# GRAVITY AND STRINGS

Self-contained and comprehensive, this definitive new edition of *Gravity and Strings* is a unique resource for graduate students and researchers in theoretical physics.

From basic differential geometry through to the construction and study of black-hole and black-brane solutions in quantum gravity – via all the intermediate stages – this book provides a complete overview of the intersection of gravity, supergravity, and superstrings.

Now fully revised, this second edition covers an extensive array of topics, including new material on non-linear electric-magnetic duality, the embedding-tensor formalism, matter-coupled supergravity, supersymmetric solutions, the geometries of scalar manifolds appearing in four- and five-dimensional supergravities, and much more. Covering reviews of important solutions and numerous solution-generating techniques, and accompanied by an exhaustive index and bibliography, this is an exceptional reference work.

TOMÁS ORTÍN is a Research Professor at the Institute for Theoretical Physics (IFT), a joint institute of the Autonomous University of Madrid and the Spanish National Research Council (UAM-CSIC). He has previously worked at the European Laboratory for Particle Physics (CERN), held postdoctoral positions at Stanford University and Queen Mary University of London, and has taught several graduate courses on advanced general relativity. His research interests include string theory, gravity, quantum gravity, and black-hole physics.

CAMBRIDGE MONOGRAPHS ON MATHEMATICAL PHYSICS

General Editors: P. V. Landshoff, D. R. Nelson, S. Weinberg

S. J. Aarseth *Gravitational N-Body Simulations: Tools and Algorithms*†
J. Ambjørn, B. Durhuus and T. Jonsson *Quantum Geometry: A Statistical Field Theory Approach*†
A. M. Anile *Relativistic Fluids and Magneto-fluids: With Applications in Astrophysics and Plasma Physics*
J. A. de Azcárraga and J. M. Izquierdo *Lie Groups, Lie Algebras, Cohomology and Some Applications in Physics*†
O. Babelon, D. Bernard and M. Talon *Introduction to Classical Integrable Systems*†
F. Bastianelli and P. van Nieuwenhuizen *Path Integrals and Anomalies in Curved Space*†
V. Belinski and E. Verdaguer *Gravitational Solitons*†
J. Bernstein *Kinetic Theory in the Expanding Universe*†
G. F. Bertsch and R. A. Broglia *Oscillations in Finite Quantum Systems*†
N. D. Birrell and P. C. W. Davies *Quantum Fields in Curved Space*†
K. Bolejko, A. Krasiński, C. Hellaby and M.-N. Célérier *Structures in the Universe by Exact Methods: Formation, Evolution, Interactions*
D. M. Brink *Semi-Classical Methods for Nucleus-Nucleus Scattering*†
M. Burgess *Classical Covariant Fields*†
E. A. Calzetta and B.-L. B. Hu *Nonequilibrium Quantum Field Theory*
S. Carlip *Quantum Gravity in 2+1 Dimensions*†
P. Cartier and C. DeWitt-Morette *Functional Integration: Action and Symmetries*†
J. C. Collins *Renormalization: An Introduction to Renormalization, the Renormalization Group and the Operator-Product Expansion*†
P. D. B. Collins *An Introduction to Regge Theory and High Energy Physics*†
M. Creutz *Quarks, Gluons and Lattices*†
P. D. D'Eath *Supersymmetric Quantum Cosmology*†
J. Dereziński and C. Gérard *Mathematics of Quantization and Quantum Fields*
F. de Felice and D. Bini *Classical Measurements in Curved Space-Times*
F. de Felice and C. J. S. Clarke *Relativity on Curved Manifolds*†
B. DeWitt *Supermanifolds, $2^{nd}$ edition*†
P. G. O. Freund *Introduction to Supersymmetry*†
F. G. Friedlander *The Wave Equation on a Curved Space-Time*†
J. L. Friedman and N. Stergioulas *Rotating Relativistic Stars*
Y. Frishman and J. Sonnenschein *Non-Perturbative Field Theory: From Two Dimensional Conformal Field Theory to QCD in Four Dimensions*
J. A. Fuchs *Affine Lie Algebras and Quantum Groups: An Introduction, with Applications in Conformal Field Theory*†
J. Fuchs and C. Schweigert *Symmetries, Lie Algebras and Representations: A Graduate Course for Physicists*†
Y. Fujii and K. Maeda *The Scalar-Tensor Theory of Gravitation*†
J. A. H. Futterman, F. A. Handler and R. A. Matzner *Scattering from Black Holes*†
A. S. Galperin, E. A. Ivanov, V. I. Ogievetsky and E. S. Sokatchev *Harmonic Superspace*†
R. Gambini and J. Pullin *Loops, Knots, Gauge Theories and Quantum Gravity*†
T. Gannon *Moonshine beyond the Monster: The Bridge Connecting Algebra, Modular Forms and Physics*†
M. Göckeler and T. Schücker *Differential Geometry, Gauge Theories, and Gravity*†
C. Gómez, M. Ruiz-Altaba and G. Sierra *Quantum Groups in Two-Dimensional Physics*†
M. B. Green, J. H. Schwarz and E. Witten *Superstring Theory Volume 1: Introduction*
M. B. Green, J. H. Schwarz and E. Witten *Superstring Theory Volume 2: Loop Amplitudes, Anomalies and Phenomenology*
V. N. Gribov *The Theory of Complex Angular Momenta: Gribov Lectures on Theoretical Physics*†
J. B. Griffiths and J. Podolský *Exact Space-Times in Einstein's General Relativity*†
S. W. Hawking and G. F. R. Ellis *The Large Scale Structure of Space-Time*†
F. Iachello and A. Arima *The Interacting Boson Model*†
F. Iachello and P. van Isacker *The Interacting Boson-Fermion Model*†
C. Itzykson and J. M. Drouffe *Statistical Field Theory Volume 1: From Brownian Motion to Renormalization and Lattice Gauge Theory*†
C. Itzykson and J. M. Drouffe *Statistical Field Theory Volume 2: Strong Coupling, Monte Carlo Methods, Conformal Field Theory and Random Systems*†
C. V. Johnson *D-Branes*†
P. S. Joshi *Gravitational Collapse and Spacetime Singularities*†
J. I. Kapusta and C. Gale *Finite-Temperature Field Theory: Principles and Applications, $2^{nd}$ edition*†
V. E. Korepin, N. M. Bogoliubov and A. G. Izergin *Quantum Inverse Scattering Method and Correlation Functions*†
M. Le Bellac *Thermal Field Theory*†
Y. Makeenko *Methods of Contemporary Gauge Theory*†
N. Manton and P. Sutcliffe *Topological Solitons*†
N. H. March *Liquid Metals: Concepts and Theory*†
I. Montvay and G. Münster *Quantum Fields on a Lattice*†
L. O'Raifeartaigh *Group Structure of Gauge Theories*†
T. Ortín *Gravity and Strings, $2^{nd}$ edition*
A. M. Ozorio de Almeida *Hamiltonian Systems: Chaos and Quantization*†
L. Parker and D. Toms *Quantum Field Theory in Curved Spacetime: Quantized Fields and Gravity*

R. Penrose and W. Rindler *Spinors and Space-Time Volume 1: Two-Spinor Calculus and Relativistic Fields*†
R. Penrose and W. Rindler *Spinors and Space-Time Volume 2: Spinor and Twistor Methods in Space-Time Geometry*†
S. Pokorski *Gauge Field Theories, 2nd edition*†
J. Polchinski *String Theory Volume 1: An Introduction to the Bosonic String*†
J. Polchinski *String Theory Volume 2: Superstring Theory and Beyond*†
J. C. Polkinghorne *Models of High Energy Processes*†
V. N. Popov *Functional Integrals and Collective Excitations*†
L. V. Prokhorov and S. V. Shabanov *Hamiltonian Mechanics of Gauge Systems*
A. Recknagel and V. Schiomerus *Boundary Conformal Field Theory and the Worldsheet Approach to D-Branes*
R. J. Rivers *Path Integral Methods in Quantum Field Theory*†
R. G. Roberts *The Structure of the Proton: Deep Inelastic Scattering*†
C. Rovelli *Quantum Gravity*†
W. C. Saslaw *Gravitational Physics of Stellar and Galactic Systems*†
R. N. Sen *Causality, Measurement Theory and the Differentiable Structure of Space-Time*
M. Shifman and A. Yung *Supersymmetric Solitons*
H. Stephani, D. Kramer, M. MacCallum, C. Hoenselaers and E. Herlt *Exact Solutions of Einstein's Field Equations, 2nd edition*†
J. Stewart *Advanced General Relativity*†
J. C. Taylor *Gauge Theories of Weak Interactions*†
T. Thiemann *Modern Canonical Quantum General Relativity*†
D. J. Toms *The Schwinger Action Principle and Effective Action*†
A. Vilenkin and E. P. S. Shellard *Cosmic Strings and Other Topological Defects*†
R. S. Ward and R. O. Wells, Jr *Twistor Geometry and Field Theory*†
E. J. Weinberg *Classical Solutions in Quantum Field Theory: Solitons and Instantons in High Energy Physics*
J. R. Wilson and G. J. Mathews *Relativistic Numerical Hydrodynamics*†

†Available in paperback

# Gravity and Strings

TOMÁS ORTÍN
*Spanish National Research Council*
*(CSIC)*

CAMBRIDGE
UNIVERSITY PRESS

CAMBRIDGE
UNIVERSITY PRESS

University Printing House, Cambridge CB2 8BS, United Kingdom

Cambridge University Press is part of the University of Cambridge.

It furthers the University's mission by disseminating knowledge in the pursuit of education, learning and research at the highest international levels of excellence.

www.cambridge.org
Information on this title: www.cambridge.org/9780521768139



First published 2004
Second edition 2015

Printed in the United Kingdom by CPI Group Ltd, Croydon CR0 4YY

*A catalog record for this publication is available from the British Library*

ISBN 978-0-521-76813-9 Hardback

To Marimar, Diego, and Tomás, the sweet strings that tie me to the real world

# Contents

# Preface to the second edition

In spite (or because) of its relentless progress, science is a perpetually unfinished work and so must be a description of any field of research at a given time. The first edition of this book tried to review the foundations and main achievements of the field that we called *semiclassical string gravity* covering the basics of general relativity, supergravity, and superstring theory[1] aiming to provide a complete and self-consistent introduction to the effective field theory description and the black-hole and black-brane solutions of the latter (ten-dimensional supergravity and some of its compactifications). However, many interesting topics and results had to be omitted then due to lack of space and many others have emerged in the following years and I started feeling quite soon that the book was not complete and the goals I had set forth had not been reached.

Of course, for the aforementioned reasons, it is intrinsically impossible to give a complete and final description of this field in the absolute sense, but I think (the reader will be the judge) that the inclusion of a reasonable number of new topics was necessary and will make the book much more useful. The second edition is the result of trying to cover that necessity while preserving the self-consistency of the book by adding background and complementary material.

The two main gaps I have tried to close are the lack of a complete discussion of the black-hole attractor mechanism and a description of the classification/characterization of the supersymmetric solutions of general (matter-coupled) four-dimensional supergravities.[2] These two subjects are linked by the original discovery of the attractor mechanism in supersymmetric extremal black-hole solutions of $N = 2, d = 4$ supergravity coupled to vector supermultiplets.

[1] This field, lying at the triple intersection of gravity, supergravity, and superstring theory, could well be named by the acronym *GRASS*.

[2] There are gaps in many other directions that could have been completed as well. For instance, a chapter on higher-derivative modifications of GR ($f(R)$ theories in particular), a deeper discussion on the definition of conserved charges in gauge theories (including gravity and supergravity) and the relation with the symmetry groups of given boundary conditions (for Kerr/CFT duality purposes), an introduction to AdS/CFT correspondence, the inclusion of asymptotically AdS and stationary solutions etc. could have been found useful by many readers. The final choice is quite subjective and associated to the author's own taste and limits.

A self-consistent description of these two subjects has required, first, the addition of several new chapters (Chapters 6–8) on matter-coupled $N = 1$ to $N = 8$ four-dimensional supergravities, including detailed descriptions of the gaugings of the $N = 1$ and $N = 2$ theories. Due to the relation via KK dimensional reduction between $N = 1, d = 5$ coupled to vector multiplets and the cubic models of $N = 2, d = 4$ supergravity, a chapter on the former (Chapter 9) has also been included, and the dimensional reduction has been performed in Chapter 15. Again, several appendices (Appendices E–J) describing the geometries of the scalar manifolds of these supergravities and the gauging of their isometries have been added for the sake of self-consistency. Furthermore, since the description of those supergravities makes heavy use of the results by Gaillard and Zumino on the general duality symmetries of (the equations of motion of) four-dimensional field theories, a section (Section 2.6) has been added describing them and their extension to higher dimensions.

With this background at hand we have been able to address the classification/characterization of the supersymmetric solutions of those supergravity theories using the Killing spinor bilinear method in Chapter 18, extending the results on the maximally supersymmetric ones of the first edition, and we have applied it in Chapter 19 to the construction of general families of supersymmetric black-hole solutions including multi-black-hole solutions and five-dimensional supersymmetric black rings.

The attractor mechanism has been explained in Chapter 27 in the framework of the Ferrara–Gibbons–Kallosh formalism and its (spacetime and worldvolume) higher-dimensional extension. Finally, the H-FGK formalism connects the results on supersymmetric black-hole solutions of Chapter 19 with the results of the FGK formalism.

There many other minor additions: an introduction to the embedding tensor formalism (Section 2.7), a review of non-linear electric–magnetic duality within Section 2.6, the algebra of four- and five-dimensional spinor bilinears (Section D.3), etc.

With the addition of all this new and highly correlated material, the organization of the book has become quite non-linear. For instance, general duality (Gaillard–Zumino) symmetries (Section 2.6) are described long before the simplest electric–magnetic duality transformations are introduced (Section 12.7). These non-linearities have no easy and economical solution, but, hopefully, they can be sorted out thanks to the cross-references provided in the main text. The index should also be helpful to those searching for specific theories, solutions, and results.

Since the publication of the first edition, several excellent books on gravity [1284, 557], supergravity [564], and superstrings [111, 860, 1248] have appeared. They deal with the basics of gravity, supergravity, and superstrings in much more depth, but I think the interdisciplinary topics studied in this book (whose contents do not fit in a nutshell, not even in a coconut shell!) provide a useful complement not specifically covered by any of them.

Just as new material had to be added to this edition, I must also add the names of people to whom I am grateful as a scientist, as a person, or both. First and foremost, I have to thank my family (Marimar, Tomás, and Diego) for their understanding and support, because nothing would have been possible without them. My students Jorge Bellorín, Pablo Bueno, Wissam Chemissany, Mechthild Hübscher, Carlos Shahbazi, and Simone Sorgato, and young collaborators Pietro Galli, José Juan Fernández-Melgarejo, Jelle Hartong, Jan Perz, Diederick Roest (now not so young!), and Silvia Vaulà helped and pushed me into new directions and taught me many things which are now in this book. I have also learned many new things from Eric Bergshoeff, Renata Kallosh, and Roberto Emparan that have found a place here. Their support, as well as that of Enrique Álvarez, Luis Álvarez-Gaumé, José Adolfo de Azcárraga, Igor Bandos, Yolanda Lozano, and Emilio Torrente-Luján, has been essential.

My long-time collaborator Patrick Meessen deserves a special mention, and he has my long-lasting gratitude for his many direct and indirect contributions to this book, for the time and energy spent in our common projects, and for his friendship. Joaquim Gomis believed in this project and shared with me his courage and wisdom. I have learned many useful things from him *¡Moltes gràcies Quim!*

The hospitality and financial support of the CERN Theory Division and the Instituto Balseiro in Bariloche have provided the calm and positive working environment that I badly needed to conclude the book. Thank you very much.

I would also like to thank Irene Pizzie for her thorough review of the manuscript. She has eliminated most inconsistencies and has made the book much more readable. Whatever defects remain are my sole responsibility.

Finally, I must thank Simon Capelin from Cambridge University Press for suggesting, encouraging, and allowing me to write this second edition to my entire satisfaction (so I am the only one to blame for its shortcomings), showing he has boundless patience.

Comments and notifications of misprints can be sent to the e-mail address `Tomas.Ortin@csic.es`. The *errata* will be posted in `http://ramon.ift.uam-csic.es/prc/misprints.html`.

# Preface to the first edition

String theory has lived for the past few years during a golden era in which a tremendous upsurge of new ideas, techniques, and results has proliferated. In what form they will contribute to our collective enterprise (theoretical physics) only time can tell, but it is clear that many of them have started to have an impact on closely related areas of physics and mathematics, and, even if string theory does not reach its ultimate goal of becoming a theory of everything, it will have played a crucial, inspiring role.

There are many interesting things that have been learned and achieved in this field that we feel can (and perhaps should) be taught to graduate students. However, we have found that this is impossible without the introduction of many ideas, techniques, and results that are not normally taught together in standard courses on general relativity, field theory, or string theory, but which have become everyday tools for researchers in this field: black holes, strings, membranes, solitons, instantons, unbroken supersymmetry, Hawking radiation.... They can, of course, be found in various textbooks and research papers, presented from various viewpoints, but not in a single reference with a consistent organization of the ideas (not to mention a consistent notation).

These are the main reasons for the existence of this book, which tries to fill this gap by covering a wide range of topics related, in one way or another, to what we may call *semiclassical string gravity*. The selection of material is according to the author's taste and personal preferences with the aim of self-consistency and the ultimate goal of creating a basic, pedagogical, reference work in which all the results are written in a consistent set of notations and conventions. Some of the material is new and cannot be found elsewhere.

Precisely because of the blend of topics we have touched upon, although a great deal of background material is (briefly) reviewed here, this cannot be considered a textbook on general relativity, supergravity, or string theory. Nevertheless, some chapters can be used in graduate courses on these matters, either providing material for a few lectures on a selected topic or combined (as the author has done with the first part, which is self-contained) into an advanced (and a bit eclectic) course on gravity.

It has not been too difficult to order logically the broad range of topics that had to be discussed, however. We can view string theory as the summit of a pyramid whose building blocks are the theories, results, and data that become more and more fundamental and basic the more we approach the base of the pyramid. At the very bottom (Part I) one can find tools such as differential geometry and the use of symmetry in physics and fundamental theories of gravity such as general relativity and extensions to accommodate fermions such as the CSK theory and supergravity. The rest of the book is supported by it. In particular, we can see string theory as the culmination of long-term efforts to construct a theory of quantum gravity for a spin-2 particle (the graviton), and our approach to general relativity as the only self-consistent classical field theory of the graviton is intended to set the ground for this view.

Part II investigates the consequences, results, and extensions of general relativity through some of its simplest and most remarkable solutions, which can be regarded as point-particle like: the Schwarzschild and Reissner–Nordström solutions, gravitational waves, and the Taub–NUT solution. In the course of this study we introduce the reader to black holes, "no-hair theorems," black-hole thermodynamics, Hawking radiation, gravitational instantons, charge quantization, electric–magnetic duality, the Witten effect, etc. We will also explain the essentials of dimensional reduction and will obtain black-hole solutions of the dimensionally reduced theory. To finish Part II we introduce the reader to the idea and implications of residual supersymmetry. We will review all our results on black-hole thermodynamics and other black-hole properties in the light of unbroken supersymmetry.

Part III introduces strings and the string effective action as a particular extension of general relativity and supergravity. String dualities and extended objects will be studied from the string-effective-action (spacetime) point of view, making use of the results of Parts I and II and paying special attention to the relation between worldvolume and spacetime phenomena. This part, and the book, closes with an introduction to the calculation of black-hole entropies using string theory.

During these years, I have received the support of many people to whom this book, and I personally, owe much: Enrique Álvarez, Luis Álvarez-Gaumé, and my long-time collaborators Eric Bergshoeff and Renata Kallosh encouraged me and gave me the opportunity to learn from them. My students Natxo Alonso-Alberca, Ernesto Lozano-Tellechea, and Patrick Meessen used and checked many versions of the manuscript they used to call the *PRC*. Their help and friendship in these years has been invaluable. Roberto Emparan, José Miguel Figueroa-O'Farrill, Yolanda Lozano, Javier Más, Alfonso Vázquez-Ramallo, and Miguel Ángel Vázquez-Mozo read several versions of the manuscript and gave me many valuable comments and advice, which contributed to improving it. I am indebted to Arthur Greenspoon for making an extremely thorough final revision of the manuscript.

Nothing would have been possible without Marimar's continuous and enduring support.

If, in spite of all this help, the book has any shortcomings, the responsibility is entirely mine. Comments and notifications of misprints can be sent to the e-mail address `tomas.ortin@uam.es`. The *errata* will be posted in `http://gesalerico.ft.uam.es/prc/misprints.html`.

This book started as a written version of a review talk on string black holes prepared for the first String Theory Meeting of the Benasque Center for Theoretical Physics, back in 1996; parts of it made a first public appearance in a condensed form as lectures for the charming Escuela de Relatividad, Campos y Cosmología "La Hechicera" organized by the Universidad de Los Andes (Mérida, Venezuela); and it was finished during a long-term visit to the CERN Theory Division. I would like to thank the organizers and members of these institutions for their invitations, hospitality, and economic support.

# Part I
# Introduction to gravity and supergravity

*Let no one ignorant of Mathematics enter here.*

*Inscription above the doorway of Plato's Academy*

# 1

# Differential geometry

The main purpose of this chapter is to fix our notation and to review the ideas and formulae of differential geometry we will make heavy use of. There are many excellent physicist-oriented references on differential geometry. Two that we particularly like are Refs. [481] and [972]. Our approach here will be quite pragmatic, ignoring many mathematical details and subtleties that can be found in the many excellent books on the subject.

## 1.1 World tensors

A *manifold* is a topological space that looks (i.e. it is homeomorphic to) locally (i.e. in a *patch*) like a piece of $\mathbb{R}^d$. $d$ is the dimension of the manifold and the correspondence between the patch and the piece of $\mathbb{R}^n$ can be used to label the points in the patch by Cartesian $\mathbb{R}^n$ coordinates $x^\mu$. In the overlap between different patches the different coordinates are consistently related by a *general coordinate transformation* (GCT) $x'^\mu(x)$. Only objects with good transformation properties under GCTs can be defined globally on the manifold. These objects are *tensors*.

A *contravariant vector field* (or $(1,0)$-type tensor or just "vector") $\xi(x) = \xi^\mu(x)\partial_\mu$ is defined at each point on a $d$-dimensional smooth manifold by its action on a function

$$\xi : f \longrightarrow \xi f = \xi^\mu \partial_\mu f, \tag{1.1}$$

which defines another function. These objects span a $d$-dimensional linear vector space at each point of the manifold called the *tangent space* $\mathrm{T}_p^{(1,0)}$. The $d$ functions $\xi^\mu(x)$ are the vector components with respect to the *coordinate basis* $\{\partial_\mu\}$.

A *covariant vector field* (or $(0,1)$-type tensor or *differential 1-form*) is an element of the dual vector space (sometimes called the *cotangent space*) $\mathrm{T}_p^{(0,1)}$ and therefore transforms vectors into functions. The elements of the basis dual to the coordinate basis of contravariant vectors are usually denoted by $\{dx^\mu\}$ and, by definition,

$$\langle dx^\mu | \partial_\nu \rangle \equiv \delta^\mu{}_\nu, \tag{1.2}$$

which implies that the action of a form $\omega = \omega_\mu dx^\mu$ on a vector $\xi(x) = \xi^\mu(x)\partial_\mu$ gives the function[1]

$$\langle \omega | \xi \rangle = \omega_\mu \xi^\mu . \tag{1.3}$$

Under a GCT, vectors and forms transform as functions, i.e. $\xi'(x') = \xi(x(x'))$ etc., which means for their components in the associated coordinate basis

$$\frac{\partial x'^\rho}{\partial x^\mu}\, \xi^\mu(x(x')) = \xi'^\rho(x'), \qquad \omega_\mu(x(x'))\, \frac{\partial x^\mu}{\partial x'^\rho} = \omega'_\rho(x'). \tag{1.4}$$

More general tensors of type $(q, r)$ can be defined as elements of the space $\mathrm{T}_p^{(q,r)}$, which is the tensor product of $q$ copies of the tangent space and $r$ copies of the cotangent space. Their components $\mathrm{T}^{\mu_1\cdots\mu_q}{}_{\nu_1\cdots\nu_r}$ transform under GCTs in the obvious way.

It is also possible to define *tensor densities of weight* $w$ whose components in a coordinate basis change under a GCT with an extra factor of the Jacobian raised to the power $w/2$. Thus, for weight $w$, the vector density components $\mathfrak{v}^\mu$ and the form density components $\mathfrak{w}_\mu$ transform according to

$$\begin{aligned} \left|\frac{\partial x'}{\partial x}\right|^{w/2} \frac{\partial x'^\rho}{\partial x^\mu}\, \mathfrak{v}^\mu(x(x')) &= \mathfrak{v}'^\rho(x'), \\ \mathfrak{w}_\mu(x(x'))\, \frac{\partial x^\mu}{\partial x'^\rho} \left|\frac{\partial x'}{\partial x}\right|^{w/2} &= \mathfrak{w}'_\rho(x'), \end{aligned} \tag{1.5}$$

where for the Jacobian we use the notation

$$\left|\frac{\partial x'}{\partial x}\right| \equiv \det\left(\frac{\partial x'^\rho}{\partial x^\mu}\right). \tag{1.6}$$

An infinitesimal GCT[2] can be written as follows:

$$\delta x^\mu = x'^\mu - x^\mu = \epsilon^\mu(x). \tag{1.7}$$

The corresponding infinitesimal transformations of scalars $\phi$ and contravariant and covariant *world vectors* (an alternative name for components in the coordinate basis) are:[3]

$$\begin{aligned} \delta\phi &= -\epsilon^\lambda \partial_\lambda \phi && \equiv -\mathcal{L}_\epsilon \phi, \\ \delta\xi^\mu &= -\epsilon^\lambda \partial_\lambda \xi^\mu + \partial_\nu \epsilon^\mu \xi^\nu && \equiv -\mathcal{L}_\epsilon \xi^\mu \equiv -[\epsilon, \xi]^\mu, \\ \delta\omega_\mu &= -\epsilon^\lambda \partial_\lambda \omega_\mu - \partial_\mu \epsilon^\nu \omega_\nu && \equiv -\mathcal{L}_\epsilon \omega_\mu, \end{aligned} \tag{1.8}$$

[1] Summation over repeated indices in any position will always be assumed, unless they are in parentheses.

[2] This is an element of a one-parameter group of GCTs (the unit element corresponding to the value 0 of the parameter) with a value of the parameter much smaller than 1.

[3] We use the functional variations $\delta\phi \equiv \phi'(x) - \phi(x)$ which refer to the value of the field $\phi$ at two different points whose coordinates are equal in the two different coordinate systems. They are denoted in Ref. [1068] by $\delta_0$. They should be distinguished from the total variations $\tilde{\delta} = \phi'(x') - \phi(x)$ which refer to the values of the field $\phi$ at the same point in two different coordinate systems. The relation between the two is $\delta\phi = \tilde{\delta}\phi - \epsilon^\mu \partial_\mu \phi$. The piece $-\epsilon^\lambda \partial_\lambda \phi$ that appears in $\delta$ variations is the "transport term," which is not present in other kinds of infinitesimal variations. The transformations $\delta$ do enjoy a group property (their commutator is another $\delta$ transformation), whereas the transformations $\tilde{\delta}$ or the transport terms by themselves do not.

and, for weight-$w$ scalar densities $\mathfrak{f}$, vector density components $\mathfrak{v}^\mu$, and the form density components $\mathfrak{w}_\mu$,

$$\begin{aligned}
\delta\mathfrak{f} &= -\epsilon^\lambda\partial_\lambda\mathfrak{f} - w\partial_\lambda\epsilon^\lambda\mathfrak{f} &&\equiv -\mathcal{L}_\epsilon\mathfrak{f},\\
\delta\mathfrak{v}^\mu &= -\epsilon^\lambda\partial_\lambda\mathfrak{v}^\mu + \partial_\nu\epsilon^\mu\mathfrak{v}^\nu - w\partial_\lambda\epsilon^\lambda\mathfrak{v}^\mu &&\equiv -\mathcal{L}_\epsilon\mathfrak{v}^\mu,\\
\delta\mathfrak{w}_\mu &= -\epsilon^\lambda\partial_\lambda\mathfrak{w}_\mu - \partial_\mu\epsilon^\nu\mathfrak{w}_\nu - w\partial_\lambda\epsilon^\lambda\mathfrak{w}_\mu &&\equiv -\mathcal{L}_\epsilon\mathfrak{w}_\mu,
\end{aligned} \tag{1.9}$$

where $\mathcal{L}_\epsilon$ is the *Lie derivative* with respect to the vector field $\epsilon$ and $[\epsilon,\xi]$ is the *Lie bracket* of the vectors $\epsilon$ and $\xi$. The definition of the Lie derivative can be extended to tensors or weight-$w$ tensor densities of any type:

$$\begin{aligned}
\mathcal{L}_\epsilon T^{\mu_1\cdots\mu_p}{}_{\nu_1\cdots\nu_q} &= -\delta_\epsilon T^{\mu_1\cdots\mu_p}{}_{\nu_1\cdots\nu_q}\\
&= \epsilon^\rho\partial_\rho T^{\mu_1\cdots\mu_p}{}_{\nu_1\cdots\nu_q} - \partial_\rho\epsilon^{\mu_1}T^{\rho\mu_2\cdots\mu_p}{}_{\nu_1\cdots\nu_q} + \cdots\\
&\quad + \partial_{\nu_1}\epsilon^\rho T^{\mu_1\cdots\mu_p}{}_{\rho\nu_2\cdots\nu_q} - w\partial_\lambda\epsilon^\lambda T^{\mu_1\cdots\mu_p}{}_{\nu_1\cdots\nu_q}.
\end{aligned} \tag{1.10}$$

In particular the metric (a symmetric $(0,2)$-type tensor to be defined later) and $r$-forms (a fully antisymmetric $(0,r)$-type tensor) transform as follows:

$$\begin{aligned}
\delta g_{\mu\nu} &= -\epsilon^\lambda\partial_\lambda g_{\mu\nu} - 2g_{\lambda(\mu}\partial_{\nu)}\epsilon^\lambda &&= -\mathcal{L}_\epsilon g_{\mu\nu},\\
\delta B_{\mu_1\cdots\mu_r} &= -\epsilon^\lambda\partial_\lambda B_{\mu_1\cdots\mu_r} - r(\partial_{[\mu_1|}\epsilon^\lambda)B_{\lambda|\mu_2\cdots\mu_r]} &&= -\mathcal{L}_\epsilon B_{\mu_1\cdots\mu_r}.
\end{aligned} \tag{1.11}$$

The main properties of the Lie derivative are that it transforms tensors of a given type into tensors of the same given type, it obeys the Leibniz rule $\mathcal{L}_\epsilon(T_1T_2) = (\mathcal{L}_\epsilon T_1)T_2 + T_1\mathcal{L}_\epsilon T_2$, it is connection independent, and it is linear with respect to $\epsilon$. Furthermore, it satisfies the *Jacobi identity*

$$[\mathcal{L}_{\xi_1},[\mathcal{L}_{\xi_2},\mathcal{L}_{\xi_3}]] + [\mathcal{L}_{\xi_2},[\mathcal{L}_{\xi_3},\mathcal{L}_{\xi_1}]] + [\mathcal{L}_{\xi_3},[\mathcal{L}_{\xi_1},\mathcal{L}_{\xi_2}]] = 0, \tag{1.12}$$

where the brackets stand for commutators of differential operators. The relation between the commutator $[\mathcal{L}_\xi,\mathcal{L}_\epsilon]$ and the Lie bracket $[\xi,\epsilon]$ is

$$[\mathcal{L}_\xi,\mathcal{L}_\epsilon] = \mathcal{L}_{[\xi,\epsilon]}. \tag{1.13}$$

Thus, the Lie bracket is an antisymmetric, bilinear product in tangent space that also satisfies the Jacobi identity

$$[\xi_1,[\xi_2,\xi_3]] + [\xi_2,[\xi_3,\xi_1]] + [\xi_3,[\xi_1,\xi_2]] = 0, \tag{1.14}$$

which one can use to give it the structure of *Lie algebra*.

## 1.2 Affinely connected spacetimes

The *covariant derivative* of world tensors is defined by

$$\begin{aligned}
\nabla_\mu\phi &= \partial_\mu\phi,\\
\nabla_\mu\xi^\nu &= \partial_\mu\xi^\nu + \Gamma_{\mu\rho}{}^\nu\xi^\rho,\\
\nabla_\mu\omega_\nu &= \partial_\mu\omega_\nu - \omega_\rho\Gamma_{\mu\nu}{}^\rho,
\end{aligned} \tag{1.15}$$

and on weight-$w$ tensor densities by

$$\begin{aligned}\nabla_\mu \mathfrak{f} &= \partial_\mu \mathfrak{f} - w\Gamma_{\mu\rho}{}^{\rho}\mathfrak{f},\\ \nabla_\mu \mathfrak{v}^\nu &= \partial_\mu \mathfrak{v}^\nu + \Gamma_{\mu\rho}{}^{\nu}\mathfrak{v}^\rho - w\Gamma_{\mu\rho}{}^{\rho}\mathfrak{v}^\nu,\\ \nabla_\mu \mathfrak{w}_\nu &= \partial_\mu \mathfrak{w}_\nu - \mathfrak{w}_\rho\Gamma_{\mu\nu}{}^{\rho} - w\Gamma_{\mu\rho}{}^{\rho}\mathfrak{w}_\nu,\end{aligned} \tag{1.16}$$

where $\Gamma$ is the *affine connection*, and is added to the partial derivative so that the covariant derivative of a tensor transforms as a tensor in all indices. This requires the affine connection to transform under infinitesimal GCTs as follows:

$$\delta\Gamma_{\mu\nu}{}^{\rho} = -\mathcal{L}_\epsilon\Gamma_{\mu\nu}{}^{\rho} - \partial_\mu\partial_\nu\epsilon^\rho, \tag{1.17}$$

and therefore it is not a tensor. In principle it can be any field with the above transformation properties and should be understood as structure added to our manifold. A $d$-dimensional manifold equipped with an affine connection is sometimes called an *affinely connected space* and is denoted by $\mathrm{L}_d$.

The definition of a covariant derivative can be extended to tensors of arbitrary type in the standard fashion. Its main properties are that it is a linear differential operator that transforms type-$(p, q)$ tensors into $(p, q+1)$ tensors (hence the name covariant) and obeys the Leibniz rule and the Jacobi identity.

Let us now decompose the connection into two (symmetric and antisymmetric) pieces under the exchange of the covariant indices:

$$\Gamma_{\mu\nu}{}^{\rho} = \Gamma_{(\mu\nu)}{}^{\rho} + \Gamma_{[\mu\nu]}{}^{\rho}. \tag{1.18}$$

The antisymmetric part is called the *torsion* and it is a tensor (which the connection is not)

$$T_{\mu\nu}{}^{\rho} = -2\Gamma_{[\mu\nu]}{}^{\rho}. \tag{1.19}$$

As we have said, the Lie derivative transforms tensors into tensors in spite of the fact that it is expressed in terms of partial derivatives. We can rewrite it in terms of covariant derivatives and torsion terms to make evident the fact that the result is indeed a tensor:

$$\begin{aligned}\mathcal{L}_\epsilon\phi &= \epsilon^\lambda\nabla_\lambda\phi,\\ \mathcal{L}_\epsilon\xi^\mu &= \epsilon^\lambda\nabla_\lambda\xi^\mu - \nabla_\nu\epsilon^\mu\xi^\nu + \epsilon^\lambda T_{\lambda\rho}{}^{\mu}\xi^\rho,\\ \mathcal{L}_\epsilon\omega_\mu &= \epsilon^\lambda\nabla_\lambda\omega_\mu + \nabla_\mu\epsilon^\nu\omega_\nu - \epsilon^\lambda\omega_\rho T_{\lambda\mu}{}^{\rho},\end{aligned} \tag{1.20}$$

etc. It should be stressed that this is just a rewriting of the Lie derivative, which is independent of any connection. There are other connection-independent derivatives. Particularly important is the *exterior derivative* defined on *differential forms* (completely antisymmetric tensors) which we will study later in Section 1.7.

The additional structure of an affine connection allows us to define *parallel transport*. In a generic spacetime there is no natural notion of parallelism for two vectors defined at two different points. We need to transport one of them keeping it "parallel to itself" to the point at which the other is defined. Then we can compare the two vectors at the same point. Using the affine connection, we can define an infinitesimal *parallel displacement* of a covariant vector $\omega_\mu$ in the direction of $\epsilon^\mu$ by

$$\delta_{\mathrm{P}_\epsilon}\omega_\mu = \epsilon^\nu\Gamma_{\nu\mu}{}^{\rho}\omega_\rho. \tag{1.21}$$

If $\omega_\mu(x)$ is a vector field, we can compare its value at a given point $x^\mu + \epsilon^\mu$ with the value obtained by parallel displacement from $x^\mu$. The difference is precisely given by the covariant derivative in the direction $\epsilon^\mu$:

$$\omega_\mu(x') - (\omega_\mu + \delta_{\mathrm{P}_\epsilon}\omega_\mu)(x) = \epsilon^\nu \nabla_\nu \omega_\mu. \tag{1.22}$$

A vector field whose value at every point coincides with the value one would obtain by parallel transport from neighboring points is a *covariantly constant* vector field, $\nabla_\nu \omega_\mu = 0$.

If the vector tangential to a curve[4] $v^\mu = dx^\mu/d\xi \equiv \dot{x}^\mu$ is parallel to itself along the curve (as a straight line in flat spacetime) then

$$v^\nu \nabla_\nu v^\mu = \ddot{x}^\mu + \dot{x}^\rho \dot{x}^\sigma \Gamma_{\rho\sigma}{}^\mu = 0, \tag{1.23}$$

which is the *autoparallel equation*. This is the equation satisfied by an *autoparallel curve*, which is the generalization of a straight line to a general affinely connected spacetime. There is a second possible generalization based on the property of straight lines of being the shortest possible curves joining two given points (*geodesics*), but it requires the notion of length and we will have to wait until the introduction of metrics.

We can understand the meaning of torsion using parallel transport: let us consider two vectors $\epsilon_1^\mu$ and $\epsilon_2^\mu$ at a given point of coordinates $x^\mu$. Let us now consider at the point of coordinates $x^\mu + \epsilon_1^\mu$ the vector $\epsilon_2'^\mu$ obtained by parallel-transporting $\epsilon_2^\mu$ in the direction $\epsilon_1^\mu$ and, at the point of coordinates $x^\mu + \epsilon_2^\mu$, the vector $\epsilon_1'^\mu$ obtained by parallel-transporting $\epsilon_1^\mu$ in the direction $\epsilon_2^\mu$. In flat spacetime, the vectors $\epsilon_1, \epsilon_2, \epsilon_1'$, and $\epsilon_2'$ form an infinitesimal parallelogram since $x^\mu + \epsilon_1^\mu + \epsilon_2'^\mu = x^\mu + \epsilon_2^\mu + \epsilon_1'^\mu$. In a general affinely connected spacetime, the infinitesimal parallelogram does not close and

$$\left(x^\mu + \epsilon_1^\mu + \epsilon_2'^\mu\right) - \left(x^\mu + \epsilon_2^\mu + \epsilon_1'^\mu\right) = \epsilon_1^\rho \epsilon_2^\sigma T_{\rho\sigma}{}^\mu. \tag{1.24}$$

Finite parallel transport along a curve $\gamma$ depends on the curve, not only on the initial and final points, so, if the curve is closed, the original and the parallel-transported vectors do not coincide. The difference is measured by the *(Riemann) curvature tensor* $R_{\mu\nu\rho}{}^\sigma$: let us consider two vectors $\epsilon_1^\mu$ and $\epsilon_2^\mu$ at a given point $x^\mu$ and let us parallel-transport the vector $\omega_\mu$ from $x^\mu$ to $x^\mu + \epsilon_1^\mu$ and then to $x^\mu + \epsilon_1^\mu + \epsilon_2^\mu$. The result is

$$\omega_\mu + (\epsilon_1^\nu + \epsilon_2^\nu)\Gamma_{\nu\mu}{}^\rho \omega_\rho + \epsilon_1^\lambda \epsilon_2^\nu \left(\partial_\lambda \Gamma_{\nu\mu}{}^\rho + \Gamma_{\lambda\delta}{}^\rho \Gamma_{\nu\mu}{}^\delta\right)\omega_\rho + \mathcal{O}(\epsilon^3). \tag{1.25}$$

If we go to the same point along the route $x^\mu$ to $x^\mu + \epsilon_2^\mu$ and then to $x^\mu + \epsilon_1^\mu + \epsilon_2^\mu$ we obtain a different value, and the difference between the parallel-transported vectors is

$$\Delta\omega_\mu = \epsilon_1^\lambda \epsilon_2^\nu R_{\lambda\nu\mu}{}^\rho \omega_\rho, \tag{1.26}$$

where

$$R_{\mu\nu\rho}{}^\sigma(\Gamma) = 2\partial_{[\mu}\Gamma_{\nu]\rho}{}^\sigma + 2\Gamma_{[\mu|\lambda}{}^\sigma \Gamma_{|\nu]\rho}{}^\lambda. \tag{1.27}$$

[4] Here we use the mathematical concept of a curve: a map from the real line $\mathbb{R}$ (or an interval) given as a function of a real parameter $x^\mu(\xi)$, rather than the *image* of the real line in the spacetime. Thus, after a reparametrization $\xi'(\xi)$, we obtain a different curve, although the image is the same and physically we would say that we have the same curve.

We can also define the curvature tensor (and the torsion tensor) through the *Ricci identities* for a scalar $\phi$, a vector $\xi^\mu$, and a 1-form $\omega_\mu$:

$$\begin{aligned}[\nabla_\mu, \nabla_\nu]\,\phi &= T_{\mu\nu}{}^\sigma \nabla_\sigma \phi,\\ [\nabla_\mu, \nabla_\nu]\,\xi^\rho &= R_{\mu\nu\sigma}{}^\rho \xi^\sigma + T_{\mu\nu}{}^\sigma \nabla_\sigma \xi^\rho,\\ [\nabla_\mu, \nabla_\nu]\,\omega_\rho &= -\omega_\sigma R_{\mu\nu\rho}{}^\sigma + T_{\mu\nu}{}^\sigma \nabla_\sigma \omega_\rho,\end{aligned} \tag{1.28}$$

or, for a general tensor,

$$[\nabla_\alpha, \nabla_\beta]\,\xi_{\mu_1\cdots}{}^{\nu_1\cdots} = -R_{\alpha\beta\mu_1}{}^\gamma \xi_{\gamma\cdots}{}^{\nu_1\cdots} - \cdots + R_{\alpha\beta\gamma}{}^{\nu_1} \xi_{\mu_1\cdots}{}^{\gamma\cdots} + \cdots + T_{\alpha\beta}{}^\gamma \nabla_\gamma \xi_{\mu_1\cdots}{}^{\nu_1\cdots}, \tag{1.29}$$

and, using the antisymmetry of the commutators of covariant derivatives and the fact that the covariant derivative satisfies the Jacobi identity, one can derive the following *Bianchi identities*:

$$\begin{aligned} R_{(\alpha\beta)\gamma}{}^\delta &= 0,\\ R_{[\alpha\beta\gamma]}{}^\delta + \nabla_{[\alpha} T_{\beta\gamma]}{}^\delta + T_{[\alpha\beta}{}^\rho T_{\gamma]\rho}{}^\delta &= 0,\\ \nabla_{[\alpha} R_{\beta\gamma]\rho}{}^\sigma + T_{[\alpha\beta}{}^\delta R_{\gamma]\delta\rho}{}^\sigma &= 0.\end{aligned} \tag{1.30}$$

(The last two identities are derived from the Jacobi identity of covariant derivatives acting on a scalar and a vector, respectively.)

In general, if we modify the affine connection by adding an arbitrary tensor[5] $\tau_{\mu\nu}{}^\rho$,

$$\Gamma_{\mu\nu}{}^\rho \rightarrow \tilde{\Gamma}_{\mu\nu}{}^\rho = \Gamma_{\mu\nu}{}^\rho + \tau_{\mu\nu}{}^\rho, \tag{1.31}$$

the curvature is modified as follows:

$$R_{\mu\nu\rho}{}^\sigma(\tilde{\Gamma}) = R_{\mu\nu\rho}{}^\sigma(\Gamma) - T_{\mu\nu}{}^\lambda \tau_{\lambda\rho}{}^\sigma + 2\nabla_{[\mu}\tau_{\nu]\rho}{}^\sigma + 2\tau_{[\mu|\lambda}{}^\sigma \tau_{|\nu]\rho}{}^\lambda. \tag{1.32}$$

The *Ricci tensor* is defined by

$$R_{\mu\nu} = R_{\mu\rho\nu}{}^\rho = \partial_\mu \Gamma_{\rho\nu}{}^\rho - \partial_\rho \Gamma_{\mu\nu}{}^\rho + \Gamma_{\mu\lambda}{}^\rho \Gamma_{\rho\nu}{}^\lambda - \Gamma_{\rho\lambda}{}^\rho \Gamma_{\mu\nu}{}^\lambda. \tag{1.33}$$

In general it is not symmetric, but, according to the second Bianchi identity,

$$R_{[\mu\nu]} = \tfrac{1}{2}\overset{*}{\nabla}_\rho \overset{*}{T}_{\mu\nu}{}^\rho + \tfrac{1}{2} R_{\mu\nu\rho}{}^\rho, \tag{1.34}$$

where we have used the *modified divergence* $\overset{*}{\nabla}_\mu$ and the *modified torsion tensor* $\overset{*}{T}_{\mu\nu}{}^\rho$,

$$\overset{*}{\nabla}_\mu = \nabla_\mu - T_{\mu\rho}{}^\rho, \qquad \overset{*}{T}_{\mu\nu}{}^\rho = T_{\mu\nu}{}^\rho - 2T_{[\mu|\sigma}{}^\sigma \delta_{|\nu]}{}^\rho. \tag{1.35}$$

If we modify the connection as in Eq. (1.31), the Ricci tensor is also modified:

$$R_{\mu\rho}(\tilde{\Gamma}) = R_{\mu\rho} - T_{\mu\nu}{}^\lambda \tau_{\lambda\rho}{}^\nu + 2\nabla_{[\mu}\tau_{\nu]\rho}{}^\nu + 2\tau_{[\mu|\lambda}{}^\nu \tau_{|\nu]\rho}{}^\lambda. \tag{1.36}$$

Another useful formula is the Lie derivative of the torsion tensor which, using the first two Bianchi identities, can be rewritten in the form

$$\mathcal{L}_\xi T_{\mu\nu}{}^\rho = \nabla_\mu\left(\xi^\lambda T_{\lambda\nu}{}^\rho\right) + \nabla_\nu\left(\xi^\lambda T_{\mu\lambda}{}^\rho\right) - \nabla_\lambda\left(\xi^\rho T_{\mu\nu}{}^\lambda\right) - 3\xi^\lambda R_{[\lambda\mu\nu]}{}^\rho + \xi^\sigma \nabla_\sigma T_{\mu\nu}{}^\rho. \tag{1.37}$$

[5] Only if $\tau$ transforms as a tensor can $\tilde{\Gamma}$ transform as a connection.

## 1.3 Metric spaces

To go further we need to add structure to a manifold: a *metric* in tangent space, i.e. an inner product for tangent-space vectors (symmetric, bilinear) associating a function $g(\xi, \epsilon)$ with any pair of vectors $(\xi, \epsilon)$. This corresponds to a symmetric $(0, 2)$-type tensor $g$ symmetric in its two covariant components $g_{\mu\nu} = g_{(\mu\nu)}$:

$$\xi \cdot \epsilon \equiv g(\xi, \epsilon) = \xi^\mu \epsilon^\nu g_{\mu\nu}. \tag{1.38}$$

The norm squared of a vector is just the product of the vector with itself, $\xi^2 = \xi \cdot \xi$. The metric will be required to be non-singular, i.e.

$$g \equiv \det(g_{\mu\nu}) \neq 0, \tag{1.39}$$

and locally diagonalizable into $\eta_{\mu\nu} = \mathrm{diag}(+ - \cdots -)$ for physical and conventional reasons. Thus, in $d$ dimensions

$$\mathrm{sign}\, g = \frac{g}{|g|} = (-1)^{d-1}. \tag{1.40}$$

As usual, a metric can be used to establish a correspondence between a vector space and its dual, i.e. between vectors and 1-forms: with each vector $\xi^\mu$ we associate a 1-form $\omega_\mu$ whose action on any other vector $\eta^\mu$ is the product of $\xi$ and $\eta$, $\omega(\eta) = \xi^\mu \eta^\nu g_{\mu\nu}$, which means the relation between components $\omega_\nu = \xi^\mu g_{\mu\nu}$. It is customary to denote this 1-form by $\xi_\mu$, and the transformation from vector to 1-form is represented by lowering the index.

The inverse metric can be used as a metric in cotangent space, and its components are those of the inverse matrix and are denoted with upper indices. The operation of raising indices can be similarly defined, and the consistency of all these operations is guaranteed because the dual of the dual is the original vector space. The extension to tensors of higher ranks is straightforward.

The determinant of the metric can also be used to relate tensors and weight $w$ tensor densities, since it transforms as a density of weight $w = 2$ and the product of a tensor and $g^{w/2}$ transforms as a density of weight $w$.

Furthermore, with a metric we can define the *Ricci scalar* $R$ and the *Einstein tensor* $G_{\mu\nu}$,

$$R = R_\mu{}^\mu, \qquad G_{\mu\nu} = R_{\mu\nu} - \tfrac{1}{2} g_{\mu\nu} R, \tag{1.41}$$

which needs not be symmetric (just like the Ricci tensor).

So far we have two independent fields defined on our manifold: the metric and the affine connection. An $\mathrm{L}_d$ spacetime equipped with a metric is sometimes denoted by $(\mathrm{L}_d, g)$. The affine connection and the metric are related by the *non-metricity tensor* $Q_{\mu\nu\rho}$,

$$Q_{\mu\nu\rho} \equiv -\nabla_\mu g_{\nu\rho}. \tag{1.42}$$

If we take the combination $\nabla_\mu g_{\rho\sigma} + \nabla_\rho g_{\sigma\mu} - \nabla_\sigma g_{\mu\rho}$ and expand it, we find that the connection can be written as follows:

$$\Gamma_{\mu\nu}{}^\rho = \left\{ {\rho \atop \mu\,\nu} \right\} + K_{\mu\nu}{}^\rho + L_{\mu\nu}{}^\rho, \tag{1.43}$$

where

$$\left\{ {\rho \atop \mu\nu} \right\} = \tfrac{1}{2} g^{\rho\sigma} \{ \partial_\mu g_{\nu\sigma} + \partial_\nu g_{\mu\sigma} - \partial_\sigma g_{\mu\nu} \} \tag{1.44}$$

are the *Christoffel symbols*, which are completely determined by the metric, and $K$ is called the *contorsion tensor* and is given in terms of the torsion tensor by

$$\begin{aligned} K_{\mu\nu}{}^{\rho} &= \tfrac{1}{2} g^{\rho\sigma} \{ T_{\mu\sigma\nu} + T_{\nu\sigma\mu} - T_{\mu\nu\sigma} \} \,, \\ K_{[\mu\nu]}{}^{\rho} &= -\tfrac{1}{2} T_{\mu\nu}{}^{\rho}, \qquad K_{\mu\nu\rho} = -K_{\mu\rho\nu}. \end{aligned} \tag{1.45}$$

Finally

$$L_{\mu\nu}{}^{\rho} = \tfrac{1}{2} \{ Q_{\mu\nu}{}^{\rho} + Q_{\nu\mu}{}^{\rho} - Q^{\rho}{}_{\mu\nu} \} \,. \tag{1.46}$$

Observe that the contorsion tensor depends on the metric whereas the torsion tensor does not. Furthermore, observe that, since the contorsion and non-metricity tensors transform as tensors, the piece responsible for the non-homogeneous term in the transformation of the affine connection is the Christoffel symbol.

With a metric it is also possible to define the length of a curve $\gamma$, $x^\mu(\xi)$, by the integral

$$s = \int_\gamma d\xi \, \sqrt{g_{\mu\nu}(x) \dot{x}^\mu \dot{x}^\nu}. \tag{1.47}$$

If we consider the above expression as a functional in the space of all curves joining two given points, we can ask which of those curves minimizes it. The answer is given by the Euler–Lagrange equations, which take the simple form

$$\ddot{x}^\mu + \dot{x}^\rho \dot{x}^\sigma \left\{ {\mu \atop \rho\sigma} \right\} = 0, \tag{1.48}$$

if we parametrize the curve by its proper length $s$. This is the *geodesic equation*, and is different from the autoparallel equation (1.23) whenever there is torsion and non-metricity.

In the standard theory of gravity metric and affine connection are not independent variables since we want to describe only the degrees of freedom corresponding to a massless spin-2 particle. To relate these two fields one imposes the *metric postulate*

$$Q_{\mu\rho\sigma} = -\nabla_\mu g_{\rho\sigma} = 0, \tag{1.49}$$

which makes the operations of raising and lowering of indices commute with the covariant derivative. A connection satisfying the above condition is said to be *metric compatible* and a spacetime $(\mathrm{L}_d, g)$ with a metric-compatible connection is called a *Riemann–Cartan spacetime* and is denoted by $\mathrm{U}_d$.

Sometimes a weaker condition is required: the vanishing of the trace-free part of the non-metricity tensor $\hat{Q}$

$$\hat{Q}_{\mu\nu\rho} \equiv Q_{\mu\nu\rho} - \tfrac{1}{d} Q_{\mu\sigma}{}^{\sigma} g_{\nu\rho} = 0. \tag{1.50}$$

In this case, the non-metricity must take the form

$$Q_{\mu\nu\rho} = -A_\mu g_{\nu\rho}, \quad \Rightarrow \quad L_{\mu\nu}^{\rho} = A_{(\mu} g_{\nu)}{}^{\rho} - \tfrac{1}{2} g_{\mu\nu} A^{\rho}, \tag{1.51}$$

where $A_\mu = -\frac{1}{d}Q_{\mu\sigma}{}^{\sigma}$ is known as the *Weyl connection* and the corresponding space is a *Cartan–Weyl* space. The presence of a Weyl connection makes the spacetime invariant under conformal rescalings of the metric

$$g'_{\mu\nu} \equiv \Omega^2 g_{\mu\nu}, \qquad \nabla'_\mu g'_{\nu\rho} = \Omega^2 \nabla_\mu g_{\nu\rho}, \tag{1.52}$$

assuming that the Weyl connection transforms as

$$A'_\mu = A_\mu - \partial_\mu \ln \Omega^2, \tag{1.53}$$

that is, as a gauge connection with gauge group $\mathbb{R}$ (see Appendix A).[6]

Still, even if the Weyl connection vanishes, the metric postulate leaves the torsion undetermined. If we want to have a connection completely determined by the metric, left as the only independent field, one has to impose the vanishing of the torsion tensor. The torsionless, metric-compatible connection is called *Levi-Civita connection* and its components are given by the Christoffel symbols.[7] A Riemann–Cartan spacetime $\mathrm{U}_d$ with vanishing torsion is a *Riemann spacetime* $\mathrm{V}_d$. If the torsion vanishes but there is a Weyl connection, the spacetime is an *Einstein–Weyl spacetime*.

There is another way of reducing the number of independent fields: by imposing the vanishing of the curvature tensor. In this case (for vanishing Weyl connection), both the metric and the connection are completely determined by the *Vielbein* (to be defined latter). The connection is called *Weitzenböck connection* [1243, 1244] and has torsion (also determined by the Vielbein). A Riemann–Cartan spacetime with Weitzenböck connection is a *Weitzenböck spacetime* $\mathrm{A}_d$; if one includes a Weyl connection, then it is called a *Weitzenböck–Weyl space*.

If both torsion and curvature vanish, the space has to be *Minkowski spacetime* $\mathrm{M}_d$ since the *Minkowski metric* $g_{\mu\nu} = \eta_{\mu\nu}$ is the only one that makes the full Riemann tensor vanish in the absence of torsion. Again, adding a Weyl connection, we have a *Minkowski–Weyl space*.

The diagram in Fig. 1.1, a more complete version of those in Refs. [736, 737] taken essentially from Ref. [927], summarizes the different particular structures that we can have on an affinely connected manifold equipped with a metric. We have introduced some changes in the diagram of Ref. [927] to indicate that Weyl spaces have a Weyl connection and not just arbitrary non-metricity.

In the rest of this section we are going to study the particular properties of some of these spacetimes. The Weitzenböck spacetime will be studied after the introduction of Vielbeins in Section 1.4.

### 1.3.1 *Riemann–Cartan spacetime* $\mathrm{U}_d$

As has been said, this is an affinely connected metric spacetime with a metric-compatible connection, so the non-metricity tensor vanishes, $Q_{\mu\nu\rho} = 0$. According to the general result,

[6] This is, as a matter of fact, how gauge symmetries were introduced in physics (see Ref. [1004] and references therein.)

[7] Sometimes (especially in the supergravity context) the Levi-Civita connection is written $\Gamma(g)$ to stress the fact that it is a function of the metric in order to distinguish it from arbitrary connections that are independent of the metric. We will do so only when necessary.

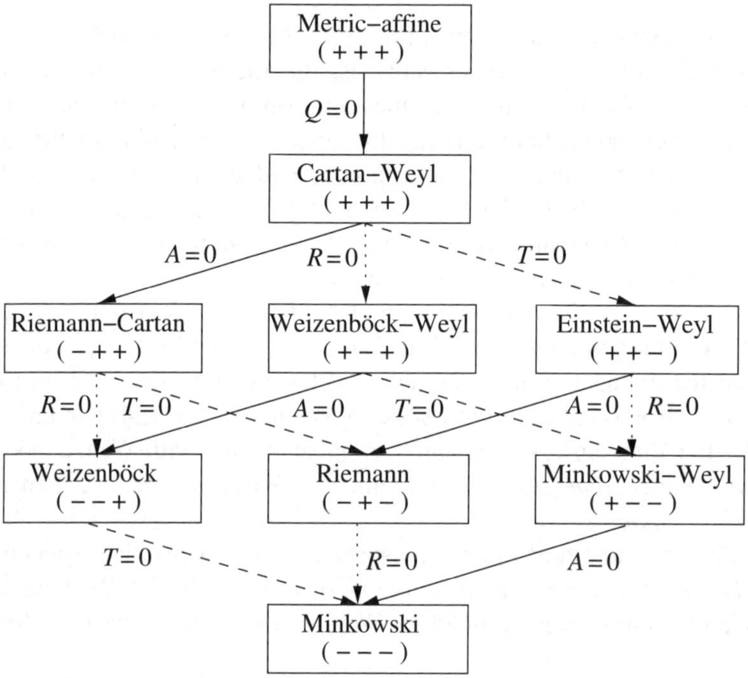


Fig. 1.1. Particular structures in an affinely connected spacetime equipped with a metric $(\mathrm{L}_d, g)$ classified by their vanishing $(-)$ or non-vanishing $(+)$ Weyl connection $A$, curvature $R$, and torsion $T$ expressed as the triplet $(A, R, T)$. Only Weyl connections with vanishing trace-free non-metricity $\hat{Q}$ are usually considered and they are called Cartan–Weyl spacetimes. They can have arbitrary Weyl connection, Riemann curvature, and torsion, and they are denoted by $(+\ +\ +)$. A Cartan–Weyl spacetime with vanishing Weyl connection is a Riemann–Cartan spacetime $(-\ +\ +)$. If the curvature vanishes instead, we have a Weitzenböck–Weyl spacetime $(+\ -\ +)$, and, if it is the torsion that vanishes, we have an Einstein–Weyl spacetime $(+\ +\ -)$. The rest of the diagram follows the same pattern.

a metric-compatible connection of a Riemann–Cartan spacetime always has the form

$$\Gamma_{\mu\nu}{}^{\rho} = \left\{ {\rho \atop \mu\,\nu} \right\} + K_{\mu\nu}{}^{\rho}. \tag{1.54}$$

Observe that the symmetric part of the contorsion tension does not vanish, but

$$K_{(\mu\nu)\,\rho} = \tfrac{1}{2}\left(T_{\mu\rho\nu} + T_{\nu\rho\mu}\right) \neq 0. \tag{1.55}$$

This means that the presence of torsion implies not only that the connection has a non-vanishing antisymmetric part, but also that the symmetric part is not fully determined by the metric but

$$\Gamma_{(\mu\nu)}{}^{\rho} = \left\{ {\rho \atop \mu\,\nu} \right\} + K_{(\mu\nu)}{}^{\rho} \neq \left\{ {\rho \atop \mu\,\nu} \right\}. \tag{1.56}$$

The curvature, Ricci, and Einstein tensors of a metric-compatible connection satisfy further identities. On contracting the $\gamma$ and $\sigma$ indices in the third Bianchi identity Eqs. (1.30) and using the metric postulate, we find the so-called *contracted Bianchi identity*

$$\nabla_{\alpha} G_{\mu}{}^{\alpha} + T_{\mu\alpha\beta} R^{\beta\alpha} - \tfrac{1}{2} T_{\alpha\beta\gamma} R_{\mu}{}^{\gamma\alpha\beta} = 0. \tag{1.57}$$

Furthermore, by applying the Ricci identity to the metric and using the metric postulate, one can prove a fourth Bianchi identity:

$$R_{\alpha\beta(\gamma\delta)} = 0. \tag{1.58}$$

If we modify the connection according to Eq. (1.31) and $\Gamma$ is metric-compatible, the Ricci scalar is

$$R(\tilde{\Gamma}) = R(\Gamma) - T_{\mu\nu}{}^{\rho}\tau_{\rho}{}^{\mu\nu} + 2\nabla_{\mu}\tau_{\nu}{}^{\mu\nu} + \tau_{\mu}{}^{\mu\lambda}\tau_{\nu}{}^{\nu}{}_{\lambda} + \tau_{\nu}{}^{\mu\rho}\tau_{\mu\rho}{}^{\nu}. \tag{1.59}$$

If $\tilde{\Gamma}$ is not a metric-compatible connection, then $\tau$ contains all the contributions of the non-metricity tensor $\tau_{\mu\nu}{}^{\rho} = L_{\mu\nu}{}^{\rho}$ and the above formula allows us to work in the framework of a Riemann–Cartan spacetime with non-metric-compatible connections. We will use this formula when we deal with Einstein–Weyl spacetimes.

If both $\tilde{\Gamma}$ and $\Gamma$ are metric-compatible connections and $\tilde{\Gamma}$ has torsion but $\Gamma = \Gamma(g)$, then $\tau = \tilde{K}$, the contorsion tensor of $\tilde{\Gamma}$, and the above formula takes a simpler form:

$$R(\tilde{\Gamma}) = R[\Gamma(g)] + 2\nabla_{\mu}\tilde{K}_{\nu}{}^{\mu\nu} + (\tilde{K}_{\mu}{}^{\mu\lambda})^2 + \tilde{K}_{\nu}{}^{\mu\rho}\tilde{K}_{\mu\rho}{}^{\nu}. \tag{1.60}$$

Now, this formula allows us to work with torsion in a Riemann spacetime. Particularly interesting is the case in which the contorsion $\tilde{K}_{\mu\nu\rho}$ is a completely antisymmetric tensor (proportional to the Kalb–Ramond field strength $H_{\mu\nu\rho}$, for instance). Then we have, if

$$\tilde{K}_{\mu\nu\rho} = \frac{1}{\sqrt{12}} H_{\mu\nu\rho}, \tag{1.61}$$

$$\int d^{d}x \sqrt{|g|}\, R(\tilde{\Gamma}) = \int d^{d}x \sqrt{|g|} \left\{ R[\Gamma(g)] + \frac{1}{2\cdot 3!} H_{\mu\nu\rho} H^{\mu\nu\rho} \right\}. \tag{1.62}$$

### 1.3.2 Einstein–Weyl spacetime $\mathrm{EW}_d$

This is defined by the conditions $\hat{Q} = T = 0$ which determine the connection $\Gamma$ to be the sum of the Levi-Civita connection and a Weyl connection so it is invariant under GCTs and Weyl transformations. If we compute the commutator of two covariant derivatives on the metric and use the Ricci identity, we find that the curvature tensor is not antisymmetric in the second pair of indices, and instead of Eq. (1.58) we have this Bianchi identity

$$R_{\mu\nu(\rho\sigma)} = -\tfrac{1}{2}F_{\mu\nu}(A)g_{\rho\sigma}, \quad \text{where} \quad F_{\mu\nu}(A) = 2\partial_{[\mu}A_{\nu]}. \tag{1.63}$$

Therefore, $R_{\mu\nu(\rho\sigma)} = 0$ if $A$ is *pure gauge* $A_\mu = \partial_\mu \ln \Omega^2$.

The complete curvature and Ricci tensors and the Ricci scalar can be easily computed using Eqs. (1.31), (1.32), (1.36), and (1.59) with $T = 0$ and $\tau_{\mu\nu}{}^{\rho} = L_{\mu\nu}{}^{\rho}$:

$$\begin{aligned} R_{\mu\nu\rho}{}^{\sigma}(\Gamma) = {} & R_{\mu\nu\rho}{}^{\sigma}(g) + \tfrac{1}{2}F_{\mu\nu}(A)g_{\rho}{}^{\sigma} + \nabla_{[\mu|}A_{\rho}g_{|\nu]}{}^{\sigma} - \nabla_{[\mu}A^{\sigma}g_{\nu]\rho} \\ & + \tfrac{1}{2}\left\{A_{[\mu}A^{\sigma}g_{\nu]\rho} - A_{[\mu|}A_{\rho}g_{|\nu]}{}^{\sigma} - \tfrac{1}{2}A^2 g_{[\mu}{}^{\sigma}g_{\nu]\rho}\right\}, \end{aligned} \tag{1.64}$$

$$\begin{aligned} R_{\mu\nu}(\Gamma) = {} & R_{\mu\nu}(g) + \tfrac{1}{4}(d-2)F_{\mu\rho}(A) + \tfrac{1}{2}(d-2)\left[\nabla_{(\mu}A_{\rho)} + \frac{1}{(d-2)}\nabla_\lambda A^\lambda g_{\mu\rho}\right] \\ & - \tfrac{1}{4}(d-2)\left[A_\mu A_\rho - A^2 g_{\mu\rho}\right], \end{aligned} \tag{1.65}$$

$$R(\Gamma) = R(g) + (d-1)\nabla_\lambda A^\lambda + \tfrac{1}{4}(d-2)(d-1)A^2. \tag{1.66}$$

For $A_\mu = \partial_\mu \ln \Omega^2$ we recover the relations between the curvature and Ricci tensors of a metric before and after a Weyl rescaling in Appendix M. The Ricci scalar has an overall factor $\Omega^{-2}$ because its definition uses an inverse metric.

### 1.3.3 Riemann spacetime $\mathrm{V}_d$

It is defined by the conditions $Q = T = 0$ which determine the connection to be the Levi-Civita connection $\Gamma(g)$ whose components in a coordinate basis are given by the Christoffel symbols. In a Riemann spacetime one can construct infinitesimal parallelograms, and autoparallel curves are also geodesics (as in flat spacetime). There are also additional interesting properties. To start with, we can write the transformation of tensors under infinitesimal GCTs (Lie derivatives) in terms of covariant derivatives alone (all torsion terms vanish). In particular, for the metric and $r$-forms we can write

$$\begin{aligned} \delta_\xi g_{\mu\nu} &= -2\nabla_{(\mu}\xi_{\nu)}, \\ \delta_\xi B_{\mu_1\cdots\mu_r} &= -\xi^\lambda \nabla_\lambda B_{\mu_1\cdots\mu_r} - r(\nabla_{[\mu_1|}\xi^\lambda)B_{\lambda|\mu_2\cdots\mu_r]}. \end{aligned} \tag{1.67}$$

Furthermore, we have the usual identity

$$\Gamma_{\rho\mu}{}^{\rho} = \partial_\mu \ln\left(\sqrt{|g|}\right), \tag{1.68}$$

which allows us to write the Laplacian of a scalar function $f$ in this way:

$$\nabla^2 f = \frac{1}{\sqrt{|g|}}\partial_\mu\Big(\sqrt{|g|}\,\partial^\mu f\Big), \tag{1.69}$$

and the divergence of a completely antisymmetric tensor ($k$-form) in this way:[8]

$$\nabla_{\mu_1} F^{\mu_1\mu_2\cdots\mu_k} = \frac{1}{\sqrt{|g|}}\partial_{\mu_1}\Big(\sqrt{|g|}\,F^{\mu_1\mu_2\cdots\mu_k}\Big). \tag{1.70}$$

The Bianchi identities take the form

$$R_{(\alpha\beta)\gamma}{}^{\delta} = 0, \qquad R_{[\alpha\beta\gamma]}{}^{\delta} = 0, \qquad \nabla_{[\alpha} R_{\beta\gamma]\rho}{}^{\sigma} = 0, \qquad R_{\alpha\beta(\gamma\delta)} = 0. \tag{1.71}$$

The first and fourth identities imply together

$$R_{\alpha\beta\gamma\delta} = R_{\gamma\delta\alpha\beta}, \tag{1.72}$$

which in turn implies that the Ricci and Einstein tensors are symmetric. The contracted Bianchi identity says now that the Einstein tensor is divergence free:

$$\nabla_\mu G^{\mu\nu} = 0, \tag{1.73}$$

which is a crucial identity in the development of general relativity.

The number of independent components of the curvature in $d$ dimensions after taking into account all these Bianchi identities is $(1/12)\,d^2(d^2-1)$.

The four-dimensional curvature tensor can be split into different pieces which transform irreducibly under the Lorentz group: a scalar piece $D(0,0)$, which is nothing but the Ricci scalar $R$, a two-index symmetric, a traceless piece $R_{\mu\nu} - \frac{1}{4}g_{\mu\nu}R$ (corresponding to the representation $D(1,1)$), and a four-index tensor with the same symmetries as the Riemann tensor but traceless: the Weyl tensor $C_{\mu\nu\rho}{}^{\sigma}$ with $C_{\mu\sigma\rho}{}^{\sigma} = 0$,

$$R_{\mu\nu}{}^{\rho\sigma} = C_{\mu\nu}{}^{\rho\sigma} + 2\Big(R_{[\mu}{}^{[\rho} - \tfrac{1}{4}Rg_{[\mu}{}^{[\rho}\Big)g_{\nu]}{}^{\sigma]} + \tfrac{1}{6}Rg_{[\mu}{}^{\rho}g_{\nu]}{}^{\sigma}. \tag{1.74}$$

The Weyl tensor can be decomposed into its self-dual and anti-self-dual parts (with respect to the last two indices). These two complex tensors transform in the $D(2,0)$ and $D(0,2)$ representations, respectively.

In $d$ dimensions the Weyl tensor is defined by

$$C_{\mu\nu}{}^{\rho\sigma} = R_{\mu\nu}{}^{\rho\sigma} - \frac{4}{d-2}R_{[\mu}{}^{[\rho}g_{\nu]}{}^{\sigma]} + \frac{2}{(d-1)(d-2)}R\,g_{[\mu}{}^{[\rho}g_{\nu]}{}^{\sigma]}. \tag{1.75}$$

The main property of the Weyl tensor $C_{\mu\nu\rho}{}^{\sigma}$ with the indices in these positions is that it is left invariant by Weyl rescalings of the metric (see Appendix M). Furthermore, just as the Riemann curvature vanishes only for Minkowski spacetime, the Weyl tensor vanishes only for conformally flat (Minkowski) spacetimes, i.e. spacetimes that are related to Minkowski's by a given conformal transformation.

[8] Observe that this implies that the second term on the r.h.s. of Eq. (1.60) times $\sqrt{|g|}$ is a total derivative.

A final property of the Levi-Civita connection that is worth mentioning is the form of its variation under an arbitrary variation of the metric:

$$\delta\Gamma_{\mu\nu}{}^{\rho}(g) = \tfrac{1}{2}g^{\rho\sigma}\{\nabla_\mu \delta g_{\nu\sigma} + \nabla_\nu \delta g_{\mu\sigma} - \nabla_\sigma \delta g_{\mu\nu}\}\,. \tag{1.76}$$

Since $\delta g_{\mu\nu}$ is a tensor, $\delta\Gamma$ is a tensor even though $\Gamma$ is not.

## 1.4 Tangent space

So far we have considered, for a given coordinate system, only one basis in tangent space: the coordinate basis. We are now going to consider an arbitrary basis in tangent space. Such a basis is defined by a set of $d$ contravariant vectors labeled by a *tangent-space index* $a$ $\{e_a = e_a{}^\mu \partial_\mu\}$ and is also referred to as a *frame* or, generically, *Vielbein* basis.[9] The coordinate basis is now a particular case in which $e_a{}^\mu = \delta_a{}^\mu$. Now we can express any vector in this basis $\xi = \xi^a e_a$ and its components $\xi^a$ will be related to the coordinate basis components by

$$\xi^\mu = \xi^a e_a{}^\mu. \tag{1.77}$$

We can immediately define the dual basis of 1-forms $\{e^a = e^a{}_\mu dx^\mu\}$ defined by

$$\langle e^a | e_b \rangle = \delta^a{}_b, \tag{1.78}$$

which implies that the matrix of components $e^a{}_\mu$ of the 1-forms in the coordinate basis is the inverse, transposed, of that of the vectors:

$$e^a{}_\mu e_b{}^\mu = \delta^a{}_b, \quad \Rightarrow e_a{}^\mu e^a{}_\nu = \delta^\mu{}_\nu. \tag{1.79}$$

We can now relate frame and world indices of any tensor using these two matrices. In particular, we can use the frame components $g_{ab}$ of the metric,

$$g_{ab} = e_a{}^\mu e_b{}^\nu g_{\mu\nu}, \tag{1.80}$$

that can also be interpreted as the matrix of inner products of the Vielbein basis $g(e_a, e_b) = g_{ab}$. An orthonormal Vielbein basis leads to $g_{ab} = \eta_{ab}$. Frames are usually chosen in such a way as to obtain a particular $g_{ab}$ and orthonormal frames will be particularly important in what follows.

It is easy to see that $g_{ab}$ and its inverse $g^{ab}$ can be consistently used to raise and lower frame indices. In particular,

$$e^a{}_\mu = g_{\mu\nu} e_b{}^\nu g^{ab}, \qquad g_{\mu\nu} = e^a{}_\mu e^b{}_\nu g_{ab}. \tag{1.81}$$

A frame is invariant under GCTs (only the components in the coordinate basis change) and, thus, frame components of any tensor are also invariant. However, we can make a change of basis. Any two Vielbein bases are related by a $\mathrm{GL}(d,\mathbb{R})$ transformation $\Lambda^a{}_b$ in tangent space at a given point of the manifold. This transformation can in fact be different

[9] *Einbein* for $d=1$, *Zweibein* for $d=2$, *Dreibein* for $d=3$, *Vierbein* for $d=4$, etc. In four dimensions it is also called a *tetrad*.

at each point and thus we have to consider local frame transformations $\Lambda^a{}_b(x)$. We write their action on vectors and forms as follows:

$$e'_a = e_b\left(\Lambda^{-1}\right)^b{}_a, \qquad e'^a = \Lambda^a{}_b e^b. \tag{1.82}$$

The *Ricci rotation coefficients* (or *anholonomy coefficients*) $\Omega_{ab}{}^c$ are the Lie brackets

$$[e_a, e_b] = -2\Omega_{ab}{}^c e_c, \qquad \Omega_{ab}{}^c = e_a{}^\mu e_b{}^\nu \partial_{[\mu} e^c{}_{\nu]}. \tag{1.83}$$

A *non-holonomic* frame is one with non-vanishing $\Omega$s. Observe that, given a basis of vectors $\{e_a\}$, we could try to find a new set of coordinates $y^a(x^\mu)$ such that

$$e_a y^b = e_a{}^\mu \partial_\mu y^b = \delta_a{}^b. \tag{1.84}$$

The integrability condition for the system of partial differential equations $[e_c, e_a]y^b = 0$ is precisely the vanishing of the anholonomy coefficients $\Omega_{ab}{}^c$. A holonomic basis of vectors $\{e_a\}$ is one for which these coefficients vanish and then we can trivialize them ($e_a{}^\mu = \delta_a{}^\mu$) by a change of coordinates.

Just as we defined a covariant derivative transforming world tensors into world tensors we are now going to define a derivative that transforms tangent-space tensors into tangent-space tensors transforming well under local $\mathrm{GL}(d,\mathbb{R})$ transformations associated with a connection $\omega$. Its action on scalars, vectors, and forms is[10]

$$\begin{aligned}
\mathcal{D}_a\phi &= \partial_a\phi, && (= e_a{}^\mu \partial_\mu \phi),\\
\mathcal{D}_a\xi^b &= \partial_a\xi^b + \omega_{ac}{}^b \xi^c, && (= e_a{}^\mu \mathcal{D}_\mu \xi^b),\\
\mathcal{D}_a\varepsilon_b &= \partial_a\varepsilon_b - \varepsilon_c\,\omega_{ab}{}^c, && (= e_a{}^\mu \mathcal{D}_\mu \varepsilon_b).
\end{aligned} \tag{1.85}$$

Local $\mathrm{GL}(d,\mathbb{R})$ covariance implies the inhomogeneous transformation law for the connection:

$$\omega'_{ab}{}^c = \left[\Lambda^c{}_d \omega_{ef}{}^d (\Lambda^{-1})^f{}_b - (\Lambda^{-1})^d{}_b \partial_e \Lambda^c{}_d\right](\Lambda^{-1})^e{}_a. \tag{1.86}$$

The curvature of this connection can be defined through the Ricci identities in the standard fashion (observe that there are no torsion terms here):

$$\begin{aligned}
[\mathcal{D}_\mu, \mathcal{D}_\nu]\phi &= 0,\\
[\mathcal{D}_\mu, \mathcal{D}_\nu]\xi^a &= R_{\mu\nu b}{}^a \xi^b,\\
[\mathcal{D}_\mu, \mathcal{D}_\nu]\varepsilon_a &= -\varepsilon_b\, R_{\mu\nu a}{}^b,
\end{aligned} \tag{1.87}$$

and then the curvature is given by[11]

$$R_{\mu\nu a}{}^b = 2\partial_{[\mu}\,\omega_{\nu]a}{}^b - 2\omega_{[\mu|a}{}^c\,\omega_{|\nu]c}{}^b. \tag{1.89}$$

[10] Of course, the formalism we are developing is just that of a $\mathrm{GL}(d,\mathbb{R})$ gauge theory and our notation is basically identical to that of Appendix A. Here we are dealing with vector representations of $\mathrm{GL}(d,\mathbb{R})$. The $d^2$ generators of its Lie algebra can be labeled by a pair of vector indices $ab$ and they are given, for instance, by $(T_{ab})^c{}_d = -\eta_{ad}\eta_b{}^c$. Thus $A_\mu{}^I \Gamma_{\mathrm{v}}(T_I) = \omega_\mu{}^{ab}(T_{ab})^c{}_d = -\omega_{\mu\,d}{}^c$. The subgroup $\mathrm{SO}(1, d-1, \mathbb{R})$ will be treated in more detail.

[11] Observe that, with all Latin indices, $R_{abc}{}^d = e_a{}^\mu e_b{}^\nu R_{\mu\nu c}{}^d$ and, therefore,

$$R_{abc}{}^d = 2\partial_{[a}\,\omega_{b]c}{}^d - 2\omega_{[a|c}{}^e\,\omega_{|b]e}{}^d + 2\Omega_{ab}{}^e\omega_{ec}{}^d. \tag{1.88}$$

At this point we have introduced a new connection $\omega$ that is independent of the metric. In the previous section we managed to relate the connection $\Gamma$ to the metric via the metric postulate. Here we are going to generalize the metric postulate first to relate the two connections (the first Vielbein postulate) and then to relate them to the metric (the second Vielbein postulate). Before we enunciate these postulates we introduce the *total covariant derivative*, covariant with respect to all the indices of the object it acts on. We denote it by $\nabla$ again, and, for instance, acting on Vielbeins it is

$$\nabla_\mu e_a{}^\nu = \partial_\mu e_a{}^\nu + \Gamma_{\mu\rho}{}^\nu e_a{}^\rho - e_b{}^\nu \omega_{\mu a}{}^b. \tag{1.90}$$

We can motivate the first Vielbein postulate as follows: we would like to be able to convert tangent into world indices and vice versa inside the total covariant derivative, so $e^a{}_\nu \nabla_\mu \xi^\nu = \mathcal{D}_\mu \xi^a$ and $\mathcal{D}$ is just the projection of $\nabla$ onto the Vielbein basis. To have this property we impose the *first Vielbein postulate*,

$$\nabla_\mu e_a{}^\nu = 0. \tag{1.91}$$

It is worth stressing that this does not imply the covariant constancy of the metric $\nabla_\mu g_{\nu\rho} = 0$. The above postulate implies the following relation between the connections:

$$\omega_{\mu a}{}^b = \Gamma_{\mu a}{}^b - e_a{}^\nu \partial_\mu e^b{}_\nu. \tag{1.92}$$

Furthermore, the curvatures of the two connections are now related by

$$R_{\mu\nu\rho}{}^\sigma(\Gamma) = e^a{}_\rho e_b{}^\sigma R_{\mu\nu a}{}^b(\omega). \tag{1.93}$$

The first Vielbein postulate also gives an important relation between the torsion and the Vielbein: on taking the antisymmetric part of $\nabla_\mu e^a{}_\nu = 0$, we obtain

$$2\mathcal{D}_{[\mu} e^a{}_{\nu]} = 2\left(\partial_{[\mu} e^a{}_{\nu]} - \omega_{[\mu}{}^a{}_{\nu]}\right) = -T_{\mu\nu}{}^a. \tag{1.94}$$

The significance of the torsion in this formalism, from the point of view of the gauge theory of $\mathrm{GL}(d,\mathbb{R})$, is unclear. We can provide an interpretation in the framework of the gauge theory of the affine group $\mathrm{IGL}(d,\mathbb{R})$ but we will do it in the more restricted context of the Lorentz and Poincaré groups in Section 4.5.

The first Vielbein postulate has allowed us to recover the structure of affinely connected spacetime $(\mathrm{L}_d, g)$ with only one (independent) connection, generalized to allow the use of an arbitrary basis in tangent space. Furthermore, we can recover the different particular structures that we defined in the previous section, also generalized to allow the use of arbitrary basis in tangent space. First, if $\Gamma$ is a completely general connection, it is given by Eq. (1.43) and then $\omega$ (which is related to $\Gamma$ by the first Vielbein postulate) is given by

$$\omega_{ab}{}^c = \omega_{ab}{}^c(e) + K_{ab}{}^c + L_{ab}{}^c, \tag{1.95}$$

where $\omega(e)$ is the (*Cartan* or even *Levi-Civita*) connection related to the Levi-Civita connection $\Gamma(g)$ by Eq. (1.92). It is completely determined by the Vielbeins through the anholonomy coefficients defined in Eq. (1.83):

$$\omega_{ab}{}^c(e) = \left\{ {c \atop a\,b} \right\} + \{-\Omega_{ab}{}^c + \Omega_b{}^c{}_a - \Omega^c{}_{ab}\}, \tag{1.96}$$

where

$$\left\{ {c \atop a\,b} \right\} = \tfrac{1}{2} g^{cd} \{ \partial_a g_{bd} + \partial_b g_{ad} - \partial_d g_{ab} \}. \tag{1.97}$$

$K_{ab}{}^c$ is nothing but the contorsion tensor expressed in a tangent-space basis, i.e. $K_{ab}{}^c = e_a{}^\mu e_b{}^\nu e^c{}_\rho K_{\mu\nu}{}^\rho$ and, similarly, $L_{ab}{}^c = e_a{}^\mu e_b{}^\nu e^c{}_\rho L_{\mu\nu}{}^\rho$.

Observe that

$$\omega_{a(bc)} = \tfrac{1}{2}(Q_{abc} + \partial_a g_{bc}). \tag{1.98}$$

We can impose the metric-compatibility condition Eq. (1.49), which in this context is known as the *second Vielbein postulate*, and we have a Riemann–Cartan spacetime $\mathrm{U}_d$. The result is that $\Gamma$ is again given by Eqs. (1.54), (1.44), and (1.45), and $\omega$ (which is related to $\Gamma$ by the first Vielbein postulate) is given by

$$\omega_{ab}{}^c = \omega_{ab}{}^c(e) + K_{ab}{}^c. \tag{1.99}$$

If we now impose the vanishing of torsion, we obtain the Levi-Civita and Cartan connections $\Gamma(g)$ and $\omega(e)$ and we recover a Riemann spacetime $\mathrm{V}_d$.

The two most important cases to which we can apply this general formalism are the following.

1. The case in which we use a coordinate basis $e_a{}^\mu = \delta_a{}^\mu$, so $g_{ab} = g_{\mu\nu}$, $\Omega = 0$, and the connections $\Gamma$ and $\omega$ are identical.
2. The case in which we use an orthonormal basis $g_{ab} = \eta_{ab}$ in which

$$\left\{ {c \atop a\,b} \right\} = 0$$

and

$$\omega_{abc} = \omega_{abc}(e) + K_{abc} + L_{abc}, \qquad \omega_{abc}(e) = -\Omega_{abc} + \Omega_{bca} - \Omega_{cab}. \tag{1.100}$$

In the second case we would like to restrict ourselves to those changes of frame that preserve the form of the metric in tangent-space indices (here usually referred to as *flat indices* because they are raised and lowered with the flat space metric). By definition, these are transformations of the $d$-dimensional Lorentz group $\mathrm{SO}(1, d-1)$ whose gauge theory we are now led to consider. This gauge theory is developed in Section A.2.3 of Appendix A and the spinorial representations of the Lorentz group are studied in Appendix D. We are simply going to rewrite here the main formulae we have obtained, adapted to the Lorentz subgroup of $\mathrm{GL}(d, \mathbb{R})$.

The main justification for making this step is that the Lorentz group admits spinorial representations, which are necessary in order to describe fermions, whereas the diffeomorphism group of a manifold does not. This is the only known method by which to describe spinors in curved spacetime in arbitrary coordinates and, thus, the only method known to couple fermions to gravity. This formalism was pioneered by Weyl [1252].

First of all, the generators $M_I$ of the Lorentz subgroup of $\mathrm{GL}(d,\mathbb{R})$ are just the antisymmetric combinations of those of $\mathrm{GL}(d,\mathbb{R})$ and can be labeled by two antisymmetric vector indices, i.e. $M_{ab}$. In this notation every generator appears twice and factors of $\frac{1}{2}$ have to be included in the right places. However, in general, the connection $\omega_\mu{}^{ab}$ is not antisymmetric in the "gauge" indices $ab$ unless it is also metric compatible ($\mathcal{D}\eta_{ab}=0$), according to Eq. (1.98). We are going to consider only metric-compatible connections that are fully antisymmetric in the gauge indices and we will call them *spin connections*.[12]

Using the explicit form of the infinitesimal Lorentz generators in the vector representation $\Gamma_{\rm v}(M_{bc})^a{}_d$ given in Eq. (A.60) and in the spinorial representation $\Gamma_{\rm s}(M_{ab})^\alpha{}_\beta$ (we use temporarily the first few Greek letters $\alpha,\beta,\ldots$ as spinorial indices) given in Eq. (D.3), we find the following expressions for the (total) covariant derivatives of contravariant and covariant vectors and spinors:

$$
\begin{aligned}
\nabla_\mu \xi^a &= \partial_\mu \xi^a - \tfrac{1}{2}\omega_\mu{}^{bc}\Gamma_{\rm v}(M_{bc})^a_d \xi^d = \partial_\mu \xi^a + \omega_{\mu b}{}^a \xi^b,\\
\nabla_\mu \varepsilon_a &= \partial_\mu \varepsilon_a + \varepsilon_d \tfrac{1}{2}\omega_\mu{}^{bc}\Gamma_{\rm v}(M_{bc})^d_a = \partial_\mu \varepsilon_a - \varepsilon_b\, \omega_{\mu a}{}^b,\\
\nabla_\mu \psi^\alpha &= \partial_\mu \psi^\alpha - \tfrac{1}{2}\omega_\mu{}^{ab}\Gamma_{\rm s}(M_{ab})^\alpha{}_\beta \psi^\beta = \partial_\mu \psi^\alpha - \tfrac{1}{4}\omega_\mu{}^{ab}(\Gamma_{ab})^\alpha{}_\beta \psi^\beta,\\
\nabla_\mu \varphi_\alpha &= \partial_\mu \varphi_\alpha + \varphi_\beta \tfrac{1}{2}\omega_\mu{}^{ab}\Gamma_{\rm s}(M_{ab})^\beta{}_\alpha = \partial_\mu \varphi_\alpha + \varphi_\beta \tfrac{1}{4}\omega_\mu{}^{ab}(\Gamma_{ab})^\beta{}_\alpha.
\end{aligned}
\tag{1.101}
$$

These definitions, once we impose the Vielbein postulates, are consistent with the raising and lowering of vector indices with the Minkowski metric and with Dirac conjugation of the spinors. With the postulates, the spin connection is given by Eqs. (1.100).

The Ricci identities can now be written for the total covariant derivative in this form:

$$
\begin{aligned}
[\nabla_\mu,\nabla_\nu]\,\phi &= T_{\mu\nu}{}^\rho \nabla_\rho \phi,\\
[\nabla_\mu,\nabla_\nu]\,\xi^a &= R_{\mu\nu b}{}^a(\omega)\xi^b + T_{\mu\nu}{}^\rho \nabla_\rho \xi^a,\\
[\nabla_\mu,\nabla_\nu]\,\varepsilon_a &= -\varepsilon_b R_{\mu\nu a}{}^b(\omega) + T_{\mu\nu}{}^\rho \nabla_\rho \varepsilon_a,\\
[\nabla_\mu,\nabla_\nu]\,\psi &= -\tfrac{1}{4}R_{\mu\nu}{}^{ab}(\omega)\Gamma_{ab}\psi + T_{\mu\nu}{}^\rho \nabla_\rho \psi,\\
[\nabla_\mu,\nabla_\nu]\,\varphi &= +\tfrac{1}{4}\varphi R_{\mu\nu}{}^{ab}(\omega)\Gamma_{ab} + T_{\mu\nu}{}^\rho \nabla_\rho \varphi.
\end{aligned}
\tag{1.102}
$$

For more general tensors one has to add a curvature ($\omega$) term for each flat index and a curvature ($\Gamma$) term for each world index. The curvatures have the same form as in Eqs. (1.27) and (1.89) but now $R_{\mu\nu}{}^{ab}$ is antisymmetric in $ab$.

The following expression is sometimes used:

$$
R_{ab} = -\partial_a \omega_c{}^c{}_b - \partial_c \omega_{ab}{}^c + \omega_{cda}\omega^{dc}{}_b + \omega_{abd}\omega_c{}^{cd}.
\tag{1.103}
$$

The Vielbein formalism allows us to study the Weitzenböck spacetime defined on p. 11.

### 1.4.1 Weitzenböck spacetime $\mathrm{A}_d$

This spacetime is defined by a metric-compatible connection that we denote by $W_{\mu\nu}{}^\rho$ and call the *Weitzenböck connection* [1243, 1244] whose Riemann curvature is identically zero,

[12] If we wanted to have a non-metric-compatible spin connection, we would have to modify the first Vielbein postulate.

$R_{\mu\nu\rho}{}^{\sigma}(W) = 0$. Trying to solve this equation directly for $W \neq 0$ is a very difficult task. However, we can use the Vielbein formalism to find a solution. Let us denote by $W_{\text{s}\,\mu}{}^{ab}$ the tangent-space connection associated with $W$ via the first Vielbein postulate

$$\nabla_\mu e^a{}_\nu = \partial_\mu e^a{}_\nu - W_{\mu\nu}{}^a + W_{\text{s}\,\mu\nu}{}^a = 0. \tag{1.104}$$

The curvature of $W_\text{s}$ is obviously zero on account of Eq. (1.93). Now, however, we can use the trivial solution to the equation $R_{\mu\nu}{}^{ab}(W_\text{s}) = 0$, namely $W_\text{s} = 0$, because, according to the above relation, $W_\text{s} = 0$ does not imply $W = 0$ but

$$W_{\mu\nu}{}^\rho = e_a{}^\rho \partial_\mu e^a{}_\nu. \tag{1.105}$$

This is the Weitzenböck connection whose curvature vanishes identically. It cannot be rewritten in terms of the metric: it is necessary to use the Vielbein formalism. Observe that, using this connection, we can write the relation between any two connections $\Gamma$ and $\omega$ satisfying the first Vielbein postulate in the form

$$\Gamma_{\mu\nu}{}^\rho = W_{\mu\nu}{}^\rho + \omega_{\mu\nu}{}^\rho. \tag{1.106}$$

$\omega_{\mu\nu}{}^\rho$ is a tensor, but $\Gamma_{\mu\nu}{}^\rho$ is not (it is an affine connection), and responsible for this is the Weitzenböck connection $W_{\mu\nu}{}^\rho$. We can also write

$$\omega_\mu{}^{ab} = \Gamma_\mu{}^{ab} - W_\mu{}^{ab}, \qquad W_\mu{}^{ab} = e^{a\nu} \partial_\mu e^b{}_\nu. \tag{1.107}$$

Now $\Gamma_\mu{}^{ab}$ is a $\text{GL}(d,\mathbb{R})$ tensor in the upper two indices whereas $\omega_\mu{}^{ab}$ is not (because it is a $\text{GL}(d,\mathbb{R})$ connection). Again, the Weitzenböck connection $W_\mu{}^{ab}$ is responsible for this.

Even though we have to search explicitly for a metric-compatible connection to find $W$, it is easy to check that it is indeed metric compatible. Then, it can be decomposed into the sum of the Levi-Civita connection and the contorsion tensor. The torsion tensor is

$$T_{\mu\nu}{}^\rho = -2\Omega_{\mu\nu}{}^\rho, \tag{1.108}$$

and, therefore, the contorsion tensor is given by

$$K_{\mu\nu\rho}(W) = \Omega_{\mu\nu\rho} - \Omega_{\nu\rho\mu} + \Omega_{\rho\mu\nu} = -\omega_{\mu\nu\rho}(e), \tag{1.109}$$

where $\omega(e)$ is, as usual, the Cartan connection (which is associated via the first Vielbein postulate with the Levi-Civita connection $\Gamma(g)$).

Now, if we use Eqs. (1.31) and (1.32) for $\tilde{\Gamma} = W, \Gamma = \Gamma(g)$, and $\tau = K$, we find an expression for the Riemann curvature tensor of the Levi-Civita connection in terms of the contorsion tensor of the Weitzenböck connection:

$$R_{\mu\nu\rho}{}^\sigma[\Gamma(g)] = -2\nabla_{[\mu} K_{\nu]\rho}{}^\sigma - 2K_{[\mu|\lambda}{}^\sigma K_{|\nu]\rho}{}^\lambda. \tag{1.110}$$

On contracting indices, and eliminating a total derivative, we find that

$$\int d^dx\,\sqrt{|g|}\,R(g) = -\int d^dx\,\sqrt{|g|}\,\left\{K_\mu{}^{\mu\lambda} K_\mu{}^\mu{}_\lambda + K_\nu{}^{\mu\rho} K_{\mu\rho}{}^\nu\right\}, \tag{1.111}$$

which can be expressed entirely in terms of the anholonomy coefficients $\Omega_{\mu\nu\rho}$, providing an alternative form of the Einstein–Hilbert action. This is, in fact, an alternative way of deriving Palatini's identity Eq. (L.4).

It is worth stressing here that the building blocks of the Riemann curvature tensor of the Levi-Civita connection in the above expression (the anholonomy coefficients/torsion and contorsion) are tensors, whereas in the standard expression for the curvature the building blocks are the Christoffel symbols, which are not tensors.

The main property of the Weitzenböck spacetime (the vanishing of the curvature) implies that parallel transport is path independent and it is possible to define parallelism of vectors *at different spacetime points*:[13] two vectors $v^\mu(x_1), w^\mu(x_2)$ are parallel if their components in the Vielbein basis $\{e_a{}^\mu\}$ are proportional. This is a consistent definition because the components in the Vielbein basis are invariant under the parallel transport defined by the $W$ connection associated with that Vielbein basis. Indeed, the vector $v^\mu(x)$, parallel-transported to $x^\mu + \epsilon^\mu$ is

$$v^\mu(x+\epsilon) = v^\mu(x) - \epsilon^\nu v^\rho(x) W_{\nu\rho}{}^\mu(x), \tag{1.112}$$

and its tangent-space components can be found with the inverse Vielbein basis at $x^\mu + \epsilon^\mu$:

$$\begin{aligned} e^a{}_\mu(x+\epsilon)v^\mu(x+\epsilon) &= [e^a{}_\mu(x) + \epsilon^\nu \partial_\nu e^a{}_\mu(x)][v^\mu(x) - \epsilon^\nu v^\rho(x) W_{\nu\rho}{}^\mu(x)] \\ &= e^a{}_\mu(x) v^\mu(x). \end{aligned} \tag{1.113}$$

Also, it can be shown that the vanishing of the curvature is equivalent to the existence of $d$ vector fields (the Vielbeins) covariantly constant with respect to the $W$ connection,

$$\overset{W}{\nabla}_\mu e^a{}_\nu = \partial_\mu e^a{}_\nu - W_{\mu\nu}{}^\rho e^a{}_\rho = 0. \tag{1.114}$$

## 1.5 Killing vectors

If, given a metric $g_{\mu\nu}$, there exists a vector field $k^\mu$ such that the Lie derivative of $g_{\mu\nu}$ with respect to it vanishes,

$$\mathcal{L}_k g_{\mu\nu} = 2\nabla_{(\mu} k_{\nu)} = 0, \tag{1.115}$$

we say that $g_{\mu\nu}$ admits the *Killing vector* $k^\mu$. The above equation is the *Killing equation*. It means that the metric does not change along the integral curves of $k^\mu$ and it is also said that the metric possesses an *isometry* in the direction $k^\mu$. If the metric does not change along the integral curves of a Killing vector, and we use as a coordinate the parameter of those integral curves (*adapted coordinates*), then the metric does not depend on that coordinate.

The Ricci identity implies the following consistency condition:[14]

$$\nabla_\alpha \nabla_\beta k_\nu = -R^\lambda{}_{\alpha\beta\nu} k_\lambda, \quad \Rightarrow \nabla^2 k^\mu = R^\mu{}_\nu k^\nu. \tag{1.116}$$

---

[13] Also known as *teleparallelism* or *absolute parallelism*.

[14] These equations can be obtained as follows: from the Killing equation we have

$$\nabla_\alpha \nabla_{(\beta} k_{\gamma)} = 0, \qquad \text{and} \qquad \nabla_\alpha \nabla_{(\beta} k_{\gamma)} + \nabla_\beta \nabla_{(\gamma} k_{\alpha)} - \nabla_\gamma \nabla_{(\alpha} k_{\beta)} = 0.$$

The last equation can be rewritten in the form

$$\nabla_{(\alpha}\nabla_{\beta)} k_\gamma = -\tfrac{1}{2}[\nabla_\alpha, \nabla_\gamma] k_\beta - \tfrac{1}{2}[\nabla_\beta, \nabla_\gamma] k_\alpha,$$

A weaker but also interesting property that a metric can have is a *conformal isometry*. This happens when there is a vector field $c^\mu$ along whose integral curves the metric changes only by a conformal factor,

$$\mathcal{L}_c g_{\mu\nu} = 2\nabla_{(\mu}c_{\nu)} = 2\lambda g_{\mu\nu}. \tag{1.117}$$

On taking the trace of the above equation, we find in $d$ dimensions

$$\lambda = \frac{1}{d}\nabla_\mu c^\mu, \quad \Rightarrow \nabla_{(\mu}c_{\nu)} - \frac{1}{d}g_{\mu\nu}\nabla_\rho c^\rho = 0, \tag{1.118}$$

called the *conformal Killing equation*; $c^\mu$ is then known as a *conformal Killing vector*.

## 1.6 Duality operations

The antisymmetric Levi-Civita tensor is defined in $d$ dimensions in tangent space by

$$\epsilon^{01\cdots(d-1)} = +1, \qquad \Rightarrow \epsilon_{01\cdots(d-1)} = (-1)^{d-1}, \tag{1.119}$$

and in curved indices by

$$\epsilon^{\mu_1\cdots\mu_d} = \sqrt{|g|}\, e^{\mu_1}{}_{a_1}\cdots e^{\mu_d}{}_{a_d}\epsilon^{a_1\cdots a_d}, \tag{1.120}$$

so, with upper indices, it is independent of the metric and, in curved indices, which we underline to distinguish them from the tangent-space ones,

$$\epsilon^{\underline{0\cdots(d-1)}} = +1, \qquad \epsilon_{\underline{0\cdots(d-1)}} = g = (-1)^{d-1}|g|. \tag{1.121}$$

The contraction of $n$ indices of two $\epsilon$ symbols gives

$$\epsilon^{\mu_1\cdots\mu_n\rho_1\cdots\rho_{(d-n)}}\epsilon_{\mu_1\cdots\mu_n\sigma_1\cdots\sigma_{(d-n)}} = n!(d-n)!\, g\, g^{\rho_1\cdots\rho_{(d-n)}}{}_{\sigma_1\cdots\sigma_{(d-n)}}, \tag{1.122}$$

where

$$\begin{aligned} g^{\rho_1\cdots\rho_{(d-n)}}{}_{\sigma_1\cdots\sigma_{(d-n)}} &= g_{[\rho_1}{}^{\sigma_1}\cdots g_{\rho_{(d-n)}]}{}^{\sigma_{(d-n)}} \\ &= \frac{1}{(d-n)!}\begin{vmatrix} g_{\rho_1}{}^{\sigma_1} & \cdots & g_{\rho_1}{}^{\sigma_{(d-n)}} \\ \vdots & \vdots & \vdots \\ g_{\rho_{(d-n)}}{}^{\sigma_1} & \cdots & g_{\rho_{(d-n)}}{}^{\sigma_{(d-n)}} \end{vmatrix}. \end{aligned} \tag{1.123}$$

---

and, using the Ricci identities for the l.h.s.,

$$\nabla_{(\alpha}\nabla_{\beta)}k_\gamma = \tfrac{1}{2}k_\lambda R_{\alpha\gamma\beta}{}^\lambda + \tfrac{1}{2}k_\lambda R_{\beta\gamma\alpha}{}^\lambda.$$

Now, using the Bianchi identities for the metric-compatible torsionless curvature we get

$$\nabla_{(\alpha}\nabla_{\beta)}k_\gamma = -k_\lambda R^\lambda{}_{\alpha\beta\gamma} + \tfrac{1}{2}k_\lambda R^\lambda{}_{\gamma\beta\alpha}.$$

Combining this equation with the Ricci identity

$$\nabla_{[\alpha}\nabla_{\beta]}k_\gamma = -\tfrac{1}{2}k_\lambda R_{\alpha\beta\gamma}{}^\lambda = \tfrac{1}{2}k_\lambda R^\lambda{}_{\gamma\alpha\beta},$$

we obtain Eq. (1.116).

We define the *dual* (or the *Hodge dual*) of a completely antisymmetric tensor of rank $k$ (a *differential form of rank* $k$ or $k$-*form*[15]) $F_{(k)}$ as the completely antisymmetric tensor of rank $d-k$, which we denote by $\star F_{(d-k)}$ and whose components are given by

$$\star F_{(k)}{}^{\mu_1\cdots\mu_{(d-k)}} = \frac{1}{k!\sqrt{|g|}}\epsilon^{\mu_1\cdots\mu_{(d-k)}\mu_{(d-k+1)}\cdots\mu_d} F_{(k)\mu_{(d-k+1)}\cdots\mu_d}. \tag{1.124}$$

The dual of the dual is the original tensor up to a sign that depends both on the dimension and on the rank of the tensor,

$$\star\star F_{(k)} = (-1)^{(d-1)+k(d-k)} F_{(k)}. \tag{1.125}$$

An important case is when the spacetime dimension is even and $k=d/2$, so $\star$ (the *Hodge star*) is an operator on the space of rank $d/2$ tensors. Then, we have

$$\begin{aligned} \star\star F_{(d/2)} &= +F_{(d/2)}, \qquad d=4n+2, \\ \star\star F_{(d/2)} &= -F_{(d/2)}, \qquad d=4n, \end{aligned} \tag{1.126}$$

for $n$ an integer. In the former case $\star$ has eigenvalues $+1$ and $-1$ and in the latter $+i$ and $-i$, and any rank $d/2$ tensor can be decomposed into the sum of its self-dual and anti-self-dual parts $F^+$ and $F^-$. For $d=4n+2$

$$\begin{aligned} F^{\pm}_{(d/2)} &= \tfrac{1}{2}\left(F_{(d/2)} \pm \star F_{(d/2)}\right), \\ \star F^{\pm}_{(d/2)} &= \pm F^{\pm}_{(d/2)}, \end{aligned} \tag{1.127}$$

and, for $d=4n$,

$$\begin{aligned} F^{\pm}_{(d/2)} &= \tfrac{1}{2}\left(F_{(d/2)} \mp i \star F_{(d/2)}\right), \\ \star F^{\pm}_{(d/2)} &= \pm i F^{\pm}_{(d/2)}. \end{aligned} \tag{1.128}$$

Real (as opposed to complex) (anti-)self-duality $\star F = (-)+F$ is therefore consistent only in $d=4n+2$ dimensions.

Another important case is when $k=p+2$ and $F_{(k)}$ is the field strength of the potential $A_{(p+1)}$, so $F_{(p+2)\mu_1\cdots\mu_{(p+2)}} = (p+2)\partial_{[\mu_1}A_{(p+1)\mu_2\cdots\mu_{(p+2)}]}$. The kinetic term of its action is normalized as follows:

$$S_{(p)}[A_{(p+1)}] = \int d^dx\sqrt{|g|}\left[\frac{(-1)^{p+1}}{2\cdot(p+2)!}F^2_{(p+2)}\right], \tag{1.129}$$

and its energy–momentum tensor is given by

$$\begin{aligned} T^{A_{(p+1)}}_{\mu\nu} = \frac{-2}{\sqrt{|g|}}\frac{\delta S_{(p)}}{\delta g^{\mu\nu}} = \frac{(-1)^p}{(p+1)!}\Big[ & F_{(p+2)\mu}{}^{\rho_1\cdots\rho_{(p+1)}}F_{(p+2)\nu\rho_1\cdots\rho_{(p+1)}} \\ & -\frac{1}{2(p+2)}g_{\mu\nu}F^2_{(p+2)}\Big]. \end{aligned} \tag{1.130}$$

[15] We will introduce the notation specific for differential forms in the next section.

The rank of its dual tensor is $\tilde{p}+2$, where $\tilde{p}=d-p-4$, and we are interested in rewriting the action and energy–momentum tensor in terms of the dual. We immediately find

$$S_p[A_{(p+1)}] = -\int d^dx\sqrt{|g|}\left[\frac{(-1)^{\tilde{p}+1}}{2\cdot(\tilde{p}+2)!}F^2_{(\tilde{p}+2)}\right] = -S_{\tilde{p}}[\tilde{A}_{(\tilde{p}+1)}], \tag{1.131}$$

which would be the action of a dual vector field $\tilde{A}_{(\tilde{p}+1)}$ such that

$$\star F_{\mu_1\cdots\mu_{(\tilde{p}+2)}} = (\tilde{p}+2)\partial_{[\mu_1}\tilde{A}_{(\tilde{p}+1)\,\mu_2\cdots\mu_{(\tilde{p}+2)}]}.$$

Using

$$\begin{aligned}\star\, F_{(\tilde{p}+2)\mu}{}^{\rho_1\cdots\rho_{(\tilde{p}+1)}} \star F_{(\tilde{p}+2)\nu\rho_1\cdots\rho_{(\tilde{p}+1)}} &= \frac{(-1)^{d-1}(\tilde{p}+1)!}{(p+2)!}g_{\mu\nu}F^2_{(p+2)} \\ &+ \frac{(-1)^{d}(\tilde{p}+1)!}{(p+1)!}F_{(p+2)\mu}{}^{\sigma_1\cdots\sigma_{(p+1)}}F_{(p+2)\nu\sigma_1\cdots\sigma_{(p+1)}},\end{aligned} \tag{1.132}$$

we obtain

$$T_{\mu\nu}^{A_{(p+1)}} = T_{\mu\nu}^{\tilde{A}_{(\tilde{p}+1)}}. \tag{1.133}$$

A useful expression for the energy–momentum tensor is

$$\begin{aligned}T_{\mu\nu}^{A_{(p+1)}} = \frac{1}{2}\Bigg\{&\frac{(-1)^p}{(p+1)!}F_{(p+2)\mu}{}^{\rho_1\cdots\rho_{(p+1)}}F_{(p+2)\nu\rho_1\cdots\rho_{(p+1)}} \\ &+ \frac{(-1)^{\tilde{p}}}{(\tilde{p}+1)!}\star F_{(\tilde{p}+2)\mu}{}^{\rho_1\cdots\rho_{(\tilde{p}+1)}}\star F_{(\tilde{p}+2)\nu\rho_1\cdots\rho_{(\tilde{p}+1)}}\Bigg\}.\end{aligned} \tag{1.134}$$

## 1.7 Differential forms and integration

As we have said before, a *differential form of rank* $k$, or $k$*-form* for short, is nothing but a totally antisymmetric tensor field $\omega_{\mu_1\cdots\mu_k} = \omega_{[\mu_1\cdots\mu_k]}$. We write all $k$-forms in this way:

$$\omega = \frac{1}{k!}\omega_{\mu_1\cdots\mu_k}dx^{\mu_1}\wedge\cdots\wedge dx^{\mu_k}, \tag{1.135}$$

so the action of the *exterior derivative* $d$ on the components is defined by

$$(d\omega)_{\mu_1\cdots\mu_{k+1}} = (k+1)\partial_{[\mu_1}\omega_{\mu_2\cdots\mu_{k+1}]} = (k+1)(\partial\omega)_{\mu_1\cdots\mu_{k+1}}. \tag{1.136}$$

It is not difficult to check using the definitions that the exterior derivative and the Lie derivative commute:

$$\mathcal{L}_\xi d\omega = d(\mathcal{L}_\xi\omega). \tag{1.137}$$

The *Hodge dual* is defined by[16]

$$(\star\omega)_{\mu_1\cdots\mu_{n-k}} = \frac{1}{k!\sqrt{|g|}}\epsilon_{\mu_1\cdots\mu_{n-k}\nu_1\cdots\nu_k}\omega^{\nu_1\cdots\nu_k}, \tag{1.138}$$

[16] Observe that we need a metric to do it and that the dual depends explicitly on that metric.

and, as before,

$$(\star)^2 = (-1)^{k(d-k)} \operatorname{sign} g = (-1)^{k(d-k)+d-1}. \tag{1.139}$$

The adjoint of $d$ with respect to the inner product of $k$-forms defined on manifolds M without border $\partial\mathrm{M} = \phi$

$$(\alpha_k|\beta_k) = \int_{\mathrm{M}} \alpha_k \wedge \star\beta_k, \tag{1.140}$$

is defined by

$$(\alpha_k|d\beta_{k-1}) = (\delta\alpha_k|\beta_{k-1}), \quad \Rightarrow \delta = (-1)^{d(k-1)-1} \operatorname{sign} g \ \star d\star. \tag{1.141}$$

Since

$$(\star d \star\omega)_{\rho_1\cdots\rho_{k-1}} = (-1)^{k(d-k+1)-1} \operatorname{sign} g\, \nabla_\mu \omega^\mu{}_{\rho_1\cdots\rho_{k-1}}, \tag{1.142}$$

we find that the relation between $\delta$ and the divergence is

$$(\delta\omega)_{\rho_1\cdots\rho_{k-1}} = (-1)^d \nabla_\mu \omega^\mu{}_{\rho_1\cdots\rho_{k-1}}. \tag{1.143}$$

Only $k$-forms can be integrated on $k$-dimensional manifolds. If $\omega$ is a $(d-1)$-form defined on a $d$-dimensional manifold M with boundary $\partial$M, then *Stokes' theorem* states that

$$\int_{\mathrm{M}} d\omega = \int_{\partial\mathrm{M}} \omega. \tag{1.144}$$

It is convenient to define *volume forms* for a manifold and its lower-dimensional submanifolds. Their contraction with other tensors results in differential forms that can be integrated. Thus, we define in a $d$-dimensional manifold, for $(d-n)$-dimensional submanifolds $\mathrm{M}^{d-n}$, $0 \leq n \leq d$, the volume forms

$$d^{d-n}\Sigma_{\mu_1\cdots\mu_n} \equiv dx^{\nu_1}\cdots dx^{\nu_{d-n}} \frac{1}{(d-n)!\sqrt{|g|}} \epsilon_{\nu_1\cdots\nu_{d-n}\mu_1\cdots\mu_n}. \tag{1.145}$$

Observe that the standard invariant-volume form for the total manifold $\mathrm{M}^d$ is just $d^d\Sigma$ up to a sign (we now use the signature $(+-\cdots-)$):

$$d^d\Sigma = (-1)^{d-1} dx^1 \wedge \cdots \wedge\ dx^d \sqrt{|g|} \equiv (-1)^{d-1} d^dx \sqrt{|g|}. \tag{1.146}$$

Now, if we have a rank-$n$ completely antisymmetric contravariant tensor $T^{\mu_1\cdots\mu_n}$ and contract it with the volume element $d^{d-n}\Sigma_{\mu_1\cdots\mu_n}$, we have constructed a $(d-n)$-form that can be integrated over a $(d-n)$-dimensional submanifold. Up to numerical factors, that form is the Hodge dual of the $n$-form that one gets by lowering the indices of $T_{\mu_1\cdots\mu_n}$:

$$\frac{1}{n!} d^{d-n}\Sigma_{\mu_1\cdots\mu_n} T^{\mu_1\cdots\mu_n} = \star T. \tag{1.147}$$

We can also take the divergence of the tensor and contract it with the volume element $d^{d-n+1}\Sigma_{\mu_1\cdots\mu_{n-1}}$. The result is

$$\frac{(-1)^{d-n}}{(n-1)!} d^{d-n+1}\Sigma_{\mu_1\cdots\mu_{n-1}} \nabla_\rho T^{\rho\mu_1\cdots\mu_{n-1}} = d \star T. \tag{1.148}$$

Stokes' theorem for the exterior derivative of form $\star T$ integrated over a $(d-n+1)$-dimensional submanifold $\mathrm{M}^{d-n+1}$ with $(d-n)$-dimensional boundary $\partial\mathrm{M}^{d-n+1}$ is now

$$\begin{aligned}&\int_{\mathrm{M}^{d-n+1}} d^{d-n+1}\Sigma_{\mu_1\cdots\mu_{n-1}}\nabla_\rho T^{\rho\mu_1\cdots\mu_{n-1}}\\&=\frac{(-1)^{d-n}}{n}\int_{\partial\mathrm{M}^{d-n+1}} d^{d-n}\Sigma_{\mu_1\cdots\mu_n}T^{\mu_1\cdots\mu_n}.\end{aligned} \tag{1.149}$$

The $n=1$ case is the *Gauss–Ostrogradski theorem,*

$$\int_{\mathrm{M}^d} d^dx\sqrt{|g|}\,\nabla_\mu v^\mu=(-1)^{d-1}\int_{\partial\mathrm{M}} d^{d-1}\Sigma_\mu v^\mu. \tag{1.150}$$

The Vielbein and spin-connection 1-forms and the torsion 2-form are

$$e^a=e_\mu{}^a dx^\mu,\qquad \omega^{ab}=\omega_\mu{}^{ab}dx^\mu,\qquad T^a=\tfrac{1}{2}T_{\mu\nu}{}^a dx^\mu\wedge dx^\nu. \tag{1.151}$$

These 1-forms are related by the *first structure equation*

$$\mathcal{D}e^a+T^a=de^a+\omega_b{}^a\wedge e^b+T^a=0, \tag{1.152}$$

which is nothing but the first Vielbein postulate Eq. (1.94) expressed in differential-form language. We have defined implicitly the Lorentz-covariant exterior derivative $\mathcal{D}$ acting on the Vielbein 1-form. The generalization to other cases is obvious.

This equation gives a convenient way of finding $\omega$. The curvature 2-form and the Ricci-tensor 1-form are given by

$$\begin{aligned}R_a{}^b&=\tfrac{1}{2}R_{\mu\nu a}{}^b dx^\mu\wedge dx^\nu=d\omega_a{}^b-\omega_a{}^c\wedge\omega_c{}^b,\\R^a&=R_\mu{}^a dx^\mu=R_{\mu\lambda}{}^{ab}e^\lambda{}_b dx^\mu.\end{aligned} \tag{1.153}$$

The *second structure equation* is obtained by applying the Lorentz-covariant exterior derivative $\mathcal{D}$ to the first structure equation and using the second Ricci identity in Eq. (1.87):

$$\mathcal{D}\left[\mathcal{D}e^a+T^a\right]=R_b{}^a\wedge e^b+\mathcal{D}T^a=0. \tag{1.154}$$

This equation contains the second of the Bianchi identities in Eq. (1.30). Acting again with $\mathcal{D}$ on the second structure equation we get

$$\mathcal{D}\left[R_b{}^a\wedge e^b+\mathcal{D}T^a\right]=\mathcal{D}R_b{}^a\wedge e^b-2R_b{}^aT^b=0, \tag{1.155}$$

which contains the third of the Bianchi identities in Eq. (1.30).

## 1.8 Extrinsic geometry

Let us consider a hypersurface $\Sigma$ embedded in a $d$-dimensional spacetime with metric $g_{\mu\nu}$ and with normal unit vector $n^\mu$:

$$n^\mu n_\mu=\varepsilon,\qquad \begin{cases}\varepsilon=+1, & \Sigma\text{ spacelike},\\ \varepsilon=-1, & \Sigma\text{ timelike}.\end{cases} \tag{1.156}$$

The metric induced on $\Sigma$ by $g_{\mu\nu}$ is defined by

$$h_{\mu\nu} = g_{\mu\nu} - \varepsilon n_\mu n_\nu. \tag{1.157}$$

$h_{\mu\nu}$ has $(d-1)$-dimensional character but it is written in $d$-dimensional form and it is evidently singular and cannot be inverted. Its indices are raised and lowered with $g$. Observe that $h_{\mu\nu}n^\nu = 0$ and thus $h$ can be used to project tensors onto the hypersurface $\Sigma$.

A way to measure how $\Sigma$ is curved inside the spacetime would be to measure the variation of the normal unit vector along it. Mathematically this would be expressed by

$$\mathcal{K}_{\mu\nu} \equiv h_\mu{}^\alpha h_\nu{}^\beta \nabla_{(\alpha} n_{\beta)}, \tag{1.158}$$

where $\mathcal{K}_{\mu\nu}$ is the *extrinsic curvature* or *second fundamental form*.

We can consider a field of unit vectors $n^\mu$ defined in the whole spacetime determining a family of hypersurfaces. Then we can calculate the Lie derivative of the induced metrics in the direction of the normal unit vectors. We find that this is twice the extrinsic curvature,

$$\mathcal{K}_{\mu\nu} = \tfrac{1}{2}\mathcal{L}_n h_{\mu\nu}. \tag{1.159}$$

The trace of the extrinsic curvature is denoted by $\mathcal{K}$, and given by

$$\mathcal{K} = h^{\mu\nu}\mathcal{K}_{\mu\nu} = h^{\mu\nu}\nabla_\mu n_\nu. \tag{1.160}$$

# 2

# Symmetries and Noether's theorems

In Chapter 3, we are going to introduce general relativity as the result of the construction of a self-consistent special-relativistic field theory (SRFT) of gravity. In this construction, gauge symmetry and the energy–momentum tensor will play a key role. In this chapter we want to review Noether's theorems, the relation between global symmetries and conserved charges, and the relation between local symmetries and gauge identities. We will define the canonical energy–momentum tensor as the conserved Noether current associated with the invariance under constant translations and we will review several ways of improving it that are associated with invariance under other spacetime transformations (Lorentz rotations and rescalings). Finally, we will relate these improved energy–momentum tensors to the energy–momentum tensor used in general relativity.

## 2.1 Equations of motion

Let us consider an action $S[\varphi]$ for a generic field $\varphi$, which may have (spacetime or internal) indices that we do not exhibit for the sake of simplicity. Allowing for Lagrangians containing higher derivatives of $\varphi$, we write the action as follows:

$$S[\varphi] = \int_{\Sigma} d^d x \, \mathcal{L}(\varphi, \partial\varphi, \partial^2\varphi, \ldots). \tag{2.1}$$

In most cases, $\mathcal{L}$ is a scalar density under the relevant spacetime transformations (Poincaré transformations in SRFTs and general coordinate transformations in general-covariant theories). It is also possible to use a Lagrangian that is a scalar density up to a total derivative,[1] and thus we will make absolutely no assumptions about the transformation properties of the Lagrangian $\mathcal{L}$.

Under arbitrary infinitesimal variations of the field variable $\delta\varphi$

$$\delta S = \int_{\Sigma} d^d x \, \delta\mathcal{L} = \int_{\Sigma} d^d x \left\{ \frac{\partial \mathcal{L}}{\partial \varphi}\delta\varphi + \frac{\partial \mathcal{L}}{\partial \partial_\mu \varphi}\delta\partial_\mu\varphi + \frac{\partial \mathcal{L}}{\partial \partial_\mu \partial_\nu \varphi}\delta\partial_\mu\partial_\nu\varphi + \cdots \right\}. \tag{2.2}$$

[1] For instance, in general relativity one may want to eliminate the piece of the Lagrangian with second derivatives, which is a total derivative, but then the rest is not a scalar density.

The variation of the coordinates is zero by hypothesis. Then the variation of the field commutes with the derivatives. On integrating by parts to obtain an overall factor of $\delta\varphi$, we find

$$\delta S = \int_\Sigma d^d x \left\{ \frac{\delta S}{\delta\varphi}\delta\varphi + \partial_\mu \left[ \left( \frac{\partial \mathcal{L}}{\partial\partial_\mu\varphi} - \partial_\nu \frac{\partial \mathcal{L}}{\partial\partial_\mu\partial_\nu\varphi} \right) \delta\varphi + \frac{\partial \mathcal{L}}{\partial\partial_\mu\partial_\nu\varphi} \partial_\nu \delta\varphi + \cdots \right] \right\}, \tag{2.3}$$

where we have defined the first variation of the action $\delta S/\delta\varphi$,

$$\frac{\delta S}{\delta\varphi} \equiv \frac{\partial \mathcal{L}}{\partial\varphi} - \partial_\mu \frac{\partial \mathcal{L}}{\partial\partial_\mu\varphi} + \partial_\mu\partial_\nu \frac{\partial \mathcal{L}}{\partial\partial_\mu\partial_\nu\varphi} + \cdots. \tag{2.4}$$

We now use Stokes' theorem Eq. (1.150) to reexpress the integral of the total derivative as an integral over the boundary $\partial\Sigma$:

$$\delta S = \int_\Sigma d^d x \frac{\delta S}{\delta\varphi}\delta\varphi + (-1)^{d-1} \int_{\partial\Sigma} d^{d-1}\Sigma_\mu \left\{ \left( \frac{\partial \mathcal{L}}{\partial\partial_\mu\varphi} - \partial_\nu \frac{\partial \mathcal{L}}{\partial\partial_\mu\partial_\nu\varphi} \right) \delta\varphi \right.$$
$$\left. + \frac{\partial \mathcal{L}}{\partial\partial_\mu\partial_\nu\varphi} \partial_\nu \delta\varphi + \cdots \right\}. \tag{2.5}$$

In theories without higher derivatives $\mathcal{L}(\varphi, \partial\varphi)$ it is enough to impose that the field variations vanish over the boundary $\delta\varphi|_{\partial\Sigma} = 0$, to see that the boundary term vanishes. Then, requiring that the action is stationary, $\delta S = 0$, under those variations we obtain the usual *Euler–Lagrange equations*

$$\frac{\delta S}{\delta\varphi} = \frac{\partial \mathcal{L}}{\partial\varphi} - \partial^\mu \left( \frac{\partial \mathcal{L}}{\partial\partial^\mu\varphi} \right) = 0. \tag{2.6}$$

If the Lagrangian contains higher derivatives of the field, it is necessary either to impose boundary conditions for derivatives of the variation of the field or to introduce (if possible) into the action boundary terms that do not change the equations of motion but eliminate the $\partial\delta\varphi$ term in the total derivative. In any of these cases we obtain the equations of motion

$$\frac{\delta S}{\delta\varphi} = \frac{\partial \mathcal{L}}{\partial\varphi} - \partial^\mu \left( \frac{\partial \mathcal{L}}{\partial\partial^\mu\varphi} \right) + \partial^\nu\partial^\mu \left( \frac{\partial \mathcal{L}}{\partial\partial^\nu\partial^\mu\varphi} \right) - \cdots = 0. \tag{2.7}$$

As we can see, the equations of motion are of degree higher than 2 in derivatives of the field. Thus, to solve them completely it is also necessary to give boundary conditions for the field, and for its first and higher derivatives.

If we add a total derivative term $\partial_\mu \mathfrak{k}^\mu(\varphi)$ to the Lagrangian, it is clear that the equations of motion will not be modified *as long as the boundary conditions for $\delta\varphi$ and its derivatives make $\mathfrak{k}^\mu(\varphi) = 0$ on the boundary.*

## 2.2 Noether's theorems

Let us now consider the infinitesimal transformations of the coordinates and fields $\tilde{\delta}x^\mu$ and $\tilde{\delta}\varphi$:

$$\tilde{\delta}x^\mu = x'^\mu - x^\mu,$$
$$\tilde{\delta}\varphi(x) \equiv \varphi'(x') - \varphi(x), \tag{2.8}$$

where $x'$ and $x$ stand for the coordinates of the same point in the two different coordinate systems. The transformation of the fields may contain terms associated with the coordinate transformations and also with other "internal" transformations (see note 3 in Chapter 1).

We want to find the consequences of the invariance, possibly up to a total derivative that depends on the variations, of the action Eq. (2.1) under the above infinitesimal changes of the field and the coordinates (which are, then, symmetry transformations). We express this invariance as follows:

$$\tilde{\delta} S = \int_\Sigma d^d x\, \partial_\mu \mathfrak{s}^\mu(\tilde{\delta}). \tag{2.9}$$

Let us now perform directly the variation of the action explicitly,[2]

$$\tilde{\delta} S = \int_\Sigma \left[\tilde{\delta} d^d x\, \mathcal{L} + d^d x\, \tilde{\delta}\mathcal{L}\right]. \tag{2.10}$$

We have

$$\begin{aligned}
\tilde{\delta} d^d x &= d^d x\, \partial_\mu \tilde{\delta} x^\mu, \\
\tilde{\delta}\mathcal{L} &= \delta\mathcal{L} + \tilde{\delta} x^\mu \partial_\mu \mathcal{L}, \\
\delta\mathcal{L} &= \frac{\partial\mathcal{L}}{\partial\varphi}\delta\varphi + \frac{\partial\mathcal{L}}{\partial\partial_\mu\varphi}\delta\partial_\mu\varphi + \frac{\partial\mathcal{L}}{\partial\partial_\mu\partial_\nu\varphi}\delta\partial_\mu\partial_\nu\varphi + \cdots,
\end{aligned} \tag{2.11}$$

where $\delta$ stands for the variation of the field at two different points whose coordinates are the same in the two different coordinate systems considered,

$$\delta\varphi(x) \equiv \varphi'(x) - \varphi(x), \tag{2.12}$$

and we have used the field-operator identity

$$\tilde{\delta} = \delta + \tilde{\delta} x^\mu \partial_\mu. \tag{2.13}$$

$\delta$ and $\partial_\mu$ commute since $\delta$ does not involve any change of coordinates. Thus

$$\tilde{\delta} S = \int_\Sigma d^d x \left\{\partial_\mu \tilde{\delta} x^\mu \mathcal{L} + \tilde{\delta} x^\mu \partial_\mu \mathcal{L} + \frac{\partial\mathcal{L}}{\partial\varphi}\delta\varphi + \frac{\partial\mathcal{L}}{\partial\partial_\mu\varphi}\partial_\mu\delta\varphi + \frac{\partial\mathcal{L}}{\partial\partial_\mu\partial_\nu\varphi}\partial_\mu\partial_\nu\delta\varphi + \cdots\right\}. \tag{2.14}$$

On integrating by parts as many times as necessary, we obtain

$$\tilde{\delta} S = \int_\Sigma d^d x \left\{\partial_\mu\left[\mathcal{L}\tilde{\delta} x^\mu + \left(\frac{\partial\mathcal{L}}{\partial\partial_\mu\varphi} - \partial_\nu \frac{\partial\mathcal{L}}{\partial\partial_\mu\partial_\nu\varphi}\right)\delta\varphi + \frac{\partial\mathcal{L}}{\partial\partial_\mu\partial_\nu\varphi}\partial_\nu\delta\varphi + \cdots\right] + \frac{\delta S}{\delta\varphi}\delta\varphi\right\}. \tag{2.15}$$

On reexpressing $\delta\varphi$ in terms of $\tilde{\delta}\varphi$ inside the total derivative, and equating the result with Eq. (2.9), we arrive at

$$\int_\Sigma d^d x \left\{\partial_\mu \mathfrak{j}^\mu_{\mathrm{N1}}(\tilde{\delta}) + \frac{\delta S}{\delta\varphi}\delta\varphi\right\} = 0, \tag{2.16}$$

[2] We follow here Ref. [1068].

where

$$\mathrm{j}^{\mu}_{\mathrm{N}1}(\tilde{\delta}) = -\mathfrak{s}^{\mu}(\tilde{\delta}) + T_{\mathrm{can}}{}^{\mu}{}_{\nu}\tilde{\delta}x^{\nu} - \frac{\partial\mathcal{L}}{\partial\partial_{\mu}\partial_{\nu}\varphi}\partial_{\rho}\varphi\partial_{\nu}\tilde{\delta}x^{\rho} + \left[\frac{\partial\mathcal{L}}{\partial\partial_{\mu}\varphi} - \partial_{\nu}\left(\frac{\partial\mathcal{L}}{\partial\partial_{\mu}\partial_{\nu}\varphi}\right)\right]\tilde{\delta}\varphi + \frac{\partial\mathcal{L}}{\partial\partial_{\mu}\partial_{\nu}\varphi}\partial_{\nu}\tilde{\delta}\varphi + \cdots, \quad (2.17)$$

where, in turn,

$$T_{\mathrm{can}}{}^{\mu}{}_{\nu} = \eta^{\mu}{}_{\nu}\mathcal{L} - \frac{\partial\mathcal{L}}{\partial\partial_{\mu}\varphi}\partial_{\nu}\varphi - \frac{\partial\mathcal{L}}{\partial\partial_{\mu}\partial_{\rho}\varphi}\partial_{\nu}\partial_{\rho}\varphi + \partial_{\rho}\left(\frac{\partial\mathcal{L}}{\partial\partial_{\mu}\partial_{\rho}\varphi}\right)\partial_{\nu}\varphi + \cdots. \quad (2.18)$$

$T_{\mathrm{can}}{}^{\mu}{}_{\nu}$ is the *canonical energy–momentum tensor* and is the only piece of $\mathrm{j}^{\mu}_{\mathrm{N}1}$ that survives (apart from $\mathfrak{s}^{\mu}$) when we consider constant $\tilde{\delta}x^{\mu}$s.

It is worth stressing that the total-derivative term will not vanish in general after use of Stokes' theorem because the variations $\tilde{\delta}x^{\mu}$ and $\tilde{\delta}\varphi$ do not vanish on the boundary.

Now we want to derive conservation laws from this identity. We see that, in the general case, if the equations of motion $\delta S/\delta\varphi = 0$ are satisfied, then we can conclude that $\mathrm{j}^{\mu}_{\mathrm{N}1}(\tilde{\delta})$ is a *conserved vector current* (*Noether current*), i.e. satisfies the *continuity equation*

$$\partial_{\mu}\mathrm{j}^{\mu}_{\mathrm{N}1}(\tilde{\delta}) = 0. \quad (2.19)$$

Thus, for a theory that is exactly invariant under constant translations, the canonical energy–momentum tensor is the associated Noether conserved current.

Strictly speaking $\mathrm{j}^{\mu}_{\mathrm{N}1}(\tilde{\delta})$ is a vector density. In the presence of a metric, we can define a vector current $\mathrm{j}^{\mu}_{\mathrm{N}1}(\tilde{\delta}) = \sqrt{|g|}j^{\mu}_{\mathrm{N}1}(\tilde{\delta})$ and write the continuity equation in general-covariant form:

$$\nabla_{\mu}j^{\mu}_{\mathrm{N}1}(\tilde{\delta}) = 0. \quad (2.20)$$

In Minkowski spacetime this distinction is unnecessary. Such terms are called "conserved" because they are used to define quantities (*charges*) that are conserved in time, as we will see next.

This is the best we can do if the transformations are global, i.e. when they take the form

$$\tilde{\delta}x^{\mu} \equiv \sigma^{A}\tilde{\delta}_{A}x^{\mu}, \qquad \tilde{\delta}\varphi \equiv \sigma^{A}\tilde{\delta}_{A}\varphi, \quad (2.21)$$

where $\tilde{\delta}_{A}x^{\mu}$ and $\tilde{\delta}_{A}\varphi$ are given functions of the coordinates and $\varphi$ and the $\sigma^{A}$, $A = 1,\ldots,n$, are the constant transformation parameters. Then, we find $n$ on-shell conserved currents $\mathrm{j}^{\mu}_{N1\,A}$ independent of the parameters $\sigma^{A}$ and they are given by

$$\mathrm{j}^{\mu}_{\mathrm{N}1\,A} = -s^{\mu}(\tilde{\delta}_{A}) + T_{\mathrm{can}}{}^{\mu}{}_{\nu}\tilde{\delta}_{A}x^{\nu} - \frac{\partial\mathcal{L}}{\partial\partial_{\mu}\partial_{\nu}\varphi}\partial_{\rho}\varphi\partial_{\nu}\tilde{\delta}_{A}x^{\rho} + \left[\frac{\partial\mathcal{L}}{\partial\partial_{\mu}\varphi} - \partial_{\nu}\left(\frac{\partial\mathcal{L}}{\partial\partial_{\mu}\partial_{\nu}\varphi}\right)\right]\tilde{\delta}_{A}\varphi + \frac{\partial\mathcal{L}}{\partial\partial_{\mu}\partial_{\nu}\varphi}\partial_{\nu}\tilde{\delta}_{A}\varphi + \cdots. \quad (2.22)$$

If the transformations are local, i.e. they depend on $n$ local parameters $\sigma^{A}(x)$, the generic result of the on-shell conservation of the current $\mathrm{j}^{\mu}_{\mathrm{N1}}$ is still true,[3] but we can do more. First, observe that, in general, in the local case, the transformations contain derivatives of the local parameters. We eliminate these derivatives of the transformation parameters by integration by parts in Eq. (2.16). We obtain an identity of the form[4]

$$\int_{\Sigma} d^d x \left\{ \partial_\mu \mathrm{j}^{\mu}_{\mathrm{N2}}(\sigma) + \sigma^A D_A \frac{\delta S}{\delta \varphi} \right\} = 0, \tag{2.23}$$

where $D_A$ are operators containing derivatives acting on the equations of motion. This identity is true for arbitrary parameters. We can choose parameters such that $\mathrm{j}^{\mu}_{\mathrm{N2}}(\sigma)$ vanishes on the boundary. Then, we obtain the off-shell identities that do not involve the transformation parameters

$$D_A \frac{\delta S}{\delta \varphi} = 0, \tag{2.24}$$

that relate the equations of motion, so not all of them are independent. These identities are called *Noether*, *gauge*, or *Bianchi identities*. Since they are identically true for arbitrary values of the parameters, we obtain the off-shell "conservation law"[5]

$$\partial_\mu \mathrm{j}^{\mu}_{\mathrm{N2}}(\sigma) = 0. \tag{2.25}$$

Since this is an identity that holds independently of the equations of motion, it follows that the current density $\mathrm{j}^{\mu}_{\mathrm{N2}}(\sigma)$ can always be written as the divergence of a two-index antisymmetric tensor, usually called the *superpotential*, that is,

$$\mathrm{j}^{\mu}_{\mathrm{N2}}(\sigma) = \partial_\nu \mathrm{j}^{\nu\mu}_{\mathrm{N2}}(\sigma), \qquad \mathrm{j}^{\nu\mu}_{\mathrm{N2}}(\sigma) = -\mathrm{j}^{\mu\nu}_{\mathrm{N2}}(\sigma). \tag{2.26}$$

This identity for the vector densities is written in terms of the vectors $\mathrm{j}^{\mu}_{\mathrm{N2}} = \sqrt{|g|}\, j^{\mu}_{\mathrm{N2}}$:

$$j^{\mu}_{\mathrm{N2}}(\sigma) = \nabla_\nu j^{\nu\mu}_{\mathrm{N2}}(\sigma), \qquad j^{\nu\mu}_{\mathrm{N2}}(\sigma) = -j^{\mu\nu}_{\mathrm{N2}}(\sigma). \tag{2.27}$$

Observe that the difference between $\mathrm{j}^{\mu}_{\mathrm{N1}}$ and $\mathrm{j}^{\mu}_{\mathrm{N2}}$ is always a term proportional to the equations of motion, i.e. it vanishes on-shell. Thus, these two currents are identical on-shell. In general we are free to add any term that vanishes on-shell to the current $j^{\mu}_{\mathrm{N1}}$ since it is conserved only on-shell. We have just seen that there is a specific on-shell-vanishing term that relates $j^{\mu}_{\mathrm{N1}}$ to $j^{\mu}_{\mathrm{N2}}$. $j^{\mu}_{\mathrm{N2}}$ cannot be modified in this way because its defining property is that it is conserved off-shell. However, we could add to both currents terms of the form $\partial_\nu \Psi^{\nu\mu}$, where $\Psi^{\mu\nu} = \Psi^{[\mu\nu]}$, which would change the superpotential. If $\Psi^{\nu\mu}$ is of the form $\partial_\rho U^{\rho\nu\mu}$, with $U^{\rho\nu\mu} = U^{[\rho\nu]\mu}$, then $\partial_\nu \Psi^{\nu\mu} = 0$ and the change in the superpotential will not change the Noether current.

[3] In principle there is one current for each value of the local parameters. This gives an infinite number of on-shell conserved currents. However, only for a certain asymptotic behavior of the $\sigma^A(x)$s will the integrals defining the conserved charges converge. These asymptotic behaviors are usually associated with the global invariances of the vacuum configuration.

[4] This expression is just symbolic. We need a more explicit form of the infinitesimal transformations in order to obtain more explicit expressions. We will find several examples in the following chapters.

[5] *Strong conservation law* in the language of Ref. [147].

It is easy to see that Noether currents are sensitive to the addition of total derivatives to the Lagrangian even if these do not modify the equations of motion: on adding to the action (2.1)

$$\Delta S = \int_{\Sigma} d^d x \, \partial_\mu \mathcal{L}^\mu, \tag{2.28}$$

which is also invariant up to a total derivative

$$\tilde{\delta} \Delta S = \int_{\Sigma} d^d x \partial_\mu \Delta \mathfrak{s}^\mu(\tilde{\delta}), \tag{2.29}$$

and, repeating the same steps as those we followed to find the Noether currents, we find a correction to the Noether current Eq. (2.17):

$$\begin{aligned} \Delta \mathrm{j}^\mu_{\mathrm{N1}}(\tilde{\delta}) &= -\Delta \mathfrak{s}^\mu(\tilde{\delta}) + \tilde{\delta}\mathcal{L}^\mu + \partial_\rho \Psi^{\rho\mu}{}_\nu \tilde{\delta} x^\nu, \\ \Psi^{\rho\mu}{}_\nu &= 2\mathcal{L}^{[\rho}\eta^{\mu]}{}_\nu. \end{aligned} \tag{2.30}$$

If we consider only constant spacetime translations and $\mathcal{L}^\mu$ is a vector density, then $\tilde{\delta}\mathcal{L}^\mu = \Delta \mathfrak{s}^\mu(\tilde{\delta})$ and we simply find a correction to the canonical energy–momentum tensor with the form of a superpotential.

We end this section with an important remark: no change in the superpotential can be related to the addition of a total derivative to the Lagrangian.

## 2.3 Conserved charges

Given a conserved current (density) $\mathrm{j}^\mu$, by taking the integral of its time component $\mathrm{j}^0$ over a piece $\mathrm{V}_t$ of a constant-time hypersurface we can define a quantity (*charge*) $Q(\mathrm{V}_t)$,

$$Q(\mathrm{V}_t) = \int_{\mathrm{V}_t} d^{d-1}x \, \mathrm{j}^0. \tag{2.31}$$

If we take the total time derivative of $Q(\mathrm{V}_t)$, since the volume of $\mathrm{V}_t$ does not depend on time (the subindex $t$ indicates only that it is in a given constant-$t$ hypersurface, but it is the same spatial volume for all $t$) the total time derivative "goes through the integral symbol" and becomes a partial time derivative of $\mathrm{j}^0$ ($c = 1$):

$$\frac{d}{dt} Q(\mathrm{V}_t) = \int_{\mathrm{V}_t} d^{d-1}x \, \partial_0 \mathrm{j}^0. \tag{2.32}$$

The continuity equation for the current and Stokes' theorem imply that

$$\frac{d}{dt} Q(\mathrm{V}_t) = \int_{\mathrm{V}_t} d^{d-1}x \, \partial_i \mathrm{j}^i = \int_{\partial \mathrm{V}_t} d^{d-2}\Sigma_i \, j^i, \tag{2.33}$$

which is interpreted as the flux of charge across the boundary of the volume of $\mathrm{V}_t$. Observe that the last integral is performed over $j^i$ rather than over $\mathrm{j}^i$.

This is a *local charge-conservation law*: the charge contained in the volume of $\mathrm{V}_t$ is only lost (or gained) by the interchange of charge with the exterior; it does not disappear into nothing and it is not created from nothing. This is what we mean by *conserved charge*.

If we take the boundary of the volume to spatial infinity, and we assume that the currents go to zero at infinity (there are no sources at infinity for the charges), then the flux integral over the boundary vanishes and we see that the total charge contained in space at a given time is conserved in absolute terms. It is usually denoted by $Q$ (all reference to time dependence has been eliminated).

Sometimes it is convenient to use a more-covariant expression for the charge:

$$Q(\mathrm{V}_t) = \int_{\mathrm{V}_t} d^{d-1}\Sigma_\mu j^\mu. \tag{2.34}$$

If the current can be expressed as the divergence of an antisymmetric two-index tensor $j^\mu = \nabla_\nu j^{\nu\mu}, j^{\nu\mu} = -j^{\mu\nu}$, then we can again use Stokes' theorem to express the charge as an integral over the boundary of $\mathrm{V}_t$:

$$Q(\mathrm{V}_t) = \frac{(-1)^{d-2}}{2} \int_{\partial \mathrm{V}_t} d^{d-2}\Sigma_{\mu\nu} j^{\mu\nu}. \tag{2.35}$$

The total charge is found by integrating over the boundary of a constant-time slice, which in general has the topology of an $\mathrm{S}^{d-2}$ sphere and lies at spatial infinity. Then, the general expression for the total conserved charge associated with a gauge symmetry is

$$Q = \frac{(-1)^{d-2}}{2} \int_{\mathrm{S}^{d-2}_\infty} d^{d-2}\Sigma_{\mu\nu} j^{\mu\nu}. \tag{2.36}$$

A change in the superpotential $\Psi^{\mu\nu}$ will also change the conserved charge unless the change in the potential vanishes at infinity or unless the change in the superpotential is also of the form $\partial_\rho U^{[\rho\mu]\nu}$ because we can use again Stokes' theorem and reduce the above integral to an integral over the boundary of $\mathrm{V}_t$, which is zero.

## 2.4 The special-relativistic energy–momentum tensor

In special-relativistic field theories the Lagrangian is, by hypothesis, a scalar under Poincaré transformations, i.e. $\tilde{\delta}\mathcal{L} = 0$. These are translations $a^\mu$ and Lorentz transformations $\Lambda^\mu{}_\nu$,

$$x'^\mu = \Lambda^\mu{}_\nu x^\nu + a^\mu, \qquad \Lambda^\mu{}_\rho \eta_{\mu\nu} \Lambda^\nu{}_\sigma = \eta_{\rho\sigma}, \tag{2.37}$$

or, infinitesimally,

$$\tilde{\delta} x^\mu = \sigma^\mu{}_\nu x^\nu + \sigma^\mu, \qquad \sigma^{\mu\nu} = -\sigma^{\nu\mu}. \tag{2.38}$$

The Minkowskian volume element $d^d x$ is also invariant under these transformations $\tilde{\delta} d^d x = 0$ and so the action is also exactly invariant, $\tilde{\delta} S = 0$ ($\mathfrak{s}^\mu = 0$).

Let us first consider infinitesimal translations. In SRFTs all fields are scalars under them, i.e. $\tilde{\delta}\varphi = 0$. Following the standard Noether procedure, we obtain $d$ conserved Noether currents (one for each independent translation) that can be labeled by a subindex $(\nu)$,

$$j^\mu_{\mathrm{N1}\,(\nu)} = T_{\mathrm{can}}{}^\mu{}_\rho \tilde{\delta}_\nu x^\rho = T_{\mathrm{can}}{}^\mu{}_\nu, \tag{2.39}$$

since $\tilde{\delta}_\nu x^\rho = \delta_\nu{}^\rho$. The $d$ conserved currents transform as a contravariant vector with respect to the label $(\nu)$ and thus they are put together into the canonical energy–momentum tensor given by Eq. (2.18) for higher-derivative theories.

Let us take for example a real scalar field $\varphi$. The Lagrangian and equation of motion are

$$\mathcal{L}(\varphi) = \tfrac{1}{2}(\partial\varphi)^2, \qquad \partial^2\varphi = 0, \tag{2.40}$$

and the canonical energy–momentum tensor resulting from the use of the general formula in this case is symmetric and conserved using the above equation of motion:

$$\begin{aligned} T_{\mu\nu}(\varphi) &= -\partial_\mu\varphi\partial_\nu\varphi + \tfrac{1}{2}\eta_{\mu\nu}(\partial\varphi)^2, \\ \partial^\mu T_{\text{matter}\,\mu\nu}(\varphi) &= -\partial_\nu\varphi\partial^2\varphi\big|_{\text{on-shell}} = 0. \end{aligned} \tag{2.41}$$

If we add a total derivative $\partial_\rho(\varphi\partial^\rho\varphi)$ to the above Lagrangian, the equations of motion do not change, as can be seen by using the Euler–Lagrange equations for higher-derivative theories (2.7). According to Eq. (2.18), the energy–momentum tensor acquires the extra term

$$+\partial_\rho\Psi^{\rho\mu\nu}, \qquad \Psi^{\rho\mu\nu} = 2\eta^{\nu[\mu}\varphi\partial^{\rho]}\varphi, \tag{2.42}$$

which is also symmetric but contains second derivatives of the field.

Although the canonical energy–momentum tensor arises as the Noether current associated with invariance under constant translations, we are going to see that it is a much richer object and contains information on the response of a theory to spacetime transformations.

Observe that the canonical energy–momentum tensor is not symmetric in general. In fact, it is symmetric only for scalar fields. However, it can be symmetrized, as we are going to explain when we study the conservation of angular momentum.

For each vector current, we can define the charge $Q_{(\nu)}$,

$$Q_{(\nu)} = \int_{V_t} d^{d-1}x\, j^0_{(\nu)} = \int_{V_t} d^{d-1}x\, T_{\text{can}}{}^0{}_\nu. \tag{2.43}$$

The $d$ conserved charges associated with the energy–momentum tensor are the $d$ components of a contravariant Lorentz vector, which is nothing but the momentum vector, and thus we have derived the local conservation laws of energy and momentum. It is customary to write $P^\nu = Q_{(\nu)}$.

### 2.4.1 Conservation of angular momentum

Let us now consider the infinitesimal Lorentz transformations. The fields appearing in SRFTs transform covariantly or contravariantly in definite representations of the Lorentz group. Let us take, for instance, a field $\varphi^\alpha$ transforming contravariantly in the representation $r$ of the Lorentz group. The index $\alpha$ goes from 1 to $d_r$, the dimension of the representation $r$. If, in the representation $r$, the generators of the Lorentz group are the $d_r \times d_r$ matrices $\Gamma_r\,(M_{\mu\nu})^\alpha{}_\beta$, then an infinitesimal Lorentz transformation of the field $\varphi$ can be written in the form

$$\tilde{\delta}\varphi^\alpha = \tfrac{1}{2}\sigma^{\mu\nu}\Gamma_r(M_{\mu\nu})^\alpha{}_\beta\varphi^\beta = \tfrac{1}{2}\sigma^{\mu\nu}\tilde{\delta}_{(\mu\nu)}\varphi^\alpha. \tag{2.44}$$

Observe that we can write

$$\tilde{\delta}_{(\rho\sigma)}x^\mu = \Gamma_{\text{v}}(M_{\rho\sigma})^\mu{}_\nu x^\nu, \tag{2.45}$$

where $\Gamma_{\text{v}}$ is the vector representation given in Eq. (A.60).

According to the general result Eq. (2.22), we find the following set of $d(d-1)/2$ conserved currents labeled by a pair of antisymmetric indices:[6]

$$j_{N1\,(\rho\sigma)}{}^{\mu} = T_{\rm can}{}^{\mu}{}_{\lambda}\tilde{\delta}_{(\rho\sigma)}x^{\lambda} + \frac{\partial\mathcal{L}}{\partial\partial_{\mu}\varphi^{\alpha}}\tilde{\delta}_{(\rho\sigma)}\varphi^{\alpha} = 2T_{\rm can}{}^{\mu}{}_{[\rho}x_{\sigma]} + \frac{\partial\mathcal{L}}{\partial\partial_{\mu}\varphi^{\alpha}}\Gamma_r(M_{\rho\sigma})^{\alpha}{}_{\beta}\varphi^{\beta}. \tag{2.46}$$

The first contribution to this current is the *orbital-angular-momentum tensor* and the second is the *spin-angular-momentum tensor*, $S^{\mu}{}_{\rho\sigma}$,

$$S^{\mu}{}_{\rho\sigma} \equiv \frac{1}{2}\frac{\partial\mathcal{L}}{\partial\partial_{\mu}\varphi^{\alpha}}\Gamma_r(M_{\rho\sigma})^{\alpha}{}_{\beta}\varphi^{\beta}. \tag{2.47}$$

Only the total angular-momentum current is conserved.

The $d(d-1)/2$ conserved charges are the components of a two-index antisymmetric tensor: the angular-momentum tensor $M_{\mu\nu}$,

$$M_{\mu\nu} = Q_{(\mu\nu)} = \int_{V_t} d^{d-1}x\, j^{0}_{N1\,(\mu\nu)}. \tag{2.48}$$

It is instructive to take the divergence of the above current. Since in the theories we are dealing with we always have $\partial_{\mu}T_{\rm can}{}^{\mu}{}_{\nu} = 0$, one finds

$$\partial_{\mu}j_{N1\,(\rho\sigma)}{}^{\mu} = -2T_{{\rm can}\,[\rho\sigma]} + \partial_{\mu}\left(\frac{\partial\mathcal{L}}{\partial\partial_{\mu}\varphi^{\alpha}}\Gamma_r(M_{\rho\sigma})^{\alpha}{}_{\beta}\varphi^{\beta}\right), \tag{2.49}$$

which should vanish on-shell according to the general formalism. This means that, except for scalars, $T_{{\rm can}\,\mu\nu}$ is not symmetric and the antisymmetric part is given by

$$T_{{\rm can}\,[\rho\sigma]} = \partial_{\mu}S^{\mu}{}_{\rho\sigma}, \tag{2.50}$$

*up to terms vanishing on-shell*. This formula suggests that we can symmetrize the canonical energy–momentum tensor, exploiting the ambiguities of Noether currents mentioned earlier, i.e. adding to it a term of the form

$$\partial_{\mu}\Psi^{\mu\rho}{}_{\sigma}, \qquad \Psi^{\mu\rho}{}_{\sigma} = -\Psi^{\rho\mu}{}_{\sigma}, \tag{2.51}$$

whose divergence is automatically zero, which in this case would be given by the *spin–energy potential*

$$\Psi^{\mu\rho}{}_{\sigma} = -S^{\mu\rho}{}_{\sigma} + S^{\rho\mu}{}_{\sigma} + S_{\sigma}{}^{\mu\rho}, \tag{2.52}$$

and also removing all the antisymmetric terms that vanish on-shell. The resulting symmetric energy–momentum tensor is usually considered as the energy–momentum tensor to which gravity couples[7] [1241] and we will denote it simply by $T^{\mu}{}_{\nu}$. It is also called the *Belinfante tensor* [125]. Using it, the conserved current associated with Lorentz rotations is

$$j_{N1\,(\rho\sigma)}{}^{\mu} = 2T^{\mu}{}_{[\rho}x_{\sigma]} + \partial_{\lambda}\left(\Psi^{\lambda\mu}{}_{[\rho}x_{\sigma]}\right), \tag{2.53}$$

[6] Here we concentrate on theories without higher derivatives.

[7] This can be justified in the framework of the Cartan–Sciama–Kibble (CSK) theory of gravity. As we will see in Section 4.4, the Belinfante tensor has to coincide with the Rosenfeld energy–momentum tensor, whose definition is based precisely on the coupling to gravity. It is also worth mentioning that, in the CSK theory, the spin–energy potential also couples to gravity through the torsion (the energy–momentum tensor couples through the metric).

again, up to terms that vanish on-shell. The second term in this expression can be eliminated by the usual procedure. The spin-angular-momentum tensor has been absorbed into the new angular-momentum tensor. We are left with the following conserved on-shell currents associated with translations and Lorentz rotations, both of them expressed in terms of the same energy–momentum tensor (the Belinfante tensor):

$$\begin{aligned} j_{\mathrm{N1}\,(\nu)}{}^{\mu} &= T^{\mu}{}_{\rho}\tilde{\delta}_{(\nu)}x^{\rho} &&= T^{\mu}{}_{\rho}, \\ j_{\mathrm{N1}\,(\rho\sigma)}{}^{\mu} &= T^{\mu}{}_{\lambda}\tilde{\delta}_{(\rho\sigma)}x^{\lambda} &&= 2T^{\mu}{}_{[\rho}x_{\sigma]}. \end{aligned} \tag{2.54}$$

It is worth stressing that the existence of these conserved currents is primarily due to the invariance of the Minkowski metric that enters into special-relativistic Lagrangians and of the Minkowski volume element under the Poincaré group or, in other words, to the existence of $d(d+1)/2$ Killing vectors precisely of the form

$$\tilde{\delta}_{(\nu)}x^{\mu}\partial_{\mu} = \partial_{\nu}, \qquad \tilde{\delta}_{(\rho\sigma)}x^{\mu}\partial_{\mu} = -2x_{[\rho}\partial_{\sigma]}. \tag{2.55}$$

A couple of simple examples of the symmetrization of the canonical energy–momentum tensor are in order here.

*The energy–momentum tensor of a vector field.* The Lagrangian and canonical energy–momentum tensor are given by

$$\mathcal{L} = -\tfrac{1}{4}F^2, \qquad F_{\mu\nu} = 2\partial_{[\mu}A_{\nu]}, \qquad T_{\mathrm{can}}{}^{\mu}{}_{\nu} = F^{\mu\rho}\partial_{\nu}A_{\rho} - \tfrac{1}{4}\eta^{\mu}{}_{\nu}F^2. \tag{2.56}$$

Under Lorentz rotations we have

$$\tilde{\delta}A_{\mu} = -A_{\nu}\sigma^{\nu}{}_{\mu} \Rightarrow S^{\mu}{}_{\rho\sigma} = F^{\mu}{}_{[\rho}A_{\sigma]} \Rightarrow \Psi^{\rho\mu}{}_{\nu} = F^{\rho\mu}A_{\nu}, \tag{2.57}$$

and, using the equations of motion $\partial_{\rho}F^{\rho\mu} = 0$,

$$T^{\mu}{}_{\nu} = T_{\mathrm{can}}{}^{\mu}{}_{\nu} + \partial_{\rho}\Psi^{\rho\mu}{}_{\nu} = F^{\mu\rho}F_{\nu\rho} - \tfrac{1}{4}\eta^{\mu}{}_{\nu}F^2, \tag{2.58}$$

which is the standard, gauge-invariant, energy–momentum tensor of a vector field, coinciding with the one derived via Rosenfeld's prescription, which we are going to introduce in Section 2.4.3, inspired by general relativity.

There is yet another way to obtain this energy–momentum tensor that is worth pointing out: let us consider the transformations

$$\tilde{\delta}x^{\mu} = \epsilon^{\mu}, \qquad \tilde{\delta}A_{\mu} = \epsilon^{\lambda}\partial_{\lambda}A_{\mu} - \mathcal{L}_{\epsilon}A_{\mu} = -\partial_{\mu}\epsilon^{\lambda}A_{\lambda}. \tag{2.59}$$

Following the same steps as those we followed to prove the Noether theorem, we find now

$$\tilde{\delta}S = \int d^dx\,\partial_{\mu}\epsilon^{\nu}T^{\mu}{}_{\nu}, \tag{2.60}$$

with $T^{\mu}{}_{\nu}$ as above (the Belinfante tensor). This variation vanishes if

$$\partial_{(\mu}\epsilon_{\lambda)} = 0, \tag{2.61}$$

which is the Killing equation in Minkowski spacetime. Then, there is invariance under Poincaré transformations whose generators are $\tilde{\delta}x^\lambda = \tilde{\delta}_{(\nu)}x^\lambda$ and $\tilde{\delta}x^\lambda = \tilde{\delta}_{(\rho\sigma)}x^\lambda$.

On integrating the above variation by parts, using the fact that it vanishes for Poincaré transformations, and using the equation of motion (which implies that $\partial_\mu T^\mu{}_\lambda = 0$), we find

$$\int d^dx\, \partial_\mu\left(\tilde{\delta}x^\lambda T^\mu{}_\nu\right) = 0, \tag{2.62}$$

and we find automatically the above Noether currents.

This method is clearly inspired by general relativity. We will find more applications for it soon.

*The energy–momentum tensor of a Dirac spinor.* The Lagrangian of a massive Dirac spinor is[8]

$$\mathcal{L} = \tfrac{1}{2}(i\bar{\psi}\not{\partial}\psi - i\bar{\psi}\overleftarrow{\not{\partial}}\psi) - m\bar{\psi}\psi, \qquad \bar{\psi}\overleftarrow{\not{\partial}} \equiv \partial_\mu\bar{\psi}\gamma^\mu. \tag{2.63}$$

It is customary to vary $\psi$ and $\bar{\psi}$ as if they were independent. This simplifies somewhat the calculations but we have to bear in mind that they are not independent. The equations of motion of $\psi$ and $\bar{\psi}$ are the Dirac conjugates of each other:

$$(i\not{\partial} - m)\psi = 0, \qquad \bar{\psi}\left(i\overleftarrow{\not{\partial}} + m\right) = 0. \tag{2.64}$$

Acting with $\not{\partial}$ on the first equation, we find that $\psi$ satisfies the Klein–Gordon equation

$$(\partial^2 + m^2)\psi = 0. \tag{2.65}$$

The canonical energy–momentum tensor is

$$\begin{aligned} T_{\text{can}\,\lambda}{}^\mu &= -\frac{\partial\mathcal{L}}{\partial\partial_\mu\psi}\partial_\lambda\psi - \partial_\lambda\bar{\psi}\frac{\partial\mathcal{L}}{\partial\partial_\mu\bar{\psi}} + \eta_\lambda{}^\mu\mathcal{L} \\ &= -\frac{i}{2}\bar{\psi}\gamma^\mu\partial_\lambda\psi + \frac{i}{2}\partial_\lambda\bar{\psi}\gamma^\mu\psi + \eta_\lambda{}^\mu\left[\frac{1}{2}(i\bar{\psi}\not{\partial}\psi - i\bar{\psi}\overleftarrow{\not{\partial}}\psi) - 2m\bar{\psi}\psi\right], \end{aligned} \tag{2.66}$$

and it is clearly not symmetric. The spin-angular-momentum tensor is

$$\begin{aligned} S^\mu{}_{\rho\sigma} &= \frac{1}{2}\frac{\partial\mathcal{L}}{\partial\partial_\mu\psi}\Gamma_{\text{s}}(M_{\rho\sigma})\,\psi + \frac{1}{2}\bar{\psi}\Gamma_{\bar{\text{s}}}(M_{\rho\sigma})\,\frac{\partial\mathcal{L}}{\partial\partial_\mu\bar{\psi}} \\ &= \frac{1}{2}\frac{i}{2}\bar{\psi}\gamma^\mu\left(\frac{1}{2}\gamma_{\nu\rho}\right)\psi + \frac{1}{2}\bar{\psi}\left(-\frac{1}{2}\gamma_{\nu\rho}\right)\left(-\frac{i}{2}\gamma^\mu\psi\right) \\ &= \frac{i}{4}\bar{\psi}\gamma^\mu{}_{\nu\rho}\psi, \end{aligned} \tag{2.67}$$

and it is totally antisymmetric. The spin–energy potential is just

$$\Psi^{\mu\nu}{}_\rho = -S^{\mu\nu}{}_\rho, \tag{2.68}$$

[8] Our conventions for spinors and gamma matrices are explained in Appendix D.

and, after use of the equations of motion, we find the Belinfante tensor

$$T_\lambda{}^\mu = \frac{i}{4}\partial_\nu\bar{\psi}\,(\gamma^\mu\eta^\nu{}_\lambda + \gamma_\lambda\eta^{\nu\mu})\psi - \frac{i}{4}\bar{\psi}(\gamma^\mu\eta^\nu{}_\lambda + \gamma_\lambda\eta^{\nu\mu})\partial_\nu\psi$$
$$+ \eta_\lambda{}^\mu\left[\tfrac{1}{2}(i\bar{\psi}\not{\partial}\psi - i\bar{\psi}\overleftarrow{\not{\partial}}\,\psi) - 2m\bar{\psi}\psi\right]. \tag{2.69}$$

In the case of the vector field, we managed to find the Belinfante tensor by a method based on the vector transformation law under GCTs. However, it is not clear how to use this method in the present case. The spinorial character is associated only with Lorentz transformations and it is not clear what the spinor transformation law should be for other GCTs. In fact, the only consistent form of dealing with spinors on curved spacetime is to treat them as scalars under GCTs and to associate the spinorial character with the Lorentz group that acts on the tangent space at each given point. This is the formalism invented by Weyl in Ref. [1252] which we will study later on.

### 2.4.2 Dilatations

Let us consider now constant rescalings (dilatations) by a factor $\Omega = e^\sigma$:

$$\begin{array}{r} x'^\mu = \Omega x^\mu, \\ \varphi'(x') = \Omega^\omega\varphi(x), \end{array} \quad \Rightarrow \quad \left\{\begin{array}{l} \tilde{\delta}x^\mu = \sigma x^\mu \equiv \sigma\tilde{\delta}_{\rm D}x^\mu, \\ \tilde{\delta}\varphi = \omega\sigma\varphi. \end{array}\right. \tag{2.70}$$

The associated conserved current is

$$j_{\rm N1\,D}{}^\mu = T_{\rm can}{}^\mu{}_\nu x^\nu + J^\mu, \qquad\qquad J^\mu \equiv \omega\frac{\partial\mathcal{L}}{\partial\partial_\mu\varphi}\varphi. \tag{2.71}$$

If we take the divergence of this current and set it equal to zero, we obtain the identity

$$T_{\rm can}{}^\mu{}_\mu + \partial_\mu J^\mu = 0. \tag{2.72}$$

It is always possible to find a redefinition of the canonical energy–momentum tensor that is symmetric, divergenceless, and, furthermore, traceless if there is scale invariance (see e.g. Refs. [286, 365, 697, 1050] and references therein). This redefined energy–momentum tensor is called the *improved energy–momentum tensor* and can be constructed systematically: on rewriting the dilatation current in the form

$$j_{\rm N1\,D}{}^\mu = \left[T_{\rm can}{}^\mu{}_\nu + \frac{2}{d-1}\partial_\rho\Big(J^{[\mu}\eta^{\rho]}{}_\nu\Big)\right]x^\nu - \frac{2}{d-1}\partial_\nu\Big(J^{[\mu}x^{\nu]}\Big), \tag{2.73}$$

we observe that

$$T^{\mu\nu} = T_{\rm can}{}^{\mu\nu} + \frac{2}{d-1}\partial_\rho\Big(J^{[\mu}\eta^{\rho]\nu}\Big), \tag{2.74}$$

is on-shell traceless on account of the identity Eq. (2.72) and also on-shell divergenceless since the piece that we add to the canonical energy–momentum tensor is of the form $\partial_\rho\Psi^{[\rho\mu]\nu}$. Observe that this term can also be obtained directly from the action if we add to it a total derivative term of the form

$$\Delta S = \frac{\omega}{d-1}\int d^dx\,\partial_\rho\left(\frac{\partial\mathcal{L}}{\partial\partial_\rho\varphi}\varphi\right). \tag{2.75}$$

Furthermore, the second term in the dilatation current is also of the form $\partial_\rho \Psi^{[\rho\mu]}$ and, then, up to this term we can write, as in Eqs. (2.54),

$$j_{\mathrm{N1\,D}}{}^{\mu} = T^{\mu}{}_{\nu}\tilde{\delta}_{\mathrm{D}} x^{\nu} = T^{\mu}{}_{\nu} x^{\nu}. \tag{2.76}$$

This result, and the analogous result for Lorentz rotations, suggest the following general picture: for *any given spacetime symmetry* generated by $\tilde{\delta}x^\mu$ it always seems possible to find a redefinition of the canonical energy–momentum tensor $T^{\mu\nu}$ such that it is symmetric and on-shell divergenceless and such that the conserved current associated with the spacetime symmetry is given, up to terms of the form $\partial_\rho \Psi^{[\rho\mu]}$, by

$$j_{\mathrm{N1}}{}^{\mu} = T^{\mu}{}_{\nu}\tilde{\delta} x^{\nu}. \tag{2.77}$$

It is to this energy–momentum tensor that gravity (a gauge theory for all spacetime transformations) couples. This immediately suggests Rosenfeld's prescription for finding the energy–momentum tensor. Before we study it, let us work out a couple of simple examples. First, let us consider a free scalar field $\varphi$ in $d$ dimensions with Lagrangian

$$\mathcal{L} = \tfrac{1}{2}(\partial\varphi)^2. \tag{2.78}$$

The action is invariant if $\omega = -(d-2)/2$. The canonical energy–momentum tensor and dilatation current are in this case

$$\begin{aligned} T_{\mathrm{can}}{}^{\mu}{}_{\nu} &= -\partial^\mu\varphi\partial_\nu\varphi + \tfrac{1}{2}\eta^{\mu}{}_{\nu}(\partial\varphi)^2, \\ j_{\mathrm{N1\,D}}{}^{\mu} &= T_{\mathrm{can}}{}^{\mu}{}_{\nu}x^{\nu} + \frac{\omega}{2}\partial^\mu\varphi^2. \end{aligned} \tag{2.79}$$

The improved energy–momentum tensor is written in a form in which it is clear that we are adding a total derivative:

$$T^{\mu\nu} = T_{\mathrm{can}}{}^{\mu\nu} - \frac{\omega}{2(d-1)}\partial_\rho\left(\eta^{\rho(\mu}\partial^{\nu)}\varphi^2 - \eta^{\mu\nu}\partial^\rho\varphi^2\right). \tag{2.80}$$

Using the improved energy–momentum tensor, the dilatation current can be written as expected:

$$j_{\mathrm{N1\,D}}{}^{\mu} = T^{\mu}{}_{\nu}x^{\nu} + \frac{\omega}{2(d-1)}\partial_\nu\left(x^\nu\partial^\mu\varphi^2 - x^\mu\partial^\nu\varphi^2\right). \tag{2.81}$$

Our second example is a $d$-dimensional vector field whose action is invariant for the same value[9] of $\omega$. In and only in $d=4$ is the Belinfante tensor traceless. By the same procedure, we find the on-shell traceless, conserved energy–momentum tensor, in any dimension:

$$T^{\mu}{}_{\nu} = \frac{d}{2(d-1)}F^{\mu\rho}\partial_\nu A_\rho - \frac{1}{4(d-1)}\eta^{\mu}{}_{\nu}F^2 - \frac{d-2}{2(d-1)}\partial_\nu F^{\mu\sigma}A_\sigma. \tag{2.82}$$

This energy–momentum tensor changes under gauge transformations of the vector field.

[9] This is true for any free-field theory described by a Lagrangian quadratic in first derivatives of the field.

### 2.4.3 Rosenfeld's energy–momentum tensor

Rosenfeld's prescription [1086] is precisely based on the minimal coupling to gravity postulated by general relativity that we will study later on: place the matter fields in a curved background substituting everywhere the flat Minkowski metric $\eta_{\mu\nu}$ by a general background metric $\gamma_{\mu\nu}$, partial derivatives by covariant derivatives compatible with the background metric, and the flat volume element $d^dx$ by $d^dx\sqrt{|\gamma|}$. Then the energy–momentum tensor is given by[10]

$$T_{\text{matter}}^{\mu\nu} = 2\frac{\delta S_{\text{matter}}}{\delta \gamma_{\mu\nu}}\bigg|_{\gamma_{\mu\nu}=\eta_{\mu\nu}}. \tag{2.83}$$

Of course, one has to define first which fields are independent of the metric. For instance, if we have a vector field $A_\mu$, we have to decide which of $A_\mu$ and $A^\mu$ is fundamental. The other field then depends on the metric used to raise or lower the index. Furthermore, we have to decide whether the fields are tensors or tensor densities, and, depending on our choice, we may have to add factors proportional to the determinant of the auxiliary metric or not and we may have to add additional connection terms in the covariant derivatives or not.

This energy–momentum tensor is symmetric by construction and conserved on-shell due to the Bianchi identity[11] associated with the invariance under GCTs of the action written in the background metric $\gamma_{\mu\nu}$. Furthermore, it can be shown to be always identical up to a term of the form $\partial_\rho \Psi^{[\rho\mu]\nu}$ to the canonical one under very general assumptions [77]. For a scalar and a vector field, the energy–momentum tensor found via Rosenfeld's prescription (the Rosenfeld or *metric* energy–momentum tensor) is identical to the canonical tensor and the Belinfante energy–momentum tensor, respectively. We will see in Chapter 3 that the same is true for a spin-2 field, and later on we will see that the same is true in a generalized sense for a Dirac spinor. This identity is not a mere coincidence but it can be justified, as has already been pointed out, in the framework of the Cartan–Sciama–Kibble theory of gravity that we will review in Section 4.4. In general, it is easier to compute the energy–momentum tensor using Rosenfeld's prescription than using the canonical one, especially if we are interested in a *symmetric* energy–momentum tensor. The Rosenfeld energy–momentum tensor has the required properties.[12,13]

---

[10] Observe that the variation here is defined with respect to the metric $\gamma_{\mu\nu}$ and not with respect to the inverse metric $\gamma^{\mu\nu}$, in which case we would have to add an overall minus sign to the definition.

[11] This will be explained and proven later on in Chapter 3.

[12] However, this is still confusing because we have two different symmetric, on-shell divergenceless energy–momentum tensors for a scalar field (the canonical and the improved, which is traceless), and Rosenfeld's procedure seems to give a unique energy–momentum tensor. This is not true, though: when we covariantize a special-relativistic action introducing a metric the result is unique up to curvature terms that vanish in Minkowski spacetime. In the case of the scalar, a covariantization that preserves the scaling invariance is

$$S[\varphi,\gamma] = \int d^dx\sqrt{|\gamma|}\left[\frac{1}{2}(\partial\varphi)^2 + \frac{\omega}{4(d-1)}\varphi^2 R(\gamma)\right], \tag{2.84}$$

where $R(\gamma)$ is the Ricci scalar of the background metric. This action is invariant, in fact, under local Weyl rescalings of the metric and local rescalings of the scalar, leaving the coordinates untouched:

$$\varphi' = \Omega^{(2-d)/2}(x)\varphi, \qquad \gamma'_{\mu\nu} = \Omega^2(x)\gamma_{\mu\nu}. \tag{2.85}$$

To illustrate this point, let us go back to the massless vector field of the previous section. Let us consider the effect of conformal transformations on its action. The conformal group consists of transformations that leave the Minkowski metric invariant up to a global (possibly local) factor: (infinitesimal) constant translations $\tilde{\delta}x^\mu = \xi^\mu$, Lorentz rotations $\tilde{\delta}x^\mu = \sigma^\mu{}_\nu x^\nu \equiv \sigma^\mu$ (these two generate the Poincaré group), dilatations $\tilde{\delta}x^\mu = \sigma x^\mu \equiv w^\mu$, and special conformal transformations (or conformal boosts) $\tilde{\delta}x^\mu = 2(\zeta \cdot x)x^\mu - x^2\zeta^\mu \equiv v^\mu$. The vector field transforms under these coordinate transformations according to the general rule for (world) vectors (2.59) with $\epsilon^\mu = \xi^\mu + \sigma^\mu + w^\mu + v^\mu$. The variation of the action is, again, given by Eq. (2.60). Conformal transformations are generated by conformal Killing vectors of Minkowski spacetime that satisfy

$$\partial_{(\mu}\epsilon_{\lambda)} \propto \eta_{\mu\lambda}. \tag{2.87}$$

The proportionality factor is zero for Poincaré transformations but non-zero for dilatations and conformal boosts. Then, the variation of the action will be zero only if the energy–momentum tensor is traceless. This happens only in $d = 4$ dimensions. On integrating by parts, etc., we find that the Noether current has the form Eq. (2.77), always with the same (Rosenfeld's) energy–momentum tensor.

We may expect that this is completely general. We need to know only how the fields transform under general coordinate transformations,[14] which determines completely the coupling to gravity in general relativity.

Just as it is possible to give a prescription for how to find the energy–momentum tensor on the basis of its coupling to gravity through the metric in general relativity, it is possible to give a definition of the spin–energy potential $\Psi^{\mu\nu}{}_\rho$ based on its coupling to gravity (maybe we should say geometry instead of gravity) through the torsion tensor in the framework of the Cartan–Sciama–Kibble (CSK) theory:

$$\Psi^{\mu\nu}{}_\rho = -\frac{2}{\sqrt{|g|}}\frac{\delta S}{\delta T_{\mu\nu}{}^\rho}\bigg|_{\gamma=T=0}. \tag{2.88}$$

The equivalence of this definition and the definition we gave in terms of the spin-angular-momentum tensor $S^\mu{}_{\rho\sigma}$ can also be proven in the CSK theory. In fact, the above

Using the results of Section 4.2, we find

$$2\frac{\delta S[\varphi,\gamma]}{\delta\gamma_{\mu\nu}} = T_{\text{can}}{}^{\mu\nu} - \frac{\omega}{2(d-1)}\left(\nabla^\mu\partial^\nu\varphi^2 - \gamma^{\mu\nu}\nabla^2\varphi\right) + \frac{\omega}{2(d-1)}\varphi G^{\mu\nu}(\gamma), \tag{2.86}$$

where $G^{\mu\nu}(\gamma)$ is the Einstein tensor of the background metric. On setting $\gamma_{\mu\nu} = \eta_{\mu\nu}$, we find precisely the improved energy–momentum tensor Eq. (2.80). Something similar can be said of the vector field in $d \neq 4$. If, in the presence of a curved metric, the vector field scales as in Minkowski spacetime, the vector field is really a vector density and then its covariantization is different from the standard one and should lead to a Rosenfeld energy–momentum tensor identical to the improved one.

[13] When the field theory has a symmetry, it is desirable or necessary to have an energy–momentum tensor that is also invariant under the same transformations. For instance, the Belinfante energy–momentum tensor for the Maxwell field is gauge invariant, as is the Maxwell action. It can be shown that, in general, symmetries of a theory are also symmetries of the Rosenfeld energy–momentum tensor if the symmetries are also symmetries of the same theory covariantized with an arbitrary background metric. The Maxwell action in a curved background is still gauge invariant, and the gauge invariance of the Belinfante–Rosenfeld energy–momentum tensor follows.

[14] A conformal scalar of weight $\omega$ is nothing but a scalar density of weight $\omega/d$.

definition is the main characteristic of that theory in which intrinsic (i.e. not orbital) angular momentum is the source of another field that has a geometrical interpretation (torsion).

## 2.5 The Noether method

There is a useful recipe for how to find the Noether current associated with global symmetry transformations of the fields $\delta\varphi$: if the action is invariant under transformations with constant parameters, then, if we use local parameters, upon use of the equations of motion, the variation of the action would be proportional to the derivative of the parameters:

$$\delta S = -\int d^d x \partial_\mu \sigma^A \, j_A^\mu, \tag{2.89}$$

because, by hypothesis, it has to vanish for constant $\sigma^A$. Up to a total derivative, this is

$$\delta S = \int d^d x \sigma^A \, \partial_\mu j_A^\mu, \tag{2.90}$$

that vanishes for constant $\sigma^A$ only if $\partial_\mu j_A^\mu = 0$. Thus the currents $j_A^\mu$ are the Noether currents associated with the global symmetry.

The observation that the variation of the action must be of the above form is the basis of the so-called *Noether method* which is used to couple fields in a symmetric way. The simplest example of how this method works is the coupling of a complex scalar field $\Phi$ to the electromagnetic field $A_\mu$. The Lagrangian of the electromagnetic field Eq. (2.56) is invariant under the transformations with local parameter $\Lambda$,

$$\delta A_\mu = \partial_\mu \Lambda, \tag{2.91}$$

while the Lagrangian for the complex scalar,

$$\mathcal{L} = \tfrac{1}{2}\partial_\mu \Phi \partial^\mu \bar{\Phi}, \tag{2.92}$$

is invariant under phase transformations with a constant parameter $\sigma$ and a constant $g$ that infinitesimally look like this:

$$\delta\Phi = ig\sigma\Phi. \tag{2.93}$$

These transformations constitute a $\mathrm{U}(1)$ symmetry group. $g$ labels the representation of $\mathrm{U}(1)$ corresponding to $\Phi$. If $\sigma$ takes values in the interval $[0, 2\pi]$, then $g$ can be any integer.

If the conserved current of the scalar Lagrangian is seen as an electric current, it is natural to couple it to the electromagnetic vector field to obtain the Maxwell equation with sources:

$$\partial_\nu F^{\nu\mu} = g j_{\mathrm{N}}^\mu. \tag{2.94}$$

From a Lagrangian point of view, this equation can be obtained by adding to the free Lagrangians of the vector and scalar a coupling of the form $gA_\mu j_{\mathrm{N}}^\mu$. However, this term modifies the equation of motion of the scalar, so the electric current $j_{\mathrm{N}}^\mu$ is not conserved on-shell. This renders the above equation inconsistent since the l.h.s. is automatically divergenceless. Clearly, the addition of a new term to the Lagrangian modifies the Noether current. The modified Noether current should be conserved on-shell upon use of the modified equations of motion. It is easy to see that the vector field contributes to it. This is the Noether current that we should use in the Lagrangian now, and this induces new modifications. This may go on indefinitely until the new correction does not contribute to the new Noether current. Observe that the modified Noether current is found using a local phase transformation according to the above general observation. It should also be stressed that the physical reason why there was inconsistency is that we did not take into account the contribution of the vector field to the electric current. Only the total electric current should be consistently conserved.

The Noether method is essentially a systematic way of performing these iterations emphasizing the role of symmetry. In the case at hand, the basic idea is that one has to identify $\sigma$ with $\Lambda$ and one has to make the whole system invariant under transformations of the same form with $\Lambda$ local. We start by calling $\mathcal{L}_0$ the Lagrangian which is the sum of the free electromagnetic and scalar Lagrangians and using the above general observation: under a local $\Lambda$ transformation ($\sigma = \Lambda$), and up to total derivatives,

$$\delta\mathcal{L}_0 = g\Lambda\partial_\mu j_{\mathrm{N}}^\mu, \qquad j_{\mathrm{N}}^\mu = -\frac{i}{2}\left(\Phi\partial^\mu\bar{\Phi} - \bar{\Phi}\partial^\mu\Phi\right). \tag{2.95}$$

$j^\mu$ is the on-shell conserved current associated with the global invariance of the Lagrangian.

The Noether method consists in the addition to $\mathcal{L}_0$ of terms that will be of higher order in the constant $g$ to compensate for the above non-vanishing variation. Typically the first correction will be of the form

$$\mathcal{L}_1 = \mathcal{L}_0 + gA_\mu j_{\mathrm{N}}^\mu. \tag{2.96}$$

The additional term cancels out the variation of $\mathcal{L}_0$ but generates, due to the variation of the Noether current itself, another term of order $\mathcal{O}(g^2)$. Up to total derivatives

$$\delta\mathcal{L}_1 = -g^2|\Phi|^2 A_\mu\partial^\mu\Lambda. \tag{2.97}$$

This variation can be exactly canceled out by

$$\mathcal{L}_2 = \mathcal{L}_1 + \tfrac{1}{2}g^2|\Phi|^2A^2, \tag{2.98}$$

which can be rewritten in the standard, manifestly gauge-invariant form

$$\begin{aligned}\mathcal{L}_2 &= -\tfrac{1}{4}F^2 + \tfrac{1}{2}\mathcal{D}_\mu\Phi\mathcal{D}^\mu\bar{\Phi},\\ \mathcal{D}_\mu\Phi &= (\partial_\mu - igA_\mu)\Phi.\end{aligned} \tag{2.99}$$

A more interesting example is provided by a set of $r$ vector fields $A^A{}_\mu$ with Lagrangian

$$\mathcal{L}_0 = +\tfrac{1}{4}g_{AB}f^A{}_{\mu\nu}f^{B\,\mu\nu}, \qquad f^A{}_{\mu\nu} = 2\partial_{[\mu}A^A{}_{\nu]}. \tag{2.100}$$

Here $g_{AB}$ is a negative-definite constant metric. This Lagrangian is evidently invariant under $r$ local gauge transformations,

$$\delta_\Lambda A^A{}_\mu = \partial_\mu\Lambda^A, \tag{2.101}$$

because the field strengths are. It is less evident, but equally true, that the above Lagrangian is invariant under global transformations that form a group G of dimension $d$ such that the Killing metric of the associated Lie algebra[15] is precisely $g_{AB}$. Under these transformations, the vector fields transform in the adjoint representation, that is, infinitesimally:

$$\delta_\sigma A^A{}_\mu = g\sigma^C \Gamma_{\text{Adj}}(T_C)^A{}_B A^B{}_\mu, \qquad \Gamma_{\text{Adj}}(T_C)^A{}_B = f_{CB}{}^A. \tag{2.102}$$

These two symmetries form a closed symmetry algebra:

$$[\delta_\Lambda, \delta_\sigma] = \delta_{\Lambda'}, \qquad \Lambda^{A\prime} = \sigma^C \Gamma_{\text{Adj}}(T_C)^A{}_B \Lambda^B. \tag{2.103}$$

There are $d$ conserved Noether currents $j^\mu_{\text{N}\,A}$ associated with the global invariance of G. Following the general argument, they can be found by performing a local G transformation:

$$\begin{aligned} \delta_{\sigma(x)} \mathcal{L}_0 &= g\sigma^A \partial_\mu j^\mu_{\text{N}\,A}, \\ j^\mu_{\text{N}\,A} &= f_{ABC} f^C{}_\mu{}^\nu A^B{}_\nu, \quad f_{ABC} \equiv f_{AB}{}^D g_{DC}. \end{aligned} \tag{2.104}$$

Let us now consider the coupling of $d$ conserved currents $j_A{}^\mu$ associated with some other set of matter fields invariant under global G transformations to the vector fields. As in the Maxwell case, we add to the action the terms $gA^A{}_\mu j_A{}^\mu$ and find that the currents are no longer conserved on-shell because the equations of motion of the fields have changed due to the new coupling term. As in the Maxwell case, the problem is that we have not taken into account all the sources of charge, since only the total charge associated with invariance of G will be conserved once the coupling has been introduced. Thus, we should couple the vector fields to their own Noether currents. We can forget about the matter fields now and try to solve the self-consistency problem of the coupling of the vector fields to themselves by use of the Noether method.

Since Noether currents were found via local G transformations, we look for invariance under local G transformations of $\mathcal{L}_0$. To cancel out $\delta_{\sigma(x)}\mathcal{L}_0$ we have to do two things: first, we have to identify $\Lambda^A = \sigma^A$ and then we have to introduce a correction that is of first order in $g$ into the Lagrangian that takes the characteristic form

$$\mathcal{L}_1 = \mathcal{L}_0 + \frac{g}{2} A^A{}_\mu j^\mu_{\text{N}\,A}. \tag{2.105}$$

In this way, by enforcing local symmetry we arrive at the same conclusion as before by physical arguments: we have to add the self-coupling term. This makes sense if the algebra of the new transformations,

$$\delta_\sigma A^A{}_\mu = \partial_\mu \sigma^A + g\sigma^C \Gamma_{\text{Adj}}(T_C)^A{}_B A^B{}_\mu, \tag{2.106}$$

closes, as is the case. The new term in the Lagrangian produces a new term of second order in $g$ in the transformation:

$$\delta_\sigma \mathcal{L}_1 = g^2 f_{ABC} f_{ED}{}^C A^A_\mu A^B{}_\nu A^{D\,\nu} \partial^\mu \sigma^E, \tag{2.107}$$

[15] See Appendix A for notation and conventions.

which can be exactly canceled out by the addition of an $\mathcal{O}(g^2)$ term that finishes the iterative procedure:

$$\mathcal{L}_2 = \mathcal{L}_1 - \frac{g^2}{4} f_{ABC} f_{ED}{}^{C} A^{A}{}_{\mu} A^{B}{}_{\nu} A^{D\,\nu} A^{E\,\mu}. \tag{2.108}$$

This Lagrangian can be written in the standard, manifestly gauge-invariant form

$$\begin{aligned} \mathcal{L}_2 &= \tfrac{1}{4} g_{AB} F^{A}{}_{\mu\nu} F^{B\,\mu\nu}, \\ F^{A}{}_{\mu\nu} &= f^{A}{}_{\mu\nu} + g f_{BC}{}^{A} A^{B}{}_{\mu} A^{C}{}_{\nu}. \end{aligned} \tag{2.109}$$

It is customary to use dimensionless gauge parameters. On rescaling $\sigma^A \to \sigma^A/g$ we recover the gauge transformations in the conventions of Appendix A.

In more complicated cases the Noether procedure will require the addition of more corrections both to the Lagrangian and to the field-transformation rules (see e.g. Ref. [983]). The procedure is simplified considerably by using first-order actions [430, 431]. Only in this way is it possible to find all the corrections to the Fierz–Pauli Lagrangian. This is explained in Section 3.2.7.

In the recipe that we have given here we have added to the original Lagrangian the exact number of vector fields needed to gauge the existing symmetry, but in many cases the field content of the theory cannot be changed and we have a given number of vector fields transforming in a given representation of the global symmetry group. The possible gaugings of the global symmetry group are usually strongly constrained and a method to explore systematically all the possibilities is needed. The *embedding tensor formalism* is such a method, and it will be reviewed in Section 2.7 after we study which are the possible global symmetries that a field theory can have.

## 2.6 Generic symmetries of field theories

In this section we are going to review the results obtained by Gaillard and Zumino in their study of the general symmetries of a four-dimensional theory [570]. In the symmetries of the theory are included electric–magnetic duality rotations[16] that will be studied in more detail in Chapter 12. These transformations are only symmetries of the equations of motion of the theory and they do not leave invariant the action (not even up to total derivatives), and the standard Noether procedure cannot be used to construct a conserved current, which, nevertheless, turns out to exist. But this is not the only result of the investigation: a condition for the general transformations that we are going to discuss to be symmetries

[16] Electric–magnetic duality transformations were first introduced by Schrödinger in Ref. [1110] in the context of the Born–Infeld theory of non-linear electrodynamics. They were extended to curved spacetime by Misner and Wheeler in Ref. [953], and their presence in supergravity theories was first realized and used in Refs. [985, 682] for $N = 1$ Maxwell–Einstein and pure $N = 2$ supergravity, respectively. The fact that these dualities can be extended in four dimensions to U($N$) was first observed in [539, 388]. The first general and systematic treatment of duality symmetries (in supergravity but also in general field theories) was made by Gaillard and Zumino [570] and we will study it here.

of the equations of motion[17] will be found and the largest possible group of symmetries of the equations of motion will be determined, which is $\mathrm{Sp}(2\bar{n}, \mathbb{R})$ for theories containing $\bar{n}$[18] Abelian vector fields (1-forms); the symmetry group of the equations of motion of any such theory will always be a subgroup of $\mathrm{Sp}(2\bar{n}, \mathbb{R})$.

This result, which can be generalized to other dimensions and form fields of ranks other than 1, has an extraordinary importance. We will construct entire classes of theories (the supergravities reviewed in Chapters 6–9) making the $\mathrm{Sp}(2\bar{n}, \mathbb{R})$ invariance manifest even if only a certain subgroup leaves invariant a given theory, and we will be able to obtain very general results valid for the entire class of theories. Furthermore, the derivation of these results is a very good exercise in classical field theory.

Instead of following the original reference [570] we are going to follow the functional method of Ref. [67], which is valid in more general cases. Thus, we are going to consider an action $S$ which is a functional of the $\bar{n}$ field strengths of as many Abelian 1-form fields[19] $F^{\Lambda} = dA^{\Lambda}$, and an indeterminate number of other fields $\varphi^i$. For the time being, we do not exclude the presence of higher-derivative terms $\partial^p F, \partial^q \varphi$ in the action. As a warm-up, we start with a simple case.

### 2.6.1 *Single vector field*

Let us consider functionals of a single Abelian field strength[20] $S[F]$. Due to its definition $F = dA$, the field strength $F$ always satisfies the *Bianchi identity* $dF = 0$, which can also be written in the form

$$\partial_\mu \star F^{\mu\nu} = 0. \tag{2.110}$$

It is useful to define the *dual ("magnetic") field strength* $G$ by

$$\star G^{\mu\nu} \equiv 2\frac{\delta S}{\delta F_{\mu\nu}}. \tag{2.111}$$

With this definition, the equations of motion of the 1-form (the "Maxwell equations") can always be written in the form

$$\partial_\mu \star G^{\mu\nu} = 0, \tag{2.112}$$

which is that of a "Bianchi identity" for $G$. Indeed, for the Maxwell action

$$S_{\text{Maxwell}}[F] \equiv \int d^4x \left[-\tfrac{1}{4}F^2\right], \tag{2.113}$$

17 Another way of expressing this invariance of the equations of motion is to say that the theory is *self-dual* under the transformations [626, 627], that is, the transformed equations of motion can be derived from an action that has the same form as the original one.

18 We reserve $n_V$ for the number of vector supermultiplets of a given supergravity theory. For general $N, d = 4$ SUEGRAs $\bar{n} = N(N-1)/2 + n_V$.

19 Here we use capital Greek indices $\Lambda, \Sigma, \Delta, \Gamma$, etc. to label the vector fields.

20 For simplicity, the coupling to a metric is not included in the problem, but the results can be immediately covariantized.

the dual field strength is given by

$$G = \star F, \tag{2.114}$$

and Eq. (2.112) is the standard Maxwell equation.

In all cases, the similarity between Eqs. (2.110) and (2.112) written in these variables leads to their invariance under global linear transformations of $F$ and $G$ of the form

$$\begin{pmatrix} F' \\ G' \end{pmatrix} = \begin{pmatrix} A & B \\ C & D \end{pmatrix} \begin{pmatrix} F \\ G \end{pmatrix} \longrightarrow \delta \begin{pmatrix} F \\ G \end{pmatrix} = \begin{pmatrix} a & b \\ c & d \end{pmatrix} \begin{pmatrix} F \\ G \end{pmatrix}. \tag{2.115}$$

In general, the action $S[F]$ is not invariant under all these transformations; not even up to total derivatives, as can be easily seen in the Maxwell case. This is not surprising since we are mixing the (Maxwell) equations of motion with other equations (the Bianchi identities) that have a completely different status. There is, however, a more basic issue: by definition, $G$ is a functional of $F$, $G(F)$, which is just $G(F) = \star F$ in the Maxwell case. It is natural to expect that $G'$ is another functional of $F'$, $G'(F')$. Indeed, in the Maxwell case,

$$G'(F') = \frac{DB + AC}{A^2 + B^2} F' + \frac{AD - BC}{A^2 + B^2} \star F. \tag{2.116}$$

It is natural to ask when $G'(F')$ is the same functional as $G(F)$, that is

$$G'(F') = G(F'), \tag{2.117}$$

which is equivalent to the requirement that the action $S'[F']$ from which one can derive $G'(F')$ using the definition (2.111) is the same as the original one, that is

$$S'[F'] = S[F']. \tag{2.118}$$

This property is known in the literature as *self-duality*, and it is a property both of the duality transformations and of the theory $S[F]$. In the Maxwell case, Eq. (2.116) tells us that self-duality ($G'(F') = \star F'$) requires $A = D$ and $B = -C$. This group of transformations is the product $\mathbb{R}^+ \times \mathrm{SO}(2)$. Only the second factor leaves invariant the energy–momentum tensor (which is necessary to have a symmetry of the equations of motion when we couple the theory to Einstein's gravity), and, therefore, the electric–magnetic duality group of Maxwell's theory is just $\mathrm{SO}(2)$.

We would like to characterize all the possible theories which are self-dual and the corresponding duality groups. We are going to require consistency of the definition of the dual vector field strength in Eq. (2.111) before and after an infinitesimal duality transformation (2.115) with the self-duality condition (2.118):

$$\Delta \star G^{\mu\nu} = 2\Delta \frac{\delta S[F]}{\delta F_{\mu\nu}} = 2 \left\{ \frac{\delta S'[F']}{\delta F'_{\mu\nu}} - \frac{\delta S[F]}{\delta F_{\mu\nu}} \right\} = 2 \left\{ \frac{\delta S[F']}{\delta F'_{\mu\nu}} - \frac{\delta S[F]}{\delta F_{\mu\nu}} \right\}, \tag{2.119}$$

by hypothesis. Here we have used capital $\Delta$ to denote the infinitesimal duality transformation to distinguish it from the functional variation of the fields. Furthermore, we consider all fields as functions of $x$, but we will not write this dependence explicitly unless it is necessary. Using the (functional) chain rule,

$$\begin{aligned}
\frac{\delta S[F']}{\delta F'_{\mu\nu}} &= \int d^4y \frac{\delta S[F']}{\delta F_{\alpha\beta}(y)} \frac{\delta F_{\alpha\beta}(y)}{\delta F'_{\mu\nu}} \\
&= \int d^4y \frac{\delta S[F']}{\delta F_{\alpha\beta}(y)} \frac{\delta}{\delta F'_{\mu\nu}} \left[F'_{\alpha\beta}(y) - aF'_{\alpha\beta}(y) - bG'_{\alpha\beta}(y)\right] \\
&= (1-a)\frac{\delta S[F']}{\delta F_{\mu\nu}} - b\int d^4y \frac{\delta S[F']}{\delta F_{\alpha\beta}(y)} \frac{\delta G'_{\alpha\beta}(y)}{\delta F'_{\mu\nu}}.
\end{aligned} \tag{2.120}$$

In the terms that are multiplied by an infinitesimal parameter (to first order in those parameters), we can replace primed by unprimed fields, obtaining

$$\begin{aligned}
\frac{\delta S[F']}{\delta F'_{\mu\nu}} &= \frac{\delta S[F']}{\delta F_{\mu\nu}} - a\frac{\delta S[F]}{\delta F_{\mu\nu}} - b\int d^4y \frac{\delta S[F]}{\delta F_{\alpha\beta}(y)} \frac{\delta G_{\alpha\beta}(y)}{\delta F_{\mu\nu}} \\
&= \frac{\delta S[F']}{\delta F_{\mu\nu}} - \frac{a}{2} \star G^{\mu\nu} - b\int d^4y \star G^{\alpha\beta}(y)\frac{\delta G_{\alpha\beta}(y)}{\delta F_{\mu\nu}} \\
&= \frac{\delta S[F']}{\delta F_{\mu\nu}} - \frac{a}{2} \star G^{\mu\nu} - \frac{b}{2}\frac{\delta}{\delta F_{\mu\nu}} \int d^4y \star G^{\alpha\beta}(y) G_{\alpha\beta}(y),
\end{aligned} \tag{2.121}$$

where we have used the definition of the dual vector field strength in the second line. Substituting this result into Eq. (2.119) we get

$$\Delta \star G^{\mu\nu} = 2\frac{\delta}{\delta F_{\mu\nu}} \left\{S[F'] - S[F]\right\} - \frac{a}{2} \star G^{\mu\nu} - \frac{b}{2}\frac{\delta}{\delta F_{\mu\nu}} \int d^4y \star G^{\alpha\beta}(y) G_{\alpha\beta}(y), \tag{2.122}$$

and, using Eq. (2.115), we get

$$\begin{aligned}
\frac{\delta \Delta S[F]}{\delta F_{\mu\nu}} &= \frac{c}{2} \star F^{\mu\nu} + \frac{(d+a)}{2} \star G^{\mu\nu} + \frac{b}{4}\frac{\delta}{\delta F_{\mu\nu}} \int d^4y \star G^{\alpha\beta}(y) G_{\alpha\beta}(y) \\
&= (d+a)\,\frac{\delta S[F]}{\delta F_{\mu\nu}} + \tfrac{1}{4}\frac{\delta}{\delta F_{\mu\nu}} \int d^4y \left[c\, F^{\alpha\beta} \star F_{\alpha\beta} + b\, G^{\alpha\beta} \star G_{\alpha\beta}\right],
\end{aligned} \tag{2.123}$$

which we rewrite in the form

$$\frac{\delta}{\delta F_{\mu\nu}} \left\{\Delta S[F] - (d+a)S[F] - \tfrac{1}{4}\int d^4y \left[c\, F^{\alpha\beta} \star F_{\alpha\beta} + b\, G^{\alpha\beta} \star G_{\alpha\beta}\right]\right\} = 0. \tag{2.124}$$

Finally, using

$$\Delta S[F] = \int d^4y\, \frac{\delta S[F]}{\delta F_{\alpha\beta}(y)} \Delta F_{\alpha\beta}(y) = \int d^4y\, \frac{\delta S[F]}{\delta F_{\alpha\beta}} \left[aF_{\alpha\beta} + bG_{\alpha\beta}\right], \tag{2.125}$$

we arrive at the final result, which is the condition that the action, the dual vector field strengths, and the duality parameters must satisfy in a self-dual theory:

$$\frac{\delta}{\delta F_{\mu\nu}} \left\{(d+a)S[F] + \tfrac{1}{4}\int d^4y \left[c\, F^{\alpha\beta} \star F_{\alpha\beta} - b\, G^{\alpha\beta} \star G_{\alpha\beta} - 2a\, F^{\alpha\beta} \star G_{\alpha\beta}\right]\right\} = 0, \tag{2.126}$$

sometimes called the *Noether–Gaillard–Zumino (NGZ) identity.*

Since the infinitesimal parameter $d$ only occurs when multiplying the action, it is clear that the above identity can only be satisfied if $d$ is related to the other parameters. The duality group can only be, at most, three dimensional. Requiring the invariance of the energy–momentum tensor restricts this group to a one-dimensional one with $a = d = 0$ and $b = \pm c$. However, since we are usually interested in non-linear higher-derivative extensions of the Maxwell theory, it is customary to focus only on the SO(2) duality group $a = d = 0$, $b = -c$, and we will do so in the following. In that case, the NGZ identity takes the following simple form (integrating the functional derivative and eliminating the integration constant [626]):

$$F^{\alpha\beta} \star F_{\alpha\beta} + G^{\alpha\beta}(F) \star G_{\alpha\beta}(F) = 0. \tag{2.127}$$

The Bianchi identity implies the local existence of a 1-form such that $F = dA$. The Maxwell equation has the form of a Bianchi identity for the dual vector field strength and, by the same token, implies the local existence of a dual 1-form potential $B$ such that $G = dB$. Then, on-shell we can write

$$\begin{aligned} \star F^{\alpha\beta} F_{\alpha\beta} + \star G^{\alpha\beta} G_{\alpha\beta} &= 2\left[\star F^{\alpha\beta}\partial_\alpha A_\beta + \star G^{\alpha\beta}\partial_\alpha B_\beta\right] \\ &= \partial_\alpha J^{\mu}_{\text{NGZ}} - 2\left[\partial_\alpha \star F^{\alpha\beta} A_\beta + \partial_\alpha \star G^{\alpha\beta} B_\beta\right], \end{aligned} \tag{2.128}$$

where we have defined the *NGZ current*

$$J^{\alpha}_{\text{NGZ}} \equiv 2\left[\star F^{\alpha\beta} A_\beta + \star G^{\alpha\beta} B_\beta\right]. \tag{2.129}$$

The Maxwell equation and Bianchi identity imply the on-shell vanishing of the second term in Eq. (2.128), and then the NGZ identity implies the on-shell conservation of the NGZ current.

Furthermore, the NGZ identity implies that, in all self-dual theories,

$$S_{\text{inv}}[F] \equiv S[F] - \tfrac{1}{4}\int d^4x F_{\alpha\beta} \star G^{\alpha\beta} \tag{2.130}$$

is automatically duality invariant. $S_{\text{inv}}[F]$ vanishes identically for the Maxwell theory but it is non-trivial in more complicated cases [306]. Under infinitesimal duality transformations,

$$\begin{aligned}\Delta S_{\text{inv}}[F] &= \int d^4x \left\{ \frac{\delta S[F]}{\delta F_{\alpha\beta}} \Delta F_{\alpha\beta} - \tfrac{1}{4} \star G^{\alpha\beta} \Delta F_{\alpha\beta} - \tfrac{1}{4} \star F_{\alpha\beta} \Delta G^{\alpha\beta} \right\} \\ &= \int d^4x \left\{ \tfrac{1}{4} \star G^{\alpha\beta} \Delta F_{\alpha\beta} - \tfrac{1}{4} \star F_{\alpha\beta} \Delta G^{\alpha\beta} \right\} \\ &= \tfrac{1}{4} b \int d^4x \left\{ \star G^{\alpha\beta} G_{\alpha\beta} + \star F_{\alpha\beta} F^{\alpha\beta} \right\}, \end{aligned} \tag{2.131}$$

which vanishes by virtue of the NGZ identity.

Finally, if the action $S[\lambda, F]$ depends on another parameter (or field) $\lambda$ which is not affected by the duality transformations, then [346]

$$\boxed{S_{\text{inv}}[\lambda, F] = -\lambda \frac{d}{d\lambda} S[\lambda, F].} \tag{2.132}$$

The proof of the invariance of this expression is a particular case of the theorem proven in the appendices of Ref. [570]:

$$\begin{aligned}\Delta S_{\text{inv}}[\lambda, F] &= -\lambda \frac{d}{d\lambda} \Delta S[\lambda, F] = -\lambda \frac{d}{d\lambda} \int d^4x \frac{\delta S[F]}{\delta F_{\alpha\beta}} \Delta F_{\alpha\beta} \\ &= -\lambda \frac{d}{d\lambda} \int d^4x \, \frac{b}{2} \star G^{\alpha\beta} G_{\alpha\beta}. \end{aligned} \tag{2.133}$$

Since $F$ is $\lambda$ independent, we can add a term $\star F^{\alpha\beta} F_{\alpha\beta}$ to the integrand, which then vanishes by virtue of the NGZ identity. Then, the identification of the above expression with $S_{\text{inv}}$ defined in Eq. (2.130) comes from the uniqueness of the invariant combination and the comparison with known explicit cases [346].

If we know $G(\lambda, F)$, the above property can be used to reconstruct $S[\lambda, F]$: from Eq. (2.130) and the above property,

$$S[\lambda, F] + \lambda \frac{d}{d\lambda} S[\lambda, F] = \tfrac{1}{4} \int d^4x F_{\alpha\beta} \star G^{\alpha\beta}. \tag{2.134}$$

The left-hand side is just $d(\lambda S)/d\lambda$, and, integrating both sides with respect to $\lambda$, we get

$$\boxed{S[\lambda, F] = \frac{1}{4\lambda} \int d\lambda d^4x F_{\alpha\beta} \star G^{\alpha\beta}(\lambda, F).} \tag{2.135}$$

*Example: the Born–Infeld theory.* The most famous non-trivial example of self-dual theory is the non-linear model of Born and Infeld proposed in Refs. [235, 237] (see also Dirac's contribution in Ref. [456]), and reviewed in Ref. [236], which was shown to be self-dual by Schrödinger in Ref. [1110]. The action of this model is

$$S_{\text{Born-Infeld}}[g^2, F] \equiv \frac{1}{g^2}\int d^4x \left\{1 - \sqrt{-\det(\eta_{\mu\nu} + gF_{\mu\nu})}\right\}, \tag{2.136}$$

where the determinant in the action has the explicit expression

$$-\det(\eta_{\mu\nu} + gF_{\mu\nu}) = 1 + \tfrac{1}{2}g^2F^2 + \tfrac{1}{16}g^4(F \star F)^2 \equiv \Delta. \tag{2.137}$$

Then, for weak fields, on expanding the square root we get the action given by an infinite power series in the two Lorentz invariants

$$\alpha \equiv \tfrac{1}{4}F^{\alpha\beta}F_{\alpha\beta}, \qquad \beta \equiv \tfrac{1}{4}F^{\alpha\beta} \star F_{\alpha\beta} \;\Rightarrow\; \Delta = 1 + 2g^2\alpha + g^4\beta^2, \tag{2.138}$$

whose first terms are

$$S_{\text{Born-Infeld}}[g^2, F] \equiv \frac{1}{g^2}\int d^4x \left\{-\alpha + \tfrac{1}{2}g^2(\alpha^2 + \beta^2) + \mathcal{O}(g^4)\right\}. \tag{2.139}$$

The action contains only even powers of the coupling constant $g^2$, and for $g^2 = 0$ we recover Maxwell's theory.

The dual vector field strength $G(g^2, F)$ is, for this theory, given by

$$G(g^2, F) = \Delta^{-1/2}(\star F + g^2\beta F), \tag{2.140}$$

and it is immediate to see that the NGZ identity Eq. (2.127) is satisfied (so the theory is self-dual) and that

$$\begin{aligned} -g^2\frac{\partial S[g^2, F]}{\partial g^2} &= S[g^2, F] - \tfrac{1}{4}\int d^4x\, F \star G = S_{\text{inv}}[g^2, F] \\ &= \frac{1}{g^2}\int d^4x \left\{1 - \frac{(1 + g^2\alpha)}{\Delta^{1/2}}\right\}. \end{aligned} \tag{2.141}$$

This is not the only self-dual theory of an Abelian vector field; the problem of finding the most general theory (with no higher derivatives) satisfying the NGZ identity (Eq. 2.127) has been treated and solved in Refs. [626, 571].

### 2.6.2 The general case

Let us now consider the general case. For each of the fundamental (or "electric") vector field strengths $F^\Lambda$ we can define a dual (or "magnetic") vector field strength $G_\Lambda(F, \varphi)$,

$$\star G_\Lambda{}^{\mu\nu} \equiv 2\frac{\delta S[F, \varphi]}{\delta F^\Lambda{}_{\mu\nu}}, \tag{2.142}$$

so that the Maxwell equations for each fundamental vector $A^\Lambda$ take the form of a Bianchi identity for the dual vector field strength

$$\partial_\mu \star G_\Lambda{}^{\mu\nu} = 0, \tag{2.143}$$

implying the local existence of as many dual (or "magnetic") vector fields $A_\Lambda$ such that $G_\Lambda = dA_\Lambda$.

Just as in the Abelian case, the similarity between the Bianchi identities and the Maxwell equations raises the possibility of mixing them linearly. Thus, we construct $2\bar{n}$-component vectors of the fundamental and dual vector field strengths and consider the linear transformations with constant matrices

$$\begin{pmatrix} F' \\ G' \end{pmatrix} = S \begin{pmatrix} F \\ G \end{pmatrix}, \qquad \Delta \begin{pmatrix} F \\ G \end{pmatrix} = \mathcal{T} \begin{pmatrix} F \\ G \end{pmatrix}, \tag{2.144}$$

where we have defined the matrices $S$ and $T$ (see also Eq. (C.6)) as

$$S \equiv \begin{pmatrix} A & B \\ C & D \end{pmatrix}, \qquad \mathcal{T} \equiv \begin{pmatrix} a & b \\ c & d \end{pmatrix}. \tag{2.145}$$

Now, however, consistency will require that these transformations are accompanied by (generically non-linear) transformations of the other fields

$$\varphi'^{\,i} = f^i(\varphi), \qquad \Delta\varphi^i = \xi^i(\varphi). \tag{2.146}$$

We are not going to attack the problem of characterizing the most general theory and duality group.[21] Instead, we will content ourselves with the study of the linear (in vector field strengths) theories. These have the generic form[22]

$$S[F,\varphi] = \int d^4x \sqrt{|g|}\,\{R + \mathcal{G}_{ij}\partial_\mu\varphi^i\partial^\mu\varphi^j$$
$$+ 2\Im\mathrm{m}\mathcal{N}_{\Lambda\Sigma}F^{\Lambda\,\mu\nu}F^{\Sigma}{}_{\mu\nu} - 2\Re\mathrm{e}\mathcal{N}_{\Lambda\Sigma}F^{\Lambda\,\mu\nu} \star F^{\Sigma}{}_{\mu\nu}\}, \tag{2.147}$$

where $\mathcal{G}_{ij}(\varphi)$ is a scalar-dependent, positive-definite matrix that can be viewed as a metric in a (*target*) space in which the scalar fields play the role of coordinates, and $\mathcal{N}_{\Lambda\Sigma}(\varphi)$ is the symmetric, complex, scalar-dependent, $\bar{n} \times \bar{n}$ *period matrix* that codifies the couplings between the scalars and the vector fields, whose imaginary part must be negative definite (so the vector's kinetic term has the correct sign).

The bosonic sectors of all the four-dimensional ungauged supergravities can be written in this form, and therefore the above action is general enough to cover many four-dimensional cases of interest. The normalizations are those used in Ref. [522]. An important term that

[21] This problem has been studied in Ref. [67].

[22] For simplicity, all the fields $\varphi^i$ are assumed to be scalars in this example. We also change the normalization of the vector fields for convenience, since this is the normalization we will use in matter-coupled $N = 1, 2, d = 4$ supergravities in Chapters 6 and 7; we also introduce a general spacetime metric $g_{\mu\nu}$ and the Einstein–Hilbert term for convenience, although they play no role in the discussion of duality and we have not yet introduced the Einstein–Hilbert action. A scalar potential $V(\varphi)$ (such as the one that can be generated in $N = 1, d = 4$ theories by adding a superpotential or in $N = 1$ and $N = 2, d = 4$ theories via Fayet–Iliopoulos terms associated to Abelian gaugings) could be added to this action but it would not change the conclusions.

could be added to it is a scalar potential such as those that occur in gauged supergravities. Its only effect on our conclusions is to restrict the possible symmetries of the theory to those that leave it invariant.

The dual vector field strengths are defined, for convenience, with a different normalization

$$\star G_\Lambda{}^{\mu\nu} \equiv \frac{1}{4\sqrt{|g|}} \frac{\delta S}{\delta F^\Lambda{}_{\mu\nu}} \quad \Rightarrow \quad G_\Lambda = \Re\mathfrak{e}\mathcal{N}_{\Lambda\Sigma} F^\Sigma + \Im\mathfrak{m}\mathcal{N}_{\Lambda\Sigma} \star F^\Sigma, \tag{2.148}$$

which can also be written in the equivalent form

$$\boxed{G_\Lambda{}^+ = \mathcal{N}^*_{\Lambda\Sigma} F^{\Sigma +}.} \tag{2.149}$$

The relation between $G$ and $F$ in the Maxwell case ($G = \star F$) is sometimes called the *linear self-duality constraint* (if we do not impose it, we deal with the fundamental and dual vector fields as independent variables). The above relation between the $G_\Lambda$ and the fundamental $F^\Lambda$ is sometimes called the *linear twisted self-duality constraint.*

Self-duality means that, after the duality transformations Eq. (2.144), $G'_\Lambda$ is related to $F'^\Lambda$ by the same linear, twisted, self-duality constraint Eq. (2.149), where now the period matrix may also need to be transformed:

$$G'_\Lambda{}^+ = \mathcal{N}'^*_{\Lambda\Sigma} F'^{\Sigma +}. \tag{2.150}$$

Expressing the primed vector field strengths in terms of the unprimed ones using Eq. (2.144) and expressing the dual field strengths in terms of the fundamental ones using the constraint Eq. (2.149), we get

$$\{(C + D\mathcal{N}^*) - \mathcal{N}'^*(A + B\mathcal{N}^*)\} F^+ = 0, \tag{2.151}$$

which implies the transformation rule for the period matrix,

$$\boxed{\mathcal{N}' = (C + D\mathcal{N})(A + B\mathcal{N})^{-1},} \tag{2.152}$$

where we have used the reality of the matrices $A, B, C$, and $D$. These transformations must preserve the essential properties of the period matrix: $\mathcal{N}'$ must be symmetric and its imaginary part must be negative definite. The condition $\mathcal{N}'^T = \mathcal{N}'$ and the above transformation rule lead to

$$C^T A - (A^T D - C^T B)\mathcal{N} + \mathcal{N}(D^T B)\mathcal{N} - \text{transposed} = 0, \tag{2.153}$$

which is solved by

$$C^T A = A^T C, \qquad B^T D = D^T B, \qquad A^T D - C^T B = \kappa \mathbb{1}_{\bar{n}\times\bar{n}}, \tag{2.154}$$

for some real constant $\kappa$ determined to be $\kappa = +1$ if we want to leave invariant the energy–momentum tensor (see the discussion that follows Eq. (2.163)). From the above transformation rules and properties of the matrices $A, B, C$, and $D$, it follows that the imaginary part of the period matrix transforms as

$$\Im\mathrm{m}\mathcal{N}' = \kappa(A^T + \mathcal{N}^\dagger B^T)^{-1}\Im\mathrm{m}\mathcal{N}(A + B\mathcal{N})^{-1}, \tag{2.155}$$

and the negative definiteness is preserved for $\kappa > 0$ and, in particular, for $\kappa = +1$, the only value that we are going to consider.[23]

These conditions identify the duality group as a subgroup[24] of $\mathrm{Sp}(2\bar{n}, \mathbb{R})$, which is defined as the group of transformations $S$ that preserve the symplectic "metric" $\Omega$[25]

$$S^T\Omega S = \Omega, \qquad \Omega \equiv \begin{pmatrix} 0 & \mathbb{1} \\ -\mathbb{1} & 0 \end{pmatrix}. \tag{2.160}$$

See Appendix C for further information on the symplectic group.

It is sometimes useful to introduce symplectic indices $M, N, \ldots$, each of them equivalent to one upper index and one lower index $\Lambda, \Sigma, \ldots$. The $2\bar{n}$-component vector of fundamental and dual vector field strengths transforms a symplectic vector:

$$(\mathcal{F}^M) \equiv \begin{pmatrix} F^\Lambda \\ G_\Lambda \end{pmatrix}, \qquad \mathcal{F}'^M = S^M{}_N \mathcal{F}^N, \tag{2.161}$$

and the symplectic metric can be used to raise and lower symplectic indices according to the convention

$$\mathcal{F}_N \equiv \Omega_{NM}\mathcal{F}^M, \qquad \mathcal{F}^N = \mathcal{F}_M \Omega^{MN}, \qquad \Omega^{MN} = -(\Omega^{-1})^{MN} = \Omega_{MN}, \tag{2.162}$$

so the components of the covariant symplectic vector are related to those of the contravariant one by

$$(\mathcal{F}_M) = (G_\Lambda, -F^\Lambda). \tag{2.163}$$

This notation is very useful. For example, let us express the energy–momentum tensor of the vectors in the action (2.147) in this language. According to the Rosenfeld prescription (Section 2.4.3), it is given by

$$T^{\text{vect}}_{\mu\nu} = -8\Im\mathrm{m}\mathcal{N}_{\Lambda\Sigma}\left[F^\Lambda{}_\mu{}^\rho F^\Sigma{}_{\nu\rho} - \tfrac{1}{4}g_{\mu\nu}F^{\Lambda\,\rho\sigma}F^\Sigma{}_{\rho\sigma}\right], \tag{2.164}$$

---

23 We can also check these properties in the infinitesimal transformations, which take the form

$$\boxed{\delta\mathcal{N} = c - \mathcal{N}a + d\mathcal{N} - \mathcal{N}b\mathcal{N}.} \tag{2.156}$$

The preservation of the symmetry requires

$$(c^T - c) + (d + a^T)\mathcal{N} + \mathcal{N}(d^T + a) - \mathcal{N}(b^T - b)\mathcal{N} = 0, \tag{2.157}$$

which, for arbitrary period matrices, implies that the generators of the duality transformations satisfy

$$c_{[\Lambda\Sigma]} = b^{[\Lambda\Sigma]} = 0, \qquad a^\Lambda{}_\Sigma = -d_\Sigma{}^\Lambda. \tag{2.158}$$

24 There are more conditions to be taken into account, which we will discuss shortly.

25 If we allow for $\kappa \neq 1$, the transformations only preserve $\Omega$ up to a scalar factor $\kappa$,

$$S^T\Omega S = \kappa\Omega, \tag{2.159}$$

and the duality group is $\mathbb{R}^+ \times \mathrm{Sp}(2\bar{n}, \mathbb{R})$.

and can be rewritten, using Eq. (1.132), in the form

$$T_{\mu\nu}^{\text{vect}} = -4\Im\mathfrak{m}\mathcal{N}_{\Lambda\Sigma}\left[F^{\Lambda}{}_{\mu}{}^{\rho}F^{\Sigma}{}_{\nu\rho} + \star F^{\Lambda}{}_{\mu}{}^{\rho} \star F^{\Sigma}{}_{\nu\rho}\right]. \tag{2.165}$$

Using the symplectic vector of vector field strengths Eq. (2.161)

$$\boxed{T_{\mu\nu}^{\text{vect}} = -4\mathcal{M}_{MN}(\mathcal{N})\mathcal{F}^{M}{}_{\mu}{}^{\rho}\mathcal{F}^{N}{}_{\nu\rho} = -4\Omega_{MN} \star \mathcal{F}^{M}{}_{\mu}{}^{\rho}\mathcal{F}^{M}{}_{\nu\rho},} \tag{2.166}$$

where we have defined the symmetric $2\bar{n} \times 2\bar{n}$ matrix $\mathcal{M}(\mathcal{N})$ by

$$\boxed{(\mathcal{M}_{MN}(\mathcal{N})) \equiv \begin{pmatrix} I_{\Lambda\Sigma} + R_{\Lambda\Gamma}I^{\Gamma\Omega}R_{\Omega\Sigma} & -R_{\Lambda\Gamma}I^{\Gamma\Sigma} \\ -I^{\Lambda\Omega}R_{\Omega\Sigma} & I^{\Lambda\Sigma} \end{pmatrix},} \tag{2.167}$$

where, in turn,

$$I_{\Lambda\Sigma} \equiv \Im\mathfrak{m}\mathcal{N}_{\Lambda\Sigma}, \qquad R_{\Lambda\Sigma} \equiv \Re\mathfrak{e}\mathcal{N}_{\Lambda\Sigma}, \qquad I_{\Lambda\Omega}I^{\Omega\Sigma} = \delta_{\Lambda}{}^{\Sigma}. \tag{2.168}$$

The symmetric matrix $\mathcal{M}_{MN}(\mathcal{N})$ satisfies

$$\mathcal{M}_{MP}(\mathcal{N})\Omega^{PQ}\mathcal{M}_{QN}(\mathcal{N}) = \Omega_{MN} \;\Rightarrow\; (\mathcal{M}^{-1}(\mathcal{N}))^{MN} = \Omega^{MP}\mathcal{M}_{PQ}(\mathcal{N})\Omega^{QN}, \tag{2.169}$$

and it is therefore a symplectic matrix whose block components satisfy Eq. (2.154) with $\kappa = +1$. The energy–momentum tensor will be invariant under duality transformations $\mathcal{F}' = S\mathcal{F}$ if the matrix $\mathcal{M}_{MN}(\mathcal{N})$ transforms according to

$$\mathcal{M}'_{MN}(\mathcal{N}) = (S^{-1})^{P}{}_{M}\mathcal{M}_{PQ}(\mathcal{N})(S^{-1})^{Q}{}_{N}. \tag{2.170}$$

This transformation rule is consistent with that of the period matrix, that is

$$\mathcal{M}'_{MN}(\mathcal{N}) = \mathcal{M}_{MN}(\mathcal{N}'), \tag{2.171}$$

where $\mathcal{M}'$ is given by Eq. (2.170) and $\mathcal{N}'$ is given by Eq. (2.152). However, $\mathcal{M}$ will remain a symplectic matrix only if $\kappa = +1$.[26]

Observe that, in terms of $\mathcal{F}^M$, the linear twisted self-duality constraint takes the form

$$\boxed{\star\mathcal{F}^{M} = \Omega^{MN}\mathcal{M}_{NP}(\mathcal{N})\mathcal{F}^{P}.} \tag{2.172}$$

Now we have to focus our attention on the duality transformation rule of the period matrix Eq. (2.152). The functional dependence of the transformed period matrix on the

[26] The transformation rule of the period matrix is insensitive to rescalings of the matrix $S$, but that of $\mathcal{M}(\mathcal{N})$ is not.

scalar fields will, in general, be different from that of the original one. This is inconsistent unless we can recover the same dependence in terms of some transformed scalars, i.e. if we can define some $\varphi'^{\,i}$ such that

$$\mathcal{N}'(\varphi) = [C + D\mathcal{N}(\varphi)][A + B\mathcal{N}(\varphi)]^{-1} = \mathcal{N}(\varphi'). \tag{2.173}$$

In other words, the linear transformation is equivalent to a reparametrization of the scalars. These transformed $\varphi'^{\,i}$ will only exist for some subgroup of $\mathrm{Sp}(2\bar{n},\mathbb{R})$, restricting the possible duality group. On top of this restriction, if we accept the transformation of the scalars, we have to take into account the transformation of the scalars' kinetic term. Infinitesimally,

$$\begin{aligned}\Delta \int d^4x \mathcal{G}_{ij}\partial_\mu\varphi^i\partial^\mu\varphi^j &= \int d^4x \left\{\partial_k\mathcal{G}_{ij}\partial_\mu\varphi^i\partial^\mu\varphi^j\xi^k + 2\mathcal{G}_{kj}\partial_\mu\xi^k\partial^\mu\varphi^j\right\} \\ &= \int d^4x \left\{\xi^k\partial_k\mathcal{G}_{ij} + 2\mathcal{G}_{kj}\partial_i\xi^k\right\}\partial_\mu\varphi^i\partial^\mu\varphi^j \\ &= \int d^4x \mathcal{L}_\xi\mathcal{G}_{ij}\partial_\mu\varphi^i\partial^\mu\varphi^j,\end{aligned} \tag{2.174}$$

which vanishes if $\xi^i(\varphi)$ is a Killing vector of the metric $\mathcal{G}_{ij}(\varphi)$, i.e. if the transformation from the $\varphi^i$ to the $\varphi'^{\,i}$ is an isometry of $\mathcal{G}_{ij}(\varphi)$.

To summarize, we have shown that the most general symmetry of the equations of motion of theories of the generic form 2.147 acts linearly on the vector fields as a subgroup of $\mathrm{Sp}(2\bar{n},\mathbb{R})$ and, simultaneously, on the scalars as a reparametrization that leaves the scalar metric invariant and produces on the period matrix the same effect as the linear transformations. The most general duality group would then be $\mathrm{Sp}(2\bar{n},\mathbb{R})$. We must be aware, though, that the reasoning leading to this result is based on the dependence of the period matrix on the scalars. There may be scalars that do not occur on the period matrix.[27] In those cases the duality group contains, as a factor additional to the subgroup of $\mathrm{Sp}(2\bar{n},\mathbb{R})$, the group of isometries of the scalar metric that only act on those scalars.

Another instance in which the general analysis is not valid is the $R = \Re\mathfrak{e}\mathcal{N} = 0$ case. Splitting the period matrix transformation rule Eq. (2.152) into its real and imaginary parts, we get two different equations for the transformation of $I = \Im\mathfrak{m}\mathcal{N}$:

$$AI' = ID, \qquad I'BI = -C. \tag{2.175}$$

There are three different possibilities. If we set $A = D = 0$ or $B = C = 0$ we get symplectic matrices with two vanishing submatrices. When the four submatrices are non-zero, we get two transformation rules that must be satisfied simultaneously. They both must preserve the symmetry of $I$ and the energy–momentum tensor (which implies that the transformation must be symplectic). All this implies that

$$C^TA = A^TC, \qquad B^TD = D^TB, \qquad D^TA = \alpha, \qquad C^TB = \alpha - 1, \tag{2.176}$$

[27] An example is provided by the scalars in hypermultiplets of $N = 2, d = 4$ supergravity theories or the scalars in chiral multiplets of $N = 1, d = 4$ supergravities when the period matrix (which is only required to be holomorphic in them) is chosen not to depend on them at all.

with $\alpha > 1$ to preserve the negative definiteness. The compatibility of the two transformation rules is only possible if $I$ enjoys the special property

$$\Im\mathrm{m}\mathcal{N}^{-1} = \frac{\alpha}{\alpha-1} E^T I E, \qquad E \equiv A^{-1}E, \tag{2.177}$$

for some matrix $E = E^T$ (for the transformation to be symplectic). If that property holds, then $B = AE$, and all the submatrices of $S$ are given in terms of $A$ and $E$.

### 2.6.3 Extension to higher dimensions and ranks

It is natural to investigate what the possible duality groups are if we work in dimensions other than four and/or if we deal with $n$ higher-rank form potentials, which are ubiquitous in higher-dimensional supergravities. To this end, we consider the following generalization of the action in Eq. (2.147):

$$\mathcal{I}[g, A^{\Lambda}_{(p+1)}, \phi^i] = \int d^dx\sqrt{|g|}\left\{R + \mathcal{G}_{ij}(\phi)\partial_\mu\phi^i\partial^\mu\phi^j + 4\tfrac{(-1)^p}{(p+2)!} I_{\Lambda\Sigma}(\phi) F^{\Lambda}_{(p+2)} \cdot F^{\Sigma}_{(p+2)}\right\}, \tag{2.178}$$

where the scalar fields $\phi^i$ now couple through the scalar-dependent, symmetric, negative-definite, kinetic matrix $I_{\Lambda\Sigma}(\phi)$ to the $n$ $(p+2)$-form field strengths $F^{\Lambda}_{(p+2)}$ of as many $(p+1)$-form potentials[28] $A^{\Lambda}_{(p+1)}$. In differential-form and in components languages, the field strengths are defined in terms of the potentials, respectively, by

$$F^{\Lambda}_{(p+2)} \equiv dA^{\Lambda}_{(p+1)}, \qquad F^{\Lambda}_{(p+2)\,\mu_1\cdots\mu_{p+2}} = (p+2)\partial_{[\mu_1|}A^{\Lambda}{}_{(p+1)\,|\mu_2\cdots\mu_{p+2}]}. \tag{2.179}$$

Furthermore, in the above action, we use the notation

$$F^{\Lambda}_{(p+2)} \cdot F^{\Sigma}_{(p+2)} \equiv F^{\Lambda}_{(p+2)\,\mu_1\cdots\mu_{p+2}} F^{\Sigma}_{(p+2)}{}^{\mu_1\cdots\mu_{p+2}}. \tag{2.180}$$

This definition implies that the $(p+2)$-form field strengths satisfy the Bianchi identities

$$dF^{\Lambda}_{(p+2)} = 0. \tag{2.181}$$

The normalizations of this action have been chosen so as to recover the $d=4$, $p=0$ action Eq. (2.147), considered in Ref. [522], if we identify $I_{\Lambda\Sigma}$ with $\Im\mathrm{m}\mathcal{N}_{\Lambda\Sigma}$ and the action studied in Ref. [939] for general $d$ and $p=0$ with the original normalizations.

The action Eq. (2.147) has one additional term that we have not included here, however, because, for generic values $d$ and $p$, it does not exist. It only exists when $p = \tilde{p} = (d-4)/2$ and the Hodge dual of a $(p+2)$-form is another $(p+2)$-form. The cases relevant for string theory or supergravity are $p=0$ (associated to charged black holes) when $d=4$, $p=1$ (associated to black strings) in $d=6$, $p=2$ (associated to membranes) in $d=8$, and $p=3$ (associated to black 3-branes) in $d=10$. When it does exist, the additional term has the generic form

$$+4\xi^2\tfrac{(-1)^p}{(p+2)!} R_{\Lambda\Sigma}(\phi) F^{\Lambda}_{(p+2)} \cdot \star F^{\Sigma}_{(p+2)}, \tag{2.182}$$

[28] The parametrization of the rank of the form potentials by $p+1$ is related to the fact that $(p+1)$-form potentials couple to $p$-branes, as will be explained in Chapter 24.

where $R_{\Lambda\Sigma}(\phi)$ is another scalar-dependent matrix. In general, this will not be a symmetric matrix because, in general, $F^{\Lambda}_{(p+2)} \cdot \star F^{\Sigma}_{(p+2)}$ is not symmetric either, and $R_{\Lambda\Sigma}$ must have the same symmetry for the new term to be non-vanishing. The symmetry of that term is related to the value of the square of the Hodge star when it acts on a $(p+2)$-form: $\star^2 = \xi^2$, which is, according to the conventions and general results in Section 1.6, $\xi^2 = -(-1)^{d/2} = (-1)^{p+1}$. The conclusion is that

$$R_{\Lambda\Sigma} = -\xi^2 R_{\Sigma\Lambda}, \qquad \xi^2 = -(-1)^{d/2} = (-1)^{p+1}. \tag{2.183}$$

The value of $\xi$ ($+1$ or $+i$) will determine the duality group.

We will work with the complete generic action

$$\mathcal{I}[g, A^{\Lambda}_{(p+1)}, \phi^i] = \int d^dx \sqrt{|g|} \left\{ R + \mathcal{G}_{ij}(\phi)\partial_\mu \phi^i \partial^\mu \phi^j + 4\tfrac{(-1)^p}{(p+2)!} I_{\Lambda\Sigma}(\phi) F^{\Lambda}_{(p+2)} \cdot F^{\Sigma}_{(p+2)} \right.$$
$$\left. + 4\xi^2 \tfrac{(-1)^p}{(p+2)!} R_{\Lambda\Sigma}(\phi) F^{\Lambda}_{(p+2)} \cdot \star F^{\Sigma}_{(p+2)} \right\}, \tag{2.184}$$

with the understanding that we must set in the results $R_{\Lambda\Sigma} = 0$ whenever $p \neq \tilde{p} = (d-4)/2$.

The discussion now proceeds along the same lines as in the $d=4$, $p=0$ case. We start by defining the dual (*magnetic*) $(\tilde{p}+2)$-form field strengths $G_{(\tilde{p}+2)\,\Lambda}$ through the variation of the action with respect to the *electric* $(p+2)$-form field strengths. With a convenient normalization they are given by

$$G_{(\tilde{p}+2)\,\Lambda} \equiv R_{\Lambda\Sigma} F^{\Sigma}_{(p+2)} + I_{\Lambda\Sigma} \star F^{\Sigma}_{(p+2)}, \tag{2.185}$$

and the equations of motion ("Maxwell equations") of the $(p+1)$-form potentials take the form of Bianchi identities with respect to them:

$$dG_{(\tilde{p}+2)\,\Lambda} = 0. \tag{2.186}$$

In general, it does not make sense to arrange the electric and magnetic field strengths in a single $2n$-component vector, because they are differential forms of different ranks that cannot be rotated into each other, but we can do it anyway if we keep in mind that only in the $p = \tilde{p}$ case can we mix the upper and lower components. Thus, we define

$$(\mathcal{F}^M) \equiv \begin{pmatrix} F^{\Lambda}_{(p+2)} \\ G_{(\tilde{p}+2)\,\Lambda} \end{pmatrix}, \tag{2.187}$$

and write the Maxwell equations (2.186) and Bianchi identities (2.181) in the unified form

$$d\mathcal{F}^M = 0. \tag{2.188}$$

We consider now the linear transformations of $\mathcal{F}^M$:

$$\mathcal{F}'^M = S^M{}_N \mathcal{F}^N, \qquad S \equiv \begin{pmatrix} A & B \\ C & D \end{pmatrix}, \tag{2.189}$$

bearing in mind that when $p \neq \tilde{p}$ we have to set $B = C = 0$.

The consistency of these linear transformations with the definitions of the magnetic field strengths requires that the matrices $R, I$ transform according to

$$\mathcal{N}' = (C + D\mathcal{N})\,(A + B\mathcal{N})^{-1}\,, \tag{2.190}$$

where we have defined the *generalized period matrix*

$$\mathcal{N} \equiv R + \xi I. \tag{2.191}$$

If $p \neq \tilde{p}$, setting $R = B = C = 0$, the general expression reduces to the transformation for $I$:

$$I' = DIA^{-1}. \tag{2.192}$$

On the other hand, the contribution of the $(p+1)$-form potentials to the energy–momentum tensor,

$$T^{A_{(p+1)}}_{\mu\nu} = 8\frac{(-1)^{p+1}}{(p+1)!} I_{\Lambda\Sigma} \left\{ F^{\Lambda}_{(p+2)\,\mu\rho_1\cdots\rho_{p+1}} F^{\Sigma}_{(p+2)\,\nu}{}^{\rho_1\cdots\rho_{p+1}} - \frac{1}{2(p+2)} g_{\mu\nu} F^{\Lambda}_{(p+2)} \cdot F^{\Sigma}_{(p+2)} \right\}. \tag{2.193}$$

can be written, with the help of Eq. (1.134), in the form

$$T^{A_{(p+1)}}_{\mu\nu} = (-1)^{p+1} 4 I_{\Lambda\Sigma} \left\{ \frac{1}{(p+1)!} F^{\Lambda}_{(p+2)\,\mu}{}^{\rho_1\cdots\rho_{p+1}} F^{\Sigma}_{(p+2)\,\nu\rho_1\cdots\rho_{p+1}} \right.$$
$$\left. + \frac{\xi^4}{(\tilde{p}+1)!} \star F^{\Lambda}_{(\tilde{p}+2)\,\mu}{}^{\rho_1\cdots\rho_{\tilde{p}+1}} \star F^{\Sigma}_{(\tilde{p}+2)\,\nu\rho_1\cdots\rho_{\tilde{p}+1}} \right\}, \tag{2.194}$$

where we can take in all cases ($p = \tilde{p}$ or not) $\xi^4 = (-1)^d$. This coefficient disappears when $p = \tilde{p}$. Next, we can rewrite the energy–momentum tensor:

$$T^{A_{(p+1)}}_{\mu\nu} = \frac{4(-1)^{p+1}}{(p+1)!} \mathcal{M}_{MN}(\mathcal{N}) \mathcal{F}^{M}{}_{\mu}{}^{\cdots} \mathcal{F}^{N}{}_{\nu\cdots}, \tag{2.195}$$

where the dots stand for the rest of the indices, which are contracted, and a normalization factor of $n!$ is introduced when $n$ indices are contracted and where we have introduced the symmetric matrix

$$(\mathcal{M}_{MN}(\mathcal{N})) \equiv \begin{pmatrix} I - \xi^2 R I^{-1} R & \xi^2 R I^{-1} \\ -I^{-1} R & \xi^4 I^{-1} \end{pmatrix};$$
$$(\mathcal{M}^{MN}(\mathcal{N})) = \begin{pmatrix} I^{-1} & -\xi^2 I^{-1} R \\ R I^{-1} & \xi^4 (I - \xi^2 R I^{-1} R) \end{pmatrix} = (\mathcal{M}_{NP}(\mathcal{N}))^{-1}. \tag{2.196}$$

When $p = \tilde{p}$, all forms have the same rank and the above expression is always consistent. When $p \neq \tilde{p}$, $R = 0$ and $\mathcal{M}_{MN}$ is diagonal, and only the indices forms of the same rank are contracted in each term.

Finally, using the relation

$$\mathcal{M}_{MN}(\mathcal{N})\mathcal{F}^N = \xi^2\Omega_{MN} \star \mathcal{F}^M, \tag{2.197}$$

where we have defined the metric

$$(\Omega_{MN}) \equiv \begin{pmatrix} 0 & \mathbb{I} \\ \xi^2\mathbb{I} & 0 \end{pmatrix}, \tag{2.198}$$

which will be used to raise and lower $M, N$ indices, we can put the energy–momentum tensor in the following final form:

$$T_{\mu\nu}^{A_{(p+1)}} = 4(-1)^{p+1}\xi^4\Omega_{MN} \star \mathcal{F}^M{}_\mu{}^{\cdots}\mathcal{F}^N{}_{\nu\cdots}. \tag{2.199}$$

We can conclude that the only transformations that leave the energy–momentum tensor invariant are those that leave $\Omega$ invariant, that is, for $p=\tilde{p}$, $\mathrm{O}(n,n)$, when $\xi^2=+1$ and $\mathrm{Sp}(2n,\mathbb{R})$ when $\xi^2=-1$. For $p\neq\tilde{p}$ there is no constraint on the symmetry group at all and we can have $\mathrm{GL(n)}$. This is the case of the vector fields ($p=0$) of the $N=1, d=5$ supergravities that will be studied in Chapter 9.

The same interpretation, caveats, and exceptions discussed in the $d=4, p=0$ case apply to the general case.

## 2.7 The embedding tensor formalism

The embedding tensor formalism,[29] introduced in Refs. [371, 1260, 1261, 1259, 1262] allows the study of the most general gaugings of a field theory with given field content and global symmetry group, such as supergravity theories. Actually, it allows us to explore the most general deformations of a theory which are compatible with gauge invariance. The simplest deformations are directly related to the gauging of global symmetries and are parametrized by the *embedding tensor*, but massive (Stückelberg) deformations can (and sometimes must) be considered as well. These deformations are parametrized by objects which are, in principle, independent of the embedding tensor. In the maximally and half-maximally extended supergravities in various dimensions studied in Refs. [1263, 1101, 1108, 1264, 157, 1258, 520], all the deformation parameters are related to the embedding tensor due to the high amount of (super)symmetry of the theories, but in general matter-coupled $N=1, d=4$ supergravity that is not the case [712].

Another interesting aspect of the formalism is that the consistency of the deformation procedure requires the existence of a hierarchy of tensor (differential-form)[30] fields called the *tensor hierarchy* of the theory [1259, 1262, 1258, 159, 1268]. The tensor hierarchy must be compatible with the original field content of the theory, and therefore all the new fields that need to be introduced[31] must be dual to the original ones or must carry no continuous degrees of freedom (like $d$ or $(d-1)$-forms in $d$ dimensions). Since higher-rank form fields are associated with extended objects of the theory (see Chapter 24), the

[29] For recent reviews see Refs. [1201, 1237, 1100].

[30] We are going to use differential-form language and exterior derivatives throughout this section.

[31] Typically, a supergravity theory is given in terms of the *fundamental* or *electric* fields of the theory, with the lowest ranks. The tensor hierarchy contains fields of all ranks up to $d$.

embedding tensor formalism provides us with an indirect way of determining the possible *branes* of the theory.

The deformation procedure has also been applied to generic theories like the one given in Eq. (2.147) in four dimensions,[32] and in $d = 4, 5, 6$ in Refs. [159, 713] determining the most general field theory in those dimensions compatible with gauge symmetry and the corresponding tensor hierarchies. Very interesting patterns emerge from these works that explain the meaning of the fields of rank $d$ and $(d-1)$ which cannot be dual to any of the fundamental fields [157, 159]. As we shall explain, they are dual to the deformation parameters themselves and to the constraints that they must obey.

In what follows we review the formalism performing the deformation of a generic theory in an arbitrary number of dimensions. The deformation procedure starts with the lowest-rank fields and the gauging of their global symmetries. We assume that the theory has some scalar fields $\phi^i$ parametrizing a target space with metric $\mathcal{G}_{ij}(\phi)$, 1-form fields $A^I$, 2-form fields $B_x$, etc. Furthermore, we assume that the equations of motion of the theory are invariant under a global symmetry group with infinitesimal generators $\{T_A\}$ satisfying the algebra

$$[T_A, T_B] = f_{AB}{}^C T_C. \tag{2.200}$$

The group acts linearly on all the forms of rank $\geq 1$, and the corresponding matrices that represent the generators are $\{T_A{}^I{}_J\}, \{T_A{}^x{}_y\}$ etc., so we have[33]

$$\delta_\alpha A^I = \alpha^A T_A{}^I{}_J A^J \equiv \alpha^A \delta_A A^I, \qquad \delta_\alpha B_x = -\alpha^A T_A{}^y{}_x B_y \equiv \alpha^A \delta_A B_x, \tag{2.201}$$

etc. The indices carried by these fields have special properties in each dimension.

- In $d = 4$, as we have shown in Section 2.6, $A^I$ must include the $n$ fundamental 1-forms and their $n$ duals and the matrices $T_A{}^I{}_J \in \mathfrak{sp}(2n, \mathbb{R})$ so $T_A{}^K{}_{[J}\Omega_{I]K} = 0$. Following our conventions, we would use indices $M, N, \ldots$. We also expect the 2-forms to be dual to the Noether current 1-forms $j_A$,[34] and therefore the index $x$ should be an adjoint index. We expect the $(d-2)$-forms of all theories to share this property. We will discuss the indices of the 3- and 4-forms ($(d-1)$- and $d$-forms) later.

- In $d = 5$, the 2-forms are dual to the 1-forms so the index $x$ should be a (covariant) $I$ and we could denote them by $\tilde{A}_I$. We expect the 3-forms to carry a covariant adjoint index $A$.

- In $d = 6$, according to the results of Section 2.6, $B_x$ must include the $n$ fundamental and $n$ dual 2-forms and the matrices $T_A{}^y{}_x \in \mathfrak{so}(n, n)$. The 3-forms are dual to the 1-forms and should carry a covariant index $I$; we can denote them by $\tilde{A}_I$. The 4-forms should carry a covariant adjoint index $A$, as usual.

[32] Equation (2.178) is not the most general action that one can have in $d \geq 4$ dimensions because we are considering only $(p+1)$-form fields.

[33] For objects transforming in the adjoint representation $T_A{}^B{}_C = f_{AC}{}^B$.

[34] They are not directly dual to the scalars, as one may have naively expected. In the end, though, not all 2-forms will be independent fields, and the number of independent 2-forms will equal the number of scalars.

- The pattern is clear and repeats itself, in a more complicated way, in higher dimensions. The $(d-1)$- and $d$-forms also obey a general pattern, as we will explain later.

The scalars transform non-linearly, and their transformations are isometries of $\mathcal{G}_{ij}(\phi)$ generated by the Killing vectors $K_A{}^i$,

$$\delta_\alpha \phi^i = \alpha^A K_A{}^i \equiv \alpha^A \delta_A \phi^i. \tag{2.202}$$

The Killing vectors have the Lie brackets

$$[K_A, K_B] = -f_{AB}{}^C K_C. \tag{2.203}$$

Some of the matrices or Killing vectors may vanish if parts of the total symmetry group do not act on some of the fields.

Finally, we assume that each of the differential-form fields is invariant under Abelian gauge transformations,

$$\delta_\sigma A^I = d\sigma^I, \qquad \delta_\sigma B_x = d\sigma_x, \tag{2.204}$$

etc., where the $\sigma^I$ are arbitrary functions, the $\sigma_x$ are arbitrary 1-forms, etc.

As explained in Section 2.5 and in Appendix J,[35] if we want the theory to be invariant under the above transformation, with the constant parameters $\alpha^A$ promoted to arbitrary spacetime functions $\alpha^A(x)$, we must identify them with some of the gauge parameters of the 1-forms, $\sigma^I(x)$, or, equivalently, some of the 1-forms $A^I$ must be identified with the gauge fields $A^A$ associated with the local transformations $\alpha^A$. The most general way of making this identification is through the *embedding tensor* $\vartheta_I{}^A$:

$$\boxed{\alpha^A(x) = \vartheta_I{}^A \sigma^I(x), \qquad A^A = \vartheta_I{}^A A^I.} \tag{2.205}$$

The transformations of the scalars which we want the theory to be invariant under are

$$\boxed{\delta_\sigma \phi^i = \sigma^I \vartheta_I{}^A K_A{}^i.} \tag{2.206}$$

In general, the embedding tensor is not invertible. It has to be an invariant tensor, that is

$$\delta_\sigma \vartheta_I{}^A = \sigma^J \vartheta_J{}^B \left(-T_B{}^K{}_I \vartheta_K{}^A + f_{BC}{}^A \vartheta_I{}^C\right) \equiv \sigma^J \mathcal{Q}_{JI}{}^A = 0. \tag{2.207}$$

The condition

$$\boxed{\mathcal{Q}_{IJ}{}^A = -X_I{}^K{}_J \vartheta_K{}^A + \vartheta_I{}^B \vartheta_J{}^C f_{BC}{}^A = 0, \quad \text{where} \quad X_I{}^K{}_J \equiv \vartheta_I{}^B T_B{}^K{}_J,} \tag{2.208}$$

[35] The conventional gauging of the isometries of different spaces has been reviewed in Appendix J. In particular, we are going to use some of the results of Section J.1, but in the general conventions of this book for the gauge fields, given in Appendix A; in Appendix J (as well as in Chapters 6 and 7) the gauge fields and their field strengths have the opposite sign. We also use $\mathcal{D}$ for the covariant derivative.

is known as the *quadratic constraint* of the embedding tensor formalism. When we introduce more deformation tensors, we get more constraints quadratic in deformation tensors (including the embedding tensor). Observe that this constraint implies that the matrices $X_I$ defined above satisfy

$$X_{(I}{}^{K}{}_{J)}\vartheta_K{}^A = 0, \tag{2.209}$$

although, in general $X_{(I}{}^{K}{}_{J)} \neq 0$. These matrices play the role of generators and structure constants. Now we can check that, if the above quadratic constraint is satisfied, their commutators and the algebra of gauge transformations on the scalars Eq. (2.206) close:

$$\begin{aligned} [X_I, X_J] &= X_{[I}{}^{K}{}_{J]}X_K + \mathcal{Q}_{IJ}{}^A T_A, \\ [\delta_{\sigma_1}, \delta_{\sigma_2}]\,\phi^i &= \sigma_1^I\sigma_2^J\left[X_{[I}{}^{K}{}_{J]}\delta_I\phi^i + \mathcal{Q}_{IJ}{}^A\delta_A\phi^i\right]. \end{aligned} \tag{2.210}$$

We assume that this constraint is always satisfied.

The next step is to construct a gauge-covariant field strength, which for the scalars reduces to the covariant derivative $\mathcal{D}$. Using the general recipe Eq. (J.16) we find that it must have the form[36]

$$\boxed{\mathcal{D}\phi^i = d\phi^i - A^I\delta_I\phi^i = d\phi^i - A^I\vartheta_I{}^A K_A{}^i,} \tag{2.211}$$

and transforms covariantly, i.e.

$$\boxed{\delta_\sigma\mathcal{D}\phi^i = \sigma^I\vartheta_I{}^A\partial_j K_A{}^i\mathcal{D}\phi^j,} \tag{2.212}$$

provided the 1-forms transform as follows:

$$\delta_\sigma A^I = \mathcal{D}\sigma^I + \Delta A^I, \tag{2.213}$$

where we have defined the covariant derivative of the gauge parameters by

$$\boxed{\mathcal{D}\sigma^I \equiv d\sigma^I - A^J X_J{}^I{}_K\sigma^K, \quad \text{because} \quad \delta_\sigma\mathcal{D}\sigma^I = \sigma^J X_J{}^I{}_K\mathcal{D}\sigma^K,} \tag{2.214}$$

and where $\Delta A^I$ is a possible additional term which is annihilated by $\vartheta_I{}^A$. This term plays no role so far, but it will soon become crucial.

Now we have to construct the gauge-covariant field strengths of the 1-forms. The most efficient way to do this is via the Ricci identities

$$\mathcal{D}^2\phi^i = -F^I\delta_I\phi^i, \tag{2.215}$$

[36] Observe that we are not introducing a coupling constant because it is included in the embedding tensor. Actually $\vartheta$ can contain several coupling constants associated with different gauge groups.

using the general rule $\mathcal{D}^2\phi^i = d\mathcal{D}\phi^i - A^I\delta_I\mathcal{D}\phi^i$. The result is

$$F^I = dA^I - \tfrac{1}{2}X_J{}^I{}_K A^J \wedge A^K + \Delta F^I, \tag{2.216}$$

where, again, $\Delta F^I$ is a possible additional term annihilated by $\vartheta_I{}^A$.

By construction, $F^I$ should be gauge covariant, but, since there are undetermined terms in the gauge transformations and in $F^I$ itself, we must check it directly. The result is

$$\delta_\sigma F^I = \sigma^J X_J{}^I{}_K F^K - 2X_{(J}{}^I{}_{K)}\left(\sigma^J F^K - \tfrac{1}{2}A^J \wedge \delta_\sigma A^K\right) + \mathcal{D}\Delta A^I + \delta_\sigma \Delta F^I, \tag{2.217}$$

and in order to have covariance we must add a non-trivial $\Delta F^I$ whose gauge transformation annihilates the unwanted terms. $\Delta F^I$ can only be a linear combination of the 2-forms $B_x$,

$$\Delta F^I = Z^{Ix}B_x, \tag{2.218}$$

where $Z^{Ix}$ is a new deformation tensor which must be invariant, which implies the quadratic constraint

$$\boxed{\mathcal{Q}_I{}^{Jx} \equiv X_I{}^J{}_K Z^{Kx} + X_I{}^x{}_y Z^{Ky} = 0, \quad \text{where} \quad X_I{}^x{}_y \equiv \vartheta_I{}^B T_B{}^x{}_y,} \tag{2.219}$$

and must also be annihilated by the embedding tensor. This is expressed by the quadratic *orthogonality* constraint

$$\boxed{\mathcal{Q}^{Ax} \equiv Z^{Ix}\vartheta_I{}^A = 0.} \tag{2.220}$$

Each new deformation tensor will generate a quadratic constraint of the first kind, and a few of them will satisfy quadratic orthogonality constraints as well.

This is not enough: the unwanted terms in $\delta_\sigma F^I$ must be proportional to the new tensor $Z^{Ix}$. Since $B_x$ has its own Abelian gauge transformations with 1-form parameter $\sigma_x$, we must set

$$\Delta A^I = -Z^{Ix}\sigma_x. \tag{2.221}$$

We must also demand the existence of some tensor fields $d_{xIJ}$ such that

$$X_{(J}{}^I{}_{K)} = Z^{Ix}d_{xJK}. \tag{2.222}$$

Summarizing, the 2-form field strengths and the gauge transformations of the 1- and 2-forms take the form

$$\boxed{\begin{aligned} F^I &= dA^I - \tfrac{1}{2}X_J{}^I{}_K A^J \wedge A^K + Z^{Ix}B_x, \\ \delta_\sigma A^I &= \mathcal{D}\sigma^I - Z^{Ix}\sigma_x, \\ \delta_\sigma B_x &= \mathcal{D}\sigma_x + 2d_{xJK}\left(\sigma^J F^K - \tfrac{1}{2}A^J \wedge \delta_\sigma A^K\right) + \Delta B_x, \end{aligned}} \tag{2.223}$$

where $\Delta B_x$ is a possible new term annihilated by $Z^{Ix}$.

The vector gauge transformations generated by the terms $Z^{Ix}\sigma_x$ are known as *massive* or *Stückelberg* gauge transformations or *shifts*, and can be used to eliminate or fix to some expected value some of the 1-forms which are "eaten" by the 2-forms. The 1-forms being eaten are called *Stückelberg fields*. The terms $Z^{Ix}B_x$ in the 2-form field strengths, known as *Stückelberg couplings*, become mass terms for the 2-forms.[37] The tensors $Z$ are sometimes called *massive deformations*. A well-known example is Romans' massive deformation of $N=2A, d=10$ supergravity, studied in Section 22.2.

In some dimensions the tensors $Z^{Ix}$ and $d_{xIJ}$ have special properties.

- In $d=4$ ($x \to A$, $I,J \to M,N$, and $T_A{}^P{}_N \in \mathfrak{sp}(2n,\mathbb{R})$) it is customary[38] to choose (up to normalization)

$$Z^{MA} = \Omega^{NM}\vartheta_N{}^A, \qquad d_{AMN} = -\tfrac{1}{2}T_{AMN} = -\tfrac{1}{2}T_A{}^P{}_N\Omega_{MP}. \tag{2.224}$$

  The choice for $d_{AMN}$ implies that the tensors $X_{MNP} = X_M{}^Q{}_P\Omega_{QN}$ are completely symmetric in the three indices. This is usually expressed as a new (*linear* or *representation*) constraint [1262],

$$\boxed{L_{MNP} \equiv X_{(MNP)} = 0,} \tag{2.225}$$

  satisfied in all four-dimensional supergravity theories that are free of gauge anomalies [426]. This constraint is related to the quadratic constraints by

$$3L_{MNP}\vartheta^{MA} - \mathcal{Q}^{AB}T_{BNP} + 2\mathcal{Q}_{(NP)}{}^A = 0. \tag{2.226}$$

- In $d=5$ dimensions ($x \to I$), the tensor $d_{IJK}$ can be chosen as the totally symmetric tensor $C_{IJK}$ that appears in the Lagrangian of $N=1, d=5$ supergravities [713] (see Chapter 9).

We can address the next problem in the construction of the tensor hierarchy: finding the covariant field strength of the 2-forms $B_x$. We can repeat the steps exactly: if we act with the covariant derivative on $F^I$, we must get zero up to a term annihilated by $\vartheta_I{}^A$. This term must be a 3-form transforming covariantly, and must be proportional to $Z^{Ix}$. We arrive at the Bianchi identity

$$\mathcal{D}F^I - Z^{Ix}H_x = 0, \tag{2.227}$$

where $H_x \sim dB_x$ is the field strength we were looking for. A direct computation of $\mathcal{D}F^I$ gives $H_x$ up to a term $\Delta H_x$ which is annihilated by $Z^{Ix}$ (as $\Delta B_x$). It must be a linear

[37] Actually, the terms $A^I\vartheta_I{}^AK_A{}^i$ in the covariant derivatives of the scalars can play exactly the same role: some scalars can be gauged away or given an expected value, and those terms become mass terms for the 1-forms in the Lagrangian (see Section 12.2.2).

[38] These are the most economical choices since they do not require the introduction of new independent tensors beyond $\vartheta_M{}^A$ and $T_A{}^M{}_N$.

combination of the 3-forms of the theory $Z_{xa}C^a$, where $Z^{Ix}Z_{xa} = 0$ and where the gauge transformations of the $C^a$ can be found by requiring gauge covariance of $H_x$ up to terms annihilated by the new deformation tensor. The Bianchi identity for $H_x$ has the form

$$\mathcal{D}H_x + d_{xIJ}F^I \wedge F^J - Z_{xa}G^a = 0, \tag{2.228}$$

where $G^a$ are the 4-form field strengths, etc.

At some point, the dual potentials will appear and we will be able to use some of the deformation tensors we introduced before; later we shall deal with the duals of the 1-forms $A^I$, $(d-3)$-forms denoted by $\tilde{A}_I$ with $(d-2)$-form field strengths $\tilde{F}_I$.[39] These field strengths will contain a Stückelberg coupling to the $(d-2)$-form potentials that we can denote by $C_A$, and the deformation tensor will be the embedding tensor itself:

$$\tilde{F}_I \sim \mathcal{D}\tilde{A}_I + \cdots + \vartheta_I{}^A C_A. \tag{2.229}$$

The field strength of $C_A$, $G_A$ will be determined by the usual method up to terms annihilated by $\vartheta_I{}^A$ and which will be linear combinations of the $(d-1)$-form potentials. The latter are dual to all the deformation tensors that we denote collectively by $c^\sharp$, $\sharp$ standing for all the corresponding indices. This duality is different from the general Poincaré duality reviewed in Section 12.7.1. To make sense of it, the deformation parameters have to be promoted to fields $c^\sharp(x)$ constrained to be constant. The constraint is implemented in the action by a term of the form

$$\int D_\sharp \wedge dc^\sharp, \tag{2.230}$$

where the Lagrange multipliers $D_\sharp$ are $(d-1)$-form potentials.[40]

Following the general pattern, the field strength $G_A$ will contain a Stückelberg term $Y_A{}^\sharp D_\sharp$:

$$\boxed{G_A \sim \mathfrak{D}C_A + \cdots + Y_A{}^\sharp D_\sharp.} \tag{2.232}$$

The deformation tensors $Y_A{}^\sharp$ can be found as follows [713]: as we have stressed, there is a quadratic constraint $\mathcal{Q}_I{}^\sharp = 0$ associated with the invariance of each deformation tensor $c^\sharp$:

$$\delta_\sigma c^\sharp = \sigma^I \delta_I c^\sharp = \sigma^I \vartheta_I{}^A \delta_A c^\sharp = 0. \tag{2.233}$$

[39] This happens very soon in four dimensions, and these fields are taken into account from the beginning, but the following discussion applies to that case too.

[40] Actually, the constraint can be solved by piecewise constant $c^\sharp(x)$s, and the discontinuities can be understood to be caused by the domain-wall defects associated with the $(d-1)$-form potentials (see Chapter 24). This is how Romans' mass parameter in $N = 2A, d = 10$ SUEGRA is associated with the $C^{(9)}$ RR potential of D8-brane (see p. 668) [186, 170], and in this way one can also understand domain walls in $N = 1, d = 4$ supergravity [796]. Observe that, if we promote the deformation tensors to fields, there are new equations of motion associated with their variations. Taking into account Eq. (2.230),

$$K_\sharp \sim \star \frac{\partial V}{\partial c^\sharp}, \tag{2.231}$$

plus terms that vanish when the rest of the duality relations are taken into account [159].

We can define the deformation tensors and constraints by

$$\boxed{Y_A{}^\sharp \equiv \delta_A c^\sharp, \qquad \mathcal{Q}_I{}^\sharp = -\vartheta_I{}^A Y_A{}^\sharp = -\delta_I c^\sharp,} \tag{2.234}$$

and then the orthogonality constraint $\vartheta_I{}^A Y_A{}^\sharp = 0$ is identical to the quadratic constraints $\mathcal{Q}_I{}^\sharp = 0$.

The field strengths of the $(d-1)$-form potentials $D_\sharp$, $K_\sharp$ will have Stückelberg couplings to top-form potentials $E_\flat$,

$$\boxed{K_\sharp \sim \mathcal{D}D_\sharp + \cdots + \textstyle\sum_\flat W_\sharp{}^\flat E_\flat,} \tag{2.235}$$

where the new deformation tensors $W_\sharp{}^\flat$ are annihilated by the $Y_A{}^\sharp$. These new deformation potentials can be found as follows: the top-form potentials are associated with all the constraints $\mathcal{Q}^\flat$ (not just the $\mathcal{Q}_I{}^\sharp$); they can be understood as the Lagrange multipliers enforcing them in the action via terms of the form

$$\mathcal{Q}^\flat E_\flat. \tag{2.236}$$

Of course, to introduce these terms in the action we have to work without assuming that the constraints are solved. This can be done, but it makes the whole procedure much harder (just like working with local deformation parameters $c^\sharp(x)$). We will not do it here, but it is the possibility of doing it that helps us to make sense of the top-form potentials $E_\beta$. Let us consider, then, the above combinations which vanish identically because we assume the constraints to be satisfied.

The infinitesimal transformations $\delta_A$ of this term also vanish because they are linear in the constraints $\mathcal{Q}^\flat$. Using the chain rule for the variations we get

$$\delta_A\left(\mathcal{Q}^\flat E_\flat\right) = \delta_A c^\sharp \frac{\partial \mathcal{Q}^\flat}{\partial c^\sharp} E_\flat = \left(Y_A{}^\sharp \frac{\partial \mathcal{Q}^\flat}{\partial c^\sharp}\right) E_\flat = 0 \tag{2.237}$$

for arbitrary $E_\flat$, which implies the identities

$$Y_A{}^\sharp \frac{\partial \mathcal{Q}^\flat}{\partial c^\sharp} = 0. \tag{2.238}$$

This leads to the identification

$$\boxed{W_\sharp{}^\flat \equiv \frac{\partial \mathcal{Q}^\flat}{\partial c^\sharp}.} \tag{2.239}$$

All the tensor hierarchies can be built along these lines. Imposing duality relations between fundamental and dual fields, Noether currents and $(d-2)$-forms, deformation parameters and $(d-1)$-forms, and constraints and top-forms, one recovers the equations of motion of a field theory.

# 3

# A perturbative introduction to general relativity

The standard approach to general relativity (GR) is purely geometrical: spacetime is curved by its energy content according to Einstein's equation and test particles move along geodesics. This point of view is what makes GR a theory completely different from the theories that describe all the other known interactions that are special-relativistic field theories (SRFTs) that, after quantization, explain the interaction between two charged bodies as the interchange of quanta of the field.

The enormous success of relativistic quantum field theories with a gauge principle made it unavoidable to try to find a theory of that kind to describe gravitational interactions at both the classical and quantum levels. This path was followed by many people, and it was found that such a theory, whose starting point is the linear perturbation theory of GR (the Fierz–Pauli theory for a free, massless spin-2 particle), would be self-consistent only after the introduction of an infinite number of non-linear terms whose summation should be equivalent to the full non-linear GR theory.[1] Thus, this approach may lead to a different justification of Einstein's theory and provides an alternative interpretation of it that is worth studying.[2] Some of the predictions of GR can be obtained at leading or next to leading order in this approach. Since this is not the standard approach, there are only a few complete treatments in the literature: the book based on Feynman's lectures on gravitation [542], which also contains many references, some of which we will follow in Section 3.2; and also Deser's lectures on the gravitational field [431]. Reference [32] is also an excellent review with many references.

In this chapter, as a warm-up exercise, we are first going to study the construction of SRFTs of gravity based on a scalar field. This is the simplest way of searching for a SRFT of gravitational interaction, and it offers us the possibility of studying, in a simple setting, problems that we shall meet later.

As is well known, scalar theories of gravity predict no global bending of light rays (in contrast to observation) and a value for the precession of the perihelion of Mercury which

[1] There are other alternative special-relativistic field theories for spin-2 particles. See, for example, Ref. [899], in which gravity is based on a *massive* (with extremely small mass) spin-2 field.

[2] Some string theories have a massless spin-2 particle in their spectra. If these string theories are consistent, the argument we will develop will imply that they contain gravity, which, to the lowest order, will be described by Einstein's theory.

is wrong (in magnitude and sign), and thus we have to consider the next logical possibility: a spin-2 field. First, we have to find a SRFT (the Fierz–Pauli theory) for the free spin-2 field. Gauge invariance plays a crucial role in the construction of this theory and we will emphasize it. We then proceed to introduce the interaction with matter fields and find the gravitational field produced by a massive point particle. We will immediately show that the interacting theory is consistent (at the classical level) only if the gravitational field couples to itself in the same form as that in which it couples to matter: through the energy–momentum tensor. Making this self-coupling consistent requires an infinite number of corrections to the Fierz–Pauli theory. We will try to find the first correction via the Noether method, at which point we encounter difficulties that arise in the definition of the gravitational energy–momentum tensor, about which we will have more to say in Chapter 10. The choice of energy–momentum tensor, which is usually defined up to the divergence of an antisymmetric tensor or up to the addition of on-shell-vanishing terms, is crucial in this context, because different choices lead to different theories with different predictions of the value for the precession of the perihelion of Mercury [986].

These problems are avoided by the use of Deser's argument, which allows one to find in just one step both the right energy–momentum tensor for the gravitational field at lowest order and all the corrections to the Fierz–Pauli theory that convert it into a self-consistent theory for a self-interacting massless spin-2 particle. This theory is just GR. We will discuss whether this is the only possible solution to our problem, since Deser's result shows the existence of a solution but not its uniqueness.

In any case, this is how we are going to introduce the Einstein equations and the Einstein–Hilbert action that will be studied in more detail in Chapter 4 and also the action for point-particles moving in a curved background. We will conclude the chapter by studying the perturbative expansion of GR (i.e. the interacting Fierz–Pauli theory consistent to a certain order in the coupling constant) in flat and curved backgrounds for later use.

## 3.1 Scalar SRFTs of gravity

If we were particle physicists in the pre-Yang–Mills[3] era wanting to describe gravity, we would certainly try to do it (Feynman in Ref. [542] or Thirring in Ref. [1177]) with a relativistic field theory of a bosonic massless particle (to provide long-range interactions) propagating in Minkowski spacetime whose interchange would be responsible for the gravitational interaction between massive bodies. Which particle? The simplest possibility is that of a scalar particle (after all, in Newtonian physics, gravity is described by the Newtonian gravitostatic potential $\phi$ alone and there was no hint of the existence of any *gravitomagnetic* field). For this reason, and considering the attractive nature of scalar-mediated interactions (see, for instance, Ref. [1155]), scalar SRFTs were the first candidates used to describe relativistic gravitation.[4]

[3] A different approach to gravity based on the gauge theory of the Poincaré and (anti-)de Sitter groups is also possible and is described in Section 4.5.

[4] Scalar theories of gravity were first proposed by Abraham [4, 5, 6, 7, 8, 9, 10, 11], Nordström [989, 990, 991, 992, 993, 995], and Einstein [486, 487, 488]. (Some old reviews are in Refs. [12, 873, 885], and a modern review is in Ref. [997].) They played an important role in the developments that led Einstein to GR. Our interest in them is purely pedagogical.

A free scalar propagating in Minkowski spacetime is described by the action

$$S = \int d^d x \ \tfrac{1}{2}(\partial\phi)^2, \qquad (\partial\phi)^2 \equiv \eta^{\mu\nu}\partial_\mu\phi\partial_\nu\phi, \tag{3.1}$$

and has as its equation of motion

$$\partial^2\phi = 0, \qquad \partial^2 \equiv \eta^{\mu\nu}\partial_\mu\partial_\nu. \tag{3.2}$$

The source for the Newtonian gravitational field is the *gravitational mass of matter* which is experimentally found to be proportional (equal in appropriate units) to the inertial mass for all material bodies. In special relativity, the inertial mass, the energy, and the momentum of a physical system are combined into the energy–momentum tensor $T^{\mu\nu}$, and therefore the source for the gravitational field will be the matter energy–momentum tensor. This is an object of utmost importance and was studied in some detail in Chapter 2.

### 3.1.1 Scalar gravity coupled to matter

From our previous discussion, the source of the scalar gravitational field (the r.h.s. of Eq. (3.2)) must be a scalar built out of the energy–momentum tensor of the matter fields. The simplest scalar is the trace $T_{\text{matter}} \equiv T_{\text{matter}}{}^{\mu}{}_{\mu}$; using it, taking into account all factors of $c$, we arrive at the action for matter coupled to scalar gravity

$$S = \frac{1}{c}\int d^d x \left\{ \frac{1}{2Cc^2}(\partial\phi)^2 + \frac{\phi}{c^2}T_{\text{matter}} + \mathcal{L}_{\text{matter}} \right\}, \tag{3.3}$$

where $C$ is a proportionality constant to be determined. From this action we can derive the equation of motion for the scalar gravitational field,

$$\partial^2\phi = C T_{\text{matter}}, \tag{3.4}$$

and the equation of motion for matter in the gravitational field.

Observe that the conservation of the matter energy–momentum tensor plays no role whatsoever in the construction of this theory. In fact, if it was required in some sense for consistency, we would be in trouble because, after the coupling to the gravitational field, the matter energy–momentum tensor is no longer conserved: only the *total* energy–momentum tensor of the above Lagrangian (the matter energy–momentum tensor, plus the gravitational energy–momentum tensor, plus an interaction term) is conserved. However, the equation of motion that we have obtained is perfectly consistent as it stands.

Observe also that nowhere is it required that the energy–momentum tensor is symmetric (although only its symmetric part contributes to the trace). In fact, there are no conditions that we can impose on the energy–momentum tensor to select only one out of the infinitely many possible energy–momentum tensors that we can obtain by adding terms proportional to the equations of motion or superpotential terms. We can view this as a weakness of scalar SRFTs of gravity. In the cases that we are going to consider, we will simply take the canonical energy–momentum tensor obtained from the matter action in its simplest form.

Now, to determine the constant $C$, we can require $\phi$ to be identical to the Newtonian gravitational potential in the static, non-relativistic limit in which only the $T_{\text{matter}\,00} =$

$-\rho c^2$ component contributes to the trace, $\rho$ being the mass density.[5] In this case, Eq. (3.4) becomes the Poisson equation,

$$\partial_i \partial_i \phi = C c^2 \rho \quad \Rightarrow C = \frac{(d-3) 8\pi G_{\rm N}^{(d)}}{(d-2)c^2}, \tag{3.5}$$

where $G_{\rm N}^{(d)}$ is the $d$-dimensional Newton constant.[6] For a point-particle of mass $M$ at rest at the origin,

$$\rho = M \delta^{(d-1)}(\vec{x}_{d-1}) \tag{3.6}$$

and

$$\phi = -\frac{16\pi G_{\rm N}^{(d)} M}{2(d-2)\omega_{(d-2)}} \frac{1}{|\vec{x}_{d-1}|^{d-3}}, \tag{3.7}$$

where $\omega_{(n)}$ is the area of the round $n$-sphere of unit radius (Eq. (K.11)).

This identification will be completely justified if, in the limit considered, $\phi$ affects the motion of matter just as the Newtonian gravitational potential does. Let us consider the motion of a massive particle in the gravitational field $\phi$. The coupling is given by the above action. All we need is the action for the free special-relativistic massive point-particle. Since we are going to make extensive use of this action, we start by reviewing it.

### 3.1.2 *The action for a relativistic massive point-particle*

The special-relativistic action for a point-particle of mass $M$ can be written as follows:

$$\boxed{S_{\rm pp}[X^\mu(\xi)] = -Mc \int d\xi \sqrt{\eta_{\mu\nu} \dot{X}^\mu \dot{X}^\nu}, \qquad \dot{X}^\mu \equiv \frac{dX^\mu}{d\xi},} \tag{3.8}$$

where $\xi$ is a general parameter for the particle's worldline. The reality of the action is related to the fact that usual massive particles move along timelike curves, $\dot{X}^\mu \dot{X}_\mu > 0$. The equations of motion that one derives from it simply express the conservation of the $d$ components of the linear momentum:

$$\frac{dP_\mu}{d\xi} = 0, \qquad P_\mu \equiv \frac{\partial L}{\partial \dot{X}^\mu} = -Mc \frac{\eta_{\mu\nu} \dot{X}^\nu}{\sqrt{\eta_{\rho\sigma} \dot{X}^\rho \dot{X}^\sigma}}. \tag{3.9}$$

The conservation of the $d(d-1)/2$ components of the angular momentum,

$$M_{\mu\nu} = 2X_{[\mu} P_{\nu]}, \tag{3.10}$$

follows. The $d(d+1)/2$ conserved quantities are, as is well known, associated with the invariance of the action under global Poincaré transformations of the spacetime coordinates

$$x'^\mu = \Lambda^\mu{}_\nu x^\nu + a^\mu, \qquad \Lambda^\mu{}_\alpha \Lambda^\nu{}_\beta \eta_{\mu\nu} = \eta_{\alpha\beta}, \tag{3.11}$$

[5] Note the minus sign in our conventions.

[6] This is an unfortunate convention in the literature in which the factor $4\pi$, which is appropriate for rationalized units in four dimensions, is indiscriminately used in all dimensions.

via the Noether theorem for global transformations: using the infinitesimal form of the Poincaré transformations,

$$\delta x^\mu = \sigma^\mu{}_\nu x^\nu + \sigma^\mu, \qquad \sigma^{\mu\nu} = -\sigma^{\nu\mu}, \tag{3.12}$$

we obtain the conservation law

$$\frac{dJ(\sigma)}{d\xi} = 0, \qquad J(\sigma) = P_\mu \delta X^\mu. \tag{3.13}$$

The conserved quantity associated with translations is the linear momentum $J(\sigma^\mu) \sim P^\mu$ and the conserved quantity associated with Lorentz transformations is the angular momentum $J(\sigma^{\mu\nu}) \sim M^{\mu\nu}$.

Observe that the invariance of the action is due to the fact that it depends only on the derivatives of the coordinates. In particular, the Minkowski metric does not depend on the coordinates. A better way to express this fact is to say that the Minkowski metric has $d(d+1)/2$ independent isometries that generate the $d$-dimensional Poincaré group. This association between spacetime isometries and conserved quantities will still hold in more complicated spacetimes.

This action is also invariant under non-singular reparametrizations of the worldline $\xi'(\xi)$. These are local (gauge) transformations that infinitesimally can be written $\delta\xi = \epsilon(\xi)$. Taking into account that the $X^\mu$s are scalars with respect to these transformations, we find

$$\delta\xi = \epsilon(\xi), \qquad \delta d\xi = \dot{\epsilon} d\xi, \qquad \tilde{\delta} X^\mu = 0, \qquad \delta \dot{X}^\mu = -\dot{\epsilon}\dot{X}^\mu, \tag{3.14}$$

and it is a simple exercise to check that $\tilde{\delta} S = 0$ identically. If we now consider the variation of the action under just

$$\delta X^\mu = -\epsilon \dot{X}^\mu, \tag{3.15}$$

we find that it is invariant only up to a total derivative

$$\delta S = \int d\xi \, \frac{d}{d\xi}\left( Mc\epsilon\sqrt{\eta_{\mu\nu}\dot{X}^\mu \dot{X}^\nu}\right). \tag{3.16}$$

On varying the action with respect to general variations of the coordinates first and integrating by parts, we obtain

$$\delta S = \int d\xi \left\{ \epsilon \frac{\delta S}{\delta X^\nu}\dot{X}^\nu + \frac{d}{d\xi}\left( Mc\epsilon\sqrt{\eta_{\mu\nu}\dot{X}^\mu \dot{X}^\nu}\right)\right\}. \tag{3.17}$$

By equating the two results and taking into account that the equation is valid for arbitrary functions $\epsilon(\xi)$, we obtain the gauge identity

$$\frac{\delta S}{\delta X^\nu}\dot{X}^\nu = 0 \Rightarrow \dot{P}_\nu \dot{X}^\nu = 0, \tag{3.18}$$

which is satisfied off-shell (trivially on-shell). Since $\dot{X}^\nu$ is proportional to the momentum, this identity is proportional to

$$\frac{d(P^\mu P_\mu)}{d\xi} = 0. \tag{3.19}$$

Indeed, $P^\mu P_\mu$ is a constant: using the definition of momentum, we find, without using the equations of motion, the mass-shell condition

$$P^\mu P_\mu = M^2 c^2, \tag{3.20}$$

and we have just shown that this constraint can be understood as a consequence of reparametrization invariance.

There are two special parameters one can use.[7] One is the particle's proper time (or length) $\xi = s$, defined by the property

$$\eta_{\mu\nu} \dot{X}^\mu \dot{X}^\nu = 1. \tag{3.21}$$

Owing to this definition, the action is usually written as

$$S_{\mathrm{pp}}[X(s)] = -Mc \int ds. \tag{3.22}$$

Although this form is unsuitable for finding the equations of motion, it tells us that the action of a massive point-particle is proportional to its worldline's proper length, and the minimal-action principle tells us that the particle moves along worldlines of minimal proper length. Observe that, from the quantum mechanics point of view, since the measure in the path integral is the exponential of

$$\frac{i}{\hbar} S = i \frac{Mc}{\hbar} \int ds = \frac{i}{\lambda\!\!\!^{-}_{\mathrm{Compton}}} \int ds, \tag{3.23}$$

the proper length is measured in units of the particle's reduced Compton wavelength.

The second special parameter that we can use is the coordinate time $\xi = X^0 = cT$. This choice of gauge fixes one of the particle's coordinates $X^0(\xi) = \xi$. In this gauge (the *physical* or *static gauge*) one can study the non-relativistic limit $\dot{X}^i \dot{X}^i = (v/c)^2 \ll 1$. In this limit the action Eq. (3.8) becomes, up to a total derivative, the non-relativistic action of a particle:

$$S[X^i(t)] = \int dt \left[\tfrac{1}{2} M v^2 - Mc^2\right]. \tag{3.24}$$

### 3.1.3 The massive point-particle coupled to scalar gravity

The coupling to the scalar gravitational field is dictated by the action Eq. (3.3). We compute the energy–momentum tensor using Rosenfeld's prescription (Section 2.4.3):

$$T_{\mathrm{pp}}^{\mu\nu}(x) = -Mc^2 \int d\xi \frac{\dot{X}^\mu \dot{X}^\nu}{\sqrt{\eta_{\rho\sigma} \dot{X}^\rho \dot{X}^\sigma}} \delta^{(d)}[X(\xi) - x], \tag{3.25}$$

which is conserved, as one can prove by using the equations of motion. The trace is identical to the Lagrangian,[8] and thus the action for the coupled particle-plus-gravity system

[7] Purists call the same curve with two different parametrizations different curves, but from a physical point of view they are clearly the same object.

[8] Observe that, in the static gauge, the 00 component of this tensor gives Eq. (3.6).

Eq. (3.3) becomes

$$S[\phi(x), X^\mu(\xi)] = \frac{1}{Cc^3}\int d^dx\,\tfrac{1}{2}(\partial\phi)^2 - Mc\int d\xi\left(1+\frac{\phi(X)}{c^2}\right)\sqrt{\eta_{\mu\nu}\dot{X}^\mu\dot{X}^\nu}. \tag{3.26}$$

For low speeds, in the static gauge, the second term is

$$\sim \int dt\,\{\tfrac{1}{2}Mv^2 - M\phi - Mc^2\}, \tag{3.27}$$

which confirms the consistency of our identification of $\phi$ with the Newtonian potential in this limit. The complete relativistic action predicts corrections to the Newtonian theory. The next two terms in the expansion of the relativistic action are

$$\int dt\,\{-\tfrac{1}{4}Mv^2(v/c)^2 + \tfrac{1}{2}Mv^2\phi/c^2\}. \tag{3.28}$$

The first term is there also for free particles, but the second represents a relativistic correction to the Newtonian coupling to the gravitational field. Owing to its sign, if the particle that acts as source for the scalar gravitational field moves, the kinetic energy contributes to $T_{\mathrm{pp}}$ with sign opposite to the rest mass, and a particle in motion produces (and therefore feels) a weaker gravitational field than when it is at rest. The gravitational field, in fact, would vanish in the limit in which the particle moves at the speed of light. This also means that the gravitational field will not affect the motion of particles moving at the speed of light.

Let us now consider the motion of a second massive particle in the scalar gravitational field produced by the first particle. Although $\phi$ is identical to the Newtonian potential, the action (just the last term in Eq. (3.26) with $\phi$ given by Eq. (3.7)) also predicts corrections to the Newtonian motion. We will not enter into details, but it can be shown [145] that the lowest-order correction to the Newtonian orbits of planets is a precession of their perihelion which is a factor $-\frac{1}{6}$ of that predicted by GR (which is experimentally confirmed). This is a clear drawback for the scalar SRFT of gravity.

With a SRFT of gravity we can also study the effect of gravity on massless particles or the gravitational field produced by massless particles, which is impossible in Newtonian gravity. Thus, there is no non-relativistic limit for this problem. First, we need to find an action for a massless particle.

### *3.1.4 The action for a massless point-particle*

Clearly, the action (3.8) (from now on referred to as a *Nambu–Goto-type action*[9]) is not well suited to take the $M \to 0$ limit. Furthermore, in spite of the straightforward physical interpretation of the Nambu–Goto-type action, the square root makes it highly non-linear and it would be desirable to have a different, more linear, action giving the same equations of motion.

---

[9] The origin of this action can be traced back to Planck. However, the generalization of this action to one-dimensional objects was proposed by Nambu (in lectures given at the 1970 Copenhagen Symposium) and Goto [643], respectively, and has inspired further generalizations for higher-dimensional objects. Hence it has become customary to refer to these kinds of actions as *Nambu–Goto-type* actions.

Thus, we are going to propose an equivalent action that we will call a *Polyakov-type action* with a new, independent, dimensionless, auxiliary "field"[10] $\gamma$ that can be interpreted as a metric on the worldline. This action is

$$\boxed{S_{\rm pp}[X^\mu(\xi),\gamma(\xi)] = -\tfrac{1}{2}Mc\int d\xi\sqrt{\gamma}\left[\gamma^{-1}\eta_{\mu\nu}\dot{X}^\mu\dot{X}^\nu+1\right]} \tag{3.29}$$

and is, yet again, invariant under Poincaré transformations of the spacetime coordinates and invariant under reparametrizations of the worldline $\xi \to \xi'(\xi)$ under which $\gamma$ transforms as follows:

$$\gamma(\xi) = \gamma'\left[\xi'(\xi)\right]\left(\frac{d\xi'}{d\xi}\right)^2. \tag{3.30}$$

The equation of motion of $\gamma$ is a constraint that simply tells us that $\gamma$ is, on-shell, the *induced metric* on the worldline,

$$\gamma = \eta_{\mu\nu}\dot{X}^\mu\dot{X}^\nu. \tag{3.31}$$

This equation is purely algebraic and can be substituted into the action to eliminate[11] $\gamma$, resulting in the Nambu–Goto-type action Eq. (3.8).

Although equivalent, this action is, however, more versatile: we can obtain from it an action for a massless particle. For this we first have to rescale $\gamma$ to $\gamma' = M^{-2}c^{-2}\gamma$ and then we can take the limit $M \to 0$. We rescale back to obtain a dimensionless worldline metric $\gamma' = p^{-2}\tilde{\gamma}$ (obviously $\tilde{\gamma}$ cannot be identified with the original $\gamma$), giving

$$\boxed{S[X^\mu(\xi),\tilde{\gamma}(\xi)] = -\frac{p}{2}\int d\xi\sqrt{\tilde{\gamma}}\,\tilde{\gamma}^{-1}\eta_{\mu\nu}\dot{X}^\mu\dot{X}^\nu,} \tag{3.32}$$

where $p$ is a constant with dimensions of momentum. In the path integral the action (which is no longer the proper length) is now measured in de Broglie's wavelength units $p/\hbar = 1/\lambda\!\!\!^{-}_{\rm deBroglie}$ associated with the characteristic momentum $p$.

The equation of motion for $\tilde{\gamma}$ states that the particle's worldline is light like:

$$\eta_{\mu\nu}\dot{X}^\mu\dot{X}^\nu = 0, \tag{3.33}$$

but this equation cannot be used to eliminate $\tilde{\gamma}$ from the action as in the massive case.

By definition, the proper length of a massless particle's worldline is always zero and cannot be used to parametrize it, but the time coordinate can be used for this purpose.

---

[10] Just a dynamical variable (not a field) of the worldline parameter in the zero-dimensional (point-like) case. This was first done in Refs. [259, 450] for strings. Our discussion follows closely that of standard string-theory references. See e.g. [670, 838, 915, 40, 1049] and also Section 20.1.

[11] It is guaranteed that, under these conditions, the equations of motion derived from the resulting action are the same equations as those one would obtain from the elimination of $\gamma$ from the original equations of motion.

### 3.1.5 The massless point-particle coupled to scalar gravity

We can now try to couple this action to gravity, which is impossible in the Newtonian theory. The energy–momentum tensor is

$$T_{\rm pp}^{\mu\nu} = -pc \int d\xi \sqrt{\gamma}\,\gamma^{-1} \dot{X}^{\mu} \dot{X}^{\nu} \delta^{(d)}[X(\xi) - x]. \tag{3.34}$$

On taking the trace and substituting into Eq. (3.3), we immediately realize that we can make the coupling to gravity disappear by rescaling the worldline auxiliary metric $\gamma$ with a factor $(1 + \phi(X)/c^2)^{-\frac{1}{2}}$. In other words: there is no coupling of a massless particle to scalar gravity. This was to be expected: we have already mentioned the weakening of the scalar gravitational interaction of a massive particle when we increase the speed. On the other hand, the trace of the energy–momentum tensor of a massless particle vanishes on-shell.

We know, however, that the light of stars passing near the Sun is bent by its gravitational field. This is the second drawback of this theory.

We could also have used the Maxwell action and the energy–momentum tensor,

$$S_{\rm matter} = \frac{1}{c} \int d^d x \, \{ -\tfrac{1}{4} F^2 \}, \qquad T_{{\rm matter}\,\mu\nu} = F_{\mu}{}^{\rho} F_{\nu\rho} - \tfrac{1}{4} \eta_{\mu\nu} F^2, \tag{3.35}$$

to study the coupling of the scalar gravitational field to massless particles (fields). On taking the trace and substituting into Eq. (3.3), we find the action

$$S = \frac{1}{c} \int d^d x \left\{ \frac{1}{2Cc^2} (\partial\phi)^2 - \frac{1}{4} \left[ 1 + \frac{d-4}{4} \phi/c^2 \right] F^2 \right\}. \tag{3.36}$$

In $d = 4$ (but only in $d = 4$!) the Maxwell energy–momentum tensor is traceless and there is no coupling to the scalar gravitational field, as expected. In other dimensions, though, there is interaction, in contradiction to the absence of gravitational interaction for massless particles. This apparent paradox can be avoided by the use of the traceless energy–momentum tensor Eq. (2.82). This energy–momentum tensor is not invariant under gauge transformations of the vector field, but, since only its trace enters the Lagrangian, the whole theory is gauge invariant and, simply, there is no interaction.

### 3.1.6 Self-coupled scalar gravity

So far, we have found several serious problems hindering this theory from describing gravity realistically, and we could simply abandon scalar theories of gravity as hopeless and try the next candidate for a SRFT of gravity. However, before we do, we want to introduce, for illustrative purposes, a possible modification of this theory that cannot fix most of the problems encountered, but is the answer to a legitimate question: does gravity couple to all forms of matter/energy including gravitational energy or only to non-gravitational energies? In the theory we have constructed, gravity does not couple to itself. However, since gravitational energy can be transformed into other forms of energy and vice versa, it would be reasonable to expect that gravity couples to all forms of energy equally. Can we modify our theory so as to fulfill this expectation?

We are looking for a theory with the equation of motion

$$\partial^2\phi = CT, \tag{3.37}$$

where $T$ is the trace of the *total* energy–momentum tensor, which should include contributions from the scalar gravitational field, matter fields, and interaction terms. The energy–momentum tensor of $\phi$ in the free theory is quadratic in $\partial\phi$. To obtain it on the r.h.s. of the equation of motion, we must add to the Lagrangian a term of the form $\phi(\partial\phi)^2$. However, this term will also contribute to the new energy–momentum tensor, and, to produce it on the r.h.s. of the new equation of motion, we need a term $\phi^2(\partial\phi)^2$ in the Lagrangian, and so on. Thus, we need to introduce an infinite number of corrections to the scalar Lagrangian.

As for the interaction terms, they contain the trace of the matter energy–momentum tensor, and thus we need to make some assumption about the form of the matter Lagrangian in order to make some progress: we will take it to be of the form

$$\mathcal{L}_{\text{matter}} = K - V, \tag{3.38}$$

where $K$ is quadratic in the first partial derivatives of the matter fields and $V$ is just a function of the fields. This implies that

$$T_{\text{matter}} = (d-2)K - dV, \tag{3.39}$$

and the action Eq. (3.3), which we can consider the lowest order in an expansion in small $\phi$, takes the form

$$S = \frac{1}{c}\int d^dx \left\{\frac{1}{2Cc^2}(\partial\phi)^2 + \left(1 + \frac{d-2}{c^2}\phi\right)K - \left(1 + \frac{d\phi}{c^2}\right)V\right\}. \tag{3.40}$$

It is reasonable to expect that the full action, with all the $\phi$ corrections, takes the form

$$S = \frac{1}{c}\int d^dx \left\{\frac{1}{2Cc^2}f(\phi)(\partial\phi)^2 + g(\phi)K - h(\phi)V\right\}, \tag{3.41}$$

where $f, g$, and $h$ are functions of $\phi$ to be found by imposing the condition that the equation of motion of $\phi$ can be written in the form Eq. (3.37), where $T$ is the trace of the total energy–momentum tensor of the above Lagrangian, which is easily found to be

$$T = (d-2)\frac{1}{2Cc^2}f(\phi)(\partial\phi)^2 + (d-2)g(\phi)K - dh(\phi)V. \tag{3.42}$$

The $\phi$ equation of motion coming from Eq. (3.41) is

$$\partial^2\phi = -\tfrac{1}{2}(f'/f)(\partial\phi)^2 + Cg'/(fK) - Ch'/(fV), \tag{3.43}$$

and, on comparing this with Eqs. (3.37) and (3.42), one finds

$$f = \frac{1}{a + [(d-2)/c^2]\phi}, \qquad g = f/b, \qquad h = (f/e)^{\frac{d}{d-2}}, \tag{3.44}$$

where $a, b$, and $e$ are integration constants. If we want to recover Eq. (3.40) in the weak-field limit, we have to take $a = b = e = 1$. Then, we have succeeded and we have found the action

$$S = \frac{1}{c}\int d^dx \left\{ \frac{1}{2Cc^2} \frac{(\partial\phi)^2}{1 + [(d-2)/c^2]\phi} + \left[1 + \frac{d-2}{c^2}\phi\right] K - \left[1 + \frac{d-2}{c^2}\phi\right]^{\frac{d}{d-2}} V \right\} \tag{3.45}$$

that gives rise to the equation of motion Eq. (3.37), with $T$, the trace of the total energy–momentum tensor corresponding to the above action, given by

$$T = \frac{d-2}{2Cc^2} \frac{(\partial\phi)^2}{1 + [(d-2)/c^2]\phi} + (d-2)\left[1 + \frac{d-2}{c^2}\phi\right] K - d\left[1 + \frac{d-2}{c^2}\phi\right]^{\frac{d}{d-2}} V. \tag{3.46}$$

This result was presented in Refs. [566] and [439], but the theory obtained is the one proposed by Nordström back in 1913 in Refs. [990, 991] in terms of different variables: on introducing

$$\Phi \equiv c^2 \left[1 + \frac{d-2}{c^2}\phi\right]^{\frac{1}{2}}, \tag{3.47}$$

the action Eq. (3.45) takes the form

$$S = \frac{1}{c}\int d^dx \left\{ \frac{2}{(d-2)^2 Cc^2} (\partial\Phi)^2 + [\Phi/c^2]^2 K - [\Phi/c^2]^{\frac{2d}{d-2}} V \right\}. \tag{3.48}$$

In the case in which $V = 0$, taking into account Eq. (3.5), the equation of motion can be written in the standard form

$$\partial^2\Phi = \frac{(d-3)4\pi G_N^{(d)}}{c^2} \Phi\, T^{(0)}_{\text{matter}}, \tag{3.49}$$

where $T^{(0)}_{\text{matter}}$ is the trace of the matter energy–momentum tensor obtained from the uncoupled $\mathcal{L}_{\text{matter}}$. In Nordström's theory, this is the equation valid in all cases ($V \neq 0$).

In this form it is very difficult to see that the theory has the property we wanted (that the source for the gravitational scalar field is the trace of the total energy–momentum tensor).

There is yet another way of rewriting this theory, which was found by Einstein and Fokker [499]. This was one of Einstein's first attempts at building a relativistic theory of gravity in which the gravitational field is represented by a metric, as suggested by Grossmann.

### 3.1.7 The geometrical Einstein–Fokker theory

The Einstein–Fokker theory is based on a conformally flat metric,

$$g_{\mu\nu} \equiv [\Phi/c^2]^{\frac{4}{d-2}} \eta_{\mu\nu}. \tag{3.50}$$

Only the conformal factor $\Phi$ is dynamical. The equation of motion for the metric (i.e. for $\Phi$) is

$$R(g) = \frac{(d-1)(d-3)}{d-2} \frac{16\pi G_{\mathrm{N}}^{(d)}}{c^2} T_{\mathrm{matter}}, \tag{3.51}$$

where $R(g)$ is the Ricci scalar for the metric $g_{\mu\nu}$ and $T_{\mathrm{matter}}$ is calculated from the canonical, special-relativistic fully covariant energy–momentum tensor $T_{\mathrm{matter}\,\mu\nu}$, by contracting both indices with $g^{\mu\nu}$.

Alternatively, the Einstein–Fokker theory can be formulated by giving the above equation for an arbitrary metric, but adding another equation,

$$C_{\mu\nu}{}^{\rho\sigma}(g) = 0, \tag{3.52}$$

where $C_{\mu\nu}{}^{\rho\sigma}$ is the Weyl tensor. This equation implies that the metric is conformally flat and can be written, in appropriate coordinates, in the form given in Eq. (3.50).

Using the formulae in Appendix M, we find

$$R(g) = \frac{4(d-1)}{d-2} [\Phi/c^2]^{-\frac{d+2}{d-2}} \partial^2 [\Phi/c^2]. \tag{3.53}$$

This, together with

$$T_{\mathrm{matter}} = [\Phi/c^2]^{-\frac{4}{d-2}} T^{(0)}_{\mathrm{matter}}, \tag{3.54}$$

gives Eq. (3.49).

Einstein and Fokker did not give a Lagrangian for gravity coupled to matter, and therefore they had to postulate how gravity affects the motion of matter. Here, the power of the Einstein–Fokker formulation of Nordström's theory becomes manifest: Einstein and Fokker suggested replacing the flat spacetime metric $\eta_{\mu\nu}$ by the conformally flat metric $g_{\mu\nu}$ everywhere in the matter Lagrangian. This prescription can be used in most matter Lagrangians (not involving spinors). For instance, for the massive particle, it leads to

$$\begin{aligned} S_{\mathrm{pp}}[X^\mu(\xi)] &= -Mc \int d\xi \sqrt{g_{\mu\nu}(X) \dot{X}^\mu \dot{X}^\nu} \\ &= -Mc \int d\xi \, [\Phi(X)/c^2]^{\frac{2}{d-2}} \sqrt{\eta_{\mu\nu} \dot{X}^\mu \dot{X}^\nu} \\ &\sim -Mc \int d\xi \, [1 + \phi(X)/c^2 + \cdots] \sqrt{\eta_{\mu\nu} \dot{X}^\mu \dot{X}^\nu}, \end{aligned} \tag{3.55}$$

which is, to lowest order in $\phi$, our old result. In general, the equation of motion simply tells us that massive particles move along timelike geodesics with respect to the metric $g_{\mu\nu}$. This is a very powerful statement that goes far beyond Nordström's original theory.

For the massless particle, we also find that the coupling can again be absorbed into the worldline auxiliary metric. There is no bending of light in this theory. However, one can argue [483] that, although there is no *global* bending, there is *local* bending of light rays. As explained in Ref. [483], local bending is a kinematical effect associated with accelerating reference frames and occurs, via Einstein's equivalence principle of gravitation and inertia (to be discussed in Section 3.3), in any theory, independently of any equation of motion. Global bending is an integral of local bending, depending on the conformal spacetime structure, which depends on the specific equations of motion of each theory. The contribution of local bending to global bending is just half the value predicted by GR and is experimentally confirmed. In scalar gravity, this contribution is canceled out.

## 3.2 Gravity as a self-consistent massless spin-2 SRFT

In the previous section we have seen that the simplest possible SRFT of gravity, scalar gravity, is not a good candidate since it does not pass two of the classical tests: bending of light and precession of the perihelion of Mercury. Apart from this, the theory did not have consistency problems regarding coupling to matter[12] or to the gravity field itself, but, precisely because of this, there was a lot of freedom in choosing the energy–momentum tensor which could be the matter energy–momentum tensor or the total energy–momentum tensor. We argued that this could be considered a weakness of the theory.

Now we have to try the next simplest possibility. Excluding a vector field (a spin-1 particle) because it leads to repulsion between like charges, the next possibility is that gravity is mediated by a massless spin-2 particle (the graviton).

The field that describes a spin-2 particle is a symmetric two-index *Lorentz* tensor $h_{\mu\nu}$ whose indices are raised and lowered with the Minkowski metric $\eta_{\mu\nu}$ (this is a SRFT). For the free field $h_{\mu\nu}$, one can try the equation of motion [213, 214] (see also [100, 101, 215, 216, 965])

$$\partial^2 h^{\mu\nu} = 0. \tag{3.56}$$

Things are, however, not that simple. On the one hand, this theory does not have positive-definite energy unless one imposes a consistency condition:

$$\partial^\mu \left( h_{\mu\nu} - \tfrac{1}{2}\eta_{\mu\nu} h^\rho{}_\rho \right) = 0, \tag{3.57}$$

as pointed out by Weyl [1253]. On the other hand, the field $h_{\mu\nu}$ describes many more helicity states than those of a massless spin-2 particle (a symmetric $h_{\mu\nu}$ has $d(d+1)/2$ independent components, some of which describe spin-1 and spin-0 helicity states) and therefore the equations of motion of this field should be such that, on-shell, it describes only the $d(d-3)/2$ helicity states that a massless spin-2 particle has in $d$ dimensions (two in four dimensions: $s_z = -2, +2$).

These two problems are related since the negative contribution to the energy comes precisely from some of the unwanted helicities which are eliminated when one imposes the above condition (which we will later call the *De Donder*[13] gauge condition [425]). To eliminate all the helicities not corresponding to the spin-2 particle we want to describe, we have to impose another condition,

$$h^\mu{}_\mu = 0. \tag{3.58}$$

[12] We saw, however, that there was some disagreement between the effect of gravity on massless fields and the effect on massless particles.

[13] Also known in the literature as the *harmonic* or *Hilbert* [1177], *Hilbert–Lorentz* [1000], or *Einstein* gauge condition.

Actually, the correct way of arriving at these two conditions is to introduce into the theory some kind of gauge freedom so that $\partial^\mu\big(h_{\mu\nu} - \frac{1}{2}\eta_{\mu\nu}h^\rho{}_\rho\big)$ and $h^\mu{}_\mu$ can take arbitrary values, in particular zero. However, let us accept for the moment the theory given by Eq. (3.56) supplemented by the conditions Eqs. (3.57) and (3.58) and let us now consider the coupling to matter. As in any SRFT of gravitation, matter must couple to gravity through the energy–momentum tensor. The l.h.s. of the equation of motion has two free indices and, therefore, it is natural to expect the matter energy–momentum tensor on the r.h.s., that is[14]

$$\partial^2 h^{\mu\nu} = \chi T^{\mu\nu}_{\text{matter}}, \tag{3.59}$$

where $\chi$ is a coupling constant whose dimensions and value we will discuss later. As opposed to the scalar case, this equation (which still has to be supplemented by Eqs. (3.57) and (3.58)) does impose consistency conditions on the matter energy–momentum tensor. First, it has to be symmetric because the l.h.s. is. Second, it has to be divergence free (conserved), because the l.h.s. is, as a result of the supplementary conditions imposed on $h_{\mu\nu}$. Both conditions are satisfied by the Belinfante or Rosenfeld energy–momentum tensors and by an infinite number of tensors obtained from these by adding a superpotential correction that does not modify their symmetry. Nevertheless, it is clear that this is a theory with a structure tighter than the scalar one, and it is encouraging to find that the consistency of the theory imposes physically meaningful conditions on the energy–momentum tensor. All this makes it worth studying.

Of course, we want to find the gauge-invariant equations of motion (or Lagrangian) and the gauge transformations which allow us to impose the conditions Eqs. (3.57) and (3.58) and arrive at Eq. (3.59). These equations of motion must necessarily be of the form

$$\mathcal{D}^{\mu\nu}(h) = \chi T^{\mu\nu}_{\text{matter}}, \tag{3.60}$$

where, now, by consistency with the conservation of the matter energy–momentum tensor, the wave operator $\mathcal{D}^{\mu\nu}(h)$ should also be divergenceless, namely

$$\partial_\mu \mathcal{D}^{\mu\nu}(h) = 0, \tag{3.61}$$

*off-shell*, i.e. independently of the equations of motion (which, in vacuum, should have the form $\mathcal{D}_{\mu\nu}(h) = 0$). In other words, the theory has to have the above property as a *Bianchi* or *gauge identity*. This kind of identity can be derived from theories with a gauge symmetry according to the general procedure outlined in Chapter 2, and, if we obtain a theory with this property (which is easier to do), we will most surely have obtained a theory with the gauge symmetry needed to remove the unwanted degrees of freedom.

The problem of finding a theory with these properties, a theory for a massless spin-2 particle, was solved by Fierz and Pauli in Ref. [543] and it was studied again by Ogievetsky and Polubarinov in Ref. [1000] in a more general setting, including possible self-interactions of the gravitational field.

The matter energy–momentum tensor in Eq. (3.60) is calculated from the free-matter field theory. When it is coupled to gravity, only the *total* (matter plus gravity)

[14] Certainly, there are other possibilities: we can add to the r.h.s. terms like $\eta^{\mu\nu}T^{\rho}_{\text{matter}\,\rho}$. However, these possibilities are inconsistent with the supplementary conditions Eqs. (3.57) and (3.58).

energy–momentum tensor is conserved. This is the inconsistency problem of this SRFT of gravity (see, for instance, Ref. [952]). Then, we should add, at least, the gravitational energy–momentum tensor calculated from the Lagrangian from which we derived Eq. (3.60) to the r.h.s. of Eq. (3.60), for consistency. However, if we want to derive the new equation of motion from a Lagrangian, we need to add to the old Lagrangian a cubic term, which, in turn, will introduce a correction to the gravitational energy–momentum tensor. If we add this correction to the r.h.s. of Eq. (3.60), we will have to add a further correction to the Lagrangian, and so on. The coupling to matter requires an infinite number of corrections to the free spin-2 (Fierz–Pauli) theory.

The problem of consistent self-interaction of the gravitational field is of great importance and was pointed out for the first time by Gupta [701, 702] and Kraichnan [876, 877]; in the classical works of Feynman [542] and Thirring [1177], in which the first correction to the free equation of motion was found and used to calculate the precession of the perihelion of Mercury;[15] in Ref. [1280]; in the works of Weinberg [1238, 1239], in which it was shown that a quantum theory of a massless spin-2 particle can have a Lorentz-invariant quantum S matrix only if it couples to the total energy–momentum tensor; in Deser's paper [430], in which it was shown that GR can be seen as the result of adding this infinite number of corrections;[16] in Boulware and Deser's paper [244], in which Weinberg's result was completed by a determination of the form of the gravitational energy–momentum tensor to which gravity itself would couple in a consistent quantum theory, which was found to be, *in the long-wavelength limit*, the one predicted by GR; in Refs. [393, 517, 739, 740, 1230, 1231], in which general, consistent, non-linear theories of a spin-2 particle were investigated with the conclusion that the only possible symmetries of these theories were "normal spin-2 gauge invariance" (to be defined later) and general covariance and, more recently, in Ref. [243], in which an alternative theory for a $d = 3$ spin-2 particle was found.

In this section we are going to study the Fierz–Pauli theory and its gauge symmetry. Then, we will couple it to matter and find the predictions for the bending of light by gravity and the precession of the perihelion of Mercury. The latter will come out as the wrong value and we will see the need to introduce corrections into the theory, as the inconsistency problem suggests. We will try to envisage a systematic way of introducing these corrections on the basis of the Noether method explained in Chapter 2. Then, we will spend some time trying to find the first correction (i.e. the gravitational energy–momentum tensor) for various methods and we will calculate the corresponding correction to the precession of the perihelion of Mercury, discovering that the Belinfante–Rosenfeld energy–momentum tensor (employed by Thirring in [1177]) does not give the right result, whereas the one used in GR does. We will then use Deser's procedure to find a theory that is consistent to all orders. This theory will turn out to be GR, which we introduce in Section 3.3.

Before proceeding to the construction of the Fierz–Pauli theory, it is worth studying a simpler example of the relation among gauge symmetry, Bianchi (gauge) identities, and conserved charges in the SRFT of a spin-1 particle.

---

[15] Thirring's result is actually wrong [986], as we will see.

[16] This result was extended in Ref. [433] to general vacua.

### 3.2.1 Gauge invariance, gauge identities, and charge conservation in the SRFT of a spin-1 particle

A massive or massless spin-1 particle is described by a vector field $A^\mu$. The simplest relativistic wave equation we could imagine for it would be

$$\left(\partial^2 + m^2\right) A^\mu = 0. \tag{3.62}$$

However, the energy density of this theory is not positive-definite unless one imposes the *Lorentz* or *transversality* condition

$$\partial_\mu A^\mu = 0. \tag{3.63}$$

Furthermore, just as in the spin-2-particle case, the vector $A^\mu$ describes spin-1 helicity states but also spin-0 helicity states. A $d$-dimensional vector field has $d$ independent components, but a massive spin-1 particle in $d$ dimensions has $d-1$ states (three in $d=4$: $s_z = -1, 0, 1$) and a massless spin-1 particle has $d-2$ helicity states (two in $d=4$: $s_z = -1, +1$). It is precisely the unwanted spin-0 helicity states that contribute negatively to the energy and the Lorentz condition projects them out.

If we couple the massless theory to charged matter, by Lorentz covariance, this has to be described by a vector current $j^\mu$, so we have

$$\partial^2 A^\mu = j^\mu \tag{3.64}$$

and, by consistency with the Lorentz condition, the vector current has to be conserved, $\partial_\mu j^\mu = 0$, which is, again, a physically meaningful condition that coincides with our experience with electric charges and currents.

We would like to construct a theory in which the Lorentz condition arises as a consequence of the equation of motion in the massive case and in which $\partial_\mu A^\mu$ is completely arbitrary in the massless case. These conditions guarantee the removal of the unwanted helicities. We expect the equation of motion to be of the form

$$\mathcal{D}^\mu(A) + m^2 A^\mu = j^\mu, \tag{3.65}$$

where, now, by consistency, the massless wave operator $\mathcal{D}^\mu(A)$ has to satisfy off-shell the identity

$$\partial_\mu \mathcal{D}^\mu(A) = 0, \tag{3.66}$$

which should arise as the gauge identity associated with some gauge symmetry.

We could proceed as in Ref. [1000], translating these conditions into a gauge identity for a general Lagrangian and then trying to find, with as much generality as possible, a gauge symmetry (forming a group) leading to that gauge identity. As is well known, the result is the *Proca Lagrangian* and the equation of motion

$$\begin{aligned} S[A] &= \int d^d x \left[-\tfrac{1}{4} F^2 + \frac{m^2}{2} A^2\right], \\ \mathcal{D}^\mu_{(m)}(A) &= \mathcal{D}^\mu(A) + m^2 A^\mu = 0, \end{aligned} \tag{3.67}$$

where

$$\mathcal{D}^\mu(A) \equiv \partial_\mu F^{\mu\nu}, \qquad F_{\mu\nu} = 2\partial_{[\mu} A_{\nu]}, \tag{3.68}$$

which, in the $m=0$ limit, reduce to Maxwell's Lagrangian and Maxwell's equation. Owing to the antisymmetry of $F_{\mu\nu}$, the massless wave operator does indeed have the off-shell property Eq. (3.66), which implies, in turn, Eq. (3.63) in the massive case, as we needed in order to obtain a positive-definite energy and to eliminate the spin-0 degree of freedom. In turn, the massless theory is easily seen to be invariant under gauge transformations,

$$\delta A_\mu = \partial_\mu \Lambda(x). \tag{3.69}$$

Given any $A^\mu$, we can gauge-transform it into another $A'^\mu$ satisfying Lorentz's condition: it is enough to choose a gauge parameter $\Lambda$ that is a solution of $\partial^2\Lambda = -\partial_\mu A^\mu$. Lorentz's condition does not completely fix the gauge: there are many potentials $A_\mu$ that satisfy that gauge condition and are related by non-trivial gauge transformations (those with parameters satisfying $\partial^2\Lambda = 0$). To fix the gauge invariance completely, it is necessary to impose another gauge condition. This is why this gauge symmetry reduces to $d-2$ the number of degrees of freedom described by a massless vector field, as required in order to describe just the spin-1 case.

Furthermore, as expected, the identity Eq. (3.66) is related to the above gauge symmetry via Noether's theorems. Let us follow Chapter 2: we know that the Maxwell action

$$S[A] = \int d^dx \, \{-\tfrac{1}{4}F^2\} \tag{3.70}$$

is exactly invariant under gauge transformations because $F_{\mu\nu}$ is. Thus,[17]

$$\begin{aligned} \delta S &= \int d^dx \, \{-F^{\mu\nu}\partial_\mu \delta A_\nu\} \\ &= \int d^dx \, \{\mathcal{D}^\mu(A)\delta A_\nu - \partial_\mu(F^{\mu\nu}\delta A_\nu)\} \\ &= \int d^dx \, \{-\partial_\mu \mathcal{D}^\mu(A)\Lambda - \partial_\mu(F^{\mu\nu}\partial_\nu\Lambda - \mathcal{D}^\mu(A)\Lambda)\}\,. \end{aligned} \tag{3.71}$$

Now we argue as follows: if the gauge parameter $\Lambda(x)$ and its derivatives vanish on the boundary, the integral of the total derivative term is zero. Since the variation is zero for any $\Lambda$, then $\partial_\mu \mathcal{D}^\mu(A) = 0$. This is the gauge identity. Now that we know it always holds, we can consider more general gauge parameters and the invariance of the action implies that

$$\partial_\mu j^\mu_{\rm N2}(\Lambda) = 0, \qquad j^\mu_{\rm N2}(\Lambda) = j^\mu_{\rm N1}(\Lambda) - \mathcal{D}^\mu(A)\Lambda, \qquad j^\mu_{\rm N1}(\Lambda) = F^{\mu\nu}\partial_\nu\Lambda, \tag{3.72}$$

where $j^\mu_{\rm N1}(\Lambda)$ and $j^\mu_{\rm N2}(\Lambda)$ are *Noether currents* associated with the gauge parameter $\Lambda$; $j^\mu_{\rm N1}(\Lambda)$ is conserved only on-shell, but $j^\mu_{\rm N2}(\Lambda)$ is automatically conserved (i.e. off-shell). On-shell they are evidently identical. Furthermore, as can easily be checked in this case, the Noether current $j^\mu_{\rm N2}(\Lambda)$ associated with a gauge symmetry enjoys another property [147]: it is always the divergence of an antisymmetric tensor. In this case

$$j^\mu_{\rm N2}(\Lambda) = \partial_\nu j^{\nu\mu}_{\rm N2}(\Lambda), \qquad j^{\nu\mu}_{\rm N2}(\Lambda) = -F^{\nu\mu}\Lambda. \tag{3.73}$$

---

17 We write here $\delta$ instead of $\tilde{\delta}$ because these transformations do not involve any coordinate transformation and the two variations are identical.

The conserved charge, which can be written covariantly, up to normalization, as

$$q(\Lambda) \sim \int_{\mathrm{V}_t} d^{d-1}\Sigma_\mu j^\mu_{\mathrm{N2}}(\Lambda), \tag{3.74}$$

where $\mathrm{V}_t$ is a spacelike hypersurface (a constant time slice for some time coordinate), can be reexpressed as an integral over the boundary of $\mathrm{V}_t$, i.e. a surface integral over an $\mathrm{S}^{d-2}$ sphere at infinity in $d$-dimensional Minkowski spacetime:

$$q(\Lambda) \sim \frac{1}{2} \int_{\mathrm{S}^{d-2}_\infty} d^{d-2}\Sigma_{\mu\nu} j^{\nu\mu}_{\mathrm{N2}}(\Lambda) = -\tfrac{1}{2} \int_{\mathrm{S}^{d-2}_\infty} d^{d-2}\Sigma_{\mu\nu} F^{\nu\mu}\Lambda. \tag{3.75}$$

For $\Lambda = 1$ or any $\Lambda(x)$ that goes to 1 at spatial infinity, $q(\Lambda)$ is just the electric charge. In differential-form language

$$q = \int_{\partial\Sigma} \star F. \tag{3.76}$$

For later use, it should be noted that, as a matter of fact, the massless theory could have been found by this simple procedure: write the most general Lorentz-invariant Lagrangian quadratic in derivatives of $A^\mu$ with arbitrary coefficients $a$ and $b$:

$$S[A] = \int d^dx \, \{a\partial_\mu A_\nu \partial^\mu A^\nu + b\partial_\mu A_\nu \partial^\nu A^\mu\}, \tag{3.77}$$

and impose on the equations of motion the gauge identity Eq. (3.66). This fixes $a = -b$ and, on choosing the overall normalization suitably, one obtains Maxwell's Lagrangian. Then we can immediately find the gauge symmetry that leaves it invariant.

How would the presence of sources modify these results? Essentially in no way, but we have to be a bit more careful. First of all, under a gauge transformation, the first variation of the action with sources

$$S_j[A] = \int d^dx \, \{-\tfrac{1}{4}F^2 - A_\mu j^\mu\} \tag{3.78}$$

is

$$\delta S_j = \int d^dx \, \{\Lambda \partial_\mu j^\mu - \partial_\mu (\Lambda j^\mu)\}, \tag{3.79}$$

and we have invariance up to a total derivative only if the source current is conserved. Conservation is also required by consistency of the equation of motion

$$\mathcal{D}^\mu(A) = j^\mu. \tag{3.80}$$

On the other hand, we can vary the action as before: first under a general variation $\delta A_\mu$ and then using the form of the gauge transformation:

$$\begin{aligned}
\delta S_j &= \int d^dx \, \{-F^{\mu\nu}\partial_\mu \delta A_\nu - \delta A_\nu j^\nu\} \\
&= \int d^dx \, \{[\mathcal{D}^\mu(A) - j^\mu]\,\delta A_\nu - \partial_\mu(F^{\mu\nu}\delta A_\nu)\} \\
&= \int d^dx \, \{-\partial_\mu[\mathcal{D}^\mu(A) - j^\mu]\Lambda - \partial_\mu\{F^{\mu\nu}\partial_\nu\Lambda - [\mathcal{D}^\mu(A) - j^\mu]\Lambda\}\}.
\end{aligned} \tag{3.81}$$

The two forms of the variation of the action are identical. By identifying them we arrive at the same results as in the sourceless case because the source terms cancel each other out.

### *3.2.2 Gauge invariance, gauge identities, and charge conservation in the SRFT of a spin-2 particle*

Inspired by the lessons learned in finding the SRFT of a spin-1 particle, we return to the spin-2 theory. We could follow Ref. [1000] and try to determine the most general theory with the required properties, including non-linear couplings and transformations. Instead, since we want to start with a linear theory (which will be adequate for a free spin-2 particle), we are going to use the shortcut we used in the massless spin-1 case: construct the most general (up to total derivatives) Lorentz-invariant action that is quadratic in $\partial_\rho h_{\mu\nu}$ and impose the gauge identity Eq. (3.61). This should determine the action for the massless theory up to total derivatives and overall normalization, and we can then search for the gauge invariance which the theory surely enjoys and prove that it is enough to eliminate the unwanted degrees of freedom. Then we can add terms polynomial in $h_{\mu\nu}$ in order to find the action for the massive theory.

There are only four different possible terms in the Lagrangian up to total derivatives. We can write all of them with unknown coefficients,

$$S = \int d^d x \left\{ a\partial^\rho h^{\mu\nu}\partial_\rho h_{\mu\nu} + b\partial^\mu h^{\nu\rho}\partial_\nu h_{\mu\rho} + c\partial^\mu h \partial^\lambda h_{\lambda\mu} + d\partial^\mu h \partial_\mu h \right\}, \tag{3.82}$$

where we use the standard notation $h$ for the trace of $h_{\mu\nu}$,

$$h \equiv h_\mu{}^\mu . \tag{3.83}$$

We normalize the kinetic term canonically[18] by setting $a = +\frac{1}{4}$, and then easily find that the equations of motion will satisfy Eq. (3.61) if $b = -\frac{1}{2}$, $c = \frac{1}{2}$, and $d = -\frac{1}{4}$, so the action we are looking for is the Fierz–Pauli action [543]

$$\boxed{S = \int d^d x \left\{ \tfrac{1}{4}\partial^\mu h^{\nu\rho}\partial_\mu h_{\nu\rho} - \tfrac{1}{2}\partial^\mu h^{\nu\rho}\partial_\nu h_{\mu\rho} + \tfrac{1}{2}\partial^\mu h \partial^\lambda h_{\lambda\mu} - \tfrac{1}{4}\partial^\mu h \partial_\mu h \right\}.} \tag{3.84}$$

We want the above action to be dimensionless in natural units $\hbar = c = 1$. The field $h_{\mu\nu}$ has to have the dimensions of $L^{-\frac{d-2}{2}}$. Then, since the energy–momentum tensor has the same dimensions as the Lagrangian, Eq. (3.60) implies that $\chi$ has the inverse dimensions of $h_{\mu\nu}$, so $\chi h_{\mu\nu}$ is dimensionless.

The corresponding divergenceless equations of motion are

$$\begin{aligned} \frac{\delta S}{\delta h_{\mu\nu}} &\equiv -\tfrac{1}{2}\mathcal{D}^{\mu\nu}(h), \\ \mathcal{D}_{\mu\nu}(h) &= \partial^2 h_{\mu\nu} + \partial_\mu\partial_\nu h - 2\partial^\lambda\partial_{(\mu}h_{\nu)\lambda} - \eta_{\mu\nu}\left(\partial^2 h - \partial_\lambda\partial_\sigma h^{\lambda\sigma}\right) = 0. \end{aligned} \tag{3.85}$$

---

[18] That is, $S_{\text{FP}} = \int d^d x \left\{ +\frac{1}{4}\partial_t h_{ij}\partial_t h_{ij} + \cdots \right\}$.

By subtracting the trace of this equation we can simplify it without any loss of information:

$$\hat{\mathcal{D}}_{\mu\nu}(h) \equiv \mathcal{D}_{\mu\nu}(h) - \frac{1}{d-2}\eta_{\mu\nu}\mathcal{D}_\rho{}^\rho(h) = \partial^2 h_{\mu\nu} + \partial_\mu\partial_\nu h - 2\partial^\lambda\partial_{(\mu}h_{\nu)\lambda} = 0. \quad (3.86)$$

Sometimes the equation of motion (3.85) is written in terms of the convenient variable $\bar{h}_{\mu\nu}$:

$$\mathcal{D}_{\mu\nu}(\bar{h}) = \partial^2\bar{h}_{\mu\nu} - 2\partial^\lambda\partial_{(\mu}\bar{h}_{\nu)\lambda} + \eta_{\mu\nu}\partial_\lambda\partial_\sigma\bar{h}^{\lambda\sigma} = 0, \quad (3.87)$$

where

$$\bar{h}_{\mu\nu} \equiv h_{\mu\nu} - \tfrac{1}{2}\eta_{\mu\nu}h. \quad (3.88)$$

Finally, we can write the Fierz–Pauli wave operator as the divergence of a tensor $\eta^{\mu\nu\rho}$,

$$\mathcal{D}^{\nu\rho}(h) = 2\partial_\mu\eta^{\nu\rho\mu}, \quad (3.89)$$

but the tensor $\eta^{\mu\nu\rho}$ is not uniquely defined. Some possible candidates are

$$\begin{aligned} \eta_{\mathrm{T}}^{\nu\rho\mu} &= \eta_{\mathrm{T}}^{(\nu\rho)\mu} &&= -\partial_\sigma H^{\mu\sigma\nu\rho}, \\ \eta_{\mathrm{LL}}^{\nu\rho\mu} &= \eta_{\mathrm{LL}}^{\nu[\rho\mu]} &&= -\partial_\sigma K^{\nu\sigma\rho\mu}, \\ \eta_{\mathrm{AD}}^{\nu\rho\mu} &= \eta_{\mathrm{AD}}^{[\nu|\rho|\mu]} &&= -\partial_\sigma K^{\nu\mu\rho\sigma}, \end{aligned} \quad (3.90)$$

where

$$\begin{aligned} K^{\mu\nu\rho\sigma} &= \tfrac{1}{2}\left[\eta^{\mu\sigma}\bar{h}^{\nu\rho} + \eta^{\nu\rho}\bar{h}^{\mu\sigma} - \eta^{\mu\rho}\bar{h}^{\nu\sigma} - \eta^{\nu\sigma}\bar{h}^{\mu\rho}\right], \\ H^{\mu\sigma\nu\rho} &= \tfrac{1}{2}\left[\eta^{\sigma\rho}\bar{h}^{\mu\nu} + \eta^{\sigma\nu}\bar{h}^{\mu\rho} - \eta^{\nu\rho}\bar{h}^{\mu\sigma} - \eta^{\mu\sigma}\bar{h}^{\nu\rho}\right]; \end{aligned} \quad (3.91)$$

$H$ is symmetric in the last two indices and $K$ is antisymmetric. In fact, $K$ has exactly the same symmetries as the Riemann tensor (in the Levi-Civita case).

On the other hand, $\eta_{\mathrm{T}}^{\mu\nu\rho}$ has the defining property

$$\frac{\partial\mathcal{L}_{\mathrm{FP}}}{\partial\partial_\mu h_{\nu\rho}} = \eta_{\mathrm{T}}^{\nu\rho\mu}, \quad (3.92)$$

for the Fierz–Pauli Lagrangian written in Eq. (3.84).

Using any of the last two $\eta^{\nu\rho\mu}$s, the fact that the Fierz–Pauli wave operator $\mathcal{D}^{\mu\nu}(h)$ is divergenceless becomes manifest.

Let us now determine the gauge symmetry of the Fierz–Pauli Lagrangian. Under a general variation of $h_{\mu\nu}$, the variation of the action is, up to a total derivative,

$$\delta S_{\mathrm{FP}} = -\tfrac{1}{2}\int d^dx\, \mathcal{D}^{\mu\nu}\delta h_{\mu\nu}. \quad (3.93)$$

If $\delta h_{\mu\nu}$ is a gauge transformation, we know that, up to total derivatives, the integrand of the variation of the action has to be proportional to the gauge identity Eq. (3.61), i.e.

$$\int d^dx\, \mathcal{D}^{\mu\nu}\delta h_{\mu\nu} \sim \int d^dx\, \partial_\mu\mathcal{D}^{\mu\nu}\epsilon_\nu, \quad (3.94)$$

(the gauge parameter $\epsilon_\mu(x)$ has to be a local Lorentz vector). On integrating the r.h.s. by parts, and choosing a convenient normalization, we find the gauge transformation

$$\delta_\epsilon h_{\mu\nu} = -2\partial_{(\mu}\epsilon_{\nu)}. \tag{3.95}$$

We can now check directly that the Fierz–Pauli Lagrangian is invariant under these transformations:

$$\delta S_{\rm FP} = \int d^dx \frac{\partial \mathcal{L}_{\rm FP}}{\partial\partial_\mu h_{\nu\rho}} \partial_\mu \delta h_{\nu\rho} = -\int d^dx \partial_\sigma H^{\mu\sigma\nu\rho} \partial_\mu \delta h_{\nu\rho}. \tag{3.96}$$

Here we have used Eqs. (3.90) and (3.92). On integrating by parts and using the explicit form of the variation, we have

$$\delta S_{\rm FP} = \int d^dx \frac{\partial \mathcal{L}_{\rm FP}}{\partial\partial_\mu h_{\nu\rho}} \partial_\mu \delta h_{\nu\rho} = \int d^dx \left\{\partial_\sigma [2H^{\mu\sigma\nu\rho}\partial_\mu\partial_\nu\epsilon_\rho] - 2H^{\mu\sigma\nu\rho}\partial_\sigma\partial_\mu\partial_\nu\epsilon_\rho\right\}. \tag{3.97}$$

The second term vanishes identically and the action turns out to be invariant up to a total derivative (the first term).

To complete our program for the massless spin-2 theory, it remains only to show that, using this gauge symmetry, we can remove $2d$ of the $d(d+1)/2$ independent components of $h_{\mu\nu}$ to leave only the $d(d-3)/2$ degrees of freedom of a massless spin-2 particle in $d$ dimensions. The counting of degrees of freedom in a gauge theory is not straightforward. See e.g. Ref. [745] for simple rules, but one can show that, using the gauge transformations (3.95), one can indeed eliminate $2d$ components (set them to a given value by fixing the gauge).

There are two popular gauges: the *transverse, traceless gauge*

$$\partial_\mu h^{\mu\nu} = h = 0, \tag{3.98}$$

which automatically leads to the equation of motion

$$\mathcal{D}_{\mu\nu}(h) = \partial^2 h_{\mu\nu} = 0 \tag{3.99}$$

typical of a massless field, and the *De Donder* or *harmonic* gauge

$$\partial_\mu \bar{h}^{\mu\nu} = 0, \tag{3.100}$$

which leads to

$$\mathcal{D}_{\mu\nu}(h) = \partial^2 \bar{h}_{\mu\nu} = 0. \tag{3.101}$$

The traceless transverse gauge implies the De Donder gauge, but not conversely. The transverse, traceless condition does not completely fix the gauge, since it is preserved by gauge transformations with $\epsilon^\mu = \partial^\mu \epsilon$ and $\partial^2\epsilon = 0$.

After the identification of the gauge symmetry of the massless theory, the next step in our program is finding the massive theory. We need to modify the massless equation of motion so that it gives the equation of motion

$$(\partial^2 + m^2)h^{\mu\nu} = 0, \tag{3.102}$$

plus the De Donder and traceless conditions. These $d+1$ constraints leave only the $(d-2)(d+1)/2$ degrees of freedom of the massive spin-2 particle in $d$ dimensions (five in $d=4$: $s_s=-2,-1,0,+1,+2$). We know that the massless wave operator is transverse due to a Bianchi identity. Thus, we know that we have to add a term $-m^2 h_{\mu\nu}$ to it. This is not enough, though: if we take the trace we find

$$\partial^2 h - \frac{m^2}{d-2} h - \partial_\mu \partial_\nu h^{\mu\nu} = 0. \tag{3.103}$$

This equation would give $h=0$ if, instead of having just transversality, we had $\partial^\mu h_{\mu\nu} = \partial_\nu h$ and then we would recover transversality. Thus, we add a term $+m^2(h_{\mu\nu} - \eta_{\mu\nu} h)$ and obtain the massive Fierz–Pauli action and equation [543]

$$\boxed{\begin{aligned} S = \int d^d x \, \{ & \tfrac{1}{4}\partial^\rho h^{\mu\nu} \partial_\rho h_{\mu\nu} - \tfrac{1}{2}\partial^\rho h^{\mu\nu} \partial_\mu h_{\rho\nu} + \tfrac{1}{2}\partial^\mu h \partial^\lambda h_{\lambda\mu} - \tfrac{1}{4}\partial^\mu h \partial_\mu h \\ & - \tfrac{1}{4} m^2 (h^{\mu\nu} h_{\mu\nu} - h^2) \} , \end{aligned}} \tag{3.104}$$

$$\mathcal{D}^{\mu\nu}_{(m)}(h) = \mathcal{D}^{\mu\nu}(h) + m^2 (h_{\mu\nu} - \eta_{\mu\nu} h) = 0, \tag{3.105}$$

from which we obtain, as expected,

$$h = 0, \qquad \partial_\mu h^{\mu\nu} = 0, \qquad (\partial^2 + m^2) h_{\mu\nu} = 0. \tag{3.106}$$

To finalize our study of the free Fierz–Pauli theory, we can use the gauge symmetry to derive conserved currents along the path set out in Chapter 2. We have already calculated the direct variation of the Fierz–Pauli action under gauge transformations and have found invariance up to a total derivative. We now calculate the variation of the Fierz–Pauli action by performing first a general variation $\delta h_{\mu\nu}$, obtaining (after integration by parts) a total derivative term and the term proportional to $\delta h_{\mu\nu}$ whose coefficient is the equation of motion:

$$\begin{aligned} \delta S_{\rm FP} &= \int d^d x \left\{ \partial_\mu \left[ \frac{\partial \mathcal{L}_{\rm FP}}{\partial \partial_\mu h_{\nu\rho}} \delta h_{\nu\rho} \right] - \partial_\mu \frac{\partial \mathcal{L}_{\rm FP}}{\partial \partial_\mu h_{\nu\rho}} \delta h_{\nu\rho} \right\} \\ &= \int d^d x \, \{ \partial_\mu [ 2 \partial_\sigma H^{\mu\sigma\nu\rho} \partial_\nu \epsilon_\rho ] - \mathcal{D}^{\nu\rho}(h) \partial_\nu \epsilon_\rho \} , \end{aligned} \tag{3.107}$$

where we have used the explicit form of the gauge transformation. On integrating again by parts, we obtain the second form of the variation of the action,

$$\delta S_{\rm FP} = \int d^d x \, \{ \partial_\mu \left[ 2 \partial_\sigma H^{\mu\sigma\nu\rho} \partial_\nu \epsilon_\rho + \mathcal{D}^{\mu\rho}(h) \epsilon_\rho \right] - \partial_\nu \mathcal{D}^{\nu\rho}(h) \epsilon_\rho \} . \tag{3.108}$$

By identifying the two forms of the variation of the action and reasoning as in the Maxwell theory, we find the Bianchi identity (the terms proportional to the gauge-transformation parameter) Eq. (3.61) and the conserved current:

$$\begin{aligned} j^\mu_{\rm N2}(\epsilon) &= j^\mu_{\rm N1}(\epsilon) + \mathcal{D}^{\mu\nu}(h) \epsilon_\nu, \\ j^\mu_{\rm N1}(\epsilon) &= 2 \partial_\sigma H^{\mu\sigma\nu\rho} \partial_\nu \epsilon_\rho - 2 H^{\sigma\mu\nu\rho} \partial_\sigma \partial_\nu \epsilon_\rho . \end{aligned} \tag{3.109}$$

Using $\eta_{\rm AD}^{\mu\nu\rho}$ in Eq. (3.90), we can write the Fierz–Pauli wave operator as

$$\mathcal{D}^{\mu\nu}(h) = -2\partial_\sigma\partial_\lambda K^{\mu\lambda\rho\sigma}, \tag{3.110}$$

and, on substituting this into $j^{\mu}_{\rm N2}(\epsilon)$ above and making the obvious manipulations, it takes the form

$$\begin{aligned} j^{\mu}_{\rm N2}(\epsilon) &= \partial_\nu\{-2\partial_\sigma K^{\mu\nu\rho\sigma}\epsilon_\rho + 2(H^{\mu\nu\sigma\rho} + K^{\mu\sigma\rho\nu})\partial_\sigma\epsilon_\rho\} \\ &\quad - 2\Big\{K^{\mu\nu\rho\delta} + H^{\mu\delta\nu\rho} + H^{\delta\mu\nu\rho}\Big\}\,\partial_\delta\partial_\nu\epsilon_\rho. \end{aligned} \tag{3.111}$$

The second term in brackets is antisymmetric in $\delta\nu$ and vanishes. Then, we can write[19]

$$\begin{aligned} j^{\mu}_{\rm N2}(\epsilon) &= \partial_\nu j^{\nu\mu}_{\rm N2}(\epsilon), \\ j^{\nu\mu}_{\rm N2}(\epsilon) &= -2\partial_\sigma K^{\mu\nu\rho\sigma}\epsilon_\rho + 2(H^{\mu\nu\sigma\rho} + K^{\mu\sigma\rho\nu})\partial_\sigma\epsilon_\rho. \end{aligned} \tag{3.112}$$

We can now use this expression to calculate conserved charges associated with the gauge parameters $\epsilon_\mu$. Observe that the term proportional to $H$ vanishes for $\epsilon$ that are Killing vectors of the Minkowski spacetime. We will come back to this point in Chapter 10.

The interpretation of the corresponding conserved charges is more complicated. In the cases in which $\epsilon$ is a Killing vector, a symmetry of Minkowski spacetime, we can associate these charges with momenta in the directions associated with those Killing vectors (linear or angular momenta). We will discuss this point also in Chapter 10.

### *3.2.3 Coupling to matter*

As we discussed at the beginning of this section, the coupling of the Fierz–Pauli theory to matter is described by Eq. (3.60). To obtain this equation of motion from a Lagrangian, we will have to add to the Fierz–Pauli Lagrangian $\mathcal{L}_{\rm FP}(h)$ the matter Lagrangian $\mathcal{L}_{\rm matter}(\varphi)$ and a coupling term weighted by the gravitational coupling constant $\chi$ combined into a modified matter Lagrangian $\mathcal{L}_{\rm matter}(\varphi, h)$:

$$\begin{aligned} \mathcal{L} &= \mathcal{L}_{\rm FP}(h) + \mathcal{L}_{\rm matter}(\varphi, h), \\ \mathcal{L}_{\rm matter}(\varphi, h) &= \mathcal{L}_{\rm matter}(\varphi) + \tfrac{1}{2}\chi h_{\mu\nu}T^{\mu\nu}_{\rm matter}(\varphi). \end{aligned} \tag{3.113}$$

From this Lagrangian, Eq. (3.60) follows. We also obtain an equation of motion for $\varphi$ modified by the coupling to $h_{\mu\nu}$. The gauge identity implies that $T^{\mu\nu}_{\rm matter}(\varphi)$ has to be conserved, $\partial_\mu T^{\mu\nu}_{\rm matter}(\varphi) = 0$, for consistency. Furthermore, this Lagrangian is invariant (up to total derivatives) under the gauge transformations $\delta_\epsilon h_{\mu\nu} = -2\partial_{(\mu}\epsilon_{\nu)}$ only if $\partial_\mu T^{\mu\nu}_{\rm matter}(\varphi) = 0$.

Two questions now arise:

1. Which $T^{\mu\nu}_{\rm matter}(\varphi)$ should we use?
2. Is the conservation of $T^{\mu\nu}_{\rm matter}(\varphi)$ consistent with the modifications to the $\varphi$ equations of motion introduced by the coupling to $h_{\mu\nu}$?

[19] As was explained in Chapter 2, this rewriting is not unique.

Let us address the first question. The energy–momentum tensor on the r.h.s. of Eq. (3.60) has to be symmetric and divergenceless. These two properties are enjoyed by the *Belinfante energy–momentum tensor* of the *free*-matter field theory, which, as explained in Chapter 2, is a symmetrization of the canonical energy–momentum tensor obtained by the addition of superpotential terms (which are identically divergenceless) and on-shell-vanishing terms. The Belinfante energy–momentum tensor is generally considered as the energy–momentum tensor to which gravity couples minimally (see e.g. Ref. [1241]).

There are many other symmetric energy–momentum tensors (in fact, an infinite number of them), such as the *improved* energy–momentum tensor associated with some scale-invariant theories. It can be argued that the improved energy–momentum tensor is in general associated with non-minimal couplings to gravity. The example discussed in Chapter 2 (a conformal scalar) should illuminate this point. In the simplest cases (scalar and vector field) the canonical and the Belinfante tensor are just what we need.[20] In Chapter 2 we also discussed an alternative prescription for how to find a symmetric, conserved, energy–momentum tensor that does not consist in finding some symmetric modification of the canonical energy–momentum tensor, Rosenfeld's. In the scalar, vector, and symmetric-tensor cases that we are going to consider, the Rosenfeld and Belinfante energy–momentum tensors are identical, and, therefore, this is the energy–momentum tensor that we shall use.

Although the ultimate justification for Rosenfeld's prescription, whose logical connection to the physical concept of an energy–momentum tensor is obscure, relies on the final formulation of GR we are tied to, we can already see that the inclusion of the coupling to gravity in the matter action,

$$S_{\rm matter}[\varphi,\eta_{\mu\nu}] + \int d^dx \chi h_{\mu\nu} \left.\frac{\delta S_{\rm matter}[\varphi,\gamma_{\mu\nu}]}{\delta\gamma_{\mu\nu}}\right|_{\gamma_{\mu\nu}=\eta_{\mu\nu}}, \tag{3.114}$$

suggests that this is the beginning of a functional series expansion of the action functional $S_{\rm matter}[\varphi,\gamma_{\mu\nu}]$ of a metric $\gamma_{\mu\nu} = \eta_{\mu\nu} + \chi h_{\mu\nu}$ around the vacuum metric $\eta_{\mu\nu}$,

$$\begin{aligned} S_{\rm matter}[\varphi,\gamma_{\mu\nu}] &= S_{\rm matter}[\varphi,\eta_{\mu\nu}] + \int d^dx \chi h_{\mu\nu} \left.\frac{\delta S_{\rm matter}[\varphi,\gamma_{\mu\nu}]}{\delta\gamma_{\mu\nu}}\right|_{\gamma_{\mu\nu}=\eta_{\mu\nu}} \\ &+ \int d^dx d^dx' \chi^2 h_{\mu\nu}(x) h_{\rho\sigma}(x') \left.\frac{\delta^2 S_{\rm matter}[\varphi,\gamma_{\mu\nu}]}{\delta\gamma_{\mu\nu}\delta\gamma_{\rho\sigma}}\right|_{\gamma_{\mu\nu}=\eta_{\mu\nu}} + \cdots, \end{aligned} \tag{3.115}$$

truncated at first order.

As to the answer to the second question, we postpone it until we work out a simple example to show that the theory we have obtained indeed describes a SRFT of gravity that is compatible with our experience.

*The gravitational field of a massive point-particle.* Just as we did to derive the simplest predictions of the scalar SRFT of gravity, we are going to find the gravitational field

[20] A vector field in four dimensions is also invariant under dilatations and, in fact, under the whole conformal group. The improved energy–momentum tensor is, however, nothing but the Belinfante tensor.

produced by a massive point-particle of mass $M$ placed at rest at the origin of coordinates in some inertial frame. In this calculation we write all the factors of $c$ that we usually omit in order to find the value of $\chi$, and perhaps therefore generate expressions that are more familiar.

The action and energy–momentum tensor for a massive point-particle are given, respectively, by Eqs. (3.8) and (3.25), and the modified action that includes the coupling to gravity is, after the mutual elimination of the spacetime integral and the $d$-dimensional Dirac delta function,

$$S_{\rm pp}[X^\mu] = -Mc\int d\xi \frac{1}{\sqrt{\eta_{\rho\sigma}\dot{X}^\rho\dot{X}^\sigma}}(\eta_{\mu\nu} + \tfrac{1}{2}\chi h_{\mu\nu}(X))\dot{X}^\mu\dot{X}^\nu. \tag{3.116}$$

Before solving any equations, we want to make the following two important observations [1177]. First, as happens in general, this action is not invariant under the gauge transformations unless $\partial_\mu T^{\mu\nu}_{\rm matter}(\varphi) = 0$. However, it is invariant to lowest order in the coupling constant $\chi$ without this assumption if we transform the particle coordinates according to

$$\delta_\epsilon X^\mu = \chi\epsilon^\mu(X), \tag{3.117}$$

which is precisely the form of an infinitesimal GCT. This is the first sign of a relation between the gauge symmetry of the Fierz–Pauli field and spacetime transformations.

Second, there are fields $h_{\mu\nu}$ that are gauge-equivalent to zero, for instance [1177]

$$h_{\mu\nu} = b_{\mu\nu} + a_{\mu\nu\rho}x^\rho, \tag{3.118}$$

with $b_{\mu\nu}$ and $a_{\mu\nu\rho}$ constants, which can be canceled out by a gauge transformation,

$$\epsilon_\mu = \tfrac{1}{2}(b_{\mu\nu}x^\nu + a_{\mu\nu\rho}x^\nu x^\rho). \tag{3.119}$$

Combined with the previous observation, this means that, by a change of coordinates, we can remove certain gravitational fields. This fact is contained in the principle of equivalence of gravitation and inertia that was one of the basic postulates on which Einstein founded GR.

Now, let us consider the gravitational field equation[21]

$$\mathcal{D}^{\mu\nu}(h) = (\chi/c)T^{\mu\nu}_{\rm pp}. \tag{3.120}$$

The energy–momentum tensor has to be evaluated on a solution of the equations of motion of a free particle $\dot{P}^\mu = 0$ plus $P_\mu P^\mu = M^2c^2$. A solution describing the particle at rest at the origin of coordinates is given by $X^i = 0$ and $\xi = X^0 = cT$. We can perform the integral over $\xi$ eliminating the $\delta(X^0 - x^0)$. The energy–momentum tensor becomes[22]

$$T^{\mu\nu}_{\rm pp} = -Mc^2\eta^\mu{}_0\eta^\nu{}_0\delta^{(d-1)}(\vec{x}_{d-1}), \qquad \vec{x}_{d-1} = (x^1,\ldots,x^{d-1}), \tag{3.121}$$

and the gravitational field equations are

$$\mathcal{D}^{00}(h) = -\chi Mc\delta^{(d-1)}(\vec{x}_{d-1}), \qquad \mathcal{D}^{ij}(h) = 0. \tag{3.122}$$

21 Let us recall that $S = (1/c)\int d^dx\,\mathcal{L}$. The Fierz–Pauli action does not acquire any factor of $c$; that is, $c^{-1}\mathcal{L}_{\rm FP} = \frac{1}{4}\partial_\mu h_{\nu\rho}\partial^\mu h^{\nu\rho} - \cdots$.

22 We stress again that, with our conventions, $T^{00}$ is negative-definite (it is minus the energy).

It is convenient to use the variable $\bar{h}_{\mu\nu}$ and the De Donder gauge[23] $\partial^\mu \bar{h}_{\mu\nu} = 0$. Then a solution can be immediately obtained for $d \geq 4$:

$$\bar{h}_{\mu\nu} = -\eta_{\mu 0}\eta_{\nu 0}\frac{\chi M c}{(d-3)\omega_{(d-2)}}\frac{1}{|\vec{x}_{d-1}|^{d-3}}, \tag{3.123}$$

and the non-vanishing components of $h_{\mu\nu}$ are[24]

$$h_{00} \equiv \frac{2}{\chi c^2}\phi, \quad h_{ii} = \frac{2}{(d-3)\chi c^2}\phi, \quad \phi = -\frac{\chi^2 M c^3}{2(d-2)\omega_{(d-2)}}\frac{1}{|\vec{x}_{d-1}|^{d-3}}. \tag{3.124}$$

The notation we have chosen suggests, correctly, that $\phi$ can be identified with the Newtonian potential as in the scalar SRFT of gravity (Eq. (3.7)). Also, as in the case of the scalar SRFT of gravity, we have to see how it affects the motion of test particles in order to confirm it.

*The gravitational field of a massless point-particle.* The action and energy–momentum tensor for a free massless particle moving in Minkowski spacetime are given, respectively, by Eqs. (3.32) and (3.34). After coupling to the gravitational field $h_{\mu\nu}$, the modified action is

$$S[X^\mu(\xi), \gamma(\xi)] = -\frac{p}{2}\int d\xi \sqrt{\gamma}\,\gamma^{-1}[\eta_{\mu\nu} + \chi h_{\mu\nu}(X)]\dot{X}^\mu \dot{X}^\nu. \tag{3.125}$$

This time the gravitational field cannot be absorbed into a redefinition of the worldline metric $\gamma$ (unless $h_{\mu\nu} \propto \eta_{\mu\nu}$) and a massless particle interacts with the gravitational field.

Let us first find the gravitational field produced by a massless particle by solving the equation $\mathcal{D}^{\mu\nu}(h) = (\chi/c)T^{\mu\nu}_{\text{pp}}$, where the energy–momentum tensor has to be calculated for a solution of the equations of motion of the free massless particle $\dot{P}^\mu = P^\mu P_\mu = 0$. It is convenient to use light-cone coordinates $u, v$, and $\vec{x}_{d-2}$ defined by

$$u = \frac{1}{\sqrt{2}}(t-z), \qquad v = \frac{1}{\sqrt{2}}(t+z), \qquad (\vec{x}_{d-2}) = (x^1, \ldots, x^{d-2}), \tag{3.126}$$

where $z \equiv x^{d-1}$, in which the Minkowski metric takes the form

$$(\eta_{\mu\nu}) = \left(\begin{array}{cc|c} 0 & 1 & \\ 1 & 0 & \\ \hline & & -\mathbb{I}_{(d-2)\times(d-2)} \end{array}\right). \tag{3.127}$$

A solution describing the particle moving at the speed of light along the $z$ axis toward $+\infty$ is given by

$$U = \vec{X}_{d-2} = 0, \qquad V = \xi, \qquad \gamma = 1. \tag{3.128}$$

[23] In this case we cannot impose a traceless gauge because the particle's energy–momentum tensor itself is not traceless.

[24] To compare this with Thirring's results [1177] it has to be taken into account that Thirring's energy–momentum tensor is twice ours and that its coupling constant $f = \chi/2$.

For this solution, the energy–momentum tensor (3.34) takes, after integration of one of the Dirac delta-function components, the form

$$T^{\mu\nu}_{\text{pp}} = -\sqrt{2}pc\ell^\mu\ell^\nu \int d\xi\delta(\xi - v)\delta(u)\delta^{(d-2)}(x^i), \qquad \ell^\mu = \delta^\mu{}_v. \tag{3.129}$$

On integrating over $\xi$ and substituting this into the gravitational equation with $h_{\mu\nu}$ in the transverse, traceless gauge, we arrive at the equation

$$\partial^2 h^{\mu\nu} = -\sqrt{2}p\chi\ell^\mu\ell^\nu\delta(u)\delta^{(d-2)}(x^i). \tag{3.130}$$

Only one component of $h^{\mu\nu}$ will be non-trivial. We define the function $K(u, \vec{x}_{d-2})$ by

$$\chi h^{\mu\nu} = 2K(u, \vec{x}_{d-2})\ell^\mu\ell^\nu, \tag{3.131}$$

which satisfies

$$\vec{\partial}^{\,2}_{d-2}K(u, \vec{x}_{d-2}) = \frac{p\chi^2}{\sqrt{2}}\delta(u)\delta^{(d-2)}(\vec{x}_{d-2}). \tag{3.132}$$

A solution can immediately be found. For $d \geq 5$ we have

$$K(u, \vec{x}_{d-2}) = \frac{p\chi^2}{\sqrt{2}(d-4)\omega_{(d-3)}}\frac{1}{|\vec{x}_{d-2}|^{d-4}}\delta(u), \tag{3.133}$$

and, for $d = 4$,

$$K(u, \vec{x}_2) = -\frac{p\chi^2}{\sqrt{2}2\pi}\ln|\vec{x}_2|\delta(u). \tag{3.134}$$

This solution describes a sort of gravitational shock wave. We will see in Chapter 14 that this result, which was found in a linear theory, is actually exact in GR and corresponds to the Aichelburg–Sexl solution found in Ref. [24] by completely different means.

*Motion of massive and massless test particles in a gravitational field.* We can now plug any of the two solutions we have found into the actions (3.116) and (3.125) to find the dynamics of a second test particle of mass $m$ or of a second test massless particle in the gravitational field created by the first particle.[25] Clearly, the most important case is the one corresponding to motion in the field of a massive particle, whose mass we will denote by $M$. We first study the massive case, since it is the one that has a non-relativistic limit. Using the static gauge $\xi = X^0 = cT$ (we write $t$ instead of $T$), we find

$$S_{\text{pp}}[X] = -mc^2\int dt\left\{\sqrt{1-(v/c)^2} + \frac{1}{\sqrt{1-(v/c)^2}}\left[1 + \frac{1}{d-3}\left(\frac{v}{c}\right)^2\right]\frac{\phi}{c^2}\right\}, \tag{3.135}$$

and, in the non-relativistic limit in which we ignore terms of order higher than $\mathcal{O}[(v/c)^4]$ and the constant term, we find

$$S_{\text{pp}}[X] = \int dt\left\{\tfrac{1}{2}mv^2 - m\phi - \tfrac{1}{4}mv^2\left(\frac{v}{c}\right)^2 + \frac{d-1}{2(d-3)}mv^2\frac{\phi}{c^2}\right\}. \tag{3.136}$$

[25] Test particle meaning that the effect of its own gravitational field on the first particle (and, correspondingly, on the gravitational field created by the first particle) can be ignored.

The first term is the kinetic energy of a particle of inertial mass $m$ and the second term is (minus) the potential energy of a particle of *gravitational mass* $m$ moving in a Newtonian gravitational potential $\phi$ (confirming the definition of $\phi$). In this scheme the gravitational and inertial masses of a particle are identical. This is essentially the content of the *principle of equivalence of gravitation and inertia* in its weak form, as we will see in Section 3.3. This was also the case for the scalar SRFT of gravity and it is the consequence of taking the energy–momentum tensor (or its trace) as the source for the gravitational field.

There are also two correction terms. One is the standard relativistic correction to the kinetic energy of a free particle and the other correction represents the contribution of the kinetic energy to the gravitational interaction. A similar term was present in the scalar SRFT of gravity (compare with Eq. (3.28)), but with a coefficient that is different in absolute value and sign. Thus, in this case, all gravitational effects will not vanish in the $v \to c$ limit. On the contrary, we see that, due to the sign of the fourth term, the kinetic energy feels, and also is a source of, gravity, just like the (inertial/gravitational) rest mass.

We can now check that the value of $\phi$ that we have obtained from our relativistic gravitational theory is correct (i.e. coincides with the Newtonian potential created by a mass $M$). In $d = 4$

$$\phi = -\frac{\chi^2 c^3}{16\pi}\frac{M}{|\vec{x}_3|} \quad \Rightarrow \chi^2 = \frac{16\pi G_N^{(4)}}{c^3}, \tag{3.137}$$

where $G_N^{(4)}$ is the Newton constant. The force between the masses $m$ and $M$ is then

$$\vec{F} = -m\vec{\nabla}\phi = -G_N^{(4)} mM \frac{\vec{x}_3}{|\vec{x}_3|^3}. \tag{3.138}$$

For higher dimensions the functional form of $\phi$ is correct. It is (unfortunately) customary in the literature to define in any dimension $d$

$$\chi^2 = 16\pi G_N^{(d)}/c^3, \tag{3.139}$$

even though the rational definition would have been

$$\chi^2 = 2(d-2)\omega_{(d-2)} G_N^{(d)}/c^3.$$

With these conventions the force between the masses $m$ and $M$ is

$$\vec{F} = -m\vec{\nabla}\phi = -\frac{8(d-3)\pi G_N^{(d)} mM}{(d-2)\omega_{(d-2)}} \frac{\vec{x}_{d-1}}{|\vec{x}_{d-1}|^{d-1}}. \tag{3.140}$$

Before we use the fully relativistic action to find corrections to Keplerian orbits, etc., there is one more point worth discussing. We have learned how the Newtonian gravitational field is encoded in the relativistic field $h_{\mu\nu}$. Of course, the relativistic field has more components and at least one more degree of freedom. We can compare this situation with that of the electrostatic field: to build a relativistic theory of the electrostatic field we would have had to use a vector field (with a scalar field we would never have been able to describe attraction between opposite charges and repulsion between like charges) that has more components. Then we could have discovered the magnetic field as part of the electromagnetic field and we would have discovered electromagnetic radiation. Thus, just to see what other non-relativistic terms the full action for general $h_{\mu\nu}$ produces, let us go back to the action Eq. (3.125), choose the static gauge again, and, instead of substituting the $h_{\mu\nu}$ we obtained for a static point-like charge, let us consider a general background gravitational field and define

$$h_{0i} = \frac{1}{\chi c^2} A_i. \tag{3.141}$$

Then, in the non-relativistic limit and ignoring $\mathcal{O}(hv^2)$ terms, we have

$$S_{\rm pp} \sim \int dt \left\{ \tfrac{1}{2} m v^2 - m\phi + \frac{m}{c} \vec{A} \cdot \vec{v} - mc^2 \right\}. \tag{3.142}$$

The new term is a non-Newtonian velocity-dependent interaction. The whole action is identical to the action of a charged particle in an electromagnetic field (see Eq. (12.55)). Then, by analogy, the last term describes the interaction of the particle with the *gravitomagnetic* field, whose existence is one of the main predictions of any relativistic theory of gravitation (including GR) but which has not yet been detected (see e.g. Ref. [360]). The Newtonian term is also called, by analogy, the *gravitostatic* potential.

We are now ready to calculate the corrections to the Keplerian orbits of planets as predicted by this theory. The main effect will be the precession of the perihelion of planets, a secular, cumulative effect that was known before Einstein's construction of GR and whose explanation by this theory was one of its early successes.

Our starting point will be Eq. (3.116) (with $M$ replaced by $m$). We consider only the $d = 4$ case. First, we rewrite this action in terms of an action for a particle moving in the background of an effective metric field $g_{\mu\nu}$:

$$S_{\rm pp}[X^\mu] = -mc \int d\xi \sqrt{g_{\mu\nu}(X) \dot{X}^\rho \dot{X}^\sigma}, \qquad g_{\mu\nu} \equiv \eta_{\mu\nu} + \chi h_{\mu\nu}(X), \tag{3.143}$$

which is equivalent to our original action Eq. (3.116) to first order in $h$. As we explained, this can always be done and it is the basis of Rosenfeld's prescription for calculating a symmetric energy–momentum tensor. The Hamilton–Jacobi equation associated with this action is [883]

$$g^{\mu\nu}(X) \frac{\partial S_{\rm pp}}{\partial X^\mu} \frac{\partial S_{\rm pp}}{\partial X^\nu} - m^2 c^2 = 0, \tag{3.144}$$

and, to first order in $\chi$, it is valid also for our original action. Let us now consider a general static, spherically symmetric metric written as follows:

$$ds^2 = \lambda(r) c^2 dt^2 - \mu(r) dr^2 - R^2(r) d\Omega^2, \qquad d\Omega^2 = d\theta^2 + \sin^2\theta \, d\varphi^2, \tag{3.145}$$

and, knowing that all the dynamics will take place in a plane, let us set $\theta = \pi/2$ from now on. The Hamilton–Jacobi equation takes the form

$$\frac{1}{\lambda c^2} (\partial_t S_{\rm pp})^2 - \frac{1}{\mu} (\partial_r S_{\rm pp})^2 - \frac{1}{\mu R^2} (\partial_\varphi S_{\rm pp})^2 - m^2 c^2 = 0; \tag{3.146}$$

$S_{\rm pp}$ has the form

$$S_{\rm pp} = -Et + l\varphi + W, \tag{3.147}$$

where $W$ is a function only of $r$. On substituting into the above equation, we find that $W$ is given by

$$W = \int dr \sqrt{\mu\lambda^{-1}\left(\frac{E}{c}\right)^2 - \frac{l^2}{R^2} - m^2c^2\mu}. \tag{3.148}$$

In the absence of the gravitational field, $\lambda = \mu = 1$, and $R = r$. On defining the non-relativistic energy $E' = E - mc^2$, assuming that $E' \ll mc^2$ so that

$$\left(\frac{E}{c}\right)^2 - m^2c^2 = m^2c^2\left[\left(\frac{E}{mc^2}\right)^2 - 1\right] = m^2c^2\left[\left(\frac{E'}{mc^2}\right)^2 + 2\frac{E'}{mc^2}\right] \sim 2mE', \tag{3.149}$$

and substituting in the integrand, we obtain $W$ for a classical free particle of energy $E'$. In the presence of a spherically symmetric gravitational field, vanishing at infinity, on making the same approximation $E' \ll mc^2$, expanding

$$\begin{aligned} &\lambda \sim 1 + \frac{\lambda_1}{r} + \frac{\lambda_2}{r^2} + \cdots, \qquad \mu \sim 1 + \frac{\mu_1}{r} + \frac{\mu_2}{r^2} + \cdots, \\ &R^2 \sim r^2\left(1 + \frac{R_1}{r} + \cdots\right), \end{aligned} \tag{3.150}$$

and expanding the expression under the square root to order $\mathcal{O}(1/r^2)$, we find

$$W \sim \int dr \sqrt{2mE' - \frac{\lambda_1 m^2c^2}{r} - \frac{l^2 - [\lambda_1(\lambda_1 - \mu_1) - \lambda_2]m^2c^2}{r^2}}. \tag{3.151}$$

For the solution Eq. (3.124),

$$\mu_1 = -\lambda_1 = R_S \equiv 2MG_N^{(4)}/c^2, \qquad R_1 = 0, \tag{3.152}$$

where we have introduced $R_S$, the *Schwarzschild* or *gravitational* radius of an object of mass $M$, and we obtain from Eq. (3.151)

$$W \sim \int dr \sqrt{2mE' + \frac{R_S m^2c^2}{r} - \frac{l^2 - 2R_S^2 m^2c^2}{r^2}}. \tag{3.153}$$

We should first compare this expression with the Newtonian expression[26]

$$W_{\text{Newtonian}} = \int dr \sqrt{2mE' + \frac{R_S m^2c^2}{r} - \frac{l^2}{r^2}}. \tag{3.154}$$

The second term is the Newtonian potential energy. We see in Eq. (3.153) that there is an $\mathcal{O}(1/r^2)$ relativistic correction to the Newtonian potential. The main consequence will be that the orbits will not be closed and the perihelions will shift. To evaluate the angular difference between two consecutive perihelions we reason, following Ref. [883], as follows. The equation for the orbit can be found from

$$\varphi = \beta_\varphi - \frac{\partial W}{\partial l}. \tag{3.155}$$

[26] We assume that $M \gg m$ so that the reduced mass can be approximated by $m$.

In a complete revolution

$$\Delta\varphi = -\frac{\partial}{\partial l}\Delta W. \tag{3.156}$$

By expanding $W$ around the Newtonian $W_{\text{Newtonian}}$ as a power series in the relativistic correction $\delta = 2R_{\text{S}}^2 m^2 c^2$ and observing that

$$\left.\frac{\partial W}{\partial \delta}\right|_{\delta=0} = -\frac{\partial W}{\partial l^2}, \tag{3.157}$$

we obtain

$$\begin{aligned} W \sim W|_{\delta=0} + \delta \left.\frac{\partial W}{\partial \delta}\right|_{\delta=0} &= W_{\text{Newtonian}} - \delta \frac{1}{2l}\frac{\partial W_{\text{Newtonian}}}{\partial l} \\ &= W_{\text{Newtonian}} - \frac{R_{\text{S}}^2 m^2 c^2}{l}\frac{\partial W_{\text{Newtonian}}}{\partial l}, \end{aligned} \tag{3.158}$$

and

$$\begin{aligned} \Delta W = \Delta W_{\text{Newtonian}} &- \frac{R_{\text{S}}^2 m^2 c^2}{l}\frac{\partial \Delta W_{\text{Newtonian}}}{\partial l} \\ = \Delta W_{\text{Newtonian}} &+ \frac{R_{\text{S}}^2 m^2 c^2}{l}\Delta\varphi_{\text{Newtonian}}, \end{aligned} \tag{3.159}$$

where we have used Eq. (3.156) for $W_{\text{Newtonian}}$. On substituting this into Eq. (3.156) we find

$$\Delta\varphi = \Delta\varphi_{\text{Newtonian}} + \frac{R_{\text{S}}^2 m^2 c^2}{l^2}\Delta\varphi_{\text{Newtonian}}. \tag{3.160}$$

Newtonian orbits are closed, so in one revolution $\Delta\varphi_{\text{Newtonian}} = 2\pi$ and the deviation from the Newtonian value is, according to this theory,

$$\delta\varphi = \frac{2\pi R_{\text{S}}^2 m^2 c^2}{l^2}. \tag{3.161}$$

This result is $\frac{4}{3}$ of the actual value; that is, it is close (better than the value given by the scalar SRFT of gravity) but not quite right. We will have to find a correction to our theory in order to obtain the right value.

The second effect that we want to calculate is the deflection of a light ray (or a massless particle) by the central gravitational field of a massive body, given by Eq. (3.124). To first order in $\chi$ we can simply take the Hamilton–Jacobi equation for a relativistic massive particle, Eq. (3.144), and set $m = 0$ [883]. The resulting equation can be solved as in the massive case with the replacement of $E = -\partial_t S$ by $\omega = -\partial_t S$. For $W$ we obtain the equation

$$W = \int dr \sqrt{\mu\lambda^{-1}\left(\frac{\omega}{c}\right)^2 - \frac{l^2}{R^2}}. \tag{3.162}$$

On expanding $\mu$ and $\lambda$ in powers of $1/r$, we obtain, for the solution Eq. (3.124),

$$W \sim \int dr \sqrt{\left(\frac{\omega}{c}\right)^2 + 2R_{\text{S}}\left(\frac{\omega}{c}\right)^2\frac{1}{r} - \frac{l^2}{r^2}}. \tag{3.163}$$

The $1/r$ term is not present[27] in the Newtonian case[28] and, as we did before, we expand $W$ around its Newtonian value,

$$\begin{aligned} W &\sim \int dr \sqrt{\left(\frac{\omega}{c}\right)^2 - \frac{l^2}{r^2}} + 2R_{\rm S}\frac{\partial}{\partial x}\int dr\sqrt{\left(\frac{\omega}{c}\right)^2 + \left(\frac{\omega}{c}\right)^2\frac{x}{r} - \frac{l^2}{r^2}}\Bigg|_{x=0} \\ &\sim W_{\rm Newtonian} + \frac{R_{\rm S}\omega}{c}\int dr\frac{1}{\sqrt{r^2-\rho^2}} \\ &\sim W_{\rm Newtonian} + \frac{R_{\rm S}\omega}{c}\text{arccosh}\left(\frac{r}{\rho}\right), \end{aligned} \tag{3.164}$$

where $\rho = cl/\omega$ is clearly the minimal value of $r$ in the path of the massless particle. Following Ref. [883], the variation of $W$ when the particle starts from $r = R \gg \rho$, goes through $r = \rho$, and again reaches $r = R$ is

$$\Delta W \sim \Delta W_{\rm Newtonian} + \frac{2R_{\rm S}\omega}{c}\,\text{arccosh}\left(\frac{R}{\rho}\right), \tag{3.165}$$

and, according to Eq. (3.156),

$$\Delta\varphi \sim \frac{\partial}{\partial l}\Delta W_{\rm Newtonian} + \frac{2R}{\rho}\frac{1}{\sqrt{1-\rho/R}} \stackrel{R\rightarrow\infty}{\longrightarrow} \pi + \frac{2R}{\rho}, \tag{3.166}$$

and we find that the deviation from the Newtonian value $\Delta\varphi = \pi$ (which means simply no bending of the light ray) is $\delta\varphi = 2R/\rho$, in good agreement with observation. This is an encouraging result, which indicates that we have found a reasonable relativistic theory of gravitation worth studying in more detail.

At this point, we remember that we still have to answer the second question posed on p. 92. The answer will prompt us to seek and introduce into our theory corrections that will make the prediction for the precession of the perihelion of Mercury agree completely with observations.

### 3.2.4 The consistency problem

The answer to the second question formulated on p. 92 is that, in general, the matter energy–momentum tensor derived from the free-matter Lagrangian is no longer conserved. As explained in Chapter 2, the divergence of the energy–momentum tensor is proportional to the equations of motion derived from the same Lagrangian, but the coupling to gravity changes these equations. This can be seen in the modified massive-particle action of the above example but the real scalar field which we studied in Chapter 2, however, makes a better example.

[27] There are also $1/r^2$ corrections, but we take only the most important one.

[28] The Newtonian case corresponds to a *free* massive particle (i.e. vanishing gravitational potential energy) moving at the speed of light with $2mE' = (\omega/c)^2$.

The modified matter Lagrangian and equation of motion are

$$\begin{aligned}\mathcal{L}_{\text{matter}}(\varphi,h) &= \tfrac{1}{2}(\partial\varphi)^2 + \tfrac{1}{2}\chi h_{\mu\nu}T_{\text{matter}}{}^{\mu\nu}(\varphi)\\ &= \tfrac{1}{2}\left(\eta^{\mu\nu}-\chi\bar{h}^{\mu\nu}\right)\partial_\mu\varphi\partial_\nu\varphi,\\ 0 &= \partial_\mu\left[\left(\eta^{\mu\nu}-\chi\bar{h}^{\mu\nu}\right)\partial_\nu\varphi\right].\end{aligned} \tag{3.167}$$

Using the new equation of motion

$$\partial_\mu T_{\text{matter}}{}^{\mu\nu}(\varphi) = -\partial_\mu\left(\bar{h}^{\mu\rho}\partial_\rho\varphi\right)\partial^\nu\varphi, \tag{3.168}$$

which is not zero, implying that the first-order matter–gravity coupled system is inconsistent. This is the essence of the consistency problem of the Fierz–Pauli theory.

How can we overcome this problem? One solution is to modify the equation of motion Eq. (3.60) by adding a term on the r.h.s. to make it divergenceless again, consistent with the new equation of motion for matter.[29] In fact, since we have modified the matter Lagrangian to include the coupling, the energy–momentum tensor has also been modified and we should replace $T^{\mu\nu}_{\text{matter}}(\varphi)$ by $T^{\mu\nu}_{\text{matter}}(\varphi,h)$ calculated from $\mathcal{L}_{\text{matter}}(\varphi,h)$. This, however, does not work because, if we include the coupling term in the calculation of the energy–momentum tensor, we should also include the Fierz–Pauli Lagrangian: only the *total* energy–momentum tensor (matter plus gravity plus interactions) is conserved. Clearly this is the physical principle behind our problem.

The situation is not too different from the ones encountered in Section 2.5 in the coupling of Abelian and non-Abelian vector fields to matter. There one also has to take into account the contribution of the vector fields themselves to the full Noether currents, since only then are these conserved.

It is reasonable to expect that full consistency can be achieved only if we can derive the new equation of motion from a Lagrangian. However, to make the correction to the energy–momentum tensor appear in the equation of motion, we have to add new terms to the Lagrangian, which introduce new modifications into the energy–momentum tensor, and so on. This problem is present in the pure-gravity system once we accept that it has to

[29] There is another possibility, proposed and studied in Ref. [441], in which consistency without addition of extra terms is recovered at the expense of locality: use on the r.h.s. of the gravitational equation a divergence-free projection of the matter energy–momentum tensor $J_{\mu\nu}$ obtained by applying the manifestly divergence-free Lorentz-covariant projection operator

$$P_{\mu\nu} \equiv \eta_{\mu\nu} - \frac{\partial_\mu\partial_\nu}{\partial^2}. \tag{3.169}$$

The most general divergence-free definition of $J_{\mu\nu}$ is

$$J_{\mu\nu} = (P_{\mu\alpha}P_{\nu\beta} + pP_{\mu\nu}\eta_{\alpha\beta} + qP_{\mu\nu}P_{\alpha\beta})T^{\alpha\beta}_{\text{matter}}. \tag{3.170}$$

Thus, the gravitational field couples only to matter, but in this consistent way. The constants $p$ and $q$ are fixed so as to obtain the right predictions for the classical tests of GR. Only two sets of values of $p$ and $q$ are admissible (all the classical tests are passed by the theory) and for one of them, $q = -p = 1$, the theory can be written in a local form with the introduction of auxiliary fields. In this form, it is shown that there are propagating spin-0 degrees of freedom in the theory. Clearly, this theory cannot pass tests in which the self-coupling of the gravitational field (the *strong form of the principle of equivalence*) is probed and it will predict, for instance, a finite value for the Nordtvedt effect (see e.g. Chapter 3 in Ref. [360] and references therein).

couple to itself through its own energy–momentum *tensor*,[30] customarily denoted by $t^{\mu\nu}$, in the same form and with the same strength as it does to matter. This coupling encodes the *strong form of the principle of equivalence*.

In conclusion, we can say that we have achieved consistency if the equations of motion

$$\mathcal{D}^{\mu\nu}(h) = \chi[T^{\mu\nu}_{\text{matter}}(\varphi, h) + t^{\mu\nu}] \tag{3.171}$$

are consistent with the equations of motion for matter, i.e.

$$\partial_\mu[T^{\mu\nu}_{\text{matter}}(\varphi, h) + t^{\mu\nu}] = 0, \tag{3.172}$$

on-shell. Equivalently, we can say that the corrected theory is consistent if we can derive the above equation of motion from a Lagrangian and derive the total energy–momentum tensor $T^{\mu\nu}_{\text{matter}}(\varphi, h) + t^{\mu\nu}$ from the same Lagrangian.

It is interesting to try to find at least the first correction.[31] We can follow an iterative procedure that stresses the importance of symmetry: the Noether method, explained in Section 2.5 and applied there to the problem of finding consistent coupling of Abelian and non-Abelian vector fields to charged matter. This case will be much more complex but what we will learn will be worth the effort. In the following section we will give a very elegant and economic argument due to Deser [430] to prove that GR is a self-consistent extension of the Fierz–Pauli theory. In this setup, only one iteration will be necessary.

When a solution to a problem is found, the problem of the uniqueness of that solution arises. The results of Weinberg [1238, 1239] and Boulware and Deser [244], mentioned at the beginning of this section, indicate that a quantum massless spin-2 theory can have a Lorentz-invariant quantum S matrix only if it couples to the *total* energy–momentum tensor, including the gravitational energy–momentum tensor whose form, *in the long-wavelength limit*, is the one predicted by GR, Eq. (3.200). Thus, *any interacting quantum theory of a spin-2 particle coincides with GR in the infrared limit*.[32]

The approach that we are going to follow stresses the importance of the conservation of the total energy–momentum tensor and its relation to gauge symmetry, and it is motivated by the hypothesis of the coupling of the spin-2 field to the matter energy–momentum tensor.[33] Other approaches have tried to determine the most general self-interacting classical SRFT of a spin-2 particle, not using as input the coupling to matter and trying to derive the gauge invariance from the requirement of self-consistency of the equations of motion. We will discuss this approach and its results at the end.

### *3.2.5 The Noether method for gravity*

We start with the Fierz–Pauli Lagrangian Eq. (3.84) plus the Lagrangian for a real scalar:

$$\mathcal{L}^{(0)} = \mathcal{L}_{\text{FP}} + \mathcal{L}_{\text{matter}}(\varphi), \qquad \mathcal{L}_{\text{matter}}(\varphi) = \tfrac{1}{2}(\partial\varphi)^2. \tag{3.173}$$

[30] It is a Lorentz tensor. In the full GR theory it will still be a Lorentz tensor but not a tensor under GCTs, and that is why it will be called in that context the *gravity energy–momentum pseudotensor.*

[31] It is enough to obtain the correct value for the precession of the perihelion of Mercury. The derivation of all the corrections is sometimes called *Gupta's program* [517].

[32] Actually, string theory contains corrections to GR in the ultraviolet limit.

[33] The non-interacting theory is perfectly consistent as it stands.

This Lagrangian is invariant under the local gauge transformations given in Eq. (3.95) with parameter $\epsilon^\mu(x)$ and *global* translations with constant parameter $\xi^\mu$ (just like any SRFT):

$$\begin{aligned}\tilde{\delta}x^\mu &= \chi\xi^\mu,\\ \delta h_{\mu\nu} &= -2\partial_{(\mu}\epsilon_{\nu)} - \chi\xi^\lambda\partial_\lambda h_{\mu\nu},\\ \delta\varphi &= -\chi\xi^\lambda\partial_\lambda\varphi.\end{aligned} \tag{3.174}$$

Both symmetries are Abelian. The conserved current associated with the global symmetry can be found by performing a local transformation of the same form in the Lagrangian, as explained in Section 2.5. Up to total derivatives,

$$\begin{aligned}\delta_{\xi(x)}\mathcal{L}_{\text{matter}}(\varphi) &= -\chi\xi^\sigma(x)\partial_\mu T_{\text{can}}{}^\mu{}_\sigma(\varphi),\\ \delta_{\xi(x)}\mathcal{L}_{\text{FP}} &= -\chi\xi^\sigma(x)\partial_\mu t^{(0)}_{\text{can}}{}^\mu{}_\sigma(h),\\ T_{\text{can}}{}^\mu{}_\sigma(\varphi) &= -\partial^\mu\varphi\partial_\sigma\varphi + \tfrac{1}{2}\eta^\mu{}_\sigma(\partial\varphi)^2,\\ t^{(0)}_{\text{can}}{}^\mu{}_\sigma(h) &= -\tfrac{1}{2}\partial^\mu h^{\nu\rho}\partial_\sigma h_{\nu\rho} + \partial^\nu h^{\mu\rho}\partial_\sigma h_{\nu\rho} - \tfrac{1}{2}\partial_\lambda h^{\lambda\mu}\partial_\sigma h\\ &\quad - \tfrac{1}{2}\partial_\sigma h^{\mu\rho}\partial_\rho h + \tfrac{1}{2}\partial^\mu h\partial_\sigma h + \eta^\mu{}_\sigma\mathcal{L}_{\text{FP}}.\end{aligned} \tag{3.175}$$

Here $T_{\text{can}}{}^\mu{}_\sigma(\varphi)$ is the canonical energy–momentum tensor of the real scalar field, and $t^{(0)}_{\text{can}}{}^\mu{}_\sigma(h)$ is that of the gravitational field. The latter is not symmetric. Both are separately conserved on-shell. In particular,

$$\partial_\mu t^{(0)}_{\text{can}}{}^\mu{}_\sigma(h) = -\tfrac{1}{2}\partial_\sigma h_{\nu\rho}\mathcal{D}^{\nu\rho}(h). \tag{3.176}$$

Our physical problem is to couple consistently these two fields, which requires the self-coupling of the gravity field. From the symmetry point of view, following the Noether philosophy, we will have a consistent theory if we manage to construct a theory that is invariant under the *local* versions of these two symmetries. Since, under local transformations, the Lagrangian transforms as above (up to total derivatives that we will systematically ignore here), it is reasonable to expect that we will have invariance to first order in the coupling constant $\chi$ if we introduce an interaction term of the typical form

$$\mathcal{L}^{(1)} = \mathcal{L}^{(0)} + \tfrac{1}{2}\chi h^{\mu\sigma}\Big[T_{\text{can}\,\mu\sigma}(\varphi) + t^{(0)}_{\text{can}\,\mu\sigma}(h)\Big] \tag{3.177}$$

and we identify the two local parameters $\xi^\mu(x) = \epsilon^\mu(x)$. This identification is also suggested by the observation that the point-particle action coupled to gravity is gauge invariant only if we complement the gauge transformation of $h_{\mu\nu}$ with a local transformation of the particle's coordinates. It is clear, however, that this is too naive: from the above Lagrangian one cannot obtain the consistent equation of motion (3.171) because the variation of the interaction term with respect to $h_{\mu\nu}$ does give $\chi T^{(0)}_{\text{can}\,\mu\sigma}(\varphi)$ on the r.h.s. but not the corresponding term for the gravitational field (unless some miracle happens, which it does not). Thus, we will have to look for a term quadratic in derivatives of $h$, symbolically $\mathcal{L}^{(1)}{}_{\mu\nu}(\partial h\partial h)$, and different from the energy–momentum tensor such that the Lagrangian

$$\mathcal{L}^{(1)} = \mathcal{L}^{(0)} + \tfrac{1}{2}\chi h^{\mu\sigma}[T_{\text{can}\,\mu\sigma}(\varphi) + \mathcal{L}^{(1)}{}_{\mu\sigma}] \tag{3.178}$$

produces the wanted equations of motion and is invariant up to $\mathcal{O}(\chi^2)$ under the corrected transformations with local parameter $\epsilon^\mu$,

$$\begin{aligned} \delta_\epsilon^{(1)} h_{\mu\nu} &= -2\partial_{(\mu}\epsilon_{\nu)} - \chi\epsilon^\lambda\partial_\lambda h_{\mu\nu}, \\ \delta_\epsilon^{(1)}\varphi &= -\chi\epsilon^\lambda\partial_\lambda\varphi. \end{aligned} \tag{3.179}$$

This is just the simplest possibility. Clearly there are infinitely many local transformations that reduce to some given global transformations, all of them different by terms proportional to the derivatives of the gauge parameters. We need additional criteria in order to find the right $\partial\epsilon$ terms here. The main property that gauge transformations have to enjoy is that they must generate one and the same algebra both on $\varphi$ and on $h_{\mu\nu}$. Then, given two transformations $\delta^{(1)}$ in Eqs. (3.179) with infinitesimal parameters $\epsilon_1$ and $\epsilon_2$, their commutator, applied to $\varphi$ and $h_{\mu\nu}$, must give another transformation $\delta^{(1)}$ with an infinitesimal parameter $\epsilon_3$ that should be a function of $\epsilon_1$ and $\epsilon_2$; that is,

$$[\delta_{\epsilon_1}^{(1)}, \delta_{\epsilon_2}^{(1)}] = \delta_{\epsilon_3(\epsilon_1,\epsilon_2)}^{(1)}. \tag{3.180}$$

The simple transformations Eqs. (3.179) do not have this property. The problem of finding the most general gauge transformations which have this property, and reduce at order zero in $\chi$ to the normal spin-2 gauge transformations of $h_{\mu\nu}$, was considered by Ogievetsky and Polubarinov in Ref. [1000]. Their conclusion, which is similar (in spite of the different setup) to Wald's in Ref. [1230], is that, apart from the $\chi = 0$ Abelian transformations, the only gauge transformations with the required properties are

$$\begin{aligned} \delta_\epsilon^{(1)} h_{\mu\nu} &= -2\partial_{(\mu}\epsilon_{\nu)} - \chi\left[\epsilon^\lambda\partial_\lambda h_{\mu\nu} + 2\partial_{(\mu}\epsilon^\lambda h_{\nu)\lambda}\right] = -2\partial_{(\mu}\epsilon_{\nu)} - \chi\mathcal{L}_\epsilon h_{\mu\nu}, \\ \delta_\epsilon^{(1)}\varphi &= -\chi\epsilon^\lambda\partial_\lambda\varphi = -\chi\mathcal{L}_\epsilon\varphi, \end{aligned} \tag{3.181}$$

and similarly for matter tensor fields of other ranks. These transformations have the algebra of infinitesimal GCTs

$$[\delta_{\epsilon_1}^{(1)}, \delta_{\epsilon_2}^{(1)}] = \delta_{[\epsilon_1,\epsilon_2]}^{(1)}, \tag{3.182}$$

where $[\epsilon_1, \epsilon_2]$ is the Lie bracket of the two vector fields.

For a scalar field, the Noether current associated with these transformations (which are not symmetries of the action) is the canonical energy–momentum tensor, as in Eqs. (3.175). However, for a vector field with action Eq. (2.56) the Noether current is not the canonical energy–momentum tensor, but the symmetric Belinfante–Rosenfeld energy–momentum tensor Eq. (2.58),

$$\delta_\epsilon^{(1)}\mathcal{L}_{\text{matter}}(A) = -\chi\epsilon^\sigma\partial_\mu T^\mu{}_\sigma(A). \tag{3.183}$$

This sounds promising, because we need symmetric energy–momentum tensors. For the gravitational field, we have (as usual, up to total derivatives)

$$\delta_\epsilon^{(1)}\mathcal{L}_{\text{FP}} = -\chi\epsilon^\sigma\partial_\mu\left[t^{(0)}_{\text{can}}{}^\mu{}_\sigma(h) + \mathcal{D}^\mu{}_\rho(h)h^\rho{}_\sigma\right]. \tag{3.184}$$

The additional term that we obtain vanishes on-shell. In general it is possible to add to a Noether current any term that vanishes on-shell and so we may understand the additional term as a redefinition of the canonical energy–momentum tensor. This redefinition is, however, important. On the one hand, the equations of motion are going to be corrected and therefore the addition of terms vanishing on-shell to first order is going to become meaningful at higher orders and should be considered with care. On the other hand, if we obtain an action that is invariant under the above gauge symmetry, the equations of motion are going to satisfy a gauge identity that is, at the same time, the condition for the invariance of the action. By varying directly the Lagrangian Eq. (3.178) under $\delta_\epsilon^{(1)}$, we find that it will be invariant up to $\mathcal{O}(\chi^2)$ if the gravitational energy–momentum tensor that appears in the equations of motion (3.171) satisfies, to first order in $\chi$,

$$\partial_\mu t^{(0)\,\mu}{}_\sigma(h) = \partial_\mu \left[ t^{(0)\,\mu}_{\text{can}\,\sigma}(h) + \mathcal{D}^\mu{}_\rho(h) h^\rho{}_\sigma \right]. \tag{3.185}$$

On taking explicitly the derivative on the r.h.s., we obtain the gauge identities[34] associated with invariance under $\delta_\epsilon^{(1)}$:

$$\partial_\mu t^{(0)\mu}{}_\sigma(h) = \gamma_{\nu\rho\sigma} \mathcal{D}^{\nu\rho}(h), \qquad \gamma_{\nu\rho\sigma} = \tfrac{1}{2}\{\partial_\nu h_{\rho\sigma} + \partial_\rho h_{\nu\sigma} - \partial_\sigma h_{\nu\rho}\}. \tag{3.186}$$

Thus, if we look for invariance under the gauge transformations $\delta_\epsilon^{(1)}$, the gravitational energy–momentum tensor that we will put on the r.h.s. of Eq. (3.171) has to be of the form

$$t^{(0)}{}_{\mu\sigma}(h) = t^{(0)}_{\text{can}\,\mu\sigma}(h) + \mathcal{D}^\mu{}_\rho(h) h^\rho{}_\sigma + \partial_\rho \Psi^\rho{}_{\mu\sigma}, \tag{3.187}$$

but we can no longer add on-shell-vanishing terms proportional to $\mathcal{D}^\mu{}_\rho(h)$ because then the above gauge identities would not be satisfied. Here we see how the requirement of gauge symmetry constrains the possible energy–momentum tensors. Comparing this situation with our construction of the scalar theory of gravity in which the energy–momentum tensor could be asymmetric and did not have to satisfy any kind of conditions, we are much better off.

Still, the redefined canonical energy–momentum tensor

$$t^{(0)}_{\text{can}\,\mu\sigma}(h) + \mathcal{D}^\mu{}_\rho(h) h^\rho{}_\sigma$$

is not symmetric as we had hoped and we have to find additional terms $\partial_\rho \Psi^\rho{}_{\mu\sigma}$ that cancel out exactly the antisymmetric part of our energy–momentum tensors. There is only one systematic procedure for doing this and only for the canonical one: the Belinfante method explained in Chapter 2, which, unfortunately, requires the addition of on-shell-vanishing terms. Let us, nevertheless, see where we are taken by this method. It is straightforward (but long and tedious) to find

$$\Psi^{\rho\mu}{}_\sigma = -2\partial^{[\rho} h^{\mu]}{}_\beta h^\beta{}_\sigma - 2\partial_\beta h^{[\rho}{}_\sigma h^{\mu]\beta} + \partial^{[\rho} h h^{\mu]}{}_\sigma + \eta_\sigma{}^{[\rho} \partial_\beta h h^{\mu]\beta}. \tag{3.188}$$

The antisymmetric part of the modified canonical tensor $t^{(0)}_{\text{can}\,\mu\sigma} + \partial_\rho \Psi^\rho{}_{\mu\sigma}$ is $-\mathcal{D}_{\rho[\mu} h^\rho{}_{\sigma]}$,

[34] We are using the zeroth-order gauge identity $\partial_\mu \mathcal{D}^{\mu\nu}(h) = 0$, which is, obviously, valid.

and, therefore, on discarding it, we obtain the Belinfante tensor

$$
\begin{aligned}
t^{(0)}_{\text{Bel}\,\mu\sigma} &\equiv t^{(0)}_{\text{can}\,\mu\sigma} + \partial_\rho \Psi^\rho{}_{\mu\sigma} + \mathcal{D}_{\rho[\mu}(h)h^\rho{}_{\sigma]} \\
&= -\tfrac{1}{2}\partial_\mu h^{\nu\rho}\partial_\sigma h_{\nu\rho} - \partial_\nu h_{\rho\mu}\partial^\nu h^\rho{}_\sigma - \partial_\nu h_{\rho\mu}\partial^\rho h^\nu{}_\sigma + 2\partial_{(\mu|}h_{\nu\rho}\partial^\nu h^\rho{}_{|\sigma)} \\
&\quad + \partial_\nu h_{\mu\sigma}\partial_\rho h^{\rho\nu} - \partial_\nu h^\nu{}_{(\mu}\partial_{\sigma)}h + \tfrac{1}{2}\partial^\nu h_{\mu\sigma}\partial_\nu h + \tfrac{1}{2}\partial_\mu h\partial_\sigma h \\
&\quad + \tfrac{1}{2}\eta_{\mu\sigma}\left[\tfrac{1}{2}\partial_\lambda h_{\nu\rho}\partial^\lambda h^{\nu\rho} - \partial_\lambda h_{\nu\rho}\partial^\nu h^{\lambda\rho} - \tfrac{1}{2}(\partial h)^2\right] \\
&\quad + h_{(\mu}{}^\nu\partial_{\sigma)}\partial_\rho h^\rho{}_\nu - h^\nu{}_{(\mu|}\partial_\nu\partial_\rho h_{|\sigma)}{}^\rho - h^\nu{}_{(\mu}\partial^2 h_{\sigma)\nu} \\
&\quad + h^{\lambda\nu}\partial_\lambda\partial_\nu h_{\mu\sigma} - \tfrac{1}{2}\eta_{\mu\sigma}h^{\lambda\nu}\partial_\lambda\partial_\nu h + \tfrac{1}{2}h_{\mu\sigma}\partial^2 h.
\end{aligned}
\tag{3.189}
$$

As expected, this tensor does not satisfy the gauge identities required because of the addition of on-shell-vanishing terms. However,

$$
t^{(0)}_{\text{can}\,\mu\sigma} + \mathcal{D}_{\rho\mu}(h)h^\rho{}_\sigma + \partial_\rho\Psi^\rho{}_{\mu\sigma} = t^{(0)}_{\text{Bel}\,\mu\sigma} + \mathcal{D}_{\rho(\mu}(h)h^\rho{}_{\sigma)} \tag{3.190}
$$

is symmetric and evidently satisfies the gauge identities associated with $\delta^{(1)}_\epsilon$.

Thus, using (more or less) the Belinfante method, we have been able to symmetrize the energy–momentum tensor associated with the gauge transformations $\delta^{(1)}_\epsilon$. This is basically the energy–momentum tensor used by Thirring in Ref. [1177], although he expressed it in the harmonic gauge. As we shall see, it is unacceptable from several points of view.

We can now try to find the Lagrangian correction from which to derive the above energy–momentum tensor as the r.h.s. of the gravitational equation of motion. It should be a term linear in $h_{\mu\nu}$ and quadratic in $\partial_\mu h_{\nu\rho}$. Unfortunately no such term can be found.[35] This means that further modifications $\partial_\rho\Psi^\rho{}_{\mu\sigma}$ are required, but this time they have to be symmetric in the two free indices and they have to lead to a term derivable from a Lagrangian, which is a difficult problem with no guaranteed unique solution.

As an act of desperation we can try to see whether Rosenfeld's energy–momentum tensor has the properties that we are looking for (even if it is not evidently associated with any Noether current). We first rewrite the Fierz–Pauli action in a background metric $\gamma_{\mu\nu}$:

$$
S = \int d^dx\sqrt{|\gamma|}\,\left\{\tfrac{1}{4}\nabla^\rho h^{\mu\nu}\nabla_\rho h_{\mu\nu} - \tfrac{1}{2}\nabla^\rho h^{\mu\nu}\nabla_\mu h_{\rho\nu} + \tfrac{1}{2}\nabla^\mu h\nabla^\lambda h_{\lambda\mu} - \tfrac{1}{4}\nabla^\mu h\nabla_\mu h\right\}, \tag{3.191}
$$

where $\gamma = \det(\gamma_{\mu\nu})$ and $\nabla_\mu$ is the covariant derivative with respect to the Levi-Civita connection $C_{\mu\nu}{}^\rho(\gamma)$ associated with $\gamma_{\mu\nu}$. Now we vary this with respect to the background metric, taking into account that $h_{\mu\nu}$ is assumed to be metric independent. By varying separately

[35] It is easy to see that, to reproduce the term $\eta_{\mu\sigma}(\partial_\rho h\partial^\rho h)$ in the energy–momentum tensor, we need a term of the form $h(\partial_\rho h\partial^\rho h)$ in the Lagrangian, but this term produces another term of the form $\eta_{\mu\sigma}h(\partial_\rho\partial^\rho h)$, which is not present in the energy–momentum tensor.

the terms without and with partial derivatives of the background metric, we obtain

$$\begin{aligned}\delta S = \int d^dx\sqrt{|\gamma|}\delta\gamma_{\alpha\beta}\Big\{ & -\tfrac{1}{4}\nabla^\alpha h_{\nu\rho}\nabla^\beta h^{\nu\rho} - \tfrac{1}{2}\nabla_\nu h_\rho{}^\alpha\nabla^\nu h^{\rho\beta} \\ & + \nabla^{(\alpha|}h_{\nu\rho}\nabla^\nu h^{\rho|\beta)} - \tfrac{1}{2}\nabla_\nu h^{\nu\rho}\nabla_\rho h^{\alpha\beta} - \tfrac{1}{2}\nabla^{(\alpha}h^{\beta)\rho}\nabla_\rho h \\ & - \tfrac{1}{2}\nabla_\nu h^{\nu(\alpha}\nabla^{\beta)}h + \tfrac{1}{2}\nabla^\nu h^{\alpha\beta}\nabla_\nu h + \tfrac{1}{4}\nabla^\alpha h\nabla^\beta h + \tfrac{1}{2}\gamma^{\alpha\beta}\mathcal{L}_{\rm FP} + \frac{\delta C_{\mu\nu}{}^\lambda}{\delta\gamma_{\alpha\beta}}f_\lambda{}^{\mu\nu}\Big\},\end{aligned} \tag{3.192}$$

$$f_\lambda{}^{\mu\nu} = h_{\lambda\rho}\nabla^\rho h^{\mu\nu} - h_\lambda{}^{(\mu}\nabla_\sigma h^{\nu)\sigma} + \tfrac{1}{2}h_\lambda{}^{(\mu}\nabla^{\nu)}h - \tfrac{1}{2}\gamma^{\mu\nu}h_{\lambda\rho}\nabla^\rho h.$$

Using now

$$\delta C_{\mu\nu}{}^\lambda = \tfrac{1}{2}\gamma^{\lambda\tau}\{\nabla_\mu\delta\gamma_{\nu\tau} + \nabla_\nu\delta\gamma_{\mu\tau} - \nabla_\tau\delta\gamma_{\mu\nu}\} \tag{3.193}$$

in the last term and integrating it by parts, it becomes

$$\int d^dx\sqrt{|\gamma|}\,\delta\gamma_{\alpha\beta}\Big\{-\tfrac{1}{2}\nabla_\mu f^{(\alpha|\mu|\beta)} - \tfrac{1}{2}\nabla_\nu f^{(\alpha\beta)\nu} + \tfrac{1}{2}\nabla_\tau f^{\tau\alpha\beta}\Big\}. \tag{3.194}$$

By expanding all the terms and setting $\gamma_{\alpha\beta} = \eta_{\alpha\beta}$, we obtain the Rosenfeld energy–momentum tensor, which turns out to be identical to the symmetrized one in Eq. (3.190).

At this point it looks impossible, without any other guiding principle, to find the right symmetric energy–momentum tensor satisfying the gauge identities and leading to an equation of motion derivable from a Lagrangian. However, we can try to solve our problem starting from the end; that is, by writing down the most general $\mathcal{L}^{(1)}{}_{\mu\sigma}$ quadratic in $\partial_\alpha h_{\beta\gamma}$ and imposing gauge invariance of the total Lagrangian $\mathcal{L}^{(1)} = \mathcal{L}_{\rm FP} + \frac{1}{2}\chi h^{\mu\sigma}\mathcal{L}^{(1)}{}_{\mu\sigma}$ to first order in $\chi$, or, equivalently, using the fact that the equation of motion derived from it satisfies the gauge identities Eq. (3.186). Up to total derivatives, the most general $\mathcal{L}^{(1)}{}_{\mu\sigma}$ is

$$\begin{aligned}\mathcal{L}^{(1)}{}_{\alpha\beta} = {} & a\partial_\alpha h_{\lambda\delta}\partial_\beta h^{\lambda\delta} + b\partial_{(\alpha|}h_{\lambda\delta}\partial^\lambda h^\delta{}_{|\beta)} + c\partial_\lambda h_{\delta\alpha}\partial^\lambda h^\delta{}_\beta \\ & + q\partial_\lambda h^\lambda{}_\alpha\partial_\delta h^\delta{}_\beta + d\partial_\lambda h_{\delta\alpha}\partial^\delta h^\lambda{}_\beta + e\partial_{(\alpha}h_{\beta)\lambda}\partial_\delta h^{\delta\lambda} \\ & + f\partial_\lambda h_{\alpha\beta}\partial_\delta h^{\delta\lambda} + g\partial_{(\alpha}h_{\beta)\lambda}\partial^\lambda h + i\partial_\lambda h_{\alpha\beta}\partial^\lambda h + j\partial_\lambda h^\lambda{}_{(\alpha}\partial_{\beta)}h \\ & + m\partial_\alpha h\partial_\beta h + \eta_{\alpha\beta}\Big[k\partial_\gamma h_{\delta\lambda}\partial^\gamma h^{\delta\lambda} + l\partial_\gamma h_{\delta\lambda}\partial^\delta h^{\gamma\lambda} \\ & + r\partial_\lambda h^{\lambda\delta}\partial_\gamma h^\gamma{}_\delta + n\partial_\gamma h^{\gamma\delta}\partial_\delta h + p(\partial h)^2\Big].\end{aligned} \tag{3.195}$$

We substitute this expression into $\mathcal{L}^{(1)}$, find the equation of motion, identify the gravitational energy–momentum tensor

$$t^{(0)\,\alpha\beta} = \mathcal{L}^{(1)\,\alpha\beta} - \partial_\lambda\left(h^{\mu\sigma}\frac{\partial\mathcal{L}^{(1)}{}_{\mu\sigma}}{\partial\partial_\lambda h_{\alpha\beta}}\right), \tag{3.196}$$

and substitute this into the gauge identity Eq. (3.186) to arrive at the condition

$$\partial_\alpha\mathcal{L}^{(1)\,\alpha\beta} - \partial_\alpha\partial_\lambda\left(h^{\mu\sigma}\frac{\partial\mathcal{L}^{(1)}{}_{\mu\sigma}}{\partial\partial_\lambda h_{\alpha\beta}}\right) = \gamma_{\mu\sigma}{}^\beta\mathcal{D}^{\mu\sigma}(h). \tag{3.197}$$

This is an equation in the constant coefficients $a, b, c, d, \ldots$. To solve it, we first observe that all the terms with the structure $h\partial\partial\partial h$ on the l.h.s. must vanish because they do not occur on the r.h.s. Then, we also impose the vanishing of all the terms with the structure $\partial h(\partial\partial h)^\beta$ on the l.h.s. for the same reason. Finally, we identify the terms with structures $\partial^\beta h(\partial\partial h)$ and $\partial h^\beta(\partial\partial h)$ on both sides of Eq. (3.197). The result can be expressed in terms of two parameters $x$ and $y$, which are left undetermined:

$$\begin{aligned} &a=-\tfrac{1}{2},\ b=2-y,\ c=-1,\ d=1-y,\ e=y,\ f=-1,\\ &g=-1-x,\ i=1,\ j=-1+x,\ k=\tfrac{1}{4},\ l=-\tfrac{1}{2}-x,\\ &m=\tfrac{1}{2},\ n=\tfrac{1}{2},\ p=-\tfrac{1}{4},\ q=y,\ r=x. \end{aligned} \tag{3.198}$$

On substituting these into the general expression for $\mathcal{L}^{(1)}{}_{\mu\sigma}$ and collecting all the terms proportional to the two parameters $x$ and $y$, we obtain

$$\begin{aligned} \mathcal{L}^{(1)}{}_{\mu\sigma} &= \mathcal{L}^{(1)}_{\mathrm{GR}\,\mu\sigma} + \text{total derivatives after contraction with } h^{\mu\sigma},\\ \mathcal{L}^{(1)}_{\mathrm{GR}\,\mu\sigma} &= -\tfrac{1}{2}\partial_\mu h^{\nu\rho}\partial_\sigma h_{\nu\rho} - \partial^\nu h^\rho{}_\mu\partial_\nu h_{\rho\sigma} + \partial^\nu h^\rho{}_{(\mu|}\partial_\rho h_{\nu|\sigma)}\\ &\quad + 2\partial^\nu h^\rho{}_{(\mu}\partial_{\sigma)}h_{\nu\rho} - \partial_{(\mu}h_{\sigma)}{}^\nu\partial_\nu h - \partial^\nu h_{\mu\sigma}\partial_\rho h^\rho{}_\nu\\ &\quad - \partial_\nu h^\nu{}_{(\mu}\partial_{\sigma)}h + \partial_\nu h_{\mu\sigma}\partial^\nu h + \tfrac{1}{2}\partial_\mu h\partial_\sigma h + \eta_{\mu\sigma}\mathcal{L}_{\mathrm{FP}}, \end{aligned} \tag{3.199}$$

an unambiguous, unique, answer (up to total derivatives), which leads to a Lagrangian $\mathcal{L}^{(1)}$ that is invariant to first order in $\chi$ under the gauge transformations Eq. (3.181). The equations of motion are fully determined and the gravitational energy–momentum tensor is the piece of these equations of motion that is proportional to $\chi$, given by Eq. (3.196), or, more explicitly, by

$$\begin{aligned} t^{(0)\,\mu\sigma}_{\mathrm{GR}} &= \tfrac{1}{2}\partial^\mu h_{\lambda\delta}\partial^\sigma h^{\lambda\delta} + \partial_\lambda h_\delta{}^\mu\partial^\lambda h^{\delta\sigma} - \partial_\lambda h_\delta{}^\mu\partial^\delta h^{\lambda\sigma} + \partial_\lambda h^{\mu\sigma}\partial_\delta h^{\delta\lambda}\\ &\quad - 2\partial^{(\mu}h^{\sigma)}{}_\delta\partial_\lambda h^{\lambda\delta} - \tfrac{1}{2}\partial_\lambda h^{\mu\sigma}\partial^\lambda h + \partial^{(\mu}h^{\sigma)\lambda}\partial_\lambda h\\ &\quad + \eta^{\mu\sigma}\Big[-\tfrac{3}{4}\partial_\alpha h_{\beta\gamma}\partial^\alpha h^{\beta\gamma} + \tfrac{1}{2}\partial_\alpha h_{\beta\gamma}\partial^\beta h^{\alpha\gamma} + \partial_\lambda h^{\lambda\alpha}\partial_\delta h^\delta{}_\alpha\\ &\quad - \partial_\lambda h^{\lambda\alpha}\partial_\alpha h + \tfrac{1}{4}\partial_\lambda h\partial^\lambda h\Big]\\ &\quad + h^{\alpha\beta}\Big[\partial_\alpha\partial_\beta h^{\mu\sigma} - 2\partial_\alpha\partial^{(\mu}h^{\sigma)}{}_\beta + \partial^\mu\partial^\sigma h_{\alpha\beta} + 2\eta^{(\sigma}{}_\alpha\hat{\mathcal{D}}^{\mu)}{}_\beta(h)\\ &\quad - \tfrac{1}{2}\eta^\mu{}_\alpha\eta^\sigma{}_\beta\hat{\mathcal{D}}^\rho{}_\rho(h) - \tfrac{1}{2}\eta_{\alpha\beta}\mathcal{D}^{\mu\sigma}(h) - \eta^{\mu\sigma}\hat{\mathcal{D}}_{\alpha\beta}(h)\Big]. \end{aligned} \tag{3.200}$$

This is clearly the energy–momentum tensor we were looking for. It is related to the Rosenfeld energy–momentum tensor Eq. (3.190) by

$$\begin{aligned} t^{(0)\,\mu\sigma}_{\mathrm{GR}} - (t^{(0)\,\mu\sigma}_{\mathrm{can}} + \mathcal{D}^{\rho\mu}(h)h_\rho{}^\sigma + \partial_\rho\Psi^{\rho\mu\sigma}) &\equiv \partial_\rho\Psi^{\rho\mu\sigma}_{\mathrm{GR-Ros}},\\ \Psi^{\rho\mu\sigma}_{\mathrm{GR-Ros}} = \partial_\nu\Big[\eta^{\sigma[\rho}\eta^{\mu]\nu}h^{\lambda\delta}h_{\lambda\delta} + 2\eta^{\nu[\rho}h^{\mu]}{}_\lambda h^{\lambda\sigma} &- 2\eta^{\sigma[\rho}h^{\mu]\lambda}h_\lambda{}^\nu\\ + \eta^{\sigma[\rho}h^{\mu]\nu}h + \eta^{\nu[\mu}h^{\rho]\sigma}h &- \tfrac{1}{2}\eta^{\nu[\mu}\eta^{\rho]\sigma}h^2\Big]. \end{aligned} \tag{3.201}$$

Summarizing: the Noether procedure allows us to find corrections to the free Fierz–Pauli theory order by order in the parameter $\chi$, making it self-consistent to that given order. The procedure seems to be unambiguous and not complicated but is tedious and time-consuming since there is no systematic way of finding the next correction for the energy–momentum tensor (the Belinfante and Rosenfeld prescriptions have proved to be inadequate in this problem). For instance, at second order, we would have to find the second-order corrections to the gauge-transformation rules Eq. (3.181) (quadratic in $h_{\mu\nu}$, linear in the gauge parameter $\epsilon^{\mu}$, with two partial derivatives and satisfying the group property), the second-order gauge identities associated with the invariance of the Lagrangian under those gauge transformations at the given order, and the second-order corrections to the Lagrangian (these would be of the form $h^{\mu\nu}h^{\rho\sigma}\mathcal{L}^{(2)}{}_{\mu\nu\rho\sigma}(\partial h\partial h)$), and then we would have to write the most general $\mathcal{L}^{(2)}{}_{\mu\nu\rho\sigma}(\partial h\partial h)$ symmetric in the pairs $(\mu\nu)$ and $(\rho\sigma)$ and then impose on the corresponding term in the Lagrangian the second-order gauge identity. A more efficient way of finding these corrections, like Deser's, is necessary, but, before we study it, it is worth checking that the correction to the equations of motion implied by this gravitational energy–momentum tensor leads to the right value of the precession of the perihelion of Mercury. We also study some other properties of $t_{\mathrm{GR}}^{(0)\,\mu\sigma}$.

### 3.2.6 Properties of the gravitational energy–momentum tensor $t_{\mathrm{GR}}^{(0)\,\mu\sigma}$

Our first observation concerns the gauge-transformation properties of $t_{\mathrm{GR}}^{(0)\,\mu\sigma}$. The first-order Lagrangian $\mathcal{L}^{(1)}$ is invariant under the gauge transformations Eq. (3.181) to first order in $\chi$. This implies the invariance of the first-order equations of motion

$$\mathcal{D}^{\mu\sigma}(h) = \chi t_{\mathrm{GR}}^{(0)\,\mu\sigma}(h). \tag{3.202}$$

The l.h.s. is invariant under the zeroth-order gauge transformations, and this implies that the zeroth-order variation of the r.h.s. does not vanish and is identical to the first-order variation of the l.h.s.

From the point of view of the linear (Fierz–Pauli) theory, we can say that the energy–momentum tensor $t_{\mathrm{GR}}^{(0)\,\mu\sigma}$ is not invariant under the same (zeroth-order) gauge transformations as those that leave the Lagrangian invariant. Rosenfeld's [58] and other energy–momentum tensors defined in the literature also lack this invariance. In the case of the Rosenfeld energy–momentum tensor, it can be shown that it is not gauge invariant because the Fierz–Pauli theory is not invariant under the zeroth-order gauge transformations (or their covariantization) when it is written in an arbitrary curved background as in Eq. (3.191). This invariance cannot be recovered by adding terms proportional to the Riemann tensor of the background metric [59].

This lack of gauge invariance is in contrast to the invariance of the Rosenfeld energy–momentum tensor of other gauge fields under the relevant gauge transformations. However, while the lack of gauge invariance of the energy–momentum tensors of other gauge theories would be a serious problem in its coupling to gravity, it is not a problem for gravity itself since, as we have seen, only in this way can the full equation of motion be gauge invariant under the full gauge transformations.

On the other hand, the situation is not too different from the one encountered in the Noether procedure for $n$ vector fields in which the Noether current associated with the lowest-order gauge transformations is not invariant under them, and we have to add further corrections.

Once this point has been clarified, we proceed to evaluate the correction to the linear solution for the gravitational field of a point-like massive particle Eq. (3.124) and the gravitational field of a point-like massless particle Eqs. (3.131), (3.133), and (3.134). The general setup used to calculate corrections is the following. From the first-order Lagrangian,[36]

$$\mathcal{L}^{(1)} = \mathcal{L}_{\rm FP} + \mathcal{L}_{\rm matter}(\varphi) + \tfrac{1}{2}(\chi/c)h^{\mu\nu}\Big[\mathcal{L}^{(1)}{}_{\mu\nu}(h) + T_{{\rm matter}\,\mu\nu}(\varphi)\Big], \tag{3.203}$$

we obtain the equations of motion

$$\begin{aligned} \mathcal{D}_{\mu\nu}(h) - (\chi/c)\left[t^{(0)}{}_{\mu\nu}(h) + T_{{\rm matter}\,\mu\nu}(\varphi)\right] &= 0, \\ \mathcal{D}^{(0)}(\varphi) + (\chi/c)\mathcal{D}^{(1)}(\varphi, h) &= 0. \end{aligned} \tag{3.204}$$

To find solutions to these equations, we expand the gravitational and matter fields

$$h_{\mu\nu} = h^{(0)}{}_{\mu\nu} + \chi h^{(1)}{}_{\mu\nu} + \cdots, \qquad \varphi = \varphi^{(0)} + \chi\varphi^{(1)} + \cdots, \tag{3.205}$$

around a solution $(h^{(0)}, \varphi^{(0)})$ of the equations

$$\begin{aligned} \mathcal{D}_{\mu\nu}(h^{(0)}) - (\chi/c)T_{{\rm matter}\,\mu\nu}(\varphi^{(0)}) &= 0, \\ \mathcal{D}^{(0)}(\varphi^{(0)}) &= 0. \end{aligned} \tag{3.206}$$

On substituting the expansion into the first-order equations of motion, taking into account that $t^{(0)}{}_{\mu\nu}(h)$ is quadratic in $h$, $\mathcal{D}^{(0)}(\varphi)$ is linear in $\varphi$, and $\mathcal{D}^{(1)}(h,\varphi)$ is linear both in $h$ and in $\varphi$, and using the above zeroth-order equations, we find, to lowest order in $\chi$,

$$\begin{aligned} \mathcal{D}_{\mu\nu}(h^{(1)}) - \frac{1}{c}t^{(0)}{}_{\mu\nu}(h^{(0)}) &= 0, \\ \mathcal{D}^{(0)}(\varphi^{(1)}) + \frac{1}{c}\mathcal{D}^{(1)}(h^{(0)}, \varphi^{(0)}) &= 0. \end{aligned} \tag{3.207}$$

We are interested in $h^{(1)}$ in $d=4$ and we calculate it by using the Rosenfeld energy–momentum tensor Eq. (3.190) on the linear solution $t^{(0)}_{{\rm Ros}\,\mu\nu}(h^{(0)})$ and the energy–momentum tensor Eq. (3.200) we found by imposing $\delta^{(1)}_{\epsilon}$ gauge invariance on the linear solution $t^{(0)}_{{\rm GR}\,\mu\nu}(h^{(0)})$.

In $d=4$ the solution Eq. (3.124) for a massive particle can be written in the simple form

$$h^{(0)}{}_{\mu\nu} = \delta_{\mu\nu}k, \qquad k = -\frac{\chi M c}{8\pi}\frac{1}{|\vec{x}_3|}. \tag{3.208}$$

On substituting this expression into the energy–momentum tensors, we find

$$\begin{aligned} \frac{1}{c}t^{(0)}_{{\rm Ros}\,\mu\nu}(h^{(0)}) &= -\partial_\mu k\partial_\nu k - \left(\tfrac{3}{2}\eta_{\mu\nu} + 2\delta_{\mu\nu}\right)(\partial k)^2 - (\eta_{\mu\nu} + \delta_{\mu\nu})k\partial^2 k, \\ \frac{1}{c}t^{(0)}_{{\rm GR}\,\mu\nu}(h^{(0)}) &= \partial_\mu k\partial_\nu k - \tfrac{3}{2}(\partial k)^2 + 2k\partial_\mu\partial_\nu k - (\eta_{\mu\nu} - \delta_{\mu\nu})k\partial^2 k. \end{aligned} \tag{3.209}$$

[36] We again restore all factors of $c$.

There are two types of terms: (i) those of the forms $\partial k \partial k$ and $k\partial_\mu\partial_\nu k$, which give rise to finite contributions, and (ii) those of the form $k\partial^2 k$, which give rise to singular contributions ($\partial^2 k \sim \delta^{(3)}(\vec{x}_3)$), but only at the origin $\vec{x}_3 = \vec{0}$, and they have to be absorbed into a renormalization of the source. In the Rosenfeld case, it is just a renormalization of the mass, but in the second case the mass is not renormalized and, instead, the source's energy–momentum tensor has singular terms $T_{\text{source } ij} \sim \delta_{ij}\delta^{(3)}(\vec{x}_3)$, which do not fit within the concept of a point-particle. Since we are mainly interested in obtaining corrections to the gravitational field of massive, finite-sized bodies of spherical symmetry (the Sun, for instance), we opt for hiding this problem in the closet with the other skeletons for the moment and simply ignore these terms.

By taking the derivatives on the r.h.s. of the above expressions, we find

$$\begin{aligned}
\frac{1}{c}t^{(0)}_{\text{Ros }00}(h^{(0)}) &= \frac{7}{2}\frac{R_S^2}{\chi^2}\frac{1}{|\vec{x}_3|^4},\\
\frac{1}{c}t^{(0)}_{\text{Ros }ij}(h^{(0)}) &= -\frac{R_S^2}{\chi^2}\left(\frac{x^i x^j}{|\vec{x}_3|^6} - \frac{1}{2}\delta_{ij}\frac{1}{|\vec{x}_3|^4}\right),\\
\frac{1}{c}t^{(0)}_{\text{GR }00}(h^{(0)}) &= \frac{3}{2}\frac{R_S^2}{\chi^2}\frac{1}{|\vec{x}_3|^2},\\
\frac{1}{c}t^{(0)}_{\text{GR }ij}(h^{(0)}) &= 7\frac{R_S^2}{\chi^2}\left(\frac{x^i x^j}{|\vec{x}_3|^6} - \frac{1}{2}\delta_{ij}\frac{1}{|\vec{x}_3|^4}\right).
\end{aligned} \tag{3.210}$$

To solve these equations, we could try to eliminate all the off-diagonal terms in the energy–momentum tensor by a gauge transformation, as Thirring did in Ref. [1177]. However, as observed in Ref. [986], the gauge transformation that we have to use is $\epsilon_\mu \sim \partial_\mu \ln r$, which does not go to zero at infinity and, furthermore, takes us out of the De Donder gauge in which we want to solve the equation. This clearly invalidates Thirring's results.

We can, however, solve directly the first of Eqs. (3.207) in the De Donder gauge: observe that the r.h.s. of this equation,

$$\partial^2 \bar{h}^{(1)}{}_{\mu\nu} = \frac{1}{c}t^{(0)}{}_{\mu\nu}(h^{(0)}), \tag{3.211}$$

is divergence free. For the Rosenfeld energy–momentum tensor we obtain [986]

$$\bar{h}^{(1)}{}_{00} = -\frac{7}{4}\frac{R_S^2}{\chi^2}\frac{1}{|\vec{x}_3|^2}, \qquad \bar{h}^{(1)}{}_{ij} = -\frac{1}{4}\frac{R_S^2}{\chi^2}\frac{x^i x^j}{|\vec{x}_3|^4}, \tag{3.212}$$

and, by combining this correction and the linear term into $g_{\mu\nu} = \eta_{\mu\nu} + \chi h^{(0)}{}_{\mu\nu} + \chi^2 h^{(1)}{}_{\mu\nu}$, we obtain the spherically symmetric metric

$$ds^2_{\text{Ros}} = \left(1 - \frac{R_S}{r} - \frac{R_S^2}{r^2}\right)c^2dt^2 - \left(1 + \frac{R_S}{r} + \frac{R_S^2}{r^2}\right)dr^2 - \left(1 + \frac{R_S}{r} + \frac{3}{4}\frac{R_S^2}{r^2}\right)r^2 d\Omega^2_{(2)}, \tag{3.213}$$

where we have defined $r = |\vec{x}_3|$ and used $dr = x^i dx^i/|\vec{x}_3|$, $dx^i dx^i = dr^2 + r^2 d\Omega^2$, etc.

For the GR energy–momentum tensor we obtain [986]

$$\bar{h}^{(1)}{}_{00} = -\frac{3}{4}\frac{R_S^2}{\chi^2}\frac{1}{|\vec{x}_3|^2}, \qquad \bar{h}^{(1)}{}_{ij} = \frac{7}{4}\frac{R_S^2}{\chi^2}\frac{x^i x^j}{|\vec{x}_3|^4}, \tag{3.214}$$

and the metric

$$ds^2_{\rm GR} = \left(1 - \frac{R_S}{r} + \frac{1}{2}\frac{R_S^2}{r^2}\right)c^2dt^2 - \left(1 + \frac{R_S}{r} - \frac{1}{2}\frac{R_S^2}{r^2}\right)dr^2 - \left(1 + \frac{R_S}{r} + \frac{5}{4}\frac{R_S^2}{r^2}\right)r^2 d\Omega^2_{(2)}. \tag{3.215}$$

It is, however, more convenient to perform a gauge transformation with parameter

$$\epsilon_i = -R_S x^i/r^2, \tag{3.216}$$

which changes the gauge of $h^{(1)}{}_{\mu\nu}$ and leaves the metric in the form

$$ds^2_{\rm GR} = \left(1 - \frac{R_S}{r} + \frac{1}{2}\frac{R_S^2}{r^2}\right)c^2dt^2 - \left(1 + \frac{R_S}{r} + \frac{1}{2}\frac{R_S^2}{r^2}\right)dr^2 - \left(1 + \frac{R_S}{r} + \frac{1}{4}\frac{R_S^2}{r^2}\right)r^2 d\Omega^2_{(2)}, \tag{3.217}$$

which we will be able to compare later on with the expansion of an exact solution of general relativity (hence the subscript "GR"), Eq. (11.34). In any case, this gauge transformation does not change the coefficient $\lambda_2$ in the expansion Eq. (3.150), which is all we need to recalculate the precession of the perihelion of Mercury. Taking into account the values of $\lambda_2$ obtained, and substituting into Eq. (3.151), we obtain

$$\delta\varphi_{\rm Ros} = 3\pi R_S^2 m^2 c^2/l^2, \qquad \delta\varphi_{\rm GR} = \tfrac{3}{2}\pi R_S^2 m^2 c^2/l^2. \tag{3.218}$$

The second is in agreement with observations. This result gives us more confidence in the self-consistent spin-2 theory that we are constructing and confirms the importance of gauge symmetry, which is a property not enjoyed by the theory built on Rosenfeld's energy–momentum tensor.

Now we can do the same for the massless point-like-particle gravitational field given in Eqs. (3.131), (3.133), and (3.134). We can write the solution in this form:

$$h^{\mu\nu} = k\ell^\mu \ell^\nu, \qquad k = k(u, x). \tag{3.219}$$

It is easy to see that all these terms vanish identically:

$$h = 0, \quad h_{\mu\rho}h^{\mu\nu} = 0, \quad \partial_\mu h^{\mu\nu} = 0, \quad h^{\mu\nu}\partial_\nu h_{\alpha\beta} = 0, \tag{3.220}$$

and all terms in $t^{(0)}_{{\rm GR}\,\mu\nu}(h^{(0)})$ identically vanish. There is neither renormalization of the source nor corrections to the lowest-order solution. The same must also be true if we consider higher-order corrections to the equations of motion, and, therefore, we expect the solution Eqs. (3.131), (3.133), and (3.134) to be an exact solution of the full theory, whatever it is. Actually, we will study this solution in Chapter 14, and we can compare the present solution with the one in Eqs. (14.23) and (14.26).

Now that we have convinced ourselves that the self-consistent spin-2 theory is a good candidate for a theory of gravitation, but is at the same time hard to obtain in a perturbative series, we are prepared to use Deser's argument, which shows that GR is precisely the resummation of the perturbative series we were generating in such a painful way.

### 3.2.7 Deser's argument

In Ref. [430] Deser presented an argument that allows one to see GR as the self-consistent SRFT of a spin-2 particle we were looking for in the sense that, in GR, the gravitational field couples to its own energy–momentum tensor, at least for a certain choice of field variables, Lagrangian, and energy–momentum tensor. The emphasis is on physical consistency rather than on gauge invariance, and, therefore, the choice of energy–momentum tensor is not based on that criterion, as in our previous discussions about the Noether method. These would be weak points if we wanted to take this work as proof of the uniqueness of GR as a solution to our initial problem, but we should understand Deser's work as a proof that GR is *a* solution to our problem from the physical standpoint.

The starting point in Deser's argument is a first-order version of the Fierz–Pauli action that uses two (off-shell) independent fields $\varphi^{\mu\nu}$ and $\Gamma_{\mu\nu}{}^{\rho}$ (see Ref. [1122] for a construction of this action),

$$\boxed{S^{(1)}_{\text{FP}}[\varphi^{\mu\nu},\Gamma_{\mu\nu}{}^{\rho}] = \frac{1}{\chi^2}\int d^dx\,\{-\chi\varphi^{\mu\nu}2\partial_{[\mu}\Gamma_{\rho]\nu}{}^{\rho} + \eta^{\mu\nu}2\Gamma_{\lambda[\mu}{}^{\rho}\Gamma_{\rho]\nu}{}^{\lambda}\}\,,} \tag{3.221}$$

which are Lorentz tensors symmetric in the pair of indices $\mu\nu$. This action is invariant up to a total derivative under the gauge transformations

$$\delta_\epsilon\varphi_{\mu\nu} = -2\partial_{(\mu}\epsilon_{\nu)} + \eta_{\mu\nu}\partial_\rho\epsilon^\rho, \qquad \delta_\epsilon\Gamma_{\mu\nu\rho} = -\chi\partial_\mu\partial_\nu\epsilon_\rho, \tag{3.222}$$

and it is equivalent on-shell to the Fierz–Pauli action because it gives the same equations of motion: the equations of motion of the fields $\varphi^{\mu\nu}$ and $\Gamma_{\mu\nu}{}^{\rho}$ are

$$\begin{aligned}
\chi\frac{\delta S^{(1)}}{\delta\varphi^{\mu\nu}} &= -\partial_{(\mu}\Gamma_{\nu)\rho}{}^{\rho} + \partial_\rho\Gamma_{\mu\nu}{}^{\rho} = 0,\\
\chi^2\frac{\delta S^{(1)}}{\delta\Gamma_{\mu\nu}{}^{\rho}} &= 2\Gamma_{\rho}{}^{(\mu\nu)} - \eta^{\mu\nu}\Gamma_{\rho\lambda}{}^{\lambda} - \eta^{\tau\sigma}\Gamma_{\tau\sigma}{}^{(\mu}\eta^{\nu)}{}_{\rho} - \chi\partial_\rho\varphi^{\mu\nu} + \chi\eta_\rho{}^{(\mu}\partial_\sigma\varphi^{\nu)\sigma} = 0.
\end{aligned} \tag{3.223}$$

The second equation is just a constraint for $\Gamma_{\mu\nu}{}^{\rho}$. On contracting it with $\eta^{\rho\sigma}$, we obtain

$$\eta^{\rho\sigma}\Gamma_{\rho\sigma}{}^{\nu} = \chi\partial_\lambda\varphi^{\lambda\nu}, \tag{3.224}$$

and, on contracting instead with $\eta_{\mu\nu}$ and using the last result, we find

$$\Gamma_{\rho\lambda}{}^{\lambda} = -\frac{1}{d-2}\chi\partial_\rho\varphi, \qquad \varphi = \varphi_\mu{}^{\mu}. \tag{3.225}$$

Using now these two last equations in the equation for $\Gamma_{\mu\nu}{}^{\rho}$, we obtain

$$\Gamma_{\rho\mu\nu} + \Gamma_{\nu\rho\mu} = \chi\partial_\rho h_{\mu\nu}, \qquad h_{\mu\nu} = \varphi_{\mu\nu} - \frac{1}{d-2}\eta_{\mu\nu}\varphi. \tag{3.226}$$

In order to solve for $\Gamma$, we add to this equation ($\rho\mu\nu$) the permutation $\mu\nu\rho$ and subtract the permutation $\nu\rho\mu$, obtaining, finally,

$$\Gamma_{\rho\mu\nu} = \tfrac{1}{2}\chi\{\partial_\rho h_{\mu\nu} + \partial_\mu h_{\nu\rho} - \partial_\nu h_{\rho\mu}\}. \tag{3.227}$$

On substituting this into the equation of motion for $\varphi^{\mu\nu}$, we find that, in terms of the variable $h_{\mu\nu}$, it takes the form

$$\frac{\delta S^{(1)}}{\delta\varphi^{\mu\nu}} = -\tfrac{1}{2}\left[\mathcal{D}_{\mu\nu}(h) - \frac{1}{d-2}\eta_{\mu\nu}\mathcal{D}_\rho{}^\rho(h)\right] = 0, \tag{3.228}$$

which is equivalent to the Fierz–Pauli equation.

Now we want to find a correction $S^{(2)}$ such that the equation of motion becomes

$$\mathcal{D}_{\mu\nu}(h) = \chi t_{\mu\nu}, \tag{3.229}$$

for the total action $S^{(1)} + S^{(2)}$, i.e. we have to obtain

$$\frac{\delta S^{(2)}}{\delta\varphi^{\mu\nu}} = \tfrac{1}{2}\chi\left(t_{\mu\nu} - \frac{1}{d-2}\eta_{\mu\nu}t_\rho{}^\rho\right) \equiv \tau_{\mu\nu}, \tag{3.230}$$

where $t_{\mu\nu}$ is the energy–momentum tensor of $\varphi^{\mu\nu}$ in $S^{(1)}$. We first calculate $t_{\mu\nu}$ using Rosenfeld's prescription. In writing the action $S^{(1)}$ in the background metric $\gamma_{\mu\nu}$, we will assume (and this is one of the key points of this argument) that $\varphi^{\mu\nu}$ is a *tensor density* of weight $w = 1$, i.e. it transforms as $\sqrt{|\gamma|}\, f^{\mu\nu}$, where $f^{\mu\nu}$ is an ordinary tensor. Thus, there is no need to introduce a $\sqrt{|\gamma|}$ factor in front of $\varphi^{\mu\nu}$ and, furthermore, $\varphi^{\mu\nu}$ is independent of the background metric. By expanding the covariant derivatives[37] of $\Gamma_{\mu\nu}{}^{\rho}$, we obtain

$$\begin{aligned} S^{(1)}[\varphi^{\mu\nu}, \Gamma_{\mu\nu}{}^{\rho}, \gamma_{\mu\nu}] = \frac{1}{\chi^2}\int d^dx\, \{&-\chi\varphi^{\mu\nu}\left[2\partial_{[\mu}\Gamma_{\rho]\nu}{}^{\rho} - 2C_{\nu[\mu}{}^{\sigma}\Gamma_{\rho]\sigma}{}^{\rho} + 2C_{\sigma[\mu}{}^{\rho}\Gamma_{\rho]\nu}{}^{\sigma}\right] \\ &+ \sqrt{|\gamma|}\,\gamma^{\mu\nu}2\Gamma_{\lambda[\mu}{}^{\rho}\Gamma_{\rho]\nu}{}^{\lambda}\}. \end{aligned} \tag{3.231}$$

A long calculation gives

$$\begin{aligned} \chi^2 t_{\alpha\beta} = {}& -\frac{2\chi^2}{\sqrt{|\gamma|}}\frac{\delta S^{(1)}}{\delta\gamma^{\alpha\beta}}\bigg|_{\gamma_{\alpha\beta}=\eta_{\alpha\beta}} \\ = {}& -4\Gamma_{\lambda[\alpha}{}^{\rho}\Gamma_{\rho]\beta}{}^{\lambda} + 2\eta_{\alpha\beta}\eta^{\kappa\delta}\Gamma_{\lambda[\kappa}{}^{\rho}\Gamma_{\rho]\delta}{}^{\lambda} \\ & -\chi\partial_\tau\big\{\eta_{\alpha\beta}\varphi^{\mu\nu}\Gamma_{\mu\nu}{}^{\tau} + 2\varphi^{\tau}{}_{(\alpha}\Gamma_{\beta)\rho}{}^{\rho} + \varphi_{\alpha\beta}\Gamma^{\tau}{}_{\rho}{}^{\rho} \\ & -2\varphi^{\tau\mu}\Gamma_{\mu(\alpha\beta)} - 2\varphi^{\mu}_{(\alpha}\left[\Gamma_{\mu|\beta)}{}^{\tau} - \Gamma_{\mu}{}^{\tau}{}_{|\beta)}\right]\big\}, \end{aligned} \tag{3.232}$$

[37] In the end $\Gamma_{\mu\nu}{}^{\rho}$ will not be a general-covariant tensor. However, it is a Lorentz tensor, and Rosenfeld's prescription tells us to replace its partial derivatives by covariant derivatives in order to find Lorentz energy–momentum tensors.

and, thus,

$$\begin{aligned}\tau_{\alpha\beta} &= -2\chi^{-1}\Gamma_{\lambda[\alpha}{}^{\rho}\Gamma_{\rho]\beta}{}^{\lambda} \\ &+ \tfrac{1}{2}\partial_\tau\left\{\frac{1}{d-2}\eta_{\alpha\beta}(\varphi^{\mu\nu}\Gamma_{\mu}{}^{\tau}{}_{\nu} - \tfrac{1}{2}\varphi\Gamma^{\tau}{}_{\rho}{}^{\rho}) - (\varphi^{\tau}{}_{\alpha}\Gamma_{\beta\rho}{}^{\rho} + \varphi^{\tau}{}_{\beta}\Gamma_{\alpha\rho}{}^{\rho} - \varphi_{\alpha\beta}\Gamma^{\tau}{}_{\rho}{}^{\rho})\right. \\ &\left. + \varphi^{\tau\mu}(\Gamma_{\mu\alpha\beta} + \Gamma_{\mu\beta\alpha}) + \varphi^{\mu}{}_{\alpha}(\Gamma_{\mu\beta}{}^{\tau} - \Gamma_{\mu}{}^{\tau}{}_{\beta}) + \varphi^{\mu}{}_{\beta}(\Gamma_{\mu\alpha}{}^{\tau} - \Gamma_{\mu}{}^{\tau}{}_{\alpha})\right\}. \end{aligned} \tag{3.233}$$

The correction to the action with the property (3.230) is precisely

$$S^{(2)} = \frac{1}{\chi^2}\int d^dx\left\{-2\chi\varphi^{\alpha\beta}\Gamma_{\lambda[\alpha}{}^{\rho}\Gamma_{\rho]\beta}{}^{\lambda}\right\}. \tag{3.234}$$

One could naively think that, with this correction, we can obtain only the first term (that quadratic in $\Gamma_{\mu\nu}{}^{\rho}$) in $\tau_{\alpha\beta}$. However, we have to take into account that the equation for $\Gamma_{\mu\nu}{}^{\rho}$ changes and, hence, substituting its solution into the equation for $\varphi^{\mu\nu}$ will give us all the terms we need. Observe also that this correction is cubic in fields whereas the action we started from is quadratic. Finally, observe that this term will not contribute to the energy–momentum tensor: there are no Minkowski metrics here to be replaced by the background metric and there is no need to introduce $\sqrt{|\gamma|}$ because $\varphi^{\mu\nu}$ is, by hypothesis, a tensor density. Thus, if this term really works, we will not need to introduce any more corrections.

For the total action

$$\boxed{S^{(1)} + S^{(2)} = \frac{1}{\chi^2}\int d^dx\,\{-\chi\varphi^{\mu\nu}2\partial_{[\mu}\Gamma_{\rho]\nu}{}^{\rho} + (\eta^{\mu\nu} - \chi\varphi^{\mu\nu})2\Gamma_{\lambda[\mu}{}^{\rho}\Gamma_{\rho]\nu}{}^{\lambda}\},} \tag{3.235}$$

we find the following equations of motion:

$$\begin{aligned}\chi\frac{\delta(S^{(1)}+S^{(2)})}{\delta\varphi^{\mu\nu}} &= -R_{\mu\nu}(\Gamma) = 0, \\ \chi^2\frac{\delta(S^{(1)}+S^{(2)})}{\delta\Gamma_{\mu\nu}{}^{\rho}} &= 2\Gamma_{\rho}{}^{(\mu\nu)} - \eta^{\mu\nu}\Gamma_{\rho\lambda}{}^{\lambda} - \eta^{\lambda\sigma}\Gamma_{\lambda\sigma}{}^{(\mu}\eta^{\nu)}{}_{\rho} - \chi\partial_\rho\varphi^{\mu\nu} + \chi\eta_{\rho}{}^{(\mu}\partial_\sigma\varphi^{\nu)\sigma} \\ &- 2\chi\varphi^{\delta(\mu}\Gamma_{\rho\delta}{}^{\nu)} + \chi\varphi^{\mu\nu}\Gamma_{\rho\sigma}{}^{\sigma} + \chi\varphi^{\lambda\sigma}\Gamma_{\lambda\sigma}{}^{(\mu}\eta^{\nu)}{}_{\rho} = 0,\end{aligned} \tag{3.236}$$

where $R_{\mu\nu}(\Gamma)$ is nothing but the Ricci tensor associated with the connection $\Gamma_{\mu\nu}{}^{\rho}$ given in Eq. (1.33). By defining

$$\mathfrak{g}^{\mu\nu} = \eta^{\mu\nu} - \chi\varphi^{\mu\nu} \tag{3.237}$$

and its inverse $\mathfrak{g}^{\mu\rho}\mathfrak{g}_{\rho\nu} = \mathfrak{g}^{\mu}{}_{\nu} = \delta^{\mu}{}_{\nu}$, which we are going to use as a metric to raise and lower indices, we can write

$$\chi^2\frac{(\delta S^{(1)}+S^{(2)})}{\delta\Gamma_{\mu\nu}{}^{\rho}} = 2\mathfrak{g}^{\delta(\mu}\Gamma_{\rho\delta}{}^{\nu)} - \mathfrak{g}^{\mu\nu}\Gamma_{\rho\delta}{}^{\delta} - \mathfrak{g}^{\lambda\sigma}\Gamma_{\lambda\sigma}{}^{(\mu}\mathfrak{g}^{\nu)}{}_{\rho} + \partial_\rho\mathfrak{g}^{\mu\nu} - \mathfrak{g}_{\rho}{}^{(\mu}\partial_\sigma\mathfrak{g}^{\nu)\sigma} = 0. \tag{3.238}$$

Now we proceed as before: we contract this equation of motion with $\mathfrak{g}_\mu{}^\rho$, giving

$$\Gamma_\lambda{}^{\lambda\nu} = -\partial_\sigma \mathfrak{g}^{\sigma\nu}, \tag{3.239}$$

and then contract with $\mathfrak{g}_{\mu\nu}$, using the last equation, giving

$$\Gamma_{\rho\lambda}{}^\lambda = \frac{1}{d-2}\mathfrak{g}_{\mu\nu}\partial_\rho \mathfrak{g}^{\mu\nu} = \frac{1}{d-2}\partial_\rho \ln|\mathfrak{g}|, \qquad |\mathfrak{g}| \equiv \det \mathfrak{g}^{\mu\nu}. \tag{3.240}$$

We already see here that the expression for $\Gamma_{\mu\nu}{}^\rho$ in terms of $\varphi^{\mu\nu}$ involves an infinite series of terms. This is the reason why one iteration will be enough even though we had expected an infinite series of corrections.

On substituting the last two results into the equation for $\Gamma_{\mu\nu}{}^\rho$, we find

$$\Gamma_{\rho\mu}{}^\sigma \mathfrak{g}_{\sigma\nu}|\mathfrak{g}|^{\frac{1}{d-2}} + \Gamma_{\rho\nu}{}^\sigma \mathfrak{g}_{\sigma\mu}|\mathfrak{g}|^{\frac{1}{d-2}} = \partial_\rho\left(|\mathfrak{g}|^{\frac{1}{d-2}}\mathfrak{g}_{\mu\nu}\right). \tag{3.241}$$

We see that, again, it is convenient to make the following definition:

$$g_{\mu\nu} \equiv |\mathfrak{g}|^{\frac{1}{d-2}}\mathfrak{g}_{\mu\nu} \qquad \Rightarrow \mathfrak{g}^{\mu\nu} = \sqrt{|g|}\, g^{\mu\nu}. \tag{3.242}$$

In terms of the variable $g_{\mu\nu}$, the above equation can be solved using the same procedure as before. The result is that $\Gamma_{\mu\nu}{}^\rho$ is given by the Christoffel symbols associated with the metric $g_{\mu\nu}$ (1.44). The two equations of motion can now be combined into one:

$$R_{\mu\nu}(g) = 0, \tag{3.243}$$

where $R_{\mu\nu}(g)$ is the Ricci tensor associated with the Levi-Civita connection of the metric $g_{\mu\nu}$. This is the *vacuum Einstein equation*, the equation of motion of GR, as we will see.

So far we have not shown that the corrected action has the required self-consistency property. We are now going to do this, and this will allow us to claim that the vacuum Einstein equation is the self-consistent extension of the Fierz–Pauli theory we were looking for, written in terms of the new variable $g_{\mu\nu}$, which turns out to have a geometrical meaning that is really unexpected, given our starting point of view.

We turn back to the equation for $\Gamma_{\mu\nu}{}^\rho$ and try to solve it without the use of $\mathfrak{g}^{\mu\nu}$ and its inverse, by raising and lowering indices with the Minkowski metric again. First, we contract it with $\eta_\mu{}^\rho$, giving

$$(\eta^{\rho\sigma} - \chi\varphi^{\rho\sigma})\Gamma_{\rho\sigma}{}^\nu = \chi\partial_\sigma\varphi^{\sigma\nu}. \tag{3.244}$$

Contracting now with $\eta_{\mu\nu}$ and substituting into it the last result, we obtain

$$\Gamma_{\rho\delta}{}^\delta = -\frac{1}{d-2}\chi\left[-\partial_\rho\varphi + 2\varphi^\delta{}_\mu\Gamma_{\rho\delta}{}^\mu - \varphi\Gamma_{\rho\delta}{}^\delta\right], \tag{3.245}$$

and, on plugging these results into the full equation, we arrive at

$$\begin{aligned}\Gamma_{\rho\mu\nu} + \Gamma_{\nu\rho\mu} &= \chi\partial_\rho h_{\mu\nu} + f_{\rho\mu\nu},\\ f_{\rho\mu\nu} &= 2\chi\varphi^\delta{}_{(\mu|}\Gamma_{\rho\delta|\nu)} - \chi\varphi_{\mu\nu}\Gamma_{\rho\delta}{}^\delta - \frac{1}{d-2}\chi\eta_{\mu\nu}\left[2\varphi^\delta{}_\lambda\Gamma_{\rho\delta}{}^\lambda - \varphi\Gamma_{\rho\delta}{}^\delta\right],\end{aligned} \tag{3.246}$$

which can be "solved" in exactly the same way, giving

$$\Gamma_{\rho\mu\nu} = \tfrac{1}{2}\chi\{\partial_\rho h_{\mu\nu} + \partial_\mu h_{\nu\rho} - \partial_\nu h_{\rho\mu}\} + \tfrac{1}{2}\{f_{\rho\mu\nu} + f_{\mu\nu\rho} - f_{\nu\rho\mu}\}. \tag{3.247}$$

There are Γs on the r.h.s. of this equation, but we do not need anything better (neither can we obtain it without inverting the matrix $\varphi^{\mu\nu}$). On substituting into the equation for $\varphi^{\mu\nu}$, we find

$$\begin{aligned} -\frac{1}{\chi}R_{\mu\nu}(\Gamma) = &-\tfrac{1}{2}\left[\mathcal{D}_{\mu\nu}(h) - \frac{1}{d-2}\eta_{\mu\nu}\mathcal{D}_\rho{}^\rho(h)\right] - 2\chi^{-1}\Gamma_{\lambda[\mu}{}^\rho\Gamma_{\rho]\nu}{}^\lambda \\ &+ \frac{1}{2\chi}\partial_\tau\left\{f_{\nu\mu}{}^\tau + f_\mu{}^\tau{}_\nu - f^\tau{}_{\nu\mu} + \frac{2\chi}{d-2}\eta^\tau{}_{(\nu|}\left[2\varphi^\delta{}_\lambda\Gamma_{|\mu)\delta}{}^\lambda - \varphi\Gamma_{|\mu)\delta}{}^\delta\right]\right\}. \end{aligned} \tag{3.248}$$

On expanding the last term we find agreement with Eqs. (3.228), (3.230), and (3.233).

Let us review this result: we have obtained a first-order action for $\varphi^{\mu\nu}$, which, to lowest order in an expansion in the parameter $\chi$, is equivalent to the free Fierz–Pauli action. The full equation of motion is the equation of motion of GR in vacuum, and we have shown that it is equivalent to the Fierz–Pauli equation with a source that is precisely the conserved energy–momentum tensor of the $\varphi^{\mu\nu}$ field that one derives directly from the action using the Rosenfeld prescription and without having to add any $\partial_\rho\Psi^{\mu\nu\rho}$ term. The action Eq. (3.235) satisfies the physical criterion of self-consistency we asked for and is the action of GR. We have, though, not checked that the Rosenfeld energy–momentum tensor is the Noether current associated with the symmetry of the problem, and we have not discussed the gauge invariance of the result.

In this construction we have found that the objects that appear in the self-consistent action have a simple geometrical interpretation: there is a non-linear function of the field $\varphi^{\mu\nu}$, $g_{\mu\nu}(\varphi)$ that we can interpret as a metric tensor, and the other field in the first-order action $\Gamma_{\mu\nu}{}^\rho$ is the associated Levi-Civita connection on-shell. The equation of motion (the vacuum Einstein equation) states that the metric is Ricci-flat. This equation is covariant under GCTs.

This geometrical interpretation is very powerful because all the infinite non-linear terms that the theory would have when written in terms of $\varphi^{\mu\nu}$ are packaged into objects that can be easily manipulated. However, this new interpretation also goes far beyond the original theory, which was a SRFT in Minkowski spacetime (that is, $\mathbb{R}^d$ equipped with the Minkowski metric). In the original SRFT of gravity, any gravitational field is always defined on $\mathbb{R}^d$ and the Minkowski metric is always there. However, in GR, in many cases it is not possible to find or define a Minkowski metric in the whole spacetime. Furthermore, many metric fields that solve the equations of motion of GR cannot be interpreted as metric fields defined on the whole $\mathbb{R}^d$ but demand spacetime manifolds with different topology. This is particularly true when there are submanifolds on which the metric field is singular. The geometrical theory is therefore much richer because non-trivial topology and causal structures (as in black-hole spacetimes) are the origin of very interesting phenomena (such as Hawking radiation).

Another strong point of the geometrical interpretation is that it provides us with a simple principle to couple matter to gravity: that of *covariance under general coordinate transformations* ("general covariance"), which encodes the equivalence principle in its stronger form. The matter action has to be a scalar under GCTs and this is achieved by introducing the metric field in the right places (precisely as in Rosenfeld's prescription for how to calculate the energy–momentum tensor). A general-covariant matter energy–momentum tensor arises from this formalism in a natural way. However, in the SRFT approach we would have to find, case by case, the corrections to the lowest-order coupled system, which we know is inconsistent. A weakness of the geometrical point of view is that there is no general-covariant energy–momentum tensor of the gravitational field itself. As we have seen, there is a Lorentz-covariant energy–momentum tensor (or pseudotensor) of the gravitational field embedded in the Ricci tensor together with the wave operator, but it cannot be promoted to a general-covariant tensor, as can be understood from the equivalence principle. This obscures the physical interpretation of vacuum solutions of the geometrical theory, which are not strictly speaking *vacuum* solutions since the whole spacetime is filled by a non-trivial gravitational field that acts as a source for itself. We will come back to this point in Chapter 10.

Where does this principle of general covariance come from? We started from a theory with an Abelian gauge symmetry[38] $\delta_\epsilon^{(0)} h_{\mu\nu} = -2\partial_{(\mu}\epsilon_{\nu)}$. We argued that this symmetry was necessary in order to have a consistent theory of free massless spin-2 particles. Then we coupled this free theory to the conserved energy–momentum tensor of the matter fields, saw the need to introduce a self-coupling of the spin-2 field, and argued that the form of the coupling should be dictated by gauge invariance with respect to the corrected transformations $\delta_\epsilon^{(1)} h_{\mu\nu} = -2\partial_{(\mu}\epsilon_{\nu)} - \chi\mathcal{L}_\epsilon h_{\mu\nu}$, which combined the Abelian gauge symmetry we started from and "localized" translations in such a way that the commutator of two $\delta_\epsilon^{(1)}$ infinitesimal transformations gives another $\delta_\epsilon^{(1)}$ transformation. This is the only possible extension of the Abelian $\delta_\epsilon^{(0)}$ transformations [1000, 1230] and the algebra is the algebra of infinitesimal GCTs.[39] In fact, we can easily see how the full gauge

---

[38] Any two of these gauge transformations commute because $\delta_{\xi_1}^{(0)}\delta_{\xi_2} h_{\mu\nu} = \delta_{\xi_1+\xi_2}^{(0)} h_{\mu\nu}$.

[39] As shown in Ref. [1230], it is possible to have a self-coupled spin-2 theory with only "normal spin-2 gauge symmetry" ($\delta_\epsilon^{(0)}$). For instance, we can add to the Fierz–Pauli Lagrangian a term proportional to some (for instance, the third) power of the linearized Ricci scalar

$$\partial^2 h - \partial_\mu\partial_\nu h^{\mu\nu}, \tag{3.249}$$

which is exactly invariant under $\delta_\epsilon^{(0)}$. Of course, the resulting higher-derivative theory cannot have the same interpretation, since the r.h.s. of the equation of motion is not the gravitational energy–momentum tensor. Also, we can couple the linear theory to matter and obtain an interacting theory that is invariant under $\delta_\epsilon^{(0)}$: we just have to add to the free-matter Lagrangian and the Fierz–Pauli Lagrangian an interaction term of the form

$$\int d^dx\, h^{\mu\nu} J_{\mu\nu}(\varphi), \tag{3.250}$$

where $J_{\mu\nu}$ is any symmetric, identically conserved tensor built out of $\varphi$ and its derivatives. This excludes the matter energy–momentum tensor, which is conserved only on-shell. Since $J_{\mu\nu}$ is identically conserved, the modification introduced into the equations of motion for matter by the coupling to gravity is immaterial. Local $J_{\mu\nu}$s can be constructed from local four-index tensors with the symmetries $J_{\mu\rho\nu\sigma} = J_{[\mu\rho][\nu\sigma]} = J_{\nu\sigma\mu\rho}$, defining

$$J_{\mu\nu} = \partial^\rho\partial^\sigma J_{\mu\rho\nu\sigma}. \tag{3.251}$$

transformation $\delta_\epsilon^{(1)} h_{\mu\nu}$ arises from the effect of a GCT on the metric $g_{\mu\nu} = \eta_{\mu\nu} + \chi h_{\mu\nu}$, just by substituting and expanding in powers of $\chi$ the infinitesimal GCT

$$\delta_\epsilon g_{\mu\nu} = -\mathcal{L}_\epsilon g_{\mu\nu} = -2\nabla_{(\mu}\epsilon_{\nu)}.$$

We can consider, then, that the gauge transformations that we found in the Noether method are just the perturbative expansion of GCTs. The self-consistent Fierz–Pauli theory can be considered as a perturbative expansion of the geometrical theory (GR) either in powers of a weak field $\varphi^{\mu\nu}$ or in powers of the dimensional coupling constant $\chi$, which we know from experience is extremely small. From this point of view, the geometrical action is extremely non-perturbative. Thus, the free Fierz–Pauli theory has been the starting point of any attempt to quantize the gravitational interaction in the standard sense (that is, in perturbation theory), as a special-relativistic quantum field theory (SRQFT).[40] Although they were unsuccessful,[41, 42] these attempts have rendered many benefits to the general theory of covariant quantization of gauge field theories,[43] leading, for instance, to Feynman's discovery of ghosts [541].

We know that the theory we have obtained is experimentally correct, although most experiments probe the perturbative regime only to a very low order in $\chi$. However, we have two very different interpretations. Which is the right one? This is a very difficult question which is still open. For many years, the geometrical form of the theory of *general relativity*, which was the first to be obtained (it is clearly easier to obtain) and was proposed by Einstein himself, was accepted as the only possible one. On the other hand, the SRFT form of the theory is necessary in order to study aspects such as the self-coupling of gravity and gravitational waves. Also, any *standard* quantization of GR[44] has to go through the identification of the particles which are going to be the gravitational-field quanta and this takes us to the SRFT. However, the quantization of this theory has been unsuccessful.[45]

---

These $J_{\mu\nu}$s are called *Pauli terms* in Ref. [1240]. It is also possible to define identically conserved non-local $J_{\mu\nu}$s, for instance the non-local projection of the energy–momentum tensor Eq. (3.170). In all these cases, we see that the spin-2 field does not couple to the total energy–momentum tensor and the quantum theories are not consistent, according to Ref. [1239].

[40] Classical references on this approach are Refs. [541, 452, 451, 1216]. More can be found in Refs. [32, 542].

[41] Pure gravity, perturbatively derived from GR, is one-loop convergent but it is divergent at the same order when coupled to matter [1179, 1181, 442, 443, 446, 447] and at two-loop order without coupling to matter [642, 1218].

[42] The situation can change dramatically when the matter coupled to gravity includes fermions and there is (necessarily local) supersymmetry; in other words, in supergravity theories, which are introduced in Chapter 5. In spite of many generic arguments (and others, more specific, based on superspace) the group led by Bern, Dixon, and Kosower showed in Ref. [199] that the supergravity theory with the largest amount of supersymmetry, which is $N = 8$ in $d = 4$, is ultraviolet finite. This surprising result was extended to the four-loop order (and also to $d = 5$) in Ref. [200]. There are cancelations between the contributions of different Feynman diagrams whose origin is unknown (they are not simply due to supersymmetry) and it has been conjectured that the theory may be ultraviolet finite at all orders in perturbation theory (see Ref. [201] for a recent review). Additional evidence has been provided by a similar computation in the (less supersymmetric) $N = 4, d = 4$ theory performed in Ref. [202]. In this case there are ultraviolet divergences but their origin can be traced to an anomaly of the U(1) duality of the theory, which has no analogies in the $N = 8$ theory.

[43] See e.g. Ref. [982].

[44] Other proposals, such as Euclidean quantum gravity and loop quantization, which we may consider less standard, do not need the identification of gravitons.

[45] In view of the fact that the self-consistency of the theory requires the inclusion of an infinite number of

We are tempted to say that any theory of gravity with the same weak-field limit that we could quantize should be the true theory. Actually, this is the main argument in favor of string theory. Meanwhile, it is probably healthy to use both aspects of the theory in the appropriate realms. This is what we intend to do here.

There is a final detail we should comment upon: we have obtained geometrical equations of motion, but the action Eq. (3.235) is not fully geometrical in the sense that it is not invariant under GCTs. We need to add a total derivative term to it:

$$S^{(0)} = \frac{1}{\chi^2}\int d^dx \left\{2\eta^{\mu\nu}\partial_{[\mu}\Gamma_{\rho]\nu}{}^{\rho}\right\}. \tag{3.252}$$

Then, written in terms of $g_{\mu\nu}$ and $\Gamma_{\mu\nu}{}^{\rho}$, the total action $S^{(0)} + S^{(1)} + S^{(2)}$ becomes the first-order Einstein–Hilbert action

$$\boxed{S_{\rm EH}[g_{\mu\nu}, \Gamma_{\mu\nu}{}^{\rho}] = \frac{1}{\chi^2}\int d^dx\sqrt{|g|}\, g^{\mu\nu}R_{\mu\nu}(\Gamma),} \tag{3.253}$$

which can be taken as the starting point of GR. Observe that the equation of motion of $g_{\mu\nu}$ looks different from that of $\mathfrak{g}^{\mu\nu}$, although it is completely equivalent. In Chapter 4 we will see in detail that the equation of motion is

$$G^{\mu\nu} = 0, \tag{3.254}$$

where $G^{\mu\nu} = R^{\mu\nu} - \frac{1}{2}g^{\mu\nu}R$ is the Einstein tensor.

To end this section, we would like to remark that the addition of the total derivative changes the gravity energy–momentum tensor by a $\partial_\rho\Psi^{\rho\mu\nu}$ term. In any case, we are going to need to add total derivatives to this action for various reasons. The issue of the gravitational-field energy–momentum tensor will be studied in Chapter 10.

## 3.3 General relativity

The search for self-consistency of the Fierz–Pauli theory has led us to the Einstein–Hilbert action, Eq. (3.253), which has the property of invariance under GCTs (*general covariance*). This property, elevated to the rank of the *principle of general covariance of relativity* (PGR) is the basis of the theory of general relativity which we want to review here in an extremely condensed way.

The PGR can be considered as the generalization of the principle of (special) relativity and states that all laws of physics should be form invariant (or covariant) under arbitrary changes of reference frame. Since any SRFT requires the use of the standard constant Minkowski metric $(\eta_{\mu\nu}) = \mathrm{diag}(+ - - \cdots -)$, which is invariant only under transformations between inertial frames related by Poincaré transformations, general covariance requires its substitution by a metric field $g_{\mu\nu}(x)$ behaving as a tensor under all GCTs and

higher-order terms, it is legitimate to wonder whether the lack of success is due to the theory itself, to the method of quantization, or just to our inability to quantize in the standard manner a theory with an infinite number of terms without making truncations that would render it inconsistent even if only to some order in $\chi$.

the substitution of all partial derivatives by (general-)covariant derivatives. If the metric field $g_{\mu\nu}$ is simply the Minkowski metric in a non-Cartesian, non-inertial reference frame, then it will always be possible to perform a GCT to a Cartesian, inertial reference frame in which the metric $g_{\mu\nu}$ takes the constant standard form $\eta_{\mu\nu}$. Later on we will extend this property to more general metrics in a local form. Finally, if we do not want to introduce any new fields in using covariant derivatives, we have to use the Levi-Civita connection $\Gamma(g)$.

To see how far we are taken by this principle, we first apply it to point-particles.

*Point-particle actions.* Actions for free point-particles moving in spacetime that are consistent with the PGR and reduce to the special-relativistic action can be readily written by replacing $\eta_{\mu\nu}$ by a general metric $g_{\mu\nu}$ in Eqs. (3.8), (3.29), and (3.32). In this way we obtain the Nambu–Goto-type action for a massive particle in a general background metric $g_{\mu\nu}(x)$,

$$\boxed{S[X^\mu(\xi)] = -Mc\int d\xi\, \sqrt{g_{\mu\nu}(X)\dot{X}^\mu\dot{X}^\nu},} \tag{3.255}$$

which is still proportional to the particle's proper time $s$ as in Eq. (3.22), where the proper time is now defined by

$$\frac{ds}{d\xi} = \sqrt{g_{\mu\nu}(X)\dot{X}^\mu\dot{X}^\nu}. \tag{3.256}$$

The Polyakov-type action for a massive particle is

$$\boxed{S[X^\mu(\xi),\gamma(\xi)] = -\frac{Mc}{2}\int d\xi\sqrt{\gamma}\left[\gamma^{-1}g_{\mu\nu}(X)\dot{X}^\mu\dot{X}^\nu + 1\right],} \tag{3.257}$$

which is also equivalent to the Nambu–Goto-type action upon elimination of the worldline metric $\gamma(\xi)$ through its own equation of motion and the Polyakov-type action for a massless particle:

$$\boxed{S[X^\mu(\xi),\gamma'(\xi)] = -\frac{p}{2}\int d\xi\sqrt{\gamma'}\,\gamma'^{\,-1}g_{\mu\nu}(X)\dot{X}^\mu\dot{X}^\nu.} \tag{3.258}$$

These three actions are manifestly invariant under reparametrizations of the worldline as in the Minkowski case. Thus, there is going to be a constraint associated with this invariance and it is going to coincide with the mass-shell condition in each case:

$$P^\mu P_\mu = M^2c^2, \qquad P_\mu \equiv \frac{\partial\mathcal{L}}{\partial\dot{X}^\mu} \tag{3.259}$$

(evidently $M = 0$ in the massless case).

Furthermore, they all are invariant[46] under spacetime GCTs $X^{\mu} \to X^{\mu\,\prime}(X)$ under which the metric transforms as follows:

$$g_{\mu\nu}(X) = g'_{\rho\sigma}\left[X'(X)\right] \frac{\partial X'^{\,\rho}}{\partial X^{\mu}} \frac{\partial X'^{\,\sigma}}{\partial X^{\nu}}, \tag{3.260}$$

so the combination $g_{\mu\nu}(X)\dot{X}^{\mu}\dot{X}^{\nu}$ is invariant.

Since the Polyakov-type action is equivalent to the Nambu–Goto-type one, let us find the equations of motion derived from the Nambu–Goto-type action. These are

$$\ddot{X}^{\lambda} + \Gamma_{\rho\sigma}{}^{\lambda}\dot{X}^{\rho}\dot{X}^{\sigma} - \frac{d}{d\xi}\left(\ln \gamma^{\frac{1}{2}}\right)\dot{X}^{\lambda} = 0, \tag{3.261}$$

where we have introduced the *induced metric* on the worldline $\gamma$

$$\gamma(\xi) \equiv g_{\mu\nu}(X)\dot{X}^{\mu}\dot{X}^{\nu}. \tag{3.262}$$

We use for it the same symbol as for the auxiliary metric of the Polyakov-type action because the equation of motion for $\gamma$ says that $\gamma$ is the induced worldline metric.

We can easily recognize in Eq. (3.261) the *geodesic equation* written in terms of an arbitrary parameter. Curves obeying that equation are called *geodesics* and are the curves of minimal (occasionally maximal) proper length between two given points. When $\xi = s$ (the proper time), $\gamma = 1$, and the third term in Eq. (3.261) vanishes; then the standard form of the geodesic equation is recovered:

$$\ddot{X}^{\lambda} + \Gamma_{\rho\sigma}{}^{\lambda}\dot{X}^{\rho}\dot{X}^{\sigma} = 0. \tag{3.263}$$

If the metric is $\eta_{\mu\nu}$, it is clear that we recover all the special-relativistic results. Furthermore, if the metric is related to $\eta_{\mu\nu}$ through a GCT, it is clear that we will be describing the same motion (straight lines in spacetime) in some system of curvilinear coordinates. Thus, even though it is difficult to see, the dynamics of the particle will have the same $d(d+1)/2$ conserved quantities associated with the invariances of the Minkowski metric in Cartesian coordinates. Now that we are dealing with general curvilinear coordinates, it is good to have a better characterization of the invariances of a metric and how they are associated with conserved quantities in the dynamics of a particle.

Let us consider the effect of infinitesimal transformations of the form

$$\begin{aligned} \delta X^{\mu} &= \epsilon^{\mu}(X), \\ \delta g_{\mu\nu} &= \epsilon^{\lambda}\partial_{\lambda} g_{\mu\nu}. \end{aligned} \tag{3.264}$$

It is worth stressing that these transformations are not GCTs in spacetime (the metric does not transform in the required way). We know that the action (3.255) is invariant under arbitrary GCTs. However, under the above transformations

$$\delta S_{\rm pp} = -Mc \int d\xi \frac{1}{2\sqrt{g_{\mu\nu}\dot{X}^{\mu}\dot{X}^{\nu}}} \dot{X}^{\rho}\dot{X}^{\sigma}\mathcal{L}_{\epsilon} g_{\rho\sigma}, \tag{3.265}$$

and is invariant only if $\epsilon^{\mu} = \epsilon k^{\mu}$, where $\epsilon$ is an infinitesimal constant parameter and $k^{\mu}$ is a Killing vector satisfying the Killing equation (1.115).

---

[46] By invariant we mean "form invariant" or, as it is sometimes put, *covariant*. This is all that the PGR requires.

These transformations can be exponentiated, giving a one-dimensional group (for one Killing vector) that leaves the action invariant. There is a conserved quantity associated with it via the Noether theorem for global symmetries,[47]

$$P(k) = -\frac{Mc}{\sqrt{g_{\mu\nu}\dot{X}^{\mu}\dot{X}^{\nu}}} k_{\rho}\dot{X}^{\rho}, \tag{3.266}$$

which can be interpreted as the components of the momentum vector in the direction of the Killing vector.

This general framework can be applied to any metric in any coordinate system. We can use it to recover the conserved quantities of a free particle moving in Minkowski spacetime. First of all, observe that we can always use coordinates adapted to a given Killing vector $k^{\mu}$: there is a coordinate $z$ such that $k^{\mu}\partial_{\mu} = \partial_z$ and $\partial_z g_{\mu\nu} = 0$. Then, there is always a coordinate system in which the action does not depend on the variable $Z(\xi)$ and hence the momentum associated with it is conserved as usual. Thus, we are simply encoding known facts in coordinate-independent form. Second, we can check that the above general expression gives the usual linear- and angular-momentum components when we use the Killing vectors of the Minkowski metric:

$$k^{(\mu)\,\rho} = \eta^{\mu\rho}, \qquad k^{([\mu\nu])\,\rho} = 2\eta^{\rho[\mu}x^{\nu]}, \tag{3.267}$$

where $(\mu)$ and $([\mu\nu])$ are labels for the $d$ translational and $d(d-1)/2$ rotational isometries.

To finish this digression, let us mention that the Polyakov-type actions (3.257) and (3.258) are one-dimensional examples of what is called a *non-linear $\sigma$-model*.[48] The non-linearity is associated with the dependence of the metric on the coordinates, which are the dynamical degrees of freedom.

*The principle of equivalence.* Accepting that, according to the PGR, the action (3.255) gives the dynamics of a massive particle in the background given by the metric $g_{\mu\nu}$, we are led to the discovery of the *principle of equivalence of gravitation and inertia* (PEGI) formulated by Einstein in Refs. [484, 485]: consider a near-Minkowskian metric $g_{\mu\nu} = \eta_{\mu\nu} + \chi h_{\mu\nu}$ with $\chi h_{\mu\nu} \ll 1$. It is easy to see that, up to second-order terms, the action is precisely the one given by Eq. (3.116). In particular, we studied the low-velocity (non-relativistic) limit in order to show that the field $h_{\mu\nu}$ describes a gravitational special-relativistic field and how in the non-relativistic limit that action can be interpreted as the non-relativistic action of a particle with potential energy $Mc^2\chi h_{00}/2$ proportional to its inertial mass. This potential energy can be interpreted as a gravitational potential energy, identifying in this way inertial and gravitational masses and $\chi h_{00}$ with $2\phi/c^2$, where $\phi$ is the Newtonian gravitational potential.

Thus, a GCT that, applied to an inertial frame, generates a non-trivial $h_{00}$ can be seen as generating a gravitational field. We are identifying the so-called inertial forces with a gravitational field, and we are saying that we cannot distinguish between them.

---

[47] The components of $k^{\mu}$ are fixed functions of the spacetime coordinates, and the parameters of the group have to be constant over the worldline; they cannot be arbitrary functions of $\xi$. Thus, this is a group of global transformations. These transformations can be gauged by the standard method of introducing a gauge vector and a covariant derivative, as will be seen in due course.

[48] Two useful references on $\sigma$-models are Refs. [802, 289].

Furthermore, all the effects of the gravitational field can be eliminated by going to an inertial frame. This is the essence of the PEGI, which we will refine later. One can distinguish among *weak* (or *Galilean*), medium-strong (or *Einstein's*), and *strong* forms of the PEGI [360].

The weak form applies to the dynamics of one particle (precisely our case): we cannot distinguish whether we are describing its motion in a non-inertial frame or whether there is a gravitational field present. This implies that the inertial and gravitational masses of any particle are always proportional, with a universal proportionality constant that, in carefully chosen units, can be made 1. We have seen that, in the action Eq. (3.116), the inertial and gravitational masses of the particle are identical. We can certainly say that the PGR implies the weak form of the PEGI.

The medium-strong form extends the rank of applicability from the dynamics of one particle to all non-gravitational laws of physics. The introduction of the curved metric $g_{\mu\nu}$ into the actions of all known interactions guarantees that it is also a consequence of the PGR.

The strong form applies to all laws of physics, including gravity itself. There is nothing we can say about this form of the PEGI for the moment, although we already mentioned in the previous section that GR satisfies it, but let us mention that it is not a direct consequence of general covariance, for we can write SRFTs in Minkowski spacetime in general-covariant form.

So far we have considered only $g_{\mu\nu}$s that can be generated by GCTs from $\eta_{\mu\nu}$. Our experience tells us that there are non-trivial gravitational fields in what we would previously have called inertial frames. These gravitational fields must be described by the metric, too. To incorporate them into the theory, we are forced to allow for all kinds of metrics $g_{\mu\nu}$ that cannot be transformed into $\eta_{\mu\nu}$ by a GCT. However, for any arbitrary spacetime metric *at a given point*, there will always be coordinate systems defining *local inertial frames* in which $g_{\mu\nu}$ is equal to $\eta_{\mu\nu}$ at that given spacetime point P and in which the first derivatives of $g_{\mu\nu}$ vanish at that given point[49] and so all the components of the Levi-Civita connection $\Gamma_{\mu\nu}{}^{\rho}(g)$ also vanish at P. One such system is provided by the *Riemann normal coordinates at the point* P (see e.g. Ref. [952]), which have the following properties:

$$\begin{aligned} g_{\mu\nu}(\mathrm{P}) &= \eta_{\mu\nu}, & \partial_\rho g_{\mu\nu}(\mathrm{P}) &= 0, \\ \partial_\rho \partial_\sigma g_{\mu\nu}(\mathrm{P}) &= \tfrac{2}{3} R_{\mu(\rho\sigma)\nu}(\mathrm{P}), & R_{\mu\nu\rho\sigma}(\mathrm{P}) &= 2\partial_\mu \partial_{[\rho} g_{\sigma]\nu}(\mathrm{P}). \end{aligned} \tag{3.268}$$

In this coordinate system, although the first derivatives vanish, the second derivatives do not. In fact, in general, there is no coordinate system in which both first and second derivatives at P vanish, because, otherwise, the Riemann tensor would also vanish at P, which is possible only if it vanishes at P in any coordinate system. This reflects the fact that, although the gravitational field is encoded in the metric tensor, it is actually characterized by the Riemann curvature tensor. The two tensors play a role similar in this respect to those of the vector potential and the field strength in Maxwell electrodynamics. Then, if there is

[49] Any real non-singular metric can be diagonalized at a given point using the appropriate coordinate system, the non-vanishing components being $+1$s and $-1$s. The number of $-1$s minus the number of $+1$s cannot be changed by a further coordinate transformation and is an intrinsic property of the metric, an invariant called the *signature*. Continuity of the metric implies that the signature is the same at all points of spacetime. We consider only metrics of signature $d-2$, the signature of $\eta_{\mu\nu}$ in our conventions.

a non-trivial gravitational field at P, the curvature tensor will not vanish at that point and the same will be true in any coordinates, including Riemann normal coordinates. Thus, to what extent is it true that all gravitational effects can be eliminated in the neighborhood of a point as the PEGI states? The point is that *observable gravitational effects* depend on the product of Riemann tensor components and spacetime coordinate intervals that can be made arbitrarily small, and the upshot of this discussion is that the equivalence between gravitation and inertia will work only *locally* and for observable effects. The PEGI is only local, and we can say that observable effects of the gravitational field can be eliminated locally in a small enough neighborhood of a given point. A longer discussion with examples can be found in Ref. [360].

There is an ongoing debate on the validity and interpretation of the PEGI into which we will not enter. Some interesting criticisms can be found in Ref. [899].

So far we have seen that the PGR forces us to use general spacetime metrics $g_{\mu\nu}$ and that these encode gravitational and inertial forces on the same footing, implying the PEGI in its medium-strong form. Any theory making use of a metric in this way would do the same. Now we want to find an equation of motion for the metric field which determines the dynamics of the gravitational field.

The PGR tells us that the equation of motion of the metric field must be a general tensor equation, $A^{\alpha\beta} = T^{\alpha\beta}_{\text{matter}}$. We have to find a suitable two-index, symmetric, tensor $A^{\alpha\beta} = A^{(\alpha\beta)}$ that is a function only of the metric and its first and second derivatives, $A^{\alpha\beta} = A^{\alpha\beta}(g_{\mu\nu}, \partial_\rho g_{\mu\nu}, \partial_\sigma\partial_\rho g_{\mu\nu})$. Now comes a very important point: in special relativity the matter energy–momentum tensor is always conserved: $\partial_\mu T^{\mu\nu}_{\text{matter}} = 0$. Now we require that the covariant generalization (as required by the PGR) of this equation

$$\nabla_\alpha T^{\alpha\beta}_{\text{matter}} = 0 \tag{3.269}$$

also holds. The connection is the Levi-Civita connection. It has to be stressed that this equation is no longer a conservation equation, as we will explain in detail in Chapter 10. However, it is the covariant generalization of the special-relativistic continuity equation and reduces to it in locally inertial frames and it seems a plausible requirement. Thus, we have to ask that $A^{\alpha\beta}$ be covariantly divergence free.

The problem of finding the most general tensor $A^{\alpha\beta}$ satisfying these conditions was solved by Lovelock in Ref. [903], and the solution is

$$A^\alpha{}_\beta = \sum_{p=1}^{p=[\frac{d+1}{2}]} c_p g^{\alpha\gamma_1\cdots\gamma_{2p}}{}_{\beta\delta_1\cdots\delta_{2p}} R_{\gamma_1\gamma_2}{}^{\delta_1\delta_2} \cdots R_{\gamma_{2p-1}\gamma_{2p}}{}^{\delta_{2p-1}\delta_{2p}} + c_0 g^\alpha{}_\beta, \tag{3.270}$$

where the $c$s are arbitrary constants and the Riemann tensor is the one associated with the Levi-Civita connection. If we also want to recover the Fierz–Pauli equation in the linear limit $g_{\mu\nu} = \eta_{\mu\nu} + \chi h_{\mu\nu}$, $A^{\alpha\beta}$ has to be linear in second derivatives of the metric. In that case, the only possibility is, as originally proven in [1220, 307, 1251],

$$A^{\alpha\beta} = aG^{\alpha\beta} + bg^{\alpha\beta}, \tag{3.271}$$

where $G^{\alpha\beta}$ is the Einstein tensor. This is also the only possibility in $d = 4$ even if we do not impose the requirement of linearity in second derivatives of the metric. The vanishing of its covariant divergence is due to the contracted Bianchi identity $\nabla_\mu G^{\mu\nu} = 0$, when the connection is the Levi-Civita connection as we have assumed, and to the metric compatibility of the same connection.

In the Fierz–Pauli theory there is no room for the constant $b$. Thus, let us set it to zero for the moment. Now we have only to fix the proportionality constant $a$, which can be inferred from the linearized (Fierz–Pauli) theory. We obtain the *Einstein equation*

$$G_{\mu\nu} = \frac{8\pi G_{\rm N}^{(d)}}{c^4} T_{\text{matter}\,\mu\nu}. \tag{3.272}$$

As we will see in detail in Chapter 4, this equation can be derived from the following action principle (up to boundary terms that we will find then):

$$\boxed{S[g_{\mu\nu}, \varphi] = \frac{c^3}{16\pi G_{\rm N}^{(d)}} \int d^d x \sqrt{|g|}\, R(g) + S_{\text{matter}}[g, \varphi],} \tag{3.273}$$

where $R(g)$ is the Ricci scalar for the Levi-Civita connection[50] and the matter energy–momentum tensor is defined by

$$\boxed{T_{\text{matter}}^{\mu\nu} = \frac{2c}{\sqrt{|g|}} \frac{\delta S_{\text{matter}}}{\delta g_{\mu\nu}},} \tag{3.274}$$

justifying Rosenfeld's definition of the energy–momentum tensor.

We may wonder whether the contracted Bianchi identity that supports the above equations of motion[51] is associated with some sort of gauge symmetry. Indeed, the group of GCTs can be understood as an infinite-dimensional continuous group of local transformations, and from the invariance of the action under this group we will derive a gauge identity (the contracted Bianchi identity) and conserved currents in Chapter 10.

In this quick review we have seen how to use the PGR to construct the theory of GR. We have introduced the minimal number of elements necessary for a general-covariant theory, but there are additional objects that one can introduce. One of them is torsion. We will see that it can be introduced consistently in the presence of fermions without adding further degrees of freedom to the theory. Another object compatible with general covariance that we can add to the theory is a *cosmological constant*, which is basically the constant $b$ that we discarded on the basis of its absence from the Fierz–Pauli theory. It occurs as the

---

[50] Thus, this is the second-order Einstein–Hilbert action that one obtains from the first-order action (3.253) by eliminating $\Gamma_{\mu\nu}{}^{\rho}$ through its equation of motion. This action is quadratic in first-order derivatives of the metric but contains second-order derivatives, which, however, appear in total derivatives.

[51] Einstein himself proposed first $R_{\mu\nu} = (8\pi G_{\rm N}^{(d)}/c^4) T_{\text{matter}\,\mu\nu}$ until he realized the inconsistency of this equation with the covariant "conservation" of the energy–momentum tensor.

constant $\Lambda$ in the action[52]

$$S[g_{\mu\nu}, \varphi] = \frac{c^3}{16\pi G_N^{(d)}} \int d^dx \sqrt{|g|}\, [R(g) - (d-2)\Lambda] + S_{\text{matter}}[\varphi], \tag{3.275}$$

leading to the *cosmological Einstein equation*

$$G_{\mu\nu} + \frac{d-2}{2}\Lambda g_{\mu\nu} = \frac{8\pi G_N^{(d)}}{c^4} T_{\text{matter}\,\mu\nu}. \tag{3.276}$$

This constant can be understood in various ways: first of all, we may think of some kind of matter distributed in spacetime in such a way that its energy–momentum tensor is precisely $T_{\mu\nu} = -[(d-2)/16\pi G_N^{(d)}]\Lambda g_{\mu\nu}$. It is commonly accepted that the vacuum energy of the quantum fields gives $\Lambda$. The value of $\Lambda$ obtained according to this prescription is many orders of magnitude bigger than the experimental upper bound. This huge disagreement is known as the "cosmological-constant problem" (see e.g. Ref. [1242]).

We can also understand $\Lambda$ as a fundamental constant of Nature. Then, the question of its smallness (if it is not zero) need not be such a big problem, at least not bigger than the question of why the values of the other fundamental constants of Nature are what they are, some of them being really small (such as the Planck length).

The main effect of the cosmological constant is to change the vacuum of the theory, which in this context we can define as the maximally symmetric solution of the classical equations of motion with all matter fields set to zero. In the presence of a cosmological constant, Minkowski spacetime is no longer a vacuum solution and the new maximally symmetric solutions are *de Sitter* ($dS_d$) spacetime for positive $\Lambda$ and *anti-de Sitter* ($AdS_d$) spacetime for negative $\Lambda$. Now, in the weak-field limit, we should be considering perturbations around the new vacuum $\bar{g}_{\mu\nu}$ as follows: $g_{\mu\nu} = \bar{g}_{\mu\nu} + \chi h_{\mu\nu}$. The theory that one obtains by linearizing the cosmological Einstein theory is not the Fierz–Pauli theory in Minkowski spacetime. This is why there was no room for the constant $b$ in considering that limit. In Section 3.4 we study precisely the linearized theory obtained by expanding the cosmological Einstein equation around a general vacuum metric $\bar{g}_{\mu\nu}$ that can be curved or can even be the Minkowski metric in arbitrary coordinates.

## 3.4 The Fierz–Pauli theory in a curved background

In the preceding sections we have constructed a theory of spin-2 particles moving in the background of Minkowski spacetime in Cartesian coordinates (constant, diagonal $\eta_{\mu\nu}$). In this section we want to try to extend this construction to other backgrounds. As we have seen, the Fierz–Pauli theory can also be considered as the lowest-order perturbation theory of GR over Minkowski spacetime. Here we will construct extensions of the Fierz–Pauli theory by constructing the lowest-order perturbation theory of GR over a given background spacetime metric that is a vacuum solution of the full GR theory.

[52] The dimension-dependent factor has been chosen in order to have the equation $R_{\mu\nu} = \Lambda g_{\mu\nu}$ in vacuum.

We may wonder whether it is possible to write the Fierz–Pauli theory (or a generalization thereof) in an arbitrary curved background metric. Such a construction would be necessary, for instance, in order to couple a spin-2 particle to GR in the same way as we couple scalars or vector fields. Such a theory would necessarily contain the same terms as the flat spacetime one but covariantized so that it has the right flat-spacetime limit and can also contain additional terms proportional to the curvature of the background metric that vanish in that limit. The guiding principle determining whether to introduce these terms is gauge invariance: the theory should be invariant under the general-covariantized gauge transformations $\delta^{(0)} h_{\mu\nu} = -2\bar{\nabla}_{(\mu}\epsilon_{\nu)}$. However, it can be shown that it is not possible to write this gauge-invariant theory, no matter what curvature terms one introduces [59]. This is one of the indications of the problems one encounters in trying to couple spin-2 particles to (GR) gravity.

While a Fierz–Pauli theory in a general curved background does not exist, such a theory does exist in backgrounds that solve the vacuum (cosmological) Einstein equations, and this is the theory we are going to obtain here. Its construction is useful for many purposes. We will use it in constructing conserved quantities in spacetimes with arbitrary asymptotics and we can use it to work with the Minkowskian Fierz–Pauli theory in arbitrary coordinates. However, apart from these prosaic applications it will also teach us interesting things, e.g. how to define masslessness in curved backgrounds.

To be as general as possible we will include a cosmological-constant term from the beginning as in Eq. (3.275).

### *3.4.1 Linearized gravity*

Let us first describe the setup: we consider a spacetime metric $g_{\mu\nu}$ that solves the $d$-dimensional cosmological Einstein equations for some matter energy–momentum tensor $T^{\mu\nu}_{\text{matter}}$ (here we set $c = 1$, as usual),

$$G_c{}^{\mu\nu} = 8\pi G_{\text{N}}^{(d)} T^{\mu\nu}_{\text{matter}}, \tag{3.277}$$

where $G_c{}^{\mu\nu}$ is the *cosmological Einstein tensor*,

$$G_c{}^{\mu\nu} \equiv G^{\mu\nu} + \frac{d-2}{2}\Lambda g^{\mu\nu}. \tag{3.278}$$

The metric $g_{\mu\nu}$ must be such that we can consider it as produced by a small perturbation of the background metric $\bar{g}_{\mu\nu}$, i.e. we can write

$$g_{\mu\nu} = \bar{g}_{\mu\nu} + h_{\mu\nu}, \tag{3.279}$$

where the perturbation $h_{\mu\nu}$ goes to zero at infinity fast enough that the metric $g_{\mu\nu}$ is asymptotically $\bar{g}_{\mu\nu}$. Furthermore, $h_{\mu\nu}$ and its derivatives are assumed to be small enough that we can ignore higher-order terms.[53]

Usually, the background metric $\bar{g}_{\mu\nu}$ will be the vacuum metric, i.e. a maximally symmetric solution of the vacuum Einstein equations

$$\bar{G}_c{}^{\mu\nu} = 0. \tag{3.280}$$

---

[53] Here we have absorbed the coupling constant $\chi = \sqrt{16\pi G_{\text{N}}^{(d)}}$ into $h_{\mu\nu}$.

Therefore, the metrics $g_{\mu\nu}$ that we consider describe, in gravitational language, isolated systems. There are no matter sources of the gravitational field at infinity. In the absence of a cosmological constant, the vacuum metric $\bar{g}_{\mu\nu} = \eta_{\mu\nu}$, the Minkowski metric, and the metrics $g_{\mu\nu}$ will be *asymptotically flat*. With positive (negative) cosmological constant, the (maximally symmetric) vacuum solution is the *(anti-)de Sitter* ((A)dS$_d$) spacetime and the metrics $g_{\mu\nu}$ will be *asymptotically (anti-)de Sitter*. However, we will keep the background metric completely general in order to cover other interesting cases in which a solution $g_{\mu\nu}$ goes asymptotically to a $\bar{g}_{\mu\nu}$ that is not the vacuum solution or even a solution of the vacuum Einstein equations. Thus, we will use only Eqs. (3.277) and (3.279) to find the equation satisfied by the perturbation $h_{\mu\nu}$. Later on, we will impose the condition that the background metric solves the Einstein equation (3.280).

The first thing we have to do is to expand this equation in powers of the perturbation $h_{\mu\nu}$. The perturbation can be treated as a tensor on the background manifold. Then, it is natural to lower and raise its indices (and those of all tensors) with the background metric $\bar{g}_{\mu\nu}$ and its inverse $\bar{g}^{\mu\nu}$. In particular, $h^{\mu\nu} = \bar{g}^{\mu\rho}\bar{g}^{\nu\sigma}h_{\rho\sigma}$ is not the inverse of $h_{\mu\nu}$ (which need not exist) and we also define $h = \bar{g}^{\mu\nu}h_{\mu\nu}$. All barred covariant derivatives are also taken with respect to the background metric's Levi-Civita connection $\bar{\Gamma}_{\mu\nu}{}^{\rho}$. We find

$$\begin{aligned} g^{\mu\nu} &= \bar{g}^{\mu\nu} - h^{\mu\nu} + \mathcal{O}(h^2), \\ \Gamma_{\mu\nu}{}^{\rho} &= \bar{\Gamma}_{\mu\nu}{}^{\rho} + \gamma_{\mu\nu}{}^{\rho} + \mathcal{O}(h^2), \\ R_{\mu\nu\rho}{}^{\sigma} &= \bar{R}_{\mu\nu\rho}{}^{\sigma} + 2\bar{\nabla}_{[\mu}\gamma_{\nu]\rho}{}^{\sigma} + \mathcal{O}(h^2), \end{aligned} \tag{3.281}$$

with

$$\gamma_{\mu\nu}{}^{\rho} = \tfrac{1}{2}\bar{g}^{\rho\sigma}\{\bar{\nabla}_{\mu}h_{\sigma\nu} + \bar{\nabla}_{\nu}h_{\mu\sigma} - \bar{\nabla}_{\sigma}h_{\mu\nu}\}. \tag{3.282}$$

This equation is essentially the equation that gives the variation of the Levi-Civita connection $\delta\Gamma_{\mu\nu}{}^{\rho}$ ($\equiv \gamma_{\mu\nu}{}^{\rho}$) under an arbitrary variation of the metric[54] $\delta g_{\mu\nu}$ ($\equiv h_{\mu\nu}$),

$$\delta\Gamma_{\mu\nu}{}^{\rho} = \tfrac{1}{2}g^{\rho\sigma}\{\nabla_{\mu}\delta g_{\sigma\nu} + \nabla_{\nu}\delta g_{\sigma\rho} - \nabla_{\sigma}\delta g_{\mu\nu}\}. \tag{3.284}$$

Now we can find the expansion of $R_{\mu\nu\rho}{}^{\sigma}$ to first order in $h_{\mu\nu}$ using *Palatini's identity*, which gives the variation of the curvature tensor under an arbitrary variation of the connection,

$$\delta R_{\mu\nu\rho}{}^{\sigma} = +2\nabla_{[\mu}\delta\Gamma_{\nu]\rho}{}^{\sigma}. \tag{3.285}$$

Palatini's identity follows from Eqs. (1.31) and (1.36), on setting the torsion equal to zero, identifying $\tau_{\mu\nu}{}^{\rho}$ with $\delta\Gamma_{\mu\nu}{}^{\rho}$, and keeping only the linear terms. We stress that, unlike $\Gamma_{\mu\nu}{}^{\rho}$, the variation $\delta\Gamma_{\mu\nu}{}^{\rho}$ is a true tensor and its covariant derivative is well defined.[55]

---

[54] For further use we quote here the generalization of this equation when there is torsion present:

$$\begin{aligned} \delta\Gamma_{\alpha\beta}{}^{\gamma} = {} & \tfrac{1}{2}g^{\gamma\delta}\{\nabla_{\alpha}\delta g_{\beta\delta} + \nabla_{\beta}\delta g_{\alpha\delta} - \nabla_{\delta}\delta g_{\alpha\beta}\} \\ & + \tfrac{1}{2}\left\{g^{\delta\gamma}g_{\sigma\beta}\delta T_{\alpha\delta}{}^{\sigma} + g^{\delta\gamma}g_{\sigma\alpha}\delta T_{\beta\delta}{}^{\sigma} - \delta T_{\alpha\beta}{}^{\gamma}\right\}. \end{aligned} \tag{3.283}$$

[55] Also for further use, here we quote the formula valid for a general connection:

$$\delta R_{\mu\rho} = \nabla_{\mu}\delta\Gamma_{\nu\rho}{}^{\nu} - \nabla_{\nu}\delta\Gamma_{\mu\rho}{}^{\nu} - T_{\mu\nu}{}^{\lambda}\delta\Gamma_{\lambda\rho}{}^{\nu}. \tag{3.286}$$

For the variation of $\Gamma_{\mu\nu}{}^{\rho}$ that we have just found, we obtain

$$R_{\mu\nu\rho}{}^{\sigma} = \bar{R}_{\mu\nu\rho}{}^{\sigma} + \bar{g}^{\sigma\lambda}\{\bar{\nabla}_{[\mu}\bar{\nabla}_{\nu]}h_{\lambda\rho} + \bar{\nabla}_{[\mu|}\bar{\nabla}_{\rho}h_{|\nu]\lambda} - \bar{\nabla}_{[\mu|}\bar{\nabla}_{\lambda}h_{|\nu]\rho}\} + \mathcal{O}(h^2), \quad (3.287)$$

and, on contracting the indices $\sigma$ and $\nu$, we find[56]

$$R_{\mu\rho} = \bar{R}_{\mu\rho} + \tfrac{1}{2}\left\{\bar{\nabla}^2 h_{\mu\rho} - 2\bar{\nabla}^{\lambda}\bar{\nabla}_{(\mu}h_{\rho)\lambda} + \bar{\nabla}_{\mu}\bar{\nabla}_{\rho}h\right\} + \mathcal{O}(h^2). \quad (3.288)$$

On contracting with $g^{\mu\rho} = \bar{g}^{\mu\rho} - h^{\mu\rho}$, we find that the Ricci scalar is given by

$$R = \bar{R} - \bar{R}_{\lambda\sigma}h^{\lambda\sigma} + \bar{\nabla}^2 h - \bar{\nabla}_{\lambda}\bar{\nabla}_{\sigma}h^{\lambda\sigma} + \mathcal{O}(h^2). \quad (3.289)$$

To find the cosmological Einstein tensor we use

$$G_{\rm c}{}^{\alpha\beta} = \left(g^{\alpha\mu}g^{\beta\rho} - \tfrac{1}{2}g^{\alpha\beta}g^{\mu\rho}\right)R_{\mu\rho} + \frac{d-2}{2}\Lambda g^{\alpha\beta}, \quad (3.290)$$

obtaining

$$\begin{aligned}
G_{\rm c}{}^{\alpha\beta} &= \bar{G}_{c}{}^{\alpha\beta} + G_{\rm cL}{}^{\alpha\beta} + \mathcal{O}(h^2),\\
G_{\rm cL}{}^{\alpha\beta} &= G_{\rm cL1}{}^{\alpha\beta} + G_{\rm cL2}{}^{\alpha\beta},\\
G_{\rm cL1}{}^{\alpha\beta} &= \tfrac{1}{2}\left\{\bar{\nabla}^2 h^{\alpha\beta} - 2\bar{\nabla}^{\lambda}\bar{\nabla}^{(\alpha}h_{\lambda}{}^{\beta)} + \bar{\nabla}^{\alpha}\bar{\nabla}^{\beta}h\right\} - \tfrac{1}{2}\bar{g}^{\alpha\beta}\{\bar{\nabla}^2 h - \bar{\nabla}^{\mu}\bar{\nabla}^{\nu}h_{\mu\nu}\},\\
G_{\rm cL2}{}^{\alpha\beta} &= -\left\{h^{\alpha\mu}\bar{g}^{\beta\rho} + \bar{g}^{\alpha\mu}h^{\beta\rho} - \tfrac{1}{2}h^{\alpha\beta}\bar{g}^{\mu\rho} - \tfrac{1}{2}\bar{g}^{\alpha\beta}h^{\mu\rho}\right\}\bar{R}_{\mu\rho} - \frac{d-2}{2}\Lambda h^{\alpha\beta}.
\end{aligned} \quad (3.291)$$

On substituting into the cosmological Einstein equation (3.277), we find

$$\bar{G}_{\rm c}{}^{\mu\nu} + G_{\rm cL}{}^{\mu\nu} = 8\pi G_{\rm N}^{(d)}(T^{\mu\nu}_{\rm matter} + t^{\mu\nu}), \quad (3.292)$$

where the l.h.s. contains terms up to first order in $h_{\mu\nu}$, and $8\pi G_{\rm N}^{(d)}t^{\mu\nu}$ stands for all the second- and higher-order terms in $h_{\mu\nu}$ and is referred to as the *gravitational energy–momentum (pseudo-)tensor*. This is the definition we will use in Section 6.1.2, and it is clearly justified by our previous results.

Now we can particularize to the case in which the background metric satisfies the vacuum cosmological Einstein equation (3.280), which, upon subtraction of the trace, implies

$$\bar{R}_{\mu\nu} = \Lambda\bar{g}_{\mu\nu}. \quad (3.293)$$

We find the same expressions as before for $R_{\mu\rho}$ and $G_{\rm cL1}{}^{\alpha\beta}$, but the expression for $G_{\rm cL2}{}^{\alpha\beta}$ is considerably simpler, i.e.

$$G_{\rm cL2}{}^{\alpha\beta} = -\Lambda\left(h^{\alpha\beta} - \tfrac{1}{2}\bar{g}^{\alpha\beta}h\right), \quad (3.294)$$

---

[56] Sometimes the subindex L is used to indicate that the object is the part linear in $h_{\mu\nu}$ of the corresponding tensor with the indices in the same position. Observe that for any tensor $T_{\rm L}{}^{\mu} \neq \bar{g}^{\mu\nu}T_{\rm L\,\nu}$, and for this reason we try to avoid this notation.

and the l.h.s. of the cosmological Einstein equation is purely linear in $h_{\mu\nu}$:

$$G_{\text{cL}}{}^{\mu\nu} = 8\pi G_{\text{N}}^{(d)}(T_{\text{matter}}^{\mu\nu} + t^{\mu\nu}). \tag{3.295}$$

This l.h.s. gives us the generalization of the Fierz–Pauli equations wave operator in curved spacetime we were looking for:

$$\begin{aligned}\bar{\mathcal{D}}^{\alpha\beta}(h) = 2G_{\text{cL}}{}^{\alpha\beta} &= \bar{\nabla}^2 h^{\alpha\beta} - 2\bar{\nabla}^\lambda\bar{\nabla}^{(\alpha}h_\lambda{}^{\beta)} + \bar{\nabla}^\alpha\bar{\nabla}^\beta h \\ &- \bar{g}^{\alpha\beta}\{\bar{\nabla}^2 h - \bar{\nabla}^\mu\bar{\nabla}^\nu h_{\mu\nu}\} - 2\Lambda\left(h^{\alpha\beta} - \tfrac{1}{2}\bar{g}^{\alpha\beta}h\right) \\ &= 16\pi G_{\text{N}}^{(d)}\left(T_{\text{matter}}^{\alpha\beta} + t^{\alpha\beta}\right),\end{aligned} \tag{3.296}$$

which justifies the present definition of $t^{\mu\nu}$, which coincides with the one we have used before. In fact, in the preceding sections we found the lowest-order term (quadratic in $h$) of $t^{\mu\nu}$ in the case $\bar{\mu\nu} = \eta_{\mu\nu}$ ($\Lambda = 0$) which we denoted by $t_{\text{GR}}^{(0)\,\mu\nu}$.

This equation, with the r.h.s. set to zero, is the equation of motion of a massless spin-2 field moving on a background spacetime $\bar{g}_{\mu\nu}$, which we are going to study in Section 3.4.2. We can already see that this equation does not look like the typical wave equation for a massless field because it has mass-like terms proportional to the cosmological constant. However, we are going to argue that precisely those terms are necessary in order to describe massless fields in a spacetime with $\bar{R}_{\mu\nu} = \Lambda\bar{g}_{\mu\nu}$.

Observe that, since $\zeta^{\mu\nu} = h^{\mu\nu} + \mathcal{O}(h^2)$ and $h^{\mu\nu} = \zeta^{\mu\nu} + \mathcal{O}(\zeta^2)$, we could have arrived at the same linear-order results by expanding around the inverse metric

$$g^{\mu\nu} = \bar{g}^{\mu\nu} - \zeta^{\mu\nu}. \tag{3.297}$$

We would like to have an action from which to derive the above equation of motion with vanishing r.h.s. Instead of guessing, we simply expand the integrand of the Einstein–Hilbert action to second order in $h_{\mu\nu}$. Using the matrix identity

$$\sqrt{|M|} = \exp(\tfrac{1}{2}\text{Tr}\ln M) \tag{3.298}$$

and the expansions

$$\begin{aligned}(1+x)^{-1} &= 1 - x + x^2 - x^3 + \cdots, \\ \ln(1+x) &= x - \tfrac{1}{2}x^2 + \tfrac{1}{3}x^3 - \tfrac{1}{4}x^4 + \cdots, \\ \exp y &= 1 + y + \frac{1}{2!}y^2 + \frac{1}{3!}y^3 + \cdots,\end{aligned} \tag{3.299}$$

we can easily calculate second- and higher-order terms:

$$\begin{aligned}g_{\mu\nu} &= \bar{g}_{\mu\nu} + h_{\mu\nu}, \\ g^{\mu\nu} &= \bar{g}^{\mu\nu} - h^{\mu\nu} + h^\mu{}_\sigma h^{\sigma\nu} - h^\mu{}_\sigma h^{\sigma\rho}h_\rho{}^\nu + \mathcal{O}(h^4), \\ \sqrt{|g|} &= \sqrt{|\bar{g}|}\left(1 + \frac{1}{2}h + \frac{1}{8}h^2 - \frac{1}{4}h_{\mu\nu}h^{\mu\nu} + \frac{1}{6}h_\mu{}^\nu h_\nu{}^\rho h_\rho{}^\mu \right. \\ &\quad \left. - \frac{1}{8}hh_{\mu\nu}h^{\mu\nu} + \frac{1}{48}h^3\right) + \mathcal{O}(h^4).\end{aligned} \tag{3.300}$$

For the Levi-Civita connection we can write the exact expression,

$$\Gamma_{\mu\nu}{}^{\rho} = \bar{\Gamma}_{\mu\nu}{}^{\rho} + g^{\rho\sigma}\gamma_{\mu\nu\sigma}, \qquad \gamma_{\mu\nu\sigma} = \tfrac{1}{2}\{\bar{\nabla}_{\mu}h_{\nu\sigma} + \bar{\nabla}_{\nu}h_{\mu\sigma} - \bar{\nabla}_{\sigma}h_{\mu\nu}\}, \tag{3.301}$$

and we just have to substitute the above expansion of $g^{\rho\sigma}$ to the desired order. For the Riemann curvature tensor and the Ricci tensor we can also write exact expressions,

$$\begin{aligned} R_{\mu\nu\rho}{}^{\sigma} &= \bar{R}_{\mu\nu\rho}{}^{\sigma} + 2\bar{\nabla}_{[\mu}(g^{\sigma\lambda}\gamma_{\nu]\rho\lambda}) + 2g^{\sigma\delta}g^{\lambda\epsilon}\gamma_{[\mu|\lambda\delta}\gamma_{|\nu]\rho\epsilon}, \\ R_{\mu\rho} &= \bar{R}_{\mu\rho} + \bar{\nabla}_{\mu}(g^{\sigma\lambda}\gamma_{\sigma\rho\lambda}) - \bar{\nabla}_{\sigma}(g^{\sigma\lambda}\gamma_{\mu\rho\lambda}) + g^{\sigma\delta}g^{\lambda\epsilon}(\gamma_{\mu\lambda\delta}\gamma_{\sigma\rho\epsilon} - \gamma_{\sigma\lambda\delta}\gamma_{\mu\rho\epsilon}), \end{aligned} \tag{3.302}$$

on which, again, we simply have to expand the inverse metric. A similar expression can immediately be found for the Ricci scalar $R$ and for the scalar density $\sqrt{|g|}\,R$. Then, up to total derivatives and $\mathcal{O}(h^3)$ terms, and using the equations of motion for the background $\bar{R}_{\mu\nu} = \Lambda\bar{g}_{\mu\nu}$ that we have not used so far, the Einstein–Hilbert action (3.275) becomes the Fierz–Pauli action in a curved background:

$$\boxed{\begin{aligned} S = \frac{1}{\chi^2}\int d^dx\sqrt{|\bar{g}|}\ \{&\tfrac{1}{4}\bar{\nabla}_{\mu}h_{\rho\lambda}\bar{\nabla}^{\mu}h^{\rho\lambda} - \tfrac{1}{2}\bar{\nabla}_{\mu}h_{\rho\lambda}\bar{\nabla}^{\rho}h^{\mu\lambda} + \tfrac{1}{2}\bar{\nabla}_{\mu}h^{\mu\nu}\bar{\nabla}_{\nu}h \\ &- \tfrac{1}{4}\bar{\nabla}_{\mu}h\bar{\nabla}^{\mu}h + \tfrac{1}{2}\Lambda(h^{\mu\nu}h_{\mu\nu} - \tfrac{1}{2}h^2)\}. \end{aligned}} \tag{3.303}$$

In the Minkowski background $\bar{g}_{\mu\nu} = \eta_{\mu\nu}$ $(\Lambda = 0)$ it is easier to find higher corrections both to the action and to the equations of motion. A long but straightforward calculation yields the cubic term in the action (up to total derivatives)

$$S^{(3)} = \frac{1}{\chi^2}\int d^dx\ \tfrac{1}{2}h^{\mu\sigma}\mathcal{L}^{(1)}_{\mu\sigma}, \tag{3.304}$$

where $\mathcal{L}^{(1)}_{\mu\sigma}$ is written in Eq. (3.199).[57] The equation of motion that one obtains from the variation of the vacuum Einstein–Hilbert action is

$$\frac{\delta S}{\delta g_{\mu\nu}} - \frac{1}{\chi^2}\sqrt{|g|}\,G^{\mu\nu} = 0. \tag{3.305}$$

Therefore, the linear equation of motion (the Fierz–Pauli equation) is (restoring everywhere $\chi$) obtained from the quadratic term in the action and the quadratic energy–momentum tensor from the cubic term in the action:

$$\begin{aligned} \frac{\delta S^{(2)}}{\delta h_{\mu\nu}} &= -G^{(1)\,\mu\nu} &&= -\tfrac{1}{2}\mathcal{D}^{\mu\nu}(h), \\ \frac{\delta S^{(3)}}{\delta h_{\mu\nu}} &= \chi(G^{(2)\,\mu\nu} + \tfrac{1}{2}hG^{(1)\,\mu\nu}) &&= \tfrac{1}{2}\chi t_{\rm GR}^{(0)\,\mu\nu}(h). \end{aligned} \tag{3.306}$$

This is the $t_{\rm GR}^{(0)\,\mu\nu}(h)$ given in Eq. (3.200). The physical consistency of these results has been discussed at length in the preceding sections.

[57] To recover the factors of $\chi$ we have to rescale $h_{\mu\nu} \to \chi h_{\mu\nu}$.

### 3.4.2 Massless spin-2 particles in curved backgrounds

We have obtained a generalization of the Fierz–Pauli action for curved backgrounds that are solutions of the vacuum Einstein equations and we want to see whether the theory can describe massless spin-2 particles in those backgrounds.

We should start by saying that the concepts of mass and angular momentum (spin) are in principle associated exclusively with the Poincaré group, which is the isometry group of Minkowski spacetime. In more general spaces one has to study the representations of the isometry group and, in general, there will be no obvious generalizations of these concepts that work in all cases.

Instead of proceeding case by case, trying to give definitions of the mass of a field, we adopt a general point of view and give a characterization of the masslessness of a field. The main observation is that massless fields have, as a rule, fewer degrees of freedom (DOFs) than do massive fields, the extra DOF being removed by gauge symmetries that appear when the mass parameters are set to zero. At the beginning of this chapter we studied two cases in Minkowski spacetime: a massive vector field has $d-1$ DOFs and no gauge symmetries. When we set the mass parameter to zero, the theory has a gauge symmetry and we can remove one more DOF (a total of two) so there are only the $d-2$ DOFs of a massless vector. In the spin-2 case, in the presence of mass the field describes $(d-2)(d+1)/2$ DOFs. When we switch off the mass parameter, there appears a gauge symmetry that allows us to remove $d-1$ DOFs more (a total of $2d$) and we are left with the $d(d-3)/2$ DOFs of a massless spin-2 particle.

In conclusion, we characterize masslessness by the occurrence of new gauge symmetries that appear when we switch off the mass parameter.

We have obtained a generalization of the Fierz–Pauli theory to curved backgrounds given by the action Eq. (3.303) and equation of motion Eq. (3.296) (with vanishing r.h.s.). In this theory there are terms proportional to the cosmological constant $\Lambda$ that have the form of mass terms. To see whether they really are mass terms according to our definition, we look for gauge symmetries. The obvious candidate is the linearization of the invariance under GCTs that generalizes Eq. (3.95) to curved backgrounds:

$$\delta_\epsilon h_{\mu\nu} = -2\bar{\nabla}_{(\mu}\epsilon_{\nu)}, \tag{3.307}$$

Let us first check the invariance of the action under these transformations. First we vary the action as usual. We obtain two types of terms: $\bar{\nabla}h\bar{\nabla}^2\epsilon$ and $\Lambda h\bar{\nabla}\epsilon$ (these arise from the variation of the "mass terms"). We want to move all the derivatives so they act over $\epsilon$. Thus, we integrate by parts all the terms of the first kind, obtaining $h\bar{\nabla}^3\epsilon$-type terms and a total derivative. These terms can be combined into terms of the forms $h\bar{\nabla}[\bar{\nabla},\bar{\nabla}]\epsilon$ and $h[\bar{\nabla},\bar{\nabla}]\bar{\nabla}\epsilon$. Then, the commutators of covariant derivatives can be replaced by curvature terms using the Ricci identity and all these terms become terms of the type $h\bar{R}\bar{\nabla}\epsilon$ and $h\bar{\nabla}\bar{R}\epsilon$. The first cancel out, *upon use of the vacuum cosmological Einstein equation for the background metric* $\bar{R}_{\mu\nu} = \Lambda\bar{g}_{\mu\nu}$, the $\Lambda h\bar{\nabla}\epsilon$ terms. The second cancel out upon use of the background Bianchi identity $\bar{\nabla}_{[\mu}\bar{R}_{\nu\rho]\sigma}{}^{\lambda} = 0$, and we are left with the total derivative:

$$\begin{aligned}\delta_\epsilon S = \frac{1}{\chi^2}\int d^dx\sqrt{|\bar{g}|}\,\bar{\nabla}_\mu\Big\{\tfrac{1}{2}h_{\rho\sigma}\Big[4\bar{\nabla}^{[\mu}\bar{\nabla}^{\rho]}\epsilon^\sigma - 2\bar{\nabla}^\rho\bar{\nabla}^\sigma\epsilon^\mu \\ + \bar{g}^{\rho\sigma}\Big(\bar{\nabla}^2\epsilon^\mu - \bar{\nabla}_\lambda\bar{\nabla}^\mu\epsilon^\lambda\Big) + 2\bar{g}^{\mu\rho}\bar{\nabla}^\sigma\bar{\nabla}_\lambda\epsilon^\lambda - 2\bar{g}^{\rho\sigma}\bar{\nabla}^\mu\bar{\nabla}_\lambda\epsilon^\lambda\Big]\Big\} \\ \equiv \int d^dx\sqrt{|\bar{g}|}\,\bar{\nabla}_\mu s^\mu(\epsilon).\end{aligned} \tag{3.308}$$

The Fierz–Pauli equation of motion Eq. (3.296) is, therefore, invariant for the backgrounds considered. The proof makes crucial use of the Einstein equation satisfied by the background metric. As we remarked in the introduction to this section, in general backgrounds there is no way to construct a gauge-invariant theory by adding curvature terms [59].

Furthermore, the presence of the cosmological constant terms is also crucial in the proof of invariance of the action. Had we tried to prove the invariance of the equation of motion directly, we would have seen the necessity for these terms to cancel out curvature terms coming from the commutators of covariant derivatives. We can conclude that the theory, with those terms, is massless.

It is interesting to see what kind of gauge identity and conserved current we obtain from this invariance. We proceed as usual. We first find the variation of the action under an arbitrary infinitesimal transformation of $\delta h_{\mu\nu}$:

$$\begin{aligned}\delta S_{\rm FP} &= \frac{1}{\chi^2}\int d^dx\sqrt{|\bar{g}|}\,\left\{\frac{S_{\rm FP}}{\delta h_{\alpha\beta}}\delta h_{\alpha\beta} + \bar{\nabla}_\mu\left(l^{\mu(\alpha\beta)}\delta h_{\alpha\beta}\right)\right\}, \\ \frac{S_{\rm FP}}{\delta h_{\alpha\beta}} &= -\tfrac{1}{2}\bar{\mathcal{D}}^{\alpha\beta}(h), \\ l^{\mu\alpha\beta} &= \tfrac{1}{2}\bar{\nabla}^\mu h^{\alpha\beta} - \bar{\nabla}^\alpha h^{\beta\mu} + \tfrac{1}{2}\bar{g}^{\mu\alpha}\bar{\nabla}^\beta h + \tfrac{1}{2}\bar{g}^{\alpha\beta}\bar{\nabla}_\nu h^{\mu\nu} - \tfrac{1}{2}\bar{g}^{\alpha\beta}\bar{\nabla}^\mu h.\end{aligned} \tag{3.309}$$

Using now the particular form of the gauge transformation $\delta_\epsilon h_{\mu\nu}$ in the above equation and integrating by parts, we obtain

$$\delta_\epsilon S = \int d^dx\sqrt{|\bar{g}|}\,\left\{-\frac{1}{\chi^2}\epsilon_\beta\bar{\nabla}_\alpha\bar{\mathcal{D}}^{\alpha\beta}(h) + \bar{\nabla}_\mu\left[\frac{1}{\chi^2}\bar{\mathcal{D}}^{\mu\beta}\epsilon_\beta - \frac{2}{\chi^2}l^{\mu(\alpha\beta)}\bar{\nabla}_\alpha\epsilon_\beta\right]\right\}, \tag{3.310}$$

and, on comparing this with the first form of the variation of the action that we found, we arrive finally at the identity, which is valid for arbitrary $\epsilon^\mu$s and without the use of any equations of motion:

$$\begin{aligned}0 &= \int d^dx\sqrt{|\bar{g}|}\,\left\{-\frac{1}{\chi^2}\epsilon_\beta\bar{\nabla}_\alpha\bar{\mathcal{D}}^{\alpha\beta}(h) + \bar{\nabla}_\mu j^\mu_{\rm N2}(\epsilon)\right\}, \\ j^\mu_{\rm N2}(\epsilon) &= j^\mu_{\rm N1}(\epsilon) + \frac{1}{\chi^2}\bar{\mathcal{D}}^{\mu\beta}(h)\epsilon_\beta, \\ j^\mu_{\rm N1}(\epsilon) &= -\frac{2}{\chi^2}l^{\mu(\alpha\beta)}\bar{\nabla}_\alpha\epsilon_\beta - s^\mu(\epsilon).\end{aligned} \tag{3.311}$$

From this identity we derive the gauge identity,

$$\bar{\nabla}_\alpha\bar{\mathcal{D}}^{\alpha\beta}(h) = 0, \tag{3.312}$$

and the off-shell covariant conservation of the above Noether current,

$$\bar{\nabla}_\mu j^\mu_{\rm N2}(\epsilon) = 0 \qquad \left(\partial_\mu \mathfrak{j}^\mu_{\rm N2}(\epsilon) = 0\right). \tag{3.313}$$

We know that this Noether current can always be written as $\mathfrak{j}^\mu_{\rm N2}(\epsilon) = \partial_\nu \mathfrak{j}^{\nu\mu}_{\rm N2}(\epsilon)$ with $\mathfrak{j}^{\nu\mu}_{\rm N2}(\epsilon) = -\mathfrak{j}^{\mu\nu}_{\rm N2}(\epsilon)$. Finding this antisymmetric tensor in the general case is complicated and we do so only for the most interesting case, in which $\epsilon^\mu$ is a Killing vector of the background metric $\epsilon^\mu \equiv \bar{\xi}^\mu$ with $\bar{\nabla}_{(\mu}\bar{\xi}_{\nu)} = 0$. In this case, $s^\mu(\xi)$ has to vanish identically, because the variations of $h_{\mu\nu}$ also vanish identically, and the first term of $j^\mu_{\rm N1}(\epsilon)$ also vanishes because of the Killing equation. Then, only the second term in the expression for $j^\mu_{\rm N2}(\bar{\xi})$ Eq. (3.311) survives and we are left with

$$j^\mu_{\rm N2}(\bar{\xi}) = \frac{1}{\chi^2}\bar{\mathcal{D}}^{\mu\nu}(h)\bar{\xi}_\nu. \tag{3.314}$$

The conservation of this current is easy to check using the Bianchi identity and the Killing equation. To find $j^{\mu\nu}_{\rm N2}(\bar{\xi}) = (1/\sqrt{|g|})\mathfrak{j}^{\mu\nu}_{\rm N2}(\bar{\xi})$ we follow Abbott and Deser in Ref. [1]. First we separate $\bar{\mathcal{D}}^{\mu\nu}$ into two pieces:

$$\begin{aligned}\bar{\mathcal{D}}^{\mu\nu}(h) = {} & \text{curvature terms} + [\bar{\nabla}^\nu, \bar{\nabla}_\lambda]h^{\lambda\mu} \ (\equiv 2X^{\mu\nu}) \\ & + \text{the rest} - [\bar{\nabla}^\nu, \bar{\nabla}_\lambda]h^{\lambda\mu} \quad (\equiv 2Y^{\mu\nu}).\end{aligned} \tag{3.315}$$

$Y^{\mu\nu}$ can be written in the form

$$Y^{\mu\nu} = -\bar{\nabla}_\alpha\bar{\nabla}_\beta K^{\mu\alpha\nu\beta}, \tag{3.316}$$

where $K^{\mu\alpha\nu\beta}$ is as defined in Eq. (3.91) but with a general background metric instead of the Minkowski metric, i.e.

$$K^{\mu\alpha\nu\beta} = \tfrac{1}{2}\left\{\bar{g}^{\mu\beta}\bar{h}^{\nu\alpha} + \bar{g}^{\nu\alpha}\bar{h}^{\mu\beta} - \bar{g}^{\mu\nu}\bar{h}^{\alpha\beta} - \bar{g}^{\alpha\beta}\bar{h}^{\mu\nu}\right\}. \tag{3.317}$$

This tensor has the same symmetries as the Riemann tensor and is sometimes called the *superpotential*. Using $\bar{R}_{\mu\nu} = \Lambda\bar{g}_{\mu\nu}$, we find

$$X^{\mu\nu} = \tfrac{1}{2}[\bar{\nabla}^\nu, \bar{\nabla}_\lambda]\bar{h}^{\lambda\mu} - \Lambda\bar{h}^{\mu\nu}, \tag{3.318}$$

and, using the Ricci identity, it can be rewritten as follows:

$$X^{\mu\nu} = \tfrac{1}{2}\left[\bar{R}^\mu{}_{\lambda\sigma}{}^\nu\bar{h}^{\lambda\sigma} - \Lambda\bar{h}^{\mu\nu}\right]. \tag{3.319}$$

Finally, we can also rewrite it as follows:

$$X^{\mu\nu} = \tfrac{1}{2}\bar{R}^\nu{}_{\alpha\beta\gamma}K^{\mu\alpha\beta\gamma}. \tag{3.320}$$

Using the expression for $Y$ in terms of the superpotential $K$,

$$Y^{\mu\nu}\bar{\xi}_\nu = -\bar{\nabla}_\alpha\left[\left(\bar{\nabla}_\beta K^{\mu\alpha\nu\beta}\right)\bar{\xi}_\nu - K^{\mu\beta\nu\alpha}\bar{\nabla}_\beta\bar{\xi}_\nu\right] - K^{\mu\beta\nu\alpha}\bar{\nabla}_\alpha\bar{\nabla}_\beta\bar{\xi}_\nu. \tag{3.321}$$

Using the Killing vector identity Eq. (1.116) for the background Killing vectors and the definition of the superpotential $K$, we see that

$$Y^{\mu\nu}\bar{\xi}_\nu = -\bar{\nabla}_\alpha\Big[\Big(\bar{\nabla}_\beta K^{\mu\alpha\nu\beta}\Big)\bar{\xi}_\nu - K^{\mu\beta\nu\alpha}\bar{\nabla}_\beta\bar{\xi}_\nu\Big] - X^{\mu\nu}\bar{\xi}_\nu, \tag{3.322}$$

and, therefore,

$$j^{\alpha\mu}_{\rm N2}(\bar{\xi}) = -\frac{2}{\chi^2}\Big[\Big(\bar{\nabla}_\beta K^{\mu\alpha\nu\beta}\Big)\bar{\xi}_\nu - K^{\mu\beta\nu\alpha}\bar{\nabla}_\beta\bar{\xi}_\nu\Big]. \tag{3.323}$$

### *3.4.3 Self-consistency*

In this chapter we have seen how the consistency of the Fierz–Pauli theory in Minkowski spacetime coupled to matter requires the introduction of an infinite series of higher-order terms whose resummation leads to GR without a cosmological constant. This is evidently consistent with the derivation of the Fierz–Pauli theory from GR as the linear perturbation theory around Minkowski spacetime.

We have found a generalization of the Fierz–Pauli theory in an arbitrary background satisfying the cosmological vacuum Einstein equation $\bar{R}_{\mu\nu} = \Lambda\bar{g}_{\mu\nu}$ as the linear perturbation theory around that background, and it is natural to ask ourselves whether, by requiring consistency in the coupling of this theory to matter, we are going to arrive at GR with a cosmological constant. (The linear theory coupled to matter is inconsistent for exactly the same reasons as in the Minkowski case.)

As shown by Deser in Ref. [433], the answer to this question is affirmative. We are not going to give here all the details of the proof, which follows closely the proof in the Minkowski case, but it is, however, interesting to see the first-order form of the Fierz–Pauli action in curved background that constitutes its starting point:

$$\boxed{\begin{aligned} S^{(1)}_{\rm FP}[\varphi^{\mu\nu},\Gamma_{\mu\nu}{}^\rho] = \frac{1}{\chi^2}\int d^dx\,\{&\chi\Gamma_{\mu\nu}{}^\rho\big(\delta_\rho{}^\mu\bar{\nabla}_\sigma\varphi^{\sigma\nu} - \bar{\nabla}_\rho\varphi^{\mu\nu}\big) \\ &+\bar{\mathfrak{g}}^{\mu\nu}2\Gamma_{\lambda[\mu}{}^\rho\Gamma_{\rho]\nu}{}^\lambda + \tfrac{1}{2}\Lambda\big(h^{\mu\nu}h_{\mu\nu} - \tfrac{1}{2}h^2\big)\}\,. \end{aligned}} \tag{3.324}$$

Here both $\varphi^{\mu\nu}$ and $\mathfrak{g}^{\mu\nu}$ are tensor densities.

## 3.5 Final comments

In this chapter we have found a SRFT of gravity (GR) that is very satisfactory from many points of view. First of all, it describes extremely well what is observed. Second, it is a theory with a high degree of internal self-consistency that can be obtained from very few principles (either the principle of equivalence and general covariance or consistent interaction of a massless spin-2 particle).

It also has some drawbacks, however: we wanted to follow the steps that led to the development of the SRFTs like quantum electrodynamics that we know so well, but we found at the end that the quantum theory based on this consistent classical theory is not consistent. Thus, at the microscopic level, the answer we have obtained is not satisfactory. In fact, at the microscopic level there arise questions, for instance the coupling of gravity to fermions, that have no answer in the formalism we have developed.

How GR should be modified in order to obtain a consistent quantum theory is a question that has received many tentative answers, the latest being string theory. In string theory, as in some of the alternative theories that have been proposed, there are additional fields, in the presence of which the proofs of uniqueness and self-consistency of GR are no longer valid. Furthermore, there is a prescription for the coupling of all those fields to fermions, and some of the additional gravitational fields can be interpreted as torsion. We want to gain some understanding of all these elements that enter into the gravitational part of string theory as well as other alternative theories of gravity. Some of these elements are more or less trivial extensions of GR (for instance, its reformulation in the Vielbein formalism which allows the coupling to spinors), and, in fact, it is always (or usually) possible to see the new theories as GR coupled to different fields. In the next few chapters we review these elements and theories that contain them: the Cartan–Sciama–Kibble theory, non-symmetric theories of gravity, theories of teleparallelism, and supergravity theories. The simplest way to introduce most of them is through a minimal-action principle, and the formulation of the minimal-action principle for GR will be the first step in this direction.

# 4
# Action principles for gravity

A minimal-action principle is a basic ingredient of any field theory. With it (with an action) we can systematically find conserved currents and charges, canonically conjugate momenta, and a Hamiltonian (which is necessary for canonical quantization), etc. On the other hand, it is easier to deal with actions than with equations of motion; it is easier to include new fields and couplings in the action respecting certain symmetries than to invent new consistent equations of motion for them and modifications of the equations of motion of the old fields.

In this chapter we are going to study in detail several action principles for GR and for more general theories that we will be concerned with later on. First, we will study the standard second-order Einstein–Hilbert action that we found as the result of imposing self-consistency on the Fierz–Pauli theory coupled to matter. We will derive the Einstein equations from it and we will find the right boundary term that will allow us to impose boundary conditions on the variations of the metric $\delta g_{\mu\nu}$ only, not on its derivatives. We will do the same for theories including a scalar and in a conformal frame that is not Einstein's. In these theories, an extra scalar factor $K$ (which could be $e^{-2\phi}$ in the string effective action) multiplies the Ricci scalar and obtaining the gravitational equations becomes more involved.

We are also going to study the behavior of the Einstein–Hilbert action under GCTs; we will obtain the Bianchi (gauge) identity and Noether current associated with them and see how they are modified by the addition of boundary terms to the action.

Then we will study the first-order formalism, in which the metric $g_{\mu\nu}$ and the connection $\Gamma_{\mu\nu}{}^{\rho}$ are considered as independent variables, and the first-order formalism for the Vielbein $e_{\mu}{}^{a}$ and the spin connection $\omega_{\mu}{}^{ab}$, with and without fermions, which will be seen to induce torsion. There is also a *purely affine* formulation of GR in which the only variable is the (symmetric) affine connection $\Gamma_{\mu\nu}{}^{\rho}$, and we will review it briefly.

The first-order formalism and the purely affine formulation are very useful for formulating Einstein's "unified theory," which is based on a non-symmetric "metric" tensor. We take the opportunity to revisit this and other non-symmetric gravity theories (NGTs).

Motivated by the success of the first-order formalism with Vielbein and spin connection, we will review the MacDowell–Mansouri formulation of four-dimensional gravity as the gauge theory of the four-dimensional Poincaré group, which we will obtain by

Wigner–Inönü contraction from the $AdS_4$ case.

Finally, we will briefly review teleparallel formulations and generalizations of GR.

## 4.1 The Einstein–Hilbert action

In $d$ dimensions, the Einstein–Hilbert action [752] is

$$\boxed{S_{\text{EH}}[g] = \frac{c^3}{16\pi G_{\text{N}}^{(d)}} \int_{\mathcal{M}} d^d x \sqrt{|g|}\, R(g),} \tag{4.1}$$

where $R(g)$ is the Ricci scalar of the metric $g_{\mu\nu}$, $G_{\text{N}}^{(d)}$ is the $d$-dimensional Newton constant, and $\mathcal{M}$ is the $d$-dimensional manifold we are integrating over. Since we have obtained this action by imposing consistent coupling of the special-relativistic field theory, we know that it is canonically normalized and we also know which expression for the force between two particles it leads to (see Eq. (3.140)). We have introduced here the speed of light in order to find the dimensions of $G_{\text{N}}^{(d)}$ in "unnatural units": $\text{M}^{-1}\text{L}^{d-1}\text{T}^{-2}$. Recall that the metric $g_{\mu\nu}$ is dimensionless in our conventions. Recall also that the factor of $16\pi$ is associated with rationalized units only in $d = 4$.

Observe that what will appear in the path integral

$$\mathcal{Z} = \int Dg\, e^{+iS_{\text{EH}}/\hbar} \tag{4.2}$$

is the dimensionless combination

$$\frac{S_{\text{EH}}}{\hbar} = \frac{16\pi}{\ell_{\text{Planck}}^{d-2}} \int d^d x \cdots, \tag{4.3}$$

where[1]

$$\frac{\ell_{\text{Planck}}^{d-2}}{16\pi} = \frac{\hbar G_{\text{N}}^{(d)}}{c^3} \tag{4.4}$$

is the $d$-dimensional *Planck length*.[2] In the absence of any other dimensional quantity this is the only combination of the constants $\hbar$, $c$, and $G_{\text{N}}^{(d)}$ with dimensions of length. However, if there is an object of mass $M$, there are two more combinations with dimensions of length: the *Compton wavelength* associated with the object,

$$\lambda\!\!\!^{-}_{\text{Compton}} = \frac{\hbar}{Mc}, \tag{4.6}$$

[1] This normalization is, unfortunately, different from the one used in the string theory literature for the 11-dimensional Planck length; see Eq. (22.43).

[2] Sometimes the *reduced Planck length*

$$\ell\!\!\!^{-}_{\text{Planck}} = \ell_{\text{Planck}}/(2\pi). \tag{4.5}$$

We have also been using the constant $\chi$ defined by $\chi^2 = 16\pi G_{\text{N}}^{(d)}/c^3$.

which is of purely quantum-mechanical nature, and the $d$-dimensional *Schwarzschild radius*,

$$R_{\rm S} = \left( \frac{16\pi M G_{\rm N}^{(d)} c^{-2}}{(d-2)\omega_{(d-2)}} \right)^{\frac{1}{d-3}}, \tag{4.7}$$

which is of purely classical, gravitational nature. It occurs naturally in the gravitational field of a massive point-like particle, Eq. (3.124).

With the constants $\hbar$, $c$, and $G_{\rm N}^{(d)}$ one can also build a combination with units of mass: the *Planck mass*,

$$M_{\rm Planck} = \left( \frac{\hbar^{d-3}}{G_{\rm N}^{(d)} c^{d-5}} \right)^{\frac{1}{d-2}}, \tag{4.8}$$

so the prefactor of the action in the path integral is

$$\frac{c^3}{G_{\rm N}^{(d)}\hbar} = \left( \frac{M_{\rm Planck} c}{\hbar} \right)^{d-2}. \tag{4.9}$$

If we consider objects whose masses are of the order of the Planck mass, then it is immediately obvious that their Compton wavelengths become of the order of their Schwarzschild radii, which are of the order of the Planck length:

$$M \sim M_{\rm Planck} \Rightarrow \lambda\!\!\!^{-}_{\rm Compton} \sim R_{\rm S} \sim \ell_{\rm Planck}. \tag{4.10}$$

At that point, quantum-mechanical effects become important.

If we naively try to quantize GR by standard methods (starting from its perturbative expansion), we find that the quantum gravitational coupling constant (Planck length) is dimensionful and, by standard arguments, we expect to obtain a non-renormalizable theory. This is indeed the case. As we will see, in string theory there is no unique constant that plays the role of length scale and coupling constant as does the Planck length in GR; there are two constants with dimensions of length: Planck's constant and the *string length* $\ell_{\rm s}$. The dimensionless quotient is essentially the *string coupling constant* $g_{\rm s}$. In that context the Schwarzschild radius has to be compared with $\ell_{\rm s}$ in order to see when (string) quantum-gravity effects become important. On the other hand, we can have better expectations about the perturbative renormalizability of the theory since the expansion is made in the dimensionless parameter $g_{\rm s}$, instead of $\ell_{\rm Planck}$ or $\ell_{\rm s}$.

The Einstein–Hilbert action Eq. (4.1) contains second derivatives of the metric. However, the terms with second derivatives take the form of a total derivative,[3] symbolically

$$S_{\rm EH}[g] = \frac{c^3}{16\pi G_{\rm N}^{(d)}} \int_{\mathcal{M}} d^d x \sqrt{|g|}\, (\partial g)^2 + \frac{c^3}{16\pi G_{\rm N}^{(d)}} \int_{\mathcal{M}} d^d x\, \partial_\mu \omega^\mu(\partial g). \tag{4.11}$$

This means that the original action Eq. (4.1) can in principle be used to obtain equations of motion that are of second order in derivatives of the metric. However, we would have to impose conditions on the derivatives of the metric on the boundary. Furthermore, observe

[3] See, for instance, Ref. [883] and Appendix L.

that the "vector" $\omega^\mu(\partial g)$ does not transform as such under GCTs. The solution to these problems consists in adding a general-covariant boundary term to the original Einstein–Hilbert action. Next we see how to find the equations of motion and the right boundary term.

### 4.1.1 Equations of motion

Let us vary the original Einstein–Hilbert action with respect to the metric. For simplicity we temporarily set $\chi = 1$. Bearing in mind that $R(g) = g^{\mu\nu} R_{\mu\nu}(\Gamma(g))$ and $R_{\mu\nu}(\Gamma(g))$ depends on $g$ only through the Levi-Civita connection $\Gamma(g)$, so we can use Palatini's identity Eq. (3.285),

$$\delta R_{\mu\nu} = \nabla_\mu \delta\Gamma_{\rho\nu}{}^{\rho} - \nabla_\rho \delta\Gamma_{\mu\nu}{}^{\rho}, \tag{4.12}$$

and using the identities

$$\delta g^{\mu\nu} = -g^{\nu\alpha} g^{\mu\beta} \delta g_{\alpha\beta}, \qquad \delta g = g\, g^{\alpha\beta} \delta g_{\alpha\beta}, \tag{4.13}$$

we immediately find

$$\delta S_{\text{EH}} = \int d^d x \sqrt{|g|}\, \{-G^{\mu\nu} \delta g_{\mu\nu} + g^{\mu\nu} [\nabla_\mu \delta\Gamma_{\rho\nu}{}^{\rho} - \nabla_\rho \delta\Gamma_{\mu\nu}{}^{\rho}]\}\,. \tag{4.14}$$

Since our covariant derivative is metric compatible, we can absorb the metric in the last term and combine the two terms into a single total derivative,

$$\delta S_{\text{EH}} = -\int_{\mathcal{M}} d^d x \sqrt{|g|}\, G^{\mu\nu} \delta g_{\mu\nu} + \int_{\mathcal{M}} d^d x \sqrt{|g|}\, \nabla_\rho v^\rho, \tag{4.15}$$

where

$$v^\rho = g^{\rho\mu} \delta\Gamma_{\mu\nu}{}^{\nu} - g^{\mu\nu} \delta\Gamma_{\mu\nu}{}^{\rho}. \tag{4.16}$$

We now have to use the equation that expresses the variation of the Levi-Civita connection with respect to a variation of the metric in order to find the variation of the action as a function of the variation of the metric. That expression was given in Eq. (3.282), and with it we find

$$v^\rho = g^{\rho\mu} g^{\sigma\nu} (\nabla_\mu \delta g_{\sigma\nu} - \nabla_\sigma \delta g_{\mu\nu}). \tag{4.17}$$

Using now Stokes' theorem Eq. (1.150), we reexpress the integral of the total derivative terms as an integral over the boundary,

$$\int_{\mathcal{M}} d^d x \sqrt{|g|}\, \nabla_\rho v^\rho = (-1)^{d-1} \int_{\partial\mathcal{M}} d^{d-1}\Sigma_\rho v^\rho = (-1)^{d-1} \int_{\partial\mathcal{M}} d^{d-1}\Sigma\, n_\rho v^\rho, \tag{4.18}$$

where $d^{d-1}\Sigma_\rho$ is defined in Chapter 1,

$$d^{d-1}\Sigma \equiv n^2 d^{d-1}\Sigma_\rho n^\rho, \tag{4.19}$$

and $n^\mu$ is the unit vector normal to the boundary hypersurface $\partial\mathcal{M}$ ($n^2 = +1$ for spacelike hypersurfaces with timelike normal unit vector and $n^2 = -1$ for timelike hypersurfaces with spacelike normal unit vector). Finally, we expand the integrand

$$n_\rho v^\rho = n^\mu g^{\sigma\nu} (\nabla_\mu \delta g_{\sigma\nu} - \nabla_\sigma \delta g_{\mu\nu}) = n^\mu h^{\sigma\nu} (\nabla_\mu \delta g_{\sigma\nu} - \nabla_\sigma \delta g_{\mu\nu}), \tag{4.20}$$

where $h_{\mu\nu} = g_{\mu\nu} - n^2 n_\mu n_\nu$ is the induced metric on the hypersurface $\partial\mathcal{M}$ (see Section 1.8). Thus, we arrive at

$$\begin{aligned}\delta S_{\mathrm{EH}} = &- \int_{\mathcal{M}} d^d x \sqrt{|g|}\, G^{\mu\nu} \delta g_{\mu\nu} + (-1)^{d-1} \int_{\partial\mathcal{M}} d^{d-1}\Sigma\, n^\mu h^{\sigma\nu} \nabla_\mu \delta g_{\sigma\nu} \\ &- (-1)^{d-1} \int_{\partial\mathcal{M}} d^{d-1}\Sigma\, n^\mu h^{\sigma\nu} \nabla_\sigma \delta g_{\mu\nu}.\end{aligned} \tag{4.21}$$

This is the final form of the variation of the action we were after. Now, we would like to be able to obtain the Einstein equation by requiring the action to be stationary (so $\delta S_{\mathrm{EH}} = 0$) under arbitrary variations of the metric *vanishing on the boundary*:

$$\delta g_{\mu\nu}\,|_{\partial\mathcal{M}} = 0. \tag{4.22}$$

If $\delta g_{\mu\nu}$ is constant on the boundary, then its covariant derivative projected onto the boundary directions with $h^{\mu\nu}$ must vanish:

$$h^{\sigma\nu} \nabla_\sigma \delta g_{\mu\nu} = 0, \tag{4.23}$$

and the second of the two boundary terms vanishes. However, the first does not vanish unless we impose boundary conditions for the covariant derivative of the variation of the metric. In order to obtain the Einstein equation we must cancel out that boundary term with the variation of another boundary term added to the Einstein–Hilbert action. This boundary term is nothing but the integral over the boundary of the trace of the extrinsic curvature of the boundary given in Eq. (1.160). Observe that

$$\delta\mathcal{K} = \delta h^\mu{}_\nu \nabla_\mu n^\nu + h^\mu{}_\nu \delta\Gamma_{\mu\rho}{}^\nu n^\rho. \tag{4.24}$$

The first term vanishes on the boundary due to our boundary condition (4.22). Using Eq. (3.282) for $\delta\Gamma$, we find

$$\delta\mathcal{K}|_{\partial\mathcal{M}} = \tfrac{1}{2} n^\rho h^{\mu\sigma} \nabla_\rho \delta g_{\mu\sigma}. \tag{4.25}$$

In conclusion, the action that one should use is the following [612, 1229]:

$$\boxed{S_{\mathrm{EH}}[g] = \frac{1}{\chi^2} \int_{\mathcal{M}} d^d x \sqrt{|g|}\, R + (-1)^d \frac{2}{\chi^2} \int_{\partial\mathcal{M}} d^{d-1}\Sigma\, \mathcal{K}.} \tag{4.26}$$

Under otherwise arbitrary variations of the metric satisfying Eq. (4.22), we have shown that the variation of the Einstein–Hilbert action with boundary term (4.26) is just

$$\delta S_{\mathrm{EH}} = -\frac{1}{\chi^2} \int_{\mathcal{M}} d^d x \sqrt{|g|}\, G^{\mu\nu} \delta g_{\mu\nu}, \tag{4.27}$$

and then the vacuum Einstein equation follows, as we wanted.

### 4.1.2 Gauge identity and Noether current

The Einstein–Hilbert action is invariant under GCTs and we can write $\tilde{\delta}_\xi S_{\rm EH} = 0$. For variations at the same point the action transforms into the integral of a total derivative ($\chi = 1$ again):

$$\delta_\xi S_{\rm EH} = \int_{\mathcal{M}} d^d x \delta_\xi \hat{\mathcal{L}} = -\int_{\mathcal{M}} d^d x \mathcal{L}_\xi \hat{\mathcal{L}} = -\int_{\mathcal{M}} d^d x \partial_\mu (\xi^\mu \hat{\mathcal{L}}), \tag{4.28}$$

because $\hat{\mathcal{L}}$ is a scalar density. This result will be valid for any general-covariant action.

To find the gauge identity associated with the invariance under GCTs we have to find the variation of the action under variations of the metric and then use the explicit form of the variation of the metric under GCTs. For simplicity we will use the original Einstein–Hilbert action with no boundary terms and then we will discuss the effect of the addition of boundary terms. The variation of the action is given by Eqs. (4.15) and (4.16):

$$\delta_\xi S_{\rm EH} = \int_{\mathcal{M}} d^d x \sqrt{|g|} \left\{ -G^{\mu\nu} \delta_\xi g_{\mu\nu} + \nabla_\rho (2 g^{\mu\sigma,\rho\nu} \nabla_\mu \delta_\xi g_{\sigma\nu}) \right\}, \tag{4.29}$$

and, using the expression for $\delta_\xi g_{\mu\nu}$ in Eq. (1.67) and integrating once by parts, we obtain

$$\delta_\xi S_{\rm EH} = \int_{\mathcal{M}} d^d x \sqrt{|g|} \left\{ -2(\nabla_\mu G^{\mu\nu})\, \xi_\nu + \nabla_\rho 2 \big( G^{\rho\sigma} \xi_\sigma - 2 g^{\mu\sigma,\rho\nu} \nabla_\mu \nabla_{(\sigma} \xi_{\nu)} \big) \right\}. \tag{4.30}$$

On comparing this with the first form of the variation (4.28) with $\hat{\mathcal{L}} = \sqrt{|g|}\, R$, we obtain the identity

$$\int_{\mathcal{M}} d^d x \sqrt{|g|} \left\{ -2(\nabla_\mu G^{\mu\nu}) \xi_\nu + \nabla_\rho \big( 2 R^{\rho\sigma} \xi_\sigma - 4 g^{\mu\sigma,\rho\nu} \nabla_\mu \nabla_{(\sigma} \xi_{\nu)} \big) \right\} = 0. \tag{4.31}$$

This equation is true for arbitrary infinitesimal GCTs. If we take $\xi^\mu$s such that the total derivative term vanishes on the boundary, then we obtain the contracted Bianchi identity $\nabla_\mu G^{\mu\nu} = 0$ as associated gauge identity. We know that this identity is always true in this context. This, in turn, implies that the total derivative term vanishes identically, i.e. the Noether current

$$j^\rho_{\rm N}(\xi) = 2 R^{\rho\sigma} \xi_\sigma - 4 g^{\mu\sigma,\rho\nu} \nabla_\mu \nabla_{(\sigma} \xi_{\nu)} \tag{4.32}$$

is covariantly conserved, $\nabla_\rho j^\rho_{\rm N} = 0$. By massaging this expression a bit, we can rewrite it in the form

$$j^\rho_{\rm N}(\xi) = \nabla_\mu j^{\mu\rho}_{\rm N}(\xi), \qquad j^{\mu\rho}_{\rm N}(\xi) = 2 \nabla^{[\mu} \xi^{\rho]}, \tag{4.33}$$

as is always expected in gauge theories. In Chapter 10 we will study the use of this current to define conserved quantities in GR.

Now we want to see the effect of additional total derivatives in the Einstein–Hilbert action

$$\Delta S_{\rm EH} = \int_{\mathcal{M}} d^d x \sqrt{|g|} \nabla_\mu k^\mu. \tag{4.34}$$

We just have to vary this additional piece in two different ways. One of the variations has the general form of the variation of any general-covariant action (4.28), that is,

$$\delta_\xi \Delta S_{\rm EH} = \int_{\mathcal{M}} d^d x \sqrt{|g|} \nabla_\mu (-\xi^\mu \nabla_\rho k^\rho). \tag{4.35}$$

The variation through the equation of motion gives

$$\delta_\xi \Delta S_{\text{EH}} = \int_{\mathcal{M}} d^d x \sqrt{|g|} \nabla_\mu [k^\rho \nabla_\rho \xi^\mu - \nabla_\rho (\xi^\rho k^\mu)]. \tag{4.36}$$

On combining these two results we find the additional terms in the Noether current,

$$\Delta j_{\text{N}}^{\mu}(\xi) = \nabla_\rho \left(2k^{[\rho}\xi^{\mu]}\right), \qquad \Delta j_{\text{N}}^{\mu\rho}(\xi) = 2k^{[\rho}\xi^{\mu]}. \tag{4.37}$$

### 4.1.3 Coupling to matter

As required by the PEGI, to couple matter to the gravitational field we first rewrite the Minkowskian matter action in the background of the metric that appears in the Einstein–Hilbert action, replacing everywhere $\eta_{\mu\nu}$ by $g_{\mu\nu}$, the volume element $d^d x$ by the GCT-invariant volume element $d^d x \sqrt{|g|}$, and, if necessary (in the most important cases it is not), partial derivatives by covariant derivatives with the Levi-Civita connection. The total action for the gravity–matter system is simply the sum of the Einstein–Hilbert action and the rewritten matter action Eq. (3.273). It is clear that, in general, we will not have to modify the boundary conditions for $\delta g_{\mu\nu}$ due to the addition of the matter action. Thus, the same boundary term as in the vacuum case should work. By varying this with respect to the metric, we obtain the Einstein equation (3.272), where the energy–momentum tensor is defined in Eq. (3.274), which we rewrite here for convenience:

$$T_{\text{matter}}^{\mu\nu} = \frac{2c}{\sqrt{|g|}} \frac{\delta S_{\text{matter}}}{\delta g_{\mu\nu}}. \tag{4.38}$$

First of all, we may ask ourselves about the consistency of Einstein's equation: we know that the (covariant) divergence of the l.h.s. (Einstein's tensor) vanishes due to the contracted Bianchi identity, which can be seen as a consequence of (or a condition for) the invariance of the Einstein–Hilbert action under GCTs. The r.h.s. (the energy–momentum tensor) should also be covariantly divergenceless. In fact, given *any* general-covariant action $S[\phi, g_{\mu\nu}]$, under a general variation of the fields, up to total derivatives,

$$\delta S = \int d^d x \left\{ \frac{\delta S}{\delta \phi} \delta\phi + \frac{\delta S}{\delta g_{\mu\nu}} \delta g_{\mu\nu} \right\}. \tag{4.39}$$

If the field equations of motion are satisfied and the variations are infinitesimal GCTs, then, on integrating by parts, we immediately realize that the gauge identity associated with the invariance under GCTs is always

$$\nabla_\mu \left( \frac{1}{\sqrt{|g|}} \frac{\delta S}{\delta g_{\mu\nu}} \right) = 0. \tag{4.40}$$

If the action is the Einstein–Hilbert action, this is the contracted Bianchi identity. If it is a matter action, this is the general covariantization of the Minkowskian energy–momentum conservation law $\partial_\mu T_{\text{matter}}^{\mu\nu} = 0$, namely

$$\nabla_\mu T_{\text{matter}}^{\mu\nu} = 0. \tag{4.41}$$

This equation ensures the consistency of the Einstein equations. However, it is not a conservation law. We will explain and discuss this problem in Chapter 10.

## 4.2 The Einstein–Hilbert action in different conformal frames

The simplest field that a matter Lagrangian added to the Einstein–Hilbert action can have is a scalar. Matter Lagrangians containing scalars appear in many theories, particularly extended $N > 2$ supergravity theories, Kaluza–Klein theories, and string theory. The scalars' kinetic term usually has the form of a non-linear $\sigma$-model in which the (real) scalars can be understood as coordinates in some target space, which usually is a homogeneous space. Hence, real scalars can take values in different ranges. If a particular scalar (that we will denote by $K$) takes values in $\mathbb{R}^+$, we can always rescale the metric in the Einstein–Hilbert action (which we will henceforth refer to as the *Einstein metric*) via a *Weyl* or *conformal* transformation

$$g_{\mu\nu} \rightarrow K^{\alpha} g_{\mu\nu}, \tag{4.42}$$

where $\alpha$ is some number. Sometimes this transformation is called a change of *conformal frame*. The Einstein–Hilbert action is written in the *Einstein (conformal) frame*. The new metric has the same signature and its equation of motion can be derived from the rescaled action (see Appendix M), which we will generically write in the following form, ignoring the matter Lagrangian:

$$S[g, K] \sim \int d^d x \sqrt{|g|}\, K R(g). \tag{4.43}$$

In the context of string theory, $K = e^{-2\phi}$, where $\phi$ is the *dilaton field*; then the metric is called the *string metric* and it is usually said that the action is written in the *string (conformal) frame*. In the context of Kaluza–Klein theory, if we reduce over a circle, and $K = k$, where $k$ is the Kaluza–Klein scalar and, in more general compactifications, $K$ is a scalar that measures the volume of the internal manifold, then the metric is called the *Kaluza–Klein metric* and we say that the action is written in the *Kaluza–Klein (conformal) frame*. We will define other conformal frames (p-brane frames, etc.) later on.

One important detail that has to be taken into account is the possibility that the vacuum value of the scalar $K$ is not just 1 but some number $K_0$. In that case, the vacuum of the metric $g_{\mu\nu}$ is rescaled by $K_0^{\alpha}$, which is not permissible. We will discuss this important issue at length in Section 15.2.2. In this section we are simply going to explain in detail how to obtain the metric equation of motion by direct variation of the above action (it is obvious that one can always perform the rescaling in the Einstein equation, but, as usual, we expect to obtain more information from the variation of the action).

Using Palatini's identity Eq. (3.285) and Eqs. (4.13), we find

$$\delta S[g, K] = -\int_{\mathcal{M}} d^d x \sqrt{|g|}\, \{K[G^{\mu\nu}\delta g_{\mu\nu} - \nabla_{\mu} v^{\mu}] - R\delta K\}, \tag{4.44}$$

where $v$ is given by Eq. (4.16). We ignore the part proportional to $\delta K$ because in general there will be more terms containing $K$ in the full action. Integrating by parts once gives

$$\delta S[g,K] = -\int_{\mathcal{M}} d^d x\sqrt{|g|}\,\{K[G^{\mu\nu}\delta g_{\mu\nu} + v^\mu \nabla_\mu \ln K] - \nabla_\rho(K v^\rho)\}\,. \tag{4.45}$$

Writing $v^\mu$ as

$$v^\mu = 2g^{\mu\nu,\rho\sigma}\nabla_\rho \delta g_{\sigma\nu} \tag{4.46}$$

and integrating by parts again gives

$$\delta S[g,K] = -\int_{\mathcal{M}} d^d x\sqrt{|g|}K\{G^{\mu\nu} - 2\,[\nabla_\rho \ln K \nabla_\sigma \ln K + \nabla_\rho\nabla_\sigma \ln K]g^{\sigma\mu,\rho\nu}\}\,\delta g_{\mu\nu}$$

$$+\int_{\mathcal{M}} d^d x\sqrt{|g|}\,\nabla_\lambda\Big\{2\Big[K g^{\lambda\nu,\rho\sigma}\nabla_\rho \delta g_{\sigma\nu} - \nabla_\rho K g^{\rho\nu,\lambda\sigma}\delta g_{\sigma\nu}\Big]\Big\}\,. \tag{4.47}$$

If we add the boundary term

$$-2\int d^{d-1}\Sigma\, K\mathcal{K} \tag{4.48}$$

to the action, it is clear that, on imposing the boundary condition Eq. (4.22), we will obtain the following equation for the metric:

$$G^{\alpha\beta} + \Big[\partial^\alpha \ln K \partial^\beta \ln K - g^{\alpha\beta}(\partial \ln K)^2\Big] + \Big[\nabla^\alpha\nabla^\beta \ln K - g^{\alpha\beta}\nabla^2 \ln K\Big] = 0. \tag{4.49}$$

Observe that we obtain a non-trivial equation of motion for the scalar $K$ (or $\log K$) even though there is (apparently) no kinetic term for it in the action we have considered. This is

$$\left(\nabla^2 + \frac{d-2}{2(d-1)}R\right)K = 0. \tag{4.50}$$

Otherwise, by going to a conformal frame in which the kinetic term explicitly disappears, we could eliminate a scalar degree of freedom that would be present in any other frame.

Observe also that this scalar $K$ is not a *conformal scalar*. A conformal scalar $K_{\rm c}$ has the equation of motion

$$\left(\nabla^2 + \frac{d-2}{4(d-1)}R\right)K_{\rm c} = 0, \tag{4.51}$$

which, under simultaneous Weyl rescalings of the metric and the scalar,

$$\tilde{g}_{\mu\nu} = \Omega^2 g_{\mu\nu}, \qquad \tilde{K}_{\rm c} = \Omega^{\frac{2-d}{2}} K_{\rm c}, \tag{4.52}$$

also rescales (i.e. it is invariant),

$$\left(\tilde{\nabla}^2 + \frac{2-d}{4(1-d)}\tilde{R}\right)\tilde{K} = \Omega^{-\frac{(2+d)}{2}}\left(\nabla^2 + \frac{(2-d)}{4(1-d)}R\right)K_{\rm c} = 0. \tag{4.53}$$

To construct an action for a conformal scalar, we have to add to the above action a kinetic term with the right coefficient:

$$S_{\rm c} \sim \int d^d x\sqrt{|g|}K\left[R - \frac{d-1}{d-2}(\partial \ln K)^2\right], \tag{4.54}$$

and then we find that $K = K_{\rm c}^2$, so the action written in terms of the conformal scalar is

$$S[K_{\rm c}] \sim \int d^dx \sqrt{|g|}\, K_{\rm c}^2 \left[ R - \frac{4(d-1)}{d-2} (\partial \ln K_{\rm c})^2 \right]. \tag{4.55}$$

Both the trace of the variation with respect to the metric and the variation with respect to $K_{\rm c}$ lead to the above equation of motion.

When we studied vector and tensor fields living on a general background, we adopted as a sign of their masslessness the existence of gauge transformations that left their equations of motion invariant. If we interpret Eq. (4.51) as that of a scalar field living on a background metric $g_{\mu\nu}$, we may wonder how we can tell whether the scalar field is massless. The only kind of local transformations that we can define for a scalar field are the Weyl transformations given above, and we can define as a massless field one whose equation of motion is invariant under them. Therefore we could consider the conformal scalar as a massless scalar. This means, in particular, that the equation of motion of a massless scalar in a spacetime satisfying $R_{\mu\nu} = \Lambda g_{\mu\nu}$ is

$$\left( \nabla^2 + \frac{d(d-2)}{4(d-1)} \Lambda \right) K_{\rm c} = 0, \tag{4.56}$$

and, as usual, the $\Lambda$ term is not a mass term; on the contrary, its presence ensures the masslessness of the scalar field.

## 4.3 The first-order (Palatini) formalism

This formalism [1016] consists in writing an action in which the metric and the connection (which contains the dependence on the derivatives of the metric) are considered independent variables. The connection is, therefore, not the Levi-Civita connection. It is assumed to be torsion free, i.e. $\Gamma_{[\mu\nu]}{}^{\rho} = 0$, but no other properties (metric compatibility, for example) are assumed. The first-order action contains only derivatives of the connection and it is linear in them. To obtain the equations of motion, we now have to vary the metric and the connection independently. The connection equation of motion gives us the standard relation between the connection and the metric, and the metric equation is, after substitution of the solution to the other equation, nothing but the Einstein equation.

The first-order action turns out to be essentially the Einstein–Hilbert action:[4]

$$\boxed{S[g_{\mu\nu}, \Gamma_{\mu\nu}{}^{\rho}] = \int d^dx \sqrt{|g|}\, g^{\mu\nu} R_{\mu\nu}(\Gamma).} \tag{4.57}$$

All the dependence on the metric is concentrated in the factor $\sqrt{|g|}\, g^{\mu\nu}$ since the Ricci tensor depends only on the connection and its derivatives, as shown in Eq. (1.33).

We stress that, since the connection is a variable, and it is not the Levi-Civita connection, one cannot use the standard property

$$\int d^dx \sqrt{|g|}\, \nabla_\mu \xi^\mu = \int d^dx\, \partial_\mu \left( \sqrt{|g|}\, \xi^\mu \right). \tag{4.58}$$

[4] We set $\chi = 1$ throughout this section.

The calculations are simplified if we use the density

$$\mathfrak{g}^{\mu\nu} = \sqrt{|g|}\, g^{\mu\nu} \tag{4.59}$$

as a variable. Furthermore, we are not going to assume in our derivation of the equations of motion either the symmetry of the connection or the symmetry of the "metric," which we will impose at the very end. In this way, we can obtain with minimum extra work the equations of the Einstein–Straus–Kaufman [492, 498, 501, 503] non-symmetric gravity theory (NGT), which was (unsuccessfully) proposed as a unified relativistic theory of gravitation and electromagnetism in which the antisymmetric part of the "metric" $g^{[\mu\nu]}$ should be identified with the electromagnetic-field-strength tensor[5] $F^{\mu\nu}$.

In the NGT the inverse "metric" is also denoted by $g^{\mu\nu}$ and satisfies

$$g^{\mu\nu} g_{\nu\rho} = \delta^{\mu}{}_{\rho}, \qquad g_{\alpha\beta} g^{\beta\gamma} = \delta_{\alpha}{}^{\gamma}, \tag{4.60}$$

but $g^{\mu\nu} g_{\mu\rho} \neq \delta^{\nu}{}_{\rho}$. Also, we cannot use it to lower or raise indices.

Let us now vary the above action with respect to the metric and connection. By using Palatini's identity Eq. (3.286), we find[6]

$$\begin{aligned} \delta S &= \int d^d x \left\{ \delta \mathfrak{g}^{\alpha\beta} R_{\alpha\beta}(\Gamma) + \mathfrak{g}^{\alpha\beta} [\nabla_\alpha \delta\Gamma_{\rho\beta}{}^{\rho} - \nabla_\rho \delta\Gamma_{\alpha\beta}{}^{\rho} - T_{\alpha\rho}{}^{\sigma} \delta\Gamma_{\sigma\beta}{}^{\rho}] \right\} \\ &= \int d^d x \left\{ \delta \mathfrak{g}^{\alpha\beta} R_{\alpha\beta}(\Gamma) + \nabla_\rho \left[ \left( \mathfrak{g}^{\rho\beta} \delta_\sigma{}^{\alpha} - \mathfrak{g}^{\alpha\beta} \delta_\sigma{}^{\rho} \right) \delta\Gamma_{\alpha\beta}{}^{\sigma} \right] \right. \\ &\qquad \left. + \left[ \nabla_\sigma \mathfrak{g}^{\alpha\beta} - \nabla_\rho \mathfrak{g}^{\rho\beta} \delta_\sigma{}^{\alpha} - \mathfrak{g}^{\lambda\beta} T_{\lambda\sigma}{}^{\alpha} \right] \delta\Gamma_{\alpha\beta}{}^{\sigma} \right\}. \end{aligned} \tag{4.61}$$

Using now the identity for vector densities of unit weight

$$\nabla_\mu \mathfrak{v}^\mu = \partial_\mu \mathfrak{v}^\mu + \mathfrak{v}^\mu T_{\mu\rho}{}^{\rho}, \tag{4.62}$$

and integrating by parts, we obtain, up to a total derivative

$$\begin{aligned} \delta S = \int d^d x \, \{ &\delta \mathfrak{g}^{\alpha\beta} R_{\alpha\beta}(\Gamma) + [T_{\rho\delta}{}^{\delta} ( \mathfrak{g}^{\rho\beta} \delta_\sigma{}^{\alpha} - \mathfrak{g}^{\alpha\beta} \delta_\sigma{}^{\rho} ) \\ &+ \nabla_\sigma \mathfrak{g}^{\alpha\beta} - \nabla_\rho \mathfrak{g}^{\rho\beta} \delta_\sigma{}^{\alpha} - \mathfrak{g}^{\lambda\beta} T_{\lambda\sigma}{}^{\alpha} ] \, \delta\Gamma_{\alpha\beta}{}^{\sigma} \}. \end{aligned} \tag{4.63}$$

Since the metric and the connection are independent, we obtain two equations from the minimal-action principle:

$$\begin{aligned} \frac{\delta S}{\delta \mathfrak{g}^{\alpha\beta}} &= R_{\alpha\beta}(\Gamma) = 0, \\ \frac{\delta S}{\delta \Gamma_{\alpha\beta}{}^{\gamma}} &= \nabla_\gamma \mathfrak{g}^{\alpha\beta} - \nabla_\rho \mathfrak{g}^{\rho\beta} \delta_\gamma{}^{\alpha} - \mathfrak{g}^{\lambda\beta} T_{\lambda\gamma}{}^{\alpha} + \mathfrak{g}^{\rho\beta} \delta_\gamma{}^{\alpha} T_{\rho\delta}{}^{\delta} - \mathfrak{g}^{\alpha\beta} T_{\gamma\delta}{}^{\delta} = 0. \end{aligned} \tag{4.64}$$

[5] See also Refs. [1111, 1112, 894, 1185]. A more recent NGT that reinterprets Einstein's theory was proposed in Ref. [954]. In it the antisymmetric part of the metric is also considered as a sort of new gravitational interaction. Clearly, the weak-field limit cannot be the Fierz–Pauli theory but contains another field corresponding to the antisymmetric part of the metric. While this suggests a relation with string theory, which also contains a rank-2 antisymmetric tensor (the Kalb–Ramond field), these two fields appear in quite different ways: the Kalb–Ramond field has an extra gauge symmetry, which allows it to be consistently quantized, whereas the antisymmetric part of the NGT "metric" transforms only under GCTs. See Refs. [372, 414, 415, 849].

[6] For the sake of generality, we allow for the possibility of having torsion.

The first equation would be the Einstein equation if the connection were the Levi-Civita connection. Observe that, if we couple bosonic (scalar or vector) matter minimally to this action, we do not have to introduce any terms containing the connection. Thus, the equation for the connection would not change and the equation for the metric would become the Einstein equation with non-vanishing energy–momentum tensor (again, if the connection were the Levi-Civita connection).

To find the relation between the connection and the metric, we have to solve the second equation. It is convenient to use a new connection $\tilde{\Gamma}$, defined by

$$\tilde{\Gamma}_{\mu\nu}{}^{\rho} = \Gamma_{\mu\nu}{}^{\rho} + \frac{1}{d-1} T_{\mu\sigma}{}^{\sigma} \delta_{\nu}{}^{\rho}. \tag{4.65}$$

Observe that the new connection $\tilde{\Gamma}$ does not completely determine the old one, $\Gamma$. In fact, if we shift $\Gamma$ by an arbitrary vector $f_\mu$ according to

$$\Gamma_{\mu\nu}{}^{\rho} \to \Gamma_{\mu\nu}{}^{\rho} + f_\mu \delta_\nu{}^{\rho}, \tag{4.66}$$

the connection $\tilde{\Gamma}$ is not modified. Thus, the expression for $\Gamma$ in terms of $\tilde{\Gamma}$ is

$$\Gamma_{\mu\nu}{}^{\rho} = \tilde{\Gamma}_{\mu\nu}{}^{\rho} + f_\mu \delta_\nu{}^{\rho}, \tag{4.67}$$

where $f_\mu$ cannot be determined from $\tilde{\Gamma}$. The new connection allows us to rewrite the second equation in the form

$$\partial_\sigma \mathfrak{g}^{\alpha\beta} + \tilde{\Gamma}_{\delta\sigma}{}^{\alpha} \mathfrak{g}^{\delta\beta} + \mathfrak{g}^{\alpha\delta} \tilde{\Gamma}_{\sigma\delta}{}^{\beta} - \mathfrak{g}^{\alpha\beta} \tilde{\Gamma}_{\sigma\delta}{}^{\delta} = 0. \tag{4.68}$$

On contracting in the above equation the indices $\sigma$ with $\alpha$ and $\sigma$ with $\beta$, taking the difference, and using the property

$$\tilde{\Gamma}_{\mu\rho}{}^{\rho} = \tilde{\Gamma}_{\rho\mu}{}^{\rho}, \tag{4.69}$$

we arrive at the Maxwell-like equation for the antisymmetric part of $\mathfrak{g}$:

$$\partial_\alpha \mathfrak{g}^{[\alpha\beta]} = 0. \tag{4.70}$$

By contracting now Eq. (4.68) with $g_{\alpha\beta}/\sqrt{|g|}$, we obtain

$$\partial_\sigma \ln \sqrt{|g|} = \tilde{\Gamma}_{\sigma\alpha}{}^{\alpha}, \tag{4.71}$$

and, on plugging this back into Eq. (4.68), we obtain an equation for the inverse metric:

$$\partial_\sigma g^{\alpha\beta} + \tilde{\Gamma}_{\sigma\delta}{}^{\beta} g^{\alpha\delta} + \tilde{\Gamma}_{\delta\sigma}{}^{\alpha} g^{\delta\beta} = 0. \tag{4.72}$$

We now multiply by the inverse "metrics" $g_{\gamma\alpha}$ and $g_{\beta\varphi}$ to obtain, at last,

$$\partial_\sigma g_{\gamma\varphi} - \tilde{\Gamma}_{\sigma\gamma}{}^{\beta} g_{\beta\varphi} - \tilde{\Gamma}_{\varphi\sigma}{}^{\alpha} g_{\gamma\alpha} = 0. \tag{4.73}$$

Although we have started with the connection $\Gamma$, the above equation allows us only to solve for the connection $\tilde{\Gamma}$ in terms of the metric.

It is easy to particularize this general setup for the case that interests us: a symmetric metric $g^{[\mu\nu]} = 0$ and a torsion-free connection $\Gamma_{[\mu\nu]}{}^{\rho} = 0$. In this case, $R_{\mu\nu}(\Gamma)$ is automatically symmetric, $\Gamma = \tilde{\Gamma}$, and Eq. (4.73) is the metric-compatibility equation $\nabla_\sigma g_{\gamma\varphi} = 0$ whose solution is (see Chapter 1) the Levi-Civita connection (Christoffel symbols). Then we recover the vacuum Einstein equation.

In the presence of matter, this formalism leads to the standard Einstein equation if the affine connection does not occur in the matter action, which is the case for scalars and gauge fields. Otherwise, the equation for the connection is modified and, in general, the connection has torsion. Actually, this can turn into an advantage of this formalism in certain cases (e.g. supergravity theories), although we develop a formalism to couple fermions to gravity in Section 4.4.

### *4.3.1 The purely affine theory*

We have seen two action principles leading to the Einstein equations. In the first one, the fundamental variables were the components of the metric tensor. In the second one, the fundamental variables were both the components of the metric tensor and the components of the affine connection. For completeness, we are going to see briefly that it is actually possible to write an action leading to the vacuum Einstein equations in the presence of a cosmological constant that is a functional of the components of the affine connection alone.

The simplest tensors that one can construct from the affine connection and its first derivatives are the curvature and Ricci tensors. To write an action, we need to integrate a density. The simplest density constructed from these two tensors alone that we can think of is the square root of the determinant of the Ricci tensor, so

$$S \sim \int d^d x \sqrt{|R_{\mu\nu}(\Gamma)|} \;\; \Rightarrow \delta S = \int d^d x \, \frac{\delta S}{\delta R_{\mu\nu}} \delta R_{\mu\nu}(\Gamma). \tag{4.74}$$

The crucial point in this formalism is the definition

$$\frac{\delta S}{\delta R_{\mu\nu}} \equiv \frac{\alpha}{2} \mathfrak{g}^{\mu\nu}, \tag{4.75}$$

where $\alpha$ is some constant and the metric density is $\sqrt{|g|} g^{\mu\nu}$, which does not need to be symmetric. Actually, it has the same symmetry as the Ricci tensor. Thus, if we want to have a symmetric metric, we have to take the determinant of the symmetric part of the Ricci tensor in the action, but the connection is arbitrary. From Eq. (4.75) we find an equation with the structure of the cosmological Einstein equation:

$$R_{\mu\nu}(\Gamma) = \Lambda g_{\mu\nu}, \qquad \Lambda = \alpha^{\frac{2}{d-2}}. \tag{4.76}$$

On substituting this into the variation of the action, we obtain

$$\delta S = \frac{\Lambda^{\frac{d-2}{2}}}{2} \int d^d x \, \mathfrak{g}^{\mu\nu} \delta R_{\mu\nu}(\Gamma), \tag{4.77}$$

and, using Palatini's identity, we find the same equation of motion for the connection (4.64) as in the NGT theory. If the metric is symmetric, this equation tells us that the connection is the Levi-Civita connection.

To obtain the Einstein equations in the presence of matter in this formalism, we must use more complicated techniques.[7]

## 4.4 The Cartan–Sciama–Kibble theory

The formalism developed so far can be used to couple matter fields that behave as tensors under GCTs. In general, the tensorial character of the matter fields under GCTs is determined from their behavior under Poincaré transformations and the only possible ambiguity is whether the field is just a tensor or a tensor density. However, this identification does not work for spinor fields, because it is based on a relation that exists only between the tensor representations of the Poincaré group and tensor representations of the diffeomorphism group. Thus, to couple fermions to gravity, we must first find out how to define spinors in a general curved spacetime.

In a classical paper,[8] [1252], Weyl proposed to define spinors in *tangent space*, using an orthonormal Vielbein basis $\{e^a{}_\mu\}$ as fundamental fields instead of the metric, and developed a formalism that is invariant under Lorentz transformations of this Vielbein basis, even if we perform a different Lorentz transformation in (the tangent space associated with) every spacetime point. Thus, in $d$ spacetime dimensions, the $d(d+1)/2$ *off-shell* degrees of freedom of the metric (the number of independent components of a $d \times d$ symmetric matrix) are replaced by the same number of off-shell degrees of freedom of the Vielbein (the number of independent components of a generic $d \times d$ matrix minus the $d(d-1)/2$ independent local Lorentz transformations). In modern language,[9] this is a gauge theory of the Lorentz group $\mathrm{SO}(1, d-1)$ and requires the introduction of a Lorentz covariant derivative $\mathcal{D}_\mu$ and a Lorentz (spin) connection $\omega_\mu{}^{ab}$. Otherwise, the Vielbeins will describe more degrees of freedom than the metric.

If we want to recover GR, however, we do not want to introduce new fields apart from the metric (Vielbeins), and thus we have to relate the spin connection to the Vielbeins, destroying the similarity to a standard Yang–Mills theory in which the connection is the dynamical field. The natural way to relate connection and Vielbeins is through the first Vielbein postulate Eqs. (1.91), which connects the spin and the affine connections by Eq. (1.92). This does not seem to help much, because the affine connection is completely undetermined. However, metric compatibility is automatic for spin and affine connections satisfying the first Vielbein postulate, because, by assumption, the spin connection $\omega_\mu{}^{ab}$ is antisymmetric in the indices $ab$, which implies $\nabla_\mu \eta_{ab} = 0$, which, with the first Vielbein postulate, implies $\nabla_\mu g_{\rho\sigma} = 0$. Therefore, the first Vielbein postulate determines the connection in terms of the Vielbein up to the torsion term. Now if we want to have as fundamental fields the Vielbeins alone, we need to impose the vanishing of torsion. In that case, the affine connection is the Levi-Civita connection $\Gamma(g)$ whose components

---

[7] See e.g. Ref. [924] and references therein. Further generalizations of the Einstein–Hilbert action are also reviewed there.

[8] A guide to the old literature on this formalism and its generalizations to include torsion can be found in Ref. [737]. A pedagogical introduction to this formalism is given in Ref. [1090] (see also Ref. [1091]). A more recent reference is Ref. [1133].

[9] The basic formalism of Yang–Mills gauge theories is developed in Appendix A and, for the Lorentz group in particular, for the present application, in Section 1.4.

are the Christoffel symbols Eq. (1.44) and then the relation Eq. (1.92) implies that the spin connection is the Cartan spin connection $\omega(e)$ given in Eq. (1.100). This case will be treated in Section 4.4.1. The possibility of including torsion will be studied in Section 4.4.2.

The first Vielbein postulate can be imposed from the beginning (the second-order formalism in which the only fundamental fields are the Vielbein components) or via the spin-connection equation of motion (the first-order formalism in which both the Vielbein and the spin-connection components are independent, fundamental fields). In the first-order formalism the theory resembles more a standard Yang–Mills theory, as we will discuss in Section 4.4.4.

### *4.4.1 The coupling of gravity to fermions*

In this section, as a warm-up exercise, we want to study the coupling of fermions to gravity using the torsionless Cartan (Levi-Civita) connection (see e.g. Ref. [257]).

Let us first summarize Weyl's recipe: to couple spinors to gravity we replace all partial derivatives in the special-relativistic action for Lorentz (or total-)covariant derivatives by the Cartan–Levi-Civita derivatives and the Minkowski metric $\eta_{\mu\nu}$ by the general metric $g_{\mu\nu}$ or by the Vielbeins $e^a{}_\mu$ if necessary.[10] Since the Cartan spin connection cannot be expressed in terms of the metric, it is clear that the fundamental variables in this formalism will be the Vielbeins. This does not require any change in the Einstein–Hilbert action since we simply have to use

$$\frac{\delta S_{\rm EH}[e]}{\delta e^a{}_\mu} = 2\frac{\delta S_{\rm EH}[g]}{\delta g_{\rho\sigma}} e_{a\,(\rho} g_{\sigma)}{}^\mu = -\frac{2}{\chi^2} e G_a{}^\mu, \tag{4.78}$$

and, correspondingly, redefine the matter energy–momentum tensor

$$T_{\text{matter}\,a}{}^\mu = \frac{c}{e}\frac{\delta S_{\text{matter}}[\varphi, e]}{\delta e^a{}_\mu}, \qquad e = \det(e^a{}_\mu) = \sqrt{|g|}. \tag{4.79}$$

Observe that, with this new definition, the energy–momentum tensor (that we can call the *Vielbein energy–momentum tensor*) does not have to be symmetric. However, we can prove that it is symmetric when the matter equations of motion hold: let us consider the variation of the matter action under a local Lorentz transformation with parameter $\sigma^{ab}(x)$, which we know leaves the Lagrangian invariant. Up to a total derivative,

$$\delta_\sigma S_{\text{matter}}[\varphi, e] = \int d^dx \left\{ \frac{\delta S_{\text{matter}}}{\delta\varphi}\delta_\sigma\varphi + \frac{\delta S_{\text{matter}}}{\delta e^a{}_\mu}\delta_\sigma e^a{}_\mu \right\}. \tag{4.80}$$

Using the definition of the energy–momentum tensor and the transformation rules (assuming that $\varphi$ transforms in the representation $r$ of the Lorentz group)

$$\delta_\sigma\varphi^\alpha = \tfrac{1}{2}\sigma^{ab}\Gamma_r(M_{ab})^\alpha{}_\beta\varphi^\beta, \qquad \delta_\sigma e^a{}_\mu = \tfrac{1}{2}\sigma^{cd}\Gamma_{\rm v}(M_{cd})^a{}_b e^b{}_\mu = \sigma^a{}_\mu, \tag{4.81}$$

[10] We could ask whether this formalism should also be applied to other fields: for instance, whether we should consider the Maxwell field as a tangential vector field. The answer is that we can do it and the choice of torsionless connection that we have made ensures that there is no difference, although we gain more insight if we consider the Maxwell field as a tangential vector field. In the presence of torsion and in more general contexts this will be impossible.

we find the Bianchi identity

$$T_{\text{matter}\,[ab]} = -\frac{1}{2e}\frac{\delta S_{\text{matter}}}{\delta\varphi^\alpha}\Gamma_r(M_{ab})^\alpha{}_\beta\varphi^\beta, \tag{4.82}$$

which vanishes on-shell.

We can also use the invariance under reparametrizations of the matter action to show that the Vielbein energy–momentum tensor is covariantly conserved on-shell:

$$\nabla_\mu T_a{}^\mu = 0. \tag{4.83}$$

As for the Vielbein energy–momentum tensor, we can try to determine its form by assuming the validity of a more or less standard matter Lagrangian, namely a standard Lagrangian whose dependence on the Vielbeins comes from two sources: the spin connection and the rest. It is easy to convince oneself by looking at simple examples that "the rest," which depends only algebraically on the Vielbeins, yields $e$ times the canonical energy–momentum tensor when $e^a{}_\mu = \delta^a{}_\mu$:

$$eT_{\text{can}\,a}{}^\mu = -\frac{\partial\mathcal{L}_{\text{matter}}}{\partial\nabla_\mu\varphi}\nabla_a\varphi + e_a{}^\mu\mathcal{L}_{\text{matter}}. \tag{4.84}$$

The dependence of the Lagrangian through the spin connection can be computed by observing that the matter Lagrangian depends on derivatives of the Vielbeins only through the Cartan spin connection which appears in covariant derivatives of the field $\varphi^\alpha$,

$$\mathcal{D}_\mu\varphi^\alpha = \partial_\mu\varphi^\alpha - \tfrac{1}{2}\omega_\mu{}^{ab}(e)\Gamma_r(M_{ab})^\alpha{}_\beta\varphi^\beta. \tag{4.85}$$

The contribution of these terms to the energy–momentum tensor is given by

$$\frac{\partial\mathcal{L}_{\text{matter}}}{\partial\omega_{\rho bc}}\frac{\partial\omega_{\rho bc}}{\partial e^a{}_\mu} - \partial_\nu\left(\frac{\partial\mathcal{L}_{\text{matter}}}{\partial\omega_{\rho bc}}\frac{\partial\omega_{\rho bc}}{\partial\partial_\nu e^a{}_\mu}\right). \tag{4.86}$$

Using[11]

$$\omega_{\rho bc} = 2\Delta_{\rho bc}{}^{\sigma\tau d}\Omega_{\sigma\tau d}, \qquad \Delta_{\rho bc}{}^{\sigma\tau d} = \tfrac{1}{2}\left\{\delta_\rho{}^\sigma e_c{}^\tau\delta_b{}^d + e_b{}^\sigma e_c{}^\tau e^d{}_\rho - \delta_\rho{}^\sigma e_b{}^\tau\delta_c{}^d\right\}, \tag{4.88}$$

we find that the contribution to the energy–momentum tensor of the spin connection is given by

$$2\frac{\partial\mathcal{L}_{\text{matter}}}{\partial\omega_{\rho bc}}\frac{\partial\Delta_{\rho bc}{}^{\sigma\tau d}}{\partial e^a{}_\mu}\Omega_{\sigma\tau d} - 2\partial_\nu\left(\frac{\partial\mathcal{L}_{\text{matter}}}{\partial\omega_{\rho bc}}\Delta_{\rho bc}{}^{\sigma\tau d}\frac{\partial\Omega_{\sigma\tau d}}{\partial\partial_\nu e^a{}_\mu}\right). \tag{4.89}$$

[11] Two similar useful relations are

$$K_{\mu ab} = \Delta_{\mu ab}{}^{\sigma\tau d}T_{\sigma\tau d}, \qquad \left\{ {\sigma \atop \mu\nu} \right\} g_{\sigma\rho} = -\Delta_{\nu\rho\mu}{}^{\alpha\beta\gamma}\partial_\alpha g_{\beta\gamma}. \tag{4.87}$$

Since the spin connection occurs in the matter Lagrangian only via covariant derivatives of the matter field $\varphi$, it is easy to see that

$$\frac{\partial \mathcal{L}_{\text{matter}}}{\partial \omega_{\rho bc}} = -e S^{\rho bc}, \tag{4.90}$$

and, using

$$S^{\rho bc} \Delta_{\rho bc}{}^{\sigma\tau d} = \tfrac{1}{2} \Psi^{\sigma\tau d}, \tag{4.91}$$

we obtain

$$\begin{aligned} 2 \frac{\partial \mathcal{L}_{\text{matter}}}{\partial \omega_{\rho bc}} \frac{\partial \Delta_{\rho bc}{}^{\sigma\tau d}}{\partial e^a{}_\mu} \Omega_{\sigma\tau d} &= -e \Psi^{\rho\mu}{}_b \omega_{\rho a}{}^b, \\ -2\partial_\nu \left( \frac{\partial \mathcal{L}_{\text{matter}}}{\partial \omega_{\rho bc}} \Delta_{\rho bc}{}^{\sigma\tau d} \frac{\partial \Omega_{\sigma\tau d}}{\partial \partial_\nu e^a{}_\mu} \right) &= \partial_\nu (e \Psi^{\nu\mu}{}_a), \end{aligned} \tag{4.92}$$

which add up to

$$e \nabla_\nu \Psi^{\nu\mu}{}_a, \tag{4.93}$$

and so we have (observe the order of indices)

$$T_a{}^\mu = T_{\text{can}\, a}{}^\mu + \nabla_\nu \Psi^{\nu\mu}{}_a, \tag{4.94}$$

which is the relation between the Vielbein energy–momentum tensor and the canonical one. If we subtract the antisymmetric part of the Vielbein energy–momentum tensor, we obtain a symmetric tensor that is conserved when the matter equations of motion hold. When $e^a{}_\mu = \delta^a{}_\mu$, this symmetric tensor becomes the Belinfante tensor, proving the relation between the Belinfante tensor and the metric (Rosenfeld) energy–momentum tensor that we mentioned in Section 2.4.1.

*Example: a Dirac spinor.* Let us now apply this recipe to a Dirac spinor.[12] A Dirac spinor $\psi^\alpha$ has only a spinorial index (which we usually hide). Thus, we are going to assume that it transforms as a spinor in tangent space and as a scalar under GCTs. Thus, the total covariant derivative $\nabla_\mu$ coincides with the Lorentz-covariant derivative $\mathcal{D}_\mu$ acting on it:

$$\nabla_\mu \psi = \mathcal{D}_\mu \psi = \left( \partial_\mu - \tfrac{1}{4} \omega_\mu{}^{ab} \gamma_{ab} \right) \psi. \tag{4.95}$$

In the special-relativistic Lagrangian of the Dirac spinor Eq. (2.63) the partial derivative appears contracted with a constant gamma matrix. Now we have to distinguish between the derivative index, which is a world-tensor index, and the gamma-matrix index, which is a Lorentz (tangent-space) index, and to contract both indices we have to use a Vielbein

$$\nabla\!\!\!\!/\, \psi = e_a{}^\mu \gamma^a \nabla_\mu \psi. \tag{4.96}$$

Finally, we also need the covariant derivative on the Dirac conjugate. The Dirac conjugate $\bar{\psi}_\alpha$ transforms covariantly (as opposed to the spinor $\psi^\alpha$, which transforms contravariantly). Then, applying the definitions in Section 1.4,

$$\bar{\psi} \overleftarrow{\nabla}_\mu \equiv \nabla_\mu \bar{\psi} = \partial_\mu \bar{\psi} + \tfrac{1}{4} \omega_\mu{}^{ab} \bar{\psi} \gamma_{ab}. \tag{4.97}$$

[12] The special-relativistic Dirac spinor was studied in Section 2.4.1.

With all these elements we can immediately write the action

$$S_{\text{matter}} = \int d^d x\, e \Big\{ \tfrac{1}{2}(i\bar{\psi} \nabla\!\!\!/ \psi - i\bar{\psi} \overleftarrow{\nabla\!\!\!/} \psi) - m\bar{\psi}\psi \Big\} . \tag{4.98}$$

The equations of motion are the evident covariantization of the flat-space ones:

$$(i \nabla\!\!\!/ - m)\psi = 0, \tag{4.99}$$

and the spin-angular-momentum tensor $S^{\mu}{}_{ab}$ and spin–energy potential $\Psi^{\mu\nu}{}_{a}$ are identical to those calculated in Section 2.4.1. By varying with respect to the Vielbeins, we find the Vielbein energy–momentum tensor, which has the general form Eq. (4.94) with

$$T_{\text{can}\,a}{}^{\mu} = -\frac{i}{2}\bar{\psi}\gamma^{\mu}\nabla_a\psi + \frac{i}{2}\nabla_a\bar{\psi}\gamma^{\mu}\psi + e_a{}^{\mu}\mathcal{L}_{\text{matter}}, \tag{4.100}$$

giving

$$\begin{aligned} T_a{}^{\mu} = {} & -\frac{i}{2}\bar{\psi}(\gamma^{\mu}e_a{}^{\nu} + \gamma_a g^{\mu\nu})\nabla_{\nu}\psi + \frac{i}{2}\nabla_{\nu}\bar{\psi}(\gamma^{\mu}e_a{}^{\nu} + \gamma_a g^{\mu\nu})\psi \\ & + e_a{}^{\mu}\mathcal{L}_{\text{matter}} - \frac{i}{2}\bar{\psi}\gamma^{\mu}{}_a \nabla\!\!\!/ \psi + \frac{i}{2}\bar{\psi} \overleftarrow{\nabla\!\!\!/} \gamma_a{}^{\mu}\psi, \end{aligned} \tag{4.101}$$

which is not symmetric because of the last two terms, which vanish on-shell, as expected. This is what saves the consistency of the Einstein equation

$$G_a{}^{\mu} = \frac{\chi^2}{2} T_a{}^{\mu}, \tag{4.102}$$

whose l.h.s. is symmetric in the absence of torsion. This is not too different from the way in which consistency is achieved in the standard GR theory in which the l.h.s. is divergenceless (due to the contracted Bianchi identity) and the r.h.s. is divergenceless only when the matter equations of motion are satisfied.

### *4.4.2 The coupling to torsion: the CSK theory*

Perhaps the simplest generalization of GR one can think of is the use of a (still metric-compatible) connection with non-vanishing torsion $T_{\mu\nu}{}^{\rho}$. Now, the torsion is a new field whose value we have to determine. The simplest possibility is to consider it a fundamental field and just include it in a generalized Einstein–Hilbert action and in the covariant derivatives acting on matter fields (minimal coupling). Then its equation of motion is determined, as usual, by varying the action with respect to it and imposing the vanishing of the variation. As we are going to see, the resulting equation of motion is algebraic and simply gives the torsion as a function of other fields. In fact, in the torsion equation of motion one can see the matter spin–energy potential $\Psi^{\mu\nu}{}_{a}$ as the source for torsion $T_{\mu\nu}{}^{a}$. This is essentially the definition of the Cartan–Sciama–Kibble (CSK) theory (reviewed in Ref. [737], in a more pedagogical form in Ref. [1090], and in the Newman–Penrose formalism in Ref. [1032]).

Why should we couple intrinsic spin to torsion? The CSK theory is based on Weyl's Vielbein formalism in which there are two distinct gauge symmetries: reparametrizations and local Lorentz transformations in tangent space. Reparametrization invariance leads to the coupling of the energy–momentum tensor to the metric and, similarly, local Lorentz invariance leads to the coupling of the spin–energy potential to torsion.

In the CSK theory, torsion is not a propagating new field. Furthermore, there is no way to couple it to vector gauge potentials without breaking the gauge symmetry, which is inadmissible. However, it is possible to generalize the theory further in such a way as to have propagating torsion. The most popular way of doing this, which occurs naturally in supergravity and string theory [1104], is to consider torsion as the 3-form field strength of a 2-form (*Kalb–Ramond*) field $B_{\mu\nu}$:

$$T_{\mu\nu\rho} = 3\partial_{[\mu}B_{\nu\rho]} \equiv H_{\mu\nu\rho}. \tag{4.103}$$

This particular form of torsion can be consistently coupled to gauge vector fields through the addition to the field strength of the gauge-field Chern–Simons 3-form $\omega_3$, Eq. (A.50),

$$H_{\mu\nu\rho} = 3\partial_{[\mu}B_{\nu\rho]} + \omega_{3\,\mu\nu\rho}, \tag{4.104}$$

and modifying the gauge-transformation rule for $B_{\mu\nu}$ to make $H_{\mu\nu\rho}$ gauge invariant. Since we will encounter this propagating torsion later on, we postpone its discussion until then. One of the reasons why we are reviewing the CSK theory here is precisely that it constitutes an important link in the evolutionary chain that goes from GR to supergravity and superstring theories. The next link in the chain will be the gauge theories of the Poincaré and (anti-)de Sitter groups that we will also study in this chapter.

Let us first consider the generalization of the Einstein–Hilbert action in the CSK theory,

$$S_{\text{CSK}}[e^a{}_\mu, T_{\mu\nu}{}^a] = \frac{1}{\chi^2}\int d^dx\, eR(e,T), \tag{4.105}$$

where $R(e,T)$ is the Ricci scalar constructed from the curvature associated with the metric-compatible torsionful spin connection Eq. (1.100) or its associated affine connection given in Eq. (1.54) and is, therefore, a function of the Vielbeins and torsion. We have chosen the Vielbeins instead of the metric as the fundamental fields since the CSK theory is relevant only in the coupling of gravity to fermions because, as we have already said, the coupling of torsion to vector fields by substitution of partial derivatives for covariant derivatives necessarily breaks their gauge invariance.

We now vary the above action with respect to the Vielbeins and torsion. First, we vary with respect to the metric and connection. Using Palatini's identity Eq. (3.286), we find

$$\delta S_{\text{CSK}} = \frac{1}{\chi^2}\int d^dx\, e\{-G^{\alpha\beta}\delta g_{\alpha\beta} + g^{\alpha\beta}[\nabla_\alpha \delta\Gamma_{\rho\beta}{}^\rho - \nabla_\rho\delta\Gamma_{\alpha\beta}{}^\rho - T_{\alpha\rho}{}^\sigma\delta\Gamma_{\sigma\beta}{}^\rho]\}\,. \tag{4.106}$$

The covariant derivatives can be split into Levi-Civita covariant derivatives $\overset{\{\}}{\nabla}_\mu$, which can be integrated away, and contorsion pieces. After some calculations, we find

$$\delta S_{\text{CSK}} = \frac{1}{\chi^2}\int d^dx\, e\Big\{-G^{\alpha\beta}\delta g_{\alpha\beta} + \delta\Gamma_{\alpha\beta}{}^\gamma g^{\beta\delta}\, \overset{*}{T}_{\gamma\delta}{}^\alpha\Big\}, \tag{4.107}$$

where $\overset{*}{T}$ is the modified torsion tensor defined in Eq. (1.35). Using now Eq. (3.283), we find, at last,

$$\delta S_{\rm CSK} = \frac{1}{\chi^2}\int d^dx\, e\Big\{-\Big[G^{\alpha\beta} - \overset{*}{\nabla}_\mu \overset{*}{T}{}^{\mu\alpha\beta}\Big]\,\delta g_{\alpha\beta} + \tfrac{1}{2}\Big[\overset{*}{T}_\gamma{}^{\alpha\beta} - \overset{*}{T}_\gamma{}^{\beta\alpha} - \overset{*}{T}{}^{\alpha\beta}{}_\gamma\Big]\,\delta T_{\alpha\beta}{}^\gamma\Big\}, \tag{4.108}$$

where we have also used the modified divergence $\overset{*}{\nabla}_\mu$ defined in Eq. (1.35).

Now we couple the pure gravity Lagrangian to the matter Lagrangian and use the definition of the Vielbein energy–momentum tensor Eq. (4.79) and the following definition of the spin–energy potential, which generalizes Eq. (2.88),

$$\Psi_{\rm matter}{}^{\mu\nu}{}_a = -\frac{2c}{e}\frac{\delta S_{\rm matter}}{\delta T_{\mu\nu}{}^a}, \tag{4.109}$$

to obtain the equations of the CSK theory:

$$\begin{aligned} G^{(\alpha\beta)} - \overset{*}{\nabla}_\mu \overset{*}{T}{}^{\mu(\alpha\beta)} &= \frac{\chi^2}{2} T_{\rm matter}{}^{\alpha\beta}, \\ \tfrac{1}{2}\Big[\overset{*}{T}_\gamma{}^{\alpha\beta} - \overset{*}{T}_\gamma{}^{\beta\alpha} - \overset{*}{T}{}^{\alpha\beta}{}_\gamma\Big] &= \frac{\chi^2}{2}\Psi_{\rm matter}{}^{\alpha\beta}{}_\gamma. \end{aligned} \tag{4.110}$$

We have taken into account in the l.h.s. of the first equation that only the symmetric part contributes to it, even though the r.h.s. (the Vielbein energy–momentum tensor) is not symmetric in general (we have seen that the antisymmetric part vanishes on-shell).

These equations can be rewritten in a more suggestive form: taking the modified divergence of the second equation, we find the equation

$$\overset{*}{\nabla}_\mu \overset{*}{T}{}^{\mu(\alpha\beta)} - \tfrac{1}{2}\overset{*}{\nabla}_\mu \overset{*}{T}{}^{\alpha\beta\mu} = \frac{\chi^2}{2}\overset{*}{\nabla}_\mu \Psi_{\rm matter}{}^{\mu\alpha\beta}, \tag{4.111}$$

which, when subtracted from the first Eq. (4.110), yields a more elegant equation,

$$G^{\alpha\beta} = \frac{\chi^2}{2} T_{\rm can}{}^{\alpha\beta}, \tag{4.112}$$

where we have used Eq. (1.34) and have defined the canonical energy–momentum tensor here by

$$T_{\rm can}{}^{\beta\alpha} = T_{\rm matter}{}^{\alpha\beta} - \overset{*}{\nabla}_\mu \Psi_{\rm matter}{}^{\mu\alpha\beta}. \tag{4.113}$$

This identification is evidently based on the definition of the Belinfante tensor, but we will prove that this tensor is indeed given by Eq. (4.84).

The second Eq. (4.110) can be simplified by raising the index $\gamma$ and antisymmetrizing it with $\beta$:

$$\overset{*}{T}{}^{\alpha\beta\gamma} = \chi^2 S^{\gamma\beta\alpha}. \tag{4.114}$$

Now we can use this equation to rewrite the Vielbein equation (the first of Eqs. (4.110)) in a general-relativistic form. First, we take the symmetric part of the equation that relates the Einstein tensor of the torsionful connection $\Gamma$ to the Einstein tensor of the Levi-Civita connection $\Gamma$, which is

$$G_{\alpha\beta}(\Gamma) = G_{\alpha\beta}[\Gamma(g)] - \tfrac{1}{2}\overset{*}{\nabla}_{\mu}\left[\overset{*}{T}_{\alpha}{}^{\mu}{}_{\beta} + \overset{*}{T}_{\beta}{}^{\mu}{}_{\alpha} - \overset{*}{T}_{\alpha\beta}{}^{\mu}\right] - f(T^2), \tag{4.115}$$

where $f(T^2)$ is a complicated expression that is quadratic in the torsion whose explicit form we do not need.[13] Then the Vielbein equation takes the form

$$G^{\alpha\beta}[\Gamma(g)] = \frac{\chi^2}{2} T_{\text{matter}}{}^{\alpha\beta} + f(T^2). \tag{4.116}$$

Then, by substituting Eq. (4.114) into this, we obtain

$$G^{\alpha\beta}[\Gamma(g)] = \frac{\chi^2}{2} T_{\text{matter}}{}^{\alpha\beta} + \mathcal{O}(\chi^4), \tag{4.117}$$

which coincides with Einstein's equation to order $\chi^2$. In fact, taking into account that the order-$\chi^4$ correction is associated with the density of intrinsic spins, only under the most extreme macroscopic conditions [737] can the CSK theory give predictions different from Einstein's, which is good. At the microscopic level, the CSK theory gives different predictions: for instance, it predicts contact interactions between fermions. These have two origins: the term quadratic in the torsion in the CSK gravity action[14] and the covariant derivatives in the matter action. All of them are of higher order in $\chi$.

Conceptually, the CSK theory offers clear advantages over Einstein's. It allows the coupling to fermions and the relation between the canonical and Vielbein energy–momentum tensors is clarified. As we are going to see, the simplest supergravity theory ($N = 1, d = 4$) has the structure of the CSK theory for a Rarita–Schwinger spinor coupled to gravity (and torsion). Finally, we are going to see that the separation between GCTs (which can be seen as the local generalization of translations) and local Lorentz transformations suggests a reinterpretation of gravity as a gauge theory (in the Yang–Mills sense) of the Poincaré group.

Before we move on to these developments, we want to derive the complete gauge identities and Noether currents for matter coupled to gravity in the CSK theory and study the first-order formalism for it.

### 4.4.3 Gauge identities and Noether currents

Let us consider the action of matter minimally coupled to Vielbein and torsion $e^a{}_\mu$ and $T_{\mu\nu}{}^a$:

$$S_{\text{matter}} = \frac{1}{c}\int d^dx\, \mathcal{L}_{\text{matter}}(\varphi, \nabla\varphi, e) = \frac{1}{c}\int d^dx\, \mathcal{L}_{\text{matter}}(\varphi, \partial\varphi, e, \partial e, T). \tag{4.119}$$

---

[13] Actually, the second term on the l.h.s. of this equation also contains terms quadratic in the torsion that we can include in $f(T^2)$ by replacing the modified divergence $\overset{*}{\nabla}_\mu$ by the Levi-Civita covariant derivative $\overset{\{\}}{\nabla}_\mu$.

[14] Using Eq. (1.60), we can split the CSK action into a standard Einstein–Hilbert action and a port quadratic in the torsion plus a total derivative that we can ignore:

$$S_{\text{CSK}}[e^a{}_\mu, T_{\mu\nu}{}^a] = \frac{1}{\chi^2}\int d^dx\, e\left\{R(e) + K_\mu{}^{\mu\lambda}K_\nu{}^\nu{}_\lambda + K_{\nu\mu\rho}K^{\mu\rho\nu}\right\}. \tag{4.118}$$

(According to the minimal coupling prescription, the dependence on torsion is only through the covariant derivative.) We assume that our matter fields, generically denoted by $\varphi$, have only Lorentz indices and that only their first derivatives occur in the action. Furthermore, the fundamental fields are assumed to be $e^a{}_\mu$ and $T_{\mu\nu}{}^a$ (not $T_{\mu\nu}{}^\rho$).

By construction, the action is exactly invariant under local Lorentz transformations and GCTs. Let us now compute the variation of the action through the variation of the fundamental fields. Following the standard procedure developed in Chapter 2, we find[15]

$$\tilde{\delta} S_{\text{matter}} = \frac{1}{c}\int d^dx \left\{\partial_\mu\left[\epsilon^a\left(\mathcal{L}_{\text{matter}} e_a{}^\mu - \frac{\partial \mathcal{L}_{\text{matter}}}{\partial \partial_\mu \varphi}\partial_a \varphi\right) + \frac{\partial \mathcal{L}_{\text{matter}}}{\partial \partial_\mu \varphi}\tilde{\delta}\varphi \right.\right.$$
$$\left.\left. + \frac{\partial \mathcal{L}_{\text{matter}}}{\partial \partial_\mu e^a{}_\nu}\tilde{\delta} e^a{}_\nu\right] + c\frac{\delta S_{\text{matter}}}{\delta \varphi}\delta\varphi + c\frac{\delta S_{\text{matter}}}{\delta e^a{}_\mu}\delta e^a{}_\mu + c\frac{\delta S_{\text{matter}}}{\delta T_{\mu\nu}{}^a}\delta T_{\mu\nu}{}^a\right\}. \tag{4.120}$$

The variations of the matter action with respect to the matter fields are the matter equations of motion. The variations of the matter action with respect to the geometric fields are source terms. Now, with our choice of fundamental fields, we define the spin-angular-momentum tensor $S^\mu{}_{ab}$, the spin–energy-potential tensor $\Psi^{\mu\nu}{}_a$, and the Vielbein energy–momentum tensor $T_a{}^\mu$ by

$$\frac{c}{e}\frac{\delta S_{\text{matter}}}{\delta K_\mu{}^{ab}} = -S^\mu{}_{ab}, \qquad \frac{c}{e}\frac{\delta S_{\text{matter}}}{\delta T_{\mu\nu}{}^a} = -\tfrac{1}{2}\Psi^{\mu\nu}{}_a, \qquad \frac{c}{e}\frac{\delta S_{\text{matter}}}{\delta e^a{}_\mu} = T_a{}^\mu. \tag{4.121}$$

The canonical energy–momentum tensor $T_a{}^\mu$ has an extra term due to our choice of fundamental fields:

$$T_{\text{can}\,a}{}^\mu = T_a{}^\mu - \stackrel{*}{\nabla}_\rho \Psi^{\rho\mu}{}_a - \tfrac{1}{2}\Psi^{\nu\rho}{}_a T_{\nu\rho}{}^\mu. \tag{4.122}$$

Now we substitute the explicit form of the variations of the fundamental fields under GCTs and local Lorentz transformations rewritten in a convenient form,

$$\begin{aligned}
\delta e^a{}_\mu &= -\mathcal{D}_\mu \epsilon^a + 2\epsilon^\nu \mathcal{D}_{[\mu} e^a{}_{\nu]} + \sigma'^a{}_b e^b{}_\mu,\\
\delta T_{\mu\nu}{}^a &= -\nabla_\mu\left(\epsilon^\lambda T_{\lambda\nu}{}^a\right) - \nabla_\nu\left(\epsilon^\lambda T_{\mu\lambda}{}^a\right) - \epsilon^\lambda\left[3R_{[\mu\nu\lambda]}{}^a + T_{\mu\nu}{}^\rho T_{\lambda\rho}{}^a\right] + \sigma'^a{}_b T_{\mu\nu}{}^b,\\
\tilde{\delta}\varphi &= \tfrac{1}{2}\sigma'^{ab}\Gamma_r(M_{ab})\varphi + \tfrac{1}{2}\epsilon^\lambda \omega_\lambda{}^{ab}\Gamma_r(M_{ab})\varphi,
\end{aligned} \tag{4.123}$$

where

$$\sigma'^{ab} = \sigma^{ab} - \epsilon^\mu \omega_\mu{}^{ab}. \tag{4.124}$$

---

[15] Taking into account $\tilde{\delta} x^\mu = \epsilon^\mu$ and that local Lorentz transformations with parameter $\sigma^{ab}$ act only on fields.

After some massaging, using the Bianchi identities for the curvature, we arrive at

$$\tilde{\delta} S = \frac{1}{c}\int \Bigg\{ \partial_\mu \Bigg\{ \epsilon^a \Bigg[ \left( \mathcal{L}_{\text{matter}} e_a{}^\mu - \frac{\partial \mathcal{L}_{\text{matter}}}{\partial \nabla_\mu \varphi} \nabla_a \varphi \right) - e\, T_{\text{can}\, a}{}^\mu \Bigg]$$

$$- e\left( \overset{\{\}}{\nabla}_\rho\, \epsilon_\lambda - \epsilon^\sigma K_{\sigma\rho\lambda} \right)\left( \Psi^{\mu\rho\lambda} - \underline{\Psi}^{\mu\rho\lambda} \right) \Bigg\}$$

$$+ e\epsilon^\lambda \left[ \overset{*}{\nabla}_\mu\, T_{\text{can}\, \lambda}{}^\mu + T_{\lambda\mu}{}^a T_{\text{can}\, a}{}^\mu + S^\mu{}_{ab} R_{\lambda\mu}{}^{ab} - \frac{\delta S_{\text{matter}}}{\delta \varphi} \nabla_\lambda \varphi \right]$$

$$+ e\sigma'^{ab} \left[ T_{ab} - \frac{1}{2} \Psi^{\rho\sigma}{}_a T_{\rho\sigma b} + \frac{1}{2} \frac{\delta S_{\text{matter}}}{\delta \varphi} \Gamma_r \left( M_{ab} \right) \varphi \right] \Bigg\}, \tag{4.125}$$

where we are using the notation

$$\underline{\Psi}^{\mu\rho}{}_a = -\underline{S}^{\mu\rho}{}_a + \underline{S}^{\rho\mu}{}_a + \underline{S}_a{}^{\mu\rho}, \qquad \underline{S}^\mu{}_{ab} = \frac{\partial \mathcal{L}_{\text{matter}}}{\partial \partial_\mu \varphi} \Gamma_r (M_{ab})\, \varphi. \tag{4.126}$$

Since the above variation of the action vanishes identically for arbitrary GCTs and local Lorentz transformations, we obtain four identities. The first identity just gives the expression for the canonical covariant energy–momentum tensor Eq. (4.84). The second gives the expression for the spin–energy-potential tensor Eq. (2.52). The third is the Bianchi identity associated with the invariance under GCTs,

$$\overset{*}{\nabla}_\mu\, T_{\text{can}\, \lambda}{}^\mu + T_{\lambda\mu}{}^a T_{\text{can}\, a}{}^\mu + S^\mu{}_{ab} R_{\lambda\mu}{}^{ab} - \frac{\delta S_{\text{matter}}}{\delta \varphi} \nabla_\lambda \varphi = 0, \tag{4.127}$$

that in flat, torsionless spacetime is the on-shell conservation of the energy–momentum tensor. The fourth is the Bianchi identity associated with the invariance under local Lorentz transformations,

$$T_{[ab]} - \tfrac{1}{2} \Psi^{\rho\sigma}{}_{[a|} T_{\rho\sigma|b]} + \tfrac{1}{2} \frac{\delta S_{\text{matter}}}{\delta \varphi} \Gamma_r \left( M_{ab} \right) \varphi = 0, \tag{4.128}$$

which tells us that the Vielbein energy–momentum tensor in flat, torsionless, spacetime is symmetric on-shell.

As an example, we will study a Dirac spinor coupled to the Vielbein and torsion in the CSK theory, but in first-order form (Section 4.4.4).

### 4.4.4 The first-order Vielbein formalism

As we have seen, the Einstein action written in terms of Vielbeins and the spin connection with the spin connection considered as a function of the Vielbeins provides a second-order action functional of the Vielbeins that is fully equivalent to the one written in terms of the metric.

There is also a first-order action for Vielbeins and the spin connection considered as independent variables. In differential-form language it takes the form

$$\boxed{S[e^a, \omega^{ab}] = \frac{(-1)^{d-1}}{(d-2)!} \int R^{a_1 a_2}(\omega) \wedge e^{a_3} \wedge \cdots \wedge e^{a_d} \epsilon_{a_1 \cdots a_d},} \tag{4.129}$$

where $R^{a_1 a_2}$ is the curvature 2-form associated with the spin connection $\omega$ defined in Eqs. (1.89) and (1.153).

This action is equivalent[16] to the first-order Einstein–Hilbert action for the metric and an affine connection $\Gamma$ related to this spin connection $\omega$ via the first Vielbein postulate. This equivalence can be seen by expanding the curvature 2-form in a Vielbein 1-form basis,

$$R^{a_1 a_2} = \tfrac{1}{2} R_{b_1 b_2}{}^{a_1 a_2} e^{b_1} \wedge e^{b_2}, \tag{4.130}$$

and using

$$e^{b_1} \wedge \cdots \wedge e^{b_d} = d^d x \sqrt{|g|}\, \epsilon^{b_1 \cdots b_d} \tag{4.131}$$

and the relation between the curvatures of $\omega$ and $\Gamma$, Eq. (1.93).

As mentioned before, this theory has some of the elements of a Yang–Mills gauge theory of the Lorentz group $\mathrm{SO}(1, d-1)$ introduced in Appendix A.2.3.

1. There is an independent gauge field (the spin connection).
2. The gauge field appears through its gauge field strength (the curvature).

These are also very important differences, however, which make it completely different from a standard Yang–Mills theory.

1. The action is not quadratic in the field strength. Therefore, the equation of motion of the gauge field will be a constraint, as we are going to see. This is necessary in order to obtain Einstein's gravity theory in which the connection is not dynamical and the only degrees of freedom are those contained in the metric (or the Vielbein) which describe a spin-2 particle.
2. It is not clear how the Vielbeins should be considered. They are in principle matter in the vector representation but they do not have a standard kinetic term.
3. To recover the Einstein–Hilbert action, we have assumed the invertibility of the Vielbeins. This geometrical property cannot be explained from the gauge-theory point of view.

It is clear that gravity cannot be considered a pure gauge theory of the Lorentz group. At most, it would be a gauge theory containing "matter," which is conceptually hard to understand. Later on we will see how to overcome some of these problems by considering the gauge theory of the Poincaré group.

[16] In our conventions that action is exactly equivalent to $+\int d^d x \sqrt{|g|}\, R$.

It is possible to find the equations of motion using differential-form language (as in Ref. [315]). However, we prefer to reexpress the above action in components:

$$S[e^a{}_\mu, \omega_\mu{}^{ab}] = \frac{(-1)^{d-1}}{2\cdot(d-2)!}\int d^dx\, R_{\mu_1\mu_2}{}^{a_1a_2}(\omega)e^{a_3}{}_{\mu_3}\cdots e^{a_d}{}_{\mu_d}\epsilon_{a_1\cdots a_d}\epsilon^{\mu_1\cdots\mu_d}. \tag{4.132}$$

On varying this action, taking into account the analog of Palatini's identity Eq. (3.285) for the Lorentz covariant derivative $\mathcal{D}_\mu$,

$$\delta R_{\mu\nu}{}^{ab} = 2\mathcal{D}_{[\mu}\delta\omega_{\nu]}{}^{ab}, \qquad \mathcal{D}_\mu\delta\omega_\nu{}^{ab} = \partial_\mu\delta\omega_\nu{}^{ab} - \omega_\mu{}^a{}_c\delta\omega_\nu{}^{cb} - \omega_\mu{}^b{}_c\delta\omega_\nu{}^{ac}, \tag{4.133}$$

we find

$$\begin{aligned}\delta S = \frac{(-1)^{d-1}}{2\cdot(d-2)!}\int d^dx\, [2\mathcal{D}_{\mu_1}\delta\omega_{\mu_2}{}^{a_1a_2}e^{a_3}{}_{\mu_3}\\ + (d-2)R_{\mu_1\mu_2}{}^{a_1a_2}\delta e^{a_3}{}_{\mu_3}]e^{a_4}{}_{\mu_4}\cdots e^{a_d}{}_{\mu_d}\epsilon_{a_1\cdots a_d}\epsilon^{\mu_1\cdots\mu_d}.\end{aligned} \tag{4.134}$$

We first analyze the second term:

$$\begin{aligned}&(d-2)R_{\mu_1\mu_2}{}^{a_1a_2}\delta e^{a_3}{}_{\mu_3}e^{a_4}{}_{\mu_4}\cdots e^{a_d}{}_{\mu_d}\epsilon_{a_1\cdots a_d}\epsilon^{\mu_1\cdots\mu_d}\\ &= (-1)^{d-1}3!(d-2)!\sqrt{|g|}\,R_{\mu_1\mu_2}{}^{a_1a_2}\delta e^{a_3}{}_{\mu_3}e_{a_1a_2a_3}{}^{\mu_1\mu_2\mu_3}\\ &= (-1)^d 4\cdot(d-2)!\sqrt{|g|}\,G_a{}^\mu\delta e^a{}_\mu.\end{aligned} \tag{4.135}$$

Now we consider the second term. We have to integrate by parts without the use of any special properties of the connection $\omega$. We find

$$\begin{aligned}&2\mathcal{D}_{\mu_1}\delta\omega_{\mu_2}{}^{a_1a_2}e^{a_3\cdots a_d}{}_{\mu_3\cdots\mu_d}\epsilon_{a_1\cdots a_d}\epsilon^{\mu_1\cdots\mu_d}\\ &= [2(d-2)\delta\omega_{\mu_1}{}^{a_1a_2}\partial_{\mu_2}e^{a_3}{}_{\mu_3} - 4\delta\omega_{\mu_1}{}^{a_1c}\omega_{\mu_2c}{}^{a_2}e^{a_3}{}_{\mu_3}]e^{a_4\cdots a_d}{}_{\mu_4\cdots\mu_d}\epsilon_{a_1\cdots a_d}\epsilon^{\mu_1\cdots\mu_d}\\ &\quad + \partial_{\mu_1}[2\delta\omega_{\mu_2}{}^{a_1a_2}e^{a_3\cdots a_d}{}_{\mu_3\cdots\mu_d}\epsilon_{a_1\cdots a_d}\epsilon^{\mu_1\cdots\mu_d}]\\ &= (-1)^{d-1}12\cdot(d-2)!\sqrt{|g|}\,e_{a_1a_2a_3}{}^{\mu_1\mu_2\mu_3}\delta\omega_{\mu_1}{}^{a_1a_2}\mathcal{D}_{\mu_2}e^{a_3}{}_{\mu_3}\\ &\quad + \partial_{\mu_1}\left[(-1)^{d-1}4(d-2)!\sqrt{|g|}\,\delta\omega_{\mu_2}{}^{a_1a_2}e_{a_1a_2}{}^{\mu_1\mu_2}\right],\end{aligned} \tag{4.136}$$

where we have used the identities

$$\begin{aligned}e_{\mu_3\cdots\mu_d}{}^{a_3\cdots a_d}\epsilon_{a_1\cdots a_d}\epsilon^{\mu_1\cdots\mu_d} &= (-1)^{d-1}2\cdot(d-2)!\sqrt{|g|}\,e_{a_1a_2}{}^{\mu_1\mu_2},\\ e_{\mu_4\cdots\mu_d}{}^{a_4\cdots a_d}\epsilon_{a_1\cdots a_d}\epsilon^{\mu_1\cdots\mu_d} &= (-1)^{d-1}3!(d-3)!\sqrt{|g|}\,e_{a_1a_2a_3}{}^{\mu_1\mu_2\mu_3},\\ 2e_{[a_3}{}^{\mu_3}e_{a_1]a_2}{}^{\mu_1\mu_2} &= 3e_{a_1a_2a_3}{}^{\mu_1\mu_2\mu_3} - 2e_{[a_2}{}^{\mu_3}e_{a_3]a_1}{}^{\mu_1\mu_2}.\end{aligned} \tag{4.137}$$

Assuming that the variations $\delta e^a{}_\mu$ and $\delta\omega_\mu{}^{ab}$ vanish on the boundary, we obtain the equations of motion

$$G_a{}^\mu = 0, \qquad \mathcal{D}_{[\mu}e^a{}_{\nu]} = 0. \tag{4.138}$$

Now we introduce a connection $\Gamma_{\mu\nu}{}^{\rho}$ such that the total covariant derivative satisfies the first Vielbein postulate Eq. (1.91). As we stressed before, the connection is automatically metric compatible and is the sum of a (Cartan) Levi-Civita part that depends only on the Vielbeins and a contorsion part. On comparing this now with Eq. (1.94), we conclude that the connection equation tells us that the torsion vanishes, which implies that the connection is just the (Cartan) Levi-Civita connection $\omega_{\mu}{}^{ab}(e)$ given by the standard expression Eq. (1.100). On substituting this spin connection into the Einstein tensor, we obtain the standard Einstein equations.

An interesting thing happens in $d = 4$: if we replace the connection $\omega$ in the action by its self-dual part, one still obtains Einstein's equation. This observation allows us to find new variables (*Ashtekar variables*), which are used in loop quantization of gravity [68, 587].

In coupling bosonic matter (including a cosmological constant) minimally to this action we would use only Vielbeins, but it is usually not necessary to write any term containing spin connections. Therefore, only the Einstein equation would be modified in the expected way. However, if we coupled fermions, we would necessarily have to introduce terms containing the spin connection, and its equation would be modified. On applying the definition of torsion, we would find that fermions generate torsion, and the solution for the spin connection would be the standard spin connection plus the corresponding contorsion tensor that would be a function of the fermions. This is exactly what happens in the CSK theory[17] and in supergravity theories (see e.g. Refs. [983], [315], where the so-called *rheonomic approach* for constructing supergravity theories which makes use of the first-order formalism is explained), for which the first-order formalism seems especially well suited since it leads to much simpler actions. Furthermore, in the first-order formalism, there is an independent connection, and a relation of gravity with Yang–Mills theories and a relation of supergravity with gauge theories based on supergroups can be established (see Section 4.5 and Chapter 5).

Now we will study a simple example: a Dirac spinor coupled to gravity in the first-order formalism. We are going to see that the resulting equations of motion are the same as those we would have obtained from the second-order CSK theory.

*Example: a Dirac spinor.* The action for a Dirac spinor coupled to gravity in the first-order formalism is the sum of Eq. (4.132) and Eq. (4.98),

$$S[e,\omega,\psi] = \frac{(-1)^{d-1}}{2\cdot(d-2)!\chi^2}\int d^dx\, R_{\mu_1\mu_2}{}^{a_1a_2}(\omega)e^{a_3}{}_{\mu_3}\cdots e^{a_d}{}_{\mu_d}\epsilon_{a_1\cdots a_d}\epsilon^{\mu_1\cdots\mu_d}$$

$$+\int d^dx\, e\Big\{\tfrac{1}{2}(i\bar\psi \not{\mathcal{D}}\psi - i\bar\psi \overleftarrow{\not{\mathcal{D}}}\psi) - m\bar\psi\psi\Big\}, \tag{4.139}$$

where $\mathcal{D}$ stands for the Lorentz covariant derivative.

[17] Observe that, in the first-order formalism, the Vielbein equation is the full Einstein tensor, whereas in the second-order formalism it is only the symmetric part of the Einstein tensor. The variation of the matter action will give automatically the canonical energy–momentum tensor, since there will be no contributions from the spin connection. Thus, the first-order formalism gives us the equation $G_a{}^{\mu} = (\chi^2/2)T_{\text{can}\,a}{}^{\mu}$ in just one shot.

By varying the Vierbein, spin connection, and spinor independently in the action, we find, after the use of our previous results, up to total derivatives,

$$\delta S = \frac{2}{\chi^2}\int d^d x\, e\Bigg\{-\left[G_a{}^{\mu} - \frac{\chi^2}{2}T_{\text{can}\,a}{}^{\mu}\right]\delta e^a{}_{\mu} + 3e_{abc}{}^{\mu\nu\rho}\left[\mathcal{D}_{\nu}e^c{}_{\rho} - \frac{\chi^2}{2}S^c{}_{\nu\rho}\right]\delta\omega_{\mu}{}^{ab}$$
$$+\frac{\chi^2}{2}\delta\bar{\psi}\left[i\nabla\!\!\!\!/\,\psi - \frac{i}{2}\left(\Gamma_{\mu\nu}{}^{\mu} - \left\{{}^{\;\mu}_{\mu\nu}\right\}\right)\gamma^{\nu}\psi - m\psi\right]$$
$$+\frac{\chi^2}{2}\left[-i\bar{\psi}\overleftarrow{\mathcal{D}\!\!\!\!/} + \frac{i}{2}\bar{\psi}\gamma^{\nu}\left(\Gamma_{\mu\nu}{}^{\mu} - \left\{{}^{\;\mu}_{\mu\nu}\right\}\right) - m\bar{\psi}\right]\delta\psi\Bigg\}, \tag{4.140}$$

where we have introduced an affine connection $\Gamma$ such that the total covariant derivative $\nabla$ satisfies the first Vielbein postulate, which means that it is also metric compatible, as we have explained before. Then,

$$\Gamma_{\mu\nu}{}^{\mu} - \left\{{}^{\;\mu}_{\mu\nu}\right\} = K_{\mu\nu}{}^{\mu} = T_{\nu\mu}{}^{\mu}. \tag{4.141}$$

$T_{\text{can}\,a}{}^{\mu}$ is the Dirac-spinor covariant canonical energy–momentum tensor. It has the same form as in Eq. (4.100) but now the total covariant derivative uses the general connections considered here. As we have already pointed out, in the first-order formalism, the covariant canonical energy–momentum tensor is obtained by direct variation with respect to the Vielbeins:

$$\frac{\delta S_{\text{matter}}}{\delta e^a{}_{\mu}} = e\,T_{\text{can}\,a}{}^{\mu}. \tag{4.142}$$

Finally, $S^{\mu}{}_{ab}$ is the spin-angular-momentum tensor, which is totally antisymmetric and given by Eq. (2.67).

The equations of motion are

$$G_a{}^{\mu} = \frac{\chi^2}{2}T_{\text{can}\,a}{}^{\mu}, \quad \mathcal{D}_{\nu}e_{\rho}{}^{c} = \frac{\chi^2}{2}S^c{}_{\nu\rho}, \quad i\nabla\!\!\!\!/\,\psi - m\psi = \frac{i}{2}T_{\nu\mu}{}^{\mu}\gamma^{\nu}\psi. \tag{4.143}$$

The second equation has the solution

$$T_{\mu\nu}{}^{a} = -\chi^2 S^a{}_{\mu\nu}, \tag{4.144}$$

as in the CSK theory. On account of the complete antisymmetry of $S$, this equation implies that the r.h.s. of the third Eq. (4.143) vanishes identically, so we are left with

$$i\nabla\!\!\!\!/\,\psi - m\psi = 0. \tag{4.145}$$

Finally, the first equation is just the Einstein equation one obtains in the CSK theory after several manipulations. We can split it into a Riemannian part and the torsion contributions, which we know are of quartic order in $\chi$.

As we have stressed before, the simplicity of the first-order formalism is related to the previously mentioned fact that this kind of action makes contact with the formulation of gravity as the gauge theory of the Poincaré group, which we are going to study next.

## 4.5 Gravity as a gauge theory

In Ref. [916] MacDowell and Mansouri formulated gravity as the gauge theory of the Poincaré group and supergravity as the gauge theory of the super-Poincaré group.[18] This approach was later extended successfully to many other situations, and it is interesting enough to review it briefly here because the similarities with and differences of gravity from the gauge theories of internal symmetries (some of which we have already mentioned) are manifest in this formulation. Here we will loosely follow Refs. [983, 565].

One of the differences we observed in Section 4.4 between the first-order formalism for gravity using Vielbeins and spin connection and a pure gauge theory is that we did not have an interpretation of the Vielbeins as gauge fields. Furthermore, our intuition tells us that, if gravity can be interpreted as a gauge theory at all, it cannot be a gauge theory of the Lorentz group alone and that at least gauge translations should be introduced into the game. We should then consider the gauging of the Poincaré group. It is worth stressing here that we are talking about the "Poincaré group of the tangent space." That is, at each point in the base manifold, which may but need not be invariant under any translational isometry, we consider inhomogeneous transformations of Lorentz vectors preserving the Minkowski metric. The relation between these gauge transformations and GCTs is one of the subtle points of this formulation of gravity.

To find the generators of the Poincaré group and their commutation relations, we can use the representation in position space (as differential operators) or alternatively we can use the following representation by $(d+1) \times (d+1)$ matrices of Poincaré transformations composed of a translation $a^a$ and a Lorentz transformation $\Lambda^a{}_b$:

$$\begin{pmatrix} 1 \\ v'^a \end{pmatrix} = \left(\begin{array}{c|c} 1 & 0 \\ \hline a^a & \Lambda^a{}_b \end{array}\right) \begin{pmatrix} 1 \\ v^b \end{pmatrix}. \tag{4.146}$$

This representation is suggestive because of its $(d+1)$-dimensional homogeneous form. We will see later that there is a reason for its existence.

We give here again the non-vanishing commutators of the generators $\{M_{ab}, P_a\}$:

$$\begin{aligned} [M_{ab}, M_{cd}] &= -M_{eb}\Gamma_{\text{v}}(M_{cd})^e{}_a - M_{ae}\Gamma_{\text{v}}(M_{cd})^e{}_b, \\ [P_c, M_{ab}] &= -P_d\Gamma_{\text{v}}(M_{ab})^d{}_c. \end{aligned} \tag{4.147}$$

Here $\Gamma_{\text{v}}(M_{ab})^d{}_c$ is the matrix corresponding to the generator $M_{ab}$ in the vector representation of the Lorentz group. The last commutator says that the $d$ generators of translations $P_a$ can be understood as the components of a Lorentz vector. Observe that $P_a$ acts trivially on objects with Lorentz indices. It would act non-trivially on objects with a non-trivial "$(d+1)$th" index in the above representation, but by construction they do not exist.

For each generator we would introduce a gauge field: the spin connection $\omega_\mu{}^{ab}$ for the Lorentz subalgebra plus $d$ new gauge fields for the translation subalgebra. Our theory has $d$ Vielbein fields with Lorentz-vector indices and it is natural to try to interpret them as the gauge fields of translations; the gauge field of the Poincaré group would, tentatively, be, in some representation $\Gamma$,

$$A_\mu = \tfrac{1}{2}\omega_\mu{}^{ab}\Gamma(M_{ab}) + e_\mu{}^a\Gamma(P_a)\,. \tag{4.148}$$

[18] The earliest work on this subject is in Ref. [1207].

Observe that, since $P_a$ does not act on objects with Lorentz indices, the covariant derivative contains in practice only the spin connection.

If we can reproduce Einstein's theory with these elements, we could say that Einstein's theory is the pure gauge theory of the Poincaré group. We are going to see whether this is possible. First we determine the effect of gauge transformations using the standard formalism of Appendix A. If $\sigma^{ab}$ and $\xi^a$ are the infinitesimal gauge parameters of Lorentz rotations and translations, then

$$\delta\omega_\mu{}^{ab} = -\mathcal{D}_\mu\sigma^{ab}, \qquad \delta e_\mu{}^a = -\mathcal{D}_\mu\xi^a + \sigma^a{}_b e_\mu{}^b. \tag{4.149}$$

In both cases $\mathcal{D}$ stands for the gauge-covariant derivative (no Levi-Civita connection is contained in it because, for the moment, we have no metric but a gauge field $e_\mu{}^a$). It is useful to compare the second Eq. (4.149) with the effect of an infinitesimal GCT generated by the world vector $\xi^\mu$ (unrelated in principle to the Lorentz vector $\xi^a$):

$$\begin{aligned}
&\delta_\xi x^\mu = \xi^\mu, \\
&\delta_\xi e^a{}_\mu = -\xi^\nu\partial_\nu e^a{}_\mu - \partial_\mu\xi^\nu e^a{}_\nu = -\mathcal{D}_\mu(\xi^\nu e^a{}_\nu) + 2\xi^\nu\mathcal{D}_{[\mu}e^a{}_{\nu]} - (\xi^\nu\omega_\nu{}^a{}_b)e^b{}_\mu.
\end{aligned} \tag{4.150}$$

The covariant derivative is, again, the Poincaré (Lorentz) gauge one. The effect of an infinitesimal reparametrization is identical to the effect of a $P_a$ gauge transformation with parameter $\xi^a = \xi^\mu e^a{}_\mu$ plus a local Lorentz transformation with parameter $\sigma^{ab} = \xi^\mu\omega_\mu{}^{ab}$ *if* $\mathcal{D}_{[\mu}e^a{}_{\nu]}$ *vanishes.*

We know that this condition is equivalent to the vanishing of torsion and we know that this constraint allows us to express $\omega_\mu{}^{ab}$ in terms of $e^a{}_\mu$. If we implement this constraint in our gauge theory, it will automatically become invariant under reparametrizations.[19] It is implemented in the first-order formalism of Section 4.4, where it appears as the equation of motion of $\omega_\mu{}^{ab}$.

The next step is to construct the gauge field strength:

$$\begin{aligned}
R_{\mu\nu} &= \tfrac{1}{2}R_{\mu\nu}{}^{ab}\Gamma(M_{ab}) + R_{\mu\nu}{}^a\Gamma(P_a)\,, \\
R_{\mu\nu}{}^{ab} &= 2\partial_{[\mu}\omega_{\nu]}{}^{ab} - 2\omega_{[\mu}{}^a{}_c\omega_{\nu]}{}^{cb}, \\
R_{\mu\nu}{}^a &= 2\mathcal{D}_{[\mu}e^a{}_{\nu]}.
\end{aligned} \tag{4.151}$$

The last line is identically equal to $-T_{\mu\nu}{}^a$. Thus, we have just learned that torsion can be interpreted in this formalism as the part of the gauge field strength that is associated with translations.

Now the moment to construct the action arrives. As already mentioned, in order to recover the constraint $R_{\mu\nu}{}^a$, the action has to be linear in the curvature components $R_{\mu\nu}{}^{ab}$. The requirement of Lorentz invariance also makes it very difficult to build quadratic actions (different from $\mathrm{Tr}(R \wedge \star R)$, which is wrong for gravity) that are not trivial (i.e. they do not correspond to topological invariants). We are then led to the action Eq. (4.129), which we know is correct.

---

[19] In supergravity formulated as the gauge theory of a supergroup the problem is how to relate supersymmetry transformations (in general super-reparametrizations in superspace) to gauge transformations associated with the supersymmetry and translation generators.

What have we learned by considering the gauge theory of the Poincaré group? Essentially we have given a gauge-field interpretation to Vielbeins (although we have not justified why they have to be invertible) and we have found that constraints are necessary in order to relate Poincaré gauge invariance to reparametrization invariance. The construction of the action is still rather ad hoc.

A slight improvement of the situation was achieved by MacDowell and Mansouri [916] (see also Refs. [339, 330, 1153]), who used it to construct supergravity actions [565]. Working in four dimensions, they considered the anti-de Sitter group $SO(2,3)$. Upon performing a Wigner–Inönü contraction [817] (which is essentially the zero-cosmological-constant limit), this group becomes the Poincaré group $ISO(1,3)$ and one recovers our previous results.

More precisely, we introduce $SO(2,3)$ indices $\hat{a},\hat{b},\cdots = -1,0,1,2,3$. The metric is $\hat{\eta}^{\hat{a}\hat{b}} = \mathrm{diag}(++---)$, and the algebra $\mathfrak{so}(2,3)$ can be written in the general form

$$\left[\hat{M}_{\hat{a}\hat{b}},\hat{M}_{\hat{c}\hat{d}}\right] = -\hat{\eta}_{\hat{a}\hat{c}}\hat{M}_{\hat{b}\hat{d}} - \hat{\eta}_{\hat{b}\hat{d}}\hat{M}_{\hat{a}\hat{c}} + \hat{\eta}_{\hat{a}\hat{d}}\hat{M}_{\hat{b}\hat{c}} + \hat{\eta}_{\hat{b}\hat{c}}\hat{M}_{\hat{a}\hat{d}}. \tag{4.152}$$

To perform the contraction, we need to introduce a dimensional parameter. This is, naturally, $g$, the gauge coupling constant in gauged $d=4$, $N=2$ supergravity; $g$ is related to the $AdS_4$ radius $R$ and to the cosmological constant $\Lambda$ by

$$R = 1/g = \sqrt{-3/\Lambda}. \tag{4.153}$$

We can now perform a $1+4$ splitting of the indices $\hat{a} = (-1,a)$, $a = 0,1,2,3$, to interpret this algebra from the point of view of the Lorentz subalgebra $\mathfrak{so}(1,3)$. On defining

$$\hat{M}_{ab} = M_{ab}, \qquad \hat{M}_{a-1} = -g^{-1}P_a, \tag{4.154}$$

we can rewrite the $AdS_4$ algebra as follows:

$$\begin{aligned} [M_{ab},M_{cd}] &= -\eta_{ac}M_{bd} - \eta_{bd}M_{ac} + \eta_{ad}M_{bc} + \eta_{bc}M_{ad}, \\ [P_c,M_{ab}] &= -2P_{[a}\eta_{b]c}, \qquad [P_a,P_b] = -g^2 M_{ab}. \end{aligned} \tag{4.155}$$

Taking the limit $g \to 0$, we recover the Poincaré algebra.

We could equally well have started with the four-dimensional de Sitter group $SO(1,4)$. The difference is that, instead of having an extra timelike direction (which we have denoted with a $-1$ index), we have an extra spacelike direction (which we would denote with a 4 index). The two spaces (and groups) are related by analytic continuation $x^{-1} \to x^4$, and, in the contraction of the extra dimension, we would find that the sign of the cosmological constant is reversed ($g \to ig$). We will use a general notation and point out where differences between the two groups could arise. However, one should keep in mind that only the anti-de Sitter space is a good background for QFT and only its group can consistently be supersymmetrized.

The gauge theory of the $AdS_4$ group is just a particular case of the general construction in Appendix A.2.3. We can also perform the contraction in the gauge field and curvature. First, we split the indices in the connection and then we rescale the gauge fields inversely to the generators:

$$\hat{\omega}_\mu = \tfrac{1}{2}\hat{\omega}_\mu{}^{\hat{a}\hat{b}}\Gamma\left(\hat{M}_{\hat{a}\hat{b}}\right) = \tfrac{1}{2}\hat{\omega}_\mu{}^{ab}\Gamma\left(\hat{M}_{ab}\right) + \hat{\omega}_\mu{}^{a,-1}\Gamma\left(\hat{M}_{a,-1}\right)$$
$$= \tfrac{1}{2}\omega_\mu{}^{ab}\Gamma(M_{ab}) + e^a{}_\mu\Gamma(P_a)\,, \tag{4.156}$$

where

$$\hat{\omega}_\mu{}^{ab} = \omega_\mu{}^{ab}, \qquad \hat{\omega}_\mu{}^{a,-1} = -ge^a{}_\mu. \tag{4.157}$$

In this scheme, linear momentum and Vierbeins are on the same footing as the rest of the generators and gauge fields. This is obviously due to the semisimple nature of the $\mathrm{AdS}_4$ group. There is some resemblance between this structure and the idea of grand unification in particle physics, although there are also obvious differences.

We can also split and rescale the curvature components, expressing everything in terms of Lorentz tensors:

$$\hat{R}_{\mu\nu}{}^{ab} = R_{\mu\nu}{}^{ab} + 2g^2 e^{[a}{}_\mu e^{b]}{}_\nu, \qquad \hat{R}_{\mu\nu}{}^{a,-1} = 2g\mathcal{D}_{[\mu}e^a{}_{\nu]}. \tag{4.158}$$

Now we can address again the construction of a quadratic action for this group. To have diffeomorphism invariance, the Lagrangian has to be a 4-form that we can integrate over a four-dimensional manifold and, therefore, the exterior product of two curvature terms $R \wedge R$. We now have to saturate the $\mathfrak{so}(2,3)$ indices. If we did it with the Killing metric, we would manifest SO(2, 3) invariance but the Lagrangian would be a total derivative, as in any Yang–Mills theory. Thus, we have to give up explicit SO(2, 3) invariance. We have to keep Lorentz invariance, though, and with the only two invariant tensors of the Lorentz group ($\eta_{ab}$, $\epsilon_{abcd}$) we can build two terms:

$$\hat{R}^{ab} \wedge \hat{R}^{cd}\epsilon_{abcd}, \qquad \hat{R}^{a,-1} \wedge \hat{R}^{b,-1}\eta_{ab}. \tag{4.159}$$

The second term is not invariant under parity and for this reason it is discarded. The inclusion of this term would also introduce torsion and it would also lead to the existence of non-invertible Vierbeins (see the discussion in Ref. [565]).

The first term can also be given an SO(2,3)-invariant origin [1153]: by introducing a constant vector $V^{\hat{a}} = \eta^{\hat{a}}{}_{-1}$ it can be written using the invariant tensor $\hat{\epsilon}$, and we obtain the action

$$S = \alpha \int \hat{R}^{\hat{a}\hat{b}} \wedge \hat{R}^{\hat{c}\hat{d}} V^{\hat{e}} \hat{\epsilon}_{\hat{a}\hat{b}\hat{c}\hat{d}\hat{e}}. \tag{4.160}$$

This is only formally SO(2, 3) invariant because the vector would change under $\mathrm{AdS}_4$ transformations. Nevertheless, this form of the action is very suggestive.

On expanding this action in terms of Lorentz tensors, we have

$$S = \alpha \int d^4x\, R_{\mu\nu}{}^{ab} R_{\rho\sigma}{}^{cd}\epsilon_{abcd}\epsilon^{\mu\nu\rho\sigma} - 16g^2\alpha \int d^4x\, e\big[R(e,\omega) + 6g^2\big]. \tag{4.161}$$

The first term is a total derivative (proportional to the Euler characteristic, a topological invariant) that does not contribute to the equations of motion, and the second term is the first-order Einstein–Hilbert action with cosmological constant $\Lambda = -3g^2$. In the

$g \to 0$ limit (provided that $\alpha \sim g^{-2}$) we recover the usual Einstein–Hilbert action plus a topological term. Observe that the variation of the action under $P_a$ gauge transformations is proportional to torsion terms and, thus, vanishes on-shell.

This is a very attractive result, which, however, leaves some questions unanswered, such as the reason for the invertibility of the Vierbein and the value of the vector $V^{\hat{a}}$. A possible solution has been proposed by Wilczek [1255].

To finish this section, we should mention that the gauge approach has been extended to larger groups, such as the full $d$-dimensional affine group. A comprehensive review on these developments may be found in Ref. [738].

## 4.6 Teleparallelism

In this section we provide a short introduction to relativistic theories of gravity based on *teleparallelism*, i.e. theories in which there is a well-defined notion of parallelism of vectors defined at different points. In GR and other generalizations based on the Riemannian or Riemann–Cartan geometry, gravity, described by the metric or Vielbein fields, is characterized by a curvature; therefore, parallel transport is path dependent and there is no such well-defined (path-independent) notion of parallelism. Teleparallelism is based on the Weitzenböck geometry and the Weitzenböck connection $W_{\mu\nu}{}^{\rho}$ described in Section 1.4.1, which has identically vanishing curvature (but non-vanishing torsion[20]).

These theories are interesting for several reasons: first of all, GR can be viewed as a particular theory of teleparallelism and, thus, teleparallelism could be considered at the very least as a different point of view that can lead to the same results. Of course, there are teleparallel theories different from and even inconsistent with GR. Second, in this framework, one can define an energy–momentum tensor for the gravitational field that is a true tensor under all GCTs. This is the reason why teleparallelism was reconsidered[21] by Møller in 1961 [962] when he was studying the problem of defining an energy–momentum *tensor* for the gravitational field [960, 961]. The idea was taken over by Pellegrini and Plebański in Ref. [1025], which constructed the general Lagrangian for these theories. The third reason why these theories are interesting is that they can be seen as gauge theories of the translation group [735, 351] (not the full Poincaré group) and, thus, they give an alternative interpretation of GR.

The basic field in these theories is the Vielbein $e^a{}_\mu$. This field has $d^2$ independent components, while the metric has only $d(d+1)/2$. The extra independent components that the Vielbein field has are those of an antisymmetric $d \times d$ tensor, such as the electromagnetic-field-strength tensor $F_{\mu\nu}$, and that is why Einstein thought that these theories could describe gravitation and electromagnetism in a unified way. In the standard Vielbein formalism (Weyl's), the extra $d(d-1)/2$ independent components of the Vielbein field are removed by introducing local Lorentz invariance, with a Lorentz connection that is not an independent field but is built out of the Vielbeins. Here, we are not interested a

[20] It is possible to have a non-trivial theory with vanishing curvature and torsion if the non-metricity tensor does not vanish. In Ref. [977] a theory equivalent to GR based on this geometry was constructed.

[21] Teleparallelism had originally been considered by Einstein, who studied it as a unified theory of gravitation and electromagnetism in Refs. [493, 494, 495, 496, 497] (see also Ref. [891]) until Ref. [502] showed that the particular theory considered by him was inconsistent.

priori in having this local invariance, and, in principle, we will construct only theories that are invariant under GCTs and global Lorentz transformations. Thus, as we will see, these theories describe in general more degrees of freedom than just those of a graviton.

The construction of the Lagrangian of these theories is fairly simple: we look for terms that have the required invariances and are, at most, quadratic in derivatives of the Vielbeins. The elementary building blocks are the Ricci rotation coefficients $\Omega_{\mu\nu}{}^{a} = \partial_{[\mu}e^{a}{}_{\nu]}$ that transform as tensors (2-forms) under GCTs and as vectors under (global) Lorentz transformations. In the context of the Weitzenböck geometry, the $-2\Omega_{\mu\nu}{}^{a}$s are the components of the torsion of Weitzenböck connection and, since there is no curvature tensor available, it is only natural to construct the Lagrangian using them.

In any dimension there are three terms with the required properties (they transform as densities under GCTs and as scalars under Lorentz transformations, being quadratic in first partial derivatives of the Vielbeins): the *Weitzenböck invariants* $I_1, \ldots, I_3$,

$$I_1 = e\,\Omega_{\mu\nu\rho}\Omega^{\mu\nu\rho}, \qquad I_2 = e\,\Omega_{\mu\nu\rho}\Omega^{\rho\nu\mu}, \qquad I_3 = e\,\Omega_{\mu\rho}{}^{\rho}\Omega^{\mu}{}_{\sigma}{}^{\sigma}. \tag{4.162}$$

There is another invariant $I_4$, which is quadratic only in $d=4$:

$$I_4 = \epsilon^{\mu_1\cdots\mu_{d-3}\nu_1\nu_2\nu_3}\Omega_{\mu_1\rho_1}{}^{\rho_1}\Omega_{\mu_2\rho_2}{}^{\rho_2}\cdots\Omega_{\mu_{d-3}\rho_{d-3}}{}^{\rho_{d-3}}\Omega_{\nu_1\nu_2\nu_3}, \tag{4.163}$$

but it is not invariant under parity transformations (a further requirement) and it is usually not considered. Also, $e$ by itself is another invariant (a cosmological-constant term) that we will not consider. Observe that all the Weitzenböck invariants involve the inverse Vielbeins $e_a{}^{\mu}$ and are, therefore, highly non-linear in the Vielbeins.

The general teleparallel Lagrangian of Pellegrini and Plebański [1025] is the integral of a linear combination of the Weitzenböck invariants with arbitrary coefficients:

$$\mathcal{L}_{\mathrm{T}} = \sum_{i=1}^{3} c^i I_i. \tag{4.164}$$

Only two of them are really independent since we can choose the overall normalization.

This general Lagrangian, written in differential-form language to relate it to the Poincaré gauge theory of gravity which is customarily written in it (see e.g. Ref. [738]), is known as the *Rumpf Lagrangian* [1088] (see also Refs. [683, 967]).

There are other ways to parametrize this Lagrangian, for instance by splitting $\Omega_{abc}$ into several pieces ${}^{(1)}\Omega$, ${}^{(2)}\Omega$, and ${}^{(3)}\Omega$ (*tentor, trator, and axitor*, respectively [683]). First we define

$$\begin{aligned} v_a &\equiv \Omega_{ab}{}^{b}, & {}^{(2)}\Omega_{abc} &= \frac{2}{1-d}\eta_{a[b}v_{c]}, \\ {}^{(3)}\Omega_{abc} &= \Omega_{[abc]}, & {}^{(1)}\Omega_{abc} &= \Omega_{abc} - {}^{(2)}\Omega_{abc} - {}^{(3)}\Omega_{abc}. \end{aligned} \tag{4.165}$$

Then, a Lagrangian equivalent to Pellegrini and Plebański's is [683]

$$\mathcal{L}_{\mathrm{T}} = e\,\Omega^{abc}\sum_{i=1}^{3} a_i\,{}^{(i)}\Omega_{abc}. \tag{4.166}$$

The relation between these two parametrizations is

$$a_1 = c_1 + \tfrac{1}{2}c_2, \qquad a_2 = c_1 + \tfrac{1}{2}c_2 + \frac{d-1}{2}c_3, \qquad a_3 = c_1 - c_2. \tag{4.167}$$

Another parametrization based on $v_a$, the tensors

$$a^{a_1\cdots a_{d-3}} = \frac{1}{3!}\epsilon^{a_1\cdots a_{d-3}b_1b_2b_3}\Omega_{b_1b_2b_3}, \qquad t_{abc} = \Omega_{a(bc)} - {}^{(2)}\Omega_{a(bc)}, \tag{4.168}$$

and the invariants $v^2, t^2$, and $a^2$ can be found in Ref. [736].

### 4.6.1 The linearized limit

Our goal now is to try to understand which kind of theories are those defined by the Lagrangian Eq. (4.164). First, we observe with Møller [962] that, for $c_1 = 1, c_2 = 2$, and $c_4 = -4$, this Lagrangian is identical (up to total derivatives) to the Einstein–Hilbert Lagrangian, and, therefore, gives the vacuum Einstein equations.[22] The Lagrangian turns out to be invariant under not just global but also local Lorentz transformations, and the only degrees of freedom left are (we know) those of the graviton. For general values of the parameters, the analysis is more complicated and it is convenient to start by studying the linear limit. To this end, we split the Vielbeins into their vacuum (Minkowski) values plus perturbations. Working in Cartesian coordinates for simplicity, we write

$$e^a{}_\mu = \delta^a{}_\mu + A^a{}_\mu. \tag{4.169}$$

For the inverse Vielbeins, we have

$$e_a{}^\mu = \delta_a{}^\mu - \delta_b{}^\mu \delta_a{}^\nu A^b{}_\nu + \mathcal{O}(A^2). \tag{4.170}$$

To this order we can unambiguously trade curved and flat indices, and the above formula can be rewritten:

$$e_a{}^\mu = \delta_a{}^\mu - A^\mu{}_a + \mathcal{O}(A^2), \qquad A^\mu{}_a \equiv \delta_b{}^\mu \delta_a{}^\nu A^b{}_\nu. \tag{4.171}$$

The metric perturbation that we have called $h_{\mu\nu}$ in previous chapters is given by the symmetric part of $A$ at lowest order:

$$g_{\mu\nu} = \eta_{\mu\nu} + h_{\mu\nu} + \mathcal{O}(A^2), \qquad h_{\mu\nu} \equiv 2A_{(\mu\nu)}, \qquad b_{\mu\nu} \equiv 2A_{[\mu\nu]}, \qquad A_{\mu\nu} \equiv \delta_{a\mu}A^a{}_\nu. \tag{4.172}$$

With these definitions it is straightforward to obtain, up to total derivatives, the linear limit of the action for the Lagrangian density Eq. (4.164):

$$\begin{aligned} S_{\rm T}[h,b] = \int d^dx \Big\{ & \frac{1}{16}(2c_1 + c_2)\partial_\mu h_{\nu\rho}\partial^\mu h^{\nu\rho} - \frac{1}{16}(2c_1 + c_2 - c_3)\partial_\mu h_{\nu\rho}\partial^\nu h^{\mu\rho} \\ & - \frac{1}{8}c_3 \partial_\mu h \partial_\nu h^{\nu\mu} + \frac{1}{16}c_3(\partial h)^2 - \frac{1}{16}[4c_1 + 2(c_2 + c_3)]\partial_\mu h_{\nu\rho}\partial^\rho b^{\nu\mu} \\ & + \frac{1}{16}\partial_\mu b_{\nu\rho}\partial^\mu b^{\nu\rho} - \frac{1}{16}(2c_1 - 3c_2 - c_3)\partial_\mu b_{\nu\rho}\partial^\rho b^{\nu\mu} \Big\}. \end{aligned} \tag{4.173}$$

[22] In fact, this theory is sometimes referred to as the *teleparallel equivalent of GR*.

Table 4.1. Values of the parameters $c_i$ in the general Lagrangian of Pellegrini and Plebański Eq. (4.164) for several theories: GR, the viable models, the Yang–Mills-type model (YM), and the von der Heyde model (vdH) [751] (which is one of the viable ones with $\lambda = 0$)

| | GR | Viable | YM | vdH |
|---|---|---|---|---|
| $c_1$ | 1 | $2-\lambda$ | 2 | 2 |
| $c_2$ | 2 | $2\lambda$ | 0 | 0 |
| $c_3$ | $-4$ | $-4$ | 0 | $-4$ |

The first four terms are familiar to us: up to coefficients, they are the same terms as those that appear in the Fierz–Pauli Lagrangian Eq. (3.84). The last two terms are also well known: up to coefficients, they are exactly those that appear in the Lagrangian of the Kalb–Ramond 2-form field, which we still have not seen. The third term in the second line represents a coupling (already at the linear level) between these two fields.

Now, it is clear that it is not possible to recover solutions of the vacuum Fierz–Pauli theory if the coupling terms have a non-zero coefficient: a non-vanishing $h$ field is a source for a non-vanishing $b$ field and vice versa. Thus, the only theories which we expect to be phenomenologically viable are those in the family

$$2c_1 + c_2 + c_3 = 0. \tag{4.174}$$

Furthermore, both in the Kalb–Ramond and in the Fierz–Pauli cases, for certain choices of the coefficients, the action has a gauge invariance,

$$\delta_\epsilon h_{\mu\nu} = -2\partial_{(\mu}\epsilon_{\nu)}, \qquad \delta_\eta b_{\mu\nu} = 2\partial_{[\mu}\eta_{\nu]}, \tag{4.175}$$

whose existence is crucial for its consistent quantization. We also expect that only when these gauge invariances are present will the theory be consistent. It turns out that all these conditions are simultaneously met: let us eliminate $c_3$ using Eq. (4.174) and then, calling $c_2 = 2\lambda$, the action can be rewritten in the form

$$S_{\rm T}[h,b] = \frac{c_1+\lambda}{2} S_{\rm FP}[h] + \frac{c_1-\lambda}{2} S_{\rm KR}[b], \tag{4.176}$$

where $S_{\rm FP}[h]$ is the Fierz–Pauli action given in Eq. (3.84) and $S_{\rm KR}[b]$ is the Kalb–Ramond action

$$S_{\rm KR}[b] = \int d^dx\, \frac{1}{12} H^2, \qquad H_{\mu\nu\rho} \equiv 3\partial_{[\mu}b_{\nu\rho]}, \qquad H^2 = H_{\mu\nu\rho}H^{\mu\nu\rho}. \tag{4.177}$$

For $c_1 = \lambda$ ($c_2 = 2\lambda$, $c_3 = -4\lambda$), the Kalb–Ramond Lagrangian disappears. Up to an overall normalization constant and a total derivative, this teleparallel Lagrangian is completely equivalent to the Einstein–Hilbert Lagrangian, as we mentioned before. For $c_1 = -\lambda$ the Fierz–Pauli Lagrangian disappears and only the Kalb–Ramond Lagrangian remains. This theory does not describe gravity. If we are always going to keep the Fierz–Pauli Lagrangian, then it makes sense to set $c_1 = 2 - \lambda$ ($c_3 = -4,\ c_2 = 2\lambda$) and keep the one-parameter family of actions

$$S_{\mathrm{T}}[h, b] = S_{\mathrm{FP}}[h] + (1 - \lambda) S_{\mathrm{KR}}[b], \tag{4.178}$$

which represent viable models of gravity (in the sense that they fulfill the above requirements) based on teleparallelism (see Table 4.1). The case $\lambda = 0$ is the model proposed in Ref. [751].

Of course, we know that the full non-linear theory will be consistent only if additional conditions are satisfied. In particular, we know from our results in Chapter 3 that the quantization of the spin-2 field $h_{\mu\nu}$ will be consistent only if it couples to the total energy–momentum tensor, the sum of the spin-2 energy–momentum tensor and the Kalb–Ramond energy–momentum tensor, although the presence of the Kalb–Ramond field could modify this result. Checking that this is (or not) the case in the above family of theories requires an expansion to order $\mathcal{O}(A^3)$ that would be interesting to do.

It is amusing to compare these results with the linearized limit of the low-energy string effective action (see Chapter 21). The linearized actions are identical, except for the presence of the dilaton in the string case. However, the non-linear actions are quite different: in the string case, we simply have standard gravity coupled to matter (the Kalb–Ramond field) that appears only quadratically (at least to lowest order in $\alpha'$), whereas, in the teleparallel case, the Kalb–Ramond field should also appear non-linearly in the full action.

It is possible to view the theories of teleparallelism as gauge theories of the group of translations [735, 351], with the Vielbeins playing the role of gauge vectors, but we will not enter into this interesting aspect.

# 5
# Pure $N = 1, 2, d = 4$ supergravities

In the previous chapter, we introduced increasingly complex theories of gravity, starting from GR, to accommodate fermions, and we saw that the generalizations of GR that we had to use could be thought of as gauge theories of the symmetries of flat spacetime.

A very important development of the past few decades has been the discovery of supersymmetry and its application to the theory of fundamental particles and interactions. This symmetry relating bosons and fermions can be understood as the generalization of the Poincaré or AdS groups, which are the symmetries of our background spacetime to the *super-Poincaré* or *super-*AdS (super-)groups, which are the symmetries of our background *superspacetime*, a generalization of standard spacetime that has fermionic coordinates.

It is natural to construct generalizations of the standard gravity theories that can be understood as gauge theories of the (super-)symmetries of the background (vacuum) superspacetime. These generalizations are the *supergravity* (SUGRA) theories. Given that the kind of fermions that we can have depends critically on the spacetime dimension, the SUGRA theories that we can construct also depend critically on the spacetime dimension. Furthermore, we can extend the standard bosonic spacetime in different ways by including more than one ($N$) set of fermionic coordinates. This gives rise to additional supersymmetries relating them and, therefore, to supersymmetric field theories and SUGRA theories with $N$ supersymmetries. The latter are also known as *extended* SUGRAs (SUEGRAs). The SUEGRAs can be further extended by coupling them to supersymmetric matter (matter fields that fill complete representations of the supersymmetry algebra) and they can be deformed by introducing new couplings among the fields depending on new parameters. The main deformation procedure is that of gauging global symmetries, explained in Section 2.5, but it is not the only one: some supergravities also admit the so-called *massive* deformations, the main example being Romans' massive $N = 2A, d = 10$ theory [1083], which is reviewed in Section 22.2. All the possible deformations can, apparently, be taken into account by the *embedding tensor formalism*, reviewed in Section 2.7.

There is, thus, a large variety of supergravities, but not infinitely large, because the gauging of supersymmetries with $N > 8$ in $d = 4$ dimensions or $N = 1$ in $d = 11$ needs the inclusion of more than one graviton and/or fields of spin higher than 2, which we do not know how to couple consistently.

We are going to study SUGRA theories because they provide an interesting extension of the ideas we have reviewed so far and because the effective field theories that describe the behavior of superstrings at low energies are SUGRA theories. But these theories are also interesting by themselves: recent results suggest that $N = 8, d = 4$ supergravity may provide the first example of a finite quantum field theory (QFT) of gravity, unconnected to its being the effective field theory of the type-II supergravities compactified on a 6-torus. See footnote 42 on p. 120 for more details.

Supersymmetry and SUGRA have been developed over the last several years and are currently the object of extensive work, so we cannot give here a complete review of any of these subjects. There are excellent books and reviews that cover most of the basic aspects, e.g. Refs. [983, 565, 1246, 1245, 211, 564, 1208, 1210]. Reference [1097] contains reprints of many of the original articles on SUGRA.

Our goal in this chapter is to introduce some of the concepts that we will use later on, profiting from and extending the material we have studied so far. Our method will be to construct the simplest, pure (that is, with no matter supermultiplets), four-dimensional SUGRA theories ($N = 1, d = 4$ Poincaré and AdS supergravities) by gauging the corresponding supergroups and studying them separately. We will then study the two simplest pure four-dimensional SUEGRA theories ($N = 2, d = 4$ Poincaré and AdS supergravities) since they illustrate important ideas we will make use of later. Our conventions for tensors, gamma matrices, and spinors are explained in Chapter 1 and Appendix D, respectively. In Chapters 6–9 we will study the supersymmetric coupling of matter to the theories that we have introduced here and the possible (electric) gaugings of their global symmetries, giving the complete bosonic Lagrangians and supersymmetry transformation rules (for vanishing fermions).

## 5.1 Gauging $N = 1, d = 4$ superalgebras

Just as the $d = 4$ Poincaré group can be constructed by exponentiation of the Poincaré algebra, the $N = 1, d = 4$ super-Poincaré group can be constructed by exponentiation of the $N = 1, d = 4$ super-Poincaré superalgebra. This superalgebra is an extension of the Poincaré algebra with (bosonic) generators $P_a$ and $M_{ab}$ by one set of anti-Hermitian fermionic generators $Q^\alpha$ (the *supersymmetry generators* or *supersymmetry charges*) that transform as Majorana[1] spinors under Poincaré transformations, so they have four components and

$$[Q^\alpha, M_{ab}] = \Gamma_s(M_{ab})^\alpha{}_\beta Q^\beta, \tag{5.1}$$

while the commutator with $P_a$ vanishes. To complete all the relations of the superalgebra, we need to give the commutator of two $Q^\alpha$s. Actually, in a superalgebra, one has to give the *anticommutator* of fermionic generators (that is, the difference from the bosonic ones), and the (anti)commutation relations have to satisfy a super-Jacobi identity, which takes the same form as the standard Jacobi identity but with commutators replaced by anticommutators whenever two fermionic generators are involved and with a relative sign between the

[1] The need for Majorana representations is associated with the anti-Hermiticity of the generators. In $d = 4$ Majorana and Weyl spinors are equivalent and the superalgebra can be written in terms of Weyl spinors only. Since the complete R-symmetry of the four-dimensional superalgebras and of the corresponding supergravities is only manifest when we write them in terms of Weyl spinors, we will do so in Chapters 6–9.

terms related to the permutation of fermionic generators. The anticommutation relation that satisfies the super-Jacobi identities is[2]

$$\{Q^\alpha, Q^\beta\} = i\left(\gamma^a \mathcal{C}^{-1}\right)^{\alpha\beta} P_a. \tag{5.2}$$

The non-vanishing commutation relations for the $N=1, d=4$ superalgebra are

$$\boxed{\begin{aligned}
[M_{ab}, M_{cd}] &= -M_{eb}\Gamma_{\mathrm{v}}(M_{cd})^e{}_a - M_{ae}\Gamma_{\mathrm{v}}(M_{cd})^e{}_b, \\
[P_a, M_{bc}] &= -P_e\Gamma_{\mathrm{v}}(M_{bc})^e{}_a, \\
[Q^\alpha, M_{ab}] &= \Gamma_{\mathrm{s}}(M_{ab})^\alpha{}_\beta Q^\beta, \\
\{Q^\alpha, Q^\beta\} &= i\left(\gamma^a \mathcal{C}^{-1}\right)^{\alpha\beta} P_a.
\end{aligned}} \tag{5.3}$$

This is the superalgebra that one has to gauge in order to construct $N=1, d=4$ Poincaré supergravity. However, to follow Section 4.5, we prefer to start from the supersymmetrized version of the $\mathrm{AdS}_4$ algebra and then perform a Wigner–Inönü contraction. To supersymmetrize it, we need to add consistently a set of fermionic supersymmetry generators to those of the bosonic algebra $\hat{M}_{\hat{a}\hat{b}}$. To have consistency, the fermionic generators have to transform as $\mathrm{AdS}_4$ Majorana spinors, which, as discussed in Appendix D, have four real (or purely imaginary) components. Denoting them by $\hat{Q}^\alpha$, we find the following (anti)commutation relations for the $\mathrm{AdS}_4$ superalgebra:

$$\boxed{\begin{aligned}
\left[\hat{M}_{\hat{a}\hat{b}}, \hat{M}_{\hat{c}\hat{d}}\right] &= -\hat{M}_{\hat{e}\hat{b}}\Gamma_{\mathrm{v}}\left(\hat{M}_{\hat{c}\hat{d}}\right)^{\hat{e}}{}_{\hat{a}} - \hat{M}_{\hat{a}\hat{e}}\Gamma_{\mathrm{v}}\left(\hat{M}_{\hat{c}\hat{d}}\right)^{\hat{e}}{}_{\hat{b}}, \\
\left[\hat{Q}^\alpha, \hat{M}_{\hat{a}\hat{b}}\right] &= \Gamma_{\mathrm{s}}\left(\hat{M}_{\hat{a}\hat{b}}\right)^\alpha{}_\beta \hat{Q}^\beta, \\
\{Q^\alpha, Q^\beta\} &= \left[\Gamma_{\mathrm{s}}\left(\hat{M}^{\hat{a}\hat{b}}\right)\hat{\mathcal{C}}^{-1}\right]^{\alpha\beta} \hat{M}_{\hat{a}\hat{b}}.
\end{aligned}} \tag{5.4}$$

An infinitesimal transformation generated by this superalgebra is

$$\hat{\Lambda} \equiv \tfrac{1}{2}\hat{\sigma}^{\hat{a}\hat{b}}\hat{M}_{\hat{a}\hat{b}} + \bar{\hat{\epsilon}}_\alpha \hat{Q}^\alpha, \tag{5.5}$$

where $\hat{\sigma}^{\hat{a}\hat{b}} = -\hat{\sigma}^{\hat{b}\hat{a}}$ is the infinitesimal parameter of an $\mathrm{SO}(2,3)$ transformation and $\hat{\epsilon}^\alpha$, an anticommuting Majorana spinor, is the infinitesimal parameter of a supersymmetry transformation. The bar indicates Dirac conjugation.

To construct theories that are invariant under *local* infinitesimal transformations ($\hat{\sigma}^{\hat{a}\hat{b}} = \hat{\sigma}^{\hat{a}\hat{b}}(x)$, $\bar{\hat{\epsilon}}_\alpha = \bar{\hat{\epsilon}}_\alpha(x)$), we need to introduce a gauge field $\hat{A}_\mu$,

$$\hat{A}_\mu \equiv \tfrac{1}{2}\hat{\omega}_\mu{}^{\hat{a}\hat{b}}\hat{M}_{\hat{a}\hat{b}} + \bar{\hat{\psi}}_{\mu\,\alpha}\hat{Q}^\alpha, \tag{5.6}$$

[2] Since our convention for Hermitian conjugation of fermionic objects is $(ab)^\dagger = +b^\dagger a^\dagger$, the structure constants have to be purely imaginary here. We are using a purely imaginary representation of the gamma matrices with a purely imaginary charge-conjugation matrix, hence the factor $i$.

whose components are the standard bosonic $\mathrm{SO}(2,3)$ connection $\hat{\omega}_\mu{}^{\hat{a}\hat{b}}$ from which we will obtain the Lorentz connection $\omega_\mu{}^{ab}$ and the Vierbein $e^a{}_\mu$ that will describe the graviton. It also contains a new fermionic field: the Rarita–Schwinger field, $\bar{\hat{\psi}}_{\mu\,\alpha}$, which has a vector index and a spinor index. This field describes a particle of spin $\frac{3}{2}$, the *gravitino*, which is the supersymmetric partner of the graviton, related to it by supersymmetry transformations, and other excitations, which should be eliminated if there is enough gauge symmetry in its action (as is the case).

By construction, the action of an infinitesimal transformation of the gauge field is the supercovariant derivative of $\hat{\Lambda}(x)$,

$$\delta \hat{A}_\mu = \partial_\mu \hat{\Lambda} + [\hat{\Lambda}, \hat{A}_\mu]. \tag{5.7}$$

On expanding the commutator (which should be understood as the anticommutator between the fermionic generators), we find the following transformation laws for the component fields under local $\mathrm{SO}(2,3)$ transformations and supersymmetry transformations:

$$\begin{aligned}
\delta_{\hat{\sigma}}\hat{\omega}_\mu{}^{\hat{a}\hat{b}} &= \hat{\mathcal{D}}_\mu \hat{\sigma}^{\hat{a}\hat{b}}, & \delta_{\hat{\sigma}}\bar{\hat{\psi}}_\mu &= -\bar{\hat{\psi}}_\mu \left[\tfrac{1}{2}\hat{\sigma}^{\hat{a}\hat{b}}\Gamma_{\mathrm{s}}\left(\hat{M}_{\hat{a}\hat{b}}\right)\right], \\
\delta_{\hat{\epsilon}}\hat{\omega}_\mu{}^{\hat{a}\hat{b}} &= -2\bar{\hat{\epsilon}}\Gamma_{\mathrm{s}}\left(\hat{M}_{\hat{a}\hat{b}}\right)\hat{\psi}_\mu, & \delta_{\hat{\epsilon}}\bar{\hat{\psi}}_\mu &= \hat{\mathcal{D}}_\mu \hat{\epsilon},
\end{aligned} \tag{5.8}$$

where $\mathcal{D}$ is the Lorentz SO(1,3) covariant derivative defined in Chapter 1.

The supercurvature is defined by

$$\hat{R}_{\mu\nu}(\hat{A}) \equiv 2\partial_{[\mu}\hat{A}_{\nu]} - [\hat{A}_\mu, \hat{A}_\nu], \tag{5.9}$$

and, by expanding it and decomposing it into bosonic and fermionic components, we find

$$\hat{R}_{\mu\nu}{}^{\hat{a}\hat{b}}(\hat{A}) = \hat{R}_{\mu\nu}{}^{\hat{a}\hat{b}}(\hat{\omega}) - 2\bar{\hat{\psi}}_{[\mu}\Gamma_{\mathrm{s}}\left(\hat{M}_{\hat{a}\hat{b}}\right)\hat{\psi}_{\nu]}, \qquad \bar{\hat{R}}_{\mu\nu\,\alpha}(\hat{A}) = 2\hat{\mathcal{D}}_{[\mu}\bar{\hat{\psi}}_{\nu]\,\alpha}. \tag{5.10}$$

Having the supercurvature components, we can now proceed to construct an action that has to be invariant under GCTs, local Lorentz transformations, parity transformations, and local supersymmetry transformations without the use of any metric. The requirement of invariance under local supersymmetry transformations is more difficult to impose and we will have to check it explicitly afterwards. The other requirements imply that the action has to be of the form

$$S[\hat{A}] = \alpha \int d^4x \left[\hat{R}_{\mu\nu}{}^{ab}\hat{R}_{\rho\sigma}{}^{cd}\epsilon_{abcd} + \beta \bar{\hat{R}}_{\mu\nu\,\alpha}\left(\gamma_5\right)^\alpha{}_\beta \hat{R}_{\rho\sigma}{}^\beta\right]\epsilon^{\mu\nu\rho\sigma}. \tag{5.11}$$

We now want to rewrite this action in terms of component Poincaré fields and in terms of the parameter $g$ whose zero limit gives the Wigner–Inönü contraction. First we study it in the superalgebra. Defining

$$\hat{M}_{ab} \equiv M_{ab}, \quad \hat{M}_{a,-1} \equiv -g^{-1}P_a, \quad \hat{Q}^\alpha \equiv g^{-\frac{1}{2}}Q^\alpha, \tag{5.12}$$

the $AdS_4$ superalgebra takes the form

$$
\begin{aligned}
[M_{ab}, M_{cd}] &= -M_{eb}\Gamma_{\rm v}(M_{cd})^e{}_a - M_{ae}\Gamma_{\rm v}(M_{cd})^e{}_b, \\
[P_a, M_{bc}] &= -P_e\Gamma_{\rm v}(M_{bc})^e{}_a, \qquad [P_a, P_b] = -g^2 M_{ab}, \\
\{Q^\alpha, Q^\beta\} &= -2\Big[\Gamma_{\rm s}\Big(\hat{M}^{a,-1}\Big)\hat{\mathcal{C}}^{-1}\Big]^{\alpha\beta} P_a + g\Big[\Gamma_{\rm s}\Big(\hat{M}^{ab}\Big)\Big]^{\alpha\beta} M_{ab}, \\
[Q^\alpha, M_{ab}] &= \Gamma_{\rm s}(M_{ab})^\alpha{}_\beta Q^\beta, \qquad [Q^\alpha, P_a] = -g\Gamma_{\rm s}\Big(\hat{M}_{a,-1}\Big)^\alpha{}_\beta Q^\beta.
\end{aligned}
\tag{5.13}
$$

In the $g\rightarrow 0$ limit we recover the $N=1, d=4$ Poincaré superalgebra using, for instance, the representation of $AdS_4$ gamma matrices

$$
\hat{\gamma}_a = i\gamma_a\gamma_5, \quad \hat{\gamma}_{-1} = \gamma_5, \quad \hat{\mathcal{C}} = \mathcal{C} = i\gamma_0. \tag{5.14}
$$

The infinitesimal transformation parameters and gauge fields are also split and rescaled as follows:

$$
\begin{aligned}
\hat{\omega}_\mu{}^{ab} &= \omega_\mu{}^{ab}, \quad \hat{\omega}_\mu{}^{a,-1} = g e^a{}_\mu, \quad \hat{\psi}_\mu = g^{\frac{1}{2}}\psi_\mu, \\
\hat{\sigma}^{ab} &= \sigma^{ab}, \qquad \hat{\sigma}^{a,-1} = g\sigma^a, \qquad \hat{\epsilon} = g^{\frac{1}{2}}\epsilon.
\end{aligned}
\tag{5.15}
$$

In terms of these variables, the $SO(2,3)$ and supersymmetry transformations take the forms

$$
\begin{aligned}
\delta_\sigma e^a{}_\mu &= \mathcal{D}_\mu\sigma^a + \sigma^a{}_b e^b{}_\mu, & \delta_\epsilon e^a{}_\mu &= -i\bar{\epsilon}\gamma^a\psi_\mu, \\
\delta_\sigma\omega_\mu{}^{ab} &= \mathcal{D}_\mu\sigma^{ab} + 2g^2 e^{[a}{}_\mu\sigma^{b]}, & \delta_\epsilon\omega_\mu{}^{ab} &= -2g\bar{\epsilon}\gamma^{ab}\psi_\mu, \\
\delta_\sigma\bar{\psi}_\mu &= -\bar{\psi}_\mu\big(\tfrac{1}{4}\sigma^{ab}\gamma_{ab}\big) - \frac{ig}{2}\bar{\psi}_\mu\sigma^a\gamma_a, & \delta_\epsilon\bar{\psi}_\mu &= \mathcal{D}_\mu\epsilon - \frac{ig}{2}\gamma_\mu\epsilon,
\end{aligned}
\tag{5.16}
$$

and the components of the supercurvature are given by

$$
\begin{aligned}
\hat{R}_{\mu\nu}{}^{ab} &= R_{\mu\nu}{}^{ab}(\omega) + 2g^2 e^{[a}{}_\mu e^{b]}{}_\nu + g\bar{\psi}_{[\mu}\gamma^{ab}\psi_{\nu]}, \\
\hat{R}_{\mu\nu}^{a,-1} &= -g\big(T_{\mu\nu}{}^a - i\bar{\psi}_{[\mu}\gamma^a\psi_{\nu]}\big), \\
\hat{R}_{\mu\nu} &= 2g^{\frac{1}{2}}\Big(\mathcal{D}_{[\mu}\psi_{\nu]} - \frac{ig}{2}\gamma_{[\mu}\psi_{\nu]}\Big).
\end{aligned}
\tag{5.17}
$$

On substituting these components into the action, we find that the right normalization of the Einstein–Hilbert term in the action requires $\alpha = -1/(16g^2\chi^2)$. Furthermore, the explicit[3] terms quartic in fermions drop out from the action (after *Fierzing*[4] and massaging some terms) if $\beta = -8i$. This is the value that will also make the action supersymmetry invariant. The result is the action for $N=1, d=4$ $AdS_4$ SUGRA,

$$
\boxed{S[e^a{}_\mu, \omega_\mu{}^{ab}, \psi_\mu] = \frac{1}{\chi^2}\int d^4x\, e\,\Big[R(e,\omega) + 6g^2 + 2e^{-1}\epsilon^{\mu\nu\rho\sigma}\bar{\psi}_\mu\gamma_5\gamma_\nu\hat{\mathcal{D}}_\rho\psi_\sigma\Big],} \tag{5.18}
$$

[3] Later we will see that the on-shell spin connection contains terms quadratic in the fermions, so the action contains implicitly terms quartic in fermions, just as in the CSK theory.

[4] That is, using Fierz's identities.

which, in the $g \to 0$ limit, gives the action for $N{=}1, d{=}4$ Poincaré SUGRA [449, 563]:

$$S[e^a{}_\mu, \omega_\mu{}^{ab}, \psi_\mu] = \frac{1}{\chi^2}\int d^4x\, e\big[R(e,\omega) + 2e^{-1}\epsilon^{\mu\nu\rho\sigma}\bar{\psi}_\mu\gamma_5\gamma_\nu\mathcal{D}_\rho\psi_\sigma\big]. \tag{5.19}$$

These are first-order actions in which, as indicated, the fundamental variables are the Vielbein, spin connection, and gravitino field. Thanks to our experience with the CSK theory,[5] we know that, when we solve the spin-connection equation of motion, which is purely algebraic, we are going to find that there is torsion proportional to some expression quadratic in fermions, making the $\hat{R}_{\mu\nu}{}^{a,-1}(\hat{A})$ components of the supercurvature vanish. Substituting the torsion into the action will give rise to terms that are quartic in fermions.

In what follows we are going to study these actions, their equations of motion, and their symmetries separately. The most efficient way to do this is to treat them in the so-called 1.5-order formalism: we consider that we have solved the equation of motion of the spin connection and we have substituted its solution back into the action, but we do not do it explicitly, keeping the action in its first-order form. Then we use the chain rule, varying over the two remaining fundamental fields (the Vierbein and gravitino), and over the spin connection first. That variation is its equation of motion, which has been solved, and simply vanishes. In this way, many calculations are greatly simplified.

We are going to make this study as self-contained as possible and, thus, we will repeat some of the general points explained in this introductory section.

## 5.2 $N=1, d=4$ (Poincaré) supergravity

The fields of $N=1, d=4$ supergravity are the Vierbein and the gravitino $\{e^a{}_\mu, \psi_\mu\}$. The gravitino is a vector of Majorana (real) spinors. The action is written in a first-order form, in which the spin connection $\omega_\mu{}^{ab}$ is also considered as an independent field and the action contains only first derivatives. We rewrite the action here for convenience, setting $\chi=1$:

$$S[e^a{}_\mu, \omega_\mu{}^{ab}, \psi_\mu] = \int d^4x\, e\big[R(e,\omega) + 2e^{-1}\epsilon^{\mu\nu\rho\sigma}\bar{\psi}_\mu\gamma_5\gamma_\nu\mathcal{D}_\rho\psi_\sigma\big]. \tag{5.20}$$

[5] We can interpret these actions as the CSK theory coupled to gravitino fields. However, there is more to it, because the consistency of the gravitino field theory requires its action to be invariant under the gauge transformations (in flat spacetime) $\delta\psi_\mu = \partial_\mu\epsilon(x)$ in order to decouple unwanted spins. When we couple the gravitino to gravity, consistency requires that the Vierbeins also transform under these fermionic transformations (otherwise, that gauge symmetry is broken), which become the local supersymmetry transformations. In this way, local supersymmetry does not reduce any further the number of degrees of freedom (graviton plus gravitino). The non-trivial part is the transformation of the Vierbeins under supersymmetry. We could have tried to arrive at the $N=1, d=4$ supergravity action from the linearized action, which is just the sum of the Fierz–Pauli action and the Rarita–Schwinger action, decoupled, by asking for consistent interaction and following the Noether method as we did in Chapter 3. Then, the full supersymmetry transformations should arise as the consistency requirement.

Here $\mathcal{D}_\mu$ is the Lorentz-covariant derivative (rather than the completely covariant derivative, which we denote as usual by $\nabla_\mu$),

$$\mathcal{D}_\mu \psi_\nu = \partial_\mu \psi_\nu - \tfrac{1}{4}\omega_\mu{}^{ab}\gamma_{ab}\psi_\nu, \qquad \nabla_\mu \psi_\nu = \mathcal{D}_\mu \psi_\nu - \Gamma_{\mu\nu}{}^\rho \psi_\rho, \tag{5.21}$$

and

$$R(e,\omega) = e_a{}^\mu e_b{}^\nu R_{\mu\nu}{}^{ab}(\omega), \tag{5.22}$$

where $R_{\mu\nu}{}^{ab}(\omega)$ is the Lorentz curvature of the Lorentz connection $\omega_\mu{}^{ab}$, Eq. (1.89).

As usual, to obtain the second-order action we solve the spin-connection equation of motion and substitute the solution for $\omega_\mu{}^{ab}$ in terms of $e^a{}_\mu$ and $\psi_\mu$ back into the first-order action. The spin-connection equation of motion is

$$\frac{\delta S}{\delta \omega_\mu{}^{ab}} = 3! e_{abc}{}^{\mu\nu\rho}\left(\mathcal{D}_\nu e^c{}_\rho + \frac{i}{2}\bar{\psi}_\nu \gamma^c \psi_\rho\right) = 0. \tag{5.23}$$

This equation implies that the expression in brackets, antisymmetrized in $\nu$ and $\rho$, is zero. Looking at Eq. (1.94), we see that there is torsion in this theory and it is given by[6]

$$T_{\mu\nu}{}^a = i\bar{\psi}_\mu \gamma^a \psi_\nu. \tag{5.24}$$

Furthermore, we see that the solution to the new equation is just that the Lorentz connection consists of two pieces: the one that solves the standard equation $\mathcal{D}_{[\mu} e^a{}_{\nu]} = 0$, which we denote by $\omega_\mu{}^{ab}(e)$ because it is completely determined by the Vierbein, and the contorsion tensor $K_\mu{}^{ab}$, which depends on the gravitino through the torsion. It is convenient to write the solution as follows:

$$\omega_{abc} = -\Omega_{abc} + \Omega_{bca} - \Omega_{cab}, \quad \Omega_{\mu\nu}{}^a = \Omega_{\mu\nu}{}^a(e) + \tfrac{1}{2}T_{\mu\nu}{}^a, \quad \Omega_{\mu\nu}{}^a(e) = \partial_{[\mu} e^a{}_{\nu]}. \tag{5.25}$$

The other two equations of motion that the first-order action yields are

$$\begin{aligned}
\frac{\delta S}{\delta e^a{}_\mu} &= -2e\left[G_a{}^\mu - 2T_{\text{can}\,a}{}^\mu(\psi)\right] = 0,\\
T_{\text{can}\,a}{}^\mu(\psi) &= \frac{1}{2e}\epsilon^{\rho\mu\sigma\nu}\bar{\psi}_\rho \gamma_5 \gamma_a \mathcal{D}_\sigma \psi_\nu,\\
\frac{\delta S}{\delta \bar{\psi}_\mu} &= 4\epsilon^{\mu\nu\rho\sigma}\left[\gamma_5\gamma_\nu \mathcal{D}_\rho \psi_\sigma + \tfrac{1}{4}T_{\nu\rho}{}^a \gamma_5 \gamma_a \psi_\sigma\right] = 0,
\end{aligned} \tag{5.26}$$

where we have used

$$\mathcal{D}_{[\mu}\gamma_{\nu]} = -\tfrac{1}{2}T_{\mu\nu}{}^a\gamma_a. \tag{5.27}$$

The second-order equations of motion follow from the substitution of Eq. (5.25) into the first-order ones.

The action Eq. (5.20) and equations of motion are manifestly invariant under

**general coordinate transformations,**

$$\delta_\xi x^\mu = \xi^\mu, \qquad \delta_\xi e^a{}_\mu = -\xi^\nu \partial_\nu e^a{}_\mu - \partial_\mu \xi^\nu e^a{}_\nu, \qquad \delta_\xi \psi_\mu = -\xi^\nu \partial_\nu \psi_\mu - \partial_\mu \xi^\nu \psi_\nu, \tag{5.28}$$

[6] The bilinear $\bar{\psi}_\mu \gamma^a \psi_\nu$ is automatically antisymmetric in $\mu\nu$.

**and local Lorentz transformations,**

$$\delta_\sigma e^a{}_\mu = \sigma^a{}_b e^b{}_\mu, \qquad \delta_\sigma \psi_\mu = \tfrac{1}{2}\sigma^{ab}\gamma_{ab}\psi_\mu, \tag{5.29}$$

where $\sigma^{ab} = -\sigma^{ba}$. On top of this, if we eliminate the spin connection as an independent field by substituting the solution of its equation of motion, there is invariance under

**local $N=1$ supersymmetry transformations:**

$$\delta_\epsilon e^a{}_\mu = -i\bar{\epsilon}\gamma^a\psi_\mu, \qquad \delta_\epsilon \psi_\mu = \mathcal{D}_\mu \epsilon. \tag{5.30}$$

This requires some explanation. The first-order action is also invariant under the same transformations supplemented by the supersymmetry transformation of the spin connection. In the second-order formalism, the supersymmetry variation of the spin connection is completely different and can be found by varying Eq. (5.25) with respect to the Vierbein and gravitino:

$$\delta_\epsilon \omega_\mu{}^{ab} = -i\bar{\epsilon}\gamma_\mu\psi^{ab} + i\bar{\epsilon}\gamma^a\psi^b{}_\mu - i\bar{\epsilon}\gamma^b\psi_\mu{}^a, \qquad \psi_{\mu\nu} \equiv \mathcal{D}_{[\mu}\psi_{\nu]}. \tag{5.31}$$

One may think that the gauging of the supersymmetry algebra should give us the first-order supersymmetry transformation rule for the spin connection, but it does not: it just gives $\delta_\epsilon \omega_\mu{}^{ab} = 0$. Nevertheless, to check the invariance of the action in the 1.5-order formalism we do not need this variation, as we are going to see.

Let us check the invariance of the action Eq. (5.20) under these transformations in the 1.5-order formalism. This is not a complicated calculation if we construct the right setup, which is the general setup explained in Chapter 2 for theories that are invariant under local symmetries. There we showed that a given theory would be invariant up to total derivatives under a local transformation if a certain *gauge identity* was satisfied by its equations of motion. Thus, all we have to do is to identify the gauge identity that has to be satisfied in this case by the Vierbein and gravitino equations of motion.

Under a general variation of the fields, the $N=1, d=4$ SUGRA action Eq. (5.20) transforms as follows:

$$\delta S = \int d^4x \left[ \frac{\delta S}{\delta e^a{}_\mu}\delta e^a{}_\mu + \frac{\delta S}{\delta \omega_\mu{}^{ab}}\delta\omega_\mu{}^{ab} + \delta\bar{\psi}_\mu \frac{\delta S}{\delta \bar{\psi}_\mu} \right]. \tag{5.32}$$

Here the variations are only with respect to explicit appearances of each field in the first-order action. The variation of the second-order action would be obtained by applying the chain rule to the variation with respect to the spin connection, using Eq. (5.25). However, these additional terms are proportional to the equation of motion of the spin connection $\delta S/\delta\omega_\mu{}^{ab}$, which we have assumed is satisfied (the 1.5-order formalism). Thus, the term containing $\delta\omega_\mu{}^{ab}$ will always vanish (for any kind of variation) because it is proportional to that equation of motion and we need only vary *explicit* appearances of the Vierbein and gravitino in the first-order action Eq. (5.20),

$$\delta S = \int d^4x \left[ \frac{\delta S}{\delta e^a{}_\mu}\delta e^a{}_\mu + \delta\bar{\psi}_\mu \frac{\delta S}{\delta \bar{\psi}_\mu} \right]. \tag{5.33}$$

Consider now the local supersymmetry transformations Eq. (5.30). On substituting into the above the explicit form of these transformations and integrating by parts the partial derivative in

$$\mathcal{D}_\mu \bar{\epsilon} = \overline{\mathcal{D}_\mu \epsilon} = \partial_\mu \bar{\epsilon} + \tfrac{1}{4}\bar{\epsilon}\omega_\mu{}^{ab}\gamma_{ab}, \tag{5.34}$$

we obtain, up to total derivatives,

$$\delta_\epsilon S = \int d^4x\, \bar{\epsilon}\left[-i\frac{\delta S}{\delta e^a{}_\mu}\gamma^a\psi_\mu - \mathcal{D}_\mu \frac{\delta S}{\delta \bar{\psi}_\mu}\right]. \tag{5.35}$$

The theory will be locally supersymmetric, then, if

$$\mathcal{D}_\mu \frac{\delta S}{\delta \bar{\psi}_\mu} = -i\frac{\delta S}{\delta e^a{}_\mu}\gamma^a\psi_\mu, \tag{5.36}$$

which will be, at the same time, the supersymmetry gauge identity. Let us prove it:

$$\begin{aligned}\mathcal{D}_\mu \frac{\delta S}{\delta \bar{\psi}_\mu} &= 4\epsilon^{\mu\nu\rho\sigma}\gamma_5(\mathcal{D}_\mu\gamma_\nu)\mathcal{D}_\rho\psi_\sigma + 4\epsilon^{\mu\nu\rho\sigma}\gamma_5\gamma_\nu\mathcal{D}_\mu\mathcal{D}_\rho\psi_\sigma \\ &+ \epsilon^{\mu\nu\rho\sigma}\gamma_5\gamma_a\mathcal{D}_\mu T_{\nu\rho}{}^a\psi_\sigma + \epsilon^{\mu\nu\rho\sigma}\gamma_5\gamma_a T_{\nu\rho}{}^a\mathcal{D}_\mu\psi_\sigma,\end{aligned} \tag{5.37}$$

where we have used $\mathcal{D}_\mu\gamma_a = 0$. Using Eq. (5.27) in the first term on the r.h.s. of the above equation, we obtain minus two times the last term. In the second term we first use the Ricci identity for the anticommutator of Lorentz-covariant derivatives, then expand the product of gammas in antisymmetrized products $\gamma^{(3)}$ and $\gamma^{(1)}$, reexpress the $\gamma^{(3)}$ in terms of $\gamma^{(1)}\gamma_5$ and the antisymmetric symbol, and, finally, use the identity

$$G_a{}^\mu = -\tfrac{3}{2}g_{abc}{}^{\mu\nu\rho}R_{\nu\rho}{}^{bc}. \tag{5.38}$$

We keep the third term as it is and obtain the total result

$$\mathcal{D}_\mu \frac{\delta S}{\delta \bar{\psi}_\mu} = 2eiG_a{}^\mu\gamma^a\psi_\mu - \epsilon^{\mu\nu\rho\sigma}\gamma_5\gamma_a T_{\mu\nu}{}^a\mathcal{D}_\rho\psi_\sigma + \epsilon^{\mu\nu\rho\sigma}[R_{\mu\nu\rho}{}^a + \mathcal{D}_\mu T_{\nu\rho}{}^a]\gamma_5\gamma_a\psi_\sigma. \tag{5.39}$$

The first term is one of the two we want. The second term is equal to the other term we want, due to the Fierz identity

$$(\bar{\psi}_\nu\gamma_5\gamma_a\mathcal{D}_\rho\psi_\sigma)(\gamma^a\psi_\mu) = -\tfrac{1}{2}(\bar{\psi}_\nu\gamma^a\psi_\mu)(\gamma_a\gamma_5\mathcal{D}_\rho\psi_\sigma). \tag{5.40}$$

The expression in brackets vanishes due to the Bianchi identity[7]

$$R_{[\mu\nu\rho]}{}^a + \mathcal{D}_{[\mu}T_{\nu\rho]}{}^a = 0, \tag{5.44}$$

and this proves the supersymmetry gauge identity.

---

[7] This identity can be related to the standard Bianchi identity as follows. First,

$$\mathcal{D}_\mu T_{\nu\rho}{}^a = \nabla_\mu T_{\nu\rho}{}^a - \Gamma_{\mu\nu}{}^\lambda T_{\rho\lambda}{}^a + \Gamma_{\mu\rho}{}^\lambda T_{\nu\lambda}{}^a. \tag{5.41}$$

Antisymmetrizing and using the definition of torsion $\Gamma_{[\mu\nu]}{}^\rho = -\tfrac{1}{2}T_{\mu\nu}{}^\rho$ gives

$$\mathcal{D}_\mu T_{\nu\rho}{}^a = \nabla_{[\mu}T_{\nu\rho]}{}^a + T_{[\mu\nu}{}^\lambda T_{\rho]\lambda}{}^a. \tag{5.42}$$

Finally,

$$R_{[\mu\nu\rho]}{}^a + \mathcal{D}_{[\mu}T_{\nu\rho]}{}^a = R_{[\mu\nu\rho]}{}^a + \nabla_{[\mu}T_{\nu\rho]}{}^a + T_{[\mu\nu}{}^\lambda T_{\rho]\lambda}{}^a, \tag{5.43}$$

which vanishes on account of the usual Bianchi identity Eq. (1.30).

### 5.2.1 Local supersymmetry algebra

An important check to be performed is the confirmation that we have on-shell closure of the $N=1$ supersymmetry algebra on the fields. Let us first consider the Vierbein. Using the supersymmetry rules ($\mathcal{D}_\mu\epsilon=\nabla_\mu\epsilon$), it is easy to obtain

$$[\delta_{\epsilon_1},\delta_{\epsilon_2}]e^a{}_\mu=-\nabla_\mu\xi^a, \tag{5.45}$$

where $\xi^a$ is the bilinear

$$\xi^a=-i\bar{\epsilon}_1\gamma^a\epsilon_2. \tag{5.46}$$

The effect of the GCT generated by $\xi^\mu=\xi^a e_a{}^\mu$ can be rewritten in this form:

$$\delta_\xi e^a{}_\mu=-\nabla_\mu\xi^a-\xi^\nu T_{\mu\nu}{}^a-\xi^\nu\omega_\nu{}^a{}_b e^b{}_\mu. \tag{5.47}$$

Thus, using the value of the torsion field in this theory, we find

$$[\delta_{\epsilon_1},\delta_{\epsilon_2}]e^a{}_\mu=(\delta_\xi+\delta_\sigma+\delta_\epsilon)e^a{}_\mu, \tag{5.48}$$

where

$$\sigma^a{}_b=\xi^\nu\omega_\nu{}^a{}_b, \qquad \epsilon=\xi^\mu\psi_\mu. \tag{5.49}$$

The same algebra is realized on all the fields of the theory.

## 5.3 $N=1,d=4$ AdS supergravity

The simplest $N=1,d=4$ Poincaré supergravity theory that we have just described can be generalized in essentially two ways: adding $N=1$ supersymmetric matter or generalizing the Lorentz connection. Adding certain matter supermultiplets sometimes produces enhancement of supersymmetry and in this way one obtains extended supergravities. We will review $N=2,d=4$ (*gauged* and *ungauged*) supergravity later.

The only generalizations of the four-dimensional Poincaré group which are usually studied are the four-dimensional (anti-)de Sitter groups $\mathrm{dS}_4=\mathrm{SO}(1,4)$ and $\mathrm{AdS}_4=\mathrm{SO}(2,3)$. Of these, only $\mathrm{AdS}_4$ is compatible with consistent supergravity. We have obtained at the beginning of this chapter the action for $N=1,d=4$ AdS supergravity in the first-order form[8]

$$\boxed{S[e^a{}_\mu,\omega_\mu{}^{ab},\psi_\mu]=\int d^4x\, e\Big[R(e,\omega)+6g^2+2e^{-1}\epsilon^{\mu\nu\rho\sigma}\bar{\psi}_\mu\gamma_5\gamma_\nu\hat{\mathcal{D}}_\rho\psi_\sigma\Big],} \tag{5.50}$$

where

$$\hat{\mathcal{D}}_\mu=\mathcal{D}_\mu-\frac{ig}{2}\gamma_\mu \tag{5.51}$$

[8] This theory was first constructed by Townsend in Ref. [1186]. Freedman constructed a consistent $N=1,d=4$ supergravity theory with a de Sitter-type cosmological constant [562], but this theory includes a vector supermultiplet whose 1-form gauges the global $\mathrm{U}(1)_\mathrm{R}$ symmetry or the pure, ungauged, supergravity theory. The difference is discussed in Chapter 6. See also Ref. [938] and references therein.

is the AdS$_4$-covariant derivative and $\mathcal{D}_\mu$ is the Lorentz-covariant derivative in the spinor representation.

This theory contains a negative cosmological constant proportional to the square of the Wigner–Inönü parameter $g$, $\Lambda = -3g^2$. The vacuum will be anti-de Sitter spacetime.

The equation of motion for $\omega_\mu{}^{ab}$ takes the same form as in the $g=0$ (Poincaré) case and therefore has the same solution, Eq. (5.25). The other two equations of motion suffer $g$-dependent modifications:

$$\begin{aligned}
\frac{\delta S}{\delta e^a{}_\mu} &= -2e\big[G_a{}^\mu - 3g^2 e_a{}^\mu - 2T_{\text{can}\,a}{}^\mu\big] = 0,\\
T_{\text{can}\,a}{}^\mu &= \frac{1}{2e}\epsilon^{\rho\mu\sigma\nu}\bar{\psi}_\rho\gamma_5\gamma_a\hat{\mathcal{D}}_\sigma\psi_\nu - \frac{ig}{2e}\epsilon^{\mu\nu\rho\sigma}\bar{\psi}_\nu\gamma_5\gamma_{\rho a}\psi_\sigma, \\
\frac{\delta S}{\delta\bar{\psi}_\mu} &= 4\epsilon^{\mu\nu\rho\sigma}\Big[\gamma_5\gamma_\nu\hat{\mathcal{D}}_\rho\psi_\sigma + \tfrac{1}{4}T_{\nu\rho}{}^a\gamma_5\gamma_a\psi_\sigma\Big] = 0.
\end{aligned} \tag{5.52}$$

The torsion term can be shown to vanish on-shell using Fierz identities.[9]

This theory is invariant under local Lorentz transformations and GCTs. Furthermore, it is invariant under local supersymmetry transformations,

$$\delta_\epsilon e^a{}_\mu = -i\bar{\epsilon}\gamma^a\psi_\mu, \qquad \delta_\epsilon\psi_\mu = \hat{\mathcal{D}}_\mu\epsilon. \tag{5.53}$$

To prove it, one has to prove the corresponding generalization of the Poincaré supersymmetry gauge identity

$$\hat{\mathcal{D}}_\mu\frac{\delta S}{\delta\bar{\psi}_\mu} = -i\frac{\delta S}{\delta e^a{}_\mu}\gamma^a\psi_\mu. \tag{5.54}$$

We find

$$\begin{aligned}
\hat{\mathcal{D}}_\mu\frac{\delta S}{\delta\bar{\psi}_\mu} &= \mathcal{D}_\mu\left(\frac{\delta S}{\delta\bar{\psi}_\mu}\right)_{g=0} - \frac{ig}{2}\gamma_\mu\left(\frac{\delta S}{\delta\bar{\psi}_\mu}\right)_{g=0}\\
&\quad + \mathcal{D}_\mu(-2ig\epsilon^{\mu\nu\rho\sigma}\gamma_5\gamma_{\nu\rho}\psi_\sigma) - \frac{ig}{2}\gamma_\mu(-2ig\epsilon^{\mu\nu\rho\sigma}\gamma_5\gamma_{\nu\rho}\psi_\sigma),
\end{aligned} \tag{5.55}$$

where we have simplified the gravitino equation of motion by using the fact that, on-shell (Fierzing),

$$\epsilon^{\mu\nu\rho\sigma}\gamma_5\gamma_a T_{\nu\rho}{}^a\psi_\sigma = 0. \tag{5.56}$$

The $g=0$ supersymmetry gauge identity can be used for the first term. The last term gives the cosmological-constant term in the Einstein equation. Thus, we need only check that the second and third terms (linear in $g$) give the two $g$-dependent pieces of the gravitino energy–momentum tensor, which can be combined into a single term. By expanding the third term we obtain a term that cancels out the second, a torsion term that vanishes due to the above identity, and a term

$$ig\epsilon^{\mu\nu\rho\sigma}T_{\mu\nu}{}^a\gamma_5\gamma_a\gamma_\rho\psi_\sigma, \tag{5.57}$$

which, upon Fierzing, gives the right result.

[9] This is also true in the Poincaré case.

*5.3.1 Local supersymmetry algebra*

On-shell we find

$$[\delta_{\epsilon_1}, \delta_{\epsilon_2}] = \delta_\xi + \delta_\sigma + \delta_\epsilon, \tag{5.58}$$

with

$$\xi^a = -i\bar{\epsilon}_1\gamma^a\epsilon_2, \qquad \sigma^a{}_b = \xi^\nu\omega_\nu{}^a{}_b + g\bar{\epsilon}_1\gamma^a{}_b\epsilon_2, \qquad \epsilon = \xi^\mu\psi_\mu. \tag{5.59}$$

## 5.4 Extended supersymmetry algebras

As we said in the introduction, one can generalize spacetime by adding one or more sets of fermionic coordinates. The corresponding supersymmetry algebras have one or more $(N)$ sets of supersymmetry generators that we denote by adding an index $i=1,\ldots, N$, $Q^{i\,\alpha}$. For $N>1$ they are called *extended supersymmetry algebras*. In this section we are going to introduce them in $d=4$, and in Sections 5.5 and 5.6 we will study two SUEGRA theories based on the simplest extended superalgebras.

It is convenient for our purposes to start by generalizing the $N=1, d=4$ AdS superalgebra to $N>1$. It turns out that, to have a consistent superalgebra, we are *forced* to introduce further bosonic generators $T^{ij}=-T^{ji}$, which generate SO$(N)$ rotations between the $N$ supersymmetry charges $\hat{Q}^{i\,\alpha}$ and which are part (but not all) of the so-called *R-symmetry* group that leaves invariant the superalgebra. More precisely, the R-symmetry group is the group of automorphisms of the superalgebra, sometimes denoted by $\mathrm{H_{aut}}$. In fact, consistency requires these generators to appear in the anticommutator of two supercharges. The complete superalgebra has the non-vanishing (anti) commutation relations

$$\boxed{\begin{aligned}
\left[T^{ij}, T^{kl}\right] &= \Gamma\left(T^{kl}\right)^i{}_m T^{mj} + \Gamma\left(T^{kl}\right)^j{}_m T^{im},\\
\left[\hat{Q}^{k\,\alpha}, T^{ij}\right] &= \Gamma\left(T^{ij}\right)^k{}_m \hat{Q}^{m\,\alpha},\\
\left[\hat{M}_{\hat{a}\hat{b}}, \hat{M}_{\hat{c}\hat{d}}\right] &= -\hat{M}_{\hat{e}\hat{b}}\,\Gamma_{\mathrm{v}}\left(\hat{M}_{\hat{c}\hat{d}}\right)^{\hat{e}}{}_{\hat{a}} - \hat{M}_{\hat{a}\hat{e}}\,\Gamma_{\mathrm{v}}\left(\hat{M}_{\hat{c}\hat{d}}\right)^{\hat{e}}{}_{\hat{b}},\\
\left[\hat{Q}^{i\,\alpha}, \hat{M}_{\hat{a}\hat{b}}\right] &= \Gamma_{\mathrm{s}}\left(\hat{M}_{\hat{a}\hat{b}}\right)^\alpha{}_\beta \hat{Q}^{i\,\beta},\\
\left\{\hat{Q}^{i\,\alpha}, \hat{Q}^{j\,\beta}\right\} &= \delta^{ij}\left[\Gamma_{\mathrm{s}}\left(\hat{M}^{\hat{a}\hat{b}}\right)\mathcal{C}^{-1}\right]^{\alpha\beta}\hat{M}_{\hat{a}\hat{b}} - (\mathcal{C}^{-1})^{\alpha\beta}T^{ij}.
\end{aligned}} \tag{5.60}$$

The new SO$(N)$ generators $T^{ij}$ play a very interesting role. If we gauge the algebra to obtain a supergravity theory based on this algebra, we first have to construct the superconnection $\hat{A}_\mu$, which will have the form

$$\hat{A}_\mu = \tfrac{1}{2}\hat{\omega}_\mu{}^{\hat{a}\hat{b}}\hat{M}_{\hat{a}\hat{b}} + \bar{\psi}^i_\mu\hat{Q}^i + \tfrac{1}{2}A^{ij}{}_\mu T^{ij}. \tag{5.61}$$

Thus, on general grounds, we expect the supergravity theory to have a Vierbein, $N$ gravitinos, and an SO$(N)$ connection $A^{ij}{}_\mu$, and the theory to be invariant under SO$(N)$

*gauge* transformations. Moreover, since the $T^{ij}$s rotate the supercharges, we expect the gravitinos to transform under SO($N$) gauge transformations and be charged with respect to the SO($N$) gauge field. For this reason, these theories are also called *gauged supergravities.* Since they are generalizations of the $N = 1$ case, they should also contain a negative cosmological constant and the vacuum will be anti-de Sitter spacetime. If the theory has scalar fields, the cosmological constant will be replaced by a scalar potential that can have many extrema, some corresponding to anti-de Sitter, de Sitter, or Minkowski spacetime, but also to less symmetric spacetimes, such as the D8-brane solution of the $N = 2A, d = 10$ SUEGRA that can be found in Section 25.2.6. These theories have, therefore, many potential vacua, with different symmetry and supersymmetry properties.

In $d = 4$ dimensions, the R-symmetry group $\mathrm{H}_{\mathrm{aut}}$ of $N$-extended SUGRAs is, actually, U($N$), but all this symmetry only becomes manifest when the theory is written in terms of Weyl spinors that transform in the fundamental representation of U($N$). The scalars of these theories parametrize spaces whose holonomy is precisely that of the R-symmetry group. For $N = 1$ these must be Kähler spaces (of a special kind called *Kähler–Hodge* manifolds, see Appendix E.3); for $N = 2$ they are the products of a Kähler space (of a particular kind called *special Kähler* manifolds, see Appendix F) and a quaternionic-Kähler manifold (see Appendix G); for $N \geq 3$ it is one of the non-compact symmetric spaces listed in Table 5.1. We will discuss this point in the following chapters.

It is important to realize that, while the R-symmetry group $\mathrm{H}_{\mathrm{aut}}$ will always act on the fermions of the theory, it may or may not act on the bosons. If we denote by G the global symmetry group of the ungauged theories and by $\mathrm{G}_{\mathrm{bos}}$ the subgroup of G that acts on the bosons, in the maximal ($N = 8$) or half-maximal ($N = 4$) supergravities the situation is

$$\mathrm{H}_{\mathrm{aut}} \subset \mathrm{G}_{\mathrm{bos}} = \mathrm{G}. \tag{5.62}$$

In particular, the scalars parametrize the coset $\mathrm{G}/(\mathrm{H}_{\mathrm{aut}} \times \mathrm{H}_{\mathrm{matter}})$, where $\mathrm{H}_{\mathrm{matter}}$ is related to the matter multiplets and it is trivial in maximally extended supergravities.

In $N = 1, 2, d = 4$ supergravities the situation is totally different:

$$\mathrm{G} = \mathrm{G}_{\mathrm{bos}} \times \mathrm{H}_{\mathrm{aut}}. \tag{5.63}$$

Further, in theories with low amounts of supersymmetry there may exist symmetries that act only on the vectors (and spinors) but not on the scalars. We will discuss these points in more detail in the chapters that follow.

For $N > 1$ the procedure of gauging superalgebras is no longer straightforward and more fields usually occur in the theories, but the basic general facts we have just discussed remain true. The *rheonomic approach* explained in Ref. [315] takes care of all the subtleties that arise in the gauging of general superalgebras in a rigorous way.

To obtain $N$-extended Poincaré superalgebras, we simply have to perform the Wigner–Inönü contraction Eq. (5.12) supplemented with

$$T^{ij} = g^{-1} Z^{ij}. \tag{5.64}$$

The effect of this rescaling (which is the only one that leads to a consistent superalgebra) is that these $Z^{ij}$s commute with every other generator in the superalgebra and become in fact a set of $N(N-1)/2$ SO(2) generators. Generators of this kind are called *central*

Table 5.1. List of the symmetric spaces parametrized by the scalars of $N \geq 3$ supergravity coupled to $n_V$ vector multiplets

G is the duality group that leaves invariant the equations of motion of the theory (but not the action, in general) and H = $\mathrm{H}_{\mathrm{aut}} \times \mathrm{H}_{\mathrm{matter}}$, where $\mathrm{H}_{\mathrm{aut}}=\mathrm{U}(N)$ is the group of automorphisms of the superalgebra. (Observe that $\mathrm{SO}(6) \sim \mathrm{SU}(4)$.)

| $N$ | G/H | Rep |
|---|---|---|
| 3 | $\frac{\mathrm{SU}(3,n_V)}{\mathrm{SU}(3)\times\mathrm{U}(n_V)}$ | $(\mathbf{3}+\mathbf{n_V})_\mathbf{c}$ |
| 4 | $\frac{\mathrm{SL}(2,\mathbb{R})}{\mathrm{U}(1)} \times \frac{\mathrm{SO}(6,n_V)}{\mathrm{SO}(6)\times\mathrm{SO}(n_V)}$ | $(\mathbf{2},\mathbf{6}+\mathbf{n_V})$ |
| 5 | $\frac{\mathrm{SU}(1,5)}{\mathrm{U}(5)}$ | **20** |
| 6 | $\frac{\mathrm{SO}^*(12)}{\mathrm{U}(6)}$ | **32** |
| 8 | $\frac{\mathrm{E}_{7(7)}}{\mathrm{SU}(8)/\mathbb{Z}_2}$ | **56** |

*charges* and we could forget about them if they did not occur in the anticommutator of the supercharges. Before we write the resulting superalgebra, it is instructive to make some general considerations. Now we expect the theory to have $N(N-1)/2$ SO(2) gauge fields that we can still label $A^{ij}{}_{\mu}$. Since the $Z^{ij}$s are central, we do not expect the gravitinos to be charged under the gauge fields, although they will be invariant under some sort of constant SO($N$) rotations. One may want to make the theory invariant under the local version of these SO($N$) rotations, *gauging* them, and then one would recover the *gauged supergravities* (hence the name) we obtained by gauging the $N$-extended AdS superalgebra.

Now, to perform the Wigner–Inönü contraction, we need to choose a spinor representation of SO(2, 3). There are two such representations, which are called *electric* and *magnetic* representations, which are explicitly worked out in Appendix D.2.1. They are equivalent in the sense that they are related by a similarity transformation and, obviously, they are just two of an infinite family of equivalent representations. These two are, however, of special interest. If we contract using the electric representation, we obtain, for the anticommutator of two supercharges,

$$\{Q^{\alpha\, i}, Q^{\beta\, j}\} = i\delta^{ij}\left(\gamma^a \mathcal{C}^{-1}\right)^{\alpha\beta} P_a - i\left(\mathcal{C}^{-1}\right)^{\alpha\beta} Z^{ij}, \tag{5.65}$$

whereas, if we contract using the magnetic representation, we obtain

$$\{Q^{\alpha\, i}, Q^{\beta\, j}\} = i\delta^{ij}\left(\gamma^a \mathcal{C}^{-1}\right)^{\alpha\beta} P_a - \gamma_5\left(\mathcal{C}^{-1}\right)^{\alpha\beta} Z^{ij}. \tag{5.66}$$

As we advanced, the first surprise is that the central charges occur in this anticommutator, but nowhere else. The second surprise is that the central charges occur in two different ways. From the Poincaré point of view, in the electric case the $Z^{ij}$s are scalars whereas in the magnetic case they are pseudoscalars. How should we interpret these charges? If we construct supergravity theories gauging the "electric" superalgebra, we will have to associate gauge potentials with the $Z^{ij}$s, which will be, then, interpreted as electric charges, in agreement with their scalar nature. In the magnetic case, the $Z^{ij}$s should be interpreted as magnetic charges. The similarity transformation that relates the electric and magnetic $\mathrm{AdS}_4$ representations becomes a *chiral–dual* transformation that rotates electric into magnetic charges and vice versa. In fact, we can write the most general anticommutator of the supercharges including both kinds of charges of the most general $N$-extended Poincaré superalgebra,[10]

$$\boxed{\begin{aligned}
[M_{ab}, M_{cd}] &= -M_{eb}\Gamma_{\mathrm{v}}(M_{cd})^{e}{}_{a} - M_{ae}\Gamma_{\mathrm{v}}(M_{cd})^{e}{}_{b},\\
[P_a, M_{bc}] &= -P_e\Gamma_{\mathrm{v}}(M_{bc})^{e}{}_{a},\\
\left[Q^{\alpha\, i}, M_{ab}\right] &= \Gamma_{\mathrm{s}}(M_{ab})^{\alpha}{}_{\beta} Q^{\beta\, i},\\
\{Q^{\alpha\, i}, Q^{\beta\, j}\} &= i\delta^{ij}(\gamma^a\mathcal{C}^{-1})^{\alpha\beta} P_a - i(\mathcal{C}^{-1})^{\alpha\beta} Q^{ij} - \gamma_5(\mathcal{C}^{-1})^{\alpha\beta} P^{ij},
\end{aligned}} \tag{5.67}$$

[10] When the superalgebra is written in terms of Weyl spinors, the electric and magnetic charges are combined into a single complex central-charge matrix $Z^{IJ} \equiv Q^{ij} + iP^{ij}$. The chiral–dual (electric–magnetic-duality) transformations become part of the U($N$) R-symmetry group.

and this anticommutator (and the full superalgebra) will be invariant under the chiral–dual (electric–magnetic-duality) transformations which we expect to be symmetries of the $N$-extended Poincaré supergravity theories, but not of the $N$-extended AdS supergravities. The main reason for this is that we do not know how to generalize electric–magnetic-duality transformations to the non-Abelian setting and also that, in the gauged supergravity theories, the gravitinos are electrically charged with respect to the gauge vectors but there are no additional fields magnetically charged with respect to them.

The above result opens up the possibility that there are more general central charges in the anticommutator of two supercharges that we have not considered at the beginning. We consider this interesting possibility in Section 5.4.1.

### 5.4.1 Central extensions

According to the Haag–Lopuszański–Sohnius theorem [707], the above anticommutator is the most general allowed if we impose the condition that our theory is Poincaré invariant. Let us, therefore, not require Poincaré invariance. It turns out that any (Poincaré or AdS) superalgebra can be extended by including "central charges" with $n$ antisymmetric Lorentz indices and two SO$(N)$ indices $Z^{ij}_{a_1\cdots a_n}$ [755]. Generically, they appear in the anticommutator of two supercharges in the form

$$\frac{1}{n!}\left(\gamma^{a_1\cdots a_n}\mathcal{C}^{-1}\right)^{\alpha\beta} Z^{ij}_{a_1\cdots a_n}, \tag{5.68}$$

with the factor being necessary in order to have the right Hermiticity properties (which can be a $\gamma_5$ only in Poincaré superalgebras). These are not central charges in the strict sense because they do not commute with the Lorentz generators. In fact, consistency implies

$$\left[Z^{kl}_{c_1\cdots c_n}, M_{ab}\right] = -n\Gamma_{\rm v}(M_{ab})^{e}{}_{[c_1} Z^{kl}_{|e|c_2\cdots c_n]}. \tag{5.69}$$

The new central charge will be symmetric or antisymmetric in the SO$(N)$ indices depending on whether $\left(\gamma^{a_1\cdots a_n}\mathcal{C}^{-1}\right)^{\alpha\beta}$ is symmetric or antisymmetric in $\alpha\beta$ since the full anticommutator has to be symmetric under the simultaneous interchange of $\alpha\beta$ and $ij$.

In four dimensions (and similarly in any dimensionality) it is easy to determine the symmetry of the possible terms:

$$\mathcal{C}^{-1},\quad \gamma_5\mathcal{C}^{-1},\quad \gamma_5\gamma_a\mathcal{C}^{-1},\quad \gamma_{abc}\mathcal{C}^{-1},\quad \gamma_{abcd}\mathcal{C}^{-1}, \tag{5.70}$$

are antisymmetric. In fact, the second and the fifth and the third and the fourth matrices are related by Eq. (D.104). The symmetric matrices are

$$\gamma_a\mathcal{C}^{-1},\quad \gamma_{ab}\mathcal{C}^{-1},\quad \gamma_5\gamma_{ab}\mathcal{C}^{-1},\quad \gamma_5\gamma_{abc}\mathcal{C}^{-1}. \tag{5.71}$$

The first and the fourth and the second and the third matrices are related by Eq. (D.104).

The most general anticommutator of the two central charges in $d = 4$ will, therefore, be

$$\begin{aligned}\{Q^{\alpha i}, Q^{\beta j}\} = {} & i\delta^{ij}\left(\gamma^a\mathcal{C}^{-1}\right)^{\alpha\beta} P_a + i\left(\mathcal{C}^{-1}\right)^{\alpha\beta} Z^{[ij]} + \gamma_5\left(\mathcal{C}^{-1}\right)^{\alpha\beta}\tilde{Z}^{[ij]} \\ & + \left(\gamma^a\mathcal{C}^{-1}\right)^{\alpha\beta} Z^{(ij)}_a + i\left(\gamma_5\gamma^a\mathcal{C}^{-1}\right)^{\alpha\beta} Z^{[ij]}_a \\ & + i\left(\gamma^{ab}\mathcal{C}^{-1}\right)^{\alpha\beta} Z^{(ij)}_{ab} + \left(\gamma_5\gamma^{ab}\mathcal{C}^{-1}\right)^{\alpha\beta}\tilde{Z}^{(ij)}_{ab}.\end{aligned} \tag{5.72}$$

It is equally easy to determine the most general anticommutator of two supercharges in the AdS case. It takes the form

$$\left\{\hat{Q}^{\alpha\, i}, \hat{Q}^{\beta\, j}\right\} = \delta^{ij}\left[\Gamma_{\rm s}\left(\hat{M}^{\hat{a}\hat{b}}\right)\mathcal{C}^{-1}\right]^{\alpha\beta}\hat{M}_{\hat{a}\hat{b}} - (\mathcal{C}^{-1})^{\alpha\beta}T^{ij} + i\left(\hat{\gamma}^{\hat{a}}\mathcal{C}^{-1}\right)^{\alpha\beta}\hat{Z}^{[ij]}_{\hat{a}}$$
$$+\left[\Gamma_{\rm s}\left(\hat{M}^{\hat{a}\hat{b}}\right)\mathcal{C}^{-1}\right]^{\alpha\beta}\hat{Z}^{(ij)}_{\hat{a}\hat{b}}. \qquad (5.73)$$

We already know that the $T^{ij}$ are not central extensions, but $\mathrm{SO}(N)$ generators that rotate the supercharges whose presence is required by the Jacobi identities . It is evident that, in order to satisfy these identities, the new extensions must have non-vanishing commutators with both the AdS generators as in Eq. (5.69) and with the $\mathrm{SO}(N)$ generators, that is

$$\left[\hat{Z}^{kl}_{\hat{c}_1\cdots\hat{c}_n}, \hat{M}_{\hat{a}\hat{b}}\right] = -n\Gamma_{\rm v}\left(\hat{M}_{\hat{a}\hat{b}}\right)^{\hat{e}}{}_{[\hat{c}_1}\hat{Z}^{kl}_{|\hat{e}|\hat{c}_2\cdots\hat{c}_n]}, \qquad (5.74)$$

$$\left[\hat{Z}^{kl}_{\hat{c}_1\cdots\hat{c}_n}, T^{ij}\right] = \Gamma(T^{ij})^{k}{}_{m}\hat{Z}^{ml}_{\hat{c}_1\cdots\hat{c}_n} + \Gamma(T^{ij})^{l}{}_{m}\hat{Z}^{km}_{\hat{c}_1\cdots\hat{c}_n}. \qquad (5.75)$$

One should now check all the Jacobi identities, which is left as an exercise for the reader. A consistency check of this proposal comes from the contraction to the Poincaré case using

$$\hat{Z}^{ij}_{c_1\cdots c_n} = g^{-1}Z^{ij}_{c_1\cdots c_n}, \qquad \hat{Z}^{ij}_{c_1\cdots c_{n-1}-1} = g^{-1}Z^{ij}_{c_1\cdots c_{n-1}}, \qquad (5.76)$$

which shows how all the Poincaré central charges arise from the AdS ones. In particular, it is interesting to see how the magnetic pseudoscalar charges $\tilde{Z}^{[ij]}$ arise in the contraction using the electric representation from the $-1$ component of the vector $\hat{Z}^{[i]}_{\hat{a}}$.

We are now going to study the two simplest examples of extended Poincaré and AdS supergravity.

## 5.5 $N = 2, d = 4$ (Poincaré) supergravity

As mentioned before, the $N = 1, d = 4$ Poincaré supergravity theory can also be generalized by adding supersymmetric matter, giving, in some cases (when the additional matter includes gravitinos), theories that are invariant under more supersymmetry transformations.

The simplest case in which this happens is the addition of a supermultiplet containing a second gravitino $\psi^2_\mu$ and a vector field $A_\mu$ (the original gravitino in the $N = 1$ supergravity multiplet is now denoted by $\psi^1_\mu$) and was studied by Ferrara and van Nieuwenhuizen in Ref. [537]. This theory is invariant under the original $N = 1$ local supersymmetry transformation with a parameter that we denote now by $\epsilon^1$ and under a new independent local supersymmetry transformation with parameter $\epsilon^2$. This theory is called for obvious reasons $N = 2, d = 4$ (Poincaré) supergravity and it is sometimes qualified as *ungauged* because it does not contain matter charged under the vector field.

From a different point of view, this SUEGRA is based on the $N = 2$ Poincaré superalgebra which we have just studied and could be derived by a generalization of the gauging of the algebraic procedure that worked for the $N = 1$ case (see Ref. [315]). Therefore, the fact that it has an $\mathrm{SO}(2)$ gauge vector field under which the gravitinos are not charged fits in the general scheme, according to which we also expect the theory to be invariant under some sort of chiral–dual (electric–magnetic-duality) symmetry.

Forgetting the historical way in which the theory was constructed, it can now be described by treating on an equal footing both gravitinos and supersymmetries as follows: the $N=2, d=4$ supergravity multiplet consists of the Vierbein, a couple of real gravitinos, and a vector field

$$\left\{e^a{}_\mu, \psi_\mu = \begin{pmatrix} \psi^1_\mu \\ \psi^2_\mu \end{pmatrix}, A_\mu\right\}, \tag{5.77}$$

respectively. The SO(2) indices $i=1,2$ that the fermions (and Pauli matrices) have in this theory will not be shown explicitly unless necessary and will be assumed to be contracted in obvious ways.[11]

The action for $N=2, d=4$ Poincaré supergravity is, in the first-order formalism [537],

$$\boxed{\begin{aligned} S = \int d^4x\, e\big\{&R(e,\omega) + 2e^{-1}\epsilon^{\mu\nu\rho\sigma}\bar{\psi}_\mu\gamma_5\gamma_\nu\mathcal{D}_\rho\psi_\sigma - \mathcal{F}^2 \\ &+ \mathcal{J}_{(\mathrm{m})}{}^{\mu\nu}(\mathcal{J}_{(\mathrm{e})\mu\nu} + \mathcal{J}_{(\mathrm{m})\mu\nu})\big\}, \end{aligned}} \tag{5.78}$$

where $\mathcal{D}$ is, as before, the Lorentz-covariant derivative, and

$$\mathcal{F}_{\mu\nu} = \tilde{F}_{\mu\nu} + \mathcal{J}_{(\mathrm{m})\mu\nu}, \quad \tilde{F}_{\mu\nu} = F_{\mu\nu} + \mathcal{J}_{(\mathrm{e})\mu\nu}, \quad F_{\mu\nu} = 2\partial_{[\mu}A_{\nu]}, \tag{5.79}$$

and

$$\mathcal{J}_{(\mathrm{e})\mu\nu} = i\bar{\psi}_\mu\sigma^2\psi_\nu, \qquad \mathcal{J}_{(\mathrm{m})\mu\nu} = -\frac{1}{2e}\epsilon^{\mu\nu\rho\sigma}\bar{\psi}_\rho\gamma_5\sigma^2\psi_\sigma. \tag{5.80}$$

$F$ is the standard vector field strength, $\tilde{F}$ is the *supercovariant* field strength,[12] and, in terms of $\mathcal{F}$, the ("Maxwell") equation of the vector field is simply

$$\frac{\delta S}{\delta A_\nu} = 4e\nabla_\mu(e)\mathcal{F}^{\mu\nu} = 0. \tag{5.81}$$

The divergences of $\mathcal{J}_{\mathrm{e}}$ and $\mathcal{J}_{\mathrm{m}}$ are two topologically conserved currents that appear as electric-like and magnetic-like sources for the vector field:

$$\partial_\mu(eF^{\mu\nu}) = +\partial_\mu(e\mathcal{J}_{\mathrm{e}}^{\nu\mu}) + \partial_\mu(e\mathcal{J}_{\mathrm{m}}^{\nu\mu}). \tag{5.82}$$

They are naturally associated with the electric and magnetic central charges of the $N=2, d=4$ Poincaré supersymmetry algebra.

The equation of motion for $\omega_\mu{}^{ab}$ is the same as in the $N=1$ case (except for the SO(2) indices, which we do not show explicitly) and, thus, the solution is the same; in particular, the torsion is given in terms of the gravitinos by

$$T_{\mu\nu}{}^a = i\bar{\psi}_\mu\gamma^a\psi_\nu \;\; (\equiv i\bar{\psi}_{j\,\mu}\gamma^a\psi^j_\nu). \tag{5.83}$$

[11] It is possible to combine the two real gravitinos into a single complex gravitino. This has some advantages: the theory looks simpler because there is no need to use Pauli matrices. However, the structure of the supergravity theory is somewhat obscured, and we choose the real form for pedagogical reasons.

[12] Whose supersymmetry transformation rule does not contain any derivatives of the gauge parameters.

The remaining two equations of motion are

$$\begin{aligned}
\frac{\delta S}{\delta e^a{}_\mu} &= -2e\Big[G_a{}^\mu - 2T(\psi)_a{}^\mu - 2\tilde{T}(A)_a{}^\mu\Big], \\
\frac{\delta S}{\delta \bar{\psi}_\mu} &= 4\epsilon^{\mu\nu\rho\sigma}\gamma_5\gamma_\nu\hat{\mathcal{D}}_\rho\psi_\sigma - 4i\Big(\tilde{F}^{\mu\nu} + i\star\tilde{F}^{\mu\nu}\gamma_5\Big)\sigma^2\psi_\nu,
\end{aligned} \tag{5.84}$$

where the equation of motion for $\omega_\mu{}^{ab}$ has been used, and

$$T(\psi)_a{}^\mu = -\frac{1}{2e}\epsilon^{\mu\nu\rho\sigma}\bar{\psi}_\nu\gamma_5\gamma_a\mathcal{D}_\rho\psi_\sigma, \qquad \tilde{T}(A)_a{}^\mu = \tilde{F}_a{}^\rho\tilde{F}^\mu{}_\rho - \tfrac{1}{4}e_a{}^\mu\tilde{F}^2. \tag{5.85}$$

The action and equations of motion are invariant under

**general coordinate transformations,**

$$\begin{aligned}
\delta_\xi x^\mu &= \xi^\mu, & \delta_\xi e^a{}_\mu &= -\xi^\nu\partial_\nu e^a{}_\mu - \partial_\mu\xi^\nu e^a{}_\nu, \\
\delta_\xi\psi_\mu &= -\xi^\nu\partial_\nu\psi_\mu - \partial_\mu\xi^\nu\psi_\nu, & \delta_\xi A_\mu &= -\xi^\nu\partial_\nu A_\mu - \partial_\mu\xi^\nu A_\nu,
\end{aligned} \tag{5.86}$$

**local Lorentz transformations,**

$$\delta_\sigma e^a{}_\mu = \sigma^a{}_b e^b{}_\mu, \qquad \delta_\sigma\psi_\mu = \tfrac{1}{2}\sigma^{ab}\gamma_{ab}\psi_\mu, \tag{5.87}$$

$\mathrm{U}(1)$ **gauge transformations,**

$$\delta_\chi A_\mu = \partial_\mu\chi, \tag{5.88}$$

**and internal SO$(2)$ rotations of the gravitinos,**

$$\psi'_\mu = e^{i\varphi\sigma^2}\psi_\mu, \tag{5.89}$$

where $\varphi$ is a constant (not spacetime-dependent) parameter.

The equations of motion (but not the action) are invariant under

**chiral–dual (electric–magnetic-duality) SO$(2)$ transformations,**

$$\tilde{F}'_{\mu\nu} = \cos\theta\,\tilde{F}_{\mu\nu} + \sin\theta\star\tilde{F}_{\mu\nu}, \qquad \psi'_\mu = e^{\frac{i}{2}\theta\gamma_5}\psi_\mu. \tag{5.90}$$

These transformations rotate electric into magnetic components of the supercovariant field strength and, at the same time, multiply by opposite phases the two chiral components of spinors (hence the name):

$$\psi'_\mu = \Big[\tfrac{1}{2}e^{\frac{i}{2}\theta}(1+\gamma_5) + \tfrac{1}{2}e^{-\frac{i}{2}\theta}(1-\gamma_5)\Big]\psi_\mu. \tag{5.91}$$

These transformations also rotate the two topologically conserved currents,

$$\begin{aligned}
\mathcal{J}'_{(\mathrm{e})} &= \cos\theta\,\mathcal{J}_{(\mathrm{e})} - \sin\theta\star\mathcal{J}_{(\mathrm{m})}, \\
\mathcal{J}'_{(\mathrm{m})} &= -\sin\theta\star\mathcal{J}_{(\mathrm{e})} + \cos\theta\,\mathcal{J}_{(\mathrm{m})},
\end{aligned} \tag{5.92}$$

which helps to prove that these transformations also rotate the Maxwell equation into the Bianchi identity

$$\partial_\mu(e \star F^{\mu\nu}) = 0, \tag{5.93}$$

since they are equivalent to

$$\mathcal{F}'_{\mu\nu} = \cos\theta\, \mathcal{F}_{\mu\nu} + \sin\theta \, \star F_{\mu\nu}. \tag{5.94}$$

This is, of course, the same rotation as that which takes place between the two central charges in the $N=2, d=4$ Poincaré supersymmetry algebra.

The total global symmetry group of this theory is the R-symmetry group $\mathrm{U}(2) = \mathrm{U}(1) \times \mathrm{SU}(2)$, but, using Majorana spinors, only the two $\mathrm{U}(1)$ groups described above are (more or less) manifest. The chiral–dual transformations correspond to the overall $\mathrm{U}(1)$ factor, while the internal $\mathrm{SO}(2)$ rotations that act on the spinor indices are a $\mathrm{U}(1)$ subgroup of the $\mathrm{SU}(2)$ factor. It is a fact that the only possible Abelian gaugings of the theory involve the second $\mathrm{U}(1)$ factor, even if we add matter to the theory. The simplest case (without matter) is considered in Section 5.6. More general, non-Abelian, gaugings, which are only possible in presence of matter, are studied in Chapter 7.

Finally, the theory is invariant under

**local** $N=2, d=4$ **supersymmetry transformations,**

$$\delta_\epsilon e^a{}_\mu = -i\bar{\epsilon}\gamma^a\psi_\mu, \qquad \delta_\epsilon A_\mu = -i\bar{\epsilon}\sigma^2\psi_\mu, \qquad \delta_\epsilon\psi_\mu = \tilde{\mathcal{D}}_\mu\epsilon, \tag{5.95}$$

where

$$\tilde{\mathcal{D}}_\mu = \mathcal{D}_\mu + \tfrac{1}{4}\, \tilde{\not{F}}\gamma_\mu\sigma^2, \tag{5.96}$$

is the supercovariant derivative acting on $\epsilon$.

It is instructive to check the invariance of the action under the above transformations. On varying the whole action, using the equation of motion of the spin connection, the specific form of the supersymmetry transformation rules, and

$$\tilde{\mathcal{D}}_\mu\bar{\epsilon} = \overline{\tilde{\mathcal{D}}_\mu\epsilon} = \overline{\mathcal{D}_\mu\epsilon} - \tfrac{1}{4}\bar{\epsilon}\gamma_\mu\, \tilde{\not{F}}\sigma^2, \tag{5.97}$$

and integrating by parts the partial derivative, we find that the invariance of the action depends on the $N=2, d=4$ Poincaré supersymmetry gauge identity

$$\tilde{\mathcal{D}}_\mu \frac{\delta S}{\delta\bar{\psi}_\mu} = -i\left(\frac{\delta S}{\delta e^a{}_\mu}\gamma^a + \frac{\delta S}{\delta A_\mu}\sigma^2\right)\psi_\mu, \tag{5.98}$$

where here the supercovariant derivative takes the form

$$\tilde{\mathcal{D}}_\mu \frac{\delta S}{\delta\bar{\psi}_\mu} = \left[\mathcal{D}_\mu + \tfrac{1}{4}\gamma_\mu\, \tilde{\not{F}}\sigma^2\right]\frac{\delta S}{\delta\bar{\psi}_\mu}. \tag{5.99}$$

To prove this gauge identity, we need to use some of the results we used to prove the $N=1$ gauge identity, the Bianchi identity for $F_{\mu\nu}$, and the $N=2$ Fierz identities, with which it is possible to prove two main identities (see Section 5.7):

$$e^{-1}\epsilon^{\mu\nu\rho\sigma}T_{\nu\rho}{}^a\gamma_5\gamma_a\psi_\sigma = 2\left(\star\mathcal{J}^{\mu\nu}_{(\mathrm{e})}\gamma_5 + i\mathcal{J}^{\mu\nu}_{(\mathrm{m})}\right)\sigma^2\psi_\nu \tag{5.100}$$

and

$$-e^{-1}\epsilon^{\mu\nu\rho\sigma}T_{\nu\rho}{}^{a}\gamma_5\gamma_a\mathcal{D}_\rho\psi_\sigma = 2iT(\psi)_a{}^{\mu}\gamma^a\psi_\mu - 2\mathcal{D}_\mu\left(\star\mathcal{J}^{\mu\nu}_{(\mathrm{e})}\gamma_5 + i\mathcal{J}^{\mu\nu}_{(\mathrm{m})}\right)\sigma^2\psi_\nu$$
$$+ 2\left(\star\mathcal{J}^{\mu\nu}_{(\mathrm{e})}\gamma_5 + i\mathcal{J}^{\mu\nu}_{(\mathrm{m})}\right)\sigma^2\mathcal{D}_\mu\psi_\nu. \tag{5.101}$$

### 5.5.1 *The local supersymmetry algebra*

The commutator of two supersymmetry variations closes on-shell with

$$[\delta_{\epsilon_1}, \delta_{\epsilon_2}] = \delta_\xi + \delta_\sigma + \delta_\chi + \delta_\epsilon, \tag{5.102}$$

where

$$\xi^\mu = -i\bar{\epsilon}_1\gamma^\mu\epsilon_2, \sigma^{ab} = \xi^\mu\omega_\mu{}^{ab} - i\bar{\epsilon}_2\left(\tilde{F}^{ab} - i\gamma_5\star\tilde{F}^{ab}\right)\sigma^2\epsilon_1,$$
$$\chi = -i\bar{\epsilon}_2\sigma^2\epsilon_1 + \xi^\nu A_\nu, \qquad \epsilon = \xi^\mu\psi_\mu. \tag{5.103}$$

## 5.6 $N=2, d=4$ "gauged" (AdS) supergravity

There are two main ways to arrive at this theory, apart from the algebra-gauging procedure. First, we could simply add supersymmetric matter to the $N=1, d=4$ AdS supergravity theory. Consistency requires that the pair of gravitinos are charged under the vector field with a coupling constant that is equal to the Wigner–Inönü parameter $g$. For this reason, the theory was first found from the $N=2, d=4$ Poincaré theory by a gauging procedure: the internal SO(2) symmetry that rotates the two real gravitinos can be gauged [554, 420], the gauge field being the vector field already present in the theory (the field content is, therefore, the same). As mentioned before, the U(1) gauge group is a subgroup of the SU(2) factor of the R-symmetry group. The pair of real gravitinos transforms as a complex, charged gravitino with a gauge parameter $\varphi$, and we have to relate this parameter to the gauge parameter of U(1) transformations of the vector field according to

$$\varphi = -g\chi, \tag{5.104}$$

where $g$ is the gauge coupling constant. The introduction of the minimal coupling between gravitinos and vector field requires, in order to preserve supersymmetry, the introduction of several other $g$-dependent terms, which can be absorbed into a change of connection from the Lorentz one to the anti-de Sitter one. In the end, the result is obviously the same as that obtained by adding supersymmetric matter to the $N=1, d=4$ AdS supergravity theory.

In any case, the two main characteristics of the theory are the presence of a negative cosmological constant $\Lambda = -3g^2$ and the fact that the gravitinos are minimally coupled to the vector field with coupling constant $g$.

We anticipate that there is going to be a third source term in the Maxwell equation, which is going to break the invariance under chiral–dual transformations of the "ungauged" (Poincaré) theory.

The gauged $N=2,d=4$ "gauged" supergravity action for these fields in the first-order formalism is, thus,

$$S=\int d^4x\, e\Big\{R(e,\omega)+6g^2+2e^{-1}\epsilon^{\mu\nu\rho\sigma}\bar{\psi}_\mu\gamma_5\gamma_\nu\Big(\hat{\mathcal{D}}_\rho+igA_\rho\sigma^2\Big)\psi_\sigma \\ -\mathcal{F}^2+\mathcal{J}_{(\mathrm{m})}{}^{\mu\nu}(\mathcal{J}_{(\mathrm{e})\mu\nu}+\mathcal{J}_{(\mathrm{m})\mu\nu})\Big\}\,, \tag{5.105}$$

where again $\hat{\mathcal{D}}$ is the SO(2,3) (AdS) gauge-covariant derivative

The symmetries of this action are essentially the same as in the ungauged case: GCTs, local Lorentz transformations,[13] U(1) gauge transformations, which now take the form

$$A'_\mu=A_\mu+\partial_\mu\chi,\qquad \psi'_\mu=e^{-ig\chi\sigma^2}\psi_\mu, \tag{5.106}$$

and local supersymmetry transformations, which take the same form as in the Poincaré case, but with the new supercovariant derivative

$$\tilde{\hat{\mathcal{D}}}_\mu=\hat{\mathcal{D}}_\mu+igA_\mu\sigma^2+\tfrac{1}{4}\,\tilde{\not{F}}\gamma_\mu\sigma^2. \tag{5.107}$$

As mentioned before, the chiral–dual invariance of the ungauged theory is broken by the minimal coupling between gravitinos and vector field, which results in the new Maxwell equation with a new Noether current,

$$\partial_\nu(e\mathcal{F}^{\nu\mu})-\frac{ig}{2}\epsilon^{\mu\nu\rho\sigma}\bar{\psi}_\nu\gamma_5\gamma_\rho\sigma^2\psi_\sigma. \tag{5.108}$$

For the sake of completeness, we give the remaining equations of motion:

$$\begin{aligned}0&=G_a{}^\mu-3g^2e_a{}^\mu-2T(\psi)_a{}^\mu-2\tilde{T}(A)_a{}^\mu,\\ 0&=e^{-1}\epsilon^{\mu\nu\rho\sigma}\gamma_5\gamma_\nu\Big(\hat{\mathcal{D}}_\rho+igA_\rho\sigma^2\Big)\psi_\sigma-i\Big(\tilde{F}^{\mu\nu}+i\star\tilde{F}^{\mu\nu}\gamma_5\Big)\sigma^2\psi_\nu,\end{aligned} \tag{5.109}$$

where the equation of motion for $\omega_\mu{}^{ab}$ has been used, and where

$$\begin{aligned}T(\psi)_a{}^\mu&=-\frac{1}{2e}\epsilon^{\mu\nu\rho\sigma}\bar{\psi}_\nu\gamma_5\gamma_a\Big(\hat{\mathcal{D}}_\rho+igA_\rho\sigma^2\Big)\psi_\sigma-\frac{ig}{2e}\epsilon^{\mu\nu\rho\sigma}\bar{\psi}_\nu\gamma_5\gamma_{\rho a}\psi_\sigma,\\ \tilde{T}(A)_a{}^\mu&=\tilde{F}_a{}^\rho\tilde{F}^\mu{}_\rho-\tfrac{1}{4}e_a{}^\mu\tilde{F}^2.\end{aligned} \tag{5.110}$$

To prove the invariance of the action under the local supersymmetry transformations, we check the $N=2,d=4$ AdS gauge identity

$$\tilde{\hat{\mathcal{D}}}_\mu\frac{\delta S}{\delta\bar{\psi}_\mu}=-i\Big(\frac{\delta S}{\delta e^a{}_\mu}\gamma^a+\frac{\delta S}{\delta A_\mu}\sigma^2\Big)\psi_\mu, \tag{5.111}$$

where, here,

$$\tilde{\hat{\mathcal{D}}}_\mu\frac{\delta S}{\delta\bar{\psi}_\mu}=\Big[\hat{\mathcal{D}}_\mu+\tfrac{1}{4}\gamma_\mu\,\tilde{\not{F}}\sigma^2\Big]\frac{\delta S}{\delta\bar{\psi}_\mu}. \tag{5.112}$$

[13] There is no invariance under the full SO(2,3).

To prove this identity we need only check the $g$-dependent terms (the $g$-independent ones work, as we checked in Section 5.5). To check the $g$-dependent terms, we need only the additional identities (see Section 5.7)

$$(\bar{\psi}_{[\nu|}\gamma_a\psi_{|\mu|})\gamma_5\gamma^a\sigma^2\psi_{|\rho]} = (\bar{\psi}_{[\nu|}\gamma_a\gamma_5\psi_{|\mu|})\gamma^a\sigma^2\psi_{|\rho]} \tag{5.113}$$

and

$$\begin{aligned}(\bar{\psi}_{[\nu|}\gamma_a\psi_{|\mu})\gamma_5\gamma^a\gamma_\rho\psi_{\sigma]} &+ (\bar{\psi}_{[\nu}\sigma^2\psi_\mu)\gamma_5\gamma_\rho\sigma^2\psi_{\sigma]} - (\bar{\psi}_{[\nu|}\gamma_5\sigma^2\psi_{|\mu})\gamma_\rho\sigma^2\psi_{\sigma]} \\ &= -2(\bar{\psi}_{[\nu|}\gamma_5\gamma_{a|\rho}\psi_\mu)\gamma^a\psi_{\sigma]} - 2(\bar{\psi}_{[\nu|}\gamma_5\gamma_{|\rho}\sigma^2\psi_\mu)\sigma^2\psi_{\sigma]}.\end{aligned} \tag{5.114}$$

### 5.6.1 The local supersymmetry algebra

The commutator of two supersymmetry variations closes on-shell with the same parameters as in the ungauged case except for

$$\sigma^{ab} = \xi^\mu\omega_\mu{}^{ab} - g\bar{\epsilon}_2\gamma^{ab}\epsilon_1 - i\bar{\epsilon}_2\left(\tilde{F}^{ab} - i\gamma_5 \star \tilde{F}^{ab}\right)\sigma^2\epsilon_1. \tag{5.115}$$

From the point of view of the supersymmetry algebra, we are going from Poincaré supersymmetry to AdS supersymmetry in which the generator of SO(2) rotations has to appear in the anticommutator of two supersymmetry charges, for consistency. Although it appears in the same position as a central charge, it should be stressed that it is not a central charge because it does not commute with the supercharges.

## 5.7 Proofs of some identities

Using the $N=2$ Fierz identities Eq. (D.67) we immediately find, for any spinor $\lambda$, the following two identities:

$$\begin{aligned}(\bar{\psi}_{[\nu|}\gamma_5\gamma_a\lambda)\gamma^a\psi_{|\mu]} = &-\tfrac{1}{2}(\bar{\psi}_{[\nu}\gamma_5\sigma^2\psi_{\mu]})\sigma^2\lambda - \tfrac{1}{4}(\bar{\psi}_{[\nu|}\gamma_a\gamma_5\sigma^2\psi_{|\mu]})\gamma^a\sigma^2\lambda \\ &+ \tfrac{1}{2}(\bar{\psi}_{[\nu}\sigma^2\psi_{\mu]})\gamma_5\sigma^2\lambda \\ &- \tfrac{1}{4}\left(\bar{\psi}_{[\nu|}\gamma_a\begin{pmatrix}\sigma^0\\ \sigma^1\\ \sigma^3\end{pmatrix}^{\mathrm{T}}\psi_{|\mu]}\right)\gamma^a\gamma_5\begin{pmatrix}\sigma^0\\ \sigma^1\\ \sigma^3\end{pmatrix}\lambda\end{aligned} \tag{5.116}$$

and

$$\begin{aligned}(\bar{\psi}_{[\nu|}\gamma_a\chi)\gamma_5\gamma^a\psi_{|\mu]} = &\frac{1}{2}(\bar{\psi}_{[\nu}\gamma_5\sigma^2\psi_{\mu]})\sigma^2\chi - \frac{1}{4}(\bar{\psi}_{[\nu|}\gamma_a\gamma_5\sigma^2\psi_{|\mu]})\gamma^a\sigma^2\chi \\ &- \frac{1}{2}(\bar{\psi}_{[\nu}\sigma^2\psi_{\mu]})\gamma_5\sigma^2\chi \\ &- \frac{1}{4}\left(\bar{\psi}_{[\nu|}\gamma_a\begin{pmatrix}\sigma^0\\ \sigma^1\\ \sigma^3\end{pmatrix}^{\mathrm{T}}\psi_{|\mu]}\right)\gamma^a\gamma_5\begin{pmatrix}\sigma^0\\ \sigma^1\\ \sigma^3\end{pmatrix}\chi.\end{aligned} \tag{5.117}$$

We can take $\lambda = \chi$ and subtract Eq. (5.117) from Eq. (5.116), giving

$$\begin{aligned}(\bar{\psi}_{[\nu|}\gamma_5\gamma_a\lambda)\gamma^a\psi_{|\mu]} &- (\bar{\psi}_{[\nu|}\gamma_a\lambda)\gamma_5\gamma^a\psi_{|\mu]} \\ &= -(\bar{\psi}_{[\nu|}\gamma_5\sigma^2\lambda)\sigma^2\psi_{|\mu]}\sigma^2\lambda + (\bar{\psi}_{[\nu|}\sigma^2\lambda)\gamma_5\sigma^2\psi_{|\mu]}\gamma_5\sigma^2\lambda.\end{aligned} \tag{5.118}$$

We can take $\lambda = \psi_\rho$ and antisymmetrize in $\nu\rho\mu$, giving

$$(\bar{\psi}_{[\nu|}\gamma_a\psi_{|\mu|})\gamma_5\gamma^a\psi_{|\rho]} = -(\bar{\psi}_{[\nu|}\gamma_5\sigma^2\psi_{|\mu})\sigma^2\psi_{\rho]} + (\bar{\psi}_{[\nu}\sigma^2\psi_\mu)\gamma_5\sigma^2\psi_{\rho]}, \tag{5.119}$$

from which Eq. (5.100) follows.

If we act with $\mathcal{D}_\mu$ on Eq. (5.100) and use Eq. (5.118) with $\lambda = \mathcal{D}_\mu\psi_\rho$ to relate $\mathcal{D}_\mu T_{\nu\rho}{}^a$ to $T(\psi)_a{}^\mu$, we obtain Eq. (5.101).

On substituting $\lambda = \sigma^2\psi_\rho$ into Eq. (5.116) and multiplying the result by an overall $\sigma^2$ and adding to it Eq. (5.117) with $\chi = \psi_\rho$, we obtain Eq. (5.113).

By combining Eqs. (5.116) and (5.117) with $\lambda = \gamma_\rho\psi_\sigma$ and $\chi = \sigma^2\gamma_\rho\psi_\sigma$ in several different ways, one obtains Eq. (5.114).

# 6
# Matter-coupled $N = 1, d = 4$ supergravity

Just as the simplest extension of general relativity that respects its general covariance consists in its coupling to general-covariant matter fields (that is, tensor or spinor fields), the simplest way to extend a supergravity theory respecting its invariance under local supersymmetry transformation is by coupling it to supersymmetric matter fields, that is, matter fields that form complete supermultiplets that transform irreducibly under supersymmetry. The matter-coupled theory can then be deformed by gauging its global symmetries, for instance. The supergravities that arise as effective string theories contain in general matter and gauged symmetries and, moreover, any potential application to real-world phenomenology would require these ingredients.

In this chapter we are going to study the general coupling of supersymmetric matter to the pure, ungauged, $N = 1$ supergravity theory that we constructed in Chapter 5 and the gauging of its global symmetries. Our goal will be to write the most general bosonic action of $N = 1, d = 4$ supergravity and its supersymmetry transformation rules for vanishing fermions[1] in full detail. By bosonic action we mean the action that results from setting to zero all the fermionic fields. This is always a consistent truncation: all the solutions of the equations of motion of the bosonic action are automatically solutions of the equations of motion of the complete action with all the fermions set to zero.[2] We are interested in bosonic field configurations that represent the long-range fields of black holes, waves, etc., and, therefore, we only need to know the bosonic action. We will need the supersymmetry transformation rules of all the fields with the fermions set to zero to study the unbroken supersymmetry of the bosonic solutions, a concept that will be explained in Chapter 17.

A much more rigorous treatment of the coupling of supergravity to supersymmetric matter can be found in the classic literature and also in the recent book [564].

One of the motivations to include this chapter and Chapters 7–9 is that the original

[1] The complete theory was originally constructed in Refs. [82, 83, 384].

[2] All truncations associated to $\mathbb{Z}_2$ symmetries of an action (i.e. truncations that eliminate the fields which are odd and preserve the fields which are even (invariant) under the $\mathbb{Z}_2$ symmetry) are consistent because the equations of motion of the odd (even) fields are odd (even) (and, hence, all their terms have odd (even) powers of them). Therefore, setting to zero the odd fields automatically satisfies their equations of motion, leaving no constraints.

works on $N = 1, d = 4$ supergravity do not cover the most general case in the language of Kähler geometry in arbitrary coordinates, and the standard reference [1245], apart from using two-component spinors, which obscures the relation to other theories, does not use the momentum map, which clarifies the geometric structures used. We have found it most convenient to find the general form of these theories from their truncation from $N = 2, d = 4$ theories [48, 49]. We shall use the conventions of Refs. [1011, 712], which were derived from those used in the study of $N = 2, d = 4$ theories in Refs. [131, 937, 795, 797, 798]. The same conventions will be used again in Chapter 7. This will allow us to relate the theories, solutions, Killing spinors, etc., to those of our previous works.

The theories that we are going to review are the most general available in the standard literature, and they only include *electric* gaugings[3] of symmetries of the action. Using the embedding tensor formalism, briefly reviewed in Section 2.7, it is possible to construct more general deformations and include higher-rank potentials. So far, this has only been done for general $N = 1, d = 4$ SUGRAs [712] and pure $N = 2, d = 4, 5, 6$ SUGRAs (to lowest order in fermions) [799], for $N = 2, d = 4$ conformal SUGRAs [1269] and some maximal and half-maximal SUGRAs in various dimensions [1263, 1101, 1108, 1264, 157, 1258].

## 6.1 The matter supermultiplets

Before we discuss the supermultiplets, we must explain an important notational issue. In Chapter 5 we used Majorana spinors throughout (for gravitinos, supercharges, and infinitesimal supersymmetry parameters), but we mentioned that the $\mathrm{U}(N)$ R-symmetry of the four-dimensional $N$-extended supergravities (SUEGRAs) only becomes manifest when we use Weyl spinors. Having complete control of the R-symmetry is vital to describe the gaugings and, therefore, in this chapter and in Chapters 7 and 8 we are going to switch from Majorana to (four-component) Weyl spinors. We will continue to use the Majorana representation for the gamma matrices, however, which is why we have to keep the four complex components of the Weyl spinor (although only two of them will be independent).

Let us explain the notation in the general $N$ case in which there are $N$ Majorana gravitinos $\psi^i{}_\mu$ and infinitesimal supersymmetry parameters $\epsilon^i$, $i = 1, \ldots, N$. We decompose each of them in the sum of two Weyl spinors of opposite chiralities denoted by the position of the index $I, J = 1, \ldots, N$:

$$\begin{aligned} \psi^i{}_\mu &= \psi^I{}_\mu + \psi_{I\,\mu}, & \gamma_5\psi^I{}_\mu &= +\psi^I{}_\mu, & \gamma_5\psi_{I\,\mu} &= -\psi_{I\,\mu}, \\ \epsilon^i &= \epsilon^I + \epsilon_I, & \gamma_5\epsilon^I &= +\epsilon^I, & \gamma_5\epsilon_I &= -\epsilon_I. \end{aligned} \tag{6.1}$$

Since $\gamma_5^* = -\gamma_5$, the two Weyl spinors are each other's complex conjugates:

$$(\psi_{I\,\mu})^* = \psi^I{}_\mu, \qquad (\epsilon_I)^* = \epsilon^I, \tag{6.2}$$

which guarantees the reality of the original Majorana spinors. The relation between chirality and the position of the index may differ in other spinors of the theory and will have to be given when those spinors are introduced.

[3] These use the fundamental (electric) 1-forms of the theory as gauge fields.

The change of chirality with the complex conjugation has to be taken into account in the Dirac conjugation. Thus,

$$\bar{\epsilon}^I = i(\epsilon_I)^\dagger \gamma_0, \qquad \bar{\epsilon}^I \gamma_5 = \bar{\epsilon}^I, \qquad \bar{\epsilon}_I \gamma_5 = -\bar{\epsilon}_I. \tag{6.3}$$

Observe that complex conjugation also interchanges the position of the indices of the Dirac conjugates:

$$(\bar{\epsilon}_I)^* = \bar{\epsilon}^I. \tag{6.4}$$

These Weyl spinors transform in the fundamental (or antifundamental, depending on the chirality) representation of the R-symmetry group $\mathrm{H_{aut}} = \mathrm{U}(N) = \mathrm{U}(1) \times \mathrm{SU}(N)$. The overall $\mathrm{U}(1)$ factor, which we will sometimes denote by $\mathrm{U}(1)_\mathrm{R}$, multiplies, by a constant phase $e^{-iq\theta}$, the Weyl spinors. Their infinitesimal transformations are

$$\delta_{\mathrm{U}(1)_\mathrm{R}} \psi = -iq\theta\psi. \tag{6.5}$$

The weight (which is identical to the Kähler weight to be discussed later), $q$, of all the spinors of the theory is $-\frac{1}{2}$ times the spinor's chirality (so they have opposite weights for opposite chiralities). In the formulation that we are going to use, R-symmetry acts only on fermions and not on bosons. Other symmetries may, however, induce R-symmetry transformations, as we will explain.

In the $N = 1$ case we can use a dot in a lower or upper position instead of the $\mathrm{SU}(N)$ index $I, J$: $\psi^\bullet{}_\mu, \psi_{\bullet\,\mu}$, etc., but it is simpler to suppress the dot, defining

$$\psi_\mu \equiv \psi_{\bullet\,\mu}, \qquad \epsilon \equiv \epsilon_\bullet, \tag{6.6}$$

with negative chirality and $q = \frac{1}{2}$, and using the complex conjugate for the other chirality.

Now we are ready to proceed with the matter supermultiplets. The two most important types of $N = 1, d = 4$ matter multiplets[4] that can be coupled to the supergravity multiplet (the one that contains the graviton and gravitino $\{e^a{}_\mu, \psi_\mu\}$) are the chiral and the gauge supermultiplets. We can couple an arbitrary number of them to the supergravity one.

Each of the $n_C$ chiral supermultiplets, which we will label with $i, j, \ldots = 1, \ldots, n_C$, contains a complex scalar field $Z^i$ and a spinor of positive chirality $\chi^i$ (*chiralino*). The complex conjugates of these fields will carry indices $i^*, j^*$: $\{Z^{*\,i^*}, \chi^{*\,i^*}\}$.

Each of the $n_V = \bar{n}$ vector or gauge supermultiplets, which we will label with $\Lambda, \Sigma, \ldots = 1, \ldots, \bar{n}$, contains a 1-form field $A^\Lambda{}_\mu$ and a spinor of negative chirality (*gaugino*) $\lambda^\Lambda$.

Summarizing: in our conventions the Weyl spinors

$$\psi_\mu, \epsilon, \chi^{*\,i^*}, \lambda^\Lambda \tag{6.7}$$

have negative chirality and (Kähler) weight $\frac{1}{2}$, and their complex conjugates have opposite chirality and weight.

---

[4] We do not consider supermultiplets containing spin-$\frac{3}{2}$ fields as these would increase the number of independent supersymmetries of the theory, as we showed in Chapter 5. There are other supermultiplets which contain fields dual to the ones of the supermultiplets that we are going to discuss, in particular higher-rank differential forms.

## 6.2 The ungauged theory

Supersymmetry strongly constrains the possible couplings of the matter supermultiplets to the supergravity one and of the matter supermultiplets among themselves. This is, actually, one of the attractive features of supersymmetric theories. As we are going to see, an ungauged $N=1, d=4$ supergravity theory is completely determined by three functions: the Kähler potential $\mathcal{K}(Z, Z^*)$, the holomorphic *superpotential* $\mathcal{W}(Z)$, and the arbitrary holomorphic kinetic matrix with positive-definite imaginary part $f_{\Lambda\Sigma}(Z)$. The holomorphic superpotential $\mathcal{W}(Z)$ is sometimes replaced by the covariantly holomorphic section of Kähler weight 1 $\mathcal{L}(Z, Z^*)$, which we will define shortly.

Before we describe each of these functions and their role in the theory, it is important to realize that the $N=1$ theories are the ones that allow us more freedom in choosing these functions. Increasing the number of supersymmetries decreases the number of arbitrary functions. Thus, in $N=2, d=4$ supergravity an ungauged theory, for a given matter content, is completely determined by the so-called prepotential function (which determines the Kähler potential and the *period matrix*, which is the $N=2$ equivalent of $f_{\Lambda\Sigma}(Z)$) and the choice of quaternionic Kähler manifold. For $N>2$ the scalars in the matter multiplets always parametrize coset spaces which are completely determined by $N$, and the number of matter supermultiplets and the kinetic matrices are also completely determined by $N$ and the matter content. For $N>4$ there are no matter multiplets and the theories are unique (up to gaugings).

*The Kähler potential $\mathcal{K}$.* The complex scalars in the chiral multiplets parametrize a $\sigma$-model with Hermitian metric $\mathcal{G}_{ij^*}$, and the kinetic term in the action has the form

$$+2\mathcal{G}_{ij^*}\partial_\mu Z^i \partial^\mu Z^{*j^*}. \tag{6.8}$$

Global $N=1$ supersymmetry requires the metric to be, actually, Kähler, so it is completely determined by the Kähler potential[5] through Eq. (E.45) $\mathcal{G}_{ij^*} = \partial_i \partial_{j^*}\mathcal{K}$. The Kähler potential, in turn, is only defined up to the Kähler transformations Eq. (E.47)

$$\delta_\lambda \mathcal{K}(Z, Z^*) = \lambda(Z) + \lambda^*(Z^*); \tag{6.9}$$

a fundamental feature of these theories is that all the spinors (but none of the scalars) transform under the Kähler transformations with a given weight. This Kähler weight, as we have mentioned before, coincides with the R-symmetry weight, which for the spinors $\psi_\mu, \lambda^\Sigma, \chi^{*i^*}, \epsilon$ is $\frac{1}{2}$. That is: under the above Kähler transformations Eq. (E.47),

$$\delta_\lambda \psi_\mu = -\tfrac{1}{4}(\lambda - \lambda^*)\psi_\mu, \qquad \delta_\lambda \lambda^\Sigma = -\tfrac{1}{4}(\lambda - \lambda^*)\lambda^\Sigma, \qquad \delta_\lambda \chi^i = +\tfrac{1}{4}(\lambda - \lambda^*)\chi^i. \tag{6.10}$$

The $\mathrm{H_{aut}} = \mathrm{U}(1)_{\mathrm{R}}$ transformations (6.5) can be understood as constant Kähler transformations with

$$\theta = \Im\mathfrak{m}\lambda. \tag{6.11}$$

[5] Kähler and Kähler–Hodge manifolds and the gauging of their isometries, which we will use later, are reviewed in Appendix E.

It is worth stressing that the Kähler potential is not a field and that a constant Kähler transformation is not the transformation of any scalar field. Hence, these $\mathrm{U}(1)_{\mathrm{R}}$ symmetry transformations act only on fermionic fields even if they change the Kähler potential.

Local $N=1$ supersymmetry requires the Kähler manifold to be a Hodge manifold, which amounts to saying that there is a complex line bundle defined over the Kähler manifold whose connection is, precisely, the Kähler connection 1-form Eq. (E.48) with holomorphic components:

$$\mathcal{Q}_i = \frac{1}{2i}\partial_i \mathcal{K}. \tag{6.12}$$

This technical requirement ensures the global compatibility between Kähler transformations and complex coordinate transformations in the Kähler manifold and, in particular, the consistency of the Kähler-covariant derivatives of these fields given in Eqs. (E.59) and (E.60). These are the derivatives that have to be used in the fermions' kinetic terms, which we will not consider, and that also appear in the supersymmetry transformation rule of the gravitino action on the infinitesimal supersymmetry parameter $\epsilon$.

It is important to stress that the invariance under Kähler transformations is not a local symmetry even though the phase is $Z(x)$-dependent since the transformation does not contain any arbitrary local parameters. Therefore there are no constraints (gauge identities) associated with it.

On the other hand, we have stressed the fact that R-symmetry only acts on the fermions of the theory, but the relation between R-symmetry transformations and Kähler transformations indicates that the Kähler potential (a bosonic object) transforms under R-symmetry transformations. The Kähler potential, however, is not a physical field and, as long as it only occurs in the theory via its derivatives (Kähler metric and connection), we do not need to relate its transformation to any transformation of the scalars. As we are going to see, the introduction of a superpotential will force us to do this. Without a superpotential, though, R-symmetry is a consistent symmetry of $N=1, d=4$ supergravity, as expected.

*The superpotential $\mathcal{W}$ (or its equivalent $\mathcal{L}$).* The second arbitrary function that defines an ungauged $N=1, d=4$ theory is a holomorphic section $\mathcal{W}(Z)$ known as the *superpotential*. It has Kähler weight $(2,0)$, which means that its infinitesimal Kähler transformation rule is

$$\delta_\lambda \mathcal{W}'(Z) = -\lambda(Z)\mathcal{W}(Z), \tag{6.13}$$

which preserves its holomorphicity. It is convenient to work with the covariantly holomorphic superpotential $\mathcal{L}(Z,Z^*)$, which is related to $\mathcal{W}$ by

$$\mathcal{L}(Z,Z^*) \equiv \mathcal{W}(Z)e^{\mathcal{K}/2}. \tag{6.14}$$

Since $e^{\mathcal{K}}$ has weight $(-2,-2)$, $\mathcal{L}$ has weight $(1,-1)$ (1, for short). Its covariant holomorphicity follows from the holomorphicity of $W$:

$$\mathcal{D}_{i^*}\mathcal{L} = (\partial_{i^*} + i\mathcal{Q}_{i^*})\mathcal{L} = e^{\mathcal{K}/2}\partial_{i^*}(e^{-\mathcal{K}/2}\mathcal{L}) = e^{\mathcal{K}/2}\partial_{i^*}\mathcal{W} = 0. \tag{6.15}$$

$\mathcal{L}$ appears in the bosonic action of the theory only through the scalar potential, which we denote by $V_u$ to distinguish it from that of the gauged theory:

$$\boxed{V_u(Z, Z^*) = 8e^{\mathcal{K}} \left[\mathcal{G}^{ij^*}\partial_i \mathcal{W} \partial_{j^*}\mathcal{W}^* - 3|W|^2\right],} \tag{6.16}$$

or

$$\boxed{V_u(Z, Z^*) = -24|\mathcal{L}|^2 + 8\mathcal{G}^{ij^*}\mathcal{D}_i\mathcal{L}\mathcal{D}_{j^*}\mathcal{L}^*.} \tag{6.17}$$

This definition is Kähler invariant and the fact that $\mathcal{L}$ (or $\mathcal{W}$) is bosonic but not Kähler neutral may seem innocuous for R-symmetry. However, $\mathcal{L}$ also occurs in couplings to fermions which are only Kähler covariant. Therefore, in general, those couplings, when expressed in terms of the scalar fields, will change with Kähler transformations, breaking R-symmetry, unless the transformation can be compensated by a transformation of the scalars (a coordinate transformation on the Kähler manifold). This can only be done if that coordinate transformation leaves invariant the Kähler metric and any other coupling involving the scalars. In other words: the coordinate transformation has to be another global symmetry of the theory. However, when this can be done, the R-symmetry and the coordinate transformation must be related and only one of them will remain independent. Since the main characteristic of R-symmetry is that it leaves the bosons invariant, it is natural to consider the coordinate transformation to be the one that survives and R-symmetry to be the one that is broken: R-symmetry is, therefore, always broken by a non-vanishing superpotential.

*The kinetic matrix* $f_{\Lambda\Sigma}$. Finally, the kinetic matrix $f_{\Lambda\Sigma}(Z)$, which is constrained to be holomorphic in the scalar fields, describes the coupling between these scalar fields and the 1-form potentials $A^{\Lambda}{}_{\mu}$. The imaginary part of $f_{\Lambda\Sigma}$ has to be positive-definite for the kinetic terms of the 1-forms to have the right sign. The real part generalizes the theta angle that one can add to the Maxwell action Eq. (12.180). In that case, the theta term is a total derivative because the theta angle is constant. In this case, if $\Re\mathfrak{e} f_{\Lambda\Sigma}$ is not constant, the term will not be a total derivative.

Having introduced all the ingredients, we are ready to introduce the bosonic action of the most general $N = 1, d = 4$ SUGRA:

$$\boxed{\begin{aligned} S = \int d^4x \sqrt{|g|}\, \big[R + 2\mathcal{G}_{ij^*}\partial_\mu Z^i \partial^\mu Z^{*\,j^*} - \Im\mathfrak{m} f_{\Lambda\Sigma} F^{\Lambda\,\mu\nu} F^{\Sigma}{}_{\mu\nu} \\ - \Re\mathfrak{e} f_{\Lambda\Sigma} F^{\Lambda\,\mu\nu} \star F^{\Sigma}{}_{\mu\nu} - V_u\big]\,, \end{aligned}} \tag{6.18}$$

where the vector field strengths are given by the standard expression $F^{\Lambda} \equiv dA^{\Lambda}$.

The complete theory is invariant under $N = 1$ local supersymmetry transformations. The infinitesimal transformations are generated by the local fermionic parameter $\epsilon$. For the bosons, they take the form

$$\boxed{\begin{aligned} \delta_\epsilon e^a{}_\mu &= -\tfrac{i}{4}\bar{\psi}_\mu\gamma^a\epsilon^* + \text{c.c.}, \\ \delta_\epsilon A^\Lambda{}_\mu &= \tfrac{i}{8}\bar{\lambda}^\Lambda\gamma_\mu\epsilon^* + \text{c.c.}, \\ \delta_\epsilon Z^i &= \tfrac{1}{4}\bar{\chi}^i\epsilon. \end{aligned}} \tag{6.19}$$

For the fermions, when they vanish identically, they take the form

$$\boxed{\begin{aligned} \delta_\epsilon\psi_\mu &= \mathcal{D}_\mu\epsilon + i\mathcal{L}\gamma_\mu\epsilon^* = \left[\nabla_\mu + \tfrac{i}{2}\mathcal{Q}_\mu\right]\epsilon + i\mathcal{L}\gamma_\mu\epsilon^*, \\ 2\delta_\epsilon\lambda^\Lambda &= \not{F}^{\Lambda+}\epsilon, \\ \delta_\epsilon\chi^i &= i\,\not{\partial}Z^i\epsilon^*, \end{aligned}} \tag{6.20}$$

where

$$\mathcal{Q}_\mu \equiv \mathcal{Q}_i\partial_\mu Z^i + \mathcal{Q}_{i^*}\partial_\mu Z^{*\,i^*} \tag{6.21}$$

is the pullback of the Kähler 1-form connection to the spacetime.

On the r.h.s. of the supersymmetry transformation rule of the gauginos $\lambda^\Lambda$ we have written $F^{\Lambda+}$ to emphasize that only the self-dual part of the 2-form $F^\Lambda$ contributes, because, due to the definite chirality of $\epsilon$,

$$\not{F}^\Lambda\epsilon = -\not{F}^\Lambda\gamma_5\epsilon = i\star\not{F}^\Lambda\epsilon = \tfrac{1}{2}\left(\not{F}^\Lambda + i\star\not{F}^\Lambda\right)\epsilon, \tag{6.22}$$

where we have used the gamma-matrix identity Eq. (D.104). These types of relations are common also in higher dimensions with higher-rank field strengths.

It is convenient to introduce the dual (or "magnetic") field strengths $G_{\Lambda\,\mu\nu}$ defined in Eq. (2.148). Since the period matrix and the normalization of the vector fields in the $N = 1, d = 4$ action differ from those of Eq. (2.147), we define

$$\star G_\Lambda{}^{\mu\nu} \equiv \frac{1}{2\sqrt{|g|}}\frac{\delta S}{\delta F^\Lambda{}_{\mu\nu}} \quad\Rightarrow\quad G_\Lambda = \Re\mathfrak{e} f_{\Lambda\Sigma}F^\Sigma - \Im\mathfrak{m} f_{\Lambda\Sigma}\star F^\Sigma, \tag{6.23}$$

and construct the symplectic vector of electric and magnetic vector field strengths $\mathcal{F}^M$ defined in Eq. (2.161), whose use simplifies considerably the equations of motion.

We can denote (the l.h.s. of) the classical equations of motion that follow from the above action by

$$\mathcal{E}_a{}^\mu \equiv -\frac{1}{2\sqrt{|g|}}\frac{\delta S}{\delta e^a{}_\mu}, \qquad \mathcal{E}^i \equiv -\frac{\mathcal{G}^{ij^*}}{2\sqrt{|g|}}\frac{\delta S}{\delta Z^{*j^*}}, \qquad \mathcal{E}_\Lambda{}^\mu \equiv \frac{1}{4\sqrt{|g|}}\frac{\delta S}{\delta A^\Lambda{}_\mu}, \tag{6.24}$$

so that all the equations of motion take the form $\mathcal{E}=0$. We can supplement the equations of motion with the Bianchi identities of the fundamental vector field strengths,

$$\mathcal{E}^{\Lambda\,\mu} \equiv \nabla_\nu \star F^{\Lambda\,\nu\mu}, \tag{6.25}$$

combine these with the Maxwell equations $\mathcal{E}_\Lambda{}^\mu$ in a symplectic vector of equations,

$$(\mathcal{E}^{M\,\mu}) \equiv \begin{pmatrix} \mathcal{E}^{\Lambda\,\mu} \\ \mathcal{E}_\Lambda{}^\mu \end{pmatrix}, \tag{6.26}$$

and write them in the symplectic-invariant form

$$\mathcal{E}^{M\,\mu} = \nabla_\nu \star \mathcal{F}^{M\,\nu\mu}. \tag{6.27}$$

Using Eqs. (2.166) (taking into account the changes in normalization) to express in a manifestly symplectic-invariant way the energy–momentum tensor of the vector fields, the Einstein equations $\mathcal{E}_{\mu\nu} \equiv \mathcal{E}_\mu{}^a e_{a\,\nu}$ and scalar equations of motion take the form

$$\begin{aligned}\mathcal{E}_{\mu\nu} = {} & G_{\mu\nu} + 2\mathcal{G}_{ij^*}[\partial_\mu Z^i \partial_\nu Z^{*\,j^*} - \tfrac{1}{2}g_{\mu\nu}\partial_\rho Z^i \partial^\rho Z^{*\,j^*}] \\ & + 2\mathcal{M}_{MN}(f^*)\mathcal{F}^M{}_\mu{}^\rho \mathcal{F}^N{}_{\nu\rho} + \tfrac{1}{2}V_{\rm u},\end{aligned} \tag{6.28}$$

$$\begin{aligned}\mathcal{E}^i = {} & \left\{\nabla_\mu \partial^\mu Z^i + \partial_\mu Z^j \Gamma_{jk}{}^i \partial^\mu Z^k\right\} + \tfrac{i}{2}\partial^i f^*_{\Lambda\Sigma} F^{\Lambda\,-}{}_{\mu\nu} F^{\Sigma\,-\,\mu\nu} \\ & + \tfrac{1}{2}g_{\mu\nu}\partial^i V_{\rm u}.\end{aligned} \tag{6.29}$$

### 6.2.1 Examples

*No chiral multiplets.* A non-vanishing superpotential can be introduced even in the absence of scalar fields. The only possibility is a constant superpotential $\mathcal{L} = \mathcal{W} = w \in \mathbb{C}$, which leads uniquely to the potential

$$V_{\rm u} = -24|w|^2. \tag{6.30}$$

The bosonic action for this theory is just

$$S = \int d^4x \sqrt{|g|}\, \left[R + 24|w|^2\right], \tag{6.31}$$

so we see that $w$ contributes to an anti-de Sitter-type cosmological constant. The supersymmetry transformations of the Vierbein are those of the general case, and, for the gravitino, they are

$$\delta_\epsilon \psi_\mu = \nabla_\mu \epsilon + i w \gamma_\mu \epsilon^*. \tag{6.32}$$

It is not difficult to recognize this theory as the $N=1, d=4$ AdS supergravity of Townsend [1186], described in Chapter 5, with cosmological constant $\Lambda = -12|w|^2$.

The de Sitter-type cosmological constant of the $N=1, d=4$ theory of Freedman [562] has a completely different origin: it arises in the gauging of the R-symmetry to be discussed later on.

Let us now consider a theory with one chiral multiplet. There are two Kähler potentials that often occur in the literature. Calling the complex scalars, respectively $\Phi$ and $\tau$, they are given by one chiral multiplet with $\mathcal{K}=\frac{1}{2}|\Phi|^2$ and one chiral multiplet with $\mathcal{K}=-\log \Im\mathrm{m}\tau$ (the axion–dilaton model).

*One chiral multiplet with* $\mathcal{K}=\frac{1}{2}|\Phi|^2$. The corresponding Kähler metric and 1-form connection are

$$\mathcal{G}_{\Phi\Phi^*}=\tfrac{1}{2}, \qquad \mathcal{Q}_\Phi=\tfrac{1}{4i}\Phi^*, \quad \mathcal{Q}^*_{\Phi^*}=(\mathcal{Q}_\Phi)^*. \tag{6.33}$$

The metric is just that of Euclidean space, invariant under a complex translation and multiplication of $\Phi$ by a phase. Depending on the applications, the complex scalar is rewritten in terms of real scalars as $\Phi=X+iY$ or $\Phi=\rho e^{i\varphi}$. The coordinates $X$ and $Y$ are shifted by the complex translations and the coordinate $\varphi$ is shifted by the phase.

There are many possible choices of superpotential and (if we add vector multiplets) of kinetic matrices. Their choice will be associated to the symmetries that we want to preserve and also to the kind of vacuum (associated with the extrema of the scalar potential) that we are after. The simplest choice for the kinetic matrices would be $f_{\Lambda\Sigma}=-i\delta_{\Lambda\Sigma}$, and is preserved by the subgroup of $\mathrm{Sp}(2\bar{n},\mathbb{R})$ matrices Eq. (2.144) with $D=A$ and $C=-B$. The simplest choice for a holomorphic superpotential is the constant

$$\mathcal{W}=w \quad \Rightarrow \quad V_{\mathrm{u}}=-24|w|^2 e^{\frac{1}{2}|\Phi|^2}. \tag{6.34}$$

If we want vacua (classical solutions of the equations of motion with maximal number of symmetries) with constant scalars (which do not always exist), we have to find extrema of the scalar potential. These constant values of the scalars automatically solve the scalar equations of motion, and their contribution to the energy–momentum tensor reduces to the value of the scalar at the extremum. The above potential has an extremum for $\Phi=0$ (actually, a maximum), and the value of the potential for $\Phi=0$ is that of an AdS cosmological constant $\Lambda=-12|w|^2$.

Another simple choice would be

$$\mathcal{W}=w\Phi^2 \quad \Rightarrow \quad V_{\mathrm{u}}=8|w|^2 e^{\frac{1}{2}|\Phi|^2}\left[8|\Phi|^2-3|\Phi|^4\right], \tag{6.35}$$

which has one extremum at $\Phi=0$ for which the potential vanishes (so we would have a Minkowski vacuum) and a one-parameter family of extrema with $|\Phi|^2=8/3$ which also have vanishing cosmological constant.

All these superpotentials break (apart from R-symmetry) the invariance under complex translations $\delta\Phi=c$ but preserve the invariance under multiplications by a phase $\delta\Phi=i\alpha\Phi$. If we choose the above $\Phi$-independent kinetic matrix, the symmetries acting on the scalars and on the vector fields are independent and the choice of superpotential does not constrain the symmetry group any further.

*The axion–dilaton model; i.e. one chiral multiplet with* $\mathcal{K} = -\log \Im\mathrm{m}\tau$. The corresponding Kähler metric and 1-form connection are

$$\mathcal{G}_{\tau\tau^*} = \frac{1}{4(\Im\mathrm{m}\tau)^2}, \qquad \mathcal{Q}_\tau = \frac{1}{4\Im\mathrm{m}\tau}, \quad \mathcal{Q}^*_{\tau^*} = (\mathcal{Q}_\tau)^*. \tag{6.36}$$

The Kähler metric describes the coset space SL$(2,\mathbb{R})$/SO$(2)$. A real parametrization in which $\Im\mathrm{m}\tau > 0$ (so $\tau$ parametrizes the upper complex half plane) is provided by $\tau = a + ie^{-\varphi}$ (sometimes $\tau = a + ie^{-2\varphi}$). The metric is invariant under real shifts of $\tau$ (which become shifts of $a$) and real rescalings of $\tau$ (which shift the scalar $\varphi$). The complete symmetry group is, of course, SL$(2,\mathbb{R})$. Later we will study the symmetries of this model, which is ubiquitous in supergravity (see e.g. Section 16.2 and Chapter 23), in full detail.

We could choose a constant kinetic matrix, as in the previous example, but a more interesting choice, consistent with the holomorphicity of $f_{\Lambda\Sigma}$ in $\tau$ and with the fact that $\Im\mathrm{m} f_{\Lambda\Sigma}$ must be positive-definite is

$$f_{\Lambda\Sigma}(\tau) = +\tau\delta_{\Lambda\Sigma}. \tag{6.37}$$

With this choice and parametrizing $\tau$ by $\tau = a + ie^{-2\varphi}$, the action becomes that of the axion–dilaton model Eq. (2.144) and its extension to $n$ vector fields Eq. (16.58) up to the global sign of $a$. That is, the bosonic action of pure, ungauged, $N=4$, $d=4$ supergravity for $n=6$.

Another choice of kinetic matrix, related to the former, is (for just two vector fields)

$$(f_{\Lambda\Sigma}(\tau)) = \begin{pmatrix} \tau & 0 \\ 0 & -1/\tau \end{pmatrix}. \tag{6.38}$$

As we are going to see, there is an $N=2$, $d=4$ theory with a very similar form. If we replace the second vector field strength $F^2$ by its dual $\Im\mathrm{m}\tau \star F_2 + \Re\mathrm{e}\tau\, F_2$ (which amounts to a field redefinition), the theory takes the form of the previous one (a truncation of pure $N=4$, $d=4$ supergravity to just two vector fields, up to the global sign of the axion field $a$).

As for the superpotential and potential, the simplest choices yield

$$\begin{aligned} \mathcal{W} &= w, & V_{\mathrm{u}} &= -\frac{24|w|^2}{\Im\mathrm{m}\tau}, \\ \mathcal{W} &= w\sqrt{\tau}, & V_{\mathrm{u}} &= \frac{8|w|^2}{\Im\mathrm{m}\tau|\tau|}\left[(\Im\mathrm{m}\tau)^2 - 3|\tau|^2\right]. \end{aligned} \tag{6.39}$$

Only the first preserves the invariance under shifts of $\tau$ by a real constant, and only the second preserves the invariance under rescalings of $\tau$.

## 6.3 The gauged theory

In order to gauge a generic $N=1, d=4$ theory, we must first have a complete understanding of all the global symmetries that it may have.

### 6.3.1 The global symmetries

As discussed in Chapter 5 (for details, see Ref. [712]), for four-dimensional $N=1$ and $N=2$ SUGRAs the global symmetry group G is the direct product

$$\mathrm{G} = \mathrm{G}_{\mathrm{bos}} \times \mathrm{U}(1)_{\mathrm{R}}, \tag{6.40}$$

where $\mathrm{G}_{\mathrm{bos}}$ is the part of G that acts on the bosonic fields (the R-symmetry group does not). In the $N=1$ case, we can split $\mathrm{G}_{\mathrm{bos}}$ into the symmetries that act on the scalars (holomorphic isometries, belonging to the group[6] $\mathrm{G}_{\mathrm{iso}} \subseteq \mathrm{G}_{\mathrm{bos}}$) and those that do not. The latter, as we will see, constitute the subgroup $\mathrm{G}_{\mathrm{V}} \subseteq \mathrm{G}_{\mathrm{bos}}$ of symmetries that only act on the vector (super)fields and leave invariant the kinetic matrix $f_{\Lambda\Sigma}$. We have, then, in general,

$$\mathrm{G}_{\mathrm{bos}} = \mathrm{G}_{\mathrm{iso}} \times \mathrm{G}_{\mathrm{V}}, \tag{6.41}$$

since any bosonic symmetry transformation is an element either of $\mathrm{G}_{\mathrm{iso}}$ or of $\mathrm{G}_{\mathrm{V}}$; by construction no element of $\mathrm{G}_{\mathrm{iso}}$ can also be an element of $\mathrm{G}_{\mathrm{V}}$ and vice versa.

On the other hand, we must take into account the results of Section 2.6: the symmetries that act on the vectors are a subgroup of $\mathrm{Sp}(2\bar{n},\mathbb{R})$, if there are $\bar{n}$ vector fields. Thus

$$\mathrm{G}_{\mathrm{V}} \subset \mathrm{Sp}(2\bar{n},\mathbb{R}). \tag{6.42}$$

Although one may naively expect that $\mathrm{G}_{\mathrm{iso}} \subset \mathrm{Sp}(2\bar{n},\mathbb{R})$, there can be chiral multiplets that do not couple to the vectors and do not induce $\mathrm{Sp}(2\bar{n},\mathbb{R})$ transformations of the vectors. Therefore, $\mathrm{G}_{\mathrm{iso}}$ is in general the product of two subgroups: the subgroup of transformations of the scalars that do not act on the vectors $\mathrm{G}_{\mathrm{iso,\,noV}}$ and that of the transformations that act both on scalars and vectors $\mathrm{G}_{\mathrm{iso\,,V}}$:

$$\mathrm{G}_{\mathrm{iso}} = \mathrm{G}_{\mathrm{iso,\,noV}} \times \mathrm{G}_{\mathrm{iso,\,V}}. \tag{6.43}$$

We must conclude that the global symmetry group G of a generic $N=1, d=4$ SUGRA theory can be written, in principle, as

$$\mathrm{G} = \mathrm{G}_{\mathrm{iso\,,noV}} \times \mathrm{G}_{\mathrm{iso\,,V}} \times \mathrm{G}_{\mathrm{V}} \times \mathrm{U}(1)_{\mathrm{R}}, \tag{6.44}$$

and the general results of Gaillard and Zumino (see Section 2.6) imply that

$$\mathrm{G}_{\mathrm{iso\,,V}} \times \mathrm{G}_{\mathrm{V}} \subset \mathrm{Sp}(2\bar{n},\mathbb{R}). \tag{6.45}$$

There remains some ambiguity in this splitting, since, as we are going to explain and was mentioned in Section 6.2 when we saw how R-symmetry may survive in the presence of a non-trivial superpotential, $\mathrm{U}(1)_{\mathrm{R}}$ transformations can be combined with transformations of the other two factors.

We have already described the action of the $\mathrm{U}(1)_{\mathrm{R}}$ transformations on the fields of the theory. Let us now describe the isometries and the rotations of the vectors.

The transformations that act on the scalars can always be seen as coordinate transformations (reparametrizations) $Z^{i\,\prime} = f^i(Z, Z^*)$ of the Kähler manifold. Not all reparametrizations are admissible, though, as a number of conditions have to be satisfied:

[6] Not all the isometries of the metric are symmetries of the full theory. They have to satisfy further conditions that we are going to study next. Thus, in order to simplify the notation, we denote by $\mathrm{G}_{\mathrm{iso}}$ only those isometries that really are symmetries of the full theory and not the full group of isometries of $\mathcal{G}_{ij^*}$ (although they may coincide).

1. They must be *holomorphic* reparametrizations, i.e. ${k_\Lambda}^i$, the infinitesimal generator of the reparametrization[7] defined by

$$\delta_\alpha Z^i = \alpha^\Lambda {k_\Lambda}^i(Z), \tag{6.46}$$

where $\alpha^\Lambda$ are the constant parameters of the transformations, has to be a holomorphic function of the $Z^i$:

$$\partial_{i^*} {k_\Lambda}^j = 0, \qquad \partial_i k^*_\Lambda{}^{j^*} = 0. \tag{6.47}$$

This condition is necessary to preserve the Hermiticity of the metric.[8]

2. They must leave invariant the Kähler metric (i.e. they must be isometries). This means that the ${k_\Lambda}^i(Z)$ are just the holomorphic components of real Killing vector fields

$$K_\Lambda = {k_\Lambda}^i \partial_i + k^*_\Lambda{}^{i^*} \partial_{i^*}, \quad k^*_\Lambda{}^{i^*} = ({k_\Lambda}^i)^* \tag{6.48}$$

and they satisfy Eq. (E.38). This condition guarantees the invariance under the above reparametrizations of the scalar kinetic term in the action, but it is not enough to leave the full theory invariant.

3. They must preserve the complete Kähler–Hodge structure. Locally, this means that the Kähler potential has to be invariant under these isometries up to Kähler transformations, as expressed in Eq. (E.61), which we reproduce here for convenience:

$$\mathcal{L}_\Lambda \mathcal{K} \equiv {k_\Lambda}^i \partial_i \mathcal{K} + k^*_\Lambda{}^{i^*} \partial_{i^*} \mathcal{K} = \lambda_\Lambda(Z) + \lambda^*_\Lambda(Z^*). \tag{6.49}$$

This equation also defines the holomorphic functions $\lambda_\Lambda(Z)$ when they exist. These functions satisfy the equivariance property Eq. (E.63). When the $\lambda_\Lambda$ are non-trivial, all the spinors of the theory, which have non-vanishing Kähler weight, will undergo induced, possibly scalar-dependent, transformations. Also the superpotential will be affected (see the following) and the Kähler 1-form connection also transforms as

$$\mathcal{L}_\Lambda \mathcal{Q} = -\tfrac{i}{2} d\lambda_\Lambda + \text{c.c.} \tag{6.50}$$

[7] To be consistent with the earlier literature on the subject, we use the symplectic indices $\Lambda, \Sigma, \ldots$ to label the symmetries of the theory, instead of the adjoint indices $A, B, C, \ldots$ used in Ref. [712]. It is understood that not all the objects (Killing vectors, symplectic generators, etc.) that carry this index are necessarily non-vanishing. It is not difficult to see that, at least in the context of $N=1$ supergravity, this notation is not very precise: since we can have as many chiral multiplets decoupled from the vectors as we want, we can have a global symmetry group with more generators than vectors. Therefore, we cannot, in general, label all the global symmetries with the symplectic indices. However, since we can only gauge as many symmetries as we have vectors, we can always use symplectic indices to label the global symmetry subgroup that we are going to gauge. This restriction will be understood from now on. Furthermore, observe that, by using this notation, we assume implicitly that the vector fields transform in the adjoint representation of the group. This is not necessarily true for generic symmetries, but it has to be true for those that we want to gauge, which is our final goal.

[8] This and all the issues concerning the symmetries of Kähler and Kähler–Hodge manifolds and their gauging are treated in full detail in Appendix J.

If the Kähler structure is preserved, one can show that for each generator there exists a *momentum map* $\mathcal{P}_\Lambda$, defined in Eq. (E.66). We will make heavy use of momentum maps in the gauging of the isometries, of which they are one of the cornerstones.

4. The reparametrizations must leave the superpotential invariant up to Kähler transformations. This condition is equivalent to requiring that the superpotential be an *invariant section* which satisfies Eq. (J.47), i.e.

$$\mathbb{L}_\Lambda \mathcal{L} = \{\mathcal{L}_\Lambda + \tfrac{1}{2}(\lambda_\Lambda - \lambda^*_\Lambda)\}\mathcal{L} = 0 \;\Rightarrow\; K_\Lambda \mathcal{L} = -\tfrac{1}{2}(\lambda_\Lambda - \lambda^*_\Lambda)\mathcal{L}, \tag{6.51}$$

or, in terms of the holomorphic superpotential,

$$\mathbb{L}_\Lambda \mathcal{W} = \{\mathcal{L}_\Lambda + \lambda_\Lambda\}\mathcal{W} = 0 \;\Rightarrow\; k^i_\Lambda \partial_i \mathcal{W} = -\lambda_\Lambda \mathcal{W}. \tag{6.52}$$

This implies that $\mathcal{D}_i\mathcal{L}$ is also an invariant section; these two properties not only make the potential exactly invariant, but also ensure the invariance of all the fermionic terms in which $\mathcal{L}$ or its Kähler-covariant derivative occur.

5. When the kinetic matrix $f_{\Lambda\Sigma}$ depends on the scalars, the reparametrizations must be compatible with the transformations of the vectors as well.

The transformations that act on the vectors and leave the equations of motion (supplemented by the Bianchi identities) invariant are the ones discussed in Section 2.6; we could repeat the general discussion with $\mathcal{N}_{\Lambda\Sigma}$ replaced by the holomorphic $f_{\Lambda\Sigma}(Z)$ of $N=1$ SUGRA, but it will be advantageous to be a bit more precise.

The transformations of the vectors take the general form in Eqs. (2.144) and (2.145), but we can rewrite $\mathcal{T}$ as a linear combination of the global parameters of the transformations $\alpha^\Lambda$ and the generators[9] $\mathcal{T}_\Lambda$:

$$S \sim \mathbb{1}_{2\bar{n}\times 2\bar{n}} + \alpha^\Omega \mathcal{T}_\Omega, \qquad \mathcal{T}_\Omega, \equiv \begin{pmatrix} a_\Omega{}^\Lambda{}_\Sigma & b_\Omega{}^{\Lambda\Sigma} \\ c_{\Omega\,\Lambda\Sigma} & d_{\Omega\,\Lambda}{}^\Sigma \end{pmatrix}. \tag{6.53}$$

According to the general results, $S$ is always a symplectic matrix, and the matrices[10] $\mathcal{T}_\Lambda$ generate a subgroup of $\mathrm{Sp}(2\bar{n},\mathbb{R})$. Furthermore, the kinetic matrix must transform infinitesimally under these rotations as (see Eq. (2.156))

$$\mathcal{T}_\Lambda(f_{\Sigma\Omega}) = c_{\Lambda\,\Sigma\Omega} + d_{\Lambda\,\Sigma}{}^\Gamma f_{\Gamma\Omega} - f_{\Sigma\Gamma} a_\Lambda{}^\Gamma{}_\Omega + b_\Lambda{}^{\Gamma\Delta} f_{\Sigma\Gamma} f_{\Delta\Omega}. \tag{6.54}$$

Only when this transformation can be compensated for by a reparametrization of the scalars satisfying all the other conditions will these transformations be a symmetry of the full theory. Technically, the compensation of these transformations is expressed by the condition

$$\mathbb{L}_\Lambda f_{\Sigma\Omega} \equiv [\mathcal{L}_\Lambda - \mathcal{T}_\Lambda] f_{\Sigma\Omega} = 0. \tag{6.55}$$

---

[9] Our conventions for writing the indices of symplectic matrices are different from those of Ref. [712].

[10] Some of these matrices can be identically zero if the corresponding symmetry acts only on the scalars. This is, again, a shortcoming of the notation employed.

Condition Eq. (6.55) relates the reparametrizations of the scalars that occur in the kinetic matrix to symplectic transformations. Since the symplectic transformations are, in general, only symmetries of the equations of motion plus the Bianchi identities, we must distinguish several cases.

The transformations with $b \neq 0$ are non-perturbative because they transform the electric field strengths $F^{\Lambda}$ into the magnetic ones $F_{\Lambda}$. They are not symmetries of the action Eq. (6.18) and we will not gauge them; henceforth we will set $b = 0$.

The transformations with $c \neq 0$ (generalized Peccei–Quinn shifts[11]) are symmetries of the action up to total derivatives, and we will not consider their gauging either.

If the Killing vectors satisfy the Lie algebra Eq. (E.39), which we reproduce here for convenience,

$$[K_{\Lambda}, K_{\Sigma}] = -f_{\Lambda\Sigma}{}^{\Omega} K_{\Omega}, \tag{6.56}$$

then, taking into account that the gauge fields always transform in the adjoint representation of the gauge group,[12] we can express the non-vanishing matrices $a, d$ in terms of the structure constants as

$$a_{\Lambda}{}^{\Omega}{}_{\Sigma} = f_{\Lambda\Sigma}{}^{\Omega}, \qquad d_{\Lambda\Omega}{}^{\Sigma} = -f_{\Lambda\Omega}{}^{\Sigma} \quad \Rightarrow \quad \mathcal{T}_{\Lambda} = \begin{pmatrix} f_{\Lambda\Sigma}{}^{\Omega} & \\ & -f_{\Lambda\Omega}{}^{\Sigma} \end{pmatrix}, \tag{6.57}$$

so the symplectic generators $\mathcal{T}_{\Lambda}$ automatically satisfy

$$[\mathcal{T}_{\Lambda}, \mathcal{T}_{\Sigma}] = +f_{\Lambda\Sigma}{}^{\Omega} \mathcal{T}_{\Omega}. \tag{6.58}$$

### *6.3.2 Example: symmetries of the axion–dilaton model*

We now focus on the axion–dilaton model considered above with $\mathcal{K} = -\log \Im\mathfrak{m}\tau$ and, for concreteness, $f_{\Lambda\Sigma} = -\tau\delta_{\Lambda\Sigma}$ (we consider an arbitrary number of vectors $\bar{n}$). We will analyze several superpotentials.

To find the symmetries that act on $\tau$ we simply have to find the Killing vectors of the metric, and we have to see if they can be split into holomorphic and antiholomorphic components. Since the metric is that of the coset space $\mathrm{SL}(2,\mathbb{R})/\mathrm{SO}(2)$, the metric will be invariant under the three-dimensional group $\mathrm{SL}(2,\mathbb{R})$. The corresponding three Killing vectors are

$$K_1 = \tau\partial_{\tau} + \text{c.c.}, \qquad K_2 = \tfrac{1}{2}(1-\tau^2)\partial_{\tau} + \text{c.c.}, \qquad K_3 = \tfrac{1}{2}(1+\tau^2)\partial_{\tau} + \text{c.c.}, \tag{6.59}$$

and they satisfy the Lie algebra

$$[K_m, K_n] = \epsilon_{mnq}\eta^{qp} K_p \quad \Rightarrow f_{mn}{}^{p} = -\epsilon_{mnq}\eta^{qp}, \quad m, n, \ldots = 1, 2, 3, \tag{6.60}$$

where $\eta = \mathrm{diag}(++-)$, which is that of $\mathrm{SO}(2,1) \sim \mathrm{SL}(2,\mathbb{R}) \approx \mathrm{Sp}(2,\mathbb{R})$ (see Eq. (A.83)); as required, the $\tau$-components are holomorphic in $\tau$.

---

[11] See p. 475.

[12] As we have stressed before, this is not necessary for a transformation to be a global symmetry. It is only necessary for it to be a *gaugeable* symmetry, which are the symmetries we are focusing on.

While these reparametrizations leave invariant the Kähler metric, they only leave invariant the Kähler potential up to Kähler transformations according to Eq. (6.51) with the holomorphic functions $\lambda_m(\tau)$ given by

$$\lambda_1 = -\tfrac{1}{2}, \qquad \lambda_2 = \tfrac{1}{2}\tau, \qquad \lambda_3 = -\tfrac{1}{2}\tau. \tag{6.61}$$

As we can see, they satisfy the equivariance property

$$\mathcal{L}_{k_m}\lambda_n - \mathcal{L}_{k_n}\lambda_m = -f_{mn}{}^p\lambda_p. \tag{6.62}$$

This is enough for the Kähler structure to be preserved, but the presence of non-trivial $\lambda$s has important consequences: under an infinitesimal $\mathrm{SL}(2,\mathbb{R})$ transformation

$$\delta_\alpha\tau = \alpha^m k_m{}^\tau = \tfrac{1}{2}(\alpha^2+\alpha^3) + \alpha^1\tau - \tfrac{1}{2}(\alpha^2-\alpha^3)\tau^2, \tag{6.63}$$

all the fermions transform with $\tau$-dependent phases. The gravitino, for example, transforms as

$$\delta_\alpha\psi_\mu = -\tfrac{i}{4}(\alpha^2-\alpha^3)\Im\mathrm{m}\tau\,\psi_\mu. \tag{6.64}$$

The finite $\mathrm{SL}(2,\mathbb{R})$ transformations take the form

$$\tau' = \frac{\alpha\tau+\beta}{\gamma\tau+\delta}, \qquad \psi'_\mu = e^{-\frac{i}{2}\mathrm{Arg}(\gamma\tau+\delta)}\psi_\mu, \tag{6.65}$$

where $\left(\begin{smallmatrix}\alpha & \beta\\ \gamma & \delta\end{smallmatrix}\right)$ is an $\mathrm{SL}(2,\mathbb{R})$ matrix, i.e. its components satisfy $\alpha\delta-\beta\gamma=1$. Our parametrization of this matrix will be related to that of the $\mathrm{Sp}(2\bar{n},\mathbb{R})$ matrix $S=\left(\begin{smallmatrix}A & B\\ C & D\end{smallmatrix}\right)$, which is the reason why we have chosen an unusual action of $\mathrm{SL}(2,\mathbb{R})$ on $\tau$. The choice of generators of this linear representation of $\mathrm{SL}(2,\mathbb{R})$, which is related to the above choice of Killing vectors,[13] is, in terms of the Pauli matrices,

$$\begin{pmatrix}\alpha & \beta\\ \gamma & \delta\end{pmatrix} \sim \mathbb{1}_{2\times 2} + \alpha^m T_m, \qquad T_1 = \tfrac{1}{2}\sigma^3,\; T_2 = \tfrac{1}{2}\sigma^1,\; T_3 = \tfrac{i}{2}\sigma^2, \tag{6.66}$$

and they satisfy the Lie algebra

$$[T_m, T_n] = -\epsilon_{mnq}\eta^{qp}T_p. \tag{6.67}$$

So far, $\mathrm{SL}(2,\mathbb{R})$ meets all the necessary requirements to be a global symmetry of the theory. If we couple $\bar{n}$ vector multiplets with the period matrix $f_{\Lambda\Sigma} = -\tau\delta_{\Lambda\Sigma}$, we have to check Eq. (6.55); its explicit form for this case is (lowering the indices with $\delta_{\Lambda\Sigma}$)

$$\begin{aligned}&\left\{\tfrac{1}{2}(\alpha^2+\alpha^3) + \alpha^1\tau - \tfrac{1}{2}(\alpha^2-\alpha^3)\tau^2\right\}\delta_{\Lambda\Sigma}\\ &-\alpha^a\left\{c_a + (d_a-a_a)\tau + b_a\tau^2\right\}_{\Lambda\Sigma} = 0,\end{aligned} \tag{6.68}$$

where the index $a$ includes the index $m$ and other possible values $x$ associated with the transformations of the vectors that do not act on $\tau$. Solving this equation for $a, b, c, d$ (taking

[13] Another way of putting it is that their effect is the same as that of the reparametrizations $\delta_\alpha\tau = \alpha^m k_m{}^\tau(\tau)$.

into account that they are generators of $\mathrm{Sp}(2\bar{n},\mathbb{R})$ and must satisfy Eqs. (2.158)), we obtain the following non-vanishing components:

$$c_{2\,\Lambda\Sigma} = c_{3\,\Lambda\Sigma} = -b_{2\,\Lambda\Sigma} = b_{3\,\Lambda\Sigma} = d_{1\,(\Lambda\Sigma)} = -a_{1\,(\Lambda\Sigma)} = \tfrac{1}{2}\delta_{\Lambda\Sigma}, \qquad a_{a\,[\Lambda\Sigma]} = d_{a\,[\Lambda\Sigma]}. \tag{6.69}$$

We can set $a_{m\,[\Lambda\Sigma]} = d_{m\,[\Lambda\Sigma]} = 0$, without any loss of generality (we just assign all the antisymmetric generators of the $\mathrm{O}(\bar{n})$ rotations to $\mathcal{T}_x$) and writing

$$\mathcal{T}_m = \mathbb{1}_{\bar{n}\times\bar{n}} \otimes T_m, \tag{6.70}$$

where the $T_m$ are the generators of $\mathrm{SL}(2,\mathbb{R})$ found above. We can interpret the result as follows: the rotations of the vector fields that are symmetries of the theory are the product of $\mathrm{SL}(2,\mathbb{R})$ and $\mathrm{O}(\bar{n})$. The orthogonal group does not act on the scalars. The finite transformations of the vectors have the following $\mathrm{SL}(2,\mathbb{R})\times \mathrm{O}(\bar{n}) \subset \mathrm{Sp}(2\bar{n},\mathbb{R})$ matrices:

$$S = \mathbb{1}_{\bar{n}\times\bar{n}} \otimes \begin{pmatrix} \alpha & \beta \\ \gamma & \delta \end{pmatrix} \otimes \begin{pmatrix} R & 0 \\ 0 & R^T \end{pmatrix}, \qquad \alpha\delta - \beta\gamma = 1, \qquad RR^T = \mathbb{1}, \tag{6.71}$$

and, at the same time, $\tau$ must transform according to Eq. (6.65) under $\mathrm{SL}(2,\mathbb{R})$ only.

Finally, let us consider how some of the possible superpotentials respect the $\mathrm{SL}(2,\mathbb{R})$ symmetry (it does not affect $\mathrm{O}(\bar{n})$). The equation which is satisfied for each unbroken symmetry generator is Eq. (6.52):

$$k_m{}^{\tau}\partial_\tau \mathcal{W} = -\lambda_m \mathcal{W}, \tag{6.72}$$

and we get for the transformations $K_1$, $K_2 + K_3$, and $K_2 - K_3$, respectively, the differential equations

$$\tau\partial_\tau \mathcal{W}_1 = \tfrac{1}{2}\mathcal{W}_1, \qquad \partial_\tau \mathcal{W}_{2+3} = 0, \qquad \tau^2\partial_\tau \mathcal{W}_{2-3} = \mathcal{W}_{2-3}, \tag{6.73}$$

with the different solutions

$$\mathcal{W}_1 = w_1\sqrt{\tau}, \qquad \mathcal{W}_{2+3} = w_{2+3}, \qquad \mathcal{W}_{2-3} = w_{2+3}e^{\frac{1}{\tau}}, \tag{6.74}$$

where the $\omega$s are different integration constants. Other, more complicated, superpotentials may leave unbroken different combinations of the generators of $\mathrm{SL}(2,\mathbb{R})$, but it should be clear that the only way to preserve the whole of $\mathrm{SL}(2,\mathbb{R})$ is to have no superpotential, just as with $\mathrm{U}(1)_{\mathrm{R}}$, which is a good symmetry of this theory only if there is no superpotential. Since the only constant $\lambda_m$ is real, there is no way to identify the $\mathrm{U}(1)_{\mathrm{R}}$ transformation with a subgroup of $\mathrm{SL}(2,\mathbb{R})$.

### 6.3.3 The gauging of the global symmetries

Having identified the possible global symmetries of generic matter-coupled $N = 1$, $d = 4$ supergravities, parametrized by the global parameters $\alpha^\Lambda$, now we want to modify the theory such that it is still invariant when these global parameters are promoted to arbitrary local functions (*gauge parameters*) $\alpha^\Lambda(x)$.[14] This can only be done if we relate these local functions to those of other already existing local symmetries of the theory: these can only be the gauge parameters of the Abelian vector fields of the ungauged theory $A^\Lambda{}_\mu$. As we have discussed in Chapter 2, this requires additional terms in the action, derivatives, vector field strengths, supersymmetry transformations, etc., which we are going to describe in detail.

We start by writing the gauge transformations of all the fields but the vectors:

$$\delta_\alpha Z^i = \alpha^\Lambda(x) k_\Lambda{}^i, \tag{6.75}$$

$$\delta_\alpha \psi_\mu = -\tfrac{1}{4}\alpha^\Lambda(x)(\lambda_\Lambda - \lambda^*_\Lambda)\psi_\mu, \tag{6.76}$$

$$\delta_\alpha \lambda^\Lambda = -\tfrac{1}{4}\alpha^\Sigma(x)(\lambda_\Sigma - \lambda^*_\Sigma)\lambda^\Lambda + \alpha^\Sigma(x) f_{\Sigma\Omega}{}^\Lambda \lambda^\Omega, \tag{6.77}$$

$$\delta_\alpha \chi^i = +\tfrac{1}{4}\alpha^\Lambda(x)(\lambda_\Lambda - \lambda^*_\Lambda)\chi^i + \alpha^\Lambda(x)\partial_j k_\Lambda{}^i \chi^j. \tag{6.78}$$

Observe that the transformations of the gauginos $\lambda^\Lambda$ include a rotation because they belong to the vector supermultiplets and must transform homogeneously with the vector fields.[15] The chiralino transformations also include an additional term for the same reason.

The $\mathrm{U}(1)_\mathrm{R}$ transformations can be understood to be included in these rules for some $\lambda_\Lambda = i$ not associated with an isometry.

The gauge transformations of the vector fields are a combination of the standard Abelian gauge transformations and the (gauged) rotations, and can be rewritten as the covariant derivative of the gauge parameter $\alpha^\Lambda$ understood as a field in the adjoint representation,

$$\delta_\alpha A^\Lambda{}_\mu = -\frac{1}{g}\partial_\mu \alpha^\Lambda(x) + \alpha^\Sigma(x) f_{\Sigma\Omega}{}^\Lambda A^\Omega{}_\mu = -\frac{1}{g}\mathfrak{D}_\mu \alpha^\Lambda, \tag{6.79}$$

with

$$\mathfrak{D}_\mu \alpha^\Lambda \equiv \partial_\mu \alpha^\Lambda + g A^\Sigma{}_\mu f_{\Sigma\Omega}{}^\Lambda \alpha^\Omega. \tag{6.80}$$

The next step is the construction of the covariant derivatives and field strengths for the above fields. This is explained in Appendix J: for the scalars and vector fields, we have the standard expressions that we reproduce here for convenience:

$$\mathfrak{D}_\mu Z^i = \partial_\mu Z^i + g A^\Lambda{}_\mu k_\Lambda{}^i, \tag{6.81}$$

$$F^\Lambda{}_{\mu\nu} = 2\partial_{[\mu} A^\Lambda{}_{\nu]} + g f_{\Sigma\Omega}{}^\Lambda A^\Sigma{}_{[\mu} A^\Omega{}_{\nu]}. \tag{6.82}$$

---

[14] We remind the reader that we only want to gauge the perturbative transformations that leave invariant the action, i.e. those with $b = c = 0$. The submatrices $a$ and $d$ are, then, given by Eqs. (6.57) and correspond to transformations of the symmetry group in the covariant and contravariant adjoint representations. This implicitly assumes that the vector fields transform in the adjoint representation of the symmetry group that we want to gauge. Under these conditions, any subgroup of the symmetry group can be gauged, but the reader should be aware that not all the subgroups of the global symmetry group fulfill them and are *gaugeables*.

[15] The transformations of the vector fields include the characteristic inhomogeneous term of gauge connections.

Things are more complicated for spinors. These only transform by multiplication by a field-dependent $\mathrm{U}(1)$ phase, whether the gauge group is Abelian or non-Abelian, while the gauge fields transform in the adjoint of the gauge group (up to the inhomogeneous term). In order to construct a spinorial covariant derivative with the help of the gauge fields, an intertwiner that converts one kind of transformation into another is needed. In $N = 1$, $d = 4$ theories[16] the intertwiner is the *momentum map* $\mathcal{P}_\Lambda$. This is locally defined by Eq. (E.67), which we reproduce here,

$$i\mathcal{P}_\Lambda = k_\Lambda{}^i \partial_i \mathcal{K} - \lambda_\Lambda, \tag{6.83}$$

and which determines $\mathcal{P}_\Lambda$ up to a constant. That constant is determined, for non-Abelian groups, by the equivariance property Eq. (E.71); i.e.

$$\mathcal{L}_\Lambda \mathcal{P}_\Sigma = -f_{\Lambda\Sigma}{}^\Omega \mathcal{P}_\Omega. \tag{6.84}$$

The additive constant in the Abelian directions of the gauge group can be chosen arbitrarily. In particular, we can choose one for symmetries unrelated to isometries such as $\mathrm{U}(1)_\mathrm{R}$. These constant momentum maps are called *D*- or *Fayet–Iliopoulos* terms and appear in the supersymmetry transformation rules of the gauginos and in the scalar potential.

The covariant derivatives of spinors of Kähler weight $q$ include a term of the form $-iqgA^\Lambda{}_\mu \mathcal{P}_\Lambda$. This term is combined with the pullback of the Kähler connection $\mathcal{Q}$ (all the derivatives of the spinors must be Kähler covariant) into a gauge-covariant Kähler connection $\hat{\mathcal{Q}}$:

$$\hat{\mathcal{Q}}_\mu \equiv \mathcal{Q}_\mu + gA^\Lambda{}_\mu \mathcal{P}_\Lambda = \tfrac{1}{2i}\left[\partial_i \mathcal{K} \mathfrak{D}_\mu Z^i - \partial_{i^*} \mathcal{K} \mathfrak{D}_\mu Z^{*\,i^*}\right] + gA^\Lambda{}_\mu \Im\mathrm{m}\lambda_\Lambda. \tag{6.85}$$

Taking all this into account, the covariant derivatives of the spinors are given by Eqs. (J.63)–(J.65), which we reproduce here for convenience:

$$\mathfrak{D}_\mu \psi_\nu = \left\{\nabla_\mu + \tfrac{i}{2}\hat{\mathcal{Q}}_\mu\right\}\psi_\nu, \tag{6.86}$$

$$\mathfrak{D}_\mu \lambda^\Lambda = \left\{\nabla_\mu + \tfrac{i}{2}\hat{\mathcal{Q}}_\mu\right\}\lambda^\Lambda + gA^\Sigma{}_\mu f_{\Sigma\Omega}{}^\Lambda \lambda^\Omega, \tag{6.87}$$

$$\mathfrak{D}_\mu \chi^i = \left\{\nabla_\mu - \tfrac{i}{2}\hat{\mathcal{Q}}_\mu\right\}\chi^i + \left\{\mathfrak{D}_\mu Z^k \Gamma_{kj}{}^i + gA^\Lambda{}_\mu \partial_j k_\Lambda{}^i\right\}\chi^j, \tag{6.88}$$

where $\nabla_\mu$ stands for the Lorentz- and general-covariant derivative.

As we have mentioned before, the fermion supersymmetry transformation rules are also modified by the gauging, but those for the bosons are not; these are still given by Eqs. (6.19). Apart from changing all the derivatives and field strengths by covariant derivatives, we have to add the so-called *fermion shifts*. In the $N = 1$, $d = 4$ case they are given by the functions

$$\mathcal{N}^i \equiv 2\mathcal{G}^{ij^*}\mathcal{D}_{j^*}\mathcal{L}^*, \qquad \mathcal{D}^\Lambda \equiv -g(\Im\mathrm{m}\, f)^{-1|\Lambda\Sigma}\,\mathcal{P}_\Sigma, \tag{6.89}$$

[16] Actually, they occur in all SUEGRAs, but they only receive this name in the $N = 1$ and $N = 2$ cases in which the scalar manifolds are not necessarily coset spaces G/H as for $N > 2$. For $N > 2$, a G transformation of the scalars comes with a compensating field-dependent H transformation of the spinors. The intertwiner connecting G and H transformations is, naturally, the H-compensator studied in Section A.4.

and only affect the gaugino and chiralino transformations. A a result, the fermion supersymmetry transformations become, for vanishing fermions,

$$\begin{aligned}
\delta_\epsilon \psi_\mu &= \mathfrak{D}_\mu \epsilon + i\mathcal{L}\gamma_\mu \epsilon^* = \left[\nabla_\mu + \tfrac{i}{2}\hat{\mathcal{Q}}_\mu\right]\epsilon + i\mathcal{L}\gamma_\mu\epsilon^*, \\
2\delta_\epsilon \lambda^\Lambda &= (\not{F}^{\Lambda +} + i\mathcal{D}^\Lambda)\epsilon, \\
\delta_\epsilon \chi^i &= i\not{\mathfrak{D}} Z^i \epsilon^* + \mathcal{N}^i \epsilon.
\end{aligned} \tag{6.90}$$

The invariance of the action under the new transformations requires the addition of new terms quadratic in the fermion shifts to the scalar potential:

$$V(Z, Z^*) = -24|\mathcal{L}|^2 + 2\mathcal{G}_{ij^*}\mathcal{N}^i\mathcal{N}^{*j^*} + \tfrac{1}{2}\Im\mathrm{m}\, f_{\Lambda\Sigma}\mathcal{D}^\Lambda\mathcal{D}^\Sigma, \tag{6.91}$$

or

$$V(Z, Z^*) = V_{\mathrm{u}}(Z, Z^*) + \tfrac{1}{2}g^2(\Im\mathrm{m}\, f)^{-1|\Lambda\Sigma}\mathcal{P}_\Lambda\mathcal{P}_\Sigma, \tag{6.92}$$

where $V_{\mathrm{u}}$ is the scalar potential of the ungauged theory given in Eq. (6.17).

With all these ingredients at hand, we are ready to write the action for the bosonic fields of the most general $N = 1, d = 4$ gauged supergravity of the kind considered here:

$$\begin{aligned}
S = \int d^4x \sqrt{|g|}\, \big[ R &+ 2\mathcal{G}_{ij^*}\mathfrak{D}_\mu Z^i \mathfrak{D}^\mu Z^{*j^*} - \Im\mathrm{m} f_{\Lambda\Sigma} F^{\Lambda\,\mu\nu} F^\Sigma{}_{\mu\nu} \\
&- \Re\mathrm{e} f_{\Lambda\Sigma} F^{\Lambda\,\mu\nu} \star F^\Sigma{}_{\mu\nu} - V \big].
\end{aligned} \tag{6.93}$$

We can use the same definitions for (the l.h.s. of) the bosonic equations of motion as in the ungauged case Eqs. (6.24), but we have to modify the Bianchi identities of the vector field strengths:

$$\mathcal{E}^{\Lambda\,\mu} \equiv \mathfrak{D}_\nu \star F^{\Lambda\,\nu\mu} \tag{6.94}$$

In this case, the (l.h.s. of the) equations of motion take the form

$$\begin{aligned}
\mathcal{E}_{\mu\nu} = G_{\mu\nu} &+ 2\mathcal{G}_{ij^*}[\mathfrak{D}_\mu Z^i \mathfrak{D}_\nu Z^{*j^*} - \tfrac{1}{2} g_{\mu\nu}\mathfrak{D}_\rho Z^i \mathfrak{D}^\rho Z^{*j^*}] \\
&+ 2\mathcal{M}_{MN}(f^*)\mathcal{F}^M{}_\mu{}^\rho \mathcal{F}^N{}_{\nu\rho} + \tfrac{1}{2}V,
\end{aligned} \tag{6.95}$$

$$\mathcal{E}^i = \mathfrak{D}^2 Z^i + \tfrac{i}{2}\partial^i f^*_{\Lambda\Sigma} F^{\Lambda\,-}{}_{\mu\nu} F^{\Sigma\,-\,\mu\nu} + \tfrac{1}{2} g_{\mu\nu}\partial^i V, \tag{6.96}$$

$$\mathcal{E}_\Lambda{}^\mu = \mathfrak{D}_\nu \star G_\Lambda{}^{\nu\mu} + \tfrac{1}{4} j^\mu, \tag{6.97}$$

where we have defined the current

$$j_{\Lambda\,\mu} = 2g(k_{\Lambda\,i^*}\mathfrak{D}_\mu Z^{*\,i} + \text{c.c.}). \tag{6.98}$$

The square of the covariant derivative acting on the scalars that appears in $\mathcal{E}^i$ is given in Eq. (J.38).

### 6.3.4 Examples of gauged $N=1, d=4$ supergravities

*Freedman's gauged $N=1$, $d=4$ supergravity.* The simplest example of a gauged $N=1$, $d=4$ supergravity theory is that of Freedman [562]; namely $N=1$, $d=4$ supergravity coupled to a single vector multiplet which is used to gauge the $\mathrm{U}(1)_\mathrm{R}$ symmetry group via a Fayet–Iliopoulos term. There is no superpotential, as that would break the global symmetry that we want to gauge. The single Fayet–Iliopoulos term is associated with a constant momentum map that we set arbitrarily to $\mathcal{P}=1$. We also set the single component of the kinetic matrix to $-i$ for simplicity. The momentum map only occurs in the fermion supersymmetry transformation rules via the covariant derivatives and fermion shift $\mathcal{D}=-g$ and in $V=+\frac{1}{2}g^2$. Since there are no chiral multiplets and only a single vector multiplet, the supersymmetry transformation rules of the two fermions of the theory are

$$\begin{aligned} \delta_\epsilon\psi_\mu &= \left[\nabla_\mu + \tfrac{i}{2}gA_\mu\right]\epsilon, \\ 2\delta_\epsilon\lambda &= \left[\not{F}^+ - ig\,\right]\epsilon. \end{aligned} \tag{6.99}$$

The boson supersymmetry rules are always the same. The bosonic action is just

$$S = \int d^4x\sqrt{|g|}\,\left[R - F^2 - \tfrac{1}{2}g^2\right], \tag{6.100}$$

that is, Einstein–Maxwell–de Sitter with $\Lambda = +\frac{1}{4}g^2$.

This theory should be compared to Townsend's in Eq. (6.31).

*Gaugings of the axion–dilaton model.* Although in principle one can gauge a combination of $\mathrm{U}(1)_\mathrm{R}$, $\mathrm{SL}(2,\mathbb{R})$, and $\mathrm{SO}(\bar{n})$, for simplicity we consider then as being gauged separately.

In the absence of a superpotential, the R-symmetry can be gauged exactly as in Freedman's theory, via a Fayet–Iliopoulos term that does not affect the covariant derivatives of the scalars. To continue to treat all the vectors symmetrically, we leave all the components $\mathcal{P}_\Lambda$ undetermined. If several of them are different from zero, we will use a linear combination of vectors as the gauge field, as can be seen from the additional term that appears in the fermion covariant derivative: $-iqgA^\Lambda\mathcal{P}_\Lambda$. It is not difficult to see that $g\mathcal{P}_\Lambda$ is just the set of components of the embedding tensor associated with the generator of $\mathrm{U}(1)_\mathrm{R}$ (see Section 2.7). The fermion shifts and the potential $V$ are different from Freedman's theory because we have a non-trivial kinetic matrix $f_{\Lambda\Sigma} = -\tau\delta_{\Lambda\Sigma}$:

$$\mathcal{D}^\Lambda = g\frac{\mathcal{P}^\Lambda}{\Im\mathfrak{m}\tau}, \qquad V = -g^2\frac{\mathcal{P}^\Lambda\mathcal{P}_\Lambda}{2\Im\mathfrak{m}\tau}. \tag{6.101}$$

These are the only new elements to be taken into account in the new bosonic action and fermion supersymmetry rules.

Gauging $\mathrm{O}(\bar{n})$ is a trivial exercise: only the vectors and gauginos transform under the global group and we only have to replace field strengths and derivatives by their covariant counterparts. The momentum maps associated with these symmetries must vanish, and so do the $\mathcal{D}^{\Lambda}$ fermion shifts. Thus, in the absence of a superpotential, the theory has no scalar potential. A superpotential would not break $\mathrm{O}(\bar{n})$ because this group does not act on $\tau$ and, therefore, it can always be added to the $\mathrm{O}(\bar{n})$-gauged theory generating a $\mathcal{N}^{\tau}$ fermion shift and a scalar potential.

The gauging of $\mathrm{SL}(2,\mathbb{R})$ is a bit more challenging: we can easily find the momentum maps for the three generators to be

$$\mathcal{P}_1 = \frac{\Re\mathfrak{e}\tau}{2\Im\mathfrak{m}\tau}, \qquad \mathcal{P}_2 = \frac{1-|\tau|^2}{4\Im\mathfrak{m}\tau}, \qquad \mathcal{P}_3 = \frac{1+|\tau|^2}{4\Im\mathfrak{m}\tau}, \tag{6.102}$$

but we encounter several problems. First of all, observe that we cannot gauge the whole $\mathrm{SL}(2,\mathbb{R})$ unless we use the fully fledged embedding tensor formalism because the action of the group on the vectors includes non-vanishing $b$ and $c$ submatrices (see the discussion at the end of Section 6.3.1). The submatrices $a$ and $d$ are both associated to the generator $\mathcal{T}_1$ of real rescalings of $\tau$, a non-compact $\mathbb{R}^+ \approx \mathbb{R}$ subgroup of $\mathrm{SL}(2,\mathbb{R})$, which is the only one we could gauge.[17] There is, however, an obstruction to this program: assume we want to use the vector $A^1{}_\mu$ for this Abelian gauging. All the vectors, and $A^1{}_\mu$ in particular, transform under the rescalings of $\tau$:

$$\tau' = \xi^2\tau, \qquad A^{\Lambda\prime}{}_\mu = \xi^{-1}A^{\Lambda}{}_\mu. \tag{6.103}$$

When this happens, the vector cannot be used to gauge the symmetry because no consistent gauge transformation rule exists for it and a covariant vector field strength cannot be constructed. The standard covariant derivative for $\tau$,

$$\mathfrak{D}_\mu\tau = (\partial_\mu + gA^1{}_\mu)\tau, \tag{6.104}$$

does not scale homogeneously even under constant rescalings. The conclusion is that at this level we cannot gauge any subgroup of $\mathrm{SL}(2,\mathbb{R})$.

[17] Observe that all this is a consequence of our choice of kinetic matrix. There are choices for which the action of all the generators of $\mathrm{SL}(2,\mathbb{R})$ on the vector fields is always perturbative. However, these kinetic matrices are not holomorphic in $\tau$.

# 7

# Matter-coupled $N = 2, d = 4$ supergravity

$N = 2, d = 4$ supergravity is the simplest theory of extended supergravity, and, even though it has no *direct* phenomenological relevance in particle physics due to the impossibility of reproducing the chiral structure of the standard model, it is a very interesting theory on its own. The increased amount of supersymmetry puts more constraints on possible matter couplings, as compared to the $N = 1$ case, but these are not completely determined by supersymmetry as in the $N > 2$ cases. This means that there are many possible $N = 2, d = 4$ theories with different matter couplings (even for the same matter content). It also means that all these couplings are different realizations of a common and very rich mathematical structure known as *special Kähler geometry* that governs the couplings of the vector multiplets to supergravity and of another, less interesting, structure known as *quaternionic-Kähler geometry* that governs the coupling of the *hypermultiplets* (to be defined) to supergravity.

These new structures, which we review in Appendices F and G, emerge from the additional constraints that the second supersymmetry imposes on the Kähler–Hodge geometry that dictates the matter couplings of $N = 1$ theory. In particular, special Kähler geometry contains information on the symplectic structure present in the couplings between scalars and vector fields in four-dimensional theories (see Section 2.6.2) because $N = 2$ supersymmetry, unlike $N = 1$ supersymmetry, constrains the period matrix $\mathcal{N}_{\Lambda\Sigma}$ (defined in Section 2.6.2) that codifies them.[1]

As we will see in subsequent chapters, the couplings allowed by $N = 2, d = 4$ supersymmetry to the vector supermultiplets are just right for having supersymmetric charged black-hole solutions with non-trivial scalars, which is impossible in $N = 1, d = 4$ theories. This makes these theories particularly interesting for us. Furthermore, many (but not all) of them arise from the compactification of type-II theories in Calabi–Yau 3-folds (see e.g. Refs. [111, 583] for a pedagogical introduction), which means that many of the results and solutions obtained in the framework of those $N = 2, d = 4$ theories can be embedded in full superstring theory.

Our presentation of these theories will be very concise: our goal is to give a recipe for building the most general matter-coupled $N = 2, d = 4$ supergravity theory and to write its

[1] In the $N = 1$ case the same matrix has been denoted by $f_{\Lambda\Sigma}$ and has also been referred to as the *kinetic matrix* (see Section 6.2). The precise relation is $f_{\Lambda\Sigma} = 2\mathcal{N}^*_{\Lambda\Sigma}$.

bosonic action and supersymmetry transformation rules for vanishing fermions, since these will be our main tools in the construction and study of gravitational solutions (black holes, strings, domain walls, etc.), just as in the $N=1$ case. Although we will strive to give a consistent basis for the definitions and operational properties of all the objects and terms that we are going to define, it should be clear from the beginning that the reader interested in mastering this field should study the original references, [1265, 1257], the review Ref. [44], or the recent book Ref. [564].

We first present the matter multiplets and construct the most general ungauged theories. Then, we will explain the gauging procedure (similar to that of the $N=1$ case) to construct the most general (electrically) gauged theory. The same remarks made at the end of the introduction of Chapter 6 concerning the need to use the embedding tensor formalism to find the most general gauged theory apply here. A partial answer for $N=2, d=4$ theories has been explored in Ref. [1224].

## 7.1 The matter supermultiplets

As explained in Section 6.1, some (but not all!) of the spinors of $N=2, d=4$ supergravity will be pairs of Weyl spinors labeled by SU(2) indices $I, J, \ldots = 1, 2$, whose position, associated with their chirality and Kähler weight, changes under complex conjugation. The chiralities of the gravitinos and of the supersymmetry parameters with lower SU(2) indices are negative (as in Eqs. (6.1)) and their Kähler weights are $\frac{1}{2}$. The chiralities of the remaining spinors will be given in the following.

The conventional supermultiplets of $N=2, d=4$ supergravity are the supergravity multiplet, the vector supermultiplet, and the hypermultiplet. The supergravity multiplet was introduced in Section 5.5, and, in the conventions that we are going to use here, they can be written as

$$\{e^a{}_\mu, \psi^I{}_\mu, A_\mu\}. \tag{7.1}$$

In $N$-SUEGRAS, the graviphoton fields could be labeled by an antisymmetric pair of SU($N$) indices $A^{IJ}{}_\mu$. In the $N=2$ case we just have $A^{IJ}{}_\mu = \varepsilon^{IJ} A_\mu$.

The vector supermultiplets of $N=2, d=4$ supergravity can be understood as a combination of a chiral multiplet and a vector supermultiplet of $N=1, d=4$. If we couple $n_V$ vector multiplets, we label them with indices $i, j = 1, \ldots, n_V$. The field content of the $i$th vector multiplet is

$$\{A^i{}_\mu, \lambda^{iI}, Z^i\}, \tag{7.2}$$

where $\lambda^{iI}$ is a pair of *gauginos* and $Z^i$ is a complex scalar which, in the coupled theory, will be interpreted as a complex coordinate in a *special Kähler manifold* (see Appendix F). The chirality and Kähler weights of $\lambda^{iI}$ are the opposite of that of the gravitinos with the same SU(2) indices:

$$\gamma_5 \lambda^{iI} = -\lambda^{iI}. \tag{7.3}$$

Since duality rotations will mix the graviphoton field with the matter vector fields it is convenient to use a common notation for all of them. Thus, we will add the index 0 to the graviphoton $A_\mu \to A^0{}_\mu$ and we will use capital Greek indices $\Lambda, \Sigma = 0, \ldots, n_V$ to label all $\bar{n} = n_V + 1$ of them:

$$(A^\Lambda{}_\mu) = (A^0{}_\mu, A^i{}_\mu). \tag{7.4}$$

The naive graviphoton and matter vector field strengths would simply be $F^0{}_{\mu\nu} = 2\partial_{[\mu}A^0{}_{\nu]}$ and $F^i{}_{\mu\nu} = 2\partial_{[\mu}A^i{}_{\nu]}$. However, in the fully coupled theory, the objects that play these roles[2] are linear combinations of these field strengths with scalar-dependent coefficients. They will be denoted, respectively, by $T_{\mu\nu}$ and $G^i{}_{\mu\nu}$, and they are defined in Eqs. (7.15) and (7.16).

Hypermultiplets are a special feature of $N=2, d=4$ supergravity. Each hypermultiplet consists of four real *hyperscalars* and two *hyperinos* which do not carry $\mathrm{SU}(2)$ indices and are, therefore, insensitive to this part of the R-symmetry group. They are sensitive to the $\mathrm{U}(1)$ factor, however, since they have a non-zero Kähler weight. If we couple $m$ hypermultiplets to supergravity, we can collectively label their field content using two indices, $u, v = 1, \ldots, 4m$ and $\alpha = 1, \ldots, 2m$:

$$\{q^u, \zeta_\alpha\}. \tag{7.5}$$

In the coupled theory, the $4m$ hyperscalars parametrize a quaternionic–Kähler manifold (defined and studied in Appendix G). The chirality of the hyperinos with lower index is negative and their Kähler weight is $\frac{1}{2}$ ,

$$\gamma_5\zeta_\alpha = -\zeta_\alpha, \qquad \gamma_5\zeta^\alpha \equiv \gamma_5(\zeta_\alpha)^* = +\zeta^\alpha. \tag{7.6}$$

Summarizing, the following spinors have negative chirality and Kähler weight $\frac{1}{2}$:

$$\psi_{I\mu}, \epsilon_I, \lambda^{iI}, \zeta_\alpha. \tag{7.7}$$

## 7.2 The ungauged theory

As we have mentioned repeatedly, the coupling of the vector multiplets (and, in particular, of their complex scalars $Z^i$) to supergravity in $N=2, d=4$ theories is governed by special Kähler geometry and that of the hypermultiplets (and, in particular, of their quaternionic scalars $q^u$), by quaternionic-Kähler geometry. Specifying the choice for these two geometries is enough to determine fully an ungauged $N=2, d=4$ theory.

This should be contrasted with the $N=1, d=4$ case, in which the coupling of the chiral multiplets (in particular of their complex scalars $Z^i$) to supergravity was not fully determined by the choice of Kähler–Hodge geometry (or, equivalently, of Kähler potential $\mathcal{K}$) because we have complete freedom to choose independently $\mathcal{K}$, a superpotential $\mathcal{W}$ and a kinetic (or period) matrix $f_{\Lambda\Sigma}$. There is no superpotential in $N=2, d=4$ theories, and the period matrix $\mathcal{N}_{\Lambda\Sigma}$ and the Kähler potential $\mathcal{K}$ cannot be chosen independently: they follow from the choice of special Kähler geometry.

The most economical way of showing how this comes about is perhaps to present the bosonic action and the supersymmetry transformation rules (for vanishing fermions) of the most general ungauged $N=2, d=4$ supergravity theory and then explain how all the coupling functions that appear in them fit in special Kähler and quaternionic-Kähler geometry. In a second stage, we will show how, given a special Kähler geometry and a quaternionic-Kähler geometry, one can construct the corresponding $N=2, d=4$ supergravity theory, and we provide several examples.

---

[2] The graviphoton and matter vector field strengths appear, by definition, in the gravitino and gaugino supersymmetry transformation rules.

The action of the bosonic fields of the theory is given by

$$
S = \int d^4x \sqrt{|g|}\,\big[R + 2\mathcal{G}_{ij^*}\partial_\mu Z^i \partial^\mu Z^{*\,j^*} + 2\mathsf{H}_{uv}\partial_\mu q^u \partial^\mu q^v
\\
+ 2\Im\mathrm{m}\mathcal{N}_{\Lambda\Sigma} F^{\Lambda\,\mu\nu} F^{\Sigma}{}_{\mu\nu} - 2\Re\mathrm{e}\mathcal{N}_{\Lambda\Sigma} F^{\Lambda\,\mu\nu} \star F^{\Sigma}{}_{\mu\nu}\big]\,. \tag{7.8}
$$

In this expression, $\mathcal{G}_{ij^*}$ is a Kähler metric[3] parametrized only by the complex scalars $Z^i$ of the vector supermultiplets and $\mathrm{H}_{uv}$ is the metric of a quaternionic-Kähler space parametrized by the hyperscalars $q^u$ only. The period matrix $\mathcal{N}_{\Lambda\Sigma}$ is a function of the $Z^i$ (and their complex conjugates) only and is determined by the choice of special Kähler geometry. Therefore, the hyperscalars are completely decoupled from the bosonic fields in the vector supermultiplets. Actually, this is true not just for the bosonic fields but for all the fields in these supermultiplets, which implies that in the ungauged theories[4] it is always consistent to truncate the hypermultiplets or the hyperscalars in the bosonic action, setting them to some convenient constant value. The $Z^i$s cannot be decoupled in the same way.

Observe that the bosonic action is of the general form Eq. (2.147), with the generic scalars $\varphi^i$ and its metric $\mathcal{G}_{ij}$ split into the $Z^i$ and $q^u$ and the metrics $\mathcal{G}_{ij^*}$ and $\mathrm{H}_{uv}$.

For vanishing fermions, the supersymmetry transformation rules of the fermions are

$$
\begin{aligned}
\delta_\epsilon \psi_{I\,\mu} &= \mathfrak{D}_\mu \epsilon_I + \varepsilon_{IJ}\, T^+{}_{\mu\nu}\gamma^\nu\, \epsilon^J,\\
\delta_\epsilon \lambda^{iI} &= i\, \not\partial Z^i \epsilon^I \ + \varepsilon^{IJ}\, \not G^{i\,+}\, \epsilon_J,\\
\delta_\epsilon \zeta_\alpha &= -i\mathbb{C}_{\alpha\beta}\, \mathsf{U}^{\beta I}{}_u\, \varepsilon_{IJ}\, \not\partial q^u\, \epsilon^J.
\end{aligned} \tag{7.9}
$$

Let us explain the different terms that appear in these formulae.

1. In the gravitino supersymmetry rule, $\mathfrak{D}_\mu$ is the Lorentz-, Kähler-, and SU(2)-covariant derivative, with the standard spinorial spin connection, the pullback of the Kähler 1-form $\mathcal{Q}_\mu$ defined in Eq. (E.48), and the pullback of the SU(2) connection $\mathsf{A}^x{}_\mu$ described in Appendix G. Taking into account the Kähler weight of $\epsilon_I$ and the position of its SU(2) index, its explicit form is

$$
\mathfrak{D}_\mu \epsilon_I = (\nabla_\mu + \tfrac{i}{2}\,\mathcal{Q}_\mu)\,\epsilon_I - \mathsf{A}_\mu{}^J{}_I\,\epsilon_J. \tag{7.10}
$$

$\mathsf{A}_\mu{}^J{}_I = \partial_\mu q^u \mathsf{A}_u{}^J{}_I(q)$ is the only object that depends on the hyperscalars in the gravitino and gaugino supersymmetry transformation rules.

The next term in the gravitino supersymmetry rule, $T_{\mu\nu}$, is a linear combination of the vector field strengths $F^\Lambda{}_{\mu\nu}$ with scalar-dependent coefficients such that its self-dual

[3] Not any Kähler metric, but the metric of a special Kähler manifold, of course.

[4] As we are going to see, the scalar potential contains direct couplings between the hyperscalars and the $Z^i$s.

part is given by

$$T^{+}{}_{\mu\nu} \equiv 2i\mathcal{L}^{\Sigma}\Im\mathrm{m}\,\mathcal{N}_{\Sigma\Lambda}F^{\Lambda\,+}{}_{\mu\nu}, \tag{7.11}$$

where, in turn, $\mathcal{L}^{\Sigma}$ are the upper components of the *(covariantly holomorphic) canonical symplectic section* $(\mathcal{V}^{M}) = \begin{pmatrix} \mathcal{L}^{\Lambda} \\ \mathcal{M}_{\Lambda} \end{pmatrix}$ defined in Appendix F.1. Since $T_{\mu\nu}$ is the object the gravitinos transform into under supersymmetry, it is called the *graviphoton field strength*. In Eq. (7.15) we give an equivalent symplectic-invariant definition of $T_{\mu\nu}$. The reason why only the self-dual part of $T_{\mu\nu}$ occurs in the transformation is the chirality of the spinors (as explained in more detail in Eq. (6.22) and the preceding paragraph).

2. The objects that appear in the gaugino supersymmetry transformation rules $G^{i}{}_{\mu\nu}$ are, again, linear combinations of the vector field strengths $F^{\Lambda}{}_{\mu}$ with scalar-dependent coefficients such that

$$G^{i\,+}{}_{\mu\nu} \equiv -\mathcal{G}^{ij^*}f^{*\,\Sigma}{}_{j^*}\Im\mathrm{m}\,\mathcal{N}_{\Sigma\Lambda}F^{\Lambda\,+}{}_{\mu\nu}, \tag{7.12}$$

where $f^{*\,\Sigma}{}_{j^*}$ is the upper component of the Kähler-covariant derivative of the symplectic section $\mathcal{U}^{*\,M}_{j^*} = \mathcal{D}_{j^*}\mathcal{V}^{*\,\mathcal{M}} = \begin{pmatrix} f^{*\,\Lambda}{}_{j^*} \\ h^{*}_{\Lambda\,j^*} \end{pmatrix}$.

We will give shortly a manifestly symplectic-invariant definition for $G^{i}{}_{\mu\nu}$. Meanwhile, it is worth stressing that the recombination of the vector field strengths $F^{\Lambda}$ into graviphoton and matter vector field strengths is nothing but a change of basis from one in which the duality transformations act linearly as symplectic transformations to one that diagonalizes the fermion supersymmetry transformation rules. The opposite change of basis is

$$F^{\Lambda}{}_{\mu\nu} = i\mathcal{L}^{*\,\Lambda}T_{\mu\nu} + 2f^{\Lambda}{}_{i}G^{i}{}_{\mu\nu}, \tag{7.13}$$

by virtue of the identity Eq. (F.31). This decomposition and its properties can also be studied using the projectors defined in Eqs. (F.32) and (F.33).

3. Finally $\mathsf{U}^{\beta I}{}_{u}$ in the hyperino supersymmetry transformation rule is a *Quadbein*, i.e. a Vielbein in the quaternionic-Kähler space, and $\mathbb{C}_{\alpha\beta}$ is the $\mathrm{Sp}(2m)$-invariant metric, both of which are defined in Appendix G.

The boson supersymmetry transformations are

$$\boxed{\begin{aligned}
\delta_\epsilon e^{a}{}_{\mu} &= -\tfrac{i}{4}\bar{\psi}_{I\,\mu}\gamma^{a}\epsilon^{I} + \text{c.c.},\\
\delta_\epsilon A^{\Lambda}{}_{\mu} &= \tfrac{1}{4}\varepsilon^{IJ}\left(\mathcal{L}^{*\,\Lambda}\bar{\psi}_{I\,\mu} + \tfrac{i}{2}f^{\Lambda *}{}_{i^*}\bar{\lambda}^{i^*}{}_{I}\gamma_{\mu}\right)\epsilon_{J} + \text{c.c.},\\
\delta_\epsilon Z^{i} &= \tfrac{1}{4}\bar{\lambda}^{iI}\epsilon_{I},\\
\delta_\epsilon q^{u} &= \mathsf{U}_{\alpha I}{}^{u}\bar{\zeta}^{\alpha}\epsilon^{I} + \text{c.c.}
\end{aligned}} \tag{7.14}$$

We have already described the objects that appear in them.

It is convenient to introduce the dual (or "magnetic") field strengths $G_{\Lambda\,\mu\nu}$ defined in Eq. (2.148) and the symplectic vector of electric and magnetic vector field strengths, and construct the symplectic vector of electric and magnetic vector field strengths $\mathcal{F}^M$ defined in Eq. (2.161).[5] We can use many of the results obtained in Section 2.6.2 since the vector part of the action Eq. (2.147) is identical.

In terms of this object, the graviphoton and matter vector field strengths can be written as follows:

$$T_{\mu\nu} = \langle \mathcal{V} \mid \mathcal{F}_{\mu\nu} \rangle = \mathcal{V}_M \mathcal{F}^M{}_{\mu\nu}, \tag{7.15}$$

$$G^i{}_{\mu\nu} = \tfrac{i}{2}\mathcal{G}^{ij^*} \langle \mathcal{D}_{j^*}\mathcal{V}^* \mid \mathcal{F}_{\mu\nu} \rangle = \tfrac{i}{2}\mathcal{G}^{ij^*}\mathcal{D}_{j^*}\mathcal{V}^*_M \mathcal{F}^M{}_{\mu\nu}, \tag{7.16}$$

and are, by construction, self-dual. For instance, using the above definition for the graviphoton field strength and the definition of $G_\Lambda$, we get

$$\begin{aligned} T &= \mathcal{M}_\Sigma F^\Sigma - \mathcal{L}^\Lambda G_\Lambda = (\mathcal{M}_\Sigma - \mathcal{L}^\Lambda \Re\mathrm{e}\mathcal{N}_{\Lambda\Sigma})F^\Sigma - \mathcal{L}^\Lambda \Im\mathrm{m}\mathcal{N}_{\Lambda\Sigma} \star F^\Sigma \\ &= (\mathcal{M}_\Sigma - \mathcal{L}^\Lambda \mathcal{N}_{\Lambda\Sigma})F^\Sigma + 2i\mathcal{L}^\Lambda \Im\mathrm{m}\mathcal{N}_{\Lambda\Sigma} \star F^{\Sigma\,+}. \end{aligned} \tag{7.17}$$

The first term vanishes upon use of the defining property of the period matrix Eq. (F.20), leaving us with the self-dual tensor of Eq. (7.11). Furthermore, using the completeness relation Eq. (F.13) and the above relations, we get[6]

$$\mathcal{F}^M = i\mathcal{V}^{*\,M} T^+ + 2\mathcal{U}_i G^{i\,+} + \text{c.c.} \tag{7.18}$$

$\mathcal{F}^M$ can also be used to write the equations of motion in a more compact and manifestly symplectic-invariant form. We can denote the l.h.s. of the equations of motion as follows:

$$\begin{aligned} \mathcal{E}_a{}^\mu &\equiv -\frac{1}{2\sqrt{|g|}}\frac{\delta S}{\delta e^a{}_\mu}, & \mathcal{E}^i &\equiv -\frac{1}{2\sqrt{|g|}}\mathcal{G}^{ij^*}\frac{\delta S}{\delta Z^{*\,ij^*}}, \\ \mathcal{E}_\Lambda{}^\mu &\equiv \frac{1}{8\sqrt{|g|}}\frac{\delta S}{\delta A^\Lambda{}_\mu}, & \mathcal{E}^u &\equiv -\frac{1}{4\sqrt{|g|}}\mathsf{H}^{uv}\frac{\delta S}{\delta q^v}, \end{aligned} \tag{7.19}$$

so that all the equations of motion take the form $\mathcal{E} = 0$. As we did in the $N = 1, d = 4$ case, we can supplement the equations of motion with the Bianchi identities Eqs. (6.25), combine the Maxwell equations with the Bianchi identities as in Eq. (6.26), and write them in the symplectic-invariant form Eq. (6.27).

Using Eqs. (2.166), the Einstein equations $\mathcal{E}_{\mu\nu} \equiv \mathcal{E}_\mu{}^a e_{a\,\nu}$ and scalar equations of motion

[5] Observe that in Section 2.6 these matrices are $2\bar{n} \times 2\bar{n}$ matrices, where $\bar{n}$ denotes the total number of vector fields, which in our case is $\bar{n} = n_V + 1$ and $n_V$ is just the number of vector supermultiplets.

[6] Observe that the l.h.s. of this equation is the sum of the self-dual and anti-self-dual parts, and it is real.

take the form

$$\mathcal{E}_{\mu\nu} = G_{\mu\nu} + 2\mathcal{G}_{ij^*}[\partial_\mu Z^i \partial_\nu Z^{*j^*} - \tfrac{1}{2} g_{\mu\nu} \partial_\rho Z^i \partial^\rho Z^{*j^*}]$$

$$+ 4\mathcal{M}_{MN}(\mathcal{N}) \mathcal{F}^M{}_\mu{}^\rho \mathcal{F}^N{}_{\nu\rho} + 2\mathsf{H}_{uv}\,[\partial_\mu q^u \partial_\nu q^v - \tfrac{1}{2} g_{\mu\nu} \partial_\rho q^u \partial_\rho q^v], \tag{7.20}$$

$$\mathcal{E}_i = \nabla_\mu (\mathcal{G}_{ij^*} \partial^\mu Z^{*i^*}) - \partial_i \mathcal{G}_{jk^*} \partial_\rho Z^j \partial^\rho Z^{*k^*} + \partial_i G_{\Lambda\,\mu\nu} \star F^{\Lambda\,\mu\nu}\,, \tag{7.21}$$

$$\mathcal{E}^u = \nabla_\mu \partial^\mu q^u + \Gamma_{vw}{}^u \partial^\mu q^v \partial_\mu q^w, \tag{7.22}$$

where $\Gamma_{vw}{}^u$ are the Christoffel symbols for the metric $\mathsf{H}_{uv}$.

As argued before, the equations of motion make it clear that setting the hyperscalar to an arbitrary constant value is always a consistent truncation.

Our next task consists in showing how one can fully define a $N=2, d=4$ supergravity model from the special Kähler and quaternionic-Kähler geometries. The basic recipe for the special Kähler case can be found on p. 912. Here we will illustrate it through several examples.

### 7.2.1 Examples

*Pure, ungauged $N=2$, $d=4$ supergravity.* The formalism of the special and quaternionic-Kähler geometries encompasses all $N=2, d=4$ supergravities. It is natural to start by showing how the pure, ungauged, theory that we studied in Section 5.5 fits into this framework.

The pure supergravity theory has no hypermultiplets and, therefore, there is no need to specify a quaternionic-Kähler geometry. It has no vector supermultiplets either, but the special Kähler geometry framework describes the couplings to all the vector fields, including the graviphoton, even if these couplings are trivial. It is described in terms of one projective (unphysical) coordinate $\mathcal{X}^0$, which can be eliminated from all the physical quantities using Kähler transformations, if necessary.

A special Kähler geometry can be defined via a prepotential $\mathcal{F}(\mathcal{X})$ or a canonical symplectic section $\mathcal{V}^M$. The latter can always be derived from the former. Pure supergravity corresponds to the prepotential

$$\mathcal{F} = -\tfrac{i}{4}(\mathcal{X}^0)^2, \tag{7.23}$$

from which, following the recipe on p. 912, we can obtain the holomorphic canonical section, the Kähler potential, and the period matrix:

$$\mathcal{F}_0 = -\tfrac{i}{2}\mathcal{X}^0, \qquad \Omega = \begin{pmatrix} \mathcal{X}^0 \\ -\tfrac{i}{2}\mathcal{X}^0 \end{pmatrix}, \qquad e^{-\mathcal{K}} = |\mathcal{X}^0|^2, \qquad \mathcal{N}_{00} = -\tfrac{i}{2}. \tag{7.24}$$

To obtain the Kähler potential as a function of the physical scalars only, we can just eliminate $\mathcal{X}^0$ through a Kähler transformation Eq. (E.47) with $\lambda(\mathcal{X}^0) = \log \mathcal{X}^0$. The result is

$$e^{-\mathcal{K}} = 1, \qquad \Omega = \mathcal{V} = \begin{pmatrix} 1 \\ -\tfrac{i}{2} \end{pmatrix}, \qquad \mathcal{Q} = 0. \tag{7.25}$$

Having computed all these quantities, we can just substitute into the general expressions for the action and supersymmetry transformation rules. We obtain (removing the index from the graviphoton) the bosonic (Einstein–Maxwell) action

$$\boxed{S = \int d^4x \sqrt{|g|}\,\left[R - F^2\right]} \tag{7.26}$$

and the supersymmetry transformation rules[7]

$$\boxed{\begin{aligned} \delta_\epsilon e^a{}_\mu &= -\tfrac{i}{4}\bar{\psi}_{I\,\mu}\gamma^a\epsilon^I + \text{c.c.},\\ \delta_\epsilon \psi_{I\,\mu} &= \nabla_\mu \epsilon_I - i\varepsilon_{IJ}\, F^+{}_{\mu\nu}\gamma^\nu\,\epsilon^J,\\ \delta_\epsilon A_\mu &= \tfrac{1}{4}\varepsilon^{IJ}\bar{\psi}_{I\,\mu}\,\epsilon_J + \text{c.c.} \end{aligned}} \tag{7.29}$$

*The axion–dilaton model.* In Chapter 6 (see p. 208) we defined an axion–dilaton model as an $N=1, d=4$ theory with one chiral multiplet and with the complex scalar $\tau$ parametrizing the coset space $\mathrm{SL}(2,\mathbb{R})/\mathrm{SO}(2)$. One could add vector supermultiplets coupled to arbitrary kinetic matrices $f_{\Lambda\Sigma}$ to it. The axion–dilaton (or *axidilaton*) model of $N=2, d=4$ supergravity also has one complex scalar parametrizing the space $\mathrm{SL}(2,\mathbb{R})/\mathrm{SO}(2)$ (hence the name), but the $N=2, d=4$ structure forces us to have one vector field from the same vector supermultiplet that $\tau$ comes from, and the period matrix will be completely determined by that structure.[8]

The $N=2, d=4$ axion–dilaton model is defined by the prepotential

$$\mathcal{F} = -i\mathcal{X}^0\mathcal{X}^1, \tag{7.30}$$

and the axidilaton scalar $\tau = a + ie^{-\varphi}$, $a$ being the axion and $\varphi$ being the dilaton, is conventionally defined in terms of the $\mathcal{X}^\Lambda$ as the special coordinate (up to the $i$ factor)

$$\tau \equiv i\mathcal{X}^1/\mathcal{X}^0. \tag{7.31}$$

---

[7] It is not difficult to see that these supersymmetry transformations are completely equivalent to those in Section 5.5 upon the change from Majorana to Weyl spinors defined in Eqs. (6.1) and a change in the normalization of the vector field. In the graviphoton term one has to use the identity

$$\not{F}\gamma_\mu = -2(F_{\mu\nu} - i\star F_{\mu\nu}\gamma_5)\gamma^\nu, \tag{7.27}$$

based on the gamma-matrix identities

$$\gamma^{ab}\gamma_c = \gamma^{ab}{}_c + 2\gamma^{[a}\eta^{b]}{}_c, \qquad \gamma^{ab}{}_c = -i\varepsilon^{ab}{}_{cd}\gamma^d\gamma_5. \tag{7.28}$$

[8] There are other $N=2, d=4$ models, like the $t^3$ model, in which the scalar parametrizes the coset space $\mathrm{SL}(2,\mathbb{R})/\mathrm{SO}(2)$, but with different period matrices.

We first construct the holomorphic symplectic section, the Kähler potential, and the period matrix:

$$
(\Omega^M) = \begin{pmatrix} \mathcal{X}^0 \\ \mathcal{X}^1 \\ -i\mathcal{X}^1 \\ -i\mathcal{X}^0 \end{pmatrix}, \qquad e^{-\mathcal{K}} = 4|\mathcal{X}^0|^2 \Im\mathfrak{m}\tau, \qquad (\mathcal{N}_{\Lambda\Sigma}) = \begin{pmatrix} -\tau & 0 \\ 0 & 1/\tau \end{pmatrix}. \tag{7.32}
$$

To eliminate $\mathcal{X}^0$ from the Kähler potential we can perform a Kähler transformation with $\lambda(\mathcal{X}^0) = \log -2i\mathcal{X}^0$ so that

$$
(\Omega^M) = \tfrac{1}{2} \begin{pmatrix} i \\ \tau \\ -i\tau \\ 1 \end{pmatrix}, \qquad e^{-\mathcal{K}} = \Im\mathfrak{m}\tau, \qquad (\mathcal{V}^M) = \frac{1}{2\sqrt{\Im\mathfrak{m}\tau}} \begin{pmatrix} i \\ \tau \\ -i\tau \\ 1 \end{pmatrix}. \tag{7.33}
$$

The period matrix remains invariant. Observe that we have introduced a factor $i$ in the holomorphic section, but we could have introduced any other numerical factor. The effect is analogous to that of *choosing a gauge for* $\mathcal{X}^0$ ($\mathcal{X}^0 = i/2$ in this particular case) and we will use this terminology from now onwards.

Since the Kähler potential is identical to that of the $N=1, d=4$ axidilaton model, the Kähler metric and 1-form connection are also identical, that is

$$
\mathcal{G}_{\tau\tau^*} = \frac{1}{4(\Im\mathfrak{m}\tau)^2}, \qquad \mathcal{Q}_\tau = \frac{1}{4\Im\mathfrak{m}\tau} \quad \Rightarrow \quad \mathcal{Q} = \frac{d\Re\mathfrak{e}\tau}{2\Im\mathfrak{m}\tau}. \tag{7.34}
$$

It is possible to choose the kinetic matrix of the $N=1, d=4$ theory to be similar to the period matrix of the $N=2, d=4$ theory, as in Eq. (6.38), but some field redefinitions (normalization and changing the sign of the axion $a = \Re\mathfrak{e}\tau$) are necessary in order to map one bosonic action into the other because $\mathcal{N}_{\Lambda\Sigma}$ is related to $f^*_{\Lambda\Sigma}$, which should be a function of $\tau^*$, not $\tau$. We will not delve further into this relation.

The action for the bosonic fields is

$$
\boxed{
\begin{aligned}
S = \int d^4x \sqrt{|g|} \Big\{ R + \frac{\partial_\mu\tau\partial^\mu\tau^*}{2(\Im\mathfrak{m}\tau)^2} - 2\Im\mathfrak{m}\tau \left[ (F^0)^2 + |\tau|^{-2}(F^1)^2 \right] \\
+ 2\Re\mathfrak{e}\tau \left[ F^0 \star F^0 - |\tau|^{-2} F^1 \star F^1 \right] \Big\}.
\end{aligned}
} \tag{7.35}
$$

If we replace the matter vector field $A^1$ by its dual ($G_1 = \Im\mathfrak{m}\tau \star F^1 + \Re\mathfrak{e}\tau F^1$), the action takes the more (manifestly) symmetric form

$$S = \int d^4x \sqrt{|g|} \left\{ R + \frac{\partial_\mu \tau \partial^\mu \tau}{(2\Im\mathrm{m}\tau)^2} - 2\Im\mathrm{m}\tau \left[ (F^0)^2 + (G_1)^2 \right] \right.$$
$$\left. + 2\Re\mathrm{e}\tau \left[ F^0 \star F^0 + G_1 \star G_1 \right] \right\}. \tag{7.36}$$

This action is identical (up to the normalization of the vector fields) to that of pure $N = 4, d = 4$ SUEGRA, given in Eq. (16.58). In fact, the model describes a consistent truncation of pure $N = 4, d = 4$ SUEGRA down to $N = 2$.[9]

In order to write the supersymmetry transformation rules of the vector fields, the axidilaton, and the gaugino (the Vierbein supersymmetry transformation is the same in all models) we need to compute several objects first:

$$T^+{}_{\mu\nu} = \sqrt{\Im\mathrm{m}\tau} \left[ F^{0\,+}{}_{\mu\nu} - \frac{i}{\tau^*} F^{1\,+}{}_{\mu\nu} \right],$$

$$(\mathcal{U}^M_\tau) = \tfrac{i}{4} (\Im\mathrm{m}\tau)^{-3/2} \begin{pmatrix} i \\ \tau^* \\ -i\tau^* \\ 1 \end{pmatrix}, \tag{7.37}$$

$$G^{\tau\,+} = -(\Im\mathrm{m}\tau)^{3/2} \left[ F^{0\,+}{}_{\mu\nu} + \frac{i}{\tau^*} F^{1\,+}{}_{\mu\nu} \right],$$

to obtain

$$\begin{aligned}
\delta_\epsilon A^0{}_\mu &= -\tfrac{i}{8\sqrt{\Im\mathrm{m}\tau}} \varepsilon^{IJ} \left( \bar{\psi}_{I\,\mu} + \tfrac{1}{4\Im\mathrm{m}\tau} \bar{\lambda}^{\tau^*\,I} \gamma_\mu \right) \epsilon^J + \text{c.c.}, \\
\delta_\epsilon A^1{}_\mu &= \tfrac{1}{8\sqrt{\Im\mathrm{m}\tau}} \varepsilon^{IJ} \left( \tau^* \bar{\psi}_{I\,\mu} - \tfrac{\tau}{4\Im\mathrm{m}\tau} \bar{\lambda}^{\tau^*\,I} \gamma_\mu \right) \epsilon^J + \text{c.c.}, \\
\delta_\epsilon \tau &= \tfrac{1}{4} \bar{\lambda}^{\tau\,I} \epsilon_I, \\
\delta_\epsilon \psi_{I\,\mu} &= \left[ \nabla_\mu + \tfrac{i}{4} \frac{\partial_\mu \Re\mathrm{e}\tau}{\Im\mathrm{m}\tau} \right] \epsilon_I + \sqrt{\Im\mathrm{m}\tau}\, \varepsilon_{IJ} \left[ F^{0\,+}{}_{\mu\nu} - \frac{i}{\tau^*} F^{1\,+}{}_{\mu\nu} \right] \gamma^\nu \epsilon^J, \\
\delta_\epsilon \lambda^{\tau\,I} &= i\, \partial\!\!\!/ \tau \epsilon^I - (\Im\mathrm{m}\tau)^{3/2} \varepsilon^{IJ} \left[ F\!\!\!/^{0\,+} + \frac{i}{\tau^*} F\!\!\!/^{1\,+}{}_{\mu\nu} \right] \epsilon_J.
\end{aligned} \tag{7.38}$$

The axidilaton is the simplest of several families of models. It is the simplest of the quadratic models characterized by prepotentials that are second-order polynomials in the

[9] The supersymmetric embeddings of the axidilaton model into higher-$N$ SUEGRAs have been discussed in Ref. [328].

$\mathcal{X}^\Lambda$ and, in particular, it is the simplest of the $\overline{\mathbb{CP}}^n$ models associated with the symmetric special Kähler geometry $\mathrm{U}(1,n)/(\mathrm{U}(1)\times\mathrm{U}(n))$ that we will study next. The axidilaton model is also the simplest ($n=1$) of the ST$[2,n]$ models which have the symmetric special Kähler geometries $\frac{\mathrm{SL}(2,\mathbb{R})}{\mathrm{U}(1)}\times\frac{\mathrm{SO}(n-1,2)}{\mathrm{SO}(n-1)\times\mathrm{SO}(2)}$ and which are also associated with truncations of $N=4, d=4$ SUEGRA coupled to $n-1$ vector supermultiplets down to $N=2$.[10] These models are, in turn, particular examples of a wide class of models known as *cubic models*, characterized by prepotentials that are third-order polynomials in the $\mathcal{X}^\Lambda$. We will briefly review these as well.

*The $\overline{\mathbb{CP}}^n$ models.* The $\overline{\mathbb{CP}}^n$ models are models with $n_V=n$ defined by the following series of quadratic prepotentials:

$$\mathcal{F}=-\tfrac{i}{4}\eta_{\Lambda\Sigma}\mathcal{X}^\Lambda\mathcal{X}^\Sigma, \qquad (\eta_{\Lambda\Sigma})=\mathrm{diag}(+-\cdots-). \tag{7.39}$$

The $n$ physical scalar fields can be defined as the special coordinates

$$Z^i\equiv\mathcal{X}^i/\mathcal{X}^0, \tag{7.40}$$

but it is convenient to extend the definition to $Z^0\equiv1$ in order to have more covariant-looking expressions and use $Z^\Lambda$ and $Z_\Lambda$ defined by

$$(Z^\Lambda)\equiv(\mathcal{X}^\Lambda/\mathcal{X}^0)=(1,Z^i), \qquad (Z_\Lambda)\equiv(\eta_{\Lambda\Sigma}Z^\Sigma)=(1,Z_i)=(1,-Z^i). \tag{7.41}$$

In the $\mathcal{X}^0=1$ gauge, the Kähler potential and the Kähler metric and its inverse are given by

$$\mathcal{K}=-\log(Z^{*\Lambda}Z_\Lambda), \quad \mathcal{G}_{ij^*}=-e^{\mathcal{K}}\left(\eta_{ij^*}-e^{\mathcal{K}}Z^*_iZ_{j^*}\right), \quad \mathcal{G}^{ij^*}=-e^{-\mathcal{K}}\left(\eta^{ij^*}+Z^iZ^{*j^*}\right). \tag{7.42}$$

The reality of the Kähler potential requires the strict positivity of $e^{-\mathcal{K}}=Z^{*\Lambda}Z_\Lambda$, which implies that the complex scalars have to obey the following restriction:

$$0\leq\sum_i|Z^i|^2<1. \tag{7.43}$$

The covariantly holomorphic symplectic section and its Kähler-covariant derivative are given by

$$\mathcal{V}=e^{\mathcal{K}/2}\begin{pmatrix}Z^\Lambda\\ \\ -\frac{i}{2}Z_\Lambda\end{pmatrix}, \qquad \mathcal{U}_i=e^{\mathcal{K}/2}\begin{pmatrix}-e^{\mathcal{K}}Z^*_iZ^\Lambda+\delta_i{}^\Lambda\\ \\ \frac{i}{2}(e^{\mathcal{K}}Z^*_iZ_\Lambda-\eta_{i\Lambda})\end{pmatrix}, \tag{7.44}$$

and they can be used to check that the period matrix

$$\mathcal{N}_{\Lambda\Sigma}=\frac{i}{2}\left[\eta_{\Lambda\Sigma}-2\frac{Z_\Lambda Z_\Sigma}{Z^\Gamma Z_\Gamma}\right], \qquad (\mathcal{N}^{-1})^{\Lambda\Sigma}=-2i\left[\eta^{\Lambda\Sigma}-2\frac{Z^\Lambda Z^\Sigma}{Z^\Gamma Z_\Gamma}\right], \tag{7.45}$$

[10] In all these truncations, four vectors out of the six in the $N=4, d=4$ supergravity multiplet are eliminated. One of the remaining vectors of the supergravity multiplet, plus the axidilaton scalar, become part of a matter $N=2, d=4$ vector supermultiplet.

satisfies its defining properties.

As we mentioned before, the scalars of this model parametrize the symmetric space $\mathrm{U}(1,n)/(\mathrm{U}(1)\times\mathrm{U}(n))$. We will show later that the Kähler metric is indeed invariant under the group $\mathrm{U}(1,n)$. The above metric is the standard (Bergman) metric for these symmetric spaces [206].

In the $n=1$ case, the change of variables

$$\mathcal{X}^\Lambda = \mathcal{S}^\Lambda{}_\Sigma \mathcal{Y}^\Sigma, \qquad (\mathcal{S}^\Lambda{}_\Sigma) = \sqrt{2}\begin{pmatrix} 1 & 1 \\ -1 & 1 \end{pmatrix}, \tag{7.46}$$

transforms the prepotential into that of the axidilaton model. It also transforms the Kähler potential into that of the axidilaton model via

$$Z = \frac{\tau + i}{\tau - i}, \tag{7.47}$$

but in a gauge different to the one we imposed ($\mathcal{Y}^0 = i/2$) because the above transformation is incompatible with setting $\mathcal{X}^0 = 1$ at the same time. Actually, setting $\mathcal{X}^0 = 1$ implies the gauge $\mathcal{Y}^0 = -\frac{i}{\sqrt{2}(\tau - i)}$.

This shows that we can map one model into the other so they are just the same model in two different representations. The fact that the scalars parametrize the same coset space ($\mathrm{SL}(2,\mathbb{R})/\mathrm{SO}(2) \sim \mathrm{U}(1,1)/(\mathrm{U}(1)\times\mathrm{U}(1))$) is not enough to reach that conclusion, however: we will find another model (the so-called $t^3$ model) that has the same coset space but a different period matrix.

*The cubic models.* These models, first studied in Ref. [387], are defined by a prepotential which is the quotient of a cubic polynomial and $\mathcal{X}^0$:

$$\mathcal{F} = -\frac{1}{3!}\frac{d_{ijk}\mathcal{X}^i\mathcal{X}^j\mathcal{X}^k}{\mathcal{X}^0}. \tag{7.48}$$

The tensor $d_{ijk}$ is completely symmetric in the three indices $i, j, k$, which, as usual, take values from 1 to $n_V$. It is clear that it completely defines the model, and, therefore, the corresponding special Kähler geometries are sometimes called *d-geometries* [329] and the tensor $d_{ijk}$ is sometimes known as the *d-tensor*. This class of models is very rich and interesting because $d$-geometries naturally arise in the compactification of type-II superstring theories in complex three-dimensional Calabi–Yau manifolds (or Calabi–Yau 3-fold, for short).[11] The $d$-tensor follows from the topology of the Calabi–Yau 3-fold. They also arise in the dimensional reduction over a circle of $N=1, d=5$ supergravities: as we shall see in Chapter 8, the couplings of the vector supermultiplets in these theories are entirely determined by a $d$-tensor (that we will denote by $C_{IJK}$) [693], which is not surprising since these theories can be obtained by compactification of 11-dimensional supergravity on Calabi–Yau 3-folds. The same $d$-tensor will characterize the $d$-geometry of the $N=2, d=4$ supergravity obtained after compactification.

[11] See, for example, Refs. [291, 794, 671, 69, 111] for introductory reviews.

A property that characterizes the $d$-tensors of symmetric cubic models (i.e. those with a scalar manifold which is a symmetric space) is [693]

$$d_{m(ij}d_{kl)n}d_{mnp} = \tfrac{4}{3}\delta_{p(i}d_{jkl)}. \tag{7.49}$$

Using special coordinates $Z^i \equiv \mathcal{X}^i/\mathcal{X}^0$, the canonical holomorphic section and the Kähler potential are

$$\Omega = \mathcal{X}^0 \begin{pmatrix} 1 \\ Z^i \\ -\mathcal{F}(Z) \\ -\frac{1}{2}d_{ijk}Z^j Z^k \end{pmatrix}, \qquad e^{-\mathcal{K}} = 8|\mathcal{X}^0|^2\, \tfrac{1}{3!}d_{ijk}\Im\mathrm{m}\, Z^i\, \Im\mathrm{m}\, Z^j\, \Im\mathrm{m}\, Z^k, \tag{7.50}$$

where $\mathcal{F}(Z)$ is the prepotential in the gauge $\mathcal{X}^0 = 1$. The Kähler 1-form connection, the Kähler metric, and the tensor $\mathcal{C}_{ijk}$ defined in Eq. (F.14) (computed using Eq. (F.55)) are given by

$$\begin{aligned} \mathcal{Q}_i &= e^{\mathcal{K}}d_{ijk}\Im\mathrm{m}\, Z^j\, \Im\mathrm{m}\, Z^k, \\ \mathcal{G}_{ij^*} &= 4\mathcal{Q}_i\mathcal{Q}_{j^*} - 2e^{\mathcal{K}}d_{ijk}\Im\mathrm{m}\, Z^k, \\ \mathcal{C}_{ijk} &= -e^{\mathcal{K}}d_{ijk}. \end{aligned} \tag{7.51}$$

The components of the period matrix are

$$\begin{aligned} \mathcal{N}_{00} &= -\tfrac{1}{3}d_{ijk}R^iR^jR^k - i\left[\tfrac{1}{8}e^{-\mathcal{K}} - d_{ijk}I^iR^jR^k + 2e^{\mathcal{K}}(d_{ijk}I^iI^jR^k)^2\right], \\ \mathcal{N}_{0i} &= \tfrac{1}{2}d_{ijk}R^jR^k - i\left[d_{ijk}I^jR^k - 2e^{\mathcal{K}}d_{ijk}I^jI^kd_{klm}I^kI^lR^m\right], \\ \mathcal{N}_{ij} &= -d_{ijk}R^k - i\left[2e^{\mathcal{K}}d_{ikl}I^kI^ld_{jmn}I^mI^n - d_{ijk}I^k\right], \end{aligned} \tag{7.52}$$

where we have used the shorthand notation $Z^i \equiv R^i + iI^i$. This result can be written in a more compact way by exploiting the relation between the cubic models of $N = 2, d = 4$ supergravity and $N = 1, d = 5$ supergravity coupled to vector supermultiplets, as we will see in Section 15.2.5 (see also Ref. [446]).

Some particularly interesting cubic models are the so-called *magic models* or *magic supergravities* [692, 693, 694],[12] which are associated with the four Jordan algebras of $3 \times 3$ Hermitian matrices $\mathbb{J}_3^{\mathbb{R}}, \mathbb{J}_3^{\mathbb{C}}, \mathbb{J}_3^{\mathbb{Q}}, \mathbb{J}_3^{\mathbb{O}}$, where $\mathbb{R}, \mathbb{C}, \mathbb{Q}, \mathbb{O}$ are the four division algebras (those of the real and complex numbers, quaternions, and octonions, respectively). The octonionic magic model is also called the *exceptional magic model* because, first, it is the only magic model that cannot be obtained from a truncation of $N = 8, d = 4$ supergravity and, second, its associated Jordan algebra $\mathbb{J}_3^{\mathbb{O}}$, which involves the real octonions with a norm invariant under $\mathrm{O}(8)$, is the only one of the four that cannot be obtained as a truncation of $\mathbb{J}_3^{\mathbb{O}_\mathbb{S}}$, which is the Jordan algebra associated with $\mathbb{O}_\mathbb{S}$, the split form of the octonions with a quadratic form invariant under $\mathrm{O}(4,4)$ (see e.g. Ref. [524]).

The magic models correspond to the four symmetric special Kähler spaces in Table F.1 which are not part of an infinite series (the two infinite series of models being the $\overline{\mathbb{CP}}^n$

[12] They can be obtained by dimensional reduction from magic $N = 1, d = 5$ theories.

Table 7.1. Characteristics of the exceptional symmetric special Kähler spaces (or *magic models*, see p. 232) associated with the division algebras $\mathbb{A}$.

The first column shows the corresponding coset space, and the second shows the dimensional of the symplectic representation in which the duality group G acts, for instance on the canonical symplectic section. The next two columns indicate the representations of the stability group H in which the upper components of the section $(\mathcal{X}^\Lambda) = (\mathcal{X}^0, \mathcal{X}^i)$ transform. The prepotential of these models, Eq. (7.53), is determined by a cubic polynomial of the $\mathcal{X}^i$ components, which is invariant under H and is listed in the next column. The last column contains the largest group that can be gauged in these models. The names of the representations are those used in Ref. [1142]. Adapted from Ref. [798].

| $\mathbb{A}$ | G/H | $\mathrm{G} \circ \mathcal{V}$ | $\mathrm{H} \circ \mathcal{X}^0$ | $\mathrm{H} \circ \mathcal{X}^i$ | $\mathsf{I}_3(\mathcal{X}^i)$ | $\mathrm{G}_{\text{gauge max}}$ |
|---|---|---|---|---|---|---|
| $\mathbb{R}$ | $\frac{\mathrm{Sp}(6,\mathbb{R})}{\mathrm{U}(3)}$ | $\mathbf{14}'$ | $\mathbf{1}_{-3}$ | $\mathbf{6}_{-1}$ | $\det(\mathcal{X})$ | |
| $\mathbb{C}$ | $\frac{\mathrm{U}(3,3)}{\mathrm{U}(3)\times\mathrm{U}(3)}$ | $\mathbf{20}$ | $(\mathbf{1},\mathbf{1})_{-3}$ | $(\mathbf{3},\overline{\mathbf{3}})_{-1}$ | $\det(\mathcal{X})$ | $\mathrm{SU}(3)_{\text{diag}}$ |
| $\mathbb{Q}$ | $\frac{\mathrm{SO}^*(12)}{\mathrm{U}(6)}$ | $\mathbf{32}'$ | $\mathbf{1}_{-3}$ | $\mathbf{15}_{-1}$ | $\mathrm{Pf}\,(\mathcal{X})$ | $\mathrm{SU}(4)$ |
| $\mathbb{O}$ | $\frac{\mathrm{E}_{7(-25)}}{\mathrm{E}_6\times\mathrm{SO}(2)}$ | $\mathbf{56}$ | $\mathbf{1}_3$ | $\mathbf{27}_1$ | $\mathrm{Tr}\,([\Omega\mathcal{X}]^3)/3!$ | |

and $ST[2,n]$ ones, which are the first and last in Table F.1): the duality symmetries of the real, complex, quaternionic, and octonionic magic models are, respectively, $\mathrm{Sp}(6,\mathbb{R})$, $\mathrm{SU}(3,3)$, $\mathrm{SO}^*(12)$, and $\mathrm{E}_{7(-25)}$ (see Table 7.1, which gives more precise information on the representations in which the duality group acts).

The prepotentials that characterize these models are always of the form

$$\mathcal{F} = -\frac{\mathsf{I}_3\,(\mathcal{X}^i)}{\mathcal{X}^0}, \tag{7.53}$$

where $\mathsf{I}_3$ is a cubic invariant of $\mathrm{H}'$ (the subgroup of the stability group of the cosets, H, which results from removing the $\mathrm{U}(1)$ factors) built entirely in terms of the components $\mathcal{X}^i$ of $\mathcal{X}^\Lambda$. $\mathsf{I}_3$ can be written in the standard form $\frac{1}{3!}d_{ijk}\mathcal{X}^i\mathcal{X}^j\mathcal{X}^k$, but there are better descriptions that make the invariance manifest and are given in Table 7.1. These descriptions make use of the fact that the $\mathcal{X}^i$ can be arranged in matrices with simple transformation properties under $\mathrm{H}'$: for instance, in the real case, $\mathcal{X}^i$ transforms in the $\mathbf{6}$ of U(3), that is as a symmetric $3 \times 3$ matrix whose determinant gives $\mathsf{I}_3$, etc.

The last two examples of special Kähler geometries and their associated $N=2, d=4$ supergravities are two particular examples of cubic models: the so-called STU model and the family of $\mathrm{ST}[2,n]$ models.

*The STU model.* The STU model was originally obtained in Ref. [472] as a truncation of the $N=4, d=4$ supergravity coupled to $16+6$ vector supermultiplets (which corresponds, as we will see in Chapter 22, to a toroidally compactified heterotic string or, at this level, $N=1, d=10$ supergravity coupled to 16 vector supermultiplets). This truncation is an $N=2, d=4$ supergravity with $n_V=3$ vector supermultiplets characterized by a cubic prepotential with $d_{ijk}=|\varepsilon_{ijk}|$, so

$$\mathcal{F}=-\frac{\mathcal{X}^1\mathcal{X}^2\mathcal{X}^3}{\mathcal{X}^0}. \tag{7.54}$$

It is customary to use special coordinates $Z^i=\mathcal{X}^i/\mathcal{X}^0$ and, further, to call these $S$, $T$, and $U$ for $i=1,2$, and 3, respectively. The model has an evident triality symmetry since the three scalars enter into it in exactly the same way.
From the expressions obtained for the generic cubic models we get, for $\mathcal{X}^0=1$,

$$\Omega=\begin{pmatrix}1\\ S\\ T\\ U\\ STU\\ -TU\\ -SU\\ -TU\end{pmatrix},\qquad e^{-\mathcal{K}}=8\Im\mathrm{m}\,S\,\Im\mathrm{m}\,T\,\Im\mathrm{m}\,U,\qquad \mathcal{Q}_i=\frac{1}{4\Im\mathrm{m}\,Z^i},\qquad \mathcal{G}_{ij^*}=\frac{\delta_{(i)j^*}}{4\left(\Im\mathrm{m}\,Z^{(i)}\right)^2}, \tag{7.55}$$

where there is no sum over the index $i$ in the final formula. The Kähler metric is the product of the metrics of three $\mathrm{SL}(2,\mathbb{R})/\mathrm{SO}(2)$ coset spaces, each of them identical to that of the axidilaton model studied earlier. In a sense, it can be said that the model has three axidilatons. The reality of the Kähler potential plus the triality symmetry imply that the imaginary parts of the three axidilatons must be taken to be positive.[13] The period matrix, which can be written using the general form for cubic models Eq. (7.52), replacing $d_{ijk}$ by $|\varepsilon_{ijk}|$, has manifest triality in the indices $i,j$ and triality invariance in the index 0.

There are another two cubic models related to the STU: the $\mathrm{ST}^2$, which is obtained by identifying $T=U$, and therefore with prepotential

$$\mathcal{F}=-\frac{\mathcal{X}^1(\mathcal{X}^2)^2}{\mathcal{X}^0}, \tag{7.56}$$

and the $t^3$ (typically written with lowercase $t$), which is obtained by identifying the three scalars $S=T=U$ and with prepotential

$$\mathcal{F}=-\frac{(\mathcal{X}^1)^3}{\mathcal{X}^0}. \tag{7.57}$$

[13] As in the axidilaton case, this is not an essential feature of the model but a particular choice of conventions. The opposite sign arises if we change the global sign of the prepotential. Both conventions are common in the literature.

The latter is a very interesting *1-modulus*[14] model in which the scalar parametrizes the coset space $\mathrm{SL}(2,\mathbb{R})/\mathrm{SO}(2)$, as in the axidilaton model, but with very different couplings, which leads to very different classical solutions. It is the model that arises in the reduction of pure $N=1, d=5$ supergravity studied in Section 15.2.5, and this makes it worth doing some explicit computations. Defining the complex scalar $t$ by

$$t = -\mathcal{X}^1/\mathcal{X}^0, \tag{7.58}$$

we get, in the $\mathcal{X}^0 = \sqrt{2}/4$ gauge, the Kähler potential and metric and the holomorphic canonical section

$$e^{-\mathcal{K}} = (\Im\mathrm{m}\, t)^3, \qquad \mathcal{G}_{tt^*} = \frac{3}{4(\Im\mathrm{m}\, t)^2}, \qquad \Omega = \begin{pmatrix} 1 \\ -t \\ -t^3 \\ -3t^2 \end{pmatrix}. \tag{7.59}$$

The matrices $\mathcal{F}_{\Lambda\Sigma}$ and $\mathcal{N}_{\Lambda\Sigma}$ are given by

$$\mathcal{F}_{\Lambda\Sigma} = \begin{pmatrix} -2t^3 & 3t^2 \\ 3t^2 & -6t \end{pmatrix}, \qquad \mathcal{N}_{\Lambda\Sigma} = \begin{pmatrix} \dfrac{4|t|^2 - t^2(t+t^*)^2}{2(t-t^*)} & \frac{3}{2}t(t+t^*) \\ \frac{3}{2}t(t+t^*) & -3(t+t^*) - \frac{3}{2}(t-t^*) \end{pmatrix}, \tag{7.60}$$

and, if we call $R$ and $I$ the real and imaginary parts of $t$, respectively, we can write the complete action for the bosonic fields in the form

$$\boxed{\begin{aligned} S = \int d^4x \sqrt{|g|}\, \Big\{ & R + \frac{3\partial_\mu t \partial^\mu t^*}{2(\Im\mathrm{m}\, t)^2} - 2I(I^2 + 3R^3)(F^0)^2 + 12RIF^0F^1 - 6I(F^1)^2 \\ & - 4R^3 F^0 \star F^0 + 12R^2 F^0 \star F^1 - 12RF^1 \star F^1 \Big\}. \end{aligned}} \tag{7.61}$$

*The $ST[2,n]$ models.* This is a family of cubic models with $n_V = n+1$ vector supermultiplets and complex scalars, characterized by the $d$-tensor with non-vanishing components $d_{1\alpha\beta} = \eta_{\alpha\beta}$ with $(\eta_{\alpha\beta}) = \mathrm{diag}(+-\cdots-)$ and where the indices $\alpha, \beta$ take $n$ values between 2 and $n+1$. The first scalar $Z^1 = \mathcal{X}^1/\mathcal{X}^0$ does not enter in the model in the same way as the rest; it is customarily called $S$, and parametrizes, yet again, an $\mathrm{SL}(2,\mathbb{R})/\mathrm{SO}(2)$ coset space. For this reason (and for the role it plays) $S$ is called an axidilaton. The rest

[14] This model arises in certain compactifications of quintic Calabi–Yau manifolds whose deformations are controlled by just one complex *modulus* parameter which arises in the supergravity theory as the scalar $t$ [294, 1161, 292, 672, 673, 293, 295, 296, 297]. (For a review, see the appendix in Ref. [579].)

of the special coordinates $Z^\alpha = \mathcal{X}^\alpha/\mathcal{X}^0$ parametrize an $\mathrm{SO}(2,n)/(\mathrm{SO}(2)\times\mathrm{SO}(n))$ coset space. The Kähler metric and 1-form connection are the products of those two spaces.

Using this notation and setting $\mathcal{X}^0 = 1$:

$$\Omega = \begin{pmatrix} 1 \\ S \\ Z^\alpha \\ \frac{1}{2}S\eta_{\alpha\beta}Z^\alpha Z^\beta \\ -\frac{1}{2}\eta_{\alpha\beta}Z^\alpha Z^\beta \\ -S\eta_{\alpha\beta}Z^\beta \end{pmatrix}, \qquad e^{-\mathcal{K}} = 4\Im\mathrm{m}\, S\; \eta_{\alpha\beta}\Im\mathrm{m}\, Z^\alpha\, \Im\mathrm{m}\, Z^\beta,$$

$$\mathcal{Q}_S = \frac{1}{4\Im\mathrm{m}\, S}, \qquad \mathcal{Q}_\alpha = \frac{\eta_{\alpha\beta}\Im\mathrm{m}\, Z^\beta}{2\eta_{\gamma\delta}\Im\mathrm{m}\, Z^\gamma\, \Im\mathrm{m}\, Z^\delta}, \tag{7.62}$$

$$\mathcal{G}_{SS^*} = \frac{1}{4(\Im\mathrm{m}\, S)^2}, \qquad \mathcal{G}_{\alpha\beta^*} = \frac{\eta_{\alpha\gamma}\Im\mathrm{m}\, Z^\gamma\, \eta_{\beta\delta}\Im\mathrm{m}\, Z^\delta}{[\eta_{\epsilon\varphi}\Im\mathrm{m}\, Z^\epsilon\, \Im\mathrm{m}\, Z^\varphi]^2} - \frac{\eta_{\alpha\beta}}{2\eta_{\epsilon\varphi}\Im\mathrm{m}\, Z^\epsilon\, \Im\mathrm{m}\, Z^\varphi}.$$

As we will see, the group $\mathrm{SO}(2,n)$ does not act linearly on the special coordinates $Z^\alpha$, and it is sometimes more convenient to give a different formulation of those models in which it does. The second formulation is related to this one by a symplectic transformation realized on the canonical section, which can be found, for instance, in Refs. [44, 564], and the transformed holomorphic canonical section takes the simple form [326, 327]

$$\Omega = \begin{pmatrix} \mathcal{Y}^\Lambda \\ S\eta_{\Lambda\Sigma}\mathcal{Y}^\Sigma \end{pmatrix}, \qquad (\eta_{\Lambda\Sigma}) = \mathrm{diag}(++-\cdots-), \tag{7.63}$$

where the coordinates $\mathcal{Y}^\Lambda$ are constrained to belong to the hypersurface

$$\eta_{\Lambda\Sigma}\mathcal{Y}^\Lambda\mathcal{Y}^\Sigma = 0, \tag{7.64}$$

which is evidently invariant under linear $\mathrm{SO}(2,n)$ transformations. This hypersurface can be parametrized, for instance, by the *Calabi–Visentini coordinates* $y^\alpha$ [280], which correspond to the physical scalars

$$\mathcal{Y}^\alpha \equiv y^\alpha, \qquad \mathcal{Y}^0 = \tfrac{1}{2}(1 + y^\alpha y^\alpha), \qquad \mathcal{Y}^1 = \tfrac{i}{2}(1 - y^\alpha y^\alpha), \tag{7.65}$$

which do not transform linearly under the group $\mathrm{SO}(2,n)$.

It can be easily shown that there is no prepotential from which the above holomorphic canonical section can be derived, and this formulation of the model is commonly used to illustrate how the complete $N = 2, d = 4$ supergravity theory can be obtained from the canonical section in absence of a prepotential which seemed to be necessary in earlier formulations of these theories.

*The universal hypermultiplet.* A common feature of all the $N = 2, d = 4$ SUEGRAs obtained in the compactification of type-II SUEGRAs on Calabi–Yau 3-folds is the presence of a hypermultiplet, commonly known as the *universal hypermultiplet*, whose scalars parametrize the non-compact symmetric space $\mathrm{U}(1,2)/(\mathrm{U}(1)\times\mathrm{U}(2))$ or $\overline{\mathbb{CP}}^2$ [317]. We have studied this space before as a particular example of a special Kähler manifold, and here we are going to show that it is a quaternionic-Kähler manifold as well.

The four real coordinates of the universal hypermultiplet are usually written in terms of two complex coordinates $S$ and $C$, and the metric can be conveniently written in the form [1159]

$$ds^2 = 2(uu^* + vv^*) \equiv \mathsf{H}_{uv}dq^u dq^v, \tag{7.66}$$

where $u$ and $v$ are the following 1-forms:

$$\begin{aligned} u &= e^{\phi}dC, \\ v &= e^{2\phi}(\tfrac{1}{2}S - C^* dC), \end{aligned} \tag{7.67}$$

and where $\phi$ is defined as the following real function of $S$ and $C$:

$$e^{-2\phi} = \tfrac{1}{2}(S+S^*) - CC^*. \tag{7.68}$$

This metric is a Kähler metric, $ds^2 = 2\partial_i\partial_{j^*}\mathcal{K}dZ^i dZ^{*\,j^*}$, with Kähler potential

$$e^{\mathcal{K}} = \tfrac{1}{2}e^{2\phi}, \tag{7.69}$$

and can be brought into the standard form in which we have studied the $\mathbb{CP}^n$ special Kähler geometries (now for $n=2$) by changing to the Bergman coordinates $Z^1$, $Z^2$ [851]:[15]

$$S = \frac{1-Z^1}{1+Z^1}, \qquad C = \frac{Z^2}{1+Z^1}. \tag{7.70}$$

In order to determine whether this space fulfills all the conditions to be a quaternionic-Kähler manifold, we first need to find a Quadbein basis $\mathsf{U}^{\alpha I}$. Choosing $\mathbb{C}_{\alpha\beta} \equiv -\varepsilon_{\alpha\beta}$, it is not difficult to see that the above metric Eq. (7.66) can be obtained from the Quadbein

$$\mathsf{U} \equiv (\mathsf{U}_u{}^{\alpha I}\, dq^u) = \begin{pmatrix} u & v \\ v^* & -u^* \end{pmatrix}, \tag{7.71}$$

where, as usual in this book, the index $\alpha$ labels rows and the index $I$ labels columns. In terms of the matrix and $\varepsilon = i\sigma^2$ we can write, for instance,

$$ds^2 = \mathrm{Tr}(\varepsilon\mathsf{U}\varepsilon\mathsf{U}^T) = \mathrm{Tr}(\mathsf{U}^\dagger\mathsf{U}), \tag{7.72}$$

and the Vielbein postulate Eq. (G.25), in differential-form language, becomes

$$\mathsf{DU} = d\mathsf{U} + \Delta\wedge\mathsf{U} - \mathsf{U}\wedge\mathsf{A}^T, \qquad \Delta \equiv (\Delta_u{}^{\alpha}{}_{\beta}\, dq^u), \qquad \mathsf{A} \equiv (\mathsf{A}_u{}^{I}{}_{J}\, dq^u). \tag{7.73}$$

Using the relations

$$du = -\tfrac{1}{2}(v+v^*)\wedge u, \qquad dv = u\wedge u^* + v\wedge v^*, \tag{7.74}$$

[15] It is much easier to perform the change of coordinates in the Kähler potential, but then, in order to put it in the simple form $e^{-\mathcal{K}} = 1 - |Z^1|^2 - |Z^2|^2$), we need a Kähler transformation with $f = -\log(1+Z^1)/\sqrt{2}$.

the $\mathrm{SU}(2)$ and $\mathrm{Sp}(2)$ connections, and their curvatures can be readily obtained:

$$\mathsf{A}=\begin{pmatrix} \frac{1}{4}(v-v^*) & -u \\ u^* & -\frac{1}{4}(v-v^*) \end{pmatrix}, \qquad \Delta=-\tfrac{3}{4}(v-v^*)\sigma^3,$$

$$\mathsf{F}=\begin{pmatrix} -\frac{1}{2}(u\wedge u^*-v\wedge v^*) & v^*\wedge u \\ -v\wedge u^* & \frac{1}{2}(u\wedge u^*-v\wedge v^*) \end{pmatrix}, \quad (R^\alpha{}_\beta)=-\tfrac{3}{2}(u\wedge u^*+v\wedge v^*)\sigma^3. \tag{7.75}$$

From the above form of $\mathsf{F}$ we can obtain the three $\mathfrak{su}(2)$ components $\mathsf{F}^x=i\mathrm{Tr}(\mathsf{F}\sigma^x)$. The result is

$$\mathsf{F}^2+i\mathsf{F}^1=-2v^*\wedge u, \qquad \mathsf{F}^3=i(v\wedge v^*-u\wedge u^*), \tag{7.76}$$

which can be immediately compared with the hyper-Kähler structure 2-forms, which can be constructed from the Quadbein as

$$\mathsf{K}^x=-\tfrac{i}{2}\varepsilon_{IK}\sigma^{x\,K}{}_J\mathsf{U}^{\alpha I}\wedge\mathsf{U}^{\beta J}, \tag{7.77}$$

to obtain the expected result, $\mathsf{F}^x=-\mathsf{K}^x$.

This hypermultiplet (or any other number of hypermultiplets) can be added to any of the models of ungauged $N=2, d=4$ SUEGRA with vector supermultiplets that we have described with minimal changes (the addition of the $\mathrm{SU}(2)$ connection in the spinor covariant derivatives).

## 7.3 The gauged theory

As in the $N=1, d=4$ case, we require a thorough knowledge of the structure of the global symmetries of generic $N=2, d=4$ theories in order to study the possible generic gaugings one can make. We are actually going to consider the global symmetries of the equations of motion studied in Section 2.6.2, which may or may not be symmetries of the action that is sometimes called the *duality group* because it contains as a subgroup the electric–magnetic duality rotations of the vector fields. The formalism is designed to deal with the full duality group and not just with the symmetries of the action, but we will have to restrict ourselves to the latter to gauge them because, otherwise, we would have to use the embedding tensor formalism (Section 2.7) to do it.

### *7.3.1 The global symmetries*

As in the $N=1$ case, the global symmetry group G of any $N=2, d=4$ SUEGRA is the direct product of the R-symmetry group of transformations that act only on the fermions, which in this case is $\mathrm{U}(2)_R$, and $\mathrm{G_{bos}}$, the part of G that acts on the bosonic fields. The R-symmetry group naturally splits into the product $\mathrm{SU}(2)_\mathrm{R}\times\mathrm{U}(1)_\mathrm{R}$, and, therefore, we have

$$\mathrm{G}=\mathrm{G_{bos}}\times\mathrm{SU}(2)_\mathrm{R}\times\mathrm{U}(1)_\mathrm{R}. \tag{7.78}$$

In the $N=2$ case, the vector fields are in supermultiplets that also contain scalars, and, therefore, generically[16], there is no subgroup of $\mathrm{G}_{\mathrm{bos}}$ acting exclusively on the vector fields and not on the scalars. All the symmetries in $\mathrm{G}_{\mathrm{bos}}$ act on scalars but, of course, there is a natural separation between the symmetries that act on the complex scalar fields (henceforth, on the complete vector supermultiplets), which we will denote by $\mathrm{G}_{\mathrm{V}}$, and those that act on the hyperscalars, which we will denote by $\mathrm{G}_{\mathrm{hyper}}$. The latter do not act on the vector fields for the same reason. Thus, we have

$$\mathrm{G}=\mathrm{G}_{\mathrm{V}}\times\mathrm{G}_{\mathrm{hyper}}\times\mathrm{SU}(2)_{\mathrm{R}}\times\mathrm{U}(1)_{\mathrm{R}}. \tag{7.79}$$

In both cases (with the aforementioned exceptions for $\mathrm{G}_{\mathrm{V}}$) the symmetries are necessarily isometries of the scalar manifolds.

Let us review each of these factors.

$\mathrm{U}(1)_{\mathrm{R}}$. Since this group acts just as in the $N=1$ case, we shall be brief. This group of transformations parametrized by $\theta$ multiplies the spinors by a phase constant $e^{-iq\theta}$, where $q$ coincides with the Kähler weight of the field. Thus, it can be seen as the group of constant Kähler transformations $\delta_\lambda\mathcal{K}(Z,Z^*)=\lambda(Z)+\lambda^*(Z^*)$ with $\theta=\Im\mathfrak{m}\lambda$. Kähler transformations with non-constant parameters $\lambda(Z)$ act as field-dependent $\mathrm{U}(1)_{\mathrm{R}}$ transformations on the fermions as well.

$\mathrm{SU}(2)_{\mathrm{R}}$. This group acts in the fundamental representation on the indices $I,J$ carried by all the fermion fields, but it also acts on the Quadbeins $\mathsf{U}^{I\alpha}$ and (via the adjoint action) on the $\mathrm{SU}(2)$ connection $\mathsf{A}^x{}_u$ of the quaternionic-Kähler manifold parametrized by the hyperscalars $q^u$. This action does not imply any transformation on the bosonic fields (it is just a change of basis). However, as with $\mathrm{U}(1)_{\mathrm{R}}$, the $q$-dependent $\mathrm{SU}(2)$ transformations that one can perform in the tangent bundle of that manifold also act on the fermions.

$\mathrm{G}_{\mathrm{V}}$. The discussion of this group is similar to that on p. 209 for the $N=1$ case and, therefore, the reader should find there the conditions that the transformations Eq. (6.46) must satisfy in order to preserve not just the metric but the whole Kähler–Hodge structure which underlies special Kähler geometry (see also Appendix J).[17] But, as explained in detail in Appendix F.3, we need the whole special Kähler structure to be preserved and this requires the preservation of the covariantly holomorphic symplectic section $\mathcal{V}$. This is equivalent to saying that the covariant Lie derivative of $\mathcal{V}$ must vanish in Eq. (F.59), which we reproduce here for convenience:

$$\mathbb{L}_\Lambda\mathcal{V}=\left\{K_\Lambda-\left[\mathcal{T}_\Lambda-\tfrac{1}{2}(\lambda_\Lambda-\lambda^*_\Lambda)\right]\right\}\mathcal{V}=0, \tag{7.80}$$

where $\mathcal{T}_\Lambda$ are the infinitesimal $\mathfrak{sp}(2\bar{n})$ matrices in Eq. (6.53), which satisfy Eq. (6.58) if the Killing vectors $K_\Lambda$ satisfy Eq. (6.56), and the $\lambda_\Lambda(Z)$ are the field-dependent parameters of the Kähler transformations associated with the isometries defined in Eq. (6.49). In other

[16] There are some exceptions in which a symmetry has a non-trivial action on the canonical symplectic section (and, therefore, on the vector fields) but leaves invariant the physical scalars.

[17] There is no superpotential in the $N=2$ case, though.

words, when the scalars undergo the infinitesimal transformation Eq. (6.46), the symplectic section must simultaneously undergo an infinitesimal symplectic transformation and a Kähler transformation.

This condition incorporates the general results in Section 2.6, according to which, if there are $n_V$ vector supermultiplets and, hence, $\bar{n} = n_V + 1$ vector fields, the symmetries that act on the vectors must be a subgroup of $\mathrm{Sp}(2\bar{n}, \mathbb{R})$. This fact is built into the special Kähler structure. Since all the couplings and, in particular, the period matrix $\mathcal{N}_{\Lambda\Sigma}$ can be derived from the canonical section $\mathcal{V}$, it follows that if the above condition is satisfied then all the couplings are preserved by these transformations and, in particular, that the period matrix satisfies

$$\mathbb{L}_\Lambda \mathcal{N}_{\Sigma\Omega} = \{K_\Lambda - \mathcal{T}_\Lambda\}\mathcal{N}_{\Sigma\Omega} = 0. \tag{7.81}$$

This means that, when the scalars undergo the infinitesimal transformation Eq. (6.46), the period matrix undergoes an infinitesimal symplectic transformation $\mathcal{T}_\Lambda(\mathcal{N}_{\Sigma\Omega})$ as in Eq. (2.156), which is what is needed for the equations of motion of the full theory to remain invariant.

The symplectic matrices $\mathcal{T}_\Lambda = (\mathcal{T}_\Lambda{}^M{}_N)$ are the same for all the objects with symplectic indices in the theory (which includes gauginos, for instance) and can be easily found by transforming the scalars in the canonical section $\mathcal{V}^M$, as we will see in several examples.

An important consequence of the preservation of the canonical symplectic section $\mathcal{V}$ is that in $N = 2, d = 4$ theories the holomorphic momentum maps $\mathcal{P}_\Lambda$ and Killing vectors $k_\Lambda{}^i$ can be expressed in terms of $\mathcal{V}$ and the matrices $\mathcal{T}_\Lambda$ using Eqs. (F.62) and (F.63), which we rewrite here for convenience:

$$\mathcal{P}_A = \langle\, \mathcal{V}^* \mid \mathcal{T}_A \mathcal{V}\,\rangle, \tag{7.82}$$

$$k_A{}^i = i\langle\, \mathcal{V} \mid \mathcal{T}_A \mathcal{U}^{*i}\,\rangle. \tag{7.83}$$

$G_{\text{hyper}}$. The symmetries in the hypermultiplet sector of the theory must be isometries of the metric $\mathsf{H}_{uv}$ that respect the whole quaternionic-Kähler structure and, in particular, the hyper-Kähler structure $\mathsf{K}^x{}_{uv}$. They are called *triholomorphic isometries* and are studied in Appendix G.1.

The situation is analogous to that of the isometries that preserve other structures: in general we must allow them to be invariant up to some induced or compensating symmetry of that structure (Kähler and symplectic transformations in the case of the special Kähler structure, for instance). The invariance condition must be expressed in terms of a conveniently defined Lie derivative which transforms covariantly under that symmetry. Typically these induced symmetry transformations act on other fields, and this will have to be taken into account in the gauging of the original symmetry.

In the case at hand, the hyper-Kähler structure $\mathsf{K}^x{}_{uv}$ is required to be invariant under the transformations

$$\delta_\alpha q^u = \alpha^\Lambda \mathsf{k}_\Lambda{}^u(q), \tag{7.84}$$

where the $\mathsf{k}_\Lambda{}^u$ are the isometries of $\mathsf{H}_{uv}$ only up to $q$-dependent $\mathrm{SU}(2)$ transformations characterized by the parameters $\lambda_\Lambda{}^x$, that is

$$\mathbb{L}_\Lambda \mathsf{K}^x{}_{uv} = 0, \tag{7.85}$$

where $\mathbb{L}_\Lambda$ is the $\mathrm{SU}(2)$-covariant derivative with respect to the Killing vector $\mathsf{k}_\Lambda$, and it is given in Eq. (G.43). The induced $\mathrm{SU}(2)$ transformations with parameters $\lambda_\Lambda{}^x$ act on all the fields with adjoint $x, y$ or fundamental $I, J$ $\mathrm{SU}(2)$ indices; as we have stressed before, we will have to take this into account when gauging the isometry.

The $\mathrm{SU}(2)$ parameters $\lambda_\Lambda{}^x$ are given in terms of the contraction of the Killing vectors $\mathsf{k}_\Lambda{}^u$ with the $\mathrm{SU}(2)$ connection $\mathsf{A}^x{}_u$ and the so-called *triholomorphic momentum maps* $\mathsf{P}_\Lambda{}^x$ by Eq. (G.41). These satisfy a number of properties listed in Appendix G.1 that can be used to compute and work with them.

The triholomorphic momentum maps are important ingredients in the gauging of the symmetries $G_{\text{hyper}}$, just as the holomorphic momentum maps are in the $G_{\text{V}}$ case (or in $N = 1, d = 4$ theories) and they perform identical geometrical roles, which is the reason why they have the same name. This can be made manifest by comparing the equations that define them: their relation to the relevant connection and compensating gauge transformation parameter given by Eqs. (E.69) and (G.41), and the relation between their derivatives and the contraction of the Killing vectors with the complex $\mathcal{J}$ and hyper-Kähler $\mathsf{K}^x$ structures Eqs. (E.70)[18] and (G.47):

$$\begin{cases} \mathcal{P}_A = i_{k_A}\mathcal{Q} - \frac{1}{2i}(\lambda_A - \lambda_A^*), \\ \\ \mathsf{P}_A{}^x = i_{\mathsf{k}_A}\mathsf{A}^x - \lambda_A{}^x, \end{cases} \qquad \begin{cases} \partial_{i^*}\mathcal{P}_A = k_A{}^j \mathcal{J}_{ji^*}, \\ \\ \mathsf{D}_u \mathsf{P}_A{}^x = -\varkappa\, \mathsf{k}_A{}^v \mathsf{K}^x{}_{vu}. \end{cases} \tag{7.86}$$

In both cases the Killing vectors can be obtained by taking derivatives of the momentum maps using Eqs. (E.70) and (G.48):

$$k_{A\,i^*} = i\partial_{i^*}\mathcal{P}_A, \qquad \mathsf{k}_A{}^u = -\frac{1}{3\varkappa}\,\mathsf{K}^{x\,uv}\mathsf{D}_v\mathsf{P}_A{}^x, \tag{7.87}$$

and for this reason they are called (*holomorphic* or *triholomorphic*) *Killing prepotentials.*

Observe that both the complex $\mathcal{J}$ and hyper-Kähler $\mathsf{K}^x$ structures are the field strengths of connections $\mathcal{Q}$ and $\mathsf{A}^x$ associated with the factors of the R-symmetry group $\mathrm{U}(1)_{\mathrm{R}} \times \mathrm{SU}(2)_{\mathrm{R}}$, and they can be understood as part of the compensating gauge transformations of these groups associated with the isometries.

### 7.3.2 Examples

*The global symmetries of the axidilaton model.* In Section 6.3.2 we studied the global symmetries of the $N = 1, d = 4$ axidilaton model, which, in that context, is equivalent to the isometries of the symmetric Kähler–Hodge manifold $\mathrm{SL}(2,\mathbb{R})/\mathrm{SO}(2)$. The $N = 2$ axidilaton model shares the same Kähler–Hodge geometry, and, therefore, the axidilaton transforms as in Eq. (6.65) and we have the same Killing vectors $K_m$ of Eq. (6.59), the

[18] We have used the relation between the components of the Kähler metric and those of the complex structure $\mathcal{G}_{ij^*} = i\mathcal{J}_{ij^*}$.

same parameters $\lambda_m$ of Eq. (6.61), and the same holomorphic momentum maps $\mathcal{P}_m$ of Eq. (6.102). Now, however, we have to study the preservation of the full special Kähler structure by these isometries, which translates into Eq. (7.80) or its finite version

$$\mathcal{V}'^M(\tau) \equiv \mathcal{V}^M(\tau'(\tau)) = e^{-i\Im\mathrm{m} f(\tau)}\, S^M{}_N \mathcal{V}^N, \tag{7.88}$$

where $S^M{}_N$ is the symplectic representation of the symmetry group and we have defined

$$\mathcal{K}'(\tau) \equiv \mathcal{K}(\tau'(\tau)) = \mathcal{K}(\tau) + f(\tau) + f^*(\tau^*). \tag{7.89}$$

Using the transformations of the axidilaton Eq. (6.65) in the Kähler potential $\mathcal{K} = -\log \Im\mathrm{m}\tau$ and in the symplectic section in Eq. (7.33) it is straightforward to find that Eq. (7.88) is satisfied, with

$$f(\tau) = \ln(\gamma\tau + \delta), \tag{7.90}$$

$$(S^M{}_N) = \begin{pmatrix} \delta & & -\gamma & \\ & \alpha & & \beta \\ -\beta & & \alpha & \\ & -\gamma & & \delta \end{pmatrix}. \tag{7.91}$$

In this four-dimensional representation the infinitesimal symplectic generators $\mathcal{T}_m$ defined in Eq. (6.53) are given by

$$(\mathcal{T}_1{}^M{}_N) = \tfrac{1}{2}\begin{pmatrix} \sigma^3 & \\ & -\sigma^3 \end{pmatrix}, \quad (\mathcal{T}_2{}^M{}_N) = -\tfrac{1}{2}\begin{pmatrix} & \sigma^3 \\ \sigma^3 & \end{pmatrix}, \quad (\mathcal{T}_3{}^M{}_N) = -\tfrac{1}{2}\begin{pmatrix} & \mathbb{1} \\ -\mathbb{1} & \end{pmatrix}. \tag{7.92}$$

In this representation $\mathrm{SL}(2,\mathbb{R})$ does not act irreducibly on symplectic vectors, which are the tensor product of two-dimensional vectors transforming contravariantly and covariantly in the fundamental: the action of the symplectic transformation $S^M{}_N$ on a vector $(\mathcal{Q}^M) = \begin{pmatrix} p^\Lambda \\ q_\Lambda \end{pmatrix}$ is equivalent to

$$(p^0, q_0)' = (p^0, q_0)\Lambda^{-1}, \qquad \begin{pmatrix} p^1 \\ q_1 \end{pmatrix}' = \Lambda \begin{pmatrix} p^1 \\ q_1 \end{pmatrix}, \tag{7.93}$$

where $\Lambda$ is the $\mathrm{SL}(2,\mathbb{R})$ matrix:

$$\Lambda \equiv \begin{pmatrix} \alpha & \beta \\ \gamma & \delta \end{pmatrix}, \qquad \Lambda^{-1} = \begin{pmatrix} \delta & -\beta \\ -\gamma & \alpha \end{pmatrix}. \tag{7.94}$$

We can denote these doublets with upper and lower indices $a, b = 1, 2$ transforming according to $P'^a = \Lambda^a{}_b P^b$ and $Q_a = Q_b(\Lambda^{-1})^b{}_a$ and relabel the components of the symplectic vector as components of these two doublets in an obvious way. Since the duality group of the theory is not the full $\mathrm{Sp}(4,\mathbb{R})$ but only $\mathrm{SL}(2,\mathbb{R}) \subset \mathrm{Sp}(4,\mathbb{R})$, we can use this new notation to construct duality invariants. The only one available in this simple theory is

$$P^a Q_a = p^0 p^1 + q_0 q_1. \tag{7.95}$$

There seem to be no further symmetries. There is, however, an additional $\mathrm{U}(1)$ factor in the symmetry group that only has a non-trivial action on objects with symplectic indices and non-trivial Kähler weight [535] and leaves $\tau$ and, actually, the whole canonical symplectic section invariant:

$$\mathcal{V}' \equiv e^{i\theta} \begin{pmatrix} \cos\theta & 0 & 0 & \sin\theta \\ 0 & \cos\theta & \sin\theta & 0 \\ 0 & -\sin\theta & \cos\theta & 0 \\ -\sin\theta & 0 & 0 & \cos\theta \end{pmatrix} \mathcal{V} = \mathcal{V}. \tag{7.96}$$

There is an extra symmetry of this kind in all the $\overline{\mathbb{CP}}^n$ models (this is the $n=1$ case), which is why we have written the coset spaces as $\mathrm{U}(1,n)/(\mathrm{U}(1)\times\mathrm{U}(n))$ instead of $\mathrm{SU}(1,n)/\mathrm{U}(n)$, which is geometrically equivalent but does not make explicit all the global symmetry of the model. This symmetry coincides with the continuous global Freudenthal duality transformation [530].

*The global symmetries of the $\overline{\mathbb{CP}}^n$ models.* As mentioned before, the $n$ complex scalars of the $\overline{\mathbb{CP}}^n$ model parametrize the symmetric coset space $\mathrm{U}(1,n)/\mathrm{U}(1)\times\mathrm{SU}(n)$, and the full theory is invariant under global $\mathrm{U}(1,n)=\mathrm{U}(1)\times\mathrm{SU}(1,n)$ transformations. Let us consider the $\mathrm{SU}(1,n)$ subgroup first.

$\mathrm{SU}(1,n)$ acts linearly, in the fundamental representation, over the coordinates $\mathcal{X}^\Lambda$,

$$\mathcal{X}'^\Lambda = \Lambda^\Lambda{}_\Sigma \mathcal{X}^\Sigma, \tag{7.97}$$

and, by definition of $\mathrm{SU}(1,n)$, the matrices $\Lambda = (\Lambda^\Lambda{}_\Sigma)$ preserve the Hermitian metric $\mathcal{X}^{*\Lambda}\eta_{\Lambda\Sigma}\mathcal{X}^\Sigma$ and have unit determinant, i.e.

$$\Lambda^{*\Gamma}{}_\Lambda\, \eta_{\Gamma\Delta}\, \Lambda^\Delta{}_\Sigma = \eta_{\Lambda\Sigma} \quad (\text{or}\ \ \Lambda^\dagger\eta\Lambda = \eta), \qquad \det\Lambda = 1. \tag{7.98}$$

Since the Kähler potential is given by minus the logarithm of this Hermitian metric before fixing the gauge $\mathcal{X}^0 = 1$, it is evident that this group will leave it invariant up to the compensating Kähler transformations necessary to restore that gauge choice. Indeed, the action induced on the special coordinates is

$$Z'^\Lambda = \frac{\Lambda^\Lambda{}_\Sigma Z^\Sigma}{\Lambda^0{}_\Sigma Z^\Sigma}, \qquad Z'_\Lambda = \frac{\Lambda_\Lambda{}^\Sigma Z_\Sigma}{\Lambda^0{}_\Sigma Z^\Sigma}, \tag{7.99}$$

where we have defined the components of $(\Lambda^\dagger)^{-1}$ as

$$[(\Lambda^\dagger)^{-1}]_\Lambda{}^\Sigma \equiv \Lambda_\Lambda{}^\Sigma \equiv \eta_{\Lambda\Gamma}\Lambda^\Gamma{}_\Omega \eta^{\Omega\Sigma}, \tag{7.100}$$

and the Kähler potential is invariant under these transformations up to Kähler transformations $\mathcal{K}' = \mathcal{K} + f + f^*$, with

$$f(Z) = \log\left(\Lambda^0{}_\Sigma Z^\Sigma\right). \tag{7.101}$$

The Kähler metric is, then, automatically invariant under this group.

The symplectic section $\mathcal{V}^N$ is also invariant up to a Kähler transformation with the above parameter and a symplectic transformation given by

$$(S^M{}_N) = \begin{pmatrix} \Re\mathfrak{e}\,\Lambda^\Lambda{}_\Sigma & -2\Im\mathfrak{m}\,\Lambda^{\Lambda\Sigma} \\ \\ \frac{1}{2}\Im\mathfrak{m}\,\Lambda_{\Lambda\Sigma} & \Re\mathfrak{e}\,\Lambda_\Lambda{}^\Sigma \end{pmatrix}, \tag{7.102}$$

where we have used the metric $\eta_{\Lambda\Sigma}$ and its inverse to raise and lower the indices of $\Lambda^\Lambda{}_\Sigma$. This matrix gives the embedding of the group $\mathrm{SU}(1,n)$ in the symplectic group $\mathrm{Sp}(2n+2,\mathbb{R})$[19] and expresses the action of that duality group on all objects with symplectic indices.

We can define the $n(n+2)$ infinitesimal generators $T_m$ of $\mathfrak{su}(1,n)$ in the fundamental representation as

$$\Lambda^\Lambda{}_\Sigma \sim \delta^\Lambda{}_\Sigma + \alpha^m\, T_m{}^\Lambda{}_\Sigma, \tag{7.104}$$

and, by definition, they must be such that $T_{m\,\Lambda\Sigma} \equiv \eta_{\Lambda\Gamma}\, T_m{}^\Gamma{}_\Sigma$ is anti-Hermitian; $\det\Lambda = 1$ requires tracelessness of these matrices as well. In terms of these infinitesimal generators and using the explicit expression of the symplectic $\mathrm{SU}(1,n)$ matrix Eq. (7.102), we can construct the infinitesimal generators in this representation (the $\mathcal{T}$ matrices in Eq. (6.53)):

$$(\mathcal{T}_m{}^M{}_N) = \begin{pmatrix} \Re\mathfrak{e}T_m{}^\Lambda{}_\Sigma & -2\Im\mathfrak{m}T_m{}^{\Lambda\Sigma} \\ \\ \frac{1}{2}\Im\mathfrak{m}T_{m\,\Lambda\Sigma} & \Re\mathfrak{e}T_{m\,\Lambda}{}^\Sigma \end{pmatrix}. \tag{7.105}$$

The anti-Hermiticity of $T_{m\,\Lambda\Sigma}$ is enough to show that these matrices belong to $\mathfrak{sp}(1,n)$.

Using Eq. (7.104) in the transformations of the special coordinates Eq. (7.99), we find that we can put them in the form

$$Z'^\Lambda = Z^\Lambda + \alpha^m k_m{}^\Lambda(Z), \qquad k_m{}^\Lambda(Z) = T_m{}^\Lambda{}_\Sigma\, Z^\Sigma - T_m{}^0{}_\Omega\, Z^\Omega Z^\Lambda, \tag{7.106}$$

and we can immediately identify these $k_m{}^\Lambda(Z)$ with the holomorphic part of the Killing vectors $K_m$.[20]

If the generators $T_m$ have the commutation relations $[T_m, T_n] = f_{mn}{}^p\, T_p$, then it is easy to see from the above relations Eqs. (7.106) that

$$[\mathcal{T}_m, \mathcal{T}_n] = f_{mn}{}^p\, \mathcal{T}_p, \qquad [K_m, K_n] = -f_{mn}{}^p\, K_p. \tag{7.107}$$

Let us now consider the $\mathrm{U}(1)$ subgroup. The canonical symplectic section and the special coordinates are left invariant by the combined constant Kähler and symplectic transformations:

$$\mathcal{V}' = e^{i\theta} \begin{pmatrix} \cos\theta\,\delta^\Lambda{}_\Sigma & 2\sin\theta\,\eta^{\Lambda\Sigma} \\ \\ -\frac{1}{2}\sin\theta\,\eta_{\Lambda\Sigma} & \cos\theta\,\delta_\Lambda{}^\Sigma \end{pmatrix} \mathcal{V} = \mathcal{V}. \tag{7.108}$$

---

[19] The group condition $\Lambda^\dagger\eta\Lambda = \eta$ implies for the real and imaginary parts of $\Lambda$

$$\Re\mathfrak{e}\Lambda_{\Delta\Lambda}\,\Im\mathfrak{m}\Lambda^\Delta{}_\Sigma = \Im\mathfrak{m}\Lambda_{\Delta\Lambda}\,\Re\mathfrak{e}\Lambda^\Delta{}_\Sigma, \qquad \Re\mathfrak{e}\Lambda_{\Delta\Lambda}\,\Re\mathfrak{e}\Lambda^\Delta{}_\Sigma + \Im\mathfrak{m}\Lambda_{\Delta\Lambda}\,\Im\mathfrak{m}\Lambda^\Delta{}_\Sigma = \eta_{\Lambda\Sigma}, \tag{7.103}$$

which means that the matrix $(S^M{}_N)$ above satisfies $S^T\Omega S = \Omega$.

[20] The $k_m{}^0(Z)$ component vanishes identically, as it must, but it is convenient to keep it.

There is no associated Killing vector because the scalars are inert under this symmetry.

Finally, we need to compute, for later use, the holomorphic functions $\lambda_m(Z)$ defined by $\mathcal{L}_{K_m}\mathcal{K} = \lambda_m + \lambda_m^*$ and the momentum maps defined by $i\mathcal{P}_m = k_m{}^i\partial_i\mathcal{K} - \lambda_m$. An easy direct calculation yields

$$\lambda_m = T_m{}^0{}_\Sigma Z^\Sigma, \qquad \mathcal{P}_m = ie^{\mathcal{K}} T_m{}^\Lambda{}_\Sigma Z^\Sigma Z^*_\Lambda = ie^{\mathcal{K}}\eta_{\Lambda\Omega} T_m{}^\Lambda{}_\Sigma Z^\Sigma Z^{*\,\Omega}. \tag{7.109}$$

Observe that the reality of the momentum maps is guaranteed by the anti-Hermiticity of the matrices $\eta_{\Lambda\Omega}T_m{}^\Omega{}_\Sigma$. On the other hand, the holomorphic momentum map and Killing vectors have exactly the form of Eqs. (7.82) and (7.83), which we could have used directly from the beginning.

*The global symmetries of the STU model.* The Kähler potential and metric of the $STU$ model are those of the coset space $[\mathrm{SL}(2,\mathbb{R})/\mathrm{SO}(2)]^3$, and, therefore, we expect the whole special Kähler structure to be invariant under global $[\mathrm{SL}(2,\mathbb{R})]^3$ transformations. These transformations act independently on each of the scalars $Z^i$, $i = 1, 2, 3$, according to

$$Z'^i = \frac{\alpha_i Z^i + \beta_i}{\gamma_i Z^i + \delta_i}, \qquad \Lambda_i \equiv \begin{pmatrix} \alpha_i & \beta_i \\ \gamma_i & \delta_i \end{pmatrix} \in \mathrm{SL}_i(2,\mathbb{R}). \tag{7.110}$$

Given the definition of the $Z^i$s as special coordinates, it follows that under $\mathrm{SL}_i(2,\mathbb{R})$ the components of the symplectic section $\Omega$ transform, up to Kähler transformations, as the following four doublets:

$$\begin{pmatrix} \mathcal{X}^i \\ \mathcal{X}^0 \end{pmatrix}' = \Lambda_i \begin{pmatrix} \mathcal{X}^i \\ \mathcal{X}^0 \end{pmatrix}, \quad (\mathcal{F}_i, \mathcal{F}_0)' = (\mathcal{F}_i, \mathcal{F}_0)\Lambda_i^{-1}, \quad (\mathcal{X}^i, |\varepsilon_{ijk}|\mathcal{F}_k)' = (\mathcal{X}^i, |\varepsilon_{ijk}|\mathcal{F}_k)\Lambda_j^{-1}, \tag{7.111}$$

where, in the last case, $i \neq j$. Then, using the notation introduced below Eq. (7.94) for $\mathrm{SL}(2,\mathbb{R})$ doublets, we find that we can label the components of $\Omega$ using the three indices associated with the three $\mathrm{SL}(2,\mathbb{R})$ duality groups as follows:

$$\begin{aligned} &\mathcal{X}^0 = \Omega^{222}, \quad \mathcal{X}^1 = \Omega^1{}_{11}, \quad \mathcal{X}^2 = \Omega_1{}^1{}_1, \quad \mathcal{X}^3 = \Omega_{11}{}^1, \\ &\mathcal{F}_0 = \Omega_{222}, \quad \mathcal{F}_1 = \Omega_{122}, \quad \mathcal{F}_2 = \Omega_{212}, \quad \mathcal{F}_3 = \Omega_{221}. \end{aligned} \tag{7.112}$$

The two-dimensional symplectic metric can be used to raise and lower indices of $\mathrm{SL}(2,\mathbb{R})$ doublets because $\mathrm{SL}(2,\mathbb{R}) \sim \mathrm{Sp}(2)$ so that $P^1 = -P_2$ and $P^2 = P_1$; then, if we start from a symplectic vector such as $(\mathcal{Q}^M) = \begin{pmatrix} p^\Lambda \\ q_\Lambda \end{pmatrix}$, we can rewrite it as $Q^{a_1a_2a_3}$ with

$$\begin{aligned} &Q^{111} = -q_0, \quad Q^{211} = q_1, \quad Q^{121} = q_2, \quad Q^{112} = q_3, \\ &Q^{222} = p^0, \quad Q^{122} = p^1, \quad Q^{212} = p^2, \quad Q^{221} = p^3. \end{aligned} \tag{7.113}$$

All the symplectic vectors of this model can be rewritten as a three-index object, which simplifies the construction of the duality invariants.

The $8\times 8$ symplectic matrices that embed the duality group can be easily constructed using the above doublets. For instance, for the first $\mathrm{SL}(2,\mathbb{R})$ factor that acts on $S = Z^1$ the components of any symplectic vector such as $(\mathcal{Q}^M) = \binom{p^\Lambda}{q_\Lambda}$ arrange themselves into four doublets:

$$(Q^{a11}) = \begin{pmatrix} -q_1 \\ q_0 \end{pmatrix}, \ (Q^{a12}) = \begin{pmatrix} q^3 \\ p^2 \end{pmatrix}, \ (Q^{a21}) = \begin{pmatrix} q_2 \\ p^3 \end{pmatrix}, \ (Q^{a22}) = \begin{pmatrix} p^1 \\ p^0 \end{pmatrix}, \tag{7.114}$$

and the corresponding symplectic matrix is given by

$$(S_1^M{}_N) = \left(\begin{array}{cc|cc|cc|cc} \delta_1 & \gamma_1 & & & & & & \\ \beta_1 & \alpha_1 & & & & & & \\ \hline & & \delta_1 & & & & & -\gamma_1 \\ & & & \delta_1 & & & -\gamma_1 & \\ \hline & & & & \alpha_1 & -\beta_1 & & \\ & & & & -\gamma_1 & \delta_1 & & \\ \hline & & & -\beta_1 & & & \alpha_1 & \\ & & -\beta_1 & & & & & \alpha_1 \end{array}\right). \tag{7.115}$$

It can be checked that under these transformations the Kähler potential and the canonical symplectic section transform according to

$$\mathcal{K}' = \mathcal{K} + f(Z) + f^*(Z^*), \tag{7.116}$$

$$\mathcal{V}'^M = e^{-i\Im\mathrm{m} f} S^M{}_N \mathcal{V}^N, \tag{7.117}$$

where the symplectic matrices take the form explained above and the Kähler transformations are just the sum of those of the individual $\mathrm{SL}_i(2,\mathbb{R})$ factors,

$$f(Z) = \sum_{i=1,2,3} \log\left(\gamma_i Z^i + \delta_i\right). \tag{7.118}$$

The Killing vectors and related objects are the obvious generalization of those of the axidilaton model.

These results cannot be generalized to arbitrary cubic models. However, they can be used to derive those of the $ST^2$ and $t^3$ models.

*The global symmetries of the universal hypermultiplet.* Since this quaternionic space is the same (at least locally) as the special Kähler space $\overline{\mathbb{CP}}^n$, whose symmetries we have studied before, we can use some of the results obtained there, such as the expressions for the Killing vectors in Bergman coordinates. Using these expressions demands that we rewrite the $\mathrm{SU}(2)$ 1-form connection $\mathsf{A}$ and its curvature $\mathsf{F}$ (equal to minus the hyper-Kähler structure $\mathsf{K}$) in these coordinates. It is enough to change coordinates in the 1-forms $u$ and $v$

that constitute the Quadbein:

$$u = e^{\mathcal{K}/2}\left(\frac{1+Z^{*1}}{1+Z^1}\right)^{1/2}\left[dZ^2 - \frac{Z^2 dZ^1}{1+Z^1}\right],$$
$$v = -\frac{dZ^1}{1+Z^1} + e^{\mathcal{K}} Z^*_i dZ^i. \tag{7.119}$$

The next step consists in rewriting the differential equation (G.47) in a convenient matrix form, such as

$$d\mathsf{P} + \mathsf{AP} - \mathsf{PA} = (\mathsf{F}_{st} K^s dq^t), \tag{7.120}$$

where $\mathsf{P} \equiv (\mathsf{P}^I{}_J)$. The r.h.s. of this equation is given by the matrix

$$(\mathsf{F}_{st} K^s dq^t) = \begin{pmatrix} -\frac{1}{2}[(u^* K)u - (v^* K)v - \text{c.c}] & (uK)v^* - (v^* K)u \\ -(u^* K)v + (vK)u^* & \frac{1}{2}[(u^* K)u - (v^* K)v - \text{c.c.}] \end{pmatrix}, \tag{7.121}$$

from which we can read the components in a $u, v$ basis. We obtain the following six independent equations:

$$\begin{aligned} &\partial_u \mathsf{P}^1{}_1 + (\mathsf{P}^1{}_2)^* = -\tfrac{1}{2}(u^* K), && \partial_v \mathsf{P}^1{}_1 = \tfrac{1}{2}(v^* K), \\ &\partial_u \mathsf{P}^1{}_2 + 2\mathsf{P}^1{}_1 = -(v^* K), && \partial_v \mathsf{P}^1{}_2 + \tfrac{1}{2}\mathsf{P}^1{}_2 = 0, \\ &\partial_u (\mathsf{P}^1{}_2)^* = 0, && -\partial_v (\mathsf{P}^1{}_2)^* + \tfrac{1}{2}(\mathsf{P}^1{}_2)^* = -(u^* K), \end{aligned} \tag{7.122}$$

where $(u\partial_u) = 1$, etc. These equations have to be solved for every Killing vector $K_m = k_m{}^i \partial_i + \text{c.c.}$, but we will not attempt to do that here.

### 7.3.3 The gauging of the global symmetries

We are now ready to gauge the perturbative symmetries of a generic $N=2, d=4$ supergravity model. As we have stressed before, this is a restricted set of symmetries of the theory: dualities that leave invariant the equations of motion but not the action will not be gauged and non-perturbative Peccei–Quinn-type symmetries of the action will not be gauged either. Furthermore, we will only use fundamental (or *electric*) vector fields as gauge fields. We will, therefore, restrict ourselves to the symmetries whose associated infinitesimal symplectic matrices $\mathcal{T}_\Lambda$ defined in Eq. (6.53) have $b = c = 0$ and are block-diagonal.[21]

We will deal only with the symmetries that we are going to gauge. These symmetries must necessarily act in the adjoint representation on the vector fields (and, therefore, on the whole supermultiplets) and that is why we can label them using the indices $\Lambda, \Sigma, \ldots$, which we have reserved for the vector fields. Thus, if we want to gauge symmetries of the hypermultiplet sector or a subgroup of the R-symmetry group, we must have the same symmetry in the vector supermultiplet sector acting in the adjoint representation. A more

[21] This applies to the vector supermultiplet sector only: in the hypermultiplet sector all symmetries are perturbative symmetries of the action.

general and also more transparent (but considerably more complicated) way to describe the gauging procedure would be to use the embedding tensor formalism described in Section 2.7.

These observations imply that the infinitesimal symplectic generators are given by the structure constants, as in Eq. (6.57), which we rewrite here:

$$\mathcal{T}_\Lambda = \begin{pmatrix} f_{\Lambda\Sigma}{}^{\Omega} & \\ & -f_{\Lambda\Omega}{}^{\Sigma} \end{pmatrix}. \tag{7.123}$$

This restriction further simplifies the expressions for the holomorphic momentum map and Killing vectors Eqs. (7.82) and (7.83). They can be written as in Eqs. (J.72) and (J.73), which we rewrite here for convenience:

$$\mathcal{P}_\Lambda = 2 f_{\Lambda\Sigma}{}^{\Gamma}\, \Re\mathfrak{e}\, \left(\mathcal{L}^{\Sigma}\mathcal{M}^{*}_{\Gamma}\right), \tag{7.124}$$

$$k_{\Lambda\, i^*} = i f_{\Lambda\Sigma}{}^{\Gamma} \left(f^{*\Sigma}_{i^*}\mathcal{M}_{\Gamma} + \mathcal{L}^{\Sigma} h^{*}_{\Gamma i^*}\right). \tag{7.125}$$

The gauging of the isometries of the special Kähler manifold and of those of the quaternionic-Kähler manifold are reviewed in full detail in Appendices J.4 and J.5, respectively, and, therefore, we shall be very brief here. The gauging of the R-symmetry group (or the part of it that can be gauged) can be treated as a special case of the gauging of isometries of the quaternionic-Kähler manifold. To gauge the theory we have to replace the vector field strengths by the covariant ones and the different derivatives by their gauge-covariant generalization. The holomorphic momentum map $\mathcal{P}_\Lambda(Z, Z^*)$ defined in Eqs. (E.67) and (E.70) and the triholomorphic momentum map $\mathsf{P}_\Lambda{}^{x}$ defined by Eqs. (G.41) and (G.47) play crucial roles in the construction of these gauge-covariant derivatives.

One of the main results of the detailed study in Appendices J.4 and J.5 is that, out of all the global symmetries that a generic $N = 2, d = 4$ theory has, we can only gauge the following:

$\mathrm{U}(1)_{\mathrm{R}} \subset \mathrm{SU}(2)_{\mathrm{R}}$ using an Abelian Fayet–Iliopoulos term which can be understood as a constant triholomorphic momentum map $\mathsf{P}_\Lambda{}^{x}$ which only points in one direction in $\mathfrak{su}(2)$ space. This arises as a particular solution of Eqs. (G.47) and (G.50), which define the triholomorphic momentum map in the absence of hyperscalars.

$\mathrm{SU}(2)_{\mathrm{R}}$ using a Fayet–Iliopoulos term which can be understood as a constant triholomorphic momentum map $\mathsf{P}_\Lambda{}^{x}$ of the form

$$\mathsf{P}_\Lambda{}^{x} = e_\Lambda{}^{x}\xi, \tag{7.126}$$

$\xi$ being an arbitrary constant and the $e_\Lambda{}^{x}$ being non-zero for $\Lambda$ in the range of vectors that will gauge this $\mathrm{SU}(2)$ group[22] and satisfy

$$\varepsilon^{xyz} e_\Lambda{}^{y} e_\Sigma{}^{z} = f_{\Lambda\Sigma}{}^{\Omega} e_\Omega{}^{x}. \tag{7.127}$$

[22] The gauge fields $A^x$ can be linear combinations of several $A^\Lambda$. These combinations are dictated by the embedding tensor in general, and, actually, these constant triholomorphic momentum maps can be understood as components of the embedding tensor $\vartheta_\Lambda{}^{A}$, with $A = x$ (see Section 2.7).

If we use three vectors $A^x{}_\mu$ to gauge this symmetry, $\Lambda = x$ and $f_{\Lambda\Sigma}{}^\Omega = \varepsilon^{xyz}$, and the equation is trivialized by taking $e_x{}^y = \delta_x{}^y$. Thus we have three vectors that point in three independent directions in $\mathfrak{su}(2)$ space. Again, this arises as a particular solution of the equations that define the triholomorphic momentum map in the absence of hyperscalars.

$\mathrm{G_{NA}}$ non-Abelian, such that, simultaneously, $\mathrm{G_{NA}} \subset \mathrm{G_V}$ and $\mathrm{G_{NA}} \subset \mathrm{G_{hyper}}$, and it acts on the vector supermultiplets in the adjoint representation. These symmetries must be symmetries of the prepotential, when available, as a consequence of Eq. (J.75).

*A product* of the first or the second with the third.

The formalism that we use encompasses all these possibilities. An important condition that the gauge group has to satisfy is that it has to be a symmetry of the prepotential, according to Eq. (J.75), which follows from the preservation of the canonical section and the restriction to non-perturbative symmetries and electric gaugings.

The gauging of the supergravity theory for the gauge groups considered requires the following.

1. The replacement of the vector field strengths and standard (partial, general-, Lorentz-, Kähler-, and SU(2)-covariant) derivatives by their gauge-covariant deformations in the action and in the supersymmetry transformation rules.

2. The addition to the fermion supersymmetry transformation rules of *fermion shifts*, as in the $N=1, d=4$ case. This is required by supersymmetry, not (only) by gauge symmetry. The supersymmetry transformation rules of the bosonic fields are not modified at all by the gauging, and are still given by Eqs. (7.14).

3. The addition of a scalar potential quadratic in the fermion shifts to the action.

The gauge-covariant derivatives of the scalars and vector field strengths are given by the standard expressions

$$\mathfrak{D}_\mu Z^i = \partial_\mu Z^i + g A^\Lambda{}_\mu k_\Lambda{}^i, \tag{7.128}$$

$$\mathfrak{D}_\mu q^u = \partial_\mu q^u + g A^\Lambda{}_\mu \mathsf{k}_\Lambda{}^u, \tag{7.129}$$

$$F^\Lambda{}_{\mu\nu} = 2\partial_{[\mu} A^\Lambda{}_{\nu]} + g f_{\Sigma\Omega}{}^\Lambda A^\Sigma{}_{[\mu} A^\Omega{}_{\nu]}. \tag{7.130}$$

The gauge-covariant derivatives of the fermions are given in Eqs. (J.91)–(J.94), which we rewrite here for convenience:

$$\mathfrak{D}_\mu \epsilon_I = \{\partial_\mu - \tfrac{1}{4}\omega_\mu{}^{ab}\gamma_{ab} + \tfrac{i}{2}\hat{\mathcal{Q}}_\mu\}\epsilon_I - \hat{\mathsf{A}}_\mu{}^J{}_I \epsilon_J, \tag{7.131}$$

$$\mathfrak{D}_\mu \psi_{I\,\nu} = \left\{\partial_\mu - \tfrac{1}{4}\omega_\mu{}^{ab}\gamma_{ab} + \tfrac{i}{2}\hat{\mathcal{Q}}_\mu\right\}\psi_{I\,\nu} - \Gamma_{\mu\nu}{}^\rho \psi_{I\,\rho} - \hat{\mathsf{A}}_\mu{}^J{}_I \psi_{J\,\nu}, \tag{7.132}$$

$$\mathfrak{D}_\mu \lambda^{Ii} = \left\{\partial_\mu - \tfrac{1}{4}\omega_\mu{}^{ab}\gamma_{ab} + \tfrac{i}{2}\hat{\mathcal{Q}}_\mu\right\}\lambda^{Ii}$$

$$+\left\{\mathfrak{D}_\mu Z^k \Gamma_{kj}{}^i + gA^\Lambda{}_\mu \partial_j k_\Lambda{}^i\right\}\lambda^{Ij} + \hat{\mathsf{A}}_\mu{}^I{}_J \lambda^{Ji}, \tag{7.133}$$

$$\mathfrak{D}_\mu \zeta_\alpha = \left\{\partial_\mu - \tfrac{1}{4}\omega_\mu{}^{ab}\gamma_{ab} + \tfrac{i}{2}\hat{\mathcal{Q}}_\mu\right\}\zeta_\alpha - \Delta_{\mu\,\alpha}{}^\beta \zeta_\beta. \tag{7.134}$$

In these expressions, $\omega_\mu{}^{ab}$ is the spin connection that gauges the Lorentz group, $\Gamma_{\mu\nu}{}^\rho$ is the Levi-Civita connection that gauges GCTs, $\Delta_{\mu\,\alpha}{}^\beta$ is the pullback of the $\mathrm{Sp}(2m)$ part of the spin connection of the quaternionic-Kähler manifold (for $m$ hypermultiplets), and the hatted objects $\hat{\mathcal{Q}}_\mu$ and $\hat{\mathsf{A}}_\mu{}^I{}_J$ are the following deformations of the pullbacks of the Kähler ($\mathcal{Q}$) and $\mathrm{SU}(2)$ ($\mathsf{A}$) connections of the special and quaternionic-Kähler manifolds:

$$\hat{\mathcal{Q}}_\mu = \mathcal{Q}_\mu + gA^\Lambda{}_\mu \mathcal{P}_\Lambda, \tag{7.135}$$

$$\hat{\mathsf{A}}_\mu{}^I{}_J = -\tfrac{i}{2}\{\mathsf{A}^x{}_\mu + gA^\Lambda{}_\mu \mathsf{P}_\Lambda{}^x\}\sigma^{x\,I}{}_J. \tag{7.136}$$

The fermion shifts $S^x, W^i, W^{i\,x}$, and $N_\alpha{}^I$ are given by

$$S^x = \tfrac{1}{2} g\mathcal{L}^\Lambda \mathsf{P}_\Lambda{}^x, \tag{7.137}$$

$$W^i = \tfrac{1}{2} g\mathcal{L}^{*\Lambda} k_\Lambda{}^i = -\tfrac{i}{2} g\mathcal{G}^{ij^*} f^{*\Lambda}{}_{j^*}\mathcal{P}_\Lambda, \tag{7.138}$$

$$W^{i\,x} = g\mathcal{G}^{ij^*} f^{*\Lambda}{}_{j^*}\mathsf{P}_\Lambda{}^x, \tag{7.139}$$

$$N_\alpha{}^I = g\mathsf{U}_\alpha{}^I{}_u \mathcal{L}^{*\Lambda}\mathsf{k}_\Lambda{}^u, \tag{7.140}$$

and they enter the fermion supersymmetry transformation rules (for vanishing fermions) as follows:

$$\begin{aligned}
\delta_\epsilon \psi_{I\,\mu} &= \mathfrak{D}_\mu \epsilon_I + \left[T^+{}_{\mu\nu}\varepsilon_{IJ} - \tfrac{1}{2}S^x \eta_{\mu\nu}\varepsilon_{IK}(\sigma^x)^K{}_J\right]\gamma^\nu \epsilon^J,\\
\delta_\epsilon \lambda^{Ii} &= i\not{\mathfrak{D}} Z^i \epsilon^I + \left[\left(\not{G}^{i+} + W^i\right)\varepsilon^{IJ} + \tfrac{i}{2}W^{i\,x}(\sigma^x)^I{}_K \varepsilon^{KJ}\right]\epsilon_J,\\
\delta_\epsilon \zeta_\alpha &= i\mathsf{U}_{\alpha I\,u}\not{\mathfrak{D}} q^u \epsilon^I + N_\alpha{}^I \epsilon_I.
\end{aligned} \tag{7.141}$$

Here $T_{\mu\nu}$ and $G^i{}_{\mu\nu}$ are, respectively, the graviphoton and matter vector field strengths defined in Eqs. (7.15) and (7.16).

The invariance of the action under these modified supersymmetry transformation rules requires the introduction of a potential for the scalar fields which is quadratic in the fermion shifts:

$$\boxed{V(Z,Z^*,q) = -6S^{*\,x}S^x + 2\mathcal{G}_{ij^*}W^iW^{*j^*} + \tfrac{1}{2}\mathcal{G}_{ij^*}W^{i\,x}W^{*j^*\,x} + 2N_\alpha{}^I N^\alpha{}_I.} \tag{7.142}$$

Replacing the fermion shifts by their values, the scalar potential takes this form:

$$\boxed{\begin{aligned} V(Z,Z^*,q) = g^2\,\big[&-\tfrac{1}{4}\Im\mathrm{m}\mathcal{N}^{\Lambda\Sigma}\mathcal{P}_\Lambda\mathcal{P}_\Sigma + \tfrac{1}{2}\mathcal{L}^{*\,\Lambda}\mathcal{L}^\Sigma(4\mathsf{H}_{uv}\mathsf{k}_\Lambda{}^u\mathsf{k}_\Sigma{}^v - 3\mathsf{P}_\Lambda{}^x\mathsf{P}_\Sigma{}^x) \\ &+\tfrac{1}{2}\mathcal{G}^{ij^*}f^\Lambda{}_i f^{*\,\Sigma}{}_{j^*}\mathsf{P}_\Lambda{}^x\mathsf{P}_\Sigma{}^x\,\big]\,. \end{aligned}} \tag{7.143}$$

Finally, the action for the bosonic fields takes the form [44]

$$\boxed{\begin{aligned} S = \int d^4x\sqrt{|g|}\,\big[&R + 2\mathcal{G}_{ij^*}\mathfrak{D}_\mu Z^i\mathfrak{D}^\mu Z^{*\,j^*} + 2\mathsf{H}_{uv}\mathfrak{D}_\mu q^u\mathfrak{D}^\mu q^v \\ &+ 2\Im\mathrm{m}\mathcal{N}_{\Lambda\Sigma}F^{\Lambda\,\mu\nu}F^\Sigma{}_{\mu\nu} - 2\Re\mathrm{e}\mathcal{N}_{\Lambda\Sigma}F^{\Lambda\,\mu\nu}\star F^\Sigma{}_{\mu\nu} - V(Z,Z^*,q)\big]\,. \end{aligned}} \tag{7.144}$$

As we said before, in the ungauged theory the hyperscalars do not couple directly to any of the fields of the vector supermultiplets. As we can see, in the gauged theory the hyperscalars couple directly to the vectors in the covariant derivatives of the hyperscalars and to the complex scalars in the scalar potential.

Let us now consider the equations of motion of the bosonic fields using the same notation as in the ungauged case defined in Eqs. (7.19). The Einstein equation is identical to that of the ungauged case with the replacement of vector field strengths and partial derivatives by their gauge-covariant deformations. The equations of motion of the scalars have the derivatives replaced by the gauge-covariant ones and take a simpler form because our definition of the second gauge-covariant derivative includes the pullback of the Levi-Civita connection in the target space (see Eq. (J.38)). These equations also have an additional term: the derivative of the scalar potential. As for the Maxwell equations and Bianchi identities, the gauging that we have performed, using only "electric" (or fundamental) vector fields as gauge fields, breaks the electric–magnetic duality of the equations of motion of the ungauged theory: the Maxwell equations now will have sources and the Bianchi identities will remain sourceless.

Their explicit form is

$$\mathcal{E}_{\mu\nu} = G_{\mu\nu} + 4\mathcal{M}_{MN}(\mathcal{N})\mathcal{F}^{M}{}_{\mu}{}^{\rho}\mathcal{F}^{N}{}_{\nu\rho} + 2\mathcal{G}_{ij^*}[\mathfrak{D}_{(\mu}Z^{i}\mathfrak{D}_{\nu)}Z^{*j^*} - \tfrac{1}{2}g_{\mu\nu}\mathfrak{D}_{\rho}Z^{i}\mathfrak{D}^{\rho}Z^{*j^*}]$$

$$+ 2\mathsf{H}_{uv}\,[\mathfrak{D}_{\mu}q^{u}\mathfrak{D}_{\nu}q^{v} - \tfrac{1}{2}g_{\mu\nu}\mathfrak{D}_{\rho}q^{u}\mathfrak{D}_{\rho}q^{v}] + \tfrac{1}{2}g_{\mu\nu}V(Z,Z^*,q), \tag{7.145}$$

$$\mathcal{E}^{i} = \mathfrak{D}^{2}Z^{i} + \partial^{i}G_{\Lambda\,\mu\nu} \star F^{\Lambda\,\mu\nu} + \tfrac{1}{2}\partial^{i}V(Z,Z^*,q), \tag{7.146}$$

$$\mathcal{E}^{u} = \mathfrak{D}^{2}q^{u} + \tfrac{1}{4}\partial^{u}V(Z,Z^*,q), \tag{7.147}$$

$$\mathcal{E}_{\Lambda}{}^{\mu} = \mathfrak{D}_{\nu} \star G_{\Lambda}{}^{\nu\mu} + \tfrac{1}{4}j_{\Lambda}{}^{\mu}, \tag{7.148}$$

$$\mathcal{E}^{\Lambda\,\mu} = \mathfrak{D}_{\nu} \star F^{\Lambda\,\nu\mu}, \tag{7.149}$$

where we have defined the current

$$j_{\Lambda\,\mu} \equiv g(k_{\Lambda\,i^*}\mathfrak{D}_{\mu}Z^{*i^*} + k^{*}_{\Lambda\,i}\mathfrak{D}_{\mu}Z^{i}) + \tfrac{1}{2}g\mathsf{k}_{\Lambda\,u}\mathfrak{D}_{\mu}q^{u}. \tag{7.150}$$

### *7.3.4 Examples of gauged $N=2, d=4$ supergravities*

*Pure $N=2, d=4$ "gauged" (AdS) supergravity.* Our first example will be provided by the theory we constructed and studied in Section 5.6, which we will recover using the general formalism we just introduced. The starting point is the pure, ungauged, theory as we studied it on p. 226, that is determined by the prepotential $\mathcal{F} = -\frac{i}{4}(\mathcal{X}^{0})^{2}$. In the absence of scalars there are no isometries and we can only gauge the R-symmetry group, and, with just one vector field, we can gauge $\mathrm{U}(1)_{\mathrm{R}} \subset \mathrm{SU}(2)_{\mathrm{R}}$ only, introducing a constant triholomorphic momentum map $\mathsf{P}_{0}^{x} = \xi e^{x}$, where $\xi$ is a normalization constant and $e^{x}$ is a constant $\mathfrak{su}(2)$ vector that can be taken to have unit norm, $e^{x}e^{x} = 1$. This vector selects the particular $\mathrm{U}(1)_{\mathrm{R}} \subset \mathrm{SU}(2)_{\mathrm{R}}$ that is going to be gauged using the graviphoton, whose index we remove.

The only effect of the gauging in the bosonic action is the addition of a potential which, according to the general formula Eq. (7.143), must be given by

$$V = -\tfrac{3}{2}g^{2}|\mathcal{L}^{0}|^{2}\mathsf{P}_{0}{}^{x}\mathsf{P}_{0}{}^{x} = -\tfrac{3}{2}g^{2}\xi^{2} = -6g^{2}, \tag{7.151}$$

where we have set $\xi = 2$ to get the action

$$\boxed{S = \int d^{4}x\sqrt{|g|}\,[R - F^{2} + 6g^{2}],} \tag{7.152}$$

which is the bosonic sector of the action in Eq. (5.105).

Only the gravitino supersymmetry rule is deformed by the gauging: the general- and Lorentz-covariant derivative has to be replaced by the gauge-covariant one in Eq. (J.91), which has the additional connection $\hat{A}^x{}_\mu$ whose value is just $2gA_\mu e^x$ in this case. Furthermore, we have to add the shift $S^x$, whose value here is just $ge^x$. The final result is

$$\boxed{\begin{aligned}\delta_\epsilon \psi_{I\,\mu} &= \{\partial_\mu - \tfrac{1}{4}\omega_\mu{}^{ab}\gamma_{ab}\}\epsilon_I + igA_\mu e^x \sigma^{x\,J}{}_I \epsilon_J \\ &\quad + \left[T^+{}_{\mu\nu}\varepsilon_{IJ} - \tfrac{1}{2}ge^x\eta_{\mu\nu}\varepsilon_{IK}\sigma^{x\,K}{}_J\right]\gamma^\nu\epsilon^J.\end{aligned}} \tag{7.153}$$

If we compare this expression with Eq. (5.107), we see that the only difference (apart from the spinor conventions and normalization) is that there a particular direction $e^x = \delta^x{}_2$ was picked. For that particular case, $e^x\varepsilon_{IK}\sigma^{x\,K}{}_J = i\delta_{IJ}$ and the Pauli matrix disappears from the fermion shift.

*The $N=2, d=4$ Einstein–Yang–Mills supergravities* The theories of gauged $N=2, d=4$ supergravity which have no hypermultiplets can be seen as the simplest $N=2$ supersymmetric generalization of Einstein–Yang–Mills theories: the $N=2$ supersymmetrization of a Yang–Mills theory minimally coupled to gravity requires the addition of fermions (gravitinos and gauginos) and complex scalars $Z^i$ to complete the supergravity and vector multiplets, but not hypermultiplets. If we do not allow for Fayet–Iliopoulos terms, these theories have a very simple scalar potential which is positive semi-definite:[23]

$$V(Z,Z^*) = -\tfrac{1}{4}g^2\Im\mathfrak{m}\mathcal{N}^{\Lambda\Sigma}\mathcal{P}_\Lambda\mathcal{P}_\Sigma \geq 0. \tag{7.154}$$

When the scalars take constant values, this potential contributes to the equations of motion as a non-negative cosmological constant $\Lambda = V(Z,Z^*)/2 \geq 0$, which can lead to Minkowski ($\Lambda = 0$) or de Sitter ($\Lambda > 0$) but not anti-de Sitter ($\Lambda < 0$) vacua. As the previous example shows, to have the latter we need Fayet–Iliopoulos terms or hypermultiplets. Thus, without these multiplets and terms, these theories admit, potentially, at least some of the asymptotically flat bosonic solutions of the non-supersymmetric Einstein–Yang–Mills theories.

For all these reasons, these gauged $N=2, d=4$ supergravities with no hypermultiplets or Fayet–Iliopoulos terms were called $N=2, d=4$ *Einstein–Yang–Mills supergravities* in Ref. [797]. Here we are going to study in detail a possible gauging of one of the models that we have studied before: an $\mathrm{SO}(3)$ gauging of the $\overline{\mathbb{CP}}^3$ model. We are also going to study the possibility of gauging more complicated models (the magic models studied on p. 232) along the lines of Ref. [798]. The action, fermionic supersymmetry transformation rules (the bosonic ones are always the same), and equations of motion of these theories can be easily obtained from those of the general case by setting to zero all hyperinos, hyperscalars, $\mathrm{SU}(2)$ connections, and triholomorphic momentum maps (which includes possible Fayet–Iliopoulos terms), and, therefore, we will not write them explicitly.

The scalars of the $\overline{\mathbb{CP}}^n$ models, defined on p. 230, parametrize the symmetric space $\mathrm{U}(1,n)/(\mathrm{U}(1)\times\mathrm{U}(n))$, and, as shown on p. 243, they are symmetric under the whole

---

[23] Remember that the holomorphic momentum map $\mathcal{P}_\Lambda(Z,Z^*)$ is real and that the imaginary part of the period matrix $\Im\mathfrak{m}\mathcal{N}_{\Lambda\Sigma}$ and its inverse (denoted by $\Im\mathfrak{m}\mathcal{N}^{\Lambda\Sigma}$) are negative-definite.

$\mathrm{U}(1,n)$ group, which acts linearly in the fundamental (vector) representation on the $\mathcal{X}^\Lambda$:

$$\mathcal{X}'^\Lambda = \Lambda^\Lambda{}_\Sigma \mathcal{X}^\Sigma. \tag{7.155}$$

By definition, this group preserves the Hermitian form $\mathcal{X}^{*\Lambda}\eta_{\Lambda\Sigma}\mathcal{X}^\Sigma$, but the prepotential does not depend on it but rather on $\eta_{\Lambda\Sigma}\mathcal{X}^\Lambda\mathcal{X}^\Sigma$. Therefore, the prepotential is left invariant by the subgroup of $\mathrm{U}(1,n)$ with real matrices, which is, of course, $\mathrm{O}(1,n)$. Since we can only gauge symmetries of the prepotential, $\mathrm{SO}(1,n)$ is the largest group that we may gauge. However, the gauge group has to act on the gauge vectors (which transform as the $\mathcal{X}^\Lambda$) in the adjoint representation, and only for the rank-3 groups $\mathrm{SO}(1,2)$ and $\mathrm{SO}(3)$ is the vector representation isomorphic to the adjoint representation. The latter possibility (compact) is the most interesting; thus, we choose the subgroup $\mathrm{SO}(3)$ as the gauge group and we will consider it in $\overline{\mathbb{CP}}^3$, which is the simplest model in this class that includes it.

$\mathrm{SO}(3)$ will act only on the indices $\Lambda = i$, leaving $\Lambda = 0$ inert. The infinitesimal transformations can be written in the form

$$\delta_\alpha Z^i = \alpha^j \mathcal{T}_j{}^i{}_k Z^k, \tag{7.156}$$

where

$$\mathcal{T}_j{}^i{}_k = f_{jk}{}^i = -\epsilon_{jki} \quad \Rightarrow \quad \mathcal{T}_i = \begin{pmatrix} 0 & & \\ & f_{ij}{}^k & \\ & & -f_{ik}{}^j \\ & & & 0 \end{pmatrix}. \tag{7.157}$$

The holomorphic momentum map and compensating Kähler transformations can be read from Eq. (7.109). The latter simply vanish and the holomorphic Killing vectors can be read from the $\delta_\alpha Z^i$ or computed by taking the derivatives of the momentum map. We get

$$\mathcal{P}_i = -ie^{\mathcal{K}}\epsilon_{ijk}Z^j Z^{*k}, \qquad k_i{}^j = -\epsilon_{ijk}Z^k, \qquad \lambda_i = 0. \tag{7.158}$$

These are all the ingredients we need to construct the gauged theory. The gauge-covariant derivative of the scalars is just that of a complex adjoint $\mathrm{SO}(3)$ scalar,

$$\mathfrak{D}_\mu Z^i = \partial_\mu Z^i + g\epsilon_{ijk}A^j{}_\mu Z^k, \tag{7.159}$$

and the hatted Kähler 1-form connection that enters in the spinor covariant derivatives is given by

$$\hat{\mathcal{Q}}_\mu = e^{\mathcal{K}}\left\{\frac{1}{2i}(Z_{i^*}\partial_\mu Z^\mu - Z_i^*\partial_\mu Z^i) - ig\epsilon_{ijk}A^i{}_\mu Z^j Z^{*k^*}\right\} = \mathcal{Q}_i\mathfrak{D}_\mu Z^i + \mathcal{Q}_{i^*}\mathfrak{D}_\mu Z^{*i^*}, \tag{7.160}$$

because in this particular case there are no compensating Kähler transformations, which means that the spinors of the theory are gauge invariant: only the scalars and the vector fields feel the gauge transformations.

The scalar potential can be constructed with the inverse of the imaginary part of the period matrix, given in Eq. (7.45):

$$\Im\mathrm{m}\mathcal{N}_{\Lambda\Sigma} = \frac{1}{2}\left[\eta_{\Lambda\Sigma} - \frac{Z_\Lambda Z_\Sigma}{Z^\Gamma Z_\Gamma} - \frac{Z^*_\Lambda Z^*_\Sigma}{Z^{*\Gamma} Z^*_\Gamma}\right] \Rightarrow \Im\mathrm{m}\mathcal{N}^{\Lambda\Sigma} = 2\left[\eta^{\Lambda\Sigma} - 2e^{\mathcal{K}} Z^{(\Lambda} Z^{*\Sigma)}\right], \tag{7.161}$$

and reads

$$V(Z,Z^*) = -\tfrac{1}{2} g^2 e^{\mathcal{K}} \epsilon_{ijk}\epsilon_{imn} Z^j Z^{*k^*} Z^m Z^{*n^*}, \tag{7.162}$$

or, equivalently,

$$V(Z,Z^*) = 2g^2 e^{\mathcal{K}}\left[(\Re\mathrm{e}Z^i \Re\mathrm{e}Z^i)(\Im\mathrm{m}Z^j \Im\mathrm{m}Z^j) - (\Re\mathrm{e}Z^i \Im\mathrm{m}Z^i)(\Re\mathrm{e}Z^j \Im\mathrm{m}Z^j)\right]. \tag{7.163}$$

Is this the only possible gauging of the $\overline{\mathbb{CP}}^n$ models? The following observation proves that basically any group can be gauged in a $\overline{\mathbb{CP}}^n$ model for sufficiently large $n$ [798] thanks to the existence of an $\mathrm{SO}(n)$ subgroup of its symmetry group: a given compact simple Lie algebra $\mathfrak{g}$ of a group G is always a subalgebra of $\mathfrak{so}(\dim(\mathfrak{g}))$. Furthermore, the vector representation of $\mathfrak{so}(\dim(\mathfrak{g}))$ branches into the adjoint representation of $\mathfrak{g}$.

The same ideas can be applied to $ST[2,n]$ models: an $\mathrm{SO}(3)$ subgroup can be gauged for $n=3$ or we can gauge an arbitrary group G as long as we take $n \geq \dim(\mathfrak{g})$.

Let us now move to the magic models studied on p. 232. By definition, the prepotentials of these models are automatically invariant under the subgroup $\mathrm{H}' \subset \mathrm{H}$, which results from removing the $\mathrm{U}(1)$ factors from the latter, and this implies that the gauge group $\mathrm{G}_{\mathrm{gauge}}$ can only be a subgroup of $\mathrm{H}'$. There are two additional constraints: we can only gauge groups whose associated infinitesimal symplectic transformations satisfy the constraint $b=c=0$, and we can only gauge groups whose action on vector fields is the adjoint (and nothing else, apart from possible singlets). Taking into account all these constraints, the maximal gaugings of the magic models were found in Ref. [798] and are given in the last column of Table 7.1.

# 8
# A generic description of all the $N \geq 2, d = 4$ SUEGRAs

Since supergravity can be seen as a theory of supergeometry, a supersymmetric generalization of GR, one may wonder why theories with different $d$ or $N$ look so different whereas the Einstein–Hilbert action takes exactly the same form in any dimension and we only need to change the range of values of the indices. It is clear that the spinor representations are different in different dimensions but, still, for a given $d$, theories with different $N$ look terribly different. There is no deep reason for this and, indeed, in Ref. [47] it was shown that all $N \geq 2, d = 4$ supergravities coupled to vector multiplets can be described in an essentially uniform fashion using the fact that all the supergravity multiplets and all the vector multiplets for all $N = 1, \ldots, 8$ can be written in a uniform way.

This construction, enlightening and useful, which we are going to review in this chapter for $d = 4$ adapted to our conventions, will allow us to complete the description of all the ungauged supergravities since only $N = 2$ theories admit hypermultiplets. A uniform study of all the possible gaugings of these theories is still lacking but could, in principle, be achieved by using the embedding tensor formalism (Section 2.7).

### 8.1 Generic supermultiplets

We start by introducing the generic gravity and vector supermultiplets which are valid for all $N$-extended $d = 4$ supergravities if we just take into account that in each case the $\mathrm{U}(N)$ R-symmetry indices $I, J$ take values from 1 to $N$ and that fields or tensors that carry more than $N$ antisymmetric $\mathrm{U}(N)$ indices simply vanish in that theory. We also adopt the convention that terms with Levi-Civita symbols $\epsilon^{I_1 \cdots I_M}$ should only be considered when $M = N$.

The generic supergravity multiplet in four dimensions has the following field content, ordered by an increasing number of $\mathrm{U}(N)$ indices:

$$\left\{ e^a{}_\mu, \psi_{I\,\mu}, A^{IJ}{}_\mu, \chi_{IJK}, P_{IJKL\,\mu}, \chi^{IJKLM} \right\}. \tag{8.1}$$

Here $e^a{}_\mu$ is the Vierbein, $\psi_{I\,\mu}$ are the $N$ gravitinos, $A^{IJ}{}_\mu$ are the $N(N-1)/2$ graviphotons, $\chi_{IJK}$ are the $N(N-1)(N-2)$ dilatinos, $P_{IJKL\,\mu}$ is the pullback over the spacetime of a Vielbein in the scalar manifold,[1] and $\chi^{IJKLM}$ are another set of $N!/(5! \cdot 3!)$ dilatinos.

[1] It is not possible to derive a generic description of all the supergravities if we use particular coordinates in

According to to our conventions, $\chi_{IJK}$ occurs only in $N \geq 3$, the scalars $P_{IJKL\,\mu}$ occur only in $N \geq 4$, and $\chi^{IJKLM}$ occurs only in $N \geq 5$.

The generic vector multiplets (labeled by $i = 1, \ldots, n_V$) have the field content

$$\{A_{i\,\mu}, \lambda_{iI}, P_{iIJ\,\mu}, \lambda_i{}^{IJK}\}. \tag{8.2}$$

Here $A_{i\,\mu}$ are the $n_V$ matter vector fields, $P_{iIJ\,\mu}$ is the pullback of the Vielbein of the matter scalar manifold, and $\lambda_{iI}$ and $\lambda_i{}^{IJK}$ are gauginos.

All the spinors are Weyl spinors, and $\psi_{I\,\mu}, \chi_{IJK}, \chi^{IJKLM}, \lambda_{iI}$, and $\lambda_i{}^{IJK}$ have positive chirality with these positions of the $\mathrm{U}(N)$ indices. These are exactly the same conventions used in Chapter 7 in the $N = 2$ case for $\psi_{I\,\mu}$ and $\lambda_{iI}$ (the other spinors do not occur in $N = 2$).

The generic fields of these supermultiplets are subject to the following constraints:

$$P_{i\,IJ} = \tfrac{1}{2}\varepsilon_{IJKL}P^{*\,i\,KL}, \tag{8.3}$$

$$\lambda_{i\,I} = \tfrac{1}{3!}\varepsilon_{IJKL}\lambda^{i\,JKL}, \tag{8.4}$$

$$\lambda_I = \tfrac{1}{5!}\varepsilon_{IJKLMN}\chi^{JKLMN}, \tag{8.5}$$

$$\chi_{IJK} = \tfrac{1}{3!}\varepsilon_{IJKLMN}\lambda^{LMN}, \tag{8.6}$$

$$P_{IJKL\,\mu} = \tfrac{1}{2}\varepsilon_{IJKLMN}P^{*\,MN}{}_{\mu}, \tag{8.7}$$

$$P_{IJKL\,\mu} = \tfrac{1}{4!}\varepsilon_{IJKLMNOP}P^{*\,MNOP}{}_{\mu}, \tag{8.8}$$

$$\chi_{IJK} = \tfrac{1}{5!}\varepsilon_{IJKLMNOP}\chi^{LMNOP}, \tag{8.9}$$

which are all the possible dualities in $\mathrm{U}(N)$ indices between fields of the same kind. According to our convention, these constraints must be disregarded when the number of indices of the Levi-Civita symbol is not exactly $N$. It follows that the first two constraints are only relevant for $N = 4$, the next two for $N = 4$, the next three for $N = 6$ (where, as we are going to see, there is just one "vector multiplet" which carries no index[2]), and the last two for $N = 8$.

---

the scalar manifolds. These Vielbeins and other objects associated with the scalar manifold are determined by a generalization of the canonical symplectic section of special geometry.

[2] Chiral multiplets only exist in $N = 1$ theories, hypermultiplets in $N = 2$, and vector supermultiplets in $N \leq 4$ theories. However, in the $N = 6$ case the supergravity multiplet does not coincide with the generic supermultiplet in Eq. (8.1) but turns out to be equivalent to a generic supergravity multiplet plus one generic vector supermultiplet. The reason behind this anomaly is that the decomposition of the duality group of the spinorial representation of $\mathrm{SO}^*(12)$ (**32**) with respect to $\mathrm{SU}(6)$ produces a singlet (this is the "practical reason" why Eq. (8.1) is not enough). Observe that $\mathrm{SO}^*(12)/\mathrm{U}(6)$ is the coset space of the quaternionic magic model of $N = 2$ supergravity in Table 7.1 and that the number and representations of vectors and scalars of the $N = 2$ and $N = 6$ supergravities coincide. Their bosonic sectors are identical, and the theories only differ in the additional fermions that $N = 6$ has.

The following tables, taken from Refs. [47, 943], give the field content of each $N > 3$ theory with $n_V$ vector multiplets; $\sharp$ stands for the number of fields of that particular kind; $N = 7$ is not included in the analysis because it is $N = 8$.

| $N = 3$ | | | | | | | | |
|---|---|---|---|---|---|---|---|---|
| | $e^a{}_\mu$ | $\psi_{I\mu}$ | $A^{IJ}{}_\mu$ | $\chi_{IJK}$ | $A^i{}_\mu$ | $\lambda_{iI}$ | $\lambda_{iIJK}$ | $P_{iIJ\mu}$ |
| $\sharp$ | 1 | 3 | 3 | 1 | $n$ | $3n_V$ | $n_V$ | $(3+3)n_V$ |

| $N = 4$ | | | | | | | | | |
|---|---|---|---|---|---|---|---|---|---|
| | $e^a{}_\mu$ | $\psi_{I\mu}$ | $A^{IJ}{}_\mu$ | $\chi_{IJK}$ | $P_{IJKL\mu}$ | $A^i{}_\mu$ | $\lambda_{iI}$ | $\lambda_{iIJK}$ | $P_{iIJ\mu}$ |
| $\sharp$ | 1 | 4 | 6 | 4 | 1+1 | $n_V$ | $4n_V$ | $4n_V$ | $(6+6)n_V$ |

| $N = 5$ | | | | | | |
|---|---|---|---|---|---|---|
| | $e^a{}_\mu$ | $\psi_{I\mu}$ | $A^{IJ}{}_\mu$ | $\chi_{IJK}$ | $\chi^{IJKLM}$ | $P_{IJKL\mu}$ |
| $\sharp$ | 1 | 5 | 10 | 10 | 1 | $5+5$ |

| $N = 6$ | | | | | | | | | | |
|---|---|---|---|---|---|---|---|---|---|---|
| | $e^a{}_\mu$ | $\psi_{I\mu}$ | $A^{IJ}{}_\mu$ | $\chi_{IJK}$ | $\chi^{IJKLM}$ | $P_{IJKL\mu}$ | $A$ | $\lambda_I$ | $\lambda_{IJK}$ | $P_{IJ}$ |
| $\sharp$ | 1 | 6 | 15 | 20 | 6 | $15+15$ | 1 | 6 | 20 | $15+15$ |

| $N = 8$ | | | | | | |
|---|---|---|---|---|---|---|
| | $e^a{}_\mu$ | $\psi_{I\mu}$ | $A^{IJ}{}_\mu$ | $\chi_{IJK}$ | $\chi^{IJKLM}$ | $P_{IJKL\mu}$ |
| $\sharp$ | 1 | 8 | 28 | 56 | 56 | $70+70$ |

As we have seen, the scalars of these theories are described by the generic Vielbeins $P^{IJKL}$ and $P_{iIJ}$, but there is a more fundamental description studied in Appendix I. This description is based on a generalization of the canonical covariantly holomorphic symplectic section $\mathcal{V}^M$ of the $N = 2$ theories multiplied by $\varepsilon_{IJ}$, $\mathcal{V}^M \varepsilon_{IJ}$ to the object $\mathcal{V}^M{}_{IJ}$ (which cannot be factored for $N > 2$), and on a generalization of the Kähler-covariant derivative of that section $\mathcal{U}^{*}_{\hat{i}^*}{}^M = \mathcal{D}_{\hat{i}^*} \mathcal{V}^{*\,M}$ to another object[3] $\mathcal{V}^M{}_i$, which is completely independent of $\mathcal{V}^M{}_{IJ}$ for $N > 2$. The $N = 2$ is a sort of degenerate particular case. Observe that the symplectic indices take $2\bar{n}$ values, where $\bar{n}$ is the total number of vector fields. As explained in Appendix I, these two symplectic objects, whose components are denoted

[3] The indices $\hat{\imath}, \hat{\jmath}$ that we use here are, actually, tangent-space indices of the scalar manifold, not world indices as in $\mathcal{U}_i{}^M = \mathcal{D}_i \mathcal{V}^M$, but this should only be taken into account if we want to recover in detail the $N = 2$ theories from the general setup. See Ref. [528] for more details.

by $(\mathcal{V}^M{}_{IJ}) = \begin{pmatrix} f^\Lambda{}_{IJ} \\ h_{\Lambda IJ} \end{pmatrix}$, $(\mathcal{V}^M{}_i) = \begin{pmatrix} f^\Lambda{}_i \\ h_{\Lambda i} \end{pmatrix}$, satisfy the constraints

$$\mathcal{V}_{M\,IJ}\mathcal{V}^{*\,M\,KL} = -2i\delta^{KL}{}_{IJ}, \qquad \mathcal{V}_{M\,i}\mathcal{V}^{*\,M\,j} = -i\delta_i{}^j, \tag{8.10}$$

with the rest of the symplectic products vanishing. These constraints (which generalize the special geometry properties Eqs. (F.5), (F.7), (F.9), and (F.10) to $N > 2$) have the following meaning: with the two square matrices $f \equiv (f^\Lambda{}_{IJ}, f^\Lambda{}_i)$ and $h \equiv (h_{\Lambda\,IJ}, h_{\Lambda\,i})$ one can construct the $2\bar{n} \times 2\bar{n}$ $\mathrm{USp}(\bar{n},\bar{n})$ matrix $U$ in Eq. (I.2). The scalar manifolds of all the $N \geq 2$ SUEGRAs coupled to vector multiplets can be embedded in this group, and $U$ provides us with a generic "coset representative" from which we can obtain, via the Maurer–Cartan 1-form, the scalar Vielbeins $P_{IJKL}, P_{iIJ}$ (plus $P_{ij}$, which does not describe independent degrees of freedom) as well as the connections $\Omega^{IJ}{}_{LK}, \Omega^i{}_j$ (the $\Omega^i{}_{IJ}$ components vanish). Via the Maurer–Cartan equations, the curvatures of the scalar manifolds, defined in terms of the $\Omega$ connections, can be written entirely in terms of the Vielbeins, as in Eqs. (I.36) and (I.38).

According to this, $\mathcal{V}^M{}_{IJ}$ and $\mathcal{V}^M{}_i$ provide us with the Vielbeins and connections necessary for the kinetic terms of the scalars, and the covariant derivatives of the spinors[4] and other objects with $\mathrm{U}(N)$ indices. They also provide us with the period matrix $\mathcal{N}_{\Lambda\Sigma}$ of the theory that determines the couplings between the vector field strengths and the scalars. It is given by Eq. (I.10), which we reproduce here for convenience

$$\mathcal{N} = hf^{-1} = \mathcal{N}^T, \tag{8.11}$$

and can be used to raise or lower the indices $\Lambda, \Sigma$ as in the $N = 2$ theories (see Eq. (I.11)).

This information suffices to describe the bosonic action and the supersymmetry transformation rules (the fermionic ones, for vanishing fermions, as usual). We present them in the following section.

## 8.2 The theories

The vector fields that appear in the generic supermultiplets that we have presented, namely the graviphotons $A^{IJ}{}_\mu$ and the matter vector fields $A^i{}_\mu$, do not appear directly in the supergravity action. As in the $N = 2$ case, the action is *democratically* written in terms of the vectors $A^\Lambda{}_\mu$, which are mixed by the duality transformations. The relation between these[5] is (compare with the $N = 2$ relation Eq. (7.13))

$$A^\Lambda{}_\mu \equiv \tfrac{1}{2} f^\Lambda{}_{IJ} A^{IJ}{}_\mu + f^\Lambda{}_i A^i{}_\mu, \tag{8.12}$$

The inverse relations can be found using Eq. (I.13) and the constraints that define $\mathcal{V}_{IJ}$ and $\mathcal{V}_i$. In order to write these expressions in the simplest way, we introduce the dual vector field strengths

$$G_{\Lambda\,\mu\nu} \equiv -\frac{1}{4\sqrt{|g|}}\frac{\delta S}{\delta \star F^\Lambda{}_{\mu\nu}} = 2\Re\mathrm{e}(\mathcal{N}_{\Lambda\Sigma}F^{\Sigma\,+}) = \Re\mathrm{e}\mathcal{N}_{\Lambda\Sigma}F^\Sigma{}_{\mu\nu} + \Im\mathrm{m}\mathcal{N}_{\Lambda\Sigma} \star F^\Sigma{}_{\mu\nu}, \tag{8.13}$$

[4] They carry a single $\mathrm{U}(N)$ index and the connection that acts on them is $\Omega^I{}_J$, which is such that $\Omega^{IJ}{}_{LK} = \Omega^{[I}{}_{[L}\delta^{J]}{}_{K]}$.

[5] Strictly speaking, this should be a relation between field strengths, as in the $N = 2$ case but this is not important because $A^{IJ}{}_\mu$ and $A^i{}_\mu$ will never be used and the field strengths $T^{IJ}{}_{\mu\nu}$ and $G^i{}_{\mu\nu}$ will appear as linear combinations of the $F^\Lambda{}_{\mu\nu}$.

and define the symplectic vector of the vector field strengths $(\mathcal{F}^M) = \begin{pmatrix} F^\Lambda \\ G_\Lambda \end{pmatrix}$. Then, for both the graviphoton $\mathbf{a} = IJ$ and the matter vector field strengths $\mathbf{a} = i$ we get

$$T_{\mathbf{a}\,\mu\nu} = \langle \mathcal{V}_{\mathbf{a}} \mid \mathcal{F} \rangle = 2if^\Lambda{}_{\mathbf{a}}\, \Im\mathrm{m}\mathcal{N}_{\Lambda\Sigma}\, F^{\Sigma+}{}_{\mu\nu} = T^+{}_{\mathbf{a}\,\mu\nu}, \tag{8.14}$$

as in the $N=2$ case.

In writing the action we must also take into account the existence of $\mathrm{U}(N)$ duality constraints obeyed by the scalar Vielbeins for $N=4,6,8$. Actually, the best way to construct an action in these cases is to ignore these constraints and assume that, in order to derive the equations of motion, they have to be imposed on the Euler–Lagrange equations that follow from the generic action.

Then, a generic bosonic action that can be used as explained above is

$$\boxed{\begin{aligned} S = \int d^4x\sqrt{|g|}\, \big[&R + 2\Im\mathrm{m}\mathcal{N}_{\Lambda\Sigma}F^{\Lambda\,\mu\nu}F^\Sigma{}_{\mu\nu} - 2\Re\mathrm{e}\mathcal{N}_{\Lambda\Sigma}F^{\Lambda\,\mu\nu}\star F^\Sigma{}_{\mu\nu} \\ &+\tfrac{2}{4!}\alpha_1 P^{*\,IJKL}{}_\mu P_{IJKL}{}^\mu + \alpha_2 P^{*\,iIJ}{}_\mu P_{iIJ}{}^\mu\big], \end{aligned}} \tag{8.15}$$

where the parameters $\alpha_1, \alpha_2$ are equal to 1 in all cases except for those in which the Vielbeins satisfy $\mathrm{U}(N)$ duality constraints, that is $N=4,6,8$. For $N=4$ and $N=8$, $\alpha_2 = \frac{1}{2}$, and, for $N=6$, $\alpha_1 + \alpha_2 = 1$ (the simplest choice being $\alpha_2 = 0$).

The Einstein and Maxwell equations are just the Euler–Lagrange equations that follow from the generic action, but the $\mathrm{U}(N)$ duality constraints must be taken into account in the scalar equations of motion. Calling the physical scalar fields $\phi^A$ and defining, as usual, the (l.h.s. of the) equations of motion by

$$\begin{aligned} \mathcal{E}_a{}^\mu &\equiv -\frac{1}{2\sqrt{|g|}}\frac{\delta S}{\delta e^a{}_\mu}, & \mathcal{E}^{IJKL} &\equiv -\frac{1}{2\sqrt{|g|}}\left(\frac{\delta S}{\delta U}U\right)^{IJKL} = -\frac{1}{2\sqrt{|g|}}P^{*\,IJKL\,A}\frac{\delta S}{\delta\phi^A}, \\ \mathcal{E}_\Lambda{}^\mu &\equiv \frac{1}{8\sqrt{|g|}}\frac{\delta S}{\delta A^\Lambda{}_\mu}, & \mathcal{E}^{iIJ} &\equiv -\frac{1}{2\sqrt{|g|}}\left(\frac{\delta S}{\delta U}U\right)^{iIJ} = -\frac{1}{2\sqrt{|g|}}P^{*\,iIJ\,A}\frac{\delta S}{\delta\phi^A}, \end{aligned} \tag{8.16}$$

where $P^{*\,IJKL\,A}$ and $P^{*\,iIJ\,A}$ are the inverse Vielbeins, we find that the Einstein and Maxwell equations (plus the Bianchi identities) are given by the generic expressions

$$\begin{aligned} \mathcal{E}_{\mu\nu} = G_{\mu\nu} &+ \tfrac{1}{12}\alpha_1\left[P^{*\,IJKL}{}_{(\mu|}P_{IJKL\,|\nu)} - \tfrac{1}{2}g_{\mu\nu}P^{*\,IJKL}{}_\rho P_{IJKL}{}^\rho\right] \\ &+ \alpha_2 P^{*\,iIJ}{}_{(\mu|}P_{iIJ\,|\nu)} - \tfrac{1}{2}g_{\mu\nu}P^{*\,iIJ}{}_\rho P_{iIJ}{}^\rho \\ &+ 4\mathcal{M}_{MN}(\mathcal{N})\mathcal{F}^M{}_\mu{}^\rho\mathcal{F}^N{}_{\nu\rho}, \end{aligned} \tag{8.17}$$

$$\mathcal{E}^{M\,\mu} = \nabla_\nu \star\mathcal{F}^{M\,\nu\mu}. \tag{8.18}$$

After imposing the constraints, and using Eqs. (I.32) and (I.33), the scalar equations of motion take different forms for different $N$:

$N = 2$

$$\mathcal{E}^{iIJ} = \mathfrak{D}^{\mu} P^{*\, iIJ}{}_{\mu} + 2T^{i\,-}{}_{\mu\nu} T^{IJ\,-\,\mu\nu} + P^{*\, iIJ\, A} P^{*\, jk}{}_{A} T_j{}^{+}{}_{\mu\nu} T_k{}^{+\,\mu\nu}; \tag{8.19}$$

$N = 3$

$$\mathcal{E}^{iIJ} = \mathfrak{D}^{\mu} P^{*\, iIJ}{}_{\mu} + 2T^{i\,-}{}_{\mu\nu} T^{IJ\,-\,\mu\nu}; \tag{8.20}$$

$N = 4$

$$\mathcal{E}^{IJKL} = \mathfrak{D}^{\mu} P^{*\, IJKL}{}_{\mu} + 6T^{[IJ|-}{}_{\mu\nu} T^{|KL]-\,\mu\nu} + P^{*\, IJKL\, A} P^{*\, ij}{}_{A} T_i{}^{+}{}_{\mu\nu} T_j{}^{+\,\mu\nu}, \tag{8.21}$$

$$\mathcal{E}^{iIJ} = \mathfrak{D}^{\mu} P^{*\, iIJ}{}_{\mu} + T^{i\,-}{}_{\mu\nu} T^{IJ\,-\,\mu\nu} + \tfrac{1}{2}\varepsilon^{IJKL} T_i{}^{+}{}_{\mu\nu} T_{KL}{}^{+\,\mu\nu}; \tag{8.22}$$

$N = 5$

$$\mathcal{E}^{IJKL} = \mathfrak{D}^{\mu} P^{*\, IJKL}{}_{\mu} + 6T^{[IJ|-}{}_{\mu\nu} T^{|KL]-\,\mu\nu}; \tag{8.23}$$

$N = 6$

$$\mathcal{E}^{IJKL} = \mathfrak{D}^{\mu} P^{*\, IJKL}{}_{\mu} + 6T^{[IJ|-}{}_{\mu\nu} T^{|KL]-\,\mu\nu} + \varepsilon^{IJKLMN} T^{+}{}_{\mu\nu} T_{MN}{}^{+\,\mu\nu}; \tag{8.24}$$

$N = 8$

$$\mathcal{E}^{IJKL} = \mathfrak{D}^{\mu} P^{*\, IJKL}{}_{\mu} + 6T^{[IJ|-}{}_{\mu\nu} T^{|KL]-\,\mu\nu} + \tfrac{1}{4}\varepsilon^{IJKLMNPQ} T_{MN}{}^{+}{}_{\mu\nu} T_{PQ}{}^{+\,\mu\nu}. \tag{8.25}$$

The supersymmetry transformations of the bosonic fields take the generic form

$$\delta_\epsilon e^a{}_\mu = -i\bar{\psi}_{I\mu}\gamma^a\epsilon^I - i\bar{\psi}^I{}_\mu\gamma^a\epsilon_I, \tag{8.26}$$

$$\begin{aligned}\delta_\epsilon A^\Lambda{}_\mu = {} & f^\Lambda{}_{IJ}\bar{\psi}^I{}_\mu\epsilon^J + f^{*\Lambda IJ}\bar{\psi}_{I\mu}\epsilon_J - \tfrac{i}{2}(f^\Lambda{}_i\bar{\lambda}^{iI}\gamma_\mu\epsilon_I + f^{*\,\Lambda i}\bar{\lambda}_{iI}\gamma_\mu\epsilon^I) \\ & - \tfrac{i}{4}(f^\Lambda{}_{IJ}\bar{\chi}^{IJK}\gamma_\mu\epsilon_K + f^{*\Lambda IJ}\bar{\chi}_{IJK}\gamma_\mu\epsilon^K),\end{aligned} \tag{8.27}$$

$$(U^{-1}\delta_\epsilon U)_{IJKL} = 4\bar{\chi}_{[IJK}\epsilon_{L]} + \bar{\chi}_{IJKLM}\epsilon^M, \tag{8.28}$$

$$(U^{-1}\delta_\epsilon U)_{iIJ} = 2\bar{\lambda}_{i[I}\epsilon_{J]} + \tfrac{1}{2}\bar{\lambda}_{iIJK}\epsilon^K, \tag{8.29}$$

and the supersymmetry transformations of the fermions, for vanishing fermions, are given by

$$\delta_\epsilon \psi_{I\mu} = \mathfrak{D}_\mu \epsilon_I + T_{IJ}{}^+{}_{\mu\nu}\gamma^\nu \epsilon^J, \tag{8.30}$$

$$\delta_\epsilon \chi_{IJK} = -\tfrac{3i}{2}\, \not{T}_{[IJ}{}^+ \epsilon_{K]} + i\, \not{P}_{IJKL}\epsilon^L, \tag{8.31}$$

$$\delta_\epsilon \lambda_{iI} = -\tfrac{i}{2}\, \not{T}_i{}^+ \epsilon_I + i\, \not{P}_{iIJ}\epsilon^J, \tag{8.32}$$

$$\delta_\epsilon \chi_{IJKLM} = -5i\, \not{P}_{[IJKL}\epsilon_{M]} + \tfrac{i}{2}\varepsilon_{IJKLMN}\, \not{T}^- \epsilon^N + \tfrac{i}{4}\varepsilon_{IJKLMNOP}\, \not{T}^{NO-}\epsilon^P, \tag{8.33}$$

$$\delta_\epsilon \lambda_{iIJK} = -3i\, \not{P}_{i[IJ}\epsilon_{K]} + \tfrac{i}{2}\varepsilon_{IJKL}\, \not{T}_i{}^- \epsilon^L + \tfrac{i}{4}\varepsilon_{IJKLMN}\, \not{T}^{LM-}\epsilon_N, \tag{8.34}$$

and where the covariant derivative acting on the fermions is

$$\mathfrak{D}_\mu \epsilon_I \equiv \nabla_\mu \epsilon_I - \epsilon_J \Omega_\mu{}^J{}_I. \tag{8.35}$$

Observe that in some cases the fields do not exist for a given $N$ or they are related by duality constraints. Therefore, Eqs. (8.33) and (8.34) only need to be considered in the cases $N = 3$ and 5, and, furthermore, only the first term on the l.h.s. is non-vanishing.

# 9

# Matter-coupled $N = 1, d = 5$ supergravity

So far we have studied four-dimensional supergravities. There are several reasons why higher-dimensional supergravities are also interesting, but, from our point of view, the most important reasons are that they naturally arise in superstring or M-theory compactifications and that their own compactification gives rise to the four-dimensional supergravities that we have studied. This will allow us to rewrite many theories and their classical solutions in a higher-dimensional language, geometrizing many quantities and objects. In turn, this will allow us to generate new four-dimensional or higher-dimensional solutions using the solution-generation techniques explained in Chapter 15.

In this chapter we are going to study the simplest example of higher-dimensional supergravity: the minimal five-dimensional supergravity (the one with the minimal spinors) and some of its couplings to matter multiplets. Since the supersymmetry parameter, the gravitino, and other fermions are given by just one minimal spinor (eight real components), these theories deserve to be called $N = 1, d = 5$ supergravities. This is what we will do here. However, in the literature they are very often called $N = 2, d = 5$ supergravities, the reason being their close relation to the $N = 2, d = 4$ theories: the compactification of all these theories in a circle gives rise to an $N = 2, d = 4$ supergravity. Not all the four-dimensional $N = 2$ theories can be obtained in this way, but for the many that can (most of them very interesting from several points of view) there is an alternative and most useful rewriting in five-dimensional language. In five dimensions the scalars are real and life is usually much simpler. All this, together with the fact that many of them can be obtained from compactifications of 11-dimensional supergravity on Calabi–Yau 3-folds [274], makes the study of these theories very attractive.

Our presentation will be very concise, since many of the concepts necessary to understand the ingredients that enter in the construction of these theories are very similar to those that have been discussed at length in Chapters 6 and 7: the combination of the graviphoton field with the vector fields from the vector supermultiplets to make manifest the maximal symmetry available (now $\mathrm{GL}(\bar{n})$, according to the results in Section 2.6.3), the description of the scalars using functions that transform linearly, like the vector fields, etc. A more detailed presentation can be found in the recent book, Ref. [564]. We will follow essentially the notation of Ref. [153] with the changes made in Ref. [129].[1] The original references on matter-coupled $N=1, d=5$ SUGRA are Refs. [693, 694].

## 9.1 The matter supermultiplets

Dirac spinors in $d=5$ have four complex components, as in $d=4$. As $d=5$ is odd, the only possibility to have smaller spinors would be to impose a Majorana (reality) condition, but, as explained in Appendix D.1.10, it is impossible to do that on single spinors. Thus, the minimal spinors in $d=5$ dimensions are pairs of spinors in which one spinor is related to the other by complex conjugation. In other words, one member of the pair is an independent Dirac spinor and the other one is basically its complex conjugate. Then, at first sight, it seems that we could do with only one member of each pair, but, as we are going to see, this doubling and the symplectic indices $i, j = 1, 2$ and $A, B = 1, \ldots, 2n_H$ that come with it are related to the action of the R-symmetry group ($\mathrm{USp}(2N)$ for $N$-SUEGRA in $d=5$ and therefore $\mathrm{USp}(2)$ for the case at hand) and necessary for the different couplings.

The supergravity multiplet consists of the graviton $e^a{}_\mu$, the graviphoton vector fields $A_\mu$, and the gravitino $\psi^i{}_\mu$, which is one (pair of) symplectic-Majorana spinors

$$\{e^a{}_\mu, \psi^i{}_\mu, A_\mu\}, \qquad (\psi^i{}_\mu)^* = \varepsilon_{ij}\psi^j{}_\mu\gamma_4. \tag{9.1}$$

The vector supermultiplets consist of a vector field $A_\mu$, a (symplectic-Majorana) *gaugino* $\lambda^i$, and a real scalar $\phi$. If we couple $n_V$ vector supermultiplets, and label them with $x = 1, \ldots, n_V$ the $x$th will have the field content

$$\{A^x{}_\mu, \lambda^{xi}, \phi^x\}, \qquad (\lambda^{xi})^* = \varepsilon_{ij}\lambda^{xj}\gamma_4. \tag{9.2}$$

In $d=5$ it is possible to dualize the vector fields into 2-form fields, but it is not possible to eliminate completely the vectors from the theory, due to the Chern–Simons-type couplings that these five-dimensional theories have. In the ungauged theories this dualization amounts, then, to an unnecessary doubling of some fields that can always be undone. However, after gauging this doubling cannot be undone in general, and the most general $d=5$ gauged supergravity theories need to be described using the 2-forms as well, as in Ref. [153].[2] We will not consider that possibility here.

On the other hand, according to the results in Section 2.6.3, in $d=5$ the most general symmetries of the equations of motion will be a subgroup of $\mathrm{GL}(\bar{n})$, $\bar{n} = n_V + 1$, which in general rotates the graviphoton into the matter vector fields and vice versa. It is convenient to use a notation that treats all vector fields on equal footing: add an index 0 to the graviphoton $A_\mu \to A^0{}_\mu$ and use indices $I, J = 0, \ldots, n_V$ to label all the vectors:

$$(A^I{}_\mu) = (A^0{}_\mu, A^x{}_\mu). \tag{9.3}$$

[1] The sign of the metric (to have mostly minus signature), multiplying all $\gamma^a$s by $+i$ and all $\gamma_a$s by $-i$ and setting $\kappa = 1/\sqrt{2}$. In particular, all the relevant definitions and conventions for gamma matrices and spinors and for real special geometry are collected in Appendices D.1.10 and H (see also Appendix G.1.1). Observe that the real special geometry identities in Appendix C of Ref. [153] are only valid for $\kappa = 1$. Furthermore, we will not consider 2-forms as in Ref. [153].

[2] See also Ref. [713] for a study of the most general couplings compatible with gauge symmetry in a generic (not necessarily supersymmetric) theory.

In general, the scalars $\phi^x$, which parametrize a real special manifold (see Appendix H) with $\sigma$-model metric $g_{xy}(\phi)$, transform non-linearly under the symmetries of the theory, and, as in the $N=2, d=4$ case, it is convenient to introduce $\bar{n}$ functions of the scalars transforming as the vector fields, i.e. in the vector representation of $\mathrm{GL}(\bar{n})$. These functions are called $h^I(\phi)$ and can be seen as coordinates in an $\bar{n}$-dimensional Riemannian space with a metric $a_{IJ}(\phi)$ which occurs in the theory as the kinetic matrix of the vector fields. Within this space, the real special manifold parametrized by the physical scalars $\phi^x$ can be described as the codimension-1 hypersurface defined by the equation

$$C_{IJK}h^I h^J h^K = 1, \tag{9.4}$$

where $C_{IJK}$ is a fully symmetric real constant tensor which characterizes completely the couplings in the vector sector: given this tensor, the functions $h^I(\phi)$ are just a set of solutions to the above constraint, and, with them and $C_{IJK}$, one can construct the functions $h_I(\phi)$ and the metrics $a_{IJ}(\phi)$ and $g_{xy}(\phi)$ using Eqs. (H.2), (H.4), and (H.11). The tensor $C_{IJK}$ itself determines the Chern–Simons terms.

Finally, as in the $N=2, d=4$ theories, there are hypermultiplets whose content is identical in five and four dimensions. We use a different notation, though: if there are $n_H$ hypermultiplets, the hyperscalars will be denoted by $q^X$, $X=1,\ldots,4n_H$, and the hyperinos by $\zeta^A$, $A=1,\ldots,2n_H$, where the $\mathrm{Sp}(2n_H)$ index $A$ is used in this five-dimensional case in the symplectic-Majorana condition

$$\{q^X, \zeta^A\}, \qquad (\zeta^A)^* = \mathbb{C}_{AB}\zeta^B\gamma_4. \tag{9.5}$$

The $\sigma$-model metric will be denoted by $g_{XY}(q)$.

As in the four-dimensional case, the pullback of the $\mathrm{SU}(2)$ connection of the quaternionic-Kähler manifold will act on objects with $\mathrm{SU}(2)$ indices (such as the gravitino and gaugino), and the pullback of the $\mathrm{Sp}(2n_H)$ component will act on objects with that kind of indices (the hyperino, the Quadbein, etc.).

## 9.2 The ungauged theory

An $N=1, d=5$ (minimal) ungauged supergravity coupled to $n_V$ vector multiplets and $n_H$ hypermultiplets is completely determined by the choice of real special and quaternionic-Kähler geometries. In turn, the former is completely determined by the tensor $C_{IJK}$, as we have discussed previously. Given these inputs, the bosonic part of the corresponding action is

$$\boxed{\begin{aligned} S = \int d^5x\sqrt{g}\,\Big\{ & R + \tfrac{1}{2}g_{xy}\partial_\mu\phi^x\partial^\mu\phi^y + \tfrac{1}{2}g_{XY}\partial_\mu q^X\partial^\mu q^Y \\ & - \tfrac{1}{4}a_{IJ}F^{I\,\mu\nu}F^J{}_{\mu\nu} + \tfrac{1}{12\sqrt{3}}C_{IJK}\frac{\varepsilon^{\mu\nu\rho\sigma\alpha}}{\sqrt{g}}F^I{}_{\mu\nu}F^J{}_{\rho\sigma}A^K{}_\alpha\Big\}. \end{aligned}} \tag{9.6}$$

We have already described all the objects that appear in this expression. The supersymmetry transformation rules for the fermionic fields, for vanishing fermions, are

$$\begin{aligned}
\delta_\epsilon \psi^i_\mu &= \mathsf{D}_\mu \epsilon^i - \tfrac{1}{8\sqrt{3}} h_I F^{I\,\alpha\beta}\left(\gamma_{\mu\alpha\beta} - 4 g_{\mu\alpha}\gamma_\beta\right)\epsilon^i,\\
\delta_\epsilon \lambda^{ix} &= \tfrac{1}{2}\left(\not\partial\phi^x - \tfrac{1}{2} h^x_I \not F^I\right)\epsilon^i,\\
\delta_\epsilon \zeta^A &= \tfrac{1}{2} f_X{}^{iA}\, \not\partial q^X \epsilon_i,
\end{aligned} \tag{9.7}$$

where $\mathsf{D}_\mu$ is the Lorentz- and $\mathrm{SU}(2)$-covariant derivative

$$\mathsf{D}_\mu \epsilon^i \equiv \{\partial_\mu - \tfrac{1}{4}\omega_\mu{}^{ab}\gamma_{ab}\}\epsilon^i + \epsilon^j \mathsf{A}_j{}^i{}_\mu, \tag{9.8}$$

and the $\mathfrak{su}(2)$ connection is the pullback of the $\mathfrak{su}(2)$ connection of the quaternionic-Kähler space

$$\mathsf{A}^r{}_\mu \equiv \partial_\mu q^X\, \omega_X{}^r, \qquad \mathsf{A}_j{}^i = i\mathsf{A}^r\, \sigma^r{}_j{}^i. \tag{9.9}$$

The supersymmetry transformation rules of the bosonic fields are

$$\begin{aligned}
\delta_\epsilon e^a{}_\mu &= \tfrac{i}{2}\bar\epsilon_i \gamma^a \psi^i_\mu,\\
\delta_\epsilon A^I{}_\mu &= -\tfrac{i\sqrt{3}}{2} h^I \bar\epsilon_i \psi^i_\mu + \tfrac{i}{2} h^I_x \bar\epsilon_i \gamma_\mu \lambda^{x\,i},\\
\delta_\epsilon \phi^x &= \tfrac{i}{2}\bar\epsilon_i \lambda^{x\,i},\\
\delta_\epsilon q^X &= -i f_{iA}{}^X \bar\epsilon^i \zeta^A.
\end{aligned} \tag{9.10}$$

We use the following notation for the (l.h.s. of the) equations of motion:

$$\mathcal{E}_a{}^\mu \equiv -\frac{1}{2\sqrt{g}}\frac{\delta S}{\delta e^a{}_\mu}, \quad \mathcal{E}^x \equiv -\frac{g^{xy}}{\sqrt{g}}\frac{\delta S}{\delta \phi^y}, \quad \mathcal{E}^X \equiv -\frac{g^{XY}}{\sqrt{g}}\frac{\delta S}{\delta q^Y}, \quad \mathcal{E}_I{}^\mu \equiv \frac{1}{\sqrt{g}}\frac{\delta S}{\delta A^I{}_\mu}. \tag{9.11}$$

They are given by

$$\begin{aligned}
\mathcal{E}_{\mu\nu} = G_{\mu\nu} &- \tfrac{1}{2}a_{IJ}\left(F^I{}_\mu{}^\rho F^J{}_{\nu\rho} - \tfrac{1}{4}g_{\mu\nu}F^{I\,\rho\sigma}F^J{}_{\rho\sigma}\right) + \tfrac{1}{2}g_{xy}\left(\partial_\mu\phi^x\partial_\nu\phi^y - \tfrac{1}{2}g_{\mu\nu}\partial_\rho\phi^x\partial^\rho\phi^y\right)\\
&+ \tfrac{1}{2}g_{XY}\left(\partial_\mu q^X\partial_\nu q^Y - \tfrac{1}{2}g_{\mu\nu}\partial_\rho q^X\partial^\rho q^Y\right),
\end{aligned} \tag{9.12}$$

$$\mathcal{E}^x = \mathfrak{D}_\mu\partial^\mu\phi^x + \tfrac{1}{4}g^{xy}\partial_y a_{IJ}F^{I\,\rho\sigma}F^J{}_{\rho\sigma}, \tag{9.13}$$

$$\mathcal{E}^X = \mathfrak{D}_\mu\partial^\mu q^X, \tag{9.14}$$

$$\mathcal{E}_I{}^\mu = \nabla_\nu(a_{IJ}F^{J\,\nu\mu}) + \tfrac{1}{4\sqrt{3}}C_{IJK}\frac{\varepsilon^{\mu\nu\rho\sigma\alpha}}{\sqrt{g}}F^J{}_{\nu\rho}F^J{}_{\sigma\alpha}. \tag{9.15}$$

In these equations $\mathfrak{D}_\mu$ is the covariant derivative in the spacetime and in the corresponding scalar manifold. Equation (9.14) can be interpreted as stating that $q$ is a harmonic map from spacetime to the quaternionic-Kähler manifold.

We could have added to these equations the Bianchi identities of the vector field strengths

$$\mathcal{E}^I{}_{\mu\nu\rho} \equiv 3\nabla_{[\mu}F^I{}_{\nu\rho]}. \tag{9.16}$$

Observe that, as in the four-dimensional case, the hyperscalars do not couple to any of the fields in the vector supermultiplets in the action. Furthermore, they only appear in the gravitino and gaugino supersymmetry transformation rules through the pullback $\mathfrak{su}(2)$ connection. It is clear that they can always be eliminated from these ungauged theories by setting them to some constant value.

### 9.2.1 Examples

*Pure, ungauged, $N=1, d=5$ supergravity.* To obtain the simplest $N=1, d=5$ supergravity, with no matter multiplets and no gaugings, we simply have to truncate all the hypermultiplets from the general framework, setting them to zero, and set the indices $I, J, K$ to their only allowed value, 0. The only component tensor $C_{IJK}$, $C_{000}$, can be normalized to $C_{000}=1$ so $h^0 = a_{00} = 1$ as well. Substituting these values into the general action and supersymmetry rules Eq. (9.6), (9.7), and (9.10), we get the bosonic action and supersymmetry transformation rules of pure $N=1, d=5$ supergravity [382]:

$$\boxed{S = \int d^5x\sqrt{g}\left\{R - \tfrac{1}{4}F^{\mu\nu}F_{\mu\nu} + \tfrac{1}{12\sqrt{3}}\frac{\varepsilon^{\mu\nu\rho\sigma\alpha}}{\sqrt{g}}F_{\mu\nu}F_{\rho\sigma}A_\alpha\right\},} \tag{9.17}$$

and

$$\boxed{\begin{aligned} \delta_\epsilon e^a{}_\mu &= \tfrac{i}{2}\bar{\epsilon}_i\gamma^a\psi^i_\mu, \\ \delta_\epsilon\psi^i_\mu &= \nabla_\mu\epsilon^i - \tfrac{1}{8\sqrt{3}}F^{\alpha\beta}\left(\gamma_{\mu\alpha\beta} - 4g_{\mu\alpha}\gamma_\beta\right)\epsilon^i, \\ \delta_\epsilon A_\mu &= -\tfrac{i\sqrt{3}}{2}\bar{\epsilon}_i\psi^i_\mu, \end{aligned}} \tag{9.18}$$

and the equations of motion reduce to

$$\mathcal{E}_{\mu\nu} = G_{\mu\nu} - \tfrac{1}{2}\left(F_\mu{}^\rho F_{\nu\rho} - \tfrac{1}{4}g_{\mu\nu}F^{\rho\sigma}F_{\rho\sigma}\right), \tag{9.19}$$

$$\mathcal{E}^\mu = \nabla_\nu F^{\nu\mu} + \tfrac{1}{4\sqrt{3}}\frac{\varepsilon^{\mu\nu\rho\sigma\alpha}}{\sqrt{g}}F_{\nu\rho}F_{\sigma\alpha}. \tag{9.20}$$

The dimensional reduction of this action to an $N=2, d=4$ SUEGRA with one vector multiplet (the $t^3$ mode, reviewed on p. 235) and the truncation of this vector multiplet to recover pure $N=2, d=4$ SUEGRA are studied in Section 15.2.5.

*An $n_V = 1$, ungauged, $N=1$, $d=5$ supergravity.* To add some more juice to the pure supergravity theory we can couple it to a single vector supermultiplet, so the indices $I, J, K = 0, 1$. There are several inequivalent choices of $C_{IJK}$ tensor (and, correspondingly, of real special geometry) even if all $\sigma$-models with one-dimensional target space are equivalent up to field redefinitions. Of course, these choices differ in the couplings of the single real scalar to the vector fields. We are going to consider a tensor whose only non-vanishing components are $C_{011} = \frac{1}{3}$ (and permutations). This model, considered in Ref. [939], is interesting because it can be obtained by dimensional reduction of minimal $N=(1,0), d=6$ supergravity.

The first step in the construction of the model consists in solving the constraint

$$C_{IJK}h^I h^J h^K = h^0(h^1)^2 = 1. \tag{9.21}$$

This equation has two solutions which correspond to two different patches of the real special manifold that need to be covered with different coordinates. Each patch gives rise to a different *branch* of the same theory. In terms of a physical, unconstrained, scalar (which we call $\phi$ in both cases), to simplify the notation the two solutions, labeled by $\sigma = \pm 1$, are parametrized as follows:

$$\begin{aligned} h^0_{(\sigma)} &= e^{\sqrt{\frac{2}{3}}\phi}, & h^1_{(\sigma)} &= \sigma\, e^{-\frac{1}{\sqrt{6}}\phi}, \\ h_{(\sigma)\,0} &= \tfrac{1}{3}e^{-\sqrt{\frac{2}{3}}\phi}, & h_{(\sigma)\,1} &= \tfrac{2}{3}\sigma\, e^{\frac{1}{\sqrt{6}}\phi}. \end{aligned} \tag{9.22}$$

The scalar metric $g_{\phi\phi}$ and the vector field strengths metric $a_{IJ}$ take exactly the same values in both branches:

$$g_{\phi\phi} = 1, \qquad a_{IJ} = \frac{1}{3}\begin{pmatrix} e^{-2\sqrt{\frac{2}{3}}\phi} & 0 \\ 0 & 2e^{\sqrt{\frac{2}{3}}\phi} \end{pmatrix}, \tag{9.23}$$

which means that the bosonic action takes exactly the same form in both branches, namely

$$\boxed{\begin{aligned} S = \int d^5x\sqrt{g}\,\Big\{ & R + \tfrac{1}{2}(\partial\phi)^2 - \tfrac{1}{12}e^{-2\sqrt{\frac{2}{3}}\phi}(F^0)^2 - \tfrac{1}{6}e^{\sqrt{\frac{2}{3}}\phi}(F^1)^2 \\ & + \frac{1}{12\sqrt{3}}\frac{\varepsilon^{\mu\nu\rho\sigma\alpha}}{\sqrt{g}} F^1{}_{\mu\nu}F^1{}_{\rho\sigma}A^0{}_{\alpha} \Big\}, \end{aligned}} \tag{9.24}$$

where we have performed several integrations by parts to simplify the Chern–Simons term.

The supersymmetry rules are different, though. For the fermions in particular,

$$\begin{aligned}
\delta_\epsilon \psi^i_\mu &= \nabla_\mu \epsilon^i - \tfrac{1}{24\sqrt{3}} \left[ e^{-\sqrt{\frac{2}{3}}\phi} F^{0\,\alpha\beta} + 2\sigma\, e^{\frac{1}{\sqrt{6}}\phi} F^{1\,\alpha\beta} \right] (\gamma_{\mu\alpha\beta} - 4 g_{\mu\alpha}\gamma_\beta)\, \epsilon^i, \\
\delta_\epsilon \lambda^i &= \tfrac{1}{2} \left( \not{\partial}\phi + \tfrac{1}{3\sqrt{2}} \left[ e^{-\sqrt{\frac{2}{3}}\phi} \not{F}^0 - \sigma\, e^{\frac{1}{\sqrt{6}}\phi} \not{F}^1 \right] \right) \epsilon^i.
\end{aligned} \tag{9.25}$$

The dimensional reduction of this model in a circle gives the $ST^2$ model mentioned on p. 234: the reduction of the $N=1, d=5$ supergravity multiplet gives the supergravity multiplet and a vector supermultiplet of $N=2, d=4$ (see the previous example). It is not difficult to see that the reduction of an $N=1, d=5$ vector supermultiplet gives exactly one $N=2, d=4$ vector supermultiplet,[3] and, therefore, the reduction of an $N=1, d=5$ supergravity coupled to $n_V$ vector supermultiplets gives an $N=2, d=4$ supergravity coupled to $n_V+1$. If the five-dimensional theory is characterized by the symmetric tensor $C_{IJK}$, it is clear that the four-dimensional theory must also be determined by it. In fact, as we will see at the end of Section 15.2.5, the four-dimensional theory is a cubic model (see p. 231) characterized by the same tensor:

$$d_{ijk} \sim 6 C_{i-1\, j-1\, k-1}. \tag{9.26}$$

According to this rule, pure $N=1, d=5$ supergravity gives rise to the $t^3$ model, characterized by a single component, $d_{111}$, which corresponds to the single component $C_{000}$. The model that we are considering gives rise to the cubic model with only non-vanishing component $d_{122}$ (and permutations) which is the $\mathrm{ST}^2$. The $N=2, d=4$ STU model studied on p. 234 originates in a five-dimensional model characterized by a tensor with only non-zero component $C_{012}$ (up to permutations). This model is sometimes referred to as the five-dimensional STU model and will be our next (and final) example.

The special Kähler geometries that can be obtained from (real) very special geometries by dimensional reduction on a circle are called complex very special geometries. A list of the symmetric ones is given in Table H.1. In all these cases, the symmetric tensor that defines the geometry satisfies the property Eq. (7.49). The last four rows contain the *magic models* discussed on p. 232 for the four-dimensional case.

*The five-dimensional $STU$ model.* This model has two vector multiplets and is defined by $C_{IJK} = \frac{1}{3!}|\varepsilon_{IJK}|$ or, equivalently, by the polynomial

$$h^0 h^1 h^2 = 1. \tag{9.27}$$

[3] Using the machinery that we will develop in Chapter 15 it is easy to see that a five-dimensional real scalar gives a four-dimensional real scalar, whereas a five-dimensional vector gives a four-dimensional vector and a real scalar which can be combined with the other one to give a complex scalar with the same index. Each pair of five-dimensional symplectic-Majorana spinors is equivalent to a five-dimensional Dirac spinor, which is also a four-dimensional Dirac spinor (four independent complex components in both cases) and can be decomposed in a pair of independent Weyl spinors.

This model also has several branches associated with the possible signs of the $h^I$: all positive or one positive and the other two negative (four in total). We will consider the first only, for the sake of simplicity. Using the above constraint, the dual variables and the metric are given by

$$h_I = \frac{1}{3h^I}, \qquad a_{IJ} = 3\delta_{I(J)}(h_{(J)})^2 \quad \text{(no sum over } J\text{)}. \tag{9.28}$$

A convenient parametrization of the scalar manifold that gives $g_{xy} = \delta_{xy}$ is

$$h^0 = e^{\sqrt{6}\phi^1}, \qquad h^1 = e^{-\sqrt{\frac{3}{2}}\phi^1 - \frac{3}{\sqrt{2}}\phi^2}, \qquad h^2 = e^{-\sqrt{\frac{3}{2}}\phi^1 + \frac{3}{\sqrt{2}}\phi^2}. \tag{9.29}$$

The action takes the explicit form

$$\boxed{\begin{aligned} S = \int d^5x\sqrt{g}\,\Big\{ & R + \tfrac{1}{2}\partial_\mu\phi^x\partial^\mu\phi^x - \tfrac{1}{12}e^{-2\sqrt{6}\phi^1}(F^0)^2 - \tfrac{1}{12}e^{\sqrt{6}\phi^1+3\sqrt{2}\phi^2}(F^1)^2 \\ & - \tfrac{1}{12}e^{\sqrt{6}\phi^1-3\sqrt{2}\phi^2}(F^2)^2 + \frac{1}{12\sqrt{3}}\frac{\varepsilon^{\mu\nu\rho\sigma\alpha}}{\sqrt{g}}F^0{}_{\mu\nu}F^1{}_{\rho\sigma}A^2{}_\alpha \Big\}. \end{aligned}} \tag{9.30}$$

The supersymmetric transformation rules can be easily obtained using the above parametrization and the general expressions.

## 9.3 The gauged theory

As usual, we must start with a study of the global symmetries of generic $N=1, d=5$ supergravities.

### 9.3.1 The global symmetries

Since in $d=5$ dimensions there are no fields whose electric–magnetic dual is another field of the same kind, there are no symmetries of the equations of motion (dualities) which are not symmetries of the action, and these will be either R-symmetries or isometries of the scalar manifolds preserving the relevant structures (real special and quaternionic-Kähler geometry). Thus, we have

$$\mathrm{G} = \mathrm{G_V} \times \mathrm{G_{hyper}} \times \mathrm{SU}(2)_\mathrm{R}. \tag{9.31}$$

Let us review each of these factors. We shall be brief, since their nature and actions are similar to those that occur in the four-dimensional theories.

$\mathrm{SU}(2)_\mathrm{R}$. As in the $N=2, d=4$ case, this group acts in the fundamental representation on the indices $i, j$ carried by all the fermion fields, on the Quadbeins $f^{iA}{}_X$, and (via the adjoint action) on the $\mathrm{SU}(2)$ connection $\mathsf{A}^r{}_X$ of the quaternionic-Kähler manifold but not on any of the physical bosonic fields.

$G_V$. The preservation of the real special structure by the isometries of the target-space metric $g_{xy}(\phi)$, studied in Appendix H.1, requires two conditions:

1. The functions $h^I(\phi)$ must be invariant fields annihilated by the covariant Lie derivative

$$\mathbb{L}_A h^I \equiv (\mathcal{L}_A - T_A)h^I = k_A{}^x \partial_x h^I - T_A{}^I{}_J h^J = 0, \tag{9.32}$$

where the $k_A{}^x(\phi)$ are the Killing vectors and the $T_A$s are, according to the general results in Section 2.6.3, a matrix representation of the Lie algebra isometry group $\mathfrak{g}_V \subset \mathfrak{so}(\bar{n})$ (they are antisymmetric matrices),

$$[k_A, k_B] = -f_{AB}{}^C k_C, \qquad [T_A, T_B] = f_{AB}{}^C T_C. \tag{9.33}$$

2. The symmetric tensor $C_{IJK}$ must be invariant under the compensating $\mathrm{GL}(\bar{n})$ transformations

$$\mathbb{L}_A C_{IJK} = 3T_A{}^L{}_{(I} C_{JK)L} = 0. \tag{9.34}$$

These conditions imply, in particular, that the kinetic matrix $a_{IJ}(\phi)$ is also an invariant field and they lead to the simple expression Eq. (H.23) for the Killing vectors, which we reproduce here for convenience:

$$k_A{}^x = -\sqrt{3} T_A{}^I{}_J h_I^x h^J. \tag{9.35}$$

$G_{\mathrm{hyper}}$. The description of this group is identical to the four-dimensional case with some changes in the notation and we will not repeat the discussion here. Let's just mention that Fayet–Iliopoulos terms that gauge the whole R-symmetry group or a $\mathrm{U}(1)_{\mathrm{R}}$ subgroup can also be introduced via constant triholomorphic momentum maps.

### 9.3.2 Examples

*The global symmetries of pure, ungauged, $N=1$, $d=5$ supergravity.* In the absence of scalars, the only symmetry of the theory is $\mathrm{SU}(2)_{\mathrm{R}}$, which acts on the corresponding indices of the fermions in the standard form.

*The global symmetries of the $n_V=1$, ungauged, $N=1$, $d=5$ supergravity.* Apart from R-symmetry, the one-dimensional $\sigma$-model is invariant under constant ($\mathbb{R}$ or, more commonly in the literature, $\mathrm{SO}(1,1)$) shifts of the single scalar, and it is not difficult to see by simple inspection that the effect of these shifts can be compensated for by rescalings of the vector fields, which just happen to leave the Chern–Simons term invariant. The Killing vector and the associated $\mathfrak{gl}(2)$ matrix that generate this $\mathrm{SO}(1,1)$ symmetry are

$$k^\phi = 1, \qquad T = \sqrt{\tfrac{2}{3}} \begin{pmatrix} 1 & 0 \\ 0 & -\tfrac{1}{2} \end{pmatrix}. \tag{9.36}$$

Since the two vectors scale under this symmetry we will not be able to use them to gauge it. Unfortunately, more complicated models are necessary to provide examples of gaugings of the isometries of the real special geometry.

### 9.3.3 The gauging of the global symmetries

The procedure used to gauge the global symmetries of $N=1, d=5$ supergravities[4] does not differ substantially from the one followed in the previous cases (the discussion of the $N=2, d=4$ case applies almost without any change to the present case[5]) and we shall be extremely brief. The construction of the covariant derivatives of all the fields when the isometries of the real special geometry are gauged is reviewed in Appendix J.6. The construction of the covariant derivatives of all the fields when the isometries of the quaternionic-Kähler manifold are gauged is reviewed in Appendix J.5 with the change in notation explained in Appendices G.1.1 and J.5.1.

The gauge-covariant derivatives of the scalars and vector field strengths are given by the standard expressions

$$\mathfrak{D}_\mu \phi^x = \partial_\mu \phi^x + gA^I{}_\mu k_I{}^x, \tag{9.37}$$

$$\mathfrak{D}_\mu q^X = \partial_\mu q^X + gA^I{}_\mu \mathsf{k}_I{}^X, \tag{9.38}$$

$$F^I{}_{\mu\nu} = 2\partial_{[\mu} A^I{}_{\nu]} + gf_{JK}{}^I A^J{}_{[\mu} A^K{}_{\nu]}, \tag{9.39}$$

and those of the fermions are

$$\mathfrak{D}_\mu \epsilon^i = \nabla_\mu \epsilon^i + \hat{\mathsf{A}}_{\mu\, j}{}^i \epsilon^j, \tag{9.40}$$

$$\mathfrak{D}_\mu \psi^i_\nu = \nabla_\mu \psi^i_\nu + \hat{\mathsf{A}}_{\mu\, j}{}^i \psi^j_\nu, \tag{9.41}$$

$$\mathfrak{D}_\mu \lambda^{ix} = \nabla_\mu \lambda^{ix} + \left\{\mathfrak{D}_\mu \phi^z \Gamma_{zy}{}^x + gA^I{}_\mu \partial_y k_I{}^x\right\} \lambda^{iy} + \hat{\mathsf{A}}_{\mu\, j}{}^i \lambda^{jx}, \tag{9.42}$$

$$\mathfrak{D}_\mu \zeta^A = \nabla_\mu \zeta^A + \Delta_{\mu\, B}{}^A \zeta^B. \tag{9.43}$$

In these expressions, $\nabla_\mu$ is the Lorentz- and general-covariant derivative and includes the spin connection (when acting on the gravitino, also the Levi-Civita connection), $\Delta_{\mu\, A}{}^B$ the pullback of the $\mathrm{Sp}(2n_H)$ part of the spin connection of the quaternionic-Kähler manifold (for $n_H$ hypermultiplets), and $\hat{\mathsf{A}}_{\mu\, i}{}^j$ is the following deformation of the pullback of the $\mathrm{SU}(2)$ ($\mathsf{A}$) connections of the quaternionic-Kähler manifolds:

$$\hat{\mathsf{A}}_{\mu\, i}{}^j = \mathsf{A}_{\mu\, i}{}^j + \tfrac{1}{2} gA^I{}_\mu \mathsf{P}_{I\, i}{}^j = i(\mathsf{A}_\mu{}^r + \tfrac{1}{2} gA^I{}_\mu \mathsf{P}^r_I)\sigma^r{}_i{}^j. \tag{9.44}$$

[4] The gauging of the global symmetries of $N=1, d=5$ supergravity theories coupled to vector supermultiplets was first considered in Refs. [693, 694], and it was generalized to theories with tensors (2-forms, which we are not considering here) in Refs. [695, 696]. Hypermultiplets and their gaugings were first considered in Ref. [321]. More general gaugings of the tensor multiplets and $N=1, d=5$ supergravities that do not admit actions were considered in Refs. [152, 153].

[5] A difference from the four-dimensional case is the presence of Chern–Simons terms, which contain not just vector field strengths (which are simply replaced by their gauge-covariant deformations) but also the vector fields without any derivatives. The gauge covariantization of the Chern–Simons term is more complicated and requires the addition of several terms, as we shall see.

The supersymmetry transformation rules of the bosonic fields are exactly the same as in the ungauged case but the supersymmetry transformation rules for the fermionic fields include fermion shifts to which we do not give any specific names. Evaluated on vanishing fermions, they take the form[6]

$$\begin{aligned}
\delta_\epsilon \psi^i_\mu &= \mathfrak{D}_\mu \epsilon^i - \tfrac{1}{8\sqrt{3}} h_I F^{I\,\alpha\beta}\left(\gamma_{\mu\alpha\beta} - 4 g_{\mu\alpha}\gamma_\beta\right)\epsilon^i + \tfrac{1}{2\sqrt{3}} g h^I \gamma_\mu \epsilon^j \mathsf{P}_{I\,j}{}^i, \\
\delta_\epsilon \lambda^{i\,x} &= \tfrac{1}{2}\left(\not{\mathfrak{D}}\phi^x - \tfrac{1}{2} h^x_I \not{F}^I\right)\epsilon^i + g h^{Ix}\epsilon^j \mathsf{P}_{I\,j}{}^i, \\
\delta_\epsilon \zeta^A &= \tfrac{1}{2} f_X{}^{iA}\left(\not{\mathfrak{D}} q^X + \sqrt{3} g h^I \mathsf{k}_I{}^X\right)\epsilon_i.
\end{aligned} \tag{9.45}$$

In order for the action to be invariant under the new supersymmetry transformation rules with fermion shifts, a scalar potential quadratic in the fermion shifts must be added:

$$\begin{aligned}
V(\phi, q) = -g^2 \Big[&\left(4h^I h^J - 2g^{xy} h^I_x h^J_y\right) \vec{\mathsf{P}}_I \cdot \vec{\mathsf{P}}_J \\
&- \tfrac{3}{2} h^I h^J \left(k_I{}^x k_J{}^y g_{xy} + \mathsf{k}_I{}^X \mathsf{k}_J{}^Y g_{XY}\right)\Big],
\end{aligned} \tag{9.46}$$

and, finally, the bosonic action of $N = 1, d = 5$ gauged supergravity is given by

$$\begin{aligned}
S = \int d^5x \sqrt{g}\, \Big\{ &R + \tfrac{1}{2} g_{xy} \mathfrak{D}_\mu \phi^x \mathfrak{D}^\mu \phi^y + \tfrac{1}{2} g_{XY} \mathfrak{D}_\mu q^X \mathfrak{D}^\mu q^Y - V(\phi, q) \\
&- \tfrac{1}{4} a_{IJ} F^{I\,\mu\nu} F^J{}_{\mu\nu} + \tfrac{1}{12\sqrt{3}} C_{IJK} \frac{\varepsilon^{\mu\nu\rho\sigma\alpha}}{\sqrt{g}} \Big[ F^I{}_{\mu\nu} F^J{}_{\rho\sigma} A^K{}_\alpha \\
&- \tfrac{1}{2} g f_{LM}{}^I F^J{}_{\mu\nu} A^K{}_\rho A^L{}_\sigma A^M{}_\alpha + \tfrac{1}{10} g^2 f_{LM}{}^I f_{NP}{}^J A^K{}_\mu A^L{}_\nu A^M{}_\rho A^N{}_\sigma A^P{}_\alpha \Big] \Big\}.
\end{aligned} \tag{9.47}$$

The equations of motion, for which we use the same notation as in the ungauged case, Eq. (9.11), are

$$\begin{aligned}
\mathcal{E}_{\mu\nu} = G_{\mu\nu} &- \tfrac{1}{2} a_{IJ}\left(F^I{}_\mu{}^\rho F^J{}_{\nu\rho} - \tfrac{1}{4} g_{\mu\nu} F^{I\,\rho\sigma} F^J{}_{\rho\sigma}\right) + \tfrac{1}{2} g_{xy}\left(\mathfrak{D}_\mu \phi^x \mathfrak{D}_\nu \phi^y - \tfrac{1}{2} g_{\mu\nu} \mathfrak{D}_\rho \phi^x \mathfrak{D}^\rho \phi^y\right) \\
&+ \tfrac{1}{2} g_{XY}\left(\mathfrak{D}_\mu q^X \mathfrak{D}_\nu q^Y - \tfrac{1}{2} g_{\mu\nu} \mathfrak{D}_\rho q^X \mathfrak{D}^\rho q^Y\right) + \tfrac{1}{2} g_{\mu\nu} V,
\end{aligned} \tag{9.48}$$

[6] The fermion shift $\frac{\sqrt{3}}{2} g h^I K_I{}^x$ in $\delta_\epsilon \lambda^{i\,x}$ vanishes identically for the gaugings considered due to the property $h^I k_I{}^x = 0$.

$$\mathcal{E}^x = \mathfrak{D}_\mu \mathfrak{D}^\mu \phi^x + \tfrac{1}{4} g^{xy} \partial_y a_{IJ} F^{I\,\rho\sigma} F^J{}_{\rho\sigma} + g^{xy} \partial_y V, \tag{9.49}$$

$$\mathcal{E}^X = \mathfrak{D}_\mu \mathfrak{D}^\mu q^X + g^{XY} \partial_Y V, \tag{9.50}$$

$$\mathcal{E}_I{}^\mu = \mathfrak{D}_\nu F_I{}^{\nu\mu} + \tfrac{1}{4\sqrt{3}} \frac{\varepsilon^{\mu\nu\rho\sigma\alpha}}{\sqrt{g}} C_{IJK} F^J{}_{\nu\rho} F^k{}_{\sigma\alpha} + g\left(k_{I\,x} \mathfrak{D}^\mu \phi^x + \mathsf{k}_{I\,X} \mathfrak{D}^\mu q^X\right). \tag{9.51}$$

### 9.3.4 Examples of gauged $N=1, d=5$ supergravities

*Pure $N=1, d=5$ "gauged" (AdS) supergravity.* With one vector field we can only gauge a $\mathrm{U}(1)_{\mathrm{R}}$ subgroup of the R-symmetry group by introducing a constant triholomorphic momentum map $\mathsf{P}_0^r$ that we can normalize: $\mathsf{P}_0^r \mathsf{P}_0^r = 1$. $\mathsf{P}_0^r$ is just an $\mathfrak{su}(2)$ vector that selects the $\mathfrak{u}(1)$ direction to be gauged using $A^0{}_\mu$ as a vector field.

Since no boson is charged under this symmetry, the only change that the gauging introduces in the bosonic action of the pure, ungauged, theory is the addition of a negative (anti-de Sitter-type) cosmological constant $\Lambda = V/3 = -\frac{4}{3} g^2$. Since all the fermions are charged, their covariant derivatives are modified and fermion shifts appear in the gravitino supersymmetry transformation rule. Therefore (suppressing the graviphoton index 0 for simplicity) we have

$$\boxed{S = \int d^5x \sqrt{g}\left\{R - \tfrac{1}{4} F^{\mu\nu} F_{\mu\nu} + \tfrac{1}{12\sqrt{3}} \frac{\varepsilon^{\mu\nu\rho\sigma\alpha}}{\sqrt{g}} F_{\mu\nu} F_{\rho\sigma} A_\alpha + 4g^2\right\}} \tag{9.52}$$

and

$$\boxed{\delta_\epsilon \psi^i_\mu = \nabla_\mu \epsilon^i + \tfrac{i}{2} g A_\mu \mathsf{P}^r \sigma^r{}_j{}^i \epsilon^j - \tfrac{1}{8\sqrt{3}} F^{\alpha\beta}\left(\gamma_{\mu\alpha\beta} - 4 g_{\mu\alpha}\gamma_\beta\right)\epsilon^i + \tfrac{i}{2\sqrt{3}} g \gamma_\mu \mathsf{P}^r \sigma^r{}_j{}^i \epsilon^j.} \tag{9.53}$$

(The boson supersymmetry transformation rules remain as in Eq. (9.53).) The only modification to the bosonic equations of motion comes from the cosmological constant.

# 10
# Conserved charges in general relativity

The definition of conserved charges in GR (and, in general, in non-Abelian gauge theories) is a very important and rather subtle subject, which is related to the definition of the energy–momentum tensor of the gravitational field. As we saw in the construction of the SRFT of gravity, perturbatively (that is, for asymptotically flat, well-behaved gravitational fields), GR gives a unique energy–momentum (Poincaré) tensor. It is natural to ask whether there is a fully general-covariant energy–momentum tensor for the gravitational field that would reduce to this in the weak-field limit. Many people (starting from Einstein himself) have unsuccessfully tried to find such a tensor, the current point of view being that it does not exist and that we have to content ourselves with energy–momentum *pseudotensors* for the gravitational field, which are covariant only under a restricted group of coordinate transformations (in most cases, Poincaré's). This, in fact, would be one of the characteristics of the gravitational field tied to the PEGI (see e.g. the discussion in Section 2.7 of Ref. [360]) that says that all the physical effects of the gravitational field (and one should include amongst them its energy density) can be locally eliminated by choosing a locally inertial coordinate system.[1]

The most important consequence of the absence of a fully general-covariant energy–momentum tensor for the gravitational field is the non-localizability of the gravitational energy: only the total energy of a spacetime is well defined (and conserved) because the integral of the energy–momentum pseudotensor over a finite volume would be dependent on the choice of coordinates. Some people find this unacceptable and, thus, the search for the general-covariant tensor goes on.[2]

Apart from the problem of the gravitational energy–momentum tensor, the definition of conserved quantities in GR has many interesting points. Several approaches have been

[1] It must be mentioned that, in spite of all these considerations, the teleparallel approach to GR gives a local expression for the energy. This is, in fact, the reason why Møller [962] was led to the study of this class of theories.

[2] For instance, one can argue that the gravity field is actually characterized by the curvature tensor, not by the metric tensor. Even if we locally make the metric tensor flat by a coordinate transformation, we cannot do the same with the Riemann tensor. Thus one could look for energy–momentum tensors for the gravitational field constructed from the Riemann tensor. These are usually called "super-energy–momentum tensors" and an example of them is the Bel–Robinson tensor.

proposed and here we are going to study two: the construction of an energy–momentum pseudotensor for the gravitational field and the Noether approach. In both approaches there is a great deal of arbitrariness, and in Section 10.1 we will study and compare several different results given in the literature in the weak-field limit, finding complete agreement with and a deep relation to the massless spin-2 relativistic field theory studied in Chapter 3.

## 10.1 The traditional approach

As we have stressed several times, the metric (or Rosenfeld) energy–momentum tensor of any general-covariant Lagrangian always satisfies (on-shell) the equation

$$\nabla_\mu T_{\text{matter}}{}^{\mu\nu} = 0, \tag{10.1}$$

as a direct consequence of general covariance. This equation is crucial for the consistency of the theory. Furthermore, it is the covariantization of the Minkowskian energy–momentum-conservation equation

$$\partial_\mu T_{\text{matter}}{}^{\mu\nu} = 0, \tag{10.2}$$

which is discussed at length in Chapter 2, and from which we can derive local conservation laws of the mass, momentum, and angular momentum, and, in general, of those charges related to the invariance of a theory under certain coordinate transformations.

In curved spacetime, however, Eq. (10.1) is not equivalent to a continuity equation for the tensor density $\sqrt{|g|}\, T_{\text{matter}}{}^{\mu\nu}$ that holds in Minkowski spacetime. Actually, we can rewrite Eq. (10.1) in the form

$$\partial_\mu \left( \sqrt{|g|}\, T_{\text{matter}}{}^{\mu\nu} \right) = -\Gamma_{\rho\sigma}{}^{\nu} T_{\text{matter}}{}^{\rho\sigma}, \tag{10.3}$$

and, in general, the r.h.s. of this equation does not vanish. From this equation we cannot derive any local conservation law.

In a sense this was to be expected: only the *total* (matter plus gravity) energy and momentum should be conserved[3] and, therefore, we can only hope to be able to find local conservation laws for the total energy–momentum tensor. Now, how is the gravity energy–momentum tensor defined in GR? This is an old problem of GR.[4] It is clear that we cannot use the same definition (Rosenfeld's) as for the matter energy–momentum tensor because that leads to a total energy–momentum tensor that vanishes identically on-shell. On the other hand, if we found a covariantly divergenceless gravitational energy–momentum tensor, the total energy–momentum tensor would have the same problem as the matter one.

In fact, it can be argued, on the basis of the PEGI, that it is impossible to define a fully general-covariant gravitational energy–momentum tensor: according to the PEGI we can remove all the physical effects of a gravitational field locally, at any given point, by using an appropriate (free-falling) reference frame. This means that we could make the gravitational energy–momentum tensor vanish at any given point. However, that would mean that the

[3] Actually, the coupling of gravity to the *total*, conserved, energy–momentum tensor was the main principle leading in Chapter 3 to GR.

[4] Some early references on the energy–momentum tensor of the gravitational field are Refs. [489, 901, 490, 491, 1109, 109, 861].

energy–momentum tensor vanishes at any given point in any reference frame and, therefore, it is identically zero.[5]

Instead of being a problem, the lack of a gravitational energy–momentum (general-covariant) *tensor* really tells us that we should not be looking for such a tensor: after all, what we want is a total energy–momentum tensor satisfying the continuity equation $\partial_\mu T_{\text{total}}{}^{\mu\nu} = 0$, which is not a tensor equation. At most, it is a tensor equation with respect to the Poincaré group, if $T_{\text{total}}{}^{\mu\nu}$ behaves as a Lorentz tensor. Then we should simply be looking for a gravitational energy–momentum *pseudotensor* $t^{\mu\nu}$ transforming as a Lorentz tensor but not as a general-covariant tensor and such that

$$\partial_\mu \sqrt{|g|}\,(T_{\text{matter}}{}^{\mu\nu} + t^{\mu\nu}) = 0. \tag{10.4}$$

This should remind the reader of the self-consistency problem of the SRFT of gravitation that we studied in Chapter 3 in which we wanted to find the energy–momentum tensor of the gravitational field with respect to the vacuum which was Minkowski spacetime.

Another point to be stressed is that it looks as if we are forced to abandon general covariance to define conserved quantities. This is not so surprising: conserved quantities are in general naturally associated with the symmetries of the vacuum, not with the full symmetry of the theory. The vacuum is generically invariant under a finite-dimensional global symmetry group, in this case the Poincaré group. The conserved quantities we are after (momentum and angular momentum) are associated with that group.[6] In asymptotically flat spacetimes, only the infinity will have the invariances of the energy–momentum pseudotensor, and, thus, only integrals over the boundary of (timelike hypersurfaces of) the whole spacetime will give well-defined conserved quantities. This implies the non-localizability of the energy mentioned in the introduction.

Our task now will be to find the gravitational energy–momentum pseudotensor and use it to define conserved quantities. Many candidates for a gravitational energy–momentum pseudotensor have been proposed in the literature. We are going to review just two of them that are physically very appealing: the Landau–Lifshitz pseudotensor [883], for asymptotically flat spacetimes, and the Abbott–Deser pseudotensor [1], for spacetimes with general asymptotics.

---

[5] This argument is not completely correct, though. In GR we can make the metric flat and its first derivatives vanishing at any given point, but not the second derivatives of the metric (i.e. the curvature). Although one can argue that, at one point (or any small enough neighborhood of a point), these non-vanishing derivatives will produce no observable physical effect (for instance, we need spatially separated test particles in order to measure tidal forces), this is not enough to say that the gravitational energy–momentum tensor will vanish identically at that point. In fact, the Landau–Lifshitz energy–momentum pseudotensor that we are going to study is precisely identified with the non-vanishing piece of the Einstein tensor at a point in a free-falling reference frame in which the metric is Minkowski's. The real problem, which is at the very foundations of GR, is that we do not have a good description of gravity in free-falling reference frames. That description could be covariantized (as happens with most other fields whose Lagrangians are well known in free-falling frames) and an energy–momentum tensor of the gravitational field and its coupling to itself could be found.

[6] We will see later what can be done for vacua different from Minkowski spacetime and for conserved quantities that are not necessarily associated with symmetries of the vacuum.

### 10.1.1 The Landau–Lifshitz pseudotensor

The main physical idea behind the definition of the Landau–Lifshitz energy–momentum pseudotensor is precisely that gravity can be locally eliminated at point P by using a free-falling coordinate system at P. Then, the starting point is to choose, for instance, Riemann normal coordinates described by Eq. (3.268) at the given point P where we want to define the energy–momentum pseudotensor. In this coordinate system, at point P the equation satisfied by the matter energy–momentum tensor takes the form

$$\nabla_\mu T_{\text{matter}}{}^{\mu\nu} = \partial_\mu T_{\text{matter}}{}^{\mu\nu} = 0, \tag{10.5}$$

and the matter energy–momentum tensor is conserved in the usual sense there because we have eliminated the gravitational field, its interaction with matter, and its own energy–momentum pseudotensor through the choice of coordinates. Thus, in this coordinate system, the gravitational energy–momentum pseudotensor vanishes.

Technically, this equation is satisfied identically due to the Bianchi identity of the r.h.s. of Einstein's equation. This means that, in this coordinate system, at the point P in question, Einstein's equation must be of the form (taking into account that the determinant of the metric can go through partial derivatives taken at P in this coordinate system)

$$\frac{1}{|g|}\partial_\rho \eta^{\mu\nu\rho} = T_{\text{matter}}{}^{\mu\nu}, \qquad \eta^{\mu\nu\rho} = -\eta^{\mu\rho\nu}. \tag{10.6}$$

Actually, it can be checked that, in this coordinate system, at the point in question, Einstein's equations take precisely the above form, with

$$\eta^{\mu\nu\rho} = -\frac{2}{\chi^2}\partial_\sigma \mathfrak{g}^{\mu\sigma,\nu\rho}, \quad \mathfrak{g}^{\mu\nu} = \sqrt{|g|}g^{\mu\nu}, \quad \mathfrak{g}^{\mu\sigma,\nu\rho} = \tfrac{1}{2}(\mathfrak{g}^{\mu\nu}\mathfrak{g}^{\sigma\rho} - \mathfrak{g}^{\mu\rho}\mathfrak{g}^{\sigma\nu}). \tag{10.7}$$

Now, in any coordinate system we can define the *Landau–Lifshitz energy–momentum pseudotensor* by

$$t_{\text{LL}}{}^{\mu\nu} = \frac{1}{|g|}\partial_\rho \eta^{\mu\nu\rho} - T_{\text{matter}}{}^{\mu\nu}, \tag{10.8}$$

so, due to the symmetries of $\eta^{\mu\nu\rho}$,

$$\partial_\mu\{|g|(T_{\text{matter}}{}^{\mu\nu} + t_{\text{LL}}{}^{\mu\nu})\} = 0, \tag{10.9}$$

which is essentially what we wanted.

To determine the explicit form of $t_{\text{LL}}{}^{\mu\nu}$ we use Einstein's equation

$$t_{\text{LL}}{}^{\mu\nu} = -\frac{2}{\chi^2}(\partial_\rho\partial_\sigma \mathfrak{g}^{\mu\sigma,\nu\rho} + G^{\mu\nu}), \tag{10.10}$$

and by expanding both terms we obtain a very complicated and not very illuminating expression that is quadratic in the metric and quadratic in connections that can be found in most standard gravity textbooks [952, 883, 360]. Since this depends on connections, it does not transform as a world tensor, but it does transform as a Lorentz tensor (affine connections transform as Lorentz tensors), just as expected.

With the total conserved energy–momentum pseudotensor obeying the local continuity equation, we go on to define the conserved charges (momentum and angular momentum) by the volume integrals[7]

$$\begin{aligned} P^{\mu} &= \int_{\Sigma} d^{d-1}\Sigma_{\nu}\sqrt{|g|}(T_{\text{matter}}{}^{\mu\nu} + t_{\text{LL}}{}^{\mu\nu}), \\ M^{\mu\alpha} &= \int_{\Sigma} d^{d-1}\Sigma_{\nu}\sqrt{|g|}\left[2x^{[\alpha}(T_{\text{matter}}{}^{\mu]\nu} + t_{\text{LL}}{}^{\mu]\nu})\right], \end{aligned} \tag{10.11}$$

where it is assumed that one integrates over a timelike hypersurface $\Sigma$.

Two shortcomings of this approach are (i) it is not clear why these are the (only) conserved charges and (ii) how we can generalize it to other spacetimes in which these are not necessarily the conserved charges. The Abbott–Deser approach will solve this problem.

A second shortcoming is the large number of terms that have to be calculated in order to find the conserved quantities. In practice, though, one uses Eq. (10.8) to rewrite, using Stokes' theorem,

$$P^{\mu} = \tfrac{1}{2}\int_{\partial\Sigma} d^{d-2}\Sigma_{\nu\rho}\frac{\eta^{\mu\nu\rho}}{\sqrt{|g|}}, \tag{10.12}$$

and similarly for $M^{\mu\alpha}$. This is an interesting expression that has to be evaluated at the boundary of the hypersurface $\Sigma$, which is, typically for asymptotically flat spacetimes, a $(d-2)$-sphere at spatial infinity, $\mathrm{S}^{d-2}_{\infty}$. We could integrate over the boundary of smaller regions of the spacetime. However, the integrand is not a general-covariant tensor, and the result of the integral would be coordinate dependent and the momentum would not be well defined. Only when we integrate over $\mathrm{S}^{d-2}_{\infty}$ in asymptotically flat spacetimes does the integral transform as a Poincaré tensor. This is the common behavior of most superpotentials used to define conserved quantities in GR, except for Møller's [962], which is a true tensor.

For asymptotically flat spacetimes, we can use the weak-field expansion[8] $g_{\mu\nu} = \eta_{\mu\nu} + h_{\mu\nu}$. In this limit, we see that

$$\mathfrak{g}^{\mu\sigma,\nu\rho} = K^{\mu\sigma\nu\rho} + \mathcal{O}(h^2), \tag{10.13}$$

and

$$\eta^{\mu\nu\rho} = 2\eta_{\text{LL}}^{\mu\nu\rho} + \mathcal{O}(h^2), \qquad 2\partial_{\rho}\eta_{\text{LL}}^{\mu\nu\rho} = \mathcal{D}^{\mu\nu}(h), \tag{10.14}$$

where $\eta_{\text{LL}}^{\mu\nu\rho}$ was defined in Eqs. (3.90) and $\mathcal{D}^{\mu\nu}(h)$ is the Fierz–Pauli wave operator.

Thus, in practice, all we have to do is to integrate the Fierz–Pauli wave operator over the volume $\Sigma$ or $\eta_{\text{LL}}$ over the boundary $\partial\Sigma$, if the asymptotic weak-field expansion of the metric is well defined. Many different gravity energy–momentum pseudotensors have been proposed in the literature but, in the end, one never uses them directly. Instead one integrates over $\Sigma$, using the equations of motion, an expression that, in the weak-field limit, is equivalent to the Fierz–Pauli wave operator. Usually this expression is rewritten as an integral over the boundary using Stokes' theorem. This can be done in many different ways, as we discussed in Chapter 3, and here is where the differences arise.[9]

---

[7] Observe the "extra" factors of $\sqrt{|g|}$.

[8] Observe that the $h_{\mu\nu}$ that we are using in this chapter is the $\chi h_{\mu\nu}$ of Chapter 3.

[9] Some expressions may be better suited for certain boundary conditions. When we compare the weak-field limits of the various expressions proposed in the literature, we have to bear in mind that the expansions used are valid only under certain asymptotic conditions.

With the zeroth component of Eq. (10.12) in $d=4$ for a stationary asymptotically flat metric in Cartesian coordinates, we obtain the Arnowitt–Deser–Misner (ADM) mass formula, which was first derived by canonical methods in Ref. [66]:

$$M_{\rm ADM} = \frac{2}{\chi^2}\int_{S^2_\infty} d^2 S_k (\partial_k h_{ll} - \partial_l h_{lk}), \tag{10.15}$$

where

$$dS_k \equiv \tfrac{1}{2}\epsilon_{ijk} dx^i \wedge dx^j. \tag{10.16}$$

We can immediately apply this formula to the simplest spacetime: Schwarzschild spacetime. The four-dimensional Schwarzschild solution in Schwarzschild coordinates is

$$ds^2 = \left(1-\frac{k}{r}\right)dt^2 - \left(1-\frac{k}{r}\right)^{-1} dr^2 - r^2 d\Omega^2_{(2)}, \tag{10.17}$$

where $k$ is the integration constant. To apply the ADM mass formula Eq. (10.15), we first rewrite the metric in *isotropic coordinates*:

$$r = (\rho + k/4)^2/\rho, \tag{10.18}$$

obtaining

$$ds^2 = \left(\frac{\rho - k/4}{\rho + k/4}\right)^2 dt^2 - \left(1+\frac{k/4}{\rho}\right)^4 d\vec{x}_3^{\,2}, \qquad \rho = |\vec{x}_3|. \tag{10.19}$$

With $\chi^2 = 16\pi G_{\rm N}^{(4)}$, the ADM mass formula gives, in agreement with our results of Chapter 3,

$$k = 2G_{\rm N}^{(4)} M. \tag{10.20}$$

### *10.1.2 The Abbott–Deser approach*

Abbott and Deser [1] proposed a general definition for spacetimes of arbitrary asymptotic behavior associating conserved charges with isometries of the asymptotic geometry which is supposed to be the vacuum (so we are physically calculating the conserved charges of an isolated system). This definition is very useful and can be extended to more complicated cases in which some dimensions are compactified [444] (see also Ref. [228]), other contexts such as supercharges in supersymmetric theories [1] (associated with Killing spinors of the vacuum, which will be studied in Chapter 17), and charges in non-Abelian gauge theories (associated with gauge Killing vectors) [2].

In this section we essentially repeat and extend the calculations of Abbott and Deser [1] in our conventions, comparing the result with that in Section 10.1.1. As an example of its usefulness, we will also use it to calculate the mass of a spacetime that is not asymptotically flat.

The first step in this approach is the expansion of the gravitational field around an arbitrary background metric $\bar{g}_{\mu\nu}$ that solves the vacuum cosmological Einstein equations, and the derivation of the linearized Einstein equations in that background. We already did this in Section 3.4.1, where we also gave a definition of the gravitational energy–momentum pseudotensor that was different from Landau and Lifshitz's. Here we use the notation and definitions of that section.

The second step consists in the construction of a conserved quantity. First, we observe that the linearized cosmological Einstein tensor satisfies the Bianchi identity with respect to the background metric:

$$\bar{\nabla}_\mu G_{c\,\mathrm{L}}{}^{\mu\nu} = 0. \tag{10.21}$$

This can be proven either by direct calculation or by taking the divergence of the cosmological Einstein tensor,

$$\nabla_\mu G_c{}^{\mu\nu} = \bar{\nabla}_\mu \bar{G}_c{}^{\mu\nu} + \gamma_{\mu\rho}{}^{\mu} \bar{G}_c{}^{\rho\nu} + \gamma_{\mu\rho}{}^{\nu} \bar{G}_c{}^{\mu\rho} + \bar{\nabla}_\mu G_{c\,\mathrm{L}}{}^{\mu\nu} + \mathcal{O}(h^2), \tag{10.22}$$

and observing that, by hypothesis, $\bar{G}_c{}^{\mu\nu} = 0$, and also that the Bianchi identity has to be satisfied order by order in $h$.

Using now Eq. (3.292) and Eq. (10.21), we find

$$\bar{\nabla}_\mu T^{\mu\nu}_{\mathrm{total}} = 0, \qquad T^{\mu\nu}_{\mathrm{total}} = T^{\mu\nu}_{\mathrm{matter}} + t^{\mu\nu}_{\mathrm{AD}}. \tag{10.23}$$

Finally, if $\bar{\xi}_\mu$ is a Killing vector field of the background metric, the above equation implies

$$\bar{\nabla}_\mu (T^{\mu\nu}_{\mathrm{total}} \bar{\xi}_\nu) = 0, \tag{10.24}$$

and, from this, we find that the quantity

$$E(\bar{\xi}) \equiv \int_\Sigma d^{d-1}x \sqrt{|\bar{g}|}\, T^{0\nu}_{\mathrm{total}} \bar{\xi}_\nu, \tag{10.25}$$

where the integral is performed over a constant time slice $\Sigma$, is a conserved quantity, namely the conserved quantity associated with the background Killing vector $\bar{\xi}^\mu$. If the Killing vector generates translations in time in the background, the conserved quantity is the energy (or mass). (In general, if $\bar{\xi}$ is a timelike Killing vector, $E(\bar{\xi})$ is called the *Killing energy*.) For Killing vectors that generate rotations, we obtain components of the angular momentum etc.

The covariant form of the above expression is

$$E(\bar{\xi}) \equiv \int_\Sigma d^{d-1}\Sigma_\mu T^{\mu\nu}_{\mathrm{total}} \bar{\xi}_\nu, \tag{10.26}$$

where

$$d^{d-1}\Sigma_\mu = \frac{1}{(d-1)!\sqrt{|\bar{g}|}} \epsilon_{\mu\rho_1\cdots\rho_{d-1}} dx^{\rho_1} \wedge \cdots \wedge dx^{\rho_{d-1}}. \tag{10.27}$$

This equation can be seen as a generalization of Landau and Lifshitz's Eq. (10.11), which is valid for any background metric $\bar{g}_{\mu\nu}$ and any of its Killing vectors $\bar{\xi}_\mu$, with a different definition of the gravitational energy–momentum pseudotensor. Indeed, Landau and Lifshitz's Eq. (10.11) can be written in the form

$$E(\xi) = \int_\Sigma d^{d-1}\Sigma_\mu \sqrt{|g|} (T_{\mathrm{matter}}{}^{\mu\nu} + t_{\mathrm{LL}}{}^{\mu\nu}) \xi_\nu, \tag{10.28}$$

where the Minkowski spacetime Killing vectors $\xi^{(\mu)\,\nu} = \eta^{\mu\nu}$ that generate constant translations are used to obtain the $P^\mu$s and those which generate Lorentz transformations $\xi^{(\mu\alpha)\,\nu} = -2x^{[\mu}\eta^{\alpha]\nu}$ are used to obtain the $M^{\mu\alpha}$s. The different definition of $t^{\mu\nu}$ is responsible for the extra factor of $\sqrt{|g|}$ in this formula compared with Abbott and Deser's. On the other hand, in the Landau–Lifshitz approach we are forced not only to work with asymptotically flat spacetimes, but also to use Cartesian coordinates. The Abbott–Deser approach can be used for any spacetime in any coordinate system.

The main problem with Eq. (10.26) is that the expression for $t^{\mu\nu}$ is very complicated; it is, in fact, an infinite series in $h$. The solution is, again, to use the equation of motion Eq. (3.292) to rewrite it. The new integrand is, as we argued it would in general be, just the covariantized Fierz–Pauli wave operator $\bar{\mathcal{D}}^{\mu\nu}(h)$ contracted with a background Killing vector, that is

$$E(\bar{\xi}) = \frac{2}{\chi^2}\int_\Sigma d^{d-1}\Sigma_\mu \bar{\mathcal{D}}^{\mu\nu}(h)\bar{\xi}_\nu. \tag{10.29}$$

At this point we note that the integrand of this expression is nothing but the conserved Noether current $j^\mu_{\mathrm{N}}(\bar{\xi})$ in Eq. (3.314), and we can use the results of Section 3.4.1 to rewrite it as a total derivative and then use Stokes' theorem to rewrite it as a $(d-2)$-surface integral:

$$\boxed{E(\bar{\xi}) = -\frac{2}{\chi^2}\int_{\partial\Sigma = \mathrm{S}^{d-2}_\infty} d^{d-2}\Sigma_{\mu\alpha}\Big[\Big(\bar{\nabla}_\beta K^{\mu\alpha\nu\beta}\Big)\bar{\xi}_\nu - K^{\mu\beta\nu\alpha}\bar{\nabla}_\beta\bar{\xi}_\nu\Big],} \tag{10.30}$$

where

$$d^{d-2}\Sigma_{\mu\alpha} = \frac{1}{(d-2)!\sqrt{|\bar{g}|}}\epsilon_{\mu\alpha\rho_1\cdots\rho_{d-2}}dx^{\rho_1}\wedge\cdots\wedge dx^{\rho_{d-2}}. \tag{10.31}$$

This is essentially Abbott and Deser's final result, although one can massage the above expression further to make it useful in specific situations. For instance, the following alternative expression is noteworthy. We first observe the identity

$$\bar{\nabla}_\beta K^{\mu\alpha\nu\beta} = 3\bar{g}^{\lambda\mu\alpha,\,\nu}{}_{\rho\sigma}\gamma_\lambda{}^{\rho\sigma}. \tag{10.32}$$

We can replace $\gamma_{\mu\nu}{}^\rho$ by $\Delta\Gamma_{\mu\nu}{}^\rho = \Gamma_{\mu\nu}{}^\rho - \bar{\Gamma}_{\mu\nu}{}^\rho$ because the difference is quadratic and higher in $h_{\mu\nu}$, which is assumed to go to zero at infinity fast enough. Then

$$E(\bar{\xi}) = -\frac{2}{\chi^2}\int_{\mathrm{S}^{d-2}_\infty} d^{d-2}\Sigma_{\mu\alpha}\Big[3\bar{g}^{\lambda\mu\alpha,\,\nu}{}_{\rho\sigma}\,\Delta\Gamma_\lambda{}^{\rho\sigma}\bar{\xi}_\nu - K^{\mu\beta\nu\alpha}\bar{\nabla}_\beta\bar{\xi}_\nu\Big]. \tag{10.33}$$

Furthermore, in Minkowski spacetime in Cartesian coordinates the generators of translations are covariantly constant and the second term can be dropped, so we obtain for any component of the momentum (and, in particular, for the energy) of asymptotically flat spacetimes the expression

$$\boxed{E(\bar{\xi}) = -\frac{2}{\chi^2}\int_{\partial\Sigma} d^{d-2}\Sigma_{\mu\alpha}3\bar{g}^{\lambda\mu\alpha,\,\nu}{}_{\rho\sigma}\,\Delta\Gamma_\lambda{}^{\rho\sigma}\bar{\xi}_\nu,} \tag{10.34}$$

which was used first in Ref. [976] and henceforth in all proofs of the positivity of the mass or Bogomol'nyi bounds based on Nester's construction ([821, 618, 825], etc.).

It is also interesting to compare Eq. (10.30) with Landau and Lifshitz's result. In flat spacetime, with Cartesian coordinates, for translational Killing vectors (which are covariantly constant) Eq. (10.30) simplifies to

$$P^\mu = E(\bar{\xi}^{(\mu)}) = -\frac{2}{\chi^2}\int_{\mathrm{S}^{d-2}_\infty} d^{d-2}\Sigma_{\nu\alpha}\partial_\beta K^{\nu\alpha\mu\beta}, \tag{10.35}$$

and we see that the integrand is nothing but $\eta_{\mathrm{AD}}^{\nu\mu\alpha}$ as defined in Eqs. (3.90). The difference from $\eta_{\mathrm{LL}}^{\nu\mu\alpha}$ is just

$$\eta_{\mathrm{LL}}^{\nu\mu\alpha} - \eta_{\mathrm{AD}}^{\nu\mu\alpha} = \partial_\beta K^{\nu\mu\alpha\beta}, \tag{10.36}$$

so

$$\partial_\alpha\left(\eta_{\mathrm{LL}}^{\nu\mu\alpha} - \eta_{\mathrm{AD}}^{\nu\mu\alpha}\right) = 0, \tag{10.37}$$

and the difference should not contribute to the conserved charges.

To end this section, let us apply these results to a simple example: the four-dimensional Reissner–Nordström–de Sitter spacetime. First, for any static, spherically symmetric metric

$$ds^2 = g_{tt}(r)dt^2 + g_{rr}(r)dr^2 - r^2 d\Omega^2_{(2)} \tag{10.38}$$

and backgrounds

$$d\bar{s}^2 = \bar{g}_{tt}(r)dt^2 + \bar{g}_{rr}(r)dr^2 - r^2 d\Omega^2_{(2)}, \tag{10.39}$$

and for the obvious timelike Killing vector $\bar{\xi}_\nu = \delta_{0\nu}\bar{g}_{tt}$, we obtain the mass formula

$$M = -\frac{1}{2G_{\mathrm{N}}^{(4)}}\frac{|\bar{g}_{tt}|^{1/2}}{|\bar{g}_{rr}|^{3/2}} r(g_{rr} - \bar{g}_{rr}). \tag{10.40}$$

This formula can be directly applied to the Schwarzschild metric given in Eq. (10.17), and it gives the correct result. It can also be applied to asymptotically (anti-)de Sitter spacetimes. We can apply it, for instance, to the Reissner–Nordström–(anti-)de Sitter metric in static coordinates,

$$\begin{aligned} ds^2 &= V dt^2 - V^{-1}dr^2 - r^2 d\Omega^2_{(2)}, \\ V &= 1 - \frac{k}{r} + \frac{Z^2}{4r^2} - \frac{1}{3}\Lambda r^2. \end{aligned} \tag{10.41}$$

We obtain again $M = k/(2G_N^{(4)})$.

## 10.2 The Noether approach

The standard method used to obtain the conserved charge is that through the Noether current. We have seen that, in fact, the Abbott–Deser formula for conserved charges can be seen from this point of view as the integral of the conserved Noether current associated with a background Killing vector of linearized gravity. Here we are going to investigate this point of view further, since there is a Noether current $j^\mu_{\mathrm{N2}}(\xi)$ associated with *any* vector field

$\xi^\mu$, Killing or otherwise, as we proved in Section 3.4.1, and we could simply calculate the superpotential $j_{N2}^{\mu\nu}(\xi)$ associated with an arbitrary vector field by generalizing Abbott and Deser's result.[10] We will, however, content ourselves with reviewing some known results.

For GR we found the Noether current $j_N^\mu(\xi)$ for any vector $\xi^\mu$ in Section 4.1.2 and we saw how it could be rewritten as the divergence of the antisymmetric superpotential tensor $j_N^{\mu\nu}(\xi) = 2\nabla^{[\mu}\xi^{\nu]}$. Using it, we can define a conserved charge for each vector $\xi^\mu$:

$$\boxed{E(\xi) = -\frac{2}{\chi^2}\int_{\partial\Sigma} d^{d-2}\Sigma_{\mu\alpha}\nabla^\mu\xi^\alpha.} \tag{10.42}$$

If $\xi^\mu$ is timelike, then $E(\xi)$ is the energy and the above formula is *Komar's formula* [869]. This formula can also be obtained from physical principles, as in Section 11.2 of Ref. [1229].

We can compare Komar's formula with Abbott and Deser's Eq. (10.30). On rewriting it as

$$E(\xi) = -\frac{2}{\chi^2}\int_{\partial\Sigma} d^{d-2}\Sigma_{\mu\alpha} g^{\mu\alpha,\nu\beta}\nabla_\nu\xi_\beta, \tag{10.43}$$

and using the weak-field expansion $g_{\mu\nu} = \bar{g}_{\mu\nu} + h_{\mu\nu}$, we find that it reproduces exactly the first term in Eq. (10.30) (after integrating by parts) but not the second one. This difference is probably responsible for one of the known drawbacks of Komar's formula: it gives the wrong value for the angular momentum of the Kerr solution.

Komar's formula can be modified by adding to the Einstein–Hilbert action total derivative terms that modify the Noether current, as explained in Section 4.1.2. The problem now lies in determining which total derivative should be added. In Ref. [1038] some examples of total derivative terms that have been added in the literature can be found and a new one is proposed. Using it and also, basically, Eq. (4.125) in the absence of torsion, the authors propose a new superpotential whose integral (if it is convergent) gives a conserved charge for any vector field $\xi$. In the weak-field limit, it can be written in the form

$$E(\bar{\xi}) = -\frac{2}{\chi^2}\int_{\partial\Sigma} d^{d-2}\Sigma_{\mu\alpha}\left[\left(\bar{\nabla}_\beta K^{\mu\alpha\nu\beta}\right)\xi_\nu - \bar{h}^{\sigma[\mu}\bar{\nabla}_\sigma\xi^{\alpha]}\right]. \tag{10.44}$$

The first term is identical to the first term in Eq. (10.30) and the second is identical to the second if $\xi^\mu = \bar{\xi}^\mu$, a background Killing vector, but the formula can be applied to more general cases. In fact, the complete formula in Ref. [1038] gives correct results in the presence of radiation, whereas Abbott and Deser's does not, probably because the weak-field expansion is not consistent in those spacetimes.

## 10.3 The positive-energy theorem

Now that we know how to define conserved quantities in GR, and, in particular, the mass (total energy), we are going to prove that the mass of an asymptotically flat spacetime that

[10] Of course, not for every vector will the integral defining the corresponding conserved charge converge, but we will not deal with this problem here.

solves the Einstein equations

$$G_{\mu\nu} = \frac{\chi^2}{2} T_{\text{matter}\,\mu\nu}, \tag{10.45}$$

with a matter energy–momentum tensor satisfying the *dominant energy condition*

$$T_{\text{matter}}{}^{\mu\nu} k_\mu n_\nu \geq 0, \qquad \forall\, n_\mu,\, k_\mu \text{ non-spacelike}, \tag{10.46}$$

is always non-negative, vanishing only for flat spacetime. This result was first obtained by Schoen and Yau [1107]. A new proof based on spinor techniques inspired by SUGRA was afterwards presented by Witten [1272], and subsequently by Nester [976] and Israel and Nester [821]. Previously, the positivity of mass in SUGRA and GR (as the bosonic part of $N = 1$ SUGRA) had been established in Refs. [445, 680]. Here we are going to use this Witten–Nester–Israel (WNI) technique because (i) it can be generalized to more complicated cases and (ii) it has a strong relation to supergravity that we will use in Chapter 17.

The positive-energy theorem is a very important result associated with the *cosmic-censorship conjecture*: in the gravitational collapse of a star, the gravitational binding energy, which is negative, grows in absolute value. If the process were to continue indefinitely, the total energy of the collapsing star would become negative. However, according to the positive-mass theorem, this cannot happen and we expect a black-hole horizon to appear before the mass becomes negative.

The WNI technique starts with the construction of the Nester 2-form $E^{\mu\nu}$. In this case (pure $d = 4$ gravity; the extension to higher dimensions is straightforward) it is simply

$$E^{\mu\nu}(\epsilon) = +\frac{i}{2}\bar{\epsilon}\gamma^{\mu\nu\rho}\nabla_\rho\epsilon + \text{c.c.}, \tag{10.47}$$

where $\epsilon$ is a *commuting* Dirac spinor. The Nester form is manifestly real. We then define the integral $I$,

$$I(\epsilon) = \int_{\partial\Sigma} \star E(\epsilon), \tag{10.48}$$

where $\Sigma$ is a three-dimensional spacelike hypersurface (for instance, a constant time slice) whose boundary $\partial\Sigma$ is a 2-sphere at infinity, $S^2_\infty$. Observe that the Nester form can be rewritten in the form

$$E^{\mu\nu}(\epsilon) = +i\bar{\epsilon}\gamma^{\mu\nu\rho}\nabla_\rho\epsilon + \nabla_\rho\Big(-\frac{i}{2}\bar{\epsilon}\gamma^{\mu\nu\rho}\epsilon\Big), \tag{10.49}$$

and only the first term contributes to $I$.

The proof has two parts.

1. Prove that, for suitably chosen spinors $\epsilon$ and $T_{\text{matter}}{}^{\mu\nu}$ satisfying the dominant energy condition, $I(\epsilon) \geq 0$.
2. Relate $I(\epsilon)$ to conserved charges.

1. We use Stokes' theorem

$$I(\epsilon) = \int_{\partial\Sigma} \star E(\epsilon) = \int_\Sigma d\star E(\epsilon) = -\tfrac{1}{2}\int_\Sigma d^3\Sigma_\nu\{-i\nabla_\mu\bar{\epsilon}\gamma^{\mu\nu\rho}\nabla_\rho\epsilon - i\bar{\epsilon}\gamma^{\mu\nu\rho}\nabla_\mu\nabla_\rho\epsilon\}, \tag{10.50}$$

where

$$d^3\Sigma_\mu = \frac{1}{3!\sqrt{|g|}} dx^\rho \wedge dx^\sigma \wedge dx^\lambda \epsilon_{\rho\sigma\lambda\mu}. \tag{10.51}$$

The second term in the integral is proportional to the Lorentz curvature tensor due to the Ricci identities Eqs. (1.102). Expanding the product of the $\gamma^{\mu\nu\rho}$ and the $\gamma_{ab}$ gives

$$-i\bar{\epsilon}\gamma^{\mu\nu\rho}\nabla_\mu\nabla_\rho\epsilon = \frac{i}{2}\bar{\epsilon}G_\mu{}^\nu\gamma^\mu\epsilon, \tag{10.52}$$

and, since the spacetime we are considering satisfies the Einstein equations,

$$-i\bar{\epsilon}\gamma^{\mu\nu\rho}\nabla_\mu\nabla_\rho\epsilon = \frac{\chi^2}{4} T_{\text{matter}\,\mu}{}^\nu k^\mu, \tag{10.53}$$

where we have defined the vector $k^a$ as the following real bilinear of the spinor $\epsilon$:

$$k^a = i\bar{\epsilon}\gamma^a\epsilon. \tag{10.54}$$

Now we want to show that $k^\mu$ is a non-spacelike vector by calculating $k^2$ directly. Using the $d = 4$ Fierz identities *for commuting spinors*, we obtain

$$k^2 = 2(i\bar{\epsilon}\epsilon)^2 + 2(\bar{\epsilon}\gamma_5\epsilon)^2 + \ell^2, \tag{10.55}$$

where we have also defined the real pseudovector $\ell^a$:

$$\ell^a = \bar{\epsilon}\gamma^a\gamma_5\epsilon. \tag{10.56}$$

Calculating now $\ell^2$ using the Fierz identities again, we obtain

$$\ell^2 = -\tfrac{2}{3}(i\bar{\epsilon}\epsilon)^2 + \tfrac{1}{3}k^2 + \tfrac{2}{3}(\bar{\epsilon}\gamma_5\epsilon)^2, \tag{10.57}$$

from which we obtain

$$k^2 = 2(i\bar{\epsilon}\epsilon)^2 + 4(\bar{\epsilon}\gamma_5\epsilon)^2, \tag{10.58}$$

which is manifestly non-negative because the bilinears $i\bar{\epsilon}\epsilon$ and $\bar{\epsilon}\gamma_5\epsilon$ are real.[11]

On collecting these results and writing $d^3\Sigma_\mu = d^3\Sigma n_\mu$, where $n_\mu$ is the non-spacelike unit vector normal to the hypersurface $\Sigma$, we find that the integral of the second term in $I(\epsilon)$ is

$$\int_\Sigma d^3\Sigma_\nu\{-i\bar{\epsilon}\gamma^{\mu\nu\rho}\nabla_\mu\nabla_\rho\epsilon\} = \frac{\chi^2}{4}\int_\Sigma d^3\Sigma\, T_{\text{matter}}{}^{\mu\nu}k_\mu n_\nu. \tag{10.59}$$

The dominant energy condition, Eq. (10.46), implies that the second term in $I(\epsilon)$ is non-negative.

[11] If $\epsilon$ were a Majorana spinor, we would have $k^2 = 0$.

As for the second term, let us use a coordinate system in which $n_\mu = \delta_{\mu 0}$ ($\mu = 0, i$). Then, it can be rewritten in the form

$$\int_\Sigma d^3\Sigma (\nabla_i \epsilon)^\dagger_\alpha (\nabla_i \epsilon)^\alpha - \int_\Sigma d^3\Sigma \left(i\gamma^i \nabla_i \epsilon\right)^\dagger_\alpha \left(i\gamma^j \nabla_j \epsilon\right)^\alpha. \tag{10.60}$$

These two terms are manifestly positive. The second one vanishes if we use spinors satisfying the *Witten condition*

$$\gamma^i e_i{}^\mu \nabla_\mu \epsilon = 0. \tag{10.61}$$

Thus, we have proven that, if the dominant energy condition is satisfied and we use spinors satisfying the Witten condition, $I(\epsilon)$ is non-negative.

2. We rewrite $I(\epsilon)$ as follows:

$$I(\epsilon) = \tfrac{1}{2} \int_{\partial\Sigma} d^2\Sigma_{\mu\nu} \epsilon^{\mu\nu\rho\sigma} \bar{\epsilon} \gamma_5 \gamma_\sigma \nabla_\rho \epsilon, \tag{10.62}$$

and expand the integrand around the vacuum $\bar{g}_{\mu\nu}$ (Minkowski spacetime) to which the solution asymptotically tends. We also impose on the chosen spinors that they admit the expansion

$$\epsilon = \epsilon_0 + \mathcal{O}\left(\frac{1}{r}\right), \tag{10.63}$$

where $r \to \infty$ at spatial infinity and

$$\bar{\nabla}_\mu \epsilon_0 = 0. \tag{10.64}$$

A spinor satisfying this condition in $N = 1$ SUGRA is a *Killing spinor* of the solution $\bar{g}_{\mu\nu}$. Since $\nabla_\mu = \bar{\nabla}_\mu - \frac{1}{4} \Delta\omega_\mu{}^{ab} \gamma_{ab}$ and the integral is taken at spatial infinity,

$$\begin{aligned} I(\epsilon) &= -\tfrac{1}{8} \int d^2\Sigma_{\mu\nu} \epsilon^{\mu\nu\rho\sigma} \bar{\epsilon}_0 \gamma_5 \gamma_\sigma \nabla_\rho \epsilon_0 = \tfrac{1}{4} \int d^2\Sigma_{\mu\nu} \left[-3 g^{\rho\mu\nu,\gamma}{}_{\alpha\beta} \Delta\omega_\rho{}^{\alpha\beta} k_{0\,\gamma}\right] \\ &= \tfrac{1}{4} E(k_0), \end{aligned} \tag{10.65}$$

where we have used Eq. (10.34) and the fact that

$$k_0^a = i \bar{\epsilon}_0 \gamma^a \epsilon_0 \tag{10.66}$$

is, trivially, a Killing vector of the vacuum $\bar{g}_{\mu\nu}$. When it is timelike, $k_0$ is the generator of translations in time and $E(k_0)$ is just the mass.

This proves that $M \geq 0$ and $M = 0$ for Minkowski spacetime.

The relation Eq. (10.66) between Killing spinors and Killing vectors is quite generic and, with minor adaptations, is true in most SUGRAs. Just as the Killing vectors of a metric constitute a Lie algebra that generates its isometry group, the Killing spinors *and* Killing vectors of a solution of a SUGRA theory, which may involve other fields apart from the metric, constitute a superalgebra that generates a supergroup that leaves the solution invariant. The simplest case is Minkowski spacetime, whose invariance supergroup is the super-Poincaré one.

# Part II
# Gravitating point-particles

*[The Universe] cannot be read until we have learnt the language and become familiar with the characters in which it is written. It is written in mathematical language, and the letters are triangles, circles and other geometrical figures, without which means it is humanly impossible to comprehend a single word.*

*Galileo Galilei*

# 11
# The Schwarzschild black hole

In this chapter we start the study of a number of important classical solutions of GR.[1] There is no doubt that the most important solution is that due to Schwarzschild, which describes the static, spherically symmetric gravitational field in the absence of matter that one finds outside any static, spherically symmetric object (star, planet, etc.). It is this, the simplest non-trivial solution, that leads to the concept of a black hole (BH), which affords a privileged theoretical laboratory for *Gedankenexperiment* in classical and quantum gravity.

It is, in fact, a firmly established belief in the scientific community that macroscopic BHs (of the size studied by astrophysicists) are the endpoints of the *gravitational collapse* of stars, which, after a long time, gives rise to Schwarzschild BHs if the stars do not rotate. There should be many macroscopic Schwarzschild BHs in our Universe, since many stars have enough mass to undergo gravitational collapse, and there is evidence of supermassive BHs in the centers of galaxies.[2] It has been suggested that smaller BHs could have been produced in the Big Bang. Here we are going to be interested in BHs of all sizes, independent of their origin (primordial, quantum-mechanical, astrophysical, . . . ).

We begin by deriving the Schwarzschild solution and studying its classical properties in order to find its physical interpretation. The physical interpretation of vacuum solutions of the Einstein equations is a most important and complicated point [232, 233] since the source, located by definition in the region in which the vacuum Einstein equations are not solved, is unknown. In the case of the Schwarzschild solution, we will be led to the new concepts of the *event horizon* and BHs. Some of the classical properties of BHs can be formulated as laws of thermodynamics, but, classically, the analogy cannot be complete. It is the existence of Hawking radiation, a quantum phenomenon, that makes the analogy complete and allows us to take it seriously, raising at the same time the problem of the statistical interpretation of the BH (Bekenstein–Hawking) entropy and the BH information problem. Finally, we rederive the expression for the BH entropy in the Euclidean quantum-gravity approach and generalize our previous results and the Schwarzschild solution to higher dimensions.

---

[1] Some of these solutions (Schwarzschild, Reissner–Nordström, Taub–NUT, etc.) are also reviewed in Ref. [209] from a different perspective and emphasize different properties.

[2] For general references on astrophysical evidence for the existence of BHs see, for instance, Refs. [305, 1073, 945].

There are many excellent books and reviews on these subjects. We would like to mention Novikov and Frolov's book [998], which is the most complete reference on BH physics, Townsend's lectures [1194], the books on quantum-field theory (QFT) on curved spacetimes [218, 1233], and the review articles [607, 1235].

## 11.1 The Schwarzschild solution

To solve the vacuum Einstein equations

$$R_{\mu\nu} - \tfrac{1}{2}g_{\mu\nu}R = 0 \quad \Rightarrow R_{\mu\nu} = 0, \tag{11.1}$$

it is necessary to make a simplifying *ansatz* for the metric. The ansatz must, at the same time, reflect the physical properties that we want the solution to enjoy. In this case we want to obtain the metric in the spacetime outside a massive spherically symmetric body that is at rest in a given coordinate system. The latter property is contained in the assumption of *staticity*[3] of the metric and the first is contained in the assumption of spherical symmetry.[4] Under these assumptions, the most general metric can always be cast in the form

$$ds^2 = W(r)(dct)^2 - W^{-1}(r)dr^2 - R^2(r)d\Omega^2_{(2)}, \tag{11.2}$$

where $W(r)$ and $R(r)$ are two undetermined functions of the coordinate $r$ and $d\Omega^2_{(2)}$ is the metric on the unit 2-sphere $S^2$ (see Appendix K). On substituting this ansatz into the equations of motion one finds (e.g. Ref. [1229]) a general solution for $W$ and $R$,

$$W = 1 + \omega/r, \qquad R^2 = r^2, \tag{11.3}$$

with one integration constant $\omega$. We see that the solution is asymptotically flat; i.e. that, as the coordinate $r$ approaches infinity, the metric approaches Minkowski's. Physically, the requirement of asymptotic flatness means that we are dealing with an isolated system, with a source of gravitational field confined in a finite volume. The constant $\omega$ has dimensions of length and we will study its meaning in a moment.

The result is the Schwarzschild solution [1118][5] in *Schwarzschild coordinates* $\{t, r, \theta, \varphi\}$:

$$\boxed{ds^2 = W(dct)^2 - W^{-1}dr^2 - r^2 d\Omega^2_{(2)}, \qquad W = 1 + \omega/r.} \tag{11.4}$$

[3] That is, the metric admits a timelike Killing vector with the property of *hypersurface orthogonality*: the space can be foliated by a family of spacelike hypersurfaces that are orthogonal to the orbits of the timelike Killing vector, and can be labeled by the parameter of these orbits, which takes the same value at any point of each of these hypersurfaces. If the space does not have this property, the explicit dependence of the metric on the associated time coordinate can always be avoided, but there will always be non-vanishing off-diagonal terms in the metric, mixing time components with space components, breaking at the same time spherical symmetry: all stationary, spherically symmetric spacetimes are also static.

[4] Invariance under the group $SO(3)$ of spatial rotations in $d = 4$.

[5] An English translation of the original paper by Karl Schwarzschild is available (see Ref. [1118]). As pointed out in Ref. [55], Karl Schwarzschild did not solve the standard Einstein equations but rather the earliest version of them in which the coordinate condition $\sqrt{|g|} = 1$ (invariant only under volume-preserving diffeomorphisms) had been imposed. This coordinate choice was motivated by convenience; it had no physical meaning, and could be replaced by any other.

Let us now review the properties of this solution.

### 11.1.1 General properties

1. *Schwarzschild's is the only spherically symmetric solution of $R_{\mu\nu} = 0$ (static or not).* This is Birkhoff's theorem [212] (first obtained by Jebsen [831] and, later, independently, by other authors [25, 504]). A simple proof can be found in Refs. [360, 952] (see also Ref. [437]).
2. The Schwarzschild solution is stable under small perturbations, gravitational or those associated with external fields [340]: the perturbations in the geometry grow small with time, being carried by waves to either $r \to \infty$ or $r \to 0$.
3. The integration constant $\omega$ is, in principle, arbitrary. It has the following meaning: for large values of $r$, where the gravitational field is weak, the trajectories of massive test particles (geodesics) approach the Keplerian orbits that they would describe if they were subject to the Newtonian gravitational field produced by a spherically symmetric object of total mass
$$M = -\frac{\omega c^2}{2G_{\mathrm{N}}^{(4)}}, \tag{11.5}$$
centered at $r = 0$. Then we can identify $M$ with the mass of the object we are describing in GR and $-\omega$ is the Schwarzschild radius defined in Eq. (4.7) associated with such an object,
$$\omega = -R_{\mathrm{S}}. \tag{11.6}$$
We can arrive at the same conclusion by using the ADM mass formula Eq. (10.15), which we rewrite here for convenience,
$$M = \frac{c^2}{8\pi G_{\mathrm{N}}^{(4)}} \int d^2 S_i \, (\partial_j h_{ij} - \partial_i h_{jj}). \tag{11.7}$$
Therefore, $M$ is the (ADM) mass of the Schwarzschild solution and it is taken to be positive for two reasons: first, nobody has seen an object with negative gravitational mass; and second, the Schwarzschild solution with negative mass has a *naked singularity*, which we will explain later, which is thought to be unacceptable on physical grounds.
4. We conclude that, as we wanted, the Schwarzschild metric describes the gravitational field created by a spherically symmetric, massive object as seen from far away (in the vacuum region) by a static observer to whom the above (Schwarzschild) coordinates $\{t, r, \theta, \varphi\}$ are adapted.[6]

[6] Actually, the coordinate $r$ does not have, a priori, the meaning of a radius, even though we have been referring to it as the *radial coordinate*. There are smooth, topologically non-trivial solutions with spherical symmetry but no center [880]. In the Schwarzschild solution $r$ has the meaning of a radius only asymptotically, as we are going to see. Sometimes it is called the *area radius* because its meaning, anywhere, is that surfaces of constant $t$ and $r$ are spheres with area $4\pi r^2$.

5. Usually, the Schwarzschild solution is used from $r = \infty$ to some finite value $r = r_\mathrm{E} > R_\mathrm{S}$ (later we will see why we have to have $r_\mathrm{E} > R_\mathrm{S}$), where it is glued to another static, spherically symmetric metric that is a solution of the Einstein equations for some matter energy–momentum tensor appropriate to describe a static, spherically symmetric star[7] or any other body whose surface is at $r = r_\mathrm{E}$. These metrics, called *Schwarzschild interior solutions*,[8] describe the spacetime in the interiors of stars and Schwarzschild's metric describes all their exteriors (by virtue of Birkhoff's theorem).
6. The Schwarzschild metric is singular (i.e. det $g_{\mu\nu} = 0$ or certain components of the metric blow up) at $r = 0, R_\mathrm{S}$. We know that the Schwarzschild metric is physically sensible for large values of $r$, but we cannot take it seriously for $r \leq R_\mathrm{S}$ because we have to go through a singularity.

   The singularity at $r = R_\mathrm{S}$ can be physical or merely the result of a bad choice of coordinates (just like the singularity at the origin in the Euclidean plane in polar coordinates). If the singularity is physical, then the region $r \leq R_\mathrm{S}$ has nothing to do with the region $r > R_\mathrm{S}$ that describes the exterior of massive bodies. However, if the singularity at $r = R_\mathrm{S}$ is just a coordinate singularity, we can use another coordinate system that is related to the Schwarzschild system in the region $r > R_\mathrm{S}$ by a standard coordinate change but such that the metric is regular at $r = R_\mathrm{S}$. The *analytic extension* of the Schwarzschild metric obtained in this way will also cover the region $r < R_\mathrm{S}$.

   To find the nature of the singularities it is necessary to perform an analysis of the curvature invariants and of the geodesics.[9]

   - Obviously $R = 0$ and $R_{\mu\nu}R^{\mu\nu} = 0$ because the Schwarzschild metric solves the equations of motion $R_{\mu\nu} = 0$. However, other higher-order curvature invariants do not vanish, for instance, the *Kretschmann invariant*

     $$R^{\mu\nu\rho\sigma}R_{\mu\nu\rho\sigma} = \frac{48M^2\cos^2\theta}{r^6} + \cdots. \tag{11.8}$$

     By examining all of them, one concludes that there is a curvature singularity at $r = 0$ but not at $r = R_\mathrm{S}$. This means that the singularity at $r = R_\mathrm{S}$ could be a coordinate singularity, but the singularity at $r = 0$ is certainly a physical singularity that will be present in any coordinate system.
   - If an observer[10] with rest mass $m$ moves in the Schwarzschild field, its equation of motion obeys the general mass-shell constraint Eq. (3.259),

     $$g_{\alpha\beta}p^\alpha p^\beta = m^2c^2, \tag{11.9}$$

---

[7] It is possible to prove that all stellar models describing isolated stars in equilibrium have spherically symmetric metrics [116, 896].

[8] For more details see, for instance, Ref. [1229].

[9] A general reference on the analysis of singularities is Ref. [361].

[10] Traditionally, in this *Gedankenexperiment* the observer sent to probe the Schwarzschild gravitational field at $r = R_\mathrm{S}$ is a graduate student who periodically sends reports to his/her advisor, who sits comfortably away from that point. If the gravitational field at the advisor's position is weak enough, the proper time will be well approximated by the Schwarzschild time $t$. We will break this cruel custom by referring to the former as a free-falling observer and to the latter as the Schwarzschild observer.

where $p^\alpha = -m dx^\alpha/d\tau$ is the observer's four-momentum, $\tau$ is the observer's proper time, and we have set $\xi = c\tau$. On the other hand, since the Schwarzschild metric admits a timelike Killing vector $k^\mu = \delta^{\mu 0}$, the observer's motion has an associated conserved momentum $p(k) \equiv p^0$ given by Eq. (3.266) that we can identify with the observer's total energy

$$E = -p^0 c. \tag{11.10}$$

To simplify the calculations, we assume that the observer has only radial motion (i.e. zero angular momentum) so $p^\theta = p^\varphi = 0$. Then, using the conservation of the energy, the mass-shell constraint becomes a simple equation for $p^r$, which is a differential equation for $r$,

$$\left(\frac{dr}{cd\tau}\right)^2 = \left(\frac{E}{mc^2}\right)^2 - W, \tag{11.11}$$

which can be integrated to give the total proper time:

$$\tau = \frac{1}{c}\int_{r_1}^{r_2} dr \left(\frac{R_S}{r} - \frac{R_S}{R_0}\right)^{-\frac{1}{2}}, \tag{11.12}$$

where

$$R_0 = \frac{R_S}{1 - [E/(mc^2)]^2} \tag{11.13}$$

is the value of the radial coordinate $r$ for which the speed of the observer is zero and $E = mc^2$.

We can use Eq. (11.12) to calculate how long it takes for the free-falling observer to go from $r = R_0 > R_S$ to the curvature singularity at $r = 0$, going through the surface $r = R_S$. The answer is, surprisingly, finite:

$$\Delta\tau = \frac{\pi}{2c} R_0 \left(\frac{R_0}{R_S}\right)^{1/2}. \tag{11.14}$$

This confirms that nothing unphysical happens at $r = R_S$ and that the singularity is only a problem of the Schwarzschild coordinates. It should, then, be possible to find a coordinate system which is not singular there.[11]

This is essentially the idea on which the Eddington–Finkelstein coordinates $\{v, r, \theta, \varphi\}$ are based [479, 550]. In these coordinates the Schwarzschild solution takes the form

$$ds^2 = W dv^2 - 2dvdr - r^2 d\Omega_{(2)}^2, \tag{11.15}$$

[11] There is, of course, another issue: whether the tidal forces at the horizon are big or small. For big enough Schwarzschild BHs they are small, but this might not be a universal behavior of BHs [772, 773].

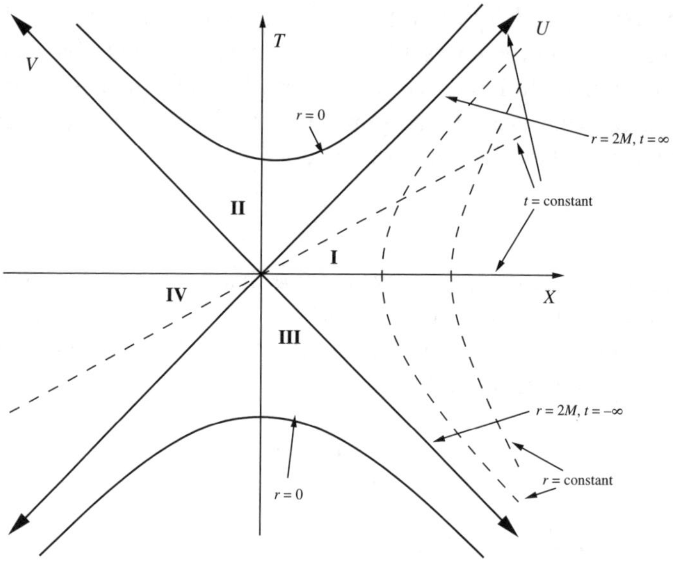


Fig. 11.1. Kruskal coordinates for the Schwarzschild solution. We have set $c = G_{\rm N}^{(4)} = 1$ and each point corresponds to a 2-sphere of radius $r(T, X)$, where $U = X + T$ and $V = X - T$.

where the coordinate $v$ is related to $t$ and $r$ in the region $r > R_{\rm S}$ by

$$v = ct + r + R_{\rm S} \ln |W|, \tag{11.16}$$

and is constant for lightlike radial geodesics (the worldlines of free-falling photons). This metric is regular everywhere except at $r = 0$, as expected, and analytically extends the original solution to the region $r < R_{\rm S}$, allowing its study. Observe that, for the Schwarzschild observer, however, things look quite different. The proper time of the Schwarzschild observer (equal to the Schwarzschild time $t$) is related to the proper time of the free-falling observer $\tau$ by

$$\frac{dt}{d\tau} = \frac{E/mc^2}{1 - R_{\rm S}/r}, \tag{11.17}$$

and will approach infinity when $r$ approaches $r = R_{\rm S}$. This infinite redshift factor is related to the singularity of the Schwarzschild metric in Schwarzschild coordinates. This seemingly paradoxical disagreement between the two observers is, however, not inconsistent, because, as we are going to see, the two observers cannot compare their observations.

7. To study the region $r < R_{\rm S}$ it is more convenient to use the Kruskal–Szekeres [879, 1167] coordinates $\{T, X, \theta, \varphi\}$ that provide the maximal analytic extension of the Schwarzschild metric, describing regions not covered by the Eddington–Finkelstein coordinates (see Fig. 11.1). The region covered by the original Schwarzschild coordinates is just the first quadrant in the figure, whereas the Eddington–Finkelstein coordinates cover the first two quadrants, separated by the $r = R_{\rm S}$ ($r = 2M$ in Fig. 11.1) line. There are two additional regions in quadrants III and IV. Of course, the curvature singularity at $r = 0$ is also present in these new coordinates.

The Schwarzschild metric in Kruskal–Szekeres coordinates takes the form

$$ds^2 = \frac{4R_S^3 e^{\frac{-r}{R_S}}}{r}\left[(dcT)^2 - dX^2\right] - r^2 d\Omega_{(2)}^2, \tag{11.18}$$

where $r$ is a function of $T$ and $X$ that is implicitly given by the coordinate transformations between the pairs $t$ and $r$ and $T$ and $X$:

$$\begin{aligned} \left(\frac{r}{R_S} - 1\right) e^{\frac{r}{R_S}} &= X^2 - c^2T^2, \\ \frac{ct}{R_S} = \ln\left(\frac{X + cT}{X - cT}\right) &= 2\,\text{arctanh}(cT/X), \end{aligned} \tag{11.19}$$

so the Schwarzschild time $t$ is an angular coordinate and the constant-$r$ lines are similar to hyperbolas that asymptotically approach the $X = \pm cT$ lines.

A convenient feature of the Kruskal–Szekeres coordinates is that the $T, X$ part is conformally flat and at each point in the $T, X$ plane the light cones have the same form as in Minkowski spacetime and no particle can have a worldline forming an angle smaller than $\pi/4$ with the $X$ axis. The $r = R_S$ lightlike hypersurface that separates these two quadrants (I, the *exterior*, and II, the *interior*) is called the *event horizon*. It is then clear that particles or signals can go from the exterior to the interior but no signal or particle (including light signals) can go from the interior to the exterior. For this reason, the object described by the full Schwarzschild metric (with no star at $r_E > R_S$) is called a *black* hole.

The existence of an event horizon has very important consequences. First, the free-falling observer can never come back from the BH and cannot send any information that contradicts the Schwarzschild observer's experiences. In this way, the two different observations are made compatible, completely against our classical intuition. Second, it is impossible for the Schwarzschild observer to have any experience of the physical singularity at $r = 0$. This is pictorially expressed by saying that "the singularity is covered by the event horizon."

8. There is another kind of diagram that can be useful for studying the causal structure of the spacetime: Penrose diagrams (see e.g. Ref. [728]) are obtained by performing a conformal transformation of the metric (that preserves the light-cone structure) such that the infinity is brought to a finite distance in the new metric. A Penrose diagram of the Schwarzschild spacetime is drawn in Fig. 11.2.

   Apart from the existence of an event horizon, we also see clearly in this diagram that the fate of the free-falling observer will always be to reach the singularity $r = 0$, which is now a spacelike hypersurface in which he/she will be crushed by infinite tidal forces.

9. We know that there are many objects in the Universe whose gravitational fields are very well described by a region $r > r_E > R_S$ of the Schwarzschild spacetime, but what kind of object gives rise to the $r < R_S$ region, that is, to the BH metric? This

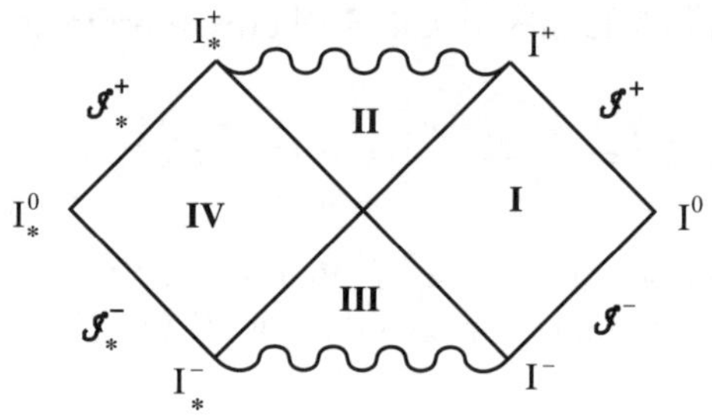


Fig. 11.2. Penrose diagram of the Schwarzschild spacetime.

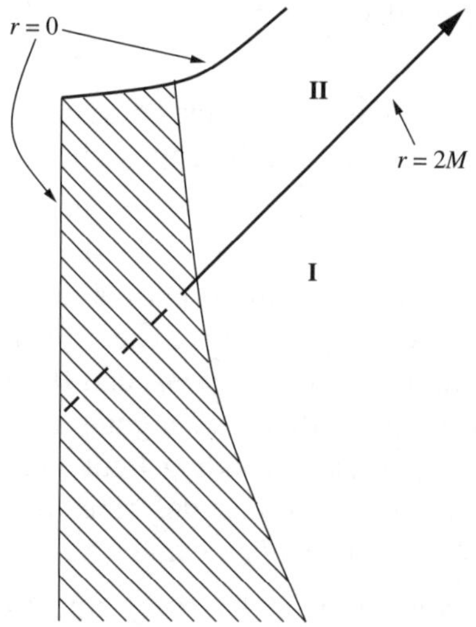


Fig. 11.3. The spacetime corresponding to the gravitational collapse of a star.

question can be answered only by inventing a new kind of object (the BH), which is, by definition, an object giving rise to a spacetime with an event horizon.

How are BHs created in the Universe? In Thorne's book [1182] the story is told of how, in a process that took almost 50 years, the scientific community arrived at the conclusion that BHs could originate from the *gravitational collapse* of very massive stars and, furthermore, that the gravitational collapse would be unavoidable if the star had a mass a few times that of the Sun.

It is evident that the spacetime described by the maximally extended Schwarzschild solution cannot originate from a gravitational collapse (there is no star in the past). Instead it describes an *eternal BH*. In Fig. 11.3 the spacetime corresponding to the spherically symmetric gravitational collapse of a star has been represented in Kruskal–Szekeres-like coordinates. The shaded region represents the star's interior, and the exterior is just the Schwarzschild spacetime in Kruskal coordinates. The BH appears when the collapsing star has a radius smaller than $R_S$.

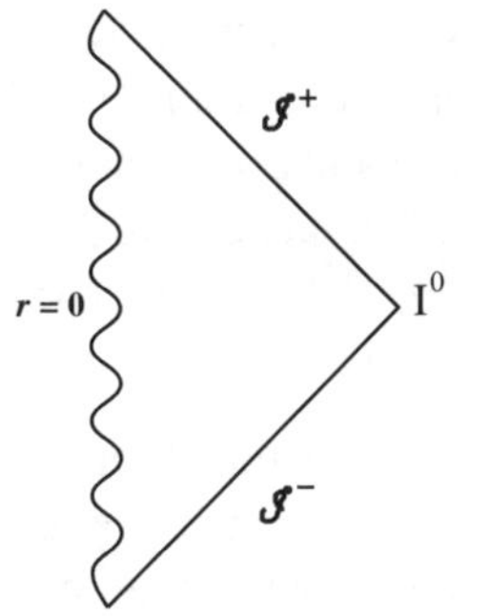


Fig. 11.4. The Penrose diagram of the typical spacetime of a naked singularity.

At this point it may seem an exaggeration to assume that the collapse of any star, in any initial state, is going to give rise to a Schwarzschild-like BH. We will elaborate on this crucial point in a moment.

10. When $M$ is negative, there is no horizon covering the singularity at $r = 0$ and it could be "seen" by all observers (see the Penrose diagram in Fig. 11.4) that could be causally affected by it.

    This can be the source of many problems; to avoid them, we can then argue that such a metric, with a singularity that can be seen from infinity, will never be the endpoint of the gravitational collapse of an ordinary star (or any kind of matter with a physically acceptable energy–momentum tensor). This is the essence of Penrose's *cosmic-censorship hypothesis*, which he first suggested in Ref. [1027] (see also Refs. [1029, 1031] and the reviews [362, 1139, 1234]), in its weak form.[12] There is a strong relation between cosmic censorship and the positivity of energy. Since the gravitational binding energy is negative, when a cloud of self-gravitating matter starts compressing itself, the total energy diminishes more and more and eventually it would become negative. Before that happens an event horizon should appear.

11. There must be other BHs apart from Schwarzschild ones: those corresponding to intermediate states of the gravitational collapse of a star, those that result from perturbing a Schwarzschild BH, or those that describe the gravitational collapse of an electrically charged star. Furthermore, a star can be in many possible states, and it is reasonable to think that they will give rise to many different BHs (or to the same BH in many different states).

    On the contrary, the analysis of the perturbations of the Schwarzschild BH [1061, 1062] shows that all perturbations decay and, after a time of the order of the Schwarzschild radius, the perturbed BH will be Schwarzschild ones,[13] determined solely by $M$, independent of the initial state of the gravitational collapse that

[12] In its strong form, the cosmic-censorship hypothesis states that, in physically acceptable spacetimes, no singularity, except for initial (Big-Bang) singularities, is ever visible to any observer. Rigorous formulations can be found in Ref. [1229].

[13] Or, in general, the Kerr–Newman BH, which is entirely determined by the mass $M$, the electric charge $Q$, and the angular momentum $J$. For simplicity we ignore angular momentum.

originated it and of how it was perturbed. All the higher multipole momenta of the gravitational field (quadrupolar and higher[14]) and of the electromagnetic field (dipole and higher[15]) [118, 1171, 1227] and all momenta of any scalar field are radiated away to infinity so the resulting BH is always a Schwarzschild BH ($M \neq 0,\ Q, J = 0$), a Kerr BH ($M, J \neq 0,\ Q = 0$), a Reissner–Nordström BH ($M, Q \neq 0,\ J = 0$), or a Kerr–Newman BH ($M, J, Q \neq 0$).

Nevertheless, it is conceivable that there might be BH solutions with higher momenta of the gravitational and electromagnetic fields or with a non-trivial scalar field that is not created by perturbations or by gravitational collapse. However, it can be shown (*uniqueness theorems*[16]) that the only BH in the absence of angular momentum and other fields is Schwarzschild [818], that with electric charge is the Reissner–Nordström BH [819], and that with mass and angular momentum is a Kerr BH [309, 1226]. Furthermore, there are no BHs with a non-constant scalar field[17] [342, 122, 933, 1163].

This does not mean that there are no solutions with the forbidden momenta: they actually exist, but they are not BHs, they do not have an event horizon, and they have naked singularities. A simple example is the family of static, spherically symmetric solutions with a non-trivial scalar field discussed in Section 12.1. A more complex example is provided by Bonnor's magnetic-dipole solution[18] [230].

We conclude that there cannot be BHs with other characteristics (*hairs*) different from $M$, $J$, and $Q$ (and, in general, other locally conserved charges). Although this has not been fully proven in all cases [358, 748, 123], this suggests that stationary BHs "have no hair" [1087] (the *no-hair conjecture*, which has a somewhat imprecise formulation).

We would like to make two comments about this conjecture.

(a) Given that the presence of hair is associated with the absence of an event horizon, the no-hair conjecture is intimately related to cosmic censorship: for an event horizon to form in gravitational collapse, all the higher momenta of the fields have to be radiated away. Cosmic censorship is related to the positivity of energy[19] and, thus, so is the no-hair conjecture. Non-stationary BHs with scalar hair and positive energy are also known to exist [1006], but the cosmic censorship and "baldness" conjectures tell us that the hair must disappear in the evolution of the BH toward a stationary state. This is possible because the "scalar charge" is not a locally conserved charge.

[14] The monopole momentum is the mass $M$ and the dipole momentum is determined by the angular momentum $J$.

[15] The monopole momentum of the electromagnetic field is the electric charge.

[16] Two reviews on uniqueness theorems containing many references are Refs. [747, 749].

[17] This statement will be made more precise in the following chapters.

[18] For a physical interpretation, see Ref. [510]; for generalizations, see Ref. [424].

[19] If negative kinetic energies are allowed, BHs with non-trivial scalar fields are possible.

(b) Since the gravitational collapse of many different systems always gives rise to the same BHs, characterized by a very small number of parameters, it is natural to wonder what has happened to all the information about the original state. This is essentially the *BH information problem*, which can be stated more precisely in quantum-mechanical language. Furthermore, it is also natural to attribute to the BHs a very big entropy, which we should be able to compute using standard statistical methods if we knew all the BH microstates that a BH characterized by $M, Q$, and $J$ can be in. This is the essence of the *BH entropy problem*. To solve these two problems, we need a theory of quantum gravity.

12. The event horizons of stationary BHs are usually *Killing horizons*, hypersurfaces that are invariant under one isometry wherein the modulus of the corresponding Killing vector $k^\mu$ of the metric vanishes, $k^2 = 0$. In the Schwarzschild case, $k^\mu = \delta^{\mu t}$ generating translations in time: $k^2\big|_{r=R_S} = g_{tt}\big|_{r=R_S} = 0$. Furthermore, the horizon hypersurface $r = R_S$ is, as a whole, time-translation invariant.

    Killing horizons (and, hence, event horizons) are *null hypersurfaces*.[20] Furthermore, for each value of $t$, the Killing horizon is a 2-sphere of radius $R_S$. This is the only topology allowed according to the *topological-censorship* theorems [721, 568]. Like many other important results in GR, these theorems depend heavily on energy-positivity conditions, and, thus, it is not surprising that they break down in the presence of a negative cosmological constant; then it is possible to find *topological black holes* [792, 888, 889, 43, 279, 890, 1145, 256, 925, 926, 1214, 217, 863, 864, 281] whose event horizons can have the topology of any compact Riemann surface. In particular, generalizations of the asymptotically anti-de Sitter Schwarzschild BH with horizons with the topology of Riemann surfaces of arbitrary genus were given in Ref. [1214].

13. The area of the event horizon is given by

$$A = \int_{r=R_S} d\Omega^2\, r^2 = 4\pi R_S^2. \tag{11.20}$$

    Hawking proved in Ref. [725] that the Einstein equations imply that the area $A$ of the event horizon of a BH never decreases with time. On top of this, if two BHs coalesce to form a new BH, the area of the horizon of this final BH is larger than the sum of the areas of the horizons of the initial BHs. (This result holds for more general kinds of BHs having electric charge and angular momentum.)

    There is a clear analogy between the area $A$ of a BH event horizon and the entropy of a thermodynamical system as never decreasing quantities [117, 119, 120, 357] which deserves to be investigated further.

14. For Killing horizons, one can define, following Boyer [246], the quantity known as the *surface gravity* $\kappa$, given by

$$\kappa^2 = -\tfrac{1}{2}(\nabla^\mu k^\nu)(\nabla_\mu k_\nu)\big|_{\text{horizon}}\,. \tag{11.21}$$

[20] That is, the vector field normal to a Killing horizon is a null vector field. This vector field, due to the Lorentzian signature, always belongs to the tangent space of the null hypersurface.

If $\kappa \neq 0$, the Killing horizon is part of a *bifurcate horizon*, whereas if $\kappa = 0$ it is a *degenerate Killing horizon.*

In the particular case of static, spherically symmetric metrics, which can always be written in the form

$$ds^2 = g_{tt}(r)dt^2 + g_{rr}(r)dr^2 - r^2 d\Omega^2_{(2)}, \tag{11.22}$$

the Killing vector $k^\mu$ is just $\delta^{\mu t}$, and the surface gravity takes the value

$$\kappa = \tfrac{1}{2}\frac{\partial_r g_{tt}}{\sqrt{-g_{tt}g_{rr}}}c, \tag{11.23}$$

which, for the Schwarzschild BH, is the non-vanishing constant

$$\kappa = \frac{c^4}{4G_{\rm N}^{(4)}M}. \tag{11.24}$$

It can be shown that the surface gravity is also constant over the horizon in more general cases [102, 310, 722]. This is analogous to the fact that the temperature is the same at any point of a system in thermodynamical equilibrium, and it constitutes the first analogy between the surface gravity and the BH temperature (and the second between a BH and a thermodynamical system). Physically, the surface gravity is the force that must be exerted at infinity to hold a unit mass in place when $r \to R_{\rm S}$ and has dimensions of acceleration, $\mathrm{LT}^{-2}$.

15. Another set of coordinates that is useful in some problems is the *isotropic coordinates* $\{t, \vec{x}_3\}$ with $\vec{x}_3 = (x^1, x^2, x^3)$, in which the three-dimensional constant-time slices are conformally flat and isotropic. The change of coordinates is given by

$$r = \left(\rho - \frac{\omega}{4}\right)^2 / \rho, \tag{11.25}$$

and the metric takes the form

$$\boxed{ds^2 = \left(1 + \frac{\omega/4}{\rho}\right)^2 \left(1 - \frac{\omega/4}{\rho}\right)^{-2} dt^2 - \left(1 - \frac{\omega/4}{\rho}\right)^4 d\vec{x}_3^{\,2},} \tag{11.26}$$

where $d\rho^2 + \rho^2 d\Omega^2_{(2)} \equiv d\vec{x}_3^2$ and $\rho = |\vec{x}_3|$. In this coordinate system the horizon is located at $\rho = -\omega/4$.

16. In order to study the possibility of some attractor behavior on the horizon, it is convenient to use a radial coordinate $\tau$ such that $\tau \to -\infty$ on the horizon. There is no such behavior in the Schwarzschild BH, but, nevertheless, a coordinate with this property does exist [608] and we will use it in Chapter 27 to describe more general families of BH solutions and how the *attractor mechanism* works in them. This coordinate is related to the $\rho$ in Eq. (11.26). by

$$\rho = -\frac{\omega}{4\tanh\frac{\omega}{4}\tau}, \tag{11.27}$$

and in terms of it the Schwarzschild metric takes the general form

$$\boxed{\begin{aligned} ds^2 &= e^{2U}dt^2 - e^{-2U}\gamma_{\underline{mn}}dx^{\underline{m}}dx^{\underline{n}}, \\ \gamma_{\underline{mn}}dx^{\underline{m}}dx^{\underline{n}} &= \frac{r_0^4}{\sinh^4 r_0\tau}d\tau^2 + \frac{r_0^2}{\sinh^2 r_0\tau}d\Omega^2_{(2)}, \end{aligned}} \tag{11.28}$$

with

$$r_0 = -\omega/2 = M, \qquad e^{-2U} = e^{-2r_0\tau}. \tag{11.29}$$

This is a *conformastationary* metric like the one in Appendix N.2.4, with $|M|$ replaced by $e^U$ and with a particular three-dimensional (*spatial, transverse*) metric $\gamma_{\underline{mn}}$. Many BH solutions can be written in the above general form with different values of the function $U(\tau)$ and of the *non-extremality parameter* $r_0$. The *exterior* of those BHs corresponds to the interval $\tau \in (-\infty, 0)$ (0 corresponding to spatial infinity). The above metric, for $\tau \in (0, +\infty)$, describes a Schwarzschild solution with negative mass: the naked singularity lies at $\tau \to +\infty$, and $\tau \to 0^+$ describes another asymptotic region. In more general solutions, such as those in Chapter 27, that interval will describe other patches of the same BH spacetime.

17. Yet another system of coordinates: let us consider some arbitrary coordinate system $\{y^\alpha\}$ and let us take four scalar functions labeled by $\mu = 0, 1, 2, 3$ of the coordinates $y^\alpha$ and $H^\mu(y)$, which we require to be harmonic:

$$\nabla^2 H^\mu = \frac{1}{\sqrt{|g|}}\partial_\alpha\left(\sqrt{|g|}\, g^{\alpha\beta}\partial_\beta H^\mu\right) = 0. \tag{11.30}$$

Now we can define new coordinates $x^\mu \equiv H^\mu(y)$, which are called *harmonic coordinates*. In the system of harmonic coordinates, the above equation takes the form of a condition on the metric:

$$\partial_\alpha\left(\sqrt{|g|}\, g^{\alpha\mu}\right) = 0. \tag{11.31}$$

If we expand the metric in a perturbation series around flat spacetime,

$$
\begin{aligned}
g_{\mu\nu} &= \eta_{\mu\nu} + \chi h^{(0)}{}_{\mu\nu} + \chi^2 h^{(1)}{}_{\mu\nu} + \cdots, \\
g^{\mu\nu} &= \eta^{\mu\nu} - \chi h^{(0)\,\mu\nu} + \chi^2\left(h^{(0)\,\mu\rho}h^{(0)}{}_\rho{}^\nu - h^{(1)\,\mu\nu}\right), \\
g &= 1 + \chi h^{(0)} + \chi^2\left[h^{(1)} + \tfrac{1}{2}\left(h^{(0)\,2} - h^{(0)\,\mu\nu}h^{(0)}{}_{\mu\nu}\right)\right], \\
\sqrt{|g|} &= 1 + \tfrac{1}{2}\chi h^{(0)} + \tfrac{1}{4}\chi^2\left[2h^{(1)} + h^{(0)\,2} - 2h^{(0)\,\mu\nu}h^{(0)}{}_{\mu\nu}\right], \\
\sqrt{|g|}\,g^{\mu\nu} &= \eta^{\mu\nu} - \chi\bar{h}^{(0)\,\mu\nu} + \chi^2\left[-h^{(1)}{}_{\mu\nu}h^{(0)\,\mu\rho}h^{(0)}{}_\rho{}^\nu - h^{(0)}h^{(0)\,\mu\nu}\right. \\
&\quad \left. + \tfrac{1}{4}\left(2h^{(1)} + h^{(0)\,2} - 2h^{(0)\,\alpha\beta}h^{(0)}{}_{\alpha\beta}\right)\eta^{\mu\nu}\right],
\end{aligned}
\tag{11.32}
$$

where, as usual, $h \equiv h^\rho{}_\rho$ and $\bar{h}^{(0)}{}_{\mu\nu} \equiv h^{(0)}{}_{\mu\nu} - \frac{1}{2}\eta_{\mu\nu}h^{(0)}$. On substituting these into Eq. (11.32), we find that the linear perturbation $h^{(0)}{}_{\mu\nu}$ of the metric in harmonic coordinates is in the *harmonic gauge*, Eq. (3.57), but the next order is not.

To set the Schwarzschild solution in a harmonic coordinate system, it turns out that we just have to shift the Schwarzschild radial coordinate $r \equiv r_{\rm h} - \omega/2$ to obtain

$$
ds^2 = \left(\frac{r_{\rm h} + \omega/2}{r_{\rm h} - \omega/2}\right)(dct)^2 - \left(\frac{r_{\rm h} - \omega/2}{r_{\rm h} + \omega/2}\right)dr_{\rm h}^2 + (r_{\rm h} - \omega/2)^2\,d\Omega_{(2)}^2, \tag{11.33}
$$

and reexpress the metric in terms of coordinates $\vec{x}_3$ (having nothing to do with the isotropic coordinates introduced before) such that $r_{\rm h} = |\vec{x}_3|$ using $r_{\rm h}dr_{\rm h} = \vec{x}_3 \cdot d\vec{x}_3$ and $d\vec{x}_3^{\,2} = dr_{\rm h}^2 + r_{\rm h}^2 d\Omega_{(2)}^2$:

$$
\boxed{
\begin{aligned}
ds^2 &= \left(\frac{r_{\rm h} + \omega/2}{r_{\rm h} - \omega/2}\right)(dct)^2 - \left(1 - \frac{\omega/2}{r_{\rm h}}\right)^2 d\vec{x}_3^{\,2} \\
&\quad - \left(\frac{r_{\rm h} - \omega/2}{r_{\rm h} + \omega/2}\right)\frac{(\omega/2)^2}{r_{\rm h}^4}(\vec{x}_3 \cdot d\vec{x}_3)^2.
\end{aligned}
}
\tag{11.34}
$$

This is the metric whose first two non-trivial terms in a perturbative series expansion, Eq. (3.217), we obtained in Chapter 3 by imposing self-consistency of the SRFT of a spin-2 particle. Observe that the metric Eq. (3.217) has no event horizon. Only if we calculated all the higher-order corrections and summed them to obtain an exact solution of Einstein's equations could we obtain an event horizon. In this sense, BHs are a highly non-perturbative phenomenon.

The differences between GR and the SRFT of the spin-2 particle also become manifest when one compares the causal structures and the asymptotic behaviors.

It seems that it is possible to make compatible either of them but not both for the Schwarzschild spacetime and the Minkowski spacetime in which the SRFT of gravity is defined [1030]. This may be a serious problem for any SRFT of gravity.

The fact that we obtained the first approximation to the Schwarzschild solution by solving the Fierz–Pauli equation in the presence of a massive point-like source may lead us to think that the full solution also corresponds to a point-like source. This is an interesting point that we are going to discuss in the following section.

## 11.2 Sources for the Schwarzschild solution

We would like to identify the object which is the source of the full Schwarzschild gravitational field (with no interior solution). Although we have found it by solving the vacuum Einstein equations, it has a singularity ($r = 0$) where the sourceless Einstein equations are not solved and we can proceed by analogy with the Maxwell case: if we solve Maxwell's equations in vacuum, imposing spherical symmetry and staticity, we find the Coulomb solution $A_\mu = \delta_{t\mu} q/(4\pi r)$, which is singular at $r = 0$, and there the equations are not solved. However, one can add at $r = 0$ a singular source corresponding to a point-like electric charge. The Maxwell equations are then solved everywhere by the Coulomb solution and one can say that the source of the Coulomb field is a point-like electric charge. The solution, however, is not completely consistent since the equations of motion of the charged particle in its own electric field are not solved because this diverges at the position of the particle. This is a well-known problem of the classical model of the electron[21] that is solved by the quantum theory.

There are several reasons why we can expect a negative result: first of all, if the source for the Schwarzschild field were a massive point-particle, it would give rise to a timelike singularity along its worldline, but we know that the Schwarzschild singularity is spacelike. Second, the source for the gravitational field is not just mass, but any kind of energy, including the gravitational field itself. Thus, even if we have a mass distribution confined to a finite region of space (in an idealized case, a point), the gravitational field that it generates will fill the whole space and the source (mass and field) will not be confined to that region. In a sense this is already taken care of by Einstein's equations: in our construction of the self-consistent spin-2 theory we saw that the Einstein tensor contains the "gravitational energy–momentum (pseudo)tensor" and only the matter sources are on the r.h.s. of Einstein's equations.

Anyway, we are going to check explicitly that the massive point-particle cannot be the source for the Schwarzschild metric. This calculation will prepare us for future calculations of the same kind, which, in contrast, will be successful and will help us to understand the reason why.

We consider the action for a massive particle coupled to gravity (we ignore boundary terms):

$$S[g_{\mu\nu}, X^\mu(\xi)] = \frac{c^3}{16\pi G_N^{(4)}} \int d^4x \sqrt{|g|}\, R - Mc \int d\xi \sqrt{|g_{\mu\nu}(X)\dot{X}^\mu \dot{X}^\nu|}. \qquad (11.35)$$

[21] For a review see e.g. Ref. [1175].

The equations of motion of $g_{\mu\nu}(x)$ and $X^\mu(\xi)$ are, respectively,

$$G_{\mu\nu}(x) + \frac{8\pi M G_N^{(4)} c^{-2}}{\sqrt{|g|}} \int d\xi \frac{g_{\mu\rho}(X) g_{\nu\sigma}(X) \dot{X}^\rho \dot{X}^\sigma}{\sqrt{|g_{\lambda\tau}(X) \dot{X}^\lambda \dot{X}^\tau|}} \delta^{(4)}[X(\xi) - x] = 0, \tag{11.36}$$

$$\gamma^{\frac{1}{2}} M \nabla^2(\gamma) X^\lambda + M \gamma^{-\frac{1}{2}} \Gamma_{\rho\sigma}{}^\lambda \dot{X}^\rho \dot{X}^\sigma = 0, \tag{11.37}$$

where

$$\gamma = g_{\mu\nu}(X) \dot{X}^\mu \dot{X}^\nu. \tag{11.38}$$

In the physical system that we are considering, the Schwarzschild gravitational field is produced by a point-particle that is at rest in the frame that we are going to use (Schwarzschild coordinates). Then, we expect the solution for $X^\mu(\xi)$ to be

$$X^\mu(\xi) = \delta^\mu{}_0 \xi. \tag{11.39}$$

However, the $X^\mu$ equations of motion are not satisfied because the component $\Gamma_{00}{}^r$ does not vanish at the origin. Actually, it diverges, and we face here the problem of the infinite force that the gravitational field exerts over the source itself, which is similar to the infinite-self-energy problem of the classical electron mentioned at the beginning of this section. We will see that, in certain situations (in the presence of unbroken supersymmetry), this problem does not occur because the divergent gravitational field is canceled out by another divergent field (electromagnetic, scalar, ...) and the equation of motion of the particle (or *brane*) can be solved exactly.

The above solution for $X^\mu(\xi)$ leads to an energy–momentum tensor whose only non-vanishing component is $T_{00} \sim \delta^{(3)}(\vec{x})$. However, on recalculating carefully[22] the components of the Einstein tensor for the Schwarzschild metric, we find that all the diagonal components, not only $G_{00}$, are different from zero at the origin:

$$\begin{aligned} G_{00} &= -\frac{W}{\sin^2\theta} 4\pi R_S \delta^{(3)}(r), & G_{rr} &= \frac{W^{-1}}{\sin^2\theta} 4\pi R_S \delta^{(3)}(r), \\ G_{\theta\theta} &= -\frac{r^2}{\sin^2\theta} 2\pi R_S \delta^{(3)}(r), & G_{\varphi\varphi} &= \sin^2\theta \, G_{\theta\theta}. \end{aligned} \tag{11.43}$$

[22] In this calculation one has to be careful to keep singular ($\delta$-like) contributions that are non-zero only at a certain point. These contributions come in two forms. One is the standard four-dimensional identity

$$\partial_i \partial_i \frac{1}{|\vec{x}|} = -4\pi \delta^{(3)}(\vec{x}), \qquad i = 1, 2, 3, \tag{11.40}$$

adapted to spherical coordinates,

$$\partial_r \left[ r^2 \partial_r \frac{1}{r} \right] = -\frac{4\pi}{\sin\theta} \delta^{(3)}(r), \tag{11.41}$$

and the other one is

$$\partial_r \left( r \frac{1}{r} \right) = \frac{4\pi}{\sin\theta} \delta^{(3)}(r), \tag{11.42}$$

both of which can be checked by partial integration. Here $\delta^{(3)}(\vec{x}) \neq \delta^{(3)}(r)$. The latter is defined by $\int dr d\varphi d\theta \delta^{(3)}(r) = 1$. The result obtained coincides with that obtained by more rigorous methods in Ref. [88, 89].

This is related to the spacelike nature of the Schwarzschild singularity, as expected. In the cases in which we will be able to identify the source of a solution with a particle (or a brane) the singularity of the metric will be non-spacelike.

## 11.3 Thermodynamics

We have seen in the preceding sections that, classically, according to the Einstein equations, there are two magnitudes in a Schwarzschild BH, the area $A$ and the surface gravity $\kappa$, that behave in some respects like the entropy $S$ and the temperature $T$ of a thermodynamical system. From this point of view, the constancy of $\kappa$ over the event horizon would be the "*zeroth law of BH thermodynamics*" and the never-decreasing nature of $A$ would be the "*second law of BH thermodynamics*." In a thermodynamical system, $S, T$, and the energy $E$ are related by the *first law of thermodynamics*:

$$dE = TdS. \tag{11.44}$$

To take the thermodynamical analogy any further, it is necessary to prove that $\kappa$ and $A$ are also related to the analog of the energy $E$ by a similar equation. The natural analog for the energy is the BH mass $M$ (times $c^2$), and thus it is necessary to have (the factor of $G_{\text{N}}^{(4)}$ appears for dimensional reasons)

$$dM \sim \frac{1}{G_{\text{N}}^{(4)}} \kappa dA. \tag{11.45}$$

This relation turns out to be true. The coefficient of proportionality can be determined [117, 357, 1143], and the *first law of BH thermodynamics* takes the form

$$\boxed{dM = \frac{1}{8\pi G_{\text{N}}^{(4)}} \kappa dA.} \tag{11.46}$$

There is an integral version of this relation that can be checked immediately (the *Smarr formula* [1143]) by simple substitution of the values of $\kappa$ and $A$ for the Schwarzschild BH:

$$M = \frac{1}{4\pi G_{\text{N}}^{(4)}} \kappa A. \tag{11.47}$$

These relations (conveniently generalized to include other conserved quantities such as the electric charge and the angular momentum) seem to hold under very general conditions [102] (see also Refs. [622, 750, 1232]).

This surprising set of analogies suggests the identification between the area of the BH horizon $A$ and the BH entropy, and between the surface gravity $\kappa$ and the BH temperature. Stimulated by these ideas, the authors of Ref. [102] conjectured, giving some plausibility arguments, a "*third law of BH thermodynamics*," namely that "it is impossible by any procedure, no matter how idealized, to reduce $\kappa$ to zero by a finite sequence of operations."

Several specific examples were studied by Wald [1228]. We will comment more on this in the case of the Reissner–Nordström BH.

The analogy is, though, not sufficient to make a full identification. Indeed, as the authors of Ref. [102] say,

> It can be seen that $\kappa/(8\pi)$ is analogous to the temperature in the same way that $A$ is analogous to the entropy. It should, however, be emphasized that $\kappa/(8\pi)$ and $A$ are distinct from the temperature and entropy of the BH.
>
> In fact the effective temperature of a BH is absolute zero. One way of seeing this is to note that a BH cannot be in equilibrium with black-body radiation at any non-zero temperature, because no radiation could be emitted from the hole whereas some radiation would always cross the horizon into the BH.

On the other hand, in the identification $A \sim S,\ \kappa \sim T$ it is not clear what the proportionality constants should be (apart from what the dimensional analysis dictates).

Hawking's discovery [723, 724] that, when the quantum effects produced by the existence of an event horizon are taken into account,[23] BHs radiate as if they were black bodies with temperature[24]

$$\boxed{T = \frac{\hbar\kappa}{2\pi c},} \tag{11.48}$$

dramatically changed this situation. On the one hand, it removed the last obstruction to a complete identification of BHs as thermodynamical systems. On the other, the coefficient of proportionality between $\kappa$ and $T$ was completely determined, which determined, in turn, that between $A$ and $S$:

$$\boxed{S = \frac{Ac^3}{4\hbar G_{\rm N}^{(4)}}.} \tag{11.49}$$

Observe that this relation can be rewritten in the following way:

$$S = \frac{1}{4}\frac{A}{\ell_{\rm Planck}^2}, \tag{11.50}$$

that is, one-quarter of the area of the horizon measured in Planckian units, a huge number for astrophysical-size BHs, in agreement with our discussions about the no-hair conjecture. Observe also that the appearance of $\hbar$ in the equation for $T$ makes manifest its quantum-mechanical origin.

---

[23] This was originally done in a semiclassical calculation in which the background geometry is classical and fixed and there are quantum fields around the BH. The existence of an event horizon gives rise to the Hawking radiation, but the effect of the Hawking radiation on the BH horizon (*backreaction*) is not taken into account. A pedagogical review of this calculation can be found in Ref. [1199].

[24] In our units Boltzmann's constant is 1 and dimensionless so $T$ has dimensions of energy, $\mathrm{ML^2T^{-2}}$ or $\mathrm{L^{-1}}$ in natural units, and the entropy is dimensionless.

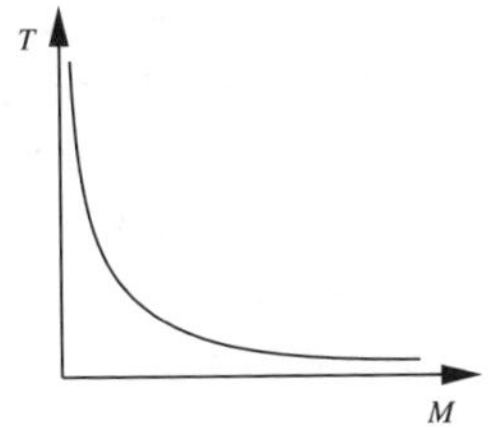


Fig. 11.5. Temperature $T$ versus mass $M$ of a Schwarzschild black hole.

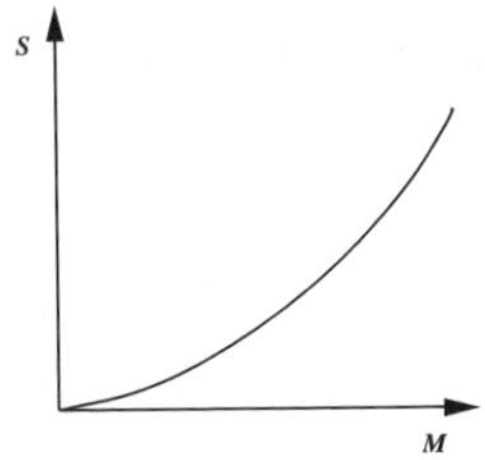


Fig. 11.6. Entropy $S$ versus mass $M$ of a Schwarzschild black hole.

In particular, for a Schwarzschild BH we have (see Figs. 11.5 and 11.6)

$$T = \frac{\hbar c^3}{8\pi G_{\rm N}^{(4)} M}, \qquad S = \frac{4\pi G_{\rm N}^{(4)} M^2}{\hbar c}, \tag{11.51}$$

and so the first law of BH thermodynamics and Smarr's formula take the forms

$$dMc^2 = TdS, \qquad Mc^2 = 2TS. \tag{11.52}$$

How can a BH from which nothing can ever escape (classically) radiate? The physical mechanism behind the Hawking radiation seems to be the process of Schwinger-pair creation in strong background fields [998, 265], which was originally discovered for electric fields [1119], rather than quantum tunneling across the horizon, which would violate causality. In the electric-field case, the background field gives energy to the particles of a virtual pair, separating them. In the BH case, the pair is produced outside but close to the event horizon and one of the particles of the pair falls into the BH while the other one escapes to infinity. The net effect is a loss of BH mass and the "emission" of radiation by the BH.

The same effect causes the spontaneous discharge of charged bodies (such as a positively charged sphere, say) left in vacuum: if the electric field is strong enough, the electron and positron of a virtual pair can be separated. The electron will move toward the sphere, being captured by it, while the positron will be accelerated to infinity. From far away, one would observe a radiation of positrons coming from the charged sphere, whose charge would diminish little by little. In fact, this process is believed to cause the discharge of Reissner–Nordström BHs [929, 311, 418, 605, 973, 1014, 998] and was discovered before the publication of Hawking's results.[25] The energy spectrum of the charged pairs produced

[25] It should also be pointed out that the production of particles in the gravitational field of a rotating BH was also discovered before [1285, 951, 1149, 1205], but this is not a purely quantum-mechanical effect, but the quantum translation of the well-known classical *super-radiance* effect.

in an electric field is also thermal [1154], but only charged particles are produced and the temperature is different depending on the kind of charged particles considered (electron–positron, proton–antiproton, etc.), whereas in the gravitational case, due to the universal coupling of gravity to all forms of energy, all kinds of particles are produced with thermal spectra with a common Hawking temperature.

The thermodynamics of BHs has several problems or peculiarities.

1. The temperature of a Schwarzschild BH (and of all known BHs far from the *extreme* limit, which we will define and discuss later) decreases as the mass (the energy) increases (see Fig. 11.5), and therefore a Schwarzschild BH has a negative specific heat (Fig. 11.7)

$$C^{-1} = \frac{\partial T}{\partial M} = -\frac{\hbar c^3}{8\pi G_{\rm N}^{(4)} M^2} < 0, \tag{11.53}$$

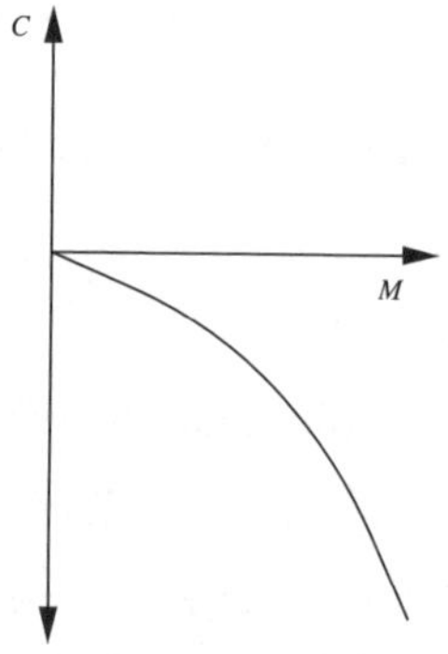


Fig. 11.7. Specific heat $C$ versus mass $M$ of a Schwarzschild black hole.

and becomes colder when it absorbs matter instead of when it radiates (as do ordinary thermodynamical systems). Thus, a BH cannot be put into equilibrium with an infinite heat reservoir because it would absorb the energy and grow without bounds.

2. The temperature grows when the mass decreases (in the evaporation, for instance) and diverges near zero mass.[26] At the same time, the specific heat becomes bigger

[26] Precisely when the metric becomes (apparently, smoothly) Minkowski. The temperature of the Minkowski spacetime is zero, rather than being infinite like the $M \to 0$ limit of the BH temperature. This result is, at first sight, paradoxical, but similar results are very frequent and we will soon meet another one (see footnote 2 on p. 321). *We will very often find that physical properties of a family of metrics parametrized by a number of continuous parameters are not themselves continuous functions of those parameters.* There is no paradox, though, because metrics in that family given by infinitesimally different values of the parameters are not always infinitely close in the space of metrics. Thus, the distance between the Minkowski metric and the Schwarzschild metric with an infinitesimal mass is not infinitesimal. Physically, this is easy to see: no matter how small the mass is, the Schwarzschild spacetime has an event horizon and does not look at all like the Minkowski spacetime.

in absolute value and stays negative. If these formulae remained valid all the way to $M = 0$, the final stage of the Hawking evaporation of a BH would be a violent explosion in which the BH would disappear. However, when $R_S$ becomes of the order of the BH's Compton wavelength (this happens when $M \sim M_{\text{Planck}}$ and implies that $R_S \sim \ell_{\text{Planck}}$), quantum-gravity effects should become important and should determine (we do not know how) the fate of the BH.

3. If a BH can radiate, its entropy can diminish. This is against the second law of BH thermodynamics (which is purely classical). However, the analogy with the second law of thermodynamics can still be preserved because it can be proven that the total entropy (BH plus radiation) never decreases. This is sometimes called the *generalized second law of BH thermodynamics* [121].

4. Returning to the BH information problem, the Hawking radiation seems to carry no more information about the BH than $M$, $J$, and $Q$ (just like the metric itself, so it is not so surprising), but we can ask ourselves whether, in the real world, beyond the approximations made, it would carry more information, and we may be able to see it in a full quantum computation of the gravitational collapse of matter in a well-defined quantum state and the subsequent evaporation of the resulting BH. For 't Hooft, Susskind, and many others, the answer is a definite "yes," namely a BH is just another (peculiar) quantum system and all the information that comes in should unitarily come out: the theory of quantum gravity is unitary. From this point of view, the absorption and radiation of matter by a BH is similar to any standard scattering experiment.

   If no information is carried by the Hawking radiation and the BH evaporates indefinitely, the information about the initial state from which the BH originated is completely lost forever and the theory of quantum gravity governing all these processes is non-unitary, in contrast to all the other physical theories. This is, for instance, Hawking's own viewpoint.

   There is a third group that proposes that the information is not carried out of the BH by Hawking radiation but rather that the evaporation process stops at some point, leaving a *BH remnant* containing that information.

   There is a little-explored fourth possibility, which is consistent with the classical results on stability of BHs and the no-hair conjecture; namely that the information never enters BHs.

   There is, however, no conclusive solution for the BH information problem. In the models based on string theory that we will explain here, BHs are standard quantum-mechanical systems and information is always recovered (even if after a long time).

5. Concerning the BH entropy problem, the statistical-mechanical entropy of systems of fixed energy $E$ is given by

$$S(E) = \ln \rho(E), \tag{11.54}$$

where $\rho(E)$ is the density of states of the system whose energy is $E$. If a BH is just another quantum-mechanical system with $E = M$, a good theory of quantum gravity should allow us to calculate the Bekenstein–Hawking entropy $S$ from knowledge of the density of BH microstates $\rho(M)$. Also, if that theory exists and the above relation is justified, our knowledge of the Bekenstein–Hawking entropy can be used to find $\rho(M)$ for large values of $M$ (when the quantum corrections are small):

$$\rho(M) \sim \exp M^2. \tag{11.55}$$

We see that the number of BH states with a given mass must grow extremely fast if it is to explain the BH's huge entropy (for a solar-mass BH, $\rho \sim 10^{10^{76}}$). The thermodynamical description of systems whose densities of states grow so fast with the energy is, however, very complicated: the canonical partition function

$$\mathcal{Z}(T) \sim \int dE \rho(E) e^{-E/T} \tag{11.56}$$

diverges whenever $\rho(E)$ grows like $e^E$ or faster. For instance, the density of states of any string theory grows exponentially with the mass, and the partition function diverges above *Hagedorn's temperature* (see e.g. Ref. [31]). For $p$-branes [38]

$$\rho(M) \sim \exp\left(\lambda M^{\frac{2p}{p+1}}\right), \tag{11.57}$$

and for $p > 1$ the partition function diverges already at zero temperature. The density of states of BHs must grow faster than that of any of these theories.

As we are going to see, string theory allows us to calculate the entropy and temperature of certain BHs, for which this theory provides quantum-mechanical models, from the density of the associated microstates. In this way string theory seems to solve (at least to some extent) the BH entropy and information problems by treating BHs as ordinary quantum-mechanical systems.

## 11.4 The Euclidean path-integral approach

It is desirable to have an independent and more direct calculation of the BH entropy and temperature. This can be achieved by using the Euclidean path integral, as suggested by Gibbons and Hawking [612, 727].

The thermodynamical study of a statistical-mechanical system starts with the calculation of a thermodynamical potential. If there are certain conserved charges $C_i$ (their related potentials being $\mu_i$), it is convenient to work in the grand canonical ensemble, where the fundamental object is the grand partition function

$$\mathcal{Z} = \mathrm{Tr}\, e^{-\beta(H - \mu_i C_i)}, \tag{11.58}$$

and the thermodynamic potential

$$W = E - TS - \mu_i C_i \tag{11.59}$$

is related to the grand partition function by

$$e^{-W/T} = \mathcal{Z}. \tag{11.60}$$

All thermodynamic properties of the system can be obtained from knowledge of $\mathcal{Z}$. In particular, the entropy is given by

$$S = (E - \mu_i C_i)/T + \ln \mathcal{Z}. \tag{11.61}$$

The idea is to calculate the *thermal* grand partition function of quantum gravity through the path integral of a Euclidean version of the Einstein–Hilbert action Eq. (4.26), $\tilde{S}_{\text{EH}}$,

$$\mathcal{Z} = \int Dg \, e^{-\tilde{S}_{\text{EH}}/\hbar}, \tag{11.62}$$

where one has to sum over all metrics with period[27] $\beta = \hbar c/T$. The only modification that has to be made to the Einstein–Hilbert action is the addition of a surface term to normalize the action so that the on-shell Euclidean action vanishes for flat Euclidean spacetime (the vacuum). The Einstein–Hilbert action becomes [612]

$$S_{\text{EH}}[g] = \frac{c^3}{16\pi G_{\text{N}}^{(4)}} \int_{\mathcal{M}} d^4x \sqrt{|g|}\, R + \frac{c^3}{8\pi G_{\text{N}}^{(4)}} \int_{\partial\mathcal{M}} d^3\Sigma \,(\mathcal{K} - \mathcal{K}_0), \tag{11.63}$$

where $\mathcal{K}_0$ is calculated by substituting the vacuum metric into the expression for $\mathcal{K}$.

The path integral is now to be calculated in the (semiclassical) saddle-point approximation (from now on we set $\hbar = c = G_{\text{N}}^{(4)} = 1$ for simplicity):

$$\mathcal{Z} = e^{-\tilde{S}_{\text{EH}}(\text{on-shell})}. \tag{11.64}$$

The classical solution used to calculate the on-shell Euclidean action is the Euclidean Schwarzschild solution, which we now discuss.

### *11.4.1 The Euclidean Schwarzschild solution*

The Euclidean Schwarzschild solution solves the Einstein equations with the Euclidean metric (in our case $(-,-,-,-)$). It can be obtained by performing a Wick rotation $\tau = it$ of the Lorentzian Schwarzschild solution. If we use Kruskal–Szekeres (KS) coordinates $\{T, X, \theta, \varphi\}$, we have to define the Euclidean KS time $\mathcal{T} = iT$. This Wick rotation has important effects. The relation between the Schwarzschild coordinate $r$ and the $T, X$ coordinates was

$$(r/R_{\text{S}} - 1)e^{r/R_{\text{S}}} = X^2 - T^2. \tag{11.65}$$

[27] $\beta$ has dimensions of length if $T$ has dimensions of energy.

The l.h.s. is bigger than $-1$ and that is why the $T, X$ coordinates also cover the BH interior. However, in terms of $\mathcal{T}$,

$$(r/R_S - 1)e^{r/R_S} = X^2 + \mathcal{T}^2 > 0, \tag{11.66}$$

and the interior $r < R_S$ of the BH is not covered by the Euclidean KS coordinates. On the other hand, the relation between the Schwarzschild time $t$ and $X, \mathcal{T}$,

$$\frac{X+T}{X-T} = e^{t/R_S}, \tag{11.67}$$

becomes

$$\frac{X - i\mathcal{T}}{X + i\mathcal{T}} = e^{-2i\,\mathrm{Arg}(X+i\mathcal{T})} = e^{-i\tau/R_S}. \tag{11.68}$$

Since $\mathrm{Arg}(X + i\mathcal{T})$ takes values between 0 and $2\pi$ (which should be identified), for consistency (to avoid conical singularities) $\tau$ must take values in a circle of length $8\pi M$ [612, 1286]. The period of the Euclidean time can be interpreted as the inverse temperature $\beta$, which coincides with the known Hawking temperature. This is the reason why we can use this metric to calculate the thermal partition function.

The result is a Euclidean metric with periodic time that covers only the exterior of the BH (region I of the KS diagram, Fig. 11.2). The $X, \mathcal{T}$ part of the metric describes a semi-infinite "cigar" (times a 2-sphere) that goes from the horizon to infinity with topology $\mathbb{R}^2 \times \mathrm{S}^2$.

Knowing the result beforehand, we could just as well have used Schwarzschild coordinates, which cover smoothly the BH exterior, and proceeded in this much more economical way [763]: given a static, spherically symmetric BH with regular horizon at $r = 0$, the $r$–$\tau$ part of its Euclidean metric can always be put in the form

$$-d\sigma^2 = f(r)d\tau^2 + f^{-1}(r)dr^2 \sim f'(0)rd\tau^2 + \frac{1}{f'(0)r}dr^2, \tag{11.69}$$

near the horizon. Defining another radial coordinate $\rho$ such that $g_{\rho\rho} = 1$, we obtain

$$-d\sigma^2 \sim \left(\frac{f'(0)}{2}\rho\right)^2 d\tau^2 + d\rho^2 \equiv \rho^2 d\tau'^2 + d\rho^2. \tag{11.70}$$

Now this metric is just the 2-plane metric in polar coordinates[28] if $\tau' \in [0, 2\pi]$. Otherwise it is the metric of a cone and has a conical singularity at $\rho = 0$ (the horizon). Then, $\tau \in [0, \beta = 4\pi/f'(0)]$.

---

[28] The Lorentzian version of this metric is the *Rindler metric*. This metric is equivalent to the wedge $X \in (0, \infty)$, $T \in (-X, X)$ of the Minkowski metric $ds^2 = dT^2 - dX^2 - \cdots$, and it can be obtained from it by transforming the coordinates to those adapted to an observer moving with constant proper acceleration $a$ in the direction $X$:

$$T = \rho \sinh at, \qquad X = \rho \cosh at \quad \Rightarrow \quad ds^2 = a^2\rho^2 dt^2 - d\rho^2 = \cdots. \tag{11.71}$$

Comparing with the above metric, the acceleration of the Rindler observer is $a = \frac{f'(0)}{2}$, and, following the argument, $a = 2\pi T$, where $T$ is the temperature and therefore $a = \kappa$, the surface gravity. Thus, we arrive at an important result: the near-horizon geometry of Schwarzschild-like BHs is described by a Rindler space with a proper acceleration equal to the surface gravity of the event horizon.

In practice we do not even need the Euclidean Schwarzschild metric. We need only the information about the period of the Euclidean time (temperature) and the fact that the BH interior disappears (the integration region), and we can simply replace $-\tilde{S}_{\rm EH}(\text{on-shell})$ by $+iS_{\rm EH}(\text{on-shell})$ because it gives the same result once we take into account the above two points. Thus, in our calculation we will use the Lorentzian Schwarzschild metric in Schwarzschild coordinates using these observations.

### 11.4.2 The boundary terms

The Euclidean Schwarzschild solution, being a solution of the vacuum Einstein equations, has $R=0$ everywhere (the singularity $r=0$ together with the whole BH interior is not included) and only the boundary term contributes to the on-shell action. We now calculate its value.

The only boundary of the Euclidean Schwarzschild metric, with the time compactified on a circle of length $\beta$, is $r\rightarrow\infty$. (If we gave the Euclidean time a different periodicity, there would be another boundary at the horizon, but there is no reason to do this.) This boundary is then the hypersurface $r=r_{\rm c}$ when the constant $r_{\rm c}$ goes to infinity. A vector normal to the hypersurface $r-r_{\rm c}=0$ is $n_\mu\sim\partial_\mu(r-r_{\rm c})=\delta_{\mu r}$; normalized to unity ($n_\mu n^\mu=-1$ because it is spacelike) with the right sign to make it outward-pointing, for a generic spherically symmetric metric Eq. (11.22), it is given by

$$n_\mu=-\frac{\delta_{\mu r}}{\sqrt{-n^2}}=-\sqrt{-g_{rr}}\,\delta_{\mu r}. \tag{11.72}$$

The four-dimensional metric $g_{\mu\nu}$ induces the following metric $h_{\mu\nu}$ on the hypersurface $r-r_{\rm c}=0$:

$$ds^2_{(3)}=h_{\mu\nu}dx^\mu dx^\nu=g_{tt}dt^2-r^2d\Omega^2_{(2)}\Big|_{r=r_{\rm c}}. \tag{11.73}$$

The covariant derivative of $n_\mu$ is

$$\nabla_\mu n_\nu=-\sqrt{-g_{rr}}\left\{\delta_{\mu r}\delta_{\nu r}\partial_r\ln\sqrt{-g_{rr}}-\Gamma_{\mu\nu}{}^{r}\right\}, \tag{11.74}$$

and the trace of the extrinsic curvature of the $r-r_{\rm c}=0$ hypersurface is (the Christoffel symbols can be found in Appendix N.2)

$$\mathcal{K}=h^{\mu\nu}\nabla_\mu n_\nu=\frac{1}{\sqrt{-g_{rr}}}\left\{\tfrac{1}{2}\partial_r\ln g_{tt}+2/r\right\}\Big|_{r=r_{\rm c}}. \tag{11.75}$$

The regulator $\mathcal{K}_0$ can be found from this expression to be

$$\mathcal{K}_0=(2/r)|_{r=r_0}. \tag{11.76}$$

On the other hand, for any static, spherically symmetric, asymptotically flat metric we must have for large $r$

$$g_{tt}\sim 1-\frac{2M}{r},\qquad g_{rr}\sim-\left(1+\frac{2M}{r}\right)\quad\Rightarrow(\mathcal{K}-\mathcal{K}_0)|_{r=r_{\rm c}}\sim-M/r_{\rm c}^2. \tag{11.77}$$

Finally, we have

$$\frac{i}{8\pi}\int_{r_0\to\infty} d^3x\sqrt{|h|}\,(\mathcal{K}-\mathcal{K}_0) = \lim_{r_c\to\infty}\frac{i}{8\pi}\int_0^{-i\beta} dt \int_{S^2} d\Omega^2 r_c^2\sqrt{g_{tt}(r_c)}\,(\mathcal{K}-\mathcal{K}_0)$$

$$= \lim_{r_0\to\infty}\frac{\beta}{2}r_c^2(\mathcal{K}-\mathcal{K}_0) = -\frac{\beta M}{2}. \tag{11.78}$$

For Schwarzschild $\beta = 8\pi M$, and Eqs. (11.61) and (11.64) lead to the expected result

$$S = \beta M + \ln\mathcal{Z} = \beta M/2 = 4\pi M^2. \tag{11.79}$$

## 11.5 Higher-dimensional Schwarzschild metrics

If we consider the $d$-dimensional vacuum Einstein equations, it is natural to look for the generalization of the Schwarzschild solution: static, spherically symmetric metrics. Here, spherical symmetry means invariance under global SO$(d-1)$ transformations. The appropriate ansatz that generalizes Eq. (11.2) is

$$ds^2 = W(r)(dct)^2 - W^{-1}(r)dr^2 - R^2(r)d\Omega^2_{(d-2)}, \tag{11.80}$$

where $d\Omega^2_{(d-2)}$ is the metric element on the $(d-2)$-sphere $\mathrm{S}^{d-2}$ (see Appendix K).

One finds the following generalization of the Schwarzschild solution [1168, 968]:

$$\boxed{ds^2 = W(dct)^2 - W^{-1}dr^2 - r^2 d\Omega^2_{(d-2)}, \qquad W = 1+\omega/r^{d-3},} \tag{11.81}$$

where $d \geq 4$: there are no Schwarzschild BHs in fewer than four dimensions.[29]

The integration constant $\omega$ is related to the $d$-dimensional analog of the Schwarzschild radius. To establish the above relation between the Schwarzschild radius and the mass, we can use, for instance, Komar's formula Eq. (10.42) correctly normalized [968]:

$$Mc^2 = -\frac{1}{16\pi G_N^{(d)}}\frac{d-2}{d-3}\int_{S^{d-2}_\infty} d^{d-2}\Sigma_{\mu\nu}\nabla^\mu k^\nu. \tag{11.82}$$

The result of the integral is $(d-3)\omega_{(d-2)}\omega c$, with $\omega_{(d-2)}$ given in Eq. (K.11), and thus

$$\omega = -R_S^{d-3} = -\frac{16\pi G_N^{(d)} Mc^{-2}}{(d-2)\omega_{(d-2)}}. \tag{11.83}$$

The solutions Eq. (11.81) are almost straightforward generalizations of the four-dimensional Schwarzschild solution in every sense. Their most interesting property is the

[29] In the presence of a negative cosmological constant there is, however, an asymptotically $\mathrm{AdS}_3$ three-dimensional solution that can be identified with a BH: the BH of Bañados, Teitelboim, and Zanelli (BTZ) [95].

existence of event horizons at $r = R_S$ in all of them, with properties that generalize those of the $d = 4$ ones and lead us to the study of their thermodynamics. The uniqueness of these (static BH) solutions was proved in Refs. [814, 620]. There is no uniqueness for stationary BHs in higher dimensions, as the existence of the rotating black ring of Ref. [512] shows.

### 11.5.1 Thermodynamics

In $d$ dimensions, the first law of BH thermodynamics and Smarr's formula are [968]

$$dMc^2 = \frac{d-2}{2(d-3)} TdS, \qquad Mc^2 = \frac{d-2}{d-3} TS, \tag{11.84}$$

where the temperature $T$ is now given in terms of the surface gravity $\kappa$ by the same expression as in four dimensions, Eq. (11.48), while $\kappa$ is defined by the same formula, Eq. (11.21), in any dimension. The entropy is given in terms of the volume of the $(d-2)$-dimensional constant-time slices of the event horizon $V^{(d-2)}$ by

$$\boxed{S = \frac{c^3 V^{(d-2)}}{4\hbar G_N^{(d)}}.} \tag{11.85}$$

The volume and surface gravity of the event horizon are

$$V^{(d-2)} = R_S^{d-2} \omega_{(d-2)}, \qquad \kappa = \frac{(d-3)c^2}{2R_S}, \tag{11.86}$$

and therefore

$$T = \frac{(d-3)\hbar c}{4\pi R_S}, \qquad S = \frac{R_S^{d-2} \omega_{(d-2)} c^3}{4\hbar G_N^{(d)}}. \tag{11.87}$$

Smarr's formula can be easily checked using these results.

The temperature of the higher $d$-dimensional BHs can also be calculated in the Euclidean formalism with the criterion of avoiding conical singularities of the $\tau$–$r$ part of the metric on the event horizon. A Euclidean calculation of the entropy may also be done.

# 12

# The Reissner–Nordström black hole

In the preceding chapter we obtained and studied the Schwarzschild solution of the vacuum Einstein equations and arrived at the BH concept. However, many of the general features of BHs that we discussed, such as the no-hair conjecture, make reference to BHs in the presence of matter fields. In this chapter we are going to initiate the study and construction of BH solutions of the Einstein equations in the presence of matter fields, starting with the simplest ones: massless scalar and vector fields.

The (unsuccessful) search for BH solutions of gravity coupled to a scalar field will allow us to deepen our understanding of the no-hair conjecture.

The (successful) search for BH solutions of gravity coupled to a vector field will allow us to find the simplest BH solution different from the Schwarzschild solution: the Reissner–Nordström (RN) solution. Simple as it is, it has very interesting features, in particular the existence of an extreme limit with a regular horizon and zero Hawking temperature that will be approached with positive specific heat, as in standard thermodynamical systems. Later on we will relate some of these properties to the unbroken supersymmetry of the extreme RN (ERN) solution, which will allow us to reinterpret it as a self-gravitating supersymmetric soliton interpolating between two vacua of the theory.

The ERN BH is the archetype of the more complicated self-gravitating supersymmetric solitons that we are going to encounter later on in the context of superstring low-energy effective actions (actually, one of our goals will be to recover it as a superstring solution), and many of its properties will be shared by them. Furthermore, the four-dimensional Einstein–Maxwell system exhibits *electric–magnetic duality* in its simplest form. Electric–magnetic duality will play a crucial role in many of the subsequent developments either as a classical solution-generating tool or as a tool that relates the weak- and strong-coupling regimes of QFTs.

It is, therefore, very important to study all these properties in this simple system.

In this chapter we are first going to study the coupling of a free massless real scalar to gravity, discussing the (non-)existence of BH solutions and its relation to the no-hair conjecture. Then, we will study the coupling of a massless vector field to gravity (the Einstein–Maxwell system), its gauge symmetry, and the notion and definition of electric charge and its conservation law. Immediately afterwards we will introduce and study the electrically charged RN BH and its sources, thermodynamics, and Euclidean action. Once we are done with the electrically charged RN BH, we will introduce electric–magnetic duality, the notion and definition of magnetic charge, and the Dirac–Schwinger–Zwanziger quantization condition. Using electric–magnetic duality, we will construct magnetically charged and dyonic RN BHs. Finally, we will consider higher-dimensional RN BH solutions.

## 12.1 Coupling a scalar field to gravity and no-hair theorems

The simplest field to which we can couple gravity is a free (vanishing potential) massless real scalar field $\varphi$. The action of this system is (choosing the simplest normalization)

$$S[g_{\mu\nu}, \varphi] = S_{\rm EH} + \frac{c^3}{8\pi G_{\rm N}^{(4)}} \int d^4x \sqrt{|g|}\, \partial_\mu\varphi\partial^\mu\varphi. \tag{12.1}$$

The equations of motion for the metric and the scalar are

$$G_{\mu\nu} + 2\left[\partial_\mu\varphi\partial_\nu\varphi - \tfrac{1}{2}g_{\mu\nu}(\partial\varphi)^2\right] = 0, \qquad \nabla^2\varphi = 0. \tag{12.2}$$

If we take the divergence of the Einstein equation above and use the contracted Bianchi identity $\nabla^\mu G_{\mu\nu} = 0$, we obtain

$$\nabla^2\varphi\nabla_\nu\varphi = 0, \tag{12.3}$$

which implies the equation of motion for the scalar field $\varphi$ if $\nabla_\nu\varphi \neq 0$. If $\nabla_\nu\varphi = 0$ the scalar equation of motion is automatically solved and, thus, we can say that the Einstein equations imply the scalar field equation of motion and we only have to solve the former. If we subtract its trace, we are left with

$$R_{\mu\nu} + 2\partial_\mu\varphi\partial_\nu\varphi = 0 \tag{12.4}$$

as the only set of equations that we really need to solve.

One can then proceed by trying to find a BH-type solution (i.e. one with a metric similar to that of the Schwarzschild solution, possessing an event horizon) of the equation of motion of this system. It is clear that any solution of the vacuum Einstein equations (in particular, Schwarzschild's) will be a solution of these equations with a constant scalar $\varphi = \varphi_0$, but we are really interested only in solutions with a non-trivial $\varphi$. How could we characterize the non-triviality of $\varphi$? By analogy with other fields, we could consider multipole expansions of $\varphi$. The monopole momentum of $\varphi$ (the coefficient of the $1/r$ term), which is the only one that respects spherical symmetry, could be understood as the "scalar charge" and we could characterize the simplest BH-type solutions (the static and spherically symmetric ones) by the mass (the monopole momentum of the gravitational field) and the "scalar charge."

We would like to have, though, a more physical definition of the "scalar charge." The first definition of "scalar charge" we could try is suggested by the form of a possible source for $\varphi$: it would have to be a scalar $\rho$ satisfying $\nabla^2\varphi = \rho$, corresponding to a coupling of the form $\varphi\rho$ in the action. Then, the integral over some spatial volume (let us say a constant-time slice of the whole spacetime) of the source would give the charge, and, using the equation of motion, we could define

$$\int_{\Sigma} d^3\Sigma_\mu n^\mu \nabla^2 \varphi, \tag{12.5}$$

where $n^\mu$ is the unit vector normal to the spacelike hypersurface $\Sigma$. This integral is indeed proportional to the coefficient of $1/r$ in the multipole expansion of $\varphi$. However, there is no way to show that this "charge" is conserved using the scalar equation of motion. Nothing prevents this kind of "charge" from disappearing, and, in fact, according to the results on gravitational collapse and perturbations[1] of the Schwarzschild solution upon which the no-hair conjecture is based, this is actually what happens in the gravitational collapse, although no complete proof is available.

Still, one could conceive of a situation in which not all the "scalar charge" disappears and after a long time the system settles into a static, spherically symmetric state with non-vanishing scalar charge. The no-hair conjecture asserts that the solution describing this state will not be a BH, which in general means that it will have naked singularities. The cosmic-censorship conjecture then tells us that this state could not have been produced in the gravitational collapse of well-behaved matter with physically admissible initial conditions, in complete agreement with the no-hair conjecture.

Now we can put to the test the no-hair and cosmic-censorship conjectures, either by trying to find static, spherically symmetric solutions with non-trivial scalar fields, or by evolving initial data sets describing one or several regular BHs with mass and scalar charge that are not in equilibrium, such as those in Ref. [1006]. This has not yet been done and, therefore, we will concentrate on finding scalar BH solutions. It is worth mentioning that some exceptions to the cosmic-censorship conjecture are known, especially in Einstein–Yang–Mills systems (see Section 19.3.2), and only by evolving the initial data can one really find out whether the same will happen here.

To find static, spherically symmetric solutions we make the ansatz ($c = G_{\rm N}^{(4)} = 1$)

$$ds^2 = \lambda(r)dt^2 - \lambda^{-1}(r)dr^2 - R^2(r)d\Omega^2_{(2)}, \qquad \varphi = \varphi(r), \tag{12.6}$$

and, using the formulae in Appendix N.2.2, we find the Janis–Newman–Winicour (JNW) solutions [827, 18]

$$\boxed{\begin{aligned} ds^2 &= W^{\frac{2M}{\omega}-1} W dt^2 - W^{1-\frac{2M}{\omega}}\left[W^{-1}dr^2 + r^2 d\Omega^2_{(2)}\right], \\ \varphi &= \varphi_0 + \frac{\Sigma}{\omega}\ln W, \\ W &= 1 + \tfrac{\omega}{r}, \qquad \omega = \pm 2\sqrt{M^2+\Sigma^2}. \end{aligned}} \tag{12.7}$$

[1] See e.g. Ref. [367], in which the wave equation for a scalar field on a Schwarzschild BH background is analyzed and it is shown that it has no physically acceptable solutions, the conclusion being that a BH cannot act as a source for the scalar field and that there will be no BH solutions with non-trivial scalar hair.

The three fully independent parameters that characterize each solution are the mass $M$, the "scalar charge" $\Sigma$, and the value of the scalar at infinity $\varphi_0$. As expected, only when the "scalar charge" vanishes ($\Sigma = 0$) does one have a regular solution (Schwarzschild's).[2] In all other cases there is a singularity at $r = r_0$, when $r_0 > 0$, or at $r = 0$.[3]

Although a regular BH cannot act as a source for scalar charge, other fields can. This is what happens in the "a-model" (also known as Einstein–Maxwell-dilaton (EMD) gravity, see Section 16.1) in which the scalar ("dilaton") equation of motion is roughly of the form

$$\nabla^2 \varphi = \tfrac{1}{8} a e^{-2a\varphi} F^2. \tag{12.11}$$

In this theory we can expect BHs with non-trivial scalar fields. However, the scalar charge will be completely determined by the mass and electric and magnetic charges of the electromagnetic field, according to a certain formula. This kind of hair, which does depend on the mass, angular momentum, and conserved charges, is called *secondary hair* [367]. If the scalar charge does not have the value dictated by the formula, then there is another source for the scalar field apart from the electromagnetic field as in the solutions of Ref. [19], so the BH would also have *primary hair*. This is the only kind of hair that the solutions Eqs. (12.8) have and is the kind forbidden by the no-hair conjecture.

At this point it is worth mentioning that there are other kinds of scalar charges that are locally conserved. This discussion anticipates concepts that we will encounter in Part III. First, the equation of motion $\nabla^2 \varphi = 0$ can be rewritten in the form $\partial_\mu \left( \sqrt{|g|}\, F^\mu \right) = 0$,

---

[2] This is another example (see footnote 26 on p. 310) of a family of metrics parametrized by a continuous parameter whose physical properties are not continuous functions of those parameters.

[3] Observe that the given family of solutions includes a non-trivial *massless* solution. On setting $M = 0$, we find

$$\boxed{\begin{aligned} ds^2 &= dt^2 - dr^2 - W r^2 d\Omega^2_{(2)}, \\ \varphi &= \varphi_0 + \tfrac{1}{2} \ln W, \\ W &= 1 + \tfrac{\omega}{r}. \end{aligned}} \tag{12.8}$$

This solution is related to Schwarzschild's (with positive or negative mass) by a Buscher "T-duality" (to be explained later on) transformation on the time direction. It is still singular for any value of $\omega$ different from zero. This is perhaps best seen after the coordinate change

$$r = \frac{1}{\rho}\left(\rho - \frac{\omega}{4}\right)^2, \tag{12.9}$$

which allows us to rewrite the metric in the isotropic form

$$\begin{aligned} ds^2 &= dt^2 - \left(1 + \frac{\omega/4}{\rho}\right)^2 \left(1 - \frac{\omega/4}{\rho}\right)^2 d\vec{x}_3^{\,2}, \\ \varphi &= \varphi_0 - \tfrac{1}{2} \ln \left[ \left(1 - \frac{\omega/4}{\rho}\right)^2 \left(1 + \frac{\omega/4}{\rho}\right)^{-2} \right], \qquad \rho = |\vec{x}_3|. \end{aligned} \tag{12.10}$$

The interpretation of these static, massless solutions is not easy. Since the mass of a spacetime is its total energy and the scalar field must contribute a positive amount to the total energy, we have to admit that the gravitational field contributes a negative amount to it. Here we see again the relation among the no-hair conjecture, the cosmic censorship, and the positivity of the energy.

where $F^{\mu} = \nabla^{\mu}\varphi$. As will be explained later for the electric charge, this is just the continuity equation for the current $F^{\mu}$ and suggests the following definition of scalar charge:

$$\int_V d^3\Sigma_\mu \nabla^\mu \varphi, \tag{12.12}$$

which will be locally conserved. The conservation of this current is associated via Noether's theorem with the invariance of the action under constant shifts of the scalar.

Second, the Bianchi-type identity $\partial_{[\mu}\partial_{\nu]}\varphi = 0$ can be rewritten in the form $\nabla_\mu F^{\mu\nu\rho} = 0$, where we have defined the completely antisymmetric tensor $F^{\mu\nu\rho} = (1/\sqrt{|g|})\epsilon^{\mu\nu\rho\sigma}\partial_\sigma\varphi$. With this definition it is possible to show that the line integral

$$\frac{1}{3!}\oint_\gamma d^1\Sigma_{\mu\nu\rho}F^{\mu\nu\rho} = \oint_\gamma d\varphi \tag{12.13}$$

along the curve $\gamma$ is conserved. Observe that, if $\gamma$ is closed, the integral will only be different from zero if $\varphi$ is multivalued, for instance if $\varphi$ is an axion (a pseudoscalar) that takes values in a circle.

How should we interpret these charges? We will see later in this chapter that the electromagnetic field $A_\mu$ has a natural coupling to the worldline of a particle with electric charge $q$ given by Eq. (12.53). The particle's electric charge is given by the surface integral over a sphere $S^2$ of the Hodge dual of the electromagnetic-field-strength 2-form $F_{\mu\nu}$. The particle's magnetic charge is given by the surface integral over a sphere $S^2$ of the electromagnetic-field-strength 2-form. The electric charge is conserved due to the equation of motion and the magnetic charge is conserved due to the Bianchi identity. A topologically non-trivial configuration of the field is needed in order to have magnetic charge.

Potentials that are differential forms of higher rank couple to the worldvolumes of extended objects: a $(p+1)$-form potential $A_{(p+1)}$ naturally couples to $p$-dimensional objects with a $(p+1)$-dimensional worldvolume (we will explain how this comes about in Chapter 24). The electric charge is the integral over the sphere $S^{d-(p+2)}$ transverse to the object's worldvolume of the Hodge dual of the $(p+2)$-form field strength $F_{(p+2)} = dA_{(p+1)}$. The magnetic charge would be the electric charge of the dual $(d-p-4)$-dimensional object, charged under the dual potential whose field strength is the Hodge dual of $F_{(p+2)}$.

Looking now at the above charges, we immediately realize that the charge defined in Eq. (12.13) is the charge of a one-dimensional object (string) and that the charge defined in Eq. (12.12) is that of a "$-1$-dimensional object." Such an object would be an instanton, defined in Euclidean space and with zero-dimensional worldvolume. Then, "charge conservation" is not a concept to be applied to it. In both cases $\varphi$ has to be a pseudoscalar.

Observe that, indeed, a line integral such as Eq. (12.13) cannot measure a point-like charge because we could continuously contract the loop $\gamma$ to a point without meeting the singularity at which the charge rests. The line integral has to have a non-vanishing linking number with the one-dimensional object, which has to have either infinite length or the topology of $S^1$; otherwise the integral would be zero by the same argument. The behavior of the scalar field has to be $\varphi \sim \ln\rho$, where $\rho$ measures the distance to the one-dimensional object in the two-dimensional plane orthogonal to it.

Similar arguments apply to the definition Eq. (12.12) and $\varphi \sim 1/\rho^2$, where now $\rho$ measures the distance to the instanton in the four-dimensional Euclidean space.

From this point of view, if BHs can be understood as particle-like objects, looking for BHs with a well-defined scalar charge is utterly hopeless. One should look instead for "black strings" and instantons, and in due time we will do so and find them.[4]

There is another point of view concerning scalar fields: in some cases they should be interpreted not as matter fields but as "local coupling constants" (as in the case of the string-theory dilaton) or, more generally, as *moduli fields*, which we will define in Chapter 15, in which case they should be treated as backgrounds and there would be no room for the notion of scalar charge.

In conclusion, if we want to find new BH solutions, we need to couple the Einstein–Hilbert action to matter fields that have associated conserved charges. The charges must be those of point-particles or we will naturally obtain solutions describing extended objects instead of black holes. Thus, we have to consider vector fields, and the simplest one is an Abelian vector field $A_\mu$. We are going to study in some detail the resulting system because later we will find generalizations of all the concepts and formulae developed here.

## 12.2 The Einstein–Maxwell system

The action for gravity coupled to an Abelian vector field $A_\mu$ is the so-called Einstein–Maxwell action[5] obtained by adding the Einstein–Hilbert and the Maxwell action with $\eta_{\mu\nu}, \partial_\mu$, and $d^4x$ replaced by $g_{\mu\nu}, \nabla_\mu$, and $d^4x$:

$$\boxed{S_{\rm EM}[g_{\mu\nu}, A_\mu] = S_{\rm EH}[g] + \frac{1}{c}\int d^4x\sqrt{|g|}\,\left[-\tfrac{1}{4}F^2\right].} \tag{12.15}$$

$F_{\mu\nu}$ is the field strength of the electromagnetic vector field $A_\mu$ and is again given by

$$F_{\mu\nu} = 2\partial_{[\mu}A_{\nu]}, \qquad F^2 = F_{\mu\nu}F^{\mu\nu}, \tag{12.16}$$

since, in the absence of torsion, $\nabla_{[\mu}A_{\nu]} = \partial_{[\mu}A_{\nu]}$. The components of $A_\mu$ and $F_{\mu\nu}$ in a given coordinate system are customarily split in this way,

$$(A_\mu) = (\phi, -\vec{A}), \qquad (F_{\mu\nu}) = \begin{pmatrix} 0 & E_1 & E_2 & E_3 \\ -E_1 & 0 & -B_3 & B_2 \\ -E_2 & B_3 & 0 & -B_1 \\ -E_3 & -B_2 & B_1 & 0 \end{pmatrix}, \tag{12.17}$$

[4] This argument really applies to pseudoscalar fields.

[5] In this section we work in the Heaviside system of units, so the Coulomb force between two charges is

$$\frac{1}{4\pi}\frac{q_1 q_2}{r_{12}^2}. \tag{12.14}$$

In the Gaussian system we should replace $1/(4c)$ by $1/(16\pi c)$ and the factor of $4\pi$ disappears from the Coulomb force. The dimensions of the vector field $A_\mu$ are $\mathrm{M}^{1/2}\mathrm{L}^{1/2}\mathrm{T}^{-1}$ (that is, $\mathrm{L}^{-1}$ in natural units $\hbar = c = 1$) and the electric charge's units are $M^{1/2}L^{3/2}T^{-1}$, so it is dimensionless in natural units. At the end we will introduce another system of units, which will be the one we more often will work with, taking $c = 1$ and replacing the factor of $1/(4c)$ in front of $F^2$ by $1/(64G_{\rm N}^{(4)})$.

where $\vec{E} = (E_1, E_2, E_3)$ and $\vec{B} = (B_1, B_2, B_3)$ are the electric and magnetic 3-vector fields in that coordinate system, and, thus, with $\vec{\nabla} = (\partial_{\underline{1}}, \partial_{\underline{2}}, \partial_{\underline{3}})$,

$$\begin{cases} E_i = F_{\underline{0i}}, \\ B_i = -\frac{1}{2}\epsilon_{ijk}F_{\underline{kl}}, \end{cases} \Leftrightarrow \begin{cases} \vec{E} = -\vec{\nabla}\phi - \frac{1}{c}\frac{\partial}{\partial t}\vec{A}, \\ \vec{B} = \vec{\nabla} \times \vec{A}. \end{cases} \tag{12.18}$$

The field strength (and the action) are invariant under the Abelian gauge transformations

$$A'_\mu = A_\mu + \partial_\mu \Lambda \tag{12.19}$$

with smooth, gauge parameter $\Lambda$. Depending on which gauge group we consider ($\mathbb{R}$ or $U(1)$), $\Lambda$ must be a single-valued or multivalued function.[6] In differential-form language

$$A = A_\mu dx^\mu, \qquad A' = A + d\Lambda, \qquad F = \tfrac{1}{2}F_{\mu\nu}dx^\mu \wedge dx^\nu = dA, \tag{12.20}$$

and the gauge invariance of $F$ is a consequence of $d^2 = 0$. Using these differential forms, the Maxwell action can be rewritten as follows:

$$S_{\rm M}[A] = \frac{1}{8c}\int F \wedge \star F. \tag{12.21}$$

Observe that there is no matter charged with respect to $A_\mu$ in this system. This is analogous to the presence of no matter fields in the Einstein–Hilbert action. However, the Einstein–Hilbert action contains the self-coupling of gravity and therefore the presence of a coupling constant in it makes sense, whereas in the Maxwell theory there are no *direct* interactions between photons and, in principle, there is neither an electromagnetic coupling constant nor a unit of electric charge. We will see that things are a bit more complicated in the presence of gravity, through which photons do interact.

The equations of motion of $g_{\mu\nu}$ and $A_\mu$ are

$$G_{\mu\nu} - \frac{8\pi G_{\rm N}^{(4)}}{c^4}T_{\mu\nu} = 0, \tag{12.22}$$

$$\nabla_\mu F^{\mu\nu} = 0 \quad \text{(Maxwell's equation)}, \tag{12.23}$$

where

$$T_{\mu\nu} = \frac{-2c}{\sqrt{|g|}}\frac{\delta S_{\rm M}[A]}{\delta g^{\mu\nu}} = F_{\mu\rho}F_\nu{}^\rho - \tfrac{1}{4}g_{\mu\nu}F^2 \tag{12.24}$$

is the energy–momentum tensor of the vector field, which is traceless[7] in $d = 4$. The

[6] If the gauge group is $\mathbb{R}$, the elements of the group will be $e^{\Lambda/L}$, whereas, if it is $U(1)$, they will be $e^{i\Lambda/L}$, where $L$ is a constant introduced to make the exponent dimensionless because $\Lambda$ is dimensionful. In the second case $\Lambda$ will have to be identified with $\Lambda + 2\pi L$. When there is a unit of charge, $L$ is related to it.

[7] This property is associated with the invariance of the Maxwell Lagrangian in curved spacetime under Weyl rescalings of the metric,

$$g'_{\mu\nu} = \Omega^2(x)g_{\mu\nu}. \tag{12.25}$$

In fact, if $\Omega = e^\sigma$, then for infinitesimal transformations $\delta_\sigma g_{\mu\nu} = 2\sigma(x)g_{\mu\nu}$ we have

$$\delta_\sigma S_{\rm M} = \frac{\delta S_{\rm M}}{\delta g_{\mu\nu}}\delta_\sigma g_{\mu\nu} \sim \sigma T^{\mu\nu}g_{\mu\nu} = 0. \tag{12.26}$$

tracelessness of the electromagnetic energy–momentum tensor implies that $R=0$ and the Einstein equation takes the simpler form

$$R_{\mu\nu} = \frac{8\pi G_{\rm N}^{(4)}}{c^4} T_{\mu\nu}. \tag{12.27}$$

On taking the divergence of the Einstein equation and using the contracted Bianchi identity for the Einstein tensor $\nabla_\mu G^{\mu\nu}=0$, we find

$$F_{\nu\rho}\nabla_\mu F^{\mu\rho} - \tfrac{3}{2}F^{\mu\rho}\nabla_{[\mu}F_{\rho\nu]} = 0. \tag{12.28}$$

Since the Levi-Civita connection is symmetric,

$$\nabla_{[\mu}F_{\rho\nu]} = \partial_{[\mu}F_{\rho\nu]} = 0 \quad \text{(the Bianchi identity)} \tag{12.29}$$

identically, using the definition of $F_{\mu\nu}$, and then we see that the Einstein equation implies generically the Maxwell equation. Using Eq. (1.70), the Maxwell equation can also be written in a simpler, equivalent, form:

$$\partial_\mu\left(\sqrt{|g|}F^{\mu\nu}\right) = 0. \tag{12.30}$$

The equations are written in terms of the field strength $F$ and usually they are solved in terms of it. However, we are ultimately interested in the vector field $A$ itself and we have to make sure that the $F$ we obtain is such that it is related to some vector field by Eq. (12.16) or Eq. (12.20). It turns out that, locally, $A$ exists if the electromagnetic Bianchi identity Eq. (12.29) is satisfied.[8]

[8] In fact, if we are given $F$ and the Bianchi identity is satisfied, we can always find the corresponding vector potential by using, e.g., the formula

$$\boxed{A_\mu(x) = -\int_0^1 d\lambda\lambda x^\nu F_{\mu\nu}(\lambda x).} \tag{12.31}$$

To check this formula it is necessary to use the Bianchi identity: taking the curl of the l.h.s.,

$$\partial_{[\rho}A_{\mu]}(x) = -\int_0^1 d\lambda\lambda\partial_{[\rho}[x^\nu F_{\mu]\nu}(\lambda x)], \tag{12.32}$$

and operating,

$$\partial_{[\rho}\left(F_{\mu]\nu}(\lambda x)\right) = \lambda\partial_{[\rho}F_{\mu]\nu}(\lambda x) - \tfrac{1}{2}\lambda(\partial_\nu F_{\rho\mu})(\lambda x), \tag{12.33}$$

where the Bianchi identity Eq. (12.29) has been used in the last identity, we obtain

$$\begin{aligned}\partial_{[\rho}A_{\mu]}(x) &= \tfrac{1}{2}\int_0^1 d\lambda\{\lambda^2 x^\nu\partial_\nu F_{\rho\mu}(\lambda x) - \lambda F_{\mu\rho}(\lambda x)\} = \tfrac{1}{2}\int_0^1 d\lambda\frac{d}{d\lambda}[\lambda^2 F_{\rho\mu}(\lambda x)]\\ &= \frac{1}{2}F_{\rho\mu}(x).\end{aligned} \tag{12.34}$$

This means that, up to gauge transformations, the field strength determines completely the potential in the Maxwell theory, which one can see as the simplest Abelian Yang–Mills theory. Interestingly, this is no longer true in the non-Abelian Yang–Mills case, as first noticed in Refs. [1278, 448].

The Bianchi identity can also be written in this form (by contracting Eq. (12.29) with $\epsilon^{\mu\nu\rho\sigma}$, introducing it into the partial derivative (because it is constant), and using the definition of the Hodge dual and Eq. (12.30) for the divergence):

$$\nabla_\mu \star F^{\mu\sigma} = 0. \tag{12.35}$$

In the language of differential forms, the Maxwell equation and Bianchi identity are

$$d \star F = 0, \tag{12.36}$$

$$dF = 0, \tag{12.37}$$

and the Bianchi identity is just a consequence of the definition Eq. (12.20) and $d^2 = 0$.

Then, if we work with the field strength, we find that there are two pairs of equations, (12.23) and (12.35) and (12.36) and (12.23), which are (as pairs) invariant if one replaces $F$ by $\star F$ (by virtue of $\star\star F = -F$). This is an *electric–magnetic-duality* transformation. The name is due to the fact that this transformation interchanges the electric and magnetic fields in any given coordinate system according to

$$\vec{E}\,' = \vec{B}, \qquad \vec{B}\,' = -\vec{E}. \tag{12.38}$$

Actually, this pair of homogeneous equations (the Maxwell equation and the Bianchi identity) would be invariant under the (invertible) substitution for $F$ of any linear combination of $F$ and $\star F$. We would have a symmetry of all the equations of motion if the Einstein equation were also invariant under this replacement. We will see in Section 12.7 when this is the case and how the Einstein–Maxwell theory is invariant under electric–magnetic duality.

The four Maxwell equations in Minkowski spacetime can be deduced from the Maxwell equation and the Bianchi identity (two of them imply the existence of the potential $A_\mu$ and are equivalent to the latter). We have

$$\begin{aligned} \partial_\mu F^{\mu\nu} = 0 \quad &\Leftrightarrow \quad \begin{cases} \vec{\nabla}\cdot\vec{E} = 0, \\ \vec{\nabla}\times\vec{B} - \dfrac{1}{c}\dfrac{\partial}{\partial t}\vec{E} = 0, \end{cases} \\ \partial_\mu \star F^{\mu\nu} = 0 \quad &\Leftrightarrow \quad \begin{cases} \vec{\nabla}\cdot\vec{B} = 0, \\ \vec{\nabla}\times\vec{E} + \dfrac{1}{c}\dfrac{\partial}{\partial t}\vec{B} = 0. \end{cases} \end{aligned} \tag{12.39}$$

### 12.2.1 Electric charge

The electric charge can be defined in terms of a source coupled to the electromagnetic field (this is analogous to the energy–momentum-pseudotensor approach for the gravitational field) or in terms of the Noether current associated with the gauge invariance (the approach that leads to Komar's formula and its generalizations for the gravitational field). The two definitions are equivalent and are very closely related to each other because the gauge invariance of the free theory imposes strong constraints on the possible couplings.

Let us first introduce the electric charge using sources. A source for the Maxwell field is described by a current $j^\mu$, which naturally couples to the vector field through a term in the action of the form

$$\frac{1}{c^2}\int d^4x\sqrt{|g|}\,[-A_\mu j^\mu]. \tag{12.40}$$

This additional interaction term spoils the action's gauge invariance unless the source $j^\mu$ is divergence free,

$$\nabla_\mu j^\mu = 0 \;\Leftrightarrow\; d\star j = 0 \qquad (j \equiv j_\mu dx^\mu), \tag{12.41}$$

which implies the *continuity equation* for the vector density $\mathfrak{j}^\mu \equiv \sqrt{|g|}\,j^\mu$,

$$\partial_\mu \mathfrak{j}^\mu = 0. \tag{12.42}$$

The continuity equation can be used to establish the local conservation of the electric charge, as explained in Section 2.3, if the electric charge contained in a three-dimensional volume at a given time $t$, $\mathrm{V}^3_t$, is defined by[9]

$$q(t) = -\frac{1}{c}\int_{\mathrm{V}^3_t} d^3x\,\mathfrak{j}^0, \tag{12.43}$$

or, in a more covariant form,

$$q(t) = \frac{1}{c}\int_{\mathrm{V}^3_t} \star j. \tag{12.44}$$

As explained in Section 2.3, this quantity is not constant: its variation is related to the flux of charge through the boundary of $\mathrm{V}^3_t$. If $\mathrm{V}^3_t$ is a constant-time slice of the whole spacetime with no boundary, then the above integrals give the total charge, which will be constant in time. If we can foliate our spacetime with constant-time hypersurfaces, then we take the four-dimensional spacetime $\mathrm{V}^4$ contained in between two constant-time slices $\mathrm{V}^3_{t_1}$ and $\mathrm{V}^3_{t_2}$, integrate the continuity equation over it, and use Stokes' theorem. The boundary of the four-dimensional region we have proposed is made up of the two constant-time slices with opposite orientations, so

$$0 = \int_{\mathrm{V}^4} d\star j = \int_{\mathrm{V}^3_{t_1}} \star j - \int_{\mathrm{V}^3_{t_2}} \star j, \tag{12.45}$$

and the total electric charge is constant in time.

Thus, gauge invariance of the action implies that the source is divergence free, and from this the local conservation of the electric charge (and the global conservation of the total electric charge) follows.

On the other hand, in the presence of the source, the Maxwell equation is modified into

$$\nabla_\mu F^{\mu\nu} = \frac{1}{c}j^\nu, \tag{12.46}$$

[9] The sign is conventional.

or, equivalently,

$$d \star F = \frac{1}{c} \star j, \tag{12.47}$$

and, using the antisymmetry of $F_{\mu\nu}$ or $d^2 = 0$, it is trivial to see that, since the l.h.s. of the equation is divergence free, the r.h.s. of the equation is also, for consistency, divergence free, as we knew it had to be in order to preserve the gauge invariance of the action. This is no coincidence: the fact that the r.h.s. of the Maxwell equation is divergence free is in fact the gauge identity associated with the invariance under $\delta A_\mu = \partial_\mu \Lambda$, as we are going to see.

Finally, using the Maxwell equation (12.47), we can rewrite the definition of the total electric charge Eq. (12.44) in terms of the field strength and again use Stokes' theorem. If the boundary of a constant-time slice has the topology of a sphere, $S^2$, at infinity, we obtain

$$\boxed{q = \int_{S^2_\infty} \star F} \tag{12.48}$$

which is a useful definition of the total electric charge of a spacetime in terms of the field strength (the electric flux) and which we will generalize further in Part III.

This is the kind of formula that we will use because in the Einstein–Maxwell system there are no fields explicitly written that act as sources for $A_\mu$. Just as in the case of the Maxwell equations in vacuum, we can obtain solutions describing the field of charges. These solutions are singular near the place where the charge ought to be and the solution is not a solution there (there are no charges explicitly included in the system). However, the above expression allows us to calculate the charge that ought to be placed there to produce the flux of electromagnetic field that we observe.[10] We have introduced sources as a device for understanding the definition.

We could also have used the invariance of the Einstein–Maxwell action to find the conserved Noether current and define the electric charge through it.

We studied the invariance of the Maxwell action and found the corresponding Noether current in Minkowski spacetime in Section 3.2.1. The coupling to gravity introduces only minor changes and the conclusion is, again, that the electric charge can be defined by Eq. (12.48).

It is useful to consider a simple example of a source: the current associated with a particle of electric charge $q$ and worldline $\gamma$ parametrized by $X^\mu(\xi)$. In a manifestly covariant form it is given by

$$j^\mu(x) = qc \int_\gamma dX^\mu \frac{1}{\sqrt{|g|}} \delta^{(4)}[x - X(\xi)], \tag{12.49}$$

where $dX^\mu = d\xi dX^\mu/d\xi$. On making the choice $\xi = X^0$ and integrating over $X^0$, we

[10] Of course, this is just a covariant generalization of the Gauss theorem that relates the flux of electric field through a closed surface to the charge enclosed by it.

obtain

$$
\begin{aligned}
j^\mu(x^0,\vec{x}) &= qc\int dX^0 \frac{dX^\mu}{dX^0}\frac{1}{\sqrt{|g|}}\delta^{(3)}(\vec{x}-\vec{X})\delta(x^0-X^0) \\
&= -qc\frac{dX^\mu}{dx^0}\frac{\delta^{(3)}[\vec{x}-\vec{X}(x^0)]}{\sqrt{|g|}}.
\end{aligned} \tag{12.50}
$$

If the particle is at rest at the origin in the chosen coordinate system, the current is

$$
j^\mu(x^0,\vec{x}) = -qc\delta^{\mu 0}\frac{\delta^{(3)}(\vec{x})}{\sqrt{|g|}}, \tag{12.51}
$$

and it is easy to see that $q$ is indeed the electric charge according to the above definitions. The current $j^\mu$ is conserved:

$$
\begin{aligned}
\nabla_\mu j^\mu &\sim \frac{\partial}{\partial x^\mu}\left\{\sqrt{|g(x)|}\int d\xi \dot{X}^\mu \frac{1}{\sqrt{|g(X)|}}\delta^{(4)}[x-X(\xi)]\right\} \\
&= \int d\xi \dot{X}^\mu \frac{\partial}{\partial x^\mu}\delta^{(4)}[x-X(\xi)] = -\int d\xi \dot{X}^\mu \frac{\partial}{\partial X^\mu}\delta^{(4)}[x-X(\xi)] \\
&= -\int d\xi \frac{d}{d\xi}\delta^{(4)}[x-X(\xi)] = -\delta^{(4)}[x-X(\xi)]\Big|_{\xi_1}^{\xi_2} = 0,
\end{aligned} \tag{12.52}
$$

generically, except for the initial and final positions of the particle $X^\mu(\xi_1)$ and $X^\mu(\xi_2)$, which look like a one-particle source and a sink and can be taken to infinity.

Observe that, for the current Eq. (12.49), the interaction term Eq. (12.40) becomes the integral of the 1-form $A$ over the worldline $\gamma$:

$$
-\frac{q}{c}\int_{\gamma(\xi)} A_\mu \dot{x}^\mu d\xi = -\frac{q}{c}\int_\gamma A. \tag{12.53}
$$

This term has to be added to the action of the particle, Eq. (3.255), (3.257), or (3.258), in order to obtain the worldline action of a massive electrically charged particle,

$$
\boxed{S[X^\mu(\xi)] = -Mc\int d\xi\,\sqrt{g_{\mu\nu}(X)\dot{X}^\mu\dot{X}^\nu} - \frac{q}{c}\int d\xi A_\mu \dot{X}^\mu,} \tag{12.54}
$$

or that of a massless one. That kind of term is known as a *Wess–Zumino* (WZ) term. In this form it is easy to see that, under a gauge transformation, the action changes by a total derivative. The integral of the total derivative vanishes exactly only for special boundary conditions, though.

This action can be used as a source, but it also describes the motion of a charged particle in a gravitational/electromagnetic background. In the special-relativistic limit, taking $\xi = X^0 = ct$, the action takes the standard form

$$
S \sim \int dt\left\{-Mc^2 + \tfrac{1}{2}Mv^2 - q\phi + \tfrac{q}{c}\vec{A}\cdot\vec{v}\right\}. \tag{12.55}
$$

If there is a point-like charge $q$ at rest at the origin the only non-vanishing components of $F$ are $F_{\underline{0}r}$ and they should depend only on $r$ because of the spherical symmetry of the problem. Using the above definition of charge and working in general static spherical coordinates Eq. (11.22), we find

$$q = \int_{S^2_\infty} \frac{\epsilon_{\mu\nu\rho\sigma}}{4\sqrt{|g|}} F^{\rho\sigma} dx^\mu \wedge dx^\nu = \int_{S^2_\infty} d\Omega^2 r^2 F_{\underline{0}r} = \omega_{(2)} \lim_{r\to\infty} \left(r^2 F_{\underline{0}r}\right), \tag{12.56}$$

where $\omega_{(2)}$ is the volume of the 2-sphere $4\pi$. Then, the electromagnetic field of a point-like charge must behave for large $r$ as follows:

$$E_r = F_{\underline{0}r} \sim +\frac{1}{4\pi}\frac{q}{r^2}, \qquad \phi \sim +\frac{1}{4\pi}\frac{q}{r}. \tag{12.57}$$

(Of course, this result is exact in the absence of gravity, in Minkowski spacetime.) On plugging this result into Eq. (12.55), we find that the electrostatic force between two particles is, in this unit system, $q_1 q_2/(4\pi r^2)$, as we said.

In the units that we are using, $M$ appears multiplied by $G_N^{(4)}$ in the metric (as in the Schwarzschild solution) and $q$ does not. Some simplification is achieved by using the following normalization and units that are standard in this field; we set $c = 1$ and rewrite the Einstein–Maxwell action as follows:

$$\boxed{S_{\text{EM}}[g, A] = \frac{1}{16\pi G_N^{(4)}} \int d^4x \sqrt{|g|}\, \left[R - \tfrac{1}{4}F^2\right].} \tag{12.58}$$

In these units both $A_\mu$ and $g_{\mu\nu}$ are dimensionless. The factor $16\pi G_N^{(4)}$ disappears from the equations of motion. Furthermore, if we keep (by definition) the WZ term as in Eq. (12.54) without any additional normalization factor, the electric charge is now

$$\boxed{q = \frac{1}{16\pi G_N^{(4)}} \int_{S^2_\infty} \star F,} \tag{12.59}$$

and has dimensions of mass (energy). Finally, for a point-like charge we expect, for large $r$,

$$E_r = F_{0r} \sim \frac{4G_N^{(4)} q}{r^2}, \tag{12.60}$$

which implies that the force between two charges is

$$F_{12} = 4G_N^{(4)} \frac{q_1 q_2}{r_{12}^2}. \tag{12.61}$$

### 12.2.2 Massive electrodynamics

Before concluding this section it is worth considering which facts would be modified if the vector field were massive. A massive vector field in Minkowski spacetime is described by the Proca Lagrangian Eq. (3.67) and its generalization to curved spacetime is straightforward. The equation of motion is

$$\nabla_\nu F^{\nu\mu} + m^2 A^\mu = 0. \tag{12.62}$$

We immediately see that this equation is completely different from the Bianchi identity Eq. (12.35), which is also valid in the massive case, which implies that massive electrodynamics, apart from gauge invariance, has no electric–magnetic duality. This implies that, in principle, there will be no Dirac magnetic monopoles dual to the electric ones, which explains the results of Ref. [816].

If we take the divergence of this equation, we find the integrability condition

$$\nabla_\nu A^\nu = 0, \tag{12.63}$$

which removes one of the degrees of freedom described by the vector field, leaving only three that correspond to the three possible helicities of a massive spin-1 particle $(-1, 0, +1)$.

The quanta of the Proca field, being massive, will propagate at a speed smaller than 1 $(c)$ and the interaction they mediate will be short ranged. We can see this by finding the static, spherically symmetric solution that describes the field of an electric monopole in this theory in Minkowski spacetime. On substituting the ansatz

$$A_\mu = \delta_{\mu 0} \frac{f(r)}{r} \tag{12.64}$$

into the equation of motion, we obtain the differential equation

$$f'' - m^2 f = 0, \tag{12.65}$$

whose solution is (with the boundary condition $A_\mu \to 0$ when $r \to \infty$)

$$A_\mu = Q \delta_{\mu 0} \frac{e^{-mr}}{r}, \tag{12.66}$$

where $Q$ is an integration constant that is somehow related to the "electric charge." However, the lack of gauge invariance suggests that the "electric charge" is not conserved in this system. In fact, it is not easy to define what is meant by electric charge here. It is then useful to consider a slightly more general system with the following classically equivalent action for $A_\mu$ and a scalar auxiliary field $\phi$:

$$S[A_\mu, \phi] = \int d^4x \sqrt{|g|} \left[ -\tfrac{1}{4} F^2 + \tfrac{1}{2} (\partial\phi + mA)^2 \right]. \tag{12.67}$$

This action is invariant under the following *massive gauge transformations*:

$$\delta A_\mu = \partial_\mu \Lambda, \qquad \delta\phi = -m\Lambda. \tag{12.68}$$

Observe that, for consistency, the scalar $\phi$ has to live in the gauge group manifold: either $\mathbb{R}$ or $S^1$ (if the gauge group is U(1)). On fixing the gauge $\phi = 0$ we recover the Proca Lagrangian, and any solution of the equations of motion of the original system is also a solution of this one in this gauge.

It is sometimes said that the scalar $\phi$ is "eaten" by the vector field, which acquires a mass in the process; $\phi$ is then referred to as a *Stückelberg field* [1162]. Observe that the number of degrees of freedom before and after the gauge fixing are the same. Observe also that this procedure for obtaining a massive vector field is different from the standard spontaneous symmetry-breaking mechanism. There are two main differences: the scalar is real and carries no charge with respect to the vector field and there is no potential for the scalar. (Actually, there is no way to write a gauge-invariant potential with only one real scalar.)

The equations of motion corresponding to the new Lagrangian are

$$\nabla_\nu F^{\nu\mu} + m\left(\nabla^\mu \phi + m A^\mu\right) = 0, \qquad \nabla^2 \phi + m \partial_\mu A^\mu = 0. \tag{12.69}$$

To define a conserved charge, we can either introduce a source $j^\mu$ into the first equation or use the conserved Noether current associated with *constant*[11] shifts of $\phi$:

$$j^\mu_{\mathrm{N}} = \partial_\nu F^{\nu\mu} + m\left(\partial^\mu \phi + m A^\mu\right). \tag{12.70}$$

The source $j^\mu$ is conserved, but only on-shell (upon use of the $\phi$ equation of motion) and the same applies to the Noether current, which is associated with a global symmetry. In both cases we can define the electric charge in this system by

$$q = \int d^3x \sqrt{|g|}\, j^0_{\mathrm{N}} = \int d^3x \sqrt{|g|} \left[\nabla_\nu F^{\nu 0} + m\left(\nabla^0 \phi + m A^0\right)\right]. \tag{12.71}$$

On applying this definition to the electric monopole solution Eq. (12.66) and using

$$\vec{\nabla}^2 \left(\frac{e^{-mr}}{r}\right) = -4\pi \delta^{(3)}(\vec{x}) + \frac{m^2 e^{-mr}}{r}, \tag{12.72}$$

we find $q = 4\pi Q$, as we naively expected. It should be stressed, though, that this charge is of a completely different nature from the usual electric charge since it is associated with a *global* symmetry of a different field. In principle, the no-hair conjecture should apply (negatively) to charges of this kind associated with short-range interactions and global (rather than local) symmetries.

Finally, note that neither the original Proca action nor the new one with the Stückelberg field $\phi$ has any duality *symmetry*. However, the new action can be dualized (i.e. written in *dual variables*), as we will see in Section 12.7.5.

## 12.3 The electric Reissner–Nordström solution

We are now ready to find BH-type solutions of the equations of motion derived from the Einstein–Maxwell action normalized as in Eqs. (12.58). Since the Maxwell equation is

[11] If we use the full gauge invariance of the theory, we recover exactly the same Noether current and Bianchi identity as in the massless case. The definition Eq. (12.59) then gives zero charge because $F$ goes to zero too fast at infinity.

satisfied if the Einstein equation is, we only have to solve the latter with the trace subtracted,

$$R_{\mu\nu} = \tfrac{1}{2}\left[F_{\mu}{}^{\rho}F_{\nu\rho} - \tfrac{1}{4}g_{\mu\nu}F^2\right], \tag{12.73}$$

plus the Bianchi identity. We are looking for a static, spherically symmetric solution and, therefore, as usual, we make the ansatz Eq. (11.2) for the metric. This time we also have to make an ansatz for the electromagnetic field. If we are looking for a point-like electrically charged object at rest, taking into account Eq. (12.60), an appropriate ansatz that is readily seen to satisfy the Maxwell equation and the Bianchi identity for the metric Eq. (11.2) is

$$F_{tr} \sim \pm\frac{1}{R^2(r)}. \tag{12.74}$$

The $\pm$ corresponds to the two possible signs of the electric charge. The metric cannot depend on this sign because the action is invariant under the (admittedly rather trivial) duality symmetry $F \to -F$, $g \to g$. The solution obtained in this way is the Reissner–Nordström (RN) solution[12] [996, 1074] and can be conveniently written as follows:

$$\boxed{\begin{aligned} ds^2 &= f(r)dt^2 - f^{-1}(r)dr^2 - r^2 d\Omega^2_{(2)}, \\ F_{tr} &= \frac{4G_{\rm N}^{(4)}q}{r^2}, \\ f(r) &= \frac{(r-r_+)(r-r_-)}{r^2}, \\ r_\pm &= G_{\rm N}^{(4)}M \pm r_0, \qquad r_0 = G_{\rm N}^{(4)}\left(M^2 - 4q^2\right)^{\frac{1}{2}}, \end{aligned}} \tag{12.75}$$

where $q$ is the electric charge, normalized as in Eq. (12.59), and $M$ is the ADM mass.

Some remarks are necessary.

1. The solution describes the gravitational and electromagnetic fields created by a spherical (or point-like), electrically charged object of total mass $M$ and electric charge $q$, as seen from far away by a static observer to which the coordinates $\{t, r, \theta, \varphi\}$ (that we can keep calling the "Schwarzschild coordinates") are adapted. The Schwarzschild solution is contained as the special case $q = 0$.

   Included in the (total) mass is the energy associated with the presence of an electromagnetic field. We cannot covariantly separate the energy associated with "matter" from the energy associated with the electromagnetic field and the gravitational field, but we must keep in mind that the mass of the spacetime contains all these contributions.

[12] The RN solution is also a particular case (the spherically symmetric case) of the general static axisymmetric electrovacuum solutions discovered independently by Weyl [1249, 1250] and should also bear his name.

2. The vector field that gives the above field strength and whose local existence is guaranteed by the fact that $F$ satisfies the Bianchi identity is

$$A_\mu = \delta_{\mu t} \frac{4G_{\rm N}^{(4)} q}{r}. \tag{12.76}$$

3. There is a generalization of Birkhoff's theorem for RN BHs (see exercise 32.1 of Ref. [952]): RN is the only spherically symmetric family of solutions (that includes Schwarzschild's) of the Einstein–Maxwell system.

4. The above solution is valid for any values of the parameters $M$ and $q$ and, therefore, of $r_\pm$, including complex ones.

5. The metric is singular at $r=0$ and also at $r_-$ and $r_+$, if $r_+$ and $r_-$ are real. At $r=r_\pm$ the signature changes and, in the region between $r_+$ and $r_-$, $r$ is timelike and $t$ is spacelike, and in that region the metric is not static as in the Schwarzschild horizon interior. To find the nature of these singularities, we calculate curvature invariants and study the geodesics: $R=0$ due to $T^\mu{}_\mu = 0$, but other curvature invariants (and $F^2$ as well) tell us that there is a curvature singularity at $r=0$ but not at $r=r_\pm$. In fact, an analysis similar to the one made in the Schwarzschild case shows that, when it is real and positive, $r_+$ is an event horizon of area

$$A = 4\pi r_+^2, \tag{12.77}$$

surrounding the curvature singularity, in agreement with the weak form of the cosmic-censorship conjecture, whereas $r_-$ is a *Cauchy horizon*: in the RN spacetime there is no Cauchy hypersurface on which we can give initial data for arbitrary fields and predict their evolution in the whole spacetime. By definition, we can have a Cauchy hypersurface only for the region outside the Cauchy horizon. This horizon seems to be unstable under small perturbations [1026] associated with the infinite blueshift that incoming radiation suffers in its neighborhood (opposite to the infinite redshift that incoming radiation suffers in the neighborhood of the event horizon), and it is conjectured that a spacelike singularity should appear in its place [998, 269].

Both horizons exist when $M > 2|q|$ and then the RN metric describes a BH. In Fig. 12.1 we have represented part of its Penrose diagram, based on the maximal analytic extension of the RN metric Eq. (12.75) found in Ref. [661]. In this diagram there are two "universes" (quadrants I and IV, which have asymptotically flat regions), as in Schwarzschild's case, but the complete diagram consists of an infinite number of pairs of "universes" arranged periodically. The singularities are timelike, not spacelike like the Schwarzschild singularities, and can be avoided by observers that enter the BH. In fact, there are timelike geodesics that, starting in a certain "universe," enter the BH crossing the event horizon $r_+$ and, after crossing two Cauchy horizons $r_-$, emerge in a different "universe." Analogous effects take place in the gravitational collapse of spherically symmetric shells of electrically charged matter [245]: depending on the characteristics of the shell, the gravitational collapse can end in a singularity, or the shell can stop contracting and start to expand in a different "universe."

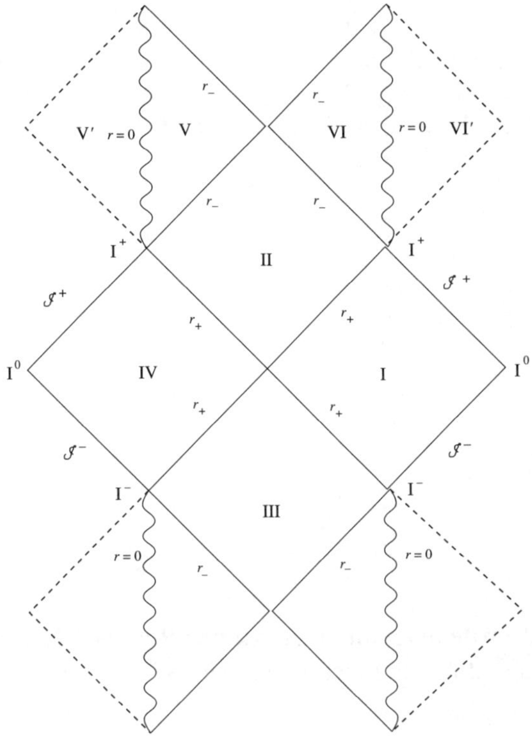


Fig. 12.1. Part of the Penrose diagram of a Reissner–Nordström black hole $M > 2|q|$. Only two "universes" are shown. The complete diagram repeats periodically the part shown.

Observe that, although the cosmic-censorship conjecture is obeyed by the RN spacetime in its weak form, it is violated in its strong form: an observer that takes the inter-"universe" trip will see the singularity.[13] However, if the Cauchy horizon indeed became a spacelike singularity, such a problem would not arise.

6. When $M < -2|q|$ (negative), $r_\pm$ are real and negative, and there is no horizon surrounding the curvature singularity at $r = 0$. The Penrose diagram of this spacetime is the one in Fig. 11.4. This case could be excluded by invoking cosmic censorship, which is violated in its weak form by this metric. It is reasonable to think (and the positive-energy theorem proves it) that, if we start with physically reasonable initial conditions, we will not end up with a negative mass.
7. When $-2|q| < M < 2|q|$, the constants $r_\pm$ are complex and there are no horizons; the only singularity left is the one at $r = 0$, and it is naked, the Penrose diagram being again Fig. 11.4. Again, cosmic censorship should exclude this range of values of $M$. This includes the special case $M = 0$. Observe that, otherwise, we would have a massless, charged object at rest, which is a rather strange object. The mass is the

[13] An observer falling into a Schwarzschild BH cannot see the singularity, which always lies in its future, until he/she actually crashes onto it. This has to do with the spacelike nature of the Schwarzschild singularity.

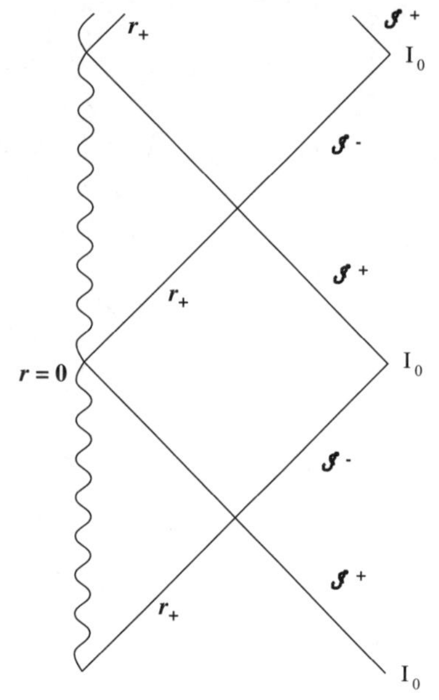


Fig. 12.2. Part of the Penrose diagram of an extreme Reissner–Nordström black hole. The complete diagram has an infinite number of "universes."

total energy of the spacetime. A non-trivial electromagnetic field such as the one produced by a point-like charge is a source of (positive) energy. Thus, our physical intuition tells us that, in order to have non-zero charge and at the same time zero mass, there must be some "negative energy density" present. It is thought that the same should happen in the other $-2|q| < M < 2|q|$ cases.

Negative energies always seem to be at the heart of naked singularities, and, in the spirit of cosmic censorship, if negative energies are not allowed initially, no naked singularities will appear in the evolution of the system.

Thus, cosmic censorship restricts the possible values of $M$ to the range $M \geq 2|q|$. What happens if we now throw into a regular RN BH charged matter with mass $M'$ and charge $q'$ such that $M + M' < 2|q + q'|$? In Ref. [1228] it was proven that, if $M = 2|q|$ (an extreme RN BH), particles whose absorption by the BH would take it into the region of forbidden parameters are not captured by the BH. However, it seems that it is possible to "overcharge" a non-extremal ($M < 2|q|$) RN BH by sending into it a charged test particle (but not by using a charged collapsing shell of charged matter) [793], although the effects of the absorption of the particle on the BH geometry (which are assumed to be small) have not yet been worked out.

We see that the RN BH provides a very interesting playground in which to test cosmic censorship. We will see that the relation between cosmic censorship and positivity of the energy can be translated into supersymmetry (BPS) bounds.

8. The limiting case $M = 2|q|$ between the naked singularity and the regular BH is very special. When $M = 2|q|$ the two horizons coincide, $r_+ = r_- = G_N^{(4)} M$, and there is

no change of signature across the resulting horizon (which is a degenerate Killing horizon), which still has a non-vanishing area given by

$$A_{\text{extreme}} = 4\pi r_+^2 = 4\pi \left(G_N^{(4)} M\right)^2. \tag{12.78}$$

This object is an *extreme* RN (ERN) BH and it will play a central role in much of what follows. Some of the properties of ERN BHs are the following.

(a) The proper distance to the horizon along radial directions at constant time,

$$\lim_{r_2 \to r_+} \int_{r_1}^{r_2} ds = \lim_{r_2 \to r_+} \int_{r_1}^{r_2} dr \left(1 - \frac{r_+}{r}\right)^{-1} = \infty, \tag{12.79}$$

diverges. This does not happen along timelike or null directions, and an observer can cross it in a finite proper time.

(b) The Penrose diagram is drawn in Fig. 12.2. As we can see, the causal structure is completely different from that of any regular RN BH no matter how close to the extreme limit it is. Thus, we can expect physical properties of the family of RN BHs to be discontinuous at the extreme limit.

(c) The relative values of their charge and mass are such that, if we have two of them, $M_1 = 2|q_1|$ and $M_2 = 2|q_2|$, it will always happen that

$$G_N^{(4)} M_1 M_2 = 4 G_N^{(4)} |q_1 q_2|, \tag{12.80}$$

and, *if both charges have the same sign* and we divide by the relative distance between them, we obtain

$$F_{12} = -G_N^{(4)} \frac{M_1 M_2}{r_{12}^2} + 4 G_N^{(4)} \frac{q_1 q_2}{r_{12}^2} = 0. \tag{12.81}$$

This is nothing but the force between two point-like, massive, charged, non-relativistic objects on account of Eqs. (3.140) and (12.61), and it vanishes, so they will be in equilibrium. Then, this suggests that it should be possible to find static solutions describing two (or many) ERN BHs in equilibrium.

(d) On shifting the radial coordinate $r = \rho + G_N^{(4)} M$ of the ERN metric, it becomes

$$ds^2 = \left(1 + \frac{G_N^{(4)} M}{\rho}\right)^{-2} dt^2 - \left(1 + \frac{G_N^{(4)} M}{\rho}\right)^2 \left(d\rho^2 + \rho^2 d\Omega_{(2)}^2\right). \tag{12.82}$$

On defining new Cartesian coordinates $\vec{x}_3 = (x^1, x^2, x^3)$ such that $|\vec{x}_3| = \rho$ and $d\vec{x}_3^{\,2} = d\rho^2 + \rho^2 d\Omega_{(2)}^2$, we obtain a new form of the ERN solution:

$$\boxed{\begin{aligned} ds^2 &= H^{-2} dt^2 - H^2 d\vec{x}_3^{\,2}, \\ A_\mu &= -2\delta_{\mu t}\, \mathrm{sign}(q) \left(H^{-1} - 1\right), \\ H &= 1 + \frac{2 G_N^{(4)} |q|}{|\vec{x}_3|} = 1 + \frac{G_N^{(4)} M}{|\vec{x}_3|}. \end{aligned}} \tag{12.83}$$

Observe that, in this case, due to the shift in the radial coordinate, the event horizon is placed at $\vec{x}_3 = \vec{0}$, which in flat Minkowski spacetime is just a point. It is, though, easy to see that the surface labeled by $\vec{x}_3 = \vec{0}$ is not just a point but rather a sphere of finite area because in the limit $\rho \to 0$ one has to take into account the $\rho^2$ factor of $d\Omega^2_{(2)}$ that cancels out the poles in $H^2$, so the induced metric in the $\rho = 0, t = \text{constant}$ hypersurface is, indeed,

$$ds^2 = -(G_{\rm N}^{(4)} M)^2 d\Omega^2_{(2)}. \tag{12.84}$$

$H$ is a harmonic function in the three-dimensional Euclidean space spanned by the coordinates $\vec{x}_3$, i.e. it satisfies

$$\partial_{\underline{i}}\partial_{\underline{i}} H = 0. \tag{12.85}$$

This fact could just be a coincidence, but, if we use Eqs. (12.83) as an ansatz in the equations of motion without imposing any particular form for $H$, we find that they are solved for *any harmonic function* $H$, not just for the one in Eqs. (12.83). We have obtained in this way the *Majumdar–Papapetrou* (MP) family of solutions [918, 1019]:

$$\boxed{\begin{aligned} ds^2 &= H^{-2}dt^2 - H^2 d\vec{x}_3^{\,2}, \\ A_\mu &= \delta_{\mu t}\alpha\left(H^{-1} - 1\right), \qquad \alpha = \pm 2, \\ \partial_{\underline{i}}\partial_{\underline{i}} H &= 0. \end{aligned}} \tag{12.86}$$

If we want to find solutions describing several ERN BHs in static equilibrium, it is, therefore, natural to search amongst this class of solutions.[14]

Maxwell's theory in Minkowski spacetime is a linear theory and it obeys the superposition principle. It is possible to find a solution describing an arbitrary number of electric charges at rest in arbitrary positions by adding the corresponding Coulomb solutions. With our normalizations we would have

$$A_\mu = -\delta_{\mu t} \sum_{i=1}^{N} \frac{2G_{\rm N}^{(4)} q_i}{|\vec{x}_3 - \vec{x}_{3,i}|}, \tag{12.88}$$

in a certain gauge. As we have stressed before, Maxwell's theory in Minkowski spacetime does not know about interactions, and this is why we can have a

[14] We could also try to look for solutions of this form in the "scalar coupled to gravity" system. Since the force between two objects with "scalar charge" is always attractive, we do not expect on physical grounds to find any. In fact, it is possible to find such solutions if we pay the price of having purely imaginary "scalar charges" (which repel each other). The solutions have the following form:

$$\left\{\begin{aligned} ds^2 &= e^{2H}dt^2 - e^{-2H}d\vec{x}_3^{\,2}, \\ \varphi &= c \pm iH, \end{aligned}\right. \tag{12.87}$$

where $c$ is any constant and $H$ is any harmonic function $\partial_i\partial_i H = 0$.

static solution, which we know would be possible in the real world only if there were another force holding the charges in place. If we introduce source terms for the charges (massive or massless point-like particles of electric charges $q_i$) then we will have to solve a (non-linear) coupled system of equations: the Maxwell field equations and the equations of motion for the particles. The solutions will be, in general, time dependent (and realistic).

Newtonian gravity is another linear theory, and, thus, there are static solutions corresponding to arbitrary mass distributions even if we know that external forces are needed to hold the masses in place. Again, on introducing sources, the solutions become realistic (and, in general, time dependent).

Now, if we again introduce sources interacting both gravitationally and electrostatically, we can have static solutions describing particles with masses and charges $M_i = 2|q_i|$ in equilibrium. Newtonian gravity is insensitive to the electrostatic interaction energy and to the gravitational interaction energy.

In GR, a non-linear (non-Abelian, self-coupling) theory, things are quite different. There is no need to introduce sources: the theory *knows* that two Schwarzschild BHs, for instance, cannot be in static equilibrium and the corresponding solution does not exist. The coupling to gravity makes the electromagnetic interaction effectively non-Abelian, and it does not need the introduction of sources to *know* that only ERN BHs can be in static equilibrium[15] [255]. This coupling gives rise to many other interesting phenomena in RN backgrounds, such as the conversion of electromagnetic into gravitational waves [1001].

Since the horizon of a single ERN BH looks like a point in isotropic coordinates, we can try harmonic functions with several point-like singularities:

$$H(\vec{x}_3) = 1 + \sum_{i=1}^{N} \frac{2G_{\rm N}^{(4)}|q_i|}{|\vec{x}_3 - \vec{x}_{3,i}|}. \tag{12.89}$$

The overall normalization is chosen so as to obtain an asymptotically flat solution and the coefficients of each pole are taken positive so that $H(\vec{x}_3)$ is nowhere vanishing and the metric is non-singular. Also this choice gives a potential like the one in Eq. (12.88) for large values of $|\vec{x}_3|$.

It can be seen [711] that each pole of $H$ indeed corresponds to a BH horizon. In fact, to see that there is a surface of finite area at $\vec{x}_{3,i}$, we simply have to shift the origin of coordinates to that point and then examine the $\rho \to 0$ limit as in the single-BH case. The charge of each BH can be calculated most simply using Eq. (12.43), where the volume encloses only one singularity (the current is nothing but a collection of Dirac-delta terms). The charges turn out to be $\text{sign}(-\alpha)|\,q_i|$, i.e. all the charges have the same sign.

In GR it is, however, impossible to calculate the *mass* of each BH because there is no local conservation law for the mass and there is no such concept as the

[15] As a matter of fact, the identity $M_1 M_2 = 4|q_1 q_2|$ does not imply that both objects are ERN BHs. It can be satisfied by a non-extremal RN BH with $M_1 > 2|q_1|$ and a naked singularity with $M_2 < 2|q_2|$, but the corresponding static solutions (if any) are not known.

mass of some region of the spacetime. Only one mass can be defined, and that is the total mass of the spacetime, $M = 2\sum_{i=1}^{N}|q_i|$. However, the equilibrium of forces existing between the BHs suggests that the electrostatic and gravitational interaction energies (to which GR gravity is sensitive) cancel out everywhere. If that were true, the masses and charges would be *localized* at the singularities and then we could assign a mass $M_i = 2|q_i|$ to each BH [255]. It is, perhaps, this localization of the mass of ERN BHs that will allow us to find sources for them, something that turned out to be impossible for Schwarzschild BHs. This is physically a very appealing idea, but it is certainly not a rigorous proof.

If we do not care about singularities, we can also take some coefficients of the poles of the harmonic function to be negative. In this way it is possible to obtain solutions with vanishing total mass. Here, it is intuitively clear that a negative coefficient is associated with some "negative mass density" and cosmic censorship should eliminate these solutions.

(e) If we take the *near-horizon limit* $\rho \to 0$ in the ERN metric Eqs. (12.83), the constant 1 can be ignored and we find another MP solution with harmonic function $H = 2G_N^{(4)}|q|/\rho$:

$$\boxed{\begin{aligned} ds^2 &= \left(\frac{\rho}{2G_N^{(4)}|q|}\right)^2 dt^2 - \left(\frac{\rho}{2G_N^{(4)}|q|}\right)^{-2} d\rho^2 - \left(2G_N^{(4)}|q|\right)^2 d\Omega_{(2)}^2, \\ A_t &= -\frac{\rho}{G_N^{(4)} q}, \qquad F_{t\rho} = \frac{1}{G_N^{(4)} q}. \end{aligned}} \tag{12.90}$$

This exact solution is the Robinson–Bertotti (RB) solution [205, 1077] and it describes the ERN metric near the horizon. It is the only solution of the Einstein–Maxwell equations which is homogeneous and has a homogeneous non-null electromagnetic field (Theorem 10.3 in Ref. [878]). It is the direct product of two two-dimensional spaces of constant curvature: a two-dimensional anti-de Sitter ($AdS_2$) spacetime with "radius" $R_{AdS} = 2G_N^{(4)}|q|$, and therefore with two-dimensional scalar curvature $R^{(2)} = -1/[2(G_N^{(4)}|q|)^2]$, in the $t-\rho$ part of the metric, and a 2-sphere $S^2$ of radius $R_S = 2G_N^{(4)}|q|$ and curvature $R^{(2)} = +1/[2(G_N^{(4)}|q|)^2]$ in the $\theta$–$\varphi$ part of the metric. The sum of the two-dimensional scalar curvatures vanishes, as it should, because all solutions of the Einstein–Maxwell system have $R = 0$. Evidently, it is not asymptotically flat.

$AdS_2$ is invariant under the isometry group SO(1, 2) (which is also called $AdS_2$) and $S^2$ under SO(3). If we compare the RB isometry group with the ERN isometry group (SO(1, 1) $\times$ SO(3) and SO(1, 1) $\sim \mathbb{R}^+ \times \mathbb{Z}_2$ are shifts in time and time inversions) we see that there is an enhancement of symmetry when we approach the horizon. As we will see in Chapter 17, there is also an enhancement of unbroken supersymmetry, which is maximal in this limit. This is enough to consider the RB solution as a vacuum of the theory alternative to Minkowski.

In turn, this allows us to view the ERN solution as interpolating between the Minkowski vacuum (which is at infinity) and the RB solution (which is at the horizon), and then we can interpret it as a *gravitational soliton* [609].

(f) There are many other solutions in the MP class. However, it has been argued in Ref. [711], and proven under certain assumptions [359], that the only BH solutions in this class (and in a bigger class that we will study in Chapter 13, the IWP class) are the ones we have written above. One could look for solutions describing extended objects by allowing the harmonic function $H$ to have one- or two-dimensional singularities. They are not asymptotically flat and they are not natural, so we will not consider them.

9. If we shift the radial coordinate by $r = \rho + r_\pm$ in the RN solution Eqs. (12.75), it takes the following form:

$$\begin{aligned} ds^2 &= \left(1 + \frac{r_\pm}{\rho}\right)^{-2}\left(1 + \frac{\pm 2r_0}{\rho}\right)dt^2 \\ &\quad - \left(1 + \frac{r_\pm}{\rho}\right)^{2}\left[\left(1 + \frac{\pm 2r_0}{\rho}\right)^{-1} d\rho^2 + \rho^2 d\Omega^2_{(2)}\right], \\ A'_\mu &= -\delta_{\mu t}\frac{4G_N^{(4)}q}{r_\pm}\left[\left(1 + \frac{r_\pm}{\rho}\right)^{-1} - 1\right]. \end{aligned} \tag{12.91}$$

The RN metric looks in this form (taking the minus sign) like a Schwarzschild metric with mass $r_0/G_N^{(4)}$ "dressed" with some factors related to the gauge potentials or, alternatively, as the ERN solution dressed with some Schwarzschild-like factors. The Schwarzschild component of this metric completely disappears in the extreme limit, leaving an ERN isotropic metric. This form of charged BH metric is quite common and occurs, as we will see, in various contexts, rewritten in this way:

$$\boxed{\begin{aligned} &ds^2 = H^{-2}W dt^2 - H^2\left[W^{-1}d\rho^2 + \rho^2 d\Omega^2_{(2)}\right], \\ &A_\mu = \delta_{\mu t}\alpha\left(H^{-1} - 1\right), \\ &H = 1 + \frac{h}{\rho}, \qquad W = 1 + \frac{\omega}{\rho}, \qquad \omega = h\left[1 - (\alpha/2)^2\right]. \end{aligned}} \tag{12.92}$$

We will obtain many solutions in this form. Afterwards, we will identify the integration constants that appear in them in terms of the physical constants

$$\alpha = -4G_N^{(4)}q/r_\pm, \qquad h = r_\pm, \qquad \omega = \pm 2r_0. \tag{12.93}$$

10. The metric of the RN BH can also be rewritten in the general form Eqs. (11.28) [608] in which the outer (event) horizon is located at $\tau \to -\infty$ and spatial infinity is

located at $\tau \to 0^-$. Transforming the coordinate $\rho$,

$$\rho = -\frac{r_0 e^{-r_0\tau}}{\sinh(r_0\tau)}, \tag{12.94}$$

so the metric takes the form of Eqs. (11.28), with the metric function $e^{-2U}$ given by $(G_{\rm N}^{(4)} = 1)$

$$e^{-2U} = \left(\frac{r_+}{2r_0}e^{-r_0\tau} - \frac{r_-}{2r_0}e^{r_0\tau}\right)^2. \tag{12.95}$$

In this case, the same metric also describes the interior of the inner (Cauchy) horizon when $\tau \in (\frac{2}{r_0}\text{arctanh}\sqrt{\frac{M-2|q|}{M+2|q|}}, +\infty)$. The limits of this interval correspond to the singularity and the inner horizon. This can be checked using the relation between the radial coordinates $\tau$ and $r$. In the two intervals we are dealing with, it is given by [578]

$$\begin{aligned} \tau &= \frac{2}{r_0}\text{arctanh}\left\{\frac{-r_0}{(r-M)+\sqrt{(r-M)^2-r_0^2}}\right\}, \qquad r \in (r_+, +\infty), \\ \tau &= \frac{2}{r_0}\text{arctanh}\left\{\frac{-r_0}{(r-M)-\sqrt{(r-M)^2-r_0^2}}\right\}, \qquad r \in (0, r_-); \end{aligned} \tag{12.96}$$

$e^{2U}$ vanishes in the two limits $\tau \to \pm\infty$ (the tell-tale sign of a horizon). The coefficient that multiplies $d\Omega^2_{(2)}$ in this limit, which is the square radius of the horizons, is $r_\pm^2$, as expected.

11. Finally, BH solutions for an action containing several different vector fields $A^\Lambda{}_\mu$, $\Lambda = 1, \ldots, N$, can easily be found. Let us consider the action

$$S[g_{\mu\nu}, A^\Lambda{}_\mu] = \frac{1}{16\pi G_{\rm N}^{(4)}}\int d^4x\sqrt{|g|}\left[R - \tfrac{1}{4}\sum_{\Lambda=1}^{\Lambda=N}\left(F^\Lambda\right)^2\right]. \tag{12.97}$$

This action is invariant under global $\mathrm{O}(N)$ rotations of the $N$ vector field strengths. This is a simple example of *duality symmetry*. Now, any solution of the Einstein–Maxwell theory (one vector field) is a solution of this theory with the remaining $N-1$ vector fields equal to zero, and, by performing general $\mathrm{O}(N)$ rotations, we can generate new solutions in which the $N$ vector fields are non-trivial. It is clear that, if the original solution had the electric charge $q_1$, the electric charges of the new solution $q_\Lambda$ will satisfy $\sum_{\Lambda=1}^N q'^2_\Lambda = q_1^2$. This duality symmetry does not act on the metric and, therefore, all we have to do is to replace $q_1^2$ by $\sum_{\Lambda=1}^N q'^2_\Lambda$ in it.

For example, had we started from the RN solution Eqs. (12.92), we would have obtained by this procedure a RN solution with many Abelian electric charges:

$$\boxed{\begin{aligned}
ds^2 &= H^{-2} W dt^2 - H^2 \Big[W^{-1} d\rho^2 + \rho^2 d\Omega^2_{(2)}\Big],\\
A^{\Lambda}{}_{\mu} &= \delta_{\mu t}\alpha^{\Lambda}(H^{-1} - 1),\\
H &= 1 + \frac{h}{\rho}, \qquad W = 1 + \frac{\omega}{\rho},\\
\omega &= h\left[1 - \tfrac{1}{4}\sum_{\Lambda=1}^{N}(\alpha^{\Lambda})^{2}\right],
\end{aligned}} \tag{12.98}$$

and

$$\alpha^{\Lambda} = -4G_{\rm N}^{(4)} q^{\Lambda}/r_{\pm}, \qquad h = r_{\pm}, \qquad \omega = \pm 2r_0, \tag{12.99}$$

where now

$$r_{\pm} = G_{\rm N}^{(4)} M \pm r_0, \qquad r_0 = G_{\rm N}^{(4)}\left(M^2 - 4\sum_{\Lambda=1}^{\Lambda=N} q_{\Lambda}^2\right)^{\frac{1}{2}}. \tag{12.100}$$

This is the first and simplest example of the use of duality symmetries as solution-generating symmetries. We will find more complex examples later on, but the main ideas are the same.

Observe that, in this procedure of generating new solutions out of known ones, the new solutions are expressed at the beginning in terms of the old physical parameters and the parameters of the duality transformation (in this case, $\mathrm{O}(N)$ and the sines and cosines of angles). Then one has to identify those constants in terms of the physical parameters of the new solution. This is usually quite a painful calculation (sometimes, in cases more complicated than this one, it is impossible to do) unless we use invariance properties such as the invariance of $\sum_{\Lambda=1}^{\Lambda=N} q_{\Lambda}^2$ under $\mathrm{O}(N)$ transformations.

In the end, we should obtain a general duality-invariant family of solutions such that a further duality transformation takes us to another member of the family but the form of the general solution no longer changes. These families of solutions reflect many of the symmetries of the theory and depend only on duality-invariant combinations of charges and moduli.

The family we have obtained is duality invariant: the effect of a further duality transformation is just to replace all charges by primed charges, but the general form of the solution does not change.

## 12.4 Sources of the electric RN black hole

Just as we did with the Schwarzschild solution, we want to try to find a source for the RN solution such that it becomes a solution everywhere, including at the singularity $r = 0$. Our candidate source will be a point-like particle at rest at $r=0$ whose mass and electric charge match those of the RN BH. As in the Schwarzschild case, our expectations are not good because most of the reasons why we were unsuccessful (delocalization of the gravitational energy and the infinite self-force of the particle) are valid also in the general RN case. The only change is the causal nature of the singularity: spacelike in the Schwarzschild case, timelike in the RN case. However, we could argue that the gravitational and electromagnetic energy densities in the ERN BH cancel each other out everywhere so they are somehow *localized* at the origin $r=0$, and, thus, in this particular case we have some hope.

Our starting point is, therefore, the action of the Einstein–Maxwell system Eq. (12.58) coupled to the action of a massive, charged particle ($c=1$):

$$S=\frac{1}{16\pi G_{\rm N}^{(4)}}\int d^4x\sqrt{|g|}\,\left[R-\tfrac{1}{4}F^2\right]-M\int d\xi\,\sqrt{g_{\mu\nu}(X)\dot{X}^\mu\dot{X}^\nu}-q\int d\xi A_\mu\dot{X}^\mu. \tag{12.101}$$

The equations of motion of the dynamical fields $g_{\mu\nu}$, $A_\mu$, and $X^\mu$ are, respectively,

$$G_{\mu\nu}-8\pi G_{\rm N}^{(4)}T_{\mu\nu}^{(A)}+\frac{8\pi G_{\rm N}^{(4)}M}{\sqrt{|g|}}\int d\xi\frac{g_{\mu\rho}g_{\nu\sigma}\dot{X}^\rho\dot{X}^\sigma}{\sqrt{|g_{\lambda\tau}\dot{X}^\lambda\dot{X}^\tau|}}\delta^{(4)}[X(\xi)-x]=0, \tag{12.102}$$

$$\partial_\mu\left(\sqrt{|g|}F^{\mu\nu}\right)-16\pi G_{\rm N}^{(4)}q\int d\xi\dot{X}^\nu\delta^{(4)}[X(\xi)-x]=0, \tag{12.103}$$

$$\gamma^{1/2}M\nabla^2(\gamma)X^\lambda+M\gamma^{-1/2}\Gamma_{\rho\sigma}{}^\lambda\dot{X}^\rho\dot{X}^\sigma-qF^\lambda{}_\rho\dot{X}^\rho=0, \tag{12.104}$$

where

$$\gamma=g_{\mu\nu}(X)\dot{X}^\mu\dot{X}^\nu. \tag{12.105}$$

Let us first consider the Einstein equation. We use the RN solution in the coordinates Eqs. (12.92) with the upper sign and, following the same steps as in the Schwarzschild case, we obtain the following non-vanishing components of the Einstein tensor:[16]

$$\begin{aligned}
G_{00}&=g_{00}H^{-2}\{-\delta(W)+2(WH^{-1})\delta(H)+(WH^{-1})'H'\},\\
G_{\rho\rho}&=g_{\rho\rho}H^{-2}\{-\delta(W)+(WH^{-1})'H'\},\\
G_{\theta\theta}&=g_{\theta\theta}H^{-2}\{\tfrac{1}{2}\delta(W)-(WH^{-1})'H'\},\\
G_{\varphi\varphi}&=\sin^2\theta\,G_{\theta\theta},
\end{aligned} \tag{12.106}$$

[16] Needless to say, the mathematical rigor in all these manipulations is scarce. For instance, we feel free to multiply delta functions by functions that may diverge or be zero if we integrated the product. Some of these manipulations could possibly be justified by working with tensor densities instead of tensors, etc. The ultimate justification for presenting these calculations is the result, which allows us to match physical parameters such as mass and electric charge with integration constants of solutions.

where we are using the notation

$$\delta(W) = -\frac{4\pi}{\sin\theta}\omega\delta^{(3)}(\rho), \qquad \delta(H) = -\frac{4\pi}{\sin\theta}h\delta^{(3)}(\rho). \tag{12.107}$$

The electromagnetic energy–momentum tensor does not need to be calculated explicitly. It does not have any distributional term (delta function) and we know that it cancels out exactly the finite terms of the Einstein tensor. Thus, in the Einstein equation, we need only focus on the distributional terms coming from the Einstein tensor and the particle's energy–momentum tensor, which depends on our ansatz for $X^\mu$. For a particle at rest at $\vec{x}_3 = \vec{0}$, we must set[17]

$$X^\mu = \delta^{\mu 0}\xi, \tag{12.108}$$

but with this choice only the 00 component of the particle's energy–momentum tensor is non-vanishing, as in the Schwarzschild case. However, in the extreme case $\omega = 0$ only the 00 component of the Einstein has a distributional term that matches exactly the particle's energy–momentum tensor

$$8\pi G_N^{(4)} M H^{-5}\delta^{(3)}(\vec{x}_3) \tag{12.109}$$

(after integration over $\xi$).

The Maxwell equation is also satisfied (even in the non-extreme case). Let us turn to the particle's equation of motion. The time component is just $dg_{00}^{-1/2}/d\xi = dH/d\xi$ in the extreme case. $H$ diverges on the particle's path, and, even though it is independent of $\xi$, we cannot say that this equation is truly solved. The radial component can be put in the form

$$-M\partial_r g_{00}^{\frac{1}{2}} - q\partial_r A_0 = 0, \tag{12.110}$$

and it is satisfied identically by the ERN solution.[18] If we considered the motion of any other particle with $M' = 2|q'|$, we would see that it can be at rest anywhere in the ERN solution.

These kinds of cancelations are indications of supersymmetry, which, as we will see, is present in the ERN solution.

## 12.5 Thermodynamics of RN black holes

As we noted during the discussion of Schwarzschild BH thermodynamics, most of the results can be generalized to BHs containing charges or angular momentum. In particular, the zeroth and second laws of BH thermodynamics take exactly the same form, and so do the identifications between the surface gravity and temperature of the horizon, Eq. (11.48), and between area and entropy, Eq. (11.49). The first law requires the addition of a new term

[17] Observe that, in this coordinate system, $\vec{x}_3 = \vec{0}$ is the event horizon! The delta functions that we obtain have support only there and we are forced to make this ansatz if we want the particle's energy–momentum tensor and electric current to reproduce the singularities of the Einstein tensor and the Maxwell equation.

[18] Again, we manipulate $g_{00}$ etc., not taking into account that they are zero or diverge along the particle's path.

that takes into account the possible changes in the BH mass due to changes in the charge ($\hbar = c = 1$),

$$dM = \frac{1}{8\pi G_{\rm N}^{(4)}} \kappa dA + \phi^{\rm h} dq, \tag{12.111}$$

where $\phi^{\rm h}$ is the electrostatic potential on the horizon. In this case

$$\begin{aligned} T &= \frac{r_0}{2\pi r_+^2} = \frac{1}{2\pi G_{\rm N}^{(4)}} \frac{\sqrt{M^2 - 4q^2}}{\left(M + \sqrt{M^2 - 4q^2}\right)^2}, \\ S &= \frac{\pi r_+^2}{G_{\rm N}^{(4)}} = \pi G_{\rm N}^{(4)} \left(M + \sqrt{M^2 - 4q^2}\right)^2, \\ \phi^h &= \phi(r_+) = \frac{4G_{\rm N}^{(4)} q}{r_+}, \end{aligned} \tag{12.112}$$

and the Smarr formula takes the form

$$M = 2TS + q\phi^{\rm h} \text{ or } r_0 = 2TS. \tag{12.113}$$

It is worth stressing that the above formulae have been obtained using a generic RN metric (i.e. non-extremal). However, we know that the limit in which we approach the ERN solution with $M = 2|q|$ is not continuous: the topology of the ERN, its causal structure, is different from that of any non-extreme RN BH, no matter how close to the extreme limit it is. Furthermore, it seems that the extreme limit cannot be approached by a finite series of physical processes (the third law of BH thermodynamics), and it has also been argued that the thermodynamical description of the RN BH breaks down when we approach the extreme limit [1060] (see also Ref. [893]): close enough to the extreme limit, the emission of a single quantum with energy equal to the Hawking temperature would take the mass of the RN BH *beyond* the extreme limit. Then, the change in the spacetime metric caused by Hawking radiation would be very big and Hawking's calculation, in which backreaction of the metric to the radiation is ignored, becomes inconsistent.

For all these reasons we may expect surprises if we naively take the limit $M \to 2|q|$ in the above formulae, but this seems not to happen: in that limit the temperature vanishes and the entropy remains finite, and, if we calculate both directly on the ERN solution, we find the same result. In any case, this is a very important issue because essentially these are the only BHs for which a statistical computation of the entropy based on string theory has been performed, and we should try to compute both by other methods, for instance using the Euclidean path-integral formalism. Before we do so, let us mention other remarkable aspects of the RN BH thermodynamics.

We have drawn the behavior of the RN BH temperature for fixed charge in Fig. 12.3. In it we see that, for large values of the mass, the temperature diminishes when the mass

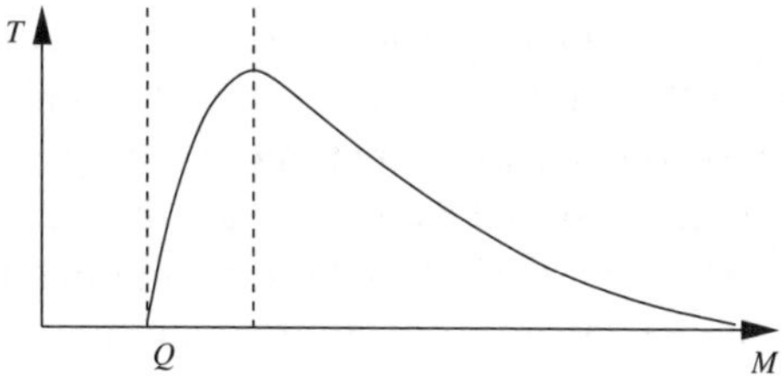


Fig. 12.3. The temperature $T$ versus the mass $M$ of a Reissner–Nordström black hole of charge $Q = 2q$.

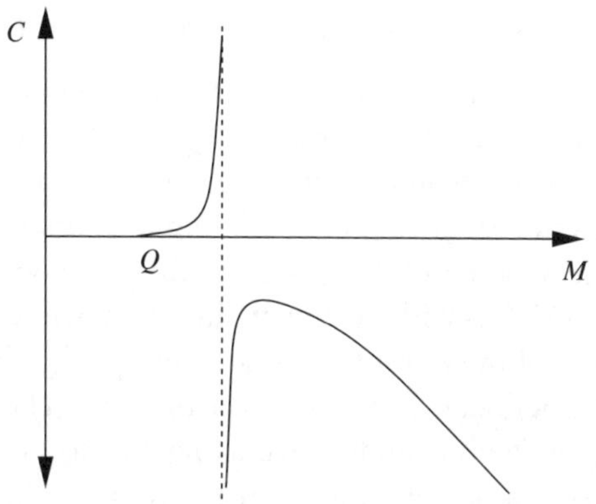


Fig. 12.4. The specific heat $C$ at constant charge versus the mass $M$ of a Reissner–Nordström black hole of charge $Q = 2q$ (not to scale).

grows, just as in the Schwarzschild BH, but, for values of the mass comparable to the charge, close to the extreme limit, the temperature grows with the mass, as in any ordinary thermodynamical system. There is a maximum temperature for RN BHs (for constant charge), which is reached for $M^\star = 4|q|/\sqrt{3}$. The maximum value for the temperature is given by $T^\star$:

$$T^\star = T(M^\star, q) = \frac{1}{12\sqrt{3}\pi G_{\rm N}^{(4)}|q|}. \tag{12.114}$$

In the plot of the specific heat at constant charge $C$ of Fig. 12.4 we clearly see the two regions in which the thermodynamical behavior is "standard" (positive specific heat) and "Schwarzschild like" (negative specific heat). At the point $M^\star$ at which the temperature reaches its maximum value, $\partial T/\partial M = 0$ and the specific heat diverges. It is tempting to associate that divergence with a phase transition between the two kinds of behavior. It is also tempting to associate the success of the statistical calculation of the ERN BH entropy with its standard thermodynamical behavior in its neighborhood.

What would be the endpoint of the Hawking evaporation of a RN BH? As we mentioned before, the electric charge is lost faster than the mass and, before the extreme limit is reached, we should have an uncharged Schwarzschild BH, whose fate we have already discussed. We can, however, speculate what would happen if the charge of the RN were of a kind not carried by any elementary particle[19] so that it could not be lost by Hawking radiation, or if the carriers of that kind of charge were extremely heavy particles[20] (unlike electrons) so that the BH would discharge much more slowly than it would lose mass. In these cases, assuming that nothing special happens when the mass is such that $\partial T/\partial M = 0$, one would expect the RN BH to approach the extreme limit in a very long-lasting (perhaps eternal) process in which the BH loses mass and temperature at lower and lower rates. It has been conjectured that the ERN BH could be a BH remnant, storing all the information contained in the original BH that is not radiated away.

## 12.6 The Euclidean electric RN solution and its action

The Euclidean (non-extreme) RN solution has a structure identical to that of the Euclidean Schwarzschild solution; in particular, it covers only the BH exterior, and it also requires the compactification of the Euclidean time in order to eliminate a conical singularity. This allows us to calculate the temperature again by finding the period of the Euclidean time that makes the metric on the horizon regular. If we use spherical coordinates with origin on the horizon (like those of Eqs. (12.92) with the upper sign) then we see that $T = g'_{\tau\tau}(0)/(4\pi) = \omega/(4\pi h^2)$. With this period of the Euclidean time, the topology is $\mathbb{R}^2 \times S^2$.

Let us now study the Euclidean ERN solution directly. The interesting region is the neighborhood of the horizon, and to study it we expand as usual the metric components of Eqs. (12.92), (with $\omega = 0$) in a power series in the inverse of the radial coordinate around the origin and keep the lowest-order terms; instead of an approximate solution, we have the Euclidean continuation of the RB metric Eqs. (12.90). This metric is completely regular for any periodicity of the Euclidean time. It is convenient to use the coordinate $r = R\ln(\rho/R)$ with $R = 2G_N^{(4)}|q|$, in which the Lorentzian RB solution becomes

$$
\boxed{\begin{aligned}
&ds^2 = e^{2r/R}dt^2 - dr^2 - R^2 d\Omega^2_{(2)},\\
&A_t = -2e^{r/R}, \qquad F_{tr} = \frac{2}{R}e^{\frac{r}{R}}, \qquad R = 2G_N^{(4)}|q|.
\end{aligned}} \tag{12.115}
$$

In these coordinates it is evident that the horizon $r = -\infty$ is at an infinite distance in the $r$ direction, and a constant-time slice of this spacetime looks like an infinite tube whose $r = \text{constant}$ sections are 2-spheres of constant radius $R$. It does not make much sense to talk about the period of $\tau$ that makes the Wick-rotated metric on the horizon regular because it is regular for any period. The same applies to flat Euclidean spacetime. The temperature cannot be uniquely assigned in this formalism. The reason could be the fact that both Minkowski spacetime and the RB solution can be considered vacua of the theory.

[19] This BH could not have been created by standard gravitational collapse. Instead, a process like quantum pair creation has to be invoked to justify its existence.

[20] For instance, the carriers of Kaluza–Klein charges and the massive modes in string theory are usually assigned very large masses (of the order of the Planck mass) in order to explain why they have not yet been observed.

As a conclusion of this discussion, then, if we compactify the Euclidean time with some arbitrary period $\beta$, the topology is not $\mathbb{R}^2 \times \mathrm{S}^2$ as in the non-extreme case, but $\mathbb{R} \times \mathrm{S}^1 \times \mathrm{S}^2$. The factor $\mathbb{R} \times \mathrm{S}^1$, with the topology of a cylinder, corresponds to $\mathbb{R}^2 - \{0\}$, the $\tau$–$r$ plane with the point at the origin (the event horizon, which is at an infinite distance) removed. The Euclidean RN solution has, therefore, two boundaries: at infinity (as in the non-extreme case) and at the horizon. One way to check this fact is to calculate the Euler characteristic of the Euclidean ERN solution using the Gauss–Bonnet theorem adapted to manifolds with boundaries. The Euler characteristic $\chi$ is a topological invariant whose value is an integer, and the Gauss–Bonnet theorem states that the integral of the 4-form

$$\frac{1}{32\pi^2}\epsilon_{abcd}R^{ab} \wedge R^{cd} \tag{12.116}$$

over a four-dimensional compact manifold M is precisely $\chi$. If the manifold has a boundary $\partial$M, then $\chi$(M) is given by the integral over M of the above 4-form plus the integral over the boundary of a 3-form [350, 481],

$$\chi(\mathrm{M}) = \frac{1}{32\pi^2}\int_{\mathrm{M}} \epsilon_{abcd}R^{ab} \wedge R^{cd} - \frac{1}{32\pi^2}\int_{\partial\mathrm{M}} \epsilon_{abcd}\left[2\theta^{ab} \wedge R^{cd} - \tfrac{4}{3}\theta^{ab} \wedge \theta^{c}{}_{e} \wedge \theta^{ed}\right], \tag{12.117}$$

where $\theta^{ab}$ is the *second fundamental 1-form* on $\partial$M, which can be constructed as explained in Ref. [481]. The contribution of the boundary integral is crucial in order to have $\chi = 2$ in the non-extreme case, corresponding to the topology $\mathbb{R}^2 \times \mathrm{S}^2$. In the extreme case, only by taking into account the boundary at the horizon does one obtain $\chi = 0$, the correct value for the topology $\mathbb{R} \times \mathrm{S}^1 \times \mathrm{S}^2$ [621].

This is going to have important consequences in what follows.

Once we have determined the period, we are ready to calculate the partition function using the Euclidean path-integral formalism in the saddle-point approximation. We are going to do it as in the Schwarzschild case, using the Lorentzian action and solution but taking into account the periodicity of the Euclidean time and the fact that the Euclidean solution covers only the exterior of the horizon.

In $\hbar = c = G_{\mathrm{N}}^{(4)} = 1$, the Einstein–Maxwell system with boundary terms is

$$S_{\mathrm{EM}}[g_{\mu\nu}, A_\mu] = \frac{1}{16\pi}\int d^4x\,\sqrt{|g|}\left[R - \tfrac{1}{4}F^2\right] + \frac{1}{8\pi}\int d^3\Sigma(\mathcal{K} - \mathcal{K}_0), \tag{12.118}$$

and, using the definition of $F_{\mu\nu}$ and integrating by parts, we rewrite it in this form

$$\begin{aligned} S_{\mathrm{EM}}[g_{\mu\nu}, A_\mu] = {}& \frac{1}{16\pi}\int d^4x\,\sqrt{|g|}[R + \tfrac{1}{2}A_\nu\nabla_\mu F^{\mu\nu}] \\ &+ \frac{1}{8\pi}\int d^3\Sigma[(\mathcal{K} - \mathcal{K}_0) + \tfrac{1}{4}n_\mu F^{\mu\nu}A_\nu]. \end{aligned} \tag{12.119}$$

Only the boundary term contributes to the action because the volume term vanishes on-shell. Furthermore, for a generic non-extreme RN BH there is only one boundary at infinity. The contribution of the extrinsic curvature terms for large values of $r_{\mathrm{c}}$ is always given by

Eqs. (11.77) for spherically symmetric, static, asymptotically flat metrics such as the RN metric. The electromagnetic boundary term has to be computed in the gauge in which $A_\mu$ vanishes on the horizon, i.e. using $A'_\mu \equiv A_\mu - A_\mu(r_+)$, because the Killing vector $\partial/\partial\tau$ is singular on the event horizon (which is a Killing horizon). We find

$$\tfrac{1}{4} n_\mu F'^{\,\mu\nu} A'_\nu = -\frac{q}{r^2\sqrt{-g_{rr}g_{tt}}}\left(\frac{4q}{r} - \frac{4q}{r_+}\right)\bigg|_{r=r_{\mathrm{c}}\to\infty} \sim \frac{4q^2}{r_{\mathrm{c}}^2 r_+} + \mathcal{O}(r_{\mathrm{c}}^{-3}). \tag{12.120}$$

Finally, taking into account that

$$d^3\Sigma = dt d\Omega^2 r^2 \sqrt{g_{tt}}\big|_{r=r_{\mathrm{c}}\to\infty} \sim dt d\Omega^2 r_{\mathrm{c}}^2, \tag{12.121}$$

we find in the limit $r_{\mathrm{c}} \to \infty$ for the Euclidean action

$$-\tilde{S}_{\mathrm{EM}} = -\frac{\beta}{2}[M - q\phi(r_+)] = -\frac{\beta}{2} r_0, \tag{12.122}$$

where $\phi(r_+) = A_t(r_+)$ is the electrostatic potential on the horizon. The entropy is

$$S = \beta[M - q\phi(r_+)] + \ln \mathcal{Z} = +\frac{\beta}{2}[M - q\phi(r_+)] = \frac{\beta}{2} r_0 = \pi r_+^2, \tag{12.123}$$

that is, one quarter of the area of the horizon.

This calculation is valid for generic non-extreme RN BHs. We should now repeat the calculation directly for ERN BHs. There are two important differences.

1. The period $\beta$ of the Euclidean time is not determined.
2. The Euclidean ERN solution has another boundary at the horizon, and the action contains the contribution of the boundary at infinity, given in Eq. (12.122), and the contribution from the new boundary that we can calculate straight away:

$$-\frac{i}{8\pi}\int_0^{-i\beta} dt \int_{S^2} d\Omega^2 r^2 \sqrt{g_{tt}} \left\{ \frac{1}{\sqrt{-g_{rr}}}\left[\tfrac{1}{2}\partial_r \ln g_{tt} + \frac{2}{r}\right] - \frac{2}{r} \right.$$
$$\left. - \frac{q}{r^2\sqrt{-g_{rr}g_{tt}}}\left(\frac{4q}{r} - \frac{4q}{r_+}\right)\right\}, \tag{12.124}$$

where we have to substitute $r = r_+$. The result is $-\beta r_0/2$, and, thus, we have

$$-\tilde{S}_{\mathrm{EM}} = -\beta r_0, \tag{12.125}$$

which gives[21] identically [621, 1174, 730]

$$S = \beta[M - q\phi(r_+)] + \ln \mathcal{Z} = 0. \tag{12.126}$$

[21] Of course, $\beta r_0$ would be identically zero for ERN BHs ($r_0 = 0$) if $\beta$ were taken finite.

It has been suggested that the same is true for any extreme charged BH, not just ERN BHs, and also that the Bekenstein–Hawking entropy formula Eq. (11.49) should be [892]

$$S = \frac{\chi A c^3}{8 G_{\rm N}^{(4)} \hbar}. \tag{12.127}$$

Since one of the main successes of string theory has been the calculation of the (finite!) entropy of the ERN BH, this result is a bit disturbing. Actually it implies that string theory and the Euclidean path-integral approach to quantum gravity give different predictions for the entropy of the ERN BH. It has been argued by Sen (see e.g. Ref. [765]) that the near-horizon ERN geometry suffers important corrections in string theory. The reason would be that, although the topology is that of a cylinder, the geometry is rather that of a *pipette*, with a radius that tends to zero at infinity when we asymptotically approach the horizon. String theory compactified on a circle undergoes a phase transition when the radius reaches the *self-dual* value. Thus, beyond the point of the pipette at which the radius has that value, the geometry may indeed change,[22] although no precise calculations have been done so far.

## 12.7 Electric–magnetic duality

As we explained in Section 12.2,[23] the full set of sourceless Maxwell equations (the Maxwell equation plus the Bianchi identity) is invariant (up to signs) under the replacement of the field strength $F$ by its dual $\tilde{F} = \star F$

$$F \to \tilde{F} = \star F. \tag{12.128}$$

This is true in flat as well as in curved spacetime. In a given frame, this transformation corresponds to the interchange of electric and magnetic fields according to Eqs. (12.38), hence the name *electric–magnetic duality*. This transformation squares to (minus) the identity and, therefore, it generates a $\mathbb{Z}_2$ electric–magnetic-duality group.

The $\mathbb{Z}_2$ can easily be extended to a continuous symmetry group:[24]

$$\tilde{F} = aF + b \star F, \quad \Rightarrow \quad \star\tilde{F} = -bF + a \star F \quad a^2 + b^2 \neq 0, \tag{12.129}$$

is an invertible transformation that leaves the set of the two equations invariant (up to factors). It is convenient to define the *duality vector*

$$\vec{F} \equiv \begin{pmatrix} F \\ \star F \end{pmatrix}. \tag{12.130}$$

It is subject to the constraint

$$\star\vec{F} = \begin{pmatrix} 0 & 1 \\ -1 & 0 \end{pmatrix} \vec{F}, \tag{12.131}$$

[22] See analogous discussions on p. 785 about the *correspondence principle*.

[23] This topic is also covered in a complementary way in Section 2.6. The reader is encouraged to study the results explained in that section, in particular the interesting case in which there are several Maxwell fields.

[24] For the moment, all these are classical considerations. We will see that quantum effects (in particular, charge quantization) break the continuous symmetry to a discrete subgroup.

with which the Maxwell equations can be written as

$$\nabla_\mu \vec{F}^{\mu\nu} = 0, \tag{12.132}$$

and it transforms in the vector representation of the duality group, a subgroup of $\mathrm{GL}(2,\mathbb{R})$:

$$\tilde{\vec{F}} = M\vec{F}, \qquad M = \begin{pmatrix} a & b \\ -b & a \end{pmatrix} = \pm\lambda \begin{pmatrix} \cos\xi & \sin\xi \\ -\sin\xi & \cos\xi \end{pmatrix}. \tag{12.133}$$

In this form we see that the duality group consists of rescalings and $\mathrm{O}(2)$ rotations of $\vec{F}$.

Observe that, if we integrate the Hodge dual of the duality vector $\star\vec{F}$ over a 2-sphere at infinity, we obtain a *charge vector* whose first component is $16\pi G_{\mathrm{N}}^{(4)} q$, in our conventions. The second component will be, by definition, the *magnetic charge* $p$:

$$\int_{\mathrm{S}^2_\infty} \star\vec{F} = \begin{pmatrix} 16\pi G_{\mathrm{N}}^{(4)} q \\ p \end{pmatrix} \equiv 16\pi G_{\mathrm{N}}^{(4)} \vec{q}, \qquad \vec{q} = \begin{pmatrix} q \\ p/(16\pi G_{\mathrm{N}}^{(4)}) \end{pmatrix}. \tag{12.134}$$

Although this transformation looks very simple written in terms of the electromagnetic field strength $F_{\mu\nu}$, it is very non-local in terms of the true field variable $A_\mu$. To see this, we simply have to use Eq. (12.31) to obtain an explicit relation between $A_\mu$ and the dual vector field $\tilde{A}_\mu$:

$$\tilde{A}_\mu(x) = -\int_0^1 d\lambda \lambda x^\nu \frac{\epsilon_{\mu\nu}{}^{\rho\sigma}}{\sqrt{|g|}} \partial_\rho A_\sigma(\lambda x). \tag{12.135}$$

This non-locality is, at the same time, what makes this duality transformation interesting and the source of problems. To start with, the replacement of $F$ by $\star F$ is not a symmetry of the Maxwell action because $(\star F)^2 = -F^2$. The reason for this is that the transformation should be done on the right variable, namely the vector field, but this is difficult to do. Another possibility is to write an action that really is a functional of the field strength. On this action, the above replacement can be performed and gives the right results. This procedure is called *Poincaré duality* and we explain it in detail in Section 12.7.1.

Let us now see what modifications the coupling to gravity Eq. (12.58) produces. The main difference is that we now have one more equation (Einstein's). For our purposes, it is useful to rewrite it in this form (see Section 1.6 and Eq. (1.134)):

$$G_{\mu\nu} - \left(F_\mu{}^\rho F_{\nu\rho} + \star F_\mu{}^\rho \star F_{\nu\rho}\right) = 0, \tag{12.136}$$

or, using the duality vector,

$$G_{\mu\nu} - \left(\vec{F}_\mu{}^\rho\right)^{\mathrm{T}} \vec{F}_{\nu\rho} = 0, \tag{12.137}$$

which makes it clear that only the O(2) subgroup leaves the Einstein equation invariant. Out of this O(2) group, the parity transformation clearly belongs to a different class (if we had $N$ vector fields, it would belong to the $\mathrm{O}(N)$ group that rotates the vectors amongst themselves). Thus, the classical electric–magnetic-duality group of the Einstein–Maxwell theory is actually SO(2).

We are studying an Abelian theory without matter and therefore it has no coupling constant. However, we could think of this U(1) gauge symmetry as part of a bigger, non-Abelian, broken symmetry group and introduce a (dimensionless in natural units in $d = 4$) coupling constant $g$ that appears as a $g^{-2}$ factor in front of $F^2$ in the action and that we will not reabsorb into a rescaling of the vector field. The appropriate duality vector, the integral of whose dual over $S^2_\infty$ is $16\pi G_N^{(4)}\vec{q}$, is now

$$\vec{F} \equiv \begin{pmatrix} g^{-2}F \\ \star F \end{pmatrix}. \tag{12.138}$$

In terms of this duality vector, the Einstein equation can be rewritten as follows:

$$G_{\mu\nu} + \left(\vec{F}_\mu{}^\rho\right)^{\mathrm{T}} \eta \star \vec{F}_{\nu\rho} = 0, \qquad \eta \equiv \begin{pmatrix} 0 & 1 \\ -1 & 0 \end{pmatrix}, \tag{12.139}$$

and it is invariant under $\mathrm{Sp}(2,\mathbb{R}) \sim \mathrm{SL}(2,\mathbb{R})$. Now, it can be checked that, out of the full group, and allowing for transformations of $g$, only the following transformations (rescalings and $\mathbb{Z}_2$ duality rotations and their products) are consistent with the duality-vector constraint:

$$\begin{aligned} M &= \begin{pmatrix} a & 0 \\ 0 & 1/a \end{pmatrix}, \qquad & g' &= a^{-1}g, \\ M &= \begin{pmatrix} 0 & 1 \\ -1 & 0 \end{pmatrix}, \qquad & g' &= 1/g. \end{aligned} \tag{12.140}$$

Now we see the main reason why this duality is interesting: if the coupling constant $g$ of the original theory is large so perturbation theory cannot be used and non-perturbative states become light, then the coupling constant of the dual theory $g' = 1/g$ is small and can be used to do perturbative expansions, and the dual theory gives a better description of the same phenomena and states. In particular, magnetic monopoles are typical non-perturbative states of gauge theories with masses proportional to $1/g^2$ and become perturbative, electrically charged states of the dual theory.

Although, originally, electric–magnetic duality arose as a symmetry of the theory, a better point of view is that it is a relation, a mapping, between two theories that describe the same degrees of freedom in different ways. One of them can describe better one region of the *moduli space*[25] than can the other. Dualities in which the coupling constant is inverted and perturbative (weak-coupling) and non-perturbative (strong-coupling) regimes are related go by the name of *S dualities*. Electric–magnetic duality in the Maxwell theory is the simplest example. Perturbative dualities such as the $\mathrm{O}(N)$ rotation between the $N$ vector fields that we considered in Section 12.3 go by the name of *T dualities*, at least in the string-theory context. In some string theories (type II) the two kinds of dualities are part of a bigger duality group (which is not just the direct product of the S and T duality groups), which is called the *U duality* group [808].

[25] The coupling constant $g$ and other parameters necessary to describe completely a theory are usually called *moduli*. The space in which they take values is the *moduli space* of the theory.

A last comment on semantics: when talking about duality, there are always certain ambiguities in the use of the word "theory." Two theories that are dual are two different descriptions of the same physical system, and many physicists would say that they are, therefore, the same "theory" written in different variables. We would like to call them different "theories" describing the same reality. Both points of view are legitimate and are similar to the active and passive points of view in symmetry transformations.

### *12.7.1 Poincaré duality*

One of the peculiarities of the electric–magnetic-duality transformation is that it does not leave the Einstein–Maxwell action invariant: the direct replacement of $F$ by its dual in the action changes the sign of the kinetic terms $F^2$. The reason is that the action Eq. (12.58) is actually a functional of the vector potential. To be able to replace $F$ by $\star F$ we need an action that is a functional of $F$. The so-called Poincaré-dualization procedure provides a systematic way of finding actions that are functionals of the field strengths and on which we can perform electric-duality transformations, obtaining the correct dual action. Furthermore, this procedure can be generalized to other $k$-form potentials and dimensions.

Since the metric does not play a role, we consider only the vector-field kinetic term in Eq. (12.58). From that action one obtains only half of the Maxwell equations: the Bianchi identity has been solved and it is assumed that $F = dA$. Thus, if we want to have a functional of $F$ that produces all the Maxwell equations (sometimes called a first-order action), it has to give also the Bianchi identity $dF = 0$. This action can be constructed simply by adding to the standard Einstein–Maxwell action a Lagrange-multiplier term enforcing the Bianchi identity. $dF$ is a 3-form and so the Lagrange multiplier has to be a 1-form $\tilde{A} = \tilde{A}_\mu dx^\mu$ (which will become the dual potential) and then the term to be added to the action is $\sim \int \tilde{A} \wedge dF$. Integrating by parts, this term is rewritten as $\sim \int d\tilde{A} \wedge F$. More explicitly, in component language, the action with the Lagrange-multiplier term is

$$S[F_{\mu\nu}, \tilde{A}_\mu] = \frac{1}{16\pi G_{\rm N}^{(4)}} \int d^4x \sqrt{|g|}\,[-\tfrac{1}{4}F^2] - \frac{1}{16\pi G_{\rm N}^{(4)}} \int d^4x\, \tfrac{1}{2}\epsilon^{\mu\nu\rho\sigma}\partial_\mu \tilde{A}_\nu F_{\rho\sigma}. \tag{12.141}$$

This action gives rise to the same equations of motion as the original action $S[A]$: the equation of motion of $F$ is

$$F = \star\tilde{F}, \tag{12.142}$$

where we have defined

$$\tilde{F} = d\tilde{A}. \tag{12.143}$$

The Bianchi identity $d\tilde{F} = 0$, which is a consequence of its definition, becomes the Maxwell equation $d\star F = 0$ by virtue of Eq. (12.42). Furthermore, by construction, the equation of motion of $\tilde{A}$ is nothing but the Bianchi identity $dF = 0$ that implies the existence of the original vector field $A_\mu$.

Since the equation of motion of $F$ is purely algebraic, we can use it in the above action to eliminate it. The result is an action that is a functional of the dual potential $\tilde{A}$ and is identical to the original Einstein–Maxwell action (with the right sign):

$$S[\tilde{A}_\mu] = \frac{1}{16\pi G_{\rm N}^{(4)}} \int d^4x \sqrt{|g|} \left[ -\tfrac{1}{4}\tilde{F}^2 \right]. \tag{12.144}$$

### 12.7.2 Magnetic charge: the Dirac monopole and the Dirac quantization condition

The electric–magnetic-duality invariance of the vacuum Maxwell equations is automatically broken when one adds sources $j^\mu$. This is not surprising since $j^\mu$ describes static or dynamical electric (only) charges. It is necessary to introduce magnetic sources that can be rotated into the electric ones in order to maintain duality invariance of the Maxwell equations. We have already seen in Eq. (12.134) that electric–magnetic duality needs the introduction of magnetic charges into which electric charges can transform. By definition, then, the magnetic charge is given by:[26]

$$p \equiv \tilde{q} = \int_{S^2_\infty} d^2\vec{S} \cdot \tilde{\vec{E}} = \int_{S^2_\infty} d^2\vec{S} \cdot \vec{B} = -\int_{S^2_\infty} F. \tag{12.145}$$

The simplest electric-charge distribution is a point-like electric charge and its dual is a magnetic point-like charge, which should be given by a magnetic field obeying

$$\vec{\nabla} \cdot \vec{B} = p\, \delta^{(3)}(\vec{x}_3), \tag{12.146}$$

which is the *Dirac-monopole equation* for the vector potential.

Introducing magnetic sources to preserve electric–magnetic duality is, however, a very dangerous move: the Bianchi identity is not satisfied at the locations of the magnetic sources, and there the vector potential, the true dynamical field, cannot be defined or, more precisely, it cannot be defined everywhere: it will have singularities. This may not be as bad as it looks at first sight, because, after all, the electrostatic potential is not defined at the location of an electric point-like charge, either. It depends on how bad the singularities of the vector field are. In the electric case, it is quite benign, since the singularity affects only the particle that gives rise to the field. Let us see what happens with the vector potential of a point-like magnetic monopole. First, we have to find it.

Knowing that

$$\nabla^2 \frac{1}{|\vec{x}_3|} = -4\pi\delta^{(3)}(\vec{x}_3), \tag{12.147}$$

we find that the magnetic field is given by

$$\vec{B} = -\frac{p}{4\pi}\vec{\nabla}\frac{1}{|\vec{x}_3|}, \tag{12.148}$$

[26] We work again in the standard units of the beginning of Section 12.2.1 and in flat spacetime. At the end of this section we will say which changes have to be made when using our normalization Eq. (12.58).

which implies, due to $\vec{B} = \vec{\nabla} \times \vec{A}$, for the Dirac-monopole equation

$$\vec{\nabla} \times \vec{A} = -\frac{p}{4\pi} \vec{\nabla} \frac{1}{|\vec{x}_3|}, \tag{12.149}$$

or, defining, to simplify matters $\vec{f} = -(4\pi/p)\vec{A}$, the following, standard form:

$$\boxed{\partial_m f_n - \partial_n f_m = \epsilon_{mnp} \partial_p \frac{1}{|\vec{x}_3|}.} \tag{12.150}$$

The integrability condition for this system of coupled partial differential equations is found by rewriting it in the equivalent form

$$\epsilon_{mnp} \partial_m f_n = \partial_p \frac{1}{|\vec{x}_3|} \tag{12.151}$$

and acting with $\partial_p$,

$$\partial_p \partial_p \frac{1}{|\vec{x}_3|} = 0, \tag{12.152}$$

which is true everywhere except at the origin. Instead of $1/|\vec{x}_3|$ we could have used any other harmonic function on the three-dimensional Euclidean space.

A solution of the Dirac-monopole equation is provided by (see e.g. Refs. [638, 364, 972, 1137])

$$\vec{f}^{\,+} = -\frac{(0,0,1) \times (x,y,z)}{|\vec{x}_3|(|\vec{x}_3| + z)}. \tag{12.153}$$

This solution is singular at $|\vec{x}_3| = -z$, i.e. the whole negative $z$ axis, not just at the location of the magnetic monopole.

In spherical coordinates (Fig. 12.5)

$$\begin{aligned} x^1 &= x = r \sin\theta \sin\varphi, \\ x^2 &= y = r \sin\theta \cos\varphi, \\ x^3 &= z = r \cos\theta, \end{aligned} \tag{12.154}$$

the above solution has as its only non-vanishing component

$$f_\varphi^+ = 1 - \cos\theta. \tag{12.155}$$

In these coordinates the solution looks regular. However, one has to take into account that the unit vector orthogonal to constant $\varphi$ surfaces is singular over the $z$ axis. Over the positive $z$ axis, $\vec{f}^{\,+}$ is regular because $f_\varphi^+$ vanishes.

Owing to this singularity, $\vec{f}^{\,+}$ is not a solution of the Dirac-monopole equation (12.150) everywhere. This can be seen just as one sees that $\nabla^2|\vec{x}|^{-1} \sim \delta^{(3)}(\vec{x})$ by integrating and

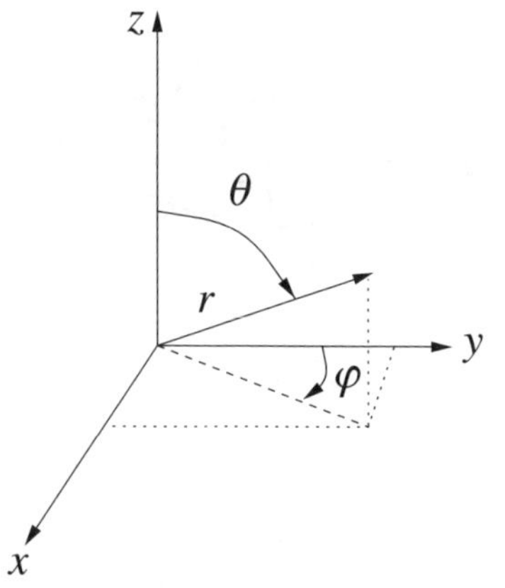


Fig. 12.5. Spherical versus Cartesian coordinates.

applying Stokes' theorem. Let us consider the integral of $\vec{\nabla} \times \vec{f}^{\,+} + \vec{x}/|\vec{x}|^3$ over a surface $\Sigma$. If $\vec{f}^{\,+}$ were a solution of Eq. (12.150) everywhere, this integral would be zero for any surface $\Sigma$. Now let us apply Stokes' theorem to the first term. We find

$$\Phi_\Sigma = \int_\Sigma d^2\vec{S} \cdot \left(\vec{\nabla} \times \vec{f}^{\,+} + \frac{\vec{x}}{|\vec{x}|^3}\right) = \oint_{\gamma=\partial\Sigma} d\vec{x} \cdot \vec{f}^{\,+} + \int_\Sigma d^2\vec{S} \cdot \frac{\vec{x}}{|\vec{x}|^3}, \qquad (12.156)$$

where $\gamma$ is the one-dimensional boundary of the two-dimensional surface $\Sigma$. Let us consider the particular surface $\Sigma_+$ (a sector of the unit sphere whose boundary is $\gamma^+ : \theta = \theta_0$ oriented in the sense of negative $\varphi$) shown in Fig. 12.6. We find

$$\left.\begin{aligned} &\int_{\Sigma+} d^2\vec{S} \cdot \frac{\vec{x}}{|\vec{x}|^3} = 2\pi(1-\cos\theta_0),\\ &\int_{\gamma+} d\vec{x} \cdot \vec{f}^{\,+} = -2\pi(1-\cos\theta_0), \end{aligned}\right\} \Rightarrow \Phi_{\Sigma+} = 0. \qquad (12.157)$$

Let us now consider a different surface $\Sigma_-$ (a sector of the unit sphere whose boundary is $\gamma^- : \theta = \theta_0$ oriented in the sense of positive $\varphi$), shown in Fig. 12.7:

$$\left.\begin{aligned} &\int_{\Sigma-} d^2\vec{S} \cdot \frac{\vec{x}}{|\vec{x}|^3} = 2\pi(1+\cos\theta_0),\\ &\int_{\gamma^-} d\vec{x} \cdot \vec{f}^{\,+} = 2\pi(1-\cos\theta_0), \end{aligned}\right\} \Rightarrow \Phi_{\Sigma-} = 4\pi. \qquad (12.158)$$

The above results are valid for any value of $\theta_0$, and we conclude that $\vec{f}^{\,+}$ does indeed solve the Dirac-monopole equation only away from the negative $z$ axis $\theta = \pi$. More precisely,

$$\vec{\nabla} \times \vec{f}^{\,+} = -\frac{\vec{x}_3}{|\vec{x}_3|^3} - 4\pi\delta(x)\delta(y)\theta(-z)\vec{u}_z, \qquad (12.159)$$

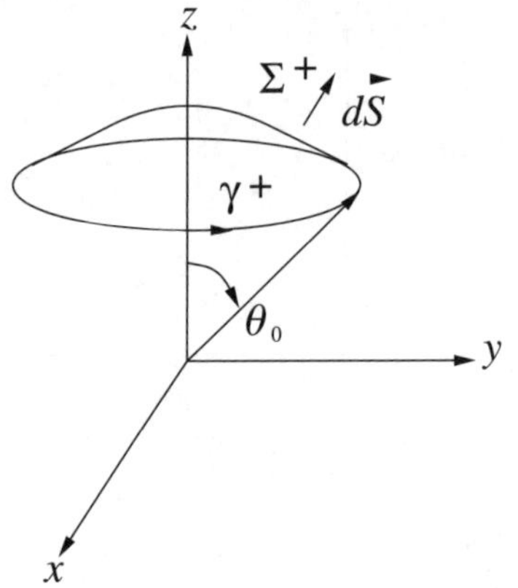


Fig. 12.6. The surface $\Sigma^+$ and its boundary $\gamma^+$.

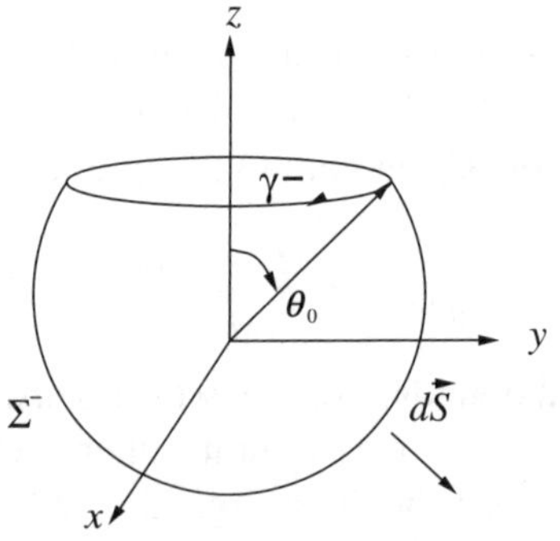


Fig. 12.7. The surface $\Sigma^-$ and its boundary $\gamma^-$.

where $\vec{u}_z$ is a unit vector along the $z$ axis.

The singularity along $\theta = \pi$ is known as the *Dirac string*. Physically, the Dirac string can be visualized as the zero-section limit of a semi-infinite tube of magnetic flux. Thus, the flux of

$$\vec{B}^{\,+} = \vec{\nabla} \times \vec{A}^{\,+} = \frac{p}{4\pi} \frac{\vec{x}_3}{|\vec{x}_3|^3} + p\delta(x)\delta(y)\theta(-z)\vec{u}_z \tag{12.160}$$

across any closed 2-surface is zero. The Dirac string appears as a singularity of the magnetic field and hence, in principle, it should be considered a physical singularity.

Thus, at first sight we have not succeeded in finding a solution of the Dirac-monopole equation. Still, we can ask ourselves whether the Dirac string is observable and has any physical effect. First of all, observe that the Dirac string can be moved (but not removed) by gauge transformations. For instance, the transformed gauge potential $\vec{A}^{\,-}$,

$$A^-_\varphi = A^+_\varphi + \partial_\varphi\left(\frac{p}{2\pi}\varphi\right) = \frac{p}{4\pi}(1 + \cos\theta), \tag{12.161}$$

is now singular over the positive $z$ axis only. We have changed the position of the Dirac string from the negative to the positive axis. From this one could naively conclude that the Dirac string is just a gauge artifact and, as such, unphysical. This is not strictly correct, though. First, $\vec{A}^+$ and $\vec{A}^-$ are related by a gauge transformation that is multivalued. Two configurations related by a multivalued gauge transformation are definitely not physically equivalent if the gauge group is $\mathbb{R}$. Second, and more important, no matter what the gauge group is, classically, $\vec{A}^+$ and $\vec{A}^-$ can be distinguished by a classical charged particle crossing the string singularity ($\vec{B}^+$ and $\vec{B}^-$ are indeed different).

However, quantum mechanically they *may* be completely equivalent if the gauge group is U(1), provided that the gauge function has the right periodicity. To analyze this problem we have to consider the quantum-mechanical coupling of the U(1) vector field to charged matter. Thus, let us consider the Schrödinger equation for a particle of mass $M$ and electric charge $q$ in an electromagnetic field:

$$H\Psi = i\hbar\frac{\partial}{\partial t}\Psi. \tag{12.162}$$

To obtain the Hamiltonian $H$ we start from the action for a massive relativistic particle in an electromagnetic background field, Eq. (12.54), which in the non-relativistic limit gives Eq. (12.55), from which, after subtracting the zero-point energy $Mc^2$, we can identify the non-relativistic Lagrangian $L$ and construct the classical Hamiltonian

$$H = \vec{P}\cdot\dot{\vec{X}} - L = \frac{1}{2M}\left[\vec{P} - \frac{q}{c}\vec{A}\right]^2 + q\phi, \qquad \vec{P} \equiv M\dot{\vec{X}} + \frac{q}{c}\vec{A}. \tag{12.163}$$

In the quantization of this system, the momentum $\vec{P}$ is replaced by the operator $-i\hbar\vec{\nabla}$. We obtain the Hamiltonian

$$H = -\frac{\hbar^2}{2M}\vec{D}^2 + q\phi, \tag{12.164}$$

where $\vec{D}$ is the *covariant derivative*,

$$\vec{D} = \vec{\nabla} - ie\vec{A}, \tag{12.165}$$

and the gauge coupling constant is

$$e = \frac{q}{\hbar c}. \tag{12.166}$$

Under a gauge transformation of the vector field $A'_\mu = A_\mu + \partial_\mu\Lambda$, the Hamiltonian Eq. (12.164) is not invariant, but its transformation can be compensated by the following gauge transformation of the wave function:

$$\Psi' = e^{-ie\Lambda}\Psi. \tag{12.167}$$

So the Schrödinger equation is gauge covariant (it changes by the above overall phase).

If the gauge group is $\mathbb{R}$ (i.e. $\Lambda \in \mathbb{R}$), $\Lambda$ has to be single-valued and the same must be true for the wave function. If the gauge group is U(1), however, $\Lambda$ lives in a circle, or equivalently in a lattice, and we have to identify two different values of $\Lambda$ differing by the period $T$,

$$\Lambda \sim \Lambda + T. \tag{12.168}$$

In a topologically trivial spacetime any closed path is contractible to a point. This implies that the wave function has to be single-valued around any closed path. This implies in turn that only gauge transformations such that the gauge phase $e^{-ie\Lambda}$ is single-valued around any closed path are allowed. Since we just admitted that $\Lambda$ can be multivalued with period $T$, we conclude that the only $T$s allowed are those satisfying

$$Te = 2\pi n, \qquad n \in \mathbb{Z}. \tag{12.169}$$

For application to the Dirac-monopole case, in which the space is topologically non-trivial ($\mathbb{R}^3$ minus the positive or negative $z$ axis) but has no non-contractible closed paths, we conclude that the gauge transformation that moves the Dirac string relates two quantum-mechanically equivalent configurations in which the wave function is single-valued if the gauge parameter has the right periodicity. If the two configurations are equivalent, in spite of the fact that they have Dirac strings in different places, then the Dirac strings have no physical effect. Going around the $z$ axis once gives

$$\Lambda(\varphi + 2\pi) = \Lambda + p, \tag{12.170}$$

so we find that we can do consistent quantum mechanics ignoring the Dirac string if the magnetic charge is related to the electric charge by the *Dirac quantization condition*[27] [455],

$$\boxed{qp = n2\pi\hbar c.} \tag{12.171}$$

It is worth remarking that this formula is invariant (up to a global sign) under electric–magnetic-duality transformations $q \to p, p \to -q$.

Using the normalization of Eq. (12.58), the definitions of electric and magnetic charge that satisfy the Dirac quantization condition in the above form (without any extra factors) are

$$\boxed{q \equiv \frac{1}{16\pi G_{\rm N}^{(4)}} \int_{S^2_\infty} \star F, \qquad p \equiv -\int_{S^2_\infty} F, \qquad pq = 2\pi n.} \tag{12.172}$$

In a non-simply connected spacetime there will be closed paths that are not contractible to a point (that is, it will have a non-trivial $\pi_1$). The wave function will not in general be single-valued around those closed paths but will pick up a phase, the *Aharonov–Bohm*

[27] There are other ways of finding this condition, such as studying the quantization of the angular momentum of the electromagnetic field created by the electric and magnetic particles. See, for instance, Ref.[638].

*phase* [20, 21], which can be detected by interference experiments. The Dirac quantization condition can be considered as the condition of cancelation of a would-be Aharonov–Bohm phase around the Dirac string, which physically is unacceptable. The concept of the Aharonov–Bohm phase is, however, much more general and deals with the non-triviality of the topology of the gauge field itself when it is seen as a section of a fiber bundle. To study the Aharonov–Bohm phase, thus, we first reformulate the Dirac monopole in this language.

### 12.7.3 The Wu–Yang monopole

Wu and Yang [1279] were the first to reformulate the Dirac monopole in the modern language. The basic idea is to generalize the basic concepts of tensors in manifolds to gauge fields:[28] a manifold is a topological space that in general is not isomorphic to $\mathbb{R}^n$. Thus it needs to be covered by patches that are isomorphic to parts of $\mathbb{R}^n$. Each patch provides a local coordinate system. Neighboring patches must overlap and the two different coordinates of points in the overlaps are related by diffeomorphisms. Now one can define tensor fields on a manifold. A given well-defined tensor field will have different components in the overlaps, corresponding to the different coordinate systems that are defined there, but they will be related by the tensor-transformation laws corresponding to the diffeomorphisms that relate the different coordinate systems.

Now, a gauge field defined on a manifold is a 1-form field, and its definitions in different patches will be related by the standard transformation rules of 1-forms under diffeomorphisms. The new freedom that we have in fiber bundles is that these different definitions can also be related by gauge transformations. The most basic example is precisely the Dirac monopole. The space manifold is just $\mathbb{R}^3$ and, in principle, we need only one patch to cover it. However, we are going to use two, because the topology of the gauge-field configuration requires it. The two patches will be the two halves of $\mathbb{R}^3$ with $z \geq 0$ and $z \leq 0$, which overlap over the plane $z = 0$. The two coordinate systems that we are going to use are trivially related and we will not distinguish them. The U(1) gauge field in the first patch $z \geq 0$ will be $A^+$, which is completely regular there (except at the origin) because its Dirac string lies in the second patch. In the second patch $z \leq 0$ the gauge field will be $A^-$, which is also regular there for analogous reasons. In the overlap $z = 0$ we have two different values of the gauge field, but they are related (by construction) by the gauge transformation Eq. (12.161). The discussion of which gauge transformations are allowed is still valid here and we arrive at the same Dirac quantization condition.

One of the advantages of this formulation is that, at the expense of introducing non-trivial topology for the gauge field, we have eliminated completely the Dirac string and have a completely regular gauge field (except at the origin). The magnetic field $\vec{B}$ is only singular at the origin, too. A calculation of the magnetic charge through the magnetic flux should now give the right result. First we rewrite the flux in differential-form language:

$$\int_{S^2} d\vec{S} \cdot \vec{B} = \int_{S^2} d\vec{S} \cdot \vec{\nabla} \times \vec{A} = \int_{S^2} dA. \tag{12.173}$$

[28] For a less-pedestrian explanation, there are many reviews and textbooks that the interested reader can consult; for instance Refs. [868, 356, 481, 972, 974].

We cannot use Stokes' theorem here because $A$ is multivalued. We divide the 2-sphere into two halves $\Sigma^{\pm}$ overlapping at the equator $z=0$. In each of these two halves, $A$ is single-valued and Stokes' theorem can be applied:

$$\int_{S^2} d\vec{S}\cdot\vec{B} = \int_{S^2} dA = \int_{\Sigma^+} dA^+ + \int_{\Sigma^-} dA^- = \int_{\gamma^+} A^+ + \int_{\gamma^-} A^-, \qquad (12.174)$$

where $\gamma^{\pm}$ are the boundaries of $\Sigma^{\pm}$: equatorial circumferences are oriented in the negative and positive $\varphi$ directions, so $\gamma^+ = -\gamma^-$. Using the relation between $A^+$ and $A^-$, we find

$$\int_{S^2} d\vec{S}\cdot\vec{B} = -\int_{\gamma^+} \partial_\varphi\left(\frac{p}{2\pi}\varphi\right) = p. \qquad (12.175)$$

The magnetic charge is given by the non-trivial monodromy of the gauge parameter.

The topology of gauge fields (fiber bundles) such as the monopole can be characterized by the values of topological invariants. In the case of the Abelian monopole it is the *first Chern class*,

$$c_1 = -\frac{1}{2\pi}\int_{S^2} F, \qquad (12.176)$$

which is nothing but the magnetic charge $p/(2\pi)$ and should be an integer $n$, according to general arguments. This result is stated in units in which $q=\hbar=c=1$ and then we see that this is nothing but the Dirac quantization condition.

### 12.7.4 *Dyons and the DSZ charge-quantization condition*

If objects with electric charge and objects with magnetic charge exist, then objects with both kinds of charges, called *dyons*, may exist. The electromagnetic field they produce is just a linear superposition of those produced by electric and magnetic monopoles.

Considering the quantum evolution of one dyon in the field of another dyon, it is found that consistency requires the four charges of these objects to obey the *Dirac–Schwinger–Zwanziger (DSZ) quantization condition* [1120, 1121, 1287, 1288]:

$$\boxed{q_1 p_2 - q_2 p_1 = n 2\pi\hbar c.} \qquad (12.177)$$

With the normalization and units of Eqs. (12.58) and (12.172) the condition takes the same form but with no $\hbar c$ constants.

Now, this condition is completely invariant under $\mathbb{Z}_2$ electric–magnetic-duality transformations. This can be more easily seen if we rewrite it in this very suggestive form using the charge vectors we introduced before:

$$\vec{q}_1^{\,\mathrm{T}}\eta\,\vec{q}_2 = \frac{n}{8G_N^{(4)}}, \qquad \vec{q} = \begin{pmatrix} q \\ p/(16\pi G_N^{(4)}) \end{pmatrix}. \qquad (12.178)$$

We saw that the Einstein equation could also be written using duality vectors and the matrix $\eta = i\sigma^2$ (Eq. (12.139)), which is the *metric* preserved by the symplectic group. The presence of that matrix implied that the duality group was a subgroup of $\mathrm{SL}(2,\mathbb{R}) \sim \mathrm{Sp}(2,\mathbb{R})$. Now we obtain the same result from the DSZ quantization condition, which is invariant under $\mathrm{Sp}(2,\mathbb{R})$ transformations.[29]

This condition does not take into account all the quantum effects, such as the quantization of electric charge (independently of any magnetic-monopole charge). These effects will break the classical duality group to some discrete subgroup, but will not change the DSZ quantization condition.

*Inclusion of a theta angle and the Witten effect.* The Einstein–Maxwell action can be modified by the addition of a *topological term* of the form

$$-\frac{\theta}{8\pi c}\int d^4x\,\sqrt{|g|}\;F\;\star F, \tag{12.180}$$

or, in differential-form language,

$$-\frac{\theta}{4\pi c}\int F\wedge F, \tag{12.181}$$

where we see that the metric does not appear in it, which is the reason why it is called topological.

On the other hand, using the Bianchi identity, the integrand of this term can be shown to be the total derivative of the *Chern–Simons 3-form* $F\wedge A$,

$$F\wedge F = d(A\wedge F), \tag{12.182}$$

and therefore it does not contribute to the classical equations of motion. However, a change in the Lagrangian produces a change in the Noether current and in the definition of the corresponding conserved charge. To make this more concrete, let us consider the Einstein–Maxwell Lagrangian Eq. (12.58) with the $\theta$-term and coupling constant $g$ with our conventions and units:

$$S[g_{\mu\nu},A_\mu] = \frac{1}{16\pi G_{\mathrm{N}}^{(4)}}\int d^4x\sqrt{|g|}\left[R - \frac{1}{4g^2}F^2 - \frac{\theta}{8\pi}F\;\star F\right]. \tag{12.183}$$

The Noether current associated with the gauge transformations of the vector field is now

$$j_{\mathrm{N}}^{\mu} = \frac{1}{16\pi G_{\mathrm{N}}^{(4)}}\nabla_\nu\left[\frac{1}{g^2}F^{\nu\mu} + \frac{\theta}{2\pi}\star F^{\nu\mu}\right]. \tag{12.184}$$

---

[29] If there are $\bar{n}$ Maxwell fields $A^{\Lambda}{}_{\mu}$, the electric $q_\Lambda$ and magnetic $p^\Lambda$ charges (setting $/16\pi G_{\mathrm{N}}^{(4)} = 1$) can be arranged in a vector $(\mathcal{Q}^M) = \begin{pmatrix} p^\Lambda \\ q_\Lambda \end{pmatrix}$ that transforms exactly as the vector $(\mathcal{F}^M)$ of field strengths introduced in Eq. (2.161), that is, linearly under $\mathrm{Sp}(2\bar{n},\mathbb{R})$. Then, the DSZ quantization condition can be written as the manifestly symplectic-invariant form

$$\boxed{\mathcal{Q}_{1\,M}\mathcal{Q}_2^M = q_{1\,\Lambda}p_2^\Lambda - q_{2\,\Lambda}p_1^\Lambda = n2\pi\hbar c.} \tag{12.179}$$

This result takes exactly the same form in the presence of scalar fields, using the definitions in Section 2.6.

In this simple Abelian case that we are considering, the second term vanishes by virtue of the Bianchi identity. Still we will keep it and, after using Stokes' theorem in the definition of electric charge, the second term gives a net contribution

$$q \equiv \frac{1}{16\pi G_{\rm N}^{(4)}} \int_{S^2_\infty} \left[\frac{1}{g^2} \star F - \frac{\theta}{2\pi} F\right]. \tag{12.185}$$

The magnetic charge is still given by Eq. (12.172).

If we start with a magnetic monopole in a vacuum with $\theta = 0$ and then "switch on" $\theta$, we see in the above formulae that the magnetic monopole acquires an electric charge proportional to $\theta$ and becomes a dyon. This is the *Witten effect* [1270].

We studied the classical duality group when we introduced the coupling constant $g$ and allowed it to transform under it. It is interesting to see what happens after we introduce $\theta$ and we allow it to transform as well. This can be seen more easily if we redefine the duality vector

$$\vec{F} \equiv \begin{pmatrix} \frac{1}{g^2} F + \frac{\theta}{2\pi} \star F \\ \star F \end{pmatrix}, \tag{12.186}$$

whose two components $F^1$ and $F^2$ are subject to the constraint

$$F^1 = \frac{\theta}{2\pi} F^2 - \frac{1}{g^2} \star F^2, \tag{12.187}$$

and define the complexified coupling constant

$$\tau = \frac{\theta}{2\pi} + \frac{i}{g^2}. \tag{12.188}$$

In terms of the new duality vector, the Einstein equation still has the form (12.139) and we see that any $\mathrm{SL}(2,\mathbb{R})$ transformation leaves it invariant. Furthermore, we can see that the $\mathrm{SL}(2,\mathbb{R})$-transformed duality vector has the same form as the original one but with $\tau$ transformed as follows: if the $\mathrm{SL}(2,\mathbb{R})$ transformation is

$$\tilde{\vec{F}} = \begin{pmatrix} \alpha & \beta \\ \gamma & \delta \end{pmatrix} \vec{F}, \qquad \alpha\delta - \beta\gamma = 1, \tag{12.189}$$

then the complexified coupling constant transforms simultaneously as follows:

$$\tilde{\tau} = \frac{\alpha\tau + \beta}{\gamma\tau + \delta}. \tag{12.190}$$

So, the full set of equations of motion is invariant under $\mathrm{SL}(2,\mathbb{R})$-duality transformations.

Furthermore, in the presence of a $\theta$-term, the electric and magnetic charges naturally fit into a duality vector, which is the integral of the Hodge dual of the duality vector of the 2-form field strengths defined above, with a $1/(16\pi G_{\rm N}^{(4)})$ normalization factor. The DSZ quantization condition still takes the form Eq. (12.178) (the $\theta$-dependent terms drop out from it) and we see that it is fully form invariant under $\mathrm{SL}(2,\mathbb{R})$ transformations. If the electric charge were quantized, $q \in \mathbb{Z}$, it is clear that only the discrete subgroup $\mathrm{SL}(2,\mathbb{Z})$ would preserve its quantization and the spectrum of charged particles (generically dyons characterized by their charge vectors). This is the general S duality group, and it will appear in different forms in many places in what follows.

### 12.7.5 Duality in massive electrodynamics

To acquire some training in the use of the Poincaré-duality procedure explained in Section 12.7.1 in more general settings than that of Maxwell's theory, it is interesting to consider the dualization of the Proca Lagrangian rewritten using the Stückelberg scalar in Eq. (12.67), which seems to have no electric–magnetic-duality *symmetry*. We first rewrite it in this form:

$$S[A_\mu, \phi] = \int d^4x \sqrt{|g|}\left[\tfrac{1}{2}G^2 - \tfrac{1}{4}F^2\right], \tag{12.191}$$

where $G$ and $F$ are the scalar and vector gauge-invariant field strengths

$$G_\mu = \partial_\mu \phi + mA_\mu, \qquad F_{\mu\nu} = 2\partial_{[\mu}A_{\nu]}. \tag{12.192}$$

To the equations of motion of this system one can now add a Bianchi identity for $G_\mu$:

$$\partial_{[\mu}(G_{\nu]} - mA_{\nu]}) = 0. \tag{12.193}$$

However, there is no duality *symmetry* because the dual of a 1-form field strength is a 3-form field strength. Nevertheless, we can perform a duality *transformation* to an equivalent system with a 3-form field strength. In other words, we can apply the Poincaré-duality procedure to the scalar (only its derivatives appear in the action), replacing it by a 2-form potential. Following the general dualization procedure, we want to find an equivalent action that is a functional of the field strength $G_\mu$ instead of the scalar $\phi$. Thus, we add to the above action a Lagrange-multiplier term enforcing the Bianchi identity for $G$,

$$\tfrac{1}{2}\int dx^4 \epsilon^{\mu\nu\rho\sigma}\partial_\mu B_{\nu\rho}(G_\sigma - mA_\sigma), \tag{12.194}$$

where we have already integrated by parts. The new action is a functional of $G_\mu$, $A_\mu$, and $B_{\mu\nu}$. The equation of motion for $G_\mu$ is

$$G = \star H, \qquad H_{\mu\nu\rho} = 3\partial_{[\mu}B_{\nu\rho]}, \tag{12.195}$$

where $H$ is the field strength of the 2-form $B$, which is invariant under the gauge transformations $\delta B_{\mu\nu} = \partial_{[\mu}\Lambda_{\nu]}$. On substituting this back into the action and integrating again by parts, we obtain an action that is a functional of the fields $A_\nu$ and $B_{\mu\nu}$:

$$S[A_\mu, B_{\mu\nu}] = \int dx^4 \sqrt{|g|}\left[\frac{1}{2\cdot 3!}H^2 - \tfrac{1}{4}F^2 + \frac{m}{4}\frac{\epsilon}{\sqrt{|g|}}FB\right]. \tag{12.196}$$

We have completely dualized $\phi$ into $B_{\mu\nu}$. Now, $A_\mu$ does not occur explicitly any longer in this action, but only through its field strength, and thus we can now Poincaré-dualize with respect to it. By adding a term

$$\int d^4x \, \tfrac{1}{2}\epsilon \, \partial\tilde{A}F, \tag{12.197}$$

and eliminating $F$ through its equation of motion

$$F = \star\tilde{F}, \qquad \tilde{F} = 2\Big(\partial\tilde{A} + \frac{m}{2}B\Big), \tag{12.198}$$

we obtain the action dual to the original:

$$S[\tilde{A}_\mu, B_{\mu\nu}] = \int d^4x \sqrt{|g|}\left[\frac{1}{2\cdot 3!}H^2 - \tfrac{1}{4}\tilde{F}^2\right]. \tag{12.199}$$

The dual vector field strength is now invariant under *dual massive gauge transformations*

$$\delta B_{\mu\nu} = \partial_{[\mu}\Lambda_{\nu]}, \qquad \delta\tilde{A}_\mu = -\frac{m}{2}\Lambda_\mu, \tag{12.200}$$

which allow us to eliminate $\tilde{A}$ completely, leaving us with a massive 2-form. $\tilde{A}$ now plays the role of the Stückelberg field for $B$.

The relation between the dual and original variables is

$$\begin{aligned} H &= -\star G, \\ \tilde{F} &= -(\star F + mB). \end{aligned} \tag{12.201}$$

## 12.8 Magnetic and dyonic RN black holes

We have seen that the full set of equations of motion of the Einstein–Maxwell system without a $\theta$-term and without the introduction of any coupling constant is invariant under the SO(2) group of electric–magnetic duality,

$$\tilde{F} = \cos(\xi)F + \sin(\xi) \star F, \qquad \star\tilde{F} = -\sin(\xi)F + \cos(\xi) \star F. \tag{12.202}$$

Duality symmetries can be used as solution-generating transformations. For instance, we generated new solutions for a theory with $N$ vector fields from the 1-vector RN solution using the O($N$) duality symmetry that rotates the vector fields. We can now do the same to generate new solutions with both electric and magnetic charges out of the purely electric RN or MP solutions. Let us take the single electric RN BH solution as given in Eqs. (12.75). Trivially we obtain a new solution with the same metric and with

$$\tilde{F}_{tr} = \frac{4G_{\rm N}^{(4)}\cos(\xi)q}{r^2}, \qquad \tilde{F}_{\theta\varphi} = 4G_{\rm N}^{(4)}\sin(\xi)q\,\sin\theta. \tag{12.203}$$

After the new solution has been found, it has to be expressed in terms of the new physical parameters $\tilde{q}$ and $\tilde{p}$, which turn out to be related to the old ones as follows:

$$\tilde{q} = \cos(\xi)q, \quad \tilde{p} = -16\pi G_{\rm N}^{(4)}\sin(\xi)q, \quad \Rightarrow |\tilde{\vec{q}}|^2 = \tilde{q}^2 + \left(\frac{\tilde{p}}{16\pi G_{\rm N}^{(4)}}\right)^2 = q^2. \tag{12.204}$$

The last equation is due to the fact that SO(2) leaves the norm of the charge vector $\vec{q}$ invariant. In the metric, we need only replace $q^2$ everywhere by $|\tilde{\vec{q}}|^2$. In the vector-field strength the other two equations have to be used to replace $q$ and $\xi$ by $\tilde{q}$ and $\tilde{p}$. The result is (now suppressing tildes)

$$
\begin{aligned}
ds^2 &= f(r)dt^2 - f^{-1}(r)dr^2 - r^2 d\Omega^2_{(2)},\\
F_{tr} &= \frac{4G_{\rm N}^{(4)} q}{r^2}, \qquad F_{\theta\varphi} = -\frac{1}{4\pi} p \sin\theta,\\
f(r) &= \frac{(r-r_+)(r-r_-)}{r^2},\\
r_\pm &= G_{\rm N}^{(4)} M \pm r_0, \qquad r_0 = G_{\rm N}^{(4)} \sqrt{M^2 - 4|\vec{q}|^2}.
\end{aligned}
\tag{12.205}
$$

This is a dyonic RN BH. The metric is essentially the same as that of the purely electric one with the replacement $q^2 \to |\vec{q}|^2$ and most of its properties are also essentially identical.

Starting with the MP solutions, we find the dyonic MP solutions

$$
\begin{aligned}
ds^2 &= H^{-2} dt^2 - H^2 d\vec{x}_3^{\,2},\\
F_{t\underline{i}} &= -2\cos\alpha\, \partial_{\underline{i}} H^{-1}, \qquad F_{\underline{i}\underline{j}} = 2\sin\alpha\, \epsilon_{ijk} \partial_{\underline{k}} H,\\
\partial_{\underline{i}}\partial_{\underline{i}} H &= 0.
\end{aligned}
\tag{12.206}
$$

From the point of view of finding new solutions, the important lesson to be learned is that we have generated a new solution with one more physical parameter (the magnetic charge) using a one-parameter solution-generating transformation group. Observe that, in the MP case, the family of solutions depends on only one arbitrary real harmonic function.

We could view these solutions and, in particular, the dyonic RN solution, as solutions of the more general theory with $g = 1$ and $\theta = 0$ and we can try to generate solutions of the more general theory using general $\mathrm{SL}(2,\mathbb{R})$ transformations. These have three independent parameters, but we have already used the one corresponding to SO(2). The other two parameters would precisely generate non-trivial values of $g$ and $\theta$. Let us obtain these RN solutions.

First we use on the dyonic RN solution Eq. (12.205) $\mathrm{SL}(2,\mathbb{R})$ rescalings, corresponding to matrices of the form

$$
\begin{pmatrix} a & 0 \\ 0 & 1/a \end{pmatrix}.
\tag{12.207}
$$

They generate a non-trivial $\tilde{g} = a$ and rescale the electric and magnetic charges and field strength. The transformed solution, written in terms of the transformed parameters $q, p$, and

$g$ (without the tildes), takes the form

$$\begin{aligned}
ds^2 &= f(r)dt^2 - f^{-1}(r)dr^2 - r^2 d\Omega^2_{(2)}, \\
F_{tr} &= \frac{4G_N^{(4)} gq}{r^2}, \qquad F_{\theta\varphi} = -\frac{1}{4\pi g} p \sin\theta, \\
f(r) &= \frac{(r-r_+)(r-r_-)}{r^2}, \\
r_\pm &= G_N^{(4)} M \pm r_0, \qquad r_0 = G_N^{(4)} \sqrt{M^2 - 4\vec{q}^{\,\mathrm{T}} \mathcal{M}^{-1} \vec{q}},
\end{aligned} \tag{12.208}$$

where $\mathcal{M}$ is the matrix

$$\mathcal{M} = \begin{pmatrix} 1/g^2 & 0 \\ 0 & g^2 \end{pmatrix}, \qquad \mathcal{M}^{-1} = \begin{pmatrix} g^2 & 0 \\ 0 & 1/g^2 \end{pmatrix}. \tag{12.209}$$

Now we use the $\mathrm{SL}(2,\mathbb{R})$ transformations that shift the $\theta$-parameter from its zero value, corresponding to matrices of the form

$$\begin{pmatrix} 1 & b \\ 0 & 1 \end{pmatrix}. \tag{12.210}$$

They generate a non-trivial $\theta/(2\pi) = b$ and mix different components of the field strength and the electric and magnetic charges (the Witten effect). The transformed solution, written in terms of the transformed parameters $q, p, g$, and $\theta$ (without the tildes), takes the form[30]

$$\begin{aligned}
ds^2 &= f(r)dt^2 - f^{-1}(r)dr^2 - r^2 d\Omega^2_{(2)}, \\
\vec{F}_{tr} &= \frac{4G_N^{(4)} \vec{q}}{gr^2}, \\
f(r) &= \frac{(r-r_+)(r-r_-)}{r^2}, \\
r_\pm &= G_N^{(4)} M \pm r_0, \qquad r_0 = G_N^{(4)} \sqrt{M^2 - 4\vec{q}^{\,\mathrm{T}} \mathcal{M}^{-1} \vec{q}},
\end{aligned} \tag{12.211}$$

where $\mathcal{M}$ is now the matrix

$$\mathcal{M} = g^2 \begin{pmatrix} |\tau|^2 & \theta/(2\pi) \\ \theta/(2\pi) & 1 \end{pmatrix}, \qquad \mathcal{M}^{-1} = g^2 \begin{pmatrix} 1 & -\theta/(2\pi) \\ -\theta/(2\pi) & |\tau|^2 \end{pmatrix}, \tag{12.212}$$

which has interesting properties: it belongs to $\mathrm{SL}(2,\mathbb{R})$ but it is symmetric. It can be seen that it parametrizes the $\mathrm{SL}(2,\mathbb{R})/\mathrm{SO}(2)$ cosets. Furthermore, under an $\mathrm{SL}(2,\mathbb{R})$ transformation $\Lambda$, it transforms (due to the transformation of $g$ and $\theta$) according to

$$\mathcal{M} = \Lambda \mathcal{M} \Lambda^{\mathrm{T}}, \tag{12.213}$$

[30] Giving the $tr$ components of the two components of the duality vector is equivalent to, but much simpler than, giving the $tr$ and $\theta\varphi$ components of $F$.

so $\vec{q}^{\,\mathrm{T}}\mathcal{M}^{-1}\vec{q}$ is form invariant under $\mathrm{SL}(2,\mathbb{R})$ transformations. Thus, using $\mathrm{SL}(2,\mathbb{R})$ duality, we cannot generate any new solutions not yet contained in the above family.

The result is that we have generated a family of solutions that contains three parameters more than the initial one by using a three-dimensional duality group. The solutions are expressed in the simplest form when one uses objects that have good transformation properties under the duality group: duality vectors and matrices. On the other hand, the family covers the most general BH-type solution of the Einstein–Maxwell theory that one can have according to the no-hair conjecture: the BH solution depends on only two conserved charges (electric and magnetic) and two moduli parameters, which are not really characteristic of the BH but rather of the vacuum of the theory.

This example may look quite simple, but it has the same features as some more complicated and juicy cases.

To end this section, let us comment on a couple of subtle points.

- Electric–magnetic-duality rotations and the Wick rotation do not commute. Although we did not stress it, the Euclidean electric RN solution has a purely imaginary electromagnetic field. Electric–magnetic-duality rotations of the Euclidean purely electric RN solution generate a Euclidean solution with imaginary magnetic charge that remains imaginary when we Wick-rotate back to the Lorentzian signature. If we Wick-rotate the dyonic RN solution, we obtain a Euclidean solution with real magnetic charge. This gives rise to problems in the calculation of the entropy in the Euclidean path-integral formalism,[31] but they can be dealt with, as shown in Refs. [266, 734, 440, 434].
- In the extreme magnetic RN BH case, we could also try to look for a source. However, the only thing that works is to view the magnetic charge as the electric charge of the dual vector field.

## 12.9 Higher-dimensional RN solutions

Just as there are higher-dimensional analogs of the Schwarzschild BH, there are also higher-dimensional analogs of the electric RN BH, which are solutions of the equations of motion obtained from considering the Einstein–Maxwell action in $d$ dimensions.

Let us first consider the higher-dimensional generalization of the Einstein-scalar system that we considered at the beginning of this chapter:

$$\boxed{S[g_{\mu\nu},\varphi] = \frac{c^3}{16\pi G_{\mathrm{N}}^{(d)}}\int d^dx\sqrt{|g|}\,\left[R + 2\partial_\mu\varphi\partial^\mu\varphi\right].} \tag{12.214}$$

It is natural to ask whether the no-hair conjecture that says that there are no regular BH-type solutions of that system in four dimensions holds in more than four dimensions. Thus, we

[31] The thermodynamical quantities that one derives from the Lorentzian metric of the dyonic RN solution are clearly S-duality invariant.

can try to find static, spherically symmetric BH solutions of this system. For the metric, we will use the straightforward generalization of the ansatz of the "dressed Schwarzschild metric" form Eq. (12.91) that we found for the four-dimensional RN solution (and that is also valid for the four-dimensional solutions of this system, Eqs. (12.7)):

$$ds^2 = H^{2x} W dt^2 - H^{-2y}\left[W^{-1}dr^2 + r^2 d\Omega^2_{(d-2)}\right], \tag{12.215}$$

where $W$ will be a function of the form

$$W = 1 + \frac{\omega}{r^{d-3}}, \tag{12.216}$$

and where $H$ is related to the scalar by

$$\varphi = \varphi_0 + z \ln H, \tag{12.217}$$

$z$ being a constant, so, when the scalar becomes constant, the above metric is just the higher-dimensional Schwarzschild metric. It is easy to see that we are forced to set $H = W$ and $y \neq 0$ in order to have a solution. This implies that the would-be "horizon" is always singular, except when the scalar is constant. For the sake of completeness we give the form of these solutions, which generalize those obtained in Refs. [827, 18]

$$\begin{aligned}
ds^2 &= W^{\frac{M}{\omega}-1} W dt^2 - W^{\frac{1}{d-3}\left(1-\frac{M}{\omega}\right)}\left[W^{-1}dr^2 + r^2 d\Omega^2_{(d-2)}\right],\\
\varphi &= \varphi_0 \pm \frac{\Sigma}{\omega}\ln W,\\
W &= 1 + \frac{\omega}{r^{d-3}}, \qquad \omega = \pm 2\sqrt{M^2 + 2\left(\frac{d-3}{d-2}\right)\Sigma^2}.
\end{aligned} \tag{12.218}$$

For $\Sigma = 0$ we recover the $d$-dimensional Schwarzschild solution. In all other cases we have metrics with naked singularities at $r = 0$ or at $r^{d-3} = -\omega$ (if possible).

Now let us return to the higher-dimensional Einstein–Maxwell system, normalized as in Eq. (12.58) ($c = 1$),

$$S[g_{\mu\nu}, A_\mu] = \frac{1}{16\pi G_N^{(d)}}\int d^d x\sqrt{|g|}\,\left[R - \tfrac{1}{4}F^2\right]. \tag{12.219}$$

The Einstein and Maxwell equations are

$$G_{\mu\nu} - \tfrac{1}{2}T_{\mu\nu} = 0, \qquad \nabla_\mu F^{\mu\nu} = 0, \tag{12.220}$$

where the electromagnetic energy–momentum tensor $T_{\mu\nu}$ is again given by Eq. (12.24).

There are a few differences from the four-dimensional case. First we observe that, in more than four dimensions, the energy–momentum tensor of the Maxwell field is no longer traceless because the Maxwell action is not invariant under Weyl rescalings of the metric. This implies that, in general, the curvature scalar is not zero on solutions, but, instead

$$R = \frac{d-4}{4(d-2)} F^2, \tag{12.221}$$

and, thus, on subtracting the trace in the Einstein equation, we are now left with the equation

$$R_{\mu\nu} - \tfrac{1}{2}\left[F_\mu{}^\rho F_{\nu\rho} - \frac{1}{2(d-2)} g_{\mu\nu} F^2\right] = 0 \tag{12.222}$$

plus the Maxwell equation to solve.

The second difference is the definition of the electric charge (we treated the definition of the mass in higher-dimensional spaces in the Schwarzschild case). If we follow exactly the same steps as in the four-dimensional case, we arrive at

$$\boxed{q = (-1)^d \frac{1}{16\pi G_N^{(d)}} \int_{S^{(d-2)}_\infty} \star F,} \tag{12.223}$$

where $\star F$ is now a $(d-2)$-form and $S^{(d-2)}_\infty$ is a $(d-2)$-sphere at spatial infinity (constant $t$, $r \to \infty$). This means that, if there is a charge $q$ at the origin in an asymptotically flat spacetime, the asymptotic behavior of $F$ and the vector $A$ is given by

$$F_{tr} \sim \frac{16\pi G_N^{(d)} q}{\omega_{(d-2)}} \frac{1}{r^{d-2}}, \qquad A_\mu \sim -\delta_{\mu t} \frac{16\pi G_N^{(d)} q}{(d-3)\omega_{(d-2)}} \frac{1}{r^{d-3}}, \tag{12.224}$$

where $\omega_{(d-2)}$ is the volume of the unit $(d-2)$-sphere (see Appendix K). In $d \neq 4$ one can perform an electric–magnetic-duality transformation, replacing $F$ by its Hodge dual $\tilde{F} = \star F$, which is a $(d-2)$-form field strength for a $(d-3)$-form potential $\tilde{F} = d\tilde{A}$. This transformation is not a symmetry. Now, we can define the electric charge associated with the dual $(d-3)$-form potential, which is what we would define as magnetic charge, by analogy with the four-dimensional case. However, the carrier of the electric charge of the dual $(d-3)$-form potential cannot be a point-like particle, but has to be a $(d-4)$-dimensional extended object (*brane*). Thus, a standard BH of the kind we are interested in now cannot carry that kind of charge and we will not consider it here, although we will in Part III.

Our immediate goal is, then, to find $d$-dimensional analogs of the RN BH. Again, we use an ansatz of the "dressed Schwarzschild metric" form:

$$\begin{aligned} ds^2 &= H^{2a} W dt^2 - H^{-2b}\left[W^{-1} dr^2 + r^2 d\Omega^2_{(d-2)}\right], \\ A_\mu &= \alpha \delta_{\mu t}(H^{-1} - 1), \qquad H = 1 + \frac{h}{r^{d-3}}, \end{aligned} \tag{12.225}$$

where $a, b, h,$ and $\alpha$ are constants to be found. The electric charge is proportional to $h$ and, thus, we expect that, when it vanishes, $h$ becomes zero ($H = 1$) and we recover the higher-dimensional Schwarzschild metric Eq. (11.81), so we can guess that

$$W = 1 + \frac{\omega}{r^{d-3}}. \tag{12.226}$$

On the other hand, with this ansatz, the metric will have two horizons at $r = -h, -\omega$ when both $h$ and $\omega$ are non-vanishing. When $\omega$ vanishes ($W = 1$) there is only one horizon and this should correspond to the extreme limit. In this case, the above metric becomes isotropic and we should be able to find whether $H$ becomes a harmonic function and multi-BH solutions exist.

On substituting into the equations of motion, we find the $d$-dimensional RN solutions:

$$\boxed{\begin{aligned} ds^2 &= H^{-2} W dt^2 - H^{\frac{2}{d-3}} \left[W^{-1} dr^2 + r^2 d\Omega^2_{(d-2)}\right], \\ A_\mu &= \delta_{\mu t} \alpha (H^{-1} - 1), \\ H &= 1 + \tfrac{h}{r^{d-3}}, \qquad W = 1 + \tfrac{\omega}{r^{d-3}}, \\ \omega &= h\left[1 - \tfrac{d-3}{2(d-2)} \alpha^2\right]. \end{aligned}} \tag{12.227}$$

By examining the asymptotic behavior of the metric and vector field, we can relate the integration constants $h, \omega$, and $\alpha$ to the mass $M$ and the electric charge $q$ as follows:

$$\alpha = \frac{16\pi G_N^{(d)}}{(d-3)\omega_{(d-2)}} \frac{q}{r_\pm^{d-3}}, \qquad h = r_\pm^{d-3}, \qquad \omega = \pm r_0^{d-3}, \tag{12.228}$$

where now

$$r_\pm^{d-3} = \frac{8\pi G_N^{(d)}}{(d-2)\omega_{(d-2)}} M \pm r_0^{d-3}, \qquad r_0^{d-3} = \frac{8\pi G_N^{(d)}}{(d-2)\omega_{(d-2)}} \sqrt{M^2 - \frac{2(d-2)}{d-3} q^2}. \tag{12.229}$$

If we take the lower signs, we obtain a BH solution very similar to the four-dimensional RN solution: if $r_0^{d-3}$ is real and finite (we take it positive) and $M$ is positive, there is an event horizon at $r = r_0$ and a Cauchy horizon at $r = 0$. The reality of $r_0^{d-3}$ implies a lower bound for the mass,

$$M \geq \sqrt{\frac{2(d-2)}{d-3}} |q|. \tag{12.230}$$

When this bound is saturated, $r_0 = 0$ ($\omega = 0$) and there is only one horizon, which is regular, and we are in the extreme limit. Furthermore, if we set $W = 1$ in the ansatz, the equations of motion are solved by any arbitrary harmonic function in $(d-1)$-dimensional Euclidean space $H$:

$$\boxed{\begin{aligned} ds^2 &= H^{-2} dt^2 - H^{\frac{2}{d-3}} d\vec{x}^{\,2}_{d-1}, \\ A_\mu &= \delta_{\mu t} \alpha (H^{-1} - 1), \qquad \alpha = \pm 2, \\ \partial_{\underline{i}} \partial_{\underline{i}} H &= 0. \end{aligned}} \tag{12.231}$$

These solutions are the generalization of the MP solutions [969]. The $d$-dimensional ERN solution is a particular case of the $d$-dimensional MP family, which, evidently, contains also multi-BH solutions. If we take the near-horizon limit of the $d$-dimensional ERN solution, we find, after the coordinate change $r^{d-3} = (d-3)h\rho,\ t \to t/h^{\frac{1}{d-3}}$, a generalization of the RB solution,

$$\boxed{\begin{aligned} ds^2 &= \left(\frac{\rho}{R}\right)^2 dt^2 - \left(\frac{R}{\rho}\right)^2 d\rho^2 - h^{\frac{2}{d-3}} d\Omega^2_{(d-2)}, \\ A_\mu &= \delta_{\mu t}\alpha\rho, \end{aligned}} \tag{12.232}$$

where

$$R = \frac{h^{\frac{1}{d-3}}}{d-3}. \tag{12.233}$$

The metric is that of the direct product $\mathrm{AdS}_2 \times \mathrm{S}^{d-2}$.

# 13
# The Taub–NUT solution

The asymptotically flat, static, spherically symmetric Schwarzschild and RN BH solutions that we have studied in Chapters 11 and 12 were the only solutions of the Einstein and Einstein–Maxwell equations with those properties. To find more solutions, we have to relax these conditions or couple to gravity more general types of matter, as we will do later on. If we stay with the Einstein(–Maxwell) theory, one possibility is to look for static, axially symmetric solutions and another possibility is to relax the condition of staticity and only ask that the solution be stationary, which implies that we have to relax the condition of spherical symmetry as well and look for stationary, axisymmetric spacetimes. In the first case one finds solutions like those in Weyl's family [1249, 1250], which can be interpreted as describing the gravitational fields of axisymmetric sources with arbitrary multipole momenta[1], or Melvin's solution [944] (which has cylindrical symmetry and was constructed earlier by Bonnor [229] via a Harrison transformation [710] of the vacuum), among many others. In the second case, we find the Kerr–Newman BHs [850, 980] with angular momentum and electric or magnetic charge and also the Taub–Newman–Unti–Tambourino (Taub–NUT) solution [1170, 981], which may but need not include charges. The Taub–NUT metric does not describe a BH because it is not asymptotically flat. In fact, the only stationary, axially symmetric BHs of the Einstein–Maxwell theory belong to the Kerr–Newman family of solutions (see e.g. Refs. [749, 748]).

The Taub–NUT solution has a number of features that are particularly interesting for us, which we are going to discuss in this chapter. In particular, it carries a new type of charge (*NUT charge*), which is of topological nature and can be viewed as "gravitational magnetic charge," so the solution is a sort of *gravitational dyon* and its Euclidean continuation (for certain values of the mass and NUT charge) is the solution known in other contexts as a *Kaluza–Klein (KK) monopole*. This is a very important solution with interesting properties such as the self-duality of its curvature and its relation to the Belavin–Polyakov–Schwarz–Tyupkin (BPST) SU(2) instanton and the Wu–Yang SU(2) monopole. In Chapter 15 we will study how it arises in KK theory. Here we will describe it as a self-dual gravitational instanton and we will take the opportunity to mention other gravitational instantons.

[1] For a review, see Ref. [1064].

The charged Taub–NUT solutions will help us to introduce a very large and interesting family of solutions: the Israel–Wilson–Perjés (IWP) solutions, which have very important properties from the point of view of supersymmetry and duality.

## 13.1 The Taub–NUT solution

General stationary, axially symmetric metrics have only two Killing vectors, $k = \partial_t$ and $m = \partial_\varphi$, that generate time translations and rotations around the symmetry axis ($z$). These two Killing vectors are not mutually orthogonal, which implies that the off-diagonal component of the metric $g_{t\varphi} = k^\mu m_\mu$ does not vanish (otherwise we would have a static spacetime). Furthermore, the components of the metric can depend on the other coordinates, which we call $r$ and $\theta$ in $d = 4$. A general ansatz for these spacetimes has the form

$$ds^2 = g_{tt}dt^2 + 2g_{t\varphi}dtd\varphi + g_{rr}dr^2 + g_{\theta\theta}d\theta^2 + g_{\varphi\varphi}d\varphi^2, \tag{13.1}$$

where all the components may depend on $r$ and $\theta$. The new interesting ingredient is the component $g_{t\varphi}(r,\theta)$. If the metric is asymptotically flat for $r \to \infty$ and $g_{t\varphi}(r,\theta)$ has the asymptotic behavior

$$g_{t\varphi} \sim 2J\frac{\sin^2\theta}{r}, \tag{13.2}$$

then the solution describes a spacetime with angular momentum $J$ in the direction of the $z$ axis. The only vacuum solution of this kind is Kerr's [850], which in Boyer–Lindquist coordinates takes the form

$$\boxed{\begin{aligned} ds^2 &= \left(1 - \frac{2Mr}{\Sigma}\right)dt^2 + 2\frac{2aMr\sin^2\theta}{\Sigma}dtd\varphi - \frac{\Sigma}{\Delta}dr^2 - \Sigma d\theta^2 - \frac{\mathcal{A}}{\Sigma}\sin^2\theta\, d\varphi^2, \\ \mathcal{A} &= \Sigma(r^2 + a^2) + 2Mra^2\sin^2\theta, \\ \Sigma &= r^2 + a^2\cos^2\theta, \qquad \Delta = r^2 - 2Mr + a^2, \end{aligned}} \tag{13.3}$$

where $a = J/M$. If $M^2 \geq a^2$ this solution describes rotating BHs with mass $M$ and angular momentum $J = Ma$. The event horizon is placed at $r = r_+ = M + \sqrt{M^2 - a^2}$ (the larger value of $r$ for which $\Delta = 0$). When $a = 0$ we recover the Schwarzschild solution. Observe that, if we take $M \to 0$, keeping $a$ finite, we also obtain Minkowski spacetime, as opposed to the limit $M \to 0$ with finite $q$ in the RN case. If $M^2 < a^2$ the solution describes naked singularities. This resembles what happens in the RN case. Here we can think that a star with a large enough angular momentum cannot undergo spontaneous gravitational collapse and so the Kerr solutions with $M^2 < a^2$ and naked singularities never arise, according to the cosmic-censorship conjecture.

The Kerr solution for $r > r_+$ is not the metric of any known rotating body: there is no known "interior Kerr solution" as in the Schwarzschild case. Instead, such spacetimes are produced by certain rotating-disk sources (see Section 6.2 of Ref. [209] for a short review with references). However, the Kerr solution describes *all* isolated, rotating, uncharged BHs.

More details on the Kerr solutions can be found in most standard textbooks on GR and in the monograph [1002]. Our subject now is the Taub–NUT solution.

If, asymptotically,

$$g_{t\varphi} \sim 2N\cos\theta, \tag{13.4}$$

the solution describes an object with *NUT charge* $N$. We will discuss soon the meaning of this new charge, for which there is no Newtonian analog. The simplest vacuum solution with this kind of charge is the Taub–NUT solution [1170, 981]

$$\boxed{\begin{aligned} ds^2 &= f(r)(dt + 2N\cos\theta\, d\varphi)^2 - f^{-1}(r)dr^2 - \left(r^2+N^2\right)d\Omega^2_{(2)}, \\ f(r) &= \frac{(r-r_+)(r-r_-)}{r^2+N^2}, \\ r_\pm &= M \pm r_0, \qquad r_0^2 = M^2 + N^2, \end{aligned}} \tag{13.5}$$

which is a generalization of the Schwarzschild solution with NUT charge, and reduces to it when $N = 0$.

Let us list some immediate properties of this spacetime.

1. The solution is non-trivial in the $M \to 0$ limit, in which it may be interpreted as the gravitational field of a pure "spike" of spin [463, 231].

2. The mass of the solution can be found by standard methods and it is $M$. In particular, we know that we can determine the mass by studying the weak-field expansion and making contact with the Newtonian limit. The Newtonian gravitational potential is given in this approximation by $\phi \sim (g_{tt} - 1)/2 = -M/r$. The Taub–NUT solution has other non-vanishing components of the metric. The diagonal components are still related to the gravitostatic Newtonian potential $\phi$, but the off-diagonal ones $g_{t\underline{i}}$ are related to a gravitomagnetic potential $\vec{A}$ according to Eq. (3.141). In the coordinates that we are using, we see that the Taub–NUT gravitational field has, as the non-vanishing component of the gravitomagnetic potential,

$$A_\varphi = g_{t\varphi} = 2N\cos\theta. \tag{13.6}$$

   This is essentially the electromagnetic field of a magnetic monopole of charge proportional to $N$. Thus, the NUT charge $N$ can be considered as a sort of "magnetic mass" [428], and so the Taub–NUT solution can be interpreted as a gravitational dyon [462].

3. This metric is not asymptotically flat but defines its own class of asymptotic behavior (*asymptotically Taub–NUT spacetimes*) labeled by $N$, which is associated with the non-vanishing at infinity of the off-diagonal $g_{t\varphi}$ component of the metric and, as we are going to see, with the periodicity of the time coordinate. The reason for this periodicity is the desire to avoid certain singularities and to have a spherically symmetric solution. Thus, let us first study the singularities.

4. This metric does not have curvature singularities and is perfectly regular at $r = 0$. However, it has the so-called "wire singularities" at $\theta = 0$ and $\theta = \pi$ where the metric fails to be invertible. These coordinate singularities cannot be cured simultaneously. Misner [950] found a way to make the metric regular everywhere by introducing two coordinate patches.

   (a) One patch covers the region $\theta \geq \pi/2$ around the north pole. In this region we change the time coordinate from $t$ to $t^{(+)}$ defined by

   $$t = t^{(+)} - 2N\varphi, \tag{13.7}$$

   so

   $$ds^2_{(+)} = f(r)\left[dt^{(+)} - 2N(1 - \cos\theta)d\varphi\right]^2 - f^{-1}(r)dr^2 - \left(r^2 + N^2\right) d\Omega^2_{(2)}. \tag{13.8}$$

   (b) The second patch covers the region $\theta \leq \pi/2$ around the south pole. In this region we change the time coordinate from $t$ to $t^{(-)}$ defined by

   $$t = t^{(-)} + 2N\varphi, \tag{13.9}$$

   so

   $$ds^2_{(-)} = f(r)\left[dt^{(-)} + 2N(1 + \cos\theta)d\varphi\right]^2 - f^{-1}(r)dr^2 - \left(r^2 + N^2\right) d\Omega^2_{(2)}. \tag{13.10}$$

   In the overlap region $t^{(+)} = t^{(-)} + 4N\varphi$ and, since $\varphi$ is compact with period $2\pi$, then both of $t^{(\pm)}$ have to be compact with period $8\pi N$.

5. The metric admits three Killing vectors whose Lie brackets are those of the $\mathfrak{so}(3)$ Lie algebra. When the period of the time coordinates is precisely $8\pi N$ this local symmetry can be integrated to give a global SO(3) symmetry and the metric is indeed spherically symmetric [812]. Furthermore, the Taub–NUT spacetime now has a very different topology: the hypersurfaces of constant $r$ are 3-spheres $S^3$ constructed as a Hopf fibration of $S^2$, the fiber being the time $S^1$. Thus, Taub–NUT has the topology of $\mathbb{R}^4$.

6. This way of eliminating the wire singularities is identical to the way in which we eliminated the string singularity in the vector field of the Dirac monopole because the mathematical problem is identical. The Dirac quantization condition translates into a relation between the periodicity of the time coordinate and the NUT charge.

   This relation is more than just a coincidence: in Chapter 15 we will generate by compactification of the Euclidean time of the Euclidean version of the Taub–NUT solution a magnetically charged BH. For this reason, the Euclidean Taub–NUT solution, which we will study later, is also known as the *Kaluza–Klein monopole*.

7. The metric function $f(r)$ has two zeros at $r = r_\pm$ and the metric has coordinate singularities there. For $r > r_+$ and $r < r_-$ (where $t$ is timelike and $r$ spacelike) the metric has closed timelike curves. Thus, although the form of the metric is similar to the Reissner–Nordström metric, no BH interpretation is possible. Furthermore, the "extremality parameter" $r_0$ vanishes only for $M = N = 0$.
8. In the region $r_- < r < r_+$, the coordinate $t$ is spacelike and $r$ is timelike. This region describes a non-singular, anisotropic, closed cosmological model. It can be thought of as a closed universe containing gravitational radiation having the longest possible wavelength [254].
9. There is no known generalization to higher dimensions. It can be embedded in higher-dimensional spacetimes but always as a product metric. The NUT charge seems to be an intrinsically four-dimensional charge (see, however, Ref. [804]).
10. There are interior Taub–NUT solutions [247].

## 13.2 The Euclidean Taub–NUT solution

The Euclidean Taub–NUT metric is interesting in itself, as we are going to see. We obtain it by Wick-rotating the time, which also has to be accompanied by a Wick rotation of the NUT charge $N$ in order to keep the metric real. We denote the Euclidean time by $\tau$. The result is (taking into account the two patches)

$$\boxed{\begin{aligned} -d\sigma_\pm^2 &= f(r)\left[d\tau^{(\pm)} \mp 2N\,(1 \mp \cos\theta)\,d\varphi\right]^2 + f^{-1}(r)dr^2 + \left(r^2 - N^2\right)d\Omega_{(2)}^2, \\ f(r) &= \frac{(r - r_+)(r - r_-)}{r^2 - N^2}, \\ r_\pm &= M \pm r_0, \qquad r_0^2 = M^2 - N^2. \end{aligned}} \tag{13.11}$$

We see that, in the Euclidean case, there is an extreme limit[2] $r_0 = 0$, which corresponds to $M = |N|$. In this case, after shifting the radial coordinate by $M$, we find that the solution can be written in isotropic coordinates in the following way (we suppress the $^+$; it is understood that $\tau$ is a compact coordinate with period $8\pi N$; and the 1-form $A$ is defined by patches so it is regular everywhere):

$$\boxed{\begin{aligned} -d\sigma^2 &= H^{-1}(d\tau + A)^2 + H d\vec{x}_3^{\,2}, \\ H &= 1 + \frac{2|N|}{|\vec{x}_3|}, \\ A &= A_i dx^i, \qquad \epsilon_{ijk}\partial_i A_j = \text{sign}(N)\,\partial_k H. \end{aligned}} \tag{13.12}$$

[2] In the literature it is the extreme limit that usually receives the name of Euclidean Taub–NUT solution.

This solution is known as the (Sorkin–Gross–Perry) *Kaluza–Klein (KK) monopole* [1148, 684]. The 1-form $A$ satisfies the Dirac-monopole equation (12.150), which we know has to be solved in two different patches.

### 13.2.1 Self-dual gravitational instantons

If we use the form of the solution Eqs. (13.12) as an ansatz in the vacuum Einstein equations, we find that we have a solution for every function $H$ that is harmonic in three-dimensional space:

$$\begin{aligned} -d\sigma^2 &= H^{-1}(d\tau + A)^2 + H d\vec{x}_3^{\,2}, \\ A &= A_i dx^i, \qquad \epsilon_{ijk}\partial_i A_j = \pm \partial_k H, \qquad \partial_{\underline{i}}\partial_{\underline{i}} H = 0. \end{aligned} \tag{13.13}$$

In fact, we know that the Laplace equation is the integrability condition of the Dirac-monopole equation, ensuring that it can be (locally) solved. Now it is possible to have solutions with several KK monopoles in equilibrium by taking a harmonic function $H$ with several point-like singularities (*Gibbons–Hawking multicenter metrics* [613]):

$$H = \epsilon + \sum_{I=1}^{k} \frac{2|N_I|}{|\vec{x}_3 - \vec{x}_{3\,I}|}. \tag{13.14}$$

If we choose $\epsilon = 1$, we have the multi-Taub–NUT metric. If all the NUT charges $N_I$ are equal to $N$, then all the wire singularities associated with each pole can be removed simultaneously by taking the period of $\tau$ equal to $8\pi N$. Asymptotically the topology is that of a *lens space*: an $S^3$ in which $k$ points have been identified, and so they are not asymptotically flat in general.

If we choose $\epsilon = 0$, the wire singularities can be eliminated by the same procedure, but the $N_I$s can all be made equal by a rescaling of the coordinates. The topology is the same as in the $\epsilon = 1$ case, but the metrics are *asymptotically locally Euclidean* (ALE), i.e. they are asymptotic to the quotient of Euclidean space by a discrete subgroup of SO(4). The $k = 1$ solution is just flat space. The $k = 2$ solution is equivalent [1057] to the Eguchi–Hanson solution [482], which is usually written in the form

$$-d\sigma^2 = \left(1 - \frac{a^4}{\rho^4}\right)\frac{\rho^2}{4}(d\tau + \cos\theta d\varphi)^2 + \left(1 - \frac{a^4}{\rho^4}\right)^{-1} d\rho^2 + \frac{\rho^2}{4} d\Omega_{(2)}^2. \tag{13.15}$$

This solution has an apparent singularity at $\rho = a$ that can be removed by identifying $\tau \sim \tau + 2\pi$. With this identification, all the $\rho > a$ constant hypersurfaces are $\mathbb{RP}^3$ ($S^3$ with antipodal points identified).

All these solutions are *gravitational instantons*, the gravitational analog of the $SU(2)$ BPST Yang–Mills (YM) instantons discovered in Ref. [124], i.e. non-singular solutions of the Euclidean Einstein equations with finite action, i.e. local minima of the Euclidean Einstein action that can be used to compute the partition function in the saddle-point approximation[3] [726]. This definition also applies to the Euclidean Schwarzschild and RN solutions, of course. It also applies to the general Euclidean Taub–NUT Eq. (13.11), which, for the particular value $M = \frac{5}{4}|N|$, is known [481] as the *Taub-bolt solution* [1015]. However, the gravitational instantons with the Gibbons–Hawking metric Eq. (13.13) have a very special property that brings them closer to their YM counterparts: the SU(2) YM instantons have an (anti-)self-dual field strength[4]

$$\star F_{\mu\nu} = \pm F_{\mu\nu}, \tag{13.16}$$

and the above gravitational instantons have an (anti-)self-dual Lorentz (SO(4)) curvature

$$R_{\mu\nu}{}^{ab}(\omega) = \pm \star R_{\mu\nu}{}^{ab}(\omega). \tag{13.17}$$

The (anti-)self-duality of the YM field strength implies, upon use of the Bianchi identity Eq. (A.43), the YM equations of motion Eq. (A.45). The (anti-)self-duality of the Lorentz curvature[5] implies, via the Bianchi identity $R_{[\mu\nu\rho]}{}^{\sigma} = 0$, the vanishing of the Ricci tensor and the Einstein equations. Both in the YM case and in the gravitational case, (anti-)self-duality is also related to special supersymmetry properties (see Chapter 17).

Four-dimensional SU(2) YM instantons can be characterized by topological invariants such as the *second Chern class*,

$$c_2 = \frac{1}{16\pi^2}\int d^4x \, \mathrm{Tr}(F \star F). \tag{13.18}$$

Then, the manifestly positive integrals

$$\int d^4x \, (F \mp \star F)^2 = 2\int d^4x \, (F^2 \mp F \star F) = 8S_{\mathrm{EYM}} \mp 32\pi^2 c_2 \geq 0 \tag{13.19}$$

can be used to obtain a bound for the Euclidean YM action $S_{\mathrm{EYM}}$:

$$S_{\mathrm{EYM}} \geq 4\pi^2|c_2|. \tag{13.20}$$

(Anti-)self-dual YM field configurations are the solutions that minimize the Euclidean action in a sector characterized by the given topological number $c_2$.

---

[3] A table with the properties of these and other gravitational instantons can be found in Appendix D of Ref. [481]. A calculation of the Euclidean actions based on the isometries of the instantons was done in Refs. [614, 662] (for more recent references, see Refs. [732, 733, 729, 731, 809]).

[4] (Anti-)self-duality can be consistently imposed only in even dimensions and depending on the signature: with Lorentzian signature, only for $d = 4n + 2$; and with Euclidean signature, only in $d = 4n$.

[5] Observe that, in Riemannian spaces, the symmetry property (Bianchi identity) $R_{\mu\nu\rho\sigma} = R_{\rho\sigma\mu\nu}$ implies that the Lorentz curvature 2-form $R_{\mu\nu}{}^{ab}$is also (anti-)self-dual in the Lorentz indices $ab$. Furthermore, if the SO(4) curvature is (anti-)self-dual, there is always a gauge (a frame $e_a{}^{\mu}$) in which the connection $\omega_{\mu}{}^{ab}$ is also (anti-)self-dual in the Lorentz indices $ab$ [482]. The Gibbons–Hawking multicenter metric has an (anti-) self-dual connection in the frame Eq. (13.68), but not in the frame Eq. (13.75). This property of (anti-)self-dual curvatures is a particular case of a more general property: as we are going to see in the next section, an object with (anti-)self-dual SO(4) indices is in fact an object with SU(2) indices embedded in SO(4), and therefore (anti-)self-dual SO(4) curvatures are SU(2) curvatures or curvatures of *special* SU(2) *holonomy*. The "reduction theorem" (Section II.7 of Vol. 1 of Ref. [868]) states that there is always a frame in which the spin connection has the same holonomy as the curvature.

Four-dimensional gravitational instantons are characterized by two topological invariants: the *Hirzebruch signature* $\tau(\mathrm{M})$, which is a third of the integral of the *first Pontrjagin class* $p_1$,

$$\tau(\mathrm{M}) = \tfrac{1}{3}\int_{\mathrm{M}} p_1 = -\frac{1}{24\pi^2}\int_{\mathrm{M}} \mathrm{Tr}_{\mathrm{v}}(R \wedge R) = \frac{1}{96\pi^2}\int_{\mathrm{M}} d^4x\sqrt{|g|}\epsilon^{\mu\nu\rho\sigma} R_{\mu\nu}{}^{ab} R_{\rho\sigma\, ab}, \tag{13.21}$$

where $\mathrm{Tr}_{\mathrm{v}}$ denotes the trace in the vector representation, and $\chi(\mathrm{M})$, the *Euler characteristic*, given in Eq. (12.117), but there is no obvious direct relation between these invariants and the Einstein–Hilbert action (nevertheless, see the discussion in Section 12.6).

The relation between these YM and gravitational configurations is worth investigating a little bit further. Let us first review the BPST SU(2) instanton in the form known as the *'t Hooft ansatz* [826].

### 13.2.2 The BPST instanton

The most transparent way to construct the BPST SU(2) instanton is through the embedding of the group SU(2) in SO(4)[6] using the fact that the algebra $\mathfrak{so}(4) \cong \mathfrak{su}(2)_+ \oplus \mathfrak{su}(2)_-$. If we denote by $M_{mn}$ the SO(4) generators, the two $\mathfrak{su}(2)$ subalgebras correspond to their three self- and three anti-self-dual projections $\{M^{(+)}_{mn}\}$ and $\{M^{(-)}_{mn}\}$ defined by

$$M^{(\pm)}_{mn} \equiv \tfrac{1}{2}\left(\delta_{mn}{}^{rs} \pm \tfrac{1}{2}\epsilon_{mn}{}^{rs}\right) M_{rs}. \tag{13.22}$$

Let us see exactly how the two subalgebras can be identified. In the fundamental (vector) representation that we are going to use (as different from Ref. [826], for instance) the generators of SO(4) are, according to the conventions in Section A.2.3, the $4 \times 4$ matrices

$$\Gamma_{\mathrm{v}}(M_{mn})^{pq} \equiv (M_{mn})^{\,pq} = +2\delta_{mn}{}^{pq}, \tag{13.23}$$

and their (anti-)self-dual parts are

$$\Gamma_{\mathrm{v}}(M^{(\pm)}_{mn})^{pq} \equiv (M^{(\pm)}_{mn})^{pq} \equiv \tfrac{1}{2}\left(\delta_{mn}{}^{rs} \pm \tfrac{1}{2}\epsilon_{mn}{}^{rs}\right)(M_{rs})^{pq} = \delta_{mn}{}^{pq} \pm \tfrac{1}{2}\epsilon_{mn}{}^{pq}. \tag{13.24}$$

These matrices have the same duality properties in the Lie algebra indices $mn$ and in the representation indices $pq$ because they have the interchange property $(M^{(\pm)}_{mn})^{pq} = (M^{(\pm)}_{pq})^{mn}$. An explicit calculation using the commutator Eq. (A.64) gives

$$\left[M_{mn}, M^{(\pm)}_{pq}\right] = \left[M^{(\pm)}_{mn}, M^{(\pm)}_{pq}\right] = -M^{(\pm)}_{rn}(M^{(\pm)}_{pq})^{r}{}_{m} - M^{(\pm)}_{mr}(M^{(\pm)}_{pq})^{r}{}_{n}. \tag{13.25}$$

(Anti-)self-duality relates the $0i$ and $ij$ generators:

$$M^{(\pm)}_{mn} = \pm\tfrac{1}{2}\epsilon_{mn}{}^{pq} M^{(\pm)}_{mn} \quad\Rightarrow\quad M^{(\pm)}_{ij} = \pm\epsilon_{ijk} M^{(\pm)}_{0k}, \tag{13.26}$$

[6] Here we are in flat four-dimensional Euclidean space with mostly plus signature and we use non-underlined Latin indices $m, n, p, q = 0, 1, 2, 3$ and $i, j, k = 1, 2, 3$ for convenience both for space coordinates and for the vector representation of SO(4). We are setting the YM coupling constant $g = 1$ for simplicity, too.

so there are only three independent generators in each set $\{M^{(+)}_{mn}\}$ and $\{M^{(-)}_{mn}\}$. We can take as the independent generators of the $\mathfrak{su}(2)_+$ and $\mathfrak{su}(2)_-$ subalgebras $\{M^{(+)}_{0i}\}$ and $\{M^{(-)}_{i0}\}$, respectively, because their Lie brackets coincide with our convention Eq. (A.83), that is

$$[M^{(+)}_{0i}, M^{(+)}_{0j}] = -\epsilon_{ijk} M^{(+)}_{0k}, \qquad [M^{(-)}_{i0}, M^{(-)}_{j0}] = -\epsilon_{ijk} M^{(-)}_{k0}. \tag{13.27}$$

These generators are usually called *'t Hooft symbols*:

$$\eta^i{}_{pq} \equiv (M^{(+)}_{0i})^{pq}, \qquad \overline{\eta}^i{}_{pq} \equiv (M^{(-)}_{0i})^{pq}. \tag{13.28}$$

They can be used to construct the gauge field of the SU(2) monopoles. We will use directly the $(\mathrm{M}^{(\pm)}_{mn})^{pq}$, which is, of course, completely equivalent. Of course, these two SU(2) groups are only distinguished as subgroups of SO(4). We are working with only one SU(2) group using two different representations.

After these preliminaries, let us consider the ansatz for the SU(2) instanton connection 1-form $A_m$

$$A^{(\pm)}_m = M^{(\pm)}_{mn} V_n, \tag{13.29}$$

where we have already defined the SU(2) generators and where $V_n$ is a vector field to be determined by requiring the 2-form field strength $F^{(\pm)} = dA^{(\pm)} - A^{(\pm)} \wedge A^{(\pm)}$ to be (anti-)self-dual in the spatial indices:

$$\star F^{(\pm)}_{mn} = \alpha F^{(\pm)}_{mn}, \qquad \alpha = \pm 1. \tag{13.30}$$

Observe that we are not assuming the sign of $\alpha$ to be correlated with the signs in the gauge field strengths. We will try both possibilities.

A straightforward calculation gives

$$F^{(\pm)} = \left\{ \tfrac{1}{2} M^{(\pm)}_{mn} V_p V_p - M^{(\pm)}_{mp} V_{np} \right\} dx^m \wedge dx^n, \qquad V_{np} \equiv \partial_n V_p + V_n V_p. \tag{13.31}$$

If we assume spherical symmetry and set $V_m = f(\rho) x^m$ with $\rho^2 = x^m x^m$, it is easy to see that $\star F^{(\pm)} = \pm F^{(\pm)}$ if $f(\rho) = 2(\rho^2 + \lambda^2)^{-1}$, and the solution is given by

$$A^{(\pm)} = -\frac{2}{\rho^2 + \lambda^2} M^{(\pm)}_{mn} x^m dx^n, \qquad F^{(\pm)} = \frac{-2\lambda^2}{(\rho^2 + \lambda^2)} M^{(\pm)}_{mn} dx^m \wedge dx^n, \tag{13.32}$$

which is the BPST instanton solution in its original form [124]. (Anti-)self-duality implies that the YM equations are satisfied. The converse is, obviously, not true. For instance, in Ref. [373] it was found that the above field strength satisfies the YM equations for $V_m = -\partial_m \ln H$, where $H$ satisfies the equation $\partial_m \partial_m H = C H^3$ for any constant $C$ (the BPST being the solution for $C = -8\lambda^2$), but it is easy to see that in general (especially for $C = 0$) they have no definite self-duality properties.

Generically, we have

$$F^{(\pm)} \mp \star F^{(\pm)} = \left\{ \tfrac{1}{2} M^{(\pm)}_{mn} V_{pp} - M^{(\pm)}_{mp} (V_{pn} + V_{np}) \right\} dx^m \wedge dx^n, \tag{13.33}$$

$$F^{(\pm)} \pm \star F^{(\pm)} = \left\{ -\tfrac{1}{2} M^{(\pm)}_{mn} (V_{pp} - 2V_p V_p) \right. \tag{13.34}$$

$$\left. + M^{(\pm)}_{mp} (V_{pn} - V_{np}) \right\} dx^m \wedge dx^n. \tag{13.35}$$

The first combination is very difficult to cancel generically (it does cancel for the BPST instanton). The second can be set to zero (so $F^+$ is anti-self-dual and $F^{(-)}$ is self-dual) if we demand

$$\star f_{mn} = \mp f_{mn}, \quad \text{where} \quad f_{mn} \equiv \partial_m V_n - \partial_n V_m, \tag{13.36}$$

and

$$\partial_m V_m - V_m V_m = 0. \tag{13.37}$$

The first condition can be satisfied by choosing $V_m = -\partial_m \ln H$. Then, substituting in the second condition we find

$$H^{-1} \partial_m \partial_m H = 0, \tag{13.38}$$

and we can construct instanton solutions using arbitrary harmonic functions in $\mathbb{R}^4$, although not all of them will be regular and have finite action. The general form of these solutions is (using now the $\pm$ labels in correspondence with the self-duality properties and not with the original ansatz)

$$\boxed{\begin{aligned} A^{(\pm)} &= M^{(\mp)}_{mn} \partial_m \ln H \, dx^n, \\ F^{(\pm)} &= \left\{ \tfrac{1}{2} M^{(\mp)}_{mn} (\partial \ln H)^2 - 2 M^{(\mp)}_{mp} \partial_p \ln H \partial_n \ln H \right\} dx^m \wedge dx^n. \end{aligned}} \tag{13.39}$$

It is useful to translate the components of the gauge fields in Eq. (13.39) to components of a conventional $\mathrm{SU}(2)$ gauge field not embedded in $\mathrm{SO}(4)$. Taking into account the convention of the $\mathfrak{su}(2)$ structure constants that forced us to select as generators $\{M^{(+)}_{0i}\}$ and $\{M^{(-)}_{i0}\}$, we find that those instanton solutions give rise to the following $\mathrm{SU}(2)$ gauge fields:

$$\boxed{\begin{aligned} A^{(\pm)\,i} &= \pm \partial_i \ln H dx^0 \mp [\delta^i{}_j \partial_0 \ln H \pm \epsilon_{ijk} \partial_k \ln H] dx^j, \\ F^{(\pm)\,i} &= \mp [\delta^i{}_j (\partial \ln H)^2 - 2\partial_i \ln H \partial_j \ln H] dx^0 \wedge dx^j \mp 2\partial_0 \ln H \partial_m \ln H dx^i \wedge dx^m. \end{aligned}} \tag{13.40}$$

The choice of harmonic function

$$H = 1 + \frac{\lambda^2}{\rho^2} \quad \Rightarrow \quad \partial_m \ln H = \frac{-2\lambda^2 x^m}{\rho^2(\rho^2 + \lambda^2)} \tag{13.41}$$

gives a solution which is gauge equivalent to the BPST instanton. In this gauge the gauge potential is singular, and it is sometimes referred to as the *singular Landau gauge* to distinguish it from the one in which the gauge potential is regular.[7] The main advantage of the singular gauge (that is, of the use of harmonic function) is that it allows us to write solutions with many singularities that describe many instantons:

$$H = \epsilon + \sum_{a=1}^{k} \frac{\lambda_a^2}{|\vec{x}_4 - \vec{x}_{4\,a}|^2}, \qquad \epsilon = 1, 0. \tag{13.42}$$

### 13.2.3 Instantons and monopoles

There is an interesting relation between instantons and certain monopoles in spite of their different (Euclidean, Lorentzian) natures. Let us restrict ourselves to YM field configurations that do not depend on the coordinate $x^0 = \tau$. The restricted theory is, thus, effectively three dimensional. The component $A_0$ now has the interpretation of a three-dimensional scalar in the adjoint representation that we denote by $\Phi$, while the other three components become the components of the three-dimensional YM vector field. The $F_{i0}$ components of the field strength are

$$F_{i0} = \partial_i \Phi - [A_i, \Phi] = \mathcal{D}_i \Phi, \tag{13.43}$$

i.e. the three-dimensional YM covariant derivative of the scalar $\Phi$. After integrating over the redundant coordinate $\tau$ (which we take to be periodic with period $2\pi$), the Euclidean YM action becomes

$$S_{\text{EYM}} = 2\pi \int d^3x \, \text{Tr}\left[\tfrac{1}{4} F_{ij} F_{ij} + \tfrac{1}{2} \mathcal{D}_i \Phi \mathcal{D}_i \Phi\right], \tag{13.44}$$

and the (anti-)self-duality[8] equation for $F$ becomes the *Bogomol'nyi equation* [227, 366]

$$\boxed{F_{ij} = \mp \epsilon_{ijk} \mathcal{D}_k \Phi.} \tag{13.45}$$

It is interesting to see how this first-order equation guarantees that the field configuration that satisfies it also satisfies the second-order equations of motion. Let us first rewrite it in three-dimensional differential-form language:

$$F = \mp \star_{(3)} \mathcal{D}\Phi. \tag{13.46}$$

If we now act with the exterior covariant derivative on both sides, using the Bianchi identity for $F$, $\mathcal{D}F = 0$, we get

$$\mathcal{D} \star_{(3)} \mathcal{D}\Phi = 0, \tag{13.47}$$

which is equivalent to the Higgs equation of motion for a time-independent field configuration. If, instead, we act with $\star_{(3)} \mathcal{D} \star_{(3)}$ (the covariant divergence) we get

$$\star_{(3)} \mathcal{D} \star_{(3)} F = [\Phi, \mathcal{D}\Phi], \tag{13.48}$$

which is equivalent, for a time-independent field configuration, to the YM equation sourced by the Higgs current.

[7] In both, $\partial_m A_m^{(\pm)} = 0$.

[8] Self-duality (the upper, $+$, sign) corresponds to the upper, $-$, sign in this equation.

Let us now consider the (four-dimensional, Lorentzian) Georgi–Glashow model [601], which consists of an SU(2) gauge field $A$ coupled to a triplet of Higgs fields $\Phi$ with a potential $V(\Phi) = \frac{1}{2}\lambda[\mathrm{Tr}(\Phi^2) - 1]^2$:

$$S_{\mathrm{GG}} = \int d^4x \left\{-\tfrac{1}{4}\mathrm{Tr}\, F^2 + \tfrac{1}{2}\mathrm{Tr}(\mathcal{D}\Phi)^2 - \tfrac{1}{2}\lambda[\mathrm{Tr}(\Phi^2) - 1]^2\right\}. \tag{13.49}$$

't Hooft [1180] and Polyakov [1053] found a magnetic-monopole solution of this model that generalizes Dirac's magnetic-monopole solution. In the $\lambda = 0$ limit (the *Bogomol'nyi–Prasad–Sommerfield (BPS) limit*), the solution takes a particularly simple form [1058] and has special properties that can also be related to supersymmetry (see Chapter 17).

Let us focus on purely magnetic (i.e. $A_0 = 0$) and static ($\partial_0 A_\mu = \partial_0 \Phi = 0$) field configurations. Their energy (taking $\lambda = 0$) is given precisely by $[1/(2\pi)]S_{\mathrm{EYM}}$ in Eq. (13.44). It is not surprising, therefore, that the energy of these configurations is bounded: the manifestly positive integral

$$\int d^3x\, \mathrm{Tr}(F_{ij} \pm \epsilon_{ijk}\mathcal{D}_k\Phi)^2 = 8E \pm \int d^3x \epsilon_{ijk}\mathrm{Tr}(F_{ij}\mathcal{D}_k\Phi) \geq 0. \tag{13.50}$$

On integrating by parts and using the three-dimensional Bianchi identity, we find that

$$\int d^3x \epsilon_{ijk}\mathrm{Tr}(F_{ij}\mathcal{D}_k\Phi) = \int d^3x \partial_i(\epsilon_{ijk}\Phi F_{ij}) = 4\int_{S^2_\infty} \mathrm{Tr}(\Phi F) = -4p, \tag{13.51}$$

where we have used Stokes' theorem and where $p$ is the SU(2) magnetic charge. Thus,

$$E \geq \tfrac{1}{2}|p|, \tag{13.52}$$

which is the *Bogomol'nyi* or *BPS bound*. We know that $p$ is quantized (for $g = 1$), $p = 2\pi n$. Using this fact and the relation $E = [1/(2\pi)]S_{\mathrm{EYM}}$, this relation is completely equivalent to Eq. (13.20). On the other hand, the configurations that minimize the energy $E = \frac{1}{2}|p|$ (saturate the BPS bound) are those satisfying the first-order Bogomol'nyi equation, and it is easy to prove that these configurations also solve all the (second-order) equations of motion of the $\lambda = 0$ Georgi–Glashow model.

The immediate conclusion of this discussion is that, if we take SU(2) (anti-)self-dual instantons that do not depend on the $\tau$ coordinate, we automatically have a magnetic-monopole solution of the Georgi–Glashow model with $\lambda = 0$ satisfying the Bogomol'nyi bound.[9] In particular, the non-Abelian Wu–Yang SU(2) monopole [1277] is obtained using the 't Hooft ansatz with the harmonic function

$$H = \frac{p}{r}, \tag{13.53}$$

where $r \equiv |\vec{x}_3|$ for any value of the constant $p$, which does not occur in the physical fields.

It is also possible to search for and find solutions of the Bogomol'nyi equations directly in three spatial dimensions. An ansatz frequently used to solve these equations for the gauge group $\mathrm{SU}(2)$ is the so-called *hedgehog ansatz*

$$\boxed{\Phi^i = \mp\delta^i{}_m f(r)x^m, \qquad A^i{}_m = -\epsilon^i{}_{mn}x^n h(r).} \tag{13.54}$$

[9] The normalization of the SU(2) generators that we are using here is that of Eq. (A.83).

This ansatz can be related to the instanton solutions Eq. (13.40) when the harmonic function $H$ is independent of the coordinate $\tau = x^0$. Under these conditions, the $\mathrm{SU}(2)$ gauge field in the instanton solution Eq. (13.40) becomes

$$A^{(\pm)\,i}{}_0 = \pm\partial_i \ln H, \qquad A^{(\pm)\,i}{}_j = -\epsilon_{ijk}\partial_k \ln H. \tag{13.55}$$

Identifying the Higgs field $\Phi^i$ with the components $A^{(\pm)\,i}{}_0$ and the vector field $A^i{}_m$ with $A^{(\pm)\,i}{}_j$ (relabeling now $i, j$ as $m = 1, 2, 3$ where necessary) we get a family of solutions of the Bogomol'nyi equation that depends on a harmonic function $H$ and can describe many monopoles:

$$\boxed{\Phi^i = \pm\delta^i{}_m\partial_m \ln H, \qquad A^i{}_m = -\epsilon^i{}_{mn}\partial_n \ln H.} \tag{13.56}$$

If we assume that $H$ is a function of $r$ only, then these solutions fit into the hedgehog ansatz for

$$-fr = hr = \partial_r \ln H. \tag{13.57}$$

Since the only spherically symmetric harmonic function available is, up to an additive constant, $H = p/r$, the only monopole solution which lies at the intersection of these families of solutions (hedgehog ansatz and dimensionally reduced 't Hooft ansatz) has the gauge and Higgs fields

$$\boxed{\Phi^i = \mp\delta^i{}_m\frac{x^m}{r^2}, \qquad A^i{}_m = \epsilon^i{}_{mn}\frac{x^n}{r^2}.} \tag{13.58}$$

The gauge field coincides with that of the Wu–Yang $\mathrm{SU}(2)$ monopole [1277]. However, that is known to be a solution of the pure YM theory while our solution has a non-trivial Higgs field. There is, however, no contradiction because the above Higgs field generates a vanishing Higgs current (the r.h.s. of Eq. (13.48)) and the YM and YM–Higgs equations of motion are equivalent in this case. The Higgs field is not trivial (covariantly constant), though.

The Wu–Yang monopole has a singularity at $r = 0$, something to be expected since there are no static globally regular solutions of the YM equations in $d = 4$ [432].[10] It is related to the embedding of the Dirac monopole in $\mathrm{SU}(2)$ by a singular gauge transformation that removes the Dirac-string singularity (see e.g. Ref. [1137]): let us call $B^{(s)}$ the field of a Dirac monopole with the Dirac-string singularity lying in the direction of the constant vector $s^m$, which in Cartesian coordinates is given by

$$B^{(s)} = \frac{1}{2}\left(1 - \frac{s^m}{s}\frac{x^m}{r}\right)^{-1} \varepsilon_{mnp}\frac{s^m}{s}\frac{x^m}{r} d\frac{x^p}{r}, \qquad s^2 \equiv s^m s^m. \tag{13.59}$$

[10] The coupling to the Higgs field is crucial to get a regular solution.

We can always embed this solution in $\mathrm{SU}(2)$ by selecting a $\mathfrak{u}(1)$ direction within $\mathfrak{su}(2)$ and identifying the corresponding component of the $\mathrm{SU}(2)$ potential with the above $\mathrm{U}(1)$ field. Explicitly, if $T_i = \frac{i}{2}\sigma^i$ are the $\mathrm{SU}(2)$ generators in the conventions of Eq. (A.83), the $\mathrm{SU}(2)$ gauge field and gauge field strength are given by

$$A^{(s)} \equiv -2B^{(s)}\frac{s^m}{s}\delta_m{}^i T_i, \qquad F(A^{(s)}) = -2F(B^{(s)})\frac{s^m}{s}\delta_m{}^i T_i; \tag{13.60}$$

then, the gauge transformation with

$$\Gamma[g^{(s)}] \equiv \frac{1}{\sqrt{2\left(1-\frac{s^m}{s}\frac{u^m}{u}\right)}}\left[1-\frac{s^m}{s}\frac{u^m}{u}-2\varepsilon_{mnp}\frac{s^m}{s}\frac{u^n}{u}T_p\right] \tag{13.61}$$

transforms $A^{(s)}$ into the field of the Wu–Yang $\mathrm{SU}(2)$ monopole Eq. (13.58). Note, however, that $\Gamma[g^{(s)}]$ has singularities, and one cannot conclude that the Wu–Yang $\mathrm{SU}(2)$ monopole is gauge equivalent to the non-Abelian embedding of the Dirac monopole.

Let us go back to the hedgehog ansatz and let us try to find other spherically symmetric solutions which do not fit into the 't Hooft ansatz. Substituting it directly into the Bogomol'nyi equations Eq. (13.45) with the positive sign we get a system of differential equations for $f(r)$ and $h(r)$:

$$\begin{cases} r\partial_r h + 2h - f(1+gr^2h)=0, \\ r\partial_r(h+f) - gr^2h(h+f)=0. \end{cases} \tag{13.62}$$

This system was analyzed in Ref. [1063] by Protogenov, who found all the solutions with finite energy. They form a 2-parameter and a 1-parameter family. The 2-parameter family has the form

$$\boxed{\begin{aligned} -rf &= \frac{1}{r}\left[\mu r \coth(\mu r + s) - 1\right], \\ rh &= -\frac{1}{r}\left[\frac{\mu r}{\sinh(\mu r + s)} - 1\right], \end{aligned}} \tag{13.63}$$

where the parameter $s$ has been named in the context of BHs as the *Protogenov hair parameter* [935]. The only globally regular solution of this family corresponds to the value $s=0$ and it is the 't Hooft–Polyakov monopole in the BPS limit.

In the $s \to \infty$ limit the general solution takes the form

$$\begin{aligned} -rf &= \mu - \frac{1}{r}, \\ rh &= -\frac{1}{r}, \end{aligned} \tag{13.64}$$

which, only for $\mu = 0$, corresponds to the Wu–Yang monopole.

One of the main drawbacks of the Wu–Yang $\mathrm{SU}(2)$ monopole as a solution of the pure YM theory is that it has zero magnetic charge if one uses the naive definition of magnetic charge (the integral over a 2-sphere at spatial infinity of the 2-form field strength divided by $4\pi$) (see, for instance, Ref. [1137]). However, in the YM–Higgs theory one can use the Higgs field to construct gauge-invariant 2-forms such as the 't Hooft tensor [1180] or just

$$\mathrm{Tr}(\hat{\Phi}F_{\mu\nu}), \qquad \text{where} \qquad \hat{\Phi} \equiv \frac{\Phi}{\sqrt{\mathrm{Tr}(\Phi^2)}}. \tag{13.65}$$

Applying these definitions to the Wu–Yang $\mathrm{SU}(2)$ monopole as a solution of the YM–Higgs theory we find that it describes a monopole of unit magnetic charge ($1/g$ if we restore the YM coupling constant). The same value is obtained for all the solutions in the same family, independently of $\mu$ and $s$.

The 1-parameter family of solutions is given by [935]

$$\boxed{rf = -rh = \frac{1}{r}\left[\frac{1}{1+\lambda^2 r}\right],} \tag{13.66}$$

and it has zero magnetic charge.

### 13.2.4 The BPST instanton and the KK monopole

We are now ready to establish a relation between the Euclidean Taub–NUT solution (KK monopole) and the BPST instanton. We are going to see that the spin-connection frame components $\omega_m{}^{np}$ of the KK monopole are identical to the SO(4)-embedded components of the BPST instanton connection $A_m^{(\pm)np}$ with the two harmonic functions $H$ identical and depending on just three coordinates $\vec{x}_3$. The latter is given in these conditions by

$$\begin{aligned} A_0^{(\pm)0i} &= -\tfrac{1}{2}\partial_i \ln H, & A_i^{(\pm)0j} &= \mp\tfrac{1}{2}\epsilon_{ijk}\partial_k \ln H, \\ A_0^{(\pm)ij} &= \pm\tfrac{1}{2}\epsilon_{ijk}\partial_k \ln H, & A_i^{(\pm)jk} &= -\delta_{i[j}\partial_{k]} \ln H. \end{aligned} \tag{13.67}$$

In the simplest frame,

$$\begin{aligned} e^0 &= H^{-\frac{1}{2}}[d\tau + A_{\underline{i}}dx^i], & e_0 &= H^{\frac{1}{2}}\partial_\tau, \\ e^i &= H^{\frac{1}{2}}dx^i, & e_i &= H^{-\frac{1}{2}}[\partial_{\underline{i}} - A_{\underline{i}}\partial_\tau], \end{aligned} \tag{13.68}$$

the frame components of the spin connection (which is just an SO(4) connection) are

$$\begin{aligned} \omega_0{}^{0i}(e) &= \tfrac{1}{2}\partial_i \ln H, & \omega_i{}^{0j}(e) &= H^{-1}\partial_{[\underline{i}}A_{\underline{j}]}, \\ \omega_0{}^{ij}(e) &= -H^{-1}\partial_{[\underline{i}}A_{\underline{j}]}, & \omega_i{}^{jk}(e) &= \delta_{i[j}\partial_{k]} \ln H. \end{aligned} \tag{13.69}$$

Here it is important to observe that all partial derivatives in these expressions have frame indices. Using the Dirac-monopole equation for the 1-form $A$,

$$\epsilon_{ijk}\partial_{[\underline{i}}A_{\underline{j}]} = \pm\partial_{\underline{k}}H, \tag{13.70}$$

the KK-monopole spin connection becomes

$$\begin{aligned} \omega_0{}^{0i}(e) &= \tfrac{1}{2}\partial_i \ln H, & \omega_i{}^{0j}(e) &= \pm\tfrac{1}{2}\epsilon_{ijk}\partial_k \ln H, \\ \omega_0{}^{ij}(e) &= \mp\tfrac{1}{2}\epsilon_{ijk}\partial_k \ln H, & \omega_i{}^{jk}(e) &= \pm\delta_{i[j}\partial_{k]} \ln H, \end{aligned} \tag{13.71}$$

which is identical to the instanton connection up to an overall minus sign, which can be corrected by changing the sign of all the frame vectors. It is, therefore, (anti-)self-dual and has SU(2) holonomy.

### 13.2.5 Bianchi IX gravitational instantons

In Ref. [625] the class of gravitational instantons with an SU(2) or SO(3) isometry group acting transitively (Bianchi IX metrics) was studied, with special emphasis on those with self-dual curvature. This class includes some of the gravitational instantons that we have studied, namely Taub–NUT, Taub-bolt, and Eguchi–Hanson instantons, and its discussion will provide us with some further interesting examples.

All Ricci-flat ($R_{\mu\nu} = 0$) Bianchi IX metrics can locally be written in the form

$$d\sigma^2 = (a_1 a_2 a_3) d\eta^2 + \sum_{i=1,2,3} (a_i \sigma^i)^2, \tag{13.72}$$

where the $a_i$s depend only on $\eta$ and the $\sigma^i$s are the $\eta$-independent SU(2) Maurer–Cartan 1-forms denoted by $e^i$ in Appendix A.3.1.

A simple solution of the Einstein equations with $a_1^2 = a_2^2$ is given by the Euclidean Taub–NUT solution ($M \neq N$),

$$\begin{aligned} a_1^1 = a_2^2 &= \tfrac{1}{4} q \sinh[q(\eta - \eta_2)] \,\mathrm{cosech}^2[q(\eta - \eta_1)], \\ q(\eta - \eta_2) a_3^2 &= \mathrm{cosech}[q(\eta - \eta_2)], \end{aligned} \tag{13.73}$$

where $q, \eta_1$, and $\eta_2$ are integration constants. The relation to the standard integration constants and coordinates is

$$\begin{aligned} N^2 &= -\tfrac{1}{4} q \,\mathrm{cosech}[q(\eta_2 - \eta_1)], \\ M &= N \cosh[q(\eta_2 - \eta_1)], \\ r &= \frac{q}{4N}\{\coth[\tfrac{1}{2} q(\eta - \eta_1)] - \coth[q(\eta_2 - \eta_1)]\}, \\ \tau &= 4N\psi. \end{aligned} \tag{13.74}$$

On taking the limit $q \to 0$ we obtain the $M = |N|$ Taub–NUT metric with self-dual curvature. With the obvious frame choice

$$e^0 = a_1 a_2 a_3 d\eta, \qquad e^i = a_i \sigma^i, \tag{13.75}$$

its connection is *not* (anti-)self-dual. With $\eta_1 = \eta_2$ we obtain the Eguchi–Hanson metric Eq. (13.15) with

$$M = N + \frac{a^4}{128 N^3}, \qquad r = M + \frac{\rho^2}{8N}, \tag{13.76}$$

after taking the $N \to \infty$ limit. This metric has self-dual curvature and connection (using the above frame). On setting $M = \frac{5}{4}|N|$ we obtain the Taub-bolt metric.

If we impose the condition that the Lorentz curvature is self-dual in the above frame, one obtains, after one integration, the equations

$$2\frac{d}{d\eta}\ln a_1 = \sum_{i=1,2,3} a_i^2 - 2a_1^2 - 2\lambda_1 a_2 a_3, \qquad \lambda_1 = \lambda_2\lambda_3, \tag{13.77}$$

and the equations one obtains from these by cyclic permutations of the indices $i = 1, 2, 3$. The algebraic equations for the constants $\lambda_i$ admit three possible solutions:

$$(\lambda_1, \lambda_2, \lambda_3) = (0, 0, 0),\ (1, 1, 1),\ (-1, -1, 1). \tag{13.78}$$

The first solution corresponds to metrics whose connection is self-dual and can be completely integrated. The general solution is [126]

$$d\sigma^2 = (f_1 f_2 f_3)^{-\frac{1}{2}} d\eta^2 + (f_1 f_2 f_3)^{\frac{1}{2}} \left[\frac{\rho^2}{4}\sum_{i=1,2,3}(f_i^{-\frac{1}{2}}\sigma^i)^2\right], \qquad f_i = 1 - \frac{b_i^4}{\rho^4}. \tag{13.79}$$

$b_1 = b_2 = a$, $b_3 = 0$ is the Eguchi–Hanson metric Eq. (13.15). Solutions of the second class have not been obtained, except for the special case $a_1 = a_2$ that gives the self-dual Taub–NUT metric. The third case is not equivalent to the second and corresponds to the Atiyah–Hitchin metric [70], which governs the interaction of two slowly moving BPS SU(2) monopoles.

## 13.3 Charged Taub–NUT solutions and IWP solutions

Let us consider stationary, axially symmetric solutions of the Einstein–Maxwell system. Some of them are the result of adding electric or magnetic charges to vacuum solutions.

The charged version of the Kerr solution was found in Ref. [980] and is known as the Kerr–Newman solution, which takes the form

$$\begin{aligned}
ds^2 &= \left(1 - \frac{2Mr - 4q^2}{\Sigma}\right)dt^2 + 2\frac{a(2Mr - 4q^2)\sin^2\theta}{\Sigma}dt d\varphi \\
&\quad - \frac{\Sigma}{\Delta}dr^2 - \Sigma d\theta^2 - \frac{\mathcal{A}}{\Sigma}\sin^2\theta\, d\varphi^2, \\
\Sigma &= r^2 + a^2\cos^2\theta, \qquad \Delta = r^2 - 2Mr + 4q^2 + a^2, \\
\mathcal{A} &= \Sigma(r^2 + a^2) + (2Mr - 4q^2)a^2\sin^2\theta, \\
A_\mu &= \frac{4qr}{\Sigma}[\delta_{\mu t} - \delta_{\mu\varphi} a \sin^2\theta].
\end{aligned} \tag{13.80}$$

Again, if $M^2 \geq 4q^2 + a^2$, this solution describes BHs with mass $M$, angular momentum $J = Ma$, and electric charge $q$, with the event horizon at $r = r_+ = M + \sqrt{M^2 - 4q^2 - a^2}$ (the larger value of $r$ for which $\Delta = 0$).

Observe that, although the solution is only electrically charged, the rotation induces a magnetic dipole moment and the $A_\varphi$ component of the vector field is non-zero.

The electrically charged Taub–NUT solution was found by Brill in Ref. [254]

$$
\boxed{
\begin{aligned}
ds^2 &= f(r)(dt + 2N\cos\theta\, d\varphi)^2 - f^{-1}(r)dr^2 - \left(r^2 + N^2\right) d\Omega^2_{(2)}, \\
F_{tr} &= \frac{4q(r^2 - N^2)}{(r^2 + N^2)^2}, \qquad (\star F)_{tr} = \frac{8qNr}{(r^2 + N^2)^2}, \\
f(r) &= \frac{(r - r_+)(r - r_-)}{r^2 + N^2}, \\
r_\pm &= M \pm r_0, \qquad r_0^2 = M^2 + N^2 - 4q^2.
\end{aligned}
}
\tag{13.81}
$$

It reduces to the RN solution when we set the NUT charge to zero. It is trivial to generalize these solutions to the magnetic and dyonic cases.

In contrast to the Taub–NUT solution, the charged Taub–NUT solution does have an extremal limit $M^2 + N^2 = 4q^2$ in which the extremality parameter $r_0$ vanishes and the two zeros of the metric function $f(r)$ coincide. In this case, by shifting the radial coordinate to $\rho = r - M$ and defining Cartesian coordinates such that $\rho = |\vec{x}_3|$, we find a simple form of the solution,[11]

$$
\boxed{
\begin{aligned}
ds^2 &= |\mathcal{H}|^{-2}\,(dt + A)^2 - |\mathcal{H}|^2 d\vec{x}_3^{\,2}, \\
A_t &= 2\mathrm{Re}(e^{i\alpha}\mathcal{H}^{-1}), \qquad \tilde{A}_t = 2\mathrm{Im}(e^{i\alpha}\mathcal{H}^{-1}), \\
\mathcal{H} &= 1 + \frac{M + iN}{|\vec{x}_3|}, \\
A &= A_{\underline{i}} dx^i, \qquad \epsilon_{ijk}\partial_i A_j = \pm\mathrm{Im}(\overline{\mathcal{H}}\partial_k \mathcal{H}).
\end{aligned}
}
\tag{13.82}
$$

As in some of the other "extreme" solutions that we have found so far,[12] it turns out that we obtain a solution for *any complex harmonic function* $\mathcal{H}(\vec{x}_3)$. By absorbing the complex

[11] Here we are actually taking the extreme limit of the dyonic solution, which indeed has a simpler form. The information on the electric and magnetic charges is contained in the SO(2) electric–magnetic-duality phase $e^{i\alpha}$.

[12] But not in all of them. In particular, not in the Kerr BH.

phase $e^{i\alpha}$ into $\mathcal{H}$, we can write the general solution in the following form:

$$\boxed{\begin{aligned} ds^2 &= |\mathcal{H}|^{-2}(dt+A)^2 - |\mathcal{H}|^2 d\vec{x}_3^{\,2}, \\ A_t &= 2\mathrm{Re}\,\mathcal{H}^{-1}, \qquad \tilde{A}_t = -2\mathrm{Re}(i\mathcal{H}^{-1}), \\ A &= A_{\underline{i}}dx^i, \qquad \epsilon_{ijk}\partial_i A_j = \pm\mathrm{Im}(\overline{\mathcal{H}}\partial_k\mathcal{H}), \\ \partial_{\underline{i}}\partial_{\underline{i}}\mathcal{H} &= 0. \end{aligned}} \tag{13.83}$$

Metrics of the above form are known as *conformastationary metrics* [878]. Observe that the integrability condition of the equation for the 1-form $A$ is the Laplace equation for $\mathcal{H}$. This big family of solutions is known as the *Israel–Wilson–Perjés (IWP) solutions* [822, 1033], although they were first discovered by Neugebauer [978]. This family contains all the "extreme" solutions (RN, charged Taub–NUT, and their multicenter generalizations) that we have found so far, plus many others that may have mass, electric, and magnetic charges, NUT charge, and also angular momentum. In particular, the $M^2 = 4q^2$ Kerr–Newman solutions, for arbitrary angular momentum, belong to this family; their complex harmonic function is

$$\mathcal{H} = 1 + \frac{M}{\sqrt{x^2+y^2+(z-ia)^2}}. \tag{13.84}$$

In terms of more suitable oblate spheroidal coordinates,

$$\begin{aligned} x+iy &= [(r-M)^2+a^2]^{\frac{1}{2}}\sin\theta\, e^{i\varphi}, \\ z &= (r-M)\cos\theta, \end{aligned} \tag{13.85}$$

the function $\mathcal{H}$ takes the form

$$\mathcal{H} = 1 + \frac{M}{r-M-ia\cos\theta}, \tag{13.86}$$

and the Euclidean three-dimensional metric becomes

$$d\vec{x}_3^{\,2} = \left[(r-M)^2+a^2\cos^2\theta\right]\left[\frac{dr^2}{(r-M)^2+a^2}+d\theta^2\right] + \left[(r-M)^2+a^2\right]\sin^2\theta\, d\varphi^2. \tag{13.87}$$

Furthermore, the 1-form $A$ is given by

$$A = \frac{(2Mr-M^2)a\sin^2\theta}{(r-M)^2+a^2\cos^2\theta}d\varphi, \tag{13.88}$$

and

$$|\mathcal{H}|^2 = \frac{(r-m)^2-a^2\cos^2\theta}{r^2+a^2\cos^2\theta}, \tag{13.89}$$

and we recover the Kerr–Newman solutions with $M^2 = 4q^2$. These solutions are not BHs because they violate the bound $M^2 - 4q^2 - a^2 \geq 0$. In fact, it has been argued by Hartle and Hawking that the only BH-type solutions in the IWP family of metrics are the multi-ERN solutions.

For us, one of the main interests of this family is that it is electric–magnetic-duality invariant and it is the most general family that we can have with the above charges always satisfying the identity $M^2 = 4|\vec{q}|^2$. An electric–magnetic-duality transformation is nothing but a change in the phase of $\mathcal{H}$. Non-extreme solutions can be constructed from the IWP class by adding a "non-extremality function" $W$, as in the RN case [906]. We will study them in Chapter 16 as a subfamily of the most general BH-type solutions of pure $N = 4, d = 4$ SUEGRA.

# 14
# Gravitational pp-waves

As we saw in Part I, the weak-field limit of GR is just a relativistic field theory of a massless spin-2 particle propagating in Minkowski spacetime. In the absence of sources, by choosing the De Donder gauge Eq. (3.100), it can be shown that the gravitational field $h_{\mu\nu}$ satisfies the wave equation (3.101) and, correspondingly, there are wave-like solutions of the weak-field equations like the one we found in Section 3.2.3 associated with a massless point-particle moving at the speed of light.

GR is, however, a highly non-linear theory and it is natural to wonder whether there are exact wave-like solutions of the full Einstein equations. The answer is definitely yes and in this chapter we are going to study some of them, the so-called pp-waves, which are especially interesting for us. In particular we are going to see that the linear solution we found in Section 3.2.3 is an exact solution of the full Einstein equations that has the same interpretation. We will use this solution many times in what follows to describe the gravitational field of Kaluza–Klein momentum modes, for instance.

## 14.1 pp-waves

In GR, pp-waves (shorthand for *plane-fronted waves with parallel rays*) are metrics that, by definition, admit a covariantly constant null Killing vector field $\ell_\mu$:

$$\nabla_\mu \ell_\nu = 0, \qquad \ell^2 = \ell_\mu \ell^\mu = 0. \tag{14.1}$$

The first spacetimes with this property were discovered by Brinkmann [263]. To describe pp-wave metrics, we define light-cone coordinates $u$ and $v$ in terms of the usual Cartesian coordinates

$$u = \frac{1}{\sqrt{2}}(t - z), \qquad v = \frac{1}{\sqrt{2}}(t + z), \tag{14.2}$$

which are related to the null Killing vector by

$$\ell_\mu = \partial_\mu u, \qquad \ell^\mu \partial_\mu v = 1, \tag{14.3}$$

i.e. $v$ is the coordinate we can make the metric independent of, the only non-vanishing components of $\ell$ are $\ell_u = \ell^v = 1$, and the metric describes a gravitational wave propagating in the positive direction of the $z$ axis. The most general metric admitting a covariantly constant null Killing vector in $d$ dimensions [264] takes the form

$$ds^2 = 2du(dv + Kdu + A_{\underline{i}}dx^i) + \tilde{g}_{\underline{ij}}dx^i dx^j, \tag{14.4}$$

where $i, j = 1, 2, \ldots, d-2$ and the vector (the *Sagnac connection* [616]) $A_{\underline{i}}$, and the metric $\tilde{g}_{\underline{ij}}$ in the transverse space do not depend on $v$. The connection and curvature for this metric are given in Appendix N.3.6. It is possible to eliminate either $K$ or the $A_{\underline{i}}$s by performing a GCT $(u, v, x^i) \to (u, v', x^{i\prime})$ that preserves the above form of the metric. Under

$$x^i = x^i(u, x'), \qquad v = v' + f(u, x'), \tag{14.5}$$

we obtain a metric of the same form but with

$$\begin{aligned}
A'_{\underline{i}} &= A_{\underline{j}}M^{\underline{j}}{}_{\underline{i}} + \tilde{g}_{\underline{kj}}\partial_{\underline{u}}x^k M^{\underline{j}}{}_{\underline{i}} + \frac{\partial f}{\partial x^{i\prime}}, \\
K' &= K + A_{\underline{i}}\partial_{\underline{u}}x^i + \tfrac{1}{2}\tilde{g}_{\underline{ij}}\partial_{\underline{u}}x^i\partial_{\underline{u}}x^j + \partial_{\underline{u}}f, \\
\tilde{g}'_{\underline{ij}} &= \tilde{g}_{\underline{kl}}M^{\underline{k}}{}_{\underline{i}}M^{\underline{l}}{}_{\underline{j}}, \\
M^{\underline{i}}{}_{\underline{j}} &\equiv \frac{\partial x^j}{\partial x^{i\prime}}.
\end{aligned} \tag{14.6}$$

It is now possible to solve the equation $A'_{\underline{i}} = 0$ with $f = 0$ and the $x^{i\prime}$ given by the solutions of the first-order differential equation

$$\partial_{\underline{u}}x^i = -\tilde{g}^{\underline{ij}}A_{\underline{j}}, \qquad \tilde{g}^{\underline{ij}}\tilde{g}_{\underline{jk}} = \delta^{\underline{i}}{}_{\underline{k}}, \tag{14.7}$$

if the matrix $M^{\underline{i}}{}_{\underline{j}}$ can be inverted. The equation $K' = 0$ can also be solved with

$$x^{i\prime} = x^i, \qquad \partial_{\underline{u}}f = -K. \tag{14.8}$$

### 14.1.1 Hpp-waves

A family of pp-waves known as *homogeneous pp-waves* or Hpp-waves was constructed by Cahen and Wallach as *symmetric* (not just homogeneous) Lorentzian spacetimes [273]. Some of these spacetimes (in $d = 4$ [875], $d = 6$ [934], $d = 10$ [221], and $d = 11$ [874, 548]) are *maximally supersymmetric*, as we will explain in Chapter 17, and are, therefore, vacua of the corresponding supersymmetric theory, just as the RB solution is another vacuum of $N = 2, d = 4$ SUGRA. In fact, the maximally supersymmetric Hpp-waves are the *Penrose limits* [1028, 700] of RB-type ($AdS_n \times S^{d-n}$) vacua, which also occur in $d = 4, 6, 10$, and 11 [223, 222]. This makes them particularly interesting. Here we review their construction following Ref. [548] and using Appendix A.

First, we need some definitions: the *Heisenberg algebra* $H(2n + 1)$ is the Lie algebra generated by $\{q_i, p_j, V\}$, $i, j = 1, \ldots, n$, with the only non-vanishing Lie brackets

$$[q_i, p_i] = \delta_{ij}V. \tag{14.9}$$

The *Heisenberg algebra* $H(2n+2)$ is the semidirect sum of $H(2n+1)$ and the Lie algebra generated by the automorphism $U$ whose action is determined by the new non-vanishing Lie brackets

$$[U, q_i] = p_i, \qquad [U, p_i] = -q_i. \tag{14.10}$$

In the complex basis

$$\alpha_i = \frac{1}{\sqrt{2}}(q_i + ip_i), \qquad I = iV, \qquad N = -iU, \tag{14.11}$$

the Lie brackets take the form

$$[\alpha_i, \alpha_j^\dagger] = \delta_{ij} I, \qquad [N, \alpha_i] = -\alpha_i, \qquad [N, \alpha_i^\dagger] = +\alpha_i^\dagger, \tag{14.12}$$

in which we recognize $N$ as the *number operator*.

All the Heisenberg algebras are solvable and have a singular Killing metric.[1] $V$ ($I$) is always central.

The Heisenberg algebras can be *deformed* as follows: let us denote by $x_r$, $r = 1, \ldots, 2n$, the column vector formed by the $q_i$s and $p_i$s. The Lie brackets can be written in this form:

$$[x_r, x_s] = \eta_{rs} V, \qquad [U, x_r] = \eta_{rs} x_s, \qquad (\eta_{rs}) = \begin{pmatrix} 0 & \mathbb{I}_{n\times n} \\ -\mathbb{I}_{n\times n} & 0 \end{pmatrix}. \tag{14.13}$$

Now, we can define a new (solvable) Lie algebra with brackets

$$[x_r, x_s] = M_{rs} V, \qquad [U, x_r] = N_{rs} x_s, \qquad MN^{\mathrm{T}} - NM^{\mathrm{T}} = 0. \tag{14.14}$$

In some cases, but not always, this algebra is equivalent to the original Heisenberg algebra up to a $\mathrm{GL}(2n)$ transformation.

The $(n+2)$-dimensional Hpp-wave spacetimes are constructed starting from a $(2n+2)$-dimensional algebra of the above form with

$$(M_{rs}) = \begin{pmatrix} 0 & -2A \\ 2A & 0 \end{pmatrix}, \qquad (N_{rs}) = \begin{pmatrix} 0 & \mathbb{I}_{n\times n} \\ 2A & 0 \end{pmatrix}, \qquad A_{ij} = A_{ji}, \tag{14.15}$$

which is inequivalent to the original Heisenberg algebra $H(2n+2)$. In the coset construction, $\mathfrak{h}$ will be the Abelian subalgebra generated by the $p_i \equiv M_i$, and its orthogonal complement $\mathfrak{k}$ is generated by $q_i \equiv P_i$, $V \equiv P_v$, and $U \equiv P_u$; $\mathfrak{h}$ and $\mathfrak{k}$ are a symmetric pair.

Using the coset representative

$$u = e^{vP_v} e^{uP_u} e^{x^i P_i}, \tag{14.16}$$

we obtain the 1-forms

$$\begin{aligned} e^u &= -du, & e^i &= -dx^i, \\ e^v &= -(dv + A_{ij} x^i x^j du), & \vartheta^i &= -x^i du. \end{aligned} \tag{14.17}$$

[1] Actually, the algebras $H(2n+1)$ are nilpotent, which implies an identically vanishing Killing metric.

To construct an invariant Riemannian metric, we use the $H$-invariant metric[2] $B_{uv} = +1$, $B_{ij} = +\delta_{ij}$ on $\mathfrak{k}$, and the result is a pp-wave of the form

$$ds^2 = 2du(dv + A_{ij}x^i x^j du) + d\vec{x}_n^{\,2}. \tag{14.18}$$

These Hpp-waves are characterized by the eigenvalues of $A$. They are invariant under the $(2n+2)$-dimensional Heisenberg group but also under the rotations of the wavefront coordinates that preserve the eigenspaces.

## 14.2 Four-dimensional pp-wave solutions

In four dimensions it is useful to define complex coordinates on the (plane) wavefront $\xi, \bar{\xi}$,

$$\xi = \frac{1}{\sqrt{2}}(x + iy), \tag{14.19}$$

so, using the fact that any two-dimensional metric is conformally equivalent to flat space, the four-dimensional metric can always be written in the form

$$ds^2 = 2du[dv + K(u,\xi,\bar{\xi})du] - 2P(u,\xi,\bar{\xi})d\xi d\bar{\xi}. \tag{14.20}$$

The Einstein vacuum equations are solved if $K$ is a harmonic function on the wavefront,

$$\partial_\xi \partial_{\bar{\xi}} K = 0, \tag{14.21}$$

and $P$ is a function of $u$ alone, and then we can absorb it into a redefinition of $\xi$ that does not change the form of the metric. The only non-trivial element of the metric in this adapted coordinate system is, therefore, $g_{\underline{uu}} = K(u,\xi,\bar{\xi})$. Observe that this function $K$ has exactly the form of a perturbation of the gravitational field about the vacuum (flat Minkowski space with metric $\eta_{\mu\nu}$) since

$$2dudv - 2d\xi d\bar{\xi} = \eta_{\mu\nu}dx^\mu dx^\nu, \tag{14.22}$$

and the metric Eq. (14.20) can also be written in the form

$$ds^2 = \eta_{\mu\nu}dx^\mu dx^\nu + 2K(u,\xi,\bar{\xi})du^2, \qquad h_{uu} = 2K. \tag{14.23}$$

The most general pp-wave solutions of the four-dimensional Einstein–Maxwell theory Eq. (12.58) are also known (see Ref. [878]), and take the form

$$\begin{aligned} ds^2 &= 2du(dv + Kdu) - 2d\xi d\bar{\xi}, \\ F_{\xi u} &= \partial_\xi C, \qquad K = \mathrm{Re}\, f + \tfrac{1}{4}|C|^2, \qquad \partial_{\bar{\xi}} f = \partial_{\bar{\xi}} C = 0. \end{aligned} \tag{14.24}$$

[2] This metric has mostly plus signature, because $B_{uv} = +1, B_{ij} = -\delta_{ij}$ is not $H$-invariant. We have to perform Wick rotations to obtain a mostly minus metric.

The specific properties of each pp-wave solution depend on the form of the function $K$,[3] which has two different terms. The first is independent of the electromagnetic field; only the second depends on it. The first term (the real part of the analytic $f(u,\xi)$) is just any harmonic function $H(u,\vec{x}_2)$ in the wavefront Euclidean two-dimensional space and it provides a purely gravitational solution. It represents a sort of perturbation of the electromagnetic and gravitational background described by the second term of $K$.

A particularly interesting type of pp-wave is a *shock* or *impulse wave*, with the first term of $K$ given by

$$K(u,\xi,\bar{\xi}) = \delta(u)K(\xi,\bar{\xi}). \tag{14.25}$$

An example of a gravitational shock wave is provided by the purely gravitational *Aichelburg–Sexl solution* [24]

$$K = H(u,\vec{x}_2) = \delta(u)\ln|\xi|, \tag{14.26}$$

which describes the gravitational field of a massive point-like particle boosted to the speed of light. In Ref. [24] this metric was obtained by performing an infinite boost in the $z$ direction to a Schwarzschild BH. This method for generating impulsive waves also works in (anti-)de Sitter spacetimes [782] using the Schwarzschild–(anti-)de Sitter solution and has also been applied to the Kerr–Newman solution [540, 902, 90, 91] and to Weyl's axisymmetric vacuum solutions [1045].[4] However, in Section 14.3 we will identify $d$-dimensional Aichelburg–Sexl (AS) shock waves as the gravitational field produced by a massless particle moving at the speed of light, checking explicitly that AS shock waves satisfy the equations of motion of Einstein's action coupled to a massless particle.

This interpretation will turn out to be very useful. In Chapter 15 we will be interested in the gravitational field produced by massless particles moving at the speed of light in compact dimensions. These particles appear as massive and charged in the non-compact dimensions, and their gravitational field (a charged extreme BH) can be derived from the massless-particle gravitational field. Then, we will simply have to adapt the AS shock-wave solution to a spacetime with compact dimensions.

Another example, this time with the first term of $K$ vanishing, is provided by a solution with Hpp-wave-type metrics Eq. (14.18). A particular case is the *four-dimensional Kowalski–Glikman solution* KG4 [875],

$$\boxed{\begin{aligned} ds^2 &= 2du(dv + \tfrac{1}{8}\lambda^2|\vec{x}_2|^2 du) - d\vec{x}_2^{\,2}, \\ F_{\underline{u1}} &= \lambda, \end{aligned}} \tag{14.27}$$

which is a maximally supersymmetric solution of the $d=4$ Einstein–Maxwell theory that is the Penrose limit of the RB solution. We will study the (super)symmetries of these vacua in Chapter 17.

Before studying shock-wave sources, we consider the higher-dimensional generalization of the pp-wave solutions Eq. (14.24).

[3] A detailed classification and description of metrics of this kind that are solutions of the Einstein–Maxwell equations can be found in Ref. [878].

[4] For further results and references on impulse waves see e.g. Refs. [1150, 1044].

### 14.2.1 Higher-dimensional pp-waves

A general pp-wave solution of the $d$-dimensional Einstein–Maxwell theory Eq. (12.219) is given by [905]

$$\boxed{\begin{aligned} ds^2 &= 2du(dv + K du) - \tilde{g}_{\underline{i}\underline{j}}(\vec{x}_{d-2})dx^i dx^j, \\ F_{u\underline{i}} &= C_{\underline{i}}, \qquad \tilde{\nabla}^2 K = \tfrac{1}{4}\tilde{C}_{\underline{i}}\tilde{C}^{\underline{i}}, \qquad \tilde{d}C = \tilde{d}\star C = 0, \qquad \tilde{R}_{\underline{i}\underline{j}} = 0, \end{aligned}} \tag{14.28}$$

i.e. $C(u, \vec{x}_{d-2})_{\underline{i}}dx^i$ is a harmonic 1-form in the Ricci-flat wavefront space and $K$ satisfies the above differential equation that can be integrated if the Green function for the Laplacian on the wavefront space is known. If the wavefront space is flat, $\tilde{g}_{\underline{i}\underline{j}} = -\delta_{\underline{i}\underline{j}}$, and we take the $\vec{x}_{d-2}$-independent harmonic 1-form $C_{\underline{i}}(u)$, $K$ is given by

$$K = H(u, \vec{x}_{d-2}) + \tfrac{1}{4}C_{\underline{i}}C^{\underline{i}}(u)M_{ij}(u)x^i x^j, \qquad \mathrm{Tr}(M) = 1, \qquad \partial_{\underline{i}}\partial_{\underline{i}}H = 0. \tag{14.29}$$

Again, $K$ consists of two terms: the first is a harmonic function on the Euclidean wavefront space $H(u, \vec{x}_{d-2})$. This is the part of $K$ that can be related to singular sources (massless particles), as we are going to see in Section 14.3. The second term in $K$ describes the gravitational and electromagnetic background. The solutions with $H = 0$ and $C_{\underline{i}}$ and $M_{ij}$ constant have, again, Hpp-wave metrics:

$$\boxed{\begin{aligned} ds^2 &= 2du(dv + A_{ij}x^i x^j du) - d\vec{x}^{\,2}_{d-2}, \\ F_{u\underline{i}} &= C_{\underline{i}}, \qquad \mathrm{Tr}(A) = \tfrac{1}{4}C_{\underline{i}}C^{\underline{i}}. \end{aligned}} \tag{14.30}$$

One particular case is the KG4 solution Eq. (14.27). Another interesting case is the five-dimensional Kowalski–Glikman solution KG5 [934], which is also maximally supersymmetric in $N = 1, d = 5$ SUGRA [382]:

$$\boxed{\begin{aligned} ds^2 &= 2du\left[dv + \frac{\lambda_5^2}{24}(4z^2 + x^2 + y^2)du\right] - dx^2 - dy^2 - dz^2, \\ F &= \lambda_5 du \wedge dz. \end{aligned}} \tag{14.31}$$

## 14.3 Sources: the AS shock wave

We consider a massless particle moving in $d$-dimensional curved space coupled to the Einstein action for the gravitational field. This coupled system is described by the following action (see Section 3.3, where, in particular, the action for a massless particle Eq. (3.258) was derived) with $c = 1$:

$$S = \frac{1}{16\pi G_{\mathrm{N}}^{(d)}} \int d^d x \sqrt{|g|}\, R - \frac{p}{2} \int d\xi \sqrt{\gamma}\gamma^{-1} g_{\mu\nu}(X) \dot{X}^\mu \dot{X}^\nu . \tag{14.32}$$

The equations of motion for $g_{\mu\nu}(x)$, $X^\mu(\xi)$, and $\gamma(\xi)$ are, respectively,

$$\begin{aligned}
\frac{16\pi G_{\mathrm{N}}^{(d)}}{\sqrt{|g|}} \frac{\delta S}{\delta g^{\mu\nu}} &= G_{\mu\nu} + \frac{8\pi G_{\mathrm{N}}^{(d)} p}{\sqrt{|g|}} \int d\xi \sqrt{\gamma}\gamma^{-1} g_{\mu\rho} g_{\nu\sigma} \dot{X}^\rho \dot{X}^\sigma \delta^{(d)}(x - X) = 0, \\
\frac{\gamma^{\frac{1}{2}}}{p} g^{\sigma\rho} \frac{\delta S}{\delta X^\rho} &= \ddot{X}^\sigma + \Gamma_{\rho\nu}{}^\sigma \dot{X}^\rho \dot{X}^\nu - \frac{d}{d\xi} (\ln \gamma)^{\frac{1}{2}} \dot{X}^\sigma = 0, \\
\frac{4\gamma^{\frac{3}{2}}}{p} \frac{\delta S}{\delta \gamma} &= g_{\mu\nu} \dot{X}^\mu \dot{X}^\nu = 0.
\end{aligned} \tag{14.33}$$

Since the particle is massless, it must move at the speed of light (this is the content of the equation of motion of $\gamma$). If it moves in the direction of the $x^{d-1} \equiv z$ axis, one can use the light-cone coordinates $u$ and $v$.

If the particle moves in the sense of increasing $z$ at the speed of light, its equation of motion is $U(\xi) = 0$. We can set $V(\xi) = \sqrt{2}\,\xi$. Thus, our ansatz for the $X^\mu(\xi)$ is

$$U(\xi) = 0, \qquad V(\xi) = \sqrt{2}\xi, \qquad \vec{X} \equiv (X^1, \ldots, X^{d-2}) = \vec{0}. \tag{14.34}$$

A gravitational wave moves at the speed of light, and thus our ansatz for the spacetime metric is that of a gravitational pp-wave moving in the same direction (i.e. with null Killing vector $\ell_\mu = \delta_{\mu u}$ so, in particular, nothing depends on $v$):

$$ds^2 = 2du dv + 2K(u, \vec{x}_{d-2}) du^2 - d\vec{x}^{\,2}_{d-2}, \qquad \vec{x}_{d-2} = \left(x^1, \ldots, x^{d-2}\right). \tag{14.35}$$

Now we plug our ansatz into the above equation of motion. First, we immediately see that the equation for $\gamma$ is satisfied because $\dot{X}^\mu = \sqrt{2}\delta^\mu{}_v$ and $g_{vv} = 0$. The equation of motion for $X^\mu$ is also satisfied by taking a constant worldline metric $\gamma = 1$ because $\Gamma_{vv}{}^\sigma = 0$.

Only one equation remains to be solved. On substituting our ansatz for the coordinates and $\gamma$ plus $|g| = 1$ (which holds for the above pp-waves), we find

$$G_{\mu\nu} + 8\pi G_{\mathrm{N}}^{(d)} p \int d\xi \delta_{\mu u} \delta_{\nu u} \delta(u) \delta(v - \sqrt{2}\xi) \delta^{(d-2)}(\vec{x}_{d-2}) = 0. \tag{14.36}$$

For the pp-wave metric Eq. (14.35) we also have *exactly* (that is, without using any property of the metric apart from the lightlike character of $\ell^\mu$)

$$G_{\mu\nu} = -\delta_{\mu u} \delta_{\nu u} \vec{\partial}^{\,2}_{d-2} K(u, \vec{x}_{d-2}). \tag{14.37}$$

Then, on integrating over $\xi$ and substituting the above result, the Einstein equation reduces to the following equation for $K(u, \vec{x}_{d-2})$:

$$\vec{\partial}^{\,2}_{d-2} K(u, \vec{x}_{d-2}) = -\sqrt{2}\, 8\pi G_{\mathrm{N}}^{(d)} p \delta(u) \delta^{(d-2)}(\vec{x}_{d-2}). \tag{14.38}$$

We found precisely the same equation in Section 3.2.3, and it has the same solution, Eqs. (3.133) and (3.134). Thus, we have found the solution

$$
\boxed{
\begin{aligned}
ds^2 &= 2dudv + 2K(u, \vec{x}_{d-2})du^2 - d\vec{x}^2_{d-2}, \\
K(u, \vec{x}_{d-2}) &= \frac{\sqrt{2}\, p 8\pi G_{\rm N}^{(d)}}{(d-4)\omega_{(d-3)}} \frac{1}{|\vec{x}_{d-2}|^{d-4}}\, \delta(u), \qquad d \geq 5, \\
K(u, \vec{x}_2) &= -\sqrt{2}\, p 4 G_{\rm N}^{(4)} \ln|\vec{x}_2|\, \delta(u), \qquad d = 4.
\end{aligned}
}
\tag{14.39}
$$

The $d = 4$ solution is the AS shock wave found in Ref. [24]. Observe that this solution is exactly the same as that obtained in Section 3.2.3 by solving the linear-order theory. There are no higher-order corrections to the first-order solution, which is not *renormalized*. This is due to the special structure of the linear solution and can be related to supersymmetry as well.

There is another useful way to rewrite the pp-wave metrics that we have found. Defining the function

$$
H \equiv 1 - K, \tag{14.40}
$$

the solution takes the form

$$
\boxed{
\begin{aligned}
ds^2 &= H^{-1}dt^2 - H\left[dz - \alpha(H^{-1} - 1)dt\right]^2 - d\vec{x}^2_{d-2}, \qquad \alpha = \pm 1, \\
H &= 1 - \frac{\sqrt{2}\, p 8\pi G_{\rm N}^{(d)}}{(d-4)\omega_{(d-3)}} \frac{1}{|\vec{x}_{d-2}|^{d-4}}\, \delta\left[\frac{1}{\sqrt{2}}(t - \alpha z)\right], \qquad d \geq 5, \\
H &= 1 + \sqrt{2}\, p 4 G_{\rm N}^{(4)} \ln|\vec{x}_2|\, \delta\left[\frac{1}{\sqrt{2}}(t - \alpha z)\right], \qquad d = 4,
\end{aligned}
}
\tag{14.41}
$$

where we have introduced the constant $\alpha = \pm 1$ to take care of the two possible directions of propagation toward $z = \alpha\infty$.

Had we tried to solve the vacuum Einstein equations with the ansatz Eq. (14.35), we would have arrived at the conclusion that any function $K$ (or $H$) harmonic in $(d-2)$-dimensional Euclidean space transverse to $z$ provides a solution. Thus, we obtain a family of pp-wave solutions of the form

$$
\boxed{
\begin{aligned}
ds^2 &= H^{-1}dt^2 - H[dz - \alpha(H^{-1} - 1)dt]^2 - d\vec{x}^2_{d-2}, \\
\vec{\partial}^2_{(d-2)} H &= 0, \qquad \alpha = \pm 1.
\end{aligned}
}
\tag{14.42}
$$

# 15
# The Kaluza–Klein black hole

Kaluza's [848] and Nordström's [994] original idea/observation that electromagnetism could be seen as part of five-dimensional gravity, combined with Klein's curling up of the fifth dimension in a tiny circle [862], constitutes one of the most fascinating and recurring themes of modern physics. Kaluza–Klein (KK) theories[1] are interesting both in their own right (in spite of their failure to produce realistic four-dimensional theories [1274], at least when the internal space is a manifold) and because of the usefulness of the techniques of dimensional reduction for treating problems in which the dynamics in one or several directions is irrelevant. We saw an example in Chapter 9, when we related four-dimensional instantons to monopoles.

On the other hand, the effective field theories of some superstring theories (which are supergravity theories) can be obtained by dimensional reduction of 11-dimensional supergravity, which is the low-energy effective field theory of (there is no real consensus on this point) M theory or one of its dual versions. In turn, string theory needs to be "compactified" to take a four-dimensional form and, to obtain the four-dimensional low-energy effective actions, one can apply the dimensional-reduction techniques.

Here we want to give a simple overview of the physics of compact dimensions and the techniques used to deal with them (dimensional reduction, etc.) in a non-stringy context. We will deal only with the compactification of pure gravity and vector fields, leaving aside compactification in the presence of more general matter fields (including fermions) until Part III. We will also leave aside many subjects such as spontaneous compactification and the issue of constructing realistic KK theories, which are covered elsewhere [476, 1271]. In addition to establishing the basic results, we want to study classical solutions of the original and dimensionally reduced theories and to see how KK techniques can be used to generate new solutions of both of them.

This chapter is organized as follows. We first study in Section 15.1 the classical and quantum mechanics of a massless particle in flat spacetime with a compact spacelike dimension. We find that the spectrum consists of an infinite tower of massive states

[1] Reference [57] contains many reprints of the most influential papers on the subject. Two old textbooks that describe the classical KK theory are Refs. [146, 894]. More recent accounts can be found in Refs. [1071, 1176, 225]. Even more recent reviews are Refs. [465, 476]. A book that describes the geometrical foundations is Ref. [370].

and explain the full spectrum of the compactified theory. Next, we perform the simplest dimensional reduction of pure gravity in $\hat{d}$ dimensions to $d = \hat{d} - 1$ dimensions using Scherk and Schwarz's formalism in Section 15.2. We find the action, equations of motion, and symmetries for the massless fields and study various choices of conformal frame. In Section 15.2.3 we study the ("direct") dimensional reduction of the effective action of a massless particle moving in curved spacetime with one compact dimension using the Scherk–Schwarz formalism. We recover the known results about the spectrum of the KK theory in the following form: the massless $\hat{d}$-dimensional particle effective action reduces to the action of a massive, charged, particle moving in $(\hat{d} - 1)$-dimensional space, with mass and charge proportional to the momentum in the compact direction. In Section 15.2.4 we obtain the S dual of the reduced KK theory by the procedure of Poincaré duality explained in Section 12.7.1. In Section 15.2.5 we first reduce the Einstein–Maxwell theory in an arbitrary dimension and then we use the result to reduce the bosonic sector of the $N = 1, d = 5$ SUGRA action Eq. (15.98) (which is a modification of the Einstein–Maxwell action). This will allow us to reduce the solutions of that theory studied earlier.

Once the reduction of theories has been established, in Section 15.3 we study the reduction of particular solutions of the Einstein–Maxwell theory and the "oxidation" of particular solutions of the dimensionally reduced Einstein–Maxwell theory. We will reduce ERN BHs in Section 15.3.1 and the AS shock-wave solution (obtaining in this way the electrically charged KK BH) in Section 15.3.2, and study the possible reduction of Schwarzschild and non-extreme RN BHs in Section 15.3.3. Finally we will see some examples of the use of KK reduction and oxidation combined with dualities to generate new solutions in Section 15.3.4. In particular, exploiting the four-dimensional S-duality symmetry studied in Section 15.2.4, we will obtain the magnetically charged KK BH that becomes, after oxidation to five dimensions, the (Sorkin–Gross–Perry) KK monopole [684, 1148] studied in Chapter 9.

In the remaining sections we give an overview of more general dimensional-reduction techniques: toroidal in Section 15.4, the Scherk–Schwarz generalized dimensional reduction in Section 15.5, and orbifold compactification in Section 15.6.

## 15.1 Classical and quantum mechanics on $\mathbb{R}^{1,3} \times S^1$

The main idea of all KK theories can be stated as follows.

> *KK principle:* our spacetime may have extra dimensions, and spacetime symmetries in those dimensions are seen as internal (gauge) symmetries from the four-dimensional point of view. All symmetries could then be unified.

There are several versions of the extra dimensions (brane-worlds, etc.), and here we will consider only the "standard" extra dimensions which are curled up in a very small compact manifold, the simplest case which we are going to study (and the one originally considered by Kaluza and Klein) being a circle. The motion of particles in this dimension should not be observable in the usual sense by (empirically well-established) assumption and that is why it is considered compact and small.

Spacetime symmetries are associated with the graviton. It is, thus, natural to start by studying the classical and quantum kinematics of a free massless particle representing a graviton in flat five-dimensional spacetime with a compact fifth dimension of length equal to $2\pi R_z$ and parametrized by the *periodic coordinate* $x^4 = z$, which takes values in $[0, 2\pi\ell]$,

$$z \sim z + 2\pi\ell, \tag{15.1}$$

that can be seen as the vacuum of the full KK theory just as Minkowski spacetime is the vacuum of GR. $\ell$ is some fundamental length unit (the Planck length $\ell_{\rm Planck}$, in string theory the string length $\ell_s = \sqrt{\alpha'}$, etc.) and $R_z$ is a fundamental *datum* defining our KK vacuum spacetime, the simplest example of *moduli*. The choice of vacuum in KK theory is, however, arbitrary, and one of the main objections to KK theories is that no dynamical mechanisms explaining why one dimension is compact and has the size indicated by the modulus are provided. This is generically known as the *moduli problem*.

The five-dimensional metric of this spacetime is, then, in these coordinates[2],

$$d\hat{s}^2 = \eta_{\mu\nu}dx^\mu dx^\nu - (R_z/\ell)^2\, dz^2. \tag{15.2}$$

We can already see that the assumption that the fifth dimension is compact has an immediate and important consequence: *five-dimensional Poincaré invariance of the KK vacuum is spontaneously broken*,

$$\mathrm{ISO}(1,4) \to \mathrm{ISO}(1,3) \times \mathrm{U}(1).$$

The five-dimensional Lorentz transformations that mix the compact and non-compact dimensions are not symmetries of the metric (they leave it formally invariant if we set $\ell = R_z$ but they change the periodicity properties of the coordinates). Amongst the five-dimensional Poincaré transformations that do not mix compact and non-compact coordinates, clearly Poincaré transformations in the four non-compact dimensions are a symmetry of the theory and constant shifts in the internal coordinate $z$ are also a $\mathrm{U}(1)$ symmetry of the theory. These are the symmetries of the KK vacuum.

The rescalings of the compact coordinate rescale $\ell$, but not $R_z$, unless we choose to ignore the rescaling of the period of $z$, which is the point of view that is usually adopted. In this case, the rescalings are not a symmetry of the theory because they change the modulus $R_z$ which is part of our definition of the (vacuum of the) theory. This is a *duality transformation* that takes us from one theory to another one (albeit of the same class).

We assume that the kinematics in the fifth dimension are the most straightforward generalization of the four-dimensional ones.[3] Thus, we assume that a free, massless particle moving in a flat five-dimensional spacetime always satisfies[4]

$$\hat{p}^{\hat{\mu}}\hat{p}_{\hat{\mu}} = 0. \tag{15.3}$$

---

[2] Usually in the literature $\ell = R_z$. We prefer this parametrization which emphasizes the distinction between $R_z$, which is a physical parameter, and $\ell$, the range of $z$ which is unphysical. One could also normalize $\ell = 1/(2\pi)$ but coordinates have dimensions of length and it is useful to keep their dependence on $\ell$. In some cases it is easier to take $\ell = R_z$ and we will do so by indicating it explicitly.

[3] It is always implicitly assumed that fundamental constants such as the speed of light $c$ and Planck constant $h$ have the same value in the five-dimensional world and the extra dimension is always taken to be spacelike. These assumptions are completely *ad hoc* and should be taken as *minimal* assumptions, although it is known that extra timelike dimensions give fields with kinetic terms with the wrong sign in lower dimensions and this justifies the assumption.

[4] As we will always do in this and other chapters, we denote five- or, in general, higher-dimensional objects and indices with a hat. Therefore $(\hat{p}^{\hat{\mu}}) = (p^\mu, \hat{p}^z)$ and $(p^\mu) = (p^0, p^1, p^2, p^3)$.

If we separate the four- and five-dimensional pieces of the above equation, it takes the form of a four-dimensional mass-shell condition:

$$p^\mu p_\mu = (p^z R_z/\ell)^2, \tag{15.4}$$

and we see that the momentum in the fifth dimension is "seen" as a four-dimensional mass,

$$M = |\hat{p}^z| R_z/\ell. \tag{15.5}$$

We can now consider the quantum-mechanical side of the problem. A free-particle wave function is a momentum eigenmode

$$\hat{P}_{\hat{\mu}}\hat{\Psi} \equiv -i\hbar\partial_{\hat{\mu}}\hat{\Psi} = \hat{p}_{\hat{\mu}}\hat{\Psi} \quad \Rightarrow \hat{\Psi} = e^{\frac{i}{\hbar}\hat{p}_{\hat{\mu}}\hat{x}^{\hat{\mu}}} \tag{15.6}$$

with $\hat{p}^2 = 0$. The wave function is supposed to be single-valued (periodic) in the compact dimension. For the above wave function, however, we have

$$\hat{\Psi}(x^\mu, z + 2\pi\ell) = e^{-\frac{i}{\hbar}\left(\frac{R_z}{\ell}\right)^2 2\pi\ell\hat{p}^z}\,\hat{\Psi}(x^\mu, z), \tag{15.7}$$

and therefore the momentum in the internal dimension can only take the values

$$\hat{p}^z = n\ell\hbar/R_z^2, \quad \hat{p}_z = n\hbar/\ell, \quad n \in \mathbb{Z}, \tag{15.8}$$

and, on account of Eq. (15.5), the spectrum of four-dimensional masses is given by

$$\boxed{M = \frac{|n|\hbar}{R_z}, \qquad n \in \mathbb{Z}.} \tag{15.9}$$

This is the first prediction of the KK theory: the five-dimensional graviton momentum modes give rise to a discrete spectrum of massive four-dimensional particles plus some massless ones related to $n = 0$. The mass of these *KK modes* is inversely proportional to the size of the internal dimension. If the size of the internal dimension is of the order of the Planck length, these particles will have masses that are multiples of the Planck mass, which would account for the fact that they are not observed.

Observe that $M$ does not depend on $\ell$, but only on the modulus $R_z$.

Let us now move to field theory and consider a five-dimensional, massless, complex scalar field $\hat{\varphi}$ satisfying the five-dimensional sourceless Klein–Gordon equation

$$\hat{\Box}\hat{\varphi} = 0. \tag{15.10}$$

It is natural to Fourier-expand the field:

$$\hat{\varphi}(\hat{x}) = \sum_{n\in\mathbb{Z}} e^{\frac{inz}{\ell}}\varphi^{(n)}(x). \tag{15.11}$$

On substituting Eq. (15.11) into Eq. (15.10), we see that each Fourier mode satisfies the Klein–Gordon equation for massive fields ($\hbar = 1$),

$$\left[\Box - (n/R_z)^2\right]\varphi^{(n)}(x) = 0, \tag{15.12}$$

Table 15.1. Decomposition of the five-dimensional graviton in four-dimensional fields and the physical spectrum.

As explained in the main text, the three four-dimensional fields $g^{(n)}_{\mu\nu}$, $A^{(n)}_{\mu}$, and $k^{(n)}$ for each $n \neq 0$ combine via the Higgs mechanism and represent a massive spin-2 particle (massive graviton) with mass $m = |n|/R_z$, which has five degrees of freedom (DOFs). There are no massive scalars or vectors in the spectrum.

| $n$ | $\hat{d} = 5$ | DOF | $d = 4$ fields | DOF | Physical spectrum |
|---|---|---|---|---|---|
| 0 | $\hat{g}^{(0)}_{\hat{\mu}\hat{\nu}}$ | 5 | $g_{\mu\nu}$ | 2 | graviton $m = 0$ |
| | | | $A_\mu$ | 2 | vector $m = 0$ |
| | | | $k$ | 1 | scalar $m = 0$ |
| $n \neq 0$ | $\hat{g}^{(n)}_{\hat{\mu}\hat{\nu}}$ | 5 | $g^{(n)}_{\mu\nu}$ | 2 | |
| | | | $A^{(n)}_{\mu}$ | 2 | graviton $m = \|n\|/R_z$ |
| | | | $k^{(n)}$ | 1 | |

and, therefore, each Fourier mode corresponds to a scalar KK mode. Dimensional reduction amounts to taking the zero mode alone. If $\hat{\varphi}$ is to be interpreted as a "relativistic wave function," this is all we need to know. However, if we want to do field theory, we are interested in the Green function for the Klein–Gordon equation. For instance, for time-independent sources we are interested in the Laplace equation

$$\Delta_{(4)}\hat{\varphi} = \delta^{(4)}(\vec{x}_4), \qquad \vec{x}_4 = (x^1, \ldots, x^4), \tag{15.13}$$

and we want to know which kind of equations it implies for each KK mode and what its solution is. That is, we want to know the harmonic function $H_{\mathbb{R}\times S^1}$ in $\mathbb{R}^3 \times S^1$ and its relation to harmonic functions in $\mathbb{R}^3$. We will deal with this problem in Appendix O.

The same analysis cannot be naively applied to the five-dimensional metric field $\hat{g}_{\hat{\mu}\hat{\nu}}$. The Fourier modes of a five-dimensional scalar field can be interpreted as scalar fields in four dimensions, but the Fourier modes of the five-dimensional metric cannot be interpreted as four-dimensional metrics because they are $5 \times 5$ matrices. The same applies to vector or spinor fields. We have to decompose the fields with respect to the four-dimensional Poincaré group.

For the graviton, the result is represented schematically in Table 15.1. Let us first focus on the Fourier zero mode, which is a $5 \times 5$ symmetric matrix. It can be decomposed (in several ways) into a $4 \times 4$ symmetric matrix that can be interpreted as the four-dimensional metric (graviton), a four-dimensional vector, and a scalar. We will see in detail in Section 15.2 how this four-dimensional massless mode of the five-dimensional graviton $\hat{g}^{(0)}_{\hat{\mu}\hat{\nu}}$ (five helicity states) can be decomposed into one massless graviton $g_{\mu\nu}$ (two helicity states), one massless vector $A_\mu$ (two helicity states), which we will call a *KK vector*, and one massless scalar $k$ (one helicity state), which we will call a *KK scalar*. The number of helicity states (degrees of freedom) is conserved in this decomposition. The massless spectrum is, thus,

$$\{g_{\mu\nu}, A_\mu, k\}, \tag{15.14}$$

and its symmetries are the local version of symmetries of the KK vacuum determined by the metric Eq. (15.2) plus a vanishing vacuum expectation value for the vector field $\langle A_\mu \rangle = 0$, i.e. four-dimensional GCTs times local U(1) whose gauge field is $A_\mu$.

The infinite tower of four-dimensional massive modes is constituted by spin-2 particles (*massive gravitons*) [1098]. They appear as interacting massless[5] gravitons, vectors, and scalars labeled by an integer:

$$\{g_{\mu\nu}^{(n)}, A_\mu^{(n)}, k^{(n)}\}. \tag{15.15}$$

As for the $n \neq 0$ modes, we will see that these fields are related by an infinite symmetry group that contains the Virasoro group [458]. These symmetries are spontaneously broken in the above KK vacuum, and the fields $A_\mu^{(n)}$ and $k^{(n)}$ are the corresponding Goldstone bosons. Owing to the Higgs mechanism, a massless vector and scalar are "eaten" by each massless graviton, giving rise to the massive gravitons [458, 353, 354]. Observe that the number of helicity states is also preserved.[6]

A brief and approximate description of how the Higgs mechanism works in this case is worth giving. Some of the symmetries acting on the $n \neq 0$ sector are *massive gauge transformations*, which include shifts of the scalars $k^{(n)}$ by arbitrary functions that are also standard gauge parameters for the vectors ${A_\mu}^{(n)}$ and shifts of the vectors ${A_\mu}^{(n)}$ by arbitrary vectors that are standard gauge transformations for the $g_{\mu\nu}^{(n)}$s. This means that the gauge-invariant field strengths of the scalars and vectors have, very roughly, the structure

$$\partial_\mu k^{(n)} + nA_\mu^{(n)}, \qquad \partial_\mu A_\nu^{(n)} + ng_{\mu\nu}^{(n)}. \tag{15.16}$$

[5] Strictly speaking, one cannot speak about the mass of these fields since, due to the interactions, neither of them is a mass eigenstate [353, 354]. By massless here we simply mean that they enjoy gauge invariances analogous to those of the massless fields.

[6] More generally, in $\hat{d}$ dimensions the graviton (spin 2) has $\hat{d}(\hat{d}-3)/2$ helicity states and a massless $(p+1)$-form potential has $(\hat{d}-2)!/[(p+1)!(\hat{d}-p-3)!]$ helicity states. In particular, a massless spin-1 particle (vector, $p=0$) has $\hat{d}-2$ and a spin-0 particle (scalar $p=-1$) always has one. A *massive graviton* (spin-2 particle) has $\hat{d}(\hat{d}-1)/2-1$ helicity states and a massive $(p+1)$-form potential has $(\hat{d}-1)!/[(p+1)!(\hat{d}-p-2)!]$ helicity states. In particular, a massive spin-1 particle (a massive vector, $p=0$) has $\hat{d}-1$ helicity states and a massive spin-0 particle (a massive scalar, $p=-1$) has just one. Thus, just on the basis of counting helicity states, the $\hat{d}$-dimensional graviton can always be decomposed into a $(\hat{d}-1)$-dimensional massless graviton, vector, and scalar, and, if the interactions allow it, via the Higgs mechanism, these massless particles can combine into a $(\hat{d}-1)$-dimensional *massive graviton*, which has the same number of helicity states as the massless $\hat{d}$-dimensional one. Analogously, a massless $\hat{d}$-dimensional $(p+1)$-form potential gives rise to massless $(\hat{d}-1)$-dimensional $(p+1)$- and $p$-form potentials. If the interactions allow it, these two potentials can combine via the Higgs mechanism into a $(\hat{d}-1)$-dimensional massive $(p+1)$-form potential that has the same number of helicity states as the massless $\hat{d}$-dimensional one. Since invariance under GCTs is (see Appendix 3.2) nothing but the gauge symmetry of the massless spin-2 particle, the theory of the *massive graviton* cannot have it. However, in the description of the *massive graviton* as a coupled system of massless graviton, vector, and scalar field, it is possible to have invariance under GCTs that is spontaneously broken by the Higgs mechanism.

These field strengths appear squared in the action. Using the massive gauge transformations, $k^{(n)}$ and $A_\mu{}^{(n)}$ can be gauged away, leaving mass terms for the $g^{(n)}_{\mu\nu}$s. Thus, $k^{(n)}$ and $A_\mu{}^{(n)}$ play the role of Stückelberg fields, like the scalar that one can introduce in massive electrodynamics to preserve a (formal) gauge invariance (see Section 12.2.2). More examples of massive gauge transformations can be found in Section 15.5.

In the above vacuum, the masslessness of the KK scalar is associated with this being the Goldstone boson of dilatations of the compact coordinate (under which it scales).

In the full KK theory ("compactification") all modes should be taken into account. More often, though, all massive modes (all KK modes) are ignored and only the massless spectrum is kept. This is equivalent to ignoring all dynamics in the internal dimensions and it is called *dimensional reduction*. This is the only consistent truncation of the full theory. It is, on the other hand, the effective theory which describes the low-energy behavior of the full theory and contains a good deal of information about the full theory. In particular, the massive modes reappear in it as solitonic solutions: extreme electrically charged KK BHs. This non-trivial fact makes the truncated action even more interesting.

In the "decompactification" limit $R_z \to \infty$ the difference between the masses of the $n$th and $(n+1)$th modes goes to zero and the spectrum becomes continuous, just like the usual momentum spectrum in a non-compact direction.

To complete our description of the KK spectrum, we should mention that, as we will see later, the KK modes also carry electric charge with respect to the massless KK vector field $A_\mu$. However (as with the details on the spectrum that we have just given), this cannot be seen in flat spacetime. In fact, now we see only that they have a certain rest mass. We know that the gravitational field will couple to it, and we know this even if we do not introduce the gravitational field. However, we can see the electric charge only in the presence of an electromagnetic field. Both the gravitational field and the electric field originate from the five-dimensional gravitational field, which we have not included so far. We will show this in Section 15.2, and we will show that KK modes carry electric charge with respect to this field in Section 15.2.3.

To end this section, let us mention that it has been argued that the KK vacuum is quantum mechanically unstable [1273].

## 15.2 KK dimensional reduction on a circle $S^1$

In this section we are going to perform the dimensional reduction of $\hat{d}$-dimensional gravity to $d \equiv \hat{d}-1$ dimensions in the formalism developed by Scherk and Schwarz in Ref. [1106]. Thus, here we are going to consider only the massless modes of the graviton field. By definition they do not depend on the compact ("internal") spacelike coordinate $\hat{x}^{\hat{d}-1}$, which we denote by $z$ and which is periodically identified with period $2\pi\ell$, where $\ell$ is some fundamental length in the theory. We are also going to see how the graviton field splits into $d$-dimensional fields.

At this point we would like to stress that, in KK theory, the use of distinguished coordinates is unavoidable: up to constant shifts, there is only one coordinate $z$ that is periodic with period $2\pi\ell$ and the Fourier mode expansion has to be done with respect to that coordinate. The metric zero mode is defined by the fact that it does not depend on that coordinate. Furthermore, technically, the dimensional-reduction procedure requires that we use the coordinate $z$.

Our starting point, therefore, is a $\hat{d}$-dimensional[7] metric $\hat{g}_{\hat{\mu}\hat{\nu}}$ independent of $z$.

It is sometimes convenient to give a coordinate-independent characterization of the metrics we are going to deal with. These are metrics admitting a spacelike Killing vector $\hat{k}^{\hat{\mu}}$. If the metric admits the Killing vector $\hat{k}^{\hat{\mu}}$ then its Lie derivative with respect to it vanishes:

$$\mathcal{L}_{\hat{k}}\hat{g}_{\hat{\mu}\hat{\nu}} = 2\hat{\nabla}_{(\hat{\mu}}\hat{k}_{\hat{\nu})} = 0 \tag{15.17}$$

(this is just the Killing equation, see Section 1.5) and this is the condition we would impose on other fields, if we had them.

To this local condition we have to add a global condition: that the integral curves of the Killing vector are closed. $z$ will be the coordinate parametrizing those integral curves (the "adapted coordinate") and it can be rescaled to make it have period $2\pi\ell$. This global condition will not be explicitly used in most of what follows, but only this condition guarantees consistency. In adapted coordinates, $\hat{k}^{\hat{\mu}} = \delta_{\underline{z}}{}^{\hat{\mu}}$.

It is reasonable to think of the hypersurfaces orthogonal to the Killing vector as the $d$-dimensional spacetime of the lower-dimensional theory. Then, the first object of interest is the metric induced on them. This is

$$\hat{\Pi}_{\hat{\mu}\hat{\nu}} \equiv \hat{g}_{\hat{\mu}\hat{\nu}} + k^{-2}\hat{k}_{\hat{\mu}}\hat{k}_{\hat{\nu}}, \qquad k^2 \equiv -\hat{k}^{\hat{\mu}}\hat{k}_{\hat{\mu}}. \tag{15.18}$$

$\hat{\Pi}^{\hat{\mu}}{}_{\hat{\nu}} = \hat{g}^{\hat{\mu}\hat{\rho}}\hat{\Pi}_{\hat{\rho}\hat{\nu}}$ can be used to project onto directions orthogonal to the Killing vector and $-k^{-2}\hat{k}^{\hat{\mu}}\hat{k}_{\hat{\nu}}$ can be used to project onto directions parallel to it. In adapted coordinates, due to the orthogonality of $\hat{\Pi}$ and $\hat{k}$, we have

$$k = |\hat{k}^{\hat{\mu}}\hat{k}_{\mu}|^{\frac{1}{2}} = |\hat{g}_{\underline{zz}}|, \qquad \hat{\Pi}_{\hat{\mu}\underline{z}} = 0. \tag{15.19}$$

The remaining components define the $(\hat{d}-1)$-dimensional metric

$$g_{\mu\nu} \equiv \hat{\Pi}_{\mu\nu}. \tag{15.20}$$

To understand why this is the right definition of the $(\hat{d}-1)$-dimensional metric instead of just $\hat{g}_{\mu\nu}$ (apart from the reason to do with orthogonality to the Killing vector), we need to examine the effect of $\hat{d}$-dimensional GCTs on it. Under the infinitesimal GCTs $\delta_{\hat{\epsilon}}\,\hat{x}^{\hat{\mu}} = \hat{\epsilon}^{\hat{\mu}}(\hat{x})$, the $\hat{d}$-dimensional metric transforms as follows:

$$\delta_{\hat{\epsilon}}\,\hat{g}_{\hat{\mu}\hat{\nu}} = -\hat{\epsilon}^{\hat{\lambda}}\partial_{\hat{\lambda}}\hat{g}_{\hat{\mu}\hat{\nu}} - 2\hat{g}_{\hat{\lambda}(\hat{\mu}}\partial_{\hat{\nu})}\hat{\epsilon}^{\hat{\lambda}}. \tag{15.21}$$

For the moment, we are interested only in $\hat{d}$-dimensional GCTs that respect the KK ansatz, i.e. that do not introduce any dependence on the internal coordinate $z$. These fall into two classes: those with infinitesimal generator $\hat{\epsilon}^{\hat{\mu}}$ independent of $z$ and those generated by a $z$-dependent $\hat{\epsilon}^{\hat{\mu}}$. The latter act only on $z$ and they are found to be only

$$\delta z = az, \qquad a \in \mathbb{R}, \tag{15.22}$$

[7] All $\hat{d}$-dimensional objects carry a hat, whereas $d = (\hat{d}-1)$-dimensional ones do not. The $\hat{d}$-dimensional indices split as follows: $\hat{\mu} = (\mu, \underline{z})$ (curved) and $\hat{a} = (a, z)$ (tangent-space indices).

which can be integrated to give global rescalings plus shifts of the coordinate $z$:

$$z' = az + b, \qquad a, b \in \mathbb{R}. \tag{15.23}$$

The former can be projected onto the directions orthogonal or parallel to the Killing vector. In orthogonal directions they are just $(\hat{d}-1)$-dimensional GCTs,

$$\delta_\epsilon x^\mu = \epsilon^\mu, \qquad \epsilon^\mu = \hat{\Pi}^\mu{}_{\hat{\nu}}\hat{\epsilon}^{\hat{\nu}} = \hat{\epsilon}^\mu. \tag{15.24}$$

In parallel directions they act only on $z$,

$$\delta_\Lambda z = -\Lambda, \qquad \Lambda = k^{-2}\hat{k}_{\hat{\nu}}\hat{\epsilon}^{\hat{\nu}} = \hat{\epsilon}^{\underline{z}}, \tag{15.25}$$

which must correspond to some local internal symmetry of the lower-dimensional theory. As we argued before, the $\hat{d}$-dimensional metric is going to give rise to the massless $(\hat{d}-1)$-dimensional fields (15.14). These fields should have good transformation properties under this internal symmetry. In particular, the metric must be invariant under it and the vector must transform under it in the standard way (because it is massless):

$$\delta_\Lambda A_\mu = \partial_\mu \Lambda. \tag{15.26}$$

Observe that the periodicity of $\Lambda$ has to be the same as the periodicity of $z$ in order for it to be a well-defined coordinate transformation. We know that the period of the $\mathrm{U}(1)$ gauge parameters is related to the unit of electric charge, and we will see that this is also the case in KK theories.

Using the above transformation law for the various components of the $\hat{d}$-dimensional metric, we arrive at the conclusion that the lower-dimensional fields are the following natural combinations of them:

$$g_{\mu\nu} = \hat{g}_{\mu\nu} - \hat{g}_{\underline{z}\mu}\hat{g}_{\underline{z}\nu}/\hat{g}_{\underline{zz}}, \qquad A_\mu = \hat{g}_{\mu\underline{z}}/\hat{g}_{\underline{zz}}, \qquad k = |\hat{g}_{\underline{zz}}|^{\frac{1}{2}} = |\hat{k}^{\hat{\mu}}\hat{k}_{\hat{\mu}}|^{\frac{1}{2}}. \tag{15.27}$$

Equivalently, we can say that the higher-dimensional metric decomposes as follows:

$$\hat{g}_{\mu\nu} = g_{\mu\nu} - k^2 A_\mu A_\nu, \qquad \hat{g}_{\mu\underline{z}} = -k^2 A_\mu, \qquad \hat{g}_{\underline{zz}} = -k^2. \tag{15.28}$$

Furthermore, under the global transformations of the internal space Eq. (15.23), the metric is invariant and only $A_\mu$ and $k$ transform. The shifts of $z$ have no effect on them and we are left with a multiplicative $\mathbb{R}$ duality group that can be split according to $\mathbb{R} = \mathbb{R}^+ \times \mathbb{Z}_2$. Only $\mathbb{R}^+$ acts on $k$,

$$A'_\mu = aA_\mu, \qquad k' = a^{-1}k, \qquad a \in \mathbb{R}^+, \tag{15.29}$$

and only $A_\mu$ transforms under the $\mathbb{Z}_2$ factor,

$$A'_\mu = -A_\mu. \tag{15.30}$$

It is a general rule that, in dimensional reductions, global internal transformations give rise to non-compact global symmetries of the lower-dimensional-theory action which generally rescale and/or rotate the fields among themselves. In particular, they act on scalars, and thus scalars naturally parametrize a $\sigma$-model. In this case $k$ parametrizes a $\sigma$-model with target space $\mathbb{R}^+$. As we explained before, these transformations should not be understood as symmetries but as *dualities* relating different theories.[8]

Observe that, in Section 15.1, the radius of the compact dimension $R_z$ appeared explicitly in the metric. In curved spacetime and at each point of the lower-dimensional spacetime we can define a local radius of the compact dimension $R_z(x)$,

$$2\pi R_z(x^\mu) = \int_0^{2\pi\ell} dz|\hat{g}_{\underline{zz}}|^{\frac{1}{2}} = \int_0^{2\pi\ell} kdz. \tag{15.31}$$

Thus, we see that the KK scalar measures the local size of the internal dimension. We should require that, asymptotically, our five-dimensional metric approaches that of the vacuum Eq. (15.2). Then, we find the following relation among the modulus $R_z$, the fundamental scale length $\ell$, and the asymptotic value of the KK scalar $k_0$:

$$\boxed{R_z = \ell k_0, \qquad k_0 = \lim_{r\to\infty} k.} \tag{15.32}$$

Sometimes the word modulus is used for the full scalar $k$. However, only its value at infinity, which we will see is not determined by the equations of motion and thus has to be set by hand as a datum defining the theory, really deserves that name.

Since masses are measured at infinity and, in KK theory, we know that these depend on the radius of the compact dimension through Eq. (15.9), we expect that the masses will depend on the value at infinity of the radius of the compact dimension $R_z$ (which is why we have used the same symbol to denote them).

### 15.2.1 The Scherk–Schwarz formalism

Having determined the relations Eqs. (15.28) and (15.27) between the lower- and higher-dimensional fields, one can simply plug them into the equations of motion of the higher-dimensional fields (here just Einstein's equations) and obtain equations for the lower-dimensional ones. This procedure automatically ensures that any field configuration that solves the lower-dimensional equations of motion also solves (when it is translated to higher-dimensional fields) the higher-dimensional equations of motion.

In this way one can see that it is not correct to set the KK scalar to a constant as was usually done in the very early KK literature. As was first realized in Ref. [1178], the KK scalar has a non-trivial equation of motion, which we will find later, and, if one sets it to a constant, this equation of motion transforms into a constraint for the vector field strength. This constraint is not generically satisfied and, therefore, solutions with $k = k_0$ that do not satisfy this constraint are not solutions of the original theory.

As a general rule, one cannot naively truncate actions by setting some fields to specific values. Doing this in the equations of motion (the correct procedure) would leave us with constraints that must be satisfied and cannot be obtained from the truncated actions. In other words, we cannot reproduce all the truncated equations of motion from a truncated action.

[8] Observe that $\ell$ is fixed. Dualities change $R_z$ only.

When will a truncation in the action be consistent? Also as a general rule, if there is a discrete symmetry in the action, eliminating only the fields which are not invariant under it will always be consistent. From this point of view, since there is no discrete symmetry acting on $k$, the inconsistency of its elimination is not surprising. On the other hand, there is a $\mathbb{Z}_2$ symmetry that acts only on $A_\mu$ and it is easy to see that it is consistent to eliminate only this field. For instance, this truncation is used to obtain $N=1, d=10$ supergravity from $N=1, \hat{d}=11$ supergravity (or the heterotic string from M theory) and can be related to dimensional reduction over the orbifold $\mathrm{S}^1/\mathbb{Z}_2$ (a segment of a line, with two boundaries) instead of on the circle $\mathrm{S}^1$.

Performing the dimensional reduction on the equations of motion is in general a quite lengthy calculation. Furthermore, the above decomposition of higher-dimensional fields into lower-dimensional ones cannot be used in the presence of fermions.

In Ref. [1106] Scherk and Schwarz described a systematic procedure for performing the dimensional reduction in the action and using the Vielbein formalism so it can also be applied to fermions. Another advantage of using Vielbeins is that we can work with objects that have only Lorentz indices and are, therefore, scalars under GCTs. Since some of the GCTs become internal gauge transformations, those objects are automatically GCT-scalars and gauge invariant.

The first thing to do is to reexpress the relations Eqs. (15.27) and (15.28) in terms of Vielbeins. Using local Lorentz rotations, we can always choose an upper-triangular Vielbein basis of the form

$$\left(\hat{e}_{\hat{\mu}}{}^{\hat{a}}\right) = \begin{pmatrix} e_\mu{}^a & kA_\mu \\ 0 & k \end{pmatrix}, \qquad \left(\hat{e}_{\hat{a}}{}^{\hat{\mu}}\right) = \begin{pmatrix} e_a{}^\mu & -A_a \\ 0 & k^{-1} \end{pmatrix}, \tag{15.33}$$

where $A_a = e_a{}^\mu A_\mu$ and we will assume that all $d$-dimensional fields with Lorentz indices have been contracted with the $d$-dimensional Vielbeins.

This choice of Vielbein basis breaks the $\hat{d}$-dimensional local Lorentz invariance to the $d=(\hat{d}-1)$-dimensional one, which is the subgroup that preserves our choice. If there were other symmetries (such as supersymmetry) acting on the Vielbeins, we would have to add to them compensating Lorentz transformations in order to preserve the choice of Vielbeins.

Next, we find the non-vanishing components of $\hat{\Omega}_{\hat{a}\hat{b}\hat{c}}$,

$$\hat{\Omega}_{abc} = \Omega_{abc}, \qquad \hat{\Omega}_{abz} = -\tfrac{1}{2}kF_{ab}, \qquad \hat{\Omega}_{azz} = -\tfrac{1}{2}\partial_a \ln k, \tag{15.34}$$

where

$$F_{ab} = e_a{}^\mu e_b{}^\nu F_{\mu\nu}, \qquad F_{\mu\nu} = 2\partial_{[\mu}A_{\nu]}, \tag{15.35}$$

is the vector field strength. With these we find the non-vanishing components of the spin connection $\hat{\omega}_{\hat{a}\hat{b}\hat{c}}$:

$$\begin{aligned} \hat{\omega}_{abc} &= \omega_{abc}, & \hat{\omega}_{abz} &= \tfrac{1}{2}kF_{ab}, \\ \hat{\omega}_{zbc} &= -\tfrac{1}{2}kF_{bc}, & \hat{\omega}_{zbz} &= -\partial_b \ln k. \end{aligned} \tag{15.36}$$

Now, instead of calculating the Ricci scalar, which involves derivatives of the spin connection, we use the following simplifying trick: we first eliminate the derivatives of the spin connection from the action by integration by parts. The result is known as *Palatini's identity* and it is derived in Appendix L for a more general case. On plugging the above results plus

$$\sqrt{|\hat{g}|} = \sqrt{|g|}\, k \tag{15.37}$$

into the $\hat{d}$-dimensional Palatini's identity Eq. (L.4) with $K = 1$, we immediately find that the $\hat{d}$-dimensional Einstein–Hilbert action can be reexpressed, up to total derivatives, in $(\hat{d} - 1)$-dimensional language as follows:

$$\begin{aligned}\int d^{\hat{d}}\hat{x}\, \sqrt{|\hat{g}|}\, \hat{R} = \int dz \int d^{\hat{d}-1}x\, \sqrt{|g|}\, k\Big\{ & -\omega_b{}^{ba}\omega_c{}^c{}_a - \omega_a{}^{bc}\omega_{bc}{}^a \\ & + 2\omega_b{}^{ba}\partial_a \ln k - \tfrac{1}{4}k^2 F^2\Big\}.\end{aligned} \tag{15.38}$$

Nothing depends on the internal coordinate $z$ and we can integrate over it, obtaining a factor of $2\pi\ell$. Using now "backwards" the $(\hat{d} - 1)$-dimensional Palatini's identity with $K = k$, we find at last

$$\boxed{\hat{S} = \frac{1}{16\pi G_N^{(\hat{d})}} \int d^{\hat{d}}\hat{x}\, \sqrt{|\hat{g}|}\, \hat{R} = \frac{2\pi\ell}{16\pi G_N^{(\hat{d})}} \int d^{\hat{d}-1}x \sqrt{|g|}\, k\big[R - \tfrac{1}{4}k^2F^2\big].} \tag{15.39}$$

This result is correct up to total derivatives (the ones ignored in applying Palatini's identity). In particular, let us stress that there was not a scalar $\hat{K}$ as in Eq. (4.43) in the original action, because objects that were total derivatives in the previous case would not be so in this case, and in the various integrations by parts factors of $\partial\hat{K}$ would be picked up. These factors are taken into account in the generalized Palatini's identity Eq. (L.4). We will often deal with this kind of Lagrangian in Part III.

Another important point is to realize that this action rescales under the global rescalings Eq. (15.29). This happens, though, only because we have chosen to ignore the effect of the rescalings on the period of $z$. On taking that effect into account, the action would be a scalar, as is the original action.

The KK scalar appears in a strange way because it does not seem to have a kinetic term, so one would say that it has no dynamics. However, one has to remember that, in deriving the Einstein equations of motion, one has to integrate several times by parts. In these integrations, derivatives of $k$ are picked up and one can see that $k$ has standard equations of motion that are implicit in Einstein's equations.

The equations of motion are[9]

$$\begin{aligned}\frac{16\pi G_N^{(\hat{d})}}{2\pi\ell}\frac{1}{k\sqrt{|g|}}\frac{\delta S}{\delta g^{\alpha\beta}} = {} & G_{\alpha\beta} + \big[\partial_\alpha \ln k\, \partial_\beta \ln k - g_{\alpha\beta}(\partial \ln k)^2\big] \\ & + \big[\nabla_\alpha\partial_\beta \ln k - g_{\alpha\beta}\nabla^2 \ln k\big] - \tfrac{1}{2}k^2\big[F_\alpha{}^\mu F_{\beta\mu} - \tfrac{1}{4}g_{\alpha\beta}F^2\big] = 0,\end{aligned} \tag{15.40}$$

[9] See Section 4.2 for a detailed derivation of Einstein's equations in the presence of an overall scalar factor.

$$\frac{16\pi G_{\rm N}^{(\hat{d})}}{2\pi\ell}\frac{1}{\sqrt{|g|}}\frac{\delta S}{\delta k} = R - \tfrac{3}{4}k^2F^2 = 0, \tag{15.41}$$

$$\frac{16\pi G_{\rm N}^{(\hat{d})}}{2\pi\ell}\frac{1}{\sqrt{|g|}}\frac{\delta S}{\delta A_\alpha} = \nabla_\beta\left(k^3F^{\beta\alpha}\right) = 0. \tag{15.42}$$

On combining the KK scalar equation with the trace of the Einstein equation, we find a standard equation of motion for $k$,

$$\nabla^2 k = -\frac{\hat{d}-2}{4}k^3F^2. \tag{15.43}$$

Setting $k = k_0$ is consistent only if $F^2 = 0$, which is not true in general. As we explained before, the KK scalar cannot simply be ignored, as was first realized in Ref. [1178]. The truncation $A_\mu = 0$ is, nevertheless, consistent.

Another way to see that the KK scalar is dynamical is to rescale the metric to the so-called *Einstein conformal frame*. By definition, this frame is the one in which the Einstein–Hilbert action has the standard form (without the factor of $k$). The rescaled metric is the *Einstein metric* $g_{{\rm E}\,\mu\nu}$. In the context of Jordan–Brans–Dicke theories, the metric $g_{\mu\nu}$ is sometimes called the *Jordan metric*, but we will call it the *KK metric* and we will refer to the corresponding conformal frame as the *KK conformal frame*.

Using the formulae of Appendix M, we find that the conformal factor is[10] (for $\hat{d} \neq 3$)

$$\Omega = k^{\frac{-1}{d-2}}, \qquad g_{\mu\nu} = k^{\frac{-2}{d-2}}g_{{\rm E}\,\mu\nu}, \tag{15.44}$$

and with it we obtain

$$\boxed{S_{\rm E} = \frac{2\pi\ell}{16\pi G_{\rm N}^{(d+1)}}\int d^dx\sqrt{|g_{\rm E}|}\left[R_{\rm E} + \frac{d-1}{d-2}k^{-2}(\partial k)^2 - \tfrac{1}{4}k^{2\frac{d-1}{d-2}}F^2\right].} \tag{15.45}$$

This action is *not* invariant under the global rescalings Eq. (15.29) because the Einstein metric also rescales under them. Rather, it rescales by a global factor that could be absorbed into the rescaling of $\ell$ (which we have chosen not to do).

However, we can combine these rescalings with a rescaling of the $\hat{d}$-dimensional metric that rescales the $\hat{d}$- and $d$-dimensional actions in such a way that the Einstein metric is invariant and only the KK scalar and vector field rescale:

$$k' = ck, \qquad A'_\mu = c^{-\frac{d-1}{d-2}}A_\mu, \qquad c \in \mathbb{R}^+. \tag{15.46}$$

[10] We replace $\hat{d}-1$ by $d$ to avoid confusion, since we are going to use these actions very often.

The KK action in the Einstein frame exhibits manifest invariance under these global rescalings which are, together with the $\mathbb{Z}_2$ transformations, the duality group of the theory. This is a standard feature of KK and supergravity theories in the Einstein frame: they are manifestly invariant under duality symmetries. In particular, the scalars that appear in these theories parametrize some $\sigma$-model. In this case, the kinetic term for $k$ is the $\mathbb{R}^+$ $\sigma$-model. It is sometimes convenient to use a scalar with a standard kinetic term $\varphi$,

$$k = e^{\pm\sqrt{2\frac{d-2}{d-1}}\varphi}, \tag{15.47}$$

in terms of which the action takes the form

$$\boxed{S_{\rm E} = \frac{2\pi\ell}{16\pi G_{\rm N}^{(d+1)}}\int d^dx\sqrt{|g_{\rm E}|}\left[R_{\rm E} + 2(\partial\varphi)^2 - \tfrac{1}{4}e^{\pm 2\sqrt{2\frac{d-1}{d-2}}\varphi}F^2\right].} \tag{15.48}$$

$\varphi$ transforms under the global rescalings Eq. (15.46) by constant shifts,

$$\varphi' = \varphi \pm \sqrt{\frac{d-1}{2(d-2)}}\ln c. \tag{15.49}$$

The redefinition of the field above is just a change of variables; $\varphi$ parametrizes $\mathbb{R}$. The two group manifolds are isomorphic, one as a multiplicative group and the other as an additive group.

Owing to its behavior under dilatations, the KK scalar is sometimes called the *dilaton*. We reserve this name for the string-theory dilaton. However, in Section 22.1 we will see that the KK scalar one obtains in the reduction of $\hat{d} = 11$ supergravity to $N = 2A, d = 10$ supergravity can be interpreted as the type-IIA string-theory dilaton.

In fact, the action Eq. (15.48) is an example of the general class of actions described by the "$a$-model" whose action Eq. (16.1) depends on a continuous parameter $a$. In this case

$$a = \pm\sqrt{\frac{2d-1}{d-2}}. \tag{15.50}$$

In Chapter 16 we will find BH-type solutions of the $a$-model for any value $a$ and here we will simply use those results for the given specific value of $a$.

### 15.2.2 Newton's constant and masses

In the presence of gravity, masses are measured at infinity in asymptotically flat spacetimes. When one dimension is compact, we can speak only about asymptotic flatness in the non-compact directions.[11] In particular, the diagonal component of the metric in the compact dimension does not have to go to $-1$ at infinity but can be any real negative number. If the metric is asymptotically flat in the non-compact directions then the dimensionally reduced

[11] The definition of mass in spacetimes with compact dimensions has also been discussed in Refs. [228, 444].

metric (assuming that the compact dimension is isometric) will be asymptotically flat in the KK conformal frame and the value of $k$ at infinity will be some positive real number $k_0$.

When we rescale the metric to go to the Einstein conformal frame, the metric does not look asymptotically flat any longer, but

$$\lim_{r\to\infty} g_{\mathrm{E}\,\mu\nu} = k_0^{\frac{2}{d-2}} \eta_{\mu\nu}, \tag{15.51}$$

and a change of coordinates is necessary:

$$x^\mu \to x'^\mu = k_0^{\frac{1}{d-2}} x^\mu \Rightarrow g_{\mathrm{E}\,\mu\nu} \to g'_{\mathrm{E}\,\mu\nu} = k_0^{-\frac{2}{d-2}} g_{\mathrm{E}\,\mu\nu} \overset{r'\to\infty}{\longrightarrow} \eta_{\mu\nu}. \tag{15.52}$$

Thus, if we start with $\hat{d}$-dimensional metrics that are asymptotically flat in the non-compact dimensions, we are forced to perform a rescaling of the coordinates, which is, at the very least, quite unusual. Of course, this change of coordinates does not modify the action Eq. (15.45).

We could have decided to start with $\hat{d}$-dimensional metrics, which naturally lead to asymptotically flat Einstein metrics with no need for changes of coordinates, but this looks rather artificial.

As we pointed out before, a very interesting aspect of the massless sector of the KK theory is that the truncated massive modes reappear as solitonic solutions. A further problem of the standard Einstein conformal frame is that the masses found for solitons are not the ones expected in the spectrum of Kaluza–Klein theories. We are going to check this explicitly in Section 15.2.3.

The prescription we have used to go to the Einstein frame is not canonical, though. We just wanted to eliminate the unconventional (local) factor of $k$ in front of the curvature scalar, and the conformal factor that does the job is unique only up to an overall constant factor. In particular, we could have rescaled the KK metric by the factor $\tilde{\Omega} = (k/k_0)^{-\frac{1}{d-2}}$ which defines the *modified Einstein conformal frame*

$$g_{\mu\nu} = (k/k_0)^{-\frac{2}{d-2}} \tilde{g}_{\mathrm{E}\,\mu\nu}. \tag{15.53}$$

One of the main characteristics of this metric is that it is invariant under the scale transformations Eq. (15.29). It is appropriate to use with it fields that are also invariant under those rescalings:

$$\tilde{A}_\mu = k_0 A_\mu, \qquad \tilde{k} = k/k_0. \tag{15.54}$$

In terms of these scale-invariant fields, the action takes the form

$$\boxed{\tilde{S}_{\mathrm{E}} = \frac{2\pi\ell k_0}{16\pi G_{\mathrm{N}}^{(\hat{d})}} \int d^dx \sqrt{|\tilde{g}_{\mathrm{E}}|} \left[ \tilde{R}_{\mathrm{E}} + \frac{d-1}{d-2}\tilde{k}^{-2}\left(\partial\tilde{k}\right)^2 - \tfrac{1}{4}\tilde{k}^{2\frac{d-1}{d-2}}\tilde{F}^2 \right],} \tag{15.55}$$

which is identical to the action in the original "Einstein frame" Eq. (15.45) except for the overall factor.

This is the frame that leads to correct results.[12] The main difference is the overall factor $k_0$ which modifies the effective value of the $d$-dimensional Newton constant which is given by (recall Eq. (15.32))

$$\boxed{G_{\rm N}^{(d)} = \frac{G_{\rm N}^{(\hat{d})}}{2\pi R_z} = \frac{G_{\rm N}^{(\hat{d})}}{V_z}.} \tag{15.56}$$

Here $V_z$ stands for the volume of the compact dimension.

Now, in $d$ dimensions, in the Einstein frame, with the action normalized,

$$S = \frac{1}{16\pi G_{\rm N}^{(d)}} \int d^dx \sqrt{|g_{\rm E}|}\, R_{\rm E}, \tag{15.57}$$

the mass $M_{\rm E}$ of a given asymptotically flat solution can be read off from $g_{{\rm E}\,tt}$:

$$g_{{\rm E}\,tt} \sim 1 - \frac{16\pi G_{\rm N}^{(d)} M_{\rm E}}{(d-2)\omega_{(d-2)}} \frac{1}{r^{d-3}}. \tag{15.58}$$

This definition can be used to find the mass in the modified Einstein frame, which we denote by $M$, or in the Einstein frame (after rescaling the coordinates so the metric is asymptotically flat), which we denote by $M_{\rm E}$. The relation between these two masses for the same spacetime can easily be computed:

$$\begin{aligned} g'_{{\rm E}\,t't'} &\sim 1 - \frac{16\pi G_{\rm N}^{(\hat{d})} M_{\rm E}}{2\pi\ell(d-2)\omega_{(d-2)}} \frac{1}{r'^{\,d-3}}, \\ \tilde{g}_{{\rm E}\,tt} &\sim 1 - \frac{16\pi G_{\rm N}^{(\hat{d})} M}{2\pi\ell k_0(d-2)\omega_{(d-2)}} \frac{1}{r^{d-3}}, \end{aligned} \tag{15.59}$$

and, using the relation between primed and unprimed coordinates, we find

$$\boxed{M_{\rm E} = k_0^{-\frac{1}{d-2}} M.} \tag{15.60}$$

It is also handy to have the definition of the electric and magnetic charges $\tilde{q}$ and $\tilde{p}$ to which the scale-invariant KK vector $\tilde{A}_\mu$ couples. To define the charge, we first find the

[12] The names "Einstein frame" and "modified Einstein frame" are a bit confusing and we keep them just because they are standard names in the literature. Both are Einstein frames in the sense that there is no scalar factor in the action in front of the Ricci scalar. However, there is an infinite number of conformal frames with that property, related by constant rescalings. Among that infinite number there is only one in which we recover what we knew about the spectrum: the "modified Einstein frame" which is related to the asymptotically flat $\hat{d}$-dimensional metric by a conformal factor that goes to 1 at infinity. The "Einstein frame" is just the simplest rescaling.

Noether current associated with $\mathrm{U}(1)$ gauge transformations in the modified Einstein-frame action:

$$\tilde{\jmath}^{\mu}_{\mathrm{N}} = \frac{1}{16\pi G_{\mathrm{N}}^{(d)}} \nabla_{\nu}\left[\tilde{k}^{2\frac{d-1}{d-2}} \tilde{F}^{\nu\mu}\right]. \tag{15.61}$$

We define

$$\tilde{q} = \int_{\mathrm{B}_{(d-1)}} d\Sigma_{\mu} \tilde{\jmath}^{\mu}_{\mathrm{N}}, \tag{15.62}$$

where $\mathrm{B}_{d-1}$ is a $(d-1)$-dimensional $t = \text{constant}$ hypersurface with boundary $\partial\mathrm{B}_{d-1} = \mathrm{S}^{d-2}$ at infinity. Using Stokes' theorem, we end up with the following definition of electric charge, which we write together with the definition of magnetic charge:

$$\boxed{\tilde{q} = \frac{1}{16\pi G_{\mathrm{N}}^{(d)}} \int_{\mathrm{S}^{d-2}_{\infty}} \tilde{k}^{2\frac{d-1}{d-2}} \star \tilde{F}, \qquad \tilde{p} = -\int_{\mathrm{S}^{2}_{\infty}} \tilde{F}.} \tag{15.63}$$

These charges have the right normalization and so the Dirac quantization condition can be written in terms of them:

$$\tilde{q}\tilde{p} = 2\pi n. \tag{15.64}$$

Observe that the period of the gauge parameter of the rescaled vector field $\tilde{A}$,

$$\delta_{\tilde{\Lambda}} z = k_0^{-1}\tilde{\Lambda}, \qquad \delta_{\tilde{\Lambda}} \tilde{A}_{\mu} = \partial_{\mu}\tilde{\Lambda}, \tag{15.65}$$

has to be $2\pi R_z$, in agreement with the unit of electric charge $1/R_z$ and Eqs. (12.168) and (12.169).

### 15.2.3 KK reduction of sources: the massless particle

One of the most interesting things we have learned so far, in several different ways, is that gravitons (or any other massless particles) traveling at the speed of light in the compact dimension look like massive, electrically charged particles in one dimension fewer.

In this section we are going to recover this result in yet another, particularly useful, way. We are going to see that the action for a massless particle moving in a $\hat{d}$-dimensional spacetime, given in Eq. (3.258), becomes that of a massive, charged "K particle" moving in $d = (\hat{d} - 1)$-dimensional spacetime when the $\hat{d}$-dimensional spacetime has an isometry. Furthermore, the mass and electric charge are both proportional to the momentum in the isometric direction and, if we assume that this dimension is compact, we recover exactly the results about the KK spectrum of Section 15.1.

By a "K particle" we mean a slight generalization of the standard massive particle with an extra coupling to a scalar, which we denote generically by $K$. The Nambu–Goto-type action takes the form

$$\boxed{S = -MK_0^{-1} \int d\xi K(X) \sqrt{\left|g_{\mu\nu}\dot{X}^{\mu}\dot{X}^{\nu}\right|}.} \tag{15.66}$$

The scalar cannot appear anywhere else. In particular it cannot appear in the Wess–Zumino term which describes the coupling of the particle to an electromagnetic field,

$$WZ = -q \int d\xi A_\mu \dot{X}^\mu, \tag{15.67}$$

because that would spoil U(1) gauge invariance. The scalar acts as a sort of local coupling constant. In particular, its presence modifies the mass of the particle, which is no longer the coefficient in front of the action: if the metric is asymptotically flat and $K_0$ is the constant value at infinity of $K$, then the mass is the coefficient in front of the action times $K_0$. We have already taken this into account in writing Eq. (15.66).

KK modes, and also string-theory objects called "winding modes" and "D0-branes" that we will study later, are examples of "K particles." The former couple to the inverse of the KK scalar, i.e. $K = k^{-1}$, as we are immediately going to see. Winding modes couple to $k$ directly, $K = k$, and D0-branes couple to the dilaton $e^{-\phi}$ in string theory.

Although we are going to explain this procedure (called *direct dimensional reduction*) in full detail, it is worth stressing that we are not going to prove that the two actions are completely equivalent. Rather, what we are going to prove is that all the solutions of the first action are of the form of those of the second one for some value of the mass and charge. If we take only one specific value of the mass and charge, we are reducing the system to some sector with a given, fixed, momentum in the internal direction.

Our starting point is the action of a point-like massless particle given in Eq. (3.258), which we rewrite here for convenience:

$$\hat{S}[\hat{X}^{\hat{\mu}}(\xi), \gamma(\xi)] = -\frac{p}{2} \int d\xi\, \gamma^{-\frac{1}{2}} \hat{g}_{\hat{\mu}\hat{\nu}}(\hat{X}) \dot{\hat{X}}^{\hat{\mu}} \dot{\hat{X}}^{\hat{\nu}}. \tag{15.68}$$

This action is usually said to be *invariant* under GCTs. In fact it is just *covariant*, since one goes from one metric to a different (even if physically equivalent) one. This happens typically when the action depends on potentials instead of field strengths. The infinitesimal transformations giving $\tilde{\delta} S = 0$ are

$$\begin{aligned} \tilde{\delta}\hat{X}^{\hat{\mu}} &= \hat{X}'^{\hat{\mu}} - \hat{X}^{\hat{\mu}} &&= \hat{\epsilon}^{\hat{\mu}}(\hat{X}), \\ \tilde{\delta}\hat{g}_{\hat{\mu}\hat{\nu}} &= \hat{g}'_{\hat{\mu}\hat{\nu}}(\hat{X}') - \hat{g}_{\hat{\mu}\hat{\nu}}(\hat{X}) &&= -2\hat{g}_{\hat{\lambda}(\hat{\mu}}\partial_{\hat{\nu})}\hat{\epsilon}^{\hat{\lambda}}. \end{aligned} \tag{15.69}$$

Let us now consider infinitesimal displacements in the direction $\hat{\epsilon}^{\hat{\mu}}$,

$$\begin{aligned} \delta_{\hat{\epsilon}}\hat{X}^{\hat{\mu}} &= \hat{\epsilon}^{\hat{\mu}}, \\ \delta_{\hat{\epsilon}}\hat{g}_{\hat{\mu}\hat{\nu}} &= \hat{g}_{\hat{\mu}\hat{\nu}}(\hat{X}') - \hat{g}_{\hat{\mu}\hat{\nu}}(\hat{X}) = \hat{\epsilon}^{\hat{\lambda}}\partial_{\hat{\lambda}}\hat{g}_{\hat{\mu}\hat{\nu}}. \end{aligned} \tag{15.70}$$

Using the formulae in Chapter 1, we find that the change of the action is now

$$\delta_{\hat{\epsilon}}\hat{S} = -\frac{p}{2} \int d\xi \gamma^{-\frac{1}{2}} \left[\mathcal{L}_{\hat{\epsilon}}\, \hat{g}_{\hat{\mu}\hat{\nu}}\right] \dot{\hat{X}}^{\hat{\mu}} \dot{\hat{X}}^{\hat{\nu}}. \tag{15.71}$$

Thus, the action is invariant if and only if $\hat{\epsilon}^{\hat{\mu}} = \hat{\epsilon}\hat{k}^{\hat{\mu}}$, $\hat{\epsilon}$ being an infinitesimal constant parameter and $\hat{k}^{\hat{\mu}}$ being a Killing vector. In other words, if the metric admits an isometry, the above action is invariant under the above symmetry and there is a conserved quantity, namely the momentum in the $\hat{k}^{\hat{\mu}}$ direction:

$$\hat{P} = -p\gamma^{-\frac{1}{2}}\hat{k}_{\hat{\mu}}\dot{\hat{X}}^{\hat{\mu}}, \qquad \dot{\hat{P}} = 0. \tag{15.72}$$

What one would like to do now is to fix the value of this momentum, which completely determines the dynamics in the isometry direction, and find the effective dynamics in the remaining directions. Doing this in a general coordinate system is very complicated (if it is possible at all) and hence we have to work in adapted coordinates as before. We will use, then, all the machinery and notation developed in this section.

In adapted coordinates the fact that there is a conserved momentum becomes evident since the action no longer depends on the isometric coordinate $z$.

To simplify the problem further, we split the $\hat{d}$-dimensional fields and coordinates in terms of the $d$-dimensional ones according to Eq. (15.28), obtaining

$$\hat{S}[X^{\mu}(\xi), Z(\xi), \gamma(\xi)] = -\frac{p}{2}\int d\xi\, \gamma^{-\frac{1}{2}}\left[g_{\mu\nu}\dot{X}^{\mu}\dot{X}^{\nu} - k^2F^2(Z)\right], \tag{15.73}$$

where the combination

$$F(Z) = \dot{Z} + A_{\mu}\dot{X}^{\mu} \tag{15.74}$$

that naturally appears in the action is the "field strength" of the extra worldline scalar $Z$, which now does not have a coordinate interpretation.

As we explained, the original action Eq. (15.68) is covariant under target-space diffeomorphisms and so must the action Eq. (15.73) be, since it is a simple rewriting of the former. In particular, it must be covariant under $X^{\mu}$-dependent shifts of the redundant coordinate $Z$,

$$\delta_{\Lambda}Z = -\Lambda(X^{\mu}), \tag{15.75}$$

which do not take us out of our choice of coordinates (i.e. coordinates adapted to the isometry) either. As discussed before, these transformations generate gauge transformations of the U(1) gauge potential,

$$\delta_{\Lambda}A_{\mu} = \partial_{\mu}\Lambda. \tag{15.76}$$

The field strength of $Z$ is covariant under this transformation, which justifies its definition.

Related to the constant shifts of $Z$ (which is an invariance) is the conservation of the momentum conjugate to $Z$,

$$P_z \equiv \frac{\partial\mathcal{L}}{\partial\dot{Z}} = p\gamma^{-\frac{1}{2}}F(Z), \qquad \dot{P}_z = 0. \tag{15.77}$$

Now we want to eliminate $Z$ from the action completely, using its equation of motion ($\dot{P}_z = 0$), and thus obtain the action that governs the effective $d$-dimensional dynamics. However, we cannot simply substitute into the action Eq. (15.73) $P_z = p\gamma^{\frac{1}{2}}F(Z) =$ constant because from the resulting action we do not obtain the same equations of motion as we would from making the substitution into the equations of motion. The reason for this is that the equation of motion of $Z$ is not algebraic because $\dot{Z}$ occurs in the action.

A consistent procedure by which to eliminate $Z$ is to perform first the Legendre transformation of the Lagrangian with respect to the redundant coordinate $Z$, just as we would to find the Hamiltonian if the Lagrangian depended only on $Z$. We express $\dot{Z}$ in terms of $X, \dot{X}$, and $P_z$ by using the definition of the latter and then define the Legendre transform

$$\mathcal{H}_z(X,\dot{X},P_z) \equiv -P_z\dot{Z}(X,\dot{X},P_z) + \mathcal{L}[X,\dot{X},\dot{Z}(X,\dot{X},P_z)]. \tag{15.78}$$

After the Legendre transform has been performed, the action that gives the corresponding equations of motion is

$$\begin{aligned}\hat{S}_z[X,\dot{X},Z,P_z,\gamma] &= \int d\xi \left(-\dot{P}_z Z + \mathcal{H}_z\right)\\ &= -\frac{p}{2}\int d\xi\gamma^{-\frac{1}{2}}\left[g_{\mu\nu}\dot{X}^\mu\dot{X}^\nu + \gamma k^{-2}(P_z/p)^2\right] + \int d\xi P_z F(Z).\end{aligned} \tag{15.79}$$

By explicit calculation we can now see that the equation for $Z$ (which now appears explicitly in the first term of the action) is just $\dot{P}_z = 0$ and that the equation for $P_z$ is trivially satisfied. Nothing wrong happens, then, on using the equation of motion of $Z$ in the action and replacing the variable $P_z$ by the constant $-p_z$, giving

$$S[X,\gamma] = -\frac{p}{2}\int d\xi\gamma^{-\frac{1}{2}}\left[g_{\mu\nu}\dot{X}^\mu\dot{X}^\nu + \gamma k^{-2}(p_z/p)^2\right] - p_z\int d\xi\left(\dot{Z} + A_\mu\dot{X}^\mu\right). \tag{15.80}$$

Here $\dot{Z}$ still occurs, but in a total derivative term that we can eliminate. Otherwise, we can keep it as an auxiliary scalar, which maintains explicit covariance under gauge transformations. Eliminating this term may give rise to boundary terms under gauge transformations, and thus we prefer to keep it, although it is, admittedly, unusual.

For $p_z \neq 0$, this is the action of a massive charged "K particle" in $(\hat{d}-1)$-dimensional spacetime. For $p_z = 0$, this is, again, the action of a massless particle moving in a $(\hat{d}-1)$-dimensional spacetime. To rewrite the $p_z \neq 0$ action in the usual Nambu–Goto form, we eliminate $\gamma$ directly from the action (no derivatives of $\gamma$ occur in it) by using its equation of motion:

$$\gamma = (p/p_z)^2 k^2 g_{\mu\nu}\dot{X}^\mu\dot{X}^\nu, \tag{15.81}$$

obtaining

$$\boxed{S = -|p_z|\int d\xi k^{-1}\sqrt{\left|g_{\mu\nu}\dot{X}^\mu\dot{X}^\nu\right|} - p_z\int d\xi\left(\dot{Z} + A_\mu\dot{X}^\mu\right),} \tag{15.82}$$

or, ignoring the total derivative and using the scale-invariant (with the tildes) fields,

$$\boxed{S = -|p_z|k_0^{-1}\int d\xi\tilde{k}^{-1}\sqrt{\left|g_{\mu\nu}\dot{X}^\mu\dot{X}^\nu\right|} - p_z k_0^{-1}\int d\xi\tilde{A}_\mu\dot{X}^\mu.} \tag{15.83}$$

This is a remarkable result. For a given momentum in the internal dimension, the massless particle looks like a "K particle" (in fact, a KK mode) with $K = k^{-1}$, mass

$$M = |p_z| k_0^{-1}, \tag{15.84}$$

and charge[13]

$$\tilde{q} = p_z k_0^{-1}. \tag{15.85}$$

The following identity, known as a *Bogomol'nyi identity*, is satisfied:[14]

$$\boxed{M = |\tilde{q}|.} \tag{15.86}$$

This is similar to the identity satisfied by the electric charge and mass of an ERN BH, between the mass and the NUT charge of an extreme Taub–NUT solution, or between the action and the second Chern class of instantons. We will see that this is not a coincidence. In Chapter 17 we will see that all of them are Bogomol'nyi identities (saturated *Bogomol'nyi bounds*) signaling the presence of residual supersymmetries in the background.

If $Z$ is a compact coordinate with period $2\pi\ell$ then the single-valuedness of the wave function implies that the momentum $p_z$ would be quantized,

$$p_z = n/\ell \tag{15.87}$$

(in natural units), and so would the mass and charge of the corresponding KK mode, as we know. Actually, since $k_0 = R_z/\ell$, we find

$$M = |n|/R_z, \qquad \tilde{q} = n/R_z. \tag{15.88}$$

To finish this section we can try to see how far we can go without assuming that there is an isometry in the direction of the compact coordinate $z$. Using the split Eq. (15.28), we can equally well arrive at the action Eq. (15.73) but now with the fields having periodic dependences on $z$. Now we should proceed to Fourier-expand all of them. This is not trivial, however, since we do not know how to expand $Z$ because it is not a periodic function of $Z$ (although $\dot{Z}$ is).

### 15.2.4 Electric–magnetic duality and the KK action

As in the case of the four-dimensional Einstein–Maxwell theory, the four-dimensional KK theory has an electric–magnetic symmetry, but, instead of being a continuous symmetry (at the classical level), it is a discrete $\mathbb{Z}_2$ symmetry. The duality transformation has to be defined very carefully in order to give consistent results. When this is done, the duality can be used to construct new solutions of the *same* theory. In general the duality transformation is not a symmetry, but relates two different theories or different degrees of freedom of the same theory.

[13] $q = p_z$ for the $A_\mu$ (without the tilde) field.
[14] $M = |q| k_0^{-1}$ for the $A_\mu$ (without the tilde) field.

We start by performing a Poincaré-duality transformation on the (modified-Einstein-frame) KK action. We remind the reader that the replacement of $\tilde{F}$ by its dual in the action leads in general to an action with the wrong sign for the kinetic term, which does not give rise to the dual equations of motion. This is why one has to follow the Poincaré-duality procedure explained in Section 12.7.1. Only the term involving the vector field in Eq. (15.55) is important here. We want to replace the vector (1-form) potential $\tilde{A}$ by its dual $(d-3)$-form potential $\tilde{A}_{(d-3)}$ and for this we have to rewrite the action in terms of the 2-form field strength $\tilde{F}$. We need to add a Lagrange-multiplier term to enforce the Bianchi identity and in order to be able to recover the equation $\tilde{F} = d\tilde{A}$. The Lagrange multiplier is the dual potential. The action is, therefore,

$$\begin{aligned} S[\tilde{F}, \tilde{A}_{(d-3)}] = & \frac{1}{16\pi G_N^{(d)}} \int d^d x \sqrt{|\tilde{g}_E|} \left[ -\tfrac{1}{4} \tilde{k}^{2\frac{d-1}{d-2}} \tilde{F}^2 \right] \\ & - \frac{1}{16\pi G_N^{(d)}} \int d^d x \frac{1}{2 \cdot (d-3)!} \epsilon^{\mu_1 \cdots \mu_{d-2} \nu_1 \nu_2} \partial_{\mu_1} \tilde{A}_{(d-3)\ \mu_2 \cdots \mu_{d-2}} \tilde{F}_{\nu_1 \nu_2}. \end{aligned} \tag{15.89}$$

This action gives rise to the same equations of motion as does the original action $S[\tilde{A}]$. The equation of motion of $\tilde{F}$ is

$$\tilde{F} = \tilde{k}^{-2\frac{d-1}{d-2}} \star \tilde{F}_{(d-2)}, \qquad \tilde{F}_{(d-2)\mu_1 \cdots \mu_{d-2}} = (d-2) \partial_{[\mu_1} \tilde{A}_{(d-3)\ \mu_2 \cdots \mu_{d-2}]}, \tag{15.90}$$

and, on substituting this into the action, we obtain the dual action, which we rewrite here in full:

$$\boxed{\tilde{S}_{\text{dualE}} = \frac{1}{16\pi G_N^{(d)}} \int d^d x \sqrt{|\tilde{g}_E|} \left[ \tilde{R}_E + \frac{d-1}{d-2} \tilde{k}^{-2} (\partial \tilde{k})^2 + \frac{(-1)^{d-3}}{2 \cdot (d-2)!} \tilde{k}^{-2\frac{d-1}{d-2}} \tilde{F}^2_{(d-2)} \right].} \tag{15.91}$$

This transformation has a chance of being a symmetry of the same theory only in four dimensions. However, even in four dimensions it is not a symmetry because the prefactor of the $\tilde{F}^2$ term was inverted in the transformation.[15] If we interpret the KK scalar as a sort of local coupling constant then we can say that the electric–magnetic-duality transformation relates two different regimes (strong and weak coupling) of the same theory. This can be made explicit by supplementing the electric–magnetic-duality transformation with an inversion of the "coupling constant." This does give us a transformation that leaves invariant the action (via the Poincaré-duality procedure) and the full set of equations of motion (including Bianchi identities),

$$\tilde{F}' = \tilde{k}^{+2\frac{d-1}{d-2}} \star \tilde{F}, \qquad \tilde{k}' = \tilde{k}^{-1}. \tag{15.93}$$

[15] This is another particular example of the "$a$-model" action (see Eq. (16.1)) with the opposite value of $a$,

$$a = \mp \sqrt{\frac{2(d-1)}{d-2}}. \tag{15.92}$$

Observe that this transformation does not involve any transformation of the modulus $k_0$ which defines our theory. (We have stressed several times that a theory is defined also by the expectation values of the moduli, in this asymptotically flat gravitational context by their constant values at infinity.) Thus, we can truly say that the above transformation is a *symmetry* of the theory.

We could have considered similar transformations for the fields without tildes. For instance, the following transformation leaves the equations of motion invariant:

$$F' = k_0^{-2}(k/k_0)^{+2\frac{d-1}{d-2}} \star F, \qquad k' = k^{-1}. \tag{15.94}$$

However, this is *not* a symmetry of the theory. The above transformation inverts $k_0$. If we went back to the $\hat{d}$ theory, we would find that the radius of the compact dimension is inverted and that the $\hat{d}$-dimensional Newton constant does not have the same value.

The transformation Eqs. (15.93) is going to relate electric and magnetic objects in the same theory. If a quantum theory with electrically and magnetically charged states is going to make sense, all the possible pairs of electric and magnetic charges must satisfy the Dirac quantization condition Eq. (12.171). The electric–magnetic-duality symmetry allows us to generate magnetic charges from electric charges, and we want the magnetic charges created to be compatible with the original electric charges that we have shown the KK theory to have. We defined the electric and magnetic charges of a solution in Eq. (15.63).

If we start with a field $\tilde{F}$ with electric charge $\tilde{q}$ and perform the electric–magnetic-duality transformation above, we generate the following magnetic charge:

$$\tilde{p}' = -\int_{S^2_\infty} \tilde{F}' = -\int_{S^2_\infty} \tilde{k}^{-2\frac{d-1}{d-2}} \star \tilde{F} = -16\pi G_N^{(4)} \tilde{q}, \tag{15.95}$$

where we have used the definition of $\tilde{q}$ in Eqs. (15.63). Then (ignoring the sign)

$$\tilde{p}' \tilde{q} = 16\pi G_N^{(4)} \tilde{q}^2 = 16\pi G_N^{(4)} n^2/R_z^2, \tag{15.96}$$

on account of Eqs. (15.88) and (15.32). This quantity will be an integer multiple of $2\pi$ if

$$\boxed{R_z = \sqrt{8G_N^{(4)}/|m|}, \qquad m \in \mathbb{Z}.} \tag{15.97}$$

*The existence of electric–magnetic-duality symmetry (so that each object and its dual can coexist) requires the radius of the internal dimension to be of the order of the Planck length.*

Similar constraints on the sizes of the internal dimensions or the values of other moduli can be found in string theory, requiring that each object and its U dual can coexist. A non-trivial check of U duality is that the constraints on moduli obtained from different dual-object pairs are consistent. We will see in Section 25.3, for instance, that the coexistence of all ten-dimensional D-$p$-branes and their electric–magnetic duals implies the same condition on the value of the ten-dimensional Newton constant.

We can say that, for values of the compactification radius, the theory can undergo a duality transformation into another theory, but, for the "self-dual compactification radius," the theory enjoys an additional symmetry. U duality will become a symmetry for the "self-dual values of the moduli." In this language, there is an enhancement of symmetry at the self-dual point in moduli space, a well-known phenomenon in the context of T duality, in which there is an enhancement of gauge symmetry at the self-dual points.

We should also stress that the electric–magnetic-duality transformation acts on the KK frame and $\hat{d}$ metric in a highly non-trivial way. Also, since it is only a discrete $\mathbb{Z}_2$ transformation even at the classical level, we cannot use it to construct dyonic solutions, although some dyonic solutions can be found.

A final remark: the dual KK action Eq. (15.91) in $d=5$ is identical to the five-dimensional string effective action up to $k_0$ factors, see Eq. (21.13), with the identification $\tilde{k}=e^{\phi}$. Evidently, in the Einstein frame the two actions would be absolutely identical with the identification $k=e^{\phi}$. Then, if we are careful enough with factors of $k_0$, we can identify any solution of the five-dimensional string effective action involving only the dilaton, the Kalb–Ramond 2-form (these fields are introduced in Part III), and the metric in six-dimensional pure gravity.

### 15.2.5 Reduction of the Einstein–Maxwell action and $N=1, d=5$ SUGRAs

Although the beauty of Kaluza–Klein theories is that they geometrize other interactions, unifying all of them in gravity, it is possible, and sometimes necessary, to introduce other fields in $\hat{d}$ dimensions. For instance, in the compactification of supergravity theories we have to include at the very least all the fields that enter into the supermultiplet in which the graviton lies. In higher dimensions, apart from gravitinos, the minimal supergravity multiplet necessarily contains other fermions plus scalars and $k$-form fields. In Part III we are going to reduce several of these supergravity theories, but now we want to see in a simple example ($N=1, d=5$ SUGRA) how the Scherk–Schwarz formalism works in the presence of matter fields.

In $\hat{d}=5$ the minimal SUGRA[16] [382] has a metric, a vector field, and a pair of symplectic-Majorana gravitinos that are associated with eight real supercharges. The action of the bosonic sector is essentially the Einstein–Maxwell action with an extra *topological* (in the sense of metric-independent) cubic *Chern–Simons* term:

$$\boxed{\hat{S}=\int d^5\hat{x}\sqrt{|\hat{g}|}\left[\hat{R}-\tfrac{1}{4}\hat{G}^2+\frac{1}{12\sqrt{3}}\frac{\hat{\epsilon}}{\sqrt{|\hat{g}|}}\hat{G}\hat{G}\hat{V}\right],} \tag{15.98}$$

where $\hat{G}=2\partial\hat{V}$ is the 2-form field strength of the vector $\hat{V}$. The field strength and the action (up to a total derivative) are invariant under the gauge transformations $\delta_{\hat{\chi}}\hat{V}=\partial\hat{\chi}$.

We want to reduce this theory on a circle, but with the same effort we can first perform the reduction of the $\hat{d}$-dimensional Einstein–Maxwell theory (without any topological term) on a circle and then apply the results to our case.

[16] This theory has been obtained on p. 267 as a particular example of the general matter-coupled, ungauged, $N=1, d=5$ SUGRA. We rewrite it here using a slightly different notation for convenience.

Before we dimensionally reduce the action of the $\hat{d}$-dimensional Einstein–Maxwell theory, it is convenient to know the spectrum of new states that appear when we consider a massless spin-1 particle on a circle. According to general arguments, we expect an infinite tower of states with masses proportional to the inverse of the compactification radius. Furthermore, we know that these massive states will be electrically charged under the massless KK vector that arises from the metric. On the other hand, we have to take into account that the $\hat{d}$-dimensional vector representation of $\mathrm{SO}(1,\hat{d}-1)$ gives rise to a vector and a scalar of $\mathrm{SO}(1,d-1)$ at each mass level:

$$\hat{V}^{(n)}_{\hat{\mu}} \rightarrow V^{(n)}_{\mu},\; l^{(n)}. \tag{15.99}$$

Our previous experience tells us that, in the $n \neq 0$ levels, the scalars $l^{(n)}$ will act as Stückelberg fields for the vectors $V^{(n)}_{\mu}$, giving rise to the mass terms for them that we expect according to the general KK arguments. For $n=0$ we obtain a massless vector and a massless scalar, $V_{\mu}$ and $l$. These are the only ones we keep in the dimensional reduction of the theory. The massless scalar is associated with the spontaneous breaking of the $\hat{d}$-dimensional gauge transformations $\delta_{\hat{\chi}}\hat{V}_{\hat{\mu}} = \partial_{\hat{\mu}}\hat{\chi}$ that depend on the coordinate $z$. In fact, the only $z$-dependent gauge transformations that preserve the KK ansatz are those linear in $z$ that shift the component $\hat{V}_{\underline{z}}$, and they give rise to a global, non-compact symmetry (duality) of the reduced theory.

Of course, we need to identify the lower-dimensional fields that transform correctly under all the gauge symmetries in order to see all these arguments working. The action is

$$\hat{S}[\hat{g}_{\hat{\mu}\hat{\nu}}, \hat{V}_{\hat{\mu}}] = \frac{1}{16\pi G_{\mathrm{N}}^{(\hat{d})}} \int d^{\hat{d}}\hat{x}\sqrt{|\hat{g}|}\left[\hat{R} - \tfrac{1}{4}\hat{G}^2\right]. \tag{15.100}$$

The reduction of the Einstein–Hilbert term goes exactly as before. We need only take care of the Maxwell term. In accord with the Scherk–Schwarz formalism, we use flat indices to identify fields that are invariant under the KK U(1) gauge transformations. Thus, the massless $d$-dimensional vector field $V_{\mu}$ is, using the Vielbein ansatz Eq. (15.33),

$$e_a{}^{\mu}V_{\mu} \equiv \hat{e}_a{}^{\hat{\mu}}\hat{V}_{\hat{\mu}} = \left(\hat{V}_{\mu} - \hat{V}_{\underline{z}}A_{\mu}\right)e_a{}^{\mu} \Rightarrow V_{\mu} = \hat{V}_{\mu} - \hat{V}_{\underline{z}}A_{\mu}. \tag{15.101}$$

The $\hat{V}_{\underline{z}}$ component becomes automatically the $d$-dimensional massless scalar $l$, and, thus, we have the decomposition

$$\begin{aligned} \hat{V}_{\underline{z}} &= l, & l &= \hat{V}_{\underline{z}}, \\ \hat{V}_{\mu} &= V_{\mu} + lA_{\mu}. & V_{\mu} &= \hat{V}_{\mu} - \hat{V}_{\underline{z}}\hat{g}_{\mu\underline{z}}/\hat{g}_{\underline{zz}}. \end{aligned} \tag{15.102}$$

It is easy to check that the $d$-dimensional scalar and vector fields obtained in this way are invariant under the KK U(1) $\delta_{\Lambda}$ transformations. Under the $z$-independent $\hat{d}$-dimensional transformations, only $V_{\mu}$ transforms,

$$\delta_{\chi}V_{\mu} = \partial_{\mu}\chi, \qquad \chi = \hat{\chi}(x), \tag{15.103}$$

and, under the linear gauge transformations $\hat{\chi} = mz$,

$$\delta_m l = m, \qquad \delta_m V_{\mu} = -mA_{\mu}. \tag{15.104}$$

Finally, under the rescalings of the $z$ coordinate that rescale $k$ and $A_\mu$, only $l$ transforms:

$$l' = a^{-1} l. \tag{15.105}$$

Now we need to identify the $d$-dimensional field strength. This is going to be related to $\hat{G}_{ab}$, which is invariant under $\delta_\Lambda$ and $\delta_{\hat{\chi}}$ transformations (including the linear ones $\delta_m$):

$$\hat{G}_{ab} = e_a{}^\mu e_b{}^\nu (2\partial_{[\mu} V_{\nu]} + 2l\partial_{[\mu} A_{\nu]}), \tag{15.106}$$

and we define the gauge-invariant $G_{\mu\nu}$ and the gauge-plus-global-invariant $\mathcal{G}_{ab} = \hat{G}_{ab}$:

$$G_{\mu\nu} = 2\partial_{[\mu} V_{\nu]}, \qquad \mathcal{G} = G + lF. \tag{15.107}$$

On the other hand,

$$\hat{G}_{az} = k^{-1}\partial_a l \tag{15.108}$$

is also invariant under $\delta_m$, and, therefore,

$$\hat{G}^2 = \hat{G}_{ab}\hat{G}^{ab} - 2\hat{G}_{az}\hat{G}^a{}_z = \mathcal{G}^2 - 2k^{-2}(\partial l)^2, \tag{15.109}$$

and the full dimensionally reduced Einstein–Maxwell action is

$$\boxed{\hat{S} = \frac{2\pi\ell}{16\pi G_N^{(\hat{d})}} \int d^{\hat{d}-1}x \sqrt{|g|}\, k \left[ R + \tfrac{1}{2}k^{-2}(\partial l)^2 - \tfrac{1}{4}k^2 F^2 - \tfrac{1}{4}\mathcal{G}^2 \right].} \tag{15.110}$$

Let us now go back to $\hat{d} = 5$ and let us reduce the Chern–Simons term. First, we convert the Chern–Simons term into an expression with only Lorentz indices,

$$\hat{\epsilon}^{\hat{\mu}_1\cdots\hat{\mu}_5} \hat{G}_{\hat{\mu}_1\hat{\mu}_2} \hat{G}_{\hat{\mu}_3\hat{\mu}_4} \hat{V}_{\hat{\mu}_5} = \sqrt{|\hat{g}|}\hat{\epsilon}^{\hat{a}_1\cdots\hat{a}_5} \hat{G}_{\hat{a}_1\hat{a}_2} \hat{G}_{\hat{a}_3\hat{a}_4} \hat{V}_{\hat{a}_5}, \tag{15.111}$$

and then we use the relation

$$\hat{\epsilon}^{abcdz} = \epsilon^{abcd} \tag{15.112}$$

between the five- and four-dimensional Levi-Civita symbols:

$$\sqrt{|\hat{g}|}\hat{\epsilon}\hat{G}\hat{G}\hat{\hat{V}} = k\sqrt{|g|}\epsilon(\hat{G}\hat{G}\hat{V}_z - 4\hat{G}\hat{G}_z\hat{V}) = \sqrt{|g|}\epsilon(\mathcal{G}\mathcal{G}l - 4\mathcal{G}\partial l V). \tag{15.113}$$

On turning back to curved indices and integrating by parts, the action takes the form

$$\boxed{\begin{aligned} S = \frac{2\pi\ell}{16\pi G_N^{(\hat{d})}} \int d^4x \sqrt{|g|}\, k \Bigg\{ & R + \tfrac{1}{2}k^{-2}(\partial l)^2 - \tfrac{1}{4}k^2 F^2(A) - \tfrac{1}{4}\mathcal{G}^2 \\ & + \frac{k^{-1}}{12\sqrt{3}\sqrt{|g|}}\epsilon \left[ 3lGG + 3l^2 GF + l^3 FF \right] \Bigg\}. \end{aligned}} \tag{15.114}$$

This theory is a four-dimensional SUGRA theory that is invariant under eight independent local $z$-independent supersymmetry transformations. Thus, it is an $N=2, d=4$ SUGRA theory. Pure $N=2, d=4$ SUGRA was described in Section 5.5 and its only bosonic fields are the metric and a vector. Therefore, the extra vector and two scalars that we obtain must be matter fields, actually the bosonic fields of an $N=2, d=4$ vector supermultiplet [334]. This reducibility of the gravity supermultiplet after dimensional reduction is a general characteristic of non-maximal SUGRAs. The matter and supergravity vector fields are combinations of the two vectors $A$ and $V$. To identify them, we can use the fact that eliminating a matter supermultiplet is always a consistent truncation of the theory. The equations of motion of $k$ and $l$ after setting $k=1$ and $l=0$ (their truncation values) give the constraints

$$\begin{aligned} 3F^2(A) + G^2(V) &= 0, \\ \sqrt{3}F(A) - \star G(V) &= 0. \end{aligned} \tag{15.115}$$

The second constraint implies the first and tells us that, with $k=1$ and $l=0$, the matter vector's field strength is, precisely, the combination[17] $(\sqrt{3}/2)F(A) - \frac{1}{2} \star G(V)$ that has to be set to zero for the truncation of the (matter) scalars to be consistent. The orthogonal combination $\frac{1}{2} \star F(A) - (\sqrt{3}/2)G(V)$ is the supergravity vector field. On setting the matter scalars and vector field to zero, we obtain the action of pure $N=2, d=4$ SUGRA (Einstein–Maxwell) with the normalization of Eq. (15.100).

We find the following relations between four-dimensional Einstein–Maxwell fields $g_{\mu\nu}$ and $A_\mu$ and five-dimensional fields satisfying the truncation condition:

$$\begin{aligned} \hat{g}_{\underline{z}\underline{z}} &= -1, & \hat{V}_{\underline{z}} &= 0, \\ 2\partial_{[\mu}\hat{g}_{\nu]\underline{z}} &= -\frac{1}{4\sqrt{|g|}}\epsilon_{\mu\nu\rho\sigma}F^{\rho\sigma}(A), & \hat{V}_\mu &= -\frac{\sqrt{3}}{2}A_\mu, \\ \hat{g}_{\mu\nu} &= g_{\mu\nu} - \hat{g}_{\mu\underline{z}}\hat{g}_{\nu\underline{z}}, \end{aligned} \tag{15.116}$$

which can be used to uplift *any* $N=2, d-4$ SUGRA (Einstein–Maxwell) solution to a $N=1, d=5$ SUGRA solution preserving the supersymmetry properties. Similar results can be found in the reduction of the minimal $N=(1,0), d=6$ SUGRA (which also has eight supercharges) to $d=5$ [905], and we will make use of them in Section 17.4 to relate maximally supersymmetric solutions of these three theories by dimensional reduction.

Still, it would be interesting to know the complete $N=2, d=4$ theory one obtains in the dimensional reduction. We can profit from our study of the general matter-coupled $N=2, d=4$ theories in Chapter 7 to identify it completely. In order to do this, we first rescale the metric in the action Eq. (15.114) to have it in the Einstein frame.[18] Expanding

[17] Actually, its Hodge dual. The supersymmetry transformation rules have to be examined in order to determine these ambiguities [905].

[18] We will ignore the constant value of the KK scalar at infinity since our only goal here is to identify the four-dimensional theory.

the Chern–Simons term and integrating by parts when necessary, we get

$$S \sim \int d^4x\sqrt{|g|}\,\Big\{R + \tfrac{1}{2}k^{-2}(\partial l)^2 + \tfrac{3}{2}k^{-2}(\partial k)^2 - \tfrac{1}{4}k(k^2+l^2)F^2 - \tfrac{1}{4}kG^2$$
$$- \tfrac{1}{2}klGF + \tfrac{1}{6\sqrt{3}}l^3F\star F + \tfrac{1}{2\sqrt{3}}lG\star G + \tfrac{1}{6\sqrt{3}}lF\star G\Big\}. \quad (15.117)$$

The kinetic term of the scalars is very similar to that of the axidilaton model discussed on p. 208 and 227. This suggests the definition of the complex scalar

$$t \equiv \tfrac{1}{\sqrt{3}}l + ik, \quad (15.118)$$

so the kinetic term takes the form

$$\frac{3\partial_\mu t\partial^\mu t^*}{2(\Im\mathrm{m}t)^2}, \quad (15.119)$$

which is three times that of the axidilaton, corresponding to the coset space $\mathrm{SL}(2,\mathbb{R})/\mathrm{SO}(2)$. We have studied two $N=2, d=4$ models with one vector supermultiplet and this symmetry: the axidilaton model on p. 227 and the $t^3$ model on p. 235. The normalization of the scalar kinetic term (which cannot be easily changed) suggests that the above action could correspond indeed with that of the latter, given in Eq. (7.61). It is easy to see that, indeed, both actions are completely equivalent with the identifications

$$F(A) = 2\sqrt{2}F^0, \qquad G(V) = -2\sqrt{6}F^1. \quad (15.120)$$

This allows us to rewrite any solution of the $t^3$ model as a solution of pure $N=1, d=5$ supergravity and vice versa, if the five-dimensional solution has an isometry.

It is not difficult to extend this result to general theories of $N=1, d=5$ supergravity coupled to $n_V$ vector supermultiplets since all the vector fields are dimensionally reduced in exactly the same way. This is an interesting exercise that gives the precise relation between those theories and the cubic models of $N=2, d=4$ supergravity discussed on p. 231.

The starting point is the five-dimensional action in Eq. (9.6), with hats added to the five-dimensional fields to avoid confusion with the four-dimensional ones, and we assume that they all are independent of $z$. We do not need to repeat the reduction of the Einstein–Hilbert term (identical in all theories), so we move to the next term: the non-linear $\sigma$-model with metric $\hat{g}_{xy}(\hat{\phi})$ parametrized by the scalars $\hat{\phi}^x$, $x=1,\ldots,n_V$. It is convenient to rewrite this term using the properties of real special geometry reviewed in Appendix H as

$$\tfrac{1}{2}\hat{g}_{xy}\partial_{\hat{\mu}}\hat{\phi}^x\partial^{\hat{\mu}}\hat{\phi}^y = \tfrac{3}{2}\hat{a}_{IJ}\partial_{\hat{\mu}}\hat{h}^I\partial^{\hat{\mu}}\hat{h}^J, \quad (15.121)$$

where $\hat{h}^I(\hat{\phi})$, $I=0,\ldots,n_V$ are $n_V+1$ functions of the scalars constrained by the relation

$$C_{IJK}\hat{h}^I\hat{h}^J\hat{h}^K = 1, \quad (15.122)$$

and $\hat{a}_{IJ}$ is the scalar-dependent kinetic matrix of the vector fields defined by Eqs. (H.2) and (H.4). Each of these functions $\hat{h}^I$ gives rise, upon dimensional reduction, to a four-dimensional (constrained) scalar field $h^I = \hat{h}^I$ (also $a_{IJ} \equiv \hat{a}_{IJ}$). For the derivatives, we have

$$\partial_z\hat{h}^I = 0, \qquad \partial_a\hat{h}^I = \partial_a\hat{h}^I \;\Rightarrow\; \hat{a}_{IJ}\partial_{\hat{\mu}}\hat{h}^I\partial^{\hat{\mu}}\hat{h}^J = a_{IJ}\partial_\mu h^I\partial^\mu h^J. \quad (15.123)$$

As we have said, all the vector fields are reduced in exactly the same way. We only need to add indices $I, J, K$:

$$\hat{A}^{I}{}_{\underline{z}} = l^{I}, \qquad l^{I} = \hat{A}^{I}{}_{\underline{z}},$$
$$\hat{A}^{I}{}_{\mu} = A^{I}{}_{\mu} + l^{I} A_{\mu}. \qquad A^{I}{}_{\mu} = \hat{A}^{I}{}_{\mu} - \hat{A}^{I}{}_{\underline{z}} \hat{g}_{\mu\underline{z}} / \hat{g}_{\underline{zz}}, \tag{15.124}$$

so that we have, for the field strengths

$$\begin{aligned} \hat{F}^{I}{}_{ab} &= F^{I}{}_{ab} + l^{I} F_{ab}, \\ \hat{F}^{I}{}_{az} &= k^{-1}\partial_{a} l^{I}, \\ \hat{F}^{I}{}_{\hat{\mu}\hat{\nu}} \hat{F}^{J\,\hat{\mu}\hat{\nu}} &= (F^{I} + l^{I}F)_{\mu\nu}(F^{J} + l^{J}F)^{\mu\nu} - 2k^{-2}\partial_{\mu} l^{I}\partial^{\mu} l^{J}, \\ C_{IJK}\hat{\epsilon}\hat{F}^{I}\hat{F}^{J}\hat{A}^{K} &= C_{IJK}\epsilon\left[3l^{I}F^{J}F^{K} + 3l^{I}l^{J}F^{K}F + l^{I}l^{J}l^{K}FF\right], \end{aligned} \tag{15.125}$$

where $F$ is the field strength of the KK vector field $A$ and we have performed several integrations by parts in the Chern–Simons term in order to put it in a more symmetric form.

Collecting all these partial results, rescaling the action to have a canonical Einstein–Hilbert term, and absorbing the $\epsilon$ symbol into the definition of Hodge dual field strengths, we arrive at the following four-dimensional action:

$$\begin{aligned} S \sim \int d^{4}x\sqrt{|g|}\,\Big\{ & R + \tfrac{1}{2}\left[3k^{-2}(\partial k)^{2} + 3a_{IJ}\partial h^{I}\partial h^{J} + k^{-2}a_{IJ}\partial l^{I}\partial l^{J}\right] \\ & -\tfrac{1}{4}k a_{IJ}(F^{I} + l^{I}F)(F^{J} + l^{J}F) - \tfrac{1}{4}k^{3}F^{2} \\ & + \tfrac{1}{2\sqrt{3}}C_{IJK}\left[l^{I}F^{J}\star F^{K} + l^{I}l^{J}F^{K}\star F + \tfrac{1}{3}l^{I}l^{J}l^{K}F\star F\right]\Big\}. \end{aligned} \tag{15.126}$$

To show that this is the bosonic action of an $N=2, d=4$ supergravity we have to rewrite it in the form Eq. (7.8), identifying the complex scalars $Z^{i}$, the Kähler metric $\mathcal{G}_{ij^{*}}$, the vector fields $A^{\Lambda}$, and the period matrix $\mathcal{N}_{\Lambda\Sigma}(Z, Z^{*})$. This is not enough: we need to show that $\mathcal{G}_{ij^{*}}$ and $\mathcal{N}_{\Lambda\Sigma}(Z, Z^{*})$ correspond to a special geometry. Let us start with the scalar fields. We can rewrite the kinetic terms as follows:

$$\tfrac{3}{2}k^{-2}a_{IJ}\left(\tfrac{1}{\sqrt{3}}l^{I} + ikh^{I}\right)\left(\tfrac{1}{\sqrt{3}}l^{J} - ikh^{J}\right), \tag{15.127}$$

which suggests the identification

$$Z^{i} \equiv \tfrac{1}{\sqrt{3}}l^{i-1} + ikh^{i-1}, \qquad \mathcal{G}_{ij^{*}} \equiv \tfrac{3}{4}k^{-2}a_{i-1\,j^{*}-1}. \tag{15.128}$$

It can be shown that this is a Kähler metric and that it is, in fact, identical to that of a cubic model with $d_{ijk} = 6C_{i-1\,j-1\,k-1}$ as defined on p. 231. Observe that the Kähler potential is completely determined by the KK scalar:

$$e^{-\mathcal{K}} = 8\tfrac{1}{3!}d_{ijk}\Im\mathrm{m}Z^{i}\,\Im\mathrm{m}Z^{j}\,\Im\mathrm{m}Z^{k} = 8k^{3}C_{IJK}h^{I}h^{J}h^{K} = 8k^{3}. \tag{15.129}$$

The period matrix can be immediately read from the above action. It would be tempting to identify the KK vector $A$ with $A^{0}$ and the vector fields $A^{I}$ with $A^{I+1}$ in $A^{\Lambda}$ (and, actually, one can always do that), but the same $N=2, d=4$ model can be written in different symplectic frames, with different canonical symplectic sections, and the coincidence of the Kähler metrics does not imply the identity of the canonical symplectic sections. It is, however, not difficult to check that the period matrix is identical to that in Eq. (7.52), derived from the prepotential Eq. (7.48), if we make the identifications

$$A^0 = \frac{1}{2\sqrt{2}}A, \qquad A^i = -\frac{1}{2\sqrt{6}}A^{i-1}, \tag{15.130}$$

with the components of $A^\Lambda$ on the l.h.s. and the vector fields in the above action on the r.h.s.

Summarizing, the relation between the fields of the cubic model of $N=2, d=4$ supergravity and the corresponding model of $N=1, d=5$ supergravity is

$$\begin{aligned} g_{\mu\nu} &= |\hat{g}_{\underline{zz}}|^{\frac{1}{2}}\left(\hat{g}_{\mu\nu} - \hat{g}_{\mu\underline{z}}\hat{g}_{\nu\underline{z}}/\hat{g}_{\underline{zz}}\right), \qquad d_{ijk} = 6C_{i-1\,j-1\,k-1}, \\ A^0 &= \frac{1}{2\sqrt{2}}\hat{g}_{\mu\underline{z}}/\hat{g}_{\underline{zz}}, \qquad A^i = -\frac{1}{2\sqrt{6}}\left(\hat{A}^{i-1}{}_\mu - \hat{A}^{i-1}{}_{\underline{z}}\hat{g}_{\mu\underline{z}}/\hat{g}_{\underline{zz}}\right), \\ Z^i &= \frac{1}{\sqrt{3}}\hat{A}^{i-1}{}_{\underline{z}} + i|\hat{g}_{\underline{zz}}|^{\frac{1}{2}}\hat{h}^{i-1}, \end{aligned} \tag{15.131}$$

and the inverse relations are

$$\begin{aligned} \hat{g}_{\underline{zz}} &= -k^2, & \hat{A}^I{}_{\underline{z}} &= \sqrt{3}\Re\mathrm{e}Z^{I+1}, \\ \hat{g}_{\mu\underline{z}} &= -2\sqrt{2}k^2A^0{}_\mu, & \hat{A}^I{}_\mu &= -2\sqrt{6}\left(A^{I+1}{}_\mu - \Re\mathrm{e}Z^{I+1}A^0{}_\mu\right), \\ \hat{g}_{\mu\nu} &= k^{-1}g_{\mu\nu} - 8kA^0{}_\mu A^0{}_\nu, & \hat{h}^I &= k^{-1}\Im\mathrm{m}Z^{I+1}. \end{aligned} \tag{15.132}$$

## 15.3 KK reduction and oxidation of solutions

We have learned how to perform dimensional reduction for the action of pure gravity and the Einstein–Maxwell theory. In particular, we have learned to relate the fields in lower and higher dimensions, and the main property of these reductions is that any solution of the lower-dimensional theory is automatically a solution of the higher-dimensional theory that does not depend on one coordinate and vice versa: any solution of the higher-dimensional theory that does not depend on a certain coordinate is a solution of the dimensionally reduced theory, *even if the coordinate is not periodic*.

This puts in our hands an incredibly powerful tool for generating new solutions both of the higher- and of the lower-dimensional theories.

The simplest use consists in taking a solution of the higher-dimensional theory that does not depend on one coordinate, which we identify as the compact one, and reducing it using the relations between higher- and lower-dimensional fields; or taking a solution of the lower-dimensional theory and uplifting (*oxidizing*) it to a solution of the higher-dimensional theory.

A more sophisticated use combines reduction and oxidation with a duality transformation of the lower-dimensional solution or a GCT of the higher-dimensional solution.

In this section we are going to see the most important examples of these techniques.

### 15.3.1 ERN black holes

*Periodic arrays and reduction.* Let us consider the Einstein–Maxwell theory in $\mathbb{R}^d \times S^1$. The action is given in Eq. (15.100) and is no different from the action in $\mathbb{R}^{d+1}$ and, thus, the equations of motion admit the same solutions, but now we have to impose different boundary conditions, namely periodicity in the coordinate $z$. Obviously, solutions that do not depend on the coordinate $z$ are trivially periodic, but we are interested primarily in solutions that do depend on $z$.

The Einstein–Maxwell theory has MP-type solutions, Eq. (12.231), in any dimension, which depend on a completely arbitrary harmonic function $H$. Harmonic functions with a point-like singularity that tend to 1 at infinity give asymptotically flat ERN BHs. We can also require the harmonic function to be periodic in the coordinate $z$ in order to obtain an ERN solution in $\mathbb{R}^d \times S^1$. There is a systematic way to construct a harmonic function periodic in $z$ with a point-like singularity [969] that makes use of the fact that we can construct solutions with an arbitrary number of ERN BHs by taking harmonic functions with that many point-like singularities. The idea is to place an infinite number of ERN BHs with identical masses at regular intervals along the $z$ axis. The corresponding solution is physically equivalent to one with a single ERN BH and a periodic $z$ coordinate. The harmonic function is given by the series

$$H = 1 + h \sum_{n=-\infty}^{n=+\infty} \frac{1}{\left(|\vec{x}_{\hat{d}-2}|^2 + (z + 2\pi n R_z)^2|\right)^{\frac{\hat{d}-3}{2}}}, \tag{15.133}$$

where we have assumed for simplicity that $z \in [0, 2\pi R_z]$; it is (if it converges[19]) a periodic function of $z$ with a pole in $\vec{x}_{\hat{d}-2} = z = 0$ in the interval $[0, 2\pi R_z]$, as we wanted.

Now that we have a solution of the Einstein–Maxwell theory in $\mathbb{R}^d \times S^1$, we can follow the standard procedure: expand in a Fourier series, take the $z$-independent zero mode, and use the relation between higher- and lower-dimensional fields to obtain a $d$-dimensional solution of the action Eq. (15.110). For $\hat{d} = 5$, this is done in Appendix O, but for general $\hat{d}$ it is unnecessary to sum the infinite series and then calculate the zero mode [969]: it is possible to approximate the infinite sum by an integral. First we change variables,

$$u_n = \frac{z - 2\pi n R_z}{|\vec{x}_{\hat{d}-2}|}, \qquad u_n \in \left[\frac{2\pi n R_z}{|\vec{x}_{\hat{d}-2}|}, \frac{2\pi(n+1)R_z}{|\vec{x}_{\hat{d}-2}|}\right], \tag{15.134}$$

and we have

$$\begin{aligned} H &= 1 + \frac{h}{|\vec{x}_{\hat{d}-2}|^{\hat{d}-3}} \sum_{n=-\infty}^{n=+\infty} \frac{1}{(1+u_n^2)^{\frac{\hat{d}-3}{2}}} \\ &\sim 1 + \frac{h}{|\vec{x}_{\hat{d}-2}|^{\hat{d}-3}} \frac{1}{\frac{2\pi R_z}{|\vec{x}_{\hat{d}-2}|}} \int_{-\infty}^{+\infty} \frac{du}{(1+u^2)^{\frac{\hat{d}-3}{2}}} = 1 + \frac{h'}{|\vec{x}_{\hat{d}-2}|^{\hat{d}-4}}, \end{aligned} \tag{15.135}$$

---

[19] It certainly does converge for $\hat{d} = 5$. In fact, this procedure was first developed in Ref. [709] in order to obtain harmonic functions on $\mathbb{R}^3 \times S^1$ and periodic SU(2) instanton solutions using the 't Hooft ansatz Eq. (13.29). Some related calculations can be found in Appendix O.

with

$$h' = \frac{h\omega_{(\hat{d}-4)}}{2\pi R_z \omega_{(\hat{d}-5)}}. \tag{15.136}$$

It is clear that this approximation is valid if $|\vec{x}_{\hat{d}-2}| \gg R_z$ and for $\hat{d} \geq 5$. For $\hat{d} = 4$ the series does not converge. In fact, defining now

$$u_n = (z - 2\pi n R_z) \in [2\pi n R_z, 2\pi(n+1)R_z], \tag{15.137}$$

we have

$$\begin{aligned} H &= 1 + \frac{h}{|\vec{x}_2|} \sum_{n=-\infty}^{n=+\infty} \frac{1}{(|\vec{x}_2|^2 + u_n^2)^{\frac{1}{2}}} \sim 1 + \frac{h}{2\pi R_z} \int_{-\infty}^{+\infty} \frac{du}{(|\vec{x}_2|^2 + u^2)^{\frac{1}{2}}} \\ &= 1 + \lim_{v \to +\infty} \frac{h}{\pi R_z} \ln \left\{ \frac{v}{|\vec{x}_2|} + \sqrt{1 + \left( \frac{v}{|\vec{x}_2|} \right)^2} \right\} \\ &\sim -\frac{h}{\pi R_z} \ln |\vec{x}_2| + D, \end{aligned} \tag{15.138}$$

where $D$ is a divergent constant. The solution to this problem [969] is to redefine each term in the $H$ series with a constant chosen so as to cancel $D$ out:

$$H = h \sum_{n=-\infty}^{n=+\infty} \frac{1}{(|\vec{x}_2|^2 + (z + 2\pi n R_z)^2|)^{\frac{1}{2}}} - 2h \sum_{n=1}^{n=+\infty} \frac{1}{2\pi n R_z}. \tag{15.139}$$

The solution is not asymptotically flat, but this is to be expected on physical grounds.

These are very useful formulae that we are going to use many times and they deserve to be rewritten and framed. For $n > 1$ and $n = 1$, respectively,

$$\boxed{\begin{aligned} H &= 1 + h \sum_{m=-\infty}^{m=+\infty} \frac{1}{[|\vec{x}_{n+1}|^2 + (z + 2\pi m R_z)^2]^{\frac{n}{2}}} \\ &\sim 1 + \frac{h\omega_{(n-1)}}{2\pi R_z \omega_{(n-2)}} \frac{1}{|\vec{x}_{n+1}|^{n-1}}, \\ H &= h \sum_{m=-\infty}^{m=+\infty} \frac{1}{[|\vec{x}_2|^2 + (z + 2\pi m R_z)^2]^{\frac{1}{2}}} - 2h \sum_{m=1}^{m=+\infty} \frac{1}{2\pi m R_z} \\ &\sim -\frac{h}{\pi R_z} \ln |\vec{x}_2|. \end{aligned}} \tag{15.140}$$

Using this approximated $H$ (the zero mode of the periodic one) in the $\hat{d}$-dimensional MP solution, we obtain a solution that does not depend on the periodic coordinate $z$, and now we can rewrite the solution in terms of the $d = (\hat{d} - 1)$-dimensional fields:[20]

$$\boxed{\begin{aligned} ds^2_{\rm KK} &= H^{-2}dt^2 - H^{\frac{2}{d-2}}d\vec{x}^{\,2}_{d-1}, \\ ds^2_{\rm E} &= H^{-\frac{5}{2}}dt^2 - H^{-\frac{d-6}{d-2}}d\vec{x}^{\,2}_{d-1}, \\ V_\mu &= \delta_{\mu t}\alpha(H^{-1} - 1), \qquad \alpha = \pm 2, \\ k &= H^{\frac{1}{d-2}}, \qquad V = V_0. \qquad \partial_{\underline{i}}\partial_{\underline{i}}H = 0, \end{aligned}} \tag{15.141}$$

where we have included a possible constant value for $\hat{V}_{\underline{z}}$. This form is valid for any $\hat{d}$-dimensional MP solution with a $z$-independent harmonic function, and, in particular, for the above $H$ that corresponds to the zero mode of the $\hat{d}$-dimensional periodic ERN BH.

Why have we gone through the long procedure of finding periodic ERN BH solutions and finding their zero modes when we could simply reduce the whole MP family assuming independence of $z$? The reason is that, in the cases that will interest us, we will have a well-defined $\hat{d}$-dimensional source that will determine the coefficient $h$ of the $\hat{d}$-dimensional harmonic function and only by going through this procedure can we relate it to the coefficient of the $d$-dimensional harmonic function.

The dimensionally reduced ERN solution does not have a regular horizon: near the origin (the only place where the horizon could be placed), using spherical coordinates $r = |\vec{x}_{d-1}|$,

$$\left(1 + h'/r^{d-3}\right)^{-\frac{d-6}{d-2}} r^2 d\Omega^2_{(d-2)} \sim h' r^{\frac{(d-3)(d-6)}{d-2}} d\Omega^2_{(d-2)}, \tag{15.142}$$

which never goes to a $(d-2)$-sphere with finite radius. However, we *know* that this solution corresponds to a solution with a regular horizon in $d + 1$ dimensions! One possible way to explain what is happening here is the following: the results of the dimensional-reduction procedure are meaningful within certain approximations. In particular, we assume that the massive modes can be ignored because their masses are very large, which means that the compactification radius is small. In this geometry, the compactification radius, measured by the modulus $k$, is not constant over the space but depends on $r$, blowing up when $r \to 0$ (the locus of the putative horizon). Thus, near this point, there are KK modes whose masses become small enough to be taken into account, but we have not done this and the solution cannot be considered valid near $r = 0$. Near $r = 0$ the solution is indeed $\hat{d}$-dimensional and regular. Similar mechanisms have been proposed in other cases and in the context of string theory to show how some singularities disappear when we take into account the higher-dimensional origin of the solution [617].

*Oxidation.* Dimensional oxidation is in general a much simpler operation than reduction: we simply take a solution of the lower-dimensional theory and rewrite it in terms of the $\hat{d}$-dimensional fields, obtaining a solution of the higher-dimensional theory that does not

[20] Since we have absorbed the asymptotic value of the KK scalar into the period of the coordinate $z$, $k = \tilde{k}$ and there is no difference between the Einstein and modified Einstein frames.

depend on the compact coordinate. However, this solution may (but need not) be the zero mode of a solution that does depend periodically on the compact coordinate, and in general we cannot know which of these possibilities is true.

In any case, the first step consists in having a solution of the lower-dimensional theory and our problem is that the $d$-dimensional ERN solution (in general, the MP solutions) is not a solution of the dimensionally reduced $\hat{d} = (d+1)$-dimensional Einstein–Maxwell theory. Let us examine the KK scalar equation of motion in the Einstein frame. It takes the form

$$\nabla^2 \ln k \sim c_1 k^{a_1} F^2 + c_2 k^{a_2} \mathcal{G}^2, \tag{15.143}$$

and requires a non-trivial $k$ if $F^2 \neq 0$ or $\mathcal{G} \neq 0$, as is the case here. Thus, the MP solutions cannot, in general, be considered solutions of the reduced Einstein–Maxwell equations and, thus, cannot be dimensionally oxidized.

There are, however, exceptions.

1. Solutions with $F^2 = 0$ satisfy the KK scalar equation of motion and thus can be oxidized to a purely gravitational solution. One example is the dyonic ERN BH with electric and magnetic charges related by $p = \pm 16\pi G_{\rm N}^{(4)} q$ (see p. 445). Another example is provided by electromagnetic pp-waves.

2. We have seen in Section 15.2.5 that any solution of the four-dimensional Einstein–Maxwell theory ($N = 2, d = 4$ SUGRA) can be oxidized to a solution of $N = 2, d = 5$ SUGRA using Eqs. (15.116) and we have mentioned that solutions of the latter can be further oxidized to $N = (1,0), d = 6$ SUGRA.

Observe that we can oxidize the four-dimensional Einstein–Maxwell solutions with $F^2 = 0$ in two different ways to $d = 5$. The second form makes use of the supersymmetric structure of the theory and ensures that the supersymmetry properties will be preserved in the oxidation, whereas in the first case they will not.

### 15.3.2 *Dimensional reduction of the AS shock wave: the extreme electric KK black hole*

Now we are going to consider the dimensional reduction of the AS shock-wave solution Eq. (14.41). We must distinguish between two cases: when the wave propagates in the compact coordinate and when it propagates in an orthogonal direction. The second case is simpler, and we study it first.

To avoid confusion, we are going to call $y$ the direction in which the wave propagates and $z$ the compact direction. The AS solution depends on a harmonic function of the transverse coordinates $H(\vec{x}_{\hat{d}-2})$ and on a delta function $\delta[(1/\sqrt{2})(t - \alpha y)]$. If the compact coordinate is $x^{\hat{d}-2} \equiv z$, we split the transverse-coordinates vector into $\vec{x}_{\hat{d}-2} = (\vec{x}_{d-2}, z)$. We know that any harmonic function $H$ provides a solution, and hence we can repeat the construction of a harmonic function of $(\vec{x}_{d-2}, z)$ that has a single point-like pole and is periodic in the coordinate $z$ by constructing a periodic array and taking the average. For $\hat{d} \geq 5$ the reduced solution is another AS shock wave with the same metric in one dimension fewer and with the coefficients of the harmonic functions related as above.

The case in which the wave propagates in the compact direction is far more interesting. We should be able to guess the result, since we have reduced the source of the AS shock wave (the massless point-particle action) in Section 15.2.3 and found the action of a massive KK mode that is electrically charged with respect to the KK vector field and with charge and mass equal to the momentum in the compact direction. We expect, then, that the reduction in the direction in which the wave propagates should give a metric describing a massive, electrically charged object which will be "extreme" in some sense, corresponding to the special relation between its mass and charge.

First, we adapt the solution Eq. (14.41) to the compactness of $z$, rescaling it to $k_0 z$, and at the same time rescaling $\ell$ so the periodicity of $z$ is always $2\pi\ell$. These rescalings introduce $k_0$ factors in several places. The solution we are going to start with is

$$\begin{aligned}
d\hat{s}^2 &= H^{-1}dt^2 - H\left[k_0 dz - \alpha(H^{-1}-1)dt\right]^2 - d\vec{x}^{\,2}_{\hat{d}-2}, \qquad \alpha = \pm 1, \\
H &= 1 - \frac{\sqrt{2}\,p 8\pi G_{\rm N}^{(\hat{d})}}{(\hat{d}-4)\omega_{(\hat{d}-3)}} \frac{1}{|\vec{x}_{\hat{d}-2}|^{\hat{d}-4}}\, \delta\left[\frac{1}{\sqrt{2}}(t-\alpha k_0 z)\right], \qquad \hat{d} \geq 5, \\
H &= 1 + \sqrt{2}\,p 4 G_{\rm N}^{(4)} \ln|\vec{x}_2|\, \delta\left[\frac{1}{\sqrt{2}}(t-\alpha k_0 z)\right], \qquad \hat{d} = 4.
\end{aligned} \tag{15.144}$$

Before we proceed, it is necessary to identify the constant $p$. In asymptotically flat cases, $p$ is just the absolute value of the momentum carried by the massless particle. In the present case, the momentum of the massless particle in the $z$ direction is given by (just take the KK vacuum spacetime limit)

$$p_z = \alpha p k_0, \tag{15.145}$$

and we should replace $p$ by $|p_z|/k_0$ accordingly in the above harmonic functions.

Now, we should Fourier-expand all the components of the metric, but we are going to content ourselves with taking the zero mode, which will be a solution of the KK-theory action Eq. (15.39). We expand

$$\delta\left[\frac{1}{\sqrt{2}}(t-\alpha k_0 z)\right] = -\frac{\sqrt{2}}{k_0}\sum_n \frac{1}{2\pi\ell} e^{in\left(z-\frac{t}{k_0}\right)/\ell}, \tag{15.146}$$

and keep only the zero mode $-1/(\sqrt{2}\pi\ell k_0)$. The replacement of the delta function by its constant zero mode gives us the $z$-independent harmonic functions and metric, which can be immediately rewritten in terms of $d$-dimensional fields that we express both in the KK frame and in the modified Einstein frame for the interesting, asymptotically flat $d > 4$

cases:

$$
\begin{aligned}
&ds^2_{\rm KK} = H^{-1}dt^2 - d\vec{x}^{\,2}_{d-1},\\
&d\tilde{s}^2_{\rm E} = H^{-\frac{d-3}{d-2}}dt^2 - H^{\frac{1}{d-2}}d\vec{x}^{\,2}_{d-1},\\
&\tilde{A}_t = \alpha(H^{-1}-1), \qquad \tilde{k} = H^{\frac{1}{2}}, \qquad \alpha = \pm 1,\\
&H = 1 + \frac{h}{|\vec{x}_{d-1}|^{d-3}}, \qquad h = \frac{16\pi G_{\rm N}^{(\hat{d})} p_z}{2\pi\ell k_0^2 (d-3)\omega_{(d-2)}}.
\end{aligned}
\tag{15.147}
$$

This is the $d$-dimensional *extreme electric KK BH* solution. As expected, it describes a massive, electrically charged object that should be a KK mode. It does not have a regular horizon. It is clear that, had we started from the general family of pp-wave solutions Eqs. (14.42), we would have obtained a family of solutions of the same form but with arbitrary harmonic functions. Thus, we can construct solutions of the KK action Eq. (15.39) with several of these objects with charges of the same sign in static equilibrium by the standard procedure. Now, the equilibrium is more difficult to describe because a third interaction, mediated by the KK scalar $k$, comes into play. On the other hand, in the reduction of the ERN solution we also found a solution describing a massive object charged with respect to a vector field and with a non-trivial scalar, but different from this one. The reason is that they obey different equations of motion, the difference being the strength with which the KK scalar couples to the vector field. We will study these *dilaton "BHs"* in more detail in Section 16.1.

We can calculate the mass and charge of the above solutions to check that they do indeed correspond to those of a KK mode. From

$$
\tilde{g}_{{\rm E}\,tt} = H^{-\frac{d-3}{d-2}} \sim 1 - \frac{d-3}{d-2}\frac{h}{|\vec{x}_{d-1}|^{d-3}},
\tag{15.148}
$$

and the definition of the mass $M$,

$$
\tilde{g}_{{\rm E}\,tt} \sim 1 - \frac{16\pi G_{\rm N}^{(d)} M}{(d-2)\omega_{(d-2)}}\frac{1}{|\vec{x}_{d-1}|^{d-3}},
\tag{15.149}
$$

we find $M = p_z k_0^{-1}$, as expected.

The electric charge can be calculated using the definition in Eqs. (15.63), finding first

$$
\tilde{k}^{2\frac{d-1}{d-2}} \star \tilde{F} = \pm(d-3)h\, d\Omega^{d-2},
\tag{15.150}
$$

where $d\Omega^{d-2}$ is the unit $(d-2)$-sphere volume form, whose integral over the sphere just gives $\omega_{(d-2)}$ (see Appendix K). The final result is $\tilde{q} = \pm p_z k_0^{-1}$ ($p_z$ was taken to be positive), also as expected.

We conclude that the extreme electric KK BH solution does indeed describe the long-range fields of a KK mode.

The name "extreme BH" for a solution that does not have a regular event horizon needs some justification: the reason is that this solution belongs to a larger family of BH solutions with regular event horizons and also with Cauchy horizons, which we will construct in Section 15.3.4. When the mass and electric charge are equal (the "extreme limit"), the event and Cauchy horizons coincide and become singular. The general families of non-extreme dilaton BHs will be studied in Section 16.1. Those with the right dilaton coupling can be oxidized to one dimension more.

Finally, observe that purely gravitational pp-waves can always be oxidized to one dimension more by taking the product with the metric of a flat line. We know that the dependence of the harmonic functions can be extended to that coordinate. The first observation is also true for any purely gravitational solution, which is always a solution of the KK action Eq. (15.39). However, the dependences of the functions in the metric cannot always be extended to the new compact coordinate. This is the case for the Schwarzschild BH solution, as we are going to see.

### *15.3.3 Non-extreme Schwarzschild and RN black holes*

*Dimensional reduction.* Paradoxically, the simplest and most fundamental BH solutions are also the most difficult to reduce because it is also more difficult to generalize them to the case in which one coordinate is compact. We certainly cannot construct, in a simple and naive way, infinite periodic arrays of Schwarzschild and non-extreme RN BHs because it is not at all clear how to construct solutions for more than one non-extreme BH, and, on physical grounds, one does not expect them even to exist because the interaction between non-extreme BHs is not balanced and they cannot be in static equilibrium.

Nevertheless, there are solutions describing an arbitrary number of aligned Schwarzschild BHs: the Israel–Khan solutions [820]. They belong to Weyl's family of axisymmetric vacuum solutions [1249, 1250, 878] and, thus, they have a metric that, in Weyl's canonical coordinates $\{t, \rho, z, \varphi\}$, takes the form

$$ds^2 = e^{2U} dt^2 - e^{-2U} \left[e^{2k}(d\rho^2 + dz^2) + \rho^2 d\varphi^2\right], \tag{15.151}$$

where $U$ is a harmonic function in three-dimensional Euclidean space that is independent of $\varphi$ (because of axisymmetry) and $k$ depends on $U$ through two first-order differential equations that can be integrated straight away:

$$\begin{aligned} \partial_{\underline{i}}\partial_{\underline{i}} U &= 0, \\ \partial_\rho k &= \rho[(\partial_\rho U)^2 - (\partial_z U)^2], \\ \partial_z k &= 2\rho \partial_\rho U \partial_z U. \end{aligned} \tag{15.152}$$

The simplest choice of $U$ is, in spherical coordinates $r^2 = \rho^2 + z^2$,

$$U = -\frac{G_{\rm N}^{(4)} M}{r}, \qquad k = -\frac{(G_{\rm N}^{(4)} M)^2 \sin^2\theta}{2r^2}, \tag{15.153}$$

and gives the Chazy–Curzon solution [343, 392]. In spite of the spherically symmetric $U$, the solution is only axisymmetric. The Schwarzschild solution corresponds to a $U$ equal to the Newtonian gravitational potential for an ideal homogeneous rod of finite length $2G_{\rm N}^{(4)}M$ and total mass $M$,

$$U = \tfrac{1}{2}\ln\left(\frac{r_+ + z_+}{r_- + z_-}\right) = \tfrac{1}{2}\ln\left(\frac{r_+ + r_- + (z_+ - z_-)}{r_+ + r_- - (z_+ - z_-)}\right),$$
$$k = \tfrac{1}{2}\ln\left(\frac{r_+r_- + z_+z_- + \rho^2}{2r_+r_-}\right), \tag{15.154}$$

where

$$r_\pm \equiv \sqrt{\rho^2 + z_\pm^2}, \qquad z_\pm \equiv z - \left(z_0 \pm G_{\rm N}^{(4)}M\right), \tag{15.155}$$

and $z_0$ is the value of $z$ at the center of the rod. The two very different-looking forms of the function $U$ are completely equivalent. The coordinate transformation

$$\rho = \sqrt{r\left(r - 2G_{\rm N}^{(4)}M\right)}, \qquad z - z_0 = \left(r - 2G_{\rm N}^{(4)}M\right)\cos\theta, \tag{15.156}$$

gives back the Schwarzschild metric in Schwarzschild coordinates.

This metric is singular at the position of the rod over the $z$ axis ($M > 0$): $\rho = 0$, $z_0 - G_{\rm N}^{(4)}M < z < z_0 + G_{\rm N}^{(4)}M$, where $U$ diverges, but the singularity can be removed by a coordinate transformation and indicates only the presence of the event horizon [820]. On the other hand, $k = 0$ on the axis, so there are no conical singularities there, as we are going to explain.

Since $U$ satisfies a linear equation, we can linearly superpose the potentials of $N$ separated rods with masses $M_i$ and lengths $2G_{\rm N}^{(4)}M_i$ to give a solution that, in principle, can describe several Schwarzschild BHs in static equilibrium. We just have to calculate $k$:

$$U = \sum_{i=1}^{N} U_i, \quad U_i = \tfrac{1}{2}\ln\left(\frac{r_{+i} + r_{-i} + (z_{+i} - z_{-i})}{r_{+i} + r_{-i} - (z_{+i} - z_{-i})}\right),$$
$$k = \sum_{i,j=1}^{N} k_{ij}, \quad k_{ij} = \tfrac{1}{4}\ln\left(\frac{r_{+i}r_{-j} + z_{+i}z_{-j} + \rho^2}{r_{+i}r_{+j} + z_{+i}z_{+j} + \rho^2} + (+\leftrightarrow -)\right), \tag{15.157}$$

where now

$$r_{\pm i} \equiv \sqrt{\rho^2 + z_{\pm i}^2}, \qquad z_{\pm i} \equiv z - (z_{0\,i} \pm G_{\rm N}^{(4)}M_i), \tag{15.158}$$

and the centers of the rods are at $z_{0\,i}$. These are Israel–Khan solutions [820]. Since, physically, we did not expect these solutions to exist, where is the catch? These solutions have additional conical singularities over the $z$ in between the rods (BHs): $e^{\pm 2U}$ is completely regular in between the axes because the $U_i$s are. The $k_{ij}$s vanish when $z$ is not in between the rods $i$ and $j$, and, in between the rods $i$ and $j$ (that is, assuming that $z_{0\,i} < z_{0\,j}$, $z_{\pm i} > 0$ and $z_{\pm j} < 0$), on taking the $\rho \to 0$ limit carefully, we obtain

$$k_{ij}^0 \equiv \lim_{\rho\to 0} k_{ij} = \tfrac{1}{2}\ln\left|\frac{(z_{+i} - z_{-j})(z_{-i} - z_{+j})}{(z_{+i} - z_{+j})(z_{-i} - z_{-j})}\right|, \tag{15.159}$$

which is constant and proportional to the Newtonian gravitational force between the rods $i$ and $j$. Thus, in general $k$ will be a constant $k^0 = \sum_{i.j} k^0_{ij}$ that differs from zero over the $z$ axis. This implies the existence of conical singularities over the axis: when $\rho \to 0$ the spatial part of the metric Eq. (15.151) takes the form

$$-ds^2_{(3)} \sim e^{-2(U-k)}\Big[d\rho^2 + dz^2 + \rho^2 e^{-2k^0} d\varphi^2\Big]. \tag{15.160}$$

The metric in brackets is the Euclidean metric in cylindrical coordinates if $e^{-k^0}\varphi$ has period $\Delta e^{-k^0}\varphi = 2\pi$; otherwise, there is a conical singularity with a deficit angle $\delta = 2\pi - \Delta e^{-k^0}\varphi$. However, for analogous reasons, the period of $\varphi$ has to be $2\pi$ if the metric is to be asymptotically flat rather than asymptotically conical and, in general, there is a defect angle $\delta = 2\pi(1 - e^{-k^0})$. For instance, for two rods separated by a coordinate distance $\Delta z$ (so, $z_{0\,1} - z_{0\,2} = \Delta z + G^{(4)}_{\rm N}(M_1 + M_2)$) the deficit angle is (see e.g. Ref. [377])

$$\delta = -2\pi \frac{4G^{(4)\,2}_{\rm N} M_1 M_2}{\Delta z[\Delta z + 2G^{(4)}_{\rm N}(M_1 + M_2)]}. \tag{15.161}$$

This conical singularity can be considered as a *strut* that holds the two BHs in place in spite of their gravitational attraction. The BH horizons are deformed by all these interactions [377]. The conical singularities are unavoidable: it can be shown that the only non-singular solution is the one with a single BH [966, 604]. In fact, in Ref. [606] it is shown that the only static, axisymmetric, asymptotically flat solutions with many BHs are the MP solutions. Nevertheless, the Euclidean action is well defined even in the presence of conical singularities [624].

It is clear that the Israel–Khan solution can be used to construct an infinite periodic array of identical Schwarzschild BHs of mass $M$ whose rod centers are separated by a coordinate distance $2\pi R_z$. This construction was made in Ref. [969]. The series $U = \sum_{n=-\infty}^{n=\infty} U_n$ diverges (asymptotically, it is similar to the second series $H$ in Eq. (15.140)) and we need to redefine it:

$$U = \sum_{n=-\infty}^{n=\infty} U_n \quad \sum_{n=1}^{+\infty} \ln\left(\frac{1 - G^{(4)}_{\rm N} M/(n2\pi R_z)}{1 + G^{(4)}_{\rm N} M/(n2\pi R_z)}\right). \tag{15.162}$$

The same is true for the $k_{nm}$s:

$$k = \sum_{n,m=-\infty}^{n,m=+\infty} k_{nm} - \sum_{n,m=0}^{n,m=+\infty} \ln\left[1 - \frac{4G^{(4)\,2}_{\rm N} M^2}{(n+m+1)^2(2\pi R_z)^2}\right]. \tag{15.163}$$

When the number of BHs is infinite, we expect the total force exerted on each BH by all the others (an infinite number to its left and to its right) to vanish and, indeed, one finds $k^0 = 0$, a total absence of conical singularities in between the BHs.

This solution can now be considered as a Schwarzschild BH with a compact dimension, asymptotically $\mathbb{R}^3 \times \mathrm{S}^1$. We could extract its Fourier zero mode and then dimensionally reduce it to three dimensions using the standard procedure. This construction can be generalized to other non-extreme BHs such as the RN BH with like [377] or opposite charges [513], and the generalization to $d = 5$ dimensions can be performed using the higher-dimensional generalization of the Weyl class recently found in Ref. [511], although for higher dimensions there are still problems.

*Oxidation: black branes.* We have already mentioned that any purely gravitational solution is automatically a solution of the KK action given in Eq. (15.39) with constant KK scalar $k = k_0$ and, therefore, it is also a solution of the higher-dimensional purely gravitational theory. The procedure can be repeated as many times as we want ($p$, say) and the result is a purely gravitational solution with a metric that is the direct product of the original metric and the metric of $p$ circles (a $p$-torus $\mathrm{T}^p$).

This remark applies in particular to $d = (\hat{d} - p)$-dimensional Schwarzschild BHs. On oxidizing them to $\hat{d}$ dimensions and adding $p$ coordinates $\vec{y}_p = (y^1, y^2, \ldots, y^p)$ with $y^i \in [0, 2\pi R_i]$, we obtain the following metric [775]:

$$\boxed{\begin{aligned} d\hat{s}^2 &= W dt^2 - d\vec{y}_p^{\,2} - W^{-1} dr^2 - r^2 d\Omega^2_{[\hat{d}-(p+2)]}, \\ W &= 1 - \frac{16\pi G_{\mathrm{N}}^{(\hat{d}-p)} M}{[\hat{d} - (p+2)]\omega_{[\hat{d}-(p+2)]}} \frac{1}{r^{\hat{d}-p-3}}. \end{aligned}} \tag{15.164}$$

These solutions are known as *(Schwarzschild) black p-brane* solutions and represent the gravitational field of massive, extended objects of $p$ spatial dimensions (*$p$-branes*), which are parametrized by the coordinates $\vec{y}_p$. They are asymptotically flat only in the directions orthogonal (or transverse) to the *worldvolume* directions $t$ and $\vec{y}_p$ (even in the $R_i \to \infty$ limit). Since the mass is the same, even for infinite compactification radii, it is clear that these objects are really characterized by a mass density per unit $p$-volume (in the $\vec{y}_p$ directions), which is called the *brane tension* $T_{(p)}$, rather than by $M$, which is the mass of the point-like object they give rise to after compactification on $\mathrm{T}^p$. To calculate the tension $T_{(p)}$ we use $p$ times the relation between the Newton constants in different dimensions,

$$G_{\mathrm{N}}^{(\hat{d}-p)} = G_{\mathrm{N}}^{(\hat{d})}/V_p, \qquad V_p = (2\pi)^p R_1 \cdots R_p, \tag{15.165}$$

and define

$$G_{\mathrm{N}}^{(\hat{d}-p)} M = G_{\mathrm{N}}^{(\hat{d})} T_{(p)} \Rightarrow M = V_p T_{(p)}. \tag{15.166}$$

Solutions like this are going to be studied in detail in Part III.

It is also possible to oxidize to purely gravitational solutions the solutions Eqs. (12.218) of the Einstein-scalar theory, but the resulting solutions, a sort of generalized black $p$-branes, do not have a clear interpretation.

### 15.3.4 Simple KK solution-generating techniques

KK oxidation and reduction can be used to generate new solutions. In general, the procedure consists in using a well-defined symmetry of the higher- or lower-dimensional theory as an intermediate step between oxidation and reduction or reduction and oxidation. Let us study some examples.

*Generation of charged solutions by higher-dimensional boosts.* The first example consists of three steps.

1. Oxidation of the Schwarzschild solution to $\hat{d} = d + 1$ dimensions.
2. Lorentz-boosting the Schwarzschild black 1-brane solution in the compact direction.
3. Reduction in the same direction.

We have already performed the first operation in Section 15.3.3. The $\hat{d}$-dimensional solution is

$$d\hat{s}^2 = W dt^2 - dz^2 - W^{-1}dr^2 - r^2 d\Omega^2_{[\hat{d}-3]}, \qquad W = 1 + \frac{\omega}{r^{\hat{d}-4}}, \tag{15.167}$$

and we are ready to perform a Lorentz boost in the positive- or negative-$z$ direction, which evidently transforms a solution of the $\hat{d}$-dimensional Einstein equations into another one:

$$\begin{pmatrix} t \\ z \end{pmatrix} \rightarrow \begin{pmatrix} \cosh\gamma & \pm\sinh\gamma \\ \pm\sinh\gamma & \cosh\gamma \end{pmatrix} \begin{pmatrix} t \\ z \end{pmatrix}, \qquad \gamma > 0. \tag{15.168}$$

The new solution can be rewritten in the form

$$\boxed{\begin{aligned} &d\hat{s}^2 = H^{-1}W dt^2 - H\left[dz - \alpha(H^{-1} - 1)dt\right]^2 - W^{-1}dr^2 - r^2 d\Omega^2_{[\hat{d}-3]}, \\ &W = 1 + \frac{\omega}{r^{\hat{d}-4}}, \qquad H = 1 + \frac{h}{r^{\hat{d}-4}}, \qquad \omega = h(1-\alpha^2), \end{aligned}} \tag{15.169}$$

if we parametrize $\alpha = \pm\coth\gamma$, which is a sort of "black pp-wave" metric. The non-extremality function $W$ disappears when we boost at the speed of light $\alpha = \pm 1$ and then we recover exactly the pp-wave solutions Eq. (14.42), for which $H$ can be any general harmonic function in $(\hat{d} - 2)$-dimensional Euclidean space.

Now, the third step gives a new $d$-dimensional class of solutions whose existence we announced:

$$\boxed{\begin{aligned} &ds^2_{\text{KK}} = H^{-1}W dt^2 - W^{-1}dr^2 - r^2 d\Omega^2_{(d-2)}, \\ &d\tilde{s}^2_{\text{E}} = H^{-\frac{d-3}{d-2}} W dt^2 - H^{\frac{1}{(d-2)}} \left(W^{-1}dr^2 - r^2 d\Omega^2_{(d-2)}\right), \\ &\tilde{A}_t = \alpha(H^{-1} - 1), \qquad \tilde{k} = H^{\frac{1}{2}}, \\ &W = 1 + \frac{\omega}{r^{d-3}}, \qquad H = 1 + \frac{h}{r^{d-3}}, \qquad \omega = h(1-\alpha^2). \end{aligned}} \tag{15.170}$$

These are the *non-extreme electric KK BHs*. They have regular event horizons and Cauchy horizons (for negative $\omega$) and, in the extreme limit $\omega = 0$, they become the extreme electric KK BHs, Eq. (15.147).

The same procedure can be used with higher-$p$ branes and also with "charged $p$-branes."

*Lower-dimensional S dualities and generation of KK branes.* In this example, we are also going to study a three-step mechanism for generating new solutions that exploits the existence of an S-duality symmetry in the four-dimensional KK theory, as discussed in Section 15.2.4.

1. Reduction of a purely gravitational five-dimensional solution to a four-dimensional KK-theory solution.
2. S dualization of the four-dimensional KK-theory solution.
3. Oxidation of the S-dualized KK-theory solution to a new purely gravitational five-dimensional solution.

In particular, we are going to apply this recipe to the "black pp-wave" solutions given in Eqs. (15.169). Using the transformation Eq. (15.93) on the four-dimensional solution Eq. (15.170), we immediately obtain

$$\boxed{\begin{aligned} ds^2_{\mathrm{KK}} &= W dt^2 - H\Big[W^{-1}dr^2 + r^2 d\Omega^2_{(2)}\Big], \\ d\tilde{s}^2_{\mathrm{E}} &= H^{-\frac{1}{2}} W dt^2 - H^{\frac{1}{2}}\Big[W^{-1}dr^2 + r^2 d\Omega^2_{(2)}\Big], \\ \tilde{F} &= \alpha h d\Omega^2, \qquad \tilde{k} = H^{-\frac{1}{2}}, \\ H &= 1 + \tfrac{h}{r}, \qquad W = 1 + \tfrac{\omega}{r}, \qquad \omega = h(1-\alpha^2). \end{aligned}} \tag{15.171}$$

The solution is naturally given in terms of the field strength $\tilde{F}$. Finding the potential is equivalent to solving the Dirac-monopole problem, which we already solved in Section 12.7.2. Here we simply quote the result: in spherical coordinates the non-vanishing components of $\tilde{F}$ are

$$\tilde{F}_{\theta\varphi} = \alpha h \sin\theta = \partial_\theta \tilde{A}_\varphi \Rightarrow \tilde{A}_\varphi = -\alpha h \cos\theta, \tag{15.172}$$

up to gauge transformations. This potential is singular at $\theta = 0$ and $\theta = \pi$, and the solution to this problem is to define the potential in two different patches $\tilde{A}^{(\pm)}_\varphi$ related by a gauge transformation:

$$\tilde{A}^{(\pm)}_\varphi = \pm\alpha h(1 \mp \cos\theta). \tag{15.173}$$

It is useful to rewrite the equation that the $A$ without the tilde has to satisfy (the Dirac-monopole equation) in a coordinate-independent way that will allow us to generalize the solutions,

$$\partial_{[\underline{i}} A_{\underline{j}]} = \alpha k_0^{-1} \tfrac{1}{2} \epsilon_{ijk} \partial_{\underline{k}} H. \tag{15.174}$$

All the properties that depend only on the modified Einstein-frame metric (singularities, horizons, causality, extremality, thermodynamics, etc.) are the same as in the electric case. The characteristic features of the magnetic BH appear in the KK frame and in $\hat{d}$ dimensions.

Using the relations Eqs. (15.28), we can immediately find the $\hat{d}$-dimensional metric which gives rise to the fields of the magnetic solution given in Eqs. (15.171):

$$\begin{aligned} d\hat{s}^2 &= W dt^2 - H\left[W^{-1}dr^2 + r^2 d\Omega^2_{(2)}\right] - k_0^2 H^{-1}W^{-1}[dz + A]^2, \\ A &= A_{\underline{i}}dx^i, \qquad \partial_{[\underline{i}}A_{\underline{j}]} = \alpha k_0^{-1}\tfrac{1}{2}\epsilon_{ijk}\partial_{\underline{k}}H, \\ H &= 1 + \tfrac{h}{r}, \qquad W = 1 + \tfrac{\omega}{r}, \qquad \omega = h(1-\alpha^2). \end{aligned} \tag{15.175}$$

This solution has no simple interpretation. The extreme $\omega = 0$ case is particularly interesting because the metric becomes a product of time and a non-trivial four-dimensional Euclidean manifold:

$$\boxed{\begin{aligned} d\hat{s}^2 &= dt^2 - H d\vec{x}_3^{\,2} - k_0^2 H^{-1}[dz + A]^2, \\ A &= A_{\underline{i}}dx^i, \qquad \partial_{[\underline{i}}A_{\underline{j}]} = \alpha k_0^{-1}\tfrac{1}{2}\epsilon_{ijk}\partial_{\underline{k}}H, \\ H &= 1 + \frac{|\tilde{p}|}{4\pi}\frac{1}{|\vec{x}_3|}, \qquad \alpha = \pm 1, \end{aligned}} \tag{15.176}$$

where we have identified the constant $h$ in $H$ in terms of the four-dimensional magnetic charge $\tilde{p}$. As usual, in the extreme case the function $H$ can be any harmonic function in flat three-dimensional space. The non-trivial four-dimensional manifold is nothing but the Euclidean Taub–NUT solution[21] Eq. (13.12) up to a rescaling

$$z = k_0^{-1}\tau, \qquad A_{\mathrm{KK}} = k_0^{-1}A_{\mathrm{TN}}. \tag{15.177}$$

This is the reason why the latter is called the (Sorkin–Gross–Perry) *Kaluza–Klein monopole* [1148, 684]. We have to identify the magnetic charge and the Taub–NUT charge,

$$|\tilde{p}|/(4\pi) = 2|N|. \tag{15.178}$$

$N$ is related to the period of $\tau$, $8\pi|N|$, which, upon making the identification $\tau = k_0 z$, implies $4|N| = R_z$ and $|\tilde{p}| = 2\pi R_z$, which is consistent with the known quantization of the KK modes' electric charge $\tilde{q} = n/R_z$ and the Dirac quantization condition Eq. (15.64).

Summarizing, we have performed a purely gravitational five dimensional duality transformation that interchanges momentum in the direction $z$ with (Euclidean) NUT charge. These two purely gravitational charges are seen in four dimensions as electric and magnetic U(1) charges.[22] This mechanism can be used in more general contexts, whenever the dimensionally reduced theory has an S-duality symmetry (see, for instance, Ref. [907]).

The KK S-duality symmetry is just a discrete $\mathbb{Z}_2$ transformation and it is natural to wonder whether there are dyonic solutions, even if they cannot be generated by continuous S-duality transformations. There is, to the best of our knowledge, only one dyonic KK BH solution that is also a dyonic ERN BH.

---

[21] But the solution Eq. (15.175) is not the Euclidean continuation of the non-extreme Taub–NUT solution, which is four-dimensional.

[22] For a discussion of the geometrical NUT charge and its representation as a $(d-3)$-form potential in $d$ dimensions, see Ref. [804].

*The RN–KK dyon.* Let us consider the dyonic MP solutions Eqs. (12.206). A quick calculation gives

$$F^2 = 8(\cos^2\alpha - \sin^2\alpha)\partial_{\underline{i}} H^{-1}\partial_{\underline{i}} H^{-1}, \tag{15.179}$$

which vanishes for $\alpha = \pm\pi/4$. For this value of the charges (whose signs we can still change, preserving $F^2 = 0$) the dyonic MP solutions are also solutions of the KK action with constant KK scalar $k = k_0$ (=1 for simplicity) and can be uplifted to a purely gravitational five-dimensional solution [855]:

$$\boxed{\begin{array}{l} d\hat{s}^2 = H^{-2}dt^2 - H^2 d\vec{x}_3^2 - \left[dz + \sqrt{2}\left(\alpha_q H^{-1}dt + \alpha_p A_{\underline{i}} dx^i\right)\right]^2, \\ \partial_{[\underline{i}} A_{\underline{j}]} = \alpha_p \tfrac{1}{2}\epsilon_{ijk}\partial_{\underline{k}} H, \qquad \partial_{\underline{k}}\partial_{\underline{k}} H = 0, \qquad \alpha_q^2 = \alpha_p^2 = 1, \end{array}} \tag{15.180}$$

where $\alpha_q$ and $\alpha_p$ are the possible signs of the electric and magnetic charges.

*Skew KK reduction and generation of fluxbranes.* Our last example, "skew KK reduction" [460, 461], shows the power of the KK techniques to generate new solutions from "almost nothing." The general setup is the following.[23] Let us consider a metric that admits two isometries, one compact (a U(1)), associated with the coordinate $\theta$, and one non-compact (an $\mathbb{R}$), associated with the coordinate $z$ with a metric of the product form

$$d\hat{s}^2 = -dz^2 - f^2[d\theta + f_m dx^m]^2 + f_{mn}dx^m dx^n, \tag{15.181}$$

where we have normalized the period of $\theta \in [0, 2\pi]$. We want now to construct a new spacetime by identifying points in the above spacetime according to

$$\begin{pmatrix} z + 2\pi R_z \\ \theta \end{pmatrix} \sim \begin{pmatrix} z \\ \theta - 2\pi B \end{pmatrix}. \tag{15.182}$$

To apply the standard Scherk–Schwarz formalism, we need to use a coordinate independent of $z$ and thus we define a new coordinate $\theta'$ adapted to the above identifications,

$$\theta' = \theta - \frac{B}{R_z} z, \tag{15.183}$$

adapted to the Killing vector $R_z\partial_{\underline{z}} - B\partial_{\underline{\theta}}$, and rewrite the metric, adapting it to KK reduction in the direction $z$. The lower-dimensional fields are

$$\begin{aligned} ds^2_{\rm KK} &= -\frac{R_z}{B}k^{-2}\left(d\theta' + f_m dx^m\right)^2 + f_{mn}dx^m dx^n, \\ A_{\theta'} &= \frac{B}{R_z}k^{-2}, \qquad A_{\underline{m}} = \frac{B}{R_z}k^{-2}f_m, \\ k^2 &= 1 + \frac{B^2}{R_z^2}f^2. \end{aligned} \tag{15.184}$$

[23] Here we follow Ref. [459], where more uses of this technique to construct new solutions can be found.

If we start from flat spacetime in polar coordinates [459],

$$d\hat{s}^2 = -dz^2 - (d\rho^2 + \rho^2 d\theta^2) + dt^2 - d\vec{y}^2_{(\hat{d}-4)}, \tag{15.185}$$

$f = \rho,\ f_m = 0$, and we obtain

$$\boxed{\begin{aligned} ds^2_{\mathrm{KK}} &= dt^2 - d\vec{y}^2_{(d-3)} - \frac{R_z}{B} k^{-2} d\theta'^2, \\ A_{\theta'} &= \frac{B}{R_z} k^{-2}, \qquad\qquad k^2 = 1 + \frac{B^2}{R_z^2}\rho^2. \end{aligned}} \tag{15.186}$$

This is the *Kaluza–Klein Melvin solution* [459]. It generalizes the original Melvin solution that *describes a parallel bundle of magnetic flux held together by its own gravitational pull* [944]. These solutions are also known as $(d-3)$-fluxbranes since they have a magnetic flux orthogonal to a $(d-3)$-dimensional spacelike submanifold that is invariant under all possible translations, just like the Schwarzschild black $p$-branes which we saw have a $p$-dimensional spacelike translation-invariant submanifold.

## 15.4 Toroidal (Abelian) dimensional reduction

The next simplest case we can consider is a $\hat{d}$-dimensional spacetime that locally (and asymptotically) is the product of $d$-dimensional Minkowski spacetime and $n$ circles ($\hat{d} = d + n$). The product of $n$ circles is *topologically* an $n$-torus $\mathrm{T}^n$ and this case, which is a trivial generalization of the single-circle case, is called *toroidal compactification. Metrically*, the relative sizes and angles of the circles define the torus.

A useful way to characterize tori is the following: a circle of length $2\pi R$ can be considered as a coset manifold, namely the quotient of the group of continuous translations $\mathbb{R}$ by the subgroup of discrete translations of size $2\pi R$, which we can denote by $2\pi R\mathbb{Z}$. Thus, $\mathrm{S}^1 = \mathbb{R}/(2\pi R\mathbb{Z})$.

A torus $\mathrm{T}^n$ can be similarly considered as the quotient of the group of $n$-dimensional translations $\mathbb{R}^n$ by a discrete $n$-dimensional subgroup called an $n$-dimensional *lattice*[24] $\Gamma^n$, $\mathrm{T}^n = \mathbb{R}^n/\Gamma^n$. The information about sizes and angles is evidently contained in $\Gamma^n$.

The quotient affects only the global properties of the torus, which locally is just $\mathbb{R}^n$, and therefore it has $n$ independent translational isometries. We can choose $n$ independent mutually commuting Killing vectors in the directions of the lattice generators. The $n$ adapted coordinates $z^m$ that parametrize their integral curves will then be periodic coordinates that can be used simultaneously.

The analysis of the theory in these spaces proceeds along the same lines as in the case of a single compact dimension. First, to find the spectrum, one performs an $n$-dimensional Fourier-mode expansion in the vacuum. The single zero mode will be the only massless mode and all the other modes will be massive. The massless $\hat{d}$-dimensional graviton mode

[24] A lattice $\Gamma^n$ is generated by linear combinations with integer coefficients of $n$ linearly independent vectors of $\mathbb{R}^n$, $\{\vec{u}_i\}$, $i = 1, \ldots, n$. Thus, a generic element $\vec{u} \in \Gamma^n$ can be written as $\vec{u} = n^i \vec{u}_i$, $n^i \in \mathbb{Z}$.

has to be decomposed into $d$-dimensional fields. It takes little effort to see that one obtains a $d$-dimensional graviton, $n$ $d$-dimensional vectors, and $(n+1)n/2$ scalars. The graviton and the $n$ vectors gauge the unbroken symmetries of the vacuum: $\mathrm{ISO}(1, d-1) \times \mathrm{U}(1)^n$ ($d$-dimensional Poincaré times the $n$ periodic isometries of the torus $\mathrm{T}^n$). The (asymptotic values of the) scalars are *moduli*: they appear naturally arranged in an $n$-dimensional metric, which is the metric of the internal space $\mathrm{T}^n$ and they carry the information about circle sizes and relative angles. Evidently they generalize $k$, which contains only information about the size of the single internal circle.

On the moduli will act the global symmetries of the torus: the affine group $\mathrm{IGL}(n, \mathbb{R})$,

$$z^{m\prime} = (R^{-1\,\mathrm{T}})^m{}_n z^n + a^m, \qquad R \in \mathrm{GL}(n, \mathbb{R}), \quad a^m \in \mathbb{R}^n, \tag{15.187}$$

which will give rise to the duality symmetries of the lower-dimensional theory.

It makes sense again to perform dimensional reduction of the theory, keeping only the massless mode. Our goal in this section will therefore be to perform the dimensional reduction of the $\hat{d}$-dimensional Einstein–Hilbert action to $d = \hat{d} - n$ dimensions.

The setup is the following: since we keep only the zero mode of the $\hat{d}$-dimensional metric, in practice we will be considering a metric that does not depend on the $n$ coordinates $z^m$ which parametrize the torus.[25] This is equivalent to saying that our metric does admit $n$ mutually commuting, translational, and periodic spacelike Killing vectors $\hat{k}_{(m)}^{\hat{\mu}}$, which we identify with those of the internal torus. We assume that all the internal coordinates have the same period $2\pi\ell$.

We can find the right definitions of the $d$-dimensional fields as in the $n=1$ case. There is not much new to be learned there, so we start by performing the following decomposition of the $\hat{d}$-dimensional Vielbeins $\hat{e}_{\hat{\mu}}{}^{\hat{a}}$ (KK ansatz) into $d$-dimensional Vielbeins $e_\mu{}^a$, vector fields $A^m{}_\mu$, and the $n$-dimensional internal metric Vielbeins $e_m{}^i$, which become scalars of the $(\hat{d}-n)$-dimensional theory:

$$\left(\hat{e}_{\hat{\mu}}{}^{\hat{a}}\right) = \begin{pmatrix} e_\mu{}^a & A^m{}_\mu e_m{}^i \\ 0 & e_m{}^i \end{pmatrix}, \qquad \left(\hat{e}_{\hat{a}}{}^{\hat{\mu}}\right) = \begin{pmatrix} e_a{}^\mu & -A^m{}_a \\ 0 & e_i{}^m \end{pmatrix}. \tag{15.188}$$

This ansatz is always possible because there always is a Lorentz rotation of the Vielbeins that brings them into this upper-triangular form. As usual, the $d$-dimensional metric is built out of the Vielbeins in this way,

$$g_{\mu\nu} = e_\mu{}^a e_\nu{}^b \eta_{ab}, \tag{15.189}$$

and we use them to trade curved and flat lower-dimensional indices, so, for instance,

$$A^m{}_a = A^m{}_\mu e_a{}^\mu. \tag{15.190}$$

We also have for the internal metric scalars (recall our mostly minus signature)

$$G_{mn} = -e_m{}^i e_n{}^j \delta_{ij}. \tag{15.191}$$

[25] We split coordinates and indices as follows: $(\hat{x}^{\hat{\mu}}) = (x^\mu, z^m)$ and, for Lorentz indices, $(\hat{a}) = (a, i)$.

The relation between $\hat{d}$- and $d$-dimensional fields is

$$\begin{aligned}\hat{g}_{\mu\nu} &= g_{\mu\nu} + A^m{}_\mu A^n{}_\nu G_{mn},\\ \hat{g}_{\mu n} &= A^m{}_\mu G_{mn} = \hat{k}_{(n)\,\mu},\\ \hat{g}_{mn} &= G_{mn} = \hat{k}_{(m)}{}^{\hat{\mu}}\hat{k}_{(n)\,\hat{\mu}}.\end{aligned} \tag{15.192}$$

These fields transform correctly as tensors, vectors, and scalars under $\hat{d}$-dimensional GCTs in the non-compact dimensions ($\hat{\epsilon}^\mu \equiv \epsilon^\mu$). Furthermore, under $\hat{d}$-dimensional GCTs in the internal dimensions ($\hat{\epsilon}^m \equiv -\Lambda^m$), the vectors undergo standard U(1) transformations,

$$\delta_{\Lambda^m} A^n{}_\mu = \delta_m{}^n \partial_\mu \Lambda^m. \tag{15.193}$$

The constant shifts of the internal coordinates have no effect whatsoever on the $d$-dimensional fields. Furthermore, under the $\mathrm{GL}(n,\mathbb{R})$ transformations only objects with internal indices transform. Thus, the $d$-dimensional metric is invariant and, in matrix notation, the internal metric and vectors transform according to

$$G' = R\,G\,R^{\mathrm{T}}, \qquad \vec{A}'_\mu = R^{-1\,\mathrm{T}}\vec{A}_\mu. \tag{15.194}$$

The group $\mathrm{GL}(n,\mathbb{R})$ can be decomposed into $\mathrm{SL}(n,\mathbb{R}) \times \mathbb{R}^+ \times \mathbb{Z}_2$, the $\mathbb{R}^+$ factor corresponding to rescalings analogous to those of the $n=1$ case, that change the determinant of the internal metric, and later we will want to redefine the fields so they transform well under those factors.

To calculate now the components of the spin connection in the above Vielbein basis, we first calculate the Ricci rotation coefficients $\hat{\Omega}_{\hat{a}\hat{b}\hat{c}}$; the non-vanishing ones are

$$\hat{\Omega}_{abc} = \Omega_{abc}, \qquad \hat{\Omega}_{abi} = \tfrac{1}{2} e_{mi} F^m{}_{ab}, \qquad \hat{\Omega}_{ibj} = -\tfrac{1}{2} e_i{}^m \partial_b e_{mj}. \tag{15.195}$$

They give

$$\begin{aligned}\hat{\omega}_{abc} &= \omega_{abc}, & \hat{\omega}_{abi} &= -\tfrac{1}{2} e_{im} F^m{}_{ab},\\ \hat{\omega}_{ibc} &= -\hat{\omega}_{bci}, & \hat{\omega}_{aij} &= -e_{[i|}{}^m \partial_a e_{m|j]},\\ \hat{\omega}_{ibj} &= \tfrac{1}{2} e_i{}^m e_j{}^n \partial_b G_{mn},\end{aligned} \tag{15.196}$$

where we have used

$$e_{(i|}{}^m \partial_a e_{|m|j)} = \tfrac{1}{2} e_i{}^m e_j{}^n \partial_a G_{mn}, \tag{15.197}$$

and we have defined

$$F^m{}_{\mu\nu} \equiv 2\partial_{[\mu} A^m{}_{\nu]}, \qquad F^m{}_{ab} = e_a{}^\mu e_b{}^\nu F^m{}_{\mu\nu}. \tag{15.198}$$

Next, we plug this result into the Ricci scalar term in the action expressed in terms of the spin-connection coefficients with the help of Palatini's identity Eq. (L.4) to obtain

$$\begin{aligned}\int d^{\hat{d}}\hat{x}\sqrt{|\hat{g}|}\,\hat{R} = \int d^n z \int d^d x \sqrt{|g|}\,K\Big\{&-\omega_b{}^{ba}\omega_c{}^c{}_a - \omega_a{}^{bc}\omega_{bc}{}^a + 2\omega_b{}^{ba}\partial_a \ln K\\ &- (\partial \ln K)^2 + \tfrac{1}{4}F^2 - \tfrac{1}{4}\partial_a G_{mn}\partial^a G^{mn}\Big\},\end{aligned} \tag{15.199}$$

where

$$K^2 \equiv |\det G_{mn}|, \qquad F^2 \equiv F^{m\,\mu\nu} F^{n}{}_{\mu\nu} G_{mn}, \tag{15.200}$$

so

$$G^{mn}\partial G_{mn} = 2\partial \ln K, \qquad \sqrt{|\hat{g}|} = \sqrt{|g|}\, K. \tag{15.201}$$

The sign of $F^2$ looks wrong, but one has to take into account the internal metric $G_{mn}$ which is negative-definite.

Using again Palatini's identity (but now in $d$ dimensions) and integrating over the internal coordinates, we find

$$\boxed{S = \frac{(2\pi\ell)^n}{16\pi G_{\rm N}^{(\hat{d})}} \int d^d x \sqrt{|g|}\, K \Big\{ R - (\partial \ln K)^2 + \tfrac{1}{4} F^2 - \tfrac{1}{4} \partial_a G_{mn} \partial^a G^{mn} \Big\}.} \tag{15.202}$$

We now want to use variables that are invariant under the $\mathbb{R}^+$ subgroup of rescalings, just as in the $n = 1$ case (the variables with tildes). First, we observe that any transformation $R \in \mathrm{GL}(n, \mathbb{R})$ can be written as follows:

$$R = a^{\frac{1}{n}} S X, \qquad a = |\det R| \in \mathbb{R}^+, \qquad S \in \mathrm{SL}(n, \mathbb{R}), \qquad X^2 = \mathbb{I}_{n\times n}. \tag{15.203}$$

Second, we define the modulus $K_0$ as the value of the scalar $K$ at infinity. According to its definition, $K$ is nothing but the volume element of the internal torus and it generalizes the scalar $k$ of the $n = 1$ case. The volume of the internal torus at a point $x$ of the $d$-dimensional space is

$$V_n(x) = \int_{\mathrm{T}^n} d^n z \sqrt{|\det G_{mn}(x)|} = K(x) \int_{\mathrm{T}^n} d^n z = (2\pi\ell)^n K(x), \tag{15.204}$$

and its value at infinity $V_n$ is measured in terms of the modulus $K_0$:

$$V_n = \lim_{r\rightarrow\infty} V_n(x) = (2\pi\ell)^n K_0. \tag{15.205}$$

If the torus were made up of orthogonal circles of local radii $R_m(x)$, then the internal metric would be diagonal,

$$G_{mn} = -\delta_{(m)n} K_{(m)}, \qquad K_m = (R_m(x)/\ell)\,, \tag{15.206}$$

and the volume would factorize into the product of the volumes of the circles. We would have

$$V_n = \prod_{m=1}^{m=n} (2\pi\ell K_{m\,0}) = \prod_{m=1}^{m=n} (2\pi R_m), \tag{15.207}$$

but it is worth stressing that this is not the case in general.

Under the transformation $R \in \mathrm{GL}(n, \mathbb{R})$ decomposed as above, the scalar $K$ and the modulus $K_0$ transform only under the $\mathbb{R}^+$ factor,

$$K' = a^{-1} K, \qquad K_0' = a^{-1} K_0, \tag{15.208}$$

and thus we can use them to define fields that are invariant under this factor:

$$\begin{aligned} \tilde{K} &= K/K_0, & \tilde{g}_{E\,\mu\nu} &= \tilde{K}^{\frac{2}{d-2}} g_{\mu\nu}, \\ \tilde{A}^m{}_\mu &= K_0^{\frac{1}{n}} A^m{}_\mu, & \mathcal{M}_{mn} &= -K^{-\frac{2}{n}} G_{mn}. \end{aligned} \tag{15.209}$$

$\mathcal{M}$ and $\tilde{A}_\mu$ transform only under the $SX \in \mathrm{SL}(n,\mathbb{R}) \times \mathbb{Z}_2$ factor as expected:

$$\mathcal{M}' = S\mathcal{M}S^{\mathrm{T}}, \qquad \vec{\tilde{A}}'_\mu = S^{-1\,\mathrm{T}} \vec{\tilde{A}}_\mu. \tag{15.210}$$

The metric is the "modified Einstein-frame metric" and the action takes the form

$$\boxed{\begin{aligned} S = \frac{V_n}{16\pi G_{\mathrm{N}}^{\hat{d}}} \int d^d x \sqrt{|\tilde{g}_{\mathrm{E}}|} \Big\{ & \tilde{R}_{\mathrm{E}} + \frac{\hat{d}-2}{n(d-2)} (\partial \ln \tilde{K})^2 - \tfrac{1}{4} \partial_\mu \mathcal{M}_{mn} \partial^\mu \mathcal{M}^{mn} \\ & - \tfrac{1}{4} \tilde{K}^{2\frac{\hat{d}-2}{n(d-2)}} \mathcal{M}_{mn} \tilde{F}^{m\,\mu\nu} \tilde{F}^{n}{}_{\mu\nu} \Big\}. \end{aligned}} \tag{15.211}$$

In this action, $\tilde{K}$ parametrizes an $\mathbb{R}^+$ $\sigma$-model, but what about $\mathcal{M}_{mn}$? This is a unimodular $n \times n$ matrix and, therefore, it belongs to $\mathrm{SL}(n,\mathbb{R})$ itself. Furthermore, it is symmetric and, therefore, it is not the most general $\mathrm{SL}(n,\mathbb{R})$ matrix we can find and it does not parametrize $\mathrm{SL}(n,\mathbb{R})$. In fact, with its $n(n+1)/2 - 1$ degrees of freedom, it parametrizes the coset space $\mathrm{SL}(n,\mathbb{R})/\mathrm{SO}(n,\mathbb{R})$. This can be seen as follows: we can view $\mathcal{M}$ as the product of two unimodular $n$-beins $\mathcal{V}_m{}^i$,

$$\mathcal{M}_{mn} = \mathcal{V}_m{}^i \mathcal{V}_n{}^j \delta_{ij}, \qquad \mathcal{V}_m{}^i = K^{-\frac{1}{n}} \hat{e}_m{}^i. \tag{15.212}$$

These unimodular $n$-beins transform under global $S \in \mathrm{SL}(n,\mathbb{R})$ transformations and local $\Lambda(x) \in \mathrm{SO}(n,\mathbb{R})$ transformations according to

$$\mathcal{V}' = S\mathcal{V}\Lambda^{\mathrm{T}}(x). \tag{15.213}$$

We can now choose $\mathcal{V}$ to be upper triangular. This can always be achieved by a suitable local $\mathrm{SO}(n,\mathbb{R})$ rotation. That matrix contains $n(n+1)/2 - 1$ degrees of freedom and parametrizes the coset space $\mathrm{SL}(n,\mathbb{R})/\mathrm{SO}(n,\mathbb{R})$ because it is an $\mathrm{SL}(n,\mathbb{R})$ matrix generated by the exponentiation of all the generators of that group except for those of the $\mathrm{SO}(n,\mathbb{R})$ subgroup, which necessarily generate non-upper-triangular matrices.[26] We can see our choice of upper-triangular matrices as a coset-representative or gauge choice. $S$ transformations take us out of our gauge choice but we can implement an $S$-dependent compensating $\Lambda$ transformation to restore the upper-triangular form.

The constant value of $\mathcal{M}$ at infinity, $\mathcal{M}_0$, contains the modular parameters of the torus (relative sizes and angles of the circles).

[26] The transpose of an upper-triangular matrix with all terms above and on the diagonal non-vanishing can never be the inverse of that matrix.

### 15.4.1 The 2-torus and the modular group

In our study of the global transformations of the internal torus we have not yet taken into account the periodic boundary conditions of the coordinates, which have to be preserved by the diffeomorphisms in the KK setting. Clearly the rescalings $R$ do not respect the torus boundary conditions, but they rescale $\ell$. The rotations $S$ respect the boundary conditions only if $S^{-1}\vec{n} \in \mathbb{Z}^n$; the matrix entries are integers, i.e. $S \in \mathrm{SL}(n, \mathbb{Z})$.

The case $n = 2$ is particularly interesting because it occurs in many instances,[27] some (but not all of them) associated with S dualities. In the case $n = 2$, up to a reflection $S = -\mathbb{I}_{2\times 2}$, these diffeomorphisms are known as *Dehn twists* and are not connected to the identity (in fact, they constitute the mapping class group of torus diffeomorphisms) and they constitute the *modular group* $\mathrm{PSL}(2, \mathbb{Z}) = \mathrm{SL}(2, \mathbb{Z})/\{\pm\mathbb{I}_{2\times 2}\}$. This is the group that acts on $\mathcal{M}$.

It is convenient to relate $\mathcal{M}$ to the complex modular parameter $\tau$ of the torus. We start by defining a complex modular-invariant coordinate $\omega$ on $\mathrm{T}^2$ by

$$\omega = \frac{1}{2\pi\ell}\vec{\omega}^{\mathrm{T}} \cdot \vec{z}, \qquad \vec{\omega} \in \mathbb{C}^2, \tag{15.214}$$

where, under $\mathrm{PSL}(2, \mathbb{Z})$ modular transformations, we assume that the complex vector $\vec{\omega}$ transforms according to

$$\vec{\omega}' = S\vec{\omega}. \tag{15.215}$$

The periodicity of $\omega$ is

$$\omega \sim \omega + \vec{\omega}^{\mathrm{T}} \cdot \vec{n}, \qquad \vec{n} \in \mathbb{Z}^2. \tag{15.216}$$

The lattice generated in the $\omega$ plane by $\vec{\omega}$ is represented in Fig. 15.1. In terms of the modular-invariant complex coordinate, the torus metric element

$$ds^2_{\mathrm{Int}} = d\vec{z}^{\mathrm{T}} G d\vec{z} \tag{15.217}$$

takes the form

$$ds^2_{\mathrm{Int}} = K^{\frac{1}{2}} \frac{1}{\mathrm{I_m}(\omega_1\bar{\omega}_2)} d\omega d\bar{\omega}. \tag{15.218}$$

(Observe that $\mathrm{I_m}(\omega_1\bar{\omega}_2)$ is a modular-invariant term, and a quite important one.)

What we have just done is to transfer the information contained in the metric (more precisely, in $\mathcal{M}$) into the complex periods $\vec{\omega}$. The relation between these two is

$$\mathcal{M} = \frac{1}{\mathrm{I_m}(\omega_1\bar{\omega}_2)} \begin{pmatrix} |\omega_1|^2 & \mathrm{R_e}(\omega_1\bar{\omega}_2) \\ \mathrm{R_e}(\omega_1\bar{\omega}_2) & |\omega_2|^2 \end{pmatrix}. \tag{15.219}$$

We can check that the transformation rules for the complex periods Eq. (15.215) and for the matrix $\mathcal{M}$ Eq. (15.210) are perfectly compatible.

[27] Owing to the isomorphisms $\mathrm{SL}(2, \mathbb{R}) \sim \mathrm{Sp}(2, \mathbb{R}) \sim \mathrm{SU}(1, 1)$ it takes several different, but equivalent, forms.

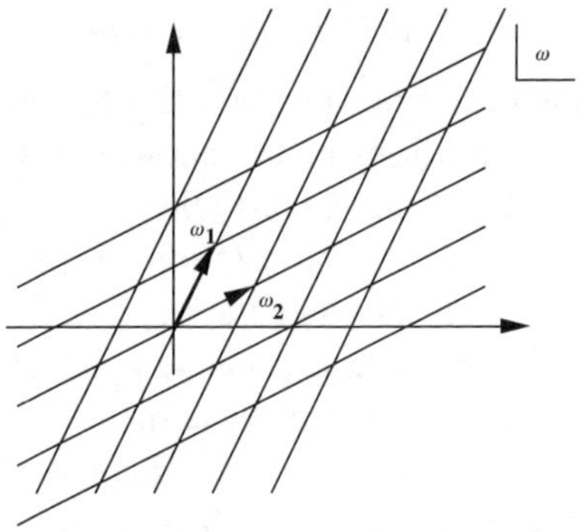


Fig. 15.1. The lattice generated in the $\omega$ plane by $\omega_1$ and $\omega_2$.

It should be clear that not all pairs of complex periods characterize different tori. Recall that $\mathcal{M}$ has only two independent entries whereas $\vec{\omega}$ contains four real independent quantities. In particular, we can see that multiplying $\vec{\omega}$ by any complex number leaves the matrix $\mathcal{M}$ invariant. It is customary to define the complex modulus parameter $\tau$,

$$\tau = \omega_1/\omega_2, \tag{15.220}$$

that can always be chosen to belong to the upper half of the complex plane $\mathbb{H}$, $\Im\mathrm{m}(\tau) \geq 0$ ($-\omega_1$ defines the same torus as $\omega_1$).

Under a modular transformation with $S$ parametrized by

$$S = \begin{pmatrix} \alpha & \beta \\ \gamma & \delta \end{pmatrix}, \tag{15.221}$$

with $\alpha\delta - \beta\gamma = 1$, the modular parameter $\tau$ undergoes a fractional-linear transformation:

$$\tau' = \frac{\alpha\tau + \beta}{\gamma\tau + \delta}. \tag{15.222}$$

Finally, in terms of $\tau$, the matrix $\mathcal{M}$ reads

$$\mathcal{M} = \frac{1}{\Im\mathrm{m}(\tau)} \begin{pmatrix} |\tau|^2 & \Re\mathrm{e}(\tau) \\ \Re\mathrm{e}(\tau) & 1 \end{pmatrix}. \tag{15.223}$$

The linear transformation of the matrix $\mathcal{M}$ Eq. (15.210) and the (non-linear) fractional-linear transformation Eq. (15.222) are completely equivalent.

The parametrization of the unimodular $\mathcal{V}$, in terms of $\tau$, is

$$\mathcal{V} = \begin{pmatrix} [\Im\mathrm{m}(\tau)]^{\frac{1}{2}} & [\Im\mathrm{m}(\tau)]^{-\frac{1}{2}}\,\Re\mathrm{e}(\tau) \\ 0 & [\Im\mathrm{m}(\tau)]^{-\frac{1}{2}} \end{pmatrix}, \tag{15.224}$$

and the $SL(2,\mathbb{R})/SO(2)$ $\sigma$-model action takes the form

$$\int d^d x \sqrt{|\tilde{g}_E|}\left[-\tfrac{1}{4}\partial_\mu \mathcal{M}_{mn}\partial^\mu \mathcal{M}^{mn}\right] = \int d^d x \sqrt{|\tilde{g}_E|}\left[\tfrac{1}{2}\frac{\partial_\mu \tau \partial^\mu \bar{\tau}}{(\Im\mathrm{m}(\tau))^2}\right]. \tag{15.225}$$

As said, this $\sigma$-model and the global symmetry group $SL(2,\mathbb{R})$ (broken by boundary conditions or quantum effects to $SL(2,\mathbb{Z})$) appear in many instances, apart from $T^2$ compactifications. To start with, $SL(2,\mathbb{Z})$ is the S-duality group and, under it, the complexified coupling constant that we also called $\tau$ transforms in the same way as the modular parameter of a torus (see Eq. (12.190)), but not every $SL(2,\mathbb{Z})$ is an S duality.

Another important example is provided by $N=2B, d=10$ SUGRA (the effective field theory of the type-IIB superstring), which we will review in Chapter 23, in which there is an $SL(2,\mathbb{R})/SO(2)$ $\sigma$-model[28] with $\Re\mathrm{e}(\tau)=\hat{C}^{(0)}$, the RR (pseudo)scalar, and $\Im\mathrm{m}(\tau)=e^{-\hat{\varphi}}$, $\hat{\varphi}$ being the dilaton, and invariance under global $SL(2,\mathbb{R})$ duality transformations that are interpreted as an S duality that rotates perturbative into non-perturbative states of the theory. In this case, the scalar $\sigma$-model does not arise from compactification. However, as we will explain in detail there, in $d=9$ dimensions it can be identified with the $\sigma$-model that arises in the compactification of $N=1, d=11$ supergravity to $d=9$ on $T^2$.

In the compactification of $N=1, d=11$ supergravity to $d=8$ dimensions on $T^3$, there is an $SL(3,\mathbb{R})/SO(3)$ $\sigma$-model that naturally contains the $SL(2,\mathbb{R})/SO(2)$ $\sigma$-model we just mentioned, but there is another $SL(2,\mathbb{R})/SO(2)$ $\sigma$-model that arises because the 11-dimensional 3-form gives rise to an eight-dimensional pseudoscalar [29]. Similar effects give rise to many $SL(2,\mathbb{R})$ subgroups of the total (U) duality group in various compactifications of 11- and ten-dimensional SUGRAs [907].

In $N=4, d=4$ SUGRA (the theory which results from the compactification on $T^6$ of $N=1, d=10$ supergravity, the effective field theory of the heterotic and type-I strings), which we will review in Section 16.2, $\Re\mathrm{e}(\tau)=a$ is a pseudoscalar in the heterotic case, $a$ is the Hodge dual of the dimensionally reduced Neveu–Schwarz 2-form and in the type-I case it is the Hodge dual of the dimensionally reduced Ramond–Ramond 2-form. In both cases, $\Im\mathrm{m}(\tau)=e^{-2\phi}$, where $\phi$ is the four-dimensional dilaton.[29] In this case, the scalar $SL(2,\mathbb{R})/SO(2)$ $\sigma$-model does not arise from compactification on $T^2$ either.

[28] In [1114] this coset was described in the form SU(1,1)/U(1). This is natural if one wants to construct the supergravity theory from scratch, using complex fields, but, from the point of view of string theory, the natural parametrization is the real one $SL(2,\mathbb{R})/SO(2)$. The relation between the SU(1,1)/U(1) variable $S$ and the $SL(2,\mathbb{R})$ parameter $\tau$ is

$$\tau = i\frac{1-S}{1+S},$$

and the relation between the kinetic terms is

$$\tfrac{1}{2}\frac{\partial_\mu \tau \partial^\mu \bar{\tau}}{(\Im\mathrm{m}(\tau))^2} = 2\frac{\partial_\mu S \partial^\mu \overline{S}}{(1-S\overline{S})^2}.$$

[29] In Ref. [389] this coset space was also described in the form SU(1,1)/U(1).

### *15.4.2 Masses, charges, and Newton's constant*

In the scale-invariant variables with tildes that we have defined we can immediately see that the $d$-dimensional Newton constant is given by

$$\boxed{G_{\rm N}^{(d)} = G_{\rm N}^{(\hat{d})}/V_n.} \tag{15.226}$$

To find the right definitions for the $n$ electric charges, we need the Noether currents. These are

$$j_n^{\nu} = \frac{1}{16\pi G_{\rm N}^{(d)}} \tilde{\nabla}_{\mu}\left(\tilde{K}^{2\frac{\hat{d}-2}{n(d-2)}} \mathcal{M}_{nm} \tilde{F}^{m\,\mu\nu}\right), \tag{15.227}$$

and then the electric and magnetic charges of the vector fields are defined by

$$\tilde{q}_n = \frac{1}{16\pi G_{\rm N}^{(d)}} \int_{S_{\infty}^{d-2}} \tilde{K}^{2\frac{(\hat{d}-2)}{n(d-2)}} \mathcal{M}_{nm} \star \tilde{F}^{m\,\mu\nu}, \qquad \tilde{p}^n = -\int_{S_{\infty}^2} \tilde{F}^n. \tag{15.228}$$

With these definitions, the electric and magnetic charges of the vector field $\tilde{A}^n_{\mu}$ satisfy the Dirac quantization condition

$$\tilde{q}_n \tilde{p}^n = 2\pi m, \qquad m \in \mathbb{Z}. \tag{15.229}$$

## 15.5 Generalized dimensional reduction

In Refs. [1105, 1106] Scherk and Schwarz introduced the idea of *generalized dimensional reduction* (GDR) and developed a general formalism. Here we want to explain the principle underlying the idea of GDR.

We can understand GDR[30] as the answer to the question "How do we dimensionally reduce multivalued fields?" There are at least two types of multivalued fields: fields that take values in some topologically non-trivial space (e.g. a circle) and fields that are defined up to some kind of local transformation (e.g. gauge vector fields, spinors (defined up to local Lorentz transformations) etc.). Let us take the simplest: a real scalar field $\hat{\varphi}$ taking values on a circle of radius $m$ (like an axion, which is, as a matter of fact, a pseudoscalar). In practice, to represent a multivalued field one takes a field living on the real line and then identifies

$$\hat{\varphi} \sim \hat{\varphi} + 2\pi m. \tag{15.230}$$

A single-valued field has to be a strictly periodic function of the compact coordinate: on going once around the compact dimension, we return to the same point and there the field has to have the same value. However, a multivalued field such as $\hat{\varphi}$ is allowed to take a

[30] Originally, GDR was introduced as just a generalized KK ansatz in which the $\hat{d}$-dimensional fields were allowed to depend on the internal coordinates $z^m$ in such a way that the lower-dimensional fields did not depend on them and, at the same time, some symmetries were broken. Here, we prefer to take the view that GDR is *the* KK ansatz for multivalued fields and it is not an option or just a clever trick.

different value as long as it is a multiple of $2\pi m$ because the two values of the field are assumed to be physically equivalent. Thus, in general, we can have

$$\hat{\varphi}(x, z + 2\pi\ell) = \hat{\varphi}(x, z) + 2\pi N m \sim \hat{\varphi}(x, z). \tag{15.231}$$

The Fourier expansion of such a multivalued field in $z$ is now

$$\hat{\varphi}^{(N)}(x, z) = \frac{mN}{\ell} z + \sum_{n \in \mathbb{Z}} e^{\frac{2\pi i n z}{\ell}} \hat{\varphi}^{(n)}(x). \tag{15.232}$$

The extra term linear in $z$ is responsible for the multivaluedness. This term is clearly non-dynamical, unlike the KK modes $\hat{\varphi}^{(n)}(x)$ which are dynamical, which means that the value of $N$ cannot change (at least, classically). $N$ is chosen once and for all and its value defines the vacuum. Therefore, it is a (discrete) modulus of the theory.

It should be obvious that the above field configurations are topologically non-trivial: the field is "wound" $N$ times around the compact dimension. The topological number that characterizes these configurations is the winding number $N$,

$$N = \frac{1}{2\pi\ell m} \oint \hat{\varphi}. \tag{15.233}$$

The choice of vacuum is also a choice of topological sector in the space of configurations.

It should be stressed that all this makes sense if there are solutions of the form

$$\hat{\varphi} = \frac{mN}{\ell} z \tag{15.234}$$

compatible with the vacuum configurations of the other fields. Otherwise, one cannot talk about those new vacua labeled by $N$.

How do we perform the dimensional reduction of this field in the vacuum $N$? The logic is always the same: we simply ignore the massive modes and keep the massless ones. This means that, to carry out dimensional reduction of the above field, we should consider the KK ansatz

$$\hat{\varphi} = \frac{mN}{\ell} z + \hat{\varphi}^{(0)}(x) = \frac{mN}{\ell} z + \varphi(x). \tag{15.235}$$

Now the question of how we are supposed to obtain a truly $d = (\hat{d} - 1)$-dimensional theory if we start with a field that depends on the internal coordinate $z$ arises. We can argue that the dependence on $z$ will always disappear in the lower-dimensional theory: a field that lives on a circle necessarily appears in the action in a form that is invariant under *arbitrary constant shifts*. This means that the action can always be rewritten in terms of derivatives of $\hat{\varphi}$. Then, the linear term will either completely disappear (if it is hit by the derivative with respect to $x^\mu$) or remain without the $z$ (if it is hit by the derivative with respect to $z$). The surviving term will play the role of a mass term in general, as we will see.

This argument leads us to make three observations.

1. The rule of thumb for how to perform GDR in this context is to implement a $z$-dependent shift in the scalar field's standard KK ansatz. If we consider more general multivalued fields $\Phi$, which are identified by

$$\hat{\Phi} \sim e^{i\omega Q} \hat{\Phi}, \tag{15.236}$$

where $\mathcal{Q}$ is some symmetry of the theory, then the generalized KK ansatz is, ignoring higher KK modes,

$$\hat{\Phi}(\hat{x}) \sim e^{\frac{i\omega \mathcal{Q} z}{2\pi\ell}} \Phi(x). \tag{15.237}$$

The symmetry generated by $\mathcal{Q}$ is generically broken.

2. The converse is not true: the invariance of the action under constant shifts of $\hat{\varphi}$ does not mean that the field lives on a circle and GDR makes sense. Formally, the GDR procedure can be performed, but the result could be meaningless since no vacuum solution associated with the GDR ansatz is guaranteed to exist. We are going to see an example of this fact in Section 15.5.1.
3. Under U(1) gauge transformations

$$\delta_\Lambda z = -\Lambda(x), \qquad \delta_\Lambda A_\mu = \partial_\mu \Lambda(x), \qquad \delta_\Lambda \varphi = \frac{mN}{\ell}\Lambda(x), \tag{15.238}$$

i.e. the lower-dimensional scalar field transforms by shifts of the gauge parameter! This kind of gauge transformation is called a *massive gauge transformation* and allows us to eliminate $\varphi$ completely by fixing the gauge. $\varphi$ plays the role of the Stückelberg field for $A_\mu$ [1162]. KK gauge invariance is broken after this gauge fixing and this is reflected, as we will see, in a new mass term for the vector field. It is usually said that the vector has "eaten" the scalar, becoming massive. This is a sort of Higgs phenomenon, the difference being that there is no scalar potential. Observe that $\varphi$ can be removed consistently by a gauge transformation if both $\Lambda$ and $\varphi$ live in circles, as we have assumed.

In the following sections we are going to see some examples of GDR that illustrate these ideas. In the first example we perform the complete GDR of the real scalar field that we have discussed above and give an alternative interpretation.

### 15.5.1 Example 1: a real scalar

Let us consider the simple model

$$\hat{S} = \int d^{\hat{d}}\hat{x}\sqrt{|\hat{g}|}\left[\hat{R} + \tfrac{1}{2}(\partial\hat{\varphi})^2\right], \tag{15.239}$$

where $\hat{\varphi}$ is a real scalar field. This action is invariant under constant shifts of the scalar and therefore it is possible to use the standard recipe for GDR: we perform now a $z$-dependent shift of the usual $z$-independent ansatz $\hat{\varphi}(x,z) = \varphi(x) + mNz/\ell$, which will lead us to a $d$-dimensional theory with no dependence on $z$.

As we have stressed repeatedly, this recipe makes real sense only if the scalar field lives in a circle and is identified periodically, $\hat{\varphi} \sim \hat{\varphi} + 2\pi m$. Although it looks as if we can simply decree that identification, the above action does not contain enough structure to enforce it and we will see that, in particular, there is no vacuum solution with $\hat{\varphi}(x,z) = mNz/\ell$. This example is therefore just an academic exercise.

Using the standard ansatz for the Vielbein Eq. (15.33) but adding a subscript (1) to the KK scalar field, we find

$$S = \int d^dx \sqrt{|g|}\, k \left[ R - \tfrac{1}{4}k^2 F_{(2)}^2 + \tfrac{1}{2}(\mathcal{D}\phi)^2 - \tfrac{1}{2}\left(\frac{mN}{\ell}\right)^2 k^{-2} \right], \tag{15.240}$$

where the field strengths are defined by

$$F_{(2)\,\mu\nu} = 2\partial_{[\mu}A_{(1)\,\nu]}, \qquad \mathcal{D}_\mu\varphi = \partial_\mu\varphi - \frac{mN}{\ell}A_{(1)\,\mu}, \tag{15.241}$$

and are invariant under the massive gauge transformations

$$\delta_\Lambda z = -\Lambda, \qquad \delta_\Lambda\varphi = \frac{mN}{\ell}\Lambda, \qquad \delta A_{(1)\,\mu} = \partial_\mu\Lambda. \tag{15.242}$$

As we expected from our general discussion, $\varphi$ is a Stückelberg field for $A_{(1)\,\mu}$, which becomes massive by "eating" it, and the KK U(1) symmetry is broken by our choice of vacuum if $N \neq 0$.

Now, let us try to find a vacuum solution of the reduced theory. It will correspond to the gauge-breaking vacuum of the $\hat{d}$-dimensional theory. We can assume that the vacuum solutions will have $A_\mu = 0$ and $\varphi = \varphi_0$, a constant. Solutions of this kind can be derived consistently from the above action by setting those fields to zero. On going to the Einstein frame and redefining $k$ as in Eq. (15.47) but now calling $\chi$ the new scalar, we find the action of a real scalar with an unbounded potential coupled to gravity:

$$S = \int d^dx \sqrt{|g|} \left[ R + 2(\partial\chi)^2 - \tfrac{1}{2}\left(\frac{mN}{\ell}\right)^2 e^{-2\sqrt{2\frac{d-1}{d-2}}\chi} \right]. \tag{15.243}$$

Since the potential has no minima, there are no vacuum solutions with constant $\chi$ equal to some minimum of the potential and a Minkowski metric. The vacuum has to have a non-trivial metric.

Typical solutions of actions of this kind, with generic potentials, are *domain-wall* solutions that interpolate between two asymptotic regions in which the scalar field takes the value of a different minimum of the potential, i.e. two vacua in which the scalar has a constant value equal to the minimum of the potential.[31] The region in which the value of the scalar switches from one vacuum value to another is the domain wall. It is a $(d-2)$-dimensional region (plus the time) that is orthogonal to the coordinate on which the scalar typically depends. In fact, it can have zero thickness or some finite thickness in the direction of the transverse coordinate.

Although this potential has no minima, there might be some domain-wall-type solution since potentials like this one admit them: $(d-2)$-branes. However, precisely for the above potential, the generic solution given in Ref. [913] breaks down. Although this is far from a proof, it seems plausible that no such solution exists, confirming our suspicion that the GDR that we have performed is not consistent because it is based on a non-existent vacuum.

[31] A general reference for domain-wall solutions in $d = 4$ dimensions is given in Ref. [400].

*GDR and* $(\hat{d}-3)$*-branes.* We have mentioned that actions such as Eq. (15.240) generically admit $(d-2)$-brane solutions. However, we have said that $p$-branes couple to a $(p+1)$-form potential with a $(p+2)$-form field strength and there is no $d$-form field strength in that action, only a potential proportional to the square of the mass parameter. However, terms of this kind, which are typical of massive supergravities, should not naively be interpreted as potentials. Instead, we should compare such a term with the kinetic term for the 1-form, which is also multiplied by a power of $k$. The analogy (and the fact that the sign is the correct one) suggests that we should interpret that term as a sort of "kinetic" term for a 0-form field strength (the mass constant), which happens to be the Hodge dual of the $d$-form field strength associated with the $(d-2)$-brane solutions.

There is another way, different from GDR, to see how the $(d-2)$-brane solutions arise: before performing the reduction of the scalar, we could have Hodge-dualized the $\hat{d}$-dimensional scalar into a $(\hat{d}-2)$-form potential by the Poincaré-duality procedure explained in Section 12.7.1,

$$\star d\hat{\varphi} = d\hat{A}_{(\hat{d}-2)} \equiv F^2_{(\hat{d}-1)}, \tag{15.244}$$

obtaining the (on-shell) equivalent model

$$\tilde{\hat{S}} = \int d^{\hat{d}}\hat{x}\sqrt{|\hat{g}|}\left[\hat{R} + \frac{(-1)^{\hat{d}-2}}{2\cdot(\hat{d}-1)!}\hat{F}^2_{(\hat{d}-1)}\right]. \tag{15.245}$$

The KK dimensional reduction of $p$-forms follows the pattern of the reduction of the Maxwell vector field performed in Section 15.2.5: a $p$-form in $\hat{d}$ dimensions gives rise to a $p$-form and a $(p-1)$-form in $d$ dimensions. The potentials and gauge-invariant field strengths are identified using tangent-space indices. In this case, we obtain a $(d-2)$-form potential and a $(d-1)$-form potential, and the action

$$\tilde{S} = \int d^{\hat{d}-1}x\sqrt{|g|}\,k\left[R - \tfrac{1}{4}k^2F^2_{(2)} + \frac{(-1)^{d-1}}{2\cdot d!}F^2_{(d)} + \frac{(-1)^{d-2}}{2\cdot(d-1)!}k^{-2}F^2_{(d-1)}\right], \tag{15.246}$$

where[32]

$$F_{(d)} = d\partial A_{(d-1)} + (-1)^d A_{(1)}F_{(d-1)}, \qquad F_{(d-1)} = (d-1)\partial A_{(d-2)} \tag{15.247}$$

are the field strengths. We can now dualize the potentials. A $(d-1)$-form potential in $d$ dimensions has a $d$-form field strength whose Hodge dual is some function $f = \star F_{(d)}$. The equation of motion of the $(d-1)$-form potential $d\star F_{(d)} = 0$ becomes the Bianchi identity for the dual $df = 0$, which implies that $f$ is a constant that we call $Nm/\ell$. On adding the term

$$-\frac{1}{d!}\int d^dx\,\frac{mN}{\ell}\epsilon\Big[F_{(d)} + (-1)^{d+1}dA_{(1)}F_{(d-1)}\Big] \tag{15.248}$$

to the action and eliminating $F_{(d)}$ using its equation of motion,

$$mN/\ell = k\star F_{(d)}, \tag{15.249}$$

[32] When indices are not explicitly shown, we assume all indices to be antisymmetrized with weight unity.

in the action, we obtain

$$\tilde{S} = \int d^d x \sqrt{|g|} \left\{ k \left[ R - \tfrac{1}{4} k^2 F_{(2)}^2 + \frac{(-1)^{d-2}}{2 \cdot (d-1)!} k^{-2} F_{(d-1)}^2 - \tfrac{1}{2} \left( \frac{mN}{\ell} \right)^2 k^{-2} \right] - \frac{1}{(d-1)!} \left( \frac{mN}{\ell} \right) \frac{\epsilon}{\sqrt{|g|}} F_{(d-1)} A_{(1)} \right\}. \tag{15.250}$$

Now we dualize into a scalar field the $(d-2)$-form potential: we add the term

$$\frac{1}{(d-1)(d-1)!} \int d^d x \, \epsilon F_{(d-1)} \partial \varphi \tag{15.251}$$

and eliminate $F_{(d-1)}$ by substituting into the action its equation of motion

$$F_{(d-1)} = (-1)^{(d-1)} k \star \mathcal{D} \varphi, \tag{15.252}$$

obtaining the same result as with GDR. The two possible routes by which to arrive at the same $d$-dimensional theory are represented in Fig. 15.2.

Thus, the standard recipe for GDR is just a way to take into account all the fields and degrees of freedom that can arise in the dimensional reduction. The new degrees of freedom are discrete degrees of freedom described by a $(d-1)$-form potential or by the dual variable that can take the values $Nm/\ell, N \in \mathbb{Z}$, and are associated with a choice of vacuum.

Now, with the form $\hat{A}_{(\hat{d}-2)}$ we can associate a $(\hat{d}-3)$-brane. If one dimension is compact, there are two possibilities: either one of the dimensions of the brane is wrapped around the compact dimension or none is. From the $d$-dimensional point of view, the first configuration looks like a $(\hat{d}-4) = (d-3)$-brane and the second like a $(\hat{d}-3) = (d-2)$-brane. The $(\hat{d}-3) = (d-2)$-brane has no dynamics and has only one degree of freedom: its charge (or mass, which is usually proportional), which is the mass parameter that appears in the $d$-dimensional action. The mass parameters are to be considered fields, although one can equally consider them as expectation values of those fields. In this language we can say that our vacuum contains a $(d-2)$-brane.[33]

The charge of the $(\hat{d}-3)$-brane can be associated with the monodromy of $\hat{\varphi}$:

$$q \sim \oint \star \hat{F}_{(\hat{d}-1)} \sim \oint d\hat{\varphi} \sim mN. \tag{15.253}$$

### 15.5.2 Example 2: a complex scalar

The simplest example just considered failed GDR because we did not really have a multivalued scalar field. Let us consider a more interesting model with a complex scalar $\hat{\Phi}$:

$$\hat{S} = \int d^{\hat{d}} \hat{x} \sqrt{|\hat{g}|} \left[ \hat{R} + \tfrac{1}{2} \partial \hat{\Phi} \partial \hat{\Phi}^* \right]. \tag{15.254}$$

[33] We have said that it actually does not in this academic example, although it will in more general cases.

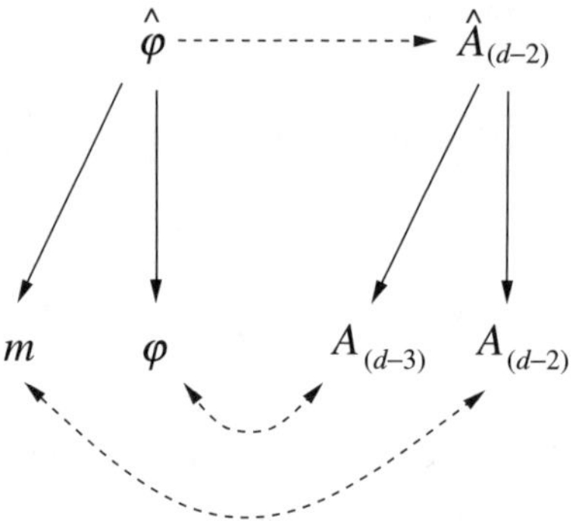


Fig. 15.2. This diagram represents two different ways of obtaining the same result: generalized dimensional reduction and "dual" standard dimensional reduction.

It is invariant under phase shifts of the scalar. Actually, we must consider as equivalent $\hat{\Phi}$ and $e^{2\pi i}\hat{\Phi}$. If we split it into its modulus $\hat{\rho}$ and phase $\hat{\sigma}$, $\hat{\Phi} = \hat{\rho} e^{\frac{i\hat{\sigma}}{m}}$, $\hat{\sigma}$ must be identified with $\hat{\sigma} + 2\pi m$ and we can say that it lives in a circle of radius $m$. Following the rule of thumb of GDR, our ansatz now has to be

$$\hat{\sigma}(x, z) = \hat{\sigma}(x) + \frac{Nm}{\ell} z \quad \Rightarrow \quad \hat{\Phi}(\hat{x}) = e^{\frac{iNz}{\ell}} \Phi(x). \tag{15.255}$$

We obtain the action

$$S = \int d^d x \sqrt{|g|}\, k \Big[ R - \tfrac{1}{4} k^2 F^2 + \tfrac{1}{2} \mathcal{D}\Phi \mathcal{D}\Phi^* - \tfrac{1}{2} \left(\frac{N}{\ell}\right)^2 k^{-2} |\Phi|^2 \Big], \tag{15.256}$$

where the field strengths are now given by

$$F_{(2)\,\mu\nu} = 2\partial_{[\mu} A_{\nu]}, \qquad \mathcal{D}_\mu \Phi = \partial_\mu \Phi + i \frac{N}{\ell} A_\mu \Phi, \tag{15.257}$$

and are invariant under the massive U(1) gauge transformations

$$\delta_\Lambda A_\mu = \partial_\mu \Lambda, \qquad \delta_\Lambda \Phi = e^{\frac{iN\Lambda}{\ell}} \Phi. \tag{15.258}$$

This is (ignoring $k$) the Lagrangian for a complex massive scalar field with U(1) charge $N/\ell$. In this case, the massive gauge transformations are simply the standard gauge transformations for a charged scalar field.

In terms of the real fields $\Phi = \rho e^{\frac{i\sigma}{m}}$ we find

$$\mathcal{D}\Phi \mathcal{D}\Phi^* = (\partial\rho)^2 + \frac{1}{m^2} \rho^2 (\mathcal{D}\sigma)^2, \qquad \mathcal{D}_\mu \sigma = \partial_\mu \sigma + \frac{mN}{\ell} A_\mu. \tag{15.259}$$

$\rho$ is invariant, but $\sigma$ transforms under massive gauge transformations,

$$\delta_\Lambda \sigma = \frac{mN}{\ell} \Lambda, \tag{15.260}$$

and is a Stückelberg field for $A_\mu$ and can be gauged away, leaving a mass term for $A_\mu$. U(1) can be spontaneously broken. A $|\hat{\Phi}|^4$ potential would produce a gravity-coupled version of the *Ginzburg–Landau Lagrangian*.

This model has an obvious solution $\rho = A_\mu = 0$ with the Minkowski metric. For $\rho = 0$ the $\sigma$-model defined by the scalars' kinetic terms is singular and we cannot distinguish between the different vacua labeled by $N$. Thus, this is also a failed example of GDR.

### 15.5.3 Example 3: an $\mathrm{SL}(2,\mathbb{R})/\mathrm{SO}(2)$ σ-model

This is given by the action

$$\hat{S} = \int d^{\hat{d}}\hat{x}\sqrt{|\hat{g}|}\left[\hat{R} + \tfrac{1}{2}\frac{\partial\hat{\tau}\partial\hat{\tau}^*}{(\Im\mathrm{m}(\tau))^2}\right]. \tag{15.261}$$

This action is invariant under global $\mathrm{SL}(2,\mathbb{R})$ fractional-linear transformations of $\hat{\tau} = \hat{a} + ie^{-\hat{\varphi}}$ and, in particular, under constant shifts $\hat{\tau} \to \hat{\tau} + b$, which act only on the real part. Furthermore, we have argued that, in many cases, we should consider as equivalent two values of $\tau$ related by $\mathrm{SL}(2,\mathbb{Z})$ transformations, which in particular means that $a$ lives in a circle of unit length. In this case, the $\sigma$-model metric is regular for finite values of $\hat{\varphi}$.

The general recipe of GDR tells us to use the ansatz

$$\hat{\tau}(\hat{x}) = \tau(x) + \frac{N}{2\pi\ell}z, \tag{15.262}$$

and we obtain the action

$$S = \int d^d x\sqrt{|g|}\,k\left[R - \tfrac{1}{4}k^2F^2 + \tfrac{1}{2}(\partial\varphi)^2 + \tfrac{1}{2}(\mathcal{D}a)^2 - \tfrac{1}{2}\left(\frac{N}{2\pi\ell}\right)^2 k^{-2}e^{-2\varphi}\right], \tag{15.263}$$

where

$$\mathcal{D}_\mu a = \partial_\mu a + \frac{N}{2\pi\ell}A_\mu, \tag{15.264}$$

and there is invariance under the massive U(1) gauge transformations

$$\delta_\Lambda A_\mu = \partial_\mu\Lambda, \qquad \delta_\Lambda a = \frac{N}{2\pi\ell}\Lambda. \tag{15.265}$$

Global $\mathrm{SL}(2,\mathbb{R})$ invariance is now clearly broken and the KK U(1) gauge invariance is also broken by the standard Stückelberg mechanism.

Let us look for a vacuum solution that will have $A_\mu = 0$ and $a = a_0$, a constant. The action for the remaining fields, in the Einstein frame, is

$$S = \int d^d x\sqrt{|g|}\left[R + 2(\partial\chi)^2 + \tfrac{1}{2}(\partial\varphi)^2 - \tfrac{1}{2}\left(\frac{N}{2\pi\ell}\right)^2 e^{-2\sqrt{2\frac{d-1}{d-2}}\chi - 2\varphi}\right]. \tag{15.266}$$

The potential for the remaining scalars has no lower bound, but we can still look for $(d-2)$-brane solutions. First, we diagonalize the potential by redefining the scalars:

$$\begin{aligned}\chi' &= \sqrt{\frac{d-1}{3d-5}}\chi + \frac{2}{\sqrt{\frac{2(3d-5)}{d-2}}}\frac{\varphi}{2},\\ \frac{\varphi'}{2} &= -\frac{2}{\sqrt{\frac{2(3d-5)}{d-2}}}\chi + \sqrt{\frac{d-1}{3d-5}}\frac{\varphi}{2},\end{aligned} \tag{15.267}$$

leaving the action in the form

$$S = \int d^dx\sqrt{|g|}\left[R + 2(\partial\chi')^2 + \tfrac{1}{2}(\partial\varphi')^2 - \tfrac{1}{2}\left(\frac{N}{2\pi\ell}\right)^2 e^{-2\sqrt{\frac{2(3d-5)}{d-2}}\chi'}\right]. \tag{15.268}$$

There are $(d-2)$-brane solutions with $\varphi' = 0$ in any dimension $d > 2$ [913]. These solutions are associated with $(\hat{d}-3)$-brane solutions in $\hat{d}$ dimensions of the $\mathrm{SL}(2,\mathbb{R})$ $\sigma$-model that, in the context of the ten-dimensional type-IIB superstring theory, are known as D-7-branes [186, 611]. We can say that the above action is the result of compactifying in a vacuum that contains $N$ $(\hat{d}-3)$-branes.

At last we have a successful realization of GDR and its relation to $(\hat{d}-3)$-branes.

### *15.5.4 Example 4: Wilson lines and GDR*

Another simple and interesting example is provided by a Dirac spinor $\hat{\psi}$ coupled to a U(1) gauge field $\hat{A}_{\hat{\mu}}$ in flat (for simplicity) $\hat{d} = 5$ spacetime,

$$\hat{S} = \int d^5\hat{x}\left\{\frac{i}{2}\bar{\hat{\psi}}\left(\hat{\not\partial} - ig\hat{\not A}\right)\hat{\psi} + \text{c.c.}\right\}. \tag{15.269}$$

This action is invariant under local U(1) transformations,

$$\hat{A}'_{\hat{\mu}} = \hat{A}_{\hat{\mu}} + \partial_{\hat{\mu}}\hat{\chi}, \qquad \hat{\psi}' = e^{ig\hat{\chi}}\hat{\psi}, \tag{15.270}$$

where the period of $\hat{\chi}$ has to be $2\pi/g$, and also under global phase shifts of the Dirac spinor (gauge transformations with constant $\hat{\chi}$). Thus, we can use the GDR ansatz

$$\hat{\psi}(\hat{x}) = e^{\frac{iNgz}{2\pi\ell}}\psi(x). \tag{15.271}$$

The dependence on the coordinate $z$ can be eliminated by a gauge transformation[34] with $\hat{\chi} = Nz/(2\pi\ell)$, but then the $z$ component of the gauge vector acquires a constant value (a non-vanishing vacuum expectation value, VEV)

$$\hat{A}'_{\hat{\mu}} = \hat{A}_{\hat{\mu}} + \frac{N}{2\pi\ell}\delta_{\hat{\mu}z}. \tag{15.272}$$

[34] Observe that the gauge parameter does not have the right periodicity.

The line integral of the vector field $\hat{A}_{\hat{\mu}}$ around the compact dimension is finite:

$$\oint_{\gamma} \hat{A} = N. \tag{15.273}$$

This configuration is said to have a U(1) *Wilson line*. The effect of the Wilson line (or the non-trivial dependence of the spinor on $z$) is to give a mass to the Dirac fermion. This is known as the *Hosotani or Wilson-line mechanism* [778, 779, 780] and we see that it can be transformed into Scherk–Schwarz GDR.

## 15.6 Orbifold compactification

Sometimes it is possible to compactify on spaces that are not manifolds. The prototypes of these spaces are *orbifolds*. These can be constructed as the quotients of manifolds by discrete symmetries. The simplest case is the segment, which can be constructed as the quotient $S^1/\mathbb{Z}_2$. To describe the quotient we need to define the action of $\mathbb{Z}_2$, and for this it is convenient to describe the circle itself as the quotient of the real line parametrized by $z$ by the group $\mathbb{Z}$ of discrete translations $z \to z + 2\pi n\ell$. There are no fixed points of the real line under this group and, therefore, we obtain a non-singular manifold.

Now, in terms of this coordinate $z$, $\mathbb{Z}_2$ acts by $z \to -z$. The result is the segment of line that goes from $z = 0$ to $z = 2\pi$. There are two fixed points under this group $z = 0$, for obvious reasons, and $z = \pi\ell$, since $-\pi\ell \sim -\pi\ell + 2\pi\ell = \pi\ell$, and they are the singular endpoints of the segment, which is not a manifold.[35]

The description of orbifolds as quotients is very convenient because in general the discrete symmetries have a well-defined action on the fields of the theory. In standard KK theory there are only tensor fields and their behavior under $z$ reflections depends on the number of $z$ indices they have: they acquire a minus sign for each index $z$. Only the KK vector has an odd number of $z$ indices, $A_\mu = \hat{g}_{\mu\underline{z}}/\hat{g}_{\underline{zz}}$, and thus it reverses its sign while the metric and KK scalar remain invariant.

The rule is that the spectrum of the KK theory on an orbifold can contain only fields that are invariant under the discrete symmetry. The reason is that odd fields will be given in solutions by odd functions of $z$ on the circle and they would be double valued (i.e. not well defined) on the orbifold. Thus, in the standard KK theory the KK vector is projected out of it. It is precisely this mechanism that was used by Hořava and Witten in Refs. [760, 761] to eliminate the RR 1-form $\hat{C}^{(1)}$ in the reduction of one-dimensional supergravity (the effective field theory of "M theory" in some corner of moduli space) to obtain chiral $N = 1, d = 10$ supergravity (the effective field theory of the heterotic string) instead of unchiral $N = 2A, d = 10$ supergravity (see Section 22.4).

In supersymmetric KK theory one has to define the action of the $\mathbb{Z}_2$ group on fermions. In odd dimensions one typically defines

$$\hat{\psi}' = \pm\hat{\Gamma}_z\hat{\psi}, \tag{15.274}$$

where $\Gamma_z$ is the gamma matrix associated with the direction $z$ and is proportional to the chirality matrix in one dimension fewer. Then, in the orbifold compactification only one chiral half of the spinors survives the projection.

35 The corresponding spacetime, taking into account the metric, would have a size of $\pi R_z$.

# 16

# Dilaton and dilaton/axion black holes

In Chapter 15 we have seen how scalar fields coupled to gravity arise naturally in KK compactification. In Part III we are going to see that scalar fields are also present, even before compactification, in some higher-dimensional supergravity theories that are the low-energy effective field theories of certain superstring theories. In all these examples the scalar fields couple in a characteristic way to vector (or $p$-form in higher dimensions) field strengths. In this chapter we are going to study first, in Section 16.1, a simple model that synthesizes the main features of those theories.

The $a$-model describes a real scalar coupled to gravity and to a vector field strength. The coupling is exponential and depends on a parameter $a$ (hence the name "$a$-model" that we are giving it here). Since the scalar can be identified in some cases with the string dilaton (or with the KK scalar, which is also called the dilaton sometimes), these models are also generically referred to as *dilaton gravity*. We will be able to obtain BH-type solutions for general values of $a$ and in any dimension $d \geq 4$; however, only a handful of values of $a$ actually occur in the theories of interest, although they occur in many different ways (*embeddings* [854]).

After studying the main properties of these *dilaton BHs*, we are going to study in Section 26.1 a more complex (four-dimensional) model that involves several scalar and vector fields. We are going to obtain extreme BH solutions that can be understood as *composite* BHs. This interpretation will open the door to the construction of four-dimensional extreme BH solutions in string theory as composite objects, the building blocks being $p$-branes and other extended objects that we will study in Chapter 26.

In Section 16.2 we add to the $a$-model with $a = 1$ and $d = 4$ a second scalar that couples not to the vector field kinetic term $F^2$ but to $F \star F$ and also couples to the dilaton. This kind of scalar (actually, a pseudoscalar, to preserve invariance of the action under parity) is called an *axion*. The model obtained has equations of motion that are invariant under global $\mathrm{SL}(2, \mathbb{R})$ (S) duality transformations (the dilaton and the axion parametrize an $\mathrm{SL}(2,\mathbb{R})/\mathrm{SO}(2)$ coset space) and it is sometimes called *axion–dilaton* gravity. The S-duality transformations can be used to obtain new solutions from known solutions.

As a matter of fact, the axion–dilaton-gravity model is a truncation of the bosonic sector of pure, ungauged, $N = 4, d = 4$ SUGRA which has five additional vector fields. This theory is a consistent truncation of the effective field theory of the heterotic superstring compactified on $T^6$, as we will see in Chapter 22, and we will therefore study its BH solutions. The most general solution (compatible with the no-hair conjecture) takes a very interesting duality-invariant form. The most general BH solution of the heterotic superstring compactified on $T^6$ should have a similarly duality-invariant form, but it is, unfortunately, unknown.

## 16.1 Dilaton black holes: the $a$-model

The $d$-dimensional "$a$-model" action is[1]

$$S = \frac{1}{16\pi G_{\rm N}^{(\rm d)}} \int d^d x \sqrt{|g|} \left[R + 2(\partial\varphi)^2 - \tfrac{1}{4} e^{-2a\varphi} F^2\right], \tag{16.1}$$

where, as usual, $F_{\mu\nu} = 2\partial_{[\mu} A_{\nu]}$ (or the equations of motion have to be supplemented by the Bianchi identity for $F$).

Our goal in this section is to find and study BH-type solutions of this model. Since there is a scalar, we might think that the only BH solutions are those with a trivial (constant) scalar field, because those are the only ones with no scalar hair. However, as we discussed in Section 12.1, we should distinguish between primary and secondary scalar hair. Secondary scalar hair is related to other conserved charges by a certain, fixed, formula, and is compatible with the existence of event horizons. We have already met in the preceding chapter some examples of dilaton BHs with non-trivial scalar fields and regular horizons. We do not know *a priori* the formula that relates the allowed, secondary, scalar "charge" to the conserved charges (mass and electric or magnetic charges), but we can deduce it from explicit BH solutions, if we find them.

The equations of motion are

$$G_{\mu\nu} + 2T^{\varphi}_{\mu\nu} - \tfrac{1}{2} e^{-2a\varphi} T^{A}_{\mu\nu} = 0, \qquad \nabla^2\varphi - \tfrac{1}{8} a e^{-2a\varphi} F^2 = 0, \qquad \nabla_\mu\left(e^{-2a\varphi} F^{\mu\nu}\right) = 0, \tag{16.2}$$

where

$$T^{\varphi}_{\mu\nu} = \partial_\mu\varphi\partial_\nu\varphi - \tfrac{1}{2} g_{\mu\nu}(\partial\varphi)^2, \qquad T^{A}_{\mu\nu} = F_\mu{}^\rho F_{\nu\rho} - \tfrac{1}{4} g_{\mu\nu} F^2. \tag{16.3}$$

Observe that, when $a = 0$, this is the Einstein–Maxwell system with an uncoupled scalar, which we can take to be constant. For $a \neq 0$ the only solutions that have a trivial dilaton (and are, therefore, solutions of the Einstein–Maxwell system) are those with $F^2 = 0$. We have already made use of this observation to embed solutions of the Einstein–Maxwell theory (dyonic RN BHs) into the KK theory (p. 445) to obtain the RN–KK dyon.

In Section 15.2 we showed that, in the KK reduction of pure $(d + 1)$-dimensional gravity on a circle, we always obtain an $a$-model with $a$ given by

$$a_{\rm KK} = \pm\sqrt{\frac{2(d-1)}{d-2}}. \tag{16.4}$$

[1] Generalizations with a massive dilaton have been studied in Refs. [675, 762].

(The two signs are related by the transformation $\varphi \to -\varphi$.) We will see that, in the reduction of the heterotic string on $T^6$, we naturally obtain $a = 1$.

If we take the divergence of the Einstein equation and use both the Bianchi identity for the metric and the Bianchi identity for $F$, we obtain

$$\left[\nabla^2\varphi - \tfrac{1}{8}ae^{-2a\varphi}F^2\right]\partial_\nu\varphi = 0, \tag{16.5}$$

which implies the scalar equation of motion *provided that the scalar is not constant*. For a non-constant scalar the only equations that we have to solve are, then, the Maxwell equation and the Einstein equation (with the trace subtracted for convenience):

$$\begin{aligned} R_{\mu\nu} + 2\partial_\mu\varphi\partial_\nu\varphi - \tfrac{1}{2}e^{-2a\varphi}\left[F_\mu{}^\rho F_{\nu\rho} - \frac{1}{2(d-2)}g_{\mu\nu}F^2\right] &= 0, \\ \nabla_\mu\left(e^{-2a\varphi}F^{\mu\nu}\right) &= 0. \end{aligned} \tag{16.6}$$

We want to find solutions of these equations describing electrically charged BHs that have to have a non-trivial scalar field. The solutions will reduce to the RN BH when $a = 0$, and when $F = 0$ we expect to recover the solutions of Refs. [827, 18] and the higher-dimensional analogs presented in Eq. (12.218). On the basis of our previous experience and discussions, it is natural to make an ansatz for the (static, spherically symmetric) metric that generalizes the "dressed Schwarzschild" metric in such a way that we can call it the "dressed RN" metric:

$$\begin{aligned} ds^2 &= f^{2x}H^{-2}W dt^2 - f^{-2y}H^{\frac{2}{d-3}}\left[W^{-1}dr^2 + r^2 d\Omega^2_{(d-2)}\right], \\ A_\mu &= \alpha\delta_{\mu t}(H^{-1} - 1), \qquad e^{-2a\varphi} = f^z, \end{aligned} \tag{16.7}$$

where $f = HW^b$ and

$$H = 1 + \frac{h}{r^{d-3}}, \qquad W = 1 + \frac{\omega}{r^{d-3}}, \tag{16.8}$$

and $x, y, z, b, h, \omega$, and $\alpha$ are constants to be found. Observe that the action is invariant under constant shifts of $\varphi$ accompanied by rescalings of the vector field. We can use this symmetry later to add a constant value at infinity to $\varphi$ and have

$$A_\mu = e^{a\varphi_0}\alpha\delta_{\mu t}(H^{-1} - 1), \qquad e^{-2a\varphi} = e^{-2a\varphi_0}f^z. \tag{16.9}$$

On substituting into the above equations of motion, one finds that there are two families of solutions, one with $b = 0$ and another one with $b = -1$. All the constants are identical in the two families, so all the fields (except for the metric) are identical. The first family contains regular BHs, but the second does not. The difference between the two families is the relation between the scalar charge and the mass and electric charge. We can view this difference as the presence of secondary scalar hair in the $b = 0$ family and of primary scalar hair in the $b = -1$ family. We are primarily interested in regular BHs and, therefore, we write only the $b = 0$ family of *dilaton-BH* solutions in its final form:

$$
\boxed{
\begin{aligned}
ds^2 &= e^{-2a(\varphi-\varphi_0)}H^{-2}W dt^2 - \left(e^{-2a(\varphi-\varphi_0)}H^{-2}\right)^{-\frac{1}{d-3}}\left[W^{-1}dr^2 + r^2 d\Omega^2_{(d-2)}\right],\\
A_\mu &= \alpha e^{a\varphi_0}\delta_{\mu t}(H^{-1}-1), \qquad e^{-2a\varphi} = e^{-2a\varphi_0}H^{2x},\\
H &= 1+\frac{h}{r^{d-3}}, \qquad W = 1+\frac{\omega}{r^{d-3}}, \qquad \omega = h\left[1-\frac{a^2}{4x}\alpha^2\right],\\
x &= \frac{(a^2/2)c}{1+(a^2/2)c}, \qquad c = \frac{d-2}{d-3}.
\end{aligned}
}
\tag{16.10}
$$

On adding the corresponding factors of $W$ to this solution, one obtains the $b = -1$ family. Here $a$ and $d$ are given parameters that determine our theory, and $\alpha$, $\varphi_0$ (the value of the dilaton at infinity), and $h$ (the coefficient of $r^{-(d-3)}$ in $H$) are the independent parameters. The relation between $\omega$ and $h$ is valid only for $h \neq 0$. If $h = 0$, $\omega$ is an arbitrary constant, there is no electromagnetic field, and we recover the solutions (12.218) which have primary scalar hair (the scalar charge is unrelated to the conserved charges) and are singular except for $b = 0$ or for $a = 0$ (which implies that $x = 0$), which is the Schwarzschild solution. For $a = 0$ we recover the RN solution, as we wanted.

Furthermore, for all values of $d, a$, and $b$, when $\omega = 0$ (*extreme dilaton BHs*) $H$ can be any arbitrary harmonic function in the transverse $(d-2)$-dimensional Euclidean space. This allows us to construct multi-BH (in general, multicenter) solutions, as in the MP family (which is included in this one with $a = 0$).

The $b = 0$ solutions were first obtained and studied in Refs. [608, 623]. The $d = 4$ solutions were rediscovered from a string-theory point of view in Ref. [589], and those with arbitrary $a$ were also studied in Ref. [756]. The multicenter solutions were found in Ref. [1135] (see also Ref. [1005]), and the solutions with $b = -1$ are presented for the first time here.

Let us now study the properties and the geometry of the $b = 0$ family. First, we want to relate the integration constants to the physical parameters: mass, electric charge, and "scalar charge." Only the first two are independent. For the sake of clarity we omit most numerical factors and define these charges by the asymptotic expansions of the fields:

$$
g_{tt} \sim 1 - \frac{\mathcal{M}}{r^{d-3}}, \qquad A_t \sim -\frac{\mathcal{Q}}{r^{d-3}}, \qquad \varphi \sim \varphi_0 - \frac{\mathcal{S}}{r^{d-3}}. \tag{16.11}
$$

These charges are related to the integration constants by

$$
\mathcal{M} = 2(1-x)h - \omega, \qquad \mathcal{Q} = \alpha e^{a\varphi_0}h, \qquad \mathcal{S} = xh. \tag{16.12}
$$

The inverse relations are, for $x \neq \frac{1}{2}$,

$$
\begin{aligned}
h &= \frac{\mathcal{M} \pm \sqrt{\mathcal{M}^2 - \frac{1-2x}{x} a^2 e^{-2a\varphi_0} \mathcal{Q}^2}}{2(1-2x)}, \\
\alpha &= \frac{2(1-2x)e^{-a\varphi_0}\mathcal{Q}}{\mathcal{M} \pm \sqrt{\mathcal{M}^2 - \frac{1-2x}{x} a^2 e^{-2a\varphi_0} \mathcal{Q}^2}}, \\
\omega &= \frac{x}{1-2x}\mathcal{M} \pm \frac{1-x}{1-2x}\sqrt{\mathcal{M}^2 - \frac{1-2x}{x} a^2 e^{-2a\varphi_0} \mathcal{Q}^2},
\end{aligned}
\tag{16.13}
$$

and they give us the expression of the "scalar charge" $\mathcal{S}$ in terms of $\mathcal{M}$ and $\mathcal{Q}$:

$$
\mathcal{S} = \frac{a^2 e^{-2a\varphi_0}\mathcal{Q}^2}{2\left(\mathcal{M} \mp \sqrt{\mathcal{M}^2 - \frac{1-2x}{x} a^2 e^{-2a\varphi_0}\mathcal{Q}^2}\right)}.
\tag{16.14}
$$

We see that it vanishes for vanishing $a$ (the RN solution) or vanishing $\mathcal{Q}$. This does not happen in the $b=-1$ family. If $\mathcal{S}$ has a different value (as in the $b=-1$ family) then there is primary scalar hair and we have solutions without regular event horizons.

The integration constants $h, \omega$, and $\alpha$ are real only when

$$
\mathcal{M}^2 \geq \frac{1-2x}{x} a^2 e^{-2a\varphi_0}\mathcal{Q}^2.
\tag{16.15}
$$

This is a constraint on $\mathcal{M}$ and $\mathcal{Q}$ only for $x < \frac{1}{2}$, that is, for $(a^2/2)c < 1$. This includes, as we know, the RN case.

When $x = \frac{1}{2}$, that is, $a = \pm\sqrt{2(d-3)/(d-2)}$, we find

$$
h = \frac{d-3}{d-2}\frac{e^{-2a\varphi_0}\mathcal{Q}^2}{\mathcal{M}}, \qquad \alpha = \frac{d-2}{d-3}\frac{e^{a\varphi_0}\mathcal{M}}{\mathcal{Q}}, \qquad \omega = -\frac{\mathcal{M}^2 - \frac{d-3}{d-2}e^{-2a\varphi_0}\mathcal{Q}^2}{\mathcal{M}},
\tag{16.16}
$$

and the "scalar charge" is given by

$$
\mathcal{S} = \pm\frac{d-3}{2(d-2)}\frac{e^{-2a\varphi_0}\mathcal{Q}^2}{\mathcal{M}}.
\tag{16.17}
$$

These metrics are a generalization of the RN metric and they have two horizons, at $r=0$ and $r=-\omega$. If we take the lower sign, $r=-\omega$ is the (regular in all cases with $\omega \neq 0$) event horizon, but the "horizon" at $r=0$ is generically singular, except for $a=0$. When $\omega=0$ (the extremal limit) the two horizons coincide. This happens when

$$
\mathcal{M} = \frac{2(1-x)}{x}\mathcal{S} = \frac{1-x}{\sqrt{x}} a e^{-a\varphi_0}|\mathcal{Q}|.
\tag{16.18}
$$

However, in this limit we have a regular BH only for $a=0$ (the ERN BH). All the other extreme dilaton BHs have a singular "horizon," i.e. a naked singularity.

There is a better way to express the extremality condition, using the scalar charge:

$$\omega^2 = \mathcal{M}^2 + 4\left(\frac{1}{x} - 1\right)\mathcal{S}^2 - a^2\left(\frac{1}{x} - 1\right)e^{-2a\varphi_0}\mathcal{Q}^2 = 0. \tag{16.19}$$

This form suggests that, as in the ERN case, the extremality condition can be viewed as a no-force condition. The difference here is that the dilaton field carries an additional interaction proportional to the "scalar charge" $\mathcal{S}$. This gives a physical explanation for the existence of regular multi-dilaton BH solutions. A pair of dilaton BHs with charges $(\mathcal{M}_i, \mathcal{Q}_i)$, satisfying separately the extremality condition, will also satisfy the no-force condition,

$$\mathcal{M}_1\mathcal{M}_2 + 4\left(\frac{1}{x} - 1\right)\mathcal{S}_1\mathcal{S}_2 - a^2\left(\frac{1}{x} - 1\right)e^{-2a\varphi_0}\mathcal{Q}_1\mathcal{Q}_2 = 0. \tag{16.20}$$

### 16.1.1 The a-model solutions in four dimensions

The general solution for the four-dimensional $a$-model is

$$\boxed{\begin{aligned} ds^2 &= H^{-\frac{2}{1+a^2}}W dt^2 - H^{-\frac{2}{1+a^2}}\left[W^{-1}dr^2 + r^2 d\Omega^2_{(2)}\right], \\ A_\mu &= \alpha e^{a\varphi_0}\delta_{\mu t}(H^{-1} - 1), \qquad e^{-2\varphi} = e^{-2\varphi_0}H^{\frac{2a}{1+a^2}}, \\ H &= 1 + \frac{h}{r}, \qquad W = 1 + \frac{\omega}{r}, \qquad \omega = h\left[1 - (1 + a^2)(\alpha/2)^2\right]. \end{aligned}} \tag{16.21}$$

Among all the possible values of $a$, only a few are relevant, at least in SUGRA theories (and hence for string theory). The most important value is $a = \sqrt{3}$. This is the value that we obtain in KK compactification from five to four dimensions, but it also appears in many other ways. The metric and dilaton field are[2]

$$\boxed{\begin{aligned} ds^2 &= H^{-\frac{1}{2}}W dt^2 - H^{\frac{1}{2}}\left[W^{-1}dr^2 + r^2 d\Omega^2_{(2)}\right], \\ e^{-2\varphi} &= e^{-2\varphi_0}H^{\frac{\sqrt{3}}{2}}, \qquad \omega = h\left[1 - \alpha^2\right]. \end{aligned}} \tag{16.22}$$

All the extended objects of type-II string theory compactified on tori give rise precisely to this Einstein metric (see, for instance, Eqs. (26.7) and (26.9)). This illustrates the comment we made in the introduction about the many possible embeddings of the $a$-model solutions into SUGRA theories.[3]

[2] The vector field and the functions $H$ and $W$ always have the same form as in Eqs. (16.21).

[3] A systematic study of embeddings of these four-dimensional dilaton BHs in the effective field theory of the heterotic string ($N = 1, d = 10$ SUGRA plus 16 vector multiplets) and their unbroken supersymmetries was presented in Ref. [854].

The next values of interest from the string-theory supergravity point of view are $a = 1$,

$$\boxed{\begin{aligned} ds^2 &= H^{-1}W dt^2 - H\left[W^{-1}dr^2 + r^2 d\Omega^2_{(2)}\right], \\ e^{-2\varphi} &= e^{-2\varphi_0}H, \qquad \omega = h\left[1 - (\alpha/\sqrt{2})^2\right], \end{aligned}} \tag{16.23}$$

and $a = 1/\sqrt{3}$,

$$\boxed{\begin{aligned} ds^2 &= H^{-\frac{3}{2}}W dt^2 - H^{\frac{3}{2}}\left[W^{-1}dr^2 + r^2 d\Omega^2_{(2)}\right], \\ e^{-2\varphi} &= e^{-2\varphi_0}H^{\frac{\sqrt{3}}{2}}, \qquad \omega = h\left[1 - \alpha^2/3\right]. \end{aligned}} \tag{16.24}$$

Four-dimensional stringy solutions with these metrics will appear in the compactification of solutions that describe the intersection of two and three extended objects, respectively, instead of just one as in the previous case. We are going to see how this comes about in Section 26.1.[4]

Finally, we have $a = 0$, the RN BH:

$$\boxed{\begin{aligned} ds^2 &= H^{-2}W dt^2 - H^{2}\left[W^{-1}dr^2 + r^2 d\Omega^2_{(2)}\right], \\ e^{-2\varphi} &= e^{-2\varphi_0}, \qquad \omega = h\left[1 - (\alpha/2)^2\right]. \end{aligned}} \tag{16.25}$$

This case will be seen to arise from the intersection of four extended objects in higher dimensions.

In four dimensions, we can define the mass $M$, electric charge $q$, and "scalar charge" $\Sigma$ more precisely by the asymptotic expansions

$$g_{tt} \sim 1 - \frac{2G_{\rm N}^{(4)}M}{r}, \qquad A_t \sim \frac{4G_{\rm N}^{(4)}e^{2a\varphi_0}q}{r}, \qquad \varphi \sim \varphi_0 + \frac{G_{\rm N}^{(4)}\Sigma}{r}. \tag{16.26}$$

The dilaton-dependent factor $e^{2a\varphi_0}$ in the definition of the electric charge is related to the integral definition

$$q = \frac{1}{16\pi G_{\rm N}^{(4)}}\int_{S^2_\infty} e^{-2a\varphi} \star F, \tag{16.27}$$

which is in turn related to the modification of the Gauss law introduced by the dilaton.

[4] The embedding of the model studied there into the STU model of $N = 2, d = 4$ supergravity is explained on p. 578.

For $a \neq 1$ the integration constants in the solutions are given by

$$\begin{aligned} h &= \frac{1+a^2}{1-a^2} G_{\rm N}^{(4)} \left( M \pm \sqrt{M^2 - 4(1-a^2)e^{2a\varphi_0}q^2} \right), \\ \alpha &= \frac{1-a^2}{1+a^2} \frac{4e^{a\varphi_0}q}{M \pm \sqrt{M^2 - 4(1-a^2)e^{2a\varphi_0}q^2}}, \\ \omega &= \frac{2a^2}{1-a^2} G_{\rm N}^{(4)} M \pm \frac{2}{1-a^2} G_{\rm N}^{(4)} \sqrt{M^2 - 4(1-a^2)e^{2a\varphi_0}q^2}, \end{aligned} \tag{16.28}$$

and $\Sigma$ is related to the conserved charges by

$$\Sigma = \frac{4a^2 e^{2a\varphi_0}q^2}{M \pm \sqrt{M^2 - 4(1-a^2)e^{2a\varphi_0}q^2}}. \tag{16.29}$$

For $a = 1$

$$h = \frac{4G_{\rm N}^{(4)} e^{2\varphi_0}q^2}{M}, \qquad \alpha = -\frac{M}{e^{\varphi_0}q}, \qquad \omega = -2G_{\rm N}^{(4)} \frac{M^2 - 2e^{2\varphi_0}q^2}{M}, \tag{16.30}$$

and $\Sigma$ is related to the conserved charges by

$$\Sigma = -\frac{2G_{\rm N}^{(4)} e^{2\varphi_0}q^2}{M}. \tag{16.31}$$

The extremality condition always takes the form

$$\left( \frac{\omega}{2G_{\rm N}^{(4)}} \right)^2 = M^2 + \frac{1}{a^2}\Sigma^2 - 4e^{2\varphi_0}q^2 = 0. \tag{16.32}$$

*Thermodynamics of four-dimensional dilaton BHs.* The coupling to scalar fields requires a modification of the first law of BH thermodynamics, which has to include a new term [380, 381, 622] in order to take into account possible variations of the energy due to variations of the scalar fields. This term is proportional to the "scalar charges" and to the variations of the values of the scalar fields at infinity (*moduli*) that characterize the vacuum of the theory,

$$\Sigma_a d\varphi^a. \tag{16.33}$$

Apart from this new term, the temperature and the entropy of dilaton BHs are related to the area and surface gravity of the event horizon by the standard formulae. When $\omega \leq 0$ (as we will assume) the event horizon is placed at $r = -\omega$. Its area is given by

$$A = 4\pi H^{\frac{2}{1+a^2}} r^2 \Big|_{r=-\omega}, \tag{16.34}$$

and so the entropy is given by

$$S = \pi (h + |\omega|)^{\frac{2}{1+a^2}} |\omega|^{\frac{2a^2}{1+a^2}}. \tag{16.35}$$

The temperature is given by

$$T = \frac{1}{4\pi}(h + |\omega|)^{-\frac{2}{1+a^2}} |\omega|^{\frac{1-a^2}{1+a^2}}. \tag{16.36}$$

These expressions can be compared with those in Ref. [756], where the thermodynamics of four-dimensional dilaton BHs was studied, with $|\omega| = r_+ - r_-$ and $h + |\omega| = r_+$.

The behavior of $T$ and $S$ in the extreme limit depends on the value of the parameter $a$:

$$\lim_{\omega \to 0} \begin{cases} T \to 0, & S \to \pi h^{\frac{2}{1+a^2}}, & 0 \leq a < 1, \\ T \to h, & S \to 0, & a = 1, \\ T \to \infty, & S \to 0, & a > 1. \end{cases} \tag{16.37}$$

Below $a = 1$ the behavior is similar to that of the RN BH, which means that near the extreme limit the specific heat is positive. Above $a = 1$ the behavior is similar to that of the Schwarzschild BH in the zero-mass limit and the specific heat is negative.

*Electric–magnetic duality in the four-dimensional a-model.* In four dimensions the $a$-model has electric–magnetic duality: the equations of motion are invariant under the discrete transformation

$$F' = \tilde{F} \equiv e^{-2a\varphi} \star F, \qquad \varphi' = \tilde{\varphi} \equiv -\varphi. \tag{16.38}$$

The fields with tildes are, by definition, the S-dual fields. This symmetry allows us to transform the above electrically charged solutions into magnetic solutions that have the same (Einstein-frame) metric. All the properties that depend on the Einstein metric (for instance, thermodynamical properties) are not affected by this transformation. However, in some cases we are interested in properties that depend on the metric given in a different frame (such as the string frame, which we will study in Chapter 21, or the KK frame that we studied in Chapter 15) that is related to Einstein's by a conformal rescaling by a function of the dilaton. Since the dilaton changes in this electric–magnetic transformation, so does the (KK or stringy) metric. A good example is provided by the electric–magnetic-duality rotation of the electrically charged KK BH studied on p. 443.

For special values of the parameter $a$ there are also dyonic dilaton BH solutions, carrying both electric and magnetic charges.[5] This happens trivially for $a = 0$, the Einstein–Maxwell plus uncoupled scalar case, because in this case (as we have already seen) the electric–magnetic-duality symmetry is a continuous symmetry and one can continuously rotate the purely electric solution into the purely magnetic one. In the case $a = 1$ there is no obvious reason for this to happen. However, the $a = 1$ model is a truncation of the $N = 4, d = 4$ SUEGRA action that we are going to see next, which does have a continuous electric–magnetic-duality symmetry.

[5] There is no theorem ensuring this, but all attempts to build dyonic solutions for other values of $a$ have been unsuccessful.

The dyonic solutions take the form[6] [608, 623, 841]

$$
\begin{aligned}
ds^2 &= (H_1H_2)^{-1}W dt^2 - H_1H_2\Big[W^{-1}dr^2 + r^2 d\Omega^2_{(2)}\Big],\\
A_t &= \frac{-4G_N^{(4)}e^{\varphi_0}q}{r_- - G_N^{(4)}\Sigma}(H_1^{-1}-1), \qquad \tilde{A}_t = \frac{1}{4\pi}\frac{e^{-\varphi_0}p}{r_- + G_N^{(4)}\Sigma}(H_2^{-1}-1),\\
e^{-2\varphi} &= e^{-2\varphi_0}H_1/H_2,\\
H_1 &= 1 + \frac{r_- - G_N^{(4)}\Sigma}{r}, \qquad H_2 = 1 + \frac{r_- + G_N^{(4)}\Sigma}{r},\\
W &= 1 - \frac{2r_0}{r}, \qquad r_\pm = M \pm r_0,\\
r_0^2 &= M^2 + \Sigma^2 - 4\left[e^{2\varphi_0}q^2 + e^{-2\varphi_0}\left(\frac{p}{16\pi G_N^{(4)}}\right)^2\right],\\
\Sigma &= \frac{2}{M}\left[e^{2\varphi_0}q^2 - e^{-2\varphi_0}\left(\frac{p}{16\pi G_N^{(4)}}\right)^2\right].
\end{aligned}
\tag{16.39}
$$

Here we have used the S-dual potential $\tilde{A}_\mu$ which is the potential related to the S-dual field strength,

$$\tilde{F}_{\mu\nu} = 2\partial_{[\mu}\tilde{A}_{\nu]}, \tag{16.40}$$

whose existence is ensured by the equation of motion of $A_\mu$, which is just the Bianchi identity for $\tilde{F}_{\mu\nu}$. Knowledge of the electric components $F_{tr}$ and $\tilde{F}_{tr}$ and of the metric and dilaton is enough to find all the components $F_{\mu\nu}$, but this form of presenting the result is more elegant and convenient since it exhibits the symmetries of the theory acting on the solution. In particular, we see that S duality interchanges $A_\mu$ and $\tilde{A}_\mu$ and $q$ and $p/(16\pi G_N{}^{(4)})$, and takes $\varphi_0$ to $-\varphi_0$, which also takes $\Sigma$ to $-\Sigma$.

The purely electric dilaton BH solutions with $a=1$ are recovered when $H_2 = 1$, and the purely magnetic ones are recovered when $H_1 = 1$. When $H_1 = H_2 = H$ the scalar becomes trivial and we recover the RN solutions. Thus, these solutions are the most general from the point of view of electric–magnetic duality.

As usual, when $W = 1$, $H_1$ and $H_2$ can be arbitrary harmonic functions in three-dimensional Euclidean space. They may but need not have coincident poles and, thus, the solutions describe electric and magnetic monopoles and dyons in static equilibrium.

Solutions of the four-dimensional $(a = 1)$-model with primary scalar hair and electric charge have been presented in Ref. [19] and probably can be generalized to all values of $a$ and to higher dimensions. We will not pursue this issue any further.

[6] Solutions with additional scalar hair are also possible, but we will not consider them further.

## 16.2 Dilaton/axion black holes

The $a$-model is a good starting point from which to study BH solutions of supergravity/superstring theories, but it is clearly too simple. It is natural to introduce successive generalizations to this model that make it closer to the real thing. In higher dimensions we can introduce differential-form potentials of higher rank, but these are associated with extended objects. In four dimensions we can introduce, as a first step, additional vector fields, all of them coupled in the same way to the scalar field. Then, we can introduce new scalars or different couplings of the scalar(s) to the vector fields. We would have an action of the form

$$S = \frac{1}{16\pi G_{\rm N}^{(4)}} \int d^4x \sqrt{|g|}\, \left[R + \tfrac{1}{2} g_{ij}\partial_\mu \varphi^i \partial^\mu \varphi^j - \tfrac{1}{4} M_{ij} F^i{}_{\mu\nu} F^{j\,\mu\nu}\right], \tag{16.41}$$

where $g^{ij}(\varphi)$ and $M_{ij}(\varphi)$ are some square matrices depending on the scalars. $g^{ij}$ can be interpreted as the inverse metric of some space of which the scalars $\varphi_i$ are the coordinates. The scalar kinetic term is a *$\sigma$-model.*

A good example of an action of this kind is provided by the four-dimensional KK action that one obtains from $\hat{d} = 4 + N$ dimensions by compactification on $\mathrm{T}^N$, Eq. (15.211). The scalars parametrize an $\mathbb{R}^+ \times \mathrm{SL}(N, \mathbb{R})/\mathrm{SO}(2)$ coset space.

There is another kind of coupling of scalars to vectors that we can introduce in four dimensions: couplings of the form

$$-\tfrac{1}{4} N_{ij}(\varphi) F^i{}_{\mu\nu} \star F^{j\,\mu\nu}. \tag{16.42}$$

As a matter of fact, the bosonic sectors of all four-dimensional SUEGRAs can be written in this form. Each of them is characterized by the number of vectors and scalars, by the $\sigma$-model metric $g_{ij}(\varphi)$, and by the matrices of couplings $M_{ij}(\varphi)$ and $N_{ij}(\varphi)$. The general case is studied in Ref. [621].

The simplest model with a coupling of the above kind is the so-called *axion–dilaton-gravity* model

$$S = \frac{1}{16\pi G_{\rm N}^{(4)}} \int d^4x \sqrt{|g|}\, \left[R + 2(\partial\varphi)^2 + \tfrac{1}{2} e^{4\varphi}(\partial a)^2 - e^{-2\varphi} F^2 + a F \star F\right]. \tag{16.43}$$

The scalar field that couples to $F \star F$ is called the *axion* and should be a pseudoscalar for the above action to be parity invariant. It plays the role of a local $\theta$-parameter (see Eq. (12.180)) just as the dilaton plays the role of local coupling constant. In fact, this model is a version of the one studied in Section 12.7.4 with local coupling constants (moduli) and, as we are going to see, it exhibits the same S-duality symmetry [1115, 1127].

1. The factor $e^{4\varphi}$ of the axion kinetic term allows us to combine the axion and the dilaton into a complex scalar field, the *axidilaton* $\tau$;

$$\tau = a + i e^{-2\varphi}, \tag{16.44}$$

and its kinetic term takes the form of an $SL(2,\mathbb{R})/SO(2)$ $\sigma$-model, Eq. (15.225). We can also use the symmetric $SL(2,\mathbb{R})$ matrix $\mathcal{M}$ defined in Eq. (15.223). As discussed in Section 15.4.1, the $\sigma$-model is invariant under global $SL(2,\mathbb{R})$ transformations that are fractional-linear transformations of $\tau$ given by Eqs. (15.221) and (15.222). This group contains three different kinds of transformations.

(a) Rescalings of $\tau$:

$$S = \begin{pmatrix} \alpha & 0 \\ 0 & \alpha^{-1} \end{pmatrix}, \qquad \tau' = \alpha^2 \tau. \tag{16.45}$$

These transformations rescale the axion and shift the value of the dilaton at infinity, $\varphi_0' = \varphi_0 - \ln \alpha$.

(b) Constant shifts:

$$S = \begin{pmatrix} 1 & \beta \\ 0 & 1 \end{pmatrix}, \qquad \tau' = \tau + \beta. \tag{16.46}$$

These transformations shift only the value of the axion at infinity, $a_0' = a_0 + \beta$.

(c) $SO(2)$ rotations:

$$S = \begin{pmatrix} \cos\theta & \sin\theta \\ -\sin\theta & \cos\theta \end{pmatrix}, \qquad \tau' = \frac{\cos\theta\,\tau + \sin\theta}{-\sin\theta\,\tau + \cos\theta}. \tag{16.47}$$

The rotation with $\theta = \pi/2$ inverts $\tau$:

$$S = \begin{pmatrix} 0 & 1 \\ -1 & 0 \end{pmatrix}, \qquad \tau' = -1/\tau. \tag{16.48}$$

When $a = 0$ this transformation is just the electric–magnetic-duality transformation of the dilaton model $\varphi' = -\varphi$.

2. The action is invariant under the first two kinds of transformations; the rescalings of $\tau$ can be compensated by opposite rescalings of $F$:

$$F' = \frac{1}{\alpha} F, \tag{16.49}$$

and the shifts of $a$ simply change the action by a total derivative $\beta\sqrt{|g|}F \star F$. This is just an Abelian version of the Peccei–Quinn symmetry. In the Euclidean non-Abelian $SU(2)$ case the total derivative is proportional to a topological invariant; namely the second Chern class defined in Eq. (13.18) that takes integer values. If the Euclidean action is properly normalized, the Peccei–Quinn transformation simply shifts it by $\beta$ times an integer, which results in a phase change in the integrand of the path integral. Thus, the classical continuous Peccei–Quinn symmetry is broken to $\mathbb{Z}$ since the only transformations that leave the path integral invariant are those with $\beta = 2\pi n, n \in \mathbb{Z}$. This is one of the quantum effects[7] that breaks $SL(2,\mathbb{R})$ to $SL(2,\mathbb{Z})$, the group of S duality.

[7] The other one is charge quantization.

3. The equations of motion (but not the action) of the whole theory are also invariant under $\mathrm{SO}(2)$ rotations. To see this (to check invariance under the whole $\mathrm{SL}(2,\mathbb{R})$), it is convenient to define the $\mathrm{SL}(2,\mathbb{R})$-dual $\tilde{F}$ of the vector field strength $F$:

$$\tilde{F}_{\mu\nu} \equiv e^{-2\varphi} \star F_{\mu\nu} + aF_{\mu\nu}. \tag{16.50}$$

The Maxwell equation is now the Bianchi identity of the S-dual field strength:

$$\nabla_\mu \star \tilde{F}^{\mu\nu} = 0. \tag{16.51}$$

It is convenient to define two S-duality vectors $\vec{F}$ and $\vec{\mathcal{F}}$,

$$\vec{F} \equiv \begin{pmatrix} \star F \\ F \end{pmatrix}, \qquad \vec{\mathcal{F}} \equiv e^{-\varphi}\mathcal{V}\vec{F} = \begin{pmatrix} \tilde{F} \\ F \end{pmatrix}, \tag{16.52}$$

where $\mathcal{V}$ is the upper-triangular unimodular matrix that we defined in Eq. (15.224) that satisfies $\mathcal{V}\mathcal{V}^{\mathrm{T}} = \mathcal{M}$. Note that $\vec{\mathcal{F}}$ transforms covariantly under $S \in \mathrm{SL}(2,\mathbb{R})$:

$$\vec{\mathcal{F}}' = S\vec{\mathcal{F}}. \tag{16.53}$$

The two components of this vector are not independent, but are related by a constraint that involves $\tau$. This constraint must be preserved by $S$ and one can check that this happens if and only if $\tau$ transforms according to Eqs. (15.221) and (15.222). The transformation $\tau' = -1/\tau$ interchanges the two components of the duality vector. For a vanishing axion field, this is the discrete electric–magnetic-duality transformation of the dilaton-gravity model.

In terms of the duality vector $\vec{\mathcal{F}}$ the Maxwell equation and the Bianchi identity take the $\mathrm{SL}(2,\mathbb{R})$-invariant form

$$\nabla_\mu \vec{\mathcal{F}}^{\mu\nu} = 0, \tag{16.54}$$

and the Einstein equation can also be written in invariant form (see Eq. (12.139)):

$$R_{\mu\nu} + \frac{\partial_\mu \tau \partial_\nu \bar{\tau}}{(\mathrm{Im}(\tau))^2} + \vec{\mathcal{F}}^{\mathrm{T}} \eta \star \vec{\mathcal{F}} = 0, \tag{16.55}$$

with $\eta = i\sigma^2$, due to the property $S\eta S^{\mathrm{T}} = \eta$ of $\mathrm{Sp}(2) \sim \mathrm{SL}(2,\mathbb{R})$ matrices $S$. The remaining two equations of motion,

$$\begin{aligned} &\nabla^2\varphi - \tfrac{1}{2}e^{4\varphi}(\partial a)^2 - \tfrac{1}{2}e^{-2\varphi}F^2 = 0, \\ &\nabla^2 a + 4\partial_\mu\varphi\, \partial^\mu a - e^{-4\varphi}F \star F = 0, \end{aligned} \tag{16.56}$$

can also be rewritten in a manifestly duality-invariant form:

$$\nabla_\mu(\partial^\mu \mathcal{M}\mathcal{M}^{-1}) + \vec{\mathcal{F}}\vec{\mathcal{F}}^{\mathrm{T}}\eta = 0. \tag{16.57}$$

The action Eq. (16.43) is a truncation of the bosonic sector of ungauged $N=4, d=4$ SUEGRA [389], that contains the metric $g_{\mu\nu}$, complex scalar $\tau$, and six Abelian vector fields $A^{\Lambda}{}_{\mu}$, $n=1,\ldots,6$. On setting $G_{\rm N}^{(4)}=1$, it takes the form

$$S = \frac{1}{16\pi}\int d^4x\sqrt{|g|}\left[R+2(\partial\varphi)^2+\tfrac{1}{2}e^{4\varphi}(\partial a)^2-e^{-2\varphi}F^{\Lambda}F^{\Lambda}+aF^{\Lambda}\,{}^{\star}F^{\Lambda}\right]. \tag{16.58}$$

This theory, in turn, can be obtained by dimensional reduction and consistent truncation from $N=1, d=10$ SUGRA, the effective theory of the heterotic string, as we will see in Chapter 22. In this context, $\varphi$ coincides with the four-dimensional string dilaton and there are many things about the general stringy case that we can learn by studying this simpler case.

Apart from S duality, this action has a trivial invariance under SO(6) (T-duality) rotations of the vector fields. This may seem to suggest that considering just one vector field would be enough to obtain the most general BH solution (up to SO(6) rotations), but we are going to see that this is not the case: at least two vectors are needed if one wants to obtain a BH solution from which we can generate the most general one[8] by more or less trivial SO(6) rotations (a *generating solution*). To explain why this is the case, we need to discuss how the conserved charges enter in the metric and scalar fields.

First we use Eq. (16.54) to define the conserved electric and magnetic charges of the six Abelian vector fields $\vec{q}^{\Lambda}$,

$$\vec{q}^{\Lambda} \equiv \frac{1}{4\pi}\int_{S^2_\infty}\vec{\mathcal{F}}^{\Lambda}, \qquad \vec{q}^{\Lambda} = \begin{pmatrix} q^{\Lambda} \\ p^{\Lambda} \end{pmatrix}, \tag{16.59}$$

that we can arrange into a 12-dimensional vector $\mathbf{q}$; $\mathbf{q}$ transforms linearly under S- and T-duality transformations $S$ and $R$:

$$\mathbf{q}' = S\otimes R\mathbf{q}. \tag{16.60}$$

The charges[9] must enter into the metric in duality invariant combinations because the metric is duality invariant. There are only two such invariants that are quadratic and quartic in the charges:

$$I_2 \equiv \mathbf{q}^{\rm T}\mathcal{M}_0^{-1}\otimes\mathbb{I}_{6\times 6}\mathbf{q}, \qquad I_4 \equiv \det\left[\sum_{n=1}^{n=6}\vec{q}^{\Lambda}\vec{q}^{\Lambda\,\rm T}\right]. \tag{16.62}$$

[8] By definition, the one with the highest possible number of charges (mass, angular momentum, and electric and magnetic charges) and moduli (the asymptotic value of $\tau$) allowed by the no-hair conjecture.

[9] Observe that, the axion being a local $\theta$-parameter, it induces a Witten effect on the charges, as explained in Section 12.7.4. Furthermore, the DSZ quantization condition takes the manifestly SL$(2,\mathbb{R})$-invariant form

$$\vec{q}_1^{\Lambda\,\rm T}\eta\,\vec{q}_2^{\Lambda} = m/2, \qquad m\in\mathbb{Z}. \tag{16.61}$$

($q^{\Lambda}$ is canonically normalized, but $p^{\Lambda}$ is $1/(4\pi)$ times the canonical magnetic charge. The product of the canonical charges is quantized in integer multiples of $2\pi$.)

Here $\mathcal{M}_0$ is the asymptotic value of the scalar matrix $\mathcal{M}$. Thus, $I_2$ is moduli dependent and $I_4$ is moduli independent. On the other hand, $I_4$ vanishes when only one vector field is non-trivial and, therefore, starting from the most general charge configuration with only one vector field and $I_4 = 0$, we cannot generate the most general charge configuration with $I_4 \neq 0$ by S- and T-duality transformations. The generating solution has to have both $I_2$ and $I_4$ generically non-vanishing.

To attain a better understanding, we can try to construct the most general solution, starting from the $d = 4, a = 1$ dilaton BH solutions we studied in Section 16.1. We simply have to observe that the equations of motion of the axion–dilaton model coincide[10] with those of the four-dimensional $a = 1$ model if the axion $a = 0$ and $F \star F = 0$. Then, the purely electric BH Eq. (16.23) provides a solution of the axion–dilaton model with one independent charge and one non-trivial modulus ($\varphi_0$). By performing a SO(2) S-duality transformation Eq. (16.47), we can generate a solution that has electric and magnetic charge. As in the Einstein–Maxwell case, the SO(2) parameter becomes a new independent charge. A non-trivial axion is generated. Further $\mathrm{SL}(2, \mathbb{R})$ transformations only shift $\varphi_0$ and add an asymptotic value to the axion $a_0$. In this way we have obtained the most general axion–dilaton BH solution with one vector field [1132], but it has the same metric as the purely dilatonic BH.

This solution is also a solution of $N = 4, d = 4$ SUEGRA with five vanishing vector fields. We could excite them by performing $\mathrm{SO}(6)/\mathrm{SO}(5)$ T-duality rotations that do not leave the charge vector invariant. However, in this way we can obtain only solutions in which all the magnetic charges are proportional to all the electric charges with the same proportionality factor. We would have added only five new independent parameters to the solution and the metric would still be the same (because $I_4 = 0$).

A more general solution with two non-vanishing charges in different vectors $q^{(1)}$ and $p^{(2)}$ was found in Ref. [608] and, later on, studied in Ref. [841]. It has a different metric (and non-vanishing $I_4$). It has a non-vanishing dilaton, a vanishing axion, and trivial moduli, which, however, could be generated by S-duality transformations. In fact, it is clear that S and T dualities suffice to generate the four possible independent charges of the two vector fields and, actually, the $2N$ independent charges of $N$ vector fields and, thus, it is the (static) generating solution of this theory.

This static generating solution is essentially the $d = 4, a = 1$ dyonic dilaton BH solution given in Eq. (16.39) but where the electric and magnetic components of the vector field belong to two different vector fields.[11] In the conventions that we are using in this section,

[10] The vector fields have a different normalization.

[11] Sometimes these solutions are called $\mathrm{U}(1)^2$ BHs.

it takes the form

$$
\begin{aligned}
ds^2 &= (H_1H_2)^{-1}W dt^2 - H_1H_2\Big[W^{-1}dr^2 + r^2 d\Omega^2_{(2)}\Big], \\
A^{(1)}{}_t &= \frac{-q}{r_- - \Sigma}(H_1^{-1}-1), \qquad \tilde{A}^{(2)}{}_t = \frac{p}{r_- + \Sigma}(H_2^{-1}-1), \\
e^{-2\varphi} &= H_1/H_2, \\
H_1 &= 1 + \frac{r_- - \Sigma}{r}, \qquad H_2 = 1 + \frac{r_- + \Sigma}{r}, \qquad W = 1 - \frac{2r_0}{r}, \\
r_\pm &= M \pm r_0, \qquad r_0^2 = M^2 + \Sigma^2 - (q^2+p^2), \qquad \Sigma = 2(q^2-p^2)/M.
\end{aligned}
\tag{16.63}
$$

It is, however, very convenient to have the most general solution written explicitly in terms of the physical charges. Moreover, the most general static solution can be immediately generalized in a natural way by adding angular momentum and NUT charge, becoming the truly most general stationary BH-type solution, which we will call the *SWIP solution*[12] [906]. It will be S- and T-duality invariant by definition, and its physical properties will be given in terms of duality-invariant combinations of charges. Ungauged $N=4, d=4$ SUEGRA is the most complicated case in which the most general solution is explicitly known and the attempts to write the most general solution of more complicated theories are inspired by it. For these reasons, it is worth studying.

### *16.2.1 The general SWIP solution*

The general solution is determined by two complex harmonic functions, $\mathcal{H}_{1,2}$, the non-extremality function, $W$, the spatial background metric, ${}^{(3)}\gamma_{ij}$, and $N$ complex constants $k^\Lambda$:

$$
\begin{aligned}
ds^2 &= e^{2U}W(dt + A_\varphi d\varphi)^2 - e^{-2U}W^{-1}\,{}^{(3)}\gamma_{ij}dx^i dx^j, \\
A^\Lambda{}_t &= 2e^{2U}\,\mathrm{Re}\big(k^\Lambda \mathcal{H}_2\big), \qquad \tilde{A}^\Lambda{}_t = 2e^{2U}\,\mathrm{Re}\big(k^\Lambda \mathcal{H}_1\big), \qquad \tau = \mathcal{H}_1/\mathcal{H}_2,
\end{aligned}
\tag{16.64}
$$

where

$$
\begin{aligned}
e^{-2U} &= 2\,\mathrm{Im}\big(\mathcal{H}_1\bar{\mathcal{H}}_2\big), \\
A_\varphi &= 2N\cos\theta + \alpha\sin^2\theta\left(e^{-2U}W^{-1} - 1\right).
\end{aligned}
\tag{16.65}
$$

[12] The construction of the most general BH-type solution was initiated in Ref. [1005] and the most general static solution was obtained in Ref. [842]. There, the solution was written in terms of two complex functions $\mathcal{H}_{1,2}$ (harmonic in the extreme limit) that obeyed a constraint. It was realized in Ref. [168] that removing the constraint in the extreme case immediately resulted in the natural inclusion of NUT charge and angular momentum. The new solutions had been obtained independently in Ref. [1184]. Finally, the general, non-extreme solution was constructed in Ref. [906]. Related work was done in Refs. [839, 582, 1079, 581, 588, 1080, 583, 363, 85, 584, 586, 1081, 1082].

The functions $\mathcal{H}_{1,2}$ take the form

$$\mathcal{H}_1 = \frac{1}{\sqrt{2}} e^{\varphi_0} e^{i\beta} \left( \tau_0 + \frac{\tau_0 \mathfrak{M} + \bar{\tau}_0 \Upsilon}{r + i\alpha\cos\theta} \right), \qquad \mathcal{H}_2 = \frac{1}{\sqrt{2}} e^{\varphi_0} e^{i\beta} \left( 1 + \frac{\mathfrak{M} + \Upsilon}{r + i\alpha\cos\theta} \right), \tag{16.66}$$

and $W$ and the background metric ${}^{(3)}\gamma_{ij}$ take the forms

$$\begin{aligned} W &= 1 - \frac{r_0^2}{r^2 + \alpha^2\cos^2\theta}, \\ {}^{(3)}\gamma_{ij} dx^i dx^j &= \frac{r^2 + \alpha^2\cos^2\theta - r_0^2}{r^2 + \alpha^2 - r_0^2} dr^2 + \left(r^2 + \alpha^2\cos^2\theta - r_0^2\right) d\theta^2 \\ &\quad + \left(r^2 + \alpha^2 - r_0^2\right)\sin^2\theta \, d\varphi^2. \end{aligned} \tag{16.67}$$

The complex constants are given by

$$k^\Lambda = -\frac{1}{\sqrt{2}} e^{-i\beta} \frac{\mathfrak{M}\Gamma^\Lambda + \Upsilon\overline{\Gamma^\Lambda}}{|\mathfrak{M}|^2 - |\Upsilon|^2}. \tag{16.68}$$

The metric can also be written in a more standard form:

$$\begin{aligned} ds^2 &= \frac{\Delta - \alpha^2\sin^2\theta}{\Sigma} dt^2 + 2\alpha\sin^2\theta \frac{\Sigma + \alpha^2\sin^2\theta - \Delta}{\Sigma} dt d\varphi \\ &\quad - \frac{\Sigma}{\Delta} dr^2 - \Sigma \, d\theta^2 - \frac{\left(\Sigma + \alpha^2\sin^2\theta\right)^2 - \Delta\alpha^2\sin^2\theta}{\Sigma} \sin^2\theta \, d\varphi^2, \\ \Delta &= r^2 - R_0{}^2 = r^2 + \alpha^2 - r_0{}^2, \\ \Sigma &= (r + M)^2 + (n + \alpha\cos\theta)^2 - |\Upsilon|^2. \end{aligned} \tag{16.69}$$

We have expressed the functions that enter the solution in terms of physical constants (charges and moduli); $\alpha = J/M$ is the angular momentum ($J$) per unit mass ($M$), and we have combined the mass and NUT charge ($N$) into the complex "mass"

$$\mathfrak{M} \equiv M + iN, \tag{16.70}$$

and the electric and magnetic charges into

$$\Gamma^\Lambda \equiv Q^\Lambda + iP^\Lambda, \qquad \vec{Q}^\Lambda \equiv \mathcal{V}_0^{-1}\vec{q}^\Lambda. \tag{16.71}$$

The (complex) axion–dilaton charge, $\Upsilon$, and $\tau_0$, its asymptotic value, are defined by

$$\tau \sim \tau_0 - ie^{-2\varphi_0}\frac{2\Upsilon}{r}. \tag{16.72}$$

In these solutions $\Upsilon$ depends on the conserved charges in this fixed way:

$$\Upsilon = -\tfrac{1}{2}\sum_n \frac{(\bar{\Gamma}^\Lambda)^2}{\mathfrak{M}}. \tag{16.73}$$

Finally, the "non-extremality" parameter $r_0$ is given by

$$r_0^2 = |\mathfrak{M}|^2 + |\Upsilon|^2 - \sum_n |\Gamma^\Lambda|^2. \tag{16.74}$$

In non-static cases when $r_0 = 0$ the solution is *supersymmetric*, but for $\alpha \neq 0$ it is not an extreme BH. A more appropriate name is *supersymmetry parameter*. The *extremality parameter* will be $R_0^2 = r_0^2 - \alpha^2$. When it is positive, we have two horizons, placed at $r_\pm = M \pm R_0$. The area of the event horizon (the one at $r_+$) is given, for BH solutions with zero NUT charge, by

$$A = 4\pi\left(r_+^2 + \alpha^2 - |\Upsilon|^2\right). \tag{16.75}$$

### 16.2.2 Supersymmetric SWIP solutions

When $r_0 = 0$ and $W = 1$ the general SWIP solution has special properties. First, the background metric ${}^{(3)}\gamma_{ij}$ is nothing but the metric of Euclidean three-dimensional space in oblate spheroidal coordinates, which are related to the ordinary Cartesian ones by

$$\begin{aligned} x &= \sqrt{r^2 + \alpha^2}\, \sin\theta \cos\varphi, \\ y &= \sqrt{r^2 + \alpha^2}\, \sin\theta \sin\varphi, \\ z &= r \cos\theta. \end{aligned} \tag{16.76}$$

On rewriting the solution Eqs. (16.64) in Cartesian coordinates, we find the solutions

$$\boxed{\begin{aligned} ds^2 &= 2\,\mathrm{Im}(\mathcal{H}_1\bar{\mathcal{H}}_2)\,(dt + A)^2 - [2\,\mathrm{Im}(\mathcal{H}_1\bar{\mathcal{H}}_2)]^{-1} d\vec{x}_3^{\,2}, \\ A^\Lambda{}_t &= 2e^{2U}\,\mathrm{Re}\big(k^\Lambda \mathcal{H}_2\big), \qquad \tilde{A}^\Lambda{}_t = 2e^{2U}\,\mathrm{Re}\big(k^\Lambda \mathcal{H}_1\big), \qquad \tau = \mathcal{H}_1/\mathcal{H}_2, \\ A &= A_{\underline{i}} dx^i, \qquad \epsilon_{ijk}\partial_i A_j = \pm \mathrm{Re}\big(\mathcal{H}_1 \partial_k \bar{\mathcal{H}}_2 - \bar{\mathcal{H}}_2 \partial_k \mathcal{H}_1\big), \\ \partial_{\underline{i}}\partial_{\underline{i}}\mathcal{H}_{1,2} &= 0, \qquad \sum_{n=1}^N (k^\Lambda)^2 = 0, \qquad \sum_{n=1}^N |k^\Lambda|^2 = \frac{1}{2}. \end{aligned}} \tag{16.77}$$

That is, for any arbitrary pair of complex harmonic functions $\mathcal{H}_{1,2}(\vec{x}_3)$ in the three-dimensional Euclidean space, it is clear that we can construct multi-BH solutions and that $r_0 = 0$ can be reinterpreted as a no-force condition between the BHs.

These solutions include the IWP metrics Eqs. (13.83) when

$$\mathcal{H}_1 = i\mathcal{H}_2 = \frac{1}{\sqrt{2}}\mathcal{H} \tag{16.78}$$

(which trivializes the axidilaton $\tau$) that in turn include the MP solutions Eqs. (12.86). These are the only BH-type solutions in the IWP family: the addition of angular momentum eliminates the event horizon and the addition of NUT charge eliminates the asymptotic flatness. Something similar is true for the supersymmetric SWIP solutions: the only supersymmetric single BHs in this family are the static ones with $A_i = 0$, which imposes a very non-trivial constraint on the harmonic functions, which was implicit in Ref. [842].

The general solutions include, for vanishing axidilaton charge $\Upsilon = 0$ (which corresponds to special choices of the electric and magnetic charges), the Kerr–Newman solution in Boyer–Lindquist coordinates Eq. (13.79).

### 16.2.3 Duality properties of the SWIP solutions

Solutions of the general and supersymmetric SWIP families are the most general BH-type solutions of $N = 4, d = 4$ SUEGRA and, therefore, an S- or T-duality transformation takes one member of the family into another member of the family. Thus, the effect of duality transformations is just to replace all the constants and functions that enter the solutions with primed constants and functions. The structure of the solutions thus reflects the $\mathrm{SL}(2,\mathbb{R})\times \mathrm{SO}(6)$ duality invariance of the equations of motion.

Let us see in a bit more detail how the charges and functions transform under duality. $\mathfrak{M}$ is obviously invariant. The complex combinations of electric and magnetic charges $\Gamma^{\Lambda}$ are $\mathrm{SO}(6)$ vectors and change by a phase under $\mathrm{SL}(2,\mathbb{R})$,

$$\Gamma^{\Lambda\,\prime} = e^{i\arg(\gamma\tau_0+\delta)}\Gamma^{\Lambda}, \tag{16.79}$$

while the axidilaton charge also changes by a phase but is an $\mathrm{SO}(6)$ scalar,

$$\Upsilon' = e^{-2i\arg(\gamma\tau_0+\delta)}\Upsilon, \tag{16.80}$$

and, therefore, its absolute value is duality invariant and can be expressed in terms of the two invariants $I_2$ and $I_4$:

$$|\Upsilon|^2 = \frac{1}{4|\mathfrak{M}|^2}(I_2^2 - 4I_4). \tag{16.81}$$

It is also easy to show that

$$\sum_n |\Gamma^{\Lambda}|^2 = I_2. \tag{16.82}$$

Since $\mathfrak{M}$ is trivially duality invariant, the last two equations imply the duality invariance of the supersymmetry parameter $r_0$, given in Eq. (16.74), and of the *supersymmetry bound* (to be defined in Chapter 17) $r_0^2 \geq 0$.

It is useful to define the two combinations of charges [166]

$$|Z_{1,2}|^2 \equiv \tfrac{1}{2}I_2 \pm I_4^{\frac{1}{2}}, \tag{16.83}$$

which are at most interchanged by duality transformations. In terms of them, the supersymmetry parameter and the BH entropy take the suggestive forms

$$\begin{aligned} r_0^2 &= \frac{1}{|\mathfrak{M}|^2}\left(|\mathfrak{M}|^2 - |Z_1|^2\right)\left(|\mathfrak{M}|^2 - |Z_2|^2\right), \\ S &= \pi\left\{(M^2 - |Z_1|^2) + (M^2 - |Z_2|^2) + 2\sqrt{(M^2 - |Z_1|^2)(M^2 - |Z_2|^2) - J^2}\right\}, \end{aligned} \tag{16.84}$$

which we will discuss in Chapter 17. Observe that $|\Upsilon|$ is given in general by $|Z_1 Z_2|^2 \mathfrak{M}^{-2}$.

The functions $\mathcal{H}_{1,2}$ transform as a doublet under $\mathrm{SL}(2, \mathbb{R})$, whereas the $k^\Lambda$s are invariant because, although they transform with the same phase as $\Gamma^\Lambda$, they can be absorbed into the arbitrary phase $\beta$ that appears in the solution. The $k^\Lambda$s are clearly $\mathrm{SO}(6)$ vectors, as are the corresponding vector potentials.

# 17

# Unbroken supersymmetry I: supersymmetric vacua

In our study of several solutions in previous chapters we have mentioned that some special properties that arise for special values of the parameters (mass, charge) are related to supersymmetry; more precisely, to the existence of (*unbroken*) supersymmetry. Those statements were a bit surprising because we were dealing with solutions of purely bosonic theories (Einstein–Maxwell, Kaluza–Klein, ...).

The goal of this chapter is to explain the concept and implications of unbroken supersymmetry and how it can be applied in purely bosonic contexts, including pure GR. Supersymmetry will be shown to have a very deep meaning, underlying more familiar symmetries that can be constructed as squares of supersymmetries. At the very least, supersymmetry can be considered as an extremely useful tool that simplifies many calculations and demonstrations of very important results in GR that are related directly or indirectly to the positivity of energy (a manifest property of supersymmetric theories).

As a further reason to devote a full chapter to this topic, unbroken supersymmetry is a crucial ingredient in the stringy calculation of the BH entropy by the counting of microstates. It ensures the stability of the solution and the calculation under classical and quantum perturbations.

To place this subject in a wider context, we will start by giving in Section 17.1 a general definition of residual (unbroken) symmetry and we will relate it to the definition of a vacuum. Vacua are characterized by their symmetries, which determine the conserved charges of point-particles moving in them and, ultimately, the spectra of quantum-field theories (QFTs) defined on them. These definitions will be applied in Section 17.2 to supersymmetry as a particular case. In this section we will have to develop a new tool, the *covariant Lie derivative*, which will be used to find the unbroken-supersymmetry algebra of any given solution according to Figueroa-O'Farrill's prescription in Ref. [546]. In Section 17.3 we will apply this prescription and the geometrical methods of Ref. [26] to the vacua of the simplest four-dimensional supergravity theories and we will try to recover the supersymmetry algebras that we gauged to construct them in Chapter 5. These vacuum superalgebras will then be used in the next chapter to understand the properties of other solutions (with or without unbroken supersymmetry) with the same asymptotic behavior. In particular, they can be used to derive *supersymmetry* or *BPS bounds*. We will also discuss the results known for minimal $d = 5, 6$ supergravities, but we will leave higher-dimensional supergravities and theories with more supercharges for Part III because these theories can be derived from ten-dimensional superstring effective theories, but we will say what can be expected from general arguments based on the structures of the respective superalgebras.

## 17.1 Vacuum and residual symmetries

The solutions of the equations of motion of a given theory usually break most (or all) of its symmetries. Sometimes a solution has (*preserves*) some of them, which receive the name of *residual* (or *unbroken*) *symmetries*, and, being symmetries, they form a symmetry group. The solution is said to be *symmetric*. The symmetries of the theory which are broken by the symmetric solution can be used to generate new solutions of the theory. Let us look at two examples.

**Classical mechanics.** The Lagrangian of a free relativistic particle moving in Minkowski spacetime is invariant under the whole Poincaré group ISO(1,3). However, every solution is a straight line, invariant only under translations parallel to it and rotations with it as the axis. These are the residual symmetries of every solution and they form a two-dimensional group $\mathbb{R} \times SO(2)$. The remaining Poincaré transformations move the line and generate other solutions.

**Field theory.** Einstein's equations are invariant under the infinite-dimensional group of GCTs. However, a given solution (metric) is invariant only under a finite-dimensional group of *isometries*. By definition, an infinitesimal isometry is an infinitesimal GCT that leaves the metric invariant, that is

$$\delta_\xi g_{\mu\nu} = -\mathcal{L}_\xi g_{\mu\nu} = -2\nabla_{(\mu}\xi_{\nu)} = 0, \tag{17.1}$$

which is known as the *Killing equation*. The solutions $\xi^\mu = \xi k^\mu$ are each the product of an infinitesimal constant $\xi$ times a *Killing vector* $k^\mu$, the generator of the isometry.

The isometries of a metric form an isometry group. This is a finite-dimensional Lie group, whose generators are Killing vectors. The finite-dimensional Lie algebra of isometries coincides with the Lie algebra of the Killing vectors with the Killing bracket by virtue of the property of the Lie derivative

$$[\mathcal{L}_{k_1}, \mathcal{L}_{k_2}] = \mathcal{L}_{[k_1,k_2]}. \tag{17.2}$$

(This structure is induced from the infinite-dimensional group of all GCTs, of which the isometry group is a subgroup.)

Formally we can associate a generator of the abstract symmetry algebra of the solution $P_A$ with each of its Killing vectors $k_A$. This abstract generator is represented on the metric by an operator, which is just minus the Lie derivative with respect to the corresponding Killing vector $P_A \sim -\mathcal{L}_{k_A}$. Then, if the Lie algebra of the isometries is $[k_A, k_B] = -f_{AB}{}^C k_C$, the abstract symmetry algebra takes the form

$$[P_A, P_B] = f_{AB}{}^C P_C. \tag{17.3}$$

What happens if there are matter fields in the theory? If they are standard tensor fields[1] $T$, infinitesimal GCTs act on them through (minus) the Lie derivative:

$$\delta_\xi T = -\mathcal{L}_\xi T. \tag{17.4}$$

Only those GCTs that leave invariant all fields of a solution will be (unbroken) symmetries of that solution. Thus, only those isometries that leave invariant the matter fields,

$$-\mathcal{L}_{k_A} T = 0, \tag{17.5}$$

generate the symmetry algebra of the solution.

Finally, GCTs that are not symmetries transform the solution into another solution, which may be physically equivalent if the boundary conditions are invariant, but will be inequivalent otherwise.

The second example is evidently richer and more interesting. In it the presence of residual symmetries has far-reaching consequences. For instance, we have proven in Section 3.3 that point-particles moving in a curved spacetime with isometries have a conserved quantity associated with every isometry. If we construct QFTs in such a spacetime, the quanta of the fields will appear in unitary representations of the symmetry (isometry) group, according to Wigner's theorem. The spectrum and the kinematics of the QFT are thus determined by the symmetry group.

The simplest and best-known example is Minkowski spacetime, whose isometry group is Poincaré's $\mathrm{ISO}(1, d-1)$: a particle moving in Minkowski spacetime has $d(d+1)/2$ conserved quantities (the $d$ components of the momentum and the $d(d-1)/2$ components of the angular momentum). QFTs in Minkowski spacetime are constructed preserving Poincaré symmetry and the quanta of the fields will be particles defined by the values of the invariants that can be constructed with the conserved quantities (mass and spin).

It is natural to associate solutions with a maximal number of unbroken symmetries with possible vacuum states of the QFT. These states will be annihilated by the operators associated with these symmetries in the quantum theory. In GR with no cosmological constant, the only maximally symmetric solution is the Minkowski spacetime (ten isometries in $d = 4$ dimensions). With a (negative) positive cosmological constant, the Minkowski metric is not a solution and the only maximally symmetric solutions are the (anti-)de Sitter spacetimes whose isometry group (SO(2,3)) SO(1,4) is also ten dimensional. These are the only maximally symmetric solutions of GR.

It is possible to define field theories in (anti-)de Sitter spacetime, but it is also possible (albeit unusual) to do it in spacetimes with fewer isometries, except in higher dimensions:

[1] If the matter fields are not standard tensor fields, i.e. if they are spinors or fields transforming covariantly under some other local symmetry of the theory (local Lorentz transformations for spinors and Vielbeins, gauge transformations for charged fields, ...), then the standard Lie derivative does not give a good representation of the infinitesimal GCTs because it is not covariant under those local symmetries and the results would depend on the frame or gauge chosen. Instead we have to use a generalized *covariant Lie derivative*, as we will see in Section 17.2, since this problem is relevant in supergravity theories.

for instance, we have studied in Chapter 15 Kaluza–Klein (KK) vacua that are the products of $d$-dimensional Minkowski spacetime and a circle whose isometry group is considerably smaller than that of $(d+1)$-dimensional Minkowski spacetime, which is spontaneously broken by the choice of vacuum. The spectrum of the KK theory is determined by the unbroken symmetry group, and it is the spectrum of a $d$-dimensional theory with gravity. The name *spontaneous compactification* could be applied to this and other cases in which there is a classical solution that we associated with a vacuum in which the spacetime is a product of a lower-dimensional spacetime and a compact space.

We can also consider other solutions of GR that asymptotically approach one of the three vacua we just mentioned. As we have stressed repeatedly, solutions of this kind represent isolated systems in GR. We can use the Abbott–Deser formalism of Section 6.1.2 to find the values of the $d(d+1)/2$ conserved quantities of those spacetimes which are associated with the isometries of the vacuum (even if the solutions themselves do not have any isometry). If we associate with the systems described by the asymptotically vacuum solutions states of a QFT built over the associated vacuum state, then the generators of the symmetry algebra have a well-defined action on them.[2] On the other hand, only the vacuum state is annihilated by all those generators, corresponding to its invariance under all the isometries. In particular, the vacuum state will be annihilated by the energy operator, and thus (if we restrict ourselves to states with non-negative energy) it will be the state with minimal energy. This point is problematic in de Sitter spacetimes, which compromises their stability.

This association of solutions that approach asymptotically a vacuum and states of a quantum theory on which the generators of the vacuum isometries act is a very fruitful point of view that we will use extensively. It can be extended to less-symmetric vacua, defining its own class of asymptotic behavior.

We are now ready to extend this concept to the supersymmetry context.

## 17.2 Supersymmetric vacua and residual (unbroken) supersymmetries

In general, the solutions of a supergravity theory are not invariant under any of the (infinite) supersymmetry transformations that leave the theory invariant. Those which are invariant under some (always a finite number of) *residual or unbroken supersymmetries* are said to be *supersymmetric*, *BPS*, or *BPS saturated*.[3]

Schematically, the local supersymmetry transformations take the form

$$\begin{aligned}\delta_\epsilon B &\sim \epsilon F,\\ \delta_\epsilon F &\sim \partial\epsilon + B\epsilon,\end{aligned}\tag{17.6}$$

[2] It should be stressed that this can be done for all the states corresponding to spacetimes with the same asymptotic behavior. We cannot compare the energies of, say, asymptotically flat and asymptotically anti-de Sitter spacetimes.

[3] We focus on local supersymmetries, although it is evidently possible to define unbroken supersymmetry in theories that are invariant only under global supersymmetry. For instance, in the context of super-Yang–Mills theory, the Bogomol'nyi–Prasad–Sommerfield (BPS) limit of the 't Hooft–Polyakov monopole discussed in Section 13.2.3 has some unbroken supersymmetries. This is why supersymmetric solutions are often called BPS solutions. Actually, even though there are non-supersymmetric BPS-type limits and solutions in the literature, BPS is synonymous with supersymmetric. The reason why they are called BPS saturated will be explained when we discuss *supersymmetry, Bogomol'nyi*, or *BPS bounds*.

for boson ($B$) and fermion ($F$) fields. We are interested in purely bosonic solutions since these are the ones that correspond to classical solutions.[4] They are also solutions of the bosonic action that one obtains by setting to zero all the fermion fields of the supergravity theory, because this is always a consistent truncation. These bosonic actions are just well-known actions of GR coupled to matter fields (for instance, the Einstein–Maxwell theory in the $N=2, d=4$ supergravity case).

According to the general definition, a bosonic solution will be supersymmetric if the above transformations vanish for some infinitesimal supersymmetry parameter $\epsilon(x)$. In the absence of fermion fields, the bosonic fields are always invariant, and it is necessary only that the supersymmetry transformations of the fermion fields vanish:

$$\delta_\kappa F \sim \partial\epsilon + B\epsilon = 0. \qquad (17.7)$$

From the superspace point of view, this can be seen as invariance under an infinitesimal super-reparametrization. Thus, by analogy with GR, this is called the *Killing spinor equation* and its solutions can be seen as the product of an infinitesimal anticommuting number $\epsilon$ and a finite commuting spinor $\kappa$ called a *Killing spinor* that also satisfies the above equation. There is a different Killing spinor equation for each supergravity theory but, since we have defined it for purely bosonic configurations, it can be used without any reference to supergravity or fermion fields.

What is the symmetry group generated by the Killing spinors? Clearly, it has to be a finite-dimensional *supergroup* of which the Killing spinors are the fermionic generators. The supergroup is part of the infinite-dimensional supergroup of superspace superreparametrizations that includes all the local supersymmetry transformations, GCTs, etc. However, where are the bosonic generators?

In the case of the isometry group of a metric, the structures of the finite-dimensional group and of the algebra of its generators are inherited from those of the infinite-dimensional group of all GCTs. In this case, the structure of the finite-dimensional supersymmetry group of a solution is inherited from that of the infinite-dimensional supergroup of all local supersymmetry transformations, GCTs, etc., of the supergravity theory. The commutator of two local supersymmetry transformations is a combination of all the symmetries of the theory: for instance, in $N=2, d=4$ Poincaré supergravity, given by Eq. (5.102), a GCT, a local Lorentz rotation, a U(1) gauge transformation, and a local supersymmetry transformation with parameters that depend on $\epsilon_{1,2}$ and the fields of the theory. Now, if $\kappa_{1,2}$ are Killing spinors of a bosonic solution, the commutator will give bosonic symmetries of the same solution. In particular, we find that the solution will be invariant[5] under GCTs generated by bilinears of the form[6]

$$k^\mu = -i\bar{\kappa}_1\gamma^\mu\kappa_2. \qquad (17.8)$$

[4] We observe only macroscopic bosonic fields in nature. However, technically, we could equally well consider non-vanishing fermionic fields. Also, we can *generate* fermionic fields by performing supersymmetry transformations on purely bosonic solutions.

[5] This statement will be made more precise shortly.

[6] If $\kappa_1$ and $\kappa_2$ are identical commuting Killing spinors, the bilinear does not vanish. Furthermore, it can be shown that $k^\mu = -i\bar{\kappa}\gamma^\mu\kappa$ is always timelike or null in $d=4$, null in $N=1, d=10$ supergravity, etc.

Other Killing spinor bilinears will be associated with generators of other (non-geometrical) symmetries of the solution. This is how the bosonic generators of the supersymmetry group of a bosonic solution arise.

Following our previous discussion of isometries in general-covariant theories, we can associate solutions admitting a maximal number of Killing spinors (maximally supersymmetric solutions) with vacua of the supergravity theory. Now, a given supergravity can have more than one maximally supersymmetric solution (vacuum). Usually, one of the vacua is also a maximally symmetric solution (Minkowski or AdS), but the other vacua are not and have non-vanishing matter fields. Each of these vacua defines a class of solutions with the same asymptotic behavior, which can be associated with states of the QFT that one would construct on the corresponding vacuum. The vacuum supersymmetry algebras can be used to define conserved quantities for those spacetimes/states. Thus, we can study the supersymmetries of these spacetimes using knowledge of their conserved charges and the superalgebra of the asymptotic vacuum spacetime or by solving the Killing spinor equation directly. We will do this in Section 18.1.

Our immediate task is to develop a method by which to find the supersymmetry algebras of the vacuum (or any other) solutions. Let us proceed by analogy with the non-supersymmetric-gravity case discussed in Section 17.1. There will be a bosonic generator $P_A$ of the abstract supersymmetry algebra for each Killing vector $k_A{}^\mu$ that generates a GCT that leaves invariant all the fields of the solution, there will be other "internal" bosonic generators $B_M$ associated with each invariance of the matter fields, and there will be a fermionic generator $Q_I$ of the abstract supersymmetry algebra for each Killing spinor $\kappa_I^\alpha$.

Now, we have to identify all the generators of the abstract supersymmetry algebra with operators acting on the supergravity fields. The (anti)commutators of these operators will give the corresponding (anti)commutators of the superalgebra generators.

Let us start with the bosonic generators $P_A$. On world tensors, each $P_A$ is represented by (minus) the standard Lie derivative with respect to the corresponding Killing vector $k_A$, which transforms world tensors into world tensors of the same rank. However, most of the fields in supergravity theories are Lorentz tensors (with vector or spinor indices), the standard Lie derivative is not covariant under local Lorentz transformations, and its action on Lorentz tensors is frame-dependent.

This has annoying consequences: for instance, the Lie derivatives of Vielbeins with respect to a Killing vector will not be zero in general, even though the same Lie derivative of the metric always will.

On the other hand, Lorentz tensors (and, in particular, spinors) in curved spaces are treated as scalars under GCTs in (Weyl's) standard formalism explained in Section 1.4. Then, if we work in Minkowski spacetime in curvilinear coordinates using Weyl's formalism and perform a Lorentz transformation, all Lorentz tensors and spinors will be invariant. This looks strange, but is not unphysical: in practice one always makes a choice of frame based on some simplicity criterion. For instance, we could always set the Vielbein matrix in an upper triangular form using local Lorentz transformations. This choice can be seen as a gauge-fixing condition that uses up all the Lorentz gauge symmetry. If we now perform a GCT (for instance, the Lorentz transformation we were discussing), it will be necessary to implement a compensating local Lorentz transformation in order to keep the Vielbein matrix upper triangular. This local Lorentz transformation will act on all Lorentz tensors and can be understood as the effect of the GCT on them.

It is necessary for our purposes to find an operator acting on Lorentz tensors that implements the adequate compensating local Lorentz transformation for each GCT. This operator is the *Lie–Lorentz derivative* [1009], which was first introduced for spinors by Lichnerowicz and Kosmann in Refs. [895, 870, 871] and used in supergravity by Figueroa-O'Farrill in Ref. [546] (see also Refs. [811, 1212, 1213]). In simple terms, it is just a Lorentz-covariant Lie derivative.

Analogous problems arise whenever there are additional local symmetries. For instance, in $N=2, d=4$ supergravity there is a local U(1) symmetry. In the Poincaré case only the gauge potential $A_\mu$ transforms under it, but in the AdS case ("gauged $N=2, d=4$ supergravity") the gravitinos and infinitesimal supersymmetry parameters transform as doublets (they are charged). A U(1)-covariant derivative (*Lie–Maxwell derivative*) is needed in order to represent infinitesimal GCTs on these fields.

Covariant Lie derivatives can be found also in the context of the geometry of reductive coset spaces $G/H$ (see Appendix A.4) on which there is a well-defined action of $H$. In fact, the Lie–Lorentz derivative coincides with it in coset spaces in which spinors can be defined and $H$ is a subgroup of the Lorentz group [26].

More generally, they can be defined in principal bundles with a reductive $G$-structure[7] [639], but here we will not make use of this formalism.

### *17.2.1 Covariant Lie derivatives*

*The Lie–Lorentz derivative.* The spinorial Lie–Lorentz derivative with respect to any vector $v$ of a Lorentz tensor $T$ transforming in the representation $r$ is given by

$$\mathbb{L}_v T \equiv v^\rho \nabla_\rho T + \tfrac{1}{2}\nabla_{[a}v_{b]}\,\Gamma_r(M^{ab})T, \tag{17.9}$$

and, on mixed world–Lorentz tensors $T_{\mu_1\cdots\mu_m}{}^{\nu_1\cdots\nu_n}$,

$$\begin{aligned}\mathbb{L}_v T_{\mu_1\cdots\mu_m}{}^{\nu_1\cdots\nu_n} \equiv{}& v^\rho \nabla_\rho T_{\mu_1\cdots\mu_m}{}^{\nu_1\cdots\nu_n} - \nabla_\rho v^{\nu_1} T_{\mu_1\cdots\mu_m}{}^{\rho\nu_2\cdots\nu_n} - \cdots \\ &+ \nabla_{\mu_1} v^\rho T_{\rho\mu_2\cdots\mu_m}{}^{\nu_1\cdots\nu_n} + \cdots + \frac{1}{2}\nabla_{[a}v_{b]}\,\Gamma_r(M^{ab})T_{\mu_1\cdots\mu_m}{}^{\nu_1\cdots\nu_n},\end{aligned} \tag{17.10}$$

where $\nabla_\mu$ is the full (affine plus Lorentz) torsionless covariant derivative satisfying the first Vielbein postulate and $\Gamma_r(M^{ab})$ are the generators of the Lorentz algebra in the representation $r$; $\nabla_{[a}v_{b]}$ is the parameter of the compensating Lorentz transformation.

This derivative enjoys certain properties only when it is taken with respect to a Killing vector or a conformal Killing vector. In particular, the property Eq. (17.12) which allows us to define a Lie algebra structure holds only for conformal Killing vectors, and we are going to restrict our study to that case.

For any two mixed tensors $T_1$ and $T_2$, any two conformal Killing vectors $k_1$ and $k_2$, and constants $a^1$ and $a^2$ we have the following.

1. $\mathbb{L}_k$ satisfies the Leibniz rule:

$$\mathbb{L}_k(T_1T_2) = \mathbb{L}_k(T_1)T_2 + T_1\mathbb{L}_k T_2. \tag{17.11}$$

[7] Recall that $G/H$ is a principal bundle over $G/H$ with structure group $H$, so this is a special case.

2. The commutator of two Lie–Lorentz derivatives

$$[\mathbb{L}_{k_1}, \mathbb{L}_{k_2}]\, T = \mathbb{L}_{[k_1,k_2]}\, T, \tag{17.12}$$

where $[k_1, k_2]$ is the Lie bracket.

3. $\mathbb{L}_k$ is linear in the vector fields

$$\mathbb{L}_{a^1 k_1 + a^2 k_2}\, T = a^1\, \mathbb{L}_{k_1} T + a^2\, \mathbb{L}_{k_2} T. \tag{17.13}$$

Thus, $\mathbb{L}_k$ is a derivative and provides a representation of the Lie algebra of conformal isometries of the manifold.

Some further properties are the following.

1. The Lie–Lorentz derivative of the Vielbein is

$$\mathbb{L}_k e^a{}_\mu = \frac{1}{d} \nabla_\rho k^\rho e^a{}_\mu, \tag{17.14}$$

and vanishes when $k$ is a Killing vector (not just conformal). In this case, we have

$$\mathbb{L}_k \xi^a = e^a{}_\mu \mathcal{L}_k \xi^\mu. \tag{17.15}$$

2. If $k^\mu = \sigma^\mu{}_\nu x^\nu$ with $\sigma^{\mu\nu} = -\sigma^{\nu\mu}$ and constant is an infinitesimal global Lorentz transformation in Minkowski spacetime with Cartesian coordinates, then, on a spinor $\psi$, as required,

$$\mathbb{L}_k \psi = k^\mu \mathcal{D}_\mu \psi + \tfrac{1}{4} \mathcal{D}_{[a} k_{b]} \gamma^{ab} \psi = k^\mu \partial_\mu \psi + \tfrac{1}{4} \sigma_{ab} \gamma^{ab} \psi. \tag{17.16}$$

3.

$$\mathbb{L}_k \gamma^a = 0. \tag{17.17}$$

4. Owing to Eqs. (17.11), (17.14), and (17.17), the Lie–Lorentz derivative with respect to Killing vectors preserves the Clifford action of vectors $v$ on spinors $\psi$, $v \cdot \psi \equiv v_a \Gamma^a \psi = \not{v}\psi$:

$$[\mathbb{L}_k, \not{v}]\, \psi = [k, v] \cdot \psi. \tag{17.18}$$

5. Also, for Killing vectors $k$ only, it preserves the covariant derivative

$$[\mathbb{L}_k, \nabla_v]\, T = \nabla_{[k,v]}\, T. \tag{17.19}$$

6. All this implies that the Lie–Lorentz derivative with respect to Killing vectors preserves the supercovariant derivative of $N = 1, 2, d = 4$ Poincaré and $N = 1, d = 4$ AdS supergravity theories,

$$[\mathbb{L}_k, \tilde{\mathcal{D}}_v]\, \psi = \tilde{\mathcal{D}}_{[k,v]}\, \psi, \tag{17.20}$$

if

$$\mathcal{L}_k F_{\mu\nu} = 0. \tag{17.21}$$

It should be clear that (minus) the Lie–Lorentz derivative with respect to the Killing vectors of the theory $-\mathbb{L}_{k_A}$ should be the operator that represents the bosonic generators $P_A$ on the Vielbein $e^a{}_\mu$ and on the infinitesimal supersymmetry parameters $\epsilon$ in $N = 1, 2, d = 4$ Poincaré and $N = 1, d = 4$ AdS theories.

*The Lie–Maxwell derivative.* How are the $P_A$s represented on the other fields $A_\mu$ and $\psi_\mu$? $A_\mu$ is defined in any solution up to U(1) gauge transformations,[8] and, even though it transforms under GCTs as a vector field, the action of the standard Lie derivative is also gauge dependent. This is similar to our problem with Lorentz tensors, but not quite the same, because $A_\mu$ is a connection and does not transform as a U(1) tensor. Thus, we do not expect to find a U(1)-covariant non-trivial generalization of the Lie derivative for it. It is easy to construct a gauge-invariant generalization of the Lie derivative by adding a compensating U(1) gauge transformation:

$$\mathcal{L}_k A_\mu - \partial_\mu (k^\nu A_\nu), \tag{17.22}$$

but it does not have the crucial Lie-algebra property. We could try to add another gauge transformation with parameter $\Omega$,[9]

$$\mathcal{L}_k A_\mu - \partial_\mu (k^\nu A_\nu + \Omega), \tag{17.23}$$

but it works only if

$$\partial_\mu \Omega = k^\lambda F_{\mu\lambda}, \tag{17.24}$$

and then Eq. (17.23) vanishes for any $A_\mu$. This is in fact how the $P_A$s are represented on $A_\mu$: on looking into the commutator Eq. (5.102), we find on the r.h.s. a GCT and a gauge transformation with parameter $\chi = k^\nu A_\nu + \Omega$ with $\Omega = -i\bar{\epsilon}_2 \sigma^2 \epsilon_1$, and it can be checked that for Killing spinors we have, precisely,

$$\partial_\mu(-i\bar{\kappa}_J \sigma^2 \kappa_I) = k_A{}^\lambda F_{\mu\lambda}, \qquad k_A{}^\mu \equiv -i\bar{\kappa}_I \gamma^\mu \kappa_J. \tag{17.25}$$

This exercise is useful because there can be other gauge-dependent fields in the supergravity theory: in $N=2, d=4$ AdS (gauged) supergravity the gravitinos $\psi_\mu$ and the supersymmetry parameters $\epsilon$ (and, therefore, the Killing spinors, if any) are electrically charged and transform according to

$$A'_\mu = A_\mu + \partial_\mu \chi, \qquad \psi'_\mu = e^{-ig\chi\sigma^2} \psi_\mu, \qquad \epsilon' = e^{-ig\chi\sigma^2} \epsilon, \tag{17.26}$$

and we need to define a U(1)- and Lorentz-covariant Lie derivative for them. For the supersymmetry parameters $\epsilon$ and Killing vectors $k$, we redefine the

$$\mathbb{L}_k \epsilon \equiv \mathbb{L}_k^{\text{Lorentz}} \epsilon + ig(k^\mu A_\mu + \Omega)\sigma^2 \epsilon, \tag{17.27}$$

where $\Omega$ has been defined in Eq. (17.23) and exists if $k$ is Killing and Eq. (17.21) is satisfied. This derivative has the Lie algebra property and also preserves the $N=2, d=4$ AdS supercovariant derivative

$$[\mathbb{L}_k, \hat{\tilde{\mathcal{D}}}_v]\, \epsilon = \hat{\tilde{\mathcal{D}}}_{[k,v]}\, \epsilon, \tag{17.28}$$

[8] Of course, the $P_A$s are represented on the field strength $F_{\mu\nu}$ by the standard Lie derivative. The condition Eq. (17.21) is a necessary condition for the corresponding $P_A$ to be a symmetry of the complete solution.

[9] In general, this $\Omega$ is known as a *momentum map*. See, e.g., Eq. (G.41) and the accompanying discussion.

under the condition Eq. (17.21), which is necessary anyway in order for the associated $P_A$ to be a symmetry of the whole solution.

A non-Abelian generalization of all these formulae can be found in Appendix A.4.1.

For the gravitinos $\psi_\mu$ we expect problems similar to those we found for $A_\mu$ since they can be considered (super) gauge fields and transform inhomogeneously under supersymmetry. The role of the supersymmetry transformation that appears in the commutators Eqs. (5.45), (5.58), and (5.102) will clearly be that of compensating the effect of the GCT.

### 17.2.2 Calculation of supersymmetry algebras

We have developed all the tools we need to calculate the symmetry superalgebra of any supergravity solution. Now we just have to follow this six-step recipe [546].

1. First we have to solve the Killing and Killing-spinor equations. We keep only the Killing vectors that leave invariant all the fields of the solution. Furthermore, we have to find any other "internal" invariance of the fields.

2. With each Killing vector $k_A{}^\mu$ we associate a bosonic generator of the superalgebra $P_A$, with any internal symmetry of the fields we associate another bosonic generator $B_M$, and with each Killing spinor $\kappa_I{}^\alpha$ we associate a fermionic generator (supercharge) $Q_I$.

   The bosonic subalgebra is in general the sum of two subalgebras generated by the $P_A$s and the $B_M$s with structure constants $f_{AB}{}^C$ and $f_{MN}{}^P$. The fermionic generators are in representations of these bosonic subalgebras. These representations are determined by the structure constants $f_{IA}{}^J$ and $f_{IM}{}^J$ that appear in

$$[Q_I, P_A] = f_{IA}{}^J Q_J, \qquad [Q_I, B_M] = f_{IM}{}^J Q_J. \tag{17.29}$$

   The superalgebra is determined by these four sets of structure constants plus the structure constants $f_{IJ}{}^A$ that appear in the anticommutators

$$\{Q_I, Q_J\} = f_{IJ}{}^A P_A. \tag{17.30}$$

3. The structure constants $f_{AB}{}^C$ of the bosonic subalgebra are simply those of the isometry Lie algebra

$$[k_A, k_B] = -f_{AB}{}^C k_C. \tag{17.31}$$

4. The commutators $[Q_I, P_A]$ can be interpreted as the action of the bosonic generators on the fermionic generators, which transform under some (spinorial) representation of the bosonic subalgebra with matrices $\Gamma_s(P_A)^J{}_I = f_{IA}{}^J$. Since the covariant Lie derivative has been defined to represent the action of infinitesimal GCTs on any kind of Lorentz tensors or spinors, and, according to Eqs. (17.19) and (17.20), transforms Killing spinors into Killing spinors, which, therefore, furnish a representation of the bosonic subalgebra, it is natural to expect that the structure constants $f_{IA}{}^J$ are given by the covariant Lie derivatives

$$\mathbb{L}_{k_A} \kappa_I \equiv f_{IA}{}^J \kappa_J. \tag{17.32}$$

5. We have mentioned that the bilinears of Killing spinors $-i\bar{\kappa}_I\gamma^\mu\kappa_J$ are Killing vectors. In fact, in the commutator of two local $N=1,2,d=4$ supersymmetry transformations with parameters $\epsilon_{1,2}$ given in Eqs. (5.45), (5.58), and (5.102) we found a GCT (more precisely, (minus) a standard Lie derivative) with parameter $-i\bar{\epsilon}_1\gamma^\mu\epsilon_2$, a local Lorentz transformation, and a gauge transformation. When we use two Killing spinors $\kappa_{I,J}$ instead, on the r.h.s. we always find $-\mathbb{L}_k$, where $k^\mu$ is the Killing vector $-i\bar{\kappa}_I\gamma^\mu\kappa_J$, which must be a linear combination of the Killing vectors $k_A$. The structure constants $f_{IJ}{}^A$ are thus given by the decomposition of the bilinears

$$-i\bar{\kappa}_I\gamma^\mu\kappa_J \equiv f_{IJ}{}^A k_A{}^\mu. \tag{17.33}$$

6. The structure constants involving the internal generators $B_M$ have to be determined case by case. They appear in extended supergravities and in general they are constant gauge transformations of vector fields that also act on the spinors.

We are now ready to apply these prescriptions to some basic examples, but it is useful to present some general considerations first.

## 17.3 $N=1,2,d=4$ vacuum supersymmetry algebras

We have defined supergravity vacua as the classical solutions that admit a maximal number of Killing spinors, i.e. four in $N=1,d=4$ theories and eight in $N=2,d=4$ theories. A necessary (and locally sufficient) condition for a solution to be maximally supersymmetric is that the integrability condition of the Killing spinor equation admit the maximal number of possible solutions.

The Killing spinor equation takes the generic form $\tilde{D}_\mu\kappa=0$, where $\tilde{D}_\mu=\partial_\mu-\Omega_\mu$ (the supercovariant derivative) can be understood as a standard covariant derivative with a connection $\Omega_\mu$ that is the combination of the spin connection and other supergravity fields contracted with gamma matrices:

$$\Omega_\mu = \Omega_\mu{}^A\Gamma_{\rm s}(T_A), \tag{17.34}$$

where $\Gamma_{\rm s}(T_A)$ stands for different antisymmetrized products of gamma matrices that constitute a (spinorial) representation of some of the generators of some algebra. Thus, the Killing spinor equation can be understood as an equation of parallelism. This is why Killing spinors are sometimes called *parallel spinors*.

The integrability condition says that the commutator of the supercovariant derivative on the Killing spinor has to be zero, that is

$$[\tilde{D}_\mu,\tilde{D}_\nu]\kappa=0 \;\Rightarrow\; R_{\mu\nu}(\Omega)\kappa=0, \tag{17.35}$$

where $R_{\mu\nu}(\Omega)$ is the curvature associated with the connection $\Omega$. This is a homogeneous equation. The space of non-trivial solutions is determined by the rank of the matrix $R_{\mu\nu}(\Omega)$, which is a linear combination of $\Gamma_{\rm s}(T_A)$s with coefficients that depend on the values of the supergravity fields in the solution. In particular, we can have maximal supersymmetry only if $R_{\mu\nu}(\Omega)=0$ identically (the connection is *flat*), which means that all the coefficients in the linear combination have to vanish.

All the known maximally supersymmetric solutions have homogeneous reductive spacetimes with invariant metrics, and the connection 1-form $\Omega$ turns out to be the Maurer–Cartan 1-form $V$ defined in Eq. (A.106) in a spinorial representation [26, 27]. In symmetric spaces, the spin connection contributes to the vertical components of $V$:

$$-\tfrac{1}{4}\omega_{ab}\gamma^{ab} = -\vartheta^i \Gamma_{\mathrm{s}}(M_i), \qquad \Gamma_{\mathrm{s}}(M_i) \equiv \tfrac{1}{4} f_{ia}{}^{b}\gamma_b{}^{a}, \tag{17.36}$$

due to Eq. (A.117) and the fact that the structure constants $f_{ia}{}^{b}$ are a representation of $\mathfrak{h}$ on $\mathfrak{k}$, which makes the above $\Gamma_{\mathrm{s}}(M_i)$ a spinorial representation of $\mathfrak{h}$.

All the horizontal components of $V$ must come from the contribution of the supergravity fields. In the non-symmetric case [27] a combination of the two contributions gives $V$.

The curvature of the 1-form $V$ is identically zero: in the language of differential forms,

$$\tilde{D} = d - V \;\Rightarrow\; R(V) = dV - V \wedge V = 0, \tag{17.37}$$

which are precisely the Maurer–Cartan equations. The Killing spinor equations admit a maximal number of solutions, and, since $V = -\Gamma_{\mathrm{s}}(u^{-1})d\Gamma_{\mathrm{s}}(u)$, where $\Gamma_{\mathrm{s}}(u)$ is the coset representative defined in Eq. (A.104) using the spinorial representation $\Gamma_{\mathrm{s}}(P_{(a)})$ dictated by the supergravity theory, the Killing spinors take the form

$$\kappa = \Gamma_{\mathrm{s}}(u^{-1})\kappa_0, \tag{17.38}$$

where $\kappa_0$ is any constant spinor. Choosing independent constant spinors, we find the following basis of Killing spinors:

$$\kappa_{(\alpha)}{}^{\beta} = \Gamma_{\mathrm{s}}(u^{-1})^{\beta}{}_{\alpha}. \tag{17.39}$$

This result reproduces the construction of Killing spinors on spheres and AdS space made in Ref. [912], but the calculations are dramatically simplified and the geometrical meaning of the construction is clearer.

On the other hand, this form of the Killing spinors is extremely useful for computing the supersymmetry algebra. First, it can be shown [26] that the Lie–Lorentz derivatives associated with the Killing vectors coincide with the $H$-covariant Lie derivatives defined in Section A.4.1. Then, using the property Eq. (A.128), we immediately find

$$\mathbb{L}_{k_A}\kappa_{(\alpha)}{}^{\beta} = -\kappa^{\beta}_{(\gamma)}\Gamma_{\mathrm{s}}(T_A)^{\gamma}{}_{\alpha} \;\Rightarrow\; f_{\alpha A}{}^{\gamma} = -\Gamma_{\mathrm{s}}(T_A)^{\gamma}{}_{\alpha}. \tag{17.40}$$

Let us now consider the Killing spinor bilinears. We consider only Majorana spinors. Then the bilinears take the form

$$-i\bar{\kappa}_{(\alpha)}\gamma^{\mu}\kappa_{(\beta)}\partial_{\mu} = -i\Gamma_{\mathrm{s}}(u^{-1})_{\alpha}{}^{\gamma}\mathcal{C}_{\gamma\delta}(\gamma^{a})^{\delta}{}_{\epsilon}\Gamma_{\mathrm{s}}(u^{-1})^{\epsilon}{}_{\beta}, \tag{17.41}$$

where $\mathcal{C}$ is the charge-conjugation matrix.

Usually, one finds that the spinorial representation is proportional to the gamma matrices,

$$\Gamma_{\mathrm{s}}(P_a) = \mathcal{S}\gamma_a. \tag{17.42}$$

When $\mathcal{S}$ is invertible and the Killing metric is also invertible[10] we can write

$$\gamma^a = \mathcal{S}\Gamma_s(P^a), \tag{17.43}$$

where $\Gamma_s(P^a)$ is a dual representation. The combination $\tilde{\mathcal{C}} \equiv \mathcal{C}\mathcal{S}$ acts as a charge-conjugation matrix in the subspace spanned by the horizontal generators in the spinorial representation

$$\tilde{\mathcal{C}}^{-1}\Gamma_s(P^a)^{\mathrm{T}}\tilde{\mathcal{C}} = -\Gamma_s(P^a), \tag{17.44}$$

so

$$\Gamma_s(u^{-1})^{\mathrm{T}}\mathcal{C}\gamma^a = \Gamma_s(u^{-1})^{\mathrm{T}}\tilde{\mathcal{C}}\Gamma_s(P^a) = \tilde{\mathcal{C}}\Gamma_s(u)\Gamma_s(P^a), \tag{17.45}$$

and, thus,

$$-i\bar{\kappa}_{(\alpha)}\gamma^\mu\kappa_{(\beta)}\partial_\mu = -i\tilde{\mathcal{C}}_{\alpha\gamma}\Gamma_s(u)^\gamma{}_\delta\Gamma_s(P^a)^\delta{}_\epsilon\Gamma_s(u^{-1})^\epsilon{}_\beta e_a. \tag{17.46}$$

In this expression we can recognize $uP^a u^{-1}$ in the spinorial representation, which is the coadjoint action of the coset element $u$ on $P^a$,

$$\begin{aligned} -i\bar{\kappa}_{(\alpha)}\gamma^\mu\kappa_{(\beta)}\partial_\mu &= -i\tilde{\mathcal{C}}_{\alpha\gamma}\Gamma_s(T^A)^\gamma{}_\beta\Gamma_{\mathrm{Adj}}(u^{-1})^a{}_A e_a = -i\tilde{\mathcal{C}}_{\alpha\gamma}\Gamma_s(T^A)^\gamma{}_\beta k_A, \\ \Rightarrow f_{\alpha\beta}{}^A &= -i\tilde{\mathcal{C}}_{\alpha\gamma}\Gamma_s(T^A)^\gamma{}_\beta, \end{aligned} \tag{17.47}$$

where we have used Eq. (A.114).

In extended supergravities, one may also have to compute other Killing spinor bilinears $-i\bar{\kappa}_{(\alpha)}\Sigma^M\kappa_{(\beta)}$, where $\Sigma^M$ acts on internal spinor indices that we are not showing. In some cases, the internal symmetries are related to the vertical generators $M^i$,

$$\Sigma^{(i)} = \mathcal{S}\Gamma_s(M^i), \tag{17.48}$$

and, using the property of the matrix $\mathcal{S}$, one finds

$$\begin{aligned} -i\bar{\kappa}_{(\alpha)}\Sigma^{(i)}\kappa_{(\beta)} &= -i\tilde{\mathcal{C}}_{\alpha\gamma}\Gamma_s(T^A)^\gamma{}_\beta\Gamma_{\mathrm{Adj}}(u^{-1})^i{}_A \\ &= i\tilde{\mathcal{C}}_{\alpha\gamma}\Gamma_s(T^A)^\gamma{}_\beta(k_A^\mu\vartheta^i{}_\mu + W^i{}_A). \end{aligned} \tag{17.49}$$

In the light of Appendix A.4.1 this means that $-i\bar{\kappa}_{(\alpha)}\Sigma^{(i)}\kappa_{(\beta)}$ gives the infinitesimal parameter of the gauge transformation of the vertical Maurer–Cartan 1-forms $\vartheta^i$ that can be compensated by a diffeomorphism.

This result can be understood only after the realization that, in all these cases, the vertical Maurer–Cartan 1-forms $\vartheta^i$ enter the solution in a non-trivial form. The simplest example is the Robinson–Bertotti solution of $N=2, d=4$ Poincaré supergravity, with $\mathrm{AdS}_2 \times \mathrm{S}^2$ geometry in which the Maxwell vector field is identical to a linear combination of the two vertical Maurer–Cartan 1-forms. The same is true for the KG4 solution. In the higher-dimensional cases that we will study in Part III (Section 25.5.1), the matter fields (differential forms of higher rank) are also given in terms of the $\vartheta^i$s, which are non-Abelian.

We are now going to focus again on the simplest $N=1,2, d=4$ cases. First, we are going to derive the Killing spinor integrability condition for $N=2, d=4$ gauged supergravity because, in the $g \to 0$ limit, we can recover the integrability condition for the ungauged case, and, on setting $A_\mu = 0$ and ignoring one fermion, we obtain the $N=1$ cases.

[10] This is usually the case. The exceptions are the Kowalski–Glikman Hpp-waves and the five-dimensional Gödel-like solution of Ref. [596].

### 17.3.1 The Killing spinor integrability condition

The integrability condition in $N=2,d=4$ gauged supergravity is

$$[\tilde{\hat{\mathcal{D}}}_\mu, \tilde{\hat{\mathcal{D}}}_\nu]\kappa = 0. \tag{17.50}$$

The supercovariant derivative $\tilde{\hat{D}}_\mu$ is given in Eq. (5.107). We immediately find

$$[\tilde{\hat{\mathcal{D}}}_\mu, \tilde{\hat{\mathcal{D}}}_\nu] = [\hat{\mathcal{D}}_\mu, \hat{\mathcal{D}}_\nu] + gF_{\mu\nu}i\sigma^2 - \frac{i}{2}\hat{\nabla}_{[\mu}\not{F}\gamma_{\nu]}i\sigma^2 + \frac{1}{8}\not{F}\gamma_{[\mu}\not{F}\gamma_{\nu]}. \tag{17.51}$$

The first term gives the SO(2,3) curvature that can be put into the form

$$[\hat{\mathcal{D}}_\mu, \hat{\mathcal{D}}_\nu] = -\tfrac{1}{4}\left(R_{\mu\nu}{}^{ab} + 2g^2 e_\mu{}^{[a}e_\nu{}^{b]}\right)\gamma_{ab}. \tag{17.52}$$

The third term can also be put into the form

$$-\frac{i}{2}\hat{\mathcal{D}}_{[\mu}\not{F}\gamma_{\nu]}i\sigma^2 = -\frac{i}{2}\nabla_{[\mu}\not{F}\gamma_{\nu]}i\sigma^2 - \frac{g}{4}\left(\gamma_{[\mu}\not{F}\gamma_{\nu]} + \not{F}\gamma_{\mu\nu}\right)i\sigma^2. \tag{17.53}$$

The $g$-dependent terms together with the second term in Eq. (17.51) combine to give

$$-\frac{g}{8}F_{ab}\left(3\gamma^{ab}\gamma_{\mu\nu} + \gamma_{\mu\nu}\gamma^{ab}\right)i\sigma^2, \tag{17.54}$$

and, with some gamma gymnastics, it is possible to rewrite it in the form

$$-\frac{i}{2}\nabla_{[\mu}\not{F}\gamma_{\nu]}i\sigma^2 = -\frac{i}{2}\not{\nabla}(F_{\mu\nu} + i\star F_{\mu\nu}\gamma_5)i\sigma^2 + \frac{i}{2}\gamma_{\mu\nu\rho}\nabla_\sigma(F^{\sigma\rho} + i\star F^{\sigma\rho}\gamma_5)i\sigma^2. \tag{17.55}$$

This term is manifestly electric–magnetic-duality invariant. Since it is not proportional to $g$, this agrees with the chiral–dual invariance of the ungauged theory. Now

$$\not{F}\gamma_{[\mu}\not{F}\gamma_{\nu]} = 8iF_{[\mu}{}^\rho \star F_{\nu]\rho}\gamma_5 + 8T(A)_{[\mu|\rho}\gamma^\rho{}_{|\nu]}. \tag{17.56}$$

The first term here can be shown to be identically zero[11] and $T(A)_{\mu\nu}$ is the bosonic part of the vector-field energy–momentum tensor Eq. (5.85). These two terms are also electric–magnetic-duality invariant and independent of $g$.

Putting everything together, we find

$$\begin{aligned}[\tilde{\hat{\mathcal{D}}}_\mu, \tilde{\hat{\mathcal{D}}}_\nu] = &-\tfrac{1}{4}R_{\mu\nu}{}^{ab}\gamma_{ab} - \tfrac{1}{2}g^2\gamma_{\mu\nu} + T(A)_{[\mu|\alpha}\gamma^\alpha{}_{|\nu]} - \frac{i}{2}\not{\nabla}(F_{\mu\nu} + i\star F_{\mu\nu}\gamma_5)i\sigma^2 \\ &+ \frac{i}{2}\gamma_{\mu\nu\rho}\nabla_\sigma(F^{\sigma\rho} + i\star F^{\sigma\rho}\gamma_5)i\sigma^2 - \frac{g}{8}F_{ab}\left(3\gamma^{ab}\gamma_{\mu\nu} + \gamma_{\mu\nu}\gamma^{ab}\right)i\sigma^2. \end{aligned} \tag{17.57}$$

[11] One has to use the self-evident four-dimensional identity

$$\eta_{a[b}\epsilon_{a_1\ldots a_4]}F^{a_1a_2}F^{a_3a_4} = 0.$$

We can now use the bosonic part of the Einstein equation rewritten in this form (substituting the value of $R = -12g^2$)

$$T(A)_{\mu\nu} = \tfrac{1}{2}R_{\mu\nu} + \tfrac{3}{2}g^2 g_{\mu\nu}, \tag{17.58}$$

plus the bosonic part of the Maxwell equation and the Bianchi identity. We obtain

$$-\tfrac{1}{4}\Big\{C_{\mu\nu}{}^{ab}\gamma_{ab} + 2i\,\not{\nabla}(F_{\mu\nu} + i\star F_{\mu\nu}\gamma_5)i\sigma^2 + \frac{g}{2}F_{ab}\Big(3\gamma^{ab}\gamma_{\mu\nu} + \gamma_{\mu\nu}\gamma^{ab}\Big)i\sigma^2\Big\}\,\kappa = 0. \tag{17.59}$$

This is a homogeneous linear equation for $\kappa$. The $8\times 8$ matrix is a linear combination of tensor products of gamma matrices and Pauli matrices, all of them linearly independent. There are terms with two gammas (and $\otimes\mathbb{I}_{2\times 2}$), whose coefficients are the components of the Weyl tensor; there are terms proportional to one gamma and $\gamma_5$ ($\otimes\sigma^2$), whose coefficients are the components of the covariant derivative of the electromagnetic tensor and its dual; and, finally, there are terms with zero, two, and four gammas ($\otimes\sigma^2$), whose coefficients are the components of the electromagnetic tensor. In order to have maximally supersymmetric solutions, each of these terms has to vanish. This imposes severe constraints on the vacuum candidates. Let us now study each case separately, and let us calculate the symmetry superalgebra using the recipe of Section 17.2.2.

### *17.3.2 The vacua of $N = 1, d = 4$ Poincaré supergravity*

On setting $g = F_{\mu\nu} = 0$ in Eq. (17.59), we find the integrability condition for the Killing spinors of $N = 1, d = 4$ Poincaré supergravity. The maximally supersymmetric solutions are those with vanishing Weyl tensor, which (since the equations of motion are $R_{\mu\nu} = 0$) implies a vanishing Riemann curvature tensor. Thus, Minkowski spacetime is the only maximally supersymmetric vacuum of $N = 1, d = 4$ Poincaré supergravity.

In this simple case it is unnecessary to use the coset description of Minkowski spacetime. We can compute Killing spinors and vectors directly. In Cartesian coordinates and with the trivial frame $e^a{}_\mu = \delta^a{}_\mu$ the spin and Levi-Civita connections vanish, the Killing spinors are just constant, and the Killing vectors are the ten known generators of the Poincaré group.

*Solution:*

$$ds^2 = \eta_{\mu\nu}dx^\mu dx^\nu, \qquad e^a{}_\mu = \delta^a{}_\mu, \qquad e_a{}^\mu = \delta_a{}^\mu. \tag{17.60}$$

*Killing spinors:*

$$\{\kappa_A{}^\beta\} = \{\kappa_\alpha{}^\beta = \delta_\alpha{}^\beta\}. \tag{17.61}$$

*Killing vectors:*

$$\{k_A{}^\mu\} = \{k_a{}^\mu = e_a{}^\mu,\ k_{ab}{}^\mu = 2e_{[a}{}^\mu e_{b]\sigma}x^\sigma\}. \tag{17.62}$$

The commutators of the bosonic generators are given by the Lie algebra of the Killing vectors. This is nothing but the Poincaré algebra, which we do not need to write explicitly.

To find the anticommutator of two supercharges, we need to decompose the Killing vectors $-i\bar{\kappa}_{(\alpha)}\gamma^{\mu}\kappa_{(\beta)}$ as a combination of the $k_A{}^{\mu}$s for each pair of indices $\alpha$ and $\beta$. This is easy:

$$-i\bar{\kappa}_{\alpha}\gamma^{\mu}\kappa_{\beta} = -i(\mathcal{C}\gamma^{a})_{\alpha\beta}k_{a}{}^{\mu} \quad \Rightarrow \quad \{Q_{\alpha}, Q_{\beta}\} = -i(\mathcal{C}\gamma^{a})_{\alpha\beta}P_{a}, \tag{17.63}$$

which we can convert into the standard form by raising the indices $\alpha$ and $\beta$ with $\mathcal{C}^{-1}$. The fact that only the translational symmetries occur could have been predicted since all the $N = 1, d = 4$ Killing vectors $-i\bar{\kappa}_{\alpha}\gamma^{\mu}\kappa_{\beta}$ are covariantly constant. These are also the ones with a vanishing compensating Lorentz transformation in the Lie–Lorentz derivative. Since the Killing spinors are constant, we find that only $\mathbb{L}_{k_{ab}}\kappa_{\alpha}{}^{\beta}$ is different from zero and, using $\nabla_{[a|}k_{(bc)|d]} = -2\eta_{ad,bc}$, we find, as expected,

$$[Q_{\alpha}, P_{ab}] = -Q_{\beta}\tfrac{1}{2}(\gamma_{ab})^{\beta}{}_{\alpha}. \tag{17.64}$$

### 17.3.3 *The vacua of* $N = 1, d = 4$ $AdS_4$ *supergravity*

The integrability condition in $N = 1, d = 4$ AdS supergravity again implies the vanishing of the Weyl tensor, but now this implies

$$R_{\mu\nu}{}^{ab} + 2g^{2}e_{[\mu}{}^{a}e_{\nu]}{}^{b} = 0, \tag{17.65}$$

i.e. the space is a maximally symmetric space of constant curvature $-2g^2$, so it is locally $AdS_4$ with *AdS radius* $R = 1/g$.

To construct the Killing spinors and find the symmetry superalgebra using the coset method, we can construct the metric and Vierbeins using the method explained in Appendix A.4,[12] or we can use another coordinate and Vielbein basis, which may be more convenient. We are going to give the metric, Killing spinors, and Killing vectors and compute the superalgebra both in holographic coordinates and using the coset space construction.

*The AdS Killing spinors and superalgebra in holographic coordinates.* In these coordinates the metric of AdS takes the form[13]

$$ds^{2} = (r/R)^{2}\eta_{ij}dy^{i}dy^{j} - (R/r)^{2}dr^{2}, \tag{17.66}$$

where $\eta_{ij}$, $i, j = 0, \ldots, d-2$, is the Minkowski metric in one dimension fewer. The simplest choice of Vielbeins is

$$e^{i}{}_{\underline{j}} = \delta^{i}{}_{j}r/R, \qquad e^{r}{}_{\underline{r}} = R/r, \tag{17.67}$$

[12] Actually, we do not even need to implement this construction explicitly. It is enough to assume that the Vierbeins and metric (and hence the spin connection) have been constructed by that procedure. We are going to construct the $AdS_4$-invariant metric just to illustrate the procedure and fix the notation.

[13] Most of these results are valid in $d$ dimensions, the only exception being the Killing spinors themselves since AdS supergravities do not exist in $d > 7$ dimensions, and we should also take into account the reality properties of spinors that vary for each dimension. Still, we can write formally the AdS Killing spinor equation $\hat{\mathcal{D}}_{\mu}\kappa = 0$ and solve it in any dimension, and later particularize those cases that really make sense. Here only $d = 4$ is interesting for us.

and the corresponding non-vanishing components of the Christoffel symbols and spin connection are

$$\Gamma_{\underline{ij}}{}^{\underline{r}} = \frac{r}{R^2} g_{ij}, \qquad \Gamma_{\underline{ri}}{}^{\underline{j}} = \delta_{\underline{i}}{}^{\underline{j}} \frac{1}{r}, \qquad \Gamma_{\underline{rr}}{}^{\underline{r}} = -\frac{1}{r}, \qquad \omega_{\underline{i}}{}^{\underline{rj}} = -\delta_{\underline{i}}{}^{\underline{j}} r/R^2. \quad (17.68)$$

The Killing spinor equation is

$$\hat{\mathcal{D}}_\mu \kappa = \left( \partial_\mu - \tfrac{1}{4} \omega_{\mu\, ab} \gamma^{ab} - \frac{ig}{2} \gamma_\mu \right) \kappa = 0. \quad (17.69)$$

Since this is the first time we meet a non-trivial Killing spinor equation, it is worth explaining in detail how it is solved. First, we simply substitute the spin connection coefficients into the $d$ equations $\hat{\mathcal{D}}_\mu \kappa = 0$, which then take the form

$$\left(\partial_{\underline{r}} - \tfrac{i}{2} r^{-1} \gamma_r\right) \kappa = 0, \qquad \left[\partial_{\underline{i}} + \tfrac{i}{2R^2} r \gamma_i (i\gamma_r - 1)\right] \kappa = 0. \quad (17.70)$$

Naively, it looks like the only solution is of the form $\kappa = r^{1/2} \kappa_0$, $(i\gamma_r - 1)\kappa_0 = 0$, where $\kappa_0$ is a constant spinor. However, the second equation is a constraint: the matrix $\gamma_r$ is unitary, traceless, and (in dimensions in which there are Majorana representations) purely imaginary. Thus, $i\gamma_r$ has an equal number of $+1$ and $-1$ eigenvalues and only eigenspinors with eigenvalue $+1$ ($1/2$ of the maximal possible number) are allowed. However, we know that AdS is maximally supersymmetric and the other half should also give solutions. It is convenient to make the following splitting:

$$\kappa = \kappa_+ + \kappa_-, \qquad \Pi_\pm \kappa_\pm = \kappa_\pm, \quad (17.71)$$

where we have introduced the following notation for the projectors:

$$\Pi_\pm \equiv \tfrac{1}{2}(1 \pm \gamma_r). \quad (17.72)$$

Substituting into the first of Eqs. (17.70), we see that the equation itself splits into two separate equations for $\kappa_+$ and $\kappa_-$, which have the solutions

$$\kappa_+ = r^{1/2} \kappa_{1\,+}, \qquad \kappa_- = R r^{-1/2} \kappa_{1\,-}, \quad (17.73)$$

where $\kappa_{1\,\pm}$ do not depend on $r$. Substituting into the remaining Eqs. (17.70), we find

$$r^{1/2} \partial_{\underline{i}} \kappa_{1\,+} + R r^{-1/2} \partial_{\underline{i}} \kappa_{1\,-} - \tfrac{i}{R} r^{1/2} \gamma_i \kappa_{1\,-} = 0, \quad (17.74)$$

which has a solution if $\kappa_{1\,-} = \kappa_{0\,-}$ (a constant spinor) and

$$\kappa_{1\,+} = \kappa_{0\,+} + \tfrac{i}{R} y^i \gamma_i \kappa_{0\,-}. \quad (17.75)$$

Thus, the full solution is

$$\kappa = r^{1/2} \left(\kappa_{0\,+} + \tfrac{i}{R} y^i \gamma_i \kappa_{0\,-}\right) + R r^{-1/2} \kappa_{0\,-}, \quad (17.76)$$

and corresponds to a four-dimensional space of solutions (in $d = 4$) spanned by the following four Killing spinors $\{\kappa_\alpha{}^\beta\}$:

$$\{\kappa_A{}^\beta\} = \left\{ \kappa_\alpha{}^\beta = \left[ r^{\frac{1}{2}} (\Pi_+ - \tfrac{i}{R} y^i \gamma_i \Pi_-) - R r^{-\frac{1}{2}} \Pi_- \right]^\beta{}_\alpha \right\}. \quad (17.77)$$

Let us now consider the Killing vectors of AdS. We can label them as follows:

$$\{k_A{}^\mu\} = \{k_{\hat{a}\hat{b}}{}^\mu\}, \quad \hat{a}, \hat{b} = -1, 0, \ldots, d-1. \tag{17.78}$$

By definition, the Lie brackets in this basis take the familiar form for the $\mathfrak{so}(2, d-1)$ algebra

$$[k_{\hat{a}\hat{b}}, k_{\hat{a}\hat{b}}] = +\hat{\eta}_{\hat{a}\hat{c}} k_{\hat{b}\hat{d}} + \cdots . \tag{17.79}$$

It is convenient to express these Killing vectors in terms of the conformal Killing vectors of constant $r$ hypersurfaces $\{k_{m_{ij}}, k_{p_i}, k_{k_i}, k_d\}$. These are given by

$$\begin{aligned} k_{m_{ij}} &= -2y_{[\underline{i}}\partial_{\underline{j}]}, & k_{p_i} &= \partial_{\underline{i}}, \\ k_{k_i} &= (2y_{\underline{i}}y^{\underline{j}} - y^2\delta_i{}^{\underline{j}})\partial_{\underline{j}}, & k_d &= -y^{\underline{i}}\partial_{\underline{i}}, \end{aligned} \tag{17.80}$$

and their Lie brackets are those of the AdS algebra

$$\begin{aligned} [k_{m_{ij}}, k_{m_{kl}}] &= +\eta_{ik} k_{m_{jl}} + \cdots , & [k_{p_i}, k_{k_j}] &= +2m_{ij} - 2\eta_{ij} d, \\ [k_{p_i}, k_{m_{jk}}] &= +k_{p_l}\eta_{[j}{}^l \eta_{k]i}, & [k_{p_i}, k_d] &= -k_{p_i}, \\ [k_{k_i}, k_{m_{jk}}] &= +k_{k_l}\eta_{[j}{}^l \eta_{k]i}, & [k_{k_i}, k_d] &= +k_{k_i}. \end{aligned} \tag{17.81}$$

With these we construct the AdS Killing vectors $\{k_{M_{ij}}, k_{P_i}, k_{K_i}, k_D\}$

$$\begin{aligned} k_{M_{ij}} &= k_{m_{ij}}, & k_{P_i} &= k_{p_i}, \\ k_{K_i} &= k_{k_i} + R^4/r^2 \partial_{\underline{i}} - 2r y_{\underline{i}} \partial_{\underline{r}}, & k_D &= k_d - r\partial_{\underline{r}} \end{aligned} \tag{17.82}$$

that have the same Lie brackets (replacing lower case by capital letters). Finally, these are related to the $\{k_{\hat{a}\hat{b}}\}$:

$$k_{ij} = k_{M_{ij}}, \quad k_{-1i} = \tfrac{1}{2}(k_{P_i} + k_{K_i}), \quad k_{ri} = \tfrac{1}{2}(k_{P_i} - k_{K_i}), \quad k_{-1r} = k_D. \tag{17.83}$$

We are now ready to recover the $N = 1, d = 4$ AdS superalgebra. The bosonic generators $P_{\hat{a}\hat{b}}$ satisfy the SO(2, 3) algebra (we usually denote these generators by $\hat{M}_{\hat{a}\hat{b}}$). The anticommutator of the fermionic generators $Q_\alpha$ is found by expanding the bilinears $-i\kappa_\alpha\gamma^\mu\kappa_\beta$ as a linear combination of the Killing vectors $k_{\hat{a}\hat{b}}{}^\mu$. For each value of $\mu$ the bilinear can be considered as a symmetric $4 \times 4$ matrix that is a linear combination of the symmetric matrices $m^{\hat{a}\hat{b}\,\alpha\beta}$ defined in Eq. (D.166), with coefficients easily found to be (using Eqs. (D.167) and (D.168)), precisely, the Killing vectors:

$$-i\kappa_\alpha\gamma^\mu\kappa_\beta = i m^{\hat{a}\hat{b}\,\alpha\beta} k_{\hat{a}\hat{b}}{}^\mu \quad \Rightarrow \{Q_\alpha, Q_\beta\} = m^{\hat{a}\hat{b}}{}_{\alpha\beta} P_{\hat{\alpha}\hat{\beta}}, \tag{17.84}$$

where we are using the charge-conjugation matrix to raise and lower spinor indices. Finally,

$$\mathbb{L}_{k_{\hat{a}\hat{b}}} \kappa_\alpha{}^\beta = -m_{\hat{a}\hat{b}\alpha}{}^\gamma \kappa_\gamma{}^\beta \quad \Rightarrow \{Q_\alpha, P_{\hat{a}\hat{b}}\} = m_{\hat{a}\hat{b}\alpha}{}^\beta Q_\beta. \tag{17.85}$$

*Coset space construction.* $AdS_4$ can be identified with the coset space $SO(2,3)/SO(1,3)$. We use the index conventions of Section 4.5. Note that $\mathfrak{g} = \mathfrak{so}(2,3)$, the Lie algebra of $G = SO(2,3)$, is given in Eq. (4.152). It is convenient to rescale and rename the generators as in Eq. (4.154), and the commutation relations take the form Eqs. (4.155).

The $M_{ab}$s generate the subalgebra $\mathfrak{h} = \mathfrak{so}(1,3)$ of the Lorentz subgroup. The orthogonal complement $\mathfrak{k}$ is generated by the $P_a$s and, on looking at the commutation relations Eqs. (4.155), we see that we have a symmetric pair.

Now we construct the coset representative $u$,

$$u(x) = e^{x^3 P_3} e^{x^2 P_2} e^{x^1 P_1} e^{x^0 P_0}, \tag{17.86}$$

and the Maurer–Cartan 1-form $V$,

$$\begin{aligned} V &= -u^{-1} du \\ &= -P_0 dx^0 - e^{-x^0 P_0} P_1 e^{x^0 P_0} dx^1 - e^{-x^1 P_1} e^{-x^0 P_0} P_2 e^{x^0 P_0} e^{x^1 P_1} dx^2 \\ &\quad - e^{-x^2 P_2} e^{-x^1 P_1} e^{-x^0 P_0} P_3 e^{x^0 P_0} e^{x^1 P_1} e^{x^2 P_2} dx^3. \end{aligned} \tag{17.87}$$

Using the definition of the adjoint action of the group on the algebra, we see that

$$e^{-x^0 P_0} P_1 e^{x^0 P_0} = T_I \Gamma_{\mathrm{Adj}}(e^{-x^0 P_0})^I{}_1, \tag{17.88}$$

etc., and, by projecting onto the horizontal subspace, we find the Vierbeins,

$$\begin{aligned} e^0 &= -dx^0, \qquad e^1 = -\cos x^0\, dx^1, \qquad e^2 = -\cos x^0 \cosh x^1\, dx^2, \\ e^3 &= -\cos x^0 \cosh x^1 \cosh x^2\, dx^2, \end{aligned} \tag{17.89}$$

which, with the Killing metric $(+---)$, give the following form of the $AdS_4$ metric:

$$ds^2 = (dx^0)^2 - \cos^2 x^0 \{(dx^1)^2 + \cosh^2 x^1 [(dx^2)^2 + \cosh^2 x^2 (dx^3)^2]\}. \tag{17.90}$$

The explicit form of the vertical 1-forms $\vartheta^{ab}$ is not necessary, but we need to know how they enter the spin connection. According to Eq. (A.117),

$$\omega^a{}_b = \tfrac{1}{2}\vartheta^{cd} f_{cd\,-1b}{}^{-1a} = \vartheta^{ac}\eta_{cb}. \tag{17.91}$$

The Killing spinor equation is

$$dx^\mu \hat{\mathcal{D}}_\mu \kappa = \left( d - \tfrac{1}{4}\omega_{ab}\gamma^{ab} - \frac{ig}{2}\gamma_a e^a \right)\kappa = 0, \tag{17.92}$$

and can be written in the form $(d - V)\kappa = 0$ with[14]

$$\Gamma_s(P_a) = \frac{ig}{2}\gamma_a, \qquad \Gamma_s(M_{ab}) = \tfrac{1}{2}\gamma_{ab}. \tag{17.93}$$

The Killing spinors are thus of the form Eqs. (17.38) and (17.39).

[14] Compare this with Eq. (D.157).

The dual generators $\Gamma_{\rm s}(P^a)$ can be defined by

$$\begin{aligned}\mathrm{Tr}\,[\Gamma_{\rm s}(P^a)\Gamma_{\rm s}(P_b)] &= \delta^a{}_b \ \Rightarrow \Gamma_{\rm s}(P^a) = -\frac{i}{2g}\gamma^a,\\ \mathrm{Tr}\,[\Gamma_{\rm s}(M^{ab})\Gamma_{\rm s}(P_{cd})] &= \delta^{ab}{}_{cd} \ \Rightarrow \Gamma_{\rm s}(M^{ab}) = -\tfrac{1}{2}\gamma^{ab}.\end{aligned} \tag{17.94}$$

We see that the matrix $\mathcal{S}$ is just the identity in this case. The bilinears give

$$\begin{aligned}-i\bar{\kappa}\gamma^a\kappa e_a &= 2g\Gamma_{\rm s}(u^{-1})^{\rm T}\mathcal{C}\Gamma_{\rm s}(P^a)\Gamma_{\rm s}(u^{-1})e_a\\ &= g\mathcal{C}\Gamma_{\rm s}(\hat{M}^{\hat{b}\hat{c}})\Gamma_{\rm Adj}(u^{-1})^a{}_{\hat{b}\hat{c}}e_a\\ &= g\mathcal{C}\Gamma_{\rm s}(\hat{M}^{\hat{b}\hat{c}})k_{\hat{b}\hat{c}},\end{aligned} \tag{17.95}$$

and we recover the standard anticommutator of the supercharges,

$$\{Q_\alpha, Q_\beta\} = g[\mathcal{C}\Gamma_{\rm s}(\hat{M}^{\hat{a}\hat{b}})]_{\alpha\beta}\hat{M}_{\hat{a}\hat{b}} = -i(\mathcal{C}\gamma^a)_{\alpha\beta}P_a - \frac{g}{2}(\mathcal{C}\gamma^{ab})_{\alpha\beta}M_{ab}. \tag{17.96}$$

The commutators $[Q_\alpha, \hat{M}_{\hat{a}\hat{b}}]$ are found using Eq. (17.40):

$$[Q_\alpha, \hat{M}_{\hat{a}\hat{b}}] = -Q_\beta\Gamma_{\rm s}(\hat{M}_{\hat{a}\hat{b}})^\beta{}_\alpha. \tag{17.97}$$

### 17.3.4 *The vacua of $N=2, d=4$ Poincaré supergravity*

The integrability condition Eq. (17.59) now gives two independent conditions for maximal supersymmetry: a vanishing Weyl tensor and a covariantly constant Maxwell field strength. Only three solutions satisfy them: Minkowski spacetime, the Robinson–Bertotti (RB) solution [205, 1077] given in Eq. (12.90), whose metric is that of the $\mathrm{AdS}_2 \times \mathrm{S}^2$ symmetric space, and the $d=4$ Kowalski–Glikman solution (KG4) [875] given in Eq. (14.27) with a Hpp-wave metric (again, a symmetric space, which we have constructed as a coset space in Section 14.1.1). The symmetry superalgebra of the Minkowski spacetime is identical to that of the $N=1$ case, with additional indices $i,j=1,2$ and, thus, we will focus on the RB and KG4 solutions since they are the simplest of a series of maximally supersymmetric solutions with metrics of the same form: $\mathrm{AdS}_m \times \mathrm{S}^n$ and Hpp whose symmetry superalgebras can be calculated in a very similar fashion [26]. The five- and six-dimensional cases will be discussed in Section 17.4 and the ten- and 11-dimensional cases in Section 25.5.1, but these four-dimensional examples already exhibit all the interesting features.

*The Robinson–Bertotti superalgebra.* The solution is given in Eq. (12.90), but we rewrite it in a more convenient form, adapted to the normalization we used for the Maxwell field of $N=2, d=4$ supergravity (a factor of two difference):

$$ds^2 = R_2^2\, d\Pi^2_{(2)} - R_2^2\, d\Omega^2_{(2)}, \qquad F = -R_2\,\omega_{\mathrm{AdS}_2}. \tag{17.98}$$

Here $d\Pi^2_{(2)}$ is the metric of the $\mathrm{AdS}_2$ spacetime of unit radius and $d\Omega^2_{(2)}$, is the metric of the 2-sphere of unit radius. That is why there are factors of $R_2^2$ in the metric. On the other hand, $\omega_{\mathrm{AdS}_2}$ is the volume 2-form of the $\mathrm{AdS}_2$ spacetime of unit radius.

Both $AdS_2$ and $S^2$ are symmetric spacetimes, $SO(2,1)/SO(2)$ and $SO(3)/SO(2)$, and their product is a symmetric space as well. We call the $SO(2,1)$ generators $\{T_I\}$, $I = 1,2,3$, with commutators given by Eq. (A.84) with $Q = \mathrm{diag}(++-)$, and the $SO(3)$ generators $\{\tilde{T}_I\}$, $I = 1,2,3$, with commutators given by Eq. (A.84) with $Q = \mathrm{diag}(+++)$.

The subalgebra $\mathfrak{h}$ is generated by $T_1$ and $\tilde{T}_3$ and $\mathfrak{k}$ is generated by $T_2, T_3, \tilde{T}_1$, and $\tilde{T}_2$. We perform the following redefinitions:

$$T_1 = M_1,\ T_2 = R_2 P_1,\ T_3 = R_2 P_0,\ \tilde{T}_1 = R_2 P_3,\ \tilde{T}_2 = R_2 P_2,\ \tilde{T}_3 = M_2. \tag{17.99}$$

The coset representative is the product of the coset representatives $u$ and $\tilde{u}$,

$$u = e^{R_2 \phi P_0} e^{R_2 \chi P_1}, \qquad \tilde{u} = e^{R_2 \varphi P_3} e^{R_2 (\theta - \frac{\pi}{2}) P_2}. \tag{17.100}$$

We obtain the Maurer–Cartan 1-forms,

$$\begin{aligned} e^0 &= -R_2 \cosh\chi\, d\phi, & e^2 &= -R_2 d\theta, & \vartheta^1 &= -\sinh\chi\, d\phi, \\ e^1 &= -R_2\, d\chi, & e^3 &= -R_2 \sin\theta d\varphi, & \vartheta^2 &= -\cos\theta\, d\varphi, \end{aligned}$$

and, with the Minkowski metric, we obtain the metric of Eq. (17.98) with

$$d\Pi^2_{(2)} \equiv \cosh^2\chi\, d\phi^2 - d\chi^2, \qquad d\Omega^2_{(2)} \equiv d\theta^2 + \sin^2\theta\, d\varphi^2. \tag{17.101}$$

Observe that we can construct not just the metric, but also the vector field of the solution, using the geometry. The RB solution Eq. (17.98) is purely electric. The gauge field is just

$$A = R_2 \vartheta^1. \tag{17.102}$$

We could also use the magnetic RB solution. The gauge field of that solution is

$$A = R_2 \vartheta^2. \tag{17.103}$$

We will work with the electric RB solution. The $N = 2, d = 4$ Killing spinor equation takes the form

$$dx^\mu \tilde{\mathcal{D}}_\mu \kappa = (d - \tfrac{1}{4}\omega_{ab}\gamma^{ab} + \tfrac{1}{4}\sigma^2 \not{F} \gamma_a e^a)\kappa = 0. \tag{17.104}$$

Equation (17.36) identifies the spin-connection term with the vertical components of the Maurer–Cartan 1-form $V$. The coefficients of the Vierbeins $e^a$ are the horizontal generators

$$\Gamma_s(P_a) = -\tfrac{1}{4}\sigma^2 \not{F} \gamma_a. \tag{17.105}$$

It can be checked that the representation explicitly given by

$$\begin{aligned} \Gamma_s(P_0) &= \frac{1}{2R_2}\gamma^1\sigma^2, & ;\Gamma_s(P_2) &= \frac{1}{2R_2}\gamma^0\gamma^1\gamma^3\sigma^2, & \Gamma_s(M_1) &= \tfrac{1}{2}\gamma^0\gamma^1, \\ & & ; & & & \\ \Gamma_s(P_1) &= -\frac{1}{2R_2}\gamma^0\sigma^2, & ;\Gamma_s(P_3) &= \frac{1}{2R_2}\gamma^0\gamma^1\gamma^2\sigma^2, & \Gamma_s(M_2) &= \tfrac{1}{2}\gamma^2\gamma^3, \end{aligned} \tag{17.106}$$

satisfies the algebra, and thus the Killing spinor equation takes the form $(d - V)\kappa = 0$ and the Killing spinors have the standard form (here $\kappa = \Gamma_s(u^{-1}\tilde{u}^{-1})\kappa_0$).

To calculate the bilinears $-i\bar{\kappa}\gamma^\mu\kappa$ we need the duals $\Gamma_s(P^a)$:

$$\gamma^a = -\frac{4}{R_2}\mathcal{S}\Gamma_s(P^a), \qquad \mathcal{S}=\gamma^0\gamma^1\sigma^2, \qquad \mathrm{Tr}[\Gamma_s(P^a)\Gamma_s(P_b)] = \delta^a{}_b. \tag{17.107}$$

The modified charge-conjugation matrix $\tilde{\mathcal{C}} = \mathcal{C}\mathcal{S}$ has the required property

$$\tilde{\mathcal{C}}^{-1}\Gamma_s(P^a)^{\mathrm{T}}\tilde{\mathcal{C}} = -\Gamma_s(P^a) \quad \Rightarrow \quad (u\tilde{u})^{-1\,\mathrm{T}}\tilde{\mathcal{C}} = \tilde{\mathcal{C}}u\tilde{u}, \tag{17.108}$$

which allows us to express the bilinears in the form

$$-i\bar{\kappa}_{\alpha i}\gamma^a\kappa_{\beta j}e_a = \frac{4i}{R_2}\left\{\tilde{\mathcal{C}}[\Gamma_s(T^A)k_A + \Gamma_s(\tilde{T}^A)\tilde{k}_A]\right\}_{\alpha i\,\beta j}, \tag{17.109}$$

where the $k_A$s are the Killing vectors of $\mathrm{AdS}_2$ and the $\tilde{k}_A$s are those of $\mathrm{S}^2$. This translates into the anticommutator

$$\{Q_{\alpha i}, Q_{\beta j}\} = -i\delta_{ij}(\mathcal{C}\gamma^a)_{\alpha\beta}P_a + \frac{i}{R_2}\mathcal{C}_{\alpha\beta}\epsilon_{ij}M_1 + \frac{1}{R_2}(\mathcal{C}\gamma_5)_{\alpha\beta}\epsilon_{ij}M_2, \tag{17.110}$$

which looks exactly like the $N=2, d=4$ Poincaré anticommutator with the two possible central charges. The difference is that now the $P_a$s commute neither with each other nor with the $M_i$s, which are no longer central.

The appearance of the two "central-charge" terms is a bit surprising. Actually, they are related to the invariance of the solution under the combined action of a gauge transformation with parameter $W_A{}^i$ and a reparametrization generated by $k_A$, due to the identification of the gauge field and the vertical 1-forms $\vartheta^i$ (see the discussion in Appendix A.4.1). This invariance does not commute with other reparametrizations and thus they no longer lead to central charges.

The "central-charge" terms can also be found by calculating the bilinears $-i\bar{\kappa}_{\alpha i}\sigma^2\kappa_{\beta j}$ and $-i\bar{\kappa}_{\alpha i}\sigma^2\gamma_5\kappa_{\beta j}$, using

$$\sigma^2 = \mathcal{S}\Gamma_s(M_1), \qquad \sigma^2\gamma_5 = \mathcal{S}\Gamma_s(M_2), \tag{17.111}$$

the definition of the adjoint action, and the expression of $W_I{}^i$, Eq. (A.114).

The commutators of the fermionic and bosonic generators are given by Eq. (17.40).

The RB solution induces spontaneous compactification on $\mathrm{S}^2$. The bosonic generators of the $\mathfrak{so}(2,1)$ subalgebra will have a spacetime interpretation, and the $\mathfrak{so}(3)$ ones will have the interpretation of internal symmetries, which is typical of extended AdS superalgebras. Since these act non-trivially on the supercharges, we would obtain a gauged supergravity with $\mathfrak{so}(3)$-charged gravitinos.

Let us consider now the contraction limit $R_2 \to \infty$. This contraction takes $\mathrm{AdS}_2$ into the two-dimensional Poincaré algebra $\mathrm{ISO}(1,1)$ ($M_1$ being the single Lorentz generator) and $\mathrm{SO}(3)$ into $\mathrm{ISO}(2)$ ($M_2$ being the generator of $\mathrm{SO}(2)$ rotations of $P_2$ and $P_3$, which now have the interpretation of two-dimensional central charges).

There is another contraction limit one can take: $R_2 \to \infty$, after redefining $M_1 \equiv R_2 Q$ and $M_2 \equiv R_2 P$. In this limit, all the bosonic generators again commute with each other and one recovers exactly the $N=2, d=4$ Poincaré superalgebra (without Lorentz generators, which can be understood here as $R$-parity generators) with $Q$ and $P$ as electric and magnetic central charges.

Generalizations of these facts will take place in other $\mathrm{AdS}_n \times \mathrm{S}^m$ solutions.

*The KG4 superalgebra.* The solution is given in Eq. (14.27), but we have to take into account the factor of two due to the normalizations being different, which changes $F_{\underline{u1}} = \frac{1}{2}\lambda$. The metric is that of a Cahen–Wallach symmetric space, whose coset construction is reviewed in Section 14.1.1, with $A_{ij} = -\frac{1}{8}\lambda^2\delta_{ij}$. There is one subtlety: in order to use the coset method, we are forced to use mostly the plus signature, in which the Killing spinor equation takes the form

$$\left(d - \tfrac{1}{4}\omega^a{}_b\gamma_a{}^b + \frac{i}{4}\sigma^2 \not{F}\gamma_a e^a\right)\kappa = 0. \tag{17.112}$$

As in the RB case, not just the metric, but also the Maxwell field, can be constructed using the vertical Maurer–Cartan 1-forms. In fact

$$A = \tfrac{1}{2}\lambda\vartheta^1. \tag{17.113}$$

An electric–magnetic-duality rotation is equivalent, in this solution, to a rotation in the wavefront plane. The new Maxwell field would be

$$A = \tfrac{1}{2}\lambda\vartheta^2. \tag{17.114}$$

On substituting the background fields into the above Killing spinor equation, we find that it takes the form $(d - V)\kappa = 0$, with the spinorial representation

$$\Gamma_s(P_a) = -\frac{i}{4}\lambda\sigma^2\gamma^u\gamma^1\gamma_a, \qquad \Gamma_s(M_i) = -\tfrac{1}{8}\lambda^2\gamma^u\gamma^i, \qquad \Gamma_s(V) = 0, \tag{17.115}$$

with $a = u, v, 1, 2$ ($\Gamma_s(P_v) = 0$) of the Heisenberg algebra:

$$[P_i, M_j] = \tfrac{1}{4}\lambda^2\delta_{ij}P_v, \qquad [P_u, P_i] = M_i, \qquad [P_u, M_i] = -\tfrac{1}{4}\lambda^2 P_i. \tag{17.116}$$

The Killing spinors can be constructed immediately, and the action of the Heisenberg-algebra generators on them is trivial to find, using Eq. (17.40). However, note the following.

1. H(6) is not the whole isometry algebra of the KG4 metric: SO(2) rotations in the wavefront plane leave the metric invariant and the full isometry group is their semidirect product $\mathrm{SO}(2) \ltimes \mathrm{H}(6)$ (because SO(2) acts on the H(6) generators). We would have to include SO(2) in the coset construction in order to obtain the commutator between the generator of SO(2) and the supercharges. However, SO(2) is not a symmetry of the full solution because it does not leave the field strength invariant.
2. Owing to the non-semisimplicity of the Heisenberg algebra, it is not possible to find a relation between $\gamma^a$ and the dual representation $\Gamma_s(P^a)$. Thus, the Killing spinor bilinears have to be calculated by brute force, but we will not do it here.

### 17.3.5 *The vacua of $N = 2, d = 4$ AdS supergravity*

The integrability condition now imposes a third constraint in order for the terms with zero, two, and four gammas to vanish: the Maxwell field-strength tensor has to vanish. Then, the only maximally supersymmetric solutions are those of $N = 1, d = 4$ AdS supergravity, i.e. $AdS_4$ spacetime, and a basis of Killing spinors will be provided by

$$\kappa_{\alpha i}{}^{\beta j} = \Gamma_s (u^{-1})^{\beta}{}_{\alpha} \delta^{j}{}_{j}. \tag{17.117}$$

The superalgebra will now include a bosonic generator associated with the constant-gauge transformations that rotate the spinor doublets and leave the vector field invariant. This generator must appear in the anticommutator of two supercharges, and the corresponding structure constants are given by the bilinears $-i\bar{\kappa}_{\alpha i}\sigma^2 \kappa_{\beta j} = \mathcal{C}_{\alpha\beta}\epsilon_{ij}$.

## 17.4 The vacua of $d = 5, 6$ supergravities with eight supercharges

To complete our overview of supergravity vacua in lower dimensions, we are going to study the vacua of the minimal supergravities in $d = 5$ and $d = 6$ dimensions that are related by dimensional reduction with $N = 2, d = 4$ Poincaré supergravity and have the same number of supercharges. Almost all the maximally supersymmetric vacua of these theories also happen to be related by dimensional reduction [905].

### *17.4.1 $N = (1, 0), d = 6$ supergravity*

The fields of this theory are the metric $\hat{\hat{e}}^{\hat{\hat{a}}}{}_{\hat{\hat{\mu}}}$, 2-form field $\hat{\hat{B}}^{-}_{\hat{\hat{\mu}}\hat{\hat{\nu}}}$ with anti-self-dual field strength $\hat{\hat{H}}^{-} = 3\partial\hat{\hat{B}}^{-}$ and positive-chirality symplectic-Majorana–Weyl gravitino $\hat{\hat{\psi}}^{+}_{\hat{\hat{\mu}}}$ [988] (we use the gamma matrices of Appendix D).

To write an action for an anti-self-dual 3-form, one has to introduce auxiliary fields. Alternatively, one can write an action for a generic 3-form,

$$\boxed{\hat{\hat{S}} = \int d^6\hat{\hat{x}}\sqrt{|\hat{\hat{g}}|}\left[\hat{\hat{R}} + \frac{1}{12}\hat{\hat{H}}^2\right],} \tag{17.118}$$

and impose the anti-self-duality constraint $\star\hat{H}^{-} = -\hat{H}^{-}$ on the equations of motion. The Killing spinor equation is

$$\left(\hat{\hat{\nabla}}_{\hat{\hat{a}}} - \frac{1}{48}\hat{\hat{\not{H}}}^{-}\hat{\hat{\gamma}}_{\hat{\hat{a}}}\right)\hat{\hat{\kappa}}^{+} = 0, \tag{17.119}$$

where $\hat{\hat{\kappa}}^{+}$ is a spinor of positive chirality.

Three maximally supersymmetric vacua of this theory are known:[15] Minkowski spacetime, the near-horizon limit of the extreme anti-self-dual string solution [617] that has an $AdS_3 \times S^3$ geometry, and the KG6 Hpp-wave solution [934]. The latter can be obtained by taking the Penrose limit of the former [1028, 700, 223].

[15] Actually, it has been shown in Ref. [333] that the maximally supersymmetric vacua of this theory and of $N = (2, 0), d = 6$ supergravity are in one-to-one correspondence, and their metrics are locally isometric to bi-invariant Lorentzian metrics of six-dimensional Lie groups with anti-self-dual parallelizing torsion. The three solutions that we present have this property and exhaust all the possibilities.

The $\mathrm{AdS}_3 \times \mathrm{S}^3$ solution can be written as follows:

$$d\hat{\hat{s}}^2 = R_3^2\, d\Pi^2_{(3)} - R_3^2\, d\Omega^2_{(3)}, \qquad \hat{\hat{H}}^- = 4R_3(\omega_{\mathrm{AdS}_3} + \omega_{\mathrm{S}^3}), \tag{17.120}$$

where $d\Pi^2_{(3)}$ is the metric of an $\mathrm{AdS}_3$ spacetime of unit radius and $\omega_{\mathrm{AdS}_3}$ is its volume 3-form. Remarkably, the electric and magnetic components of the 2-form potential can be written in terms of the vertical Maurer–Cartan 1-forms in $\mathrm{AdS}_3$ and $\mathrm{S}^3$, which are $\mathrm{SO}(2,1)$ and $\mathrm{SO}(3)$ Yang–Mills solutions (the $c_a$ are constants):

$$\hat{\hat{B}}^{-\,a} = c_{(a)}\epsilon^{(a)bc}\vartheta^i f_{ibc}. \tag{17.121}$$

The KG6 solution can be written as follows:

$$d\hat{s}^2 = 2du\left[dv + \frac{\lambda_6^2}{8}\vec{x}^{\,2}_{(4)}du\right] - d\vec{x}^{\,2}_{(4)}, \qquad \hat{B}^- = -\lambda_6 du \wedge (x^1dx^2 - x^3dx^4). \tag{17.122}$$

The potential can be also written as a Yang–Mills field:

$$\hat{\hat{B}}^{-\,i} \sim \vartheta^j f_j{}^{iu}. \tag{17.123}$$

The calculation of the Killing spinors and superalgebras can be done following the general method. It is, however, more interesting to see how the dimensional reduction of these solutions gives all the maximally supersymmetric vacua of $N=1, d=5$ supergravity.

### *17.4.2 $N=1, d=5$ supergravity*

The dimensional reduction of $N=(1,0), d=6$ supergravity gives $N=1, d=5$ supergravity (Section 15.2.5) coupled to a vector multiplet. We are interested in reducing maximally supersymmetric six-dimensional solutions preserving all their unbroken supersymmetries. When is this possible?

Let us consider the component of the gravitino in the compact direction $w$ (which gives rise to a five-dimensional spin $-\frac{1}{2}$ "gaugino") and its supersymmetry variation,

$$\delta_{\hat{\epsilon}}\hat{\hat{\psi}}^+_{\underline{w}} \sim (\partial_{\underline{w}} + M)\hat{\epsilon}^+, \tag{17.124}$$

where $M$ is a combination of gamma matrices. This equation has to vanish identically for a $w$-independent Killing spinor $\hat{\kappa}^+$ in order to have five-dimensional unbroken supersymmetry, which implies that $M$ has to vanish identically. If we reduce a maximally supersymmetric six-dimensional solution and $M$ does not vanish identically, then, since the above equation vanishes for some Killing vectors, they must depend on $w$, and five-dimensional supersymmetry is broken. The amount of supersymmetry broken depends on the rank of $M$, which tells us how many non-trivial solutions of $M\hat{\hat{\kappa}}^+ = 0$ exist and how many six-dimensional Killing spinors are independent of $w$.

Up to this point, this discussion carries over to any other case without modification. However, there are two different possibilities concerning the vanishing of $M$: in the present case, we obtain a five-dimensional reducible theory of which we can always truncate consistently the matter multiplet. Matter fields in a given multiplet are identified in the supersymmetry transformation rules because they transform among themselves. Now, $\hat{\hat{\psi}}^{+}_{\underline{w}}$ corresponds to a matter spinor and therefore $M$ consists basically of matter fields whose truncation sets $M = 0$. This is indeed possible because all the terms in $M$ contain two gammas, and a combination of them can always be set to zero. This truncation gives the pure $N = 1, d = 5$ supergravity action Eq. (15.98) and the supersymmetry transformation rule of the five-dimensional gravitino [905]:

$$\delta_{\hat{\epsilon}}\hat{\psi}_{\hat{a}} = \left\{\hat{\nabla}_{\hat{a}} - \frac{1}{8\sqrt{3}}(\hat{\gamma}^{\hat{b}\hat{c}}\hat{\gamma}_{\hat{a}} + 2\hat{\gamma}^{\hat{b}}\hat{g}^{\hat{c}}{}_{\hat{a}})\hat{G}_{\hat{b}\hat{c}}\right\}\hat{\epsilon}. \tag{17.125}$$

The relation between the six- and five-dimensional fields is

$$\hat{\hat{g}}_{\underline{ww}} = -1, \quad \hat{\hat{B}}^{-}_{\hat{\mu}\underline{w}} = \frac{1}{\sqrt{3}}\hat{V}_{\hat{\mu}}, \quad \hat{\hat{g}}_{\hat{\mu}\underline{w}} = \frac{1}{\sqrt{3}}\hat{V}_{\hat{\mu}}, \quad \hat{\hat{g}}_{\hat{\mu}\hat{\nu}} = \hat{g}_{\hat{\mu}\hat{\nu}} - \tfrac{1}{3}\hat{V}_{\hat{\mu}}\hat{V}_{\hat{\nu}}, \tag{17.126}$$

with the $\hat{B}^{-}_{\mu\nu}$ components determined by anti-self-duality.

The $AdS_3 \times S^3$ solution can be reduced preserving all the supersymmetry in two directions: the direction of the Hopf fiber of the $S^3$ when we see it as a fibration over $S^2$ (i.e. the coordinate $\psi$ in the Euler-angle parametrization Eq. (A.97)) and the analog in $AdS^3$ when we see it as a fibration over $AdS_2$, i.e. the coordinate $\eta$ in

$$d\Pi^2_{(3)} \equiv \tfrac{1}{4}\left[d\Pi^2_{(2)} - (d\eta + \sinh(\chi/2)d\phi)^2\right]. \tag{17.127}$$

Actually, it is also possible to perform a dimensional reduction in a combination of the two directions: on rotating by an angle $\xi$,

$$w = \frac{R_3}{2}(\cos\xi\,\eta + \sin\xi\,\psi), \qquad y = -\sin\xi\,\eta + \cos\xi\,\psi, \tag{17.128}$$

and reducing in the direction $w$ we obtain the two-parameter family [398]

$$\boxed{\begin{aligned} d\hat{s}^2 &= R_2^2 d\Pi^2_{(2)} - R_2^2 d\Omega^2_{(2)} - R_2^2(dy + \cos\xi\cos\theta\,d\varphi - \sin\xi\sinh\chi\,d\phi)^2, \\ \hat{G} &= \sqrt{3}R_2(\cos\xi\,\omega_{AdS_2} - \sin\xi\,\omega_{S^2}), \qquad R_2 = R_3/2. \end{aligned}} \tag{17.129}$$

It is maximally supersymmetric for any $\xi$ and can be obtained as the near-horizon limit of the $d = 5$ rotating extreme BH [249, 252, 598],[16] which, as usual, has only half of the maximal supersymmetry [846]. Note that $\sin\xi$ plays the role of the rotation parameter $\jmath < 1$ of Ref. [1197], and it is no surprise that, when $\xi = 0$, we recover the near-horizon limit of

[16] The BMPV and the near-horizon limit that gives the above metric is described on p. 603.

the static $d=5$ extreme BH, which has the geometry of $AdS_2 \times S^3$ $(y=\psi)$ [332]. It is a bit more surprising that for $\xi=\pi/2$ we recover the near-horizon limit of the $d=5$ extreme string solution [617]. Although many computations can be performed for arbitrary $\xi$, and then we can take the limits $\xi \to 0, \pi/2$, it is clear that this is yet another example of a family of solutions that seems to depend continuously on a parameter, while the physical properties (the symmetry superalgebras, for instance) do not.

The above solution admits a description as a homogeneous, reductive but non-symmetric spacetime that can be used to compute its symmetry superalgebra [27].

Before the KG6 solution can be reduced, preserving all its supersymmetries, we need to identify isometric directions satisfying the truncation condition. One of the directions is found after performing a $u$-dependent rotation,

$$x^3 = \cos\left(\frac{\lambda_6}{2}u\right)x^{3\prime} + \sin\left(\frac{\lambda_6}{2}u\right)x^{4\prime}, \; x^4 = -\sin\left(\frac{\lambda_6}{2}u\right)x^{3\prime} + \cos\left(\frac{\lambda_6}{2}u\right)x^{4\prime},$$

$$v = v' + \frac{\lambda_6}{2}x^{3\prime}x^{4\prime},$$

which leaves the solution in the form

$$\begin{aligned} d\hat{\hat{s}}^2 &= 2du\left[dv' + \frac{\lambda_6^2}{8}(x^2+y^2)du + \lambda_6 x^{3\prime}dx^{4\prime}\right] - d\vec{x}'^2_{(4)}, \\ \hat{\hat{B}}^- &= -\lambda_6 du \wedge (x^1dx^2 - x^{3\prime}dx^{4\prime}). \end{aligned} \tag{17.130}$$

On reducing now in the isometric coordinate $w \equiv x^{4\prime}$, we obtain the KG5 solution Eq. (14.31) with $\lambda_5 = -\sqrt{3}\lambda_6$. A similar rotation in $x^1$ and $x^2$ produces the same result. The isometric direction $w=(1/\sqrt{2})(u+v)$ can also be used. If we perform the two rotations and reduce in the direction $w$, we obtain a $d=5$ maximally supersymmetric Gödel-like solution [596]:

$$d\hat{s}^2 = (dt+\omega)^2 - d\vec{x}^2_{(4)}, \qquad \hat{V} = -\sqrt{3}\,\omega, \qquad \omega = \tfrac{1}{\sqrt{2}}\lambda_6(x^1dx^2 - x^3dx^4). \tag{17.131}$$

The relations among all the known vacua are represented in Fig. 17.1.

### 17.4.3 Relation to the $N=2, d=4$ vacua

The dimensional reduction of $N=1, d=5$ supergravity gives $N=2, d=4$ supergravity coupled to a vector multiplet that can be consistently truncated as shown in Section 15.2.5. The discussion at the beginning of Section 17.4.2 applies to this situation: maximal supersymmetry survives the dimensional reduction only if no matter fields are generated.

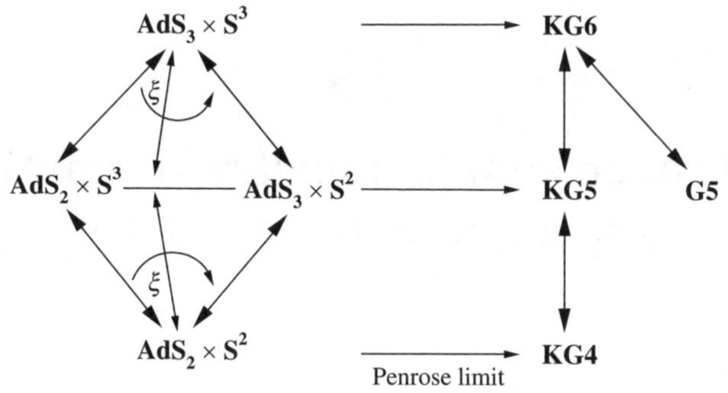


Fig. 17.1. Relations between the non-trivial $d = 4, 5, 6$ vacua with eight supercharges.

This condition cannot be satisfied for the Gödel-like solution Eq. (17.131), and only the KG5 Eq. (14.31) and the near-horizon limit of the rotating BH Eq. (17.129) and string give maximally supersymmetric four-dimensional solutions: the KG4 solution Eq. (14.27), with $\lambda_4 = (2/\sqrt{3})\lambda_5$, and the dyonic RB solution, in which the rotation parameter $\sin\xi$ now plays the role of the electric–magnetic-duality rotation parameter:

$$ds^2 = R_2^2\, d\Pi_{(2)}^2 - R_2^2\, d\Omega_{(2)}^2, \qquad F = -R_2(\cos\xi\, \omega_{\mathrm{AdS}_2} - \sin\xi\, \omega_{\mathrm{S}^2}). \tag{17.132}$$

It is amusing to observe that a spatial rotation in $d = 6$ has become an electric–magnetic-duality rotation in $d = 4$ [472]. This is one of the phenomena we will study in Part III.

# 18

# Unbroken supersymmetry II: partially supersymmetric solutions

Now that we have studied maximally supersymmetric vacua, we are ready to study solutions with fewer unbroken supersymmetries. We are particularly interested in solutions that tend asymptotically to a maximally supersymmetric solution because we can assign to them conserved charges and supercharges: as explained in Section 10.1.2, spacetimes (supersymmetric or not) with these asymptotics have well-defined conserved bosonic charges associated with the isometries of the vacua they approach asymptotically, but it is also possible to assign to them supercharges using the same formalism [1]. Then, it is possible (at least formally) to associate with these solutions states in the quantum theory transforming under the vacuum supersymmetry algebra, with well-defined quantum numbers associated with all the generators. The supersymmetry algebra satisfied by the unbroken bosonic symmetries and supersymmetries of the vacuum can then be used to find relations that have to be obeyed by the charges and supercharges of the solutions with partially unbroken supersymmetry and that imply bounds for certain functions of the charges of non-supersymmetric solutions under certain assumptions.

We will study these properties in Section 18.1, considering the solutions with partially unbroken supersymmetry not as vacua but as excitations of some vacuum to which they tend asymptotically. By associating states in a quantum theory with these solutions and using the vacuum superalgebra, general supersymmetry bounds for the mass can be derived. These bounds are saturated by (supersymmetric or "BPS") states with partially unbroken supersymmetry. The bounds can be extended to solutions of the theory, even in the absence of supersymmetry, if certain conditions on the energy–momentum tensor are imposed. These are very powerful techniques.

For these and other reasons, the solutions with partially unbroken supersymmetry (to which we will refer as *supersymmetric* or *BPS* in what follows) are very interesting, which raises the question of how to find them. One can always test the supersymmetry of a solution by checking whether or not it admits Killing spinors but it is desirable to have more general characterizations or classifications. Fortunately, the requirement that a field configuration (not necessarily a solution of the classical equations of motion) is so powerful that in all the cases studied so far the form of the field configuration is completely fixed (up to reparametrizations etc.), with all the field components depending on a very reduced number of independent functions. When these supersymmetric field configurations are plugged into the classical equations of motion one finds that they reduce to a very small number of equations for those independent functions. On solving these equations (which in some cases are very simple) one gets supersymmetric solutions to the very complicated classical equations of motion of the original supergravity theory, which is, by itself, very interesting.

There is a systematic way of finding all the supersymmetric field configurations of a supergravity theory, pioneered by Tod in Ref. [1183], which we have named, accordingly, *Tod's program* and which we are going to review in Section 18.2 in a formulation due to Gauntlett [599, 596] supplemented by the use of the *Killing spinor identities* (KSIs) [843, 130]. Tod's program is based on the exploitation of all the integrability and consistency conditions implied or required by the existence of (at least) one solution to the Killing spinor equations (KSEs) and originally made use of the Newman–Penrose formalism [1032], only valid in four dimensions. As proposed in Refs. [599, 596], these consistency and integrability conditions can also be expressed in terms of tensors constructed as bilinears of the Killing spinor (KS) whose existence has been assumed. On the other hand, the integrability conditions that relate the (left-hand sides of the off-shell) equations of motion can be derived in a very simple way using the KSIs, as shown in Refs. [843, 130]. This greatly reduces the number of independent equations of motion that need to be considered. Tod's program will be the subject of Section 18.2.

## 18.1 Partially supersymmetric solutions

By assumption, the partially supersymmetric solutions that we are going to consider in this section are invariant under the supersymmetry transformation generated by a spinor $\epsilon^{\alpha}$ that asymptotically approaches a linear combination of the vacuum KSs $\kappa_I{}^{\alpha}$:

$$\epsilon^{\alpha} \rightarrow \epsilon^{I} \kappa_I{}^{\alpha}, \tag{18.1}$$

where the $\epsilon^I$ are (commuting) constants. Then, the (BPS) state $|s\rangle$ associated with this solution will be annihilated by the supercharge $\epsilon^I Q_I$:

$$\delta_{\epsilon}|s\rangle \sim \epsilon^I Q_I|s\rangle = 0. \tag{18.2}$$

The game now will be to use results about the states obtained through the study of the superalgebra to predict properties of the associated solutions.

The superalgebra is a very powerful tool. It is a fact that, in all cases, one can construct a manifestly positive quadratic combination of the supercharges that equals a combination of the bosonic charges including the energy,[1] which turns out to bound it from below.

These *supersymmetry* or *Bogomol'nyi (B) bounds*[2] have very important consequences, for instance the positivity of the energy itself and the stability of the theory and of the BPS states that saturate the bound since these have the minimal possible energy in the given

[1] This has been studied mostly in Poincaré superalgebras that can be applied to asymptotically flat spacetimes. There are fewer general results for AdS superalgebras and none for KG Hpp-wave superalgebras.

[2] These bounds were first discovered by Bogomol'nyi [227] in a purely bosonic context (see Section 13.2.3) and then by Witten and Olive [1276] as a consequence of supersymmetry.

charge sector.[3] When the bound is saturated, the state is annihilated by some combination of supercharges and it is supersymmetric. In the supergravity theory general bounds on the energy and other charges of solutions can be established under mild assumptions using *Witten–Nester–Israel (WNI) constructions* [1286, 1002, 846, 642] (see also Ref. [825]), the simplest of which was used in Section 10.3 to prove the positive-energy theorem. The solutions associated with supersymmetric states (with the same charges/quantum numbers) are also supersymmetric (they admit KSs) and also have special stability properties.

Our goal in this section is to study these B bounds in Minkowski spacetimes and their consequences for asymptotically flat supergravity solutions.

### *18.1.1 Partially unbroken supersymmetry, supersymmetry bounds, and the superalgebra*

Let us consider a supersymmetric state $|s\rangle$ in the quantum theory associated with the $N$-extended $d=4$ Poincaré superalgebra, including central charges. The state describes a quantum object in Minkowski spacetime and will be annihilated by a combination of the Minkowski (Poincaré) supercharges (the Minkowski vacuum is annihilated by all of them):[4]

$$\delta_\epsilon |s\rangle \sim \bar{\epsilon}^{i}_{\alpha} Q^{i\,\alpha} |s\rangle = 0. \tag{18.3}$$

The anticommutator of this supercharge with itself gives (if $|s\rangle$ is normalizable)

$$\bar{\epsilon}\mathfrak{M}\epsilon = 0, \tag{18.4}$$

where, using Eq. (5.72), the matrix $\mathfrak{M}$ is given by the expression

$$\mathfrak{M} \equiv i\delta^{ij}\gamma^a P_a + iZ^{[ij]} + \gamma_5 \tilde{Z}^{[ij]} + \gamma^a Z^{(ij)}_a + i\gamma_5\gamma^a Z^{[ij]}_a + i\gamma^{ab} Z^{(ij)}_{ab} + \gamma_5\gamma^{ab}\tilde{Z}^{(ij)}_{ab}, \tag{18.5}$$

in which we have replaced the operators by their values on $|s\rangle$. This equation has $4N - \mathrm{rank}(\mathfrak{M})$ independent solutions (preserved supersymmetries) if $\det \mathfrak{M} = 0$. Finding all the values of $P_a, Z^{ij}, \tilde{Z}^{ij}, \ldots$ for which $\mathfrak{M}$ is singular would be exceedingly difficult. However, there are simple solutions with a simple physical interpretation.

Let us consider first a state describing a point-like, massless, uncharged (all $Z=0$) particle. In the reference system in which it moves in direction 1 its null momentum takes the form $(P^\mu) = (p, \pm p, 0, \ldots, 0)$ and

$$\mathfrak{M} = ip\delta^{ij}\gamma^0(1 \pm \gamma^0\gamma^1). \tag{18.6}$$

$\mathfrak{M}$ is singular: $(\gamma^0\gamma^1)^2 = 1$ and $\mathrm{Tr}(\gamma^0\gamma^1) = 0$ imply that $\frac{1}{2}(1 \pm \gamma^0\gamma^1)$ is idempotent (and, therefore, a projector) with trace equal to 2, and, thus, with eigenvalues $+1$ with multiplicity

[3] Furthermore, in the Poincaré case, there are no fields that carry that charge and the BPS state cannot perturbatively lose that charge by emitting charged particles.

[4] Here we are going to focus on supersymmetric asymptotically flat solutions associated to states in Minkowski spacetime. The supersymmetric bound for asymptotically $AdS_4$ black-hole and black-string solutions has been studied in Refs. [790, 789, 791].

2 and 0 with multiplicity 2. Half of the supersymmetries of this state will be preserved, precisely those generated by $\bar{\epsilon}^i_\alpha Q^{i\,\alpha}$, where $\gamma^{\pm}\epsilon = \frac{1}{2}(\gamma^0 \pm \gamma^1)\epsilon = 0$. This is precisely the constraint satisfied by the KSs of generic pp-wave solutions, some of which can be associated with the gravitational field of massless point-particles (see Section 14.3). All supergravity theories admit these supersymmetric states and solutions.

The next simplest case is that of a massive point-particle with mass $M$ and electric charge $Q$ (massive neutral particle states cannot be supersymmetric) in pure $N=2, d=4$ Poincaré SUEGRA ($Z^{[ij]} = Q\epsilon^{ij}$). In the particle's rest frame $(P^\mu) = (M, 0, \ldots, 0)$ and

$$\mathfrak{M} = i\gamma^0 M\big[\delta^{ij} + (Q/M)\gamma^0\epsilon^{ij}\big]. \tag{18.7}$$

The $8\times 8$ matrix $\mathfrak{M}$ is singular if and only if $Q = \pm M$, in which case it is proportional to another projector: $\frac{1}{2}(\delta^{ij} \pm \gamma^0\epsilon^{ij})$. The unbroken supersymmetries are generated by the supercharges $\bar{\epsilon}^i Q^i$, where $\epsilon^i$ satisfies the constraint

$$\tfrac{1}{2}(\delta^{ij} \pm \gamma^0\epsilon^{ij})\epsilon^j = 0. \tag{18.8}$$

The rank of this projector (and $\mathfrak{M}$) is 4: half of the supersymmetries are unbroken.

Since pure $N=2, d=4$ Poincaré SUEGRA is invariant under electric–magnetic duality transformations combined with chiral transformations of the spinors (the *chiral–dual* transformations in Eqs. (5.90)), we expect similar results for massive dyons: if the state has magnetic charge, then there is another central charge $\tilde{Z}^{[ij]} = P\epsilon^{ij}$ and, if $P^2 + Q^2 = M^2$,

$$[\delta^{ij} \pm (\cos\xi + i\sin\xi\,\gamma_5)\gamma^0\epsilon^{ij}] = [\delta^{ij} \pm e^{i\xi\gamma_5}\gamma^0\epsilon^{ij}], \tag{18.9}$$

with

$$e^{i\xi} = (Q + iP)/M \equiv \mathcal{Z}/M. \tag{18.10}$$

Sometimes it is convenient to use Weyl instead of Majorana spinors in these analyses. The change of basis is given in Eqs. (6.1), and in the new basis the matrix $\mathfrak{M}$ takes the form

$$\mathfrak{M} = i\delta^{ij}\gamma^a P_a + i\mathcal{Z}^{IJ} + \gamma^a \mathcal{Z}^{IJ}_a + i\gamma^{ab}\mathcal{Z}^{IJ}_{ab}, \tag{18.11}$$

where we have defined the new complex central charges

$$\mathcal{Z}^{IJ} = Z^{ij} + i\tilde{Z}^{ij}, \qquad \mathcal{Z}^{IJ}_a = Z^{ij}_a - i\tilde{Z}^{ij}_a, \qquad \mathcal{Z}^{IJ}_{ab} = Z^{ij}_{ab} + i\tilde{Z}^{ij}_{ab}. \tag{18.12}$$

In the pure $N=2, d=4$ SUEGRA case that we are considering

$$\mathcal{Z}^{IJ} \equiv \mathcal{Z}\epsilon^{IJ}, \qquad \mathcal{Z} = Q + iP. \tag{18.13}$$

Which solution of $N=2, d=4$ SUEGRA could we associate this supersymmetric state with? It should be a static, spherically symmetric solution (the closest to a point-like object we can have) with charge $M = |Q+iP| = 2|q+ip|$ with the conventions of Chapter 12: an extreme, dyonic charged Reissner–Nordström (RN) BH. We will see that indeed the extreme RN (ERN) BH admits KSs that satisfy the above constraint.

In more general $N=2, d=4$ theories with vector supermultiplets (see Chapter 7) the central charge $\mathcal{Z}$ is a complex combination of the graviphoton's electric and magnetic charges (the "electric charge" of $T^+$), and, therefore, a combination of the electric $q_\Lambda$ and magnetic $p^\Lambda$ charges of all the vectors $A^\Lambda{}_\mu$ and the values of the complex scalars $Z^i$ at infinity $Z^i_\infty$, although, often, the spacetime-dependent form of the scalars is used. The definition is symplectic invariant (up to a phase) so $|\mathcal{Z}|$ is always duality invariant, and it is given by

$$\boxed{\mathcal{Z}(\mathcal{Q}, Z, Z^*) \equiv \mathcal{V}_M \mathcal{Q}^M,} \tag{18.14}$$

where $\mathcal{V}^M$ is the canonical covariantly holomorphic symplectic section defined in Appendix F.1, $(\mathcal{Q}^M) \equiv \binom{p^\Lambda}{q_\Lambda}$ is the symplectic charge vector, and therefore $\mathcal{V}_M \mathcal{Q}^M = p^\Lambda \mathcal{M}_\Lambda - q_\Lambda \mathcal{L}^\Lambda$.

The states of these theories with $M = |\mathcal{Z}_\infty| = |\mathcal{Z}(\mathcal{Q}, Z_\infty, Z^*_\infty)|$, preserving half of the possible supersymmetries of the theory, are associated with solutions of the SUEGRA theory that describe static, asymptotically flat BHs with the same charges $\mathcal{Q}^M$ and with scalars with the same asymptotic values $Z^i_\infty$. These solutions can be constructed systematically, as we will show in Section 19.2 using the characterization of all the supersymmetric solutions of these theories that we will obtain in Section 18.4. These, and analogous extremal solutions which do not have supersymmetry properties, will be the subject of Chapter 27 as well.

What happens in the general case? Let us focus first on the $d=4$ $N$-extended case with strictly central charges and let us use again a (complex) Weyl representation in which the electric and magnetic central-charge matrices are combined into a single, complex, antisymmetric matrix $\mathcal{Z}_{IJ}$ [538]. As in the $N=2$ case, this matrix, which can be understood as the matrix of the electric and magnetic charges of the $N(N-1)/2$ graviphotons of the theory, is, actually, a function of the electric and magnetic charges of all the vectors of the theory (graviphotons and matter vector fields alike) contained in $(\mathcal{Q}^M) \equiv \binom{p^\Lambda}{q_\Lambda}$ and of the asymptotic values of the scalars. Using the generic scalar-dependent symplectic vectors $\mathcal{V}_{IJ}(\phi)$ defined in Section 8.1, we can write

$$\boxed{\mathcal{Z}_{IJ}(\mathcal{Q}, \phi) \equiv \mathcal{Q}^M \mathcal{V}_{M\,IJ}.} \tag{18.15}$$

This matrix has $[N/2]$ skew eigenvalues $\mathcal{Z}_a$ which also depend on the charges and moduli. For massive states, in the rest frame, $\mathfrak{M}$ is singular whenever the absolute value of one of these skew eigenvalues is equal to the mass, $|\mathcal{Z}_a| = M$. The amount of supersymmetry preserved depends on the number of skew eigenvalues whose absolute values are equal to $M$. When all of them are equal to $M \neq 0$, half of the supersymmetries will be unbroken. Let us see what happens in the most interesting cases $N = 1, 2, 4, 8, d = 4$.[5]

[5] It should be stressed that this discussion applies to massive states in Minkowski spacetime or to asymptotically flat solutions. There are supersymmetric solutions with different amounts of unbroken supersymmetry (for instance, a quarter in $N=2$), but they are not asymptotically flat as they involve non-trivial hyperscalar fields [795, 943].

**N = 1.** There are no massive supersymmetric states.

**N = 2.** Half of the supersymmetries are unbroken when[6]

$$M = |\mathcal{Z}|.$$

**N = 4.** Half of the supersymmetries are preserved when

$$M = |\mathcal{Z}_1| = |\mathcal{Z}_2|,$$

and a quarter are preserved when

$$M = |\mathcal{Z}_1| \neq |\mathcal{Z}_2|.$$

**N = 8.** Half of the supersymmetries are unbroken when

$$M = |\mathcal{Z}_1| = |\mathcal{Z}_2| = |\mathcal{Z}_3| = |\mathcal{Z}_4|,$$

a quarter when

$$M = |\mathcal{Z}_1| = |\mathcal{Z}_2| \neq |\mathcal{Z}_{3,4}|,$$

and an eighth when

$$M = |\mathcal{Z}_1| \neq |\mathcal{Z}_{2,3,4}|.$$

In all cases, the projectors on the KSs are of the form $\delta^{IJ} + \gamma^0 \alpha^{IJ}$, where the complex matrix $\alpha^{IJ}$ depends on the SUEGRA theory and the specific solution we are considering: different combinations of charges can lead to the same $|\mathcal{Z}_a|$s. These combinations are usually related by dualities of the theory (e.g. electric–magnetic duality in pure $N = 2, d = 4$ SUEGRA, which multiplies $\mathcal{Z}$ by a complex phase). There is one projector for each $\frac{1}{2}$ factor of broken supersymmetry and all the projectors must commute among themselves in order for them to be compatible. In all cases, the associated solutions are generalizations of the ERN BH like those of Section 16.2.

Just as we have defined $\mathcal{Z}_{IJ}$ to describe the electric and magnetic charges of the graviphotons, one can define charges (*matter central charges*) $\mathcal{Z}_i$, $i = 1, \cdots, n_V$, that describe the electric and magnetic charges of the matter vector fields of the theory (for $N \geq 2$) and are given as linear combinations of the electric and magnetic charges of all the $\bar{n}$ vector fields of the theory with coefficients that depend on the scalar fields:

$$\boxed{\mathcal{Z}_i(\mathcal{Q}, \phi) \equiv \mathcal{Q}^M \mathcal{V}_{M\,i},} \tag{18.16}$$

where $\mathcal{V}^M{}_i$ is the scalar-dependent symplectic vector defined in Section 8.1. The completeness relation Eq. (I.15) leads to the important identity

$$\boxed{\tfrac{1}{2}\mathcal{Z}_{IJ}\mathcal{Z}^{*\,IJ} + \mathcal{Z}_i \mathcal{Z}^{*\,i} = -\tfrac{1}{2}\mathcal{M}_{MN}(\mathcal{N})\mathcal{Q}^M \mathcal{Q}^N \equiv -V_{\mathrm{bh}}(\phi, \mathcal{Q}),} \tag{18.17}$$

[6] In $N = 2$ the only skew eigenvalue of $\mathcal{Z}_{IJ} = \mathcal{Z}\varepsilon_{IJ}$ is, evidently, $\mathcal{Z}$.

where $\mathcal{M}(\mathcal{N})$ is the symmetric symplectic matrix defined in Eq. (C.42). This object is known as the *black-hole potential*[7] of the $N$-extended supergravity theory [522] and plays a very important role in the physics of static extremal BHs.

Let us now consider the case in which there is one (real) quasi-central charge $Z^{(p)}_{a_1\cdots a_p}$. In the appropriate reference frame it is always possible to write

$$\mathfrak{M} = i\gamma^0 M\left(\delta^{ij} + (Z^{(p)}/M)\gamma^0\gamma^1\cdots\gamma^p\alpha^{ij}\right), \tag{18.18}$$

which is singular when $M = Z^{(p)}$. The projector is not one of those associated with point-particles. Actually, quasi-central charges with $p$ Lorentz indices are associated in the corresponding SUEGRA theory with $(p+1)$-form potentials[8] $A^{(p+1)}{}_{\mu_1\cdots\mu_{p+1}}$. As we will see in Chapter 24, these potentials couple naturally to the $(p+1)$-dimensional worldvolumes of $p$-dimensional extended objects (*p-branes*) just as vector fields couple to the worldlines of point-particles. Thus, the projector and the supersymmetric state correspond to a half-supersymmetric $p$-brane. We will study the SUEGRA solutions in Chapter 25.

In a more general case, there can be several non-vanishing quasi-central charges. The corresponding state is then interpreted as being composed of several supersymmetric $p$-branes that intersect (see Section 25.6), generically one for each projector. This interpretation, which can also be applied to supersymmetric point-particle states (Chapter 26), is based on the observed property that supersymmetric objects can be in equilibrium: there is a cancelation between gravitational attraction and other interactions and this allows the existence of solutions that describe several of these supersymmetric objects in equilibrium. The simplest example is provided by the Majumdar–Papapetrou solutions that describe several ERN BHs in equilibrium.

$\mathfrak{M}$ has another important property: it is roughly the square of the supercharges and, thus, we can expect its eigenvalues to have some positivity properties. Indeed, it can be shown that the Hamiltonian of supersymmetric QFTs is non-negative [445] or, equivalently, that $M \geq 0$. In four-dimensional theories with extended supersymmetry one can go even further and prove that the mass is bounded from below by the skew eigenvalues of the central-charge matrix $\mathcal{Z}^{IJ}$ [1276, 538],

$$M \geq |\mathcal{Z}_a|, \qquad \forall a = 1, \ldots, [N/2], \tag{18.19}$$

and the results can be generalized to other dimensions. These bounds are known as *Bogomol'nyi*, *BPS*, or *supersymmetry bounds* and play a crucial role in the stability of states and theories. Broken supersymmetry is restored when one of these bounds is saturated, as we have seen. Supersymmetric ("BPS") states then have the minimal masses allowed for given values of the central charges. The central charges cannot change because there are no (perturbative) states in the theory that carry them and, then, the masses of BPS states cannot diminish and the states cannot decay and are stable.

As for the associated BPS solutions and SUEGRA theories, some stability properties can also be proven, such as the positive-energy theorem of GR the proof of which, inspired

[7] Here we are using the convention that we will use in Chapter 27, opposite to that of much of the literature.

[8] One can view them as the components of the gauge superpotential associated with the quasi-central charges: $A^{(p+1)}{}_{\mu}{}^{a_1\cdots a_p} Z^{(p)}_{a_1\cdots a_p}$ if we think of SUGRA theories as gauge theories of given superalgebras, as we did in Chapter 5.

by $N=1, d=4$ SUGRA, we gave in Section 10.3. The relation between positivity of the energy and the supersymmetry algebra was studied in Ref. [801]. There are generalizations based on the WNI construction (see e.g. Ref. [618] for $N=2, d=4$). The stability of solutions manifests itself, most remarkably, in the absence of Hawking radiation ($T=0$) from ERN and other supersymmetric BHs. This establishes an interesting link between BH thermodynamics and supersymmetry (see e.g. [1008]), which we use in Chapter 26.

We are going to review the (not maximally) supersymmetric solutions of the simplest theories: pure $N=1,2,4, d=4$ SUGRA, and their relation (for asymptotically flat and asymptotically Taub–NUT solutions) to the supersymmetry bounds one can derive from the superalgebras. In Section 18.2 we will review results on the characterization and classification of supersymmetric solutions of more general supergravities.

### 18.1.2 Examples

*Pure $N=1, d=4$ Poincaré supergravity.* The only solutions with partially unbroken supersymmetry in this theory are the purely gravitational pp-waves given by Eqs. (14.24) with $C=0$. The KSE $\nabla_\mu \kappa = 0$ is solved for any constant spinor $\kappa$ satisfying the constraint $\gamma^u \kappa = 0$, which is precisely what we expected from the superalgebra. The absence of massive supersymmetric solutions can be understood as a consequence of the $N=1$ supersymmetry bound $M \geq 0$. This is in agreement with the finite temperature and entropy of all Schwarzschild BHs. The bound is also in agreement with the cosmic-censorship conjecture.

Things are different in Euclidean $N=1, d=4$ supergravity: the integrability condition of the KSE,

$$R_{\mu\nu}{}^{ab}\gamma_{ab}\kappa = 0, \tag{18.20}$$

also admits solutions when the curvature is (anti-)self-dual because then it is proportional to a projector $\frac{1}{2}(1 \pm \gamma^1\gamma^2\gamma^3\gamma^4)$,

$$R_{\mu\nu}{}^{ab}\gamma_{ab} = R_{\mu\nu}{}^{ab}\gamma_{ab}\tfrac{1}{2}(1 \pm \gamma^1\gamma^2\gamma^3\gamma^4)\kappa, \tag{18.21}$$

and half of the supersymmetries are preserved. Then all metrics (in any dimension!) with (anti-)self-dual curvature in four of them (special SU(2) holonomy; see footnote 5 on p. 380) preserve half of the supersymmetries. These are the metrics of (anti-)self-dual gravitational instantons, which were discussed in Section 13.2.1. In the frame in which the spin connection is also (anti-)self-dual, the KSE takes the form

$$\nabla_\mu \kappa = \left[\partial_\mu - \tfrac{1}{4}\omega_\mu{}^{ab}\gamma_{ab}\tfrac{1}{2}(1 \pm \gamma^1\gamma^2\gamma^3\gamma^4)\right]\kappa = \partial_\mu \kappa = 0, \tag{18.22}$$

where we have used $\frac{1}{2}(1 \pm \gamma^1\gamma^2\gamma^3\gamma^4)\kappa = 0$. Then, in this frame, the KSs are the constant spinors that satisfy that constraint.

It is actually possible to reformulate the KSE problem in terms of holonomy, which turns out to be an extremely powerful tool. See, for instance, Ref. [685].

An interesting case of a metric of special SU(2) holonomy is the Euclidean Taub–NUT solution because its supersymmetry can be understood from the point of view of

BPS saturation: it is in principle possible to include the NUT charge in the BPS bound in asymptotically Taub–NUT Lorentzian spaces: $M + |N| \geq 0$. This explains why the massless Lorentzian Taub–NUT solution is not supersymmetric. In the rotation to Euclidean time, the bound becomes $M - |N| \geq 0$, although $M$ no longer has the interpretation of mass,[9] and can be saturated for $M = |N|$, which corresponds to the self-dual Euclidean Taub–NUT solution Eqs. (13.12).

*Pure $N = 2, d = 4$ Poincaré supergravity.* All the field configurations (metric and U(1) vector field) admitting KSs in pure $N = 2, d = 4$ SUEGRA were constructed by Tod in Ref. [1183]. His construction includes field configurations that do not solve the (Einstein–Maxwell) equations of motion. The supersymmetric solutions fall into two classes: pp-waves such as those of Eqs. (14.24) (including the maximally supersymmetric KG4 solution Eqs. (14.27)) and the IWP solutions of Eqs. (13.83) (including the maximally supersymmetric RB solution Eqs. (12.90)). Generically, the solutions of both families preserve half of the supersymmetries.

The IWP family contains many different solutions. We can use the superalgebra to study the asymptotically flat ones which are the (multi-)Kerr–Newman solutions Eq. (13.83) with the saturated BPS bound $M = |Q + iP| = |\mathcal{Z}|$ for any value of the angular momentum, which does not appear in the bound, basically because the Lorentz generators do not appear in the supercharge anticommutator. All these solutions are singular for non-vanishing angular momentum. The only regular ones are the ERN solutions. These can be seen as interpolating between the Minkowski vacuum at asymptotic infinity and the RB vacuum at the near-horizon limit, which supports their interpretation as solitons [609]. This vacuum interpolation, which takes place in more general cases [629, 468], plays a very important role in many instances, such as the AdS/CFT correspondence [921] (for a review see Ref. [23]) and non-conformal generalizations [234].

Let us now consider the thermodynamics of generic dyonic RN BHs. Their temperature and entropy are given by

$$T = \frac{1}{2\pi} \frac{\sqrt{M^2 - |\mathcal{Z}|^2}}{M + \sqrt{M + \sqrt{M^2 - |\mathcal{Z}|^2}}} \stackrel{M \to |\mathcal{Z}|}{\longrightarrow} 0, \tag{18.23}$$

$$S = \pi \left[ M + \sqrt{M^2 - |\mathcal{Z}|^2} \right]^2 \stackrel{M \to |\mathcal{Z}|}{\longrightarrow} \pi M^2. \tag{18.24}$$

Observe that, like any other physical quantity that depends only on the metric, which is invariant under the electric–magnetic duality of the theory, both $T$ and $S$ are also duality invariant: they can be written in terms of the absolute value of the single mass-matrix skew eigenvalue $\mathcal{Z} = Q + iP$. On the other hand, in the extreme limit, which (in the absence of angular momentum) coincides with the supersymmetric limit[10] we find that $M \to |\mathcal{Z}|$, $T \to 0$ so there is no Hawking radiation. This is in agreement with our previous arguments about stability. Furthermore, $S \to \pi M^2 = \pi |\mathcal{Z}|^2$ is completely determined by the central

[9] Compare this bound with Eq. (13.52).

[10] In more general $N \geq 2$ supergravities there are extremal static BHs which are not supersymmetric [855]. They will be studied in more detail in Chapter 27.

charges. We discussed at length the arguments in favor of and against there being a finite value of the entropy for ERN BHs in Section 12.6. In any case, we can always say that the area of the ERN BH is finite and given by a duality-invariant combination of the central charges. This observation is relevant because we will find that, in general, charged supersymmetric spherically symmetric solutions have a regular horizon with finite area only when they preserve the same amount of supersymmetry as the ERN BH (i.e. four supersymmetries).

If we consider asymptotically Taub–NUT IWP solutions, we find that all of them satisfy the saturated bound $M+|N|=|Q+iP|=|\mathcal{Z}|$ (with or without angular momentum).

The supersymmetry of some solutions of $N=2, d=4$ AdS supergravity has been studied in Refs. [1084, 872, 281]. The supersymmetry of the largest class of solutions, given by Plebański and Demiański in Refs. [1042, 1043], was studied in Ref. [28].

$N=4, d=4$ *supergravity.* The bosonic part of this theory is described by the action Eq. (16.58). All the field configurations admitting KSs were found by Tod in Ref. [1184] (see also Refs. [131, 943]) and they include pp-waves and the SWIP solutions of Refs. [842, 166] described in Section 16.2.1. These have been studied more thoroughly in Chapter 16. There are SWIP solutions that preserve half of the supersymmetries (i.e. eight) and SWIP solutions that preserve a quarter (i.e. four) (these include the IWP solutions).

Actually, to study them, it is convenient to focus on the asymptotically flat ones and use the supersymmetry bounds of the theory. There are two BPS bounds: $M^2-|\mathcal{Z}_{1,2}|^2\geq 0$, neither of which is invariant under the dualities of the theory. However, we can construct a duality-invariant *generalized BPS bound* by taking their product and dividing by $M^2$:

$$M^2+\frac{|\mathcal{Z}_1\mathcal{Z}_2|^2}{M^2}-|\mathcal{Z}_1|^2-|\mathcal{Z}_2|^2\geq 0. \tag{18.25}$$

This bound is satisfied by all the regular static non-extreme SWIP solutions of Ref. [906] and it is saturated by all the supersymmetric SWIP solutions of Ref. [166]: the skew eigenvalues of the central-charge matrix correspond precisely to the combinations of electric and magnetic charges of Eq. (16.83), and the product of the two supersymmetry bounds gives precisely the supersymmetry parameter $r_0^2$ of the SWIP solutions, Eqs. (16.84).

The bound can be saturated in two different ways: when $M=|\mathcal{Z}_{1,2}|\neq|\mathcal{Z}_{2,1}|$, a quarter of the supersymmetries are unbroken, and when $M=|\mathcal{Z}_1|=|\mathcal{Z}_2|$, half of the supersymmetries are unbroken. The only supersymmetric solutions with regular horizons are the static ones with only a quarter of the supersymmetries preserved. All the supersymmetric solutions have zero temperature.

The entropy (see the second Eq. (16.84)) and the temperature of the BHs of $N=4, d=4$ can be expressed in a form that is manifestly invariant under the two duality groups of the theory: $\mathrm{SL}(2,\mathbb{R})$ (S duality) and $\mathrm{SO}(6)$ (T duality). The entropy of the static supersymmetric BHs that preserve a quarter of the supersymmetries is given by

$$S=4\pi\left||\mathcal{Z}_1|^2-|\mathcal{Z}_2|^2\right|=4\pi\sqrt{I_4}, \tag{18.26}$$

where $I_4$ is the quartic invariant defined in Eqs. (16.62), which depends on the charges but not on the asymptotic value of the axidilaton. In the extreme limit, i.e. when the entropy is finite, its expression in terms of the central charges is not only duality invariant but also *moduli independent*. This is a very important property that suggests that the BH entropy does indeed count microstates, since this counting would not be changed by small variations of the moduli.

In fact, this happens irrespective of the values of the scalars at infinity, which are free parameters in the general solutions, and indicates the existence of an *attractor mechanism* in these extremal supersymmetric BHs. The attractor mechanism was originally discovered in extremal, supersymmetric solutions of $N = 2, d = 4$ supergravity coupled to vector supermultiplets [531, 1157, 526, 527], but, as we will study in Chapter 27, it can be shown that it is present in very general $d = 4$ theories (including all four-dimensional ungauged supergravities) [522].

## 18.2 Tod's program

The ideas presented in Section 18.1 only allow us to expect certain asymptotically flat solutions with well-defined asymptotic charges saturating a supersymmetry bound to be supersymmetric. It would be highly desirable to be able to find, in a systematic way if possible, other supersymmetric solutions perhaps with different asymptotics or to explore the possibility of having more than one with a given set of charges and asymptotic values of the scalars (moduli). Classical solutions of GR are always interesting and difficult to construct and, as we have discussed, the supersymmetric ones are especially interesting.

Supersymmetric solutions should be especially easy to find since their form is constrained by the (super)symmetry requirements and they must depend on a reduced number of independent functions. One could substitute educated ansatzs for the metric and other bosonic fields in the KSEs and find out which additional conditions the proposed field configuration has to satisfy to be supersymmetric. Since (as we are going to show in full detail) unbroken supersymmetry does not imply that the equations of motion are satisfied, we will have to impose these on the supersymmetric field configuration. The integrability conditions of the KSEs relate different equations of motion, and only a few of them need to be actually checked.

This procedure can be and has been used successfully many times in the literature. The requirement that a field configuration be supersymmetric is, however, so powerful that the first step (making an educated ansatz) can be entirely omitted: it is possible to find all the necessary conditions that a generic field configuration has to satisfy to be supersymmetric. In all the cases studied so far, these conditions have been shown to be sufficient as well, which leads to a complete characterization of the supersymmetric field configurations of the theory under study.

The characterization of all the supersymmetric solutions of a given supergravity theory was first achieved by Tod in Ref. [1183] for pure $N = 2, d = 4$ Poincaré supergravity, following the pioneering work by Gibbons and Hull [618], by assuming that a generic field configuration (metric and vector field) admits at least one KS (so the KSEs are solved for some $\kappa$) and then studying all the consistency conditions that could be derived from the KSEs, finding all the necessary and sufficient conditions for a field configuration to be supersymmetric.

In Ref. [1183] (and later for pure $N = 4, d = 4$ SUEGRA in Ref. [1184]) Tod used the four-dimensional Newman–Penrose formalism [1032], which cannot be easily generalized to other dimensions or signatures and which is not commonly used in the supergravity community. For these reasons a different formalism based on tensors (instead of the spinors of the Newman Penrose formalism), the *spinor bilinear method*, was proposed in Ref. [599] to perform essentially the same analysis. Soon after, this method was used [596] to find all the supersymmetric solutions of pure $N=1, d=5$ supergravity. These results for $N=1, d=5$ supergravity were immediately extended: the gauged (AdS) case was solved in Ref. [593]; the ungauged theory coupled to an arbitrary number of vector multiplets was solved in [595];[11] the Abelian gaugings of the latter were considered in Refs. [705, 706]; the addition of both vector multiplets and hypermultiplets to the ungauged theory was studied in Ref. [129];[12] and the gaugings of these theories were studied in Ref. [132]. Finally, the most general case was studied in Ref. [127].

The new method was first applied to $N=2, d=4$ supergravities in Ref. [282], in which all the supersymmetric solutions of pure, gauged (AdS) $N=2, d=4$ supergravity were characterized (they were later studied in Ref. [275]). The supersymmetric solutions of ungauged $N=2, d=4$ supergravity coupled to vector supermultiplets were characterized in Ref. [937] and the ungauged case with vector multiplets and hypermultiplets was solved in Ref. [795]. The case with only vector multiplets and Abelian gaugings was completely solved in Refs. [278, 866, 867]. The timelike supersymmetric solutions[13] of theories with vector supermultiplets and non-Abelian gaugings of the isometries of the special Kähler manifold were characterized and studied in Refs. [797, 798] and the inclusion of hypermultiplets and their gaugings were studied in Ref. [940]. The null supersymmetric solutions of the most general gauged theory have not yet been characterized.

For other values of $N$ and $d$, the method has been used for pure, ungauged $N=4, d=4$ SUEGRA in Ref. [131]; for ungauged matter-coupled $N=1, d=4$ SUGRA in Ref. [1011] (without superpotential but with non-trivial kinetic matrix);[14] for minimal $d=6$ SUGRA in Ref. [704] (see also Ref. [333]); some gaugings of this theory were considered in Ref. [302]; the coupling to hypermultiplets has been fully solved in Ref. [835]. In $d=7$ it has been used in Ref. [303] and in $d=3$ in Ref. [427].

The basic idea of the spinor bilinear method is to translate the KSEs into tensor equations containing the same information and then proceed as before, finding the conditions that one has to impose on a generic field configuration for these equations to be satisfied. The tensors that appear in the new equations are constructed as bilinears of the KSs and the equations that they are assumed to satisfy follow from those assumed to be satisfied by the spinors (the KSEs).

The following recipe provides a detailed description of how this method can be carried out in any theory.

1. Write the KSEs (that is, the equations $\delta_\kappa F = 0$ for supersymmetry parameters $\kappa$ and for all the fermionic field $F$ of the theory) of the theory for a generic bosonic field configuration and assume that they are satisfied for at least one $\kappa$ (KS). In what

[11] Previous, partial, results had been obtained in Refs. [335, 1094].

[12] Previous, partial, results were obtained in Refs. [277, 320].

[13] We will soon explain what *timelike supersymmetric solution* means.

[14] The gauged case, with superpotential but with trivial kinetic matrix, has been studied by spinorial-geometry methods in Ref. [647].

follows $\kappa$ will be treated as a commuting spinor, which will not affect the analysis because the KSEs are linear in $\kappa$.

2. Construct with $\kappa$ and gamma matrices all the possible spinor bilinears $\sim \bar{\kappa}\gamma^{\mu_1\cdots\mu_n}\kappa$. We could call the resulting tensors "Killing tensors," but we will not because, in general, they are not Killing vectors or tensors in the standard sense.[15] Here it is crucial to treat $\kappa$ as a commuting variable since some important tensors automatically vanish for a single anticommuting spinor (see footnote 6 of Chapter 17).

3. The products of tensor bilinears can be decomposed in different products of tensor bilinears using Fierz identities, giving rise to an algebra that provides us with important relations between the tensor bilinears. The next step consists in computing the algebra of tensor bilinears. The algebra can force us to consider separately different cases (timelike or null vector bilinears, for instance). The four- and five-dimensional cases are worked out in detail in Appendix D.3 for an arbitrary number of minimal (Weyl) spinors in $d = 4$ and for just one (a symplectic-Majorana pair) in $d = 5$.

4. Every KSE contains, among other terms, an algebraic or differential operator $\mathcal{O}$ acting on the KS $\mathcal{O}\kappa + B\kappa = 0$. To translate that KSE into a tensor equation, we must act with that operator on the tensor bilinears $\bar{\kappa}\gamma^{\mu_1\cdots\mu_n}\kappa$. We will obtain two terms in which the operator acts on only one of the KSs of the bilinear $\sim \bar{\kappa}\gamma^{\mu_1\cdots\mu_n}\mathcal{O}\kappa$, and then we can use the KSE to replace the operator acting on the KS by some other terms linear on the KS $-B\kappa$, giving $\sim -\bar{\kappa}\gamma^{\mu_1\cdots\mu_n}B\kappa$. Typically, $B$ contains gamma matrices whose product with $\gamma^{\mu_1\cdots\mu_n}$ can always be expressed as a sum of antisymmetric products of gammas. The corresponding new terms can be written entirely in terms of tensor bilinears.

   In general one obtains several tensor equations for each KSE, but they are not independent.

5. The tensor equations have to be analyzed. We will always find that there is a combination of the vector bilinears which is always a Killing vector, and we will find equations that indicate that the whole field configuration is invariant under the associated isometry. To make progress one must introduce adapted coordinates and write the metric in those coordinates. The Killing vector will appear explicitly or implicitly in that metric, and the equations satisfied by it lead to equations for the metric components. At the end of this analysis one has a number of necessary conditions that the field configuration must satisfy for the KSEs to admit a solution $\kappa$. Typically these conditions imply that supersymmetric field configurations depend on a reduced number of independent functions, which may or may not be subject to further conditions.

6. We plug a field configuration satisfying the conditions derived in the previous step (that is, a generic supersymmetric field configuration) into the original KSEs and

[15] This seems to contradict some statements in Chapter 17, but, from one KS $\kappa$ there are several vector bilinears $\bar{\kappa}\gamma^{\mu}\kappa$ distinguished by their R-symmetry indices. Only one combination of them will always be a Killing vector, and that is the combination implicit in those statements. The examples that we are going to study should clarify this point.

check whether they are also sufficient for the KSEs to admit a solution. In all the examples that we are going to review, the necessary conditions turn out to be also sufficient. Solving the KSEs one finds explicitly the KSs and the amounts of supersymmetry preserved by the generic configuration and, sometimes, by some less generic configurations as well.

7. At this point we have identified all the supersymmetric field configurations of the theory. Only some of them satisfy the equations of motion and are supersymmetric solutions. Before plugging the general form of the supersymmetric field configurations into the equations of motion to try to solve them it is worth checking which of them are independent in order to minimize the effort.

   The integrability conditions of the KSEs are basically relations between different ("left-hand sides") of the equations of motion, which only hold for supersymmetric field configurations. Using these relations we can identify a minimal number of equations that need to be satisfied by a supersymmetric field configuration to solve all of them and be a supersymmetric solution.

   There is a shorter route to obtain the same relations between the equations of motion based on the Noether identities (see p. 33) associated to the invariance under local supersymmetry transformations of supergravity theories. The relations obtained in this way (first found in Ref. [843] and applied to this problem in Ref. [130]) are known as *Killing spinor identities* (KSIs) and we are going to derive them in Section 18.2.1.

   Deriving the KSIs for the theory at hand and finding the independent equations of motion constitutes the next step in this method. Plugging the generic supersymmetric configurations into these independent equations of motion is the final step, and the result is usually a set of simple equations that the few independent functions that enter the generic supersymmetric field configuration have to satisfy. Different solutions of these equations lead to different supersymmetric solutions of the theory that occasionally may admit more KSs than the generic configuration. Such is the case of the maximally supersymmetric solutions (vacua) studied in Chapter 17.

Before concluding this overview and jumping into the examples, it is important to mention that there is another method that can be used to characterize in a systematic way all the supersymmetric solutions of a given theory called *spinorial geometry* [631]. This method is based on the use of spinors in a preferred frame and we will not cover it here, but it gives a much more refined classification of the supersymmetric solutions of a supergravity theory because it distinguishes between solutions with different amounts of unbroken supersymmetry whereas the spinor bilinear method does not. This method has been used mostly in 10 and 11 dimensions [645, 658, 646, 651, 656, 657, 652, 653, 654, 660, 659, 655, 648, 1017, 649, 650], but also for $N = 1, d = 4$ supergravity [647] and for pure, gauged (AdS) $N = 2, d = 4$ supergravity [276].

### *18.2.1 The Killing spinor identities*

As we have stressed, a key simplification in the spinor bilinear method is the existence of a large number of relations between the (left-hand sides) of the equations of motion of the bosonic fields in supersymmetric field configurations. These relations, which we call Killing spinor identities (KSIs), can be obtained as follows [843, 130]: the action $S$ of a supergravity theory with bosonic and fermionic fields $\phi^b, \phi^f$ is, by definition, invariant under local supersymmetry transformations with arbitrary parameters $\epsilon(x)$. Up to total derivatives, it follows that

$$\delta_\epsilon S = \int d^dx \left[ S_{,b}\,\delta_\epsilon\phi^b + S_{,f}\,\delta_\epsilon\phi^f \right] = 0, \tag{18.27}$$

where $b$ and $f$ are treated as indices running, respectively, over the bosonic and fermionic fields and $S_{,b} \equiv \delta S/\delta\phi^b$ (resp. $f$) are the (l.h.s. of the) Euler–Lagrange equations of the bosonic (fermionic) fields. From the integrand of this expression we obtain the Noether identities relating these equations, but they are not useful for our purposes because they are odd in fermions and vanish identically for vanishing fermions. Therefore, we take a second functional derivative with respect to the fermionic fields only (with the index $f_1$) to obtain an expression even in them, and then we set the fermionic fields to zero:

$$\left\{ S_{,bf_1}\,\delta_\epsilon\phi^b + S_{,b}\,(\delta_\epsilon\phi^b)_{,f_1} + S_{,ff_1}\,\delta_\epsilon\phi^f + S_{,f}\,(\delta_\epsilon\phi^f)_{,f_1} \right\}\Big|_{\phi^f=0} = 0. \tag{18.28}$$

The terms $\delta_\epsilon\phi^b\big|_{\phi^f=0}$, $S_{,f}\big|_{\phi^f=0}$, and $(\delta_\epsilon\phi^f)_{,f_1}\big|_{\phi^f=0}$ vanish because they are odd in the fermion fields and we are left with an identity valid for any bosonic fields $\phi^b$ and any supersymmetry parameter $\epsilon$:

$$\left\{ S_{,b}\,(\delta_\epsilon\phi^b)_{,f_1} + S_{,ff_1}\,\delta_\epsilon\phi^f \right\}\Big|_{\phi^f=0} = 0. \tag{18.29}$$

The second term in this identity can be made to vanish if we restrict ourselves to supersymmetric field configurations and the corresponding KS $\kappa$ since, by definition, $\delta_\kappa\phi^f\big|_{\phi^f=0}$. For this case, then, we obtain the KSIs

$$\boxed{\mathcal{E}^b\,(\delta_\kappa\phi^b)_{,f_1}\big|_{\phi^f=0} = 0.} \tag{18.30}$$

These identities consist in a sum of terms linear in the KS with coefficients that contain the bosonic equations of motion. They can be conveniently translated into tensor identities by multiplying them by $\bar{\kappa}\gamma^{\mu_1\cdots\mu_n}$ from the left, as we will see in the examples. Observe that these identities depend only on the supersymmetry transformation rules of the bosonic fields, which are not modified by gaugings and other deformations of the original theories. Therefore, these identities have essentially the same form for all the theories that have the same $N, d$ and field content when we write them in terms of the $\mathcal{E}^b$ symbols. Of course, the bosonic equations represented by $\mathcal{E}^b$ that appear in the identities will be deformed.

### 18.3 All the supersymmetric solutions of ungauged $N = 1, d = 4$ supergravity

In this section we are going to review, following Ref. [1011], how the spinor bilinear method can be used to characterize all the supersymmetric solutions of a generic ungauged

$N=1, d=4$ supergravity coupled to vector and chiral multiplets, as described in Chapter 6, to which we refer the reader for notations, conventions, etc. We will not include a superpotential in our analysis. We will give many details of the procedure and calculations in this first example that we will suppress in the following ones for the sake of brevity.

### 18.3.1 Supersymmetric configurations

Following the recipe on p. 523, we start by assuming that a purely bosonic field configuration of these theories $\{g_{\mu\nu}, A^{\Lambda}{}_{\mu}, Z^i\}$ is supersymmetric and, therefore, the KSEs admit at least one solution (KS) $\kappa$. Thus, we have by hypothesis

$$\delta_\kappa \psi_\mu = \mathcal{D}_\mu \kappa = 0, \tag{18.31}$$

$$\delta_\kappa \lambda^\Lambda = \tfrac{1}{2} \not{F}^{\Lambda +} \kappa = 0, \tag{18.32}$$

$$\delta_\kappa \chi^i = i \not{\partial} Z^i \kappa = 0. \tag{18.33}$$

The second step in the recipe requires us to construct with $\kappa$ all the possible spinor bilinears. We have done this for the general four-dimensional case in Section D.3.1, and for $N=1$ we find that there is no non-vanishing scalar bilinear, which implies that the only vector bilinear that can be built with $\kappa$ is necessarily null; we denote it by $l_\mu$. There is no timelike case in $N=1, d=4$ supergravity, a fact associated with the absence of central charges in the $N=1, d=4$ Poincaré superalgebra, the consequence of which is that there are no massive point-like supersymmetric objects in these theories. Apart from the null vector, one can also construct one self-dual 2-form $\Phi_{\mu\nu}$.

As explained at the end of Section D.3.1, in the null case we must introduce an auxiliary spinor $\eta$ to construct a null tetrad $l, n, m, m^*$ defined in Eqs. (D.224) (replacing $\epsilon$ by $\kappa$). With it we can also construct additional self-dual 2-forms as in Eqs. (D.227). It is very important to define clearly the orientation of the null tetrad. Our choice [937] is that the relation between a standard Cartesian tetrad $\{e^0, e^1, e^2, e^3\}$ and the complex null tetrad $\{e^u, e^v, e^z, e^{z^*}\} = \{\hat{l}, \hat{n}, \hat{m}, \hat{m}^*\}$ is

$$\begin{pmatrix} e^u \\ e^v \\ e^z \\ e^{z^*} \end{pmatrix} = \frac{1}{\sqrt{2}} \left(\begin{array}{cc|cc} 1 & 1 & & \\ 1 & -1 & & \\ \hline & & 1 & i \\ & & 1 & -i \end{array}\right) \begin{pmatrix} e^0 \\ e^1 \\ e^2 \\ e^3 \end{pmatrix} \Rightarrow \begin{pmatrix} \gamma^u \\ \gamma^v \\ \gamma^z \\ \gamma^{z^*} \end{pmatrix} = \begin{pmatrix} \not{l} \\ \not{n} \\ \not{m} \\ \not{m}^* \end{pmatrix} = \frac{1}{\sqrt{2}} \left(\begin{array}{cc|cc} 1 & 1 & & \\ 1 & -1 & & \\ \hline & & 1 & i \\ & & 1 & -i \end{array}\right) \begin{pmatrix} \gamma^0 \\ \gamma^1 \\ \gamma^2 \\ \gamma^3 \end{pmatrix}. \tag{18.34}$$

With this choice, the chirality matrix is given by

$$\gamma_5 \equiv -i\gamma^0\gamma^1\gamma^2\gamma^3 = -\gamma^{uv}\gamma^{zz^*}. \tag{18.35}$$

Now, we have to reexpress the KSEs Eqs. (18.31)–(18.33) as equations involving the bosonic fields and the spinor bilinears. Acting on Eq. (18.32) with $\bar{\kappa}\gamma_\mu$ and $\bar{\eta}\gamma_\mu$ we get

$$F^{\Lambda +}{}_{\mu\nu} l^\nu = 0, \qquad F^{\Lambda +}{}_{\mu\nu} m^{*\nu} = 0, \tag{18.36}$$

and acting on Eq. (18.33) with $i\sqrt{2}\bar{\kappa}$ and $i\sqrt{2}\bar{\eta}$ we get

$$l^\mu \partial_\mu Z^i = 0, \qquad m^\mu \partial_\mu Z^i = 0. \tag{18.37}$$

To rewrite Eq. (18.31) in terms of spinor bilinears, we need information about the covariant derivative of the auxiliary spinor. All we have is the normalization condition Eq. (D.222) and, taking its Kähler-covariant derivative, we find that

$$\mathcal{D}_\mu \eta + a_\mu \kappa = 0, \tag{18.38}$$

for some $a_\mu$ with Kähler weight $-2$ times that of $\kappa$, that is

$$\mathcal{D}_\mu a_\nu = (\nabla_\mu - i\mathcal{Q}_\mu) a_\nu; \tag{18.39}$$

$a_\mu$ will be determined by consistency.

Now, taking the covariant derivative of the null tetrad vectors defined in Eqs. (D.224) (replacing $\epsilon$ by $\kappa$) using Eq. (18.38) and the KSE Eq. (18.31) we get

$$\mathcal{D}_\mu l_\nu = \nabla_\mu l_\nu = 0, \tag{18.40}$$

$$\mathcal{D}_\mu n_\nu = \nabla_\mu n_\nu = -a^*_\mu m_\nu - a_\mu m^*_\nu, \tag{18.41}$$

$$\mathcal{D}_\mu m_\nu = (\nabla_\mu - i\mathcal{Q}_\mu) m_\nu = -a_\mu l_\nu. \tag{18.42}$$

Equations (18.36)–(18.42) have the same information content as the original KSEs Eqs. (18.31)–(18.33) but expressed in terms of tensors. Now we have to extract it. Equations (18.36) and (18.37) imply that the vector field strengths and the derivatives of the scalars must be of the form[16]

$$F^{\Lambda +} = \tfrac{1}{2}\phi^\Lambda \hat{l} \wedge \hat{m}^*, \qquad dZ^i = A^i \hat{l} + B^i \hat{m}, \tag{18.43}$$

for some functions $\phi^\Lambda$, $A^i$, $B^i$ to be determined.

Equation (18.40) implies that the metric admits a covariantly constant null vector. The most general metric of this kind is a Brinkmann pp-wave metric [263, 264]. We have studied these metrics in Chapter 14, giving in Eq. (14.4) its most general form. In four dimensions the wavefront metric is two-dimensional and can be written using complex coordinates as in Eq. (14.20) (the one-dimensional connection having been eliminated by a change of coordinates). We are going to use essentially this form of the Brinkmann metric, with the 1-form connection[17] and with some changes in the notation: $\xi \to z$, $K \to H$, $A_i dx^i \to \hat{\omega} = \omega_{\underline{z}} dz + \omega_{\underline{z}^*} dz^*$. We conclude that Eq. (18.40) is equivalent to the statement that the metric is of the form

$$ds^2 = 2du(dv + Hdu + \hat{\omega}) - 2e^{2U} dz dz^*, \tag{18.44}$$

[16] Here we use hats to denote differential forms: $\hat{l} \equiv l_\mu dx^\mu$, $\hat{m}^* \equiv m^*_\mu dx^\mu$, etc.

[17] The coordinate transformation needed to remove it may or may not be compatible with the rest of the equations that we have to solve. Thus, we have to take the most general form of the metric and check later if and how the 1-form can be removed.

where $U$ may depend on $z, z^*$, and $u$. The coordinates $u$ and $v$ are defined in terms of $l_\mu$ by

$$\hat{l} = du, \qquad l^\mu \partial_\mu = \frac{\partial}{\partial v}, \tag{18.45}$$

and this implies that the scalars $Z^i$ and the functions $A^i$ and $B^i$ are functions of $z$ and $u$ but not of $z^*$ and $v$.

The rest of the tetrad is given by

$$\hat{n} = dv + H du + \hat{\omega}, \qquad \hat{m} = e^U dz, \tag{18.46}$$

and we have to find under which conditions Eqs. (18.41) and (18.42) are satisfied (Eq. (18.40) has been completely solved by the above choice of metric). Equation (18.42) yields two conditions:

$$\hat{a} = n^\mu(\partial_\mu U - i\mathcal{Q}_\mu)\hat{m} + D\hat{l}, \qquad m^\mu \partial_\mu(U - i\mathcal{Q}_\mu) = 0 \tag{18.47}$$

($D$ being a function to be determined), the second of which can be rewritten using the definition of $\mathcal{Q}$ and the dependence $Z^i(z,u)$ in the form

$$\partial_{\underline{z}^*}(U + \mathcal{K}/2) = 0 \ \Rightarrow \ U = -\mathcal{K}/2 + h(u), \tag{18.48}$$

where $h(u)$ can be eliminated by a coordinate redefinition that does not change the general form of the Brinkmann metric. Equation (18.41) yields another two conditions:

$$D = e^{-U}(\partial_{\underline{z}^*} H - \dot{\omega}_{\underline{z}^*}), \tag{18.49}$$

$$(d\omega)_{\underline{zz}^*} = 2ie^{2U} n^\mu \mathcal{Q}_\mu, \tag{18.50}$$

which lead to

$$\hat{a} = [\dot{U} - \tfrac{1}{2}e^{-2U}(d\omega)_{\underline{zz}^*}]\hat{m} + e^{-U}(\partial_{\underline{z}^*} H - \dot{\omega}_{\underline{z}^*})\hat{l}. \tag{18.51}$$

We have concluded the analysis of the consistency conditions derived from our initial assumption that we have a bosonic field configuration that admits (at least) one Killing spinor $\kappa$. The result is that the metric must be of the form Eqs. (18.44), where the 1-form $\hat{\omega}$ satisfies Eq. (18.50), and the function $U$ satisfies Eq. (18.48), while the field strengths and scalars must take the form given in Eqs. (18.43). Now we have to prove that these necessary conditions for the bosonic field configuration to be supersymmetric are also sufficient by solving the KSEs directly.

Assuming that the scalars and the vector field strengths are given by Eqs. (18.43), the KSE $\delta_\kappa \chi^i = 0$ becomes

$$i[A^i \not{l} + B^i \not{m}]\kappa^* = 0, \tag{18.52}$$

which can be solved by imposing two algebraic conditions on the spinors:

$$\not{l}\kappa^* = 0\,, \qquad \not{m}\kappa^* = 0\,. \tag{18.53}$$

The same two conditions ensure that the KSE $\delta_\kappa \lambda^\Lambda = 0$ is satisfied for field strengths of the form Eqs. (18.43) without any further assumptions.

These two conditions are completely equivalent [937]. It is worth reviewing the proof, since the same pair of constraints appears in many other cases. If we multiply the first condition by $\not{n}$ and the second by $\not{m}^*$ we obtain conditions which are equivalent to the ones above but resemble more a projector ($(\gamma^{uv})^2 = \mathbb{1}$, etc.):

$$\not{n}\,\not{l}\kappa^* = (1 - \gamma^{uv})\kappa^* = 0, \qquad \not{m}^*\not{m}\kappa^* = -(1 + \gamma^{zz^*})\kappa^* = 0. \tag{18.54}$$

Let us assume that $\kappa^*$ satisfies the second condition. On using Eq. (18.35) we get

$$\gamma^{zz^*}\kappa^* = -\kappa^* \Rightarrow \gamma^{uv}\gamma^{zz^*}\kappa^* = \gamma^{uv}\kappa^* \Rightarrow -\gamma^5\kappa^* = \gamma^{uv}\kappa^*, \tag{18.55}$$

which, due to the chirality of $\kappa^*$, is the same as the first condition.

The $u$ and $v$ components of the KSE $\delta_\kappa \psi_a = 0$ are automatically satisfied if we assume that $\kappa$ is $u$- and $v$-independent. Using the condition $\gamma^{zz^*} = \epsilon$ implied by the constraints Eqs. (18.53) and Eqs. (18.48), the $z$ and $z^*$ components take the simple form

$$\partial_{\underline{z}}\kappa = \partial_{\underline{z}^*}\kappa = 0. \tag{18.56}$$

The KSs are, therefore, constant spinors that satisfy Eqs. (18.53). It is not difficult to see that half of the eigenvalues of the matrices involved are zero and, thus, that they only impose two constraints on the four components of the Killing spinors. Then, generically, the field configurations that we have identified preserve at least half of the supersymmetries. Some configurations may not require Eqs. (18.53) and may have more (never less) unbroken supersymmetry, but this would require the scalars to be constant and the vanishing of vector field strengths. In presence of a superpotential and gaugings, things, however, can be quite different.

### 18.3.2 Supersymmetric solutions

Having identified the form of the supersymmetric field configurations we now have to impose the equations of motion. Following the general recipe, we start by finding the relations existing between the off-shell (l.h.s. of the) bosonic equations of motion when we substitute the supersymmetric field configurations. The off-shell (l.h.s. of the) bosonic equations of motion are defined in Eqs. (6.24) and given explicitly in Eqs. (6.27), (6.28), and (6.29). The relations we search for are encoded into the KSIs, which can be easily computed from the general formula Eq. (18.30) having the supersymmetry transformation rules of the bosonic fields Eqs. (6.19). The result is

$$\mathcal{E}^\mu{}_a\gamma^a\kappa^* = 0, \tag{18.57}$$

$$\mathcal{E}_\Lambda{}^\mu\gamma_\mu\kappa^* = 0, \tag{18.58}$$

$$\mathcal{E}_i\kappa = 0. \tag{18.59}$$

Before we translate these spinor equations into tensor equations using the spinor bilinears, we are going to generalize these identities to include in them the Bianchi identities of the vector field strengths so as to give them an electric–magnetic duality- (symplectic-) invariant form.[18] We will do the same in all the examples that do not involve gaugings, since these naturally break electric–magnetic-duality invariance.

The Bianchi identities Eq. (6.25) have implicitly been assumed to be satisfied in this derivation of the KSIs (we are using the vector potentials), but we could have obtained these as integrability equations of the KSEs and, in that case, the Bianchi identities would have occurred naturally in the KSIs (there would have been no need to assume they are satisfied), and they would have done so, necessarily, in a formally symplectic-invariant way.

The only KSI in which the Bianchi identities can occur is the second one, which involves the Maxwell equations. The natural symplectic-covariant generalization we are going to replace it with is

$$(\mathcal{E}_\Lambda{}^\mu - f^*_{\Lambda\Sigma}\mathcal{B}^{\Sigma\,\mu})\gamma_\mu\kappa^* = 0. \tag{18.60}$$

(That this combination is the one that appears naturally can be checked, for instance, by acting with $\not{\mathcal{D}}$ on the KSE $\delta_\kappa\lambda^\Lambda = 0$.)

Observe that the general form of the KSIs in $N=1, d=4$ theories is the same for theories with or without superpotential and gaugings (excluding the modification just made) because the supersymmetry transformation rules of the bosonic fields are the same in all cases. Only the explicit form of the (l.h.s. of the) equations of motion will change, of course.

Now, acting with $i\sqrt{2}\bar{\kappa}$ and $i\sqrt{2}\bar{\eta}$ on the l.h.s of Eqs. (18.57) and (18.60) we get

$$\mathcal{E}_{\mu\nu}l^\nu = \mathcal{E}_{\mu\nu}m^\nu = 0, \tag{18.61}$$

$$(\mathcal{E}_{\Lambda\,\mu} - f_{\Lambda\Sigma}\mathcal{B}^\Sigma{}_\mu)l^\mu = (\mathcal{E}_{\Lambda\,\mu} - f_{\Lambda\Sigma}\mathcal{B}^\Sigma{}_\mu)m^\mu = 0, \tag{18.62}$$

while Eq. (18.59) implies, directly,

$$\mathcal{E}_i = 0. \tag{18.63}$$

These are identities satisfied off-shell by all supersymmetric field configurations identified in Section 18.3.1 and imply that the equations of motion of the scalars are satisfied automatically and do not need to be checked. The same happens for several components of the Einstein and Maxwell equations–Bianchi identities and, as a result, the only independent equations of motion that need to be imposed on supersymmetric configurations are the following three:

$$\mathcal{E}_{\mu\nu}n^\mu n^\nu = 0, \tag{18.64}$$

$$(\mathcal{E}_{\Lambda\,\mu} - f_{\Lambda\Sigma}\mathcal{B}^\Sigma{}_\mu)n^\mu = 0, \tag{18.65}$$

$$(\mathcal{E}_{\Lambda\,\mu} - f_{\Lambda\Sigma}\mathcal{B}^\Sigma{}_\mu)m^{*\mu} = 0. \tag{18.66}$$

[18] On the other hand, apart from reasons of symmetry and beauty, we have characterized only the general form of the vector field strengths of supersymmetric field configurations, not the vector potentials, and we do need to impose the Bianchi identities.

In differential-form language, the Bianchi identities take the form

$$\hat{\mathcal{B}}^{\Sigma} = -dF^{\Lambda} = \tfrac{1}{2}d(\phi^{\Sigma}\hat{m} + \text{c.c}) \wedge \hat{l}, \tag{18.67}$$

and are solved by vector potentials that depend on holomorphic functions $\varphi^{\Lambda}(z,u)$ in this way:

$$A^{\Lambda} = \varphi^{\Lambda}(z,u)du + \text{c.c.}, \quad \text{where} \quad e^{\mathcal{K}/2}\partial_{\underline{z}}\varphi^{\Lambda}(z,u) = \phi^{*\Lambda}. \tag{18.68}$$

The Maxwell equations take the form

$$\hat{\mathcal{E}}_{\Lambda} = d(f_{\Lambda\Sigma}F^{\Lambda+} + \text{c.c.}) = -\tfrac{1}{2}d(f_{\Lambda\Sigma}\phi^{\Sigma}\hat{m}^{*} + \text{c.c}) \wedge \hat{l}, \tag{18.69}$$

and are solved by holomorphic functions $\varphi_{\Lambda}(z,u)$ such that

$$e^{\mathcal{K}/2}\partial_{\underline{z}}\varphi_{\Lambda}(z,u) = f^{*}_{\Lambda\Sigma}\phi^{*\Sigma}. \tag{18.70}$$

The holomorphic functions $\varphi^{\Lambda}(z,u)$ and $\varphi_{\Lambda}(z,u)$ are related by

$$\partial_{\underline{z}}\varphi_{\Lambda}(z,u) = f^{*}_{\Lambda\Sigma}\partial_{\underline{z}}\varphi^{\Sigma}(z,u). \tag{18.71}$$

We know that the kinetic matrix $f_{\Lambda\Sigma}$ is a holomorphic function of the $Z^i$s, which are, in turn, holomorphic functions of $z$ in supersymmetric configurations. Therefore, the above equation can be satisfied in only two ways: either the $Z^i$s are $z$-independent or the functions $\varphi_{\Lambda}$ and $\varphi^{\Lambda}$ are $z$-independent and the vector fields $A^{\Lambda}$ are pure gauge. These two cases have to be analyzed independently.

Finally, the component $\mathcal{E}_{\mu\nu}n^{\mu}n^{\nu}$ of the Einstein equations in the gauge $\omega = 0$ and taking into account Eq. (18.71) leads to an equation for the function $H$:

$$\partial_{\underline{z}}\partial_{\underline{z}^*}H - e^{-\mathcal{K}/2}\partial^2_{\underline{u}}e^{-\mathcal{K}/2} - e^{-\mathcal{K}}\mathcal{G}_{ij^*}\partial_{\underline{u}}Z^i\partial_{\underline{u}}Z^{*j^*} - \tfrac{1}{2}\Im\mathfrak{m} f_{\Lambda\Sigma}\partial_{\underline{z}}\varphi^{\Lambda}\partial_{\underline{z}^*}\varphi^{*\Sigma} = 0. \tag{18.72}$$

In the first case, with $z$-independent $Z^i$s, $H$ can be found to be of the form

$$H = \Re\mathfrak{e} f(z) + \left[e^{-\mathcal{K}/2}\partial^2_{\underline{u}}e^{-\mathcal{K}/2} + e^{-\mathcal{K}}\mathcal{G}_{ij^*}\partial_{\underline{u}}Z^i\partial_{\underline{u}}Z^{*j^*}\right]|z|^2 + \tfrac{1}{2}\Im\mathfrak{m} f_{\Lambda\Sigma}\varphi^{\Lambda}\varphi^{*\Sigma}, \tag{18.73}$$

for some arbitrary holomorphic function $f(z)$. These solutions describe gravitational, electromagnetic, and scalar pp-waves.

In the second case, when the $Z^i$s do depend on $z$ and the vector field strengths vanish, the equation for $H$ is not easy to integrate except when the $Z^i$s are $u$-independent holomorphic functions of $z$, in which case

$$H = \Re\mathfrak{e} f(z). \tag{18.74}$$

The presence of this function characterizes the gravitational pp-waves we have studied in Chapter 14 (there are no electromagnetic or scalar waves superposed in this case). On the other hand, the holomorphic scalars $Z^i$ and the metric function $U$ proportional to the Kähler potential characterize a class of solutions known as *cosmic strings*, which have been studied, for instance, in Refs. [73, 674, 161, 160] and which appear in many supergravity theories in four [1184, 130, 937, 795] as well as in other different dimensions (the metric, then, includes additional *inert* coordinates). A common example is provided by the axidilaton model, which we have studied in $N=1,2,4, d=4$ supergravity but which also occurs in type-IIB supergravity in ten dimensions (see Chapter 23). There, the cosmic string solutions have six additional *inert* spacelike coordinates and receive the name of *7-branes*. We will review them in Section 25.2.8.

The solutions corresponding to the second case are superpositions of gravitational waves and cosmic strings. These two types of solutions can exist and be supersymmetric independently. The fact that they can be superposed is characteristic of supersymmetric solutions.

## 18.4 All the supersymmetric solutions of ungauged $N=2, d=4$ supergravity

In our next example we are going to deal with the most general ungauged $N=2, d=4$ supergravity theory with $n_V$ vector supermultiplets and $n_H$ hypermultiplets and generic special and quaternionic-Kähler geometries, following Refs. [937, 795]. These theories are described in Chapter 7 and Appendix F, whose conventions we use. This is a very interesting set of theories with very interesting supersymmetric solutions (BHs, in particular) and the results we are going to review here are extremely useful by themselves, but also because the results that one obtains for higher $N$ supergravities in $d=4$ seem to follow a similar pattern (see Section 18.6). Furthermore, this is the lowest $N$ supergravity with a timelike sector of supersymmetric solutions, non-existent for $N=1$. We are going to give fewer details than in the preceding example (especially in the null case). Details can be found in the original references.

The KSEs of ungauged $N=2, d=4$ supergravity are

$$\delta_\kappa \psi_{I\,\mu} = \mathfrak{D}_\mu \kappa_I + \varepsilon_{IJ}\, T^+{}_{\mu\nu}\gamma^\nu\, \kappa^J = 0, \tag{18.75}$$

$$\delta_\kappa \lambda^{iI} = i\, \partial\!\!\!/ Z^i \kappa^I \,+\, \varepsilon^{IJ}\, G\!\!\!/^{i\,+}\, \kappa_J = 0, \tag{18.76}$$

$$\delta_\kappa \zeta_\alpha = -i\mathbb{C}_{\alpha\beta}\, \mathsf{U}^{\beta I}{}_u\, \varepsilon_{IJ}\, \partial\!\!\!/ q^u\, \kappa^J = 0, \tag{18.77}$$

(observe that $\mathfrak{D}_\mu$ contains the pullback of the $\mathrm{SU}(2)$ connection of the quaternionic-Kähler manifold, $\mathsf{A}$) and we are going to start by assuming that the above equations admit at least one solution $\kappa^I$ for the bosonic field configuration $\{g_{\mu\nu}, A^\Lambda{}_\mu, Z^i, q^u\}$.

The bilinears that can be constructed with this single, commuting, $\kappa^I$ and their properties can be found in Section D.3.1. The antisymmetric scalar matrix $M_{IJ}$ is equivalent to a single complex scalar $X$ in the $N=2$ case $M_{IJ} = X\varepsilon_{IJ}$, and the norm of the $\mathrm{U}(2)$-neutral vector $V_\mu = V^I{}_{J\,\mu}$ is $4|X|^2 \geq 0$. The timelike and null cases are, therefore, distinguished by the value of $X$ and we have to deal with them separately.

### 18.4.1 The timelike case: supersymmetric configurations

We are going to omit the details of the translation of the KSEs from their spinorial form (in terms of $\kappa^I$) to the tensor form (in terms of the spinor bilinears $X, V^I{}_{J\,\mu}, \Phi_{IJ\,\mu\nu}$), since the procedure is the same as the one we followed in the $N = 1$ case. Analyzing the tensorial KSEs, one arrives at the following results.

1. The timelike vector $V_\mu$ is a Killing vector.
2. The three spacelike vectors $V^x{}_\mu \equiv \frac{1}{\sqrt{2}}(\sigma^x)_I{}^J\, V_J{}^I{}_\mu$, $x = 1, 2, 3$, satisfy the equation (in differential-form language, hence the hats)
$$d\hat{V}^x + \varepsilon^{xyz}\, \mathsf{A}^y \wedge \hat{V}^z = 0. \tag{18.78}$$
3. Using Eqs. (D.189), which we rewrite here for convenience, the vector field strengths are given by[19]
$$F^{\Lambda +} = V^{-2}[\hat{V} \wedge \hat{C}^{\Lambda +} + i \star(\hat{V} \wedge \hat{C}^{\Lambda +})],$$
$$C^{\Lambda +}{}_\mu \equiv V^\nu F^{\Lambda +}{}_{\nu\mu} = \mathcal{L}^{*\Lambda}\mathfrak{D}_\mu X + X^*\mathfrak{D}_\mu \mathcal{L}^\Lambda. \tag{18.79}$$
4. The complex scalars from the vector multiplets must satisfy
$$V^\mu \partial_\mu Z^i = 0. \tag{18.80}$$
5. Those from the hypermultiplets must satisfy
$$\mathsf{U}^{\alpha I}{}_u V^J{}_I{}^\mu \partial_\mu q^u - \mathsf{U}^{\alpha J}{}_u V^\mu \partial_\mu q^u = 0. \tag{18.81}$$
6. The scalar bilinear satisfies the equation
$$\mathfrak{D}_\mu X = -iT^+{}_{\mu\nu}V^\nu \quad \Rightarrow \quad V^\mu \partial_\mu X = 0. \tag{18.82}$$

To make progress we must choose coordinates and determine the general form of the metric. It is natural to choose the time coordinate by (the proportionality factor is conventional)
$$V^\mu \partial_\mu \equiv \sqrt{2}\partial_t \quad \Rightarrow \quad \hat{V} = 2\sqrt{2}|X|^2(dt + \omega), \tag{18.83}$$
for some 1-form $\omega$ in $\mathbb{R}^3$ to be determined later. Then $X$ and the $Z^i$ and the metric must be time independent. In absence of hypermultiplets the 1-forms $\hat{V}^x$ would be exact and we could use them to define the three spacelike coordinates $\hat{V}^x \sim dx^x$. In the general case we cannot do that, but we can still use those vectors thanks to their mutual orthogonality and their orthogonality to $V$, as Dreibeins of the three-dimensional spacelike metric $\gamma_{\underline{mn}}$ $m, n = 1, 2, 3$,
$$\delta_{xy}\hat{V}^x \otimes \hat{V}^y \equiv \gamma_{\underline{mn}} dx^m dx^n, \tag{18.84}$$
and, then, using Eq. (D.190) we can write the metric in the form
$$ds^2 = \frac{1}{4|X|^2}\hat{V} \otimes \hat{V} - \frac{1}{2|X|^2}\delta_{xy}\hat{V}^x \otimes \hat{V}^y, \tag{18.85}$$

[19] A more refined expression will be available after we discuss the choice of coordinates.

or

$$ds^2 = e^{2U}(dt+\omega)^2 - e^{-2U}\gamma_{\underline{mn}}dx^m dx^n, \tag{18.86}$$

where the metric (sometimes called *warp*) factor $e^{2U}$ is given by

$$e^{2U} = 2|X|^2. \tag{18.87}$$

The metric of all the timelike supersymmetric solutions of four-dimensional supergravity is always of this (so-called *conformastationary*) form, studied in Section N.2.4. It reduces to the IWP-type metric of Section N.2.3 for $\gamma_{\underline{mn}} = \delta_{\underline{mn}}$. However, the metric function $e^{2U}$, the 1-form $\omega$, and the three-dimensional metric $\gamma_{\underline{mn}}$ obey slightly different equations in each case.

Let us study $\gamma_{\underline{mn}}$ first. In the gauge in which the pullback of the $\mathrm{SU}(2)$ connection vanishes, $\mathsf{A}^x{}_t = 0$, Eq. (18.78) can then be interpreted as Cartan's first structure equation Eq. (1.152) for a torsionless connection $\varpi$ in three-dimensional $t$-independent space:

$$d\hat{V}^x - \varpi^{xy}\wedge\hat{V}^y = 0, \quad \text{with} \quad \varpi_{\underline{m}}{}^{xy} = \varepsilon^{xyz}\mathsf{A}^z{}_u\partial_{\underline{m}}q^u. \tag{18.88}$$

In other words: in supersymmetric configurations the $\mathrm{SU}(2)$ connection of the quaternionic-Kähler manifold must be embedded in the spin connection of the three-dimensional manifold with metric $\gamma_{\underline{mn}}$ so their holonomies are identical when acting on the Killing spinors. In absence of hypermultiplets, or when the hyperscalars are constant, $\gamma_{\underline{mn}} = \delta_{\underline{mn}}$.

The time independence of the three-dimensional spin connection implies via the above condition the time independence of the pullback of the $\mathrm{SU}(2)$ curvature, and then Eq. (G.31) implies that the pullback of the Quadbein is also time independent; we conclude that the hyperscalars $q^u$ are also time independent. This, together with Eqs. (18.81) and the decomposition of the vector bilinears in terms of the Vierbeins and Pauli matrices, leads to the condition

$$\mathsf{U}^{\alpha J}{}_x(\sigma^x)_J{}^I = 0, \qquad \text{where} \qquad \mathsf{U}^{\alpha J}{}_x \equiv V_x{}^{\underline{m}}\partial_{\underline{m}}q^u\mathsf{U}^{\alpha J}{}_u. \tag{18.89}$$

Next let us consider the spatial 1-form $\omega$. Taking the exterior derivative of the 1-form $\hat{V}$ in Eq. (18.83) we get

$$d\omega = \tfrac{1}{2\sqrt{2}}d(|X|^{-2}\hat{V}), \tag{18.90}$$

and using the equations satisfied by the vector bilinears we find, first,

$$d\omega = -\tfrac{i}{2\sqrt{2}}\star\left[(X\mathfrak{D}X^* - X^*\mathfrak{D}X)\wedge\frac{\hat{V}}{|X|^4}\right]; \tag{18.91}$$

after some massaging we obtain the following three-dimensional formula:

$$d\omega = -\frac{i}{2|X|^4} \star_{(3)} (X^* \mathfrak{D} X - X \mathfrak{D} X^*), \tag{18.92}$$

or, in components,

$$(d\omega)_{\underline{mn}} = -\frac{i}{2|X|^4 \sqrt{|\gamma|}} \varepsilon_{\underline{mnp}} (X^* \mathfrak{D}^{\underline{p}} X - X \mathfrak{D}^{\underline{p}} X^*). \tag{18.93}$$

At this point we have determined the general form of the timelike supersymmetric configurations. Before moving forward to the next point in the recipe, we introduce two real symplectic vectors $\mathcal{I}^M$ and $\mathcal{R}^M$. They will play a very important role in what follows since they can be seen as the fundamental building blocks of the supersymmetric field configurations (at least of the vector multiplet sector):

$$\boxed{\mathcal{R}^M \equiv \Re\mathfrak{e}(\mathcal{V}^M / X), \qquad \mathcal{I}^M \equiv \Im\mathfrak{m}(\mathcal{V}^M / X),} \tag{18.94}$$

where $\mathcal{V}$ is the symplectic section defined in Section F.1. We are going to rewrite all the fields of the supersymmetric configurations (except for the hyperscalars and the three-dimensional metric) in terms of them. First of all, the metric function is given by the symplectic product

$$\boxed{e^{-2U} = \mathcal{R}_M \mathcal{I}^M,} \tag{18.95}$$

the equation for $\omega$ can be written as

$$\boxed{d\omega = 2 \star_{(3)} \mathcal{I}_M d\mathcal{I}^M,} \tag{18.96}$$

the complex scalars are given (for instance) by

$$\boxed{Z^i = \frac{\mathcal{V}^i / X}{\mathcal{V}^0 / X} = \frac{\mathcal{R}^i + i\mathcal{I}^i}{\mathcal{R}^0 + i\mathcal{I}^0},} \tag{18.97}$$

and the symplectic vector of fundamental and dual vector field strengths (which can be extracted from Eq. (D.189)) can be written as

$$\boxed{\mathcal{F} = -\tfrac{1}{2}\left\{ d(\mathcal{R}\hat{V}) + \star(\hat{V} \wedge d\mathcal{I}) \right\}.} \tag{18.98}$$

For $\omega$ to exist, the integrability condition of its defining equation (18.96),

$$\boxed{\mathcal{I}_M d\star_{(3)} d\mathcal{I}^M = 0\,,} \tag{18.99}$$

must be satisfied everywhere. If the $\mathcal{I}^M$ are harmonic functions on the three-dimensional space with metric $\gamma_{\underline{mn}}$ then the above equation is solved everywhere except at the singularities. Canceling the contributions of these singularities to the integrability equation imposes strong constraints on the coefficients of the harmonic functions, as we will study in Chapter 19 (see Refs. [429, 108, 128]).

We have found that the necessary conditions for a bosonic field configuration is that the fields are independent and take the form in the framed equations (18.86) and (18.95)–(18.98) in terms of the real symplectic vectors defined in Eqs. (18.94) and, further, that the hyperscalars satisfy Eqs. (18.89) and are related to the three-dimensional spacelike metric by the holonomy condition Eqs. (18.88). Now we have to prove that these conditions are also sufficient by finding explicitly the Killing spinors.

The KSE $\delta_\kappa \zeta_\alpha = 0$ can be solved requiring the KSs to be annihilated by a projector $\Pi^x$,

$$\Pi^x{}_I{}^J \equiv \tfrac{1}{2}\left[\, \delta_I{}^J - \gamma^{0(x)} \sigma^{(x)}{}_I{}^J \,\right], \tag{18.100}$$

for each non-vanishing component $x = 1,2,3$ of the pullback of the Quadbein $\mathsf{U}_{\alpha I\, x} = \mathsf{U}_{\alpha I\, u} V^{x\,\mu} \partial_\mu q^u$. These three projectors commute with each other, and the three possible constraints on the KS are consistent. Each of them reduces the number of independent components of the KSs by $1/2$, but it has to be taken into account that each projector is implied by the other two and, therefore, imposing three projectors amounts to only two independent conditions on the KSs and $1/4$ of their components remain independent.

As for the meaning of these projectors, observe that $\Pi^x{}_I{}^J \kappa_J = 0$ is equivalent to

$$\sigma^x{}_I{}^J \kappa_J = \gamma^{0x} \kappa_I, \tag{18.101}$$

which states that the action of the subgroup of the $\mathrm{SU}(2)$ R-symmetry group generated by the $\sigma^x$s and that of the subgroup of the Lorentz group generated by the $\gamma^{0x}$ on the KSs must be the same. Obviously, only then can the action of the holonomies of the pullback of the $\mathrm{SU}(2)$ connection $\mathsf{A}$ and of the three-dimensional spin connection $\varpi$, related by the supersymmetry condition Eqs. (18.88), on the KSs be the same.

Solving the $\delta_\kappa \lambda^{iI} = 0$ equation requires

$$\kappa_I + i\gamma_0\, (X/X^*)^{1/2} \varepsilon_{IJ} \kappa^J = 0, \tag{18.102}$$

which is compatible with and independent of the $\Pi^x$ projectors and reduces by an additional factor of $1/2$ the number of independent components of the KSs. Using this condition and the holonomy condition Eqs. (18.88), together with the $\Pi^x$ projectors in the $\delta_\kappa \psi_{I\mu} = 0$ equation, many terms disappear and we are left with a simple differential equation whose solution is that $\kappa_I = X^{1/2} \kappa_{I\,0}$, where $\kappa_{I\,0}$ is a constant spinor.

We conclude that the necessary conditions are also sufficient for the bosonic field configurations to admit KSs of the form

$$\boxed{\kappa_I = X^{1/2}\kappa_{I\,0}, \qquad \kappa_{I\,0} + i\gamma_0\varepsilon_{IJ}\kappa^J{}_0 = 0, \qquad \Pi^x{}_I{}^J\,\kappa_{J\,0} = 0,} \tag{18.103}$$

which have $1/2, 1/4$, or $1/8$ of their components unrestricted for zero, one, or two/three $\Pi^x$ conditions imposed, respectively. This is the fraction of supersymmetry generically preserved by these field configurations.

### *18.4.2 The timelike case: supersymmetric solutions*

We start by finding which equations of motion are not automatically satisfied by the supersymmetric field configurations of these theories. Applying the general formula one gets three KSIs (one for each fermion of the theory) and, introducing the Bianchi identities to make the expressions symplectic invariant as explained in the $N = 1$ example, we get

$$\mathcal{E}_a{}^\mu\gamma^a\kappa^I - 4i\varepsilon^{IJ}\mathcal{E}_M{}^\mu\mathcal{V}^M\kappa_J = 0, \tag{18.104}$$

$$\mathcal{E}^i\kappa^I - 2i\varepsilon^{IJ}\,\not{\mathcal{E}}_M\mathcal{U}^{M*i}\kappa_J = 0, \tag{18.105}$$

$$\mathcal{E}^u\mathsf{U}^{\alpha I}{}_u\,\kappa_I = 0, \tag{18.106}$$

where the (l.h.s. of the) equations of motion of these theories have been defined in Eqs. (7.19), (6.25), (6.26), and (6.27) and are given explicitly in Eqs. (7.20)–(7.22). These equations are valid also in the gauged theories that we are going to study next, but the equations of motion will be different.

We proceed as in the $N = 1$ example to translate these spinorial equations into equations involving only the tensor spinor bilinears. The timelike and null cases have to be dealt with separately. In the timelike case, with the Vierbeins chosen as in Appendix N.2.4 we find

$$\mathcal{E}^{ab} = \delta^a{}_0\delta^b{}_0\mathcal{E}^{00}, \tag{18.107}$$

$$(\mathcal{V}_M/X)\mathcal{E}^{M\,a} = \tfrac{1}{4}|X|^{-1}\mathcal{E}^{00}\delta^a{}_0, \tag{18.108}$$

$$\mathcal{U}^*_{M\,i^*}\mathcal{E}^{M\,a} = \tfrac{1}{2}(X^*/X)^{1/2}\mathcal{E}_{i^*}\delta^a{}_0, \tag{18.109}$$

$$\mathcal{E}^u = 0. \tag{18.110}$$

The last identity indicates that the equations of motion of the hyperscalars are always satisfied for supersymmetric configurations. The first identity tells us that all the components of the Einstein equations except for $\mathcal{E}^{00}$ are automatically satisfied. The second[20] and third say that $\mathcal{E}^{00}$ and the equations of motion of the complex scalars $\mathcal{E}_i$ are proportional to the only components of the Maxwell equations and Bianchi identities that are not automatically solved: $\mathcal{E}^{M\,0}$. Thus, we just need to solve $\mathcal{E}^{M\,0}=0$.

These identities have to be obeyed *everywhere* if the field configuration is going to be supersymmetric. In particular, they must be obeyed at the singularities (or apparent singularities), where the classical equations of motion are not solved and one must place "sources" of the fields, a concept which is not rigorous in GR but which has proven useful, precisely, in the realm of supersymmetric solutions (see, for instance, Section 12.4 for the ERN BH, which is supersymmetric in pure $N=2, d=4$ supergravity, the Einstein–Maxwell theory). Then, the KSIs can be seen as constraining the sources of the fields. It has been shown in Ref. [128] that they eliminate many, if not all, of the possible singular BH solutions of these theories, such as the rotating IWP solutions of the Einstein–Maxwell theory.[21] Asking for globally preserved supersymmetry amounts to cosmic censorship in these theories, much in the spirit of Ref. [841]. In a similar fashion one can associate Eq. (18.109) with the existence of an attractor mechanism for the values of the complex scalars of the vector multiplets on the horizon of a supersymmetric BH[22] that forces those values to be related to the electric and magnetic charges of the vector fields, irrespective of their values at infinity (see Chapter 27). By the same token, Eq. (18.110) means that there is no such mechanism at work for the hyperscalars, which, after all, is not surprising since they do not couple to the vector fields.

It is not hard to see that the equations $\mathcal{E}^{M\,0}=0$ imply that the $2(n_V+1)$ components of the real symplectic vectors $\mathcal{I}^M$ defined in Eqs. (18.94) have to be harmonic functions in the three-dimensional space with the metric $\gamma_{\underline{mn}}$:

$$\boxed{d\star_{(3)} d\mathcal{I}^M = 0.} \tag{18.113}$$

In Section 18.4.1 we managed to write all the fields of a supersymmetric configuration (except for the hyperscalars, which have to be found independently, which determines $\gamma_{\underline{mn}}$) in terms of both $\mathcal{I}^M$ and $\mathcal{R}^M$, but the equations of motion seem only to constrain the former. The latter cannot be chosen to be arbitrary, though, because they are not independent of the

[20] Observe that the l.h.s. of the second identity is complex whereas the r.h.s. is real and we get two independent KSIs (apart from $\mathcal{E}^{M\,m}=0$):

$$\mathcal{R}_M \mathcal{E}^{M\,0} = \tfrac{1}{4}|X|^{-1}\mathcal{E}^{00}, \tag{18.111}$$

$$\mathcal{I}_M \mathcal{E}^{M\,0} = 0. \tag{18.112}$$

The second equation is identical to the integrability condition Eq. (18.99).

[21] Because these have "sources" of angular momentum forbidden by the KSI $\mathcal{E}^{0m}=0$. In the IWP family this is the only way a solution can have angular momentum. As we will see in Chapter 19, the supersymmetric multi-BH solutions of $N=2, d=4$ supergravity can have angular momentum that does not arise from singular sources [429, 108, 128].

[22] In the BH solutions that we are discussing, the "singularity" is, actually, the position of the event horizon of an extremal supersymmetric BH.

former: the $2(n_V+1)$ complex components of $\mathcal{V}^M/X$ depend on just $n_V+1$ variables $Z^i, X$, and, therefore, they satisfy relations that, in particular, should allow us to find $\mathcal{R}^M$ as a function of $\mathcal{I}^M$, a problem commonly known in the literature as *solving the stabilization equations* [526, 527]. The solution is strongly model dependent and can be very difficult to find. We will study some examples in Chapters 19 and 27.

### 18.4.3 The null case

As discussed in Section D.3, in the null case the $N$ spinors in $\kappa_I$ are proportional and we can write $\kappa_I = \phi_I \kappa$ for a $\mathrm{U}(N)$ vector $\phi_I$ which can be normalized $\phi^I \phi_I = 1$. Substituting $\kappa_I = \phi_I \kappa$ in the KSEs we obtain another set of KSEs with $\phi_I$ and a single spinor $\kappa$, which gives rise to a null vector, and we can proceed as in the $N=1$ case, introducing an auxiliary spinor $\eta$ to construct all the possible bilinears and, in particular, the null tetrad in Eq. (D.224). The presence of $\phi_I$ gives rise to an imaginary connection

$$\zeta \equiv \phi^I \mathfrak{D} \phi_I, \tag{18.114}$$

acting on $\kappa$.

As in the $N=1$ case, the null vector $l_\mu$ is not just Killing but covariantly constant so the metric is of Brinkmann type and we can make the same choice of coordinates as before. The consistency conditions lead to the following conclusions.

1. It is possible to choose a gauge for $\zeta_\mu$ so that its only component is proportional to $l_\mu$ and, in this gauge, the metric function $U$ and the Kähler potential $\mathcal{K}$ are related, as in the $N=1$ case, by Eq. (18.48).

2. In this gauge the vector fields and complex scalars can depend on $u$ and $z$ (they are given by Eqs. (18.43) as in the $N=1$ case) but the hyperscalars are functions only of $u$.

3. The KSIs are, in the null case,

$$\mathcal{E}_{\mu\nu} l^\nu = \mathcal{E}_{\mu\nu} m^\nu = 0, \tag{18.115}$$

$$\mathcal{V}_M \mathcal{E}^{M\,\mu} = \mathcal{U}^*_{M\,i^*} \mathcal{E}^{M\,\mu} l_\mu = \mathcal{U}^*_{M\,i^*} \mathcal{E}^{M\,\mu} m^*_\mu = 0, \tag{18.116}$$

$$\mathcal{E}^i = 0, \tag{18.117}$$

$$\mathcal{E}^u \mathsf{U}^{\alpha I}{}_u \, \phi_I = 0, \tag{18.118}$$

so the only independent equations of motion that one has to check on supersymmetric configurations are $\mathcal{E}_{\mu\nu} n^\mu n^\nu = 0$, $\mathcal{U}^*_{M\,i^*} \mathcal{E}^M{}_\mu n^\mu = 0$, $\mathcal{U}^*_{M\,i^*} \mathcal{E}^M{}_\mu m^\mu = 0$, and $\mathcal{E}^u = 0$ (a projection of this one only vanishes automatically).

The hyperscalar equations of motion are satisfied for $q$s that depend only on $u$. The Maxwell equations and Bianchi identities can be analyzed exactly as in the $N=1$

case because $\mathcal{N}_{\Lambda\Sigma}$ is not holomorphic or antiholomorphic in $z$. There are, again, two separate cases with and without dependence of the $Z^i$s on $z$. The equation for the metric function $H$ is essentially the same as in the $N=1$ case, with the obvious modifications:

$$\partial_{\underline{z}}\partial_{\underline{z}^*}H - e^{-\mathcal{K}/2}\partial_{\underline{u}}^2 e^{-\mathcal{K}/2} - e^{-\mathcal{K}}\left[\mathcal{G}_{ij^*}\partial_{\underline{u}}Z^i\partial_{\underline{u}}Z^{*\,j^*} + \mathsf{H}_{vw}\partial_{\underline{u}}q^v\partial_{\underline{u}}q^w\right] + \Im\mathfrak{m}\mathcal{N}_{\Lambda\Sigma}\partial_{\underline{z}}\varphi^\Lambda\partial_{\underline{z}^*}\varphi^{*\Sigma} = 0, \tag{18.119}$$

and in the case in which the $Z^i$s only depend on $u$ it can be easily integrated.

## 18.5 The timelike supersymmetric solutions of $N=2, d=4$ SEYM theories

In this example we are going to characterize the timelike supersymmetric solutions of a class of gauged $N=2, d=4$ supergravities with no hypermultiplets and with no Fayet–Iliopoulos (FI) terms called $N=2, d=4$ super-Einstein–Yang Mill (SEYM) theories, following Ref. [798].[23] The only symmetries that have been gauged are (necessarily non-Abelian) subgroups of the isometry group of the special Kähler manifold. The presence of non-Abelian gauge symmetries makes these theories and their possible solutions much richer, while the absence of hypermultiplets and their associated gaugings keeps them simple enough. The most general case has been studied in Ref. [940], but the problem has been solved only for the timelike class of supersymmetric solutions.[24]

In this case the KSEs take the form

$$\delta_\kappa\psi_{I\,\mu} = \mathfrak{D}_\mu\kappa_I + \varepsilon_{IJ}T^+{}_{\mu\nu}\gamma^\nu\kappa^J = 0, \tag{18.120}$$

$$\delta_\kappa\lambda^{Ii} = i\not{\mathfrak{D}}Z^i\kappa^I + \varepsilon^{IJ}[\not{G}^{i\,+} + W^i]\kappa_J = 0. \tag{18.121}$$

The structure is essentially the same as in the ungauged case, Eqs. (18.75) and (18.76), but there are two important differences: all the derivatives and vector field strengths are now gauge covariant and the gauge vectors occur explicitly in them, and there is one additional term in the second – the fermion shift $\varepsilon^{IJ}W^i\kappa_J$. The structure of the equations for the bilinears will also be very similar, as we are going to see.

### 18.5.1 Supersymmetric configurations

The analysis of this case follows very closely that of the ungauged case (setting to zero the hypermultiplets) and we use exactly the same notation and definitions, in particular for the real symplectic vectors $\mathcal{R}^M$ and $\mathcal{I}^M$. The main difference (and source of difficulties) is the need to deal with the vector fields directly (not just through their field strengths). We will not be able to write the full solution in terms of $\mathcal{R}^M$ and $\mathcal{I}^M$ only, and we will need to specify explicitly at least part of the gauge vectors. Furthermore, many of the equations that one derives are gauge dependent and a proper gauge choice has to be found.

[23] The null supersymmetric solutions have also been studied there.

[24] The problem has also been solved for both the timelike and null classes in theories with only vector multiplets and Abelian gaugings (via FI terms).

The equations that the vector bilinears satisfy have the same structure as in the ungauged case and $V_\mu$ is a Killing vector as it was there. The $\hat{V}^x{}_\mu dx^\mu$ are exact 1-forms and we can define the same time coordinate and three spacelike coordinates $x^x$ by $\hat{V}^x{}_\mu dx^\mu = dx^x$. Using these coordinates, the metric can be put in the conformastationary form Eq. (18.86) with $\gamma_{\underline{mn}} = \delta_{\underline{mn}}$ (i.e. it is what we have called an IWP-type metric, studied in Section N.2.3) and with $e^{2U} = 2|X|^2 = (\mathcal{R}_M \mathcal{I}^M)^{-1}$, as in the ungauged case.

Once we have chosen coordinates, the most convenient gauge choice turns out to be

$$\boxed{A^\Lambda{}_t = -\tfrac{1}{\sqrt{2}} e^{2U} \mathcal{R}^\Lambda,} \tag{18.122}$$

because in this gauge the complex scalars $Z^i$ (which must transform in the adjoint representation of the gauge group) and the scalar bilinear $X$ are time-independent. We will discuss the form of the vector field strengths later.

The equation of the spatial 1-form $\omega$ of the metric is formally the same as in the ungauged case Eq. (18.91), but now the derivative $\mathcal{D}_\mu$ is also covariant with respect to the gauge group and contains the gauge field $A^\Lambda{}_\mu$. This equation can be rewritten in a purely three-dimensional form, performing a KK dimensional reduction in the time direction to the space $\mathbb{R}^3$ with the Euclidean metric.[25] We also have to reduce the gauge field $A^\Lambda{}_\mu$ as we did in Section 15.2.5, using the Vierbein defined in Section N.2.3 and, as we showed there, the three-dimensional gauge field is given by

$$\boxed{\tilde{A}^\Lambda{}_{\underline{m}} \equiv A^\Lambda{}_{\underline{m}} - \omega_{\underline{m}}\, A^\Lambda{}_t.} \tag{18.123}$$

The result of this process is that the three-dimensional equation for $\omega$ takes the same form as in the ungauged case, Eq. (18.96), but with the three-dimensional exterior derivative replaced by an exterior covariant derivative denoted by $\tilde{\mathfrak{D}}$, which has the same form as $\mathfrak{D}$ but with the gauge field replaced by the one with the tilde:

$$\boxed{d\omega = 2 \star_{(3)} \mathcal{I}_M \tilde{\mathfrak{D}} \mathcal{I}^M.} \tag{18.124}$$

The integrability condition of this equation now takes the form

$$\boxed{\mathcal{I}_M \tilde{\mathfrak{D}} \star_{(3)} \tilde{\mathfrak{D}} \mathcal{I}^M = 0.} \tag{18.125}$$

We can now focus on the vector field strengths $F^\Lambda$ (and their duals $G_\Lambda$). The procedure that we follow to find them is the same as in the ungauged case, and the expressions one

[25] Observe that the IWP-type and conformastationary metric Eq. (18.86) is already written in a KK form in which the KK scalar is $e^U$, $\omega$ is the KK vector, and $\gamma_{\underline{mn}}$ ($\delta_{\underline{mn}}$ in this case) is the three-dimensional metric.

obtains are formally the same (with the obvious replacement of covariant derivatives). The outcome is, not surprisingly, identical to Eq. (18.98) with that replacement:

$$\boxed{\mathcal{F} = -\tfrac{1}{2}\left\{\mathfrak{D}(\mathcal{R}\hat{V}) + \star(\hat{V} \wedge \mathfrak{D}\mathcal{I})\right\}.} \tag{18.126}$$

We conclude that the supersymmetric field configurations of these theories are characterized, in the gauge defined by Eq. (18.122), by time-independent complex scalars $Z^i, X$, an IWP-type metric with the 1-form $\omega$ and the metric function $e^{2U}$ defined in terms of the real vectors $\mathcal{R}^M$ and $\mathcal{I}^M$ by Eqs. (18.124) and (18.95), respectively, and vector field strengths of the form Eq. (18.126). It is not difficult to show that these properties are sufficient for the field configuration to admit KSs that have exactly the same form as in the ungauged case with vanishing hyperscalars, that is Eqs. (18.103) without the projectors $\Pi^x$.

### *18.5.2 Supersymmetric solutions*

Since the general form of the KSIs depends only on the supersymmetry transformation rules of the bosonic fields and these are the same in all $N=2, d=4$ theories, the conclusion we arrived at in the ungauged case remains valid: the only independent equations of motion for supersymmetric configurations are the time components of the Maxwell equations and Bianchi identities, and we need to impose only these.

In this case, though, we are going to do this in a different way. First, we observe that, according to Eq. (18.126), the spacelike components of the supersymmetric vector field strengths can be put in the following form:

$$\boxed{\tilde{F}^{\Lambda}{}_{\underline{mn}} = -\tfrac{1}{\sqrt{2}}\epsilon_{mnp}\tilde{\mathfrak{D}}_{\underline{p}}\mathcal{I}^{\Lambda}.} \tag{18.127}$$

Comparing with Eq. (13.45) this equation can be interpreted as the Bogomol'nyi equation [227] for the gauge field $\tilde{A}^{\Lambda}{}_{\underline{m}}$ and a real "Higgs" field[26] $\mathcal{I}^{\Lambda}$. The integrability condition of this equation is obtained by taking the exterior $\tilde{\mathfrak{D}}$ derivative on both sides and using the standard Yang–Mills Bianchi identity for the gauge field with tildes, which yields

$$\boxed{\tilde{\mathfrak{D}} \star_{(3)} \tilde{\mathfrak{D}}\mathcal{I}^{\Lambda} = 0.} \tag{18.128}$$

This integrability condition is equivalent to the Bianchi identity of the four-dimensional full gauge field $A^{\Lambda}{}_{\mu}$ and the vector field strength $F^{\Lambda}{}_{\mu\nu}$ (a highly non-trivial fact) and only the Maxwell equations remain to be checked. A heavy calculation leads to the following

[26] Remember that the index $\Lambda$ is an adjoint index in the gauged theory.

single equation (as predicted by the KSIs):[27]

$$\tilde{\mathfrak{D}} \star_{(3)} \tilde{\mathfrak{D}} \mathcal{I}_\Lambda = \tfrac{1}{2} g^2 \left[ f_{\Lambda(\Sigma}{}^{\Gamma} f_{\Delta)\Gamma}{}^{\Omega}\, \mathcal{I}^{\Sigma} \mathcal{I}^{\Delta} \right] \mathcal{I}_\Omega . \tag{18.129}$$

(It can be checked that Eqs. (18.128) and (18.129) imply the integrability condition of $\omega$ up to singularities.)

How should one proceed in order to construct supersymmetric solutions of an $N = 2, d = 4$ SEYM theory? The non-linearity of non-Abelian theories cannot be avoided, but there are methods for constructing solutions which are simpler than others. For instance, one could try to find a set of functions $\mathcal{I}^\Lambda$ satisfying Eqs. (18.128), but writing these equations requires our having previously made a choice of three-dimensional gauge field $\tilde{A}^{\Lambda}_{\underline{m}}$, and solving the resulting second-order equations is always complicated.

Perhaps the best way to proceed is to find directly a solution $\tilde{A}^{\Lambda}_{\underline{x}}, \mathcal{I}^\Lambda$ of the Bogomol'nyi equation (18.127). There are many solutions of these equations, and general methods for constructing them are available in the literature on monopoles in Yang–Mills theories and one can make use of them right away. Equations (18.128) are, then, automatically solved and we can focus on Eqs. (18.129). In general, this is a horribly complicated problem, but, fortunately, for compact groups there is always a solution which is very simple (albeit not the most general, by far): $\mathcal{I}_\Lambda = \alpha \mathcal{I}^\Lambda$ for some real number $\alpha$ solves those equations because the r.h.s. vanishes identically. Having $\mathcal{I}^M$ we can proceed as in the ungauged case: solving the stabilization equations for $\mathcal{R}^M$ so we can construct the metric function $e^{-2U}$ and the scalars $Z^i$ using Eqs. (18.95) and (18.97). If the integrability condition Eq. (18.125) can be solved, then we can, in turn, solve Eq. (18.124) for $\omega$ and reconstruct the full gauge field, using the gauge-fixing condition Eq. (18.122) for the time component and $\tilde{A}^{\Lambda}_{\underline{m}}$, and its definition Eq. (18.123) for the space components. We will give a complete example in Chapter 19.

## 18.6 All the supersymmetric solutions of ungauged $N \geq 2, d = 4$ supergravity

The formalism developed in Ref. [47] and reviewed in Chapter 8 that describes all $N \geq 2$ SUEGRAs coupled to vector supermultiplets in a uniform way can be used to find, also in a uniform way, all the supersymmetric solutions of these theories. So far, only the timelike case has been completely solved, in Ref. [943]. We are going to review these results here and, although they are going to take a very complicated form that does not seem useful for the systematic construction of solutions, in the following chapter we will show how one can find, at least, the general form of the metrics of the BHs of these theories.

Using the relations between the fermionic fields of the generic supermultiplet, one is left with the following independent KSEs:

$$\mathfrak{D}_\mu \kappa_I + T_{IJ}{}^{+}{}_{\mu\nu} \gamma^\nu \kappa^J = 0, \tag{18.130}$$

$$P_{IJKL} \kappa^L - \tfrac{3}{2}\, T_{[IJ}{}^{+} \kappa_{K]} = 0, \tag{18.131}$$

[27] In components, $\tilde{\mathfrak{D}} \star_{(3)} \tilde{\mathfrak{D}} \mathcal{I}_\Lambda = \tilde{\mathfrak{D}}_{\underline{m}} \tilde{\mathfrak{D}}_{\underline{m}} \mathcal{I}_\Lambda$.

$$\not{P}_{i\,IJ}\kappa^J - \tfrac{1}{2}\not{T}_i{}^+\kappa_I = 0, \tag{18.132}$$

$$\not{P}_{[IJKL}\kappa_{M]} = 0, \tag{18.133}$$

$$\not{P}_{i\,[IJ}\kappa_{K]} = 0, \tag{18.134}$$

where, according to our general conventions, the last two KSEs should be considered only for $N = 5$ and $N = 3$, respectively. These equations can be translated to tensor language by the usual method (the algebra of spinor bilinears for arbitrary $N$ is given in Section D.3.1), and the result is the following set of equations:

$$\mathfrak{D}_\mu M_{IJ} - 2iT_{K[I|}{}^+{}_{\mu\nu}V^K{}_{|J]}{}^\nu = 0, \tag{18.135}$$

$$\mathfrak{D}_\mu V^I{}_{J\nu} + i\left[M^{IK}T_{JK}{}^+{}_{\mu\nu} - \Phi^{IK}{}_{(\mu|}{}^\rho T_{KJ}{}^+{}_{|\nu)\rho} - \text{h.c.}\right] = 0, \tag{18.136}$$

$$M^{KL}P_{KLIJ\mu} + 6iT_{[IJ|}{}^+{}_{\mu\nu}V^K{}_{|K]}{}^\nu = 0, \tag{18.137}$$

$$P_{IJKL}\cdot V^L{}_M - \tfrac{3i}{2}T_{[IJ}{}^+\cdot\Phi_{K]M} = 0, \tag{18.138}$$

$$M^{IJ}P_{iIJ\mu} + 2iT_i{}^+{}_{\mu\nu}V^\nu = 0, \tag{18.139}$$

$$P_{iIJ}\cdot V^J{}_K - \tfrac{i}{2}T_i{}^+\cdot\Phi_{IK} = 0. \tag{18.140}$$

For the $N = 5$ theory we also get

$$P_{[IJKL}\cdot V^N{}_{M]} = 0, \tag{18.141}$$

$$P_{[IJKL|\,\mu}M_{|M]N} = 0, \tag{18.142}$$

and for the $N = 3$ case we get

$$P_{i[IJ}\cdot V^L{}_{K]} = 0, \tag{18.143}$$

$$P_{i[IJ|\,\mu}M_{|K]L} = 0. \tag{18.144}$$

Using the general formula and the bosonic supersymmetry transformation rules one can also derive the general KSIs:

$$\mathcal{E}_a{}^\mu\gamma^a\kappa^I - 4i\mathcal{E}_\Lambda{}^\mu f^{*\,\Lambda\,IJ}\kappa_J = 0, \tag{18.145}$$

$$\mathcal{E}_\Lambda{}^\mu f^{*\,\Lambda\,[IJ}\gamma_\mu\kappa^{K]} - \tfrac{i}{3!}\mathcal{E}^{IJKL}\kappa_L = 0, \tag{18.146}$$

$$\mathcal{E}_\Lambda{}^\mu f^{*\,\Lambda\,i}\gamma_\mu\kappa^I - \tfrac{i}{2}\mathcal{E}^{i\,IJ}\kappa_J = 0, \tag{18.147}$$

$$\mathcal{E}^{[IJKL}{}_{K}{}^{M]} = 0, \tag{18.148}$$

$$\mathcal{E}^{i\,[IJ}{}_{K}{}^{K]} = 0, \tag{18.149}$$

where, again, the last two KSIs should be considered only for $N = 5$ and $N = 3$, respectively. As usual, the Bianchi identities $\mathcal{E}^{\Lambda\,\mu}$ can be included in these identities by replacing everywhere (with lower indices $IJ$ or $i$)

$$\mathcal{E}_{\Lambda}{}^{\mu} f^{\Lambda} \longrightarrow \mathcal{E}_{M}\mathcal{V}^{M}. \tag{18.150}$$

As we have stressed several times, these identities remain formally true after the gaugings.

We are going to study only the case in which $V_{\mu} = V^{I}{}_{I\,\mu}$ is timelike ($V^{2} = 2|M|^{2} > 0$). The analysis of these theories is far more complicated than that of the $N = 2$ theories due to one main reason: the expressions we have to deal with are formally $\mathrm{U}(N)$ covariant but the theory does not really have a local $\mathrm{U}(N)$ symmetry: scalar- (and, hence, spacetime-) dependent $\mathrm{U}(N)$ transformations appear as compensating transformations for the global duality symmetries of the theory, just as the scalar-dependent $\mathrm{U}(1)$ Kähler or $\mathrm{SU}(2)$ quaternionic-Kähler transformations that act on the spinors of the $N = 2$ theory are not truly local symmetries depending on arbitrary spacetime-dependent parameters. The parameters are the global (constant) parameters of the duality transformations. An immediate consequence of this fact is that, if the analysis of the KSEs indicates that a certain $\mathrm{U}(N)$ tensor is covariantly constant, we cannot conclude from it that there is a $\mathrm{U}(N)$ gauge in which they are actually constant.

This problem affects in particular the rank-2 Hermitian projector $\mathcal{J}^{I}{}_{J}$ defined in Eq. (D.191), which can be shown to be covariantly constant. As mentioned in Section D.3.1, if it were possible to choose a gauge in which only the first two diagonal entries were different from zero and equal to $+1$, the general problem would be reduced to an effective $N = 2$ problem, and the supersymmetric solutions of these theories would be supersymmetric solutions of $N = 2$ truncations. In spite of this possibility being commonly accepted in the literature, it has not been proven in this context, and counterexamples (supersymmetric solutions of $N = 8, d = 4$ SUEGRA which are not supersymmetric in any of the possible $N = 2$ truncations of this theory of which they are, nevertheless, solutions) have been constructed in Ref. [139]. This difficulty was first noticed in Ref. [1184], in the context of pure $N = 4, d = 4$ SUEGRA.

We are going to present the final results of the analysis of the KSEs. First of all, as usual, $V_{\mu}$ is a timelike Killing vector and the metric can be put in the conformastationary form Eq. (18.86) with the metric function $e^{-2U}$ and the spatial 1-form $\omega$ being given, again, by Eqs. (18.95) and (18.96) in terms of the real symplectic vectors $\mathcal{R}^{M}, \mathcal{I}^{M}$, which are now defined by the $\mathrm{U}(N)$-invariant expressions[28]

$$\boxed{\mathcal{R}^{M} \equiv \mathfrak{Re}\left(|M|^{-2}M^{IJ}\mathcal{V}_{IJ}{}^{M}\right), \qquad \mathcal{I}^{M} \equiv \mathfrak{Im}\left(|M|^{-2}M^{IJ}\mathcal{V}_{IJ}{}^{M}\right).} \tag{18.151}$$

[28] Observe that $e^{-2U} = \mathcal{R}_{M}\mathcal{I}^{M} = |M|^{-2} = (M^{IJ}M_{IJ})^{-2}$.

The vector field strengths are also given by the $(N=2)$-like expression Eq. (18.98). Observe that the $N=2$ case with no hypermultiplets ($M_{IJ} = X\varepsilon_{IJ}$, $\mathcal{V}^M_{IJ} = \mathcal{V}^M \varepsilon_{IJ}$) is automatically contained in these expressions.

The supersymmetry conditions on the scalars (more precisely, on the Vielbeins; for general $N$ there are no simple explicit expressions such as Eq. (18.97) for the $N=2$ case) and on the three-dimensional spacelike metric require heavy use of the projector $\mathcal{J}^I{}_J$ and its complementary $\tilde{\mathcal{J}}^I{}_J$, and also of the generalized Pauli matrices $(\sigma^m)^I{}_J$ (actually, an almost hypercomplex structure) defined in Eq. (D.200). These U($N$) tensors must satisfy the supersymmetry conditions

$$\boxed{\begin{aligned} \mathfrak{D}\mathcal{J} \equiv d\mathcal{J} - [\mathcal{J}, \Omega] &= 0, \\ \mathcal{J} d\sigma^m \mathcal{J} &= 0. \end{aligned}} \tag{18.152}$$

Using the projectors, the scalar Vielbeins are split into two subsets: those that would be the complex scalars in $N=2$ vector multiplets and those that would be hyperscalars in hypermultiplets in an $N=2$ truncation (if the projectors were constant):

$$\begin{aligned} &P_{IJKL}\, \mathcal{J}^I{}_{[M} \mathcal{J}^J{}_N \tilde{\mathcal{J}}^K{}_P \tilde{\mathcal{J}}^L{}_{Q]}, \quad \text{and} \quad P_{i\,IJ}\, \mathcal{J}^I{}_{[K} \mathcal{J}^J{}_{L]}, \\ &P_{IJKL}\, \mathcal{J}^I{}_{[M} \tilde{\mathcal{J}}^J{}_N \tilde{\mathcal{J}}^K{}_P \tilde{\mathcal{J}}^L{}_{Q]}, \quad \text{and} \quad P_{i\,IJ}\, \mathcal{J}^I{}_{[K} \tilde{\mathcal{J}}^J{}_{L]}. \end{aligned} \tag{18.153}$$

These two subsets play very different roles, analogous to the role played by the complex scalars and hyperscalars in $N=2$ theories. Only the scalars in the second subset are constrained by supersymmetry; they must satisfy

$$\boxed{\begin{aligned} P_{IJKL\,m}\, \mathcal{J}^I{}_{[M} \tilde{\mathcal{J}}^J{}_N \tilde{\mathcal{J}}^K{}_P \tilde{\mathcal{J}}^L{}_{Q]} (\sigma^m)^Q{}_R &= 0, \\ P_{i\,IJ\,m}\, \mathcal{J}^I{}_{[K} \tilde{\mathcal{J}}^J{}_{L]} (\sigma^m)^L{}_M &= 0. \end{aligned}} \tag{18.154}$$

We will see that they can be consistently truncated (just as the hyperscalars in the $N=2$ case[29]) because their equations of motion are satisfied independently of the rest of the fields. Furthermore, they will not be subject to any sort of attractor mechanism in BH solutions.

It should be possible to construct the scalars in the first subset from $\mathcal{R}^M, \mathcal{I}^M$, but explicit formulae are not available. The equations of motion of these scalars are related to those of the vector fields in such a way that they cannot be truncated consistently without imposing constraints on the latter.

[29] Compare these supersymmetry constraints with Eq. (18.89).

Finally, the three-dimensional spacelike metric $\gamma_{\underline{mn}}$ is constrained by a condition that relates its spin connection to an SU(2) projection of the generic U($N$) connection $\Omega$:[30]

$$\boxed{\varpi^{mn} = i\varepsilon^{mnp}\text{Tr}\,[\sigma^p\Omega].} \tag{18.155}$$

If we set to zero the would-be hyperscalars, we can set to zero the matrices $\sigma^p$, and this condition says that the three-dimensional metric is flat.

These conditions are sufficient to ensure that the field configuration[31] admits the KSs

$$\boxed{\kappa_I + i\sqrt{2}|M|^{-1}M_{IJ}\gamma^0\kappa^J = \Pi^{m-\,I}{}_J\kappa^J = 0, \quad \mathcal{J}^I{}_J\kappa^J = \kappa^I, \quad \kappa_I = |M|^{1/2}\kappa_{I\,0},} \tag{18.156}$$

where $\kappa_{I\,0}$ is a constant spinor, and the projection $\Pi^{m-\,I}{}_J\kappa^J = 0$, where

$$\Pi^{m\pm\,I}{}_J \equiv \tfrac{1}{2}[\delta^I{}_J \pm \gamma^{0(m)}(\sigma^{(m)})^I{}_J] = 0 \tag{18.157}$$

has to be imposed for each value of $m$ for which $P_{iIJ\,m} \neq 0$. Consistency of these projections is guaranteed by the third projector. These results are almost identical to those found in the $N = 2$ case; they have the same interpretation and lead to the same (absolute) counting of independent components of the KSs.

Finally, let us see which conditions have to satisfy these supersymmetric configurations to be solutions. The generic KSIs, Eqs. (18.145)–(18.149), with the modification Eq. (18.150), imply, for the timelike case, that the following equations of motion are automatically satisfied:

$$\mathcal{E}^{0m} = \mathcal{E}^{mn} = \mathcal{E}_M{}^m = 0, \tag{18.158}$$

$$\mathcal{E}^{MNPQ}\mathcal{J}^{[I}{}_M\tilde{\mathcal{J}}^J{}_N\tilde{\mathcal{J}}^K{}_P\tilde{\mathcal{J}}^{L]}{}_Q = 0, \tag{18.159}$$

$$\mathcal{E}^{i\,MN}\mathcal{J}^{[I}{}_M\tilde{\mathcal{J}}^{J]}{}_N = 0. \tag{18.160}$$

The last two of these equations of motion are those of the would-be hyperscalars in the $N = 2, d = 4$ *truncation*. The other two projections of the scalar equations of motion are

[30] This supersymmetry constraint should be compared with that of the $N = 2$ case, Eq. (18.88).

[31] The existence of $\{\mathcal{J}, \sigma^m\}$ with the right properties is an integral part of the solutions. This is also the main problem one faces when one tries to construct explicitly the solutions. As we shall see in the next chapter, it can be bypassed in certain conditions.

related to $\mathcal{E}_M{}^0$, but in an $N$-dependent way:

$$N=2,3 \qquad \mathcal{E}^{iIJ} = 2\sqrt{2}\frac{M^{IJ}}{|M|}\mathcal{V}_M{}^{*\,i}\mathcal{E}^{M\,0}, \tag{18.161}$$

$$N=4,5 \qquad \mathcal{E}^{IJKL} = 2\sqrt{2}\frac{M^{[IJ|}}{|M|}\mathcal{V}_M{}^{*\,|KL]}\mathcal{E}^{M\,0}, \tag{18.162}$$

$$N=4 \qquad \mathcal{E}^{iIJ} = 2\sqrt{2}\left\{\frac{M^{IJ}}{|M|}\mathcal{V}_M{}^{*\,i} + \frac{\tilde{M}^{IJ}}{|M|}\mathcal{V}_{M\,i}\right\}\mathcal{E}^{M\,0}, \tag{18.163}$$

$$N=6 \qquad \mathcal{E}^{IJ} = 2\sqrt{2}\left\{\frac{M^{IJ}}{|M|}\mathcal{V}_M{}^{*\,i} + \tfrac{1}{2}\frac{\tilde{M}^{IJKL}}{|M|}\mathcal{V}_{M\,KL}\right\}\mathcal{E}^{M\,0}, \tag{18.164}$$

$$N=8 \qquad \mathcal{E}^{IJKL} = 12\sqrt{2}\left\{\frac{M^{[IJ|}}{|M|}\mathcal{V}_M{}^{*\,KL]} + \tfrac{1}{12}\frac{\tilde{M}^{IJKLPQ}}{|M|}\mathcal{V}_{M\,PQ}\right\}\mathcal{E}^{M\,0}. \tag{18.165}$$

In all cases, the index $M$ is the symplectic index and $I, J, K, L, P, Q$ are $\mathrm{U}(N)$ indices. The identity for $N=6$ could have been written in terms of the complex conjugate of the $\mathrm{SU}(6)$ dual of the $\mathcal{E}^{IJ}$, $\mathcal{E}^{IJKL}$.

The 00 component of the Einstein equations and the 0 components of the combined Maxwell equations and Bianchi identities are related by[32]

$$\mathcal{E}^{00} - 2\sqrt{2}|M|\mathcal{R}_M\mathcal{E}^{M\,0} = 0, \tag{18.166}$$

$$\mathcal{I}_M\mathcal{E}^{M\,0} = 0. \tag{18.167}$$

The first equation can be related to the (saturated) BPS bound of the solutions, and the second can be related to the absence of sources of NUT charge as in Ref. [128].

The above identities make it clear that we need to impose only $\mathcal{E}_M{}^0 = 0$ on the supersymmetric field configurations and these equations take the same form as in the $N=2$ case, Eq. (18.113). That is: the components of the real symplectic vector $\mathcal{I}^M$ must be harmonic functions in the background of the three-dimensional spatial metric $\gamma_{\underline{mn}}$.

## 18.7 All the supersymmetric solutions of ungauged $N=1, d=5$ supergravity

In this example we are going to review the results obtained in Ref. [129], which extend those of Ref. [595] by including hypermultiplets[33] and the associated $\mathrm{SU}(2)$ connection.

[32] These KSIs are identical to those of the $N=2$ case in Eqs. (18.111) and (18.112).

[33] More general results on the gauged theories have been published in Refs. [132, 127].

We are going to use the notation and conventions of those references, which are the same as those in Chapter 9 and Appendix H.

As we have emphasized in Chapters 7 and 9, the dimensional reduction of any $N = 1, d = 5$ supergravity gives rise to an $N = 2, d = 4$ supergravity whose vector multiplet sector is a *cubic model* of the kind described on p. 231. The explicit relation between the five- and four-dimensional fields of these theories (excluding the hypermultiplets) has been obtained in Section 15.2.5 as an example of KK dimensional reduction. This relation means that the reduction of any supersymmetric solution of an $N = 1, d = 5$ ungauged supergravity admitting a spacelike translational isometry[34] should give a supersymmetric solution of the corresponding $N = 2, d = 4$ ungauged supergravity. Thus, after finding the most general supersymmetric solutions, we will pay special attention to those that have one extra translational isometry.

We start by writing the KSEs and KSIs in tensor language for both the timelike and null cases. The KSEs of these theories are

$$\delta_\kappa \psi^i_\mu = \left\{ \mathsf{D}_\mu - \tfrac{1}{8\sqrt{3}} h_I F^{I\,\alpha\beta} \left(\gamma_{\mu\alpha\beta} - 4 g_{\mu\alpha}\gamma_\beta\right) \right\} \kappa^i = 0, \tag{18.168}$$

$$\delta_\kappa \lambda^{ix} = \left(\not\partial \phi^x - \tfrac{1}{2} h^x_I \not F^I\right) \kappa^i = 0, \tag{18.169}$$

$$2\delta_\kappa \zeta^A = f_X{}^{iA} \not\partial q^X \kappa_i = 0. \tag{18.170}$$

The algebra of spinor bilinears is reviewed in Section D.3.2. The bilinears that one can construct from the commuting spinor $\kappa^i$ are a real scalar $f$, a real vector $V_\mu$ (both $\mathrm{SU}(2)$ singlets), and a Hermitian and traceless matrix of 2-forms $\Phi_i{}^j{}_{\mu\nu}$, which can be decomposed as a linear combination of the three Pauli matrices as in Eq. (D.237) and can be treated as an $\mathrm{SU}(2)$ triplet. As in the $N \geq 2, d = 4$ theories, there are two different cases distinguished by the causal nature of $V_\mu$: timelike ($f \neq 0$) and null ($f = 0$).

The translation of the KSE $\delta_\kappa \psi^i_\mu = 0$ to tensor form yields the following differential equations:

$$df - \tfrac{1}{\sqrt{3}} h_I i_V F^I = 0, \tag{18.171}$$

$$\nabla_{(\mu} V_{\nu)} = 0, \tag{18.172}$$

$$dV + \tfrac{2}{\sqrt{3}} f h_I F^I + \tfrac{1}{\sqrt{3}} h_I \star \left(F^I \wedge V\right) = 0, \tag{18.173}$$

[34] If the solution does not admit an extra spacelike isometry it cannot be dimensionally reduced to four dimensions. The requirement that the isometry be translational so it does not act with fixed points is necessary for the preservation of supersymmetry, as shown in Ref. [167]: when the isometry does not act without fixed points the KSs depend on the isometric coordinate and, therefore, they have no place in the lower-dimensional theory. In KK dimensional reductions it is implicitly assumed that the isometry is translational.

$$\mathsf{D}_\alpha \Phi^r{}_{\beta\gamma} + \tfrac{1}{\sqrt{3}} h_I F^{I\,\rho\sigma} \left( g_{\rho[\beta} \star \Phi^r{}_{\gamma]\alpha\sigma} - g_{\rho\alpha} \star \Phi^r{}_{\beta\gamma\sigma} \right.$$
$$\left. - \tfrac{1}{2} g_{\alpha[\beta} \star \Phi^r{}_{\gamma]\rho\sigma} \right) = 0, \tag{18.174}$$

while the translation of $\delta_\kappa \lambda^{ix} = 0$ and $\delta_\kappa \zeta^A = 0$ gives the following constraints:[35]

$$\pounds_V \phi^x = 0, \tag{18.176}$$

where $\pounds$ denotes the same Lie derivative defined in Section 1.8,

$$h_I^x F^I{}_{\alpha\beta} \Phi^{r\,\alpha\beta} = 0, \tag{18.177}$$

$$f d\phi^x + h_I^x i_V F^I = 0, \tag{18.178}$$

$$\Phi^r{}_{\mu\nu} \partial^\nu \phi^x + \tfrac{1}{4} \epsilon_{\mu\nu\alpha\beta\gamma} h_I^x F^{I\,\nu\alpha} \Phi^{r\,\beta\gamma} = 0, \tag{18.179}$$

$$\pounds_V q^X = 0, \tag{18.180}$$

$$f \partial_\mu q^X + \Phi^r{}_\mu{}^\nu \partial_\nu q^Y J^r{}_Y{}^X = 0. \tag{18.181}$$

As we see, $V_\mu$, whether timelike or null, is always Killing, and all the scalars $f, \phi^x, q^X$ will be independent of the adapted coordinate. Furthermore, on antisymmetrizing Eq. (18.174) in all its spacetime indices, we find that

$$\mathsf{D}\Phi^r = d\Phi^r + 2\varepsilon^{rst} \mathsf{A}^s \wedge \Phi^t = 0, \tag{18.182}$$

so the triplet of 2-forms is covariantly constant with respect to the pullback of the $\mathrm{SU}(2)$ connection in five dimensions. The rest of the equations will have to be analyzed separately in the timelike and null cases.

The KSIs that one derives directly from the general formula and the supersymmetry transformation rules of the bosonic fields Eqs. (9.10) are

$$\left( \mathcal{E}_\mu{}^\nu \gamma_\nu + \tfrac{\sqrt{3}}{2} h^I \mathcal{E}_{I\,\mu} \right) \kappa^i = 0, \tag{18.183}$$

$$\left( \mathcal{E}_x - h_x^I \not{\mathcal{E}}_I \right) \kappa^i = 0, \tag{18.184}$$

$$f_{iA}{}^X \mathcal{E}_X \kappa^i = 0, \tag{18.185}$$

where the (l.h.s. of the) equations of motion are defined in Eqs. (9.11) and given explicitly in Eqs. (9.12)–(9.15).

---

[35] Here we have used complex structures of the quaternionic-Kähler manifold,

$$J^r{}_Y{}^X = f_Y{}^{iA} \mathsf{J}^r{}_{iA}{}^{jB} f_{jB}{}^X. \tag{18.175}$$

We would like to include explicitly the Bianchi identities in the KSIs, as we did in other cases. However, in five dimensions there is no electric–magnetic duality and it is not clear how this should be done. A direct calculation of the integrability conditions shows that the first two KSIs must be replaced by

$$\left[\left(\mathcal{E}_{\mu}{}^{\sigma} + \tfrac{\sqrt{3}}{2} h_I \star \mathcal{E}^I{}_{\mu}{}^{\sigma}\right)\gamma_\sigma + \tfrac{\sqrt{3}}{2} h^I \mathcal{E}_{I\,\mu}\right]\kappa^i = 0, \tag{18.186}$$

$$\left[\mathcal{E}_x - h^I_x \left(\not{\mathcal{E}}_I + \tfrac{1}{6} a_{IJ} \not{\mathcal{E}}^J\right)\right]\kappa^i = 0, \tag{18.187}$$

where the Bianchi identities $\mathcal{E}^I$ are defined by the 3-form with components

$$\mathcal{E}^I{}_{\mu\nu\rho} \equiv 3\nabla_{[\mu} F^I{}_{\nu\rho]}. \tag{18.188}$$

In tensor language these identities take the form

$$f\left(\mathcal{E}_{\mu\rho} + \tfrac{\sqrt{3}}{2} h_I {}^\star\mathcal{B}^I{}_{\mu\rho}\right) + \tfrac{\sqrt{3}}{2} h^I \mathcal{E}_{I\,\mu} V_\rho = 0, \tag{18.189}$$

$$\left(\mathcal{E}_{\mu\rho} + \tfrac{\sqrt{3}}{2} h_I {}^\star\mathcal{B}^I{}_{\mu\rho}\right) V^\rho + \tfrac{\sqrt{3}}{2} f h^I \mathcal{E}_{I\,\mu} = 0, \tag{18.190}$$

$$\mathcal{E}_x V^\rho - f h^I_x \mathcal{E}_I{}^\rho = 0, \tag{18.191}$$

$$f\mathcal{E}_x - h^I_x \mathcal{E}_{I\,\rho} V^\rho = 0, \tag{18.192}$$

$$\mathcal{E}_X = 0. \tag{18.193}$$

In the timelike case, not all these identities are independent. Let us now focus on the timelike $V^\mu V_\mu = f^2 \neq 0$ case.

### 18.7.1 The timelike case: supersymmetric configurations

We define the adapted time coordinate $t$ by $V = \partial_t$ and write the metric in the conforma-stationary form

$$\boxed{ds^2 = f^2 (dt + \omega)^2 - f^{-1} h_{\underline{mn}} dx^m dx^n,} \tag{18.194}$$

where the time-independent fields $\omega = \omega_{\underline{m}} dx^m$ and $h_{\underline{mn}}$, $m, n = 1, \ldots, 4$, are, respectively, a 1-form and a Riemannian metric in the four-dimensional transverse space.[36]

[36] This metric is studied in Section N.3.2.

In a four-dimensional Riemannian space one can define real (anti-)self-duality of 2-forms (always with respect to the metric $h$).[37] This possibility plays a crucial role in the analysis. To start with, one defines the self- and anti-self-dual parts of the 2-form $f d\omega$ [596]:

$$\boxed{f d\omega = G^+ + G^-;} \tag{18.195}$$

$\omega$ is found by solving this equation after $f$, and $G^+$ and $G^-$ are known. It is then shown that the vector field strengths must take the form

$$\boxed{F^I = -\sqrt{3}\left\{d\left[f h^I (dt+\omega)\right] + \Theta^I\right\},} \tag{18.196}$$

where the $\Theta^I$s are spatial self-dual 2-forms and

$$\boxed{G^+ = -\tfrac{3}{2} h_I \Theta^I.} \tag{18.197}$$

There are other 2-forms that are naturally anti-self-dual: from the Fierz identities Eqs. (D.245)–(D.247) it follows that, in the coordinates chosen, the 2-form bilinears $\Phi^r$ only have spacelike components, are anti-self-dual and satisfy the algebra of the imaginary unit quaternions

$$\Phi^r{}_m{}^n \Phi^s{}_n{}^p = -\delta^{rs}\delta_m{}^p + \varepsilon^{rst}\Phi^t{}_m{}^p, \tag{18.198}$$

so the spatial manifold is endowed with an *almost* quaternionic structure (see Eq. (G.2) and Appendix G). A longer calculation shows that this structure is also SU(2)- and Lorentz-covariantly constant over the four-dimensional spatial manifold, i.e.

$$\partial_m \Phi^r{}_{np} - 2\xi_{m[n|}{}^q \Phi^r{}_{q|p]} + 2\epsilon^{rst} \mathsf{A}^s{}_m \Phi^t{}_{np} = 0, \tag{18.199}$$

where $\xi$ is the four-dimensional spin connection. This equation, like Cartan's first structure equation (which occurs in the $N=2, d=4$ case in Eq. (18.88)) should be understood as an equation for the four-dimensional spin connection. Actually, it is only for the anti-self-dual part $\xi_m^- \in \mathfrak{su}(2)_-$, which is the only part that occurs in it.[38] We can always choose

[37] The self- and anti-self-dual parts play very different roles, as we are going to see. Here the conventions (the signature, the definition of the Levi-Civita symbol $\epsilon^{a_1\cdots a_5}$, and the choice of gamma matrices and spinor conventions) play a very significant role in determining which part plays which role.

[38] The split into self- and anti-self-dual parts corresponds to the algebraic split of the algebra of the holonomy group $\mathfrak{so}(4) \cong \mathfrak{su}(2)_+ \oplus \mathfrak{su}(2)_-$. This split also plays a crucial role in the construction of the BPST instanton reviewed in Section 13.2.2: the generators $(\mathbf{M}^{(-)}_{mn})_{pq}$ are equivalent to the hyper-Kähler structure $\mathsf{J}^r$. (See also Section 13.2.2 and subsequent sections for the geometrical realizations of the same ideas.)

Vierbeins and an $\mathrm{SU}(2)$ gauge such that the $\Phi^r$ are constant and equal to the hyper-Kähler structure of $\mathbb{R}^4$, denoted by $\mathsf{J}^r$, and then the above condition is solved by

$$\boxed{\xi^-_{m\,n}{}^q = -\mathsf{A}^r{}_m \mathsf{J}^r{}_n{}^q,} \tag{18.200}$$

which is the analog of the embedding condition Eq. (18.88). In the absence of hypermultiplets, $\mathsf{A}^r$ and its field strength vanish, and so does $\xi^-$. The four-dimensional manifold is, then, a hyper-Kähler manifold.

Finally, Eq. (18.181) becomes the four-dimensional equation

$$\boxed{\partial_m q^X = \mathsf{J}^r{}_m{}^n\, \partial_n q^Y\, J^r{}_Y{}^X,} \tag{18.201}$$

which states that $q$ would be a *quaternionic map* if it was a map between hyper-Kähler manifolds [349]. In this case it can be shown that the above condition implies that $q$ is a *harmonic map*, i.e. it satisfies

$$\mathfrak{D}_m \partial^m q^X = 0. \tag{18.202}$$

(Incidentally, this equation is equivalent to the equation of motion $\mathcal{E}_X = 0$ once time independence has been taken into account.)

We have shown that the supersymmetric field configurations are characterized by the time-independent scalars $f, \phi^x, q^X$, the latter being a quaternionic map satisfying Eq. (18.201), the metric given by Eq. (18.194), where the four-dimensional metric $h$ is indirectly and partially determined by the relation Eq. (18.200) and the 1-form $\omega$ is partially related to the vector field strengths (given by Eq. (18.196)) by the condition Eq. (18.197). It can be shown that these conditions are sufficient to solve the KSEs. The KSs have the form

$$\boxed{\kappa^i = f^{1/2}\kappa^i_0, \qquad R^-\kappa^i_0 = \Pi^{r\,+}{}_j{}^i \kappa^j_0 = 0,} \tag{18.203}$$

where we have defined the projectors $R^\pm, \Pi^{r\,\pm}{}_j{}^i$ as follows:

$$R^\pm \equiv \tfrac{1}{2}\left(1 \pm \gamma^0\right), \tag{18.204}$$

$$\Pi^{r\,\pm}{}_i{}^j \equiv \tfrac{1}{2}\left[\delta \pm \tfrac{i}{4}\, \not{\mathsf{J}}^{(r)}\sigma^{(r)}\right]_i{}^j. \tag{18.205}$$

Since $\gamma^0 = \gamma^{1234}$, $R^\pm$ projects onto spinors of definite chirality in the four-dimensional space and constrains half of the components of $\kappa_0$. This ensures that the differential equation $(\partial_m + \frac{1}{4}\not{\xi}^+{}_m)\kappa_0{}^i = 0$ is automatically solved for constant spinors. The $\Pi^{r+}{}_j{}^i$ are projectors only if the $R^-\kappa^i_0 = 0$ condition is satisfied. We need to use only one of these constraints for each non-trivial component $\mathsf{A}^r$. As in the four-dimensional case, this constraint states that the action of the generators of $\mathfrak{su}(2)$ and $\mathfrak{su}(2)_-$ is the same (compare with Eq. (18.101)),

$$\kappa^j\, i\sigma^r{}_j{}^i = \tfrac{1}{4}\mathsf{J}^r{}_{mn}\gamma^{mn}\kappa^i. \tag{18.206}$$

Also as in the four-dimensional case, only two of these projectors are independent, the third being implied by the other two, and each of these projectors constrains a further half of the components of the KS spinors up to a grand total of $1/8$.

### 18.7.2 The timelike case: supersymmetric solutions

In the timelike case, using the Vielbein $e^0{}_\mu = V_\mu/f$, the KSIs Eqs. (18.189)–(18.193) take the form following:

$$\mathcal{E}_{ab} + \tfrac{\sqrt{3}}{2}h^I\mathcal{E}_{I\,(a}\delta^0{}_{b)} = 0, \tag{18.207}$$

$$h_I \star \mathcal{E}^I{}_{ab} + h^I\mathcal{E}_{I\,[a}\delta^0{}_{b]} = 0, \tag{18.208}$$

$$\mathcal{E}_x - h^I_x\mathcal{E}_I{}^0 = 0, \tag{18.209}$$

$$\mathcal{E}_X = 0, \tag{18.210}$$

from which it follows that we need to impose on the supersymmetric field configurations the full Maxwell equations and Bianchi identities. However, it turns out that they reduce to the following two equations to be solved in the four-dimensional transverse space:

$$\boxed{\begin{aligned}\nabla^2_{(4)}(h_I/f) - \tfrac{1}{4}C_{IJK}\Theta^J\cdot\Theta^K &= 0,\\ d\Theta^I &= 0.\end{aligned}} \tag{18.211}$$

These are the basic equations to be solved after a four-dimensional transverse space with the right properties in relation to the hyperscalars has been identified. However, there are several issues that we have ignored, such as the integrability of the equation for $\omega$, Eq. (18.195). We will go through all of these issues when we consider the solutions that have one additional isometry in Section 18.7.5.

### 18.7.3 The null case: supersymmetric configurations

As usual, in the null case we write $l_\mu$ instead of $V_\mu$. We define a null coordinate $v$ adapted to $l$ and the other null coordinates $u$ by

$$l^\mu\partial_\mu = \partial_{\underline{v}}, \qquad l_\mu dx^\mu = f du; \tag{18.212}$$

$f$ and the scalars $\phi^x$, $q^X$ may depend on $u$ but not on $v$, and the $v$-independent metric can be put in the form[39]

$$ds^2 = 2f du(dv + H du + \omega) - f^{-2}\gamma_{\underline{rs}} dx^r dx^s, \tag{18.213}$$

for some three-dimensional 1-form $\omega = \omega_{\underline{r}} dx^r$ and spacelike metric $\gamma_{\underline{rs}}$. A Vielbein basis for this metric and the corresponding spin connection and curvature for it are given in Section N.4. The five-dimensional Vielbein and spin connection depend on the Dreibein and spin connection of $\gamma_{\underline{rs}}$, denoted by $v^r$ and $\varpi_{\underline{r}}{}^{st}$.

The KSEs imply that the spatial components of the pullback of the SU(2) connection $\mathsf{A}^p{}_X\, \partial_{\underline{r}} q^X$ are related to those of the three-dimensional spin connection $\varpi_{\underline{r}}{}^{st}$, as in the four-dimensional case Eq. (18.88):

$$\varpi_{\underline{r}}{}^{st} = 2\varepsilon^{stp} \mathsf{A}^p{}_X \partial_{\underline{r}} q^X. \tag{18.214}$$

We should expect that a condition similar to Eq. (18.101) will have to be imposed on the Killing spinors.

For the hyperscalars we find a condition identical to the one we found in four dimensions, Eq. (18.89):

$$\partial_r q^X f_X{}^{iA} \sigma^r{}_i{}^j = 0. \tag{18.215}$$

Using hats for differential operators in the three-dimensional space of metric $\gamma$ and introducing a set of $v$-independent $n_V + 1$ 1-forms $\psi^I$, the vector field strengths can be written in the form

$$F^I = \left[\tfrac{1}{\sqrt{3}} f^2 h^I \hat{\star}\hat{d}\omega - \psi^I\right] \wedge du + \sqrt{3}\, \hat{\star}\hat{d}(h^I/f), \qquad \text{where} \qquad h_I \psi^I = 0. \tag{18.216}$$

The null supersymmetric configurations of $N = 1, d = 5$ supergravity must satisfy the above framed equations. These conditions also prove to be sufficient for the solutions to admit constant KSs satisfying the constraints

$$\gamma^+ \kappa^i = \Pi_{p\,j}{}^i \kappa^j = 0 \tag{18.217}$$

(on projection $\Pi_p$ for each direction in which $\mathsf{A}^p$ does not vanish) in the gauge in which

$$\mathsf{A}^p{}_u = -\tfrac{1}{4} \varepsilon_{prs} v_r{}^{\underline{t}} \partial_u v_{s\underline{t}}, \tag{18.218}$$

where we have introduced the mutually commuting projectors

$$\Pi_p = \tfrac{1}{2}\left[1 - i\gamma^p (\sigma^{(p)})^T\right]. \tag{18.219}$$

[39] $l_\mu$ is not covariantly constant in general and this is not a Brinkmann metric.

### 18.7.4 The null case: supersymmetric solutions

In the null case ($f=0$) the KSIs Eqs. (18.189)–(18.193) reduce to

$$\mathcal{E}_{\mu\rho}l^{\rho} + \tfrac{\sqrt{3}}{2}h_I \star \mathcal{E}^{I}{}_{\mu\rho}l^{\rho} = 0, \tag{18.220}$$

$$h^I \mathcal{E}_{I\,\mu} = h^I_x \mathcal{E}_{I\,\rho} l^{\rho} = 0, \tag{18.221}$$

$$\mathcal{E}_x = \mathcal{E}_X = 0, \tag{18.222}$$

which implies that we have many more equations to check than in the timelike case. They can be put in the following form:

$$\begin{aligned}
\hat{\nabla}^2 K^I &= 0,\\
\hat{\nabla}^2 L_I &= 0,\\
\hat{\nabla}^2 N &= 0,\\
\nabla_s(\dot{\gamma})_{rs} - \partial_r(\dot{\gamma})_{ss} - 3f^3\dot{K}_I\partial_r K^I - g_{XY}\dot{q}^X\partial_r q^Y &= 0,\\
\hat{\star}\hat{d}\omega - \sqrt{3}\left(L_I dK^I + K^I dL_I\right) + 3K_I\dot{\alpha}^I &= 0,\\
C_{IJK}\left[\hat{\nabla}_r\left(K^J\dot{\alpha}^K\right)_r + \partial_r K^J\left(\dot{\alpha}^K\right)_r\right] &= 0,\\
\nabla_r(\dot{\omega})_r + 3(\dot{\omega})_r\partial_r \log f + \tfrac{1}{2}f^{-3}(\ddot{\gamma})_{rr} + \tfrac{1}{4}f^{-3}(\dot{\gamma})^2 - \tfrac{3}{2}f^{-4}\dot{f}(\dot{\gamma})_{rr}&\\
- 3C_{IJK}K^I\left(\dot{K}^J\dot{K}^K + (\dot{\alpha}^J)_r(\dot{\alpha}^K)_r + \tfrac{2}{\sqrt{3}}(\dot{\alpha}^J)_r\partial_r M^K\right)&\\
+ \tfrac{1}{2}f^{-3}g_{XY}\dot{q}^X\dot{q}^Y + 12f^3\left(K_I\dot{K}^I\right)^2 &= 0.
\end{aligned} \tag{18.223}$$

The last two Eqs. (18.223) arise as gauge-fixing conditions for the spatial 1-forms $\alpha^I$ (to be defined below) and for $\omega$ that simplify the other equations.

The functions $K^I, K_I, L_I, N$, and three-dimensional 1-forms $\alpha^I = \alpha^I_r dx^r$, in terms of which are written the above equations of motion, are defined as follows:[40]

$$K^I \equiv h^I/f, \tag{18.224}$$

[40] The equation for $\omega$ is not exactly an equation of motion, but we write it here as one of the equations that need to be solved.

$$K_I \equiv C_{IJK} K^J K^K, \tag{18.225}$$

$$\hat{d}\alpha^I = \hat{\star}\hat{d}K^I, \tag{18.226}$$

$$L_I \equiv C_{IJK} K^J M^K, \tag{18.227}$$

$$N \equiv H - \tfrac{1}{2} L_I M^I = H - \tfrac{1}{2} C_{IJK} K^I M^J M^K, \tag{18.228}$$

while the functions $M^I$ are functions of $x^r$ and $u$ which must be found by solving the equations of motion.

Once the harmonic functions $K^I, L_I, N$ are chosen (this determines directly the functions $K_I$ and indirectly the functions $M^I$), the 1-forms $\alpha^I$ are found by integrating its definition and then going to the gauge in which the sixth Eq. (18.223) is satisfied. This allows one to integrate the equation for $\gamma$ (the fourth in Eq. (18.223)) whose $u$-dependence is not determined by the holonomy condition Eq. (18.214). Finally, the equation for $\omega$ (the fifth Eq. (18.223)) can be integrated going to the gauge in which the final Eq. (18.223) is satisfied.

The functions that appear in the metric and in the vector field strengths Eqs. (18.213), (18.216) are given in terms of those building blocks by

$$f^{-3} = K_I K^I, \qquad h^I = f K^I,$$

$$H = N - \tfrac{1}{2} L_I M^I, \quad \psi^I = f^3 K^I \left( L_J \hat{d} K^J - K^J \hat{d} L_J - \sqrt{3} K_I \dot{\alpha}^I \right) + \hat{d} M^I. \tag{18.229}$$

### 18.7.5 Solutions with an additional isometry

The equations of motion obtained for the supersymmetric solutions of $N=1, d=5$ supergravity simplify greatly with the assumption of an additional isometry. The resulting solutions are interesting by themselves and because they can be reduced to solutions of $N=2, d=4$ supergravity, providing potentially interesting relations between five- and four-dimensional BH (for instance), their entropies, etc.

Again, we have to deal separately with the timelike and null cases.

*Timelike supersymmetric solutions.* In the absence of hypermultiplets, the four-dimensional Riemannian space is a hyper-Kähler manifold with self-dual spin connection (the anti-self-dual part vanishes due to Eq. (18.200)), and, if it admits a triholomorphic Killing vector (i.e. a Killing vector that also preserves the full hyper-Kähler structure), then it has been shown in Ref. [628] that it must be a Gibbons–Hawking space [613, 614] whose metric is given by Eq. (13.13) (taking the $+$ sign, since we want self-duality, not anti-self-duality) [596, 595].

In the presence of hypermultiplets it is natural to generalize the Gibbons–Hawking metric by adding a non-trivial three-dimensional metric $\gamma_{\underline{rs}}$ whose generic $\mathfrak{su}(2)$ holonomy will become the $\mathfrak{su}(2)_-$ part of the four-dimensional holonomy so that Eq. (18.200) can be satisfied [129]:

$$
\boxed{
\begin{aligned}
&h_{\underline{mn}} dx^m dx^n = H^{-1}(dz+\chi)^2 + H\gamma_{\underline{rs}} dx^r dx^s, \\
&\chi = \chi_{\underline{r}} dx^r, \qquad \hat{d}H = +\hat{\star}\hat{d}\chi, \\
&\hat{\nabla}^2 H = 0,
\end{aligned}
}
\tag{18.230}
$$

where $r, s = 1, 2, 3$, all the fields are independent of the isometric coordinate $z$, and the hatted differential operators refer to the three-dimensional background metric $\gamma_{\underline{rs}}$.

With the orientation $\varepsilon_{z123} = +1$ and the Vierbein basis

$$
V^z = H^{-1/2}(dz+\chi), \qquad V^r = H^{1/2} v^r, \tag{18.231}
$$

where the $v^r$ is the Dreibein of $\gamma_{\underline{rs}}$, we find that the anti-self-dual part of the spin connection of the four-dimensional metric $h_{\underline{mn}}$, $\xi^-$, is related to the spin connection of the three-dimensional metric $\gamma_{\underline{rs}}$, $\varpi$, by[41]

$$
\xi_{\underline{r}}^{-\,zs} = -\tfrac{1}{2}\varepsilon^{stu}\,\varpi_{\underline{r}}{}^{tu}, \tag{18.232}
$$

and supersymmetry requires that it is, in turn, related to the pullback of the SU(2) connection by

$$
\boxed{\xi_{\underline{r}}^{-\,zs} = -2\mathsf{A}^s{}_X\,\partial_{\underline{r}} q^X.} \tag{18.233}
$$

These metrics and hyperscalars are the most general timelike supersymmetric ones that have one additional isometry. The solutions are then formally identical to those found in the absence of hyperscalars in Ref. [595], the only difference being that the $2n_V+4$ functions $K^I, L_I, M, H$ on which the fields these solutions depend are harmonic in the three-dimensional metric $\gamma_{\underline{rs}}$. The explicit relation between the fields and these functions is

$$
\begin{aligned}
h_I/f &= L_I + H^{-1}C_{IJK}K^J K^K, \\
\Theta^I &= [(dz+\chi)\wedge d(K^I/H) + H\hat{\star}d(K^I/H)], \\
\omega &\equiv \omega_5(dz+\chi) + \hat{\omega}, \\
\omega_5 &= M + \tfrac{3}{2}H^{-1}L_I K^I + H^{-2}C_{IJK}K^I K^J K^K, \\
\hat{\star}d\hat{\omega} &= HdM - MdH + \tfrac{3}{2}\left(K^I dL_I - L_I dK^I\right).
\end{aligned}
\tag{18.234}
$$

[41] The self-duality condition $\hat{d}H = +\hat{\star}\hat{d}\chi$ ensures neither $H$ nor $\chi$ contribute to $\xi^-$.

The metric function $f$ can be determined using the identities of real special geometry in Appendix H, but there is no general expression available except for symmetric spaces [595]. For $n_V = 0$ it is just

$$f^{-1} = L_0 + (K^0)^2/H. \tag{18.235}$$

The complete five-dimensional metric can be cast in the form

$$\begin{aligned}
ds^2 &= -k^2[dz + B]^2 + k^{-1}\left[e^{2U}(dt+\hat{\omega})^2 - e^{-2U}\gamma_{\underline{rs}}dx^r dx^s\right], \\
e^{2U} &= \frac{fH^{-1}}{\sqrt{f^{-1}H^{-1} - f^2\omega_5^2}}, \qquad k^2 = f^{-1}H^{-1} - f^2\omega_5^2, \\
B &= \chi - f^2\omega_5 k^{-2}(dt+\hat{\omega}).
\end{aligned} \tag{18.236}$$

It is easy to see the form of the $N = 2, d = 4$ supersymmetric solution that will appear after dimensional reduction (the expression in brackets) and the KK vector field ($B$). The rest of the fields can be obtained from Eqs. (15.131). These relations and results have been used to study the relations between four- and five-dimensional supersymmetric solutions (especially BHs) in Refs. [136, 137, 574, 509, 573, 138, 114]. We will study some examples in Chapter 19.

*Null supersymmetric solutions.* In the null case, the additional isometry we are interested in is actually null and we will require all the fields to be $u$-independent. Since they are already $v$-independent, they will be independent of the spacelike coordinate $y \equiv \frac{1}{\sqrt{2}}(u - v)$.

The gauge-fixing conditions (the last two Eqs. (18.223)) as well as the equation for $\gamma$ (the fourth Eq. (18.223)) are automatically solved for $u$-independent solutions. These are, then, completely determined by the choice of harmonic functions $K^I, L_I, N$ (which implies, indirectly, that of $M^I$), and one just has to integrate the equation for $\omega$ (whose integrability equations are solved automatically by the harmonicity of $K^I$ and $L_I$ except at the singularities) and reconstruct the fields of the solution from them using Eqs. (18.229). (We are assuming that the supersymmetry constraints of the hyperscalars have been solved, of course.)

Not surprisingly, these solutions and the constraints resemble very much the ones of the $N = 2, d = 4$ timelike supersymmetric solutions. We know that these theories are related by dimensional reduction and so are the supersymmetric solutions for translational isometries. We can make the relation more explicit by rewriting the metric Eq. (18.213), for the $u$-independent case, in the following form:

$$\begin{aligned}
ds^2 &= -k^2[dy + A]^2 + k^{-1}\left[e^{2U}\left(dt + \tfrac{1}{\sqrt{2}}\omega\right)^2 - e^{-2U}\gamma_{\underline{rs}}dx^r dx^s\right], \\
e^{2U} &= \left(\frac{f^3}{1-H}\right)^{1/2}, \qquad k^2 = (1-H)f, \\
A &= -(1-H)^{-1}\left(Hdt + \tfrac{1}{\sqrt{2}}\omega\right),
\end{aligned} \tag{18.237}$$

from which the four-dimensional metric can be easily read (it is the expression in square brackets).

This four-dimensional solution should be compared to the one in Eqs. (18.236). If the four-dimensional solution is static (the three-dimensional $\omega = 0$), $A$ is purely electric in this case, while $B$ has electric and magnetic parts (via $\chi$), so these solutions are clearly less general, which we should have suspected because they depend on one harmonic function less.

# 19

# Supersymmetric black holes from supergravity

The construction of classical solutions of GR, with or without matter, is always a difficult task. In the preceding chapters we have constructed and studied the (static) black-hole solutions of some simple theories by using educated ansatz for the symmetries and general form of the metric, solution-generating transformations and other strategies, but the complexity of the solutions of the axion–dilaton model (Section 16.2) and the effort necessary to find them suggests that we must use more powerful methods or restrict ourselves to simpler kinds of solutions if we want to do the same for more complicated theories.

A combination of these two possibilities arises for the supersymmetric solutions of supergravity theories. Considering only supergravity theories is, by itself, a restriction, albeit not a very strong one, in view of the large number of possible models. Their supersymmetric solutions are usually the simplest in their class but they still have interesting physics. In any case they can be used as a first approximation to more complicated and *realistic* solutions and, sometimes, the latter can be obtained as deformations of the former. While the results obtained using supersymmetric solutions as tractable, solvable, toy models should not be extrapolated lightly to the non-supersymmetric ones,[1] it is clear that there is a great deal to be learned from them. For instance, it is in this class of solutions that the *attractor mechanism* that holds for all extremal (not necessarily supersymmetric) black-hole solutions was first discovered [531, 1157, 526, 527].

In Chapter 18 we studied a method that allows the complete characterization of those solutions to the point that, in many interesting cases, a recipe can be given to construct them all.[2] These solutions include, depending on the choices of symmetries, functions, and

[1] As we have stressed repeatedly, the physical properties of a family of solutions depending in an apparently continuous form on several parameters are not necessarily continuous functions of those parameters. For instance, the temperature of a Schwarzschild black hole diverges in the zero-mass (Minkowski) limit. Thus, we should not assume automatically that the physical properties of supersymmetric solutions are the naive supersymmetric limits of those of the general family.

[2] Some of the general classes of solutions that we are going to study were originally found by a series of ansatz of increasing generality. For instance, the supersymmetric solutions of $N = 2, d = 4$ supergravity coupled to vector supermultiplets were found in Ref. [937] (which we will follow here) through the spinor bilinear method explained in Chapter 18. However, the most general black-hole solutions of this class had already been constructed in Refs. [115, 429, 108] generalizing the partial results of Refs. [531, 168, 112, 1092, 1093].

integration constants, cosmic strings, pp waves and other interesting solutions, plus many other less interesting and typically singular solutions of difficult interpretation. In this chapter we are going to review in full detail how to construct the black-hole-type ones. We will also construct a few supersymmetric solutions of other kinds because of their intrinsic interest or because they are related to other black-hole solutions via dimensional reduction, for example.

## 19.1 Introduction

Which supergravity theories admit supersymmetric black-hole solutions (SBHSs)? Where, among all the supersymmetric solutions of these theories, do the SBHSs lie? Which properties do we expect these solutions to share?

In Chapter 18 we established a relation between supersymmetric (or BPS) solutions of a supergravity theory and supersymmetric states of the theory in representations of the supersymmetry algebra. Asymptotically flat SBHSs should be associated with massive, point-like states with the same conserved charges saturating a supersymmetry (or BPS) bound. Thus, the independent parameters of SBHSs will be their electric and magnetic charges (the latter, only in $d=4$, since there are no point-like magnetic charges in $d \neq 4$), their angular momenta, and the asymptotic values of the scalars that we will call, in this context, moduli, but not on the mass, which will be completely determined by the supersymmetry bound.

In $d=4$ there are no massive supersymmetric states for $N=1$ (there is no central charge and the BPS bound is $M=0$) and, therefore, we can only find SBHSs in $N \geq 2$ supergravities. All the five-dimensional supergravities have SBHSs but there are none (that are supersymmetric and regular[3]) in $d \geq 6$.[4]

There are several important differences between the four- and five-dimensional SBHSs to be noted:

1. The angular momentum (one independent parameter $J$ in $d=4$ and two, $J_1, J_2$, in $d=5$) does not enter in the supersymmetry algebra (neither in $d=4$ nor in $d=5$), but in $d=4$ all rotating ($J \neq 0$) single-SBHSs are singular, while in $d=5$ there is a one-parameter ($J_1 = J_2$) family of rotating single-SBHSs which are regular [598]. We have studied their maximally supersymmetric near-horizon limits in Section 17.4.2.

2. In $d=5$ the topology of the black-hole horizons is not restricted to be that of $S^3$ like the topology of the four-dimensional ones is restricted to be that of $S^2$. This was shown by the construction of a *black-ring* solution in pure five-dimensional GR by Emparan and Reall in Ref. [512]. The event horizon of this solution has the topology of $S^2 \times S^1$. A generalization with electric charge was constructed in Ref. [506] and a regular supersymmetric black-ring solution of minimal (pure $N=1$) five-dimensional supergravity was constructed in Ref. [507]. This solution was rewritten in Ref. [594] in terms of the harmonic functions that naturally occur

---

3 There are point-like supersymmetric solutions in $d \geq 6$ but they are always singular.

4 There is no propagating gravity below $d=4$ and the three-dimensional BTZ black hole [95] is an asymptotically $AdS_3$ solution (in fact, it is locally $AdS_3$). We will not study it here.

in the classification of the supersymmetric solutions that we have reviewed in Chapter 18, and choosing harmonic functions with more poles allows for the construction of concentric multi-black-ring solutions. Further generalizations have been made in Refs. [140, 508, 595, 1010]. We will review the construction of the simplest supersymmetric black-ring solutions later on, although most of the time we will focus on static SBHSs.

We have found that in $d=4$ and $d=5$ the supersymmetric solutions can be classified as timelike or null depending on the causal nature of the spinor bilinear $V_\mu$. The timelike ones can be associated to massive states and, therefore, we expect SBHSs to be timelike supersymmetric.[5] The metrics that we found for these supersymmetric solutions are of the expected form (conformastationary), as we are going to discuss.

An important consequence of the saturation of the supersymmetry bounds is the stability of the SBHSs under classical and quantum perturbations. The corollary is the absence of Hawking radiation: SBHSs must have zero Hawking temperature. This, in turn, implies that (in the $d$-dimensional static case) their metrics can always be written in the conformastatic form [939, 56] (see Chapter 27)

$$ds_d^2 = e^{2U}dt^2 - e^{-\frac{2}{d-3}U}d\vec{x}_{d-1}^2. \tag{19.1}$$

Then, the $(d-1)$-dimensional spacelike metric must be flat. Since we showed in Chapter 18 that the presence of active (non-constant) hyperscalars in supersymmetric solutions forces that metric (denoted by $\gamma_{\underline{mn}}$) to have the same holonomy as the pullback of the $\mathrm{SU}(2)$ connection $\mathsf{A}$ of the quaternionic-Kähler manifold, we conclude that SBHSs must have constant hyperscalars (or what we called *would-be hyperscalars* in Section 18.6 for $N>2$) and we will just ignore them from now on.

In the above metric there is only one independent metric function, $U$, although the full solution will depend on more functions associated with the scalar and vector fields that we can call for the moment $H^i$. Actually, $U$ can written as a function of these $H^i$, reflecting the fact that the mass of the solutions can also be written in terms of the charges and asymptotic values of the scalars (the BPS bound). In most cases these functions will be harmonic functions in Euclidean $\mathbb{E}^{d-1}$ whose poles and integration constants have to be carefully chosen for the solution to be asymptotically flat and regular. If the metric is only stationary it will include a spatial 1-form $\omega$, which will also depend, in a more complicated form, on the $H^i$.

The harmonicity of the $H^i$ opens up the possibility of having solutions describing several black holes in equilibrium, generalizations of the MP solutions of the Einstein–Maxwell theory in Eqs. (12.86). In the Einstein–Maxwell case the stationary generalization of the MP solutions (the IWP solutions, Eqs. (13.83)) are always singular, but in the supergravity theories that have scalar fields there are regular multi-black-hole solutions with non-vanishing angular momentum [429, 108, 128]. As in the case of the $d=5$ rotating SBHSs, the angular momentum is chiefly carried by the electromagnetic fields. We will study a complete example with two centers.

[5] A supersymmetric solution can be both timelike and null supersymmetric at the same time: most supersymmetric solutions have more than one Killing spinor and the $V_\mu$s constructed with them may have different causality properties. This is true for most maximally supersymmetric vacua, for instance. It is, however, immaterial for our purposes.

The entropy of SBHSs need not vanish even if the temperature must,[6] and we will see that for static black holes it is a function of the charges (which take integer values after quantization) and not of the moduli (which are continuous parameters). This property, which goes by the name of the *attractor mechanism*, suggests that the black-hole entropy has something to do with counting states and that it has a microscopic interpretation. An interpretation in the context of string theory was given by Strominger and Vafa in Ref. [1160] soon after the attractor mechanism was discovered in SBHSs of $N = 2, d = 4$ supergravity in Refs. [531, 1157, 526, 527]. It was later shown in Ref. [522] that this property holds for all the static, extremal (supersymmetric or not) black-hole solutions of $d = 4$ theories with actions of the general form Eq. (2.147). We will prove this result and we will extend it to higher dimensions in Chapter 27, where we will also study extremal non-supersymmetric solutions and their deformation to more *realistic* non-extremal solutions. In this chapter we will deal only with SBHSs.

## 19.2 The supersymmetric black holes of ungauged $N = 2, d = 4$ supergravity

### *19.2.1 The general recipe*

According to the general discussion in the introduction, the general recipe for constructing SBHSs of an ungauged $N = 2, d = 4$ supergravity characterized by the canonical covariantly holomorphic symplectic section $\mathcal{V}^M$ and some quaternionic-Kähler geometry is just the recipe for constructing timelike supersymmetric solutions with vanishing hyperscalars. It can be summarized as follows [937]:

1. Introduce the auxiliary function $X$ with the same Kähler weight as $\mathcal{V}^M$ (namely, 1) and define the Kähler-neutral real symplectic vectors $\mathcal{R}^M$ and $\mathcal{I}^M$ as follows:

$$\boxed{\mathcal{R}^M + i\mathcal{I}^M \equiv \mathcal{V}^M / X.} \tag{19.2}$$

2. The components of $\mathcal{R}^M$ and $\mathcal{I}^M$ are not independent: the former can be entirely expressed in terms of the latter, which is equivalent to solving the *stabilization equations* (also called the *Freudenthal duality equations* [268]). We will give the solutions for some models in the examples.

   The (generically non-linear) mapping $\mathcal{I}^M \longrightarrow \mathcal{R}^M(\mathcal{I})$ defines an operation called *Freudenthal duality* that can be extended to any symplectic vector in any theory of $N = 2, d = 4$ supergravity [577].[7] We will indicate the Freudenthal of any symplectic vector by a tilde, so, for instance, $\mathcal{R} = \tilde{\mathcal{I}}$. It can be shown that the Freudenthal dual of

[6] If it does, the SBHS is singular. These solutions are sometimes called *small black holes*, and it has been argued that quantum corrections can give rise to a regular event horizon (see, for instance, Ref. [411]). We will not study in detail these classically singular solutions here.

[7] This transformation was first introduced in the context of $N = 8, d = 4$ supergravity and theories associated to Freudenthal algebras in Ref. [238] and it was extended later to other $d = 4$ supergravities in Ref. [536] in a slightly different fashion.

a vector is a homogeneous function of degree 1 of the components of the original vector and that the operation is an anti-involution, i.e.[8]

$$\tilde{\tilde{\mathcal{I}}}^M = -\mathcal{I}^M. \tag{19.3}$$

The symplectic product of a symplectic vector and its Freudenthal dual is a symplectic invariant and, therefore, it is duality invariant; it is called the *Hesse potential* $\mathsf{W}$ and it is homogeneous of degree 2 in the components of the vector and characteristic of the model. For instance, for $\mathcal{I}^M$:

$$\boxed{\mathsf{W}(\mathcal{I}) = \tilde{\mathcal{I}}_M \mathcal{I}^M = \mathcal{R}_M(\mathcal{I})\mathcal{I}^M.} \tag{19.4}$$

Knowing the Hesse potential for the given model one can find the Freudenthal dual that provides a solution to the stabilization (or Freudenthal duality) equations as follows:

$$\boxed{\tilde{\mathcal{I}}_M \equiv \tfrac{1}{2}\frac{\partial \mathsf{W}}{\partial \mathcal{I}^M}.} \tag{19.5}$$

If one can prove that in a given theory there is only one duality invariant with the required degree of homogeneity in the components of the symplectic vectors, then it can be identified (up to normalization constants) with the Hesse potential and the Freudenthal duality equations can be automatically solved. This procedure works for theories with a high degree of symmetry only.

3. The components of $\mathcal{I}^M$ are real functions $H^M$ in Euclidean three-dimensional space $\mathbb{E}^3$, i.e.

$$\boxed{\partial_{\underline{p}}\partial_{\underline{p}} H^M = 0.} \tag{19.6}$$

The choice of $H^M$ determines completely the solution, since it can be written entirely in terms of $\mathcal{I}^M$ and $\mathcal{R}^M$, which is a function of $\mathcal{I}^M$. We will denote $\mathcal{R}^M(\mathcal{I})$ for $\mathcal{I}^M = H^M$ by $\tilde{H}^M$, since it is just the Freudenthal dual of $H^M$:

$$\boxed{\mathcal{I}^M = H^M, \qquad \mathcal{R}^M(\mathcal{I}) \equiv \tilde{H}^M.} \tag{19.7}$$

In the following two subsections we study carefully the choices that lead to a single static black hole and to multi-black-hole solutions. Observe that the symplectic

[8] We will prove these properties and study them in more detail in Chapter 27.

vector of functions $H^M$ transforms linearly under duality transformations, with no other factors because it is Kähler neutral.

The rest of the recipe is just an explanation of how the physical fields are constructed from $\tilde{H}^M(H)$ and $H^M$.

4. The metric has the conformastationary form

$$\boxed{ds^2 = e^{2U}(dt+\omega)^2 - e^{-2U}d\vec{x}_3^{\,2},} \tag{19.8}$$

where the metric function $e^{-2U}$ and the spatial 1-form $\omega$ are given by

$$\boxed{\begin{aligned} e^{-2U} &= \frac{1}{2|X|^2} = \mathsf{W}(H), \\ \partial_{[\underline{m}}\omega_{\underline{n}]} &= \epsilon_{mnp} H_M \partial_{\underline{p}} H^M. \end{aligned}} \tag{19.9}$$

In order to find $\omega$ we must integrate the above differential equation, which is possible only if the integrability conditions

$$\boxed{H_M \partial_{\underline{p}} \partial_{\underline{p}} H^M = 0} \tag{19.10}$$

hold. Equation (19.6) may make us naively think that the integrability condition above is always automatically satisfied, but harmonic functions generically have poles at which the Laplace equation is not satisfied and their contribution must be canceled somehow. The cancelation works in different ways and imposes different constraints on single- and multi-black-hole solutions.

5. The vector field strengths and the complex scalar fields are given by

$$\boxed{\begin{aligned} \mathcal{F}^M &= -\tfrac{1}{\sqrt{2}} d(\tilde{H}^M e^{2U}) \wedge dt - \tfrac{1}{\sqrt{2}} e^{2U} \star (dt \wedge dH^M), \\ Z^i &= \frac{\tilde{H}^i + iH^i}{\tilde{H}^0 + iH^0}. \end{aligned}}$$

Let us now see how to select the harmonic functions to construct single- and multi-SBHSs.

### 19.2.2 Single-black-hole solutions

It has been shown in Ref. [128] that the single-SBHSs of these theories are necessarily static if they are regular and, therefore, we can set $\omega = 0$. The vanishing of the r.h.s. of the equation for $\omega$ becomes a constraint that we have to take into account:

$$H_M \partial_{\underline{p}} H^M = 0. \tag{19.11}$$

The metric of a single static black hole is spherically symmetric and we can put it in these two common forms:

$$ds^2 = e^{2U}dt^2 - e^{-2U}\left[dr^2 + r^2 d\Omega^2_{(2)}\right] = e^{2U}dt^2 - e^{-2U}\frac{1}{\rho^2}\left[\frac{1}{\rho^2}d\rho^2 + d\Omega^2_{(2)}\right], \tag{19.12}$$

for two radial coordinates[9] $r = |\vec{x}| =\in (0, +\infty)$ and $\rho \in (-\infty, 0)$, which are related by $r = -1/\rho$. The left limit of these intervals corresponds to the horizon (or the singularity, if there is no regular horizon) and the right limit to spatial infinity. We will use a dot to denote derivation with respect to $\rho$.

Our experience with the MP solutions Eqs. (12.86) teaches us that, in order to have a SBHS describing a single, static, spherically symmetric object, the harmonic functions $H^M$ must be of the form

$$H^M = A^M + \tfrac{1}{\sqrt{2}}\frac{B^M}{r} = A^M - \tfrac{1}{\sqrt{2}}B^M \rho, \tag{19.13}$$

for two real constant symplectic vectors $A^M$ and $B^M$ (the $\sqrt{2}$ is a conventional normalization) which must be functions of the independent physical parameters of the black hole. These parameters are the electric $q_\Lambda$ and magnetic $p^\Lambda$ charges combined into $(\mathcal{Q}^M) = \left(\begin{smallmatrix} p^\Lambda \\ q_\Lambda \end{smallmatrix}\right)$ and the moduli $Z^i_\infty$.[10] Our goal in what follows is to find this dependence, to determine under which conditions the solution is asymptotically flat and regular, and deduce the values of the mass and the entropy in terms of the independent physical parameters of the solution.

First Eq. (19.11) becomes the following constraint for $A^M$ and $B^M$:

$$H_M \dot{H}^M = 0 \quad \Rightarrow \quad A_M B^M = 0. \tag{19.14}$$

The charges of a SBHS with the above harmonic functions can be computed by integrating the vector field strengths in Eq. (19.11) over a 2-sphere at infinity, and the result is $\mathcal{Q}^M = B^M$. This is one of the main characteristics of single-SBHSs: in Chapter 27 we will find extremal non-supersymmetric solutions with the same functional form whose main difference from the supersymmetric ones is that $B^M \neq \mathcal{Q}^M$.

To find the constants $A^M$ we start by rewriting the auxiliary function $X$ as

$$X = \tfrac{1}{\sqrt{2}} e^{U + i\alpha}, \tag{19.15}$$

[9] $\rho$ is also called $\tau$ in the literature.

[10] This is a total of $4n_V + 2$ real parameters against the $4n_V + 4$ components of $A^M$ and $B^M$, but we have to take into account the constraint Eq. (19.11) and the normalization of the metric at infinity.

for some function $\alpha$. The normalization of the metric at spatial infinity ($r \to +\infty$ or $\rho \to 0^-$),

$$e^{-2U}(A) = \mathsf{W}(A) = 1, \tag{19.16}$$

implies that at spatial infinity

$$X_\infty = \tfrac{1}{\sqrt{2}} e^{i\alpha_\infty}. \tag{19.17}$$

Taking the same limit in the definition of $\mathcal{I}^M = H^M$ in Eq. (19.2) ($H^M \to A^M$) and using Eq. (19.17) we get

$$A^M = \sqrt{2}\Im\mathfrak{m}(e^{-i\alpha_\infty}\mathcal{V}^M_\infty). \tag{19.18}$$

Using the antisymmetry of the symplectic product, the definition of $H^M = \mathcal{I}^M$, Eq. (19.15), and (18.14) we can rewrite the constraint Eq. (19.14) for $B^M = \mathcal{Q}^M$ in the form

$$A_M\mathcal{Q}^M = H_M\mathcal{Q}^M = \Im\mathfrak{m}(\mathcal{V}_M/X)\mathcal{Q}^M = \sqrt{2}e^{-U}\,\Im\mathfrak{m}(e^{-i\alpha}\mathcal{Z}) = 0. \tag{19.19}$$

Then, everywhere,

$$e^{i\alpha} = \pm\mathcal{Z}/|\mathcal{Z}|, \tag{19.20}$$

and we get the following general expressions for the integration constants $A^M$ in terms of charges $\mathcal{Q}^M$ and the asymptotic values of the scalar fields $Z^i_\infty$:

$$A^M = \pm\sqrt{2}\,\Im\mathfrak{m}\left(\frac{\mathcal{Z}^*_\infty}{|\mathcal{Z}_\infty|}\mathcal{V}^M_\infty\right). \tag{19.21}$$

So far, we have required asymptotic flatness and the fulfillment of Eq. (19.14), and we still have to require the metric to be regular. Two necessary conditions are the positivity of the mass (otherwise there is no event horizon covering the singularity) and the finiteness of the entropy (otherwise the event horizon has zero area and, again, there is a naked singularity).

The coordinates that we are using are what we called *isotropic* when we studied the ERN black hole (which is a particular example of the kind of solutions we are constructing here). Equation (12.82) tells us that in the expansion of $e^{-2U} = \mathsf{W}$ around spatial infinity the coefficient of the $1/r$ term (in $G_{\rm N}^{(4)} = 1$ units and with the standard normalization factor $(16\pi G_N^{(4)})^{-1}$ in the action) is twice the mass $M$. Equivalently, expanding in powers of $\rho$ around $\rho = 0$, the coefficient of $\rho$ is $-2M$, so

$$\boxed{M = -\tfrac{1}{2}\,\dot{\mathsf{W}}\Big|_{\rho=0}}\,. \tag{19.22}$$

Using the chain rule, the definition, Eq. (19.5), of $\tilde{H}_M$, the constraint Eq. (19.11), and the definition Eq. (19.2) we get

$$M = -\,\tilde{H}_M\dot{H}^M\Big|_{\rho=0} = -\,(\tilde{H}_M + iH_M)\dot{H}^M\Big|_{\rho=0} = -\,X^{-1}\mathcal{V}_M\dot{H}^M\Big|_{\rho=0} = \pm|\mathcal{Z}_\infty|, \tag{19.23}$$

where we have used Eqs. (19.13), (19.17), and (19.20) and the definition of central charge Eq. (18.14). Only the upper sign in Eqs. (19.20) and (19.21) leads to a positive mass and we have completely determined the harmonic functions:

$$\boxed{H^M = \sqrt{2}\,\Im\mathrm{m}\left(\frac{\mathcal{Z}^*_\infty}{|\mathcal{Z}_\infty|}\mathcal{V}^M_\infty\right) - \tfrac{1}{\sqrt{2}}\mathcal{Q}^M\rho.} \tag{19.24}$$

On the other hand, we have proven that all the single-SBHSs saturate the BPS bound of the theory, as they should.

The entropy is just $\pi$ times the radius of the 2-spheres in the $\rho \to -\infty$ limit squared. In this limit, the constants $A^M$ (which, in these SBHSs, are the only ones that depend on the moduli) in the harmonic functions $H^M$ can be ignored and, taking into account that $e^{-2U} = \mathsf{W}(H)$ is homogeneous of second order in $H$, it must behave as

$$e^{-2U} \sim \tfrac{1}{2}\mathsf{W}(\mathcal{Q})\rho^2; \tag{19.25}$$

plugging this result into the metric Eq. (19.12) we find that

$$\boxed{S/\pi = \tfrac{1}{2}\,\mathsf{W}(\mathcal{Q}),} \tag{19.26}$$

a duality-invariant expression which is totally independent of the moduli. This is the main implication of the attractor mechanism.

The regularity of the solution requires $\mathsf{W}(\mathcal{Q}) > 0$, a condition that may not be satisfied for arbitrary values of the charges. In the regions in charge space in which it is not satisfied, there are no SBHSs with those charges, but there may be other non-supersymmetric extremal black-hole solutions. It is not clear if there is always an extremal black-hole solution for any choice of charges in any theory. This question has to be studied case by case.

If we take the same limit in the physical scalars Eq. (19.11), we find

$$Z^i_{\mathrm{h}} = \frac{\tilde{\mathcal{Q}}^i + i\mathcal{Q}^i}{\tilde{\mathcal{Q}}^0 + i\mathcal{Q}^0}, \tag{19.27}$$

i.e. functions of the charges only, independent of the values at infinity (the moduli $Z^i_\infty$). Usually the scalars cannot take arbitrary values and this imposes further conditions on the possible values of the charges for which there is a well-defined SBHS.

The above near-horizon limits are called *attractors* (hence the name *attractor mechanism*) and it is this behavior that is usually understood as fundamental in the literature. However, as we will see in Chapter 27, this behavior is far from general in extremal black holes: *the values of the scalars on the horizon of an extremal static black hole generically depend on the moduli*. It is only for SBHSs and a handful of non-supersymmetric extremal black holes that they do not, but we will show that, in all cases, *the entropy of an extremal static black hole is moduli independent*.

In SBHSs the scalars always flow from the possible (but otherwise arbitrary) asymptotic values $Z^i_\infty$ to the supersymmetric attractor fixed point $Z^i_{\rm h}(\mathcal{Q})$. A possible choice of asymptotic values is $Z^i_\infty = Z^i_{\rm h}(\mathcal{Q})$ and it is achieved by setting $A^M = B^M/\beta = \mathcal{Q}^M/(\beta\sqrt{2})$ for some constant $\beta$ in all the harmonic functions. As a result, all of them are proportional,

$$H^M = \frac{1}{\sqrt{2}\beta}\mathcal{Q}^M H, \qquad H \equiv 1 - \beta\rho, \tag{19.28}$$

and, since $\mathsf{W}(H)$ is homogeneous of degree 2 in $H^M$, normalizing $\beta^2 = \mathsf{W}(\mathcal{Q})/2$ we find for the metric factor and the scalar fields

$$e^{-2U} = H^2, \qquad Z^i = Z^i_{\rm h}(\mathcal{Q}), \qquad H = 1 - \sqrt{\mathsf{W}(\mathcal{Q})/2}\rho. \tag{19.29}$$

That is: the scalars are constant and the metric is that of the ERN black hole with a charge $\sqrt{\mathsf{W}(\mathcal{Q})/2}$ which can be identified with the value of the central charge when the scalar fields take their supersymmetric attractor values $\mathcal{Z}_{\rm h}$.

These black holes are known as *double extremal black holes*.

In general, the entropy can be related to the value of the central charge on the horizon (since the scalars only depend on the charges in that limit, the central charge is moduli independent there). It is not difficult to show using the formulae of this section that

$$\boxed{S/\pi = |\mathcal{Z}_{\rm h}|^2 = \mathsf{W}(\mathcal{Q})/2, \qquad \mathcal{Z}_{\rm h} = \mathcal{Z}(Z_{\rm h}, Z^*_{\rm h}, \mathcal{Q}).} \tag{19.30}$$

### *19.2.3 Multi-black-hole solutions*

Our experience with MP solutions suggests that choosing a harmonic function with $n_C$ point-like singularities

$$H^M = A^M + \tfrac{1}{\sqrt{2}}\sum_{a=1}^{a=n_C} \frac{B^M_a}{|\vec{x} - \vec{x}_a|} \tag{19.31}$$

gives rise to a multi-SBHS describing $n_C$ black holes with their horizons placed at the singularities $\vec{x} = \vec{x}_a$. It is clear that the coefficients $B^M_a$ must be identified with charges of each black hole, and we replace $B^M_a$ by $\mathcal{Q}^M_a$ from now on. We do not need to restrict ourselves to static configurations $\omega = 0$ as in the MP case: it is possible to have non-static multi-SBHSs which are completely regular, and we are going to study this general case, which includes the static one for particular choices of the integration constants.

Our first concern is to find under which conditions the integrability condition for $\omega$ Eq. (19.10) is satisfied. For the above harmonic functions it takes the form

$$\sum_b \left\{ A_M + \tfrac{1}{\sqrt{2}} \sum_{a\neq b} \frac{\mathcal{Q}_{a\,M}}{|\vec{x}_b - \vec{x}_a|} \right\} \mathcal{Q}^M_b \delta^{(3)}(\vec{x} - \vec{x}_b) = 0, \tag{19.32}$$

where the antisymmetry of the symplectic product has eliminated the poles $1/|\vec{x}_b - \vec{x}_a|$ with $a = b$. The condition is satisfied everywhere except at the singularities. To cancel their contribution we must require each of the coefficients of the delta functions to vanish, that is [429, 108]

$$\left\{ A_M + \tfrac{1}{\sqrt{2}} \sum_{a \neq b} \frac{\mathcal{Q}_{a\,M}}{|\vec{x}_b - \vec{x}_a|} \right\} \mathcal{Q}_b^M = 0. \tag{19.33}$$

Summing these equations over $b$ we find the condition

$$A_M\, \mathcal{Q}^M = 0, \qquad \mathcal{Q}^M \equiv \sum_b \mathcal{Q}_b^M, \tag{19.34}$$

which is identical to Eq. (19.14) for the total charge of the configuration and enforces the absence of total NUT charge. Actually, each of the above conditions can be understood as the requirement that each of the possible individual sources of NUT charge vanishes [128].

From the above condition Eq. (19.34) we arrive at the expressions Eq. (19.21) and (19.23) for $A^M$ and the total mass $M$ in terms of the total charge and moduli, which we can write

$$M = \tfrac{1}{\sqrt{2}} \tilde{A}_M \mathcal{Q}^M = \sum_a \tfrac{1}{\sqrt{2}} \tilde{A}_M \mathcal{Q}_a^M \equiv \sum_a \mathsf{M}_a. \tag{19.35}$$

In spite of its appearance, this expression is not the sum of the masses that we would assign to each of the individual centers $M_a$ if they were completely isolated because $M_a \neq \mathsf{M}_a$. The reason is that $A^M$ is a non-linear function of the total charge and moduli $A^M(Z_\infty, Z^*_\infty, \mathcal{Q})$ and, if the center $a$ were isolated, we would have $M_a = \tilde{A}_{a\,M} \mathcal{Q}_a^M |\mathcal{Z}_{a\,\infty}|$, where $\tilde{A}_{a\,M} \equiv A^M(Z_\infty, Z^*_\infty, \mathcal{Q}_a)$ satisfies $A_{a\,M} \mathcal{Q}_a^M = 0$ and $\mathcal{Z}_{a\,\infty} = \mathcal{V}_{M\,\infty} \mathcal{Q}_a^M$.

From a different point of view: the total central charge is the sum of the central charges of each center $\mathcal{Z}_\infty = \sum_a \mathcal{Z}_{a\,\infty}$ but the phases of these are different and the absolute value of the total one is not the sum of the absolute values of the individual central charges.

The regularity of the metric requires that the coefficients of $1/|\vec{x} - \vec{x}_a|$ in an expansion $1/|\vec{x} - \vec{x}_a| << 1$, which are, precisely, $2\mathsf{M}_a$, are positive.

After the constants $A^M$ have been determined as a function of the moduli and total charges solving Eq. (19.34), the constraints Eqs. (19.33) become a set of $n_C - 1$ independent linear equations for the $n_C(n_C - 1)/2$ inverse distances between the centers $|\vec{x}_b - \vec{x}_a|$. The $n_C - 1$ distances of all the centers to a given one ($a = 1$, say) can be written as functions of the $(n_C - 2)(n_C - 1)/2$ remaining ones, but it is clear that these are not entirely free parameters since they must satisfy the triangular inequality. Furthermore, only for certain values of the moduli and charges do the functions that give the $n_C - 1$ distances $|\vec{x}_1 - \vec{x}_a|$ give positive values. Thus, determining the most general consistent and regular multi-SBHSs (there are yet more conditions to be met) becomes an extremely difficult problem for more than two centers even in the simplest $N = 2, d = 4$ theories.

To make further progress we must integrate the equation for $\omega$, Eq. (19.9). Assuming the integrability conditions Eqs. (19.33) are satisfied, the $\omega$ equation takes the form

$$2\partial_{[\underline{m}}\omega_{\underline{n}]} = \epsilon_{mnp} \sum_{a<b} \mathcal{Q}_{a\,M} \mathcal{Q}_b^M \left[ \frac{(x^p - x_a^p)}{|\vec{x} - \vec{x}_a|^3} \left( \frac{1}{|\vec{x} - \vec{x}_b|} - \frac{1}{|\vec{x}_a - \vec{x}_b|} \right) - (a \leftrightarrow b) \right]. \tag{19.36}$$

We can integrate separately the contributions of each pair $(a<b)$ to $\omega$, $\omega^{(a,b)}$,

$$2\partial_{[\underline{m}}\omega^{(a,b)}_{\underline{n}]} = \epsilon_{mnp}F^{(a,b)\,\underline{p}},$$
$$F^{(a,b)\,\underline{p}} \equiv \mathcal{Q}_{a\,M}\mathcal{Q}^M_b\left[\frac{(x^p-x^p_a)}{|\vec{x}-\vec{x}_a|^3}\left(\frac{1}{|\vec{x}-\vec{x}_b|}-\frac{1}{|\vec{x}_a-\vec{x}_b|}\right)-(a\leftrightarrow b)\right], \tag{19.37}$$

as follows:

1. Shift the coordinates to place the origin at the $a$th black hole:
$$\vec{y}\equiv\vec{x}-\vec{x}_a, \qquad \vec{y}_b\equiv\vec{x}_b-\vec{x}_a. \tag{19.38}$$
2. Rotate the coordinates so the $b$th black hole is on the positive third axis of the new coordinates:
$$\vec{z}\equiv R\,\vec{y}, \qquad \vec{z}_b=R\,\vec{y}_b=(0,0,l), \quad \text{where} \quad l\equiv|\vec{y}_b|=|\vec{x}_b-\vec{x}_a|. \tag{19.39}$$
3. Transform the system to spherical coordinates:
$$\begin{aligned}\partial_{[\theta}\omega^{(a,b)}_{\varphi]} &= r^2\sin\theta F^{(a,b)}_r,\\ \partial_{[r}\omega^{(a,b)}_{\varphi]} &= -\sin\theta F^{(a,b)}_\theta,\\ \partial_{[r}\omega^{(a,b)}_{\theta]} &= \sin\theta F^{(a,b)}_\varphi.\end{aligned} \tag{19.40}$$
4. $F^{(a,b)}_\varphi=0$. Taking $\omega_r=\omega_\theta=0$ and massaging the expressions a little we get the following two equations:
$$\partial_\theta\omega^{(a,b)}_\varphi = -\frac{\mathcal{Q}_{a\,M}\mathcal{Q}^M_b}{l}\sin\theta\left[1-\frac{r+l}{r_l}+\frac{rl(r+l)}{r_l^3}(1-\cos\theta)\right], \tag{19.41}$$
$$\partial_r\omega^{(a,b)}_\varphi = -\mathcal{Q}_{a\,M}\mathcal{Q}^M_b\frac{\sin^2\theta}{r_l^3}(r-l), \tag{19.42}$$
where
$$r_l\equiv|\vec{x}-\vec{x}_b|=\sqrt{r^2-2lr\cos\theta+l^2}. \tag{19.43}$$
The first equation can be integrated immediately:
$$\omega^{(a,b)}_\varphi = \frac{\mathcal{Q}_{a\,M}\mathcal{Q}^M_b}{l}\left[-C+\cos\theta+\frac{r+l}{r_l}(1-\cos\theta)+f(r)\right], \tag{19.44}$$
where $C$ is an integration constant and $f(r)$ is a function of $r$ only to be determined by substituting this result into the second equation. This turns out to be exactly solved

for a constant $f(r)$ that we set to zero. Note that $C$ has to take the value 1 in order for $\omega_{\varphi}^{(a,b)}$ to vanish when $r \to 0$ and $r_l \to 0$ (so there are no "sources of angular momentum" at the position of the black holes, which is a necessary condition for the regularity of the solution).

Then, the contribution of the pair $(a, b)$ to $\omega$ is[11]

$$\boxed{\omega_{\varphi}^{(a,b)} = \frac{\mathcal{Q}_{a\,M}\mathcal{Q}_b^M}{l}(\cos\theta - 1)\left[1 - \frac{r+l}{r_l}\right].} \tag{19.45}$$

Observe that the symplectic product $\mathcal{Q}_{a\,M}\mathcal{Q}_b^M$ is the natural generalization to many Maxwell fields of the quantity that satisfies the DSZ quantization condition discussed in Section 12.7.4. This quantity is proportional to the angular momenta of the Maxwell fields, which should be quantized in the same way. The angular momenta of the Maxwell fields are at the origin of the angular momentum of the gravitational field, since the centers are static and, as we are going to see, the horizons are static (there are no sources of angular momentum there) thanks to the boundary conditions chosen.

This contribution behaves for large values of $r$ as

$$\omega_{\varphi}^{(a,b)} \sim \mathcal{Q}_{a\,M}\mathcal{Q}_b^M \frac{\sin^2\theta}{r}, \tag{19.46}$$

and on comparing with Eq. (13.2) we see that this pair of black holes gives a contribution to the angular momentum along the axis that joins them equal to $\mathcal{Q}_{a\,M}\mathcal{Q}_b^M$, so the total angular momentum of the multi-SBHS is

$$\boxed{\vec{J} = \textstyle\sum_{a<b} \vec{J}^{(a,b)}, \qquad \vec{J}^{(a,b)} = \mathcal{Q}_{a\,M}\mathcal{Q}_b^M \dfrac{\vec{x}_b - \vec{x}_a}{|\vec{x}_b - \vec{x}_a|}.} \tag{19.47}$$

Going back to Cartesian coordinates, on undoing the rotation and the shift we find

$$\boxed{\begin{aligned} \omega &= \textstyle\sum_{a<b} \omega^{(a,b)}, \\ \omega_{\underline{m}}^{(a,b)} &= \mathcal{Q}_{a\,M}\mathcal{Q}_b^M \epsilon_{mnp}(x^n - x_a^n)(x_b^p - x_a^p) f^{(a,b)}(x), \\ f^{(a,b)}(x) &= -\frac{|\vec{x} - \vec{x}_b| - |\vec{x} - \vec{x}_a| - l}{l^2|\vec{x} - \vec{x}_a||\vec{x} - \vec{x}_b|[l|\vec{x} - \vec{x}_a| + (\vec{x}_b - \vec{x}_a)\cdot(\vec{x} - \vec{x}_a)]}. \end{aligned}} \tag{19.48}$$

[11] This is just a solution in a local patch and in a certain gauge (because $\omega$ is defined up to shifts by the total derivative of an arbitrary function). It is enough for our purposes, but in order to check the regularity of the solution one must find the global form of $\omega$, just as for the Taub–NUT solution.

It can be checked that the limits of these contributions at all of the centers are non-vanishing but finite. This is a very relevant fact when we study the behavior of the metric at each center. In the limit $\vec{x} \to \vec{x}_c$ and using the local coordinate $r \equiv |\vec{x} - \vec{x}_c| \to 0$ the term $\frac{1}{\sqrt{2}}\mathcal{Q}_c^M/r$ dominates in the harmonic functions and we have the local version of Eq. (19.25)

$$e^{-2U} \to \frac{\mathsf{W}(\mathcal{Q}_c)}{r^2}. \tag{19.49}$$

This implies that in this limit $g_{tt} = e^{2U}$ will vanish (the horizon's infinite redshift) and $g_{t\underline{m}} = e^{2U}\omega_{\underline{m}}$ and $e^{2U}\omega_{\underline{m}}\omega_{\underline{n}}$ will also vanish because $\omega$ is finite at the centers. The last term will not contribute to the near-horizon geometry, which will be that of a single-SBHS with charge vector $\mathcal{Q}_c$ (namely $\mathrm{AdS}_2 \times \mathrm{S}^2$) if $\mathsf{W}(\mathcal{Q}_c) > 0$, a condition that we must impose in order to have regularity.

The total entropy of the system will be

$$\boxed{S = \textstyle\sum_a S_a, \qquad S_a/\pi = \frac{1}{2}\mathsf{W}(\mathcal{Q}_a),} \tag{19.50}$$

and it will in general be different from the entropy of an equivalent single-SBHS with the total charge

$$S_{\text{equiv}}/\pi = \tfrac{1}{2}\mathsf{W}(\mathcal{Q}) = \tfrac{1}{2}\tilde{\mathcal{Q}}_M\mathcal{Q}^M = S/\pi + \sum_{a<b}\tilde{\mathcal{Q}}_{a\,M}\mathcal{Q}_b^M. \tag{19.51}$$

An interesting and obvious question is whether $S_{\text{equiv}}$ is larger or smaller than the actual entropy $S$ in general configurations. For an observer placed at a distance much larger than the relative distance between the centers, the whole system comprises an approximately spherically symmetric isolated object with mass $M$, charge $\mathcal{Q}^M = \sum_a \mathcal{Q}_a^M$, angular momentum $\vec{J} = \sum_{a<b}\vec{J}^{(a,b)}$, and moduli $Z^i_\infty$, which may correspond, from a *microscopical* point of view, to different multi-SBHS configurations with the same total charges, and the difference between $S_{\text{equiv}}$ and $S$ could perhaps be accounted for by this degeneracy.

Let us now study some examples.

### *19.2.4 Examples of single-SBHSs: stabilization equations*

In Section 19.2.2 we showed that, if we solve the stabilization (or Freudenthal-duality) equations of an $N=2, d=4$ theory, then we can construct algorithmically the whole solution for given values of the moduli and charges: the harmonic functions $H^M$ are given by Eq. (19.24) and with the solution of the Freudenthal-duality equations we can construct, first, their Freudenthal dual $\tilde{H}^M$ and then the Hesse potential (hence, the metric function), the vector field strengths, and the scalar fields.

We have already studied the main physical properties of the resulting SBHSs under the assumption that we have solved the Freudenthal-duality equations, and therefore here we are going to review the solutions in some of the few theories for which they are known. As we have seen, they are necessary to construct multi-SBHSs and non-Abelian solutions too. We will follow the order in which we presented the examples of ungauged $N=2, d=4$ theories in Section 7.2.1, where the canonical symplectic sections etc. of these theories can be found.

*Pure $N=2, d=4$ supergravity.* This theory was presented in the formalism of special Kähler geometry on p. 226. By definition,

$$(\mathcal{R}^M + i\mathcal{I}^M) = \begin{pmatrix} X^{-1} \\ -\frac{i}{2}X^{-1} \end{pmatrix} \quad \Rightarrow \quad \mathcal{R}_0 + i\mathcal{I}_0 = -\tfrac{i}{2}(\mathcal{R}^0 + i\mathcal{I}^0), \tag{19.52}$$

from which we find the solution

$$\boxed{(\tilde{H}^M) = \begin{pmatrix} -2H_0 \\ \frac{1}{2}H^0 \end{pmatrix}, \qquad \mathsf{W}(H) = \tfrac{1}{2}(H^0)^2 + 2(H_0)^2.} \tag{19.53}$$

The central charge is just

$$\mathcal{Z} = -(q_0 + \tfrac{i}{2}p^0), \tag{19.54}$$

and the explicit form of the harmonic functions is

$$H^M = \tfrac{1}{\sqrt{2}}\frac{\mathcal{Q}^M}{|\mathcal{Z}|}H, \qquad H \equiv 1 - |\mathcal{Z}|\rho \;\Rightarrow\; e^{-2U} = \mathsf{W}(H) = H^2, \tag{19.55}$$

so the solution is the dyonic ERN solution Eq. (12.206) (up to normalization of the charges). Observe that in this theory the charges of SBHSs can take any values (as long as they do not vanish simultaneously). Furthermore, the general formalism has shown to us that there is only one independent harmonic function, which never vanishes in the region that lies outside the event horizon $\rho \in (-\infty, 0)$ so that the metric is completely regular there.

*The axion–dilaton model.* This model was introduced on p. 227. Following the definition,

$$(\mathcal{R}^M + i\mathcal{I}^M) = \frac{1}{2X\sqrt{\Im\mathrm{m}\tau}} \begin{pmatrix} i \\ \tau \\ -i\tau \\ 1 \end{pmatrix} \quad \Rightarrow \quad \begin{cases} \mathcal{R}^0 + i\mathcal{I}^0 = i(\mathcal{R}_1 + i\mathcal{I}_1), \\ \mathcal{R}^1 + i\mathcal{I}^1 = i(\mathcal{R}_0 + i\mathcal{I}_0), \end{cases} \tag{19.56}$$

so that

$$\boxed{\tilde{H}^M = \mathcal{A}^M{}_N H^N, \quad (\mathcal{A}^M{}_N) \equiv \begin{pmatrix} 0 & -\sigma^1 \\ \sigma^1 & 0 \end{pmatrix}, \quad \mathsf{W}(H) = 2(H^0H^1 + H_0H_1).} \tag{19.57}$$

The central charge is now

$$\mathcal{Z} = \frac{1}{2\sqrt{\Im\mathrm{m}\tau}}\left[(p^1 - iq_0) - (q_1 + ip^0)\tau\right], \tag{19.58}$$

and we cannot simplify the general form of the harmonic functions. The metric function $e^{-2U} = \mathsf{W}(H)$ takes the explicit form

$$e^{-2U} = 1 + \frac{2M}{r} + \frac{S/\pi}{r^2}, \qquad M = |\mathcal{Z}_\infty|, \qquad S/\pi = \tfrac{1}{2}\mathsf{W}(\mathcal{Q}) = p^0p^1 + q_0q_1, \tag{19.59}$$

and its regularity is guaranteed by the positivity of the mass and the entropy, which is a non-trivial condition the charges must satisfy. There are extremal black-hole solutions with charges satisfying $p^0p^1 + q_0q_1 < 0$, but they are not supersymmetric and have to be obtained by other methods, explained in Chapter 27. The same condition is enough to guarantee that $\Im\mathfrak{m}\tau_{\rm h} > 0$ (which is $\tau_\infty$-independent, as expected) in this case, since[12]

$$\tau = \frac{H^1 + iH_0}{H_1 - iH^0} = \frac{H^1H_1 - H^0H_0 + \frac{i}{2}\mathsf{W}(H)}{(H_1)^2 + (H^0)^2} \longrightarrow \frac{p^1q_1 - p^0q_0 + \frac{i}{2}\mathsf{W}(\mathcal{Q})}{(q_1)^2 + (p^0)^2} = \tau_{\rm h}. \tag{19.60}$$

*The $\overline{\mathbb{CP}}^n$ models.* These models have been introduced on p. 230. Again, according to the definition,

$$(\mathcal{R}^M + i\mathcal{I}^M) = e^{\mathcal{K}/2}X^{-1}\begin{pmatrix} Z^\Lambda \\ -\frac{i}{2}Z_\Lambda \end{pmatrix} \quad \Rightarrow \quad \mathcal{R}^\Lambda + i\mathcal{I}^\Lambda = 2i\eta^{\Lambda\Sigma}(\mathcal{R}_\Sigma + i\mathcal{I}_\Sigma), \tag{19.61}$$

and we get

$$(\tilde{H}^M) = \begin{pmatrix} \tilde{H}^\Lambda \\ \tilde{H}_\Lambda \end{pmatrix} = \begin{pmatrix} -2\eta^{\Lambda\Sigma}H_\Sigma \\ \frac{1}{2}\eta_{\Lambda\Sigma}H^\Sigma \end{pmatrix}, \qquad \mathsf{W}(H) = \tfrac{1}{2}\eta_{\Lambda\Sigma}H^\Lambda H^\Sigma + 2\eta^{\Lambda\Sigma}H_\Lambda H_\Sigma. \tag{19.62}$$

The positivity of $\mathsf{W}$ is not automatic and, on the horizon, it imposes a constraint on the charges $\mathsf{W}(\mathcal{Q}) > 0$. This condition guarantees at the same time that the scalar fields satisfy the constraint $\eta_{\Lambda\Sigma}Z^\Lambda Z^{*\Sigma} > 0$ on the horizon. The metric can be written as in the axidilaton case in the form Eq. (19.59). This is obviously true for all the quadratic models.

*The cubic models.* These models were introduced on p. 231. The stabilization equations have a very complicated form, but a highly non-trivial way of solving them under certain assumptions was found by Shmakova in Ref. [1136]. The main assumption (and it is not a small one) is that one can solve the equations

$$d_{ijk}\tilde{Y}^j\tilde{Y}^k = 2p^0q_i + d_{ijk}p^jp^k \tag{19.63}$$

for the auxiliary variables $\tilde{Y}^i$. If the solutions are available, then, defining the quantities

$$\Delta \equiv \tfrac{1}{3!}d_{ijk}p^ip^jp^k, \qquad \tilde{\Delta} \equiv \tfrac{1}{3!}d_{ijk}\tilde{Y}^i\tilde{Y}^j\tilde{Y}^k, \tag{19.64}$$

[12] Remember that we have defined $\tau \equiv i\mathcal{X}^1/\mathcal{X}^0$ and we have to take the $i$ into account.

the Hesse potential for the charge vector can be written as

$$\mathsf{W}(\mathcal{Q}) = 2\sqrt{4(\tilde{\Delta}/p^0)^2 - (p^\Lambda q_\Lambda + 2\Delta/p^0)^2}, \tag{19.65}$$

and the Freudenthal dual (now a highly non-linear function) can be found using Eq. (19.5) applied to the charge vector.

The Hesse potential is in all these models twice the square root of a function which is homogeneous of order 4 in the charges and which should be, by construction, invariant under the duality transformations of the model. We will denote this invariant by $J_4(\mathcal{Q})$ and we will write

$$\mathsf{W}(\mathcal{Q}) = 2\sqrt{J_4(\mathcal{Q})} \quad \Rightarrow \quad S/\pi = \sqrt{J_4(\mathcal{Q})}. \tag{19.66}$$

In models with a high degree of symmetry, $J_4$ can be determined by invariance arguments; typically (STU $\mathrm{ST}^2$, $\mathrm{T}^3$, and other models) it is a quartic polynomial that can be written in the form

$$J_4(\mathcal{Q}) \equiv \mathbb{K}_{MNPQ}\mathcal{Q}^M\mathcal{Q}^N\mathcal{Q}^P\mathcal{Q}^Q, \tag{19.67}$$

which defines the symmetric $\mathbb{K}$-tensor introduced in Refs. [932, 50]. The use of this tensor simplifies many calculations that can be done in a model-independent way. For instance, for all the models that admit it,

$$\tilde{\mathcal{Q}}_M = \frac{4\mathbb{K}_{MNPQ}\mathcal{Q}^N\mathcal{Q}^P\mathcal{Q}^Q}{\mathsf{W}(\mathcal{Q})}. \tag{19.68}$$

It should be clear that the only SBHSs are those for which $J_4(\mathcal{Q}) > 0$. Within this orbit[13] it is not guaranteed that the scalars take only allowed values, as different from the previous cases, but it is necessary to study these properties case by case.

A well-known cubic model (and a very interesting one too) is the next one we are going to study.

*The* STU *model.* This model has been reviewed on p. 234. Equation (19.63) can be solved by

$$\tilde{Y}^i = \sqrt{\frac{\frac{1}{3!}|\varepsilon_{ijk}|\Upsilon_j\Upsilon_k}{\Upsilon_i}}, \quad \text{where} \quad \Upsilon_i \equiv 3p^0 q_i + \tfrac{3}{2}|\varepsilon_{ijk}|p^j p^k, \tag{19.69}$$

and we can readily obtain the Hesse potential[14]

$$J_4(\mathcal{Q}) = 4p^0 q_1 q_2 q_3 - 4q^0 p_1 p_2 p_3 + \textstyle\sum_{i<j} p^i q_i p^j q_j - (p^\Lambda q_\Lambda)^2. \tag{19.71}$$

[13] Different orbits of the duality group in the charge space are obviously characterized by duality invariants. There may be several orbits with $J_4(\mathcal{Q}) > 0$ and distinguished by the value of some other invariant.

[14] This system has a very high degree of symmetry that can be used to determine, up to overall normalization, the Hesse potential [203]; as we have seen, the charge vector of this theory can be rewritten following

The values of the scalars on the horizon are given by (no sum over the $i$ indices)

$$Z^i_{\rm h} = \frac{-p^\Lambda q_\Lambda + 2p^{(i)}q_{(i)} + 4i\sqrt{J_4(\mathcal{Q})}}{2q_i p^0 + |\varepsilon_{ijk}|p^j p^k}. \tag{19.72}$$

In the conventions we are using, the imaginary parts of the three scalars have to be positive and we have to require, on top of $J_4(\mathcal{Q}) > 0$, the three conditions

$$2q_i p^0 + |\varepsilon_{ijk}|p^j p^k > 0. \tag{19.73}$$

To see how these restrictions on the values of the charges for supersymmetric black holes work, it is useful to study an example with only a few non-vanishing charges and scalars. The only ones with only two non-vanishing charges (the pair $p^\Lambda, q_\Lambda$ for some value of $\Lambda$) have $J_4(\mathcal{Q}) < 0$, and therefore we see that we need at least four non-vanishing charges. Two typical choices which preserve the triality of the model are $q_0, p^1, p^2, p^3$ and $p^0, q_1, q_2, q_3$. In both cases the real parts of the attractor values of the scalars $\Re\mathrm{e} Z^i_{\rm h}$ (*axions*) vanish. Let us consider the first choice, for which $\Re\mathrm{e} Z^i_{\rm h} = 0$, $\forall i = 1,2,3$. There are two ways of satisfying all the constraints for SBHSs:

$$q_0 > 0, \;\; p^i < 0, \;\; \forall i = 1,2,3, \quad \text{or} \quad q_0 < 0, \;\; p^i > 0, \;\; \forall i = 1,2,3, \tag{19.74}$$

and they distinguish two regions in charge space separated by surfaces on which $J_4(\mathcal{Q}) = 0$.

The metric function $e^{-2U} = \mathsf{W}(H)$, the scalar fields, etc. can be obtained from these expressions by replacing the charge vector $\mathcal{Q}^M$ by the vector of harmonic functions $H^M$. The metric function does not have a finite expansion in powers of $\rho$ as in the simplest models. However, $J_4(H)$ is a polynomial of fourth degree that we can write in the obvious shorthand notation

$$\begin{aligned} J_4(H) &= \mathbb{K}A^4 - 2\sqrt{2}\,\mathbb{K}A^3\mathcal{Q}\rho + 3\mathbb{K}A^2\mathcal{Q}^2\rho^2 - \sqrt{2}\,\mathbb{K}A\mathcal{Q}^3\rho^3 + \tfrac{1}{4}\mathbb{K}\mathcal{Q}^4\rho^4 \\ &= \tfrac{1}{4} - M\rho + 3\mathbb{K}A^2\mathcal{Q}^2\rho^2 - \sqrt{2}\,\mathbb{K}A\mathcal{Q}^3\rho^3 + \tfrac{1}{4}\left(\frac{S}{\pi}\right)^2\rho^4, \end{aligned} \tag{19.75}$$

where we have used the asymptotic flatness conditions $\mathsf{W}(A) = 1$, $\tilde{A}_M\mathcal{Q}^M = \sqrt{2}M$ and the near-horizon condition $S = 2\pi\mathsf{W}(\mathcal{Q})$. The positivity of the mass and the finiteness of the entropy do not suffice to ensure that $e^{2U}$ vanishes only at the horizon $\rho = -\infty$ and we may have to impose two additional conditions on the charges and moduli:

$$\mathbb{K}A^2\mathcal{Q}^2 \geq 0, \qquad \mathbb{K}A\mathcal{Q}^3 \leq 0. \tag{19.76}$$

---

Eq. (7.113) as a tensor transforming in the product of the fundamental representations of the three $\mathrm{SL}(2,\mathbb{R}) \sim \mathrm{Sp}(2)$, $\mathcal{Q}^{a_1a_2a_3}$. Using the two-dimensional symplectic "metric" $\Omega_{ab} = \epsilon_{ab}$, we can write $J_4(\mathcal{Q})$ in the manifestly duality-invariant form

$$J_4(\mathcal{Q}) = \mathcal{Q}^{a_1a_2a_3}\mathcal{Q}^{b_1b_2b_3}\mathcal{Q}^{c_1c_2c_3}\mathcal{Q}^{d_1d_2d_3}\epsilon_{a_1b_1}\epsilon_{a_2b_2}\epsilon_{c_1d_1}\epsilon_{c_2d_2}\epsilon_{a_3c_3}\epsilon_{b_3d_3}. \tag{19.70}$$

There are no other independent invariants that one can build with a single charge vector.

These SBHSs are particularly interesting because they are solutions of $N=8, d=4$ and $N=4, d=4$ SUEGRA as well: the STU model is at the same time a consistent truncation of $N=8, d=4$ SUEGRA (the effective field theory of the type-II strings compactified on $\mathrm{T}^6$) and also of the $N=4, d=4$ theory coupled to 22 vector multiplets (the effective field theory of the heterotic string compactified on $\mathrm{T}^6$). Furthermore, the SBHSs of the STU model can generate the most general SBHS of $N=8, d=4$ SUEGRA by duality transformations [203] (something that is very hard to do in practice, though). We will show later how the metric of the most general SBHS of $N=8, d=4$ SUEGRA can be constructed in a more direct way by using the results of Chapter 18.

Related to this, it is worth mentioning that we are also going to meet again some of these SBHSs in Chapter 26: the extremal limit of the black-hole solution of $N=8, d=4$ SUEGRA given in Eq. (26.37) has essentially the same form as a solution of the STU model with only four non-vanishing charges and harmonic functions ($p^0, q_i$, and $H^0, H_i$, $i=1,2,3$, for instance, so only the imaginary parts of the three scalars $Z^i$ are active) which can be identified with $H_{\mathrm{D6}}, H_{\mathrm{D2}}, H_{\mathrm{S5}}, H_{\mathrm{W}}$ up to constants, since the latter harmonic functions are normalized to be 1 at spatial infinity. That solution has four active scalars, but only three of them are independent (the dilaton is, again up to constants, $k_1/K_6$) and they can be related to products and quotients of the three $\Im\mathfrak{m}(Z^i)$ of the solution of the STU model.

As a matter of fact, it is not difficult to see how the toy model Eq. (26.1) (an inconsistent truncation of $N=8, d=4$ and $N=4, d=4$ SUEGRA) studied in Section 26.1 coincides with a simple inconsistent truncation of the STU model in which we set to zero the real parts of all the scalars (sometimes generically called *axions*) $R^i \equiv \Re\mathfrak{e} Z^i$. By inspection of the period matrix of the model (the generic expression Eq. (7.52) with $d_{ijk}=|\epsilon_{ijk}|$), it is easy to see that the equations of motion of $R^i$ become the following constraints when $R^i=0$:

$$F^0 F^i = 0, \qquad F^i \star F^j = 0, \ \ \forall i \neq j. \tag{19.77}$$

These constraints are satisfied by the SBHSs with charges $p^0, q_i$ or $p^i, q_0$ mentioned above. Furthermore, for $R^i=0$, the period matrix becomes purely imaginary and equal to

$$\mathcal{N}_{\Lambda\Sigma} = -i\delta_{\Lambda\Sigma} I^1 I^2 I^3/(I^\Sigma)^2, \tag{19.78}$$

where $I^0 \equiv 1$. Then, defining

$$I^1 \equiv \tfrac{1}{2}e^{-2\phi}, \qquad I^2 \equiv \tfrac{1}{2}e^{-2\rho}, \qquad I^3 \equiv \tfrac{1}{2}e^{-2\sigma}, \tag{19.79}$$

the action of the STU model becomes identical to Eq. (26.1) with the following identifications:

$$F^0 = \tfrac{1}{\sqrt{2}}F^{(1)\,1}, \ \ F^1 = \tfrac{1}{\sqrt{2}}e^{-2(\phi-\rho-\sigma)} \star F^{(2)}{}_1, \ \ F^2 = \tfrac{1}{\sqrt{2}}F^{(2)}{}_2, \ \ F^3 = \tfrac{1}{\sqrt{2}}F^{(1)\,2}. \tag{19.80}$$

*The* ST[2, n] *models.* This model, studied on p. 235, is a particular case of the cubic model and we can use the general machinery to find the Freudenthal dual. Shmakova's Eqs. (19.63) are easy to solve and the Hesse potential can be written in terms of a quartic invariant:

$$\begin{aligned} \mathsf{W}(\mathcal{Q}) &= 2\sqrt{J_4(\mathcal{Q})}, \\ J_4(\mathcal{Q}) &= (p^\alpha p_\alpha + 2p^0 q_1)(q^\alpha q_\alpha - 2p^1 q_0) - (p^0 q_0 - p^1 q_1 + p^\alpha q_\alpha)^2, \end{aligned} \tag{19.81}$$

where we have used the metric $\eta_{\alpha\beta}$ to raise and lower the indices of the charges.

Defining the charge vectors $P^A, Q^A$ and the metric $\mathrm{SO}(2, n_V - 1)$ metric $\eta_{AB}$ (in a non-diagonal basis),

$$(P^A) \equiv \begin{pmatrix} p^0 \\ q_1 \\ p^\alpha \end{pmatrix}, \qquad (Q^A) \equiv \begin{pmatrix} -p^1 \\ q_0 \\ q_\alpha \end{pmatrix}, \qquad (\eta_{AB}) \equiv \left(\begin{array}{cc|c} 0 & 1 & \\ 1 & 0 & \\ \hline & & \eta_{\alpha\beta} \end{array}\right), \tag{19.82}$$

the quartic invariant can be written in a manifestly $\mathrm{SO}(2, n_V - 1)$ symmetric form:

$$J_4(\mathcal{Q}) = P^2 Q^2 - P \cdot Q. \tag{19.83}$$

### *19.2.5 Two-center SBHS of the axion–dilaton model*

The analysis performed in the generic case has effectively reduced the problem of constructing a solution of any model to the algebraic problem of finding the moduli, $n_C$ charge vectors, and the relative positions of the centers such that all the constraints required by the regularity of the solution are satisfied. Here we are going to study the simple case $n_C = 2$.

Without loss of generality, we can place one center ($a$) at the origin and the other ($b$) at a distance $l$ (the relative direction is irrelevant). We assume that the values of the constants $A^M$ have been determined as functions of the modulus $\tau_\infty$ and total charge $\mathcal{Q}^M = \mathcal{Q}_a^M + \mathcal{Q}_b^M$ so the metric is asymptotically flat, the total mass $M$ is positive, and $A_M \mathcal{Q}^M = 0$. The only independent condition in Eqs. (19.33) that remains to be satisfied can be solved by setting

$$l = \frac{J}{\sqrt{2}\, A_M \mathcal{Q}_a^M}, \qquad J = \mathcal{Q}_{a\,M} \mathcal{Q}_b^M, \tag{19.84}$$

where $J$ is the angular momentum of the system along the axis that goes from $a$ to $b$. We have assumed that $A_M \mathcal{Q}_{a,b}^M \neq 0$. If $A_M \mathcal{Q}_a^M = 0$ the same happens for the other center and we must set $J = 0$. In this kind of solution the central charges of all the centers have the same phases and $M = \sum_a M_a$. These are generalizations of the MP solutions of the Einstein–Maxwell theory.

To get a completely regular metric we must impose the positivity of the total mass and of the coefficients $\mathsf{M}_{a,b}$ of the two centers as well as the finiteness of their entropies $S_{a,b}$. A non-trivial example[15] is provided by two centers which are purely magnetic and purely

[15] A much more general treatment of this problem for all theories defined by a quadratic prepotential can be found in Ref. [521].

electric, that is

$$(\mathcal{Q}_a^M) = \begin{pmatrix} p^0 \\ p^1 \\ 0 \\ 0 \end{pmatrix}, \qquad (\mathcal{Q}_b^M) = \begin{pmatrix} 0 \\ 0 \\ q_0 \\ q_1 \end{pmatrix} \quad \Rightarrow \quad J = p^\Lambda q_\Lambda, \tag{19.85}$$

where we have suppressed the $a, b$ indices because they are unnecessary.[16] The metric function takes the simple form (with no mixed terms $\sim \frac{1}{|\vec{x}-\vec{x}_b|\,|\vec{x}-\vec{x}_a|}$)

$$e^{-2U} = 1 + \frac{2\mathsf{M}_b}{|\vec{x}-\vec{x}_b|} + \frac{2\mathsf{M}_a}{|\vec{x}-\vec{x}_a|} + \frac{1}{4}\frac{S_a/\pi}{|\vec{x}-\vec{x}_a|^2} + \frac{1}{4}\frac{S_b/\pi}{|\vec{x}-\vec{x}_b|^2}, \tag{19.86}$$

where

$$\mathsf{M}_a = A^0 p^1 + A^1 p^0, \quad \mathsf{M}_b = A_0 q_1 + A_1 q_0, \quad S_a/\pi = 2p^0 p^1, \quad S_b/\pi = 2q_0 q_1. \tag{19.87}$$

The two horizons will be regular if the two charges of each center are non-vanishing and have the same sign. This leads, automatically, to $J \neq 0$.

The central charges of each center are given by

$$\mathcal{Z}_a = \frac{1}{2\sqrt{\Im\mathfrak{m}\tau}}(p^1 - ip^0\tau), \qquad \mathcal{Z}_b = -\frac{i}{2\sqrt{\Im\mathfrak{m}\tau}}(q_0 - iq_1\tau), \qquad \mathcal{Z} = \mathcal{Z}_a + \mathcal{Z}_b. \tag{19.88}$$

The mass that each of the centers would have if it were an isolated black hole is just the absolute value of the corresponding charge, while the total mass is the absolute value of the total central charge. The regularity of the metric requires the parameters $\mathsf{M}_{a,b}$ to be positive as well. Using the general solution for $A^M$ we find that they are given by

$$\mathsf{M}_{a,b} = \Re\mathfrak{e}(e^{-i\alpha_\infty}\mathcal{Z}_{a,b\,\infty}), \quad \text{where} \quad e^{i\alpha_\infty} = \frac{\mathcal{Z}_\infty}{|\mathcal{Z}_\infty|}. \tag{19.89}$$

Operating, we get

$$\mathsf{M}_{a,b} = \frac{M_{a,b}^2 + \Delta}{M}, \quad \text{where} \quad \Delta = \Re\mathfrak{e}(\mathcal{Z}_{a\,\infty}\mathcal{Z}_{b\,\infty}^*) = (p^0 q_0 - p^1 q_1)\frac{\Re\mathfrak{e}\tau_\infty}{4\Im\mathfrak{m}\tau_\infty}. \tag{19.90}$$

Since $\Re\mathfrak{e}\tau_\infty \in \mathbb{R}$ we can always make $\Delta > 0$ so the two parameters are different from zero, and the solution is regular everywhere.

## 19.3 The supersymmetric black holes of $N = 2, d = 4$ SEYM

### 19.3.1 The general recipe

According to the general results in Section 18.5.2 we can proceed as follows:

[16] Observe that the two terms $p^0 q_0$ and $p^1 q_1$ are the non-vanishing parts of two of the six independent $\mathrm{SL}(2,\mathbb{R})$ invariants that can be built with two different charge vectors because each of them is made out of two independent $\mathrm{SL}(2,\mathbb{R})$ doublets, as explained on p. 241. The entropies are another two of those six invariants. The remaining two vanish for this particularly simple family of solutions that we have chosen. This means that it is not the most general case.

1. Find a solution of the Bogomol'nyi Eqs. (18.127) in $\mathbb{R}^3$, rewritten here for convenience,

$$\boxed{\tilde{F}^{\Lambda}{}_{\underline{mn}} = -\tfrac{1}{\sqrt{2}}\epsilon_{mnp}\tilde{\mathfrak{D}}_{\underline{p}}\mathcal{I}^{\Lambda},} \tag{19.91}$$

for a gauge field $\tilde{A}^{\Lambda}{}_{\underline{x}}$ and a real "Higgs" field $\mathcal{I}^{\Lambda}$. It should be stressed that these are nothing but the standard Bogomol'nyi equations, many of whose solutions for certain gauge groups can be found in the literature on Yang–Mills monopole solutions and used here straight away. In the Yang–Mills context only globally regular solutions were usually considered. Our experience with the Abelian case (for which the above equation is also satisfied) suggests that singular solutions of the Bogomol'nyi equations (like those associated with harmonic functions with point-like singularities) may lead to regular supergravity solutions in which the singularity corresponds to the extremal event horizon of a black hole. Nothing forbids us from using multicenter solutions of the Bogomol'nyi equations either, but we will not consider this problem here.

2. Find the lower components of $\mathcal{I}^M$ ($\mathcal{I}_{\Lambda}$) by solving Eqs. (18.129),

$$\boxed{\tilde{\mathfrak{D}} \star_{(3)}\tilde{\mathfrak{D}}\mathcal{I}_{\Lambda} = \tfrac{1}{2}g^2\left[f_{\Lambda(\Sigma}{}^{\Gamma}f_{\Delta)\Gamma}{}^{\Omega}\,\mathcal{I}^{\Sigma}\mathcal{I}^{\Delta}\right]\mathcal{I}_{\Omega}.} \tag{19.92}$$

As mentioned before, a solution which is always available for compact gauge groups is $\mathcal{I}_{\Lambda} = \alpha\mathcal{I}^{\Lambda}$ for some real constant $\alpha$ because this makes the r.h.s vanish automatically, and the l.h.s. is one of the integrability equations of the Bogomol'nyi equation.

3. The integration constants of the $\mathcal{I}^M$ must be chosen so as to satisfy the integrability conditions of the equation for $\omega$. For single-SBHSs (the only case that we are going to consider here), we expect regularity to be associated with staticity, and we must impose $\omega = 0$, which leads to the equation

$$\boxed{\mathcal{I}_M\tilde{\mathfrak{D}}\mathcal{I}^M = 0.} \tag{19.93}$$

When $\omega = 0$ we can remove the tildes from the gauge field and the covariant derivatives.

4. If the stabilization equations have been solved, we can construct the metric function $e^{-2U}$ and the complex scalars $Z^i$ as in the Abelian case and the vector field strengths using Eq. (18.126),

$$\boxed{\mathcal{F} = -\tfrac{1}{2}\left\{\mathfrak{D}(\mathcal{R}\hat{V}) + \star(\hat{V}\wedge\mathfrak{D}\mathcal{I})\right\}.} \tag{19.94}$$

### 19.3.2 Examples

Before we construct examples of SBHSs in this class, it is worth reviewing briefly some of the previous work on solutions of YM theories coupled to gravity.[17] Most of the previous work on this topic was focused on pure Einstein–Yang–Mills (EYM) theories, (the minimal non-Abelian extension of the Einstein–Maxwell theory), ignoring the possible existence of unbroken supersymmetry which is, however, one of our main concerns here.

Soon after the discovery of the 't Hooft–Polyakov monopole [1180, 1053], several groups found solutions of the pure EYM theory [1281, 352, 81, 1236, 1034][18] whose $\mathrm{SU}(2)$ gauge field is that of the Wu–Yang $\mathrm{SU}(2)$ monopole [1277]. The metric of all these solutions is that of the (dS or AdS) non-extremal Reissner–Nordström black hole, and the singularity in the gauge field (generically expected for static YM solutions [432]) is covered by an event horizon.

The coincidence of the metrics is due to the relation between the Wu–Yang $\mathrm{SU}(2)$ monopole and the non-Abelian embedding of the Dirac monopole Eq. (13.60), explained in Section 13.2.3: these two solutions give rise to exactly the same energy–momentum tensor because all the $\mathrm{SU}(2)$ matrices disappear in the trace. For this reason, these solutions have been regarded as not truly non-Abelian, even though, as we have stressed, their equivalence to the non-Abelian embedding of the Dirac monopole is up to a singular gauge transformation (see also Ref. [298]).

Finding less trivial ("genuinely or essentially non-Abelian") solutions proved much more difficult and a *non-Abelian baldness theorem* stating that the only black-hole solutions of the EYM $\mathrm{SU}(2)$ theory with a regular horizon and non-vanishing magnetic charge had to be non-Abelian embeddings of the Reissner–Nordström solution was proven in Ref. [580]. An "essentially non-Abelian" globally regular [1147] solution of the EYM theory had already been found: the Bartnik–McKinnon particle [106], but this solution, as well as the subsequent black-hole-type generalizations [1221, 219, 881] (all of them known only numerically) have vanishing gauge charges. The theorem was extended to prove the absence of regular monopole or dyon solutions of the EYM theory in Refs. [514, 220].

The classification of the possible EYM solutions for the gauge group $\mathrm{SU}(2)$ [1146] suggests that one has to add more fields to the theory in order to get "essentially non-Abelian" black-hole or gravitating monopole solutions with non-vanishing charges. In fact, globally regular gravitating monopole solutions (the Harvey–Liu [714] and Chamseddine–Volkov [337, 338] solutions) have been found in gauged $N = 4, d = 4$ SUEGRA, and, in what follows, we will find $N = 2$ versions of them. But we will also find the supersymmetric "essentially-non-Abelian" solutions with non-vanishing charges constructed by Meessen in a completely analytic form [935], and we will recover the extremal Reissner–Nordström-like solutions embedded in gauged $N = 2, d = 4$ SUEGRA.

Observe that, in order to find SBHSs of the EYM theory we must embed it in a supergravity theory. We can always embed it in gauged $N = 1, d = 4$ supergravity but, as we have emphasized repeatedly, the black-hole solutions of this theory cannot be supersymmetric and we have to work, at least, in gauged $N = 2, d = 4$ super-EYM (SEYM)

[17] For more comprehensive reviews see Refs. [1222, 1256].

[18] In some cases the EYM–Higgs theory was considered but only solutions with constant Higgs field were obtained.

to have SBHSs. These theories have scalar fields and a scalar potential, and the embedding of the known asymptotically flat pure EYM solutions is a non-trivial problem that we will, however, manage to solve for some of them, proving that they are supersymmetric and that there is an attractor mechanism at work in them. At the same time, the presence of the scalar fields (which behave as adjoint Higgs fields) will also allow us to find the above mentioned "essentially-non-Abelian" solutions with non-vanishing charges.

The only existing examples of SBHSs of $N=2, d=4$ SEYM were obtained in Refs. [797, 935, 798]. Here we will review the results of the first two of these references in which static ($\omega=0$) single, non-Abelian monopole and black-hole solutions of the SO(3)-gauged $\overline{\mathbb{CP}}^3$ model described on p. 254 were constructed and studied.

For this model, the Bogomol'nyi Eqs. (19.91) split into an Abelian part (the zeroth component) and an $\mathrm{SO}(3)$ non-Abelian part (the $i=1,2,3$ components)[19]

$$F^0{}_{mn} = -\tfrac{1}{\sqrt{2}}\epsilon_{mnp}\partial_p\mathcal{I}^0, \tag{19.95}$$

$$F^i{}_{mn} = -\tfrac{1}{\sqrt{2}}\epsilon_{mnp}\mathfrak{D}_p\mathcal{I}^i, \tag{19.96}$$

where the staticity has already been accounted for. Choosing a harmonic function $H^0 = A^0 + \frac{1}{\sqrt{2}}p^0/r$ for $\mathcal{I}^0$ is enough to satisfy the first equation and guarantee the existence of the potential $A^0$, whose explicit expression we do not need.

The second, non-Abelian, set of equations are nothing but the standard Bogomol'nyi equations for $\mathrm{SU}(2)$ monopoles Eq. (13.45) with the lower sign $(+)$ if we define

$$\Phi^i \equiv -\tfrac{1}{\sqrt{2}}\mathcal{I}^i, \tag{19.97}$$

and we can simply use the solutions obtained making the hedgehog ansatz in Section 13.2.3. The gauge and Higgs fields are given by Eqs. (13.54) in terms of $f$ and $h$, and these are in turn given by Eqs. (13.63) and (13.66) for two families of solutions which will be distinguished, respectively, using subscript $s$ and $*$. We rewrite these equations here for convenience, restoring the coupling constant $g$ which had been set to 1 there and selecting only the lower sign:

$$\boxed{\begin{aligned}
\Phi^i &= \delta^i{}_m f(r)x^m, & A^i{}_m &= -\epsilon^i{}_{mn}x^n h(r),\\
f_s &= \frac{1}{gr^2}\left[1-\mu r\coth(\mu r+s)\right], & h_s &= -\frac{1}{gr^2}\left[1-\frac{\mu r}{\sinh(\mu r+s)}\right],\\
f_* &= \frac{1}{gr^2}\left[\frac{1}{1+\lambda^2 r}\right], & h_* &= -f_*.
\end{aligned}} \tag{19.98}$$

The magnetic charge of the YM field is always the minimal $1/g$ except in the isolated case, in which it vanishes.

[19] These tensor equations are defined in $\mathbb{E}^3$ and we therefore remove the underlines that distinguish curved from flat indices.

The last building blocks of the solution are the $\mathcal{I}_\Lambda$ components, which must obey Eqs. (19.92). Choosing $\mathcal{I}_i = \alpha\mathcal{I}^i$ for some constant $\alpha$ makes the r.h.s. of that equation vanish. We will, however, choose $\alpha = 0$ to simplify the interpretation of the solutions (only one Higgs field for one gauge vector). For $\Lambda = 0$ we just have to choose an independent harmonic function $\mathcal{I}_0 = H_0 = A_0 + \frac{1}{\sqrt{2}}q_0/r$.

The integration constants $\mu, s, A^0, A_0, \alpha$ will be partially fixed by imposing Eq. (19.93),

$$A^0 q_0 - A_0 p^0 = 0, \tag{19.99}$$

and asymptotic flatness. Using the solution for the Freudenthal duality equations of this model, Eqs. (19.62), we find that the metric function and the scalar fields are given by

$$e^{-2U} = \tfrac{1}{2}(H^0)^2 + 2(H_0)^2 - (rf)^2, \tag{19.100}$$

$$Z^i = \frac{\sqrt{2}rf}{H^0 + 2iH_0}\delta^i{}_m \frac{x^m}{r}. \tag{19.101}$$

Because of the relative sign between the Abelian and non-Abelian contributions to the metric function, at least one of the two functions $H^0, H_0$ must be different from zero. For some of the non-Abelian solutions this will not be enough and at least one of the charges $p^0, q_0$ will be required to be non-zero.

Let us start by examining the case with vanishing $p^0, q_0$ and constant, non-vanishing $H^0 = A^0, H_0 = A_0$. In the $r \to 0$ limit

$$rf_{s\neq 0} \sim \frac{1}{gr}, \qquad rf_{s=0} \sim 0, \qquad rf_* \sim \frac{1}{gr}, \tag{19.102}$$

so the non-Abelian contribution grows without bound and the metric factor $e^{-2U}$ becomes negative in all cases except for the 't Hooft–Polyakov monopole $s = 0$. In fact, in this case, the metric becomes Minkowski and there are neither event horizons nor singularities in this limit. These kinds of solutions are sometimes called *global monopoles* and they are globally regular, just like the Yang–Mills monopole solution that sources the gravitational field [714, 337, 338].

The other solutions must contain a naked singularity at some finite value of $r$ and only by switching on the charges $p^0, q_0$ can the metric remain positive.

Now let us study the only non-singular solution $s = 0$ in the $r \to \infty$ limit, in which the function $f$ behaves as

$$rf_s \sim \frac{1}{g}\left(-\mu + \frac{1}{r}\right), \qquad rf_* \sim \frac{1}{g\lambda^2 r^2}. \tag{19.103}$$

To normalize $e^{-2U} = 1$ at infinity we impose

$$W_{\rm RN}(A) = 1 + (\mu/g)^2, \quad \text{where} \quad W_{\rm RN}(A) \equiv \tfrac{1}{2}(A^0)^2 + 2(A_0)^2. \tag{19.104}$$

The scalars are not asymptotically constant (after all, they transform under gauge local SU(2) transformations) but they are covariantly constant. The leading term is

$$Z^i \sim Z_\infty \delta^i{}_m \frac{x^m}{r}, \qquad Z_\infty \equiv \frac{\mu/g}{1 + (\mu/g)^2}\left(\tfrac{1}{\sqrt{2}}A^0 - \sqrt{2}iA_0\right). \tag{19.105}$$

The asymptotic value of $\lim_{r\to\infty} Z^i Z^{*i} = |Z_\infty|^2$ is gauge invariant and, using the above relations, gives an expression for $\mu$ in terms of $g$ and moduli,

$$\mu^2 = \frac{|Z_\infty|^2}{1-|Z_\infty|^2} g^2, \tag{19.106}$$

which can be used in the expression of $Z_\infty$ to find $A^0$ and $A_0$ as functions of the real and imaginary parts of $Z_\infty$ and $g$.

Taking this into account, the mass of the globally regular monopole solution is proportional to the magnetic charge $1/g$:

$$M_{\text{monopole}} = \sqrt{\frac{|Z_\infty|^2}{1-|Z_\infty|^2}}\frac{1}{g}, \tag{19.107}$$

and it is manifestly positive for $g > 0$ (as we have implicitly assumed) since the scalars of the $\overline{\mathbb{CP}}^3$ model are constrained to the values $1 < |Z|^2 \leq 0$.

Let us now consider the generic case with non-vanishing $p^0, q_0$. It is convenient to solve Eq. (19.99) by introducing a constant non-vanishing $\beta$ such that

$$\frac{A^0}{p^0/\sqrt{2}} = \frac{A_0}{q_0/\sqrt{2}} \equiv 1/\beta \;\Rightarrow\; \begin{cases} H^0 = Hp^0/(\sqrt{2}\beta) \\ H_0 = Hq_0/(\sqrt{2}\beta) \end{cases} \quad \text{where} \quad H \equiv 1 + \frac{\beta}{r}. \tag{19.108}$$

The normalization of $e^{-2U} = 1$ at infinity implies that

$$\beta^2 = \frac{W_{\text{RN}}(\mathcal{Q})/2}{1+(\mu/g)^2}, \tag{19.109}$$

where the isolated solution $f_*$ is also included with $\mu = 0$. The asymptotic behavior of the scalars is the same as in the previous case, with the complex constant $Z_\infty$ given by

$$Z_\infty \equiv \frac{\beta\mu/g}{W_{\text{RN}}(\mathcal{Q})/\sqrt{2}}\left(\tfrac{1}{\sqrt{2}}p^0 - \sqrt{2}iq_0\right), \qquad |Z_\infty|^2 \equiv \frac{\beta^2(\mu/g)^2}{W_{\text{RN}}(\mathcal{Q})/2}, \tag{19.110}$$

and the constant $\mu$ has the same value as in the previous case, while $\beta$ is given by

$$\beta^2 = (1-|Z_\infty|^2)W_{\text{RN}}(\mathcal{Q})/2. \tag{19.111}$$

Observe that this means that the isolated solution $f_*$ is characterized by $|Z_\infty|^2 = 0$.

The mass and entropy (which is moduli independent as in the ungauged cases) are given by

$$M = \sqrt{\frac{W_{RN}(\mathcal{Q})/2}{1-|Z_\infty|^2}} + M_{\text{monopole}}, \tag{19.112}$$

$$S/\pi = \tfrac{1}{2}W_{\text{RN}}(\mathcal{Q}) - \frac{1}{g^2}, \quad \text{for} \quad s \neq 0 \;\text{ and }\; |Z_\infty| = 0, \tag{19.113}$$

$$S/\pi = \tfrac{1}{2}W_{\text{RN}}(\mathcal{Q}), \quad \text{for} \quad s = 0 \tag{19.114}$$

where $M_{\text{monopole}}$ is given by Eq. (19.107) and vanishes for the isolated case because that is the $|Z_\infty| = 0$ case. The 't Hooft–Polyakov monopole does not contribute to the entropy, which suggests that it must be associated to a pure state in the quantum theory. In the $|Z_\infty| = 0$ (isolated) case, the gauge field, which is clearly non-trivial, does not contribute to the mass but it does contribute to the entropy. It represents a monopole screened at infinity. Finally, in the family $s \neq 0$, the charges $p^0, q_0$ must be such that[20]

$$W_{\text{RN}}(\mathcal{Q}) \geq \frac{1}{g^2}. \tag{19.115}$$

If the equality can be achieved (which is possible only for certain values of $g$ because the charges $p^0, q_0$ have to be quantized), the solutions are global monopoles.

The near-horizon limit of the scalars is the covariantly constant function of the charges:

$$Z^i_{\text{h}} = \frac{-1/g}{(\frac{1}{2}p^0 + iq_0)} \delta^i{}_m \frac{x^m}{r}. \tag{19.116}$$

Since the magnetic charge is $1/g$ in all cases except in the isolated one, we can say that the attractor mechanism also works here (in a covariant way) except in the isolated case.

In order to recover the solutions of Refs. [1281, 352, 81, 1236, 1034, 298] (the metric) embedded in $N = 2$ SEYM, we have to tune the parameters of the solutions so as to get covariantly constant scalars which do not contribute to the energy-momentum tensor.[21] This is possible only for the $s \to \infty$ solutions (Wu–Yang monopoles) for which $rf$ is a harmonic function, and it can be achieved by setting the constant $\beta = -1/\mu$. This corresponds to

$$Z^i = Z\delta^i{}_m \frac{x^m}{r}, \qquad Z = \frac{\sqrt{2}/g}{p^0/\sqrt{2} + i\sqrt{2}q_0} = Z_\infty. \tag{19.117}$$

Since, by design, we have constructed a double extremal black hole, the metric becomes that of the ERN black hole, as in Eqs. (19.29), with $W(\mathcal{Q})/2$ replaced by $S/\pi$. These solutions were called *black hedgehogs* in Ref. [797] and *black merons* in Ref. [298] because the gauge field of the Wu–Yang monopole can also be understood as a Lorentzian meron solution. A closely related solution with non-covariantly constant scalars was obtained in a different context in Ref. [844].

## 19.4 The supersymmetric black holes of $N = 8, d = 4$ supergravity

In Section 18.6 we found the generic form of the timelike supersymmetric solutions of all $N \geq 2, d = 4$ SUEGRAS coupled to vector supermultiplets. The SBHSs should be found among the subclass in which what we called *would-be hyperscalars* have been truncated so the transverse three-dimensional metric is that of $\mathbb{E}^3$. However, in order to construct explicitly the most general SBHSs of these theories the general result suggests that we must first construct and parametrize the most general matrices $M_{IJ}$ with the right properties, which is an extremely difficult task.

[20] Observe that neither the mass nor the entropy depend on the *Protogenov hair parameter* $s$, nor on $\lambda$.

[21] The associated Higgs field will never be covariantly constant because the Bogomol'nyi equation would lead to a trivial gauge field strength.

It was realized in Ref. [1013], however, that completing that task is unnecessary to find the metric of the SBHSs (but not the scalar fields): one can just assume that $M_{IJ}$ and some explicit parametrization of $\mathcal{V}_{IJ}^{M}$ have been used to define the two real symplectic vectors $\mathcal{R}^M$ and $\mathcal{I}^M$ as in Eq. (18.151). The components of $\mathcal{I}^M$ are harmonic functions in $\mathbb{E}^3$, $H^M$, and all we need to do is to find how to express the components of $\mathcal{R}^M$ in terms of those of $\mathcal{I}^M$.

In $N = 2, d = 4$ theories one has to face a similar problem: solving the Freudenthal duality equations, but there we had explicit expressions for $M_{IJ}$ and $\mathcal{V}_{IJ}^{M}$. However, this is not necessary if the theory has a very high degree of (duality) symmetry. In $d = 4$ the duality group acts embedded in $\mathrm{Sp}(2\bar{n}, \mathbb{R})$ if the theory contains $\bar{n}$ vector fields, and both $\mathcal{I}^M$ and $\mathcal{R}^M$ must transform in some symplectic representation of that group. In the $N = 8$ case[22] this is enough to solve the Freudenthal duality equations, as we are going to see. First, we need to review some facts about the duality group of this theory, $\mathrm{E}_{7(+7)}$.

### 19.4.1 The duality group of $N = 8, d = 4$ SUEGRA and its invariants

$N = 8, d = 4$ SUEGRA has 28 (fundamental, "electric") vector fields $A^\Lambda$, $\Lambda = 1, \ldots, 28$. The index $\Lambda$ is usually replaced by an antisymmetric pair of SO(8) indices $i, j = 1, \ldots 8$ so the vector field strengths are denoted by $F^{ij}$. The 28 dual ("magnetic") vector field strengths are denoted by $G_{ij}$, and altogether, they make a 56-dimensional symplectic vector $(\mathcal{F}^M) \equiv \begin{pmatrix} F^{ij} \\ G_{ij} \end{pmatrix}$ which transforms in the fundamental (**56**) representation of $\mathrm{E}_{7(+7)}$. This expression for the **56** is sometimes called the "real" or "SO(8)" basis[23]. The corresponding symplectic charge vector is $(\mathcal{Q}^M) \equiv \begin{pmatrix} p^{ij} \\ q_{ij} \end{pmatrix}$. The infinitesimal action of $\mathrm{E}_{7(+7)}$ on the fundamental representation is

$$\delta \mathcal{Q} = \begin{pmatrix} 2\Lambda^{[i}{}_k \delta^{j]}{}_l & \Sigma^{ijkl} \\ \Sigma_{ijkl} & 2\Lambda_{[i}{}^k \delta_{j]}{}^l \end{pmatrix} \begin{pmatrix} p^{kl} \\ q_{kl} \end{pmatrix}, \tag{19.121}$$

---

[22] The same reasoning can be used in the four-dimensional supergravities with duality groups of type E7 [529, 533, 530], leading to very similar solutions.

[23] There is another basis (the "complex" or SU(8) basis) in which the **56** is written as a complex tensor with two antisymmetric SU(8) indices $A, B = 1, \ldots 8$, $\mathcal{F}_{AB}$, so it transforms in the **28** of SU(8). The infinitesimal action of $\mathrm{E}_{7(+7)}$ on $\mathcal{F}_{AB}$ is

$$\delta \mathcal{F}_{AB} = -2\Lambda^{C|}{}_{[A|}\mathcal{F}_{C|B]} + \Sigma_{ABCD}\overline{\mathcal{F}}^{AB} \tag{19.118}$$

Here, the $\Lambda^A{}_B$ are infinitesimal transformations of the maximal, compact subgroup of $\mathrm{E}_{7(+7)}$, SU(8) (i.e. $\Lambda^A{}_B = 0$), and where the off-diagonal infinitesimal parameters $\Sigma_{ABCD}$ satisfy the complex self-duality condition

$$\overline{(\Sigma_{ABCD})} \equiv \overline{\Sigma}^{ABCD} = \tfrac{1}{4!}\varepsilon^{ABCDEFGH}\Sigma_{EFGH}. \tag{19.119}$$

The relation between the components of $\mathcal{F}$ in both bases is

$$\overline{\mathcal{F}}^{AB} \equiv \tfrac{1}{4\sqrt{2}} \left(F^{ij} - iG_{ij}\right) \Gamma^{ij}{}_{AB}, \tag{19.120}$$

where $\Gamma^{ij} = \Gamma^{[i}\Gamma^{j]}$ and the $\Gamma^i{}_{AB}$s are the SO(8) gamma matrices. It must be stressed that the SU(8) indices $A, B$ are, in principle, different from the indices $I, J$. The action of the group $\mathrm{E}_{7(+7)}$ on all the fields of $N = 8, d = 4$ has been studied in detail in Ref. [847].

where the $\Lambda^i{}_j$ are infinitesimal transformations of $\mathrm{SL}(8,\mathbb{R})$ (i.e. $\Lambda^i{}_i = 0$), the maximal, (non-compact) subgroup of $\mathrm{E}_{7(+7)}$, and where the off-diagonal infinitesimal parameters are related by

$$\Sigma^{ijkl} = \tfrac{1}{4!}\varepsilon^{ijklmnpq}\Sigma_{mnpq}. \tag{19.122}$$

$\mathrm{E}_{7(+7)}$ has two important properties:

1. The only independent $\mathrm{E}_{7(+7)}$ invariant that can be written with a single vector in the fundamental representation is Cartan's quartic invariant $J_4(\mathcal{Q})$ [308], which can be written in the real basis in the form[24,25]

$$\boxed{J_4(\mathcal{Q}) = p^{ij}q_{jk}p^{kl}q_{li} - \tfrac{1}{4}(p^{ij}q_{ji})^2 + \tfrac{1}{96}\varepsilon^{ijklmnpq}\left(q_{ij}q_{kl}q_{mn}q_{pq} + p^{ij}p^{kl}p^{mn}p^{pq}\right).} \tag{19.126}$$

2. Using a vector $\mathcal{Q}$ in the **56** there is a unique way of constructing with its components another vector transforming in the same representation: taking the *Jordan triple product* of that vector with itself three times $(\mathcal{Q},\mathcal{Q},\mathcal{Q})^M$. The definition of this product

[24] Using the complex basis one can construct the Cremmer–Julia quartic invariant $\diamondsuit(\mathcal{Q})$ [385]:

$$\diamondsuit(\mathcal{Q}) = \mathcal{Q}_{AB}\overline{\mathcal{Q}}^{BC}\mathcal{Q}_{CD}\overline{\mathcal{Q}}^{DA} - \tfrac{1}{4}\left(\mathcal{Q}_{AB}\overline{\mathcal{Q}}^{AB}\right)^2 + \tfrac{1}{96}\varepsilon^{ABCDEFGH}\left(\mathcal{Q}_{AB}\mathcal{Q}_{CD}\mathcal{Q}_{EF}\mathcal{Q}_{GH} + \text{c.c.}\right). \tag{19.123}$$

$J_4(\mathcal{Q})$ and $\diamondsuit(\mathcal{Q})$ must be proportional; indeed, it was found in Ref. [94] that

$$J_4(\mathcal{Q}) = -\diamondsuit(\mathcal{Q}). \tag{19.124}$$

For a detailed proof see Ref. [689].

[25] If one is allowed to use the moduli to construct duality invariants, there are other possibilities. The basic building block would be the central-charge matrix $\mathcal{Z}_{IJ}(\mathcal{Q},\phi)$ defined in Eq. (18.15). When acting with $\mathrm{E}_{7(+7)}$ on the charges and scalars, this matrix only transforms with (scalar-dependent) $\mathrm{SU}(8)$ compensating transformations. Any $\mathrm{SU}(8)$ invariant built with it becomes automatically an $\mathrm{E}_{7(+7)}$ invariant. For instance, the quadratic invariant

$$-V_{\mathrm{bh}}(\phi,\mathcal{Q}) \equiv \overline{\mathcal{Z}}^{IJ}\mathcal{Z}_{IJ} \tag{19.125}$$

gives the so-called *black-hole potential* of the $N=8, d=4$ theory in the FGK formalism [47, 522, 46]. There is a combination of the $\mathrm{E}_{7(+7)}$ invariants constructed from the central-charge matrix which has exactly the same form as $\diamondsuit$ in Eq. (19.123) with $\mathcal{Q}_{AB}$ replaced everywhere by $\mathcal{Z}_{IJ}$ and which we can denote by $\diamondsuit(\mathcal{Z})$, which can be shown to be moduli independent [528] and identical to $\diamondsuit(\mathcal{Q})$. In fact, it can be argued that there are values of the scalars of the $N=8$ theory for which $\mathcal{Z}_{IJ} = \delta_{IJ}{}^{AB}\mathcal{Q}_{AB}$.

is[26]

$$\boxed{\begin{aligned}(\mathcal{Q},\mathcal{Q},\mathcal{Q})^{ij} &\equiv \tfrac{1}{2}p^{ik}q_{kl}p^{lj} + \tfrac{1}{8}p^{ij}q_{kl}p^{kl} - \tfrac{1}{96}\varepsilon^{ijklmnpq}q_{kl}q_{mn}q_{pq},\\ (\mathcal{Q},\mathcal{Q},\mathcal{Q})_{ij} &\equiv -\tfrac{1}{2}q_{ik}p^{kl}q_{lj} - \tfrac{1}{8}q_{ij}q_{kl}p^{kl} + \tfrac{1}{96}\varepsilon_{ijklmnpq}p^{kl}p^{mn}p^{pq}.\end{aligned}} \tag{19.127}$$

The existence of the above Jordan triple product has profound consequences due to its properties. First of all, the symplectic product of the Jordan triple product and $\mathcal{Q}$ gives an invariant which is , precisely, $J_4(\mathcal{Q})$:

$$\boxed{(\mathcal{Q},\mathcal{Q},\mathcal{Q})_M\mathcal{Q}^M = \tfrac{1}{2}(\mathcal{Q},\mathcal{Q},\mathcal{Q})_{ij}p^{ij} - \tfrac{1}{2}(\mathcal{Q},\mathcal{Q},\mathcal{Q})^{ij}q_{ij} = J_4(\mathcal{Q}).} \tag{19.128}$$

We can multiply this vector by any invariant function of $\mathcal{Q}$, which will necessarily be a function of $J_4(\mathcal{Q})$ to obtain another vector in the $\mathbf{56}$. The combination

$$\boxed{\tilde{\mathcal{Q}}^M \equiv 2(\mathcal{Q},\mathcal{Q},\mathcal{Q})^M/\sqrt{J_4(\mathcal{Q})}} \tag{19.129}$$

defines the *Freudenthal dual* of $\mathcal{Q}^M$ [238].[27] The Freudenthal dual of $N=8$ supergravity shares the main properties of the Freudenthal dual of the $N=2$ theories given at the beginning of Section 19.2: to start with, the Freudenthal duality operation is an anti-involution

$$\tilde{\tilde{\mathcal{Q}}} = -\mathcal{Q}, \tag{19.130}$$

which implies that

$$J_4(\tilde{\mathcal{Q}}) = J_4(\mathcal{Q}). \tag{19.131}$$

If we define the Hesse potential of the $N=8$ theory by

$$\boxed{W(\mathcal{Q}) \equiv 2\sqrt{J_4(\mathcal{Q})},} \tag{19.132}$$

it is easy to see that

$$\frac{\partial W}{\partial \mathcal{Q}^M} = 2\tilde{\mathcal{Q}}_M, \tag{19.133}$$

$$W(\mathcal{Q}) = \tilde{\mathcal{Q}}_M\mathcal{Q}^M. \tag{19.134}$$

[26] The Jordan triple product of three different vectors transforming in the $\mathbf{56}$ is defined only up to terms proportional to the symplectic products of two of the three vectors. This ambiguity disappears when the three vectors are equal.

[27] A generalization of this definition to all $N \geq 2$ theories was proposed in Ref. [536] in which the Freudenthal dual is scalar dependent. For $N=2$ that definition is different from the one we have given, but they coincide for the charge vector in the near-horizon limit.

It is possible and convenient to introduce a $\mathbb{K}$-tensor [932, 50] for this theory. The $\mathbb{K}$-tensor can be found by constructing first the linearization of the Cartan invariant as explained in Ref. [518] (see also Ref. [845] for more details) $J_4'$. This is given by

$$\begin{aligned} J_4'(\mathcal{Q}_1,\mathcal{Q}_2,\mathcal{Q}_3,\mathcal{Q}_4) \equiv\ & \tfrac{1}{6}\mathrm{Tr}\left\{p_1q_2p_3q_4 + p_1q_3p_4q_2 + p_1q_4p_2q_3 + (p\leftrightarrow q)\right\} \\ & -\tfrac{1}{12}\left\{[\mathcal{Q}_1|\mathcal{Q}_2][\mathcal{Q}_3|\mathcal{Q}_4] + [\mathcal{Q}_1|\mathcal{Q}_3][\mathcal{Q}_2|\mathcal{Q}_4] + [\mathcal{Q}_1|\mathcal{Q}_4][\mathcal{Q}_2|\mathcal{Q}_3]\right\} \\ & + \tfrac{1}{96}\varepsilon_{ijklmnop}\left(p_1^{ij}p_2^{kl}p_3^{mn}p_4^{op} + q_{1\,ij}q_{2\,kl}q_{3\,mn}q_{4\,op}\right), \end{aligned} \tag{19.135}$$

where we have defined, for convenience, the symmetric product

$$[\mathcal{Q}_1 \mid \mathcal{Q}_2] \equiv -\tfrac{1}{2}\mathrm{Tr}\left(p_1q_2 + (1\leftrightarrow 2)\right). \tag{19.136}$$

$J_4'(\mathcal{Q}_1,\mathcal{Q}_2,\mathcal{Q}_3,\mathcal{Q}_4)$ is linear in each of the four entries, symmetric in them, and reduces to the Cartan invariant when the four entries are equal. The $\mathbb{K}$-tensor is implicitly defined by

$$\mathbb{K}_{MNPQ}\mathcal{Q}_1{}^M\mathcal{Q}_2{}^N\mathcal{Q}_3{}^P\mathcal{Q}_4{}^Q \equiv J_4'(\mathcal{Q}_1,\mathcal{Q}_2,\mathcal{Q}_3,\mathcal{Q}_4), \tag{19.137}$$

and, by construction,

$$\mathbb{K}_{MNPQ}\mathcal{Q}^M\mathcal{Q}^N\mathcal{Q}^P\mathcal{Q}^Q = J_4'(\mathcal{Q},\mathcal{Q},\mathcal{Q},\mathcal{Q}) = J_4(\mathcal{Q}). \tag{19.138}$$

The Jordan triple product, the Hesse potential, and the Freudenthal dual have simple expressions in terms of the $\mathbb{K}$-tensor:

$$\begin{aligned} (\mathcal{Q},\mathcal{Q},\mathcal{Q})^M &= \mathbb{K}^M{}_{NPQ}\mathcal{Q}^N\mathcal{Q}^P\mathcal{Q}^Q \equiv \mathbb{K}^M\mathcal{Q}^3, \\ W(\mathcal{Q}) &= 2\sqrt{\mathbb{K}\mathcal{Q}^4}, \\ \tilde{\mathcal{Q}}^M &= 4\mathbb{K}^M\mathcal{Q}^3/W(\mathcal{Q}), \end{aligned} \tag{19.139}$$

where we have introduced a very obvious shorthand notation.

With this machinery in place, we are ready to go back and solve our original problem.

### *19.4.2 The metric function*

We have seen that the only possibility for $\mathcal{R}^M(\mathcal{I})$ is to be proportional to the Jordan triple product of $\mathcal{I}^M$ or, equivalently, to the Freudenthal dual of $\mathcal{I}^M$. The proportionality factor can be found by requiring $\mathcal{R}^M(\mathcal{I})$ to be homogeneous and of first order in $\mathcal{I}^M$ because $e^{-2U} = \mathcal{R}_M\mathcal{I}^M$ is expected to be of second order. This means that the proportionality factor is just numerical and, since we expect the entropy to be given by $\sqrt{J_4(\mathcal{Q})}$ [527] (see Section 26.2.3, where we will discuss the string origin of these solutions), we conclude that, analogously to what happens in the $N=2$ case, the metric function is equal to the Hesse potential of the symplectic vector of harmonic functions $H^M$:

$$e^{-2U} = W(H). \tag{19.140}$$

The 1-form $\omega$ and the vector field strengths $\mathcal{F}^M$ are given by the $N = 2$ formulae Eqs. (19.9) and (19.11). The scalars, however, cannot be written explicitly without more information.

Observe that the metric function Eq. (19.140) is similar to that of many $N = 2$ theories, but there is no $N = 2$ theory with the field content and duality group of $N = 8$ SUEGRA. On the other hand, this function reduces to those found in Ref. [523] for all the magic $N = 2$ truncations of $N = 8$ supergravity. Another simple truncation gives the metric function of the STU model, whose SBHSs have been reviewed on p. 578.

### 19.4.3 Single supersymmetric black-hole solutions

We only expect regular static single-SBHSs. They are constructed as in an $N = 2$ supergravity with a Hesse potential determined by a $\mathbb{K}$-tensor, with single-pole spherically symmetric harmonic functions of the form Eq. (19.13) with $B^M = \mathcal{Q}^M$. The only problem left to solve is to find the expression of the integration constants $A^M$ as functions of the moduli and charges, because we do not have explicit expressions for the scalars. However, we can still extract some information about them by imposing the normalization of the metric at infinity and the vanishing of the 1-form $\omega$. These two conditions imply, respectively,

$$\begin{aligned} \mathbb{K}_{MNPQ} A^M A^N A^P A^Q &= \tfrac{1}{4}, \\ A_M \mathcal{Q}^M &= 0. \end{aligned} \tag{19.141}$$

In the $N \geq 2$ theories the metric function $e^{-2U}$ and $\mathcal{I}^M$ are related to the matrix $M_{IJ}$ by $e^{2U} = |M|^2 = M^{IJ} M_{IJ}$ and $\mathcal{I}^M = \Im\mathrm{m}(\mathcal{V}^M_{IJ} M^{IJ} |M|^{-2})$, and the above two constraints imply

$$\begin{aligned} |M_\infty|^2 &= 1, \\ \Im\mathrm{m}\left(\mathcal{Z}_{\infty\, IJ} M_\infty^{IJ}\right) &= 0, \end{aligned} \tag{19.142}$$

where in the second equation we have used the first and the definition of the central-charge matrix Eq. (18.15).

The projection

$$\mathcal{Z} \equiv \tfrac{1}{\sqrt{2}} \mathcal{Z}_{IJ} M^{IJ} |M|^{-2} \tag{19.143}$$

plays a role similar to the central charge in $N = 2$ theories.[28] In terms of this central charge, the second condition above states that it is real, i.e.

$$\Im\mathrm{m}\, \mathcal{Z}_\infty = 0. \tag{19.144}$$

On the other hand, the mass, which can be computed using Eq. (19.22) ($\rho = -1/r$), is just

[28] In the first-order flow equations satisfied by these SBHSs [1012] this function drives the flow of the metric function as it is the case in $N = 2$ theories [522] (but, surprisingly, not the flow of the scalars).

its real part: [29]

$$M = |\mathcal{Z}_\infty| = \Re\mathrm{e}\,\mathcal{Z}_\infty = \tfrac{1}{\sqrt{2}}\tilde{A}_M \mathcal{Q}^M = 2\sqrt{2}\mathbb{K}_{MNPQ}A^M A^N A^P \mathcal{Q}^Q. \tag{19.145}$$

It is not hard to see that the entropy is given by

$$S/\pi = \sqrt{J_4(\mathcal{Q})}, \tag{19.146}$$

as expected.

## 19.5 The supersymmetric black holes of $N=1, d=5$ supergravity

### 19.5.1 The general recipe

Following the general results in Section 18.7.2, to give a recipe for the construction of five-dimensional SBHSs we first need a convenient choice of hyper-Kähler metric $h_{\underline{mn}}$.[30] The simplest choice, $h_{\underline{mn}} = \delta_{\underline{mn}}$, which is suggested by our experience with five-dimensional ERN solutions, does not lead to a simple algorithm except for static solutions with $\Theta^I = 0,\ \forall I$. These are determined by harmonic functions in $\mathbb{E}^4$, $L_I$:[31]

$$h_I e^{-U} = L_I, \qquad \partial_{\underline{m}}\partial_{\underline{m}} L_I = 0. \tag{19.147}$$

Unlike the four-dimensional case, these functions are not subject to any constraint apart from $e^{2U} = 1$ at infinity. Generically, multi-black-hole solutions do not have angular momentum as is the case in four dimensions, basically because they are only electrically charged. The magnetically charged objects are black strings.

Given the harmonic functions $L_I$, we have to find $e^{2U}$, and for this we need to know the variables $h^I$ as a function of the $h_I = C_{IJK}h^J h^K$ that we will denote by $h^I(h_.)$. Finding these relations is equivalent to solving the Freudenthal duality equations in four dimensions. Note that $h^I(h_.)$ must be homogeneous of degree $1/2$ in $h_I$ and the fundamental constraint $C_{IJK}h^I h^J h^K = 1$ takes the form $\mathsf{W}(h_.)/2 = 1$, where $\mathsf{W}(h_.)$ is a function homogeneous of degree $3/2$ in $h_I$ that plays the same role as the Hesse potential in $d=4$. In some cases it can be found simply on symmetry grounds. Then, substituting $h_I = e^U L_I$, $\mathsf{W}(e^U L)/2 = 1$ and using homogeneity we obtain

$$e^{-\frac{3}{2}U} = \mathsf{W}(L)/2. \tag{19.148}$$

[29] Since the general form of the metric function is the same for the magic $N=2$ truncations studied in Ref. [523], it is not surprising that one finds almost identical mass formulae for the SBHSs of those theories.

[30] We have assumed that all the hyperscalars, if any, have been set to a constant value, as in the four-dimensional case.

[31] In this section we will use the notation $r = |\vec{x}_3|$, $r_4 \equiv |\vec{x}_4|$, and $\rho \equiv 1/r_4^2$. We will also denote the metric function $g_{tt} = f^2$ by $e^{2U}$ as in the four-dimensional case.

The scalar fields can be written, for instance, in the form

$$\boxed{\phi^x = h_x/h_0 = L_x/L_0.} \tag{19.149}$$

With this choice of transverse metric $h_{\underline{mn}}$, it is very difficult to find a recipe to construct stationary SBHSs. A recipe exists for the timelike supersymmetric solutions with one extra isometry discussed in Section 18.7.5, however, and it can be used for SBHSs if we choose the Gibbons–Hawking metric with

$$\boxed{H = 1/r \quad \Rightarrow \quad \chi = \cos\theta d\varphi,} \tag{19.150}$$

in standard three-dimensional spherical coordinates $r, \theta, \varphi$, because this metric,

$$h_{\underline{mn}} dx^m dx^n = H^{-1}(dz+\chi)^2 + H d\vec{x}_3^{\,2} = r(dz+\cos\theta d\varphi)^2 + \frac{1}{r}(dr^2 + r^2 d\Omega_{(2)}^2), \tag{19.151}$$

is simply the Euclidean metric in $\mathbb{R}^4$ with a non-standard radial coordinate

$$\boxed{r = r_4^2/4.} \tag{19.152}$$

Indeed, with this change of coordinates, and using $\psi$ instead of $z$, we obtain the standard metric of $\mathbb{E}^4$ in spherical coordinates,

$$h_{\underline{mn}} dx^m dx^n = dr_4^2 + r_4^2 d\Omega_{(3)}^2, \quad \text{where} \quad d\Omega_{(3)}^2 = \tfrac{1}{4}[d\Omega_{(2)}^2 + (d\psi + \cos\theta d\varphi)^2], \tag{19.153}$$

where $0 < \theta < \pi$, $0 < \varphi < 2\pi$, and $0 < \psi < 4\pi$.

One can then use the recipe of Section 18.7.5:

1. Find $2n_V + 3$ harmonic functions in $\mathbb{E}^3$ $K^I, L_I, M$ such that the integrability constraint of the 1-form $\hat{\omega}$,

$$\boxed{\frac{1}{r}\hat{\nabla}^2 M - M\hat{\nabla}^2\frac{1}{r} + \tfrac{3}{2}\left(K^I\hat{\nabla}^2 L_I - L_I\hat{\nabla}^2 K^I\right) = 0,} \tag{19.154}$$

   where $\hat{\nabla}^2$ is the Laplacian in $\mathbb{E}^3$, is satisfied everywhere.

2. Solve the equation for $\hat{\omega}$,

$$\boxed{\hat{\star} d\hat{\omega} = \frac{1}{r}dM - M d\frac{1}{r} + \tfrac{3}{2}\left(K^I dL_I - L_I dK^I\right),} \tag{19.155}$$

and construct the 1-form $\omega$:

$$\boxed{\begin{aligned}\omega &= \omega_5(d\psi + \cos\theta d\varphi) + \hat{\omega}, \\ \omega_5 &\equiv M + \tfrac{3}{2}rL_IK^I + r^2 C_{IJK}K^IK^JK^K.\end{aligned}} \tag{19.156}$$

3. The $h_I$ and the self-dual 2-forms $\Theta^I$ are given by

$$\boxed{\begin{aligned}h_Ie^{-U} &= L_I + H^{-1}C_{IJK}K^JK^K, \\ \Theta^I &= [(d\psi + \cos\theta d\varphi)\wedge d(rK^I) + \frac{1}{r}\hat{\star}d(rK^I)].\end{aligned}} \tag{19.157}$$

4. Once the $h_Ie^{-U}$ have been found, the metric function is determined in terms of W as explained above by Eq. (19.148), the vector field strengths are given by Eq. (18.196), and the scalars can be constructed, for instance, as in Eq. (19.149).

Observe that the static solutions are recovered by setting $M = K^I = 0$ and changing the radial coordinate as in Eq. (19.152), although this can be very difficult to do for multi-black-hole solutions.

### 19.5.2 Single, static, black-hole solutions

Following the general recipe we set $M = K^I = 0$. We simply need to determine the harmonic functions $L_I$ which will have a single point-like singularity at the origin (for simplicity). In this case it is convenient to use the radial coordinate $\rho = 1/r_4^2$, which vanishes at spatial infinity and tends to $+\infty$ near the horizon. The metric takes the form

$$ds^2 = e^{2U}dt^2 - e^{-U}\frac{1}{\rho}\left[\frac{1}{4\rho^2}d\rho^2 + d\Omega^2_{(3)}\right], \tag{19.158}$$

the harmonic functions $L_I$ are linear functions of $\rho$

$$L_I = A_I + B_I\rho, \tag{19.159}$$

and the problem reduces to the determination of the integration constants $A_I$ and $B_I$ in terms of the physical parameters (electric charges $q_I$ and moduli $\phi^x_\infty$) in such a way that the metric is regular and normalized at infinity.

We assume that the $h_I$ have been solved in terms of the $h^I$ and that the potential W has been found. We can immediately identify the integrations $B_I$: the vector field strengths for timelike supersymmetric configurations given in Eq. (18.196) reduce in this case to

$$F^I = -\sqrt{3}d(e^Uh^I)\wedge dt. \tag{19.160}$$

The electric charge $q_I$ is given by the generalization of Eq. (12.223),

$$q_I = -\frac{1}{16\pi G_N^{(5)}} \int_{S^3_\infty} a_{IJ} \star F^J, \tag{19.161}$$

where we have assumed that the action has the standard normalization factor $(16\pi G_N^{(5)})^{-1}$. Using the identities of real special geometry in Appendix H we can show that

$$a_{IJ} d(e^U h^I) = -e^{2U} dL_I \quad \Rightarrow \quad q_I = \frac{\sqrt{3}\pi}{4G_N^{(5)}} B_I. \tag{19.162}$$

From now on it is convenient to use units in which

$$B_I = q_I, \qquad G_N^{(5)} = \frac{\sqrt{3}\pi}{4}. \tag{19.163}$$

The scalar fields, written generically as in Eq. (19.149), are given by simple quotients of the harmonic functions $L_I$:

$$\phi^x = L_x/L_0 \quad \Rightarrow \quad \phi^x_\infty = A_x/A_0. \tag{19.164}$$

If, following Ref. [942], we introduce $\phi^0 \equiv 1$, we can write

$$A_I = A_0 \phi^I_\infty. \tag{19.165}$$

The normalization of the metric identifies $A_0$ as a function of the $\phi^I_\infty$,

$$1 = \lim_{\rho \to 0} e^{-\frac{3}{2}U} = \mathsf{W}(A)/2 = A_0^{3/2} \mathsf{W}(\phi_\infty)/2, \tag{19.166}$$

and we arrive at the general solution

$$A_I = \phi^I_\infty [\mathsf{W}(\phi_\infty)/2]^{-2/3}. \tag{19.167}$$

Having determined all the integration constants in terms of the independent physical parameters $q_I, \phi^I_\infty$ we can compute the mass, the entropy, and the values of the scalars on the horizon in terms of them.

The mass is given by

$$\boxed{M = \frac{\pi}{4G_N^{(5)}} \dot{\mathsf{W}}\Big|_{\rho=0} = \frac{1}{\sqrt{3}} \dot{\mathsf{W}}\Big|_{\rho=0},} \tag{19.168}$$

where in the last identity we have used the units in Eq. (19.163) and the overdot indicates derivation with respect to $\rho$. As we will show in Section 27.3.1,

$$\dot{\mathsf{W}} = 3(\mathsf{W}/2)^{1/3} h^I \dot{L}_I, \tag{19.169}$$

and, substituting in the above mass formula, we get

$$M = \sqrt{3}\,\mathcal{Z}_\infty, \tag{19.170}$$

where $\mathcal{Z}_\infty$ is the value of the central charge at spatial infinity:

$$\mathcal{Z}(\phi, q) \equiv h^I(\phi) q_I, \qquad \mathcal{Z}_\infty \equiv \mathcal{Z}(\phi_\mathrm{h}, q). \tag{19.171}$$

The signs of the moduli $\phi^I_\infty$ and charges must be such that the mass is positive and saturates the $N = 1, d = 5$ supersymmetry (BPS) bound

$$M \geq \sqrt{3}\,|\mathcal{Z}_\infty|, \tag{19.172}$$

which generalizes the one of the pure Einstein–Maxwell theory Eq. (12.230).

Let us now consider the near-horizon behavior of the solution. Given that $\mathsf{W}(L)$ is homogeneous of order $3/2$ in the $L_I$, which are linear in $\rho$, in the $\rho \to \infty$ limit,

$$e^{-U} \sim R_S^2 \rho, \tag{19.173}$$

where $R_\mathrm{S}$ is the Schwarzschild radius of the horizon given by

$$R_\mathrm{S} = [\mathsf{W}(q)/2]^{1/3}. \tag{19.174}$$

The full metric Eq. (19.158) becomes in this limit

$$ds^2 \sim \left(\frac{\rho}{R_\mathrm{S}/2}\right)^2 dt'^2 - \left(\frac{R_\mathrm{S}/2}{\rho}\right)^2 - R_S^2 d\Omega^2_{(3)}, \qquad t' = t/(2R_\mathrm{S}), \tag{19.175}$$

which is the product of $\mathrm{AdS}_2$ with radius $R_\mathrm{S}/2$ and $S^3$ (the constant-time sections of the event horizon, provided that $R_\mathrm{S} \neq 0$) with radius $R_\mathrm{S}$. The area of the event horizon is $2\pi^2 R_\mathrm{S}^3$ and the entropy is a quarter of that, i.e.

$$4S/\pi^2 = \mathsf{W}(q). \tag{19.176}$$

The values of the scalars on the horizon are given by

$$\phi^I_\mathrm{h} = \frac{q_I}{q_0}, \tag{19.177}$$

and they are moduli independent, which indicates the existence of an attractor mechanism similar to the one discussed in the four-dimensional case at work. Also as in the four-dimensional case the entropy can be related to the value of the central charge on the horizon $\mathcal{Z}_\mathrm{h} = \mathcal{Z}(\phi_\mathrm{h}, q)$. The precise relation can be found by considering the double extremal SBHSs of these theories: the scalars are constant and $\phi^I_\mathrm{h} = \phi^I_\infty$ when all the integration constants $A_I$ and $B_I$ ($q_I$ here) are proportional with the same proportionality constant $\beta$:

$$A_I = q_I/\beta \;\Rightarrow\; L_I = \frac{q_I}{\beta} L, \quad \text{where} \quad L = 1 + \beta\rho. \tag{19.178}$$

Again, due to the homogeneity of $\mathsf{W}(L)$ in $L$, the metric function $e^{-U}$ takes the simple form

$$e^{-U} = \frac{[\mathsf{W}(q)/2]^{2/3}}{\beta} L. \tag{19.179}$$

To normalize the metric function at spatial infinity to 1, we take $\beta = (\mathsf{W}(q)/2)^{2/3}$ so $e^{-U} = L$ and the metric becomes that of the ERN solution (Eq. (12.98)[32] with $\omega = 1$ and $H = L$). The mass of this solution is $\sqrt{3}\,\beta = \sqrt{3}\,(\mathsf{W}(q)/2)^{2/3}$. On the other hand, the general formula $M = \sqrt{3}\,|\mathcal{Z}_\infty|$ must also apply and, since for these double extremal SBHSs the scalars are constant, by construction $\mathcal{Z}_\infty = \mathcal{Z}_{\rm h} \equiv \mathcal{Z}(\phi_{\rm h}, q)$, from which it follows that

$$|\mathcal{Z}_{\rm h}| = (\mathsf{W}(q)/2)^{2/3}, \qquad 4S/\pi^2 = |\mathcal{Z}_{\rm h}|^{3/2}. \tag{19.180}$$

### 19.5.3 Examples

The general results obtained in Section 19.5.2 reduce the problem of writing all the single, static, SBHSs of a given model of $N = 1, d = 5$ supergravity to the choice of charges and the determination of the potential W for that specific model. Let us see how this is achieved by considering some simple examples.

*Pure $N = 1, d = 5$ supergravity.* This theory was described in the framework of real special geometry on p. 267. For the static metrics and purely electric vector fields that we are considering, this theory is equivalent to the Einstein–Maxwell one because the Chern–Simons term characteristic of $N = 1, d = 5$ supergravity has no practical effect. The theory is defined by the constraint $(h^0)^3 = 1$. The dual variable is $h_0 = (h^0)^2$ and, expressed in terms of it, the defining constraint becomes $(h_0)^{3/2} = 1$ so

$$\mathsf{W}(L) = 2(L_0)^{3/2}, \qquad e^{-U} = L_0 = 1 + q_0\rho. \tag{19.181}$$

The solution is just the ERN solution. Observe that here $\mathcal{Z} = q_0$.

We are going to review some stationary solutions of this theory later.

*The $d = 5$ $\mathrm{ST}^2$ model.* This theory, which is the simplest with one vector multiplet, was described in the framework of real special geometry on p. 268. The defining constraint is $h^0(h^1)^2 = 1$ and the dual variables are given by

$$h_0 = \tfrac{1}{3}(h^1)^2, \qquad h_1 = \tfrac{2}{3}h^0h^1. \tag{19.182}$$

These expressions can be inverted in two different ways labeled $\sigma = \pm 1$:

$$h^1 = \sigma\sqrt{3h_0}, \qquad h^0 = \tfrac{3}{2}\sigma\frac{h_1}{\sqrt{3h_0}}, \tag{19.183}$$

which give rise to two different *branches* of the theory that have the same bosonic action. We will find a SBHS in each branch. Two parametrizations in terms of scalar fields[33] with appropriate normalization that solve the defining constraint are given in Eqs. (9.22). These parametrizations are different from the generic one $\phi^I \equiv h_I/h_0$ but they are far better because the scalar fields chosen are totally unconstrained. Observe that, since the bosonic

---

[32] The radial coordinate $\rho$ is not the same as the one we are using here.

[33] We have denoted both of them by $\phi$ to avoid using too many indices, but they are different fields.

action is the same in both branches, each of the two SBHSs that we are going to find is a solution of the two branches, but it will be supersymmetric only in one of the two branches.

Expressing the defining constraint in terms of the dual variables we get

$$\mathsf{W}(L) = \sqrt{3^3 L_0 L_1^2}, \quad e^{-U} = 3\sqrt[3]{\frac{L_0 L_1^2}{4}}, \quad \phi = \sqrt{\tfrac{2}{3}} \log \frac{\sigma L_1}{2L_0}. \tag{19.184}$$

The signs of the integration constants must be chosen so as to make $\mathsf{W}(L)$ real and positive in each branch (i.e. $L_0 > 0$ and $\sigma L_1 > 0$) and, as in the four-dimensional case, only certain signs of the charges will be possible for SBHSs, but the specific signs will depend on the branch.

Using the normalization of the metric function at infinity and the asymptotic behavior of the scalar we find the integration constants

$$A_0 = \tfrac{1}{3} e^{-\sqrt{\frac{2}{3}}\phi_\infty} > 0, \qquad \sigma A_1 = \tfrac{2}{3} e^{\frac{1}{\sqrt{6}}\phi_\infty} > 0. \tag{19.185}$$

In order to get regular solutions we must take $q_0 > 0$ and $\sigma q_1 > 0$ in the $\sigma$ branch. Then, the central charge

$$\mathcal{Z}(\phi, q) = e^{\sqrt{\frac{2}{3}}\phi} q_0 + \sigma e^{-\frac{1}{\sqrt{6}}\phi} q_1 \tag{19.186}$$

is automatically positive in both branches and $M = \sqrt{3}|\mathcal{Z}_\infty|$. On the horizon,

$$e^{\sqrt{\frac{3}{2}}\phi_{\rm h}} = \frac{\sigma q_1}{2q_0} \quad \Rightarrow \quad \mathcal{Z}_{\rm h} = 3\sqrt[3]{\frac{q_0 q_1^2}{4}} = (\mathsf{W}(q)/2)^{2/3}, \tag{19.187}$$

and the entropy is given by

$$4S/\pi^2 = \mathsf{W}(q)/2 = \tfrac{1}{2}\sqrt{3^3 q_1^2 q_0}. \tag{19.188}$$

*The five-dimensional* STU *model.* This model was described in the framework of real special geometry on p. 269. Since $h^0 h^1 h^2 = \sqrt{3^3 h_0 h_1 h_2}$,

$$\mathsf{W}(L) = 2\sqrt{3^3 L_0 L_1 L_2}, \qquad e^{-U} = 3\sqrt[3]{L_0 L_1 L_2}, \qquad h_I = \frac{L_I}{3\sqrt[3]{L_0 L_1 L_2}}. \tag{19.189}$$

Inverting the parametrization of the $h^I$ in terms of the physical scalars $\phi^1, \phi^2$ given in Eq. (9.29) we find that

$$e^{2\sqrt{6}\phi^1} = \frac{h^1 h^2}{3h^0} = \frac{1}{9}\sqrt[3]{\frac{L_1^2 L_2^2}{L_0^4}}, \qquad e^{3\sqrt{2}\phi^2} = \frac{h_2}{h_1} = \frac{L_2}{L_1}. \tag{19.190}$$

The integration constants $A_I$ are, in terms of the moduli $\phi^1_\infty, \phi^2_\infty$,

$$A_0 = \tfrac{1}{9} e^{-\sqrt{6}\phi^1_\infty}, \quad A_1 = \frac{1}{\sqrt{3}} e^{-\sqrt{\frac{3}{2}}\phi^1_\infty - \frac{3}{\sqrt{2}}\phi^2_\infty}, \quad A_2 = \frac{1}{\sqrt{3}} e^{-\sqrt{\frac{3}{2}}\phi^1_\infty + \frac{3}{\sqrt{2}}\phi^2_\infty}. \tag{19.191}$$

The central charge of this model is

$$\mathcal{Z}(\phi, q) = e^{\sqrt{6}\phi^1} q_0 + e^{-\sqrt{\frac{3}{2}}\phi^1 - \frac{3}{\sqrt{2}}\phi^2} q_1 + e^{-\sqrt{\frac{3}{2}}\phi^1 + \frac{3}{\sqrt{2}}\phi^2} q_2, \tag{19.192}$$

and the mass is given by the standard saturated BPS bound $M=\sqrt{3}\,\mathcal{Z}_\infty$. Finally, the values of the scalars on the horizon at $\rho\to\infty$ and the entropy are given by

$$2S/\pi^2=\sqrt{3^3 q_0 q_1 q_2}, \qquad e^{2\sqrt{6}\phi^1_{\rm h}}=\tfrac{1}{9}\sqrt[3]{\frac{q_1^2 q_2^2}{q_0^4}}, \qquad e^{3\sqrt{2}\phi^2_\infty}=\frac{q_2}{q_1}. \tag{19.193}$$

In this branch (all $h^I>0$), all the charges must be positive (and so are the integration constants $A_I$). However, the solutions of the other branches will also be solutions of this branch because the bosonic action of the theory is the same in all branches. However, in each branch, only one solution (one choice of sign for the charges) will be supersymmetric.

We will meet again these SBHSs in Chapter 26: the extremal limit of the black-hole solution of maximal five-dimensional supergravity in Eq. (26.18) has exactly the same form with $L_0=H_{\rm W}$, $L_1=H_{\rm D5}$, and $L_2=H_{\rm D1}$ (up to constants, since the harmonic functions $H$ are normalized to be 1 at spatial infinity). The scalar $k_{V^4}$ in that solution can be identified with $e^{3\sqrt{2}\phi^2}$, and the dilaton $e^{-2\phi}$ and the scalar $k_1$, which are equal up to constants, can be identified with $[9e^{2\sqrt{6}\phi^1}]^{-3/8}$. The five-dimensional STU model can be obtained from a truncation of the maximal five-dimensional supergravity. As we have seen repeatedly, a truncation requires that certain fields are set to zero or identified in order to reduce the number of independent degrees of freedom. This is what happens to the scalars $e^{-2\phi}$ and the scalar $k_1$. Evidently, there are more general black-hole solutions in the maximal supergravity for which these two scalars have different spatial dependence, but they cannot be obtained from the five-dimensional STU model.

### 19.5.4 Some stationary solutions of pure $N=1, d=5$ supergravity

In this section we are going to review briefly the supersymmetric black-ring solution (SBRS) of pure $N=1, d=5$ supergravity. This solution is interesting by itself but it also provides us with a shortcut to introduce the supersymmetric rotating black holes of this theory which arise in different limits of the SBRS.

As mentioned in the introduction, a SBRS of pure $N=1, d=5$ supergravity was constructed in Ref. [507], and in Ref. [594][34] Gauntlett and Gutowski found that this SBRS is just an example of a timelike supersymmetric solution with an additional isometry ($H=1/r$), and they determined the corresponding harmonic functions in $\mathbb{E}^3$, $L_0, K^0, M$. These three depend on the single harmonic function in $\mathbb{E}^3$

$$N\equiv\frac{1}{|\vec{x}-\vec{x}_1|}, \qquad \vec{x}=(x,y,z), \qquad \vec{x}_1=(0,0,-R^2/4), \tag{19.194}$$

as follows:[35]

$$L_0=1+\frac{q_0-a^2}{4}N, \qquad K^0=-\frac{a}{2}N, \qquad M=\frac{3a}{4}\left(1-\frac{R^2}{4}N\right), \tag{19.195}$$

[34] The embedding in ten-dimensional type-IIB SUEGRA was given in Ref. [508] (see also Ref. [140]). The ten-dimensional solution is a *supertube*.

[35] In Refs. [507, 594] the parameter $q_0$ (which coincides with the electric charge in our conventions) was called $Q$ and the parameter $a$ was called $q$.

so the metric function is given by

$$e^{-U} = L_0 + (K^0)^2/H = 1 + q_0 N - a^2 N(1 - rN). \tag{19.196}$$

The 1-form $\omega$ has two parts: $\omega_5$, which can be computed immediately using the general formula Eq. (19.156), and $\hat{\omega}$, which has to be found by integrating Eq. (19.155). The result is

$$\begin{aligned}
\omega &= (F - \cos\theta G)d\varphi + (F - G)d\psi, \\
F &= \frac{3a}{4}\left[1 - \left(r + \frac{R^2}{4}\right)N\right], \\
G &= \frac{a}{16} r N^2 \left[(3q_0 - a^2) + 2a^2 rN\right],
\end{aligned} \tag{19.197}$$

where $r, \theta, \varphi$ are the usual spherical coordinates centered at $\vec{x} = 0$.

The asymptotic flatness of the solution is trivial to check. In the units defined in Eq. (19.163) $q_0$ is the electric charge of the solution and, as expected from the supersymmetry and asymptotic flatness of the solution, the mass saturates the same BPS bound as the single-SBHSs $M = \sqrt{3}\, q_0$. More interestingly (and crucial for the very existence of this solution), the two components of the angular momentum that can be independent in five dimensions are non-vanishing and different:

$$J_1 = \tfrac{1}{2\sqrt{3}} a(3q_0 - a^2), \qquad J_2 = \tfrac{1}{2\sqrt{3}} a(6R^2 + 3q_0 - a^2). \tag{19.198}$$

The three independent parameters of the solution, $q_0, a, R$ (corresponding to $q_0, J_1, J_2$ since $M$ is not independent) have to satisfy the condition $q_0 > a^2 + 2aR$ to avoid the existence of closed timelike curves.

The solution may have an event horizon at $\vec{x} = \vec{x}_1$ because $g_{tt} = e^{2U}$ vanishes in that limit, but other components of the metric are singular in that limit. However, this happens because the coordinate system is singular in that limit (especially the coordinate $\theta$). To see what the metric is like in that limit one first has to change coordinates from the spherical ones centered at $\vec{x} = 0$ $r, \theta, \varphi$ to another set of spherical coordinates centered at $\vec{x}_1$, $r', \theta', \varphi'$. We can take $\varphi' = \varphi$, and the relation between $r, \theta$ and $r', \theta'$ is implicitly given by

$$r\cos\theta = r'\cos\theta' - R^2/4, \qquad r^2 = r'^2 - \tfrac{1}{2}R^2\cos\theta' + (R^2/4)^2. \tag{19.199}$$

After this coordinate change, the metric can be expanded in powers of $r' \to 0$ and, although the behavior is better, there remain some divergent terms. These can be canceled so the metric is completely regular in that limit by making two additional transformations:

$$dt = dv + \left(\frac{b_1}{r'} + \frac{b_2}{r'^2}\right)dr', \qquad d\psi = d\varphi' + 2d\psi' + 2\frac{c_1}{r'}dr', \tag{19.200}$$

for suitably chosen constants $b_1, b_2, c_1$.[36] An event horizon with (constant-time sections of) topology $S^2 \times S^1$ and metric

$$ds^2_{(3)} = -\frac{a^2}{4}d\Omega'^2_{(2)} - l^2 d\psi'^2, \quad \text{where} \quad l \equiv \sqrt{3\left[\frac{(q_0 - a^2)^2}{4a^2} - R^2\right]}, \tag{19.201}$$

[36] Their values can be found in Ref. [594].

arises in that limit.

We can recover several interesting solutions taking different limits of the SBRS:

1. Setting $a=R=0$ we recover the ERN black hole with vanishing angular momenta that we have studied before. The limit $a\to 0,\ R\to 0$ does not commute with the near-horizon limit and, as explained before, the event horizon is a round 3-sphere.

2. Setting $R=0$ so $N=H=1/r$ we get the rotating, supersymmetric Breckenridge–Myers–Peet–Vafa (BMPV) black hole [252, 598] embedded in pure $N=1, d=5$ supergravity characterized by a mass that saturates the BPS bound $M=\sqrt{3}\,q_0$ and equal angular momenta $J_1=J_2=\frac{1}{2\sqrt{3}}a(3q_0-a^2)\equiv-\frac{1}{\sqrt{3}}J$. Since $K^0=-\frac{a}{2}H$, $e^{-U}$ is a harmonic function and the solution can be rewritten in an equivalent way with

$$L_0=1+\frac{q_0}{4}H, \qquad K^0=0, \qquad M=\tfrac{1}{8}JH, \tag{19.202}$$

so the metric can be seen as that of the ERN black hole ($e^{-U}=L_0$) deformed with [746]

$$\omega=W_M\equiv M(d\psi+\cos\theta d\varphi). \tag{19.203}$$

In the near-horizon limit $r\to 0$, defining $R_2^2=q/4$, the metric takes the form

$$\begin{aligned} ds^2 \sim\ & \frac{r^2}{R_2^4}dt^2-R_2^2\frac{dr^2}{r^2}-R_2^2d\Omega_{(2)}^2-R_2^2\left(1-\frac{J^2}{q^3}\right)(d\psi+\cos\theta d\varphi)^2 \\ & +\frac{2}{R_2}\frac{J}{q^{3/2}}r dt(d\psi+\cos\theta d\varphi). \end{aligned} \tag{19.204}$$

Rescaling the time coordinate

$$t\equiv X\phi, \qquad X/R_2\equiv\left(1-\frac{J^2}{q^3}\right)^{1/2}, \tag{19.205}$$

the metric can be rewritten in the form

$$ds^2\sim R_2^2d\Pi_{(2)}^2-R_2^2d\Omega_{(2)}^2-R_2^2\left[\frac{X}{R_2}(d\psi+\cos\theta d\varphi)-\frac{J}{q_0^{3/2}}\frac{r}{R_2^2}d\phi\right]^2, \tag{19.206}$$

where

$$d\Pi_{(2)}^2=\frac{r^2}{R_2^4}d\phi^2-\frac{dr^2}{r^2} \tag{19.207}$$

is the metric of $AdS_2$ with unit radius (the factor $R_2^{-4}$ is immaterial). Finally, defining the new parameter $\xi$ by

$$\cos\xi\equiv X/R_2 \quad\Rightarrow\quad \sin\xi=J/q_0^{3/2}, \tag{19.208}$$

and redefining the coordinates

$$y \equiv \cos\xi\,\psi, \qquad e^{\chi} \equiv r/R_2^2, \tag{19.209}$$

the metric takes the final form

$$ds^2 \sim R_2^2 d\Pi_{(2)}^2 - R_2^2 d\Omega_{(2)}^2 - R_2^2\left[dy + \cos\xi\cos\theta d\varphi - \sin\xi e^{\chi} d\phi\right]^2, \tag{19.210}$$

which should be compared with that of the maximally supersymmetric vacuum of $N=1, d=5$ supergravity in Eq. (17.129). The BMPV SBHSs have a regular horizon for $J^2 < q^3$ whose constant-time sections are squashed 3-spheres with area

$$A = 2\pi^2\sqrt{q^3 - J^2} \tag{19.211}$$

[598] and interpolate between the Minkowski vacuum and the one in Eq. (17.129), both of which are maximally supersymmetric.

3. In the $R \to \infty$ limit (see Ref. [507]) the solution becomes the black "string"[37] solution of Ref. [135]. This solution can, equivalently, be described by the choice of harmonic functions

$$H=1, \quad N \equiv \frac{1}{r}, \quad K = -\frac{a}{2}N, \quad L = 1 + \frac{q_0}{2}N, \quad M = -\frac{3a}{4}N. \tag{19.212}$$

The metric function and the 1-form $\omega$ in the metric are given by

$$e^{-U} = 1 + q_0 N + \frac{q^2}{4}N^2, \qquad \omega = -\left(\frac{3a}{2}N + \frac{3aq_0}{4}N^2 + \frac{a^3}{8}N^3\right)d\psi. \tag{19.213}$$

[37] Supersymmetric strings belong to the class of null supersymmetric solutions. This is a timelike supersymmetric solution independent of the coordinate $\psi$ so it is closer to a SBHS smeared in the $\psi$ direction. In fact, for $a=0$, $\omega$ vanishes and the solution is nothing but the ERN black hole smeared in the $\psi$ direction.

# Part III
# Gravitating extended objects of string theory

*There is geometry in the humming of the strings.*

*Pythagoras*

# 20
# String theory

In this chapter we start the study of the extended objects that appear in the non-perturbative spectrum of string theory, the subject of the third part of this book. In this part we will make use of all the techniques we have developed in the first and second parts, whose main goal was to serve as a preparation for the third.

In a certain sense, this third part also presents the synthesis and (it is hoped) culmination of the ideas presented in the previous two in the framework of string theory: on the one hand, string theory includes a presumably consistent theory of quantum gravity that contains the gravitons described at lowest order by the Fierz–Pauli theory we studied in Chapter 3 [1103, 1104]. There are two main differences from the non-renormalizable theory of GR: the presence of a dimensionless coupling constant different from the Planck length and the presence of terms of higher order in derivatives. Furthermore, consistent string theories have spacetime supersymmetry and, therefore, supergravity, which we studied in Chapters 5 and 17. On the other hand, string theory incorporates naturally extra dimensions that have to be compactified. Thus, the ideas of Kaluza and Klein studied in Chapter 15 are also integrated into the picture.

Finally, the Schwarzschild, Reissner–Nordström, pp-wave, etc. solutions studied in other chapters are also solutions of string theory and it is natural to try to use them to solve the puzzles that arise when one tries to do quantum mechanics in those backgrounds: the information and entropy problems. If string theory is really a good theory of quantum gravity, then it should help us to solve them and we will see to what extent it succeeds in Chapter 26.

The attempts to solve these long-standing problems have been made possible by recent developments in string theory (essentially dualities and D-branes) and also by a change of perspective that we could call the "spacetime approach," which is based on the effective field theories, when further advance with the "worldsheet approach" was becoming increasingly difficult and slow. Of course, the two approaches are complementary and there has been a considerable amount of feedback between them. In fact, some of the most interesting things that we have learned in this period are the relations between the two of them. The logic of these relations is represented schematically in Fig. 20.1 and it is worth pausing for a moment to describe it in detail.

The box in the center represents the worldvolume theories of all the extended objects of

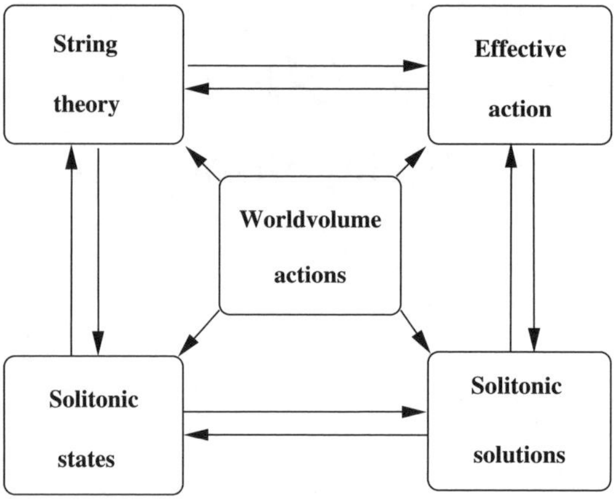


Fig. 20.1. Most of the recent progress in string theory has been based on the relations represented in this diagram between different aspects of the theory. The implications of duality symmetries in each box must be related to similar implications in other boxes.

string theory, including fundamental strings. They give, to start with, the perturbative definition of string theory (the worldsheet formulation), represented by the box in the upper-left-hand corner. This worldsheet formulation allows us to calculate the perturbative spectrum and also find some non-perturbative states (the D-branes), which are represented by the lower-left-hand corner of the diagram (their worldvolume actions are also represented by the box in the center). The low-energy effective field theories that describe the dynamics of the massless states of the perturbative spectrum are represented in the upper-right-hand corner. These theories are usually supergravity theories that have certain (on-shell) global symmetries that transform and mix the fields that represent the string massless modes. The equations of motion of these effective field theories can be derived (a most important point) from consistency conditions (conformal invariance or $\kappa$-symmetry) of worldvolume theories (the box in the center) in general backgrounds. The effective field theories also admit solitonic solutions, which are represented in the lower-right-hand corner. The solitonic solutions can be excited and deformed, and their effective dynamics is, yet again, given by worldvolume actions.

The meaning of the $\downarrow, \rightarrow$, and $\downarrow$ arrows and the arrows that come from the box in the center of this diagram is clear. The progress made in this field comes from the realization of the existence and use of the remaining arrows of the figure.

The main idea is that, generically, the global symmetries of the effective field theories correspond to *dualities* of the string theories.[1]

Some of these dualities are essentially perturbative (in the string coupling constant, which we will define later) and can be found and studied using the worldsheet approach.

[1] This is the point of view proposed in Ref. [808], but a more precise statement that includes more cases would be that the relations between different effective field theories correspond to dualities of the corresponding string theories. Some of the relations can be described as global symmetries of a single effective action, but in other cases there are relations between very different effective actions.

They are generically called *T dualities*, and the archetype of them (known long before as T duality, T from "target space," see e.g. Ref. [635]) relates string theories compactified on circles of dual radii. This is an exact symmetry at all orders in string perturbation theory [39].

On top of these, there are the so-called *S dualities*, which are non-perturbative in the string coupling constant and cannot be studied using the standard worldsheet approach. In fact, their existence in four-dimensional heterotic-string theory was suggested by the existence of the corresponding global symmetry in the effective action [553, 1075, 1124] which is nothing but that of $N = 4, d = 4$ SUGRA. As we saw in Section 16.2, the equations of motion of this theory are invariant under global $\mathrm{SL}(2, \mathbb{R})$ transformations, some of which invert the dilaton field, which can be interpreted as the string coupling constant. In general, S dualities are associated with this group[2] [808] and involve the inversion of the dilaton (string coupling constant) and the interchange of electric and magnetic fields. Another interesting example that we will study in Chapter 23 is $N = 2B, d = 10$ SUGRA, the effective field theory of the type-IIB superstring: the equations of motion of the supergravity theory are invariant under global $\mathrm{SL}(2, \mathbb{R})$ rotations that invert the dilaton, which suggests the existence of an S duality between type-IIB superstring theories.

The T- and S-duality groups are sometimes (in type-II theories) part of a bigger duality group called in Ref. [808] the U-*duality group*.

The interpretation as string dualities of the supergravity symmetries (the upper $\leftarrow$ arrow in Fig. 20.1) is possible if the string spectra reflect the same duality properties. For T dualities, this is easy to see in the perturbative spectrum, but S dualities necessarily imply the existence of new non-perturbative states. Thus, using the upper $\leftarrow$ relation, we start learning new things about string theory and its spectrum. Needless to say, the worldvolume actions for the corresponding states must also be related by the same dualities.

The symmetries of the supergravity theories also relate different solitonic solutions. In some cases this has been used to generate new solitonic solutions out of old ones. In the end one should be able to obtain complete duality-invariant families of solutions. Now, the relation between effective field theories and string theories can be used to relate solitonic supergravity solutions to perturbative and non-perturbative string states (the lower $\leftarrow$ arrow). Whereas in the supergravity theories the duality groups are continuous, in string theory quantum effects generically break them into the discrete subgroups that result from the restriction from real numbers to integers. In particular, the S-duality group is broken by charge quantization of the string states to $\mathrm{SL}(2, \mathbb{Z})$ [1125]. The full spectrum is invariant under that group [1126].

To develop these relations, we must find a dictionary to interpret, in terms of string theory, supergravity field configurations and symmetries, because only the fields associated with massless string-theory modes appear in the effective field theories, but string dualities often involve massive modes. However, these massive modes are usually charged and couple to the massless long-range potentials that appear in supergravity,[3] and we know their transformations under dualities, which helps us to know how the massive and charged string modes behave under those dualities. These couplings can be read off from the worldvolume

[2] But not the other way around: the T-duality group can contain $\mathrm{SL}(2, \mathbb{R})$ subgroups.

[3] The only bosonic fields apart from the potentials and the metric are the scalars, whose interpretation is different: they represent coupling constants or geometrical data of the compactification space, *moduli*.

actions of those states.

We encountered a similar phenomenon in KK theories in Chapter 15: the massive KK modes of the spectrum are charged with respect to the massless KK vector field that appears in the KK (effective) action. The massive KK modes reappear in the theory as massive, electrically charged solitonic solutions (extreme, electrically charged BHs). The S duality of the four-dimensional KK theory (in this case, just a $\mathbb{Z}_2$ group) could be used to generate extreme, magnetic BH solutions, which should represent non-perturbative states of the KK theory.

A final ingredient in all these relations is supersymmetry and supersymmetric ("BPS") states, which are associated with supersymmetric supergravity solutions. BPS states enter into shortened supermultiplets, which have fewer states. This and other properties (the saturation of the Bogomol'nyi bounds, which are duality invariant [1005, 1126]) do not depend on any continuous parameters and remain valid in the limits of strong and weak coupling. Therefore, these states are particularly well suited for studying, for instance, S dualities.

In this part we are going to investigate all these relations using as our main tool the effective actions, supergravity actions, in order finally to apply them to the calculation of entropies of extreme BHs. This chapter is devoted to a very brief introduction to string theory.[4] We will study in it, from the worldsheet point of view, the simplest, and most characteristic, string duality: T duality.

In Chapter 21 we study the origin and meaning of the string effective field theories which we will use to investigate string dualities in the following chapters. In this chapter we will start studying T duality in curved spaces from the point of view of effective action.

The ten-dimensional supergravities which are effective field theories of ten-dimensional superstring theories will be studied in Chapters 22 and 23. We will be able to interpret some of their symmetries in stringy terms, but many results will have to wait until Chapter 24, in which we will introduce extended objects and we will realize that the $(p+1)$-form potentials that appear in the supergravities are naturally the fields to which extended, charged $p$-dimensional objects (*$p$-branes*) couple. Then we will see the implications of the supergravity symmetries for the non-perturbative spectrum of string theory.

In Chapter 24 we are going to study generically extended objects, and in Chapter 25 we will focus on the extended objects of string theory. We will first find which these extended objects, implied by duality, are. Then, we will identify the corresponding classical solutions and, to strengthen the connection between classical solutions and string states, we will calculate the masses that duality predicts and we will use the results to identify the integration constants of the classical solutions. Here we can interpret these solutions as the long-range fields produced by one of these extended objects. We will check that they preserve the right amount of unbroken supersymmetry according to the supersymmetry algebra. Next, we will study the worldvolume effective actions of these objects. Finally, we will see solutions that describe several of these objects intersecting each other, still preserving some amount of supersymmetry. Related states must exist in string theory.

In Chapter 26 we will use many of these results to perform a microscopical computation

[4] This chapter is no substitute for the study of the many reviews and books on string theory [37, 40, 662, 670, 838, 858, 859, 915, 1003, 1049, 1113, 1102] and D-branes [1051, 832, 833]. We will simply present the concepts and results that we will use, establishing the notation and conventions.

of the entropies of some four- and five-dimensional BHs (generalizations of the extreme Reissner–Nordström BH). This is one of the main successes of string theory and has stimulated the research in this field from many different points of view.

## 20.1 Strings

Even though duality relates all the extended objects that appear in string theory, leading to the idea of *p-brane democracy* [1189], we know how to quantize only particles and strings, zero- and one-dimensional objects that have one-dimensional worldlines and two-dimensional *worldsheets*, respectively. The latter are the fundamental objects of all string theories, although some of them contain also particles (D0-*branes*). For the moment, there is no fully satisfactory formulation of a *string field theory*, so we must content ourselves with a "first quantization" and with a recipe for how to compute Feynman diagrams.

The action for a single string is the generalization, invariant under general worldsheet and target-space coordinate transformations, of the action for a point-particle Eq. (3.8) proposed by Nambu[5] and Goto [643], and measures essentially the area of the worldsheet swept out by the string when it moves in a $d$-dimensional ambient (*target*) space with a metric $g_{\mu\nu}$:

$$\boxed{S_{\rm NG}[X^{\mu}(\xi)] = -T\int_{\Sigma} d^2\xi\,\sqrt{|g_{ij}|}.} \qquad (20.1)$$

Here $\xi^i$, $i = 0, 1$, are coordinates on the two-dimensional worldvolume, $X^{\mu}(\xi)$, $\mu = 0, \ldots, d-1$ are the spacetime coordinates of the string, which give the embedding of the worldvolume into the $d$-dimensional spacetime, and $|g_{ij}|$ stands for the determinant of the induced metric $g_{ij}$ on the worldvolume (the pullback of the spacetime metric $g_{\mu\nu}$),

$$g_{ij} \equiv \partial_i X^{\mu}\partial_j X^{\nu} g_{\mu\nu}(X). \qquad (20.2)$$

$T$ is the *string tension*, a positive constant with (natural) dimensions of $\mathrm{L}^{-2}$ or $\mathrm{M}^2$, which is equivalent to mass per unit length: when the string is wrapped on a compact dimension of radius $R$, it is seen from the uncompactified dimensions as a particle with mass $2\pi RT$. It is related to the *Regge slope* $\alpha'$ by

$$T = 1/(2\pi\alpha'). \qquad (20.3)$$

The Regge slope sets the fundamental length and mass of the theory, the *string length* $\ell_{\rm s}$ and the *string mass* $m_{\rm s}$:

$$\ell_{\rm s} = \sqrt{\alpha'}, \qquad m_{\rm s} = 1/\sqrt{\alpha'}. \qquad (20.4)$$

The Nambu–Goto action is highly non-linear and very difficult to quantize [1102], even if we do it in Minkowski spacetime $g_{\mu\nu} = \eta_{\mu\nu}$, avoiding the non-linearities associated with the $X$-dependence of the spacetime metric. An action that is quadratic in derivatives of the worldsheet fields $X^{\mu}(\xi)$ can be constructed by introducing an auxiliary worldsheet metric

[5] Lectures at the 1970 Copenhagen Symposium.

$\gamma_{ij}$. The so-called *Polyakov action*[6] reads

$$S_{\rm P}[X^\mu(\xi), \gamma_{ij}(\xi)] = -\frac{T}{2}\int_\Sigma d^2\xi \sqrt{|\gamma|}\,\gamma^{ij}\partial_i X^\mu \partial_j X^\nu g_{\mu\nu}(X). \tag{20.5}$$

The general variation of this action is given by

$$\begin{aligned}\delta S_{\rm P} = {} & T\int_\Sigma d^2\xi\sqrt{|\gamma|}\,\delta X^\mu g_{\mu\nu}\left[\nabla^2 X^\nu + \gamma^{ij}\partial_i X^\rho \partial_j X^\sigma \Gamma_{\rho\sigma}{}^\nu(g)\right] \\ & - \frac{T}{2}\int_\Sigma d^2\xi\sqrt{|\gamma|}\,\delta\gamma^{ij}\left[\partial_i X^\mu \partial_j X^\nu - \frac{1}{2}\gamma_{ij}\gamma^{kl}\partial_k X^\mu \partial_l X^\nu\right] g_{\mu\nu} \\ & - T\int_{\partial\Sigma} d\Sigma^i \delta X^\mu \partial_i X^\nu g_{\mu\nu}. \end{aligned} \tag{20.6}$$

Since there is no kinetic term for the worldsheet metric, its equation of motion just gives the Rosenfeld energy–momentum tensor of the worldsheet fields $X^\mu$ and tells us that it must be zero. This equation is just a primary constraint that we can use to eliminate $\gamma_{ij}$ in the Polyakov action: on writing it in the form

$$\gamma_{kl} = 2g_{kl}/g_i{}^i, \tag{20.7}$$

and substituting this into the Polyakov action, we recover the Nambu–Goto action. Observe that the equation of motion of the worldsheet metric $\gamma_{ij}$ tells us only that it is proportional to the induced metric $g_{ij}$, but, in just two worldsheet dimensions, it is impossible to determine the proportionality coefficient $g_i{}^i = \gamma^{ij}g_{ij}$ because this equation of motion (and, hence, the energy–momentum tensor) is (off-shell) traceless. This property is related to an additional symmetry of the Polyakov action for strings: invariance under Weyl rescalings of the worldsheet metric,

$$\gamma_{ij} \to \Omega^2(\xi)\gamma_{ij}. \tag{20.8}$$

This symmetry plays a crucial role in the quantization of the Polyakov action, allowing one to gauge away the worldsheet metric completely and consistently: in two dimensions it is always possible to put the metric in the *conformal gauge* $\gamma_{ij} \propto \eta_{ij}$ by using reparametrizations. Then, with a Weyl rescaling, we can always obtain $\gamma_{ij} = \eta_{ij}$. This symmetry is, however, potentially broken by anomalies that impose further restrictions on the spacetime dimensions, metric, etc., in order to have consistent string theories (see e.g. Ref. [1102]).

Let us now consider the equation of motion of the $X^\mu$s. The boundary term can be non-vanishing if we consider the propagation of *open strings*, with free endpoints and the topology of a line segment (the alternative is to consider *closed strings*, with the topology of a circle). In order to eliminate it, we have to impose special boundary conditions for open strings. There are two main possibilities (for each disconnected piece of the boundary $\partial\Sigma^{(n)}$ and for each embedding coordinate $X^\mu$):

[6] It was Polyakov who first quantized it and its supersymmetric version (see Eq. (20.23)) using the path-integral formalism in Refs. [1054, 1055].

**Neumann (N) boundary conditions:**

$$n^i \partial_i X^\mu \big|_{\partial\Sigma^{(n)}} = 0, \tag{20.9}$$

where $n^i$ is a unit vector normal to the boundary. These conditions respect target-space Poincaré invariance when $g_{\mu\nu} = \eta_{\mu\nu}$. In the simplest situation, the propagation of a free open string, in which the worldsheet is a strip swept out with length and width parametrized by $\xi^0$ and $\xi^1$, the N boundary conditions are

$$\partial_{\underline{1}} X^\mu \big|_{\partial\Sigma^{(n)}} = 0. \tag{20.10}$$

**Dirichlet (D) boundary conditions:**

$$t^i \partial_i X^\mu \big|_{\partial\Sigma^{(n)}} = 0, \tag{20.11}$$

where $t^i$ is a unit vector tangential to the boundary. For the free open string it is just

$$\partial_{\underline{0}} X^\mu \big|_{\partial\Sigma^{(n)}} = 0. \tag{20.12}$$

This condition is equivalent to requiring that

$$X^\mu \big|_{\partial\Sigma^{(n)}} = c^{(n)\,\mu}, \tag{20.13}$$

where the $c^{(n)\,\mu}$s are constants, which restricts the endpoints of the open strings (one or both, depending on how many pieces of the boundary we impose this kind of condition on) to moving on $(p+1)$-dimensional hypersurfaces if these conditions are imposed on $(d-p-1)$ embedding coordinates. This explicitly breaks translation invariance in those $(d-p-1)$ coordinates. The hypersurfaces will later be interpreted as the worldvolumes of dynamical $p$-dimensional objects: *Dp-branes* [1047].

In Minkowski spacetime and in the conformal gauge the equation of motion of the $X^\mu$s is just the two-dimensional wave equation for free bosonic fields $\partial_i \partial^i X^\mu = 0$, whose solutions are the sum of a function $X^\mu_+(\xi^+)$ of $\xi^+ = \xi^1 + \xi^0$ (a *left-moving* wave) and a function $X^\mu_-(\xi^-)$ of $\xi^- = \xi^1 - \xi^0$ (a *right-moving* wave). In open-string worldsheets, left- and right-moving waves are related by the boundary conditions $\partial_+ X_+ = -\partial_- X_-$ for N boundary conditions and $\partial_+ X_+ = \partial_- X_-$ for D boundary conditions.

On the other hand, even if we eliminate the worldsheet metric by gauge transformations, we still have to take into account its equation of motion (the vanishing of the worldsheet energy–momentum tensor).

Apart from being open or closed, strings can be *oriented* or *unoriented*, and in each case only oriented or unoriented worldsheet surfaces are considered.

It is possible to add another Weyl-invariant term to the Polyakov action without any additional background field: a worldsheet Einstein–Hilbert term[7]

$$-\frac{\phi_0}{4\pi} \int d^2\xi \sqrt{|\gamma|}\, R(\gamma). \tag{20.14}$$

[7] In the presence of boundaries, it must be supplemented by a boundary term as in Eq. (4.26).

This term does not change the classical equations of motion because the two-dimensional Einstein–Hilbert Lagrangian density is just the curvature 2-form, which is locally a total derivative: the spin-connection 1-form is $\omega^{ab} = \epsilon^{ab}\omega$ and $R^{ab} = \epsilon^{ab}d\omega$. Actually, this term is the constant $\phi_0$ times a topological invariant: the *Euler characteristic* $\chi = 2 - 2g - b - c$, where $g$ is the *genus* (number of handles) of the two-dimensional worldsheet, $b$ is the number of boundaries, and $c$ is the number of *crosscaps* (which are present only when the worldsheet is non-orientable). As we are going to see, $\phi_0$ is the vacuum expectation value of the *dilaton*, a massless scalar field present in all string theories, and $g = e^{\phi_0}$ can be interpreted as the *string coupling constant* that counts loops in string amplitudes.

### 20.1.1 Superstrings

The string theories we have studied so far are called *bosonic* because they include only bosonic worldsheet fields and, furthermore, their spectra only contain spacetime bosons, as we will see. To construct a string theory whose spectrum includes fermions, we can generalize the action of a *spinning* point-particle (although the historical order did not follow this logic). The action for a spinning particle contains, in addition to the commuting variables $X^\mu(\xi)$ that describe the position of the particle, anticommuting variables $\psi^\mu(\xi)$ that describe the spin degrees of freedom and were first proposed in Refs. [141, 312] and studied and generalized in Refs. [142, 103, 258, 369, 260, 87, 104, 105]. The simplest action one can write has global *worldline supersymmetry* transformations that relate $X^\mu$ and $\psi^\mu$, which form a scalar supermultiplet [258, 260], and the natural generalization, which is invariant under worldline reparametrizations, is also naturally invariant under local worldline supersymmetry transformations. This generalization requires the introduction of auxiliary fields: an Einbein $e(\xi)$, as in the bosonic case ($e^2 = \gamma$), and a gravitino $\chi$, which form a one-dimensional supergravity multiplet that has no dynamics. The action for the massless case is

$$\boxed{S = -\tfrac{1}{2}\int d\xi\, e\left[e^{-2}\dot{X}^\mu \dot{X}_\mu + e^{-1}\psi^\mu \dot{\psi}_\mu - e^{-2}\chi\psi^\mu \dot{X}_\mu\right],} \tag{20.15}$$

and the supersymmetry transformations that leave it invariant are

$$\begin{aligned} \delta_\epsilon X^\mu &= \epsilon\psi^\mu, & \delta_\epsilon e &= \epsilon\chi, \\ \delta_\epsilon \psi^\mu &= -\epsilon(\dot{X}^\mu - \tfrac{1}{2}\chi\psi^\mu)e^{-1}, & \delta_\epsilon \chi &= 2\dot{\epsilon}. \end{aligned} \tag{20.16}$$

Invariance under worldline reparametrizations and supersymmetry transformations leads to constraints (gauge identities) that are necessary for consistency. Furthermore, it can be shown that the quantization of this model leads to the Dirac equation for spin-$\frac{1}{2}$ particles [103, 142, 260] (for more recent references, see Refs. [743, 600]), and this action can be used to obtain path-integral representations of propagators and Feynman diagrams (see e.g. Ref. [753] and references therein) for spin-$\frac{1}{2}$ particles. It is also remarkable that, when it is coupled to gravity [104], the action leads to the Dirac equation in curved spacetime as given in Ref. [257] and the equations of motion for a pole–dipole singularity derived by Papapetrou in Ref. [1020] using the method of Einstein, Infeld, and Hoffmann [500].[8]

The action for the spinning particle can be generalized to include other couplings to vector fields and odd-rank forms preserving worldline local supersymmetry, which can be made manifest by using *superworldsheet* coordinates and geometry (see, for instance, Ref. [754] and references therein). In particular, the coupling to a 3-form can be understood as the coupling to a completely antisymmetric torsion field, and it leads to the Dirac equation of the CSK theory and a modification of the Papapetrou equations [1089].

On the other hand, the spectrum also contains spin-0 and spin-1 particles and, thus, in spite of the worldline supersymmetry, there is no spacetime supersymmetry.

The action for a massive Dirac particle can be obtained by adding the term[9]

$$-\tfrac{1}{2}\int d\xi e[m^2 + e^{-1}\psi_5\dot{\psi}_5 + me^{-1}\chi\psi_5], \tag{20.17}$$

where $\psi_5$ is another anticommuting variable that transforms as follows:

$$\delta_\epsilon\psi_5 = m\epsilon + \frac{\epsilon}{me}\psi_5(\dot{\psi}_5 - \tfrac{1}{2}\chi). \tag{20.18}$$

By analogy, the worldsheet action for a fermionic string that can describe spacetime fermions (but, also, as we will see, spacetime bosons, depending on boundary conditions) must incorporate two-dimensional spinors. Worldsheet supersymmetry can be used as a guiding principle to introduce worldsheet fermions. Then, the theory that generalizes the Polyakov action in Minkowski spacetime is the theory of $d$ two-dimensional scalar superfields $(X^\mu, \psi^\mu)$ coupled to the two-dimensional supergravity multiplet $(e^a{}_i, \chi_i)$ with the action [259, 450]

$$\boxed{\begin{aligned} S = -\frac{T}{2}\int_\Sigma d^2\xi e\big[&\gamma^{ij}\partial_i X^\mu\partial_j X_\mu - i\bar{\psi}^\mu \not{D}\psi_\mu \\ &+ 2\bar{\chi}_i\rho^j\rho^i\psi^\mu\partial_j X_\mu + \tfrac{1}{2}(\bar{\chi}_i\rho^j\rho^i\chi_j)(\bar{\psi}^\mu\psi_\mu)\big]. \end{aligned}} \tag{20.19}$$

Both $\psi^\mu$ and $\chi_i$ are real two-dimensional spinors with hidden spinorial indices and $\rho^i = \rho^a e_a{}^i$, where $\rho^a$ are the two-dimensional gamma matrices (see Appendix D.1.7). This action is invariant under worldsheet reparametrizations and local Lorentz transformations, as usual in supergravity (see Chapter 5), and also under global Poincaré transformations of the embedding coordinates $X^{\mu\prime} = \Lambda^\mu{}_\nu X^\nu + a^\mu$. It is also invariant under local worldsheet supersymmetry transformations:

$$\begin{aligned} \delta_\epsilon X^\mu &= \bar{\epsilon}\psi^\mu, & \delta_\epsilon e^a{}_i &= -2i\bar{\epsilon}\rho^a\chi_i, \\ \delta_\epsilon\psi^\mu &= i(\partial_i X^\mu + \tfrac{1}{4}\bar{\chi}_i\psi^\mu)\rho^i\epsilon, & \delta_\epsilon\chi_i &= \tilde{\mathcal{D}}_i\epsilon, \end{aligned} \tag{20.20}$$

[8] This method was used in Ref. [552] to reobtain the equations of motion of a point-particle and in Ref. [703] to recover the equations of motion of the Nambu–Goto string.

[9] This action can be derived by dimensional reduction of the action of the massless spinning particle.

where

$$\tilde{D}_i\epsilon = (\partial_i + \tilde{\omega}_i\rho_3)\epsilon, \qquad \tilde{\omega}_i = \omega_i + i\bar{\chi}_i\rho_3\rho^j\chi_j. \tag{20.21}$$

Furthermore, and most importantly, it is also invariant under super-Weyl transformations:

$$\begin{aligned} X^{\mu\prime} &= X^\mu, & e^{a\prime}{}_i &= \Omega e^a{}_i, \\ \psi^{\mu\prime} &= \Omega^{-\frac{1}{2}}\psi^\mu & \chi_i' &= \Omega^{\frac{1}{2}}\chi_i + i\rho_i\lambda, \end{aligned} \tag{20.22}$$

where $\lambda$ is any real spinor. This symmetry allows the complete decoupling of the Zweibein $e^a$ and the worldsheet gravitino $\chi_i$, whose elimination gives the so-called *Ramond–Neveu–Schwarz (RNS)* model [979, 1067]

$$\boxed{S = -\frac{T}{2}\int_\Sigma d^2\xi[\eta^{ij}\partial_i X^\mu \partial_j X_\mu - i\bar{\psi}^\mu \partial\!\!\!/ \psi_\mu].} \tag{20.23}$$

This model is only globally supersymmetric and has to be supplemented with the equations of motion of the worldsheet metric and gravitino (even after we have eliminated them).

One can consider open, closed, oriented, and unoriented superstrings. In the open-string case, at the endpoints $\xi^1 = 0, 2\pi\ell$, N and D boundary conditions can be chosen for the $X^\mu$s, and for the $\psi^\mu$s one can choose between

**Ramond (R) boundary conditions,**

$$\psi^\mu_+(\xi^1 = 0) = \psi^\mu_-(\xi^1 = 0), \qquad \psi^\mu_+(\xi^1 = 2\pi\ell) = \psi^\mu_-(\xi^1 = 2\pi\ell), \tag{20.24}$$

**and Neveu–Schwarz (NS) boundary conditions,**

$$\psi^\mu_+(\xi^1 = 0) = \psi^\mu_-(\xi^1 = 0), \qquad \psi^\mu_+(\xi^1 = 2\pi\ell) = -\psi^\mu_-(\xi^1 = 2\pi\ell), \tag{20.25}$$

where $\psi^\mu_\pm$ are the left- and right-moving components of $\psi^\mu$.

For closed superstrings, $\xi^1 \sim \xi^1 + 2\pi\ell$ and we can have, for each component $\psi^\mu_+$ and $\psi^\mu_-$, independently,

**Ramond (R) (periodic) boundary conditions,**

$$\psi^\mu_\pm(\xi^1 = 0) = \psi^\mu_\pm(\xi^1 = 2\pi\ell), \tag{20.26}$$

**and Neveu–Schwarz (NS) (antiperiodic) boundary conditions,**

$$\psi^\mu_\pm(\xi^1 = 0) = -\psi^\mu_\pm(\xi^1 = 2\pi\ell). \tag{20.27}$$

So we can have RR, RNS, NSR, and NSNS boundary conditions.

Just as in the spinning-particle case, worldsheet supersymmetry leads to spacetime supersymmetry of the quantized theory only under quite restrictive conditions [636] and, in any case, spacetime supersymmetry is never explicit. In the alternative *Green–Schwarz (GS)* formulation [668] spacetime supersymmetry is explicit from the outset.

### 20.1.2 Green–Schwarz actions

Green–Schwarz-type actions can be constructed for particles, strings, and other extended objects, as we will see. The simplest example describes a massless particle moving in flat (target) superspace with supercoordinates $Z^M(\xi) = (X^\mu(\xi), \theta^{\alpha\, I}(\xi))$, where $I = 1, \ldots, N$ numbers the supersymmetries, $\alpha$ is a spacetime spinorial index, and the $\theta^I$ are anticommuting spacetime spinors (but worldsheet scalars) [261]:

$$S = -\frac{p}{2}\int d\xi e^{-1}(\dot{X}^\mu \delta^a{}_\mu - i\bar{\theta}^I \gamma^a \dot{\theta}^I)\eta_{ab}(\dot{X}^\nu \delta^b{}_\nu - i\bar{\theta}^I \gamma^b \dot{\theta}^I). \tag{20.28}$$

This action is invariant under worldline reparametrizations, under target-space Poincaré transformations, and also under the standard global superspace transformations

$$\delta_\epsilon \theta^I = \epsilon^I, \qquad \delta_\epsilon X^\mu = i\bar{\epsilon}^I \gamma^a \theta^I \delta_a{}^\mu. \tag{20.29}$$

In principle, there is no worldline supersymmetry, a very desirable property. A necessary condition for having linearly realized worldline supersymmetry is that the numbers of on-shell bosonic and fermionic degrees of freedom should be equal.[10] To find the numbers of on-shell degrees of freedom, it is necessary to know all the local symmetries of the action. The above action turns out to have worldline-reparametrization invariance, which can be used to gauge away one of the $X^\mu$s, and a new local symmetry generated by a fermionic infinitesimal parameter $\kappa$ (*$\kappa$-symmetry*), which halves the number of fermionic degrees of freedom [74, 75, 1138], which has already been halved by the Dirac equation. Under $\kappa$-symmetry

$$\delta_\kappa \theta^I = -i\Pi_\mu \gamma^a \delta_a{}^\mu \kappa^I, \qquad \delta_\kappa X^\mu = i\bar{\theta}^I \gamma^a \delta_a{}^\mu \delta_\kappa \theta^I, \qquad \delta_\kappa e = 4e\dot{\bar{\theta}}^I \kappa^I, \tag{20.30}$$

where $\Pi_\mu = \delta S/\delta \dot{X}^\mu$ is the momentum conjugate to $X^\mu$.

Taking into account this new symmetry, if we denote by $M$ the number of real components of the minimal spinor in the spacetime dimension $d$ considered, then the necessary condition for having worldsheet supersymmetry reads

$$NM = 4(d-1), \tag{20.31}$$

and, taking into account the values of $M$ in Table D.1, it can be satisfied for $d = 2, 3, 4, 5$, and 9 with $N = 4, 4, 3, 2$, and 2, respectively. Thus, the dimensions in which these actions can be consistent are restricted.

The condition can be generalized to objects with $p$ extended dimensions. Using worldvolume reparametrizations, we can always gauge away $p+1$ $X^\mu$s. The condition of worldvolume supersymmetry becomes now

$$NM = 4(d-p-1), \tag{20.32}$$

[10] In just one dimension, talking about degrees of freedom does not make much sense. However, the same reasoning carries over to higher-dimensional cases, which allows a classification of all the possible supersymmetric extended objects [13].

Table 20.1. A brane scan taking into account only scalar supermultiplets; $N$ is given by the quotient between the numbers given and $M$

| | | $p=-1$ | $p=0$ | $p=1$ | $p=2$ | $p=3$ | $p=4$ | $p=5$ |
|---|---|---|---|---|---|---|---|---|
| $d$ | $M$ | $d$ | $(d-1)$ | $(d-2)$ | $(d-3)$ | $(d-4)$ | $(d-5)$ | $(d-6)$ |
| 2 | 1 | **8** | **4** | | | | | |
| 3 | 2 | **12** | **8** | **4** | | | | |
| 4 | 4 | **16** | **12** | **8** | **4** | | | |
| 5 | 8 | | **16** | | **8** | | | |
| 6 | 8 | | | **16** | | **8** | | |
| 7 | 16 | | | | **16** | | | |
| 8 | 16 | **32** | | | | **16** | | |
| 9 | 16 | | **32** | | | | **16** | |
| 10 | 16 | | | **32** | | | | **16** |
| 11 | 32 | | | | **32** | | | |

and it can be solved in the cases represented in Table 20.1 [13]. There are five series of solutions. Four of them (with $NM=32,16,8,$ and 4) are associated with the four division algebras $\mathbb{O}$, $\mathbb{Q}$, $\mathbb{C}$, and $\mathbb{R}$, respectively. These series correspond to objects related by *double dimensional reduction*, as we will see.

There is another way to understand this result: if there is linearly realized worldvolume supersymmetry, the worldvolume fields must fit in $(p+1)$-dimensional scalar supermultiplets. Each solution in the table corresponds to a scalar multiplet. There is agreement with the fact that scalar multiplets exist only in up to six dimensions.

For us, it is interesting that spacetime supersymmetric string actions could in principle be constructed in $d=3,4,6,$ and 10, provided that the action has $\kappa$-symmetry. Green and Schwarz showed in Ref. [669] that the ten-dimensional action previously constructed by them in Ref. [668] has $\kappa$-symmetry. The GS action is not just a straightforward generalization of the superparticle action Eq. (20.28), because the kinetic term would not be $\kappa$-symmetric by itself. The key to $\kappa$-symmetry is the addition of a (super-)Wess–Zumino (WZ) term: the integral of a 2-form $\Omega_2$ such that $\Omega_3=d\Omega_2$ is (target) Poincaré and supersymmetry invariant [742]:[11]

$$\Omega_2=-idX^\mu\delta^a{}_\mu\wedge(\bar{\theta}^1\Gamma_a\theta^1-\bar{\theta}^2\Gamma_a d\theta^2)+(\bar{\theta}^1\Gamma_a d\theta^1)\wedge(\bar{\theta}^2\Gamma^a d\theta^2). \tag{20.33}$$

The GS action is then given by

$$\boxed{S=-\frac{T}{2}\int_\Sigma d^2\xi\sqrt{|\gamma|}\gamma^{ij}(\partial_i X^\mu\delta^a{}_\mu-i\bar{\theta}^I\gamma^a\partial_i\theta^I)(\partial_j X^\nu\delta^b{}_\nu-i\bar{\theta}^J\gamma^b\partial_j\theta^J)\eta_{ab}+T\int_\Sigma\Omega_2.} \tag{20.34}$$

[11] According to Table 20.1, this is an $N=2$ theory, with two minimal (Majorana–Weyl spinors with 16 real components) ten-dimensional spinors, $\theta^1$ and $\theta^2$, with equal or opposite chiralities: type-IIB and type-IIA strings, respectively. These theories can also be described with the RNS theory, but only after quantization.

This action, whose covariant quantization poses many problems, has very interesting features. First, it can be generalized to the other string theories with $N=2$ spacetime supersymmetry [681, 190] and other higher-dimensional objects of Table 20.1, always with a WZ term (see also Refs. [96, 97]). This was done for the super 3-brane in $d=6$ in Ref. [800] and for the super 2-brane (*supermembrane*, now *M2-brane*) of $d=11$ in Refs. [191, 192]. The M2-brane is related to the ten-dimensional string by double dimensional reduction [469, 14] (the last two members of the $NM=32$ series).

These generalizations include the coupling of the supersymmetric extended objects to supergravity fields: since GS actions are manifestly spacetime supersymmetric, the coupling to gravity implies the coupling to all the background fields of a supergravity theory.[12] The fact that it can and must be coupled to supergravity explains in part the necessity of the WZ term: supergravity theories contain potentials that are $(p+1)$-forms in superspace that can naturally couple to $p$-dimensional objects through a WZ term (the integral of the pullback of the $(p+1)$-form over the $(p+1)$-dimensional worldvolume), just like particles couple to the Maxwell 1-form Eq. (12.53) (see also Eq. (21.4)). The expansion of the $(p+1)$-form fields in components contains terms that do not vanish in flat spacetime, and $\Omega_2$ given above is one such term. The full WZ term for the superstring coupled to supergravity also contains the purely bosonic term Eq. (21.4).

$\kappa$-symmetry imposes constraints on the supergravity fields [481, 335, 137, 139, 961, 138]. In particular [191, 192], the action for the 11-dimensional supermembrane coupled to the fields of 11-dimensional supergravity can be $\kappa$ invariant only if certain constraints are solved by the equations of motion of that theory. These constraints coincide with the superspace constraints of $d=11$ supergravity [383]. Thus worldvolume $\kappa$ invariance implies the equations of motion of the spacetime supergravity fields, a highly non-trivial fact. Something similar happens in string theory coupled to background fields: by requiring invariance under worldsheet Weyl transformations in the quantum theory one obtains the equations of motion of the spacetime fields (see Section 21.1).

Although there is no clear motivation at this point, we could also include other supermultiplets in the super-$p$-brane actions [473]. The new supersymmetric extended objects include the *Dp-branes* that we will study in more detail later [1190, 786, 16, 17, 319, 318, 194, 99].

## 20.2 Quantum theories of strings

In this section we are going to overview the quantization in Minkowski spacetime of the bosonic and fermionic string actions that we have introduced in Section 20.1. We will focus on the definition of quantum string theory (in particular on string interactions) and on the results: the critical dimensions, mode expansions, and massless spectra of the simplest consistent string theories.

[12] Only recently has it been learned how to couple superstrings to RR backgrounds (to be defined in Section 20.2.2) [947, 946, 948].

### 20.2.1 Quantization of free-bosonic-string theories

Free strings can translate, rotate, and vibrate. The various allowed vibrational modes are seen as different particle states in spacetime. These particle states must fit into Poincaré multiplets characterized by mass and spin (or helicity). Failure to do so implies breaking of spacetime Poincaré invariance.

The simplest way to quantize the Polyakov or the superstring action, Eqs. (20.5) and (20.19), and obtain their spectra is to use the physical light-cone gauge in which all the gauge invariances of the action are used to eliminate unphysical degrees of freedom. This was done first for the bosonic string in Ref. [637], where it was found that Poincaré invariance is recovered only in the critical dimension $d = 26$.

Here we will follow the careful treatment of Ref. [1049] for the bosonic string, where it is shown how, using worldsheet reparametrizations and Weyl rescalings, one can always set

$$X^{+} \equiv \frac{1}{\sqrt{2}}(X^{0}+X^{1}) = \xi^{0}, \qquad \partial_{\underline{1}}\gamma_{\underline{11}} = 0, \qquad \det(\gamma_{ij}) = -1, \tag{20.35}$$

which allows the elimination of $X^{+}, X^{-}$, and $\gamma_{ij}$ and leads to the Hamiltonian

$$H = -\frac{c}{2}\int_{0}^{2\pi\ell} d\xi^{1}\left[T^{-1}\Pi^{i}\Pi^{i} + T\partial_{\underline{1}}X^{i}\partial_{\underline{1}}X^{i}\right], \qquad c = 2\pi\ell T/p^{+}, \tag{20.36}$$

where $\xi^{1} \in [0, 2\pi\ell]$, the $\Pi^{i}$ are the momenta conjugate to the $X^{i}$, and $p^{+}$ is the momentum conjugate to $x^{-}$, a cyclic variable, so $p^{+}$ is a constant of motion. The equations of motion for the $X^{i}$ that follow from this Lagrangian are

$$(\partial_{\underline{0}}^{2} - c^{2}\partial_{\underline{1}}^{2})X^{i} = 0 \;\Rightarrow X^{i} = X^{i}_{+} + X^{i}_{-}, \qquad X^{i}_{\pm} = X^{i}_{\pm}(\xi^{1} \pm c\xi^{0}). \tag{20.37}$$

The left-, $X^{i}_{+}$, and right-moving, $X^{i}_{-}$, components are related by boundary conditions in the open-string case and we consider only $X^{i}$. The boundary conditions play their role in the mode expansions of the $X^{i}$s: for open strings with N boundary conditions

$$X^{i} = x^{i} + \frac{p^{i}}{p^{+}}\xi^{0} - i\sqrt{2\alpha'}\sum_{n\neq 0}\frac{\alpha^{i}_{n}}{n}e^{\frac{icn\xi^{0}}{2\ell}}\cos\left(\frac{n\xi^{1}}{2\ell}\right). \tag{20.38}$$

If $X$ has D boundary conditions at both ends, $X(\xi^{1}=0) = x_{1}$ and $X(\xi^{1}=2\pi\ell) = x_{2}$, then

$$X = x_{1} - \frac{x_{2}-x_{1}}{2\pi\ell}\xi^{1} + \sqrt{2\alpha'}\sum_{n\neq 0}\frac{\alpha_{n}}{n}e^{\frac{icn\xi^{0}}{2\ell}}\sin\left(\frac{n\xi^{1}}{2\ell}\right). \tag{20.39}$$

For closed strings

$$X^{i} = x^{i} + \frac{p^{i}}{p^{+}}\xi^{0} - i\sqrt{\frac{\alpha'}{2}}\sum_{n\neq 0}\left[\frac{\alpha^{i}_{n}}{n}e^{\frac{in}{\ell}(\xi^{1}+c\xi^{0})} + \frac{\tilde{\alpha}^{i}_{n}}{n}e^{-\frac{in}{\ell}(\xi^{1}-c\xi^{0})}\right]. \tag{20.40}$$

Reality implies in all cases $\alpha^{i\,\dagger}_{n} = \alpha^{i}_{-n}$ and $\tilde{\alpha}^{i\,\dagger}_{n} = \tilde{\alpha}^{i}_{-n}$. On substituting the mode expansions into the equal-time commutators (using $\Pi^{i} = p^{+}\partial_{\underline{0}}X^{i}/(2\pi\ell)$),

$$[x^{-}, p^{+}] = i, \qquad [X^{i}(\xi^{1}), \Pi^{j}(\xi^{1\prime})] = i\delta^{ij}\delta(\xi^{1}-\xi^{1\prime}), \tag{20.41}$$

we find the commutation relations

$$[x^i, p^j] = i\delta^{ij}, \qquad [\alpha^i_m, \alpha^j_n] = m\delta^{ij}\delta_{m,-n}, \qquad [\tilde{\alpha}^i_m, \tilde{\alpha}^j_n] = m\delta^{ij}\delta_{m,-n}. \tag{20.42}$$

The vacuum is defined to be annihilated by all the oscillators $\alpha^i_n$ and $\tilde{\alpha}^i_n$ with $n > 0$ and states are created by acting on it with creation operators $\alpha^i_{-n}$ and $\tilde{\alpha}^i_{-n}$, with $n > 0$, on the momentum eigenstates $|0, k\rangle$

$$p^i|0, k\rangle = k^i|0, k\rangle, \qquad p^+|0, k\rangle = k^+|0, k\rangle. \tag{20.43}$$

Relative to this vacuum, the mass operator $M^2 = -2p^+H - p^ip^i$ takes the form, for open strings with only N boundary conditions,

$$M^2 = \frac{1}{\alpha'}(N + A). \tag{20.44}$$

For open strings with DD boundary conditions in one coordinate

$$M^2 = \left(\frac{x_2 - x_1}{2\pi\alpha'}\right)^2 + \frac{1}{\alpha'}(N + A), \tag{20.45}$$

and for closed strings

$$M^2 = \frac{2}{\alpha'}(N + \tilde{N} + A + \tilde{A}). \tag{20.46}$$

In all these cases

$$N = \sum_{n>0} n\alpha^i_{-n}\alpha^i_n, \qquad \tilde{N} = \sum_{n>0} n\tilde{\alpha}^i_{-n}\tilde{\alpha}^i_n \tag{20.47}$$

are the *level* operators that take only positive integer values and $A$, $\tilde{A}$ are constants that arise in the normal ordering of the Hamiltonian and take the value $A = \tilde{A} = (2 - d)/24$.

In the closed-string case there is still one constraint that has not been eliminated, which is associated with the $\xi^1$ shift symmetry:

$$N = \tilde{N}. \tag{20.48}$$

Let us now consider the lightest states of these three theories. The lightest states of the open string with N boundary conditions are the $|0, k\rangle$, whose mass is, according to Eq. (20.44), $M^2 = (2 - d)/(24\alpha')$, which is negative for $d > 2$, corresponding to a spacetime scalar tachyon and indicating the instability of the open-bosonic-string vacuum. The next lightest states are obtained by acting with the $\alpha^i_{-1}$ operators on $|0, k\rangle$ and have masses $M^2 = (26 - d)/(24\alpha')$. They fill a vector representation of SO$(d - 2)$, just like a $d$-dimensional massless spacetime vector particle. Poincaré invariance then requires the mass of these states to be zero and $d = 26$, and the spectrum contains a scalar tachyon, a massless vector, and massive and higher-spin states.

The lightest state of the DD open string is again $|0, k\rangle$, with $(d = 26)$ $M^2 = [(x_2 - x_1)/(2\pi\alpha')]^2 - 1/\alpha'$, whose value and sign depend on the distance between the hypersurfaces $x = x_1, x_2$ to which the string endpoints are attached. When $x_2 = x_1$ there are massless states $\alpha^i_{-1}|0,k\rangle$, $i \neq x$, and $\alpha^x_{-1}|0,k\rangle$ ($x$ being the DD direction), a massless vector, and a scalar in $d-1$ dimensions. When $x_2 \neq x_1$ we have the $d-2$ states of a massive $(d-1)$-dimensional vector. The scalar plays the role of a Higgs scalar.

Observe that the $\alpha_n$s would represent oscillations of the strings in directions perpendicular to the $x = x_1, x_2$ hypersurfaces, which is not possible. Hence they must represent oscillations of the $x = x_1, x_2$ hypersurfaces which, thus, are the 25-dimensional worldvolumes of dynamical 24-dimensional objects: *D24-branes*. These objects can be understood as new physical string states that must be non-perturbative since they do not appear in the perturbative spectrum. Indeed, as we will see, their tension is $\sim g^{-1}$, where $g$ is the string coupling constant that we will define later.

In the presence of two D24-branes parallel at $x = x_1, x_2$, open strings can have both ends attached to either one of them or have each end attached to each of them.[13] Since we are considering oriented strings, we must distinguish between the strings that connect 1 to 2 and those that connect 2 to 1. There are, then, four *sectors* and, according to the preceding paragraphs, two of them include two massless vectors and scalars, and the other two include two massive vectors, all of them living in the directions parallel to the D-branes. When the two 24-branes coincide, we have two extra massless vector and scalar fields. The four massless vector fields turn out to be U(2) gauge fields. In general, when $n$ D-branes coincide, there are massless $\mathrm{U}(n)$ vector fields labeled by two indices $I, J = 1, 2$ that indicate to which D-brane each string endpoint is attached, and the gauge symmetry is spontaneously broken to smaller groups when some of the D-branes become separated: the Higgs scalars give mass to the gauge fields.

It is possible to introduce labels (*Chan–Paton factors*) for open-string endpoints even if all coordinates have N boundary conditions, and the theory will have $\mathrm{U}(n)$ gauge vector fields. In this case we can imagine that the spacetime is filled with $n$ D25-branes.

Symmetry enhancements at particular values of moduli are some of the most interesting features of string theories.

Let us now consider closed strings. The lightest states obeying the constraint Eq. (20.48) after the tachyon $|0,0;k\rangle$ are of the form $\alpha^i_{-1}\tilde{\alpha}^j_{-1}|0,0;k\rangle$ and, just in $d = 26$, they fit into Poincaré representations: the part symmetric and traceless in $ij$ corresponds to a massless graviton; the trace part to a massless scalar, the *dilaton*; and the antisymmetric part to a massless 2-form field: the Kalb–Ramond (KR) field.

After introducing interactions and following the reasoning of Chapter 3, closed-string theories must contain gravity, which in a certain limit must coincide with Einstein's GR.

Finally, let us consider unoriented strings. They can be obtained from oriented strings by taking the quotient by the *worldsheet parity operator* $\Omega\colon \xi^1 \to 2\pi\ell - \xi^1$, which interchanges right- and left-moving sectors,

$$\Omega X^\mu_\pm = X^\mu_\mp, \tag{20.49}$$

and is an involution, $\Omega^2 = 1$, and generates a $\mathbb{Z}_2$ symmetry group. Taking the quotient means keeping only $\Omega$-invariant states. On level-$N$ states

$$\Omega|N;k\rangle = (-1)N|N;k\rangle, \qquad \Omega|N,\tilde{N};k\rangle = |\tilde{N},N;k\rangle. \tag{20.50}$$

[13] This system is not stable in bosonic-string theory, as the presence of tachyons indicates, but it will be in type-II superstring theories.

Thus, the KR tensor is removed from the closed-string spectrum and the massless vector is removed from the open-string spectrum unless the endpoints have Chan–Paton factors attached. In that case, only the $\mathrm{U}(n)$ vectors $V_\mu^{IJ}$ antisymmetric in $IJ$ survive, i.e. the $\mathrm{SO}(n)$ or $\mathrm{Sp}(n)$ subgroups.

In general, the quotient of a theory by a discrete symmetry of the theory (such as $\Omega$ here) is called an *orbifold* by analogy with the spacetime orbifolds discussed in Section 15.6. If in a string theory worldsheet parity is combined with a discrete symmetry of the spacetime, then the quotient is called an *orientifold* [1095, 412, 1056, 758, 759, 644, 207, 208],[14] but since orbifolds and orientifolds are often related by dualities, it is customary to call all of them orientifolds. In this language, closed-unoriented-string theory is an orientifold of closed, oriented-string theory. The hypersurfaces left invariant by the discrete spacetime symmetries are called *orientifold planes* and, although they are similar in other respects to D$p$-branes, they are not dynamical objects in the sense that they are attached to the fixed points of the spacetime orbifold and cannot translate or oscillate. However, there are dynamical fields on them. In the above case, there is no spacetime symmetry, the whole spacetime is invariant, and we can say that there is an orientifold plane of 25 spatial dimensions (O25-plane) that fills the entire spacetime, as a D25-brane does.

There is a crucial difference between orientifolds of point-particle theories and closed-string theories: in the latter one must include, for consistency, *twisted sectors*: strings that are closed up to a symmetry operation associated with the orientifold. In general, the inclusion of these twisted sectors makes the string theory non-singular at the orbifold fixed points, as distinct from point-particle theories. Twisted sectors also appear in other contexts: for instance, the *winding modes* that appear in toroidal compactifications (see Section 20.3) can be seen as strings in $\mathbb{R}^n$ closed up to an element of $\Gamma^n$, where $\Gamma^n$ is the discrete group used to define the torus: $\mathrm{T}^n \equiv \mathbb{R}^n/\Gamma^n$ ($\Gamma^n = \mathbb{Z}^n$ in the simplest case).

In bosonic-string theory we can add D-branes and O-planes more or less at will, because the theory is already inconsistent due to the tachyon. In the consistent superstring theories, D-branes and O-planes have to be introduced paying attention to anomaly and tadpole cancelations. These conditions are, in turn, related to the possibility of solving the equations of motion of the effective string theory for a background that contains those objects. In particular, one has to be able to solve the harmonic equation for $(p+1)$-form potentials in compact spaces, which is possible only if the total charge associated with the potentials vanishes. (Super-)D$p$-branes and O$p$-planes of superstring theories are charged with respect to (so-called) RR $(p+1)$-form potentials, which we will define later, and a consistent background will be one in which the sum of those charges vanishes. A very interesting example, as we will see, is the construction of the type-I $\mathrm{SO}(32)$ superstring theory by adding D9-branes and O9-planes to the type-IIB superstring theory [1095].

There is one last consideration we must make: as we are going to see, open-string interactions can produce closed strings. Thus, open strings are not fully consistent by themselves and have to be combined with a closed-string sector with the same orientability. The fields of the massless spectra of the resulting theories (without D-branes) are represented in Table 20.2. The consistency of the interacting theory also requires the addition of twisted sectors in orbifolds and orientifolds.

[14] For a review on orientifolds, see, for instance, Ref. [408] and also [54].

Table 20.2. Massless fields of the various bosonic-string theories

| Theory | Massless fields |
|---|---|
| Closed oriented | $g_{\mu\nu},\ B_{\mu\nu},\ \phi$ |
| Closed unoriented | $g_{\mu\nu},\ \phi$ |
| Open (and closed) oriented | $g_{\mu\nu},\ B_{\mu\nu},\ V_\mu^{IJ},\ \phi$ |
| Open (and closed) unoriented | $g_{\mu\nu},\ V_\mu^{[IJ]},\ \phi$ |

### 20.2.2 *Quantization of free-fermionic-string theories*

Again, the simplest way to arrive at the physical spectrum is to go to the light-cone gauge,

$$X^+ = \xi^0, \quad e^0{}_{\underline{1}} = \partial_{\underline{1}} e^1{}_{\underline{1}} = \psi^+ = \rho^i \chi_i = 0, \quad \det(e^a{}_i) = +1, \tag{20.51}$$

which allows the elimination of $e^a{}_i, \chi_i, X^\pm,$ and $\psi^\pm$ and leads to the Hamiltonian

$$H = -\frac{c}{2}\int_0^{2\pi\ell} d\xi^1 \left[T^{-1}\Pi^i_X \Pi^i_X + T\partial_{\underline{1}} X^i \partial_{\underline{1}} X^i + 2\Pi^i_\psi \rho_3 \partial_{\underline{1}} \psi^i\right], \tag{20.52}$$

and to the equations of motion (20.37) for the $X^i$, and, for the upper ($\psi^i_+$) and lower ($\psi^i_-$) components of the spinors $\psi^i$,

$$(\partial_{\underline{0}} \mp \partial_{\underline{1}})\psi^i_\pm = 0, \tag{20.53}$$

they are left- and right-moving, respectively.

The quantization proceeds essentially along the same lines as in the bosonic string, paying attention to the second-class constraints that this theory has. Here we describe just the general structure of the consistent superstring theories, leaving all details aside.

Superstring theories are Poincaré invariant only in the critical dimension $d = 10$. It is necessary to introduce the *worldvolume fermion number* $F$, defined modulo 2. The R and NS sectors are separated into $\mathrm{R}_\pm$ and $\mathrm{NS}_\pm$ subsectors with respect to the operator $e^{i\pi F}$. Then, consistency and the absence of tachyons require the combination of these subsectors (*GSO projection* [636]) in very precise ways.

**Closed strings** There are several possibilities.

- **Type-IIB$_+$ superstring** $\mathrm{R}_+\mathrm{R}_+ \oplus \mathrm{R}_+\mathrm{NS}_+ \oplus \mathrm{NS}_+\mathrm{R}_+ \oplus \mathrm{NS}_+\mathrm{NS}_+$, whose massless spectrum corresponds to an $N = 2B_+, d = 10$ supergravity multiplet with self-dual 4-form potential and chiral fermions.
- **Type-IIB$_-$ superstring** $\mathrm{R}_-\mathrm{R}_- \oplus \mathrm{R}_-\mathrm{NS}_+ \oplus \mathrm{NS}_+\mathrm{R}_- \oplus \mathrm{NS}_+\mathrm{NS}_+$, whose massless spectrum corresponds to an $N = 2B_-, d = 10$ supergravity multiplet with anti-self-dual 4-form potential and chiral fermions with the opposite chirality to the previous case.
- **Type-IIA$_{+-}$ superstring** $\mathrm{R}_+\mathrm{R}_- \oplus \mathrm{R}_+\mathrm{NS}_+ \oplus \mathrm{NS}_+\mathrm{R}_- \oplus \mathrm{NS}_+\mathrm{NS}_+$, whose massless spectrum corresponds to an $N = 2A_{+-}, d = 10$ supergravity multiplet.

Table 20.3. Massless fields of the various ten-dimensional superstring (supergravity) theories

| Theory | NSNS bosonic | RR bosonic | Chiral fermionic | Non-chiral fermionic | Vector supermultiplets |
|---|---|---|---|---|---|
| Type IIA | $\hat{g}_{\hat{\mu}\hat{\nu}}, \hat{B}_{\hat{\mu}\hat{\nu}}, \hat{\phi}$ | $\hat{C}^{(1)}{}_{\hat{\mu}}, \hat{C}^{(3)}{}_{\hat{\mu}\hat{\nu}\hat{\rho}}$ | | $\hat{\psi}_{\hat{\mu}}, \hat{\lambda}$ | |
| Type IIB | $\hat{\jmath}_{\hat{\mu}\hat{\nu}}, \hat{\mathcal{B}}_{\hat{\mu}\hat{\nu}}, \hat{\varphi}$ | $\hat{C}^{(0)}, \hat{C}^{(2)}{}_{\hat{\mu}\hat{\nu}}, \hat{C}^{(4\pm)}{}_{\hat{\mu}_1\cdots\hat{\mu}_4}$ | $\hat{\zeta}^{i\,(\mp)}_{\hat{\mu}}, \hat{\chi}^{i\,(\pm)}$ | | |
| Type I | $\hat{\jmath}_{\hat{\mu}\hat{\nu}}, \hat{\varphi}$ | $\hat{C}^{(2)}{}_{\hat{\mu}\hat{\nu}}$ | $\hat{\zeta}^{(\pm)}_{\hat{\mu}}, \hat{\chi}^{(\mp)}$ | | $(V^I{}_{\mu}, \eta^I)$ |
| Heterotic | $\hat{g}_{\hat{\mu}\hat{\nu}}, \hat{B}_{\hat{\mu}\hat{\nu}}, \hat{\phi}$ | | $\hat{\psi}^{(\pm)}_{\hat{\mu}}, \hat{\lambda}^{(\mp)}$ | | $(V^I{}_{\mu}, \eta^I)$ |

**Type-IIA$_{-+}$ superstring** $R_-R_+ \oplus R_-NS_+ \oplus NS_+R_+ \oplus NS_+NS_+$, whose massless spectrum corresponds to an $N = 2A_{-+}, d = 10$ supergravity multiplet. The sign of the Chern–Simons term of the action is different from that of the $N = 2A_{+-}$ theory.

There is a fifth possibility that gives the so-called type-0A and -0B strings, which are bosonic and have a tachyon but have sometimes been considered.

**Heterotic strings** These are constructed by combining the right-moving fields of the closed type-II superstring with the left-moving fields of the closed bosonic string. The 16 extra spacetime dimensions of the bosonic string must be compactified, which gives rise to gauge symmetry. A generic toroidal compactification gives the gauge group[15] $U(1)^{16}$, but, as explained in Section 20.3, for special values of the radii of the circles and of the angles between them, there the gauge group can be bigger and non-Abelian. In particular, one can show that the anomaly-free groups are $SO(32)$ and $E_8 \times E_8$. The massless modes are those of $N = 1_{\pm}, d = 10$ supergravity coupled to vector supermultiplets with those gauge groups. The couplings between the massless fields are, however, different from those of the type-I $SO(32)$ theory.

**Open superstrings** These result from the combination of two subsectors: $R_+ \oplus NS_+$ or $R_- \oplus NS_+$. In both cases, the massless spectrum corresponds to an $N = 1, d = 10$ vector supermultiplet $V^{IJ}_{\mu}, \chi$, where $\chi$ is a Majorana–Weyl gaugino with positive chirality in one case and negative in the other and where we have added gauge indices associated with $U(n)$ Chan–Paton factors.

Open superstrings also need a closed-superstring sector with the same spacetime supersymmetry ($N = 1$), which is constructed by taking the quotient of one type IIB by $\Omega$. Then, we have to take also open unoriented superstrings, with gauge group $SO(n)$ or $Sp(n)$. The only anomaly-free group is $SO(32)$. The result is the

**type-I$_{\pm}$ SO(32) superstring** The massless modes correspond to $N = 1_{\pm}, d = 10$ supergravity coupled to $SO(32)$ vector supermultiplets.

[15] If we had to compactify both right- and left-moving fields we would obtain $U(1)^{32}$. The explanation for this phenomenon can be found in Section 20.3.

The fields corresponding to the massless modes of all these theories are given in Table 20.3.

### 20.2.3 D-branes and O-planes in superstring theories

In the bosonic-string case we have seen how theories (oriented or unoriented) with Dirichlet boundary conditions could be constructed by adding to the simplest oriented theories D-branes or O-planes. The presence of these objects changes the boundary conditions and the orientability of the theory. In the bosonic case, consistency of the construction was not so important since the original theory was already sick, but in the construction of superstring theories we have chosen only those which are completely (self-)consistent, free of anomalies and tachyons, and we have found all of them in ten dimensions. We can, however, try to construct new theories by compactification of these or by the addition of D-branes and O-planes, breaking in general ten-dimensional Lorentz invariance and supersymmetry.

The rules for how to add D-branes and O-planes consistently to superstring theories are much more restrictive than in the bosonic case. To start with, the following facts have to be taken into account.

1. The type-IIA (B) theory admits only D$p$-branes and O$p$-planes with $p$ even (odd).

2. D$p$-branes and O$p$-planes are charged with respect to the RR $(p+1)$-form potentials of each theory [1047] (see Table 20.3) and their duals. This agrees with the fact that type IIA (B) has odd-(even)-rank RR potentials only. However, one has to introduce a 9- (10)-form potential $\hat{C}^{(9)}$ ($\hat{C}^{(10)}$) for the D8- (9)-brane and O8- (9)-plane of the type IIA (B) theory. These fields carry no local degrees of freedom and, therefore, they are not associated with states of the spectrum.

3. D$p$-branes carry a unit of positive or negative RR charge that equals its tension $q_{\mathrm{D}p} = \pm T_{\mathrm{D}p}$. O$p$-planes have charges and tensions that depend on the symmetries involved. The prototype is the type-IIB$_\pm$ O9-plane associated with worldsheet parity $\Omega$, which is a symmetry of both theories (but interchanges the two type-IIAs) and has $\hat{C}^{(10)}$ charge $q_{\mathrm{O}9} = \pm T_{\mathrm{O}9}$ and tension $T_{\mathrm{O}9} = -32 T_{\mathrm{D}9}$ (negative). The O$p$-planes related to it by T duality have

$$T_{\mathrm{O}p} = -2^{p-5} T_{\mathrm{D}p}. \tag{20.54}$$

4. The presence of a single D$p$-brane or O$p$-plane halves the supersymmetry of the theory. This is the amount preserved by those objects considered as superstring states. From this point of view they are BPS states and they must saturate some Bogomol'nyi bound, which implies a relation between their tensions and charges (identity in proper units). Then, equilibrium of forces between objects of the same kind (including spatial orientation) is to be expected, as discussed in Chapter 17, which means that we can have parallel D$p$-branes with the same charge in equilibrium and enhancement of gauge symmetry when they coincide. We will see that classical solutions that describe these systems can be found in the effective supergravity theories. The stability of these systems is reflected in the absence of tadpoles in the string theory.

5. It is possible to introduce D$p$-branes that intersect (at any angle) if the resulting system preserves supersymmetry. There are simple rules for allowed intersections at right angles, which will be studied in Section 25.6.

6. As we have mentioned, the absence of anomalies and tadpoles is related to the stability of the system of D$p$-branes and O$p$-planes. In particular, the system should be able to solve the equations of motion of the string effective theory which we are going to study in the next few chapters. The equations of motion for $(p+1)$-forms are generalizations of the harmonic equation, which, in compact spaces, can be solved only if the total charge is zero.[16]

These rules have been used extensively for building new string theories. The simplest construction leads to the type-I $SO(32)$ theory starting from type IIB: on introducing an O9-plane (i.e. taking the quotient of the type-IIB theory by $\Omega$), consistency requires the addition of 32 D9-branes in order to obtain zero total RR charge, which results in the introduction of an open, unoriented-string sector with gauge group $SO(32)$.

### 20.2.4 String interactions

Strings interact by joining and splitting. It is then easy to understand that open strings can interact to give closed strings and that consistency (unitarity) requires a closed-string sector in open-string theories.

String amplitudes are defined as path integrals over all embeddings $X^\mu$ and all worldsheet metrics $\gamma_{ij}$ with given boundaries and boundary data that determine the string states that are scattered. The boundary data are included as vertex operators in the path integral. Without vertex operators, we have vacuum amplitudes, given by the path integral

$$Z = \int \mathcal{D}X\mathcal{D}\gamma e^{-S_{\rm P}-S_{\rm Euler}}, \tag{20.55}$$

where $S_{\rm P}$ is the Euclidean Polyakov action Eq. (20.5) and $S_{\rm Euler}$ is the topological term Eq. (20.14). For closed strings we will restrict ourselves to (just oriented or oriented plus unoriented) compact surfaces. For open strings we will add surfaces with boundaries.

The sum over metrics can be decomposed into a sum of path integrals over worldsheets with given topologies. The topology of two-dimensional surfaces can be characterized completely by the numbers $g, b,$ and $c$, combined into the Euler characteristic $\chi$, which is given by the topological term Eq. (20.14) as explained on p. 614. The result takes the form

$$Z = \sum_t (e^{\phi_0})^{-\chi(t)} \int_{\{\Sigma_t\}} \mathcal{D}X\mathcal{D}\gamma e^{-S_{\rm P}\Sigma_t}, \tag{20.56}$$

where $t$ stands for given topologies and $\{\Sigma_t\}$ is the space of surfaces with topology $t$. Now, each topology can be associated with a loop order, given precisely by $-\chi(t)$, and the above

[16] The lines of force of the field can only go to sources or to infinity. In compact spacetimes, they have to start and end on sources and the total charge has to be zero.

sum can be understood as a perturbative series expansion in which $e^{\phi_0}$ plays the role of the *string coupling constant* $g$:

$$\boxed{g \equiv e^{\phi_0}.} \tag{20.57}$$

In Section 21.1 we will see that $\phi_0$ is the vacuum expectation value of the *dilaton field*.

## 20.3 Compactification on $\mathrm{S}^1$: T duality and D-branes

We can obtain four-dimensional string theories by compactification. The simplest compactification would be on a circle. Already in this case we are going to start seeing *stringy* effects (T duality, first discovered in Refs. [856, 1096], and enhancement of gauge symmetry) that we did not see in the field-theory (KK) case, which are a manifestation of the extended-object nature of strings and a suggestion that there is a minimal length in string theory. General references on T duality are Refs. [635, 35].

We are going to study first the compactification of closed bosonic strings on a circle.

### 20.3.1 Closed bosonic strings on $\mathrm{S}^1$

If $Z \equiv X^{\hat{d}-1}$ is the compact coordinate, it is convenient to identify $Z \sim Z + 2\pi R_z$, where $R_z$ is the compactification radius, and keep using the Minkowski metric. Now, in the mode expansion Eq. (20.40) of $Z$, the following apply.

1. There is another zero mode compatible with the periodicities of $\xi^1$ and $Z$:

$$\frac{R_z w}{\ell}\xi^1, \qquad w \in \mathbb{Z}. \tag{20.58}$$

When we go around the closed string once, $\xi^1 \to \xi^1 + 2\pi\ell$, we go $w$ times around the compact dimension: $Z \to Z + 2\pi R_z w$. This is a *winding mode*, a purely stringy animal that corresponds to the capacity of closed strings to be wrapped $w$ times around compact dimensions.

2. There are also string KK modes as in Chapter 15,

$$\frac{n}{R_z p^+}\xi^0, \qquad n \in \mathbb{Z}. \tag{20.59}$$

Quantization leads to the mass formula and constraint

$$M^2 = \frac{n^2}{R_z^2} + \frac{R_z^2 w^2}{\alpha'^2} + \frac{2}{\alpha'}(N + \tilde{N} - 2), \qquad N = \tilde{N} + nw. \tag{20.60}$$

Observe that the mass of the $w = 1$ mode agrees with the definition of the string tension on p. 611: it is the product of the length of the compact dimension and the string tension.

The spectrum is now that of the uncompactified theory (the $n = w = 0$ sector) plus new sectors with non-vanishing KK momentum or winding number. The spectrum of the uncompactified theory has to be interpreted now in $\hat{d} - 1$ dimensions: the $\hat{d}$-dimensional graviton gives rise to a graviton, a (KK) vector, and a (KK) scalar in $\hat{d}-1$ dimensions, while the KR 2-form gives rise to another 2-form and another (*winding*) vector, and the dilaton gives another dilaton. For generic values of the compactification radius $R_z$ there are no more massless states and the vector gauge symmetry group is $\mathrm{U}(1)^2$. As discussed in Chapter 15, KK modes are charged with respect to the KK $\mathrm{U}(1)$ vector field. We will see that winding modes are charged with respect to the winding $\mathrm{U}(1)$ vector field.

The spectrum is invariant under the *T-duality* transformation

$$\boxed{n' = w, \qquad w' = n, \qquad R'_z = \alpha'/R_z.} \tag{20.61}$$

A bosonic-string theory with one dimension compactified on a circle of radius $R_z$ and a second bosonic-string theory with one dimension compactified on a circle of radius $\alpha'/R_z$ have the same spectra, with the winding modes of one of them having the same masses as the KK modes of the other and vice versa. Not only do they have the same spectra, but also they have the same interactions and scattering amplitudes, but one has to take into account that the string coupling constants of the two theories are related by [39]

$$\boxed{g' = g\ell_s/R_z.} \tag{20.62}$$

This has very important consequences: if we diminish the size of the compactification radius beyond the *self-dual radius* $R_z = \ell_s = \sqrt{\alpha'}$, there is another completely equivalent bosonic-string theory defined on a circle of radius bigger than $\ell_s$. The self-dual radius can be interpreted as the minimal radius on which a bosonic-string theory can be compactified.

On the other hand, at the self-dual radius there are four additional massless vectors: $N = \pm n = \pm w = 1$ and $\tilde{N} = \pm n = \mp w = 1$. These, plus the KK and the winding vector, turn out to be the gauge vectors of $\mathrm{SU}(2) \times \mathrm{SU}(2)$ and we find a new enhancement of symmetry at a special point (a T-duality fixed point) of the moduli space, which is one of the most striking properties of string theory.

Is T duality related to some property of the Polyakov action that we have missed? Actually, yes: the Polyakov action is invariant under Poincaré dualization of one of the embedding coordinates, just as Maxwell's action is invariant under the dualization of the vector field (Section 12.7.1): we put $\partial_i Z \equiv F_i$ and add to the Polyakov action in Minkowski spacetime a Lagrange-multiplier term enforcing the Bianchi identity $\partial_{[i}F_{j]} = 0$:

$$S_P[X^\mu, \gamma_{ij}, Z] = -\frac{T}{2}\int d^2\xi \sqrt{|\gamma|}\,\gamma^{ij}(\partial_i X^\mu \partial_j X_\mu - F_i F_j) - T\int d^2\xi \epsilon^{ij}\partial_i Z' F_j. \tag{20.63}$$

The equations of motion of the Lagrange multiplier $Z'$ and of $F_i$ are

$$\partial_{[i}F_{j]} = 0, \qquad F^i = \epsilon^{ij}\partial_j Z', \tag{20.64}$$

and we can use the latter to eliminate $F_i$. The result is the Polyakov action with $Z$ replaced by the dual embedding coordinate $Z'$. This procedure can be used in the presence of non-trivial spacetime metrics that are independent of $z$, as we will see in Section 21.2.2.

### 20.3.2 Open bosonic strings on $\mathrm{S}^1$ and D-branes

Equations (20.64) imply $\partial^i Z = \epsilon^{ij}\partial_j Z'$, which implies that, under Poincaré duality, left-moving objects transform into left-moving objects and right-moving ones into minus right-moving objects. Then T duality can be described as the equivalent transformation

$$Z = Z_+ + Z_- \to Z' = Z_+ - Z_-. \tag{20.65}$$

This description is useful for open strings. Open strings with N boundary conditions have KK modes but no winding modes on a circle. Clearly, a T-duality transformation will not take us into another similar open-string theory. Yet, we can perform the transformation and try to identify the resulting theory. The main observation is that N boundary conditions $\partial_+ Z_+ = \partial_- Z_-$ and D boundary conditions $\partial_+ Z_+ = -\partial_- Z_-$ are interchanged by T duality. Indeed, on applying the above transformation to the open string with N boundary conditions mode expansion Eq. (20.38), which we rewrite here in the form

$$Z_\pm = \frac{z}{2} \pm \frac{p^z}{2cp^+}\xi^\pm - i\sqrt{\frac{\alpha'}{2}}\sum_{n\neq 0} e^{\pm \frac{in}{2\ell}\xi^\pm}, \tag{20.66}$$

we find

$$Z' = Z_+ - Z_- = \frac{p^z}{cp^+}\xi^1 + \sqrt{2\alpha'}\sum_{n\neq 0}\frac{\alpha_n}{n}e^{\frac{icn\xi^0}{2\ell}}\sin\left(\frac{n\xi^1}{2\ell}\right), \tag{20.67}$$

which coincides with the expansion with D boundary conditions Eq. (20.39) with $z^1 = 0$ and $z^2 = p^z/T$. If we take into account that $p^z = n/R_z$, we see that $z^2 = n2\pi R'_z$, where $R'_z = \alpha'/R_z$ is the T-dual radius. The coordinate $Z$ with N boundary conditions has been transformed into the compact coordinate $Z'$ with dual compactification radius and D boundary conditions at both ends, $z^1 = 0$ and $z^2 = 2\pi R'_z \sim 0$. The momentum mode $n$ has become a winding mode with winding number $n$: the string has one endpoint attached to the D24-brane, winds around the compact dimension $n$ times, and ends again on the same D24-brane.

We could repeat the procedure in another compact coordinate, giving a D23-brane. Thus, T duality in a direction parallel to a D$p$-brane transforms it into a D$(p-1)$-brane with a compact transverse dimension and vice versa. Furthermore, with Chan–Paton factors, we would have found endpoints with different labels on the same hypersurface. Thus, we would have overlapping D-branes and the gauge group would be preserved by T duality.

What happens when we perform T duality on two parallel non-overlapping D24-branes whose transverse dimension is compact? There are, as we discussed, four sectors labeled by pairs $ij$ with $ij = 1, 2$ indicating on which of the two D24-branes the first and second endpoints are. The spectrum is consistent with spontaneously broken $\mathrm{U}(2)$ gauge symmetry. The 11 and 22 sectors are T dual to open strings with N boundary conditions. If the D24-branes are placed at angles $\theta_1$ and $\theta_2$, the 12 and 21 sectors with winding number $w$ have expansions

$$Z^{ij}_\pm = \frac{R_z\theta_i}{2} + \frac{[2\pi w - (\theta_j - \theta_i)]R_z}{4\pi\ell}\xi^\pm \mp i\sqrt{\frac{\alpha'}{2}}\sum_{n\neq 0} e^{\pm\frac{in}{2\ell}\xi^\pm}, \tag{20.68}$$

and the T-dual expansion is that of an N open string with momentum given by $(p^{z\,ij})' = [w - (\theta_i - \theta - j)]/R'_z$. The shift with respect to the standard KK momentum $w/R'_z$ is caused by the appearance of *Wilson lines* of the $U(2)$ gauge field $V^{ij}$ of the N open strings with two Chan–Paton factors. As explained on p. 462, in a background with a compact dimension we can have topologically non-trivial configurations of the gauge fields characterized by the line integral of the gauge field around the compact direction (known as a Wilson line):

$$W[C] = \exp\left\{ i \oint_C A \right\}. \tag{20.69}$$

In this case, in the dual-string theory, we have the gauge-field configuration

$$\left(V_{\underline{z}}^{ij}\right) = -\frac{1}{2\pi R'_z} \begin{pmatrix} \theta_1 & 0 \\ 0 & \theta_2 \end{pmatrix}. \tag{20.70}$$

This non-trivial background breaks $U(2)$ down to $U(1)^2$, as can be seen in the spectrum[17] [1049]. Thus, the symmetry in the original configuration is equal to that in the dual configuration. In one case, the symmetry breaking is associated with the separation of the D24-branes and in the T dual it is associated with the presence of Wilson lines (D25-branes, being spacetime-filling branes, cannot be spatially separated), but gauge symmetry is preserved by T duality. The generalization to $n$ D-branes and $U(n)$ gauge symmetry is straightforward.

### 20.3.3 Superstrings on $S^1$

Closed and heterotic superstring theories compactified on circles also have T duals: the type-IIA and type-IIB theories compactified on circles of T-dual radii are each other's dual [412, 454] and the $SO(32)$ and $E_8 \times E_8$ heterotic strings compactified on circles of T-dual radii are each other's T-dual theory. Type-I $SO(32)$ strings also have a T-dual theory, the so-called type-I′, which has a nine-dimensional interpretation.

To make further progress in the study of string dualities, we need to study string effective actions. They will allow us to extend these results easily to more general backgrounds and they will also allow us to find possible non-perturbative dualities that the perturbative formulation of string theory that we have just sketched in this chapter is not powerful enough to exhibit.

[17] To find the spectrum, one has to take into account the coupling of the $U(2)$ gauge vector to the Polyakov action through the boundary term Eq. (21.6).

# 21
# The string effective action and T duality

After Chapter 20's brief introduction to perturbative string theory and T duality we are going to discuss how one arrives at the low-energy string effective (field theory) actions, what their meaning is, and what their limits of validity are. We are going to start exploiting them to study T duality and to find *Buscher's T-duality rules*, which relate different curved backgrounds that are equivalent from the string-theory point of view. These rules are some of the most powerful tools of string theory.

## 21.1 Effective actions and background fields

The low-energy string effective action describes the low-energy dynamics of a given string theory. Here low energy means energies lower than the relevant energy scale: the string mass $m_{\rm s}$. Thus, the low-energy limit is the $\alpha' \to 0$ limit, heuristically the limit in which the string length can be ignored and a theory of particles (a field theory) is recovered.

On the other hand, at low energies only the massless modes are relevant and their dynamics is described by a theory of the corresponding massless fields. The obvious way to find this field theory is to compute string amplitudes for the massless modes, take the $\alpha' \to 0$ limit, and then construct a field theory that reproduces these amplitudes. In principle, the effective field theory has an expansion in powers of $\alpha'$, although usually only the lowest-order terms are considered. The terms of higher order in $\alpha'$ are also of higher order in derivatives. Also, string amplitudes are calculated order by order in string perturbation theory and the effective action can also be expanded in powers of the string coupling constant, which here is the exponential of the dilaton field. Again, only the lowest orders are usually considered.

Actually, for some superstring theories, it is possible to arrive at the effective theory using (super)symmetry arguments. In particular, the massless modes of the type-II superstrings fill the supergravity multiplets of the (two only) maximal ten-dimensional supergravity theories: those of the type-IIA (non-chiral) theory fill the supergravity multiplet of $N = 2A, d = 10$ SUEGRA [290, 603, 788, 810] whose action is given in Eq. (22.38) and those of the type-IIB (chiral) theory fill the supergravity multiplet $N = 2B, d = 10$ SUEGRA [1114, 1117, 788] whose action is given in Eq. (23.4). Similarly, since there is only one $N = 1$ supergravity in $d = 10$ dimensions [636, 189, 341], coupled to vector fields with the right gauge group it must be the effective action of the heterotic and type-I strings (but here the couplings of the 2-form are different in the two theories).

The fields of these effective theories are given in Table 20.3. The NSNS fields are sometimes called the *common sector* since they occur in all of them, including the bosonic (oriented)-string theories. The fields in the common sector are the metric $g_{\mu\nu}$, associated with the graviton, the KR 2-form $B_{\mu\nu}$, and the *dilaton* $\phi$ whose vacuum expectation value $\phi_0$ gives the string coupling constant $g = e^{\phi_0}$ (see Eqs. (21.8) and (20.14)). The action for the common sector in the *string frame* to be defined in the following is given (in $d$ dimensions with $d = 10$ and 26) by

$$\boxed{S = \frac{g^2}{16\pi G_{\mathrm{N}}^{(d)}} \int d^dx \sqrt{|g|}\, e^{-2\phi} \left[ R - 4(\partial\phi)^2 + \frac{1}{2\cdot 3!} H^2 \right],} \tag{21.1}$$

where, in our notation in which indices not shown are all antisymmetrized,

$$H = 3\partial B \tag{21.2}$$

is the KR field strength, which is invariant under the gauge transformations necessary for the consistent quantization of a massless 2-form field,

$$\delta B = 2\partial\Sigma, \tag{21.3}$$

where $\Sigma_\mu$ is an arbitrary vector field. The overall factor $e^{-2\phi}$ is associated with the genus-0 (tree-level) origin of these terms and the normalization is conventional. In particular, the factor $g^2$ ($g$ is defined in Eq. (20.57)) compensates the factor $e^{-2\phi_0}$ that appears in the weak field expansion of the action around the vacuum $g_{\mu\nu} = \eta_{\mu\nu}, \phi = \phi_0$, so $G_{\mathrm{N}}^{(d)}$ can be interpreted as the $d$-dimensional Newton constant. Its value Eq. (25.26) can be determined by using duality arguments that relate it to the string coupling constant and the string length just as we determined the Newton constant in terms of the compactification radius of KK theory in Eq. (15.97), as we will see in Section 25.3.

Observe that, up to normalization, the above action is also invariant under shifts of the dilaton field that change its vacuum expectation value and thus $g$, which is a free parameter. A potential for the dilaton could give it a mass and fix $g$, solving two problems simultaneously: the determination of $g$ and the existence of a massless scalar that couples as the Jordan–Brans–Dicke field to all kinds of matter, inducing violations of the equivalence principle [416, 417]. The massless KR 2-form that also couples to matter can also be a source of violations of the equivalence principle. The KR 3-form field strength can be understood as a completely antisymmetric dynamical torsion field, as we discussed on p. 157, and, in the same spirit, the dilaton can be understood as part of a non-metricity tensor of the type considered by Weyl [1104]. The above string effective action can then be written as an Einstein–Hilbert action for a torsionful and non-metric-compatible connection using Eq. (1.59) (see also Eq. (1.62)).

The RR fields are differential forms $C^{(n)}{}_{\mu_1\cdots\mu_n}$ of even (odd) rank in the $N = 2B\,(A)$ theory and appear in the respective actions Eqs. (23.4) and Eq. (22.38) with no couplings to the dilaton in the string frame. They couple to the KR 2-form due to the definitions of the field strengths and also to the presence of Chern–Simons (CS) topological terms in the supergravity actions. These CS terms contain a great deal of information on the possible intersections of extended objects of the theory [1191].

Although the identification of the field theories on the basis of symmetry arguments is correct, the identification of the fields with the string modes is ambiguous, since the supergravity theories are unique up to field redefinitions. To establish completely the relation between supergravity fields and string modes, it is necessary to have more information. For instance, making use of the relations in Fig. 20.1, the supergravity fields must be related by dualities in the same way as the string modes are.

String effective actions also arise in a different way: string theories are usually quantized in flat spacetime, but the string worldsheet action can be written in a curved background as a non-linear $\sigma$-model, Eq. (20.5), and, furthermore, can be generalized to describe the coupling to all background fields associated with the string massless modes.[1] The coupling of the string to the Kalb–Ramond 2-form $B_{\mu\nu}$ is represented by a WZ term that generalizes the coupling of the Maxwell vector field to a charged point-particle, Eq. (12.53), i.e. it is the integral of the pullback of the 2-form over the two-dimensional worldsheet:

$$\frac{T}{2}\int_{\Sigma} B, \tag{21.4}$$

where $B$ is given by

$$B = \tfrac{1}{2}B_{ij}d\xi^i \wedge d\xi^j = d^2\xi\epsilon^{ij}B_{ij} = d^2\xi\epsilon^{ij}\partial_i X^\mu \partial_j X^\nu B_{\mu\nu}. \tag{21.5}$$

Observe that the role of the electric charge is played here by the string tension. This coefficient can be changed but we will take it as above, defining the normalization of $B_{\mu\nu}$ that we will use. This is the normalization that leads to the effective action Eq. (21.1). Observe also that the WZ term, being topological (metric independent), is automatically Weyl invariant and does not contribute to the $\gamma_{ij}$ equation of motion. Furthermore, the WZ term is invariant, up to total derivatives, under gauge transformations of the 2-form, Eq. (21.3), which means that strings are charged with respect to the KR 2-form and the charge is conserved. In open-string worldsheets the non-vanishing boundary term is canceled out by the variation of the term that represents the coupling of the open string to the 1-form $V_\mu$:

$$\int_{\partial\Sigma} V, \tag{21.6}$$

provided that the vector transforms under the KR 2-form gauge symmetry Eq. (21.3),

$$\delta V_\mu = T\Sigma_\mu. \tag{21.7}$$

Finally, this term is not parity invariant and occurs only in oriented-string theories.

---

[1] This is always true for the fields in the common sector for the bosonic and fermionic strings, although worldsheet supersymmetry has to be studied case by case. The inclusion of RR massless superstring fields in the $\sigma$-model is more complicated and how to do it is known only in certain cases.

The coupling to the dilaton is related to the topological term Eq. (20.14):

$$-\frac{1}{4\pi}\int d^2\xi\sqrt{|\gamma|}\,\phi(X)R(\gamma). \tag{21.8}$$

This term, which is of higher order in $\alpha'$ for dimensional reasons, gives no contribution to the $\gamma_{ij}$ equation of motion but breaks Weyl invariance for generic dilaton fields.

Since Weyl invariance is absolutely necessary for the consistency of string theory, the natural question to be asked is: in which backgrounds $g_{\mu\nu}$, $B_{\mu\nu}$, and $\phi$ is Weyl invariance quantum mechanically preserved? The background fields can be understood as coupling functions and then the question can be reformulated in terms of the vanishing of the beta *functionals* associated with them.

To lowest order, these beta functionals are given by[2] [285]

$$\begin{aligned}
\beta^g_{\mu\nu} &= \alpha'\left[R_{\mu\nu} - 2\nabla_\mu\nabla_\nu\phi + \tfrac{1}{4}H_\mu{}^{\alpha\beta}H_{\nu\alpha\beta}\right] + \mathcal{O}(\alpha'^2),\\
\beta^B_{\mu\nu} &= \frac{\alpha'}{2}e^{2\phi}\nabla^\rho(e^{-2\phi}H_{\rho\mu\nu}) + \mathcal{O}(\alpha'^2),\\
\beta^\phi &= \frac{d-26}{6} - \frac{\alpha'}{2}\left[\nabla^2\phi - (\partial\phi)^2 - \tfrac{1}{4}R - \tfrac{1}{48}H^2\right] + \mathcal{O}(\alpha'^2),
\end{aligned} \tag{21.9}$$

and it turns out that the vanishing of these beta functionals is equivalent to the equations of motion derived from the action Eq. (21.1) plus a term

$$\frac{g^2}{16\pi G_N^{(d)}}\int d^dx\sqrt{|g|}\,e^{-2\phi}[-2(d-2)\Lambda], \qquad \Lambda = \frac{2(d-26)}{3\alpha'(d-2)}, \tag{21.10}$$

for the bosonic string, which vanishes in the critical dimension $d = 26$ (the same happens for the fermionic string for $d = 10$). Indeed, the equations of motion are (see Section 4.2)

$$\begin{aligned}
\frac{16\pi G_N^{(d)}e^{2(\phi-\phi_0)}}{\sqrt{|g|}}\frac{\delta S}{\delta g^{\mu\nu}} &\sim \frac{1}{\alpha'}(\beta^g_{\mu\nu} - 4g_{\mu\nu}\beta^\phi) + \mathcal{O}(\alpha'),\\
\frac{16\pi G_N^{(d)}e^{2(\phi-\phi_0)}}{\sqrt{|g|}}\frac{\delta S}{\delta\phi} &\sim -\frac{16}{\alpha'}\beta^\phi + \mathcal{O}(\alpha'),\\
\frac{16\pi G_N^{(d)}e^{2(\phi-\phi_0)}}{\sqrt{|g|}}\frac{\delta S}{\delta B_{\mu\nu}} &\sim -\frac{1}{\alpha'}\beta^{B\,\mu\nu} + \mathcal{O}(\alpha').
\end{aligned} \tag{21.11}$$

Thus, we see that quantum conformal invariance leads (in the critical dimension) to the same effective action for the string common sector Eq. (21.1). The metric that appears in

[2] $\beta^\phi$ can be obtained from $\beta^g_{\mu\nu}$ using the contracted Bianchi identity. The conformal anomaly appears precisely as the integration constant. This mechanism is very reminiscent of what happens in unimodular gravity theories. These are theories with invariance only under volume-preserving diffeomorphisms. They were first considered by Einstein in 1919 and more recently by various authors in Refs. [210, 744, 1206, 1037]. See Ref. [33] for a very recent paper with many references.

that action is the same metric as that to which the string couples and therefore appears in the $\sigma$-model and is called the *string-frame metric*. A conformal rescaling,

$$g_{\mu\nu} = e^{\frac{4}{d-2}\phi} g_{\mathrm{E}\,\mu\nu}, \tag{21.12}$$

can eliminate the factor $e^{-2\phi}$ in front of the Einstein–Hilbert term (see Appendix M). The rescaled metric $g_{\mathrm{E}\,\mu\nu}$ is called the *Einstein-frame metric*. In the Einstein frame, the string action is given by

$$S = \frac{1}{16\pi G_{\mathrm{N}}^{(d)}} \int d^dx \sqrt{|g_{\mathrm{E}}|} \Big[ R_{\mathrm{E}} + \frac{4}{d-2}(\partial\phi)^2 + \frac{1}{2\cdot 3!} e^{\frac{-8}{d-2}\phi} H^2 - (d-2)\Lambda e^{\frac{4}{d-2}\phi} \Big]. \tag{21.13}$$

The solutions to these equations describe backgrounds (vacua) for bosonic-string theory in which strings can be consistently quantized, to lowest order in $\alpha'$ and the string coupling constant. The simplest is evidently ten-dimensional Minkowski spacetime, which should remain a good vacuum to all orders[3] because all fields are trivial. Other vacua can be argued to be exact and not to receive higher $\alpha'$ corrections due to their unbroken supersymmetry and/or the vanishing of their curvature invariants, as is the case with pp-wave solutions [698, 42, 774, 1215, 165, 776] and the four-dimensional solutions of [368], which are based on the classification of the four-dimensional metrics that have all the curvature invariants vanishing [1059]. The next step is to try to quantize string theory on these vacua (for instance in the KG10 solution [946, 948]).

It is amusing to see that quantizing string theory in non-trivial backgrounds amounts to finding the generalization of Pythagoras' law for vibrating strings (arguably the first law in the history of physics) and that the generalization is possible only for backgrounds that satisfy the above generalization of the Einstein equations.

The effective action has been obtained perturbatively in both $\alpha'$ and the string coupling constant $g$ and, furthermore, in the low-energy (long-distance) approximation. As a general rule, the results obtained working with it can be trusted as long as $e^{\phi} \ll 1$ and the curvature scalar $R \ll \ell_{\mathrm{s}}^{-2}$ or $R\alpha' \ll 1$. Sometimes there are lengths in the system under consideration that do not appear in the curvature, such as the radius of a circle. These distances should be bigger than the string scale $\ell_{\mathrm{s}}$. At distances of the order of $\ell_{\mathrm{s}}$, or curvatures of the order of $\ell_{\mathrm{s}}^2$, there are stringy effects that invalidate the effective action and its solutions, unless higher-order corrections to both are taken into account.

For instance, we have seen in Chapter 20 that, when the distance between two D-branes becomes of the order of $\ell_{\mathrm{s}}$ or when the compactification radius takes the self-dual value of $\ell_{\mathrm{s}}$, new massless states appear in the string spectrum that were not taken into account in the calculation of the string effective action and should be included by hand at that point. Beyond that point, the T-dual description of the effective field theory should be used. In more complex geometries one can find radii that are functions of other coordinates, so there are regions in which they are smaller than $\ell_{\mathrm{s}}$. The solution cannot be trusted in those regions. This is the basis of Sen's argument, which we mentioned on p. 351.

[3] In superstring theories. In bosonic-string theories, the 26-dimensional Minkowski spacetime is not a stable vacuum, as the presence of the tachyon suggests, and is also $\alpha'$ corrected [551], although it is not known to which solution the corrections converge.

### 21.1.1 The D-brane effective action

The effective action of a D$p$-brane (or, rather, the effective action of open-plus-closed-string theory in the presence of a D$p$-brane so the open-string sector has D boundary conditions in $25 - p$ coordinates) in a background with metric, KR 2-form, dilaton, and $(p+1)$-dimensional U(1) vector is given by the following generalization of the Nambu–Goto action [887]:

$$S = -T_{\mathrm{D}p}g \int d^{p+1}\xi \, e^{-\phi} \sqrt{|g_{ij} + B_{ij} + 2\pi\alpha' F_{ij}|}, \tag{21.14}$$

where $g_{ij}$ and $B_{ij}$ are the pullbacks over the $(p+1)$-dimensional worldvolume of the spacetime fields and $F_{ij} = 2\partial_{[i}V_{j]}$ is the standard field strength of a gauge vector that in this context is called the *Born–Infeld* (BI) vector field since the above action, with flat spacetime metric and zero dilaton and KR field, was proposed by Born and Infeld [235, 237] (see also Refs. [1110, 236]) as a non-linear model for electrodynamics: the power expansion of the action contains a Maxwell term $F^2$ and an infinite series of higher-order terms. The main property of this theory is that the spherically symmetric field was singularity free and had a *core* with characteristic size $\sim\ell_{\mathrm{s}}$ with the above normalization.[4] $T_{\mathrm{D}p}$ is the D$p$-brane tension.[5]

This action is invariant under spacetime and worldvolume reparametrizations, and also under the KR and BI vector field gauge transformations Eqs. (21.3) and (21.7) and has to be added to the *bulk* closed-string effective action Eq. (21.1). If the strings are unoriented, then there is no KR 2-form in either of them.

The fact that, in the presence of D boundary conditions, the string effective action includes the worldvolume action of an extended object, the D$p$-brane, is a final argument in favor of the interpretation of the latter as a dynamical object. On the other hand, the D$p$-brane action is a generalization of the string $\sigma$-model action and it may constitute the starting point for quantization. Actually, the D$p$-brane actions of superstring theories can (and must) couple to all the fields of the (bulk) closed-superstring effective action, which is a supergravity action. The consistent coupling to all these fields requires, first of all, a WZ term for the action to be invariant under $\kappa$-symmetry transformations. The bosonic part of the WZ term describes the coupling to a RR $(p+1)$-form potential and other RR potentials of lower rank. Thus, as first shown by Polchinski in Ref. [1047], superstring D$p$-branes carry RR charges and are sources of the RR fields of those actions. We will discuss the effective actions of these super-D-branes in Chapter 25.

## 21.2 T duality and background fields: Buscher's rules

We have introduced string effective-field-theory actions in Section 21.1, and in this section we want to show how to use them to study string dualities in the simplest case: T duality.

[4] A review of actions of this kind can be found in Ref. [1204].

[5] In the literature, the same quantity as that which we define as tension is called *effective tension*, and the coefficient in front of the D$p$-brane effective action is called *tension*. We use only the physical tension parameter defined above, which should avoid confusion.

We will consider the effective action for the string common sector, and the results will be valid only for closed bosonic strings but will later be generalized to the heterotic and type-II cases. We essentially follow Ref. [167], where the heterotic case was studied, with a few notational changes.

### *21.2.1 T duality in the bosonic-string effective action*

T duality relates closed $\hat{d}$-dimensional string theories compactified on circles of relatively dual radii. The effective field theories will be $\hat{d}-1=d$-dimensional field theories for the massless modes, and the KK formalism that was developed in Chapter 15 is perfectly suited to obtaining them from the effective actions of the uncompactified $\hat{d}$-dimensional effective theories.[6]

Our starting point is the action Eq. (21.1) with hats on every object, following the notation of Chapter 15. We denote the compact coordinate by $x^{\hat{d}-1}\equiv z\in[0,2\pi\ell_{s}]$, and assume that all fields are independent of it. We can use the standard KK ansatz Eq. (15.33) and the results concerning the spin connection, Eqs. (15.36) and (15.35), and volume element, Eq. (15.37). Before substituting in the Einstein–Hilbert part of the action, we use the $\hat{d}$-dimensional Palatini's identity Eq. (L.4) with $K=e^{-2\hat{\phi}}$ and immediately obtain

$$\int d^{\hat{d}}\hat{x}\,\sqrt{|\hat{g}|}\,e^{-2\hat{\phi}}\hat{R}=\int dz\int d^{\hat{d}-1}x\,\sqrt{|g|}\,e^{-2\hat{\phi}}k\Big\{-\omega_{b}{}^{ba}\omega_{c}{}^{c}{}_{a}-\omega_{a}{}^{bc}\omega_{bc}{}^{a} + 2\omega_{b}{}^{ba}\partial_{a}\ln\left(e^{-2\hat{\phi}}k\right)-2\partial_{a}\ln k\,\partial^{a}\ln e^{-2\hat{\phi}}-\tfrac{1}{4}k^{2}F^{2}(A)\Big\}. \tag{21.15}$$

It is evident that the combination $e^{-2\hat{\phi}}k$ now plays the role of a $d$-dimensional dilaton, and thus we define

$$\phi\equiv\hat{\phi}-\tfrac{1}{2}\ln k. \tag{21.16}$$

The kinetic term for the dilaton gives

$$\int d^{\hat{d}}\hat{x}\,\sqrt{|\hat{g}|}\,e^{-2\hat{\phi}}\Big[-4(\partial\hat{\phi})^{2}\Big]=\int dz\int d^{\hat{d}-1}x\,\sqrt{|g|}\,e^{-2\hat{\phi}}k\,\Big[-4(\partial\hat{\phi})^{2}\Big]. \tag{21.17}$$

On combining these two terms and using now the $d$-dimensional Palatini's identity with $K=e^{-2\phi}$, we obtain, straightforwardly,

$$\int d^{\hat{d}}\hat{x}\,\sqrt{|\hat{g}|}\,e^{-2\hat{\phi}}[\hat{R}-4(\partial\hat{\phi})^{2}]=\int dz\int d^{\hat{d}-1}x\,\sqrt{|g|}\,e^{-2\phi}\big[R-4(\partial\phi)^{2} + (\partial\ln k)^{2}-\tfrac{1}{4}k^{2}F^{2}(A)\big]. \tag{21.18}$$

[6] Observe that the KK formalism may describe the massive KK modes of the string but not the massive winding modes. The massless modes have zero momentum and winding number except at the self-dual radius, at which there are additional massless modes with non-trivial KK momentum and winding number. The effective action that we are going to write cannot describe the enhancement of symmetry that takes place at the self-dual radius. A direct calculation of the string effective action for that radius is necessary.

The reduction of the KR 2-form is a bit trickier. First, we reduce the field strength by identifying, in tangent-space indices,

$$\hat{H}_{abc} \equiv H_{abc}, \qquad \hat{H}_{abz} = k^{-1} F_{ab}(B), \tag{21.19}$$

where

$$F(B) = 2\partial B, \qquad B_\mu \equiv \hat{B}_{\mu\underline{z}}. \tag{21.20}$$

The above identification of $H$ does not completely determine the reduction of the KR 2-form. It simply gives

$$H_{\mu\nu\rho} = 3(\partial_{[\mu}\hat{B}_{\nu\rho]} - A_{[\mu}F(B)_{\nu\rho]}). \tag{21.21}$$

Now, we could define $B_{\mu\nu} = \hat{B}_{\mu\nu}$, but we are free to implement field redefinitions, for the sake of convenience. Here, it is convenient to define

$$\hat{B}_{\mu\nu} = B_{\mu\nu} - A_{[\mu}B_{\nu]} \;\Rightarrow H = 3[\partial B - \tfrac{1}{2}AF(B) - \tfrac{1}{2}BF(A)], \tag{21.22}$$

so $B_{\mu\nu}$ is invariant under the interchange of the two vector fields $A_\mu$ and $B_\mu$.

The presence of two additional terms in the field strength $H$ (apart from $3\partial B$), generically known as *Chern–Simons terms*, is a new feature that is quite common in higher-dimensional supergravities. $H$ is invariant under the gauge transformations of the vector fields and the 2-form, but the 2-form must also transform under gauge transformations of the vector fields (the so-called *Nicolai–Townsend transformations*):

$$\begin{aligned} \delta_\Lambda A_\mu &= \partial_\mu \Lambda, \qquad \delta_\Sigma B_\mu = \partial_\mu \Sigma, \\ \delta B_{\mu\nu} &= 2\partial_{[\mu}\Sigma_{\nu]} + B_{[\mu}\partial_{\nu]}\Lambda + A_{[\mu}\partial_{\nu]}\Sigma. \end{aligned} \tag{21.23}$$

The origin of the gauge transformation of the KK vector field $A$ is the GCTs of the compact coordinate $\delta z = -\Lambda$, while the gauge transformations of the vector field $B$ and the 2-form are the gauge transformations of the $\hat{d}$-dimensional 2-form Eq. (21.3):

$$\Sigma_\mu = \hat{\Sigma}_\mu, \qquad \Sigma = \hat{\Sigma}_{\underline{z}}. \tag{21.24}$$

After integration over the compact coordinate, the dimensionally reduced effective action takes the form

$$S \sim \int d^d x \sqrt{|g|}\, e^{-2\phi} \Big[ R - 4(\partial\phi)^2 + \frac{1}{2\cdot 3!} H^2 + (\partial \log k)^2 - \tfrac{1}{4}k^2 F^2(A) - \tfrac{1}{4}k^{-2}F^2(B) \Big]. \tag{21.25}$$

We could now rescale the $d$-dimensional fields as we did in Eq. (15.54) and rewrite the action in terms of the scale-invariant fields:

$$\tilde{k} = k/k_0, \qquad \tilde{A}_\mu = k_0 A_\mu, \qquad \tilde{B}_\mu = k_0^{-1} B_\mu. \tag{21.26}$$

The action Eq. (21.25) is manifestly invariant under the transformations

$$A_\mu \to B_\mu, \qquad B_\mu \to A_\mu, \qquad k \to k^{-1}, \tag{21.27}$$

which invert the KK scalar (and therefore the radius of compactification) and interchange the KK vector field that couples to KK modes with the vector field that, as we are going to see, couples to the winding modes. These are, evidently, part of the string T-duality transformations.

There are two ways to understand these transformations: first, we compactify a string background, T-dualize it, and decompactify it into a different background. Another way to think about them is to think of two compactifications of T-dual backgrounds: in one of them we call the KK scalar and vector field $k$ and $A$, and in the other we call them $k^{-1}$ and $B$. These two dual compactifications give the same $d$-dimensional string background. This is a better way to think about T-duality transformations because it is the one that generalizes to type-II effective actions.

In both cases, it is easy to relate the $\hat{d}$-dimensional metric, KR 2-form, and dilaton of the two dual string backgrounds (primed and unprimed):[7]

$$\boxed{\begin{array}{ll} \hat{g}'_{\underline{zz}} = 1/\hat{g}_{\underline{zz}}, & \hat{B}'_{\mu\underline{z}} = \hat{g}_{\mu\underline{z}}/\hat{g}_{\underline{zz}}, \\ \hat{g}'_{\mu\underline{z}} = \hat{B}_{\mu\underline{z}}/\hat{g}_{\underline{zz}}, & \hat{B}'_{\mu\nu} = \hat{B}_{\mu\nu} + 2\hat{g}_{[\mu|\underline{z}|}\hat{B}_{\nu]\underline{z}}/\hat{g}_{\underline{zz}}, \\ \hat{g}'_{\mu\nu} = \hat{g}_{\mu\nu} - (\hat{g}_{\mu\underline{z}}\hat{g}_{\nu\underline{z}} - \hat{B}_{\mu\underline{z}}\hat{B}_{\nu\underline{z}})/\hat{g}_{\underline{zz}}, & \hat{\phi}' = \hat{\phi} - \tfrac{1}{2}\ln|\hat{g}_{\underline{zz}}|. \end{array}} \tag{21.28}$$

These relations are known as *Buscher's rules* [270, 271, 272] and relate two backgrounds with one isometry that are completely equivalent[8] from the string-theory point of view and, in particular, are classical solutions of the string effective action Eq. (21.1). If we set $\ell_z = \ell_s$, we immediately obtain the relations Eqs. (20.61) and (20.62) between the moduli of the two dual theories.

The rules were originally derived using the string $\sigma$-model, as we are going to do in Section 21.2.2 (although at the classical level), but the effective-action method [917, 154, 166, 162] turns out to give the correct rules in a much simpler way. In Section 21.3 we will study some simple examples of string solutions and T dualization using Buscher's rules, although string solutions and their duality relations are the main theme of Part III and we will see many more examples in later chapters.

Buscher's rules refer only to solutions with an isometry.[9] However, from the string point of view, it seems that it should be possible to define T duality whenever strings can be wrapped around non-contractible cycles. However, only a (partial) realization of this more general duality has been achieved, in Ref. [676].

[7] These rules are valid only for the heterotic-string background fields (all in the NSNS sector) at lowest order in $\alpha'$. At higher orders in $\alpha'$ one has to take into account the Yang–Mills fields and also corrections to these rules [163].

[8] If the isometric direction is not compact or corresponds to an isometry with fixed points (a rotation instead of a translation) so that strings cannot wrap around it, the stringy equivalence between the two solutions related by Buscher's rules need not be true. Still, the new configuration solves the string equations of motion and it *is* another string solution [1078].

[9] It is clear, though, that they can be extended to the case of several mutually commuting symmetries (toroidal compactifications). The rules follow from the results of Section 22.5.

To end this discussion on Buscher's T-duality rules, let us make some important remarks.

1. These rules are valid only to lowest order in $\alpha'$.
2. T duality does not commute with gauge transformations (reparametrizations or gauge transformations of the KR 2-form).
3. In the presence of fermions, Buscher's rules have to be formulated in terms of the Vielbein instead of the metric. We have used the Scherk–Schwarz recipe, which employs the Vielbein formalism, to derive the rules, and one could draw the conclusion that our results automatically imply a transformation rule for the Vielbein. However, the rules involve only world tensors and they determine the transformation rules for the Vielbeins up to ($z$-independent) local Lorentz transformations, and only by considering T duality with fermions is the indeterminacy eliminated and one finds just two possible transformation rules for the Vielbein [154]:

$$\boxed{\hat{e}^{\hat{a}\,\prime}{}_{\underline{z}} = \mp\hat{e}^{\hat{a}}{}_{\underline{z}}/\hat{g}_{\underline{zz}}, \qquad \hat{e}^{\hat{a}\,\prime}{}_{\mu} = \hat{e}^{\hat{a}}{}_{\mu} - (\hat{g}_{\mu\underline{z}} \pm \hat{B}_{\mu\underline{z}})\hat{e}^{\hat{a}}{}_{\underline{z}}/\hat{g}_{\underline{zz}}.} \tag{21.29}$$

Both signs lead to the same Buscher rules for world tensors Eqs. (21.28). Now, if we start with the standard gauge choice for the Vielbein Eq. (15.33), the two possible T-dual Vielbeins are given by

$$\left(\hat{e}_{\hat{\mu}}{}^{\hat{a}\,\prime}\right) = \begin{pmatrix} e_{\mu}{}^{a} & \pm k^{-1}B_{\mu} \\ 0 & \pm k^{-1} \end{pmatrix}, \qquad \left(\hat{e}_{\hat{a}}{}^{\hat{\mu}\,\prime}\right) = \begin{pmatrix} e_{a}{}^{\mu} & -B_{a} \\ 0 & \pm k \end{pmatrix}. \tag{21.30}$$

We will see in Section 23.4 that T duality in type-II theories requires the use of the lower ("non-standard") sign for it to work in the fermionic sector.

### 21.2.2 *T duality in the bosonic-string worldsheet action*

We can also gain some insight by studying T duality from the point of view of the two-dimensional $\sigma$-model that describes the motion of a string in a $\hat{d}$-dimensional spacetime with a metric $\hat{g}_{\hat{\mu}\hat{\nu}}$ and a KR 2-form $\hat{B}_{\hat{\mu}\hat{\nu}}$:

$$\hat{S} = -\frac{T}{2}\int d^2\xi\sqrt{|\gamma|}\gamma^{ij}\hat{g}_{\hat{\mu}\hat{\nu}}(\hat{X})\partial_i\hat{X}^{\hat{\mu}}\partial_j\hat{X}^{\hat{\nu}} + \frac{T}{2}\int d^2\xi\, \epsilon^{ij}\hat{B}_{\hat{\mu}\hat{\nu}}(\hat{X})\partial_i\hat{X}^{\hat{\mu}}\partial_j\hat{X}^{\hat{\nu}}. \tag{21.31}$$

We do not include the dilaton term Eq. (21.8) since, in our purely classical approach, it is not going to play any role at all.[10] As in the effective action, we assume that the spacetime fields are independent of $z = x^{\hat{d}-1}$ and, thus, the embedding coordinate

[10] The T-duality transformation rule of the dilaton is a quantum effect. We found it in the string effective action because this action contains information about the quantum theory.

$Z$ appears only through its derivatives. We then decompose the $\hat{d}$-dimensional fields into $(\hat{d}-1)$-dimensional fields using Eqs. (15.28), (21.20), and (21.22) and, on substituting into Eq. (21.31) we obtain

$$\hat{S} = -\frac{T}{2}\int d^2\xi\sqrt{|\gamma|}\,[\gamma^{ij}g_{ij} - k^2F^2] + \frac{T}{2}\int d^2\xi\ \epsilon^{ij}[B_{ij} + A_iB_j - 2F_iB_j], \quad (21.32)$$

where $g_{ij}, B_{ij}, A_i$, and $B_i$ are the pullbacks of the $d$-dimensional metric, KR 2-form, KK vector, and winding vector, and where

$$F_i = \partial_i Z + A_i \quad (21.33)$$

is the field strength of $Z$, which is invariant under the shifts

$$\delta_\Lambda Z = -\Lambda(X), \qquad \delta_\Lambda A_\mu = \partial_\mu \Lambda. \quad (21.34)$$

There is a conserved current associated with this invariance that coincides with the momentum canonically conjugate to the cyclic coordinate $Z$:

$$P_z{}^i = \frac{1}{\sqrt{|\gamma|}}\frac{\delta\hat{S}}{\delta\partial_i Z} = \frac{1}{\sqrt{|\gamma|}}\frac{\delta\hat{S}}{\delta F_i} = T(k^2F^i - \star B^i), \qquad \nabla_i P_z{}^i = 0, \quad (21.35)$$

and there is, as usual, an associated magnetic-like (i.e. topologically) conserved current:

$$W_z{}^i = T\star F^i - \star A^i, \qquad \nabla_i W_z{}^i \sim \epsilon^{ij}\partial_i\partial_j Z = 0. \quad (21.36)$$

The charge associated with this current is the string winding number in the compact dimension: up to normalization

$$\int d\xi^1\sqrt{|\gamma|}W_z^{\underline{0}} \sim \int d\xi^1\partial_{\underline{1}}Z. \quad (21.37)$$

We will see that the charges associated with these currents are the momentum of the string in the compact dimension and the winding number, respectively.

As in the Minkowski-spacetime case, we want to perform a Poincaré-duality transformation of the action (21.32) with respect to the scalar $Z$. The Lagrange-multiplier term that we have to add to enforce the Bianchi identity is now

$$+T\int d^2\xi\ \epsilon^{ij}\partial_i Z'\ (F_j - A_j). \quad (21.38)$$

Now we want to eliminate $F_i$ by using its equation of motion, which is the constraint

$$F_i = k^{-2}\star F'_i, \qquad F'_i \equiv \partial_i Z' + B_i. \quad (21.39)$$

Since $B_\mu$ transforms under $\delta_\Sigma$ in Eq. (21.23), for $\tilde{F}_i$ to be gauge invariant (as the l.h.s. of the above equation is) $Z'$ has to transform simultaneously as follows:

$$\delta_\Sigma Z' = -\Sigma. \quad (21.40)$$

On substituting Eq. (21.39) into the action modified with the Lagrange-multiplier term, we obtain the dual action

$$\hat{S}' = -\frac{T}{2}\int d^2\xi\sqrt{|\gamma|}\left[\gamma^{ij}g_{ij} - k^{-2}F'^2\right] + \frac{T}{2}\int d^2\xi\epsilon^{ij}\left(B_{ij} + B_iA_j - 2F'_iA_j\right), \tag{21.41}$$

which has exactly the same form as the original action (21.32) with the replacements Eq. (21.27), which imply the Buscher T-duality rules Eqs. (21.28) for all fields except for the dilaton, as we have explained.

The dual action has conserved currents $P^i_{z'}$, and $W^i_{z'}$, which are related to the conserved currents of the original theory by

$$P^i_{z'} = W^i_{z}, \qquad W^i_{z'} = P^i_{z}, \tag{21.42}$$

also as expected.

To gain more insight into these transformations, we are going to find the worldline actions of the string momentum modes and winding modes and see that they have the expected dependences of the masses on the compactification radius and are interchanged by T duality, as expected. The relation between these two worldline actions is similar to the relation between T-dual D-brane worldvolume actions, as we will see in Section 21.2.3.

*Winding and momentum modes.* We want to find the action of winding modes as seen from the $(\hat{d}-1)$-dimensional point of view. We will perform a *double dimensional reduction* of the spacetime fields (as we did to obtain Eq. (21.32)) and also of the worldvolume fields, since we assume that the worldsheet coordinate $\xi^1$ is compact and that we can use the KK formalism also in the worldsheet. The result will describe a particle moving in $\hat{d}-1$ dimensions and coupled to the $(\hat{d}-1)$-dimensional fields in a specific way. Its mass will identify it as a bosonic-string winding mode.

Our starting point is thus Eq. (21.32). The next step consists in using part of the gauge freedom to set

$$Z = w\ell_z\xi^1/\ell, \tag{21.43}$$

where we take $\xi^1 \in [0, 2\pi\ell]$ and $Z \in [0, 2\pi\ell_z]$. This configuration has winding number $w$ and, indeed, on computing the conserved charge associated with $W^i_z$, we obtain a number proportional to $w$. Note that $\ell$ and $\ell_z$ can be changed at will by worldsheet and spacetime reparametrizations, and the final physical results will not depend on either of them. With this normalization, the asymptotic value of the KK scalar is $k_0 = R_z/\ell_z$. All the other worldsheet fields are taken to be independent of $\xi^1$. We split the worldsheet metric as follows:

$$\begin{aligned}
\hat{\gamma}_{\tau\tau} &= l^2(\gamma - a^2), & \hat{\gamma}^{\tau\tau} &= l^{-2}\gamma^{-1},\\
\hat{\gamma}_{\tau\sigma} &= -l^2 a_i, & \hat{\gamma}^{\tau\sigma} &= -l^{-2}\gamma^{-1}a,\\
\hat{\gamma}_{\sigma\sigma} &= -l^2, & \hat{\gamma}^{\sigma\sigma} &= -l^{-2}(1-\gamma^{-1}a^2),
\end{aligned} \tag{21.44}$$

where $\gamma$ is going to be the worldline metric after reduction.

Using this ansatz in the action Eq. (21.32) and integrating $\xi^1$, we obtain

$$S = -\pi\ell T \int d\xi^0 \left\{ \gamma^{\frac{-1}{2}} \left[ (g_{\mu\nu} - k^2 A_\mu A_\nu)\dot{X}^\mu \dot{X}^\nu + 2\frac{w\ell_z}{\ell} a k^2 A_\mu \dot{X}^\mu \right] + \left(\frac{w\ell_z}{\ell}\right)^2 \gamma^{\frac{1}{2}} (1 - \gamma^{-1} a^2) k^2 \right\} + 2\pi\ell_z w T \int d\xi^0 B_\mu \dot{X}^\mu, \tag{21.45}$$

and, on solving the equation for the metric component $a$ and substituting back into the action, we obtain the worldline action of a point-particle charged with respect to the winding vector:

$$S = -\pi\ell T \int d\xi^0 \left[ \gamma^{\frac{-1}{2}} g_{\mu\nu} \dot{X}^\mu \dot{X}^\nu + \left(\frac{w\ell_z}{\ell}\right)^2 \gamma^{\frac{1}{2}} k^2 \right] + 2\pi\ell_z w T \int d\xi^0 B_\mu \dot{X}^\mu. \tag{21.46}$$

To read off the mass, we eliminate the worldline metric $\gamma$ using its equation of motion. The final result is the worldline action for a winding mode with winding number $w$:

$$\boxed{S = -2\pi|w|\ell_z k_0 T \int d\xi^0 (k/k_0) \sqrt{|g_{\mu\nu}\dot{X}^\mu \dot{X}^\nu|} + 2\pi\ell_z w T \int d\xi^0 B_\mu \dot{X}^\mu.} \tag{21.47}$$

The mass is given by

$$M = 2\pi|w|\ell_z k_0 T = 2\pi|w|\ell_z \frac{R_z}{\ell_z}\frac{1}{2\pi\alpha'} = |w|\frac{R_z}{\alpha'}, \tag{21.48}$$

and it is equal (up to the sign) to the charge with respect to the scale-invariant winding vector $\tilde{B}_\mu$. This action thus describes a winding state $w, n = 0, N = \tilde{N} = 1$ in Eq. (20.60).

Let us now find the action for a string momentum mode. Our starting point is again Eq. (21.32). We want now to eliminate $Z$ using the conservation of the current $P_z^i$. The situation is similar to the one we found in Section 15.2.3 when we reduced the action of a massless particle moving in a space with an isometry in the direction $Z$ using the conservation of momentum in that direction. The first step is then to replace $\partial_i Z$ by $P_z^i$ in the action by performing a Legendre transformation of the Lagrangian $\hat{L}$ in Eq. (21.32), $\hat{S} = \int d^2\xi \sqrt{|\gamma|}\hat{L}$, with respect to $Z$ and only then use the equation of motion.

Therefore, we take the transformed action

$$\tilde{\hat{S}}[X^\mu, P_z^i] = \int d^2\xi \sqrt{|\gamma|}\left(-P_z{}^i \partial_i Z + L\right), \tag{21.49}$$

and eliminate $\partial_i X$ (and $F_i$) completely by using Eq. (21.35), obtaining

$$\hat{S}'[X^\mu, P_z^i] = -\frac{T}{2}\int d^2\xi \sqrt{|\gamma|}\gamma^{ij}\left[ g_{ij} + k^{-2}T^{-2}\mathcal{F}_i\mathcal{F}_j - \frac{2}{T}A_i\mathcal{F}_j \right] + \frac{T}{2}\int d^2\xi\, \epsilon^{ij}[B_{ij} - A_i B_j], \tag{21.50}$$

where

$$\mathcal{F}_i \equiv [P_{z\,i} + T \star B_i]. \tag{21.51}$$

Now we can use consistently the equations of motion and replace $P_z^i$ in Eq. (21.50) by

$$P_z{}^i = \frac{1}{\sqrt{|\gamma|}} \delta^{i\underline{0}} C, \tag{21.52}$$

which is automatically conserved. The constant $C$ is fixed by noting that the conserved charge is the momentum in the direction $z$ and is quantized in units $n/\ell_z$:

$$\int d\xi^1 \sqrt{|\gamma|} P_z^{\underline{0}} = \frac{n}{\ell_z} \quad \Rightarrow C = \frac{n}{2\pi\ell\ell_z}. \tag{21.53}$$

After elimination of all the components of the worldsheet metric, we arrive immediately at the worldline action of a KK mode with momentum number $n$, Eq. (15.83), which corresponds to the states $n, w = 0, N = \tilde{N} = 1$ in Eq. (20.60). These momentum modes are related to the winding modes whose worldline action we found before by T duality and their worldline actions are related by Buscher's T duality rules Eqs. (21.28).

### *21.2.3 T duality in the bosonic* D$p$*-brane effective action*

We have found in Section 20.3.2 how D$p$-branes transform under T duality into D$(p+1)$- or D$(p-1)$-branes depending on whether the compact coordinate is transverse or parallel to the D$p$-brane worldvolume. These relations between D$p$-branes in spacetimes with compact directions are also realized in their worldvolume effective actions Eq. (21.14), as shown in Refs. [36, 183]. Not only do the spacetime fields transform following Buscher's rules, but also the relation between components of the BI vector and embedding coordinates is realized in them. In the type-II superstring case there are WZ terms that describe the coupling to RR forms and we need the type-II Buscher rules for them. We will study them in Chapter 23 and the T-duality relations between the worldvolume actions of super-D $p$-branes in Chapter 25.

The T duality between the effective action of a D$\hat{p}$-brane ($\hat{p} = p + 1$) wrapped on a compact spacetime dimension and the effective action of a D$p$-brane can be established by performing the double dimensional reduction of the former and the direct dimensional reduction of the latter and showing that the resulting actions are identical. We follow Ref. [183].

*Double dimensional reduction of the* D$\hat{p}$*-brane effective action.* We use hats both for spacetime and for worldvolume objects, since each of them will be reduced in one dimension, parametrized, respectively, by $\hat{x}^d \equiv z \in [0, 2\pi\ell_z]$ and $\hat{\xi}^{\hat{p}} \equiv \zeta \in [0, 2\pi\ell]$. We proceed in two steps. First, we rewrite the $\hat{d}$-dimensional spacetime fields in terms of the $d$-dimensional ones, as in Section 21.2.2, giving

$$\hat{H}_{\hat{\imath}\hat{\jmath}} \equiv \hat{g}_{\hat{\imath}\hat{\jmath}} + \hat{B}_{\hat{\imath}\hat{\jmath}} + 2\pi\alpha' \hat{F}_{\hat{\imath}\hat{\jmath}} = g_{\hat{\imath}\hat{\jmath}} - k^2 F_{\hat{\imath}} F_{\hat{\jmath}} + B_{\hat{\imath}\hat{\jmath}} + A_{[\hat{\imath}} B_{\hat{\jmath}]} - 2F_{[\hat{\imath}} B_{\hat{\jmath}]} + 2\pi\alpha' \hat{F}_{\hat{\imath}\hat{\jmath}}, \tag{21.54}$$

where the pullbacks of $\hat{d}$-dimensional fields over the $(\hat{p}+1)$-dimensional worldvolume are computed with $\partial_{\hat{i}}\hat{X}^{\hat{\mu}}$, the pullbacks of $d$-dimensional fields with $\partial_{\hat{i}}X^{\mu}$, and where

$$F_{\hat{i}} = \partial_{\hat{i}} Z + A_{\hat{i}}. \tag{21.55}$$

Next, we make the gauge choice $Z = c\zeta$, with the remaining embedding coordinates independent of $\zeta$; $c$ is the constant $\ell_z/\ell$. The $(\hat{p}+1)\times(\hat{p}+1)$ matrix $\hat{H}_{\hat{i}\hat{j}}$ defined in Eq. (21.54) splits as follows:

$$\left(\hat{H}_{\hat{i}\hat{j}}\right) = \begin{pmatrix} M_{ij} & -U_i^- \\ -U_j^+ & -c^2k^2 \end{pmatrix} = \begin{pmatrix} M_{ij} & 0 \\ 0 & -c^2k^2 \end{pmatrix} \begin{pmatrix} \mathbb{I} & -(M^{-1})_{ik}U_k^- \\ c^{-2}k^{-2}U_j^+ & 1 \end{pmatrix}, \tag{21.56}$$

where

$$M_{ij} \equiv H_{ij} - k^2 A_i A_j - A_{[i}B_{j]}, \qquad U_i^{\pm} \equiv c(k^{-1}\mathcal{F}_i \pm kA_i), \qquad \mathcal{F}_i \equiv \partial_i \hat{V}_{\underline{\zeta}} + B_i. \tag{21.57}$$

The determinant of $\hat{H}_{\hat{i}\hat{j}}$ is the product of the determinants of the two matrices,

$$\det(\hat{H}_{\hat{i}\hat{j}}) = -c^2k^2 \det(M_{ij})\,[1 - U_i^+(M^{-1})_{ik}U_k^-], \tag{21.58}$$

and, on substituting this into the D$\hat{p}$-brane effective action and integrating over $\zeta$, we obtain the effective action of a D$p$-brane moving in $d$ spacetime dimensions:

$$S = -T_{\mathrm{D}\hat{p}} 2\pi\ell|c|k_0 \int d^{p+1}\xi e^{-(\phi-\phi_0)}(k/k_0)^{\frac{1}{2}}[1 - U_i^+(M^{-1})_{ik}U_k^-]^{\frac{1}{2}}\sqrt{|M|}. \tag{21.59}$$

*Direct dimensional reduction of the* D$p$*-brane effective action.* Now we use hats and primes (indicating that we are in the T-dual situation) for the spacetime fields which we are going to reduce in the direction parametrized by $z'$, and primes but no hats for the $(p+1)$-dimensional worldvolume fields. We split the spacetime fields as in Eq. (21.54),

$$\hat{H}'_{ij} \equiv \hat{g}'_{ij} + \hat{B}'_{ij} + 2\pi\alpha' F'_{ij} = M'_{ij} - U_i^{-\prime}U_j^{+\prime}, \tag{21.60}$$

where

$$M'_{ij} \equiv H_{ij} - k'^{-2}B'_iB'_j - B'_{[i}A'_{j]}, \qquad U_i^{\pm\prime} \equiv (k'\mathcal{F}'_i \pm k'^{-1}B'_i), \qquad \mathcal{F}'_i \equiv \partial_i Z' + A'_i. \tag{21.61}$$

$M'_{ij}, U_i^{\pm\prime}$, and $\mathcal{F}^i_i$ are exactly the Buscher T duals of the unprimed ones plus the relation

$$Z' = \hat{V}_{\underline{\zeta}}. \tag{21.62}$$

Now, it only remains to see that the action we obtain after this reduction is equivalent to Eq. (21.59). First, we rewrite $\hat{H}'_{ij}$ as the product of two matrices,

$$\hat{H}'_{ij} = M'_{ik}[\delta_{kj} - M'^{-1}_{ik}U_k^{-\prime}U_j^{+\prime}], \tag{21.63}$$

the second of which has $p$ times the eigenvalue $+1$ (for each of the $p$ vectors orthogonal to $U^{+\prime}$) and one time the eigenvalue $1 - U_i^{+\prime} M'^{-1}_{ij} U_j^{-\prime}$ for the eigenvector $M'^{-1}U^{-\prime}$, and

$$\det(\hat{H}'_{ij}) = \det(M'_{ij})\,[1 - U_i^{+\prime}(M'^{-1})_{ik}U_k^{-\prime}]. \tag{21.64}$$

Writing $e^{-(\hat{\phi}'-\hat{\phi}'_0)} = e^{-(\phi'-\phi'_0)}(k'/k'_0)^{-\frac{1}{2}}$, the action takes the form

$$S = -T'_{\mathrm{D}p}\int d^{p+1}\xi e^{-(\phi-\phi_0)}(k'/k'_0)^{-\frac{1}{2}}[1 - U_i^{+\prime}(M'^{-1})_{ik}U_k^{-\prime}]^{\frac{1}{2}}\sqrt{|M'|}. \tag{21.65}$$

This action is identical to Eq. (21.59) after application of Buscher's rules if we identify

$$T'_{\mathrm{D}p} = T_{\mathrm{D}(p+1)}2\pi\ell|c|k_0^{\frac{1}{2}} = 2\pi R_z T_{p+1}. \tag{21.66}$$

The D$p$-brane tension $T_p$ should be essentially independent of the compactification radius (since it takes the same value in uncompactified spacetimes). It can depend solely on $\ell_{\mathrm{s}}$ and $g$. Since the string coupling constants of T-dual theories are related by Eqs. (20.61) and (20.62), and since it has units of mass by $p$-volume, for all $p$

$$T_{\mathrm{D}p} \sim \frac{1}{\ell_{\mathrm{s}}^{p+1}g}. \tag{21.67}$$

Later, we will use other methods to find the same result and the proportionality constant. The $g^{-1}$ dependence of the tension of these objects gives them unique status, intermediate between standard solitons, whose mass is proportional to $g^{-2}$, and the *fundamental* objects that appear in the perturbative spectrum, with masses (tensions) independent of $g$, such as the *fundamental string* whose action is the string $\sigma$-model action and whose tension is $T = 1/(2\pi\ell_{\mathrm{s}}^2)$.

## 21.3 Example: the fundamental string (F1)

Many string solutions (i.e. solutions of the string effective action Eq. (21.1)) are known. For instance, all the vacuum Einstein solutions are string solutions with constant dilaton and pure gauge KR 2-form and, for each of them that admits an isometry, we can find a T-dual string solution (possibly with non-trivial dilaton and KR 2-form). To investigate T duality, however, we must choose a convenient solution to which we can give a physical and stringy interpretation, in the same spirit as that in which we chose a gravitational wave in Section 15.2.3 to illustrate KK reduction, four-dimensional electric–magnetic duality, and KK oxidation. The so-called *fundamental-string* solution [410, 409] (see also Ref. [471]) denoted by F1 represents a string at rest and can play this role.

We can understand the F1 solution as a solution of the ("bulk-plus-brane") action that results from the addition of the string effective action Eq. (21.1) to the string $\sigma$-model action Eq. (21.31) which, denoted from now on by $S_{\mathrm{F1}}$, acts as a singular one-dimensional source for the former. The equations of motion of the spacetime fields[11] are Eqs. (21.11) plus the

[11] We add hats to all $\hat{d}$-dimensional fields.

source terms

$$\begin{aligned}
\frac{16\pi G_N^{(\hat{d})} e^{2(\hat{\phi}-\hat{\phi}_0)}}{\sqrt{|\hat{g}|}} \frac{\delta \hat{S}_{F1}}{\delta \hat{g}^{\hat{\mu}\hat{\nu}}} &= +\frac{8\pi G_N^{(\hat{d})} e^{2(\hat{\phi}-\hat{\phi}_0)} T}{\sqrt{|\hat{g}|}} \int d^2\xi \sqrt{|\gamma|}\gamma^{ij} \hat{g}_{i\hat{\mu}} \hat{g}_{j\hat{\nu}} \delta^{(\hat{d})}(\hat{x}-\hat{X}), \\
\frac{16\pi G_N^{(\hat{d})} e^{2(\hat{\phi}-\hat{\phi}_0)}}{\sqrt{|\hat{g}|}} \frac{\delta \hat{S}_{F1}}{\delta \hat{\phi}} &= 0, \\
\frac{16\pi G_N^{(\hat{d})} e^{2(\hat{\phi}-\hat{\phi}_0)}}{\sqrt{|\hat{g}|}} \frac{\delta \hat{S}_{F1}}{\delta \hat{B}_{\hat{\mu}\hat{\nu}}} &= +\frac{8\pi G_N^{(\hat{d})} e^{2(\hat{\phi}-\hat{\phi}_0)} T}{\sqrt{|\hat{g}|}} \int d^2\xi \epsilon^{ij} \partial_i \hat{X}^{\hat{\mu}} \partial_j \hat{X}^{\hat{\nu}} \delta^{(\hat{d})}(\hat{x}-\hat{X}).
\end{aligned} \tag{21.68}$$

We also have to solve the equations of motion of the worldvolume fields:

$$\begin{aligned}
-\frac{2}{T\sqrt{|\gamma|}} \frac{\delta \hat{S}_{F1}}{\delta \gamma^{ij}} &= \hat{g}_{ij} - \tfrac{1}{2}\gamma_{ij} \hat{g}^k{}_k = 0, \\
\frac{1}{T\sqrt{|\gamma|}} \frac{\delta \hat{S}_{F1}}{\delta \hat{X}^{\hat{\mu}}} &= \hat{g}_{\hat{\mu}\hat{\nu}} \left[\nabla^2 \hat{X}^{\hat{\nu}} + \gamma^{ij} \hat{\Gamma}_{ij}{}^{\hat{\nu}}\right] + \frac{\epsilon^{ij}}{2\sqrt{|\gamma|}} \hat{H}_{\hat{\mu}ij} = 0.
\end{aligned} \tag{21.69}$$

We work in the *static gauge*, identifying the worldvolume coordinates with the first spacetime coordinates $\hat{X}^i = \xi^i \equiv (T, Y)$, $i = 0, 1$. The remaining spacetime coordinates are transverse to the string worldvolume and we make for them the ansatz $X^m(\xi) = 0$, $m = 1, \ldots, \hat{d} - 2$. If the solution is to describe a fundamental string at rest, it is natural to make an ansatz for the metric with Poincaré symmetry in the worldvolume directions. On the other hand, our experience with ERN BHs tells us that a full solution of the equations with sources can be expected only when there is supersymmetry/extremality and the solution depends solely on a reduced number of functions that are harmonic in transverse space, whose singularities are associated with the sources. In this case, the solution should depend on just one function, $H_{F1}(x^m)$.

A solution of all the equations that satisfy the above criteria and ansatz is the *fundamental-string solution* given, for $\hat{d} \geq 5$, by

$$\boxed{\begin{aligned}
d\hat{s}^2 &= H_{F1}^{-1}[dt^2 - dy^2] - d\vec{x}^2_{(\hat{d}-2)}, \\
\hat{B}_{t\underline{y}} &= -(H_{F1}^{-1} - 1), \\
e^{-2(\hat{\phi}-\hat{\phi}_0)} &= H_{F1}, \\
H_{F1} &= \varepsilon + \frac{h_{F1}}{|\vec{x}_{(\hat{d}-2)}|^{\hat{d}-4}}, \qquad h_{F1} = \frac{16\pi G_N^{(\hat{d})} T}{(\hat{d}-4)\omega_{(\hat{d}-3)}}.
\end{aligned}} \tag{21.70}$$

The integration constant $h_{F1}$ is completely determined by the source and the value of the Newton constant, and $\varepsilon$ has to be taken equal to 1 in order to have asymptotic flatness.

It is reasonable to assume that this solution describes a string lying at rest in the direction of the isometric coordinate $y$. Let us now take $y$ to be compact, rescaling it by $y \to k_0 y$, where $k_0 = R_y/\ell_s$, $R_y$ being the compactification radius as usual and $y \in [0, 2\pi\ell_s]$. The dimensionally reduced solution has the non-vanishing fields

$$\boxed{\begin{aligned} ds^2 &= H_{\rm F1}^{-1} dt^2 - d\vec{x}^{\,2}_{(d-1)}, \\ B_t &= -k_0 \big(H_{\rm F1}^{-1} - 1\big), \\ e^{-2(\phi-\phi_0)} &= H_{\rm F1}^{\frac{1}{2}}, \qquad k = k_0 H_{\rm F1}^{-\frac{1}{2}}, \end{aligned}} \tag{21.71}$$

with the same $H_{\rm F1}$. This is the metric of a point-like object with mass and charged with respect to the winding vector $B_\mu$, as corresponds to a string winding mode.[12] The T dual is charged with respect to the KK vector $A_\mu$, and the $\hat{d}$-dimensional T dual is the purely gravitational solution

$$\boxed{\begin{aligned} &d\hat{s}'^{\,2} = H_{\rm F1}^{-1} dt^2 - k_0^{-2}[dy' + k_0(H_{\rm F1}^{-1} - 1)dt] - d\vec{x}^{\,2}_{(\hat{d}-2)}, \\ &e^{-2\hat{\phi}'} = e^{-2\hat{\phi}'} = e^{-2\hat{\phi}_0} k_0^2, \end{aligned}} \tag{21.72}$$

which is the zero mode of the shock-wave solution Eq. (14.41) that can be shown to represent not just a point-particle, but also a fundamental string moving in the compact coordinate $y'$, as expected.

[12] To compare this with the masses predicted, we need a string value of $G_{\rm N}^{(\hat{d})}$.

# 22
# From eleven to four dimensions

In Chapter 21 we started our study of string dualities in the effective action by treating the simplest case: T duality in the string common sector. Now we are ready to handle more complicated cases: type-II T duality, type-IIB S duality, heterotic/type-I string duality, and the strong-coupling limit of the type-IIA superstring. Actually, only the first of these dualities (the only one which is perturbative) was known from the worldsheet point of view; the rest were conjectured after they had been observed in the corresponding effective actions and were interpreted in string language.

For instance, it was well known that the $N = 2A, d = 10$ supergravity theory can be obtained from $N = 1, d = 11$ supergravity [386] by dimensional reduction (i.e. compactifying on a circle and ignoring all the massive Kaluza–Klein modes). This reduction was first performed in the Einstein frame. The reduction in the string frame [162, 1275] gave new and useful information. To go to the string frame, it is necessary to identify the dilaton, which turns out to be essentially the moduli field that measures the radius of the circle in the 11th dimension, namely the KK scalar $\hat{\hat{g}}_{\underline{xx}}$. Since the dilaton is the string coupling constant, the strong-coupling limit of the type-IIA string theory corresponds to the limit of decompactification (large radius) of the 11th dimension. The surprising fact is that this statement is true, including the massive Kaluza–Klein modes and string modes, if one also includes the *solitonic modes* which appear in the non-perturbative spectrum of the string theory. These non-perturbative states at strong coupling should be identified with the ordinary Kaluza–Klein modes of 11-dimensional supergravity.

This relation between field theories is in agreement with the relation between the worldvolume action of the M2-brane wrapped on a compact dimension and that of the GS type-IIA superstring theory [469, 14], which give, in a certain sense, the 11-dimensional and $N = 2A, d = 10$ supergravity theories.

The other dualities form a chain that relates all string theories (at least under certain circumstances) to 11-dimensional supergravity. There has been conjectured the existence of the so-called *M theory* whose low-energy limit would be described by 11-dimensional supergravity which, in different limits, would give all the ten-dimensional string theories which would be different, dual, manifestations of the same unique theory.[1]

[1] A 12-dimensional origin for M theory and type-IIB superstring theory (F theory) has also been suggested.

Our interest is mainly in four-dimensional string effective field theories and classical solutions and their connection to higher-dimensional theories and solutions. It is then natural to start by introducing 11-dimensional supergravity and then performing the reduction on a circle to find the type-IIA superstring effective action. We will do this in Section 22.1. Since we are interested in classical solutions, we will study only the bosonic sectors of these theories. However, we will also need the supersymmetry transformation rules for the fermions in order to study their unbroken supersymmetries.

To study T duality between the effective actions of the type-IIA and -IIB theories, following the philosophy of Section 21.2, we will have to perform dimensional reduction of both theories to nine dimensions. The reduction of the IIA theory will be done in Section 22.3, whereas the reduction of the IIB theory will be postponed to Chapter 23, in which we will find the type-II Buscher rules.

The $E_8 \times E_8$ heterotic-string theory can be obtained by compactification of M theory on a segment (the simplest orbifold), each $E_8$ factor group living on one of the ten-dimensional boundaries. From the point of view of effective actions, we can easily obtain the heterotic-string effective action (without the gauge fields) by compactifying 11-dimensional supergravity on an orbifold, which amounts to the $S^1$ compactification which we carry out in Section 22.1 followed by a truncation that we study in Section 22.4. A similar truncation of the type-IIB theory that gives the type-I theory (again without the gauge fields) will be studied in Chapter 23 and corresponds to the O9 plus 32 D9 construction of the type-I SO(32) theory.

Further compactification increases the number of dualities: on the one hand, one can perform T dualities in more directions that can also be rotated into each other. Also, in even dimensions, new dualities that involve the Hodge-dualization of differential-form potentials appear: in $d = 4$, vectors can be dualized, in $d = 6$ 2-forms, and in $d = 8$ 3-forms. Furthermore, in odd dimensions, Hodge-dualization of differential forms can increase the number of vector fields that can be rotated into other vector fields, enhancing the duality group. Usually these dualities are manifest only in the Einstein frame and were known as *hidden symmetries* of supergravity theories. In Section 22.5 we are going to study as an example the toroidal compactification down to four dimensions of $N = 1, d = 10$ supergravity, the effective theory of the heterotic string, and we will find that, generically, the classical duality group[2] is $O(n, n + 16)$ for compactification on an $n$-torus, all of it due to T duality, but, in $d = 4$, vectors can be dualized into vectors and the symmetry is increased by the S-duality group $SL(2, \mathbb{R})$. (The duality groups that appear in toroidal compactifications of $N = 2, d = 10$ theories are given in Table 22.1.)

Finally, we are going to study the preservation of unbroken supersymmetry under duality transformations in Section 22.6.

A scheme of the dimensional reductions and truncations that we are going to study in this chapter and the next is given in Fig. 22.1. It is not a diagram of the web of dualities that relates string theories, although it is closely related to it. A general reference for this and

[2] Quantum effects such as charge quantization break the classical supergravity duality groups to discrete subgroups, typically the ones obtained by restricting the matrix entries to taking integer values [808]. On the other hand, if G is the classical duality group, the scalars parametrize a coset space G/H, where H is the maximal compact subgroup of G. For the heterotic-string case $H = O(n) \times O(n + 16)$.

Table 22.1. Hidden symmetries of toroidally compactified $N=2, d=10$ supergravities [382, 837]

$E_{3(+3)} = SL(3,\mathbb{R}) \times SL(2,\mathbb{R})$, $E_{4(+4)} = SL(5,\mathbb{R})$, and $E_{5(+5)} = SO(5,5)$. The numbers with tildes indicate that the corresponding fields are dualized by some of the duality transformations which will merely be symmetries of the equations of motion. The discrete subgroups are the U-duality groups.

| $d$ | G | H | $e^a{}_\mu$ | $C_{\mu\nu\rho}$ | $B_{\mu\nu}$ | $A_\mu$ | $\varphi$ | $\psi_\mu$ | $\lambda$ |
|---|---|---|---|---|---|---|---|---|---|
| 9 | $GL(2,\mathbb{R})$ | SO(2) | 1 | 1 | 2 | 3 | 3 | 2 | 4 |
| 8 | $E_{3(+3)}$ | SO(3) × SO(2) | 1 | $\tilde{1}$ | 3 | 6 | 7 | 2 | 6 |
| 7 | $E_{4(+4)}$ | SO(5) | 1 | 0 | 5 | 10 | 14 | 4 | 16 |
| 6 | $E_{5(+5)}$ | SO(5) × SO(5) | 1 | 0 | $\tilde{5}$ | 16 | 25 | 4 | 20 |
| 5 | $E_{6(+6)}$ | USp(8) | 1 | 0 | 0 | 27 | 42 | 8 | 48 |
| 4 | $E_{7(+7)}$ | SU(8) | 1 | 0 | 0 | $\tilde{28}$ | 70 | 8 | 56 |
| 3 | $E_{8(+8)}$ | SO(16) | 1 | 0 | 0 | 0 | 128 | 16 | 128 |

the following chapter is Ref. [1097], where much information and most original references to supergravity theories can be found.

## 22.1 Dimensional reduction from $d=11$ to $d=10$

Here we are going to obtain the bosonic sector of $N=2A, d=10$ (also known as type-IIA) supergravity and the supersymmetry transformation rules by straightforward dimensional reduction of $N=1, d=11$ supergravity using the techniques developed in Ref. [1106]. This dimensional reduction has been performed in Ref. [810], but now we will obtain the type-IIA theory directly in the string frame.

We are going to describe the procedure used to perform the dimensional reduction in some detail since our goal is to relate the various supergravity theories in different dimensions. Throughout this and the following few sections we will use double hats for 11-dimensional objects, single hats for ten-dimensional objects, and no hats for nine-dimensional objects. We first introduce the theory of 11-dimensional supergravity.

### 22.1.1 Eleven-dimensional supergravity

The fields of $N=1, d=11$ supergravity [386] are the Elfbein, a 3-form potential, and a 32-component Majorana gravitino,[3]

$$\left\{\hat{\hat{e}}^{\hat{\hat{a}}}{}_{\hat{\hat{\mu}}}, \hat{\hat{C}}_{\hat{\hat{\mu}}\hat{\hat{\nu}}\hat{\hat{\rho}}}, \hat{\hat{\psi}}_{\hat{\hat{\mu}}}\right\}. \tag{22.1}$$

[3] Our conventions for 11-dimensional gamma matrices and spinors are given in Appendix D.1.3 and are essentially identical to those of the original reference [386] except for the relation between $\hat{\hat{\Gamma}}^{10}$, the totally antisymmetric tensor, and the remaining ten gamma matrices, which is not explicitly given in that reference. In our case, that relation is given in Eq. (D.81), and in the supersymmetry transformation rules that follow the sign of the topological term in the action is the opposite to that in Ref. [386]. Observe also that our spin connection and contorsion have opposite signs, and that we have set the constant $K=\frac{1}{2}$.

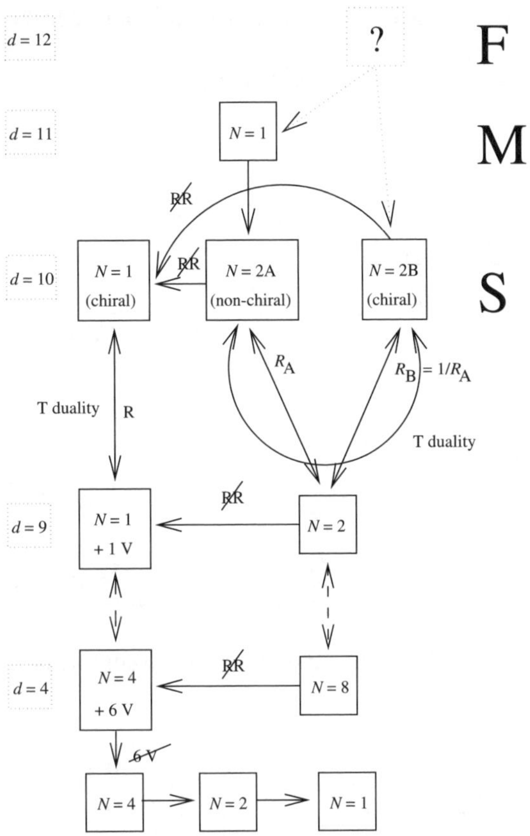


Fig. 22.1. A directory of supergravities. The relations between various supergravity theories upon compactification on circles (lines with two arrowheads) and truncation (lines with a single arrowhead) are schematically represented. The letters F, M, and S indicate that the corresponding supergravity theory is the low-energy limit of F theory, M theory, or a superstring theory.

The action for these fields is

$$\begin{aligned}\hat{\hat{S}} = \frac{1}{16\pi G_{\rm N}^{(11)}} \int d^{11}\hat{\hat{x}}\, \sqrt{|\hat{\hat{g}}|} \Bigg[ \hat{\hat{R}}(\hat{\hat{\omega}}) - \frac{i}{2} \bar{\hat{\hat{\psi}}}_{\hat{\hat{\mu}}} \hat{\hat{\Gamma}}^{\hat{\hat{\mu}}\hat{\hat{\nu}}\hat{\hat{\rho}}} \nabla_{\hat{\hat{\nu}}} \left( \frac{\hat{\hat{\omega}} + \tilde{\hat{\hat{\omega}}}}{2} \right) \hat{\hat{\psi}}_{\hat{\hat{\rho}}} - \frac{1}{2 \cdot 4!} \hat{\hat{G}}^2 \\ + \frac{1}{192} \bar{\hat{\hat{\psi}}}_{\hat{\hat{\mu}}} \hat{\hat{\Gamma}}^{[\hat{\hat{\mu}}} \hat{\hat{\Gamma}}_{\hat{\hat{\alpha}}\hat{\hat{\beta}}\hat{\hat{\gamma}}\hat{\hat{\delta}}} \hat{\hat{\Gamma}}^{\hat{\hat{\nu}}]} \hat{\hat{\psi}}_{\hat{\hat{\nu}}} \left( \hat{\hat{G}}^{\hat{\hat{\alpha}}\hat{\hat{\beta}}\hat{\hat{\gamma}}\hat{\hat{\delta}}} + \tilde{\hat{\hat{G}}}^{\hat{\hat{\alpha}}\hat{\hat{\beta}}\hat{\hat{\gamma}}\hat{\hat{\delta}}} \right) - \frac{1}{(144)^2} \frac{1}{\sqrt{|\hat{\hat{g}}|}} \hat{\hat{\epsilon}} \hat{\hat{G}} \hat{\hat{G}} \hat{\hat{C}} \Bigg].\end{aligned} \tag{22.2}$$

Let us now describe each object in this action:

$$\hat{\hat{G}} = 4\partial \hat{\hat{C}} \tag{22.3}$$

is the field strength of the 3-form and is obviously invariant under the gauge transformations

$$\delta_{\hat{\hat{\chi}}}\hat{\hat{C}} = 3\partial\hat{\hat{\chi}}, \tag{22.4}$$

where $\hat{\hat{\chi}}$ is a 2-form;

$$\tilde{\hat{\hat{G}}}_{\hat{\mu}\hat{\nu}\hat{\rho}\hat{\sigma}} = \hat{\hat{G}}_{\hat{\mu}\hat{\nu}\hat{\rho}\hat{\sigma}} - \tfrac{3}{2}\bar{\hat{\hat{\psi}}}_{[\hat{\mu}}\hat{\hat{\Gamma}}_{\hat{\nu}\hat{\rho}}\hat{\hat{\psi}}_{\hat{\sigma}]} \tag{22.5}$$

is the supercovariant field strength; and

$$\nabla_{\hat{\mu}}\left(\hat{\hat{\omega}}\right)\hat{\hat{\psi}}_{\hat{\nu}} = \partial_{\hat{\mu}}\hat{\hat{\psi}}_{\hat{\nu}} - \tfrac{1}{4}\hat{\hat{\omega}}_{\hat{\mu}}{}^{\hat{\hat{a}}\hat{\hat{b}}}\hat{\hat{\Gamma}}_{\hat{\hat{a}}\hat{\hat{b}}}\hat{\hat{\psi}}_{\hat{\nu}} \tag{22.6}$$

is the covariant derivative with

$$\begin{aligned}
\tilde{\hat{\hat{\omega}}}_{\hat{\mu}}{}^{\hat{\hat{a}}\hat{\hat{b}}} &= \hat{\hat{\omega}}_{\hat{\mu}}{}^{\hat{\hat{a}}\hat{\hat{b}}} - \frac{i}{16}\bar{\hat{\hat{\psi}}}_{\hat{\alpha}}\hat{\hat{\Gamma}}_{\hat{\mu}}{}^{\hat{\hat{a}}\hat{\hat{b}}\hat{\alpha}\hat{\beta}}\hat{\hat{\psi}}_{\hat{\beta}},\\
\hat{\hat{\omega}}_{\hat{\mu}}{}^{\hat{\hat{a}}\hat{\hat{b}}} &= \hat{\hat{\omega}}_{\hat{\mu}}{}^{\hat{\hat{a}}\hat{\hat{b}}}\left(\hat{\hat{e}}\right) + \hat{\hat{K}}_{\hat{\mu}}{}^{\hat{\hat{a}}\hat{\hat{b}}},\\
\hat{\hat{K}}_{\hat{\mu}}{}^{\hat{\hat{a}}\hat{\hat{b}}} &= -\frac{i}{16}\left[-\bar{\hat{\hat{\psi}}}_{\hat{\alpha}}\hat{\hat{\Gamma}}_{\hat{\mu}}{}^{\hat{\hat{a}}\hat{\hat{b}}\hat{\alpha}\hat{\beta}}\hat{\hat{\psi}}_{\hat{\beta}} + 2\left(\bar{\hat{\hat{\psi}}}_{\hat{\mu}}\hat{\hat{\Gamma}}^{\hat{\hat{b}}}\hat{\hat{\psi}}^{\hat{\hat{a}}} - \bar{\hat{\hat{\psi}}}_{\hat{\mu}}\hat{\hat{\Gamma}}^{\hat{\hat{a}}}\hat{\hat{\psi}}^{\hat{\hat{b}}} + \bar{\hat{\hat{\psi}}}_{\hat{\hat{b}}}\hat{\hat{\Gamma}}_{\hat{\mu}}\hat{\hat{\psi}}^{\hat{\hat{a}}}\right)\right].
\end{aligned} \tag{22.7}$$

This action is invariant under the local supersymmetry transformations with parameter $\hat{\hat{\epsilon}}$ (a Majorana spinor):

$$\boxed{\begin{aligned}
\delta_{\hat{\hat{\epsilon}}}\hat{\hat{e}}^{\hat{\hat{a}}}{}_{\hat{\mu}} &= -\frac{i}{2}\bar{\hat{\hat{\epsilon}}}\hat{\hat{\Gamma}}^{\hat{\hat{a}}}\hat{\hat{\psi}}_{\hat{\mu}},\\
\delta_{\hat{\hat{\epsilon}}}\hat{\hat{\psi}}_{\hat{\mu}} &= 2\tilde{\nabla}_{\hat{\mu}}\hat{\hat{\epsilon}} = 2\nabla_{\hat{\mu}}\left(\tilde{\hat{\hat{\omega}}}\right)\hat{\hat{\epsilon}} + \frac{i}{144}\left(\hat{\hat{\Gamma}}^{\hat{\alpha}\hat{\beta}\hat{\gamma}\hat{\delta}}{}_{\hat{\mu}} - 8\hat{\hat{\Gamma}}^{\hat{\beta}\hat{\gamma}\hat{\delta}}\hat{\hat{\eta}}_{\hat{\mu}}{}^{\hat{\alpha}}\right)\hat{\hat{\epsilon}}\,\tilde{\hat{\hat{G}}}_{\hat{\alpha}\hat{\beta}\hat{\gamma}\hat{\delta}},\\
\delta_{\hat{\hat{\epsilon}}}\hat{\hat{C}}_{\hat{\mu}\hat{\nu}\hat{\rho}} &= \tfrac{3}{2}\bar{\hat{\hat{\epsilon}}}\hat{\hat{\Gamma}}_{[\hat{\mu}\hat{\nu}}\hat{\hat{\psi}}_{\hat{\rho}]}.
\end{aligned}} \tag{22.8}$$

Observe that the "topological" *Chern–Simons* term in the action $\hat{\hat{\epsilon}}\hat{\hat{G}}\hat{\hat{G}}\hat{\hat{C}}$ seems to break parity ($\hat{\hat{\epsilon}}$ is a tensor density, or pseudotensor). Parity is, however, preserved because the 3-form $\hat{\hat{C}}$ is also a pseudotensor and so transforms with an extra sign under reflections.[4]

---

[4] The sign of the topological term can be changed by a field redefinition $\hat{\hat{C}} \to -\hat{\hat{C}}$. However, as we will see in Chapter 23, with our conventions, to establish T duality between the type-IIA theory that we will obtain by dimensional reduction to ten dimensions and the conventional type-IIB theory with *self-dual* 5-form and positive-chirality gravitinos, taking into account everything (i.e. not just the bosonic sector), we are forced to take the negative sign. The type-IIA theory that one would obtain with positive sign seems to be related to the unconventional type-IIB theory with *anti-self-dual* 5-form and negative-chirality gravitinos. In other words, there are *two* "different" 11-dimensional supergravity theories, which differ in the sign of the topological term and are related by the (rather trivial) duality transformation $\hat{\hat{C}} \to -\hat{\hat{C}}$. Each of these theories gives rise to a different type-IIA theory (type $IIA_{+-}$ and type $IIA_{-+}$), which in turn are related by T duality to the two "different" $N = 2B_+$ and $N = 2B_-$ supergravity theories. We mentioned all these theories in Section 20.2.2.

Now we are going to perform the dimensional reduction. For many purposes it is enough to reduce the bosonic fields by setting to zero all the fermions in the action and also reducing the supersymmetry transformation laws, and this is what we are going to do.

### 22.1.2 Reduction of the bosonic sector

The action for the bosonic fields is

$$\boxed{\hat{\hat{S}} = \frac{1}{16\pi G_{\rm N}^{(11)}}\int d^{11}\hat{\hat{x}}\,\sqrt{|\hat{\hat{g}}|}\left[\hat{\hat{R}} - \frac{1}{2\cdot 4!}\hat{\hat{G}}^2 - \frac{1}{(144)^2}\frac{1}{\sqrt{|\hat{\hat{g}}|}}\hat{\hat{\epsilon}}\hat{\hat{G}}\hat{\hat{G}}\hat{\hat{C}}\right],} \tag{22.9}$$

and the equations of motion are

$$\begin{aligned} \hat{\hat{R}}_{\hat{\hat{\mu}}\hat{\hat{\nu}}} - \frac{1}{12}\left[\hat{\hat{G}}_{\hat{\hat{\mu}}}{}^{\hat{\hat{\alpha}}_1\hat{\hat{\alpha}}_2\hat{\hat{\alpha}}_3}\hat{\hat{G}}_{\hat{\hat{\nu}}\hat{\hat{\alpha}}_1\hat{\hat{\alpha}}_2\hat{\hat{\alpha}}_3} - \frac{1}{12}\hat{\hat{g}}_{\hat{\hat{\mu}}\hat{\hat{\nu}}}\hat{\hat{G}}^2\right] &= 0, \\ \nabla_{\hat{\hat{\mu}}}\hat{\hat{G}}^{\hat{\hat{\mu}}\hat{\hat{\nu}}\hat{\hat{\rho}}\hat{\hat{\sigma}}} - \frac{1}{3^2\cdot 2^7}\frac{1}{\sqrt{|\hat{\hat{g}}|}}\hat{\hat{\epsilon}}^{\hat{\hat{\nu}}\hat{\hat{\rho}}\hat{\hat{\sigma}}\hat{\hat{\mu}}_1\cdots\hat{\hat{\mu}}_4\hat{\hat{\nu}}_1\cdots\hat{\hat{\nu}}_4}\hat{\hat{G}}_{\hat{\hat{\mu}}_1\cdots\hat{\hat{\mu}}_4}\hat{\hat{G}}_{\hat{\hat{\nu}}_1\cdots\hat{\hat{\nu}}_4} &= 0. \end{aligned} \tag{22.10}$$

The equation of motion of the 3-form potential can be rewritten in this way:

$$\partial\left(\star\hat{\hat{G}} + \frac{35}{2}\hat{\hat{C}}\hat{\hat{G}}\right) = 0. \tag{22.11}$$

This equation has the form of a Bianchi identity and we could identify the expression in parentheses with $7\partial\hat{\hat{\tilde{C}}}$ where $\hat{\hat{\tilde{C}}}$ is a 6-form potential that is the dual of the 3-form potential. This implies that the field strength of the dual 6-form is [22, 188, 174]

$$\star\hat{\hat{G}} = 7(\partial\hat{\hat{\tilde{C}}} - 10\hat{\hat{C}}\partial\hat{\hat{C}}) \equiv \hat{\hat{\tilde{G}}}. \tag{22.12}$$

This field strength is obviously invariant under the gauge transformations

$$\delta_{\hat{\hat{\tilde{\chi}}}}\hat{\hat{\tilde{C}}} = 6\partial\hat{\hat{\tilde{\chi}}}, \tag{22.13}$$

where $\hat{\hat{\tilde{\chi}}}$ is a 5-form. However, in its definition the 3-form appears explicitly and, to make it invariant under the 3-form gauge transformations (22.4), $\hat{\hat{\tilde{C}}}$ has to transform as follows:

$$\delta_{\hat{\hat{\chi}}}\hat{\hat{\tilde{C}}} = -30\partial\hat{\hat{\chi}}\hat{\hat{C}}. \tag{22.14}$$

This procedure for defining the dual of some field in which the original field has not been completely eliminated, by making use of the equations of motion, is usually referred to as "on-shell" dualization. The fact that the 3-form potential appears in the field strength of its dual 6-form makes it very difficult (but not impossible, see Ref. [174]) to find a formulation of 11-dimensional supergravity in terms of the 6-form alone. On-shell dualization is enough for determining the gauge-transformation laws and a field strength in which the 3-form should be interpreted as a complicated function of the dual 6-form.

We now focus on the dimensional reduction of the theory in its 3-form formulation. We assume that all fields are independent of the coordinate $z = x^{\underline{10}}$ which we choose to correspond to a spacelike direction ($\hat{\hat{\eta}}_{\underline{zz}} = -1$) and we rewrite the fields and action in a ten-dimensional form. The dimensional reduction of the metric gives rise to the ten-dimensional metric, a vector field, and a scalar (the dilaton), whereas the dimensional reduction of the 3-form potential gives rise to a ten-dimensional 3-form and a (KR) 2-form, which are the fields of the ten-dimensional $N = 2A, d = 10$ supergravity theory

$$\left\{\hat{g}_{\hat{\mu}\hat{\nu}}, \hat{B}_{\hat{\mu}\hat{\nu}}, \hat{\phi}, \hat{C}^{(3)}{}_{\hat{\mu}\hat{\nu}\hat{\rho}}, \hat{C}^{(1)}{}_{\hat{\mu}}\right\}. \tag{22.15}$$

The metric, the KR two-form, and the dilaton are NSNS fields and the 3-form and the vector are RR fields. We are going to use for RR forms the conventions proposed and used in Refs. [667, 151, 174, 936], which we will explain later. Furthermore, we are going to use the string metric, to which the NSNS sector couples as in the bosonic-string effective action Eq. (21.1). The coupling to the RR sector is then completely determined just by defining RR fields with "natural" gauge-transformation rules (with no scalars involved in them). On the other hand, we have chosen for the RR fields the simplest normalization, which is also the most common: they appear at the same order in $\alpha'$ as the NSNS ones.[5]

Using the Scherk–Schwarz procedure as explained in Chapter 15 and rescaling the metric so as to obtain the action in the string frame, we find that the fields of the 11-dimensional theory have to be expressed in terms of the ten-dimensional ones as follows:

$$\begin{aligned}
\hat{\hat{g}}_{\hat{\mu}\hat{\nu}} &= e^{-\frac{2}{3}\hat{\phi}}\hat{g}_{\hat{\mu}\hat{\nu}} - e^{\frac{4}{3}\hat{\phi}}\hat{C}^{(1)}{}_{\hat{\mu}}\hat{C}^{(1)}{}_{\hat{\nu}}, &\qquad \hat{\hat{C}}_{\hat{\mu}\hat{\nu}\hat{\rho}} &= \hat{C}^{(3)}{}_{\hat{\mu}\hat{\nu}\hat{\rho}},\\
\hat{\hat{g}}_{\hat{\mu}\underline{z}} &= -e^{\frac{4}{3}\hat{\phi}}\hat{C}^{(1)}{}_{\hat{\mu}}, & \hat{\hat{C}}_{\hat{\mu}\hat{\nu}\underline{z}} &= \hat{B}_{\hat{\mu}\hat{\nu}},\\
\hat{\hat{g}}_{\underline{zz}} &= -e^{\frac{4}{3}\hat{\phi}}. &&
\end{aligned} \tag{22.16}$$

For the Elfbeins we have

$$\left(\hat{\hat{e}}^{\hat{\hat{a}}}{}_{\hat{\hat{\mu}}}\right) = \begin{pmatrix} e^{-\frac{1}{3}\hat{\phi}}\hat{e}^{\hat{a}}{}_{\hat{\mu}} & e^{\frac{2}{3}\hat{\phi}}\hat{C}^{(1)}{}_{\hat{\mu}} \\ 0 & e^{\frac{2}{3}\hat{\phi}} \end{pmatrix}, \qquad \left(\hat{\hat{e}}_{\hat{\hat{a}}}{}^{\hat{\hat{\mu}}}\right) = \begin{pmatrix} e^{\frac{1}{3}\hat{\phi}}\hat{e}_{\hat{a}}{}^{\hat{\mu}} & -e^{\frac{1}{3}\hat{\phi}}\hat{C}^{(1)}{}_{\hat{a}} \\ 0 & e^{-\frac{2}{3}\hat{\phi}} \end{pmatrix}. \tag{22.17}$$

[5] This is not the only normalization used in the literature. For instance, in Ref. [715], all the kinetic terms of the RR fields in the action have an extra factor $\alpha'$ with respect to ours, so their RR fields have units of $L^{-1}$ (ours are dimensionless). In Refs. [1051, 1048] the RR fields have dimensions $L^{-4}$ and their kinetic terms appear in the action with a purely numerical factor instead of the factor $(16\pi G_{\rm N}^{(10)})^{-1}$ that they carry in our case. Our conventions are also those of Ref. [41], where the issue of normalizations has been studied exhaustively.

The inverse relations are

$$\begin{aligned}
\hat{g}_{\hat{\mu}\hat{\nu}} &= \left(-\hat{\hat{g}}_{\underline{zz}}\right)^{\frac{1}{2}}\left(\hat{\hat{g}}_{\hat{\mu}\hat{\nu}} - \hat{\hat{g}}_{\hat{\mu}\underline{z}}\hat{\hat{g}}_{\hat{\nu}\underline{z}}/\hat{\hat{g}}_{\underline{zz}}\right), & \hat{C}^{(3)}{}_{\hat{\mu}\hat{\nu}\hat{\rho}} &= \hat{\hat{C}}_{\hat{\mu}\hat{\nu}\hat{\rho}}, \\
\hat{C}^{(1)}{}_{\hat{\mu}} &= \hat{\hat{g}}_{\hat{\mu}\underline{z}}/\hat{\hat{g}}_{\underline{zz}}, & \hat{B}_{\hat{\mu}\hat{\nu}} &= \hat{\hat{C}}_{\hat{\mu}\hat{\nu}\underline{z}}, \\
\hat{\phi} &= \tfrac{3}{4}\ln\left(-\hat{\hat{g}}_{\underline{zz}}\right).
\end{aligned} \tag{22.18}$$

Now we can perform the reduction of the action Eq. (22.9). We first consider the Ricci scalar term. To reduce this term, we use a slight generalization of Palatini's identity Eq. (L.4). With the above ansatz for the Elfbeins, the non-vanishing components of the 11-dimensional spin connection are

$$\begin{aligned}
\hat{\hat{\omega}}_{z\hat{a}z} &= -\tfrac{2}{3}e^{\frac{1}{3}\hat{\phi}}\partial_{\hat{a}}\hat{\phi}, & \hat{\hat{\omega}}_{z\hat{a}\hat{b}} &= -\tfrac{1}{2}e^{\frac{4}{3}\hat{\phi}}\hat{G}^{(2)}{}_{\hat{a}\hat{b}}, \\
\hat{\hat{\omega}}_{\hat{a}\hat{b}z} &= \tfrac{1}{2}e^{\frac{4}{3}\hat{\phi}}\hat{G}^{(2)}{}_{\hat{a}\hat{b}}, & \hat{\hat{\omega}}_{\hat{a}\hat{b}\hat{c}} &= e^{\frac{1}{3}\hat{\phi}}\left(\hat{\omega}_{\hat{a}\hat{b}\hat{c}} + \tfrac{2}{3}\delta_{\hat{a}[\hat{b}}\partial_{\hat{c}]}\hat{\phi}\right),
\end{aligned} \tag{22.19}$$

where

$$\hat{G}^{(2)} = 2\partial\hat{C}^{(1)} \tag{22.20}$$

is the field strength of the ten-dimensional RR 1-form $\hat{C}^{(1)}{}_{\hat{\mu}}$. Using

$$\sqrt{\hat{\hat{g}}} = \sqrt{|\hat{g}|}\, e^{-\frac{8}{3}\hat{\phi}}, \tag{22.21}$$

plus Palatini's identity Eq. (L.4) for $d = 11$ and $\hat{\hat{\phi}} = 0$, plus the fact that the coordinate $z$ conventionally lives in a circle of radius equal to the reduced 11-dimensional Planck length $\ell^{(11)}_{\text{Planck}} \equiv \ell^{(11)}_{\text{Planck}}/(2\pi)$, we find

$$\begin{aligned}
\int d^{11}\hat{\hat{x}}\,\sqrt{|\hat{\hat{g}}|}\left[\,\hat{\hat{R}}\,\right] = 2\pi\ell^{(11)}_{\text{Planck}}\int d^{10}\hat{x}\,\sqrt{|\hat{g}|}\Big\{ &-e^{-2\hat{\phi}}\left[\left(\hat{\omega}_{\hat{b}}{}^{\hat{b}\hat{a}} + 2\partial^{\hat{a}}\hat{\phi}\right)^2 + \hat{\omega}_{\hat{a}}{}^{\hat{b}\hat{c}}\hat{\omega}_{\hat{b}\hat{c}}{}^{\hat{a}}\right] \\
&- \tfrac{1}{4}\left(\hat{G}^{(2)}\right)^2\Big\}.
\end{aligned}$$

Finally, using Palatini's identity Eq. (L.4) again, but now for $d = 10$ and $\phi = \hat{\phi}$, we obtain for the Ricci scalar term

$$\int d^{11}\hat{\hat{x}}\,\sqrt{|\hat{\hat{g}}|}\left[\,\hat{\hat{R}}\,\right] = 2\pi\ell^{(11)}_{\text{Planck}}\int d^{10}\hat{x}\,\sqrt{|\hat{g}|}\left\{e^{-2\hat{\phi}}\left[\hat{R} - 4\left(\partial\hat{\phi}\right)^2\right] - \tfrac{1}{4}\left(\hat{G}^{(2)}\right)^2\right\}. \tag{22.22}$$

Now we have to reduce the $\hat{\hat{G}}$-term in Eq. (22.9). We identify field strengths in 11 and ten dimensions with flat indices (this automatically ensures gauge invariance), taking into account the scaling of the ten-dimensional metric

$$\hat{G}^{(4)}{}_{\hat{a}\hat{b}\hat{c}\hat{d}} = e^{-\frac{4}{3}\hat{\phi}}\,\hat{\hat{G}}_{\hat{a}\hat{b}\hat{c}\hat{d}}, \tag{22.23}$$

which leads to

$$\hat{G}^{(4)} = 4\left(\partial \hat{C}^{(3)} - \hat{H}\hat{C}^{(1)}\right), \tag{22.24}$$

where $\hat{H}$ is the field strength of the NSNS 2-form $\hat{B}$,

$$\hat{H} = 3\partial \hat{B}. \tag{22.25}$$

The remaining components of $\hat{\hat{G}}$ are given by

$$\hat{\hat{G}}_{\hat{a}\hat{b}\hat{c}z} = e^{\frac{1}{3}\hat{\phi}} \hat{H}_{\hat{a}\hat{b}\hat{c}}, \tag{22.26}$$

and the contribution of the $\hat{\hat{G}}$-term to the ten-dimensional action becomes

$$\int d^{11}\hat{\hat{x}}\, \sqrt{|\hat{\hat{g}}|}\left[-\frac{1}{2\cdot 4!}\left(\hat{\hat{G}}\right)^2\right] = 2\pi\ell^{(11)}_{\text{Planck}} \int d^{10}\hat{x}\, \sqrt{|\hat{g}|}\left[\frac{1}{2\cdot 3!}e^{-2\hat{\phi}}\hat{H}^2 - \frac{1}{2\cdot 4!}\left(\hat{G}^{(4)}\right)^2\right]. \tag{22.27}$$

Finally, taking into account

$$\hat{\hat{\epsilon}}^{\,\hat{\mu}_0\cdots\hat{\mu}_9 \underline{z}} = \hat{\epsilon}^{\,\hat{\mu}_0\cdots\hat{\mu}_9}, \tag{22.28}$$

the third term in the $d = 11$ action Eq. (22.9) (all terms with curved indices) gives

$$\hat{\hat{\epsilon}}\hat{\hat{G}}\hat{\hat{G}}\hat{\hat{C}} = 48\hat{\epsilon}\partial\hat{C}^{(3)}\partial\hat{C}^{(3)}\hat{B} - 96\hat{\epsilon}\partial\hat{C}^{(3)}\partial\hat{B}\hat{C}^{(3)}, \tag{22.29}$$

and, on integrating by parts, we obtain

$$\int d^{11}\hat{\hat{x}}\left[-\frac{1}{(144)^2}\hat{\hat{\epsilon}}\hat{\hat{G}}\hat{\hat{G}}\hat{\hat{C}}\right] = 2\pi\ell^{(11)}_{\text{Planck}} \int d^{10}\hat{x}\left[-\frac{1}{144}\hat{\epsilon}\partial\hat{C}^{(3)}\partial\hat{C}^{(3)}\hat{B}\right]. \tag{22.30}$$

On putting all these results together, we find what is described in the literature as the bosonic part of the $N = 2A, d = 10$ supergravity action in ten dimensions in the string frame:

$$\begin{aligned}\hat{S} = \frac{2\pi\ell^{(11)}_{\text{Planck}}}{16\pi G^{(11)}_{\text{N}}} \int d^{10}\hat{x}\, \sqrt{|\hat{g}|}\Big\{ & e^{-2\hat{\phi}}\left[\hat{R} - 4\left(\partial\hat{\phi}\right)^2 + \frac{1}{2\cdot 3!}\hat{H}^2\right] \\ & - \left[\frac{1}{4}\left(\hat{G}^{(2)}\right)^2 + \frac{1}{2\cdot 4!}\left(\hat{G}^{(4)}\right)^2\right] - \frac{1}{144}\frac{1}{\sqrt{|\hat{g}|}}\,\hat{\epsilon}\partial\hat{C}^{(3)}\partial\hat{C}^{(3)}\hat{B}\Big\}.\end{aligned} \tag{22.31}$$

In the first line of Eq. (22.31) we can see the known action for the bosonic NSNS fields in the string frame (characterized by the overall factor $e^{-2\hat{\phi}}$). The second line has no dilaton factor and describes the RR sector whose truncation leaves us with the action of $N = 1, d = 10$ supergravity.

In string theory, however, we want the string metric to be asymptotically flat and we want, in 11-dimensional supergravity, to have asymptotically flat metrics. Both things cannot be true at the same time regarding the relations between 11- and ten-dimensional fields since, with $z$ compact, $\hat{\hat{g}}_{\underline{zz}}$ does not have to go to $-1$ at infinity in the non-compact directions and the ten-dimensional metric is essentially the 11-dimensional one rescaled by powers of $\hat{\hat{g}}_{\underline{zz}}$ (the dilaton). If the 11-dimensional metric is asymptotically flat and we denote by $\hat{\phi}_0$ the asymptotic value of the dilaton, then the ten-dimensional metric that we have defined behaves at infinity as follows:

$$\hat{g}_{\hat{\mu}\hat{\nu}} \to e^{\frac{2}{3}\hat{\phi}_0}\hat{\eta}_{\hat{\mu}\hat{\nu}}. \tag{22.32}$$

The exponential of the asymptotic value of the dilaton is the (type-IIA) *string coupling constant* $\hat{g}_{\mathrm{A}}$ that counts loops in string amplitudes,

$$\hat{g}_{\mathrm{A}} = e^{\hat{\phi}_0}. \tag{22.33}$$

We could repeat here the discussion of Section 15.2.2 with some minor modifications, but it should be clear that, to obtain the "right" string metric, we have to rescale the metric. To eliminate the additional factors of $e^{\hat{\phi}_0}$ that would appear, we have to rescale all the other fields, although this does not seem strictly necessary. Thus, we make the following substitutions:

$$\begin{aligned} \hat{g}_{\hat{\mu}\hat{\nu}} &\to e^{\frac{2}{3}\hat{\phi}_0}\hat{g}_{\hat{\mu}\hat{\nu}}, & \hat{C}^{(1)}{}_{\hat{\mu}} &\to e^{\frac{1}{3}\hat{\phi}_0}\hat{C}^{(1)}{}_{\hat{\mu}}, \\ \hat{B}_{\hat{\mu}\hat{\nu}} &\to e^{\frac{2}{3}\hat{\phi}_0}\hat{B}_{\hat{\mu}\hat{\nu}}, & \hat{C}^{(3)}{}_{\hat{\mu}\hat{\nu}\hat{\rho}} &\to e^{\hat{\phi}_0}\hat{C}^{(3)}{}_{\hat{\mu}\hat{\nu}\hat{\rho}}, \end{aligned} \tag{22.34}$$

after which the relation between the 11- and ten-dimensional fields is

$$\begin{aligned} \hat{\hat{g}}_{\hat{\mu}\hat{\nu}} &= e^{-\frac{2}{3}(\hat{\phi}-\hat{\phi}_0)}\hat{g}_{\hat{\mu}\hat{\nu}} - e^{\frac{4}{3}\hat{\phi}}e^{\frac{2}{3}\hat{\phi}_0}\hat{C}^{(1)}{}_{\hat{\mu}}\hat{C}^{(1)}{}_{\hat{\nu}}, & \hat{\hat{C}}_{\hat{\mu}\hat{\nu}\hat{\rho}} &= e^{\hat{\phi}_0}\hat{C}^{(3)}{}_{\hat{\mu}\hat{\nu}\hat{\rho}}, \\ \hat{\hat{g}}_{\hat{\mu}\underline{z}} &= -e^{\frac{4}{3}\hat{\phi}}e^{\frac{1}{3}\hat{\phi}_0}\hat{C}^{(1)}{}_{\hat{\mu}}, & \hat{\hat{C}}_{\hat{\mu}\hat{\nu}\underline{z}} &= e^{\frac{2}{3}\hat{\phi}_0}\hat{B}_{\hat{\mu}\hat{\nu}}, \\ \hat{\hat{g}}_{\underline{zz}} &= -e^{\frac{4}{3}\hat{\phi}}. \end{aligned} \tag{22.35}$$

For the Elfbein we have

$$\begin{aligned} \left(\hat{\hat{e}}^{\hat{\hat{a}}}{}_{\hat{\hat{\mu}}}\right) &= \begin{pmatrix} e^{-\frac{1}{3}(\hat{\phi}-\hat{\phi}_0)}\hat{e}^{\hat{a}}{}_{\hat{\mu}} & e^{\frac{2}{3}\hat{\phi}}e^{\frac{1}{3}\hat{\phi}_0}\hat{C}^{(1)}{}_{\hat{\mu}} \\ 0 & e^{\frac{2}{3}\hat{\phi}} \end{pmatrix}, \\ \left(\hat{\hat{e}}_{\hat{\hat{a}}}{}^{\hat{\hat{\mu}}}\right) &= \begin{pmatrix} e^{\frac{1}{3}(\hat{\phi}-\hat{\phi}_0)}\hat{e}_{\hat{a}}{}^{\hat{\mu}} & -e^{\frac{1}{3}\hat{\phi}}\hat{C}^{(1)}{}_{\hat{a}} \\ 0 & e^{-\frac{2}{3}\hat{\phi}} \end{pmatrix}. \end{aligned} \tag{22.36}$$

The inverse relations are

$$\begin{aligned} \hat{g}_{\hat{\mu}\hat{\nu}} &= e^{-\frac{2}{3}\hat{\phi}_0}\left(-\hat{\hat{g}}_{\underline{zz}}\right)^{\frac{1}{2}}\left(\hat{\hat{g}}_{\hat{\mu}\hat{\nu}} - \hat{\hat{g}}_{\hat{\mu}\underline{z}}\hat{\hat{g}}_{\hat{\nu}\underline{z}}/\hat{\hat{g}}_{\underline{zz}}\right), & \hat{C}^{(3)}{}_{\hat{\mu}\hat{\nu}\hat{\rho}} &= e^{-\hat{\phi}_0}\hat{\hat{C}}_{\hat{\mu}\hat{\nu}\hat{\rho}}, \\ \hat{C}^{(1)}{}_{\hat{\mu}} &= e^{-\frac{1}{3}\hat{\phi}_0}\hat{\hat{g}}_{\hat{\mu}\underline{z}}/\hat{\hat{g}}_{\underline{zz}}, & \hat{B}_{\hat{\mu}\hat{\nu}} &= e^{-\frac{2}{3}\hat{\phi}_0}\hat{\hat{C}}_{\hat{\mu}\hat{\nu}\underline{z}}, \\ \hat{\phi} &= \tfrac{3}{4}\ln\left(-\hat{\hat{g}}_{\underline{zz}}\right), \end{aligned} \tag{22.37}$$

and, finally, the action becomes

$$\boxed{\begin{aligned}\hat{S} = \frac{\hat{g}_{\rm A}^2}{16\pi G_{\rm N\,A}^{(10)}}\int d^{10}\hat{x}\,\sqrt{|\hat{g}|}\Big\{ & e^{-2\hat{\phi}}\Big[\hat{R} - 4\big(\partial\hat{\phi}\big)^2 + \frac{1}{2\cdot 3!}\hat{H}^2\Big] \\ & - \Big[\tfrac{1}{4}\big(\hat{G}^{(2)}\big)^2 + \frac{1}{2\cdot 4!}\big(\hat{G}^{(4)}\big)^2\Big] \\ & - \frac{1}{144}\frac{1}{\sqrt{|\hat{g}|}}\,\hat{\epsilon}\partial\hat{C}^{(3)}\partial\hat{C}^{(3)}\hat{B}\Big\},\end{aligned}} \tag{22.38}$$

where we have made the identification of the prefactor of the action

$$\frac{2\pi\ell_{\rm Planck}^{(11)}e^{\frac{8}{3}\hat{\phi}_0}}{16\pi G_{\rm N}^{(11)}} = \frac{\hat{g}_{\rm A}^2}{16\pi G_{\rm N\,A}^{(10)}}. \tag{22.39}$$

The factor $g_{\rm A}^2$ absorbs the asymptotic value of the dilaton in the action.

Then, we find the following relation:

$$G_{\rm N\,A}^{(10)} = \frac{G_{\rm N}^{(11)}}{2\pi\ell_{\rm Planck}^{(11)}\hat{g}_{\rm A}^{\frac{2}{3}}}. \tag{22.40}$$

Observe that we have taken the 11th coordinate $z$ to live in a circle of radius equal to the reduced 11-dimensional Planck length $\ell_{\rm Planck}^{(11)}$, which, up to numerical factors, is the only scale available in 11-dimensional supergravity. Actually, we have to distinguish between the interval in which $z$ takes values ($z \in [0, 2\pi\ell_{\rm Planck}^{(11)}]$) and the actual radius of the 11th dimension (measured at infinity), which we denote here by $R_{11}$ and which is naturally measured with the 11-dimensional metric:

$$R_{11} = \frac{1}{2\pi}\lim_{r\to\infty}\int\sqrt{|\hat{\hat{g}}_{\underline{zz}}|}dz = \ell_{\rm Planck}^{(11)}e^{\frac{2}{3}\hat{\phi}_0} = \ell_{\rm Planck}^{(11)}\hat{g}_{\rm A}^{\frac{2}{3}}. \tag{22.41}$$

By using this relation in Eq. (16.40), we find

$$G_{\rm N\,A}^{(10)} = \frac{G_{\rm N}^{(11)}}{2\pi R_{11}} = \frac{G_{\rm N}^{(11)}}{V_{11}}, \tag{22.42}$$

as usual in KK theory ($V_{11}$ is the volume of the internal space).

Now, in this case, we define the 11-dimensional Planck length[6] by

$$\boxed{16\pi G_{\rm N}^{(11)} \equiv \frac{(\ell_{\rm Planck}^{(11)})^9}{2\pi},} \tag{22.43}$$

[6] The reason for this somewhat unusual definition is explained in Section 25.1.1.

and, thus, in terms of it and $\hat{g}_A$ only, the ten-dimensional Newton constant is given by

$$G_{NA}^{(10)} = \frac{(\ell_{\text{Planck}}^{(11)})^8}{32\pi^2 \hat{g}_A^{\frac{2}{3}}}. \tag{22.44}$$

This result has to be compared with the value we will obtain in Section 25.1.1 for $G_N^{(10)}$ in terms of the type-IIA stringy variables $\ell_s$ and $\hat{g}_A$, Eq. (25.26):

$$G_{NA}^{(10)} = 8\pi^6 \hat{g}_A^2 (\alpha')^4. \tag{22.45}$$

They are consistent if

$$\ell_{\text{Planck}}^{(11)} = 2\pi \ell_s \hat{g}_A^{\frac{1}{3}}, \tag{22.46}$$

and, thus,

$$R_{11} = \ell_s \hat{g}_A. \tag{22.47}$$

These are the two main relations between 11- and ten-dimensional type-IIA constants, and they deserve to be framed together:

$$\boxed{\begin{aligned} \ell_{\text{Planck}}^{(11)} &= 2\pi \ell_s \hat{g}_A^{\frac{1}{3}}, \\ R_{11} &= \ell_s \hat{g}_A. \end{aligned}} \tag{22.48}$$

Once we have expressed $R_{11}$ in these stringy variables, it is easy to see that the strong coupling limit of the type-IIA theory ($\hat{g}_A \to \infty$) coincides with the decompactification limit $R_{11}$ in which a new dimension becomes macroscopic [1275].

Just as the KK scalar of the 11-dimensional theory gives the string-theory dilaton, the KK vector gives the RR 1-form, and the 3-form gives the RR 3-form and the NSNS 2-form. Since we know that there are D0- and D2-branes associated with the RR 1-form, and 3-form, we find that they originate, respectively, from the 11-dimensional graviton moving in the compact direction and from a two-dimensional object that couples to the 11-dimensional 3-form: the M2-brane. An M2-brane wrapped around the compact dimension gives the type-IIA string ($\hat{B}_{\hat{\mu}\hat{\nu}} = e^{-\frac{2}{3}\hat{\phi}_0} \hat{\hat{C}}_{\hat{\mu}\hat{\nu}\underline{z}}$) and, unwrapped, gives the D2-brane ($\hat{C}^{(3)}{}_{\hat{\mu}\hat{\nu}\hat{\rho}} = e^{-\hat{\phi}_0} \hat{\hat{C}}_{\hat{\mu}\hat{\nu}\hat{\rho}}$). These and more relations between extended objects are represented in Fig. 25.5 on p. 758.

### 22.1.3 Magnetic potentials

For each of the differential-form potentials $\hat{C}^{(1)}$, $\hat{C}^{(3)}$, and $\hat{B}$, a magnetic dual, $\hat{C}^{(7)}$, $\hat{C}^{(5)}$, and $\hat{B}^{(6)}$, respectively, can be introduced whose equation of motion is equivalent to the Bianchi identity of the original (electric) potential and vice versa. (In general we can dualize only on-shell since the electric potentials occur without derivatives in the action.) These potentials will be useful in studying D-brane solutions and D-brane effective actions.

Dual field strengths are defined through the Hodge dual of the original field strengths, but potentials can be defined in many ways. A way to define the dual (magnetic) potentials is to fix the form of the corresponding field strength. For RR potentials, a convenient general form for the field strengths was proposed in Refs. [667, 151] and used explicitly in Refs. [174, 936]. In differential-form and component language, it is

$$\begin{aligned} \hat{G}^{(2n)} &= d\hat{C}^{(2n-1)} - \hat{H}\hat{C}^{(2n-3)}, \\ \hat{G}^{(2n)} &= 2n\Big[\partial\hat{C}^{(2n-1)} - \tfrac{1}{2}(2n-1)(2n-2)\partial\hat{B}\hat{C}^{(2n-3)}\Big]. \end{aligned} \tag{22.49}$$

The differential-form language considerably simplifies the expressions. The gauge transformations of the RR form potentials are

$$\delta\hat{C}^{(2n-1)} = d\hat{\Lambda}^{(2n-2)} - d\hat{B} \wedge \hat{\Lambda}^{(2n-4)}. \tag{22.50}$$

This normalization is extremely useful because it can be generalized to the type-IIB and massive type-IIA fields. For massless type-II supergravities, we can also introduce the notation

$$\begin{aligned} \hat{C} &= \hat{C}^{(0)} + \hat{C}^{(1)} + \hat{C}^{(2)} + \cdots, \\ \hat{G} &= \hat{G}^{(1)} + \hat{G}^{(2)} + \cdots, \\ \hat{\Lambda}^{(\cdot)} &= \hat{\Lambda}^{(0)} + \hat{\Lambda}^{(1)} + \cdots, \end{aligned} \tag{22.51}$$

with which we can write[7] (as we are going to show)

$$\begin{aligned} \hat{H} &= d\hat{B}, & \delta\hat{B} &= d\hat{\Lambda}, \\ \hat{G} &= d\hat{C} - \hat{H} \wedge \hat{C}, & \delta\hat{C} &= d\hat{\Lambda}^{(\cdot)} - d\hat{B} \wedge \hat{\Lambda}^{(\cdot)}. \end{aligned} \tag{22.52}$$

Now we have to prove that it is indeed possible to have magnetic RR potentials with field strengths of that kind. First of all, the field strengths $\hat{G}^{(2)}$ and $\hat{G}^{(4)}$ we are already using conform to this normalization. Their Bianchi identities are

$$d\hat{G}^{(2)} = 0, \qquad d\hat{G}^{(4)} - \hat{H} \wedge \hat{G}^{(2)} = 0, \tag{22.53}$$

and, from the action, the equations of motion are found to be

$$d \star \hat{G}^{(2)} + H \wedge \star\hat{G}^{(4)} = 0, \qquad d \star \hat{G}^{(4)} - \hat{H} \wedge \hat{G}^{(4)} = 0. \tag{22.54}$$

The Bianchi identities for the dual field strengths $\hat{G}^{(8)}$ and $\hat{G}^{(6)}$ are, according to the general normalization,

$$d\hat{G}^{(8)} - \hat{H} \wedge \hat{G}^{(6)} = 0, \qquad d\hat{G}^{(6)} - \hat{H} \wedge \hat{G}^{(4)} = 0. \tag{22.55}$$

By comparison with the equations of motion for the electric potentials, we find the relations

$$\hat{G}^{(8)} = - \star \hat{G}^{(2)}, \qquad \hat{G}^{(6)} = + \star \hat{G}^{(4)}. \tag{22.56}$$

[7] The RR gauge transformations are also written, after a redefinition of the gauge parameters, in the form

$$\delta\hat{C} = d\hat{\Lambda}^{(\cdot)} e^{\hat{B}}.$$

These relations define the dual field strengths, which, together with the general form for the field strengths, define the magnetic RR potentials.

The relations between electric- and magnetic-field strengths and their Bianchi identities can be written in the general form

$$\boxed{\begin{aligned} \hat{G}^{(10-k)} &= (-1)^{[k/2]} \star \hat{G}^{(k)}, \\ d\hat{G} - \hat{H} \wedge \hat{G} &= 0, \end{aligned}} \tag{22.57}$$

which we will see apply in this form to the massive type-IIA theory in Section 22.2 and to the type-IIB theory in Chapter 23.

Observe that the set of all the Bianchi identities plus the duality relations between electric and magnetic potentials determine all the equations of motion of the RR fields. Observe also that all the "electric" RR potentials are pseudotensors and that their duals are, by definition, tensors.

Now let us dualize the NSNS 2-form potential. Using the general definitions for RR potentials, field strengths, etc., we can write

$$\boxed{\begin{aligned} \hat{H}^{(7)} &= e^{-2\hat{\phi}} \star \hat{H}, \\ dH &= 0, \\ d\hat{H}^{(7)} + \tfrac{1}{2} \star \hat{G} \wedge \hat{G} &= 0, \\ d\left(e^{-2\phi} \star \hat{H}\right) + \tfrac{1}{2} \star \hat{G} \wedge \hat{G} &= 0, \\ d\left(e^{2\phi} \star \hat{H}^{(7)}\right) &= 0, \end{aligned}} \tag{22.58}$$

which, again, apply in precisely this form to massive type-IIA and type-IIB theories. In the type-IIA case, one can take [829]

$$\hat{H}^{(7)} = d\hat{B}^{(6)} - \tfrac{1}{2} \sum_{n=1}^{n=4} \star \hat{G}^{(2n+2)} \wedge \hat{C}^{(2n-1)}. \tag{22.59}$$

It is more difficult to relate the magnetic potentials to 11-dimensional potentials since some of them can be defined only in ten dimensions. For instance, the $\hat{C}^{(7)}$ associated with the D6-brane is the dual of the $\hat{C}^{(1)}$, which in $d = 11$ is part of the metric. In fact, the D6-brane can be obtained by compactification of the 11-dimensional KK monopole (KKM), which is a purely gravitational solution that, by definition, always has a compact direction. The potential to which the 11-dimensional KKM couples has been studied in Ref. [804]. The potentials $\hat{B}^{(6)}$ and $\hat{C}^{(5)}$ can be related to $\hat{\hat{\tilde{C}}}$, and the associated extended objects (the so-called *solitonic 5-brane* S5A which, by definition, couples to $\hat{B}^{(6)}$ and the D4-brane) originate from the 11-dimensional 5-brane M5 which, by definition, is the object that couples to $\hat{\hat{\tilde{C}}}$. These relations are represented in Fig. 25.5 on p. 758.

### 22.1.4 Reduction of fermions and the supersymmetry rules

Here we want to reduce to ten dimensions the fermions and the supersymmetry transformation laws in Eq. (22.8). We will keep only the terms up to second order in fermions.

First we need to decompose the 11-dimensional gamma matrices in terms of the ten-dimensional ones. This is done in Appendix D.1.4. Next, we have to decompose the 11-dimensional spinors into ten-dimensional spinors. Eleven-dimensional Majorana spinors are also ten-dimensional Majorana spinors. However, in ten-dimensional supergravity, the elementary spinor is a Majorana–Weyl spinor. Thus, each 11-dimensional spinor can be considered as a pair of Majorana–Weyl spinors with opposite chiralities (this is why this theory is non-chiral and it is $N = 2$). In principle we could split all the spinors into their chiral halves, but it is not worth doing this for the moment. Later on, we will have to do it in order to relate the spinors to those of the type-IIB theory, which are of the same chirality and cannot be considered as the two halves of any Majorana spinor. As we will see, there are two options in the type-IIB case: either we use indices $i = 1, 2$ for the spinors or we combine them into a chiral complex spinor. We will use the first option.

We express the 11-dimensional spinors in terms of the ten-dimensional spinors (gravitino $\hat{\psi}_{\hat{\mu}}$, dilatino $\hat{\lambda}$, and the supersymmetry transformation parameter $\hat{\epsilon}$) as follows:[8]

$$\hat{\hat{\epsilon}} = e^{-\frac{1}{6}\hat{\phi}}\,\hat{\epsilon}, \quad \hat{\hat{\psi}}_{\hat{a}} = e^{\frac{1}{6}\hat{\phi}}\left(2\hat{\psi}_{\hat{a}} - \tfrac{1}{3}\hat{\Gamma}_{\hat{a}}\hat{\lambda}\right), \quad \hat{\hat{\psi}}_{z} = \frac{2i}{3}e^{\frac{1}{6}\hat{\phi}}\hat{\Gamma}_{11}\hat{\lambda}. \tag{22.60}$$

Observe that, with these definitions, the gravitino $\hat{\psi}_{\hat{\mu}}$ is real but the dilatino $\hat{\lambda}$ is purely imaginary. We could use a purely real dilatino just by multiplying by $i$, but then its supersymmetry rule would look unconventional.

We now want to use the relation between the 11- and ten-dimensional bosonic fields that we have already obtained. However, we have performed the dimensional reduction working in a special Lorentz gauge $\hat{\hat{e}}^{\hat{a}}{}_{\underline{z}} = 0$ and supersymmetry transformations do not preserve this gauge. In fact,

$$\delta_{\hat{\hat{\epsilon}}}\hat{\hat{e}}^{\hat{a}}{}_{\underline{z}} = \tfrac{1}{3}e^{\frac{1}{3}\hat{\phi}}\,\bar{\hat{\epsilon}}\hat{\Gamma}^{\hat{a}}\hat{\Gamma}_{11}\hat{\lambda}. \tag{22.61}$$

We have to introduce a compensating local Lorentz transformation in order to preserve our gauge choice. Then, the ten-dimensional supersymmetry transformation $\delta_{\hat{\epsilon}}$ will be a combination of an 11-dimensional supersymmetry transformation $\delta_{\hat{\hat{\epsilon}}}$ and an 11-dimensional compensating local Lorentz transformation $\delta_{\hat{\hat{\sigma}}}$ such that

$$\delta_{\hat{\epsilon}}\hat{\hat{e}}^{\hat{a}}{}_{\underline{z}} \equiv \left(\delta_{\hat{\hat{\epsilon}}} + \delta_{\hat{\hat{\sigma}}}\right)\hat{\hat{e}}^{\hat{a}}{}_{\underline{z}} = \tfrac{1}{3}e^{\frac{1}{3}\hat{\phi}}\,\bar{\hat{\epsilon}}\hat{\Gamma}^{\hat{a}}\hat{\Gamma}_{11}\hat{\lambda} + \tfrac{1}{2}\hat{\hat{\sigma}}^{\hat{\hat{b}}\hat{\hat{c}}}\Gamma_{\mathrm{v}}\left(\hat{\hat{M}}_{\hat{\hat{b}}\hat{\hat{c}}}\right)^{\hat{a}}{}_{\hat{\hat{d}}}\,\hat{\hat{e}}^{\hat{d}}{}_{\underline{z}} = 0. \tag{22.62}$$

Since the generators of the Lorentz group in the vector representation are given by Eq. (A.60), the parameter of the compensating Lorentz transformation is given by

$$\hat{\hat{\sigma}}^{\hat{a}}{}_{z} = -\tfrac{1}{3}e^{\frac{1}{3}\hat{\phi}}\,\bar{\hat{\epsilon}}\hat{\Gamma}^{\hat{a}}\hat{\Gamma}_{11}\hat{\lambda}. \tag{22.63}$$

[8] As a first step, we simply identify, up to factors involving the dilaton, the dilatino $\hat{\lambda}$ with the (flat) component $\hat{\hat{\psi}}_z$ and the gravitino $\hat{\psi}_{\hat{a}}$ with the (flat) components $\hat{\hat{\psi}}_{\hat{a}}$. Then we see that it is natural to combine this dilatino and this gravitino into a new one whose supersymmetry transformation rules are much simpler. The final combinations are the ones we write.

We now have to apply this definition of a ten-dimensional supersymmetry transformation to all fields, performing a compensating Lorentz transformation on all of them with the above parameter and using the explicit form of the Lorentz transformations on each individual field.

After some calculations we find the following $N = 2A, d = 10$ supersymmetry transformation laws (only for the "electric" NSNS and RR potentials):

$$
\begin{aligned}
\delta_{\hat{\epsilon}} \hat{e}^{\hat{a}}{}_{\hat{\mu}} &= -i\bar{\hat{\epsilon}}\hat{\Gamma}^{\hat{a}}\hat{\psi}_{\hat{\mu}}, \\
\delta_{\hat{\epsilon}}\hat{\psi}_{\hat{\mu}} &= \left\{\partial_{\hat{\mu}} - \tfrac{1}{4}\left(\hat{\not{\omega}}_{\hat{\mu}} + \tfrac{1}{2}\Gamma_{11}\hat{\not{H}}_{\mu}\right)\right\}\hat{\epsilon} \\
&\quad + \frac{i}{8}e^{\hat{\phi}}\sum_{n=1,2}\frac{1}{(2n)!}\hat{\not{G}}^{(2n)}\hat{\Gamma}_{\hat{\mu}}\left(-\hat{\Gamma}_{11}\right)^{n}\hat{\epsilon}, \\
\delta_{\hat{\epsilon}}\hat{B}_{\hat{\mu}\hat{\nu}} &= -2i\bar{\hat{\epsilon}}\hat{\Gamma}_{[\hat{\mu}}\hat{\Gamma}_{11}\hat{\psi}_{\hat{\nu}]}, \\
\delta_{\hat{\epsilon}}\hat{C}^{(1)}{}_{\hat{\mu}} &= -e^{-\hat{\phi}}\bar{\hat{\epsilon}}\hat{\Gamma}_{11}\left(\hat{\psi}_{\hat{\mu}} - \tfrac{1}{2}\hat{\Gamma}_{\hat{\mu}}\hat{\lambda}\right), \\
\delta_{\hat{\epsilon}}\hat{C}^{(3)}{}_{\hat{\mu}\hat{\nu}\hat{\rho}} &= 3e^{-\hat{\phi}}\bar{\hat{\epsilon}}\hat{\Gamma}_{\hat{\mu}\hat{\nu}}\left(\hat{\psi}_{\hat{\rho}]} - \frac{1}{3!}\hat{\Gamma}_{\hat{\rho}]}\hat{\lambda}\right) + 3\hat{C}^{(1)}{}_{[\hat{\mu}}\delta_{\hat{\epsilon}}\hat{B}_{\hat{\nu}\hat{\rho}]}, \\
\delta_{\hat{\epsilon}}\hat{\lambda} &= \left(\hat{\not{\partial}}\hat{\phi} + \frac{1}{12}\hat{\Gamma}_{11}\hat{\not{H}}\right)\hat{\epsilon} + \frac{i}{4}e^{\hat{\phi}}\sum_{n=1,2}\frac{5-2n}{(2n)!}\hat{\not{G}}^{(2n)}\left(-\hat{\Gamma}_{11}\right)^{n}\hat{\epsilon}, \\
\delta_{\hat{\epsilon}}\hat{\phi} &= -\frac{i}{2}\bar{\hat{\epsilon}}\hat{\lambda}.
\end{aligned}
\tag{22.64}
$$

Observe that, in principle, one obtains two additional terms in $\delta_{\hat{\epsilon}}\hat{e}_{\hat{\mu}}{}^{\hat{a}}$:

$$
\frac{i}{3}\bar{\hat{\epsilon}}\hat{\Gamma}^{\hat{a}}\hat{\Gamma}_{\hat{\mu}}\hat{\lambda} + \tfrac{1}{3}\delta_{\hat{\epsilon}}\hat{\phi}\,\hat{e}_{\hat{\mu}}{}^{\hat{a}}, \tag{22.65}
$$

which combine into an infinitesimal Lorentz transformation of the Zehnbein with parameter

$$
\hat{\sigma}^{\hat{a}\hat{b}} = -\frac{i}{6}\bar{\hat{\epsilon}}\hat{\Gamma}^{\hat{a}\hat{b}}\hat{\lambda}. \tag{22.66}
$$

The same transformation also appears in the other fields with tangent-space indices (i.e. just the fermions) but at higher orders in fermions. Thus, it can be absorbed into a redefinition of the ten-dimensional local Lorentz transformations and that is why we have ignored it.

Another point is that we have obtained these expressions without taking into account the final rescaling of the bosonic fields by powers of $e^{\hat{\phi}_0}$. It can be checked that, if we also rescale the fermions according to

$$
\hat{\psi}_{\hat{\mu}} \rightarrow e^{\frac{1}{6}\hat{\phi}_0}\hat{\psi}_{\hat{\mu}}, \quad \hat{\lambda} \rightarrow e^{-\frac{1}{6}\hat{\phi}_0}\hat{\lambda}, \quad \hat{\epsilon} \rightarrow e^{\frac{1}{6}\hat{\phi}_0}\hat{\epsilon}, \tag{22.67}
$$

the supersymmetry transformation rules remain invariant.

In order to study which solutions of this theory preserve some supersymmetry, it is desirable to include the magnetic potentials in the fermionic supersymmetry transformation rules, since many solutions are naturally expressed in terms of the magnetic variables.[9] Using the general definitions Eqs. (22.57) and (22.58) plus the identity Eq. (D.87), it is easy to prove that

$$\begin{aligned} \frac{1}{k!}\,\hat{\not{G}}^{(k)} &= \frac{(-1)^k}{(10-k)!}\,\hat{\not{G}}^{(10-k)}\hat{\Gamma}_{11} = \tfrac{1}{2}\left\{\frac{1}{k!}\,\hat{\not{G}}^{(k)} + \frac{(-1)^k}{(10-k)!}\,\hat{\not{G}}^{(10-k)}\hat{\Gamma}_{11}\right\}, \\ \frac{1}{3!}\hat{\Gamma}_{11}\,\hat{\not{H}} &= -\frac{1}{7!}e^{2\hat{\phi}}\,\hat{\not{H}}^{(7)} \qquad\qquad = \tfrac{1}{2}\left\{\frac{1}{3!}\hat{\Gamma}_{11}\,\hat{\not{H}} - \frac{1}{7!}e^{2\hat{\phi}}\,\hat{\not{H}}^{(7)}\right\}. \end{aligned} \tag{22.68}$$

Furthermore,

$$\hat{\not{H}}_{\hat{\mu}} = \frac{2}{7!}e^{2\hat{\phi}}\hat{\Gamma}_{11}\hat{\Gamma}_{\hat{\mu}\hat{\nu}_1\cdots\hat{\nu}_7}\hat{H}^{(7)\,\hat{\nu}_1\cdots\hat{\nu}_7} = \tfrac{1}{2}\left\{\hat{\not{H}}_{\hat{\mu}} + \frac{2}{7!}e^{2\hat{\phi}}\hat{\Gamma}_{11}\hat{\Gamma}_{\hat{\mu}\hat{\nu}_1\cdots\hat{\nu}_7}\hat{H}^{(7)\,\hat{\nu}_1\cdots\hat{\nu}_7}\right\}. \tag{22.69}$$

We simply have to substitute these identities into Eqs. (22.64) to obtain

$$\boxed{\begin{aligned} \delta_{\hat{\epsilon}}\hat{\psi}_{\hat{\mu}} &= \left\{\partial_{\hat{\mu}} - \tfrac{1}{4}\left(\hat{\not{\omega}}_{\hat{\mu}} + \tfrac{1}{4}\Gamma_{11}\,\hat{\not{H}}_{\mu} + \frac{1}{2\cdot 7!}e^{2\hat{\phi}}\hat{\Gamma}_{\hat{\mu}\hat{\nu}_1\cdots\hat{\nu}_7}\hat{H}^{(7)\,\hat{\nu}_1\cdots\hat{\nu}_7}\right)\right\}\hat{\epsilon} \\ &\quad + \frac{i}{16}e^{\hat{\phi}}\sum_{n=1}^{n=4}\frac{1}{(2n)!}\,\hat{\not{G}}^{(2n)}\hat{\Gamma}_{\hat{\mu}}\left(-\hat{\Gamma}_{11}\right)^n\hat{\epsilon}, \\ \delta_{\hat{\epsilon}}\hat{\lambda} &= \left[\hat{\not{\partial}}\hat{\phi} + \tfrac{1}{4}\left(\frac{1}{3!}\hat{\Gamma}_{11}\,\hat{\not{H}} - \frac{1}{7!}e^{2\hat{\phi}}\hat{\not{H}}^{(7)}\right)\right]\hat{\epsilon} \\ &\quad + \frac{i}{8}e^{\hat{\phi}}\sum_{n=1}^{n=4}\frac{5-2n}{(2n)!}\,\hat{\not{G}}^{(2n)}\left(-\hat{\Gamma}_{11}\right)^n\hat{\epsilon}. \end{aligned}} \tag{22.70}$$

When using these expressions one should always keep in mind that the magnetic potentials are *not* independent and that, if the magnetic potentials do not vanish, the electric ones will not (and vice versa) and their contributions have to be added.

Finally, it is possible to assign on-shell supersymmetry transformation rules to all the RR potentials (electric and magnetic) [187]. It should also be possible to assign on-shell supersymmetry transformation rules to the dual KR 6-form.

## 22.2 Romans' massive $N=2A, d=10$ supergravity

Romans showed in Ref. [1083] that (in contrast to 11-dimensional or $N=2B, d=10$ supergravity) $N=2A, d=10$ supergravity can be deformed by introducing a mass parameter $m$ while keeping, formally, all the supersymmetry and gauge symmetry, although

[9] It is possible to find supersymmetry transformation rules for the magnetic RR potentials for which the supersymmetry algebra is satisfied on-shell (which is not a problem since these potentials are, anyway, defined only on-shell). See e.g. Ref. [187]. The same is probably true for the NSNS magnetic potentials, but the transformation rules have not been given in the literature.

actually both are broken, as we will see. It is not known how to derive this theory from the standard 11-dimensional SUGRA and it is not possible to deform 11-dimensional SUGRA in any way to include a cosmological constant, while preserving at the same time 11-dimensional Poincaré invariance [110, 435, 436].[10]

The deformation of the $N=2A, d=10$ theory consists in a deformation of the RR field strengths and in a deformation of the Lagrangian. Both are consistent with a generalization of the systematic definition of electric and magnetic RR field strengths in Section 22.1.3 in which we replace the second and fourth of Eqs. (22.52) by[11]

$$\begin{aligned}\hat{G} &= d\hat{C} - \hat{H} \wedge \hat{C} + me^{\hat{B}},\\ \delta\hat{C} &= d\hat{\Lambda}^{(\cdot)} - d\hat{B} \wedge \hat{\Lambda}^{(\cdot)} - m\hat{\Lambda} \wedge e^{\hat{B}},\end{aligned} \tag{22.71}$$

while the equations of motion, Bianchi identities, and duality relations Eqs. (22.57) remain valid as they are. In particular, the 2- and 4-form field strengths are

$$\hat{G}^{(2)} = 2\partial\hat{C}^{(1)} + m\hat{B}. \qquad \hat{G}^{(4)} = 4\partial\hat{C}^{(3)} - 12\partial\hat{B}\hat{C}^{(1)} + 3m\hat{B}\hat{B}, \tag{22.72}$$

and the gauge transformations of the RR 1- and 3-forms are

$$\delta\hat{C}^{(1)} = \partial\hat{\Lambda}^{(0)} - m\hat{\Lambda}, \qquad \delta\hat{C}^{(3)} = 3\partial\hat{\Lambda}^{(2)} - \hat{H}\hat{\Lambda}^{(0)} - 3m\hat{\Lambda}\hat{B}. \tag{22.73}$$

The Lagrangian is deformed by replacing the RR field strengths by the deformed ones, by adding a "cosmological-constant" term proportional to $m^2$, and by adding new terms to the Chern–Simons piece of the action. The action of the bosonic sector is

$$\boxed{\begin{aligned}\hat{S} = \frac{g_A^2}{16\pi G^{(10)}_{\mathrm{NA}}}\int d^{10}\hat{x}\,\sqrt{|\hat{g}|}\Big\{e^{-2\hat{\phi}}\Big[\hat{R} - 4\big(\partial\hat{\phi}\big)^2 + \frac{1}{2\cdot 3!}\hat{H}^2\Big]\\ -\Big[\tfrac{1}{2}m^2 + \frac{1}{2\cdot 2!}\big(\hat{G}^{(2)}\big)^2 + \frac{1}{2\cdot 4!}\big(\hat{G}^{(4)}\big)^2\Big]\\ -\frac{1}{144}\frac{1}{\sqrt{|\hat{g}|}}\hat{\epsilon}\Big[\partial\hat{C}^{(3)}\partial\hat{C}^{(3)}\hat{B} + \tfrac{1}{2}m\partial\hat{C}^{(3)}\hat{B}\hat{B}\hat{B}\\ + \frac{9}{80}m^2\hat{B}\hat{B}\hat{B}\hat{B}\hat{B}\Big]\Big\}.\end{aligned}} \tag{22.74}$$

Actually, there is no cosmological-constant term: in the Einstein frame $m^2$ carries a dilaton factor and so it is a potential for the dilaton. On the other hand, the $\hat{\Lambda}$ gauge

[10] See, however, Refs. [174, 936, 29, 30]. Also, it is possible to derive other ten-dimensional massive theories from $d=11$ [783].

[11] On redefining $\hat{\Lambda}^{(\cdot)}$, the RR gauge transformations can also be rewritten in the form

$$\delta\hat{C} = (d\hat{\Lambda}^{(\cdot)} - m\hat{\Lambda}) \wedge e^{\hat{B}}.$$

transformations can be used to gauge away $\hat{C}^{(1)}$ completely, which is nothing but a Stückelberg field for the NSNS 2-form $\hat{B}$ which transforms its kinetic term in the action into a conventional mass term $m^2\hat{B}^2$, and is the reason why this theory is called *massive*.

One may wonder how some of the fields of the supergravity multiplet can become massive while the theory is formally invariant under $N=2$ local supersymmetry, since linearly realized supersymmetry implies that all states in the same supermultiplet have the same mass. The reason is not just gauge symmetry but also that supersymmetry is broken in this theory. There are two ways to see this: on the one hand, a local supersymmetry transformation can be used to gauge away one dilatino and give mass to one gravitino; on the other hand, the most (super)symmetric vacuum of this theory, which is not Minkowski spacetime but the D8-brane, breaks half of the supersymmetries[12] as well as some of the isometries of Minkowski spacetime. Romans' massive $N=2A, d=10$ supergravity can be interpreted as the effective field theory of type-IIA superstrings with an open-string sector associated with a D8-brane that breaks translation invariance and supersymmetry [1052].

There are good reasons to interpret the mass parameter $m$ as another (0-form) RR field strength,

$$\hat{G}^{(0)} \equiv m. \tag{22.75}$$

A 0-form field strength has to be constant due to the Bianchi identity $d\hat{G}^{(0)}=0$. It can be dualized (on-shell) into a 10-form field strength $\hat{G}^{(10)}$ whose equation of motion is this Bianchi identity and whose Bianchi identity also follows the general rule Eqs. (22.57) and implies the existence of a RR 9-form $\hat{C}^{(9)}$, which must be non-trivial in order to have $\star\hat{G}^{(10)}=-\hat{G}^{(0)}=-m$. A RR 9-form potential was required by string theory since the type-IIA theory admits all even $p$ D$p$-branes, which couple to RR $(p+1)$-form potentials and in massless $N=2A, d=10$ supergravity there is no potential for the D8-brane. Romans' theory describes the effective type-IIA string theory in the presence of D8-branes. The trouble with the 9-form potential is that it does not have dynamical degrees of freedom and, if we include it in the form of a mass parameter, there is a D8-brane; if we do not include it, there is not a D8-brane, whereas, for lower-rank potentials, the *same* theory admits solutions with and without branes.[13]

The 11-dimensional origin of the D8-brane and its associated mass parameter are unknown, although there are arguments based on the superalgebras of $N=2A, d=10$ and $d=11$ supergravity that support the idea that there is a nine-dimensional extended object in $d=11$ (the M9-brane discussed in Refs. [193, 155], also known as the KK9M-brane [907]), which could also be associated with the $(9+1)$-dimensional boundaries of the Hořava–Witten construction discussed in Section 22.4.

[12] Actually, the mass of the KR 2-form should be determined in the D8 background. The Stückelberg mechanism for higher-rank form potentials underlies many gauged higher-dimensional supergravities, but there are cases in which the theory admits a maximally supersymmetric AdS vacuum with respect to which the forms are massless in spite of the explicit "mass terms" they have in the action.

[13] Actually, the theory can be generalized slightly, admitting the possibility that $\hat{G}^{(0)}$ is only piecewise constant, which is equivalent to the introduction of sources for the dual $\hat{C}^{(9)}$ potential which are D8-branes placed at the discontinuities of $\hat{G}^{(0)}$ [186, 170], but we will not consider this generalization here. See the discussion on p. 68.

The supersymmetry transformation rules are given by Eqs. (22.70), where the sums have to be extended up to $n=5$ to include $\hat{G}^{(10)}$ and the new field strengths have to be used. The same is true for the expression for $\hat{H}^{(7)}$, Eq. (22.70).

## 22.3 Further reduction of $N=2A, d=10$ SUEGRA to nine dimensions

Now we consider the dimensional reduction of the action of the massless $N=2A, d=10$ supergravity Eq. (22.38) to nine dimensions. This reduction should give us the effective field theory of the type-IIA superstring compactified on a circle, which we will later compare with the effective theory of the type-IIB theory on another circle in order to find their relations under T duality.[14]

We could have reduced the 11-dimensional theory directly on a 2-torus, obtaining an equivalent result with manifest invariance under global $\mathrm{GL}(2,\mathbb{R})$ transformations (in the Einstein frame), according to the general arguments of Section 15.4. This would facilitate the comparison with the reduction of the $N=2B, d=10$ theory in its manifestly $\mathrm{SL}(2,\mathbb{R})$-invariant form [936] since these two symmetries coincide in nine dimensions [162], although they have very different (geometrical and non-geometrical) origins. However, the compactification in two steps is necessary in order to obtain the T-duality relations between the ten-dimensional fields, since these and their physical interpretation are much simpler in the string frame, with string variables.

We start by reducing the bosonic NSNS sector of the action Eq. (22.38). Apart from the fact that we are going to call $x$ the compact coordinate, $A^{(1)}$ the KK vector, and $A^{(2)}$ the winding vector, the result of this reduction was given in Eq. (21.25) and we can use it directly. The only subtlety has to do with the normalization factor: after integration of the compact coordinate $x \in [0, 2\pi\ell_{\mathrm{s}}]$, we obtain

$$\frac{2\pi\ell_{\mathrm{s}}\hat{g}_{\mathrm{A}}^{2}}{16\pi G_{\mathrm{N\,A}}^{(10)}} = \frac{2\pi\ell_{\mathrm{s}}g_{\mathrm{A}}^{2}k_{0}}{16\pi G_{\mathrm{N\,A}}^{(10)}} = \frac{g_{\mathrm{A}}^{2}}{16\pi G_{\mathrm{N\,A}}^{(9)}}, \tag{22.76}$$

where we have used

$$g_{\mathrm{A}} = \hat{g}_{\mathrm{A}}k_{0}^{-\frac{1}{2}}, \qquad G_{\mathrm{N\,A}}^{(9)} = G_{\mathrm{N\,A}}^{(10)}/(2\pi R_{x}). \tag{22.77}$$

Next, we perform the dimensional reduction of the bosonic RR sector.

### 22.3.1 Dimensional reduction of the bosonic RR sector

The task of reducing the RR field strengths is greatly simplified by using the normalization Eqs. (22.52). We find that the ten-dimensional odd-rank RR potentials split into the following nine-dimensional RR potentials of odd and even rank:

$$\begin{aligned}\hat{C}^{(2n-1)}{}_{\mu_1\cdots\mu_{2n-1}} &= C^{(2n-1)}{}_{\mu_1\cdots\mu_{2n-1}} + (2n-1)A^{(1)}{}_{[\mu_1}C^{(2n-2)}{}_{\mu_2\cdots\mu_{2n-1}]},\\ \hat{C}^{(2n-1)}{}_{\mu_1\cdots\mu_{2n-2}\underline{x}} &= C^{(2n-2)}{}_{\mu_1\cdots\mu_{2n-2}},\end{aligned} \tag{22.78}$$

[14] T duality can also be established between the massive $N=2A, d=10$ supergravity and $N=2B, d=10$ supergravity [936, 186], but, due to lack of space, we will restrict ourselves to the simplest case.

and the even-rank RR field strengths reduce to nine dimensions according to

$$\hat{G}^{(2n)}{}_{a_1\cdots a_{2n}} = G^{(2n)}{}_{a_1\cdots a_{2n}}, \qquad \hat{G}^{(2n)}{}_{a_1\cdots a_{2n-1}x} = k^{-1}G^{(2n-1)}{}_{a_1\cdots a_{2n-1}}. \tag{22.79}$$

The nine-dimensional RR field strengths are defined as follows:

$$\begin{aligned} G^{(2n+1)} &= dC^{(2n)} - HC^{(2n-2)} + F^{(2)}C^{(2n-1)}, \\ G^{(2n)} &= dC^{(2n-1)} - HC^{(2n-3)} + F^{(1)}C^{(2n-2)}, \end{aligned} \tag{22.80}$$

and the nine-dimensional gauge transformations that leave them invariant are

$$\begin{aligned} \delta A^{(i)} &= d\Sigma^{(i)}, \\ \delta B &= d\Lambda - d\Sigma^{(1)}A^{(2)} - d\Sigma^{(2)}A^{(1)}, \\ \delta C^{(2n)} &= d\Lambda^{(2n-1)} - H\Lambda^{(2n-3)} - F^{(2)}\Lambda^{(2n-2)}, \\ \delta C^{(2n+1)} &= d\Lambda^{(2n)} - H\Lambda^{(2n-2)} - F^{(1)}\Lambda^{(2n-1)}. \end{aligned} \tag{22.81}$$

Using the notation introduced in Section 22.1.3, we can write the nine-dimensional RR field strengths and gauge transformations in this way:

$$\begin{aligned} G &= dC - HC + F^{(2)}\Pi_{\text{odd}}C + F^{(1)}\Pi_{\text{even}}C. \\ \delta C &= d\Lambda^{(\cdot)} - H\Lambda^{(\cdot)} - F^{(2)}\Pi_{\text{even}}\Lambda^{(\cdot)} - F^{(1)}\Pi_{\text{odd}}\Lambda^{(\cdot)}. \end{aligned} \tag{22.82}$$

The RR kinetic terms in the action reduce as follows:

$$-\frac{\sqrt{|\hat{g}|}}{2\cdot(2n+2)!}\left(\hat{G}^{(2n+2)}\right)^2 = -\frac{\sqrt{|g|}}{2\cdot(2n+2)!}k\left(G^{(2n+2)}\right)^2 + \frac{\sqrt{|g|}}{2\cdot(2n+2)!}k^{-1}\left(G^{(2n+1)}\right)^2. \tag{22.83}$$

The reduction of the Chern–Simons term is straightforward. On putting everything together, after some integration by parts we obtain

$$\begin{aligned} S = \frac{g_{\text{A}}^2}{16\pi G_{\text{N A}}^{(9)}}\int d^9x\,\sqrt{|g|}\Bigg\{ & e^{-2\phi}\Big[R - 4(\partial\phi)^2 + \frac{1}{2\cdot 3!}H^2 + (\partial\ln k)^2 \\ & - \tfrac{1}{4}k^2\left(F^{(1)}\right)^2 - \tfrac{1}{4}k^{-2}\left(F^{(2)}\right)^2\Big] - \tfrac{1}{2}\sum_{n=1,\dots,4}\frac{(-1)^n k^{(-1)^n}}{n!}\left(G^{(n)}\right)^2 \\ & - \frac{1}{2^3\cdot 3^2}\frac{\epsilon}{\sqrt{|g|}}\Big[\partial C^{(3)}\partial C^{(3)}A^{(2)} - 3\partial C^{(3)}\partial C^{(2)}\left(B + A^{(1)}A^{(2)}\right) \\ & + 6\partial C^{(3)}\partial A^{(1)}C^{(2)}A^{(2)} + 9\partial C^{(2)}\partial A^{(1)}C^{(2)}\left(B + A^{(1)}A^{(2)}\right) \\ & + 9\partial A^{(1)}\partial A^{(1)}C^{(2)}C^{(2)}A^{(2)}\Big]\Bigg\}. \end{aligned} \tag{22.84}$$

### 22.3.2 Dimensional reduction of fermions and supersymmetry rules

We can use the decomposition of ten- into nine-dimensional gamma matrices explained in Appendix D.1.5, which corresponds to the decomposition of ten-dimensional 32-component fermions,

$$\hat{f} = \begin{pmatrix} f^1 \\ f^2 \end{pmatrix}, \tag{22.85}$$

into their two chiral 16-dimensional halves $f^1$ and $f^2$. Thus, from each ten-dimensional spinor we obtain a pair of nine-dimensional spinors with internal indices $i = 1, 2$, which we do not write explicitly in general and on which the Pauli matrices act. Now, from the two ten-dimensional spinors $\hat{\lambda}$ and $\hat{\psi}_{\hat{\mu}}$ we obtain three pairs of nine-dimensional spinors, $\rho$, $\lambda$, and $\psi_\mu$. The ten-dimensional supersymmetry parameter $\hat{\epsilon}$ gives a pair of nine-dimensional supersymmetry parameters $\epsilon$. The explicit relations are

$$\begin{aligned} \hat{\psi}_\mu &= \psi_\mu + kA^{(1)}{}_\mu \sigma^3 \rho, \qquad & \hat{\lambda} &= \sigma^2(\lambda + \rho), \\ \hat{\psi}_{\underline{x}} &= k\sigma^3 \rho, & \hat{\epsilon} &= \epsilon. \end{aligned} \tag{22.86}$$

Observe that all these nine-dimensional spinors are real (we remind the reader that $\hat{\lambda}$ was defined to be imaginary).

Observe also that, in the dimensional reduction, the Dirac conjugates acquire an extra $\sigma^2$. For instance

$$\bar{\hat{\epsilon}} = \bar{\epsilon}\sigma^2. \tag{22.87}$$

The dimensional reduction of the supersymmetry rules is a repetition of what we did in the reduction from 11 to ten dimensions, and we quote only the final results.

For the NSNS bosons:

$$\boxed{\begin{aligned} &\delta_\epsilon e^a{}_\mu = -i\bar{\epsilon}\Gamma^a \psi_\mu, \qquad \delta_\epsilon k = -ik\bar{\epsilon}\rho, \qquad \delta_\epsilon \phi = -\frac{i}{2}\bar{\epsilon}\lambda, \\ &\delta_\epsilon A^{(1)}{}_\mu = -ik^{-1}\bar{\epsilon}\sigma^3(\psi_\mu - \Gamma_\mu \rho), \quad \delta_\epsilon A^{(2)}{}_\mu = -ik\bar{\epsilon}(\psi_\mu + \Gamma_\mu \rho), \\ &\delta_\epsilon B_{\mu\nu} = -2i\bar{\epsilon}\sigma^3 \Gamma_{[\mu}\psi_{\nu]} + \delta_\epsilon A^{(1)}{}_{[\mu} A^{(2)}{}_{\nu]} + \delta_\epsilon A^{(2)}{}_{[\mu} A^{(1)}{}_{\nu]}. \end{aligned}} \tag{22.88}$$

For the RR bosons:

$$\begin{aligned}
\delta_\epsilon C^{(2n)}{}_{\mu_1\cdots\mu_{2n}} &= 2nie^{-\phi}k^{1/2}\bar{\epsilon}\mathcal{P}_{n-1}\Gamma_{[\mu_1\cdots\mu_{2n-1}}\left[\psi_{\mu_{2n}]} - \frac{1}{4n}\Gamma_{\mu_{2n}]}(\lambda-\rho)\right] \\
&\quad - i2n(2n-1)\bar{\epsilon}\sigma^3 C^{(2n-2)}{}_{[\mu_1\cdots\mu_{2n-2}}\Gamma_{\mu_{2n-1}}\psi_{\mu_{2n}]} \\
&\quad - 2n\delta_\epsilon A^{(2)}{}_{[\mu_1}C^{(2n-1)}{}_{\mu_2\cdots\mu_{2n}]}, \\
\delta_\epsilon C^{(2n-1)}{}_{\mu_1\cdots\mu_{2n-1}} & \\
= -i(2n-1)e^{-\phi}k^{-\frac{1}{2}}\bar{\epsilon}\mathcal{P}_{n-1}\Gamma_{[\mu_1\cdots\mu_{2n-2}} & \left[\psi_{\mu_{2n-1}]} - \frac{1}{2(2n-1)}\Gamma_{\mu_{2n-1}]}(\lambda+\rho)\right] \\
- i(2n-1)(2n-2)\bar{\epsilon}\sigma^3 C^{(2n-3)}{}_{[\mu_1\cdots\mu_{2n-3}}\Gamma_{\mu_{2n-2}}\psi_{\mu_{2n-1}]} & \\
- (2n-1)\delta_\epsilon A^{(1)}{}_{[\mu_1}C^{(2n-2)}{}_{\mu_2\cdots\mu_{2n-1}]}. &
\end{aligned} \tag{22.89}$$

For the fermions:

$$\begin{aligned}
\delta_\epsilon\psi_\mu &= \left\{\partial_\mu - \tfrac{1}{4}\left[\not{\omega}_\mu + \tfrac{1}{2}\not{H}_\mu\sigma^3 + \left(k\,\not{F}^{(1)}{}_\mu\sigma^3 + k^{-1}\,\not{F}^{(2)}{}_\mu\right)\right]\right\}\epsilon \\
&\quad + \frac{1}{16}e^\phi\sum_{i=1}^{8}\frac{1}{n!}k^{\frac{(-1)^n}{2}}\,\not{G}^{(n)}\Gamma_\mu\mathcal{P}_{[\frac{n}{2}]+1}\epsilon, \\
\delta_\epsilon\rho &= \left[\tfrac{1}{2}\,\not{\partial}\ln k + \tfrac{1}{8}\left(k\,\not{F}^{(1)}\sigma^3 - k^{-1}\,\not{F}^{(2)}\right)\right]\epsilon \\
&\quad + \frac{1}{16}e^\phi\sum_{i=1}^{8}\frac{1}{n!}k^{\frac{(-1)^n}{2}}\,\not{G}^{(n)}\mathcal{P}_{[\frac{n}{2}]+1}\epsilon, \\
\delta_\epsilon\lambda &= \left\{\not{\partial}\phi - \frac{1}{12}\not{H}\sigma^3 - \tfrac{1}{8}\left(k\,\not{F}^{(1)}\sigma^3 + k^{-1}\,\not{F}^{(2)}\right)\right\}\epsilon \\
&\quad + \frac{1}{16}e^\phi\sum_{i=1}^{8}(-1)^n\frac{(9-2n)}{n!}k^{\frac{(-1)^n}{2}}\,\not{G}^{(n)}\mathcal{P}_{[\frac{n}{2}]+1}\epsilon.
\end{aligned} \tag{22.90}$$

## 22.4 The effective field theory of the heterotic string

The full action and supersymmetry transformation rules of the $N=2A, d=10$ supergravity theory are invariant under the two $\mathbb{Z}_2$ transformations:

$$\hat{C}^{(2n+1)} \to -\hat{C}^{(2n+1)}, \qquad \hat{\psi}_{\hat{\mu}} \to \pm\hat{\Gamma}_{11}\hat{\psi}_{\hat{\mu}}, \qquad \hat{\lambda} \to \mp\hat{\Gamma}_{11}\hat{\lambda}, \qquad \hat{\epsilon} \to \pm\hat{\Gamma}_{11}\hat{\epsilon}. \tag{22.91}$$

These transformations correspond to the 11-dimensional transformation of the compact coordinate[15] $z \to -z$ combined with the transformation $f \to \pm f$ for all the fermions of

[15] One should remember that the 11-dimensional $\hat{\hat{C}}_{\hat{\hat{\mu}}\hat{\hat{\nu}}\hat{\hat{\rho}}}$ is a pseudotensor.

the theory, which is always a symmetry. Eliminating all the fields which are odd under these transformations is always a consistent truncation of $N = 2A, d = 10$ supergravity that is equivalent, according to the discussion in Section 15.6, to the compactification of 11-dimensional supergravity on the orbifold $S^1/\mathbb{Z}_2$. In the two possible truncations all the RR fields and half of the fermions (a chiral half) are eliminated. The result is a chiral theory that is invariant under supersymmetry transformations generated by a single Majorana–Weyl fermion, i.e. $N = 1, d = 10$ supergravity. The action for the bosonic sector of this theory is that of the common sector Eq. (21.1). As for the supersymmetry transformation rules, defining for all fermions $\hat{f}$

$$\hat{f} = \hat{f}^{(+)} + \hat{f}^{(-)}, \qquad \hat{\Gamma}_{11}\hat{f}^{(\pm)} \equiv \pm\hat{f}^{(\pm)}, \tag{22.92}$$

we obtain, for the two possible truncations,

$$\boxed{\begin{aligned} \delta_{\hat{\epsilon}}\hat{\psi}^{(\pm)}_{\hat{a}} &= \hat{\nabla}^{(\pm)}_{\hat{a}}\hat{\epsilon}^{(\pm)}, \\ \delta_{\hat{\epsilon}}\hat{\lambda}^{(\mp)} &= \left(\not{\partial}\hat{\phi} \pm \frac{1}{12}\hat{\not{H}}\right)\hat{\epsilon}^{(\pm)}, \end{aligned}} \tag{22.93}$$

where $\hat{\nabla}^{(\pm)}_{\hat{a}}$ are the covariant derivatives associated with the two torsional spin connections

$$\hat{\Omega}^{(\pm)}_{\hat{a}\hat{b}\hat{c}} = \hat{\omega}_{\hat{a}\hat{b}\hat{c}} \pm \tfrac{1}{2}\hat{H}_{\hat{a}\hat{b}\hat{c}}. \tag{22.94}$$

From the string/M-theoretical point of view, though, this is not the whole story: first of all one expects to obtain the effective field theory of some $N = 1, d = 10$ superstring theory. There are three of these: type-I SO(32), heterotic with gauge group $E_8 \times E_8$, and heterotic with gauge group SO(32). Hořava and Witten showed in Refs. [760, 761] that, on each of the $(1 + 9)$-dimensional boundaries of the compactified spacetime that correspond to the endpoints of the segment $S^1/\mathbb{Z}_2$, there is an $E_8$ vector supermultiplet, so the orbifold compactification of M theory gives the $E_8 \times E_8$ heterotic-string theory.

This yields an entirely new way to view this string theory which, as is type-IIA, is also related to M theory. For instance, the heterotic-string dilaton measures the distance between the $(1 + 9)$-dimensional boundaries. Also, the dimensional reduction of the heterotic string on a circle can be related to toroidal compactifications of M theory, which are related to type-II string theory. In the end, this will give a relation between heterotic- and type-I string theories, as we will discuss in Chapter 23.

As for the effective action of the heterotic-string theory, it is obtained by coupling the action of pure $N = 1, d = 10$ supergravity [189] whose bosonic sector is given by Eq. (21.1) to the corresponding vector supermultiplets. In the bosonic sector, this requires the addition of the Yang–Mills kinetic term to the action,

$$\boxed{\hat{S} = \frac{\hat{g}_{\mathrm{h}}^2}{16\pi G_{\mathrm{N\,h}}^{(10)}} \int d^{10}\hat{x}\,\sqrt{|\hat{g}|}\,e^{-2\hat{\phi}}\left[\hat{R} - 4\left(\partial\hat{\phi}\right)^2 + \frac{1}{2\cdot 3!}\hat{H}^2 - \tfrac{1}{4}\alpha'\hat{F}^I{}_{\hat{\mu}\hat{\nu}}\hat{F}^{I\,\hat{\mu}\hat{\nu}}\right],} \tag{22.95}$$

together with a modification of the KR field strength by the addition of the Yang–Mills Chern–Simons 3-form term defined in Eq. (A.50) [341]:

$$\hat{H} = 3\partial\hat{B} - \tfrac{1}{2}\alpha'\hat{\omega}_3, \qquad \hat{\omega}_3 = -3\hat{A}^I\hat{F}^I + 2f_{IJK}A^IA^JA^K. \tag{22.96}$$

The supersymmetry transformation rules of the gravitino and dilatino fields are still given by Eqs. (22.93), with $\hat{H}$ as defined above, but we also have to consider that of the gauginos,

$$\boxed{\delta_{\hat{\epsilon}}\hat{\chi}^I = -\frac{\sqrt{2\alpha'}}{8}\,\hat{\not{F}}^I.} \tag{22.97}$$

## 22.5 Toroidal compactification of the heterotic string

In this section we are going to study toroidal compactifications of the heterotic-string effective field theory from $\hat{d} = 10$ to $d = 10 - n$ dimensions. Our goal is to find the T and S dualities that arise in the compactification, especially in $d = 4$ dimensions. In contrast to maximal supergravities (32 supercharges, $N = 2, d = 10$ theories and their toroidal reductions, for example), after dimensional reduction on an $n$-torus the supergravity multiplet becomes reducible into a lower-dimensional supergravity multiplet and $n$ vector multiplets. We will study how to separate the two kinds of fields. This is important, since matter fields can always be consistently truncated, but they have to be correctly identified in order to preserve supersymmetry.

This dimensional reduction was first done in Ref. [331] in the Einstein frame and in Ref. [917] in the string frame, in which a stringy interpretation was given to the dualities found. Here we repeat what they did, first for pure $N = 1, d = 10$ supergravity using our own conventions and emphasizing the relations between ten-dimensional and lower-dimensional fields that will allow us to relate solutions in different dimensions. Later we will add Yang–Mills fields in order to have the complete heterotic-string effective field theory.

### 22.5.1 Reduction of the action of pure $N = 1, d = 10$ supergravity

*The Ricci scalar and dilaton terms.* We can use the notation and ansatz we made for the metric in Section 15.4 and apply immediately Eqs. (15.199)–(15.201), although we have to insert into the first of them the dilaton prefactor $e^{-2\hat{\phi}}$. On defining the $d = (\hat{d} - n)$-dimensional dilaton field by

$$e^{-2\phi} \equiv e^{-2\hat{\phi}}K, \tag{22.98}$$

integrating over the $n$ redundant coordinates, and applying again Palatini's identity to re-express the $d$-dimensional spin-connection coefficients in terms of the Ricci scalar, we obtain

$$\frac{\hat{g}^2}{16\pi G_{\rm N}^{(10)}}\int d^{10}\hat{x}\sqrt{|\hat{g}|}\,e^{-2\hat{\phi}}\left[\hat{R}-4(\partial\hat{\phi})^2\right]$$
$$=\frac{g^2}{16\pi G_{\rm N}^{(d)}}\int d^{d}x\sqrt{|g|}\,e^{-2\phi}\left[R-4(\partial\phi)^2+\tfrac{1}{4}F^2-\tfrac{1}{4}\partial_a G_{mn}\partial^a G^{mn}\right], \tag{22.99}$$

where the $d$-dimensional string coupling constant is

$$g=e^{\phi_0}=e^{\hat{\phi}_0}\frac{1}{\sqrt{K_0}}=\hat{g}\left(\frac{V_n}{(2\pi\ell_{\rm s})^n}\right)^{\frac{1}{2}}, \qquad V_n=(2\pi)^n R_9\cdots R_{(10-n)}, \tag{22.100}$$

and the $d$-dimensional Newton constant $G_{\rm N}^{(d)}$ is related to $G_{\rm N}^{(10)}$ by

$$G_{\rm N}^{(d)}=G_{\rm N}^{(10)}/V_n. \tag{22.101}$$

*The KR term.* As usual, we define the lower-dimensional KR field strength as identical to the higher-dimensional one in flat indices so the gauge invariance is automatically inherited:

$$H_{abc}\equiv\hat{H}_{abc}=e_a{}^{\mu}e_b{}^{\nu}e_c{}^{\rho}\left[\hat{H}_{\mu\nu\rho}-3A^m{}_{[\mu}\hat{H}_{\nu\rho]m}+3A^m{}_{[\mu}A^n{}_{\nu}\hat{H}_{\rho]mn}\right]. \tag{22.102}$$

On the other hand, we have another two gauge-invariant combinations,

$$\hat{H}_{abi}=e_a{}^{\mu}e_b{}^{\nu}e_i{}^{m}\left[\hat{H}_{\mu\nu m}-2A^n{}_{[\mu}\hat{H}_{\nu]mn}\right], \tag{22.103}$$

$$\hat{H}_{aij}=e_a{}^{\mu}e_i{}^{m}e_j{}^{n}\hat{H}_{\mu mn}, \tag{22.104}$$

where

$$\hat{H}_{\mu\nu m}=2\partial_{[\mu}\hat{B}_{\nu]m}, \qquad \hat{H}_{\mu mn}=\partial_{\mu}\hat{B}_{mn}. \tag{22.105}$$

$\hat{H}_{\mu mn}$ is just the field strength for the $d$-dimensional scalars,

$$B_{mn}=\hat{B}_{mn}. \tag{22.106}$$

Although $\hat{H}_{\mu\nu n}$ looks like a good vector-field strength, it is not gauge invariant. It enters into the gauge-invariant combination $\hat{H}_{abi}$, which can be rewritten in this way:

$$\hat{H}_{abi}=e_a{}^{\mu}e_b{}^{\nu}e_i{}^{m}\left[2\partial_{[\mu}\left(\hat{B}_{\nu]m}+A^n{}_{\nu]}\hat{B}_{nm}\right)-2\partial_{[\mu}A^n{}_{\nu]}\hat{B}_{nm}\right]. \tag{22.107}$$

$2\partial_{[\mu}A^n{}_{\nu]}=F^n{}_{\mu\nu}$ is gauge invariant by itself, and so the other piece on the r.h.s. must also be gauge invariant. This suggests the following (re)definition of the $d$-dimensional vector fields and their strengths:

$$\begin{aligned} A^{(1)m}{}_{\mu}&=A^m{}_{\mu}, & F^{(1)m}{}_{\mu\nu}&=2\partial_{[\mu}A^{(1)m}{}_{\nu]},\\ A^{(2)}{}_{m\,\mu}&=\hat{B}_{\mu m}-A^n{}_{\mu}\hat{B}_{nm}, & F^{(2)}{}_{m\,\mu\nu}&=2\partial_{[\mu}A^{(2)}{}_{|m|\nu]}, \end{aligned} \tag{22.108}$$

which leads to

$$\hat{H}_{abi} = e_i{}^m \left( F^{(2)}{}_{m\,ab} + F^{(1)n}{}_{ab} B_{nm} \right). \tag{22.109}$$

Using these results in the KR field strength Eq. (22.102), we arrive at the following natural definition for the $d$-dimensional axion field:

$$B = \hat{B} - A^{(1)m} \hat{B}_{mn} A^{(1)n} + A^{(1)m} A^{(2)}{}_m, \tag{22.110}$$

which implies

$$H = 3\partial B - \tfrac{3}{2} A^{(1)m} F^{(2)}{}_m - \tfrac{3}{2} A^{(2)}{}_m F^{(1)m}. \tag{22.111}$$

Finally, we have

$$\hat{H}^2 = H^2 + 3\hat{H}_{abi}\hat{H}^{abi} + 3\hat{H}_{aij}\hat{H}^{aij} = H^2 + 3\mathcal{F}^2 + 3G^{mn}G^{pq}\partial_\mu B_{mp}\partial^\mu B_{nq}, \tag{22.112}$$

where we use the shorthand notation

$$\mathcal{F}_m = F^{(2)}{}_m + F^{(1)p} B_{pm}, \qquad \mathcal{F}^2 = G^{mn}\mathcal{F}_m\mathcal{F}_n. \tag{22.113}$$

On putting together the results of this and Section 22.4, we obtain

$$\boxed{\begin{aligned} S = {} & \frac{g_{\mathrm{h}}^2}{16\pi G_{\mathrm{N}}^{(d)}} \int d^dx \sqrt{|g|}\, e^{-2\phi} \Big\{ R - 4(\partial\phi)^2 + \frac{1}{2\cdot 3!} H^2 \\ & - \tfrac{1}{4}[\partial_\mu G_{mn}\partial^\mu G^{mn} - G^{mn}G^{pq}\partial_\mu B_{mp}\partial^\mu B_{nq}] + \tfrac{1}{4}\left(F^{(1)}\right)^2 + \tfrac{1}{4}\mathcal{F}^2 \Big\}. \end{aligned}} \tag{22.114}$$

This is essentially the result we were after, although we will have to massage it a bit more and we will also have to add the fermions to the picture. We also wanted the relation between the ten- and $d$-dimensional fields in order to be able to reconstruct in ten-dimensional language four-dimensional solutions. If we have such a solution in terms of the fields $g_{\mu\nu}$, $B_{\mu\nu}$, $A_\mu^{(1)m}$, $A^{(2)}{}_{m\mu}$, $G_{mn}$, $B_{mn}$, and $\phi$, the ten-dimensional fields of the corresponding ten-dimensional solutions are given by

$$\begin{aligned} \hat{g}_{\mu\nu} &= g_{\mu\nu} + A^{(1)m}{}_\mu A^{(1)n}{}_\nu G_{mn}, & \hat{g}_{mn} &= G_{mn}, \\ \hat{B}_{\mu\nu} &= B_{\mu\nu} + A^{(1)m}{}_\mu A^{(1)n}{}_\nu B_{mn} - A^{(1)m}{}_{[\mu} A^{(2)}{}_{|m|\nu]}, & \hat{B}_{mn} &= B_{mn} \\ \hat{B}_{\mu m} &= A^{(2)}{}_{m\mu} + A^{(1)n}{}_\mu B_{nm}, & \hat{g}_{\mu m} &= A^{(1)n}{}_\mu G_{nm}, \\ \hat{\phi} &= \phi + \tfrac{1}{4}\ln(\det G). \end{aligned} \tag{22.115}$$

*Manifestly* $\mathrm{O}(n,n)$*-symmetric action.* The $d$-dimensional action that we have just obtained has a global $\mathrm{GL}(n,\mathbb{R})$ invariance that acts on the indices associated with the compact dimensions. In particular, the $\mathrm{O}(n)$ subgroup acts irreducibly on the vectors $A^{(1)m}$ and $A^{(2)}{}_m$ without mixing them. The same happens for the scalars $G_{mn}$ and $B_{mn}$. We know, though, that, in the $n = 1$ case, Eq. (21.25), the action is invariant under the $\mathbb{Z}_2$ T-duality transformations Eqs. (21.27) that interchange these fields and which, combined with the SO(1,1) rescalings of the fields, form an $\mathrm{O}(1,1)$ duality group. We expect now that all the KK and winding vectors $A^{(1)m}$ and $A^{(2)}{}_m$ can be interchanged independently and also rotated. The resulting T-duality group will be $\mathrm{O}(n,n)$ but this cannot be seen with the action as written.

To make $\mathrm{O}(n,n)$ manifest, following Ref. [917] we define the matrices $G \equiv (G_{mn})$ and $B \equiv (B_{mn})$, construct the $2n \times 2n$ symmetric matrix

$$M = \begin{pmatrix} -G^{-1} & G^{-1}B \\ -BG^{-1} & -G+BG^{-1}B \end{pmatrix}, \qquad M^{-1} = \begin{pmatrix} -G+BG^{-1}B & -BG^{-1} \\ G^{-1}B & -G^{-1} \end{pmatrix}, \tag{22.116}$$

and introduce the $2n \times 2n$ matrix $L$,

$$L = \begin{pmatrix} 0 & \mathbb{I}_{n\times n} \\ \mathbb{I}_{n\times n} & 0 \end{pmatrix} = L^{-1}, \tag{22.117}$$

which is nothing but the $\mathrm{O}(n,n)$ metric $(\mathrm{diag}(+,\cdots,+,-,\cdots,-))$ in a non-diagonal form.[16] The essential property of $M$ is that, as we can see in Eq. (22.116),

$$LML = M^{-1} \quad \Rightarrow M^T LM = L, \tag{22.118}$$

which is the definition of an $\mathrm{O}(n,n)$ matrix in this non-diagonal basis, i.e. $M \in \mathrm{O}(n,n)$. Actually, $M$ parametrizes the coset space, $\mathrm{O}(n,n)/(\mathrm{O}(n)\times\mathrm{O}(n))$, as can be seen by counting the number of independent scalars and comparing it with the dimension of the coset space, and also in the construction of Section 22.5.2.

Using the cyclic property of the trace and Eq. (22.118), we can rewrite the kinetic term of the scalars in the action Eq. (22.114) in this manifestly $\mathrm{O}(n,n)$-invariant way:

$$\tfrac{1}{4}[\partial_\mu G_{mn}\partial^\mu G^{mn} - G^{mn}G^{pq}\partial_\mu B_{mp}\partial^\mu B_{nq}] = \tfrac{1}{8}\mathrm{Tr}(\partial ML\partial ML). \tag{22.119}$$

The scalars are coupled to the vectors and we also need to rewrite their kinetic terms. Defining the $\mathrm{O}(n,n)$ column vectors

$$A^\Sigma = \begin{pmatrix} A^{(1)m}{}_\mu \\ A^{(2)}{}_{m\,\mu} \end{pmatrix}, \qquad F^\Sigma = 2\partial A^\Sigma, \tag{22.120}$$

we can rewrite the kinetic term with them as follows:

$$\tfrac{1}{4}\left(F^{(1)}\right)^2 + \tfrac{1}{4}\mathcal{F}^2 = -\tfrac{1}{4}(M^{-1})_{\Sigma\Lambda}F^\Sigma F^\Lambda = -\tfrac{1}{4}(LML)_{\Sigma\Lambda}F^\Sigma F^\Lambda, \tag{22.121}$$

and give the KR field strength in the form

$$H = 3\partial B - \tfrac{3}{2}L_{\Sigma\Lambda}A^\Sigma F^\Lambda, \tag{22.122}$$

[16] We reserve the symbol $\eta$ for the diagonal form of the $\mathrm{O}(n,n)$ metric.

to arrive at the following $O(n,n)$-invariant action:

$$S = \frac{g_{\rm h}^2}{16\pi G_{\rm N}^{(d)}} \int d^dx \sqrt{|g|}\, e^{-2\phi} \Big\{ R - 4(\partial\phi)^2 + \frac{1}{2\cdot 3!} H^2 - \tfrac{1}{8}{\rm Tr}\,(\partial M L \partial M L) - \tfrac{1}{4}(LML)_{\Sigma\Lambda} F^{\Sigma} F^{\Lambda} \Big\}. \tag{22.123}$$

The scalars and vectors transform under $\Omega \in O(n,n)$ according to

$$M' = \Omega M \Omega^T, \qquad F' = \Omega F. \tag{22.124}$$

In the supergravity literature [331, 172, 1085] $\eta$, the diagonal metric of $O(n,n)$, is used instead of $L$. The diagonalization is important since, as we are going to see, it distinguishes vector fields in the supergravity multiplet (*graviphotons*) from vectors in the matter supermultiplets. To do this, it suffices to perform a change of basis:

$$\mathcal{A} = RA, \qquad \mathcal{F} = RF, \qquad \mathcal{M} = (R^{-1})^T M R^{-1}, \tag{22.125}$$

$$\eta = (R^{-1})^T L R^{-1} = \begin{pmatrix} \mathbb{I}_{n\times n} & 0 \\ 0 & -\mathbb{I}_{n\times n} \end{pmatrix}, \tag{22.126}$$

$$R = (R^{-1})^T = \frac{1}{\sqrt{2}} \begin{pmatrix} \mathbb{I}_{n\times n} & \mathbb{I}_{n\times n} \\ -\mathbb{I}_{n\times n} & \mathbb{I}_{n\times n} \end{pmatrix}. \tag{22.127}$$

The action and the KR field strength take the same form with $L$ replaced by $\eta$ and the vector and scalar fields $A$ and $M$ replaced by $\mathcal{A}$ and $\mathcal{M}$.

*Manifestly $O(p,p)$-covariant equations of motion.* It is clear that the equations of motion for the metric, KR, dilaton, and vector fields are automatically $O(p,p)$ covariant. However, $\mathcal{M}$ is a constrained matrix and its equations of motion have to be calculated with care. It can be shown that they can be put into the manifestly $O(p,p)$-covariant form

$$\nabla_\mu \Big( e^{-2\phi} \mathcal{M}^{-1} \partial^\mu \mathcal{M} \Big) = \tfrac{1}{2} e^{-2\phi} \big( \eta \mathcal{M} \eta \mathcal{F} \mathcal{F}^T - \eta \mathcal{F} \mathcal{F}^T \eta \mathcal{M} \big). \tag{22.128}$$

This equation transforms in the adjoint representation of $O(p,p)$.

### 22.5.2 Reduction of the fermions and supersymmetry rules of $N = 1, d = 10$ SUGRA

To reduce the fermions and the supersymmetry rules, we need to construct Vielbeins for the $\sigma$-model scalars, i.e. Vielbeins in the coset space $O(n,n)/(O(n) \times O(n))$. We define the matrix $E \equiv (e^i{}_m)$ so $E^{-1} = (e^m{}_i)$ and $E^T E = -G$ and construct the $2n \times 2n$ matrix

$$V \equiv \frac{1}{\sqrt{2}} \begin{pmatrix} -E + (E^{-1})^T B & -(E^{-1})^T \\ -E - (E^{-1})^T B & (E^{-1})^T \end{pmatrix}, \tag{22.129}$$

with inverse

$$V^{-1} = -\frac{1}{\sqrt{2}}\begin{pmatrix} E^{-1} & E^{-1} \\ E^{T}+BE^{-1} & -E^{T}+BE^{-1} \end{pmatrix}. \tag{22.130}$$

This matrix satisfies

$$V^{T}V = M^{-1}, \qquad V^{-1}(V^{-1})^{T} = M, \tag{22.131}$$

and transforms under global $\Omega \in \mathrm{O}(n,n)$ transformations on the right and local $N \in \mathrm{O}(n)\times\mathrm{O}(n)$ on the left,

$$V' = NV\Omega. \tag{22.132}$$

It is natural to use indices $V = \left(V^{A}{}_{\Sigma}\right)$ and $V^{-1} = \left(V^{\Sigma}{}_{A}\right)$, where $A, B = 1,\ldots,2n$. On the other hand, the change of basis from $L$ to the diagonal metric $\eta$ is

$$\mathcal{V} \equiv VR^{-1}. \tag{22.133}$$

The combinations $\mathcal{V}\mathcal{F}$ and $\mathcal{V}\partial\mathcal{V}^{-1}$ which are invariant under global $\mathrm{O}(n,n)$ transformations appear naturally in the reduction of the fermionic supersymmetry transformation rules. We are interested only in the purely bosonic transformation rules (with all fermions set to zero). The reduction is made in two steps: first one reduces all the tensor fields, then one decomposes ten-dimensional 32-component spinors into $d$-dimensional spinors with extra internal indices and ten-dimensional gamma matrices into tensor products of $d$-dimensional gamma matrices and matrices associated with the internal symmetries. The second step depends strongly on $n$ and, thus, we are going to perform only the first step, with the ultimate goal of finding the right truncation of the matter vector fields.

It is convenient to split the $2n$-dimensional indices $A, B$ into $A_1, B_1$ running from 1 to $n$ and $A_2, B_2$ that take values between $n+1$ and $2n$. Furthermore, in order to indicate the correct contractions with the gamma matrices, we have defined the $2n$ "vector"

$$\hat{\Gamma}^{A} = (\hat{\Gamma}^{d+1},\ldots,\hat{\Gamma}^{d+p},\hat{\Gamma}^{d+1},\ldots,\hat{\Gamma}^{d+p}). \tag{22.134}$$

The result of the reduction is, then,

$$\begin{aligned}
\delta_{\hat{\epsilon}}\hat{\psi}_{a}^{(+)} &= \nabla_{a}^{(+)}\,\hat{\epsilon}^{(+)} + \frac{\sqrt{2}}{4}\hat{\Gamma}_{A_1}\mathcal{V}^{A_1}{}_{\Sigma}\,\mathcal{F}^{\Sigma}{}_{a}\hat{\epsilon}^{(+)} \\
&\quad - \tfrac{1}{4}\hat{\Gamma}_{A_1}\mathcal{V}^{A_1}{}_{\Sigma}\partial_{a}\mathcal{V}^{\Sigma}{}_{B_1}\hat{\Gamma}^{B_1}\hat{\epsilon}^{(+)}, \\
\delta_{\hat{\epsilon}}\hat{\psi}^{(+)\,A_2} &= -\frac{\sqrt{2}}{8}\mathcal{V}^{A_2}{}_{\Sigma}\,\not{\mathcal{F}}^{\Sigma}\,\hat{\epsilon}^{(+)} - \tfrac{1}{2}\mathcal{V}^{A_2}{}_{\Sigma}\,\not{\partial}\mathcal{V}^{\Sigma}{}_{A_1}\hat{\Gamma}^{A_1}\,\hat{\epsilon}^{(+)}, \\
\delta_{\hat{\epsilon}}(\hat{\lambda}^{(-)} - \hat{\Gamma}^{i}\hat{\psi}_{i}^{(+)}) &= \left(\not{\partial}\phi - \frac{1}{12}\not{H}\right)\hat{\epsilon}^{(+)} + \frac{\sqrt{2}}{8}\hat{\Gamma}_{A_1}\mathcal{V}^{A_1}{}_{\Sigma}\,\not{\mathcal{F}}^{\Sigma}\,\hat{\epsilon}^{(+)}.
\end{aligned} \tag{22.135}$$

It is clear that $\hat{\psi}_{a}^{(+)}$ will split into several $d$-dimensional gravitinos (four in $d=4$, since the ten-dimensional chiral spinors $\hat{\psi}_{a}^{(+)}$ have 16 real components for each $a$, which corresponds to $N=4, d=4$ SUEGRA) and the combination $\hat{\lambda}^{(-)} - \hat{\Gamma}^i \hat{\psi}_i^{(+)}$ will split into several $d$-dimensional dilatinos (again four in $d=4$) since they transform into the dilaton under supersymmetry. The $n$ spinors $\hat{\psi}^{(+)\,A_2}$ transform into vectors and so they will split into $d$-dimensional gauginos ($4n$ in $d=4$) of the $n$ vector supermultiplets. The supersymmetry parameter splits into as many $d$-dimensional supersymmetry parameters as the gravitino, giving the number $N$ of independent supersymmetry transformations that can be performed ($N=4$ in $d=4$).

These supersymmetry transformation rules are clearly covariant under $O(n,n)$ T-duality transformations. This means that any $d$-dimensional solution will have the same number of unbroken supersymmetries after an $O(n,n)$ rotation. The corresponding ten-dimensional solutions may but need not have the same amount of supersymmetry. This is due to the fact that unbroken supersymmetry can be broken in dimensional reduction. We are going to discuss this subtle point in Section 22.6, but obviously it applies to many other situations: for instance the relation between the unbroken supersymmetries of $N=1, d=11$ and $N=2A, d=10$ supergravity solutions.

### 22.5.3 The truncation to pure supergravity

The fields of the reduced theory correspond to pure supergravity (16 supercharges in $d$ dimensions) coupled to $n$ vector supermultiplets. The fields in the supergravity multiplet are the gravitinos $\hat{\psi}_a^{(+)}$, the dilatino $\hat{\lambda}^{(-)} - \hat{\Gamma}^i \hat{\psi}_i^{(+)}$, the graviton $e^a{}_\mu$, the dilaton $\phi$, the KR 2-form $B_{\mu\nu}$, and $n$ of the $2n$ KK and winding vectors $A^\Sigma$. The $n$ vector supermultiplets are made out of the $n^2$ scalars contained in $\mathcal{V}^A{}_\Sigma$, $n$ of the $2n$ KK and winding vectors $A^\Sigma$, and the gauginos $\hat{\psi}^{(+)\,A_2}$. Thus, we know to which supermultiplet each field belongs, except for the vector fields. These, however, can be identified by studying the truncation of the vector multiplets, which consists in

$$E = \mathbb{I}_{n\times n}, \qquad B=0, \qquad \hat{\psi}^{(+)\,A_2}=0, \tag{22.136}$$

plus the vanishing of the matter vector fields. Since the truncation has to be consistent at the level of the equations of motion, if we substitute the above values of the fields into the equations of motion of the theory, we will be forced to set to zero $n$ combinations of the $2n$ vector fields $A^\Sigma$, which are then identified with the matter vector fields. The $n$ orthogonal combinations that remain are the supergravity vector fields.

Substituting $\mathcal{M} = \mathbb{I}_{2n\times 2n}$ into Eq. (22.128) tells us only that $\mathcal{F}^{\Sigma_1}\mathcal{F}^{\Sigma_2}=0$, however, and we also have to impose consistency of the truncation of the supersymmetry transformation rules Eqs. (22.135). On substituting $\mathcal{V}=0$ and $\delta_{\hat{\epsilon}}\hat{\psi}^{(+)\,A_2}=0$, we find

$$\mathcal{F}^{\Sigma_2}=0 \quad \Rightarrow -\frac{1}{\sqrt{2}}(F^{\Sigma_1} - F^{\Sigma_2}) = 0, \tag{22.137}$$

which implies that the combinations $\mathcal{F}^{\Sigma_2}$ are the matter vector fields and the $\mathcal{F}^{\Sigma^1}$ are the supergravity ones.

The action of the truncated pure supergravity $d$-dimensional theory is

$$S = \frac{g_{\text{h}}^2}{16\pi G_{\text{N}}^{(d)}} \int d^d x \sqrt{|g|}\, e^{-2\phi} \left[ R - 4(\partial\phi)^2 + \frac{1}{2\cdot 3!} H^2 - \tfrac{1}{4}\mathcal{F}^{\Sigma_1}\mathcal{F}^{\Sigma_2} \right], \tag{22.138}$$

where

$$H = 3\partial B - \tfrac{3}{2}\mathcal{A}^{\Sigma_1}\mathcal{F}^{\Sigma_1}. \tag{22.139}$$

In $d=4$ (the $n=6$ case), as we are going to see in Section 22.5.5, the KR 2-form can be dualized into a pseudoscalar axion field $a$ and the action is exactly as we wrote it in Eq. (16.58) (up to notational details and the rescaling to the Einstein frame). Then, any solution of $N=4, d=4$ SUEGRA like those we studied in Chapter 12 can be oxidized to a solution of $N=1, d=10$ supergravity and understood as a string solution. Furthermore, if we take into account that the four-dimensional supergravity vectors are

$$\mathcal{A}^m = \frac{1}{\sqrt{2}}(A^{(1)\,m} + A^{(2)}{}_m), \tag{22.140}$$

and the truncation condition implies $A^{(1)\,m} = A^{(2)}{}_m$, we will find that all supersymmetric four-dimensional solutions are also supersymmetric using the ten-dimensional supersymmetry transformation rules.

### 22.5.4 Reduction with additional U(1) vector fields

Let us now consider the reduction of the full action Eq. (22.95). We are going to do this for generic points of the moduli space of toroidal compactifications at which, as discussed on p. 625, the symmetry group is the Abelian $\mathrm{U}(1)^{16}$, but here we will keep the number of vector fields "$p$" generic. Observe that the Yang–Mills terms appear in the heterotic-string effective action at first order in $\alpha'$ and we will explicitly exhibit this constant.

Before we work out this generic case, we should mention a particular but most interesting case: the $\mathrm{E}_8 \times \mathrm{E}_8$ and SO(32) are related by T duality after compactification on a circle.

All we need to reduce now is the KR and vector fields. We just quote the results, since the procedure followed is the same as before. The $p$ vector fields give $p$ vector fields $A^I{}_\mu$ and $p \times n$ scalars $a^I{}_m$:

$$\hat{A}^I{}_\mu = \frac{1}{\sqrt{\alpha'}} A^I{}_\mu + \hat{A}^I{}_m A^{(1)\,m}{}_\mu, \qquad \hat{A}^I{}_m = \frac{1}{\sqrt{\alpha'}} a^I{}_m. \tag{22.141}$$

The KR 2-form gives the same fields in $d$ dimensions but with different definitions:

$$\begin{aligned}
\hat{B}_{\mu\nu} &= B_{\mu\nu} + \hat{B}_{mn} A^{(1)\,m}{}_\mu A^{(1)\,n}{}_\nu - A^{(1)\,m}{}_{[\mu|} A^{(2)}{}_{m|\nu]} - a^I{}_m A^{(1)\,m}{}_{[\mu} A^I{}_{\nu]}, \\
\hat{B}_{\mu m} &= A^{(2)}{}_{m\,\mu} + A^{(1)\,n}{}_\mu B_{nm} + \tfrac{1}{2} a^I{}_m A^I{}_\mu, \\
\hat{B}_{mn} &= B_{mn}.
\end{aligned} \tag{22.142}$$

The vector-field strength decomposes as

$$\hat{F}^I{}_{ab} = \frac{1}{\sqrt{\alpha'}}\big(F^I{}_{ab} + a^I{}_m F^{(1)\,m}{}_{ab}\big),$$
$$\hat{F}^I{}_{ai} = \frac{1}{\sqrt{\alpha'}} e_i{}^m \partial_a a^I{}_m, \tag{22.143}$$

where $F^I{}_{\mu\nu} = 2\partial_{[\mu} A^I{}_{\nu]}$, whereas the KR field strength decomposes as

$$\hat{H}_{aij} = e_i{}^m e_j{}^n \big(\partial_a B_{mn} - \tfrac{1}{2} a^I{}_m \partial_a a^I{}_n + \tfrac{1}{2} a^I{}_n \partial_a a^I{}_m\big),$$
$$\hat{H}_{abi} = e_i{}^m \big(F^{(2)}{}_{m\,ab} - C_{mn} F^{(1)\,n}{}_{ab} + a^I{}_m F^I{}_{ab}\big), \tag{22.144}$$
$$\hat{H}_{abc} = H_{abc},$$

where the $d$-dimensional KR field strength and the scalars $C_{mn}$ are given by

$$H_{\mu\nu\rho} = 3\partial_{[\mu} B_{\nu\rho]} - \tfrac{3}{2} L_{\Sigma\Lambda} A^\Sigma{}_{[\mu} F^\Lambda{}_{\nu\rho]}, \qquad C_{mn} = B_{mn} - \tfrac{1}{2} a^I{}_m a^I{}_n, \tag{22.145}$$

we have defined the $(2n+p)$-dimensional vector

$$(A^\Sigma{}_\mu) = \begin{pmatrix} A^{(1)m}{}_\mu \\ A^{(2)}{}_{m\,\mu} \\ A^I{}_\mu \end{pmatrix}, \qquad F^\Sigma{}_{\mu\nu} = 2\partial_{[\mu} A^\Sigma{}_{\nu]}, \qquad \Sigma = 1, \ldots, 2n+p, \tag{22.146}$$

and $L_{\Sigma\Lambda}$ is the $\mathrm{O}(p, p+n)$ metric in a non-diagonal basis:

$$(L_{\Sigma\Lambda}) = \begin{pmatrix} 0 & \mathbb{I}_{p\times p} & 0 \\ \mathbb{I}_{p\times p} & 0 & 0 \\ 0 & 0 & -\mathbb{I}_{n\times n} \end{pmatrix}. \tag{22.147}$$

On putting everything together, we obtain an action of the form Eq. (22.123) but with the fields and $L_{\Sigma\Lambda}$ defined above and the matrix $M$ now of dimension $(2n+p)\times(2n+p)$ parametrizing an $\mathrm{O}(n, n+p)/(\mathrm{O}(n)\times\mathrm{O}(n+p))$ coset space and given by

$$M = \begin{pmatrix} -G^{-1} & G^{-1}C & G^{-1}a^T \\ C^T G^{-1} & -G - C^T G^{-1} C + a^T a & -C^T G^{-1} a^T + a^T \\ a G^{-1} & -a G^{-1} C + a & \mathbb{I}_{p\times p} - a G^{-1} a^T \end{pmatrix}. \tag{22.148}$$

It can be constructed with the Vielbein

$$V \equiv \frac{1}{\sqrt{2}} \begin{pmatrix} -E + (E^{-1})^T C & -(E^{-1})^T & -(E^{-1})^T a^T \\ -E - (E^{-1})^T C & (E^{-1})^T & (E^{-1})^T a^T \\ \sqrt{2}\, a & 0 & \sqrt{2}\, \mathbb{I}_{p\times p} \end{pmatrix}, \qquad V^T V = M^{-1}. \tag{22.149}$$

All the properties enjoyed by the old $M$ as an $\mathrm{O}(p,p)$ matrix are now enjoyed by the new $M$ as an $\mathrm{O}(p, p+n)$ matrix with the new metric $L$. This metric can also be diagonalized to $\eta = \text{diag}(\mathbb{I}_{n\times n}, -\mathbb{I}_{(n+p)\times(n+p)})$ by the same matrix $R$ given in Eq. (22.127) acting only on the first $2n$ indices. Clearly, all the vector fields associated with the $n+p$ negative eigenvalues of $\eta$ are matter vector fields. The truncation to pure supergravity now includes the conditions

$$a^I{}_m = A^I{}_\mu = 0. \tag{22.150}$$

The T-duality group is now $O(n, n+p)$ and includes the interchange of KK and winding vectors with the U(1) gauge vectors. This is not too surprising if we take into account that the gauge fields of the heterotic string originate from the compactification of the right- or left-moving part of 16 worldsheet scalars.

### 22.5.5 *Trading the KR 2-form for its dual*

As we mentioned in the introduction, in certain dimensions, the symmetry of the compactified theory can be bigger than $O(n, n+p)$, for instance due to the possibility of dualizing fields that can be rotated into other already existing fields. Here we are going to see an important example in $d=4$ dimensions, in which the heterotic-string KR 2-form can be dualized into a pseudoscalar axion field, which, together with the dilaton, parametrizes an $SL(2,\mathbb{R})/SO(2)$ coset space, Eq. (15.225) (the one present in $N=4, d=4$ supergravity, studied in Section 16.2). It turns out that the equations of motion (but not the action) of the full theory are $SL(2,\mathbb{R})$ covariant because the $SL(2,\mathbb{R})$ transformations involve the dualization of the vector fields (which are dual to vector fields precisely in $d=4$). This new hidden symmetry of the supergravity theory has been conjectured to be a non-perturbative S duality of the full heterotic-string theory [553, 1075, 1124].

We are also interested in the dualization of the KR 2-form in $d=6$, in which one obtains another 2-form potential. A transformation of the dilaton and metric brings the theory into the form of the theory that one obtains by compactifying $N=2A, d=10$ supergravity on K3, which is evidence of a strong–weak-coupling duality between the full heterotic-string theory compactified on $T^4$ and the full type-IIA string theory compactified on K3.

The SO(32) heterotic string is also related by another strong–weak-coupling duality to the type-I SO(32) superstring but the relation does not involve the dualization of the (NSNS) KR 2-form, but rather its interchange by a RR 2-form, as we will see in Section 23.5.

Then we are going to perform the dualization of the KR 2-form in arbitrary dimension $d$. The general procedure for Poincaré dualizations is explained in Section 12.7.1: we consider the action as a functional of $H$ and add a Lagrange-multiplier term to enforce the Bianchi identity

$$E^{\mu_1\cdots\mu_{6-n}} \equiv \nabla_\mu \star H^{\mu\mu_1\cdots\mu_{6-n}} + \frac{(-1)^{6-n}}{4}\eta_{\Sigma\Lambda}\mathcal{F}^\Sigma{}_{\nu\rho} \star \mathcal{F}^{\Lambda\nu\rho\mu_1\cdots\mu_{6-n}} = 0. \tag{22.151}$$

The Lagrange-multiplier term that has to be added to Eq. (22.123) (diagonalized, so $L$ is replaced by $\eta$, $F$ by $\mathcal{F}$, etc.) is

$$\frac{g_{\text{h}}^2}{16\pi G_{\text{N}}^{(d)}} \int d^dx\sqrt{|g|}\, \frac{1}{(6-n)!}\tilde{B}_{\mu_1\cdots\mu_{6-n}} E^{\mu_1\cdots\mu_{6-n}}. \tag{22.152}$$

Now we want to eliminate $H$ completely from the action by using its equation of motion:

$$H^{\nu_1\nu_2\nu_3} = \frac{1}{(7-n)!} e^{2\phi} \frac{1}{\sqrt{|g|}} \epsilon^{\mu_1\cdots\mu_{7-n}\nu_1\nu_2\nu_3} \tilde{H}_{\mu_1\cdots\mu_{7-n}}, \tag{22.153}$$

where

$$\tilde{H} = (7-n)\partial\tilde{B} \tag{22.154}$$

is the $(7-n)$-form dual to $H$.

Since $B$ appears in the action only through $H$, $B$ is completely eliminated from this action and replaced by $\tilde{B}$, and we obtain the dual action

$$\begin{aligned} S = \frac{g_{\mathrm{h}}^2}{16\pi G_{\mathrm{N}}^{(d)}} \int d^dx \sqrt{|g|}\, e^{-2\phi} \Big\{ & R - 4(\partial\phi)^2 + \frac{(-1)^d}{2\cdot(7-n)!} e^{4\phi}\tilde{H}^2 - \tfrac{1}{8}\,\mathrm{Tr}\,(\partial\mathcal{M}\eta\partial\mathcal{M}\eta) \\ & - \tfrac{1}{4}(\eta\mathcal{M}\eta)_{\Sigma\Lambda}\,\mathcal{F}^{\Sigma}\mathcal{F}^{\Lambda} \\ & + \frac{(-1)^{6-p}}{8\cdot(6-p)!}\eta_{\Sigma\Lambda}\frac{1}{\sqrt{|g|}}\epsilon\tilde{B}\mathcal{F}^{\Sigma}\mathcal{F}^{\Lambda} \Big\}. \end{aligned} \tag{22.155}$$

Now we study two special cases.

*The* $n=4, d=6$ *case.* In this case $\tilde{H}$ is just another 3-form field strength. The field content of the effective action coincides with the massless sector of the type-IIA string compactified on K3 and, in fact, there is a field redefinition that takes us from the massless fields of the heterotic string compactified on $\mathrm{T}^4$ to the massless fields of the type-IIA string compactified on K3, which supports the duality between the two theories (including the massive modes) [1123, 470, 808, 466, 1275]:

$$\tilde{H} = H_{\mathrm{IIA}}, \qquad \phi = -\phi_{\mathrm{IIA}}, \qquad g_{\mu\nu} = e^{-2\phi_{\mathrm{IIA}}} g_{\mathrm{IIA}\,\mu\nu}. \tag{22.156}$$

On performing this change of variables, we obtain

$$\begin{aligned} S = \frac{g_{\mathrm{h}}^2}{16\pi G_{\mathrm{N}}^{(6)}} \int d^6x \sqrt{|g_{\mathrm{IIA}}|}\, \Big\{ e^{-2\phi_{\mathrm{IIA}}} \Big[ & R_{\mathrm{IIA}} - 4(\partial\phi_{\mathrm{IIA}})^2 + \frac{1}{2\cdot 3!} H_{\mathrm{IIA}}^2 \\ - \tfrac{1}{8}\,\mathrm{Tr}(\partial\mathcal{M}\eta\partial\mathcal{M}\eta)\Big] & - \tfrac{1}{4}\,(\eta\mathcal{M}\eta)_{\Sigma\Lambda}\,\mathcal{F}^{\Sigma}\mathcal{F}^{\Lambda} + \frac{1}{16}\eta_{\Sigma\Lambda}\frac{1}{\sqrt{|g_{IIA}|}}\epsilon B_{IIA}\mathcal{F}^{\Sigma}\mathcal{F}^{\Lambda} \Big\}. \end{aligned} \tag{22.157}$$

Observe that, in these variables, the vector fields do not carry the factor $e^{-2\phi_{\mathrm{IIA}}}$. This is due to the fact that all of them are of RR origin.

*The $n=6, d=4$ case.* To exhibit the new symmetry, we first go to the Einstein frame by rescaling the metric,

$$g_{\mu\nu} = e^{2\phi} g_{E\mu\nu}. \tag{22.158}$$

Using the formulae in Appendix M to rescale the Ricci scalar and defining the complex scalar field $\tau$ that parametrizes the coset space $\mathrm{SL}(2,\mathbb{R})/\mathrm{SO}(2)$,

$$\tau \equiv a + ie^{-2\phi}, \qquad a \equiv \tilde{B}, \tag{22.159}$$

we obtain the action of $N=4, d=4$ SUEGRA coupled to $p$ vector multiplets:

$$\boxed{\begin{aligned} S = \frac{1}{16\pi G_{\mathrm{N}}^{(4)}} \int d^4x \sqrt{|g_{\mathrm{E}}|} \bigg\{ & R_{\mathrm{E}} - \tfrac{1}{2}\frac{\partial_\mu \tau \partial^\mu \overline{\tau}}{(\mathrm{Im}(\tau))^2} - \tfrac{1}{8}\,\mathrm{Tr}(\partial \mathcal{M}\eta\partial\mathcal{M}\eta) \\ & - \tfrac{1}{4} e^{-2\phi}\,(\eta\mathcal{M}\eta)_{\Sigma\Lambda}\,\mathcal{F}^{\Sigma}\mathcal{F}^{\Lambda} + \tfrac{1}{8}\eta_{\Sigma\Lambda}\frac{1}{\sqrt{|g|}} a\epsilon \mathcal{F}^{\Sigma}\mathcal{F}^{\Lambda} \bigg\}. \end{aligned}} \tag{22.160}$$

The truncation of the vector fields $\mathcal{A}^{\Sigma}$, $\Sigma = 7, \ldots, 12+p$ and the scalar fields $\mathcal{M} = \mathbb{I}_{(12+p)\times(12+p)}$ takes us to the pure supergravity action Eq. (16.58). The truncated action is invariant under SO(6) rotations of the vector fields, which are associated with rotations[17] in the internal $\mathrm{T}^6$, and, as discussed in Section 16.2, the equations of motion are covariant under $\mathrm{SL}(2,\mathbb{R})$ transformations of $\tau$, a non-perturbative symmetry that will not have a simple interpretation until we introduce the solitonic 5-brane.

## 22.6 T duality, compactification, and supersymmetry

The hidden symmetries of supergravity theories that we have studied can be used to transform 11- or ten-dimensional solutions with the appropriate number of isometries using one of the mechanisms we described in Chapter 15 to generate new solutions: reduce, use the $d$-dimensional hidden symmetry transformation, and oxidize again. The T-duality Buscher rules of the string common sector can be interpreted as the simplest application of this mechanism, using the $\mathbb{Z}_2$ symmetry of the theory reduced on a circle.

On the other hand, these hidden symmetries of supergravity theories are evidently symmetries of the supersymmetry transformation rules, which means that they preserve the unbroken supersymmetries of the $d$-dimensional solutions, acting covariantly on their Killing spinors.

This may lead us to think that the whole procedure of reduction–dualization–oxidation preserves the unbroken supersymmetry of the 11- or ten-dimensional solutions. There is, however, one loophole in all these arguments: unbroken supersymmetry has to be preserved by dimensional reduction and this requires that the 11- or ten-dimensional Killing

[17] This SO(6) is part of the original T-duality group $\mathrm{O}(6, 6+p)$, but it does not contain any interchanges of winding and KK vectors, which are constrained to be equal by the truncation conditions.

spinors are independent of the internal coordinates, but this need not be true *even if all the bosonic fields of the solution are independent of them.* We have indeed assumed that the supersymmetry parameters are independent of the internal coordinates to obtain the supersymmetry transformation rules of the $d$-dimensional theories, and a Killing spinor depending on the internal coordinates simply would not appear as a $d$-dimensional Killing spinor, the $d$-dimensional solution would not be supersymmetric, and, after dualization and oxidation, in general, the new 11- or ten-dimensional solution will not be supersymmetric either. In those cases, duality does not respect supersymmetry [84, 86, 167, 34, 718, 474].

This is a very interesting phenomenon with potentially important implications and it is worth studying a concrete example: the T-dualization of Minkowski spacetime [167]. We have found T duality by studying string theory in Minkowski spacetime with a compact dimension parametrized by a Cartesian coordinate with radius $R_z$, realizing that the spectrum was identical to that of another string theory in Minkowski spacetime with a compact dimension parametrized by a Cartesian coordinate with radius $\alpha'/R_z$, which is the string solution T dual to the former. Both spacetimes preserve all supersymmetries of $N=1, d=10$ supergravity. The Buscher rules allow us, however, to find T duals to Minkowski spacetime associated with any other of its isometries, not just with the translational ones.

Let us consider, then, the maximally supersymmetric ten-dimensional Minkowski spacetime with a two-dimensional subspace written in polar coordinates $\rho$ and $\theta$:

$$d\hat{s}^2 = dt^2 - d\vec{x}_7^2 - d\rho^2 - \rho^2 d\theta^2. \tag{22.161}$$

The T dual with respect to the isometry associated with shifts in the angular coordinate $\theta \in [0, 2\pi]$ is the solution

$$d\hat{s}'^2 = dt^2 - d\vec{x}_7^2 - d\rho^2 - \rho^{-2} d\theta^2, \qquad e^{-2\hat{\phi}'} = \rho^2. \tag{22.162}$$

On looking into the supersymmetry transformation rules Eqs. (22.93), it takes no time to see that this solution has no unbroken supersymmetries at all because the dilatino variation $\delta_{\hat{\epsilon}}\hat{\lambda} = \rho^{-1}\hat{\Gamma}^9\hat{\epsilon}$ will never vanish for non-trivial $\hat{\epsilon}$.

Apparently, the T-duality transformation has broken completely the supersymmetry of the original background. The technical reason can be traced to the reduction to nine dimensions: the nine-dimensional solution

$$ds^2 = dt^2 - d\vec{x}_7^2 - d\rho^2, \qquad e^{-2\phi} = k = \rho, \tag{22.163}$$

is not supersymmetric according to Eqs. (22.135). The ten-dimensional Killing spinor equations are satisfied and the nine-dimensional ones are not, and the reason why is that the ten-dimensional Killing spinor depends on the internal coordinate $\theta$. This seems to happen whenever the isometry is not translational, but rotational, acting with fixed points.

Physically, the dimensional reduction in the direction $\theta$ is not a compactification in the standard KK sense. On the one hand, there are no non-contractible loops associated with $\theta$ and there are no associated winding modes. On the other hand, the "radius" of the $\theta$ direction, $\rho$, goes to zero in one limit and to infinity in another limit, and this space is certainly not asymptotically the KK vacuum. There seems to be no clear reason to expect these two backgrounds, which are related by Buscher's rules, to be equivalent for strings in the standard sense of interchange of KK and winding modes. Nevertheless, the two backgrounds are related by Buscher's rules and there seems to be no reason why supersymmetry should disappear [34]. It has been suggested that supersymmetry is still present but realized non-locally [718]. For the moment, there is no definite physical explanation for this phenomenon.

# 23

# The type-IIB superstring and type-II T duality

In Chapter 22 we initiated the study of the 11- and ten-dimensional supergravity theories which arise in the low-energy limits of the various string theories and M theory. Our goal was to study the dualities that relate the various string theories and M theory using effective-field-theory actions as we did in Section 21.2 with T duality in the effective action of the common string sector. In the coming chapters we will study these dualities from the point of view of their effect on classical solutions of the effective actions that represent the classical long-range fields of perturbative and non-perturbative states of these theories, as we did in Section 21.3 with the solutions associated with string and winding modes.

In this chapter we are going to study the $N = 2B, d = 10$ (chiral) supergravity theory, the effective field theory of the type-IIB superstring, and how it is related to the $N = 2A$ (non-chiral) theory after compactification on a circle (type-II T duality). Furthermore, we are also going to study the truncations to $N = 1$ theories, which are the effective field theories of the type-I and heterotic superstrings, finding the field-theory version of the type-I/heterotic-string duality.

First, in Section 23.1, we will study the bosonic sector of the theory, giving a non-self-dual action from which one can derive equations of motion that have to be supplemented by the self-duality constraint of the 5-form field strength [148]. We will also give the supersymmetry transformation rules to lowest order in fermions. Then, in Section 23.2 we will study the S-duality symmetry of this theory, which becomes manifest only after several redefinitions.

Next, in Section 23.3 we will perform the dimensional reduction to nine dimensions of $N = 2B, d = 10$ supergravity compactified on a circle. As we will see, the nine-dimensional theory thus obtained is *identical* to the theory obtained by dimensional reduction of the $N = 2A, d = 10$ theory, Eq. (22.84). This will allow us to establish a correspondence between fields of the ten-dimensional $N = 2A$ and $N = 2B$ theories compactified on circles. This correspondence is part of the T duality existing between the two corresponding superstring theories and with our procedure we will have obtained the generalization of the T-duality Buscher rules to type-II theories in Ref. [162] and generalized in Ref. [936].

Finally, in Section 23.5 we will study various consistent truncations of the theory and their relations to $N = 1$ theories and the corresponding full string theories. In particular, we will find a non-perturbative strong–weak-coupling relation between the type-I SO(32) and the heterotic SO(32) superstring theories.

## 23.1 $N = 2B, d = 10$ supergravity in the string frame

The fields of $N = 2B, d = 10$ SUEGRA [1114, 1117, 788] associated with the massless modes of the type-IIB superstring are given in Table 20.2. Actually, as we have mentioned a few times, there are two $N = 2B, d = 10$ theories, with all fermionic chiralities reversed and opposite self-duality properties of the RR 5-form field strength. Here we are going to describe the theory with a RR 4-form $\hat{C}^{(4)+}$ with self-dual 5-form field strength (suppressing the upper index + for convenience),[1]

$$\hat{G}^{(5)} = + \star \hat{G}^{(5)}, \tag{23.1}$$

and negative-chirality pairs of gravitinos and supersymmetry transformation parameters $\hat{\zeta}^{i\,(-)}_{\hat{\mu}}$ and $\hat{\varepsilon}^{i\,(-)}$ and positive-chirality dilatinos $\hat{\chi}^{i\,(+)}$. Suppressing the $\pm$ and the $i = 1, 2$ indices of the SO(2) global symmetry that rotates the fermions for convenience the spinors of the theory are

$$\hat{\Gamma}_{11}\hat{\zeta}_{\hat{\mu}} = -\hat{\zeta}_{\hat{\mu}}, \qquad \hat{\Gamma}_{11}\hat{\varepsilon} = -\hat{\varepsilon}, \qquad \hat{\Gamma}_{11}\hat{\chi} = +\hat{\chi}. \tag{23.2}$$

Although the self-duality equation for $\hat{G}^{(5)}$ looks like a constraint, it is indeed one of the equations of motion of the theory [1114], and it arises as such in the superspace formalism [788]. It is not hard to see that, combined with the Bianchi identity, it gives a conventional-looking equation of motion:

$$\partial \hat{G}^{(5)} - \frac{10}{3}\hat{\mathcal{H}}\hat{G}^{(3)} = 0 \;\Rightarrow \partial \star \hat{G}^{(5)} - \frac{10}{3}\hat{\mathcal{H}}\hat{G}^{(3)} = 0. \tag{23.3}$$

It is known that it is not possible to write a covariant action for this self-duality equation of motion [928]. Nevertheless, having an action is very useful (for instance, to perform dimensional reductions) and we would like to write one. The main observation is that, if we do away with the self-duality of the 5-form, we can find an action that gives the conventional equation of motion Eq. (23.3) [148]. This equation of motion does not imply self-duality when it is combined with the Bianchi identity, but only $\partial \hat{G}^{(5)} = \partial \star \hat{G}^{(5)}$. However, it is consistent with self-duality. This should be reflected in the following property of the action: if we dualize the 4-form, we must end up with an identical action written in terms of the dual 4-form. In other words, the action of the theory of the non-self-dual (NSD) 5-form must itself be "Poincaré self-dual." In our conventions the action of such a Poincaré self-dual NSD theory is

$$\boxed{\begin{aligned} S_{\rm NSD} = {} & \frac{\hat{g}_{\rm B}^2}{16\pi G_{\rm NB}^{(10)}} \int d^{10}\hat{x}\, \sqrt{|\hat{\jmath}|}\, \Big\{ e^{-2\hat{\varphi}} \Big[ \hat{R}(\hat{\jmath}) - 4\,(\partial\hat{\varphi})^2 + \frac{1}{2\cdot 3!}\hat{\mathcal{H}}^2 \Big] \\ & + \tfrac{1}{2}\Big(\hat{G}^{(1)}\Big)^2 + \frac{1}{2\cdot 3!}\Big(\hat{G}^{(3)}\Big)^2 + \frac{1}{4\cdot 5!}\Big(\hat{G}^{(5)}\Big)^2 - \frac{1}{192}\frac{1}{\sqrt{|\hat{\jmath}|}}\hat{\epsilon}\, \partial\hat{C}^{(4)}\partial\hat{C}^{(2)}\hat{\mathcal{B}} \Big\}. \end{aligned}}$$

$$\tag{23.4}$$

[1] All the RR field strengths are normalized as explained in Section 22.1.3 according to Eqs. (22.51) and (22.52), the only difference being the use of calligraphic letters for the NSNS fields $\hat{\jmath}_{\hat{\mu}\hat{\nu}}$, $\hat{\mathcal{B}}_{\hat{\mu}\hat{\nu}}$, and $\hat{\varphi}$.

It is worth remarking that the sign of the last term is linked to the self-duality of the 5-form. If we wanted to impose consistently instead anti-self-duality of $\hat{G}^{(5)}$ and have an action for the opposite chirality $N = 2B, d = 10$ theory, the sign would have been exactly the opposite. This sign will ultimately be related, via T duality, to the sign of the Chern–Simons term of the $N = 2A, d = 10$ theory and, via dimensional oxidation, to the analogous sign of 11-dimensional supergravity.

It is also important to observe that the kinetic term for the 4-form has an extra factor of $\frac{1}{2}$ with respect to the standard normalization. In a sense, this factor takes into account that we have twice the right number of degrees of freedom. When, after dimensional reduction, we eliminate the self-duality constraint, we will obtain fields with the correct normalization thanks to this extra $\frac{1}{2}$.

Observe also that we have introduced, as in the $N = 2A$ case, a prefactor $g_{\rm B}^2$ to absorb in it the asymptotic value of the dilaton, using the definition

$$g_{\rm B} \equiv e^{\hat{\varphi}_0}. \tag{23.5}$$

Assuming, as we do here, that the metric in the string frame is asymptotically flat, then it will also be asymptotically flat in the "modified-Einstein-frame" metric, and we conclude that the Newton constant of this theory is $G_{\rm N\,B}^{(10)}$. Its value is different from that of the IIA theory, so the quotient $\hat{g}_{\rm B}^2/G_{\rm N\,B}{}^{(10)}$ does not depend on $\hat{g}_{\rm B}$ (see Eq. (25.26)). This will be important in finding a frame in which the action is invariant under S duality.

### 23.1.1 *Magnetic potentials*

On comparing the equations of motion of the RR 0- and 2-form potentials,

$$d \star \hat{G}^{(1)} - \hat{\mathcal{H}} \star \hat{G}^{(3)} = 0, \qquad d \star \hat{G}^{(3)} + \hat{\mathcal{H}}\hat{G}^{(5)} = 0, \tag{23.6}$$

with the Bianchi identities of the corresponding dual potentials,

$$d\hat{G}^{(9)} - \hat{\mathcal{H}}\hat{G}^{(7)} = 0, \qquad d\hat{G}^{(7)} - \hat{\mathcal{H}}\hat{G}^{(5)} = 0, \tag{23.7}$$

we find the relation between the original and the dual field strengths:

$$\hat{G}^{(9)} = + \star \hat{G}^{(1)}, \qquad \hat{G}^{(7)} = - \star \hat{G}^{(3)}, \tag{23.8}$$

which defines them, in complete agreement with the first of Eqs. (22.57). Observe that $\hat{C}^{(0)}$ and $\hat{C}^{(2)}$ are pseudotensors and acquire an extra minus sign under parity transformations. $\hat{C}^{(5)}$ has no definite parity properties: in fact, parity transforms the self-duality constraint into an anti-self-duality constraint and chiral spinors into anti-chiral spinors and vice versa. Thus the $N = 2B, d = 10$ supergravity is not invariant under parity, which, in fact, relates the two possible $N = 2B, d = 10$ supergravities. The magnetic potentials $\hat{C}^{(8)}$ and $\hat{C}^{(6)}$ are tensors.

There is a RR potential, electric or magnetic, for all the associated $p$ odd D$p$-branes of the type-IIB string theory, except for the D9-brane that requires a $\hat{C}^{(10)}$ potential, which can be added to the theory at no cost.

Finally, the dual of the NSNS 2-form is a 6-form whose field strength can be written in the form

$$\hat{\mathcal{H}}^{(7)} = d\hat{\mathcal{B}}^{(6)} + \tfrac{1}{2}\sum_{n=0}^{n=3} \star\hat{G}^{(2n+3)} \wedge \hat{C}^{(2n)}. \tag{23.9}$$

### 23.1.2 The type-IIB supersymmetry rules

The supersymmetry transformation rules of $N = 2B, d = 10$ supergravity, generalized to include the magnetic RR potentials and field strengths plus $\hat{C}^{(10)}$, are, suppressing the $i, j = 1, 2$ SO(2) indices in fermions and Pauli matrices [187],

$$\begin{aligned}
\delta_{\hat\varepsilon}\hat{e}_{\hat\mu}{}^{\hat a} &= -i\bar{\hat\varepsilon}\hat\Gamma^{\hat a}\hat\zeta_{\hat\mu},\\
\delta_{\hat\varepsilon}\hat\zeta_{\hat\mu} &= \nabla_{\hat\mu}\hat\varepsilon - \tfrac{1}{8}\not{\hat{\mathcal{H}}}_{\hat\mu}\sigma_3\hat\varepsilon + \frac{1}{16}e^{\hat\varphi}\sum_{n=1,\cdots,5}\frac{1}{(2n-1)!}\not{\hat{G}}^{(2n-1)}\hat\Gamma_{\hat\mu}\mathcal{P}_n\hat\varepsilon,\\
\delta_{\hat\varepsilon}\hat{\mathcal{B}}_{\hat\mu\hat\nu} &= -2i\bar{\hat\varepsilon}\sigma^3\hat\Gamma_{[\hat\mu}\hat\zeta_{\hat\nu]},\\
\delta_{\hat\varepsilon}\hat{C}^{(2n-2)}{}_{\hat\mu_1\cdots\hat\mu_{2n-2}} &= i(2n-2)e^{-\hat\varphi}\bar{\hat\varepsilon}\mathcal{P}_n\hat\Gamma_{[\hat\mu_1\cdots\hat\mu_{2n-3}}\left(\hat\zeta_{\hat\mu_{2n-2}]} - \frac{1}{2(2n-2)}\hat\Gamma_{\hat\mu_{2n-2}]}\hat\chi\right)\\
&\quad + \tfrac{1}{2}(2n-2)(2n-3)\hat{C}^{(2n-4)}{}_{[\hat\mu_1\cdots\hat\mu_{2n-4}}\delta_{\hat\varepsilon}\hat{\mathcal{B}}_{\hat\mu_{2n-3}\hat\mu_{2n-4}]},\\
\delta_{\hat\varepsilon}\hat\chi &= \left(\not{\partial}\hat\varphi - \frac{1}{12}\not{\hat{\mathcal{H}}}\sigma^3\right)\hat\varepsilon + \tfrac{1}{4}e^{\hat\varphi}\sum_{n=1,\cdots,5}\frac{(n-3)}{(2n-1)!}\not{\hat{G}}^{(2n-1)}\mathcal{P}_n\hat\varepsilon,\\
\delta_{\hat\varepsilon}\hat\varphi &= -\frac{i}{2}\bar{\hat\varepsilon}\hat\chi,
\end{aligned} \tag{23.10}$$

where

$$\mathcal{P}_n = \begin{cases}\sigma^1, & n \text{ even},\\ i\sigma^2, & n \text{ odd}.\end{cases} \tag{23.11}$$

Observe that the consistency of these supersymmetry transformations demands that the gravitinos and supersymmetry transformation parameters have the same chirality, opposite to that of the dilatinos. Observe also that, due to the self-duality of $\hat{G}^{(5)}$,

$$\not{\hat{G}}^{(5)} = \not{\hat{G}}^{(5)}\tfrac{1}{2}(1+\hat\Gamma_{11}). \tag{23.12}$$

The $\not{\hat{G}}^{(5)}$ term in $\delta_{\hat\varepsilon}\hat\zeta_{\hat\mu}$ survives due to the negative chirality of $\hat\varepsilon$ and does not survive in $\delta_{\hat\varepsilon}\hat\chi$ for the same reason. This fact plays an important role in the existence of maximally supersymmetric solutions of this theory.

## 23.2 Type-IIB S duality

In the original version of the ten-dimensional, chiral $N = 2$ supergravity [1114] the theory has a classical SU(1,1) global symmetry. The two scalars parametrize the coset space

SU(1,1)/U(1), U(1) being the maximal compact subgroup of SU(1,1), and transform under a combination of a global SU(1,1) transformation and a local U(1) transformation that depends on the global SU(1,1) transformation. They are combinations of the dilaton and the RR scalar. The group SU(1,1) is isomorphic to $\mathrm{SL}(2,\mathbb{R})$, the conjectured classical S-duality symmetry group for the type-IIB string theory [808]. A simple field redefinition [162] is enough to rewrite the action in terms of two real scalars parametrizing the coset space $\mathrm{SL}(2,\mathbb{R})/\mathrm{SO}(2)$, which can now be identified with the dilaton and the RR scalar. Now the S-duality symmetry becomes manifest only when we rescale the metric to work in the Einstein frame:

$$\hat{\jmath}_{\mathrm{E}\,\mu\nu} = e^{-\frac{\varphi}{2}} \jmath_{\mu\nu}. \tag{23.13}$$

The RR potentials we are working with here, however, are not the most appropriate to exhibit manifest $\mathrm{SL}(2,\mathbb{R})$ symmetry. In fact, they have been chosen because they are the most appropriate to study T duality and the worldvolume effective actions of D-branes. In particular, while the NSNS and RR 2-forms we are using form an $\mathrm{SL}(2,\mathbb{R})$ doublet (as we are going to see), their field strengths do not. Furthermore, our self-dual RR 4-form potential $\hat{C}^{(4)}$ is not $\mathrm{SL}(2,\mathbb{R})$ invariant. Thus, for the purpose of exhibiting the $\mathrm{SL}(2,\mathbb{R})$ symmetry it is convenient to perform the following field redefinitions:[2,3]

$$\hat{\vec{\mathcal{B}}} = \begin{pmatrix} \hat{C}^{(2)} \\ \hat{\mathcal{B}} \end{pmatrix}, \qquad \hat{D} = \hat{C}^{(4)} - 3\hat{\mathcal{B}}\hat{C}^{(2)}. \tag{23.14}$$

These new fields undergo the following gauge transformations:

$$\delta\hat{\vec{\mathcal{B}}} = 2\hat{\vec{\Sigma}}, \qquad \delta\hat{D} = 4\partial\hat{\Delta} + 2\hat{\vec{\Sigma}}^{T}\eta\,\hat{\vec{\mathcal{H}}}, \tag{23.15}$$

and have field strengths

$$\begin{aligned} \hat{\vec{\mathcal{H}}} &= 3\partial\hat{\vec{\mathcal{B}}}, \\ \hat{F} &= \hat{G}^{(5)} = +\star\hat{F} = 5\left(\partial\hat{D} - \hat{\vec{\mathcal{B}}}^{\,T}\eta\,\hat{\vec{\mathcal{H}}}\right), \end{aligned} \tag{23.16}$$

where $\eta$ is the $2\times 2$ matrix

$$\eta = i\sigma^2 = \begin{pmatrix} 0 & 1 \\ -1 & 0 \end{pmatrix} = -\eta^{-1} = -\eta^{T}. \tag{23.17}$$

Given the isomorphism $\mathrm{SL}(2,\mathbb{R}) \sim \mathrm{Sp}(2,\mathbb{R})$, it plays the role of an invariant metric:

$$S\eta S^{\mathrm{T}} = \eta \;\Rightarrow\; \eta S\eta^{T} = (S^{-1})^{T}, \qquad S \in \mathrm{SL}(2,\mathbb{R}). \tag{23.18}$$

Next, we define the complex scalar $\hat{\tau}$ that parametrizes the coset space $\mathrm{SL}(2,\mathbb{R})/\mathrm{SO}(2)$,

$$\hat{\tau} = \hat{C}^{(0)} + ie^{-\hat{\varphi}}, \tag{23.19}$$

---

[2] In our conventions all fields are either invariant or transform *covariantly* as opposed to *contravariantly*.

[3] The complete relations (including fermions) between the formulation of the $N=2B$ theory in "stringy variables" that we have introduced in Section 23.1 and the manifestly S-duality-covariant formulation that we are going to introduce in this section can be found in Ref. [602].

and the $2 \times 2$ symmetric $SL(2,\mathbb{R})$ matrix $\hat{\mathcal{M}}_{ij}$, given in Eq. (15.223) in terms of $\hat{\tau}$, that satisfies the property

$$\hat{\mathcal{M}}^{-1} = \eta \hat{\mathcal{M}} \eta^T. \tag{23.20}$$

Under $S \in SL(2,\mathbb{R})$ given by Eq. (15.221) the new variables that we have defined transform as follows:

$$\hat{\mathcal{M}}' = S\hat{\mathcal{M}}S^T, \qquad \hat{\tau}' = \frac{\alpha\hat{\tau} + \beta}{\gamma\hat{\tau} + \delta}, \qquad \hat{\vec{\mathcal{B}}}' = S\hat{\vec{\mathcal{B}}}, \tag{23.21}$$

and the 4-form $\hat{D}$ and the Einstein metric $\hat{\jmath}_{\rm E}$ are inert.

Now, it is a simple exercise to rewrite the NSD $N = 2B$ action in the following manifestly S-duality-invariant form:

$$\begin{aligned}\hat{S}_{\rm NSD} = {} & \frac{\hat{g}_{\rm B}^2}{16\pi G_{\rm NB}^{(10)}} \int d^{10}\hat{x}\, \sqrt{|\hat{\jmath}_{\rm E}|} \left\{ \hat{R}(\hat{\jmath}_{\rm E}) + \tfrac{1}{4}{\rm Tr}\left(\partial\hat{\mathcal{M}}\hat{\mathcal{M}}^{-1}\right)^2 \right. \\ & \left. + \frac{1}{2\cdot 3!}\hat{\vec{\mathcal{H}}}^{\,T}\hat{\mathcal{M}}^{-1}\hat{\vec{\mathcal{H}}} + \frac{1}{4\cdot 5!}\hat{F}^2 - \frac{1}{2^7\cdot 3^3}\frac{1}{\sqrt{|\hat{\jmath}_{\rm E}|}}\epsilon\hat{D}\hat{\vec{\mathcal{H}}}^T\eta\,\hat{\vec{\mathcal{H}}} \right\}.\end{aligned} \tag{23.22}$$

Observe that the factor $\hat{g}_{\rm B}^2/(16\pi G_{\rm NB}^{(10)})$ is S-duality invariant because it does not depend on $\hat{g}_{\rm B}$ (see Eq. (25.26)). Thus it is in the Einstein frame that the full action is invariant under S duality and masses measured in this frame are S-duality invariant. As usual, this is not the metric in which we should measure masses (at least masses that we want to compare with the string spectrum) because, if the string metric is asymptotically flat, the Einstein metric is not. We should use the modified Einstein frame. This metric will also be S-duality invariant, but the action will have the prefactor $1/(16\pi G_{\rm NB}^{(10)})$ which is not invariant and, thus, masses measured in it will not be invariant.

It is easy to find how the stringy fields $\hat{\mathcal{H}}$, $\hat{G}^{(3)}$, and $\hat{C}^{(4)}$ transform under $SL(2,\mathbb{R})$:

$$\begin{aligned}\hat{\mathcal{H}}' &= \left(\delta + \gamma\hat{C}^{(0)}\right)\hat{\mathcal{H}} + \gamma\hat{G}^{(3)}, \\ \hat{G}^{(3)\prime} &= \frac{1}{|\gamma\hat{\tau} + \delta|^2}\left[\left(\delta + \gamma\hat{C}^{(0)}\right)\hat{G}^{(3)} - \gamma e^{-2\hat{\varphi}}\hat{\mathcal{H}}\right], \\ \hat{C}^{(4)\prime} &= \hat{C}^{(4)} - 3\left(\hat{C}^{(2)}\ \hat{\mathcal{B}}\right)\begin{pmatrix}\alpha\gamma & \beta\gamma \\ \beta\gamma & \delta\beta\end{pmatrix}\begin{pmatrix}\hat{C}^{(2)} \\ \hat{\mathcal{B}}\end{pmatrix}.\end{aligned} \tag{23.23}$$

$\hat{\tau}$ transforms as above and we stress that the string metric does transform under $SL(2,\mathbb{R})$:

$$\hat{\jmath}' = |\gamma\hat{\tau} + \delta|\hat{\jmath}. \tag{23.24}$$

Some of the $SL(2,\mathbb{R})$ transformations of $N = 2B, d = 10$ SUEGRA involve an inversion of the dilaton, and, hence, of the string coupling constant $\hat{g}_{\rm B}$, just as we discussed in the case of $N = 4, d = 4$ SUEGRA, the effective theory of the heterotic string (Sections 16.2 and 22.5.5). These are, therefore, non-perturbative transformations from the string-theory point of view, and the perturbative description that we have of it gives little information about them. We can take the point of view that the existence of this symmetry of the supergravity theory indicates the existence of a similar string S duality (with $\mathrm{SL}(2,\mathbb{R})$ broken to $\mathrm{SL}(2,\mathbb{Z})$ by quantum effects such as charge quantization) that relates strongly coupled type-IIB string theory to another weakly coupled type-IIB theory (because the supergravity is invariant under these transformations), and try to check the implications.

As we learnt in Section 12.7 and applied later in the case of S duality in four-dimensional KK theory, one of the main characteristics of S duality is that it interchanges fundamental, perturbative states with solitonic, non-perturbative states. If the S dual of a theory is another theory of the same kind (as is the case here), then the full spectrum of the theory, including perturbative and non-perturbative states, must be S-duality invariant. Thus, if there is S duality in type-IIB superstring theory, there must be (or we must add) non-perturbative states in it that are interchanged with the fundamental-string states that we already know. These non-perturbative states turn out to be D1-branes, namely *D-strings*, and dyonic states known as *pq-strings*. Their presence implies the possible addition of open-string sectors to the IIB theory. Since the type-IIA and -IIB theories are T dual to each other, namely D-strings are T dual to D0- and D2-branes, we will have to admit the possibility of having the corresponding open-string sectors in the IIA theory and, again, due to T duality, D3-branes in the IIB, and so on and so forth.

The very first string solution that we studied, the F1 solution Eq. (21.70), represents the long-range fields associated with a kind of fundamental, perturbative string state. S duality requires the existence of S-dual solutions: D-string and *pq*-string solutions. We will study all these issues related to the stringy interpretation of the supergravity symmetry that we have uncovered in this section in more detail later on.

## 23.3 Dimensional reduction of $N=2B, d=10$ SUEGRA and type-II T duality

In this section we are going to study from the point of view of the type-II string effective actions the T duality between the IIA and IIB superstring theories compactified on circles discovered in Refs. [412, 454]. As in the string common sector, it is possible to find T-duality relations ("type-II Buscher rules") between fields and solutions of both theories by performing the dimensional reduction of the effective field theories ($N=2A$ and $N=2B, d=10$ SUEGRAs) to nine dimensions on a circle. The dimensional reduction of the $N=2A$ theory was performed in Section 22.3 and here we are going to do the same to the action of the $N=2B$ theory Eq. (23.4), assuming now that all the fields are independent of the dual compact coordinate that we call $y$ in this case. Since the NSD action has to be complemented by the self-duality constraint, at the end we will also have to reduce the constraint and then use it in the reduced nine-dimensional action to obtain exactly Eq. (22.84).

We start with the reduction of the NSNS sector, which we have already performed. The result is given in Eq. (21.25) but now we call $y$ the compact coordinate, $A^{(2)}$ the KK vector, $A^{(1)}$ the winding vector, and $k^{-1}$ the KK scalar (a summary of the relations between the ten- and nine-dimensional fields can be found in Section 23.3.1). The asymptotic value of this KK scalar is $R_y/\ell_{\mathrm{s}}$ but, since $k$ is to be identified with the KK scalar coming from the reduction of the IIA theory, it is also $\ell_{\mathrm{s}}/R_x$, and thus we have the T-duality relation between compactification radii

$$\frac{R_y}{\ell_s} = \frac{\ell_s}{R_x}. \tag{23.25}$$

Furthermore, the normalization factor in front of the action is now

$$\frac{2\pi\ell_s\hat{g}_B^2}{16\pi G_{NB}^{(10)}} = \frac{2\pi\ell_s g_B^2 k_0^{-1}}{16\pi G_{NB}^{(10)}} = \frac{g_B^2}{16\pi G_{NB}^{(9)}}, \tag{23.26}$$

where we have used

$$g_B = \hat{g}_B k_0^{\frac{1}{2}}, \qquad G_{NB}^{(9)} = G_{NB}^{(10)}/(2\pi R_y). \tag{23.27}$$

The normalization factor is independent both of the $d=10$ or $d=9$ string coupling constants and of the compactification radius, although the $d=9$ string coupling constant squared and the $d=9$ Newton constant do depend on them in precisely the same way.

This is the T-dual reduction to the one we performed in Section 22.3 for the IIA theory. The ten-dimensional IIB NSNS fields $\hat{\jmath}_{\hat{\mu}\hat{\nu}}, \hat{\mathcal{B}}_{\hat{\mu}\hat{\nu}}$, and $\hat{\varphi}$ decompose in terms of the same nine-dimensional IIA/B NSNS fields in the T-dual way.

As for the dimensional reduction of the RR sector, when we reduced the $N=2A$ theory to nine dimensions, we defined the field strengths and gauge transformations for the nine-dimensional RR potentials in Eqs. (22.79) and (22.80). Since we have completely determined the reduction of the NSNS fields, the reduction of the $N=2B$ RR potentials is completely determined by the requirement that they have the same nine-dimensional field strengths as in the $N=2A$ case. This is achieved by the following identifications:

$$\begin{aligned} \hat{C}^{(2n)}{}_{\mu_1\cdots\mu_{2n}} &= C^{(2n)}{}_{\mu_1\cdots\mu_{2n}} - 2nA^{(2)}{}_{[\mu_1}C^{(2n-1)}{}_{\mu_2\cdots\mu_{2n}]}, \\ \hat{C}^{(2n)}{}_{\mu_1\cdots\mu_{2n-1}\underline{x}} &= -C^{(2n-1)}{}_{\mu_1\cdots\mu_{2n-1}}. \end{aligned} \tag{23.28}$$

The field strengths reduce as follows:

$$\hat{G}^{(2n+1)}{}_{a_1\cdots a_{2n+1}} = G^{(2n+1)}{}_{a_1\cdots a_{2n+1}}, \qquad \hat{G}^{(2n+1)}{}_{a_1\cdots a_{2n}y} = -kG^{(2n)}{}_{a_1\cdots a_{2n}}, \tag{23.29}$$

and the corresponding kinetic terms in the action reduce as follows:

$$\frac{\sqrt{|\hat{\jmath}|}}{2\cdot(2n+1)!}\left(\hat{G}^{(2n+1)}\right)^2 = \frac{\sqrt{|g|}}{2\cdot(2n+1)!}k^{-1}\left(G^{(2n+1)}\right)^2 - \frac{\sqrt{|g|}}{2\cdot(2n)!}k\left(G^{(2n)}\right)^2. \tag{23.30}$$

Although only the electric RR forms appear in the action, the reduction formulae work for the magnetic ones as well. The final check for the RR fields is that the field strengths satisfy the same duality relations as one obtains in the $N=2A$ case because, since we have used the same definitions for the field strengths as in the $N=2A$ case, they satisfy the same Bianchi identities and, then, if they satisfy the same duality relations, they satisfy the same equations of motion. Indeed, the RR field strengths obtained from both theories satisfy

$$G^{(9-k)} = - \star G^{(k)}, \tag{23.31}$$

which is always consistent in $d=9$.

This is clearly enough to conclude that the reduction of the $N = 2A$ and $N = 2B$ theories to $d = 9$ gives the same nine-dimensional theory. However, just as a check, we can complete the reduction of the action of the $N = 2B$ theory and see that it coincides with Eq. (22.84). We are only going to outline how this is done: the NSD action can be reduced to $d = 9$ straightforwardly using the above formulae, but we obtain a theory with 5- and 4-form RR field strengths that originate from the ten-dimensional self-dual 5-form field strength. The nine-dimensional 5- and 4-forms are related by Eq. (23.31), which is a constraint of the action that we have to eliminate in order to arrive at the action Eq. (22.84). To eliminate consistently the constraint and the 5-form, we first Poincaré-dualize it into a second 4-form, following the standard procedure. Finally, we identify the two 4-forms and the result is Eq. (22.84) with a single 4-form and with the correct sign in the Chern–Simons term.

This result allows us to map fields of one ten-dimensional theory onto fields of the other ten-dimensional theory (which is always independent of one coordinate). This mapping is the generalization of Buscher's T-duality rules to type-II theories [162, 936] that we describe in Section 23.3.1, but it is worth making some preliminary remarks.

1. The rules reflect the T-duality rules for D-branes that we discussed on p. 630.

2. We could have reduced the manifestly S-duality invariant action Eq. (23.22) and we would have obtained a manifestly $\mathrm{SL}(2,\mathbb{R})$-invariant action in $d = 9$. As usual in KK compactification, the action would also be invariant under a group $\mathbb{R}^+$ of rescalings of the internal dimension and other $\mathbb{Z}_2$ factors which combine into $\mathrm{GL}(2,\mathbb{R})$, which is the invariance group that one obtains in the reduction from $d = 11$ to $d = 9$. The IIB S duality now has a geometrical interpretation in the IIA theory.

3. It is possible to use the full $\mathrm{SL}(2,\mathbb{R})$ invariance of the action to perform a GDR of the theory [936, 602]. The result is a family of *massive* supergravity theories that depends on three mass parameters transforming in the adjoint of $\mathrm{SL}(2,\mathbb{R})$ that fit into a symmetric *mass matrix*. One of the theories, which depends on a single parameter, is precisely the theory one would obtain by reducing Romans' theory to $d = 9$ [186],[4] but there are other theories that cannot be obtained from known 11- and ten-dimensional supergravities. Most of the $d = 9$ theories obtained in this way are gauged supergravities [378, 987] and the gauge group is determined by the conjugacy class of the chosen mass matrix [196]. While there is a simple string interpretation for Romans' theory based on the D8-brane, the remaining massive/gauged theories have a less-conventional interpretation that is based on non-conventional extended branes.

4. Although only low-rank RR potentials appear in the action, we have extended the relation to the high-rank magnetic RR potentials. This implies the existence of new high-rank RR potentials unrelated to the electric ones: the IIB $\hat{C}^{(8)}$ can be related to the IIA $\hat{C}^{(7)}$, but also to a $\hat{C}^{(9)}$ that exists in Romans' theory only since it is the

[4] This is the theory that one obtains with the GDR ansatz studied in Section 15.5.3. Observe that the mass parameter is naturally quantized since it is a winding number. This implies that the mass parameter of Romans' theory must also be quantized, if we insist on identifying these theories as string theory indicates. This GDR ansatz is related to the RR 9-form potential we are going to discuss next.

magnetic dual of the mass parameter.[5] In turn, the IIA $\hat{C}^{(9)}$ implies the existence of a IIB $\hat{C}^{(10)}$ associated with the D9-brane. On combining these results with S duality, we arrive at the possible existence of the S dual of $\hat{C}^{(10)}$, $\hat{\mathcal{B}}^{(10)}$, whose on-shell supersymmetry transformation is given in Eq. (23.41). A new T duality implies the existence of another $\hat{B}^{(10)}$ in the IIA theory. These potentials play an interesting role, as we are going to see in Section 23.5 [806, 155]. T-duality rules for $\mathcal{B}^{(10)}$ and $\mathcal{B}^{(6)}$ have been given in [515].

### 23.3.1 The type-II T-duality Buscher rules

We are now ready to relate the $N=2A, B, d=10$ fields. For the sake of completeness we summarize the relation between ten- and nine-dimensional fields for each theory first.

**Summary of the type-IIA reduction**

NSNS fields:

$$
\begin{aligned}
\hat{g}_{\mu\nu} &= g_{\mu\nu} - k^2 A^{(1)}{}_{\mu} A^{(1)}{}_{\nu}, & g_{\mu\nu} &= \hat{g}_{\mu\nu} - \hat{g}_{\mu\underline{x}}\hat{g}_{\nu\underline{x}}/\hat{g}_{\underline{xx}},\\
\hat{B}_{\mu\nu} &= B_{\mu\nu} - A^{(1)}{}_{[\mu} A^{(2)}{}_{\nu]}, & B_{\mu\nu} &= \hat{B}_{\mu\nu} + \hat{g}_{[\mu|\underline{x}|}\hat{B}_{\nu]\underline{x}}/\hat{g}_{\underline{xx}},\\
\hat{\phi} &= \phi + \tfrac{1}{2}\ln k, & \phi &= \hat{\phi} - \tfrac{1}{4}\ln|\hat{g}_{\underline{xx}}|,\\
\hat{g}_{\mu\underline{x}} &= -k^2 A^{(1)}{}_{\mu}, & A^{(1)}{}_{\mu} &= \hat{g}_{\mu\underline{x}}/\hat{g}_{\underline{xx}},\\
\hat{B}_{\mu\underline{x}} &= A^{(2)}{}_{\mu}, & A^{(2)}{}_{\mu} &= \hat{B}_{\mu\underline{x}},\\
\hat{g}_{\underline{xx}} &= -k^2, & k &= |\hat{g}_{\underline{xx}}|^{\frac{1}{2}}.
\end{aligned}
\tag{23.32}
$$

RR fields:

$$
\begin{aligned}
\hat{C}^{(2n-1)}{}_{\mu_1\cdots\mu_{2n-1}} &= C^{(2n-1)}{}_{\mu_1\cdots\mu_{2n-1}} + (2n-1)A^{(1)}{}_{[\mu_1} C^{(2n-2)}{}_{\mu_2\cdots\mu_{2n-1}]},\\
\hat{C}^{(2n+1)}{}_{\mu_1\cdots\mu_{2n}\underline{x}} &= C^{(2n)}{}_{\mu_1\cdots\mu_{2n}},\\
C^{(2n-1)}{}_{\mu_1\cdots\mu_{2n-1}} &= \hat{C}^{(2n-1)}{}_{\mu_1\cdots\mu_{2n-1}} - (2n-1)\hat{g}_{[\mu_1|\underline{x}|}\hat{C}^{(2n-1)}{}_{\mu_2\cdots\mu_{2n-1}]\underline{x}}/\hat{g}_{\underline{xx}},\\
C^{(2n)}{}_{\mu_1\cdots\mu_{2n}} &= \hat{C}^{(2n+1)}{}_{\mu_1\cdots\mu_{2n}\underline{x}}.
\end{aligned}
\tag{23.33}
$$

**Summary of the type-IIB reduction**

NSNS fields:

$$
\begin{aligned}
\hat{\jmath}_{\mu\nu} &= g_{\mu\nu} - k^{-2} A^{(2)}{}_{\mu} A^{(2)}{}_{\nu}, & g_{\mu\nu} &= \hat{\jmath}_{\mu\nu} - \hat{\jmath}_{\mu\underline{y}}\hat{\jmath}_{\nu\underline{y}}/\hat{\jmath}_{\underline{yy}},\\
\hat{\mathcal{B}}_{\mu\nu} &= B_{\mu\nu} + A^{(1)}{}_{[\mu} A^{(2)}{}_{\nu]}, & B_{\mu\nu} &= \hat{\mathcal{B}}_{\mu\nu} + \hat{\jmath}_{[\mu|\underline{y}|}\hat{\mathcal{B}}_{\nu]\underline{y}}/\hat{\jmath}_{\underline{yy}},\\
\hat{\varphi} &= \phi - \tfrac{1}{2}\ln k, & \phi &= \hat{\varphi} - \tfrac{1}{4}\ln|\hat{\jmath}_{\underline{yy}}|,\\
\hat{\jmath}_{\mu\underline{y}} &= -k^{-2} A^{(2)}{}_{\mu}, & A^{(1)}{}_{\mu} &= \hat{\mathcal{B}}_{\mu\underline{y}},\\
\hat{\mathcal{B}}_{\mu\underline{y}} &= A^{(1)}{}_{\mu}, & A^{(2)}{}_{\mu} &= \hat{\jmath}_{\mu\underline{y}}/\hat{\jmath}_{\underline{yy}},\\
\hat{\jmath}_{\underline{yy}} &= -k^{-2}, & k &= |\hat{\jmath}_{\underline{yy}}|^{-\frac{1}{2}}.
\end{aligned}
\tag{23.34}
$$

---

[5] As explained on p. 458, a non-vanishing value of the potential $\hat{C}^{(9)}$ is related to the GDR associated with the shifts of $\hat{C}^{(0)}$ that we discussed before, which are in turn related to the mass parameter of Romans' theory. Clearly, the whole picture is consistent.

RR fields:

$$
\begin{aligned}
\hat{\mathcal{C}}^{(2n)}{}_{\mu_1\cdots\mu_{2n}} &= C^{(2n)}{}_{\mu_1\cdots\mu_{2n}} - 2nA^{(2)}{}_{[\mu_1}C^{(2n-1)}{}_{\mu_2\cdots\mu_{2n}]},\\
\hat{\mathcal{C}}^{(2n)}{}_{\mu_1\cdots\mu_{2n-1}\underline{y}} &= -C^{(2n-1)}{}_{\mu_1\cdots\mu_{2n-1}},\\
C^{(2n)}{}_{\mu_1\cdots\mu_{2n}} &= \hat{\mathcal{C}}^{(2n)}{}_{\mu_1\cdots\mu_{2n}} + 2n\hat{\jmath}_{[\mu_1|\underline{y}|}\hat{\mathcal{C}}^{(2n)}{}_{\mu_2\cdots\mu_{2n}]\underline{y}}/\hat{\jmath}_{\underline{yy}},\\
C^{(2n-1)}{}_{\mu_1\cdots\mu_{2n-1}} &= -\hat{\mathcal{C}}^{(2n)}{}_{\mu_1\cdots\mu_{2n-1}\underline{y}}.
\end{aligned}
\tag{23.35}
$$

We obtain the following generalization of Buscher's rules [162, 936].

**From IIA to IIB**

$$
\begin{aligned}
\hat{\jmath}_{\mu\nu} &= \hat{g}_{\mu\nu} - \left(\hat{g}_{\mu\underline{x}}\hat{g}_{\nu\underline{x}} - \hat{B}_{\mu\underline{x}}\hat{B}_{\nu\underline{x}}\right)\Big/\hat{g}_{\underline{xx}}, &\qquad \hat{\jmath}_{\mu\underline{y}} &= \hat{B}_{\mu\underline{x}}/\hat{g}_{\underline{xx}},\\
\hat{\mathcal{B}}_{\mu\nu} &= \hat{B}_{\mu\nu} + 2\hat{g}_{[\mu|\underline{x}}\hat{B}_{\nu]\underline{x}}/\hat{g}_{\underline{xx}}, & \hat{\mathcal{B}}_{\mu\underline{y}} &= \hat{g}_{\mu\underline{x}}/\hat{g}_{\underline{xx}},\\
\hat{\varphi} &= \hat{\phi} - \tfrac{1}{2}\ln|\hat{g}_{\underline{xx}}|, & \hat{\jmath}_{\underline{yy}} &= 1/\hat{g}_{\underline{xx}},\\
\hat{\mathcal{C}}^{(2n)}{}_{\mu_1\cdots\mu_{2n}} &= \hat{C}^{(2n+1)}{}_{\mu_1\cdots\mu_{2n}\underline{x}} + 2n\hat{B}_{[\mu_1|\underline{x}|}\hat{C}^{(2n-1)}{}_{\mu_2\cdots\mu_{2n}]}\\
&\quad - 2n(2n-1)\hat{B}_{[\mu_1|\underline{x}|}\hat{g}_{\mu_2|\underline{x}|}\hat{C}^{(2n-1)}{}_{\mu_3\cdots\mu_{2n}]\underline{x}}/\hat{g}_{\underline{xx}},\\
\hat{\mathcal{C}}^{(2n)}{}_{\mu_1\cdots\mu_{2n-1}\underline{y}} &= -\hat{C}^{(2n-1)}{}_{\mu_1\cdots\mu_{2n-1}}\\
&\quad + (2n-1)\hat{g}_{[\mu_1|\underline{x}|}\hat{C}^{(2n-1)}{}_{\mu_2\cdots\mu_{2n-1}]\underline{x}}/\hat{g}_{\underline{xx}}.
\end{aligned}
\tag{23.36}
$$

**From IIB to IIA**

$$
\begin{aligned}
\hat{g}_{\mu\nu} &= \hat{\jmath}_{\mu\nu} - \left(\hat{\jmath}_{\mu\underline{y}}\hat{\jmath}_{\nu\underline{y}} - \hat{\mathcal{B}}_{\mu\underline{y}}\hat{\mathcal{B}}_{\nu\underline{y}}\right)\Big/\hat{\jmath}_{\underline{yy}}, &\qquad \hat{g}_{\mu\underline{x}} &= \hat{\mathcal{B}}_{\mu\underline{y}}/\hat{\jmath}_{\underline{yy}},\\
\hat{B}_{\mu\nu} &= \hat{\mathcal{B}}_{\mu\nu} + 2\hat{\jmath}_{[\mu|\underline{y}}\hat{\mathcal{B}}_{\nu]\underline{y}}/\hat{\jmath}_{\underline{yy}}, & \hat{B}_{\mu\underline{x}} &= \hat{\jmath}_{\mu\underline{y}}/\hat{\jmath}_{\underline{yy}},\\
\hat{\phi} &= \hat{\varphi} - \tfrac{1}{2}\ln|\hat{\jmath}_{\underline{yy}}|, & \hat{g}_{\underline{xx}} &= 1/\hat{\jmath}_{\underline{yy}},\\
\hat{C}^{(2n+1)}{}_{\mu_1\cdots\mu_{2n+1}} &= -\hat{\mathcal{C}}^{(2n+2)}{}_{\mu_1\cdots\mu_{2n+1}\underline{y}} + (2n+1)\hat{\mathcal{B}}_{[\mu_1|\underline{y}|}\hat{\mathcal{C}}^{(2n)}{}_{\mu_2\cdots\mu_{2n+1}]}\\
&\quad + 2n(2n+1)\hat{\mathcal{B}}_{[\mu_1|\underline{y}|}\hat{\jmath}_{\mu_2|\underline{y}|}\hat{\mathcal{C}}^{(2n)}{}_{\mu_3\cdots\mu_{2n+1}]\underline{y}}/\hat{\jmath}_{\underline{yy}},\\
\hat{C}^{(2n+1)}{}_{\mu_1\cdots\mu_{2n}\underline{x}} &= \hat{\mathcal{C}}^{(2n)}{}_{\mu_1\cdots\mu_{2n}} + 2n\hat{\jmath}_{[\mu_1|\underline{y}|}\hat{\mathcal{C}}^{(2n)}{}_{\mu_2\cdots\mu_{2n}]\underline{y}}/\hat{\jmath}_{\underline{yy}}.
\end{aligned}
\tag{23.37}
$$

## 23.4 Dimensional reduction of fermions and supersymmetry rules

As we discussed on p. 641, we have to take into account carefully the T-duality transformation of the Vielbeins when dealing with fermions. There are two possible rules compatible with fermions, which, combined with the standard KK ansatz Eqs. (15.33), give

Eqs. (21.30). In our case we have already reduced the $N=2A$ fermions and supersymmetry transformation rules using the standard KK ansatz (with $A_\mu$ renamed $A^{(1)}{}_\mu$) and it turns out that we can obtain agreement with those results only by using the lower sign in Eqs. (21.30) (with $B_\mu$ renamed $A^{(2)}{}_\mu$) [719, 720]. To be more explicit, the ansatz that we must use in the reduction of the $N=2B, d=10$ theory is

$$\left(\hat{e}_{\hat{\mu}}{}^{\hat{a}}\right)=\begin{pmatrix} e_\mu{}^a & -k^{-1}A^{(2)}{}_\mu \\ 0 & -k^{-1}\end{pmatrix}, \qquad \left(\hat{e}_{\hat{a}}{}^{\hat{\mu}}\right)=\begin{pmatrix} e_a{}^\mu & -A^{(2)}{}_a \\ 0 & -k\end{pmatrix}. \tag{23.38}$$

The sign is irrelevant in the reduction of the bosonic sector, as we stressed, and, thus, it does not change the type-II Buscher rules we just derived.

The $N=2B, d=10$ spinors are pairs of Majorana–Weyl spinors with only 16 real non-vanishing components out of the 32 in the chiral basis that we are using with $\hat{\Gamma}_{11}=\mathbb{I}_{16\times 16}\otimes\sigma^3$. The indices that label each pair of fermions are usually not explicitly shown. The Pauli matrices that appear in the supersymmetry rules Eqs. (23.10) act only on those indices and both survive in the nine-dimensional theory. In the decomposition of the ten-dimensional gamma matrices that we have used in the type-IIA case new Pauli matrices appear but they do not act on those indices; rather, they act on the chiral (upper and lower) components of the 32-component spinors. These Pauli matrices do not survive the reduction.

Taking all these facts into account, the ten-dimensional 32-component fermions, which, including the supersymmetry parameter, are $\hat{\zeta}^i_{\hat{\mu}}, \hat{\chi}^i$, and $\hat{\varepsilon}^i$, and the nine-dimensional, 16-component fermions $\psi^i_\mu, \lambda^i, \rho^i$, and $\epsilon^i$, are related by

$$\begin{aligned} \hat{\zeta}^i_{\underline{y}} &= \begin{pmatrix} 0 \\ -k^{-1}\rho^i \end{pmatrix}, & \hat{\zeta}^i_\mu &= \begin{pmatrix} 0 \\ \psi^i_\mu - k^{-1}A^{(2)}{}_\mu\rho^i \end{pmatrix}, \\ \hat{\chi}^i &= \sigma^2\begin{pmatrix} 0 \\ \lambda^i-\rho^i \end{pmatrix}, & \hat{\varepsilon}^i &= \begin{pmatrix} 0 \\ \epsilon^i \end{pmatrix}; \end{aligned} \tag{23.39}$$

using these relations, we obtain complete agreement with the nine-dimensional supersymmetry transformation rules that we derived from the $N=2A$ theory, Eqs. (22.88)–(22.90).

Using Eqs. (22.86) and (23.39), we can derive Buscher's rules for the fermions. They are not very interesting except for the supersymmetry parameters, since they can be used for Killing spinors, if they are independent of the compact coordinates, which is not always the case, as we discussed in Section 22.6. Recalling that the gamma matrix which points into the direction into which we T-dualize is $\hat{\Gamma}^9=\mathbb{I}_{16\times 16}\otimes i\sigma^1$, it is immediately possible to derive the two T-duality rules:

$$\boxed{\begin{aligned} &\hat{\epsilon} = \hat{\varepsilon}^2 - i\hat{\Gamma}^9\hat{\varepsilon}^1, \\ &\hat{\varepsilon}^1 = -\frac{i}{2}\hat{\Gamma}^9\left(1+\hat{\Gamma}_{11}\right)\hat{\epsilon}, \qquad \hat{\varepsilon}^2 = \tfrac{1}{2}\left(1-\hat{\Gamma}_{11}\right)\hat{\epsilon}. \end{aligned}} \tag{23.40}$$

## 23.5 Consistent truncations and heterotic/type-I duality

In Section 22.4 we saw how the $N=2A, d=10$ theory could be truncated to $N=1, d=10$ SUGRA to which vector supermultiplets could be coupled. We also studied how this truncation and the addition of $E_8 \times E_8$ vector supermultiplets could be justified from the string/M-theory point of view as arising from an orbifold compactification of 11-dimensional supergravity (M theory) with ten-dimensional $E_8$ vector supermultiplets living on the two boundaries.

In this section we are going to study the possible consistent truncations of the $N=2B, d=10$ theory and their relations to other string effective field theories. We will follow Refs. [187, 1076] and, in particular, we will include in our discussion the RR 10-form potential and its S dual $\mathcal{B}^{(10)}$, to which one can give the on-shell supersymmetry transformation rule

$$\delta_{\hat{\varepsilon}}\hat{\mathcal{B}}^{(10)} = -ie^{-2\hat{\varphi}}\bar{\hat{\varepsilon}}\sigma^3\Big(10\hat{\Gamma}_{[\hat{\mu}_1\cdots\hat{\mu}_9}\hat{\psi}_{\hat{\mu}_{10}]} - \hat{\Gamma}_{\hat{\mu}_1\cdots\hat{\mu}_{10}}\hat{\lambda}\Big). \tag{23.41}$$

Several $\mathbb{Z}_2$ symmetries of the $N=2B, d=10$ theory are known.

1. $(-1)^F$, which changes the signs of all fermions and leaves bosons invariant. It is present in all theories and it is related to the truncation to $N=0$ eliminating all fermions.

2. The symmetry associated with the worldsheet parity symmetry $\Omega$,

$$\hat{f} \to \sigma^1\hat{f}, \quad \hat{C}^{(2n-2)} \to (-1)^n\hat{C}^{(2n-2)}, \quad \hat{\mathcal{B}} \to -\hat{\mathcal{B}}, \quad \hat{\mathcal{B}}^{(10)} \to -\hat{\mathcal{B}}^{(10)}, \tag{23.42}$$

where $\hat{f}$ stands for any fermion doublet. The associated truncations are

$$\hat{C}^{(2n-2)} = 0, \quad n=1,3,5, \qquad \hat{\mathcal{B}} = 0, \qquad \hat{\mathcal{B}}^{(10)} = 0, \qquad (1+\sigma^1)\hat{f} = 0. \tag{23.43}$$

The remaining fields are those of the $N=1, d=10$ supergravity multiplet (plus $\hat{C}^{(10)}$) but they now appear as in the type-I string effective action:

$$\boxed{S = \frac{\hat{g}_{\rm I}^2}{16\pi G_{\rm NI}^{(10)}}\int d^{10}\hat{x}\,\sqrt{|\hat{\jmath}|}\,\Big\{e^{-2\hat{\varphi}}\Big[\hat{R}(\hat{\jmath}) - 4\,(\partial\hat{\varphi})^2\Big] + \frac{1}{2\cdot 3!}\Big(\hat{G}^{(3)}\Big)^2\Big\}.} \tag{23.44}$$

We know that the quotient of the type-IIB string theory by $\Omega$ is equivalent to the introduction of an O9-plane and that consistency requires the introduction of 32 D9-branes whose RR charges and tensions will cancel out exactly those of the O9-plane, introducing at the same time open strings whose massless states fill an SO(32) gauge supermultiplet. At lowest order in $\alpha'$ these vector supermultiplets will contribute to the above action with a term [78]

$$\boxed{\frac{\hat{g}_{\rm I}^2}{16\pi G_{\rm N\,I}^{(10)}}\int d^{10}\hat{x}\,\sqrt{|\hat{\jmath}|}\left\{\frac{\alpha'}{4}e^{-\hat{\varphi}}\,{\rm Tr}_{\rm Adj}\,(F^2)\right\}} \tag{23.45}$$

(with the Killing metric normalized to $K_{IJ}=-\delta_{IJ}$) plus the addition of a Chern–Simons term Eq. (A.50) to $\hat{G}^{(3)}$,

$$\hat{G}^{(3)} \to 3\partial\hat{C}^{(2)} - \tfrac{1}{2}\alpha'\hat{\omega}_3, \tag{23.46}$$

which is needed for the supersymmetric coupling. The sum of Eqs. (23.44) and (23.45) is the effective action of the type-I SO(32) superstring theory. Observe that the vector fields kinetic term carries a dilaton factor $e^{-\hat{\varphi}}$, which is associated with the fact that these terms come from different string diagrams (worldsheet topologies).

3. The product $(-1)^F\Omega$ induces a truncation of the $N=2B, d=10$ SUEGRA that consists in keeping the same bosonic fields and the combination of fermions orthogonal to that of the previous case. The result is again an $N=1, d=10$ SUGRA with bosonic action Eq. (23.44). From the string-theory point of view this truncation has been associated in Ref. [1076] with one in which the tensions of the O9-plane and the D9-branes cancel out, but the charges do not. The corresponding string theory, which has gauge group USp(32) and was constructed in Ref. [1164], is not supersymmetric, even though it is tachyon free. The supersymmetry of the supergravity theory is broken by the coupling to matter, which fails to be supersymmetric because the vector fields are not in the adjoint representation (a necessary condition for supersymmetry). The consistency of the coupling is due to the fact that supersymmetry is spontaneously broken [464, 963], which makes it a fascinating theory.

4. The S-duality transformation $S=\eta$. It does not lead to any supersymmetric truncation, but it allows us to discuss the S duals of other truncations.

5. Those that correspond to the worldsheet transformations $(-1)^{F_{\rm L}}$ and $(-1)^{F_{\rm R}}$, where $F_{\rm L(R)}$ is the spacetime fermion number coming from the left- (right-)movers. These two transformations are related by

$$\Omega(-1)^{F_{\rm L}}\Omega=(-1)^{F_{\rm R}}, \qquad (-1)^F(-1)^{F_{\rm L(R)}}=(-1)^{F_{\rm R(L)}}. \tag{23.47}$$

Their action on the supergravity fields is

$$\hat{f}\to\pm\sigma^3\hat{f}, \qquad \hat{C}^{(2n-2)}\to-\hat{C}^{(2n-2)}. \tag{23.48}$$

The truncation is

$$\hat{C}^{(2n-2)}=0, \quad n=1,\ldots,6, \qquad (1\mp\sigma^3)\hat{f}=0. \tag{23.49}$$

The remaining fields are those of pure $N=1, d=10$ supergravity just as they appear in the heterotic-string effective action Eq. (21.1) (plus $\hat{\mathcal{B}}^{(10)}$). In string theory one has to take into account the twisted sectors that arise and which have been argued to give the type-IIA superstring theory [408]. On the other hand, in Refs. [806, 155] it has been argued that $(-1)^{F_{\rm L}}$ is actually the S dual of $\Omega$,

$$S\Omega S^{-1} = (-1)^{F_{\rm L}}, \tag{23.50}$$

so one can also consider the S dual of the construction that leads to the type-I theory: an O9-plane associated with $(-1)^{F_{\rm L}}$ and 32 S duals of the D9-branes (S9-branes). The result is the heterotic SO(32) superstring theory which arises, then, as the S dual of the type-I SO(32) superstring. These theories are each other's strong-coupling limit [406, 803, 1052]. The fields of the effective theories are related by the strong–weak-coupling transformation

$$\boxed{\hat{\jmath}_{\hat{\mu}\hat{\nu}} = e^{-\hat{\phi}}\hat{g}_{\hat{\mu}\hat{\nu}}, \qquad \hat{\varphi} = -\hat{\phi}_h, \qquad \hat{C}^{(2)}{}_{\hat{\mu}\hat{\nu}} = \hat{B}_{\hat{\mu}\hat{\nu}}.} \tag{23.51}$$

Since $S(-1)^F\Omega S^{-1} = (-1)^{F_{\rm R}}$, one may also expect to find a (non-supersymmetric) heterotic dual of the USp(32) superstring.

We can combine the construction of the type-I theory and this heterotic/type-I duality with our knowledge of type-II T duality. Since we can consider the type-I theory as simply type IIB with one O9-plane and 32 D9-branes, and we know what the T dual of each of them is (type IIA with an O8-plane and 32 D8-branes), we can immediately say that the T dual of the type-I theory (called type I′ [1048]) is essentially a nine-dimensional theory with $N=1$ supersymmetry (16 supercharges) and gauge group SO(32)× $U(1)^2$. Since

$$T\Omega T^{-1} = I_x\Omega, \qquad I_x x = -x, \tag{23.52}$$

the presence of the O8-plane implies that, instead of $\mathbb{R}^9 \times S^1$, we have $\mathbb{R}^9 \times S^1/\mathbb{Z}_2$ and we actually have two O8-planes with RR charge $-16$ in the two nine-dimensional boundaries. Introduction of Wilson lines into the compactification separates the D8-branes, and one can obtain different gauge groups [964, 80].

The S-dual version of this T duality is well known to lead from the heterotic SO(32) theory to the heterotic $E_8 \times E_8$ theory (up to the possible introduction of Wilson lines) with one dimension compactified, which is associated with the Hořava–Witten scenario with one extra dimension compactified, that is, M theory on $S^1 \times S^1/\mathbb{Z}_2$. We have learned that type-IIB S duality is a rotation of the 2-torus on which we compactify M theory, and here we are seeing precisely that the type I′ theory is a rotated version of the heterotic $E_8 \times E_8$ theory compactified on a circle and both are related to M theory. Furthermore, the mysterious objects at the boundaries of the Hořava–Witten scenario, compactified on a circle, are related to the O8-planes and D8-branes. More consequences of these chains of dualities were studied in Ref. [155].

# 24
# Extended objects

## 24.1 Introduction

In the previous chapters we have studied the upper-left- and upper-right-hand boxes of Fig. 20.1 that concern the standard perturbative formulation of string theory and the effective actions of the ten-dimensional string theories (and M theory). We have also learned a bit about the existence of some non-perturbative states in the string spectrum, in particular D-branes and KK and winding modes in compactified theories (the lower-left-hand box of Fig. 20.1). We have studied in the three cases the existence of dualities that related various theories and how these dualities are realized in the worldsheet action (when this is possible, i.e. for T duality) and in the effective actions. We have also mentioned that S dualities and T dualities imply the existence of new solitonic states in the string spectrum.

In this chapter and the next we are going to study systematically the lower-right-hand and central boxes of Fig. 20.1, that is, the solitonic solutions of the string effective field theories and their worldvolume actions. We will study the implications that the various dualities have for them (which are evidently related to the effects of dualities on the effective actions) and for the non-perturbative string spectrum. This chapter will be devoted to a general introduction to extended objects, and in Chapter 25 we will deal specifically with those that occur in string/M theory.

These are subjects with many facets that are related in many ways to each other and to the subjects of the previous chapters. Therefore, it is hopeless to try to give a complete, or even half-complete, account of them in the space that we have at our disposal. Our aim will be to cover the basic material and the essential results and solutions in a unified system of conventions (like the rest of the book), giving pointers to the literature for further developments.

We start in Section 24.2 with a general introduction to the kinematics and dynamics of generic extended objects in which we will discuss various forms of the actions for these objects, their coupling to background fields (Section 24.2.1), and the generalization of the Dirac quantization condition for extended objects (Section 24.2.2).

In Section 24.3 we treat the simplest generic black and extreme solutions of the "$p$-brane $a$-model," which is itself a generalization of the "$a$-model" studied in Section 16.1. The string-theory solutions that we will study later are in general special cases of these general families of solutions.

## 24.2 Generalities

The basic extended objects are known as *$p$-branes*, objects with $p$ spatial dimensions that sweep out $(p+1)$-dimensional worldvolumes as they evolve in time in a $d$-dimensional ambient (or target) spacetime. Strings, which we have already studied, are the simplest examples of $p$-branes ($p=1$), but there are many other examples that differ by their worldvolume dimensions, their worldvolume fields, their couplings to background fields, and other characteristics such as being associated with a compact dimension (such as the KK monopole and more general KK *$p$-branes* [907]). We will discuss these variants of the basic $p$-brane in order of increasing complexity, and in Chapter 25 we will see which of them occur in string/M theory.

The dynamics of all these objects is governed by their $(p+1)$-dimensional worldvolume actions; in what follows we are going to study them and use them as tools to classify the objects. In this chapter we consider only bosonic actions, but in Chapter 25 we will briefly discuss the $\kappa$-symmetric addition of fermions which is required by the coupling to supergravity.

### 24.2.1 Worldvolume actions

The basic dynamical variables of a $p$-brane are the spacetime coordinates of the object $X^{\mu}(\xi)$, $\mu=0,\ldots,d-1$ ($\xi^{i}$, $i=0,\ldots,p$ are the worldvolume coordinates), which give the embedding of the worldvolume in the $d$-dimensional target spacetime and are worldvolume scalar fields. Some $p$-branes may have additional worldvolume fields (scalars, vectors – such as the BI vector of D$p$-branes – and tensors), whose physical meanings will be discussed in Section 25.6.

The simplest worldvolume and spacetime reparametrization-invariant action for a $p$-brane is the generalization of the Nambu–Goto action Eq. (14.1) (in the same notation),

$$\boxed{S^{(p)}_{\rm NG}[X^{\mu}(\xi)] = -T_{(p)}\int d^{p+1}\xi\,\sqrt{|g_{ij}|},} \tag{24.1}$$

which is proportional to the volume swept out by the $p$-brane. The proportionality constant $T_{(p)}$ is the $p$-brane tension and has natural dimensions of $L^{-(p+1)}$ or, equivalently, of mass per unit of spatial $p$-dimensional volume. In fact, let us consider a spacetime that is the direct product of a non-compact $(d-p)$-dimensional spacetime and a $p$-dimensional compact space so the metric of the original spacetime $\hat{g}_{\hat{\mu}\hat{\nu}}=\mathrm{diag}(g_{\mu\nu}, g_{mn})$ and a configuration in which the $p$-brane[1] wraps the $p$-dimensional space so $\hat{X}^{m}=\xi^{m}$ $m=1,\ldots,p$ and the remaining embedding coordinates are independent of the $\hat{X}^{m}=\xi^{m}$. The Nambu–Goto (NG) action becomes the action of a massive particle moving in the

[1] This is the straightforward generalization of a string winding-mode configuration.

$(d-p)$-dimensional spacetime,

$$S^{(p)}_{\rm NG}[X^{\mu}(\xi)] = -T_{(p)}V_{(p)}\int d\xi^{0}\,\sqrt{|g_{\mu\nu}\dot{X}^{\mu}\dot{X}^{\nu}|}\,, \qquad V_{(p)} = \int d\xi^{1}\cdots d\xi^{p}\sqrt{|g_{mn}|}\,, \tag{24.2}$$

where $V_{(p)}$ is the volume of the internal manifold and $T_{(p)}V_{(p)}$ is the mass of the particle if the spacetime metric is asymptotically flat in the $d-p$ directions orthogonal ("transverse") to the worldvolume. Transverse space, the space in which the wrapped $p$-brane moves as a particle, plays a very important role in the definition of mass (as we have just seen) and charge.

As we have discussed several times, on the one hand, the NG action is highly non-linear and, on the other hand, it cannot describe massless (tensionless) objects (also known as *null branes*). These problems are solved by introducing auxiliary fields. Several possibilities have been proposed in the literature. For instance, we can introduce a scalar density field $v$ and write the action

$$\boxed{S^{(p)}[X^{\mu}(\xi), v] = \int d^{p+1}\xi\,\frac{1}{2v}\big[\,|g| + v^{2}T^{2}_{(p)}\big]\,,} \tag{24.3}$$

which is equivalent to the NG action upon elimination of $v$ using its equation of motion. In this action we can take consistently the tensionless limit to obtain a null brane action [897]. Although this action is still highly non-linear, it is useful for certain purposes: we can replace the tension (a constant) by a worldvolume $p$-form potential[2] $c_{(p)\,i_1\cdots i_p}$ whose equation of motion tells us that the dual of its field strength $\mathcal{G}_{(p+1)} = (p+1)\partial c_{(p)}$ is just a constant. The action is [1187]

$$\boxed{S^{(p)}[X^{\mu}(\xi), v, c_{(p)}] = \int d^{p+1}\xi\,\frac{1}{2v}\big[\,|g| + \big(\star\,\mathcal{G}_{(p+1)}\big)^{2}\big]\,,} \tag{24.4}$$

where here $\star\mathcal{G}_{(p+1)} = [1/(p+1)!]\epsilon^{i_1\cdots i_{p+1}}\mathcal{G}_{(p+1)\,i_1\cdots i_{p+1}}$ and the equation of motion of $c_{(p)}$ has the solution

$$\mathcal{G}_{(p+1)\,i_1\cdots i_p} = \frac{T_{(p)}}{v}\epsilon_{i_1\cdots i_p}\,, \tag{24.5}$$

where $T_{(p)}$ arises as just an integration constant. On substituting this solution into the action, we recover exactly the action Eq. (24.3) and then $T_{(p)}$ is identified as the $p$-brane tension. One can also consider solutions in which $\star\mathcal{G}_{(p+1)}/v$ is only piecewise constant, the discontinuities being associated with brane intersections (see, for instance, Ref. [1193] for an application involving string and D-string junctions). On the other hand, these actions are also suitable for supersymmetric objects and can be made $\kappa$-symmetric [173, 195].

[2] This is similar to the replacement of the mass parameter by the RR 9-form potential in Romans' massive supergravity, Section 22.2.

Another, more common, possibility is to introduce an independent metric on the worldvolume $\gamma_{ij}(\xi)$ and write the following Polyakov-type action:

$$S_{\rm P}^{(p)}[X^{\mu},\gamma_{ij}] = -\frac{T_{(p)}}{2}\int d^{p+1}\xi\sqrt{|\gamma|}\left[\gamma^{ij}g_{ij} + (1-p)\right]. \tag{24.6}$$

The "cosmological-constant" term $(p-1)$ has been chosen to vanish in the string case $p=1$ (otherwise conformal invariance would be broken) and for $p \neq 1$ to give $\gamma_{ij} = g_{ij}$ identically as the solution of the equation of motion for $\gamma_{ij}$. On substituting this solution into the above action to eliminate the auxiliary worldvolume metric, one recovers the NG action with the above normalization. (The $p=1$ case was discussed in Section 20.1.)

In the $p \neq 1$ case, the "cosmological-constant" term can be considered as a mass (tension) term by following the same procedure as in the particle case. First we rescale the worldvolume metric,

$$\gamma_{ij} = T_{(p)}^{-\frac{2}{p-1}}\gamma'_{ij}, \tag{24.7}$$

giving

$$S_{\rm P}^{(p)}[X^{\mu},\gamma_{ij}] = -\tfrac{1}{2}\int d^{p+1}\xi\sqrt{|\gamma'|}\left[\gamma'^{\,ij}g_{ij} + T_{(p)}^{\frac{p+1}{p-1}}(1-p)\right]. \tag{24.8}$$

Now we can take the tensionless limit $T_{(p)} \rightarrow 0$ and then again rescale the worldvolume metric to a dimensionless metric, again obtaining an action suitable for describing null branes:

$$S_{\rm P}^{(p)}[X^{\mu},\gamma_{ij}] = -\frac{P_{(p)}}{2}\int d^{p+1}\xi\sqrt{|\gamma''|}\,\gamma''^{\,ij}g_{ij}. \tag{24.9}$$

The procedure of introducing auxiliary fields can also be used to linearize the Born–Infeld action of D-branes [3], as we will see.

*KK-brane actions.* Certain objects necessarily live in spacetimes with compact dimensions, with some of their worldvolume dimensions wrapped around them and no dynamics in those directions. Their worldvolume actions can be formulated as *gauged $\sigma$-models.* The main example is the KK monopole [164, 156, 515], whose interpretation as an extended object of string/M theory is implied by duality, as we will see. The effective actions of the M-theory objects which reduce to those of the objects of the massive type-IIA theory are also given by gauged $\sigma$-models [904, 1007, 174, 516]. The gauging of a group of isometries of a $\sigma$-model (with no Wess–Zumino (WZ) term) with Riemannian metric is reviewed in detail in Appendix J.1. Here we adapt and shorten that presentation of the subject to this context and notation,

Our starting point is the $(p+1)$-dimensional $\sigma$-model ($p$-brane Polyakov-type action)

$$S = -\frac{T_{(p)}}{2}\int d^{p+1}\xi\sqrt{|\gamma|}\left[\gamma^{ij}g_{ij} - (p-1)\right], \tag{24.10}$$

which is invariant under the GCTs

$$X^\mu \to X^{\mu\,\prime} = X^\mu + \epsilon^\mu(X),$$
$$g_{\mu\nu}(X) \to g'_{\mu\nu}(X') = g_{\mu\nu}(X) - 2\big(\partial_{(\mu|}\epsilon^\rho\big) g_{\rho|\nu)}. \tag{24.11}$$

Let us assume now that the metric admits an isometry group generated by the Killing vectors $k_A{}^\mu(X),\ \ A = 1,\ldots,r$,

$$[k_A, k_B] = -f_{AB}{}^C k_C, \tag{24.12}$$

and let us consider the following infinitesimal transformations (which are *not* infinitesimal GCTs):

$$X^\mu \to X^{\mu\,\prime} = X^\mu + \alpha^A k_A{}^\mu,$$
$$g_{\mu\nu}(X) \to g_{\mu\nu}(X') = g_{\mu\nu}(X) + \alpha^A k_A{}^\lambda \partial_\lambda g_{\mu\nu}, \tag{24.13}$$

where the $\alpha^A$s are constant infinitesimal parameters. The variation of the action vanishes if (as is assumed) the $k_A$ satisfy the Killing equation $\nabla_{(\mu|}k_{A|\nu)} = 0$ in the target space.

Now we want to gauge this symmetry promoting the $\alpha^A$s to arbitrary functions of $X^\mu$.

To make the $\sigma$-model invariant under these transformations, it suffices to replace the partial derivative of the worldvolume scalars $X^\mu$ by the covariant derivative

$$\mathfrak{D}_i X^\mu = \partial_i X^\mu + C^A{}_i k_A{}^\mu, \tag{24.14}$$

where we have introduced the non-dynamical[3] worldvolume vector fields $C^A{}_i$ which transform as standard gauge potentials:

$$\delta_\alpha C^A{}_i = -\big(\partial_i \alpha^A + f_{BC}{}^A C^B{}_i \alpha^C\big) = -\mathfrak{D}_i \alpha^A. \tag{24.15}$$

The gauged $\sigma$-model action (without WZ term) then takes the form

$$\boxed{S = -\frac{T_{(p)}}{2}\int d^{p+1}\xi \sqrt{|\gamma|}\big[\gamma^{ij}\mathfrak{D}_i X^\mu \mathfrak{D}_j X^\nu g_{\mu\nu} - (p-1)\big].} \tag{24.16}$$

Since the worldvolume vector fields $C^A{}_i$ are not dynamical (their derivatives do not occur in the action), they play the role of Lagrange multipliers and may be eliminated by using their equations of motion directly in the action (at least in this simple $\sigma$-model with no WZ term). These are

$$k_{A\,\mu}\mathfrak{D}_i X^\mu = 0, \tag{24.17}$$

and their solution is

$$C^A{}_i = -h^{AB} k_B{}^\mu \partial_i X^\nu g_{\mu\nu}, \tag{24.18}$$

where we have defined the metric $h_{AB}$ (*assumed* to be invertible, which is not true in general, but is in many cases of interest),

$$h_{AB} = k_A{}^\mu k_B{}^\nu g_{\mu\nu}, \qquad h^{AB} h_{BC} = \delta^A{}_C. \tag{24.19}$$

[3] They must be non-dynamical if we do not want to introduce additional degrees of freedom in the theory. This just means that we do not add kinetic terms for these vector fields.

In this case, we have on-shell

$$\mathfrak{D}_i X^\mu = \left(g^\mu{}_\nu - h^{AB} k_A{}^\mu k_{B\,\nu}\right) \partial_i X^\nu . \tag{24.20}$$

Observe that the matrix

$$\Pi^\mu{}_\nu \equiv \left(g^\mu{}_\nu - h^{AB} k_A{}^\mu k_{B\,\nu}\right), \qquad \Pi^\mu{}_\nu \Pi^\nu{}_\rho = \Pi^\mu{}_\rho, \tag{24.21}$$

projects onto the space orthogonal to the orbits of the isometry group:

$$\Pi^\mu{}_\nu k_A{}^\nu = 0, \qquad \forall\, I = 1, \ldots, r, \tag{24.22}$$

so $r$ directions simply disappear from the $\sigma$-model action. We will see this more clearly when we perform the target-space (*direct*) dimensional reduction of the gauged $\sigma$-model.

After eliminating the auxiliary vector fields (assuming that this was possible), the gauged $\sigma$-model takes the form

$$S = -\frac{T_{(p)}}{2} \int d^{p+1}\xi \sqrt{|\gamma|} \left[\gamma^{ij} \partial_i X^\mu \partial_j X^\nu \Pi_{\mu\nu} - (p-1)\right], \tag{24.23}$$

since $g_{\rho\sigma} \Pi^\rho{}_\mu \Pi^\sigma{}_\nu = \Pi_{\mu\nu}$. In the case of just one isometry, we have seen in Section 15.2 that, in adapted coordinates, $\hat{\Pi}_{\hat{\mu}\hat{\nu}}$ is zero except for the $(\hat{d}-1) \times (\hat{d}-1)$ submatrix $\hat{\Pi}_{\mu\nu}$ which is the metric in $\hat{d}-1$ dimensions. The above $\sigma$-model (adding hats everywhere) does not depend on the isometric coordinate $Z$ and reduces simply to a $\sigma$-model with $(\hat{d}-1)$-dimensional target space.

In this simple case, then, the gauged $\sigma$-model with $d$-dimensional target space is actually a $\sigma$-model with $(d-1)$-dimensional target space in disguise, written in $d$-dimensional covariant language. In more general cases it is not possible to eliminate completely the non-physical degrees of freedom (such as $Z$), but the $\sigma$-model still has $d-r$ degrees of freedom.

Let us now consider the coupling of $p$-branes to other spacetime fields. The simplest and more natural ones are the couplings to scalar fields (which act as local coupling "constants") and to $(p+1)$-form potentials, the fields $p$-branes can be charged with respect to. Let us start with $(p+1)$-form potentials.

### 24.2.2 *Charged branes and Dirac charge quantization for extended objects*

The Lagrangian of a $(p+1)$-form potential $A_{(p+1)\,\mu_1\cdots\mu_{p+1}}$ is usually constructed in terms of its $(p+2)$-form field strength,

$$F_{(p+2)\,\mu_1\cdots\mu_{p+2}} = (p+2)\partial_{[\mu_1} A_{(p+1)\,\mu_2\cdots\mu_{p+2}]} \tag{24.24}$$

($F_{(p+2)} = dA_{(p+1)}$ in differential-form language), which is invariant under the gauge transformations generated by a $p$-form $\Lambda_{(p)}$,

$$\delta A_{(p+1)\,\mu_1\cdots\mu_{p+1}} = (p+1)\partial_{[\mu_1} \Lambda_{(p)\,\mu_2\cdots\mu_{p+1}]} \tag{24.25}$$

($\delta A_{(p+a)} = d\Lambda_{(p)}$ in differential-form language). The Lagrangian is[4]

$$S[A_{(p+1)}] = \int d^d x \sqrt{|g|} \left[ \frac{(-1)^{(p+1)}}{2 \cdot (p+2)!} F^2_{(p+2)} \right], \tag{24.26}$$

and the equation of motion is just

$$\nabla_\mu F_{(p+2)}{}^{\mu\mu_1 \cdots \mu_{p+1}} = 0 \tag{24.27}$$

($d \star F_{(p+2)} = 0$ in differential-form language). As usual, we can work directly with the field strength, provided that we impose on it the Bianchi identity

$$\nabla_\mu \star F_{(p+2)}{}^{\mu\mu_1 \cdots \mu_{d-p-3}} = 0 \tag{24.28}$$

($dF_{(p+2)} = 0$) to ensure the local existence[5] of the $A_{(p+1)}$.

$A_{(p+1)}$ couples naturally to a current $j^{\mu_1 \cdots \mu_{p+1}}$, the coupling being represented by the Lagrangian term

$$\int d^d x \sqrt{|g|} \frac{(-1)^{(p+1)}}{(p+1)!} A_{(p+1)\mu_1 \cdots \mu_{p+1}} j^{\mu_1 \cdots \mu_{p+1}}. \tag{24.30}$$

This term is gauge invariant if the current is divergence free ("conserved"),

$$\nabla_\mu j^{\mu\mu_1 \cdots \mu_p} = 0 \tag{24.31}$$

($d \star j = 0$ in differential-form language). This condition follows from the gauge identity of the free theory $\nabla_\mu \nabla_\nu F_{(p+2)}{}^{\mu\nu\mu_1 \cdots \mu_p} = 0$ (off-shell) plus the equation of motion

$$\nabla_\mu F_{(p+2)}{}^{\mu\mu_1 \cdots \mu_{p+1}} = j^{\mu_1 \cdots \mu_{p+1}}, \tag{24.32}$$

or, in differential-form language,

$$\delta F = (-1)^d j, \quad d \star F_{(p+2)} = (-1)^{d+p} \star j. \tag{24.33}$$

The conservation law for the current $d \star j = 0$ suggests the following definition for the charge $q_p$ associated with $A_{(p+1)}$:

$$q_p \equiv \int_{\mathrm{B}^{(d-p-1)}} \star j, \tag{24.34}$$

where, by definition, $\mathrm{B}^{(d-p-1)}$ is a *capping surface* whose boundary is (topologically) $\partial \mathrm{B}^{(d-p-1)} = \mathrm{S}^{(d-p-2)}$. More precisely, this is the charge contained in the capping surface. The total charge would be calculated by integrating over a capping surface whose boundary

---

[4] We are ignoring here possible scalar couplings, Chern–Simons terms, etc.

[5] Given a field strength $F_{(p+2)}$ satisfying the Bianchi identity, $A_{(p+1)}$ is given, up to gauge transformations, by the formula that generalizes Eq. (12.31):

$$A_{\mu_1 \cdots \mu_{p+1}} = (-1)^{(p+1)} \int_0^1 d\lambda \lambda^{(p+1)} F_{(p+2)\,\mu_1 \cdots \mu_{p+2}}(\lambda x) x^{\mu_{p+2}}. \tag{24.29}$$

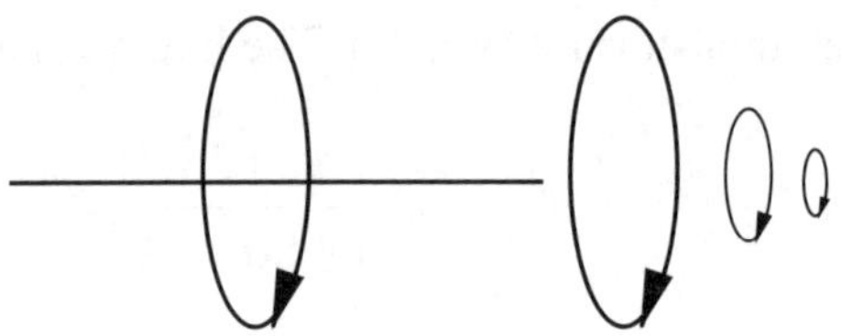

Fig. 24.1. Open strings cannot carry charge.

is the $(d-p-2)$-sphere at infinity, $S_{\infty}^{(d-p-2)}$. As usual this definition is invariant under smooth deformations of the capping surface in source-free ($j=0$) regions.

Using the generalization of the Gauss law Eq. (24.33) and Stokes' theorem, we obtain the usual definition

$$q_p = (-1)^{d+p} \int_{S^{(d-p-2)}} \star F_{(p+2)}, \tag{24.35}$$

which is also invariant under smooth deformations of the $(d-p-2)$-dimensional surface in source-free regions.

Given a $p$-brane sweeping out a $(p+1)$-dimensional worldvolume $W^{(p+1)}$, we can immediately construct a conserved current for it that generalizes Eq. (12.49) for the current of a point-like charged object:

$$\begin{aligned} j^{\mu_1\cdots\mu_{p+1}}(x) &= q_p \int_{W^{(p+1)}} dX^{\mu_1} \wedge \cdots \wedge dX^{\mu_{p+1}} \frac{\delta^{(d)}(x - X(\xi))}{\sqrt{|g(X)|}} \\ &= q_p \int_{W^{(p+1)}} d^{p+1}\xi \frac{\partial(X^{\mu_1} \cdots X^{\mu_{p+1}})}{\partial(\xi^{\mu_1} \cdots \xi^{\mu_{p+1}})} \frac{\delta^{(d)}(x - X(\xi))}{\sqrt{|g(X)|}}. \end{aligned} \tag{24.36}$$

The charge associated with this current is $q_p$, except when the $p$-brane worldvolume has boundaries, since, in that case, we can continuously deform the surface of integration and contract it to a point without meeting the $p$-brane source, obtaining zero charge. This is easy to visualize for a string of finite length in a four-dimensional target space: the string charge with respect to a 2-form potential is calculated through a closed line integral around the string that can be slid off the string and contracted to a point, as shown in Fig. 24.1. In a sense this explains why the KR 2-form does not appear in the open-string spectrum. It is clear now that $q_p \neq 0$ only when the $p$-brane has compact topology or extends to infinity. A brane with boundaries can also carry charge if the boundaries are attached to another object just as open strings are attached to D-branes. This case, which is more complicated (and interesting), will be studied in Section 25.6 since it depends strongly on the theory we are considering.

On substituting the above $p$-brane current into the interaction term Eq. (24.30), we obtain the standard form of the WZ term for $p$-branes which appears in the $p$-brane worldvolume action,

$$(-1)^{(p+1)} q_p \int_{W^{(p+1)}} A_{(p+1)}. \tag{24.37}$$

*Electric–magnetic duality for extended objects.* The electric–magnetic dual of a charged $p$-brane is, by definition, an object that couples to the electric–magnetic dual of the $(p+1)$-form potential $A_{(p+1)}$. We have seen several examples, for instance, in Section 22.5.5, in which we used Poincaré duality to replace the KR 2-form completely by its dual in various dimensions and also in Sections 22.1.3 and 23.1.1 in which we defined the on-shell duals of the RR potentials that appear in the $N=2A, B, d=10$ supergravity actions, not being able to replace the original potentials completely by their duals.

In all (massless) cases, the dual is a $(\tilde{p}+1)$-form $\tilde{A}_{(\tilde{p}+1)}$ with $\tilde{p}=d-p-4$, whose field strength is, by definition, the Hodge dual of $\tilde{F}_{\tilde{p}+2}=\star F_{(p+2)}$. The electric–magnetic dual of a $p$-brane is therefore a $\tilde{p}$-brane with $\tilde{p}\neq p$ in general, which is electrically charged with respect to $\tilde{A}_{(\tilde{p}+1)}$. Only in even dimensions do objects with $p=(d-4)/2$ have duals of the same dimension, and then we can have $p$-brane dyons.

A new feature with respect to point-particles is that there can be *self-dual p-branes*, charged with respect to a self-dual potential: point-particles couple only to vectors, which are dual to vectors in $d=4$, but, in $d=4$, self- or anti-self-duality is consistent only with a Euclidean signature. However, self-dual 2-forms in $d=6$ and 4-forms in $d=10$ can be consistently defined. They occur in $N=(1,0), d=6$ SUGRA (Section 17.4.1) and in $N=2B, d=10$ SUEGRA (Chapter 23), and are associated with chiral theories.

*Dirac charge quantization.* The charge of a $p$-brane moving in the background $A_{(p+1)}$ field sourced by a dual $\tilde{p}$-brane is quantized as in the case of point-particles [975, 1172, 1173].[6]

Let us consider the quantum propagation of a charged $p$-brane moving along a closed path[7] so its worldvolume is topologically a $(p+1)$-sphere $\mathrm{S}^{(p+1)}$. The interesting term in the path integral is the WZ term, which, using Stokes' theorem, can be written in the form

$$(-1)^{(p+1)}q_p\int_{\mathrm{B}^{(p+2)}} dA_{(p+1)}=(-1)^{(p+1)}q_p\int_{\mathrm{B}^{(p+2)}} F_{(p+2)}, \tag{24.38}$$

where $\mathrm{B}^{(p+2)}$ is one of the many possible $(p+2)$-dimensional capping surfaces with $\partial\mathrm{B}^{(p+2)}=\mathrm{S}^{(p+1)}$. To avoid having any ambiguities in the path integral, the difference between choosing two different capping surfaces $\mathrm{B}_1^{(p+2)}$ and $\mathrm{B}_2^{(p+2)}$ must be an integer multiple of $2\pi$:

$$(-1)^{(p+1)}q\int_{\mathrm{B}_1^{(p+2)}} F_{(p+2)}-(-1)^{(p+1)}q\int_{\mathrm{B}_2^{(p+2)}} F_{(p+2)}=2\pi n. \tag{24.39}$$

Now, the first capping surface plus the second (with reversed orientation) form (topologically) a $(p+2)$-sphere $\mathrm{S}^{(p+2)}$, and we arrive at the condition

$$q_p\int_{\mathrm{S}^{(p+2)}} F_{(p+2)}=(-1)^{(p+1)}2\pi n. \tag{24.40}$$

[6] The derivation of the Dirac quantization condition we are about to explain is different from that used in Section 12.7.2 for point-particles, which can also be generalized to charged $p$-branes.

[7] This will be the analog of a point-particle moving along a closed path encircling a Dirac string.

Defining now the "magnetic" charge, or, better, the electric charge of the dual $\tilde{p}$-brane, by

$$p_p = q_{\tilde{p}} = (-1)^{d+\tilde{p}} \int_{S^{p+2}} F_{(p+2)}, \tag{24.41}$$

the above result takes the form of the generalization of the Dirac quantization condition we were after:

$$q_p q_{\tilde{p}} = 2\pi n, \qquad n \in \mathbb{Z}. \tag{24.42}$$

The charge quantization condition for $p$-brane dyons was found in Ref. [438].

Observe that, in the string $\sigma$-model action Eq. (21.31), the constant that plays the role of the charge with respect to the KR 2-form is the string tension $T$. This identity, which has the form of a BPS bound, indicates that the object described by the action is a BPS object. All the extended objects we are going to deal with have the same property.

Observe also that, even though we have defined a scalar charge $q_p$ because this is a useful quantity, this number does not give all the information. Actually, the charge should be the tensor

$$Z^{\mu_1\cdots\mu_p} \sim \int_{\Sigma} d^{d-1}\Sigma_\mu j^{\mu\mu_1\cdots\mu_p} \tag{24.43}$$

(where $\Sigma$ is a spacelike hypersurface) and includes information about spatial orientation, etc. This explains why objects with the same tensions and $q_p$s can be in equilibrium if they are parallel (so their $Z^{\mu_1\cdots\mu_p}$s are identical) but not when one of them is tilted. These tensorial charges appear naturally in supersymmetry algebras, as we are going to see in Section 25.5.

### 24.2.3 The coupling of p-branes to scalar fields

A spacetime scalar $K(X)$ can be introduced in only one place in the NG action if we want to preserve reparametrization invariance and the gauge invariance of the WZ term,

$$S^{(p)}_{\rm NG}[X^\mu(\xi)] = -T_{(p)} \int d^{p+1}\xi (K/K_0)\sqrt{|g_{ij}|}. \tag{24.44}$$

We have introduced $K_0$, the asymptotic value of $K$ at infinity (assuming that the metric $g_{\mu\nu}$ is asymptotically flat, at least in the directions transverse to the $p$-brane), so $T_{(p)}$ is the physical $p$-brane tension[8] and usually is proportional to $K_0$.

The coupling to $K$ can always be changed or eliminated by a Weyl rescaling of the spacetime metric. It is important to make clear in which Weyl conformal reference frame we are writing the $p$-brane action. There are two special frames that can always be defined. We use the fundamental-string worldsheet action to illustrate the definitions.

[8] This is called in the literature *effective tension*. See footnote 5 on p. 637.

*(Fundamental) p-brane frames.* These are defined as the Weyl conformal frames in which the $p$-brane action does not couple to the scalar $K$. At the same time, all the terms in the action for the spacetime fields should carry the same $K$-dependent factor [471].

For instance, by definition, the fundamental-string worldsheet action Eq. (21.31) does not depend on the dilaton when it is written in the string conformal frame. At the same time, all the terms in the action for the spacetime fields that couple to the string Eq. (21.1) carry the same $e^{-2\phi}$ factor.

*Dual p-brane frames.* These are the frames in which the electric–magnetic dual $\tilde{p}$-brane would be fundamental. In the fundamental-string case in $d$ dimensions the KR 2-form is dual to a $(d-4)$-form potential $C$ with field strength $G=(d-3)\partial C$ and the dual object has $\tilde{p}=d-5$ (a 5-brane in $d=10$). On Poincaré-dualizing the KR 2-form, we obtain

$$S=\frac{g^2}{16\pi G_{\rm N}^{(d)}}\int d^dx\sqrt{|g|}e^{-2\phi}\left[R-4(\partial\phi)^2+\frac{(-1)^{(d-4)}}{2\cdot(d-3)!}e^{4\phi}G^2\right]. \tag{24.45}$$

Thus, $g_{\mu\nu}$, the string metric, is not the metric to which the dual $(d-5)$-brane naturally couples. On performing a conformal transformation,

$$g=\Omega_{(1)-(d-5)}g_{(d-5)}, \tag{24.46}$$

and imposing that the dilaton factor is the same for all terms in the action, one obtains

$$\Omega_{(1)-(d-5)}=e^{\frac{4}{d-4}\phi}, \tag{24.47}$$

and

$$S=\frac{g^2}{16\pi G_{\rm N}^{(d)}}\int d^dx\sqrt{|g_{(d-5)}|}e^{\frac{4}{d-4}\phi}\left[R+\frac{(-1)^{(d-4)}}{2\cdot(d-3)!}G^2\right]. \tag{24.48}$$

In the frame $g_{(d-5)}$, by definition, the NG $(d-5)$-brane action has no dilaton factors. Then, going back to the string frame, we find

$$S_{\rm NG}^{(d-5)}[X^\mu(\xi)]=-T_{(d-5)}\int d^{d-4}\xi\, e^{-2\phi}\sqrt{|g_{ij}|}. \tag{24.49}$$

The dual of the fundamental string has, then, a tension proportional to $g^{-2}$ ($g$ being the string coupling constant), as a typical solitonic object. The dual object to the fundamental string in $d=10$ dimensions is known as the *solitonic 5-brane* or S5-brane (also called *NS5-brane* to distinguish it from the D5-brane that couples to the RR 6-form potential).

It is natural to use in all cases the string frame because it is the theory of strings that we know how to quantize (even if imperfectly). From the string point of view, we can classify all the extended objects that we are going to see in terms of the scalar couplings that appear in their NG actions

**Fundamental $p$-branes.** They do not couple to the dilaton (to lowest order in $\alpha'$), which does not occur in the NG action

$$S_{\rm NG}^{(\rm pf)}[X^\mu(\xi)]=-T_{(p)}\int d^{p+1}\xi\,\sqrt{|g_{ij}|}. \tag{24.50}$$

Their mass is independent of the string coupling constant $g = e^{\phi_0}$ ($\phi_0$ being the constant value of the dilaton at infinity). In $d = 10$ there is only one: the fundamental string. In $d = 11$ the M2- and M5-branes can both be considered fundamental.

**Solitonic $p$-branes.** They couple to the dilaton as follows:

$$S_{\rm NG}^{(\rm ps)}[X^\mu(\xi)] = -T_{(p)} \int d^{p+1}\xi\, e^{-2\phi}\sqrt{|g_{ij}|}. \tag{24.51}$$

Their mass is proportional to $g^{-2}$, which is typical of standard solitons. In $d = 10$ there is only one: the S5-brane.

**Dirichlet (D) $p$-branes.** They couple to the dilaton as follows:

$$S_{\rm NG}^{(\rm D)}[X^\mu(\xi)] = -T_{(p)} \int d^{p+1}\xi\, e^{-\phi}\sqrt{|g_{ij}|}. \tag{24.52}$$

Their mass is proportional to $g^{-1}$. They are a new type of purely stringy solitons. They occur only in $d = 10$ and lower dimensions and couple to RR potentials.

**Momentum modes.** They are charged point-like objects that couple to the KK scalar $k$ as follows:

$$S_{\rm NG}[X^\mu(\xi)] = -T_{(0)} \int d\xi\, k^{-1}\sqrt{|g_{\xi\xi}|}. \tag{24.53}$$

Their mass is proportional to $k_0^{-1}$, that is, to the inverse of the radius of the compact dimension, and they couple to the KK vector. We have met them in several places. We are going to see that $d = 11$ momentum modes can be seen as $d = 10$ D0-branes, given the relation between the KK scalar $k$ and the dilaton and between the KK vector and the RR 1-form, Eqs. (22.35).

**Winding modes.** [9] They are charged point-like objects that couple to the KK scalar $k$ as follows:

$$S_{\rm NG}[X^\mu(\xi)] = -T_{(0)} \int d\xi\, k\sqrt{|g_{\xi\xi}|}. \tag{24.54}$$

Their mass is proportional to $k_0$, that is, to the radius of the compact dimension, and they couple to the winding vector.

**Kaluza–Klein (KK) branes.** They are described by gauged $\sigma$-models and couple to a positive power of the volume of the compact space associated with the gauging. In the decompactification limit the tension becomes infinite. Thus, these objects exist only when there are compact dimensions. The archetype of these objects is the KK monopole, which is described by a U(1)-gauged $\sigma$-model and coupled both to the dilaton and to the KK scalar $k$ as follows:

$$S_{\rm NG}^{(\rm KK)}[X^\mu(\xi)] = -T_{(p)} \int d^{p+1}\xi\, e^{-2\phi}k^2\sqrt{|\Pi_{ij}|}. \tag{24.55}$$

[9] There are also winding modes associated with other branes wrapped on compact spaces. Here we refer only to the string winding modes.

## 24.3 General $p$-brane solutions

In this section we are going to construct the simplest classical solutions that describe uncharged (*Schwarzschild*) and charged $p$-branes in a $d$-dimensional spacetime. They are solutions of a generalization proposed in Ref. [775] of the $a$-model discussed in Section 16.1 in which we will replace the 1-form potential adequate for BHs (which, in a sense, are point-like objects, 0-branes) by a $(p+1)$-form potential that is adequate for $p$-branes. This $p$-brane $a$-model is, for specific $a$s and $p$s, a simplified version of most of the supergravity actions we are dealing with and the solutions we obtain will be supergravity (superstring) solutions.

We are also going to see how, according to Ref. [471], it is possible to find a $p$-brane worldvolume source for the "extreme" ones, generalizing the results we obtained for ERN and KK BHs (Sections 12.4 and 15.2.3, respectively), and for the fundamental-string solution (Section 21.3), which will be particular cases of our general solution.

### 24.3.1 Schwarzschild black $p$-branes

The Schwarzschild solution describes the gravitational field of a massive, point-like object in vacuum. Is there an analogous solution of the $d$-dimensional Einstein–Hilbert action describing the gravitational field of a massive $p$-brane in vacuum?

Let us consider the simplest $p$-brane configuration (a "flat" $p$-brane with trivial topology). This configuration should give rise to an asymptotically flat spacetime characterized by $p+1$ translational isometries associated with the $p$-brane worldvolume. The requirement that the solution be asymptotically flat is essential, if we want to describe an *isolated* $p$-brane. However, we cannot impose asymptotic flatness in the direction along which the isometries act, but only in the $d-(p+1)$ spacelike transverse dimensions. In what follows, we will use the concept of asymptotic flatness in this restricted sense.

Solutions with these properties, *Schwarzschild p-branes*, were constructed using KK techniques in Section 15.3.3, and the metric is given by Eqs. (15.164), although here we are going to ignore all the hats. The first thing we can do is compute the $p$-brane tension[10] using the definition we gave in Section 24.2.1. This was done in Section 15.3.3, Eqs. (15.165) and (15.166), and the result can be expressed in this way: we can identify the $p$-brane tension as the constant $T_{(p)}$ in the expansion of the $g_{tt}$ component of the metric:

$$g_{tt} \sim 1 - \frac{16\pi G_{\rm N}^{(d)} T_{(p)}}{(\tilde{p}+2)\omega_{(\tilde{p}+2)}} \frac{1}{r^{\tilde{p}+1}}. \tag{24.56}$$

The coefficient of $r^{-(\tilde{p}+1)}$ is, by definition, the $p$-brane Schwarzschild radius to the power $\tilde{p}+1$.

In Section 27.2 we will consider more general black $p$-brane metrics, however, and a more intrinsic and general definition of tension is needed for them, such as the one proposed in Ref. [908]. Assume, following Ref. [970], that we perform a weak-field expansion of the metric around Minkowski's $g_{\mu\nu} = \eta_{\mu\nu} + h_{\mu\nu}$ and the perturbation $h_{\mu\nu}$ can be written in the

[10] The restricted asymptotic flatness of the metric does not allow us to define a finite $d$-dimensional mass.

form

$$h_{\mu\nu} = \frac{c_{\mu\nu}}{r^{\tilde{p}+1}}, \tag{24.57}$$

where $c_{\mu\nu}$ is a constant tensor and $r$ is a radial coordinate such that the slices of constant worldvolume and radial coordinate have, asymptotically, the metric $r^2 d\Omega_{\tilde{p}+3}$. Then, the $p$-brane's energy–momentum tensor $t_{ab}$ (here $a, b = 0, \ldots, p$ are curved indices) is given by

$$t_{ab} = -\frac{\omega_{\tilde{p}+2}}{16\pi G_{\rm N}^{(d)}} \left[(\tilde{p}+1)c_{ab} + \eta_{ab}\eta^{cd}c_{cd}\right], \tag{24.58}$$

where $\omega_{\tilde{p}+2}$ is the volume of the unit round $S^{\tilde{p}+2}$. The tension $T_p$ is just the $t_{00}$ component.

The constant tensor $c_{\mu\nu}$ will change slightly when we change coordinates, but the expression Eq. (24.58) will still be valid.

An important property of this definition to be remembered is that $T_p$ will coincide with the mass of the BH in $(\tilde{p}+4)$ dimensions that one obtains by dimensional reduction over the spacelike worldvolume coordinates (taking properly into account the relation between the $d$- and the $(\tilde{p}+4)$-dimensional Newton constant).

It is clear from the definition that infinite (uncompactified) $p$-branes have infinite energy.

Is this Schwarzschild $p$-brane solution analogous to the Schwarzschild BH solution in the sense that it exhibits an event horizon? In other words: is it a *black* $p$-brane solution? In the presence of the $p$ translational isometries the only sensible definition of an event horizon is equivalent to the standard definition of an event horizon in the transverse space. The event horizon thus defined becomes a $(p+2)$-dimensional extended object: the product of a BH horizon and the $p$-dimensional Euclidean space spanned by the $p$-brane. Clearly, for positive tension, the Schwarzschild $p$-brane solution has an event horizon of that form, whereas, for negative tension, the curvature singularity at $r = 0$ will be naked.

When the spacetime has $n$ compact dimensions, it is possible to have black $p$-branes, $p \leq n$, wrapping the compact dimensions with the same (finite) mass and event horizons of different topologies. Therefore, the uniqueness of four-dimensional BHs is not true in higher dimensions if some of the dimensions are compact. We know now that it is not true even in the absence of compact dimensions, as the existence of the asymptotically flat rotating black ring of Ref. [512] shows.

Black $p$-brane solutions (the Schwarzschild ones and the charged non-extreme ones that we are going to see next) are classically unstable [677, 678] (the charged, extreme ones are stable [679], as is usual in supersymmetric solutions) under linear perturbations along the worldvolume dimensions with wavelengths larger than the Schwarzschild radius. On the other hand, they are also quantum mechanically unstable, because there is Hawking radiation associated with their event horizons, but also because the area of the event horizon (entropy) of several BHs with the same total mass is in general larger than the area of the event horizon of a Schwarzschild $p$-brane. For a Schwarzschild string compactified on a circle this happens whenever the length of the circle is larger than the Schwarzschild $p$-brane radius. The two instabilities seem to be related in the sense that the classical one is present whenever the thermodynamical one is present [686, 687, 1072].

Although the thermodynamical argument seems to indicate that a black $p$-brane will break up into several BHs that will eventually merge into one, it has been argued in

Ref. [767] that, for the black string, this process cannot take place in a finite time and that, instead, the black string decays into a new non-translationally invariant ("inhomogeneous") black string. Initial data sets for inhomogeneous $p$-brane solutions have subsequently been proposed [768].

As we did in the BH case, we are going to use the Schwarzschild $p$-brane solution Eqs. (15.164) as our basic black $p$-brane solution and we are going to see that the charged $p$-brane solutions of the $p$-brane $a$-model we are about to study can be built by "dressing" it with appropriate factors.

### 24.3.2 The p-brane a-model

As we have already said, this is the simplest model that embodies the main characteristics of the string effective action (and supergravity actions): gravity coupled to one scalar and a $(p+1)$-form to which the scalar couples non-minimally. It generalizes the $a$-model for BHs that we used before, which appears here as the $p=0$ case, but here we canonically normalize the $(p+1)$-form field strengths. The action is[11]

$$S = \frac{1}{16\pi G_{\rm N}^{(d)}} \int d^d x \sqrt{|g|} \left[ R + 2(\partial\varphi)^2 + \frac{(-1)^{p+1}}{2\cdot(p+2)!} e^{-2a\varphi} F_{(p+2)}^2 \right], \tag{24.59}$$

where $F_{(p+2)} = dA_{(p+1)}$ is the field strength of the $(p+1)$-form potential $A_{(p+1)}$.

The equations of motion are

$$\begin{aligned} G_{\mu\nu} + 2T^{\varphi}_{\mu\nu} - \tfrac{1}{2} e^{-2a\varphi} T^{A_{(p+1)}}_{\mu\nu} &= 0, \\ \nabla^2\varphi + \frac{(-1)^{p+1}}{4\cdot(p+2)!} a e^{-2a\varphi} F_{(p+2)}^2 &= 0, \\ \nabla_\mu \left( e^{-2a\varphi} F_{(p+2)}{}^{\mu\nu_1\cdots\nu_{p+1}} \right) &= 0, \end{aligned} \tag{24.60}$$

where $T^{A_{(p+1)}}$ is the $(p+1)$-form energy–momentum tensor, given in Eq. (1.130). As usual, not all of them are independent in general (they can be derived from the Bianchi identities). On the other hand, the solutions we will find will be defined up to gauge transformations, including *large* gauge transformations that change the asymptotic behavior of the fields so the physics could be inequivalent. The classical equations of motion are insensitive to these subtleties.

We want to find single charged black $p$-brane solutions of an analogous nature to the black Schwarzschild $p$-brane of Section 24.2.1. The method we used to construct them in Section 15.3.3 cannot be used in the presence of $p$-forms: we cannot simply add extra dimensions to a lower-dimensional charged BH metric. For instance, if we took a

[11] This action is equivalent to the original one written in Ref. [775] (whose main results we reobtain in a different fashion here) which was given in the string frame. Here we do not want to assume that the scalar is the dilaton and thus we prefer to use an Einstein-frame action. The constant $a$ is, then, not the same as in Ref. [775], but we obtain simpler, more-symmetric expressions.

four-dimensional RN BH and added several extra dimensions, we would obtain a higher-dimensional metric with vanishing scalar curvature (as in four dimensions) but now the trace of the Maxwell energy–momentum tensor would not be zero in more than four dimensions. Another way of expressing the same fact is that, if we dimensionally reduce the above action to $d-p$ dimensions, we do not simply obtain the above action written directly in $d-p$ dimensions, but rather we find extra fields. In particular, we find extra scalars that couple to the form. The approach taken originally by Horowitz and Strominger in Ref. [775] was to perform this dimensional reduction, simultaneously reducing the $(p+2)$-form field strength to a 2-form. The resulting problem is a BH problem. In fact, we obtain the $a$-model for BHs! (The two parameters $a$ are related but different.) Then, they used the solutions found in Refs. [608, 623] and rewrote them back into $d$-dimensional language.

Our approach will be to solve the problem directly in $d$ dimensions with the ansatz

$$ds^2 = f[Wdt^2 - d\vec{y}_p^{\,2}] - g^{-1}[W^{-1}d\rho^2 + \rho^2 d\Omega^2_{(\tilde{p}+2)}], \qquad W = 1 + \frac{\omega}{\rho^{\tilde{p}+1}}, \tag{24.61}$$

which can be seen as a dressing of the Schwarzschild $p$-brane metric Eqs. (15.164) by the functions $f$ and $g$, which are related to the presence of the dilaton and the $(p+1)$-form. The natural ansatz for $A_{(p+1)}$, if the $p$-brane is electrically charged and we want $A_{(p+1)}$ to vanish at infinity, is, by analogy with the BH case,

$$A_{\underline{t}\underline{y}^1\cdots\underline{y}^p} = \alpha\big(H^{-1} - 1\big), \qquad H = 1 + \frac{h}{\rho^{\tilde{p}+1}}. \tag{24.62}$$

The dilaton and the functions $f$ and $g$ are just functions of $H$, if we do not want to consider primary scalar hair. We are going to assume that the value of the scalar field at infinity vanishes, $\varphi_0 = 0$. We can always generate a non-vanishing value by rescalings of $A_{(p+1)}$ and shifts of $\varphi$, but, since these rescalings will be different in different conformal frames, we postpone them until we specify those frames.

On substituting this ansatz into the equations of motion and using Appendix N.3.3, we arrive at the final form for the desired solutions, which is valid for $\tilde{p} \leq 0$:

$$\boxed{\begin{aligned} ds^2 &= \big(e^{-2a\varphi}H^{-2}\big)^{\frac{1}{p+1}}\big[Wdt^2 - d\vec{y}_p^{\,2}\big] \\ &\quad - \big(e^{-2a\varphi}H^{-2}\big)^{-\frac{1}{\tilde{p}+1}}\big[W^{-1}d\rho^2 + \rho^2 d\Omega^2_{(\tilde{p}+2)}\big], \\ e^{-2a\varphi} &= H^{2x}, \quad A_{\underline{t}\underline{y}^1\cdots\underline{y}^p} = \alpha(H^{-1}-1), \quad H = 1 + \frac{h}{\rho^{\tilde{p}+1}}, \quad W = 1 + \frac{\omega}{\rho^{\tilde{p}+1}}, \\ \omega &= h\Big[1 - \frac{a^2}{4x}\alpha^2\Big], \qquad x = \frac{(a^2/2)c}{1+(a^2/2)c}, \qquad c = \frac{(p+1)+(\tilde{p}+1)}{(p+1)(\tilde{p}+1)}. \end{aligned}}$$

(24.63)

For $p = 0$ these solutions reduce to those of the dilaton $a$-model Eqs. (16.10). As has happened in all the cases we have previously studied, when the extremality parameter $\omega = 0$ so the "Schwarzschild factor" $W$ becomes 1 (and $W$ disappears), $H$ becomes an arbitrary harmonic function in the $(\tilde{p}+3)$-dimensional Euclidean space transverse to the object. These will be solutions describing extreme $p$-branes. When $h=0$, so $H=1$ and disappears, we recover the Schwarzschild $p$-brane solutions. Finally, this general solution is written in such a way that it is valid for the case $a=0$ as well. In that case, the scalar decouples from $A_{(p+1)}$ and the only solutions with no primary scalar hair have a completely trivial scalar field.

In general, these solutions have a regular event horizon at $\rho=\omega^{\frac{1}{\tilde{p}+1}}$ when $h\geq 0$ and $\omega<0$.

The general solution Eq. (24.63) can be generalized further by assuming that there are $q$ additional translational isometries in the directions $z_1,\ldots,z_q$. Using as ansatz the metric and the results of Section N.3.4, we quite straightforwardly obtain

$$\boxed{\begin{aligned}
ds^2 &= \left(e^{-2a\varphi}H^{-2}\right)^{\frac{1}{p+1}}\left[Wdt^2 - d\vec{y}_p^{\,2}\right] \\
&\quad - \left(e^{-2a\varphi}H^{-2}\right)^{-\frac{1}{\tilde{p}+1}}\left[d\vec{z}_q^{\,2} + W^{-1}d\rho^2 + \rho^2 d\Omega^2_{(\delta-2)}\right], \\
e^{-2a\varphi} &= H^{2x}, \quad A_{\underline{t}\underline{y}^1\cdots\underline{y}^p} = \alpha(H^{-1}-1), \quad H = 1+\frac{h}{\rho^{\delta-3}}, \quad W = 1+\frac{\omega}{\rho^{\delta-3}}, \\
\omega &= h\left[1-\frac{a^2}{4x}\alpha^2\right], \qquad x = \frac{(a^2/2)c}{1+(a^2/2)c}, \qquad c = \frac{(p+1)+(\tilde{p}+1)}{(p+1)(\tilde{p}+1)},
\end{aligned}}$$

(24.64)

where $\delta=d-(p+q)>3$. Observe that $q$ is essentially arbitrary. This solution can be considered as the zero mode of the original solution when the $q$s of the transverse dimensions are compact. Equivalently, we can say that it is the original solution "smeared" over $q$ dimensions.

*Electric–magnetic duality in the p-brane a-model.* Except in a few particular cases, the $p$-brane $a$-model does not have any electric–magnetic-duality symmetry. Instead, in general, there is an electric–magnetic duality relating *pairs* of these models: the equations of motion (not the actions, as usual) of the $(a,p)$ model and the $(a,\tilde{p})$ models ($\tilde{p}=d-p-4$) are related by the transformation (see the formulae in Section 1.6)

$$F_{(p+2)} = e^{-2a\varphi_{\tilde{p}}}F_{(\tilde{p}+2)}, \qquad \varphi_p = -\varphi_{\tilde{p}}. \tag{24.65}$$

We have a symmetry of the same theory only when $\tilde{p}=p$ (BHs in $d=4$, strings in $d=6$, membranes in $d=8$, 3-branes in $d=10$, etc.).

At the level of solutions, these transformations allow us to rewrite an electric solution of the $(a,p)$ model as a magnetic solution of the $(a,\tilde{p})$ model and vice versa. Thus, we do not obtain, strictly speaking, more solutions by this procedure: the magnetic dual of an electric $p$-brane is the electric $\tilde{p}$-brane, and all these solutions are already contained in the general one, Eqs. (24.64). In many cases, though, the theory is written in terms of $(p+2)$-forms directly and it is useful to write these "magnetic" solutions directly in terms of them. On the other hand, since the dilaton is inverted in the magnetic solution, if we use it to reexpress the Einstein metric in a different conformal frame (the string frame, say) the metrics of the electric and magnetic solutions will be different. We straightforwardly find

$$
\boxed{
\begin{aligned}
ds^2 &= \left(e^{+2a\varphi}H^{-2}\right)^{\frac{1}{\tilde{p}+1}}\left[W dt^2 - d\vec{y}^{\,2}_{\tilde{p}}\right] \\
&\quad - \left(e^{+2a\varphi}H^{-2}\right)^{-\frac{1}{p+1}}\left[d\vec{z}^{\,2}_q + W^{-1}d\rho^2 + \rho^2 d\Omega^2_{(\tilde{\delta}-2)}\right], \\
e^{-2a\varphi} &= H^{-2x}, \qquad F_{(p+2)z_1\cdots z_q\psi_1\cdots\psi_{(\tilde{\delta}-2)}} = (\tilde{\delta}-3)\alpha h \Omega^{(\tilde{\delta}-2)}_{\psi_1\cdots\psi_{(\tilde{\delta}-2)}}, \\
H &= 1 + \frac{h}{\rho^{\tilde{\delta}-3}}, \qquad W = 1 + \frac{\omega}{\rho^{\tilde{\delta}-3}}, \\
\omega &= h\left[1 - \frac{a^2}{4x}\alpha^2\right], \qquad x = \frac{(a^2/2)c}{1+(a^2/2)c}, \qquad c = \frac{(p+1)+(\tilde{p}+1)}{(p+1)(\tilde{p}+1)},
\end{aligned}
}
\tag{24.66}
$$

where $\tilde{\delta} = d - (\tilde{p} + q)$ and $\Omega^{(n)}$ is the volume form of an $n$-sphere.

Most of the single-$p$-brane solutions we are going to deal with are included in these general solutions, except for the self-dual ones. It is easy to deal with them using solutions describing two $p$-branes and therefore we will study them after we study intersecting $p$-brane solutions in Section 25.6.

### 24.3.3 Sources for solutions of the p-brane a-model

Our experience tells us that we may find charged $p$-brane sources for the extreme, and only for the extreme ($\omega = 0$), charged $p$-brane solutions of the $a$-model that we have just found. On finding these sources, we will be able to relate the integration constants $h$ and $\alpha$ of the solution to the brane tension $T_{(p)}$ and charge parameter $\mu_p$.

We consider the following generic coupled system:

$$S = S_a + S_p, \tag{24.67}$$

where $S_a$ is the bulk $a$-model action Eq. (24.59) and $S_p$ is the charged $p$-brane action:

$$
\begin{aligned}
S_p[X^\mu, \gamma_{ij}] = &-\frac{T_{(p)}}{2}\int d^{p+1}\xi\sqrt{|\gamma|}\left[e^{-2b\varphi}\gamma^{ij}\partial_i X^\mu \partial_j X^\nu g_{\mu\nu} - (p-1)\right] \\
&+ \frac{(-1)^{p+1}\mu_p}{(p+1)!}\int d^{p+1}\xi\, \epsilon^{i_1\cdots i_{p+1}} A_{(p+1)\,\mu_1\cdots\mu_{p+1}}\partial_{i_1}X^{\mu_1}\cdots\partial_{i_{p+1}}X^{\mu_{p+1}}.
\end{aligned}
\tag{24.68}
$$

The coupling of the scalar to the $p$-brane is, in principle, arbitrary. However, in all the relevant cases the parameters $a$ and $b$ turn out to be related by

$$a = -(p+1)b, \tag{24.69}$$

and only then do we have a solution of the coupled system, as we are going to see.

The equations of motion of the spacetime fields are

$$
\begin{aligned}
& G_{\mu\nu} + 2T^{\varphi}_{\mu\nu} + \frac{(-1)^{p+1}}{2\cdot(p+1)!} e^{-2a\varphi} T^{A_{(p+1)}}_{\mu\nu} \\
& \quad - \frac{8\pi G_{\rm N}^{(d)} T_{(p)}}{\sqrt{|g|}} \int d^{p+1}\xi \sqrt{|\gamma|}\, e^{-2b\varphi} \gamma^{ij} \partial_i X^{\mu} \partial_j X^{\nu} g_{\rho\mu} g_{\sigma\nu}\, \delta^{(d)}(x - X(\xi)) = 0, \\
& \nabla^2\varphi + \frac{(-1)^{p+1} a}{4\cdot(p+2)!} e^{-2a\varphi} F^2_{(p+2)} \\
& \quad - \frac{4\pi G_{\rm N}^{(d)} T_{(p)} b}{\sqrt{|g|}} \int d^{p+1}\xi \sqrt{|\gamma|}\, e^{-2b\varphi}\, \gamma^{ij} \partial_i X^{\mu} \partial_j X^{\nu} g_{\mu\nu}\, \delta^{(d)}(x - X(\xi)) = 0, \\
& \nabla_{\mu}\left(e^{-2a\varphi} F_{(p+2)}{}^{\mu\nu_1\cdots\nu_{p+1}}\right) \\
& \quad - \frac{16\pi G_{\rm N}^{(d)} \mu_p}{\sqrt{|g|}} \int d^{p+1}\xi \epsilon^{i_1\cdots i_{p+1}} \partial_{i_1} X^{\nu_1} \cdots \partial_{i_{p+1}} X^{\nu_{p+1}} \delta^{(d)}(x - X(\xi)) = 0,
\end{aligned}
\tag{24.70}
$$

and those of the worldvolume fields are

$$
\begin{aligned}
& \gamma_{ij} - e^{-2b\varphi} g_{\mu\nu} \partial_i X^{\mu} \partial_j X^{\nu} = 0, \\
& \nabla^2(\gamma) X^{\mu} + \gamma^{ij} \partial_i X^{\rho} \partial_j X^{\sigma} \left(\Gamma_{\rho\sigma}{}^{\mu} - 2b\partial_{(\rho}\varphi g_{\sigma)}{}^{\mu}\right) \\
& \quad + \frac{(-1)^{p+1}\mu_p/T_{(p)}}{(p+1)!\sqrt{|\gamma|}} e^{2b\varphi} F_{(p+2)}{}^{\mu}{}_{\mu_1\cdots\mu_{p+1}} \partial_{i_1} X^{\mu_1} \cdots \partial_{i_{p+1}} X^{\mu_{p+1}} \epsilon^{i_1\cdots i_{p+1}} = 0.
\end{aligned}
\tag{24.71}
$$

The first of the worldvolume equations can be used immediately in all the other equations to eliminate the worldvolume metric. Furthermore, using the static gauge for the first $(p+1)$ $p$-brane embedding coordinates that we denote by $Y^i$,

$$Y^i(\xi) = \xi^i, \tag{24.72}$$

and the following ansatz for the transverse embedding coordinates:

$$X^m(\xi) = 0, \tag{24.73}$$

which corresponds to a $p$-brane at rest at $x^m = 0$, it is possible to perform the worldvolume integrals in the equations of motion of the spacetime fields and only $(\tilde{p}+3)$-dimensional Dirac delta functions remain as sources.

Our ansatz for the spacetime fields is given by the extreme $(\omega = 0)$ $p$-brane solutions Eqs. (24.63), $H$ being now a function of the transverse coordinates $x^m$ to be determined.

In the absence of sources (or outside of them) $H$ can be any harmonic function of those $\tilde{p}+3$ transverse coordinates, satisfying

$$\partial_m \partial_m H(\vec{x}_{(\tilde{p}+3)}) = 0. \tag{24.74}$$

In general, $H$, and therefore the solution, has singularities that can be understood as originated by sources that are not included explicitly in the action. When we include source terms in the action, the singularities of $H$ have to match them. In this case, the sources are $(\tilde{p}+3)$-dimensional Dirac delta functions placed at the origin in transverse space, and $H$ has to have a single pole there, with the coefficient $h$ necessary to match that of the Dirac delta functions.

We find that all the equations are solved everywhere (including at the Dirac delta-function singularity) if and only if $a$ and $b$ are related by Eq. (24.69) and the tension $T_{(p)}$ and charge parameter $\mu_{(p)}$ are related by

$$\boxed{\mu_p = (-1)^p T_{(p)}/\alpha.} \tag{24.75}$$

Then, $H$ is given by

$$\boxed{H = \begin{cases} \epsilon + \dfrac{h}{|\vec{x}_{(\tilde{p}+3)}|}, & h = \dfrac{16\pi G_{\rm N}^{(d)} T_{(p)}}{(\tilde{p}+1)\omega_{(\tilde{p}+2)}\alpha^2}, \quad \tilde{p} \geq 0, \\ \epsilon + h \ln|\vec{x}_2|, & h = -\dfrac{16\pi G_{\rm N}^{(d)} T_{(p)}}{2\pi\alpha^2}, \quad \tilde{p} = -1, \\ \epsilon + h|x|, & h = -\dfrac{16\pi G_{\rm N}^{(d)} T_{(p)}}{2\alpha^2}, \quad \tilde{p} = -2, \end{cases}} \tag{24.76}$$

where $\epsilon$ is not determined by the equations of motion alone. Asymptotic flatness in transverse space ($\tilde{p} \geq 0$) requires that $\epsilon = +1$. The solutions with $\epsilon = 0$ can sometimes be understood as the $\vec{x}_{\tilde{p}+3} \to 0$ limit of the asymptotically flat ones. Typically, the limit $\vec{x}_{\tilde{p}+3} \to 0$ is a near-horizon limit. We will see some very important examples in Section 25.5.1, but we have already studied the simplest example: the near-horizon limit of the ERN BH which corresponds to the RB solution Eq. (12.90).

Observe that, for $\tilde{p} \geq 0$ ($\tilde{p} < 0$), $h$ is naturally positive (negative) ($T_{(p)}$ being naturally positive). In general, the metrics of solutions with a single $p$-brane with $\tilde{p} < 0$ will unavoidably have singularities that are unrelated to the $p$-brane sources because $H$ will become zero or negative at some point. To obtain consistent solutions, one must combine several branes. An example of this kind of construction of a regular solution for branes with $\tilde{p} = -1$, i.e. $(d-3)$-branes, can be found in Ref. [674]. For branes with $\tilde{p} = -2$, i.e. $(d-2)$-branes (*domain walls*), a popular way of obtaining a regular metric consists in "cutting" the space before the critical distance $x = 1/|h|$ at which $H = 0$ is reached, for instance by constructing a one-dimensional orbifold with the positive-tension $(d-2)$-brane at one of the fixed points.[12] Consistency requires us to place a brane with opposite tension and charge at

[12] This is the basis of the Hořava–Witten scenario, and also of the Randall–Sundrum scenarios [1069, 1070]. See also Ref. [170].

the other fixed point so the total charge and tension are zero. For this system

$$H = 1 - |hx|, \qquad x \in [0, \pi R], \ \ \pi R < 1/|h|. \tag{24.77}$$

In stringy constructions, the negative-tension brane is an orientifold plane associated with the symmetry $x \to -x$ that gives rise to the orbifold.

# 25
# The extended objects of string theory

After the general introduction to extended objects of Chapter 24, in this chapter we are going to study specifically the extended objects that appear in string theory. The existence of these objects is implied by our previous knowledge of existing objects (strings and D$p$-branes) combined with duality. This path will be followed in Section 25.1, in which we will arrive at the diagrams in Figs. 25.5 and 25.6 that represent, respectively, more- and less-conventional extended string/M-theory objects and their duality relations. The duality relations can be used to find the masses of all these objects compactified on tori (Tables 25.1–25.3) using as input the mass of a string wound once on a circle (i.e. the mass of a winding mode). To obtain consistent results (in particular for electric–magnetic dual branes to coexist satisfying the Dirac quantization condition), the ten-dimensional Newton constant has to have a specific value in terms of the string coupling constant and the string length that we will determine.

The next step (Section 25.2) will be to identify which are, among the general solutions of the $p$-brane $a$-model, those that represent the long-range fields of the basic extended objects of string and M theory that we found before. We will first identify families of solutions and then we will study one by one the most important solutions. In Section 25.3 we will check the values of the integration constants of those solutions against the masses and charges of the extended objects that we determined using duality arguments. Then, the duality relations between the solutions will be checked in Section 25.4.

In Section 25.5 we will learn how a great deal of information about all these objects is encoded in the spacetime superalgebras of the effective (supergravity) theories. In particular, the superalgebras tell us (up to a point) which extended objects may exist and the amount of unbroken supersymmetry preserved by each of them (always half of the total), as we will check by solving explicitly the Killing spinor equations (Section 25.5.1).

In Section 25.6 we will study the possible intersections between several of these objects. The worldvolume fields of the extended objects contain a large amount of information about these intersections and we will briefly review the worldvolume theories of the extended objects of string/M theory first. We will construct solutions describing the simplest intersections, which will be used in Chapter 26 to construct four-dimensional BH solutions.

Some general references with emphasis on $p$-brane solutions of the string/M-theory effective actions are Refs. [471, 467, 1151, 1192, 590, 1152, 1282, 1144]. The standard general references on D-branes are the second volume of Polchinski's book [1049] and Refs. [1051, 78, 832] and Johnson's book [833]. In the last few years, Bergshoeff, Riccioni, and collaborators have started a classification program for the branes of the most relevant supergravity theories in different dimensions (in general those related to $N=2A,B,d=10$ supergravity via toroidal compactifications). See Refs. [177, 178, 179, 176, 175, 171, 180, 181, 150, 182].

## 25.1 String-theory extended objects from duality

We have already met some of the extended objects of string theory: (fundamental) strings (F1) and, in type-II and type-I theories, D$p$-branes (D$p$) with $p=0,2,4,6,8$ for the type-IIA theory, $p=1,3,5,7,9$ for the type-IIB theory, and $p=1,5$ for the type-I theory. Although D$p$-branes have masses proportional to the inverse string coupling constant, their existence has been inferred from the perturbative formulation of string theory, which was reviewed in Chapter 20. It is not surprising that T duality, a perturbative string duality, does not require the existence of any new extended objects in the theory: it just relates D$p$-branes to D$(p\pm 1)$-branes and fundamental strings to fundamental strings in different states.[1] These relations are represented from the viewpoint of the associated classical solutions in Fig. 25.5, which also contains many other relations and new objects required by duality.[2]

We have stressed, however, that non-perturbative S dualities require in general the existence of new non-perturbative states dual to the ones present in the perturbative spectrum. $N=2B,d=10$ SUEGRA has a global SL$(2,\mathbb{R})$ symmetry and it was proposed in Ref. [808] that this symmetry of the effective action reflects an S duality between type-IIB superstring theories, which would be related by the discrete subgroup SL$(2,\mathbb{Z})$, as discussed in Section 23.2. Let us consider systematically what the implications of the existence of this S duality, fundamental strings, and D$p$-branes (with $p$ odd) in type-IIB superstring theory are.

*Extended objects from type-IIB S duality.* Fundamental strings couple to the KR 2-form potential $\hat{\mathcal{B}}_{\hat{\mu}\hat{\nu}}$, which is interchanged with the RR 2-form $\hat{C}^{(2)}_{\hat{\mu}\hat{\nu}}$ to which D1-branes (*D-strings*) couple, by the S-duality transformation $S=\eta$. Thus, the S dual of the fundamental string is the D-string and the two objects form an S-duality doublet, as represented in Fig. 25.6. This is not new information, but it fits nicely in the conjectured S duality.

Let us now consider D$p$-branes beyond the D-string. The D3-brane couples to the 4-form potential with self-dual field strength, which transforms into itself under S duality. Thus, the D3-brane is an S-duality singlet. The D5-brane couples to $\hat{C}^{(6)}$, which is the electric–magnetic dual of $\hat{C}^{(2)}$ (D5-branes are the electric–magnetic duals of D-strings, but not their S duals). Under S duality $\hat{C}^{(6)}$ must transform into $\hat{\mathcal{B}}^{(6)}$, the electric–magnetic dual of $\hat{\mathcal{B}}$.

[1] To the standard D$p$-branes with $p\geq 0$ we can add a IIB D$(-1)$-brane, the *D-instanton*, with zero worldvolume directions and ten transverse Euclidean directions. It can be obtained by T-dualizing the D0-brane in the Euclidean time direction.

[2] Less-conventional objects whose existence is also implied by string dualities are represented in Fig. 25.6. On the other hand, the T duality between fundamental strings is represented in Fig. 25.5 as T duality between gravitational-wave solutions and fundamental-string solutions, which we know represent momentum and winding string states (Section 21.3). Waves and KK6 monopoles can be viewed as electric–magnetic duals.

We have to add to the string spectrum the 5-brane (the *solitonic 5-brane*, *NS 5-brane*, or *S5-brane*) that couples to $\hat{\mathcal{B}}^{(6)}$ that we mentioned on p. 713. The D5-brane and the S5-brane transform as an S-duality doublet. We will determine more properties of the S5-brane using S duality later on.

Next, there is the D7-brane, which couples to $\hat{C}^{(8)}$, the electric–magnetic dual of $\hat{C}^{(0)}$. This pseudoscalar field does not transform linearly under S duality and it is very difficult (if it is possible at all) to find how $\hat{C}^{(8)}$ transforms. Still, the existence of an object related to D7 by $S = \eta$ is required for the S-duality conjecture to be true. Such an object can be called an *S7-brane*. We will see that it is possible to find a classical solution that represents it and that it is related to other solutions by T duality, which makes its presence necessary from this point of view.

Finally, it has been conjectured that the D9-brane is dual to an S9-brane that couples to a 10-form potential $\hat{\mathcal{B}}^{(10)}$. The S9-brane plays an important role in the type-IIB construction of the heterotic SO(32) superstring theory that we have reviewed in Section 23.5.

*Extended objects from type-II T duality.* We have completed the string spectrum to make it consistent with type-IIB S duality. Now, the enhanced type-IIB spectrum has to be consistent with type-IIA/B T duality. First, what is the T dual of the S5-brane in a transverse direction? It must be an object that couples to the T dual of $\hat{\mathcal{B}}^{(6)}$, that is, to the electric–magnetic dual of the type-IIA metric components that give rise to the KK vector field in $d = 9$. We know of only one object of this kind: the KK monopole, Eq. (15.176), which we generated via electric–magnetic duality of the KK vector on p. 444. We will check that the classical solutions that represent the S5 and the KK monopole are indeed related by T duality. It is clear that we can add to the type-IIA theory an S5A that will be dual to the KK monopole of the type-IIB theory.

This is a remarkable relation with very important implications.

1. So far, we have seen the KK monopole just as a topologically non-trivial solution (a gravitational instanton). Now we are going to view it as a new kind of extended object, a KK$p$-brane with non-trivial worldvolume dynamics. The KK monopole of the type-IIA theory has a $(5+1)$-dimensional worldvolume with a vector and a scalar in addition to the ten embedding coordinates $\hat{X}^{\hat{\mu}}$ and we will denote it by KK6A. The worldvolume theory of the KK6B contains a self-dual 2-form potential plus two extra scalars [164].

2. The KK-monopole solution Eq. (15.176) is well defined only when the coordinate $z$ (which is the one associated with the S5–KK6 T duality) is compact and has the right periodicity, which is related to its charge and tension. Thus, we cannot take the decompactification limit.[3] KK6-branes do not exist in uncompactified theories.

3. The above two points are consistent with the requirements for a $\kappa$-symmetric worldvolume action: the bosonic-field contents of the KK6A and KK6B theories have one too many local degrees of freedom, but, as explained in Section 24.2.1,

[3] In other words, the tension and charge are proportional to the compactification radius and would diverge in the decompactification limit.

they can be eliminated by gauging one isometry, which would be associated with the coordinate $z$.

The KK6B, being purely gravitational, is an S-duality singlet and does not require the introduction of any other object. On the other hand, T duality in the worldvolume directions of the KK6- and S5-branes takes us into another KK6-brane or S5-brane.

Now, what is the T dual of an S7-brane? The actual transformation of the S7 solution shows that the T dual, in a worldvolume direction, is a solution called the $D6_1$-brane in Ref. [907]. It shares with KK-branes the requirement that a transverse direction should be compact and it shares with the D6-brane the fact that it has a non-trivial $\hat{C}^{(7)}$.

We can go on tracing all the objects related to these by S and T dualities. In general, starting with the D7/S7 doublet,[4] one finds KK-branes. Their presence is, nevertheless, required by these dualities in ten and 11 dimensions and by U duality in lower dimensions [804, 224, 805, 999], but they have not been studied thoroughly and they are often considered exotic objects.

*Extended objects from type-IIA/M-theory duality.* We have shown in Section 22.1 that the fact that the dimensional reduction of 11-dimensional supergravity on a circle gives $N = 2A, d = 10$ SUEGRA can be interpreted as a suggestion that the strong-coupling limit of type-IIA superstring theory is 11-dimensional supergravity or a theory that reduces to it at low energies (*M theory*). In that limit, then, all the type-IIA extended objects must be related to some 11-dimensional M-theory object [1188].

The identification of the $d = 11$ objects requires the oxidation of the $d = 10$ classical solutions associated with the type-IIA extended objects which will not be performed until Section 25.4, after we find such solutions in Section 25.2. Here we will only give the results, but we can check their internal consistency[5] and their consistency with $d = 10$ dualities.

It has long being known that the type-IIA fundamental string (F1A) can be seen as the 11-dimensional membrane (2-brane) (M2) wrapped on a circle [469]. However, if there was a membrane in $d = 11$, there should have been a membrane in $d = 10$ as well. Now we know that type-IIA superstring theory has a membrane: the D2-brane. We will see that the tensions, charges, and worldvolume effective actions of the fundamental string and the D2-brane can be derived from that of the M2-brane.

How about the other D$p$-branes? The D0-brane turns out to correspond to a $d = 11$ graviton KK mode, and its electric–magnetic dual, the D6-brane, is a $d = 11$ KK monopole compactified on the U(1) fiber direction ($z$ in Eq. (15.176)). The D4-brane has to originate on a $d = 11$ 4- or 5-brane, but there are no 3-branes in $d = 10$, and there is a 5-brane, the

[4] The reduction to $d = 9$ of the D7 and S7 gives a pair of objects that can be constructed by reducing the KK7M on two directions in different orders. All S-duality doublets of the IIB theory have this property (see the next footnote). The situation is, actually, more complicated, since there are an infinite number of "$pq$-7-branes" of which the D7 and S7 are just two examples that do not give rise to all the possibilities.

[5] If we assume that one $d = 11$ object exists, all the different ways of reducing that object over a circle must give different $d = 10$ objects. On the other hand, if we compactify the same object down to $d = 9$ over the same directions but in different orders, then, since these compactifications are related by an $SL(2, \mathbb{R})$ transformation in the internal torus, we must obtain $d = 9$ S-duality doublets. Many examples of this fact can be found in Fig. 25.6.

S5A. Both the D4 and the S5A come from a $d=11$ 5-brane (M5) associated with the dual 6-form potential.

The D8-brane has to be associated with a $d=11$ 8- or 9-brane, but none of these is explicitly known. As we explained in Section 22.2, the natural candidate would be the KK9M-brane, but no associated solution of $d=11$ supergravity corresponding to it is known, and possibly no such solution exists. However, it may exist as a solution of a modified theory [155, 193, 907]. Its inclusion is necessary for the consistency of the diagrams in Figs. 25.5 and 25.6 on p. 758 and 759.

### *25.1.1 The masses of string- and M-theory extended objects from duality*

We can immediately use our knowledge of the duality relations between the extended objects of string and M theories to find their masses when they are compactified on tori. All we need to know is the duality transformation rules of compactification radii, the string coupling constant, the string masses, and the mass of one of the extended objects. Let us first recall the duality transformation rules.

In T duality, the relation between the compactification radii $R_{\rm A(B)}$ and coupling constants $g_{\rm A(B)}$ of the type-IIA(B) theories (or any other pair of string theories) is given by Eqs. (20.61) and (20.62), which we rewrite here for convenience:

$$\boxed{R_{\rm A,B} = \ell_{\rm s}^2/R_{\rm B,A}, \qquad g_{\rm A,B} = g_{\rm B,A}\ell_{\rm s}/R_{\rm B,A}.} \tag{25.1}$$

In type-IIB S duality (with $S=\eta$ and vanishing $\hat{C}^{(0)}$), the transformation rules for the string coupling constant and radii can be deduced from Eqs. (23.21) and (23.24), and take the forms

$$\boxed{g'_{\rm B} = 1/g_{\rm B}, \qquad R'_i = R_i/g_{\rm B}^{\frac{1}{2}}.} \tag{25.2}$$

These rules have to be supplemented by the following transformation rule for the masses, which follows from Eq. (23.24) and the definition of mass, as we will explain in Section 25.3:

$$\boxed{M' = g_{\rm B}^{\frac{1}{2}} M.} \tag{25.3}$$

In type-IIA/M-theory duality we have to use the relations Eqs. (22.48) (rewritten in Eqs. (25.4)) between the string length and the 11-dimensional Planck length and between the string coupling constant and the compactification radius of the 11th coordinate that we call here $R_{10}$ for convenience:

$$\boxed{\ell_{\rm s} = \ell_{\rm Planck}^{(11)\,3/2}/R_{10}^{1/2}, \qquad g_{\rm A} = \left(R_{10}/\ell_{\rm Planck}^{(11)}\right)^{3/2}.} \tag{25.4}$$

*Masses from $d = 10$ string dualities.* We are going to apply these rules to the transformation of the (almost) only object whose mass we know: the (fundamental) string F1 wound once around a compact coordinate ($x^9 \in [0, 2\pi R_9]$, say). Its mass is just the mass of a winding mode with $w = 1$ in the mass formula Eq. (20.60) (which is valid for superstrings if we include left- and right-moving fermionic oscillators),

$$M_{\mathrm{F1}w} = \frac{R_9}{\ell_s^2}, \tag{25.5}$$

which is just the string tension times the volume of the compact space.

As a warm-up exercise, let us first perform a T duality in the $x^9$ direction, in which we know we should obtain an F1 with minimal momentum in the compact direction:

$$M_{\mathrm{F1}m} = M'_{\mathrm{F1}w} = \frac{R_9}{\ell_s^2} = \frac{1}{R'_9}, \tag{25.6}$$

in agreement with Eq. (20.60) with $n = 1$. In this example, it did not matter whether we were dealing with the IIA or IIB fundamental string. Let us now assume that it is the IIB one, F1B. An S-duality transformation should take us to the D-string wound once around $x^9$. Using Eqs. (25.2) and (25.3), we find

$$M_{\mathrm{D1}} = M'_{\mathrm{F1B}w} = g_{\mathrm{B}}^{\frac{1}{2}} M_{\mathrm{F1B}w} = g_{\mathrm{B}}^{\frac{1}{2}} \frac{R_9}{\ell_s^2} = \frac{R'_9}{g'_{\mathrm{B}}\ell_s^2}. \tag{25.7}$$

This mass should be equal to the D-string tension times the volume of the circle, so

$$T_{\mathrm{D1}} = \frac{M_{\mathrm{D1}}}{2\pi R'_9} = \frac{1}{(2\pi\ell_s) g'_{\mathrm{B}}\ell_s}. \tag{25.8}$$

We can now perform successive T-duality transformations to find the masses and tensions of all the D$p$-branes. A T duality in the direction $x^9$ takes us to the D0-brane, whose mass and tension are

$$T_{\mathrm{D0}} = M_{\mathrm{D0}} = M'_{\mathrm{D1}} = \frac{R_9}{g_{\mathrm{B}}\ell_s^2} = \frac{\ell_s^2/R'_9}{g'_{\mathrm{A}}\ell_s/(R'_9\ell_s^2)} = \frac{1}{g'_{\mathrm{A}}\ell_s}. \tag{25.9}$$

If we T-dualize the D-string in a transverse direction ($x^8$), we obtain instead the D2-brane:

$$M_{\mathrm{D2}} = M'_{\mathrm{D1}} = \frac{R_9}{g_{\mathrm{B}}\ell_s^2} = \frac{R_9}{g'_{\mathrm{A}}\ell_s/(R'_8\ell_s^2)} = \frac{R_8 R_9}{g'_{\mathrm{A}}\ell_s^3} \Rightarrow T_{\mathrm{D2}} = \frac{1}{(2\pi\ell_s)^2 g'_{\mathrm{A}}\ell_s}. \tag{25.10}$$

By repeating this procedure, we obtain the mass of the D$p$-brane wrapped around a $p$-torus and its tension (removing the primes):

$$M_{\mathrm{D}p} = \frac{R_{10-p}\cdots R_9}{g\ell_s^{p+1}}, \qquad T_{\mathrm{D}p} = \frac{1}{(2\pi\ell_s)^p g\ell_s}. \tag{25.11}$$

The S5B-brane is the S-dual of the D5-brane:

$$M_{\mathrm{S5B}} = g^{\frac{1}{2}} M'_{\mathrm{D5}} = g'^{-\frac{1}{2}} \frac{R'_5/g'^{\frac{1}{2}} \cdots R'_9/g'^{\frac{1}{2}}}{g'^{-1}\ell_{\mathrm{s}}^6} = \frac{R_5 \cdots R_9}{g^2 \ell_{\mathrm{s}}^6},$$
$$T_{\mathrm{S5B}} = \frac{1}{(2\pi\ell_{\mathrm{s}})^5 g^2 \ell_{\mathrm{s}}}, \tag{25.12}$$

which confirms its non-perturbative character ($M \sim g^{-2}$). On S-dualizing the D7- and D9-branes, we obtain the S7- and S9-branes,

$$M_{\mathrm{S7}} = \frac{R_9 \cdots R_3}{g_{\mathrm{B}}^3 \ell_{\mathrm{s}}^8}, \qquad T_{\mathrm{S7}} = \frac{1}{(2\pi\ell_{\mathrm{s}})^7 g_{\mathrm{B}}^3 \ell_{\mathrm{s}}}, \tag{25.13}$$

$$M_{\mathrm{S9}} = \frac{R_9 \cdots R_1}{g_{\mathrm{B}}^4 \ell_{\mathrm{s}}^{10}}, \qquad T_{\mathrm{S9}} = \frac{1}{(2\pi\ell_{\mathrm{s}})^9 g_{\mathrm{B}}^4 \ell_{\mathrm{s}}}, \tag{25.14}$$

which are even more non-perturbative ($M \sim g^{-3}, g^{-4}$).

On T-dualizing the S5B-brane in a transverse direction ($x^9$), we find the IIA KK monopole (KK6A) with the U(1) fiber in the dual $x^9$ direction. Its mass is

$$M_{\mathrm{KK6A}} = \frac{R_9^2 R_8 \cdots R_4}{g_{\mathrm{A}}^2 \ell_{\mathrm{s}}^8}, \qquad T_{\mathrm{KK6A}} = \frac{R_9}{(2\pi\ell_{\mathrm{s}})^6 g_{\mathrm{A}}^2 \ell_{\mathrm{s}}^2}. \tag{25.15}$$

This object is non-perturbative ($M \sim g^{-2}$). Furthermore, its tension is proportional to the radius of the U(1) fiber, and diverges in the decompactification limit, as we announced.

Since this is a purely gravitational object, there is an identical object in the IIB theory, KK6B, whose mass is identical with the replacement of $g_{\mathrm{A}}$ by $g_{\mathrm{B}}$. Furthermore, by virtue of T duality, the S5A-brane also has the same mass as the S5B with the obvious replacements.

If we dualize the S7-brane in the transverse direction $x^2$, we find the KK8A,

$$M_{\mathrm{KK8A}} = \frac{R_9 \cdots R_3 R_2^3}{g_{\mathrm{A}}^3 \ell_{\mathrm{s}}^{11}}, \qquad T_{\mathrm{KK8A}} = \frac{R_2^3}{(2\pi\ell_{\mathrm{s}})^8 g_{\mathrm{A}}^3 \ell_{\mathrm{s}}^3}, \tag{25.16}$$

which is highly non-perturbative and tightly wrapped around $x^2$. Similar results are obtained when we T-dualize the S9-brane in any direction (say $x^9$): we obtain the KK9A,

$$M_{\mathrm{KK9A}} = \frac{R_9^3 R_8 \cdots R_1}{g_{\mathrm{A}}^4 \ell_{\mathrm{s}}^{12}}, \qquad T_{\mathrm{KK9A}} = \frac{R_9^3}{(2\pi\ell_{\mathrm{s}})^9 g_{\mathrm{A}}^4 \ell_{\mathrm{s}}^3}. \tag{25.17}$$

All these results are collected in Tables 25.1 and 25.2.

*Masses from type-IIA/M-theory duality.* We said that the F1A is the M2 wrapped in the 11th dimension. Thus, the mass of the M2 wrapped on a torus must equal that of the F1A wrapped on a circle:

$$M_{\mathrm{M2}} = M_{\mathrm{F1A}} = \frac{R_9}{\ell_{\mathrm{s}}^2} = \frac{R_9 R_{10}}{(\ell_{\mathrm{Planck}}^{(11)})^3} \Rightarrow T_{\mathrm{M2}} = \frac{1}{(2\pi\ell_{\mathrm{Planck}}^{(11)})^2 \ell_{\mathrm{Planck}}^{(11)}}. \tag{25.18}$$

Table 25.1. Masses of the type-IIA extended objects

Masses of the various extended objects of type-IIA superstring theory are given in ten-dimensional language (the compactification radii $R_i$, the string coupling constant $g_A$, and the string length $\ell_s$) and in 11-dimensional (M-theory) language (the compactification radii $R_i$ and the reduced 11-dimensional Planck length $\ell\!\!\!/^{(11)}_{\text{Planck}} = \ell^{(11)}_{\text{Planck}}/(2\pi)$). The coordinate which is compactified to go from the 11- to the ten-dimensional theory is assumed to be $x^{10}$, so the "11th-dimensional radius" is here denoted by $R_{10} = g_A \ell_s = g_A^{2/3} \ell\!\!\!/^{(11)}_{\text{Planck}}$. Furthermore, the configurations of the various 11-dimensional objects that give rise to the ten-dimensional ones are also provided in a notation whose meaning is the following: the array corresponds to the 11 coordinates, starting from $\hat{x}^0$ up to $\hat{x}^{10}$. A plus means that one of the worldvolume directions occupies that spacetime direction. A star means that the object has a special isometry in the corresponding direction. The corresponding direction cannot be decompactified.

| Type-IIA object | Mass in $d = 10$ constants | Mass in $d = 11$ constants | 11-dimensional object |
|---|---|---|---|
| F1m | $R_9^{-1}$ | | |
| D0 | $g_A^{-1}\ell_s^{-1}$ | $R_{10}^{-1}$ | WM$(+,-^{10})$ |
| F1w | $R_9\ell_s^{-2}$ | $R_{10}R_9(\ell\!\!\!/^{(11)}_{\text{Planck}})^{-3}$ | M2$(+,-^8,+^2)$ |
| D2 | $R_9R_8g_A^{-1}\ell_s^{-3}$ | $R_9R_8(\ell\!\!\!/^{(11)}_{\text{Planck}})^{-3}$ | M2$(+,-^7,+^2,-)$ |
| D4 | $R_9\cdots R_6g_A^{-1}\ell_s^{-5}$ | $R_{10}R_9\cdots R_5(\ell\!\!\!/^{(11)}_{\text{Planck}})^{-6}$ | M5$(+,-^5,+^5)$ |
| S5A | $R_9\cdots R_5g_A^{-2}\ell_s^{-6}$ | $R_9\cdots R_5(\ell\!\!\!/^{(11)}_{\text{Planck}})^{-6}$ | M5$(+,-^4,+^5,-)$ |
| D6 | $R_9\cdots R_4g_A^{-1}\ell_s^{-7}$ | $R_{10}^2R_9\cdots R_4(\ell\!\!\!/^{(11)}_{\text{Planck}})^{-9}$ | KK7M$(+,-^3,+^6,-^\star)$ |
| KK6A | $R_9^2R_8\cdots R_4g_A^{-2}\ell_s^{-8}$ | $R_{10}R_9^2\cdots R_4(\ell\!\!\!/^{(11)}_{\text{Planck}})^{-9}$ | KK7M$(+,-^3,+^5,+^\star,+)$ |
| D8 | $R_9\cdots R_2g_A^{-1}\ell_s^{-9}$ | $R_{10}^3R_9\cdots R_2(\ell\!\!\!/^{(11)}_{\text{Planck}})^{-12}$ | KK9M$(+,-,+^8,+^\star)$ |
| KK8A | $R_9^3R_8\cdots R_2g_A^{-3}\ell_s^{-11}$ | $R_{10}R_9^3R_8\cdots R_2(\ell\!\!\!/^{(11)}_{\text{Planck}})^{-12}$ | KK9M$(+,-,+^7,+^\star,+)$ |
| KK9A | $R_9^3R_8\cdots R_1g_A^{-4}\ell_s^{-12}$ | $R_9^3R_8\cdots R_1(\ell\!\!\!/^{(11)}_{\text{Planck}})^{-12}$ | KK9M$(+,+^8,+^\star,-)$ |

Table 25.2. Masses of the type-IIB extended objects

Masses of the various extended objects of type-IIB superstring theory are given in terms of the compactification radii $R_i$, the string coupling constant $g_B$, and the string length $\ell_s$; when a radius appears with a power different from 1, it means that that is a special isometric direction of the object (a KK object).

| Type-IIB object | Mass | Type-IIB object | Mass |
|---|---|---|---|
| F1m | $R_9^{-1}$ | KK6A | $R_9^2R_8\cdots R_4g_B^{-2}\ell_s^{-8}$ |
| F1w | $R_9\ell_s^{-2}$ | D7 | $R_9\cdots R_3g_B^{-1}\ell_s^{-8}$ |
| D1 | $R_9g_B^{-1}\ell_s^{-2}$ | S7 | $R_9\cdots R_3g_B^{-3}\ell_s^{-8}$ |
| D3 | $R_9\cdots R_7g_B^{-1}\ell_s^{-4}$ | D9 | $R_9\cdots R_1g_B^{-1}\ell_s^{-10}$ |
| D5 | $R_9\cdots R_5g_B^{-1}\ell_s^{-6}$ | S9 | $R_9\cdots R_1g_B^{-4}\ell_s^{-10}$ |
| S5B | $R_9\cdots R_5g_B^{-2}\ell_s^{-6}$ | | |

Table 25.3. Masses of the M-theory extended objects

Masses of the extended objects of M theory given in terms of the compactification radii $R_i$ and the reduced 11-dimensional Planck length $\ell_{\text{Planck}}^{(11)} = \ell_{\text{Planck}}^{(11)}/(2\pi)$; when a radius appears with a power different from 1, it means that that is a special isometric direction of the object (a KK object).

| M object | Mass |
|---|---|
| WM | $0$ |
| M2 | $R_{10}R_9(\ell_{\text{Planck}}^{(11)})^{-3}$ |
| M5 | $R_{10}\cdots R_6(\ell_{\text{Planck}}^{(11)})^{-6}$ |
| KK7M | $R_{10}^2 R_9\cdots R_4(\ell_{\text{Planck}}^{(11)})^{-9}$ |
| KK9M | $R_{10}^3 R_9\cdots R_2(\ell_{\text{Planck}}^{(11)})^{-12}$ |

On the other hand, the mass of the M2 wrapped in two directions different from the one that we consider the 11th (say $x^8$ and $x^9$) must coincide with that of the D2. Indeed,

$$M_{\text{M2}} = \frac{R_8 R_9}{(\ell_{\text{Planck}}^{(11)})^3} = \frac{R_8 R_9}{g_{\text{A}}\ell_{\text{s}}^3} = M_{\text{D2}}. \tag{25.19}$$

The D4 is an M5 wrapped in the 11th dimension,

$$M_{\text{M5}} = M_{\text{D4}} = \frac{R_6\cdots R_9}{g_{\text{A}}\ell_{\text{s}}^5} = \frac{R_6\cdots R_{10}}{(\ell_{\text{Planck}}^{(11)})^6} \Rightarrow T_{\text{M5}} = \frac{1}{(2\pi\ell_{\text{Planck}}^{(11)})^5\ell_{\text{Planck}}^{(11)}}, \tag{25.20}$$

and the S5A is an M5 that is not wrapped there,

$$M_{\text{M5}} = \frac{R_5\cdots R_9}{(\ell_{\text{Planck}}^{(11)})^6} = \frac{R_5\cdots R_9}{g_{\text{A}}^2\ell_{\text{s}}^6} = M_{\text{S5A}}. \tag{25.21}$$

To complete this section, we can see that the D0 is nothing but a KK mode moving in the 11th direction,

$$M_{\text{D0}} = \frac{1}{g_{\text{A}}\ell_{\text{s}}} = \frac{1}{R_{10}}. \tag{25.22}$$

These results are collected in Tables 25.1 and 25.3.

*The Newton constant.* If we want all the extended objects just found to be quantum-mechanically compatible, we know that the charges of those pairs of extended objects which are electric–magnetic duals must satisfy the Dirac quantization condition. We do not know the charges of all of these objects, except in the case of the F1, since we know the coefficient of the WZ term in the string $\sigma$-model action Eq. (21.31); namely the string tension $T$. This coefficient coincides with the F1 charge, canonically normalized taking into account the normalization of the spacetime effective action Eq. (21.1). Then

$$q_{\text{F1}} = T = \frac{1}{(2\pi\ell_{\text{s}})\ell_{\text{s}}}. \tag{25.23}$$

This identity between the canonically normalized charge and the coefficient of the WZ term in the extended objects' effective action is completely general. Furthermore, those coefficients are always identical to the coefficients of the kinetic (NG) terms given on p. 713 for various kinds of objects. Let us, then, consider the electric–magnetic dual of the F1, the S5, whose tension is given in Eq. (25.12). Then

$$q_{\text{S5}} = T_{\text{S5}}g^2 = \frac{1}{(2\pi\ell_{\text{s}})^5\ell_{\text{s}}}. \tag{25.24}$$

With the normalization Eq. (21.1), the Dirac quantization condition for $n = 1$ reads

$$q_{\text{F1}}q_{\text{S5}} = 2\pi\frac{g^2}{16\pi G_{\text{N}}^{(10)}}, \tag{25.25}$$

which is possible only if

$$\boxed{G_{\text{N}}^{(10)} = 8\pi^6 g^2 \ell_{\text{s}}^8.} \tag{25.26}$$

We obtained a similar result in Chapter 15 in the context of $\hat{d} = 5$ KK theory.

There are more pairs of electric–magnetic duals: D$p$- and D$\tilde{p}$-branes. According to the above observations, the D$p$-brane charge is

$$q_{\text{D}p} = T_{\text{D}p}g = \frac{1}{(2\pi\ell_{\text{s}})^p\ell_{\text{s}}}, \tag{25.27}$$

and we find again ($d = 10$)

$$q_{\text{D}p}q_{\text{D}\tilde{p}} = \frac{1}{(2\pi\ell_{\text{s}})^6\ell_{\text{s}}^2} = q_{\text{F1}}q_{\text{S5}}. \tag{25.28}$$

For the two M-branes, we have

$$q_{\text{M}p} = \frac{1}{(2\pi\ell_{\text{Planck}}^{(11)})^p\ell_{\text{Planck}}^{(11)}}, \tag{25.29}$$

so, on account of the definition of $\ell_{\text{Planck}}^{(11)}$, Eq. (22.43),

$$q_{\text{M}p}q_{\text{M}\tilde{p}} = \frac{1}{(2\pi\ell_{\text{Planck}}^{(11)})^7(\ell_{\text{Planck}}^{(11)})^2} = 2\pi\frac{2\pi}{(\ell_{\text{Planck}}^{(11)})^9} = 2\pi\frac{1}{16\pi G_{\text{N}}^{(11)}}, \tag{25.30}$$

which is the correct form of Dirac's quantization condition with the standard normalization of $d = 11$ supergravity. This is the reason behind the unusual definition of $\ell_{\text{Planck}}^{(11)}$.

## 25.2 String-theory extended objects from effective-theory solutions

The results discussed in Section 25.1 and depicted in Figs. 25.5 and 25.6 are supported by (and, in many cases, based on) the study of explicit classical solutions of the string effective action that can be interpreted as the long-range fields associated with the extended objects of string theory. That interpretation is based on a comparison between the charges of the sources and those of the solutions; $p$-branes can be charged with respect to $(p+1)$-form potentials and, therefore, it is natural to start looking for $p$-brane solutions that are charged with respect to each of the $(p+1)$-form potentials of the string effective theory. The existence of these solutions is a clear argument in support of the existence of the associated string-theory extended object.

The search for these solutions can be systematized using the $p$-brane $a$-model solutions as follows: for solutions that involve only one brane, the Chern–Simons terms can be ignored and the relevant part of the string effective action is just

$$S = \frac{g^2}{16\pi G_N^{(d)}} \int d^dx \sqrt{|g|} \left\{ e^{-2\phi} \left[ R - 4(\partial\phi)^2 + \frac{(-1)^{p_1+1}}{2\cdot(p_1+2)!}(H^{(p_1+2)})^2 \right] \right.$$
$$\left. + \frac{(-1)^{p_2+1}}{2\cdot(p_2+2)!}(G^{(p_2+2)})^2 \right\}, \tag{25.31}$$

where

$$H^{(p_1+2)} = dB^{(p_1+1)}, \qquad G^{(p_2+2)} = dC^{(p_1+1)}. \tag{25.32}$$

By definition, since it is written in the string frame, it describes the potentials that couple to fundamental $p_1$-branes and D$p_2$-branes. We have to rewrite it in the *modified Einstein-frame metric*[6] $\tilde{g}_{\mathrm{E}\,\mu\nu}$ in which the $p$-brane $a$-model is given.[7] Using the relation

$$g_{\mu\nu} = e^{\frac{4}{d-2}(\phi-\phi_0)}\, \tilde{g}_{\mathrm{E}\,\mu\nu}, \tag{25.34}$$

we obtain (ignoring tildes) (see Appendix M)

$$S = \frac{1}{16\pi G_N^{(d)}} \int d^dx \sqrt{|g|} \left[ R + \frac{4}{d-2}(\partial\phi)^2 + \frac{(-1)^{p_1+1}}{2\cdot(p_1+2)!} e^{-4\frac{p_1+1}{d-2}(\phi-\phi_0)} (H^{(p_1+2)})^2 \right.$$
$$\left. + \frac{(-1)^{p_2+1}}{2\cdot(p_2+2)!} e^{2\frac{\tilde{p}_2-p_1}{d-2}(\phi-\phi_0)} g^2 (G^{(p_2+2)})^2 \right]. \tag{25.35}$$

On comparing this action with the $a$-model action Eq. (24.59), we find that the relations between the string dilaton $\phi$ and the $a$-model scalar $\varphi$, and between the string potentials

---

[6] We always take the string-frame metric $g_{\mu\nu}$ to be asymptotically flat (at least in the non-compact directions). On rescaling that metric by a power of the dilaton,

$$g_{\mu\nu} = e^{\frac{4}{d-2}\phi} g_{\mathrm{E}\,\mu\nu}, \tag{25.33}$$

we obtain the *Einstein-frame metric* $g_{\mathrm{E}\,\mu\nu}$ in which the Einstein–Hilbert term has no dilaton factors. However, this metric is not asymptotically flat and a constant rescaling by the dilaton VEV is necessary. The result is the asymptotically flat *modified Einstein-frame metric* [919] $\tilde{g}_{\mathrm{E}\,\mu\nu}$ given in Eq. (25.34).

[7] Recall that we chose $\varphi_0 = 0$.

$B^{(p_1+1)}$ and $C^{(p_2+1)}$ and the $a$-model potentials $A_{(p_{1,2}+1)}$, are

$$\phi-\phi_0=\sqrt{\frac{d-2}{2}}\varphi, \qquad B^{(p_1+1)}=A_{(p_1+1)}, \qquad C^{(p_2+1)}=g^{-1}A_{(p_2+1)}, \quad (25.36)$$

and, furthermore,

$$a_1=\frac{2(p_1+1)}{\sqrt{2(d-2)}}, \qquad a_2=\frac{-(\tilde{p}_2-p_2)}{\sqrt{2(d-2)}}. \quad (25.37)$$

We could have Poincaré-dualized the field strengths. In that case, we would have obtained for the duals of fundamental $p_1$-branes ($p_3=\tilde{p}_1$-branes) an $a$-model with

$$a_3=-\frac{2(p_3+1)}{\sqrt{2(d-2)}}, \qquad p_3=\tilde{p_1}, \qquad \tilde{B}^{(p_3+1)}=g^{-2}A_{(p_3+1)}, \quad (25.38)$$

whereas for the dual D$p_2$-branes we would have the same $a_2$.

Finally, to find solutions of $d=11$ supergravity that describe M-theory branes (M$p$-branes), we just have to set $a=0$, since there is no scalar in that theory.

The four families of solutions, which are valid for $\delta>3$, are given by the following.
(Black) M-theory $p$-branes (M$p$):

$$\begin{aligned}
&d\tilde{s}_{\rm E}^2=H_{{\rm M}p}^{-\frac{2}{p+1}}\left[Wdt^2-d\vec{y}_p^{\,2}\right]-H_{{\rm M}p}^{\frac{2}{\tilde{p}+1}}\left[d\vec{z}_q^{\,2}+W^{-1}d\rho^2+\rho^2d\Omega_{(\delta-2)}^2\right],\\
&C^{(p+1)}{}_{\underline{t}\underline{y}^1\cdots\underline{y}^p}=\alpha\left(H_{{\rm M}p}^{-1}-1\right),\\
&H_{{\rm M}p}=1+\frac{h_{{\rm M}p}}{\rho^{\delta-3}}, \qquad W=1+\frac{\omega}{\rho^{\delta-3}}, \qquad \omega=h_{{\rm M}p}\left[1-\frac{1}{2c}\alpha^2\right].
\end{aligned} \quad (25.39)$$

(Black) fundamental $p$-branes (F$p$):

$$\begin{aligned}
&d\tilde{s}_{\rm E}^2=H_{{\rm F}p}^{-\frac{2}{d-2}\frac{\tilde{p}+1}{p+1}}\left[Wdt^2-d\vec{y}_p^{\,2}\right]-H_{{\rm F}p}^{\frac{2}{d-2}}\left[d\vec{z}_q^{\,2}+W^{-1}d\rho^2+\rho^2d\Omega_{(\delta-2)}^2\right],\\
&ds_{\rm s}^2=H_{{\rm F}p}^{-\frac{2}{p+1}}\left[Wdt^2-d\vec{y}_p^{\,2}\right]-\left[d\vec{z}_q^{\,2}+W^{-1}d\rho^2+\rho^2d\Omega_{(\delta-2)}^2\right],\\
&e^{-2\phi}=e^{-2\phi_0}H_{{\rm F}p}, \qquad B^{(p+1)}{}_{\underline{t}\underline{y}^1\cdots\underline{y}^p}=\alpha\left(H_{{\rm F}p}^{-1}-1\right),\\
&H_{{\rm F}p}=1+\frac{h_{{\rm F}p}}{\rho^{\delta-3}}, \qquad W=1+\frac{\omega}{\rho^{\delta-3}}, \qquad \omega=h_{{\rm F}p}\left[1-\frac{p+1}{2}\alpha^2\right].
\end{aligned}$$

(25.40)

(Black) solitonic $p$-branes (S$p$):

$$
\begin{aligned}
d\tilde{s}_{\rm E}^2 &= H_{{\rm S}p}^{-\frac{2}{d-2}}\left[W dt^2 - d\vec{y}_p^{\,2}\right] - H_{{\rm S}p}^{\frac{2}{d-2}\frac{p+1}{\tilde{p}+1}}\left[d\vec{z}_q^{\,2} + W^{-1}d\rho^2 + \rho^2 d\Omega^2_{(\delta-2)}\right],\\
ds_{\rm s}^2 &= W dt^2 - d\vec{y}_p^{\,2} - H_{{\rm S}p}^{\frac{2}{\tilde{p}+1}}\left[d\vec{z}_q^{\,2} + W^{-1}d\rho^2 + \rho^2 d\Omega^2_{(\delta-2)}\right],\\
e^{-2\phi} &= e^{-2\phi_0}H_{{\rm S}p}^{-1}, \qquad \tilde{B}^{(p+1)}{}_{\underline{t}\underline{y}^1\cdots \underline{y}^p} = \alpha e^{-2\phi_0}\left(H_{{\rm S}p}^{-1} - 1\right),\\
H_{{\rm S}p} &= 1 + \frac{h_{{\rm S}p}}{\rho^{\delta-3}}, \qquad W = 1 + \frac{\omega}{\rho^{\delta-3}}, \qquad \omega = h_{{\rm S}p}\left[1 - \frac{\tilde{p}+1}{2}\alpha^2\right].
\end{aligned}
\tag{25.41}
$$

(Black) D$p$-branes:

$$
\begin{aligned}
d\tilde{s}_{\rm E}^2 &= H_{{\rm D}p}^{-8\frac{\tilde{p}+1}{(d-2)^2}}\left[W dt^2 - d\vec{y}_p^{\,2}\right] - H_{{\rm D}p}^{8\frac{p+1}{(d-2)^2}}\left[d\vec{z}_q^{\,2} + W^{-1}d\rho^2 + \rho^2 d\Omega^2_{(\delta-2)}\right],\\
ds_{\rm s}^2 &= H_{{\rm D}p}^{-\frac{4}{d-2}}\left[W dt^2 - d\vec{y}_p^{\,2}\right] - H_{{\rm D}p}^{\frac{4}{d-2}}\left[d\vec{z}_q^{\,2} + W^{-1}d\rho^2 + \rho^2 d\Omega^2_{(\delta-2)}\right],\\
e^{-2\phi} &= e^{-2\phi_0}H_{{\rm D}p}^{-2\frac{\tilde{p}-p}{d-2}}, \qquad C^{(p+1)}{}_{\underline{t}\underline{y}^1\cdots \underline{y}^p} = \alpha e^{-\phi_0}\left(H_{{\rm D}p}^{-1} - 1\right),\\
H_{{\rm D}p} &= 1 + \frac{h_{{\rm D}p}}{\rho^{\delta-3}}, \qquad W = 1 + \frac{\omega}{\rho^{\delta-3}}, \qquad \omega = h_{{\rm D}p}\left[1 - \frac{d-2}{8}\alpha^2\right].
\end{aligned}
\tag{25.42}
$$

### 25.2.1 Extreme $p$-brane solutions of string and M theories and sources

The four families of solutions given in Eqs. (25.39)–(25.42) contain subfamilies of extreme solutions with $\omega = 0$ in which the $H$ functions can be arbitrary functions of the transverse coordinates $\vec{x}_{(\delta-1)}$ ($\rho = |\vec{x}_{(\delta-1)}|$) for $\delta \geq 2$. These are isotropic coordinates in which the metric of the transverse space is conformally flat. As in the ERN BH case, in some cases they do not cover the whole spacetime which can be analytically extended beyond $\rho = 0$, which is only a coordinate singularity.

In Section 24.3.3 we saw that some of the extreme solutions of the $p$-brane $a$-model (with $q = 0$ and a single-pole $H$) could be matched against some charged $p$-brane sources (obeying Eqs. (24.69) and (24.75)), which allowed us to determine $h$, the coefficient of the pole of $H$, in terms of the tension and charge of the source and in terms of the Newton constant.

It turns out that the four families of objects that we are considering always satisfy Eqs. (24.69) and (24.75) and we can use those results to determine $h$ in the extreme solutions with $q = 0$ and a single pole in terms of the tensions and Newton constants. Then, for the ten- and 11-dimensional objects whose tensions we found in Section 25.1.1, we can determine $h$ as a function of $\ell_s, g,$ and $\ell_{\text{Planck}}^{(11)}$, using the values of $G_N^{(10)}$ and $G_N^{(11)}$ that we determined there. For these $d = 10, 11$ objects $\alpha = \pm 1$ is just the relative sign between the charge parameter $\mu_p$ and the tension $T_{(p)}$, which we consider positive.

All we have to do to find $h$ for the families of extreme M$p$-branes, D$p$-branes, etc. is to substitute into Eqs. (24.76) the values of $T_{(p)}$ (which are unknown in general, except in the cases studied in the previous sections) and $\alpha$, determined by setting $\omega = 0$ in the solutions;

$$\alpha_{\text{M}p}^2 = \frac{2(d-2)}{(p+1)(\tilde{p}+1)}, \quad \alpha_{\text{F}p} = \frac{2}{p+1}, \quad \alpha_{\text{S}p} = \frac{2}{\tilde{p}+1}, \quad \alpha_{\text{D}p} = \frac{8}{d-2}. \tag{25.43}$$

Observe that, indeed, for the string/M-theory branes (M2 and M5 in $d = 11$ and F1, S5, and D$p$ in $d = 10$) $\alpha^2 = +1$.

Writing the four families of extreme solutions with the right values for $h$ is straightforward, but not very interesting, except in the case of the $d = 11, 10$ string/M-theory branes, which we are going to write and study next.

### 25.2.2 *The M2 solution*

On substituting the values $d = 11$ and $p = 2$ into the general black M$p$-brane family of solutions Eqs. (25.39), we immediately find the black M2 solution [699],

$$\boxed{\begin{aligned} d\hat{\hat{s}}^2 &= H_{\text{M2}}^{-\frac{2}{3}}\left[W dt^2 - d\vec{y}_2^{\,2}\right] - H_{\text{M2}}^{\frac{1}{3}}\left[W^{-1}d\rho^2 + \rho^2 d\Omega_{(7)}^2\right], \\ \hat{\hat{C}}_{\underline{t}\underline{y}^1\underline{y}^2} &= \alpha\left(H_{\text{M2}}^{-1} - 1\right), \\ H_{\text{M2}} &= 1 + \frac{h_{\text{M2}}}{\rho^6}, \qquad W = 1 + \frac{\omega}{\rho^6}, \qquad \omega = h_{\text{M2}}\left[1 - \alpha^2\right], \end{aligned}} \tag{25.44}$$

and, in the extreme limit $\omega = 0$, $\alpha = \pm 1$, we obtain [478]

$$\boxed{\begin{aligned} d\hat{\hat{s}}^2 &= H_{\text{M2}}^{-\frac{2}{3}}\left[dt^2 - d\vec{y}_2^{\,2}\right] - H_{\text{M2}}^{\frac{1}{3}}\, d\vec{x}_8^{\,2}, \\ \hat{\hat{C}}_{\underline{t}\underline{y}^1\underline{y}^2} &= \pm\left(H_{\text{M2}}^{-1} - 1\right), \qquad H_{\text{M2}} = 1 + \frac{h_{\text{M2}}}{|\vec{x}_8|^6}, \end{aligned}} \tag{25.45}$$

which is the solution commonly called an M2-brane in the literature written in isotropic coordinates. The integration constant $h_{\text{M2}}$ can be determined by the first of Eqs. (24.76), the value of the M2 tension by Eq. (25.18), and the value of $G_N^{(11)}$ by Eq. (22.43),

$$\boxed{h_{\text{M2}} = \frac{\left(\ell_{\text{Planck}}^{(11)}\right)^6}{6\omega_{(7)}}.} \tag{25.46}$$

In Section 25.3 we will check by classical field-theory methods that this value of this integration constant and the values of other integration constants really correspond to the tension and charge of the M2-brane and the other objects.

It is clear that, if we want the solution to describe $N_{\mathrm{M2}}$ parallel M2-branes, we just have to replace $H_{\mathrm{M2}}$ by another harmonic function with as many poles as branes, each of them with the coefficient $h_{\mathrm{M2}}$. When $N_{\mathrm{M2}}$ M2s coincide, there is a single pole with coefficient $N_{\mathrm{M2}}h_{\mathrm{M2}}$. Similar observations are valid for all the extreme solutions that follow.

The black M2-brane has a regular non-degenerate event horizon at $\rho = (-\omega)^{\frac{1}{6}}$ whose constant-time sections have the topology of $\mathrm{S}^7$ and a singular (would-be) inner horizon covered by the event horizon.

In the extreme limit the event horizon does not become singular but becomes degenerate (as in the ERN BH case) and the singularity covered by it becomes timelike. The Penrose diagram of the extreme M2-brane is similar to that of the ERN BH but with only one asymptotically flat region [468, 629] and is represented in Fig. 25.1 (a more detailed diagram can be found in Ref. [1152]).

In fact, on taking the near-horizon limit $\rho = |\vec{x}_8| \to 0$ (which consists in the deletion of the constant 1 in $H_{\mathrm{M2}}$) in spherical coordinates, we obtain a solution whose metric is the direct product of those of $\mathrm{AdS}_4$ and $\mathrm{S}^7$ with radii $R_4$ and $2R_4$ (after a rescaling of the worldvolume coordinates),

$$\boxed{d\hat{\hat{s}}^2 = R_4^2\, d\Pi^2_{(4)} - (2R_4)^2\, d\Omega^2_{(7)}, \qquad \hat{\hat{C}}_{\underline{t}\underline{y}^1\underline{y}^2} = \left(\frac{r}{R_4}\right)^3, \qquad R_4 = \frac{h_{\mathrm{M2}}^{\frac{1}{2}}}{2},} \tag{25.47}$$

where we are using the following form of the metric of the $\mathrm{AdS}_n$ space with radius $R_n$:

$$R_n^2\, d\Pi^2_{(n)} \equiv \left(\frac{r}{R_n}\right)^2 \left(dt^2 - d\vec{y}^{\,2}_{n-2}\right) - \left(\frac{R_n}{r}\right)^2 dr^2. \tag{25.48}$$

The dual 7-form field strength is given by the $\mathrm{S}^7$ volume form $\tilde{\hat{\hat{G}}}{}^{(7)} = 6(2R_4)^6\omega_{(7)}$.

All the extreme string/M-theory solutions that we are going to study preserve half of the supersymmetries, but this near-horizon limit preserves all the supersymmetry and can be considered a vacuum of M theory. The M2-brane can then be seen as a soliton interpolating between two maximally supersymmetric vacua, Minkowski at infinity and $\mathrm{AdS}_4 \times \mathrm{S}^7$ at the horizon [629]. Since the $\mathrm{S}^7$ is a compact space, this vacuum can be seen to induce spontaneous compactification to $d = 4$ [567]. The compactification of $D = 11$ supergravity on $\mathrm{S}^7$ gives rise to a gauged $N = 8, d = 4$ SUEGRA with gauge group $\mathrm{SO}(8)$ (the isometry group of the compact space) and an $\mathrm{AdS}_4$ vacuum [477] (see also Ref. [476] and references therein).

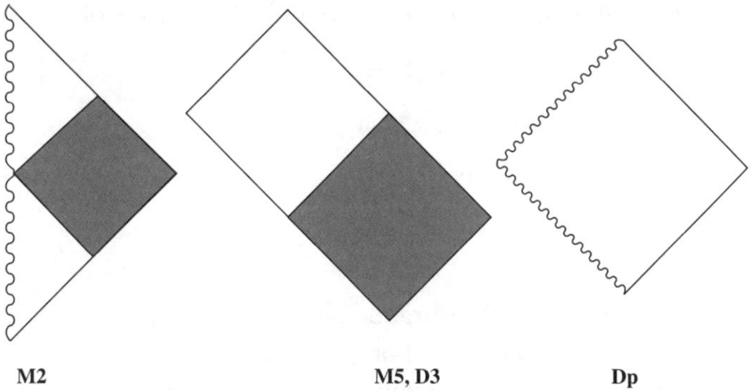


Fig. 25.1. Penrose diagrams of different (extreme) string/M-brane solutions: the M2-brane, which has a timelike singularity covered by a horizon which does not allow it to be seen from the asymptotic region covered by the isotropic coordinates (shaded); the M5- and D3-brane that are regular everywhere and have two asymptotic regions separated by the horizon; and the D$p$-branes with $p \neq 3$, which have singular horizons. In all cases the angular coordinates of the transverse spheres and the spacelike worldvolume coordinates have been ignored.

### 25.2.3 The M5 solution

On substituting the values $d = 11$ and $p = 5$ into the general black M$p$-brane family of solutions Eqs. (25.39), we obtain the black M5-brane solution [699],

$$\boxed{\begin{aligned} d\hat{\hat{s}}^2 &= H_{\mathrm{M5}}^{-\frac{1}{3}}\left[W dt^2 - d\vec{y}_5^{\,2}\right] - H_{\mathrm{M5}}^{\frac{2}{3}}\left[W^{-1}d\rho^2 + \rho^2 d\Omega_{(4)}^2\right], \\ \hat{\hat{\tilde{C}}}_{t\underline{y}^1\cdots\underline{y}^5} &= \alpha\left(H_{\mathrm{M5}s}^{-1} - 1\right), \\ H_{\mathrm{M5}} &= 1 + \frac{h_{\mathrm{M5}}}{\rho^3}, \qquad W = 1 + \frac{\omega}{\rho^3}, \qquad \omega = h_{\mathrm{M5}}[1-\alpha^2], \end{aligned}} \tag{25.49}$$

and, in the extreme limit $\omega = 0$, $\alpha = \pm 1$,

$$\boxed{\begin{aligned} d\hat{\hat{s}}^2 &= H_{\mathrm{M5}}^{-\frac{1}{3}}\left[dt^2 - d\vec{y}_5^{\,2}\right] - H_{\mathrm{M5}}^{\frac{2}{3}}\, d\vec{x}_5^{\,2}, \\ \hat{\hat{\tilde{C}}}_{t\underline{y}^1\cdots\underline{y}^5} &= \pm\left(H_{\mathrm{M5}}^{-1} - 1\right), \qquad H_{\mathrm{M5}} = 1 + \frac{h_{\mathrm{M5}}}{|\vec{x}_5|^3}, \end{aligned}} \tag{25.50}$$

referred to usually in the literature as the M5-brane solution in isotropic coordinates. This solution is regular everywhere.

Using the first of Eqs. (24.76), and Eqs. (25.20) and (22.43), we obtain

$$h_{\mathrm{M5}} = \frac{(\ell_{\mathrm{Planck}}^{(11)})^3}{3\omega_{(4)}}. \tag{25.51}$$

The event horizon of the black M5-brane, placed at $\rho = (-\omega)^{\frac{1}{3}}$, has (constant-time sections of) $\mathrm{S}^4$ topology. There is an inner horizon which is singular.[8]

As in the M2 case, the event horizon remains regular but degenerate in the extreme limit but now the singularity disappears and there is a new asymptotically flat region across the horizon (see its Penrose diagram in Fig. 25.1). Using the coordinate $r$ defined by

$$\rho = h_{\mathrm{S5}}^{\frac{1}{3}} r^2/(1-\rho^6)^{\frac{1}{3}}, \tag{25.52}$$

the metric takes the form [1152]

$$d\hat{s}^2 = r^2\left[dt^2 - d\vec{y}_5^{\,2}\right] - h_{\mathrm{S5}}^{\frac{2}{3}}\left[\frac{4}{r^2(1-r^6)^{\frac{8}{3}}}dr^2 + \frac{d\Omega_{(4)}^2}{(1-r^6)^{\frac{2}{3}}}\right] \tag{25.53}$$

and covers both sides of the event horizon $\rho = r = 0$. The spatial infinities are now at $r = \pm 1$. This metric is invariant under $r \to -r$, which allows us to identify the two asymptotic regions. There are more asymptotic regions, all of which may also be identified.

In the near-horizon limit $\rho = |\vec{x}_5| \to 0$ of the extreme M5 we obtain another maximally supersymmetric solution, whose metric is the direct product of those of $\mathrm{AdS}_7$ and $\mathrm{S}^4$ with radii $\mathrm{R}_7$ and $\mathrm{R}_7/2$,

$$d\hat{s}^2 = R_7^2\, d\Pi_{(7)}^2 - (R_7/2)^2\, d\Omega_{(4)}^2, \qquad \tilde{\hat{\hat{C}}}_{t\underline{y}^1\cdots\underline{y}^5} = \left(\frac{r}{R_7}\right)^6, \qquad R_7 = 2h_{\mathrm{M5}}^{\frac{1}{2}}, \tag{25.54}$$

where we are using again the notation of Eq. (25.48). The 4-form field strength is given by the $\mathrm{S}^4$ volume form $\hat{\hat{G}} = 3(R_7/2)^3\omega_{(4)}$ and, again, the M5 can be seen as a vacuum-interpolating soliton [629]. This vacuum induces spontaneous compactification on $\mathrm{S}^4$, which is described by a $d=7$ gauged SUEGRA with $\mathrm{SO}(5)$ gauge group (the isometry group of $\mathrm{S}^4$) and an $\mathrm{AdS}_7$ vacuum [1039].

These vacua play a crucial role in the AdS/CFT correspondence proposed by Maldacena in Ref. [921] (for a review see Ref. [23]).

[8] This is a common feature of all extreme $p$-branes with $p \geq 1$ [1152].

### 25.2.4 The fundamental string F1

The ten-dimensional black fundamental-string solution is given, in the modified Einstein frame and in the string frame, by

$$\begin{aligned}
d\tilde{\hat{s}}_{\rm E}^2 &= H_{\rm F1}^{-\frac{3}{4}}[W dt^2 - dy^2] - H_{\rm F1}^{\frac{1}{4}}\left[W^{-1}d\rho^2 + \rho^2 d\Omega_{(7)}^2\right],\\
d\hat{s}_{\rm s}^2 &= H_{\rm F1}^{-1}[W dt^2 - dy^2] - \left[W^{-1}d\rho^2 + \rho^2 d\Omega_{(7)}^2\right],\\
e^{-2\hat{\phi}} &= e^{-2\hat{\phi}_0} H_{\rm F1}, \qquad \hat{B}_{t\underline{y}} = \alpha\left(H_{\rm F1}^{-1} - 1\right),\\
H_{\rm F1} &= 1 + \frac{h_{\rm F1}}{\rho^6}, \qquad W = 1 + \frac{\omega}{\rho^6}, \qquad \omega = h_{\rm F1}[1 - \alpha^2].
\end{aligned} \tag{25.55}$$

The extreme limit $\omega = 0, \alpha = \pm 1$ is known as the *fundamental-string solution* [410, 409]:

$$\begin{aligned}
d\tilde{\hat{s}}_{\rm E}^2 &= H_{\rm F1}^{-\frac{3}{4}}\left[dt^2 - dy^2\right] - H_{\rm F1}^{\frac{1}{4}} d\vec{x}_8^2,\\
d\hat{s}_{\rm s}^2 &= H_{\rm F1}^{-1}[dt^2 - dy^2] - d\vec{x}_8^2,\\
e^{-2\hat{\phi}} &= e^{-2\hat{\phi}_0} H_{\rm F1}, \qquad \hat{B}_{t\underline{y}} = \pm\left(H_{\rm F1}^{-1} - 1\right), \qquad H_{\rm F1} = 1 + \frac{h_{\rm F1}}{|\vec{x}_8|^6}.
\end{aligned} \tag{25.56}$$

Using the first of Eqs. (24.76), and Eqs. (20.3), (20.4), and (25.26), we obtain

$$h_{\rm F1} = \frac{(2\pi \ell_{\rm s})^6 g^2}{6\omega_{(7)}}, \tag{25.57}$$

which, in the weak-coupling limit $g \to 0$, with $\ell_{\rm s}$ fixed, goes quickly to zero, giving a flat spacetime metric. The F1 solution can then be understood as the long-range fields produced by a fundamental string in the strong-coupling limit. In the weak-coupling limit, the string decouples from the supergravity fields.

The event horizon of the black solution becomes singular in the extreme limit both in the Einstein and in the string frame, and in that limit the dilaton also diverges at the horizon as $\hat{\phi} \sim \ln|\vec{x}_8|$. In the dual string frame (the frame in which the S5-brane is fundamental and there is no dilaton factor in its worldvolume action, which is related to the string frame by Eqs. (24.46) and (24.47)), ignoring the constant in $H_{\rm F1}$ leads to the solution with metric

$$d\hat{s}_{\rm S5}^2 = \frac{\rho^4}{h_{\rm F1}^{\frac{2}{3}}}\left[dt^2 - dy^2\right] - h_{\rm F1}^{\frac{1}{3}}\frac{d\rho^2}{\rho^2} - h_{\rm F1}^{\frac{1}{3}} d\Omega_{(7)}^2, \tag{25.58}$$

which is the direct product of the round $S^7$ metric with radius $h_{\rm F1}^{\frac{1}{6}}$ and the metric of a 1-brane in three dimensions (i.e. a domain wall), which is singular. This near-horizon limit is well defined in this frame, even if it leads to a metric with singularities, but, unlike the M2 and M5 cases, it is not a maximally supersymmetric vacuum. These geometries play roles analogous to the $\mathrm{AdS}_n \times \mathrm{S}^m$ geometries in non-conformal versions of the AdS/CFT correspondence [234].

### *25.2.5 The S5 solution*

The ten-dimensional solitonic 5-brane solution is given by

$$\boxed{\begin{aligned} d\tilde{\hat{s}}^2_{\mathrm{E}} &= H_{\mathrm{S5}}^{-\frac{1}{4}}\left[W dt^2 - d\vec{y}_5^{\,2}\right] - H_{\mathrm{S5}}^{\frac{3}{4}}\left[W^{-1}d\rho^2 + \rho^2 d\Omega^2_{(3)}\right], \\ d\hat{s}^2_{\mathrm{s}} &= W dt^2 - d\vec{y}_5^{\,2} - H_{\mathrm{S5}}\left[W^{-1}d\rho^2 + \rho^2 d\Omega^2_{(3)}\right], \\ e^{-2\hat{\phi}} &= e^{-2\hat{\phi}_0}H_{\mathrm{S5}}^{-1}, \qquad \hat{B}^{(6)}{}_{t\underline{y}^1\cdots\underline{y}^5} = \alpha e^{-2\hat{\phi}_0}\left(H_{\mathrm{S5}}^{-1} - 1\right), \\ H_{\mathrm{S5}} &= 1 + \frac{h_{\mathrm{S5}}}{\rho^2}, \qquad W = 1 + \frac{\omega}{\rho^2}, \qquad \omega = h_{\mathrm{S5}}\left[1 - \alpha^2\right], \end{aligned}} \tag{25.59}$$

and, in the extreme limit $\omega = 0, \alpha = \pm 1$, in which it is usually known as the *solitonic 5-brane solution* [283, 284] (also known as the *NS 5-brane*), it takes the form

$$\boxed{\begin{aligned} d\tilde{\hat{s}}^2_{\mathrm{E}} &= H_{\mathrm{S5}}^{-\frac{1}{4}}\left[dt^2 - d\vec{y}_5^{\,2}\right] - H_{\mathrm{S5}}^{\frac{3}{4}}\, d\vec{x}_4^{\,2}, \\ d\hat{s}^2_{\mathrm{s}} &= dt^2 - d\vec{y}_5^{\,2} - H_{\mathrm{S5}}\, d\vec{x}_4^{\,2}, \\ e^{-2\hat{\phi}} &= e^{-2\hat{\phi}_0}H_{\mathrm{S5}}^{-1}, \qquad \tilde{\hat{B}}^{(6)}{}_{t\underline{y}^1\cdots\underline{y}^5} = \pm e^{-2\hat{\phi}_0}\left(H_{\mathrm{S5}}^{-1} - 1\right), \\ H_{\mathrm{S5}} &= 1 + \frac{h_{\mathrm{S5}}}{|\vec{x}_4|^2}, \end{aligned}} \tag{25.60}$$

and is, like the M5-brane, regular everywhere. Using the first of Eqs. (24.76), the S5 tension, Eqs. (25.12), and $G_{\mathrm{N}}^{(10)}$, Eq. (25.26), we obtain

$$\boxed{h_{\mathrm{S5}} = \ell_{\mathrm{s}}^2,} \tag{25.61}$$

which is independent of $g$ and remains constant in the weak-coupling limit.

In the near-horizon limit $\rho = |\vec{x}_4| \to 0$ in the string frame (which is the dual S5-brane frame) we obtain a metric that is the product of Minkowski $6+1$ and that of a round $\mathrm{S}^3$,

$$d\hat{s}^2 = dt^2 - d\vec{y}_5^{\,2} - dz^2 - h_{\mathrm{S5}} d\Omega^2_{(3)}, \qquad z = h_{\mathrm{S5}}^{\frac{1}{2}} \ln\left(\frac{\rho}{h_{\mathrm{S5}}^{\frac{1}{2}}}\right). \tag{25.62}$$

The S5-brane metric interpolates, then, between Minkowski spacetime at infinity and the above regular metric at the horizon, which is at an infinite proper distance. There is no need to continue the metric analytically beyond the horizon and, actually, it would be more correct to say that, in the limit $\rho \to 0$, one finds another asymptotic region with the above metric.

### 25.2.6 *The* D$p$*-branes*

The generic solution for black D$p$-branes $N = 2A, B, d = 10$ SUEGRA with $p < 7$ is

$$
\begin{aligned}
d\tilde{\hat{s}}_{\rm E}^2 &= H_{{\rm D}p}^{-\frac{7-p}{8}}\left[W dt^2 - d\vec{y}_p^{\,2}\right] - H_{{\rm D}p}^{\frac{p+1}{8}}\left[W^{-1}d\rho^2 + \rho^2 d\Omega_{(8-p)}^2\right],\\
d\hat{s}_{\rm s}^2 &= H_{{\rm D}p}^{-\frac{1}{2}}\left[W dt^2 - d\vec{y}_p^{\,2}\right] - H_{{\rm D}p}^{\frac{1}{2}}\left[W^{-1}d\rho^2 + \rho^2 d\Omega_{(8-p)}^2\right],\\
e^{-2\hat{\phi}} &= e^{-2\hat{\phi}_0}H_{{\rm D}p}^{\frac{p-3}{2}}, \qquad \hat{C}^{(p+1)}{}_{\underline{t}\underline{y}^1\cdots\underline{y}^p} = \alpha e^{-\hat{\phi}_0}\left(H_{{\rm D}p}^{-1} - 1\right),\\
H_{{\rm D}p} &= 1 + \frac{h_{{\rm D}p}}{\rho^{7-p}}, \qquad W = 1 + \frac{\omega}{\rho^{7-p}}, \qquad \omega = h_{{\rm D}p}[1-\alpha^2].
\end{aligned}
\tag{25.63}
$$

This solution is not entirely correct for $p = 3$ since it does not take into account the self-duality of the 5-form field strength, but the only change that has to be made is in the 4-form potential: the metric and dilaton fields are correct, as we will see.

In the extreme limit $\omega = 0, \alpha = \pm 1$ the solutions are valid for all $p = 0, \ldots, 9$ (with the same caveats in the case $p = 3$) for harmonic functions with poles of the right order. These are the solutions usually known as *Dp-brane solutions* in the literature:

$$
\begin{aligned}
d\tilde{\hat{s}}_{\rm E}^2 &= H_{{\rm D}p}^{\frac{p-7}{8}}\left[dt^2 - d\vec{y}_p^{\,2}\right] - H_{{\rm D}p}^{\frac{p+1}{8}}\, d\vec{x}_{9-p}^{\,2},\\
d\hat{s}_{\rm s}^2 &= H_{{\rm D}p}^{-\frac{1}{2}}\left[dt^2 - d\vec{y}_p^{\,2}\right] - H_{{\rm D}p}^{\frac{1}{2}} d\vec{x}_{9-p}^{\,2},\\
e^{-2\hat{\phi}} &= e^{-2\hat{\phi}_0}H_{{\rm D}p}^{\frac{p-3}{2}}, \qquad \hat{C}^{(p+1)}{}_{\underline{t}\underline{y}^1\cdots\underline{y}^p} = \pm e^{-\hat{\phi}_0}\left(H_{{\rm D}p}^{-1} - 1\right),\\
H_{{\rm D}p} &= 1 + \frac{h_{{\rm D}p}}{|\vec{x}_{9-p}|^{7-p}}, \ p<7, \quad H_{{\rm D}7} = 1 + h_{{\rm D}7}\ln|\vec{x}_2|, \quad H_{{\rm D}8} = 1 + h_{{\rm D}8}|x|.
\end{aligned}
$$

Using Eqs. (24.76), the D$p$-brane tension formula, Eq. (25.11), and the value of $G_{\rm N}^{(10)}$, Eq. (25.26), we find

$$
h_{{\rm D}p} = \frac{(2\pi\ell_{\rm s})^{7-p}g}{(7-p)\omega_{(8-p)}}, \ p<7, \qquad h_{{\rm D}7} = -\frac{g}{2\pi}, \qquad h_{{\rm D}8} = -\frac{g}{4\pi\ell_{\rm s}}.
\tag{25.64}
$$

Several remarks are in order here.

1. The D-instanton ($p = -1$) solution is not included in this general case. It will be dealt with in Section 25.2.7.

2. The D-string ($p = 1$) solution is related by IIB S duality (with $S = \eta$) to the F1B solution. More general S-duality transformations generate solutions that represent bound states of $q$ F1Bs and $p$ D1s called $pq$-strings [1116]. The same can be said about the D5 and the S5B, which can be combined into $pq$ 5-branes [909]. There are also $pq$ 7-brane solutions, but they have a more complicated interpretation. We will study these solutions in Section 25.4.3.

3. The metric and dilaton of the $p = 3$ solution are those of the self-dual D3-brane solution, but the RR potential is different. The correct field strength is just the self-dual part of the field strength of the generic solution and its components are

$$\hat{G}^{(5)}{}_{\underline{m}\underline{t}\underline{y}^1\underline{y}^3\underline{y}^3} = \mp\frac{e^{-\hat{\varphi}_0}}{2} H_{\mathrm{D3}}^{-2}\partial_{\underline{m}} H_{\mathrm{D3}}, \qquad \hat{G}^{(5)}{}_{\underline{m}_1\cdots\underline{m}_5} = \pm\frac{e^{-\hat{\varphi}_0}}{2}\epsilon_{\underline{m}_1\cdots\underline{m}_5\underline{m}_6}\partial_{\underline{m}_6} H_{\mathrm{D3}}. \tag{25.65}$$

4. The solutions for $p < 7$ are well defined for all values of $|\vec{x}_{9-p}| > 0$ ($H_{\mathrm{D}p} > 0$). The D7 and D8 solutions are well defined only in certain regions of the transverse space for which $H_{\mathrm{D}p} > 0$ due to the negative signs of $h_{\mathrm{D7}}$ and $h_{\mathrm{D8}}$. To obtain solutions that are well defined everywhere in the transverse space, one has to consider configurations with several branes and compact transverse spaces. The only singularities are then at the positions of the branes.[9] We are going to study the simplest of these combinations of D7-branes in Section 25.2.8. The simplest combination of D8-branes which leads to a regular metric is the orbifold construction discussed on p. 723.

5. The Hodge dual of the 10-form field strength associated with the D8-brane solution is $\star\hat{G}^{(10)} = \pm h_{\mathrm{D8}} g^{-1} = \mp 1/(4\pi\ell_s)$. This must, then, be the value of the mass parameter $m = \hat{G}^{(0)}$ of the Romans massive $N = 2A, d = 10$ supergravity that describes the effective string theory in the presence of one D8-brane [1052, 186]. The parameter $m$, which was completely arbitrary from the supergravity point of view, must be quantized from the string-theory point of view.

6. The $p = 9$ solution is just flat spacetime.

7. In all the $p < 7$ cases, except for $p = 3$, the D$p$-brane horizon is singular and its Penrose diagram is given in Fig. 25.1. The near-horizon geometries (in the dual frame) also correspond to solutions with singularities.

8. In the $p = 3$ case the solution has a regular horizon and the analytic continuation across it is completely regular, as in the M5 case. There is again a discrete isometry that relates the old and new asymptotically flat regions. The near-horizon geometry

[9] The positions of the branes can be identified in general with the poles of the harmonic functions, although we know that, in many cases, these poles are just coordinate singularities and correspond to regular event horizons.

of the D3-brane is a maximally supersymmetric solution with the metric of the completely regular space $\mathrm{AdS}_5 \times \mathrm{S}^5$:

$$d\hat{s}_{\mathrm{s}}^2 = R_5^2\, d\Pi_{(5)}^2 - R_5^2\, d\Omega_{(5)}^2, \quad \hat{G}^{(5)} = \pm 2 e^{-\hat{\varphi}_0} R_5^4 (\omega_{\mathrm{AdS}_5} - \omega_{\mathrm{S}^5}), \quad R_5 = h_{\mathrm{D3}}^{\frac{1}{4}} \tag{25.66}$$

(compare with Eq. (17.120)), which has played a crucial role in the AdS/CFT-correspondence conjecture [921, 23]. This solution induces spontaneous compactification on $\mathrm{S}^5$. The theory is described by gauged $N = 4, d = 5$ SUEGRA with gauge group $\mathrm{SO}(6)$ [690, 857].

### 25.2.7 *The* D*-instanton*

If we extrapolate the association of $(p+1)$-forms to objects with a $(p+1)$-dimensional worldvolume to a 0-form (scalar) potential such as the type-IIB RR 0-form $\hat{C}^{(0)}$, we conclude that the IIB theory admits a "$-1$-brane," an object with a zero-dimensional worldvolume: just a point in spacetime. Such an object must be an instanton, which is a Euclidean solution. Then, associated with the type-IIB RR 0-form $\hat{C}^{(0)}$, we expect to find a *D-instanton solution* to the Euclidean equations of motion of the $N = 2B, d = 10$ supergravity theory.[10] It is not clear how to define the complete Euclidean $N = 2B, d = 10$ supergravity because it is not possible to have a *real* self-dual 5-form (a formal definition has nevertheless been given in Ref. [611]), but this problem does not arise if one ignores the 5-form field strength, as we do here.

We start with the action of the truncated $N = 2B, d = 10$ theory in the Einstein frame and with Lorentzian signature in which we keep only the metric and the dilaton and the RR 0-form combined in the complex scalar $\hat{\tau} = \hat{C}^{(0)} + ie^{-\hat{\varphi}}$:

$$\hat{S} = \frac{\hat{g}_{\mathrm{B}}^2}{16\pi G_{\mathrm{N}}^{(10)}} \int d^{10}\hat{x} \sqrt{|\hat{\jmath}_{\mathrm{E}}|} \left\{ \hat{R}_{\mathrm{E}} + \tfrac{1}{2} \frac{\partial_\mu \hat{\tau} \partial^\mu \bar{\hat{\tau}}}{(\Im\mathrm{m}\, \hat{\tau})^2} \right\}. \tag{25.67}$$

To obtain the Euclidean action we have to perform a Wick rotation, which can be understood as a coordinate redefinition $t = x^0 = i\bar{x}^{10}$, where $\tau$ is treated as real afterwards. $\hat{\varphi}$ is a scalar and transforms as such under this reparametrization. However, $\hat{C}^{(0)}$ can be treated either as a scalar (and then its Hodge dual, the RR 7-form $\hat{C}^{(7)}$, has to be treated as a pseudotensor) or as a pseudoscalar (and the RR 7-form $\hat{C}^{(7)}$ has to be treated as a tensor).[11] The second option (which is also the one we adopted for consistency when we defined the magnetic RR potentials) was chosen by the authors of Ref. [611], who performed the Wick rotation using the RR 7-form and treating it as a tensor. If $\hat{C}^{(0)}$ is a pseudoscalar then

[10] D-instantons of bosonic-string theories had been considered in Refs. [1046, 664] before the relation between RR charges and D-branes was discovered in Ref. [1047]. The IIB instanton BPS condition had been obtained in Ref. [663].

[11] The difference between these two options can be seen as the reason why Wick rotations and Hodge duality do not commute.

it acquires an extra factor of $i$ in the Wick rotation: $\hat{C}^{(0)} = i\hat{\bar{C}}^{(0)}$. The Euclidean action becomes

$$\bar{S} = \frac{\hat{g}_{\rm B}^2}{16\pi G_{\rm N}^{(10)}} \int d^{10}\bar{\hat{x}} \sqrt{|\bar{\hat{g}}_{\rm E}|} \Big\{ \bar{\hat{R}}_{\rm E} + \tfrac{1}{2} e^{2\hat{\varphi}} \Big[ (\partial e^{-\hat{\varphi}})^2 - (\partial \hat{\bar{C}}^{(0)})^2 \Big] \Big\}. \tag{25.68}$$

Observe that the two scalars contribute with different signs to the action. Their "energy–momentum" tensors will appear with opposite signs in the Einstein equation. Thus, we can obtain a solution with flat spacetime by taking the derivatives of the two scalars to be equal, up to a global sign. Then we need only solve the scalar equations. The *D-instanton solution* takes the following form in the (unmodified) Einstein and string frames [611]:

$$\boxed{\begin{array}{ll} d\bar{\hat{s}}_{\rm E}^2 = e^{-\frac{\hat{\varphi}_0}{2}} d\bar{\vec{x}}_{10}^2, & d\bar{\hat{s}}_{\rm s}^2 = H_{{\rm D}i}^{\frac{1}{2}} d\bar{\vec{x}}_{10}^2, \\ e^{-2\hat{\varphi}} = e^{-2\hat{\varphi}_0} H_{{\rm D}i}^{-2}, & \hat{\bar{C}}^{(0)} = \pm e^{-\hat{\varphi}_0} \left( H_{{\rm D}i}^{-1} - 1 \right), \\ H_{{\rm D}i} = 1 + \dfrac{h_{{\rm D}i}}{|\bar{\vec{x}}_{10}|^8}. & \end{array}} \tag{25.69}$$

The value of $h_{{\rm D}i}$ is the extrapolation to $p=-1$ of the value of $h_{{\rm D}p}$, Eq. (25.64). This value can also be obtained via a T-duality relation with the D0-brane.

At first sight there is a singularity at $\rho = |\bar{\vec{x}}_{10}| = 0$ (in the string frame). However, the string metric is invariant under the reparametrization

$$\rho = h_{{\rm D}i}^{\frac{1}{4}}/\tilde{\rho}, \tag{25.70}$$

which shows that, in the limit $\rho \to 0$, one finds another asymptotically flat region identical to the one at $\rho \to \infty$. The metric, therefore, describes a sort of Euclidean wormhole joining the two asymptotically flat regions and is regular everywhere.

The value of the Euclidean action of the D-instanton can be calculated. In the modified Einstein frame, and normalized in our conventions, it takes the value

$$I = 2\pi/g_{\rm B} = q_{{\rm D}i}/g_{\rm B}. \tag{25.71}$$

The action Eq. (25.67) appears in other contexts in $d \neq 10$ dimensions. We have met it, for instance, as a truncation of the $N=4, d=4$ SUEGRA theory that arises in the toroidal reduction of the heterotic-string effective action Eq. (22.160), but it appears in many other reductions, as shown in Ref. [907]. It is possible to find instanton solutions for all of them that are almost identical with the D-instanton solution, differing only in the harmonic function which, in $d>2$ dimensions, will be $H = 1 + h/|\vec{x}_d|^{d-2}$. The string-frame geometry will be different in each case. The $d=4$ solution associated with the heterotic string was found in Ref. [630] and also has a wormhole interpretation.

### *25.2.8 The D7-brane and holomorphic $(d-3)$-branes*

The D7-brane solution Eqs. (25.64) is just the simplest of a very rich family of solutions of the action Eq. (25.67), which share many interesting properties and some pathologies that

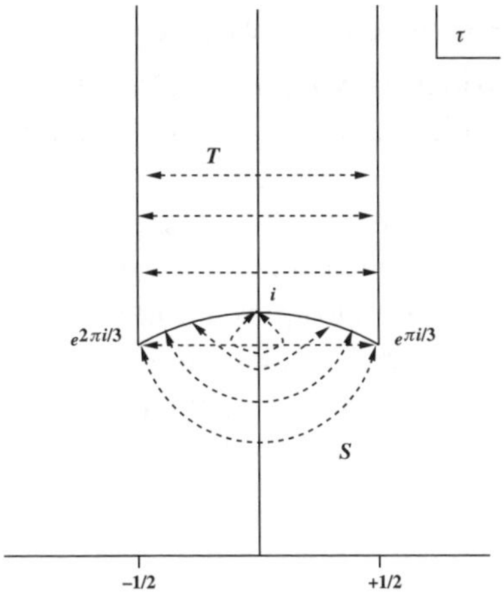


Fig. 25.2. The fundamental domain of the modular group.

can be eliminated after a careful analysis. They are the subject of this section.

The $\mathrm{SL}(2,\mathbb{R})/\mathrm{SO}(2)$ $\sigma$-model of Eq. (25.67) is invariant under transformations of the whole group $\mathrm{SL}(2,\mathbb{R})$, Eqs. (15.221) and (15.222), but only the discrete subgroup $\mathrm{SL}(2,\mathbb{Z})$ is supposed to relate equivalent (dual) type-IIB theories. On the other hand, as discussed in Section 15.4.1, only the modular group $\mathrm{G} \equiv \mathrm{PSL}(2,\mathbb{Z}) = \mathrm{SL}(2,\mathbb{Z})/\{\pm\mathbb{I}_{2\times 2}\}$ acts on $\hat{\tau}$. In conclusion, type-IIB S duality tells us that values of the $\hat{\tau}$ field that are related by modular transformations must be considered equivalent and should be *identified*. The same will be true in the cases in which $\tau$ can be viewed as the modular parameter of a torus.[12] Thus $\tau$, which in principle takes values in the whole complex upper half plane $\mathbb{H}$, can be restricted to take values in the *fundamental domain* of the modular group in $\mathbb{H}$, which we are going to discuss now.

The modular group G is generated by the elements $T$ and $S$,

$$S = \begin{pmatrix} 0 & -1 \\ 1 & 0 \end{pmatrix}, \qquad T = \begin{pmatrix} 1 & 1 \\ 0 & 1 \end{pmatrix}, \tag{25.72}$$

whose actions on $\tau$ are $T(\tau) = \tau + 1$ and $S(\tau) = -1/\tau$. Observe that $S^2 = -\mathbb{I}_{2\times 2} \sim \mathbb{I}_{2\times 2}$ in G and also $(ST)^3 = (T^{-1}S)^3 \sim \mathbb{I}$ in G. Thus $S$ and $ST$ generate two cyclic subgroups of orders 2 and 3, respectively.

The fundamental domain of G in $\mathbb{H}$ can be defined as the quotient $\mathbb{H}/\mathrm{G}$ and corresponds to the region $|\tau| \geq 1$ and $-\frac{1}{2} \leq \Re\mathfrak{e}(\tau) \leq \frac{1}{2}$ with the lines $\Re\mathfrak{e}(\tau) = -\frac{1}{2}$ and $\Re\mathfrak{e}(\tau) = +\frac{1}{2}$ identified by a $T$ transformation and with the arc of unit radius $e^{i\theta}$, $\theta \in [\pi/3, 2\pi/3]$, joining the fundamental domain corners $e^{\frac{2\pi i}{3}}$ and $e^{\frac{\pi i}{3}}$ identified with itself ("orbifolded") according to $e^{i\theta} \sim e^{i(\pi-\theta)}$ (the $S$ transformation) (see Fig. 25.2).

$\mathbb{H}/\mathrm{G}$ has, therefore, two special points associated with the two cyclic subgroups

[12] We remove all the hats henceforth, since the results of this section will be valid for many cases and dimensions apart from the $N = 2B, d = 10$ case. The results for D7-branes in $d = 10$ dimensions will be valid for $(d-3)$-branes in $d$ dimensions.

generated by $S$ and $ST$: $\tau = i$, which is invariant under $S$; and $\tau = \rho \equiv e^{\frac{2\pi i}{3}}$, which is invariant under $ST$. If we consider its compactification $\widehat{\mathbb{H}/\mathrm{G}}$ in which the point at infinity is added, then a third special point appears: $\infty$ itself, which is invariant under the infinite subgroup of integer powers of $T$ and can be understood as an infinite-order orbifold point.

Since the fundamental domain in which $\tau$ takes values is topologically non-trivial, we expect $\tau(x)$, which maps the transverse space on the fundamental domain, to be a multivalued function of $x$ whose monodromies are in G. On the other hand, the real part of $\tau$ has, therefore, the typical behavior of an axion field and takes values in a circle.

We can now interpret the D7-brane solution in the light of the preceding discussion. First, it is useful to rewrite it using $\omega = x^1 + ix^2$ in the form

$$\boxed{d\tilde{\hat{s}}_{\mathrm{E}}^2 = dt^2 - d\vec{y}_7^{\,2} - \Im\mathrm{m}(\mathcal{H})\, d\omega d\bar{\omega}, \qquad \tau = \mathcal{H},} \tag{25.73}$$

where

$$\mathcal{H} = ie^{-\hat{\varphi}_0} h_{\mathrm{D}7} \ln \omega, \quad \text{or} \quad \mathcal{H} = ie^{-\hat{\varphi}_0} h_{\mathrm{D}7} \ln \bar{\omega}, \tag{25.74}$$

for D7- and anti-D7-branes (positive and negative charge with respect to $\hat{C}^{(0)}$), respectively, where we have eliminated the constant 1 in $H_{\mathrm{D}7}$ since the solution is not asymptotically flat anyway. If we go around the origin $\omega = 0$ at which the (anti-)D7 is placed, then, according to the source calculation,

$$\omega \rightarrow e^{2\pi i}\omega \Rightarrow \tau \rightarrow \tau \pm 1 = T^{\pm 1}(\tau). \tag{25.75}$$

The D7-brane solution has, as we expected from our general discussion, non-trivial monodromy. Furthermore, the monodromy around a D7-brane with charge $n$ is $T^n$. For $(d-3)$-branes monodromy plays the role of charge (they are equivalent, when standard charge can be defined), which can be represented by a monodromy matrix.

The D7-brane solution is, however, defined only in the disk $|\omega| < 1$ due to the negative sign of $h_{\mathrm{D}7}$ and we may be interested in different transverse spaces. The corresponding solutions may be found thanks to the following observation: the ansatz Eqs. (25.73) is a solution for any $\mathcal{H}$ that is a holomorphic or antiholomorphic function of $\omega$ [674]. D7-branes will be placed at points around which the monodromy of $\tau$ is $T^n$. The restriction to (anti)holomorphicity has to do with the impossibility of having objects with opposite charges in equilibrium.

Observe that what appears in the metric is $\Im\mathrm{m}(\mathcal{H})$, *not* $\Im\mathrm{m}(\tau)$, even if they coincide in this form of the solution. Then $g_{\omega\bar{\omega}}$ does not transform under G transformations of $\tau$, but the relation $g_{\omega\bar{\omega}} = \Im\mathrm{m}(\tau)$ breaks down. In fact, the general 7-brane solution can be written in a more general form[13]

$$\boxed{d\tilde{s}_{\rm E}^2 = dt^2 - d\vec{y}_7^{\,2} - \Im\mathrm{m}(\mathcal{H})\,|f(\omega)|^2 d\omega d\bar{\omega}, \qquad \tau = \mathcal{H},} \tag{25.77}$$

where $f(\omega)$ is any holomorphic function of $\omega$, but $f(\omega)$ can always be reabsorbed (locally!) into a change of coordinates $\omega' = F(\omega)$, $dF/d\omega = f$, and $\tau(\omega') = \tau[F^{-1}(\omega')]$.

$\mathcal{H}$ is in general multivalued and hence $g_{\omega\bar{\omega}}$ could also be. When it is, we use a multivalued $f(\omega)$ in order to make $g_{\omega\bar{\omega}}$ single valued. A general solution for $f(\omega)$ can be given on the basis of the observation that $\Im\mathrm{m}(\tau)$ transforms under G (or under monodromy) as the absolute value squared of a modular form of weight $-1$ would,[14] i.e.

$$\Im\mathrm{m}(\tau') = \frac{\Im\mathrm{m}(\tau)}{|\gamma\tau + \delta|^2}, \tag{25.78}$$

and, therefore, going around closed loops in transverse ($\omega$) space $\Im\mathrm{m}(\mathcal{H})$ can transform in this way for some values of $\gamma$ and $\delta$. Then, we can build a single-valued function by multiplying $\Im\mathrm{m}(\mathcal{H})$ by the absolute value squared of a modular form of $\mathcal{H}$ of opposite weight 1, i.e. an $f[\mathcal{H}(\omega)]$ such that

$$f\left[\frac{\alpha\mathcal{H} + \beta}{\gamma\mathcal{H} + \delta}\right] = (\gamma\mathcal{H} + \delta) f(\mathcal{H}). \tag{25.79}$$

Then $g_{\omega\bar{\omega}} = \Im\mathrm{m}(\mathcal{H})\,|f(\omega)|^2$ will always be single valued. The choice of modular form is not unique, though. The choice in Ref. [674] was

$$f = \eta^2(\mathcal{H}) \prod_{n=1}^{N} (\omega - \omega_n)^{-\frac{1}{12}}, \tag{25.80}$$

where $\eta$ is Dedekind's function[15]

$$\eta(z) = q^{\frac{1}{24}} \prod_{n=1}^{\infty} (1 - q^n), \qquad q = e^{2\pi i z}, \tag{25.81}$$

---

[13] There is another form of the general solution, which is manifestly $\mathrm{SL}(2,\mathbb{R})$ invariant:

$$\boxed{d\tilde{s}_{\rm E}^2 = dt^2 - d\vec{y}_7^{\,2} - e^{-2U} d\omega d\bar{\omega}, \qquad \tau = \mathcal{H}_1/\mathcal{H}_2, \qquad e^{-2U} = \Im\mathrm{m}(\mathcal{H}_1\bar{\mathcal{H}}_2),} \tag{25.76}$$

where $\mathcal{H}_{1,2}$ are two arbitrary holomorphic functions $\omega$ transforming as a doublet under $\mathrm{SL}(2,\mathbb{R})$, both in $\tau$ and in the metric (but $e^{-2U}$ is invariant, as it must be). The structure of this family is similar to that of the SWIP solutions of $N = 4, d = 4$ SUGRA (Section 16.2.1). We can relate it either to the solution Eq. (25.73) as the particular case $\mathcal{H}_1 = \mathcal{H},\ \mathcal{H}_2 = 1$ or to the solution Eq. (25.77) as the particular case $\mathcal{H}_1/\mathcal{H}_2 = \mathcal{H},\ f = \mathcal{H}_2$ since $\Im\mathrm{m}(\mathcal{H}_1\bar{\mathcal{H}}_2) = |\mathcal{H}_2|^2 \Im\mathrm{m}(\mathcal{H}_1/\mathcal{H}_2)$.

[14] There are no modular forms of negative weight, however; see Ref. [1130].

[15] Dedekind's $\eta$ function is not, strictly speaking, a modular form: only its 24th power is a weight-12 modular form (actually, a cusp form, since it vanishes at infinity where it admits the expansion $\eta^{24} = q - 24q^2 + \cdots$) because its transform contains a phase factor that is a 24th root of unity. However, since we are taking its absolute value, this phase is immaterial.

and the $\omega_n$s are the *decompactification points*[16] at which $\mathcal{H} \sim i\ln(\omega - \omega_n)$ so $\Im\mathrm{m}(\tau)$ diverges when we approach them. Near the decompactification points, $\eta^2 \sim (\omega - \omega_i)^{\frac{1}{12}}$ and the metric would become singular without the additional factor $\Pi_{n=1}^{N}(\omega - \omega_n)^{-\frac{1}{12}}$. As a result of the presence of this factor, the spacetime will be asymptotically conical: when $|\omega| \to \infty$, $|f(\omega)| \sim |\omega|^{-\frac{N}{12}}$, and the transverse metric takes the form

$$dr^2 + (1 - N/12)^2 r^2 d\theta^2, \qquad r = |\omega|^{1-\frac{N}{12}}/(1 - N/12), \tag{25.82}$$

and it is asymptotically conical with deficit angle $N/12$ when $N < 12$. For $N = 12$ (i.e. 12 D7-branes) the space is asymptotically cylindrical; for $N > 12$ it has finite volume, but it is singular except in the exceptional case $N = 24$ [674, 611].

Let us now go back to the problem of finding a globally well-defined D7 ($(d-3)$-brane) solution. This problem was first considered and solved in Ref. [674] with the Riemann sphere as the transverse space. The crucial observation is that there is essentially a unique function that maps the fundamental region of the modular group bijectively onto the sphere [1130]: the modular invariant $j(\tau)$ given in terms of the even Jacobi $\theta$ functions $\theta_2(\tau), \theta_3(\tau),$ and $\theta_4(\tau)$ by

$$j(\tau) = \frac{(\theta_2^8 + \theta_3^8 + \theta_4^8)^3}{\eta^{24}}. \tag{25.83}$$

The simplest solution is then implicitly given by $j(\tau) = \omega$ or

$$\tau(\omega) = j^{-1}(\omega). \tag{25.84}$$

More general solutions can be obtained by replacing $\omega$ by a holomorphic function $h(\omega)$, which is often a quotient of polynomials.

$j(\tau)$ is single valued on $\widehat{\mathbb{H}/G}$ and $j^{-1}(\omega)$ is multivalued on $S^2$ with monodromy in G. The only points around which there are non-trivial monodromies are the inverse images of the orbifold points $\widehat{\mathbb{H}/G}$; namely $j(\infty) = \infty$, $j(i) \equiv \omega_i$, and $j(\rho) \equiv \omega_\rho$, and the monodromies are related to the transformations that leave them invariant. In fact we can describe the monodromies of $\tau(\omega) = j^{-1}(\omega)$ by a sphere with two branch cuts joining the points $\infty$ and $\omega_\rho$ and $\omega_\rho$ and $\omega_i$, as in Fig. 25.3. Crossing the first cut in the sense of the arrow, the function jumps from $\tau$ to $T(\tau)$; crossing the second, it jumps from $\tau$ to $S(\tau)$. To check this, we can compute the monodromy along closed paths around those points. The paths are represented in Fig. 25.3 and are the inverse images of the open paths in $\mathbb{H}$ (closed in $\mathbb{H}/G$) represented in Fig. 25.4.

Around $\infty$ the above solution admits the expansion $\tau(\omega) \sim -[1/(2\pi i)]\ln\omega + \cdots$. Evidently, on going once around infinity ($\omega \to \omega e^{-2\pi i}$), $\tau \to \tau + 1 = T(\tau)$, and we can say that this solution describes a $(d-3)$-brane (a D7-brane in the ten-dimensional string context) of unit charge at $\omega = \infty$ and, due also to some other welcome properties, this is why this solution Eq. (25.84) is generally known as *the* finite-energy $(d-3)$-brane solution.

Around $\omega_i$ the above solution admits the expansion $\tau(\omega) \sim i + \alpha(\omega - \omega_i)^{1/2} + \cdots$, and, on going once around it (i.e. $\omega - \omega_i \to (\omega - \omega_i)e^{+2\pi i}$), we see that, to leading order in $\omega - \omega_i$, $\tau \to -1/\tau = S(\tau)$. Finally, around $\omega_\rho$, $\tau(\omega) \sim \rho + \beta(\omega - \omega_\rho)^{\frac{1}{3}} + \cdots$, and, on going once around it (i.e. $\omega - \omega_\rho \to (\omega - \omega_\rho)e^{+2\pi i}$), we see that $\tau \to -(1 + 1/\tau) = T^{-1}S(\tau)$.

[16] These are the points at which there are D7-branes, associated with $T^n$ monodromy.

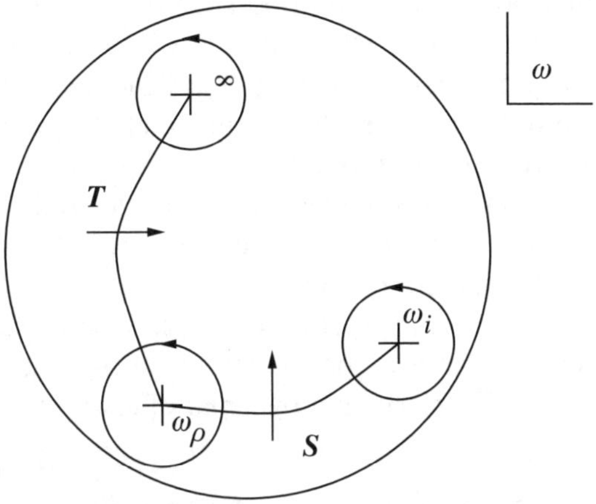


Fig. 25.3. The image of the fundamental domain of the modular group by $j(\tau)$.

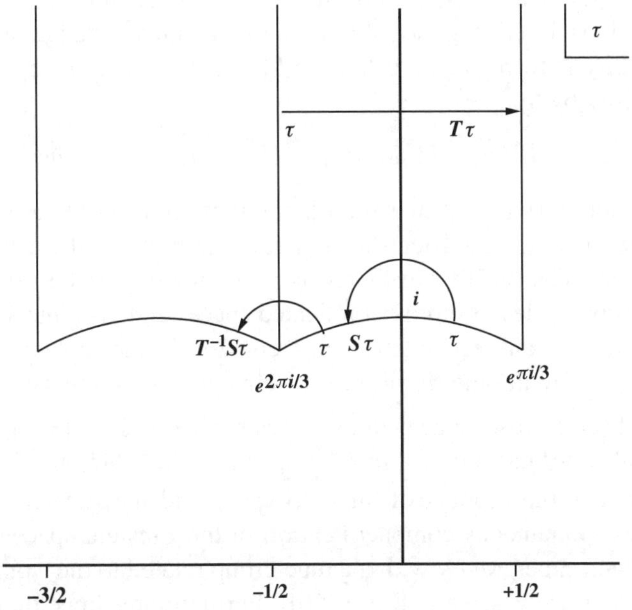


Fig. 25.4. The inverse image of the monodromy paths.

Apart from the fact that these singular points are given to us by the structure of the solution Eq. (25.84), it should be clear that the consistency of monodromy implies the existence of other singular points apart from $\infty$ such that the monodromy around all of them is $T^{-1}$ [611]. The same conclusion could have been drawn from conservation of charge. Here the consistency of monodromy plays the same role as conservation of charge: we are dealing with a compact space and, just as the total charge has to be zero in such a space, the "total monodromy" has to be trivial.[17]

It is natural to associate with each singular point characterized by a monodromy matrix in $G$ a $(d-3)$-brane. The standard IIB D7-branes are associated with $T^n$ monodromies, but a consistent solution on a sphere requires, as we have seen, 7-branes with $S$ and $ST$ monodromy that may also have negative tensions. These 7-brane solutions are generically known as *pq 7-branes*, but, as distinct from $pq$-strings or 5-branes (Section 25.4.3), they are characterized by a $\mathrm{PSL}(2,\mathbb{Z})$ matrix, not by a pair of charges.

### 25.2.9 Some simple generalizations

The extreme $p$-brane solutions admit many generalizations. The most interesting ones describe intersections of branes and we will study them in Section 25.6. For a single $p$-brane, the simplest generalizations involve either a modification of the worldvolume geometry, which we have taken so far to be flat $(p+1)$-dimensional spacetime, or a modification of the geometry of the transverse space.

Since the supergravity equations of motion are local, it is clear that global modifications of the worldvolume geometry such as imposing periodicity conditions on the $n$ coordinates will give new solutions describing the $p$-branes *wrapped* on a rectangular $n$-torus. Another interesting possibility is to replace the flat worldvolume metric $\eta_{ij}$ by $g_{ij}(t,\vec{y}_p)$ and the transverse metric $\delta_{mn}$ by $h_{mn}(x)$:

$$ds^2 = H^{\alpha}(x)g_{ij}(x)dy^i dy^j - H^{\beta}(x)h_{mn}(x)dx^m dx^n. \tag{25.85}$$

The equations of motion (with a minor modification of the $(p+1)$-form potential ansatz) are still solved if $g_{ij}$ and $h_{mn}$ are Ricci-flat and $H$ is harmonic in the new transverse space (see Refs. [591, 248, 828, 547]). This kind of solution can also be used to describe the wrapping of branes on cycles of more complicated spaces and also intersections.

A possible choice of transverse metric $h_{mn}$ consists in the replacement of the round $\mathrm{S}^{(\tilde{p}+2)}$ metric $d\Omega^2_{(\tilde{p}+2)}$ by the metric of an Einstein space with the same curvature as the round $\mathrm{S}^{(\tilde{p}+2)}$ [475] (G/H coset spaces in the cases studied in Ref. [313]) in the flat metric written in spherical coordinates $d\rho^2+\rho^2 d\Omega^2_{(\tilde{p}+2)}$. In the M2, M5, and D3 cases, the near-horizon limits are now the product of an AdS space and the Einstein space. This kind of solution induces spontaneous compactification in the Einstein space and the theory is described by a gauged supergravity with a gauge group related to the isometry group of the Einstein space (for a review, see e.g. Ref. [556]). Furthermore, since they do not preserve all the supersymmetries (unlike the $\mathrm{AdS}_n\times\mathrm{S}^m$ solutions), the supergravities will also have fewer supercharges.

[17] Just as in the D8-brane case, the $(d-3)$-branes that we have to add may have negative tensions.

It is also possible to look directly for metrics that can be understood as near-horizon limits (see, for instance, Ref. [545]).

## 25.3 The masses and charges of the $p$-brane solutions

In Section 25.1.1 we found the masses and charges of the extended objects of string/M theory using duality arguments. We matched these with the coefficients of the harmonic functions of the extreme solutions by studying the coupling of supergravity to the sources. This procedure is, however, difficult or impossible to follow for generic solutions that represent complex systems of extended objects or are "black" (non-extreme). In those cases we need a procedure by which to calculate the masses and charges of the objects described by the solutions using only the solution and the normalizations of the fields that appear in the action. This is the subject of this section. We will follow Refs. [919, 41].

### 25.3.1 Masses

We need to collect together several pieces of data that are scattered over several chapters.

1. The closed-superstring worldvolume action is given in Eq. (20.19): $T = 1/(2\pi\alpha')$ is the string tension, $\alpha' = \ell_s^2$ is the Regge slope, and $\ell_s$ is the string length. With that normalization of the worldvolume fields, the low-energy effective action of the common sector of the different superstring theories is, in the string frame, given by Eq. (21.1). The complete effective actions of the type-IIA and -IIB superstring theories are given in Eqs. (22.38) and (23.4), respectively.

2. In the normalization factor of Eq. (21.1) $\hat{g}$ is the (dimensionless) $d = 10$ string coupling constant, which is related to the dilaton vacuum expectation value by Eq. (20.57). (Sometimes we use $\hat{g}_A$ ($\hat{g}_B, \hat{g}_I, \hat{g}_h$) for the coupling constant of the type-IIA (-IIB, -I, heterotic) superstring theory, but it should be clear from the context which coupling constant we are talking about.)

   The explicit factor of $\hat{g}^2$ is meant to absorb the asymptotic value of the dilaton, so that $G_N^{(10)}$ is really the ten-dimensional Newton constant. We are assuming here that string-frame metrics are asymptotically flat.

3. When we are dimensionally reducing the above actions to $d$ dimensions, we integrate over the compact, redundant, dimensions and obtain an overall factor that is the volume of these compact dimensions, $V_{10-d}$. One finds the following relations between the $d$-dimensional string coupling constant $g$ and Newton constant $G_N^{(d)}$ and the ten-dimensional ones (see Section 15.2.2):

$$G_N^{(d)} = G_N^{(10)}/V_{10-d}^{1/2}, \qquad g = \hat{g}/V_{10-d}. \tag{25.86}$$

   Here we are going to compactify on rectangular tori, with orthogonal circles, and so the volume of the compact space is

$$V_{10-d} = (2\pi)^{10-d} R_9 \cdots R_d. \tag{25.87}$$

4. The ten-dimensional Einstein metric $\hat{g}_{\mathrm{E}\,\hat{\mu}\hat{\nu}}$ is related to the string metric $\hat{g}_{\hat{\mu}\hat{\nu}}$ by

$$\hat{g}_{\mathrm{E}\,\hat{\mu}\hat{\nu}} = e^{-\frac{\hat{\phi}}{2}}\hat{g}_{\hat{\mu}\hat{\nu}}, \tag{25.88}$$

i.e. by performing this rescaling in the above action we bring it into the canonical form because the factor $e^{-2\hat{\phi}}$ in front of the curvature disappears. However, the Einstein metric cannot be asymptotically flat if the string metric is and the factor in front of the action is $\hat{g}^2(16\pi G_{\mathrm{N}}^{(10)})^{-1}$.

Following Ref. [919], we define a *modified Einstein metric* $\tilde{g}_{\mathrm{E}\,\hat{\mu}\hat{\nu}}$ that has the same value at infinity as the string metric and is the one to use to define masses. These masses are *the same as those masses that appear in the string spectrum*, and, in particular, the mass of the fundamental string is independent of the string coupling constant:[18]

$$\tilde{\hat{g}}_{\mathrm{E}\,\hat{\mu}\hat{\nu}} \equiv e^{\frac{-\hat{\phi}-\hat{\phi}_0}{2}}\hat{g}_{\hat{\mu}\hat{\nu}} = \hat{g}^{\frac{1}{2}}\hat{g}_{\mathrm{E}\,\hat{\mu}\hat{\nu}}. \tag{25.90}$$

If we rewrite the string effective action in terms of the modified Einstein metric, we obtain the correct normalization factor $(16\pi G_{\mathrm{N}}^{(10)})^{-1}$ and no dilaton factors.

5. In $d$ dimensions, the mass $M$ of any static, asymptotically flat metric describing a point-like object can be found from its asymptotic behavior at spatial infinity [1168, 968]. In the string-theory context, the mass $M_{\mathrm{E}}$ associated with the "Einstein metric" (i.e. with the "wrong" normalization factor $g^2(16\pi G_{\mathrm{N}}^{(d)})^{-1}$ in the action) can be implicitly defined by

$$g_{\mathrm{E}\,tt} \sim 1 - \frac{16\pi G_{\mathrm{N}}^{(d)} M_{\mathrm{E}}}{g^2(d-2)\omega_{(d-2)}}\frac{1}{|\vec{x}_{d-1}|^{d-3}}, \qquad \vec{x}_{d-1} = (x^1, \ldots, x^{d-1}). \tag{25.91}$$

The mass $M$ associated with the *modified Einstein metric* (i.e. with the "right" normalization factor $(16\pi G_{\mathrm{N}}^{(d)})^{-1}$) is defined analogously by

$$\tilde{g}_{\mathrm{E}\,tt} \sim 1 - \frac{16\pi G_{\mathrm{N}}^{(d)} M}{(d-2)\omega_{(d-2)}}\frac{1}{|\vec{x}_{d-1}|^{d-3}}. \tag{25.92}$$

If both the string metric and the "modified string metric" are asymptotically flat (as we have assumed) then the Einstein metric is not, and it is necessary to rescale the coordinates with factors of $g^{\frac{1}{4}}$ in order to be able to use Eq. (25.91). Taking this into account, we find that the relation (in any dimension) between $M_{\mathrm{E}}$ and $M$ is given by

$$M_{\mathrm{E}} = g^{\frac{1}{4}} M. \tag{25.93}$$

[18] In $d$ dimensions, the relation between the modified Einstein metric and the string metric is

$$\tilde{g}^{(d)}_{\mathrm{E}\,\mu\nu} = e^{-\frac{4}{d-2}(\phi-\phi_0)} g_{\mu\nu}. \tag{25.89}$$

6. Under IIB S duality, it is the (unmodified) Einstein-frame metric that is invariant. Then $M_{\rm E}$ is S-duality invariant, which implies the following S-duality transformation rule for $M$, which was already given in Eq. (25.3):

$$M'_{\rm E} = M_{\rm E} \Rightarrow M' = g'^{-\frac{1}{4}}_{\rm B} g^{\frac{1}{4}}_{\rm B} M = g^{\frac{1}{2}}_{\rm B} M. \tag{25.94}$$

We can apply these formulae to find the masses of any of the solutions we have studied. They should coincide with the masses of the corresponding states in string/M theory. Let us take, for example, the F1 solution given in Eq. (25.56), assuming that the string is compactified on a circle and $y$ is a compact dimension so we can dimensionally reduce the above solution, and calculate the modified Einstein mass of the resulting point-like object that lives in $d=9$ by using Eq. (25.92).

First, we need the nine-dimensional dilaton (see Eq. (21.16)),

$$e^{-2(\phi-\phi_0)} = e^{-2(\hat{\phi}-\hat{\phi}_0)}\sqrt{|\hat{g}_{\underline{yy}}|} = H^{\frac{1}{2}}_{\rm F1}, \tag{25.95}$$

so, in this case, the relation between the nine-dimensional metrics is

$$\tilde{g}_{{\rm E}\mu\nu} = H^{\frac{1}{7}}_{\rm F1}\, g_{\mu\nu} \Rightarrow \tilde{g}_{{\rm E}\,tt} = H^{-\frac{6}{7}}_{\rm F1} \sim 1 - \frac{6h_{\rm F1}}{7\rho^6}, \tag{25.96}$$

which, compared with Eq. (25.92), gives the right value,

$$M_{{\rm F1}w} = \frac{6h_{\rm F1}\omega_{(7)}}{16\pi G^{(9)}_{\rm N}} = \frac{12\pi R_9 h_{\rm F1}\omega_{(7)}}{16\pi G^{(10)}_{\rm N}} = \frac{R_9}{\ell^9_{\rm s}}, \tag{25.97}$$

on account of Eqs. (25.86), (25.26), and (25.57).

For the D$p$-brane solutions ($p<7$) Eqs. (25.64) compactified on $p$ circles we calculate first the $(10-p)$-dimensional dilaton, which is given by

$$e^{-2(\phi-\phi_0)} = e^{-2(\hat{\phi}-\hat{\phi}_0)}\sqrt{\hat{g}_{\underline{y^1y^1}}\cdots\hat{g}_{\underline{y^py^p}}} = H^{\frac{p-6}{4}}_{{\rm D}p}\,. \tag{25.98}$$

Then, according to Eq. (25.89),

$$\tilde{g}_{{\rm E}\,\mu\nu} = H^{\frac{p-6}{2(8-p)}}_{{\rm D}p}\, g_{\mu\nu} \;\Rightarrow\; \tilde{g}_{{\rm E}\,tt} = H^{-\frac{p-7}{p-8}}_{{\rm D}p} \sim 1 - \frac{p-7}{p-8} h_{{\rm D}p}\frac{1}{|\vec{x}_{p-9}|^{7-p}}, \tag{25.99}$$

which, compared with Eq. (25.92), gives again the right value,

$$M_{{\rm D}p} = \frac{(7-p)h_{{\rm D}p}\omega_{(8-p)}}{16\pi G^{(10-p)}_{\rm N}} = \frac{(7-p)(2\pi)^p R_9\cdots R_{10-p} h_{{\rm D}p}\omega_{(8-p)}}{16\pi G^{(10)}_{\rm N}} = \frac{R_9\cdots R_{10-p}}{\ell^{p+1}_{\rm s} g}. \tag{25.100}$$

### *25.3.2 Charges*

With the normalization of the superstring effective actions Eqs. (22.38) and (23.4), the (electric) charges associated with the KR 2-form and the RR $(p+1)$-form potentials, which are carried, respectively, by fundamental strings and D$p$-branes, can be defined by the integrals[19]

$$q_{\rm F1} = \frac{g^2}{16\pi G_{\rm N}^{(10)}} \int_{S^7_\infty} e^{-2\hat{\phi}} \star \hat{H}, \qquad q_{{\rm D}p} = \frac{g^2}{16\pi G_{\rm N}^{(10)}} \int_{S^{8-p}_\infty} \star \hat{G}^{(p+2)}, \tag{25.101}$$

whereas the charge associated with the NSNS 6-form potential, carried by the S5, is given by

$$q_{\rm S5} = \frac{g^2}{16\pi G_{\rm N}^{(10)}} \int_{S^4_\infty} e^{2\hat{\phi}} \star \hat{H}^{(7)} = \frac{g^2}{16\pi G_{\rm N}^{(10)}} \int_{S^4_\infty} \hat{H}. \tag{25.102}$$

$q_{\rm F1}$ and $q_{\rm S5}$ are the electric–magnetic duals of each other, as are $q_{{\rm D}p}$ and $q_{{\rm D}\tilde{p}}$. With the above normalization, the generalization of the Dirac quantization condition for extended objects reads

$$q_{{\rm D}p} q_{{\rm D}\tilde{p}} = 2\pi n \frac{16\pi G_{\rm N}^{(10)}}{\hat{g}^2}, \qquad q_{\rm F1} q_{\rm S5} = 2\pi n \frac{16\pi G_{\rm N}^{(10)}}{\hat{g}^2}, \qquad n \in \mathbb{Z}. \tag{25.103}$$

It is easy to see that the values of the charges of the string/M-theory solutions coincide with the values we gave in Section 25.1.1. It should be stressed that both the masses and the charges of the extreme solutions are determined by the same $h$. This is due to the fact that the masses (tensions) and charges of these objects saturate BPS bounds. The solutions preserve half of the supersymmetries of the corresponding supergravity theory (see Section 25.5.1).

## 25.4 Duality of string-theory solutions

In Section 25.1 we used string dualities to find and relate all the extended objects of string and M theories. In the subsequent sections we have established a relation between those objects and certain classical solutions of the string effective actions and $d = 11$ supergravity using arguments based on the symmetries of the solutions which determine the dimensions of the worldvolumes of the objects they describe, on the basis of the charges they carry and the matching with $p$-brane sources.

On the other hand, in Chapters 21–23 we learned how string dualities manifest themselves in string effective actions; to close the loop, here we are going to see how the duality relations between string states are realized as relations between solutions of the effective actions. These relations are represented in Figs. 25.5 and 25.6.

The three main types of duality relations that we are going to study are (i) those between the solutions of $d = 11$ supergravity and solutions of $N = 2A, d = 10$ supergravity, via the dimensional-reduction formulae Eqs. (22.35); (ii) those between solutions of $N = 2A, d = 10$ and $N = 2B, d = 10$ supergravity, via the type-II Buscher T-duality rules Eqs. (23.36) and (23.37); and (iii) those between solutions of $N = 2B, d = 10$ supergravity, via $\mathrm{SL}(2, \mathbb{Z})$

[19] These definitions are valid for field configurations in which only one potential is non-trivial. In general, the charge is obtained by integrating the form $F$ such that $dF = 0$ is the equation of motion. (This is sometimes called the *Page charge* [930].) The presence of non-trivial Chern–Simons terms in the action implies that $F$ consists in various terms, as we will discuss in Section 25.6.1.

transformations Eqs. (23.21) or (23.23) and (23.24). We are also going to need the results of Section 15.3.1 in order to perform reductions on transverse directions.

The supergravity duality transformations can also be used to construct new solutions. We will study two families of solutions constructed in this way: $pq$-strings and $pq$-5-branes.

### 25.4.1 $N = 2A, d = 10$ SUEGRA solutions from $d = 11$ SUGRA solutions

There are two basic $p$-brane solutions of $d = 11$ SUGRA: the M2- and M5-brane solutions Eqs. (25.45) and (25.50). If they really describe the M2- and M5-brane states of M theory, their reduction must give rise to the F1A, D2, D4, and S5A solutions Eqs. (25.56), (25.64), and (25.60) of $N = 2A, d = 10$ SUEGRA [469, 14, 1190] under *double* and *direct* dimensional reductions, i.e. in a worldvolume direction (corresponding to branes wrapped in the compact dimension) or in a transverse direction.

Double dimensional reductions are, by definition, made in a direction none of the fields depends on, and one just has to rewrite the solution in $d = 10$ variables using Eqs. (22.35) in a straightforward manner. The only subtlety is that, in order to have a non-trivial value for $\hat{g} = e^{\hat{\phi}_0}$, one must first rescale the compact worldvolume coordinate (that we call here $z$) $z \rightarrow e^{\frac{2}{3}\hat{\phi}_0} z$ in Eqs. (25.45) and (25.50).

Direct dimensional reductions are made precisely in one of the directions on which the $p$-brane metric depends. We could substitute the harmonic function for another one independent of the compact direction but, in that case, we would lose the relation to the quantum object it represents. The right procedure is, as we explained in Section 15.3.1, to construct first the correct solution that describes the $p$-brane in a transverse space with a compact coordinate, which amounts to solving the Laplace equation in such a space, and then to Fourier-expand the solution, keeping only the zero mode. The solution is the same harmonic function as that which describes an infinite periodic array of parallel $p$-branes separated by a distance equal to the length of the compact direction. This harmonic function is the linear superposition of those of each $p$-brane:

$$H_p = 1 + \frac{h_p}{|\vec{x}_{\tilde{p}+3}|^{\tilde{p}+1}} \longrightarrow H_p = 1 + \sum_{m\in\mathbb{Z}} \frac{h_p}{\left[|\vec{x}_{\tilde{p}+2}|^2 + (z + 2\pi m R_z)\right]^{\frac{\tilde{p}+1}{2}}}. \tag{25.104}$$

The zero mode can then be found using Eqs. (15.140). For M2- and M5-branes with a compact transverse coordinate ($\tilde{p} = 5, 2$ and $n = 6, 3$, respectively) we find

$$H_{\text{M2}} \sim 1 + \frac{h_{\text{M2}}\omega_{(5)}}{2\pi R_z \omega_{(4)}} \frac{1}{|\vec{x}_7|^5} = H_{\text{D2}}, \qquad H_{\text{M5}} \sim 1 + \frac{h_{\text{M5}}\omega_{(2)}}{2\pi R_z \omega_{(1)}} \frac{1}{|\vec{x}_4|^2} = H_{\text{S5A}}, \tag{25.105}$$

using the actual values of the integration constants $h$. The reduction is now straightforward.

This procedure is sometimes called *smearing*. The brane is said to be *delocalized* in one dimension. Sometimes duality gives delocalized solutions (for instance, oxidizing the F1A and D4 solutions using Eqs. (22.37)) and one has to show that one can indeed add the missing coordinate. This is a clear insufficiency of these methods.

There are more extended solutions in $N = 2A, d = 10$ SUEGRA that do not originate on M2- or M5-branes: the D0, D6, and D8 solutions. The $d = 11$ origin of the D8 is

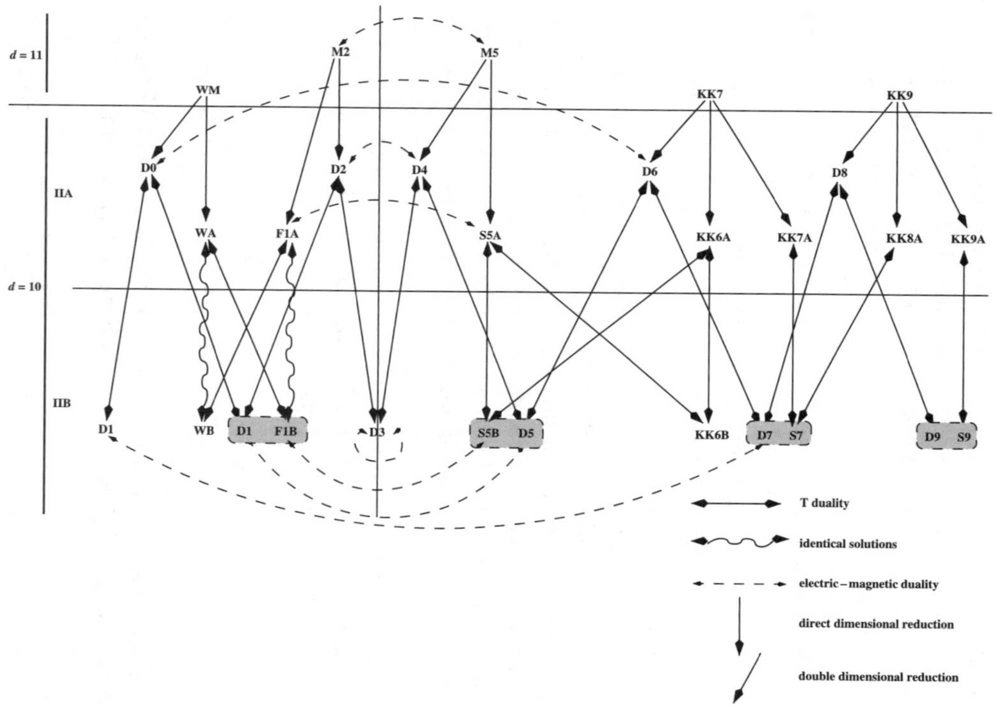


Fig. 25.5. Duality relations between classical solutions of ten- and 11-dimensional supergravity theories: $p$-branes, M-branes, D-branes, waves, and Kaluza–Klein monopoles. Lines with two arrows denote T-duality relations; dashed lines denote S-duality relations. Lines with a single arrow denote relations of dimensional reduction, either vertical (direct dimensional reduction) or diagonal (double dimensional reduction). Pairs of branes in boxes are type-IIB S-duality doublets.

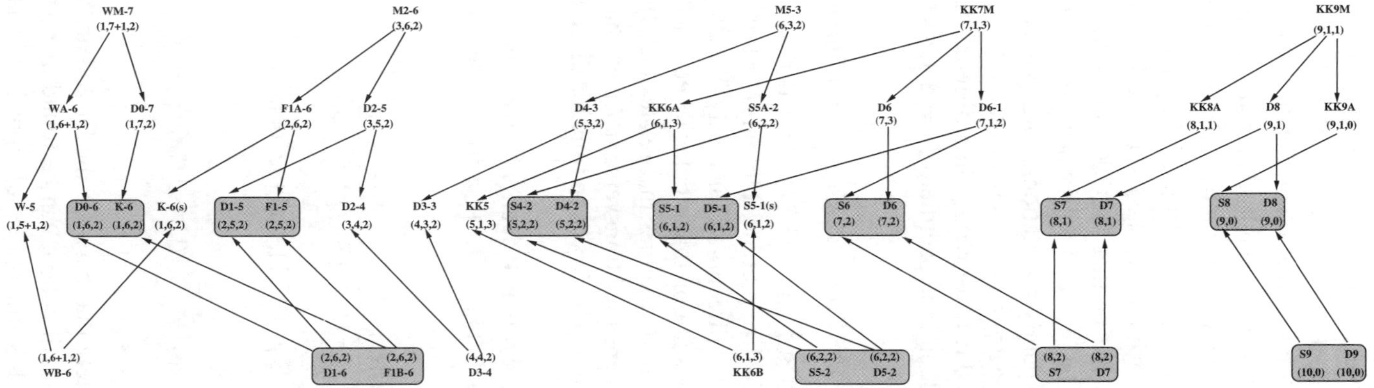


Fig. 25.6. Duality relations between KK branes. The numbers in parentheses represent the worldvolume dimension and the isometric and transverse directions. The arrows indicate dimensional reduction in the corresponding kind of direction. In the upper row we represent M-theory KK branes; immediately below are ten-dimensional type-IIA branes; and below them are nine-dimensional branes. Type-IIB KK branes are in the bottom row. Pairs of branes in boxes are S-duality doublets. Singlets are denoted with an (s).

not known. To find the origins of the D0 and D6, we can simply apply the oxidation formulae Eqs. (22.37).

For the D0, we find that the only non-trivial field is the metric, which has the form of the pp-wave Eq. (14.42) with $H$ replaced by $H_{\rm D0}$ and $z$ by $e^{\frac{2}{3}\hat{\phi}_0}$. Actually, $H_{\rm D0}$ is the zero mode of the harmonic function $H$ of an AS shock pp-wave moving in the compact 11th dimension. The reduction of such a shock wave was studied in Section 15.3.2, and, if we compare $h_{\rm D0}$ with $h$ in Eq. (15.147), we find that it corresponds to a shock wave with the minimal momentum $p_z = 1/\ell^{(11)}_{\rm Planck}$. The D0-brane is, therefore, nothing but a KK mode.

The D6 also oxidizes to a purely gravitational solution: the KK monopole Eq. (15.176) (with six extra dimensions) with $k_0$ replaced by $e^{\frac{2}{3}\hat{\phi}_0}$ and $H$ by $H_{\rm D6}$. Observe that $h_{\rm D6} = \ell_{\rm s} g/2 = R_z/2$ also has the right value, which is related to the periodicity of $z$ and diverges in the decompactification limit. This means that the KK monopole is not a solution of standard $d = 11$ supergravity and can be included only when a dimension is compact.

This exercise shows the need to add purely gravitational solutions such as KK monopoles and waves to be able to explain the spectrum of objects of type-IIA superstring theory. Of course, once they are included in $d = 11$, if we reduce in a different dimension we find a gravitational wave and a KK monopole in (compactified) $N = 2A, d = 10$ supergravity.

### *25.4.2 $N = 2A/B, d = 10$ SUEGRA T-dual solutions*

The type-II T-duality rules Eqs. (23.36) and (23.37) were derived using dimensional reduction; therefore, if we want to T-dualize $p$-brane solutions in a transverse direction, we have to *delocalize* them previously in that direction. Conversely, T duality in a worldvolume direction typically takes us to a $p$-brane solution that is delocalized in a transverse direction. The main example is the whole chain of T dualities that relates the D$p$-branes: if we start with the D0, fully localized in nine-dimensional transverse space, we have to delocalize it in one direction in order to find the D-string, which will be fully localized in eight dimensions and vice versa.

One finds a precise correspondence through T duality between the coefficients $h_{{\rm D}p}$ that we have determined before; on T-dualizing a D$p$ in a transverse direction, we find the following coefficient after smearing:

$$\frac{h_{{\rm D}p}\omega_{(6-p)}}{2\pi R\omega_{(5-p)}} = \frac{(2\pi\ell_{\rm s})^{7-p}g}{2\pi R}\frac{\omega_{(6-p)}}{(7-p)\omega_{(8-p)}\omega_{(5-p)}}. \tag{25.106}$$

Using the T-duality rules for $g$ and $R$, Eqs. (25.1), and the identity

$$\frac{\omega_{(n-1)}}{n\omega_{(n+1)}\omega_{(n-2)}} = \frac{1}{(n-1)\omega_{(n)}},$$

we find $h_{{\rm D}(p+1)}(g')$, as given in Eq. (25.64).

Another example of T duality, namely that between the F1 and an AS shock wave that describes a string moving in a compact direction, can be found in Section 21.3. It is also a simple exercise to relate the S5 and KK monopole solutions by T duality in a transverse direction of the S5, which becomes the special isometric direction of the KK monopole.

### 25.4.3 S duality of $N=2B, d=10$ SUEGRA solutions: pq-branes

Performing S-duality transformations to relate the F1B to the D1 and the S5B to the D5 poses no problems but offers some opportunities: $\mathrm{SL}(2,\mathbb{R})$ is a three-dimensional group and, after a general transformation, the new solution may have up to three new independent physical parameters. This procedure was used by Schwarz in Ref. [1116] to construct a solution with four independent parameters. In the (unmodified) Einstein frame in which $\mathrm{SL}(2,\mathbb{R})$ invariance is manifest, it takes the form

$$\boxed{\begin{array}{ll} d\hat{s}_{\mathrm{E}}^2 = H_{pq1}^{-\frac{3}{4}}[dt^2 - dy^2] - H_{pq1}^{\frac{1}{4}}d\vec{x}_8^2, & \\ \hat{\mathcal{B}}_{t\underline{y}} = \vec{a}(H_{pq1}^{-1} - 1), & \hat{\mathcal{M}} = \vec{a}\vec{a}^{\mathrm{T}}H_{pq1}^{-\frac{1}{2}} + \vec{b}\vec{b}^{\mathrm{T}}H_{pq1}^{\frac{1}{2}}, \\ H_{pq1} = 1 + \dfrac{h_{pq1}\hat{g}^{-\frac{3}{2}}}{|\vec{x}_8|^6}, & \vec{a}^{\mathrm{T}}\eta\vec{b} = 1. \end{array}} \tag{25.107}$$

Observe that $h_{pq1}$ comes with a factor $\hat{g}^{-\frac{3}{2}}$ to take into account the rescaling of coordinates necessary to relate this metric to an asymptotically flat string of modified string metric with the usual formulae. The constant vectors $\vec{a}$ and $\vec{b}$ can be seen as the two column vectors of an $\mathrm{SL}(2,\mathbb{R})$ matrix,

$$\left(\vec{a}\,\vec{b}\right) = \begin{pmatrix} \alpha & \beta \\ \gamma & \delta \end{pmatrix}, \qquad \vec{a}^{\mathrm{T}}\eta\vec{b} = \alpha\delta - \beta\gamma = 1, \tag{25.108}$$

and transform covariantly under $\mathrm{SL}(2,\mathbb{R})$, which leaves this family invariant. The four independent parameters correspond to the asymptotic values of the two scalars, combined in the matrix $\hat{\mathcal{M}}_0$, and the charges $q_{\mathrm{F1}}$ and $q_{D1}$ . The mass must be a function of those four parameters (again, a saturated BPS bound). To find the values of the physical parameters of this solution in terms of the constants $\vec{a}$, $\vec{b}$, and $h_{pq1}$, we can use an $\mathrm{SL}(2,\mathbb{R})$ definition for the charges:

$$\begin{pmatrix} q_{\mathrm{D1}} \\ q_{\mathrm{F1B}} \end{pmatrix} = \vec{q} = \frac{g^2}{16\pi G_{\mathrm{N}}^{(10)}}\int_{\mathrm{S}^7_\infty} \star\hat{\mathcal{M}}^{-1}\vec{\hat{\mathcal{H}}} = \frac{6\omega_{(7)}h_{pq1}\hat{g}^{-\frac{3}{2}}}{(2\pi\ell_{\mathrm{s}})^7\ell_{\mathrm{s}}}\hat{\mathcal{M}}_0^{-1}\vec{a}, \tag{25.109}$$

where $\hat{\mathcal{M}}_0 = \vec{a}\vec{a}^{\mathrm{T}} + \vec{b}\vec{b}^{\mathrm{T}}$. Using the property $\vec{a}^{\mathrm{T}}\hat{\mathcal{M}}_0^{-1}\vec{a} = 1$, we find the relation between $\vec{q}$ and $h_{pq1}$:

$$h_{pq1} = \frac{(2\pi\ell_{\mathrm{s}})^7\ell_{\mathrm{s}}\hat{g}^{\frac{3}{2}}}{6\omega_{(7)}}\sqrt{\vec{q}^{\,\mathrm{T}}\hat{\mathcal{M}}_0\vec{q}}. \tag{25.110}$$

We can now express the full solution in terms of the physical parameters $\mathcal{M}_0$ and $\vec{q}$.

The object described by any of these solutions (usually called the *pq-string*) is a $p=1$ object (a string) that has both $q_{\mathrm{F1B}}$ and $q_{\mathrm{D1}}$ charges in a IIB vacuum characterized by the moduli $\mathcal{M}_0$ ($e^{\hat{\varphi}_0} = \hat{g}$ and $\hat{C}_0^{(0)} = \hat{\theta}/(2\pi)$), and can be understood as the superposition of

D1s and F1Bs. The values of the charges are therefore quantized: they can only be multiples of those of one (D, F) string: $n/(2\pi\ell_s^2)$. The tension of this object is proportional to $h_{pq1}$, and, therefore, for trivial moduli, to

$$\sqrt{q_{\rm D1}^2 + q_{\rm F1B}^2} < |q_{\rm D1}| + |q_{\rm F1B}|.$$

The tension of two parallel (or coincident) strings of the same kind would be the sum of the tensions of each of them (*bound states at threshold*), which means that there is zero interaction energy and it costs zero energy to disintegrate the system. In this case, the tension is in general smaller, which means that this solution represents a bound state of F1Bs and D1s with non-zero binding energy (*non-threshold bound states*). However, the solution is stable with respect to disintegration only if the numbers of D1s and F1As are relatively prime: if they have a g.c.d. different from 1, say $N$, the tension is $N$ times that of a single $pq$-string with $n_{\rm D1}/N$ and $n_{\rm F1}/N$ strings, and it takes zero energy to disintegrate it. The $pq$-strings with coprime numbers of strings are the basic states of the theory.

A solution describing analogous bound states of D5 and S5Bs (*pq-5-branes*) was constructed in Ref. [909]:

$$\begin{aligned}
d\hat{s}_{\rm E}^2 &= H_{pq5}^{-\frac{1}{4}}[dt^2 - d\vec{y}_5^{\,2}] - H_{pq5}^{\frac{3}{4}} d\vec{x}_5^{\,2}, \\
\hat{\mathcal{B}}_{t\underline{y}^1\cdots\underline{y}^5} &= \eta\vec{b}(H_{pq5}^{-1} - 1), \qquad \hat{\mathcal{M}} = \vec{a}\vec{a}^{\rm T} H_{pq5}^{\frac{1}{2}} + \vec{b}\vec{b}^{\rm T} H_{pq5}^{-\frac{1}{2}}, \\
H_{pq5} &= 1 + h_{pq5}\frac{\hat{g}^{-\frac{1}{2}}}{|\vec{x}_4|^2}, \qquad \vec{a}^{\rm T}\eta\vec{b} = 1.
\end{aligned} \tag{25.111}$$

On T-dualizing these solutions, one obtains new bound states of F1s and D$p$s [910, 911]. These solutions describe intersecting branes with non-zero interaction energy. Other intersecting solutions with non-zero interaction energy are, for instance, the systems of D$p$-branes and D$(p+2)$-branes studied in Refs. [250, 376].

## 25.5 String-theory extended objects from superalgebras

In Chapter 5 we introduced supergravities as the gauge theories of the supersymmetry algebras. These contain a great deal of information about the global and local symmetries of each supergravity theory. When we studied four-dimensional Poincaré-extended supersymmetry algebras, we saw that, associated with each of the possible "electric" central charges $Q^{ij}$, there was an SO(2) gauge potential $A_\mu^{ij}$ whose gauge symmetry is generated by $Q^{ij}$. They contributed to the gauge superpotential with a term $\frac{1}{2}A_\mu^{ij}Q^{ij}$. The "magnetic" central charges could be associated with the electric–magnetic dual potentials, which are not independent. The central charges could be associated with electric and magnetic charges of supergravity solutions in Chapter 17, and the superalgebra could be used to see whether the solutions preserved any supersymmetries.

This correspondence between central charges and Abelian potentials holds for "quasi-central charges" with Lorentz indices as well, and with each charge $Z^{(p)}_{a_1\cdots a_p}$ we can associate in the supergravity theory a $(p+1)$-form potential $A^{(p+1)}$ that transforms under Abelian gauge transformations. They contribute to the gauge superpotential with a term $(1/p!)A^{(p+1)}{}_{\mu}{}^{a_1\cdots a_p}Z^{(p)}_{a_1\cdots a_p}$. The electric–magnetic dual $(\tilde{p}+1)$-form potential is associated with a $Z^{(\tilde{p})}_{a_1\cdots a_{\tilde{p}}}$ quasi-central charge that must also be present in the superalgebra. It is clear that these quasi-central charges must be associated with $p$-brane solutions of the supergravity theory [73] and that the superalgebra can be used to study their unbroken supersymmetries.

This is a very powerful tool that can be used to determine which objects/states may exist in a supergravity theory knowing just which quasi-central charges are algebraically allowed in the anticommutator of the supercharges of a given superalgebra.[20] Here we are going to write the superalgebras of the string/M-theory effective actions (supergravities) and we are going to study some examples, following in part Ref. [1196] and starting with the algebra of $d=11$ supergravity.

The superalgebra of $d=11$ superalgebra (also known as *M superalgebra*) admits quasi-central charges of ranks $1, 2, 5, 6, 9,$ and 10. The last three values are just the duals of the first three. Therefore, the M superalgebra is usually written in the form

$$\left\{\hat{\bar{Q}}^{\alpha}, \hat{\bar{Q}}^{\beta}\right\} = c(\hat{\Gamma}^{\hat{a}}\hat{\mathcal{C}}^{-1})^{\alpha\beta}\hat{P}_{\hat{a}} + \frac{c_2}{2!}(\hat{\Gamma}^{\hat{a}\hat{b}}\hat{\mathcal{C}}^{-1})^{\alpha\beta}\hat{\mathcal{Z}}^{(2)}_{\hat{a}\hat{b}} + \frac{c_5}{5!}(\hat{\Gamma}^{\hat{a}_1\cdots\hat{a}_5}\hat{\mathcal{C}}^{-1})^{\alpha\beta}\hat{\mathcal{Z}}^{(5)}_{\hat{a}_1\cdots\hat{a}_5}, \tag{25.112}$$

with constants $c, c_2,$ and $c_5$ that are convention dependent and immaterial for our discussion.[21] We immediately recognize the momentum and the charges associated with the M2- and M5-branes. The gravitational wave is associated with the momentum, but what is the charge associated with the KK monopole (KK7M)? Furthermore, is there a charge for the KK9M? As a matter of fact, as we have stressed repeatedly, these KK-branes are not states of the *uncompactified* $d=11$ theory and appear only after compactification. Still, we can include them in the $d=11$ superalgebra using dual charges *and* vectors $k^a$ and $l^a$ that project the charges in the compact direction. The two terms that correspond to the KK7M and the KK9M and should be added are [907]

$$+\frac{c_6}{6!}(\hat{\Gamma}^{\hat{a}_1\cdots\hat{a}_6}\hat{\mathcal{C}}^{-1})^{\alpha\beta}\hat{\mathcal{Z}}^{(7)}_{\hat{a}_1\cdots\hat{a}_7}k^{\hat{a}_7} + \frac{c_9}{9!}(\hat{\Gamma}^{\hat{a}_1\cdots\hat{a}_9}\hat{\mathcal{C}}^{-1})^{\alpha\beta}\hat{\mathcal{Z}}^{(8)}_{\hat{a}_1\cdots\hat{a}_8}l_{\hat{a}_9}. \tag{25.113}$$

The dimensional reduction of the M algebra should give the $N=2A, d=10$ superalgebra with a charge for each of the known objects of this theory. We need only reduce the vector indices (as we did in the reduction of $d=11$ supergravity). Each of the standard quasi-central charges gives two in $d=10$: $\hat{\hat{P}}_{\hat{a}} = \hat{P}_{\hat{a}}$, $\hat{\hat{P}}_z = \hat{\mathcal{Z}}^{(0)}$, $\hat{\hat{\mathcal{Z}}}^{(2)}_{\hat{a}\hat{b}} = \hat{\mathcal{Z}}^{(2)}_{\hat{a}\hat{b}}$, $\hat{\hat{\mathcal{Z}}}^{(2)}_{\hat{a}z} = \hat{\mathcal{Z}}^{(1)}_{\hat{a}}$, etc. The non-standard ones give rise to three, for instance $\hat{\hat{\mathcal{Z}}}^{(7)}_{\hat{a}_1\cdots\hat{a}_7} = \hat{\mathcal{Z}}^{(7)}_{\hat{a}_1\cdots\hat{a}_7}$,

[20] We found all the possibilities in $d=4$ in Section 5.4.1. In higher dimensions the analysis is almost identical.

[21] This formula can be interpreted as a decomposition of a symmetric bi-spinor into Lorentz tensors. A consistency check is provided by the counting of independent components on both sides of the equation: $33 \times 32/2$. Physically, this formula should be understood as an inventory of possibilities: for instance, the superalgebra of $N=1, d=10$ supergravity admits quasi-central charges of ranks 1 and 5, but, physically, we expect on the r.h.s. one rank-5 and two rank-1 charges: momentum and the string charge. The counting on the two sides gives different results, but physically it is correct.

$\hat{\mathcal{Z}}^{(7)}_{\hat{a}_1\cdots\hat{a}_6 z} = \hat{\mathcal{Z}}^{(6)}_{\hat{a}_1\cdots\hat{a}_6}$, and $\hat{\mathcal{Z}}^{(7)}_{\hat{a}_1\cdots z\hat{a}_6} = \hat{\mathcal{Z}}^{(6)}_{\hat{a}_1\cdots\hat{a}_5 z}$, corresponding, respectively, to the KK7A, the D6, and the KK6A. The KK6A is the standard KK monopole. The KK7A is the solution one obtains by reducing the KK7M (the M-theory KK monopole) in a genuine transverse direction (the harmonic function is smeared by the usual procedure and then one solves for the vector field in the metric [936]).

The result of the reduction is the $N = 2A, d = 10$ superalgebra generalized with the inclusion of KK-brane charges:

$$
\begin{aligned}
\left\{\hat{Q}^\alpha, \hat{Q}^\beta\right\} &= c(\hat{\Gamma}^{\hat{a}}\hat{\mathcal{C}}^{-1})^{\alpha\beta}\hat{P}_{\hat{a}} + \sum_{n=0,1,4,8} \frac{c_n}{n!}(\hat{\Gamma}^{\hat{a}_1\cdots\hat{a}_n}\hat{\Gamma}_{11}\hat{\mathcal{C}}^{-1})^{\alpha\beta}\mathcal{Z}^{(n)}_{\hat{a}_1\cdots\hat{a}_n} \\
&+ \sum_{n=2,5,6} \frac{c_n}{n!}(\hat{\Gamma}^{\hat{a}_1\cdots\hat{a}_n}\hat{\mathcal{C}}^{-1})^{\alpha\beta}\hat{\mathcal{Z}}^{(n)}_{\hat{a}_1\cdots\hat{a}_n} \\
&+ \frac{c_5}{5!}(\hat{\Gamma}^{\hat{a}_1\cdots\hat{a}_5}\hat{\Gamma}_{11}\hat{\mathcal{C}}^{-1})^{\alpha\beta}\hat{\mathcal{Z}}^{(6)}_{\hat{a}_1\cdots\hat{a}_5\hat{a}_6}\hat{k}^{\hat{a}_6} + \frac{c_6}{6!}(\hat{\Gamma}^{\hat{a}_1\cdots\hat{a}_6}\hat{\mathcal{C}}^{-1})^{\alpha\beta}\hat{\mathcal{Z}}^{(7)}_{\hat{a}_1\cdots\hat{a}_6\hat{a}_7}\hat{l}^{\hat{a}_7} \\
&+ \frac{c_8}{8!}(\hat{\Gamma}^{\hat{a}_1\cdots\hat{a}_8}\hat{\mathcal{C}}^{-1})^{\alpha\beta}\hat{\mathcal{Z}}^{(7)}_{\hat{a}_1\cdots\hat{a}_7}\hat{m}_{\hat{a}_8} + \frac{c_9}{9!}(\hat{\Gamma}^{\hat{a}_1\cdots\hat{a}_9}\hat{\mathcal{C}}^{-1})^{\alpha\beta}\hat{\mathcal{Z}}^{(8)}_{\hat{a}_1\cdots\hat{a}_8}\hat{n}_{\hat{a}_9}.
\end{aligned}
\tag{25.114}
$$

Let us now turn to the $N = 2B, d = 10$ superalgebra. It contains an $\mathrm{SO}(2)$ pair of chiral supercharges labeled by $i, j = 1, 2$, and the charges that appear on the r.h.s. of their anticommutator carry a pair symmetric or antisymmetric in these indices. The allowed ranks for antisymmetric indices are 3 and 7 and those for symmetric indices are 1, 5, and 9. The charges with antisymmetric indices are proportional to $\sigma^2$ and those with symmetric indices can be decomposed into a basis of symmetric $2 \times 2$ matrices: $\mathbb{I}, \sigma^1$, and $\sigma^3$. The charges proportional to $\sigma^2$ and $\mathbb{I}$ are invariant under SO(2), and charges proportional to $\sigma^1$ and $\sigma^3$ form SO(2) doublets. Combining the latter into symmetric traceless charges denoted by $(ij)$, the algebra is usually written in the form

$$
\begin{aligned}
\left\{\hat{Q}^{i\,\alpha}, \hat{Q}^{j\,\beta}\right\} &= c\delta^{ij}(\hat{\Gamma}^{\hat{a}}\hat{\mathcal{C}}^{-1})^{\alpha\beta}\hat{P}_{\hat{a}} + c_1(\hat{\Gamma}^{\hat{a}}\hat{\mathcal{C}}^{-1})^{\alpha\beta}\hat{\mathcal{Z}}^{(1)(ij)}_{\hat{a}} \\
&+ \frac{c_3}{3!}(\sigma^2)^{ij}(\hat{\Gamma}^{\hat{a}_1\hat{a}_2\hat{a}_3}\hat{\mathcal{C}}^{-1})^{\alpha\beta}\hat{\mathcal{Z}}^{(3)}_{\hat{a}_1\hat{a}_2\hat{a}_3} \\
&+ \frac{c_5}{5!}\delta^{ij}(\hat{\Gamma}^{\hat{a}_1\cdots\hat{a}_5}\hat{\mathcal{C}}^{-1})^{\alpha\beta}\hat{\mathcal{Z}}^{(5)}_{\hat{a}_1\cdots\hat{a}_5} + \frac{c_5}{5!}(\hat{\Gamma}^{\hat{a}_1\cdots\hat{a}_5}\hat{\mathcal{C}}^{-1})^{\alpha\beta}\hat{\mathcal{Z}}^{(5)(ij)}_{\hat{a}_1\cdots\hat{a}_5},
\end{aligned}
\tag{25.115}
$$

where it is understood that the r.h.s. has to be projected over the positive-chirality subspace.

$\mathrm{SL}(2, \mathbb{R})$ acts on the spinors through $\mathrm{SO}(2)$ rotations and, therefore, $\hat{\mathcal{Z}}^{(1)(ij)}$ and $\hat{\mathcal{Z}}^{(5)(ij)}$ correspond to the two doublets of strings (D1 and F1B) and 5-branes (D5 and S5B). $\hat{\mathcal{Z}}^{(3)}$ corresponds to the S-duality invariant D3. There is no invariant 5-brane and $\hat{\mathcal{Z}}^{(5)}_{\hat{a}_1\cdots\hat{a}_5}$ should be replaced by $\hat{\mathcal{Z}}^{(6)}_{\hat{a}_1\cdots\hat{a}_5\hat{a}_6}\hat{k}^{\hat{a}_6}$ associated with the KK6B. To these charges one should add $\hat{\mathcal{Z}}^{(9)(ij)}$ for the D9–S9 doublet and two $\hat{\mathcal{Z}}^{(7)}$s for the D7–S7 doublet[22] if we are to relate

[22] Each of them is $\mathrm{SO}(2)$ invariant, but we may assume that they are interchanged by S duality. The situation is still not completely clear since, as we have stressed before, there are an infinite number of *pq*-7-branes, not just a doublet, and this is difficult to reflect in the superalgebra.

this superalgebra to the $N = 2A, d = 10$ by T duality, as we should expect. As Figs. 25.5 and 25.6 show, more charges with more auxiliary vectors need to be included if one really wants to have complete agreement and consistency with all the dualities conjectured.

### 25.5.1 Unbroken supersymmetries of string-theory solutions

If we follow now the reasoning of Section 18.1.1, we arrive at the conclusion that the quantum theory based on the above superalgebras admits states with momentum $P_0 = T_{(p)}$ (the tension) and the quasi-central charge $\mathcal{Z}^{(p)}_{1\cdots p} = T_{(p)}$, corresponding to extended objects that are invariant under the supersymmetry transformations generated by spinors that satisfy constraints of the form Eq. (18.18), which we can rewrite in the form

$$(\mathbb{I} \pm \Gamma^{01\cdots p}\mathcal{O})\epsilon = 0, \tag{25.116}$$

where $\mathcal{O}$ is an operator that depends on the theory and the state. In all cases $\Gamma^{01\cdots p}\mathcal{O}$ is a traceless operator that squares to the identity; half of its eigenvalues are $+1$ and the other half are $-1$. Therefore $\frac{1}{2}(\mathbb{I} \pm \Gamma^{01\cdots p}\mathcal{O})$ is a projector that eliminates half of the components of $\epsilon$. There is, therefore, a half-supersymmetric state for each quasi-central charge in the given superalgebras, and we expect the associated solutions of the supergravity theories (which will be extreme solutions) to have unbroken supersymmetries generated by Killing spinors that satisfy the same constraints. Let us see some examples.

*Unbroken supersymmetries of the M2-brane.* We are going to work out in detail this example to illustrate how the Killing-spinor equations are usually solved. The rest of the examples follow the same pattern and we will give only the results.

The $d = 11$ SUGRA Killing spinor equations are $\delta_{\hat{\hat{\kappa}}}\hat{\hat{\psi}}_{\hat{\hat{\mu}}} = 0$ with $\delta_{\hat{\hat{\kappa}}}\hat{\hat{\psi}}_{\hat{\hat{\mu}}}$ given by Eq. (22.8). We just have to substitute into it the spin-connection and 4-form components of the M2 solution Eq. (25.45). Choosing the Elfbeins

$$\hat{\hat{e}}_{\underline{i}}{}^{j} = H_{\rm M2}^{-\frac{1}{3}}\delta_i{}^j, \qquad \hat{\hat{e}}_{\underline{m}}{}^{n} = H_{\rm M2}^{\frac{1}{6}}\delta_m{}^n, \tag{25.117}$$

and using the results of Appendix N.3.5, we find the non-vanishing components

$$\begin{aligned}
\hat{\hat{\omega}}_{\underline{m}}{}^{nl} &= -\tfrac{1}{3}H_{\rm M2}^{-1}\partial_{\underline{q}}H_{\rm M2}\eta_m{}^{[n}\eta^{l]q}, \\
\hat{\hat{\omega}}_{\underline{i}}{}^{mj} &= \tfrac{2}{3}H_{\rm M2}^{-\frac{3}{2}}\partial_{\underline{q}}H_{\rm M2}\eta_i{}^{[m}\eta^{j]q}, \\
\hat{\hat{G}}_{mijk} &= \mp\epsilon_{ijk}\, H_{\rm M2}^{-\frac{7}{6}}\partial_{\underline{m}}H_{\rm M2},
\end{aligned} \tag{25.118}$$

and substituting, and assuming that $\partial_{\underline{i}}\hat{\hat{\kappa}} = 0$, we find the equations

$$\begin{aligned}
\delta_{\hat{\hat{\kappa}}}\hat{\hat{\psi}}_{\underline{i}} &= \tfrac{1}{3}H_{\rm M2}^{-\frac{3}{2}}\partial_{\underline{n}}H_{\rm M2}\hat{\hat{\Gamma}}_{(i)}{}^{n}\left(1 \mp \frac{i}{2}\epsilon_{(i)jk}\hat{\hat{\Gamma}}^{(i)jk}\right)\hat{\hat{\kappa}} = 0, \\
\delta_{\hat{\hat{\kappa}}}\hat{\hat{\psi}}_{\underline{m}} &= 2\partial_{\underline{m}}\hat{\hat{\kappa}} - \tfrac{1}{6}H_{\rm M2}^{-1}\partial_{\underline{n}}H_{\rm M2}\left[\hat{\hat{\Gamma}}^{mn} \mp i\left(\hat{\hat{\Gamma}}^{mn} + 2\delta^{mn}\right)\hat{\hat{\Gamma}}^{012}\right]\hat{\hat{\kappa}} = 0.
\end{aligned} \tag{25.119}$$

The first equation is purely algebraic and can be solved only if $\hat{\kappa}$ satisfies the constraint

$$\left(1 \mp i\hat{\Gamma}^{\hat{0}\hat{1}\hat{2}}\right)\hat{\kappa} = 0. \tag{25.120}$$

Using this constraint in the second equation, it takes the form

$$2\left(\partial_{\underline{m}} + \tfrac{1}{6}H_{\mathrm{M2}}^{-1}\partial_{\underline{m}}H_{\mathrm{M2}}\right)\hat{\kappa} = 0, \tag{25.121}$$

whose solution is $\hat{\kappa} = H_{\mathrm{M2}}^{-\frac{1}{6}}\hat{\kappa}_0$, where $\hat{\kappa}_0$ is a constant spinor. The M2 Killing spinors are, therefore,

$$\boxed{\hat{\kappa} = H_{\mathrm{M2}}^{-\frac{1}{6}}\hat{\kappa}_0, \qquad \left(1 \mp i\hat{\Gamma}^{\hat{0}\hat{1}\hat{2}}\right)\hat{\kappa}_0 = 0.} \tag{25.122}$$

The constraint has the form Eq. (25.116) predicted by the supersymmetry algebra and therefore only half of the components of $\hat{\kappa}_0$ are independent and only half of the supersymmetries are unbroken. We have included the two possible signs of the M2 charge. They are irrelevant for a single brane but may be crucial in the presence of other branes.

Observe that the Killing spinor exists for *any* function $H_{\mathrm{M2}}$ (not necessarily harmonic!). The Killing spinor equations do not imply the equations of motion that restrict $H_{\mathrm{M2}}$ to be harmonic. On the other hand, strictly speaking, the solution is supersymmetric only if the Killing spinors have the correct asymptotic behavior: if the solution is asymptotically the vacuum, the Killing spinors have to approach the vacuum Killing spinors asymptotically. Furthermore, they have to be normalizable. These conditions are satisfied if $H_{\mathrm{M2}}$ is the harmonic function that describes parallel M2-branes. If $H_{\mathrm{M2}}$ corresponds to the $\mathrm{AdS}_4 \times \mathrm{S}^7$ solution, which is a vacuum solution itself, the asymptotic behavior is right by definition. Furthermore, the Killing spinor equation can be solved for spinors that do not satisfy the constraint by introducing dependence on the "worldvolume" coordinates and the solution is maximally supersymmetric. The group-theoretical methods explained in Chapter 17 are better suited for solving the equation.

*Unbroken supersymmetries of the* M5*-brane.* An entirely analogous calculation gives

$$\boxed{\hat{\kappa} = H_{\mathrm{M5}}^{-\frac{1}{12}}\hat{\kappa}_0, \qquad \left(1 \mp \hat{\Gamma}^{\hat{0}\hat{1}\hat{2}\hat{3}\hat{4}\hat{5}}\right)\hat{\kappa}_0 = 0.} \tag{25.123}$$

As in the M2 case, in the near-horizon limit (or, equivalently, on choosing the $H_{\mathrm{M5}}$ that describes the $\mathrm{AdS}_7 \times \mathrm{S}^4$ solution) the spinors that do not satisfy the constraint also solve the Killing spinor equation and the solution is maximally supersymmetric.

*Unbroken supersymmetries of KK monopoles.* The KK-monopole solution, in all theories, is the direct product of the $d = 4$ Euclidean Taub–NUT solution and $(d-4)$-dimensional

Minkowski spacetime and can be seen as a $(d-5)$-brane. In all cases, the Killing spinor equations reduce to

$$\nabla_{\underline{m}}\kappa = 0, \qquad m = (d-4), (d-3), (d-2), (d-1), \tag{25.124}$$

since the solution is trivial in the worldvolume directions. As discussed on p. 519, in the frame in which the spin connection is (anti-)self-dual, Eq. (13.68), the Killing spinors are constant spinors satisfying the constraint $(1 \pm \Gamma^{(d-4)(d-3)(d-2)(d-1)})\kappa = 0$, which can be rewritten in the form Eq. (25.116) with $p = d-5$ and $\mathcal{O}$ depending on the specific theory. For $N = 1, d = 11$, $N = 2A, d = 10$, and $N = 2B, d = 10$ (with self-dual RR 5-form), respectively, we have

$$\boxed{(1 \mp i\hat{\hat{\Gamma}}^{0123456})\hat{\hat{\kappa}} = 0, \qquad (1 \mp \hat{\Gamma}^{012345}\hat{\Gamma}_{11})\hat{\kappa} = 0, \qquad (1 \pm \hat{\Gamma}^{012345})\hat{\kappa} = 0.} \tag{25.125}$$

*Unbroken supersymmetries of the* D$p$*-branes.* Taking into account that the NSNS 3-form field strength is zero for these solutions, and that only the field strength $\hat{G}^{(p+2)}$ and its dual, whose combinations add up, is different from zero, the $N = 2A$ and $N = 2B$ Killing spinor equations can be written in the following unified way:

$$\begin{aligned}
\delta_{\hat{\kappa}}\hat{\psi}_{\hat{\mu}} &= \left\{\partial_{\hat{\mu}} - \tfrac{1}{4}\not{\hat{\omega}}_{\hat{\mu}} + \tfrac{1}{8}e^{\hat{\phi}}\frac{1}{(p+2)!}\,\not{\hat{G}}^{(p+2)}\hat{\Gamma}_{\hat{\mu}}\mathcal{O}_{\mathrm{D}p}\right\}\hat{\kappa}, \\
\delta_{\hat{\kappa}}\hat{\lambda} &= \left\{\not{\partial}\hat{\phi} - \tfrac{1}{4}e^{\hat{\phi}}\frac{p-3}{(p+2)!}\,\not{\hat{G}}^{(p+2)}\mathcal{O}_{\mathrm{D}p}\right\}\hat{\kappa},
\end{aligned} \tag{25.126}$$

where

$$\mathcal{O}_{\mathrm{D}p} = i(-\hat{\Gamma}_{11})^{\frac{p+2}{2}}, \quad p \text{ odd (IIA)}, \qquad \mathcal{O}_{\mathrm{D}p} = \mathcal{P}_{\frac{p+3}{2}}, \quad p \text{ even (IIB)}, \tag{25.127}$$

where $\mathcal{P}_n$ is defined in Eq. (23.11).

With this notation, the Killing spinors are given by

$$\boxed{\hat{\kappa} = H_{\mathrm{D}p}^{-\frac{1}{8}}\hat{\kappa}_0, \qquad (1 \mp \hat{\Gamma}^{01\cdots p}\mathcal{O}_{\mathrm{D}p})\hat{\kappa}_0 = 0.} \tag{25.128}$$

*Unbroken supersymmetries of the fundamental string.* Since, for this solution, the RR potentials vanish, we can write the Killing spinor equations for the $N = 2$ and $N = 1$ theories in the unified form

$$\begin{aligned}
\delta_{\hat{\kappa}}\hat{\psi}_{\hat{\mu}} &= \left\{\partial_{\hat{\mu}} - \tfrac{1}{4}\left(\not{\hat{\omega}}_{\hat{\mu}} + \tfrac{1}{2}\not{\hat{H}}_{\hat{\mu}}\mathcal{O}\right)\right\}\hat{\kappa} = 0, \\
\delta_{\hat{\kappa}}\hat{\lambda} &= \left\{\not{\partial}\hat{\phi} - (1/12)\not{\hat{H}}\mathcal{O}\right\}\hat{\kappa} = 0,
\end{aligned} \tag{25.129}$$

where $\hat{\kappa}$ is a Majorana spinor, a pair of Majorana–Weyl spinors, or a Majorana–Weyl spinor, and $\mathcal{O}_{\rm F1} = \hat{\Gamma}_{11}, \sigma^3,$ and $\mathbb{I}$, for the $N = 2A, 2B,$ and 1 theories, respectively. The solutions are given by

$$\boxed{\hat{\kappa} = H_{\rm F1}^{\frac{1}{4}}\kappa_0, \qquad \left(1 \pm \hat{\Gamma}^{01}\mathcal{O}_{\rm F1}\right)\hat{\kappa}_0 \quad = \quad 0.} \tag{25.130}$$

*Unbroken supersymmetries of the solitonic 5-brane.* The $N = 2A, 2B, 1, d = 10$ cases can also be treated in a unified way. The result is

$$\boxed{\hat{\kappa} = \hat{\kappa}_0, \qquad \left(1 \pm \hat{\Gamma}^{0\cdots 5}\mathcal{O}_{\rm S5}\right)\hat{\kappa}_0 \quad = \quad 0.} \tag{25.131}$$

where now $\mathcal{O}_{\rm S5} = \sigma^3$ for the $N = 2B$ theory and $\mathbb{I}$ in the other two cases.

Observe that, up to possible $\hat{\Gamma}_{11}$ factors, this is essentially the projector for a metric of SU(2) holonomy (see the discussion on p. 519) in the frame in which the spin connection is (anti-)self-dual (because the Killing spinor is constant). Actually, the Killing spinor equation for the S5 can be seen as the condition of covariant constancy of the spinor with respect to the torsion spin connection $\hat{\Omega}^{(\pm)}_{\hat{\mu}} = \hat{\omega}_{\hat{\mu}} \pm \hat{\not{H}}_{\hat{\mu}}$ in the proper subspaces of $\mathcal{O}$. The torsionful spin connection for the S5 is *identical* to that of the BPST instanton given in Section 13.2.2.

*T duality of the Killing spinors.* Several of the solutions whose Killing spinors we have calculated are related by T duality and so must be the Killing spinors themselves. The T-duality transformation rules for the Killing spinors are given in Eqs. (23.40), and here we are going to check them on D$p$-brane Killing spinors. Clearly, the $H$-dependent factor plays no role and we are going to focus on the projectors.

The D0-brane Killing spinor satisfies

$$\left(1 \pm i\hat{\Gamma}^0\hat{\Gamma}_{11}\right)\hat{\kappa}_{\rm D0} = 0 \Rightarrow \hat{\Gamma}_{11}\hat{\kappa}_{\rm D0} = \pm i\hat{\Gamma}^0\hat{\kappa}_{\rm D0}. \tag{25.132}$$

A T-duality transformation in the ninth direction gives the following two Majorana–Weyl Killing spinors of the $N = 2B$ theory:

$$\hat{\kappa}^1_{\rm D1} = -\tfrac{i}{2}\hat{\Gamma}^9(1 + \hat{\Gamma}_{11})\hat{\kappa}_{\rm D0}, \qquad \hat{\kappa}^2_{\rm D1} = \tfrac{1}{2}\hat{\Gamma}^9(1 - \hat{\Gamma}_{11})\hat{\kappa}_{\rm D0}. \tag{25.133}$$

Using the D0 constraint in Eq. (25.132), we can rewrite them in the form

$$\hat{\kappa}^1_{\rm D1} = \mp\hat{\Gamma}^{09}\tfrac{1}{2}(1 \mp i\hat{\Gamma}^0)\hat{\kappa}_{\rm D0}, \quad \hat{\kappa}^2_{\rm D1} = \tfrac{1}{2}(1 \mp i\hat{\Gamma}^0)\hat{\kappa}_{\rm D0} \Rightarrow (\mathbb{I} \pm \hat{\Gamma}^{09}\sigma^1)\hat{\kappa}_{\rm D1} = 0, \tag{25.134}$$

which is the constraint of the D1 Killing spinor. The dependence on $H$ is clearly correct.

Another T-duality transformation in the eighth direction gives, using the constraint of $\hat{\kappa}_{\rm D1}$,

$$\hat{\kappa}_{\rm D2} = \hat{\kappa}^2_{\rm D1} - i\hat{\Gamma}^8\hat{\kappa}^1_{\rm D1} \pm i\hat{\Gamma}^{098}\left(\hat{\kappa}^2_{\rm D1} - i\hat{\Gamma}^8\hat{\kappa}^1_{\rm D1}\right) \Rightarrow \left(1 \mp i\hat{\Gamma}^{098}\right)\hat{\kappa}_{\rm D2} = 0, \tag{25.135}$$

which is the D2-brane Killing spinor algebraic constraint.

*Maximally supersymmetric vacua of string and M theories.* We have mentioned that the $AdS_4 \times S^7$ and $AdS_7 \times S^4$ solutions of $d = 11$ supergravity Eqs. (25.54) and Eqs. (25.47) are maximally supersymmetric solutions and, therefore, vacua of the theory. The metrics of these spaces are products of those of symmetric spaces and the Killing spinors, and symmetry superalgebras can be constructed and studied using the methods of Chapter 17 (see Ref. [26]). The superalgebras are extended AdS superalgebras of the kind we studied in Section 5.4, written in 11-dimensional notation. These are also the superalgebras of the gauged SUEGRAs one obtains by compactification on $S^7$ and $S^4$.

These are not the only maximally supersymmetric solutions of $d = 11$ supergravity since we can always take the Penrose limit of any solution while preserving (or increasing) the number of supersymmetries [1028, 700, 222, 223]. The Penrose limits of the above two vacua give the same KG11 solution (first found in Ref. [874]), which has an Hpp-wave metric of the form Eq. (14.18) with

$$\hat{\hat{G}}_{\underline{ux^1\cdots x^3}} = \lambda, \qquad A_{ij} = \begin{cases} -\dfrac{1}{18}\lambda^2\delta_{ij} & i,j = 1,2,3, \\ -\dfrac{1}{72}\lambda^2\delta_{ij} & i,j = 4,\ldots,9. \end{cases} \tag{25.136}$$

The symmetry superalgebra of this solution was studied in Ref. [548]. It does not seem to be associated with any known supergravity. The same happens for the other KG solutions.

There are no more maximally supersymmetric vacua in $d = 11$ [874, 549]. Let us turn now to the ten-dimensional theories. In the $N = 2A, 1$ cases the only maximally supersymmetric vacuum is Minkowski spacetime [549]. In the $N = 2B$ theory there is, as we have seen, a maximally supersymmetric solution with the metric of $AdS_5 \times S^5$ Eq. (25.66). The superalgebra is that of gauged $N = 4, d = 5$ SUEGRA with gauge group $SO(6)$, but it is naturally written in ten-dimensional notation. The Penrose limit gives the maximally supersymmetric KG10 solution [221], which also has an Hpp-wave metric of the form Eq. (14.18) with

$$\hat{G}^{(5)}_{\underline{ux^1\cdots x^4}} = \hat{G}^{(5)}_{\underline{ux^5\cdots x^8}} = \lambda, \qquad A_{ij} = -\tfrac{1}{2}\delta_{ij}\lambda^2, \quad i,j, = 1,\ldots,8, \tag{25.137}$$

in our conventions.

There are no more maximally supersymmetric vacua in $d = 10$ [549], but there are other vacua with fewer supersymmetries, which are perhaps more interesting from a phenomenological point of view. We have mentioned some of them (those that can be obtained by replacing the spheres by other Einstein spaces). A complete classification is still lacking, but there is currently intense work in this direction (see, for instance, Refs. [807, 599]).

## 25.6 Intersections

Although now it may seem natural to look for solutions that represent simultaneously branes of different kinds in equilibrium, the first solutions of that kind were only identified in Ref. [1018] among general solutions found years before in Ref. [699]. After that, very many solutions were quickly constructed and the basic rules that govern their existence studied. Trying to review all these solutions and the various approaches in depth in the space available would be utterly hopeless. Thus, we will be pragmatic, focusing on the simplest families of solutions and the general rules. More information can be obtained from reviews such as those in Refs. [590, 1144, 813].

The solutions we have studied so far describe $p$-branes at rest, in their lower energy states in which none of their worldvolume fields is excited and has a non-trivial configuration. We have checked this by matching the solutions with $p$-brane sources in Sections 24.3.3 and 25.2.1. Excited worldvolume fields describe, first of all, deformations of the $p$-brane in the target spacetime when they involve the embedding coordinates $X^{\mu}(\xi)$ (as happens in all the supersymmetric cases). These deformations can be seen, in certain cases, as other branes that end on or intersect the original brane. The converse relation is always true: a brane that ends on or intersects another brane always corresponds to an excitation of the worldvolume fields of the latter. The dimension of the intersection, which behaves as a dynamical solitonic object in the worldvolume of the *host* brane, determines the nature of the intersecting brane. That dimension is associated with the rank of the excited worldvolume differential-form fields: $k$-brane intersections to $(k+1)$-form worldvolume fields. There are three main examples of this correspondence.

1. By definition, open strings end on D$p$-branes and their endpoints are seen as point-particles electrically charged with respect to the BI vector field (*BIons*). Worldvolume half-supersymmetric solutions of the D$p$-brane action in flat spacetime describing a point-like electric charge were found in Refs. [288, 610] (see also Refs. [886, 716]). There is always an excited embedding scalar that corresponds to a *spike* sticking out of the D$p$-brane. The energy of this solitonic worldvolume solution per unit length of the spike is precisely equal to the fundamental-string tension. Furthermore, perturbations along the spike have Dirichlet boundary conditions and one concludes that this solution represents a fundamental string attached to the D$p$-brane.

   Since these are supersymmetric worldvolume solutions, it is not surprising that there are solutions describing several parallel (or antiparallel) spikes in equilibrium.

   The BI vector field can be dualized into a $(p-2)$-form potential, which is another BI vector for the D3-brane [626, 627, 666]. The D3 electric–magnetic self-duality is related to type-IIB S duality, and the dual BIons turn out to describe D-strings ending on a D3-brane.[23] The dual $(p-3)$-BIons of other D$p$-branes are related to this one by T duality: a D$(p-2)$ ending on a D$p$ with a $(p-3)$-dimensional intersection associated with the $(p-2)$-form dual of the BI vector field.

   These intersections can be written in the form $\mathrm{F1} \perp \mathrm{D}p(0)$ and $\mathrm{D}(p{-}2) \perp \mathrm{D}p(p{-}3)$.

2. In Ref. [784] a supersymmetric worldvolume solution of the M5 equations of motion in flat spacetime [188, 787, 1021, 98, 15] in which two embedding scalars and the self-dual 2-form[24] were excited was found. It corresponds to the intersection

[23] If the D-string ended on $N$ coincident D3-branes, whose worldvolume field theory contains a non-Abelian $\mathrm{SU}(N)$ BI vector field (in fact, it is a non-linear generalization of $N=4$, $d=4$ super-Yang–Mills theory), the intersection would be seen as an $\mathrm{SU}(N)$ magnetic monopole in the worldvolume [666, 453].

[24] The bosonic worldvolume fields of the string/M-theory extended objects can be found in Table 25.4.

$\mathrm{M2} \perp \mathrm{M5}(1)$. The dimensional reduction along the intersection corresponds to $\mathrm{F1}A \perp \mathrm{D4}(0)$, which was discussed in example 1.

3. A solution describing the supersymmetric M5 worldvolume soliton associated with the intersection $\mathrm{M5} \perp \mathrm{M5}(3)$ was constructed in Ref. [785]. The worldvolume gauge field is here the dual of an embedding scalar. These are present in any $p$-brane with $p < d - 1$ and are $(p-1)$-forms. They indicate the possibility of two $p$-branes intersecting over a $(p-2)$-brane.

   Indeed, on T-dualizing the $\mathrm{D1} \perp \mathrm{D3}(0)$ in a direction parallel to the D3 and perpendicular to the D1, we find $\mathrm{D2} \perp \mathrm{D2}(0)$. T duality in directions transverse to both branes generates another sequence of possible intersections, $\mathrm{D}p \perp \mathrm{D}p(p-2)$.

   Had we dualized in a direction perpendicular to the D3 and parallel to the D1, we would have generated $\mathrm{D0} \perp \mathrm{D4}(0)$ and then further T dualities would have generated the sequence $\mathrm{D}p \perp \mathrm{D}(p+4)(p)$.

It is clear that we can go on generating new intersections via dualities. The results, in terms of supergravity solutions (including gravitational waves and KK monopoles [185, 914]), are summarized in Tables 25.5 and 25.6. Some of the intersections (named *overlaps* in Ref. [597]) cannot be associated with excited worldvolume fields. They arise, in fact, in degenerate limits of intersections involving more than two branes. For instance, the $\mathrm{M5} \perp \mathrm{M5}(1)$ intersection corresponds to an M2 ending on two M5s in a limit in which these become infinitely close and the M2 disappears.

As we have mentioned, these intersections, seen as excited worldvolume configurations (branes within branes), always preserve some supersymmetry. Actually, in general excitations,[25] no supersymmetry would be preserved, but we are interested in the cases in which some supersymmetry is preserved, in part because they are easier to deal with and we can also expect to find a classical (supersymmetric) solution associated with that brane configuration. Worldvolume supersymmetry and the worldvolume superalgebras have been used to study the possible intersections [158, 592].

What is the spacetime version of these worldvolume arguments? As we have seen, the low-energy effective actions are very powerful tools with which to study extended objects that arise as classical solutions. Supersymmetric solutions describing single branes can typically be related to elementary brane sources and we expect the same to be true for intersecting brane sources, although not many results have been obtained in this direction for specific solutions. However, one can use general arguments based on the field equations (charge conservation [1158, 1191, 65]) and spacetime supersymmetry to determine which intersections are allowed. Then, one can try to find the corresponding supergravity solutions.

Let us review these arguments.

[25] There are non-supersymmetric BIon solutions with no scalars excited, and, therefore, such solutions are not associated with deformations of the worldvolume [610].

Table 25.4. Bosonic worldvolume fields of string/M-theory branes: $S$ and $T$ are worldvolume scalars, $V_i$ is a worldvolume vector, and $V^+_{ij}$ is a 2-form with self-dual field strength

| Object | Worldvolume dimension | Worldvolume fields |
|---|---|---|
| M2 | $2+1$ | $X^\mu$ |
| M5 | $5+1$ | $X^\mu, V^+_{ij}$ |
| KK7M | $6+1$ | $X^\mu, V_i$ |
| S5A | $5+1$ | $X^\mu, V^+_{ij}, S$ |
| KK6B | $5+1$ | $X^\mu, V^+_{ij}, S, T$ |
| S5B | $5+1$ | $X^\mu, V_i$ |
| KK6A | $5+1$ | $X^\mu, V_i, S$ |
| D$p$ | $p+1$ | $X^\mu, V_i$ |

Table 25.5. Elementary intersections of ten-dimensional extended objects

| |
|---|
| F1 $\parallel$ S5, F1 $\perp$ D$p(0)$, |
| S5 $\perp$ S5$(1)$, S5 $\perp$ S5$(3)$, S5 $\perp$ D$p(p-1)$ $(p>1)$, |
| D$p$ $\perp$ D$p'(m)$, $p+p'=4+2m$, |
| W $\parallel$ F1, W $\parallel$ S5, W $\parallel$ D$p$, |
| KK6 $\perp$ D$p(p-2)$ |

Table 25.6. Elementary intersections of 11-dimensional extended objects

| |
|---|
| M2 $\perp$ M2$(0)$, M2 $\perp$ M5$(1)$, M5 $\perp$ M5$(1)$, M5 $\perp$ M5$(3)$, |
| W $\parallel$ M2, W $\parallel$ M5, |
| KK7M $\parallel$ M2, KK7M $\perp$ M2$(0)$, KK7M $\parallel$ M5, KK7M $\perp$ M5$(1)$, KK7M $\perp$ M5$(3)$, |
| W $\parallel$ KK, W $\perp$ KK7M$(2)$, W $\perp$ KK7M$(4)$ |

### 25.6.1 Brane-charge conservation and brane surgery

Following Ref. [1191], let us consider the charge carried by an F1B solution. In the absence of any other object (i.e. with only the $\hat{\mathcal{B}}$ potential excited), the charge is given by the first of Eqs. (25.101) and it is different from zero only if the string has no free endpoints at a finite distance; otherwise, we could slide the $S^7$, on which we integrate along the string beyond its endpoint, and contract it to a point[26] without encountering any singularity (source). This is because (this is just the $\hat{\mathcal{B}}$ equation of motion)

$$d\left(e^{-2\hat{\varphi}} \star \hat{\mathcal{H}}\right) = 0 \tag{25.138}$$

outside the string. In the presence of other fields, the equation of motion has additional terms, and the homotopy-invariant definition of charge is

$$q_{\rm F1} \sim \int_{S^7} \left(e^{-2\hat{\varphi}} \star \hat{\mathcal{H}} - \star\hat{G}^{(3)}\hat{C}^{(0)} - \hat{G}^{(5)}\hat{C}^{(2)}\right), \tag{25.139}$$

where $S^7$ surrounds the string. Let us consider a semi-infinite string. At a large enough distance $L$ from the endpoint, boundary effects are not important and the charge is still approximately given by the first of Eqs. (25.101). The larger $L$ is for a fixed value of the $S^7$ radius $R_7$, the better the approximation. Closer to the endpoint, the additional terms must contribute (otherwise, we are back to the previous case), but we can obtain the same value for the integral by making $R_7 \to 0$, keeping $R_7/L$ constant, until the only contribution to the integral comes from the endpoint. The degenerate $S^7$ can be decomposed, for convenience, into the product $S^5 \times S^2$, if we assume that the contribution to $q_{\rm F1B}$ comes from the last term in the integral in Eq. (25.139). The integral decomposes into a product of integrals,

$$\int_{S^5} \hat{G}^{(5)} \int_{S^2} \hat{C}^{(2)}. \tag{25.140}$$

The first integral gives the D3-brane charge (assuming, as we are doing here, that $\hat{\mathcal{H}}$ does not contribute), $\hat{G}^{(5)} = \star\hat{G}^{(5)}$, and thus the string endpoint must be at a D3-brane. If there is no D1-brane present, then $\hat{G}^{(3)} \sim d\hat{C}^{(2)} = 0$ inside the D3-brane and, *locally*, $\hat{C}^{(2)} = dV$, where $V$ is a vector that lives in the D3-brane worldvolume. Then

$$q_{\rm F1B} \sim \int_{S^2} dV. \tag{25.141}$$

The interpretation is clear: an F1B can end on a D3-brane, and at the intersection point there is an excited worldvolume vector field (the dual BI vector field) whose magnetic charge is proportional to the F1B charge. This is the same result as we obtained before.

[26] This can be visualized best in $d = 4$, in which $S^7$ is replaced by $S^1$.

A similar reasoning indicates that, if it is the second term that contributes to the charge integral, the F1B can also end on a D-string and at the intersection the BI vector field is excited so its dual field strength is a constant.[27]

These arguments that determine the opening of branes seem to depend on field redefinitions (the charge integrand is defined up to total derivatives). However, the different expressions for the charge are just choices that are more or less adequate to describe a given physical situation. The most symmetric expression for $q_{\mathrm{F1B}}$ can be obtained by using Eq. (23.9). Each of the four possible terms corresponds to the F1B ending on one of the four D$p$-branes $p = 1, 3, 5, \text{ and } 7$ and exciting the dual BI field magnetically.

### 25.6.2 Marginally bound supersymmetric states and intersections

We are considering only supersymmetric brane intersections, in which the branes that intersect do not interact and are in supersymmetric equilibrium. These intersections can be considered as bound states with zero binding energy (or *marginally bound* states), and their existence depends on whether it is possible to impose the simultaneous annihilation of that state by the supercharges that annihilate those associated with each individual brane.

As we have seen, the annihilation of a $p$-brane state by a given set of supercharges is entirely equivalent to the action of a projector $P_p$ of the generic form Eq. (25.116) on a spinor, $P_p \epsilon = 0$. Then, the existence of a supersymmetric state composed of a $p$-brane and a $p'$-brane depends on the compatibility of the respective projectors $P_p$ and $P_{p'}$: it will exist if

$$[P_p, P_{p'}] = 0, \tag{25.142}$$

and the state will preserve a quarter of the supersymmetries.

This equation depends on $p$ and $p'$ but also on the spatial orientation of the branes. A general analysis is complicated because of the different $\mathcal{O}_p$ that occur in the projectors. Let us consider a simple example first: two $p$-branes of the same kind, S5A for simplicity, extended along five Cartesian coordinates (so they are either parallel or orthogonal). It is relatively easy to see that the two associated projectors commute if the number of relative transverse dimensions (those which are parallel to one brane and transverse to the other) is $0 \bmod 4$, which leads to the allowed (supersymmetric) intersections $\mathrm{S5} \perp \mathrm{S5}(3)$ and $\mathrm{S5} \perp \mathrm{S5}(1)$, which are included in Table 25.5. For D$p$-branes $\mathcal{O}_p = i\mathbb{I}, i\Gamma_{11}, \sigma^1, i\sigma^2$ depend on $p \bmod 4$ and the analysis of intersections between D$p$-branes gives the allowed intersection $\mathrm{D}p \perp \mathrm{D}(p+4)(p)$ and, with a little more effort, the other cases in the table [64, 480].

This analysis, which is essentially based on the spacetime supersymmetry algebra, allows the study of more complicated intersections involving more branes [184] or non-orthogonal intersections (*branes at angles* [197, 251, 169, 1195, 1134]). The inclusion of another brane is allowed if its associated projector commutes with the other ones. The amount of unbroken supersymmetry is generically halved each time a brane is included, except in the case in which the projector of the additional brane does not impose any new constraint

[27] This is similar to viewing the mass parameter of Romans' $N = 2A, d = 10$ SUEGRA as the dual of the RR 10-form field strength.

on the spinor. The canonical example is that of a D5 in the directions 12345 and an S5B in the directions 12678, so they intersect in two directions. Their associated projectors $P_{\rm D5}(12345)$ and $P_{\rm S5B}(12679)$ (in the obvious notation) commute, and

$$P_{\rm D5}(12345)P_{\rm S5B}(12679)\epsilon \sim P_{\rm D3}(129)\epsilon = 0, \tag{25.143}$$

and, thus, including a D3 in the directions 129 does not break any additional supersymmetry (for any sign of the charge). This property gives rise to the phenomenon of D3-brane creation when a D5 and an S5B cross [708, 505, 419, 144, 79]. Adding a fourth brane may break all or no additional supersymmetry, depending on the sign of its charge (Section 26.1).

Another case in which no additional constraint is imposed is when we add a brane of the same kind, but rotated by a supersymmetry-preserving angle ("branes at angles") [197]. We have no space to review this important case and we refer the reader to the literature.

### *25.6.3 Intersecting-brane solutions*

The worldvolume and spacetime arguments that we have reviewed so far suggest that classical solutions of the low-energy effective string/M theories describing intersecting branes should exist. In this section we are going to study the simplest intersecting-brane solutions. These (like most of the solutions that have been found so far) are actually "imperfect" and represent intersecting branes (even in the cases in which we expect a brane ending on another brane) that are partially delocalized, smeared along the relative transverse directions. These intersecting-brane solutions have to be understood, then, as approximations to the true field configurations, but are, nevertheless, worth studying.

The construction of these solutions is surprisingly simple using the *harmonic-superposition rule* [1203, 597]. This rule can be used for marginally bound systems of supersymmetric (extreme) branes when the number of overall transverse dimensions is finite (although it gives asymptotically flat solutions only when it is $\geq 3$) and when the Chern–Simons terms in the field strengths do not contribute to the equations of motion. This rule gives an Ansatz for the metric, $(p+1)$-form potentials, and dilaton (in $d = 10$) that is based on the forms of the solutions that describe each of the extreme branes independently, namely Eqs. (25.45), (25.50), (25.56), (25.64), and (25.60). Of course, the solutions constructed using this rule have to be checked directly in the equations of motion. For all combinations of two branes this can be done using a generalization of the $p$-brane $a$-model that we will briefly study in Section 25.6.4 (see Refs. [63, 61] for further generalizations to more branes and non-extremal branes).

Basically, the harmonic-superposition rule says that the metric is diagonal and each component consists in the same factors as each of the individual solutions smeared over the relative transverse directions,[28] multiplied. The same is true for the dilaton. Finally, the differential-form potentials are the sum of those of each individual solution.

Let us use this rule to construct the solution describing the intersection of a F1 in the direction $y$ with a D$p$-brane extended in the orthogonal directions $\vec{z}_p \equiv (z^1, \ldots, z^p)$.

[28] That is, all the harmonic functions depend only on the same overall transverse directions.

By combining Eqs. (25.56) and (25.64), smearing $H_{\mathrm{F1}}$ along $\vec{z}_p$ and $H_{\mathrm{D}p}$ along $y$, we obtain

$$
\begin{array}{l}
d\hat{s}_{\mathrm{s}}^2 = H_{\mathrm{D}p}^{-\frac{1}{2}} H_{\mathrm{F1}}^{-1} dt^2 - H_{\mathrm{D}p}^{+\frac{1}{2}} H_{\mathrm{F1}}^{-1} dy^2 - H_{\mathrm{D}p}^{-\frac{1}{2}} d\vec{z}_p^{\,2} - H_{\mathrm{D}p}^{+\frac{1}{2}} d\vec{x}_{8-p}^{\,2}, \\
e^{-2\hat{\phi}} = e^{-2\hat{\phi}_0} H_{\mathrm{D}p}^{\frac{p-3}{2}} H_{\mathrm{F1}}, \qquad \hat{C}^{(p+1)}{}_{t\underline{z}^1\cdots\underline{z}^p} = \pm e^{-\hat{\phi}_0}\left(H_{\mathrm{D}p}^{-1} - 1\right), \\
\hat{B}_{t\underline{y}} = \pm\left(H_{\mathrm{F1}}^{-1} - 1\right), \qquad H_{\mathrm{D}p,\mathrm{F1}} = 1 + \dfrac{h_{\mathrm{D}p,\mathrm{F1}}}{|\vec{x}_{8-p}|^{6-p}}.
\end{array}
\tag{25.144}
$$

It is straightforward to apply the rule to other cases and we will do this in Chapter 26, where more examples can be found. The metrics of some basic intersections can be found, for instance, in Refs. [590, 1144, 813].

The main insufficiency of these solutions is the delocalization of the branes. For instance, the above solution does not tell us at which point along the coordinate $y$ and in the hyperplane $\vec{z}_p$ the string intersects the D$p$-brane. This is immaterial if at the end we want to wrap the solution in those directions and perform dimensional reduction of the solution, as we will do in Chapter 26 to construct $d = 4, 5$ BHs. Nevertheless, since we have a clear worldvolume picture of supersymmetric intersections, it is natural to try to find the solutions that would have BIons and their generalizations as sources (this was the approach of Ref. [641]). Partially localized solutions and some special fully localized solutions have been found in Refs. [717, 60, 823, 1165, 519, 898, 1283, 777], but in Refs. [931, 1024] it was argued, using AdS/CFT-correspondence arguments, that fully localized solutions might not exist in general, and these arguments seem to be confirmed by the results in Ref. [641]. If the string in the F1 $\perp$ D$p(0)$ intersection can be seen as "hair" on the D$p$-brane, then the absence of a fully localized solution for that configuration can be seen as a sort of "no-hair theorem" for D$p$-brane solutions.

Nevertheless, an ansatz for fully localized intersections of this and other kinds has been given in Ref. [1066]. The solutions depend on an unknown function that satisfies a highly non-linear differential equation, but it is not known whether this equation has solutions with the appropriate boundary conditions.

### 25.6.4 The $(a_1$–$a_2)$-model for $p_1$- and $p_2$-branes and black intersecting branes

This is a straightforward generalization of the $p$-brane $a$-model that includes $(p_1 + 1)$- and $(p_2 + 1)$-form potentials, coupled to a scalar with parameters $a_1$ and $a_2$:

$$
S = \frac{1}{16\pi G_{\mathrm{N}}^{(d)}} \int d^d x \sqrt{|g|} \left[ R + 2(\partial\varphi)^2 + \sum_{i=1,2} \frac{(-1)^{p_i+1}}{2\cdot(p_i+2)!} e^{-2a_i\varphi} F_{(p_i+2)}^2 \right], \tag{25.145}
$$

and it is just a convenient simplification of the higher-dimensional supergravity actions we are dealing with. Note, in particular, the absence of Chern–Simons terms: the solutions we will obtain will be solutions of the full supergravity action only when those terms do not contribute to the equations of motion. This condition will be fulfilled in most cases.

The equations of motion corresponding to this action are

$$
\begin{aligned}
G_{\mu\nu} + 2T^{\varphi}_{\mu\nu} + \sum_{i=1,2} \frac{(-1)^{p_i+1}}{2\cdot(p_i+1)!} e^{-2a_i\varphi} T^{A_{(p_i+1)}}_{\mu\nu} &= 0, \\
\nabla^2\varphi + \sum_{i=1,2} \frac{(-1)^{p_i+1}}{4\cdot(p_i+2)!} a_i e^{-2a_i\varphi} F^2_{(p_i+2)} &= 0, \\
\nabla_\mu (e^{-2a_i\varphi} F_{(p_i+2)}{}^{\mu\nu_1\ldots\nu_{p_i+1}}) &= 0, \quad i=1,2.
\end{aligned}
\tag{25.146}
$$

The harmonic-superposition rule and our experience indicate that an adequate ansatz for a $p_1$- and a $p_2$-brane intersecting over $r$ spatial directions may depend on three functions: $H_i$, $i=1,2$, which will be independent harmonic functions in the extreme limit and are associated with the potentials, and the "Schwarzschild factor" $W$, which becomes 1 in the extreme limit. The ansatz must be such that, when a given $H_i$ is set to 1 to recover a solution for a single $p_j$-brane, $i\neq j$. All these conditions are fulfilled by the metric ansatz

$$
\begin{aligned}
ds^2 = {} & H_1^{2z_1} H_2^{2z_2}\left[W dt^2 - d\vec{y}_r^{\,2}\right] - H_1^{2z_1} H_2^{-2y_2} d\vec{y}^{\,2}_{(p_1-r)} - H_1^{-2y_1} H_2^{2z_2} d\vec{y}^{\,2}_{(p_2-r)} \\
& - H_1^{-2y_1} H_2^{-2y_2}\left[d\vec{y}_q^{\,2} + W^{-1} d\rho^2 + \rho^2 d\Omega^2_{(\delta-2)}\right].
\end{aligned}
\tag{25.147}
$$

The coordinates $\vec{y}_r = (y_1^1,\ldots,y_1^r)$ (plus, of course, time) correspond to the common directions of the two branes relative to the worldvolume of the intersection, and the solution is assumed to be independent of them. The coordinates $\vec{y}_{(p_1-r)} = (y_1^2,\ldots,y_2^{(p_1-r)})$ and $\vec{y}_{(p_2-r)} = (y_1^3,\ldots,y_3^{(p_2-r)})$ are relative transverse coordinates (to the $p_1$- and $p_2$-branes, respectively). For simplicity, we will also assume the solution to be independent of them (i.e. it will be delocalized). The solution may depend only on the overall transverse coordinates (the rest), but, as we did for single-brane solutions, we include, for completeness, $q$ additional isometries, and the solution will not depend on $\vec{y}_q$. Finally, the parameters $z_1, z_2, y_1,$ and $y_2$ are determined by the single-brane solutions of the $a$-model, Eq. (24.64).

The dilaton is assumed to be a certain product of powers of $H_1$ and $H_2$. Finally, the ansatz for the potentials is the usual one, and we need only take into account that the $p_i$-brane lies in the directions $\vec{y}_r$ and $\vec{y}_{(p_i-r)}$:

$$
A_{(p_i+1)1_1\cdots r_1 1_2\cdots(p_i-r)_2} = \alpha_i (H_i^{-1} - 1), \qquad i=1,2.
\tag{25.148}
$$

If this ansatz is to work, then, by insisting on the independence of the two would-be harmonic functions $H_1$ and $H_2$, we should simply acquire constraints on $\omega, r, a_1,$ and $a_2$. On plugging the ansatz into the equations of motion, we find, after a long and boring calculation, that it does indeed lead to solutions under a few conditions on those constants:

$$
\begin{aligned}
ds^2 = {} & \left(e^{-2a_1\varphi_1} H_1^{-2}\right)^{\frac{1}{p_1+1}} \left(e^{-2a_2\varphi_2} H_2^{-2}\right)^{\frac{1}{p_2+1}} \left[W dt^2 - d\vec{y}_r^{\,2}\right] \\
& - \left(e^{-2a_1\varphi_1} H_1^{-2}\right)^{\frac{1}{p_1+1}} \left(e^{-2a_2\varphi_2} H_2^{-2}\right)^{-\frac{1}{\tilde{p}_2+1}} d\vec{y}^{\,2}_{(p_1-r)} \\
& - \left(e^{-2a_1\varphi_1} H_1^{-2}\right)^{-\frac{1}{\tilde{p}_1+1}} \left(e^{-2a_2\varphi_2} H_2^{-2}\right)^{\frac{1}{p_2+1}} d\vec{y}^{\,2}_{(p_2-r)} \\
& - \left(e^{-2a_1\varphi_1} H_1^{-2}\right)^{-\frac{1}{\tilde{p}_1+1}} \left(e^{-2a_2\varphi_2} H_2^{-2}\right)^{-\frac{1}{\tilde{p}_2+1}} \\
& \left[d\vec{y}_q^{\,2} + W^{-1} d\rho^2 + \rho^2 d\Omega^2_{(\delta-2)}\right], \\
& A_{(p_i+1)1_1\cdots r_1 1_2 \cdots (p_i-r)_2} = \alpha_1 (H_1^{-1} - 1), \\
e^{-2a_i\varphi_i} \equiv H_i^{2x_i}, & \qquad e^{-2a_i\varphi} = e^{-2a_i\varphi_i}\left(e^{-2a_j\varphi_j}\right)^{\frac{l_i}{x_j}}, \qquad i \neq j, \\
H_i = 1 + \frac{h_i}{\rho^{\delta-3}}, & \qquad W = 1 + \frac{\omega}{\rho^{\delta-3}}, \\
\omega = h_i \left[1 - \frac{a_i^2}{4x_i}\alpha_i^2\right], & \qquad l_i = (x_i - 1)\left[c_i(r+1) - \frac{p_j+1}{\tilde{p}_i+1}\right], \qquad i \neq j, \\
x_i = \frac{(a_i^2/2)c_i}{1 + (a_i^2/2)c_i}, & \qquad c_i = \frac{(p_i+1) + (\tilde{p}_i+1)}{(p_i+1)(\tilde{p}_i+1)}, \\
a_1 a_2 = -2(r - r_0), & \qquad r_0 = \frac{(p_1+1)(p_2+1)}{d-2} - 1.
\end{aligned}
\tag{25.149}
$$

This solution generalizes to the non-extreme regime extreme intersecting solutions obtained in Ref. [62] in another $(a_1$–$a_2)$ model (see also Refs. [1223, 63] and references therein). Some of these generalizations had already been obtained in certain cases in Refs. [401, 374, 375]. As usual, in the extremal limit $W = 1$, the $H_i$ are arbitrary independent harmonic functions of the overall transverse coordinates.

The most interesting relation that we obtain is the one in the last line, that among the $a_i$s, the dimensionality of the branes, and $r$:

$$
r = \frac{(p_1+1)(p_2+1)}{d-2} - \frac{a_1 a_2}{d-2} - 1. \tag{25.150}
$$

This equation contains the intersection rules and we can apply it to some basic examples, using the values for the $a_i$ constants appropriate for each kind of brane (see p. 735).

As an example, let us consider the case of $d = 11$ SUGRA.[29] Since there is no scalar, $a_1 = a_2 = 0$, which implies $r = r_0$. Equation (25.150) immediately gives the three

[29] Examples in $d = 10$ can be found in Chapter 26, where they are used to construct BH solutions: the black D1 || D5 is given in Eqs. (26.13) and the black D2 || S5 || D6 is given in Eqs. (26.36).

intersections $M2 \perp M2(0)$, $M2 \perp M5(1)$, and $M5 \perp M5(3)$ (but not the overlap $M5 \perp M5(1)$, which requires a different ansatz, see e.g. Ref. [590]). For example, the solution corresponding to a black intersection $M2 \perp M2(0)$ in which brane 1 lies in the directions $\vec{y}_2 = (y^1, y^2)$ and brane 2 lies in the directions $\vec{z}_2 = (z^1, z^2)$ is given by

$$
\boxed{
\begin{aligned}
d\hat{\hat{s}}^2 &= H_1^{-\frac{2}{3}} H_2^{-\frac{2}{3}} W dt^2 - H_1^{-\frac{2}{3}} H_2^{\frac{1}{3}} d\vec{y}_2^{\,2} - H_1^{\frac{1}{3}} H_2^{-\frac{2}{3}} d\vec{z}_2^{\,2} \\
&\quad - H_1^{\frac{1}{3}} H_2^{\frac{1}{3}} \left[ d\vec{w}_q^{\,2} + W^{-1} d\rho^2 + \rho^2 d\Omega^2_{(5-q)} \right], \\
\hat{\hat{C}}_{t\underline{y}^1\underline{y}^2} &= \alpha_1 (H_1^{-1} - 1), \qquad \hat{\hat{C}}_{t\underline{z}^1\underline{z}^2} = \alpha_2 (H_2^{-1} - 1), \\
H_i &= 1 + \frac{h_i}{\rho^{4-q}}, \qquad W = 1 + \frac{\omega}{\rho^{4-q}}, \qquad \omega = h_i \left[1 - \alpha_i^2\right].
\end{aligned}
}
\tag{25.151}
$$

On reducing in a relative transverse dimension, we obtain a black intersecting solution, $F1A \perp D2(0)$. On reducing in one of the extra isometric transverse directions $\vec{w}_q$, we obtain $D2 \perp D2(0)$ (this is the reason why the extra isometric $\vec{w}_q$s are introduced into the ansatz).

The solution corresponding to a black intersection $M2 \perp M5(1)$ in which the M2 lies in the directions $\vec{y}_2 = (y^1, y^2)$ and the M5 in $y^1, \vec{z}_4 = (z^1, \ldots, z^4)$ is

$$
\boxed{
\begin{aligned}
d\hat{\hat{s}}^2 &= H_1^{-\frac{2}{3}} H_2^{-\frac{1}{3}} [W dt^2 - (dy^1)^2] - H_1^{-\frac{2}{3}} H_2^{\frac{2}{3}} (dy^2)^2 - H_1^{\frac{1}{3}} H_2^{-\frac{2}{3}} d\vec{z}_4^{\,2} \\
&\quad - H_1^{\frac{1}{3}} H_2^{\frac{2}{3}} \left[ W^{-1} d\rho^2 + \rho^2 d\Omega^2_{(2)} \right], \\
\hat{\hat{C}}_{t\underline{y}^1\underline{y}^2} &= \alpha_1 (H_1^{-1} - 1), \qquad \hat{\tilde{\hat{C}}}_{t\underline{y}^1\underline{z}^1\cdots\underline{z}^4} = \alpha_2 (H_2^{-1} - 1), \\
H_i &= 1 + \frac{h_i}{\rho}, \qquad W = 1 + \frac{\omega}{\rho}, \qquad \omega = h_i \left[1 - \alpha_i^2\right].
\end{aligned}
}
\tag{25.152}
$$

# 26
# String black holes in four and five dimensions

Following our general plan, in Chapter 25 we have started to see classical solutions that describe the long-range fields generated by configurations of extended objects in string/M theory. In general, the solutions do not reflect some of the characteristics of the brane configuration which may be understood as "hair," but in many cases of interest (in general, in the presence of unbroken supersymmetry), given a classical supergravity solution, we can tell which brane configurations give rise to it. This is in itself a very interesting development, but there is more, because, if the brane configurations only involve D-branes, they can be associated with two-dimensional CFTs (string theories) over which we have good control. Furthermore, each of the branes considered here (D- or not D-) has a worldvolume supersymmetric field theory associated with it. All this allows us to relate supergravity configurations to QFTs whose degrees of freedom can be understood as the microscopical degrees of freedom of the quantum (super)gravity theory contained in string/M theory. This is, roughly speaking, the basis of the AdS/CFT correspondence and generalizations [921, 234] and also the basis for the microscopical computations of BH entropies [1160], the subject of this chapter.

In this chapter we are going to present $N = 2A/B, d = 10$ SUEGRA solutions associated with configurations of extended objects of type-II superstring theories that lead to BH solutions of maximal $d = 5, 4$ SUEGRAs ($N = 4, d = 5$ and $N = 8, d = 4$) (Section 26.2) upon toroidal compactification.[1] The association can be understood as a strong–weak-coupling limit (see Fig. 26.1). We will carefully relate the solutions' integration

[1] The compactification of the $N = 1, d = 11$ theory on Calabi–Yau 3-folds ($\mathrm{CY}_3$) gives $N = 1, d = 5$ supergravities (reviewed in Chapter 9) whose matter content and defining tensor $C_{IJK}$ are related to certain topological invariants of the $\mathrm{CY}_3$ as shown in Ref. [274]. The compactification of the $N = 2A/B, d = 10$ SUEGRAs on $\mathrm{CY}_3$ gives the $N = 2, d = 4$ SUEGRAs reviewed in Chapter 7 with analogous properties. For more details, see, for instance, Refs. [111, 564]. Therefore, the BH solutions of certain $N = 1, d = 5$ and $N = 2, d = 4$ theories are also superstring or M-theory BHs although their uplift to ten or 11 dimensions is, in general, more difficult to describe and we will not consider it here. (Some $N = 1, d = 5$ and $N = 2, d = 4$ theories arise as truncations of the toroidally compactified ten- and 11-dimensional theories, and, in those cases, the uplift is simpler.) The supersymmetric BH solutions of those theories have been reviewed in Chapter 19 profiting from the general classifications of supersymmetric solutions made in Chapter 18. We will study them together with the non-supersymmetric ones using the FGK formalism in Chapter 27.

constants to the physical parameters of the stringy sources and then, using our knowledge of the QFTs associated with those sources in the extreme and supersymmetric cases, we will count the states of these QFTs at each energy level, and the corresponding entropy will be shown to coincide with one quarter of the area of the BH's horizon (Section 26.3).

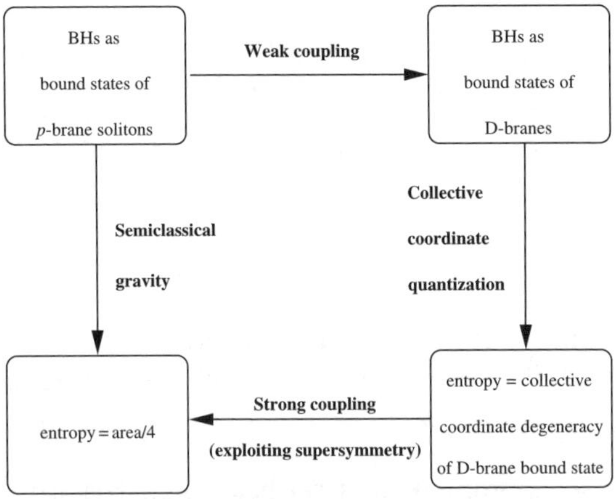


Fig. 26.1. The logic behind the string-theory calculation of extreme BH entropies is represented in this diagram [92].

Although it is quite self-contained, this chapter is far from complete due to lack of space and also to the immense amount of literature on this subject. Fortunately, there are some good reviews such as [1023, 423] and also Maldacena's Ph.D. thesis [919]. Other reviews that are interesting for their emphasis on particular aspects of the problem or as sources of bibliography are Refs. [764, 765, 920, 1022, 71, 1140, 76, 955, 1225, 971, 45].

Before we study how to construct BHs as composed of intersecting systems of branes, it is interesting to review how the idea of compositeness of BHs came about by studying a $d = 4$ model whose BH-type solutions are related to the dilaton BHs studied in Section 16.1.1.

## 26.1 Composite dilaton black holes

Let us consider the following string-inspired model (which is actually an inconsistent truncation of the heterotic- and, simultaneously, of the type II-string effective action compactified on $T^6$, so not all its solutions will be string solutions):[2]

$$\boxed{\begin{aligned} S = \int d^4x\,\sqrt{|g|}\,\{R + 2\left[(\partial\phi)^2 + (\partial\sigma)^2 + (\partial\rho)^2\right] - \tfrac{1}{4}e^{-2\phi}\big[e^{-2(\sigma+\rho)}(F^{(1)\,1})^2 \\ + e^{-2(\sigma-\rho)}(F^{(1)\,2})^2 + e^{2(\sigma+\rho)}(F^{(2)}{}_1)^2 + e^{2(\sigma-\rho)}(F^{(2)}{}_2)^2\big]\}\,. \end{aligned}}$$

Extreme BH-type solutions of this model (which are actually heterotic- (or type-II) string solutions) were found in Ref. [404], discussed further in Refs. [402, 403], and later

[2] The string effective actions can be consistently truncated to the $N = 2, d = 4$ STU model studied on p. 234. The inconsistent truncation that gives the model that we are going to study is explained on p. 578.

rediscovered and reinterpreted in Ref. [1065]. They can be written as follows:

$$
\boxed{
\begin{aligned}
ds^2 &= (H^{(1)\,1}H^{(1)\,2}H^{(2)}{}_1H^{(2)}{}_2)^{-\frac{1}{2}}dt^2 - (H^{(1)\,1}H^{(1)\,2}H^{(2)}{}_1H^{(2)}{}_2)^{-\frac{1}{2}}d\vec{x}_3^{\,2},\\
A^{(1)\,m}{}_t &= \alpha^{(1)\,m}(H^{(1)\,m}-1)^{-1}, \quad (\alpha^{(1)\,m})^2=1, \quad m=1,2,\\
\tilde{A}^{(2)}{}_{m\,t} &= \alpha^{(2)}{}_m(H^{(2)}{}_m-1)^{-1}, \quad (\alpha^{(2)}{}_m)^2=1, \quad m=1,2,\\
e^{-4\phi} &= \frac{H^{(1)\,1}H^{(2)}{}_1}{H^{(1)\,2}H^{(2)}{}_2}, \qquad e^{-4\sigma}=\frac{H^{(1)\,1}H^{(2)}{}_2}{H^{(1)\,2}H^{(2)}{}_1}, \qquad e^{-4\rho}=\frac{H^{(1)\,1}H^{(1)\,2}}{H^{(2)}{}_1H^{(2)}{}_2},
\end{aligned}
}
\tag{26.1}
$$

where the $H^{(i)}{}_m$s are independent harmonic functions in three-dimensional Euclidean space and the potentials $\tilde{A}^{(2)}{}_m$ correspond to the dual field strengths

$$\tilde{F}^{(2)}{}_1 = e^{-2(\phi+\sigma-\rho)} \star F^{(2)}{}_1, \qquad \tilde{F}^{(2)}{}_2 = e^{-2(\phi-\sigma+\rho)} \star F^{(2)}{}_2. \tag{26.2}$$

The harmonic functions appropriate to describe a single BH are

$$H^{(i)}{}_m = 1 + \frac{|q^{(i)}{}_m|}{|\vec{x}_3|}. \tag{26.3}$$

The $q^{(1)}{}_m$s are electric charges and the $q^{(2)}{}_m$s are magnetic charges. Their signs are given by the $\alpha^{(i)}{}_m$ constants. The signs in the harmonic functions are chosen in order to have a regular metric. The ADM mass and horizon are ($G_{\rm N}^{(4)}=1$)

$$M = \tfrac{1}{4}\sum_{i,m=1,2}|q^{(i)}{}_m|, \qquad A = 4\pi\sqrt{\prod_{i,m=1,2}|q^{(i)}{}_m|}. \tag{26.4}$$

The area is non-zero (and the horizon is regular) only when the four charges are finite.

The metrics can be related to those of the extreme $a$-model dilaton BHs of Section 16.1.1: when only one charge is different from zero, the metric is that of the extreme $a=\sqrt{3}$ BH, Eq. (16.22). If there are two non-vanishing charges that are equal, the metric is that of the extreme $a=1$ BH, Eq. (16.23). Three identical non-vanishing charges give the metric of the extreme $a=1/\sqrt{3}$ BH, Eq. (16.24); and four identical non-vanishing charges give the metric of the $a=0$ BH (the ERN BH), Eq. (16.25). This fact suggests the interpretation of ERN BHs as objects composed of four extreme $a=\sqrt{3}$ "BHs" [1085], each of which breaks/preserves separately half of the supersymmetries while the ERN preserves an eighth as a type-II (i.e. $N=8, d=4$ SUEGRA) solution.

It is interesting to study in a little more detail the preservation of supersymmetries in terms of BPS bounds. As we discussed in Section 18.1.1, there are four central-charge skew eigenvalues $Z_i$ in $N=8, d=4$ SUEGRA. Their absolute values are in this case [840]

$$
\begin{aligned}
|Z_1| &= \tfrac{1}{4}|q^1{}_1+q^2{}_1+q^1{}_2+q^2{}_2|, \quad |Z_2| = \tfrac{1}{4}|q^1{}_1-q^2{}_1+q^1{}_2-q^2{}_2|,\\
|Z_3| &= \tfrac{1}{4}|q^1{}_1+q^2{}_1-q^1{}_2-q^2{}_2|, \quad |Z_4| = \tfrac{1}{4}|q^1{}_1-q^2{}_1-q^1{}_2+q^2{}_2|.
\end{aligned}
\tag{26.5}
$$

If only one of the charges $q$ is different from zero (the extreme $a=\sqrt{3}$ dilaton BH), $M=|Z_i|,\ i=1,\ldots,4$, and half of the supersymmetries are preserved. If two are different from zero (the extreme $a=\sqrt{3}$ dilaton BH) (say $q^1{}_1$ and $q^2{}_1$, both positive), then $M=|Z_{1,2}|<|Z_{3,4}|$ and a quarter of the supersymmetries are preserved. For three (say $q^1{}_1, q^2{}_1$, and $q^1{}_2$, all positive), $M=|Z_1|<|Z_{2,3,4}|$ and an eighth of the supersymmetries are preserved. If we add a fourth charge $q^2{}_2$, then, if it is positive, no additional supersymmetries are broken, but all are broken if it is negative.

This discussion parallels the discussion of the addition of branes to a type-II configuration on p. 775, and we need only establish the link between the $d=4$ solutions and $d=10$ brane solutions wrapped on $\mathrm{T}^6$ to arrive at the conclusion that $d=4$ BHs can be understood as composed of wrapped branes and that, in order to obtain $d=4$ BHs with regular horizons, we need to include enough branes to break seven-eighths of the supersymmetries. After reaching this conclusion, it is natural to try the construction of $d=4$ and $d=5$ BHs directly from $d=10$ extended objects in order to identify precisely which elementary string-theory objects these BHs are made of. This information will be used later in the entropy calculation.

## 26.2 Black holes from branes

### 26.2.1 Black holes from single wrapped branes

To gain some insight, we are first going to investigate the construction of BHs in any dimension by wrapping completely $p$-brane solutions on $\mathrm{T}^p$. The harmonic functions do not change and we need only reduce the metrics. These are diagonal and lead to a metric and the KK scalar modulus associated with the volume of the torus and we just have to rescale with it the reduced metric to the modified Einstein frame. The $(p+1)$-form gives directly a vector in $10-p$ dimensions and higher forms that vanish in these simple solutions.

The simplest example is provided by the F1, Eq. (25.56), wrapped on a circle. It gives rise to a $d=9$ charged extremal BH solution:

$$
\begin{aligned}
d\tilde{s}_{\mathrm{E}}^2 &= H_{\mathrm{F1}}^{-\frac{6}{7}}dt^2 - H_{\mathrm{F1}}^{\frac{1}{7}}d\vec{x}_8^{\,2}, & A_t &= \pm\big(H_{\mathrm{F1}}^{-1}-1\big),\\
ds_{\mathrm{s}}^2 &= H_{\mathrm{F1}}^{-1}dt^2 - d\vec{x}_8^{\,2}, & e^{-2\phi} &= e^{-2\phi_0}H_{\mathrm{F1}}^{\frac{1}{2}},\\
k &= k_0 H_{\mathrm{F1}}^{-\frac{1}{2}}.
\end{aligned}
\tag{26.6}
$$

The horizon would be at the pole of $H_{\mathrm{F1}}$, $\vec{x}_8=0$. However, its area (volume) is zero and so it is singular. Furthermore, the size of the compact coordinate, measured in terms of the modulus $k$, vanishes there and the dilaton diverges.

If we reduce further on $\mathrm{T}^5$, smearing the harmonic function, we obtain the $d=4$ solution,

$$
\begin{aligned}
d\tilde{s}_{\mathrm{E}}^2 &= H_{\mathrm{F1}}^{-\frac{1}{2}}dt^2 - H_{\mathrm{F1}}^{\frac{1}{2}}d\vec{x}_3^{\,2}, & A_t &= \pm\big(H_{\mathrm{F1}}^{-1}-1\big),\\
ds_{\mathrm{s}}^2 &= H_{\mathrm{F1}}^{-1}dt^2 - d\vec{x}_3^{\,2}, & e^{-2\phi} &= e^{-2\phi_0}H_{\mathrm{F1}}^{\frac{1}{2}},\\
k &= k_0\, H_{\mathrm{F1}}^{-\frac{1}{2}}, & V_5 &= V_{5\,0},
\end{aligned}
\tag{26.7}
$$

which has the metric of the $a=\sqrt{3}$ dilaton BH, one component of the solution Eq. (26.1). The metric and moduli of this solution are also singular at the horizon.

Let us try with D$p$-brane solutions Eq. (25.64) wrapped on $T^p$ ($p \leq 6$). They give

$$
\begin{aligned}
d\tilde{s}_E^2 &= H_{Dp}^{-\frac{7-p}{8-p}} dt^2 - H_{Dp}^{\frac{1}{8-p}} d\vec{x}_{9-p}^2, & A_t &= \pm\left(H_{Dp}^{-1} - 1\right), \\
ds_s^2 &= H_{Dp}^{-\frac{1}{2}} dt^2 - H_{Dp}^{\frac{1}{2}} d\vec{x}_{9-p}^2, & e^{-2\phi} &= e^{-2\phi_0} H_{Dp}^{\frac{p-6}{4}}, \\
V_p &= V_{p,0}\, H_{Dp}^{-\frac{p}{4}}.
\end{aligned}
\tag{26.8}
$$

In all cases, the metric, the modulus that measures the volume of the torus, $V_p$, and the dilaton are singular on the horizon.[3]

If we reduce further to $d = 4$ ($p \leq 6$) on a $T^{6-p}$, we obtain the solutions

$$
\begin{aligned}
d\tilde{s}_E^2 &= H_{Dp}^{-\frac{1}{2}} dt^2 - H_{Dp}^{\frac{1}{2}} d\vec{x}_3^2, & A_t &= \pm\left(H_{Dp}^{-1} - 1\right), \\
ds_s^2 &= H_{Dp}^{-\frac{1}{2}} dt^2 - H_{Dp}^{\frac{1}{2}} d\vec{x}_3^2, & e^{-2\phi} &= e^{-2\phi_0}, \\
V_p &= V_{p0}\, H_{Dp}^{-\frac{p}{4}}, & V_{6-p} &= V_{6-p\,0}\, H_{Dp}^{\frac{6-p}{4}},
\end{aligned}
\tag{26.9}
$$

which also has the (singular) metric of the extreme $a = \sqrt{3}$ dilaton BH. The moduli are different but, clearly, D$p$-branes and F1s can also be seen as the building blocks of the $d = 4$ BHs, Eqs. (26.1). An analogous calculation can be done for S5s, with analogous results, and, if we consider KK monopoles or gravitational waves, we are going to obtain the same result. All these objects can be used to construct the $d = 4$ BHs, and we know that we need at least four of them in order to break more supersymmetries and obtain one extreme BH with a regular horizon.

There is a BH solution of the model in Eq. (26.1) that also describes non-extreme BHs with regular horizons and reduces to the solution Eqs. (26.1) in the extremal limit. It can also be seen as originating from a combination of simpler objects, which turn out to be the dimensional reductions of the non-extremal F1, D$p$, S5, etc. solutions. This is the kind of solution we will try to obtain in Section 26.2.2 in $d = 4$ and also in $d = 5$.

Now, however, we are going to discuss briefly what happens near the singularity of the extreme $d = 4$ BHs we have obtained by reducing a single $d = 10$ object. These are solutions of the string effective action and, as we discussed in Section 21.1, in general they can be trusted only as long as their curvature, measured in string units $\ell_s^{-2}$, is small. Furthermore, these are solutions of the *compactified* string effective action, and we know that the standard KK dimensional reduction of the effective action is valid when the compactification radii are larger than the self-dual radius $\ell_s$ because close to it new stringy massless degrees of freedom that were not considered in the effective action appear. Similar effects are expected to take place when the curvature approaches $\ell_s^{-2}$.

Thus, new massless stringy degrees of freedom must come into play near the singularities of the given $d = 4$ solutions since the moduli that measure the volumes of the tori in general vanish there, and the solution cannot be trusted beyond a surface of radius $\sim \ell_s$ around the singularity that is sometimes called the *stretched horizon*. For the BH obtained

[3] Except for the $p = 3$ metric in the string reference frame and the $p = 6$ dilaton, which is constant.

by compactifying the F1 on $T^6$, Sen suggested in Ref. [1128] that an entropy that would coincide with the entropy associated with the degeneracy of string states with the same mass could be associated with the stretched horizon. This suggestion is along the same lines as the idea that BHs could be identified with highly excited string states and vice versa [1166].

This correspondence between string states and BHs was made more precise in Ref. [770], taking into account the dependence of the radius of the stretched horizon on the string coupling constant $\sim g^2\ell_s$: at strong coupling the size of the stretched horizon is bigger than the string length and the solution that describes the microscopical configuration should be a BH with a regular horizon. At weak coupling, the string picture is the right one. The transition between BHs and strings takes place at the value of $g$ at which the BH's Schwarzschild radius $R_S \sim \ell_s$, and at this point the string and BH descriptions of the BH/string object of a given mass $M$ should agree.[4] The mass of a highly excited closed-string state is, according to Eqs. (20.46) and (20.48), $M \sim N^{\frac{1}{2}}/\ell_s$, while, at the correspondence point, the Schwarzschild BH's mass is $M \sim R_S/(g\ell_s)^2 \sim 1/(g^2\ell_s)$. If both are the mass of the same object then $g \sim N^{-\frac{1}{4}}$.

Let us now compute the entropy of this object in the string description and in the BH description at the correspondence point $R_S \sim \ell_s$, $g \sim N^{-\frac{1}{4}}$. The BH entropy is

$$S \sim R_S^2/G_N^{(4)} \sim g^{-2} \sim \sqrt{N}. \tag{26.10}$$

The string entropy is the logarithm of the degeneracy of states at the mass level $M$. String theories are a particular case of two-dimensional CFTs [634], which in general have an infinite spectrum of states. The degeneracy of states of a CFT characterized by a *central charge* $c$, for large values of the two-dimensional energy $E$, is given by *Cardy's formula*,

$$\rho(E) \sim e^{\sqrt{\pi(c-24E_0)EL/3}}, \tag{26.11}$$

where $E_0$ is the lowest energy and $L$ is the size of the spatial coordinate of the two-dimensional theory. For string theories $L \sim \ell_s$ and $E = M^2 \sim N/\ell_s^2$. Therefore, $\rho \sim e^{M/M_0}$ and $S = \ln \rho \sim \sqrt{N}$, in good qualitative agreement with the result in the BH picture.

The BH–string correspondence principle can be extended to higher-dimensional Schwarzschild BHs and also to charged BHs, generalizing at the same time the string picture to a string/brane picture characterized by the same conserved quantities.

Clearly, this principle underlies the logic of the calculation of entropies of stringy BHs depicted in Fig. 26.1. It works best when there is unbroken supersymmetry (extreme BHs); it can be argued that the counting of states remains unmodified when we vary $g$. In the following sections we are going to construct these $d = 4, 5$ stringy extreme supersymmetric BH solutions.

### *26.2.2 Black holes from wrapped intersecting branes*

We have seen that, in order to construct $d = 4$ extreme BH solutions with regular horizons, we need at least four extended objects that break seven-eighths of the supersymmetries.

[4] The BH–fundamental-string transition has been studied further in Refs. [771, 852, 413, 853].

In $d=5$, three are necessary and, since this case is simpler and the counting of microstates for it clearer, we are going to start with it.

There are many possible configurations of three extended objects that give rise to a regular BH in $d=5$ upon compactification on $\mathrm{T}^5$. They are related by string (U) dualities in the five compact dimensions, which appear in $d=5$ as hidden symmetries of the maximal $N=4, d=5$ SUEGRA. These symmetries do not act on the Einstein metric (although they do act on the moduli), thus any of these configurations is equally good for obtaining a BH metric (the issue of U duality will be studied in Section 26.2.3). Not all the corresponding $d=10$ configurations are equally simple to treat, basically because we do not have good string descriptions of KK monopoles of S5-branes. D-brane configurations are clearly preferred. The simplest configurations of this kind are D5 $\|$ D1 $\|$W, as proposed in Ref. [287] as a simpler alternative to the original configuration proposed by Strominger and Vafa in Ref. [1160], and the T-dual configuration W$\|$ D2 $\|$ D6 proposed in Ref. [922]. We are going to study the former.

In $d=4$, a configuration that fits the requirements, W $\|$ D2 $\|$ S5A $\|$ D6, was also proposed in Ref. [922], and two T-dual alternatives D0 $\|$ D4 and D1 $\|$ D5 plus an F1 and a KK monopole were proposed in Ref. [834]. We are going to study the configuration proposed in Ref. [922].

We will not study rotating BHs, although we have mentioned them in several places, remarking, in particular, on the non-existence of supersymmetric rotating BHs with a regular horizon in $d=4$. It is interesting to mention their existence in $d=5$, where they can also be modeled with string-theory extended objects [249, 252].

$d=5$ *Black holes from intersecting branes.* To obtain regular extreme $d=5$ BHs, it is necessary to construct them as intersections of at least three $d=10$ extended objects. A possible configuration that leads to regular extreme $d=5$ BHs is

| | 0 | 1 | 2 | 3 | 4 | 5 | 6 | 7 | 8 | 9 |
|---|---|---|---|---|---|---|---|---|---|---|
| D1 | $+$ | $+$ | $\sim$ | $\sim$ | $\sim$ | $\sim$ | $-$ | $-$ | $-$ | $-$ |
| D5 | $+$ | $+$ | $+$ | $+$ | $+$ | $+$ | $-$ | $-$ | $-$ | $-$ |
| W | $+$ | $\rightarrow$ | $\sim$ | $\sim$ | $\sim$ | $\sim$ | $-$ | $-$ | $-$ | $-$ |

(26.12)

where $+$ signs stand for worldvolume dimensions (isometric in the solutions), $-$ signs stand for overall transverse directions on which the solution depends, $\sim$ signs stand for transverse directions in which the solution has been smeared, and $\rightarrow$ indicates the direction in which the wave propagates. The direction with $+$, $\sim$, or $\rightarrow$ signs will be compactified on a $\mathrm{T}^5=\mathrm{S}^1\times\mathrm{T}^4$ with volume $V^5=2\pi R_1 V^4$, where $R_1$ is the radius of the coordinate $y^1$ and $V^4=(2\pi)^4 R_2\cdots R_5$ is the volume of the $\mathrm{T}^4$ on which the coordinates $y^2,\ldots, y^5$ are compactified. Then the solution will depend only on the overall transverse coordinates $\vec{x}_4$.

Our procedure will be to construct first the $d=10$ solution of $N=2B, d=10$ SUEGRA that describes this system and then reduce it to $d=5$ in two steps ($\mathrm{S}^1$ and $\mathrm{T}^4$). Since we will also be interested in non-extremal BHs, we are going to construct the black intersecting solution first and then we will take the extreme limit.

The harmonic-superposition rule cannot be used directly in this black intersection.

We start from the non-extreme $D1 \parallel D5$ intersection (contained in Eq. (25.149))

$$
\begin{aligned}
d\hat{s}_s^2 &= H_{D1}^{-\frac{1}{2}} H_{D5}^{-\frac{1}{2}} [W dt^2 - (dy^1)^2] - H_{D1}^{\frac{1}{2}} H_{D5}^{\frac{1}{2}} [W^{-1} dr^2 + r^2 d\Omega_{(3)}^2] \\
&\quad - H_{D1}^{\frac{1}{2}} H_{D5}^{-\frac{1}{2}} [(dy^2)^2 + (dy^3)^2 + (dy^4)^2 + (dy^5)^2], \\
\hat{C}^{(2)}{}_{\underline{t}\underline{y}^1} &= \alpha_{D1} e^{-\hat{\phi}_0} (H_{D1}^{-1} - 1), \qquad \hat{C}^{(6)}{}_{\underline{t}\underline{y}^1 \cdots \underline{y}^5} = \alpha_{D5} e^{-\hat{\phi}_0} (H_{D5}^{-1} - 1), \\
e^{-2\hat{\phi}} &= e^{-2\hat{\phi}_0} H_{D5}/H_{D1}, \qquad H_i = 1 + \frac{r_i^2}{r^2}, \qquad W = 1 + \frac{\omega}{r^2}, \\
\omega &= r_i^2 (1 - \alpha_i^2), \qquad i = D1, D5.
\end{aligned}
\tag{26.13}
$$

Now, to add a wave, we use the procedure studied on p. 442 and boost the above solution in the direction $y^1$, obtaining [287, 769]

$$
\begin{aligned}
d\hat{s}_s^2 &= H_{D1}^{-\frac{1}{2}} H_{D5}^{-\frac{1}{2}} \{ H_W^{-1} W dt^2 - H_W [dy^1 + \alpha_W (H_W^{-1} - 1) dt]^2 \} \\
&\quad - H_{D1}^{\frac{1}{2}} H_{D5}^{\frac{1}{2}} [W^{-1} dr^2 + r^2 d\Omega_{(3)}^2] \\
&\quad - H_{D1}^{\frac{1}{2}} H_{D5}^{-\frac{1}{2}} [(dy^2)^2 + (dy^3)^2 + (dy^4)^2 + (dy^5)^2], \\
\hat{C}^{(2)}{}_{\underline{t}\underline{y}^1} &= \alpha_{D1} e^{-\hat{\phi}_0} (H_{D1}^{-1} - 1), \qquad \hat{C}^{(6)}{}_{\underline{t}\underline{y}^1 \cdots \underline{y}^5} = \alpha_{D5} e^{-\hat{\phi}_0} (H_{D5}^{-1} - 1), \\
e^{-2\hat{\phi}} &= e^{-2\hat{\phi}_0} H_{D5}/H_{D1}, \qquad H_i = 1 + \frac{r_i^2}{r^2}, \qquad W = 1 + \frac{\omega}{r^2}, \\
\omega &= r_i^2 (1 - \alpha_i^2), \qquad i = D1, D5, W.
\end{aligned}
\tag{26.14}
$$

In the extreme limit $\omega = 0$ we recover a D1 $\parallel$ D5 $\parallel$ W that could have been constructed using the harmonic-superposition rule.

Let us now dimensionally reduce this solution in the direction $y^1$. In this reduction a modulus field that measures the size of that direction arises,[5]

$$
\frac{k_1}{k_{1\,0}} = |\hat{g}_{\underline{y}^1\underline{y}^1}|^{\frac{1}{2}} = \frac{H_W^{\frac{1}{2}}}{H_{D1}^{\frac{1}{4}} H_{D5}^{\frac{1}{4}}},
\tag{26.15}
$$

and $k_{1\,0} = R_1/\ell_s$. In the reduction on $T^4$ we obtain another modulus field associated with its volume,[6]

$$
k_{V^4} = |\hat{g}_{\underline{y}^2\underline{y}^2} \hat{g}_{\underline{y}^3\underline{y}^3} \hat{g}_{\underline{y}^4\underline{y}^4} \hat{g}_{\underline{y}^5\underline{y}^5}|^{\frac{1}{2}} = H_{D1}/H_{D5}.
\tag{26.16}
$$

[5] Before reducing, we can rescale $y^1$ so that it takes values in $[0, 2\pi\ell_s]$. We will do the same systematically in common worldvolume directions, but not in relative transverse directions, since, in order to apply Eqs. (15.140), the coordinates have to take values in $[0, 2\pi R]$. The value at infinity of the corresponding modulus is 1.

[6] All we are doing here is a standard toroidal compactification of the kind we have performed in Section 22.5

The $d = 5$ dilaton is given by

$$e^{-2\phi} = e^{-2\hat{\phi}}k_1 = e^{-2\phi_0}\frac{H_{\rm W}^{\frac{1}{2}}}{H_{\rm D1}^{\frac{1}{4}}H_{\rm D5}^{\frac{1}{4}}}, \qquad e^{-2\phi_0} = e^{-2\hat{\phi}_0}k_{1,0}. \tag{26.17}$$

The solution of maximal $d = 5$ SUEGRA[7] in the modified Einstein frame is

$$\begin{aligned}
&d\tilde{s}_{\rm E}^2 = (H_{\rm D1}H_{\rm D5}H_{\rm W})^{-\frac{2}{3}}W dt^2 - (H_{\rm D1}H_{\rm D5}H_{\rm W})^{\frac{1}{3}}\,[W^{-1}dr^2 + r^2 d\Omega_{(3)}^2],\\
&A^{(i)}{}_t = \alpha_i\big(H_i{}^{-1} - 1\big), \qquad H_i = 1 + \frac{r_i^2}{|\vec{x}_4|^2}, \qquad \alpha_i^2 = 1, \qquad i = \rm{D1, D5, W},\\
&k_{V^4} = \frac{H_{\rm D1}}{H_{\rm D5}}, \qquad e^{-2\phi} = e^{-2\phi_0}\frac{H_{\rm W}^{\frac{1}{2}}}{H_{\rm D1}^{\frac{1}{4}}H_{\rm D5}^{\frac{1}{4}}}, \qquad k_1 = k_{1\,0}\frac{H_{\rm W}^{\frac{1}{2}}}{H_{\rm D1}^{\frac{1}{4}}H_{\rm D5}^{\frac{1}{4}}},\\
&W = 1 + \frac{\omega}{r^2}, \qquad \omega = r_i^2(1 - \alpha_i^2), \qquad i = \rm{D1, D5, W}.
\end{aligned} \tag{26.18}$$

This is a BH solution with event horizon at $\rho = -\omega$ $(\omega < 0)$. As usual, in the extreme limit $W = 1$ disappears, we can replace $dr^2 + r^2 d\Omega_{(3)}^2$ by $d\vec{x}_4^2$, and the $H_i$ could be arbitrary harmonic functions. However, we will take them as above in that limit and, with that choice, it is easy to see that there is a regular horizon at $\vec{x}_4 = 0$, with finite area (and, hence, finite entropy) given by

$$A = \omega_{(3)}\left(\lim_{|\vec{x}_4|\to 0} |\vec{x}_4|^6\, H_{\rm D1}H_{\rm D5}H_{\rm W}\right)^{\frac{1}{2}} = 2\pi^2(r_{\rm D1}r_{\rm D5}r_{\rm W})^{\frac{1}{2}}\,. \tag{26.19}$$

Furthermore, the moduli fields are finite there, as we wanted. If $r_{\rm D1} = r_{\rm D5} = r_{\rm W}$ then all the moduli are constant[8] and the metric takes the form

$$ds^2 = H^{-2}W dt^2 - H[W^{-1}dr^2 + r^2 d\Omega_{(3)}^2], \qquad H = H_{\rm D1} = H_{\rm D5} = H_{\rm W}, \tag{26.20}$$

and studied in general in Section 15.4. In general we would obtain a bunch of moduli fields coming from the internal metric. For this and the solutions that will follow, the internal metric is proportional to the identity and there is only one non-trivial modulus: its determinant. Its square root is $k_V$. We have not performed the toroidal reduction of RR fields, but it is clear that they give rise to a series of form potentials of equal and lower ranks. Since the potentials in this and the other solutions that we are going to study have components only in compact directions (plus time), only the time component of the vector fields that originate from the reduction will be non-trivial and have the obvious value.

[7] This solution can be embedded in the five-dimensional STU model ($N = 1, d = 5$ supergravity) described on p. 269 which arises as a truncation of the maximal SUEGRA: it only has three active vector fields and two independent scalars. See the discussion on p. 601.

[8] When we relate these constants to the numbers of branes of each kind, it will become clear that, in general, it is not possible to attain this equality except for special values of the moduli $g$, $R$, and $V$, although, for large numbers of branes, we can be arbitrarily close to the equality.

which is just that of the $d = 5$ RN BH (see Eq. (12.227)). The only difference is the number of vector fields of the total solution, which is dictated in our case by the requirement that we have an $N = 2B, d = 10$ SUEGRA solution that we can relate to a type-IIB superstring configuration.

Our next task is to relate the constants $r_i$ to the physical parameters of the solution. This is very easy to do in the extreme case in which we can immediately associate the supergravity solution with a supersymmetric configuration with $N_{\rm D1}$ D-strings (all with the same kind of charge, either positive or negative, to preserve supersymmetry), $N_{\rm D5}$ D5s (again all of them with the same kind of charge), and $N_{\rm W}$ units of momentum in the direction $y^1$. Since the D5s are not smeared, using Eq. (25.64) we obtain

$$r_{\rm D5}^2 = N_{\rm D5} h_{\rm D5} = N_{\rm D5} \ell_{\rm s}^2 \hat{g}. \tag{26.21}$$

For the D-strings, which are smeared in four directions, we have to use repeatedly Eqs. (15.140):

$$r_{\rm D1}^2 = N_{\rm D1} h_{\rm D1} \frac{\omega_{(5)}}{V^4 \omega_{(1)}} = \frac{N_{\rm D1} \ell_{\rm s}^6 \hat{g}}{R_2 \cdots R_5}. \tag{26.22}$$

For the gravitational wave that propagates in a compact direction and has four compact transverse directions, we have to use first the coefficient in Eq. (15.147) which, for $N_{\rm W}$ units of momentum, takes the form

$$h_{\rm W} = \frac{8NG_{\rm N}^{(\hat{d})}}{R_z^2 (\hat{d} - 4)\omega_{(\hat{d}-3)}}. \tag{26.23}$$

Here $z = y^1$, $\hat{d} = 10$, and $G_{\rm N}^{(10)}$ is given by Eq. (25.26). Smearing in four directions, we find

$$r_{\rm W}^2 = h_{\rm W} \frac{\omega_{(5)}}{V^4 \omega_{(1)}} = \frac{N_{\rm W} \ell_{\rm s}^8 \hat{g}^2}{R_1^2 R_2 \cdots R_5}. \tag{26.24}$$

This complete identification of all the parameters of the solution in terms of stringy quantities and compactification moduli can be used in the expressions for the area of the horizon and the entropy of the extreme BH. In any dimension, the entropy is one quarter of the area of the horizon in Planck units [968], and we obtain

$$G_{\rm N}^{(5)} = \frac{G_{\rm N}^{(10)}}{V^5} = \frac{\pi}{4} \frac{\ell_{\rm s}^8 g^2}{R_1 \cdots R_5}, \qquad S = \frac{A}{4G_{\rm N}^{(5)}} \Rightarrow \boxed{S = 2\pi \sqrt{N_{\rm D1} N_{\rm D5} N_{\rm W}},} \tag{26.25}$$

a beautiful formula that does not depend either on any moduli $g$ and $R_i$ or on the string length $\ell_{\rm s}$: it depends only on integers, which suggests that it can be explained in terms of a counting of possible string states on the background of the intersecting branes that are compatible with the same supergravity solution. Observe that the mass of the extreme BH (which is typical of a marginally bound configuration) does depend on the moduli:

$$\boxed{M = \frac{N_{\rm D1} R_1}{\hat{g} \ell_{\rm s}} + \frac{N_{\rm D5} R_1 \cdots R_5}{\hat{g} \ell_{\rm s}^6} + \frac{N_{\rm W}}{R_1}.} \tag{26.26}$$

The identification of the physical parameters in the non-extreme case is just a little more complicated and the interpretation in terms of stringy objects is quite a bit more complicated. The physical parameters of the solution are the mass, D1 charge, D5 charge, and momentum, which are not proportional to the numbers of D1s, D5s, and momentum modes. The Chern–Simons terms are zero in this solution, and the charges, measured in units of the fundamental D1 and D5 charges Eq. (25.27) (actually, charge densities), are simply

$$\begin{aligned}
\frac{\mathcal{Q}_{\rm D1}}{2\pi\ell_{\rm s}^2} &= \frac{\hat{g}^2}{16\pi G_{\rm N}^{(10)}}\int_{\rm S^3\times T^4}\star\hat{G}^{(3)} = \frac{\alpha_{\rm D1}r_{\rm D1}^2 R_2\cdots R_5}{2\pi\ell_{\rm s}^8\hat{g}} && \Rightarrow \alpha_{\rm D1}r_{\rm D1}^2 = \frac{\mathcal{Q}_{\rm D1}\ell_{\rm s}^6\hat{g}}{R_2\cdots R_5},\\
\frac{\mathcal{Q}_{\rm D5}}{(2\pi)^5\ell_{\rm s}^6} &= \frac{\hat{g}^2}{16\pi G_{\rm N}^{(10)}}\int_{\rm S^3}\star\hat{G}^{(7)} = \frac{\alpha_{\rm D5}r_{\rm D5}^2}{(2\pi)^5\ell_{\rm s}^8\hat{g}} && \Rightarrow \alpha_{\rm D5}r_{\rm D5}^2 = \mathcal{Q}_{\rm D5}\ell_{\rm s}\hat{g},\\
\mathcal{Q}_{\rm W} &= \frac{\alpha_{\rm W}r_{\rm W}^2 R_1^2 R_2\cdots R_5}{\ell_{\rm s}^8\hat{g}^2} && \Rightarrow \alpha_{\rm W}r_{\rm W}^2 = \frac{\mathcal{Q}_{\rm W}\ell_{\rm s}^8\hat{g}^2}{R_1^2 R_2\cdots R_5},
\end{aligned} \tag{26.27}$$

where the last result was obtained by comparison with Eq. (26.23). The $\mathcal{Q}_i$s are integers.

The mass of this brane configuration has to be measured in $d=5$ in the modified Einstein frame. On expanding the $g_{tt}$ component of the metric, we find

$$\omega - \tfrac{2}{3}(r_{\rm D1}^2 + r_{\rm D5}^2 + r_{\rm W}^2) = -\frac{8}{3\pi}G_{\rm N}^{(5)}M = -\frac{2M\ell_{\rm s}^8\hat{g}^2}{3R_1\cdots R_5}. \tag{26.28}$$

These four relations plus the three relations among $\omega, \alpha_i$, and $r_i$ in the last of Eqs. (26.18) allow us in principle to express all the integration constants in terms of the physical charges and moduli. In practice, however, one arrives at the equation

$$\sum_{i={\rm D1,D5,W}}\sqrt{\omega^2+4\mathcal{Q}_i^2} = \frac{2M\ell_{\rm s}^8\hat{g}^2}{3R_1\cdots R_5} \equiv \mathcal{M}, \qquad \mathcal{Q}_i \equiv \alpha_i r_i^2, \tag{26.29}$$

and it is difficult in general to write $\omega(\mathcal{M}, \mathcal{Q}_i)$ in a manageable way. So it is better to express all the constants in terms of $\omega$ and $\mathcal{Q}_i$: namely, $M$ using Eq. (26.29), the $\alpha_i$s using

$$\alpha_i = \frac{2\mathcal{Q}_i}{\omega+\sqrt{\omega^2+4\mathcal{Q}_i}}, \tag{26.30}$$

and the $r_i$s using Eqs. (26.27). The horizon area takes the value

$$A = 2\pi^2\sqrt{\prod_i (r_i^2-\omega)} \sim 2\pi^2\sqrt{\mathcal{Q}_{\rm D1}\mathcal{Q}_{\rm D5}\mathcal{Q}_{\rm W}}\left(1+\tfrac{1}{2}|\omega|\right)^{\frac{1}{2}}, \qquad \omega \ll |\mathcal{Q}_i|, \tag{26.31}$$

from which we can immediately find the entropy. However, this formula is difficult to explain in terms of counting of states since the $Q_i$s are not related to the numbers of stringy objects. In Ref. [769] it was proposed that one should interpret the $Q_i$s as total charge densities associated with $N_i$ branes *and* $\bar{N}_i$ antibranes; that is, $Q_i \equiv N_i - \bar{N}_i$. Furthermore, we also assume[9] that

$$\frac{\omega^2}{4\ell_s^4\hat{g}^2} \equiv 4N_{\mathrm{D1}}\bar{N}_{\mathrm{D1}}, \qquad \frac{\omega^2 R_2^2\cdots R_5^2}{4\ell_s^{12}\hat{g}^2} \equiv 4N_{\mathrm{D5}}\bar{N}_{\mathrm{D5}}, \qquad \frac{\omega^2 R_1^4 R_2^2\cdots R_5^2}{4\ell_s^{16}\hat{g}^4} \equiv 4N_{\mathrm{W}}\bar{N}_{\mathrm{W}}. \tag{26.32}$$

The remaining physical parameters can be written in terms of the $N_i$s and $\bar{N}_i$s:

$$\boxed{\begin{array}{l} M = (N_{\mathrm{D1}}+\bar{N}_{\mathrm{D1}})\dfrac{R_1}{\hat{g}\ell_s} + (N_{\mathrm{D5}}+\bar{N}_{\mathrm{D5}})\dfrac{R_1\cdots R_5}{\hat{g}\ell_s^6} + (N_{\mathrm{W}}+\bar{N}_{\mathrm{W}})\dfrac{1}{R_1}, \\[2ex] R_1^2 = \hat{g}^{\frac{1}{2}}\ell_s^2\dfrac{N_{\mathrm{D1}}\bar{N}_{\mathrm{D1}}}{N_{\mathrm{W}}\bar{N}_{\mathrm{W}}}, \qquad R_2^2\cdots R_5^2 = \ell_s^8\dfrac{N_{\mathrm{D1}}\bar{N}_{\mathrm{D1}}}{N_{\mathrm{D5}}\bar{N}_{\mathrm{D5}}}. \end{array}} \tag{26.33}$$

The mass formula should be compared with that of the extreme BH Eq. (26.26): branes and antibranes contribute equally to it, and there seems to be no interaction/binding energy between them.

The entropy takes the form

$$\boxed{S = 2\pi\left(\sqrt{N_{\mathrm{D5}}}-\sqrt{\bar{N}_{\mathrm{D5}}}\right)\left(\sqrt{N_{\mathrm{D1}}}-\sqrt{\bar{N}_{\mathrm{D1}}}\right)\left(\sqrt{N_{\mathrm{W}}}-\sqrt{\bar{N}_{\mathrm{W}}}\right).} \tag{26.34}$$

$d=4$ *Black holes from intersecting branes.* In Ref. [766] a stringy model based on a system of intersecting D6, D2, S5A, and W in the configuration

| | 0 | 1 | 2 | 3 | 4 | 5 | 6 | 7 | 8 | 9 |
|---|---|---|---|---|---|---|---|---|---|---|
| D6 | + | + | + | + | + | + | + | − | − | − |
| S5 | + | + | + | + | + | + | ∼ | − | − | − |
| D2 | + | + | ∼ | ∼ | ∼ | − | + | − | − | − |
| W | + | → | ∼ | ∼ | ∼ | ∼ | ∼ | − | − | − |

(26.35)

was proposed in order to describe $d=4$ BHs. In the extreme limit (which had been constructed before in Refs. [834, 922]) it has a regular horizon and moduli that are regular there (if the D2 were placed in directions 012, then the moduli would be singular at the horizon). The construction of the $d=10$ solution is similar to that of the solution associated with the $d=5$ BH, which we have discussed in detail, and we will therefore omit unnecessary details here.

The black intersecting solution is

[9] This could always be seen as a change of variables, although the resulting expressions have a very appealing physical interpretation. Observe that, when all the branes of one kind have charges of the same sign, $\omega=0$ and we have an extremal BH as before.

$$
\begin{aligned}
d\hat{s}_{\rm s}^2 &= H_{\rm D6}^{-\frac{1}{2}} H_{\rm D2}^{-\frac{1}{2}} \Big\{ H_{\rm W}^{-1} W dt^2 - H_{\rm W}\big[dy^1 + \alpha_{\rm W}(H_{\rm W}^{-1} - 1)dt\big]^2 \Big\} \\
&\quad - H_{\rm D6}^{-\frac{1}{2}} H_{\rm D2}^{\frac{1}{2}} [(dy^2)^2 + (dy^3)^2 + (dy^4)^2 + (dy^5)^2] \\
&\quad - H_{\rm D6}^{-\frac{1}{2}} H_{\rm D2}^{-\frac{1}{2}} H_{\rm S5} (dy^6)^2 - H_{\rm D6}^{\frac{1}{2}} H_{\rm D2}^{\frac{1}{2}} H_{\rm S5} [W^{-1} dr^2 + r^2 d\Omega_{(2)}^2], \\
e^{-2\hat{\phi}} &= e^{-2\hat{\phi}_0} H_{\rm D6}^{\frac{3}{2}} H_{\rm D2}^{-\frac{1}{2}} H_{\rm S5}^{-1}, \qquad \hat{B}^{(6)}{}_{\underline{t}\underline{y}^1 \cdots \underline{y}^5} = \alpha_{\rm S5} e^{-2\hat{\phi}_0} \big(H_{\rm S5}^{-1} - 1\big), \\
\hat{C}^{(3)}{}_{\underline{t}\underline{y}^1\underline{y}^6} &= \alpha_{\rm D2} e^{-\hat{\phi}_0} \big(H_{\rm D2}^{-1} - 1\big), \qquad \hat{C}^{(7)}{}_{\underline{t}\underline{y}^1 \cdots \underline{y}^6} = \alpha_{\rm D6} e^{-\hat{\phi}_0} \big(H_{\rm D6}^{-1} - 1\big), \\
H_i &= 1 + \frac{r_i}{r}, \qquad \alpha_i^2 = +1, \qquad i = {\rm D6, D2, S5, W}, \\
W &= 1 + \frac{\omega}{r}, \qquad \omega = r_i(1 - \alpha_i^2), \qquad i = {\rm D6, D2, S5, W}.
\end{aligned}
\tag{26.36}
$$

We dimensionally reduce this solution in three steps: first on $S^1$ ($y^1$), then on $T^4$ ($y^2, \ldots, y^5$), and then on $S^1$ ($y^6$). In the modified Einstein frame it takes the form[10]

$$
\begin{aligned}
d\tilde{s}_{\rm E}^2 &= (H_{\rm D6} H_{\rm D2} H_{\rm W} H_{\rm S5})^{-\frac{1}{2}} W dt^2 \\
&\quad - (H_{\rm D6} H_{\rm D2} H_{\rm W} H_{\rm S5})^{\frac{1}{2}} [W^{-1} dr^2 + r^2 d\Omega_{(2)}^2], \\
A^{(i)}{}_t &= \alpha_i \big(H_i^{-1} - 1\big), \\
e^{-2\phi} &= e^{-2\phi_0} \frac{H_{\rm W}^{\frac{1}{2}}}{H_{\rm S5}^{\frac{1}{2}}}, \qquad k_1 = k_{10} \frac{H_{\rm W}^{\frac{1}{2}}}{H_{\rm D6}^{\frac{1}{4}} H_{\rm D2}^{\frac{1}{4}}}, \\
K_6 &= \frac{H_{\rm S5}^{\frac{1}{2}}}{H_{\rm D6}^{\frac{1}{4}} H_{\rm D2}^{\frac{1}{4}}}, \qquad K_{V^4} = \frac{H_{\rm D2}}{H_{\rm D6}}. \\
H_i &= 1 + \frac{r_i}{r}, \qquad \alpha_i^2 = +1, \qquad i = {\rm D6, D2, S5, W}, \\
W &= 1 + \frac{\omega}{r}, \qquad \omega = r_i(1 - \alpha_i^2), \qquad i = {\rm D6, D2, S5, W}.
\end{aligned}
\tag{26.37}
$$

This solution has a regular horizon at $r = -\omega$ ($\omega < 0$). Actually, when all the constants $r_i$ have the same value, all the $H_i = H$ and the metric is identical to that of the $d = 4$ RN BH. In all cases, the extreme-limit $W = 1$ solution (compare it with the solution Eq. (26.1)) has a regular horizon.

[10] This solution can be embedded in the STU model of $N = 2, d = 4$ supergravity described on p. 234 which arises as a truncation of $N = 8, d = 4$ SUEGRA. See the discussion on p. 580.

In this case we are going to find the physical parameters directly in the non-extreme case. Using definitions similar to those of the $d = 5$ case, we find

$$\begin{aligned} \alpha_{\mathrm{D2}} r_{\mathrm{D2}} &= \frac{Q_{\mathrm{D2}} \ell_{\mathrm{s}}^5 \hat{g}}{2R_2 \cdots R_5}, & \alpha_{\mathrm{D6}} r_{\mathrm{D6}} &= \frac{Q_{\mathrm{D6}} \ell_{\mathrm{s}} \hat{g}}{2}, \\ \alpha_{\mathrm{S5}} r_{\mathrm{S5}} &= \frac{Q_{\mathrm{S5}} \ell_{\mathrm{s}}^2}{2R_6}, & \alpha_{\mathrm{W}} r_{\mathrm{W}} &= \frac{Q_{\mathrm{W}} \ell_{\mathrm{s}}^8 \hat{g}^2}{2R_1^2 R_2 \cdots R_6}, \end{aligned} \tag{26.38}$$

and, for the mass,

$$\omega - \tfrac{1}{2} \sum_i r_i = -\frac{M \ell_{\mathrm{s}}^8 \hat{g}^2}{4R_1 \cdots R_6}. \tag{26.39}$$

We find the following equation relating charges, mass, and $\omega$:

$$\sum_i \sqrt{\omega^2 + 4\mathcal{Q}_i} = \frac{M \ell_{\mathrm{s}}^8 \hat{g}^2}{4R_1 \cdots R_6} \equiv \mathcal{M}, \qquad \mathcal{Q}_i \equiv \alpha_i r_i, \tag{26.40}$$

which is, again, very difficult to solve. We therefore use $\omega$ as a parameter and find, for the $\alpha_i$s, again Eq. (26.30), etc. The area of the horizon is given by

$$A = 4\pi \sqrt{\prod_i (r_i - \omega)}. \tag{26.41}$$

Now, introducing a parametrization similar to the five-dimensional $Q_i \equiv N_i - \bar{N}_i$ and

$$\begin{aligned} \frac{\omega^2}{\ell_{\mathrm{s}}^2 \hat{g}^2} &\equiv 4N_{\mathrm{D6}} \bar{N}_{\mathrm{D6}}, & \frac{\omega^2 R_2^2 \cdots R_5^2}{\ell_{\mathrm{s}}^{10} \hat{g}^2} &\equiv 4N_{\mathrm{D2}} \bar{N}_{\mathrm{D2}}, \\ \frac{\omega^2 R_1^4 R_2^2 \cdots R_6^2}{\ell_{\mathrm{s}}^{16} \hat{g}^4} &\equiv 4N_{\mathrm{W}} \bar{N}_{\mathrm{W}}, & \frac{\omega^2 R_6^2}{\ell_{\mathrm{s}}^2 \hat{g}^2} &\equiv 4N_{\mathrm{D6}} \bar{N}_{\mathrm{D6}}, \end{aligned} \tag{26.42}$$

we obtain for the mass and entropy

$$\boxed{\begin{aligned} M &= (N_{\mathrm{D2}} + \bar{N}_{\mathrm{D2}}) \frac{R_1 R_6}{\hat{g} \ell_{\mathrm{s}}^3} + (N_{\mathrm{D6}} + \bar{N}_{\mathrm{D6}}) \frac{R_1 \cdots R_6}{\hat{g} \ell_{\mathrm{s}}^7} \\ &\quad + (N_{\mathrm{S5}} + \bar{N}_{\mathrm{S5}}) \frac{R_1 \cdots R_5}{\hat{g}^2 \ell_{\mathrm{s}}^6} + (N_{\mathrm{W}} + \bar{N}_{\mathrm{W}}) \frac{1}{R_1}, \\ S &= 2\pi \prod_i \left( \sqrt{N_i} - \sqrt{\bar{N}_i} \right). \end{aligned}} \tag{26.43}$$

We can also express the moduli in terms of them, $\ell_{\mathrm{s}}$, and $\hat{g}$.

### 26.2.3 Duality and black-hole solutions

The solutions we have obtained are particular solutions that have only a few vectors and scalars excited in maximal $N = 4, d = 5$ and $N = 8, d = 4$ SUEGRA. These theories have, respectively, 27 and 56 $\mathrm{U}(1)$ vector fields, which are rotated among themselves by the U-duality groups $\mathrm{E}_{6(+6)}$ and $\mathrm{E}_{7(+7)}$ (see Table 22.1) and scalars that parametrize the coset spaces $\mathrm{E}_{6(+6)}/\mathrm{USp}(8)$ and $\mathrm{E}_{7(+7)}/\mathrm{SU}(8)$ [385] and are also rotated by the same U-duality groups, but the Einstein metrics are U-duality invariant and all unbroken supersymmetries are preserved.

Several questions immediately arise. How does U duality act on these solutions and on their $d = 10$ description? How does U duality act on physical parameters such as the masses and entropies? What is the more general BH-type solution of these theories?

Most U-duality rotations correspond to T and S dualities in higher dimensions, whose effects on the components are well known to us. We can use them to find configurations that are more convenient for our purposes. For instance, we can dualize the $d = 4$ BH we have obtained into one composed entirely of D-branes [93, 94], whose stringy description is much better known than that of the S5s we have used. If we denote by $\mathrm{T}^n$ a T-duality transformation in the $n$th coordinate, ignoring time and the three overall transverse coordinates, we find

| | 1 | 2 | 3 | 4 | 5 | 6 |
|---|---|---|---|---|---|---|
| D6 | $+$ | $+$ | $+$ | $+$ | $+$ | $+$ |
| S5 | $+$ | $+$ | $+$ | $+$ | $+$ | $\sim$ |
| D2 | $+$ | $\sim$ | $\sim$ | $\sim$ | $\sim$ | $+$ |
| W | $\rightarrow$ | $\sim$ | $\sim$ | $\sim$ | $\sim$ | $\sim$ |

$\xrightarrow{\mathrm{ST}^2\mathrm{T}^1\mathrm{T}^5}$

| | 1 | 2 | 3 | 4 | 5 | 6 |
|---|---|---|---|---|---|---|
| D3 | $\sim$ | $\sim$ | $+$ | $+$ | $\sim$ | $+$ |
| D5 | $+$ | $+$ | $+$ | $+$ | $+$ | $\sim$ |
| D3 | $\sim$ | $+$ | $\sim$ | $\sim$ | $+$ | $+$ |
| D1 | $+$ | $\sim$ | $\sim$ | $\sim$ | $\sim$ | $\sim$ |

(26.44)

which can be T-dualized further into a configuration involving only D3s, etc.,

$\xrightarrow{\mathrm{T}^2\mathrm{T}^3}$

| | 1 | 2 | 3 | 4 | 5 | 6 |
|---|---|---|---|---|---|---|
| D3 | $\sim$ | $+$ | $\sim$ | $+$ | $\sim$ | $+$ |
| D3 | $+$ | $\sim$ | $\sim$ | $+$ | $+$ | $\sim$ |
| D3 | $\sim$ | $\sim$ | $+$ | $\sim$ | $+$ | $+$ |
| D3 | $+$ | $+$ | $+$ | $\sim$ | $\sim$ | $\sim$ |

(26.45)

U-duality transformations do not change the $d = 5, 4$ Einstein metric of these solutions; they amount simply to a relabeling of the vector fields and to a complicated non-linear transformation of the scalars. The physical parameters defined in terms of the Einstein metric (such as the mass and entropy) do not change, which means that there must be duality-invariant expressions for them. If the mass is an independent parameter (as it would be from the supergravity point of view), this is an empty statement, but, since the mass depends on the masses of the constituents and the moduli, which are transformed by duality, it is not, as a matter of fact.

The U-duality-invariant expressions for the masses and entropies are particularly simple in the extreme limits since they are completely determined by the moduli and the vector's electric and magnetic charges (27 electric charges in $d = 5$ and 28 electric and 28 magnetic charges in $d = 4$) encoded in the superalgebra's central-charge matrix. Actually, if the entropy counts microstates, it should not depend on any moduli, but only on charges. This is what happens in the explicit solutions that we have constructed. Since, at least in this case, there is only one U-duality invariant that can be constructed from the charges alone, the U-duality invariant expressions for the entropy of extreme BHs of $N=4, d=5$ and $N=8, d=4$ SUEGRA can easily be determined [840, 527, 532]. In the $N=8, d=4$ case, the entropy must be given by the beautiful formula

$$\boxed{S = 4\pi\sqrt{|\diamondsuit|},} \tag{26.46}$$

where $\diamondsuit$ is the quartic Cremmer–Julia invariant $E_{7(+7)}$. This invariant is defined in the (more difficult to use for our purposes) $SU(8)$ basis in Eq. (19.123). However, as explained in the same footnote, there is only one quartic invariant for $E_{7(+7)}$ and $\diamondsuit$ is identical, up to the global sign, to the Cartan invariant $J_4(\mathcal{Q})$, which is defined in the $SO(8)$ basis in Eq. (19.126) and rewritten here for convenience:

$$-\diamondsuit = J_4(\mathcal{Q}) = p^{ij}q_{jk}p^{kl}q_{li} - \tfrac{1}{4}(p^{ij}q_{ji})^2 + \tfrac{1}{96}\varepsilon^{ijklmnpq}\left(q_{ij}q_{kl}q_{mn}q_{pq} + p^{ij}p^{kl}p^{mn}p^{pq}\right). \tag{26.47}$$

In this formula the $p^{ij}$ and $q_{ij}$ are the 28 magnetic and 28 electric charges of the theory; they can easily be related to the charges of extended objects of $N=2A, d=10$ SUEGRA compactified on $T^6$ [92]: if the indices $i,j,k,l,m,n = 1,\ldots,6$ then

- $p^{ij} = \dfrac{1}{4!\sqrt{2}}\epsilon^{ijklmn}\mathcal{Z}^{(4)}_{klmn}$ correspond to D4s wrapped on the $klmn$ directions;
- $q_{ij} = \dfrac{1}{\sqrt{2}}\mathcal{Z}^{(2)}_{ij}$ correspond to D2s wrapped on the $ij$ directions;
- $p^{78} = \dfrac{1}{\sqrt{2}}\mathcal{Z}^{(0)}$ corresponds to D0 charge;
- $q_{78} = -\dfrac{1}{\sqrt{2}}\mathcal{Z}^{(6)}_{123456}$ corresponds to a D6 wrapped on $T^6$;
- $p^{i7} = \dfrac{1}{5!\sqrt{2}}\epsilon^{ijklmn}\mathcal{Z}^{(5)}_{jklmn}$ correspond to S5As wrapped on $jklmn$;
- $p^{i8} = \dfrac{1}{\sqrt{2}}p^i$ correspond to KK momentum in the directions $i$;
- $q_{i7} = \dfrac{1}{5!\sqrt{2}}\epsilon^{(i)jklmn}\mathcal{Z}^{(6)}_{jklmn(i)}$ correspond to KK6As wrapped on $jklmn$ with isometric direction $i$; and
- $q_{i8} = \dfrac{1}{\sqrt{2}}\mathcal{Z}^{(1)\,i}$ correspond to F1s wrapped in the directions $i$.

The $d=4$ extremal BH that we have constructed corresponds to $p^{18} = (1/\sqrt{2})N_{\rm W}$, $q_{16} = (1/\sqrt{2})N_{\rm D2}$, $p^{67} = (1/\sqrt{2})N_{\rm S5}$, and $q_{78} = -(1/\sqrt{2})N_{\rm D6}$, and the diamond formula immediately gives the right value for the BH entropy. The dual configurations give exactly the same result. In fact, the identification between entries of the central-charge matrix and charges of extended objects is clearly not unique, but is defined only up to U-duality rotations. We can also look at it from a different point of view: the objects we have considered are all wrapped on $T^6$ and are T-dual to each other; essentially, they cannot be distinguished from the $d=4$ central-matrix point of view. Their masses may be different if they are related by S dualities, though. On the other hand, since U duality acts on the moduli, too, we have to use the description in which compactification radii are bigger than the critical self-dual radius and the coupling constant is small.

It would be very interesting to have explicitly the most general U-duality-invariant BH-type solutions of these theories consistent with the no-hair theorems, which would be similar to the SWIP solutions of pure $N=4, d=4$ SUEGRA that we studied in Section 16.2.1 to check these formulae, but obtaining them turns out to be an extremely difficult problem. The metric of the supersymmetric ones has, however, been recently obtained in Ref. [943] (see Section 19.4). A proposal for the most general non-extremal one has been recently made in Ref. [355]. A simpler problem consists in finding a generating solution that would give the general solution when we act on it with a general U-duality transformation, which would preserve the metric. Simple arguments [395, 92] tell us that such solutions must have, respectively, four and five independent charge parameters in $d=5$ and 4 dimensions. It is not hard to find brane configurations with these parameters, and generating solutions have been proposed in [203, 204].

## 26.3 Entropy from microstate counting

In Section 26.2 we constructed BH solutions of maximal $d=4,5$ SUEGRA and identified the $N=2A/B, d=10$ SUEGRA solutions they originate from. In the extreme cases, we identified unambiguously their "components," which places us in the upper-left-hand box of Fig. 26.1, and allows us to move clockwise round the figure to calculate the microscopical entropy and see that it coincides with $A/(4G_N^{(d)})$, the commonly accepted semiclassical value.[11]

From the string-theory point of view, the BH solutions that we have obtained are just vacua on which strings should be quantized taking into account the boundary conditions imposed by the presence of D-branes and other extended objects. However, this may be difficult to do since, in general, the string coupling constant $\hat{g}$ is going to be large in order for the solutions to have a macroscopical Schwarzschild radius, and we only know how to quantize perturbatively, for small $\hat{g}$. We have argued, however, that the entropy is independent of the moduli and, therefore, we can try to calculate it in the limit $\hat{g} \to 0$ in which the Schwarzschild radius goes to zero and the BH description has to be replaced by a system of branes in Minkowski spacetime, a background we do know how to quantize type-II superstrings on.[12] The microscopic entropy is calculated in this limit and its value is then extrapolated to the large-$\hat{g}$ (BH) regime. It is also important for the validity of this

[11] See, however, the discussion on p. 351, which is clearly related to the correspondence principle.

[12] We are assuming implicitly that the $r_i$ coefficients of our BH solutions are proportional to positive powers of $\hat{g}$, which implies that they are composed of D-branes, F1s, or Ws.

calculation that only BPS microstates contribute to the entropy,[13] and that the dimension of supermultiplets (and, hence, the counting of BPS states) is independent of $\hat{g}$.

All we have to do now is to identify the string theory defined by the vacuum of extended objects associated with the BH in the weak-coupling limit and find, in particular the central charge $c$, and the values of $L$ and $E_0$ that characterize it as a CFT. Then Cardy's formula Eq. (26.11) gives the state degeneracy and its logarithm gives the entropy.

The first calculation along these lines was performed in Ref. [1160], and in Ref. [287] a similar calculation with a simpler model (the extreme D1 $\|$ D5 $\|$W $d = 5$ BH that we have constructed) was performed. The background of worldsheet string theory that we have presented in this book is not sufficient to explain in detail the identification of the two-dimensional CFT associated with this BH (i.e. with $N_{\text{D1}}$ parallel D1s intersecting in one dimension, $N_{\text{D5}}$ parallel D5s, and momentum in the direction of the D1s), but its essential aspects are not too difficult to understand. First of all, since the theory will be supersymmetric, $E_0 = 0$. Then, one has to realize that the only string states that are going to contribute correspond to the open sector with one endpoint on one of the $N_{\text{D1}}$ parallel D1s and the other on one of the $N_{\text{D5}}$ parallel D5s, which have momentum in the direction $y^1$ that contributes to their mass[14] the amount $N_{\text{W}}/R_1$. The number of these string modes (states of the two-dimensional CFT) is clearly proportional to the product $N_{\text{D1}}N_{\text{D5}}$, and we expect $c$, which measures the number of local degrees of freedom of a CFT, to be proportional to it. A more precise calculation gives $c = 6N_{\text{D1}}N_{\text{D5}}$. Finally, in this case[15] $L = 2\pi R_1$ and, on substituting this into Cardy's formula, we obtain the entropy Eq. (26.34) in the extreme limit, a very interesting result whose validity has been reviewed more recently in Ref. [1225]. Similar arguments explain why the BH entropy of the $d = 4$ BH is also proportional to the square root of the product of the numbers of D-branes.

There is, unfortunately, no more space to study more general examples of these calculations (see, for instance, Ref. [92]); neither is there time to see how these string models explain qualitatively and quantitatively Hawking radiation [422, 421, 923] or how extra dimensions relate these calculations to the $d = 3$ BH of Bañados, Teitelboim, and Zanelli and non-stringy CFTs[16] [95, 815, 1131, 304]. Nevertheless, we hope that the reader will have found these introductory notes useful.

[13] There are subleading contributions from non-BPS states as well. They introduce small corrections to the entropy [301].

[14] There are no more contributions to the masses of these states, apart from the oscillators, but states with excited oscillator modes do not have the required properties of supersymmetry. The same happens to the states that start and end on branes of the same kind.

[15] Clearly, we are considering the theory that lives in the intersection of the D-branes.

[16] To what extent are strings fundamental to this result?

# 27

# The FGK formalism for (single, static) black holes and branes

In the previous chapters of this book we have reviewed several methods to construct BH and black-$p$-brane solutions of supergravity or supergravity-like theories, some of which are related to superstring theory. In Chapter 19 we showed how the results of the general classification of supersymmetric solutions in Chapter 18 can be used to construct systematically[1] all the supersymmetric black-hole solutions (SBHSs) of some general families of four- and five-dimensional supergravity theories. However, SBHSs (and supersymmetric black-$p$-brane solutions (SBBSs)) occupy a very small region in the space of all BH and black-brane solutions or even in the smaller space of extremal solutions because, as we are going to study in this chapter, generically there are many more extremal non-supersymmetric than supersymmetric solutions in a given theory. On top of this, it is expected that all the extremal solutions arise as limits of some non-extremal (and, therefore, necessarily non-supersymmetric) family of solutions. The techniques we have developed to construct SBHSs and SBBSs within the framework of solutions with unbroken supersymmetry only allow us to explore certain corners of the space of BH and black-brane solutions, the most interesting of which are in the interior.

It is, therefore, highly desirable to find another framework in which non-supersymmetric (extremal and non-extremal) solutions can be studied. In this chapter we are going to review one such framework: the so-called *FGK formalism*, which was proposed by Ferrara, Gibbons, and Kallosh in Ref. [522] for four-dimensional asymptotically flat, single, static, extremal and non-extremal BHs and has been generalized to BHs in higher dimensions [941] and to black $p$-branes in arbitrary number of dimensions [56].[2,3]

The FGK formalism will allow us to derive a number of general results that apply to all the black solutions of a large number of theories (those with actions of the generic form Eq. (2.147) in $d=4$ or Eqs. (2.178) and (2.184) in other dimensions). In particular,

[1] Strictly speaking this is only true for the single, static BH solutions: although the general form of the stationary multicenter solutions is known, the integration constants have to be determined case by case.

[2] Always single, static, spherically symmetric, and asymptotically flat in the transverse directions. This restriction will be assumed throughout the chapter.

[3] A more general approach to the construction of BH solutions can be found in Ref. [253]. The FGK formalism can be seen as a special case (static, spherically symmetric) of the former in which the electric and magnetic potentials have been eliminated using their equations of motion.

we will prove the general existence of the *attractor mechanism* [531, 1157, 526, 527] in the extremal solutions, supersymmetric or not. We have already shown the existence of this mechanism in SBHSs in Chapters 18 and 19, but the FGK formalism will allow us to extend (and prove) that observation to the non-supersymmetric cases. To a large extent, the attractor mechanism reduces the study of extremal solutions to the search of the *attractors* of a theory, which are the extrema of a function called the *black-hole potential*. This has generated a very large literature, particularly for supergravity theories (see, for instance, Refs. [956, 71, 51, 1040, 1041, 525, 133, 557] and references therein).

For pedagogical reasons, we are going to review the original four-dimensional FGK formalism in Section 27.1 and we will extend it to higher spacetime and worldvolume dimensions in Section 27.2.

For $N = 1, d = 5$ (BHs and black strings) and $N = 2, d = 4$ theories (BHs only), the FGK formalism can be reformulated (*H-FGK formalism*) in terms of variables (the *H-variables*) that transform linearly under the duality group and can be understood as the building blocks of the physical fields of the solutions.[4,5] We will review this formalism in Section 27.3 and we will show how it can be used to construct non-extremal solutions of certain (but not all!) theories using the *hyperbolic ansatz* of Refs. [578, 939, 942].

## 27.1 The $d = 4$ FGK formalism

In Ref. [522] Ferrara, Gibbons, and Kallosh considered the BH solutions of a very generic four-dimensional supergravity-like theory with scalar and vector fields coupled to gravity as dictated by the action Eq. (2.147), which we rewrite here for convenience

$$\begin{aligned} S[g, A, \varphi] = \int d^4x \sqrt{|g|} \, \{ & R + \mathcal{G}_{ij}\partial_\mu\varphi^i\partial^\mu\varphi^j \\ & + 2\Im\mathfrak{m}\mathcal{N}_{\Lambda\Sigma}F^{\Lambda\,\mu\nu}F^{\Sigma}{}_{\mu\nu} - 2\Re\mathfrak{e}\mathcal{N}_{\Lambda\Sigma}F^{\Lambda\,\mu\nu} \star F^{\Sigma}{}_{\mu\nu} \} . \end{aligned} \tag{27.1}$$

The bosonic actions of all the ungauged four-dimensional Poincaré supergravities have this generic form, as we have pointed out elsewhere.

The equations of motion of this system can be simplified by making an educated ansatz for the metric and scalar and vector fields. The ansatz should be general enough to describe BHs in all these theories, but, at the same time, it should be simple enough if we want to be able to obtain general results. To simplify the problem, we will restrict ourselves to single, static, spherically symmetric BHs.[6]

It turns out [608] that all the single, static, spherically symmetric, asymptotically flat BHs of these theories can be described by a metric of the form Eq. (11.28), which we also

[4] This reformulation of the FGK, proposed in Ref. [941], is partially based on Refs. [959, 957] in the $N = 1, d = 5$ case and on Ref. [958] in the $N = 2, d = 4$ case. The formalism of the latter three references is wider in scope and can be applied to more general solutions.

[5] In the supersymmetric limits these become the harmonic functions that are ubiquitous in the supersymmetric solutions and which are the building blocks of all the physical fields.

[6] We will make this restriction throughout this chapter, with the necessary changes when we deal with extended objects.

rewrite here for convenience replacing the radial coordinate $\tau$ by $\rho$:

$$\begin{aligned} ds^2 &= e^{2U}dt^2 - e^{-2U}\gamma_{\underline{mn}}dx^{\underline{m}}dx^{\underline{n}}, \\ \gamma_{\underline{mn}}dx^{\underline{m}}dx^{\underline{n}} &= \frac{r_0^4}{\sinh^4 r_0\rho}d\rho^2 + \frac{r_0^2}{\sinh^2 r_0\rho}d\Omega^2_{(2)}. \end{aligned} \tag{27.2}$$

The *non-extremality parameter*, $r_0$ (a function to be determined from the physical parameters of the solution) indicates how far from the extremal limit in which the Hawking temperature $T$ vanishes a regular BH solution is: since (for $r_0 \neq 0$) the spatial (transverse) metric behaves in the near-horizon limit $\rho \to -\infty$ as

$$\gamma_{\underline{mn}}dx^{\underline{m}}dx^{\underline{n}} \sim r_0^4 e^{4r_0\rho}d\rho^2 + r_0^2 e^{2r_0\rho}d\Omega^2_{(2)}, \tag{27.3}$$

the solution will have a regular horizon at $\rho \to -\infty$ if, in the same limit, the metric function behaves as

$$e^U \sim e^{C+r_0\rho}. \tag{27.4}$$

Then, the full metric behaves as

$$ds^2 \sim e^{2C+2r_0\rho}[dt^2 - r_0^4 e^{-4C}d\rho^2] - e^{-2C}r_0^2 d\Omega^2_{(2)}, \tag{27.5}$$

so the Bekenstein–Hawking entropy $S$ is always given by

$$S = \pi e^{-2C}r_0^2. \tag{27.6}$$

To find the Hawking temperature we change coordinates, $\rho = e^{2C}\varrho/r_0 - C/r_0$, so the time-radial part of the metric takes the form

$$e^{2e^{2C}\varrho/r_0}[dt^2 - d\varrho^2]. \tag{27.7}$$

This is the metric of the *Rindler spacetime* characteristic of the near-horizon geometry of non-extremal BHs.[7] The temperature can be read from it by comparing it with the expression[8]

$$e^{\frac{4\pi}{\beta}}[dt^2 - d\varrho^2] \quad \Rightarrow \quad T = \frac{e^{2C}}{2\pi r_0}. \tag{27.8}$$

Combining the expressions for $S$ and $T$, we find [622]

$$\boxed{r_0 = 2ST,} \tag{27.9}$$

and $r_0$ vanishes when $T$ does, if the BH remains regular and $S$ is finite in that limit. The extremal limit of that metric is given in Eq. (19.12). This formula reduces to Smarr's

[7] See footnote 28 on p. 314.

[8] On Wick-rotating the time, redefining $e^{\frac{2\pi}{\beta}\varrho} = \frac{2\pi}{\beta}r$, and following the reasoning on p. 314, the inverse temperature can be found to be, precisely, $\beta$.

formula Eqs. (11.52) and (12.113) for Schwarzschild and Reissner–Nordström (RN) BHs and generalizes them.

Although generic static and spherically symmetric metrics depend on two different functions, the above metric Eq. (27.7) has a single independent function of $\rho$, $e^{2U}$ (the *metric function*, also known – quite improperly – as the *warp factor*) to be found. The metric function $e^{-2U}$ and the non-extremality parameter $r_0$ are given for the Schwarzschild BH in Eqs. (11.29) and for the RN BH in Eqs. (12.73) and (12.95). This is one of the key simplifications that make this formalism so useful.

To describe our ansatz for the vector fields we will use simultaneously the fundamental (or *electric*) vector fields $A^{\Lambda}{}_{\mu}$ and the dual (*magnetic*) vector fields $A_{\Lambda\,\mu}$. The dual vector fields are associated with the dual vector field strengths $G_{\Lambda\,\mu\nu}$ defined in Eq. (2.148) by $G_{\Lambda\,\mu\nu}=2\partial_{[\mu|}A_{\Lambda\,|\nu]}$ and are not independent of the fundamental ones $A^{\Lambda}{}_{\mu}$, which are the only ones we ultimately care about: they are related by Eq. (2.148). Now, we will assume that the $t$ component of each fundamental vector $A^{\Lambda}{}_{t}$ is a function of $\rho$ that we will call $\psi^{\Lambda}(\rho)$. The remaining components are indirectly determined by the assumption that the $t$ component of each dual vector field $A_{\Lambda\,t}$ is another function of $\rho$, $\chi_{\Lambda}(\rho)$. That is:

$$A^{\Lambda}{}_{t}=\psi^{\Lambda}(\rho) \quad\Rightarrow\quad F^{\Lambda}{}_{\underline{m}t}=\partial_{\underline{m}}\psi^{\Lambda}, \qquad A_{\Lambda\,t}=\chi_{\Lambda}(\rho) \quad\Rightarrow\quad G_{\Lambda\,\underline{m}t}=\partial_{\underline{m}}\chi_{\Lambda}. \tag{27.10}$$

This ansatz requires that we check the Bianchi identities as well as the Maxwell equations. Finally, to complete the ansatz we will assume that the scalars only depend on $\rho$.

We will proceed to substitute the ansatz in the equations of motion in two steps: first, we will deal with the metric Eq. (27.2) with an unspecified (but time-independent) spatial metric[9] $\gamma_{\underline{mn}}$ assuming general spatial dependence for the fields, and later on we will substitute the metric in the second line of that equation and will restrict the field to depend only on $\rho$.

Let us start with the Maxwell equations and Bianchi identities, respectively:

$$\partial_{\nu}(\sqrt{|g|}\star G_{\Lambda}{}^{\nu\mu})=0, \qquad \partial_{\nu}(\sqrt{|g|}\star F^{\Lambda\,\nu\mu})=0. \tag{27.11}$$

The $\mu=\underline{m}$ components are automatically solved by the ansatz. The $\mu=t$ components of the equations depend on the $\underline{mn}$ components of the field strengths and we have to use the definition Eq. (2.148) to reexpress them in terms of the $\underline{m}t$ ones. More specifically, we use the relations

$$\begin{aligned} F^{\Lambda} &= I^{-1\,\Lambda\Gamma}R_{\Gamma\Sigma}\star F^{\Sigma}-I^{-1\,\Lambda\Sigma}\star G_{\Sigma},\\ G_{\Lambda} &= (I+RI^{-1}I)_{\Lambda\Sigma}\star F^{\Sigma}-R_{\Lambda\Gamma}I^{-1\,\Gamma\Sigma}\star G_{\Sigma}, \end{aligned} \tag{27.12}$$

where the matrices $I,R$ are defined in Eqs. (2.168). Then the $t$ components of those equations can be written, combined, in the three-dimensional, manifestly symplectic form

$$\nabla_{\underline{m}}\left[e^{-2U}\mathcal{M}_{MN}\partial^{\underline{m}}\Psi^{N}\right]=0, \quad \text{where} \quad (\Psi^{M})\equiv\begin{pmatrix}\psi^{\Lambda}\\ \chi_{\Lambda}\end{pmatrix}, \tag{27.13}$$

where the symplectic matrix $\mathcal{M}_{MN}=\mathcal{M}_{MN}(\mathcal{N})$ has been defined in Eq. (2.167).

[9] The connection and curvature of this metric can be found in Appendix N.2.4, with $e^{U}$ replaced by $|M|$.

Using the given relations, the scalar equations of motion

$$\frac{1}{\sqrt{|g|}}\partial_\mu(\sqrt{|g|}\mathcal{G}_{ij}\varphi^j) + \tfrac{1}{2}\partial_i\mathcal{G}_{jk}\partial_\mu\varphi^j\partial^\mu\varphi^k + \partial_i(F^\Lambda \star G_\Lambda) = 0 \tag{27.14}$$

do not put up any difficulties and take the three-dimensional form

$$\nabla_{\underline{m}}(\mathcal{G}_{ij}\partial^{\underline{m}}\varphi^j) - \tfrac{1}{2}\partial_i\mathcal{G}_{jk}\partial_{\underline{m}}\varphi^j\partial^{\underline{m}}\varphi^k - \tfrac{1}{2}\partial_i\left(4e^{-2U}\mathcal{M}_{MN}\right)\partial_{\underline{m}}\Psi^M\partial^{\underline{m}}\Psi^N = 0. \tag{27.15}$$

Finally, the Einstein equations, which, upon use of Eq. (2.166), can be written in the manifestly symplectic invariant form

$$G_{\mu\nu} + \mathcal{G}_{ij}\left[\partial_\mu\varphi^i\partial_\nu\varphi^j - \tfrac{1}{2}g_{\mu\nu}\partial_\rho\varphi^i\partial^\rho\varphi^j\right] + 4\mathcal{M}_{MN}(\mathcal{N})\mathcal{F}^M{}_\mu{}^\rho\mathcal{F}^N{}_{\nu\rho} = 0, \tag{27.16}$$

split into three sets of three-dimensional equations corresponding, respectively, to the (flat) components $00$, $0m$, and $mn$:

$$R + 2(\partial U)^2 - 4\nabla^2 U + \mathcal{G}_{ij}\partial_{\underline{m}}\varphi^i\partial_{\underline{n}}\varphi^j - 4e^{-2U}\mathcal{M}_{MN}\partial_{\underline{m}}\Psi^M\partial^{\underline{m}}\Psi^N = 0, \tag{27.17}$$

$$\partial_{[\underline{m}}\psi^\Lambda\partial_{\underline{n}]}\chi_\Lambda = 0, \tag{27.18}$$

$$G_{mn} + 2\left[\partial_m U\partial_n U - \tfrac{1}{2}\delta_{mn}(\partial U)^2\right] + \mathcal{G}_{ij}\left[\partial_m\varphi^i\partial_n\varphi^j - \tfrac{1}{2}\delta_{mn}\partial_q\varphi^i\partial^q\varphi^j\right]$$
$$+ 4e^{-2U}\mathcal{M}_{MN}\left[\partial_m\Psi^M\partial_n\Psi^N - \tfrac{1}{2}\delta_{mn}\partial_q\Psi^M\partial^q\Psi^N\right] = 0. \tag{27.19}$$

If we use the trace of Eq. (27.19) to eliminate $R$ from Eq. (27.17), all the equations we have obtained (except Eq. (27.18), which is a constraint) look like the equations of a set of scalar fields $(\phi^A) \equiv (U, \varphi^i, \Psi^M)$ coupled to gravity. In fact, defining the metric of the indefinite signature $\mathcal{G}_{AB}$

$$(\mathcal{G}_{AB}) \equiv \left(\begin{array}{c|c|c} 2 & & \\ \hline & \mathcal{G}_{ij} & \\ \hline & & 4e^{-2U}\mathcal{M}_{MN} \end{array}\right), \tag{27.20}$$

Eqs. (27.13), (27.15), (27.17) combined with Eqs. (27.19) and (27.19) can be obtained from the three-dimensional action

$$\boxed{S[\gamma,\phi] = \int d^3x\sqrt{\gamma}\left[R(\gamma) + \mathcal{G}_{AB}(\phi)\gamma^{\underline{mn}}\partial_{\underline{m}}\phi^A\partial_{\underline{n}}\phi^B\right],} \tag{27.21}$$

which is a special case of the action found in Ref. [253]. Equation (27.18) remains to be imposed as a constraint.

We are now ready to take the second step and specify the form of the three-dimensional metric, restricting all the fields to depend on $\rho$ only. We only need the Ricci scalar of $\gamma_{\underline{mn}}$, whose only non-vanishing component is $R_{\rho\rho} = -2r_0^2$. The constraint Eq. (27.18) is automatically solved and the equations of motion that follow from the above action Eq. (27.21) take the simple form

$$\frac{d}{d\rho}(\mathcal{G}_{AB}\dot{\phi}^B) - \tfrac{1}{2}\partial_A\mathcal{G}_{BC}\dot{\phi}^B\dot{\phi}^C = 0, \tag{27.22}$$

$$\mathcal{G}_{AB}\dot{\phi}^A\dot{\phi}^B - 2r_0^2 = 0, \tag{27.23}$$

where an overdot indicates derivation with respect to $\rho$. The first equation (which is nothing but the geodesic equation) is the equation that would follow from the action

$$\boxed{S[\phi] = \int d\rho\, \mathcal{G}_{AB}\dot{\phi}^A\dot{\phi}^B,} \tag{27.24}$$

which can be viewed as the action of a mechanical system in which the radial coordinate $\rho$ plays the role of evolution parameter.[10] The second equation is a constraint known as the *Hamiltonian constraint* because $\mathcal{G}_{AB}\dot{\phi}^A\dot{\phi}^B$ is proportional to the Hamiltonian of the system, which is a conserved quantity. The constraint states that the value of that conserved quantity is precisely $2r_0^2$.

The problem of finding BH solutions has been reduced to that of finding geodesics in the target space with metric $\mathcal{G}_{AB}$. There are many group-theoretical methods to integrate equations of this kind when the target space is symmetric; see, for instance, Refs. [457, 253, 585, 558, 560, 691, 572, 347, 198, 149, 561, 242, 239, 345, 348, 688, 544, 344].

The reduction of the original action to an effective mechanical system is a great simplification, but one can still do better: since the metric $\mathcal{G}_{AB}$ does not depend on the potentials $\Psi^M$ (otherwise gauge invariance would be broken) there is a conserved quantity associated with each of them that we can identify, up to a normalization constant $\alpha$ to be determined, with the electric and magnetic charges $(\mathcal{Q}^M) \equiv \begin{pmatrix} p^\Lambda \\ q_\Lambda \end{pmatrix}$:

$$\frac{d}{d\rho}(\mathcal{G}_{MN}\dot{\Psi}^N) = 0 \quad \Rightarrow \quad \mathcal{G}_{MN}\dot{\Psi}^N = 4e^{-2U}\mathcal{M}_{MN}\dot{\Psi}^N = \mathcal{Q}_M/\alpha. \tag{27.25}$$

We can invert this relation and eliminate $\dot{\Psi}^M$ in the remaining equations of motion. The equations of motion for the variables $U(\rho), \varphi^i(\rho)$ and the Hamiltonian constraint take the form [522]

$$\ddot{U} + e^{2U}V_{\text{bh}} = 0, \tag{27.26}$$

$$\frac{d}{d\rho}(\mathcal{G}_{ij}\dot{\varphi}^j) - \tfrac{1}{2}\partial_i\mathcal{G}_{jk}\dot{\varphi}^j\dot{\varphi}^k + e^{2U}\partial_i V_{\text{bh}} = 0, \tag{27.27}$$

$$\dot{U}^2 + \tfrac{1}{2}\mathcal{G}_{ij}\dot{\varphi}^i\dot{\varphi}^j + e^{2U}V_{\text{bh}} = r_0^2, \tag{27.28}$$

[10] Geodesic actions such as this one also arise in the search for other types of solutions which depend effectively on only one direction, such as cosmological solutions, instantons, etc. See, for instance, Ref. [149] and references therein.

where $V_{\rm bh} = V_{\rm bh}(\varphi, \mathcal{Q})$, known as the *black-hole potential*, is given by[11]

$$-V_{\rm bh}(\varphi, \mathcal{Q}) \equiv -\tfrac{1}{2}\mathcal{Q}^M \mathcal{M}_{MN}\mathcal{Q}^N. \tag{27.29}$$

The first two equations can be obtained from the effective action

$$S_{\rm eff}[U, \varphi^i] = \int d\rho \left[\dot{U}^2 + \tfrac{1}{2}\mathcal{G}_{ij}\dot{\varphi}^i\dot{\varphi}^j - e^{2U}V_{\rm bh}\right]. \tag{27.30}$$

The problem of finding BH solutions of the action Eq. (27.1) has finally been reduced to an equivalent mechanical problem in which the variables $U(\rho), \varphi^i(\rho)$ obey ordinary differential equations.[12] We can now proceed to derive generic results for the four-dimensional solutions, in particular for the extremal (BPS or not) ones.

### 27.1.1 FGK theorems and the attractor mechanism

Let us first consider regular extreme BHs. In the extremal limit $r_0 \to 0$, $r_0^{-1}\sinh r_0\rho \to \rho$, and the metric Eq. (27.2) takes the form

$$ds^2 = e^{2U}dt^2 - e^{-2U}\left[\frac{d\rho^2}{\rho^4} + \frac{1}{\rho^2}d\Omega^2_{(2)}\right], \tag{27.31}$$

which is equivalent to the spatially isotropic metric

$$ds^2 = e^{2U}dt^2 - e^{-2U}d\vec{x}_3^{\,2}, \quad \text{with} \quad r = |\vec{x}_3| = -1/\rho. \tag{27.32}$$

In the near-horizon limit $\rho \to -\infty$ of an extremal BH solution the metric function $e^{-2U}$ must diverge as

$$e^{-2U} \sim \frac{A}{4\pi}\rho^2, \tag{27.33}$$

where $A$ is the area of the event horizon for the solution to have a regular horizon. Then, in this limit, the metric of a regular extremal BH will always take the form

$$ds^2 \sim \frac{4\pi}{A}\frac{dt^2}{\rho^2} - \frac{A}{4\pi}\frac{d\rho^2}{\rho^2} - \frac{A}{4\pi}d\Omega^2_{(2)}, \tag{27.34}$$

which is that of the Robinson–Bertotti solution $\mathrm{AdS}_2 \times \mathrm{S}^2$ with radii $\sqrt{A/(4\pi)}$.

---

[11] Our convention for the sign of the BH potential is the opposite to that found in most of the literature on BH attractors so that it matches the conventional sign in mechanics. We have also set the normalization constant $\alpha = 1/2$ so the charges have the normalization used in Section 19.2.

[12] There is an important difference between the action Eq. (27.24) and the action Eq. (27.30) obtained after elimination of the electric and magnetic potentials: most of the symmetries present in the former are explicitly broken by the charge vector $\mathcal{Q}^M$ in the latter unless $\mathcal{Q}^M$ is duality invariant. This can only happen for particular choices of the charges $p^\Lambda, q_\Lambda$. For this reason one can use group-theoretical methods only with the former.

We are also going to assume that the scalar fields are finite on the horizon of a regular BH solution. More precisely, we will assume, following Ref. [522], that

$$\lim_{\rho\to-\infty} \mathcal{G}_{ij}\dot{\varphi}^i\dot{\varphi}^j e^{2U}\rho^4 = \lim_{\rho\to-\infty} \frac{4\pi}{A}\mathcal{G}_{ij}\dot{\varphi}^i\dot{\varphi}^j\rho^2 \equiv \xi^2 < \infty. \tag{27.35}$$

If, under these assumptions, we multiply the Hamiltonian constraint Eq. (27.28) by $e^{2U}\rho^4$ and then by $A^2/(4\pi)$ (which, by assumption, is finite), we get

$$A + \frac{A^2}{8\pi}\xi^2 + 4\pi V_{\rm bh}(\varphi_{\rm h}, \mathcal{Q}) = 0 \quad \Rightarrow \quad A \leq -4\pi V_{\rm bh}(\varphi_{\rm h}, \mathcal{Q}). \tag{27.36}$$

This bound can be refined: if we define the coordinate $\varpi \equiv -\log(-\rho)$, then we can rewrite the definition of the parameter $\xi$ in this form:

$$\xi^2 = \lim_{\varpi\to-\infty} \frac{4\pi}{A}\mathcal{G}_{ij}\frac{d\varphi^i}{d\varpi}\frac{d\varphi^j}{d\varpi}. \tag{27.37}$$

The r.h.s. of this equation is just the kinetic term of the scalar in the original action. The only way to satisfy this identity is to have

$$\lim_{\varpi\to-\infty} \frac{d\varphi^j}{d\varpi} = \lim_{\rho\to-\infty} \rho\frac{d\varphi^j}{d\rho} = 0, \tag{27.38}$$

in this limit because, if the limit was any non-vanishing constant,[13] then $\varphi$ would be linear in $\varpi$ and would diverge on the near-horizon limit. Thus, $\xi^2 = 0$ and the above bound Eq. (27.36) must be saturated, which implies for the entropy

$$\boxed{S/\pi = -V_{\rm bh}(\varphi_{\rm h}, \mathcal{Q}).} \tag{27.39}$$

This identity is interesting but useless unless we know the values of the scalars on the horizon $\varphi_{\rm h}$. To find out information about them we analyze the near-horizon limit of the scalar equations of motion Eq. (27.27) multiplied by $\rho^2$. Taking into account Eqs. (27.33) and (27.38) we get

$$\lim_{\rho\to-\infty} \rho^2\ddot{\varphi}^i = -\frac{4\pi}{A}\,\mathcal{G}^{ij}\partial_j V_{\rm bh}\big|_{\varphi=\varphi_{\rm h}}, \tag{27.40}$$

which implies that in this limit

$$\varphi^i \sim \frac{4\pi}{A}\,\mathcal{G}^{ij}\partial_j V_{\rm bh}\big|_{\varphi=\varphi_{\rm h}} \log(-\rho) + \alpha\rho + \varphi^i_{\rm h} + \mathcal{O}(1/\rho). \tag{27.41}$$

By assumption of finiteness of the scalars on the horizon, $\alpha = 0$ and

$$\boxed{\partial_i V_{\rm bh}\big|_{\varphi=\varphi_{\rm h}} = 0.} \tag{27.42}$$

In other words: the values of the scalars on the horizon are the critical points of the BH potential, whose value at the horizon determines, in turn, the entropy.

[13] It has to be a constant if the scalar metric is going to be regular on the horizon.

In principle, these equations can be solved for the $\varphi^i_{\rm h}$ entirely in terms of the charges: the scalars will always flow from their arbitrary values at spatial infinity $\varphi^i_\infty$ (the moduli) to the values $\varphi_{\rm h}(\mathcal{Q})$, called *attractors*. This is what is usually meant in the literature by *attractor mechanism*. However, $V_{\rm bh}$ may in general have *flat directions* around certain attractors and Eqs. (27.42) will be underdetermined there. The *attractor* values of the scalars ($\varphi^i_{\rm h}$) will depend on other parameters apart from the charges. These parameters can only be the moduli $\varphi^i_\infty$. We conclude that, if the BH potential has flat directions, the attractor values will keep some dependence on the moduli $\varphi^i_{\rm h} = \varphi^i_{\rm h}(\mathcal{Q}, \varphi_\infty)$ or, equivalently, the attractor mechanism as it is usually understood only works for some of the scalar fields.

The crucial point is that, even when one attractor depends on the moduli, the value of the BH potential for that attractor only depends on the charges [1129] (otherwise, the directions parametrized by the moduli would not be flat!). Thus the entropy of extremal BHs only depends on the charges, which are quantized. This is the most important consequence of the attractor mechanism. It implies that the BH entropy of an extremal BH only depends on integer numbers, which strongly suggests that an explanation in terms of the counting of states should exist (see Chapter 26 for some string-theory examples and references).

As we are going to show at the end of this section, there is at least one extremal BH for each attractor: the double extremal BH with constant scalars equal to the attractor. Other extremal BHs with the same attractor have different values of the scalars at infinity and non-constant scalars.

On the other hand, we know that in the supergravity theories that we have considered in Chapter 19 there are extremal BH solutions that are supersymmetric. Thus, the BH potential must admit at least one attractor (the "supersymmetric attractor").

Let us now consider the spatial-infinity limit ($\rho \to 0^-$). To $\mathcal{O}(\rho^2)$ the metric function and the scalars behave as

$$U \sim M\rho, \qquad \varphi^i \sim \varphi^i_\infty + \Sigma^i \rho, \tag{27.43}$$

where $M$ is the mass and $\Sigma^i$ denotes the scalar charges. Substituting into Eq. (27.28) we get a general bound which is saturated in the extremal limit:

$$M^2 + \tfrac{1}{2}\mathcal{G}_{ij}(\varphi_\infty)\Sigma^i\Sigma^j + V_{\rm bh}(\varphi_\infty, \mathcal{Q}) = r_0^2 \geq 0. \tag{27.44}$$

As explained in Chapter 16, the scalar "charges" cannot be independent parameters in regular BHs and we expect them to be some functions $\Sigma^i = \Sigma^i(\varphi_\infty, \mathcal{Q}, M)$. If we knew these functions, we could write the non-extremality function $r_0^2 = r_0^2(\varphi_\infty, \mathcal{Q}, M)$, but they are not known in general. The bound is, nevertheless, useful if we consider double extremal BHs[14] for some given attractor $\varphi_{\rm h}$: the functions $\Sigma^i(\varphi_\infty, \mathcal{Q}, M)$ must vanish for $\varphi_\infty = \varphi_{\rm h}$ because the scalars are constant. Then the bound leads to the identity

$$M^2_{\rm double\ extremal} = -V_{\rm bh}(\varphi_{\rm h}, \mathcal{Q}) = S/\pi. \tag{27.45}$$

[14] These have been defined and discussed in the supersymmetric case on p. 571.

Before we move to apply the FGK formalism to supergravity theories, it is worth working out a couple of simple examples that illustrate the power of the method: the general case with vanishing charges and the case with constant scalars.

If the charges vanish, so does $V_{\rm bh}$, and the equation of motion for $U$ Eq. (27.26) is solved by $U = a + b\rho$. The normalization of the metric at spatial infinity requires $a = 0$ and $b = M$. This is the metric of the Schwarzschild BH Eq. (11.29). We should remember that the horizon will be regular if $e^U$ behaves in the $\rho \to -\infty$ limit as in Eq. (27.4), i.e. $r_0 = M$. On the other hand, the Hamiltonian constraint takes the form $r_0^2 = M^2 + \frac{1}{2}\mathcal{G}_{ij}\dot{\varphi}^i\dot{\varphi}^j$, which brings us to the conclusion that either the scalars are constant or the metric $\mathcal{G}_{ij}(\varphi)$ is singular on the solution. The latter is a possible way to evade the no-hair theorem, but the solution will not be stable because a singular scalar metric allows for runaway perturbations of the scalars. This can be considered as a proof of the no-hair theorem for the class of theories under consideration in the absence of vector fields.

If the scalars are constant and the charges do not vanish, according to Eq. (27.27) the constant values of the scalars must extremize $V_{\rm bh}$ and must correspond to some attractor $\varphi^i = -\varphi^i_{\rm h}$ (we are dealing with a BH that becomes a double extremal BH in the extremal limit). We can integrate the equation for $U$ Eq. (27.26), multiplying it by $\dot{U}$. This gives the Hamiltonian constraint Eq. (27.28), the only equation that needs to be solved and which can be integrated immediately, giving

$$e^{-U} = -\sqrt{-V_{\rm bh}(\varphi_{\rm h}, \mathcal{Q})/r_0^2}\,\sinh(r_0\rho + a). \tag{27.46}$$

The normalization of the metric at spatial infinity requires that $\sinh a = -\sqrt{-r_0^2/V_{\rm bh}(\varphi_{\rm h}, \mathcal{Q})}$, so that

$$e^{-U} = \cosh r_0\rho - \sqrt{r_0^2 - V_{\rm bh}(\varphi_{\rm h}, \mathcal{Q})}\,\frac{\sinh r_0\rho}{r_0} = \cosh r_0\rho - M\frac{\sinh r_0\rho}{r_0}, \tag{27.47}$$

after comparing with the $\rho \to 0^-$ expansion $e^{-U} \sim 1 - M\rho$. Therefore, the non-extremality parameter is given by

$$r_0^2 = M^2 + V_{\rm bh}(\varphi_{\rm h}, \mathcal{Q}) \geq 0, \tag{27.48}$$

and we recover Eq. (27.45) in the extremal limit. The entropy is

$$S/\pi = (M + r_0)^2 \quad \Rightarrow \quad T = \frac{r_0}{2S} = \frac{r_0}{2\pi(M + r_0)^2}. \tag{27.49}$$

It is not difficult to identify the metric Eq. (27.47) with that of the RN BH given in Eq. (12.95) with $\sqrt{-V_{\rm bh}(\varphi_{\rm h}, \mathcal{Q})}$ (which is a function of the charges only) replacing $2|q|$.

The same metric should also describe the interior of the inner horizon for $\rho \in (\rho_{\rm sing}, +\infty)$. The "entropy" of this horizon is easily found to be given by $S_{\rm inner}/\pi = (M - r_0)^2$ and

$$S_{\rm inner}S_{\rm outer} = \pi^2(V_{\rm bh}(\varphi_{\rm h}, \mathcal{Q}))^2 = S^2_{\rm extremal}, \tag{27.50}$$

which is a moduli-independent quantity. This property, sometimes called the *geometric mean property*, has been shown to hold in charged, rotating, asymptotically flat or ADS

BH solutions of a wide class of theories in Ref. [394], following earlier work [405, 884, 396, 397, 399]. A similar result for the horizons of arbitrary topology has been proven in Ref. [316].

### 27.1.2 The FGK formalism for $N=2, d=4$ supergravity

We have reviewed these theories in Chapter 7, and we have explained in Chapters 18 and 19 that, in regular BH solutions, the hyperscalars have to be constant and we can safely ignore them. Thus, comparing the generic action Eq. (27.1) with that of the ungauged $N=2, d=4$ theories Eq. (7.8) and ignoring the hyperscalars, we find that the FGK action and the Hamiltonian constraint for these theories take the general form

$$\boxed{\begin{aligned} S[U,Z^i] &= \int d\rho \left[\dot{U}^2 + \mathcal{G}_{ij^*}\dot{Z}^i\dot{Z}^{*\,j^*} - e^{2U}V_{\rm bh}\right], \\ r_0^2 &= \dot{U}^2 + \mathcal{G}_{ij^*}\dot{Z}^i\dot{Z}^{*\,j^*} + e^{2U}V_{\rm bh}. \end{aligned}} \tag{27.51}$$

Using the $N=2$ version of Eq. (18.17) we can rewrite the BH potential in terms of the central charge defined in Eq. (18.14):

$$\boxed{-V_{\rm bh}(Z,Z^*,\mathcal{Q}) = |\mathcal{Z}|^2 + \mathcal{G}^{ij^*}\mathcal{D}_i\mathcal{Z}\mathcal{D}_{j^*}\mathcal{Z}^*,} \tag{27.52}$$

where the Kähler-covariant derivative of the central charge is

$$\mathcal{D}_i\mathcal{Z} = e^{-\mathcal{K}/2}\partial_i\left(e^{\mathcal{K}/2}\mathcal{Z}\right). \tag{27.53}$$

It is convenient to rewrite the covariant derivatives of $\mathcal{Z}$ in terms of partial derivatives of $|\mathcal{Z}|$. Since $\mathcal{Z}$ is covariantly holomorphic,

$$\partial_i|\mathcal{Z}| = \mathcal{D}_i|\mathcal{Z}| = \mathcal{Z}^{*\,1/2}\mathcal{D}_i\mathcal{Z}^{1/2} = \tfrac{1}{2}(\mathcal{Z}^*/\mathcal{Z})^{1/2}\mathcal{D}_i\mathcal{Z}, \tag{27.54}$$

and

$$\boxed{-V_{\rm bh}(Z,Z^*,\mathcal{Q}) = |\mathcal{Z}|^2 + 4\mathcal{G}^{ij^*}\partial_i|\mathcal{Z}|\partial_{j^*}|\mathcal{Z}|.} \tag{27.55}$$

If we consider the supersymmetric (hence, extremal) case and combine this information and the usual BPS bound with the bound for double extremal BHs Eq. (27.45), we get two important results which apply to all the SBHSs, not just to the double extremal ones:

$$\boxed{S/\pi = |\mathcal{Z}_{\rm h}|^2, \qquad \partial_i|\mathcal{Z}||_{\rm h} = 0.} \tag{27.56}$$

In other words: in the supersymmetric case, the values of the scalars on the horizon are the critical points of $|\mathcal{Z}|$ (all the critical points of the central charge are critical points of the

BH potential[15]) and the square of the value of $|\mathcal{Z}|$ on the horizon gives the entropy [522]. It can also be shown that, at the supersymmetric attractors,

$$\partial_i\partial_{j^*}V_{\rm bh}|_{Z_{\rm h\,susy}} = 2\mathcal{G}_{ij^*}V_{\rm bh}|_{Z_{\rm h\,susy}}\,, \qquad \partial_i\partial_{j^*}|\mathcal{Z}||_{Z_{\rm h\,susy}} = \tfrac{1}{2}\mathcal{G}_{ij^*}|\mathcal{Z}||_{Z_{\rm h\,susy}}\,, \tag{27.58}$$

so that $V_{\rm bh}$ has no flat directions and there is a maximum of $V_{\rm bh}$ there (which implies stability, see footnote 11 on p. 804) and a minimum of $|\mathcal{Z}|$ if the scalar metric $\mathcal{G}_{ij^*}$ is not degenerate.

For non-supersymmetric attractors these properties do not hold in general [640, 1202].

Observe that using Eq. (27.55) we can write the FGK action Eq. (27.51), up to the derivative of $\mp 2e^U|\mathcal{Z}|$, as a sum of non-negative terms, *à la Bogomol'nyi*[16] [522]:

$$S[U,Z^i] = \int d\rho\left[\left(\dot{U}\pm e^U|\mathcal{Z}|\right)^2 + \mathcal{G}_{ij^*}\left(\dot{Z}^i \pm 2e^U\partial^i|\mathcal{Z}|\right)\left(\dot{Z}^{*\,j^*} \pm 2e^U\partial^{j^*}|\mathcal{Z}|\right)\right]. \tag{27.59}$$

This implies that the configurations that make each of these terms vanish extremize the action (which vanishes) and must solve the Euler–Lagrange equations. Those terms vanish if the following first-order (*flow*) equations are satisfied:

$$\boxed{\dot{U} = \mp e^U|\mathcal{Z}|. \qquad \dot{Z}^i = \mp 2e^U\partial^i|\mathcal{Z}|.} \tag{27.60}$$

Only one of the signs is allowed if we only consider positive mass: in the $\rho\to 0^-$ limit, the l.h.s. of the first equation is the mass and only the lower sign gives $M = |\mathcal{Z}_\infty|$. These first-order equations imply the Euler–Lagrange equations that follow from the FGK action, as is easy to check by direct substitution. The Hamiltonian constraint, though, is only satisfied for the extremal case $r_0 = 0$. Actually, these BPS equations can also be obtained by requiring unbroken supersymmetry,[17] which we know requires extremality. It is important to note that, in general, this is not the unique way of rewriting the FGK action *à la Bogomol'nyi* and there is more than one set of flow equations for a given theory. Different families of solutions satisfy different first-order equations. This will be discussed in Section 27.1.3.

In the near-horizon limit, these equations give the attractor mechanism for the supersymmetric case. However, the BH attractors are not attractors of this system of ordinary differential equations because the r.h.s. of the first equation does not vanish in the near-horizon limit and only the scalars exhibit an attractive behavior.

*The FGK formalism for $N = 8, d = 4$ SUEGRA.* Let us finish by briefly mentioning that for $N \geq 2$ SUEGRAs, the BH potential can be written as in Eq. (18.17) in terms of the central-charge matrix $\mathcal{Z}_{IJ}$ and the matter charges $\mathcal{Z}_i$, defined in Eqs. (18.15) and (18.16). In this

---

15 But not the other way around! This property can be obtained from the identity

$$-\partial_i V_{\rm bh} = 2Z^*\mathcal{D}_i\mathcal{Z} + i\mathcal{C}_{ijk}\mathcal{G}^{jj^*}\mathcal{G}^{kk^*}\mathcal{D}_{j^*}\mathcal{Z}^*\mathcal{D}_{k^*}\mathcal{Z}^*. \tag{27.57}$$

16 Compare with Eq. (13.44).

17 All the static, spherically symmetric SBHSs described in Section 19.2 satisfy them.

form it can be used as a starting point to investigate the possible attractors in relation to the central-charge matrix, as it was done for the $N=2$ case in Ref. [522]. This investigation was carried out for $N=8$ theory (in which the matter central charges $\mathcal{Z}_i = 0$) in Ref. [528] using the formulae in Appendix I. For the $N=8$ case Eq. (I.23) implies for the complete $\mathcal{V}^M{}_{IJ}$

$$\mathfrak{D}\mathcal{V}^M{}_{IJ} = \tfrac{1}{2}\mathcal{P}_{IJKL}\overline{\mathcal{V}}^{MKL}, \tag{27.61}$$

where $\mathfrak{D}$ is the SU(8)-covariant derivative and $\mathcal{P}_{IJKL}$ is the (complex SU(8) self-dual) scalar Vielbein. Then, we have

$$\mathfrak{D}\mathcal{Z}_{IJ} = \tfrac{1}{2}\mathcal{P}_{IJKL}\overline{\mathcal{Z}}^{KL}, \quad \text{and} - dV_{\text{bh}} = \tfrac{1}{4}\mathcal{P}_{IJKL}\overline{\mathcal{Z}}^{IJ}\overline{\mathcal{Z}}^{KL} + \text{c.c.} \tag{27.62}$$

Using the complex SU(8) self-duality of the scalar Vielbein to write the latter expression in terms of just $\mathcal{P}_{IJKL}$ and multiplying by its inverse (which must exist for regular solutions), we get the following algebraic condition for the critical points of $V_{\text{bh}}$ [528]:

$$\overline{\mathcal{Z}}^{[IJ}\overline{\mathcal{Z}}^{KL]} + \tfrac{1}{2}\varepsilon^{IJKLMNPQ}\mathcal{Z}_{MN}\mathcal{Z}_{PQ} = 0. \tag{27.63}$$

To go further one can use SU(8) rotations so that the only non-vanishing components of $\mathcal{Z}_{IJ}$ are $\mathcal{Z}_{n\,2n} \equiv z_n$, $n = 1,\ldots,4$ [538]. In this basis Eq. (27.63) is equivalent to the following system:

$$\begin{aligned} z_1 z_2 + z_3^* z_4^* &= 0,\\ z_1 z_3 + z_2^* z_4^* &= 0,\\ z_1 z_4 + z_2^* z_3^* &= 0. \end{aligned} \tag{27.64}$$

We can still use SU(8) to make the phases of the four $z_n$ identical, that is, $z_n = \rho_n e^{i\varphi/4}$, so the central-charge matrix takes the form [538]

$$(\mathcal{Z}_{IJ}) = \begin{pmatrix} \rho_1 & & & \\ & \rho_2 & & \\ & & \rho_3 & \\ & & & \rho_4 \end{pmatrix} \otimes \begin{pmatrix} 0 & 1 \\ -1 & 0 \end{pmatrix} e^{i\varphi/4}. \tag{27.65}$$

The attractor equations change in a simple way: $\rho_1\rho_2 + e^{-i\varphi}\rho_3\rho_4 = 0$, etc. The Cremmer–Julia invariant $\Diamond(\mathcal{Q})$, defined in Eq. (19.123), in this SU(8) basis is equal to [532][18]

$$-\Diamond(\mathcal{Z}) = [(\rho_1+\rho_2)^2 - (\rho_3+\rho_4)^2][(\rho_1-\rho_2)^2 - (\rho_3-\rho_4)^2] + 8\rho_1\rho_2\rho_3\rho_4(\cos\varphi - 1). \tag{27.66}$$

The only two ways of solving the attractor equations for regular BHs ($\Diamond \neq 0$) are as follows.

1. All $\rho_n = 0$ but one ($\rho_1$, say). For the solutions corresponding to these attractors (which preserve $1/8$ of the supersymmetries) $J_4(\mathcal{Q}) = -\Diamond(\mathcal{Z}) = \rho_1^4 > 0$ and the entropy is $S/\pi = \sqrt{J_4(\mathcal{Q})} = \rho_1^2$. These are the SBHSs that we have constructed in Section 19.4.

[18] See the discussion in footnote 25 on p. 590 in which it is explained why $\Diamond(\mathcal{Z}) = \Diamond(\mathcal{Q})$.

2. $\varphi=\pi$ and all $\rho_n=\rho$ for some $\rho$. For the solutions corresponding to these attractors (which are not supersymmetric) $J_4(\mathcal{Q}) = -\lozenge(\mathcal{Z}) = -16\rho^4 < 0$ and the entropy is $S/\pi = \sqrt{-J_4(\mathcal{Q})} = 4\rho^2$. The prototype of this type of extremal non-supersymmetric black-hole solutions is the dyonic RN–KK BH constructed in Section 15.3.4[19] and first found in Ref. [855]. From the higher-dimensional point of view this is a purely gravitational solution but it can be dualized into a system of intersecting D0- and D6-branes [532].

The attractor equations Eqs. (27.64) turn out to be identical to those of the $N=2, d=4$ STU model, which can be obtained from the general $N=2$ ones Eq. (27.57) upon the identifications

$$z_1 = i\mathcal{Z}, \qquad z_2 = \mathcal{D}_{S^*}\mathcal{Z}^*, \qquad z_3 = \mathcal{D}_{T^*}\mathcal{Z}^*, \qquad z_4 = \mathcal{D}_{U^*}\mathcal{Z}^*, \tag{27.67}$$

where the derivatives have tangent-space indices. The $N=8$ BPS case with $z_1 \neq 0$ corresponds to the $N=2$ case $\mathcal{Z}\neq 0$, $\mathcal{D}_i\mathcal{Z}=0$, but the other $N=8$ BPS cases, in which the other $z_n$ are non-vanishing, correspond to three non-BPS attractors of the $N=2$ STU model with $|\mathcal{Z}|=0$. The $N=8$ non-BPS case corresponds to the cases $|\mathcal{Z}| = |\mathcal{D}_i\mathcal{Z}| \neq 0\ \forall i$.

### 27.1.3 Flow equations

As we mentioned before, there are, in general, several ways of rewriting the FGK action Eq. (27.51) as a sum of squares, so there is more than one set of first-order (*flow*) equations that imply the Euler–Lagrange ones. We have seen there is always one way of doing it, given by Eq. (27.59), from which the BPS Eqs. (27.60), associated with unbroken supersymmetry, follow.

It is obvious that, if there was another function (known in the literature as a *prepotential* and also as a *superpotential*) $W(Z, Z^*, \mathcal{Q})$ such that the BH potential took the same form Eq. (27.55) with $|\mathcal{Z}|$ replaced by $W$, that is

$$-V_{\rm bh}(Z, Z^*, \mathcal{Q}) = W^2 + 4\mathcal{G}^{ij^*}\partial_i W \partial_{j^*} W, \tag{27.68}$$

the FGK could be immediately rewritten à la Bogomol'nyi, as in Eq. (27.59) with $|\mathcal{Z}|$ replaced by $W$, and the flow equations

$$\boxed{\dot{U} = \mp e^U W, \qquad \dot{Z}^i = \mp 2e^U \partial^i W,} \tag{27.69}$$

would immediately follow. In most theories $W$s different from $|\mathcal{Z}|$ can be found and are associated with extremal non-supersymmetric BH solutions [322, 900, 52, 534, 241, 323, 324]. In all cases, the extrema of the superpotential are also extrema of the BH potential

$$\partial_i W|_{Z_{\rm h}} = 0 \ \Rightarrow \partial_i V_{\rm bh}|_{Z_{\rm h}} = 0, \tag{27.70}$$

[19] The solution presented there has the magnetic charge equal to the electric one. In the pure KK theory, with only one scalar and one vector, this is the only possibility (apart from the purely electric and magnetic ones). Dyonic solutions with independent electric and magnetic charges are possible in more complicated models (with more vector fields) with a continuous group of electric–magnetic duality.

and the mass and the scalar charges are given by the value of the superpotential and its derivatives at spatial infinity:[20]

$$M \lim_{\rho\to 0^-} W, \qquad \Sigma^i = - \lim_{\rho\to 0^-} \mathcal{G}^{ij}\partial_j W. \tag{27.71}$$

This discussion can be extended to more general models described by the generic FGK action Eq. (27.30) and to the non-extremal case. If a function (sometimes called a *generalized superpotential*) $S(U,\varphi,\rho)$ exists such that

$$-[e^{2U}V_{\rm bh}(\varphi,\mathcal{Q}) - r_0^2] = \tfrac{1}{4}(\partial_U S)^2 + \tfrac{1}{2}\mathcal{G}^{ij}\partial_i S\partial_j S, \tag{27.72}$$

then the FGK action Eq. (27.30) can be rewritten à la Bogomol'nyi up to a total derivative if

$$\frac{\partial S}{\partial \rho} = -r_0^2, \tag{27.73}$$

and we find that the first-order equations

$$\boxed{\dot{U} = \mp\tfrac{1}{2}\partial_U S, \qquad \dot{\varphi}^i = \mp\mathcal{G}^{ij}\partial_j S} \tag{27.74}$$

imply not only the Euler–Lagrange equations, but also the Hamiltonian constraint for arbitrary values of $r_0$ [949, 830, 299, 1035]. That is: non-extremal BHs also satisfy first-order equations! Furthermore, the extrema of the generalized superpotential are also extrema of the BH potential (although these extrema seem to be relevant only in the extremal BHs), and the mass and scalar charges are also given by the derivatives of $S$ at spatial infinity. Taking into account Eq. (27.73) we expect $S$ to have the general form

$$S(U,\varphi,\rho) \equiv 2\mathcal{W}(U,\varphi) - r_0^2\rho = 2e^U W(\varphi,\rho) + r_0^2\rho, \tag{27.75}$$

where $W$ reduces the superpotential in the extremal case.

The universality of this construction can be traced to the existence of another well-known universal construction for mechanical systems: the Hamilton–Jacobi formalism.[21] Defining $(\varphi^a) \equiv (U,\varphi^i)$ and the metric $(\mathcal{G}_{ab}) \equiv \begin{pmatrix} 2 & 0 \\ 0 & \mathcal{G}_{ij}\end{pmatrix}$, the definition of the generalized superpotential Eq. (27.72) and the first-order Eqs. (27.74) take the form

$$\tfrac{1}{2}\mathcal{G}^{ab}\partial_a S\partial_b S = -[e^{2U}V_{\rm bh}(\varphi,\mathcal{Q}) - r_0^2], \qquad \pi_a = \frac{\partial S}{\partial \phi^a}, \tag{27.76}$$

where $\pi_a \equiv \mathcal{G}_{ab}\dot{\phi}^b$ are the canonical momenta.

[20] The lower sign of Eqs. (27.69) has been taken here.

[21] This formalism has been used in this context to study renormalization-group flows which are dual (in the gravity/gauge sense) to domain-wall and cosmological solutions [226, 1219, 569, 1141], to extremal BHs [781], and to other solutions of higher-dimensional supergravity describing extended objects [830]. Here we follow Ref. [53]. A deep relation between the Hamilton–Jacobi formalism and pseudosupersymmetric mechanics has been discovered in Ref. [1198].

These equations should be compared with those that define Hamilton's principal function $S(\varphi, \Pi, \rho)$ ($\Pi_a$ and $\Phi^a$ being the new constant momenta and positions):

$$\pi_a = \frac{\partial S}{\partial \phi^a}, \qquad \Phi^a = \frac{\partial S}{\partial \Pi_a}, \qquad H = -\frac{\partial S}{\partial \rho}, \tag{27.77}$$

where

$$H = \tfrac{1}{2}\mathcal{G}^{ab}\pi_a\pi_b + \mathcal{V}(\phi), \quad \text{with} \quad \mathcal{V}(\phi) \equiv e^{2U} V_{\text{bh}}(\varphi), \tag{27.78}$$

is the Hamiltonian. Combining the first and third Eqs. (27.77) with the Hamiltonian constraint we find

$$S(\phi, \rho) = \mathcal{W}(\phi) - r_0^2\rho, \tag{27.79}$$

$$\pi_a = \frac{\partial \mathcal{W}}{\partial \phi^a}, \tag{27.80}$$

$$\tfrac{1}{2}\mathcal{G}^{ab}\partial_a\mathcal{W}\partial_b\mathcal{W} = -[e^{2U}V_{\text{bh}}(\varphi, \mathcal{Q}) - r_0^2], \tag{27.81}$$

where $\mathcal{W}$ is Hamilton's characteristic function. The conclusion is that, if we find a solution for Eq. (27.81), then the solutions of the mechanical system satisfy the first-order Eqs. (27.80) that can be recast in the form

$$\dot{\phi}^a = \mathcal{G}^{ab}\frac{\partial \mathcal{W}}{\partial \phi^b}. \tag{27.82}$$

All mechanical systems (and, in particular, the FGK action) admit this treatment for a number of functions $S$ and $\mathcal{W}$ associated to different solutions. The Hamilton–Jacobi theory provides methods for computing them. See, for instance, Ref. [53].

## 27.2 The general FGK formalism

It is natural to extend the FGK formalism [522] to encompass all kinds of extended objects in any number of dimensions (higher than four). This extension was worked out in Ref. [56], which we are going to follow, with some small changes in the notation.

First we have to specify the kind of theories whose solutions we want to study. If we want to study charged $p$-branes, our action will have to include $(p+1)$-form potentials with respect to which they can be electrically charged. Most higher-dimensional supergravities include potentials of different ranks, but our action will only contain, for simplicity, potentials of the same rank $A^{\Lambda}_{(p+1)}$ coupled to gravity and to scalar fields $\phi^i$ as in Eq. (2.178).[22]

The dual potentials $A_{\Lambda\,(\tilde{p}+1)}$ ($\tilde{p} \equiv d - p - 4$) have, in general, different rank and the $p$-branes that we want to study cannot be charged with respect to them. For the same reason we cannot include terms of the form $F_{(p+2)} \star F_{(p+2)}$ in the action because they make no sense. However, when $\tilde{p} = p$ (for particular values of $d$ and $p$), $p$-branes can be

[22] We are going to rely heavily on the notation and results of Section 2.6.3; the reader is encouraged to read that section before this one.

dyonic (electrically charged with respect to both $A^{\Lambda}_{(p+1)}$ and the dual potentials $A_{\Lambda\,(p+1)}$; this second coupling and the charges will be called "magnetic") and the $F_{(p+2)} \star F_{(p+2)}$ terms do make sense. In order to treat all the possible cases in a unified way we are going to introduce those terms formally on the understanding that they are to be considered only when $\tilde{p} = p$. Furthermore, when $d = 4n+2$ the dyonic $p = \tilde{p}$-branes can also be self- or anti-self-dual (with the $(p+2)$-field strengths satisfying the corresponding constraint). This can be taken into account in this framework by identifying electric and magnetic charges up to a sign. In this way we can obtain general results, from which the particular ones are deduced by setting to zero some terms and integration constants, or by making identifications among the latter.

Accordingly, we are going to consider the action Eq. (2.184), which we reproduce here for convenience:

$$S[g, A^{\Lambda}_{(p+1)}, \varphi^i] = \int d^d x \sqrt{|g|} \left\{ R + \mathcal{G}_{ij}(\varphi)\partial_\mu \varphi^i \partial^\mu \varphi^j + 4\tfrac{(-1)^p}{(p+2)!} I_{\Lambda\Sigma}(\varphi) F^{\Lambda}_{(p+2)} F^{\Sigma}_{(p+2)} \right.$$

$$\left. + 4\xi^2 \tfrac{(-1)^p}{(p+2)!} R_{\Lambda\Sigma}(\varphi) F^{\Lambda}_{(p+2)} \star F^{\Sigma}_{(p+2)} \right\}, \qquad (27.83)$$

where $\xi^2 = -(-1)^{d/2} = (-1)^{p+1}$. Whenever $\tilde{p} \neq p$ we must set $R_{\Lambda\Sigma} = 0$.

Having specified the class of theories under consideration we must now specify the class of solutions we are interested in by making an appropriate ansatz for the metric and other fields. We want to study single, charged, static, flat,[23] black $p$-branes in $d = p + \tilde{p} + 4$ dimensions using a transverse radial coordinate $\rho$ in such a way that the event horizon is at $\rho \to \infty$.[24]

Examining the metrics of the well-known families of black $p$-brane solutions reviewed in Section 24.3 (which include the original solutions of Ref. [775]) one arrives at the following ansatz for the metric [56]:[25]

$$ds^2_{(d)} = e^{\frac{2}{p+1}\tilde{U}} \left[ W^{\frac{p}{p+1}} dt^2 - W^{-\frac{1}{p+1}} d\vec{y}_p^{\,2} \right] - e^{-\frac{2}{\tilde{p}+1}\tilde{U}} d\sigma^2_{\tilde{p}+3}, \qquad (27.84)$$

$$d\sigma^2_{\tilde{p}+3} = \left( \frac{\omega/2}{\sinh\left(\frac{\omega}{2}\rho\right)} \right)^{\frac{2}{\tilde{p}+1}} \left[ \left( \frac{\omega/2}{\sinh\left(\frac{\omega}{2}\rho\right)} \right)^2 \frac{d\rho^2}{(\tilde{p}+1)^2} + d\Omega^2_{(\tilde{p}+2)} \right], \qquad (27.85)$$

where $\vec{y}_p = (y^1, \ldots, y^p)$ are the spatial worldvolume coordinates, $d\Omega^2_{(\tilde{p}+2)}$ is the metric of the round $(\tilde{p}+2)$-sphere of unit radius, $\rho$ is the radial coordinate in the $(\tilde{p}+3)$-dimensional transverse space, $\tilde{U}$ and $W$ are two functions of $\rho$ to be found by solving the equations of motion, and $\omega$ is the *non-extremality parameter* that appears in Eqs. (11.3), (11.81), (12.92), (12.227), (24.63), etc. Observe that this metric has rational powers of $\sinh \frac{\omega}{2}\rho$ and it will be ill defined for $\rho \leq 0$ if it is well defined for $\rho \geq 0$, which is the

[23] Flat in the spatial directions of its worldvolume where the metric should be Euclidean. The metric of the full worldvolume is not flat, however.

[24] The convention $\rho \to -\infty$ for the near-horizon limit used in the four-dimensional case presents some problems in higher dimensions. It is convenient to use $\rho \to \infty$ to obtain results valid in any dimension $d$.

[25] This metric has also been obtained in Ref. [830].

convention we have chosen.

In the $d=4, p=0$ case this metric reduces to that in Eq. (27.2) upon the identification $\omega=-2r_0$. However, in general $\omega$ is not a length and using $r_0$ would be misleading, hence the different notation. It also reduces to the higher-dimensional BH ($p=0$) metric used in Ref. [939] in which $W$ disappears and $\tilde{U}$ is just the $U$ used there.

The presence of two independent functions, $\tilde{U}$ and $W$, instead of just one is surprising: if we reduced the $p$-brane metric over the $p$ spatial worldvolume directions we should obtain a BH in $\tilde{p}+4$ dimensions and they only depend on $\tilde{U}$. However, $W$ cannot be "gauged away" to make it disappear from all the components of the metric by redefining $\tilde{U}$ and the transverse metric. We must accept its unavoidable presence in the ansatz and substitute it blindly in the equations of motion. The stated apparent contradiction must then be solved.

For the *electric* $(p+1)$-form potentials $A^{\Lambda}_{(p+1)}$ we are going to assume that their only non-vanishing components are the worldvolume ones

$$A^{\Lambda}_{(p+1)\,\underline{t}\underline{y}^1\cdots\underline{y}^p} = \psi^{\Lambda}(\rho). \tag{27.86}$$

When $\tilde{p}=p$, we will also assume that the worldvolume components of the dual (*magnetic*) potentials $A_{\Lambda\,(p+1)}$ are given by

$$A_{(p+1)\,\Lambda\,\underline{t}\underline{y}^1\cdots\underline{y}^p} = \chi_{\Lambda}(\rho). \tag{27.87}$$

Since the electric and magnetic potentials are related by electric–magnetic duality this last assumption implies that some of the transverse components of both the electric and magnetic potentials are non-vanishing too, as in the $d=4, p=0$ case studied before. Using the duality relations it will be enough to consider the worldvolume components.

When $\tilde{p}\neq p$ we simply have to set to zero the $\chi_{\Lambda}$s and we do not need to impose the Bianchi identities.

Just as the electric and magnetic field strengths $F_{(p+2)}{}^{\Lambda}, G_{(\tilde{p}+2)\,\Lambda}$ can be combined into a vector $\mathcal{F}^M$, Eq. (2.187), the functions $\psi^{\Lambda}$ and $\chi_{\Lambda}$ can be combined into $(\Psi^M)\equiv\binom{\psi^{\Lambda}}{\chi_{\Lambda}}$.

Finally, we will assume that all the scalars depend only on $\rho$.

Plugging this ansatz into the Maxwell equations and Bianchi identities, collectively written as

$$d\mathcal{F}^M=0, \tag{27.88}$$

we get the equations

$$\frac{d}{d\rho}\left[e^{-2\tilde{U}}\,\mathcal{M}_{MN}\,\dot{\Psi}^N\right]=0, \tag{27.89}$$

where $\mathcal{M}_{MN}=\mathcal{M}_{MN}(\mathcal{N})$ has been defined in Eq. (2.196) and the overdots indicate derivation with respect to $\rho$. (When $\tilde{p}\neq p$ we just have to set $R_{\Lambda\Sigma}=\chi_{\Lambda}=0$.) These equations can be integrated immediately

$$\dot{\Psi}^M=\alpha e^{2\tilde{U}}\mathcal{M}^{MN}\mathcal{Q}_N, \tag{27.90}$$

where the integration constants $\mathcal{Q}_M$ are the charges with respect to the electric and magnetic potentials:

$$\mathcal{Q}_M \sim \int_{S^{\tilde{p}+2}} \star\mathcal{M}_{MN}\mathcal{F}^N, \qquad (\mathcal{Q}^M)\equiv\begin{pmatrix}p^{\Lambda}\\ q_{\Lambda}\end{pmatrix}=\mathcal{Q}_N\Omega^{NM}, \tag{27.91}$$

$\alpha$ is a normalization constant, and $\Omega_{MN}$, given in Eq. (2.198), is the $\mathrm{O}(n,n)$-invariant metric when $\xi^2=+1$ and the $\mathrm{Sp}(2n,\mathbb{R})$-invariant one when $\xi^2=-1$.

Using Eq. (2.195) the Einstein equations can be written in the form

$$G_{\mu\nu}+\mathcal{G}_{ij}\left[\partial_\mu\varphi^i\partial_\nu\varphi^j-\tfrac{1}{2}g_{\mu\nu}\partial_\rho\varphi^i\partial^\rho\varphi^j\right]+\frac{4\xi^2}{(p+1)!}\mathcal{M}_{MN}\mathcal{F}^M{}_\mu{}^{\rho_1\cdots\rho_{p+1}}\mathcal{F}^N{}_{\nu\rho_1\cdots\rho_{p+1}}=0, \tag{27.92}$$

and, upon substitution of the ansatz, they split into three equations. One of them takes a very simple form and can be integrated right away:

$$\frac{d^2\ln W}{d\rho^2}=0, \qquad W=e^{\gamma\rho}, \tag{27.93}$$

if we normalize $W=1$ at spatial infinity $\rho\to 0^+$.

Before we study the other equations of motion it is worth taking a moment to study the metric Eq. (27.84) now that we have found $W(\rho)$. Substituting the second Eq. (27.93) for $W$, the metric takes the form

$$ds^2_{(d)}=e^{\frac{2}{p+1}\tilde{U}}\left[e^{\frac{p}{p+1}\gamma\rho}dt^2-e^{-\frac{1}{p+1}\gamma\rho}d\vec{y}^{\,2}_p\right]-e^{-\frac{2}{\tilde{p}+1}\tilde{U}}d\sigma^2_{(\tilde{p}+3)}, \tag{27.94}$$

with $d\sigma^2_{\tilde{p}+3}$ still given by Eq. (27.85); as expected, it only depends on one undetermined function $\tilde{U}$, but it depends on two different constants $\omega$ and $\gamma$. The same argument that we used to argue that the metric should depend on only one undetermined function indicates that it should depend on only one undetermined constant ($\omega$).

In fact, $\gamma$ can be related to $\omega$ by demanding regularity of the black-brane horizon. In the near-horizon ($\rho\to+\infty$) limit for $\omega\neq 0$,[26] the angular part of the transverse metric $d\sigma^2_{(\tilde{p}+3)}$ behaves as

$$\sim e^{\frac{1}{\tilde{p}+1}\omega\rho}(-\omega)^{\frac{2}{\tilde{p}+1}}d\Omega^2_{(\tilde{p}+2)}, \tag{27.95}$$

and this part of the complete $d$-dimensional metric will be regular only if in this limit

$$\tilde{U}\sim C+(\omega/2)\rho, \tag{27.96}$$

for some constant $C$ whose interpretation we will give later. Using this information and demanding the regularity of the worldvolume components of the metric in the same limit, we find that $\gamma=\omega$.

The constant $C$ is related to the *entropy density by unit (world-) volume* $\tilde{S}$: the sections of the event horizons of the branes covered by our ansatz have the topology $\mathbb{R}^p\times S^{\tilde{p}+2}$ and have an infinite volume because the worldvolume is not compact. The entropy per unit worldvolume is finite, though. To normalize this quotient we divide by the volume of the unit $S^{\tilde{p}+2}$, $\omega_{(\tilde{p}+2)}$ for convenience (for four-dimensional black holes $\tilde{S}=S/\pi$). Thus, $\tilde{S}$ is defined by

$$\tilde{S}\equiv\frac{A_{\mathrm{h}\,(\tilde{p}+2)}}{\omega_{(\tilde{p}+2)}}, \tag{27.97}$$

[26] This analysis is different in the extremal case $\omega=0$, which we study in Section 27.2.1.

where $A_{\mathrm{h}\,(\tilde{p}+2)}$ is the volume of the $(\tilde{p}+2)$-dimensional constant worldvolume sections of the horizon. Given the above near-horizon behavior of $\tilde{U}$ we get

$$\tilde{S} = \left(-e^{-C}\omega\right)^{\frac{\tilde{p}+2}{\tilde{p}+1}}, \quad \text{so} \quad e^{C} = -\omega \tilde{S}^{-\frac{\tilde{p}+1}{\tilde{p}+2}}. \tag{27.98}$$

Summarizing, we have found that the metric Eq. (27.94), with $\omega \neq 0$, is regular in the near-horizon limit if

$$e^{\tilde{U}} \sim (-\omega)\tilde{S}^{-\frac{\tilde{p}+1}{\tilde{p}+2}} e^{\frac{\omega}{2}\rho}, \qquad W \sim e^{\omega\rho}, \tag{27.99}$$

and the metric has to have the form

$$\begin{aligned} ds^2_{(d)} &= e^{\frac{2}{p+1}\tilde{U}}\left[e^{\frac{p}{p+1}\omega\rho}dt^2 - e^{-\frac{1}{p+1}\omega\rho}d\vec{y}^{\,2}_p\right] - e^{-\frac{2}{\tilde{p}+1}\tilde{U}}d\sigma^2_{(\tilde{p}+3)}, \\ d\sigma^2_{\tilde{p}+3} &= \left(\frac{\omega/2}{\sinh\left(\frac{\omega}{2}\rho\right)}\right)^{\frac{2}{\tilde{p}+1}}\left[\left(\frac{\omega/2}{\sinh\left(\frac{\omega}{2}\rho\right)}\right)^2 \frac{d\rho^2}{(\tilde{p}+1)^2} + d\Omega^2_{(\tilde{p}+2)}\right]. \end{aligned} \tag{27.100}$$

The near-horizon limit of the time-radial part of this metric can be rewritten in the form

$$\sim e^{\frac{2}{p+1}C}\exp\left(-\frac{(\tilde{p}+1)e^{Cc}}{(-\omega)^{\frac{1}{\tilde{p}+1}}}\varrho\right)\left[dt^2 - d\varrho^2\right], \quad \text{where} \quad c \equiv \frac{d-2}{(p+1)(\tilde{p}+1)}, \tag{27.101}$$

and, comparing with the Rindler metric $e^{-\frac{4\pi}{\beta}\rho}\left[dt^2 - d\varrho^2\right]$, where $\beta$ is the inverse Hawking temperature, we can immediately identify the inverse Hawking temperature of the $p$-brane in terms of $C$ and $\omega$. Using the relation between $C$ and the entropy density we obtain the following relation between the temperature, the entropy density, and the non-extremality parameter,

$$(-\omega)^{\frac{1}{\tilde{p}+1}} = \tfrac{4\pi}{\tilde{p}+1} T \tilde{S}^{\frac{(d-2)}{(p+1)(\tilde{p}+2)}}, \tag{27.102}$$

which generalizes for $p$-branes in $d$ dimensions Eq. (27.9) and justifies our calling $\omega$ the "non-extremality parameter" since it vanishes when $T$ vanishes (assuming $\tilde{S}$ remains finite).

Let us return to our original problem. Substituting the ansatz into the Einstein equations taking into account our results for $W$ and $\Psi^M$ Eq. (27.90), we find two equations for $\tilde{U}, \varphi^i$:

$$\ddot{\tilde{U}} + e^{2\tilde{U}} V_{\mathrm{bb}} = 0, \tag{27.103}$$

$$\dot{\tilde{U}}^2 + \tfrac{(p+1)(\tilde{p}+1)}{d-2}\mathcal{G}_{ij}\dot{\varphi}^i\dot{\varphi}^j + e^{2\tilde{U}} V_{\mathrm{bb}} = (\omega/2)^2, \tag{27.104}$$

where

$$V_{\mathrm{bb}}(\varphi, \mathcal{Q}) \equiv 2\alpha^2 \tfrac{(p+1)(\tilde{p}+1)}{(d-2)} \mathcal{M}_{MN}\mathcal{Q}^M\mathcal{Q}^N \tag{27.105}$$

is the (negative semi-definite) *black-brane potential*. Doing the same in the equations of motion of the scalars

$$\nabla^2\varphi^i + \Gamma_{jk}{}^i\dot{\varphi}^j\dot{\varphi}^k + \frac{2}{(p+2)!}\partial^i(F^\Lambda \star G_\Lambda) = 0, \quad (27.106)$$

we get a third equation:

$$\ddot{\varphi}^i + \Gamma_{jk}{}^i\dot{\varphi}^j\dot{\varphi}^k + \tfrac{d-2}{2(\tilde{p}+1)(p+1)}e^{2\tilde{U}}\partial^i V_{\rm bb} = 0. \quad (27.107)$$

As in the four-dimensional BH case, Eqs. (27.103) and (27.107) can be derived from a mechanical effective action

$$S[\tilde{U},\varphi^i] = \int d\rho \left\{\dot{\tilde{U}}^2 + \tfrac{(p+1)(\tilde{p}+1)}{d-2}\mathcal{G}_{ij}\dot{\varphi}^i\dot{\varphi}^j - e^{2\tilde{U}}V_{\rm bb}\right\}, \quad (27.108)$$

while Eq. (27.104) states that the value Hamiltonian associated with this action (which is a conserved quantity) must take the value $(\omega/2)^2$.

### 27.2.1 FGK theorems for static flat branes

The effective action allows us to generalize the FGK theorems and the attractor mechanism for extremal and non-extremal $p$-branes. Since the generalization follows the steps of the BH case we will not give many details.

In the extremal $\omega \longrightarrow 0$ limit, the general metric Eq. (27.100) becomes[27]

$$ds^2_{(d)} = e^{\frac{2\tilde{U}}{p+1}}\left[dt^2 - d\vec{y}^{\,2}_p\right] - \frac{e^{-\frac{2\tilde{U}}{\tilde{p}+1}}}{\rho^{\frac{2}{\tilde{p}+1}}}\left[\frac{1}{\rho^2}\frac{d\rho^2}{(\tilde{p}+1)^2} + d\Omega^2_{(\tilde{p}+2)}\right]. \quad (27.109)$$

This metric will be regular in the near-horizon limit if

$$e^{\tilde{U}} \sim \tilde{S}^{-\frac{\tilde{p}+1}{\tilde{p}+2}}\rho^{-1}, \quad (27.110)$$

and then it will take the form

$$ds^2_{(d)} = \rho^{\frac{-2}{p+1}}\tilde{S}^{-\frac{2(\tilde{p}+1)}{(p+1)(\tilde{p}+2)}}\left[dt^2 - d\vec{y}^{\,2}_p\right] - \tilde{S}^{\frac{2}{\tilde{p}+2}}\left[\frac{1}{\rho^2}\frac{d\rho^2}{(\tilde{p}+1)^2} + d\Omega^2_{(\tilde{p}+2)}\right], \quad (27.111)$$

which is the metric of $\mathrm{AdS}_{p+2} \times \mathrm{S}^{\tilde{p}+2}$ with radii equal to $\tilde{S}^{\frac{1}{\tilde{p}+2}}$.

Imposing a regularity condition similar to that in Ref. [522], Eq. (27.35), and demanding finiteness of the scalars on the horizon, it is not difficult to show that the entropy density is related to the black-brane potential by

$$\tilde{S} = \left[-V_{\rm bb}(\varphi_{\rm h},\mathcal{Q})\right]^{\frac{\tilde{p}+2}{2(\tilde{p}+1)}}, \quad (27.112)$$

[27] Observe that the transverse metric is the Euclidean metric in $\mathbb{R}^{\tilde{p}+3}$, as can be seen by making the coordinate change $\rho = 1/r^{\tilde{p}+1}$.

which generalizes Eq. (27.39), and also (if the scalar metric remains invertible on the horizon) that the values of the scalars on the horizon $\varphi^i_{\rm h}$ (*attractors*) extremize the black-brane potential

$$\boxed{\partial_i V_{\rm bb}|_{\varphi=\varphi_{\rm h}} = 0.} \tag{27.113}$$

This result leads to the generalization of the attractor mechanism for $p$-branes and, in particular, to the fact that the entropy density of an extremal black $p$-brane will only depend on the charges.

The generalization of the bound Eq. (27.44) for non-extremal $p$-branes turns out to be less straightforward. Defining the constant

$$\tilde{u} \equiv - \dot{\tilde{U}}\Big|_{\rho\to 0^+}, \tag{27.114}$$

and taking the spatial-infinity ($\rho \to 0^+$) limit of the Hamiltonian constraint Eq. (27.104), we get

$$\boxed{\tilde{u}^2 + \tfrac{(p+1)(\tilde{p}+1)}{d-2}\mathcal{G}_{ij}(\varphi_\infty)\Sigma^i\Sigma^j + V_{\rm bb}(\varphi_\infty, \mathcal{Q}) = (\omega/2)^2,} \tag{27.115}$$

but now $\tilde{u}$ is not the $p$-brane tension $T_p$ (nor is it simply proportional to it). Applying the definition on p. 716[28]

$$\boxed{T_p = -\tfrac{1}{(p+1)(\tilde{p}+2)}\left[(d-2)\tilde{u} + p(\tilde{p}+1)\omega/2\right],} \tag{27.117}$$

so this bound differs from Eq. (27.44) by terms proportional to $p\omega$. Observe that for uncharged (*Schwarzschild*) branes $\tilde{u} = \omega/2$ and $T_p = -\omega/2$.

### 27.2.2 Inner horizons

Experience shows that charged, black, $p$-branes usually have two horizons (an event horizon and a Cauchy horizon) but the metric Eq. (27.100) only covers the exterior of the event horizon (the region between the event horizon and spatial infinity).

In the four-dimensional BH case the same metric covers the interior of the inner (Cauchy) event horizon (the region between the singularity and the inner horizon) in a different range of values of the radial coordinate [578]:[29] $\rho \in (-\infty, \rho_{\rm sing})$, with $\rho_{\rm sing} < 0$ (see p. 807).

[28] We are using units such that

$$\omega_{\tilde{p}+2}(\tilde{p}+2) = 8\pi G_{\rm N}^{(d)} = 1, \tag{27.116}$$

where $G_{\rm N}^{(d)}$ is the $d$-dimensional Newton constant.

[29] Here we are using the convention adopted in the general case: the event horizon lies at $\rho \to +\infty$ and spatial infinity at $\rho \to 0^+$.

The range $\rho < 0$ is, however, forbidden in the general case; in order to obtain the metric that covers the interior of the inner horizon, one must follow the recipe given in Ref. [942]: take the metric of the exterior of the outer horizon (which will be a metric of the form Eq. (27.100), regular in the interval $\rho \in (0, +\infty)$) and construct from it *another metric* of the same general form by performing the following transformation (which is not a simple coordinate change)

$$\rho \longrightarrow -\varrho, \qquad e^{-\tilde{U}(\rho)} \longrightarrow -e^{-U(-\varrho)}. \tag{27.118}$$

In the new metric the interior of the inner horizon corresponds to the interval of the new radial coordinate $\varrho \in (\varrho_{\rm sing}, +\infty)$ and with it we can compute the "temperature" and "entropy densities" of the black brane's inner horizon and check the geometric mean property discussed on p. 807.

### 27.2.3 *FGK formalism for the black holes of $N = 1, d = 5$ theories*

(Ungauged) $N = 1, d = 5$ supergravity theories provide the simplest case to which the general FGK formalism can be applied, since they contain BH and their dual black-string solutions. The formalism cannot deal with both kinds of solutions simultaneously and in this section we will study the BH case.

The general metric of five-dimensional BHs is

$$ds^2 = e^{2U}dt^2 - e^{-U}\left(\frac{\omega/2}{\sinh\left(\frac{\omega}{2}\rho\right)}\right)\left[\left(\frac{\omega/2}{\sinh\left(\frac{\omega}{2}\rho\right)}\right)^2\frac{d\rho^2}{4} + d\Omega^2_{(3)}\right]. \tag{27.119}$$

These theories were reviewed in Chapter 9. The starting point is the bosonic action Eq. (9.6), but for the kind of solutions that we are considering we can safely ignore the Chern–Simons term. Comparing this action with the generic one in Eq. (27.83), we find that we must replace $p = 0$, $\tilde{p} = 1$, $\mathcal{G}_{ij}$ by $\frac{1}{2}g_{xy}$ and $I_{\Lambda\Sigma}$ by $a_{IJ}$ ($R_{\Lambda\Sigma} = 0$ here and there are no magnetic charges) in the effective action Eq. (27.108) to obtain (writing $U$ instead of $\tilde{U}$) the following effective action:

$$S[U, \phi^x] = \int d\rho\left\{\dot{U}^2 + \tfrac{1}{3}g_{xy}\dot{\phi}^x\dot{\phi}^y - e^{2U}V_{\rm bh}\right\}, \tag{27.120}$$

plus the Hamiltonian constraint

$$\dot{U}^2 + \tfrac{1}{3}g_{xy}\dot{\phi}^x\dot{\phi}^y + e^{2U}V_{\rm bh} = (\omega/2)^2. \tag{27.121}$$

The BH potential $V_{\rm bh}(\phi, q)$ is given, with the normalization $\alpha^2 = 3/32$, by the following two equivalent expressions:

$$-V_{\rm bh}(\phi, q) = a^{IJ}q_Iq_J = \mathcal{Z}_{\rm e}^2 + 3g^{xy}\partial_x\mathcal{Z}_{\rm e}\partial_y\mathcal{Z}_{\rm e}, \tag{27.122}$$

where we have used the definition of the *(electric) BH central charge*

$$\boxed{\mathcal{Z}_{\rm e}(\phi, q) \equiv h^I(\phi) q_I,} \tag{27.123}$$

and we have used Eq. (H.11). The SBHSs of these theories that we studied in Section 19.5 and are extremal BHs saturate the supersymmetry bound $M \geq \mathcal{Z}_{\rm e}(\phi_\infty, q)$, and it is not difficult to see that the values of the scalars of the SBHSs on the horizon extremize the BH central charge $\mathcal{Z}_{\rm e}$. On the other hand, the extremization of the BH central charge $\mathcal{Z}_{\rm e}$ implies the extremization of the BH potential, analogous to what happens in the $N=2, d=4$ case,

$$\partial_x \mathcal{Z}_{\rm e}|_{\phi_{\rm h}} = 0 \quad \Rightarrow \quad \partial_x V_{\rm bh}|_{\phi_{\rm h}} = 0, \tag{27.124}$$

as can be seen by direct differentiation of $V_{\rm bh}$. The converse is not true, since, as in the $N=2, d=4$ case, there are more attractors apart from the supersymmetric ones.

The effective action Eq. (27.120) can be rewritten à la Bogomol'nyi:

$$S[U, \phi^x] = \int d\rho \left\{ (\dot{U} \pm e^U \mathcal{Z}_{\rm e})^2 + \tfrac{1}{3} g_{xy} (\dot{\phi}^x \pm 3e^U \partial^x \mathcal{Z}_{\rm e})(\dot{\phi}^y \pm 3e^U \partial^y \mathcal{Z}_{\rm e}) \mp \frac{d}{d\rho}(2e^U \mathcal{Z}_{\rm e}) \right\}, \tag{27.125}$$

which leads to the flow equations

$$\boxed{\dot{U} = \mp e^U \mathcal{Z}_{\rm e}, \qquad \dot{\phi}^x = \mp 3 e^U \partial^x \mathcal{Z}_{\rm e},} \tag{27.126}$$

which are very similar to those of the $N=2, d=4$ case and which are satisfied by SBHSs (the first equation implies $M = \mathcal{Z}_{\rm e}(\phi_\infty, q)$). Non-supersymmetric extremal BHs are associated with non-supersymmetric attractors and with superpotentials different from $\mathcal{Z}_{\rm e}$.

There is another way of rewriting the action à la Bogomol'nyi using the definition of $g_{xy}$ Eq. (H.11) so the scalars' kinetic term becomes just $a^{IJ} \dot{h}^I \dot{h}^J$:

$$S[U, \phi^x] = \int d\rho \left\{ e^{2U} a^{IJ} \left[ \frac{d}{d\rho}(e^{-U} h_I) \pm q_I \right] \left[ \frac{d}{d\rho}(e^{-U} h_J) \pm q_J \right] \pm \frac{d}{d\rho}(2e^U \mathcal{Z}_{\rm e}) \right\}. \tag{27.127}$$

The associated flow equations are

$$\frac{d}{d\rho}(e^{-U} h_I) = \mp q_I \quad \Rightarrow \quad e^{-U} h_I = A_I \mp q_I \rho, \tag{27.128}$$

for some constants $A_I$, and, thus, we recover the result obtained in Section 19.5 that in the SBHSs the combinations $e^{-U} h_I$ are harmonic functions.

### 27.2.4 *FGK formalism for the black strings of $N=1, d=5$ theories*

The vector fields of the $N=1, d=5$ theories, $A^I{}_\mu$, can be dualized into 2-forms, $B_{I\,\mu\nu}$, associated with black-string solutions. Due to the Chern–Simons term in which the vector fields appear without derivatives, it is not possible to dualize completely the action eliminating all the vector fields, but that term is irrelevant for the kind of solutions that we are going to consider, including only static black strings, and we can take as starting point the bosonic action

$$S = \int \sqrt{g}\left\{R + \tfrac{1}{2}g_{xy}\partial_\mu\phi^x\partial^\mu\phi^y + \tfrac{1}{2\cdot 3!}a^{IJ}H_I H_J\right\}, \tag{27.129}$$

in which the 3-form field strengths are defined by $H_I \equiv dB_I$ and are related to the original vector field strengths by $H_I = \star a_{IJ}F^J$; $a^{IJ}$ is the inverse of $a_{IJ}$. The string charges will be denoted by $p^I$, but they are electric with respect to the 2-form potentials and play the role of the $q_\Lambda$ in the generic action.

The metric of a five-dimensional black string extending along the $y$ direction is

$$ds^2 = e^{\tilde{U}}\left[e^{\frac{\omega}{2}\rho}dt^2 - e^{-\frac{\omega}{2}\rho}dy^2\right] - e^{-2\tilde{U}}\left(\frac{\omega/2}{\sinh\left(\frac{\omega}{2}\rho\right)}\right)^2\left[\left(\frac{\omega/2}{\sinh\left(\frac{\omega}{2}\rho\right)}\right)^2 d\rho^2 + d\Omega^2_{(2)}\right], \tag{27.130}$$

and applying straightforwardly the general formalism we arrive at the effective action

$$\boxed{S[\tilde{U},\phi^x] = \int d\rho\left\{\dot{\tilde{U}}^2 + \tfrac{1}{3}g_{xy}\dot{\phi}^x\dot{\phi}^y - e^{2U}V_{\rm bs}\right\},} \tag{27.131}$$

plus the Hamiltonian constraint

$$\dot{\tilde{U}}^2 + \tfrac{1}{3}g_{xy}\dot{\phi}^x\dot{\phi}^y + e^{2U}V_{\rm bs} = (\omega/2)^2. \tag{27.132}$$

The *black-string potential* $V_{\rm bs}(\phi, p)$ is again given by two equivalent expressions:

$$\boxed{-V_{\rm bs}(\phi,p) \equiv a_{IJ}\,p^I p^J = \mathcal{Z}_{\rm m}^2 + 3\partial_x\mathcal{Z}_{\rm m}\partial^x\mathcal{Z}_{\rm m},} \tag{27.133}$$

where we have used the definition of the *(magnetic) string central charge*

$$\boxed{\mathcal{Z}_{\rm m}(\phi,p) = h_I(\phi)p^I.} \tag{27.134}$$

Note that $\mathcal{Z}_{\rm m}(\phi, p)$ plays for black strings the role played by $\mathcal{Z}_{\rm e}(\phi, q)$ for BHs, allowing the effective action to be rewritten à la Bogomol'nyi.

## 27.3 The H-FGK formalism

Throughout this book and especially in Chapter 19 we have seen that the static extremal BH and black-brane solutions of many theories can be written in terms of a number of harmonic functions. There is essentially one harmonic function for every potential (including the dual ones) and the harmonic functions transform among themselves as the potentials do: linearly under the duality group, as we have seen in Section 2.6. The metric function $e^U$ is a duality-invariant function of these harmonic functions and the scalars are non-linear combinations with the right transformation properties under duality.

This structure extends to other supersymmetric black-brane solutions: stationary, in gauged theories, etc. It is natural to wonder if this harmonic structure is valid for more general BH or black-brane solutions such as the non-extremal ones or if it is a property of the supersymmetric or the extremal solutions only.

In Refs. [578, 939] it was shown that, at least in certain models of $N = 2, d = 4$ and $N = 1, d = 5$ supergravity, the static non-extremal and extremal non-supersymmetric solutions have exactly the same structure as the supersymmetric ones, the difference being that the harmonic functions are different or no longer harmonic. Will this property hold for all possible models? If this is the case, there must be a deep reason behind this universal structure.

Such a reason exists: these supergravity theories can be formulated in terms of real variables which are directly related to the would-be harmonic functions, the building blocks of the BH and black-brane solutions [959, 957, 958, 865]. This fact can be combined with the reduction that leads to the FGK formalism, and the FGK effective actions of the $N = 2, d = 4$ and $N = 1, d = 5$ theories can be rewritten entirely in terms of the new variables, the would-be harmonic functions, which will be denoted by $H^M$ and $H_I$. The result is a new formalism, completely equivalent to the FGK one, based on these $H$-variables called the *H-FGK* formalism [941].

The H-FGK formalism presents several advantages with respect to the standard FGK formalism:

1. The equations satisfied by the $H$-variables are much simpler than those satisfied by the metric function and scalars $H, \varphi^i$. Constructing explicit BH and black-string solutions of $N = 2, d = 4$ and $N = 1, d = 5$ supergravities is much simpler.

2. The $H$-variables transform linearly under duality transformations and, therefore, the integration constants that appear in them must transform equivariantly under duality. This condition constrains the possible integration constants and helps to solve the differential equations satisfied by the $H$-variables [579, 267, 268].

3. In the $N = 2, d = 4$ case, the H-FGK effective action exhibits a very interesting local symmetry [577]. This symmetry extends the discrete *Freudenthal duality* discovered in the BH solutions of $N = 8, d = 4$ supergravity in Ref. [238] and generalized to $N = 2, d = 4$ supergravity in Ref. [536].

4. With the H-FGK formalism it is possible to address the general question of whether the $H$-variables are always harmonic functions for extremal BHs. The short answer is *no*: solutions with anharmonic $H$-variables have been known for some time [900, 633, 575, 240], but with the H-FGK formalism it is possible to study systematically when this happens.

5. In the cases in which the $H$-variables are harmonic, first-order flow equations are easy to construct [1012, 942, 579].

In what follows we are going to review the basics of the H-FGK formalism for $N = 1, d = 5$ and $N = 2, d = 4$ BHs.[30] We will not give all the details of the construction of the effective actions (they can be found in Ref. [941]); we will focus, instead, on their properties and applications.

### 27.3.1 For the black-hole solutions of $N = 1, d = 5$

One of the main results in Section 19.5 is that the static SBHSs of the $N = 1, d = 5$ supergravity theories coupled to $n_V$ vector multiplets are determined by $\bar{n} = n_V + 1$ harmonic functions $H_I$. The relation between the metric function and scalar fields and these harmonic functions is

$$H_I \equiv e^{-U} h_I(\phi), \tag{27.135}$$

a fact that we have recovered in Section 27.2.3. The idea is to replace the $n_V$ scalars $\phi^x$ and the metric function by the $\bar{n}$ variables $H_I(\rho)$, no longer assumed to be harmonic so they can describe non-supersymmetric BHs.

It is convenient to define the $\bar{n}$ dual variables $\tilde{H}^I$ as follows:

$$\tilde{H}^I \equiv e^{-U/2} h^I(\phi), \tag{27.136}$$

which can be expressed in terms of the $H_I$, finding first the relation between the $h^I$ and $h_I$ (which is equivalent to solving the stabilization equations in $N = 2, d = 4$ theories). Those relations $\tilde{H}^I(H.)$ are homogeneous functions of degree $1/2$ in the $H_I$. We also define the function

$$\mathsf{V}(\tilde{H}) \equiv C_{IJK} \tilde{H}^I \tilde{H}^J \tilde{H}^K, \tag{27.137}$$

which is homogeneous of degree 3 in the $\tilde{H}^I$ and of degree $3/2$ in the $H_I$. The $H_I$ can be introduced as the variables conjugate to the $\tilde{H}^I$ with respect to $\mathsf{V}$:

$$H_I = \tfrac{1}{3} \frac{\partial \mathsf{V}}{\partial \tilde{H}^I}, \tag{27.138}$$

and we can define the Legendre transform of $\mathsf{V}$, $\mathsf{W}$ by

$$\mathsf{W}(H) \equiv 3\tilde{H}^I H_I - \mathsf{V}(\tilde{H}) = 2\mathsf{V}. \tag{27.139}$$

The $\tilde{H}^I$ are now the variables conjugate to the $H_I$ with respect to $\mathsf{W}$:

$$\tilde{H}^I \equiv \tfrac{1}{3} \frac{\partial \mathsf{W}}{\partial H_I} \equiv \tfrac{1}{3}\, \partial^I \mathsf{W}. \tag{27.140}$$

$\mathsf{W}(H)$ plays a central role in the five-dimensional H-FGK formalism since we can derive everything from it. In particular, the physical fields are given in terms of $\mathsf{W}$ by

$$e^{-\frac{3}{2}U} = \tfrac{1}{2}\mathsf{W}(H), \qquad h_I = (\mathsf{W}/2)^{-2/3} H_I, \qquad h^I = \tfrac{1}{3}(\mathsf{W}/2)^{-1/3}\, \partial^I \mathsf{W}. \tag{27.141}$$

[30] There is a similar formalism for the black strings of $N = 1, d = 5$ theories [942].

It only remains to rewrite the FGK action Eq. (27.120) and the Hamiltonian constraint Eq. (27.121) in terms of the new variables $H_I$. Using Eq. (H.11) to express $g_{xy}$ in terms of $a^{IJ}$ and the identity

$$a^{IJ} = -\tfrac{2}{3}\,(\mathsf{W}/2)^{4/3}\,\partial^I\partial^J \log \mathsf{W}, \tag{27.142}$$

we get

$$\boxed{-\tfrac{3}{2}S_{\text{H-FGK}}[H] = \int d\rho \left\{\partial^I\partial^J \log \mathsf{W}\left(\dot{H}_I\dot{H}_J + q_I q_J\right)\right\},} \tag{27.143}$$

and

$$\mathcal{H} \equiv \partial^I\partial^J \log \mathsf{W}\left(\dot{H}_I\dot{H}_J - q_I q_J\right) = -\tfrac{3}{2}(\omega/2)^2. \tag{27.144}$$

The kinetic term in the action is that of a non-relativistic particle moving in $\bar{n}$-dimensional space with metric $\partial^I\partial^J \log \mathsf{W}$, i.e. a *Hessian manifold* with *Hessian potential* $\log \mathsf{W}$.

The equations of motion that follow from the H-FGK effective action do not look much simpler than those of the FGK formalism,

$$\partial^K\partial^I\partial^J \log \mathsf{W}\left(\dot{H}_I\dot{H}_J - q_I q_J\right) + 2\partial^K\partial^I \log \mathsf{W}\,\ddot{H}_I = 0, \tag{27.145}$$

but from them one can derive powerful results: multiplying them by $H_K$ and using the homogeneity properties of $\mathsf{W}$ and the Hamiltonian constraint we get

$$\partial^I \log \mathsf{W}\left[\ddot{H}_I - (\omega/2)^2 H_I\right] = 0, \tag{27.146}$$

which is generically solved by functions $H_I$ satisfying[31]

$$\ddot{H}_I - (\omega/2)^2 H_I = 0, \tag{27.147}$$

that is, harmonic functions in the extremal ($\omega = 0$) cases (supersymmetric or not) and linear combinations of hyperbolic sines and cosines of $\frac{\omega}{2}\rho$ in the non-extremal case.

In the supersymmetric case we know that the coefficient of $\rho$ in $H_I$ is $q_I$. In extremal non-supersymmetric cases the coefficients will be some $B_I(q, \phi_\infty)$ (that we can call *attractors* in this context but are also known as *fake charges*) extremizing the BH potential:

$$\boxed{\partial^I V_{\text{bh}}(H, q)\big|_{H_I = B_I} = 0.} \tag{27.148}$$

This equation determines the $B_I$ up to normalization, which is fixed by solving the Hamiltonian constraint.

[31] Actually, all the solutions known satisfy it.

The attractors determine the mass and BH entropy of the extremal BHs:

$$\boxed{M = \mathcal{Z}_{\rm e}(\phi_\infty, B), \qquad \tilde{S} = [-V_{\rm bh}(B,q)]^{3/4} = \mathsf{W}(B)/2,} \tag{27.149}$$

where we have defined the *fake central charges*

$$\mathcal{Z}_{\rm e}(\phi, B) \equiv h^I(\phi) B_I. \tag{27.150}$$

More general results for extremal and non-extremal have been obtained using this formalism in Ref. [942].

### *27.3.2 For $N = 2, d = 4$ black holes*

In the $N = 2, d = 4$ case[32] our objective is to replace the variables $U, Z^i$ ($2\bar{n} - 1$ real variables) of the FGK effective action by the $2\bar{n}$ variables defined in Eq. (19.2):

$$H^M \equiv \mathcal{I}^M \equiv \Im\mathfrak{m}(\mathcal{V}^M/X), \tag{27.151}$$

where $X$ is a complex variable with the same Kähler weight as $\mathcal{V}^M$ which can be written in the form

$$X = \tfrac{1}{\sqrt{2}} e^{U + i\alpha}, \tag{27.152}$$

where the phase $\alpha$ is a variable that does not occur in the original FGK formalism. Since we are increasing the number of variables, the new action must be invariant under some local symmetry. This local symmetry is responsible for the absence of $\alpha$ in the FGK action and only acts on the FGK variables by shifting $\alpha$ by an arbitrary function, leaving $U$ and $\mathcal{V}^M$ invariant.[33]

It is convenient to introduce the (*Freudenthal*) dual variables (defined on p. 565):

$$\tilde{H}^M \equiv \mathcal{R}^M \equiv \Re\mathfrak{e}(\mathcal{V}^M/X). \tag{27.153}$$

These variables can be expressed in terms of the $H^M$, and the mapping from these to the Freudenthal dual variables defines the discrete *Freudenthal duality* transformation [238, 536, 577], which is an anti-involution

$$\tilde{\tilde{H}}^M = -H^M. \tag{27.154}$$

In terms of these variables the scalar fields are given by Eq. (19.4):

$$Z^i = \frac{\tilde{H}^i + iH^i}{\tilde{H}^0 + iH^0}. \tag{27.155}$$

Finally, we define the *Hesse potential* $\mathsf{W}(H)$ as in Eq. (19.4):

$$\mathsf{W}(H) \equiv \tilde{H}_M(H) H^M = e^{-2U} = \frac{1}{2|X|^2}. \tag{27.156}$$

[32] We go back to the radial coordinate $\rho \in (-\infty, 0)$ and to the non-extremality parameter $r_0 = -\omega/2$.

[33] It cannot be interpreted as a Kähler transformation.

The main properties of this object that characterizes the theory are its homogeneity (of degree 2) in the $H^M$ and

$$\partial_M \mathsf{W} \equiv \frac{\partial \mathsf{W}}{\partial H^M} = 2\tilde{H}_M, \tag{27.157}$$

$$\partial^M \mathsf{W} \equiv \frac{\partial \mathsf{W}}{\partial \tilde{H}_M} = 2H^M, \tag{27.158}$$

$$\partial_M \partial_N \mathsf{W} = -2\mathcal{M}_{MN}(\mathcal{F}), \tag{27.159}$$

$$\mathsf{W}\partial_M \partial_N \log \mathsf{W} = 2\mathcal{M}_{MN}(\mathcal{N}) + 4\mathsf{W}^{-1} H_M H_N, \tag{27.160}$$

where we have introduced the matrix $\mathcal{M}_{MN}(\mathcal{F})$, which is indentical to the matrix $\mathcal{M}_{MN}(\mathcal{N})$ defined in Eq. (2.167), with the period matrix $\mathcal{N}_{\Lambda\Sigma}$ replaced by $\mathcal{F}_{\Lambda\Sigma} \equiv \frac{\partial^2 \mathcal{F}}{\partial \mathcal{X}^\Lambda \partial \mathcal{X}^\Sigma}$, $\mathcal{F}$ being the prepotential of the theory.

Since the scalar fields $Z^i$ and the Hesse potential $\mathsf{W}$ (and therefore $e^U$) are invariant under discrete Freudenthal duality, this transformation must be asymmetry of the system formulated in terms of the $H$-variables.

The details of the change of variables can be found in Ref. [941]. Here we will simply quote the result, which is the effective action

$$\boxed{-S_{\text{H-FGK}}[H] = \int d\rho \left\{ \tfrac{1}{2} g_{MN} \dot{H}^M \dot{H}^N - V \right\}} \tag{27.161}$$

and the Hamiltonian constraint

$$\tfrac{1}{2} g_{MN} \dot{H}^M \dot{H}^N + V + r_0^2 = 0, \tag{27.162}$$

where the metric $g_{MN}(H)$ and the potential $V(H)$ are given in terms of $\mathsf{W}(H)$ by

$$g_{MN}(H) \equiv \partial_M \partial_N \log \mathsf{W} - 2\frac{H_M H_N}{\mathsf{W}^2}, \tag{27.163}$$

$$V(H) \equiv \left\{ -\tfrac{1}{4} \partial_M \partial_N \log \mathsf{W} + \frac{H_M H_N}{\mathsf{W}^2} \right\} \mathcal{Q}^M \mathcal{Q}^N = -V_{\text{bh}}/\mathsf{W}. \tag{27.164}$$

The metric $g_{MN}(H)$ always admits a null eigenvector [958, 577]

$$\tilde{H}^M g_{MN} = 0, \tag{27.165}$$

and therefore it is non-invertible. Taking this into account, the equations of motion are

$$g_{MN} \ddot{H}^N + (\partial_N g_{PM} - \tfrac{1}{2} \partial_M g_{NP}) \dot{H}^N \dot{H}^P + \partial_M V = 0. \tag{27.166}$$

If we expand these equations we can easily show that $\dot{H}^M = \mathcal{Q}^M/\sqrt{2}$ (which corresponds to the SBHSs) solves them. This is much more difficult to do in the FGK formalism.

Mimicking what we did in the five-dimensional case, we multiply these equations by $H^M$ using the homogeneity properties and the Hamiltonian constraint. We get

$$\tilde{H}_M \left(\ddot{H}^M - r_0^2 H^M\right) + \frac{(\dot{H}^M H_M)^2}{\mathsf{W}} = 0. \tag{27.167}$$

In the supersymmetric case, staticity (in particular, the absence of the NUT charge [128]) implies the constraint (setting $\omega = 0$ in Eq. (19.9))

$$\dot{H}^M H_M = 0. \tag{27.168}$$

Using this constraint in Eq. (27.167) brings it into the form

$$\tilde{H}_M \left(\ddot{H}^M - r_0^2 H^M\right) = 0, \tag{27.169}$$

which can be solved by harmonic functions in the extremal case and by hyperbolic functions in the non-extremal one. The corresponding solutions have been studied exhaustively in Ref. [579]. However, as we have mentioned before, not all the solutions are of this form. In particular, some extremal non-supersymmetric solutions (the most general non-supersymmetric ones of the $t^3$ and STU models, for instance) have non-harmonic $H^M$s [900, 633, 575, 240]. These *unconventional solutions* [577] do not satisfy the constraint Eq. (27.168), however. We are going to show that it is always possible to impose that constraint on any solution. Still, that does not guarantee the harmonicity of hyperbolicity of the $H^M$s.

The necessity to depart from the harmonic ansatz can be understood as due to the impossibility of having all the possible values for the moduli $Z^i_\infty$ for a given attractor $B^M$ using only the constant terms in the harmonic functions [268]. This obstruction depends on the theory and the non-supersymmetric attractor under consideration. The situation in the non-extremal case is still unclear [576]. The most general non-extremal solution of the STU model is known [355] but not in terms of the $H^M$-variables.

### 27.3.3 *Freudenthal duality*

The discrete Freudenthal duality transformation $H'^M = \tilde{H}^M$ leaves the metric function (the Hesse potential) and scalar fields invariant:

$$e^{-2U}(\tilde{H}) = e^{-2U}(H), \qquad Z^i(\tilde{H}) = Z^i(H). \tag{27.170}$$

This implies that the H-FGK effective action (and, in particular, the BH potential) is also invariant.[34] The immediate consequence is that, if $B^M$ is an attractor

$$\partial_M V_{\rm bh}(H, \mathcal{Q})\|_{H^M = B^M} = 0, \tag{27.171}$$

then its Freudenthal dual $\tilde{B}^M$, defined by

$$\tilde{B}^M \equiv -\tfrac{1}{2}\,\Omega^{MN}\,\frac{\partial \mathsf{W}(B)}{\partial B^N}, \tag{27.172}$$

[34] Observe that this means invariance under the replacement of the variables $H^M$ by their Freudenthal duals $\tilde{H}^M$, but not under the replacement of the charges $\mathcal{Q}^M$ by their Freudenthal duals, defined in Eq. (27.172). In general, $V_{\rm bh}(H, \mathcal{Q}) \neq V_{\rm bh}(H, \tilde{\mathcal{Q}})$.

is also an attractor. As in the five-dimensional case the attractors are defined up to normalization because of the scale invariance of the BH potential as a function of the $H^M$. The normalization condition is

$$V_{\rm bh}(B,\mathcal{Q}) = -\tfrac{1}{2}\mathsf{W}(B)\,, \tag{27.173}$$

which leads to the following Freudenthal-duality expression for the entropy:

$$\boxed{S/\pi = \tfrac{1}{2}\mathsf{W}(B),} \tag{27.174}$$

which generalizes Eq. (19.26), valid for the supersymmetric case only. Black holes with Freudenthal-dual charges have the same BH entropy [238].

We have mentioned before that we expect the H-FGK action to have a local invariance associated with shifts of the phase of $X$, $\alpha$. The non-invertibility of the metric $g_{MN}$ also indicates that the number of independent variables is less than $2\bar{n}$. A more direct proof comes from the following identity that relates the equations of motion:

$$\tilde{H}^M \frac{\delta S_{\text{H-FGK}}}{\delta H^M} = 0. \tag{27.175}$$

This is a Noether identity such as those introduced in Section 2.2, p. 33. The symmetry associated with it can be found by multiplying it by an infinitesimal function $f(\rho)$ and integrating it over $\rho$:

$$\delta_f S_{\text{H-FGK}} = \int d\rho \delta_f H^M \frac{\delta S_{\text{H-FGK}}}{\delta H^M} = 0, \quad \text{where} \quad \delta_f H^M \equiv f(\rho)\tilde{H}^M. \tag{27.176}$$

It can be checked that these infinitesimal transformations leave invariant the full H-FGK action. The finite form of the transformations is

$$(\tilde{H}'^M + iH'^M) = e^{if(\rho)}(\tilde{H}^M + iH^M) \quad \Rightarrow \mathcal{V}^M/X' = e^{if(\rho)}\mathcal{V}^M/X, \tag{27.177}$$

which corresponds to a transformation of the phase of $X$ only:

$$\delta_f \alpha = -f, \tag{27.178}$$

and for $f = -\pi/2$ the discrete Freudenthal duality transformations are recovered. These transformations are, therefore, a continuous and local extension of them that can be called *local Freudenthal symmetry*.

The significance of this symmetry, which seems to have no higher-dimensional counterpart, is not clear. Its use is not clear either: the Noether charge associated with the invariance under the global Freudenthal rotations vanishes identically, and the equations of motion do not get any simpler to solve. Observe, however, that the constraint Eq. (27.168) is not preserved by the local Freudenthal symmetry. It is easy to show that, given a configuration $H^M$ not satisfying Eq. (27.168), a local Freudenthal transformation with parameter defined by

$$\dot{f} = \frac{\dot{H}^M H_M}{\mathsf{W}} \tag{27.179}$$

brings it into another configuration $H'^M$ with the same physical fields satisfying Eq. (27.168). The transformed variables may have very complicated expressions, however.

# Appendix A
# Lie groups, symmetric spaces, and Yang–Mills fields

In this appendix we review some basic definitions and properties of Lie groups and algebras and their use in the construction of homogeneous and symmetric spaces and in field theory. Rigorous definitions and proofs can be found, for instance, in Refs. [741, 868], and in the physicist-oriented Refs. [107, 632, 391, 315 (Vol. II), 370 (Vol. I)].

## A.1 Generalities

A Lie group G of dimension $n$ is both a group and a differential manifold of dimension $n$: the points of the manifold are the elements of the group and the maps

$$\begin{aligned} \mathrm{G}\times\mathrm{G} &\rightarrow \mathrm{G}, & \mathrm{G} &\rightarrow \mathrm{G}, \\ (g_1, g_2) &\rightarrow g_1 g_2, & g &\rightarrow g^{-1}, \end{aligned} \tag{A.1}$$

are differentiable. For each element $g \in \mathrm{G}$ there are also two natural diffeomorphisms: *left* and *right translations by* $g$, denoted by $L_g$ and $R_g$ respectively, and defined by

$$\begin{aligned} L_g : \mathrm{G} &\rightarrow \mathrm{G}, & R_g : \mathrm{G} &\rightarrow \mathrm{G}, \\ h &\rightarrow L_g(h) \equiv gh, & h &\rightarrow R_g(h) \equiv hg. \end{aligned} \tag{A.2}$$

The identity $e$ is a naturally distinguished point. The tangent space at the identity $\mathrm{T}_e^{(1,0)}$ is the *Lie algebra* $\mathfrak{g}$ of G. This name will be justified later. Each element $v(e) \in \mathfrak{g}$ can be extended to a vector field $v(g)$ defined at all points $g \in \mathrm{G}$ by taking the push-forward of the left- or the right-translation diffeomorphisms

$$v_{\mathrm{L}}(g) \equiv L_{g\,*}\, v(e), \qquad v_{\mathrm{R}}(g) \equiv R_{g\,*}\, v(e). \tag{A.3}$$

Sometimes we use the following notation for them:

$$L_{g\,*}\, v \equiv gv, \qquad R_{g\,*}\, v \equiv vg. \tag{A.4}$$

The vector fields defined in this way have the property of being, respectively, *left* and *right invariant*, i.e. they satisfy

$$L_{g\,*}\, v_{\mathrm{L}}(h) = v_{\mathrm{L}}(gh), \qquad R_{g\,*}\, v_{\mathrm{R}}(h) = v_{\mathrm{R}}(hg). \tag{A.5}$$

Similarly, we can define left- and right-invariant differential forms $\omega$ and $\eta$ of any rank using the pullbacks associated with the left- and right-translation diffeomorphisms:

$$L_g{}^* \omega(h) = \omega(gh), \qquad R_g{}^* \eta(h) = \eta(hg). \tag{A.6}$$

It is customary to work with left-invariant vector fields and 1-forms. We can always construct a basis of left-invariant vector fields $\{e_A(g)\}$ using the above procedure starting with a basis of vector fields at the identity $\{e_A(e)\}$ and a dual basis of left-invariant 1-forms $\{e^A(g)\}$. The Lie bracket of two left-invariant vector fields is another left-invariant vector field that we can write as a linear combination of elements of the $\{e_A(g)\}$ basis.[1] Thus,

$$[e_A, e_B] = -f_{AB}{}^C e_C, \tag{A.7}$$

where $f_{AB}{}^C = -f_{BA}{}^C = +(f_{AB}{}^C)^*$ are the *structure constants*. The Jacobi identity Eq. (1.14) implies

$$f_{[AB}{}^E f_{C]E]}{}^D = 0. \tag{A.8}$$

The vector fields at the identity are thus an $n$-dimensional vector space endowed with an antisymmetric, bilinear (but non-associative) product $[\cdot,\cdot]$ (the Lie bracket of the associated left-invariant vector fields) that satisfies the Jacobi identity; that is, by definition, a Lie algebra, which justifies our definition of $\mathfrak{g}$. We will denote a basis of $\mathfrak{g}$ by $\{T_A\}$ and, by convention,

$$e_A(e) \equiv -T_A \;\Rightarrow [T_A, T_B] = f_{AB}{}^C T_C. \tag{A.9}$$

The dual left-invariant 1-forms satisfy the *Maurer–Cartan equations*

$$de^A = \tfrac{1}{2} f_{BC}{}^A e^B \wedge e^C, \tag{A.10}$$

and $d^2 e^A = 0$ is equivalent to the Jacobi identity.

The exponential map provides a local parametrization of G in a neighborhood of the identity with coordinates $\sigma^A$:

$$g(\sigma) = e^{\sigma^A T_A}. \tag{A.11}$$

If the group is a connected and compact manifold, any of its elements can be expressed in this way. With this parametrization it is easy to construct a basis of left-invariant[2] 1-forms by expanding the *Maurer–Cartan 1-form* $V$,

$$V = -g^{-1} dg = e^A T_A, \tag{A.12}$$

in terms of which the Maurer–Cartan equations are $dV - V \wedge V = 0$.

For matrix groups the generators $T_A$ are just matrices and the Lie bracket is just the standard commutator. The left and right translations are just matrix multiplications from the left, $gT$, or from the right, $Tg$. We can take different sets of matrices of different dimensions or operators that satisfy the same commutation relations and provide different

[1] The same is true for right-invariant vector fields. On the other hand, the Lie bracket of any left-invariant vector field with any right-invariant vector field vanishes.

[2] Right-invariant 1-forms are provided by $dgg^{-1}$.

*representations* that we denote by the subscript $r$ in $\Gamma_r(T_A)$. Their exponentiation generates a representation of the group,

$$\Gamma_r[g(\sigma)] = e^{\sigma^A \Gamma_r(T_A)}, \qquad [\Gamma_r(T_A), \Gamma_r(T_B)] = f_{AB}{}^C \Gamma_r(T_C). \tag{A.13}$$

A Lie algebra can be complexified and we can then perform complex changes of basis that complexify the structure constants. If all the structure constants are real (as we will usually take them to be) and all the generators have the same definite Hermiticity properties, then all of them must be anti-Hermitian, $\Gamma_r(T_A)^\dagger = -\Gamma_r(T_A)$, and the elements of the group are represented by unitary operators $\Gamma_r[g(\sigma)]^\dagger = \Gamma_r[g(\sigma)]^{-1}$, i.e. we obtain a *unitary representation*.

Any representation of a *compact* Lie group is equivalent to a unitary representation. The unitary representations of compact Lie groups are also finite dimensional and the operators can be represented by their matrices in a given basis. All matrix representations automatically satisfy the Jacobi identity.

Sometimes we will be interested in non-compact groups (for instance, the Lorentz group in $d$ dimensions $SO(1, d-1)$). Their unitary representations are infinite dimensional and their finite-dimensional representations are necessarily non-unitary. Then the generators cannot be all Hermitian or all anti-Hermitian at the same time (with real structure constants). The "non-compact generators" will be represented by Hermitian operators.

In the *adjoint representation* each element $T \in \mathfrak{g}$ is represented by an operator $\Gamma_{\text{Adj}}(T)$ acting on $\mathfrak{g}$ itself according to[3]

$$\Gamma_{\text{Adj}}(T)T' \equiv [T, T'], \quad \forall\, T, T' \in \mathfrak{g}. \tag{A.14}$$

Thus, the generators themselves are represented by

$$\Gamma_{\text{Adj}}(T_A)T_B \equiv [T_A, T_B] = f_{AB}{}^C T_C \equiv T_C \Gamma_{\text{Adj}}(T_A)^C{}_B, \tag{A.15}$$

so the components of the operator $\Gamma_{\text{Adj}}(T_A)$ in the basis $\{T_A\}$ are the structure constants

$$\Gamma_{\text{Adj}}(T_A)^C{}_B = f_{AB}{}^C. \tag{A.16}$$

These matrices satisfy the Lie algebra Eq. (A.9) due to the Jacobi identity Eq. (A.8).

The adjoint representation of the Lie algebra allows us to define the adjoint representation of the group by exponentiation,

$$\Gamma_{\text{Adj}}[g(\sigma)] \equiv \exp\{\sigma^A \Gamma_{\text{Adj}}(T_A)\}, \tag{A.17}$$

and the adjoint action of the group on the algebra,

$$T'_B = T_D (\Gamma_{\text{Adj}}\,[g(\sigma)])^D{}_B. \tag{A.18}$$

An equivalent definition of the adjoint action of the group on the algebra is

$$\Gamma_{\text{Adj}}(g)T \equiv L_{g\,*}\, R_{g^{-1}\,*}\, T = gTg^{-1}. \tag{A.19}$$

[3] This convention is chosen so that the components $\varphi^A$ of an element of the Lie algebra $\varphi$ in the basis $\{T_A\}$ transform contravariantly (as defined in Section A.2.1) under group transformations.

In any representation we have[4]

$$\Gamma_r(g)\Gamma_r(T_A)\Gamma_r(g^{-1}) = \Gamma_r[\Gamma_{\text{Adj}}(g)T_A] = \Gamma_r(T_B)(\Gamma_{\text{Adj}}(g))^B{}_A. \tag{A.20}$$

We can now introduce the symmetric, bilinear *Killing form* (or *metric*) $K(\cdot,\cdot)$ into the Lie algebra:

$$K(T,T') \equiv \text{Tr}\{\Gamma_{\text{Adj}}(T)\Gamma_{\text{Adj}}(T')\}, \qquad K_{AB} = K(T_A,T_B) = f_{AC}{}^D f_{BD}{}^C. \tag{A.21}$$

The Killing metric is invariant under the adjoint action of the group on the algebra due to the cyclic property of the trace. Infinitesimally, we have

$$f_{A(B}{}^C K_{D)C} = 0, \tag{A.22}$$

and this is the condition that any other invariant metric should satisfy. We can define

$$f_{ABC} \equiv f_{AB}{}^D K_{DC}, \tag{A.23}$$

which is fully antisymmetric on account of Eq. (A.22).

It is easy to see that

$$\text{Tr}[\Gamma_{\text{Adj}}(T_{[A})\Gamma_{\text{Adj}}(T_B)\Gamma_{\text{Adj}}(T_{C]})] = -\tfrac{1}{2} f_{ABC}. \tag{A.24}$$

The Killing metric contains a great deal of information. Let us first state some definitions.

A Lie algebra $\mathfrak{g}$ is Abelian $[T,T'] = 0,\ \forall\ T,T' \in \mathfrak{g}$. This is sometimes expressed as $[\mathfrak{g},\mathfrak{g}] = 0$. Abelian Lie algebras generate by exponentiation Abelian Lie groups.

A subgroup $\mathrm{H} \subset \mathrm{G}$ is an *invariant subgroup* if $ghg^{-1} \in \mathrm{H}\ \forall\ h \in \mathrm{H},\ g \in \mathrm{G}$. A subalgebra $\mathfrak{h} \subset \mathfrak{g}$ is an *invariant subalgebra* or *ideal* if $[M,T] \in \mathfrak{h}\ \forall M \in \mathfrak{h},\ T \in \mathfrak{g}$. This is sometimes expressed as $[\mathfrak{h},\mathfrak{g}] \subset \mathfrak{h}$. The Lie algebra $\mathfrak{h}$ of an invariant subgroup $\mathrm{H} \subset \mathrm{G}$ is an invariant subalgebra of the Lie algebra $\mathfrak{g}$ of G.

A Lie group (algebra) is *simple* if it does not have any proper invariant subgroup (subalgebra). Simple Lie algebras generate simple Lie groups.

A Lie group (algebra) is *semisimple* if it does not have any non-trivial invariant Abelian subgroup (subalgebra). Semisimple Lie algebras generate semisimple Lie groups.

For any Lie algebra, the set of all possible Lie brackets of its elements $[\mathfrak{g},\mathfrak{g}] \equiv \mathfrak{g}^{(1)} \equiv \mathfrak{g}_{(1)}$ is an ideal called the *derived subalgebra*. We can define two sequences of ideals $\mathfrak{g}^{(n)}$ and $\mathfrak{g}_{(n)}$:

$$[\mathfrak{g}^{(n-1)},\mathfrak{g}^{(n-1)}] \equiv \mathfrak{g}^{(n)}, \qquad [\mathfrak{g}_{(n-1)},\mathfrak{g}] \equiv \mathfrak{g}_{(n)}. \tag{A.25}$$

A Lie algebra is *solvable* (*nilpotent*) of degree $n$ if $\mathfrak{g}^{(n)}$ ($\mathfrak{g}_{(n)}$) is just the 0 element for some $n$. Every nilpotent algebra is solvable. It can be shown that a Lie algebra is semisimple iff it does not possess any invariant solvable subalgebra.

Now the following can be shown.

**Cartan's first criterion.** $\mathfrak{g}$ is solvable if $g(T,T') = 0,\ \forall\ T,T' \in \mathfrak{g}^{(1)}$.

[4] It is very easy to check this identity infinitesimally.

**Cartan's second criterion.** $\mathfrak{g}$ is semisimple iff its Killing form is non-degenerate.

**(Weyl.)** A connected semisimple (linear) Lie group is compact iff its Lie-algebra Killing metric is definite negative.

If the Killing metric of a Lie algebra is 0, then the algebra is solvable.

The Killing metric of a nilpotent Lie algebra is 0.

We can diagonalize and normalize the Killing metric using $\mathrm{GL}(n,\mathbb{R})$ transformations so that it only has $\pm 1, 0$ in the diagonal. The zeros are associated with invariant Abelian subalgebras and the $+1$s are associated with non-compact directions.

## A.2 Yang–Mills fields

### *A.2.1 Fields and covariant derivatives*

Fields always transform in finite-dimensional representations of the symmetry group, even if they are not unitary.[5] $\Gamma_r(g)^i{}_j$ $i,j=1,\ldots,\dim(r)$ denotes the matrix corresponding to group element $g$ in the representation labeled by $r$. The indices which are also carried by the fields will in general not be shown. In any representation there are three different types of fields according to the way they transform: *contravariant fields*, represented by a column vector and transforming according to

$$\psi^{i\,\prime} = (\Gamma_r(g))^i{}_j \psi^j; \tag{A.26}$$

*covariant fields*, represented by a row vector and transforming according to

$$\xi_i' = \xi_j (\Gamma_r^{-1}(g))^j{}_i; \tag{A.27}$$

and Lie-algebra-valued fields that transform under the adjoint action of the group

$$\begin{aligned} \varphi &= \varphi^A \Gamma_r(T_A)\,, \\ \varphi' &= \Gamma_r(g)\varphi\Gamma_r^{-1}(g) \quad \Rightarrow \varphi'^A = [\Gamma_{\mathrm{Adj}}(g)]^A{}_B \varphi^B. \end{aligned} \tag{A.28}$$

The relation among the three kinds of fields depends on the group and representation we are considering. If the representation $r$ is unitary and $\psi$ is contravariant, then $\psi^\dagger$ is covariant. If the group is defined by the property that it preserves the scalar product associated with a metric $\eta$ $\langle u|v\rangle = u^\dagger \eta v$ so $u' = \Gamma_{\mathrm{v}}(g)u$ and $\Gamma_v^\dagger(g)\eta\Gamma_{\mathrm{v}}(g)$, where $\Gamma_{\mathrm{v}}(g)$ is the matrix associated with the group element $g$ in the defining fundamental or vector representation (these are the groups $\mathrm{SO}(n_+,n_-)$, $\mathrm{SU}(n_+,n_-)$, and $\mathrm{Sp}(n)$), then, given a contravariant vector field $\psi$, the row vector $\psi^\dagger\eta$ transforms as a covariant vector field. It is also possible to relate contravariant and covariant fields in the spinor representations of $\mathrm{SO}(n_+,n_-)$ groups (see Appendix D).

Since

$$\Gamma_r(g) = \exp\{\sigma^A \Gamma_r(T_A)\} \equiv \exp\{\sigma_r\}, \tag{A.29}$$

[5] Their *solutions* correspond to *states* in the quantum theory and therefore must fit into unitary representations according to Wigner's theorem, however.

for infinitesimal values of the parameters (group manifold coordinates) $\sigma^A$, i.e. for transformations near the identity or infinitesimal transformations, the various fields transform as follows:

$$\begin{aligned}\delta_\sigma \psi &= \sigma^A \Gamma_r(T_A)\,\psi = \sigma_r \psi,\\ \delta_\sigma \xi &= -\xi \sigma^A \Gamma_r(T_A) = -\xi \sigma_r,\\ \delta_\sigma \varphi &= [\sigma_r, \varphi] \;\;\Rightarrow \delta_\sigma \varphi^A = f_{BC}{}^A \sigma^B \varphi^C = \sigma^B \Gamma_{\text{Adj}}(T_B)^A{}_C \varphi^C = (\sigma_{\text{Adj}})^A{}_B \varphi^B.\end{aligned} \tag{A.30}$$

If we now consider *local* (i.e. *gauge*) transformations of the fields with $\sigma^A = \sigma^A(x)$, the derivatives[6] of the fields $\partial_\mu \psi,\ \partial_\mu \xi$, and $\partial_\mu \varphi$ do not transform as do the fields themselves, i.e. they do not transform covariantly under local transformations.[7] It is then necessary to introduce a compensating field $A_\mu$ transforming under the adjoint action of the group on the Lie algebra (like $\varphi$) to define a new *covariant derivative*. This compensating field is the *gauge field* and can be defined in any representation

$$A_{r\,\mu} = A^A{}_\mu \Gamma_r(T_A). \tag{A.31}$$

With it we define the covariant derivative $D_\mu$ by its action on the fields (here $g$ is a coupling constant):

$$\begin{aligned}D_\mu \psi &= \partial_\mu \psi - g A_{r\,\mu} \psi,\\ D_\mu \xi &= \partial_\mu \xi + g \xi A_{r\,\mu},\\ D_\mu \varphi &= \partial_\mu \varphi - g[A_{r\,\mu}, \varphi] \;\;\Rightarrow D_\mu \varphi^A = \partial_\mu \varphi^A - g f_{BC}{}^A A^B{}_\mu \varphi^C.\end{aligned} \tag{A.32}$$

The covariant derivative transforms covariantly under gauge transformations, i.e.

$$\begin{aligned}(D_\mu \psi)' &= \Gamma_r[g(x)] D_\mu \psi,\\ (D_\mu \xi)' &= (D_\mu \xi)\Gamma_r^{-1}[g(x)],\\ (D_\mu \varphi)' &= \Gamma_r[g(x)](D_\mu \varphi)(\Gamma_r[g(x)])^{-1},\end{aligned} \tag{A.33}$$

if the gauge field transforms as follows:

$$\begin{aligned}A'_{r\,\mu} &= \Gamma_r[g(x)] A_\mu (\Gamma_r[g(x)])^{-1} + \frac{1}{g}(\partial_\mu \Gamma_r[g(x)])(\Gamma_r[g(x)])^{-1},\\ \delta_\sigma A^A{}_\mu &= \frac{1}{g}\left(\partial_\mu \sigma^A - g f_{BC}{}^A A^B{}_\mu \sigma^C\right),\\ \delta_\sigma A_{r\,\mu} &= \frac{1}{g} D_\mu \sigma_r.\end{aligned} \tag{A.34}$$

Observe that it transforms as $\varphi$ or $\sigma$ up to an inhomogeneous term typical of a connection (which is its geometrical meaning). The spin connection $\omega_\mu$ defined in Chapter 1 is just the connection for the gauge group $\mathrm{SO}(1, d-1)$.

[6] Partial derivatives should be replaced by derivatives covariant with respect to GCTs.

[7] Sometimes in the literature "covariant transformation" means a transformation in which there are no derivatives of the local parameters $\sigma^A$, which is, obviously, a necessary condition.

Observe also that the covariant derivatives of contravariant and covariant fields are compatible: if the representation is unitary (and therefore all the generators are anti-Hermitian so the real gauge field is anti-Hermitian as well)

$$(D\psi)^\dagger = \partial_\mu \psi^\dagger + g\psi^\dagger A_\mu, \tag{A.35}$$

etc. If we are dealing with a metric-preserving group, then in the vector representation

$$(D\psi)^\dagger \eta = \partial_\mu \psi^\dagger \eta + g\psi^\dagger \eta A_\mu, \tag{A.36}$$

on account of

$$\Gamma_{\rm v}(T_A^\dagger)\eta = -\eta \Gamma_{\rm v}(T_A) \,. \tag{A.37}$$

### A.2.2 Kinetic terms

Now we want to build gauge-invariant actions for the fields $\psi, \varphi$, and $A_\mu$ ($\xi$ can simply be transformed into a contravariant field). Invariants can be built in several ways. The simplest is to take the product of covariant and contravariant objects that, by definition, transform oppositely. This works both for $\psi$ and for $\varphi$ (which is once covariant and once contravariant): with our convention for the signature $(+,-,\cdots,-)$, gauge-invariant kinetic terms for $\psi$ and $\varphi$ are

$$\begin{aligned} &g^{\mu\nu}\,(D_\mu\psi)^\dagger D_\nu\psi, \\ -&g^{\mu\nu}\,\mathrm{Tr}(D_\mu\varphi D_\nu\varphi) \sim K_{AB}g^{\mu\nu}D_\mu\varphi^A D_\nu\varphi^B. \end{aligned} \tag{A.38}$$

(If the group is not compact, some kinetic terms will have a wrong sign.)

There is no covariant derivative for the gauge field because it does not transform covariantly (due to the inhomogeneous term). The closest to the covariant derivative of the gauge field that we can use to construct a kinetic term is its *gauge field strength*, which is just the curvature of the connection $A_\mu$. We can define it as we defined the Riemann curvature tensor for the Levi-Civita connection in terms of the Ricci identity:

$$\begin{aligned} [D_\mu, D_\nu]\psi &= -gF_{r\,\mu\nu}\psi, \\ [D_\mu, D_\nu]\xi &= +g\xi F_{r\,\mu\nu}, \\ [D_\mu, D_\nu]\varphi &= -g[F_{r\,\mu\nu}, \varphi]. \end{aligned} \tag{A.39}$$

From these relations we find

$$F_{r\,\mu\nu} = F^I{}_{\mu\nu}\Gamma_r(T_A) = 2\partial_{[\mu}A_{r\,\nu]} - g[A_{r\,\mu}, A_{r\,\nu}], \tag{A.40}$$

or, in components,

$$F^A{}_{\mu\nu} = 2\partial_{[\mu}A^A{}_{\nu]} - gf_{BC}{}^A A^B{}_\mu A^C{}_\nu. \tag{A.41}$$

The gauge field strength transforms under the adjoint action of the group on the algebra (like $\varphi$) and, thus,

$$D_\mu F_{r\,\nu\rho} = \partial_\mu F_{r\,\nu\rho} - g[A_{r\,\mu}, F_{r\,\nu\rho}]. \tag{A.42}$$

The gauge field strength always satisfies the *Bianchi identity*[8]

$$D_{[\mu}F_{\nu\rho]} = 0. \tag{A.43}$$

Finally, the kinetic term for $A_\mu$, which is invariant, is

$$\mathrm{Tr}(F_{r\,\mu\nu}F^{r\,\mu\nu}) \sim K_{AB}F^{A}{}_{\mu\nu}F^{B\mu\nu}, \tag{A.44}$$

where there is a proportionality coefficient that depends on conventions and on the representation $r$. The Yang–Mills equation of motion is therefore

$$D_\mu F^{\mu\nu} = 0. \tag{A.45}$$

There is another invariant that we can build in four dimensions,

$$\mathrm{Tr}\left(F_{\mu\nu} \star F^{\mu\nu}\right) \sim -\frac{1}{2\sqrt{|g|}}K_{AB}\epsilon^{\mu\nu\rho\sigma}F^{A}{}_{\mu\nu}F^{B}{}_{\rho\sigma}, \tag{A.46}$$

but it is a total derivative and does not contribute to the $A_\mu$ equations of motion. In the Euclidean signature, the integral of Eq. (A.46) is (up to numerical factors) the *instanton number*, a topological invariant that does contribute to the Euclidean path integral.

Sometimes it is useful to work with differential forms. Thus, we define the Lie-algebra-valued 1- and 2-forms,

$$A \equiv A_\mu dx^\mu, \qquad F = \tfrac{1}{2}F_{\mu\nu}dx^\mu \wedge dx^\nu \equiv dA - A \wedge A. \tag{A.47}$$

The kinetic term for $A$ can now be written as the $d$-form (in $d$ dimensions)

$$\int d^dx\sqrt{|g|}\;\mathrm{Tr}_{\mathrm{Adj}}F^2 \sim \int \mathrm{Tr}_{\mathrm{Adj}}(F \wedge \star F), \tag{A.48}$$

and the four-dimensional topological term can be rewritten as

$$\int d^4x\sqrt{|g|}\;\mathrm{Tr}_{\mathrm{Adj}}(F \star F) \sim \int \mathrm{Tr}_{\mathrm{Adj}}(F \wedge F). \tag{A.49}$$

Now we define the *Chern–Simons 3-form*

$$\omega_3 = \frac{1}{3!}\omega_{3\,\mu\nu\rho}dx^\mu \wedge dx^\nu \wedge dx^\rho \equiv \mathrm{Tr}_{\mathrm{Adj}}\left(A \wedge dA - \tfrac{2}{3}A \wedge A \wedge A\right), \tag{A.50}$$

or, in components, using the property Eq. (A.24) and the normalization $K_{AB} = \delta_{AB}$ for a compact group:

$$\omega_{3\,\mu\nu\rho} = -3!\left(A^{A}{}_{[\mu}\partial_\nu A^{A}{}_{\rho]} - \tfrac{1}{3}f_{ABC}A^{A}{}_{[\mu}A^{B}{}_{\nu}A^{C}{}_{\rho]}\right). \tag{A.51}$$

The Chern–Simons 3-form has the following very important property:[9]

$$d\omega_3 = \mathrm{Tr}_{\mathrm{Adj}}(F \wedge F), \tag{A.52}$$

which makes it evident that the topological term $F \wedge F$ is a total derivative.

---

[8] This can be checked by expanding in powers of $g$. The first term (of zeroth order in $g$) vanishes in the absence of torsion, the second vanishes due to several cancelations, and the last term $\mathcal{O}(g^2)$ vanishes due to the Jacobi identity Eq. (A.8).

[9] To prove it one simply has to realize that the trace over the exterior product of four $A$s vanishes because complete antisymmetry in four indices is the opposite to cyclic symmetry. (For three indices, complete antisymmetry and cyclic symmetry are the same thing, and this is why $\omega_3$ can be defined at all.)

### A.2.3 $\mathrm{SO}(n_+, n_-)$ *gauge theory*

The group $\mathrm{SO}(n_+, n_-)$ is defined as the group of $n \times n$ (where $n = n_+ + n_-$) real matrices[10] $\hat{\Lambda}^{\hat{a}}{}_{\hat{b}}$ that act on (contravariant) $n$-dimensional vectors by

$$\hat{V}'^{\hat{a}} = \hat{\Lambda}^{\hat{a}}{}_{\hat{b}}\, \hat{V}^{\hat{b}}, \tag{A.53}$$

have determinant $+1$, and preserve the metric $\hat{\eta}_{\hat{a}\hat{b}} = \mathrm{diag}(+\cdots+-\cdots-)$ ($n_\pm$ plus (minus) signs), so

$$\hat{V}'^{\hat{a}} \hat{\eta}_{\hat{a}\hat{b}}\, \hat{V}'^{\hat{b}} = \hat{V}^{\hat{a}} \hat{\eta}_{\hat{a}\hat{b}}\, \hat{V}^{\hat{b}}. \tag{A.54}$$

This implies that $\mathrm{SO}(n_+, n_-)$ matrices satisfy the defining property

$$\hat{\eta}_{\hat{a}\hat{b}}\, \hat{\Lambda}^{\hat{b}}{}_{\hat{d}}\, \hat{\eta}^{\hat{d}\hat{c}} = (\hat{\Lambda}^{-1})^{\hat{c}}{}_{\hat{a}}, \qquad \hat{\eta}^{\hat{a}\hat{b}} \hat{\eta}_{\hat{b}\hat{c}} = \delta^{\hat{a}}{}_{\hat{c}}. \tag{A.55}$$

This generalizes to arbitrary signature the $n_- = 0$ orthogonality condition $\hat{\Lambda}^{\mathrm{T}} = \hat{\Lambda}^{-1}$.

If we consider also translations in the $n$-dimensional vector space, we obtain the group $\mathrm{ISO}(n_+, n_-)$, which acts on contravariant vectors as follows:

$$\hat{V}'^{\hat{a}} = \hat{\Lambda}^{\hat{a}}{}_{\hat{b}}\, \hat{V}^{\hat{b}} + \hat{W}^{\hat{a}}. \tag{A.56}$$

The Poincaré group in $d$ spacetime dimensions is $\mathrm{ISO}(1, d-1)$ in this notation.

We can immediately define the action of $\mathrm{SO}(n_+, n_-)$ on covariant vectors $\hat{V}_{\hat{a}}$:

$$\hat{V}'_{\hat{a}} = \hat{V}_{\hat{b}} (\hat{\Lambda}^{-1})^{\hat{b}}{}_{\hat{a}}. \tag{A.57}$$

Using the defining property of $\mathrm{SO}(n_+, n_-)$ matrices Eq. (A.55), we can relate covariant and contravariant vectors in the standard way, raising and lowering indices with $\hat{\eta}$:

$$\hat{V}_{\hat{a}} = \hat{\eta}_{\hat{a}\hat{b}}\, \hat{V}^{\hat{b}}, \qquad \hat{V}^{\hat{a}} = \hat{\eta}^{\hat{a}\hat{b}}\, \hat{V}_{\hat{b}}. \tag{A.58}$$

Let us now consider infinitesimal $\mathrm{SO}(n_+, n_-)$ transformations $\hat{\Lambda}^{\hat{a}}{}_{\hat{b}} \sim \delta^{\hat{a}}{}_{\hat{b}} + \hat{\sigma}^{\hat{a}}{}_{\hat{b}}$. The defining property of $\mathrm{SO}(n_+, n_-)$ matrices in the vector representation Eq. (A.55) implies that the infinitesimal parameters of the transformation satisfy $\hat{\sigma}^{\hat{a}\hat{b}} = \hat{\sigma}^{[\hat{a}\hat{b}]}$ and thus the group has $n(n-1)/2$ independent generators $\hat{M}_{\hat{a}\hat{b}}$ (one for each independent parameter), which are conveniently labeled by an antisymmetric pair of indices $\hat{a}\hat{b}$ (this expresses the fact that the adjoint representation is just the antisymmetric product of two vector representations). We can write

$$\hat{\sigma}^{\hat{a}}{}_{\hat{b}} = \tfrac{1}{2} \hat{\sigma}^{\hat{c}\hat{d}}\, \Gamma_{\mathrm{v}}\left(\hat{M}_{\hat{c}\hat{d}}\right)^{\hat{a}}{}_{\hat{b}}, \tag{A.59}$$

where

$$\Gamma_{\mathrm{v}}\left(\hat{M}_{\hat{c}\hat{d}}\right)^{\hat{a}}{}_{\hat{b}} = +2\hat{\eta}_{[\hat{c}}{}^{\hat{a}} \hat{\eta}_{\hat{d}]\hat{b}} \tag{A.60}$$

are the $\mathrm{SO}(n_+, n_-)$ generators in the vector representation. Observe that we need to divide by two in order to avoid counting the same generator twice. These generators are normalized so that

$$\mathrm{Tr}\left[\Gamma_{\mathrm{v}}\left(\hat{M}_{\hat{a}\hat{b}}\right) \Gamma_{\mathrm{v}}\left(\hat{M}_{\hat{c}\hat{d}}\right)\right] = -4\hat{\eta}_{[\hat{a}\hat{b}][\hat{c}\hat{d}]}. \tag{A.61}$$

[10] We use hats to avoid confusion with (Lorentzian) tangent-space indices. When $n_+ = 1$ and $n_- = d-1$ they are, of course, identical.

The infinitesimal transformations of contravariant and covariant vectors take the forms:

$$\begin{aligned}\delta_{\hat\sigma}\hat V^{\hat a} &= \tfrac{1}{2}\hat\sigma^{\hat c\hat d}\,\Gamma_{\rm v}\left(\hat M_{\hat c\hat d}\right)^{\hat a}{}_{\hat b}\hat V^{\hat b},\\ \delta_{\hat\sigma}\hat V_{\hat a} &= \hat V_{\hat b}\left[-\tfrac{1}{2}\hat\sigma^{\hat c\hat d}\,\Gamma_{\rm v}\left(\hat M_{\hat c\hat d}\right)^{\hat b}{}_{\hat a}\right].\end{aligned} \tag{A.62}$$

Using this representation of the generators, one can find the $\mathfrak{so}(n_+, n_-)$ algebra

$$\left[\hat M_{\hat a\hat b}, \hat M_{\hat c\hat d}\right] = -\hat\eta_{\hat a\hat c}\hat M_{\hat b\hat d} - \hat\eta_{\hat b\hat d}\hat M_{\hat a\hat c} + \hat\eta_{\hat a\hat d}\hat M_{\hat b\hat c} + \hat\eta_{\hat b\hat c}\hat M_{\hat a\hat d}, \tag{A.63}$$

which can be also be written

$$\left[\hat M_{\hat a\hat b}, \hat M_{\hat c\hat d}\right] = -\hat M_{\hat e\hat b}\Gamma_{\rm v}\left(\hat M_{\hat c\hat d}\right)^{\hat e}{}_{\hat a} - \hat M_{\hat a\hat e}\Gamma_{\rm v}\left(\hat M_{\hat c\hat d}\right)^{\hat e}{}_{\hat b}. \tag{A.64}$$

These commutation relations can be interpreted as the action of $\hat M_{\hat c\hat d}$ on $\hat M_{\hat a\hat b}$, which transforms as the antisymmetric product of two covariant vectors, indicating that the adjoint representation is the antisymmetric product of two vector representations. This can be seen, for instance, by using the *Eckart–Schrödinger* representation of the $\mathfrak{iso}(n_+, n_-)$ algebra[11]

$$\hat M_{\hat a\hat b} = \hat x_{\hat a}\partial_{\hat b} - \hat x_{\hat b}\partial_{\hat a} = -\hat x^{\hat c}\Gamma_{\rm v}\left(\hat M_{\hat a\hat b}\right)^{\hat d}{}_{\hat c}\partial_{\hat d}, \qquad \hat P_{\hat a} = -\partial_{\hat a}, \tag{A.67}$$

which gives the additional commutator

$$\left[\hat P_{\hat a}, \hat M_{\hat b\hat c}\right] = -\hat P_{\hat d}\Gamma_{\rm v}\left(\hat M_{\hat b\hat c}\right)^{\hat d}{}_{\hat a}, \tag{A.68}$$

indicating that the linear momentum $\hat P_{\hat a}$ is a covariant $\mathrm{SO}(n_+, n_-)$ vector.

The $\mathfrak{so}(n_+, n_-)$ structure constants are defined by

$$\left[\hat M_{\hat a\hat b}, \hat M_{\hat c\hat d}\right] = \tfrac{1}{2}f_{\hat a\hat b\,\hat c\hat d}{}^{\hat e\hat f}\hat M_{\hat e\hat f} \quad \Rightarrow \Gamma_{\rm Adj}\left(\hat M_{\hat a\hat b}\right)^{\hat e\hat f}{}_{\hat c\hat d} = f_{\hat a\hat b\,\hat c\hat d}{}^{\hat e\hat f} = 4\Gamma_{\rm v}\left(\hat M_{\hat a\hat b}\right)^{[\hat e}{}_{[\hat c}\,\hat\eta^{\hat f]}{}_{\hat d]}, \tag{A.69}$$

where, as we are going to do systematically, we have introduced an additional factor of $\frac{1}{2}$ in order to sum over each generator only once. The Killing metric is

$$\begin{aligned}\hat K_{\hat a\hat b\,\hat c\hat d} &= \mathrm{Tr}\left[\Gamma_{\rm Adj}\left(\hat M_{\hat a\hat b}\right)\Gamma_{\rm Adj}\left(\hat M_{\hat c\hat d}\right)\right]\\ &= (d-2)\,\mathrm{Tr}\left[\Gamma_{\rm v}\left(\hat M_{\hat a\hat b}\right)\Gamma_{\rm v}\left(\hat M_{\hat c\hat d}\right)\right] = 4(d-2)\hat\eta_{[\hat a\hat b]\,[\hat c\hat d]}.\end{aligned} \tag{A.70}$$

[11] If we consider infinitesimal Poincaré transformations

$$\delta x^\mu = a^\mu + \sigma^\mu{}_\nu x^\nu \tag{A.65}$$

of vector fields in a space of signature $(n_+, n_-)$, we find

$$\begin{aligned}\delta V^\mu &= \left[a^\nu P_\nu + \frac{1}{2}\sigma^{\alpha\beta}M_{\alpha\beta}\right]V^\mu + \frac{1}{2}\sigma^{\alpha\beta}\Gamma_{\rm v}(M_{\alpha\beta})^\mu{}_\rho V^\rho,\\ \delta V_\mu &= \left[a^\nu P_\nu + \frac{1}{2}\sigma^{\alpha\beta}M_{\alpha\beta}\right]V_\mu - \frac{1}{2}\sigma^{\alpha\beta}\Gamma_{\rm v}(M_{\alpha\beta})^\rho{}_\mu V_\rho.\end{aligned} \tag{A.66}$$

The generators in the Eckart–Schrödinger representation appear in the universal transport term.

Apart from the vector and adjoint representations and other tensor representations (that can be built as tensor products of a number of covariant and contravariant vector representations), $\mathrm{SO}(n_+, n_-)$ groups also admit *spinorial representations* that are complex, in general. These are $2^{[n/2]}$ dimensional and can be constructed from a representation of a Clifford algebra, as explained in Appendix D. We denote the $\mathfrak{so}(n_+, n_-)$ generators in a spinorial representation by

$$\Gamma_{\mathrm{s}}\left(\hat{M}_{\hat{a}\hat{b}}\right)^{\alpha}{}_{\beta}, \qquad \alpha, \beta = 1, \ldots, 2^{[n/2]}. \tag{A.71}$$

*Spinors* are the elements of the representation space, and they are represented by $2^{[n/2]}$-component contravariant vectors $\hat{\psi}^{\alpha}$ and covariant vectors $\hat{\xi}_{\alpha}$. They transform infinitesimally under $\mathrm{SO}(n_+, n_-)$ according to

$$\begin{aligned} \delta_{\hat{\sigma}}\hat{\psi}^{\alpha} &= \tfrac{1}{2}\hat{\sigma}^{\hat{c}\hat{d}}\,\Gamma_{\mathrm{s}}\left(\hat{M}_{\hat{c}\hat{d}}\right)^{\hat{\alpha}}{}_{\hat{\beta}}\hat{\psi}^{\beta}, \\ \delta_{\hat{\sigma}}\hat{\xi}_{\alpha} &= \hat{\xi}_{\beta}\left[-\tfrac{1}{2}\hat{\sigma}^{\hat{c}\hat{d}}\,\Gamma_{\mathrm{s}}\left(\hat{M}_{\hat{c}\hat{d}}\right)^{\hat{\beta}}{}_{\hat{\alpha}}\right]. \end{aligned} \tag{A.72}$$

Covariant and contravariant spinors are related by the operation of *Dirac conjugation*: with each contravariant spinor $\hat{\psi}^{\alpha}$ one can associate a covariant spinor (its *Dirac conjugate*) denoted by $\bar{\hat{\psi}}_{\alpha}$ and related to it by

$$\bar{\hat{\psi}}_{\alpha} = \hat{\psi}^{\beta\,\star}\,\mathcal{D}_{\beta\alpha}, \tag{A.73}$$

where $\mathcal{D}$ is the *Dirac conjugation matrix* that satisfies

$$\mathcal{D}^{-1}\Gamma_{\mathrm{s}}\left(\hat{M}_{\hat{a}\hat{b}}\right)^{\dagger}\mathcal{D} = -\Gamma_{\mathrm{s}}\left(\hat{M}_{\hat{a}\hat{b}}\right). \tag{A.74}$$

There is another conjugation operation that transforms a contravariant spinor into a covariant one: *Majorana conjugation*. More details can be found in Appendix D.

The $\mathrm{SO}(n_+, n_-)$ connection, customarily denoted by $\hat{\omega}_{\mu}$ and called *spin connection*, is

$$\hat{\omega}_{\mu} = \tfrac{1}{2}\hat{\omega}_{\mu}{}^{\hat{a}\hat{b}}\Gamma\left(\hat{M}_{\hat{a}\hat{b}}\right), \tag{A.75}$$

and the $\mathrm{SO}(n_+, n_-)$-covariant derivative acting on contravariant fields $\hat{\psi}$ is

$$\hat{\mathcal{D}}_{\mu}\hat{\psi} = \partial_{\mu}\hat{\psi} - \hat{\omega}_{\mu}\hat{\psi}, \tag{A.76}$$

whereas that acting on Lie-algebra-valued fields is

$$\hat{\varphi} = \tfrac{1}{2}\hat{\varphi}^{\hat{a}\hat{b}}\Gamma\left(\hat{M}_{\hat{a}\hat{b}}\right), \qquad \hat{\mathcal{D}}_{\mu}\hat{\varphi} = \partial_{\mu}\hat{\varphi} - \left[\hat{\omega}_{\mu}, \hat{\varphi}\right]. \tag{A.77}$$

Note that $\hat{\psi}, \hat{\varphi}$, and the connection $\hat{\omega}_{\mu}$ undergo the following infinitesimal gauge transformations with local parameter $\hat{\sigma}^{\hat{a}\hat{b}}$:

$$\begin{aligned} \delta_{\hat{\sigma}}\hat{\psi} &= \hat{\sigma}\hat{\psi}, \\ \delta_{\hat{\sigma}}\hat{\varphi} &= \left[\hat{\sigma}, \hat{\varphi}\right], \\ \delta_{\hat{\sigma}}\hat{\omega}_{\mu} &= \hat{\mathcal{D}}_{\mu}\hat{\sigma} = (\partial_{\mu}\hat{\sigma} - [\hat{\omega}_{\mu}, \hat{\sigma}]), \qquad \hat{\sigma} \equiv \tfrac{1}{2}\hat{\sigma}^{\hat{a}\hat{b}}\Gamma\left(\hat{M}_{\hat{a}\hat{b}}\right). \end{aligned} \tag{A.78}$$

As usual, the curvature is another Lie-algebra-valued field defined through the commutator of two covariant derivatives in any representation:

$$\left[\hat{\mathcal{D}}_\mu, \hat{\mathcal{D}}_\nu\right]\hat{\psi} = -\hat{R}_{\mu\nu}\hat{\psi}, \tag{A.79}$$

where

$$\hat{R}_{\mu\nu} = \tfrac{1}{2}\hat{R}_{\mu\nu}{}^{\hat{a}\hat{b}}\Gamma\left(\hat{M}_{\hat{a}\hat{b}}\right), \qquad \hat{R}_{\mu\nu}{}^{\hat{a}\hat{b}} = 2\partial_{[\mu}\hat{\omega}_{\nu]}{}^{\hat{a}\hat{b}} - 2\hat{\omega}_{[\mu|}{}^{\hat{a}\hat{c}}\,\hat{\omega}_{|\nu]\hat{c}}{}^{\hat{b}}, \tag{A.80}$$

transforms as $\hat{\sigma}$ and satisfies the Bianchi identities

$$\hat{\mathcal{D}}_{[\mu}\hat{R}_{\nu\rho]} = 0. \tag{A.81}$$

$\mathrm{SO}(3)$ *and three-dimensional real Lie algebras.* For the particular case $n = 3$ the adjoint representation coincides with the vector representation. Then we have two different notations (with one and with two antisymmetric indices) for the same representation. The relation between the two is

$$T_i = \tfrac{1}{2}\epsilon_{ijk}T_{jk}, \qquad T_{ij} = \epsilon_{ijk}T_k, \tag{A.82}$$

and the $T_i$s satisfy the algebra

$$[T_i, T_j] = -\epsilon_{ijk}\eta^{kl}T_l \quad \Rightarrow f_{ij}{}^l = -\epsilon_{ijk}\eta^{kl} \quad \Rightarrow K_{ij} = -2\eta_{ij}. \tag{A.83}$$

For $\eta^{kl} = \delta^{kl}$ (for the group $\mathrm{SO}(3)$) an explicit representation is given in Eq. (A.94). The only other possibility, $\eta = \mathrm{diag}\,(++-)$, is $\mathrm{SO}(2,1)$. They are identical as complex algebras (it suffices to multiply $T_1$ and $T_2$ by $i$), but not as real algebras.

All the possible real three-dimensional Lie algebras can be written in terms of a matrix $\mathsf{Q}^{kl}$:

$$[T_i, T_j] = -\epsilon_{ijk}\mathsf{Q}^{kl}T_l, \qquad \mathsf{Q}^{(lk)}\epsilon_{kij}\mathsf{Q}^{ij} = 0. \tag{A.84}$$

If we make the separation $\mathsf{Q}^{lk} = \mathsf{Q}^{(kl)} - \epsilon^{kli}a_i$, the constraint on $\mathsf{Q}$ is just $\mathsf{Q}^{(kl)}a_l = 0$. $\mathsf{Q}^{(kl)}$ can be diagonalized and its eigenvectors $a_i$ can be found, and all the three-dimensional Lie algebras (nine in total) can be classified. This is the Bianchi classification (see e.g. Ref. [878]). The only semisimple ones are those of $\mathrm{SO}(3)$ and $\mathrm{SO}(2,1)$.

## A.3 Riemannian geometry of group manifolds

We can define Riemannian metrics on Lie groups. The most interesting ones are those invariant under the left- and right-translation diffeomorphisms (*bi-invariant metrics*). If $B_{AB}$ are the components of a non-singular metric in the basis of left-invariant vector fields $\{e_A\}$, then

$$ds^2 = B_{AB}e^A \otimes e^B, \tag{A.85}$$

where the $e^A$ are a basis of left-invariant 1-forms constructed for instance as in Eq. (A.12), is automatically a metric invariant under the left-translation diffeomorphisms, and has an isometry group G. Under right translations $g \to gh$

$$e^A \to \Gamma_{\mathrm{Adj}}(h^{-1})^A{}_B e^B \quad \Rightarrow B_{AB} \to \Gamma_{\mathrm{Adj}}(h^{-1})^C{}_A\Gamma_{\mathrm{Adj}}(h^{-1})^D{}_B B_{CD}. \tag{A.86}$$

Infinitesimally, $B_{AB}$ will be invariant if the analog of Eq. (A.22) holds. In that case the metric will have an isomorphism group of G $\times$ G (or smaller, if some left and right actions coincide).

For semisimple groups it is natural to use the Killing metric. If we diagonalize it and normalize it so that its diagonal contains only $+1$s and $-1$s, then the metric is just $\eta_{AB}$ and a metric-compatible connection will be an $\mathrm{SO}(n_+, n_-)$ connection, $\omega^{AB} = -\omega^{BA}$. In the absence of torsion, the connection 1-form can be determined by comparing the Maurer–Cartan equations (A.10) with the structure equation Eq. (1.152),[12]

$$\omega^A{}_B = \tfrac{1}{2} e^C f_{CB}{}^A = \tfrac{1}{2} e^C \Gamma_{\mathrm{Adj}}(T_C)^A{}_B. \tag{A.87}$$

The curvature 2-form is given simply by

$$R^A{}_B = \tfrac{1}{8} e^C \wedge e^D f_{CD}{}^E f_{EB}{}^A, \tag{A.88}$$

and all its components in this basis are constant and completely determined by the structure constants. The Ricci tensor is proportional to the Killing metric. Furthermore, repeated use of the Jacobi identity shows that the curvature is covariantly constant,

$$\nabla_A R_{BCDE} = 0. \tag{A.89}$$

Let us consider a simple but useful example.

### *A.3.1 Example: the* SU(2) *group manifold*

$\mathrm{SU}(2)$ matrices $U$ ($U^\dagger = U^{-1}$, $\det U = +1$) can be parametrized by $z_0, z_1 \in \mathbb{C}$,

$$U \equiv \begin{pmatrix} z_0 & z_1 \\ -\bar{z}_1 & \bar{z}_0 \end{pmatrix}, \qquad |z_0|^2 + |z_1|^2 = 1, \tag{A.90}$$

so $\mathrm{SU}(2)$ has the topology of $\mathrm{S}^3$. We can parametrize both by the *Euler angles* $\{\theta, \varphi, \psi\}$,

$$z_0 = \cos\left(\frac{\theta}{2}\right) e^{i(\varphi+\psi)/2}, \qquad z_1 = \sin\left(\frac{\theta}{2}\right) e^{i(\varphi-\psi)/2}, \tag{A.91}$$

where[13] $\theta \in [0, \pi]$, $\varphi \in [0, 2\pi]$, and $\psi \in [0, 4\pi]$, whose main property is that we can construct any general SU(2) rotation as the product of three rotations:

$$U(\varphi, \theta, \psi) = U(\varphi, 0, 0) U(0, \theta, 0) U(0, 0, \psi). \tag{A.92}$$

The left-invariant Maurer–Cartan 1-forms $e^i$, $i = 1, 2, 3$, are

$$U^{-1} dU \equiv -e^i T_i, \tag{A.93}$$

where the $T_i$s are the anti-Hermitian generators of the $\mathfrak{su}(2)$ Lie algebra Eq. (A.83) (with $\eta^{kl} = \delta^{kl}$),

$$T_i = \frac{i}{2} \sigma^i. \tag{A.94}$$

[12] If the metric is not $\eta_{AB} = (+\cdots, -\cdots)$ then $\omega^A{}_B = e^C f_{CB}{}^A + e^C K_{CB}{}^A$, where $K_{CB}{}^A = K_{BC}{}^A$ has to be determined case by case.

[13] On imposing the restriction $\psi \in [0, 2\pi]$, we obtain $\mathbb{RP}^3$, which is homeomorphic to SO(3).

With the Euler-angles parametrization they are

$$\begin{aligned} e^1 &= \sin\psi\, d\theta - \sin\theta\cos\psi\, d\varphi, \\ e^2 &= -(\cos\psi\, d\theta + \sin\theta\sin\psi\, d\varphi), \\ e^3 &= -(d\psi + \cos\theta\, d\varphi), \end{aligned} \tag{A.95}$$

and it can easily be checked that they satisfy the Maurer–Cartan equation

$$de^i = -\tfrac{1}{2}\epsilon_{ijk} e^j \wedge e^k. \tag{A.96}$$

Using the Killing metric we can construct a bi-invariant (i.e. $\mathrm{SU}(2) \times \mathrm{SU}(2) \sim \mathrm{SO}(4)$-invariant) metric on $\mathrm{SU}(2)$ ($\mathrm{S}^3$). On normalizing to obtain the volume of $\mathrm{S}^3$, we obtain[14]

$$d\Omega^2_{(3)} = \tfrac{1}{4}\left[(e^1)^2 + (e^2)^2 + (e^3)^2\right] = \tfrac{1}{4}\left[d\Omega^2_{(2)} + (e^3)^2\right]. \tag{A.97}$$

## A.4 Riemannian geometry of homogeneous and symmetric spaces

We can define the action of a (*transformation*) group G on a space M as a continuous map

$$\begin{aligned} \mathrm{G} \times \mathrm{M} &\rightarrow \mathrm{M} \\ (g, x) &\rightarrow gx \end{aligned} \tag{A.98}$$

such that

$$g_1(g_2 x) = (g_1 g_2)x, \quad ex = x, \quad \forall\, g_1, g_2 \in G,\, x \in M. \tag{A.99}$$

Each $g \in \mathrm{G}$ induces a homeomorphism of M into M; G is said to act *transitively* on M if, given any two points of M, there is always a transformation of the group that relates them. M is then a *homogeneous space*. The subgroup $\mathrm{H} \subset \mathrm{G}$ that leaves a given point invariant is the *isotropy* group of that point. The isotropy groups of all points of M are isomorphic, and we can talk about the isotropy group of M. Then, it can be shown that M and the coset space $\mathrm{G/H}$ defined by the equivalence classes under right multiplication by elements of H, sometimes denoted by $\{g\mathrm{H}\}$, are homeomorphic. Observe that G acts from the left on these equivalence classes.

It can also be shown that, if H is topologically closed, the coset space $\mathrm{G/H}$ can be given the structure of a manifold of dimension $\dim \mathrm{G} - \dim \mathrm{H}$ on which G acts transitively. A manifold M that is a homogeneous space is always diffeomorphic to the coset manifold $\mathrm{G/H}$, which henceforth is what we mean by homogeneous space.

A very important theorem states that G can be seen as a *principal bundle* with base space $\mathrm{G/H}$, structure group H, and projection $\mathrm{G} \rightarrow \mathrm{G/H}$.

The Lie algebra of any homogeneous space $\mathrm{G/H}$ can be decomposed as the direct sum of vector spaces $\mathfrak{g} = \mathfrak{h} \oplus \mathfrak{p}$, where $\mathfrak{h}$ is the Lie subalgebra of H and $\mathfrak{k}$ is its orthogonal complement. Since $\mathfrak{h}$ is a subalgebra

$$[\mathfrak{h}, \mathfrak{h}] \subset \mathfrak{h}. \tag{A.100}$$

[14] See Appendix K for the definitions of spheres and their volumes.

G/H is *reductive* if

$$[\mathfrak{k}, \mathfrak{h}] \subset \mathfrak{k}, \tag{A.101}$$

which means that $\mathfrak{k}$ is a representation of H.

G/H is *symmetric* if it is reductive and

$$[\mathfrak{k}, \mathfrak{k}] \subset \mathfrak{h}. \tag{A.102}$$

The pair $(\mathfrak{k}, \mathfrak{h})$ is then a *symmetric pair*.[15] The two components of a symmetric pair are mutually orthogonal with respect to the Killing metric, which is block diagonal.

If G/H is reductive and $\mathfrak{k}$ is a subalgebra,

$$[\mathfrak{k}, \mathfrak{k}] \subset \mathfrak{k} \tag{A.103}$$

(so it is an ideal), then $\mathfrak{g}$ is the *semidirect sum* of $\mathfrak{h}$ and $\mathfrak{k}$, and G is the *semidirect product* of the corresponding subgroups $\mathrm{G} = \mathrm{H} \ltimes \mathrm{K}$.

By construction, there is a natural transitive action of G on homogeneous spaces, and it is natural to define on them Riemannian metrics that are invariant under the left action of G. If the isotropy group H of a homogeneous (not necessarily symmetric or reductive) space G/H is compact, then there is always at least one G-invariant Riemannian metric. If G/H is symmetric, G is connected, and H is compact, and it is equipped with the G-invariant metric, then G/H is a *(Riemannian) globally symmetric* space.

Thus, these homogeneous spaces with a G-invariant metric have an isometry group G that acts transitively from the left. If they are reductive, their Riemann curvature tensor is covariantly constant,[16] as in Eq. (A.89).

We have already seen a particular example of homogeneous reductive spaces with a G-invariant Riemannian metric: group manifolds equipped with a bi-invariant metric on $\mathfrak{g}$. They have a trivial isotropy subgroup $\mathrm{H} = e$ and are trivial coset manifolds. They are clearly reductive but not symmetric because $\mathfrak{k} = \mathfrak{g}$ and then $[\mathfrak{k}, \mathfrak{k}] = \mathfrak{k}$. The isometry group is the product of the left isometry group G and the right isometry group, which is also G.[17]

Now, we are going to show a procedure by which to construct G-invariant Riemannian metrics on homogeneous spaces. Let us introduce some notation: we denote by $\{M_i\}$ $(i, j = 1, \ldots, \dim \mathrm{H})$ a basis of $\mathfrak{h}$, and by $\{P_a\}$ $(a, b = 1, \ldots, d = \dim \mathrm{G} - \dim \mathrm{H})$ a basis of $\mathfrak{k}$. By exponentiating the generators of $\mathfrak{k}$ we can construct a coset representative $u(x) = u(x^1, \ldots, x^d)$. We can construct the coset representative as a product of generic elements of the one-dimensional subgroups generated by the $P_a$s:

$$u(x) = e^{x^1 P_1} \cdots e^{x^q P_q}. \tag{A.104}$$

Under a left transformation $g \in \mathrm{G}$, $u$ transforms into another element of G, which becomes a coset representative $u(x')$ only after a right transformation with an element $h \in \mathrm{H}$, which is a function of $g$ and $x$:

$$gu(x) = u(x')h. \tag{A.105}$$

[15] This decomposition can be characterized by an involutive ($\sigma^2 = 1$) automorphism $\sigma$ of $\mathfrak{g}$ such that $\sigma(T) = +T,\ \forall\, T \in \mathfrak{h}$, and $\sigma(T) = -T,\ \forall\, T \in \mathfrak{k}$.

[16] These spaces should not be confused with *spaces of constant curvature*, which have $R_{IJKL} = K\delta_{IJ\,KL}$. These are a particular type of symmetric space that has the maximal number of Killing vectors allowed ($\frac{1}{2}d(d+1)$). They are also known as *maximally symmetric spaces*.

[17] In general, the right isometry group of G/H will be $N(\mathrm{H})/\mathrm{H}$, where $N(\mathrm{H})$ is the normalizer of H.

Now we construct the left-invariant Maurer–Cartan 1-form and expand it in *horizontal*, $e^a$, and *vertical*, $\vartheta^i$, components:

$$V \equiv -u^{-1}du = e^a P_a + \vartheta^i M_i. \tag{A.106}$$

The horizontal components $e^a$ can be used as Vielbeins for G/H and, given any metric $B_{ab}$ on $\mathfrak{k}$, we can construct a Riemannian metric

$$ds^2 = B_{ab} e^a \otimes e^b. \tag{A.107}$$

To find under what conditions this metric will be (left) G invariant, we have to look at the transformation of the Maurer–Cartan 1-forms under left multiplication by a constant element $g \in \mathrm{G}$, $u(x') = gu(x)h^{-1}$:

$$\begin{aligned} e^a(x') &= (he(x)h^{-1})^a = \Gamma_{\mathrm{Adj}}(h)^a{}_b e^b(x), \\ \vartheta^i(x') &= (h\vartheta(x)h^{-1})^i + (h^{-1}dh)^i + (he(x)h^{-1})^i. \end{aligned} \tag{A.108}$$

The last term in the second Eq. (A.108) is zero in the reductive case and the $\vartheta^i$s transform as a connection. Furthermore, the restriction of $\Gamma_{\mathrm{Adj}}(h)$ to $\mathfrak{k}$ is a representation of $\mathfrak{h}$. The Riemannian metric will be invariant under the left action of G if

$$f_{i(a}{}^c B_{b)c} = 0, \tag{A.109}$$

which is guaranteed if we can set $B_{ab} = K_{ab}$, the projection on $\mathfrak{k}$ of the (non-singular) Killing metric. There are many important cases in which G is not semisimple but there is a non-degenerate invariant metric. For instance, we can describe Minkowski space as the quotient of the Poincaré group (which is not semisimple because it contains the Abelian invariant subgroup of translations) by the Lorentz subgroup. The Minkowski metric is a non-degenerate invariant metric for this coset. Another example is provided by the Hpp-wave spacetimes constructed in Section 14.1.1.

The resulting Riemannian metric contains G in its isometry group (generically $\mathrm{G} \times N(\mathrm{H})/\mathrm{H}$) and must admit $n$ Killing vector fields $k_{(I)}$. The Killing vectors $k_A$ and the *H-compensator* $W_A{}^i$ are defined through the infinitesimal version of $gu(x) = u(x')h$ with

$$\begin{aligned} g &= 1 + \sigma^A T_A, \\ h &= 1 - \sigma^A W_A{}^i M_i, \\ x^{\mu\prime} &= x^\mu + \sigma^A k_A{}^\mu. \end{aligned} \tag{A.110}$$

Using these equations in

$$u(x + \delta x) = u(x) + \sigma^A k_A u, \tag{A.111}$$

we obtain

$$T_A u = k_A u - u W_A{}^i M_i. \tag{A.112}$$

Acting with $u^{-1}$ on the left and using the definitions of the adjoint action and the Maurer–Cartan 1-forms, we obtain

$$T_B \Gamma_{\mathrm{Adj}}(u^{-1})^B{}_A = -k_A{}^a P_a - (k_A{}^\mu \vartheta^i{}_\mu + W_A{}^i) M_i, \tag{A.113}$$

which, projected onto the horizontal and vertical subspaces, gives

$$k_A{}^a = -\Gamma_{\text{Adj}}(u^{-1}(x))^a{}_A, \tag{A.114}$$
$$W_A{}^i = -k_A{}^\mu \vartheta^i{}_\mu - \Gamma_{\text{Adj}}(u^{-1}(x))^i{}_A. \tag{A.115}$$

The Killing vectors associated with the right isometry group $N(\mathrm{H})/\mathrm{H}$ are just the vectors $e_a$ dual to the horizontal Maurer–Cartan 1-forms in the directions of $N(\mathrm{H})/\mathrm{H}$.

These formulae simplify considerably the calculation of Killing vectors, if we construct the space with the above recipe. As in group manifolds, the spin connection[18] can easily be found: on comparing the Maurer–Cartan equations

$$de^a - \vartheta^i \wedge e^b f_{ib}{}^a - \tfrac{1}{2} e^b \wedge e^c f_{bc}{}^a = 0 \tag{A.116}$$

with the structure equation Eq. (1.152), we obtain

$$\omega^a{}_b = \vartheta^i f_{ib}{}^a + \tfrac{1}{2} e^c f_{cb}{}^a, \tag{A.117}$$

if we do not allow for torsion, or

$$\omega^a{}_b = \vartheta^i f_{ib}{}^a, \qquad T^a = -\tfrac{1}{2} e^c \wedge e^b f_{cb}{}^a. \tag{A.118}$$

It is straightforward to compute the curvature using the Maurer–Cartan equations:

$$d\vartheta^i - \tfrac{1}{2}\vartheta^j \wedge \vartheta^k f_{jk}{}^i - \tfrac{1}{2} e^a \wedge e^b f_{ab}{}^i = 0. \tag{A.119}$$

In the symmetric case ($f_{cb}{}^a = 0$), and in the reductive case with the torsionful connection Eq. (A.118), the curvature 2-form is given by

$$R^a{}_b = [d\vartheta^i - \tfrac{1}{2}\vartheta^j \wedge \vartheta^k f_{jk}{}^i] f_{ib}{}^a = \tfrac{1}{2} e^c \wedge e^d f_{cd}{}^i f_{ib}{}^a \tag{A.120}$$

(using the Maurer–Cartan equations), and it is covariantly constant. In the reductive (non-symmetric) case

$$R^a{}_b = \tfrac{1}{2} e^c \wedge e^d \big( f_{cd}{}^i f_{ib}{}^a + \tfrac{1}{2} f_{cd}{}^e f_{eb}{}^a - \tfrac{1}{2} f_{ce}{}^a f_{db}{}^e \big). \tag{A.121}$$

The reductive case (symmetric or not) is particularly interesting because, as we have said, according to Eqs. (A.108) the vertical 1-forms $\vartheta^i$ transform as a connection for the group H. Equations (A.117) and (A.118) relate this gauge connection to the spin connection (and torsion). Sometimes this is expressed by saying that the gauge group has been embedded into the tangent-space group. These relations are used very often in the construction of solutions. This suggests the following definitions.

### *A.4.1* H-*covariant derivatives*

The H-*covariant derivative* of any object that transforms contravariantly, $\phi' = \Gamma_r(h)\phi$, or covariantly, $\psi' = \psi\Gamma_r(h^{-1})$ (for instance, $u(x)$ itself), in the representation $r$ of H is

$$\mathcal{D}_\mu \phi \equiv \partial_\mu \phi - \vartheta_\mu{}^i \Gamma_r(M_i)\phi, \qquad \mathcal{D}_\mu \psi \equiv \partial_\mu \psi + \psi \vartheta_\mu{}^i \Gamma_r(M_i). \tag{A.122}$$

[18] We assume here that $B_{ab}$ is diagonal, with only $+1$s and $-1$s on the diagonal.

The curvature $F^i$ is

$$F^i = d\vartheta^i - \tfrac{1}{2}\vartheta^j \wedge \vartheta^k f_{jk}{}^i \tag{A.123}$$

and it is covariantly constant with respect to the full (Lorentz-plus-gauge) covariant derivative if we use the torsionful spin connection Eq. (A.118) and also in the symmetric case. This statement is clearly equivalent to the covariant constancy of the Lorentz curvature.

This implies that, in any reductive coset space G/H, there is a solution of the Yang–Mills equations of motion for the group H in the curved geometry associated with the (torsionful) spin connection defined earlier. This result is implicitly or explicitly used in many places. The simplest example is provided by the coset manifold $\mathrm{SU}(2)/\mathrm{U}(1)$, which gives a round 2-sphere. The U(1) connection solves the Maxwell equations and corresponds to the Dirac monopole (see the calculation of the Robinson–Bertotti superalgebra on p. 503 and Appendix K.1).

The gauge field $\vartheta^i$ is invariant under the combination of the diffeomorphisms generated by the Killing vectors $k_A$ and gauge transformations generated by $W_A{}^i$, i.e.

$$-\mathcal{L}_{k_A}\vartheta^i = \mathcal{D}_\mu W_A{}^i, \tag{A.124}$$

where $\mathcal{L}_{k_A}$ is the standard Lie derivative.

The H-*covariant Lie derivative with respect to the Killing vectors*[19] $k_A$ of contravariant ($\phi$) or covariant ($\psi$) objects in the representation $r$ of H is

$$\mathbb{L}_{k_A}\phi \equiv \mathcal{L}_{k_A}\phi + W_A{}^i\Gamma_r(M_i)\phi, \qquad \mathbb{L}_{k_A}\psi \equiv \mathcal{L}_{k_A}\psi - \psi W_A{}^i\Gamma_r(M_i). \tag{A.125}$$

This Lie derivative satisfies, among other properties,

$$[\mathbb{L}_{k_A}, \mathbb{L}_{k_B}] = \mathbb{L}_{[k_A,k_B]}, \tag{A.126}$$

$$\mathbb{L}_{k_A}e^a = 0, \tag{A.127}$$

$$\mathbb{L}_{k_A}u = \mathcal{L}_{k_A}u - uW_A{}^iM_i = T_A u, \tag{A.128}$$

where the last property follows from Eqs. (A.115) and (A.112). The connection 1-forms $\vartheta^i$ are not covariant or contravariant objects and this definition does not apply to them. The best one can do for them is to combine the standard Lie derivative and a compensating gauge transformation. The resulting operator acting on $\vartheta^i$ is identically zero, due to Eq. (A.124).

### *A.4.2 Example: round spheres*

The $n$-dimensional sphere $\mathrm{S}^n$ (see Appendix K) is a homogeneous topological space on which the orthogonal group $\mathrm{SO}(n+1)$ acts transitively.[20] Any point is invariant under

---

[19] H-covariant Lie derivatives can be defined with respect to any vector, but the property Eq. (A.126) holds only for Killing vectors. The spinorial Lie derivative [871, 870, 895], and the more general Lie–Lorentz and Lie–Maxwell derivatives that appear in calculations of supersymmetry algebras [546, 1009] discussed in Section 17.2.1, can actually be seen as particular examples of this more general operator (see e.g. Ref [639]) and, actually, are identical objects when they are acting on Killing spinors of maximally supersymmetric spacetimes [26].

[20] $\mathrm{SO}(n+1)$ rotates in the standard form the coordinates of the ambient space $\mathbb{R}^{n+1}$, respecting the defining equation $(x^1)^2 + \cdots + (x^{n+1})^2 = 1$.

rotations around the axis that crosses that point: the isotropy group is thus $\mathrm{SO}(n)$, and $\mathrm{S}^n$ is therefore homeomorphic to $\mathrm{SO}(n+1)/\mathrm{SO}(n)$. If the $\mathrm{SO}(n+1)$ generators are $\{\hat{M}_{\hat{a}\hat{b}}\}$, $\hat{a},\hat{b}=1,\ldots,n+1$, the generators of $\mathrm{SO}(n)$ can be chosen as $\{M_{ab}\equiv\hat{M}_{ab}\}$ $a,b=1,\ldots,n$ and those of the orthogonal complement as $P_a\equiv\hat{M}_{n+1\,a}$. We immediately see that $\mathrm{S}^n$ is a symmetric space:

$$[P_a, M_{bc}] = 2\delta_{a[c}P_{b]} = -P_d\Gamma_{\mathrm{v}}(M_{bc})^d{}_a, \qquad [P_a, P_b] = M_{ab}. \tag{A.129}$$

To construct an $\mathrm{SO}(n+1)$-symmetric Riemannian metric,[21] we first construct a coset representative $u$ as before and then the Maurer–Cartan 1-form $V$:

$$\begin{aligned}-V = {} & P_n dx^n + e^{-x^n P_n} P_{n-1} e^{x^n P_n} dx^{n-1} \\ & + e^{-x^n P_n} e^{x^{n-1}P_{n-1}} P_{n-2} e^{x^n P_{n-1}} e^{x^n P_n} dx^{n-2} \cdots .\end{aligned} \tag{A.130}$$

Here we can use repeatedly the formula

$$[X,Y]=Z,\ \ [Y,Z]=X,\ \ [Z,X]=Y \ \ \Rightarrow e^{aX}Ye^{-aX} = \cos(a)\,Y + \sin(a)\,Z \tag{A.131}$$

for the triplets $P_a$, $P_b$, and $M_{ab}$, or, far better, the definition of the adjoint action:

$$e^{-x^n P_n} P_{n-1} e^{x^n P_n} = \tfrac{1}{2}\hat{M}_{\hat{a}\hat{b}}\Gamma_{\mathrm{Adj}}(e^{-x^n P_n})^{\hat{a}\hat{b}}{}_{n-1}. \tag{A.132}$$

In both cases, the result is

$$\begin{aligned}-e &= P_n dx^n + P_{n-1}\cos x^n\, dx^{n-1} + P_{n-2}\cos x^n \cos x^{n-1}\, dx^{n-2} + \cdots, \\ -\vartheta &= \sin x^n \sum_{a=1}^{a=n-1} M_{n\,a} dx^a + \cos x^n \sin x^{n-1} \sum_{a=1}^{a=n-2} M_{n-1\,a} dx^a \\ &\quad + \cos x^n \cos x^{n-1} \sin x^{n-2} \sum_{a=1}^{a=n-3} M_{n-2\,a} dx^a + \cdots.\end{aligned} \tag{A.133}$$

Using the $\mathrm{SO}(n)$-invariant metric[22] $\delta_{ab}$, we obtain the $\mathrm{SO}(n+1)$-invariant metric

$$ds^2 = (dx^n)^2 + \cos^2 x^n \left[(dx^{n-1})^2 + \cos^2 x^{n-1}\left[(dx^{n-2})^2 + \cdots\right.\right. . \tag{A.134}$$

On comparing this with the metric Eq. (K.4) in standard spherical coordinates, we see that the coset coordinates that we have used are related to them by $x^n=\theta_{n-1}+\pi/2,\ldots,x^1=\varphi$. The spin connection is given by

$$\omega^a{}_b = \tfrac{1}{2}\vartheta^{cd} f_{cd\,n+1b}{}^{n+1a} = \tfrac{1}{2}\vartheta^{cd}\Gamma_{\mathrm{v}}(M_{cd})^a{}_b = \vartheta^a{}_b, \tag{A.135}$$

and the curvature, which corresponds to a maximally symmetric space, by

$$R_{cd}{}^a{}_b = \tfrac{1}{2} f_{n+1c\,n+1d}{}^{ef} f_{ef\,n+1b}{}^{n+1a} = \tfrac{1}{2}\delta_{cd}{}^{ef}\Gamma_{\mathrm{v}}(M_{ef})^a{}_b = \delta_{[c}{}^a\delta_{d]b}. \tag{A.136}$$

We can construct $\mathrm{AdS}_d$ spacetimes in an almost identical way as the coset manifolds $\mathrm{SO}(2,d-1)/\mathrm{SO}(1,d-1)$ with $P_a\equiv\hat{M}_{-1\,a}$, $a=0,1,\ldots,d-1$, and $\{M_{ab}\equiv\hat{M}_{ab}\}$ using the metric $\eta_{ab}$ on $\mathfrak{k}$. The $d=4$ case is worked out in Section 17.3.3 and the $d=2$ case is in Section 17.3.4, p. 503.

---

[21] With $\frac{1}{2}n(n+1)$ isometries, spheres with this metric (*round spheres*) are also maximally symmetric spaces.

[22] On rescaling some entries of $\delta_{ab}$, one obtains *squashed spheres* (see Appendix K.2) that have less symmetry.

# Appendix B
# The irreducible, non-symmetric Riemannian spaces of special holonomy

**Theorem (Berger)** [143, 1099, 836] Let us consider simply connected manifolds of dimension $n$ with Riemannian metric $g$ that are irreducible and non-symmetric. Then exactly one of the following seven cases holds:[1]

1. $\mathrm{Hol}(g) = \mathrm{SO}(n)$. This is the generic holonomy group of Riemannian spaces.
2. $\mathrm{Hol}(g) = \mathrm{U}(\frac{n}{2}) \subset \mathrm{SO}(n)$ with $n \geq 4$. This is the holonomy group of Kähler spaces (see Appendix E).
3. $\mathrm{Hol}(g) = \mathrm{SU}(\frac{n}{2}) \subset \mathrm{SO}(n)$ with $n \geq 4$. This is the holonomy group of Calabi–Yau spaces, which are, evidently, Kähler spaces since $\mathrm{SU}(m) \subset \mathrm{U}(m)$. The necessary and sufficient condition for a Kähler space to be a Calabi–Yau space is that it is Ricci-flat.
4. $\mathrm{Hol}(g) = \mathrm{Sp}(\frac{n}{4}) \subset \mathrm{SO}(n)$ with $n \geq 8$. This is the holonomy group of hyper-Kähler spaces, which are Kähler and Ricci-flat because $\mathrm{Sp}(\frac{n}{4}) \subseteq \mathrm{SU}(\frac{n}{2}) \subseteq \mathrm{U}(\frac{n}{2})$. Sometimes they are considered as a special sub-case of quaternionic-Kähler manifolds (see Appendix G) and sometimes they are not.
5. $\mathrm{Hol}(g) = \mathrm{Sp}(\frac{n}{4}) \times \mathrm{Sp}(1) \subset \mathrm{SO}(n)$. This is the holonomy group of quaternionic-Kähler spaces which are Einstein, non Ricci-flat, and not Kähler (see Appendix G).
6. $n = 7$ and $\mathrm{Hol}(g) = \mathrm{G}_2 \subset \mathrm{SO}(7)$.
7. $n = 8$ and $\mathrm{Hol}(g) = \mathrm{Spin}_7 \subset \mathrm{SO}(8)$. This and case 6 are the special holonomy groups and the corresponding spaces are Ricci-flat.

The constraints on $n$ in the above classification are meant to avoid repetitions. For instance, Calabi–Yau spaces with $n = 2$ are always hyper-Kähler spaces because $\mathrm{Sp}(1) \sim \mathrm{SU}(2)$. Hyper-Kähler 4-manifolds are included in the standard Riemannian case because $\mathrm{Sp}(1) \times \mathrm{Sp}(1) \sim \mathrm{SO}(4)$. (On the other hand, it has been shown by Kodaira that any compact hyper-Kähler 4-manifold is either a K3 surface or a compact torus $\mathrm{T}^4$.)

This classification is summarized in Fig. B.1.

[1] The case $n = 16$ with $\mathrm{Hol}(g) = \mathrm{Spin}_9 \subset \mathrm{SO}(16)$ originally found by Berger has subsequently been proven always to correspond to symmetric spaces.

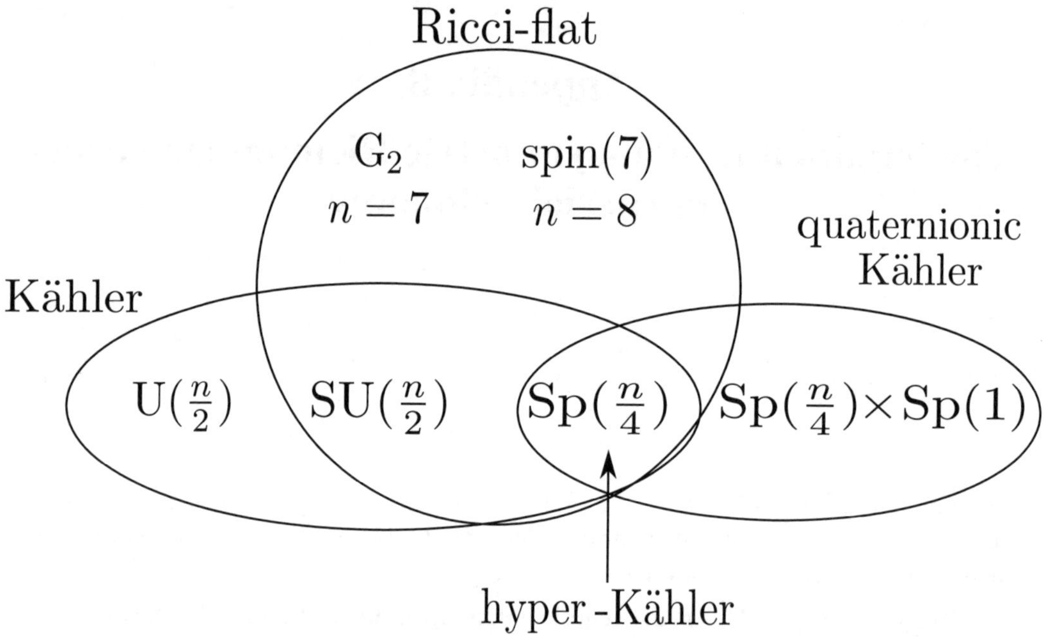


Fig. B.1. The irreducible, non-symmetric Riemannian spaces of dimensions $n$ of special holonomy of Berger's classification according to Ref. [1099]. They belong to three broad categories: Ricci-flat, Kähler, and quaternionic Kähler.

# Appendix C
# Miscellanea on the symplectic group

In this appendix we collect together a number of formulae and properties of the symplectic group and symplectic matrices which occur *partout* in four-dimensional SUEGRAs. We are going to follow very closely an appendix of Ref. [67], adapted to our notation and needs.

## C.1 The symplectic group

The symplectic group $\mathrm{Sp}(2n, \mathbb{R})$ is the group of transformations $S$ that preserve the symplectic "metric" $\Omega$:

$$(\Omega_{MN}) \equiv \begin{pmatrix} 0 & \mathbb{1}_{n\times n} \\ -\mathbb{1}_{n\times n} & 0 \end{pmatrix}, \qquad S \equiv \begin{pmatrix} A & B \\ C & D \end{pmatrix}, \qquad S^T \Omega S = \Omega. \tag{C.1}$$

This condition on $S$, which defines the fundamental representation, is equivalent to the following conditions on the $n \times n$ submatrices $A, B, C$, and $D$:[1]

$$C^T A = A^T C, \qquad B^T D = D^T B, \qquad A^T D - C^T B = \mathbb{1}. \tag{C.2}$$

If $S$ is symplectic, so is $S^T$, from which the equivalent conditions

$$AB^T = BA^T, \qquad DC^T = CD^T, \qquad AD^T - CB^T = \mathbb{1}_{n\times n}. \tag{C.3}$$

Then, if $A$ and $D$ are invertible, the matrices

$$A^T C, \;\; B^T D, \;\; CA^{-1}, \;\; BD^{-1}, \;\; A^{-1}B, \;\; D^{-1}C, \;\; AB^T, \;\; DC^T \tag{C.4}$$

are symmetric.

[1] In what follows we will write $\mathbb{1}$ instead of $\mathbb{1}_{n\times n}$.

If $D^{-1}$ exists, we can factorize the symplectic matrix as follows:

$$S = \begin{pmatrix} A & B \\ C & D \end{pmatrix} = \begin{pmatrix} \mathbb{1} & BD^{-1} \\ 0 & \mathbb{1} \end{pmatrix} \begin{pmatrix} (D^T)^{-1} & 0 \\ 0 & D \end{pmatrix} \begin{pmatrix} \mathbb{1} & 0 \\ D^{-1}C & \mathbb{1} \end{pmatrix}, \tag{C.5}$$

where $A = (D^T)^{-1} + BD^{-1}C$ follows from $BD^{-1} = (D^T)^{-1}B^T$.

If we expand the symmetric matrix $S$ infinitesimally away from the unit

$$S \sim \mathbb{1}_{2n\times 2n} + \epsilon\mathcal{T}, \qquad \mathcal{T} \equiv \begin{pmatrix} a & b \\ c & d \end{pmatrix}, \tag{C.6}$$

we find that the $n \times n$ submatrices $a, b, c$, and $d$ of $\mathcal{T}$ satisfy

$$\boxed{c = c^T, \quad b = b^T, \quad a^T = -d.} \tag{C.7}$$

*The real basis.* The symplectic indices $M, N, \ldots$ take $2n$ values and are equivalent to one pair of indices $\Lambda, \Sigma, \ldots$ (one in the upper position and one in the lower position) which take only $n$ values. Thus, a real symplectic vector $\mathcal{Q}^M$ transforms as

$$\mathcal{Q}'^M = S^M{}_N \mathcal{Q}^N \tag{C.8}$$

and contains two real $\mathrm{SO}(N)$ vectors $p^\Lambda$ and $q_\Lambda$,

$$(\mathcal{Q}^M) = \begin{pmatrix} p^\Lambda \\ q_\Lambda \end{pmatrix}, \tag{C.9}$$

which are transformed and mixed by the $A, B, C, D$ submatrices. Therefore, the submatrices carry the indices

$$(S^M{}_N) = \begin{pmatrix} A^\Lambda{}_\Sigma & B^{\Lambda\Sigma} \\ C_{\Lambda\Sigma} & D_\Lambda{}^\Sigma \end{pmatrix}, \tag{C.10}$$

and

$$\left\{ \begin{array}{rcl} p'^\Lambda & = & A^\Lambda{}_\Sigma \, p^\Sigma + B^{\Lambda\Sigma} \, q_\Sigma, \\ \\ q'_\Lambda & = & C_{\Lambda\Sigma} \, p^\Sigma + D_\Lambda{}^\Sigma \, q_\Sigma. \end{array} \right. \tag{C.11}$$

The conditions Eqs. (C.2) read in this notation

$$A^\Lambda{}_\Omega B^{\Sigma\Omega} = B^{\Lambda\Omega} A^\Sigma{}_\Omega, \quad D_\Lambda{}^\Omega C_{\Sigma\Omega} = C_{\Lambda\Omega} D_\Sigma{}^\Omega, \quad A^\Lambda{}_\Omega D_\Sigma{}^\Omega - C_{\Lambda\Omega} B^{\Sigma\Omega} = \delta_{\Lambda\Sigma}. \tag{C.12}$$

The matrix of infinitesimal generators $\mathcal{T}$ carries the same indices as those of $S$ and the submatrices $a, b, c, d$ carry the same indices as those of the submatrices $A, B, C, D$. The conditions they satisfy, Eqs. (C.7), read in this notation

$$c_{[\Lambda\Sigma]} = b^{[\Lambda\Sigma]} = 0, \qquad a^\Lambda{}_\Sigma = -d_\Sigma{}^\Lambda. \tag{C.13}$$

The symplectic "metric" $\Omega_{MN}$ and its inverse can be used to raise and lower symplectic indices according to the convention

$$\mathcal{Q}_N \equiv \Omega_{NM}\mathcal{Q}^M, \qquad \mathcal{Q}^N = \mathcal{Q}_M \Omega^{MN}, \qquad \Omega^{MN} = -(\Omega^{-1})^{MN} = \Omega_{MN}, \tag{C.14}$$

so the components of the covariant symplectic vector are related to those of the contravariant one by

$$(\mathcal{Q}_M) = (q_\Lambda, -p^\Lambda). \tag{C.15}$$

*The complex basis.* Symplectic vectors can also be expressed in an alternative complex basis in which the components are

$$(\tilde{Q}^M) = \frac{1}{\sqrt{2}} \begin{pmatrix} p + iq \\ p - iq \end{pmatrix}, \tag{C.16}$$

which are related to the components of the real basis by

$$\tilde{\mathcal{Q}} = \mathcal{A}^{-1}\mathcal{Q}, \tag{C.17}$$

where $\mathcal{A}$ is a symplectic and unitary matrix given by

$$\mathcal{A} \equiv \frac{1}{\sqrt{2}} \begin{pmatrix} \mathbb{1} & \mathbb{1} \\ -i\mathbb{1} & i\mathbb{1} \end{pmatrix}. \tag{C.18}$$

The symplectic matrix in the complex basis, $\tilde{S}$, is related to the one in the real basis, $S$, by

$$\tilde{S} = \mathcal{A}^{-1} S \mathcal{A}, \tag{C.19}$$

and always belongs to $\mathrm{U}(n,n) \cap \mathrm{Sp}(2n,\mathbb{C})$, i.e.

$$\tilde{S}^\dagger \begin{pmatrix} \mathbb{1} & 0 \\ 0 & -\mathbb{1} \end{pmatrix} \tilde{S} = \begin{pmatrix} \mathbb{1} & 0 \\ 0 & -\mathbb{1} \end{pmatrix}, \qquad \tilde{S}^T \Omega \tilde{S} = \Omega. \tag{C.20}$$

The opposite is also true: any $\mathrm{U}(n,n) \cap \mathrm{Sp}(2n,\mathbb{C})$ matrix is related to an $\mathrm{Sp}(2n,\mathbb{R})$ matrix by Eq. (C.19).

Equations (C.20) define a representation of $\mathrm{Sp}(2n,\mathbb{R})$ on $\mathbb{C}^{2n}$ which is the direct sum of the representations

$$\begin{pmatrix} \psi \\ \psi^* \end{pmatrix} \quad \text{and} \quad \begin{pmatrix} \psi \\ -\psi^* \end{pmatrix}, \tag{C.21}$$

which consist of all the linear combinations with real coefficients of all vectors with the prescribed form.

Expanding infinitesimally in the neighborhood of the unit in the complex basis Eq. (C.19)

$$\tilde{S} \sim \mathbb{1}_{2n\times 2n} + \epsilon\tilde{\mathcal{T}}, \qquad \tilde{\mathcal{T}} \equiv \begin{pmatrix} \mathsf{a} & \mathsf{b} \\ \mathsf{c} & \mathsf{d} \end{pmatrix}, \tag{C.22}$$

we find that the generators $\mathsf{a}, \mathsf{b}, \mathsf{c}, \mathsf{d}$ must satisfy simultaneously the conditions that the generators of $\mathrm{U}(n,n)$ and $\mathrm{Sp}(2n,\mathbb{C})$ satisfy, which are

$$\mathsf{a}^\dagger = -\mathsf{a}, \quad \mathsf{b} = \mathsf{c}^\dagger, \quad \mathsf{d} = -\mathsf{d}^\dagger, \quad \text{and} \quad \mathsf{a} = -\mathsf{d}^T, \quad \mathsf{b} = \mathsf{b}^T, \quad \mathsf{c} = \mathsf{c}^T, \tag{C.23}$$

respectively, and which are summarized by

$$\mathsf{a}^\dagger = -\mathsf{a}, \quad \mathsf{b} = \mathsf{b}^T, \quad \mathsf{c} = \mathsf{b}^*, \quad \mathsf{d} = \mathsf{a}^*. \tag{C.24}$$

The maximal compact subgroup of $\mathrm{U}(n,n)$ is $\mathrm{U}(n)\times\mathrm{U}(n)$, and, taking into account the second Eq. (C.20), this implies that the maximal compact subgroup of $\mathrm{Sp}(2n,\mathbb{R})$ is $\mathrm{U}(n)$.

The usual embedding of the fundamental representation of $\mathrm{U}(n)$ into the complex and the fundamental representations of $\mathrm{Sp}(2n,\mathbb{R})$ are, respectively,

$$\begin{pmatrix} u & 0 \\ 0 & u^* \end{pmatrix}, \qquad \begin{pmatrix} \Re\mathrm{e}\, u & -\Im\mathrm{m}\, u \\ \Im\mathrm{m}\, u & \Re\mathrm{e}\, u \end{pmatrix}, \tag{C.25}$$

where $u$ belongs to the fundamental of $\mathrm{U}(n)$.

The Lie subalgebra of $\mathrm{U}(n)$ in this complex basis is given by the matrices Eqs. (C.22) and (C.24) with $\mathsf{b}=0$. The Lie subalgebra of $\mathrm{U}(n)$ in the fundamental representation of $\mathrm{Sp}(2n,\mathbb{R})$ is given by the matrices of the form Eqs. (C.6) and (C.7) with $c=-b$.

*The matrices $f$, $h$ and $\mathcal{V}$.* The $f$ and $h$ matrices are $n\times n$ complex matrices that satisfy the two conditions

$$(f^\dagger, h^\dagger)\,\Omega \begin{pmatrix} f \\ h \end{pmatrix} = -i\mathbb{1} \quad\Rightarrow\quad -f^\dagger h + h^\dagger f = i\mathbb{1} \tag{C.26}$$

and

$$(f^T, h^T)\,\Omega \begin{pmatrix} f \\ h \end{pmatrix} = 0 \quad\Rightarrow\quad -f^T h + h^T f = 0, \tag{C.27}$$

which are equivalent to the requirement that the real matrix

$$S = \begin{pmatrix} A & B \\ C & D \end{pmatrix} = \sqrt{2}\begin{pmatrix} \Re\mathrm{e}\, f & -\Im\mathrm{m}\, f \\ \Re\mathrm{e}\, h & -\Im\mathrm{m}\, h \end{pmatrix} \tag{C.28}$$

belongs to the fundamental representation of $\mathrm{Sp}(2n,\mathbb{R})$. That is: they are equivalent to Eqs. (C.2). Then, given any symplectic matrix we can define matrices $f$ and $h$ satisfying the relations Eqs. (C.26) and (C.27) by

$$\begin{pmatrix} f \\ h \end{pmatrix} = \tfrac{1}{\sqrt{2}}\begin{pmatrix} A - iB \\ C - iD \end{pmatrix}. \tag{C.29}$$

In the complex basis, the matrix $\tilde{S}$ can be written in terms of the matrices $f$ and $h$ as follows:

$$\tilde{S} = \mathcal{A}^{-1} S \mathcal{A} = \tfrac{1}{\sqrt{2}}\begin{pmatrix} f + ih & f^* + ih^* \\ f - ih & f^* - ih^* \end{pmatrix}. \tag{C.30}$$

The matrix $\mathcal{V}$ is defined by

$$\mathcal{V} \equiv \begin{pmatrix} f \\ h \end{pmatrix}. \tag{C.31}$$

Defining the *symplectic product*

$$\langle\, A \mid B \,\rangle \equiv -A^M \Omega_{MN} B^N = A_M B^M, \tag{C.32}$$

we can derive the following relations from the defining relations of the $f$ and $h$ matrices Eqs. (C.26) and (C.26):

$$\langle\, \mathcal{V} \mid \mathcal{V}^* \,\rangle = -i\mathbb{1}, \qquad \langle\, \mathcal{V} \mid \mathcal{V} \,\rangle = 0. \tag{C.33}$$

*The V matrix.* With the complex $n \times n$ matrices $f$ and $h$ one can define another complex $2n \times 2n$ matrix

$$V \equiv S\mathcal{A} = \begin{pmatrix} f & f^* \\ h & h^* \end{pmatrix} = (\mathcal{V}, \mathcal{V}^*), \tag{C.34}$$

which transforms from the left in the fundamental representation of $\mathrm{Sp}(2n, \mathbb{R})$ and from the right in the complex representation of $\mathrm{Sp}(2n, \mathbb{R})$ just introduced. Since $\mathcal{A}$ is a symplectic matrix, $V$ and $V^T$ are symplectic matrices too, i.e.

$$V^P{}_M \, \Omega_{PQ} \, V^Q{}_N = \Omega_{MN}, \qquad V^M{}_P \, \Omega^{PQ} \, V^N{}_Q = \Omega^{MN}. \tag{C.35}$$

This implies that both its rows and columns are *mutually symplectic* vectors, i.e. defining by $V_\xi$ the symplectic vector with components given by the $\xi$th column of $V^M{}_\xi$, then

$$\langle V_\xi \mid V_\eta \rangle = -\Omega_{\xi\eta}. \tag{C.36}$$

*The coset space* $\mathrm{Sp}(2n, \mathbb{R})/\mathrm{U}(n)$. Let us denote by $\mathcal{S}$ an arbitrary symplectic matrix $S$ of the form Eq. (C.1) which is also symmetric and positive-definite and by $\{\mathcal{S}\}$ the space of all such matrices. The positive-definiteness of $\mathcal{S}$ in all subspaces implies that $D$ must be invertible and positive-definite as well, so we can factorize $\mathcal{S}$ according to Eq. (C.5). Furthermore, since $D$ and $D^{-1}$ are symmetric and positive-definite by hypothesis we can define their unique positive-definite square roots $\sqrt{D}$ and $\sqrt{D^{-1}}$ and we can factorize further

$$\begin{pmatrix} D^{-1} & 0 \\ 0 & D \end{pmatrix} = \begin{pmatrix} \sqrt{D^{-1}} & 0 \\ 0 & \sqrt{D} \end{pmatrix} \begin{pmatrix} \sqrt{D^{-1}} & 0 \\ 0 & D \end{pmatrix}, \tag{C.37}$$

and we find that

$$\mathcal{S} = gg^T, \quad \text{with} \quad g = \begin{pmatrix} \mathbb{1} & BD^{-1} \\ 0 & \mathbb{1} \end{pmatrix} \begin{pmatrix} \sqrt{D^{-1}} & 0 \\ 0 & \sqrt{D} \end{pmatrix} \in \mathrm{Sp}(2n, \mathbb{R}). \tag{C.38}$$

This result allows us to prove that $\{\mathcal{S}\}$ is the coset space $\mathrm{Sp}(2n, \mathbb{R})/\mathrm{U}(n)$: the maximal compact subgroup of $\mathrm{Sp}(2n, \mathbb{R})$ is $\mathrm{H} := \{g \in \mathrm{Sp}(2n, \mathbb{R}); gg^T = \mathbb{1}\}$, and, according to Eq. (C.25), it is $\mathrm{U}(n)$.

*The matrices $\mathcal{M}$ and $\mathcal{N}$.* The $n \times n$ matrices $f = (f^\Lambda{}_a)$, $a = 1, \ldots, n$, can be shown to be invertible. Let us assume they are not. Then, $f^\Lambda{}_a \, \psi^a = 0$ for some $\psi \neq 0$ and acting on Eq. (C.26) with $\psi^\dagger$ from the left and $\psi$ from the right we would get

$$-(f\psi)^\dagger h \, \psi + \psi^\dagger h^\dagger f \psi = i\psi^\dagger \psi \neq 0, \tag{C.39}$$

which is absurd. One can prove analogously that the matrices $h = (h_{\Lambda\, a})$ are also invertible.

The invertibility of $f$ allows us to define the invertible $n \times n$ matrix

$$\boxed{\mathcal{N} = hf^{-1},} \tag{C.40}$$

which is symmetric due to Eq. (C.27) and has a negative-definite imaginary part due to Eq. (C.26):

$$\mathcal{N} = \mathcal{N}^T, \qquad \Im\mathrm{m}\,\mathcal{N} = -\tfrac{i}{2}(\mathcal{N} - \mathcal{N}^\dagger) = -\tfrac{1}{2}(ff^\dagger)^{-1}. \tag{C.41}$$

$\mathcal{N}^{-1}$, however, has a positive-definite imaginary part $\mathcal{N}^{-1} - (\mathcal{N}^{-1})^\dagger = i(hh^\dagger)^{-1}$.

Any symmetric matrix with a negative-definite imaginary part (such as the period matrix of four-dimensional SUEGRAs) can be written in the form Eq. (C.40) for some matrices $(f, h)$ satisfying Eqs. (C.26) and (C.27). To prove this it is enough to consider an $f$ satisfying Eq. (C.41).

There is also a one-to-one correspondence between symmetric complex matrices $\mathcal{N}$ with a negative-definite imaginary part and symplectic, symmetric, negative-definite matrices $\mathcal{M}$. For each $\mathcal{N}$ we can define $\mathcal{M}(\mathcal{N})$ by

$$\boxed{\mathcal{M}(\mathcal{N}) \equiv \begin{pmatrix} I + RI^{-1}R & -RI \\ -IR & I^{-1} \end{pmatrix}, \qquad I \equiv \Im\mathrm{m}\,\mathcal{N}, \qquad R \equiv \Re\mathrm{e}\,\mathcal{N}.} \tag{C.42}$$

This matrix can be factorized and expressed in different ways:

$$\begin{aligned} \mathcal{M}(\mathcal{N}) &= \begin{pmatrix} \mathbb{1} & -R \\ 0 & \mathbb{1} \end{pmatrix} \begin{pmatrix} I & 0 \\ 0 & I^{-1} \end{pmatrix} \begin{pmatrix} \mathbb{1} & 0 \\ -R & \mathbb{1} \end{pmatrix} \\ &= i\Omega + \begin{pmatrix} \mathcal{N} I^{-1} \mathcal{N}^\dagger & -\mathcal{N} I^{-1} \\ -I^{-1}\mathcal{N}^\dagger & I^{-1} \end{pmatrix} \\ &= i\Omega - 2 \begin{pmatrix} hh^\dagger & -hf^\dagger \\ -fh^\dagger & ff^\dagger \end{pmatrix} \\ &= i\Omega - 2 \begin{pmatrix} -h \\ f \end{pmatrix} (-h^\dagger, f^\dagger) \\ &= -2\,\Re\mathrm{e} \left[ \begin{pmatrix} -h \\ f \end{pmatrix} (-h^\dagger, f^\dagger) \right]. \end{aligned} \tag{C.43}$$

The final expression is equivalent to

$$\mathcal{M}_{MN}(\mathcal{N}) = -\mathcal{V}_M \mathcal{V}_N^\dagger - \mathcal{V}_n \mathcal{V}_M^\dagger\,. \tag{C.44}$$

Since the symplectic symmetric positive-definite matrices, and hence the negative-definite ones, parametrize the coset space $\mathrm{Sp}(2n, \mathbb{R})/\mathrm{U}(n)$, the matrices $\mathcal{N}$ also parametrize this coset space.

The matrix $\mathcal{V}$ (equivalently, the pair $(f, h)$) transforms contravariantly under symplectic rotations:

$$\mathcal{V}' = S\mathcal{V} = \begin{pmatrix} A & B \\ C & D \end{pmatrix} \mathcal{V}; \tag{C.45}$$

then, by definition, the matrix $\mathcal{N}$, its imaginary part, and $\mathcal{M}(\mathcal{N})$ transform as

$$\mathcal{N}' = (C + D\mathcal{N})(A + B\mathcal{N})^{-1}, \tag{C.46}$$

$$I' = [(A + B\mathcal{N})^{-1}]^{\dagger} I (A + B\mathcal{N})^{-1}, \tag{C.47}$$

$$\mathcal{M}(\mathcal{N}') = (S^T)^{-1}\, \mathcal{M}(\mathcal{N})\, S^{-1}. \tag{C.48}$$

The last transformation rule can be obtained from Eq. (C.43).

The relation between the negative-definite symmetric matrix $\mathcal{M}$ defined in Eq. (C.42) and $\mathcal{S}$ defined in Eq. (C.38) can be obtained from their transformation properties under $\mathrm{Sp}(2n, \mathbb{R})$,

$$\mathcal{M} = -\mathcal{S}^{-1} = \Omega^{-1}\mathcal{S}\Omega^{-1}. \tag{C.49}$$

Finally,

$$\mathcal{M} = -(V^{-1})^{\dagger} V^{-1}. \tag{C.50}$$

# Appendix D
# Gamma matrices and spinors

In this appendix we explain our conventions for gamma matrices in diverse dimensions and their relations (via dimensional reduction). We start by reviewing some basic facts about spinors and gamma matrices in diverse dimensions. Next, we review spinors and gamma matrices in spaces of arbitrary dimensions and signatures. At the end, we review the algebra of commuting spinor bilinears in four and five dimensions which is used in the *spinor bilinear method* explained in Section 18.2.

## D.1 Generalities

Let us first review some facts about gamma matrices.[1] Gamma matrices are the generators of the $d$-dimensional *Clifford algebra* associated with the metric $\eta_{ab} = \text{diag}(+ - \cdots -)$, $a, b = 0, \ldots, d-1$, and, therefore, satisfy the anticommutation relations

$$\{\Gamma_a, \Gamma_b\} = +2\eta_{ab}. \tag{D.1}$$

Any other element of the Clifford algebra can be constructed as a linear combination of the gamma matrices and their products.

Clifford algebras are relevant in physics due to the fact that a representation of the $d$-dimensional Clifford algebra for the metric $\eta_{ab}$ can be used to construct a representation of the $d$-dimensional Lorentz algebra $\mathfrak{so}(1, d-1)$,

$$[M_{ab}, M_{cd}] = -\eta_{ac}M_{bd} - \eta_{bd}M_{ac} + \eta_{ad}M_{bc} + \eta_{bc}M_{ad}, \tag{D.2}$$

that we denote by $\Gamma_{\text{s}}$, by taking antisymmetric products of two gamma matrices:

$$\Gamma_{\text{s}}(M_{ab}) = \tfrac{1}{2}\Gamma_{ab}, \qquad \Gamma_{ab} \equiv \Gamma_{[a}\Gamma_{b]}. \tag{D.3}$$

(We use the notation

$$\Gamma^{a_1 \cdots a_n} = \Gamma^{[a_1}\Gamma^{a_2} \cdots \Gamma^{a_n]} \tag{D.4}$$

for the antisymmetrized (with weight unity) product of $n$ gamma matrices.)

[1] We will follow Ref. [984] but using our mostly minus-signature metric. See also Refs. [1217, 315, 1247].

Lorentz transformations in this representation are constructed by exponentiation[2]

$$\Gamma_s(\Lambda) = \exp\left\{\tfrac{1}{2}\sigma^{ab}\Gamma_s(M_{ab})\right\}. \tag{D.5}$$

How many different representations can be built in this way? It can be shown that

**Theorem D.1** *There is only one (physically[3]) inequivalent irreducible representation of the Clifford algebra in d dimensions and it is* $2^{[d/2]}$ *dimensional.*

The corresponding representation of the Lorentz algebra $\Gamma_s$ is a *spinorial representation* and the elements of the complex $2^{[d/2]}$-dimensional vector representation space are called (Dirac) *spinors*. We use the first Greek letters as indices in this vector space: $\alpha, \beta = 1, \ldots, 2^{[d/2]}$.

It is worth stressing that, even if we use an irreducible representation of the Clifford algebra, the corresponding representation of the Lorentz group is reducible for $d$ even. In that case, the representation space is the direct sum of two subspaces of dimension $2^{[d/2]-1}$ whose elements are called *Weyl spinors* and will be discussed later.

We consider only unitary representations. The definition of the algebra means that, if the gamma matrices are unitary, $\Gamma^0$ is Hermitian and the rest of the $\Gamma^i$ are anti-Hermitian:

$$\Gamma^{0\,\dagger} = +\Gamma^0, \qquad \Gamma^{i\,\dagger} = -\Gamma^i, \quad i = 1, \ldots, d-1. \tag{D.6}$$

This implies

$$\Gamma^0\Gamma^a\Gamma^0 = \Gamma^{a\,\dagger}. \tag{D.7}$$

In any representation, all gamma matrices are *traceless*. This can be seen by considering $\mathrm{Tr}\left(\Gamma^a\Gamma^b\Gamma^a\right)$ with $a \neq b$ and using the anticommutators and the cyclic property of the trace.

We can prove the existence of a $2^{[d/2]}$-dimensional representation of the Clifford algebra by explicit construction as in Ref. [315]. This will provide us with gamma matrices in any dimension. Let us first consider the case for $d$ even. We can proceed by induction: if we assume that a representation of the $(d-2)$-dimensional gamma matrices $\{\Gamma^a_{(d-2)}\}, a = 0, \ldots, d-3$, exists and that it is $2^{[(d-1)/2]}$ dimensional, then

$$\Gamma^a_{(d)} = \Gamma^a_{(d-2)} \otimes \sigma^1, \qquad \Gamma^{d-2}_{(d)} = \mathbb{I} \otimes i\sigma^2, \qquad \Gamma^{d-1}_{(d)} = \mathbb{I} \otimes i\sigma^3, \tag{D.8}$$

[2] The Lorentz group $O(1, d-1)$ is neither connected nor simply connected for $d \geq 3$. Exponentiation of the generators gives only transformations in the component of the group manifold connected with the identity (the proper (determinant $+1$), orthochronous (${\Lambda^1}_1 > 1$) Lorentz group). To obtain all the Lorentz transformations, one needs to multiply by discrete transformations such as time reversal and parity (depending on the dimension $d$). More precisely, the exponentiation of the generators of the Lorentz algebra $\mathfrak{so}(1, d-1)$ in this representation $\Gamma_s$ gives elements of the simply connected group which is locally isomorphic to $SO(1, d-1)$ (and, therefore, has the same Lie algebra) and which is called, by definition, $\mathrm{Spin}(1, d-1)$. The Lorentz group is doubly connected and, thus, some representations of $\mathrm{Spin}(1, d-1)$ (in particular the spinorial $\Gamma_s$) are double valued as representations of the Lorentz group. Spin$(1, d-1)$ is the double cover of $SO(1, d-1)$, in other words.

[3] By physically equivalent we mean representations of the Clifford algebra that give, with the construction of the generators of the Lorentz group in terms of the gamma matrices Eq. (D.3), equivalent representations (i.e. that are representations related by similarity transformations) of the Lorentz group. Two physically equivalent representations of the Clifford algebra may but need not be strictly equivalent.

where $\sigma^{1,2,3}$ are the (Hermitian, unitary, $2 \times 2$) Pauli matrices

$$\sigma^1 = \begin{pmatrix} 0 & 1 \\ 1 & 0 \end{pmatrix}, \qquad \sigma^2 = \begin{pmatrix} 0 & -i \\ i & 0 \end{pmatrix}, \qquad \sigma^3 = \begin{pmatrix} 1 & 0 \\ 0 & -1 \end{pmatrix} \tag{D.9}$$

that satisfy

$$(\sigma^i)^I{}_J (\sigma^j)^J{}_K = \delta^{ij} \delta^I{}_K + i\varepsilon^{ijk} (\sigma^k)^I{}_K \,, \tag{D.10}$$

$$\delta^K{}_J \delta^L{}_I = \tfrac{1}{2} \delta^K{}_I \delta^L{}_J + \tfrac{1}{2} (\sigma^i)^K{}_I (\sigma^i)^L{}_J \,, \tag{D.11}$$

$$\varepsilon^{IJ} \varepsilon_{KL} = -\tfrac{2}{3} (\sigma^i)^{[I}{}_{[K} (\sigma^i)^{J]}{}_{L]} \,, \tag{D.12}$$

$$(\sigma^{[i|})^I{}_J (\sigma^{|j]})^K{}_L = -\tfrac{i}{2} \varepsilon^{ijk} [\delta^I{}_L (\sigma^k)^K{}_J - (\sigma^k)^I{}_L \delta^K{}_J] \,, \tag{D.13}$$

$$\varepsilon_{K[I} (\sigma^i)^K{}_{J]} = (\sigma^i)^{[I}{}_K \varepsilon^{J]K} = 0 \,, \tag{D.14}$$

$$\varepsilon_{LI} (\sigma^i)^I{}_J \varepsilon^{JK} = \left[ (\sigma^i)^K{}_L \right]^* \,, \tag{D.15}$$

$$\left[ (\sigma^i)^I{}_J \varepsilon^{JK} \right]^* = -\varepsilon_{IJ} (\sigma^i)^J{}_K \,, \tag{D.16}$$

is a $2^{[d/2]}$-dimensional representation of the $d$-dimensional Clifford algebra. A $d=2$ representation of the two-dimensional Clifford algebra is provided by (for instance) $\{\mathbb{I}_{2\times 2}, i\sigma^2\}$, and this completes the proof for $d$ even.

Now, if $d$ is even and $\{\Gamma^0, \ldots, \Gamma^{d-1}\}$ are $2^{d/2} \times 2^{d/2}$ gamma matrices satisfying the $d$-dimensional Clifford algebra, then the gamma matrices $\{\Gamma^0, \ldots, \Gamma^{d-1}, \Gamma^d\}$ with

$$\Gamma^d \equiv -i\varphi(d) \Gamma^0 \cdots \Gamma^{d-1}, \qquad \varphi(d) = (-1)^{\frac{1}{4}(d-2)+1}, \tag{D.17}$$

satisfy the $(d+1)$-dimensional Clifford algebra.[4] Thus, the even $d$ irreducible representations determine the $d+1$ irreducible representations, and this completes the proof. Observe that this matrix is different from the chirality matrix $\mathcal{Q} = \Gamma_{d+1}$ ($\gamma_5$ in $d=4$):

$$\begin{aligned} &\Gamma_{d+1} = i\Gamma^d = \varphi(d) i \Gamma^0 \cdots \Gamma^{d-1}, \\ &\Gamma_{d+1}^2 = +1, \qquad \Gamma_{d+1}^\dagger = +\Gamma_{d+1}, \qquad \Gamma^0 \Gamma_{d+1} \Gamma^0 = -\Gamma_{d+1}^\dagger. \end{aligned} \tag{D.19}$$

Observe also that, in odd dimensions, by construction, the product of all gamma matrices is proportional to a constant whose sign can be chosen at will (by changing the sign of $\Gamma^d$).

[4] By putting together

$$\begin{aligned} \Gamma^0 \cdot \Gamma^{d-1} &= (-1)^{[d/2]} \Gamma^{d-1} \cdot \Gamma^0, \\ \left(\Gamma^0 \cdot \Gamma^{d-1}\right)\left(\Gamma^{d-1} \cdot \Gamma^0\right) &= (-1)^{d-1}, \end{aligned} \tag{D.18}$$

it is easy to check that $\Gamma^d$ anticommutes with all the other gammas, squares to $-1$, and is anti-Hermitian.

The two possible signs give inequivalent representations of the Clifford algebra (which are, nevertheless, physically equivalent).

Let us now consider equivalent representations of the Clifford algebra, related by a similarity transformation

$$\Gamma^{a\,\prime} = S\Gamma^a S^{-1}. \tag{D.20}$$

If $d$ is even, and if we change the sign of all the gamma matrices, we obtain an equivalent representation with $S = \mathcal{Q}$, the chirality matrix. If $d$ is odd, changing the signs of all the gamma matrices does not provide an equivalent representation because it changes the sign of the product of all the gamma matrices.

For both even and odd $d$ the Hermitian conjugates of the gamma matrices constitute another representation related to the original one by $S = \mathcal{D}$, where $\mathcal{D}$ is the *Dirac conjugation matrix* and can be taken to be $\mathcal{D} = i\Gamma^0$. We will also call this matrix $\mathcal{D}_+$ to distinguish it from $\mathcal{D}_-$, used in other dimensions and/or signatures which is defined to satisfy

$$\mathcal{D}_-\Gamma^a\mathcal{D}_-^{-1} = -\Gamma^{a\,\dagger}. \tag{D.21}$$

In even $d$ the transposed gamma matrices also provide another equivalent representation of the Clifford algebra. In that case, by definition, $S = \mathcal{C}$, the *charge-conjugation matrix*, which we will use later. For $d$ odd, sometimes it is the transposed gamma matrices that provide a representation and sometimes it is the transposed gamma matrices with the sign reversed that provide an equivalent representation. Therefore, one can also define

$$\mathcal{C}_\pm\Gamma^a\mathcal{D}_\pm^{-1} = \pm\Gamma^{a\,T}. \tag{D.22}$$

Clearly, for even $d$ (and for odd $d$, up to a sign) the (plus and minus) complex conjugates of gamma matrices also give an equivalent representation and, by definition, $\mathcal{S} = \mathcal{B}_\pm$. The matrix $\mathcal{B}$ is related to $\mathcal{D}$ and $\mathcal{C}_\pm$ by

$$\mathcal{C}_\pm = \mathcal{B}_\pm^T\mathcal{D}, \tag{D.23}$$

and we will not use it for the moment.

Gamma matrices carry a vector Lorentz index that is raised and lowered with $\eta$. They are invariant under Lorentz transformations that act on their three (two spinorial, one vector) indices:

$$\left(\Gamma'^{\,a}\right)^\alpha{}_\beta = \Lambda^a{}_b\Lambda^\alpha{}_\gamma\left(\Gamma^b\right)^\gamma{}_\delta\left(\Lambda^{-1}\right)^\delta{}_\beta = (\Gamma^a)^\alpha{}_\beta, \qquad \Lambda^\alpha{}_\gamma = \Gamma_{\rm s}(\Lambda)^\alpha{}_\gamma. \tag{D.24}$$

If we consider infinitesimal transformations, we find that gamma matrices and generators of the Lorentz group in the spinorial representation must obey the commutation relations[5]

$$[\Gamma^a, \Gamma_{\rm s}(M_{bc})] = \Gamma_{\rm v}(M_{bc})^a{}_d\Gamma^d. \tag{D.25}$$

These commutation relations are identical to those of the momentum $P^a$ and the Lorentz generators in the Poincaré algebra. Thus, they indicate that the gamma matrices transform as vectors in the spinorial representation.[6]

---

[5] This consistency check is satisfied in our conventions.

[6] Observe, though, that gamma matrices do not commute and therefore they do not provide a spinorial representation of the translation generators of the Poincaré algebra.

Let us now study spinors. Dirac spinors transform under the Lorentz group as expected (contravariantly),

$$\psi'^{\alpha} = \Lambda^{\alpha}{}_{\beta}\psi^{\beta} = \Gamma_{\rm s}(\Lambda)^{\alpha}{}_{\beta}\psi^{\beta},$$
$$\Gamma_{\rm s}(\Lambda) = \exp\left\{\tfrac{1}{2}\sigma^{ab}\Gamma_{\rm s}(M_{ab})\right\} = \exp\left\{\tfrac{1}{4}\sigma^{ab}\Gamma_{ab}\right\}, \tag{D.26}$$

or, infinitesimally,

$$\delta_{\sigma}\psi^{\alpha} = \tfrac{1}{2}\sigma^{ab}\Gamma_{\rm s}(M_{ab})^{\alpha}{}_{\beta}\psi^{\beta}. \tag{D.27}$$

In flat Minkowski spacetime, Lorentz transformations are global. In curved spacetime Lorentz transformations make sense only at one point in tangent space and therefore Lorentz transformations are naturally local ($\sigma^{ab} = \sigma^{ab}(x)$). Hence theories containing spinors are required to be invariant under local Lorentz transformations, and are naturally gauge theories of the Lorentz group. The Lorentz covariant derivative acting on a (contravariant) spinor is, according to the results of Appendix A.2.3 and Eq. (D.3),

$$\nabla_{\mu}\psi = \left(\partial_{\mu} - \tfrac{1}{4}\omega_{\mu}{}^{ab}\Gamma_{ab}\right)\psi. \tag{D.28}$$

In field theory, a Dirac spinor $\psi^{\alpha}$ is a field $\psi^{\alpha}(x)$ that satisfies the massive or massless ($m = 0$), charged or uncharged ($e = 0$) $d$-dimensional *Dirac equation*

$$(i\not{\nabla} - m + e\not{A})\psi = 0, \qquad \not{\nabla} \equiv \Gamma^{a}e_{a}{}^{\mu}\nabla_{\mu}, \qquad \not{A} \equiv \Gamma^{a}e_{a}{}^{\mu}A_{\mu}. \tag{D.29}$$

We can also define spinors $\xi_{\alpha}$ transforming covariantly,

$$\xi'_{\alpha} = \xi_{\beta}\left(\Lambda^{-1}\right)^{\beta}{}_{\alpha} = \xi_{\beta}\Gamma_{\rm s}\left(\Lambda^{-1}\right)^{\beta}{}_{\alpha},$$
$$\Gamma_{\rm s}\left(\Lambda^{-1}\right) = \exp\left\{-\tfrac{1}{2}\sigma^{ab}\Gamma_{\rm s}(M_{ab})\right\} = \exp\left\{-\tfrac{1}{4}\sigma^{ab}\Gamma_{ab}\right\}, \tag{D.30}$$

or, infinitesimally,

$$\delta_{\sigma}\xi_{\alpha} = \xi_{\beta}\left[-\tfrac{1}{2}\sigma^{ab}\Gamma_{\rm s}(M_{ab})^{\beta}{}_{\alpha}\right]. \tag{D.31}$$

The Lorentz covariant derivative acts on covariant spinors according to

$$\nabla_{\mu}\xi = \partial_{\mu}\xi - \xi\left(-\tfrac{1}{4}\omega_{\mu}{}^{ab}\Gamma_{ab}\right) \equiv \xi\overleftarrow{\nabla}_{\mu}. \tag{D.32}$$

Just as we can transform contravariant vectors into covariant vectors by "lowering the index" with the metric, we can transform contravariant spinors into covariant spinors by *conjugation*. Given a Dirac spinor transforming contravariantly, there are two kinds of conjugate spinors that transform covariantly.

**The Dirac conjugate.** $\bar{\psi}$ of a spinor $\psi$ is a new spinor that transforms covariantly and whose components $\bar{\psi}_{\alpha}$ are linear combinations of those of $(\psi^{\alpha})^{*}$:

$$\bar{\psi}_{\alpha} = \left(\psi^{\dagger}\mathcal{D}\right)_{\alpha} = (\psi^{\beta})^{*}\mathcal{D}_{\beta\alpha}, \tag{D.33}$$

where $\mathcal{D}$ is the *Dirac conjugation matrix*. According to this definition

$$\mathcal{D}\,\Gamma_{ab}\,\mathcal{D}^{-1} = -\Gamma^{\dagger}_{ab}. \tag{D.34}$$

Taking the Hermitian conjugate of Eq. (D.26) and using $\Gamma^{ab\,\dagger} = -\Gamma^0\Gamma^{ab}\Gamma^0$, we find that, up to a phase, the Dirac conjugation matrix is given by $\Gamma^0$. Our convention is $\mathcal{D} = i\Gamma^0$, so

$$\bar{\psi} \equiv i\psi^\dagger\Gamma^0. \tag{D.35}$$

Obviously, by definition, the product $\bar{\psi}_\alpha\psi^\alpha$ is a Lorentz invariant. Also $\bar{\bar{\psi}} = \psi$.

The Dirac conjugate satisfies the conjugate Dirac equation

$$i\nabla_a\bar{\psi}\Gamma^a + m\bar{\psi} - e\bar{\psi}\Gamma^a A_a = \bar{\psi}(i\overleftarrow{\not{\nabla}} + m - e\not{A}) = 0. \tag{D.36}$$

**The Majorana conjugate.** $\psi^c$ of a spinor $\psi$ is a new spinor that transforms covariantly and whose components $\psi^c_\alpha$ are linear combinations of $\psi^\alpha$,

$$\psi^c_\alpha = \left(\psi^T\mathcal{C}\right)_\alpha = \psi^\beta\mathcal{C}_{\beta\alpha} \tag{D.37}$$

(not $\psi^\star$ as in the Dirac conjugate), and that transform as the Dirac conjugate under Lorentz transformations. Here $\mathcal{C}$ is the *charge-conjugation matrix*. By transposing Eq. (D.26) and using the definition of $\psi^c$, we find that $\mathcal{C}$ must satisfy

$$\mathcal{C}\,\Gamma_{ab}\,\mathcal{C}^{-1} = -\Gamma^T_{ab}. \tag{D.38}$$

The matrices $-\frac{1}{2}\Gamma^T_{ab}$ also satisfy the Lorentz algebra. The charge-conjugation matrix $\mathcal{C}$ relates this representation and the standard one. It is natural to look for a charge-conjugation matrix that also relates a representation of the Clifford algebra and the representation obtained by transposing all gamma matrices. There are two possibilities, $\mathcal{C}_+$ and $\mathcal{C}_-$, defined by

$$\mathcal{C}_\pm\Gamma^a\mathcal{C}_\pm^{-1} = \pm\Gamma^{a\,T}. \tag{D.39}$$

When $d$ is even, both matrices exist because $+\Gamma^{a\,T}$ and $-\Gamma^{a\,T}$ generate the same finite group as $\Gamma^a$. When $d$ is odd, only one of them exists because of the definition of $\Gamma^{d-1}$ in terms of $\Gamma^0, \ldots, \Gamma^{d-2}$. Furthermore, both matrices, when they exist, are either symmetric or antisymmetric.

When $d$ is even, the charge-conjugation matrices act on $\Gamma_{d+1}$ as follows:

$$\mathcal{C}_\pm\Gamma_{d+1}\mathcal{C}_\pm^{-1} = \varphi^2\Gamma^T_{d+1}. \tag{D.40}$$

Using the $\mathcal{C}_\pm$ charge-conjugation matrices and taking the Majorana conjugate of the Dirac equation, we find that the Majorana conjugate satisfies the following equation:

$$\psi^c(i\overleftarrow{\not{\nabla}} \mp m + e\not{A}) = 0, \tag{D.41}$$

which implies that $\psi^c$ has charge opposite to that of $\bar{\psi}$ (hence the name "charge-conjugation matrix"). It is obviously desirable that both $\psi^c$ and $\bar{\psi}$ have the same mass and, thus, in the massive case the only acceptable charge-conjugation matrix is $\mathcal{C}_-$.

By construction $\psi^c\psi$ is Lorentz invariant and $(\psi^c)^c = \psi$.

We can now study various types of spinors that are in general associated with special representations of gamma matrices.

**Weyl spinors.** (Also called *chiral* spinors.) For even $d$ it is possible to define as before the *chirality matrix* $\Gamma_{d+1}$ which anticommutes with all the gamma matrices and therefore commutes with the generators of the Lorentz group $\Gamma_s(M_{ab})$ and with their exponentials, which span the $\mathrm{Spin}(1,d-1)$ group. Thus (Schur's lemma) this representation of the $\mathrm{Spin}(1,d-1)$ group, and Dirac spinors, are reducible even if the gamma matrices provide an irreducible representation of the $d$-dimensional Clifford algebra of $\eta_{ab}$. The chirality matrix is traceless and squares to unity and therefore half of its eigenvalues are $+1$s and the other half are $-1$s. It is natural to split the space of Dirac spinors into the direct sum of the subspaces of spinors with eigenvalues $+1$ and $-1$. The elements of each of these subspaces are called *Weyl spinors* and, by definition, satisfy the Weyl or chirality condition

$$\tfrac{1}{2}(1 \pm \Gamma_{d+1})\psi = \psi. \tag{D.42}$$

For the positive sign, the spinors are called left-handed (negative chirality); for the negative sign they are called right-handed (positive chirality). A Weyl spinor describes half the degrees of freedom of a Dirac spinor.

Observe that, although Weyl spinors are irreducible representations of the $\mathrm{Spin}(1,d-1)$ group, they are not irreducible representations of the Lorentz group $\mathrm{SO}(1,d-1)$ because this group contains discrete transformations that interchange the two subspaces of opposite chiralities. In particular, the parity transformation is implemented by $P = i\Gamma^0$, which does not commute but anticommutes with the chirality matrix, switching the chirality of the spinors.

Using the Dirac equation, it can be seen that the Weyl condition is preserved in time for $m = 0$.

Associated with Weyl spinors there are *Weyl (or chiral) representations* of gamma matrices. In a Weyl representation the generators of the Lorentz group are diagonal and the chirality matrix is $\Gamma_{(d+1)} = \mathbb{I} \otimes \sigma^3$. In a chiral basis of gamma matrices, half of the components of a Weyl spinor are zero, and it is sometimes advantageous to use half-size spinors.

**Majorana spinors.** are spinors whose Majorana conjugate is proportional to their Dirac conjugate (the Majorana condition):

$$\hat{\lambda} = \alpha\,\bar{\lambda}. \tag{D.43}$$

This is a reality condition because it relates the components of $\lambda$ to those of its complex conjugate. Thus, it describes half the degrees of freedom of a Dirac spinor. Using our definitions of Majorana and Dirac conjugates in the definition of a Majorana spinor, we find that it implies

$$|\alpha|^2 \Gamma^{0\star}(\mathcal{C}^{-1})^{\star}\,\Gamma^0 \mathcal{C}^{-1} = +1. \tag{D.44}$$

This condition cannot be fulfilled in all dimensions and this is the reason why Majorana spinors exist only in certain dimensions. This condition and the (anti)symmetry of $\mathcal{C}$ do not depend on the representation, and the results found in any representation are valid in general.[7]

Associated with Majorana spinors are *Majorana representations* in which all $\Gamma$s are purely imaginary.[8] If a Majorana representation exists, then the condition for the existence of Majorana spinors is automatically satisfied by the choice

$$\mathcal{C} = i\alpha\Gamma^0. \tag{D.45}$$

With any other representation we have to check explicitly whether this equation holds.

Below $d = 11$, there are Majorana spinors in all but $d = 5, 6$, and 7.

**Majorana–Weyl spinors.** satisfy both Majorana and Weyl conditions. They exist in even dimensions if, in addition to the Majorana condition, one can satisfy a compatibility condition between Majorana and Weyl conditions:

$$\mathcal{C}^{-1}\,\Gamma_{d+1}^{\mathrm{T}}\,\mathcal{C} = \Gamma^0\,\Gamma_{d+1}^{\dagger}\,\Gamma^0. \tag{D.46}$$

This condition is representation dependent and, again, can be satisfied only in certain dimensions. Using the definition of $\Gamma_{d+1}$ we have chosen and its properties, we see that the condition is satisfied whenever $\eta^2(d) = -1$ ($d$ even); that is, when

$$d = 2(\mathrm{mod}\ 4) = 2, 6, 10, \ldots. \tag{D.47}$$

Majorana spinors do not exist in $d = 6$ and so Majorana–Weyl spinors exist only in $d = 2$ and 10 (at least in our representation and in fewer than 11 dimensions). It can be shown that these are also the only dimensions in which they exist.

**Symplectic-Majorana spinors.** When defining Majorana spinors (i.e. spinors satisfying a reality condition) is not possible, one can take an even number of Dirac spinors labeled by $i = 1, \ldots, 2n$ and impose a reality condition on the whole set:

$$\bar{\psi}^i = {\psi_i}^c \equiv \Omega_{ij}\psi^{i\,c}, \tag{D.48}$$

where $\Omega$ is real and satisfies

$$\Omega_{ij}\Omega_{jk} = -\delta_{ik}. \tag{D.49}$$

Below $d = 11$ this can be done consistently in $d = 4, 5, 6, 7$, and 8. In $d = 6$ we can impose simultaneously the symplectic-Majorana and Weyl conditions.

In many cases, $n = 2$ symplectic-Majorana spinors appear combined in a single, unconstrained, Dirac spinor that contains the same number of degrees of freedom.

See Table D.1 for a summary.

---

[7] This condition and similar conditions will be studied in detail, for arbitrary signature, in Section D.2. For the moment we will simply quote the results for Lorentzian signature and $d \leq 11$.

[8] With mostly plus signature all $\Gamma$s are purely real (essentially the same matrices multiplied by $i$). If we used the Pauli metric, so that $\{\Gamma^a, \Gamma^b\} = +2\delta^{ab}$, all $\Gamma$s would be real except for $\Gamma^4$, which would be imaginary.

Table D.1. Spinors that exist in various dimensions with signature $(1, d-1)$: $M$ is the number of real independent components of the smaller spinorial representation in the given dimension (W, Weyl; M, Majorana; M–W, Majorana–Weyl; and S-M, symplectic-Majorana spinors)

| $d$ | W | M | M–W | S-M | S-M and W | $M$ |
|---|---|---|---|---|---|---|
| 2 | x | x | x | | | 1 |
| 3 | | x | | | | 2 |
| 4 | x | x | | x | | 4 |
| 5 | | | | x | | 8 |
| 6 | x | | | x | x | 8 |
| 7 | | | | x | | 16 |
| 8 | x | x | | x | | 16 |
| 9 | | x | | | | 16 |
| 10 | x | x | x | | | 16 |
| 11 | | x | | | | 32 |

### D.1.1 Useful identities

Most of the gamma identities (for up to four gammas) that we need can be obtained from the following products by symmetrization, antisymmetrization, etc.:

$$\Gamma^a\Gamma^b = \Gamma^{ab} + \eta^{ab}, \tag{D.50}$$

$$\Gamma^a\Gamma^b\Gamma^c = \Gamma^{abc} + \eta^{ab}\Gamma^c - \eta^{ca}\Gamma^b + \eta^{bc}\Gamma^a, \tag{D.51}$$

$$\begin{aligned}\Gamma^a\Gamma^b\Gamma^c\Gamma^d = {} & \Gamma^{abcd} + \eta^{ab}\Gamma^{cd} - \eta^{cb}\Gamma^{da} + \eta^{cd}\Gamma^{ab} + \eta^{da}\Gamma^{bc} \\ & - \eta^{ac}\Gamma^{bd} - \eta^{bd}\Gamma^{ac} + \eta^{ab}\eta^{cd} - \eta^{ac}\eta^{bd} + \eta^{ad}\eta^{bc}.\end{aligned} \tag{D.52}$$

For instance, we can obtain

$$\Gamma^{ab}\Gamma_{cd} = \Gamma^{ab}{}_{cd} + 4\Gamma^{[a}{}_{[d}\eta^{b]}{}_{c]} + 2\eta^a{}_{[d}\eta^b{}_{c]}, \tag{D.53}$$

from which we can derive the Lorentz algebra. With more than four, we use repeatedly

$$\Gamma^a\Gamma^{b_1\cdots b_n} = \Gamma^{ab_1\cdots b_n} + n\eta^{a[b_1}\Gamma^{b_2\cdots b_n]}, \tag{D.54}$$

$$\Gamma^{b_1\cdots b_n}\Gamma^a = \Gamma^{b_1\cdots b_n a} + n\Gamma^{[b_1\cdots b_{n-1}}\eta^{b_n]a}. \tag{D.55}$$

From these equations we get the useful formula

$$\Gamma^a\Gamma^{b_1\cdots b_n}\Gamma_a = (-1)^n(d-2n)\Gamma^{b_1\cdots b_n}, \tag{D.56}$$

the (anti)commutator

$$\left[\Gamma^a, \Gamma^{b_1\cdots b_n}\right]_\pm = [1 \pm (-1)^n]\Gamma^{ab_1\cdots b_n} + n[1 \mp (-1)^n]\eta^{a[b_1}\Gamma^{\cdots b_n]}, \tag{D.57}$$

and the general formula

$$\Gamma^{b_1\cdots b_n}\Gamma_{a_1\cdots a_m} = \sum_{p=0}^{\min(n,m)} \frac{n!m!}{(n-p)!(m-p)!p!} \Gamma^{[b_1\cdots b_{n-p}}{}_{[a_{p+1}\cdots a_m}\eta^{b_{n-p+1}\cdots b_n]}{}_{a_{m-p}a_{m-p-1}\cdots a_1]}$$
$$= \Gamma^{b_1\cdots b_n}{}_{a_1\cdots a_m} + nm\Gamma^{[b_1\cdots b_{n-1}}{}_{[a_2\cdots a_m}\eta^{b_n]}{}_{a_1]}$$
$$+ \frac{n(n-1)m(m-1)}{2!}\Gamma^{[b_1\cdots b_{n-2}}{}_{[a_3\cdots a_m}\eta^{b_{n-1}b_n]}{}_{a_2 a_1]} + \cdots . \tag{D.58}$$

### *D.1.2 Fierz identities*

These identities are used very often in supergravity theories. To derive these identities, we first need a basis $\{\mathcal{O}_I\}$ of the vector space of $2^{[d/2]} \times 2^{[d/2]}$ matrices. This basis can be built out of the gamma matrices

$$\{\mathcal{O}^I\} = \left\{\mathbb{I}, \Gamma^a, i\Gamma^{ab}, i\Gamma^{abc}, \Gamma^{abcd}, \ldots\right\}. \tag{D.59}$$

(Observe that there are $2^{2[d/2]}$ matrices in this basis. Furthermore, this is why a decomposition like Eq. (D.58) is always possible.) All these matrices are linearly independent except for the last one (the product of all gamma matrices) in odd dimensions. Now we construct a dual basis orthogonal to the one in Eq. (D.59):

$$\{\mathcal{O}_I\} = \{\mathbb{I}, \Gamma_a, i\Gamma_{ab}, i\Gamma_{abc}, \Gamma_{abcd}, \ldots\}, \qquad \mathcal{O}_I(\mathcal{O}^J) \equiv \mathrm{tr}(\mathcal{O}_I\mathcal{O}_J) = 2^{[d/2]}\delta_{IJ}, \tag{D.60}$$

as we can easily check:

$$\begin{aligned}
\mathrm{Tr}\,\mathbb{I}^2 &= 2^{[d/2]},\\
\mathrm{Tr}\left(\tilde{\Gamma}^a\tilde{\Gamma}^b\right) &= \mathrm{Tr}\left(\tilde{\Gamma}^{(a}\tilde{\Gamma}^{b)}\right) = \mathrm{Tr}\left(\mathbb{I}\delta^{ab}\right) = 2^{[d/2]}\delta^{ab},\\
\mathrm{Tr}\left(\mathbb{I}\tilde{\Gamma}^a\right) &= 0,
\end{aligned} \tag{D.61}$$

etc. Any $2^{[d/2]} \times 2^{[d/2]}$ matrix $P$ is a linear combination of the $\{\mathcal{O}^I\}$:

$$\begin{aligned}
P &= p_I\mathcal{O}^I,\\
\mathrm{Tr}(\mathcal{O}_I P) &= 2^{[d/2]}p_I,\\
P^\alpha{}_\beta &= p_I\mathcal{O}^{I\,\alpha}{}_\beta = 2^{-[d/2]}\sum_I P^\gamma{}_\delta\mathcal{O}^\delta_{I\,\gamma}\mathcal{O}^{I\,\alpha}_\beta .
\end{aligned} \tag{D.62}$$

Let us now consider the product of bilinears of spinors,

$$\mathcal{Q} = \left(\bar{\lambda}M\chi\right)\left(\bar{\psi}N\varphi\right) = \bar{\lambda}_\alpha\chi^\beta\bar{\psi}_\gamma\varphi^\delta M^\alpha{}_\beta N^\gamma{}_\delta. \tag{D.63}$$

For fixed indices $\alpha$ and $\delta$, $M^\alpha{}_\beta N^\gamma{}_\delta$ is just a matrix $P^\gamma{}_\beta(\alpha,\delta)$ to which we can apply Eq. (D.63), obtaining

$$M^\alpha{}_\beta N^\gamma{}_\delta = 2^{-[d/2]}\sum_I (M\mathcal{O}_I N)^\alpha{}_\delta\mathcal{O}^{I\,\gamma}_\beta . \tag{D.64}$$

On substituting this identity into $\mathcal{Q}$ and taking into account the necessary permutation of the spinors, we obtain the general *Fierz identities*

$$\boxed{\begin{aligned} &\mathcal{Q} = p\, 2^{-[d/2]} \sum_I \left(\bar{\lambda} M \mathcal{O}^I N \varphi\right) \left(\bar{\psi} \mathcal{O}_I \chi\right), \\ &p = \begin{cases} -1 & \text{anticommuting spinors,} \\ +1 & \text{commuting spinors.} \end{cases} \end{aligned}} \tag{D.65}$$

Now, depending on the dimensions and the particular properties of the spinors, this expression can be further simplified. For instance, if we are dealing with Majorana–Weyl spinors, then terms with $n$ and $d-n$ gammas will be related.

In some cases ($N=2$ SUGRA theories) spinors appear in SO(2) doublets $\psi^i,\ i=1,2$. It is then convenient to arrange them in a vector,

$$\psi \equiv \begin{pmatrix} \psi^1 \\ \psi^2 \end{pmatrix}, \tag{D.66}$$

and it is useful to have Fierz identities for these vectors.[9] In the space of $2\times 2$ matrices a convenient basis is provided by the Pauli matrices and the identity. It is easy to arrive at the following $N=2$ Fierz identities:

$$\boxed{\begin{aligned} &\mathcal{Q} = p\, 2^{-[d/2]-1} \sum_{I,A} \left(\bar{\lambda} M \mathcal{O}^I N \sigma^A \varphi\right) \left(\bar{\psi} \mathcal{O}_I \sigma^A \chi\right), \\ &p = \begin{cases} -1 & \text{anticommuting spinors,} \\ +1 & \text{commuting spinors,} \end{cases} \end{aligned}} \tag{D.67}$$

with $A=0,1,2,3$ and $\sigma^0 = 1$.

### *D.1.3 Eleven dimensions*

Our 11-dimensional gamma matrices satisfy the anticommutation relations

$$\{\hat{\hat{\Gamma}}^{\hat{\hat{a}}}, \hat{\hat{\Gamma}}^{\hat{\hat{b}}}\} = +2\hat{\hat{\eta}}^{\hat{\hat{a}}\hat{\hat{b}}}. \tag{D.68}$$

It is possible to choose (in a way consistent with all the properties that we are going to enumerate) the 11th gamma matrix $\hat{\hat{\Gamma}}^{10}$ to be

$$\hat{\hat{\Gamma}}^{\hat{\hat{10}}} = i\hat{\hat{\Gamma}}^{\hat{\hat{0}}} \cdots \hat{\hat{\Gamma}}^{\hat{\hat{9}}} \equiv -i\hat{\Gamma}_{11}, \tag{D.69}$$

where $\hat{\Gamma}_{11}$ will be the ten-dimensional chirality matrix $\{\hat{\Gamma}_{11}, \hat{\Gamma}^{\hat{a}}\} = 0$.

[9] In higher-$N$ supergravity theories, spinors come in higher-dimensional multiplets and further generalizations exist.

They are in a purely imaginary (i.e. Majorana) representation, i.e. $\hat{\hat{\Gamma}}^{\hat{\hat{a}}\,\star} = -\hat{\hat{\Gamma}}^{\hat{\hat{a}}}$. They are all anti-Hermitian, except for $\hat{\hat{\Gamma}}^{\hat{\hat{0}}}$, which is Hermitian:

$$\begin{aligned} \hat{\hat{\Gamma}}^{\hat{\hat{0}}\,\dagger} &= +\hat{\hat{\Gamma}}^{\hat{\hat{0}}}, \\ \hat{\hat{\Gamma}}^{\hat{\hat{\imath}}\,\dagger} &= -\hat{\hat{\Gamma}}^{\hat{\hat{\imath}}}, \qquad \hat{\hat{\imath}} = 1, \ldots, 10. \end{aligned} \tag{D.70}$$

We have the property

$$\hat{\hat{\Gamma}}^{\hat{\hat{0}}}\,\hat{\hat{\Gamma}}^{\hat{\hat{a}}}\,\hat{\hat{\Gamma}}^{\hat{\hat{0}}} = \hat{\hat{\Gamma}}^{\hat{\hat{a}}\,\dagger}. \tag{D.71}$$

The Dirac conjugation matrix $\hat{\hat{\mathcal{D}}}$ is the real antisymmetric matrix

$$\hat{\hat{\mathcal{D}}} = i\hat{\hat{\Gamma}}^{0}, \tag{D.72}$$

and thus we have

$$\hat{\hat{\mathcal{D}}}\hat{\hat{\Gamma}}^{\hat{\hat{a}}_1\cdots\hat{\hat{a}}_n}\,\hat{\hat{\mathcal{D}}}^{-1} = (-1)^{[n/2]}\left(\hat{\hat{\Gamma}}^{\hat{\hat{a}}_1\cdots\hat{\hat{a}}_n}\right)^{\dagger}. \tag{D.73}$$

Their Hermiticity properties combined with their imaginary nature mean that all are symmetric except for $\hat{\hat{\Gamma}}^{\hat{\hat{0}}}$, which is antisymmetric:

$$\hat{\hat{\Gamma}}^{\hat{\hat{0}}\,T} = -\hat{\hat{\Gamma}}^{\hat{\hat{0}}}, \qquad \hat{\hat{\Gamma}}^{\hat{\hat{\imath}}\,T} = +\hat{\hat{\Gamma}}^{\hat{\hat{\imath}}}, \quad \hat{\hat{\imath}} = 1, \ldots, 10. \tag{D.74}$$

We choose a charge-conjugation matrix equal to the Dirac conjugation matrix,

$$\hat{\hat{\mathcal{C}}} = \hat{\hat{\mathcal{D}}} = i\hat{\hat{\Gamma}}^{0}, \tag{D.75}$$

which satisfies

$$\hat{\hat{\mathcal{C}}}^{T} = \hat{\hat{\mathcal{C}}}^{\dagger} = \hat{\hat{\mathcal{C}}}^{-1} = -\hat{\hat{\mathcal{C}}}, \qquad \hat{\hat{\mathcal{C}}}\hat{\hat{\Gamma}}^{\hat{\hat{a}}}\,\hat{\hat{\mathcal{C}}}^{-1} = -\hat{\hat{\Gamma}}^{\hat{\hat{a}}\,T}. \tag{D.76}$$

The last property implies

$$\hat{\hat{\mathcal{C}}}\hat{\hat{\Gamma}}^{\hat{\hat{a}}_1\cdots\hat{\hat{a}}_n}\,\hat{\hat{\mathcal{C}}}^{-1} = (-1)^{n+[n/2]}\left(\hat{\hat{\Gamma}}^{\hat{\hat{a}}_1\cdots\hat{\hat{a}}_n}\right)^{T}. \tag{D.77}$$

The standard definitions of the Dirac conjugates and Majorana conjugates and our specific choice of Dirac and charge-conjugation matrices $\hat{\hat{\mathcal{C}}} = \hat{\hat{\mathcal{D}}}$ imply that the Majorana condition

$$\bar{\hat{\hat{\lambda}}} = \hat{\hat{\lambda}}^{c} \tag{D.78}$$

is equivalent to requiring that all components of a Majorana spinor are real. Using the property Eq. (D.77) and the definition of (*anticommuting*) Majorana spinors, one finds

$$\bar{\hat{\hat{\epsilon}}}\,\hat{\hat{\Gamma}}^{\hat{\hat{a}}_1\cdots\hat{\hat{a}}_n}\,\hat{\hat{\psi}} = (-1)^{n+[n/2]}\,\bar{\hat{\hat{\psi}}}\,\hat{\hat{\Gamma}}^{\hat{\hat{a}}_1\cdots\hat{\hat{a}}_n}\,\hat{\hat{\epsilon}}, \tag{D.79}$$

so the preceding bilinear is symmetric for $n = 0, 3, 4, 7$, and 8 and antisymmetric for $n = 1, 2, 5, 6, 9$, and 10.

On the other hand, taking the Hermitian conjugate[10] and using Eq. (D.73), we find

$$\left(\bar{\hat{\hat{\epsilon}}}\hat{\hat{\Gamma}}^{\hat{a}_1\cdots\hat{a}_n}\hat{\hat{\psi}}\right)^\dagger = (-1)^{[n/2]}\bar{\hat{\hat{\psi}}}\hat{\hat{\Gamma}}^{\hat{a}_1\cdots\hat{a}_n}\hat{\hat{\epsilon}}, \tag{D.80}$$

which implies, on comparison with Eq. (D.79), that the preceding bilinear is real for even $n$ and imaginary for odd $n$.

Finally, we have the useful identity

$$\hat{\hat{\Gamma}}^{\hat{a}_1\cdots\hat{a}_n} = i\frac{(-1)^{[n/2]+1}}{(11-n)!}\hat{\hat{\epsilon}}^{\hat{a}_1\cdots\hat{a}_n\hat{\hat{b}}_1\cdots\hat{\hat{b}}_{11-n}}\hat{\hat{\Gamma}}_{\hat{\hat{b}}_1\cdots\hat{\hat{b}}_{11-n}}. \tag{D.81}$$

### D.1.4 Ten dimensions

The 11-dimensional Majorana representation of gamma matrices can be constructed from the ten-dimensional Majorana (purely imaginary) representation, according to

$$\begin{aligned}\hat{\hat{\Gamma}}^{\hat{a}} &= \hat{\Gamma}^{\hat{a}}, \quad \hat{a} = 0,\ldots,9,\\ \hat{\hat{\Gamma}}^{10} &= +i\hat{\Gamma}^0\cdots\hat{\Gamma}^9.\end{aligned} \tag{D.82}$$

Ten-dimensional Majorana spinors are identical to 11-dimensional spinors and the same definitions and identities apply to them. However, in ten dimensions we can also have Weyl spinors that satisfy many additional identities. They are defined in terms of the chirality matrix $\hat{\Gamma}_{11}$,

$$\hat{\Gamma}_{11} = -\hat{\Gamma}^0\cdots\hat{\Gamma}^9 = i\hat{\hat{\Gamma}}^{10}, \tag{D.83}$$

so $\hat{\Gamma}_{11}$ is Hermitian and satisfies $(\hat{\Gamma}_{11})^2 = +1$. Spinors of positive, $\hat{\psi}^{(+)}$, and negative, $\hat{\psi}^{(-)}$, chiralities are defined as usual:

$$\hat{\Gamma}_{11}\hat{\psi}^{(\pm)} = \pm\hat{\psi}^{(\pm)}. \tag{D.84}$$

Furthermore, in $d = 10$ we can define Majorana–Weyl fermions. It is useful to work in a Majorana–Weyl representation of the gamma matrices in which, in addition to having imaginary gamma matrices, the chirality matrix $\hat{\Gamma}_{11}$ has the form

$$\hat{\Gamma}_{11} = \mathbb{I}_{16\times 16}\otimes\sigma^3 = \begin{pmatrix}\mathbb{I}_{16\times 16} & 0\\ 0 & -\mathbb{I}_{16\times 16}\end{pmatrix}. \tag{D.85}$$

We will see explicitly that it is possible to have $\hat{\Gamma}_{11}$ defined as before in terms of the ten gamma matrices and at the same time having precisely that form. The sign of $\hat{\Gamma}_{11}$ is chosen in order to have that relation with positive sign and to have also

$$\hat{\Gamma}_{11} = \frac{1}{10!}\hat{\epsilon}_{\hat{a}_1\cdots\hat{a}_{10}}\hat{\Gamma}^{\hat{a}_1\cdots\hat{a}_{10}} = \frac{1}{10!\sqrt{|\hat{g}|}}\hat{\epsilon}_{\hat{\mu}_1\cdots\hat{\mu}_{10}}\hat{\Gamma}^{\hat{\mu}_1\cdots\hat{\mu}_{10}}, \tag{D.86}$$

---

[10] We use the convention $(ab)^\star = +a^\star b^\star$ for anticommuting numbers. This is the convention used in Refs. [386, 824, 1068] etc. The opposite convention is used in Refs. [1246, 1245] etc.

which leads to

$$\Gamma_{11}\hat{\Gamma}^{\hat{a}_1\cdots\hat{a}_n} = \frac{(-1)^{[(10-n)/2]+1}}{(10-n)!}\hat{\epsilon}^{\hat{a}_1\cdots\hat{a}_n\hat{b}_1\cdots\hat{b}_{10-n}}\hat{\Gamma}_{\hat{b}_1\cdots\hat{b}_{10-n}}. \tag{D.87}$$

In the Majorana–Weyl representation each 32-component real Majorana spinor $\hat{\psi}$ can be constructed from one positive-chirality and one negative-chirality 16-component spinor:

$$\hat{\psi} = \begin{pmatrix} \hat{\psi}^{(+)} \\ \hat{\psi}^{(-)} \end{pmatrix}. \tag{D.88}$$

### *D.1.5 Nine dimensions*

We have chosen a Majorana–Weyl representation for the ten-dimensional gamma matrices. They can be constructed from a purely real representation of the nine-dimensional ones:

$$\hat{\Gamma}^a = \Gamma^a \otimes \sigma^2, \quad a = 0, \ldots, 8, \qquad \hat{\Gamma}^9 = \mathbb{I}_{16\times 16} \otimes i\sigma^1, \tag{D.89}$$

where $\Gamma^8$ satisfies

$$\Gamma^8 = \Gamma^0 \cdots \Gamma^7. \tag{D.90}$$

As usual, it will be proportional to the eight-dimensional chiral matrix $\Gamma_{(8)\,9}$ (see Section D.1.6). One can explicitly check that, with these definitions, the ten-dimensional representation of the gamma matrices is indeed chiral and $\hat{\Gamma}_{11} = \mathbb{I}_{16\times 16} \otimes \sigma^3$.

### *D.1.6 Eight dimensions*

The purely nine-dimensional gamma matrices we are using can be constructed in the standard way from a purely real eight-dimensional representation (which is not chiral):

$$\Gamma^a = \Gamma^a_{(8)}, \quad a = 0, \ldots, 7, \qquad \Gamma^8 = \Gamma^0 \cdots \Gamma^7. \tag{D.91}$$

The chirality matrix is defined by

$$\Gamma_{(8)\,9} = i\Gamma^8 = i\Gamma^0 \cdots \Gamma^7. \tag{D.92}$$

We will not be able to decompose this representation in terms of a seven-dimensional representation. There are no purely real or imaginary (in Lorentzian signature) representations of the gamma matrices in seven dimensions. Thus, we cannot decompose $\Gamma^a_{(8)} = \Gamma^a_{(7)} \otimes A_{2\times 2}$ with the same factor matrix $A$ for all $a = 0, \ldots, 6$. Thus, it is impossible to use this representation to perform a dimensional reduction from $d = 8$ to $d = 7, 6, 5$ dimensions because we would break Lorentz invariance.

It is possible, however, to reduce directly to four dimensions. If $\{\gamma^a\}$ is a Majorana (purely imaginary) representation of the four-dimensional gamma matrices, the purely real representation of the eight-dimensional ones can be constructed in this way:

$$\begin{aligned}
\Gamma^a_{(8)} &= \gamma^a \otimes \sigma^2 \otimes \sigma^1, \quad a = 0, 1, 2, 3, & \Gamma^4_{(8)} &= \Gamma^4_{(5)} \otimes \sigma^3 \otimes \sigma^3, \\
\Gamma^5_{(8)} &= \Gamma^4_{(5)} \otimes \sigma^1 \otimes \sigma^3, & \Gamma^6_{(8)} &= \mathbb{I}_{4\times 4} \otimes \mathbb{I}_{2\times 2} \otimes i\sigma^2, \\
\Gamma^7_{(8)} &= \mathbb{I}_{4\times 4} \otimes i\sigma^2 \otimes \sigma^3, & &
\end{aligned} \tag{D.93}$$

where

$$\Gamma^{4}_{(5)} = \gamma^0\gamma^1\gamma^2\gamma^3. \tag{D.94}$$

Since a four-dimensional Majorana representation exists, this proves the existence of the nine-, ten-, and 11-dimensional representations we are using. For the purpose of dimensional reduction of the gamma matrices from $d=10$ to $d=4$ this is not the best representation. It is more convenient to use those of Refs. [262, 636].

### *D.1.7 Two dimensions*

A (purely imaginary) Majorana–Weyl representation of the two-dimensional Clifford algebra is given by

$$\gamma^0_{(2)} = \sigma^2, \qquad \gamma^1_{(2)} = i\sigma^1. \tag{D.95}$$

The chiral matrix is, as expected in a Weyl representation,

$$\gamma_{(2)\,3} = \gamma^0_{(2)}\gamma^1_{(2)} = \sigma^3. \tag{D.96}$$

The two-dimensional gamma matrices are sometimes denoted by $\rho^a$.

### *D.1.8 Three dimensions*

We can build a purely imaginary Majorana representation from the two-dimensional (purely imaginary) Majorana–Weyl representation:

$$\gamma^a_{(3)} = \gamma^a_{(2)}, \; a=0,1, \qquad \gamma^2_{(3)} = -i\gamma^0_{(2)}\gamma^1_{(2)} = -i\sigma^3. \tag{D.97}$$

### *D.1.9 Four dimensions*

Given the Majorana representation in Eq. (D.97) (purely imaginary) of the three-dimensional gamma matrices, we can build a Majorana representation (purely imaginary) of the four-dimensional gamma matrices:

$$\gamma^a = \gamma^a_{(3)} \otimes \sigma^3, \; a=0,1,2, \qquad \gamma^3 = \mathbb{I}_{2\times 2} \otimes i\sigma^1. \tag{D.98}$$

The chiral matrix is

$$\gamma_5 = -i\gamma^0\gamma^1\gamma^2\gamma^3 = \frac{i}{4!}\epsilon_{abcd}\gamma^{abcd}, \tag{D.99}$$

and, using the explicit form of the three-dimensional gamma matrices,

$$\gamma_5 = \mathbb{I}_{2\times 2} \otimes \sigma^2. \tag{D.100}$$

It is obviously Hermitian and imaginary (and therefore antisymmetric). In this representation we can use as charge-conjugation matrix $\mathcal{C}_- = i\gamma^0$, which is real and antisymmetric. The Majorana condition says that Majorana spinors are purely real spinors, $\psi = \psi^*$.

There is another possible choice; namely $\mathcal{C}_+ = \gamma_5\gamma^0$, which is also real and antisymmetric and would impose the following condition on Majorana spinors: $\psi = -i\gamma_5\psi^*$. This is inconsistent (just take the complex conjugate of this relation) and so with it we can define only symplectic-Majorana spinors.

We can also build a Weyl representation that is complex:

$$\gamma^a = \gamma^a_{(3)} \otimes \sigma^2, \;\; a = 0, 1, 2, \qquad \gamma^3 = \mathbb{I}_{2\times 2} \otimes i\sigma^1. \tag{D.101}$$

With the Majorana representation Eq. (D.97) of three-dimensional gamma matrices, we find

$$\gamma_5 = \mathbb{I}_{2\times 2} \otimes \sigma^3. \tag{D.102}$$

There are no Majorana–Weyl fermions in four dimensions and there are no Majorana–Weyl representations of the gamma matrices. The Weyl and Majorana representations given here are related by the similarity transformation (which is valid also for $\gamma_5$)

$$\gamma^a_{\mathrm{M}} = S\gamma^a_{\mathrm{W}} S^{-1}, \qquad S = \mathbb{I}_{2\times 2} \otimes \tfrac{1}{\sqrt{2}}(\mathbb{I}_{2\times 2} - i\sigma^1). \tag{D.103}$$

We also have the identity

$$\gamma^{a_1\cdots a_n} = \frac{(-1)^{[n/2]} i}{(4-n)!} \epsilon^{a_1\cdots a_n b_1 \cdots b_{4-n}} \gamma_{b_1\cdots b_{4-n}} \gamma_5. \tag{D.104}$$

Using this identity the $d = 4$ Fierz identities for anticommuting spinors take the form

$$\begin{aligned}(\bar{\lambda} M \chi)(\bar{\psi} N \varphi) = &-\tfrac{1}{4}(\bar{\lambda} M N \varphi)(\bar{\psi}\chi) - \tfrac{1}{4}(\bar{\lambda} M \gamma^a N \varphi)(\bar{\psi}\gamma_a \chi)\\ &+ \tfrac{1}{8}(\bar{\lambda} M \gamma^{ab} N \varphi)(\bar{\psi}\gamma_{ab}\chi) + \tfrac{1}{4}(\bar{\lambda} M \gamma^a \gamma_5 N \varphi)(\bar{\psi}\gamma_a \gamma_5 \chi)\\ &- \tfrac{1}{4}(\bar{\lambda} M \gamma_5 N \varphi)(\bar{\psi}\gamma_5\chi).\end{aligned} \tag{D.105}$$

(For commuting spinors it is enough to change the global sign of the r.h.s.)

In our review of four-dimensional supergravities in the main text we always use a Majorana representation of the gamma matrices. As for the spinors, we use Majorana spinors in our review of pure $N = 1, 2, d = 4$ supergravities in Chapter 5 and Weyl spinors in our review of the matter-coupled theories in Chapters 6, 7, and 8. The Majorana spinors of four-dimensional SUEGRAs (for example, the supersymmetry parameter) carry an R-symmetry index $i = 1, \ldots, N$ which we place, conventionally, in an upper position: $\epsilon^i$. The transition to Weyl spinors is explained at the beginning of Section 6.1, but we review it here for convenience: each Majorana spinor $\epsilon^i$ is decomposed as the sum of two Weyl spinors of opposite chiralities. The chiralities denoted by the position of the index $I, J = 1, \ldots, N$ and the convention can be different for different fields. For the supersymmetry parameter

$$\epsilon^i = \epsilon^I + \epsilon_I, \qquad \gamma_5 \epsilon^I = +\epsilon^I, \qquad \gamma_5 \epsilon_I = -\epsilon_I. \tag{D.106}$$

Complex conjugation interchanges these two Weyl spinors ($\gamma_5^* = -\gamma_5$) as required by the reality of $\epsilon^i$:

$$(\epsilon_I)^* = \epsilon^I. \tag{D.107}$$

The convention for the Dirac conjugates takes into account the following fact:

$$\bar{\epsilon}^I = i(\epsilon_I)^\dagger \gamma_0, \qquad \bar{\epsilon}^I \gamma_5 = \bar{\epsilon}^I, \qquad \bar{\epsilon}_I \gamma_5 = -\bar{\epsilon}_I, \qquad (\bar{\epsilon}_I)^* = \bar{\epsilon}^I. \tag{D.108}$$

### D.1.10 Five dimensions

There are no Majorana representations in $d=5$, but only pairs of (complex) symplectic-Majorana spinors that can be combined into a single unconstrained Dirac spinor. Doing this, however, hides this structure and makes it more difficult (or impossible) to construct five-dimensional supergravities with the most general couplings. We will show how to deal with these spinors after we construct the five-dimensional gamma matrices.

Using any representation of the four-dimensional gamma matrices $\gamma^a$, $a=0,1,2,3$, we can construct a five-dimensional representation (which is necessarily complex, even if the four-dimensional gamma matrices are purely imaginary, as we take them to be in Chapter 9)

$$\hat{\gamma}^a = \gamma^a, \quad a=0,1,2,3, \qquad \hat{\gamma}^4 = -i\gamma_5 = \gamma_0\gamma_1\gamma_2\gamma_3. \tag{D.109}$$

When we use the purely imaginary four-dimensional gamma matrices $\hat{\gamma}^4$ is purely real. On the other hand, $\hat{\gamma}^0$ is Hermitian and the other gammas are anti-Hermitian.

To simplify the notation, we remove all the hats in the gamma matrices and indices, which will be understood to be five dimensional. The product of all the five-dimensional gammas is

$$\gamma^0 \cdots \gamma^4 = +1 \quad \Rightarrow \quad \gamma^{a_1\cdots a_5} = +\epsilon^{a_1\cdots a_5}, \tag{D.110}$$

and the duality formula takes the form

$$\gamma^{a_1\cdots a_n} = \frac{(-1)^{[n/2]}}{(5-n)!}\epsilon^{a_1\cdots a_n b_1\cdots b_{5-n}}\gamma_{b_1\cdots b_{5-n}}. \tag{D.111}$$

To explain our convention for symplectic-Majorana spinors, let us start by defining the Dirac, complex, and charge-conjugation matrices. In this case they are $\mathcal{D}_+, \mathcal{B}_+, \mathcal{C}_+$ ($\mathcal{B}_-$ and $\mathcal{C}_-$ do not exist in this case) and are given by the antisymmetric matrices

$$\mathcal{D} \equiv i\gamma^0, \qquad \mathcal{C} \equiv i\gamma^{04}, \qquad \mathcal{B} \equiv \gamma^4, \tag{D.112}$$

so that the relation Eq. (D.23) is satisfied and, further,

$$\mathcal{B}^*\mathcal{B} = -1. \tag{D.113}$$

With these choices, the Dirac and Majorana conjugates are defined, respectively, by[11]

$$\psi^\dagger\mathcal{D} = i\psi^\dagger\gamma^0, \qquad \psi^T\mathcal{C} = i\psi^T\gamma^{04}, \tag{D.114}$$

and the Majorana condition (Dirac conjugate = Majorana conjugate) would be equivalent to the reality condition

$$\psi^* = \gamma^4\psi. \tag{D.115}$$

Taking the complex conjugate of this equation and multiplying both sides by $\gamma^4$ we get

$$\gamma^4\gamma^{*4}\psi^* = \gamma^4\psi, \tag{D.116}$$

[11] We do not use the standard notation $\bar{\psi}$ for the Dirac conjugate here. We will use this notation for the Majorana conjugate which will be equivalent to the Dirac one up to an operation to be defined.

which is not consistent with the previous formula because $\mathcal{B}\mathcal{B}^* = \gamma^4\gamma^{*\,4} \neq +1$. Then, the Majorana condition cannot be consistently imposed. The solution to this problem is to consider $n$ pairs of spinors $\psi^i$, $i = 1, \ldots, 2n$, introduce a $2n \times 2n$ matrix $\Omega_{ij}$, and require the modified Majorana condition

$$\psi^{*\,i} = \gamma^4 \psi_i, \qquad \psi_i \equiv \Omega_{ij}\psi^j, \tag{D.117}$$

which will be consistent if

$$\Omega_{ij}\Omega_{jk} = -\delta_{ik}. \tag{D.118}$$

We can take this matrix to be the symplectic metric $(\Omega_{ij}) = \left(\begin{smallmatrix} 0 & \mathbb{1}_{n\times n} \\ -\mathbb{1}_{n\times n} & 0 \end{smallmatrix}\right)$ and operate with it using the conventions in Appendix C to raise and lower $i, j$ (symplectic) indices. For $n = 1$, $\Omega_{ij} = \varepsilon_{ij}$, and since this matrix is also preserved by the group $\mathrm{SU}(2)$ we will also refer to these indices as $\mathrm{SU}(2)$ indices.

The notation that we are going to use for the Majorana conjugate is meant to preserve the position of these indices and the corresponding transformation properties:

$$\bar{\psi}^i \equiv \psi^{i\,T}\mathcal{C}, \qquad \bar{\psi}_i \equiv \Omega_{ij}\bar{\psi}^j. \tag{D.119}$$

The symplectic-Majorana condition says that $\bar{\psi}_i$ is equal to the Dirac conjugate of $\psi^i$:

$$\psi^{i\dagger}\mathcal{D} = \bar{\psi}_i, \tag{D.120}$$

which justifies this notation.

Under complex conjugation,

$$\left(\bar{\psi}^i\right)^* = \bar{\psi}_i\gamma^4. \tag{D.121}$$

The Fierz identities for generic anticommuting spinors with or without indices are

$$\begin{aligned}(\bar{\lambda}M\varphi)\,(\bar{\psi}N\chi) = &-\frac{1}{4}\,(\bar{\lambda}MN\chi)\,(\bar{\psi}\varphi) - \frac{1}{4}\,(\bar{\lambda}M\gamma^a N\chi)\,(\bar{\psi}\gamma_a\varphi) \\ &+\frac{1}{8}\left(\bar{\lambda}M\gamma^{ab}N\chi\right)(\bar{\psi}\gamma_{ab}\varphi)\,.\end{aligned} \tag{D.122}$$

(The global sign of the r.h.s. must be reversed for commuting spinors.)

### *D.1.11 Six dimensions*

With the representation Eq. (D.109) of the five-dimensional gamma matrices, we can construct a six-dimensional representation

$$\hat{\hat{\gamma}}^{\hat{a}} = \hat{\gamma}^{\hat{a}} \otimes \sigma^1, \qquad \hat{a} = 0, 1, 2, 3, 4, \qquad \hat{\hat{\gamma}}^5 = \mathbb{I}_{4\times 4} \otimes i\sigma^2, \tag{D.123}$$

which is a Weyl representation, since the chirality matrix $\hat{\hat{\gamma}}_7$ is

$$\hat{\hat{\gamma}}_7 = \hat{\hat{\gamma}}_0 \cdots \hat{\hat{\gamma}}_5 = \mathbb{I}_{4\times 4} \otimes \sigma^3. \tag{D.124}$$

A useful formula is

$$\hat{\hat{\gamma}}^{\hat{\hat{a}}_1\cdots\hat{\hat{a}}_n} = \frac{(-1)^{[n/2]}}{(6-n)!}\,\hat{\hat{\epsilon}}^{\hat{\hat{a}}_1\cdots\hat{\hat{a}}_n\hat{\hat{b}}_1\cdots\hat{\hat{b}}_{6-n}}\,\hat{\hat{\gamma}}_{\hat{\hat{b}}_1\cdots\hat{\hat{b}}_{6-n}}\,\hat{\hat{\gamma}}_7. \tag{D.125}$$

## D.2 Spaces with arbitrary signatures

We now want to generalize our results on spinors and gamma matrices to $d$-dimensional spaces with signatures $(+^t, -^s)$, where $t$ is the number of timelike dimensions and $s$ is the number of spacelike dimensions. The essential reference is Ref. [882] and other useful references are Refs. [565, 1097, 1169], which we roughly follow.

The general setup is the same as in the signature $(1, d-1)$ case: we consider the generators of the Clifford algebra associated with the metric $\eta_{ab} = \mathrm{diag}(+^t, -^s)$, where the indices are $a, b = -(t-1), -(t-2), \ldots, 0, 1, \ldots, s$, which is the metric of $\mathrm{SO}(t, s)$. These are the $2^{[d/2]} \times 2^{[d/2]}$ gamma matrices $\Gamma^a$ which satisfy the usual anticommutation relations and out of which one can build the generators of $\mathfrak{so}(t, s)$ in the spinorial representation in the usual form. They are unique up to similarity transformations. The complex $2^{[d/2]}$-component vectors in the representation space are Dirac spinors. We consider unitary representations and, therefore, all timelike (spacelike) gamma matrices are Hermitian (anti-Hermitian):

$$\Gamma^{a\,\dagger} = +\Gamma^a, \quad a \leq 0, \qquad \Gamma^{a\,\dagger} = -\Gamma^a, \quad a > 0. \tag{D.126}$$

A representation can be constructed by "Wick-rotating" the signature $(1, d-1)$ matrices, multiplying them by factors of $i$ if necessary. Given a $d$-even representation, one can construct the chirality matrix $\mathcal{Q} = \Gamma_{d+1}$,

$$\Gamma_{d+1} = \varphi(s,t)\Gamma^{-(t-1)}\Gamma^{-(t-2)}\cdots\Gamma^{-1}\Gamma^0\Gamma^1\cdots\Gamma^s, \qquad \varphi(s,t) = -e^{\frac{\pi i}{4}(s-t)}, \tag{D.127}$$

which is unitary and Hermitian and anticommutes with all the $\Gamma^a$s. Using it, we can construct a representation of the $(d+1)$-dimensional gamma matrices: if the signature is $(t, s+1)$ we define

$$\Gamma^{s+1} = -i\Gamma_{d+1}, \tag{D.128}$$

and if the signature is $(t+1, s)$ we simply define

$$\Gamma^{-t} = \Gamma_{d+1}. \tag{D.129}$$

Thus, in even dimensions the gamma matrices are independent and in odd dimensions they are not: the product of all the gamma matrices is a power of the imaginary unit.

In even dimensions, $\{\pm\Gamma^a\}$, $\{\pm\Gamma^{a\,\dagger}\}$, $\{\pm\Gamma^{a\,T}\}$, and $\{\pm\Gamma^{a\,*}\}$ generate equivalent representations. In odd dimensions, however, due to the above-mentioned constraint, only one sign gives an equivalent representation. The matrices of the corresponding similarity transformations are the chirality matrix $\mathcal{Q}$ ($\Gamma_{d+1}$), the Dirac matrix $\mathcal{D}_\pm$, the charge-conjugation matrix $\mathcal{C}_\pm$, and the $\mathcal{B}_\pm$ matrix:

$$\begin{aligned} \mathcal{Q}\,\Gamma^a\,\mathcal{Q}^{-1} &= -\Gamma^a, & \mathcal{D}_\pm\,\Gamma^a\,\mathcal{D}_\pm^{-1} &= \pm\Gamma^{a\,\dagger}, \\ \mathcal{C}_\pm\,\Gamma^a\,\mathcal{C}_\pm^{-1} &= \pm\Gamma^{a\,\mathrm{T}}, & \mathcal{B}_\pm\,\Gamma^a\,\mathcal{B}_\pm^{-1} &= \pm\Gamma^{a\,*}. \end{aligned} \tag{D.130}$$

In even dimensions all these matrices exist and, evidently,

$$\mathcal{D}_\pm = \mathcal{D}_\mp\mathcal{Q}, \qquad \mathcal{C}_\pm = \mathcal{C}_\mp\mathcal{Q}, \qquad \mathcal{B}_\pm = \mathcal{B}_\mp\mathcal{Q}. \tag{D.131}$$

Table D.2. Possible values of $\varepsilon_\pm(\pm)$

| | $s-t$ | | | | | | | |
|---|---|---|---|---|---|---|---|---|
| | 1 | 2 | 3 | 4 | 5 | 6 | 7 | 8 |
| $\varepsilon_+$ | 0 | $-1$ | $-1$ | $-1$ | 0 | $+1$ | $+1$ | $+1$ |
| $\varepsilon_-$ | $+1$ | $+1$ | 0 | $-1$ | $-1$ | $-1$ | 0 | $+1$ |

In odd dimensions $\mathcal{Q}$ does not exist and only one of the $\mathcal{C}_\pm$ and $\mathcal{B}_\pm$ exists.

In general, $\mathcal{D}$ is defined (up to a phase $\alpha$) by

$$\mathcal{D} = \alpha\Gamma^0\Gamma^{-1}\cdots\Gamma^{-(t-1)}. \tag{D.132}$$

In our conventions we find

$$\mathcal{D}\,\Gamma^a\,\mathcal{D}^{-1} = (-1)^{t+1}\Gamma^{a\,\dagger}, \tag{D.133}$$

and thus

$$\mathcal{D} = \mathcal{D}_+, \quad \text{for odd } t, \qquad \mathcal{D} = \mathcal{D}_-, \quad \text{for even } t, \tag{D.134}$$

and then one has the relations

$$\mathcal{C}_\pm = \mathcal{B}_\pm^T\mathcal{D}, \quad \text{for odd } t, \qquad \mathcal{C}_\pm = \mathcal{B}_\mp^T\mathcal{D}, \quad \text{for even } t, \tag{D.135}$$

so the existence and properties of $\mathcal{C}_\pm$ are determined by the existence and properties of $\mathcal{B}_\pm$. The main result is[12]

$$\mathcal{B}_\pm^T = \varepsilon_\pm(t,s)\mathcal{B}_\pm, \qquad \varepsilon_\pm(t,s) = \text{sqcos}\left[\frac{\pi}{4}(s-t\pm 1)\right]. \tag{D.136}$$

When $\varepsilon = \pm 1$, $\mathcal{B}$ is symmetric or antisymmetric. When $\varepsilon_\pm = 0$, $\mathcal{B}_\pm$ does not exist. The value depends on $(s-t) \bmod 8$ and it is represented in Table D.2, from Ref. [1169]. Observe that, since these matrices are assumed to be unitary, we also have

$$\mathcal{B}_\pm^*\mathcal{B}_\pm = \varepsilon_\pm(t,s). \tag{D.137}$$

Thus, for instance, when $s=t$ only $\mathcal{B}_-$ exists and it is symmetric, whereas for $s=t+1$ both $\mathcal{B}_+$ and $\mathcal{B}_-$ exist and are, respectively, antisymmetric and symmetric.

The symmetry of $\mathcal{B}_\pm$ determines that of the charge-conjugation matrix $\mathcal{C}_\pm$:

$$\begin{aligned} \mathcal{C}_\pm^T &= \varepsilon_\pm(\pm 1)^t(-1)^{\frac{t(t-1)}{2}}\mathcal{C}_\pm, \quad \text{for odd } t, \\ \mathcal{C}_\pm^T &= \varepsilon_\mp(-1)^{\frac{t(t-1)}{2}}\mathcal{C}_\pm, \quad \text{for even } t. \end{aligned} \tag{D.138}$$

We can now define the Dirac $\bar\psi$ and Majorana $\psi^c$ conjugates of a spinor $\psi$:

$$\bar\psi = \psi^\dagger\mathcal{D}, \qquad \psi^c = \psi^T\mathcal{C}. \tag{D.139}$$

[12] The function sqcos $\theta$ is defined as the projection on the $x$ axis of the line that forms an angle $\theta$ with the $x$ axis and joins the origin to a square centered on the origin and with sides of length 2. Then $\text{sqcos}(-\pi/4, 0, \pi/4) = +1$, $\text{sqcos}(3\pi/4, \pi, 5\pi/4) = -1$, and $\text{sqcos}(\pi/2, 3\pi/2) = 0$.

The existence of these conjugation operations is due to the equivalence of the Hermitian conjugate and transposed representations of the gamma matrices.

Now we can proceed to define various types of constrained spinors.

**Weyl spinors.** In any even dimension these are eigenspinors of the chirality matrix, which has only eigenvalues $+1$ and $-1$ because $\mathcal{Q}^2 = 1$. Since it is traceless, half of the eigenvalues are $+1$ and half are $-1$. $\mathcal{Q}$ commutes with all the $\mathfrak{so}(t, s)$ generators, which means that the (Dirac) spinorial representation is reducible to the direct sum of the two Weyl spinorial representations.

**(Pseudo-)Majorana spinors.** They satisfy the reality constraint

$$\bar{\psi} = \psi^c, \tag{D.140}$$

which, using the relation among $\mathcal{D}, \mathcal{C}$, and $\mathcal{B}$, can be rewritten in the form

$$\psi^* = \mathcal{B}\psi. \tag{D.141}$$

On taking the complex conjugate of this equation, we find the consistency condition

$$\mathcal{B}\mathcal{B}^* = \varepsilon = +1. \tag{D.142}$$

Only in this case is it possible to define Majorana spinors. If the equation is satisfied by $\mathcal{B}_-$, the spinors are called Majorana spinors. If it is satisfied by $\mathcal{B}_+$, they are called pseudo-Majorana spinors. Thus, Table D.2 can be reinterpreted in terms of the existence of Majorana (M) or pseudo-Majorana (pM) spinors, as in Table D.3.

**(Pseudo-)Majorana–Weyl spinors.** The Majorana and Weyl conditions are compatible if

$$\mathcal{D}^{-1}\mathcal{Q}^\dagger\mathcal{D} = \mathcal{C}^{-1}\mathcal{Q}^T\mathcal{C}, \tag{D.143}$$

which, using the relation among $\mathcal{D}, \mathcal{C}$, and $\mathcal{B}$, can be simplified to

$$\mathcal{Q}^* = \mathcal{B}\mathcal{Q}\mathcal{B}^{-1}, \tag{D.144}$$

which is satisfied for $s - t = 0 \bmod 4$.

**(Pseudo-)symplectic-Majorana spinors.** When $\mathcal{B}\mathcal{B}^* = -1$ the Majorana reality condition cannot be consistently imposed. However, then one can introduce an even number of Dirac spinors labeled by $i = 1, \ldots, 2n$ and impose the reality condition

$$\bar{\psi}^i = \psi_i{}^c \equiv \Omega_{ij}\psi^{j\,c}, \tag{D.145}$$

where $\Omega$ is real and satisfies

$$\Omega_{ij}\Omega_{jk} = -\delta_{ik}. \tag{D.146}$$

This condition can be rewritten in the more transparent form

$$\psi^{i\,*} = \Omega_{ij}\mathcal{B}\psi^j, \tag{D.147}$$

which is consistent if $\mathcal{B}\mathcal{B}^* = -1$. The cases in which these spinors can be defined are represented in Table D.3.

Now we are going to use these results in several examples of interest.

Table D.3. Possible spinors in $d = t + s$ dimensions with signatures $(+^t, -^s)$

M stands for Majorana, pM for pseudo-Majorana, SM for symplectic-Majorana, pSM for pseudo-symplectic-Majorana, MW for Majorana–Weyl, and pMW for pseudo-Majorana–Weyl; asterisks mean that $d$ has to be even. In addition to this, Weyl spinors are possible for any even $d$.

| | | | $s-t$ | | | | |
|---|---|---|---|---|---|---|---|
| 1 | 2 | 3 | 4 | 5 | 6 | 7 | 8 |
| | pSM | pSM | pSM | | | | |
| | | | | | pM | pM | pM |
| M | M | | | | | | M |
| | | | SM | SM | SM | | |
| | | | | | | | pMW* |
| | | | | | | | MW* |

### *D.2.1* $AdS_4$ *gamma matrices and spinors*

The spinor representations of $SO(2,3)$ (which we also refer to as $AdS_4$) have the same dimension (four) as those of $SO(1,3)$. The corresponding gamma matrices, which we write with hats, are $4 \times 4$ matrices, and any representation of them includes a representation of the $SO(1,3)$ (unhatted) gamma matrices. Furthermore, it is clear that $AdS_4$ spinors transform as Lorentz spinors under the Lorentz subgroup. Our goal now will be to construct an explicit representation of the gamma matrices and the generators of the $AdS_4$ group. Since $s - t = 1$ we expect only a Majorana representation. It appears in two forms, which are equivalent through similarity transformations. We call them *electric* and *magnetic* and denote them by $-$ or $+$ subscripts or superscripts, respectively.

**The magnetic representation.** It is built by the standard procedure from the Clifford algebra associated with the $SO(2,3)$ metric $\hat{\eta}_{\hat{a}\hat{b}}$: first we construct five gamma matrices satisfying

$$\left\{\hat{\gamma}_{\hat{a}}, \hat{\gamma}_{\hat{b}}\right\} = +2\hat{\eta}_{\hat{a}\hat{b}}. \tag{D.148}$$

These matrices can be constructed by using the four $SO(1,3)$ Dirac matrices:

$$\hat{\gamma}_{+\,-1} = \gamma_5 = -i\gamma^0 \cdots \gamma^3, \qquad \hat{\gamma}_{+\,a} = \gamma_a; \tag{D.149}$$

using purely imaginary Dirac matrices, we obtain a purely imaginary representation of the $SO(2,3)$ Clifford algebra.

The $SO(2,3)$ generators in the magnetic spinorial representation are constructed from the Clifford algebra in the usual fashion,

$$\Gamma_+\left(\hat{M}_{\hat{a}\hat{b}}\right) = \tfrac{1}{2}\hat{\gamma}_{+\,\hat{a}\hat{b}}, \tag{D.150}$$

and they automatically satisfy the $\mathfrak{so}(2,3)$ algebra Eq. (4.152).

SO(2, 3) spinors $\hat{\psi}_+{}^\alpha$ transform with the exponential of all these generators and are, in particular, Lorentz spinors.

Since $t = 2$, we know that $\mathcal{D} = \mathcal{D}_-$. Furthermore, the only $\mathcal{B}$ leading to consistent Majorana spinors is $\mathcal{B}_-$, and thus we can use only $\mathcal{C}_+$, which is antisymmetric. We can take

$$\mathcal{C}_+ = \mathcal{D}_- = \hat{\gamma}_+^0 \hat{\gamma}_+^{-1} = \gamma^0 \gamma_5. \tag{D.151}$$

It is easy to check that the charge-conjugation matrix $\mathcal{C}_+$ satisfies

$$\mathcal{C}_+ \hat{\gamma}^{\hat{a}} \mathcal{C}_+^{-1} = +\hat{\gamma}^{\hat{a}\,T} = -\hat{\gamma}^{\hat{a}\,\dagger}. \tag{D.152}$$

Since the Dirac conjugation and charge-conjugation matrices are identical, $\mathcal{B}_- = \mathbb{I}_{4\times 4}$, the Majorana condition $\bar{\hat{\psi}}_+ = \hat{\psi}_+^c$ implies that Majorana spinors are purely real spinors in this representation, $\hat{\psi}_+^* = \hat{\psi}_+$.

In this representation the ten matrices

$$\left[\Gamma_+\left(\hat{M}^{\hat{a}\hat{b}}\right)\mathcal{C}_+^{-1}\right]^{\alpha\beta} \tag{D.153}$$

are real and symmetric. This is necessary in order to build the $\mathrm{osp}(N/4)$ supersymmetry algebra. The six matrices

$$(\mathcal{C}_+^{-1})^{\alpha\beta}, \qquad \left(i\hat{\gamma}_+^{\hat{a}}\mathcal{C}_+^{-1}\right)^{\alpha\beta} \tag{D.154}$$

are real and antisymmetric and we will use them to add other ("central") charges in the anticommutator $\{Q^{\alpha\, i}, Q^{\beta\, j}\}$ supersymmetry algebra.

These 16 matrices are a basis of the linear space of real $4 \times 4$ matrices. Later we will work out the details simultaneously with analogous matrices of the electric representation. For the moment it suffices to observe that other antisymmetrized products of gamma matrices can be related to the antisymmetrized products of zero, one, and two gamma matrices via

$$\hat{\gamma}_+^{\hat{a}_1\cdots\hat{a}_n} \sim \frac{1}{(5-n)!}\hat{\epsilon}^{\hat{a}_1\cdots\hat{a}_5}\hat{\gamma}_{+\,\hat{a}_{n+1}\cdots\hat{a}_5}, \tag{D.155}$$

with $\hat{\epsilon}^{-10123} = +1$. In particular,

$$\hat{\gamma}_+^{\hat{a}\hat{b}\hat{c}\hat{d}\hat{e}} = i\hat{\epsilon}^{\hat{a}\hat{b}\hat{c}\hat{d}\hat{e}}. \tag{D.156}$$

**The electric representation.** This is the representation that is used most often. The SO(2, 3) generators in the *electric* spinorial representation $\Gamma_-$ are built directly from the SO(1, 3) Dirac gamma matrices $\gamma^a$:

$$\Gamma_-\left(\hat{M}_{ab}\right) = \tfrac{1}{2}\gamma^{ab}, \qquad \Gamma_-\left(\hat{M}_{a-1}\right) = \frac{i}{2}\gamma^a. \tag{D.157}$$

It can be checked that these matrices satisfy the $\mathfrak{so}(2,3)$ algebra Eq. (4.152).

In this representation

$$i\gamma^0\Gamma_-\left(\hat{M}_{\hat{a}\hat{b}}\right)\left(i\gamma^0\right)^{-1} = -\Gamma_-\left(\hat{M}_{\hat{a}\hat{b}}\right)^\dagger = -\Gamma_-\left(\hat{M}_{\hat{a}\hat{b}}\right)^T, \tag{D.158}$$

which implies that we can take as Dirac and charge-conjugation matrices

$$\mathcal{D} = \mathcal{D}_+ = \mathcal{C}_- = i\gamma^0, \tag{D.159}$$

which coincide with the ones we used in $d = 4$ dimensions with signature $(1,3)$. This may seem contradictory. However, we have not yet identified from which representation of the Clifford algebra the $\mathfrak{so}(2,3)$ representation arises. Actually, it can be constructed by the standard procedure from the following hatted gamma matrices:

$$\hat{\gamma}_{-\,-1} = \gamma_5, \qquad \hat{\gamma}_{-\,a} = i\gamma_a\gamma_5, \tag{D.160}$$

which provide a purely imaginary representation of the Clifford algebra associated with $\hat{\eta}_{\hat{a}\hat{b}}$. Then, we can see that the Dirac conjugation and charge-conjugation matrices are given by the product of the two timelike gammas, as in the *magnetic* case:

$$\mathcal{D}_+ = \mathcal{C}_- = \hat{\gamma}_-^0\hat{\gamma}_-^{-1} = i\gamma^0. \tag{D.161}$$

As is needed for supersymmetry, in the electric representation the ten matrices

$$\left[\Gamma_-\left(\hat{M}^{\hat{a}\hat{b}}\right)\mathcal{C}_-^{-1}\right]^{\alpha\beta} \tag{D.162}$$

are also real and symmetric, and the six matrices

$$\left(\mathcal{C}_-^{-1}\right)^{\alpha\beta}, \quad \left(i\hat{\gamma}_-^{\hat{a}}\mathcal{C}_-^{-1}\right)^{\alpha\beta} \tag{D.163}$$

are real and antisymmetric.

Although these two representations are built in different ways, they are equivalent: they are related by a complete chiral–dual-type change of basis in spinor space:[13]

$$\begin{aligned} \hat{\psi}_+ &= S\hat{\psi}_-, \\ \Gamma_+\left(\hat{M}_{\hat{a}\hat{b}}\right) &= S\Gamma_-\left(\hat{M}_{\hat{a}\hat{b}}\right)S^{-1}, \\ \mathcal{C}_+ &= (S^{-1})^T\mathcal{C}_-S^{-1}, \\ S &= \frac{1}{\sqrt{2}}(1 + i\gamma_5). \end{aligned} \tag{D.164}$$

Observe that, given that $\mathcal{C}_\pm$ is always real and antisymmetric and squares to minus the identity, the condition (from now on we suppress the $+$ and $-$ subindices)

$$\mathcal{C}\Gamma\left(\hat{M}^{\hat{a}\hat{b}}\right)\mathcal{C}^{-1} = -\Gamma\left(\hat{M}^{\hat{a}\hat{b}}\right)^T \tag{D.165}$$

[13] By this we mean a change of basis, not just a rotation of the spinors as in Eqs. (5.90).

is equivalent to the statement that the $4 \times 4$ matrices $\Gamma\left(\hat{M}^{\hat{a}\hat{b}}\right)^{\alpha}{}_{\beta}$ are at the same time a spinorial representation of the algebra $\mathfrak{so}(2,3)$ and a fundamental representation of the algebra $\mathfrak{sp}(4,\mathbb{R})$. Using $\mathcal{C}_{\alpha\beta}$ as a metric to raise and lower indices,[14] we can construct real, symmetric representations of $\mathfrak{sp}(4,\mathbb{R})$ [619] $m^{\hat{a}\hat{b}\,\alpha\beta}$, where

$$m^{\hat{a}\hat{b}\,\alpha\beta} = \left[\Gamma\left(\hat{M}^{\hat{a}\hat{b}}\right)\mathcal{C}^{-1}\right]^{\alpha\beta}. \tag{D.166}$$

These objects satisfy the identity

$$m^{\hat{a}\hat{b}\,\alpha\beta} m_{\hat{c}\hat{d}\,\alpha\beta} = 2\delta^{[\hat{a}\hat{b}]}{}_{[\hat{c}\hat{d}]}, \tag{D.167}$$

which simply states that these matrices are an orthonormal basis in the ten-dimensional space of $4 \times 4$ real symmetric matrices with the trace of the standard product of matrices as scalar product and, therefore, for any symmetric matrix $O_{\alpha\beta}$,

$$O_{\alpha\beta} = \tfrac{1}{2} m^{\hat{a}\hat{b}\,\gamma\delta} m_{\hat{a}\hat{b}\,\alpha\beta} O_{\gamma\delta} \;\Rightarrow m^{\hat{a}\hat{b}\,\gamma\delta} m_{\hat{a}\hat{b}\,\alpha\beta} = 2\delta^{(\gamma\delta)}{}_{(\alpha\beta)}, \tag{D.168}$$

and, by definition of $m^{\hat{a}\hat{b}}$, we obtain the identity

$$m^{\hat{a}\hat{b}\,\alpha\beta} m_{\hat{a}\hat{b}}{}^{\gamma\delta} = (\mathcal{C}^{-1})^{\alpha\gamma}(\mathcal{C}^{-1})^{\beta\delta} + (\mathcal{C}^{-1})^{\alpha\delta}(\mathcal{C}^{-1})^{\beta\gamma}, \tag{D.169}$$

which is crucial for the consistency of the $\mathrm{osp}(4/N)$ superalgebra.

The matrices $m^{\hat{a}\hat{b}\,\alpha\beta}$ can also be used to convert objects in the adjoint of $\mathfrak{so}(2,3)$ into objects in the fundamental of $\mathfrak{sp}(4,\mathbb{R})$, which are somewhat easier to deal with.

Let us now consider the six real, antisymmetric matrices $n^{\hat{\hat{a}}\,\alpha\beta}$,

$$\begin{aligned} n^{\hat{a}\,\alpha\beta} &= \frac{1}{\sqrt{2}}\left(i\hat{\gamma}^{\hat{a}}\mathcal{C}^{-1}\right)^{\alpha\beta}, \\ n^{4\,\alpha\beta} &= \frac{1}{\sqrt{2}}(\mathcal{C}^{-1})^{\alpha\beta}, \end{aligned} \tag{D.170}$$

labeled by the index $\hat{\hat{a}} = (\hat{a}, 4)$ which we raise and lower with the $\mathrm{SO}(2,4)$ metric $\hat{\hat{\eta}}_{\hat{\hat{a}}\hat{\hat{b}}} = \mathrm{diag}(++----)$. It can be proved that these matrices are an orthonormal basis in the space of real $4 \times 4$ antisymmetric matrices with the trace as scalar product (raising and lowering indices with $\mathcal{C}$):

$$\begin{aligned} n^{\hat{\hat{a}}\,\alpha\beta} n_{\hat{\hat{b}}\,\beta\alpha} &= 2\delta^{\hat{\hat{a}}}{}_{\hat{\hat{b}}}, \\ n^{\hat{\hat{a}}\,\gamma\delta} n_{\hat{\hat{a}}\,\alpha\beta} &= -2\delta^{[\gamma\delta]}{}_{[\alpha\beta]}. \end{aligned} \tag{D.171}$$

[14] Upper-left indices are contracted with adjacent lower-right indices: $\xi_{\alpha} = \xi^{\beta}\mathcal{C}_{\beta\alpha} = -\mathcal{C}_{\alpha\beta}\xi^{\beta}$.

## D.3 The algebra of commuting spinor bilinears

The bilinears constructed from commuting spinors and the algebra they satisfy are the essential ingredients of the spinor bilinear method described in Section 18.2. Here we are going to review their computation in four and five dimensions for a generic $N$, and, as explained in the preceding sections, the spinor will carry an index $I=1,\dots,N$ or $i=1,\dots,2N$.

The first step consists in identifying the different bilinears that can be constructed with commuting spinors and studying their reality and symmetry (in $IJ$ or $ij$ indices) properties. After this is done, we just have to decompose the products of any two of them using the appropriate Fierz identities.

### *D.3.1 Four-dimensional case*

Before we start studying the bilinears that we can construct with $N$ four-dimensional Weyl spinors $\epsilon^I$, let us observe that the maximal number of independent chiral spinors at any given point is just two, which has important consequences for $N>2$.

The bilinears have the general form $\bar{\epsilon}^I\gamma^{a_1\cdots a_n}\epsilon^J$ and $\bar{\epsilon}^I\gamma^{a_1\cdots a_n}\epsilon_J$ with $n\leq 4$. The bilinears with indices in the opposite positions can be related to these by complex conjugation. The bilinears with $n=3,4$ can be related to those with, respectively, $n=1,0$ using Eq. (D.104). Finally, introducing a $\gamma_5$ between the spinors and commuting it with the gammas one finds that $\bar{\epsilon}^I\gamma^{a_1\cdots a_n}\epsilon^J$ vanishes for $n$ odd and $\bar{\epsilon}^I\gamma^{a_1\cdots a_n}\epsilon_J$ for $n$ even. Thus, we only need to study the reality and symmetry properties of $\bar{\epsilon}^I\epsilon^J$, $\bar{\epsilon}^I\gamma^a\epsilon_J$, and $\bar{\epsilon}^I\gamma^{ab}\epsilon^J$ separately. We are going to give names to these different tensor bilinears, introducing a convenient normalization.

1. The complex matrix of scalars $M_{IJ}$ is defined by

$$M_{IJ}\equiv\bar{\epsilon}_I\epsilon_J, \qquad M^{IJ}\equiv\bar{\epsilon}^I\epsilon^J=(M_{IJ})^*, \tag{D.172}$$

because $(\epsilon_I)^*=\epsilon^I$ and $(\bar{\epsilon}_I)^*=\bar{\epsilon}^I$. This matrix is antisymmetric: if we transpose the bilinear in the spinor indices, it should not change since all of them are contracted. We can then transpose the expression using the fact that the spinors are commuting variables and $\gamma_0=\gamma_0^{-1}=\gamma_0^\dagger=-\gamma_0^T$ is antisymmetric:

$$M_{IJ}=i[(\epsilon^I)^\dagger\gamma_0\epsilon_J]^T=i(\epsilon_J)^T\gamma_0^T(\epsilon^I)^*=i(\epsilon^J)^\dagger\gamma_0\epsilon_I=-M_{JI}. \tag{D.173}$$

2. The complex matrix of vectors $V^I{}_{Ja}$ and its complex conjugate $V_I{}^J{}_a$ are defined by

$$V^I{}_{Ja}\equiv i\bar{\epsilon}^I\gamma_a\epsilon_J, \qquad V_I{}^J{}_a\equiv i\bar{\epsilon}_I\gamma_a\epsilon^J=(V^I{}_{Ja})^*. \tag{D.174}$$

This is Hermitian in the R-symmetry indices, owing to the fact that $V_I{}^J{}_a=V^J{}_{Ia}$ for commuting spinors (which can be proven by transposing the expression in the spinorial indices, as we have done already):

$$(V^I{}_{Ja})^*=V_I{}^J{}_a=V^J{}_{Ia}=(V^I{}_{Ja})^T. \tag{D.175}$$

3. The complex matrix of 2-forms $\Phi_{IJ\,ab}$ and its complex conjugate $\Phi^{IJ}{}_{ab}$ defined by

$$\Phi_{IJ\,ab} \equiv \bar{\epsilon}_I \gamma_{ab} \epsilon_J, \qquad \Phi^{IJ}{}_{ab} \equiv \bar{\epsilon}^I \gamma_{ab} \epsilon^J = (\Phi_{IJ\,ab})^*. \tag{D.176}$$

It can be checked by transposition of the spinorial indices that it is symmetric in the $\mathrm{U}(N)$ indices $\Phi_{IJ\,ab} = \Phi_{JI\,ab}$, and, using Eq. (D.104), that it is self-dual:

$$\star\Phi_{IJ\,ab} = -i\Phi_{IJ\,ab} \quad \Rightarrow \quad \Phi_{IJ\,ab} = \Phi_{IJ}{}^{+}{}_{ab}. \tag{D.177}$$

Now we have to consider all the possible products between these bilinears, using the Fierz identity Eq. (D.105) (with the overall sign corresponding to commuting spinors). The calculations are straightforward and simplified by the vanishing of some of the bilinears (such as $\bar{\epsilon}^I \epsilon_J$, etc.)

There are two different products of scalar bilinears:

$$M_{IJ} M_{KL} = \tfrac{1}{2} M_{IL} M_{KJ} - \tfrac{1}{8} \Phi_{IL} \cdot \Phi_{KJ}, \tag{D.178}$$

$$M_{IJ} M^{KL} = -\tfrac{1}{2} V^L{}_I \cdot V^K{}_J, \tag{D.179}$$

where we use the notation $\Phi_{IL} \cdot \Phi_{KJ} \equiv \Phi_{IL}{}^{ab} \Phi_{KJ\,ab}$, etc. The symmetry of the 2-forms in Eq. (D.178) implies that

$$M_{I[J} M_{KL]} = 0, \tag{D.180}$$

which is a Plücker identity and implies that $\mathrm{rank}(M_{IJ}) \leq 2$. This identity is related to the above-mentioned fact that at most there are two independent spinors at a given point.

The $\mathrm{U}(N)$-dual of $M_{IJ}$ is defined by

$$\tilde{M}^{I_1 \cdots I_{N-2}} \equiv \tfrac{1}{2} \varepsilon^{I_1 \cdots I_{N-2} KL} M_{KL}, \qquad \varepsilon^{1 \cdots N} = \varepsilon_{1 \cdots N} = +1. \tag{D.181}$$

Let us now consider the products of the vector bilinears contracting the vector indices. From Eq. (D.179) and the antisymmetry of $M_{IJ}$ we get

$$V^I{}_L \cdot V^K{}_J = -V^I{}_J \cdot V^K{}_L = -V^K{}_L \cdot V^I{}_J, \tag{D.182}$$

from which it follows that all the vector bilinears $V^I{}_{J\,a}$ are null:

$$|V^I{}_J|^2 = V^I{}_J \cdot V^I{}_J = 0, \tag{D.183}$$

where there is no sum in the repeated $\mathrm{U}(N)$ indices.

The $\mathrm{U}(N)$-invariant combination of vectors $V_a \equiv V^I{}_{I\,a}$ plays a crucial role in the spinor bilinear method. It is real and can be shown to be non-spacelike:

$$V^2 = -V^I{}_J \cdot V^J{}_I = 2M^{IJ} M_{IJ} \equiv 2|M|^2 \geq 0, \tag{D.184}$$

where we have used Eqs. (D.182) and (D.179). When $\epsilon_I$ is the Killing spinor of a supersymmetric field configuration, the said configuration is called timelike (respectively null) supersymmetric when this vector bilinear $i\bar{\kappa}^I \gamma^a \kappa_I$ is timelike (respectively null). In the spinor bilinear method, both cases have to be analyzed separately.

Other relevant products of $M_{IJ}$ with the other bilinears are

$$M_{IJ}V^K{}_{La} = \tfrac{1}{2}M_{IL}V^K{}_{Ja} + \tfrac{1}{2}\Phi_{IL\,ba}V^K{}_J{}^b, \tag{D.185}$$

$$M_{IJ}\Phi^{KL}{}_{ab} = V^L{}_{I\,[a|}V^K{}_{J\,|b]} - \tfrac{i}{2}\epsilon_{ab}{}^{cd}V^L{}_{Ic}V^K{}_{Jd}. \tag{D.186}$$

Let us consider now the uncontracted product of two vectors with the same kind of indices:

$$V^I{}_{Ja}V^K{}_{Lb} = \tfrac{i}{2}\epsilon_{ab}{}^{cd}V^I{}_{Lc}V^K{}_{Jd} + V^I{}_{L\,(a|}V^K{}_{J\,|b)} - \tfrac{1}{2}g_{ab}V^I{}_L \cdot V^K{}_J. \tag{D.187}$$

Let us now focus on the timelike case $V^2 > 0$. Setting $I = J$ and $K = L$ in this identity we get an expression for the spacetime metric[15,16] in terms of the vectors (and, hence, in terms of the spinors):

$$g_{ab} = 2V^{-2}[V_aV_b - V^I{}_{Ja}V^J{}_{Ib}]. \tag{D.190}$$

This expression suggests that in the timelike case the vector bilinears can be used as a Vielbein basis, but, for $N > 2$, there are far too many and we must find the right combinations. To this end, we will use a generalization of the Pauli matrices.

We introduce a new Hermitian matrix $\mathcal{J}^I{}_J$ defined by Ref. [1184]

$$\mathcal{J}^I{}_J \equiv \frac{2M^{IK}M_{JK}}{|M|^2} = \frac{2V \cdot V^I{}_J}{V^2} \quad \Rightarrow \quad \mathcal{J}^I{}_I = +2. \tag{D.191}$$

$\mathcal{J}^I{}_J$ is idempotent, and therefore a projector, as can be shown using Eq. (D.178):

$$\mathcal{J}^I{}_J\mathcal{J}^J{}_K = \mathcal{J}^I{}_K. \tag{D.192}$$

$\mathcal{J}^I{}_J$ can be diagonalized by local U($N$) transformations[17] so that it has only two non-vanishing entries equal to $+1$ in the diagonal, which would constitute a sort of $N = 2$ subsector. However, the theories we are dealing with have no true local U($N$) symmetry (that is, with local parameters with arbitrary spacetime dependence). U($N$) transformations appear as compensating transformations of the global duality symmetry transformations and depend on the spacetime coordinates only through the scalar fields. Thus the general $N$ case cannot be reduced to an effective $N = 2$ case with more spinors unless an identity derived from the KSEs required this to be the case, but this does not seem to happen.

---

[15] Here $g_{ab} = \eta_{ab}$. We need Vielbeins to construct the full spacetime metric, and these must be determined by the KSEs or other conditions.

[16] A property that is used very often is that, given any 2-form $F = \frac{1}{2}F_{\mu\nu}dx^\mu \wedge dx^\nu$ and a non-null 1-form $\hat{V} = V_\mu dx^\mu$, $F$ can be written as

$$F = -V^{-2}[E \wedge \hat{V} - \star(B \wedge \hat{V})], \qquad E_\mu \equiv V^\nu F_{\nu\mu}, \qquad B_\mu \equiv \star V^\nu F_{\nu\mu}, \tag{D.188}$$

and for the complex combinations $F^\pm$ we have

$$F^\pm = -V^{-2}[C^\pm \wedge \hat{V} \pm i \star (C^\pm \wedge \hat{V})], \qquad C^\pm{}_\mu \equiv V^\nu F^\pm{}_{\nu\mu}. \tag{D.189}$$

There is a similar property in the null case. See footnote 23 on p. 889.

[17] This is unnecessary in the $N = 2$ case in which $\mathcal{J}^I{}_J = \delta^I{}_J$ always.

Using the Fierz identities we can show that

$$\mathcal{J}^I{}_J\epsilon^J = \epsilon^I, \qquad \epsilon_I\mathcal{J}^I{}_J = \epsilon_J, \tag{D.193}$$

so $\epsilon^I$ is an eigenvector with eigenvalue $+1$. For $N > 2$ these formulae say that the $N$ spinors labeled by $I$, $\epsilon^I$ are not linearly independent. They are independent only for $N = 2$ (and only in the timelike case that we are discussing now).

The identities Eqs. (D.193) imply that the contraction of $\mathcal{J}$ with any of the bilinears is the identity. This fact and Eq. (D.186) allow us to write the 2-forms entirely in terms of the scalar and vector bilinears:

$$\Phi^{KL}{}_{ab} = \frac{2M^{IK}M_{IJ}}{|M|^2}\Phi^{JL}{}_{ab} = \frac{2M^{IK}}{|M|^2}V^L{}_{I\,[a}V_{b]} - i\frac{M^{IK}}{|M|^2}\epsilon_{ab}{}^{cd}V^L{}_{I\,c}V_d. \tag{D.194}$$

Other useful identities involving the projector $\mathcal{J}^I{}_J$ are

$$\mathcal{J}^K{}_{[I}\mathcal{J}^L{}_{J]} = |M|^{-2}M_{IJ}M^{KL}, \tag{D.195}$$

$$\mathcal{J}^I{}_J = \delta^I{}_J - \tilde{\mathcal{J}}^I{}_J, \tag{D.196}$$

where the complementary projector $\tilde{\mathcal{J}}^I{}_J$ is defined by

$$\begin{aligned}\tilde{\mathcal{J}}^I{}_J &\equiv (N-2)|\tilde{M}|^{-2}\tilde{M}^{IK_1\cdots K_{N-3}}\tilde{M}_{JK_1\cdots K_{N-3}},\\ |\tilde{M}|^2 &\equiv \tilde{M}^{I_1\cdots I_{N-2}}\tilde{M}_{I_1\cdots I_{N-2}} = \tfrac{1}{2}(N-2)!|M|^2.\end{aligned} \tag{D.197}$$

Let us now consider the construction of a Vierbein basis $\{e^a\}$ using the matrix of vector bilinears in the timelike case. Since, by assumption, $V^a$ is a timelike vector, it can always be used to construct the zeroth component:

$$e^0 \equiv \tfrac{1}{\sqrt{2}}|M|^{-1}\hat{V}, \qquad \hat{V} \equiv V_\mu dx^\mu. \tag{D.198}$$

By construction, $e^0{}_\mu e^0{}_\nu g^{\mu\nu} = +1$. Let us choose[18] three generic auxiliary spacelike Vielbeins $e^m$, $m = 1, 2, 3$, and let us define with them the three 1-forms

$$\hat{V}^m = V^m{}_\mu dx^\mu \equiv |M|e^m, \tag{D.199}$$

and the three spacetime-dependent Hermitian matrices

$$(\sigma^m)^I{}_J \equiv -\sqrt{2}\,V^{m\,\mu}V^I{}_{J\,\mu}, \tag{D.200}$$

which are defined only up to local SO(3) rotations of the $e^m$. Then we can decompose the 1-forms $\hat{V}^I{}_J = V^I{}_{J\,\mu}dx^\mu$ in terms of the 1-forms defined in Eqs. (D.198) and (D.199)

$$\hat{V}^I{}_J = \tfrac{1}{2}\mathcal{J}^I{}_J\hat{V} + \tfrac{1}{\sqrt{2}}(\sigma^m)^I{}_J\hat{V}^m \quad\Rightarrow\quad V^I{}_{J\,a} = \tfrac{1}{\sqrt{2}}|M|\left[\delta_a{}^0\mathcal{J}^I{}_J + \delta_a{}^m(\sigma^m)^I{}_J\right]. \tag{D.201}$$

[18] The KSEs impose conditions on these Vielbeins.

The properties satisfied by the vector bilinears $V^I{}_{Ja}$ imply the following properties for the three $(\sigma^m)^I{}_J$ matrices:

$$\sigma^m\sigma^n = \delta^{mn}\mathcal{J} + i\varepsilon^{mnp}\sigma^p, \tag{D.202}$$

$$\mathcal{J}\sigma^m = \sigma^m\mathcal{J} = \sigma^m, \tag{D.203}$$

$$(\sigma^m)^I{}_I = 0, \tag{D.204}$$

$$\mathcal{J}^K{}_J\mathcal{J}^L{}_I = \tfrac{1}{2}\mathcal{J}^K{}_I\mathcal{J}^L{}_J + \tfrac{1}{2}(\sigma^m)^K{}_I(\sigma^m)^L{}_J, \tag{D.205}$$

$$M_{K[I}(\sigma^m)^K{}_{J]} = 0, \tag{D.206}$$

$$2|M|^{-2}M_{LI}(\sigma^m)^I{}_J M^{JK} = (\sigma^m)^K{}_L, \tag{D.207}$$

$$|M|^{-2}M^{IJ}M_{KL} = -\tfrac{1}{3}(\sigma^m)^{[I}{}_{[K}(\sigma^m)^{J]}{}_{L]}, \tag{D.208}$$

$$(\sigma^{[m|})^I{}_J(\sigma^{|n]})^K{}_L = -\tfrac{i}{2}\varepsilon^{mnp}[\mathcal{J}^I{}_L(\sigma^p)^K{}_J - (\sigma^p)^I{}_L\mathcal{J}^K{}_J]. \tag{D.209}$$

In order to interpret these results, let us consider the $N=2$ case. If $N=2$ then we can write

$$M_{IJ} = X\varepsilon_{IJ} \quad \Rightarrow \quad V^2 = 4|X|^2 \geq 0 \tag{D.210}$$

for some complex scalar $X$, from which it follows that

$$\mathcal{J}^I{}_J = \delta^I{}_J, \tag{D.211}$$

and the above properties are the properties satisfied by the Pauli matrices Eqs. (D.10)–(D.16). We conclude that the four spacetime-dependent Hermitian matrices $\{(\sigma^a)^I{}_J\}$, $a = 0, m$, with $\sigma^0 \equiv \mathcal{J}$ are, at each spacetime point, the four generators of a $\mathfrak{u}(2)$ subalgebra of $\mathfrak{u}(N)$ in the eigenspace of $\mathcal{J}$ of eigenvalue $+1$.[19] The $\{(\sigma^a)^I{}_J\}$ also provide a basis for all the Hermitian matrices $A$ in this subspace (i.e. those satisfying $\mathcal{J}A\mathcal{J} = A$) because Eq. (D.209) is a completeness relation in that subspace. Finally, the properties in Eqs. (D.202)–(D.209) can be used to express the Vierbein in terms of the vector bilinears:

$$e^a{}_\mu = \frac{1}{\sqrt{2}|M|}V^I{}_{J\mu}(\sigma^a)^J{}_I. \tag{D.212}$$

In the $N=2$ case we can take the $\sigma^m$ to be the standard, constant, Pauli matrices, and then this equation gives the corresponding auxiliary spacelike Vierbeins.

[19] The comments made about the diagonalization of $\mathcal{J}$ also apply to the $\sigma^m$. Even though these also transform under local $\mathrm{SO}(3)$ transformations (the local changes of basis of the auxiliary spacelike Vierbeins), it is unclear if these are enough to render them constant while Eq. (D.202) still holds with a non-constant $\mathcal{J}$.

Let us now focus on the the null case $V^2 = 0$. It is customary to write $l^a \equiv V^a$ in this case. Equation (D.184) implies that $M_{IJ} = 0$ for all $I, J$, which means that all the spinors $\epsilon_I$ are parallel and one can write

$$\epsilon_I = \phi_I \epsilon, \tag{D.213}$$

for some complex functions $\phi_I$ which transform as an $\mathrm{U}(N)$ vector,[20] and some negative-chirality spinor $\epsilon$. These are defined up to a complex factor which can be chosen so as to have the following normalization:

$$\phi_I \phi^I = 1, \quad \text{where} \quad \phi^I \equiv (\phi_I)^*. \tag{D.214}$$

The only remaining freedom in the definition of $\phi^I$ and $\epsilon$ is a change by a local phase $\theta(x)$:

$$\phi_I \to e^{i\theta}\phi_I, \qquad \epsilon \to e^{-i\theta}\epsilon. \tag{D.215}$$

In the null case one can construct another Hermitian projector $\mathcal{K}^I{}_J$ that plays a role analogous to that of $\mathcal{J}^I{}_J$ in the non-null case:

$$\mathcal{K}^I{}_J \equiv \phi^I \phi_J. \tag{D.216}$$

It satisfies the properties

$$\mathcal{K}^I{}_J \mathcal{K}^J{}_K = \mathcal{K}^I{}_K, \tag{D.217}$$

$$\mathcal{K}^I{}_I = +1, \tag{D.218}$$

$$\mathcal{K}^I{}_J \epsilon^J = \epsilon^I, \tag{D.219}$$

$$\epsilon_I \mathcal{K}^I{}_J = \epsilon_J, \tag{D.220}$$

$$V^I{}_{J\,a} = \mathcal{K}^I{}_J l_a, \tag{D.221}$$

which express the known fact that only one spinor is linearly independent in this case and the vector bilinears are all proportional, so we cannot use them to construct a Vielbein basis (tetrad).

In order to construct a tetrad we need to introduce an auxiliary spinor with the same chirality and opposite $\mathrm{U}(1)$ Kähler weight as $\epsilon$ (i.e. $-1/2$) and normalized against $\epsilon$ by the Kähler-invariant condition

$$i\epsilon^T \gamma_0 \eta = -i\eta^T \gamma_0 \epsilon = \tfrac{1}{2}. \tag{D.222}$$

With these two spinors we can construct a complex null tetrad $\{l, n, m, m^*\}$ with metric[21]

$$(\eta_{ab}) = \begin{pmatrix} 0 & 1 & 0 & 0 \\ 1 & 0 & 0 & 0 \\ 0 & 0 & 0 & -1 \\ 0 & 0 & -1 & 0 \end{pmatrix}, \quad \text{or} \quad ds^2 = 2\hat{l} \otimes \hat{n} - 2\hat{m} \otimes \hat{m}^*, \tag{D.223}$$

[20] In the $N = 1$ case, $\phi = 1$.

[21] Observe that in this basis $\epsilon^{lnmm^*} = i$.

as follows:

$$l_\mu = i\sqrt{2}\bar{\epsilon}\gamma_\mu\epsilon, \qquad n_\mu = i\sqrt{2}\bar{\eta}\gamma_\mu\eta,$$
$$m_\mu = i\sqrt{2}\bar{\epsilon}\gamma_\mu\eta, \qquad m^*_\mu = i\sqrt{2}\bar{\epsilon}^*\gamma_\mu\eta. \tag{D.224}$$

Note that $l$ and $n$ have no $\mathrm{U}(1)$ charges but $m$ ($m^*$) has charge $-1$ ($+1$).

With $\epsilon$ and $\eta$ we can also construct three independent self-dual 2-forms which can also be expressed in terms of the 1-forms in Eqs. (D.224):[22,23]

$$\begin{aligned} \Phi^{(1)}{}_{\mu\nu} &= \bar{\epsilon}^*\gamma_{\mu\nu}\epsilon = \hat{l}\wedge\hat{m}^*, \\ \Phi^{(2)}{}_{\mu\nu} &= \bar{\eta}^*\gamma_{\mu\nu}\epsilon = \tfrac{1}{2}[\hat{l}\wedge\hat{n} + \hat{m}\wedge\hat{m}^*], \\ \Phi^{(3)}{}_{\mu\nu} &= \bar{\eta}^*\gamma_{\mu\nu}\eta = -\hat{n}\wedge\hat{m}. \end{aligned} \tag{D.227}$$

Finally, let us remark that the normalization condition (D.214) leaves two freedoms in the choice of $\eta$ that become freedoms of the null tetrad: the already discussed $\mathrm{U}(1)$ freedom Eq. (D.215) and the possibility of shifting $\eta$ by terms proportional to $\epsilon$:

$$\eta' = \eta + \delta\epsilon, \tag{D.228}$$

under which the null tetrad transforms as follows:

$$l' = l, \qquad n' = n + \delta^* m + \delta m^* + |\delta|^2 l, \qquad m' = m + \delta l. \tag{D.229}$$

### *D.3.2 Five-dimensional case*

We are going to study the spinor bilinears that can be constructed with $N$ pairs of symplectic-Majorana spinors $\epsilon^i$, $i = 1, \ldots, 2N$, using the conventions described in Section D.1.10. In the end, we will focus on the $N = 1$ case, which is the only one whose algebra has been given in the literature [596, 129].

First, observe that the position of the symplectic spinor indices, which are simply raised or lowered with symplectic metric $\Omega_{ij}$, is irrelevant in the construction of the bilinears. None of the bilinears that one can construct with $n$ gamma matrices vanishes identically because the spinors are not chiral. Those with $n \geq 3$ can be related to the ones with $n = 0, 1, 2$ using the identity Eq. (D.111). Therefore, we will only consider the following three.

---

22 The identification of these 2-forms in terms of the vectors is found by studying the contractions between the 2-forms and vectors using the Fierz identities.

23 A useful property is that any self-dual 2-form $F^+$ can be written as a linear combination of these, with complex coefficients $c_i$

$$F^+ = c_i\hat{\Phi}^{(i)}, \tag{D.225}$$

which can be found by contracting $F^+$ with the null tetrad basis:

$$l^\nu F^+{}_{\nu\mu} = -\tfrac{1}{2}c_2 l_\mu - c_3 m_\mu, \qquad n^\nu F^+{}_{\nu\mu} = c_1 m^*_\mu + \tfrac{1}{2}c_2 n_\mu,$$
$$m^\nu F^+{}_{\nu\mu} = c_1 l_\mu + \tfrac{1}{2}c_2 m_\mu, \qquad m^{*\nu} F^+{}_{\nu\mu} = -\tfrac{1}{2}c_2 m^*_\mu - c_3 n_\mu. \tag{D.226}$$

This is analogous to the property in footnote 16 on p. 885.

1. The complex matrix $M_{ij}$ defined by

$$M_{ij} \equiv i\bar{\epsilon}_i \epsilon_j. \tag{D.230}$$

($M^i{}_j = \Omega^{ki} M_{kj}$ etc.) It is easy to show that this matrix satisfies the following two properties:

$$M_{ij} = -M_{ji}, \qquad (M_i{}^j)^\dagger \equiv (M_j{}^i)^* = M_i{}^j. \tag{D.231}$$

2. The complex matrix of vectors $V_{ij}{}^a$ defined by

$$V_{ij}{}^a \equiv i\bar{\epsilon}_i \gamma^a \epsilon_j, \tag{D.232}$$

which satisfies the same properties as the matrix of scalars $M_{ij}$, namely it is antisymmetric with both indices down and Hermitian with one down and one up:

$$V_{ij}{}^a = -V_{ji}{}^a, \qquad (V_i{}^{j\,a})^\dagger \equiv (V_j{}^{i\,a})^* = V_i{}^{j\,a}. \tag{D.233}$$

3. The complex matrix of 2-forms $\Phi_i{}^{j\,a}$ defined by

$$\Phi_{ij}{}^a \equiv \bar{\epsilon}_i \gamma^{ab} \epsilon_j, \tag{D.234}$$

which satisfies the two properties

$$\Phi_{ij}{}^{ab} = +\Phi_{ji}{}^{ab}, \qquad (\Phi_i{}^{j\,ab})^\dagger \equiv (\Phi_j{}^{i\,ab})^* = \Phi_i{}^{j\,a}. \tag{D.235}$$

For the minimal case $N = 1$, the antisymmetric matrices $M_{ij}$ and $V_{ij}{}^a$ must be proportional to $\Omega_{ij} = \varepsilon_{ij}$, and, therefore, $M_i{}^j$ and $V_i{}^{j\,a}$ are proportional to $\delta_i{}^j$ and, being Hermitian, the proportionality constants must be real. Therefore, one can define the following unique R-invariant (SU(2)-invariant) bilinears $f$ and $V^a$ by

$$f \equiv M_i{}^i, \qquad V^a \equiv V_i{}^{i\,a}. \tag{D.236}$$

For $N = 1$ the matrix $\Phi_i{}^{j\,ab}$ is Hermitian and traceless, and, therefore, it can be expanded in terms of the Pauli matrices[24] with coefficients $\Phi^{r\,ab}$ that transform in the adjoint of SU(2) as follows:

$$\Phi_i{}^{j\,ab} = \tfrac{1}{2}\Phi^{r\,ab}\sigma^r{}_i{}^j, \qquad \Phi^{r\,ab} = \Phi_i{}^{j\,ab}\sigma^r{}_i{}^j. \tag{D.237}$$

For $N > 1$ there are another two R-symmetry invariants, which cannot be expressed in terms of products of $f$ and $V^a$:

$$M^2 \equiv M_{ij}M^{ij} = M_i{}^j M_j{}^i, \qquad W^a \equiv M^{ij}V_{ij}{}^a = M_i{}^j V_j{}^{i\,a}. \tag{D.238}$$

[24] Our conventions for the positions of the indices of the Pauli matrices in Chapter 9 are different from those in the rest of the book. They are explained in Appendix G.1.1.

For general values of $N$, the products of these bilinears take a complicated form:

$$M_{ij}M_{kl} = \tfrac{1}{4}M_{il}M_{kj} + \tfrac{1}{4}V_{il}\cdot V_{kj} + \tfrac{1}{8}\Phi_{il}\cdot\Phi_{kj}, \tag{D.239}$$

$$\begin{aligned} V_{ij}{}^{a}V_{kl}{}^{b} = {} & +\tfrac{1}{2}V_{il}{}^{(a}V_{kj}{}^{b)} + \tfrac{1}{2}\Phi_{il}{}^{(a|c}\Phi_{kj\,c}{}^{|b)} \\ & + \tfrac{1}{4}\eta^{ab}\left(M_{il}M_{kj} - V_{il}\cdot V_{kj} + \tfrac{1}{2}\Phi_{il}\cdot\Phi_{kj}\right) \\ & + \tfrac{i}{4}\left(\Phi_{il}{}^{ab}M_{kj} + M_{il}\Phi_{kj}{}^{ab}\right) \\ & + \tfrac{i}{8}\varepsilon^{ab}{}_{cde}\left(\Phi_{il}{}^{cd}V_{kj}{}^{e} - V_{il}{}^{e}\Phi_{kj}{}^{cd}\right), \end{aligned} \tag{D.240}$$

$$\begin{aligned} M_{ij}V_{kl}{}^{a} = {} & \tfrac{1}{4}\left(V_{il}{}^{a}M_{kj} + M_{il}V_{kj}{}^{a}\right) - \tfrac{1}{16}\varepsilon^{a}{}_{bcde}\Phi_{il}{}^{bc}\Phi_{kj}{}^{de} \\ & + \tfrac{i}{4}\left(V_{il}{}^{b}\Phi_{kj\,b}{}^{a} - V_{kj}{}^{b}\Phi_{il\,b}{}^{a}\right), \end{aligned} \tag{D.241}$$

etc.

As mentioned before, the analysis of these identities has not yet been performed for $N > 1$, and, therefore, we will restrict ourselves to the $N = 1$ case. It is convenient to use a form of the Fierz identities Eq. (D.67) in which the spinor and $\mathrm{SU}(2)$ indices are contracted with those of the matrices $M$ and $N$ in the obvious way (e.g. $\bar{\lambda}M\varphi = \bar{\lambda}_i M_j{}^i\varphi^j$):

$$\begin{aligned} (\bar{\lambda}M\varphi)(\bar{\psi}N\chi) = {} & -\tfrac{1}{8}\left(\bar{\lambda}M\sigma^{\hat{r}}N\chi\right)\left(\bar{\psi}\sigma^{\hat{r}}\varphi\right) - \tfrac{1}{8}\left(\bar{\lambda}M\gamma^{a}\sigma^{\hat{r}}N\chi\right)\left(\bar{\psi}\gamma_{a}\sigma^{\hat{r}}\varphi\right) \\ & + \tfrac{1}{16}\left(\bar{\lambda}M\gamma^{ab}\sigma^{\hat{r}}N\chi\right)\left(\bar{\psi}\gamma_{ab}\sigma^{\hat{r}}\varphi\right), \end{aligned} \tag{D.242}$$

where $\hat{r} = 0, 1, 2, 3$ and $\sigma^0 = \mathbb{1}$, and the global sign must be reversed for commuting spinors. We obtain

$$V^2 = f^2, \tag{D.243}$$

$$V_aV_b = \eta_{ab}f^2 + \tfrac{1}{3}\Phi^{r}{}_{a}{}^{c}\Phi^{r}{}_{cb}, \tag{D.244}$$

$$V^a\Phi^{r}{}_{ab} = 0, \tag{D.245}$$

$$V^a(^\star\Phi^r)_{abc} = -f\Phi^{r}{}_{bc}, \tag{D.246}$$

$$\Phi^{r}{}_{a}{}^{c}\Phi^{s}{}_{cb} = -\delta^{rs}(\eta_{ab}f^2 - V_aV_b) - \varepsilon^{rst}f\Phi^{t}{}_{ab}, \tag{D.247}$$

$$\Phi^{r}{}_{[ab}\Phi^{s}{}_{cd]} = -\tfrac{1}{4}f\delta^{rs}\varepsilon_{abcde}V^e, \tag{D.248}$$

and

$$V_a\gamma^a\epsilon^i = f\epsilon^i, \tag{D.249}$$

$$\Phi^r{}_{ab}\gamma^{ab}\epsilon^i = 4if\epsilon^j\sigma^r{}_j{}^i. \tag{D.250}$$

These latter two identities only involve three spinors, but they can be obtained from the general case by removing the leftmost spinor $\bar{\lambda}$ (or replacing it by a convenient set of Kronecker deltas).

# Appendix E
# Kähler geometry

In this appendix we are going to give a brief review of Kähler manifolds oriented to their use in $N = 1$ and $N = 2$, $d = 4$ supergravities and their gauged versions. For the sake of completeness we start by reviewing the fundamentals of complex differential geometry.

## E.1 Complex manifolds

A *complex manifold*[1] of (complex) dimension $d$ is a topological space that looks locally like $\mathbb{C}^d$, and, therefore, like $\mathbb{R}^{2d}$. There is a homeomorphism between each of the open subsets (*patches)* that covers the manifold to an open subset of $\mathbb{C}^d$ which provides a set of complex and simultaneously real coordinates $z^i = x^i + iy^i$. The index $i$ (not to be confused with the imaginary unit $i^2 = -1$) takes values $i = 1, \ldots, d$. We will denote collectively the $2n$ real coordinates $(x^i, y^i)$ by $x^m$, $m = 1, \ldots, 2d$. In the overlaps between patches the coordinates are not related by simple diffeomorphisms but by diffeomorphisms which are given by *holomorphic functions*, i.e. functions $F^i(z) = f^i + ig^i$ that satisfy the *Cauchy–Riemann* equations

$$\frac{\partial f^i}{\partial x^j} = \frac{\partial g^i}{\partial y^j}, \qquad \frac{\partial f^i}{\partial y^j} = -\frac{\partial g^i}{\partial x^j}, \tag{E.1}$$

defining a *complex* (or *holomorphic*) *structure*. Since this is a more restrictive structure, every $d$-dimensional complex manifold is a $2d$ differential manifold. The converse is not true and we will be interested in finding out under which conditions this happens.

The Cauchy–Riemann equations can be rewritten in real matrix notation as

$$J_n{}^m \frac{\partial f^n}{\partial x^p} = \frac{\partial f^m}{\partial x^n} J_p{}^n, \tag{E.2}$$

where $\frac{\partial f^n}{\partial x^p}$ is the Jacobian of the coordinate transformation $((f^m) \equiv (f^i, g^i))$ and

$$(J_m{}^n) = \begin{pmatrix} 0 & \mathbb{1}_{d\times d} \\ -\mathbb{1}_{d\times d} & 0 \end{pmatrix} \tag{E.3}$$

[1] Standard references in the strings/supergravity context are Refs. [291, 972, 1211].

are the components of a tensor

$$J \equiv J_m{}^n dx^m \otimes \frac{\partial}{\partial x^n} \tag{E.4}$$

(it is trivial to check that the components are the same in any coordinate system if we only allow for holomorphic changes of coordinates) which is globally defined over the complex manifold and that can also be seen as a linear map in the standard real tangent space $\xi'^m = \xi^n J_n{}^m$.

The tangent space of a $d$-dimensional complex manifold is just its tangent space as a $2d$-dimensional (real) differential manifold, but the presence of a complex structure gives us more possibilities. First, we can *complexify* it, allowing for combinations $\xi + i\eta$, where $\xi$ and $\eta$ are real vectors. The natural coordinate basis for the complexified tangent space is

$$\partial_i \equiv \frac{\partial}{\partial z^i} = \frac{1}{2}\left(\frac{\partial}{\partial x^i} - i\frac{\partial}{\partial y^i}\right), \qquad \partial_{i^*} \equiv \frac{\partial}{\partial z^{*\,i^*}} = \frac{1}{2}\left(\frac{\partial}{\partial x^i} + i\frac{\partial}{\partial y^i}\right), \tag{E.5}$$

and the dual basis of 1-forms is

$$dz^i \equiv dx^i + idy^i, \qquad dz^{*\,i^*} \equiv dx^i - idy^i, \tag{E.6}$$

and it satisfies

$$\langle dz^i|\partial_j\rangle = \delta^i{}_j, \quad \langle dz^{*\,i^*}|\partial_{j^*}\rangle = \delta^{i^*}{}_{j^*}, \quad \langle dz^i|\partial_{j^*}\rangle = \langle dz^{i^*}|\partial_j\rangle = 0. \tag{E.7}$$

All tensors can be expressed as complex linear combinations of the tensor products of the elements of these bases. In particular, the tensor $J$ is given by

$$J = idz^i \otimes \frac{\partial}{\partial z^i} - idz^{*\,i^*} \otimes \frac{\partial}{\partial z^{*\,i^*}}, \tag{E.8}$$

from which we can read off the non-vanishing components in the complex basis:

$$J_i{}^j = i\delta_i{}^j, \quad J_{i^*}{}^{j^*} = -i\delta_{i^*}{}^{j^*}, \quad J = \begin{pmatrix} i\mathbb{1}_{d\times d} & 0 \\ 0 & -i\mathbb{1}_{d\times d} \end{pmatrix}. \tag{E.9}$$

In both bases it is easy to see that $J$ has the characteristic property

$$J^2 = -1. \tag{E.10}$$

This globally defined tensor $J$ contains all the information about the complex structure and is usually given that name.

The complex structure $J$ can be used to construct two projectors,

$$P_{\mp m}{}^n \equiv \tfrac{1}{2}(\delta_m{}^n \mp J_m{}^n), \quad P_{\mp}^2 = P_{\mp}, \quad P_{\mp}P_{mp} = 0, \quad P_{\mp} + P_{\mp} = \mathbb{1}_{2d\times 2d}, \tag{E.11}$$

that project out, respectively, the holomorphic and antiholomorphic components of tensors. Given, for instance, a 1-form $\omega = \omega_m dx^m = \rho_i dx^i + \sigma_i dy^i$, the 1-form constructed with the components of $\omega$ projected with $P_{\mp}$ are

$$\begin{aligned} P_{-m}{}^n\omega_n dx^m &= \tfrac{1}{2}(\rho_i - i\sigma_i)dz^i = \omega_i dz^i, \\ P_{+m}{}^n\omega_n dx^m &= \tfrac{1}{2}(\rho_i + i\sigma_i)dz^{*\,i^*} = \omega_{i^*} dz^{*\,i^*}. \end{aligned} \tag{E.12}$$

A Riemannian metric $g_{mn}$ defined in a complex manifold with the property

$$J_m{}^p J_n{}^q g_{pq} = g_{mn} \tag{E.13}$$

is a *Hermitian metric* with respect to the complex structure $J$, and the complex manifold is a *Hermitian manifold*. This property implies that

$$g_{mn} = (P_{-\,m}{}^p P_{+\,n}{}^q + P_{+\,m}{}^p P_{-\,n}{}^q)\, g_{pq}, \tag{E.14}$$

which implies that, out of all the four possible kinds of components that a metric has in the complex basis $g_{ij}, g_{ij^*}, g_{i^*j}, g_{i^*j^*}$, a Hermitian metric has only two non-vanishing types of components: $g_{ij^*}$ and $g_{i^*j}$. These two components are related by

$$(g_{ij^*})^* = g_{i^*j} = g_{ji^*}, \tag{E.15}$$

and, therefore, a Hermitian metric can always be put in the form

$$ds^2 = g_{mn} dx^m dx^n = 2 g_{ij^*} dz^i dz^{*\,j^*}. \tag{E.16}$$

All Riemannian complex manifolds admit a Hermitian metric: if $g_{mn}$ is the Riemannian metric, then

$$h_{mn} = \tfrac{1}{2}(g_{mn} + J_m{}^p J_n{}^q g_{pq}) \tag{E.17}$$

is automatically Hermitian.

All the components of a differential $k$-form $\omega$ in a complex manifold can be projected with $P_{\mp}$ into holomorphic and antiholomorphic components and the $k$-form itself can be split into a sum of $k$-forms with $p$ holomorphic and $q$ antiholomorphic indices $\omega^{(p,q)}$, with $p + q = k$, called $(p, q)$-forms:

$$\omega = \sum_{p+q=k} \omega^{(p,q)}, \qquad \omega^{(p,q)} = \frac{1}{p!q!}\omega_{i_1\cdots i_p i_1^*\cdots i_q^*} dz^{i_1} \wedge \cdots \wedge dz^{i_p} \wedge dz^{*\,i_1^*} \wedge \cdots \wedge dz^{*\,i_q^*}. \tag{E.18}$$

Observe that the property (E.13) can be rewritten as $J_{mn} = -J_{nm}$ and is equivalent to the existence of a globally defined $(1,1)$-form

$$\mathcal{J} \equiv \tfrac{1}{2} J_{mn} dx^m \wedge dx^n = J_{ij^*} dz^i \wedge dz^{*\,j^*} = 2i g_{ij^*} dz^i \wedge dz^{*\,j^*}, \tag{E.19}$$

known as the *fundamental 2-form* and, sometimes, as the *Kähler 2-form*, although this name is more appropriate in the context of Kähler manifolds. The $d$th exterior power of the fundamental 2-form in a complex manifold of complex dimension $d$ is proportional to the manifold's volume form:

$$\begin{aligned}
\mathcal{J} \wedge \cdots \wedge \mathcal{J} &= (-1)^{[\frac{d}{2}]} J_{i_1 j_1^*} \cdots J_{i_d j_d^*}\, dz^{i_1} \wedge \cdots \wedge dz^{i_d} \wedge dz^{*\,i_1^*} \wedge \cdots \wedge dz^{*\,i_d^*} \\
&= (2i)^d (-1)^{[\frac{d}{2}]} g_{i_1 j_1^*} \cdots g_{i_d j_d^*}\, dz^{i_1} \wedge \cdots \wedge dz^{i_d} \wedge dz^{*\,i_1^*} \wedge \cdots \wedge dz^{*\,i_d^*} \\
&= (2i)^d (-1)^{[\frac{d}{2}]} \epsilon^{i_1\cdots i_d} \epsilon^{i_1^*\cdots i_d^*} g_{i_1 j_1^*} \cdots g_{i_d j_d^*}\, d^d z d^d z^* \\
&= (2i)^d (-1)^{[\frac{d}{2}]} d! \det(g_{ij^*})\, d^d z d^d z^*, \\
&= (2i)^d (-1)^{[\frac{d}{2}]+1} d! \sqrt{|g|}\, d^d z d^d z^*,
\end{aligned} \tag{E.20}$$

where we have taken into account that $g_{ij^*}$ is only one of the two off-diagonal blocks of the full Hermitian metric whose determinant we denote by $|g|$, as before.

The exterior derivative acting on a complex $(p,q)$-form $\omega^{(p,q)}$ gives just two terms,

$$d\omega = \omega^{(p+1,q)} + \omega^{(p,q+1)}, \tag{E.21}$$

where

$$\begin{aligned}
\omega^{(p+1,q)} &= \frac{1}{p!q!}\partial_i \omega^{(p,q)}_{i_1\cdots i_p i_1^*\cdots i_q^*} dz^i \wedge dz^{i_1} \wedge \cdots \wedge dz^{i_p} \wedge dz^{*\,i_1^*} \wedge \cdots \wedge dz^{*\,i_q^*}, \\
\omega^{(p,q+1)} &= \frac{(-1)^p}{p!q!}\partial_{i^*} \omega^{(p,q)}_{i_1\cdots i_p i_1^*\cdots i_q^*} dz^{i_1} \wedge \cdots \wedge dz^{i_p} \wedge dz^{*\,i^*} \wedge dz^{*\,i_1^*} \wedge \cdots \wedge dz^{*\,i_q^*},
\end{aligned} \tag{E.22}$$

instead of what one would get if the complex structure components $J_m{}^n$ were not constant and did not commute with the partial derivatives:

$$d\omega = \omega^{(p+2,q-1)} + \omega^{(p+1,q)} + \omega^{(p,q+1)} + \omega^{(p-1,q+2)}. \tag{E.23}$$

Therefore, in a complex manifold the exterior derivative can be split into the two *Dolbeault operators* $\partial, \partial^*$,

$$d = \partial + \partial^*, \tag{E.24}$$

which are such that

$$\partial\omega^{(p,q)} = \omega^{(p+1,q)}, \qquad \partial^*\omega^{(p,q)} = \omega^{(p,q+1)}, \tag{E.25}$$

where $\omega^{(p+1,q)}$ and $\omega^{(p,q+1)}$ are given by Eqs. (E.22). The main property of these exterior derivatives is

$$\partial^2 = \partial\partial^* + \partial^*\partial = \partial^{*\,2} = 0, \tag{E.26}$$

which allows the definition of the $(p,q)-$*Dolbeault cohomology classes*.

### *E.1.1 Hermitian connections*

On complex manifolds, due to the split of the tangent space into a holomorphic and an antiholomorphic subspace with basis $\{\partial_i\}$ and $\partial_{i^*}$, respectively, we can introduce a parallel transport rule[2] that preserves the holomorphicity or antiholomorphicity of the vectors. This is equivalent to the requirement that the complex structure be covariantly constant, i.e.

$$\nabla_m J_n{}^p = 0. \tag{E.27}$$

This condition implies the covariant constancy of the projectors $P_\mp$, which leads to the vanishing of several mixed components of the connection $\Gamma_{mn}{}^p$. The rest of the mixed components are set to zero by requiring that the torsion tensor is pure in its lower indices:

$$\Gamma_{[ij^*]}{}^m = 0, \tag{E.28}$$

[2] The conventions for the affine connection, curvature, and torsion are the same as in the real case, and the torsion and curvature satisfy the same Ricci identities Eqs. (1.28).

and this leaves us with a connection whose only non-vanishing components are ${\Gamma_{ij}}^k$ and ${\Gamma_{i^*j^*}}^{k^*}$.

A third condition, namely the covariant constancy (or preservation) of a Hermitian metric $g_{ij^*}$,

$$\nabla_i g_{jk^*} = \partial_i g_{jk^*} - {\Gamma_{ij}}^l g_{lk^*} = 0, \qquad \nabla_{i^*} g_{jk^*} = \partial_{i^*} g_{jk^*} - {\Gamma_{i^*k^*}}^{l^*} g_{jl^*} = 0, \tag{E.29}$$

determines uniquely a *Hermitian connection*, whose only non-vanishing components are given by

$${\Gamma_{ij}}^k = g^{kl^*}\partial_i g_{jl^*}, \qquad {\Gamma_{i^*j^*}}^{k^*} = g^{lk^*}\partial_{i^*} g_{lj^*} = ({\Gamma_{ij}}^k)^*. \tag{E.30}$$

Observe that, in general, this connection is not just the Levi-Civita connection expressed in the complex coordinate basis, as in particular it has a non-vanishing torsion which is pure in all its indices:

$${T_{ij}}^k = -2{\Gamma_{[ij]}}^k = -2g^{kl^*}\partial_{[i}g_{j]\,l^*}, \qquad {T_{i^*j^*}}^{k^*} = ({T_{ij}}^k)^*. \tag{E.31}$$

The only non-vanishing components of the curvature tensor, given by the general expression Eq. (1.27), are

$${R_{ij^*k^*}}^{l^*} = \partial_i {\Gamma_{j^*k^*}}^{l^*}, \qquad {R_{i^*jk}}^l = ({R_{ij^*k^*}}^{l^*})^*, \tag{E.32}$$

plus those obtained by the symmetry relations

$${R_{j^*ik^*}}^{l^*} = -{R_{ij^*k^*}}^{l^*}, \qquad {R_{ji^*k}}^l = -{R_{i^*jk}}^l. \tag{E.33}$$

The *Ricci 2-form* is defined by

$$\mathfrak{R} \equiv i{R_{ij^*k^*}}^{k^*} dz^i \wedge dz^{*j^*} = i\partial_i\partial_{j^*} \ln\sqrt{|g|} dz^i \wedge dz^{*j^*} = i\partial\partial^* \ln\sqrt{|g|}. \tag{E.34}$$

It can be shown that the Ricci 2-form is globally well defined even though $|g|$ is not. On the other hand, since $\partial\partial^* = -\frac{1}{2}d(\partial - \partial^*)$, the Ricci 2-form is always closed,

$$d\mathfrak{R} = 0, \tag{E.35}$$

and therefore locally exact, but generically not globally exact. The Ricci 2-form defines the first Chern cohomology class (of the Levi-Civita connection of the tangent bundle), which is invariant under smooth changes of complex structure.

### *E.1.2 Holomorphic isometries of complex manifolds*

Given a Hermitian manifold it is natural to consider the holomorphic coordinate transformations $z'^i = f^i(z)$ that leave invariant the Hermitian metric. These transformations are generated by holomorphic vectors $k^i(z)$,

$$\delta_\epsilon z^i = \epsilon k^i(z), \qquad \partial_{j^*} k^i(z) = 0, \tag{E.36}$$

which are the holomorphic components of the vector $K(z, z^*)$ that generates the transformations of the real coordinates

$$K = K^m\partial_m = k^i(z)\partial_i + {k^*}^{i^*}(z^*)\partial_{i^*}; \tag{E.37}$$

$\epsilon$ is an infinitesimal constant parameter.

Of course, any real vector $V$ can be decomposed into components $V^i, V^{i^*}$, but these components need not be holomorphic and antiholomorphic in the coordinates $z^i$. However, only those whose components are holomorphic and antiholomorphic generate coordinate transformations that preserve the complex structure.

The preservation of the metric by the infinitesimal holomorphic reparametrization Eq. (E.36) is expressed by the Killing equation Eq. (1.115). Taking (anti)holomorphicity into account, the only non-trivial components of the Killing equation are just

$$\tfrac{1}{2}\pounds_k g_{ij^*} = \nabla_{i^*} k^*_j + \nabla_j k_{i^*} = 0. \tag{E.38}$$

If the Hermitian metric $g_{ij^*}$ admits a set of Killing vectors $\{K_A = k_A{}^i\partial_i + k^*_A{}^{i^*}\partial_{i^*}\}$ satisfying the Lie algebra

$$[K_A, K_B] = -f_{AB}{}^C K_C, \tag{E.39}$$

it is easy to see that the the components $k_A{}^i$ and $k^*_A{}^{i^*}$ of the Killing vectors satisfy, separately, the same Lie algebra.

## E.2 Almost complex structures and manifolds

In a (real) differential manifold of dimension $2d$ a globally defined tensor $J_m{}^n$ having the property Eq. (E.10),

$$J_m{}^n J_n{}^p = -\delta_m{}^p, \tag{E.40}$$

and is characteristic of a complex structure, is called an *almost complex structure*, and the manifold that admits it is an *almost complex manifold*.[3] A Riemannian metric satisfying Eq. (E.13), where $J$ is an almost complex structure, is an *almost Hermitian metric*, and a manifold equipped with such a metric is an *almost Hermitian manifold*. Almost Hermitian manifolds with closed fundamental 2-form $\mathcal{J}$ are called *almost Kähler manifolds*. Obviously, all complex manifolds are also almost complex manifolds. All almost complex manifolds are orientable, and so are all complex manifolds.

Given an almost complex structure, we can complexify the tangent space as in a complex manifold. Then, defining the projectors $P_\mp$, we can split the tangent space as the direct sum of a holomorphic and an antiholomorphic subspace, and we can split differential forms as sums of $(p,q)$-forms, etc. However, the exterior derivative is not just the sum of the two Dolbeault operators as in a complex manifold because, in general, the almost complex structure has no constant components. Even if there is a local coordinate system in which it is constant, it is not clear whether there are local choices in all the manifold's patches such that the coordinate transformations in the overlaps are holomorphic (which is necessary to keep the constant form of the tensor). When all these conditions are met, the almost complex structure is a complex structure and the almost complex manifold is a complex manifold.

We would like to find a criterion to determine when an almost complex structure is a complex structure (and when the differential manifold is a complex manifold).

[3] We have assumed from the start that the dimension of the manifold is even, but an almost complex structure can only be defined in manifolds of even dimensions: in $r$ dimensions $0 \leq (\det J)^2 = \det J^2 = \det(-\mathbb{1}_{r\times r}) = (-1)^r$, which is only consistent if $r$ is even.

First, we say that an almost complex structure is *integrable* if the Lie bracket of two (anti)holomorphic vector fields (where holomorphicity is defined with respect to that structure) is always another (anti)holomorphic vector field. Then the Newlander–Nirenberg theorem states that if the almost complex structure is integrable, it is a complex structure (and the differential manifold is a complex manifold). The converse is easily found to be true.

A practical criterion of integrability of the almost complex structure is provided by the *Nijenhuis tensor* $N_{mn}{}^p$:

$$N_{mn}{}^p \equiv -\nabla_{[m} J_{n]}{}^p + J_{[m}{}^q J_{n]}{}^r \nabla_q J_r{}^p. \tag{E.41}$$

This expression, involving (almost Hermitian) covariant derivatives, makes it clear that it is a tensor. However, there is no dependence on the connection, and the covariant derivatives can be replaced by partial derivatives.

It can be shown (see, for instance, Ref. [1211] or Ref. [291]) that the vanishing of the Nijenhuis tensor is a necessary and sufficient condition for the integrability of the almost complex structure and, therefore, for it to be a complex structure.

## E.3 Kähler manifolds

A *Kähler manifold* is a complex Hermitian manifold whose fundamental 2-form $\mathcal{J}$, defined in Eq. (E.19) and called here a *Kähler 2-form*, is closed:

$$d\mathcal{J} = 0. \tag{E.42}$$

Expanding $dJ$ we get[4]

$$dJ = 2i\partial_k \mathcal{G}_{ij^*} dZ^k \wedge dZ^i \wedge dZ^{*j^*} - 2i\partial_{k^*} \mathcal{G}_{ij^*} dZ^{*k^*} \wedge dZ^i \wedge dZ^{*j^*} = 0, \tag{E.43}$$

which leads to the two equations

$$\partial_{[k} \mathcal{G}_{i]j^*} = 0, \qquad \partial_{[k^*|} \mathcal{G}_{i|j^*]} = 0. \tag{E.44}$$

First, these equations imply the vanishing of the torsion and the identification of the Hermitian and the Levi-Civita connections. Second, in a local coordinate patch $U_{(x)}$ these equations are solved by

$$\mathcal{G}_{ij^*} = \partial_i \partial_{j^*} \mathcal{K}_{(x)} \quad \text{or} \quad \mathcal{J} = 2i\partial\partial^* \mathcal{K}_{(x)} \tag{E.45}$$

for some real function $\mathcal{K}_{(x)}(Z, Z^*)$, called the *Kähler potential*.[5] The Kähler potential in each coordinate patch is not unique: it is defined only up to *Kähler transformations* of the form

$$\mathcal{K}'_{(x)}(Z, Z^*) = \mathcal{K}_{(x)}(Z, Z^*) + \lambda(Z) + \lambda^*(Z^*), \tag{E.47}$$

---

[4] In Kähler manifolds we denote the metric by $\mathcal{G}_{ij^*}$ instead of $g_{ij^*}$ and we capitalize the complex coordinates $z^i \to Z^i$.

[5] Sometimes Eq. (E.45) is used as the definition of a Kähler manifold. Riemannian manifolds whose metric $g_{\mu\nu}$ is related to a real function of the coordinates $H(x)$ called the *Hessian potential* by the analogous relation

$$g_{\mu\nu} = \partial_\mu \partial_\nu H \tag{E.46}$$

are called *Hessian manifolds*.

where $\lambda(Z)$ is any holomorphic function of the complex coordinates $Z^i$.

We define in each coordinate patch $U_{(x)}$ the *Kähler (connection) 1-form*

$$\mathcal{Q}_{(x)} \equiv \tfrac{1}{2i}(\partial - \partial^*)\mathcal{K}_{(x)}, \tag{E.48}$$

which is such that, in each patch,

$$\mathcal{J} \equiv 2d\mathcal{Q}_{(x)}. \tag{E.49}$$

By definition, the Kähler transformations leave invariant the Kähler metric, but the Kähler 1-form transforms according to

$$\mathcal{Q}' = \mathcal{Q} + \tfrac{1}{2i}(\partial\lambda - \partial^*\lambda^*) \quad \Rightarrow \quad \mathcal{Q}'_i = \mathcal{Q}_i + \tfrac{1}{2i}\partial_i\lambda. \tag{E.50}$$

In compact manifolds, the Kähler potentials defined on two different patches $U_{(x)}, U_{(y)}$ cannot be identical on the overlap $U_{(x)} \cap U_{(y)}$; they must be related by a Kähler transformation

$$\mathcal{K}_{(x)} = \mathcal{K}_{(y)} + \lambda_{xy}(Z) + \lambda^*_{xy}(Z^*). \tag{E.51}$$

If the Kähler 1-form were globally defined and given by $\mathcal{Q}$ in all patches, and $\mathcal{J} = 2d\mathcal{Q}$, then, according to Eq. (E.20), the volume form would be a total derivative and its integral, upon use of Stokes' theorem, would vanish in a compact manifold. Therefore, in compact Kähler manifolds, a globally defined $\mathcal{Q}$ cannot exist and the 1-forms defined in different patches must be related by Kähler transformation.

The curvature tensor in a Kähler manifold has additional symmetry properties due to the vanishing of the torsion. The only non-vanishing components are

$$R_{i^*jk}{}^l = \partial_{i^*}\Gamma_{jk}{}^l \quad \Rightarrow \quad R_{i^*jkl^*} = \partial_{i^*}\partial_j\mathcal{G}_{kl^*} - \partial_{i^*}\mathcal{G}_{l^*m}\mathcal{G}^{mn^*}\partial_j\mathcal{G}_{kn^*}, \tag{E.52}$$

satisfying

$$R_{ij^*\,kl^*} = R_{kl^*\,ij^*} = R_{il^*\,kj^*} = R_{kj^*\,il^*}, \tag{E.53}$$

and the non-vanishing components of the Ricci tensor are given by

$$R_{ij^*} \equiv R_{ij^*k^*}{}^{k^*} = -R_{ij^*kl^*}\mathcal{G}^{kl^*} \quad \Rightarrow \quad \mathfrak{R} = iR_{ij^*}dZ^i \wedge dZ^{*\,j^*}. \tag{E.54}$$

In a Kähler manifold one can define objects (for instance, tensors) which also transform under the Kähler transformations Eq. (E.47). By definition, we say that such an object, which we denote by $\Psi(Z, Z^*)$ not showing the possible tensor indices, has Kähler weight $(q, \bar{q})$ if, under Eq. (E.47),

$$\Psi' = e^{-(q\lambda + \bar{q}\lambda^*)/2}\Psi. \tag{E.55}$$

According to this definition, $e^{\mathcal{K}}$ has Kähler weight $(-2, -2)$. The Kähler-covariant derivative $\mathcal{D}$ acting on these objects is given by

$$\mathcal{D}_i \equiv \nabla_i + iq\mathcal{Q}_i = \nabla_i + \tfrac{1}{2}q\partial_i\mathcal{K}, \qquad \mathcal{D}_{i^*} \equiv \nabla_{i^*} - i\bar{q}\mathcal{Q}_{i^*} = \nabla_{i^*} + \tfrac{1}{2}\bar{q}\partial_{i^*}\mathcal{K}, \tag{E.56}$$

where $\nabla$ is the standard covariant derivative associated with the Hermitian (Levi-Civita) connection associated with the tensorial nature of $\Psi$.

The Ricci identity for this covariant derivative acting on $\Psi$ is

$$[\mathcal{D}_i, \mathcal{D}_{j^*}] = [\nabla_i, \nabla_{j^*}] - \tfrac{1}{2}\,(q - \bar{q})\,\mathcal{G}_{ij^*}. \tag{E.57}$$

An especially interesting case is that of fields with $\bar{q} = -q$ (with weight $q$, for short), whose Kähler transformations are just $Z$-dependent $\mathrm{U}(1)$ transformations:

$$\Psi' = e^{-iq\Im\mathrm{m}\lambda(Z)}\Psi. \tag{E.58}$$

The structure that supports fields with these properties for $q = 1$ is that of a $\mathrm{U}(1)$ bundle associated to a complex line bundle $L^1 \to \mathcal{M}$ over the Kähler manifold $\mathcal{M}$, and the consistency of this construction requires that the first Chern class of the $L^1 \to \mathcal{M}$ bundle (given by the Ricci 2-form $\mathfrak{R}$ of the fiber's Hermitian metric) equals the Kähler 2-form $\mathcal{J}$. These Kähler manifolds are known as *Kähler–Hodge (KH) manifolds*. The manifolds parametrized by the complex scalars of the chiral multiplets of $N = 1, d = 4$ supergravity must be KH manifolds, as discussed in Chapter 6. In those theories, objects such as the superpotential and all the spinors of the theory have a well-defined Kähler weight and transform as sections of the bundle. The manifolds parametrized by the complex scalars of the vector multiplets of $N = 2, d = 4$ supergravity are also KH manifolds but must satisfy further constraints that define what is known as *special Kähler geometry*, described in Appendix F.

The spacetime pullback of the Kähler-covariant derivative on tensor fields with Kähler weight $q$ is often used and takes the simple form

$$\mathfrak{D}_\mu = \nabla_\mu + iq\mathcal{Q}_\mu, \tag{E.59}$$

where $\nabla_\mu$ is the standard spacetime-covariant derivative plus possibly the pullback of the Levi-Civita connection and $\mathcal{Q}_\mu$ is the pullback of the Kähler 1-form:

$$\mathcal{Q}_\mu = \tfrac{1}{2i}(\partial_\mu Z^i \partial_i \mathcal{K} - \partial_\mu Z^{*\,i^*} \partial_{i^*}\mathcal{K}). \tag{E.60}$$

### *E.3.1 Holomorphic isometries of Kähler manifolds*

The transformations generated by the Killing vectors will preserve the Kähler structure if they leave the Kähler potential invariant up to Kähler transformations, i.e. if, for each Killing vector $K_A$, the Lie derivative of the Kähler potential with respect to $K_A$ takes the value

$$\mathcal{L}_A\mathcal{K} \equiv k_A{}^i \partial_i\mathcal{K} + k^*_A{}^{i^*}\partial_{i^*}\mathcal{K} = \lambda_A(Z) + \lambda^*_A(Z^*) \tag{E.61}$$

for certain characteristic holomorphic functions $\lambda_A(Z)$ which are defined by this equation. This condition can be rewritten in terms of a *covariant Lie derivative* which expresses invariance up to some transformation (a Kähler transformation in this case) in this way:

$$\mathbb{L}_A\mathcal{K} \equiv \mathcal{L}_A\mathcal{K} - (\lambda_A + \lambda^*_A). \tag{E.62}$$

We will very often use generalizations of this concept in what follows.

The equivariance property

$$\mathcal{L}_A\lambda_B - \mathcal{L}_B\lambda_A = -f_{AB}{}^C\lambda_C \tag{E.63}$$

follows from Eq. (E.62).

Observe that this implies that all the objects defined on the complex manifold $\Phi(Z, Z*)$ transforming as in Eq. (E.55) and which respect this symmetry will undergo a Kähler transformation when one performs an isometric change of coordinates:

$$\mathbb{L}_A\Phi \equiv \{\mathcal{L}_A + \tfrac{1}{2}(q\lambda_A + \bar{q}\lambda^*_A)\}\Phi = 0. \tag{E.64}$$

Fields with this property (or generalizations thereof) will be called *invariant fields* (or *sections*), and it should be clear that the isometries under consideration will be symmetries of the whole Kähler structure if the fields that characterize this structure are invariant fields in the above sense.

Thus, the preservation of the Kähler structure implies the conservation of the Kähler 2-form $\mathcal{J}$:

$$\mathbb{L}_A\mathcal{J} = \mathcal{L}_A\mathcal{J} = 0. \tag{E.65}$$

The closedness of $\mathcal{J}$ implies that $\mathcal{L}_A\mathcal{J} = d(i_{k_A}\mathcal{J})$ locally, and therefore the preservation of the Kähler structure implies the local existence of a set of real 0-forms $\mathcal{P}_A$ known as *momentum maps* such that

$$i_{k_A}\mathcal{J} = d\mathcal{P}_A. \tag{E.66}$$

A local solution for this equation is provided by

$$i\mathcal{P}_A = k_A{}^i\partial_i\mathcal{K} - \lambda_A, \tag{E.67}$$

which, on account of Eq. (E.61), is equivalent to

$$i\mathcal{P}_A = -(k^*_A{}^{i^*}\partial_{i^*}\mathcal{K} - \lambda^*_A) \tag{E.68}$$

or

$$\mathcal{P}_A = i_{k_A}\mathcal{Q} - \tfrac{1}{2i}(\lambda_A - \lambda^*_A). \tag{E.69}$$

The momentum maps can be used as a prepotential from which the Killing vectors can be derived:

$$k_{A\,i^*} = i\partial_{i^*}\mathcal{P}_A. \tag{E.70}$$

Due to this property they are sometimes called *Killing prepotentials*.

The momentum maps are defined, in principle, up to an additive real constant.[6] Using Eqs. (E.39), (E.61), and (E.63) one finds

$$\mathcal{L}_A\mathcal{P}_B = 2ik_{[A}{}^i k^*_{B]}{}^{j^*}\mathcal{G}_{ij^*} = -f_{AB}{}^C\mathcal{P}_C. \tag{E.71}$$

This equation fixes the additive constant of the momentum map along directions in which the isometry group is a non-Abelian group.

Furthermore, observe that this equation can also be read as invariance up to a rotation in the adjoint representation of the symmetry group and can be written in terms of a new, more general, covariant Lie derivative as

$$\mathbb{L}_A\mathcal{P}_B \equiv \{\mathcal{L}_A - T_A\}\mathcal{P}_B, \tag{E.72}$$

[6] In $N = 1, d = 4$ theories (but not in $N = 2, d = 4$) it is possible to have non-vanishing, constant, momentum maps with $i\mathcal{P} = -\lambda =$ constant which are called *D*- or *Fayet–Iliopoulos* terms, as explained in Chapter 6.

where the matrices $T_A$ are a linear representation of the generators of the symmetry group with the commutation relations

$$[T_A, T_B] = +f_{AB}{}^C T_C. \tag{E.73}$$

For general fields $\Phi(Z, Z^*)$ with tensor indices, Kähler weight $(q, \bar{q})$, and transforming in the linear representation of the symmetry group generated by the matrices $T_A$ we can give the general definition of the covariant Lie derivative

$$\boxed{\mathbb{L}_A \Phi \equiv \{\pounds_A - [T_A - \tfrac{1}{2}(q\lambda_A + \bar{q}\lambda_A^*)]\}\Phi,} \tag{E.74}$$

and an invariant field or section will be one annihilated by the above operator.

The objects defining the special Kähler geometries that we are going to study in Appendix F transform in the fundamental representation of a symplectic group, and the matrices $T_A$ will necessarily be generators of that symplectic group (see Section F.3). We will then use a slightly different notation $\mathcal{T}_A$.

It is perhaps useful to give a slightly more general definition of the covariant derivative that can be adapted to other situations: if $\Phi$ is a field which, under the isometry generated by $K_A$, transforms according to some rule $\delta_A\Phi$, then

$$\boxed{\mathbb{L}_A \Phi \equiv \{\pounds_A - \delta_A\}\Phi,} \tag{E.75}$$

and is, by construction, an invariant field. Evidently, for the fields that we have considered so far,

$$\delta_A \Phi = [T_A - \tfrac{1}{2}(q\lambda_A + \bar{q}\lambda_A^*)]\Phi. \tag{E.76}$$

For the sake of completeness, we give the list of the irreducible symmetric Kähler spaces taken from Ref. [280] in Table E.1.

Table E.1. Irreducible symmetric Kähler spaces

This list is identical to the list of irreducible symmetric spaces G/H with a $\mathrm{U}(1) \sim \mathrm{SO}(2)$ factor in the stability group H. Only some of them are also symmetric special Kähler manifolds (which are listed in Table F.1).

| Type | Coset | Complex dimension |
|---|---|---|
| $\mathrm{I}_{m,m'}$ $1 \leq m \leq m'$ | $\frac{\mathrm{U}(m,m')}{\mathrm{U}(m)\times\mathrm{U}(m')}$ | $mm'$ |
| $\mathrm{II}_{m}$ $2 \leq m$ | $\frac{\mathrm{SO}^*(2m)}{\mathrm{U}(m)}$ | $m(m-1)/2$ |
| $\mathrm{III}_{m}$ $1 \leq m$ | $\frac{\mathrm{Sp}(2n,\mathbb{R})}{\mathrm{U}(m)}$ | $m(m+1)/2$ |
| $\mathrm{IV}_{m}$ $3 \leq 3m$ | $\frac{\mathrm{SO}(m,2)}{\mathrm{SO}(m)\times\mathrm{SO}(2)}$ | $m$ |
| V | $\frac{\mathrm{E}_{6,-14}}{\mathrm{SO}(10)\times\mathrm{SO}(2)}$ | 16 |
| VI | $\frac{\mathrm{E}_{7,-25}}{\mathrm{E}_6\times\mathrm{SO}(2)}$ | 27 |

# Appendix F
# Special Kähler geometry

Special Kähler geometry is the structure underlying the couplings between the fields of $n_V$ vector supermultiplets in $N = 2$, $d = 4$ supergravity: complex scalars parametrizing a Kähler–Hodge manifold, 1-form fields, and pairs of Weyl spinors called gauginos. These couplings are encoded in several functions of the complex scalars: the Kähler potential $\mathcal{K}$ and its derivates (metric and Kähler 1-form), the *period matrix* $\mathcal{N}_{\Lambda\Sigma}$ defined on p. 54, the *canonical symplectic section* $\mathcal{V}$, which we are about to define, and its Kähler-covariant derivative $\mathcal{U}_i$. Furthermore, if the theory is gauged, some couplings will be dictated by the holomorphic Killing vectors and their associated momentum maps. Supersymmetry requires that all these objects are related in a very specific way. These relations are the essential content of special Kähler geometry.

We are going to introduce the subject in a coordinate-independent way.[1] Later on, in Section F.2, we will introduce the so-called *prepotential* and make contact with the original formulation of de Wit, Lauwers, and Van Proeyen [1257].

## F.1 Special Kähler manifolds

Special Kähler geometry can be formally defined as follows: consider a Kähler–Hodge manifold $\mathcal{M}$ of complex dimension $n_V$ and a flat $2\bar{n}$-dimensional ($\bar{n} \equiv n_V + 1$) vector bundle $E \to \mathcal{M}$ with structure group $\mathrm{Sp}(2\bar{n};\mathbb{R})$. Let us consider the product bundle $E \otimes L^1 \to \mathcal{M}$. This product bundle will be a *special Kähler manifold* if there is a section $\mathcal{V}$ of this bundle, called the *(covariantly holomorphic) canonical symplectic section*, which satisfies certain properties. Denoting the holomorphic Kähler-covariant derivative of the

[1] Some basic references are Refs. [325, 327, 379] and the review Refs. [1209, 579]. The definition of special Kähler manifold was first made in Ref. [1156], formalizing the original results of Ref. [1265].

symplectic section by[2]

$$\mathcal{U}_i \equiv \mathcal{D}_i \mathcal{V}, \qquad (\mathcal{U}_i)^* \equiv \mathcal{U}^*_{i^*}, \tag{F.2}$$

we can write the properties that define a special Kähler geometry as follows:[3]

$$\langle \mathcal{V} \mid \mathcal{V}^* \rangle = -i, \tag{F.5}$$

$$\mathcal{D}_{i^*} \mathcal{V} = 0, \tag{F.6}$$

$$\langle \mathcal{U}_i \mid \mathcal{V} \rangle = 0. \tag{F.7}$$

The first condition is just a normalization convention, the fact that the product is purely imaginary being a consequence of the antisymmetry of the symplectic product. The second (covariant holomorphicity of the canonical section $\mathcal{V}$) and third conditions constrain the canonical section which should only describe the $n_V$ complex degrees of freedom corresponding to the complex scalars $Z^i$. Indeed, the canonical symplectic section $\mathcal{V}$ can be viewed as a highly redundant description of those $n_V$ complex scalars through $2n_V + 2$ complex functions. However, its structure carries a great deal of information about the $N = 2, d = 4$ supergravity model under consideration. In fact, $\mathcal{V}$ completely defines such a model: all the couplings of the theory in which the complex scalars are involved can be derived from $\mathcal{V}$.

From the fundamental properties listed in Eqs. (F.5)–(F.7) and the properties of the Kähler-covariant derivative, it follows that

$$\mathcal{D}_{i^*} \mathcal{U}_i = \mathcal{G}_{ii^*} \mathcal{V}, \tag{F.8}$$

$$\langle \mathcal{U}_i \mid \mathcal{U}^*_{i^*} \rangle = i\mathcal{G}_{ii^*}, \tag{F.9}$$

$$\langle \mathcal{U}_i \mid \mathcal{V}^* \rangle = 0. \tag{F.10}$$

---

[2] Note that being a section of the product bundle $E \otimes L^1 \to \mathcal{M}$ means that $\mathcal{V}$ has Kähler weight $q = 1$ and, therefore, according to the definition Eq. (E.56),

$$\begin{aligned} \mathcal{D}_i \mathcal{V} &= (\partial_i + \tfrac{1}{2}\partial_i \mathcal{K})\mathcal{V} = e^{-\mathcal{K}/2}\partial_i \left(e^{\mathcal{K}/2}\mathcal{V}\right), \\ \mathcal{D}_{i^*} \mathcal{V} &= (\partial_{i^*} - \tfrac{1}{2}\partial_{i^*} \mathcal{K})\,\mathcal{V} = e^{\mathcal{K}/2}\partial_{i^*} \left(e^{-\mathcal{K}/2}\mathcal{V}\right). \end{aligned} \tag{F.1}$$

[3] We use symplectic indices $M, N = 1, \ldots, 2\bar{n}$, or, equivalently, pairs of one upper and one lower index $\Lambda, \Sigma = 1, \ldots, \bar{n}$. The notation is described in Section 2.6.2. The symplectic product is sometimes denoted by brackets and the equivalence with the index notation is

$$\langle \mathcal{A} \mid \mathcal{B} \rangle \equiv \mathcal{A}_M \mathcal{B}^M = -\mathcal{A}^N \Omega_{NM} \mathcal{B}^M = \mathcal{B}^\Lambda \mathcal{A}_\Lambda - \mathcal{B}_\Lambda \mathcal{A}^\Lambda. \tag{F.3}$$

The components of the canonical section and its holomorphic covariant derivative are often denoted by

$$\mathcal{V} \equiv \begin{pmatrix} \mathcal{L}^\Lambda \\ \mathcal{M}_\Lambda \end{pmatrix}, \qquad \mathcal{U}_i \equiv \begin{pmatrix} f^\Lambda{}_i \\ h_{\Lambda\, i} \end{pmatrix}. \tag{F.4}$$

The covariant derivative of the identity $\langle \mathcal{U}_i \mid \mathcal{V} \rangle = 0$ gives

$$\langle \mathcal{D}_i \mathcal{U}_j \mid \mathcal{V} \rangle = -\langle \mathcal{U}_j \mid \mathcal{U}_i \rangle. \tag{F.11}$$

The r.h.s. of this equation is antisymmetric because the symplectic product is, while the l.h.s. is symmetric because it contains two covariant derivatives acting on $\mathcal{V}$. Therefore, we must conclude that

$$\langle \mathcal{D}_i \mathcal{U}_j \mid \mathcal{V} \rangle = \langle \mathcal{U}_j \mid \mathcal{U}_i \rangle = 0. \tag{F.12}$$

The latter equation allows us to write a completeness relation for symplectic sections $\mathcal{A}$:

$$\mathcal{A} = i\langle \mathcal{A} \mid \mathcal{V}^* \rangle \mathcal{V} - i\langle \mathcal{A} \mid \mathcal{V} \rangle \mathcal{V}^* + i\langle \mathcal{A} \mid \mathcal{U}_i \rangle \mathcal{G}^{ii^*} \mathcal{U}^*_{i^*} - i\langle \mathcal{A} \mid \mathcal{U}^*_{i^*} \rangle \mathcal{G}^{ii^*} \mathcal{U}_i. \tag{F.13}$$

The symplectic product of the symmetric tensor $\mathcal{D}_i\, \mathcal{U}_j$ with $\mathcal{V}^*$ and $\mathcal{U}^*_{k^*}$ vanishes identically due to the basic properties, but the product with $\mathcal{U}_k$ does not. We can, then, define a Kähler-weight 2, 3-index tensor

$$\mathcal{C}_{ijk} \equiv \langle \mathcal{D}_i\, \mathcal{U}_j \mid \mathcal{U}_k \rangle, \tag{F.14}$$

which is completely symmetric in the three indices because of the orthogonality of the $\mathcal{U}$s.[4] This tensor appears in the curvature tensor of special Kähler manifolds [390]:

$$R_{ij^*kl^*} = -\mathcal{G}_{ij^*}\mathcal{G}_{kl^*} - \mathcal{G}_{il^*}\mathcal{G}_{kl^*} + \mathcal{C}_{ikm}\mathcal{C}^*_{j^*l^*n^*}\mathcal{G}^{mn^*}. \tag{F.16}$$

One can also show that

$$\mathcal{D}_{i^*}\, \mathcal{C}_{jkl} = 0, \qquad \mathcal{D}_{[i}\, \mathcal{C}_{j]kl} = 0, \tag{F.17}$$

which implies the existence of a function $\mathcal{S}$, such that

$$\mathcal{C}_{ijk} = \mathcal{D}_i \mathcal{D}_j \mathcal{D}_k\, \mathcal{S}. \tag{F.18}$$

This function is given by

$$\mathcal{S}^{-1} = i2^7 \cdot 3\mathcal{L}^{*\,\Lambda}\mathcal{L}^{*\,\Sigma}\, \Im\mathrm{m}\,(\mathcal{N})_{\Lambda\Sigma}\,, \tag{F.19}$$

where $\mathcal{N}(Z, Z^*)_{\Lambda\Sigma}$ is the period matrix, which in special Kähler geometry is defined by the following two properties:

$$\mathcal{M}_\Lambda = \mathcal{N}_{\Lambda\Sigma}\mathcal{L}^\Sigma, \qquad h_{\Lambda\, i} = \mathcal{N}^*_{\Lambda\Sigma} f^\Sigma{}_i. \tag{F.20}$$

Then, Eq. (F.10) implies that $\mathcal{N}_{\Lambda\Sigma}$ is symmetric, which is necessary for the object we just defined in this geometry to be the period matrix that appears in the kinetic term of the vector fields in the supergravity actions. In turn, the symmetry of $\mathcal{N}_{\Lambda\Sigma}$ trivializes the relation $\langle \mathcal{U}_i \mid \mathcal{U}_j \rangle = 0$. Furthermore, $\Im\mathrm{m}\, \mathcal{N}_{\Lambda\Sigma}$ must be invertible and negative-definite for the kinetic term of the supergravity vector fields to be well defined. These properties always

[4] A useful relation can be found by using Eq. (F.13) and the definition of the tensor $\mathcal{C}_{ijk}$

$$\mathcal{D}_i\, \mathcal{U}_j = i\mathcal{C}_{ijk}\mathcal{G}^{kl^*}\mathcal{U}^*_{l^*}. \tag{F.15}$$

hold for the period matrix defined in special Kähler geometry by the relations in Eqs. (F.20) [384] and thanks to them we can use $\Im\mathrm{m}\,\mathcal{N}_{\Lambda\Sigma}$ and its inverse to raise and lower the $\Lambda$-indices.

The other defining properties, Eqs. (F.8) and (F.9), lead to the following relations:

$$\Im\mathrm{m}\,\mathcal{N}_{\Lambda\Sigma}\,\mathcal{L}^{\Lambda}\mathcal{L}^{*\Sigma} = -\tfrac{1}{2}, \tag{F.21}$$

$$\Im\mathrm{m}\,\mathcal{N}_{\Lambda\Sigma}\,\mathcal{L}^{\Lambda}f^{\Sigma}{}_{i} = 0\,, \tag{F.22}$$

$$\Im\mathrm{m}\,\mathcal{N}_{\Lambda\Sigma}\,\mathcal{L}^{\Lambda}f^{*\Sigma}{}_{i^*} = 0\,, \tag{F.23}$$

$$\Im\mathrm{m}\,\mathcal{N}_{\Lambda\Sigma}\,f^{\Lambda}{}_{i}\,f^{*\Sigma}{}_{j^*} = -\tfrac{1}{2}\mathcal{G}_{ij^*}. \tag{F.24}$$

Other useful identities that involve $\mathcal{N}_{\Lambda\Sigma}$ are

$$(\partial_i\mathcal{N}_{\Lambda\Sigma})\mathcal{L}^{\Sigma} = -2i\Im\mathrm{m}\,\mathcal{N}_{\Lambda\Sigma}\,f^{\Sigma}{}_{i}, \tag{F.25}$$

$$\partial_i\mathcal{N}^{*}{}_{\Lambda\Sigma}\,f^{\Sigma}{}_{j} = -2\mathcal{C}_{ijk}\mathcal{G}^{kk^*}\Im\mathrm{m}\,\mathcal{N}_{\Lambda\Sigma}f^{*\Sigma}{}_{k^*}, \tag{F.26}$$

$$\mathcal{C}_{ijk} = \partial_i\mathcal{N}^{*}_{\Lambda\Sigma}\,f^{\Lambda}{}_{j}f^{\Sigma}{}_{k}, \tag{F.27}$$

$$\partial_{i^*}\mathcal{N}_{\Lambda\Sigma}\,\mathcal{L}^{\Sigma} = 0, \tag{F.28}$$

$$\partial_{j^*}\mathcal{N}^{*}{}_{\Lambda\Sigma}\,f^{\Sigma}{}_{i} = 2i\mathcal{G}_{ij^*}\Im\mathrm{m}\,\mathcal{N}_{\Lambda\Sigma}\mathcal{L}^{\Sigma}, \tag{F.29}$$

$$f^{\Lambda}{}_{i}\mathcal{G}^{ij^*}f^{*\Sigma}{}_{j^*} = -\tfrac{1}{2}\Im\mathrm{m}\,\mathcal{N}^{-1|\Lambda\Sigma} - \mathcal{L}^{*\Lambda}\mathcal{L}^{\Sigma}. \tag{F.30}$$

The last identity implies

$$-2\Im\mathrm{m}\,\mathcal{N}_{\Lambda\Gamma}\left(\mathcal{L}^{*\Gamma}\mathcal{L}^{\Sigma} + f^{\Gamma}{}_{i}\mathcal{G}^{ij^*}f^{*\Sigma}{}_{j^*}\right) = \delta_{\Lambda}{}^{\Sigma}. \tag{F.31}$$

We can define two projectors that select, out of the $n_V + 1$ vector field strengths of the supergravity theory, the (single) graviphoton and the $n_V$ matter vector field strengths:[5]

$$\mathcal{T}_{\Lambda} \equiv 2i\mathcal{L}_{\Lambda} = 2i\mathcal{L}^{\Sigma}\Im\mathrm{m}\,\mathcal{N}_{\Sigma\Lambda}, \tag{F.32}$$

$$\mathcal{T}^{i}{}_{\Lambda} \equiv -f^{*i}_{\Lambda} = -\mathcal{G}^{ij^*}f^{*\Sigma}{}_{j^*}\Im\mathrm{m}\,\mathcal{N}_{\Sigma\Lambda}. \tag{F.33}$$

[5] In this formalism all the $n_V + 1$ vector fields are treated on the same footing to make all the symmetries manifest. The distinction between these two kinds of (complex) combinations of (real) vector field strengths resides in the place they occupy in the supersymmetry transformation rules: the graviphoton appears, by definition, on the r.h.s. of the supersymmetry transformation of the gravitinos and the matter vector field strengths are, again by definition, the combinations that appear on the r.h.s. of the gauginos' supersymmetry transformations.

They satisfy the properties[6]

$$\Im\mathrm{m}\,\mathcal{N}^{-1|\Sigma\Lambda}\mathcal{T}_\Lambda\mathcal{T}^{*\,i^*}{}_\Sigma = \Im\mathrm{m}\,\mathcal{N}^{-1|\Sigma\Lambda}\mathcal{T}_\Lambda\mathcal{T}^i{}_\Sigma = 0, \tag{F.34}$$

$$\mathcal{T}^*_\Lambda\mathcal{T}_\Sigma + 4\mathcal{G}_{j^*i}\mathcal{T}^{*\,j^*}{}_\Lambda\mathcal{T}^i{}_\Sigma = \Im\mathrm{m}\,\mathcal{N}_{\Sigma\Lambda}. \tag{F.35}$$

Using these definitions and the above properties one can show the following identities for the derivatives of the period matrix:

$$\partial_i\mathcal{N}_{\Lambda\Sigma} = 4\mathcal{T}_{i(\Lambda}\mathcal{T}_{\Sigma)}, \tag{F.36}$$

$$\partial_{i^*}\mathcal{N}_{\Lambda\Sigma} = 4\mathcal{C}^*{}_{i^*j^*k^*}\mathcal{T}^{i^*}{}_{(\Lambda}\mathcal{T}^{j^*}{}_{\Sigma)}. \tag{F.37}$$

## F.2 The prepotential

As we have said, a special Kähler geometry is completely specified by giving its canonical symplectic section $\mathcal{V}$. We have not yet explained how the Kähler potential and the period matrix can be derived from $\mathcal{V}$, though. In order to do that it is useful to introduce another section, $\Omega$, which is by definition holomorphic. The covariant holomorphicity of $\mathcal{V}$, Eqs. (F.6) and (F.1), can be written in the form

$$\mathcal{D}_{i^*}\mathcal{V} = e^{\mathcal{K}/2}(\partial_{i^*}e^{-\mathcal{K}/2}\mathcal{V}) = 0, \tag{F.38}$$

and, therefore,

$$\Omega \equiv e^{-\mathcal{K}/2}\mathcal{V} \equiv \begin{pmatrix} \mathcal{X}^\Lambda \\ \mathcal{F}_\Sigma \end{pmatrix} \tag{F.39}$$

is a holomorphic symplectic section of Kähler weight $(2,0)$ because $e^{-\mathcal{K}/2}$ and $\mathcal{V}$ have weights $(1,1)$ and $(1,-1)$, respectively. In terms of $\Omega$, the properties that define a special Kähler geometry, Eqs. (F.5)–(F.7), take the form

$$\langle \Omega \mid \Omega^* \rangle = -ie^{-\mathcal{K}}, \tag{F.40}$$

$$\partial_{i^*}\Omega = 0, \tag{F.41}$$

$$\langle \partial_i\Omega \mid \Omega \rangle = 0. \tag{F.42}$$

Equation (F.40), together with the definition of the period matrix $\mathcal{N}_{\Lambda\Sigma}$, Eq. (F.20), lead to the following expression for the Kähler potential:

$$e^{-\mathcal{K}} = -2\Im\mathrm{m}\,\mathcal{N}_{\Lambda\Sigma}\,\mathcal{X}^\Lambda\mathcal{X}^{*\,\Sigma}. \tag{F.43}$$

[6] Often, we will remove the $-1$ and denote the inverse of $\Im\mathrm{m}\,\mathcal{N}$ by using just upper indices, i.e. $\Im\mathrm{m}\,\mathcal{N}^{\Sigma\Lambda} \equiv \Im\mathrm{m}\,\mathcal{N}^{-1|\Sigma\Lambda}$.

As we discussed before, these symplectic sections are very redundant descriptions of the $n_V$ complex coordinates, and, therefore, their components should satisfy a large number of relations. In particular, we can assume, for instance, that all the lower components of $\Omega$, $\mathcal{F}_\Lambda$, depend on the complex coordinates $Z^i$ only through the upper components $\mathcal{X}$. In that case, the expression for the Kähler potential Eq. (F.43) leads to the following equation:

$$\partial_i \mathcal{X}^\Lambda \left[ 2\mathcal{F}_\Lambda - \frac{\partial \left(\mathcal{X}^\Sigma \mathcal{F}_\Sigma\right)}{\partial \mathcal{X}^\Lambda} \right] = 0, \tag{F.44}$$

the *generic* solution of which is

$$\mathcal{F}_\Lambda = \frac{\partial \mathcal{F}}{\partial \mathcal{X}^\Lambda}, \tag{F.45}$$

where we have defined the function of $\mathcal{X}^\Lambda$

$$\mathcal{F}(\mathcal{X}) \equiv \tfrac{1}{2} \mathcal{X}^\Sigma \mathcal{F}_\Sigma(\mathcal{X}), \tag{F.46}$$

which is called the *prepotential*. Plugging Eq. (F.45) into this definition we find that

$$\mathcal{X}^\Lambda \frac{\partial \mathcal{F}}{\partial \mathcal{X}^\Lambda} = 2\mathcal{F}, \tag{F.47}$$

which implies (Euler's theorem) that $\mathcal{F}(\mathcal{X})$ is homogeneous of second degree in the $\mathcal{X}^\Lambda$s.

Of course, the *generic* solution and, therefore, the prepotential may not exist for a given holomorphic section $\Omega$. However, it was proven in Ref. [379] that there is always a symplectic $\mathrm{Sp}(2\bar{n}, \mathbb{R})$ transformation of $\Omega$ such that a prepotential exists. This symplectic transformation will not be a duality transformation and will not leave the action invariant (only, formally, the equations of motion) and it amounts to a change of coordinates in the scalar manifold. Thus, for most purposes, the existence of a prepotential can always be assumed.

The prepotential provides us with an alternative to the canonical section ($\mathcal{V}$ or $\Omega$) to characterize a given special Kähler geometry and the vector multiplet sector of a given $N = 2, d = 4$ supergravity. All we need is a choice of coordinates $Z^i$ to express the $\mathcal{X}^\Lambda$ as holomorphic functions of them.[7] A common choice is given, implicitly, by the so-called *special coordinates*

$$Z^i \equiv \mathcal{X}^i / \mathcal{X}^0, \qquad \mathcal{X}^0 = 1. \tag{F.48}$$

A Kähler transformation would leave the first equation invariant while transforming $\mathcal{X}^0$ into an arbitrary holomorphic function. We will use this choice unless stated otherwise.

The recipe to reconstruct the whole special Kähler geometry and the vector multiplet sector of a given $N = 2, d = 4$ supergravity starting from any given homogeneous function of degree 2 of the $n_V + 1$ variables $\mathcal{X}^\Lambda$, $\mathcal{F}(\mathcal{X})$ goes as follows.

1. Construct the lower components of the holomorphic section $\Omega$ using Eq. (F.45), copied here for convenience:

$$\mathcal{F}_\Lambda = \frac{\partial \mathcal{F}}{\partial \mathcal{X}^\Lambda} \quad \Rightarrow \quad \Omega = \begin{pmatrix} \mathcal{X}^\Lambda \\ \frac{\partial \mathcal{F}}{\partial \mathcal{X}^\Lambda} \end{pmatrix}. \tag{F.49}$$

[7] When the geometry is defined through the canonical sections, these are given in some coordinates and there is no need to define them.

2. Construct the Kähler potential using Eq. (F.40):

$$e^{-\mathcal{K}} = i\langle\Omega \mid \Omega^*\rangle = i\Omega_M\Omega^{*\,M} = i\left(\mathcal{X}^{*\,\Lambda}\mathcal{F}_\Lambda - \mathcal{F}^*_\Lambda\mathcal{X}^\Lambda\right). \tag{F.50}$$

3. Given $\Omega$ and $\mathcal{K}$ we can immediately construct the covariantly holomorphic canonical section $\mathcal{V}$ using Eq. (F.39):

$$\mathcal{V} = e^{\mathcal{K}/2}\Omega. \tag{F.51}$$

4. The period matrix can be computed from

$$\mathcal{N}_{\Lambda\Sigma} = \mathcal{F}^*_{\Lambda\Sigma} + 2i\frac{\Im\mathfrak{m}\mathcal{F}_{\Lambda\Lambda'}\mathcal{X}^{\Lambda'}\Im\mathfrak{m}\mathcal{F}_{\Sigma\Sigma'}\mathcal{X}^{\Sigma'}}{\mathcal{X}^{\Omega}\Im\mathfrak{m}\mathcal{F}_{\Omega\Omega'}\mathcal{X}^{\Omega'}}, \tag{F.52}$$

where we have defined, for all $n$, the symmetric tensors

$$\mathcal{F}_{\Lambda_1\cdots\Lambda_n} \equiv \frac{\partial^n\mathcal{F}}{\partial\mathcal{X}^{\Lambda_1}\cdots\partial\mathcal{X}^{\Lambda_n}}. \tag{F.53}$$

Having the explicit form of $\mathcal{N}$, we can also compute the Kähler potential using Eq. (F.43) or, equivalently,

$$e^{-\mathcal{K}} = -2\Im\mathfrak{m}\,\mathcal{F}_{\Lambda\Sigma}\,\mathcal{X}^\Lambda\mathcal{X}^{*\,\Sigma}. \tag{F.54}$$

5. Using Eq. (F.28) we can also compute the tensor $\mathcal{C}$:

$$\mathcal{C}_{ijk} = e^{\mathcal{K}}\partial_i\mathcal{X}^\Lambda\partial_j\mathcal{X}^\Sigma\partial_k\mathcal{X}^\Omega\mathcal{F}_{\Lambda\Sigma\Omega}, \tag{F.55}$$

which, in special coordinates Eq. (F.48), gives

$$\mathcal{C}_{ijk} = e^{\mathcal{K}}\mathcal{F}_{ijk}. \tag{F.56}$$

## F.3 Holomorphic isometries of special Kähler manifolds

As in previous examples, the symmetries of special Kähler manifolds are isometries generated by Killing vectors $K_A$ satisfying the Lie algebra

$$[K_A, K_B] = -f_{AB}{}^C K_C, \tag{F.57}$$

which preserve not only the metric but also the complete special Kähler structure, which is basically contained in the covariantly holomorphic canonical section $\mathcal{V}$. It is obvious that Hermiticity and the Kähler–Hodge structure must also be preserved, so the results obtained in Appendix E.3.1 hold here, and, in particular, the existence and properties of the holomorphic functions $\lambda_A$ associated with compensating Kähler transformations and the momentum maps $\mathcal{P}_A$.

Since the fields that define a special Kähler structure transform linearly under symplectic transformations as well as under Kähler transformations, we need the covariant Lie derivative defined in Eq. (E.74) with the generators $\mathcal{T}_A$ being elements of $\mathfrak{sp}(2\bar{n})$ satisfying the commutation relations of the symmetry group

$$[\mathcal{T}_A, \mathcal{T}_B] = +f_{AB}{}^C\mathcal{T}_C. \tag{F.58}$$

Now, the preservation of the special Kähler structure requires the canonical weight $(1,-1)$ section $\mathcal{V}$ to be an invariant section, that is that it be annihilated by the symplectic and Kähler-covariant Lie derivative

$$\mathbb{L}_A \mathcal{V} = K_A \mathcal{V} - [\mathcal{T}_A - \tfrac{1}{2}(\lambda_A - \lambda^*_A)]\mathcal{V} = 0. \tag{F.59}$$

Using the covariant holomorphicity of $\mathcal{V}$ one can write

$$K_A \mathcal{V} = k_A{}^i \mathcal{U}_i - i\mathcal{P}_A \mathcal{V} - \tfrac{1}{2}(\lambda_A - \lambda^*_A)\mathcal{V}, \tag{F.60}$$

and, comparing with Eq. (F.59), we get

$$k_A{}^i \mathcal{U}_i \;=\; (\mathcal{T}_A + i\mathcal{P}_A)\mathcal{V}. \tag{F.61}$$

The symplectic product of this equation with $\mathcal{V}^*$ gives an alternative expression for the momentum map,

$$\mathcal{P}_A = \langle\, \mathcal{V}^* \mid \mathcal{T}_A \mathcal{V}\,\rangle, \tag{F.62}$$

which leads, via Eq. (E.70), to an alternative expression for the Killing vectors:

$$k_A{}^i = i\partial^i \mathcal{P}_A = i\langle\, \mathcal{V} \mid \mathcal{T}_A \mathcal{U}^{*i}\,\rangle. \tag{F.63}$$

On the other hand, the symplectic product of Eq. (F.61) with $\mathcal{V}$ gives the following condition:

$$\langle\, \mathcal{V} \mid \mathcal{T}_A \mathcal{V}\,\rangle = 0. \tag{F.64}$$

Using the same identity and Eq. (F.9) one can also show that

$$k_A{}^i k^*_B{}^{j^*} \mathcal{G}_{ij^*} = \mathcal{P}_A \mathcal{P}_B - i\langle\, \mathcal{T}_A \mathcal{V} \mid \mathcal{T}_B \mathcal{V}^*\,\rangle. \tag{F.65}$$

It follows that

$$\langle\, \mathcal{T}_{[A} \mathcal{V} \mid \mathcal{T}_{B]} \mathcal{V}^*\,\rangle \;=\; -\tfrac{1}{2} f_{AB}{}^C \mathcal{P}_C. \tag{F.66}$$

Again, for the sake of completeness, we give in Table F.1 the list of the symmetric special Kähler spaces taken from Refs. [390, 44].

Table F.1. List of symmetric special Kähler spaces

The types correspond to the list of irreducible Kähler spaces in Table E.1, but this list also includes the few symmetric special Kähler spaces which are not irreducible (the last row). On the other hand, except for the type-$I_{1,n}$ ones, these spaces are *very special Kähler* spaces, which can be obtained by dimensional reduction of the symmetric real special spaces discussed in Appendix H. The four cases which do not form an infinite family (2nd to 5th rows) correspond to the *magic models* discussed on p. 232. The coset spaces of the type $I_{1,n}$, which are the spaces $\overline{\mathbb{CP}}^n$, are usually written in the equivalent form $\frac{SU(1,n)}{U(n)}$, but the global symmetry group is the full $U(1,n)$. The case $n=1$ is also written in the equivalent form $SL(2,\mathbb{R})/SO(2)$ and corresponds to the axion–dilaton model studied on p. 227 (also in Chapter 16 from the $N=4, d=4$ perspective). By virtue of the isomorphisms $SO(1,2) \sim SL(2,\mathbb{R}) \sim SU(1,1)$ and $SO(2,2) \sim SU(1,1)\times SU(1,1)$, the last case can be rewritten for $n=2$ and $n=3$ in the equivalent forms $\frac{SU(1,1)}{U(1)} \times \frac{SU(1,1)}{U(1)}$ and $\frac{SU(1,1)}{U(1)} \times \frac{SU(1,1)}{U(1)} \times \frac{SU(1,1)}{U(1)}$, respectively. The $N=2, d=4$ supergravity theory described by the latter is the so-called "STU model" [472, 113, 134].

| Type | Coset | Complex dimension | $Sp(2n_V+2,\mathbb{R})$ | Symplectic representation of G |
|---|---|---|---|---|
| $I_{1,n}$ | $\frac{U(1,n)}{U(1)\times U(n)}$ | $n$ | $Sp(2n+2,\mathbb{R})$ | $(\mathbf{n+1})\oplus(\mathbf{n+1})$ |
| $I_{3,3}$<br>$1\leq n$ | $\frac{U(3,3)}{U(3)\times U(3)}$ | 9 | $Sp(20,\mathbb{R})$ | **20** |
| $II_6$ | $\frac{SO^*(12)}{U(6)}$ | 15 | $Sp(32,\mathbb{R})$ | **32** |
| $III_3$ | $\frac{Sp(6,\mathbb{R})}{U(3)}$ | 6 | $Sp(14,\mathbb{R})$ | **14** |
| VI | $\frac{E_{7,(-25)}}{E_6\times SO(2)}$ | 27 | $Sp(56,\mathbb{R})$ | **56** |
| | $\frac{SL(2,\mathbb{R})}{U(1)} \times \frac{SO(n-1,2)}{SO(n-1)\times SO(2)}$ | $n$ | $Sp(2n+2,\mathbb{R})$ | $\mathbf{2}\otimes[(\mathbf{n+1})\oplus(\mathbf{n+1})]$ |

# Appendix G
# Quaternionic-Kähler geometry

A *quaternionic-Kähler manifold* can be defined as a real $4m$-dimensional Riemannian manifold that satisfies the following properties.

1. It is endowed with a triplet of complex structures[1] $\mathsf{J}^x$, $x = 1, 2, 3$. In a local patch with coordinates $q^u$, $u = 1, \ldots, 4m$, we denote their components by $\mathsf{J}^x{}_u{}^v(q)$ and, by definition of complex structure, they satisfy (no sum over $x$)
$$\mathsf{J}^{(x)}{}_u{}^v \mathsf{J}^{(x)}{}_v{}^w = -\delta_u{}^w, \qquad ((\mathsf{J}^x)^2 = -1). \tag{G.1}$$
2. The three complex structures satisfy the quaternionic algebra[2]
$$\mathsf{J}^x \mathsf{J}^y = -\delta^{xy} + \varepsilon^{xyz} \mathsf{J}^z \tag{G.2}$$
(which also includes the above property for $x = y$). The indices $x, y, z = 1, 2, 3$ can and will be interpreted as $\mathfrak{su}(2)$ indices.
3. The Riemannian metric, denoted by $\mathsf{H}_{uv}$, is Hermitian with respect to each of the three complex structures (again, no sum over $x$)
$$\mathsf{J}^{(x)}{}_u{}^w \mathsf{J}^{(x)}{}_v{}^q \mathsf{H}_{wq} = \mathsf{H}_{uv}. \tag{G.3}$$
As in any Hermitian manifold we can define, for each of the complex structures, a fundamental or Kähler 2-form (see Eq. (E.19))
$$\mathsf{K}^x{}_{uv} = \mathsf{J}^x{}_{uv} \equiv \mathsf{H}_{vw} \mathsf{J}^x{}_u{}^w. \tag{G.4}$$
The triplet of Kähler forms is known in this context as the ($\mathfrak{su}(2)$-valued) *hyper-Kähler 2-form*.

[1] Complex structures were defined in Appendix E, but we use a slightly different notation here, both for them and for other objects (such as the Hermitian metric). The use of these two different notations is useful in the $N = 2, d = 4$ supergravity context in which one has, simultaneously, scalars that parametrize a special Kähler and scalars that parametrize a quaternionic-Kähler space.

[2] This is the algebra satisfied by ($-i$ times) the Pauli matrices $-i\sigma^x$; see Eq. (D.10).

4. There is an $\mathrm{SU}(2)$ bundle whose base is the hyper-Kähler manifold with connection 1-form $\mathsf{A}^x$ that allows us to define an $\mathrm{SU}(2)$-covariant derivative $\mathsf{D}$. It is required that the hyper-Kähler 2-form is covariantly constant[3] with respect to it:

$$\mathsf{D}_u \mathsf{K}^x{}_{vw} \equiv \nabla_u \mathsf{K}^x{}_{vw} + \varepsilon^{xyz}\, \mathsf{A}^y{}_u \mathsf{K}^z{}_{vw} = 0, \tag{G.5}$$

where $\nabla_u$ is the covariant derivative with Levi-Civita connection.

5. Finally, the curvature of the $\mathrm{SU}(2)$ connection, which in differential-form language takes the form[4]

$$\mathsf{F}^x \equiv d\mathsf{A}^x + \tfrac{1}{2}\varepsilon^{xyz}\mathsf{A}^y \wedge \mathsf{A}^z, \tag{G.7}$$

must be proportional to the hyper-Kähler 2-form

$$\mathsf{F}^x = \varkappa\, \mathsf{K}^x, \tag{G.8}$$

for some proportionality constant $\varkappa \neq 0$ (if $\varkappa = 0$, the manifold is a *hyper-Kähler manifold* instead, which is sometimes considered as a special example of quaternionic-Kähler). In $N=2, d=4$ supergravity the value of the proportionality constant has to be, with these conventions, $\varkappa = -1$.

One of the main properties of quaternionic-Kähler spaces, which can be taken as the defining one, is that the holonomy of their Riemannian (Levi-Civita) connection (generically $\mathrm{SO}(4m)$) is $\mathrm{SU}(2)\times \mathrm{Sp}(2m)$. Therefore, it is useful to decompose the tangent indices with respect to that group. In particular, the Vielbein's "flat" indices can be written as a pair $(\alpha I)$ consisting of one fundamental $\mathrm{SU}(2)$ (or $\mathrm{Sp}(2)$) index $I = 1,2$ and one $\mathrm{Sp}(2m)$ index $\alpha = 1,\ldots,2m$. The chiral spinors of the hypermultiplets of $N=2, d=4$ theories will carry this kind of indices. We will call the Vielbein of these spaces *Quadbein*, and we will denote it by

$$\mathsf{U}^{\alpha I}{}_u. \tag{G.9}$$

The standard antisymmetric symplectic *metrics* in $\mathrm{Sp}(2)$ and $\mathrm{Sp}(2m)$ are denoted, respectively, by $\varepsilon_{IJ}$ and $\mathbb{C}_{\alpha\beta}$. They have the same values with upper and lower indices, and, therefore, they satisfy

$$\varepsilon^{KI}\varepsilon_{KJ} = \delta^I{}_J, \qquad \mathbb{C}^{\gamma\alpha}\mathbb{C}_{\gamma\beta} = \delta^\alpha{}_\beta. \tag{G.10}$$

The metric in the tangent space of the quaternionic-Kähler space is the tensor product of these two, i.e.

$$\eta_{\alpha I\,\beta J} \equiv \varepsilon_{IJ}\mathbb{C}_{\alpha\beta}, \tag{G.11}$$

[3] Not just covariantly closed.

[4] As usual, this curvature can be defined via the Ricci identity

$$\mathsf{DDK}^x = \varepsilon^{xyz}\mathsf{F}^y \wedge \mathsf{K}^z. \tag{G.6}$$

It takes the same form for any object with adjoint $\mathrm{SU}(2)$ indices $x,y,z = 1,2,3$. The convention used in this particular case for the connection and the corresponding field strength differs by a minus sign with the general definition given in Appendix A.

and, then, the Quadbein is related to the metric $\mathsf{H}_{uv}$ by

$$\mathsf{H}_{uv} = \mathsf{U}^{\alpha I}{}_u\, \mathsf{U}^{\beta J}{}_v \varepsilon_{IJ} \mathbb{C}_{\alpha\beta}, \tag{G.12}$$

and it follows that

$$2\, \mathsf{U}^{\alpha I}{}_{(u}\, \mathsf{U}^{\beta J}{}_{v)}\, \mathbb{C}_{\alpha\beta} = \mathsf{H}_{uv} \varepsilon^{IJ}. \tag{G.13}$$

Our convention for raising and lowering these indices is the same as in Eq. (C.14):

$$\begin{aligned} \chi^I &= \chi_J \varepsilon^{JI}, & \xi_I &= \varepsilon_{IJ} \xi^J, \\ \chi^\alpha &= \chi_\beta \mathbb{C}^{\beta\alpha}, & \xi_\alpha &= \mathbb{C}_{\alpha\beta} \xi^\beta. \end{aligned} \tag{G.14}$$

It must also be taken into account that complex conjugation transforms covariant into contravariant symplectic indices and vice versa,

$$(\chi^I)^* = \chi^*_I, \qquad (\xi_I)^* = \xi^{*\,I}, \tag{G.15}$$

etc., but this operation is different from and gives a different result than lowering or raising the indices with the metrics $\varepsilon, \mathbb{C}$ (for instance $(\chi^I)^* = \chi^*_I \neq \varepsilon_{IJ}\chi^J$). The requirement that the result is the same for a given object is equivalent to imposing a reality condition on that object. The Quadbein must satisfy a reality condition of the following kind:

$$\mathsf{U}_{\alpha I\, u} \equiv (\mathsf{U}^{\alpha I}{}_u)^* = \varepsilon_{IJ} \mathbb{C}_{\alpha\beta}\, \mathsf{U}^{\beta J}{}_u. \tag{G.16}$$

By definition, the inverse Quadbein $\mathsf{U}^u{}_{\alpha I}$ satisfies

$$\mathsf{U}_{\alpha I}{}^u\, \mathsf{U}^{\alpha I}{}_v = \delta^u{}_v, \tag{G.17}$$

and, therefore,

$$\mathsf{U}_{\alpha I}{}^u = \mathsf{H}^{uv}\, \varepsilon_{IJ} \mathbb{C}_{\alpha\beta}\, \mathsf{U}^{\beta J}{}_v, \tag{G.18}$$

and

$$\mathsf{H}^{uv} \mathsf{U}^{\alpha I}{}_u \mathsf{U}^{\beta J}{}_v = \mathbb{C}^{\alpha\beta} \varepsilon^{IJ}. \tag{G.19}$$

Choosing as SU(2) generators in the fundamental representation $-\frac{i}{2}(\sigma^x)^I{}_J$, where $\sigma^x$ are the Pauli sigma matrices of Eq. (D.9), the SU(2)-covariant derivative of objects transforming contravariantly in the fundamental representation of SU(2) (that is, with upper indices $I, J, \ldots$) is given by

$$\mathsf{D}\chi^I \equiv d\chi^I + \mathsf{A}^I{}_J \chi^J, \quad \text{with} \quad \mathsf{A}^I{}_J \equiv -\tfrac{i}{2} \mathsf{A}^x (\sigma^x)^I{}_J, \tag{G.20}$$

and the corresponding field strength $\mathsf{F}^I{}_J = -\frac{i}{2}\mathsf{F}^x(\sigma^x)^I{}_J$, found through the Ricci identities, is

$$\mathsf{F}^I{}_J = d\mathsf{A}^I{}_J + \mathsf{A}^I{}_K \wedge \mathsf{A}^K{}_J, \tag{G.21}$$

with $\mathsf{F}^x$ given by Eq. (G.7).

The next step consists in defining the SU(2)-covariant derivative acting on objects with covariant indices in a way compatible with the rule Eq. (G.14) for raising and lowering SU(2) indices. The convention that we will use is

$$\mathsf{D}\xi_I \equiv d\xi_I + \mathsf{A}_I{}^J \xi_J, \quad \text{where} \quad \mathsf{A}_I{}^J = (\mathsf{A}^I{}_J)^* = -\varepsilon_{IK} \mathsf{A}^K{}_L \varepsilon^{LJ}. \tag{G.22}$$

The consistency of the definition relies on the property Eq. (D.15) of the Pauli matrices. Sometimes, the indices of the Pauli matrices are also raised and lowered:

$$(\sigma^x)_I{}^J = \left[(\sigma^x)^I{}_J\right]^* = \varepsilon_{IK}(\sigma^x)^K{}_L\varepsilon^{LJ}, \tag{G.23}$$

although this can lead to confusion. With this notation, we have

$$\mathsf{A}_I{}^J = \tfrac{i}{2}\mathsf{A}^x(\sigma^x)_I{}^J, \qquad \mathsf{F}_I{}^J = d\mathsf{A}_I{}^J + \mathsf{A}_I{}^K \wedge \mathsf{A}_K{}^J = \tfrac{i}{2}\,\mathsf{F}^x\,(\sigma^x)_I{}^J. \tag{G.24}$$

At this point we have introduced an SU(2) connection $\mathsf{A}_u{}^I{}_J$ and we are going to introduce an $\mathrm{Sp}(2m)$ connection $\Delta_u{}^\alpha{}_\beta$. These can be seen as the only non-vanishing components of the spin connection, due to the special holonomy of these spaces. We also have the Levi-Civita connection, but these three connections must be related by the *Vielbein postulate* Eq. (1.91), which in the notation employed here takes the form

$$\mathsf{D}_u\,\mathsf{U}^{\alpha I}{}_v = \partial_u\mathsf{U}^{\alpha I}{}_v - \Gamma_{uv}{}^w\,\mathsf{U}^{\alpha I}{}_w + \mathsf{A}_u{}^I{}_J\,\mathsf{U}^{\alpha J}{}_v + \Delta_u{}^\alpha{}_\beta\,\mathsf{U}^{\beta I}{}_v = 0. \tag{G.25}$$

As in the general case, this postulate also relates the three respective curvatures

$$R_{ts}{}^{uv}\mathsf{U}^{\alpha I}{}_u\,\mathsf{U}^{\beta J}{}_v = -\mathsf{F}_{ts}{}^{IJ}\,\mathbb{C}^{\alpha\beta} - \overline{\mathsf{R}}_{ts}{}^{\alpha\beta}\,\varepsilon^{IJ}, \tag{G.26}$$

where

$$\overline{R}_{ts}{}^\alpha{}_\beta = 2\partial_{[t}\Delta_{s]}{}^\alpha{}_\beta + 2\Delta_{[t|}{}^\alpha{}_\gamma\Delta_{|s]}{}^\gamma{}_\beta \tag{G.27}$$

is the curvature of the $\mathrm{Sp}(2m)$ connection. This relation is equivalent to the statement that the holonomy of these spaces is contained in $\mathrm{Sp}(2) \times \mathrm{Sp}(2m)$.

The covariant constancy of the Pauli matrices and symplectic metric together with the covariant constancy of the Quadbeins suggests that it should be possible to express the hyper-Kähler 2-forms in terms of them. One can check that

$$\mathsf{K}^x{}_{uv} = -i\varepsilon_{IK}\sigma^{x\,K}{}_J\mathsf{U}^{\alpha I}{}_u\mathsf{U}^{\beta J}{}_v\mathbb{C}_{\alpha\beta} \tag{G.28}$$

satisfies the quaternionic algebra Eq. (G.2) and is covariantly constant, as required. It then follows that

$$\mathsf{U}^{\alpha I}{}_u\,\mathsf{U}^{\beta J}{}_v\mathbb{C}_{\alpha\beta} = \tfrac{1}{2}\mathsf{H}_{uv}\varepsilon^{IJ} - \tfrac{i}{2}\mathsf{K}^x{}_{uv}\sigma^{x\,I}{}_K\varepsilon^{KJ}. \tag{G.29}$$

The symmetric part of this equation is just Eq. (G.13) and the antisymmetric part of this equation leads to

$$\mathsf{K}^{IJ}{}_{uv} = \tfrac{i}{2}\mathsf{K}^x{}_{uv}\sigma^{x\,I}{}_K\varepsilon^{KJ} = -\mathsf{U}^{\alpha I}{}_{[u}\,\mathsf{U}^{\beta J}{}_{v]}\mathbb{C}_{\alpha\beta}, \tag{G.30}$$

from which we get the useful relation

$$\mathsf{F}_{\mu\nu}{}^{IJ} = -\varkappa\,\mathbb{C}_{\alpha\beta}\mathsf{U}^{\alpha I}{}_u\mathsf{U}^{\beta J}{}_v\partial_{[\mu}q^u\partial_{\nu]}q^v. \tag{G.31}$$

## G.1 Triholomorphic isometries of quaternionic-Kähler spaces

We want to study coordinate transformations that leave the geometries invariant (as has been the case for other geometries we have studied). These symmetries of the geometry will leave the metric invariant, and, therefore, they will generically be called isometries, but they must leave invariant all the other structures that define the geometry. When the structures that define a quaternionic-Kähler geometry are preserved by the transformation, it is called a *triholomorphic isometry* (just as they are called *holomorphic isometries* in complex Kähler spaces).

In order to study under which conditions the quaternionic-Kähler structures are preserved, we start by assuming that the metric $\mathsf{H}_{uv}$ admits Killing vectors $\mathsf{k}_A{}^u$ satisfying the Lie algebra

$$[\mathsf{k}_A, \mathsf{k}_B] = -f_{AB}{}^C \mathsf{k}_C. \tag{G.32}$$

The metric and the ungauged sigma model are invariant under the global transformations (with constant parameters $\alpha^A$)

$$\delta_\alpha q^u = \alpha^A \mathsf{k}_A{}^u(q). \tag{G.33}$$

In order to discuss the preservation of the remaining structures under these transformations we need to define SU(2)-covariant Lie derivatives. Let $\psi^x(q)$ be a field on the quaternionic-Kähler manifold transforming under local $\mathrm{SU}(2)$ transformations with parameter $\lambda^x(q)$ in the adjoint representation, infinitesimally

$$\delta_\lambda \psi^x = -\varepsilon^{xyz}\lambda^y\psi^z. \tag{G.34}$$

Its $\mathrm{SU}(2)$-covariant derivative is given by

$$\mathsf{D}\psi^x = d\psi^x + \varepsilon^{xyz}\mathsf{A}^y\psi^z, \tag{G.35}$$

where the $\mathrm{SU}(2)$ connection 1-form $\mathsf{A}^x$ transforms as

$$\delta_\lambda \mathsf{A}^x = \mathsf{D}\lambda^x. \tag{G.36}$$

In order to define an $\mathrm{SU}(2)$-covariant Lie derivative with respect to the Killing vector $\mathsf{k}_A$, which we will denote by $\mathbb{L}_A$, we have to add to the standard Lie derivative $\mathcal{L}_A$ a local $\mathrm{SU}(2)$ transformation whose transformation parameter is given by a compensator field[5] $\lambda_A{}^x$,

$$\mathbb{L}_A\psi^x \equiv \mathcal{L}_A\psi^x - \varepsilon^{xyz}\lambda_A{}^y\psi^z, \tag{G.37}$$

which, by definition, is linear in the Killing vector $\mathsf{k}_A$ and transforms according to the rule

$$\delta_\lambda\lambda_A{}^x = -\mathbb{L}_A\lambda^x = -(\mathcal{L}_A\lambda^x - \varepsilon^{xyz}\lambda_A{}^y\lambda^z). \tag{G.38}$$

$\mathbb{L}_A$ is a linear operator which satisfies the Leibniz rule for scalar and vector products of $\mathrm{SU}(2)$ vectors. The Lie derivative must also satisfy the crucial property

$$[\mathbb{L}_A, \mathbb{L}_B] = \mathbb{L}_{[\mathsf{k}_A,\mathsf{k}_A]} = -f_{AB}{}^C\mathbb{L}_C, \tag{G.39}$$

[5] This field is the $\mathrm{SU}(2)$ analog of the functions $\lambda_A$ that arise in the treatment of the symmetries of Kähler manifolds: both are parameters of compensating transformations of the respective symmetry groups associated with a given Killing vector labeled $A$.

which implies the Jacobi identity for the SU(2)-covariant Lie derivative. This requirement turns into the following property of the compensator field:

$$\mathcal{L}_A\lambda_B{}^x - \mathcal{L}_B\lambda_A{}^x - \varepsilon^{xyz}\lambda_A{}^y\lambda_B^z = f_{AB}{}^C\lambda_C{}^x, \qquad (G.40)$$

where we have used the linearity of $\lambda_A$ on $\mathsf{k}_A$ to replace $\lambda_{[\mathsf{k}_A,\,\mathsf{k}_B]}$ by $-f_{AB}{}^C\lambda_C$.

In order to find a solution to Eq. (G.40) we introduce the SU(2) vector $\mathsf{P}_A{}^x$, related to the compensator field $\lambda_A{}^x$ by[6]

$$\lambda_A{}^x \equiv \mathsf{k}_A{}^u\mathsf{A}^x{}_u - \mathsf{P}_A{}^x. \qquad (G.41)$$

$\mathsf{P}_A{}^x$ is going to be the *triholomorphic momentum map* when we impose the preservation of the hyper-Kähler structure $\mathsf{K}^x$ by the global transformations Eq. (G.33). Substituting this definition into Eq. (G.40) we find that $\mathsf{P}_A{}^x$ must satisfy the equivariance condition

$$\mathsf{D}_A\mathsf{P}_B{}^x - \mathsf{D}_B\mathsf{P}_A{}^x - \varepsilon^{xyz}\mathsf{P}_A{}^y\mathsf{P}_B{}^z - \varkappa\,\mathsf{k}_A{}^u\mathsf{k}_B{}^v\,\mathsf{K}^x{}_{uv} = -f_{AB}{}^C\mathsf{P}_C{}^x, \qquad (G.42)$$

where $\mathsf{D}_A \equiv \mathsf{k}_A{}^u\mathsf{D}_u$ and we have made use of Eq. (G.8).

Using the SU(2)-covariant Lie derivative $\mathbb{L}_A$ we can express the preservation of the hyper-Kähler structure $\mathsf{K}^x$ by the global transformations Eq. (G.33) simply as

$$\begin{aligned}
\mathbb{L}_A\mathsf{K}^x{}_{uv} &= \mathcal{L}_A\mathsf{K}^x{}_{uv} - \varepsilon^{xyz}(\mathsf{k}_A{}^w\mathsf{A}^y{}_w - \mathsf{P}_A{}^y)\mathsf{K}^z{}_{uv} \\
&= -\mathsf{k}_A^w\mathsf{D}_w\mathsf{K}^x{}_{uv} - 2(\nabla_{[u|}\mathsf{k}_A{}^w)\mathsf{K}^x{}_{w|v]} + \varepsilon^{xyz}\mathsf{P}_A{}^y\mathsf{K}^z{}_{uv} \\
&= -2(\nabla_{[u|}\mathsf{k}_A{}^w)\mathsf{K}^x{}_{w|v]} + \varepsilon^{xyz}\mathsf{P}_A{}^y\mathsf{K}^z{}_{uv} \\
&= 0,
\end{aligned} \qquad (G.43)$$

where we have used the covariant constancy of the hyper-Kähler structure. Contracting the whole equation with $\mathsf{K}^{y\,uv}$ we find

$$\mathsf{K}^{x\,uv}\nabla_u\mathsf{k}_{A\,v} = -2m\mathsf{P}_A{}^x. \qquad (G.44)$$

Now, acting on both sides of this equation with $\mathsf{D}_w$ and using the identity Eq. (1.116) for the Killing vector $\mathsf{k}_A{}^u$ we get

$$\mathsf{k}_A{}^r R_{wruv}\mathsf{K}^{x\,uv} = -2m\mathsf{D}_w\mathsf{P}_A{}^x. \qquad (G.45)$$

Using Eq. (G.28) in Eq. (G.26) we get

$$R_{wruv}\mathsf{K}^{x\,uv} = -2m\,\mathsf{F}^x{}_{wr} = -2m\varkappa\,\mathsf{K}^x{}_{wr}, \qquad (G.46)$$

and substituting, finally, we arrive at the following defining equation for the triholomorphic momentum map:

$$\boxed{\mathsf{D}_u\mathsf{P}_A{}^x = \varkappa\,\mathsf{K}^x{}_{uv}\mathsf{k}_A{}^v.} \qquad (G.47)$$

[6] Observe that this equation is the exact analog of Eq. (E.67) (or the following equations) that defines the momentum map of Kähler manifolds in terms of the Kähler connection contracted with the Killing vector and the parameters of the compensating Kähler transformation.

Contracting this equation with $\mathsf{K}^{x\,wu}$ (summing over $u$ and $x$) we get an equivalent equation

$$\boxed{\mathsf{k}_A{}^u = -\tfrac{1}{3\varkappa}\,\mathsf{K}^{x\,uv}\mathsf{D}_v\mathsf{P}_A{}^x,} \tag{G.48}$$

which says that the triholomorphic Killing vectors can be obtained from the triholomorphic momentum map, which is called, for this reason, the *triholomorphic Killing prepotential.*

Contracting again Eq. (G.47) with $\mathsf{k}_B{}^u$ we find

$$\mathsf{D}_B\mathsf{P}_A{}^x = \varkappa\,\mathsf{k}_B{}^u\mathsf{k}_A{}^v\mathsf{K}^x{}_{uv}, \tag{G.49}$$

which, when substituted directly into the equivariance equation Eq. (G.42), leads to

$$\mathbb{L}_A\mathsf{P}_B{}^x = -\mathsf{D}_A\mathsf{P}_B{}^x + \varepsilon^{xyz}\mathsf{P}_A{}^y\mathsf{P}_B{}^z - f_{AB}{}^C\mathsf{P}_C{}^x = 0, \tag{G.50}$$

which states that the triholomorphic momentum map is an invariant field and

$$\varepsilon^{xyz}\mathsf{P}_A{}^y\mathsf{P}_B{}^z - \varkappa\,\mathsf{k}_A{}^u\mathsf{k}_B{}^v\mathsf{K}^x{}_{uv} = f_{AB}{}^C\mathsf{P}_C{}^x. \tag{G.51}$$

This equation can also be understood as a realization of the Lie algebra of the Killing vectors in terms of $\mathsf{P}_A{}^x$, $\mathsf{K}^x{}_{uv}$ being the symplectic structure used to define the Poisson brackets which are the l.h.s.

There is a simpler way to arrive at Eq. (G.47) which can help us to understand better where the triholomorphic momentum maps come from; it consists in focusing on the transformation properties of the $\mathrm{SU}(2)$ connection 1-form. The discussion is similar to the one on p. 492 concerning the construction of the Lie–Maxwell derivative (a $\mathrm{U}(1)$-covariant Lie derivative).

Let us consider the standard Lie derivative of the $\mathrm{SU}(2)$ connection 1-form $\mathsf{A}_u{}^I{}_J$ with respect to the Killing vectors $\mathsf{k}_A{}^u$. Using the definition and adding and subtracting terms, we can rewrite it as follows:

$$\begin{aligned}\mathcal{L}_A\mathsf{A}_u{}^I{}_J &= \mathsf{k}_A{}^v\partial_v\mathsf{A}_u{}^I{}_J + \partial_u\mathsf{k}_A{}^v\mathsf{A}_v{}^I{}_J\\ &= \mathsf{k}_A{}^v 2\partial_{[v}\mathsf{A}_{u]}{}^I{}_J + \partial_u(\mathsf{k}_a{}^v\mathsf{A}_v{}^I{}_J)\\ &= \mathsf{k}_A{}^v\mathsf{F}_{vu}{}^I{}_J + \mathsf{D}_u(\mathsf{k}_A{}^v\mathsf{A}_v{}^I{}_J).\end{aligned} \tag{G.52}$$

If the quaternionic-Kähler structure is invariant, the $\mathrm{SU}(2)$ connection 1-form $\mathsf{A}_u{}^I{}_J$ must be invariant up to $\mathrm{SU}(2)$ gauge transformations for some gauge transformation parameter. This is possible only if the first term in Eq. (G.52) is the $\mathrm{SU}(2)$-covariant derivative of some parameter that we define as (minus) the triholomorphic momentum map:

$$\mathsf{k}_A{}^v\mathsf{F}_{vu}{}^I{}_J = \mathsf{D}_u(-\mathsf{P}_A{}^I{}_J), \tag{G.53}$$

or

$$\boxed{\mathsf{k}_A{}^v\mathsf{F}_{vu}{}^x = -\mathsf{D}_u\mathsf{P}_A{}^x.} \tag{G.54}$$

This formula is equivalent to Eq. (G.47) thanks to the defining property Eq. (G.8). Using it in the preceding equation, we get

$$\mathcal{L}_A \mathsf{A}_u{}^I{}_J = \mathsf{D}_u \lambda_A{}^I{}_J, \quad \text{with} \quad \lambda_A{}^I{}_J = \mathsf{k}_A{}^v \mathsf{A}_v{}^I{}_J - \mathsf{P}_A{}^I{}_J. \tag{G.55}$$

Again, for the sake of completeness, we give in Table G.1 the list of the symmetric quaternionic-Kähler spaces taken from Ref. [44].

### *G.1.1 Alternative notation for the $d = 5$ case*

In order to conform to the conventions of Refs. [152, 153], our notation for the quaternionic-Kähler geometry of $N = 1, d = 5$ supergravities differs from the one used in the $N = 2, d = 4$ theories in a couple of points:[7] the scalars are labeled by $X, Y, \ldots$ instead of $u, v, \ldots$ and the scalar metric is $g_{XY}(q)$ instead of $\mathsf{H}_{uv}$. Adjoint $\mathrm{SU}(2)$ indices are written in vector form or with an index $r, s = 1, 2, 3$, so an $\mathfrak{su}(2)$ vector $\psi^x$ is now written in the form $\vec{\psi}$ or $\psi^r$ and the hyper-Kähler structure is $\vec{J}_{XY}$. The $\mathrm{Sp}(2n_H)$ indices will be $A, B, \ldots$ instead of $\alpha, \beta, \ldots$ and the Quadbeins $\mathsf{U}^{\alpha I}{}_u$ now take the form $f^{iA}{}_X$.

Accordingly, we use vector products $\vec{\psi} \times \vec{\eta}$ instead of $\epsilon^{xyz}\psi^y\eta^z$. The $\mathrm{SU}(2)$ connection $\mathsf{A}^x{}_u$ is now $\vec{\omega}_X$, but its pullback over the spacetime $\vec{\omega}_X \partial_\mu q^X$ is denoted by $\vec{\mathsf{A}}_\mu$. The generators of $\mathrm{SU}(2)$ in the fundamental representation are the three Pauli matrices multiplied by $i$, but the indices are written in a different position (with respect to our conventions): $\vec{\sigma}_i{}^j$. The action of infinitesimal $\mathrm{SU}(2)$ transformations with parameter $\vec{\lambda}$ on an object with an upper fundamental index is defined by

$$\delta_\lambda \psi^i = i\psi^j \vec{\sigma}_j{}^i \cdot \vec{\lambda}, \tag{G.56}$$

and the $\mathrm{SU}(2)$-covariant derivative is

$$\mathsf{D}_\mu \psi^i = \partial_\mu \psi^i + \psi^j \mathsf{A}_{\mu\, j}{}^i, \qquad \mathsf{A}_{\mu\, j}{}^i = i\mathsf{A}_\mu{}^r \sigma^r{}_j{}^i. \tag{G.57}$$

This definition induces two changes in the sign of the commutator with respect to the four-dimensional case. Then the $\mathfrak{su}(2)$ structure constants are twice those of the four-dimensional case, so infinitesimal $\mathrm{SU}(2)$ transformations with parameter $\vec{\lambda}$ on objects transforming in the adjoint representation take the form

$$\delta_\lambda \vec{\psi} = -2\vec{\lambda} \times \vec{\psi}, \qquad \delta_\lambda \vec{\omega} = -2\vec{\lambda} \times \vec{\omega} + d\vec{\lambda}. \tag{G.58}$$

The compensating $\mathrm{SU}(2)$ gauge transformation associated with the Killing vector now denoted by $k_A{}^X$ is written as $\vec{\eta}_A$, and the triholomorphic momentum map $\mathsf{P}^x_A$ is now $\frac{1}{2}\vec{P}_A$. Its defining equations take the form

$$\vec{J}_X{}^Y \nabla_Y k_A{}^X = 2n_H \vec{P}_A, \tag{G.59}$$

$$\mathfrak{D}_X \vec{P}_A = -\tfrac{1}{2}\vec{J}_{XY} k_A{}^Y, \tag{G.60}$$

and the equivariance condition takes the form

$$\vec{P}_A \times \vec{P}_B + \tfrac{1}{2} k_A{}^X \vec{J}_{XY} k_B{}^Y = f_{AB}{}^C \vec{P}_C. \tag{G.61}$$

[7] For more details the reader can consult the appendix of Ref. [132].

Table G.1. List of symmetric quaternionic-Kähler spaces: $m$ is the quaternionic dimension (the number of hypermultiplets) and the real dimension is $4m$

| $m$ | Coset G/H |
|---|---|
| $m$ | $\frac{\mathrm{Sp}(2m+2,\mathbb{R})}{\mathrm{Sp}(2,\mathbb{R})\times\mathrm{Sp}(2m,\mathbb{R})}$ |
| $m$ | $\frac{\mathrm{SU}(m,2)}{\mathrm{SU}(m)\times\mathrm{SU}(2)\times\mathrm{U}(1)}$ |
| $m$ | $\frac{\mathrm{SO}(4,m)}{\mathrm{SO}(4)\times\mathrm{SO}(m)}$ |
| 2 | $\frac{\mathrm{G}_2}{\mathrm{SO}(4)}$ |
| 7 | $\frac{\mathrm{F}_4}{\mathrm{Sp}(6,\mathbb{R})\times\mathrm{Sp}(2,\mathbb{R})}$ |
| 10 | $\frac{\mathrm{E}_6}{\mathrm{SU}(6)\times\mathrm{U}(1)}$ |
| 16 | $\frac{\mathrm{E}_7}{\mathrm{SO}(12)\times\mathrm{SU}(2)}$ |
| 28 | $\frac{\mathrm{E}_8}{\mathrm{E}(7)\times\mathrm{SU}(2)}$ |

# Appendix H
# Real special geometry

Real special geometry is the geometry that describes the couplings of the real scalars of the vector supermultiplets in minimal five-dimensional supergravity (that we call $N=1$ since the supersymmetry parameter is just one minimal five-dimensional spinor and is very often called $N=2$ in the literature).

Just as special Kähler geometry arises from the need (imposed by supersymmetry) to integrate in a single structure the Kähler geometry of the $\sigma$-model parametrized by the $n_V$ complex scalars $Z^i$, $x=i,\ldots,n_V$ with the $\mathrm{Sp}(2n_V+2,\mathbb{R})$ structure that controls their coupling to the vector fields, via the period matrix $\mathcal{N}_{\Lambda\Sigma}(Z,Z^*)$, $\Lambda,\Sigma=0,1,\ldots,n_V$, so real special geometry arises from the need to integrate in a single structure the Riemannian metric of the $\sigma$-model parametrized by the $n_V$ real scalars $\phi^x$, $x=1,\ldots,n_V$, with the $\mathrm{GL}(n_V+1)$ structure that controls their coupling to the vector fields, via the kinetic matrix $a_{IJ}(\phi)$, $I,J=0,1,\ldots,n_V$. In the special Kähler case, $\Im\mathrm{m}\mathcal{N}_{\Lambda\Sigma}$ shares many of the properties of a metric and the same happens here with $a_{IJ}$. On the other hand, the need to make the $\mathrm{GL}(n_V+1)$ symmetry manifest on the scalars requires the introduction of a redundant description in terms of a set of constrained functions $h^I(\phi)$, just as in the special Kähler case a similar need requires the introduction of the canonical symplectic section $\mathcal{V}^M(Z,Z^*)$ which satisfies a number of constraints as well. Note that $h^I(\phi)$ will transform contravariantly in the vector representation of $\mathrm{GL}(n_V+1)$.

The constraint satisfied by the $n_V+1$ functions $h^I(\phi)$ always has the same form[1],

$$C_{IJK}h^I h^J h^K = 1, \tag{H.1}$$

where $C_{IJK}$ is a constant real symmetric tensor. The space in which the $n_V$ scalars $\phi^x$ "live" (as coordinates) is a hypersurface in an $(n_V+1)$-dimensional Riemannian space defined by this constraint. Another set of variables can also be defined as follows:

$$h_I \equiv \frac{1}{3}\frac{\partial}{\partial h^I}C_{JKL}h^J h^K h^L = C_{IJK}h^J h^K \quad \Rightarrow h_I h^I = 1. \tag{H.2}$$

By construction, these variables will transform covariantly, and it is natural to define a metric $a_{IJ}$ that relates these two sets of variables $h^I$ and $h_I$ ("raising" and "lowering" the

[1] We use the conventions of Refs. [152, 153] with $\kappa=1/\sqrt{2}$. Observe that the real special geometry identities in Appendix C of Ref. [153] are valid only for $\kappa=1$.

$\mathrm{GL}(n_V + 1)$ indices):

$$h_I \equiv a_{IJ} h^J, \qquad h^I \equiv a^{IJ} h_J, \qquad a_{IJ} a^{JK} = \delta^I{}_K. \tag{H.3}$$

This metric can be understood as the metric of the $(n_V + 1)$-dimensional Riemannian space. It can be checked that a matrix $a_{IJ}$ with the required properties is given by

$$a_{IJ} = -2C_{IJK} h^K + 3h_I h_J. \tag{H.4}$$

Next, we define the derivatives of the functions $h^I$ and $h_I$ with respect to the coordinates $\phi^x$, with a normalization factor

$$h^I_x \equiv -\sqrt{3} h^I{}_{,x} \equiv -\sqrt{3} \frac{\partial h^I}{\partial \phi^x}, \tag{H.5}$$

$$h_{Ix} \equiv +\sqrt{3} h_{I,x}. \tag{H.6}$$

These two objects are related by the metric $a_{IJ}$:

$$h_{Ix} = a_{IJ} h^J_x, \tag{H.7}$$

and, furthermore, due to Eq. (H.1), they satisfy the orthogonality relations

$$h_I h^I_x = 0, \qquad h^I h_{Ix} = 0. \tag{H.8}$$

We can combine $h^I$ and $h^I_x$ in a single object that satisfies the following orthonormality and completeness properties:

$$\begin{pmatrix} h^I \\ h^I_x \end{pmatrix} (h_I, h^y_I) = \begin{pmatrix} 1 & 0 \\ 0 & \delta^y_x \end{pmatrix}, \qquad (h_I, h^x_I) \begin{pmatrix} h^J \\ h^J_x \end{pmatrix} = \delta^J_I, \tag{H.9}$$

which allows us to decompose any object with a $\mathrm{GL}(n_V + 1)$ index $I$ as

$$A^I = (h_J A^J)\, h^I + (h^x_J A^J)\, h^I_x. \tag{H.10}$$

The metric of the hypersurface $C_{IJK} h^I h^J h^K = 1$ in which the physical scalars live, $g_{xy}(\phi)$, which will appear as the $\sigma$-model metric, is defined as the pullback of $a_{IJ}$:

$$g_{xy} \equiv a_{IJ} h^I_x h^J_y = -2C_{IJK} h^I_x h^J_y h^K, \tag{H.11}$$

and can be used to raise and lower $x, y$ indices. Other useful expressions are

$$a_{IJ} = h_I h_J + h^x_I h_{Jx}, \tag{H.12}$$

$$C_{IJK} h^K = h_I h_J - \tfrac{1}{2} h^x_I h_{Jx}, \tag{H.13}$$

$$h_I h_J = \tfrac{1}{3} a_{IJ} + \tfrac{2}{3} C_{IJK} h^K, \tag{H.14}$$

$$h^x_I h_{Jx} = \tfrac{2}{3} a_{IJ} - \tfrac{2}{3} C_{IJK} h^K. \tag{H.15}$$

The Levi-Civita connection associated with the scalar metric $g_{xy}$ is denoted by $\Gamma_{xy}{}^{z}$ and the associated general-covariant derivative (in target space) is defined according to the general conventions in Chapter 1. For instance:

$$h_{Ix;y} \equiv h_{Ix,y} - \Gamma_{xy}{}^{z} h_{Iz}. \tag{H.16}$$

We define with it the $T_{xyz}$ tensor:

$$T_{xyz} \equiv \sqrt{3} h_{Ix;y} h^{I}_{z} = -\sqrt{3} h_{Ix} h^{I}_{y;z}. \tag{H.17}$$

It can be shown that

$$h_{Ix;y} = \tfrac{1}{\sqrt{3}}(h_I g_{xy} + T_{xyz} h^{z}_{I}), \tag{H.18}$$

$$h^{I}_{x;y} = -\tfrac{1}{\sqrt{3}}(h^I g_{xy} + T_{xyz} h^{Iz}), \tag{H.19}$$

$$\Gamma_{xy}{}^{z} = h^{Iz} h_{Ix,y} - \tfrac{1}{\sqrt{3}} T_{xy}{}^{w} = 8 h^{z}_{I} h^{I}_{x,y} + \tfrac{1}{\sqrt{3}} T_{xy}{}^{w}. \tag{H.20}$$

## H.1 The isometries of real special manifolds

As in other cases, we must require the Killing vectors of the $\sigma$-model metric $g_{xy}(\phi)$, $k_A{}^x(\phi)$, to respect the real special structure; this is equivalent to the following two requirements.

1. The functions $h^I(\phi)$ must be *invariant functions*: invariant under the isometries generated by the $k_A{}^x(\phi)$ up to $\mathrm{GL}(n_V+1)$ rotations with matrices $(T_A{}^I{}_J)$ satisfying

$$[k_A, k_B] = -f_{AB}{}^{C} k_C, \qquad [T_A, T_B] = f_{AB}{}^{C} T_C. \tag{H.21}$$

This condition can be written in terms of a covariant Lie derivative as

$$\mathbb{L}_A h^I \equiv (\mathcal{L}_A - T_A) h^I = k_A{}^x \partial_x h^I - T_A{}^I{}_J h^J = 0, \tag{H.22}$$

and, multiplying by $-\sqrt{3} h^{y}_{I}$ and using $h^{I}_{x} h^{y}_{I} = \delta^{y}_{x}$, this implies that the Killing vectors must be given by

$$k_A{}^x = -\sqrt{3} T_A{}^I{}_J h^{x}_{I} h^J. \tag{H.23}$$

Furthermore, multiplying by $h_I$ and using $h_I h^{I}_{x} = 0$, we get the property

$$T_A{}^I{}_J h_I h^J = 0, \tag{H.24}$$

which can be read as an orthogonality condition with respect to the metric $a_{IJ}$:

$$T_A{}^I{}_{(J} a_{K)I} = 0. \tag{H.25}$$

2. The tensor $C_{IJK}$ must be invariant under the compensating $\mathrm{GL}(n_V+1)$ transformations with matrices $(T_A{}^I{}_J)$:

$$\delta_A C_{IJK} = -3T_A{}^L{}_{(I}C_{JK)L} = 0. \tag{H.26}$$

These two conditions imply the invariance of $h^I_x$ and of the kinetic matrix $a_{IJ}$ under the isometries

$$\mathbb{L}_A h^I_x = (\mathcal{L}_A - T_A)h^I_x = k_A{}^y\partial_y h^I_x + \partial_x k_A{}^y h^I_y - T_A{}^I{}_J h^I_x = 0, \tag{H.27}$$

$$\mathbb{L}_A a_{IJ} = (\mathcal{L}_A - T_A)a_{IJ} = k_A{}^x\partial_x a_{IJ} + 2T_A{}^K{}_{(I}a_{J)K} = 0. \tag{H.28}$$

Again, for the sake of completeness, we give in Table H.1 the list of the symmetric special real spaces and other very special manifolds taken from Ref. [564].

Table H.1. Symmetric very special spaces

Taken from Ref. [564] and classified by the integers $q = -1, 0, 1, 2, \ldots$ and $P = 0, 1, 2, \ldots$ according to Ref. [1266]. In that reference, as in a large part of the literature, the real special manifolds we have discussed are referred to as *very special manifolds*. The symmetric ones are given in the third column of this table and correspond to special choices of the tensor $C_{IJK}$. Upon dimensional reduction of the supergravity to $d = 4$, all these manifolds, except for those in the second row of the table, become special Kähler manifolds of an $N = 2, d = 4$ supergravity, the relation being known as *r-map*. The corresponding special Kähler geometries are known as very special Kähler geometries and are given in the fourth column of the table. The dimensional reduction of the supergravity to $d = 3$ gives a supergravity characterized by a quaternionic-Kähler geometry known as very special quaternionic-Kähler geometry. They are given in the last column. The real dimensions of these spaces for a given $n$ is $n - 1$ in the real case, $2n$ in the complex case, and $4(n + 1)$ in the quaternionic one.

| $(q, P)$ | $n$ | Real | Kähler | Quaternionic-Kähler |
|---|---|---|---|---|
| $(-1, 0)$ | 2 | $\mathrm{SO}(1,1)$ | $\left[\frac{\mathrm{SU}(1,1)}{\mathrm{U}(1)}\right]^2$ | $\frac{\mathrm{SO}(3,4)}{(\mathrm{SU}(2))^2}$ |
| $(-1, P)$ | $2 + P$ | $\frac{\mathrm{SO}(P+1,1)}{\mathrm{SO}(P+1)}$ | | |
| $(0, P)$ | $3 + P$ | $\mathrm{SO}(1,1) \times \frac{\mathrm{SO}(P+1,1)}{\mathrm{SO}(P+1)}$ | $\frac{\mathrm{SU}(1,1)}{\mathrm{U}(1)} \times \frac{\mathrm{SO}(P+2,2)}{\mathrm{SO}(P+2)\times\mathrm{SO}(2)}$ | $\frac{\mathrm{SO}(P+4,4)}{\mathrm{SO}(P+4)\times\mathrm{SO}(4)}$ |
| $(1, 1)$ | 6 | $\frac{\mathrm{SL}(3,\mathbb{R})}{\mathrm{SU}(3)}$ | $\frac{\mathrm{Sp}(6,\mathbb{R})}{\mathrm{U}(3)}$ | $\frac{\mathrm{F}_{4(4)}}{\mathrm{USp}(6)\times\mathrm{SU}(2)}$ |
| $(2, 1)$ | 9 | $\frac{\mathrm{SL}(3,\mathbb{C})}{\mathrm{SU}(3)}$ | $\frac{\mathrm{U}(3,3)}{\mathrm{U}(3)\times\mathrm{U}(3)}$ | $\frac{\mathrm{E}_{6(2)}}{\mathrm{SU}(6)\times\mathrm{SU}(2)}$ |
| $(4, 1)$ | 15 | $\frac{\mathrm{SU}^*(6)}{\mathrm{USp}(6)}$ | $\frac{\mathrm{SO}^*(12)}{\mathrm{U}(6)}$ | $\frac{\mathrm{E}_{7(-5)}}{\mathrm{S})(12)\times\mathrm{SU}(2)}$ |
| $8, 1$ | 27 | $\frac{\mathrm{E}_{6,(-26)}}{\mathrm{F}_4}$ | $\frac{\mathrm{E}_{7,(-25)}}{\mathrm{E}_6\times\mathrm{SO}(2)}$ | $\frac{\mathrm{E}_{8,(-24)}}{\mathrm{E}_7\times\mathrm{SO}(2)}$ |

# Appendix I
# The generic scalar manifolds of $N \geq 2, d = 4$ SUEGRAs

In this appendix we review the formalism used to describe in a unified way all the scalar manifolds of all the $N \geq 2, d = 4$ SUEGRAs coupled to $n_V$ vector multiplets proposed in Ref. [47]. This formalism is used in Chapter 8 to give a unified description of all those SUEGRAs since the main qualitative difference between them lies in the scalar sector. The original Ref. [47] contains more details, examples, and also the extension of this formalism to higher dimensions.

The total number of vector fields, $\bar{n}$ (labeled, as usual, by $\Lambda, \Sigma, \ldots$) for a given $N$ is always given by the sum of the number of graviphotons, $N(N-1)/2$, and the number of generic vector multiplets of this formalism, $n_V$ (labeled by $i, j, \ldots$) according to Table I.1. As explained in the main text, in this formalism, the supergravity multiplet of the $N = 6$ theory is described by one generic supergravity multiplet plus one generic vector multiplet and, therefore, $n_V = 1$ even though there are no true $N \geq 5$ vector supermultiplets.

Table I.1.

| $N$ | 3 | 4 | 5 | 6 | 8 |
|---|---|---|---|---|---|
| $n_V$ | $n_V$ | $n_V$ | 0 | 1 | 0 |
| $\bar{n}$ | $3 + n_V$ | $6 + n_V$ | 10 | 16 | 28 |

The formalism is based on the fact that all the scalar manifolds[1] can be described by a $\mathrm{USp}(\bar{n}, \bar{n})$ matrix $U$. This matrix is constructed using two $\bar{n} \times \bar{n}$ matrices $f$ and $h$ with components

$$f \equiv (f^{\Lambda}{}_{IJ}, f^{\Lambda}{}_{i}), \qquad h \equiv (h_{\Lambda\, IJ}, h_{\Lambda\, i}), \tag{I.1}$$

where $I, J = 1, \ldots, N$ are $\mathrm{U}(N)$ indices and $f^{\Lambda}{}_{IJ}$ and $h_{\Lambda\, IJ}$ are antisymmetric in the pair $IJ$, as follows:

$$U \equiv \frac{1}{\sqrt{2}} \begin{pmatrix} f + ih & f^* + ih^* \\ f - ih & f^* - ih^* \end{pmatrix}. \tag{I.2}$$

[1] The list for $N \geq 3$ is in Table 5.1.

The condition that $U \in \mathrm{USp}(\bar{n}, \bar{n})$ is

$$U^{-1} = \begin{pmatrix} 1 & 0 \\ 0 & -1 \end{pmatrix} U^{\dagger} \begin{pmatrix} 1 & 0 \\ 0 & -1 \end{pmatrix} = \begin{pmatrix} 0 & 1 \\ -1 & 0 \end{pmatrix} U^{T} \begin{pmatrix} 0 & -1 \\ 1 & 0 \end{pmatrix}$$

$$= \tfrac{1}{\sqrt{2}} \begin{pmatrix} f^{\dagger} - ih^{\dagger} & -(f^{\dagger} + ih^{\dagger}) \\ -(f - ih) & f + ih \end{pmatrix}, \tag{I.3}$$

which implies that the matrices $f$ and $h$ satisfy the constraints[2]

$$i(f^{\dagger}h - h^{\dagger}f) = 1, \qquad f^{T}h - h^{T}f = 0. \tag{I.4}$$

These constraints are satisfied for all $N \geq 2$ SUEGRAs using appropriate definitions of $f$ and $h$ [47, 528], which justifies this formalism.

The matrices $f$ and $h$ can also be used to build two sets of symplectic vectors $\mathcal{V}^{M}{}_{IJ}$ and $\mathcal{V}^{M}{}_{i}$ defined by

$$\mathcal{V}_{IJ} \equiv \begin{pmatrix} f^{\Lambda}{}_{IJ} \\ h_{\Lambda IJ} \end{pmatrix}, \qquad \mathcal{V}_{i} \equiv \begin{pmatrix} f^{\Lambda}{}_{i} \\ h_{\Lambda i} \end{pmatrix}, \tag{I.5}$$

and, in terms of the symplectic vectors, the constraints satisfied by $f$ and $h$ can be written in the form[3]

$$\begin{aligned} \langle \mathcal{V}_{IJ} \mid \mathcal{V}^{*\,KL} \rangle &= -2i\delta^{KL}{}_{IJ}, \\ \langle \mathcal{V}_{i} \mid \mathcal{V}^{*\,j} \rangle &= -i\delta_{i}{}^{j}, \end{aligned} \tag{I.6}$$

with the rest of the symplectic products vanishing.

The indices $IJ$ and $i$ can be seen as indices in the holonomy group $\mathrm{H} = \mathrm{H}_{\mathrm{aut}} \times \mathrm{H}_{\mathrm{matter}}$, where $\mathrm{H}_{\mathrm{aut}} = \mathrm{U}(N)$ and the symplectic indices $M, N$ can be seen as indices of a symplectic representation of the duality group G, the matrix $U$ can be seen as a sort of coset representative of G/H, and one can use the formalism studied in Appendix A.4, defining a left-invariant Maurer–Cartan 1-form that can be written as a linear combination of Vielbein $P$ and 1-form connection $\Omega$ (with components that have different combinations of the indices $IJ$ and $i$). In matrix form, this combination takes the form

$$\Gamma \equiv U^{-1}dU = \begin{pmatrix} \Omega & P^{*} \\ P & \Omega^{*} \end{pmatrix}. \tag{I.7}$$

The different components of the connection are given explicitly by

$$\Omega = \begin{pmatrix} \Omega^{KL}{}_{IJ} & \Omega^{j}{}_{IJ} \\ \Omega^{KL}{}_{i} & \Omega^{j}{}_{i} \end{pmatrix} = \begin{pmatrix} i\langle d\mathcal{V}_{IJ} \mid \mathcal{V}^{*\,KL} \rangle & i\langle d\mathcal{V}_{IJ} \mid \mathcal{V}^{*\,j} \rangle \\ i\langle d\mathcal{V}_{i} \mid \mathcal{V}^{*\,KL} \rangle & i\langle d\mathcal{V}_{i} \mid \mathcal{V}^{*\,j} \rangle \end{pmatrix}, \tag{I.8}$$

and those of the Vielbein are

$$P = \begin{pmatrix} P_{KLIJ} & P_{jIJ} \\ P_{KLi} & P_{ij} \end{pmatrix} = \begin{pmatrix} -i\langle d\mathcal{V}_{IJ} \mid \mathcal{V}_{KL} \rangle & -i\langle d\mathcal{V}_{IJ} \mid \mathcal{V}_{j} \rangle \\ -i\langle d\mathcal{V}_{i} \mid \mathcal{V}_{KL} \rangle & -i\langle d\mathcal{V}_{i} \mid \mathcal{V}_{j} \rangle \end{pmatrix}. \tag{I.9}$$

[2] When we multiply these matrices we must include a factor $1/2$ for each contraction of pairs of antisymmetric indices $IJ$.

[3] Complex conjugation only raises or lowers the $i, j$ indices in this formalism.

The definition of the period matrix $\mathcal{N}_{\Lambda\Sigma}$ in this framework,

$$\mathcal{N} = hf^{-1} = \mathcal{N}^T, \tag{I.10}$$

implies the following properties that generalize those satisfied by the period matrix and the components of the $N = 2$ canonical section and its Kähler-covariant derivative:[4]

$$h_{\Lambda\,\cdot} = \mathcal{N}_{\Lambda\Sigma} f^{\Sigma}{}_{\cdot}, \tag{I.11}$$

$$\mathfrak{D}h_{\Lambda\,\cdot} = \mathcal{N}^*_{\Lambda\Sigma}\mathfrak{D}f^{\Lambda}{}_{\cdot}, \tag{I.12}$$

$$-\tfrac{1}{2}\Im\mathrm{m}\mathcal{N}^{\Lambda\Sigma} = \tfrac{1}{2}f^{\Lambda}{}_{IJ}f^{*\Sigma IJ} + f^{\Lambda}{}_{i}f^{*\Sigma i}, \tag{I.13}$$

$$i = \tfrac{1}{2} \mid \mathcal{V}_{IJ}\rangle\langle\mathcal{V}^{*\,IJ} \mid \; + \mid \mathcal{V}_i\rangle\langle\mathcal{V}^{*\,i} \mid \; - \text{ c.c.} \tag{I.14}$$

The latter two identities can be combined (using the first) into the identity of symplectic matrices

$$\tfrac{1}{2}\mathcal{V}_{M\,IJ}\mathcal{V}^{*\;IJ}_{M} + \mathcal{V}_{M\,i}\mathcal{V}^{*\;i}_{M} = -\tfrac{1}{2}(\mathcal{M}(\mathcal{N}) + i\Omega)_{MN}, \tag{I.15}$$

where $\mathcal{M}(\mathcal{N})$ is the symmetric symplectic matrix defined in Eq. (C.42) and $\Omega$ is the symplectic "metric" in Eq. (C.1).

The exterior $\mathrm{H}_{\mathrm{Aut}} \times \mathrm{H}_{\mathrm{Matter}}$-covariant derivative of the symplectic vectors $\mathcal{V}_{IJ}, \mathcal{V}_i$ can be defined using the connection $\Omega$ (with the right indices) by

$$\mathfrak{D}\mathcal{V} \equiv d\mathcal{V} - \mathcal{V}\Omega, \tag{I.16}$$

that is

$$\mathfrak{D}\mathcal{V}_i = d\mathcal{V}_i - \mathcal{V}_j\Omega^j{}_i - \tfrac{1}{2}\mathcal{V}_{IJ}\Omega^{IJ}{}_i, \tag{I.17}$$

$$\mathfrak{D}\mathcal{V}_{IJ} = d\mathcal{V}_{IJ} - \mathcal{V}_j\Omega^j{}_{IJ} - \tfrac{1}{2}\mathcal{V}_{KL}\Omega^{KL}{}_{IJ}. \tag{I.18}$$

Taking the symplectic product of these derivatives with $\mathcal{V}^{*j}$ and $\mathcal{V}^{*\,KL}$ and using Eqs. (I.6) and the definition Eq. (I.8) we find that the mixed components of the connection vanish:

$$\Omega^{KL}{}_i = \Omega^j{}_{IJ} = 0. \tag{I.19}$$

Furthermore, taking the symplectic product of these derivatives with $\mathcal{V}_j$ and $\mathcal{V}_{KL}$ and using Eqs. (I.6), the definition Eq. (I.9), and the property Eq. (I.11) we get the following expressions for the Vielbeins:

$$P_{IJKL} = -2f^{\Lambda}{}_{IJ}\Im\mathrm{m}\mathcal{N}_{\Lambda\Sigma}\,\mathfrak{D}f^{\Sigma}{}_{KL}, \tag{I.20}$$

$$P_{iIJ} = -2f^{\Lambda}{}_{i}\Im\mathrm{m}\mathcal{N}_{\Lambda\Sigma}\,\mathfrak{D}f^{\Sigma}{}_{IJ}, \tag{I.21}$$

$$P_{ij} = -2f^{\Lambda}{}_{i}\Im\mathrm{m}\mathcal{N}_{\Lambda\Sigma}\,\mathfrak{D}f^{\Sigma}{}_{j}. \tag{I.22}$$

[4] We use the notation $\Im\mathrm{m}\mathcal{N}^{\Lambda\Sigma} \equiv (\Im\mathrm{m}\mathcal{N})^{-1|\Lambda\Sigma}$.

These equations can be inverted using Eq. (I.13) to give

$$\mathfrak{D} f^{\Lambda}{}_{IJ} = f^{*\,\Lambda i} P_{iIJ} + \tfrac{1}{2} f^{*\Lambda KL} P_{IJKL}, \tag{I.23}$$

$$\mathfrak{D} f^{\Lambda}{}_{i} = f^{*\,\Lambda j} P_{ij} + \tfrac{1}{2} f^{*\Lambda IJ} P_{iIJ}. \tag{I.24}$$

Then, for any choice of indices, we have the identities

$$\langle\, \mathfrak{D}\mathcal{V} \mid \mathcal{V}^* \,\rangle = 0, \qquad \langle\, \mathfrak{D}\mathcal{V} \mid \mathcal{V} \,\rangle = \langle\, d\mathcal{V} \mid \mathcal{V} \,\rangle = iP. \tag{I.25}$$

It is useful to introduce the inverse Vielbeins $P^{*\,IJKL}, P^{*\,iIJ}, P^{*\,ij}$. If we label the physical scalar fields (the coordinates of the scalar manifold) by $\phi^A$, these are defined by

$$P^{*\,IJKL\,A} P_{MNOP\,A} = 4!\delta^{IJKL}{}_{MNOP}, \qquad P^{*\,iIJ\,A} P_{jKL\,A} = 2\delta^{i}{}_{j}\delta^{IJ}{}_{KL}. \tag{I.26}$$

Observe that $P^{*\,IJKL\,A} P_{iMN,A} = 0$ but $P^{*\,IJKL\,A} P_{ij,A} \neq 0$ in general.

Then, using the completeness relation and the definitions of the Vielbeins and their inverses, we find

$$\langle\, \mathfrak{D}_A \mathcal{V}_{IJ} \mid \mathfrak{D}_B \mathcal{V}^{*\,KL} \,\rangle = \tfrac{i}{2} P_{IJMN\,A} P^{*\,KLMN}{}_{B} + i P_{iIJ\,A} P^{*\,iKL}{}_{B}, \tag{I.27}$$

$$\langle\, \mathfrak{D}_A \mathcal{V}_{IJ} \mid \mathfrak{D}_B \mathcal{V}^{*\,i} \,\rangle = \tfrac{i}{2} P_{IJKL\,A} P^{*\,iKL}{}_{B} + i P_{jIJ\,A} P^{*\,ij}{}_{B}, \tag{I.28}$$

$$\langle\, \mathfrak{D}_A \mathcal{V}_{i} \mid \mathfrak{D}_B \mathcal{V}^{*\,j} \,\rangle = \tfrac{i}{2} P_{iIJ\,A} P^{*\,iIJ}{}_{B} + i P_{ik\,A} P^{*\,jk}{}_{B}. \tag{I.29}$$

The symplectic products of covariant derivatives with all indices up or down vanish identically.

In order to deal with the scalar equations of motion it is necessary to have identities relating the derivatives of the period matrix with respect to the scalars to the vectors $\mathcal{V}$ and the Vielbeins. Using the definition of the period matrix Eq. (I.10), Eqs. (I.11) and (I.12), and the first of Eqs. (I.4) we get

$$d\mathcal{N} = 4i\Im\mathrm{m}\mathcal{N}\, \mathfrak{D} f f^{\dagger}\, \Im\mathrm{m}\mathcal{N}. \tag{I.30}$$

The covariant derivative on the r.h.s. can be expressed in terms of the Vielbeins using Eqs. (I.23) and (I.24):

$$d\mathcal{N}_{\Lambda\Sigma} = i\Im\mathrm{m}\mathcal{N}_{\Gamma(\Lambda}\Im\mathrm{m}\mathcal{N}_{\Sigma)\Omega}\left[P_{IJKL} f^{*\,\Gamma IJ} f^{*\,\Omega KL} + 4P_{iIJ} f^{*\,\Gamma i} f^{*\,\Omega IJ} + 4P_{ij} f^{*\,\Gamma i} f^{*\,\Omega j}\right], \tag{I.31}$$

and now we can use the inverse Vielbeins to get

$$P^{*\,IJKL\,A} \frac{\partial}{\partial\phi^A} \mathcal{N}_{\Lambda\Sigma} = 4! i\Im\mathrm{m}\mathcal{N}_{\Omega(\Lambda}\Im\mathrm{m}\mathcal{N}_{\Sigma)\Delta} f^{*\,\Omega[IJ|} f^{*\,\Delta|KL]}, \tag{I.32}$$

$$P^{*\,iIJ\,A} \frac{\partial}{\partial\phi^A} \mathcal{N}_{\Lambda\Sigma} = 8i\Im\mathrm{m}\mathcal{N}_{\Omega(\Lambda}\Im\mathrm{m}\mathcal{N}_{\Sigma)\Delta} f^{*\,\Omega i} f^{*\,\Delta IJ}, \tag{I.33}$$

$$P^{*\,IJKL\,A} \frac{\partial}{\partial\phi^A} \mathcal{N}^*_{\Lambda\Sigma} = -4i\Im\mathrm{m}\mathcal{N}_{\Omega(\Lambda}\Im\mathrm{m}\mathcal{N}_{\Sigma)\Delta} P^{*\,IJKL\,A} P^{*\,ij}{}_{A} f^{\Omega}{}_{i} f^{\Delta}{}_{j}, \tag{I.34}$$

$$P^{*\,iIJ\,A} \frac{\partial}{\partial\phi^A} \mathcal{N}^*_{\Lambda\Sigma} = -4i\Im\mathrm{m}\mathcal{N}_{\Omega(\Lambda}\Im\mathrm{m}\mathcal{N}_{\Sigma)\Delta} P^{*\,iIJ\,A} P^{*\,jk}{}_{A} f^{\Omega}{}_{i} f^{\Delta}{}_{j}. \tag{I.35}$$

Having studied the Vielbeins and the connection, we must now finally study the curvature. We can use the Maurer–Cartan equations $d\Gamma + \Gamma \wedge \Gamma = 0$ for the Maurer–Cartan 1-form in Eq. (I.7) to compute in a simple way the curvatures of the non-vanishing components of the connection $\Omega$ in terms of the Vielbeins. We find

$$R^{KL}{}_{IJ} = d\Omega^{KL}{}_{IJ} + \tfrac{1}{2}\Omega^{KL}{}_{MN} \wedge \Omega^{MN}{}_{IJ}$$

$$= -\tfrac{1}{2}P^{*KLMN} \wedge P_{MNIJ} - P^{*iKL} \wedge P_{iIJ} \tag{I.36}$$

$$= -i\langle \mathfrak{D}\mathcal{V}_{IJ} \,|\, \mathfrak{D}\mathcal{V}^{*\,KL} \rangle, \tag{I.37}$$

$$R^{j}{}_{i} = d\Omega^{j}{}_{i} + \Omega^{j}{}_{k} \wedge \Omega^{k}{}_{i} = -\tfrac{1}{2}P^{*\,jIJ} \wedge P_{iIJ} - P^{*ik} \wedge P_{ik} \tag{I.38}$$

$$= -i\langle \mathfrak{D}\mathcal{V}_{i} \,|\, \mathfrak{D}\mathcal{V}^{*\,j} \rangle. \tag{I.39}$$

On the other hand, the vanishing of the mixed components of the connection $\Omega^{i}{}_{IJ}$ implies the following relation between the different components of the Vielbein:

$$\tfrac{1}{2}P_{IJKL} \wedge P^{*\,iKL} + P_{jIJ} \wedge P^{*\,ij} = -i\langle \mathfrak{D}\mathcal{V}_{IJ} \,|\, \mathfrak{D}\mathcal{V}^{*\,i} \rangle = 0. \tag{I.40}$$

# Appendix J
# Gauging isometries of non-linear $\sigma$-models

Non-linear $\sigma$-models are a common element of many of the actions considered in the main text that contain scalar fields: in the $N = 1, d = 4$ matter-coupled supergravities studied in Chapter 6 we find $\sigma$-models which correspond to Kähler–Hodge manifolds; in the $N = 2, d = 4$ theories studied in Chapter 7 we find special Kähler and quaternionic-Kähler manifolds, and the latter and real special manifolds naturally arise in the $\sigma$-models of the $N = 1, d = 5$ matter-coupled supergravities studied in Chapter 9. There are many other examples in the text, and that is why generic non-linear $\sigma$-models have been included in the generic actions Eqs. (2.147) and (2.178).

In many cases the metrics of these $\sigma$-models have isometries and the $\sigma$-model action is invariant under an associated global symmetry. Furthermore, there are many instances in which we want to *gauge* one or several of those symmetries, deforming the action so that it becomes invariant under the local (gauge) version of those symmetries. For example, gauging this kind of symmetry in the above-mentioned supergravities one obtains theories with non-Abelian Yang–Mills fields and symmetries and a scalar potential with many potential uses. Gauged $\sigma$-models also arise in the construction of KK-brane effective actions studied on p. 706.

Due to the couplings of the scalars to other fields, the symmetries of a $\sigma$-model are not automatically symmetries of the full theory. For general kinds of couplings to certain kinds of bosonic fields (differential forms of various ranks) the situation has been studied in Section 2.6.2. In the above-mentioned supergravity theories the coupling to supergravity (hence, of the $\sigma$-model scalars to vectors and spinors) requires the geometries of their $\sigma$-models to be Kähler–Hodge, special Kähler, quaternionic-Kähler, or real special. There, the Riemannian structure (the metric) is not the only structure that needs to be preserved for a transformation to be called a symmetry. These geometries and their symmetries have been reviewed in the previous appendices, and here we want to study the gauging of all these symmetries in order of increasing complexity and with an (essentially) homogeneous notation that is used in Chapters 6 and 7 to describe the complete theories.

A warning regarding the notation and conventions is necessary: in this appendix (as well as in Chapters 6 and 7) the connections and their field strengths have the opposite sign to those used in the rest of the book and, in particular, to those in Appendix A. The generators, structure constants, etc. follow the general convention. Furthermore, we use

$\mathfrak{D}$ instead of $\mathcal{D}$, reserved here to denote the Kähler-covariant derivative) to denote the covariant derivatives that we will construct by gauging. Finally, we will use $\pounds$ to denote the standard Lie derivative where necessary to avoid confusion with the upper components of the covariantly holomorphic canonical section $\mathcal{L}^\Lambda$.

We start by reviewing the case of a real $\sigma$-model completely determined by a Riemannian metric.

## J.1 Introduction: gauging isometries of Riemannian manifolds

Our starting point is the following generic non-linear $\sigma$-model action for $n$ real scalar fields $\phi^i(x)$ in a $d$-dimensional spacetime with coordinates and metric $x^\mu$, $g_{\mu\nu}(x)$:

$$S[\phi] = \int d^dx\,\sqrt{|g|}\,\left[\,\mathcal{G}_{ij}g^{\mu\nu}\partial_\mu\phi^i\partial_\nu\phi^j\,\right]. \tag{J.1}$$

$\mathcal{G}_{ij}(\phi)$ is a set of real functions of the scalars that can be interpreted as a Riemannian metric in some $n$-dimensional space parametrized by the real scalars $\phi^i$; it is sometimes called *target space*. This means that under general infinitesimal redefinitions of the scalars

$$\tilde{\delta}\phi^i = \epsilon^i(\phi), \tag{J.2}$$

the *target-space metric* $\mathcal{G}_{ij}(\phi)$ transforms according to

$$\tilde{\delta}_\epsilon\mathcal{G}_{ij} = \epsilon^k\partial_k\mathcal{G}_{ij} \tag{J.3}$$

and the variation of the action is

$$\tilde{\delta}_\epsilon S = \int d^dx\,\sqrt{|g|}\,\left[\,\mathcal{L}_\epsilon\mathcal{G}_{ij}g^{\mu\nu}\partial_\mu\phi^i\partial_\nu\phi^j\,\right], \tag{J.4}$$

where $\mathcal{L}_\epsilon\mathcal{G}_{ij}$ is the standard Lie derivative of the target-space metric with respect to $\epsilon^i(\phi)$, which can be seen as a vector field in the target space. Thus, it will not vanish unless $\epsilon^i$ is a Killing vector of $\mathcal{G}_{ij}$. In other words, to be a symmetry of the $\sigma$-model action, the infinitesimal transformation must preserve the Riemannian structure of the target space.

If we denote by $K_A{}^i(\phi)$ the Killing vectors of $\mathcal{G}_{ij}$, i.e. those satisfying

$$\mathcal{L}_A\mathcal{G}_{ij} \equiv \mathcal{L}_{K_A}\mathcal{G}_{ij} = K_A{}^k\partial_k\mathcal{G}_{ij} + 2\partial_{(i|}K_A{}^k\mathcal{G}_{|j)} = 0, \tag{J.5}$$

the infinitesimal transformations that leave the above action invariant will be linear combinations of these vectors with constant infinitesimal parameters that we will denote by $\alpha^A$:

$$\tilde{\delta}_\alpha\phi^i = \alpha^A K_A{}^i(\phi), \qquad \tilde{\delta}_\alpha\mathcal{G}_{ij} = \alpha^A K_A{}^k\partial_k\mathcal{G}_{ij} = -\alpha^A(\mathcal{L}_A - K_A)\mathcal{G}_{ij}. \tag{J.6}$$

If the Killing vectors satisfy the Lie algebra,

$$[K_A, K_B] = -f_{AB}{}^C K_C, \tag{J.7}$$

then the variation of the Killing vectors is given by

$$\tilde{\delta}_\alpha K_A{}^i = \alpha^B K_B{}^j\partial_j K_A{}^i = -\alpha^B f_{BA}{}^C K_C + \alpha^B K_A{}^j\partial_j K_B{}^i. \tag{J.8}$$

Let us promote the constant parameters $\alpha^A$ to arbitrary infinitesimal spacetime functions $\alpha^A(x)$. It is clear that the $\sigma$-model action Eq. (J.1) will fail to be invariant because now $\tilde{\delta}_\alpha \partial_\mu \phi^i = (\partial_\mu \alpha^A) K_A{}^i + \alpha^A \partial_j K_A{}^i \partial_\mu \phi$ and the first term does not vanish; in other words, the action is not invariant because the derivative of the scalar fields does not transform *covariantly*, which in a general context means without derivatives of the parameters.

The result of the variation of the action, using the fact that the $K_I$ are Killing vectors, is

$$\tilde{\delta}_\alpha S = \int d^d x \left[ -\alpha^A \partial_\mu \left( 2\sqrt{|g|} K_{Ai} \partial^\mu \phi^i \right) \right] \neq 0, \tag{J.9}$$

where we have used the target-space metric to lower the index of the Killing vector. Using the standard argument in Section 2.5,

$$\mathrm{j}_A{}^\mu \equiv 2\sqrt{|g|} K_{Ai} \partial^\mu \phi^i \tag{J.10}$$

is the Noether current (density) associated with the symmetry generated by $K_A$.

In this discussion we have used $\tilde{\delta}_A$ to denote the infinitesimal transformations. Since they involve *target-space* but not *spacetime* coordinate transformations, these transformations are identical to the ones without tildes according to the definitions in footnote 3, Section 2.2. We will remove the tildes from now onwards, but, when studying the transformations of fields $\Phi$ with target-space tensor indices, we must avoid the mistake of writing $\delta_\Lambda \Phi = -\mathcal{L}_A \Phi$ following (wrongly!) the transformation rules Eq. (1.8).

We would like to modify the $\sigma$-model action so as to make it invariant under these local transformations, or under a subgroup of them. This deformation is known as *gauging* the original action, and the preceding discussion suggests that it requires us to modify the derivative $\partial_\mu$ of the scalars so as to make it transform *covariantly* under the local transformations.

To construct the covariant derivative $\mathfrak{D}_\mu \phi^i$ it is necessary to use vector fields $A^\Lambda{}_\mu$ (called, in this context, gauge fields). These fields may already be present in the theory in a sector omitted in the action Eq. (J.1) or we may just add them without a kinetic term. In this case the introduction of additional gauge symmetries will allow us to eliminate some of the scalars, which become redundant (see, for instance, the discussion on p. 706). This is precisely why we need to gauge an isometry to construct the effective action of the KK-monopole, as discussed in Ref. [164].

If the gauge fields are already present in the theory before gauging the global symmetry of the $\sigma$-model, the vector fields are invariant under the $\mathrm{U}(1)$ gauge transformations

$$\delta_\Lambda A^\Lambda{}_\mu = -\partial_\mu \sigma^\Lambda. \tag{J.11}$$

The gauging procedure requires the identification of some of the local parameters $\sigma^\Lambda$ with some of the parameters $\alpha^A$ which are going to be promoted to local parameters. With this identification, the presence of the gauge field (we will give the details soon) in the covariant derivative can cancel the unwanted term $\partial_\mu \alpha^A$. Furthermore, and most importantly, this identification avoids the introduction of additional gauge symmetry in the theory in the gauging procedure, which means that the gauge symmetry cannot be used to eliminate any scalars (comparing with the discussion on p. 706, the equations of motion of the gauge fields are no longer constraints, but dynamical equations that cannot be used to eliminate any fields from the action). Summarizing: the couplings will change but the number of degrees of freedom will not. This is the case we are interested in in this appendix since our final goal is to gauge isometries in supergravity theories in which we cannot easily change the numbers of degrees of freedom without breaking supersymmetry.

The identification of some of the $\sigma^\Lambda$s with some of the $\alpha^A$s is made through the embedding tensor $\vartheta_\Lambda{}^A$ introduced in Section 2.7:[1]

$$\alpha^A = \sigma^\Lambda \vartheta_\Lambda{}^A, \qquad A^A{}_\mu = A^\Lambda{}_\mu \vartheta_\Lambda{}^A. \tag{J.12}$$

Doing this systematically for an arbitrary set of vectors and symmetries requires the introduction of a hierarchy of higher-rank tensor fields, as discussed in Section 2.7. We will not do it here. Instead, we are going to assume that the number of vector fields at our disposal is larger than the rank of the symmetry group that we want to gauge, ignoring other possible symmetries, and we will label all the vector fields and symmetries by the same indices $\Lambda, \Sigma, \ldots$ $A^\Lambda{}_\mu, K_\Lambda, \alpha^\Lambda, \sigma^\Lambda$, identifying $\alpha^\Lambda = g\sigma^\Lambda$, where $g$ is the gauge coupling constant. We will assume that some of the Killing vectors $K_\Lambda$, structure constants $f_{\Lambda\Sigma}{}^\Omega$, etc. simply vanish. This is the notation used in Chapters 6 and 7.

Thus, the symmetries that we want to gauge are

$$\delta_\alpha \phi^i = \alpha^\Lambda K_\Lambda{}^i(\phi), \tag{J.13}$$

where

$$[K_\Lambda, K_\Sigma] = -f_{\Lambda\Sigma}{}^\Omega K_\Omega, \tag{J.14}$$

and where the $\alpha^\Lambda$s are the parameters of the $\mathrm{U}(1)$ gauge transformations of the vector fields times $g$, i.e. $\delta_\alpha A^\Lambda{}_\mu = -\frac{1}{g}(\partial_\mu \alpha^\Lambda + \mathcal{O}(g))$.

The general structure of the covariant derivative of an object $\Phi(\phi)$ transforming covariantly under the local transformations Eq. (J.13) as

$$\delta_\alpha \Phi = \alpha^\Lambda \delta_\Lambda \Phi \tag{J.15}$$

is

$$\boxed{\mathfrak{D}_\mu \Phi \equiv \nabla_\mu \Phi + g A^\Lambda{}_\mu \delta_\Lambda \Phi,} \tag{J.16}$$

where $\nabla_\mu \Phi$ is spacetime and target-space covariant if we want this covariant derivative to be covariant under spacetime and target-space GCTs. This is optional (not our original goal) but certainly convenient. We will give its general form shortly, but we need to find first the covariant derivative of the scalar fields.

This structure guarantees (apart from general spacetime and target-space covariance) the cancelation of the $\partial_\mu \alpha$ terms in the gauge transformations, but we need the exact transformation rule of the gauge fields in order to give a general transformation rule for

[1] Actually, one could also use the magnetic duals of the vector fields $A_{\Lambda\,\mu}$ as gauge fields, and the embedding tensor must carry in general a symplectic index $M$ and a Lie-algebra index $A$: $\vartheta_M{}^i$. We will not consider this most general possibility here.

these covariant derivatives. We can find it by studying an example. For the scalar fields, according to the general rule, we have

$$\mathfrak{D}_\mu\phi^i \equiv \partial_\mu\phi^i + gA^\Lambda{}_\mu K_\Lambda{}^i, \tag{J.17}$$

and, under the gauge transformations Eq. (J.13) and upon use of Eq. (J.8), it transforms as

$$\delta_\alpha\mathfrak{D}_\mu\phi^i = \alpha^\Lambda\partial_j K_\Lambda{}^i\mathfrak{D}_\mu\phi^j, \tag{J.18}$$

if the gauge fields transform in the usual way,

$$\delta_\alpha A^\Lambda{}_\mu = -\frac{1}{g}\mathfrak{D}_\mu\alpha^\Lambda = -\frac{1}{g}\left(\partial_\mu\alpha^\Lambda + gf_{\Sigma\Omega}{}^\Lambda A^\Sigma{}_\mu\alpha^\Omega\right), \tag{J.19}$$

and their field strengths have the standard form

$$F^\Lambda{}_{\mu\nu} = 2\partial_{[\mu}A^\Lambda{}_{\nu]} + gf_{\Sigma\Omega}{}^\Lambda A^\Sigma{}_\mu A^\Omega{}_\nu, \tag{J.20}$$

and transform covariantly as

$$\delta_\alpha F^\Lambda{}_{\mu\nu} = \alpha^\Omega f_{\Omega\Sigma}{}^\Lambda F^\Sigma{}_{\mu\nu}. \tag{J.21}$$

This implies that, if the gauge fields are already present in the original (*ungauged*) theory, the theory must be invariant under the global rotations obtained from the expression Eq. (J.19) for constant $\alpha^\Lambda$

$$\delta_\alpha A^\Lambda{}_\mu = \alpha^\Omega f_{\Omega\Sigma}{}^\Lambda A^\Sigma{}_\mu. \tag{J.22}$$

In other words: as we should have suspected from the beginning, the vector field sector must be invariant under the same subgroup of global symmetries of the $\sigma$-model and it must transform in the adjoint representation of that subgroup if we want to gauge it.

Observe that if we were also working with the dual (*magnetic*) fields $A_{\Lambda\,\mu}$ these would automatically transform in the form

$$\delta_\alpha A_{\Lambda\,\mu} = -\alpha^\Omega f_{\Omega\Lambda}{}^\Sigma A_{\Sigma\,\mu}, \tag{J.23}$$

corresponding to an $\mathfrak{sp}(2n_V)$ matrix Eq. (6.57)

$$\mathcal{T}_\Omega = \begin{pmatrix} f_{\Omega\Sigma}{}^\Lambda & \\ & -f_{\Omega\Lambda}{}^\Sigma \end{pmatrix}, \tag{J.24}$$

but we could also consider more general possibilities, mixing electric and magnetic vector fields.

With this result, we find that the covariant derivative transforms in general as

$$\boxed{\delta_\alpha\mathfrak{D}_\mu\Phi = \alpha^\Lambda\mathfrak{D}_\mu\delta_\Lambda\Phi = \alpha^\Lambda\left(\nabla_\mu\delta_\Lambda\Phi + gA^\Sigma{}_\mu\delta_\Sigma\delta_\Lambda\Phi\right).} \tag{J.25}$$

Now, for $\nabla_\mu\Phi$ to have the properties that we have assumed, it has to be the spacetime-covariant pullback (that is, using $\mathfrak{D}_\mu\phi^i$) of the covariant derivative in target space to take into account any spacetime or target indices that $\Phi$ may have. For instance, if $\Phi = \mathfrak{D}_\mu\phi^i$,

$$\nabla_\mu\mathfrak{D}_\nu\phi^i = \partial_\mu\mathfrak{D}_\nu\phi^i - \Gamma_{\mu\nu}^\rho\mathfrak{D}_\rho\phi^i + \Gamma_{jk}{}^i\mathfrak{D}_\mu\phi^j\mathfrak{D}_\nu\phi^k, \tag{J.26}$$

where $\Gamma_{jk}{}^{i}$ is the Levi-Civita connection of the target-space metric $\mathcal{G}_{ij}$, and the second covariant derivative of the scalars is and transforms as

$$\begin{aligned}\mathfrak{D}_\mu\mathfrak{D}_\nu\phi^i &= \nabla_\mu\partial_\nu\phi^i + gA^\Lambda{}_\mu\partial_j K_\Lambda{}^i\mathfrak{D}_\mu\phi^j \\ &= \partial_\mu\mathfrak{D}_\nu\phi^i - \Gamma^\rho_{\mu\nu}\mathfrak{D}_\rho\phi^i + \Gamma_{jk}{}^i\mathfrak{D}_\mu\phi^j\mathfrak{D}_\nu\phi^k + gA^\Lambda{}_\mu\partial_j K_\Lambda{}^i\mathfrak{D}_\mu\phi^j, \\ \delta_\alpha\mathfrak{D}_\mu\mathfrak{D}_\nu\phi^i &= \alpha^\Lambda\partial_j K_\Lambda{}^i\mathfrak{D}_\mu\mathfrak{D}_\nu\phi^j.\end{aligned} \tag{J.27}$$

To obtain this result explicitly we have used the transformation rule of the target-space connection under infinitesimal GCTs Eq. (1.17) with $\epsilon = \alpha^\Lambda K_\Lambda$, and we have used the fact that the isometries of the metric must also leave invariant the Levi-Civita connection.

The Ricci identity for the covariant derivative acting on any function of the scalars $\Phi(\phi)$ follows from Eq. (J.27):[2]

$$\boxed{[\mathfrak{D}_\mu, \mathfrak{D}_\nu]\Phi = gF^\Lambda{}_{\mu\nu}\delta_\Lambda\Phi.} \tag{J.28}$$

The gauged $\sigma$-model action is obtained by simply replacing the partial derivatives of the scalars by the covariant ones:

$$S_{\text{gauged}}[\phi, A_\mu] = \int d^dx\,\sqrt{|g|}\,\left[\mathcal{G}_{ij}\mathfrak{D}_\mu\phi^i\mathfrak{D}^\mu\phi^j\right], \tag{J.29}$$

and, with the definitions in Eqs. (J.17), (J.27), and (J.29), the equations of motion of the scalars can be written in the simple and manifestly spacetime and target-space covariant form

$$\mathfrak{D}_\mu\mathfrak{D}^\mu\phi^i = 0. \tag{J.30}$$

Note that, if we expand the gauged $\sigma$-model action in powers of the coupling constant $g$, the first-order term is precisely the one we would have added by following step by step the Noether method explained in Section 2.5:

$$g\jmath_A{}^\mu A^A{}_\mu. \tag{J.31}$$

In a generic theory there are additional terms depending on the scalars and which may have to be modified to make the theory gauge invariant. The treatment of the kinetic (or period) matrices that couple the scalars to the vector fields is studied in detail on p. 211, for instance. Other objects define additional geometric structures of the target space that have to be preserved by the isometries of the $\sigma$-model metric. In the following sections, which will deal with $\sigma$-models with complex, Kähler–Hodge, special Kähler, quaternionic-Kähler, and real special target spaces, we will find several examples. The invariance of the corresponding structures under the isometries of the metric has been studied in other appendices; here we will simply have to use those results to construct the covariant derivatives of the fields of the theory etc.

[2] If $\Phi$ carries more indices, then the commutator of $\nabla_\mu$ will produce additional curvature terms.

## J.2 Gauging holomorphic isometries of complex manifolds

In matter-coupled $N = 1, 2$, $d = 4$ supergravities (see Chapters 6 and 7) there are complex scalar fields $Z^i$ whose kinetic terms $2\mathcal{G}_{ij^*}\partial_\mu Z^i \partial^\mu Z^{*j^*}$ define a $\sigma$-model with a complex target space with Hermitian metric $\mathcal{G}_{ij^*}(Z, Z^*)$. The metric has to satisfy additional properties (it has to be Kähler–Hodge or special Kähler), but we will leave these aside for the moment, considering only the Hermitian structure.

Only holomorphic coordinate transformations Eq. (E.36) preserve the Hermitian structure. Therefore, we will require the Killing vectors $K_\Lambda{}^m$ to split into holomorphic and antiholomorphic components:

$$K_\Lambda = K_\Lambda{}^m \partial_m = k_\Lambda{}^i \partial_i + k_\Lambda{}^{*i^*} \partial_{i^*}, \qquad \partial_j k_\Lambda{}^{*i^*} = \partial_{j^*} k_\Lambda{}^i = 0. \tag{J.32}$$

As mentioned in Section E.1.2, these holomorphic and antiholomorphic components satisfy the same Lie algebra as the real Killing vectors

$$[K_\Lambda, K_\Sigma] = -f_{\Lambda\Sigma}{}^\Gamma K_\Gamma, \quad [k_\Lambda, k_\Sigma] = -f_{\Lambda\Sigma}{}^\Gamma k_\Gamma, \quad [k^*_\Lambda, k^*_\Sigma] = -f_{\Lambda\Sigma}{}^\Gamma k^*_\Gamma. \tag{J.33}$$

The global holomorphic isometries that leave the $\sigma$-model invariant take, then, the form

$$\delta_\alpha Z^i = \alpha^\Lambda k_\Lambda{}^i(Z). \tag{J.34}$$

We can follow the general recipe to gauge the Hermitian $\sigma$-model, introducing gauge fields $A^\Lambda{}_\mu$ and using them to construct the covariant derivatives of the complex scalar fields $Z^i$ following the general recipe Eq. (J.16). They take the form

$$\mathfrak{D}_\mu Z^i = \partial_\mu Z^i + g A^\Lambda{}_\mu k_\Lambda{}^i, \tag{J.35}$$

and transform as

$$\delta_\alpha \mathfrak{D}_\mu Z^i = \alpha^\Lambda \mathfrak{D}_\mu k_\Lambda{}^i = \alpha^\Lambda \partial_j k_\Lambda{}^i \mathfrak{D}_\mu Z^j, \tag{J.36}$$

provided that the gauge potentials transform as usual Eq. (J.19).

Now, to make the $\sigma$-model kinetic term gauge invariant it is enough to replace the partial derivatives by covariant derivatives:

$$\mathcal{G}_{ij^*}\partial_\mu Z^i \partial^\mu Z^{*j^*} \quad \longrightarrow \quad \mathcal{G}_{ij^*}\mathfrak{D}_\mu Z^i \mathfrak{D}^\mu Z^{*j^*}. \tag{J.37}$$

The covariant derivatives of more general objects are (anti)holomorphic versions of the general formulae obtained in Section J.1. For instance,

$$\mathfrak{D}_\mu \mathfrak{D}_\nu Z^i = \partial_\mu \mathfrak{D}_\nu Z^i - \Gamma_{\mu\nu}{}^\rho \mathfrak{D}_\rho Z^i + \Gamma_{jk}{}^i \mathfrak{D}_\mu Z^j \mathfrak{D}_\nu Z^k + g A^\Lambda{}_\mu \partial_j k_\Lambda{}^i \mathfrak{D}_\nu Z^j, \tag{J.38}$$

$$[\mathfrak{D}_\mu, \mathfrak{D}_\nu]\, Z^i = g F^\Lambda{}_{\mu\nu} k_\Lambda{}^i. \tag{J.39}$$

## J.3 Kähler–Hodge manifolds

These are the manifolds that arise in $N = 1$, $d = 4$ supergravity. The covariant derivatives of the scalars are the same as in Section J.2, but this case presents the following important novelties with respect to the previous one.

- The need to preserve the Kähler–Hodge structure.
- The presence of additional scalar-dependent objects in the theory, like the holomorphic kinetic matrices $f_{\Lambda\Sigma}(Z)$ and the superpotential $W(Z)$ (or, equivalently, of the covariantly holomorphic section $\mathcal{L}(Z, Z^*)$, related to $W(Z)$ by Eq. (6.14)) which have to be "preserved" in some sense. We will have to define under which conditions they are preserved and we will have to construct covariant derivatives for them, since being "preserved" is not equivalent to being inert.
- The presence of spinor fields with non-zero Kähler weight, which will not be invariant under the isometries and for which we will have to construct gauge-covariant derivatives.

The preservation of the Kähler–Hodge structure under the holomorphic isometries of the metric generated by the $k_\Lambda{}^i(Z)$ has been studied in Section E.3.1, and we saw that generically it is preserved up to Kähler transformations $\lambda_\Lambda(Z)$ which act on all the fields with non-zero Kähler weight ($\mathcal{L}$ and the spinors). We also found that it is possible to find some real functions $\mathcal{P}_\Lambda(Z, Z^*)$ related to the isometries whose properties we are going to use to construct covariant derivatives of general fields.

$\mathcal{L}$ and $f_{\Lambda\Sigma}(Z)$ will be "preserved" by the holomorphic isometries if they are invariant up to Kähler transformations (in the case of $\mathcal{L}$) or up to "rotations" (in the case of $f_{\Lambda\Sigma}(Z)$, which is Kähler neutral) that can be compensated by rotations of the $n_V$ vector field strengths. The latter must belong to $\mathrm{Sp}(2n_V, \mathbb{R})$ for the equations of motion to be preserved, according to the general results in Section 2.6. The generators of the symplectic transformations $\mathcal{T}_\Lambda$ are defined by

$$S \sim \mathbb{1}_{2n_V\times 2n_V} + \alpha^\Omega \mathcal{T}_\Omega, \qquad \mathcal{T}_\Omega \equiv \begin{pmatrix} a_\Omega{}^\Lambda{}_\Sigma & b_\Omega{}^{\Lambda\Sigma} \\ c_{\Omega\,\Lambda\Sigma} & d_{\Omega\,\Lambda}{}^\Sigma \end{pmatrix}, \tag{J.40}$$

and must provide a representation of the Lie algebra of the gauge group:

$$[\mathcal{T}_\Lambda, \mathcal{T}_\Sigma] = +f_{\Lambda\Sigma}{}^\Omega \mathcal{T}_\Omega. \tag{J.41}$$

These matrices act linearly on the vector fields, but on the kinetic matrix they act according to Eq. (2.156) (with $\mathcal{N}_{\Lambda\Sigma}$ replaced by $f_{\Lambda_\Sigma}$). The explicit expression is given in Eq. (6.54), which we rewrite here for convenience:

$$\mathcal{T}_\Lambda(f_{\Sigma\Omega}) = c_{\Lambda\,\Sigma\Omega} + d_{\Lambda\,\Sigma}{}^\Gamma f_{\Gamma\Omega} - f_{\Sigma\Gamma} a_\Lambda{}^\Gamma{}_\Omega + b_\Lambda{}^{\Gamma\Delta} f_{\Sigma\Gamma} f_{\Delta\Omega}. \tag{J.42}$$

To deal with these and other fields which depend on the spacetime coordinates only through the complex scalars, it is convenient to use the symplectic and Kähler-covariant Lie derivative with respect to the Killing vector $K_\Lambda$, denoted by $\mathbb{L}_\Lambda = \mathbb{L}_{K_\Lambda}$, which is given by the general definition Eq. (E.75) for fields $\Phi$ transforming according to

$$\delta_\Lambda \Phi = [\mathcal{T}_\Lambda - \tfrac{1}{2}(q\lambda_\Lambda + \bar{q}\lambda^*_\Lambda)]\Phi \tag{J.43}$$

by

$$\mathbb{L}_\Lambda \Phi \equiv \{\pounds_\Lambda - [\mathcal{T}_\Lambda - \tfrac{1}{2}(q\lambda_\Lambda + \bar{q}\lambda^*_\Lambda)]\}\Phi. \tag{J.44}$$

Generic covariantly holomorphic sections $\mathcal{L}(Z,Z^*)$ and kinetic matrices $f_{\Lambda\Sigma}(Z)$ will not transform under the holomorphic isometries of the Kähler metric according to Eq. (J.43), but only if they do, i.e. only if they are *invariant fields* annihilated by the above covariant Lie derivative, will the isometries be symmetries of the full theory that can be gauged. Taking into account the lack of complex indices, the Kähler weights of these fields, and their (sought after) behavior under symplectic transformations, these fields should behave as

$$\delta_\Lambda \mathcal{L} = -\tfrac{1}{2}(\lambda_\Lambda - \lambda^*_\Lambda)\}\mathcal{L}, \tag{J.45}$$

$$\delta_\Lambda f_{\Sigma\Gamma} = c_{\Lambda\,\Sigma\Omega} + d_{\Lambda\,\Sigma}{}^{\Gamma} f_{\Gamma\Omega} - f_{\Sigma\Gamma} a_\Lambda{}^{\Gamma}{}_{\Omega} + b_\Lambda{}^{\Gamma\Delta} f_{\Sigma\Gamma} f_{\Delta\Omega}, \tag{J.46}$$

and the condition that they are invariant fields takes the explicit form

$$\mathbb{L}_\Lambda \mathcal{L} = \{K_\Lambda + \tfrac{1}{2}(\lambda_\Lambda - \lambda^*_\Lambda)\}\mathcal{L} = 0, \tag{J.47}$$

$$\begin{aligned}\mathbb{L}_\Lambda f_{\Sigma\Gamma} &= \{K_\Lambda - \mathcal{T}_\Lambda\} f_{\Sigma\Gamma} \\ &= K_\Lambda f_{\Sigma\Omega} - (c_{\Lambda\,\Sigma\Omega} + d_{\Lambda\,\Sigma}{}^{\Gamma} f_{\Gamma\Omega} - f_{\Sigma\Gamma} a_\Lambda{}^{\Gamma}{}_{\Omega} + b_\Lambda{}^{\Gamma\Delta} f_{\Sigma\Gamma} f_{\Delta\Omega}) = 0.\end{aligned} \tag{J.48}$$

For the reasons explained in Section 6.3.1 we will restrict our study to symmetries whose symplectic generators have vanishing $b$ and $c$ submatrices, and, since the gauge fields must transform in the adjoint representation, the submatrices $a$ and $d$ are given by the structure constants as in Eq. (6.57), which we reproduce here for convenience:

$$a_\Lambda{}^{\Omega}{}_{\Sigma} = f_{\Lambda\Sigma}{}^{\Omega}, \qquad d_{\Lambda\Omega}{}^{\Sigma} = -f_{\Lambda\Omega}{}^{\Sigma} \;\Rightarrow\; \mathcal{T}_\Lambda = \begin{pmatrix} f_{\Lambda\Sigma}{}^{\Omega} & \\ & -f_{\Lambda\Omega}{}^{\Sigma} \end{pmatrix}. \tag{J.49}$$

Let us now assume that conditions (J.47) and (J.48) are met so we have a global symmetry $\delta_\alpha \equiv \alpha^\Lambda \delta_\Lambda$ of the theory, where $\delta_\Lambda$ equals $k_\Lambda{}^i$ for the complex scalars $Z^i$, is given by Eqs. (J.45) and (J.46) for $\mathcal{L}$ and $f_{\Sigma\Gamma}$, by Eq. (E.71) for the momentum maps $\mathcal{P}_\Lambda$, and can be read from Eqs. (6.76)–(6.78) for the spinors of $N=1, d=4$ supergravity:[3]

$$\delta_\Lambda \psi_\mu = -\tfrac{1}{4}(\lambda_\Lambda - \lambda^*_\Lambda)\psi_\mu, \tag{J.50}$$

$$\delta_\Lambda \lambda^\Sigma = -\tfrac{1}{4}(\lambda_\Lambda - \lambda^*_\Lambda)\lambda^\Sigma + f_{\Lambda\Omega}{}^{\Sigma}\lambda^\Omega, \tag{J.51}$$

$$\delta_\Lambda \chi^i = +\tfrac{1}{4}(\lambda_\Lambda - \lambda^*_\Lambda)\chi^i + \partial_j k_\Lambda{}^i \chi^j. \tag{J.52}$$

Let us proceed to gauge this symmetry. In the kind of theories that we are considering it is enough to covariantize the vector field strengths and the derivatives of the fields

[3] Observe that, without the restriction to infinitesimal transformations with $b=c=0$, the momentum map would be an invariant field because it is just part of a symplectic vector whose "upper components" $\mathcal{P}^\Lambda$ are missing. A more general formalism is necessary to deal with the $b\neq 0, c\neq 0$ cases. Something similar can be said about the transformations of the gauginos changing upper to lower components.

that feel the global transformations. The covariant vector field strength takes the standard form Eq. (J.20) and the covariant derivatives of the rest of the fields can, in principle, be constructed using the general recipe Eq. (J.16), where $\nabla_\mu$ is general covariant and the gauge-covariant pullback (i.e. with $\mathfrak{D}_\mu Z^i$) of the (Kähler, for instance) covariant derivative of the field. For fields which depend only on the spacetime coordinates through the complex scalars the result is that the gauge-covariant derivative is the gauge-covariant pullback of the original covariant derivative of the field.

- This recipe gives for the scalars and their first covariant derivatives Eqs. (J.35) and (J.38).
- For the momentum maps the general prescription gives

$$\begin{aligned} \delta_\Sigma \mathcal{P}_\Lambda &= -f_{\Sigma\Lambda}{}^{\Omega}\mathcal{P}_\Omega, \\ \mathfrak{D}_\mu \mathcal{P}_\Lambda &= \partial_\mu \mathcal{P}_\Lambda - g f_{\Sigma\Lambda}{}^{\Omega} A^{\Sigma}{}_\mu \mathcal{P}_\Omega, \end{aligned} \tag{J.53}$$

and it is easy to see that it is the gauge-covariant pullback of the original derivative:

$$\mathfrak{D}_\mu \mathcal{P}_\Lambda = \mathfrak{D}_\mu Z^i \partial_i \mathcal{P}_\Lambda + \mathfrak{D}_\mu Z^{*i^*} \partial_{i^*} \mathcal{P}_\Lambda. \tag{J.54}$$

- For the covariantly holomorphic section $\mathcal{L}$, the recipe gives

$$\begin{aligned} \mathfrak{D}_\mu \mathcal{L} &= \left\{\partial_\mu + i(\mathcal{Q}_i \mathfrak{D}_\mu Z^i + \text{c.c.}) - \tfrac{1}{2} g A^\Lambda{}_\mu (\lambda_\Lambda - \text{c.c.})\right\} \mathcal{L} \\ &= \left\{\partial_\mu + i\mathcal{Q}_\mu + ig A^\Lambda{}_\mu [(k_\Lambda{}^i \mathcal{Q}_i - \tfrac{1}{2i}\lambda_\Lambda) + \text{c.c.}]\right\} \mathcal{L} \\ &= \left\{\partial_\mu + i(\mathcal{Q}_\mu + g A^\Lambda{}_\mu \mathcal{P}_\Lambda)\right\} \mathcal{L} \\ &= \{\partial_\mu + i\hat{\mathcal{Q}}_\mu\}\mathcal{L}, \end{aligned} \tag{J.55}$$

where we have used the definition of $\mathfrak{D}_\mu Z^i$ in the second line ($\mathcal{Q}_\mu$ is the standard pullback of the Kähler 1-form connection); we have used Eq. (E.69) in the third line; and we have defined

$$\hat{\mathcal{Q}}_\mu \equiv \mathcal{Q}_\mu + g A^\Lambda{}_\mu \mathcal{P}_\Lambda. \tag{J.56}$$

Observe that this 1-form is, in general, different from the "covariant pullback" of the Kähler 1-form:

$$\mathfrak{D}_\mu Z^i \mathcal{Q}_i + \text{c.c.} \tag{J.57}$$

The difference between this and the correct one is

$$\hat{\mathcal{Q}}_\mu - (\mathfrak{D}_\mu Z^i \mathcal{Q}_i + \text{c.c.}) = -g A^\Lambda{}_\mu \Im\text{m}\lambda_\Lambda, \tag{J.58}$$

which only vanishes when the isometries that have been gauged leave the Kähler potential exactly invariant (i.e. $\lambda_\Lambda = 0$). The complete covariant derivative is, nevertheless, the covariant pullback of the standard one:

$$\mathfrak{D}_\mu \mathcal{L} = \{\mathfrak{D}_\mu Z^i \mathcal{D}_i + \mathfrak{D}_\mu Z^{*i^*} \mathcal{D}_{i^*}\}\mathcal{L} = \mathfrak{D}_\mu Z^i \mathcal{D}_i \mathcal{L}, \tag{J.59}$$

where we have taken the covariant holomorphicity of $\mathcal{L}$.

- According to the general recipe, the covariant derivative of the kinetic matrix is given by[4]

$$\mathfrak{D}_\mu f_{\Sigma\Gamma} = \partial_\mu f_{\Sigma\Gamma} - 2gA^\Lambda f_{\Lambda(\Sigma}{}^\Omega f_{\Gamma)\Omega} = \mathfrak{D}_\mu Z^i \partial_i f_{\Lambda\Sigma}, \tag{J.61}$$

which can be rewritten as the gauge-covariant pullback of the holomorphic partial derivative,

$$\mathfrak{D}_\mu f_{\Sigma\Gamma} = \mathfrak{D}_\mu Z^i \partial_i f_{\Lambda\Sigma}, \tag{J.62}$$

on account of its holomorphicity.

- Finally, let us consider the spinors of the theory. They are not invariant fields, as they do not depend only on the complex scalars $Z^i$, and their gauge-covariant derivatives cannot be the gauge-covariant pullback of their usual derivatives. They have a non-vanishing Kähler weight which is $(-1/2, 1/2)$ times their chirality, but they have additional indices and associated transformations (see Eqs. (J.50)–(J.52)); we have to take all this into account to construct their gauge-covariant derivatives. Applying in a straightforward manner the general prescription, we get

$$\mathfrak{D}_\mu \psi_\nu = \left\{\nabla_\mu + \tfrac{i}{2}\hat{\mathcal{Q}}_\mu\right\}\psi_\nu, \tag{J.63}$$

$$\mathfrak{D}_\mu \lambda^\Lambda = \left\{\nabla_\mu + \tfrac{i}{2}\hat{\mathcal{Q}}_\mu\right\}\lambda^\Lambda + gA^\Sigma{}_\mu f_{\Sigma\Omega}{}^\Lambda \lambda^\Omega, \tag{J.64}$$

$$\mathfrak{D}_\mu \chi^i = \left\{\nabla_\mu - \tfrac{i}{2}\hat{\mathcal{Q}}_\mu\right\}\chi^i + \left\{\mathfrak{D}_\mu Z^k \Gamma_{kj}{}^i + gA^\Lambda{}_\mu \partial_j k_\Lambda{}^i\right\}\chi^j, \tag{J.65}$$

where $\nabla_\mu$ is the general- and Lorentz-covariant derivative[5] and $\Gamma_{ij}{}^k$ are the components of the target-space Levi-Civita connection.

## J.4 Gauging holomorphic isometries of special Kähler manifolds

Since special Kähler manifolds are Kähler–Hodge, this case can be regarded basically as an extension of the previous one (without superpotential) in which we simply require the preservation of an additional structure: the covariantly holomorphic canonical section $\mathcal{V}$. The expression of this condition is Eq. (F.59), and a number of properties have been derived from it in Section F.3, among which we would like to mention the relations of the momentum maps and holomorphic Killing vectors with the canonical section and the symplectic generators (Lie algebra structure constants upon the restriction Eq. (J.49) to symplectic generators with $b = c = 0$) Eqs. (F.62) and (F.63). These relations indicate that *only non-Abelian isometry groups can be gauged in this case.*

[4] With the restriction to infinitesimal symplectic transformations with $b = c = 0$, the gauging will only be possible if $\pounds_\Lambda f_{\Sigma\Gamma} = \delta_\Lambda f_{\Sigma\Gamma}$ with

$$\delta_\Lambda f_{\Sigma\Gamma} \equiv \mathcal{T}_\Lambda(f_{\Sigma\Gamma}) = -2f_{\Lambda(\Sigma}{}^\Omega f_{\Gamma)\Omega}. \tag{J.60}$$

[5] The gauginos $\lambda^\Lambda$ carry upper $\Lambda$ indices and should not be confused with the infinitesimal Kähler transformations $\lambda_\Lambda$.

The preservation of the canonical section implies the conservation of all the quantities that can be derived from it (Kähler potential, period matrix, momentum maps, etc.), and, therefore, it is a sufficient condition for the whole theory to be invariant under the isometry group.

The relations Eqs. (F.62) and (F.63) give new identities for the case $b = c = 0$, Eq. (J.49), which we want to explore before studying the gauging. First, observe that the condition Eq. (F.64) takes the form

$$f_{\Lambda\Sigma}{}^{\Omega}\mathcal{L}^{\Sigma}\mathcal{M}_{\Omega} = 0, \tag{J.66}$$

and the covariant derivative of Eq. (F.64) is given by

$$f_{\Lambda\Sigma}{}^{\Omega}(f^{\Sigma}{}_{i}\mathcal{M}_{\Omega} + h_{\Omega\, i}\mathcal{L}^{\Sigma}) = 0. \tag{J.67}$$

Then, using Eqs. (F.62) and (F.63) and Eqs. (F.64), (J.66), and (J.67) we find that

$$\mathcal{L}^{\Lambda}\mathcal{P}_{\Lambda} = 0, \tag{J.68}$$

$$\mathcal{L}^{\Lambda}k_{\Lambda}{}^{i} = 0, \tag{J.69}$$

$$\mathcal{L}^{*\Lambda}k_{\Lambda}{}^{i} = -if^{*\Lambda\, i}\mathcal{P}_{\Lambda}. \tag{J.70}$$

From Eqs. (J.68) and (J.69) it follows that

$$\mathcal{L}^{\Lambda}\lambda_{\Lambda} = 0. \tag{J.71}$$

Some further useful equations that can be derived are explicit versions of Eqs. (F.62) and (F.63) for this restricted class of symmetries, i.e.

$$\mathcal{P}_{\Lambda} = 2f_{\Lambda\Sigma}{}^{\Gamma}\Re\mathfrak{e}\left(\mathcal{L}^{\Sigma}\mathcal{M}^{*}_{\Gamma}\right), \tag{J.72}$$

$$k_{\Lambda\, i^*} = if_{\Lambda\Sigma}{}^{\Gamma}\left(f^{*\Sigma}_{i^*}\mathcal{M}_{\Gamma} + \mathcal{L}^{\Sigma}h^{*}_{\Gamma i^*}\right). \tag{J.73}$$

Another interesting identity is

$$k_{\Lambda\, i^*}\mathfrak{D}Z^{*i^*} - k^{*}_{\Lambda i}\mathfrak{D}Z^{i} = i\mathfrak{D}\mathcal{P}_{\Lambda} = i(d\mathcal{P}_{\Lambda} + f_{\Lambda\Sigma}{}^{\Omega}A^{\Sigma}\mathcal{P}_{\Omega}). \tag{J.74}$$

Finally, let us make an important remark: if we have constructed our theory from a prepotential $\mathcal{F}(\mathcal{X})$, then Eq. (F.64) leads to

$$f_{\Lambda\Sigma}{}^{\Gamma}\,\mathcal{X}^{\Sigma}\partial_{\Gamma}\,\mathcal{F} = 0, \tag{J.75}$$

which means that the symmetries of a special Kähler geometry are necessarily invariances of the prepotential.

To gauge the theory we just have to replace the vector field strengths and the derivatives of fields by their gauge-covariant counterparts, which are constructed by the usual procedure. The gauge-covariant derivatives of $\mathcal{V}$ and $\mathcal{U}_i = \mathcal{D}_i\mathcal{V}$ are simply given by

$$\mathfrak{D}_{\mu}\mathcal{V} = \mathfrak{D}_{\mu}Z^{i}\mathcal{D}_{i}\mathcal{V} = \mathfrak{D}_{\mu}Z^{i}\mathcal{U}_{i}, \tag{J.76}$$

$$\mathfrak{D}_{\mu}\mathcal{U}_{i} = \mathfrak{D}_{\mu}Z^{j}\mathcal{D}_{j}\mathcal{U}_{i} + \mathfrak{D}_{\mu}Z^{*j^*}\mathcal{D}_{j^*}\mathcal{U}_{i} = i\mathcal{C}_{ijk}\mathcal{U}^{*j}\mathfrak{D}_{\mu}Z^{k} + \mathcal{G}_{ij^*}\mathfrak{D}_{\mu}Z^{*j^*}\,\mathcal{V}. \tag{J.77}$$

The gauge-covariant derivatives of the gravitinos $\psi_{I\,\mu}$, the supersymmetry parameters $\epsilon_I$, and the gauginos $\lambda^{Ii}$ take the obvious form (as in the preceding case, with the additional pullback of the $\mathrm{SU}(2)$ connection $\mathsf{A}_\mu{}^I{}_J$ if there are hypermultiplets). The hyperinos $\zeta_\alpha$ have Kähler weight $1/2$, and therefore, as for the other spinors, their gauge-covariant derivatives include the pullback of the modified Kähler connection $\hat{\mathcal{Q}}_\mu$.

## J.5 Gauging isometries of quaternionic-Kähler manifolds

As we saw in Section G.1, the preservation of the quaternionic-Kähler structure under the isometries of the metric $\mathsf{H}_{uv}$ (an invariance up to $\mathrm{SU}(2)$ compensating transformations) leads to the existence of the triholomorphic momentum maps $\mathsf{P}_\Lambda{}^I{}_J = -\frac{i}{2}\mathsf{P}^x\sigma^{x\,I}{}_J$, which are also preserved by the same transformations Eq. (G.50). The construction of the covariant derivatives of the fields feeling these transformations (all the fields which are functions of the hyperscalars $q^u$ and have $\mathrm{SU}(2)$ indices but not those which have $\mathrm{Sp}(2m)$ indices) relies heavily on their properties. On the other hand, the construction can be made applying the general prescription Eq. (J.16) including the gauge-covariant pullback of the $\mathrm{SU}(2)$ connection. Needless to say, the candidates for the gauge fields are the same as in the previous case and they must transform under the isometry group of the quaternionic-Kähler manifold that we want to gauge. But, if the vectors transform under that group, all the fields in the vector supermultiplets must also transform in the same way. This means that we can only gauge a group which is simultaneously a subgroup of the two isometry groups and which acts in the adjoint representation on the vector fields. This makes the two simultaneous statements

$$[\mathsf{k}_\Lambda, \mathsf{k}_\Sigma] = -f_{\Lambda\Sigma}{}^\Omega \mathsf{k}_\Omega, \qquad [k_\Lambda, k_\Sigma] = -f_{\Lambda\Sigma}{}^\Omega k_\Omega \tag{J.78}$$

compatible.

There is one particular case which does not fit this pattern: it turns out that in the total absence of hypermultiplets (hence with no quaternionic-Kähler manifold whatsoever) the momentum map $\mathsf{P}_\Lambda{}^x$ can still be defined in two cases, in which they are equivalent to a set of constant Fayet–Iliopoulos terms.

1. In the first case, the gauge group contains an $\mathrm{SU}(2)$ factor and

$$\mathsf{P}_\Lambda{}^x = e_\Lambda{}^x \xi, \tag{J.79}$$

where $\xi$ is an arbitrary constant and the $e_\Lambda{}^x$ are constants that are non-zero for $\Lambda$ in the range of the $\mathrm{SU}(2)$ factor and satisfy

$$\varepsilon^{xyz} e_\Lambda{}^y e_\Sigma{}^z = f_{\Lambda\Sigma}{}^\Omega e_\Omega{}^x. \tag{J.80}$$

2. In the second case, the gauge group contains a $\mathrm{U}(1)$ factor and

$$e_\Lambda{}^x = e^x\, \xi_\Lambda, \tag{J.81}$$

where $e_\Lambda{}^x$ is an arbitrary $\mathfrak{su}(2)$ vector and $\xi_\Lambda$ is an arbitrary constant that is nonzero for $\Lambda$ corresponding to the $\mathrm{U}(1)$ factor (this can be a linear combination in "$\Lambda$ space").

In these two cases there are no isometries of the quaternionic-Kähler manifold. Actually, they are associated with the $\mathrm{SU}(2)_{\mathrm{R}}$ factor of the R-symmetry group of the theory and to a $\mathrm{U}(1)_{\mathrm{R}} \subset \mathrm{SU}(2)_{\mathrm{R}}$. The formalism allows us to treat them together with the general case.

- The covariant derivatives of the hyperscalars have the standard form and transform in the standard way, with the obvious changes in notation:

$$\mathfrak{D}_\mu q^u \equiv \partial_\mu q^u + gA^\Lambda{}_\mu \mathsf{k}_\Lambda{}^u, \qquad \delta_\alpha \mathfrak{D}_\mu q^u = \alpha^\Lambda \partial_v \mathsf{k}_\Lambda{}^u \mathfrak{D}_\mu q^v. \tag{J.82}$$

- Equation (G.50) can be put in the equivalent form

$$\delta_\Lambda \mathsf{P}_\Sigma{}^x = k_\Lambda{}^u \partial_u \mathsf{P}_\Sigma{}^x = -\varepsilon^{xyz} \lambda_\Lambda{}^y \mathsf{P}_\Sigma{}^z - f_{\Lambda\Sigma}{}^\Omega \mathsf{P}_\Omega{}^x, \tag{J.83}$$

  where $\lambda_\Lambda{}^x$ is the parameter of the $\mathrm{SU}(2)$ compensating transformation associated with $\mathsf{k}_\Lambda{}^x$ and given by Eq. (G.41), which we reproduce here for convenience:

$$\lambda_\Lambda{}^x = \mathsf{k}_\Lambda{}^u \mathsf{A}^x{}_u - \mathsf{P}_\Lambda{}^x. \tag{J.84}$$

  Using this $\delta_\Lambda \mathsf{P}_\Sigma{}^x$ we get the the following expression for the gauge-covariant derivative of the triholomorphic momentum map:

$$\mathfrak{D}_\mu \mathsf{P}_\Lambda{}^x = \partial_\mu \mathsf{P}_\Lambda{}^x + \varepsilon^{xyz} \hat{\mathsf{A}}^y{}_\mu \mathsf{P}_\Lambda{}^z + g f_{\Lambda\Sigma}{}^\Omega A^\Sigma{}_\mu \mathsf{P}_\Omega{}^x, \tag{J.85}$$

  where we have defined

$$\hat{\mathsf{A}}^x{}_\mu \equiv \partial_\mu q^u \mathsf{A}^x{}_u + gA^\Lambda{}_\mu \mathsf{P}_\Lambda{}^x. \tag{J.86}$$

  As usual, the gauge-covariant derivative is the gauge-covariant pullback of the standard ($\mathrm{SU}(2)$-covariant) derivative:

$$\mathfrak{D}_\mu \mathsf{P}_\Lambda{}^x = \mathfrak{D}_\mu q^u \mathsf{D}_u \mathsf{P}_\Lambda{}^x. \tag{J.87}$$

- Under Eq. (G.33), spinors with lower and upper $\mathrm{SU}(2)$ indices undergo the following transformations with parameter $\lambda_\Lambda{}^x$ given by Eq. (J.84):

$$\delta_\Lambda \psi_I = -\lambda_\Lambda{}^x \tfrac{i}{2} \sigma^{x\,J}{}_I \psi_J, \qquad \delta_\Lambda \psi^I = \lambda_\Lambda{}^x \tfrac{i}{2} \sigma^{x\,I}{}_J \psi^J. \tag{J.88}$$

  Then, using the general formula, their covariant derivative is given by

$$\mathfrak{D}_\mu \psi_I = \nabla_\mu \psi_I - \hat{\mathsf{A}}_\mu{}^J{}_I \psi_J, \qquad \mathfrak{D}_\mu \psi^I = \nabla_\mu \psi^I + \hat{\mathsf{A}}_\mu{}^I{}_J \psi^J, \tag{J.89}$$

  where

$$\hat{\mathsf{A}}_\mu{}^I{}_J = -\tfrac{i}{2} \hat{\mathsf{A}}^x{}_\mu \sigma^{x\,I}{}_J, \tag{J.90}$$

  and $\nabla_\mu$ is the covariant derivative with respect to the remaining relevant local symmetries (Lorentz etc.).

Thus, if we take into account their Kähler weight and gaugings of the isometries of the special Kähler manifold,[6] we have for the spinors of $N = 2, d = 4$ supergravity

$$\mathfrak{D}_\mu \epsilon_I = \{\partial_\mu - \tfrac{1}{4}\omega_\mu{}^{ab}\gamma_{ab} + \tfrac{i}{2}\hat{\mathcal{Q}}_\mu\}\epsilon_I - \hat{\mathsf{A}}_\mu{}^J{}_I \epsilon_J, \tag{J.91}$$

$$\mathfrak{D}_\mu \psi_{I\,\nu} = \left\{\partial_\mu - \tfrac{1}{4}\omega_\mu{}^{ab}\gamma_{ab} + \tfrac{i}{2}\hat{\mathcal{Q}}_\mu\right\}\psi_{I\,\nu} - \Gamma_{\mu\nu}{}^\rho \psi_{I\,\rho} - \hat{\mathsf{A}}_\mu{}^J{}_I \psi_{J\,\nu}, \tag{J.92}$$

$$\begin{aligned}\mathfrak{D}_\mu \lambda^{Ii} = {} & \left\{\partial_\mu - \tfrac{1}{4}\omega_\mu{}^{ab}\gamma_{ab} + \tfrac{i}{2}\hat{\mathcal{Q}}_\mu\right\}\lambda^{Ii} \\ & + \left\{\mathfrak{D}_\mu Z^k \Gamma_{kj}{}^i + g A^\Lambda{}_\mu \partial_j k_\Lambda{}^i\right\}\lambda^{Ij} + \hat{\mathsf{A}}_\mu{}^I{}_J \lambda^{Ji},\end{aligned} \tag{J.93}$$

$$\mathfrak{D}_\mu \zeta_\alpha = \left\{\partial_\mu - \tfrac{1}{4}\omega_\mu{}^{ab}\gamma_{ab} + \tfrac{i}{2}\hat{\mathcal{Q}}_\mu\right\}\zeta_\alpha - \Delta_{\mu\,\alpha}{}^\beta \zeta_\beta. \tag{J.94}$$

### *J.5.1 Alternative notation for the $d = 5$ case*

As we have explained in Appendix G.1.1, we use a slightly different notation for the quaternionic-Kähler manifolds that arise in $N = 1, d = 5$ supergravities, to conform to the conventions of Refs. [152, 153] (see also Ref. [132]).

Here we just need to give the gauge-covariant derivatives on objects with fundamental $\mathrm{SU}(2)$ indices whose standard $\mathrm{SU}(2)$-covariant derivative is given in Eq. (G.57). It has the same form but with a deformed $\mathrm{SU}(2)$ connection denoted by $\hat{\mathsf{A}}_{\mu\,i}{}^j$:

$$\mathfrak{D}_\mu \psi^i = \partial_\mu \psi^i + \psi^j \hat{\mathsf{A}}_{\mu\,j}{}^i, \qquad \hat{\mathsf{A}}_{\mu\,j}{}^i = i\hat{\mathsf{A}}_\mu{}^r \sigma^r{}_j{}^i, \tag{J.95}$$

where

$$\hat{\mathsf{A}}_\mu{}^r \equiv \mathsf{A}_\mu{}^r + \tfrac{1}{2} g A^I{}_\mu P_I{}^r, \tag{J.96}$$

identical to the $\hat{\mathsf{A}}$ of the four-dimensional case up to the factor of $1/2$ which is due to the different normalization of the triholomorphic momentum map. Observe that now the triholomorphic momentum map carries the $\mathrm{SO}(n_V + 1)$ index $I$ carried by the gauge vectors.

## J.6 Gauging isometries of real special manifolds

In the gauging of the isometries of the real special manifolds of the five-dimensional theories, we use a notation analogous to that of the four-dimensional case, labeling the symmetries to be gauged with the same indices as the vector fields $I, J$ ($\Lambda, \Sigma$ in the four-dimensional case). Then, consistency requires that the matrices that we now call $T_I$ are just the structure constants:

$$T_I{}^J{}_K = f_{IK}{}^J. \tag{J.97}$$

[6] Non-Abelian gauge groups, as we have stressed, must act simultaneously on both scalar manifolds. The only exception is the Fayet–Iliopoulos terms that gauge $\mathrm{SU}(2)_\mathrm{R}$ or $\mathrm{U}(1)_\mathrm{R} \subset \mathrm{SU}(2)_\mathrm{R}$. The $\mathrm{U}(1)_\mathrm{R}$ associated with Kähler transformations can be gauged via Fayet–Iliopoulos terms in $N = 1, d = 4$ supergravity, but it cannot be gauged in $N = 2, d = 4$ supergravity.

Then, Eqs. (H.23)–(H.26) take the form

$$k_I{}^x = -\sqrt{3} f_{IJ}{}^K h^x_K h^J, \tag{J.98}$$

$$f_{IJ}{}^K h_K h^J = 0, \tag{J.99}$$

$$\delta_I C_{JKL} = -3 f_{I(J}{}^M C_{KL)M} = 0, \tag{J.100}$$

and from Eqs. (J.98) and (J.99) we can obtain an alternative form for the Killing vectors:

$$k_I{}^x = -\sqrt{3} f_{AB}{}^K h^{Jx} h_K. \tag{J.101}$$

The convention used in this case to define the gauge transformations of the scalars is a bit different because the gauge parameters are multiplied by the coupling constant carrying the opposite sign to that of our generic conventions:

$$\delta_\Lambda \phi^x = -g\alpha^I k_I{}^x. \tag{J.102}$$

The gauge fields also have the opposite sign, so $gA^I_\mu$ follows the general convention of this book.

The functions $h^I$ must transform as the vectors, in the adjoint representation:

$$\delta_\alpha h^I = -g f_{JK}{}^I \alpha^J h^K. \tag{J.103}$$

Observe that, when the theory has vector supermultiplets and hypermultiplets, the Killing vectors of the isometries of the real special and quaternionic-Kähler manifolds $k_I{}^x(\phi)$ and $k_I{}^X(q)$ that we are gauging must satisfy the same Lie algebra.

The gauge-covariant derivatives on the scalars are

$$\mathfrak{D}_\mu h^I = \partial_\mu h^I + g f_{JK}{}^I A^J{}_\mu h^K, \tag{J.104}$$

$$\mathfrak{D}_\mu \phi^x = \partial_\mu \phi^x + g A^I{}_\mu k_I{}^x, \tag{J.105}$$

and we also have

$$\mathfrak{D}_\mu h_I = \partial_\mu h_I + g f_{AB}{}^K A^J{}_\mu h_K, \tag{J.106}$$

$$\mathfrak{D}_\mu C_{IJK} = 0. \tag{J.107}$$

# Appendix K
# $n$-spheres

An $n$-dimensional unit-radius[1] sphere $S^n$ is the hypersurface of $\mathbb{R}^{n+1}$ defined by $(x^1)^2 + \cdots + (x^{n+1})^2 = 1$. It is usually parametrized in terms of spherical coordinates $\{r, \varphi, \theta_1, \ldots, \theta_{n-1}\}$,

$$
\begin{aligned}
x^1 &= \rho_{n-1} \sin\varphi, \\
x^2 &= \rho_{n-1} \cos\varphi, \\
x^3 &= \rho_{n-2} \cos\theta_1, \\
\vdots &\quad \vdots \\
x^k &= \rho_{n-k+1} \cos\theta_{k-2}, \qquad 3 \leq k \leq n+1,
\end{aligned}
\tag{K.1}
$$

where

$$
\begin{aligned}
\rho_l &= [(x^1)^2 + \cdots + (x^{n+1-l})^2]^{\frac{1}{2}} = r \prod_{m=1}^{l} \sin\theta_{n-m}, \\
\rho_0 &= r = [(x^1)^2 + \cdots + (x^{n+1})^2]^{\frac{1}{2}},
\end{aligned}
\tag{K.2}
$$

and $\varphi \in [0, 2\pi]$, $\theta_i \in [0, \pi]$, setting $r = 1$. The metric induced on $S^n$ in spherical coordinates is denoted by $d\Omega^2_{(n)}$ and is implicitly defined in

$$
d\vec{x}^{\,2}_{(n+1)} = d\rho_0^2 + \rho_0^2 d\theta^2_{n-1} + \cdots + \rho^2_{n-2} d\theta_1^2 + \rho^2_{n-1} d\varphi^2 \equiv dr^2 + r^2 d\Omega^2_{(n)}. \tag{K.3}
$$

In practice, it is convenient to use the recursive formula

$$
d\Omega^2_{(n)} = d\theta^2_{n-1} + \sin^2\theta_{n-1}\, d\Omega^2_{(n-1)}, \qquad d\Omega^2_{(1)} = d\varphi^2. \tag{K.4}
$$

The spheres equipped with this metric, which is clearly SO$(n+1)$ invariant, are called *round spheres* (see Appendix A.4.2). Other metrics with less symmetry on the same $S^n$ manifolds are possible, but sometimes a different notation is used to denote the corresponding Riemannian spaces.

[1] For a topological space, the radius is irrelevant, but it becomes relevant when we consider the metric induced from the Euclidean metric of $\mathbb{R}^{n+1}$.

For some purposes, such as the calculation of the curvature in spacetimes with spherical symmetry, it is convenient to rename the coordinates $\varphi$ and $\theta_k$ and use $\psi_i,\ i = 1,\ldots,n$, with

$$\psi_i = \theta_{n-i}, \quad i = 1,\ldots,n-1, \qquad \psi_n = \varphi, \tag{K.5}$$

and define

$$q_i = r^2 \prod_{k=1}^{i-1} \sin\psi_i, \quad i = 1,\ldots,n, \tag{K.6}$$

so the metric takes the form

$$d\vec{x}^{\,2}_{(n+1)} = q_0 dr^2 + \sum_{i=1}^{n} q_i d\psi_i^2 = dr^2 + r^2 d\Omega^2_{(n)}. \tag{K.7}$$

The volume form on $S^n$ is, in spherical coordinates,

$$d\Omega^n \equiv d\varphi \prod_{i=1}^{n-1} \sin^i\theta_i \, d\theta_i. \tag{K.8}$$

In Cartesian coordinates in the embedding $(n+1)$-dimensional space it takes the form

$$d\Omega^n = \frac{1}{n! r^{n+1}} \epsilon_{\mu_1\cdots\mu_{n+1}} x^{\mu_{n+1}} dx^{\mu_1} \cdots dx^{\mu_n}. \tag{K.9}$$

Other useful identities are

$$d^{n+1}x = r^n dr d\Omega^n, \qquad r^n d\Omega^n = d^n y \sqrt{|g|}, \tag{K.10}$$

where the $y$ are coordinates on the $n$-sphere.

The volume of the unit $n$-sphere $S^n$ is given by

$$\omega_{(n)} = \int_{S^n} d\Omega^n = \frac{2\pi^{\frac{n+1}{2}}}{\Gamma\left(\frac{n+1}{2}\right)}. \tag{K.11}$$

Using

$$\Gamma(x+1) = x\Gamma(x), \quad \Gamma(1) = 1, \quad \Gamma\left(\frac{1}{2}\right) = \pi^{1/2}, \tag{K.12}$$

one obtains $\omega_{(1)} = 2\pi$, $\omega_{(2)} = 4\pi$, $\omega_{(3)} = 2\pi^2$, etc.

The round $n$-spheres are globally symmetric spaces $SO(n+1)/SO(n)$ (Appendix A.4.2). There is also a description of the round spheres $S^3$ and $S^7$ as principal (Hopf) bundles in which both the base space and the fiber are spheres. The $S^3$ case is based on the description of $S^2$ as the coset manifold $SO(3)/SO(2) \sim SU(2)/U(1)$ and the general theorem (see p. 843) that ensures that G ($SU(2) \sim S^3$) is a principal bundle with base G/H ($S^2$) and structure group H ($SO(2) \sim S^1$). Let us now study these Hopf fibrations.

## K.1 $S^3$ and $S^7$ as Hopf fibrations

There is a natural action of U(1) on SU(2)

$$\mathrm{U} \to \mathrm{U}\begin{pmatrix} u & 0 \\ 0 & \bar{u} \end{pmatrix}, \qquad |u|^2 = 1 \tag{K.13}$$

(i.e. through shifts of $\psi$), that allows us to take the quotient $\mathrm{SU}(2)/\mathrm{U}(1)$ that can be identified with $\mathrm{S}^2$. This is why the metric on $\mathrm{S}^2$ is the metric on the coset manifold $\mathrm{SU}(2)/\mathrm{U}(1)$:

$$d\Omega^2_{(2)} = (e^1)^2 + (e^2)^2. \tag{K.14}$$

We can then view SU(2) ($\mathrm{S}^3$) as a fiber bundle with fiber U(1) ($\mathrm{S}^1$) and base space $\mathrm{S}^2$. From this point of view $e^3$ is a U(1) connection in the bundle and its curvature coincides with that of the Dirac magnetic monopole [1200] (see Section 12.7.2).

This is the simplest case where $n = 1$ in the first sequence of Hopf principal fiber bundles [757],

$$\mathrm{S}^{2n+1} \overset{\mathrm{U}(1)}{\to} \mathbb{CP}^n, \tag{K.15}$$

since $\mathbb{CP}^1$ is nothing but the Riemann sphere $\mathrm{S}^2$. Here $\mathrm{S}^{2n+1}$ is described by the equation in $\mathbb{C}^n$, $\bar{z}_0 z_0 + \cdots + \bar{z}_n z_n = 1$. There is another infinite sequence of Hopf fiberings,[2]

$$\mathrm{S}^{4n+3} \overset{\mathrm{SU}(2)}{\to} \mathbb{HP}^n, \tag{K.16}$$

where $\mathbb{H}$ is the field of quaternions. Here $\mathrm{S}^{4n+3}$ is described by the equation in $\mathbb{H}^n$, $\bar{z}_0 z_0 + \cdots + \bar{z}_n z_n = 1$. The first member in this series describes $\mathrm{S}^7$ as a fiber bundle with SU(2) as fiber and $\mathrm{S}^4$ ($\mathbb{HP}^1$) as base space. The $\mathrm{S}^7$ metric can be similarly constructed [1200],

$$d\Omega^2_{(7)} = \tfrac{1}{4}\left[d\Omega^2_{(4)} + \sum_{i=1}^{3}(e^i + \mathcal{A}^i)^2\right], \tag{K.17}$$

where the $e^i$ are the SU(2) Maurer–Cartan 1-forms and $d\Omega^2_{(4)}$ is the metric on $\mathrm{S}^4$, which we construct as before,

$$d\Omega^2_{(4)} = d\chi^2 + \sin^2\chi\, d\Omega^2_{(3)}, \qquad d\Omega^2_{(3)} = \tfrac{1}{4}\sum_{i=1}^{3}(E^i)^2, \tag{K.18}$$

where the $E^i$ are a second set of SU(2) Maurer–Cartan 1-forms and (in different coordinates) the 1-form with $\mathfrak{su}(2)$ indices

$$\mathcal{A}^i = -\sin^2(\chi/2)\, E^i \tag{K.19}$$

coincides with the gauge connection of the BPST instanton. This metric is also maximally symmetric (SO(8) invariant).

[2] The last "sequence" can be defined analogously using octonions, but only the first element is well defined.

## K.2 Squashed $S^3$ and $S^7$

The metrics of the round $S^3$ and $S^7$ associated with their description as Hopf fibrations can easily be deformed to those of *squashed spheres* by introducing a parameter $\lambda$:

$$\begin{aligned} d\tilde{\Omega}^2_{(3)} &= \tfrac{1}{4}\left[d\Omega^2_{(2)} + \lambda^2 (e^3)^2\right], \\ d\tilde{\Omega}^2_{(7)} &= \tfrac{1}{4}\left[d\Omega^2_{(4)} + \lambda^2 \sum_{i=1}^{3} (e^i + \mathcal{A}^i)^2\right]. \end{aligned} \tag{K.20}$$

Only for certain values of $\lambda$ does one obtain Einstein metrics: $\lambda = 1$, the *round* spheres (i.e. SO(4) and SO(8) invariant), and, for the $S^7$ case only, $\lambda = 1/\sqrt{5}$. The metric of this squashed $S^7$ is only $SO(5) \times SO(3)$ invariant, which makes it interesting in Kaluza–Klein compactifications [72].

# Appendix L
# Palatini's identity

This identity allows us to express the Einstein–Hilbert action in terms of the spin-connection coefficients alone, with no partial derivatives, which are eliminated upon integrating by parts. On substituting the expression for the Ricci scalar,

$$R = 2e_a{}^\mu e_b{}^\nu \partial_{[\mu}\omega_{\nu]}{}^{ab} + \omega_a{}^{ac}\omega_b{}^b{}_c + \omega_b{}^{ac}\omega_{ac}{}^b, \tag{L.1}$$

into the Einstein–Hilbert action and integrating by parts, using the relation between the Levi-Civita connection and the spin connection

$$\partial_a \ln\sqrt{|g|} = \Gamma_{ba}{}^b = \omega_{ba}{}^b + e_a{}^\mu \partial_b e_\mu{}^b, \tag{L.2}$$

we obtain

$$\begin{aligned}\int d^dx\,\sqrt{|g|}\,KR = \int d^dx\,\sqrt{|g|}\,K\Big\{&-2\partial_{[\mu|}\left(e_a{}^\mu e_b{}^\nu\right)\omega_{|\nu]}{}^{ab} + 2\omega_a{}^{ab}(\partial_b \ln K)\\ &+ 2e_b{}^\mu\partial_c e_\mu{}^c\omega_a{}^{ab} - \omega_b{}^{ba}\omega_c{}^c{}_a - \omega_a{}^{bc}\omega_{cb}{}^a\Big\}.\end{aligned} \tag{L.3}$$

Simple manipulations of the two terms with explicit Vielbeins lead us to the following generalization of Palatini's identity which is often used:

$$\int d^dx\,\sqrt{|g|}\,KR = \int d^dx\,\sqrt{|g|}\,K\Big\{-\omega_b{}^{ba}\omega_c{}^c{}_a - \omega_a{}^{bc}\omega_{bc}{}^a + 2\omega_b{}^{ba}(\partial_a \ln K)\Big\}. \tag{L.4}$$

Observe [1106] that the integrand is a scalar under reparametrizations (there are no world indices at all).

# Appendix M
# Conformal rescalings

If we make the local scale transformation in $d$ dimensions

$$\tilde{g}_{\mu\nu} = \Omega^2 g_{\mu\nu}, \tag{M.1}$$

the determinant of the metric and the Christoffel symbols transform as follows:

$$\sqrt{|\tilde{g}|} = \Omega^d \sqrt{|g|}, \quad \tilde{\Gamma}_{\mu\nu}{}^{\rho} = \Gamma_{\mu\nu}{}^{\rho} + (\delta_\nu{}^\rho \delta_\mu{}^\alpha + \delta_\mu{}^\rho \delta_\nu{}^\alpha - g_{\mu\nu} g^{\rho\alpha})\, \partial_\alpha \ln \Omega. \tag{M.2}$$

So the covariant derivative of a vector (defined to be invariant with index down) and the Laplacian of a scalar transform as follows:

$$\begin{aligned} \tilde{\nabla}_\mu A_\nu &= \nabla_\mu A_\nu - 2A_{(\mu}\partial_{\nu)} \ln \Omega + g_{\mu\nu} A_\rho\, \partial^\rho \ln \Omega, \\ \tilde{\nabla}^2 s &= \Omega^{-2}\big[\nabla^2 s + (d-2)\, \partial_\mu \ln \Omega\, \partial^\mu s\big], \end{aligned} \tag{M.3}$$

where the formulae are written using only $\tilde{g}$ on the l.h.s. and only $g$ on the r.h.s. The completely antisymmetric tensor $\epsilon^{\mu_1 \cdots \mu_d}$ is scale invariant with our conventions.

The Ricci tensor and scalar and the Einstein tensor transform as follows:

$$\begin{aligned} \tilde{R}_{\mu\nu} = R_{\mu\nu} &- (d-2)\big[\partial_\mu \ln \Omega\, \partial_\nu \ln \Omega - g_{\mu\nu} (\partial \ln \Omega)^2\big] \\ &+ (d-2)\left[\nabla_\mu \partial_\nu \ln \Omega + \frac{1}{d-2} g_{\mu\nu} \nabla^2 \ln \Omega\right], \end{aligned} \tag{M.4}$$

$$\tilde{R} = \Omega^{-2}\big[R + (d-1)(d-2)(\partial \ln \Omega)^2 + 2(d-1)\nabla^2 \ln \Omega\big], \tag{M.5}$$

$$\begin{aligned} \tilde{G}_{\mu\nu} = G_{\mu\nu} &- (d-2)\left[\partial_\mu \ln \Omega\, \partial_\nu \ln \Omega + \frac{d-3}{2} g_{\mu\nu} (\partial \ln \Omega)^2\right] \\ &+ (d-2)\big[\nabla_\mu \partial_\nu \ln \Omega - g_{\mu\nu} \nabla^2 \ln \Omega\big]. \end{aligned} \tag{M.6}$$

# Appendix N
# Connections and curvature components

## N.1 For a $d = 3$ metric

The metric

$$ds^2 = \frac{r_0^4}{\sinh^4 r_0\tau} d\tau^2 + \frac{r_0^2}{\sinh^2 r_0\tau} d\Omega_{(2)}^2 \tag{N.1}$$

has the following non-vanishing components of the Levi-Civita connection:

$$\begin{aligned}
&\Gamma_{\tau\tau}{}^{\tau} = -2r_0 \coth r_0\tau, && \Gamma_{\tau\theta}{}^{\theta} = -r_0 \coth r_0\tau, && \Gamma_{\tau\varphi}{}^{\varphi} = \Gamma_{\tau\theta}{}^{\theta}, \\
&\Gamma_{\theta\theta}{}^{\tau} = \frac{1}{r_0} \sinh r_0\tau \cosh r_0\tau, && \Gamma_{\varphi\varphi}{}^{\tau} = \sin^2\theta\, \Gamma_{\theta\theta}{}^{\tau}, && \Gamma_{\varphi\varphi}{}^{\theta} = -\sin\theta\cos\theta, \\
&\Gamma_{\theta\varphi}{}^{\varphi} = \operatorname{cotan}\theta,
\end{aligned} \tag{N.2}$$

of the Riemann tensor,

$$R_{\tau\theta\tau}{}^{\theta} = R_{\tau\varphi\tau}{}^{\varphi} = -r_0^2, \qquad R_{\theta\varphi\theta}{}^{\varphi} = \sinh^2 r_0\tau, \tag{N.3}$$

and of the Ricci tensor,

$$R_{\tau\tau} = -2r_0^2. \tag{N.4}$$

## N.2 For some $d = 4$ metrics

### *N.2.1 General static, spherically symmetric metrics (I)*

The metric

$$ds^2 = g_{tt}(r)dt^2 + g_{rr}(r)dr^2 - r^2 d\Omega_{(2)}^2 \tag{N.5}$$

leads to the Levi-Civita connection components

$$\begin{aligned}
&\Gamma_{tt}{}^{r} = -\tfrac{1}{2}\partial_r g_{tt}/g_{rr}, && \Gamma_{tr}{}^{t} = \tfrac{1}{2}\partial_r g_{tt}/g_{tt}, && \Gamma_{rr}{}^{r} = \tfrac{1}{2}\partial_r g_{rr}/g_{rr}, \\
&\Gamma_{r\theta}{}^{\theta} = 1/r, && \Gamma_{r\varphi}{}^{\varphi} = 1/r, && \Gamma_{\theta\theta}{}^{r} = r/g_{rr}, \\
&\Gamma_{\theta\varphi}{}^{\varphi} = \cos\theta/\sin\theta, && \Gamma_{\varphi\varphi}{}^{r} = \sin^2\theta\, \Gamma_{\theta\theta}{}^{r}, && \Gamma_{\varphi\varphi}{}^{\theta} = -\sin\theta\cos\theta,
\end{aligned} \tag{N.6}$$

and the Ricci tensor,

$$
\begin{aligned}
R_{tt} &= -\frac{\sqrt{g_{tt}}\kappa'}{\sqrt{-g_{rr}}} + \frac{g'_{tt}}{r g_{rr}}, & R_{rr} &= \frac{\sqrt{-g_{rr}}\kappa'}{\sqrt{g_{tt}}} - \frac{g'_{tt}}{r g_{rr}}, \\
R_{\theta\theta} &= -\frac{r g'_{tt}}{2 g_{rr} g_{tt}} + \frac{r g'_{rr}}{2 g_{rr}^2} - \left(1 + \frac{1}{g_{rr}}\right), & R_{\varphi\varphi} &= \sin^2\theta\, R_{\theta\theta},
\end{aligned}
\tag{N.7}
$$

where the prime indicates partial derivation with respect to $r$, and

$$
\kappa = \tfrac{1}{2}\frac{g'_{tt}}{\sqrt{-g_{rr} g_{tt}}}. \tag{N.8}
$$

The Ricci scalar is

$$
R = 2\frac{\kappa'}{\sqrt{-g_{rr} g_{tt}}} - \frac{2}{r g_{rr}}\left[\ln\left(-\frac{g_{tt}}{g_{rr}}\right)\right]' + \frac{2}{r^2}\left(1 + \frac{1}{g_{rr}}\right). \tag{N.9}
$$

If we choose the Vierbein basis

$$
e_t{}^0 = \sqrt{g_{tt}}, \qquad e_r{}^1 = \sqrt{-g_{rr}}, \qquad e_\theta{}^2 = r, \qquad e_\varphi{}^3 = r\sin\theta, \tag{N.10}
$$

the non-vanishing components of the spin-connection 1-form are

$$
\omega_t{}^{01} = \kappa, \quad \omega_\theta{}^{12} = -\frac{1}{\sqrt{-g_{rr}}}, \quad \omega_\varphi{}^{13} = -\frac{\sin\theta}{\sqrt{-g_{rr}}}, \quad \omega_\varphi{}^{23} = -\cos\theta. \tag{N.11}
$$

The non-vanishing components of the curvature 2-form are

$$
\begin{aligned}
R_{tr}{}^{01} &= -\kappa', & R_{t\theta}{}^{02} &= \tfrac{1}{2} g'_{tt}/(g_{rr}\sqrt{g_{tt}}), \\
R_{t\varphi}{}^{03} &= \sin\theta\, R_{t\theta}{}^{02}, & R_{r\theta}{}^{12} &= -\tfrac{1}{2} g'_{rr}/(-g_{rr})^{\frac{3}{2}}, \\
R_{r\varphi}{}^{13} &= \sin\theta\, R_{r\theta}{}^{12}, & R_{\theta\varphi}{}^{23} &= \sin\theta\left(1 + \frac{1}{g_{rr}}\right).
\end{aligned}
\tag{N.12}
$$

### N.2.2 *General static, spherically symmetric metrics (II)*

These can be written in the form

$$
ds^2 = \lambda(r)dt^2 - \lambda^{-1}(r)dr^2 - R^2(r)d\Omega^2, \tag{N.13}
$$

and lead to the non-vanishing Levi-Civita connection coefficients

$$
\begin{aligned}
\Gamma_{tt}{}^r &= \tfrac{1}{2}\lambda\lambda', & \Gamma_{tr}{}^t &= \tfrac{1}{2}\lambda^{-1}\lambda', & \Gamma_{rr}{}^r &= -\tfrac{1}{2}\lambda^{-1}\lambda', \\
\Gamma_{r\theta}{}^\theta &= (\ln R)', & \Gamma_{r\varphi}{}^\varphi &= (\ln R)', & \Gamma_{\theta\theta}{}^r &= -\tfrac{1}{2}\lambda(R^2)', \\
\Gamma_{\theta\varphi}{}^\varphi &= \cos\theta/\sin\theta, & \Gamma_{\varphi\varphi}{}^r &= -\tfrac{1}{2}\lambda(R^2)'\sin^2\theta, & \Gamma_{\varphi\varphi}{}^\theta &= -\cos\theta\sin\theta.
\end{aligned}
\tag{N.14}
$$

The components of the Ricci tensor are

$$
\begin{aligned}
R_{tt} &= -\frac{\lambda}{2R^2}(R^2\lambda')', & R_{rr} &= -\lambda^{-2}R_{tt} + 2\frac{R''}{R}, \\
R_{\theta\theta} &= \tfrac{1}{2}\left[\lambda(R^2)'\right]' - 1, & R_{\varphi\varphi} &= \sin^2\theta\, R_{\theta\theta},
\end{aligned}
\tag{N.15}
$$

and the Ricci scalar is

$$R = -\frac{1}{R^2}\left[(R^2\lambda)'' - 2 + 2\lambda R R''\right]. \tag{N.16}$$

If we choose the Vielbein 1-form basis

$$e_t{}^0 = \lambda^{\frac{1}{2}}, \quad e_r{}^1 = \lambda^{-\frac{1}{2}}, \quad e_\theta{}^2 = R, \quad e_\varphi{}^3 = R\sin\theta, \tag{N.17}$$

we obtain the spin-connection 1-form with components

$$\omega_0{}^{01} = \tfrac{1}{2}\lambda^{-\frac{1}{2}}\lambda', \quad \omega_2{}^{21} = (\ln R)'\lambda^{\frac{1}{2}}, \quad \omega_3{}^{31} = (\ln R)'\lambda^{\frac{1}{2}}, \quad \omega_3{}^{32} = (1/R)\cot\theta, \tag{N.18}$$

and with them the curvature 2-form components

$$\begin{aligned} R_{01}{}^{01} &= -\tfrac{1}{2}\lambda'', & R_{02}{}^{02} &= -\tfrac{1}{2}(\ln R)'\lambda', \\ R_{03}{}^{03} &= -\tfrac{1}{2}(\ln R)'\lambda', & R_{12}{}^{12} &= -\frac{1}{2R}(2R''\lambda + R'\lambda'), \\ R_{13}{}^{13} &= -\frac{1}{2R}(2R''\lambda + R'\lambda'), & R_{23}{}^{23} &= -\frac{1}{R^2}\left[(R')^2\lambda - 1\right]. \end{aligned} \tag{N.19}$$

The components of the Ricci-tensor 1-form are

$$\begin{aligned} R_0{}^0 &= \tfrac{1}{2}\nabla^2 \ln\lambda, & R_1{}^1 &= \tfrac{1}{2}\nabla^2 \ln\lambda - 2\frac{R''\lambda}{R}, \\ R_2{}^2 &= -\tfrac{1}{2}\nabla^2 \ln\lambda - \frac{(R^2\lambda)'' - 2}{2R^2}, & R_3{}^3 &= -\tfrac{1}{2}\nabla^2 \ln\lambda - \frac{(R^2\lambda)'' - 2}{2R^2}, \end{aligned} \tag{N.20}$$

where the form of the Laplacian of a scalar function $f(r)$ in these coordinates is

$$\nabla^2 f(r) = -R^{-2}\left(R^2\lambda f'\right)'. \tag{N.21}$$

The latter two components in Eqs. (N.20) can also be written in this simpler form:

$$R_2{}^2 = R_3{}^3 = \nabla^2 \ln R + 1/R^2. \tag{N.22}$$

### N.2.3 $d = 4$ *IWP-type metrics*

These are stationary (not necessarily axially symmetric) metrics of the form

$$ds^2 = e^{2U}(dt + \omega)^2 - e^{-2U} d\vec{x}^{\,2}, \qquad \omega = \omega_{\underline{i}} dx^{\underline{i}} \quad \Rightarrow \sqrt{|g|} = e^{-2U}, \tag{N.23}$$

where $U$ and $\omega_{\underline{i}}$ are functions of $\vec{x}$ only. The components of the inverse metric are

$$g^{tt} = e^{-2U}\left(1 - e^{4U}\omega^2\right), \quad g^{t\underline{i}} = e^{2U}\omega_{\underline{i}}, \quad g^{\underline{ij}} = -e^{2U}\delta^{ij}. \tag{N.24}$$

A convenient Vierbein 1-form basis and its dual vector basis are provided by

$$e^0 = e^U(dt + \omega), \quad e^i = e^{-U}dx^{\underline{i}}, \quad e_0 = e^{-U}\partial_t, \quad e_i = e^U\left(-\omega_{\underline{i}}\partial_t + \partial_{\underline{i}}\right), \tag{N.25}$$

and the corresponding spin-connection 1-forms are given by

$$\omega^{0i} = \partial_{\underline{i}} e^U e^0 + e^{3U} \partial_{[\underline{i}}\omega_{\underline{k}]} e^k, \qquad \omega^{ij} = e^{3U}\left(\partial_{[\underline{i}}\omega_{\underline{j}]} e^0 - \partial_{[\underline{i}} e^{-2U} \delta_{j]\,k}\right) e^k. \tag{N.26}$$

The self-dual combinations (in the upper indices) take the form

$$\omega^{+\,0i} = \frac{i}{4} e^{3U}\left[\partial_{\underline{i}} V e^0 - i\epsilon_{ijk}\partial_{\underline{j}} V e^k\right], \quad \omega^{+\,ij} = -\tfrac{1}{4} e^{3U}\left[\epsilon_{ijk}\partial_{\underline{k}} V e^0 - 2i\partial_{[\underline{i}} V \delta_{j]\,k} e^k\right], \tag{N.27}$$

where

$$V = b + ie^{-2U}, \qquad \partial_{[\underline{i}}\omega_{\underline{j}]} = -\tfrac{1}{2}\epsilon_{ijk}\partial_{\underline{k}} b. \tag{N.28}$$

The components of the Ricci tensor are

$$\begin{aligned} R_{tt} &= e^{8U}\partial_{[\underline{i}}\omega_{\underline{j}]}\partial_{[\underline{i}}\omega_{\underline{j}]} + e^{4U}\partial^2 U = \tfrac{1}{2}(\mathrm{Im}\,V)^{-4}\partial V \partial \bar{V} - \tfrac{1}{2}(\mathrm{Im}\,V)^{-3}\partial^2 \mathrm{Im}\,V, \\ R_{t\underline{i}} &= -\tfrac{1}{2} e^{8U}\omega_{\underline{i}}\partial V \partial \bar{V} + \tfrac{1}{2} e^{6U}\omega_{\underline{i}}\partial^2 e^{-2U} + e^{6U}\epsilon_{ijk}\partial_j e^{-2U}\partial_k b, \\ R_{\underline{ij}} &= -\tfrac{1}{4} e^{4U}\left[\left(e^{4U}\omega_{\underline{i}}\omega_{\underline{j}} + \delta_{ij}\right)\partial V \partial \bar{V} - \partial_{(\underline{i}} V \partial_{\underline{j})} \bar{V}\right] \\ &\quad + \tfrac{1}{2} e^{2U}\left(e^{4U}\omega_{\underline{i}}\omega_{\underline{j}} + \delta_{ij}\right)\partial^2 e^{-2U} + 4e^{6U}\omega_{(\underline{i}}\epsilon_{j)kl}\partial_{\underline{k}} e^{-2U}\partial_{\underline{l}} b, \end{aligned} \tag{N.29}$$

and the Ricci scalar is given by

$$R = \tfrac{1}{2}\frac{\partial V \partial \bar{V}}{(\mathrm{Im}\,V)^3} + \frac{\partial^2 \mathrm{Im}\,V}{(\mathrm{Im}\,V)^2}. \tag{N.30}$$

### *N.2.4 The d = 4 conformastationary metric*

This metric is a generalization of the preceding one with a non-trivial three-dimensional metric. Our notation differs a little: we call $e^U$, $|M|$. Then,

$$ds^2 = |M|^2(dt + \omega)^2 - |M|^{-2}\gamma_{\underline{mn}} dx^m dx^n, \qquad m, n = 1, 2, 3. \tag{N.31}$$

All components are $t$-independent. We choose the Vierbein basis

$$(e^a{}_\mu) = \begin{pmatrix} |M| & |M|\omega_{\underline{m}} \\ 0 & |M|^{-1} v_{\underline{m}}{}^n \end{pmatrix}, \qquad (e^\mu{}_a) = \begin{pmatrix} |M|^{-1} & -|M|\omega_m \\ 0 & |M| v_m{}^{\underline{n}} \end{pmatrix}, \tag{N.32}$$

where the $v_{\underline{m}}{}^p$ are the Dreibeins of $\gamma_{\underline{mn}}$ which has $(+\,+\,+)$ signature, i.e.

$$\gamma_{\underline{mn}} = v_{\underline{m}}{}^p v_{\underline{n}}{}^q \delta_{pq}, \qquad v_m{}^{\underline{p}} v_{\underline{p}}{}^n v_n, \qquad \omega_m = v_m{}^{\underline{n}}\omega_{\underline{n}}. \tag{N.33}$$

The non-vanishing components of the spin-connection components are

$$\begin{aligned} \omega_{00m} &= -\partial_m |M|, & \omega_{0mn} &= \tfrac{1}{2}|M|^3 f_{mn}, \\ \omega_{m0n} &= \omega_{0mn}, & \omega_{mnp} &= -|M|\varpi_{mnp} - 2\delta_{m[n}\partial_{p]}|M|, \end{aligned} \tag{N.34}$$

where $\varpi_m{}^{np}$ is the three-dimensional spin connection associated with the Dreibeins $v_{\underline{m}}{}^{p}$ and where

$$\partial_m \equiv v_m{}^{\underline{n}}\partial_{\underline{n}}, \qquad f_{mn} = v_m{}^{\underline{p}} v_n{}^{\underline{q}} f_{\underline{pq}}, \qquad f_{\underline{mn}} \equiv 2\partial_{[\underline{m}}\omega_{\underline{n}]}. \tag{N.35}$$

The non-vanishing components of the Riemann tensor are

$$\begin{aligned}
R_{0m0n} &= \tfrac{1}{2}\nabla_m\partial_n|M|^2 + \partial_m|M|\partial_n|M| - \delta_{mn}(\partial|M|)^2 + \tfrac{1}{4}\nabla m|M|^6 f_{mp}f_{np},\\
R_{0mnp} &= -\tfrac{1}{2}\nabla_m(|M|^4 f_{np}) + \tfrac{1}{2}f_{m[n}\partial_{p]}|M|^4 - \tfrac{1}{4}\delta_{m[n}f_{p]l}\partial_q|M|^4,\\
R_{mnpq} &= -|M|^2 R_{mnpq} + \tfrac{1}{2}|M|^6(f_{mn}f_{pq} - f_{p[m}f_{n]q})\\
&\quad -2\delta_{mn,pq}(\partial|M|)^2 + 4|M|\delta_{[m}{}^{[p}\nabla_{n]}\partial^{q]}|M|,
\end{aligned} \tag{N.36}$$

where we have adopted the convention that all the objects on the r.h.s. of the equations (covariant derivatives, tensor contractions, etc.) refer to the three-dimensional spatial metric $\gamma$ and the three-dimensional spin connection $\varpi$.

The non-vanishing components of the Ricci tensor are

$$\begin{aligned}
R_{00} &= -|M|^2\nabla^2\log|M| - \tfrac{1}{4}|M|^6 f^2,\\
R_{0m} &= \tfrac{1}{2}\nabla_n(|M|^4 f_{nm}),\\
R_{mn} &= |M|^2\{R_{mn} + 2\partial_m\log|M|\partial_n\log|M| - \delta_{mn}\nabla^2\log|M| - \tfrac{1}{2}|M|^4 f_{mp}f_{np}\},
\end{aligned} \tag{N.37}$$

and the Ricci scalar is

$$R = -|M|^2\{R - \tfrac{1}{4}|M|^4 f^2 - 2\nabla^2\log|M| + 2(\partial\log|M|)^2\}. \tag{N.38}$$

## N.3 For some $d > 4$ metrics

### N.3.1 $d > 4$ general static, spherically symmetric metrics

We are going to use the metric

$$ds^2 = \lambda(\rho)dt^2 - \mu^{-1}(\rho)d\rho^2 - R^2(\rho)d\Omega^2_{(d-2)}, \tag{N.39}$$

where $d\Omega^2_{(d-2)}$, implicitly defined in Eq. (K.7), is the metric on $\mathrm{S}^{(d-2)}$ (we use the $\psi_i$ coordinates). To consider the most general higher-dimensional static, spherically symmetric metrics it would suffice to take $\mu(\rho) = \lambda(\rho)$ or $R(\rho) = \rho$. We prefer this, however, because it covers both cases and sometimes the components of the metric are simpler if we do not force $R$ to be $\rho$ or $\mu$ to be $\lambda$. This is the class of metrics we used for single (extreme or non-extreme) BHs (point-like objects with $d-2$ asymptotically flat directions).

The non-vanishing components of the Levi-Civita connection are

$$\begin{aligned}
&\Gamma_{tt}{}^{\rho} = \tfrac{1}{2}\mu\lambda', \qquad \Gamma_{t\rho}{}^{t} = \tfrac{1}{2}\lambda^{-1}\lambda', \qquad \Gamma_{\rho\rho}{}^{r} = -\tfrac{1}{2}\mu^{-1}\mu',\\
&\Gamma_{\rho p}{}^{r} = \delta_q{}^{r}(\ln R)', \qquad \Gamma_{qr}{}^{\rho} = -\tfrac{1}{2}\delta_{qr}\mu(R^2)'q_{(r)}/R^2,\\
&\Gamma_{qr}{}^{s} = \left\{\theta_{rq}\delta^s_{(r)}\cot\psi_{(q)} + \theta_{qr}\delta^s_{(q)}\cot\psi_{(r)} - \theta_{qs}\delta_{(q)r}\cot\psi_{(s)}\,q^{-1}_{(s)}q_{(q)}\right\},
\end{aligned} \tag{N.40}$$

where

$$\theta_{rq} = \begin{cases} 1 \; r > q, \\ 0 \; r \leq q, \end{cases} \tag{N.41}$$

and where $q, r, s = 1, \ldots, d-2$ label the angular coordinates and, here,

$$g_{qr} = -\delta_{qr} q_{(r)}. \tag{N.42}$$

Using the Laplacian of a scalar function of $\rho$ in this coordinate system,

$$\nabla^2 f(\rho) = -\frac{\left[(\lambda\mu)^{\frac{1}{2}} R^{d-2} f'\right]'}{(\lambda/\mu)^{\frac{1}{2}} R^{d-2}}, \tag{N.43}$$

we can write the components of the Ricci tensor in their simplest form as follows:

$$\begin{aligned} R_{tt} &= \tfrac{1}{2}\lambda\nabla^2 \ln\lambda, \qquad R_{qr} = g_{qr}\left[\nabla^2 \ln R + \frac{d-3}{R^2}\right], \\ R_{\rho\rho} &= -\tfrac{1}{2}\mu^{-1}\nabla^2 \ln\lambda + \frac{d-2}{R}\left(\frac{\lambda}{\mu}\right)^{\frac{1}{2}}\left[R'\left(\frac{\lambda}{\mu}\right)^{-\frac{1}{2}}\right]', \end{aligned} \tag{N.44}$$

and the Ricci scalar is

$$R = \nabla^2 \ln\left(\lambda R^{d-2}\right) + (d-2)(d-3)\frac{1}{R^2} - \frac{d-2}{R}(\lambda\mu)^{\frac{1}{2}}\left[R'\left(\frac{\lambda}{\mu}\right)^{-\frac{1}{2}}\right]'. \tag{N.45}$$

If we are interested in finding singular contributions to the curvature, these formulae are not completely appropriate, because in obtaining them we have performed operations in which singular contributions are ignored. The unsimplified formulae are

$$R_{qr} = -\frac{1}{d-2} g_{qr}\left\{\nabla^2 \ln\mu + \frac{1}{R^{d-2}}\left(\frac{\lambda}{\mu}\right)^{-\frac{1}{2}}\left[\left(\frac{\lambda}{\mu}\right)^{\frac{1}{2}}\left(R^{d-2}\mu\right)'\right]' - \frac{(d-2)(d-3)}{R^2}\right\}, \tag{N.46}$$

$$R = \nabla^2 \ln\left(\frac{\lambda}{\mu}\right) - \frac{1}{R^{d-2}}\left(\frac{\lambda}{\mu}\right)^{-\frac{1}{2}}\left[\left(\frac{\lambda}{\mu}\right)^{\frac{1}{2}}\left(R^{d-2}\mu\right)'\right]' - \frac{d-2}{R}(\lambda\mu)^{\frac{1}{2}}\left[R'\left(\frac{\lambda}{\mu}\right)^{\frac{1}{2}}\right]' - \frac{(d-2)(d-3)}{R^2}. \tag{N.47}$$

### *N.3.2 The $d = 5$ conformastationary metric*

The five-dimensional conformastationary metric has the form[1]

$$ds^2 = f^2\,(dt + \omega)^2 - f^{-1} h_{\underline{mn}} dx^m dx^n, \qquad \omega = \omega_{\underline{m}} dx^m, \qquad m, n = 1, \ldots, 4. \tag{N.48}$$

[1] The notation here is adapted to that of Section 18.7.1 in the main body of the text. The metric function $f$ is often called $e^U$ in the literature (as it is Chapter 27).

We choose the Vielbein basis

$$(e^a{}_\mu) = \begin{pmatrix} f & f\omega_{\underline{m}} \\ 0 & f^{-1/2}V^n{}_{\underline{m}} \end{pmatrix}, \qquad (e^\mu{}_a) = \begin{pmatrix} f^{-1} & -f^{1/2}\omega_m \\ 0 & f^{1/2}V^{\underline{n}}{}_m \end{pmatrix}, \tag{N.49}$$

where the $V_{\underline{m}}{}^p$s are a Vierbein for the four-dimensional Euclidean metric $h_{\underline{mn}}$ that we use to convert curved into flat four-dimensional spacelike indices, that is

$$h_{\underline{mn}} = V_{\underline{m}}{}^p V_{\underline{n}}{}^q \delta_{pq}, \qquad V_m{}^{\underline{p}} V_n{}^{\underline{q}} h_{\underline{pq}} = \delta_{mn}, \qquad \omega_m = V_m{}^{\underline{n}}\omega_{\underline{n}}. \tag{N.50}$$

With this choice of Vielbein, the non-vanishing components of the spin connection are

$$\begin{aligned} \omega_{00m} &= -2\partial_m f^{1/2}, & \omega_{0mn} &= \tfrac{1}{2}f^2\,(d\omega)_{mn}\,, \\ \omega_{m0n} &= \tfrac{1}{2}f^2\,(d\omega)_{mn}\,, & \omega_{mnp} &= -f^{1/2}\xi_{mnp} - 2\delta_{m[n}\partial_{p]}f^{1/2}, \end{aligned} \tag{N.51}$$

where we adopt the convention that the objects (covariant derivatives, index contractions, curvature tensors, etc.) on the r.h.s. of all the equations refer to the four-dimensional metric $h_{\underline{mn}}$ and the Vierbein basis $\{V^p\}$ and where we are denoting by $\xi_{mnp}$ the four-dimensional spin connection.

The non-vanishing components of the Ricci tensor are

$$\begin{aligned} R_{00} &= -\nabla^2 f + f^{-1}(\partial f)^2 - \tfrac{1}{4}f^4(d\omega)^2, \\ R_{0m} &= -\tfrac{1}{2}f^{-1/2}\nabla_n[f^3(d\omega)_{nm}], \\ R_{mn} &= fR_{mn} - \tfrac{1}{2}(d\omega)_{mp}(d\omega)_{np} + \tfrac{3}{2}f^{-1}\partial_m f\partial_n f - \tfrac{1}{2}\delta_{mn}[\nabla^2 f - f^{-1}(\partial f)^2], \end{aligned} \tag{N.52}$$

and the Ricci scalar is given by

$$R = -fR + \tfrac{1}{4}(d\omega)^2 + \nabla^2 f - \tfrac{5}{2}f^{-1}(\partial f)^2. \tag{N.53}$$

### *N.3.3 A general metric for (single, black) p-branes*

This metric can be understood as a generalization of the previous one with translational isometries in $p$ dimensions, and it is adequate for describing the gravitational fields of $p$-branes. Therefore, in general, it is not asymptotically flat in those $p$ dimensions. It is, roughly speaking, the result of adding those $p$ dimensions to the general, static, spherically symmetric $(d-p)$-dimensional metric of Section N.2. Thus, it has the general form

$$ds^2 = \lambda(\rho)dt^2 - f(\rho)d\vec{y}_p^{\,2} - \mu^{-1}(\rho)dr^2 - R^2(\rho)d\Omega^2_{(\tilde{p}+2)}, \tag{N.54}$$

where $\vec{y}_p = (y_p^1, \ldots, y_p^p)$ are the coordinates *on the p-brane* that we denote with the indices $i, j, k = 1, \ldots, p$, $\rho^2 = (x^{p+1})^2 + \cdots + (x^{d-1})^2$ is the radial coordinate in the $(d-p-1)$-dimensional, asymptotically flat space transverse to the $p$-brane, the $d-p-2$ angular coordinates are labeled by $q, r, s = 1, \ldots, d-p-2$, and $\tilde{p} \equiv d-p-4$ is the dimension of the object that is the electric–magnetic dual to the $p$-brane.

The non-vanishing components of the Levi-Civita connection are

$$\begin{aligned}
&\Gamma_{tt}{}^{\rho} = \tfrac{1}{2}\mu\lambda', && \Gamma_{t\rho}{}^{t} = \tfrac{1}{2}\lambda^{-1}\lambda', \quad \Gamma_{\rho\rho}{}^{\rho} = -\tfrac{1}{2}\mu^{-1}\mu', \\
&\Gamma_{\rho q}{}^{r} = \delta_{q}{}^{r}(\ln R)', && \Gamma_{ij}{}^{\rho} = -\tfrac{1}{2}\delta_{ij}\mu f', \quad \Gamma_{i\rho}{}^{j} = \tfrac{1}{2}\delta_{i}{}^{j} f^{-1} f', \\
&\Gamma_{qr}{}^{\rho} = -\tfrac{1}{2}\delta_{qr}\mu(R^2)' q_{(r)}/R^2, \\
&\Gamma_{qr}{}^{s} = \left\{\theta_{rq}\delta^{s}_{(r)} \cot\psi_{(q)} + \theta_{qr}\delta^{s}_{(q)} \cot\psi_{(r)} - \theta_{qs}\delta_{(q)r} \cot\psi_{(s)}\, q^{-1}_{(s)} q_{(q)}\right\}.
\end{aligned} \tag{N.55}$$

The non-vanishing components of the Ricci tensor are

$$\begin{aligned}
R_{tt} &= R_{tt}^{(d-p)} - \tfrac{1}{4}p\mu(\ln f)'\lambda', \quad R_{\rho\rho} = R_{\rho\rho}^{(d-p)} + \tfrac{1}{2}p(\mu f)^{-\frac{1}{2}}\left[(\mu f)^{\frac{1}{2}}(\ln f)'\right]', \\
R_{ij} &= -\tfrac{1}{2}\delta_{ij} f\nabla^2 \ln f, \qquad\qquad R_{qr} = R_{qr}^{(d-p)} - \tfrac{1}{2}p g_{qr}\mu(\ln f)'(\ln R)',
\end{aligned} \tag{N.56}$$

where we have indicated with the superscript $(d-p)$ the components of the curvature of the $(d-p)$-dimensional metric that one obtains if the $p$ coordinates $\vec{y}_p$ are suppressed and which are given in Appendix N.3.1.

The Ricci scalar is

$$R = R^{(d-p)} + \tfrac{1}{2}p\left\{\nabla^2_{(d-p)} \ln f + f^{-1}\nabla^2 f + \mu\left[(\ln f)'\right]^2\right\}. \tag{N.57}$$

### *N.3.4 A general metric for (composite, black) p-branes*

This metric is a generalization of the general higher-dimensional, static, spherically symmetric metric with translational isometries in $\sum_{n=1}^{N} r_n$ dimensions that split into $N$ blocks. The difference from the metric of Section N.3.3 is that the previous one also had spherical symmetry $SO(p)$ in the $p$ directions associated with the isometries, but in the present metric the spherical symmetry is split into $N$ groups and is, thus, $\prod_{n=1}^{N} SO(r_n)$. This metric is adequate for describing the gravitational field of composite (intersecting etc.) $p$-branes. It has the general form

$$ds^2 = \lambda(\rho)dt^2 - \sum_{n=1}^{N} f_n(\rho)d\vec{y}_n^{\,2} - \mu^{-1}(\rho)dr^2 - R^2(r)d\Omega^2_{(\delta-2)}, \tag{N.58}$$

where $\vec{y}_n = (y_n^1, \ldots, y_n^{r_n})$ are the coordinates of the $n$th "block" that we denote with the indices $i_n, j_n, k_n = 1, \ldots, r_n$, $\rho^2 = (x^{p+1})^2 + \cdots + (x^{d-1})^2$ is the radial coordinate in the $(d - \sum_n r_n - 1)$-dimensional, asymptotically flat space transverse to the $p$-branes, the $\delta - 2$ angular coordinates are labeled by $q, r, s = 1, \ldots, \delta - 2$, and $\delta$ is defined by

$$\delta = d - \sum_{n=1}^{N} r_n. \tag{N.59}$$

The non-vanishing components of the Levi-Civita connection are

$$
\begin{aligned}
\Gamma_{tt}{}^{\rho} &= \tfrac{1}{2}\mu\lambda', & \Gamma_{t\rho}{}^{t} &= \tfrac{1}{2}\lambda^{-1}\lambda', \\
\Gamma_{\rho\rho}{}^{r} &= -\tfrac{1}{2}\mu^{-1}\mu', & \Gamma_{\rho p}{}^{r} &= \delta_q{}^{r}(\ln R)', \\
\Gamma_{i_n j_m}{}^{\rho} &= -\tfrac{1}{2}\delta_{ij}\delta_{nm}\mu f'_{(n)}, & \Gamma_{i_n\rho}{}^{j_m} &= \tfrac{1}{2}\delta_i{}^{j}\delta_n{}^{m} f^{-1} f'_{(n)}, \\
\Gamma_{qr}{}^{\rho} &= -\tfrac{1}{2}\delta_{qr}\mu(R^2)' q_{(r)}/R^2, \\
\Gamma_{qr}{}^{s} &= \Big\{\theta_{rq}\delta^{s}_{(r)}\cot\psi_{(q)} + \theta_{qr}\delta^{s}_{(q)}\cot\psi_{(r)} - \theta_{qs}\delta_{(q)r}\cot\psi_{(s)}\, q^{-1}_{(s)} q_{(q)}\Big\}.
\end{aligned}
\tag{N.60}
$$

The non-vanishing components of the Ricci tensor are

$$
\begin{aligned}
R_{tt} &= R^{(\delta)}_{tt} - \tfrac{1}{4}\mu\textstyle\sum_{n=1}^{N} r_n(\ln f_n)'\lambda', \\
R_{i_n j_m} &= -\tfrac{1}{2}\delta_{ij}\delta_{nm}\Big\{\nabla^2 f_n + \mu f_{(n)}\big[(\ln f_n)'\big]^2\Big\}, \\
R_{\rho\rho} &= R^{(\delta)}_{\rho\rho} + \tfrac{1}{2}\textstyle\sum_{n=1}^{N} r_n(\mu f_{(n)})^{-\frac{1}{2}}\Big[(\mu f_{(n)})^{\frac{1}{2}}\big(\ln f_{(n)}\big)'\Big]', \\
R_{qr} &= R^{(\delta)}_{qr} - \tfrac{1}{2}g_{qr}\mu(\ln R)'\sum_{n=1}^{N} r_n(\ln f_n)',
\end{aligned}
\tag{N.61}
$$

where we have indicated with the superscript $(\delta)$ the components of the curvature of the $\delta$-dimensional metric that one obtains if the coordinates $\vec{y}_n$ are suppressed.

The Ricci scalar is

$$
R = R^{(\delta)} + \tfrac{1}{2}\sum_{n=1}^{N} r_n\Big\{\nabla^2_{(\delta)}\ln f_{(n)} + f^{-1}_{(n)}\nabla^2 f_{(n)} + \mu\big[(\ln f_{(n)})'\big]^2\Big\}. \tag{N.62}
$$

### N.3.5 A general metric for extreme $p$-branes

$d$-dimensional metrics of the general form

$$
ds^2 = H^{2x}\eta_{ij}dy^{\underline{i}}dy^{\underline{j}} + H^{-2y}\eta_{mn}dx^{\underline{m}}dx^{\underline{n}}, \tag{N.63}
$$

where $i,j = 0,1,\ldots,p$, $m,n = p+1,\ldots,d-1$, and $H$ is a function solely of the $x^{\underline{m}}$s often occur in the study of $p$-branes. The coordinates $y^{\underline{i}}$ correspond to the $p$-brane worldvolume and the coordinates $x^{\underline{m}}$ are transverse to the $p$-brane. Observe that, with our conventions, $\eta_{mn} = -\delta_{mn}$. The non-vanishing components of the Levi-Civita connection are

$$
\begin{aligned}
\Gamma_{\underline{ij}}{}^{\underline{m}} &= x\eta_{ij}H^{2(x+y)-1}\partial_{\underline{m}}H, \qquad \Gamma_{\underline{im}}{}^{\underline{j}} = x\delta_i{}^{j}H^{-1}\partial_{\underline{m}}H, \\
\Gamma_{\underline{mn}}{}^{\underline{p}} &= -yH^{-1}\Big\{\delta_{pm}\partial_{\underline{n}}H + \delta_{pn}\partial_{\underline{m}}H - \delta_{mn}\partial_{\underline{p}}H\Big\}.
\end{aligned}
\tag{N.64}
$$

The non-vanishing components of the Ricci tensor are

$$
\begin{aligned}
R_{\underline{ij}} &= g_{\underline{ij}}\nabla^2\ln H^{x}, \\
R_{\underline{mn}} &= g_{\underline{mn}}\nabla^2\ln H^{-y} + zH^{-1}\partial_{\underline{m}}\partial_{\underline{n}}H \\
&\quad + H^{-2}\partial_{\underline{m}}H\partial_{\underline{n}}H\big\{x^2(p+1) + y^2(\tilde{p}+1) + (2y-1)z\big\},
\end{aligned}
\tag{N.65}
$$

where we have used the fact that

$$\sqrt{|g|} = H^{z-2y}, \qquad z = x(p+1) - y(\tilde{p}+1), \tag{N.66}$$

and the fact that, for a scalar function of the $x^{\underline{m}}$s in this metric,

$$\begin{aligned} \nabla^2 f(x^m) &= -H^{2y-1}\left[z\partial_{\underline{m}}H\partial_{\underline{m}}f + H\partial^2 f\right], \qquad \partial^2 \equiv +\partial_{\underline{m}}\partial_{\underline{m}}, \\ \nabla^2 \ln H &= (1-z)H^{2y-2}(\partial H)^2 - H^{2y-1}\partial^2 H. \end{aligned} \tag{N.67}$$

The Ricci scalar is

$$R = \nabla^2 \ln H^{2(z-y)} - H^{2(y-1)}(\partial H)^2\left[x^2(p+1) + y^2(\tilde{p}+1) - z(z-2y)\right]. \tag{N.68}$$

The simplest choice of Vielbein is

$$e_{\underline{i}}{}^{j} = \delta_i{}^j H^x, \qquad e_{\underline{m}}{}^{n} = \delta_m{}^n H^{-y}, \tag{N.69}$$

and it gives the following non-vanishing components of the spin connection:

$$\omega_{\underline{m}}{}^{np} = 2yH^{-1}\partial_{\underline{q}}H\eta_m{}^{[n}\eta^{p]q}, \qquad \omega_{\underline{i}}{}^{mj} = -2xH^{(x+y)-1}\partial_{\underline{q}}H\eta_i{}^{[m}\eta^{j]q}. \tag{N.70}$$

### N.3.6 Brinkmann metrics

The metric of any spacetime admitting a covariantly constant null Killing vector can always be put into the Brinkmann-metric form Eq. (14.4), which we rewrite here for convenience:

$$ds^2 = 2du(dv + Kdu + A_{\underline{i}}dx^i) + \tilde{g}_{\underline{ij}}dx^i dx^j, \tag{N.71}$$

where all the functions in the metric are independent of $v$. Either $K$ or $A_{\underline{i}}$ can be removed by a coordinate transformation that preserves the above form of the metric, but here we work with the most general form.

Using also light-cone coordinates in tangent space, a natural Vielbein basis is

$$\begin{aligned} e^u &= du, \qquad & e^v &= dv + Kdu + A_{\underline{i}}dx^i, \qquad & e^i &= \tilde{e}_{\underline{j}}{}^i dx^j, \\ e_u &= \partial_{\underline{u}} - K\partial_{\underline{v}}, \qquad & e_v &= \partial_{\underline{v}}, \qquad & e_i &= \tilde{e}_i{}^{\underline{j}}\left[\partial_{\underline{j}} - A_{\underline{j}}\partial_{\underline{v}}\right], \end{aligned} \tag{N.72}$$

where the $\tilde{e}^i{}_{\underline{j}}$ are Vielbeins in the $(d-2)$-dimensional wavefront space. The associated components of the spin connection are

$$\begin{aligned} \omega_{uiu} &= \tilde{e}_i{}^{\underline{j}}\left[\partial_{\underline{j}}K - \partial_{\underline{u}}A_{\underline{j}}\right], \qquad & \omega_{uij} &= \tfrac{1}{2}\tilde{F}_{ij} - \tilde{e}_{[i|}{}^{\underline{k}}\partial_{\underline{u}}\tilde{e}_{a|\underline{k}]}, \\ \omega_{ijk} &= \tilde{\omega}_{ijk}, \qquad & \omega_{iju} &= -\tfrac{1}{2}\tilde{F}_{ij} - \tilde{e}_{(i|}{}^{\underline{k}}\partial_{\underline{u}}\tilde{e}_{\underline{k}|j)}, \end{aligned} \tag{N.73}$$

where $F_{\underline{ij}} = 2\partial_{[\underline{i}}A_{\underline{j}]}$.

The components of the Ricci tensor are

$$\begin{aligned} R_{\underline{ij}} &= \tilde{R}_{\underline{ij}}, \\ 2R_{\underline{iu}} &= \tilde{\nabla}^{\underline{j}}F_{\underline{ji}} + \tilde{\nabla}_{\underline{i}}\left(\tilde{g}^{\underline{jk}}\partial_{\underline{u}}\tilde{g}_{\underline{jk}}\right) - \tilde{\nabla}_{\underline{j}}\left(\tilde{g}^{\underline{jk}}\partial_{\underline{u}}\tilde{g}_{\underline{ki}}\right), \\ R_{\underline{uu}} &= \tilde{\nabla}_{\underline{i}}\partial^{\underline{i}}K - \tfrac{1}{4}\tilde{F}^2 + \tfrac{1}{2}\tilde{g}^{\underline{ij}}\partial^2_{\underline{u}}\tilde{g}_{\underline{ij}} + \tfrac{1}{4}\partial_{\underline{u}}\tilde{g}^{\underline{ij}}\partial_{\underline{u}}\tilde{g}_{\underline{ij}} - \tilde{g}^{\underline{ij}}\tilde{\nabla}_{\underline{i}}\left(\partial_{\underline{u}}A_{\underline{j}}\right), \end{aligned} \tag{N.74}$$

where all the objects with tildes are calculated from the metric $\tilde{g}_{\underline{ij}}$, treating $u$ as some constant. The Ricci scalar is just $R = \tilde{R}$.

## N.4 A five-dimensional metric with a null Killing vector

If the null Killing vector admitted by a certain metric is not covariantly constant, the metric cannot be written as in Section N.3.6. Here we give a Vielbein basis and the corresponding spin connection and curvature for the following five-dimensional metric of this kind shared by the null supersymmetric configurations of $N = 1, d = 5$ supergravity studied in Section 18.7.3:

$$ds^2 = 2f du(dv + H du + \omega) - f^{-2}\gamma_{\underline{rs}} dx^r dx^s, \qquad r, s = 1, 2, 3, \tag{N.75}$$

where $\omega = \omega_{\underline{r}} dx^r, f, H, \gamma_{\underline{rs}}$ are $v$ independent.

The Vielbein basis and its inverse are given by

$$\begin{aligned} e^+ &= f du, & e_+ &= f^{-1}(\partial_{\underline{u}} - H\partial_{\underline{v}}), \\ e^- &= dv + H du + \omega, & e_- &= \partial_{\underline{v}}, \\ e^r &= f^{-1} v^r, & e_r &= f(v_r - \omega_r \partial_{\underline{v}}), \end{aligned} \tag{N.76}$$

where $v^r = v^r{}_{\underline{s}} dx^s$ is a Dreibein basis for the spatial three-dimensional Riemannian metric $\gamma_{\underline{rs}}$: $\delta_{rs} v^r{}_{\underline{t}} v^s{}_{\underline{q}} = \gamma_{\underline{tq}}$, etc. The non-vanishing components of the spin connection are

$$\begin{aligned} \omega_{+r+} &= \partial_r H - \partial_{\underline{u}} \omega_{\underline{s}} v_r{}^{\underline{s}}, & \omega_{rs+} &= -\tfrac{1}{2} f^2 F_{rs} - f^{-2} \partial_{\underline{u}} f \delta_{rs} - f^{-1} v_{(r|}{}^{\underline{t}} \partial_{\underline{u}} v_{|s)\underline{t}}, \\ \omega_{+r-} &= \tfrac{1}{2} \partial_r f, & \omega_{rst} &= f \varpi_{rst} - 2\delta_{r[s} \partial_{t]} f, \\ \omega_{+rs} &= \tfrac{1}{2} f^2 F_{rs} - f^{-1} v_{[r|}{}^{\underline{t}} \partial_{\underline{u}} v_{|s]\underline{t}}, & \omega_{r+-} &= -\tfrac{1}{2} \partial_r f \\ \omega_{-r+} &= \tfrac{1}{2} \partial_r f, \end{aligned} \tag{N.77}$$

where, as usual, all the quantities on the r.h.s. of all these equations refer to the three-dimensional metric and Dreibein, $\varpi_{rst}$ being the corresponding three-dimensional spin connection, and $F$ being defined by

$$F_{rs} = v_r{}^{\underline{t}} v_s{}^{\underline{p}} F_{\underline{tp}}, \qquad F_{\underline{rs}} \equiv 2\partial_{[\underline{r}} \omega_{\underline{s}]}. \tag{N.78}$$

The non-vanishing components of the Ricci tensor are

$$\begin{aligned}
R_{++} &= -f\nabla^2 H - \tfrac{1}{4} f^4 F^2 + f\nabla^{\underline{r}}\dot{\omega}_{\underline{r}} + 3\dot{\omega}_{\underline{r}}\partial^{\underline{r}} f + \tfrac{1}{2} f^{-2}\gamma^{\underline{rs}}\ddot{\gamma}_{\underline{rs}} + \tfrac{1}{4} f^{-2}\dot{\gamma}^{\underline{rs}}\dot{\gamma}_{\underline{rs}} \\
&\quad - \tfrac{3}{2} f^{-3}\dot{f}\gamma^{\underline{rs}}\dot{\gamma}_{\underline{rs}} - 3f^{-2}\left[\partial_{\underline{u}}^2 \log f - 2\left(\partial_{\underline{u}} \log f\right)^2\right], \\
R_{+-} &= -\tfrac{1}{2} f^2 \nabla^2 \log f, \qquad\qquad (\text{N.79}) \\
R_{+r} &= -\tfrac{1}{2}\nabla_s\left(f^3 F_{sr}\right) - \tfrac{1}{2} v_r{}^{\underline{r}}\gamma^{\underline{st}}\nabla_{\underline{s}}\dot{\gamma}_{\underline{rt}} + \tfrac{1}{2} v_r{}^{\underline{r}}\partial_{\underline{u}}\left(\gamma^{\underline{st}}\partial_{\underline{r}}\gamma_{\underline{st}}\right) + \tfrac{3}{2} v_r{}^{\underline{r}}\dot{\gamma}_{\underline{rt}}\partial^{\underline{t}} \log f \\
&\quad - \tfrac{3}{2}\partial_r\partial_{\underline{u}} \log f - \tfrac{3}{4}\gamma^{\underline{st}}\dot{\gamma}_{\underline{st}}\partial_r \log f + \tfrac{3}{2}\partial_{\underline{u}} \log f \,\partial_r \log f, \\
R_{rs} &= f^2 R_{rs}(\gamma) - \delta_{rs} f^2 \nabla^2 \log f + \tfrac{3}{2}\partial_r f \partial_s f,
\end{aligned}$$

and the Ricci scalar is

$$R = -f^2 R(\gamma) + 2f^2\nabla^2 \log f - \tfrac{3}{2}\left(\partial f\right)^2. \qquad (\text{N.80})$$

# Appendix O
# The harmonic operator on $\mathbb{R}^3 \times \mathrm{S}^1$

This section is based on Ref. [676].

We want to relate the solutions of the Laplace equation in $\mathbb{R}^3 \times \mathrm{S}^1$ and in $\mathbb{R}^3$. We denote the corresponding Laplacians by $\Delta_{(4)}$ and $\Delta_{(3)}$, and we have

$$\Delta_{(4)} = \Delta_{(3)} + \partial_z^2. \tag{O.1}$$

The Laplacian, being a local operator, has the same form in $\mathbb{R}^3 \times \mathrm{S}^1$ and in $\mathbb{R}^4$. Clearly, the difference is in the periodicity conditions that the solutions must satisfy in the first case. This observation will help us to construct them starting with harmonic functions on $\mathbb{R}^4$.

The solution of the Laplace equation (more precisely, it is the Green function of the Laplacian) in $\mathbb{R}^4$ is $1/|\vec{x}_4 - \vec{x}_{4(0)}|^2$, where $\vec{x}_4 = (x^1, x^2, x^3, x^4)$. In particular, it satisfies

$$\Delta_{(4)} \frac{1}{|\vec{x}_4 - \vec{x}_{4(0)}|^2} = -4\pi^2 \delta^{(4)}(\vec{x}_4 - \vec{x}_{4(0)}). \tag{O.2}$$

This harmonic function has a singularity at $\vec{x}_4 = \vec{x}_{4(0)}$. We are in general interested in harmonic functions that go to 1 at infinity and with a different coefficient for the pole ($h$):

$$H_{\mathbb{R}^4} = 1 + \frac{h}{|\vec{x}_4 - \vec{x}_{4(0)}|^2}. \tag{O.3}$$

Since the Laplacian is linear, we can combine linearly harmonic functions to construct one with singularities placed at regular intervals on the $x^4$ axis. The resulting harmonic function will have the periodicity required for it to be a harmonic function on $\mathbb{R}^3 \times \mathrm{S}^1$. More explicitly,

$$H_{\mathbb{R}^3 \times \mathrm{S}^1} = 1 + \sum_{n \in \mathbb{Z}} \frac{h}{|\vec{x}_3 - \vec{x}_{3(0)}|^2 + (z - z_{(0)} - 2\pi n \ell)^2}. \tag{O.4}$$

This series can be summed:

$$H_{\mathbb{R}^3 \times \mathrm{S}^1} = 1 + \frac{h}{2\ell |\vec{x}_3 - \vec{x}_{3(0)}|^2} \frac{\sinh |\vec{x}_3 - \vec{x}_{3(0)}|/\ell}{\cosh |\vec{x}_3 - \vec{x}_{3(0)}|/\ell - \cos(z - z_{(0)})/\ell}. \tag{O.5}$$

In this form it is evident that we have a function with the right periodicity and one can also immediately check that it is a solution of the Laplace equation. On the other hand, near the singularity $\vec{x}_3 \to \vec{x}_{3(0)}$, $z \to z_{(0)}$, or, in the equivalent limit $\ell \to 0$ in which the periodicity of the fourth coordinate is irrelevant, $H_{\mathbb{R}^3\times \mathrm{S}^1}$ becomes exactly $H_{\mathbb{R}^4}$ plus subdominant terms.

Now we want to expand as a Fourier series the periodic harmonic function $H_{\mathbb{R}^3\times \mathrm{S}^1}$:

$$H_{\mathbb{R}^3\times \mathrm{S}^1} = \sum_{n\in\mathbb{Z}} H_{\mathbb{R}^3\times \mathrm{S}^1,n}(\vec{x}_3 - \vec{x}_{3(0)})e^{\frac{inz}{\ell}}. \tag{O.6}$$

The Fourier modes are

$$H_{\mathbb{R}^3\times \mathrm{S}^1,n}(\vec{x}_3 - \vec{x}_{3(0)}) = \delta_{n,0} + \frac{h/(2\ell)}{|\vec{x}_3 - \vec{x}_{3(0)}|}e^{-\frac{|n|(|\vec{x}_3 - \vec{x}_{3(0)}| - inz_{(0)})}{\ell}}. \tag{O.7}$$

If we consider only the zero mode, we find

$$H_{\mathbb{R}^3\times \mathrm{S}^1,0} = 1 + \frac{h/(2\ell)}{|\vec{x}_3 - \vec{x}_{3(0)}|}, \tag{O.8}$$

which is a harmonic function on $\mathbb{R}^3$, satisfying

$$\Delta_{(3)} H_{\mathbb{R}^3\times \mathrm{S}^1,0} = -\frac{2\pi h}{\ell}\delta^{(3)}(\vec{x}_3 - \vec{x}_{3(0)}). \tag{O.9}$$

Using

$$\Delta_{(3)}\frac{1}{|\vec{x}_3 - \vec{x}_{3(0)}|} = -4\pi\delta^{(3)}(\vec{x}_3 - \vec{x}_{3(0)}), \tag{O.10}$$

it is easy to see that the higher (KK) modes satisfy the massive three-dimensional Laplace equation

$$\left[\Delta_{(3)} - \frac{|n|^2}{\ell^2}\right] H_{\mathbb{R}^3\times \mathrm{S}^1,n} = -\frac{2\pi h}{\ell}\delta^{(3)}(\vec{x}_3 - \vec{x}_{3(0)})e^{\frac{inz_{(0)}}{\ell}}. \tag{O.11}$$

# References


[1] L. F. Abbott and S. Deser, Stability of Gravity with a Cosmological Constant, *Nucl. Phys.* **B195** (1982) 76.

[2] L. F. Abbott and S. Deser, Charge Definition in Non-Abelian Gauge Theories, *Phys. Lett.* **116B** (1982) 259.

[3] M. Abou Zeid and C. M. Hull, Geometric Actions for D-Branes and M-Branes, *Phys. Lett.* **B428** (1998) 277.

[4] M. Abraham, *Lincei Atti* **20** (1911) 678.

[5] M. Abraham, *Phys. Z.* **13** (1912) 1.

[6] M. Abraham, *Phys. Z.* **13** (1912) 4.

[7] M. Abraham, *Phys. Z.* **13** (1912) 176.

[8] M. Abraham, *Phys. Z.* **13** (1912) 310.

[9] M. Abraham, *Phys. Z.* **13** (1912) 311.

[10] M. Abraham, *Phys. Z.* **13** (1912) 793.

[11] M. Abraham, *Nuovo Cim.* **4** (1912) 459.

[12] M. Abraham, *Jahrbuch der Radioaktivität und Elektronik* **11** (1914) 470.

[13] A. Achúcarro, J. Evans, P. K. Townsend, and D. Wiltshire, Super $p$-Branes, *Phys. Lett.* **B198** (1987) 441.

[14] A. Achúcarro, P. Kapusta, and K. S. Stelle, Strings from Membranes: The Origin of Conformal Invariance, *Phys. Lett.* **B232** (1989) 302.

[15] M. Aganagic, J. Park, C. Popescu, and J. H. Schwarz, World-Volume Action of the M-Theory Five-Brane, *Nucl. Phys.* **B496** (1997) 191.

[16] M. Aganagic, C. Popescu, and J. H. Schwarz, D-Brane Actions with Local Kappa Symmetry, *Phys. Lett.* **B393** (1997) 311.

[17] M. Aganagic, C. Popescu, and J. H. Schwarz, Gauge-Invariant and Gauge-Fixed D-Brane Actions, *Nucl. Phys.* **B495** (1997) 99.

[18] A. G. Agnese and M. La Camera, Gravitation Without Black Holes, *Phys. Rev.* **D31** (1985) 1280.

[19] A. G. Agnese and M. La Camera, General Spherically Symmetric Solutions in Charged Dilaton Gravity, *Phys. Rev.* **D49** (1994) 2126.

[20] Y. Aharonov and D. Bohm, Significance of Electromagnetic Potentials in the Quantum Theory, *Phys. Rev.* **115** (1959) 485.

[21] Y. Aharonov and D. Bohm, Further Considerations on Electromagnetic Potentials in the Quantum Theory, *Phys. Rev.* **D3** (1961) 1511.

[22] O. Aharony, String Theory Dualities From M Theory, *Nucl. Phys.* **B476** (1996) 470.

[23] O. Aharony, S. S. Gubser, J. Maldacena, H. Ooguri, and Y. Oz, Large N Field Theories, String Theory and Gravity, *Phys. Rep.* **323** (2000) 183.

[24] P. Aichelburg and R. Sexl, On the Gravitational Field of a Massless Particle, *Gen. Relat. Gravit.* **2** (1971) 303.

[25] W. Alexandrow, *Ann. Phys.* **72** (1923) 141.

[26] N. Alonso-Alberca, E. Lozano-Tellechea, and T. Ortín, Geometric Construction of Killing Spinors and Supersymmetry Algebras in Homogeneous Spacetimes, *Classical Quant. Grav.* **19** (2002) 6009.

[27] N. Alonso-Alberca, E. Lozano-Tellechea, and T. Ortín, The Near Horizon Limit of the Extreme Rotating $d = 5$ Black Hole as a Homogeneous Space-Time, *Classical Quant. Grav.* **20** (2003) 423.

[28] N. Alonso-Alberca, P. Meessen, and T. Ortín, Supersymmetry of Topological Kerr-Newman-Taub-NUT-aDS Spacetimes, *Classical Quant. Grav.* **17** (2000) 2783.

[29] N. Alonso-Alberca, P. Meessen, and T. Ortín, An $\mathrm{SL}(3, Z)$ Multiplet of 8-Dimensional Type II Supergravity Theories and the Gauged Supergravity Inside, *Nucl. Phys.* **B602** (2001) 329.

[30] N. Alonso-Alberca and T. Ortín, Gauged / Massive Supergravities in Diverse Dimensions, *Nucl. Phys.* **B651** (2003) 263.

[31] E. Álvarez, Strings at Finite Temperature, *Nucl. Phys.* **B269** (1986) 596.

[32] E. Álvarez, Quantum Gravity: A Pedagogical Introduction to some Recent Results, *Rev. Mod. Phys.* **61** (1989) 561.

[33] E. Álvarez, Can One Tell Einstein's Unimodular Theory from Einstein's General Relativity?, *JHEP* **0503** (2005) 002.

[34] E. Álvarez, L. Álvarez-Gaumé, and I. Bakas, T Duality and Space-Time Supersymmetry, *Nucl. Phys.* **B457** (1995) 3.

[35] E. Álvarez, L. Álvarez-Gaumé, and Y. Lozano, An Introduction to T Duality in String Theory, *Nucl. Phys. Proc. Suppl.* **41** (1995) 1.

[36] E. Álvarez, J. L. F. Barbón, and J. Borlaf, T-Duality for Open Strings, *Nucl. Phys.* **B479** (1996) 218.

[37] E. Álvarez and P. Meessen, String Primer, *JHEP* **9902** (1999) 015.

[38] E. Álvarez and T. Ortín, Asymptotic Density of States of $p$-Branes, *Mod. Phys. Lett.* **A7** (1992) 2889.

[39] E. Álvarez and M. Á. Osorio, Duality is an Exact Symmetry of String Perturbation Theory, *Phys. Rev.* **D40** (1989) 1150.

[40] L. Álvarez-Gaumé and M. A. Vázquez-Mozo, Topics in String Theory and Quantum Gravity, Les Houches Summer School on Gravitation and Quantizations, Session 57. Amsterdam: North-Holland (1995).

[41] S. P. de Alwis, Coupling of Branes and Normalization of Effective Actions in String/M-Theory, *Phys. Rev.* **D56** (1997) 7963.

[42] D. Amati and C. Klimcik, Nonperturbative Computation of the Weyl Anomaly for a Class of Nontrivial Backgrounds, *Phys. Lett.* **B219** (1989) 443.

[43] S. Åminneborg, I. Bengtsson, S. Holst, and P. Peldán, Making Anti-de Sitter Black Holes, *Classical Quant. Grav.* **13** (1996) 2707.

[44] L. Andrianopoli, M. Bertolini, A. Ceresole, R. D'Auria, S. Ferrara, P. Fré, and T. Magri, $N = 2$ Supergravity and $N = 2$ Super Yang-Mills Theory on General Scalar Manifolds: Symplectic Covariance, Gaugings and the Momentum Map, *J. Geom. Phys.* **23** (1997) 111.

[45] L. Andrianopoli and R. D'Auria, Extremal Black Holes in Supergravity and the Bekenstein-Hawking Entropy, *Entropy* **4** (2002) 65.

[46] L. Andrianopoli, R. D'Auria, and S. Ferrara, U Invariants, Black Hole Entropy and Fixed Scalars, *Phys. Lett.* **B403** (1997) 12.

[47] L. Andrianopoli, R. D'Auria, and S. Ferrara, U Duality and Central Charges in Various Dimensions Revisited, *Int. J. Mod. Phys.* **A13** (1998) 431.

[48] L. Andrianopoli, R. D'Auria, and S. Ferrara, Supersymmetry Reduction of N-extended Supergravities in Four Dimensions, *JHEP* **0203** (2002) 025.

[49] L. Andrianopoli, R. D'Auria, and S. Ferrara, Consistent Reduction of $N = 2 \rightarrow N = 1$ Four Dimensional Supergravity Coupled to Matter, *Nucl. Phys.* **B628** (2002) 387.

[50] L. Andrianopoli, R. D'Auria, S. Ferrara, A. Marrani, and M. Trigiante, Two-Centered Magical Charge Orbits, *JHEP* **1104** (2011) 041.

[51] L. Andrianopoli, R. D'Auria, S. Ferrara, and M. Trigiante, Extremal Black Holes in Supergravity, *Lect. Notes Phys.* **737** (2008) 661.

[52] L. Andrianopoli, R. D'Auria, E. Orazi, and M. Trigiante, First Order Description of Black Holes in Moduli Space, *JHEP* **0711** (2007) 032.

[53] L. Andrianopoli, R. D'Auria, E. Orazi, and M. Trigiante, First Order Description of $D = 4$ static Black Holes and the Hamilton-Jacobi Equation, *Nucl. Phys.* **B833** (2010) 1.

[54] C. Angelantonj and A. Sagnotti, Open Strings, *Phys. Rep.* **371** (2002) 1.

[55] S. Antoci, David Hilbert and the Origin of the "Schwarzschild Solution", `physics/0310104`.

[56] A. de Antonio Martín, T. Ortín, and C. S. Shahbazi, The FGK Formalism for Black p-Branes in $d$ Dimensions, *JHEP* **1205** (2012) 045.

[57] T. Applequist, A. Chodos, and P. G. O. Freund, *Modern Kaluza–Klein* Theories. Menlo Park, CA: Addison-Wesley (1987).

[58] C. Aragone and S. Deser, Constraints on Gravitationally Coupled Tensor Fields, *Nuovo Cim.* **3A** (1971) 709.

[59] C. Aragone and S. Deser, Consistency Problems of Spin-2 Gravity Coupling, *Nuovo Cim.* **57B** (1980) 33.

[60] I. Ya. Aref'eva, M. G. Ivanov, O. A. Rytchkov, and I. V. Volovich, Non-Extremal Localized Branes and Vacuum Solutions in M-Theory, *Classical Quant. Grav.* **15** (1998) 2923.

[61] I. Ya. Aref'eva, M. G. Ivanov, and I. V. Volovich, Non-Extremal Intersecting $p$-Branes in Various Dimensions, *Phys. Lett.* **B406** (1997) 44.

[62] I. Ya. Aref'eva, K. S. Viswanathan, and I. V. Volovich, $p$-Brane Solutions in Diverse Dimensions, *Phys. Rev.* **D55** (1997) 4748.

[63] I. Ya. Aref'eva and A. Volovich, Composite $p$-Branes in Diverse Dimensions, *Classical Quant. Grav.* **14** (1997) 2991.

[64] R. Argurio, F. Englert, and L. Houart, Intersection Rules for $p$-Branes, *Phys. Lett.* **B398** (1997) 61.

[65] R. Argurio, F. Englert, L. Houart, and P. Windey, On the Opening of Branes, *Phys. Lett.* **B408** (1997) 151.

[66] R. Arnowitt, S. Deser, and C. Misner, The Dynamics of General Relativity, in *Gravitation: An Introduction to Current Research*, ed. L. Witten. New York: Wiley (1962), p. 227. (Reprinted in *Gen. Relativ. Grav.* **40** (2008) 1997.)

[67] P. Aschieri, S. Ferrara, and B. Zumino, Duality Rotations in Nonlinear Electrodynamics and in Extended Supergravity, *Riv. Nuovo Cim.* **31** (2008) 625.

[68] A. Ashtekar, *Lectures on Non-perturbative Quantum Gravity.* Singapore: World Scientific (1991).

[69] P. S. Aspinwall, Compactification, Geometry and Duality: $N = 2$, in *Strings, Branes and Gravity*, eds. J. A. Harvey, S. Kachruand, and E. Silverstein. River Edge, NJ: World Scientific (2001), p. 723.

[70] M. F. Atiyah and N. J. Hitchin, Low-Energy Scattering of Non-Abelian Monopoles, *Phys. Lett.* **107A** (1985) 21. (Reprinted in Ref. [615].)

[71] R. D'Auria and P. Fré, BPS Black Holes in Supergravity: Duality Groups, $p$-Branes, Central Charges and the Entropy, in *Classical and Quantum Black Holes*, eds. P. Fré, V. Gorini, G. Magli, and U. Moschella. Bristol: Insitute of Physics Publishing (1999), p. 137.

[72] M. A. Awada, M. J. Duff, and C. N. Pope, $N = 8$ Supergravity Breaks down to $N = 1$, *Phys. Rev. Lett.* **50** (1983) 294.

[73] J. A. de Azcárraga, J. P. Gauntlett, J. M. Izquierdo, and P. K. Townsend, Topological Extensions of the Supersymmetry Algebra for Extended Objects, *Phys. Rev. Lett.* **63** (1989) 2443.

[74] J. A. de Azcárraga and J. Lukierski, Supersymmetric Particles with Internal Symmetries and Central Charges, *Phys. Lett.* **B113** (1982) 170.

[75] J. A. de Azcárraga and J. Lukierski, Supersymmetric Particles in $N = 2$ Superspace: Phase Space Variables and Hamiltonian Dynamics, *Phys. Rev.* **D28** (1983) 1337.

[76] B. E. Baaquie and L. C. Kwek, Superstrings, Gauge Fields and Black Holes, *Int. J. Mod. Phys.* **A16** (2001) 2605.

[77] S. V. Babak and L. P. Grishchuk, The Energy-Momentum Tensor for the Gravitational Field, *Phys. Rev.* **D61** (2000) 024038.

[78] C. Bachas, Lectures on D-branes, in Cambridge 1997, Duality and Supersymmetric Theories, eds. D. I. Olive and P. C. West. Cambridge: Cambridge University Press (1999), p. 414.

[79] C. P. Bachas, M. R. Douglas, and M. B. Green, Anomalous Creation of Branes, *JHEP* **9707** (1997) 002.

[80] C. P. Bachas, M. B. Green, and A. Schwimmer, $(8, 0)$ Quantum Mechanics and Symmetry Enhancement in Type I′ Superstrings, *JHEP* **9801** (1998) 006.

[81] F. A. Bais and R. J. Russell, Magnetic Monopole Solution of Nonabelian Gauge Theory in Curved Space-Time, *Phys. Rev.* **D11** (1975) 2692. [Erratum *ibid.* **D12** (1975) 3368.]

[82] J. A. Bagger, Coupling The Gauge Invariant Supersymmetric Nonlinear Sigma Model To Supergravity, *Nucl. Phys.* **B211** (1983) 302.

[83] J. Bagger and E. Witten, The Gauge Invariant Supersymmetric Nonlinear Sigma Model, *Phys. Lett.* **B118** (1982) 103.

[84] I. Bakas, Space-Time Interpretation of S Duality and Supersymmetry Violations of T Duality, *Phys. Lett.* **B343** (1995) 103.

[85] I. Bakas, Solitons of Axion – Dilaton Gravity, *Phys. Rev.* **D54** (1996) 6424.

[86] I. Bakas and K. Sfetsos, T Duality and World Sheet Supersymmetry, *Phys. Lett.* **B349** (1995) 448.

[87] A. P. Balachandran, P. Salomonson, B. S. Skagerstam, and J. O. Winnberg, Classical Description of Particle Interacting with Nonabelian Gauge Field, *Phys. Rev.* **D15** (1977) 2308.

[88] H. Balasin and H. Nachbagauer, On the Distributional Nature of the Energy-Momentum Tensor of Black Hole or What Curves the Schwarzschild Geometry?, *Classical Quant. Grav.* **10** (1993) 2271.

[89] H. Balasin and H. Nachbagauer, Distributional Energy-Momentum Tensor of the Kerr-Newman Space-Time Family, *Classical Quant. Grav.* **11** (1994) 1453.

[90] H. Balasin and H. Nachbagauer, The Ultrarelativistic Kerr Geometry and its Energy-Momentum Tensor, *Classical Quant. Grav.* **12** (1995) 707.

[91] H. Balasin and H. Nachbagauer, Boosting the Kerr Geometry in an Arbitrary Direction, *Classical Quant. Grav.* **13** (1996) 731.

[92] V. Balasubramanian, How to Count the States of Extremal Black Holes in $N = 8$ Supergravity, in *Cargese 1997, Strings, Branes and Dualities,* eds. L. Baulieu, P. Di Francesco, M. Douglas, V. Kazakov, M. Picco, and P. Windey. Dordrecht: Kluwer (1999), pp. 399–410.

[93] V. Balasubramanian and F. Larsen, On D-Branes and Black Holes in Four Dimensions, *Nucl. Phys.* **B478** (1996) 199.

[94] V. Balasubramanian, F. Larsen, and R. G. Leigh, Branes at Angles and Black Holes, *Phys. Rev.* **D57** (1998) 3509.

[95] M. Bañados, C. Teitelboim, and J. Zanelli, Black Hole in Three-Dimensional Spacetime, *Phys. Rev. Lett.* **69** (1992) 1849.

[96] I. A. Bandos, J. A. de Azcárraga, J. M. Izquierdo, and J. Lukierski, An Action for Supergravity Interacting with Super-$P$-Brane Sources, *Phys. Rev.* **D65** (2002) 021901.

[97] I. A. Bandos, J. A. de Azcárraga, J. M. Izquierdo, and J. Lukierski, On Dynamical Supergravity Interacting with Super-$P$-Brane Sources, Invited talk at 3rd Int. Sakharov Conf. on Physics, Moscow, Russia, 24–29 June 2002, `hep-th/0211065`.

[98] I. Bandos, K. Lechner, A. Nurmagambetov, P. Pasti, D. P. Sorokin, and M. Tonin, Covariant Action for the Super-Five-Brane of M-Theory, *Phys. Rev. Lett.* **78** (1997) 4332.

[99] I. A. Bandos, D. P. Sorokin, and M. Tonin, Generalized Action Principle and Superfield Equations of Motion for $D = 10$ $Dp$-branes, *Nucl. Phys.* **B497**, (1997) 275.

[100] A. Barajas, Birkhoff's Theory of Gravitation and Einstein's for Weak Fields, *Proc. Nat. Acad. Sci.* **30** (1944) 54.

[101] A. Barajas, G. D. Birkhoff, C. Graef, and M. Sandoval Vallarta, On Birkhoff's New Theory of Gravitation, *Phys. Rev.* **66** (1944) 138.

[102] J. M. Bardeen, B. Carter, and S. W. Hawking, The Four Laws of Black Hole Mechanics, *Commun. Math. Phys.* **31** (1973) 161.

[103] A. Barducci, R. Casalbuoni, and L. Lusanna, Supersymmetries and the Pseudoclassical Relativistic Electron, *Nuovo Cim.* **A35** (1976) 377.

[104] A. Barducci, R. Casalbuoni, and L. Lusanna, Classical Scalar and Spinning Particles Interacting with External Yang-MIlls Fields, *Nucl. Phys.* **B124** (1977) 93.

[105] A. Barducci, R. Casalbuoni, and L. Lusanna, Classical Spinning Particles Interacting with External Gravitational Fields, *Nucl. Phys.* **B124** (1977) 521.

[106] R. Bartnik and J. McKinnon, Particle-Like Solutions of the Einstein Yang-Mills Equations, *Phys. Rev. Lett.* **61** (1988) 141.

[107] A. O. Barut and R. Raczka, *Theory of Group Representations and Applications*. Singapore: World Scientific (1986).

[108] B. Bates and F. Denef, Exact Solutions for Supersymmetric Stationary Black Hole Composites, Preprint `hep-th/0304094`.

[109] H. Bauer, *Phys. Z.* **19** (1918) 163.

[110] K. Bautier, S. Deser, M. Henneaux, and D. Seminara, No Cosmological $D = 11$ Supergravity, *Phys. Lett.* **B406** (1997) 49.

[111] K. Becker, M. Becker, and J. H. Schwarz, *String Theory and M-Theory: A Modern Introduction*. Cambridge: Cambridge University Press (2007).

[112] K. Behrndt, G. L. Cardoso, B. de Wit, R. Kallosh, D. Lüst, and T. Mohaupt, Classical and Quantum $N = 2$ Supersymmetric Black Holes, *Nucl. Phys.* **B488** (1997) 236.

[113] K. Behrndt, R. Kallosh, J. Rahmfeld, M. Shmakova, and W. K. Wong, STU Black Holes and String Triality, *Phys. Rev.* **D54** (1996) 6293.

[114] K. Behrndt, G. Lopes Cardoso, and S. Mahapatra, Exploring the Relation between *4D* and *5D BPS* Solutions, *Nucl. Phys.* **B732** (2006) 200.

[115] K. Behrndt, D. Lüst, and W. A. Sabra, Stationary Solutions of $N = 2$ Supergravity, *Nucl. Phys.* **B510** (1998) 264.

[116] R. Beig and W. Simon, On the Uniqueness of Static Perfect-Fluid Solutions in General Relativity, *Commun. Math. Phys.* **144** (1992) 373.

[117] J. D. Bekenstein, Baryon Number, Entropy, and Black Hole Physics Ph.D. Thesis, Princeton University (1972, unpublished).

[118] J. D. Bekenstein, Nonexistence of Baryon Number for Static Black Holes, *Phys. Rev.* **D5** (1972) 1239.

[119] J. D. Bekenstein, Black Holes and the Second Law, *Lett. Nuovo Cim.* **4** (1972) 737.

[120] J. D. Bekenstein, Black Holes and Entropy, *Phys. Rev.* **D9** (1973) 2333.

[121] J. D. Bekenstein, Generalized Second Law of Thermodynamics in Black Hole Physics, *Phys. Rev.* **D9** (1974) 3292.

[122] J. D. Bekenstein, Novel "No-Scalar-Hair" Theorem for Black Holes, *Phys. Rev.* **D51** (1995) R6608.

[123] J. D. Bekenstein, Black Hole Hair: 25 Years After, in *2nd Int. Sakharov Conf. on Physics*, eds. I. M. Dremin and A. M. Semikhatov. Singapore: World Scientific (1997), p. 216.

[124] A. A. Belavin, A. M. Polyakov, A. S. Schwarz, and Yu. S. Tyupkin, Pseudoparticle Solutions of the Yang-Mills Equations, *Phys. Lett.* **59B** (1975) 85.

[125] F. J. Belinfante, On the Spin Angular Momentum of Mesons, *Physica* **VI** (1939) 887.

[126] V. A. Belinskii, G. W. Gibbons, D. N. Page, and C. N. Pope, Asymptotically Euclidean Bianchi IX Metrics and Quantum Gravity, *Phys. Lett.* **B76** (1978) 433.

[127] J. Bellorín, Supersymmetric Solutions of Gauged Five-dimensional Supergravity with General Matter Couplings, *Classical Quant. Grav.* **26** (2009) 195012.

[128] J. Bellorín, P. Meessen, and T. Ortín, Supersymmetry, Attractors and Cosmic Censorship, *Nucl. Phys.* **B762** (2007) 229.

[129] J. Bellorín, P. Meessen, and T. Ortín, All the Supersymmetric Solutions of $N = 1$, $d = 5$ Ungauged Supergravity, *JHEP* **0701** (2007) 020.

[130] J. Bellorín and T. Ortín, A Note on Simple Applications of the Killing Spinor Identities, *Phys. Lett.* **B616** (2005) 118.

[131] J. Bellorín and T. Ortín, All the Supersymmetric Configurations of $N = 4$, $d = 4$ Supergravity, *Nucl. Phys.* **B726** (2005) 171.

[132] J. Bellorín and T. Ortín, Characterization of all the supersymmetric solutions of gauged $N = 1$, $d = 5$ Supergravity, *JHEP* **0708** (2007) 096.

[133] S. Bellucci, S. Ferrara, M. Gunaydin, and A. Marrani, SAM Lectures on Extremal Black Holes in $d = 4$ Extended Supergravity, *Springer Proc. Phys.* **134** (2010) 1.

[134] S. Bellucci, S. Ferrara, A. Marrani, and A. Yeranyan, stu Black Holes Unveiled, *Entropy* **10** (2008) 507.

[135] I. Bena, Splitting Hairs of the Three Charge Black Hole, *Phys. Rev.* **D70** (2004) 105018.

[136] I. Bena and P. Kraus, Microscopic Description of Black Rings in AdS/CFT, *JHEP* **0412** (2004) 070.

[137] I. Bena and P. Kraus, Microstates of the D1-D5-KK System, *Phys. Rev.* **D72** (2005) 025007.

[138] I. Bena, P. Kraus, and N. P. Warner, Black Rings in Taub-NUT, *Phys. Rev.* **D72** (2005) 084019.

[139] I. Bena, H. Triendl, and B. Vercnocke, Camouflaged Supersymmetry in Solutions of Extended Supergravities, *Phys. Rev.* **D86** (2012) 061701.

[140] I. Bena and N. P. Warner, One Ring to Rule Them All ... And in the Darkness Bind Them?, *Adv. Theor. Math. Phys.* **9** (2005) 667.

[141] F. A. Berezin and M. S. Marinov, *JETP Lett.* **21** (1975) 320.

[142] F. A. Berezin and M. S. Marinov, Particle Spin Dynamics as the Grassmann Variant of Classical Mechanics, *Ann. Phys.* **104** (1977) 336.

[143] M. Berger, Sur les groupes d'holonomie homogène des variétés à connexion affines et des variétés riemanniennes, *Bull. Soc. Math. Fran.* **83** (1955) 279.

[144] O. Bergman, M. R. Gaberdiel, and G. Lifschytz, Branes, Orientifolds and the Creation of Elementary Strings, *Nucl. Phys.* **B509** (1998) 194.

[145] O. Bergmann, Scalar Theory as a Theory of Gravitation. I, *Am. J. Phys.* **24** (1956) 38.

[146] P. G. Bergmann, *Introduction to the Theory of Relativity*. New York: Prentice-Hall (1942); New York: Dover (1976).

[147] P. G. Bergmann and R. Schiller, Classical and Quantum Theories in the Lagrangian Formalism, *Phys. Rev.* **89** (1953) 4.

[148] E. Bergshoeff, H.-J. Boonstra, and T. Ortín, $S$ Duality and Dyonic $p$-Brane Solutions in Type II String Theory, *Phys. Rev.* **D53** (1996) 7206.

[149] E. Bergshoeff, W. Chemissany, A. Ploegh, M. Trigiante, and T. Van Riet, Generating Geodesic Flows and Supergravity Solutions, *Nucl. Phys.* **B812** (2009) 343.

[150] E. A. Bergshoeff, C. Condeescu, G. Pradisi, and F. Riccioni, Heterotic-Type II Duality and Wrapping Rules, *JHEP* **1312** (2013) 057.

[151] E. Bergshoeff, P. M. Cowdall, and P. K. Townsend, Massive IIA Supergravity from the Topologically Massive $D$-2-Brane, *Phys. Lett.* **410B** (1997) 13.

[152] E. Bergshoeff, S. Cucu, T. de Wit, J. Gheerardyn, R. Halbersma, S. Vandoren, and A. Van Proeyen, Superconformal $N = 2$, $D = 5$ Matter With and Without Actions, *JHEP* **0210** (2002) 045.

[153] E. Bergshoeff, S. Cucu, T. de Wit, J. Gheerardyn, S. Vandoren, and A. Van Proeyen, $N = 2$ Supergravity in Five Dimensions Revisited, *Classical Quant. Grav.* **21** (2004) 3015.

[154] E. Bergshoeff, I. Entrop, and R. Kallosh, Exact Duality in the String Effective Action, *Phys. Rev.* **D49** (1994) 6663.

[155] E. Bergshoeff, E. Eyras, R. Halbersma, J. P. van der Schaar, C. M. Hull, and Y. Lozano, Space-Time Filling Branes and Strings with Sixteen Supercharges, *Nucl. Phys.* **B564** (2000) 29.

[156] E. Bergshoeff, E. Eyras, and Y. Lozano, The Massive Kaluza-Klein Monopole, *Phys. Lett.* **B430** (1998) 77.

[157] E. A. Bergshoeff, J. Gomis, T. A. Nutma, and D. Roest, Kac-Moody Spectrum of (Half-)Maximal Supergravities, *JHEP* **0802** (2008) 069.

[158] E. Bergshoeff, J. Gomis, and P. K. Townsend, $M$-Brane Intersections from Worldvolume Superalgebras, *Phys. Lett.* **B421** (1998) 109.

[159] E. A. Bergshoeff, J. Hartong, O. Hohm, M. Hübscher, and T. Ortín, Gauge Theories, Duality Relations and the Tensor Hierarchy, *JHEP* **0904** (2009) 123.

[160] E. A. Bergshoeff, J. Hartong, M. Hübscher, and T. Ortín, Stringy Cosmic Strings in Matter Coupled $N = 2$, $d = 4$ Supergravity, *JHEP* **0805** (2008) 033.

[161] E. A. Bergshoeff, J. Hartong, T. Ortín, and D. Roest, Seven-branes and Supersymmetry, *JHEP* **0702** (2007) 003.

[162] E. Bergshoeff, C. M. Hull, and T. Ortín, Duality in the Type II Superstring Effective Action, *Nucl. Phys.* **B451** (1995) 547.

[163] E. Bergshoeff, B. Janssen, and T. Ortín, Solution-Generating Transformations and the String Effective Action, *Classical Quant. Grav.* **13** (1996) 321.

[164] E. Bergshoeff, B. Janssen, and T. Ortín, Kaluza-Klein Monopoles and Gauged Sigma-Models, *Phys. Lett.* **B410** (1997) 131.

[165] E. Bergshoeff, R. Kallosh, and T. Ortín, Supersymmetric String Waves, *Phys. Rev.* **D47** (1993) 5444.

[166] E. Bergshoeff, R. Kallosh, and T. Ortín, Black Hole–Wave Duality in String Theory, *Phys. Rev.* **D50**, (1994) 5188.

[167] E. Bergshoeff, R. Kallosh, and T. Ortín, Duality Versus Supersymmetry and Compactification, *Phys. Rev.* **D51** (1995) 3009.

[168] E. Bergshoeff, R. Kallosh, and T. Ortín, Stationary Axion/Dilaton Solutions and Supersymmetry, *Nucl. Phys.* **B478** (1996) 156.

[169] E. Bergshoeff, R. Kallosh, T. Ortín, and G. Papadopoulos, $\kappa$-Symmetry, Supersymmetry and Intersecting Branes, *Nucl. Phys.* **B502** (1997) 149.

[170] E. Bergshoeff, R. Kallosh, T. Ortín, D. Roest, and A. Van Proeyen, New Formulations of $D = 10$ Supersymmetry and $D8$ - $O8$ Domain Walls, *Classical Quant. Grav.* **18** (2001) 3359.

[171] E. A. Bergshoeff, A. Kleinschmidt, and F. Riccioni, Supersymmetric Domain Walls, *Phys. Rev.* **D86** (2012) 085043

[172] E. Bergshoeff, I. G. Koh, and E. Sezgin, Coupling of Yang-Mills to $N = 4$, $D = 4$ Supergravity, *Phys. Lett.* **B155** (1985) 71.

[173] E. Bergshoeff, L. A. J. London, and P. K. Townsend, Spacetime Scale Invariance and the Super-$p$-Brane, *Classical Quant. Grav.* **9** (1992) 2545.

[174] E. Bergshoeff, Y. Lozano, and T. Ortín, Massive Branes, *Nucl. Phys.* **B518** (1998) 363.

[175] E. A. Bergshoeff, A. Marrani, and F. Riccioni, Brane Orbits, *Nucl. Phys.* **B861** (2012) 104.

[176] E. A. Bergshoeff, T. Ortin, and F. Riccioni, Defect Branes, *Nucl. Phys.* **B856** (2012) 210.

[177] E. A. Bergshoeff and F. Riccioni, $D$-Brane Wess-Zumino Terms and $U$-Duality, *JHEP* **1011** (2010) 139.

[178] E. A. Bergshoeff and F. Riccioni, Branes and Wrapping Rules, *Phys. Lett.* **B704** (2011) 367.

[179] E. A. Bergshoeff and F. Riccioni, The $D$-brane U-scan, `arXiv:1109.1725`.

[180] E. A. Bergshoeff and F. Riccioni, Heterotic Wrapping Rules, *JHEP* **1301** (2013) 005.

[181] E. A. Bergshoeff, F. Riccioni, and L. Romano, Branes, Weights and Central Charges, *JHEP* **1306** (2013) 019.

[182] E. A. Bergshoeff, F. Riccioni, and L. Romano, Towards a Classification of Branes in Theories with Eight Supercharges, *JHEP* **1405** (2014) 070.

[183] E. Bergshoeff and M. de Roo, D-Branes and T-Duality, *Phys. Lett.* **B380** (1996) 265.

[184] E. Bergshoeff, M. de Roo, E. Eyras, B. Janssen, and J. P. van der Schaar, Multiple Intersections of D-Branes and M-Branes *Nucl. Phys.* **B494** (1997) 119.

[185] E. Bergshoeff, M. de Roo, E. Eyras, B. Janssen, and J. P. van der Schaar, Intersections Involving Monopoles and Waves in Eleven Dimensions, *Classical Quant. Grav.* **14** (1997) 2757.

[186] E. Bergshoeff, M. de Roo, M. B. Green, G. Papadopoulos, and P. K. Townsend, Duality of Type II 7-Branes and 8-Branes, *Nucl. Phys.* **B470** (1996) 113.

[187] E. Bergshoeff, M. de Roo, B. Janssen, and T. Ortín, The Super D9-Brane and its Truncations, *Nucl. Phys.* **B550** (1999) 289.

[188] E. Bergshoeff, M. de Roo, and T. Ortín, The Eleven-Dimensional Five-Brane, *Phys. Lett.* **B386** (1996) 85.

[189] E. Bergshoeff, M. de Roo, B. de Wit, and P. van Nieuwenhuizen, Ten-Dimensional Maxwell-Einstein Supergravity, its Currents, and the Issue of its Auxiliary Fields, *Nucl. Phys.* **B195** (1982) 97.

[190] E. Bergshoeff, E. Sezgin, and P. K. Townsend, Superstring Actions in $D = 3, 4, 6, 10$ Curved Superspace, *Phys. Lett.* **B169** (1986) 191.

[191] E. Bergshoeff, E. Sezgin, and P. K. Townsend, Supermembranes and Eleven-Dimensional Supergravity, *Phys. Lett.* **189B** (1987) 75.

[192] E. Bergshoeff, E. Sezgin, and P. K. Townsend, Properties of the Eleven-Dimensional Super Membrane Theory, *Ann. Phys.* **185** (1988) 330.

[193] E. Bergshoeff and J. P. van der Schaar, On M-9-Branes, *Classical Quant. Grav.* **16** (1999) 23.

[194] E. Bergshoeff and P. K. Townsend, Super-D-Branes, *Nucl. Phys.* **B490** (1997) 145.

[195] E. Bergshoeff and P. K. Townsend, Super-D-Branes Revisited, *Nucl. Phys.* **B531** (1998) 226.

[196] E. Bergshoeff, T. de Wit, U. Gran, R. Linares, and D. Roest, (Non-)Abelian Gauged Supergravities in Nine Dimensions, *JHEP* **10** (2002) 61.

[197] M. Berkooz, M. R. Douglas, and R. G. Leigh, Branes Intersecting at Angles, *Nucl. Phys.* **B480** (1996) 265.

[198] M. Berkooz and B. Pioline, 5D Black Holes and Non-linear Sigma Models, *JHEP* **0805** (2008) 045.

[199] Z. Bern, J. J. Carrasco, L. J. Dixon, H. Johansson, D. A. Kosower, and R. Roiban, Three-Loop Superfiniteness of $N = 8$ Supergravity, *Phys. Rev. Lett.* **98** (2007) 161303.

[200] Z. Bern, J. J. Carrasco, L. J. Dixon, H. Johansson, and R. Roiban, The Ultraviolet Behavior of $N = 8$ Supergravity at Four Loops, *Phys. Rev. Lett.* **103** (2009) 081301.

[201] Z. Bern, J. J. Carrasco, L. J. Dixon, H. Johansson, and R. Roiban, Amplitudes and Ultraviolet Behavior of $N = 8$ Supergravity, *Fortsch. Phys.* **59** (2011) 561.

[202] Z. Bern, S. Davies, T. Dennen, A. V. Smirnov, and V. A. Smirnov, The Ultraviolet Properties of N = 4 Supergravity at Four Loops, *Phys. Rev. Lett.* **111** (2013) 231302.

[203] M. Bertolini, P. Fré, and M. Trigiante, The Generating Solution of Regular $N = 8$ BPS Black Holes, *Classical Quant. Grav.* **16** (1999) 2987.

[204] M. Bertolini and M. Trigiante, Regular BPS Black Holes: Macroscopic and Microscopic Description of the Generating Solution, *Nucl. Phys.* **B582** (2000) 393.

[205] B. Bertotti, Uniform Electromagnetic Field in the Theory of GR, *Phys. Rev.* **116** (1959) 1331.

[206] A. L. Besse, *Einstein Manifolds*. Berlin: Springer Verlag (1987).

[207] M. Bianchi and A. Sagnotti, On the Systematics of Open String Theories, *Phys. Lett.* **B247** (1990) 517.

[208] M. Bianchi and A. Sagnotti, Twist Symmetry and Open String Wilson Lines, *Nucl. Phys.* **B361** (1991) 519.

[209] J. Bičák, *Selected Solutions of Einstein's Field Equations: Their Role in General Relativity and Astrophysics*. Heidelberg: Springer-Verlag (2000).

[210] J. J. van der Bij, H. van Dam, and Y. J. Ng, Theory Of Gravity and the Cosmological Term: The Little Group Viewpoint, *Physica* **116A** (1982) 307.

[211] A. Bilal, Introduction to Supersymmetry, Lectures given at Summer School *Gif 2000*, `hep-th/0101055`.

[212] G. D. Birkhoff, *Relativity and Modern Physics*. Cambridge, MA: Harvard University Press (1923).

[213] G. D. Birkhoff, Matter, Electricity and Gravitation in Flat Space-Time, *Proc. Nat. Acad. Sci.* **29** (1943) 231. (Reprinted in Ref. [216].)

[214] G. D. Birkhoff, Flat Space-Time and Gravitation, *Proc. Nat. Acad. Sci.* **30** (1944) 324. (Reprinted in Ref. [216].)

[215] G. D. Birkhoff, El Concepto de Tiempo y la Gravitación, *Bol. Soc. Mat. Mexicana* **1** (nos. 4, 5) (1944) 1. (Reprinted in Ref. [216].)

[216] G. D. Birkhoff, Collected Mathematical Papers, 3 vol. New York: American Mathematical Society (1950).

[217] D. Birmingham, Topological Black Holes in Anti-de Sitter Space, Report `hep-th/9808032`.

[218] N. D. Birrell and P. C. W. Davies, *Quantum Fields in Curved Space*. Cambridge: Cambridge University Press (1989).

[219] P. Bizon, Colored Black Holes, *Phys. Rev. Lett.* **64** (1990) 2844.

[220] P. Bizon and O. T. Popp, No Hair Theorem for Spherical Monopoles and Dyons in SU(2) Einstein Yang-Mills Theory, *Classical Quant. Grav.* **9** (1992) 193.

[221] M. Blau, J. M. Figueroa-O'Farrill, C. M. Hull, and G. Papadopoulos, A New Maximally Supersymmetric Background of IIB Superstring Theory, *JHEP* **0201** (2002) 047.

[222] M. Blau, J. M. Figueroa-O'Farrill, C. M. Hull, and G. Papadopoulos, Penrose Limits and Maximal Supersymmetry, *Classical Quant. Grav.* **19** (2002) L87.

[223] M. Blau, J. M. Figueroa-O'Farrill, and G. Papadopoulos, Penrose Limits, Supergravity and Brane Dynamics, *Classical Quant. Grav.* **19** (2002) 4753.

[224] M. Blau and M. O'Loughlin, Aspects of U-Duality in Matrix Theory, *Nucl. Phys.* **B525** (1998) 182.

[225] M. Blau, W. Thirring, and G. Landi, Introduction to Kaluza-Klein Theories, in 25th Schladmig Conf., *Concepts and Trends in Particle Physics*, eds. H. Latal and H. Mitter. Berlin: Springer-Verlag (1987), p. 1.

[226] J. de Boer, E. P. Verlinde, and H. L. Verlinde, On the Holographic Renormalization Group, *JHEP* **0008** (2000) 003.

[227] E. Bogomol'nyi, Stability of Clasical Solutions, *Sov. J. Nucl. Phys.* **24** (1976) 449 [*Yad. Fiz.* **24** (1976) 861].

[228] L. Bombelli, R. K. Koul, G. Kunstatter, J. Lee, and R. D. Sorkin, On Energy in 5-Dimensional Gravity and the Mass of the Kaluza-Klein Monopole, *Nucl. Phys.* **B299** (1987) 735.

[229] W. B. Bonnor, Static Magnetic Fields in General Relativity, *Proc. Phys. Soc. London* **A67** (1954) 225.

[230] W. B. Bonnor, An Exact Solution of the Einstein-Maxwell Equations Referring to a Magnetic Dipole, *Z. Phys.* **190** (1966) 444.

[231] W. B. Bonnor, A New Interpretation of the NUT Metric in General Relativity, *Proc. Camb. Phil. Soc.* **66** (1969) 145.

[232] W. B. Bonnor, Physical Interpretation of Vacuum Solutions of Einstein's Equations, Part I. Time-Independent Solutions, *Gen. Relativ. Gravit.* **24** (1992) 551.

[233] W. B. Bonnor, J. B. Griffiths, and M. A. H. MacCallum, Physical Interpretation of Vacuum Solutions of Einstein's Equations, Part II. Time-Dependent Solutions, *Gen. Relativ. Gravit.* **26** (1994) 687.

[234] H. J. Boonstra, K. Skenderis, and P. K. Townsend, The Domain-Wall/QFT Correspondence, *JHEP* **9901** (1999) 003.

[235] M. Born, On the Quantum Theory of the Electromagnetic Field, *Proc. Roy. Soc. London* **A143** (1934) 410.

[236] M. Born, Théorie non-linéare du champ électromagnétique, *Ann. Inst. Poincaré* **7** (1939) 155.

[237] M. Born and L. Infeld, Foundations of the New Field Theory, *Proc. Roy. Soc. London* **A144** (1934) 425.

[238] L. Borsten, D. Dahanayake, M.J. Duff, and W. Rubens, Black Holes Admitting a Freudenthal Dual, *Phys. Rev.* **D80** (2009) 026003

[239] G. Bossard, The Extremal Black Holes of $N = 4$ Supergravity from $\mathfrak{so}(8, 2+n)$ Nilpotent Orbits, *Gen. Relativ. Grav.* **42** (2010) 539.

[240] G. Bossard and S. Katmadas, Duality Covariant Non-BPS First Order Systems, *JHEP* **1209** (2012) 100.

[241] G. Bossard, Y. Michel, and B. Pioline, Extremal Black Holes, Nilpotent Orbits and the True Fake Superpotential, *JHEP* **1001** (2010) 038.

[242] G. Bossard, H. Nicolai, and K. S. Stelle, Universal BPS Structure of Stationary Supergravity Solutions, *JHEP* **0907** (2009) 003.

[243] N. Boulanger and L. Gualtieri, An Exotic Theory of Massless Spin-Two Fields in Three Dimensions, *Classical Quant. Grav.* **18** (2001) 1485.

[244] B. D. Boulware and S. Deser, Classical General Relativity Derived from Quantum Gravity, *Ann. Phys.* **89** (1975) 193.

[245] D. G. Boulware, Naked Singularities, Thin Shells and the Reissner Nordström Metric, *Phys. Rev.* **D8** (1973) 2363.

[246] R. H. Boyer, Geodesic Killing Orbits and Bifurcate Killing Horizons, *Proc. Roy. Soc. London* **A311** (1969) 245.

[247] M. Bradley, G. Fodor, L. Á. Gergely, M. Marklund, and Z. Perjés, Rotating Perfect Fluid Sources of the NUT Metric, *Classical Quant. Grav.* **16** (1999) 1667.

[248] D. Brecher and M. J. Perry, Ricci-Flat Branes, *Nucl. Phys.* **B566** (2000) 151.

[249] J. C. Breckenridge, D. A. Lowe, R. C. Myers, A. W. Peet, A. Strominger, and C. Vafa, Macroscopic and Microscopic Entropy of Near-Extremal Spinning Black Holes, *Phys. Lett.* **B381** (1996) 423.

[250] J. C. Breckenridge, G. Michaud, and R. C. Myers, More D-Brane Bound States, *Phys. Rev.* **D55** (1997) 6438.

[251] J. C. Breckenridge, G. Michaud, and R. C. Myers, New Angles on D-Branes, *Phys. Rev.* **D56** (1997) 5172.

[252] J. C. Breckenridge, R. C. Myers, A. W. Peet, and C. Vafa, D-Branes and Spinning Black Holes, *Phys. Lett.* **B391** (1997) 93.

[253] P. Breitenlohner, D. Maison, and G. W. Gibbons, Four-Dimensional Black Holes from Kaluza-Klein Theories, *Commun. Math. Phys.* **120** (1988) 295.

[254] D. Brill, Electromagnetic Fields in a Homogeneous, Nonisotropic Universe, *Phys. Rev.* **133** (1964) B845.

[255] D. R. Brill and R. W. Lindquist, Interaction Energy in Geometrostatics, *Phys. Rev.* **131** (1963) 471.

[256] D. Brill, J. Louko, and P. Peldán, Thermodynamics of (3+1)-Dimensional Black Holes with Toroidal or Higher Genus Horizons, *Phys. Rev.* **D56** (1997) 3600.

[257] D. Brill and J. A. Wheeler, Interaction of Neutrinos and Granvitational Fields, *Rev. Mod. Phys.* **29** (1957) 465.

[258] L. Brink, S. Deser, B. Zumino, P. Di Vecchia, and P. S. Howe, Local Supersymmetry for Spinning Particles, *Phys. Lett.* **B64** (1976) 435.

[259] L. Brink, P. Di Vecchia, and P. Howe, A Locally Supersymmetric and Reparametrization Invariant Action for the Spinning String, *Phys. Lett.* **65B** (1976) 471.

[260] L. Brink, P. Di Vecchia, and P. Howe, A Lagrangian Formulation of the Classical and Quantum Dynamics of Spinning Particles, *Nucl. Phys.* **B118** (1977) 76.

[261] L. Brink and J. H. Schwarz, Quantum Superspace, *Phys. Lett.* **B100** (1981) 310.

[262] L. Brink, J. H. Schwarz, and J. Scherk, Supersymmetric Yang-Mills Theories, *Nucl. Phys.* **B121** (1977) 77.

[263] H. W. Brinkmann, *Proc. Nat. Acad. Sci.* **9** (1923) 1.

[264] H. W. Brinkmann, Einstein Spaces which are Mapped Conformally on Each Other, *Math. Annal.* **94** (1925) 119.

[265] R. Brout, S. Massar, R. Parentani, and P. Spindel, A Primer for Black Hole Quantum Physics, *Phys. Rep.* **260** (1995) 329.

[266] J. D. Brown, Black Hole Pair Creation and the Entropy Factor, *Phys. Rev.* **D51** (1995) 5725.

[267] P. Bueno, R. Davies, and C. S. Shahbazi, Quantum Black Holes in Type-IIA String Theory, *JHEP* **1301** (2013) 089.

[268] P. Bueno, P. Galli, P. Meessen, and T. Ortín, Black Holes and Equivariant Charge Vectors in $N = 2$, $d = 4$ Supergravity, *JHEP* **1309** (2013) 010.

[269] L. Burko and A. Ori, Introduction to the Internal Structure of Black Holes, in *Internal Structure of Black Holes and Spacetime Singularities*, eds. L. Burko and A. Ori. Bristol: Institute of Physics Publishing, and Jerusalem: The Israel Physical Society (1997).

[270] T. Buscher, Quantum Corrections and Extended Supersymmetry in New Sigma Models, *Phys. Lett.* **159B** (1985) 127.

[271] T. Buscher, A Symmetry of the String Background Field Equations, *Phys. Lett.* **194B** (1987) 59.

[272] T. Buscher, Path Integral Derivation of Quantum Duality in Non-Linear Sigma Models, *Phys. Lett.* **201B** (1988) 466.

[273] M. Cahen and N. Wallach, Lorentzian Symemtric Spaces, *Bull. Am. Math. Soc.* **76** (1970) 585.

[274] A. C. Cadavid, A. Ceresole, R. D'Auria, and S. Ferrara, Eleven-dimensional Supergravity Compactified on Calabi-Yau Threefolds, *Phys. Lett.* **B357** (1995) 76.

[275] S. L. Cacciatori, M. M. Caldarelli, D. Klemm, and D. S. Mansi, More on BPS Solutions of $N = 2$, $D = 4$ Gauged Supergravity, *JHEP* **0407** (2004) 061.

[276] S. L. Cacciatori, M. M. Caldarelli, D. Klemm, D. S. Mansi, and D. Roest, Geometry of Four-dimensional Killing Spinors, *JHEP* **0707** (2007) 046.

[277] S. L. Cacciatori, A. Celi, and D. Zanon, BPS Equations in $N = 2$, $D = 5$ Supergravity with Hypermultiplets, *Classical Quant. Grav.* **20** (2003) 1503.

[278] S. L. Cacciatori, D. Klemm, D. S. Mansi, and E. Zorzan, All Timelike Supersymmetric Solutions of $N = 2$, $D = 4$ Gauged Supergravity Coupled to Abelian Vector Multiplets, *JHEP* **0805** (2008) 097.

[279] R.-G. Cai and Y.-Z. Zhang, Black Plane Solutions in Four-Dimensional Space-Times, *Phys. Rev.* **D54** (1996) 4891.

[280] E. Calabi and E. Visentini, *Ann. Math.* **71** (1960) 472.

[281] M. M. Caldarelli and D. Klemm, Supersymmetry of Anti-De Sitter Black Holes, *Nucl. Phys.* **B545** (1999) 434.

[282] M. M. Caldarelli and D. Klemm, All Supersymmetric Solutions of $N = 2$, $D = 4$ Gauged Supergravity, *JHEP* **0309** (2003) 019.

[283] C. G. Callan, Jr. A. Harvey, and A. Strominger, World-Sheet Approach to Heterotic Instantons and Solitons, *Nucl. Phys.* **B359** (1991) 611.

[284] C. G. Callan, J. A. Harvey, and A. Strominger, Supersymmetric String Solitons, in *String Theory and Quantum Gravity '91*, ed. J. A. Harvey. Singapore: World Scientific (1992), p. 208.

[285] C. G. Callan, E. J. Martinec, M. J. Perry, and D. Friedan, Strings In Background Fields, *Nucl. Phys.* **B262** (1985) 593.

[286] C. G. Callan Jr., S. Coleman, and R. Jackiw, A New Improved Energy-Momentum Tensor, *Ann. Phys.* **59** (1970) 42.

[287] C. G. Callan Jr. and J. M. Maldacena, *D*-Brane Approach to Black Hole Quantum Mechanics, *Nucl. Phys.* **B472** (1996) 591.

[288] C. G. Callan Jr. and J. M. Maldacena, Brane Dynamics from the Born-Infeld Action, *Nucl. Phys.* **B513** (1998) 198.

[289] C. Callan and L. Thorlacius, Sigma Models and String Theory, in *Particles, Strings and Supernovae*, vol. 2, eds. A. Jevicki and C. I. Tan. Singapore: World Scientific (1989), p. 795.

[290] I. C. G. Campbell and P. C. West, $N = 2, d = 10$ Non chiral Supergravity and its Spontaneous Compactification, *Nucl. Phys.* **B243** (1984) 112.

[291] P. Candelas, Lectures on Complex Manifolds, in *Superstrings'87*. Singapore: World Scientific, (1988).

[292] P. Candelas, Yukawa Couplings Between (2,1) Forms, *Nucl. Phys.* **B298** (1988) 458.

[293] P. Candelas, P. S. Green, and T. Hubsch, Rolling Among Calabi-Yau Vacua, *Nucl. Phys.* **B330** (1990) 49.

[294] P. Candelas, G. T. Horowitz, A. Strominger, and E. Witten, Vacuum Configurations for Superstrings, *Nucl. Phys.* **B258** (1985) 46.

[295] P. Candelas and X. C. de la Ossa, Comments on Conifolds, *Nucl. Phys.* **B342** (1990) 246.

[296] P. Candelas and X. de la Ossa, Moduli Space Of Calabi-Yau Manifolds, *Nucl. Phys.* **B355** (1991) 455.

[297] P. Candelas, X. C. de la Ossa, P. S. Green, and L. Parkes, An Exactly Soluble Superconformal Theory from a Mirror Pair of Calabi-Yau Manifolds, *Phys. Lett.* **B258** (1991) 118.

[298] F. Canfora, F. Correa, A. Giacomini, and J. Oliva, Exact Meron Black Holes in Four Dimensional SU(2) Einstein-Yang-Mills Theory, *Phys. Lett.* **B722** (2013) 364.

[299] G. L. Cardoso and V. Grass, On Five-dimensional Non-extremal Charged Black Holes and FRW Cosmology, *Nucl. Phys.* **B803** (2008) 209.

[300] G. L. Cardoso, B. de Wit, J. Käppeli, and T. Mohaupt, Stationary BPS Solutions in $N = 2$ Supergravity with $R^2$ Interactions, *JHEP* **0012** (2000) 019.

[301] G. L. Cardoso, B. de Wit, and T. Mohaupt, Area Law Corrections from State Counting and Supergravity, *Classical Quant. Grav.* **17** (2000) 1007.

[302] M. Cariglia and O.A.P. Mac Conamhna, The General Form of Supersymmetric Solutions of $N = (1,0)$ U(1) and SU(2) Gauged Supergravities in Six Dimensions, *Classical Quant. Grav.* **21** (2004) 3171.

[303] M. Cariglia and O. A. P. Mac Conamhna, Timelike Killing Spinors in Seven Dimensions, *Phys. Rev.* **D70** (2004) 125009.

[304] S. Carlip, Black Hole Entropy from Horizon Conformal Field Theory, *Nucl. Phys. Proc. Suppl.* **88** (2000) 10.

[305] B. J. Carr, Black Holes in Cosmology and Astrophysics, in *General Relativity*, Eds. G. S. Hall and J. R. Pulham. London: Institute of Physics Publishing (1996), p. 143.

[306] J. J. M. Carrasco, R. Kallosh, and R. Roiban, Covariant Procedures for Perturbative Non-linear Deformation of Duality-invariant Theories, *Phys. Rev.* **D85** (2012) 025007.

[307] E. Cartan, Sur les équations de la gravitation de Einstein, *J. Math. Pure Appl.* **1** (1922) 141.

[308] E. Cartan, *Oeuvres complètes*. Paris: Editions du Centre National de la Recherche Scientifique (1984).

[309] B. Carter, Axisymmetric Black Hole Has Only Two Degrees of Freedom, *Phys. Rev. Lett.* **26** (1971) 331.

[310] B. Carter, Properties of the Kerr Metric, in *Black Holes*, eds. C. Dewitt and B. S. DeWitt. New York: Gordon and Breach (1973).

[311] B. Carter, Charge and Particle Conservation in Black-Hole Decay, *Phys. Rev. Lett.* **33** (1974) 558.

[312] R. Casalbuoni, Relativity and Supersymmetries, *Phys. Lett.* **B62** (1976) 49.

[313] L. Castellani, A. Ceresole, R. D'Auria, S. Ferrara, P. Fré, and M. Trigiante, G/H M-Branes and AdS(p+2) Geometries, *Nucl. Phys.* **B527** (1998) 142.

[314] L. Castellani, R. D'Auria, and S. Ferrara, Special Geometry Without Special Coordinates, *Classical Quant. Grav.* **1** (1990) 317.

[315] L. Castellani, R. D'Auria, and P. Fré, Supergravity and Superstrings, A Geometric Perspective, 3 vol. Singapore: World Scientific (1991).

[316] A. Castro and M. J. Rodríguez, Universal Properties and the First Law of Black Hole Inner Mechanics, *Phys. Rev.* **D86** (2012) 024008.

[317] S. Cecotti, S. Ferrara, and L. Girardello, Geometry of Type II Superstrings and the Moduli of Superconformal Field Theories, *Int. J. Mod. Phys.* **A4** (1989) 2475.

[318] M. Cederwall, A. von Gussich, B. E. Nilsson, P. Sundell, and A. Westerberg, The Dirichlet Super-*p*-Branes in Ten-Dimensional Type IIA and IIB Supergravity, *Nucl. Phys.* **B490** (1997) 179.

[319] M. Cederwall, A. von Gussich, B. E. Nilsson, and A. Westerberg, The Dirichlet Super-Three-Brane in Ten-Dimensional Type IIB Supergravity, *Nucl. Phys.* **B490** (1997) 163.

[320] A. Celi, Toward the Classification of BPS Solutions of $N = 2$, $d = 5$ Gauged Supergravity with Matter Couplings, Ph.D. Thesis, hep-th/0405283.

[321] A. Ceresole and G. Dall'Agata, General Matter Coupled $N = 2$, $D = 5$ Gauged Supergravity, *Nucl. Phys.* **B585** (2000) 143.

[322] A. Ceresole and G. Dall'Agata, Flow Equations for Non-BPS Extremal Black Holes, *JHEP* **0703** (2007) 110.

[323] A. Ceresole, G. Dall'Agata, S. Ferrara and A. Yeranyan, Universality of the Superpotential for $d = 4$ Extremal Black Holes, *Nucl. Phys.* **B832** (2010) 358.

[324] A. Ceresole, G. Dall'Agata, S. Ferrara and A. Yeranyan, First Order Flows for $N = 2$ Extremal Black Holes and Duality Invariants, *Nucl. Phys.* **B824** (2010) 239.

[325] A. Ceresole, R. D'Auria, and S. Ferrara, The Symplectic Structure of $N = 2$ Supergravity and its Central Extension, *Nucl. Phys. Proc. Suppl.* **46** (1996) 67.

[326] A. Ceresole, R. D'Auria, S. Ferrara, and A. Van Proeyen, On Electromagnetic Duality in Locally Supersymmetric $N = 2$ Yang-Mills Theory, in *Physics from Planck Scale to Electroweak Scale*, eds. P. Nath, T. Taylor, and S. Pokorski. River Edge, NJ: World Scientific (1995).

[327] A. Ceresole, R. D'Auria, S. Ferrara, and A. Van Proeyen, Duality Transformations in Supersymmetry Yang-Mills Theories Coupled to Supergravity, *Nucl. Phys.* **444** (1995) 92.

[328] A. Ceresole, S. Ferrara, A. Gnecchi, and A. Marrani, More on $N = 8$ Attractors, *Phys. Rev.* **D80** (2009) 045020.

[329] A. Ceresole, S. Ferrara, A. Gnecchi, and A. Marrani, *d*-Geometries Revisited, *JHEP* **1302** (2013) 059.

[330] A. H. Chamseddine, Massive Supergravity from Spontaneously Breaking Orthosymplectic Gauge Symmetry, *Ann. Phys.* **113** (1978) 219.

[331] A. Chamseddine, $N = 4$ Supergravity Coupled to $N = 4$ Matter and Hidden Symmetries, *Nucl. Phys.* **B185** (1981) 403.

[332] A. H. Chamseddine, S. Ferrara, G. W. Gibbons, and R. Kallosh, Enhancement of Supersymmetry Near 5d Black Hole Horizon, *Phys. Rev.* **D55** (1997) 3647.

[333] A. H. Chamseddine, J. M. Figueroa-O'Farrill, and W. Sabra, Six-Dimensional Supergravity Vacua and Anti-Selfdual Lorentzian Lie Groups, hep-th/0306278.

[334] A. H. Chamseddine and H. Nicolai, Coupling the SO(2) Supergravity through Dimensional Reduction, *Phys. Lett.* **B96** (1980) 89.

[335] A. H. Chamseddine and W. A. Sabra, Metrics Admitting Killing Spinors in Five Dimensions, *Phys. Lett.* **B426** (1998) 36.

[336] A. H. Chamseddine and W. A. Sabra, Calabi-Yau Black Holes and Enhancement of Supersymmetry in Five-dimensions, *Phys. Lett.* **B460** (1999) 63.

[337] A. H. Chamseddine and M. S. Volkov, NonAbelian BPS Monopoles in $N = 4$ Gauged Supergravity, *Phys. Rev. Lett.* **79** (1997) 3343.

[338] A. H. Chamseddine and M. S. Volkov, NonAbelian Solitons in $N = 4$ Gauged Supergravity and Leading Order String Theory, *Phys. Rev.* **D57** (1998) 6242.

[339] A. H. Chamseddine and P. C. West, Supergravity as a Gauge Theory of Supersymmetry, *Nucl. Phys.* **B129** (1977) 39.

[340] S. Chandrasekhar, *The Mathematical Theory of Black Holes*. Oxford: Clarendon Press (1983).

[341] G. F. Chapline and N. S. Manton, Unification of Yang-Mills Theory and Supergravity in Ten Dimensions, *Phys. Lett.* **B120** (1983) 105.

[342] J. E. Chase, *Commun. Math. Phys.* **19** (1970) 276.

[343] J. Chazy, Sur le champ de gravitation de deux masses fixes dans la théorie de la relativité, *Bull. Soc. Math. France* **52** (1924) 17.

[344] W. Chemissany, P. Fré, J. Rosseel, A. S. Sorin, M. Trigiante, and T. Van Riet, Black Holes in Supergravity and Integrability, *JHEP* **1009** (2010) 080.

[345] W. Chemissany, P. Fré, and A. S. Sorin, The Integration Algorithm of Lax Equation for both Generic Lax Matrices and Generic Initial Conditions, *Nucl. Phys.* **B833** (2010) 220.

[346] W. Chemissany, R. Kallosh, and T. Ortín, Born-Infeld with Higher Derivatives, *Phys. Rev.* **D85** (2012) 046002.

[347] W. Chemissany, A. Ploegh, and T. Van Riet, A Note on Scaling Cosmologies, Geodesic Motion and Pseudo-susy, *Classical Quant. Grav.* **24** (2007) 4679.

[348] W. Chemissany, J. Rosseel, M. Trigiante and T. Van Riet, The Full Integration of Black Hole Solutions to Symmetric Supergravity theories, *Nucl. Phys.* **B830** (2010) 391.

[349] J. Chen and J. Li, Quaternionic Maps Between HyperKähler Manifolds, *J. Diff. Geom.* **55** (2000) 355.

[350] S. S. Chern, On the Curvature Integral in a Riemannian Manifold, *Ann. Math.* **46** (1945) 674.

[351] Y. M. Cho, Einstein Lagrangian as the Translational Yang-Mills Lagrangian, *Phys. Rev.* **D14** (1976) 2521.

[352] Y. M. Cho and P. G. O. Freund, Gravitating 't Hooft Monopoles, *Phys. Rev.* **D12** (1975) 1588. [Erratum *ibid.* **D13** (1976) 531.]

[353] Y. M. Cho and S. W. Zoh, Explicit Construction of Massive Spin-Two Fields in Kaluza-Klein Theory, *Phys. Rev.* **D46** (1992) R2290.

[354] Y. M. Cho and S. W. Zoh, Virasoro Invariance and Theory of Internal String, *Phys. Rev.* **D46** (1992) 3483.

[355] D. D. K. Chow and G. Compère, Seed for General Rotating Non-extremal Black Holes of N = 8 Supergravity, *Classical Quant. Grav.* **31** (2014) 022001.

[356] Y. Choquet-Bruhat, C. DeWitt-Morette, and M. Dillard-Bleick, *Analysis, Manifolds and Physics*. Amsterdam: Elsevier (1977).

[357] D. Christodoulou, Investigation in Gravitational Collapse and the Physics of Black Holes, Ph.D. Thesis, Princeton University (1971, unpublished).

[358] P. T. Chruściel, "No-Hair" Theorems: Folklore, Conjectures, Results, *Contemp. Math.* **170** (1994) 23.

[359] P. T. Chruściel, H. S. Reall, and P. Tod, On Israel-Wilson-Perjes Black Holes, *Classical Quant. Grav.* **23** (2006) 2519.

[360] J. Ciufolini and J. A. Wheeler, *Gravitation and Inertia*. Princeton, NJ: Princeton University Press (1995).

[361] C. J. S. Clarke, *The Analysis of Space-Time Singularities*. Cambridge: Cambridge University Press (1993).

[362] C. J. S. Clarke, A Review of Cosmic Censorship, *Classical Quant. Grav.* **10** (1993) 1375.

[363] G. Clément and D. V. Gal'tsov, Stationary BPS Solutions to Dilaton-Axion Gravity, *Phys. Rev.* **D54** (1996) 6136.

[364] S. Coleman, in *Gauge Theories in High Energy Physics*, eds. M. K. Gaillard and R. F. Stora. Amsterdam: North-Holland (1983).

[365] S. Coleman and R. Jackiw, Why Dilatation Generators do not Generate Dilatations?, *Ann. Phys.* **67** (1971) 552.

[366] S. Coleman, S. Parke, A. Neveu, and C. M. Sommerfield, Can One Dent a Dyon?, *Phys. Rev.* **D15** (1977) 544.

[367] S. Coleman, J. Preskill, and F. Wilczek, Quantum Hair on Black Holes, *Nucl. Phys.* **B378** (1992) 175.

[368] A. A. Coley, A Class of Exact Classical Solutions to String Theory, *Phys. Rev. Lett.* **89** (2002) 281601.

[369] P. A. Collins and R. W. Tucker, An Action Principle for the Neveu-Schwarz-Ramond String and Other Systems Using Supernumerary Variables, *Nucl. Phys.* **B121** (1977) 307.

[370] R. Coquereaux and A. Jadczyk, *Riemannian Geometry, Fiber Bundles, Kaluza–Klein Theories and all that*. Singapore: World Scientific (1988).

[371] F. Cordaro, P. Fré, L. Gualtieri, P. Termonia, and M. Trigiante, $N = 8$ Gaugings Revisited: An Exhaustive Classification, *Nucl. Phys.* **B532** (1998) 245.

[372] N. J. Cornish and J. W. Moffat, Remarks on Theoretical Problems in Nonsymmetric Gravitational Theory, *Phys. Rev.* **D47** (1993) 4421.

[373] E. Corrigan and D. B. Fairlie, Scalar Field Theory and Exact Solutions to a Classical SU(2) Gauge Theory, *Phys. Lett.* **B67** (1977) 69.

[374] M. S. Costa, Composite M-Branes, *Nucl. Phys.* **B490** (1997) 202.

[375] M. S. Costa, Black Composite M-Branes, *Nucl. Phys.* **B495** (1997) 195.

[376] M. S. Costa and G. Papadopoulos, Superstring Dualities and $p$-Brane Bound States, *Nucl. Phys.* **B510** (1998) 217.

[377] M. S. Costa and M. J. Perry, Interacting Black Holes, *Nucl. Phys.* **B591** (2000) 469.

[378] P. M. Cowdall, Novel Domain Wall and Minkowski Vacua of $D = 9$ Maximal SO(2) Gauged Supergravity, `hep-th/0009016`.

[379] B. Craps, F. Roose, W. Troost, and A. Van Proeyen, What is Special Kaehler Geometry?, *Nucl. Phys.* **B503** (1997) 565.

[380] J. D. E. Creighton and R. B. Mann, Quasilocal Thermodynamics of Dilaton Gravity Coupled to Gauge Fields, *Phys. Rev.* **D52** (1995) 4569.

[381] J. D. E. Creighton and R. B. Mann, `gr-qc/9511012`.

[382] E. Cremmer, Supergravities in Five Dimensions, in *Superspace & Supergravity*, eds. S. W. Hawking and M. Roček. Cambridge: Cambridge University Press (1981), p. 267.

[383] E. Cremmer and S. Ferrara, Formulation of Eleven-Dimensional Supergravity in Superspace, *Phys. Lett.* **B91** (1980) 61.

[384] E. Cremmer, S. Ferrara, L. Girardello, and A. Van Proeyen, Yang-Mills Theories With Local Supersymmetry: Lagrangian, Transformation Laws And SuperHiggs Effect, *Nucl. Phys.* **B212** (1983) 413.

[385] E. Cremmer and B. Julia, The SO(8) Supergravity, *Nucl. Phys.* **B159** (1979) 141.

[386] E. Cremmer, B. Julia, and J. Scherk, Supergravity Theory in 11 Dimensions, *Phys. Lett.* **76B** (1978) 409.

[387] E. Cremmer, C. Kounnas, A. Van Proeyen, J. P. Derendinger, S. Ferrara, B. de Wit, and L. Girardello, Vector Multiplets Coupled to $N = 2$ Supergravity: SuperHiggs Effect, Flat Potentials and Geometric Structure, *Nucl. Phys.* **250** (1985) 385.

[388] E. Cremmer, J. Scherk, and S. Ferrara, U(n) Invariance in Extended Supergravity, *Phys. Lett.* **B68** (1977) 234.

[389] E. Cremmer, J. Scherk, and S. Ferrara, $\mathrm{SU}(4)$ Invariant Supergravity Theory, *Phys. Lett.* **74B** (1978) 64.

[390] E. Cremmer and A. Van Proeyen, Classification Of Kähler Manifolds in $N = 2$ Vector Multiplet Supergravity Couplings, *Classical Quant. Grav.* **2** (1985) 445.

[391] J. F. Cornwell, *Group Theory in Physics*, vols. 1, 2, and 3. London: Academic Press (1989).

[392] H. E. J. Curzon, Cylindrical Solutions of Einstein's Gravitational Equations, *Proc. London Math. Soc.* **23** (1924) 477.

[393] C. Cutler and R. M. Wald, A New Type of Gauge Invariance for a Collection of Massless Spin-2 Fields. 1. Existence and Uniqueness, *Classical Quant. Grav.* **4** (1987) 1267.

[394] M. Cvetič, G. W. Gibbons, and C. N. Pope, Universal Area Product Formulae for Rotating and Charged Black Holes in Four and Higher Dimensions, *Phys. Rev. Lett.* **106** (2011) 121301.

[395] M. Cvetič and C. M. Hull, Black Holes and U-Duality, *Nucl. Phys.* **B480** (1996) 296.

[396] M. Cvetič and F. Larsen, General Rotating Black Holes in String Theory: Grey Body Factors and Event Horizons, *Phys. Rev.* **D56** (1997) 4994.

[397] M. Cvetič and F. Larsen, Grey Body Factors for Rotating Black Holes in Four-dimensions, *Nucl. Phys.* **B506** (1997) 107.

[398] M. Cvetič and F. Larsen, Near Horizon Geometry of Rotating Black Holes in Five Dimensions, *Nucl. Phys.* **B531** (1998) 239.

[399] M. Cvetič and F. Larsen, Greybody Factors and Charges in Kerr/CFT, *JHEP* **0909** (2009) 088.

[400] M. Cvetič and H. Soleng, Supergravity Domain Walls, *Phys. Rep.* **282** (1997) 159.

[401] M. Cvetič and A. A. Tseytlin, Non-Extreme Black Holes from Non-Extreme Intersecting M-Branes, DAMTP Report DAMTP-R-96-27 and `hep-th/9606033`.

[402] M. Cvetič and A. A. Tseytlin, General Class of BPS Saturated Dyonic Black Holes as Exact Superstring Solutions, *Phys. Lett.* **B366** (1996) 95.

[403] M. Cvetič and A. A. Tseytlin, Solitonic Strings and BPS Saturated Dyonic Black Holes, *Phys. Rev.* **D53** (1996) 5619.

[404] M. Cvetič and D. Youm, Singular BPS Saturated States and Enhanced Symmetries of Four-Dimensional $N = 4$ Supersymmetric String Vacua, *Phys. Lett.* **B359** (1995) 87.

[405] M. Cvetič and D. Youm, Entropy of Nonextreme Charged Rotating Black Holes in String Theory, *Phys. Rev.* **D54** (1996) 2612.

[406] A. Dabholkar, Ten-Dimensional Heterotic String as a Soliton, *Phys. Lett.* **B357** (1995) 307.

[407] A. Dabholkar, Microstates of Non-supersymmetric Black Holes, *Phys. Lett.* **B402** (1997) 53.

[408] A. Dabholkar, Lectures on Orientifolds and Duality, in *High Energy Physics and Cosmology 1997*, eds. E. Gava, A. Masiero, K. S. Narain *et al.* Singapore: World Scientific (1998), p. 128.

[409] A. Dabholkar, G. W. Gibbons, J. Harvey, and F. Ruiz-Ruiz, Superstrings and Solitons, *Nucl. Phys.* **B340** (1990) 33.

[410] A. Dabholkar and J. Harvey, Non-renormalization of the Superstring Tension, *Phys. Rev. Lett.* **63** (1989) 478.

[411] A. Dabholkar, R. Kallosh, and A. Maloney, A Stringy Cloak for a Classical Singularity, *JHEP* **0412** (2004) 059.

[412] J. Dai, R. G. Leigh, and J. Polchinski, New Connections between String Theories, *Mod. Phys. Lett.* **A4** (1989) 2073.

[413] T. Damour, Strings and Black Holes, *Ann. Phys.* **11** (2000) 1.

[414] T. Damour, S. Deser, and J. McCarthy, Theoretical Problems in Non-symmetric Gravitational Theory, *Phys. Rev.* **D45** (1992) R3289.

[415] T. Damour, S. Deser, and J. McCarthy, Nonsymmetric Gravity Theories: Inconsistencies and a Cure, *Phys. Rev.* **D47** (1993) 1541.

[416] T. Damour and A. M. Polyakov, The String Dilaton and a Least Coupling Principle, *Nucl. Phys.* **B423** (1994) 532.

[417] T. Damour and A. M. Polyakov, String Theory and Gravity, *Gen. Relativ. Gravit.* **26** (1994) 1171.

[418] T. Damour and R. Ruffini, Quantum Electrodynamical Effects in Kerr-Newman Geometries, *Phys. Rev. Lett.* **35** (1975) 463.

[419] U. Danielsson, G. Ferretti, and I. R. Klebanov, Creation of Fundamental Strings by Crossing *D*-Branes, *Phys. Rev. Lett.* **79** (1997) 1984.

[420] A. Das and D. Z. Freedman, Gauge Internal Symmetry in Extended Supergravity, *Nucl. Phys.* **B120** (1977) 221.

[421] S.R. Das, G.W. Gibbons, and S.D. Mathur, Universality of Low Energy Absorption Cross Sections for Black Holes, *Phys. Rev. Lett.* **78** (1997) 417.

[422] S. R. Das and S. D. Mathur, Comparing Decay Rates for Black Holes and *D*-Branes, *Nucl. Phys.* **B478** (1996) 561.

[423] S. R. Das and S. D. Mathur, The Quantum Physics of Black Holes: Results from String Theory, *Ann. Rev. Nucl. Part. Sci.* **50** (2000) 153.

[424] A. Davidson and E. Gedalin, Finite Magnetic Flux Tube as a Black and White Dihole, *Phys. Lett.* **B339** (1994) 304.

[425] T. De Donder, *La gravique Einsteinienne*. Paris: Gauthier-Villars (1921).

[426] J. De Rydt, T. T. Schmidt, M. Trigiante, A. Van Proeyen, and M. Zagermann, Electric/Magnetic Duality for Chiral Gauge Theories with Anomaly Cancellation, *JHEP* **0812** (2008) 105.

[427] N. S. Deger, H. Samtleben, and O. Sarioglu, On The Supersymmetric Solutions of $D = 3$ Half-maximal Supergravities, *Nucl. Phys.* **B840** (2010) 29.

[428] M. Demiański and E. T. Newman, *Bull. Acad. Pol. Sci. Ser. Sci. Math. Astron. Phys.* **14** (1966) 653.

[429] F. Denef, Supergravity Flows and *D*-brane Stability, *JHEP* **0008** (2000) 050.

[430] S. Deser, Self-Interaction and Gauge Invariance, *Gen. Relativ. Gravit.* **1** (1970) 9.

[431] S. Deser, The Gravitational Field, Lectures given at Brandeis University 1971–1972 (unpublished).

[432] S. Deser, Absence of Static Solutions in Source-Free Yang-Mills Theory, *Phys. Lett.* **B64** (1976) 463.

[433] S. Deser, Gravity from Self-Interaction on a Curved Background, *Classical Quant. Grav.* **4** (1987) L99.

[434] S. Deser, Black-Hole Electromagnetic Duality, in *Proc. 7th Mexican School of Particles and Fields and 1st Latin American Symp. High-Energy Physics,* eds. J.C. D'Oliva, M. Klein-Kreisler, and H. Méndez. New York: American Institute of Physics (1997), p. 437.

[435] S. Deser, $D = 11$ Supergravity Revisited, in College Station 1998, Relativity, Particle Physics and Cosmology, ed. R.E. Allen. Singapore: World Scientific (1999), p. 1.

[436] S. Deser, Uniqueness of $D = 11$ Supergravity, in Santiago 1997, Black Holes and the Structure of the Universe, eds. C. Teitelboim and J. Zanelli. Singapore: World Scientific (2000), p. 70.

[437] S. Deser and J. Franklin, Schwarzschild and Birkhoff a la Weyl, *Am. J. Phys.* **73** (2005) 261.

[438] S. Deser, A. Gomberoff, M. Henneaux, and C. Teitelboim, *p*-Brane Dyons and Electric-Magnetic Duality, *Nucl. Phys.* **B520** (1998) 179.

[439] S. Deser and L. Halpern, Self-Coupled Scalar Gravitation, *Gen. Relativ. Gravit.* **1** (1970) 131.

[440] S. Deser, M. Hennaaux, and C. Teitelboim, Electric-Magnetic Black-Hole Duality, *Phys. Rev.* **D55** (1997) 826.

[441] S. Deser and B. E. Laurent, Gravitation without Self-Interaction, *Ann. Phys.* **50** (1968) 76.

[442] S. Deser and P. van Nieuwenhuizen, One Loop Divergences of Quantized Einstein-Maxwell Fields, *Phys. Rev.* **D10** (1974) 401.

[443] S. Deser and P. van Nieuwenhuizen, Nonrenormalizability of the Quantized Dirac-Einstein System, *Phys. Rev.* **D10** (1974) 411.

[444] S. Deser and M. Soldate, Gravitational Energy in Spaces with Compactified Dimensions, *Nucl. Phys.* **B311** (1988/89) 739.

[445] S. Deser and C. Teitelboim, Supergravity has Positive Energy, *Phys. Rev. Lett.* **39** (1977) 249.

[446] S. Deser, H.-S. Tsao, and P. van Nieuwenhuizen, Nonrenormalizability of Einstein Yang-Mills Interactions at the One Loop Level, *Phys. Lett.* **B50** (1974) 491.

[447] S. Deser, H.-S. Tsao, and P. van Nieuwenhuizen, One Loop Divergences of the Einstein Yang-Mills System, *Phys. Rev.* **D10** (1974) 3337.

[448] S. Deser and F. Wilczek, Nonuniqueness of Gauge Field Potentials, *Phys. Lett.* **D65** (1976) 391.

[449] S. Deser and B. Zumino, Consistent Supergravity, *Phys. Lett.* **62B** (1976) 335.

[450] S. Deser and B. Zumino, A Complete Action for the Spinning String, *Phys. Lett.* **65B** (1976) 369.

[451] B. S. DeWitt, Quantum Theory of Gravity, II, *Phys. Rev.* **162** (1967) 1195.

[452] B. S. DeWitt, Quantum Theory of Gravity, III, *Phys. Rev.* **162** (1967) 1239.

[453] D. E. Diaconescu, *D*-Branes, Monopoles and Nahm Equations, *Nucl. Phys.* **B503** (1997) 220.

[454] M. Dine, P. Huet, and N. Seiberg, Large and Small Radius in String Theory, *Nucl. Phys.* **B322** (1989) 301.

[455] P. A. M. Dirac, Quantised Singularities in the Electromagnetic Field, *Proc. Roy. Soc. London* **A113** (1931) 60.

[456] P. A. M. Dirac, An Extensible Model of the Electron, *Proc. Roy. Soc. London* **A268** (1962) 57.

[457] P. Dobiasch and D. Maison, Stationary, Spherically Symmetric Solutions of Jordan's Unified Theory of Gravity and Electromagnetism, *Gen. Relativ. Grav.* **14** (1982) 231.

[458] L. Dolan and M. J. Duff, Kac-Moody Symmetries of Kaluza-Klein Theories, *Phys. Rev. Lett.* **52** (1984) 14.

[459] F. Dowker, J. P. Gauntlett, G. W. Gibbons, and G. T. Horowitz, Nucleation of $p$-Branes and Fundamental Strings, *Phys. Rev.* **D53** (1996) 7115.

[460] F. Dowker, J. P. Gauntlett, S. B. Giddings, and G.T. Horowitz, On Pair Creation of Extremal Black Holes and KK Monopoles, *Phys. Rev.* **D50** (1994) 2662.

[461] F. Dowker, J. P. Gauntlett, D. A. Kastor, and J. Traschen, Pair Creation of Dilaton Black Holes, *Phys. Rev.* **D49** (1994) 2909.

[462] J. S. Dowker, The NUT Solution as a Gravitational Dyon, *Gen. Relativ. Gravit.* **5** (1974) 603.

[463] J. S. Dowker and J. A. Roche, *Proc. Phys. Soc. London* **92** (1967) 1.

[464] E. Dudas and J. Mourad, Consistent Gravitino Couplings in Non-Supersymmetric Strings, *Phys. Lett.* **B514** (2001) 173.

[465] M. J. Duff, Kaluza-Klein Theory in Perspective, in *Oskar Klein Centenary Nobel Symposium*, `hep-th/9410046`.

[466] M. J. Duff, Strong/Weak Coupling Duality from the Dual String, *Nucl. Phys.* **B442** (1995) 47.

[467] M. J. Duff, Supermembranes, `hep-th/9611203`.

[468] M. J. Duff, G. W. Gibbons, and P. K. Townsend, Macroscopic Superstrings as Interpolating Solitons, *Phys. Lett.* **B332** (1994) 321.

[469] M. J. Duff, P. S. Howe, T. Inami, and K. S. Stelle, Superstrings in $D = 10$ From Supermembranes in $D = 11$, *Phys. Lett.* **B191** (1987) 70.

[470] M. J. Duff and R. R. Khuri, *Four Dimensional String/String Duality*, *Nucl. Phys.* **B411** (1994) 473.

[471] M. J. Duff, R. R. Khuri, and J. X. Lu, String Solitons, Phys. Rep. **259** (1995) 213.

[472] M. J. Duff, J. T. Liu, and J. Rahmfeld, Four-Dimensional String-String-String Triality, *Nucl. Phys.* **B459** (1996) 125.

[473] M. J. Duff and J. X. Lu, Type II p-Branes: The Brane Scan Revisited, *Nucl. Phys.* **B390** (1993) 276.

[474] M. J. Duff, H. Lü, and C. N. Pope, Supersymmetry Without Supersymmetry, *Phys. Lett.* **B409** (1997) 136.

[475] M. J. Duff, H. Lü, C. N. Pope, and E. Sezgin, Supermembranes with Fewer Supersymmetries, *Phys. Lett.* **B371** (1996) 206.

[476] M. J. Duff, B. E. W. Nilsson, and C. N. Pope, Kaluza-Klein Supergravity, *Phys. Rep.* **130** (1986) 1.

[477] M. J. Duff and C. N. Pope, Kaluza-Klein Supergravity and the Seven Sphere, in *School on Supergravity and Supersymmetry*, eds. S. Ferrara, J. G. Taylor, and P. van Nieuwenhuizen. Singapore: World Scientific (1983).

[478] M. J. Duff and K. S. Stelle, Multi-Membrane Solutions of $D = 11$ Supergravity, *Phys. Lett.* **B253** (1991) 113.

[479] A. S. Eddington, *Nature* **113** (1924) 192.

[480] J. D. Edelstein, L. Tataru, and R. Tatar, Rules for Localized Overlappings and Intersections of $p$-Branes, *JHEP* **9806** (1998) 003.

[481] T. Eguchi, P. B. Gilkey, and A. J. Hanson, Gravitation, Gauge Theories and Differential Geometry, *Phys. Rep.* **66** (1980) 213.

[482] T. Eguchi and A. J. Hanson, Asymptotically Flat, Self-Dual Solutions to Euclidean Gravity, *Phys. Lett.* **74B** (1978) 249.

[483] J. Ehlers and W. Rindler, Local and Global Light Bending in Einstein's and other Gravitational Theories, *Gen. Relativ. Gravit.* **29** (1997) 519.

[484] A. Einstein, Über das Relativitätsprinzip und die aus demslben gezogenen Folgerungen, *Jahrbuch der Radioaktivität und Elektronik* **4** (1908) 411.

[485] A. Einstein, Über den Einfulss der Schwerkraft auf die Ausbreitung des Lichtes, *Ann. Phys.* **35** (1911) 898.

[486] A. Einstein, *Ann. Phys.* **35** (1912) 355.

[487] A. Einstein, *Ann. Phys.* **35** (1912) 443.

[488] A. Einstein, *Phys. Z.* **14** (1913) 1251.

[489] A. Einstein, *Sitzungsber. Preuß. Akad. Wiss., phys.-math. Kl.* (1915) 778.

[490] A. Einstein, *Sitzungsber. Preuß. Akad. Wiss., phys.-math. Kl.* (1918) 154.

[491] A. Einstein, *Sitzungsber. Preuß. Akad. Wiss., phys.-math. Kl.* (1918) 448.

[492] A. Einstein, *Sitzungsber. Preuß. Akad. Wiss.* (1925) 414.

[493] A. Einstein, Riemann-Geometrie mit Aufrechterhaltung des Begriffes des Fernparallelismus, *Sitzungsber. Preuß. Akad. Wiss., phys.-math. Kl.* (1928) 217.

[494] A. Einstein, Zur einheitliche Feldtheorie, *Sitzungsber. Preuß. Akad. Wiss., phys.-math. Kl.* (1929) 2.

[495] A. Einstein, Einheiliche Feldtheorie und Hamiltonsches Prinzip, *Sitzungsber. Preuß. Akad. Wiss., phys.-math. Kl.* (1929) 156.

[496] A. Einstein, Zur theorie der Räume mit Riemann-Metrik und Fernparallelismus, *Sitzungsber. Preuß. Akad. Wiss., phys.-math. Kl.* (1930) 401.

[497] A. Einstein, Auf die Riemann-Metrik und den Fern-Parallelismus gegründete einheitliche Feldtheorie, *Math. Ann.* **102** (1930) 685.

[498] A. Einstein, *The Meaning of Relativity. Including the Relativistic Theory of the Non-symmetric Field*, 5th edn. Princeton, NJ: Princeton University Press (1955).

[499] A. Einstein and A. D. Fokker, Die Nordströmsche Gravitationstheories vom Standpunkt des absolunten Differentialkalküls, *Ann. Phys.* **44** (1914) 321.

[500] A. Einstein, L. Infeld, and B. Hoffmann, *Ann. Math.* **39** (1938) 65.

[501] A. Einstein and B. Kaufman, *Ann. Math.* **62** (1955) 128.

[502] A. Einstein and W. R. P. Mayer, Zwei strenge statische Lösungen der Feldgleichungen der einheitlichen Feldtheorie, *Sitzungsber. Preuß. Akad. Wiss., phys.-math. Kl.* (1930) 110.

[503] A. Einstein and E. G. Straus, *Ann. Math.* **47** (1946) 731.

[504] J. Eisland, *Trans. A. M. S.* **27** (1925) 213.

[505] S. Elitzur, A. Giveon, and D. Kutasov, Branes and $N = 1$ Duality in String Theory, *Phys. Lett.* **B400** (1997) 269.

[506] H. Elvang, A Charged Rotating Black Ring, *Phys. Rev.* **D68** (2003) 124016.

[507] H. Elvang, R. Emparan, D. Mateos, and H. S. Reall, A Supersymmetric Black Ring, *Phys. Rev. Lett.* **93** (2004) 211302.

[508] H. Elvang, R. Emparan, D. Mateos, and H. S. Reall, Supersymmetric Black Rings and Three-charge Supertubes, *Phys. Rev.* **D71** (2005) 024033.

[509] H. Elvang, R. Emparan, D. Mateos, and H.S. Reall, Supersymmetric 4D Rotating Black Holes from 5D Black Rings, *JHEP* **0508** (2005) 042.

[510] R. Emparan, Black Diholes, *Phys. Rev.* **D61** (2000) 104009.

[511] R. Emparan and H. S. Reall, Generalized Weyl Solutions, *Phys. Rev.* **D65** (2002) 084025.

[512] R. Emparan and H. S. Reall, A Rotating Black Ring in Five Dimensions, *Phys. Rev. Lett.* **88** (2002) 101101.

[513] R. Emparan and E. Teo, Macroscopic and Microscopic Description of Black Diholes, *Nucl. Phys.* **B610** (2001) 190.

[514] A. A. Ershov and D. V. Gal'tsov, Nonexistence of Regular Monopoles and Dyons in the SU(2) Einstein Yang-Mills Theory, *Phys. Lett.* **A150** (1990) 159.

[515] E. Eyras, B. Janssen, and Y. Lozano, Five-Branes, KK Monopoles and T Duality, *Nucl. Phys.* **B531** (1998) 275.

[516] E. Eyras and Y. Lozano, The Kaluza-Klein Monopole in a Massive IIA Background, *Nucl. Phys.* **B546** (1999) 197.

[517] J. Fang and C. Fronsdal, Deformation of Gauge Groups. Gravitation, *J. Math. Phys.* **20** (1979) 2264.

[518] J. Faulkner, A Construction of Lie Algebras from a Class of Ternary Algebras, *Trans. Am. Math. Soc.* **155** (1971) 397.

[519] A. Fayyazuddin and D. J. Smith, Localized Intersections of M5-Branes and Four-Dimensional Superconformal Field Theories, *JHEP* **9904** (1999) 030.

[520] J. J. Fernández-Melgarejo, T. Ortín, and E. Torrente-Luján, The general gaugings of maximal $d = 9$ supergravity, *JHEP* **1110** (2011).

[521] J. J. Fernández-Melgarejo and E. Torrente-Luján, $N = 2$ SUGRA BPS Multi-center Solutions, Quadratic Prepotentials and Freudenthal Transformations, `arXiv:1310.4182`.

[522] S. Ferrara, G. W. Gibbons, and R. Kallosh, Black Holes and Critical Points in Moduli Space, *Nucl. Phys.* **B500** (1997) 75.

[523] S. Ferrara, E. G. Gimon, and R. Kallosh, Magic Supergravities, $N = 8$ and Black Hole Composites, *Phys. Rev.* **D74** (2006) 125018.

[524] S. Ferrara and M. Günaydin, Orbits of Exceptional Groups, Duality and BPS States in String Theory, *Int. J. Mod. Phys.* **A13** (1998) 2075.

[525] S. Ferrara, K. Hayakawa, and A. Marrani, Lectures on Attractors and Black Holes, *Fortsch. Phys.* **56** (2008) 993.

[526] S. Ferrara and R. Kallosh, Supersymmetry and Attractors, *Phys. Rev.* **D54** (1996) 1514.

[527] S. Ferrara and R. Kallosh, Universality of Supersymmetric Attractors, *Phys. Rev.* **D54** (1996) 1525.

[528] S. Ferrara and R. Kallosh, On $N = 8$ attractors, *Phys. Rev.* **D73** (2006) 125005.

[529] S. Ferrara and R. Kallosh, Creation of Matter in the Universe and Groups of Type $E7$, *JHEP* **1112** (2011) 096.

[530] S. Ferrara, R. Kallosh, and A. Marrani, Degeneration of Groups of Type E7 and Minimal Coupling in Supergravity, *JHEP* **1206** (2012) 074.

[531] S. Ferrara, R. Kallosh, and A. Strominger, $N = 2$ Extremal Black Holes, *Phys. Rev.* **D52** (1995) 5412.

[532] S. Ferrara and J. M. Maldacena, Branes, Central Charges and U-Duality Invariant BPS Conditions, *Classical Quant. Grav.* **15** (1998) 749.

[533] S. Ferrara and A. Marrani, Black Holes and Groups of Type $\mathrm{E}_7$, *Pramana* **78** (2012) 893.

[534] S. Ferrara, A. Marrani, and E. Orazi, Maurer-Cartan Equations and Black Hole Superpotentials in $N = 8$ Supergravity, *Phys. Rev.* **D81** (2010) 085013.

[535] S. Ferrara, A. Marrani, and E. Orazi, Split Attractor Flow in $N = 2$ Minimally Coupled Supergravity, *Nucl. Phys.* **B846** (2011) 512.

[536] S. Ferrara, A. Marrani, and A. Yeranyan, Freudenthal Duality and Generalized Special Geometry, *Phys. Lett.* **B701** (2011) 640.

[537] S. Ferrara and P. van Nieuwenhuizen, Consistent Supergravity with Complex Spin 3/2 Gauge Fields, *Phys. Rev. Lett.* **37** (1976) 1669.

[538] S. Ferrara, C. A. Savoy, and B. Zumino, General Massive Multiplets in Extended Supersymmetry, *Phys. Lett.* **100B** (1981) 393.

[539] S. Ferrara, J. Scherk, and B. Zumino, Algebraic Properties of Extended Supergravity Theories, *Nucl. Phys.* **B121** (1977) 393.

[540] V. Ferrari and P. Pendenza, Boosting the Kerr Metric, *Gen. Relativ. Gravit.* **22** (1990) 1105.

[541] R. P. Feynman, Quantum Theory of Gravitation, *Acta Phys. Polon.* **24** (1963) 697.

[542] R. P. Feynman, Feynman Lectures on Gravitation, eds. F. B. Morinigo, W. G. Wagner, and B. Hatfield. Reading, MA: Addison-Wesley (1995).

[543] M. Fierz and W. Pauli, Relativistic Wave Equations for Particles of Arbitrary Spin in an Electromagnetic Field, *Proc. Roy. Soc. London* **A173** (1939) 211.

[544] P. Figueras, E. Jamsin, J. V. Rocha, and A. Virmani, Integrability of Five Dimensional Minimal Supergravity and Charged Rotating Black Holes, *Classical Quant. Grav.* **27** (2010) 135011.

[545] J. M. Figueroa-O'Farrill, Near-Horizon Geometries of Supersymmetric Branes, `hep-th/9807149`.

[546] J. M. Figueroa-O'Farrill, On the Supersymmetries of Anti-de Sitter Vacua, *Classical Quant. Grav.* **16** (1999) 2043.

[547] J. M. Figueroa-O'Farrill, More Ricci-Flat Branes, *Phys. Lett.* **B471** (1999) 128.

[548] J. M. Figueroa-O'Farrill and G. Papadopoulos, Homogeneous Fluxes, Branes and a Maximally Supersymmetric Solution of M Theory, *JHEP* **0108** (2001) 036.

[549] J. M. Figueroa-O'Farrill and G. Papadopoulos, Maximally Supersymmetric Solutions of Ten- And Eleven-Dimensional Supergravities, *JHEP* **0303** (2003) 048.

[550] D. Finkelstein, Past-Future Asymmetry of the Gravitational Field of a Point Particle, *Phys. Rev.* **110** (1958) 965.

[551] W. Fischler and L. Susskind, Dilaton Tadpoles, String Condensates and Scale Invariance, *Phys. Lett.* **B171** (1986) 383.

[552] V. A. Fock, *J. Phys. USSR* **1** (1939) 81.

[553] A. Font, L. E. Ibáñez, D. Lüst, and F. Quevedo, Strong-Weak Coupling Duality and Nonperturbative Effects in String Theory, *Phys. Lett.* **B249** (1990) 35.

[554] E. S. Fradkin and M. A. Vasiliev, Model of Supergravity with Minimal Electromagnetic Interaction, Lebedev Institute preprint N 197 (1976).

[555] P. Fré, Lectures on Special Kähler Geometry and Electric-magnetic Duality Rotations, *Nucl. Phys. Proc. Suppl.* **45BC** (1996) 59.

[556] P. Fré, Gaugings and Other Supergravity Tools of $p$-Brane Physics, Lectures given at the school Recent Advances in M-Theory, `hep-th/0102114`.

[557] P. G. Fré, Gravity, a Geometrical Course. Vol. 2: Black Holes, Cosmology and Introduction to Supergravity. Dordrecht: Springer Verlag (2013).

[558] P. Fré, V. Gili, F. Gargiulo, A. S. Sorin, K. Rulik, and M. Trigiante, Cosmological Backgrounds of Superstring Theory and Solvable Algebras: Oxidation and Branes, *Nucl. Phys.* **B685** (2004) 3.

[559] P. Fré and P. Soriani, *The $N = 2$ Wonderland: From Calabi–Yau Manifolds to Topological Field Theories*. Singapore: World Scientific (1995).

[560] P. Fré and A. S. Sorin, Integrability of Supergravity Billiards and the Generalized Toda Lattice Equation, *Nucl. Phys.* **B733** (2006) 334.

[561] P. Fré and A. S. Sorin, The Weyl Group and Asymptotics: All Supergravity Billiards have a Closed Form General Integral, *Nucl. Phys.* **B815** (2009) 430.

[562] D. Z. Freedman, Supergravity with Axial-gauge Invariance, *Phys. Rev.* **D15** (1977) 1173.

[563] D. Z. Freedman, P. van Nieuwenhuizen, and S. Ferrara, Progress Toward a Theory of Supergravity, *Phys. Rev.* **D13** (1976) 3214.

[564] D. Z. Freedman and A. Van Proeyen, *Supergravity*. Cambridge: Cambridge University Press (2012).

[565] P. G. O. Freund, *Introduction to Supersymmetry*. Cambridge: Cambridge University Press (1986).

[566] P. G. O. Freund and Y. Nambu, Scalar Fields Coupled to the Trace of the Energy-Momentum Tensor, *Phys. Rev.* **174** (1968) 1741.

[567] P. O. Freund and M. A. Rubin, Dynamics of Dimensional Reduction, *Phys. Lett.* **B97** (1980) 233.

[568] J. L. Friedman, K. Schleich, and D. M. Witt, Topological Censorship, *Phys. Rev. Lett.* **71** (1993) 1486. [Erratum *ibid.* **75** (1995) 1872.]

[569] M. Fukuma, S. Matsuura, and T. Sakai, Holographic Renormalization Group, Prog. Theor. Phys. **109** (2003) 489.

[570] M. K. Gaillard and B. Zumino, Duality Rotations for Interacting Fields, *Nucl. Phys.* **B193** (1981) 221.

[571] M.K. Gaillard and B. Zumino, Selfduality in Nonlinear Electromagnetism, in *Supersymmetry and Quantum Field Theory: Proceedings of the D. Volkov Memorial Seminor held in Kharkov, Ukraine 5–7 January 1997*, eds. J. Wess and V.P. Akulov, Lecture Notes in Physics.

[572] D. Gaiotto, W. Li, and M. Padi, Non-Supersymmetric Attractor Flow in Symmetric Spaces, *JHEP* **0712** (2007) 093.

[573] D. Gaiotto, A. Strominger, and X. Yin, 5D Black Rings and 4D Black Holes, *JHEP* **0602** (2006) 023.

[574] D. Gaiotto, A. Strominger, and X. Yin, New Connections between 4D and 5D Black Holes, *JHEP* **0602** (2006) 024.

[575] P. Galli, K. Goldstein, S. Katmadas, and J. Perz, First-order Flows and Stabilisation Equations for Non-BPS Extremal Black Holes, *JHEP* **1106** (2011) 070.

[576] P. Galli, K. Goldstein, and J. Perz, On Anharmonic Stabilisation Equations for Black Holes, *JHEP* **1303** (2013) 036.

[577] P. Galli, P. Meessen, and T. Ortín, The Freudenthal Gauge Symmetry of the Black Holes of $N = 2$, $d = 4$ Supergravity, *JHEP* **1305** (2013) 011.

[578] P. Galli, T. Ortín, J. Perz, and C. S. Shahbazi, Non-extremal Black Holes of $N = 2$, $d = 4$ Supergravity, *JHEP* **1107** (2011) 041.

[579] P. Galli, T. Ortín, J. Perz, and C. S. Shahbazi, Black Hole Solutions of $N = 2$, $d = 4$ Supergravity with a Quantum Correction, in the H-FGK formalism, arXiv:1212.0303 [hep-th].

[580] D.V. Gal'tsov and A.A. Ershov, Nonabelian Baldness of Colored Black Holes, *Phys. Lett.* **A138** (1989) 160.

[581] D. V. Gal'tsov, A. A. Garcia, and O. V. Kechkin, Symmetries of the Stationary Einstein-Maxwell Dilaton-Axion Theory, *J. Math. Phys.* **36** (1995) 5023.

[582] D. V. Gal'tsov and O.V. Kechkin, Ehlers-Harrison-Type Transformations in Dilaton-Axion Gravity, *Phys. Rev.* **D50** (1994) 7394.

[583] D. V. Gal'tsov and O. V. Kechkin, U Duality and Symplectic Formulation of Dilaton-Axion Gravity, *Phys. Rev.* **D54** (1996) 1656.

[584] D. V. Gal'tsov and P. S. Letelier, Ehlers-Harrison Transformations and Black Holes in Dilaton-Axion Gravity with Multiple Vector Fields, *Phys. Rev.* **D55** (1997) 3580.

[585] D. V. Gal'tsov and O. A. Rytchkov, Generating Branes via Sigma Models, *Phys. Rev.* **D58** (1998) 122001.

[586] D. V. Gal'tsov and S. A. Sharakin, Matrix Ernst Potentials for EMDA with Multiple Vector Fields, *Phys. Lett.* **B399** (1997) 250.

[587] R. Gambini and J. Pullin, *Loops, Knots, Gauge Theories and Quantum Gravity*. Cambridge: Cambridge University Press (1996).

[588] A. Garcia, D. V. Gal'tsov, and O. V. Kechkin, Class of Stationary Axisymmetric Solutions of the Einstein-Maxwell Dilaton-Axion Field Equations, *Phys. Rev. Lett.* **74** (1995) 1276.

[589] D. Garfinkle, G. Horowitz, and A. Strominger, Charged Black Holes in String Theory, *Phys. Rev.* **D43** (1991) 3140. [Erratum *ibid.* **D45** (1992) 3888.]

[590] J. P. Gauntlett, Intersecting Branes, in *Dualities of Gauge and String Theories*, eds. Y. M. Cho and S. Nam. Singapore: World Scientific (1998), p. 146.

[591] J. P. Gauntlett, G. W. Gibbons, G. Papadopoulos, and P. K. Townsend, Hyper-Kaehler Manifolds and Multiply Intersecting Branes, *Nucl. Phys.* **B500** (1997) 133.

[592] J. P. Gauntlett, J. Gomis, and P. K. Townsend, BPS Bounds for Worldvolume Branes, *JHEP* **9801** (1998) 003.

[593] J. P. Gauntlett and J. B. Gutowski, All Supersymmetric Solutions of Minimal Gauged Supergravity in Five Dimensions, *Phys. Rev.* **D68** (2003) 105009. [Erratum *ibid.* D **70** (2004) 089901.]

[594] J. P. Gauntlett and J. B. Gutowski, Concentric Black Rings, *Phys. Rev.* **D71** (2005) 025013.

[595] J. P. Gauntlett and J. B. Gutowski, General Concentric Black Rings, *Phys. Rev.* **D71** (2005) 045002.

[596] J. P. Gauntlett, J. B. Gutowski, C. M. Hull, S. Pakis, and H. S. Reall, All Supersymmetric Solutions of Minimal Supergravity in Five- Dimensions, *Classical Quant. Grav.* **20** (2003) 4587

[597] J. P. Gauntlett, D. A. Kastor, and J. Traschen, Overlapping Branes in M Theory, *Nucl. Phys.* **B478** (1996) 544.

[598] J. P. Gauntlett, R. C. Myers, and P. K. Townsend, Supersymmetry of Rotating Branes, *Phys. Rev.* **D59** (1999) 025001.

[599] J. P. Gauntlett and S. Pakis, The Geometry of $D = 11$ Killing Spinors, *JHEP* **0304** (2003) 039.

[600] S. P. Gavrilov and D. M. Gitman, Quantization of Point-Like Particles and Consistent Relativistic Quantum Mechanics, *Int. J. Mod. Phys.* **A15** (2000) 4499.

[601] H. Georgi and S. L. Glashow, Unified Weak and Electromagnetic Interactions without Neutral Currents, *Phys. Rev. Lett.* **28** (1972) 1494.

[602] J. Gheerardyn and P. Meessen, Supersymmetry of Massive $D = 9$ Supergravity, *Phys. Lett.* **B525** (2002) 322.

[603] F. Giani and M. Pernici, $N = 2$ Supergravity in Ten Dimensions, *Phys. Rev.* **D30** (1984) 325.

[604] G. W. Gibbons, The Motion of Black Holes, *Commun. Math. Phys.* **35** (1974) 13.

[605] G. W. Gibbons, Vacuum Polarization and the Spontaneous Loss of Charge by Black Holes, *Commun. Math. Phys.* **44** (1975) 245.

[606] G. W. Gibbons, *Proc. Roy. Soc. London* **A372** (1980) 535.

[607] G. W. Gibbons, An Introduction to Black Hole Thermodynamics, Notes of Lectures given at the 1980 Summer Institute of the Ecole Normale Supérieure, LPT, Paris, France, August 4–23, 1980 (unpublished).

[608] G. W. Gibbons, Antigravitating Black Hole Solitons with Scalar Hair in $N = 4$ Supergravity, *Nucl. Phys.* **B207** (1982) 337.

[609] G. W. Gibbons, Aspects of Supergravity Theories (three lectures), in Supersymmetry, Supergravity and Related Topics, eds. F. del Águila, J. de Azcárraga, and L. Ibáñez. Singapore: World Scientific (1985), p. 147.

[610] G. W. Gibbons, Born-Infeld Particles and Dirichlet $p$-Branes, *Nucl. Phys.* **B514** (1998) 603.

[611] G. W. Gibbons, M. B. Green, and M. J. Perry. Instantons and Seven-Branes in Type IIB Superstring Theory, *Phys. Lett.* **B370** (1996) 37.

[612] G. W. Gibbons and S. W. Hawking, Action Integrals and Partition Functions in Quantum Gravity, *Phys. Rev.* **D15** (1977) 2752. (Reprinted in Ref. [615].)

[613] G. W. Gibbons and S. W. Hawking, Gravitational Multi-instantons, *Phys. Lett.* **78B** (1978) 430. (Reprinted in Ref. [615].)

[614] G. W. Gibbons and S. W. Hawking, Classification of Gravitational Instanton Symmetries, *Commun. Math. Phys.* **66** (1979) 291. (Reprinted in Ref. [615].)

[615] G. W. Gibbons and S. W. Hawking (eds.), *Euclidean Quantum Gravity*. Singapore: World Scientific (1993).

[616] G. W. Gibbons and C. A. R. Herdeiro, Supersymmetric Rotating Black Holes and Causality Violation, *Classical Quant. Grav.* **16** (1999) 3619.

[617] G. W. Gibbons, G. T. Horowitz, and P. K. Townsend, Higher-Dimensional Resolution of Dilatonic Black Hole Singularities, *Classical Quant. Grav.* **12** (1995) 297.

[618] G. W. Gibbons and C. M. Hull, A Bogomol'nyi Bound for General Relativity and Solitons in $N = 2$ Supergravity, *Phys. Lett.* **109B** (1982) 190.

[619] G. W. Gibbons, C. M. Hull, and N. P. Warner, The Stability of Gauged Supergravity, *Nucl. Phys.* **B218** (1983) 173.

[620] G. W. Gibbons, D. Ida, and T. Shiromizu, Uniqueness and Non-Uniqueness of Static Vacuum Black Holes in Higher Dimensions, *Phys. Rev. Lett.* **89** (2002) 041101.

[621] G. W. Gibbons and R. E. Kallosh, Topology, Entropy and Witten Index of Dilaton Black Holes, *Phys. Rev.* **D51** (1995) 2839.

[622] G. W. Gibbons, R. Kallosh, and B. Kol, Moduli, Scalar Charges and the First Law of Black Hole Thermodynamics, *Phys. Rev. Lett.* **77** (1996) 4992.

[623] G. W. Gibbons and K. Maeda, Black Holes and Membranes in Higher Dimensional Theories with Dilaton Fields, *Nucl. Phys.* **B298** (1988) 741.

[624] G. W. Gibbons and M. J. Perry, New Gravitational Instantons and their Interactions, *Phys. Rev.* **D22** (1980) 313.

[625] G. W. Gibbons and C. N. Pope, The Positive Action Conjecture and Asymptotically Euclidean Metrics in Quantum Gravity, *Commun. Math. Phys.* **66** (1979) 267. (Reprinted in Ref. [615].)

[626] G. W. Gibbons and D. A. Rasheed, Electric-Magnetic Duality Rotations in Nonlinear Electrodynamics, *Nucl. Phys.* **B454** (1995) 185.

[627] G. W. Gibbons and D. A. Rasheed, SL(2,R) Invariance of Non-Linear Electrodynamics Coupled to an Axion and a Dilaton, *Phys. Lett.* **B365** (1996) 46.

[628] G.W. Gibbons and P.J. Ruback, The Hidden Symmetries Of Multicenter Metrics, *Commun. Math. Phys.* **115** (1988) 267.

[629] G. W. Gibbons and P. K. Townsend, Vacuum Interpolation in Supergravity via Super $p$-Branes, *Phys. Rev. Lett.* **71** (1993) 3754.

[630] S. B. Giddings and A. Strominger, String Wormholes, *Phys. Lett.* **B230** (1989) 46.

[631] J. Gillard, U. Gran, and G. Papadopoulos, The Spinorial Geometry of Supersymmetric Backgrounds, *Classical Quant. Grav.* **22** (2005) 1033.

[632] R. Gilmore, *Lie Groups, Lie Algebras and Some of Their Applications*. New York: Wiley (1974).

[633] E. G. Gimon, F. Larsen, and J. Simon, Constituent Model of Extremal non-BPS Black Holes, *JHEP* **0907** (2009) 052.

[634] P. Ginsparg, Applied Conformal Field Theory, in *Fields, Strings and Critical Phenomena*, eds. E. Brézin and J. Zinn-Justin. Amsterdam: North Holland (1990), pp. 1–168.

[635] A. Giveon, M. Porrati, and E. Rabinovici, Target Space Duality in String Theory, *Phys. Rep.* **244** (1994) 77.

[636] F. Gliozzi, J. Scherk, and D. Olive, Supersymmetry, Supergravity Theories and the Dual Spinor Model, *Nucl. Phys.* **B122** (1977) 253.

[637] P. Goddard, J. Goldstone, C. Rebbi, and C. B. Thorn, Quantum Dynamics of a Massless Relativistic String, *Nucl. Phys.* **B56** (1973) 109.

[638] P. Goddard and D. I. Olive, Magnetic Monopoles in Gauge Field Theories, *Rep. Prog. Phys.* **41** (1978) 91.

[639] M. Godina and P. Matteucci, Reductive G-structures and Lie Derivatives, `math.DG/0201235`.

[640] K. Goldstein, N. Iizuka, R. P. Jena, and S. P. Trivedi, Non-supersymmetric Attractors, *Phys. Rev.* **D72** (2005) 124021.

[641] A. Gomberoff, D. Kastor, D. Marolf, and J. Traschen, Fully Localized Brane Intersections: The Plot Thickens, *Phys. Rev.* **D61** (2000) 024012.

[642] M. H. Goroff and A. Sagnotti, The Ultraviolet Behavior of Einstein Gravity, *Nucl. Phys.* **B266** (1986) 709.

[643] T. Goto, Relativistic Quantum Mechanics of One-Dimensional Mechanical Continuum and Subsidiary Condition of Dual Resonance Model, *Prog. Theor. Phys.* **46** (1971) 1560.

[644] J. Govaerts, Quantum Consistency of Open String Theories, *Phys. Lett.* **B220** (1989) 77.

[645] U. Gran, J. Gutowski, and G. Papadopoulos, The Spinorial Geometry of Supersymmetric IIb Backgrounds, *Classical Quant. Grav.* **22** (2005) 2453.

[646] U. Gran, J. Gutowski, and G. Papadopoulos, The G(2) Spinorial Geometry of Supersymmetric IIB Backgrounds, *Classical Quant. Grav.* **23** (2006) 143.

[647] U. Gran, J. Gutowski, and G. Papadopoulos, Geometry of all Supersymmetric Four-dimensional $N = 1$ Supergravity Backgrounds, *JHEP* **0806** (2008) 102.

[648] U. Gran, J. Gutowski, and G. Papadopoulos, Invariant Killing Spinors in 11D and Type II Supergravities, *Classical Quant. Grav.* **26** (2009) 155004.

[649] U. Gran, J. Gutowski, and G. Papadopoulos, Classification of IIB Backgrounds with 28 Supersymmetries, *JHEP* **1001** (2010) 044.

[650] U. Gran, J. Gutowski, and G. Papadopoulos, M-theory Backgrounds with 30 Killing Spinors are Maximally Supersymmetric, *JHEP* **1003** (2010) 112.

[651] U. Gran, J. Gutowski, G. Papadopoulos, and D. Roest, Systematics of IIB Spinorial Geometry, *Classical Quant. Grav.* **23** (2006) 1617.

[652] U. Gran, J. Gutowski, G. Papadopoulos, and D. Roest, Maximally Supersymmetric G-Backgrounds of IIB Supergravity, *Nucl. Phys.* **B753** (2006) 118.

[653] U. Gran, J. Gutowski, G. Papadopoulos, and D. Roest, $N = 31$ is not IIB, *JHEP* **0702** (2007) 044.

[654] U. Gran, J. Gutowski, G. Papadopoulos, and D. Roest, Aspects of Spinorial Geometry, *Mod. Phys. Lett.* **A22** (2007) 1.

[655] U. Gran, J. Gutowski, G. Papadopoulos, and D. Roest, IIB solutions with $N = 28$ Killing Spinors are Maximally Supersymmetric, *JHEP* **0712** (2007) 070.

[656] U. Gran, P. Lohrmann, and G. Papadopoulos, The Spinorial Geometry of Supersymmetric Heterotic String Backgrounds, *JHEP* **0602** (2006) 063.

[657] U. Gran, P. Lohrmann, and G. Papadopoulos, Geometry of Type II Common sector $N = 2$ Backgrounds, *JHEP* **0606** (2006) 049.

[658] U. Gran, G. Papadopoulos, and D. Roest, Systematics of M-Theory Spinorial Geometry, *Classical Quant. Grav.* **22** (2005) 2701.

[659] U. Gran, G. Papadopoulos, and D. Roest, Supersymmetric Heterotic String Backgrounds, *Phys. Lett.* **B656** (2007) 119.

[660] U. Gran, G. Papadopoulos, D. Roest, and P. Sloane, Geometry of all Supersymmetric Type I Backgrounds, *JHEP* **0708** (2007) 074.

[661] J. C. Graves and D. Brill, Oscillatory Character of Reissner-Nordström Metric for an Ideal Charged Wormhole, *Phys. Rev.* **120** (1960) 1507.

[662] M. B. Green, Introduction to string and superstring Theory 1, in *From the Planck Scale to the Weak Scale: Towards a Theory of the Universe*, ed. H. E. Haber. Singapore: World Scientific (1987).

[663] M. B. Green, Point-like States for Type IIB Superstrings, *Phys. Lett.* **B329** (1994) 435.

[664] M. B. Green, A Gas of D Instantons, *Phys. Lett.* **B354** (1995) 271.

[665] M. E. Peskin, Introduction to String and Superstring Theory 2, in *From the Planck Scale to the Weak Scale: Towards a Theory of the Universe*, ed. H. E. Haber, Singapore: World Scientific (1987).

[666] M. B. Green and M. Gutperle, Comments on Three-Branes, *Phys. Lett.* **B377** (1996) 28.

[667] M. B. Green, C. M. Hull, and P. K. Townsend, D-$p$-brane Wess-Zumino Actions, T-Duality and the Cosmological Constant, *Phys. Lett.* **B382** (1996) 65.

[668] M. B. Green and J. H. Schwarz, Covariant Description of Superstrings, *Phys. Lett.* **B136** (1984) 367.

[669] M. B. Green and J. H. Schwarz, Properties of the Covariant Formulation of Superstring Theories, *Nucl. Phys.* **B243** (1984) 285.

[670] M. B. Green, J. H. Schwarz, and E. Witten, *Superstring Theory*, 2 vol. Cambridge: Cambridge University Press (1987).

[671] B. R. Greene, String theory on Calabi-Yau manifolds, in *Fields, strings and duality*, eds. C. Efthimiou and B. R. Greene, Singapore, Singapore: World Scientific (1997).

[672] B. R. Greene and M. R. Plesser, (2,2) and (2,0) Superconformal Orbifolds, Harvard University Preprint HUTP-89/A043 (1989).

[673] B. R. Greene and M. R. Plesser, Duality In Calabi-yau Moduli Space, *Nucl. Phys.* **B338** (1990) 15.

[674] B. R. Greene, A. Shapere, C. Vafa, and S.-T. Yau, Stringy Cosmic Strings and Noncompact Calabi-Yau Manifolds, *Nucl. Phys.* **B337** (1990) 1.

[675] R. Gregory and J. A. Harvey, Black Holes with a Massive Dilaton, *Phys. Rev.* **D47** (1993) 2411.

[676] R. Gregory, J. A. Harvey, and G. Moore, Unwinding Strings and T-duality of Kaluza-Klein and H-Monopoles, *Adv. Theor. Math. Phys.* **1** (1997) 283.

[677] R. Gregory and R. Laflamme, Black Strings and $P$-Branes are Unstable, *Phys. Rev. Lett.* **70** (1993) 2837.

[678] R. Gregory and R. Laflamme, The Instability of Charged Black Strings and $p$-Branes, *Nucl. Phys.* **B428** (1994) 399.

[679] R. Gregory and R. Laflamme, Evidence for Stability of Extremal Black $p$-Branes, *Phys. Rev.* **D51** (1995) 305.

[680] M. Grisaru, Positivity of the Energy in Einstein's Theory, *Phys. Lett.* **73B** (1978) 207.

[681] M. T. Grisaru, P. S. Howe, L. Mezincescu, B. Nilsson, and P. K. Townsend, $N = 2$ Superstrings in a Supergravity Background, *Phys. Lett.* **B162** (1985) 116.

[682] M. T. Grisaru, P. van Nieuwenhuizen, and J. A. M. Vermaseren, One Loop Renormalizability of Pure Supergravity and of Maxwell-Einstein Theory in Extended Supergravity, *Phys. Rev. Lett.* **37** (1976) 1662.

[683] F. Gronwald and F. W. Hehl, On the Gauge Aspects of Gravity, *Proc.* 14th Course of the *School of Cosmology and Gravitation on Quantum Gravity*, eds. P. G. Bergmann, V. de Sabbata, and H.-J. Treder. Singapore: World Scientific (1996), p. 148.

[684] D. J. Gross and M. J. Perry, Magnetic Monopoles in Kaluza-Klein Theories, *Nucl. Phys.* **B226** (1983) 29.

[685] S. S. Gubser, Special Holonomy in String Theory and M-Theory, Lectures given at TASI (2001), `hep-th/0201114`.

[686] S. S. Gubser and I. Mitra, Instability of Charged Black Holes in Anti-De Sitter Space, `hep-th/0009126`.

[687] S. S. Gubser and I. Mitra, The Evolution of Unstable Black Holes in Anti-De Sitter Space, *JHEP* **0108** (2001) 018.

[688] M. Günaydin, Lectures on Spectrum Generating Symmetries and U-duality in Supergravity, Extremal Black Holes, Quantum Attractors and Harmonic Superspace, `arXiv:0908.0374`.

[689] M. Günaydin, K. Koepsell, and H. Nicolai, Conformal and Quasiconformal Realizations of Exceptional Lie Groups, *Commun. Math. Phys.* **221** (2001) 57.

[690] M. Günaydin and N. Marcus, The Spectrum of the $S^5$ Compactification of the Chiral $N = 2$, $D = 10$ Supergravity and the Unitary Supermultiplets of U(2, 2/4), *Classical Quant. Grav.* **2** (1985) L11.

[691] M. Günaydin, A. Neitzke, B. Pioline, and A. Waldron, Quantum Attractor Flows, *JHEP* **0709** (2007) 056.

[692] M. Günaydin, G. Sierra, and P. K. Townsend, Exceptional Supergravity Theories and the MAGIC Square, *Phys. Lett.* **B133** (1983) 72.

[693] M. Günaydin, G. Sierra, and P. K. Townsend, The Geometry of $N = 2$ Maxwell-Einstein Supergravity and Jordan Algebras, *Nucl. Phys.* **B242** (1984) 244.

[694] M. Günaydin, G. Sierra, and P. K. Townsend, Gauging The $D = 5$ Maxwell-Einstein Supergravity Theories: More On Jordan Algebras, *Nucl. Phys.* **B253** (1985) 573.

[695] M. Günaydin and M. Zagermann, The Gauging of Five-dimensional, $N = 2$ Maxwell-Einstein Supergravity Theories Coupled to Tensor Multiplets, *Nucl. Phys.* **B572** (2000) 131.

[696] M. Günaydin and M. Zagermann, Gauging the Full R-symmetry Group in Five-dimensional, $N = 2$ Yang-Mills/Einstein/Tensor Supergravity, *Phys. Rev.* **D63** (2001) 064023.

[697] F. Gürsey, *Ann. Phys.* **24** (1963) 211.

[698] R. Güven, Plane Waves in Effective Field Theories of Superstrings, *Phys. Lett.* **B191** (1987) 275.

[699] R. Güven, Black $p$-Brane Solutions of $D = 11$ Supergravity Theory, *Phys. Lett.* **276B** (1992) 49.

[700] R. Güven, Plane Wave Limits and T Duality, *Phys. Lett.* **B482** (2000) 255.

[701] S. N. Gupta, Gravitation and Electromagnetism, *Phys. Rev.* **96** (1954) 1683.

[702] S. N. Gupta, Einstein's and Other Theories of Gravitation, *Rev. Mod. Phys.* **29** (1957) 334.

[703] M. Gurses and F. Gursey, Derivation of the String Equation of Motion in General Relativity, *Phys. Rev.* **D11** (1975) 967.

[704] J. B. Gutowski, D. Martelli, and H.S. Reall, All Supersymmetric Solutions of Minimal Supergravity in Six Dimensions, *Classical Quant. Grav.* **20** (2003) 5049.

[705] J. B. Gutowski and H. S. Reall, General Supersymmetric AdS(5) Black Holes, *JHEP* **0404** (2004) 048.

[706] J. B. Gutowski and W. Sabra, General Supersymmetric Solutions of Five-dimensional Supergravity, *JHEP* **0510** (2005) 039.

[707] R. Haag, J. T. Lopuszański, and M. Sohnius, All Possible Generators of Supersymmetries of the S Matrix, *Nucl. Phys.* **B88** (1975) 257.

[708] A. Hanany and E. Witten, Type IIB Superstrings, BPS Monopoles and 3-Dimensional Gauge Dynamics, *Nucl. Phys.* **B492** (1997) 152.

[709] B. J. Harrington and H. K. Shepard, Periodic Euclidean Solutions and the Finite Temperature Yang-Mills Gas, *Phys. Rev.* **D17** (1978) 2122.

[710] B. K. Harrison, New Solutions of the Einstein-Maxwell Equations from Old, *J. Math. Phys.* **9** (1968) 1744.

[711] J. B. Hartle and S. W. Hawking, Solutions of the Einstein-Maxwell Equations with Many Black Holes, *Commun. Math. Phys.* **26**, (1972) 87.

[712] J. Hartong, M. Hübscher, and T. Ortín, The Supersymmetric Tensor Hierarchy of N = 1, d = 4 supergravity, *JHEP* **0906** (2009) 090.

[713] J. Hartong and T. Ortín, Tensor Hierarchies of 5- and 6-Dimensional Field Theories, *JHEP* **0909** (2009) 039.

[714] J. A. Harvey and J. Liu, Magnetic Monopoles in N = 4 Supersymmetric Low-energy Superstring Theory, *Phys. Lett.* **B268** (1991) 40.

[715] J. A. Harvey and A. Strominger, The Heterotic String is a Soliton, *Nucl. Phys.* **B449** (1995) 535. [Erratum *ibid.* **B458** (1996) 456.]

[716] A. Hashimoto, The Shape of Branes Pulled by Strings, *Phys. Rev.* **D57** (1998) 6441.

[717] A. Hashimoto, Supergravity Solutions for Localized Intersections of Branes, *JHEP* **9901** (1999) 018.

[718] S. F. Hassan, T Duality and Nonlocal Supersymmetries, *Nucl. Phys.* **B460** (1996) 362.

[719] S. F. Hassan, T-Duality, Space-Time Spinors and R-R Fields in Curved Backgrounds, *Nucl. Phys.* **B568** (2000) 145.

[720] S. F. Hassan, SO(d,d) Transformations of Ramond-Ramond Fields and Space-Time Spinors, *Nucl. Phys.* **B583** (2000) 431.

[721] S. W. Hawking, *Commun. Math. Phys.* **25** (1972) 152.

[722] S. W. Hawking, The Event Horizon, in *Black Holes*, eds. C. DeWitt and B. S. DeWitt. New York: Gordon and Breach (1973).

[723] S. W. Hawking, *Nature* **248** (1974) 30.

[724] S. W. Hawking, Particle Creation by Black Holes, *Commun. Math. Phys.* **43** (1975) 199. (Reprinted in Ref. [615].)

[725] S. W. Hawking, Black Holes and Thermodynamics, *Phys. Rev.* **D13** (1976) 191.

[726] S. W. Hawking, Gravitational Instantons, *Phys. Lett.* **60A** (1977) 81.

[727] S. W. Hawking, The Path-integral Approach to Quantum Gravity, in *General Relativity, An Einstein Centenary Survey*, eds. S. W. Hawking and W. Israel. Cambridge: Cambridge University Press (1979), p. 746.

[728] S. W. Hawking and G. F. R. Ellis, *The Large Scale Structure of Space-Time*. Cambridge: Cambridge University Press (1973).

[729] S. W. Hawking and G. T. Horowitz, Gravitational Hamiltonian, Action, Entropy and Surface Terms, *Classical Quant. Grav.* **13** (1996) 1487.

[730] S. W. Hawking, G. T. Horowitz, and S. F. Ross, Entropy, Area and Black Hole Pairs, *Phys. Lett.* **B383** (1996) 383.

[731] S. W. Hawking and C. J. Hunter, The Gravitational Hamiltonian in the Presence of Non-Orthogonal Boundaries, *Classical Quant. Grav.* **13** (1996) 2735.

[732] S. W. Hawking and C. J. Hunter, Gravitational Entropy and Global Structure, *Phys. Rev.* **D59** (1999) 044025.

[733] S. W. Hawking, C. J. Hunter, and D. N. Page, Nut Charge, Anti-de Sitter Space and Entropy, `hep-th/9809035`.

[734] S. W. Hawking and S. F. Ross, Duality Between Electric and Magnetic Black Holes, *Phys. Rev.* **D52** (1995) 5865.

[735] K. Hayashi and T. Nakano, Extended Translation Invariance and Associated Gauge Fields, *Prog. Theor. Phys.* **38** (1967) 491.

[736] K. Hayashi and T. Shirafuji, New General Relativity, *Phys. Rev.* **D19** (1979) 3524.

[737] F. W. Hehl, P. von der Heyde, G. D. Kerlick, and J. M. Nester, General Relativity with Spin and Torsion: Foundations and Prospects, *Rev. Mod. Phys.* **48** (1976) 393.

[738] F. W. Hehl, J. D. McCrea, E. W. Mielke, and Y. Ne'eman, Metric-Affine Gauge Theory of Gravity: Field Equations, Noether Identities, World Spinors, and Breaking of Dilation Invariance, *Phys. Rep.* **258** (1995) 1.

[739] K. R. Heiderich and W. G. Unruh, Spin-two Fields, General Covariance and Conformal Invariance, *Phys. Rev.* **D38** (1988) 490.

[740] K. R. Heiderich and W. G. Unruh, Nonlinear, Noncovariant Spin-Two Theories, *Phys. Rev.* **D42** (1990) 2057.

[741] S. Helgason, *Differential Geometry, Lie Groups and Symmetric Spaces*. New York: Academic Press (1978).

[742] M. Henneaux and L. Mezincescu, A Sigma Model Interpretation of Green-Schwarz Covariant Superstring Action, *Phys. Lett.* **B152** (1985) 340.

[743] M. Henneaux and C. Teitelboim, Relativistic Quantum Mechanics of Supersymmetric Particles, *Ann. Phys.* **143** (1982) 127.

[744] M. Henneaux and C. Teitelboim, The Cosmological Constant and General Covariance, *Phys. Lett.* **B222** (1989) 195.

[745] M. Henneaux, C. Teitelboim, and J. Zanelli, Gauge Invariance and Degree of Freedom Count, *Nucl. Phys.* **B332** (1990) 169.

[746] C. A. R. Herdeiro, Spinning Deformations of the D1-D5 System and a Geometric Resolution of Closed Timelike Curves, *Nucl. Phys.* **B665** (2003) 189.

[747] M. Heusler, *Black Holes Uniqueness Theorems*. Cambridge: Cambridge University Press (1996).

[748] M. Heusler, No-Hair Theorems and Black Holes with Hair, *Helv. Phys. Acta* **69** (1996) 501.

[749] M. Heusler, Stationary Black Holes: Uniqueness and Beyond, *Living Rev. Relativ.* **1** (1998) 6.

[750] M. Heusler and N. Straumann, The First Law of Black Hole Physics For a Class of Nonlinear Matter Models, *Classical Quant. Grav.* **10** (1993) 1299.

[751] P. von der Heyde, Is Gravitation Mediated by the Torsion of Spacetime?, *Z. Naturforsch.* **31a** (1976) 1725.

[752] D. Hilbert, Die Grundlagen der Physik (Erste Mitteilung), *Königl. Gesell. Wiss. Göttingen. Math.-phys. Klasse. Nachr.* (1915) 395.

[753] J. W. van Holten, Propagators and Path Integrals, *Nucl. Phys.* **B457** (1995) 375.

[754] J. W. van Holten, $D = 1$ Supergravity and Spinning Particles, in *From Field Theory to Quantum Groups*, eds. B. Jancewicz and J. Sobczyk. Singapore: World Scientific (1996), p. 173.

[755] J. W. Van Holten and A. Van Proeyen, $N = 1$ Supersymmetry Algebras in $D = 2$, $D = 3$, $D = 4$ Mod 8, *J. Phys.* **A15** (1982) 3763.

[756] C. F. Holzhey and F. Wilczek, Black Holes as Elementary Particles, *Nucl. Phys.* **B380** (1992) 447.

[757] H. Hopf, *Math. Ann.* **104** (1931) 637.

[758] P. Hořava, Strings on World Sheet Orbifolds, *Nucl. Phys.* **B327** (1989) 461.

[759] P. Hořava, Background Duality of Open String Models, *Phys. Lett.* **B231** (1989) 251.

[760] P. Hořava and E. Witten, Heterotic and Type I String Dynamics from Eleven Dimensions, *Nucl. Phys.* **B460** (1996) 506.

[761] P. Hořava and E. Witten, Eleven-Dimensional Supergravity on a Manifold with Boundary, *Nucl. Phys.* **B475** (1996) 94.

[762] J. H. Horne and G. T. Horowitz, Black Holes Coupled to a Massive Dilaton, *Nucl. Phys.* **B399** (1993) 169.

[763] G. T. Horowitz, The Dark Side of String Theory: Black Holes and Black Strings, in *String Theory and Quantum Gravity*, eds. J. A. Harvey, R. Iengo, K. S. Narain, S. Randjbar-Daemi, and H. L. Verlinde. Singapore: World Scientific (1993), p. 55.

[764] G. T. Horowitz, Quantum States of Black Holes, in *Black Holes and Relativistic Stars*. Chicago, IL: University of Chicago Press (1998), p. 241.

[765] G. T. Horowitz, The Origin of Black Hole Entropy in String Theory, in *Proc. Pacific Conf. Gravitation and Cosmology*, eds. Y. M. Cho, C. H. Lee, and S. W. Kim. Singapore: World Scientific (1999), p. 46.

[766] G. T. Horowitz, D. A. Lowe, and J. M. Maldacena, Statistical Entropy of Nonextremal Four-Dimensional Black Holes and U Duality, *Phys. Rev. Lett.* **77** (1996) 430.

[767] G. T. Horowitz and K. Maeda, Fate of the Black String Instability, *Phys. Rev. Lett.* **87** (2001) 131301.

[768] G. T. Horowitz and K. Maeda, Inhomogeneous Near-Extremal Black Branes, *Phys. Rev.* **D65** (2002) 104028.

[769] G. T. Horowitz, J. M. Maldacena, and A. Strominger, Nonextremal Black Hole Microstates and U Duality, *Phys. Lett.* **B383** (1996) 151.

[770] G. T. Horowitz and J. Polchinski, A Correspondence Principle for Black Holes and Strings, *Phys. Rev.* **D55** (1997) 6189.

[771] G. T. Horowitz and J. Polchinski, Self-Gravitating Fundamental Strings, *Phys. Rev.* **D57** (1998) 2557.

[772] G. T. Horowitz and S. F. Ross, Naked Black Holes, *Phys. Rev.* **D56** (1997) 2180.

[773] G. T. Horowitz and S. F. Ross, Properties of Naked Black Holes, *Phys. Rev.* **D57** (1998) 1098.

[774] G. T. Horowitz and A. R. Steif, Space-Time Singularities in String Theory, *Phys. Rev. Lett.* **64** (1990) 260.

[775] G. T. Horowitz and A. Strominger, Black Strings and $p$-Branes, *Nucl. Phys.* **B360** (1991) 197.

[776] G. T. Horowitz and A. A. Tseytlin, A New Class of Exact Solutions in String Theory, *Phys. Rev.* **D51** (1995) 2896.

[777] K. Hosomichi, On Branes Ending on Branes in Supergravity, *JHEP* **0006** (2000) 004.

[778] Y. Hosotani, Dynamical Mass Generation by Compact Extra Dimensions, *Phys. Lett.* **B126** (1983) 309.

[779] Y. Hosotani, Dynamical Gauge Symmetry Breaking as the Casimir Effect, *Phys. Lett.* **B129** (1983) 193.

[780] Y. Hosotani, Dynamics of Nonintegrable Phases and Gauge Symmetry Breaking, *Ann. Phys.* **190** (1989) 233.

[781] K. Hotta, Holographic RG Flow Dual to Attractor Flow in Extremal Black Holes, *Phys. Rev.* **D79** (2009) 104018.

[782] M. Hotta and M. Tanaka, Shock-wave Geometry with Non-Vanishing Cosmological Constant, *Classical Quant. Grav.* **10** (1993) 307.

[783] P. S. Howe, N. D. Lambert, and P. C. West, A New Massive Type IIA Supergravity from Compactification, *Phys. Lett.* **B416** (1998) 303.

[784] P. S. Howe, N. D. Lambert, and P. C. West, The Self-Dual String Soliton, *Nucl. Phys.* **B515** (1998) 203.

[785] P. S. Howe, N. D. Lambert, and P. C. West, The Threebrane Soliton of the M-Fivebrane, *Phys. Lett.* **B419** (1998) 79.

[786] P. S. Howe and E. Sezgin, Superbranes, *Phys. Lett.* **B390** (1997) 133.

[787] P. S. Howe and E. Sezgin, $D = 11, p = 5$, *Phys. Lett.* **B394** (1997) 62.

[788] P. Howe and P. C. West, The Complete $N = 2, d = 10$ Supergravity, *Nucl. Phys.* **B238** (1984) 181.

[789] K. Hristov, On BPS Bounds in $D = 4$ $N = 2$ Gauged Supergravity II: General Matter Couplings and Black Hole Masses, *JHEP* **1203** (2012) 095.

[790] K. Hristov, C. Toldo, and S. Vandoren, On BPS Bounds in $D = 4$ $N = 2$ Gauged Supergravity, *JHEP* **1112** (2011) 014.

[791] K. Hristov, C. Toldo, and S. Vandoren, Black Branes in AdS: BPS Bounds and Asymptotic Charges, *Fortsch. Phys.* **60** (2012) 1057.

[792] C.-G. Huang and C.-B. Liang, A Torus Like Black Hole, *Phys. Lett.* **A201** (1995) 27.

[793] V. E. Hubený, Overcharging a Black Hole and Cosmic Censorship, *Phys. Rev.* **D59** (1999) 064013.

[794] T. Hübsch, Calabi-Yau Manifolds: A Bestiary for Physicists, 2nd edn. Singapore: World Scientific (1994).

[795] M. Hübscher, P. Meessen, and T. Ortín, Supersymmetric Solutions of $N = 2$ $d = 4$ SUGRA: The Whole Ungauged Shebang, *Nucl. Phys.* **B759** (2006) 228.

[796] M. Hübscher, P. Meessen, and T. Ortín, Domain Walls and Instantons in $N = 1, d = 4$ Supergravity, *JHEP* **1006** (2010) 001.

[797] M. Hübscher, P. Meessen, T. Ortín, and S. Vaulà, Supersymmetric $N = 2$ Einstein-Yang-Mills Monopoles and Covariant Attractors, *Phys. Rev.* **D78** (2008) 065031.

[798] M. Hübscher, P. Meessen, T. Ortín, and S. Vaulà, $N = 2$ Einstein-Yang-Mills's BPS Solutions, *JHEP* **0809** (2008) 099.

[799] M. Hübscher, T. Ortín, and C. S. Shahbazi, The Tensor Hierarchies of Pure $N = 2, d = 4, 5, 6$ Supergravities, *JHEP* **1011** (2010) 130.

[800] J. Hughes, J. Liu, and J. Polchinski, Supermembranes, *Phys. Lett.* **B180** (1986) 370.

[801] C. M. Hull, The Positivity of Gravitational Energy and Global Supersymmetry, *Commun. Math. Phys.* **90** (1983) 545.

[802] C. M. Hull, Lectures on Non-Linear Sigma Models and Strings, in Super Field Theories, eds. H. C. Lee, V. Elias, G. Kunstatter, R. B. Mann, and K. S. Viswanathan. New York: Plenum Press (1987).

[803] C. M. Hull, String-String Duality in Ten-Dimensions, *Phys. Lett.* **B357** (1995) 545.

[804] C. M. Hull, Gravitational Duality, Branes and Charges, *Nucl. Phys.* **B509** (1998) 216.

[805] C. M. Hull, U-Duality and BPS Spectrum of Super Yang-Mills Theory and M-Theory, *JHEP* **9807** (1998) 018.

[806] C. M. Hull, The Nonperturbative SO(32) Heterotic String, *Phys. Lett.* **B462** (1999) 271.

[807] C. M. Hull, Holonomy and Symmetry in M-Theory, `hep-th/0305039`.

[808] C. M. Hull and P. K. Townsend, Unity of Superstring Dualities, *Nucl. Phys.* **B438** (1995) 109.

[809] C. J. Hunter, The Action of Instantons with NUT Charge, *Phys. Rev.* **D59** (1999) 024009.

[810] M. Huq and M. A. Namazie, Kaluza-Klein Supergravity in Ten Dimensions, *Classical Quant. Grav.* **2** (1985) 293.

[811] D. J. Hurley and M. A. Vandyck, On the Concepts of Lie and Covariant Derivatives of Spinors. Part 1, *J. Phys.* **A27** (1994) 4569.

[812] C. A. Hurst, *Ann. Phys.* **50** (1968) 51.

[813] T. Z. Husain, If I Only Had a Brane!, Ph.D. Thesis, Stockholm University, `hep-th/0304143`.

[814] S. Hwang, *Geometriae Dedicata* **71** (1998) 5.

[815] S. Hyun, U-Duality Between Three and Higher Dimensional Black Holes, `hep-th/9704055`.

[816] A. Yu. Ignatev and G. C. Joshi, Massive Electrodynamics and the Magnetic Monopoles, *Phys. Rev.* **D53** (1996) 984.

[817] E. Inönü and E. P. Wigner, *Proc. Nat. Acad. Sci.* **39** (1953) 510.

[818] W. Israel, Event Horizons in Static Vacuum Space-Times, *Phys. Rev.* **164** (1967) 1776.

[819] W. Israel, *Commun. Math. Phys.* **8** (1968) 245.

[820] W. Israel and K. A. Khan, Collinear Particles and Bondi Dipoles in General Relativity, *Nuovo Cim.* **33** (1964) 331.

[821] W. Israel and J. M. Nester, Positivity of the Bondi Gravitational Mass, *Phys. Lett.* **85A**, (1981) 259.

[822] W. Israel and G. A. Wilson, A Class of Stationary Electromagnetic Vacuum Fields, *J. Math. Phys.* **13**, (1972) 865.

[823] N. Itzhaki, A. A. Tseytlin, and S. Yankielowicz, Supergravity Solutions for Branes Localized Within Branes, *Phys. Lett.* **B432** (1998) 298.

[824] C. Itzykson and B. Zuber, Quantum Field Theory. New York: McGraw-Hill (1980).

[825] J. M. Izquierdo, N. D. Lambert, G. Papadopoulos, and P. K. Townsend, Dyonic Membranes, *Nucl. Phys.* **B460** (1996) 560.

[826] R. Jackiw, C. Nohl, and C. Rebbi, Conformal Properties of Pseudoparticle Configurations, *Phys. Rev.* **D15** (1977) 1642.

[827] A. I. Janis, E. T. Newman, and J. Winicour, Reality of the Schwarzschild Singularity, *Phys. Rev. Lett.* **20** (1968) 878.

[828] B. Janssen, Curved Branes and Cosmological $(a,b)$-Models, *JHEP* **0001** (2000) 044.

[829] B. Janssen, P. Meessen, and T. Ortín, The D8-Brane Tied Up: String and Brane Solutions in Massive Type IIIA Supergravity, *Phys. Lett.* **B453** (1999) 229.

[830] B. Janssen, P. Smyth, T. Van Riet and B. Vercnocke, A First-Order Formalism for Timelike and Spacelike Brane Solutions, *JHEP* **0804**, (2008) 007.

[831] J. T. Jebsen, *Ark. f Mat. Astron. och Fys.* **15**(18) (1921) 9 S.

[832] C. V. Johnson, D-Brane Primer, Lectures given at ICTP, TASI, and BUSSTEPP, `hep-th/0007170`.

[833] C. V. Johnson, *D-Branes*. Cambridge: Cambridge University Press (2002).

[834] C. V. Johnson, R. R. Khuri, and R. C. Myers, Entropy of 4-d Extremal Black Holes, *Phys. Lett.* **B378** (1996) 78.

[835] D. C. Jong, A. Kaya, and E. Sezgin, 6D Dyonic String With Active Hyperscalars, *JHEP* **0611** (2006) 047.

[836] D.D. Joyce, *Compact Manifolds with Special Holonomy*. Oxford: Oxford University Press (2000).

[837] B. Julia, Group Disintegrations, in *Superspace and Supergravity*, eds. S. W. Hawking and M. Roček. Cambridge: Cambridge University Press (1981), p. 331.

[838] M. Kaku, *Introduction to Superstrings*. New York: Springer-Verlag (1988).

[839] R. Kallosh, D. Kastor, T. Ortín, and T. Torma, Supersymmetry and Stationary Solutions in Dilaton-Axion Gravity, *Phys. Rev.* **D50** (1994) 6374.

[840] R. Kallosh and B. Kol, $E_7$ Symmetric Area of the Black Hole Horizon, *Phys. Rev.* **D53** (1996) 5344.

[841] R. Kallosh, A. Linde, T. Ortín, A. Peet, and A. Van Proeyen, Supersymmetry as a Cosmic Censor, *Phys. Rev.* **D46** (1992) 5278.

[842] R. Kallosh and T. Ortín, Charge Quantization of Axion–Dilaton Black Holes, *Phys. Rev.* **D48** (1993) 742.

[843] R. Kallosh and T. Ortín, Killing Spinor Identities, `hep-th/9306085`.

[844] R. Kallosh and T. Ortín, Exact SU(2) x U(1) Stringy Black Holes, *Phys. Rev.* **D50** (1994) 7123.

[845] R. Kallosh and T. Ortín, New E77 Invariants and Amplitudes, *JHEP* **1209** (2012) 137.

[846] R. Kallosh, A. Rajaraman, and W. K. Wong, Supersymmetric Rotating Black Holes and Attractors, *Phys. Rev.* **D55** (1997) 3246.

[847] R. Kallosh and M. Soroush, Explicit Action of E7(7) on $\mathcal{N}=8$ Supergravity Fields, *Nucl. Phys.* **B801** (2008) 25.

[848] T. Kaluza, Zum Unitätsproblem der Physik, *Sitzungsber. Preuß. Akad. Wiss., phys.-math. kl.* (1921) 966 (translated into English in Ref. [57]).

[849] P. F. Kelly, Expansions of Non-Symmetric Gravitational Theories about a GR Background, *Classical Quant. Grav.* **6** (1991) 1217. [Erratum *ibid.* **8** (1992) 1423(E).]

[850] R. P. Kerr, Gravitational Field of a Spinning Mass as an Example of Algebraically Special Metrics, *Phys. Rev. Lett.* **11** (1963) 237.

[851] S. V. Ketov, Universal Hypermultiplet Metrics, *Nucl. Phys.* **B604** (2001) 256.

[852] R. R. Khuri, Self-Gravitating Strings and String/Black Hole Correspondence, *Phys. Lett.* **B470** (1999) 73.

[853] R. R. Khuri, Entropy and String/Black-Hole Correspondence, *Nucl. Phys.* **B588** (2000) 253.

[854] R. R. Khuri and T. Ortín, Supersymmetric Black Holes in $N = 8$ Supergravity, *Nucl. Phys.* **B467** (1996) 355.

[855] R. R. Khuri and T. Ortín, A Non-Supersymmetric Dyonic Extreme Reissner-Nordström Black Hole, *Phys. Lett.* **B373** (1996) 56.

[856] K. Kikkawa and M. Yamasaki, Casimir Effects in Superstring Theories, Phys. Lett. **B149** (1984) 357.

[857] H. J. Kim, L. J. Romans, and P. van Nieuwenhuizen, The Mass Spectrum of Chiral $N = 2$ $D = 10$ Supergravity on $S^5$, *Phys. Rev.* **D32** (1985) 389.

[858] E. Kiritsis, Introduction to Non-perturbative String Theory, `hep-th/9708130`.

[859] E. Kiritsis, Introduction to Superstring Theory, Leuven Notes in Mathematical and Theoretical Physics **B9**. Leuven: Leuven University Press (1998).

[860] E. Kiritsis, *String Theory in a Nutshell*. Princeton, NJ: Princeton University Press (2007).

[861] F. Klein, *Konigl. Gesell. Wiss. Göttinger Nachr., math.-phys. Kl.* (1918) 394.

[862] O. Klein, Quantum Theory and Five Dimensional Theory of Relativity, *Z. Phys.* **37** (1926) 895.

[863] D. Klemm, V. Moretti, and L. Vanzo, Rotating Topological Black Holes, *Phys. Rev.* **D57** (1998) 6127.

[864] D. Klemm and L. Vanzo, Quantum Properties of Topological Black Holes, *Phys. Rev.* **D58** (1998) 104025.

[865] D. Klemm and O. Vaughan, Nonextremal Black Holes in Gauged Supergravity and the Real Formulation of Special Geometry, *JHEP* **1301** (2013) 053.

[866] D. Klemm and E. Zorzan, All Null Supersymmetric Backgrounds of $N = 2$, $D = 4$ Gauged Supergravity Coupled to Abelian Vector Multiplets, *Classical Quant. Grav.* **26** (2009) 145018.

[867] D. Klemm and E. Zorzan, The Timelike Half-supersymmetric Backgrounds of $N = 2$, $D = 4$ Supergravity with Fayet-Iliopoulos Gauging, *Phys. Rev.* **D82** (2010) 045012.

[868] S. Kobayashi and K. Nomizu, *Foundations of Differential Geometry*, vol. 1 and 2. New York: Interscience (1969).

[869] A. Komar, Covariant Conservation Laws in General Relativity, *Phys. Rev.* **113** (1954) 934.

[870] Y. Kosmann, Dérivées de Lie des spineurs, *C. R. Acad. Sci. Paris Sér. A.* **262** (1966) A289.

[871] Y. Kosmann, Dérivées de Lie des spineurs, *Annali Mat. Pura Appl.* **(IV) 91** (1972) 317.

[872] V. A. Kostelecky and M. J. Perry, Solitonic Black Holes in Gauged $N = 2$ Supergravity, *Phys. Lett.* **B371** (1996) 191.

[873] F. Kottler, *Encyklopädie der mathematikalischen Wissenschaft*, vol. 6, part II. Leipzig: B. G. Teubner (1922).

[874] J. Kowalski-Glikman, Vacuum States in Supersymmetric Kaluza-Klein Theory, *Phys. Lett.* **B134** (1984) 194.

[875] J. Kowalski-Glikman, Positive Energy Theorem and Vacuum States for the Einstein-Maxwell System, *Phys. Lett.* **B150** (1985) 125.

[876] R.H. Kraichnan, Special Relativity Derivation of Generally Covariant Gravitation Theory, *Phys. Rev.* **98** (1955) 1118. [Erratum *ibid.* **99** (1955) 1906.]

[877] R. H. Kraichnan, Possibility of Unequal Gravitational and Inertial Masses, *Phys. Rev.* **107** (1957) 1485.

[878] D. Kramer, H. Stephani, M. MacCallum, and E. Herlt, *Exact Solutions of Einstein's Field Equations*. Cambridge: Cambridge University Press (1980).

[879] M. Kruskal, Maximal Extension of the Schwarzschild Metric, *Phys. Rev.* **119** (1960) 1743.

[880] H. P. Künzle, Construction of Singularity-Free Spherically Symmetric Spacetime Manifolds, *Proc. Roy. Soc. London* **A297** (1967) 244.

[881] H.P. Künzle and A. K. M. Masood-ul-Alam, Spherically Symmetric Static SU(2) Einstein Yang-Mills Fields, *J. Math. Phys.* **31** (1990) 928.

[882] T. Kugo and P. K. Townsend, Supersymmetry and the Division Algebras, *Nucl. Phys.* **B221** (1983) 357.

[883] L. D. Landau and E. M. Lifshitz, *The Classical Theory of Fields*. Oxford: Pergamon Press (1975).

[884] F. Larsen, A String Model of Black Hole Microstates, *Phys. Rev.* **D56** (1997) 1005.

[885] M. von Laue, *Jahrbuch der Radioaktivität und Elektronik* **14** (1917) 263.

[886] S. M. Lee, A. Peet, and L. Thorlacius, Brane-Waves and Strings, *Nucl. Phys.* **B514** (1998) 161.

[887] R. G. Leigh, Dirac-Born-Infeld Action from Dirichlet Sigma Model, *Mod. Phys. Lett.* **A4** (1989) 2767.

[888] J. P. S. Lemos, Two-Dimensional Black Holes and Planar General Relativity, *Classical Quant. Grav.* **12** (1995) 1081.

[889] J. P. S. Lemos, Cylindrical Black Hole in General Relativity, *Phys. Lett.* **B353** (1995) 46.

[890] J. P. S. Lemos and V. T. Zanchin, Rotating Charged Black String And Three-Dimensional Black Holes, *Phys. Rev.* **D54** (1996) 3840.

[891] T. Levi-Cività, Vereinfachte Herstellung der Einsteinschen einheitlichen Feldgleichungen, *Sitzungsber. Preuß. Akad. Wiss., phys.-math. Kl.* (1929) 137.

[892] S. Liberati and G. Pollifrone, Entropy and Topology of Gravitational Instantons, *Phys. Rev.* **D56** (1997) 6558.

[893] S. Liberati, T. Rothman, and S. Sonego, Nonthermal Nature of Incipient Extremal Black Holes, *Phys. Rev.* **D62** (2000) 024005.

[894] A. Lichnerowicz, *Théories relativistes de la gravitation et de l'électromagnétisme*. Paris: Masson (1955).

[895] A. Lichnerowicz, Spineurs harmoniques, *C. R. Acad. Sci. Paris* **257** (1963) 7.

[896] L. Lindblom and A. K. M. Masood-ul-Alam, On the Spherical Symmetry of Static Stellar Models, *Commun. Math. Phys.* **162** (1994) 123.

[897] U. Lindström and R. von Unge, A Picture of D-Branes at Strong Coupling, *Phys. Lett.* **B403** (1997) 233.

[898] A. Loewy, Semi Localized Brane Intersections in SUGRA, *Phys. Lett.* **B463** (1999) 41.

[899] A. A. Logunov, The Relativistic Theory of Gravity and Mach's Principle, *Phys. Part. Nucl.* **29** (1) (1998) 1–32. (*Fiz. Élem. Chast. Atom Yadra* **29** (1998) 5.)

[900] G. Lopes Cardoso, A. Ceresole, G. Dall'Agata, J. M. Oberreuter, and J. Perz, First-order Flow Equations for Extremal Black Holes in Very Special Geometry, *JHEP* **0710** (2007) 063.

[901] H. A. Lorentz, *Amsterd. Versl.* **25** (1916) 468.

[902] C. O. Lousto and N. Sánchez, The Ultrarrelativistic Limit of the Boosted Kerr-Newman Geometry and the Scattering of Spin $1/2$ Particles, *Nucl. Phys.* **B383** (1992) 377.

[903] D. Lovelock, The Einstein Tensor and Its Generalizations, *J. Math. Phys.* **12** (1971) 498.

[904] Y. Lozano, Eleven Dimensions from the Massive D-2-Brane, *Phys. Lett.* **B414** (1997) 52.

[905] E. Lozano-Tellechea, P. Meessen, and T. Ortín, On d = 4, d = 5, d = 6 Vacua with Eight Supercharges, *Classical Quant. Grav.* **19** (2002) 5921.

[906] E. Lozano-Tellechea and T. Ortín, The General, Duality-Invariant Family of Non-BPS Black-Hole Solutions of $N = 4$, $d = 4$ Supergravity, *Nucl. Phys.* **B569** (2000) 435.

[907] E. Lozano-Tellechea and T. Ortín, 7-Branes and Higher Kaluza-Klein Branes, *Nucl. Phys.* **B607** (2001) 213.

[908] J. X. Lu, ADM Masses for Black Strings and p-Branes, *Phys. Lett.* **B313** (1993) 29.

[909] J. X. Lu and Shibaji Roy, An Si(2, *Z*) Multiplet of Type IIB Super Five-Branes, *Phys. Lett.* **B428** (1998) 289.

[910] J. X. Lu and Shibaji Roy, Nonthreshold (F, D*p*) Bound States, *Nucl. Phys.* **B560** (1999) 181.

[911] J. X. Lu and Shibaji Roy, (F, D5) Bound State, Si(2, *Z*) Invariance and the Descendant States in Type IIB/A String Theory, *Phys. Rev.* **D60** (1999) 126002.

[912] H. Lü, C. N. Pope, and J. Rahmfeld, A Construction of Killing Spinors on $S^n$, *J. Math. Phys.* **40** (1999) 4518.

[913] H. Lü, C. N. Pope, E. Sezgin, and K. S. Stelle, Stainless Super $p$-Branes, *Nucl. Phys.* **B456** (1995) 669.

[914] H. Lü, C. N. Pope, T. A. Tran, and K. W. Xu, Classification of $p$-Branes, NUTs, Waves and Intersections, *Nucl. Phys.* **B511** (1998) 98.

[915] D. Lüst and S. Theisen, Lectures on String Theory. Heidelberg: Springer-Verlag (1989).

[916] S. W. MacDowell and F. Mansouri, Unified Geometric Theory of Gravity and Supergravity, *Phys. Rev. Lett.* **38** (1977) 739. [Erratum *ibid.* **38** (1977) 1376.]

[917] J. Maharana and J. H. Schwarz, Non-compact Symmetries in String Theory, *Nucl. Phys.* **B390** (1993) 3.

[918] S. D. Majumdar, A Class of Exact Solutions of Einstein's Field Equations, *Phys. Rev.* **72** (1947) 390.

[919] J. M. Maldacena, Black Holes in String Theory, Ph.D. Thesis, Princeton University, `hep-th/9607235`.

[920] J. M. Maldacena, Black Holes and D-Branes, *Nucl. Phys. Proc. Suppl.* **61A** (1998) 111.

[921] J. M. Maldacena, The Large *N* Limit of Superconformal Field Theories and Supergravity, *Adv. Theor. Math. Phys.* **2** (1998) 231.

[922] J. M. Maldacena and A. Strominger, Statistical Entropy of Four-Dimensional Extremal Black Holes, *Phys. Rev. Lett.* **77** (1996) 428.

[923] J. M. Maldacena and A. Strominger, Black Hole Greybody Factors and D-Brane Spectroscopy, *Phys. Rev.* **D55** (1997) 861.

[924] G. Mangano, Are there Metric Theories of Gravity other than General Relativity?, `gr-qc/9511027`.

[925] R. B. Mann, Pair Production of Topological Anti-de Sitter Black Holes', *Classical Quant. Grav.* **14** (1997) L109.

[926] R. B. Mann, Topological Black Holes: Outside Looking In, in *Internal Structure of Black Holes and Spacetime Singularities*, eds. L. Burko and A. Ori. Bristol: Institute of Physics Publishing, and Jerusalem: The Israel Physical Society (1997), p. 311.

[927] Y. Mao, M. Tegmark, A. H. Guth, and S. Cabi, Constraining Torsion with Gravity Probe B, *Phys. Rev.* **D76** (2007) 104029.

[928] N. Marcus and J. H. Schwarz, Field Theories that have no Manifestly Lorentz Invariant Formulation, *Phys. Lett.* **115B** (1982) 111.

[929] M. A. Markov and V. P. Frolov, *Teor. Mat. Fiz.* **3** (1970) 3.

[930] D. Marolf, Chern-Simons Terms and the Three Notions of Charge, in *Proc. of the E.S. Fradkin Memorial Conf.*, hep-th/0006117.

[931] D. Marolf and A. Peet, Brane Baldness vs. Superselection Sectors, *Phys. Rev.* **D60** (1999) 105007.

[932] A. Marrani, E. Orazi, and F. Riccioni, Exceptional Reductions, *J. Phys.* **A44** (2011) 155207.

[933] A. E. Mayo and J. D. Bekenstein, No Hair for Spherical Black Holes: Charged and Nonminimally Coupled Scalar Field With Selfinteraction, *Phys. Rev.* **D54** (1996) 5059.

[934] P. Meessen, A Small Note on $pp$-Wave Vacua in Six Dimensions and Five Dimensions, *Phys. Rev.* **D65** (2002) 087501.

[935] P. Meessen, Supersymmetric Coloured/Hairy Black Holes, *Phys. Lett.* **B665** (2008) 388.

[936] P. Meessen and T. Ortín, An Sl(2, Z) Multiplet of Nine-Dimensional Type II Supergravity Theories, *Nucl. Phys.* **B541** (1999) 195.

[937] P. Meessen and T. Ortín, The Supersymmetric Configurations of $N = 2$, $D = 4$ Supergravity Coupled to Vector Supermultiplets, *Nucl. Phys.* **B749** (2006) 291.

[938] P. Meessen and T. Ortín, Ultracold Spherical Horizons in Gauged $N = 1$, $d = 4$ Supergravity, *Phys. Lett.* **B693** (2010) 358.

[939] P. Meessen and T. Ortín, Non-Extremal Black Holes of $N = 2$, $d = 5$ Supergravity, *Phys. Lett.* **B707** (2012) 178.

[940] P. Meessen and T. Ortín, Supersymmetric Solutions to Gauged $N = 2$ $d = 4$ sugra: The Full Timelike Shebang, *Nucl. Phys.* B **863** (2012) 65.

[941] P. Meessen, T. Ortín, J. Perz, and C. S. Shahbazi, H-FGK-formalism for Black-Hole Solutions of $N = 2$, $d = 4$ and $d = 5$ Supergravity, *Phys. Lett.* **B709** (2012) 260.

[942] P. Meessen, T. Ortín, J. Perz, and C. S. Shahbazi, Black Holes and Black Strings of $N = 2$, $d = 5$ Supergravity in the H-FGK formalism, *JHEP* **1209** (2012) 001.

[943] P. Meessen, T. Ortín, and S. Vaulà, All the Timelike Supersymmetric Solutions of all Ungauged $d = 4$ Supergravities, *JHEP* **1011** (2010) 072.

[944] M. A. Melvin, Pure Electric and Magnetic Geons, *Phys. Lett.* **8** (1964) 65.

[945] K. Menou, E. Quataert, and R. Narayan, Astrophysical Evidence for Black Hole Event Horizons, in *The Eighth Marcel Grossman Meeting: Proceedings*, ed. T. Piran. Singapore: World Scientific (1999), p. 204.

[946] R. R. Metsaev, Type IIB Green-Schwarz Superstring in Plane Wave Ramond-Ramond Background, *Nucl. Phys.* **B625** (2002) 70.

[947] R. R. Metsaev and A. A. Tseytlin, Superstring Action in $\mathrm{AdS}_5 \times S^5$: Kappa-Symmetry Light Cone Gauge, *Phys. Rev.* **D63** (2001) 046002.

[948] R. R. Metsaev and A. A. Tseytlin, Exactly Solvable Model of Superstring in Plane Wave Ramond-Ramond Background, *Phys. Rev.* **D65** (2002) 126004.

[949] C. M. Miller, K. Schalm, and E. J. Weinberg, Nonextremal Black Holes are BPS, *Phys. Rev.* **D76** (2007) 044001.

[950] C. Misner, The Flatter Regions of Newman, Unti and Tambourino's Generalized Schwarzschild Space, *J. Math. Phys.* **4** (1963) 924.

[951] C. Misner, Interpretation of Gravitational Wave Observations, *Phys. Rev. Lett.* **28** (1972) 994.

[952] C. W. Misner, K. S. Thorne, and J. A. Wheeler, *Gravitation*. New York: W. H. Freeman and Co. (1973).

[953] C. W. Misner and J. A. Wheeler, Classical Physics as Geometry: Gravitation, Electromagnetism, Unquantized Charge, and Mass as Properties of Curved Empty Space, *Annals Phys.* **2** (1957) 525.

[954] J. W. Moffat, New Theory of Gravitation, *Phys. Rev.* **D19** (1979) 3554.

[955] T. Mohaupt, Black Holes in Supergravity and String Theory, *Classical Quant. Grav.* **17** (2000) 3429.

[956] T. Mohaupt, Black Hole Entropy, Special Geometry and Strings, *Fortsch. Phys.* **49** (2001) 3 .

[957] T. Mohaupt and O. Vaughan, Non-extremal Black Holes, Harmonic Functions, and Attractor Equations, *Classical Quant. Grav.* **27** (2010) 235008.

[958] T. Mohaupt and O. Vaughan, The Hesse Potential, the c-map and Black Hole Solutions, *JHEP* **1207** (2012) 163.

[959] T. Mohaupt and K. Waite, Instantons, Black Holes and Harmonic Functions, *JHEP* **0910** (2009) 058.

[960] C. Møller, *Ann. Phys.* **4** (1958) 347.

[961] C. Møller, *Ann. Phys.* **12** (1961) 118.

[962] C. Møller, Conservation Laws and Absolute Parallelism in General Relativity, *Mat. Fys. Skr. Dan. Vid. Selsk.* **1** (10) (1961).

[963] S. Moriyama, USp(32) String as Spontaneously Supersymmetry Broken Theory, Phys. Lett. **B522** (2001) 177.

[964] D. R. Morrison and N. Seiberg, Extremal Transitions and Five-Dimensional Supersymmetric Field Theories, *Nucl. Phys.* **B483** (1997) 229.

[965] M. Moshinski, On the Interactions of Birkhoff's Gravitational Field with the Electromagnetic and Pair Fields, *Phys. Rev.* **80** (1950) 514.

[966] H. Müller zum Hagen and H. J. Seifert, Two Axisymmetric Black Holes Cannot Lie in Static Equilibrium, *Int. J. Theor. Phys.* **8** (1973) 443.

[967] U. Muench, F. Gronwald, and F. W. Hehl, A Small Guide to Variations in Teleparallel Gauge Theories of Gravity and the Kaniel-Itin Model, gr-qc/9801036.

[968] R. C. Myers and M. J. Perry, Black Holes in Higher Dimensional Space-Times, *Ann. Phys.* **172** (1986) 304.

[969] R. C. Myers and M. J. Perry, Higher-Dimensional Black Holes in Compactified Space, *Phys. Rev.* **D35** (1987) 455.

[970] R. C. Myers and M. J. Perry, Stress Tensors and Casimir Energies in the AdS/CFT Correspondence, *Phys. Rev.* **D60** (1999) 046002.

[971] R. C. Myers and M. J. Perry, Black Holes and String Theory, Summary of Lectures given at Fourth Mexican School on Gravitation and Mathematical Physics, `gr-qc/0107034`.

[972] M. Nakahara, *Geometry, Topology and Physics*. London: Institute of Physics Publishing (1990).

[973] T. Nakamura and H. Sato, Absorption of Massive Scalar Field by a Charged Black Hole, *Phys. Lett.* **B61** (1976) 371.

[974] C. Nash and S. Sen, *Topology and Geometry for Physicists*. London: Academic Press (1983).

[975] R. I. Nepomechie, Magnetic Monopoles from Antisymmetric Tensor Gauge Fields, *Phys. Rev.* **D31** (1985) 1921.

[976] J. M. Nester, A new Gravitational Energy Expression with a Simple Positivity Proof, *Phys. Lett.* **83A**, (1981) 241.

[977] J. M. Nester and H.-J. Yo, Symmetric Teleparallel General Relativity, `gr-qc/9809049`.

[978] G. Neugebauer, Untersuchungen zu Einstein-Maxwell-Feldern mit eindimensionaler Bewegunsgruppe, Habilitationsschrift, University of Jena (1969).

[979] A. Neveu and J. H. Schwarz, Factorizable Dual Model of Pions, *Nucl. Phys.* **D31** (1971) 86.

[980] E. T. Newman, E. Couch, K. Chinnapared, A. Exton, A. Prakash, and R. Torrence, Metric of a Rotating Charged Mass, *J. Math. Phys.* **6** (1965) 918.

[981] E. Newman, L. Tambourino, and T. Unti, Empty-space Generalization of the Scwarzschild Metric, *J. Math. Phys.* **4** (1963) 915.

[982] P. van Nieuwenhuizen, An Introduction to Covariant Quantization of Gravity, in *1975 Marcel Grossmann Meeting on General Relativity*. Oxford: Oxford University Press (1977), p. 1.

[983] P. van Nieuwenhuizen, Supergravity, *Phys. Rep.* **68** (1981) 189.

[984] P. van Nieuwenhuizen, Six Lectures at the Trieste 1981 Summer School on Supergravity, in *Supergravity '81*. Cambridge: Cambridge University Press (1982), p. 151.

[985] P. van Nieuwenhuizen and J. A. M. Vermaseren, One Loop Divergences in the Quantum Theory of Supergravity, *Phys. Lett.* **B65** (1976) 263.

[986] A. I. Nikishov, On Energy-Momentum Tensors of Gravitational Field, *Phys. Part. Nuclei* **32** (2001) 1. (*Fiz. Élem. Chast. Atom. Yadra* **32** (2001) 5.)

[987] H. Nishino and S. Rajpoot, Gauged $N = 2$ Supergravity in Nine-Dimensions and Domain-Wall Solutions, *Phys. Lett.* **B546** (2002) 261.

[988] H. Nishino and E. Sezgin, The Complete $N = 2$, $D = 6$ Supergravity With Matter And Yang-Mills Couplings, *Nucl. Phys.* **B278** (1986) 353.

[989] G. Nordström, *Phys. Z.* **13** (1912) 1126.

[990] G. Nordström, *Ann. Phys.* **40** (1913) 856.

[991] G. Nordström, Theorie der Gravitation rom Standpunkt des Relativiätsprinzips, *Ann. Phys.* **42** (1913) 533.

[992] G. Nordström, *Ann. Phys.* **43** (1914) 1101.

[993] G. Nordström, *Phys. Z.* **15** (1914) 375.

[994] G. Nordström, Über die Möglichkeit, das elektromagnetische Feld und das Gravitationsfeld zu vereinigen, *Phys. Z.* **15** (1914) 504 (translated into English in Ref. [57]).

[995] G. Nordström, *Ann. Acad. Sci. Fenn.* **57** (1914, 1915).

[996] G. Nordström, *Proc. Kon. Ned. Akad. Wet.* **20** (1918) 1238.

[997] J. D. Norton, *Arch. Hist. Exact Sciences* **45** (1993) 17.

[998] I. Novikov and V. P. Frolov, *Physics of Black Holes*. Dordrecht: Kluwer Academic Publishers (1989).

[999] N. A. Obers, B. Pioline, and E. Rabinovici, M-Theory and U-Duality on $T^d$ with Gauge Backgrounds, *Nucl. Phys.* **B525** (1998) 163.

[1000] V. I. Ogievetsky and I. V. Polubarinov, Interacting Field of Spin-2 and the Einstein Equations, *Ann. Phys.* **35** (1965) 167.

[1001] D. W. Olson and W. G. Unruh, Conversion of Electromagnetic to Gravitational Radiation by Scattering from a Charged Black Hole, *Phys. Rev. Lett.* **33** (1974) 1116.

[1002] B. O'Neill, *The Geometry of Kerr Black Holes*. Wellesley, MA: A. K. Peters (1994).

[1003] H. Ooguri and Z. Yin, TASI Lectures on Perturbative String Theories, in Fields, Strings and Duality. Proceedings, Summer School, Theoretical Advanced Study Institute in Elementary Particle Physics, TASI'96, Boulder, USA, June 2–28, 1996, eds. C. Efthimiou and B. Greene. Singapore: World Scientific (1997), p. 5. `hep-th/9612254`.

[1004] L. O'Raifeartaigh, *The Dawning of Gauge Theory*. Princeton, NJ: Princeton University Press (1997).

[1005] T. Ortín, Electric-Magnetic Duality and Supersymmetry in Stringy Black Holes, *Phys. Rev.* **D47** (1993) 3136.

[1006] T. Ortín, Time-Symmetric Initial-Data Sets in 4-D Dilaton Gravity, *Phys. Rev.* **D52** (1995) 3392.

[1007] T. Ortín, A Note on the D-2-Brane of the Massive Type IIA Theory and Gauged Sigma Models, *Phys. Lett.* **B415** (1997) 39.

[1008] T. Ortín, Non-sypersymmetric (but) Extreme Black Holes, Scalar Hair, Massless Black Holes and Other Open Problems, *Nucl. Phys. Proc. Suppl.* **61A** (1998) 131.

[1009] T. Ortín, A Note on Lie-Lorentz Derivatives, *Classical Quant. Grav.* **19** (2002) L1.

[1010] T. Ortín, A Note on Supersymmetric Godel Black Holes, Strings and Rings of Minimal $d = 5$ Supergravity, *Classical Quant. Grav.* **22** (2005) 939.

[1011] T. Ortín, The Supersymmetric Solutions and Extensions of Ungauged Matter-coupled $N = 1$, $d = 4$ Supergravity, *JHEP* **0805** (2008) 034.

[1012] T. Ortín, A Simple Derivation of Supersymmetric Extremal Black Hole Attractors, *Phys. Lett.* **B700** (2011) 261.

[1013] T. Ortín and C. S. Shahbazi, The Supersymmetric Black Holes of $N = 8$ Supergravity, *Phys. Rev.* **D86** (2012) 061702.

[1014] D. N. Page, Particle Emission Rates from a Black Hole III. Charged Leptons from a Nonrotating Black Hole, *Phys. Rev.* **D16** (1977) 2402.

[1015] D. N. Page, Taub-NUT Instanton with an Horizon, *Phys. Lett.* **B78** (1978) 249.

[1016] A. Palatini, Deduzione invariantiva della equazioni gravitazionali dal principio di Hamilton, *Red. Circ. Mat. Palermo* **43** (1919) 203.

[1017] G. Papadopoulos, New Half Supersymmetric Solutions of the Heterotic String, *Classical Quant. Grav.* **26** (2009) 135001.

[1018] G. Papadopoulos and P. K. Townsend, Intersecting M-Branes, *Phys. Lett.* **B380** (1996) 273.

[1019] A. Papapetrou, A Static Solution of the Equations of the Gravitational Field for an Arbitrary Charge-Distribution, *Proc. Roy. Irish Acad.* **A51** (1947) 191.

[1020] A. Papapetrou, *Proc. Roy. Soc. London* **209A** (1951) 248.

[1021] P. Pasti, D. P. Sorokin, and M. Tonin, Covariant Action for a $D = 11$ Five-Brane with the Chiral Field, *Phys. Lett.* **B398** (1997) 41.

[1022] A. W. Peet, The Bekenstein Formula and String Theory (N-Brane Theory), *Classical Quant. Grav.* **15** (1998) 3291.

[1023] A. W. Peet, TASI Lectures on Black Holes in String Theory, hep-th/0008241.

[1024] A. W. Peet, Baldness/Delocalization in Intersecting Brane Systems, *Classical Quant. Grav.* **17** (2000) 1235.

[1025] C. Pellegrini and J. Plebański, Tetrad Fields and Gravitational Fields, *Mat. Fys. Skr. Dan. Vid. Selsk.* **2** (4) (1963).

[1026] R. Penrose, Structure of Space-Time, in *Battelle Rencontres*, eds. C. DeWitt and J. A. Wheeler. New York: Benjamin (1968).

[1027] R. Penrose, Gravitational Collapse: The Role of General Relativity, *Riv. Nuovo Cim.* **1** (1969) 252.

[1028] R. Penrose, Any Spacetime has a Plane Wave as a Limit, in *Differential Geometry and Relativity*. Dordrecht: Reidel (1976), p. 271.

[1029] R. Penrose, Singularities and Time-Asymmetry, in *General Relativity: An Einstein Centenary Survey*, eds. S. W. Hawking and W. Israel. Cambridge: Cambridge University Press (1979), p. 581.

[1030] R. Penrose, On Schwarzschild Causality – A Problem for "Lorentz Covariant" General Relativity, in *Essays in General Relativity*, ed. F. J. Tipler. New York: Academic Press (1980), p. 1.

[1031] R. Penrose, The Question of Cosmic Censorship, in *Black Holes and Relativistic Stars*. Chicago, IL: The University of Chicago Press (1998).

[1032] R. Penrose and W. Rindler, *Spinors and Space-Time*, vol. 1 and 2. Cambridge: Cambridge University Press (1984).

[1033] Z. Perjés, Solutions of the Coupled Einstein-Maxwell Equations Representing the Fields of Spinning Sources, *Phys. Rev. Lett.* **27** (1971) 1668.

[1034] M. J. Perry, Black Holes Are Colored, *Phys. Lett.* **B71** (1977) 234.

[1035] J. Perz, P. Smyth, T. Van Riet, and B. Vercnocke, First-order Flow Equations for Extremal and Non-extremal Black Holes, *JHEP* **0903** (2009) 150.

[1036] . E. Peskin, Introduction to String and Superstring Theory 2, in *From the Planck Scale to the Weak Scale: Towards a Theory of the Universe*, ed. H. E. Haber. Singapore: World Scientific (1987).

[1037] A. N. Petrov, On the Cosmological Constant as a Constant of Integration, *Mod. Phys. Lett.* **A6** (1991) 2107.

[1038] A. N. Petrov and J. Katz, Relativistic Conservation Laws on Curved Backgrounds and the Theory of Cosmological Perturbations, *Proc. Roy. Soc. London* **458** (2002) 319.

[1039] K. Pilch, P. van Nieuwenhuizen, and P. K. Townsend, Compactification of $D = 11$ Supergravity on S(4) (or 11 = 7 + 4, too), *Nucl. Phys.* **B242** (1984) 377.

[1040] B. Pioline, Lectures on Black Holes, Topological Strings and Quantum Attractors, *Classical Quant. Grav.* **23** (2006) S981.

[1041] B. Pioline, Lectures on Black Holes, Topological Strings and Quantum Attractors (2.0), *Lect. Notes Phys.* **755** (2008) 283.

[1042] J. F. Plebański, A Class of Solutions of the Einstein-Maxwell Equations, *Ann. Phys.* **90** (1975) 196.

[1043] J. F. Plebański and M. Demiański, Rotating, Charged, and Uniformly Accelarating Mass in General Relativity, *Ann. Phys.* **98** (1976) 98.

[1044] J. Podolský, Exact Impulsive Gravitational Waves in Spacetimes of Constant Curvature, in *Gravitation: Following the Prague Inspiration*, eds. O. Semerak, J. Podolský, and M. Zefka. Singapore: World Scientific (2002), p. 205.

[1045] J. Podolský and J. B. Griffiths, Boosted Static Multipole Particles as Sources of Impulsive Gravitational Waves, *Phys. Rev.* **D58** (1998) 124024.

[1046] J. Polchinski, Combinatorics of Boundaries in String Theory, *Phys. Rev.* **D50** (1994) 6041.

[1047] J. Polchinski, Dirichlet Branes and Ramond–Ramond Charges, *Phys. Rev. Lett.* **75** (1995) 4724.

[1048] J. Polchinski, TASI Lectures on D branes, `hep-th/9611050`.

[1049] J. Polchinski, *String Theory*, vol. 1 and 2. Cambridge: Cambridge University Press (1998).

[1050] J. Polchinski, Scale and Conformal Invariance in Quantum Field Theory, *Nucl. Phys.* **B303** (1988) 226.

[1051] J. Polchinski, S. Chaudhuri, and C. V. Johnson, Notes on D branes, `hep-th/9602052`.

[1052] J. Polchinski and E. Witten, Evidence for Heterotic - Type I String Duality, *Nucl. Phys.* **B460** (1996) 525.

[1053] A. M. Polyakov, Particle Spectrum in the Quantum Field Theory, *Sov. Phys. JETP Lett.* **20** (1974) 194.

[1054] A. M. Polyakov, Quantum Geometry of Bosonic Strings, *Phys. Lett.* **B103** (1981) 207.

[1055] A. M. Polyakov, Quantum Geometry of Fermionic Strings, *Phys. Lett.* **B103** (1981) 211.

[1056] G. Pradisi and A. Sagnotti, Open String Orbifolds, *Phys. Lett.* **B216** (1989) 59.

[1057] M. K. Prasad, Equivalence of Eguchi-Hanson Metric to Two-center Gibbons-Hawking Metric, *Phys. Lett.* **B83** (1979) 310.

[1058] M. K. Prasad and C. M. Sommerfield, An Exact Classical Solution for the 't Hooft Monopole and the Julia-Zee Dyon, *Phys. Rev. Lett.* **35** (1975) 760.

[1059] V. Pravda, A. Pravdova, A. Coley, and R. Milson, All Spacetimes With Vanishing Curvature Invariants, *Classical Quant. Grav.* **19** (2002) 6213.

[1060] J. Preskill, P. Schwarz, A. Shapere, S. Trivedi, and F. Wilczek, Limitations on the Statistical Description of Black Holes, *Mod. Phys. Lett.* **A6** (1991) 2353.

[1061] R. H. Price, Non-spherical Perturbations of Gravitational Collapse. I. Scalar and Gravitational Perturbations, *Phys. Rev.* **D5** (1972) 2419.

[1062] R. H. Price, Non-spherical Perturbations of Gravitational Collapse. II. Integer-Spin, Zero-Rest-Mass-Fields, *Phys. Rev.* **D5** (1972) 2439.

[1063] A. P. Protogenov, Exact Classical Solutions Of Yang-Mills Sourceless Equations, *Phys. Lett.* **B67** (1977) 62.

[1064] H. Quevedo, *Fortsch. Phys.* **38** (1990) 733.

[1065] J. Rahmfeld, Extremal Black Holes as Bound States, *Phys. Lett.* **B372** (1996) 198.

[1066] A. Rajaraman, Supergravity Solutions for Localised Brane Intersections, *JHEP* **0109** (2001) 018.

[1067] P. Ramond, Dual Theory of Free Fermions, *Phys. Rev.* **D3** (1971) 2415.

[1068] P. Ramond, *Field Theory: A Modern Primer*. Reading, MA: Benjamin/Cummings (1981).

[1069] L. Randall and R. Sundrum, A Large Mass Hierarchy from a Small Extra Dimension, *Phys. Rev. Lett.* **83** (1999) 3370.

[1070] L. Randall and R. Sundrum, An Alternative to Compactification, *Phys. Rev. Lett.* **83** (1999) 4690.

[1071] J. Rayski, *Acta Phys. Polon.* **27** (1965) 89.

[1072] H. S. Reall, Classical and Thermodynamic Stability of Black Branes, *Phys. Rev.* **D64** (2001) 044005.

[1073] M. Rees, Astrophysical Evidence for Black Holes, in *Black Holes and Relativistic Stars*, ed. R. M. Wald. Chicago, IL: University of Chicago Press (1998).

[1074] H. Reissner, *Ann. Phys.* **50** (1916) 106.

[1075] S. J. Rey, Confining Phase of Superstrings and Axionic Strings, *Phys. Rev.* **D43**, 526 (1991).

[1076] F. Riccioni, Truncations of the D9-Brane Action and Type-I Strings, `hep-th/0301021`.

[1077] I. Robinson, A Solution of the Maxwell-Einstein Equations, *Bull. Acad. Polon. Sci.* **7** (1959) 351.

[1078] M. Rocek and E. Verlinde, Duality, Quotients, and Currents, *Nucl. Phys.* **B373** (1992) 630.

[1079] M. Rogatko, Stationary Axisymmetric Axion-Dilaton Black Holes: Mass Formulae, *Classical Quant. Grav.* **11** (1994) 689.

[1080] M. Rogatko, The Bogomolnyi-Type Bound in Axion-Dilaton Gravity, *Classical Quant. Grav.* **12** (1995) 3115.

[1081] M. Rogatko, Extrema of Mass, First Law of Black Hole Mechanics and Staticity Theorem in Einstein-Maxwell Axion Dilaton Gravity, *Phys. Rev.* **D58** (1998) 044011.

[1082] M. Rogatko, Uniqueness Theorem for Static Dilaton $U(1)^N$ Black Holes, *Classical Quant. Grav.* **19** (2002) 875.

[1083] L. J. Romans, Massive $N = 2a$ Supergravity in Ten Dimensions, *Phys. Lett.* **B169** (1986) 374.

[1084] L. J. Romans, Supersymmetric, Cold and Lukewarm Black Holes in Cosmological Einstein-Maxwell Theory, *Nucl. Phys.* **B383** (1992) 395.

[1085] M. de Roo, Matter Coupling in $N = 4$ Supergravity, *Nucl. Phys.* **B255** (1985) 515.

[1086] L. Rosenfeld, Sur le tenseur d'impulsion-énergie, *Mém. Acad. Roy. Belgique* **6** (1930) 30.

[1087] R. Ruffini and J. A. Wheeler, Introducing the Black Hole, *Phys. Today*, **24** (1971) 30.

[1088] H. Rumpf, On the Translational Part of the Lagrangian of the Poincaré Gauge Theory of Gravitation, *Z. Naturforsch.* **33a** (1978) 1224.

[1089] H. Rumpf, Supersymmetric Dirac Particles in Riemann-Cartan Space-Time, *Gen. Relativ. Gravit.* **14** (1982) 873.

[1090] V. de Sabbata and M. Gasperini, *Introduction to Gravitation*. Singapore: World Scientific (1985).

[1091] V. de Sabbata and C. Sivaram, *Spin and Torsion in Gravitation*. Singapore: World Scientific (1994).

[1092] W. A. Sabra, General Static $N = 2$ Black Holes, *Mod. Phys. Lett.* **A12** (1997) 2585.

[1093] W. A. Sabra, Black Holes in $N = 2$ Supergravity Theories and Harmonic Functions, *Nucl. Phys.* **B510** (1998) 247.

[1094] W.A. Sabra, General BPS Black Holes in Five Dimensions, *Mod. Phys. Lett.* **A13** (1998) 239.

[1095] A. Sagnotti, Open Strings and their Symmetry Groups, in *Nonperturbative Quantum Field Theory*, eds. G. 't Hooft, A. Jaffe, G. Mack, P. K. Mitter, and R. Stora. New York: Plenum Press (1988), p. 521.

[1096] N. Sakai and I. Senda, Vacuum Energies of String Compactified on Torus, *Prog. Theor. Phys.* **75** (1986) 692. [Erratum *ibid.* **77** (1987) 773.]

[1097] A. Salam and E. Sezgin, eds. *Supergravities in Diverse Dimensions*, vol. 1 and 2. Amsterdam/Singapore: North Holland/World Scientific (1989).

[1098] A. Salam and J. Strathdee, On Kaluza-Klein Theory, *Ann. Phys. (NY)* **141** (1982) 316.

[1099] S. Salamon, *Riemannian Geometry and Holonomy Groups*. Harlow: Longman Scientific and Technical (1989).

[1100] H. Samtleben, Lectures on Gauged Supergravity and Flux Compactifications, *Classical Quant. Grav.* **25** (2008) 214002.

[1101] H. Samtleben and M. Weidner, The Maximal $D = 7$ Supergravities, *Nucl. Phys.* **B725** (2005) 383.

[1102] J. Scherk, An Introduction to the Theory of Dual Models and Strings, *Rev. Mod. Phys.* **47** (1975) 123.

[1103] J. Scherk and J. H. Schwarz, Dual Models for Nonhadrons, *Nucl. Phys.* **B81** (1974) 118.

[1104] J. Scherk and J. H. Schwarz, Dual Models and the Geometry of Space-Time, *Phys. Lett.* **B52** (1974) 347.

[1105] J. Scherk and J. H. Schwarz, Spontaneous Breaking of Supersymmetry through Dimensional Reduction, *Phys. Lett.* **82B** (1979) 60.

[1106] J. Scherk and J. H. Schwarz, How to Get Masses from Extra Dimensions, *Nucl. Phys.* **B153** (1979) 61.

[1107] R. Schoen and S. -T. Yau, On the Proof of the Positive Mass Conjecture in General Relativity, *Commun. Math. Phys.* **65** (1979) 45.

[1108] J. Schon and M. Weidner, Gauged $N = 4$ Supergravities, *JHEP* **0605** (2006) 034.

[1109] E. Schrödinger, Die Energiekomponenten des Gravitationsfeldes, *Phys. Z.* **19** (1918) 4.

[1110] E. Schrödinger, Contributions to Born's New Theory of the Electromagnetic Field, *Proc. Roy. Soc. London* **A150** (1935) 465.

[1111] E. Schrödinger, *Proc. Roy Irish Acad.* **51A** (1947) 163.

[1112] E. Schrödinger, *Space-Time Structure*. Cambridge: Cambridge University Press (1954).

[1113] J. H. Schwarz, Superstring Theory, *Phys. Rep.* **89** (1982) 223.

[1114] J. H. Schwarz, Covariant Field Equations of Chiral $N = 2$, $D = 10$ Supergravity, *Nucl. Phys.* **B226** (1983) 269.

[1115] J. H. Schwarz, Dilaton-Axion Symmetry, in *String Theory, Quantum Gravity and the Unification of the Fundamental Interactions*, eds. M. Bianchi, F. Fucito, E. Marinari, and A. Sagnotti. River Edge, NJ: World Scientific (1993).

[1116] J. H. Schwarz, An SL(2, Z) Multiplet of Type IIB Superstrings, *Phys. Lett.* **B360** (1995) 13–18. [Erratum *ibid.* **B364** (1995) 252.]

[1117] J. H. Schwarz and P. C. West, Symmetries and Transformations of Chiral $N = 2$ $D = 10$ Supergravity, *Phys. Lett.* **B126** (1983) 301.

[1118] K. Schwarzschild, Über das Gravitationsfled eines Massenpunktes nach der Einsteinschen Theorie, *Sitzungsber. deutsch. Akad. Wiss. Berlin, Kl. Math. Phys. Technik* (1916) 189–196. English translation by S. Antoci and A. Loinger, available at `http://arXiv.org/pdf/physics/9905030`.

[1119] J. Schwinger, On Gauge Invariance and Vacuum Polarization, *Phys. Rev.* **82** (1951) 664.

[1120] J. Schwinger, Magnetic Charge and Quantum Field Theory, *Phys. Rev.* **144** (1966) 1087.

[1121] J. Schwinger, Sources and Magnetic Charge, *Phys. Rev.* **173** (1968) 1536.

[1122] J. Schwinger, *Particles, Sources and Fields*, vol. 1. New York: Addison-Wesley (1970).

[1123] N. Seiberg, Observations on the Moduli Space of Superconformal Field Theories, *Nucl. Phys.* **B303** (1988) 286.

[1124] A. Sen, Electric Magnetic Duality in String Theory, *Nucl. Phys.* **B404** (1993) 109.

[1125] A. Sen, Quantization of Dyon Charge and Electric Magnetic Duality in String Theory, *Phys. Lett.* **303B** (1993) 22.

[1126] A. Sen, Magnetic Monopoles, Bogomolny Bound and Sl(2,Z) Invariance in String Theory, *Mod. Phys. Lett.* **A8** (1993) 2023.

[1127] A. Sen, Strong-Weak Coupling Duality in Four-Dimensional String Theory, *Int. J. Mod. Phys.* **A9** (1994) 3707.

[1128] A. Sen, Extreme Black Holes and Elementary String States, *Mod. Phys. Lett.* **A10** (1995) 2081.

[1129] A. Sen, Entropy Function for Heterotic Black Holes, *JHEP* **0603** (2006) 008.

[1130] J.-P. Serre, *Cours d'arithmétique*. Paris: Presses Universitaires de France (1970).

[1131] K. Sfetsos and K. Skenderis, Microscopic Derivation of the Bekenstein-Hawking Entropy Formula for Non-extremal Black Holes, *Nucl. Phys.* **B517** (1998) 179.

[1132] A. Shapere, S. Trivedi, and F. Wilczek, Dual Dilaton Dyons, *Mod. Phys. Lett.* **A6**, (1991) 2677.

[1133] I. L. Shapiro, Physical Aspects of the Space-Time Torsion, *Phys. Rep* **357** (2001) 113.

[1134] M. M. Sheikh Jabbari, Classification of Different Branes at Angles, *Phys. Lett.* **B420** (1998) 279.

[1135] K. Shiraishi, Multicentered Solution for Maximally Charged Dilaton Black Holes in Arbitrary Dimensions, *J. Math. Phys.* **34** (1993) 1480.

[1136] M. Shmakova, Calabi-Yau Black Holes, *Phys. Rev.* **D56** (1997) 540.

[1137] Y. M. Shnir, *Magnetic Monopoles*. Berlin, Germany: Springer (2005).

[1138] W. Siegel, Hidden Local Supersymmetry in the Supersymmetric Particle Action, *Phys. Lett.* **B128** (1983) 397.

[1139] T. P. Singh, Gravitational Collapse, Black Holes and Naked Singularities, gr-qc/9805066.

[1140] K. Skenderis, Black Holes and Branes in String Theory, *Lect. Notes Phys.* **541** (2000) 325.

[1141] K. Skenderis and P. K. Townsend, Hamilton-Jacobi Method for Curved Domain Walls and Cosmologies, *Phys. Rev.* **D74** (2006) 125008.

[1142] R. Slansky, Group Theory For Unified Model Building, *Phys. Rept.* **79** (1981) 1.

[1143] L. Smarr, Mass Formula for Kerr Black Holes, *Phys. Rev. Lett.* **30** (1973) 71. [Erratum *ibid.* **30** (1973) 521.]

[1144] D. J. Smith, Intersecting Brane Solutions in String and M-Theory, hep-th/0210157.

[1145] W. L. Smith and R. B. Mann, Formation of Topological Black Holes from Gravitational Collapse, *Phys. Rev.* **D56** (1997) 4942.

[1146] J. A. Smoller and A. G. Wasserman, Regular Solutions of the Einstein Yang-Mills Equations, *J. Math. Phys.* **36** (1995) 4301.

[1147] J. A. Smoller, A. G. Wasserman, S.-T. Yau, and J. B. McLeod, Smooth Static Solutions of the Einstein Yang-Mills Equations, *Commun. Math. Phys.* **143** (1991) 115.

[1148] R. D. Sorkin, Kaluza-Klein Monopole, *Phys. Rev. Lett.* **51** (1983) 87.

[1149] A. A. Starobinsky, *Zh. Éksp. Teor. Fiz.* **64** (1973) 48.

[1150] R. Steinbauer, On the Geometry of Impulsive Gravitational Waves, gr-qc/9809054.

[1151] K. S. Stelle, Lectures on Supergravity $p$-Branes, in *Trieste 1996 Summer School in High Energy Physics and Cosmology*, eds. E. Gava, A. Masiero, K. S. Narain, S. Randjbar-Daemi, and Q. Shafi. Singapore: World Scientific (1997), p. 287.

[1152] K. S. Stelle, BPS Branes in Supergravity, in *Trieste 1997 Summer School in High Energy Physics and Cosmology*, eds. E. Gava, A. Masiero, K. S. Narain *et al.* Singapore: World Scientific (1998), p. 29.

[1153] K. Stelle and P. C. West, Spontaneously Broken de Sitter Symmetry and the Gravitational Holonomy Group, *Phys. Rev.* **D21** (1980) 1466.

[1154] C. R. Stephens, The Hawking Effect in Abelian Gauge Theories, *Ann. Phys.* **193** (1989) 255.

[1155] N. Straumann, Reflections on Gravity, astro-ph/0006423.

[1156] A. Strominger, Special Geometry, *Commun. Math. Phys.* **133** (1990) 163.

[1157] A. Strominger, Macroscopic Entropy of $N = 2$ Extremal Black Holes, *Phys. Lett.* **B383** (1996) 39.

[1158] A. Strominger, Open $p$-Branes, *Phys. Lett.* **B383** (1996) 44.

[1159] A. Strominger, Loop Corrections to the Universal Hypermultiplet, *Phys. Lett.* **B421** (1998) 139.

[1160] A. Strominger and C. Vafa, Microscopic Origin of the Bekenstein-Hawking Entropy, *Phys. Lett.* **B370** (1996) 99.

[1161] A. Strominger and E. Witten, New Manifolds for Superstring Compactification, *Commun. Math. Phys.* **101** (1985) 341.

[1162] E. C. G. Stückelberg, *Helv. Phys. Acta* **11** (1938) 225.

[1163] D. Sudarsky and T. Zannias, Spherical Black Holes Cannot Support Scalar Hair, *Phys. Rev.* **D58** (1998) 087502.

[1164] S. Sugimoto, Anomaly Cancellations in Type I $D9 - \bar{D}9$ System and the USp(32) String Theory, *Prog. Theor. Phys.* **102** (1999) 685.

[1165] S. Surya and D. Marolf, Localized Branes and Black Holes, *Phys. Rev.* **D58** (1998) 124013.

[1166] L. Susskind, Some Speculations About Black Hole Entropy in String Theory, hep-th/9309145.

[1167] G. Szekeres, On the Singularities of a Riemannian Manifold, *Pbl. Mat. Debrecen* **7** (1960) 285.

[1168] F. Tangherlini, *Nuovo Cim.* **77** (1963) 636.

[1169] Y. Tanii, Introduction to Supergravities in Diverse Dimensions, hep-th/9802138.

[1170] A. H. Taub, Empty Space-Times Admitting a Three-Parameter Group of Motions, *Ann. Math.* **53** (1951) 472.

[1171] C. Teitelboim, Nonmeasurability of the Quantum Numbers of a Black Hole, *Phys. Rev.* **D5** (1972) 2941.

[1172] C. Teitelboim, Gauge Invariance for Extended Objects, *Phys. Lett.* **167B** (1986) 63.

[1173] C. Teitelboim, Monopoles of Higher Rank, *Phys. Lett.* **167B** (1986) 69.

[1174] C. Teitelboim, Action and Entropy of Extreme and Non-Extreme Black Holes, *Phys. Rev.* **D51** (1995) 4315. [Erratum *ibid.*. **D52** (1995) 6201.]

[1175] C. Teitelboim, D. Villarroel, and Ch. G. van Weert, Classical Electrodynamics of Retarded Fields and Point Particles, *Riv. Nuovo Cim.* **3**, N. 9 (1980) 1.

[1176] W. E. Thirring, *Acta Phys. Austriaca Suppl.* **9** (1972) 256.

[1177] W. E. Thirring, An Alternative Approach to the Theory of Gravitation, *Ann. Phys.* **16** (1964) 96.

[1178] Y. Thiry, The Equations of Kaluza's Unified Theory, *Acad. Sci. Paris* **226** (1948) 216.

[1179] G. 't Hooft, An Algorithm for the Poles at Dimension Four in the Dimensional Regularization Procedure, *Nucl. Phys.* **B62** (1973) 444.

[1180] G. 't Hooft, Magnetic Monopoles in Unified Gauge Theories, *Nucl. Phys.* **B79** (1974) 276.

[1181] G. 't Hooft and M. Veltman, One-Loop Divergencies in the Theory of Gravitation, *Ann. Inst. Henri Poincaré* **20** (1974) 69.

[1182] K. S. Thorne, *Black Holes & Time Warps*. New York: W. W. Norton and Co. (1994).

[1183] K. P. Tod, All Metrics Admitting Supercovariantly Constant Spinors, *Phys. Lett.* **121B**, (1981) 241.

[1184] K. P. Tod, More on Supercovariantly Constant Spinors, *Classical Quant. Grav.* **12** (1995) 1801.

[1185] M. A. Tonnelat, *La théorie du champ unifié d'Einstein*. Paris: Gauthier-Villars (1955).

[1186] P. K. Townsend, Cosmological Constant In Supergravity, *Phys. Rev.* **D15** (1977) 2802.

[1187] P. K. Townsend, Worldsheet Electromagnetism and the Superstring Tension, *Phys. Lett.* **B277** (1992) 285.

[1188] P. K. Townsend, The Eleven-Dimensional Supermembrane Revisited, *Phys. Lett.* **B350** (1995) 184.

[1189] P. K. Townsend, P-Brane Democracy, in *PASCOS/Johns Hopkins 1995*, eds. J. A. Bagger, G. K. Domokos, A. F. Falk, and S. Kövesi-Domokos. Singapore: World Scientific (1996), p. 271.

[1190] P. K. Townsend, D-branes from M-branes, *Phys. Lett.* **B373** (1996) 68.

[1191] P. K. Townsend, Brane Surgery, *Nucl. Phys. Proc. Suppl.* **B475** (1996).

[1192] P. K. Townsend, Four Lectures on M Theory, in *Trieste 1996 Summer School in High Energy Physics and Cosmology*, eds. E. Gava, A. Masiero, K. S. Narain, S. Randjbar-Daemi, and Q. Shaft. Singapore: World Scientific (1997), p. 385.

[1193] P. K. Townsend, Membrane Tension and Manifest IIB S Duality, *Phys. Lett.* **B409** (1997) 131.

[1194] P. K. Townsend, Black Holes, `gr-qc/9707012`.

[1195] P. K. Townsend, M-Branes at Angles, *Nucl. Phys. Proc. Suppl.* **67** (1998) 88.

[1196] P. K. Townsend, M Theory from its Superalgebra, in *Strings, Branes and Dualities*, eds. L. Baulieu, P. Di Francesco, M. Douglas, V. Kazakov, M. Picco, and P. Windey. Dordrecht: Kluwer (1999), p. 141.

[1197] P. K. Townsend, Killing Spinors, Supersymmetries and Rotating Intersecting Branes, `hep-th/9901102`.

[1198] P. K. Townsend, Hamilton-Jacobi Mechanics from Pseudo-supersymmetry, *Classical Quant. Grav.* **25** (2008) 045017.

[1199] J. Traschen, An Introduction to Black Hole Evaporation, in *Mathematical Methods of Physics*, eds. A. Bytsenko and F. Williams. Singapore: World Scientific (2000).

[1200] A. Trautman, Solutions of the Maxwell and Yang-Mills Equations Associated with Hopf Fibrings, *Int. J. Theor. Phys.* **16** (1977) 561.

[1201] M. Trigiante, Dual Gauged Supergravities, `hep-th/0701218`.

[1202] P. K. Tripathy and S. P. Trivedi, Non-Supersymmetric Attractors in String Theory, *JHEP* **0603** (2006) 022.

[1203] A. A. Tseytlin, Harmonic Superpositions of M Branes, *Nucl. Phys.* **B475** (1996) 149.

[1204] A. A. Tseytlin, Born-Infeld Action, Supersymmetry and String Theory, in *The Many Faces of the Superworld*, ed. M. A. Shifman. Singapore: World Scientific (2000), p. 417.

[1205] W. G. Unruh, Second Quantization in the Kerr Metric, *Phys. Rev.* **D10** (1974) 3194.

[1206] W. G. Unruh, A Unimodular Theory Of Canonical Quantum Gravity, *Phys. Rev.* **D40** (1989) 1048.

[1207] R. Utiyama, Invariant Theoretical Interpretation of Interaction, *Phys. Rev.* **101** (1956) 1597.

[1208] A. Van Proeyen, Tools for Supersymmetry, in *Spring School on Quantum Field Theory: Supersymmetry and Superstrings*, `hep-th/9910030`.

[1209] A. Van Proeyen, $N = 2$ Supergravity in $d = 4, 5, 6$ and its Matter Couplings, Lectures given at the Institut Henri Poincaré, Paris, November 2000.

[1210] A. Van Proeyen, Structure of Supergravity Theories, `hep-th/0301005`.

[1211] S. Vandoren, Lectures on Riemannian Geometry, Part II: Complex Manifolds.

[1212] M. A. Vandyck, On the Problem of Space-Time Symmetries in the Theory of Supergravity, *Gen. Relativ. Gravit.* **20** (1988) 261.

[1213] M. A. Vandyck, On the Problem of Space-Time Symmetries in the Theory of Supergravity. 2: $N = 2$ Supergravity and Spinorial Lie Derivatives, *Gen. Relativ. Gravit.* **20** (1988) 905.

[1214] L. Vanzo, Black Holes with Unusual Topology, *Phys. Rev.* **D56** (1997) 6475.

[1215] H. J. de Vega, M. Ramón Medrano, and N. Sánchez, Superstring Propagation Through Supergravitational Shock Waves, *Nucl. Phys.* **B374** (1992) 425.

[1216] M. A. G. Veltman, in *Methods in Field Theory*, eds. R. Balian and J. Zinn-Justin. Amsterdam: North-Holland (1976), p. 265.

[1217] M. A. G. Veltman, Gammatrica, *Nucl. Phys.* **B319** (1989) 253.

[1218] A. E. M. van de Ven, Two Loop Quantum Gravity, *Nucl. Phys.* **B378** (1992) 309.

[1219] E. P. Verlinde and H. L. Verlinde, RG Flow, Gravity and the Cosmological Constant, *JHEP* **0005** (2000) 034.

[1220] H. Vermeil, *Nachr. Ges. Wiss. Göttingen* (1917) 334.

[1221] M. S. Volkov and D. V. Gal'tsov, NonAbelian Einstein Yang-Mills Black Holes, *JETP Lett.* **50** (1989) 346. [*Pisma Zh. Eksp. Teor. Fiz.* **50** (1989) 312.]

[1222] M.S. Volkov and D.V. Gal'tsov, Gravitating Non-Abelian Solitons and Black Holes with Yang-Mills Fields, *Phys. Rept.* **319** (1999) 1.

[1223] A. Volovich, Three-Block $p$-Branes in Various Dimensions, *Nucl. Phys.* **B492** (1997) 235.

[1224] M. de Vroome and B. de Wit, Lagrangians with Electric and Magnetic Charges of $N = 2$ Supersymmetric Gauge Theories, *JHEP* **0708** (2007) 064.

[1225] S. Wadia, Status of Microscopic Modeling of Black Holes by D1-D5 System, `hep-th/0011286`.

[1226] R. M. Wald, Final States of Gravitational Collapse, *Phys. Rev. Lett.* **26** (1971) 1653.

[1227] R. M. Wald, Electromagnetic Fields and Massive Bodies, *Phys. Rev.* **D6** (1972) 1476.

[1228] R. M. Wald, Gedankenexperiments to Destroy a Black Hole, *Ann. Phys.* **83** (1974) 548.

[1229] R. M. Wald, *General Relativity*. Chicago, IL: University of Chicago Press (1984).

[1230] R. M. Wald, Spin-Two Fields and General Covariance, *Phys. Rev.* **D33** (1986) 3613.

[1231] R.M. Wald, A New Type of Gauge Invariance for a Collection of Massless Spin-2 Fields. 2. Geometrical Interpretation, *Classical Quant. Grav.* **4** (1987) 1279.

[1232] R. M. Wald, The First Law of Black Hole Mechanics, in *College Park 1993, Directions in General Relativity*, vol. 1, eds. B. L. Hu, M. P. Ryan Jr. C. V. Vishveshwara, and T. A. Jacobson. Cambridge: Cambridge University Press (1993), p. 358.

[1233] R. M. Wald, *Quantum Field Theory in Curved Spacetime and Black Hole Thermodynamics*. Chicago, IL: University of Chicago Press (1994).

[1234] R. M. Wald, Gravitational Collapse and Cosmic Censorship, in *Black Holes, Gravitational Radiation and the Universe*, eds. B. R. Iyer, B. Bhawal, and C. V. Vishveshwara. Dordrecht: Kluwer (1998), p. 69.

[1235] R. M. Wald, The Thermodynamics of Black Holes, *Living Rev. Rel.* **4** (2001) 6.

[1236] M. Y. Wang, A Solution of Coupled Einstein SO(3) Gauge Field Equations, *Phys. Rev.* **D12** (1975) 3069.

[1237] M. Weidner, Gauged Supergravities in Various Spacetime Dimensions, *Fortsch. Phys.* **55** (2007) 843.

[1238] S. Weinberg, Derivation of Gauge Invariance and the Equivalence Principle from Lorentz Invariance in the S Matrix, *Phys. Lett.* **9** (1964) 357.

[1239] S. Weinberg, Photons and Gravitons in S Matrix Theory: Derivation of Charge Conservation and Equality of Gravitational and Inertial Mass, *Phys. Rev.* **135** (1964) B1049.

[1240] S. Weinberg, Photons and Gravitons in Perturbation Theory: Derivation of Maxwell's and Einstein's Equations, *Phys. Rev.* **138** (1965) B988.

[1241] S. Weinberg, *Gravitation and Cosmology*. New York: Wiley (1972).

[1242] S. Weinberg, The Cosmological Constant Problem, *Rev. Mod. Phys.* **61** (1989) 1.

[1243] R. Weitzenböck, *Invariantentheorie*. Groningen: Noordhoff (1923).

[1244] R. Weitzenböck, *Sitzungsber. Konigl. Preuß. Akad. Wiss., phys.-math. Kl.* (1928) 466.

[1245] J. Wess and J. Bagger, *Supersymmetry and Supergravity*. Princeton, NJ: Princeton University Press (1992).

[1246] P. C. West, *An Introduction to Supersymmetry and Supergravity*, extended 2nd ed., Singapore: World Scientific (1990).

[1247] P. C. West, Supergravity, Brane Dynamics and String Duality, in *Duality and Supersymmetric Theories*, eds. D. I. Olive and P. C. West. Cambridge: Cambridge University Press (1999), p. 147.

[1248] P. West, *Introduction to Strings and Branes*. Cambridge: Cambridge University Press (2012),

[1249] H. Weyl, Zur Gravitationstheorie, *Ann. Phys.* **54** (1917) 117.

[1250] H. Weyl, Zur Gravitationstheorie, *Ann. Phys.* **54** (1917) 185.

[1251] H. Weyl, *Space–Time–Matter*. New York: Dover (1922).

[1252] H. Weyl, Electron and Gravitation, *Z. Phys.* **56** (1929) 330. Translated in Ref. [1004].

[1253] H. Weyl, How Far can one get with a Linear Field Theory of Gravitation in Flat Space-Time?, *Am. J. Math.* **66** (1944) 591. (Reprinted in Ref. [1254].)

[1254] H. Weyl, *Gesammelte Abhandlungen*, 4 vol., Berlin: Springer-Verlag (1968).

[1255] F. Wilczek, Riemann-Einstein Structure from Volume and Gauge Symmetry, *Phys. Rev. Lett.* **80** (1998) 4851.

[1256] E. Winstanley, Classical Yang-Mills Black Hole Hair in Anti-de Sitter Space, *Lect. Notes Phys.* **769** (2009) 49.

[1257] B. de Wit, P. G. Lauwers, and A. Van Proeyen, Lagrangians of $N = 2$ Supergravity-Matter Systems, *Nucl. Phys.* **B255** (1985) 569.

[1258] B. de Wit, H. Nicolai, and H. Samtleben, Gauged Supergravities, Tensor Hierarchies, and M-Theory, *JHEP* **0802** (2008) 044.

[1259] B. de Wit and H. Samtleben, Gauged Maximal Supergravities and Hierarchies of Nonabelian Vector-Tensor Systems, *Fortsch. Phys.* **53** (2005) 442.

[1260] B. de Wit, H. Samtleben, and M. Trigiante, On Lagrangians and Gaugings of Maximal Supergravities, *Nucl. Phys.* **B655** (2003) 93.

[1261] B. de Wit, H. Samtleben, and M. Trigiante, Maximal Supergravity from IIB Flux Compactifications, *Phys. Lett.* **B583** (2004) 338.

[1262] B. de Wit, H. Samtleben, and M. Trigiante, Magnetic Charges in Local Field Theory, *JHEP* **0509** (2005) 016.

[1263] B. de Wit, H. Samtleben, and M. Trigiante, The Maximal $D = 5$ Supergravities, *Nucl. Phys.* **B716** (2005) 215.

[1264] B. de Wit, H. Samtleben, and M. Trigiante, The Maximal $D = 4$ Supergravities, *JHEP* **0706** (2007) 049.

[1265] B. de Wit and A. Van Proeyen, Potentials and Symmetries of General Gauged $N = 2$ Supergravity – Yang–Mills Models, *Nucl. Phys. B* **245** (1984) 89.

[1266] B. de Wit and A. Van Proeyen, Special Geometry, Cubic Polynomials and Homogeneous Quaternionic Spaces, *Commun. Math. Phys.* **149** (1992) 307.

[1267] B. de Wit, F. Vanderseypen, and A. Van Proeyen, Symmetry Structure of Special Geometries, *Nucl. Phys.* **B400** (1993) 463.

[1268] B. de Wit and M. van Zalk, Supergravity and M-Theory, *Gen. Relativ. Grav.* **41** (2009) 757.

[1269] B. de Wit and M. van Zalk, Electric and Magnetic Charges in $N = 2$ Conformal Supergravity Theories, *JHEP* **1110** (2011) 050.

[1270] E. Witten, Dyons of Charge $e = \theta/2\pi$, *Phys. Lett.* **86B** (1979) 283.

[1271] E. Witten, Search for a Realistic Kaluza-Klein Theory, *Nucl. Phys.* **B186** (1981) 412.

[1272] E. Witten, A New Proof of the Positive Energy Theorem, *Commun. Math. Phys.* **80**, (1981) 381.

[1273] E. Witten, Instability of the Kaluza-Klein Vacuum, *Nucl. Phys.* **195** (1982) 481.

[1274] E. Witten, Fermion Quantum Numbers in Kaluza-Klein Theory, in *Shelter Island Conf. Quantum Field Theory and the Fundamental Problems of Physics II*, eds. R. Jackiw, N. N. Khuri, S. Weinberg, and E. Witten. Cambridge, MA: MIT Press (1985), p. 227.

[1275] E. Witten, String Theory Dynamics in Various Dimensions, *Nucl. Phys.* **B443** (1995) 85.

[1276] E. Witten and D. Olive, Supersymmetry Algebras that Include Topological Charges, *Phys. Lett.* **78B** (1978) 97.

[1277] T. T. Wu and C. -N. Yang, Some Solutions of the Classical Isotopic Gauge Field Equations, in *Properties of Matter under Unusual Conditions*, eds. H. Mark and S. Fernbach. New York: Interscience (1969), p. 349.

[1278] T. T. Wu and C. -N. Yang, Some Remarks About Unquantized Nonabelian Gauge Fields, *Phys. Rev.* **D12** (1975) 3843.

[1279] T. T. Wu and C. N. Yang, Concept of Non-integrable Phase Factors and Global Formulation of Gauge Fields, *Phys. Rev.* **D12** (1975) 3845.

[1280] W. Wyss, Zur Unizität der Gravitationstheorie, *Helv. Phys. Acta* **38** (1965) 469.

[1281] P. B. Yasskin, Solutions for Gravity Coupled to Massless Gauge Fields, *Phys. Rev.* **D12** (1975) 2212.

[1282] D. Youm, Black Holes and Solitons in String Theory, *Phys. Rep.* **316** (1999) 1.

[1283] D. Youm, Partially Localized Intersecting BPS Branes, *Nucl. Phys.* **B556** (1999) 222.

[1284] A. Zee, *Einstein Gravity in a Nutshell*. Princeton, NJ: Princeton University Press (2013).

[1285] Ya. B. Zel'dovich, *Pis'ma Zh. Éskp. Teor. Fiz* **14** (1970) 270.

[1286] B. Zumino, *Ann. N.Y. Acad. Sci.* **302** (1977) 545.

[1287] D. Zwanziger, Exactly Soluble Nonrelativistic Model of Particles with Both Electric and Magnetic Charges, *Phys. Rev.* **176** (1968) 1480.

[1288] D. Zwanziger, Quantum Field Theory of Particles with Both Electric and Magnetic Charges, *Phys. Rev.* **176** (1968) 1489.

# Index

Page numbers in italic are those on which the definition or a very specific discussion of the subject can be found. "4d" can mean four dimensions or four-dimensional.